8th International Conference on Structural Health Monitoring of Intelligent Infrastructure (SHMII 2017)

AA005060

Structural Health Monitoring in Real-World Application

Brisbane, Australia
5-8 December 2017

Volume 1 of 3

Editors:

Tommy Chan **Saeed Mahini**

ISBN: 978-1-5108-6457-3

Printed from e-media with permission by:

Curran Associates, Inc.
57 Morehouse Lane
Red Hook, NY 12571

Some format issues inherent in the e-media version may also appear in this print version.

Copyright© (2017) by International Society for Structural Health Monitoring of Intelligent Infrastructure
All rights reserved.

Printed by Curran Associates, Inc. (2018)

For permission requests, please contact International Society for Structural Health Monitoring of Intelligent Infrastructure at the address below.

International Society for Structural Health Monitoring of Intelligent Infrastructure
University of Manitoba
Agricultural & Civil Engineering Bldg.
A250 - 96 Dafor Road
Winnipeg, Manitoba R3T 2N2
Canada

Fax: 204-474-7519

central@ishmii.org

Additional copies of this publication are available from:

Curran Associates, Inc.
57 Morehouse Lane
Red Hook, NY 12571 USA
Phone: 845-758-0400
Fax: 845-758-2633
Email: curran@proceedings.com
Web: www.proceedings.com

TABLE OF CONTENTS

VOLUME 1

KEYNOTE LECTURES

REGULAR SESSIONS

RS1: SMART AND OTHER ADVANCED SENSORS

RS2: WIRELESS AND OTHER ADVANCED SENSOR NETWORKS

RS3: DATA ACQUISITION, PROCESSING AND MANAGEMENT, INCLUDING BIG DATA

VOLUME 2

RS7: DAMAGE CONTROL, REPAIR AND STRENGTHENING

RS8: LIFE-CYCLE PERFORMANCE-BASED DESIGN AND PERFORMANCE MONITORING

RS10: SMART MATERIALS, STRUCTURES AND REHABILITATION

RS11: GLOBAL POSITIONING SYSTEM (GPS) AND RELATED SYSTEMS

RS12: REMOTE MONITORING SYSTEMS

RS13: INTEGRATED SYSTEMS AND IMPLEMENTATIONS OF SHM

RS15: STANDARDIZATION OF SHM SYSTEMS

RS16: CRITICAL ISSUES FOR SHM SUCH AS COST EFFECTIVENESS OF SHM SYSTEMS

RS17: RESILIENCE OF INFRASTRUCTURE SYSTEMS

MINI-SYMPOSIA

MS1: RECENT SHM ADVANCES IN AUSTRALIA

VOLUME 3

MS2: DATA-DRIVEN STRUCTURAL DAMAGE IDENTIFICATION AND PERFORMANCE ASSESSMENT

MS3: TEMPERATURE BEHAVIOUR OF STRUCTURES

SPECIAL SESSIONS

SS1: RECENT ADVANCES IN RAILWAY SYSTEM MONITORING IN CHINA

SS2: CASE STUDIES AND NEW METHODS FOR MODAL-BASED SHM OF CIVIL STRUCTURES

SS3: OPTICAL BASED MONITORING TECHNIQUES FOR INFRASTRUCTURAL MAINTENANCE AND SAFETY

International Society for Structural Health Monitoring of Intelligent Infrastructure

ISHMII

SHMII8

8th International Conference on Structural Health Monitoring of Intelligent Infrastructure

STRUCTURAL HEALTH MONITORING IN REAL-WORLD APPLICATION

Brisbane, Australia | 5 - 8 December **2017**

Proceedings

Editors: Prof Tommy Chan, Conference Chair
and Dr Saeed Mahini, Conference Co-chair

Contents

Preface

For the last three decades, Structural Health Monitoring (SHM) has been attracting enormous research efforts around the world as it targets at monitoring structural conditions to prevent catastrophic failure, and to provide quantitative data for engineers and infrastructure owners to design reliable structures and develop economical asset management plans. SHM has now been accepted as a justified effort for civil structures. It is a worldwide trend to install SHM systems on significant structures.

The International Society for Structural Health Monitoring of Intelligent Infrastructure (ISHMII) holds the SHMII official conference every two years and one or more official workshops (such as CSHM), usually in alternating years. SHMII-8 will build on the success of SHMII-1, held in Tokyo 2003, SHMII-2 in Shenzhen 2005, SHMII-3 in Vancouver 2007, SHMII-4 in Zurich 2009, SHMII-5 in Cancun 2011, SHMII-6 in Hong Kong 2013 and SHMII-7 in Turin 2015.

The three-day Conference program will include keynote lectures, invited lectures, regular sessions, special sessions, mini-symposiums, technical visit and touring activities. Over 200 delegates are invited to present and share their experience with experts, engineers, researchers and scientists from different countries and disciplines and interact with experts from the industry.

The SHMII-8 is hosted by The Australian Network of Structural Health Monitoring (ANSHM) and the Queensland University of Technology (QUT). In 2009, ANSHM was set up to raise the recognition of Structural Health Monitoring (SHM) in Australia. QUT is a university for the real world. Although the name QUT has only been used since 1989, yet it has a proud history which can be traced back as early as its establishment as the Brisbane School of Arts in 1849.

As the first time SHMII is being held in the southern hemisphere, the objectives of this SHMII-8 are to share the advancement of the SHM in Australia with other parts of the world, specifically to showcase achievements, exchange ideas and disseminate knowledge nationally and internationally, and to raise general community awareness on the need for, and value of, SHM research and application

With a theme of Structural Health Monitoring in Real-world Application aligning well with QUT being a university for the real world, the SHMII-8 Conference features a strong program of keynote lectures, invited lectures, technical visit and touring activities.

The SHMII-8 Proceedings include the keynotes lectures, and papers submitted to the conference either as regular papers or invited papers, special sessions and mini-symposiums. It has authors from 26 countries, which include Australia, Austria, Canada, China, France, Germany, Hong Kong, India, Iran, Ireland, Italy, Japan, Mexico, Pakistan, Poland, Portugal, Singapore, South Africa, South Korea, Spain, Sweden, Switzerland, Taiwan, United Kingdom, United States, and Vietnam, covering research work from six of the seven continents of the world.

The Proceedings consist of an Abstract Book and a memory stick contains the full papers in electronic format. It covers wide range of SHM topics including, smart sensors, wireless networks, data acquisition, damage assessment, reliability forecast, damage repair, performance monitoring, life cycle management of infrastructure, smart materials and rehabilitation, GPS, remote monitoring, integrated SHM, design guidelines, standardization of SHM systems, critical issues for SHM and resilience of infrastructure and other related subjects.

The Editors would like to express our sincere thanks to the contributors either as authors, reviewers, gold and silver sponsors, exhibitors, supporters, organisers of special sessions and mini-symposiums, and the Event Manager and her team. In fact, without their support and advice, a conference with this scale could not be organised.

We sincerely acknowledge the financial support from the Queensland Department of Transport and Main Roads of the Queensland Government as the gold sponsor, Mainmark, Geomotion and Worldsensing as the silver sponsors and Bestech, Engineers Enterprise, Land Solution Australia, National Instruments, SAVTek, and Strucomp P/L as the exhibitors of the conference.

The Brisbane City Council, the Queensland Department of Transport and Main Roads, the American Society of Civil Engineers - Australia Section, Engineers Australia Queensland Division and the Institution of Structural Engineers are the conference supporters and their support is gratefully acknowledged. Financial and in-kind support provided by the University of New England to the co-chair for organising the conference is also sincerely acknowledged.

Finally, we would like to express our deepest thanks to all Local Organising Committee members specially the leaders and the members of the fundraising/sponsorship and exhibitions subcommittee, publicity subcommittee and editorial subcommittees who have spent their times and efforts voluntarily and also QUT Conferences as the operational and logistic subcommittee and the related sub-contractors for their help and support throughout the organising time of the conference.

It is expected that the SHMII-8 Proceedings will serve as a valuable reference to all concerned with SHM and it will become an important knowledgebase for the advancement of this important technology.

EDITORS

Prof Tommy Chan (SHMII-8 Chair)

President of ANSHM
Professor of Civil and Structural Engineering,
Queensland University of Technology (QUT)

Dr Saeed Mahini (SHMII-8 Co-Chair)

EC member of ANSHM
Council member of ISHMII
Former Senior Lecturer in Civil Engineering
Discipline of Civil and Environmental Engineering,
University of New England (UNE)
Sessional Academic, School of Engineering, Griffith University

SHMII-8 Committees

Editorial Subcommittee

CHIEF EDITORS: Prof Tommy Chan and **Dr Saeed Mahini**

CHAIR: Dr Saeed Mahini, University of New England, New South Wales

Dr Ulrike Dackermann, University of New South Wales, New South Wales
Prof Hong Guan, Griffith University, Queensland
Dr Lei Hou, Griffith University, Queensland
Dr Jun Li, Curtin University, Western Australia
Dr Alex Ching-Tai Ng, University of Adelaide, South Australia
Dr Andy Nguyen, Queensland University of Technology, Queensland
Dr Ying Wang, University of Surrey, England
A/Prof Xinqun Zhu, Western Sydney University, New South Wales

Steering Committee

CHAIR: Prof Tommy Chan, Queensland University of Technology, Australia

CO-CHAIR: Dr Saeed Mahini, University of New England, Australia

Prof Farhad Ansari, University of Illinois, Chicago, USA
Prof Wije W Ariyaratne, Roads and Maritime Services, NSW, Australia
Dr Wolfgang Habel, Federal Institute for Materials Research and Testing (BAM), Germany
Prof Hong Hao, Curtin University, Australia
Prof Nick Haritos, University of Melbourne, Australia
Prof Sergey Kharkovsky, University of Western Sydney, Australia
Dr Yew-Chin Koay, Vic Roads, Australia
Prof Yew-Chaye Loo, Griffith University, Australia
Prof Aftab Mufti, ISIS Canada, Canada
Dr Peter Runcie, NICTA, Australia
Prof Bijan Samali, University of Western Sydney, Australia
Prof Mark Stewart, University of Newcastle, Australia
Prof David Thambiratnam, Queensland University of Technology, Australia
Prof Brian Uy, University of New South Wales, Australia
Dr Xiaoming Wang, CSIRO, Australia
Prof You-lin Xu, Hong Kong Polytechnic University, Hong Kong

Local Organising Committee

Mr Bob A Barrett, Department of Transport and Main Roads, Queensland
Prof Tommy Chan, Queensland University of Technology, Queensland (Chair)
Mr Craig Cowled, Queensland University of Technology, Queensland
Dr Ulrike Dackermann, University of New South Wales, New South Wales
Prof Hong Guan, Griffith University, Queensland
Dr Peter Ho, Queensland University of Technology, Queensland
Dr Lei Hou, Griffith University, Queensland
Mr Kenneth Jacobs, Department of Transport and Main Roads, Queensland
Prof Jianchun Li, University of Technology Sydney, New South Wales
Dr Jun Li, Curtin University, Western Australia
Dr Saeed Mahini, University of New England, New South Wales (Co-Chair)
Dr Alex Ching-Tai Ng, University of Adelaide, South Australia
Prof Tuan Duc Ngo, University of Melbourne, Victoria
Dr Andy Nguyen, Queensland University of Technology, Queensland
Mr Hung Nguyen, Department of Transport and Main Roads, Queensland
Dr Govinda Pandey, Rockfield Technologies Australia, Queensland
Dr Torill Pape, AECOM, Queensland
Mr Ben Sue, Brisbane City Council, Queensland
Dr Ying Wang, University of Surrey, England
Mr Max Willison, Brisbane City Council, Queensland
A/Prof Xinqun Zhu, Western Sydney University, New South Wales

International Scientific Committee

Prof George Akhras, Canada
Dr Sreenivas Alampalli, USA
Prof Farhad Ansari, USA
Dr Baidar Bakht, Canada
Prof Nemkumar Banthia, Canada
Prof Muhammed Basheer, UK
Prof James Brownjohn, UK
Dr Francisco Carrion-Viramontes, Mexico
Prof Joan Casas, Spain
Prof Tommy Chan, Australia
Prof Genda Chen, USA
Dr J.J. Roger Cheng, Canada
Dr Ulrike Dackermann, Australia
Prof Alessandro De Stefano, Italy
Prof Tuan Duc Ngo, Australia
Prof Branko Glisic, USA
Prof Hong Guan, Australia
Dr Wolfgang Habel, Germany
Dr Ali Hadigheh, Australia
Prof Hong Hao, Australia
Prof Nick Haritos, Australia
Dr Li Hou, Australia
Prof Vistasp Karbhari, USA
Prof Sergey Kharkovsky, Australia
Dr Yew-Chin Koay, Australia
Prof Chan Ghee Koh, Singapore
Dr Eva Lantsoght, Netherlands
Prof Hui Li, China
Prof Jianchun Li, Australia
Dr Jun Li, Australia
Prof Werner Lienhart, Austria
Prof Yew-Chaye Loo, Australia
Prof Dan M. Frangopol, USA
Dr Saeed Mahini, Australia
Prof Urs Meier, Switzerland
Dr Alexis Mendez, USA
Dr John Moore, Australia
Prof Aftab Mufti, Canada
Dr John Newhook, Canada
Dr Alex Ng, Australia
Dr Andy Nguyen, Australia
Prof Yi-Qing Ni, Hong Kong
Prof Toshiyuki Oshima, Japan
Prof Jinping Ou, China
Prof Didem Ozevin, USA
Prof Bijan Samali, Australia
Prof Ian Smith, Switzerland
Prof Mark Stewart, Australia
Prof Paul Sumitro, USA
Dr Dagmar Svecova, Canada
Prof Su Taylor, Northern Ireland
Prof David Thambiratnam, Australia
Dr Douglas Thomson, Canada
Prof Michael Todd, USA
Prof Brian Uy, Australia
Dr Xiaoming Wang, Australia
Dr Herbert Wiggenhauser, Germany
Prof Zhishen Wu, Japan
Prof You-Lin Xu, Hong Kong
Prof Ling Yu, China
Dr Jian Zhang, China
Dr Zhou Zhi, China
Dr Jing Zhou, China
A/Prof Xinqun Zhu, Australia

Sponsors and Supporters

Gold Sponsor

Queensland Government

Silver Sponsors

mainmark

Geomotion AUSTRALIA

WORLD SENSING

Supporters

Brisbane City Council

Queensland Department of Transport and Main Roads

American Society of Civil Engineers (ASCE)

Engineers Australia

The Institution of Structural Engineers

University of New England

Exhibitors

STRUCOMP PTY LTD
Proud marketers of the TechnoLab series of mechanics teaching products
TechnoLab™
www.strucomp.com

BESTECH
Sensors & Teaching Equipment

SAVTek Pty Ltd
Sound and Vibration Technology

Engineers Enterprise
Fibre optic consulting & solutions

NATIONAL INSTRUMENTS™

landsolution
AUSTRALIA
Incorporating Clem Jones Wright

The 8th International Conference on
Structural Health Monitoring of Intelligent Infrastructure
Brisbane, Australia | 5-8 December 2017

Data Science and Engineering for Structural Health Monitoring

H Li, Y. Q. Bao, S. L. Li, J. P. Ou
Key Lab of Structures Dynamic Behavior and Control of the Ministry of Education, Harbin Institute of Technology, Harbin, 150090, China, Email: lihui@hit.edu.cn
Key Lab of Intelligent Disaster Mitigation of the Ministry of Industry and Information Technology, Harbin Institute of Technology, Harbin, 150090, China
School of Civil Engineering, Harbin Institute of Technology, Harbin China, 150090

Abstract

Structural health monitoring (SHM) technology study the damage or even characteristics of disaster evolving of structures by various types of sensors installed on structures and provide scientific support for the structural design, construction, maintenance and operation safety. The significant scientific issues of SHM can be concluded as 'sensing' and 'data'. According to the widely application of SHM on civil infrastructures and accumulation of huge quantities of data, it is well-recognized that data sciences and engineering for SHM is increasingly becoming a world-wide research focus. The issue of 'data' is core target of SHM which includes the data acquisition, transmission and management, data analysis and mining, structural damage identification and structural safety assessment and so on. This paper reviews the recent research development of data sciences and engineering on our research group which includes data acquisition by compressive sensing based on group sparse optimization, computer vison and deep learning methods for data anomaly detection and structure condition assessment, the identification and modeling of spatio-temporal distribution of vehicle loads, and corrosion damage and crack identification by computer vision techniques.Finally, the artificial intelligence in civil engineering is the extension of SHM technique, and will make the civil structure have ability of "see", "sensing", "thinking" ,"talking", "action" and "adapation" (2S+2T+2A).

1. Introduction

Structural health monitoring (SHM) systems for the safety of structures have been widely investigated and installed on many civil infrastructure systems (Li et al. 2015). These systems produce huge quantities of data in each day. In recent years, with the rapid development of information technology, 'data' has been accumulated at an unprecedented speed. 'Big data', 'Artificial Intelligence' and 'Deep Learning' become one of the most important technologies in current. In 2014, Li (2014a, 2014b) proposed the research direction of data science and engineering for SHM, which including data acquisition, data transmission, data management, model updating and damage identification, data analysis and data mining, data modeling, safety evaluation and reliability prediction, data display and graphical representation. This paper presents the recently research progress of the data analysis and modeling, and the safety evaluation in our research group.

2. Data Acquisition

Traditional data acquisition must follow the Shannon-Nyquist thermo, which results in huge data. Vibration signals of most civil infrastructures have sparse characteristic, (i.e., only a few modes contribute to the vibration of the structures). Therefore, the vibration data usually have sparse representation. Additionally, the measured vibration data by the sensors placed on different locations of structure almost has the same sparse structure in the frequency domain. Based on the group sparsity of the structural vibration data, a group sparse optimization algorithm based compressive sensing (CS) for wireless sensors are proposed. The algorithm can be shortly introduced as follows.

The signal matrix $\mathbf{U}$ with considering measurement noise can be represented as

$$\mathbf{U} = \mathbf{\Psi X} + \boldsymbol{\varepsilon} \tag{1}$$

where $\boldsymbol{\varepsilon} \in R^{M\times K}$ is Gaussian noise matrix. Then, the Fourier coefficients matrix $\mathbf{X}$ is recovered by solving the following optimization problem:

$$\min_{\mathbf{X}\in C^{M\times K}} \|\mathbf{X}\|_{2,1} + \frac{\mu}{2}\left\|P_\Omega(\mathbf{\Psi X}) - P_\Omega \mathbf{U}\right\|_2^2 \tag{2}$$

Once the optimal solution $\mathbf{X}_{rec}$ is obtained, the recovered signal matrix is given by

$$\mathbf{U}_{rec} = \mathbf{\Psi X}_{rec} \tag{3}$$

The field test on a suspension bridge with wireless sensors is carried out to illustrate the effectiveness of the proposed approach. The results show that the smaller reconstruction errors can be achieved by using multiple sensors data with the group sparse optimization method than only using single sensor data. Even using only 10% random sampling data, the original data can be reconstructed by the group sparse optimization method with small reconstruction error as shown in Fig. 1. In addition, the modal parameters can also be identified from the reconstruction data with small identification errors (Bao et al. 2017a).

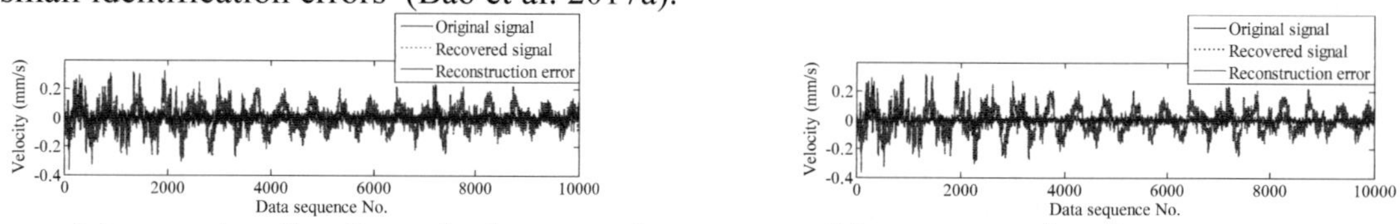

(a) reconstruction from single sensor data (b) reconstruction from group sensors data

Fig. 1. Data reconstruction results from 10% samples

3. Data Driven-based Approaches

3.1 Computer Vision and Deep Learning-based Data Anomaly Detection

Due to harsh environment of civil structures, the data measured by SHM systems are always contaminated by multiple anomalies, which will seriously affect the data analysis results and is one of main barriers to make an automatic warning in real time because it is hard to distinct the anomalies caused from either structure damage or wrong data. Inspired by manual inspection process in real world, a computer vision and deep learning-based data anomaly detection method is proposed (Bao et al. 2017b). The main idea of the approach as shown in Fig. 2, in which the data is firstly conversed by data visualization, then deep neural networks are trained for data anomaly classification. The procedures imitates the biological vision and the way of human logical thinking.

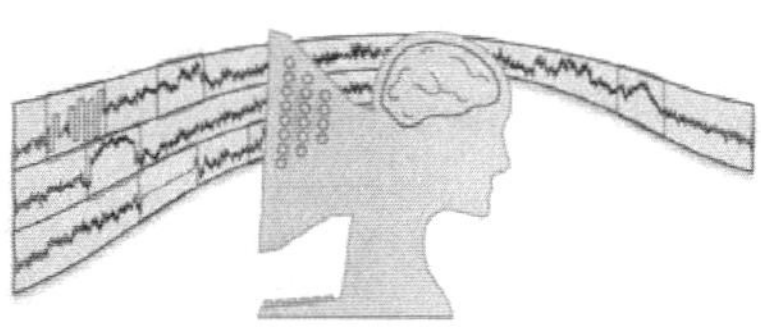

Fig. 2. Main idea of the proposed data anomaly detection method

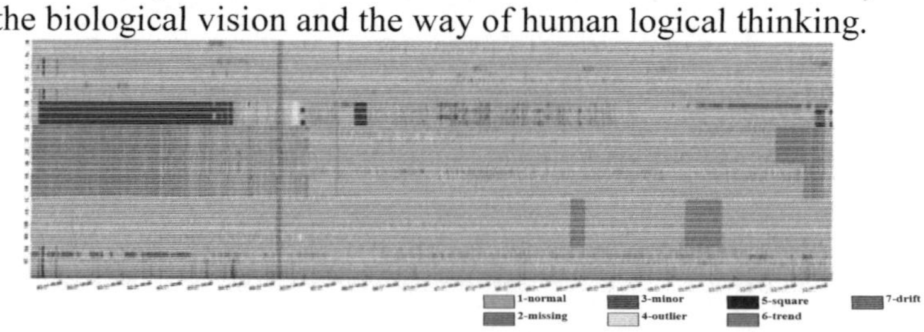

Fig. 3. Data anomaly distribution in 2012: (a) Data anomaly distribution from January to June in 2012; and (b) Data anomaly distribution from July to December in 2012

The acceleration data from the SHM system of a real long-span bridge in China is employed to present the training procedure and validate the performance of the proposed method. The data anomaly distribution in a year is shown in Fig. 3, which shows the multi-pattern anomalies of the data can be automatically detected with high accuracy.

3.2 Condition assessment of stay cables by using pattern recognition algorithms

Stay cables are critical and vulnerable type of structural components in a long-span cable-supported bridge, that suffer from cyclic fatigue and environmental corrosion. The monitoring based condition assessment of cables is of urgent needs of *in-situ* bridges. However, the raw data of cable tension can hardly be used as the condition indicator directly because cable tension is also affected by external loading, environmental effects, etc. Therefore, we proposed pattern recognition approach by using tension ratio indicator which is a new damage indicator that only related to the cable condition (Li et al. 2016).

Sample of several cables are shown in Fig. 4 with the Gaussian Mixture Model (GMM) fitted result. The recognized clusters modeled by mono-Gaussian distribution actually have a clear correspondence with the physical traffic lane: i.e., cluster 1 to 3 refers to traffic lanes 1 to 3, respectively, cluster 5 to 7 refers to traffic lanes 4 to 6, respectively, and cluster 4 refers to noise cluster. GMM parameters of some cable pairs are observed change during the monitoring period that indicate a suspected wire breakage or wire corrosion may occur to these stay cables.

(a) SA03 (b) SA09

Fig. 4 Fitted GMM model of cable Tension Ratio in May, 2006

3.3 Machine learning-based wind engineering

Conventional wind engineering relies heavily on wind tunnel experiment, which however is difficult to simulate the real full-scale structure and the complex wind condition. SHM provides an excellent opportunity to investigate the wind and wind effects on a prototype structure with long-term full-scale measurements. The mechanism and knowledge of the wind engineering is hidden in the big data of measured wind and wind effects. The machine learning-based schemes to discover the knowledge hidden in the large amount of full-scale measured data is proposed. For example, a density-based clustering method, which is an import branch of machine learning, is applied to discover the potential patterns in the wind speed field of the vortex-induced vibrations (VIVs) of a long-span bridge (Li et al., 2017). This clustering algorithm can be described by Eq. (4). The clustering result in Fig. 5 shows that each cluster in the wind speed field corresponds to a mode (frequency) of VIV, revealing the relationship between the wind field and the VIV mode for the prototype bridge.

$$\rho_i = \sum_j \chi(d_{ij} - d_c)$$
$$\delta_i = \min_{j:\rho_j > \rho_i} (d_{ij}) \tag{4}$$
$$d_{ij} = \sqrt{(U^i_{1/4} - U^j_{1/4})^2 + (U^i_{1/2} - U^j_{1/2})^2 + (U^i_{3/4} - U^j_{3/4})^2}$$

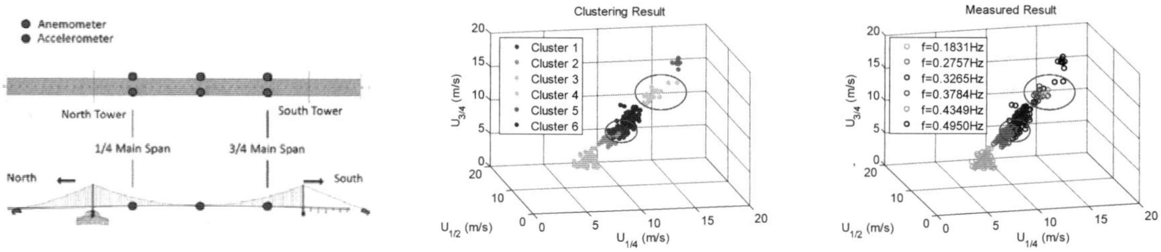

Fig. 5 Clusters hidden in the wind fields represented by the 10-min mean wind speeds at 1/4, 1/2 and 3/4 spans are found corresponding to modes of VIV.

4. Model-based Approaches

4.1 Identification of spatio-temporal distribution of vehicle loads on bridges by computer-vision technology

Identification of spatio-temporal distribution of vehicle loads is very important for understanding the exact loading conditions and behaviors of long-span bridges. Traditional moving loads identification methods for bridges are mainly based on simplified vehicle–bridge interaction models and parameter identification algorithms, which are not suitable for long-span bridges. Recent years, both WIMs and cameras have been implemented on bridges to monitor traffic conditions. The cameras can provide the vehicle status over a bridge. Inspired by this information, a computer vision-based identification approach for spatio-temporal distribution of vehicle loads over a long-span bridge is proposed (Chen et al. 2016).

The overall identification procedure of the proposed method can be summarized as follows: a background subtraction technique is used to detect whether there is a car in the detection area. Once a car is detected in the region above at the WIM location, an appropriate sub-image called template is sampled from the rear part of the car and the weight of the car measured by the WIM can be obtained based on the car pass time. The trajectory of this car can then be tracked by the template match and particle filter techniques from videos taken by cameras in the range of the bridge (see Fig. 6). Representative identified spatial samples at different instantaneous time on a real long-span bridge are presented in Fig. 7.

Fig. 6 Trajectory tracking

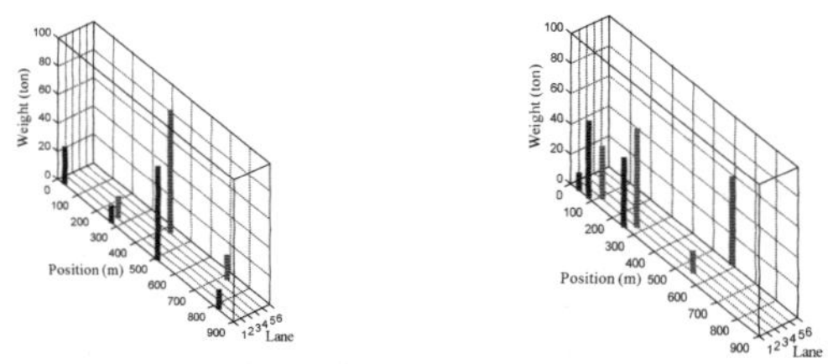

Fig. 7 Representative identified spatial distributions of vehicle loads on a long-span bridge

4.2 Markov random field model of vehicles on bridges using graphic modeling algorithm

Stochastic models of vehicle loads are of great importance not only in the design of new bridges but also in the safety assessment of existing bridges. Traditional stochastic models of vehicle loads are incapable to quantitatively model the complex spatial distribution of vehicles on the whole bridge considering multiple vehicles and multiple lanes. In the field of statistics or machine learning, graphical models are powerful tools for constructing large-scale probabilistic models for massive random variables with complex dependence patterns. Inspired by this, a graphical model-based approach is proposed to model the spatial distribution of heavy vehicle loads on long-span bridges. An example with potentials estimated from monitored samples of a cable-stayed bridge is employed to illustrate the procedure of the proposed method, corresponding graphical model and simulated random samples by this model are presented in Fig. 8 and Fig. 9, respectively. (Chen et al. 2017)

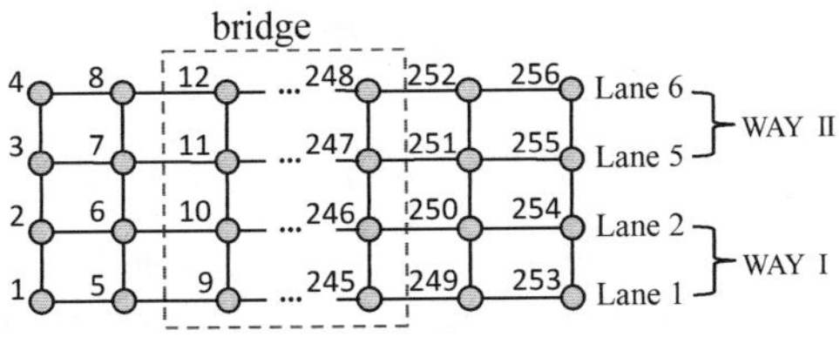

Fig. 8 Undirected graphical model

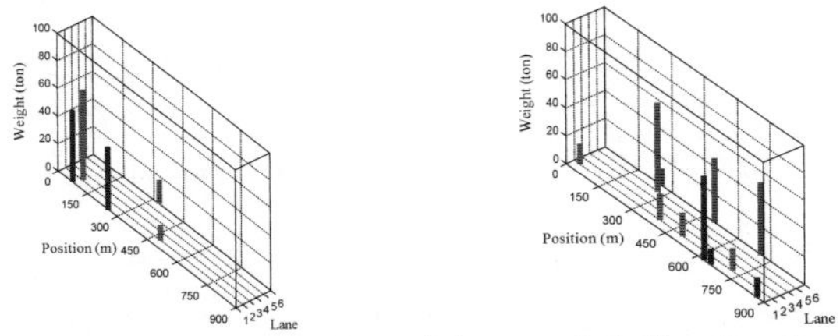

Fig. 9 Simulated samples of the spatial distribution of heavy vehicle loads based on the proposed model

4.3 Corrosion damage & modeling of steel wires in cables by computer vision technology

High-strength steel wires are critical components of cables in long-span bridges and are vulnerable under normal operational conditions. Bridge inspections reveal that for old cable-supported bridges, steel wires in the cables are heavily corroded and many of them are even broken. This part presents an approach for pit region identification and modelling of galvanized high-strength steel wires based on an image processing algorithm (Xu et al. 2016).

To extract the pit region from the background surface, the optimal entropy threshold method is employed to obtain the threshold value.

$$p_i = \frac{n_i}{n}, P_i = \sum_{k=1}^{i} p_k, H_P(t) = -\sum_{i=0}^{t} \frac{p_i}{P_t} \ln \frac{p_i}{P_t}, H_B(t) = -\sum_{i=t+1}^{L-1} \frac{p_i}{1-P_t} \ln \frac{p_i}{1-P_t}$$
$$H(t) = H_P(t) + H_B(t), T = \arg\max(H(t)) \quad (5)$$

The p-q order moment variable μ_{pq} is employed for pit corrosion shape approximation, and it can be observed that there are five depth modes according to depth distribution, shown in Fig. 10.

$$\begin{cases} \mu_{pq} = \sum\sum (x-\bar{x})^p (y-\bar{y})^q g(x,y)\Delta x \Delta y \\ a = \left(\dfrac{\mu_{20} + \mu_{02} + [(\mu_{20} - \mu_{02})^2 + 4\mu_{11}2]^{1/2}}{\mu_{00}/2}\right)^{1/2} \\ b = \left(\dfrac{\mu_{20} + \mu_{02} - [(\mu_{20} - \mu_{02})^2 + 4\mu_{11}2]^{1/2}}{\mu_{00}/2}\right)^{1/2} \\ \varphi = (1/2)\tan^{-1}[2\mu_{11}/(\mu_{20} - \mu_{02})] \end{cases} \quad (6)$$

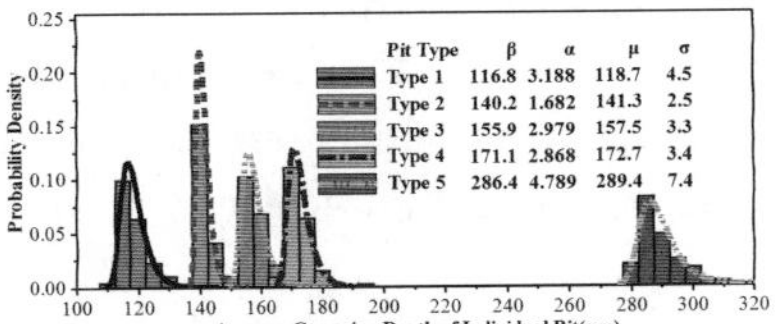

Fig. 10 Fitted Gumbel distribution of five pit groups

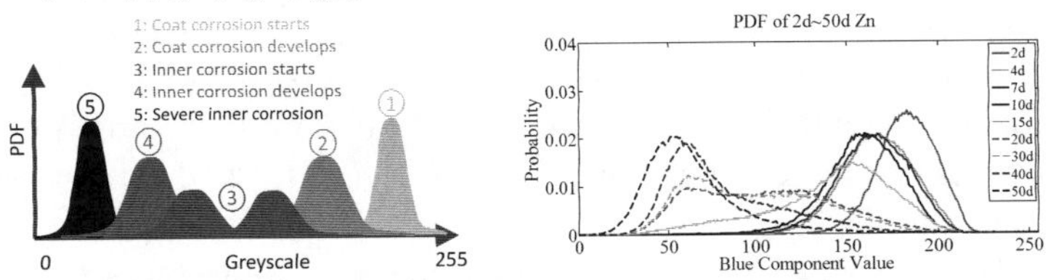

Fig. 11 Shape change of histogram plots during the corrosion process

Corroded wires under different periods are generated, and Fig. 11 shows the PDF plots under different periods. Joint Gaussian distribution is applied. The corrosion evaluation criterion is fitted by a time-variant model of 1st location parameter, and fatigue tests are conducted under different periods, and S-N curves are shown in Fig. 12 with fitted time-dependent parameters.

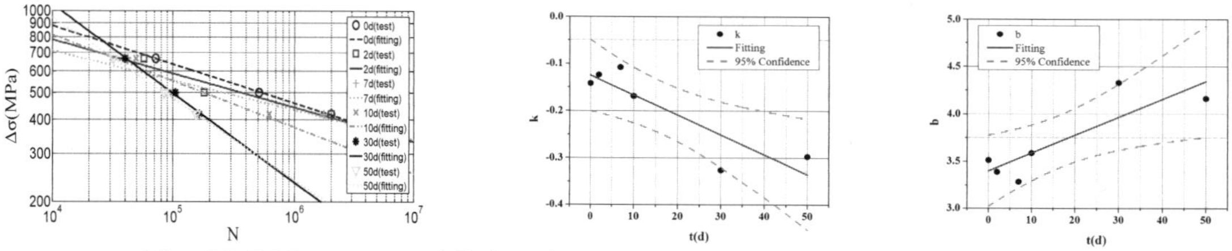

Fig. 12 S-N curves and fitting time-dependent model of S-N parameters

4.4 Crack image identification by deep Boltzmann Machine learning algorithm

The steel box girders of bridges suffer from fatigue damage caused by the initial flaws in the welding parts and living loads. These cracks will decrease the structural reliability and shorten the operational life of long-span bridges. For fatigue cracks, a conventional method is visual inspection followed by evaluation of the bridge condition based on experts' experience. To address the limitation, a framework based on deep Restricted Boltzmann Machine (RBM) learning techniques for crack identification with a complicated background in a steel box girder of bridges is proposed (Xu et al. 2017).

Sets of images are obtained by equally dividing original images into elements with pixel size of 24×24. A state representation vector is artificially labeled to every image element. With the 576 units as the input layer stretched from image elements, a three-layer Restricted Boltzmann Machine with 500, 500, and 2000 hidden units is trained in the deep learning network. The acquired test error reaches 4.8% after 1000 iterations; verifying the accuracy of the trained model, shown in Fig. 13. The results shown in Fig. 14 indicate that the proposed method is effective in identifying fatigue cracks from images containing the complicated background in a steel box girder of bridges.

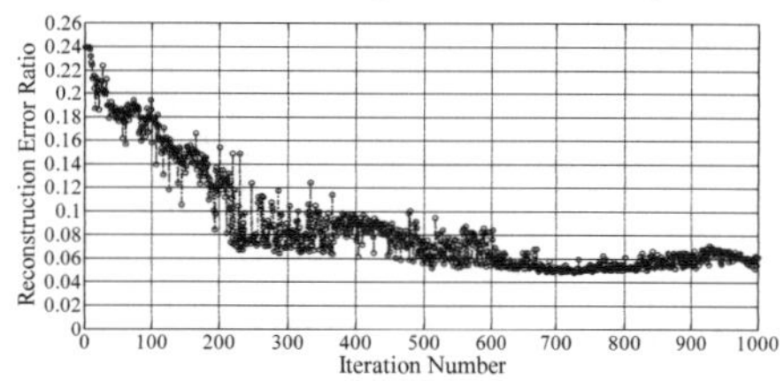

Fig. 13 Reconstruction error ratio of test samples

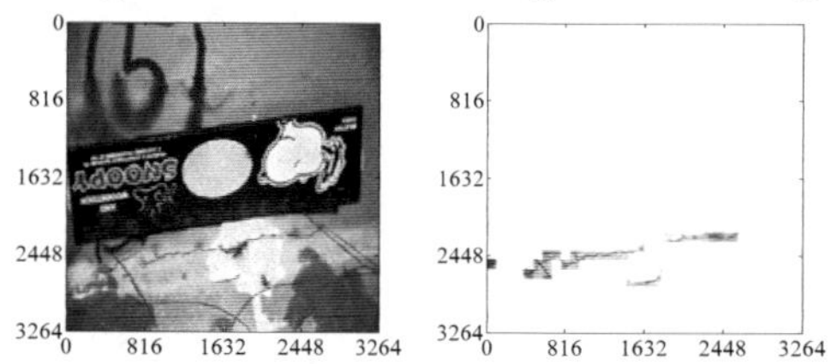

Fig. 14 Crack identification results

Conclusions

This paper presents the recently progress in our research group about the data science and engineering for SHM. Predictably, the big data of SHM in combination with artificial intelligence techniques will greatly change the traditional data analysis methods for SHM. The deep learning methods, reinforcement learning methods and the computer vision techniques etc. will provide a new perspective to understand the data of SHM and further reveal the performance degeneration of structure.

The artificial intelligence will take revolution for all aspects of civil engineering. The deep learning and reinforcement learning methods and virtual reality can be used in city planning, transportation planning, structure design and display of real world; deep learning and reinforcement learning methods, virtual reality, robotic and 3D print will make the construction and maintenance of civil structure more intelligence and automatics. All these make big revolution in theory and technology and make the structure have the ability of "see" (computer vision), "sensing" (sensing technology), "thinking" (brain) , "talk" (semantics), "act" and "adaption" (control) (2S+2T+2A), and labor free.

Acknowledgement

This research was supported by grants from the National Natural Science Foundation of China (Grant No. 51161120359, 51378154 and 51678203) the National Basic Research Program of China (Grant No.2013CB036305).

References

Li, H, Ou, J, Zhang, X, Pei, M, & Li, N 2015, 'Research and practice of health monitoring for long-span bridges in the mainland of China', *Smart Structures and Systems*, vol. 15, no. 3, pp. 555-576.

Li, H 2014a, 'SHM Data Science and Engineering', *5th Asia-Pacific Workshop on Structural Health Monitoring*, Shenzhen, China, Dec. 4-5, 2014

Li, H 2014b, 'Next Generation of SHM: Sense Everywhere and Understand Everything', *10th International Symposium on Structural Engineering for Chinese Experts*, Hefei, China, Sept. 18-20, 2014

Bao, Y, Shi, ZQ, Wang X & Li H 2017a, 'Compressive sensing of wireless sensors based on group sparse optimization for structural health monitoring', *Structural Health Monitoring*, 2017, (In press)

Bao, Y, Tang, ZY, Li, H & Zhang, Y 2017b, 'Computer Vision and Deep Learning-based Data Anomaly Detection Method for Structural Health Monitoring', *Structural Health Monitoring*, 2017 (Under review).

Li, S, Wei, S, Bao, Y & Li, H 2016. 'Condition assessment of cables by pattern recognition of vehicle-induced cable tension under normal operation', *Engineering Structures*. (Under review).

Li, SW, Laima, S & Li H 2017, 'Cluster analysis of winds and wind-induced vibrations on a long-span bridge based on long-term field monitoring data', *Engineering Structures*, vol. 138, pp. 245-259.

Chen, Z, Li, H, Bao, Y, Li, N & Jin Y. 2016, 'Identification of spatio-temporal distribution of vehicle loads on long-span bridges using computer vision technology', *Structural Control and Health Monitoring* vol. 23, no. 3, pp. 517-534.

Chen, Z, Bao, Y & Li, H 2017, 'Modeling approach of the spatial distribution of heavy vehicle loads on long-span bridges based on undirected graphical model', *Structural Control and Health Monitoring* (Under Review).

Xu, Y, Li, H, Li, S, Guan, X & Lan, C 2016, '3-D Modelling and Statistical Properties of Surface Pits of Corroded Wire Based on Image Processing Technique', *Corrosion Science* vol. 111, pp. 275-287.

Xu, Y, Li, S, Zhang, D, Jin, Y, Zhang, F, Li, N & Li, H 2017, 'Identification Framework for Cracks on a Steel Structure Surface by a Restricted Boltzmann Machines Algorithm Based on Consumer-Grade Camera Images', *Structural Control and Health Monitoring*, 2017 (Under review).

The 8th International Conference on
Structural Health Monitoring of Intelligent Infrastructure
Brisbane, Australia | 5-8 December 2017

Innovative Data Analysis Techniques for Structural Health Monitoring

Hong Hao[1], Jun Li[1]
[1]Centre for Infrastructural Monitoring and Protection, School of Civil and Mechanical Engineering, Curtin University, Perth, Australia.
Email: hong.hao@curtin.edu.au

Abstract

Many sources of uncertainties, which could be introduced into the structure during their construction and operational stages, make the reliable structural health monitoring (SHM) difficult. Environmental effect, finite element modelling error, methodology error and noise in the measurement data are the most significant types of uncertainties. The accuracy of structural condition assessment and the reliability of SHM results may be significantly affected. Besides, different data analysis techniques and damage indices may result in different structural damage identification results, even from the same measured data. The primary challenge of SHM is therefore development of methodologies that are sensitive to minor structural condition changes but insensitive to the changes by uncertain effects. This paper discusses and introduces a few signal processing techniques for SHM, including both the global vibration and local wave propagation based methods, to achieve a better accuracy and reliability of structural damage identification. The damage indices based on these new signal processing techniques are less susceptible to the uncertainties, but more sensitive to minor structural damage. These include: 1) Vibration Phase Space Topology. A damage index named change of phase space topology derived from the measured vibration responses is defined to detect the location of structural damage; 2) Phase Trajectory Change. The phase trajectories of multi-type vibration responses are obtained from a bridge under moving loads, and a damage index is defined as the separated distance between the trajectories of undamaged and damaged structures to detect the damage location; 3) Chaotic system based technique. A novel technique based on analyzing the responses of a nonlinear oscillator, i.e. Duffing-Holmes system, with the recorded ultrasonic guided wave signal as an added input is proposed for the detection of minor structural damage. Experimental verifications on various structures have been conducted to validate the accuracy and reliability of the proposed approaches to analyse the measurement data and detect the minor structural damage. The results demonstrate that these approaches are sensitive and accurate to detect the structural damage, but not sensitive to uncertainty effects.

1. Introduction

Vibration based Structural Health Monitoring (SHM) techniques have been developed to identify the structural conditions, assess and predict the structural performance under operational conditions. These techniques usually employ the measured structural vibration responses at critical locations and/or the measurements of the environmental and loading effects to detect the anomalies, local damage and continuous deterioration in structures. The general objectives of SHM are to: (1) Validate design assumptions and monitor structural performance; (2) Identify the possible damage at an early stage to ensure structural safety; (3) Provide rapid condition and safety assessment after disasters and extreme events; (4) Evaluate the current structural condition, load-carrying capacity and predict the remaining service life; (5) Support infrastructure asset management regarding the optimum inspection, rehabilitation and maintenance planning; and (6) Obtain massive amounts of in-situ data for improving the design specifications and guidelines (Ko and Ni 2005, Frangopol *et al.* 2008, Li and Hao 2016). Structural damage is usually considered as changes in the material and/or geometric properties, changes in the boundary conditions and system connectivity, which adversely affect the expected structural performance. In terms of structural condition monitoring and damage detection, periodical vibration measurements or long term monitoring data could be used for the extraction of damage-sensitive indicators to perform structural health checks and/or the quantitative assessment of structural damages.

Many sources of uncertainties, which could be introduced into the structure during their construction and operational stages, make the reliable structural condition assessment difficult. Environmental effect, finite element modelling error, methodology error and noise in the measurement data are the most significant types of uncertainties. If the effect of these uncertainties on structural vibration properties or the defined damage indicators is larger than or comparable to the effect of structural damage, false identification results may be obtained and true structural damage may not be confidentially identified. This paper presents and discusses several innovative signal processing techniques for SHM, particularly in order to achieve a better accuracy and reliability of structural damage detection. The damage indices based on these new signal processing techniques are less susceptible to the uncertainties, but more sensitive to minor structural damage.

2. Methodologies

This section will briefly review the theoretical background of the following data analysis techniques and the defined damage indices for structural damage detection. These techniques include: 1) Vibration Phase Space Topology. A damage index named change of phase space topology derived from the measured vibration responses is defined to detect the location of structural damage (Nie et al. 2013); 2) Phase Trajectory Change. The phase trajectories of multi-type vibration responses are obtained from a bridge under moving loads, and a damage index is defined as the separated distance between the trajectories of undamaged and damaged structures to detect the damage location (Zhang et al. 2017); 3) Chaotic system based technique (Zhang et al. in press). A novel technique based on analyzing the responses of a nonlinear oscillator, i.e. Duffing-Holmes system, with the recorded guided wave signal as an added input is developed for the detection of minor structural damage. Damage index is defined based on Lyapunov exponents.

2.1 Vibration Phase Space Topology

A phase space represents all the possible states of a dynamic system. Each possible state of the system corresponds to a unique point in the phase space. For mechanical systems, the phase space could include all the vibrations, i.e. displacement, velocity and acceleration. The time domain response series may reflect the changes in the physical parameters of a dynamic system. However, this information buried in the time series is not straightforward to be extracted from directly analysing the time history. The reconstructed phase space is more sensitive to detect the change in structural parameters, and could be used as a good candidate for structural damage identification.

The phase space reconstruction is to reconstruct the dynamic behaviour of a system from the measured vibration responses. The delay coordinate approach is used to reconstruct the m dimensional phase space by using the measured time responses with a proper time delay T. Consequently, a point at time n can be expressed as

$$X(n) = [x(n), x(n+T), \cdots, x(n+(m-1)T)] \tag{1}$$

where the delayed time series $x(n)$ is often named as the pseudostate vector.

A successful reconstruction requires the choice of both delay T and embedding dimension m. T must be large enough because if the delay is too small the resulting vector will be nearly the same carrying a great deal of redundant information. Methods for choosing an appropriate delay have focused on using the autocorrelation function to deduce the time at which a vector is least correlated with itself. More specifically, the first zero, or crossing of the autocorrelation function has often been used to select the delay. The choice of m is equally critical, given that too small a dimension will result in trajectories not unique. For example, if the system is embedded in one dimension (described by one variable), the phase space representation would be a line. On the other hand, too large an m will result in additional computational cost. The previous study proved that the False Nearest Neighbours approach (FNN) is a good way to proceed when choosing the embedding dimension m (Tong 1990). With appropriate T and m, the dynamics of the system can be unfolded well by a phase space.

To locate and quantify damage, an index, named *Changes of Phase Space Topology (CPST)*, is proposed. The phase trajectory as defined by Equation (1) of the responses corresponding to the undamaged structure is considered as the base trajectory. Similarly, when the structure suffers damage, a new set of data

$$Y(n)=[y(n),y(n+T),\cdots,y(n+(m-1)T)] \tag{2}$$

are collected in the same way. The object of the feature-based damage identification is to extract some metric capable of discriminating between the base trajectory (undamaged) set X and damaged dataset Y. For example, when the dimension m is selected to be 2, the coordinator of pristine phase space from time series can be described as

$$\begin{cases} X(1)=[x(1),x(2),\cdots,x(N-k)] \\ X(2)=[x(k),x(k+1),\cdots,x(N)] \end{cases} \tag{3}$$

where N is the length of time series, $k = Tf_s$ is the number of point sampled in the delay T, and f_s is the sampling frequency. A fiducial point on the trajectory corresponding to the damaged structure is selected with time index r as $Y(r)$. The nearest p neighbours of this point are then selected on the base trajectory (pristine) as

$$X(p_j): \min\|X(p_j)-Y(r)\|, \quad j=1,2,\cdots,p \tag{4}$$

where p denotes the nearest neighbours to fiducial point and the $\|\cdot\|$ operator computes the Euclidean norm. Note the time indices of the neighbours need not have any relation in time to the fiducial point as the search is made based on the Euclidean norm only. The future value of $Y(r)$ at time steps s into the future may then be forecasted using the baseline data as the model via

$$\hat{Y}(r+s)=\frac{1}{p}\sum_{j=1}^{p}X(p_j+s), \quad j=1,2,\cdots,p \tag{5}$$

where the predicted value $\hat{Y}$ is the average of the values at the neighbourhoods of time step s. Then, the damage index CPST is defined as

$$CPST = \frac{1}{p}\left\|\hat{Y}(r+s) - Y(r+s)\right\| \quad (6)$$

where $Y(r+s)$ is the averaged values for fiducial point defined in Equation (5). This calculation will be performed N-k times because the reconstructed phase trajectory has N-k points. If a structure is damaged, the value of the CPST will increase.

2.2 Phase Trajectory Change

A multi-dimensional phase space will be constructed, for example, a three dimensional one with the three axes corresponding to the displacement, velocity and acceleration respectively. The phase space trajectory can be plotted directly with the displacement, velocity and acceleration responses measured at the mid-span of a bridge under the moving load, which are the function of time t,

$$u=u(t), \quad v=v(t), \quad a=a(t) \quad (7)$$

(u_0, v_0, a_0) and (u_d, v_d, a_d) denote the spatial coordinates on the phase trajectory from the undamaged and damaged bridge models, respectively. The damage will induce the local structural stiffness change, and the distance between the corresponding points on those two trajectories of undamaged and damaged bridges subjected to a moving load will also change.

For a bridge structure subjected to a moving load, the time axis could be normalized as the location of the moving load by $x=ct$, where c and t are the travelling speed and time, respectively. In a general case, the location of the moving load is often expressed as the normalized location by $x'=x/L$, in which L is the length of the beam. The damage index is defined as the distance between the spatial coordinates of the normalized moving load locations from the undamaged and damaged phase trajectories

$$DI = \sqrt{\left(u_d^{x'} - u_0^{x'}\right)^2 + \left(v_d^{x'} - v_0^{x'}\right)^2 + \left(a_d^{x'} - a_0^{x'}\right)^2} \quad (8)$$

where $\left(u_d^{x'}, v_d^{x'}, a_d^{x'}\right)$ and $\left(u_0^{x'}, v_0^{x'}, a_0^{x'}\right)$ denote the spatial coordinates on the damaged and undamaged phase space trajectories of the normalized locations of the moving load. It is noted that by using the normalized locations of the moving load on the bridge to define the damage index, the proposed approach can be used for scenarios with different travelling speeds of the moving loads before and after damage.
Multiple types of vibration responses are used to construct the multi-dimensional vibration phase space and plot the trajectory, such as a three dimensional phase space "displacement-velocity-acceleration", or a two dimensional phase space "displacement-acceleration"

$$DI = \sqrt{\left(u_d^{x'} - u_0^{x'}\right)^2 + \left(a_d^{x'} - a_0^{x'}\right)^2} \quad (9)$$

2.3 Chaotic system based technique

Duffing-Holmes equation is a nonlinear equation describing an oscillator. It represents the motion of a damped oscillator and is an example of a dynamic system that exhibits the chaotic behavior. In mathematics, it is a second order differential equation (Jiang et al. 2016)

$$\ddot{x}+\delta\dot{x}-\alpha x^{k_1}+\beta x^{k_2}=\mathrm{F}\cos\omega t \tag{10}$$

in which δ is the damping coefficient of the system, $\left(-\alpha x^{k_1}+\beta x^{k_2}\right)$ is the nonlinear term, F is the amplitude of the periodic driving force and ω is the circular frequency of the driving force.
In most previous studies, the parameters of the nonlinear oscillator are adjusted as a critical state for state transition. The very weak signal is taken as the input into the system, which leads to the significant change in the state of the system for signal detection. The other merit of this method is the immunity to the strong random noise. The Duffing equation is becoming a very important method for detecting a weak signal with a strong noise effect.
Without considering the noise, when the guided wave signal is added to the input of the nonlinear system, the mathematical model of the Duffing-Holmes nonlinear oscillator is expressed as

$$\ddot{x}+\delta\dot{x}-x^3+x^5=\mathrm{F}\cos\omega t+A\left(1-\cos\frac{\omega_c t}{n}\right)\sin\omega_c t \tag{11}$$

In this case, the nonlinear term $\left(-x^3+x^5\right)$ is used. When $\omega_c=\omega$, after the triangular transformation, Eq. (11) can be rewritten as

$$\ddot{x}+\delta\dot{x}-x^3+x^5=\mathrm{F}'\cos\left(\omega_c t-\varphi\right) \tag{12}$$

in which $\mathrm{F}'=\sqrt{\mathrm{F}^2+A^2\left(1-\cos\frac{\omega_c t}{n}\right)^2}$ and $\tan\varphi=\frac{A}{\mathrm{F}}\left(1-\cos\frac{\omega_c t}{n}\right)$. It is difficult to have the analytical solution of such a nonlinear system. When considering the noise effect, Eq. (11) can be further written as

$$\ddot{x}+\delta\dot{x}-x^3+x^5=\mathrm{F}\cos\omega t+A\left(1-\cos\frac{\omega_c t}{n}\right)\sin\omega_c t+\sigma e(t) \tag{13}$$

Eq. (13) is the mathematical model of the nonlinear oscillator-based Ultrasonic Guided Wave (UGW) signal inspection system. The parameters of the above Duffing-Holmes nonlinear oscillator, such as δ, ω and F, need to be selected and adjusted carefully to detect the weak guided wave signal, such as a real measured reflected signal from a minor damage in structures.
Lyapunov exponent of a dynamical system is a quantity that characterizes the rate of separation of infinitesimally close trajectories. For discrete time system, Lyapunov exponent can be calculated as (Wu and Baleanu 2015)

$$\lambda\left(x_0\right)=\lim_{n\to\infty}\frac{1}{n}\sum_{i=0}^{n-1}\ln\left|\frac{\mathrm{d}f\left(x_i\right)}{\mathrm{d}x_i}\right| \tag{14}$$

The sign of Lyapunov exponents can be used to identify the state change of the system, and hence identify the weak signal in the input.

3. Experimental Investigations

In this section, experimental investigations will be presented to verify the performance and effectiveness of the above-mentioned three approaches.

3.1 Vibration Phase Space Topology

A two-span continuous reinforced concrete slab was fabricated in the laboratory for validating the presented damage detection approach based on vibration phase space topology. The dimension of the slab is designed as length l=6400mm, width w=800mm and height h=100 mm. The length of each span is 3m with 200 mm over hang at each end, as shown in Figure 1. The slab was constructed using commercially supplied grade 32 concrete, with standard cylinder tests confirming an average concrete compressive capacity of 40.2MPa after 28 days. The slab was reinforced using 11N6 reinforcements in both the positive and negative bending regions. The slab was statically loaded until failure in 12 loading steps, as shown in Figure 1. After each loading increment listed in Table 1, the damage to the slab was recorded and vibration tests performed after the static load was removed. Initially, only the first span was loaded, with small cracks appearing in the center of the span under a load of 6kN. Further loading of the first span resulted in the cracks increasing in number and severity. After extensively cracking the first span, the second span was incrementally loaded. Cracks appeared in the second span as well as at the central support under a load of 6kN. Further loading of the second span increased the damage; existing cracks widened and additional cracks formed. Ultimate failure of the slab occurred when the reinforcing steel over the central support snapped up on reloading after increment 12.

The vibration tests of the undamaged slab were also carried out to get the baseline information of the slab. Figure 2 shows the sensor placements and positions. Three sets were defined, namely Set 1 (sensors 1–9), Set 2 (sensors 10–18) and Set 3 (sensors 19–27).

Table 1 Load increments in the slab loading test

Load location	Loading increment (kN)												
	0	1	2	3	4	5	6	7	8	9	10	11	12
Load span 1 (kN)	0	3	6	12	18	18	18	18	18	25	32	35	38
Load span 2 (kN)	0	0	0	0	0	3	6	12	18	25	26	35	38

Figure 1 Experimental setup for the slab loading test

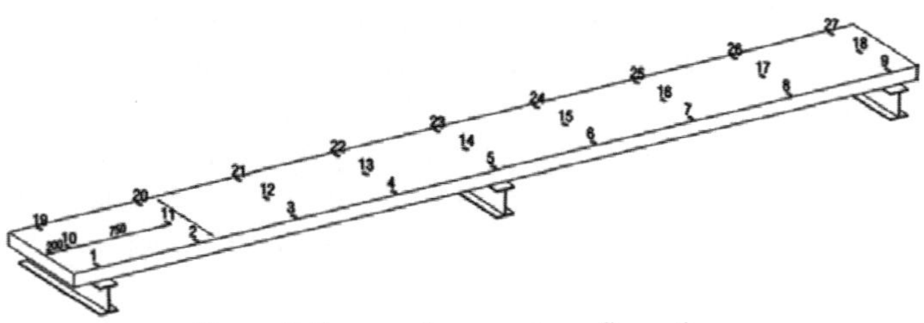

Figure 2 Sensor placement configurations

The time delay is taken as $T=63$ based on calculating the autocorrelation function of the measured responses, and the proper embedding dimension m is chosen to be 3. Figure 3 shows the typical CPST values calculated from the reconstructed phase space of dynamic responses with three embedding dimensions at three vibration measurement sets along the slab length. Except the measurement points at the supports of the slab, CPST damage values of the measured dynamic responses at all the other measurement locations increase with the load or slab damage. Larger CPST values are observed in the middle of these two spans with the loading increments, which is consistent with the observations of more cracks in the center of two spans. The results indicate that CPST damage index is very sensitive to structural damage.

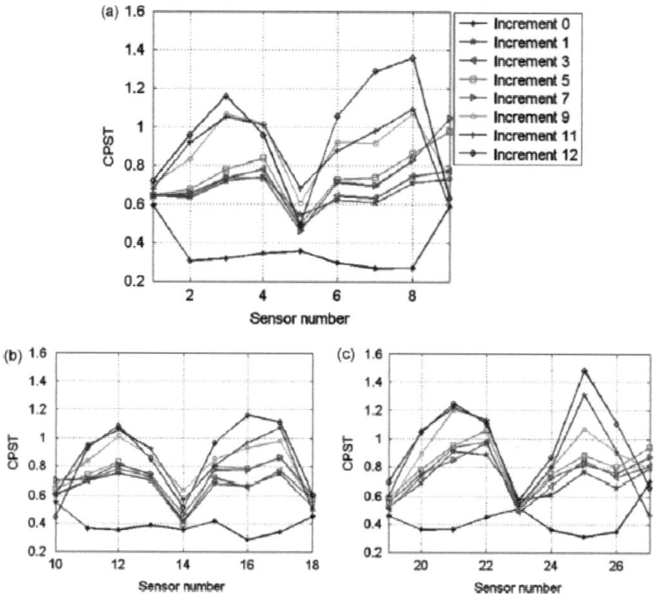

Figure 3 CPST damage index values at each loading increment: (a) Set 1; (b) Set 2 and (c) Set 3

3.2 Phase Trajectory Change of Multi-type Measurements

Figure 4 shows a composite bridge model constructed with a reinforced concrete slab supported on two steel I-type girders. Thirty-two bolts screwing into metric nuts are used to link the slab and girder together. The nuts are welded onto the reinforcement bar in the slab before pouring. Each girder has 16 connectors with an equal space of 200mm and there are 32 in total in the bridge model, which are denoted as SC1~SC16 and SC1'~SC16'. The bridge model is placed on two steel frames which are fixed to the laboratory strong floor. Figure 4 shows the experimental bridge-vehicle system. The vehicle model is simplified as a steel beam carrying two concrete blocks. The distance between the front axle and rear axle of the vehicle is 1m. A roller system is designed and the crane is used to pull the vehicle to travel on the top of the bridge model with a constant speed as the pulling force of the crane is steady and stable. Three transverse connecting plates are located at the two ends and the centre of the track to ensure the stability and rigidity of the track. Subjected to the test conditions, the velocity response is not measured in the experimental tests, and the displacement and acceleration responses are used for the analysis. In this study, only a displacement sensor and an accelerometer are used, and their locations are marked in Figure 4(c). The sampling rate is set as 2000 Hz.

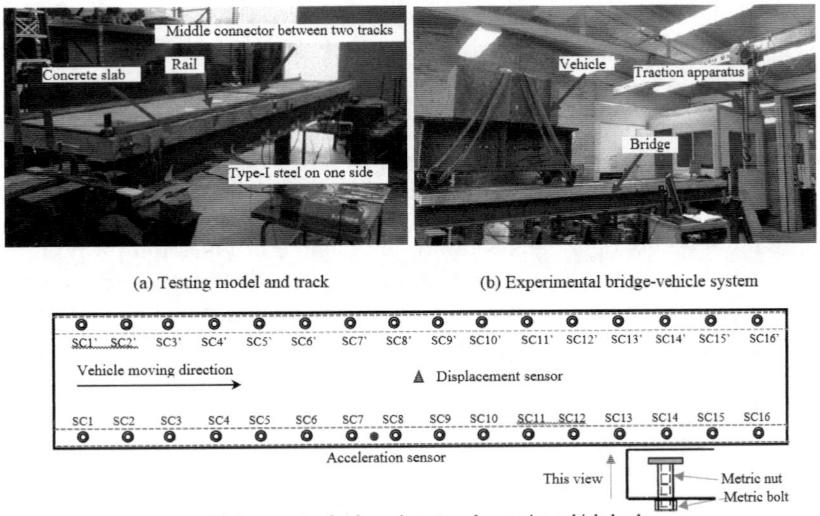

(a) Testing model and track (b) Experimental bridge-vehicle system

(c) Sensor setup for dynamic tests under moving vehicle load

Figure 4 Experimental setup of a bridge-vehicle system

When all the connectors are tight into the nut, the bridge is defined as the undamaged state. If a specific connector is removed from the nut, it indicates the failure of the particular shear connector which also shows that the bridge is under the damaged state. A single damage with shear connectors SC15 and SC15' removed is used as an example to demonstrate the method. The measured responses are pre-processed with a low pass filter with the cut-off frequency of 20Hz and then used for reconstructing the phase space and calculate the damage index.

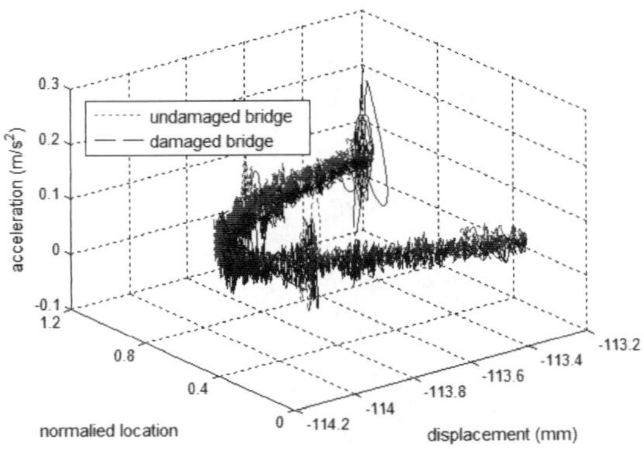

Figure 5 Comparison of the phase trajectories between the undamaged and damaged states

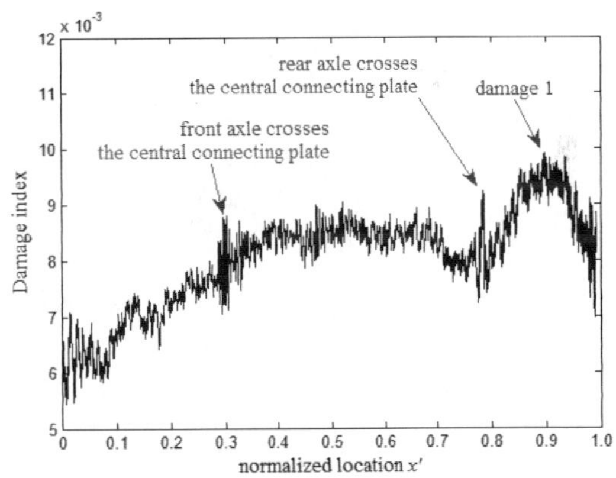

Figure 6 Damage detection results in the bridge-vehicle system

Figure 5 shows the comparison of the phase trajectories between the undamaged and damaged states. It is noted that clear difference can be observed from the undamaged and damaged phase trajectories, indicating the sensitivity of the phase trajectory on changes of structural conditions, though significant uncertainties are observed in the testing conditions, i.e. uncertain moving speed, roughness on the track and measurement noise. Since only the displacement and acceleration responses are measured, a two-dimensional phase trajectory is reconstructed and used to calculate the damage index. A Smooth Moving Average processing scheme with 250 points is used to reduce the noise effect and improve the performance of damage detection. The damage detection results for those two damage scenarios are shown in Figure 6. With the axle distance of the front and rear axles of the vehicle equal to 1m, it can be observed from Figure 4 that only the front axle will pass along the region of removed shear connectors. The highest damage index as shown in Figure 6 indicates the damage location when the front axle moves on the top of damage. It is roughly located at the normalized location 0.91, which matches well with the true introduced damaged location at 2.8m/3m=0.93, indicating that the damage can be identified effectively. The other two peak damage index values at the normalized locations of 0.30 and 0.78 correspond to the instants when the front and rear axles impact the track at the central transverse plate as described above. It should be noted that removing the nuts only causes minimum change in vibration properties of the system, therefore the conventional vibration-based approaches could not reliably detect such damages.

3.3 Detection of Minor Damage with Nonlinear Oscillator

Experimental studies on a 3m long steel pipe with minor damages are conducted to demonstrate the effectiveness and capability of this approach. The specimen of the steel pipe has the inner radius of 50.75mm and wall thickness 2.32mm. It is simply supported with two wooden supports. The experimental setup is shown in Figure 7. The generation and reception of the single mode UGW depend primarily on the transducer's spatial distribution. The testing system used 16 PZT patches on the pipe, which are evenly spaced at the pipe surface in the circumferential direction at 25mm from the excitation end. This is a pulse-echo configuration to detect the defect based on the recorded echoes. The locations of generating and receiving the guided waves are almost at the same location. With such placement for pulse-echo guided wave methods, the defect echo can be clearly distinguished from the incident wave and the end echo. 16 patches are connected in parallel and the superposed signal from the 16 patches is the final tested signal.

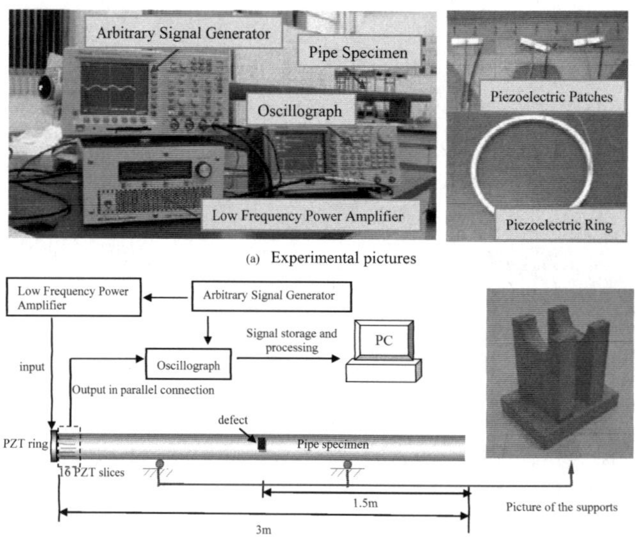

(a) Experimental pictures

(b) Schematic diagram of experiment

Figure 7 Experimental setup for the damage detection in the pipe structure

The defect introduced by a saw cut is at 1.5m from the excitation end. Five cases listed in Table 2 are considered in the tests. In experiment tests, a UGW signal with 70kHz-center frequency and 10 cycles will be generated by a signal generator, which is amplified by a power amplifier and then used to generate the guided wave in the pipe by the installed PZT ring. Both the incident and reflected wave signals are received by the same 16 PZT patches and recorded by the oscillograph.

Table 2 The crack parameters in the experimental studies

Case No.		1	2	3	4	5
Defect size	in circumference (θ)	-	π/8	π/8	π/4	π/3
	in radius (a)	-	1mm	2mm	2mm	2mm
Cross section reduction ratio		0	3%	6%	12%	20%

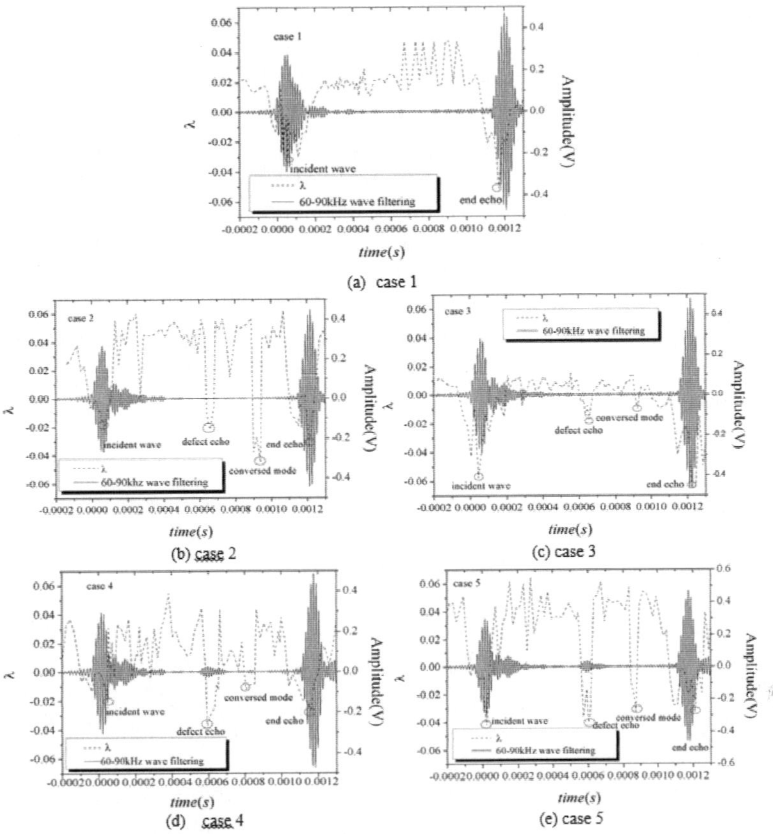

Figure 8 Damage detection with Lyapunove exponents in different damaged cases

The weak reflected wave from a small defect makes it very difficult to be reliably detected using the traditional approaches by directly analyzing the recorded waves. The nonlinear chaotic oscillator based method is used to detect these small defects in this study. For comparison, a band-pass filter with a frequency range from 60kHz - 90kHz is also used to process the tested signals. The results for the defined five cases are shown in Figure 8. For case 1 of the undamaged pipe, the Lypunov exponent λ between incident wave and end echo are all larger than zero, indicating there is no UGW signal in between and hence no structural damage. For the damaged cases, only when the damage is relatively large, i.e., Case 4 and Case 5, the reflected UGW can be distinctively observed in the filtered signal as shown in Fig. 8 (d) and (e), indicating the traditional approach can reliably identify the damage only when the damage is significant. On the other hand, using Duffing-Holmes nonlinear oscillator and Lyapunov exponent λ gives the excellent identification results. For all the cases considered in the study, as shown in Figs. 8(b)~(e), four negative peaks of Lyapunov exponents λ can be found in the λ -*t* curves. From left to right, they respectively correspond to the incident wave, the *L(0,2)* mode UGW reflected by the defect, conversion mode reflected by the defect and the pipe end echo. Clear observation of the negative Lyapunov exponent λ values indicates the existence of reflected UGW, which allows a successful identification of the damage existence. These results demonstrate that the approach can identify very minor damages that could not be identified by the traditional approach based on direct analysis of the reflected wave.

4. Conclusions

This paper summarizes several innovative data analysis techniques for structural health monitoring and damage detection, including 1) Vibration Phase Space Topology. CPST damage index derived from the measured vibration responses is defined to detect the location of structural damage; 2) Phase Trajectory Change. The phase trajectories of multi-type vibration responses are obtained from a bridge under moving loads, and the separated distance between the trajectories of undamaged and damaged structures is defined as a damage index to detect the damage location; 3) Chaotic system based technique. A novel technique based on analyzing the responses of a nonlinear oscillator, i.e. Duffing-Holmes system, with the recorded UGW signal as an added input is proposed for the detection of minor structural damage. Experimental verifications on various structures have been conducted to investigate the performance and reliability of the above three approaches to analyse the measurement data and detect the minor structural damage. The results demonstrate that these approaches are sensitive and accurate to detect the structural damage, but not sensitive to uncertainty effect and measurement noise.

Acknowledgement
The authors acknowledge the financial supports from ARC research projects.

References

Frangopol, DM, Strauss, A & Kim, S 2008, 'Use of monitoring extreme data for the performance prediction of structures: general approach', *Engineering Structures*, vol. 30, no. 12, pp. 3644-3653.

Jiang, Y, Zhu, H & Li, Z 2016, 'A new compound faults detection method for rolling bearings based on empirical wavelet transform and chaotic oscillator', *Chaos Solitons & Fractals*, vol. 89, pp. 8-19.

Ko, JM & Ni, YQ 2005, 'Technology developments in structural health monitoring of large-scale bridges', *Engineering Structures*, vol. 27, no. 12, pp. 1715-1725.

Li, J & Hao, H 2016, 'A review of recent research advances on structural health monitoring in Western Australia', *Structural Monitoring and Maintenance*, vol. 3, no. 1, pp. 33-49.

Nie, Z, Hao, H & Ma, H 2013, 'Structural damage detection based on the reconstructed phase space for reinforced concrete slab: Experimental study', *Journal of Sound and Vibration*, vol. 332, no. 4, pp. 1061-1078.

Tong, H, 1990, Non-linear time series analysis, Oxford University, UK.

Wu, GC & Baleanu, D 2015, 'Jacobian matrix algorithm for Lyapunov exponents of the discrete fractional maps', *Communications in Nonlinear Science & Numerical Simulation*, vol. 22, pp. 95-100.

Zhang, W, Hao, H, Wu, J, Li, J, Ma, H, & Li, C (in press), 'Detection of minor damage in structures with guided wave signals and nonlinear oscillator', *Measurement*.

Zhang, W, Li, J, Hao, H & Ma, H 2017, 'Damage detection in bridge structures under moving loads with phase trajectory change of multi-type vibration measurements', *Mechanical Systems and Signal Processing*, vol. 87, pp. 410-425.

The 8th International Conference on
Structural Health Monitoring of Intelligent Infrastructure
Brisbane, Australia | 5-8 December 2017

SHM-Based Life-Cycle Management of Long-Span Cable-Supported Bridges

Y.L. Xu[1]

[1] *Department of Civil and Environmental Engineering, Faculty of Construction and Environment, The Hong Kong Polytechnic University, Hong Kong, China. Email:* ceylxu@polyu.edu.hk

Abstract

Long-span cable-supported bridges are important infrastructure to our society but subjected to deterioration once they are built due to ageing effects, natural hazards and man-made extreme events. A sudden failure or loss of functionality of the bridges may have severe economic, social and enviornmental impacts. Therefore, it is crucial to implement life-cycle managment (LCM) strategises that could maintain performance of the bridges within acceptable levels in terms of cost-effective intervention actions through their life-cycle (Frangopol and Soliman, 2016). Life-cycle cost and structural performence are two basic but conflicting aspects required for serious consideration in the LCM of the bridges. A life-cycle cost cosists of not only the initial design and construction cost, but also those due to operation, inspection, maintenance and repair during a specified lifetime (Frangopol and Messervey 2007). Structural performnace can be reflected mainly in terms of fucntionality, safety, sustaibaility and reliability. Multi-objective optimization techniques along with the supplementary information, such as the cost of interventions, the status of structures, and the effect of maintenance on the structural performance, should be employed to find the optimum inspection/maintenance types and application times.

It is noted that uncertainties inevitably exist in structural models, response measurements, loading phenomena, deterioration mechanisms, and surrounding environment. These uncertainties can be classified as two broad categories: aleatory uncertainties which describe the inherent randomness of phenomenon being observed, and epistemic uncertainties which describe the errors associated with imperfect models of reality due to insufficient and inaccurate knowledge (Ang and Tang 2007). Because of these uncertainties, the reliability of the current LCM is difficult to be assessed. Structural health monitoring (SHM) technology has recently attracted increasing attention in the civil engineering community. More than a hundred long-span bridges worldwide have been equipped with SHM systems which are designed to assess the bridges' functionality and safety while tracking the symptoms of operational incidents and potential damage. SHM technology is based on a comprehensive sensory system and a sophisticated data processing system implemented with advanced information technology and supported by cultivated computer algorithms. This technology allows actual loading conditions to be monitored, various types of structural responses to be measured, and deterioration and damage to be identified. SHM technology thus can provide a promising means and solid foundation for tackling the challenging issues in the LCM of the bridges.

This paper therefore proposes an SHM-based LCM framework for long-span cable-supported bridges and their relationship is shown in Figure 1. It involves seven major tasks: (1) to integrate multiscale finite element modeling and model updating with stress analysis for predicting both global and local structural responses with external loadings, dynamic characteristics and responses measured by the SHM system; (2) to determine the optimal placement of multi-type sensors for the best global and local response reconstruction of civil structures with the input of structural responses measured by the SHM system; (3) to assess the current health status of the structures based on the previous loading histories and using the SHM-based damage detection method; (4) to perform proper inspection, maintenance and repair work on the basis of the evaluated structural health states; (5) to develop loading models based on incessant field measurement data from the SHM system so that the previous loading histories can be analyzed and future loadings can be forecast; (6) to conduct damage prognosis and predict the remaining service life of the civil structures; and (7) to develop LCM strategies for making optimal decision under multiple objective and various constraints. The seven major tasks are interconnected and will be introduced in detail in this paper.

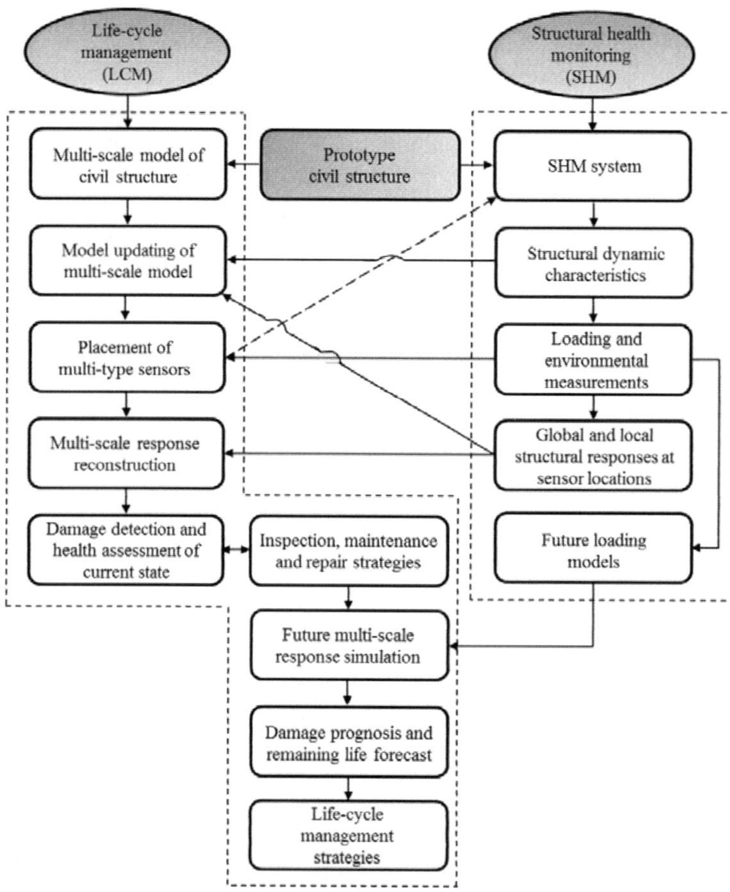

Figure 1 General components of SHM-based LCM framework

It is noted that the successful implementation of the entire SHM-based LCM framework is quite complicated and challenging, and many techniques and disciplines are involved in this framework. Although not all the seven major tasks have been completed yet in a systematic way, some relevant works which have been done by the author and his research team are presented in this paper with reference to the Tsing Ma suspension bridge and the Stonecutters cable-stayed bridge in Hong Kong. These works include (1) the Tsing Ma Bridge and its SHM system; (2) the loading assessment and loading models of the Tsing Ma Bridge based on incessant field measurement data from the SHM system; (3) the multiscale finite element modelling and model updating of the

Stonecutters Bridge; (4) the SHM-based stress analysis of the Tsing Ma Bridge due to multiple dynamic loadings; (5) the SHM-based fatigue damage prognosis and fatigue reliability analysis of the Tsing Ma Bridge; and (6) the SHM-based bridge rating system and inspection. The results from this case study indicate that the proposed SHM-based LCM framework has a potential to be used in practice for long-span cable-supported bridges with SHM systems. However, some further works shall be done following the proposed framework in order to turn the potential to the reality. These further works include (1) the determination of the optimal placement of multi-type sensors for the best global and local response reconstruction of the bridge through the input of measured responses from the SHM system; (2) the assessment of the current health status of the bridge using the SHM-based damage detection method; and (3) the development of SHM-based LCM strategies for making optimal decision under multiple objective and various constraints.

References

Ang, A.H.S. and W.H. Tang. 2007. *Probability concepts in engineering: Emphasis on applications to civil and environmental engineering. 2nd edition.* New York: Wiley.

Frangopol, D.M., and T.B. Messervey. 2007. Integrated life-cycle health monitoring, maintenance, management, and cost of civil infrastructure. In *Proceeding of 2007 International Symposium on Integrated Life-cycle Design and Management of Infrastructure*, ed. L.C. Fan, L.M. Sun, and Z. Sun, 3-20. Shanghai: Tongji University Press.

Frangopol, D.M. and Soliman, M. 2016. Life-cycle of structural systems: recent achievement and future directions. *Structure and Infrastructure Engineering,* 12(1), 1-20.

The 8th International Conference on
Structural Health Monitoring of Intelligent Infrastructure
Brisbane, Australia | 5-8 December 2017

Advanced SHM using Computer Vision and Machine Learning

Su Taylor[1] and M Lydon

[1] *Intelligent Infrastructure Group, School of Natural and Built Environment, Queen's University Belfast, Northern Ireland. Email:* s.e.taylor@qub.ac.uk

Abstract

The research presented in this paper aligns to the digital transformation of Civil Engineering and specifically Structural Health Monitoring (SHM) Systems. SHM can provide valuable information on the structural capacity and changes in structural performance, generally as an indication of damage. The applications of many SHM systems are currently limited by structure type, access for fixing of sensors, light levels and maintaining power supplies. This paper investigates the use of computer vision systems for SHM to ensure the safety and resilience of our civil infrastructure. Computer Vision is a new method of SHM which operates by recording motion pictures of a target area, or feature, on bridges and civil infrastructure. The development and validation of a contactless deflection monitoring system which tracks features to sub pixel accuracy is presented. The image is also pre-filtered for changing light levels in the environment and due to crossing freight. Machine learning is also used to identify events which provides useful data on real loading. The results of this research confirm the suitability of these systems for information to accurately determine the health of bridges.

1. Introduction

The current trend toward digitisation is having a significant impact on all levels of society and our environment. As intelligent sensing systems are being implemented across civil infrastructure, our road and rail networks are becoming smarter. Structural Health Monitoring (SHM) systems are becoming capable of providing real time information on the structural condition of our infrastructure. The implementation of SHM technologies to compliment traditional visual inspection methods provides an excellent opportunity to create a change in the current bridge assessment methods (Ojio et al, 2016; Catbas et al 2013 and Zaurin & Catbas, 2010a). The type of data used to give key information to asset owners has a vast range, from short term static deflections to longer term acceleration and strain readings which can give the dynamic modal response of a bridge. The application of such systems appears limitless as they can be tailored to monitor a simple steel beam and the most complex of long span bridges. The data for long multiple span bridges is particularly complex and, without advanced data mining and intelligent computational methodologies, the information can be difficult to interpret meaningfully. The potential of SHM data becomes exponentially more powerful through the identification, extraction and analysis of critical events. Therefore, setting appropriate limits and triggers in SHM systems is vital to ensuring structural anomalies can be detected in timely manner. This safeguards the operational life of the structure and extracts the features of data required for structural assessment thus targeting the whole life safety of the structure. Various methods exist for triggering data collection for SHM and this paper presents a method using computer vision and machine learning whereby anomalies on the data can be easily identified through a trained algorithm for vehicle detection and rating. In recent years vision based SHM has attracted considerable attention due to its practical deployment and low cost (Catbas, et al 2012,Zaurin & Catbas, 2010b). The potential of traffic vision based traffic detection systems has been successfully demonstrated within a hybrid Sensor-Camera monitoring system for damage detection and localization on a real bridge structure (Zaurin et al. 2016). The authors used a video stream of data to detect and classify vehicles; this information was simultaneously collected with strain measurements at various locations along the bridge. The hybrid system was successfully used to detect and locate damage scenarios on a real-world bascule bridge.

In this research the traffic identification method is integrated into a completely contactless vision based SHM system for deflection measurement which provides the response to known actions. The synchronised cameras simultaneously monitor traffic and measure bridge deflections, resulting in a data collection which can be stipulated by the vehicle type of interest. The system has been developed and validated through a series of laboratory trials and subsequently tested in the field. The field trials were carried out on a 30m span truss bridge in Northern Ireland. The machine learning algorithm successfully identified vehicle type from the live traffic crossing the structure and the response of the structure to this live loading was captured from an additional camera positioned adjacent to the bridge. The system has been successfully synchronised by a time stamp of millisecond accuracy which was sustained over a full day with no drift. The enabled the exact positioning of any vehicle to be connected to the dynamic deflection response of the bridge.

2. Development of contactless displacement measurement technique

In the long term, monitoring with cameras is expected to be more broadly utilised for structural health monitoring because of its potential for inexpensive deployment at any site. Additionally, due to some planning regulations and environmental protection policies, it is not always possible to monitor the condition of critical structures using conventional monitoring methods. Laboratory trials were carried out at Queens University Belfast to develop and validate a system for vision based deflection measurements and to trial a method for accurately synchronising a number of cameras to provide multipoint measurements at any point in time. Accurate camera calibration has been found to be one of the most complex aspects of this system development (Ojio et al, 2016 and Lydon et al, 2016). The accuracy of the system is determined by the intrinsic and extrinsic parameters of the camera which is used to measure translations and rotations in the structure. Lens distortion effects have to be accounted for and a scaling factor for the conversion from pixel units to equivalent translations and rotations for sub millimetre accuracy. In this research, this was successfully achieved through the adaption of the processes described in previous research by Bouguet et al (2015). There are two types of lens distortion in a standard camera, radial distortion and tangential distortion. Radial distortion occurs when light rays bend more near the edges of a lens than they do at the optical centre. The radial distortion coefficients model this type of distortion. The distorted points are denoted as ($x_{\text{distorted}}$, $y_{\text{distorted}}$):

$$x_{\text{distorted}} = x(1 + k_1 * r^2 + k_2 * r^4 + k_3 * r^6) \quad (1)$$

$$y_{\text{distorted}} = y(1 + k_1 * r^2 + k_2 * r^4 + k_3 * r^6) \quad (2)$$

- x, y — Undistorted pixel locations. x and y are in normalized image coordinates. Normalized image coordinates are calculated from pixel coordinates by translating to the optical centre and dividing by the focal length in pixels. Thus, x and y are dimensionless.
- k_1, k_2, and k_3 — Radial distortion coefficients of the lens.
- $r^2: x^2 + y^2$

Tangential Distortion occurs when the lens and the image plane are not parallel. The distorted points are denoted as $(x_{\text{distorted}}, y_{\text{distorted}})$:

$$x_{\text{distorted}} = x + [2 * p_1 * x * y + p_2 * (r^2 + 2 * x^2)] \quad (3)$$

$$y_{\text{distorted}} = y + [p_1 * (r^2 + 2 * y^2) + 2 * p_2 * x * y] \quad (4)$$

- x, y — Undistorted pixel locations. x and y are in normalized image coordinates. Normalized image coordinates are calculated from pixel coordinates by translating to the optical center and dividing by the focal length in pixels. Thus, x and y are dimensionless.
- p_1 and p_2 — Tangential distortion coefficients of the lens.
- $r^2 = x^2 + y^2$

This process is performed by taking several pictures of a checkerboard pattern at differing orientations in order to determine the intrinsic & extrinsic parameters of the camera. Once these are known, the lens distortion can be removed. The system obtains the scaling factor by determining the length of a checkerboard square in pixels from an undistorted image after converting the points used for calculation to world coordinates with the extrinsic parameters of the camera, the pattern is shown in Fig. 1. The Harris method is used for corner detection between the black and white squares (Harris & Stephens 1988). A division operation is then performed with this number of pixels and the known length of the square in mm.

Fig. 1. Checkerboard image used for camera calibration with corners identified

Following successful calibration of the camera, the algorithm was trained to identify a set of features (or single pixels) that can be reliably tracked. This process is known as feature registration, and example of identified features is shown in Fig. 2. In this research Binary Robust Invariant Scalable Keypoints (BRISK) was selected as the most accurate (Leutenegger et al. 2011). The key point detection used by BRISK is an extension of AGAST, which in turn is an extension of FAST(Rosten & Drummond, 2005). The key points are invariant of scale and are estimated in continuous scale-space and are composed of binary string, simple brightness comparison tests. Neighbourhood sampling is performed in deterministic, equally spaced concentric circles to detect features. Several long distance sample point comparisons to determine orientation are also performed and these points are returned for tracking at the next stage of the algorithm. The final stage tracked of the set of features to give calculations of translations or, if needed, rotations.

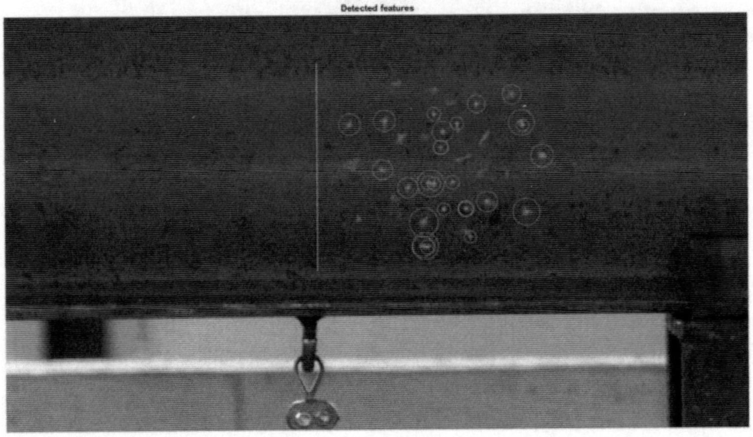

Fig. 2. Example of identified natural features for tracking displacement on test specimen

The algorithm tracks the features returned by BRISK using a Kanade-Lucas-Tomasi (KLT) (Tomasi & Kanade 1991) object tracking method. The algorithm works by generating an image pyramid, where each levels is down-sampled by a factor of two in width and height. The system begins tracking each point in the lowest resolution level, and continues tracking until all levels have been checked. This method is applicable to the recordings gathered because there will not be a large displacement between frames as the videos were captured at a rate of 60fps. The method uses a threshold value for the amount a feature can be expected to move between frames to eliminate points that cannot be reliably tracked, reducing the error in the display of results. The positions of the points in each frame were stored in a matrix in Matlab and the Euclidean distance between the points was calculated, in time, to obtain displacement measurements. Prior to site testing a series of laboratory trials were carried out to establish the accuracy of the measurements. The vision based displacements were validated against displacements using an accurately calibrated fibre optic translation sensor (FOTS) .

3. Experimental Programme

Test plan & Results for Displacement Monitoring Series

A 178mm x 102mm x19mm steel universal beam, with an effective span of 5m, was simply supported at each end and centrally loaded to induce displacement along the span of the beam. A static load of 332kg was applied to the midspan of the beam by means of a loading trolley and crane apparatus. The loading trolley was attached to the beam at midspan and weights were placed inside as shown in Fig. 3. Deflections were measured at the both sides of the flange to take into account any torsional movement in the beam but the camera was trained to investigate one point on the flange where the accurately calibrated FOTS was measuring vertical translation. The resolution of the FOTS was 0.03mm and was used to validate readings calculated using the vision based method from images. The comparison between the readings at node 4 position on the beam from the FOS and the camera is shown in Fig. 4.

Fig. 3. Crane & Trolley Apparatus

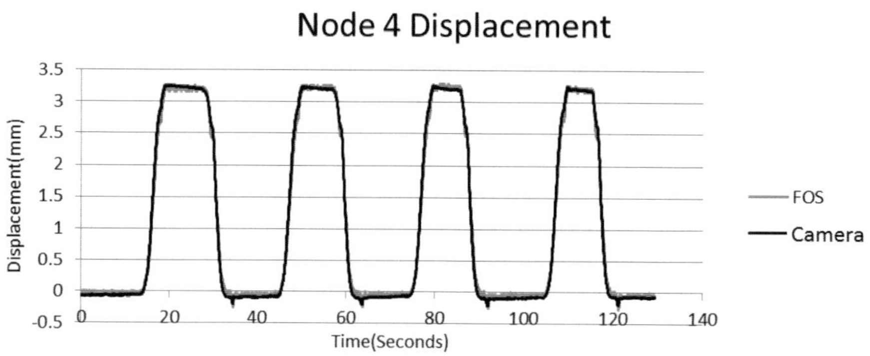

Fig. 4. Results from Node 4 Displacement Monitoring

Table 1 Results from Node 4 Displacement Monitoring

Node	RMSE(Camera vs FOS)
4	0.079

The Root Mean Square Error (RMSE) between the two readings is shown in Table 1. This shows that the camera and the FOS were synchronised successfully using a time stamp and also provided confidence in this method for monitoring deflections in the field.

4. Machine learning for vehicle identification

To process of machine learning for vehicle identification has significant potential in the deployment of autonomous vehicles and it is expected to become increasingly popular in the coming years. For this research a three step approach was adopted: 1) training data analysis, 2) Model training and 3) Eliminating false positives. In this case the training data consisted of a wide range of images containing the front of various vehicles, this singular elevation was chosen due to the camera fixed camera position for this site test. After collecting data for the training data analysis a neural network was then trained to perform recognition and classification of vehicles. This paired with camera calibration can be used to successfully identify the type and position of vehicle crossing the structure. The information is then used in conjunction with the displacement data gathered from additional cameras and allow for accurate calculation of an influence line providing data for damage detection of structures through completely contactless means.

5. Field testing

Following the successful implementation of the post-processing algorithms in the laboratory, a site investigation was carried out to determine its suitability for accurately identifying vehicles and measuring the corresponding bridge response. A single lane 30m span steel truss bridge, shown in Fig.5, was selected as a suitable structure to test this system as it was due for strengthening work so the reverse process of damage detection could be applied post strengthening work. The bridge forms part of the Tamnamore Road in Co Tyrone, Northern Ireland, and this road provides access to a busy industrial estate and is frequently used by heavy goods vehicles (HGV). Additionally the traffic on the bridge is controlled by a traffic light system which only allows for a single lane of traffic in one direction at any one time thus simplifying the case of multiple presence on the bridge. The data was collected from two cameras, one to measure bridge deflection and one to detect traffic passing over the deck.

Fig. 5. Side elevation of Verners Bridge. (Image taken for location of camera monitoring deflection)

The bridge deflection was monitored under live loading using a Nikon 810D camera which was set up on a sturdy tripod on the north west river bank at a distance of 22m from the bridge. The distance can be calibrated by using a feature on the bridge of a known length. Natural features on the midspan of the lower flange were targeted for monitoring and an 80-400mm zoom lens was used at maximum zoom to monitor these predefined points on the bridge. A pixel-mm ratio of 0.239mm was possible at this zoom, thus allowing for sub millimetre measurement of displacement. The calculated midspan deflection due to the crossing of three similar 5-axle vehicles are compared in Fig.6. Although the exact weight of the vehicles was unknown they were anticipated to be in the region of 22t to 44t based on the WIM data on a similar road in this county. A numerical analysis of the structure was carried out LUSAS software to estimate the deflections. Under approximate midspan loading of a 22t 5-axle truck, the bridge deflection was predicted to be 1.98mm which was similar to the deflections on site. Research is ongoing using a calibration truck with multiple passes and on other bridge types.

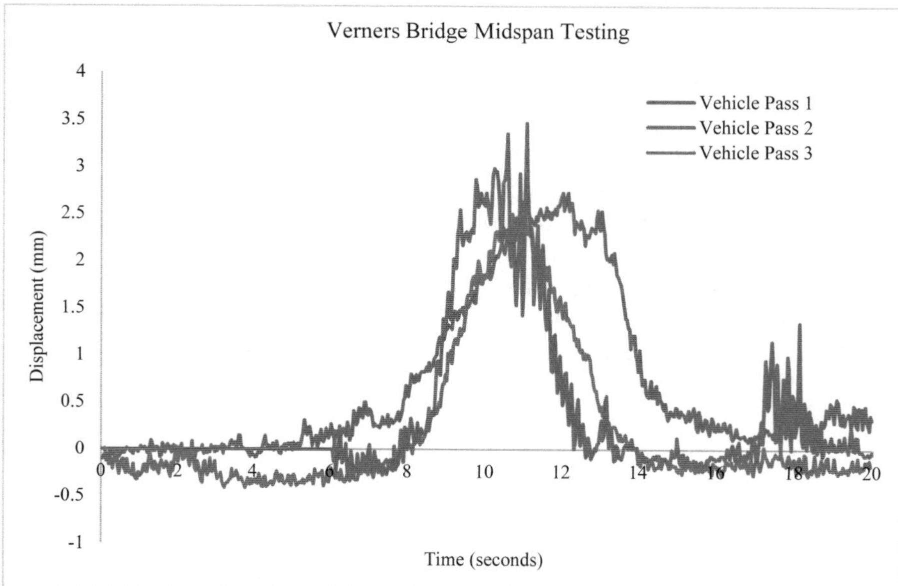

Fig. 6. Successful measurement of Midspan deflection

The success of the overall system also relies on successful vehicle identification using the second camera. As shown in Fig.7, a GoPro camera was tied to an existing stiff pole at the site. This allowed for rapid deployment of system and minimised the equipment need. An example of the successful identification of this vehicle from video of the live traffic is presented in Fig. 7. Research is ongoing to develop a fully portable non-contact monitoring system which has been . trained to recognise many different vehicle classes. It will then be possible to use the system to identify the deflections from different classes of vehicle and develop a database of bridge responses under loading. This should have the capability of detecting changes in behaviour as an indicator of damage a bridge. In this case, comparisons will be made once the bridge has had strengthening work to see the difference in behaviour as a reverse philosophy to damage.

Fig. 7. Successful classification of vehicle crossing bridge and GoPro used for Vehicle Capture

6. Conclusions

Accurate and relevant data analysis is critical in providing useful information from SHM systems. As SHM technology advances, there is a growing need to integrate smart data selection and analysis methods to provide key information on the real health of our civil infastrcture. Machine learning is a powerful tool which can bring together accurate traffic data, potentially in real time, with the real response of a bridge. This paper has presented the initial research carried out in developing a vision based SHM system which uses machine learning to identify traffic events and feature tracking to calculate the deflections under each event. There is also the potential to measure accelerations for model analysis by further post processing of the feature tracking over time by differential analysis. Ultimately this will then trigger addition data acquisition systems to obtain data. The results presented in this paper confirm the suitability of such a system for collection of data in structural health monitoring systems for bridges.

References

Bouguet, J.-Y., 2015. Camera Calibration Toolbox for Matlab. *Http://Www.Vision.Caltech.Edu/Bouguetj/Calib_Doc/Index.Html#Ref.*

Çatbaş, F. N., Kijewski-Correa, T., & Aktan, A. E. (2013). Structural identification of constructed systems. Reston (VI): American Society of Civil Engineers.

Catbas, F. N., Zaurin, R., Gul, M., & Gokce, H. B. (2012). Sensor networks, computer imaging, and unit influence lines for structural health monitoring: Case study for bridge load rating. *Journal of Bridge Engineering*, 17, 662–670.

Harris, C. & Stephens, M., 1988. A combined corner and edge detector. *Proceedings of the 4th Alvey Vision Conference*, pp.147–151.

Leutenegger, S., Chli, M. & Siegwart, R.Y., 2011. BRISK: Binary Robust invariant scalable keypoints. In *2011 International Conference on Computer Vision*. IEEE, pp. 2548–2555.

Lydon, D., Taylor, S.E., Robinson, D., Lydon, M. & Hester, D., 2016. Monitoring of Bridges using Computer Vision Methods. In *Civil Engineering Research Association of Ireland.* Galway, Ireland.

Lydon, M., Lydon, D., Taylor, S.., Hester, D. & Robinson, D., 2016. Laboratory investigation of contactless displacement measurement using computer vision systems. In *Civil Engineering Research Association of Ireland.* Galway, Ireland.

Ojio, T., Carey, C., OBrien, E., Doherty, C., and Taylor, S. (2016). Contactless Bridge Weigh-in-Motion *American Society of Civil Engineers Journal of Bridge Engineering,* 10.1061/(ASCE) BE.1943-5592.0000776 , ISSN (online): 1943-5592

Rosten, E. & Drummond, T., 2005. Fusing Points and Lines for High Performance Tracking. *Proceedings of the IEEE International Conference on Computer Vision*, 2(October), pp.1508–1511.

Tomasi, C. & Kanade, T., 1991. Detection and Tracking of Point Features. *Carnegie Mellon University Technical Report*, pp.91–132.

Zaurin, R., Khuc, T., Catbas, F.N. & Asce, F., 2016. Hybrid Sensor-Camera Monitoring for Damage Detection: Case Study of a Real Bridge. *Journal of Bridge Engineering*, 21(6).

Zaurin, R., & Catbas, F.N. (2010a). Integration of computer imaging and sensor data for structural health monitoring of bridges. *Smart Materials and Structures,* 19, 015019.

Zaurin, R., & Catbas, F. (2010b). Structural health monitoring using video stream, influence lines, and statistical analysis. *Structural Health Monitoring,* 10, 309–332.

The 8th International Conference on
Structural Health Monitoring of Intelligent Infrastructure
Brisbane, Australia | 5-8 December 2017

SHM Roles in Autonomous Inspection and Preventive Maintenance of Bridges

G.D. Chen[1], W.A. Stone[2].
[1]*Missouri University of Science and Technology – USA, Email:* gchen@mst.edu
[2]*Missouri Department of Transportation - USA*

Abstract

Missouri University of Science and Technology (Missouri S&T) is leading a ten-partner, five-year University Transportation Center with a theme of "Inspecting and Preserving Infrastructure through Robotic Exploration (INSPIRE)." The goal of the INSPIRE Center is to transform in representative demonstration cases the current labor-intensive, inconsistent, and expensive ad hoc process of bridge inspection and maintenance into an efficient, safe, reliable, and cost-effective data-driven decision-making protocol. In this presentation, the needs for an autonomous bridge inspection and maintenance program are first established. A framework of such a program integrating sensor measurement, nondestructive evaluation (NDE), and data analytics into visual inspection through a mobile platform of robotics is then envisioned to improve the reliability of inspection data and results, streamline the inspection and maintenance process, and augment the inspection of fracture critical members. Next, key to the success of the program, structural health monitoring (SHM) technologies such as lab-on-sensor innovation for the understanding of structural behaviors, microwave imaging for the internal condition monitoring of bridge structures, and adaptive wavelet transform for the extraction of damage and deterioration from images are introduced. Finally, industry-university collaborations, which is critical to realizing the envisioned integrating framework, are dicussed and the potential SHM impacts to bridge preservation are identified. In general, a global lab-on-sensor system is key to quantifying structural behaviors and, together with local NDEs, enables both qualitative and quantitative assessments of the condition of bridge systems or even transportation networks. The bridge condition identified during inspection can be used to prioritize the maintenance of bridges. An unmanned aerial vehicle with one or two remotely-controlled arms can potentially make bridge inspection and maintenance an integrated task. The autonomous inspection and maintenance program likely results in cheaper, faster, and more reliable solutions in bridge preservation.

1. Introduction

INSPIRE University Transportation Center is a diverse, experienced, and geographically distributed consortium of six major and four minor partners in order to address the labor-intensive, inconsistent, and costly management challenges of our nation's transportation system. The six major partner institutions from east to west are City College of New York, Georgia Institute of Technology, Missouri S&T, the University of Colorado, Boulder, the University of Nevada, Las Vegas, and the University of Nevada, Reno. Located in the state of Missouri, the four minor partners are Lincoln University, St. Louis Community College, East Central College, and Ozarks Technical Community College. Major partners that are widely distributed nationwide will support all research activities, whereas minor partners provide a conduit to develop the next-generation transportation workforce.

Following is a presentation of the needs and a framework of the autonomous inspection and maintenance of bridges, example SHM technologies, industry-university collaborations, and SHM impacts to bridge preservation.

2. Needs for an Autonomous Inspection and Maintenance Program

Preserving the existing transportation structures is a major task in the U.S. American Society of Civil Engineers (ASCE) rated 9.1% of the nation's 614,387 highway bridges structurally deficient in 2016 (https://www.infrastructurereportcard.org/cat-item/bridges/). Four out of ten bridges approached the end of their design life of 50 years or older. Their overall condition was rated C+ with Grade A being the best and F the worst. Therefore, cost-effectively managing the bridge inventory to ensure the safety of the public, maintain the quality of life, and meet the economic development needs is a challenge to the U.S. transportation agencies.

2.1 Accessibility for bridge inspection and maintenance

Highways and railways traverse elevated bridges. Access to these structures can be difficult, dangerous, and disruptive, as indicated in Figs. 1 and 2. Their inspection and maintenance requires the use of heavy lifting equipment (e.g. boom trucks) and access equipment (e.g. snooper trucks) or the courage of inspectors to walk on suspension cables. Traffic control is often required during the inspection and maintenance work, increasing concerns for the safety of inspectors and travellers.

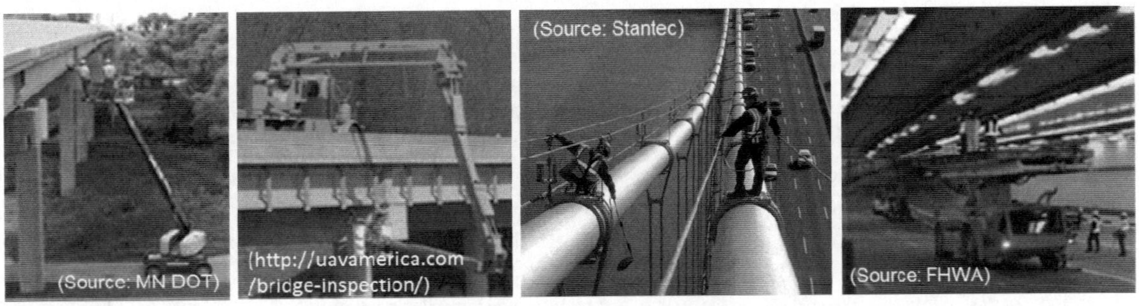

Fig. 1 Ways to access inspection areas in current practice

Fig. 2 Ways to access maintenance areas in current practice

2.2 Proximity inspection requirement for fracture critical elements

Hands-on inspection required by National Bridge Inventory Standards (NBIS) for fracture critical members, as included in the double cantilever steel truss bridge (e.g. eye bars) in Fig. 3, is particularly challenging in elevated steel bridges. For any valid inspection, an inspector must be able to lay his/her hand on the members, tap or scratch the members' surface for any potential crack and corrosion products, and even conduct a dye penetrant inspection for any surface breaking defects in metals such as fatigue cracks on in-service members.

Fig. 3 Double cantilever truss bridge with fracture critical members

2.3 Consistency of visual inspection

Visual inspection is labor intensive and subjective, resulting in inconsistent and less reliable bridge element ratings. To date, bridges have been rated mainly based on the visual observation of material and physical conditions during biennial visual inspections mandated by the federal government. For each bridge, the conditions of superstructure, substructure, and deck are reported separately. A Federal Highway Administration (FHWA) study on the reliability of visual inspection for highway bridges (Moore et al. 2001) indicated that, 45 participating inspectors assigned an average of 4 or 5 condition ratings to each element of a bridge as shown in Fig. 4, including a maximum of 6 different ratings to the superstructure and a minimum of 4 different ratings to the substructure. As such, visual inspection provided inconsistent condition ratings that have limited use going beyond a qualitative screening of bridges for federal funds allocation.

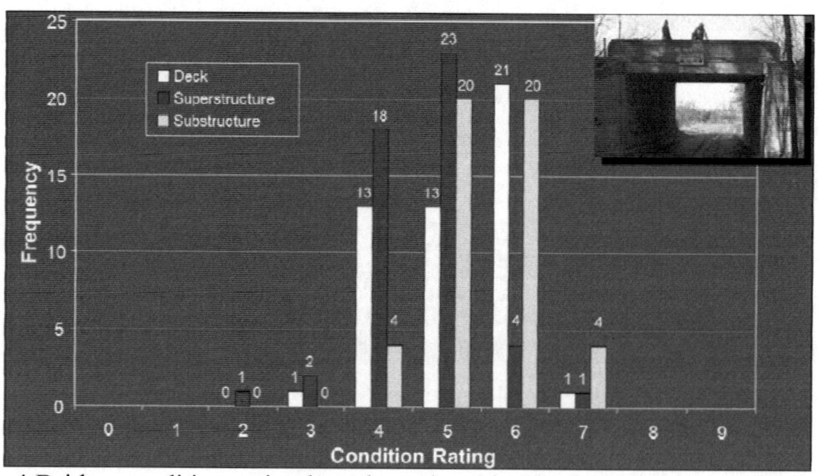

Fig. 4 Bridge condition rating based on visual inspections (Moore et al. 2001)

3. The Proposed Autonomous Bridge Inspection and Maintenance Program

To make bridge ratings more consistent and reliable as a key decision-making factor in bridge preservation, a robot-assisted bridge inspection tool (RABIT) has recently been developed and prototyped as a major deliverable of the five-year FHWA Long-term Bridge Performance Program to supplement the visual inspection with quantifiable data sets in bridge deck evaluation (https://www.fhwa.dot.gov/research/tfhrc/programs/infrastructure/structures/ltbp/ltbpresearch/rabit/index.cfm). The RABIT simultaneously implements six types of NDE technologies, including a panoramic camera to collect 360° images, a professional grade camera to capture high-resolution images of the deck surface, an electrical resistivity device to characterize the corrosive environment of the concrete, an impact echo and ultrasonic surface wave generator to evaluate concrete delamination and deck strength, a ground penetrating radar to "map" rebar and other metallic objects below the surface and qualitatively assess concrete deck deterioration, and a global positioning system to record and mark location data.

Although the invention of the RABIT is a leap towards the automated and rapid collection of data sets that are more repeatable and quantifiable, traffic control is still required during the inspection of bridge decks. More importantly, potentially conflicting data and results obtained from multiple technologies are a major concern in application. For members other than bridge decks, the accessibility and proximity requirements as discussed in Section 2 further complicate the inspection of bridges. Furthermore, inspection and maintenance works are basically not integrated into a bridge management system in that inspection data are not used to their maximum extent in the development of maintenance strategy and heavy equipment must be mobilized for inspection and maintenance separately.

Therefore, autonomous bridge inspection and preventive maintenance without affecting traffic is desirable in practice. Here, maintenance is referred to as cleansing/washing and lubrication of bearings. Fig. 5 delineates a framework of bridge network preservation with an integrated system of sensing, NDE, data analytics, and visual inspection. Take a girder bridge inventory as an example. A small number of representative bridges can be selected for monitoring due to similar materials and construction processes used. For each representative bridge, only critical elements and hot spots are instrumented for the quantitative assessment of structural behaviors. However, all bridges are inspected with cameras and other NDE devices carried by robots mainly for qualitative assessment. Bridges and members with abnormal behaviors are confirmed by visual inspection before load rating is provided. Deficient bridges are prioritized for maintenance and preventive maintenance can be done by an unmanned aerial vehicle with robotic arms.

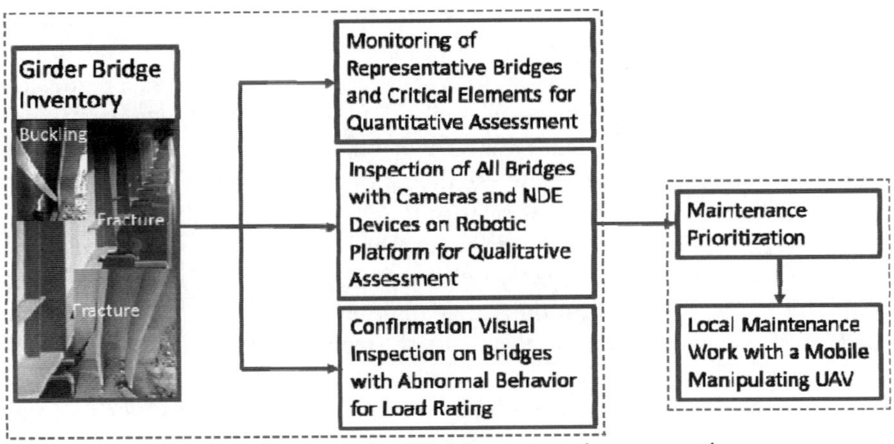

Fig. 5 A framework of bridge network preservation

3.1 Robotic platforms

In recent years, multi-modal unmanned vehicles have been developed in some applications. For example, Naviator is a hybrid flying/swimming vehicle invented by Professor Javier Diez in Mechanical and Aerospace Engineering at Rutgers University (http://mae.rutgers.edu/f-javier-diez-garias). This vehicle can provide a mobile platform for bridge inspections above and under water. SCAMP is a hybrid flying/climbing vehicle invented by Dr. Mark Cutkosky from Biomimetics and Dexterous Manipulation Laboratory at Stanford University (http://bdml.stanford.edu/).

In the INSPIRE Center, two types of structural crawlers and a mobile manipulating unmanned aerial vehicle (UAV), as shown in Fig. 6, will be developed to assist in the inspection of bridges. The structural crawlers can carry NDE devices and move underneath a bridge deck or along a bridge pier to take various types of data. The prototype in Fig. 6(a) has been demonstrated to move on a vertical masonry wall by Dr. Jizhong Xiao at City College of New York. The first design as shown in Fig. 6(b) has been manufactured and demonstrated by Dr. Hung La at the University of Nevada, Reno, to climb around a steel bar with a payload of 8 lbs. The UAV with one or two robotic arms as shown in Fig. 6(c) has been successfully demonstrated in laboratory to turn a valve by Dr. Paul Oh at the University of Nevada, Las Vegas (Korpela et al. 2014). It can be used to not only inspect fracture critical members since the robotic arms can assist in a few simple tests in addition to taking photos, but also maintain local areas of bridges such as bearing lubrication and cleansing/washing.

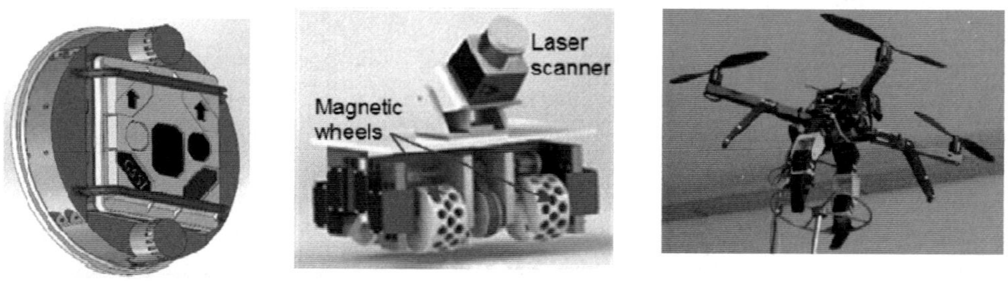

(a) Climbing robot on masonry walls (b) Climbing robot on steel members (c) Mobile manipulating UAV

Fig. 6 Current state of development in robotic platforms

3.2 Integration of NDE into SHM

Although NDE and SHM involve a research community with common interests, they are not yet integrated in a significant and meaningful way. NDE is focused more on a detailed investigation of the local changes in geometries and material properties, while SHM, as an ideal, involves an autonomous system that can collect data using sensors, cleanse and analyze data with a computer, extract mission-critical information about the global behavior of a bridge system through data mining, and automatically report findings to engineers in charge or decision makers with wireless communication. Together, structural assessment at the local and global scales can be an effective tool for bridge inspections. For example, distributed fiber optic sensors deployed along and across critical load paths can serve as a screening tool of a bridge queuing development of a prioritized inspection plan with NDE, a normalization and spatial correlation tool for vision- or image-based inspection, and a system integration tool for truck monitoring (bridge weigh-in-motion) and element-to-system rating.

To further illustrate the above point on the integration of NDE and SHM technologies, Fig. 7(a) shows the microwave synthetic aperture radar (SAR) image of a 533mm×300mm×152mm concrete slab with two embedded steel bars corroded (Ghasr et al. 2015). The two bars have a diameter of 9.5 mm and 19.1 mm, respectively. The concrete slab was partially immersed in 5% NaCl solution and the bars were subjected to impressed current to increase corrosion activity. The image clearly indicated corrosion of the embedded steel bars in concrete and its two-dimensional (2D) slice showed the two bars embedded at 25 mm deep. However, it is difficult to estimate the cross sectional loss of the steel bars, which is needed to estimate the remaining capacity of the bars in load rating. On the other hand, a Fe-C coated long period fiber grating (LPFG) sensor can be attached to the embedded bars to directly measure the steel mass loss (Chen et al. 2016a). As shown in Fig. 7(b), the calibration curve can be applied to determine the steel mass loss based on the change in resonant wavelength of the LPFG sensor as will be discussed further in Section 4.1. As such, NDE can give a qualitative evaluation of corrosion condition in a region while SHM with the corrosion sensor can be used to determine a quantitative evaluation of the steel mass loss directly.

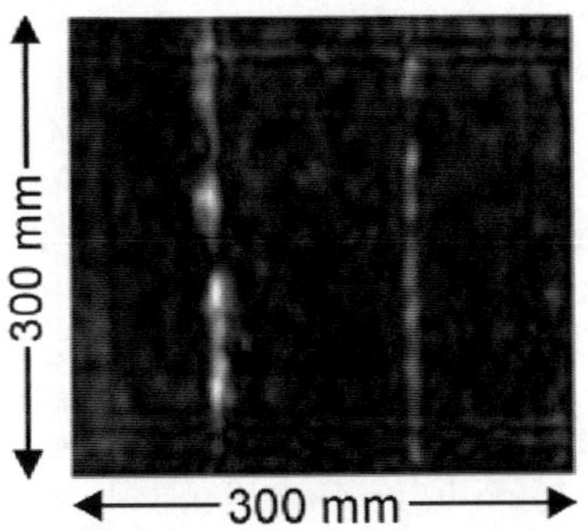

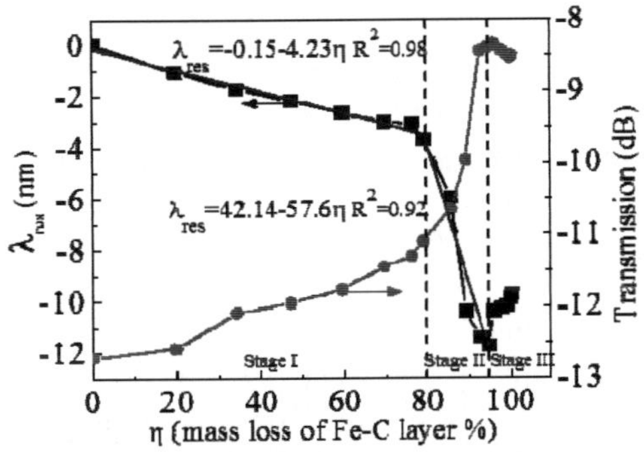

(a) Microwave SAR image (b) Calibration curve of a Fe-C coated LPFG sensor

Fig. 7 Qualitative vs. quantitative evaluation of corrosion-induced mass loss

3.3 Integration of data analytics into visual inspection for decision-making

Adaptive wavelet transform (AWT) has recently been developed by Dr. Genda Chen at Missouri S&T to extract time-varying features from a structural response over time (Qu et al. 2016). AWT is a signal processing method that adaptively adjusts itself to the time-varying signal by changing the center frequency of a Morlet mother wavelet in order to achieve the selectivity of time and frequency resolution over time. To illustrate the effectiveness of the AWT, a time function with one to three frequencies over time is considered. The continuous wavelet transform (CWT) with a center frequency of 1 Hz and the synchro-squeezing AWT (SSAWT) of the time signal are compared in Fig. 8. It can be observed from Fig. 8 that the scalogram in time-frequency domain by the CWT is blurred while that by the SSAWT is very clean. The latter can easily give an accurate evaluation of frequencies over time.

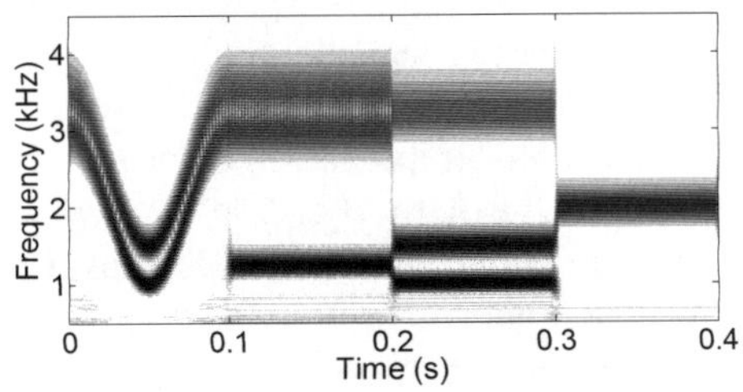

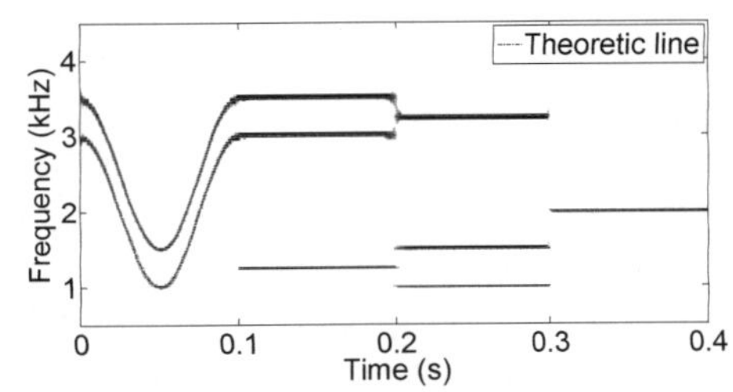

(a) CWT (F_c = 1 Hz) (b) SSAWT

Fig. 8 Time-frequency representation of a time function

In application, the results from data analysis such as the frequency identification in Fig. 8 will be combined with visual examinations of the physical objects to derive at conclusions about the actual conditions of a bridge.

4. State-of-the-art Development of Representative Technologies

In the INSPIRE Center, a number of cutting-edge technologies will be developed over a period of five years. They will be tested in laboratory and the promising ones will be demonstrated at bridge sites. Following is a brief presentation of a few technologies that will be further developed at Missouri S&T.

4.1 Lab-on-sensor innovations for structural behaviour monitoring

A lab-on-sensor design theory has been recently developed and applied to detect and assess structural behaviors such as cracking, foundation scour, and steel corrosion. For each mechanical or electrochemical behavior, the theory includes three steps: (1) Extension of the behavior from a structural element to its nearby deployed sensor with a special mechanism, (2) Calibration of the sensed parameter with the behavior of the sensor mechanism and (3) Correlation of the behavior of the sensor mechanism with the behavior of the nearby structural element.
For crack detection and assessment, coax cable sensors were designed, fabricated, calibrated and applied to an in-service low-volume bridge based on the propagation and change of electromagnetic waves in a coax cable (Chen et al. 2004). For foundation scour detection and assessment, smart rocks were designed, fabricated, calibrated and applied to an in-service high-volume bridge based on the change of magnetic fields around a smart rock near foundations (Chen et al. 2016b). For steel corrosion detection and assessment, nano iron and silica particles coated LPFG sensors were designed, fabricated, calibrated and applied to steel reinforcement corrosion in laboratory based on the change of light wavelength (Huang et al. 2013). Fig. 9 compares an LPFG corrosion sensor coated with nano iron and silica particles after being heated, a coax cable crack sensor, and a smart rock with one embedded magnet. Their diameters are in the order of 100 μm to 10 cm. Their measurement principles are quite different as briefly discussed previously.

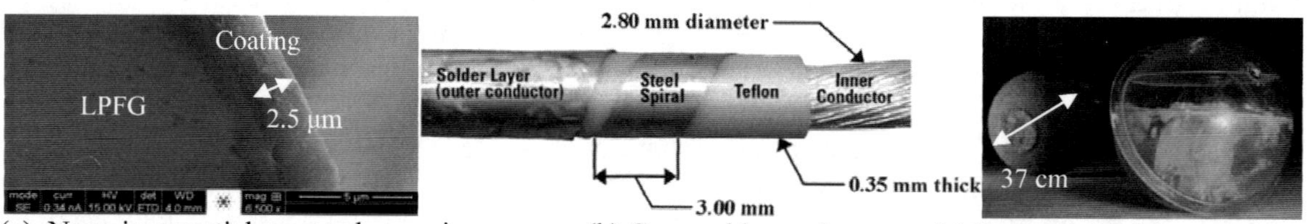

(a) Nano iron particles coated corrosion sensor (b) Coax cable crack sensor (c) Smart rock with magnet
Fig. 9 Lab-on-sensor in various scales for corrosion, crack, and scour monitoring

To further illustrate the concept of lab-on-sensor design, an LPFG sensor (cladding mode LP08) with a grating period of 387 μm and a total length of approximately 40 mm was cleansed, deposited with an approximately 0.8 μm thick silver film, and electroplated in a solution to produce an outer 20 μm thick Fe-C layer (Chen et al. 2016a). The microstructure of the Fe-C layer was examined with scanning electron microscope (SEM). As shown in Fig. 10(a), the Fe-C layer has a dendritic structure of round particles with approximately 27 nm in diameter. The dendritic structure of the outer Fe-C layer is porous. Figs. 10(b) and 10(c) show the surface morphologies of the Fe-C coated LPFG sensor prior to corrosion test and after 24 hours of immersion in 3.5 wt.% NaCl solution. On the surface of the LPFG sensor, the scattered grains as shown in Fig. 10(b) represent the pitting corrosion products formed in air during preparation for SEM imaging. After 24 hours of immersion in salt solution, the LPFG surface was covered with a significant amount of corrosion products, increasing the outer diameter from 165 μm to 400 μm as shown in Fig. 10(c).

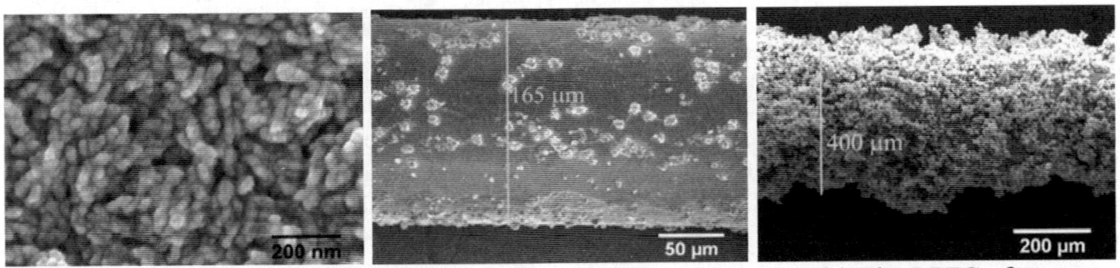

(a) Fe-C layer microstructure (b) Fe-C coated LPFG prior to test (c) The LPFG after test

Fig. 10 SEM images of Fe-C layer and LPFG sensor (Chen et al. 2016b)

When the Fe-C coated sensor was submerged in 3.5wt.% NaCl solution, both the transmission spectrum and electrochemical impedance spectroscope (EIS) were acquired up to 24 hours. The corrosion rate taken from the EIS tests was converted to steel mass loss over time. Fig. 7(b) shows the wavelength of the LPFG sensor as a function of the steel mass loss over time (normalized). The two linear relations in the wavelength-mass plot up to 0.95 steel mass loss likely suggest the chloride penetration process through Fe-C and Ag pores and the active corrosion process, respectively. The two stages could enable sensing of chloride and mass loss in application.

4.2 Microwave and hyperspectral cameras for internal condition examination

Take microwave SAR images as an example to show how effective NDE technologies are for internal condition inspection. In this example, microwave was used to detect voids behind a 1.52 m × 1.46 m × 0.178 m reinforced concrete wall panel with backfill sands, as shown in Fig. 11(a), to simulate its application in mechanically stabilized earth (MSE) walls (Bao et al. 2016). Foam is electromagnetically similar to air, and can thus properly represent the presence of an air void. Therefore, a 127 mm × 127 mm × 51 mm piece of construction foam was placed in the middle of a 610 mm × 660 mm × 300 mm wooden sandbox, as illustrated in Fig. 11(b). The sandbox was tightly pressed against the back side of the MSE wall panel by a pallet jack.

The full-size MSE wall panel tested was made of Class A-1 concrete and reinforced with #10 (SI unit) epoxy-coated steel bars. The front face of the wall panel was a 38 mm thick decoration layer. The average relative permittivity of the MSE wall panel was taken 6.0 and the operation frequency range of the imaging system was set to 1-4 GHz, resulting in image resolution of ~12 mm in depth and ~20 mm in lateral directions. Fig. 11(c) shows the 2D slice of a three-dimensional SAR image corresponding to the back side of the panel in contact with a piece of foam placed in a sandbox. This image clearly demonstrates presence of the void and the sandbox.

Fig. 11 Image slice at back side of wall with a piece of foam placed in sandbox (Bao et al. 2016)

4.3 Adaptive wavelet transform for the extraction of damage and deterioration

The SSAWT (Qu et al. 2016) was applied to a concrete slab with six embedded artificial defects (designated by A to F) as shown in Fig. 12(a) to represent delamination in bridge decks. Figs. 12(b) and 12(c) illustrate the test setup with a portable seismic property analyzer (PSPA) with one source and two receivers and the principle for delamination detection. The source generates an impact wave into the concrete slab and the receivers accept the reflected waves from the delamination or the bottom face of the slab. In addition, reflection from the side faces of the slab may also be present. A total of 40 measurements as located in Fig. 12(a) were taken. It can be seen from Fig. 12(d-f) that the scalograms for signals at Points #2, #23, and #10 are quite different, corresponding to no, shallow and deep delamination cases, respectively. From the scalograms, the long duration feature was used to separate shallow delamination from other cases and the characteristic frequencies corresponding to no and deep delamination cases can be identified (Qu et al. 2016).

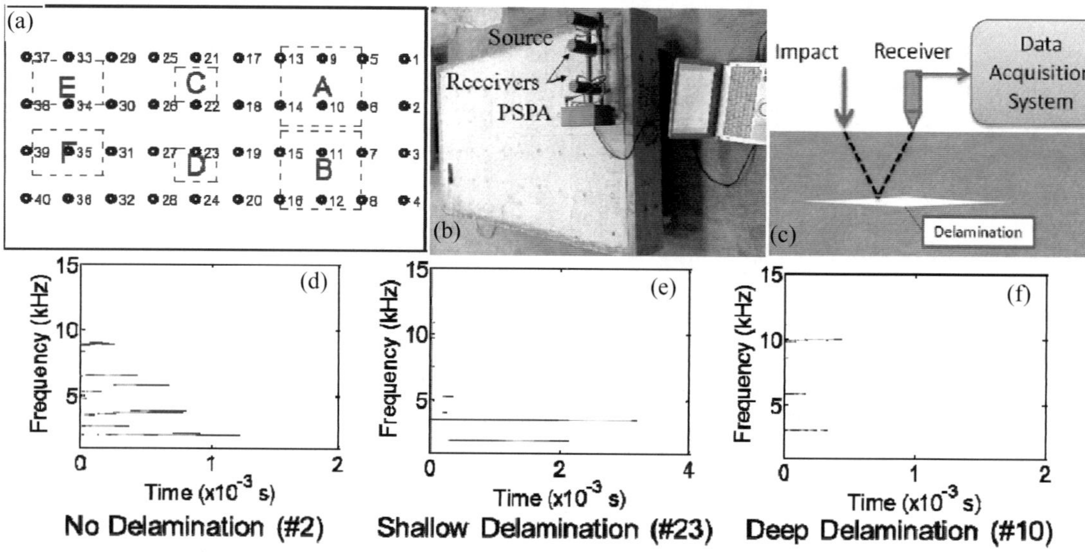

Fig. 8 Illustration of combined visual inspection and data analysis

5. Industry-University Collaborations and SHM Impacts to Bridge Preservation

The Missouri Department of Transportation (MoDOT) is collaborating with the INSPIRE Center to advance NDE and SHM technologies into practice. MoDOT is responsible for over 10,000 bridges statewide, including 53 major river crossing structures. Most of these bridges are mandated to be inspected biennially following the National Bridge Inspection Standards (NBIS). The unmanned vehicles will become the new platform that the advanced technologies can easily access to bridge elements and provide reliable inspection data that are not available today even with current access. For example, microwave SAR images can be effective for internal condition inspection. MoDOT and the INSPIRE Center is exploring a pooled fund study to engage other state DOTs interested to use unmanned aerial vehicles in their bridge inspection and preventive maintenance.

Depending on their size, age, and accessibility, most bridges can be inspected in half an hour each. In the state of Missouri, approximately 1,500 out of over 10,000 bridges require in-depth inspection in four hours each by a crew of two people. With the proposed autonomous inspection program, the time required for the in-depth inspection of each bridge can be reduced to less than one hour, amounting to over 9,000 hours in saving. When each inspector charges $70/hour, the use of the autonomous inspection program can save a direct cost of at least $630,000 each year in the state of

Missouri. If a bridge deck must be inspected in detail and surveyed, 4 workers must spend 8 hours for about $4,600 while two UAV operators can complete the same job in just over 2 hours for about $150 (according to Michigan DOT). Thus, the use of the autonomous program can save $4,450 per bridge deck or a direct cost of $6,675,000 for 1,500 bridges in the state of Missouri. If access to the problem area with traditional inspection methods must be made from bridge decks, the saving in indirect cost associated with no traffic closure with the proposed autonomous inspection program multiplies.

6. Concluding Remarks

Based on the previous discussion, the following remarks can be made:

- A global lab-on-sensor system is key to quantifying structural behaviour and, together with any local NDE, enables a more accurate assessment of the condition of bridge systems both qualitatively and quantitatively.
- A mobile manipulating UAV may meet proximity requirements on the inspection of fracture critical members and can make inspection and maintenance an integrated task.
- The autonomous inspection and maintenance program likely results in cheaper, faster, and more reliable solutions in bridge preservation.

Acknowledgements

Financial support for this study was provided by the U.S. Department of Transportation, Office of the Assistant Secretary for Research and Technology (USDOT/OST-R) under Grant No. 69A3551747126 through INSPIRE University Transportation Center (http://inspire-utc.mst.edu) at Missouri University of Science and Technology. The views, opinions, findings and conclusions reflected in this publication are solely those of the authors and do not represent the official policy or position of the USDOT/OST-R, or any State or other entity.

References

Bao, Y, Ghasr, M, Ying, K, Chen, GD & Zoughi, R 2016, 'Microwave synthetic aperture radar (SAR) imaging for nondestructive evaluation of mechanically stabilized earth walls', *Proceedings of 95th Transportation Research Board Annual Meeting*, Washington D.C., USA, 10-14 January.

Chen, GD, Mu, HM, Pommerenke, D & Drewniak, JL 2004, 'Damage detection of reinforced concrete beams with vovel distributed crack/strain sensors', *Structural Health Monitoring*, vol.3, no.3, pp. 225-243.

Chen, YZ, Tang, F, Bao, Y, Tang, Y & Chen, GD 2016a, 'A Fe-C coated long period fiber grating sensor for corrosion induced mass loss measurement', *Optics Letter*, vol. 41, no. 10, pp. 2306-2309.

Chen, GD, Tang, Y, Chen, YZ, Li, ZC, Guo, CR, Fan, L, Bao, Y, Hu, XY & Klegseth, M 2016b, *Smart rocks technology for real-time monitoring of bridge scour and riprap effectiveness – design guidelines and visualization tools*, Publication No. OASRTRS-14-H-MST, U.S. Department of Transportation Office of the Assistant Secretary for Research and Technology, Washington D.C.

Ghasr, MT, Pape, YL, Scott, DB & Zoughi, R 2015, 'Holographical microwave imaging of corroded steel bars in concrete'. *ACI Materials Journal*, vol. 112, no.1, pp. 115-123.

Huang, Y, Gao, Z, Chen, GD, and Xiao, H 2013, 'Long period fiber grating sensors coated with nano iron/silica particles for corrosion monitoring', *Smart Materials and Structures*, vol. 22, no. 7, doi:10.1088/0964-1726/22/7/075018.

Korpela, C, Orsag, M & Oh, P 2014, 'Hardware-in-the-loop verification for mobile manipulating unmanned aerial vehicles', *Journal of Intelligent Robot System*, vol. 73, pp. 725-736.

Moore, M, Phares, B, Graybeal, B, Rolander, D & Washer, G 2001, *Reliability of visual inspection for highway bridges, volume I: final report*, Report No. FHWA-RD-01-020, Federal Highway Administration, Washington, D.C.

Qu, HY, Li, MX, Chen, GD, Sneed, LH & Anderson, NL 2016, 'Multi-bandwidth wavelet transform of impact echo signals for delamination detection of thin concrete slabs', *Journal of Civil Structural Health Monitoring*, vol. 6, no. 4, pp. 649-664.

The 8th International Conference on
Structural Health Monitoring of Intelligent Infrastructure
Brisbane, Australia | 5-8 December 2017

Distributed Fibre Optic Sensing Technology in the Extreme Areas of Civil Engineering

A.Kindler[1], K.Nycz [2], S.Großwig [3], M.B.Schaller [4].
[1] *Stump Spezialtiefbau GmbH – Germany, Email:* arne.kindler@stump.de [2] *Stump Spezialtiefbau GmbH – Germany,* [3]*GESO mbH& Co. Projekt KG - Germany,* [4]*GGB mbH -Germany*

Abstract

The foundation assessment plays an equally important role in assessing the current state of the building, the current structural safety and the remaining lifespan of civil engineering structures as well as the assessment of the above-ground structure. One of the biggest problems of the continuous monitoring of subsurface foundation elements is the reliable data acquisition over a 3-year lifetime. Relevant experience of the last decades shows that the reliability of the measuring results e.g. due to the decreasing quality of the instruments is lost over time in the case of the essential long-term measuring methods in the field of special underground engineering, or an interpretation of the results is difficult. Other measurement and test methods, e.g. fibre optic Bragg grating sensors permit only point-specific statements about the load-carrying behaviour of the monitored object. Based on this, in 2016, a new innovative anchor and pile monitoring system was used in addition to a classical instrumentation in order to demonstrate and promote the applicability of fibre optic measurement technology to the construction site.
Furthermore, the distributed fibre optic sensing technology can be used as a "sleeping" sensor. Based on the measuring principle, it is now also possible to carry out measurements over the entire lifecycle of the structure and to classify the current actual state.

1. Introduction

The first fibre optic sensors came onto the market about three decades ago. Initially used as special solutions only in selected cases, now they have found use in almost all areas of technology, especially in areas, where electrical sensor systems reach their limits. The first application examples of spatially resolved fibre-optic sensors in the field of geotechnics have been in existence since the mid-1990s, in projects such as the assessment of the of radioactive repository location sites (Hurtig et al. 1994), for the monitoring of landfill sealing systems (Hurtig et al. 1995) or within the framework of the rehabilitation of mountain farming landscapes (Schreck et al. 1998).
The company's first own experience in the field of glass fibre measurement technology have been in existence since 1993. The first carried out projects were a monitoring at the Edertalsperre (Dietz et al. 2001), a monitoring with glass fibre tendon at a flood bridge in Carinthia as well as monitoring within the framework of the Storebælt crossing. In all cases, the structure was permanently monitored by means of individual-point measurements. Since 1993 various fibre-optic sensor solutions for geotechnical applications have been introduced to the market (Voet, Glötzl 1996, Hurtig et al. 1994).

2. Fibre -optic strain measurements with very high spatial resolution

Fibre optic sensors

For many years, glass fibre-based sensors have been used for detection of variety of mechanical variables. In comparison with conventional electrical sensor systems, fibre-optic sensors are advantageous due to their extremely compact design and possibility to measure over long distances. A fibre optic measuring system consists of a readout unit and passive sensor fibre. During the measurement, the read-out unit transmits a light pulse into the glass fibre, captures the reflective light signal and analyses its change. Basically, two types of fibre optic sensors are available – single-point sensors and distributed measuring sensors.

Single point measurements

In single-point measurements, the measurement is carried out by writing Fibre Bragg gratings whose reflection wavelength depends inter alia on the strain state of the fibre. Further possibility consists in measuring of optical endpoint sensors. The measurement is carried out exclusively in the area of these "mirrors" or "grids" by the changes in the wavelength caused by environmental influences such as temperature, strain or vibration. The glass fibre is used exclusively as an optical fibre and connection between the individual "mirror elements". The disadvantage of these types of fibre -optic sensors is the exclusive recording of data at discrete measuring locations, which may cause an inaccurate interpretation of the results of the measurements. A possible cause for the inaccurate interpretation may be an over-measurement of so-called "hot spots" of temperature or strain changes, which occur at locations that are not known in advance. Similar problems arise with spatially extended continuous monitoring tasks in structures such as, for example, the grout body of an anchor.

Local distributed measurement

In local distributed measurements, a standard fibre optic cable itself is the sensitive element used for the measurement. Glass fibres are integrated in special cable construction, which ensures a secure connection between the glass fibre and the outer jacket. The integration of the cable in, e.g. cement suspension or concrete is possible on the basis of a strong surface structuring of the outer mantle of the cable (see Fig. 1) or by direct connection to the structure by means of epoxy and stabilizing mats such as glass or carbon fibres.
The physical quantities acting on the fibre from the outside, such as strain, compression, transverse forces or temperature, influence the scattering of the laser light in the glass fibre via different interaction mechanisms. Local distributed fibre-optic methods are using this property of light scattering (Kindler et al. 2017).
During the measurement, short laser light pulses are coupled into the glass fibre. On their way through the fibre the light pulses are scattered on the glass molecules. A small part of the laser light is then backscattered. In addition to the Rayleigh peak, the backscatter spectrum also contains Brillouin lines as well as Raman lines. Depending on the physical quantity to be measured, either

- the Rayleigh peak for the measurement of the optical damping along the glass fibre,
- the Brillouin lines for the measurement of the strain along the glass fibre or
- the Raman lines for the measurement of the temperature along the glass fibre.

3. Case study 1: Anchor Monitoring

When constructing a prestressed structures (for example, the roof of the Olympic stadium in Munich or pylons of a bridge structures), it is necessary to permanently secure the forces in the ground by means of permanent anchors. In order to establish the load bearing capacity of these

anchors over the entire life time of such a structure, the anchors have to be specially instrument. Until now it has only been possible to do so by the means of anchor force transducer (Wichter, Meininger 2000).
Today's permanent anchor monitoring systems are based on the use of anchor force transducer. The installation of a force transducer requires additional height or supplementary structural adjustments of the anchor heads. Moreover, it is known that all force sensors, whether electrical or hydraulic, lose their reliability over time (Wichter, Meininger 2000).
Other measurement and test methods, e.g. conventional optical waveguides sensors require an adjustment with an unloaded optical waveguide and give only punctual information about the carrying behaviour. Furthermore this information has to be interpreted in order to evaluate the state of the anchor.
It is therefore necessary to find new ways to reliably monitor the component "grout anchor" over its entire life time. For this reason, attempts have been made to develop permanent anchor monitoring system on the basis of a distributed strain-sensitive sensor cable (Kindler et al., 2017c).

3.1 Measuring cable assembly in the anchoring area of the permanent anchor

For the first test phase, fife strand anchors (5-0.6 ") were instrumented with a corresponding sensor cable before their installation in the ground. Total length of the test anchors was 35 m. In the area of the grout body, the strain-sensitive sensor cable was fixed under minimal tension in the center of the individual strands (Figure 1). Thus, the strain measurement was performed in the later center of the grout body. In the area of the free steel length, the glass fibre cable was driven unstressed trough a protective tube.

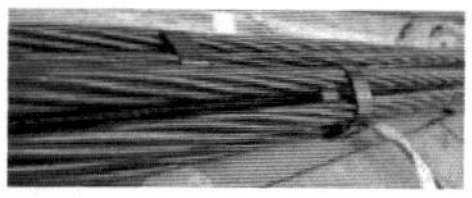

Figure 1 End position of the fibre optic cable in the area of the grout body

3.2 Measuring test sequence

The measurements in the anchor grout body were conducted with the commercial system ODiSI (optical distributed sensor interrogator). The measuring point detection was 0,005 m with a spatial resolution of approx. 0,05 m.
In order to obtain plausible measurements results all anchors were subjected to suitability test in accordance with DIN EN 1537 (2014). All anchors showed with their force-displacement line typical deformation behaviour in granular, pre-loaded soils.

3.3 Measurement results

General information
In the course of the test run, data for all the initial load stages were recorded at the beginning, in the middle (approx. after half of the stage load hold time) as well as at the end of the load stage hold

time. For the load relief and reloading stages, the measurement data acquisition for each load stage was performed only once at the end of the load stage hold time.

Overall evaluation of the results

As a result of the evaluation of all the measured anchors, a characteristic strain pattern in the grout body was established. Fig. 2 shows the measuring results of the strain measurements over all stress stages for the test anchor 5.

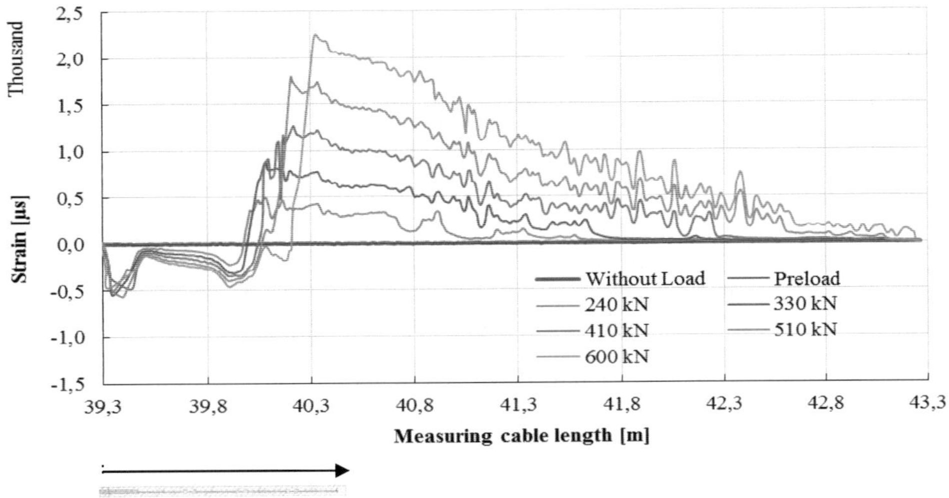

Grout body length [m]

Figure 2 Strain characteristic curves of test anchor 5 for all load levels 60, 240, 330, 420, 510 and 600 kN

The observed strain profile and, in particular, the steeply extending strain front (Fig. 2) did not correspond to the relevant technical literature but were described for the first time by Kindler et al. (2016a).

The evaluation of all test results of the individual test anchors showed that the strain behavior in the grout body can be reproduced. This leads to the question of maintaining this characteristic when the load is changed, for example, from relieving or reloading processes.

Individual evaluation of the load stages over the entire load cycle

In order to clarify the question of obtaining the characteristic, the load stages with the highest repetition frequency were considered for all test anchors. It could be shown that the load step characteristic of the maximum test load depicts in all subsequent loads and reload phases (Fig. 3).

If, as a result of the stress on the anchors, a later overload occurs, a new load step characteristic is established with the overloading. On the basis of the strain characteristic, it can be evaluated whether the grout body in its load range is. If an overload has occurred during the life time of the anchor, the value of the overload can be determined by a gradual lifting of the load during the implementation of a lifting test.

This result is the prerequisite for the application of the distributed fibre optic sensor for monitoring of the anchors bearing capacity over time.

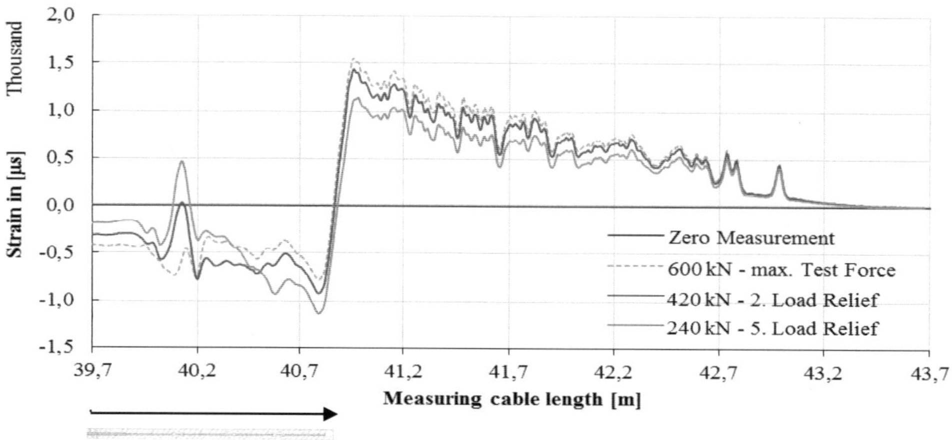

Grout body length [m]

Figure 1 Change in the strain curve upon relief to 240 kN after inducing of the maximum test load of 600 kN on the grout body of the test anchors 3

This means that as results of the suitability and compliance tests a characteristic, even "chaotic," crack pattern of the anchor grout body is obtained, which can be referred to as a load stage characteristic.
On the basis of this load stage characteristic, it is possible, according to current knowledge, for granular soils to determine whether the installed and tested anchor has been subjected to a load larger than in its test load.

4. Case study 2: Strain measurement on a bored and driven pile

In the course of the construction of a bridge structure, it is necessary to check it regard to its structural state (DIN 1076, 1999). Common bridge tests are restricted only to visible and accessible bridge components. Since bridge structures are often situated in places which are not characterized by markedly good ground conditions, it is necessary to found these bridge structures deeply in the ground.
In the year 2016 in Germany, large-bore pile for a bridge construction was instrumented with a strain-sensitive sensor cable over the entire pile length of approx. 31 m. In addition to the strain-sensitive sensor cable, the bored pile was also equipped with standard strain measuring gauges in 4 + 1 measuring levels (reference measurement). The instrumented large bore pile was designed as a floating bored pile (D = 1,2 m) in a soft sea tone.
Due to its length, the total reinforcement cage was composed of two partial reinforcement bins. Figure 4 shows a view of the instrumented lower partial reinforcing cage with strain-sensitive sensor cable (blue), conventional strain gauge (red) and cuff tubes for jacket compression.
In total, the reinforcement cage was equipped with 3 x 5 = 15 conventional strain gauges (as measuring points) and a strain sensitive fibre optic sensor cable with a total length of 4 x 30 m = 120 m (as a loop) and 600 measuring points (0,2 m).

Figure 2 Instrumented reinforcing cage of the large bored pile with fibre optic sensor cable (blue) and conventional strain gauges (red)

4.1 Measuring test sequence

During the test load, the measured data were automatically read out, saved and evaluated online in a 2-minute time intervals. Due to the lack of approval by the client, the results obtained in particular the most interesting comparison of the conventional measuring results with the results of the strain-sensitive sensor cable cannot be presented at this point.
For this reason a further project was conducted, where a driven concrete pile was equipped with conventional strain gauges (3 pieces) and a fibre optic measuring sensor cable (4 x 7.6 m). According to DIN EN 14199 (2015) the test pile with an outer diameter of 300 mm is to be classified as a micro pile. Total length of the pile was 7.6 m. The in-situ concrete pile was constructed as a friction pile

4.2 Measurement Results

Measurement of conventional strain gauges
In Figure 5, the course of the conventional strain measurements over the time of all three measurement sensors is shown.

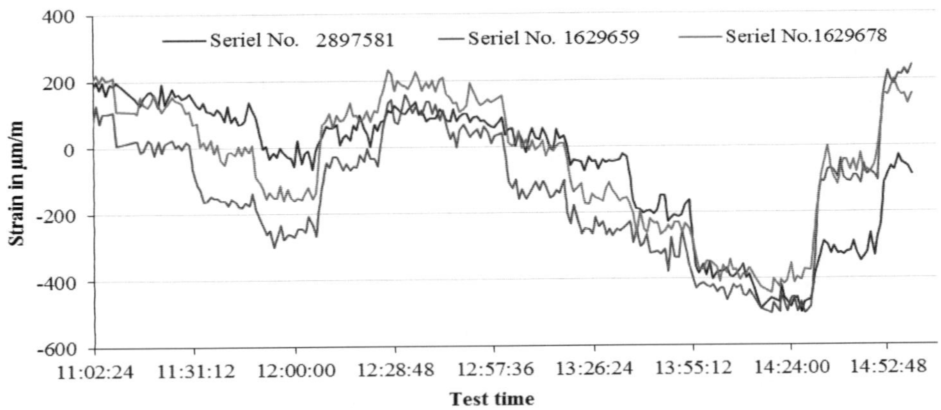

Figure 3 Measurement results of the three conventional strain gauges over the entire test period (pile test load)

The curves of the strain gauges correspond, as expected, with the individual load stages.

Measured value acquisition of the strain sensitive sensor cable
Corresponding to the measuring principle, frequencies (Brillouin frequency) are measured as the output measuring variable along the strain-sensitive sensor cable. Figure 6 shows the measured Brillouin frequency for the initial state (zero measurement) before the first load input.

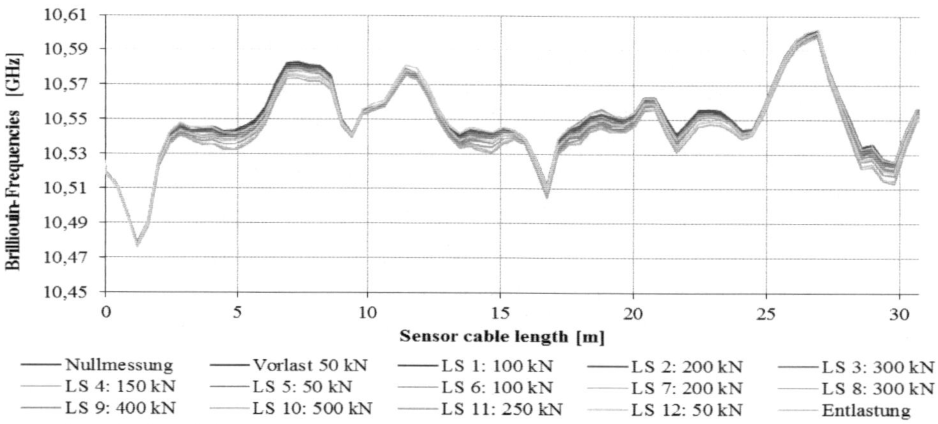

Figure 4 Distribution of the measured Brillouin frequency via the sensor cable length over all load stages

On the basis of the measuring curves, the range of the strain-sensitive sensor cable, which is integrated in the in-situ concrete pile, can be clearly distinguished from the air exposed areas (feed and return) of the sensor cable.

The measuring-relevant area of the strain-sensitive sensor cable is located between the cable positions 24 m and 72 m, as shown in Figure 6. The scattering of the measured values in this range is significantly lower than in the range of the pigtails. The positions of the plugs that connect the strain-sensitive sensor cable to the extension cable and the pigtails are characterized by the artefacts in the profile of the scattering of the Brillouin frequency.

Since the in-situ concrete pile in the soil can be considered a quasi-stationary thermal equilibrium state and the measurement time was only a period of 6.5 hours, temperature compensation could be dispensed with from the measurement point of view.

The strain changes are calculated from the measured Brillouin frequency in Figure 6 on the basis of the relationship between the Brillouin frequency shift and the strain.

Figure 7 shows the deformation profiles over the complete pile length caused by individual load stages.

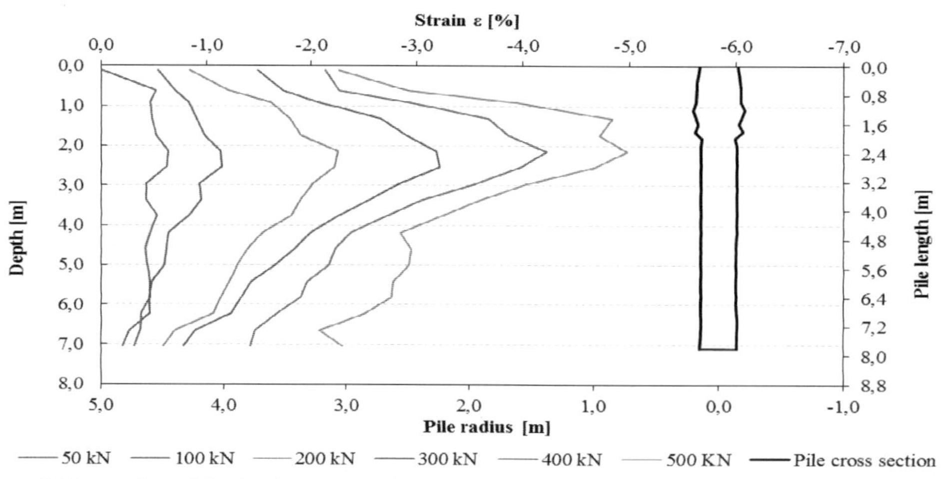

Figure 5 Mean value of the development of the deformation processes over the in-situ concrete pile depth

Overall assessment of results over the pile shaft

As a result of the evaluation of the strain profiles measured with conventional strain gauges, a typical normal force profile results over all stress load stages. Figure 8 shows the recalculated normal force curve for the load stage LS 500 kN over the pile depth.

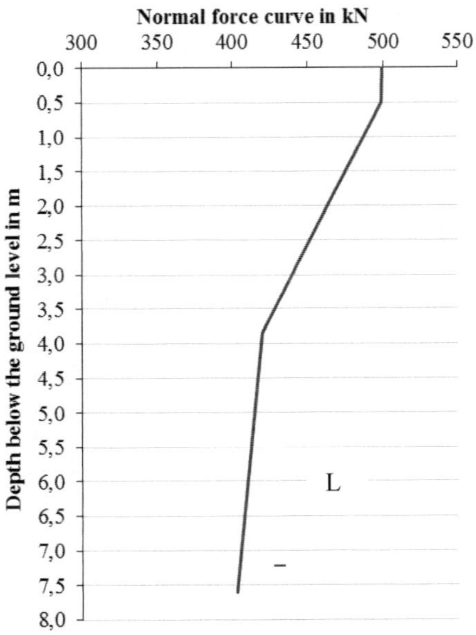

Figure 6 Recalculated normal force curve for the load stages LS = 500 kN over the pile depth based on the measurement results of three conventional strain gauges

If the evaluation is carried out for measurements conducted with the strain-sensitive sensor cable which is distributed over the entire pile length, the normal force profile over the pile shaft is as shown in Figure 9.

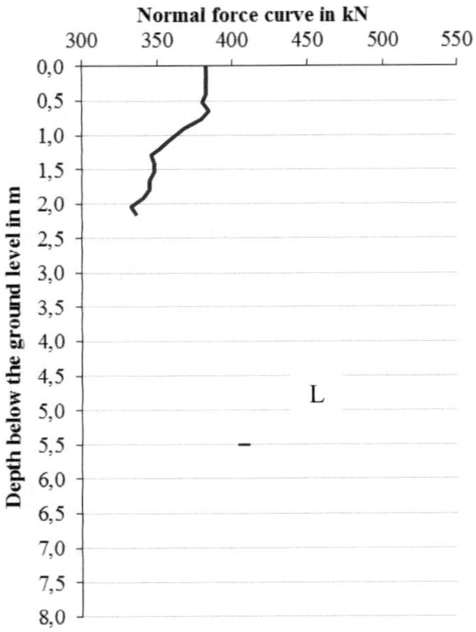

Figure 7 Recalculated normal force profile of the load stages LS = 500 kN over the depth on the basis of the measuring results of the fibre -optic strain measurement

Comparison of the measurement results

If the Figure 8 and the Figure 9 are considered, the differences in the normal force distribution over the pile length are evident as a result of the higher number of measuring points. For a better illustration, the standard force distribution curve of the load stage 10: 500 kN from Figure 8 and Figure 9 are compared as a representative of all load stages in Figure 10.

The recalculated normal force curves show significant differences in their course over the pile depth. These differences can be explained and interpreted most clearly by a consideration of the associated soil profile (see Figure 10).

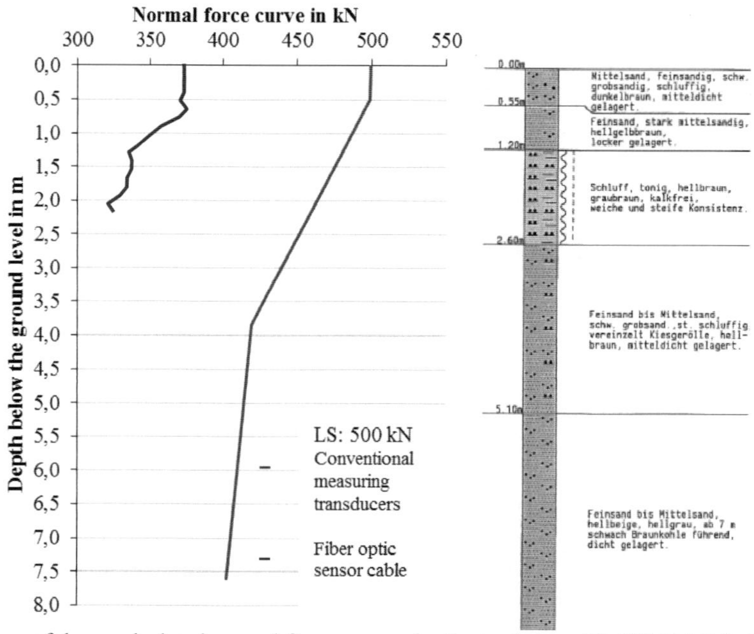

Figure 8 Comparison of the recalculated normal force curves for the load stage 10: 500 kN (LS: 10) over the depth based on conventional and fibre -optic strain measurements and the corresponding ground profile

In Figure 10, as already described, a linear curve is shown in the normal force profile of the measurements with conventional strain gauges.

Although this result is conditional to the small number of strain gauges, a larger number of conventional strain gauges for a pile of 7.6 m length is not customary or cannot be realized from a constructive point of view. This shows that a linear interpolation of the measurement results can greatly overstate or underestimate the prevailing strain conditions. On the other hand, the fibre - optic strain measurement provides much detailed measuring results. It can be concluded from Figure 10 that the strain behavior in a pile can be much more difficult than expected.

3. Conclusions

Advanced commercial systems using the Brillouin technology and the Rayleigh technology were successfully tested for geotechnical monitoring. The successful tests open a lot of new possibilities in geotechnical monitoring with the goal to create a good balance between improving safety and reducing costs.

References

Dietz, K.; Habel, W. & Feddersen, I., 2001, 'Eder Dam, stabilisation by permanent rock anchors-monitoring and long term performance', ICOLD – 69th ANNUAL MEETING, DRESDEN

Hurtig, E., Großwig, S., Jobmann, M., Kühn, K., & Marschall, P., 1994 'Fibre Optic Temperature Measurements in Shallow Boreholes: Application for Fluid Logging', Geothermics 23 (4), pp. 355 - 364.

Hurtig, E., Großwig, S., & Kühn, K., 1995 'Monitoring landfill bases and landfill base sealing with fibre -optic temperature measurements', 3rd Working Conference of the BMBF Collaborative Research Project "Further development of landfill sealing systems", BAM Federal Institute for Materials Research and Testing, Berlin, Conference Transcript, pp. 341-350.

Kindler, A., Schaller, M.-B., Glötzl, J., Brentle, J. & Lich, A, 2016, 'Nachweis der Ankertragfähigkeit am Beispiel eines innovativen Ankermonitorings', 12. Hans Lorenz Symposium, Technical University Berlin., Conference Transcript

Kindler, A., Schaller, M.-B., Glötzl, J., 2017, 'Nachweis der Ankertragfähigkeit auf Grundlage der faseroptischer Messtechnik', Bautechnik, vol. 94, no. 2, pp. 144–151

Kindler, A., Schaller, M.-B., Glötzl, J., & Großwig, 2017, 'Static Pile Load Test using Fibre -optic Strain Measurements with a Very High Local Resolution', Pile Foundation Symposium 2017, Braunschweig.

Nishigaki, M., Schreck, M., Matsuo, Y., Großwig, S.; Kühn, K., & Döring, H., 1997, 'Intrinsic Distributed Fibre Optic Measurement - A new age of sensing ?', Sensor Magazine vol. 4/97, pp. 24 – 29

Nöther, N., 2010, 'Distributed Fibre Sensors in River Embankments: Advancing and Implementing the Brillouin Optical Frequency Domain Analysis', BAM-Dissertationsreihe, Band 64

Samiec, D., 2011, ‚Verteilte faseroptische Temperatur- und Dehnungsmessung mit sehr hoher Ortsauflösung', Photonik 6 , no. 4, pp. 34–37

Schreck, M., Nishigaki, M., Christoph, G., Gläser, H., Großwig, S., Hurtig, E., Kasch, M. & Kühn, K., 1998, ‚Determination of Heat Production Zones at an Open - Cast Mine', Journal of Environmental Engineering of the American Society of Civil Engineers, vol. 124 (7), pp. 646 - 651.

Schaller, M.-B., Großwig, S., Pfeiffer, Th., 2016, 'Measuring systems and monitoring with regard to Industry 4.0, Monitoring of structures during construction', 1st Symposium "B3 - Südschnellweg Hannover Expansion", Hannover

Voet, M.R.-H.; Glötzl, F, 1996, 'Glass-fibre sensor technology and results of innovative activities in geotechnics', Specialized seminar on "Measurements in Geotechnics", Institute of Ground Engineering and Soil Mechanics, Technical University of Braunschweig.

DIN 1076:1999-11: Engineering structures in connection with roads - inspection and test, Beuth-Verlag Berlin

DIN EN 1537: 2014-07: Ausführung von Arbeiten im Spezialtiefbau - Verpressanker; Deutsche Fassung EN 1537:2013. Beuth-Verlag Berlin.

DIN EN 14199:2015-07: Ausführung von Arbeiten im Spezialtiefbau - Mikropfähle; Deutsche Fassung EN 14199:2015 Beuth-Verlag Berlin

The 8th International Conference on
Structural Health Monitoring of Intelligent Infrastructure
Brisbane, Australia | 5-8 December 2017

Intelligent Pipelines with Integrated Fibre-Optics for Single-Ended Distributed Strain Sensing

F. Kammann[1], B. Marx[2], W. Hill[3], H. Brauer[4], H. Karbasian[5], D. Krix[6].
[1] *LIOS Technology GmbH – Germany, Email:* florian.kammann@lios-tech.com, [2,3] *LIOS Technology GmbH - Germany,* [4]*Salzgitter Mannesmann Line Pipe GmbH - Germany,* [5,6]*Salzgitter Mannesmann Forschung GmbH - Germany*

Abstract

Pipelines can be exposed to considerable mechanical load during installation and in operation, e.g. by coiling on a reel or by ground movements. For a safe operation of pipelines, it is essential not to exceed certain strain levels. Plastic deformation of steel pipes starts at about 0.5 % of strain. Pipes may buckle or even tear up at higher strain loads.
Fibre-optic sensor cables mounted at pipelines or laid nearby have already been used to detect geohazards. Mostly, sensor cables were not tightly bonded to the pipelines, but embedded in the surrounding soil. This resulted in good sensitivity to ground movements. We introduce pipelines where the fibre-optic sensor cables are embedded in the cement mortar coating of pipelines. A strong mechanical coupling is achieved, and the sensor cables are capable of directly monitoring strain at the pipeline during installation and operation. A state-of-the-art Brillouin Optical Time Domain Reflectometry (BOTDR) interrogator can be used to monitor multiple pipelines with a length up to 70 km each. Different installation schemes and fibre-optic cable designs have been evaluated in a trial on a test pipeline.
We investigated the transfer of strain for different sensor cable layouts and installation schemes during deformation of a segment of an intelligent pipeline. Longitudinal installation of sensor cables parallel to the pipe resulted in high strain sensitivity. Strain readings of the BOTDR system were in good quantitative agreement with data from strain gauges. The system gives spatially resolved strain data that may also be used to detect the direction of the pipeline deformation. Negative strain (compression) of pipelines was monitored using pre-strained fibres. Polymer and FIMT sensor cables both gave similarly good strain responses.

1. Introduction

Pipelines can be exposed to considerable mechanical load during installation and in operation, e.g. by coiling on a reel or by ground movements. For a safe operation of pipelines, it is essential not to exceed certain strain levels. Plastic deformation of steel pipes starts at about 0.5 % of strain. Pipes may buckle or even tear up at higher strain loads.
Fibre-optic sensor cables mounted at pipelines or laid nearby have already been used to detect geohazards (Inaudi, D. & Glisic, B. 2010). Mostly, sensor cables were not tightly bonded to the pipelines, but embedded in the surrounding soil. This resulted in good sensitivity to ground movements. We introduce pipelines where the fibre-optic sensor cables are embedded in the cement mortar coating of pipelines. A strong mechanical coupling is achieved, and the sensor cables are capable of directly monitoring strain at the pipeline during installation and operation.

In this study we investigate the transfer of strain for different sensor cable lay-outs and installation schemes during deformation of a segment of an intelligent pipeline.

2. Assembly of the Intelligent Pipeline

The trial was conducted at Salzgitter Mannesmann Line Pipe in Siegen, Germany. The pipeline consists of twelve steel pipes that are welded to each other. The ten centre pipes are equipped with fibres. Each pipe has a length of 12 m and a diameter of 200 mm. The steel pipes are protected against corrosion and mechanical damage by a three-layer polyethylene (PE) outer coating. Three different cable types were used: one fibre-in-metal-tube (FIMT) design and two polymer cables. Tight buffers were applied to all sensor fibres for efficient strain transfer from the cable surface to the fibre itself.
Fibre-optic sensor cables were installed on the PE coating in linear and helical schemes as presented in Figure 1. The linear installation was performed manually at the following positions: 3, 6 and 12 o'clock on the pipes. Each 0.5 m the cable was fixed using adhesive tape. The cable for the helical installation was fed manually onto the pipe, which was constantly turned. The spacing length between each loop was 0.5 m. Cement mortar coating was applied over the sensor cables mounted on the PE coating. After hardening of the mortar the pipes were welded together and the optical fibres from the different pipes and installation schemes were spliced together. Strain gauges were installed as reference sensors at four of the joint locations at the 12 o'clock position of the pipes.

Figure 1: Installation of sensor cables at pipes. Left: linear installation; Centre: helical installation; Right: application of cement mortar coating

3. Experimental Set-Up

Strain was induced in the pipeline using a crane that lifted the centre part of the pipeline. The pipeline was lifted to four different height levels corresponding to strain gauge readings of 0.05 %, 0.1 %, 0.15 % and 0.2 %, at the mostly stretched location. The strain in the fibres was measured and recorded using a PRE.VENT DSM system. Readings of the strain gauges were recorded continuously during the experiments.
The PRE.VENT DSM interrogator for distributed strain sensing is based on spontaneous Brillouin scattering in standard telecommunication fibres and does not require a fibre loop or special sensor elements at the pipeline. It is cooled completely passively and does not contain any moving or wearing parts. The instrument is dedicated for use in industrial and other rough environments. With a measuring time below 10 min, the strain resolution is better than 20 $\mu\varepsilon$ for distances up to 50 km

(Marx, B., Rath, A., Kolm, F., Schröder, A., Buntebarth, C., Dreß, A. & Hill, W. 2016). For the presented investigation, a spatial resolution of 1 m was used and a strain range of up to 2 % was scanned within a measurement time of a few minutes.

4. Results and Discussion

Brillouin peak frequencies are a function of strain and temperature (Parker, T.R., Farhadiroushan, M., Handerek, V.A. & Rogers, A.J. 1997). We use the loose fibre sections in the splice cassettes as a reference for the temperature effect. This enables us to determine strain within the optical fibres from the Brillouin frequencies.
Strain patterns derived from Brillouin peak frequencies of helically installed sensor cables are shown in Figure 2 left.

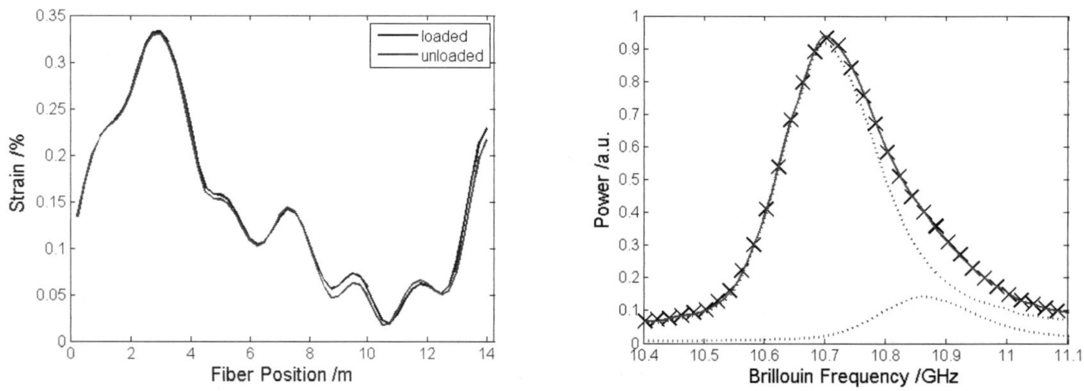

Figure 2: Results for helical cable installation; left: strain measurement over one pipe for the unloaded (red) and 0.20 % loaded (blue) state, the pipe is lifted at the right end of the graph; right: Brillouin spectrum (red) can be decomposed in a main peak and a small side peak (both dotted in blue)

Even if there was no mechanical load applied to the pipeline, the fibre showed a strong strain pattern reaching from 0.02 % to 0.33 % of strain. This result shows that the manual feeding of the sensor cable during installation results in a considerable and uneven strain load on the sensor fibre. Thus, for determining the load-induced strain, a reference strain profile has to be subtracted. Negative difference values between profiles with and without strain load point to load-induced compression.
The helical installation is favourable to detect, e.g., inner pressure of the pipeline. In this case the expansion of the pipe leads to an increase of the pipeline diameter and to a shift of the Brillouin frequency. In the experiment, the fibre distance between maximum strain on the top and the maximum compression on the bottom of the pipe was only 0.4 m, whereas the spatial resolution of the Brillouin measurements was set to 1 m. The effect of the 0.4 m being smaller than the spatial resolution of the system was observed in small changes of the measured average central frequency, because the analysed signal was averaged over domains with predominant positive or negative strain (e.g. in Figure 2 left increased strain at positions of 5 m and 9.5 m and decreased strain at position of 11.5 m).
Bending of the pipeline should lead to closely neighboured strain and compression locations resulting in inhomogeneous broadening of the Brillouin spectrum (Ravet, F., Bao, X., Ozbakkaglu, T. & Saatcioglu, M. 2006). Figure 2 right shows the peak in a loaded part of the pipe. The peak profile can be decomposed into a main peak and a side peak. However, the side peak is relatively

low. In this measurement the peak broadening could not be found to scale with the pipe load. The measurements of the longitudinal installed cable show an excellent coupling of the sensor cable to the pipeline. We assume that also the helically installed sensor cable was well coupled to the pipeline. Since the distance between maximum and minimum strain was only 0.4 m, we assume that optical fibre compensated this alternating stress by short-distance relaxation movements.
The longitudinal cable installation shows to be favourable for the sensing of the applied load. Figure 3 left presents the measurement results of four pipe segments. The cable was installed at the 12 o'clock position of each pipe. Actual maximum strain values were matched with the strain gauge. The strain profile along the pipeline is well-resolved and shows a maximum at the position of lifting at a position of 36 m, a decrease in strain going further down the pipeline, going through a minimum in an area of negative strain where the top of the pipe is compressed and finally increasing again to a value of zero. The good agreement between the results of the strain gauge and the distributed measurement proves the excellent strain transfer from the pipe to the cable and from there to the fibre itself. Besides the positive strain, negative strain can be observed as well. This is possible, because pre-strained cables were used. For higher lifting of the pipe it can be seen that the position of minimum strain moves farther away from the maximum as expected.

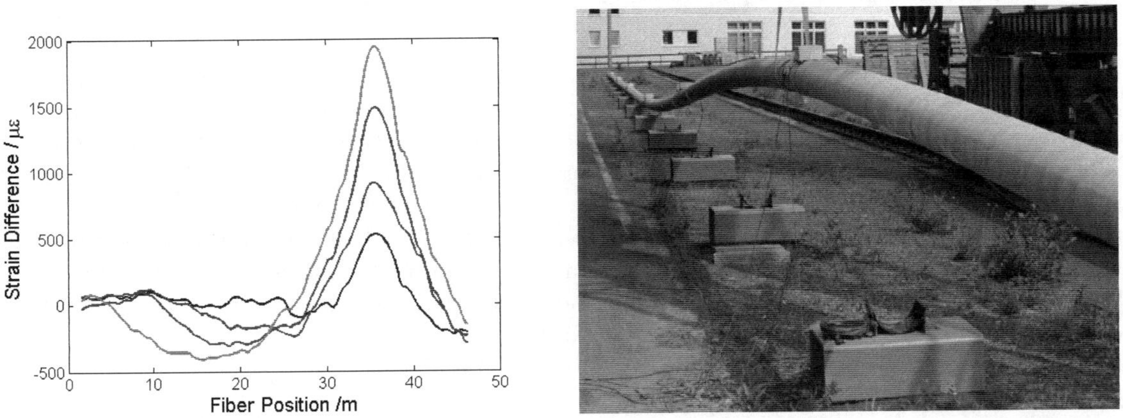

Figure 3: Results for linear cable installation; left: strain difference to reference measurement at cable position 12 o'clock for the following maximum loads: 0.5 % (blue), 1.0 % (green), 1.5 % (red), 2.0 % (turquoise); right: picture of the pipeline lifted by the crane

The exact deformation of the pipeline can be determined using several sensor cables per segment, which are positioned at complementary positions around the pipe. During the experiments fibres were positioned at three positions of the pipeline: 3, 6 and 12 o'clock. The strain at the side of the pipeline is expected to be low, because the position is close to the neutral axis of the bent pipe. Measurement results (not shown) confirm that almost no strain was detected at the side of the pipeline. Figure 4 left presents the comparison between cables at the 6 o'clock and 12 o'clock positions over the length of one pipe. When a pipeline is lifted it experiences positive strain on the top and negative strain on the bottom. The measured strain at the top of the pipeline exceeds the measured compression at the bottom. This is expected for this experimental set-up, where the tensile strain exceeds the compressive strain of the pipe. Thus one may derive the shape of the pipe from the strain pattern of the three sensor cables.

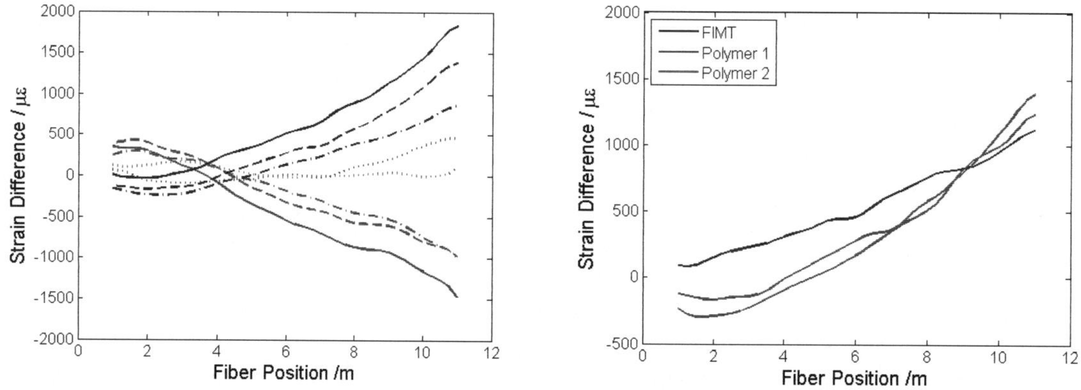

Figure 4: Results for linear cable installation; left: strain difference to reference measurement for one pipe and increasing loads 0.05 %, 0.10 %, 0.15 %, 0.20 %, blue: cable on the top of the pipe, red: cable on the bottom of the pipe; right: strain difference to reference measurement for 0.15 % maximum strain for three different cables. blue: FIMT (pipe position at end of pipeline), green: polymer 1, red: polymer 2 (for both pipes position in the middle of the pipeline)

In this study we also compare three different cable types (Figure 4 right). All cable types were carefully chosen and could easily withstand the installation procedure. Figure 4 right shows the strain measurements for a load of 0.15 % strain. Note that the results for the polymer cables are from the same pipe, while the FIMT result is from a pipe which is near the end of the pipeline. The two polymer cables show generally similar response in terms of the slope, but Polymer 1 seems to be more sensitive to tensile stress whereas Polymer 2 responds better to compressive strain. The strain in the FIMT was lower in comparison due to the position close to the open end, so that a direct comparison of the response to the polymer cables is limited, but one could conclude on a lower strain transfer due to the higher elastic modulus.

5. Conclusions

We demonstrated the strain sensing capabilities of pipelines with fibre-optic sensor cables embedded in cement mortar in combination with a state-of-the-art BOTDR interrogator. Longitudinal installation of sensor cables parallel to the pipe resulted in high strain sensitivity. Strain readings of the BOTDR system were in good quantitative agreement with data from strain gauges. The system gives spatially resolved strain data that may also be used to detect the direction of the pipeline deformation. Negative strain (compression) of pipelines was monitored using pre-strained fibres. Polymer and FIMT sensor cables both gave similarly good strain responses.

6. Outlook

Integrated distributed fibre optic sensing can be used for measurement of strain development of a complete steel pipeline or of single pipes, which might be located in critical areas like curves, slopes, river crossings or railways. Additionally, the optical fibre technology can be used for detection of third-party intrusions. Not only pipelines can be supplied with fibre optical sensing, also round, rectangular or square pipes for e.g. constructions or buildings can be monitored using this technique. Fibre optic cables can be installed in the hollows after production process in the pipe mill. In this way they are not visible from outside and do not need additional protection against mechanical or climatic impacts during service life in the topic of Structural Health Monitoring.

References

Inaudi, D. & Glisic, B. 2010, "Long-Range Pipeline Monitoring by Distributed Fiber Optic Sensing", *Journal of Pressure Vessel Technology*, vol. 132, pp. 011701-1-011707-9

Marx, B., Rath, A., Kolm, F., Schröder, A., Buntebarth, C., Dreß, A. & Hill, W. 2016, "Brillouin Distributed Temperature Sensing System for Monitoring of Submarine Export Cables of Off-Shore Wind Farms", *Proc. SPIE*, vol. 9916, pp. 99162U-1 - 99162U-4

Parker, T.R., Farhadiroushan, M., Handerek, V.A. & Rogers, A.J. 1997, "Temperature and strain dependence of the power level and frequency of spontaneous Brillouin scattering in optical fibers", *Optics Letters*, vol. 22, no. 11, pp. 787-789

Ravet, F., Bao, X., Ozbakkaglu, T. & Saatcioglu, M. 2006, "Signature of Structure Failure Using Asymmetric and Broadening Factors of Brillouin Spectrum", *IEEE Photonics Technology Letters*, vol. 18, no. 2, pp. 394-396

The 8th International Conference on
Structural Health Monitoring of Intelligent Infrastructure
Brisbane, Australia | 5-8 December 2017

Sensor development for monitoring bolt tightness in steel truss girder of a cable-stayed bridge

Zhen Sun[1], Xinhong Ouyang[1]

[1] *JSTI Group, Jiangsu, China, Email:* sunzhen08@gmail.com

Abstract

Steel truss girder system is a commonly adopted main girder type in many bridges, and bolts are often used for joint connection. During bridge operation, bolts may get loose due to wind load and truck load, which pose a danger for safety of the bridge. As loose bolts are not visible as other deterioration such as corrosion or crack, they cannot be detected by visual inspection, which is subjective and labor intensive. Structural health monitoring (SHM) gains more popularity for assessment of bridge performance. However, there is no available sensor to monitor bolt tightness. Condition assessment of loose bolts is a challenging task in bridge maintenance. In this paper, a new type of sensor is developed to monitor bolt tightness in real time. First, the mechanism of the sensor is explained and basic manufacturing characteristics are introduced. The sensor obtains the tightening force through recorded strain measurement. Second, the sensor is used on steel truss girder of a cable-stayed bridge, and data acquisition system is developed using LABVIEW software. Third, the sensor performance is evaluated with the monitored data, and the results show that the sensor can be used to monitor bolt tightness in steel truss girder bridges. Finally, conclusions are drawn for reliable application in engineering practice.

Keywords
Steel truss bridge; Bolt tightness; Structural health monitoring; data acquisition system; Sensor performance

1. Introduction

Bolt connection is widely used in steel truss bridges due to ease of implementation and high quality during the construction process. In the harsh environment, however, bolt connection may experience deterioration such as bolt loosening (Figure 1). This will reduce the load carrying capacity of bridges [1-3], and may cause severe bridge failure if not detected at an early stage. Therefore, early detection of preload change in the bolts is significant to ensure safety and reliability of bridges.

In the past few years, there have been a lot of efforts devoted to reliable monitoring of bolt tightness, which include vibration based method [4-10], ultrasonic method, piezoelectric method and impedance method. Vibration based methods use parameters as power spectral density [5], transmittance function [4] to detect damage before and after bolt loosening.

However, vibration based method require information of both undamaged and damage states, and it is easily affected by environment such as temperature, wind, etc. It is difficult to apply this method for online monitoring of bolt looseness status [11-12].

In this paper, a new type of sensor is developed to monitor bolt tightness through measuring the bolt tension with strain gauges on the washer. First, the mechanism of the sensor is explained and basic manufacturing characteristics are introduced. The sensor obtains the tightening force through recorded strain measurement. Second, the sensor is used on steel truss girder of a cable-stayed bridge, and data acquisition system is developed using LABVIEW software. Third, the sensor performance is evaluated with the monitored data, and the results show that the sensor can be used to monitor bolt tightness in steel truss girder bridges. Finally, conclusions are drawn for reliable application in engineering practice.

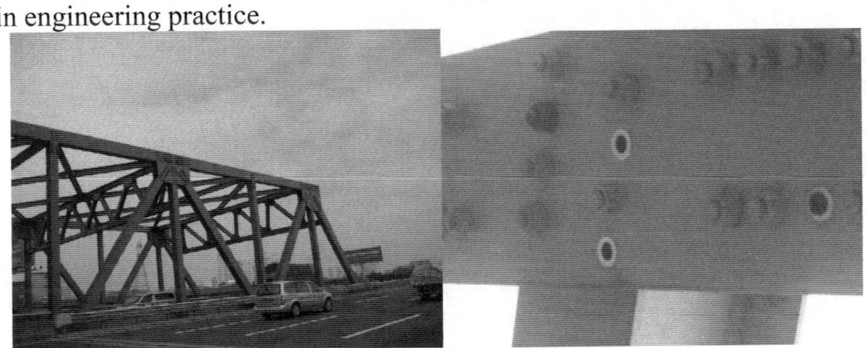

Figure 1 The bridge and lost bolts.

2. Mechanism of the sensor measurement

The sensor is a steel cylinder with 3 strain sensors put inside, as shown in Figure 2. When the cylinder is stressed longitudinally, strain gauges will measure the amplitude, and then the load can be calculated. The structure of the sensor is illustrated in Figure 3.

The strain gauges measure the vibrating frequency f. And R is calculated as follows.

$$R=f^2/1000 \tag{1}$$

The fastening load is calculated as

$$P = G \times (R_0 - R_1) + K(T_1 - T_0) \tag{2}$$

In which, G and K are factors of Loading and Temperature respectively; R0 and R1 are the values of initial and current loading; T0 and T1 are the initial and current temperature respectively.

Figure 2 Photo of the sensor

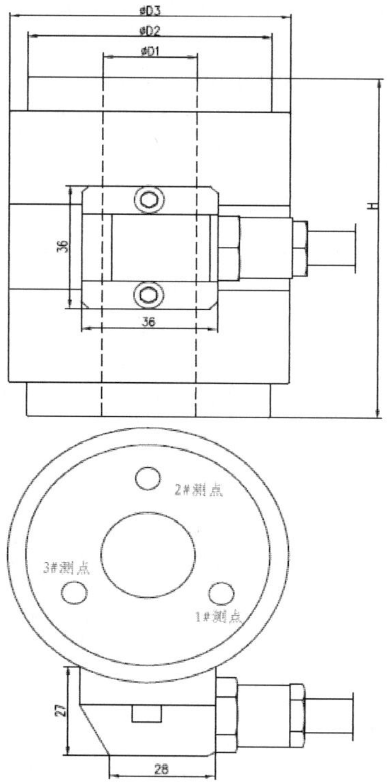

Figure 3 Structure of the sensor

The properties of the sensor are as follows.

Measurement Scope：200～999 KN

Sensitivity：0.07% FS

Working temperature：-40°C～+65°C

3. Sensor application on the bridge

Yachihe Bridge is a cable stayed bridge with a main span of 800 m (Figure 4). The pylon is an H shaped concrete pylon, and the main span is a steel truss girder. Two types of bolts were used, which are M30 and M24 in the connection of truss members.

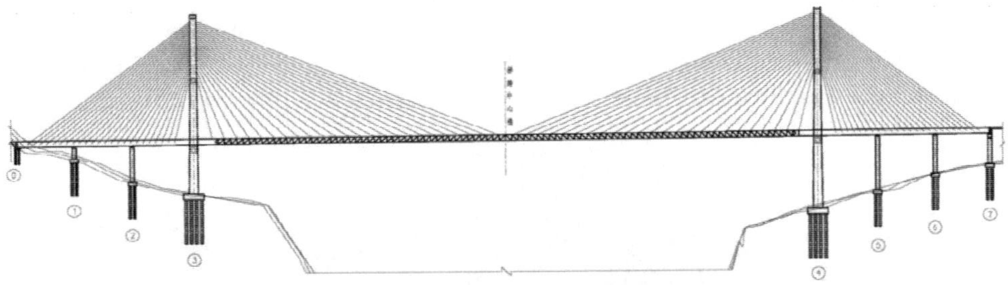

Figure 4 Yachihe Bridge

The designed preloads for M24 and M30 bolts are 240 kN and 360 kN respectively. The bolts are friction type, with a friction coefficient of not less than 0.45. The structure of the bolts is shown in Figure 5, and detailed characteristics are shown in Table 1.

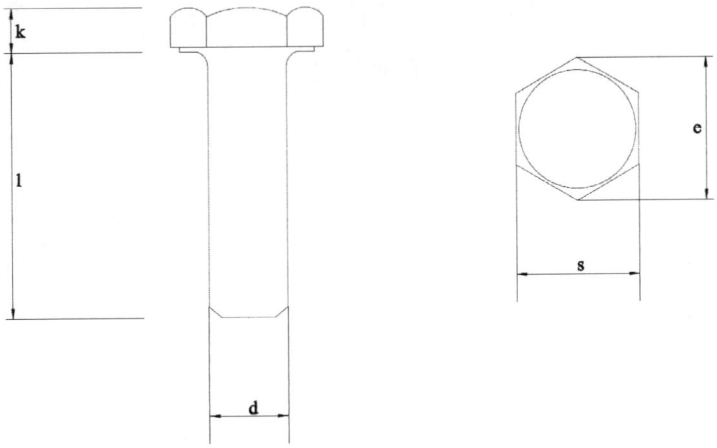

Figure 5 Bolt structure (Unit: mm)

Table 1 Characteristics of the bolts

d	k			e	s	
	Average	Max	Min	Min	Max	Min
M24	15	15.90	14.10	45.20	41	40
M30	18.7	19.75	17.65	55.37	50	49

The data acquisition system is developed with Labview, as shown in Figure 6. It consists of personal computer, machine box and data acquisition unit. The machine boxes are connected with 485 cabling, and the signal is transmitted with RS485/RS232. BGK-8001 data acquisition unit is used, which can connect with 192 sensors simultaneously.

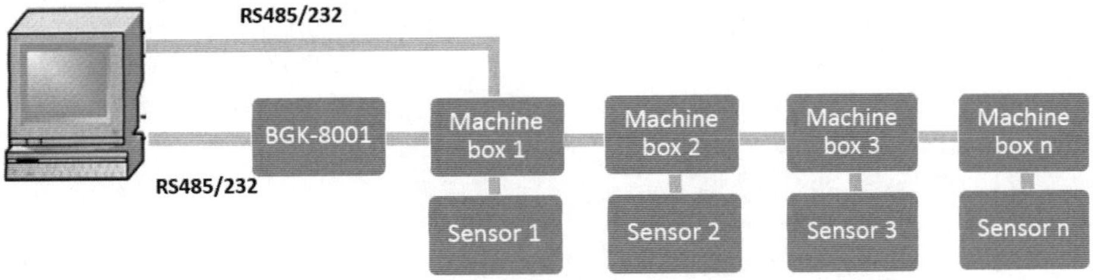

Figure 6 Data acquisition system

4. Installation and evaluation with monitored data

The targeted bolt for monitoring is selected through finite element analysis, to find the most adverse cross-section under wind and vehicle loads. The bolt is loosened with a torch wrench, and then the sensor is installed. Installed sensor is shown in Figure 7, and one day measurement is extracted for the evaluation.

Figure 7 Installed sensor on the bridge

After installation of the sensor, its efficacy and reliability are evaluated with monitored data. Design force of the bolt is 360 kN, and it is shown in Figure 8 that the actual tightening force is fluctuating around the design force. It can also be observed that there is a daily period, which is caused by temperature variation in a day.

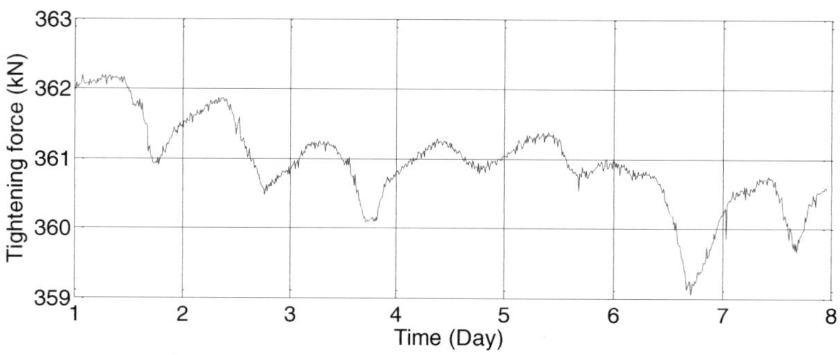

Figure 8 Tightening force variation in a week

5. Conclusions

In this paper, a new type of sensor is developed to monitor bolt tightness in a cable stayed bridge. The conclusions are shown below.

First, the mechanism of the sensor is to calculate the tightening force through recorded strain measurement.

Second, the data acquisition system is developed using LABVIEW software, which can be reliably on the bridge.

Third, the monitored data show effect of temperature variation, and the sensor can be used to monitor bolt tightness in steel truss girder bridges.

Reference

[1] S. Basava and D. P. Hess 1998, 'Bolted joint clamping force variation due to axial vibration', *Journal of Sound and Vibration*, vol. 210, no. 2, pp. 255 - 265.

[2] S. Izumi, T. Yokoyama, A. Iwasaki, and S. Sakai 2005, 'Three dimensional finite element analysis of tightening and loosening mechanism of threaded fastener' vol.12, no.4, *Engineering Failure Analysis*, 604-615.

[3] Ritdumrongkul S and Fujino Y 2006, 'Identification of the location and level of damage in multiple bolted-joint structures by PZT actuator-sensors', *Journal of Structure Engineering*, 132, pp. 304–11.

[4] Caccese V, Mewer R and Senthil S Vel 2004, 'Detection of bolt load loss in hybrid composite/metal bolted connections', *Engeering Structures,* 26, pp. 895–906.

[5] Amerini F, Barbieri E, Meo M and Polimeno U 2010, 'Detecting loosening/tightening of clamped structures using nonlinear vibration techniques', *Smart Materials and Structures,* 19, pp. 1–9.

[6] Amerini F and Meo M 2011, 'Structural health monitoring of bolted joints using linear and nonlinear acoustic/ultrasound methods', *Struc. Health Monitoring,* 10, pp. 659–72.

[7] P. Razi, R. A. Esmaeel, and F. Taheri 2013, 'Improvement of a vibration-based damage detection approach for health monitoring of bolted flange joints in pipelines', *Structural Health Monitoring*, vol. 12, no. 3, pp. 207 - 224.

[8] R. I. Zadoks and X. Yu 1997, 'An investigation of the self-loosening behavior of bolts under transverse vibration', *Journal of Sound and Vibration*, vol. 208, no. 2, pp. 189 - 209.

[9] N. G. Pai and D. P. Hess 2002, 'Experimental study of loosening of threaded fasteners due to dynamic shear loads', *Journal of Sound and Vibration*, vol. 253, no. 3, pp. 585 - 602.

[10] R. A. Ibrahim and C. L. Pettit 2005, 'Uncertainties and dynamic problems of bolted joints and other fasteners', *Journal of Sound and Vibration*, vol. 279, no. 3-5, pp. 857 - 936.

[11] H. Ahmadian and H. Jalali 2007, 'Identification of bolted lap joints parameters in assembled structures', *Mechanical Systems and Signal Processing*, vol. 21, no. 2, pp. 1041 - 1050.

[12] A. Milanese, P. Marzocca, J. M. Nichols, M. Seaver, and S.T. Trickey 2008, 'Modeling and detection of joint loosening using output-only broad-band vibration data', *Structural Health Monitoring*, vol. 7, no. 4, pp. 309 - 328.

The 8th International Conference on
Structural Health Monitoring of Intelligent Infrastructure
Brisbane, Australia | 5-8 December 2017

Leakage Monitoring for Steel Pipeline Structures Based on Piezoelectric Ceramics

S. Yan, Q.H. Hu , P.T. Zhao .
School of Civil Engineering, Shenyang Jianzhu University – China, Email: syan1962@163.com,
249076002@qq.com, *284865821@qq.com*

Abstract

The goal of the paper is to research the leakage monitoring technology for steel pipeline structures based on piezoelectric ceramics which are pasted on the surface of the structures as sensors to collect wave signals produced by the friction between the fluid and the pipe wall in the leakage time. The leakage position is identified based on time-frequency joint analysis, and time-difference of arrival (TDOA) method is used to locate the leakage position. To validate the efficiency of TDOA method, a model test is performed in a laboratory. A steel pipe of 3 meters long filled with high pressure water, where a valve is installed on the pipe to control the water leakage velocity, is tested and the passively generated signals are received by piezoelectric ceramic patches pasted on the surface of the pipe. A signal energy-based index is proposed to identify the leakage level. The experimental results show that the proposed method can successfully monitor the leakage level and position.

1. Introduction

Pipeline is an important tool for transporting fluid medium. It plays an irreplaceable role in the transportation of water, oil and natural gas, etc. However, due to a long-term use of a pipeline, dynamic loading and environmental impacts, the pipe wall may become thin and damaged, resulting in the leakage of the pipeline and may eventually cause greater economic losses, the environmental pollution and even serious accidents. In recent years, the domestic and foreign universities and research institutions are paying attention to develop pipeline leakage monitoring. The research focuses on the analysis of acoustic emission (AE) signal propagation characteristics, signal processing methods and leakage source localization and so on (Pan, 2013). Wan et al. (2012) systematically reviewed the hierarchical leak detection and localization method in natural gas pipeline monitoring sensor networks from the perspective of the signal processing, initial recognition, and final decision three phases. The properties and key techniques as well as challenges for each method or technology are reviewed in detail.

For wave propagation characteristics, many foreign scholars have carried out researches in this area. Rewerts et al (1997, 1998) studied the influence of the working medium on the attenuation of acoustic signal in the pipeline. They also investigated the dispersion characteristics of the medium wave propagation in the liquid filled pipeline, and established the dispersion model of the medium wave in the liquid filled pipeline. Wichaidit (2008) considered the influence of pipe joints to the propagation velocity of AE wave. Ahadi (2010) used the combination of short-time Fourier transform and wavelet transform to analyze the amplitude characteristics of the AE signal of the water filled plastic pipe leakage. In China, Sun et al. (2008) developed on the influence of liquid loading on the propagation characteristics of AE wave, and researched the propagation and attenuation characteristics of AE waves in liquid pipelines. Jiao (2007) proposed the comparison method to apply to the calibration of AE sensor for pipeline leakage monitoring, which provided the basis for the selection and calibration of the leak AE sensor. Wang (2010) researched the AE signal attenuation and location analysis.

For the wave processing, a lot of research fruits are obtained in the field. Pontoppidan (Pontoppidan, 2005) applied independent component analysis (ICA) method to analyze the AE energy signal, and the experimental results show that the accuracy of AE source localization accuracy of ICA is better than that of principal component analysis (PCA). Kim (2009) combined the time frequency analysis method and the boundary element method to analyze the acoustic emission signal of buried gas pipeline, and had a good effect on improving the signal to noise ratio and the detection of leakage state. In China, Gao (2009) used bispectrum analysis method to analyze the AE signal of pipeline leakage, and applied the method of bispectrum and cross Bispectrum to construct equations. Gu (2006) used empirical mode analysis to decompose the AE signal, contributing to the AE signal and noise components of phase separation, improving the result of spectrum analysis precision.

For the leakage source localization, researchers are usually focusing on the localization precision and some improvements are obtained. Lee et al. (2000) applied the AE technology in pipeline leak detection, and they used two kinds of methods of the signal amplitude attenuation and travel time on the leak source positioning. Through the comparison of the positioning effect of two kinds of methods, they found the localization method based on attenuation for leak detection is more effective. Miller (1999) established a reference standard for the construction and evaluation of AE testing equipment for pipeline leakage, which is helpful to the identification of the leakage source. Based on the analysis of the acoustic emission signals of CFRP (carbon fiber reinforced plastic) sheet, Surgeon (1999) used the AE technology to accurately locate the source of the CFRP plate, and the noise can be separated. In China, Gao (2009) used the bispectrum based statistics to realize the detection of leakage signal and location of leakage in complex urban environment noise. Jiao (2004) used the guided wave dispersion in pipelines to propose a practical method of AE leak location based on the extraction of single mode guided waves. After the empirical mode decomposition of acoustic emission signal of pipeline leakage, Sun (2008) chose the first 3 IMF components which had the characteristics of acoustic emission to reconstruct, and then carried on the correlation analysis to the acoustic emission signal which was reconstructed according to the IMF component Although some research fruits are obtained, there are also several challenges existing in the field to overcome. Most of the researches are focused on the characteristics of signal propagation and qualitative characterization of leakage state, but less research on the leakage mechanical model is reported, and the leakage localization is still a large space for promotion. Therefore, the goal of this paper is to further research the mechanical model of pipeline leakage and improve the accuracy of leakage source location. Toward this, a kind of piezoelectric ceramic passive monitoring technology is used, and through the method of simulation analysis and experimental verification, the pipeline leak monitoring problem of using passive wave technology is hopefully overcome. The paper will be focusing on the following aspects:

(1) Develop on the basic principle and mechanical model of leakage acoustic wave in pipeline structures. In this paper, the principle of pipeline leakage acoustic wave is studied by theoretical analysis, and the corresponding mechanical model is established.

(2) Research on the passive monitoring method of pipeline structure leakage based on piezoelectric ceramics. The methods of simulation analysis and experiments are used in this part, respectively. According to the analysis of pipeline leakage theory, simulation analysis of pipeline

leak based on ABAQUS is performed. Then the pipeline leakage monitoring experiment based on piezoelectric ceramic is carried out to verify the simulation analysis results.

Through this research, a mechanical model for the leakage localization by using piezoelectric ceramic passive wave based technology is proposed, which will further increase the precision of leakage localization evaluation.

2. Pipeline leakage model

When corrosion holes or cracks are produced in a pipeline, the pressure difference between the inside and outside of the pipe will leak out the liquid in the pipeline. The liquid in the pipe jets out through the corrosion holes or cracks, and the interaction between the fluid and the pipe wall produces the vibration (Feng, 2015). The vibration conveys the pipeline leakage signal, and the leakage signal propagates outside the pipe wall. The PZT sensors bonded on both sides of the leakage source can obtain the leakage signal (Wen, 2012),

$$\begin{cases} x_1(t) = s(t) + n_1(t) \\ x_2(t) = \alpha s(t - \tau_{peak}) + n_2(t) \end{cases} \tag{1}$$

where $x_1(t)$ and $x_2(t)$ are the PZT sensor output signals, $s(t)$ is the collected leakage signal, $n_1(t)$ and $n_2(t)$ are zero-mean random noises representing the ambient interference and noise, α is an attenuation factor, τ_{peak} is the time delay of the leakage signal reaching sensor A and sensor B, shown in **Figure** 1 (Bai, 2015).

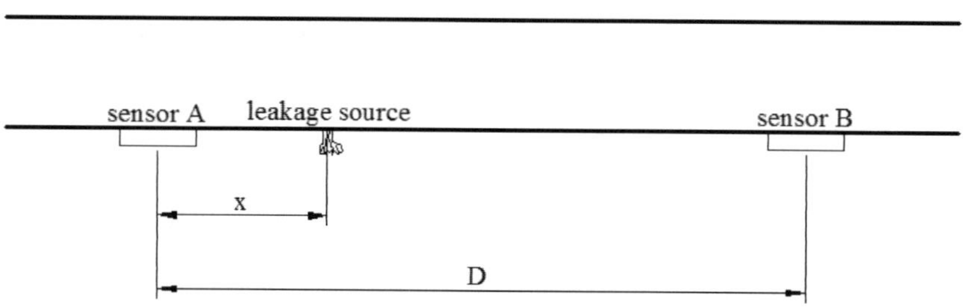

Figure 1. The schematic of pipeline leak locating.

The signals are random and non-stationary. The cross-correlation function can be used to determine the time delay for the leakage signal to reach the two sensors, and then use the TDOA method to determine the leakage source. For the leakage signal $f_1(t)$ and $f_2(t)$ the cross-correlation function is expressed as

$$R_{12}(\tau) = E[f_1(t) f_2(t - \tau_{peak})] = \lim_{T \to +\infty} \frac{1}{T} \int_0^T f_1(t) f_2(t - \tau_{peak}) dt \tag{2}$$

The TDOA method is used to locate the leakage source, expressed in **Equation** (3)

$$x = \frac{D - v\tau_{peak}}{2} \tag{3}$$

Where x is the distance of sensor A to leakage source, D is the distance between sensor A and sensor B, v is the propagation velocity of the leakage signal in the pipe wall.

3. Numerical analysis and simulation analysis

Pipeline guided wave dispersion equation is a transcendental equation. The approximate solution of the dispersion equation can be solved by numerical calculation. This is a relation between group velocity and phase velocity,

$$c_g = \frac{c_p^2}{c_p - (fd)\dfrac{dc_p}{d(fd)}} \tag{4}$$

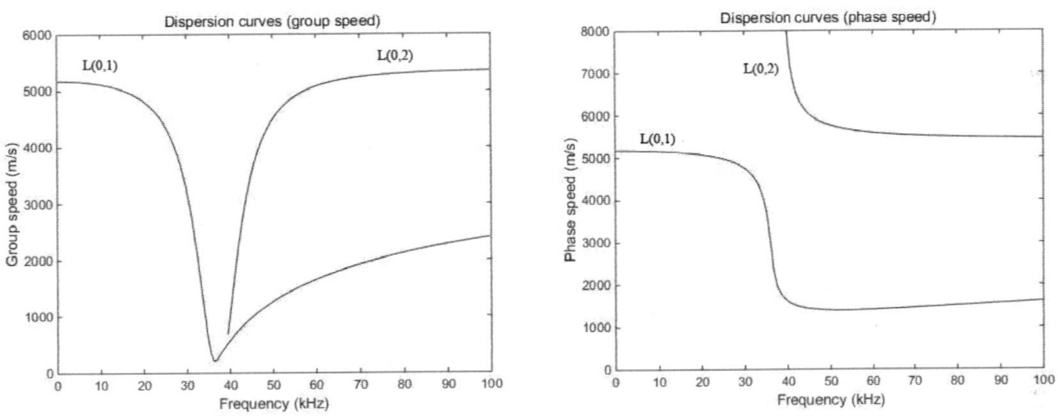

(a) group velocity dispersion curve (b) phase velocity dispersion curve
Figure 2. Hollow steel pipe dispersion curve.

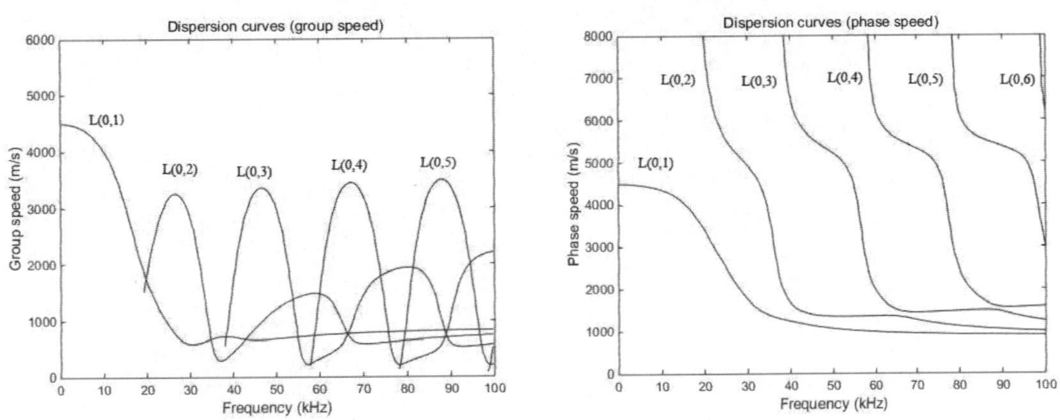

(a) group velocity dispersion curve (b) phase velocity dispersion curve
Figure 3. Liquid-filled steel pipe dispersion curve.

Taking the steel pipe with outer diameter of 48 mm, wall thickness of 3mm and density of 7850 kg / m^3 for an example, **Figure** 2 and **Figure** 3 are the dispersion curves of the hollow steel pipe and the liquid filled steel pipe, respectively. When the frequency is above 50 kHz, the group velocity of the L(0, 2) modal guided wave is much larger than the group velocity of the L(0, 1) modal guided wave. When the frequency is below 40 kHz, the hollow steel tube only has L(0, 1) mode. When the frequency is below 20 kHz, the filled steel tube only has L(0, 1) mode. Therefore, L(0, 1) mode and L(0, 2) mode guided wave can be used as a damage detection wave for the detected pipe.

4. Experimental verification

4.1 Experimental Setup

In order to locate the leakage source of the single leakage source pipeline structure, a system of leak monitoring system for the pipeline structure based on TDOA method is established. **Figure** 4 shows the experimental system. The experimental system consists of a digital oscilloscope, a computer, a piezometer, a PZT sensor for receiving signals, a water tank, a water pump and a pipe to be measured. The oscilloscope in the system is a RIGOL DS1000B four-channel digital oscilloscope, as shown in **Figure** 5.

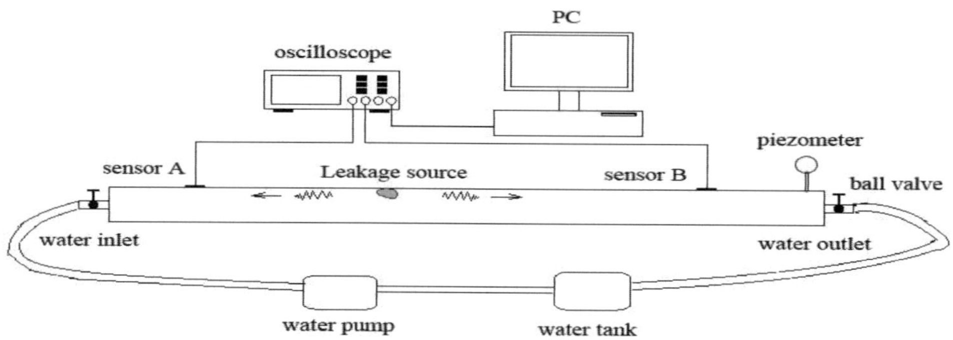

Figure 4. Schematic diagram of the experimental system.

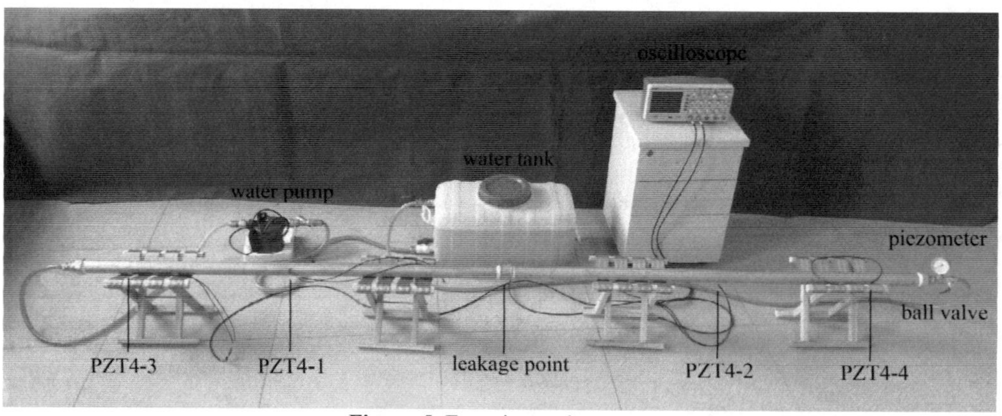

Figure 5. Experimental setup.

According to the previous section of the analysis of the pipeline model, the same material parameters, the same geometric size of the steel pipe and PZT sensor are selected to conduct experiment. The liquid in the experimental system is circulating. The liquid in the tank flows through the pump to the measured pipe and flows back to the tank from the measured pipe. In the experiment, the pressure can be applied to the measured pipe by the water pump, and then the pressure of the pipe can be adjusted by the switch of the ball valve. The tested pipe be is made of two parts of 1500 mm long pipe through the pipe joint, and there is an artificial hole of the diameter of 3 mm in the middle of the pipe joint to simulate the leakage of the pipeline leakage source. The location of the leakage source and the actual leakage sources are shown in **Figure** 6 and **Figure** 7, respectively.

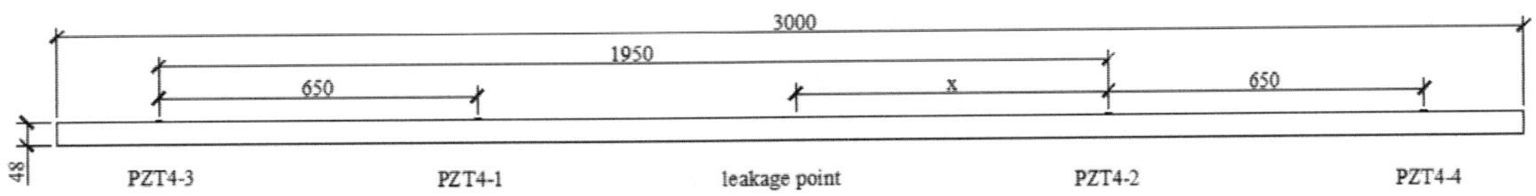

Figure 6. Diagram of the leak point.

Figure 7. Actual source of leakage.

4.2 Experimental Results Analysis

When the pipeline pressure is 0.1MPa, the sensors receive the response signal shown in **Figure** 8. **Figure** 8(a) and **Figure** 8(b) are the leak-free spectrum and the single leakage source spectrum, respectively. From **Figure** 8(a), it is known that the background noise of the pipeline operates at a maximum amplitude for the range of 0~1000 Hz. From **Figure** 8(b), when the pipeline has a single leakage source, the leakage signal is concentrated near the 1220 Hz.

In this experiment, the two groups of sensors simultaneously receive the signal, first with the PZT4-1 sensor and PZT4-2 sensor at the same time to receive the signal, and then use the PZT4-2 sensor and PZT4-3 sensor at the same time receive the signal, and finally filtered according to the frequency of the leaked sound signal. After adjusting the filter range many times, the filtering effect becomes the best when the filter range is 800Hz~1600Hz. **Figure** 8(c) and **Figure** 8(d) are time histories of filtered signals received by the PZT4-1 sensor and the PZT4-2 sensor, respectively. **Figure** 8(e) and **Figure** 8(f) are time histories of filtered signals received by the PZT4-2 sensor and the PZT4-3 sensor, respectively.

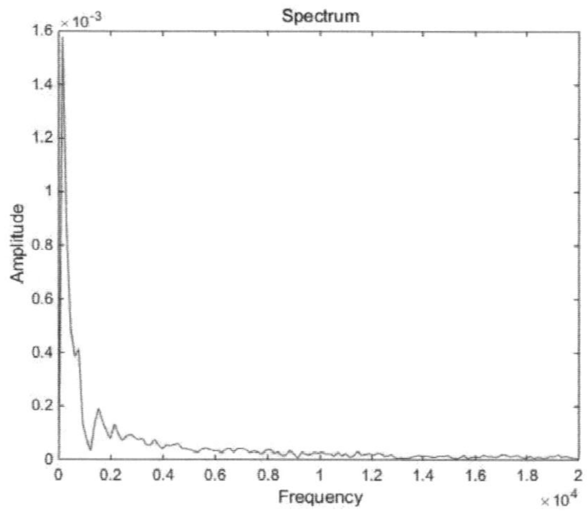

(a) The leak-free spectrum

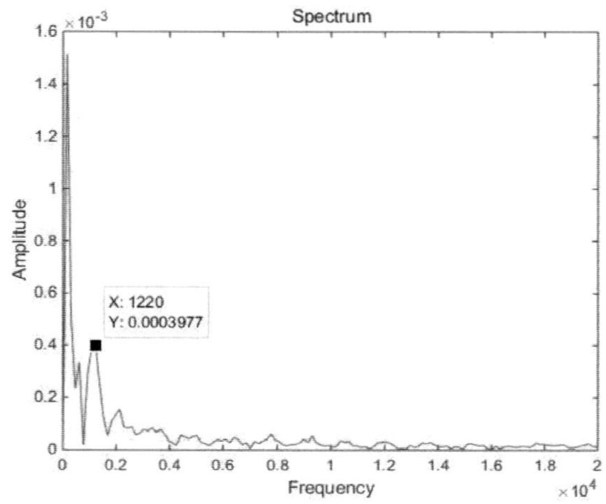

(b) The single leakage source spectrum

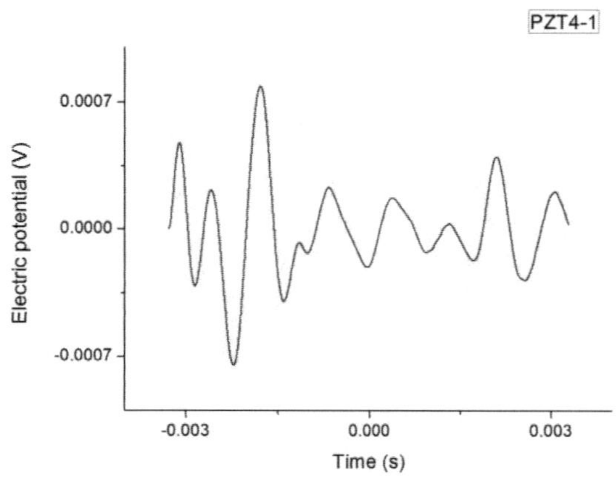

(c) Time history diagram of PZT4-1 sensor

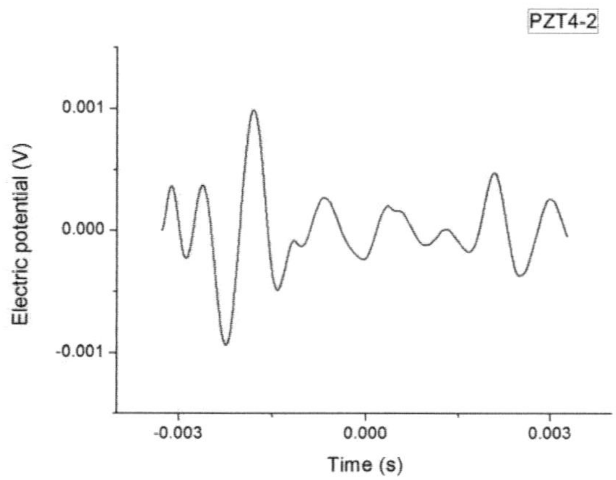

(d) Time history diagram of PZT4-2 sensor

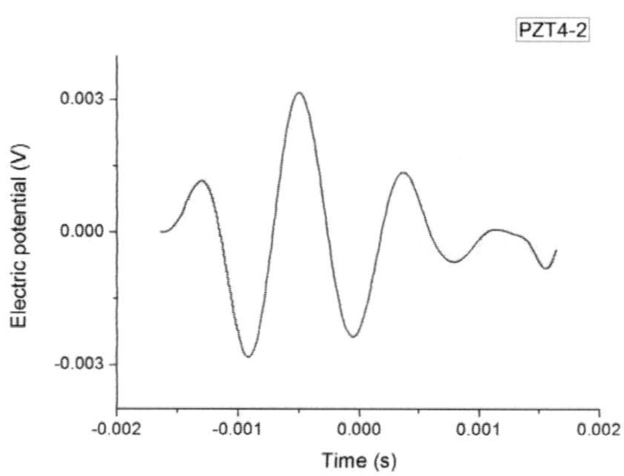

(e) Time history diagram of PZT4-2 Sensor

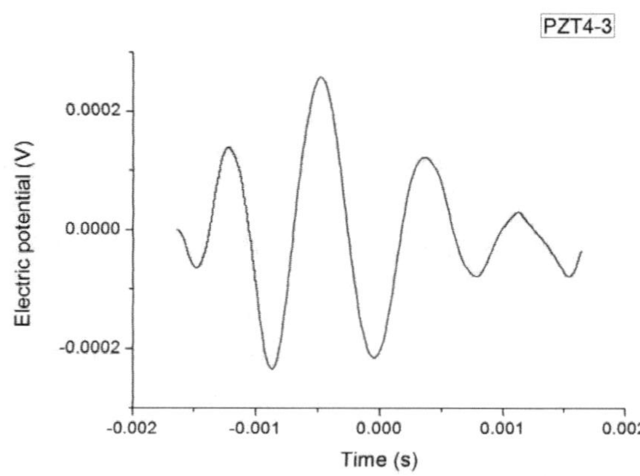

(f) Time history diagram of PZT4-3 Sensor

Figure 8. Part of the received signal graph.

The location of the leak source can be located by the TDOA method, and the time difference can be determined by solving the mutual correlation coefficients of signals received by the two groups of sensors. The time delay corresponding to the maximum value of the correlation number is the time difference between the two groups of sensors. **Figure** 9(a) is the cross-correlation coefficient of PZT4-1 sensor and PZT4-2 sensor received signals. From the figure we can see that the maximum correlation coefficient of the corresponding time delay is 2.002×10^{-5}s. Therefore, the time difference between the leakage acoustic signals of the two sensors is 2.002×10^{-5}s. **Figure** 9(b) is the cross-correlation coefficient of PZT4-2 sensor and PZT4-3 sensor received signals. From the figure we can also see that the maximum correlation coefficient of the corresponding time delay is 1.291×10^{-4}s. Therefore, the time difference between the leakage acoustic signals of the two sensors is 1.291×10^{-4}s. The frequency of the leakage acoustic signal is centered at 1220 Hz. According to the dispersion curve of the liquid filled duct in the previous section, the group velocity of the leakage acoustic signal propagating in the liquid filled pipeline is 4490m/s. According to **Equation** (3) calculation of leakage position is as follow.

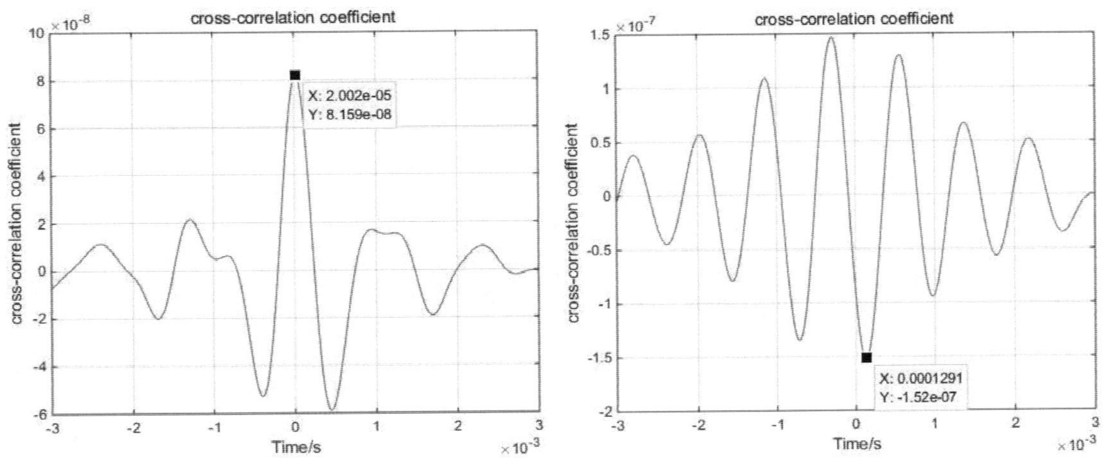

(a) The cross-correlation coefficient of PZT4-1 sensor and PZT4-2 sensor received signals

(b) The cross-correlation coefficient of PZT4-2 sensor and PZT4-3 sensor received signals

Figure 9. Part of the received signal correlation coefficient graph .

When the leak occurs between the PZT4-1 sensor and the PZT4-2 sensor, x=0.61 m. The error between the experimental calculation and the actual measurement is 6.2%. When the leak occurs between the PZT 4-2 sensor and the PZT 4-3 sensor, x=0.69 m. The error between the experimental calculation and the actual measurement is 6.0%.

The causes of the experimental errors might be as follows: the one is the actual propagation velocity of the leakage acoustic wave in the pipe is in error with the group velocity selected by the dispersion curve and the two is the actual time difference, and the time delay of the received acoustic signal received by the two sensors is in error. The distance between the PZT4-1 sensor and the PZT4-2 sensor is the same as that of the leakage source, and the energy ratio of the leakage acoustic signal received by the PZT4-1 sensor located upstream of the leakage source is smaller than that of the downstream PZT4-2 sensor. The distance of the sensor PZT4-3 from the leakage source is two times of the sensor PZT4-2, and the energy of the leakage acoustic signal received by the sensor PZT4-3 is much smaller than the energy of the sensor PZT4-2.

5. Conclusions

(1) The frequency of acoustic signals of pipeline leakage is near 1220 Hz, the frequency is low, the excitation modes are less and the frequency dispersion is smaller, so the dispersion curve can be used to determine the propagation speed of leakage sound signals in the pipeline.

(2) The leakage of acoustic signals propagates along the same distance along the pipeline, and the energy loss of the leakage acoustic signals propagating upstream is larger, and the energy loss increases as the propagation distance increases.

(3) The TDOA method can be used to approximately determine the location of the leak source, but its accuracy is closely related to the estimation of the group velocity of the leakage sound wave and the estimation of the time difference.

Acknowledgments

This work was financially supported in part by the Natural Science Foundation of Liaoning Province of China (2015020595), and the Science Foundation of Liaoning Education Department of China (LJZ2016029).

References

Ahadi, M & Bakhtiar, MS 2010, 'Leak detection in water-filled plastic pipes through the application of tuned wavelet transforms to acoustic emission signals', *Applied Acoustics*, vol. 71, pp. 634-639.

Bai, YQ, Li, K & Pei, L 2015, 'Water supply pipeline leak detection and location based on acoustic wave technology', *Water Supply and Drainage*, no. S1, pp. 340-343. (in Chinese)

Clark, MA, Reverts, LE & Roberts, Ron R 1997, 'Experimental studies on the role of backfill and pipeline characteristics in the application of acoustic leak location to underground pipeline', *Review of Progress in Quantitative Nondestructive Evaluation*, vol. 16, pp. 467-473.

Feng, XS, Wen, YM & Zhen, JP 2015, 'Analysis of acoustic vibration signal characteristics of pipeline leakage', *Acoustic Technique*, no. 05, pp. 413-418. (in Chinese)

Gao, H 2009, *Research on Key Technologies of urban underground gas pipeline leakage monitoring*, Ph.D Thesis of Tianjin University, Tianjin, China. (in Chinese).

Gao, H, Wang, BG & Du, ZH 2009, 'The application of bispectrum in the leakage location model of pipeline', *Nanotechnology and Precision Engineering*, vol. 5, pp. 445-450. (in Chinese)

Gu, XH, Hou, DB & Zhou, M 2006, 'Leak detection and location of buried pipe based on acoustic emission and EMD', *Journal of Zhejiang University (Engineering Science Edition)*, vol. 7, pp. 1105-1108. (in Chinese)

Jiao, JP, He, CF & Wu, B 2004, 'Study on acoustic emission location of pipeline leakage based on guided wave theory', *Journal of Mechanical Engineering*, vol. 10, pp. 77-81. (in Chinese)

Jiao, Y, Yang, QX & Shen, GT 2007, 'Study on calibration of acoustic emission sensor', *Nondestructive Testing*, vol. 3, pp. 139-141. (in Chinese)

Kim, MS & Lee, SK 2009, 'Detection of leak acoustic signal in buried gas pipe based on the time–frequency analysis', *Journal of Loss Prevention in the Process Industries*, vol. 22, pp. 990-994.

Lee, MR & Lee, JH 2000, 'Acoustic emission technique for pipeline leak detection', *Key Engineering Materials*, vol. 183, no. 187, pp. 888-892.

Miller, RK 1999, 'A reference for the development of acoustic emission pipeline leak detection techniques', *NDT&E International*, vol. 32, no. 1, pp. 1-8.

Pan, BX 2013, *Acoustic emission detection and location technology for pipeline leakage*, Master's Thesis of China University of Petroleum (Hua Dong), Qingdao.

Pontoppidan, NH, Sigurdsson, S & Larsen, J 2005, 'Condition monitoring with mean field independent components analysis', *Mechanical Systems and Signal Processing*, vol. 19, no. 6, pp. 1337-1347.

Rewerts, LE, Roberts, RR & Amanda, LM 1997, 'Dispersion compensation in acoustic emission pipeline leak location' *Review of Progress in Quantitative Nondestructive Evaluation*, vol. 16, pp. 427-436.

Roberts, RR, Rewerts, LE & Amanda, CM 1998, 'The role of propagation characteristics in acoustic emission', *Review of Progress in Quantitative Nondestructive Evaluation*, vol. 17, pp. 501-508.

Surgeon, M & Wevers, M 1999, 'Modal analysis of acoustic emission signals from CFRP laminates', *NDT&E International*, vol. 32, pp. 311-322.

Sun, G & Zhang, FX 1984, *Piezoelectricity*, National Defense Industry Press of China. (in Chinese)

Sun, LY, Li, YB & Han, SH 2008, 'Analysis of the influence of liquid loading on the propagation characteristics of acoustic emission wave', *Journal of Instrumentation*, vol. 8, pp. 1617-1621. (in Chinese)

Sun, LY, Li, YB & Qu, ZG 2008, 'Research on acoustic emission pipeline leak detection based on EMD', *Piezoelectrics & Acoustooptics*, vol. 2, pp. 239-241. (in Chinese)

Wan, JW, Yang, Y, Wu, YF, Feng, RJ and Yu 2012, 'Hierarchical leak detection and localization method in natural gas pipeline monitoring sensor networks', Sensors, vol.12, 1, pp. 189-214.

Wang, WY & Yang, JF 2010, 'Testing and analysis of acoustic emission signal attenuation in industrial pipeline', *Pipeline Technology and Equipment*, vol. 1, pp. 27-30. (in Chinese)

Wen, YM, Xie, RF & Li, P 2012, 'Noise adaptive suppression in pipeline leak detection', *Journal of Scientific Instrument*, no. 10, pp. 2228-2233. (in Chinese)

Wichaidit, W & Au, YHJ 2008, 'Effect of the pipe-joints on acoustic emission wave propagation velocity', *Proceeding of the 6th International Conference on Manufacturing Research (ICMR08)*, Brunel University, UK, 9-11 September, pp. 917-922.

The 8th International Conference on
Structural Health Monitoring of Intelligent Infrastructure
Brisbane, Australia | 5-8 December 2017

En"vision"ing a Novel Approach for Structural Health Monitoring – A Model for Full-Field Structural Identification using 3D-Digital Image Correlation

Mehrdad S. Dizaji[1], Devin K. Harris[2], Mohamad Alipour [3], Osman E. Ozbulut[4]

[1] University of Virginia – U.S.A., [2] University of Virginia – U.S.A., Email: dharris@virginia.edu [3] University of Virginia – U.S.A., [4] University of Virginia – U.S.A.

Abstract

Traditional structural health monitoring (SHM) measurement techniques have primarily relied on discrete sensing strategies using sensors physically attached to the structural system of interest. These sensors have proven to be effective in describing both global and local phenomena, but are limited to providing discrete response measurements of these systems. With the introduction of novel imaging tools and advances in computer vision and image analysis techniques, such as digital image correlation (DIC), the ability to measure the full-field response of these systems provides a novel approach to refining structural identification (St-Id) approaches used in SHM.

This paper explores this new concept through a case study on a series of structural test specimens analyzed using a vision-based metrology approach (3D-DIC) for St-Id. In this work, a Hybrid Genetic Algorithm (HGA), which combines the genetic algorithm as a global optimization and a gradient based method as a local optimization, was proposed for finite element model updating based on full-field measurement results of structural specimen subjected to variable loading. To minimize the error between the full field 3D-DIC measurements and FEA model updating result, an objective function was introduced that includes the contributions of both dynamic and static responses, enabling simultaneous local and global system identification. For the identification process, ABAQUS was interfaced with MATLAB to converge on the optimal unknown/uncertain system parameters (i.e. boundary conditions and constitutive relationships) of the experimental setup. The results of this study highlight the advantage of this hybrid approach and the effectiveness of deploying this vision-based strategy for large scale structural systems.

1. Introduction

Much of the physical infrastructure across the globe was built during eras of growth and with a finite intended service life, but in many cases these systems have continued to operate and have remained in service well beyond this intended period (American Society of Civil Engineers 2017). Structural health monitoring (SHM) has shown promise as a strategy for temporal condition assessment of the built environment. SHM provides a system performance evaluation strategy with the end goal of characterizing behavior and providing indications of damage and even forewarning of impending failure. To date, significant research has been performed on condition assessment (Ahlborn et al. 2010) and SHM of infrastructure with advances in novel technologies and assessment techniques (Vaghefi et al. 2012). These examples highlight the continued development in the body of work with the SHM community, much of which has included advancements in sensing and analysis techniques, but the current limitation of SHM continues to be the challenge of translating measurable phenomena for full-scale physical systems into information that describes the system's health and condition state. A comprehensive solution is still needed, one that integrates strategies for routine inspection, data management, result interpretation and decision support. From a research and development perspective, these concepts demonstrate that SHM is still in its infancy with excellent opportunities for growth and development. As we make advances towards a holistic strategy of smart and connected communities, the need for low-cost, non-invasive, and data-rich techniques is becoming paramount to the SHM community. This paper describes a preliminary experimental investigation that explores the feasibility of leveraging digital image correlation (DIC) within the SHM framework (Figure 1). Tools such as DIC have the potential to provide decision-makers with a comprehensive assessment tool to better describe the performance of this network while also being non-invasive and data-rich.

Figure 1 – Concept of Digital Image Correlation for SHM

2. Background of Digital Image Correlation

Deformation measurement has always been vital to the field of engineering mechanics; however, measurement capabilities have historically been limited to localized deformation measurement using classical tools such as strain gauges, linear variable displacement transducers and vibrating wire gauges (Webb et al. 2014). Similarly, the structural dynamics community has relied on vibration response measurement derived from accelerometers to enable the correlation between vibration response and global stiffness of structural systems (Ribeiro et al. 2012). These measurement techniques have been used extensively in SHM and St-Id in the past; however, recent advances in photogrammetric and image-based measurement techniques provide an alternative strategy for describing structural response (Vaghefi et al. 2012).

Digital image correlation (DIC) builds on the fundamental principles of photogrammetry and provides a mechanism to quantify full-field surface deformation from a series of sequential images of a specimen subjected to loading (Sutton et al. 2009). Surface displacement data, which can be transformed into strain via post-processing, is derived by correlating patterns within an image and comparing sequential pairs of digital images taken before and after the deformation (Figure 2). The DIC technique can be deployed using a single camera to quantify planar surface deformations (2D-DIC), or configured for three-dimensional surface deformation (3D-DIC) by using stereo-paired cameras. Images can be derived from a variety of sources (e.g. CCD, DSLR, etc.) with the choice of camera and lens configuration influenced by factors such as camera noise, lighting, acquisition speed, and geometric relationships between area of interest and field of view. A comprehensive treatment of DIC is not presented in this paper, but is available in the literature (Sutton et al. 2009).

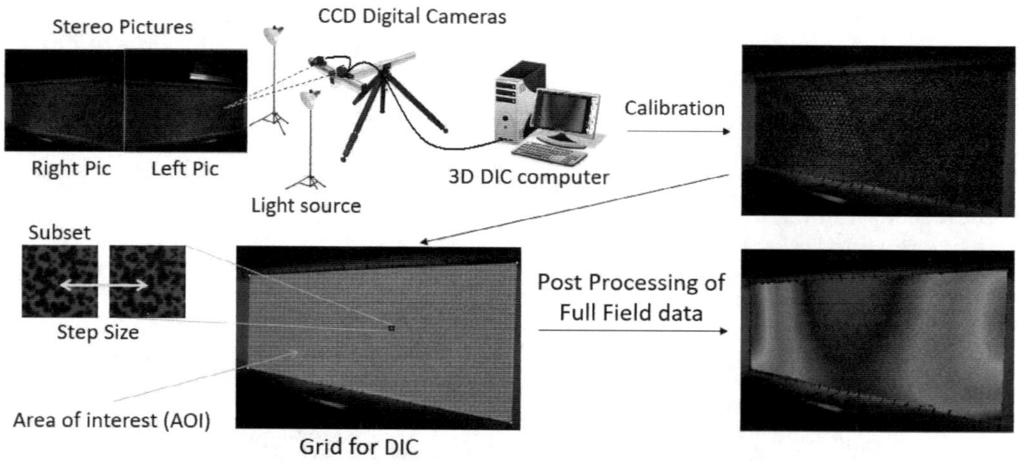

Figure 2 – Schematic of Digital Image Correlation

3. Motivation

Aside from the economic and technology adoption challenges that are projected to be overcome as technology advances, the larger issues hindering the adoption of SHM systems into practice include the challenges associated with monitoring of existing structures with no previous baseline behavior measures and the ability to translate the measurement and sensing data into meaningful, reliable and actionable information on the system's condition. Specifically, this work explores a new vision-based structural identification (St-Id) framework that allows for full-field structural behavior matching and identification. The proposed approach can result in significant enhancements compared with the use of traditional discrete-point mechanical sensors that are typically used in current SHM and St-Id applications (Dizaji et al. 2016). Figure 3 conceptualizes the potential advantage of the proposed framework for model updating and St-Id using full-field response measurements (i.e. 3D-DIC) in comparison with those obtained by discrete point sensors (e.g. strain gages, LVDTs, etc.). The dotted line depicts the predictions of an FEM model updated by matching measurements of only two discrete mechanical sensors. The dashed line is a model updated using a fine grid of points on a full-field response contour (e.g. derived from DIC measurement). In this illustration, it becomes evident that matching a complex function using a limited set of discrete mechanical sensors may provide a close agreement in the proximity of the sensors, but may not guarantee a good match in other locations on the specimen. On the other hand, matching an entire full-field response enables a more comprehensive representation of the global and local behaviors throughout the specimen. This advantage is expected to be realized in more complex structural systems, where only a few discrete points may be insufficient to describe the behavior.

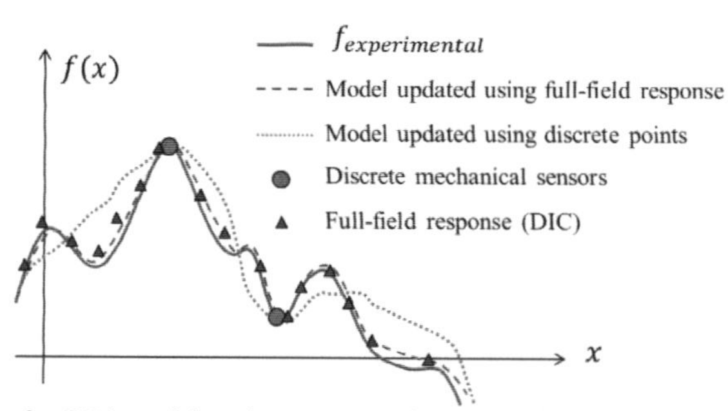

Figure 3 - FEA model updating using discrete sensors versus full-field response

An example is a structure with geometrical non-uniformities (e.g. a hole or defect) or complicated boundary conditions where the response cannot be uniquely represented by a few discrete sensors. The experimental specimen examined in this work is a relatively simple structural system, but the work presented in this paper represents a series of laboratory examples that provide the foundation for expanding this concept. Future works by the authors aim to illustrate these advantages in more complex structural systems such as full-scale highway bridge structures.

4. Experimental – Numerical Study

Experimental Investigation

An experimental program was executed to demonstrate the potential for integrating 3D-DIC into the SHM framework. The experimental program was a laboratory scale investigation of a representative steel beam subjected to various loading and boundary conditions. Multiple configurations were evaluated in the study, but only one scenario (*CF1*: simply-supported structural component subjected to a concentrated load at midspan) is presented in this paper (Figure 4).

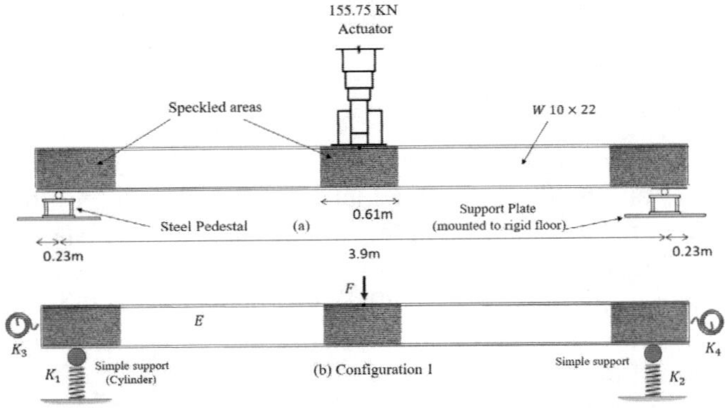

Figure 4 - Schematic of loading and boundary conditions used during experimental testing

The experimental program consisted of a series of flexural loading cycles within the elastic range (σ_{yield}= 345MPa) of a structural steel beam (ASTM A992 W10x22). The 4.37m beam was tested in a strong axis configuration in the UVA Structures Lab. The beam was also instrumented with Bridge Diagnostic Inc. (BDI) sensors at midspan and both support locations to provide a comparison between traditional SHM sensing modalities and those derived from the 3D-DIC measurements. Three paired DIC camera (Point Grey Grasshopper 2.0 CCD with 5.0MP resolution) systems were used to evaluate the midspan (Schneider 8 mm lens) and two end span (Schneider 12 mm lenses) locations. The end and midspan locations were patterned over the full depth of the beam web over 24 in. with the patterned created by applying a flat white paint base coat, followed by random speckle pattern with a permanent marker. The DIC data acquisition (DAQ) integrated output signals (load and displacement) from MTS actuators to allow for simultaneous acquisition of load, displacement, and images. The BDI DAQ system was manually synchronized at the start of each test. Figure 5 provides a basic illustration of the experimental setup used during testing. For each of the configurations, the beam was loaded monotonically under displacement-control, with the beam response kept within the elastic range. The loading sequence consisted of loading the beam to a displacement of 1.27mm at a rate of 0.051mm per second, followed by a two-cycle sinusoidal loading from 1.27mm up to a peak displacement of 7.62mm, and concluding with an unloading through the reverse of the initial loading sequence.

In addition to the quasi-static loading, the beam was subjected to an impact test using a hammer to excite the structure dynamically. BDI accelerometers were used to collect the acceleration response at a sampling rate of 1000 Hz (Figure 5b). At the time of testing, the research team did not have access to high-speed cameras for collecting dynamic DIC response; however, this equipment will be available for future comparisons.

(a) (b)

Figure 5 – Experimental set up (a) Experimental set up for DIC, 6 camera setup (3 systems or 3 camera pairs) 1 Pair at midspan (8 mm lens) and 1 pair at each support (12 mm lens) (b) Experimental set up for BDI accelerometers to collect the acceleration response.

Modal Identification

The vibration behavior of a given structure is governed by its modal properties (natural frequencies, damping ratios and mode shapes), which can be obtained by processing vibration data from a vibration test. One of the most common techniques used to estimate the modal properties of a structure is Operational Modal Analysis (OMA), which is based on using ambient vibrations under normal operating conditions (Rainieri and Fabbrocino 2014). For civil engineering structures, operational modal analysis offers an efficient method for processing vibrations of small amplitudes, which are often covered by white noise.

In this work, the vibration data was analyzed using the Enhanced Frequency Domain Decomposition (EFDD) technique to identify the modal parameters. The acceleration data from the five accelerometers were analyzed and the singular values of the spectral density function matrix were obtained as the natural frequencies. As a result, and based on the existing constraints on the sampling frequency of data acquisition, only the first two natural frequencies of the structure in strong-axis flexure could be estimated as $f_1 = 69.85\ Hz$ and $f_2 = 268.94\ Hz$. It should be noted that in this work, the modal identification was carried out using the EFDD implementation in ARTeMIS Modal software package and the reader is referred to (Rainieri and Fabbrocino 2014) for additional background on the EFDD theory and implementation.

Numerical Simulation

In this investigation, an initial finite element model of the structure was developed in ABAQUS, a robust commercially available finite element software package. This same model was later updated based on experimental response in the St-Id process. The steel beam was modeled using a total of 4,300 continuum 3D hexahedral solid elements (C3D8) with full integration. The geometry was developed from standard section properties available within the AISC Manual of Steel Construction [54]. A global view of the FEA model of the steel beam and a schematic of the experimental setup are illustrated in Figure 6. The boundary supports were modeled as a series of springs (translational and rotational) to represent the possible deviation from ideal simple conditions and to allow for updating based on experimental measurements. In addition to analyzing the model under static loading for strains and deflections, a modal analysis of the FEM model was also carried out in ABAQUS using the Lanczos Eigenvalue extraction method (Dassault Systèmes 2016). From the numerical model, natural frequencies were estimated as $f_1^{num} = 70.49\ Hz$ and $f_2^{num} = 192.75\ Hz$. To facilitate the automated iterative parameter optimization process, a direct interface was established between ABAQUS and MATLAB, a multi-paradigm numerical computing environment.

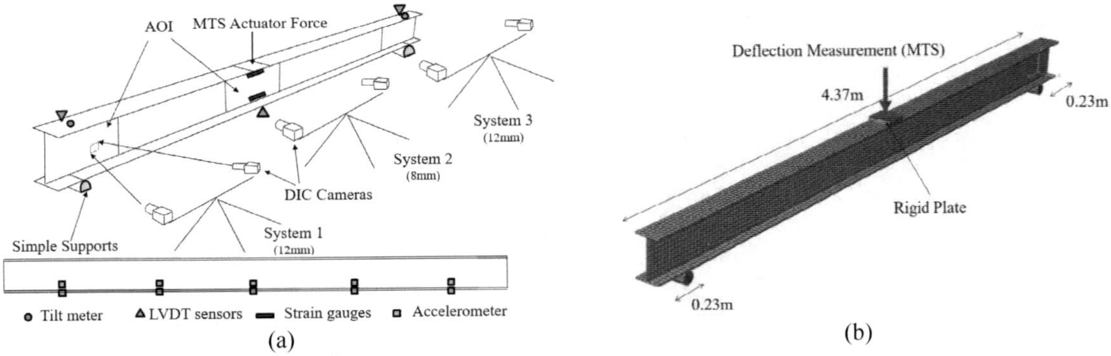

Figure 6 - Schematic of experimental setup and numerical model representation (a) DIC and instrumentation configuration (b) Representative isometric view of the finite element model of the steel beam

DIC Results Versus Reference Sensors

Results from the experimental program provided a basis for comparison of the 3D-DIC measurements with the in-place mechanical sensors that are representative of those used in traditional structural testing and SHM applications. For the comparison, a virtual gauge was selected in the DIC system to allow for local strains to be measured within both the tension and compression regions of the cross-section as shown in Figure 7. The results demonstrate that the measurements derived from both systems are comparable, but the DIC results exhibit a noisier response. This outcome is expected, but it should also be noted that the full-field measurement capability derived from DIC cannot be achieved with local sensing techniques and the full-field measurement provides a unique capability for a more robust St-Id strategy as illustrated in Figure 3. During the experiments, the DIC measurement also provided a supplemental benefit to the investigation in that vertical deflections were measured at the support locations, which were previously assumed to be fixed in this direction.

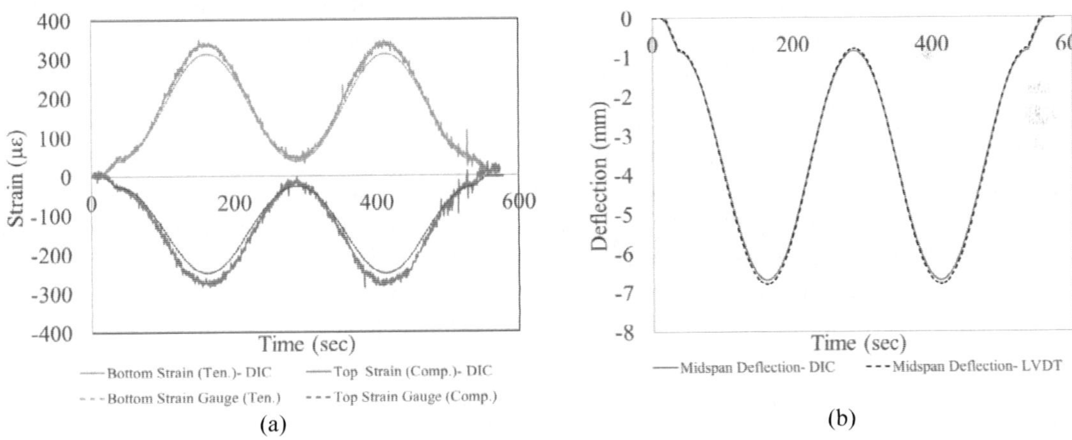

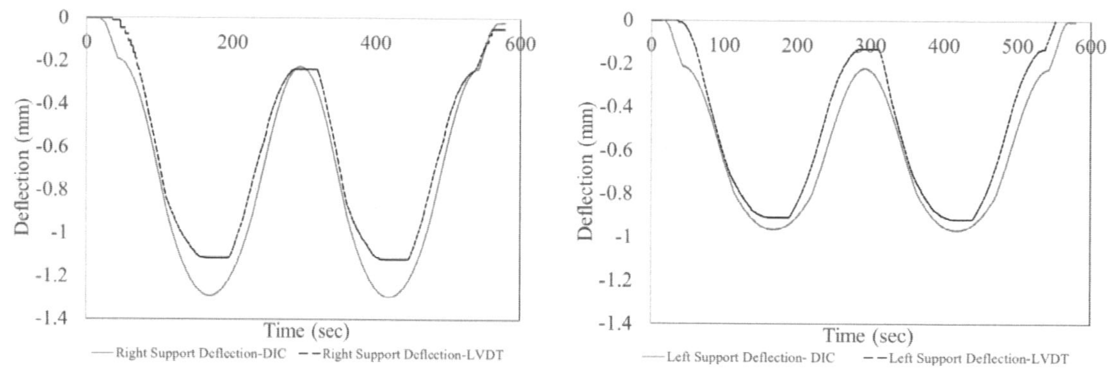

(c) (d)

Figure 7 - *Comparison of results obtained from DIC and mechanical sensors (a) midspan strain; (b) midspan deflection; (c) right support deflection; (d) left support deflection*

Definition of the Objective Function (Cost Function)

The identification problem consisted of the determination of structural parameters that minimize the difference between calculated data from a numerical model and a set of experimental data. In this research, the numerical model is a finite element model with geometry and boundary conditions defined to replicate the real conditions as closely as possible. Four different scenarios have been defined for the objective function as listed below and summarized in Table 1.

1. Full Field Model Updating (FF-MU): Full-field measurement of strains and displacements on the area of interest shown in Figure 5 was utilized in model updating.
2. Discrete Static Model Updating (DS-MU): Mechanical sensors (e.g. LVDT and Strain Gauge) were used in model updating.
3. Discrete Dynamic Model Updating (DD-MU): First two identified natural frequencies of the beam (resulting from accelerometers) were used in model updating.
4. Discrete Hybrid Model Updating (DH-MU): Combination of the second and third scenarios.

In Table 1, $\varepsilon_{xx}^{exp}, \varepsilon_{xy}^{exp}$ *and* δ_y^{exp} represent the two components of the strain tensor and displacement, respectively, that were extracted at time t and a point i of coordinates x_i (e.g. 480 measurement points for DIC results, 2 strain gauges, 3 LVDT locations and 2 natural frequencies). The values $\varepsilon_{xx}^{num}, \varepsilon_{xy}^{num}$ *and* δ_y^{num} represent the corresponding values computed from the finite element model. In this study, the data for three representative time frames, namely t=100, 150, 175 sec., have been selected to be included in the cost function to provide a representation of different stages of loading while maintaining a reasonable computational cost.

Table 1 - Different scenarios for objective function

Type	Objective Function
FFMU	$F=\sum_{i=1}^{480}\sum_{t=100,150,175s}\left(\frac{\lvert\varepsilon_{xx}^{num}(i,t)-\varepsilon_{xx}^{exp}(i,t)\rvert}{\varepsilon_{xx}^{exp}(i,t)}\right)^2+\sum_{i=1}^{480}\sum_{t=100,150,175s}\left(\frac{\lvert\varepsilon_{xy}^{num}(i,t)-\varepsilon_{xy}^{exp}(i,t)\rvert}{\varepsilon_{xy}^{exp}(i,t)}\right)^2 +\sum_{i=1}^{480}\sum_{t=100,150,175s}\left(\frac{\lvert\delta_{y}^{num}(i,t)-\delta_{y}^{exp}(i,t)\rvert}{\delta_{y}^{exp}(i,t)}\right)^2$
DSMU	$F=\sum_{i=1}^{2}\left(\frac{\lvert\varepsilon_{xx}^{num}(i)-\varepsilon_{xx}^{exp}(i)\rvert}{\varepsilon_{xx}^{exp}(i)}\right)^2+\sum_{i=1}^{3}\left(\frac{\lvert\delta_{y}^{num}(i)-\delta_{y}^{exp}(i)\rvert}{\delta_{y}^{exp}(i)}\right)^2$
DDMU	$F=\sum_{i=1}^{2}\left(\frac{\lvert f^{num}(i)-f^{exp}(i)\rvert}{f^{exp}(i)}\right)^2$
DHMU	$F=\sum_{i=1}^{2}\left(\frac{\lvert f^{num}(i)-f^{exp}(i)\rvert}{f^{exp}(i)}\right)^2+\left(\frac{\lvert\varepsilon_{xx}^{num}(i)-\varepsilon_{xx}^{exp}(i)\rvert}{\varepsilon_{xx}^{exp}(i)}\right)^2+\sum_{i=1}^{3}\left(\frac{\lvert\delta_{y}^{num}(i)-\delta_{y}^{exp}(i)\rvert}{\delta_{y}^{exp}(i)}\right)^2$

Structural Identification Process

Figure 8 provides a generalized illustration of the proposed St-Id strategy used in this investigation. In summary, numerical data from the FEM model are compared with experimental data (i.e. discrete data from mechanical sensors or full-field data from DIC) and the objective functions are evaluated for convergence. The parameters are then iteratively updated in an optimization process until convergence. This process is described in more depth in the following sections.

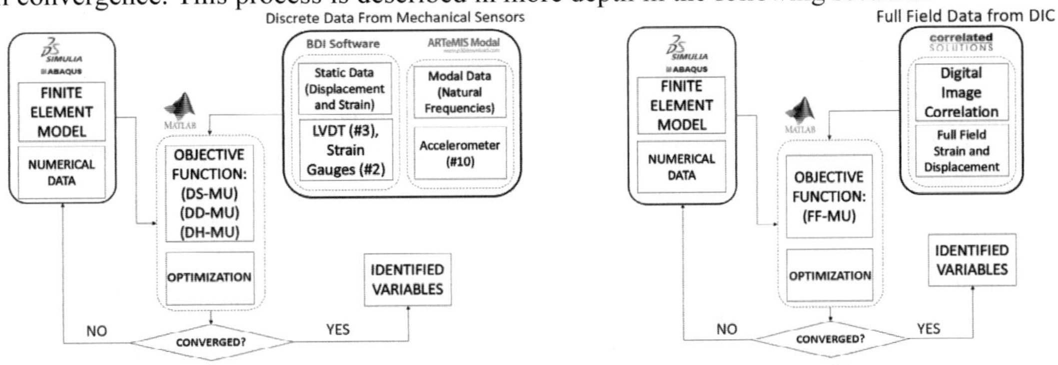

Figure 8 - Overview of the proposed full-field St-Id process (a) Model Updating using Mechanical Sensors (b) Model Updating Using Full Field DIC data

5. Results and Discussion

Optimization Process- Hybrid Genetic Algorithm (HGA)

In this study, a hybridized training algorithm was adopted to minimize the cost functions of Table 1, thus identifying the unknown parameters (E_s, K_1, K_2, K_3, K_4). The algorithm was based on the combination of a Genetic Algorithm (GA) and a gradient-based algorithm. In the first step a genetic optimization step is employed to explore the space of parameters and locate the approximate region of the optimum solution. In the second step, a gradient-based method is utilized to continue the search within the approximate region to quickly converge on the precise location of the optimum solution. As a result, the favorable characteristics of both methods namely the efficient exploration of the space by the GA and the superior convergence of the gradient-based methods are leveraged to achieve an efficient optimization. To start the process, a feasible initial guess for the parameters (Table 2) is used to generate an FEA model which upon analysis will be evaluated in the cost function. If the stopping criteria are not met, a new solution is generated through different operations in GA (e.g. selective reproduction, crossover and mutation). The new solution gives rise to a new FEA model and the process is repeated as necessary. Once the stopping criteria are satisfied, the final solution of GA is used to initiate the gradient-based scheme. This step continues until convergence criteria are satisfied when the final optimal solution is identified. The final identified parameters are presented in

Table 3 for all of the objective functions.

Table 2 - Initial, minimum and maximum values of the updating parameters

Material parameter	E_s (GPa)	K_1 (KN/mm.)	K_2 (KN/mm.)	K_3 (KN-mm/rad)	K_4 (KN-mm/rad)
Initial	172	12.26	12.26	226.06	226.06
min	138	8.75	8.75	5.65	5.65
max	276	262	262	565.15	565.15

Table 3 - Identified optimal parameters for different objective functions

Objective Functions	E_s (GPa)	K_1 (KN/mm)	K_2 (KN/mm)	K_3 (KN-mm/rad)	K_4 (KN-mm/rad)
FF-MU	200.11	17.06	15.77	9.1	6.08
DS-MU	208.78	19.54	17.11	36.39	24.97
DD-MU	267.11	27.22	23.44	169.99	238.94

DH-MU	207.98	19.14	17.24	34	25.77

The examination and comparison of the identified parameters listed in

Table 3 illustrates the efficiency of the different objective functions. It can be seen that the FF-MU has closely converged to a rational solution of E_s (assumed to be 200GPa for structural steel). Also the identified K_1 and K_2 obtained from FF-MU closely match the slope of the load-deflection response of the supports as shown in Figure 9, highlighting the rationality of the FF-MU objective function. On the other hand, while the E_s, K_1 and K_2 identified by DS-MU are still acceptable approximations of the assumed true values, the K_3 and K_4 values show a large discrepancy when compared to FF-MU. This can be attributed to the fact that the DS-MU has not utilized any rotational information from the two ends of the beam, thus rendering it incapable of successfully converging to rational rotational stiffness parameters. Interestingly, while the DD-MU results, which only use very limited information through the two available natural frequencies, are significantly far from the rational values, the addition of this new, but insufficient information to the discrete static sensors (the hybrid objective function DH-MU) provides results that are closest to the rational values of the full-field case of FF-MU, albeit still lacking in the description of the rotational restrain stiffness. This outcome is in agreement with the discussion previously presented in the motivation section and depicted in Figure 3. As was anticipated, the inclusion of more structural information in the St-Id process yields more rational and reliable identification. Specifically, full-field information yields the most efficient updating in comparison with discrete sensing information provided by mechanical sensors.

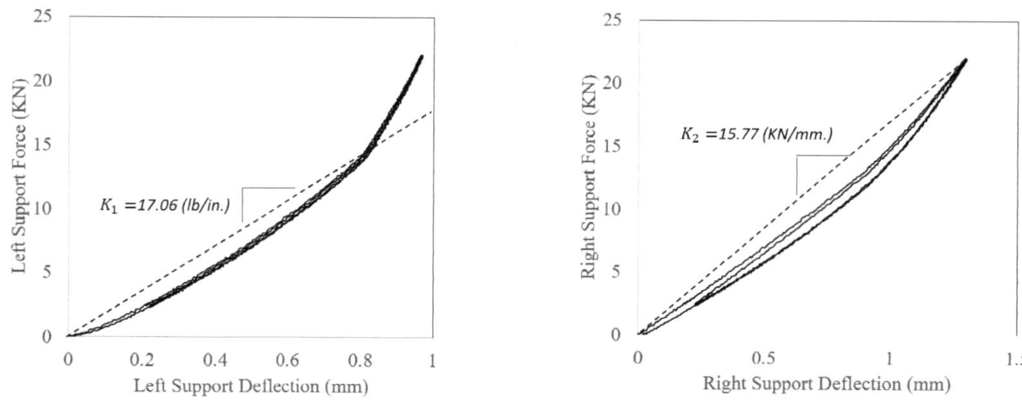

(a) *(b)*

Figure 9 - Experimental data for (a) Force versus left support deflection (b) Force versus right support deflection

6. Conclusions and Future Work

In this study, a structural identification procedure based on full-field sensing capabilities of 3D digital image correlation (3D-DIC) was developed to identify the material properties and boundary conditions of a steel beam specimen under flexural loads. Conventional mechanical sensors typically used in St-Id applications were also installed on the experimental specimen to provide context for comparison with the current practice of discrete-point matching for model updating. A series of objective functions that included the relevant components of strain, deflections and modal properties of the beam were also defined. These objective functions were then driven to a minimum by leveraging an efficient optimization algorithm consisting of a hybrid of genetic algorithm (GA) and a gradient-based optimizer. The outcomes of this study show that the St-Id results obtained using full-field image-based measurements can be successfully used as an alternative to physical in-place sensors for characterizing the response of structural systems. Also, with the rich and full-field information provided through 3D-DIC, the updating results create a more reliable and rational matching between the responses in comparison with the traditional discrete sensors.

Future work is expected to further explore the potential for using DIC for characterizing the vibration response, reducing the noise within the experiments, and evaluation of the range of applicability with respect to uncertainty in the updated parameters. These areas of focus are critical to the applicability of the proposed approach to more complex structural systems.

Acknowledgements

The authors would also like to acknowledge the contributions for Dr. Abdollah Bagheri and Mr. Abdou Ndong for their assistance with characterizing the experimental dynamic data.

References

1. American Society of Civil Engineers (2017). "Report card for America's infrastructure." <http://www.infrastructurereportcard.org/>. (June 1, 2017).
2. Ahlborn, T., Shuchman, R., Sutter, L., Brooks, C., Harris, D., Burns, J., Endsley, K., Evans, D., Vaghefi, K., and Oats, R. (2010). "The State-of-the-Practice of Modern Structural Health Monitoring for Bridges: A Comprehensive Review."
3. Vaghefi, K., Oats, R., Harris, D., Ahlborn, T., Brooks, C., Endsley, K., Roussi, C., Shuchman, R., Burns, J., and Dobson, R. (2012). "Evaluation of Commercially Available Remote Sensors for Highway Bridge Condition Assessment." *Journal of Bridge Engineering*, 17(6), 886-895.
4. Webb, G., Vardanega, P., and Middleton, C. (2014). "Categories of SHM Deployments: Technologies and Capabilities." *Journal of Bridge Engineering*, 10.1061/(ASCE)BE.1943-5592.0000735, 04014118.
5. Ribeiro, D., Calçada, R., Delgado, R., Brehm, M., and Zabel, V. (2012). "Finite element model updating of a bowstring-arch railway bridge based on experimental modal parameters." *Engineering Structures*, 40, 413-435.
6. Sutton, M. A., Orteu, J. J., and Schreier, H. (2009). *Image correlation for shape, motion and deformation measurements: basic concepts, theory and applications*, Springer Science & Business Media.
7. Dizaji, M. S., Alipour, M., and Harris, D. K. (2016). "Leveraging Vision for Structural Identification: A Digital Image Correlation Based Approach." *International Digital Imaging Correlation Society*, Springer, 121-124.
8. Rainieri, C., and Fabbrocino, G. (2014). "Operational modal analysis of civil engineering structures." *Springer, New York*, 142, 143.
9. Dassault Systèmes (2016). "ABAQUS analysis user's manual." *Abaqus*.

The 8th International Conference on
Structural Health Monitoring of Intelligent Infrastructure
Brisbane, Australia | 5-8 December 2017

Composite Impact Damage Assessment with Embedded Fiber Bragg Gratings

M. Yeager [1], C. Key [2], W. Gregory [2], M. D. Todd [1].
[1] *University of California San Diego - USA, Email:* mdtodd@ucsd.edu [2] *Applied Physical Sciences Inc. - USA*

Abstract

The use of composite materials has expanded significantly across civil, aerospace, and marine structures in both new designs and retrofits. The performance benefits from composites—typically, weight reduction with increased strength, corrosion resistance, and improved thermal and acoustic properties—are challenged by a host of failure modes whose genesis and progression are not yet well understood. As such, structural health monitoring (SHM) plays a key role for in-situ assessment for the purposes of performance/operations optimization, maintenance planning, and overall life cycle cost reduction. In this work, arrays of fiber Bragg grating optical strain sensors are embedded into a pre-preg composite specimen that was designed by surrogate finite element model simulation, and will be subjected to high-energy damaging impacts from a drop-weight impacting head. Results from SHM algorithms developed via hypothesis testing are demonstrated on blind impact tests in order to determine the efficacy of embedded fiber Bragg grating arrays for assessing structural health of such composites. Shock testing of a special fiber connector was also performed.

1. Introduction

Fiber reinforced polymer (FRP) composite material systems are being used more and more for their attractive strength and stiffness properties and corrosion resistance. On the other hand, they often have poorly-understood failure modes (e.g., delamination, disbonding, fiber breakage, matrix cracking, and bearing damage in connections) that are not well correlated with visual inspection experience. Classic nondestructive techniques like ultrasonic testing (UT) require that the structure be removed from service, precipitating considerable research and development in the implementation of structural health monitoring (SHM) systems for the purposes of performance/operations optimization, maintenance planning and overall life cycle cost reduction (Farrar & Worden 2007).

This work deploys a network of internally-embedded fiber Bragg gratings used to measure dynamic strain in a vibration context. Detailed reviews of vibration-based SHM have been performed by other researchers over many years (Sohn et al. 2003, Doebeling et al. 1998). A passive sensing scheme, such as in this work, is one that relies on ambient excitation of the structure and, therefore, generally assume that the input is unknown or otherwise immeasurable. The technical literature presents several successful implementations of passive damage detection systems using a variety of sensing modalities and damage sensitive features (Gul & Catbas 2009, Kessler et al. 2002). Fiber Bragg gratings, in particular, are an attractive sensing choice because of several advantages over traditional strain sensors including light weight, corrosion resistance, impermeability to liquid absorption, immunity to electromagnetic interference, and the ability to be multiplexed on a single fiber (Kersey et al. 1997, De Olivera et al. 2008, Majumder et al. 2012, Peters 2012). Although several studies have been performed regarding the use of optical sensors in the SHM of composites

(Jang et al. 2012, Park et al. 2011, Murayama et al. 2002), larger-scale dense deployments are quite rare in the literature. This work attempts to demonstrate an integrated data acquisition, feature selection, and decision-making process in the context of detecting impact-induced delamination damage in a composite, resulting in a full "data-to-decision" implementation of SHM.

2. Experimental Procedure

An Axiom AX-3201S/EL (woven glass fiber with toughened epoxy matrix) composite panel having uniform thickness was designed and fabricated. The target thickness was 12.7 mm (16 plies), representative of material geometry used in composite deck and masts. The panel was outfitted with a network of 40 embedded FBGs with 10 sensors multiplexed on 4 separate fibers. The FBGs were placed at 6.25% of the panel thickness, closer to the surface away from the impact face, during the prepreg layup. The panel was cured in a vacuum oven at 93.3 °C under constant pressure. Surrogate model finite element simulation was used to determine eigenmodes to aide in designing the placement of sensors within the laminate such that significant strain magnitudes would result from expected impact locations. Figure 1 (left) shows the routing paths of the FBG arrays and shows the naming convention used throughout the paper.

Figure 1. Sensor layout (left) and testing arrangement (middle and right).

A test fixture fabricated with aluminum rectangular bars was used to support the panel and introduce damage using a drop weight pendulum impact hammer. The fixture had a 111.76 cm x 111.76 cm free span and provided a clamped (fixed) boundary condition along each panel edge. The impact head was outfitted with a Piezo-based dynamic load cell for measuring impact force history and a photogate system was used to measure the actual velocity at impact. A steel cable with a pneumatic release mechanism was used to hoist the pendulum for repeatable impact energies. A surrogate, un-instrumented panel was first installed in the test fixture for system calibration and to assist in establishing damage characterization and associated energy levels.

The instrumented panel was clamped all around. On the back face, an electromechanical shaker was suspended from the frame and attached to the panel using a steel stinger rod. The shaker was used to provide pseudorandom excitation in the form of "operational" vibration loads between damaging impact tests, in which an impacting head induced delamination on the specimen. As repeated impacts were performed, the impact location was moved slightly to continue growing the delamination and avoid through-penetration of the panel. Although impact damage in a field deployment would occur at random, all impacts in this study were focused on a single location because one of the main objectives of the study is to explore the detectability threshold as a function of damage size. In between impact events, the shaker was attached to the panel and actuated for 2 minutes with band-limited white noise (BLWN) between 10-250 Hz. The upper bound was governed by the Nyquist sampling limitations of the optical interrogation hardware and the lower bound was selected to exclude very low frequency components near the oscillation frequency of the suspended shaker. Each 120-second time history was windowed and segmented into 2-second tests for a total of 58 individual tests at each of the 12 discrete damage levels. Figure 2 describes the impact protocol for incrementally growing the delamination in the panel and Table 1 describes the accumulation of delamination damage in the panel.

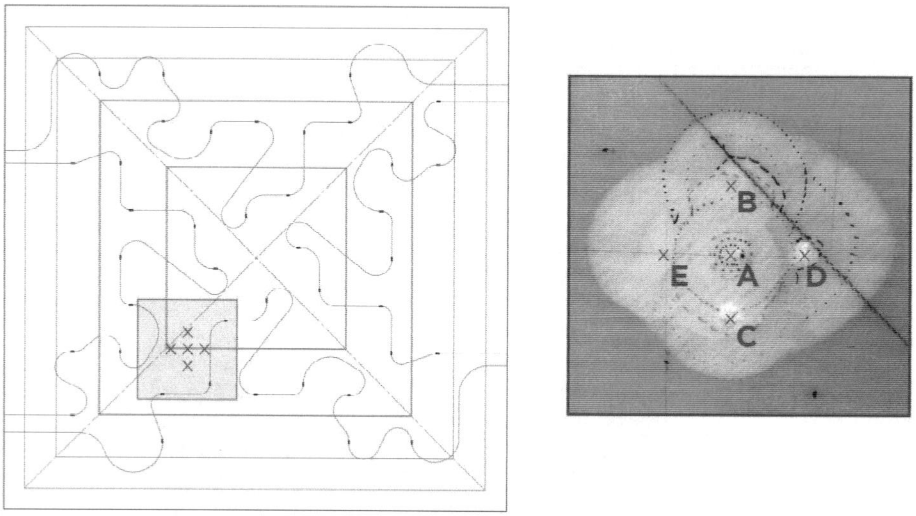

Figure 2. Impact location(s) on the test article.

2. Feature Selection

Once raw data are obtained via the sensors (FBGs here), damage-sensitive features must be mined from the raw data that will be used for detection of the damage. Spectral density estimates are a powerful frequency domain feature, where the underlying assumption is that damage will be manifested in the structure as changes in fundamental structural stiffness, and this loss in stiffness will result in a change in the fundamental vibration characteristics of the structure, observable in the spectral densities. In general for the kinds of applications in this work, the input is not assumed known or measurable, and thus it becomes difficult to discern whether vibration characteristics in the auto power spectrum are due to damage or a result of changes in the input. Motivated by previous work by the authors (Yeager et al. 2017) and Farrar et al. (Farrar et al. 1994), a feature used to detect structural changes in a bridge subjected to ambient excitation, was adapted. The feature comes from the cross power spectral density (CPSD) between *i-j* sensor pairs, quantified by

$$\hat{S}_{ij}(\omega)=(2/N)E\left[X_i^*(\omega)X_j(\omega)\right], \tag{1}$$

where the *E*[ ] operator signifies an ensemble average of the Fourier transforms X_i and X_j, and an asterisk denotes the complex conjugate. From this definition, it is evident that the quantity will be maximally correlated at global structural resonant maxima. The prominent peaks are selected from the CPSD and their shifts were tracked as a multivariate feature set i.e.

$$\mathbf{x} = \begin{bmatrix} \Delta_1 & \Delta_2 & \cdots & \Delta_k \end{bmatrix} = \left[\arg\max\left(\hat{S}^b_{ij}\left[\omega_{b,k}\right]\right)\right] - \left[\arg\max\left(\hat{S}^u_{ij}\left[\omega_{u,k}\right]\right)\right], \quad (2)$$

where $\hat{S}^b_{ij}$ and $\hat{S}^u_{ij}$ represent the CPSDs for the *i-j* th sensor pair for the baseline (*b*) and unknown (*u*) structural states, respectively. The frequency values $\omega_{b,k}$ and $\omega_{u,k}$ are the arguments that maximize $\hat{S}^b_{ij}$ and $\hat{S}^u_{ij}$, respectively. A collection of reference peaks from numerous baseline tests is established. Representative CPSD spectra for illustration purposes are shown in Figure 3, where the blue curve is the averaged CPSD of several baseline states, and its peaks serve as the reference. The red curve is a CPSD for a representative sensor pair from a test panel. The shift in each of the peaks, $\mathbf{x}_{m,p}$, becomes the damage-sensitive feature vector for the *m*-th test and the *p*-th sensor pair. The feature vectors along the *m* dimension will all be of equal dimension because a set number of peaks are tracked for a given sensor pair throughout the test, based on the number of prominent peaks in the reference CPSD. In general, however, dimension of the vectors can change along the *p* dimension because different sensor pair combinations will have CPSDs with different shape and peak prominence.

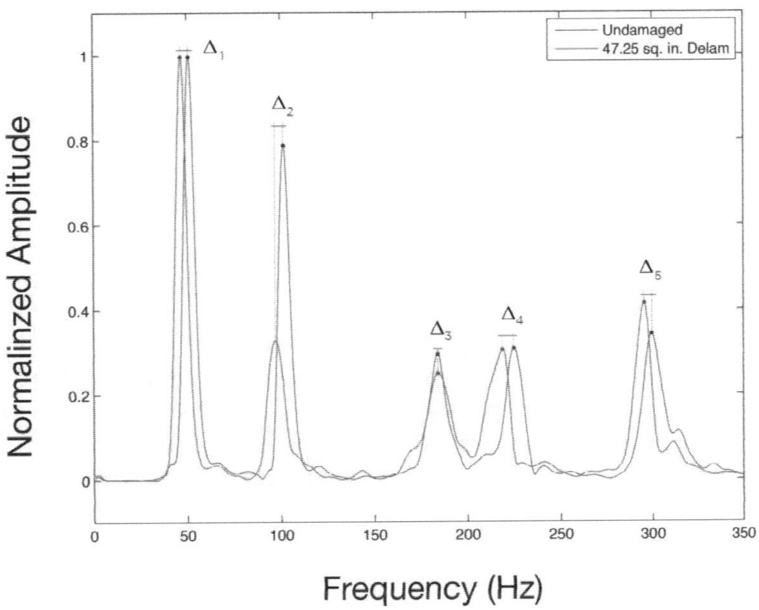

Figure 3. Representative CPSD spectra for sensor pair 1 and 5.

Finally, this multivariate feature set was reduced to the nobis covariance-weighted distance metric

$$D_{m,p} = \left(\mathbf{x}_{m,p} - \bar{\mathbf{x}}_p\right)^T \Sigma_p^{-1} \left(\mathbf{x}_{m,p} - \bar{\mathbf{x}}_p\right), \quad (3)$$

where $\mathbf{x}_{m,p}$ is a feature vector for an unknown structural state, $\bar{\mathbf{x}}_{m,p}$ is the mean of feature vectors from the baseline training feature set, and Σ_p is the covariance matrix of the baseline training feature set. The baseline training feature set is an ensemble of strain responses of the test article before any damage was introduced. Because sensor pairs are being used to generate the cross spectra, and there are 40 sensors (though only 39 are operable since one failed during fabrication), there are 780 possible sensor pair combinations. Searching all 780 sensor pairs, however, is computationally expensive and overly redundant for the purpose of this study, so a random selection of 100 sensor pairs was generated as the set of time histories that were used to generate all cross-spectra. Figure 4 presents the results for 4 randomly selected sensor pairs as a function of test number.

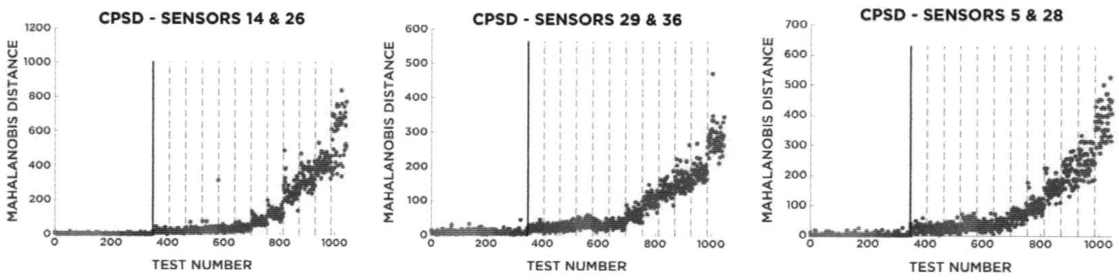

Figure 4. CPSD-based Mahalanobis distance features for three representative *i-j* sensor pairs.

In Figure 4, the data points to the left of the solid vertical line are from baseline (undamaged) tests, and the dashed vertical lines represent the different damage levels. For the purpose of this study, only the smallest delamination (1.86 cm^2) from the first impact will be considered in the damage detection scenario. It is assumed that if damage can be detected with reasonable confidence at the lowest damage level, then damage detection at higher levels is increasingly robust, under the fundamental observation that damage levels will monotonically (though not necessarily linearly) increase outlier counts.

3. Damage Detection as a Binary Hypothesis Test

Beyond raw data and feature selection, one must be able to employ some sort of outlier detection to indicate when a statistically significant change in the feature set has occurred. This fundamentally requires a model to account for the uncertainty or noise present within the feature. In this work, it is assumed that the features, particularly with all of the averaging, will tend towards a multivariate normal distribution under central limit theorem arguments. If this is true, then the squared Mahalanobis distance measure will follow a chi-square distribution with k degrees of freedom, where k is the dimension of the feature vector

$$\mathbf{x} \sim \mathbf{N}_k(\mu, \Sigma) \Rightarrow \mathbf{d}^2 \sim \chi_k^2, \tag{4}$$

where $\mathbf{x}$ is the multivariate normal feature vector with mean μ and covariance Σ and $\mathbf{d}^2$ is the squared Mahalanobis distance distributed as a chi-square distribution χ_k^2 with k degrees of freedom. This assumption is verified by plotting a histogram of the baseline Mahalanobis square distances for a given sensor pair and overlaying a chi-square distribution with the appropriate degrees of freedom. Figure 5 (left) shows this for a representative sensor pair, and the equivalence strongly supports the proposed Gaussian statistical model for the features (and chi-square model for the distances).

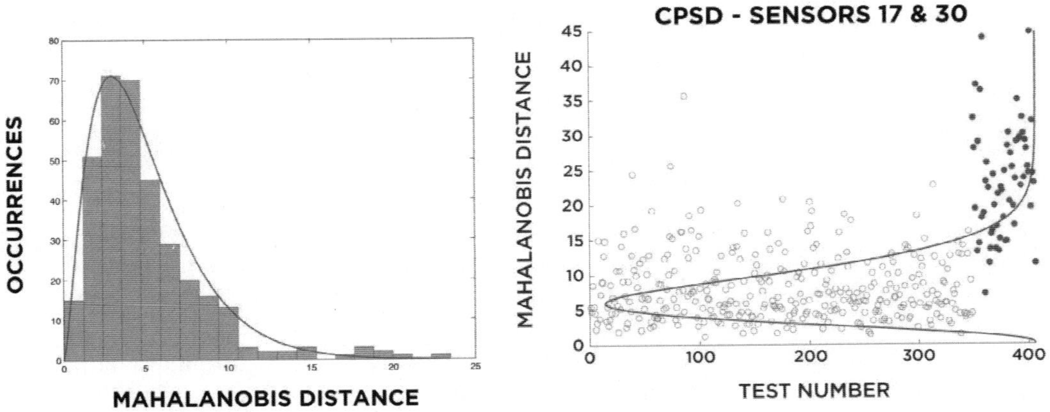

Figure 5. (left) Baseline Mahalanobis distances for sensor pair 15 & 29 compared with a chi-square distribution with 5 degrees of freedom (implying there were 5 tracked peaks in the CPSD); (right) Comparison of baseline to damaged Mahalanobis distances. The fitted chi-square distribution is given by the continuous curve.

Given this model, a detection hypothesis test may be constructed to decide "Are the features representative of the baseline or not?" This simple binary decision is the most common (and often the most useful) in SHM applications. In this work, "critical damage" was defined to be the lowest level of damage (1.86 cm^2 delamination). Figure 5 (right) shows the baseline data compared to the smallest damage level, including the chi-square distribution applied to the baseline; the filled points are being tested (compared) against the open circles via the fundamental binary test

$$\begin{aligned} H_0 &: \mathbf{d}_{ij}^2 \in \chi_p^2 \\ H_A &: \mathbf{d}_{ij}^2 \notin \chi_p^2, \end{aligned} \tag{5}$$

where "0" indicates the null hypotheses and "A" indicates the alternative hypothesis. The hypothesis test is applied to each of the 100 sensor pairs in consideration independently (since the number of degrees of freedom can change from pair to pair in accordance with cross spectral peak identification), and then a simple voting scheme may be employed across the network to arrive at a final decision.

Figure 6 shows three different sensor pair comparisons, for illustrative purposes. For the indicated sensor pair number, the histogram corresponds to the Mahalanobis distances of the baseline data for that particular sensor pair which forms the probability density function (PDF) for a chi-square distribution with k degrees of freedom. The continuous curve is the theoretical chi square distribution, and the red points represent the Mahalanobis distances of the damaged data from the first delamination. In any binary hypothesis test, the decision of which hypothesis to select depends upon a threshold that must be set, and this threshold depends on the application; in this test, a 95% confidence was selected, resulting in the individual black dashed lines drawn upon each distribution in Figure 6. This figure serves as a way to visualize the binary hypothesis test, in that any feature point to the right of the vertical dashed line would be diagnosed correctly as damaged, while any point to the left of the vertical dashed line would be diagnosed incorrectly as undamaged.

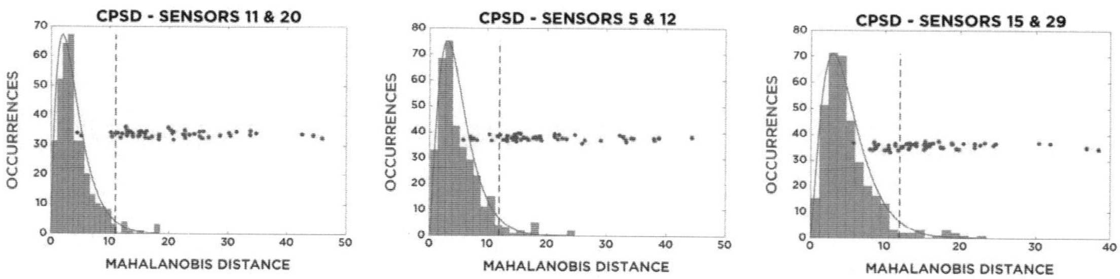

Figure 6. Specific hypothesis test visualization for three different *i-j* sensor pairs.

For each *i-j* sensor pair there will be 100 true/false test results for each candidate outlier, as the experiment was set up using 100 random sensor pair combinations. With each of the 100 *i-j* pairs thus casting a "vote", if a majority (>50) of the sensor pairs choose an outlier, then the point is labeled an outlier, and the null hypothesis is rejected. If a point, in reality, comes from a damaged structural state but is labeled undamaged, a decision-level "false-negative" is the result. A decision-level "false-positive" implies the opposite: that a data point is actually from an undamaged structure but is given a damaged assignment.

Finally, a blind test is conducted to assess the performance of the SHM algorithm. A random assortment of 290 baseline data points was used to create the baseline chi-square distribution, leaving 58 data points that are, in reality, undamaged and 58 points that came from the damaged structure. The binary hypothesis test then attempts to accurately discern between damaged and undamaged data points. The results after final sensor fusion via voting are shown in Table 1. The first row represent results for the binary test as stated, i.e., detecting the lowest damage level. No false negatives were reported, and about 11% false positives, which is consistent with the statistical model threshold. The same baseline was also applied to some higher damage levels, and the results are shown on subsequent rows. For any delamination 33.29 cm^2 and larger, there was 100% correct classification. Although this particular study employed a fairly dense sensor network, the data in the presented figures suggests that a much more sparse network of sensors could be used, as implied by Figure 6. The dense network provides more robust statistics and therefore better detection. Another important note is that the damage sensitive feature has its origins in the CPSD, which speaks to the global dynamic characteristics of the structure. This implies that the proximity of damage to any sensor in the network does not play a roll in the detection performance. Lastly, the feature is blind to the actual geometry of the structure since the feature stems from the global governing dynamics so the method should scale to more complex geometries and sparser sensor networks (with likely expected performance decreases, of course).

Table 1: Hypothesis test and sensor fusion voting scheme results for the binary detection problem.

Delamination Area (sq. cm)	Correct Prediction Percentage	False Negatives (Out of 116)	False Positives (Our of 116)
1.86	89%	0	13
5.18	92%	0	9
8.9	98%	0	2
33.29	100%	0	0

Of course, the threshold selection for this laboratory study was somewhat arbitrary; it is useful to understand how threshold selection affects both types of decision errors, false-positives and false-negatives. Figure 7 presents the prediction results along with the false-negatives and false positives for a range of potential threshold choices. According to the figure, a decision threshold of around 81%-84% would appear to be the best range of choices for a decision threshold, because that range yields the best prediction results and minimizes the occurrences both the false-negatives and false-positives. It is important to remember, however, that in many SHM applications, false-negative errors will have significantly higher consequences (e.g., life safety implications) than false positive errors (e.g., downtime for unnecessary maintenance). This practical consideration justifies the more conservative choice of 95% as the decision threshold used in this experiment.

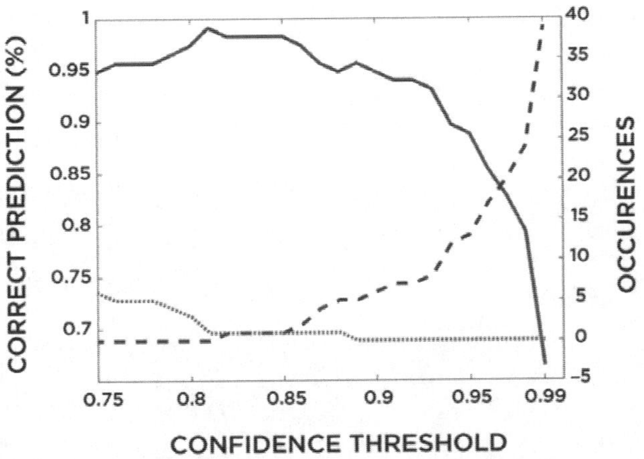

Figure 7. Detection performance as a function of threshold selection. The solid line represents overall correct detection percentage (left vertical axis), the dashed and dotted lines represent the false-positives and false-negatives respectively (right vertical axis).

4. FBG Ingress/Egress Design and Testing

Fiber ingress and egress in embedded applications is a long-standing issue. No uniformly viable commercial product to get fibers in and out of the composite specimen exists, and a design was proposed and built for testing under shock-level impacts. The housing was designed in such a way that it could undergo the fabrication process entirely under the vacuum bag, while also preventing resin infiltration into the connector. The connector was fabricated with a thin plate across the front façe that was designed to break away after the cure of the composite part. This break away piece provides a resin dam to prevent resin from flowing into the connector chamber during the cure. Figure 8 presents the basic dimensions of the Box-style housing, providing plan (left), elevation (center), and exploded (right) view of connector; all measurements are in inches. Figure 9 (left) shows the break-away detail of the round-style housing. Installing the connector under the vacuum bag requires no special procedure for bagging, which adds very little additional complexity to the fabrication procedure. Figure 9 shows the round-style housing under a vacuum bag, demonstrating the success in providing a connector design that can be incorporated under the bag with minimal additional complexity.

A separate impact system designed by Hi-Test Laboratories Inc. was used to deliver blows to the connector embedded in a separate instrumented composite specimen, very similar to the one described above. Impacting head blows up to 263.1 g peak acceleration were delivered to the specimen, and the functionality of the connector was verified. The connector functioned perfectly, even after substantial cracking occurred at the 263.1 g load level (Figure 10).

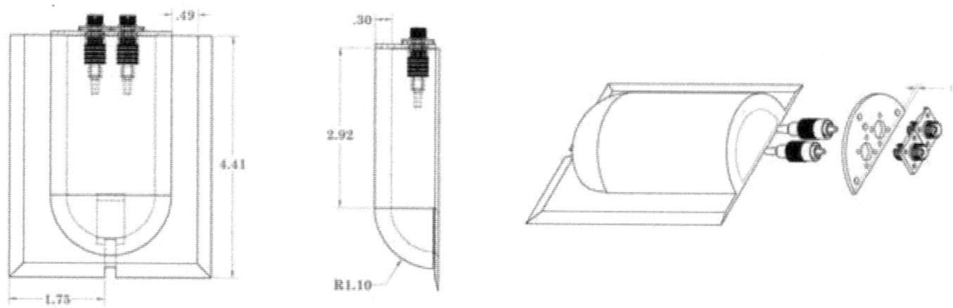

Figure 8. Housing design/geometry.

Figure 9 (left) Housing break-away front panel and (right) the housing under the vacuum bag.

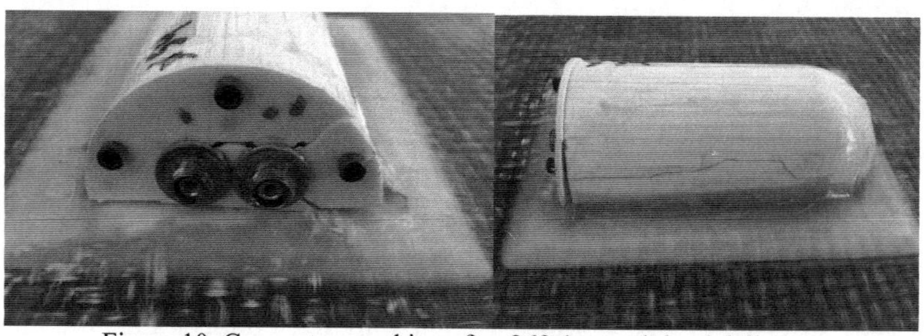

Figure 10. Connector cracking after 263.1 g peak load delivery.

5. Conclusions

This paper considered the detection of impact-induced delamination in a glass-epoxy material system using internally-embedded fiber Bragg gratings. The detection was implemented in a simulated operational condition by exciting the structure between impacts with pseudo-random noise and extracting cross-spectral density peaks estimated between sensor pairs. Using the peaks as features, the Mahalanobis distance metric was computed in order to establish an unsupervised learning application whereby baseline distances were verified to be modeled as a chi-square distribution. A blind hypothesis test on the data showed excellent statistical decision performance for detecting delamination damage as small as 1.86 cm^2, and a performance report was shown that showed the effect of decision threshold selection. A special easily-fabricated and resin-resisting ingress/egress fiber connector was also designed and shock-validated, and it survived This work represents the full system-level implementation of a “data-to-decision” SHM application.

Acknowledgements

This material is based upon work supported by the Naval Sea Systems Command under Contract No. N00024-13-c-4053. Anything expressed herein are those of the author(s) and do not necessarily reflect the views of the Naval Sea Systems Command.

References

Farrar, C. R., & Worden, K. 2007. "An introduction to structural health monitoring." *Philosophical Transactions of the Royal Society of London A: Mathematical, Physical and Engineering Sciences* vol. 365, no. 1851, pp. 303-315.

Sohn, H., et al. 2003. "A review of structural health monitoring literature: 1996–2001." *Los Alamos National Laboratory Report LA-13976-MS.*

Doebling, S. et al. 1998. "A summary review of vibration-based damage identification methods." *Shock and Vibration Digest*, vol. 30, no. 2, pp. 91-105.

Gul, M., and Catbas, F. 2009. "Statistical pattern recognition for Structural Health Monitoring using time series modeling:tTheory and experimental verifications," *Mechanical Systems and Signal Processing* vol. 23, no. 7, pp. 2192-2204.

Kessler, S., et al. 2002, "Damage detection in composite materials using frequency response methods." *Composites Part B: Engineering*, vol. 33, no. 1, pp. 87-95.

Kersey, A., et al. 1997 "Fiber grating sensors." *Journal of Lightwave Technology*, vol. 15, no. 8, pp. 1442-1463.

De Oliveira, R. et al., 2008, "Health monitoring of composite structures by embedded FBG and interferometric Fabry–Pérot sensors." *Computers & structures*, vol. 86, no. 3, pp. 340-346.

Majumder, M. et al., 2008, "Fibre Bragg gratings in structural health monitoring—Present status and applications." *Sensors and Actuators A: Physical* vol. 147, no. 1, pp. 150-164.

Peters, K. 2012, "Full-spectrum FBG analysis of inhomogeneous, fast-varying strain effects." *Asia Pacific Optical Sensors Conference*. International Society for Optics and Photonics. Sydney, Australia , January 31-February 3, 2012.

Jang, B-W. et al., 2012, "Detection of impact damage in composite structures using high speed FBG interrogator." *Advanced Composite Materials* vol 21, no. 1, pp. 29-44.

Park, S. et al., 2011, "Real-time monitoring of composite wind turbine blades using fiber Bragg grating sensors." *Advanced Composite Materials* vol 20, no. 1, pp. 39-51.

Murayama, H., et al., 2002, "Structural health monitoring of a full-scale composite structure with fiber-optic sensors." *Advanced Composite Materials*, vol 11, no. 3, pp. 287-297.

Worden, K., et al. 2007, "The fundamental axioms of structural health monitoring." *Proceedings of the Royal Society of London A: Mathematical, Physical and Engineering Sciences*. vol. 463. no. 2082, pp. 1639-1664.

Yeager, M., et al., 2017, "Assessment of embedded fiber Bragg gratings for structural health monitoring of composites." *Structural Health Monitoring*, vol. 16, no. 3, pp. 262-275.

Farrar, C., et al., 1994, “Dynamic characterization and damage detection in the I-40 bridge over the Rio Grande”, *Los Alamos National Laboratory Report LA--12767-MS.*

The 8th International Conference on
Structural Health Monitoring of Intelligent Infrastructure
Brisbane, Australia | 5-8 December 2017

Comparative Study on Two Types of Sensors for Measuring Vibration Response of Building Structures

Navid Hakimtoroghi[1], Ardalan Sabamehr[2], Mehrdad Mirshafiei[3], Rabin Raut[1], Ashutosh Bagchi[2]
[1] *Dept of Electrical and Computer Engineering, Concordia University, Montreal, Canada*
[2] *Dept of Building, Civil and Environmental Eng., Concordia University, Montreal, Canada*
[3] *Sesequake Inc., Montreal, Canada*

Abstract

Structural Health Monitoring plays a vital role in assessing in-situ the performance of a structure by use of modal properties of the structure. Modal properties can be extracted from two factors; acceleration and velocity. These two factors can be measured by a Micro Electro Mechanical Sensor (MEMS) accelerometer and geophone respectively. In this study, the response characteristics of the above two sensors are compared in the laboratory setting. For our tests, we have used analog circuitry for data acquisition part of the geophone sensor. The test has been done on the fifth floor of a building. We used a commercially proven vibration sensor as the benchmark to verify our measurements. As the frequency response of the two sensors are different, some post-compensations have been applied to the signals to match them in frequency domain.

1. Introduction

A seismic sensor is a device that measures the ground motions, and produce a physical variable related to these motions (Havskov & Alguacil 2016). Two different sensors have been used in our study; geophone and MEMS. Geophones measure the velocity by generating a voltage from a coil that freely moves in a magnetic field. (Fig. 1) shows the general structure of a geophone. We used GS-11D series of geophone from Geospace in this work.

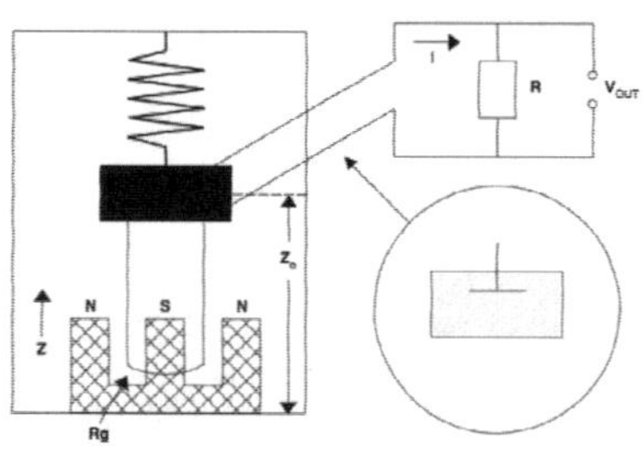

Figure 1 Geophone internal structure

MEMS measure the acceleration by a structure like geophone, but integrated inside a chip. MEMS generally generate a DC voltage and shows the changes in accelerations by changing the level of DC voltage. In this study, we have used ADXL355 MEMS form linear technology. ADXL355 provide an internal conditioning system and analog-digital convertor beside the core accelerometer (Analog Devices Inc. 2016) (Fig. 2).

Figure 2 ADXL355

2. Sensors

To compare two types of sensors, first we need to describe their intrinsic differences. The first difference between the geophones and MEMS is the type of physical parameter that the sensors are sensitive to. While MEMS are sensitive the acceleration, geophones are sensitive to velocity. As we want to compare these two sensors, in the data analysis phase, we achieved the acceleration by calculating derivative of velocity with respect to time.

The second difference is that MEMS are active elements and need an external power source for operation, while geophones are passive and the output signal voltage of the sensor is generated by the movement of the coil in a magnetic field. These two different mechanisms of generating the output voltage dictate the way we should design the conditioning circuit.

For the sensor module, we need three different geophones to detect vibrations in three perpendicular axes and there are different structures for vertical and horizontal geophones. MEMS, on the hand, generally provide three differential outputs, each represent an axis based on the physical location of the sensor.

The operating bandwidth is the other difference between these two sensors. While MEMS have flat response over a wide range of frequency, geophones because of its physical structure, cannot respond well to high frequency vibrations. Further for vibrations in the neighborhood of 10 Hz (Geospace Technologies 2012) (corner frequency of the geophone), the geophone responds like a second order high pass filter. Fig 3 shows the output frequency response of the geophone which is simulated by its equivalent circuit model. We can compensate for this feature of the geophone either adding some modules in analog conditioning circuit or using digital filters in software data analysis.

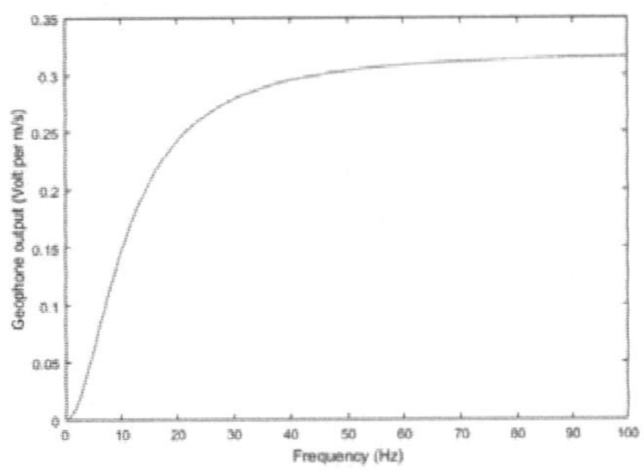

Figure 3 Geophone response curve

2.1. Geophone

For the geophone, we need three different modules before the signal is processed by the analog digital convertor (ADC). Specialty of these modules are described in the following.

For the first module, we need a low noise amplifier optimized with regards to the sensor characteristics. The considerations are that (i) the amplifier should be matched with the output impedance of the sensor, and (ii) provide enough gain over the bandwidth of the sensors. The second module is a single input to differential output amplifier stage, and the third module is a RC filter that filters out the frequencies above our frequencies of interest. The output signal from the last module is applied to the ADC module to convert the analog voltage to digital data.
The design considerations of each of these modules are explained briefly in the following sections.

2.1.1. Low Noise Amplifier (LNA)

The LNA can be implemented in three different ways i.e., using (i) transistors, (ii) integrated instrumentation amplifiers or, (iii) operational amplifier (OA). Based on the lab experiments with transistors, while each transistor may have low noise feature, addition of discrete components like capacitors and resistors to complete the circuit could increase the total noise in the circuit. It is considered better to use fewer external components in the final circuit to reduce the total noise. Instrumentation amplifiers generally need only one external resistor to set the gain. The challenge with instrumentation amplifier is that we must convert the single ended output signal to differential mode before applying to the ADC system. Since the noise level of a single OA is lower than that of an instrumentation amplifier an OAs is considered as an optimum choice for the voltage amplifier block in the overall system. OA circuits have wide diversity of applications and each circuit and application is customizable for certain usage.
We have tested several high-class OAs (i) using SPICE simulation, and (ii) using OA chips on a circuit board. In our tests, we considered current and voltage noise of each amplifier for the optimum match for the geophone output impedance.
Fig. 4 depicts comparisons of three different source of noise in different OAs current noise, voltage noise and Johnson noise. Based on the data sheet of the geophone, as the output impedance is 380 Ohms, the most important source of noise in our case is voltage noise (Horwitz & Hill 2015).

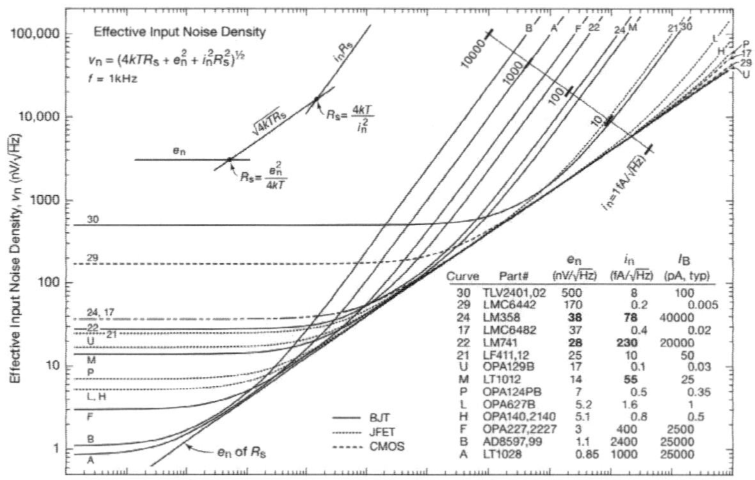

Figure 4 Effective input noise density vs. source resistance (Horwitz & Hill 2015)

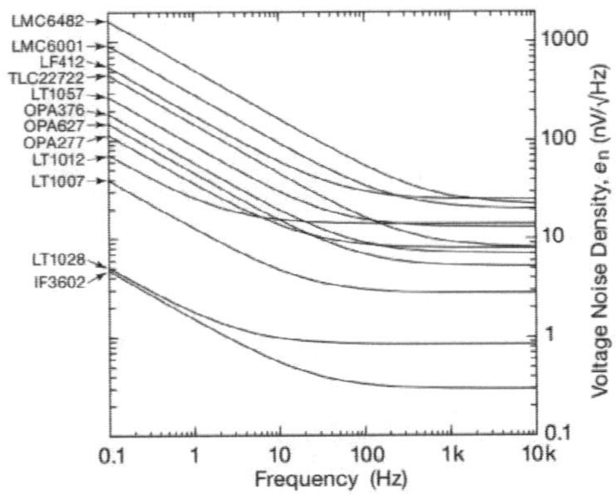

Figure 5 Voltage noise of OAs (Horwitz & Hill 2015)

Fig. 5 shows the voltage noise of different OAs and transistors at different frequencies. As IF3602 is a FET transistor, we have selected LT1028 (Linear Technology Corporation, 1992) for our circuits. We have used non-inverting structure to prevent loading the sensor by the gain resistors (W. Rodgers, 1993).

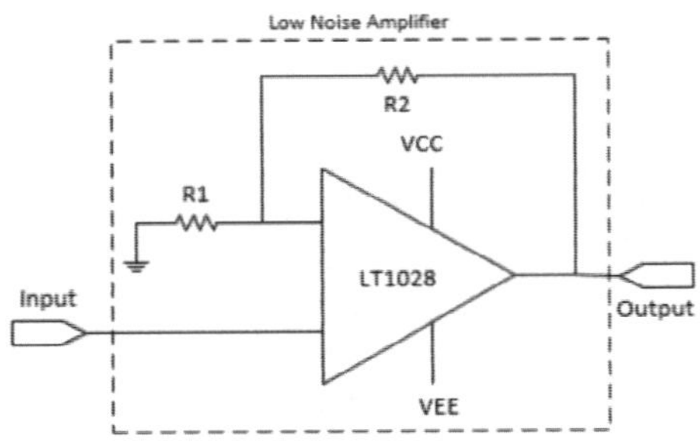

Figure 6 Non-inverting amplifier using LT1028

Design of the LNA the in this way helped to keep the noise floor of the conditioning circuit below the noise floor of the geophone (Havskov & Alguacil 2016). Therefore, the smallest possible detectable voltage at the output of the geophone will be detected by the ADC.

2.1.2. Single ended to differential amplifier

In addition to noise there is another consideration for the design of the amplifiers. For the best common mode rejection, we have chosen an ADC with differential input, while the output of our sensors is single ended. Therefore, during conditioning of the geophone signal, we must convert the single ended signal to a fully differential signal to get the best results with our ADC (Linear Technology Corporation, 2016). We used a differential amplifier, LTC6363, to form the second stage of amplification before the ADC (Fig. 7).

2.1.3. Filter

The differential amplifier is followed by a differential RC filter circuit (Fig. 7). The cut off frequency of the RC filter is set to 100 Hz to filter out high frequency noise from the analog circuitry.

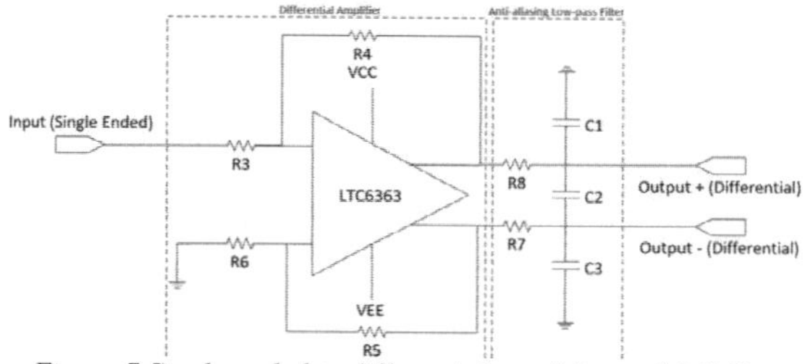

Figure 7 Single ended to differential amplifier and RC filter

2.1.4. ADC Module

After the analog conditioning, we need an ADC to convert the analog signal to digital data. We have used LTC2508-32 demo board from Linear Technology. The ADC module on this board can be controlled by a software in a host computer to analyze and save the data.
We have provided a 45MHz clock pulse for the ADC controller module. Then, based on the requirements, the controller can provide a variety of precise clock frequencies for the ADC chip. We set the ADC to get 1M samples per second with down sampling factor of 16K. Therefore, the ADC delivers 64 samples per second from the output of the analog conditioning circuit.

2.2. MEMS

ADXL355 is the MEMS sensor that have been tested in this work. The supremacy of this sensor over other MEMS is that we do not need to engage the analog conditioning part of the sensor system because it provides us the digital data over a SPI bus. To read the digital data, we need to read the SPI bus with an external microcontroller. The block diagram of the sensor is shown in Fig. 8.

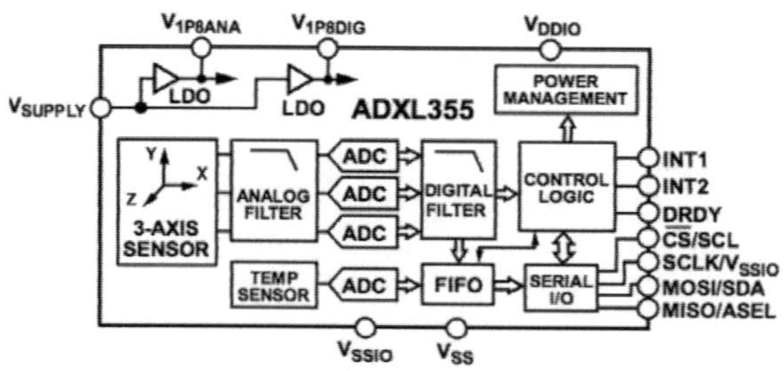

Figure 8 ADXL 355 internal block diagram (Analog Devices Inc. 2016)

The microcontroller is responsible to provide the clock pulse for the MEMS. The values from the 20-bit internal ADC are stored in the internal registers of the MEMS unit and the MEMS send them to the microcontroller in order upon request. Each phase of data transmission, includes the values of acceleration for three different axes. These data can be buffered into the microcontroller to store in the SD card as a text file

3. Experimental Results

The data has been collected from multi story building of Concordia University which is located in Montreal, Quebec, Canada. The sensor is placed in fifth floor of Engineering and Computer Science and Visual Arts Integrated Complex (EV building) in an electronic lab. The geophone has been placed in the housing that provided by the manufacturer and due to its weight, it is not necessary to fix it on the ground. The MEMS board (Fig. 2) on the other hand, needed to be fixed on the ground. So, we used wax to fix the board beside the geophone housing.

To compare the MEMS and geophone, we have used two different setups for each sensor in the same vibration environment. As mentioned, the geophone data, after three level of amplification, single ended to differential conversion and filtering is read by an ADC evaluation board. For the MEMS, we needed to program a microcontroller board to read the data from the sensor and store the data in a SD card. The text data, later is analyzed and plotted by MATLAB.

In the following plots, the frequency response of the MEMS and geophone are shown in three different axes of the test environment.

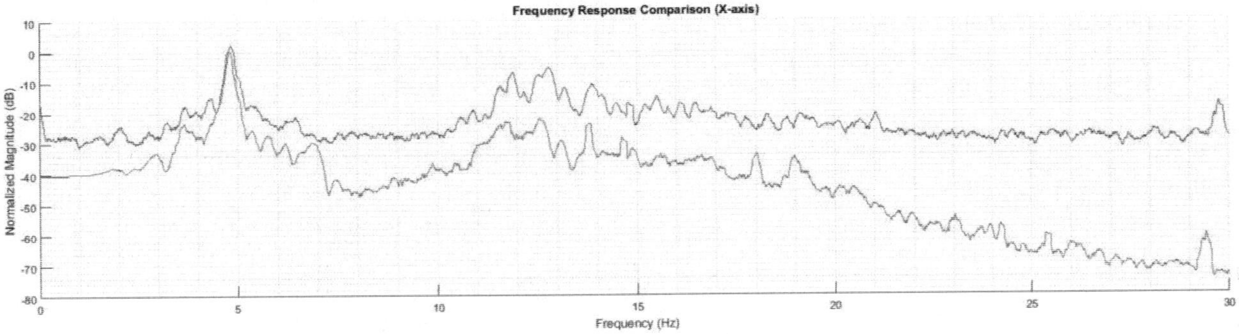

Figure 9 Normalized frequency response of the MEMS (blue) and geophone (orange) in X-axis

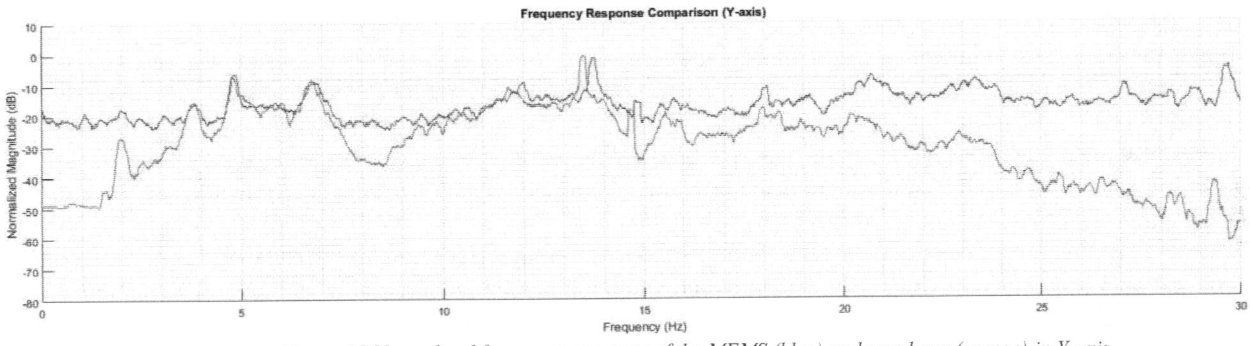

Figure 10 Normalized frequency response of the MEMS (blue) and geophone (orange) in Y-axis

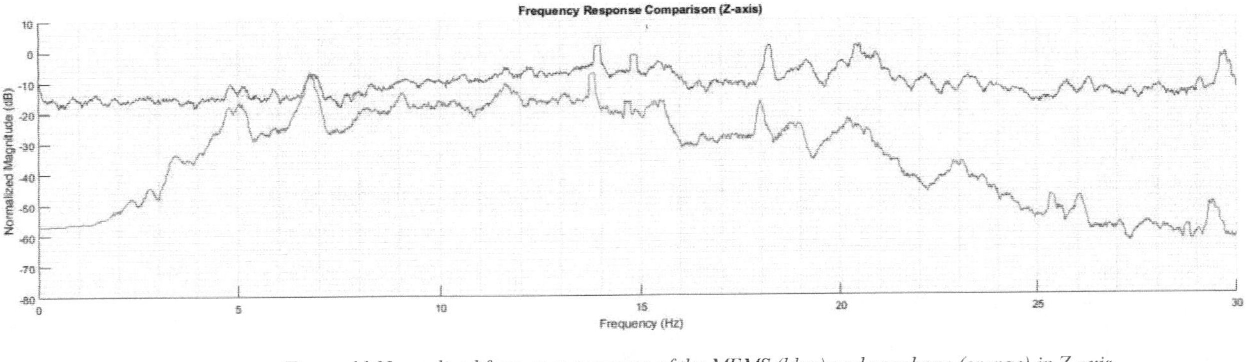

Figure 11 Normalized frequency response of the MEMS (blue) and geophone (orange) in Z-axis

Following observations can be extracted and explained from the plots:
The microcontroller that have been used for the tests generates the clock signal for the MEMS to read the values of acceleration. In this case, the internal clock of the microcontroller is responsible to provide the clock signal for the MEMS. The internal clock of the microcontroller is not precise enough to sample the data at a fixed sampling rate. On the other hand, for the geophone, we have used a 45 MHz clock pulse for the controller of the ADC. The controller then divides the frequency and provide a precise clock for the ADC which means a fixed sample rate. The variable sampling rate for the MEMS during the data acquisition phase causes a frequency shift in MEMS frequency response. For the next prototype, using external clock (crystal) and frequency divider for the microcontroller unit in hardware side or using data calibration algorithm in data analysis program in software can overcome this problem.
The next observation on the plots is the difference between the dynamic range of the MEMS and geophone response. Dynamic range in definition is the difference between the largest and smallest values that can be detected by a system. Higher dynamic range of geophone conditioning system means its noise floor is lower than that of the MEMS, and consequently, smaller vibration can be detected by the geophone.
Another observation in the plots is that the weakness of geophone in response to frequencies smaller than 10Hz is more effective in the vibrations in z-axis. This situation becomes more important when we have some mode shapes at low frequency. As an example, for the z-axis, while MEMS responded to frequencies smaller than 10 Hz, the geophone frequency response is flat in the mentioned interval.

4. Conclusion

Finally, two types of sensors have been tested in level five of a sixteen story building. The z-axis is not applicable for vibration test of building whereas it is perpendicular to the floor, the others are lateral vibration of building. Detection of the ambient or operational load is one of the main concern in vibration test. As it shows, Geophone-based sensor is more precise to have a fixed sampling rate rather than MEMS-based sensor. Apart from that, smaller vibrations can be detected by geophone-based system due to its lower noise floor in comparison with MEMS-based. It means that geophone is more preferable for ambient vibration test. However, Geophone has limitations in responding to lower frequencies. As shown in Fig. 9 to Fig. 11 it cannot detect the frequency lower that 5Hz especially along z-axis. Consequently, both sensors have some benefits and drawbacks which should be considered in vibration test.

References

Jens Havskov, Gerardo Alguacil, 2016, *Instrumentation in Earthquake Seismology*, second edition, Springer.

Analog Devices Inc., 2016, *ADXL355 Low Noise, Low Drift, Low Power, 3-Axis MEMS Accelerometers*, Available from: http://www.analog.com/en/products>. [01 June 2017].

Geospace Technologies, 2012, *GS-11D high output, rotating coil geophone,* Available from: http://www.geospacetech.ca>. [01 June 2017].

Linear Technology Corporation, 2016, *LTC2508 32-Bit Oversampling ADC with Configurable Digital Filter*, Available from: http://www.linear.com/product>. [01 June 2017].

Paul Horwitz, Winfield Hill, 2015, *The Art of Electronics*, third edition.

Linear Technology Corporation, 1992, *LT1028 Ultralow Noise Precision High Speed Op Amps,* Available from: http://www.linear.com/product>. [01 June 2017].

Peter W. Rodgers, 1993, *Maximizing the signal-to-noise ratio of the electromagnetic seismometer: The optimum coil resistance, amplifier characteristics, and circuit,* Bulletin of the Seismological Society of America April 1993vol. 83 no. 2 561-582 (Rodgers 1993)

The 8th International Conference on
Structural Health Monitoring of Intelligent Infrastructure
Brisbane, Australia | 5-8 December 2017

In-service Aqueduct Tunnel Monitoring Using Optical Fiber Sensor

M. Imai[1], K. Ikeda [2]
[1] *Kajima Technical Research Institute – Japan, Email:* michio@kajima.com
[2] *Hokkaido Electrical Power Company – Japan*

Abstract

A number of civil facilitations of water power plant are aging in Japan, thus it has to be monitored for safely supplying electrical power. Such civil infrastructures have been widely evaluated through visual inspection, but such inspection is definitely labor intensive and time consuming. Especially for aqueduct tunnel wherein water is running constantly, the plant generating electricity must shut down and stop running water for inspector to access the tunnel. Therefore, the longer the aqueduct tunnel is, the longer inspection time takes. From economical point of view, less period of power outage is demanded, of course. Instead of visual inspection, an instrumentation has been applied for structural monitoring. Unfortunately, electrical instruments are generally not designed for working underwater; on the other hand, optical fiber sensing is expected for *in-situ* monitoring even if aqueduct tunnel is filled with running water. The authors have developed a method of quick sensor installation, and are applying distributed strain measurement to in-service aqueduct tunnel. The sensor installation was completed in a couple of days, and the opening displacements of the joints of tunneling walls were successfully monitored in-service. In this paper, the sensor installation method, the monitored data and processing method are described.

1. Introduction

Civil infrastructures have been exposed to harsh environment for a long term, and large part of them had been built in economic growth period that corresponds from 1950's to 60's in Japan. Despite aging structures, a lot of tunnels and bridges are still in service nowadays. There is no alternative yet, thus replacement have difficulty in shutting it down. Especially for power utilities, continuous operation is a crucial issue for stable supply of energy. Aqueduct tunnel of hydropower plant is one of the troublesome structures, because it extends to longer area. In addition, such structures have been always applied by earth and hydraulic pressure, and exposed to abrasive stream for electric power generation. Once something happens in the tunnel, not only power outage but also big impact on the surrounding environment would occur. For safe operation, manual inspection is periodically executed when plant stops working. In general, an inspector walks along the tunnel to check whether something happens or not, also the transversal and longitudinal deformation of tunnel is surveyed along with opening displacement of the joints of tunneling wall. Moreover, damaged structure such as crack is inspected if the area is deformed. Such survey and measurement is, of course, labor intensive and time consuming. It takes more than a week in the case of larger area for inspection. Real-time monitoring on aqueduct tunnel in operation is really required for safe electrical power supply.
Instrumentation, e.g., strain gauge and linear variable differential transformer, helps us to know the useful structural information instead of visual inspection. Most of the conventional instrumentation are electrical-based, therefore they may be unable to survive for a long time due to oxidization and/or insulation failure. On the other hand, as there is no possibility for degradation, optical-based

sensor shows availability for long-term even if it is submerged. Especially for optical fiber sensor, there are many other features, such as small, light-weight, no electromagnetic immersion, no requirement of power on sensing segment, and long-distance transmission. In civil engineering field, these are the advantages of what sensing technique requires. There are many types of optical fiber sensors, however, only Brillouin-based sensor can uniquely measure strain profiles along the optical fiber (B. Glisic and D. Inaudi, 2008). Brillouin-based distributed measurement is especially valuable in monitoring a lengthwise structure such as tunneling, because the locations to be monitored are normally located on a line along the structures.
Typically, optical fiber sensor installation requires a skilled manual worker and specialized equipment for introducing uniform initial strain to sensor. Additionally, it takes time to tighten the sensor for stable and semi-permanent measurement. Just in case, temporal monitoring is needed until retrofitting. In such case, resources for monitoring are limited so, less labor, less cost, and less installation days are preferable. It can be generally stated that the cost and the quality are trade-off relationship. Some cases need optimal measurement with the expense of the best effort, but other cases need only reasonable measurement with moderate cost. Currently, there seems to be no alternative but to install optical fiber sensor with cast-in anchor on tunneling wall. Such destructive anchoring method can hold optical fiber sensor tightly for long-term, however, it has to make a hole in monitored structure partially and requires installation of time and cost.
For structural monitoring on an aging aqueduct tunnel, the authors have developed a sensor installation method which uses adhesive for quick installation and less cost. This paper presents the application of the method in the tunnel to monitor its transversal and longitudinal behavior under operation.

2. Sensor Installation

Owing to wet surface and overhead working, quick installation of optical fiber sensor on the surface of aqueduct tunnel is challenging. To overcome these obstacles, we tried to establish the installation method experimentally by a field test.

2.1 Overview

Due to its easy and quick handling, it has been decided to use adhesive to install optical fiber sensor instead of anchoring. By a preliminary test, we had already chosen four types of adhesives among commercial products for the following reasons: (1) higher adhesive stress is one of the most important requirements even if the object gets wet; (2) workability of adhesive is critical for quick installation. Throughout the year, aqueduct tunnel under the ground provides cool environment, which decreases the flexibility of adhesives when mixing; and (3) its cured behaviour should be flexible in avoiding an interfacial debonding.

2.2 Exposure Test

Prior to the start of *in-situ* monitoring, we had executed an exposure test in an aqueduct tunnel to choose better adhesives. Four types of adhesives were selected beforehand and were applied to the exposure test. Optical fiber sensor, whose size of 1.8 x 3.5 mm and covered with jacket as shown in figure 1, was attached on the surface of the tunnel with adhesives. As shown in figure 2, each length of the attachment was approximately 1.5m. The sensor is wired into a junction box, which was anchored on the tunnel, and further wired upward along a shaft toward the ground. Therefore, we could access the embedded sensor from the ground when the plant worked. When the plant was in operation, the area of the exposure test was exposed to running water for power generation.

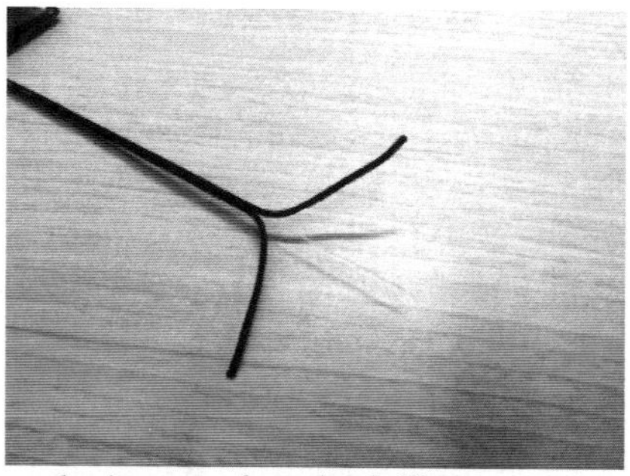

Figure 1. Armored optical fiber sensor cable

Figure 2. Exposure test in aqueduct tunnel

2.3 Test Results

By exposing in the in-service aqueduct through half a year, better installation method had been tried as a result of optical measurement. Optical loss distribution had been periodically measured by OTDR, and the loss of test area is presented in figure 3. There had been neither a trend nor changing in optical loss in this term, therefore it was suggested that no bending due to debonding of adhesives had occurred. After half a year, when the plant stopped working, it was observed that all adhesives were still bonded tightly as shown in figure 4. It could be concluded that all adhesives provided acceptable adhesivity. Based on this exposure test, the best adhesive, which mainly consisted of epoxy compound, had been experimentally determined due to its adhesivity and workability.

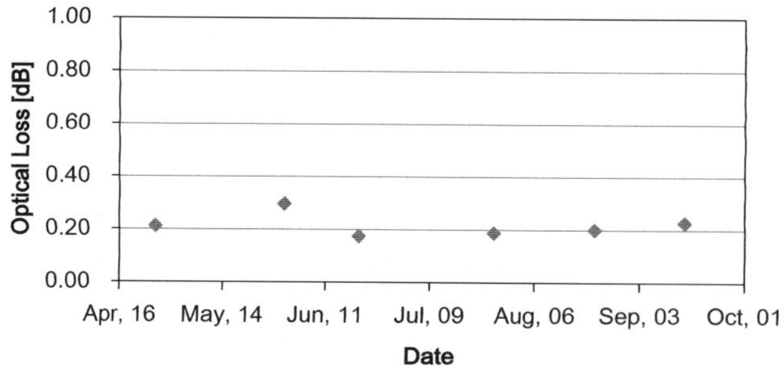

Figure 3. Measured optical loss

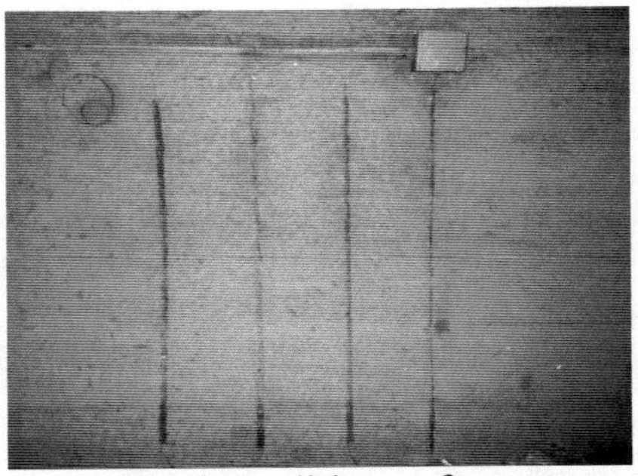
Figure 4. Final conditions of exposure test

3. In-service Monitoring

Monitored aqueduct tunnel had been constructed for hydraulic power generation in 1940's, and it has been observed that the tunnel is gradually deformed. One of the most deformed area had already been retrofitted a couple of years ago. The cause of deformation is still under investigation, making it difficult to design the retrofitting of the tunnel. Once or twice a year, an inspector periodically surveys transversal and longitudinal deformation, and measures the gap between the tunneling wall. For such regular inspection, the plant operation is required to be shutdown. Therefore, continuous in-service monitoring would provide valuable information for proper maintenance. Also, the monitoring should surely take less resources.

3.1 Overview

There are two types monitoring to be executed in this aqueduct tunnel. One is the transversal deformation of tunnel, which corresponds to the cross-section shape in the retrofitted area, and other is the longitudinal deformation of the joints of tunnelling wall, which corresponds to the opening displacement of the gap. The armored optical fiber sensor applies for both types of monitoring, and strain distribution is measured in operation. Based on the requirements of spatial resolution, transversal deformation is measured by BOCDA-based instrument, and longitudinal deformation is measured by BOTDR-based instrument. (C. K. Y. Leung, *et al.*, 2015)

3.2 Sensor Installation

As a result of the exposure test conducted in similar situation to this monitored structure, the installation method has been established. The method is very simple, i.e. the armored optical fiber sensor is embedded on the surface of the concrete structures with the use of adhesives, which is fruitful for both labor and money saving,

Figures 5 and 6 presents the installed optical fiber sensor on the transversal and the longitudinal section of the tunnel respectively. On the transversal section, the sensor is embedded all throughout the length of the inner perimeter to measure the shape of cross-section for two circles in total. On the longitudinal section, the sensor is embedded for one-meter-length adjacent to the joints for measuring the opening displacement of the gap. Totally, twelve gaps are covered by one distributed optical fiber sensor. Throughout the length of approximately 600m, both monitored sensors are wired to the end of the tunnel for connecting to the measurement instruments. Simple sensor installation method has been successfully completed within three days by four workers, including carrying materials, marking position, installing, wiring and checking the optical fiber sensor.

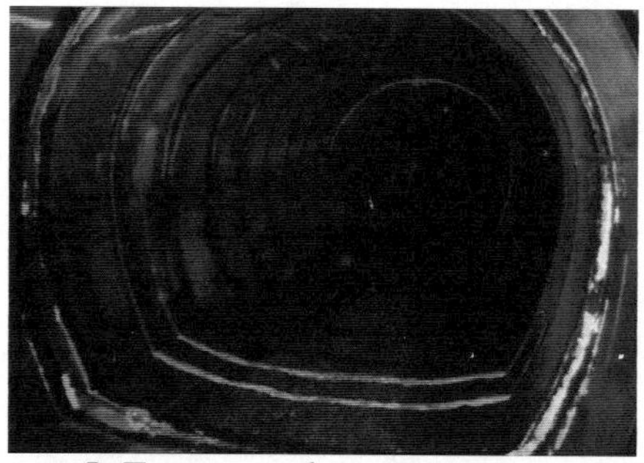

Figure 5. Transversal measurement section

Figure 6. Longitudinal measurement section

3.3 In-situ Monitoring

After a day of the sensor installation, the plant started working. Figure 7 shows the end of the aqueduct tunnel. The tunnel is always not filled with running water, therefore the pressure of the running water does not dominantly apply to the tunneling structures. When the plant works anyway, the embedded sensors are the only way to measure the behaviour of the tunnel. The result measured just after the operation is regarded with the baseline of the measurement, i.e., the initial measurement.

a) Stop operation b) In operation (Oct.) c) In operation (Mar.)

Figure 7. The end of aqueduct tunnel

3.4 Temperature Effect

It is well known that Brillouin scattering are affected by strain and temperature, thus Brillouin-based measurement has to be compensated in case of temperature change. Figure 8 shows the relationship between the measured strain by BOTDR instrument and the water temperature measured at the intake gate. This result of optical fiber sensor is measured at the location, where the strain is not considered to be applied, thus it can be said that only water temperature has an impact on the measurement. Depending on water temperature, the average of measurement at the location except for the gap shows approximately 21 [10^{-6}/deg.], which can be the coefficient of temperature.

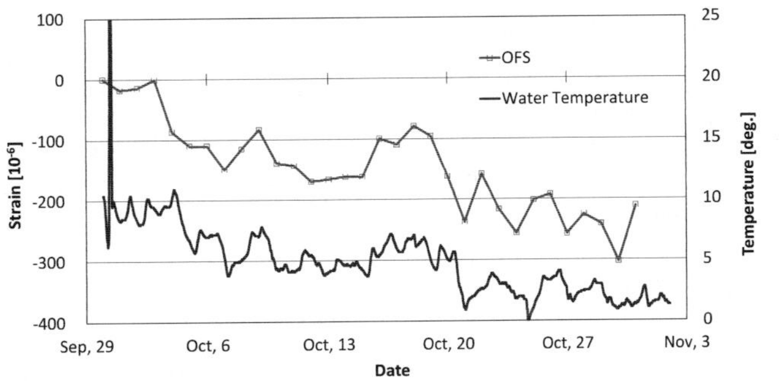

Figure 8. Measured relationship between longitudinal strain and temperature

Next, it is discussed with the temperature effect of the transversal measurement. The shape of the deformed cross section will be calculated by measuring strain distribution along the inner perimeter of the tunnel lining with some assumptions. Based on the infinitesimal deformation theory, deformation is found by integration of bending deformation. Thus, the position of neutral axis and the length of perimeter has to be assumed. Additionally, other assumption has to be introduced to decrease the integration error. For instance, the deformation at the beginning position of the integration is corresponding to those at the end of the integration, and it can be said that such assumption is physically reasonable. Also, it is assumed that the length of the perimeter remains constant, considering that no axial force is applied. In other words, this assumption can neglect temperature effect, because temperature effect acts in the direction of axial force. This assumption is physically derived from the fact that we can only measure strain along the inner perimeter in this case, otherwise we have to install the sensor inside the tunnel lining to measure along the outer perimeter. It should be noted that the full section of tunnel is assumed to show the same temperature, and this assumption is still being investigated by this *in-situ* monitoring.

3.5 Results and Discussions

As one of the longitudinal behaviours of the aqueduct tunnel, figure 11 shows the measured strain distribution at the joint between Blocks 16 and 17. In this case, the spatial resolution of BOTDR is one meter, and the measurement interval is 0.05m. It shows the generated strain distribution after one month of plant operation, and there is a peak-shaped strain distribution at 735.4m in position due to the gap opening. And, the temperature-compensated result provides almost zero strain distribution except for gap. Finally, the opening displacement of the gap is calculated by the integration of compensated strain. Figure 10 shows the measured opening displacement of the gap, putting down with water temperature. Measured opening displacement is increasing in this term, while water temperature is decreasing. It can be assumed that due to thermal expansion of the tunnelling wall, the measurement of other gaps show similar behaviour in fact.

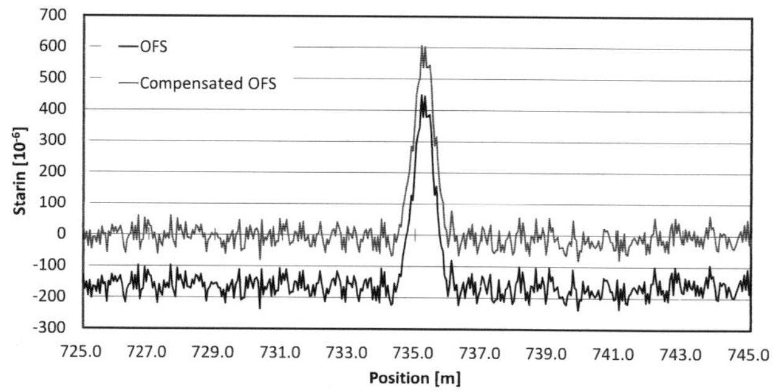

Figure 9. Measured longitudinal strain distribution at gap (BL16-17)

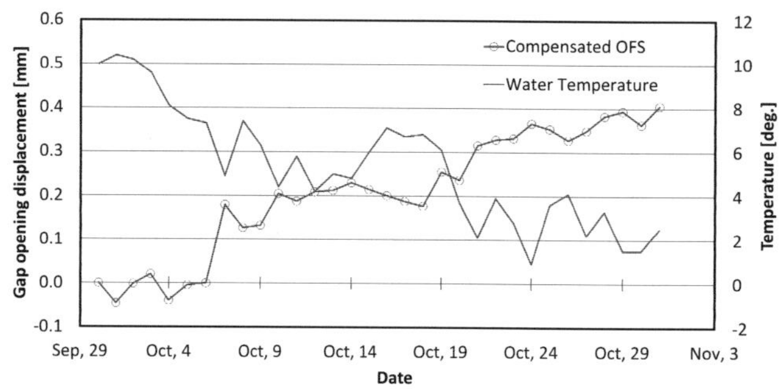

Figure 10. Measured opening displacement of gap (BL16-17)

As one of the transversal behavior of the aqueduct tunnel, Figure 11 shows the measured strain distribution around the inner perimeter of the tunnel. In this case, the spatial resolution of BOCDA is approximately 50mm, and the measurement interval of those is 25mm, respectively. It shows the generated strain distribution for almost one month of plant operation. As described earlier, transversal strain distribution is not temperature-compensated. Finally, the cross-section shape deformation of the tunnel is calculated by the integration of this strain measurement.

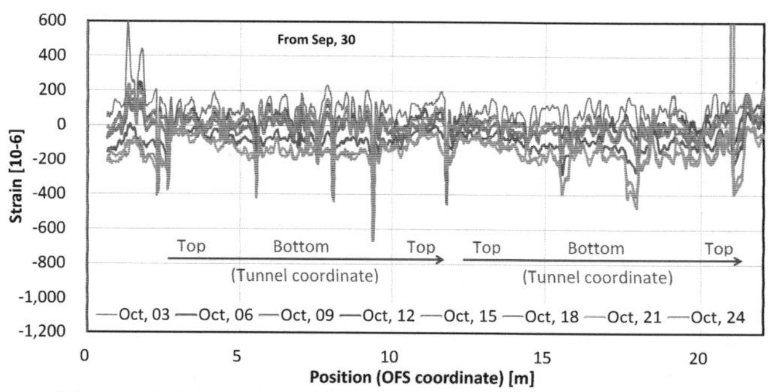

Figure 11. Measured transversal strain distribution

Figure 12 shows the schematic flow of data processing for shape deformation. Constraint 1 means the deformation at the beginning position of the integration is corresponding to those at the end of the integration. And, constraint 2 means that the length of the perimeter remains constant. To be fulfilled with two constraints, the sum of differences is distributed to all measurement. As presented in Figure 12, there are three requirements, 1) the origin of coordinates, 2) the distance from neutral axis of tunnel lining, and 3) the length of inner perimeter. Requirement 1) is determined by the measured strain distribution as shown in Figure 11, requirements 2) and 3) are determined by the designed values.

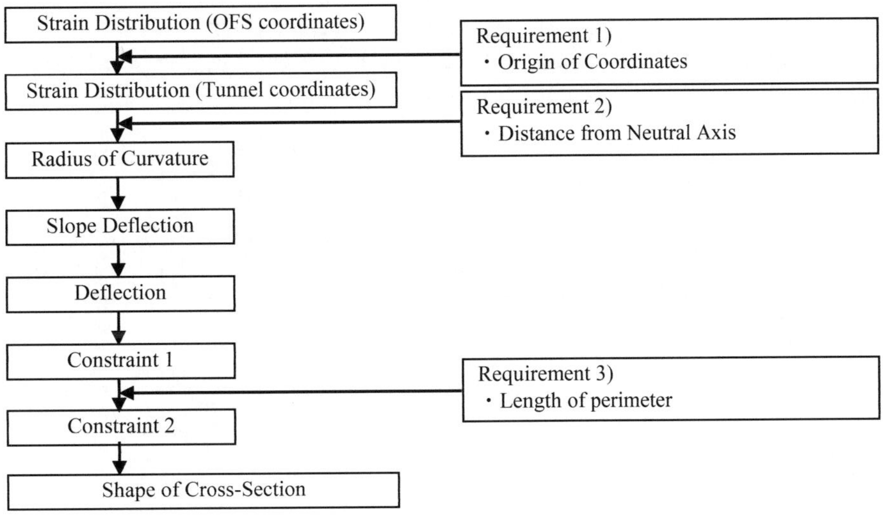

Figure 12. Schematic flow of cross-sectional shape calculation

Based on the measured shape of cross-section, Figure 13 shows the deformation of the inner diameter in four directions, converting the normalized deformation ratio by the designed diameter. Although some scattering can be seen, it can be observed that the vertical diameter is shorten and the horizontal diameter is lengthen.

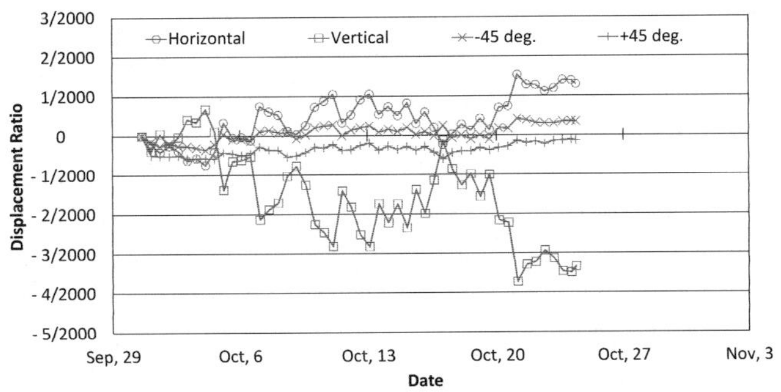

Figure 13. Measured deformation of cross-section

4. Conclusions

Brillouin-based optical fiber strain sensor has just been applied to an aqueduct tunnel, taking its advantage of distributed, water-proof and long transmittable sensor. Our main concern was to develop a monitoring method, which reveals in-service structural behaviour with less resource. Focusing on sensor installation and deformation calculation, a simple method had been proposed and applied to a real tunnel. Our method could successfully be completed to install the sensor only for a couple days for more than 600m range when the hydraulic power plant stopped working, and just started to collect monitoring data with running water for power generation.

This study has just started to monitor the aqueduct, and the measurement has never been compared with the referred data. When the next inspection will be executed, the referred data can be attained. Monitored data is expected to contribute to identify the cause of deformation and determine the method and area of retrofitting if necessary. Hereafter, strain distribution will be continuously measured, and our method will be evaluated from geotechnical point of view. Just in case, the proposed method for deformation calculation will have to be modified.

References

Glisic, B. & Inaudi, D. 2008, 'Fibre optic methods for structural health monitoring', John Wiley & Sons.

Leung, C. K.Y., Wan, K. T., Inaudi, D., Bao, X., Habel, W., Zhou, Z., Ou, J., Ghandehari, M., Wu, H. C., & Imai, M. 2015, 'Review: optical fiber sensors for civil engineering applications', *Materials and Structures*, vol.48, no.4, pp.1-36.

The 8th International Conference on
Structural Health Monitoring of Intelligent Infrastructure
Brisbane, Australia | 5-8 December 2017

In-situ Deformation Monitoring of Tunnel Segments using High-resolution Distributed Fibre Optic Sensing

C. Monsberger, W. Lienhart
Graz University of Technology, Institute of Engineering Geodesy and Measurement System – Austria
Email: christoph.monsberger@tugraz.at

Abstract

In modern tunnelling, geotechnical deformation monitoring is an important component to ensure a safe construction and a long lifetime of the tunnel. It is state of the art to measure displacements at the inner lining surface using total stations or terrestrial laser scanners and internal deformations with vibrating wire sensors. However, these measurements do not deliver a complete picture of the capacity within the tunnel lining and new methods must be evolved. In this paper, the authors report about a fibre optic sensing system that allows distributed in-situ deformation monitoring in tunnelling applications. Especially due to the high spatial resolution of about 10 mm, new information can be gathered.
The developed system was used to measure internal strains within tunnel segments during load tests at a specifically developed test rig. Sensing cables with lengths up to 60 m were embedded in several layers along the reinforcement and therefore, the utilization at about 6000 positions can be observed. Moreover, the high measurement resolution of about 1 μm/m enables the detection of cracks. The suitability of the system is proven with reference measurements, including camera-based systems for crack monitoring and internal pointwise strain measurements with vibrating wire sensors.
In addition to investigations of precast tunnel segments, a modified system for conventional tunnelling is presented, which is already installed in the shotcrete lining directly in the area of the working face in a geological interference zone. After installation, autonomous monitoring was performed during the curing of the shotcrete and the further excavation of the tunnel over several weeks. The results demonstrate the high potential of distributed fibre optic sensing systems and their capability to extend classical measurement methods in tunnelling applications.

1. Introduction

Motorway and railway tunnels are important parts of the civil infrastructure and their functionality is essential for a country's society and economy. For that reason, several large tunnel projects are currently being carried out worldwide, which will lead to shorter travel times as well as higher transport capacities in the near future. In Austria, three long railway tunnels are under construction: The Brenner Base Tunnel (length: 55 km), the Semmering Base Tunnel (length: 27 km) and the Koralm Tunnel (length: 33 km). All of these tunnel projects are part of the Trans-European Transport Network (TEN-T) and will contribute to a significant improvement of the European railway infrastructure.
However, long tunnel projects require large investments, typically more than 10 billion US-Dollars, and long construction times of 10 years and more. Hence, building owners are striving for a long service life of more than 100 years, which can only be achieved if every construction step is carried

out in accordance to specifications and maintenance work is performed at the right time to avoid severe damages.
To assess the structural behaviour of a tunnel, geotechnical monitoring systems are essential components for a safe construction and operation, see OeGG (2014). These include 3D displacement measurements at the inner lining surface with total stations (Rabensteiner, 1996) or terrestrial laser scanners (Fakete et al., 2010) as well as internal deformation measurements using vibrating wire sensors (Radončić et al., 2015) or extensometers (Barla, 2009). Nevertheless, the mentioned sensor systems do not deliver a complete picture of the internal load distribution of the tunnel section and therefore, local damages like cracks within the concrete lining might not be detected.
Distributed fibre optic measurement systems are advantageous compared to pointwise deformations sensors as the fibre optic cable itself acts as the sensing element. Thus, distributed measurements can be performed along the entire cable and a large number of measurement points can be realized. However, since an optical glass fibre is used for sensing, robust sensor cables as well as tailored installation techniques are required to protect the sensing element during the construction of the tunnel lining. Against this background, only a few approaches of fibre optic monitoring applications in tunnelling are currently known, in which mostly Brillouin analysers with a typical spatial resolution of 0.5 to 1 m are used, see e.g. Kechavarzi et al. (2016).
This paper reports about a distributed fibre optic sensing system based on Rayleigh backscattering, which allows distributed in-situ deformation monitoring in tunnel lining segments with a high spatial resolution of 10 mm. In the following, the basic principle of the measurement device is described (Sec. 2.1), sensing cables for applications in harsh environments are discussed (Sec. 2.2) and calibration results of the system in the laboratory are presented (Sec. 2.3). Afterwards, the utilization of the fibre optic system during load tests at a specifically developed test rig (Sec. 3) and autonomous monitoring of a shotcrete lining (Sec. 4) are introduced. Finally, the suitability of the system for tunnelling applications is discussed and an outlook on future research is given (Sec. 5).

2. Fibre Optic Monitoring System

In order to provide a suitable and reliable fibre optic measurement system for tunnel monitoring, various system components have to be considered. In this section, the characteristics of the sensing principle as well as different sensing cables for strain measurement in harsh environments are discussed and calibration results of the system in the IGMS (Institute of Engineering Geodesy and Measurement Systems, Graz University of Technology) laboratory are presented.

2.1 Sensing Principle

Rayleigh scattering is one of the major effects of intensity loss in optical fibres. It is mainly caused by variations of the refractive index profile along the fibre core and effects about 85 % of the natural attenuations (Wuilpart, 2011). The Rayleigh scatter amplitude has a random but static behaviour along the fibre. Changes in strain or temperature cause a spectral shift in the locally reflected Rayleigh pattern and therefore, a small, local fibre segment, which is transformed to the frequency domain, can be interpreted as a weak reflecting fibre Bragg grating (FBG) with a random period. Consequently, the distributed measurement system can be modelled by splitting the fibre into equidistant segments and calibrating the local Rayleigh shift in reference to changes in strain or temperature.
For tunnel lining monitoring, an optical backscatter reflectometer (OBR) from Luna Innovations Inc. was used, whose sensing principle is based on optical frequency domain reflectometry (OFDR). This interrogation unit is able to record sensing information with a high resolution of

about 1 µm/m for strain and about 0.1 K for temperature measurements. Moreover, a spatial resolution of about 10 mm or even better may be realized. (Luna, 2014a)
To form the distributed measurement system, the Rayleigh backscatter of the sensing fibre is recorded in an initial strain and temperature state. This measurement can be interpreted as a reference scan. Later, the fibre is scanned again, when the strain and/or the temperature state has changed. For data evaluation, the signals of both measurements are divided in equidistant segments, where the length of the segment Δz corresponds to the spatial resolution of the OBR and the spectrum of each segment is observed in the frequency domain. The wavelength spectra of an interval with a length of 10 mm for an unstrained reference scan and the spectra of the same segment of the fibre with an applied strain of about 1000 µm/m can be seen in Fig. 1a. The strain change causes a spectral shift between the reference and the influenced scan, which size can be determined through a cross correlation between the two spectra, see Fig. 1b. The resulting wavelength shift Δλ is directly proportional to the apparent strain change of this fibre segment. By calculating this shift for each segment of the sensing fibre, it is possible to realize a distributed measurement system. Detailed information on the sensing principle and the optical network may be found in e.g. Kreger et al. (2006).

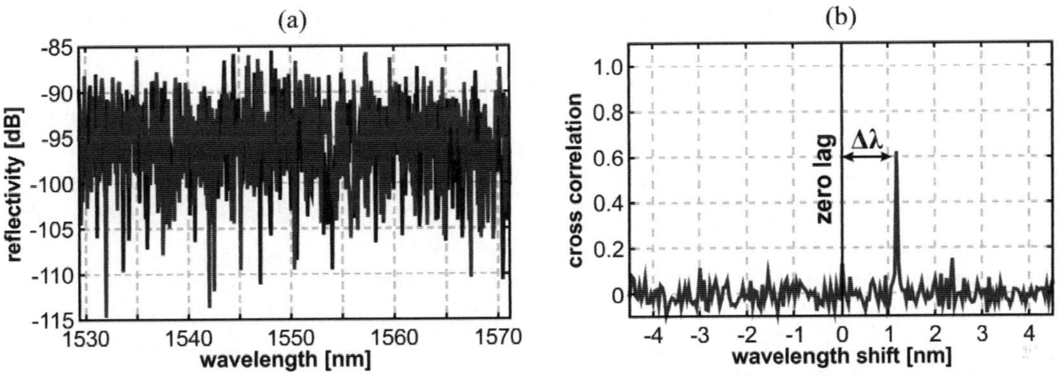

Fig. 1: Sensing principle of an optical backscatter reflectometer (Monsberger et al., 2017): (a) Wavelength spectrum for an unstrained (blue) and a strained (red) fibre segment (Δz = 10 mm) and (b) the corresponding cross correlation function

2.2 Strain Sensing Cables

As a consequence of the harsh environments for optical fibres in tunnelling applications, robust sensor cables are required to protect the sensing element during installation and monitoring. In addition, it is also important that all cable layers are interlocking to ensure the unbiased strain transfer from the outer sheath of the cable through the protection layers to the sensitive fibre core.
For fibre optic strain measurements in geotechnical projects, IGMS commonly uses strain sensing cables from Brugg Cables AG, Switzerland. Fig. 2 shows the basic structure of two selected cables for distributed strain measurements, which are especially developed for sensing in grout or concrete. Both cables protect the optical fibre through a special metal tube. Due to an interlocking multi-layer buffer, the strain transfer to the sensitive glass fibre core can be ensured. The outer protection varies depending on the cable type. Whereas type V9 (Fig. 2a) is only equipped with an outer polyimide layer, type V3 (Fig. 2b) has also a special steel armouring that additionally protects the sensing fibre against mechanical impacts. Although this armouring makes the cable more robust, the flexibility of this type (minimal bending radius, etc.) is limited and thus, the cable selection must be well adapted to the monitoring conditions. Moreover, both cables have a structured outer surface, which guarantees a solid bond with the surrounding material.

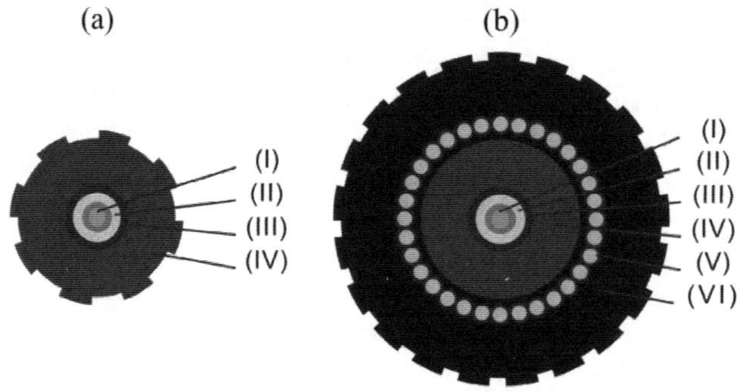

Fig. 2: Structure of strain sensing cables from Brugg Cables AG: (a) Type V9 and (b) type V3 with (I) strain sensing single mode fibre (∅ 250 μm), (II) multi-layer buffer with strain transfer layer, (III) metal tube, (IV) polyimide protection layer, (V) special steel armouring and (VI) polyimide outer sheath

2.3 System Calibration

In order to prevent systematic errors in the conversion from raw measurements to strain or temperature changes, individual calibration has to be considered as an important part within the development of a fibre optic measurement system. Many manufactures of fibre optic sensors do not specify individual calibration parameters and refer to literature values instead, which might result in errors up to 10 % (Luna, 2014b). For that reason, IGMS developed a unique calibration device within the last years, see Fig. 3a. Key components of this facility are the stable setup on a 30 m concrete bench, a linear translation stage that allows sensor elongations up to 300 mm and a laser interferometer that is used for a precise reference measurement. For details, reference is given to Woschitz et al. (2015).
As discussed in Sec. 2.1, the local Rayleigh shift between a reference scan and subsequent measurements depends on both, strain ε and temperature changes ΔT. The transfer from the measured wavelength shift Δλ [nm] or the frequency shift Δν [GHz] to these quantities may be approximated by the linear function

$$\frac{\Delta\lambda}{\lambda} = \frac{-\Delta\nu}{\nu} = K_\varepsilon \varepsilon + K_T \Delta T \tag{1}$$

with the normalized sensitivity coefficients K_ε and K_T and the centre wavelength λ or centre frequency ν of the scan. This equation is identically to the response of a fibre Bragg grating (Kreger et al., 2006). As the calibration facility is set up in the IGMS temperature and humidity controlled laboratory, the temperature is constant and the measurement signal only depends on strain:

$$\frac{\Delta\lambda}{\lambda} = \frac{-\Delta\nu}{\nu} = K_\varepsilon \varepsilon \Big|_{\Delta T = const.} \tag{2}$$

Prior to the installation within the tunnel lining segments, individual calibrations of samples of different sensing cables were carried out on the IGMS calibration facility. Regarding the experience from past calibrations, it is important to slightly pre-strain the cables to avoid non-linearities in the lower strain regions. In addition, calibrations are usually performed in full cycles to evaluate possible hysteresis effects. To determine the different strain sensing coefficients K_ε, the distributed strain values over the strained section can be averaged and a linear regression analysis can be performed.

(a) 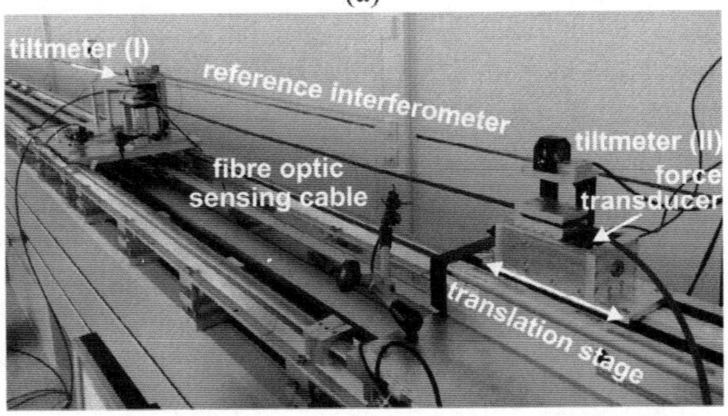

(b) 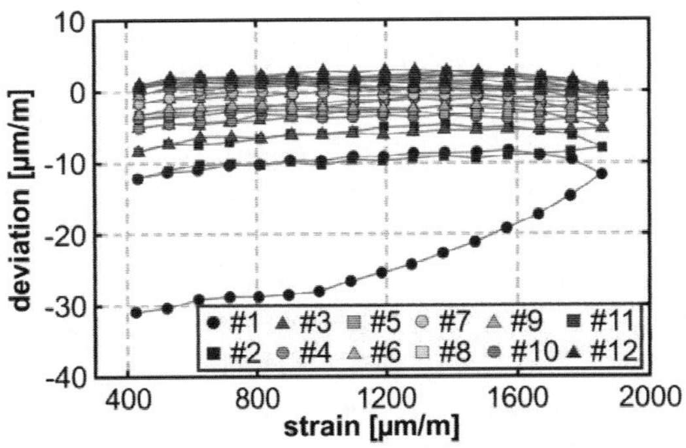

Fig. 3: (a) IGMS calibration facility for fibre optic strain sensors and (b) residuals of the linear regression analysis of cable type V3

As an example of a typical calibration result, Fig. 3b shows the residuals of the linear regression analysis of cable type V3. In the first cycle, significant drift and hysteresis effects with deviations of more than 30 µm/m become visible. Also cycle #02 does not fit exactly to the other calibration data, even if its deviations are much smaller. It can be assumed that this effect might be related to an interaction between the protecting elements of the cable during the first strain increase after production of the cable. Therefore, the first two cycles were excluded from the determination of K_ε. The estimated strain coefficient of cable type V3 can be specified with 0.748, which basically agrees to literature values, e.g. 0.780 (Luna2014b). Further information on strain calibration of sensing cables for geotechnical applications may be found in Monsberger et al. (2017).
Temperature compensation of the derived strain values usually depends on the monitoring application. In case of load tests, e.g. indoors or inside the tunnel, the concrete temperature will be almost constant over a relatively short testing period of some hours and the remaining influences may be neglected. In other applications, especially in long-term monitoring, an unstrained fibre installed nearby the strain sensing cable can be used to eliminate raising temperature effects.

3. Load Tests of Precast Tunnel Segments

When using tunnel boring machines (TBMs) for excavation work, concrete lining segments are applied, which are usually set up in a ring of four to eight elements. For a verification of the load-bearing and deformation behaviour of these precast segments, the Austrian Federal Railways (ÖBB-Infrastruktur AG) developed in cooperation with the Montanuniversität Leoben (Chair of Subsurface Engineering) a special test rig (see Fig. 4b), which allows biaxial testing of full scale precast tunnel segments under exact known loading conditions (Gehwolf et al., 2016). The basic measurement equipment of the rig includes wire-rope sensors to measure the vertical deformations as well as strain gauges and extensometers on the segment surface for crack detection. In addition, tests with camera-based systems for crack monitoring are carried out (Pogats, 2016). To gather information about the internal load distribution, IGMS developed a fibre optic sensing system to measure internal strains within the tunnel segments. This approach is also patented, see Lienhart & Galler (2016). Subsequently, the sensing fibre installation is discussed and some results of the load tests are presented, which demonstrate the capabilities of the system.

3.1 Production of Tunnel Segment and Fibre Optic Instrumentation

In the course of the project, tunnel segments are produced with classical reinforcement elements. For that reason, the sensing cable can be individually guided along the reinforcement bars (see Fig. 4a) without any supporting structure and thus, a grid may be built up with complete coverage of each segment. Although harsh environments are prevalent for the optical sensing fibre, the less protected cable type V9 is used to provide more flexibility. During the installation, the cable is mounted on the reinforcement using cable ties to guarantee a fixed position inside the segment. Later, a solid bond between the concrete and the cable is given due to the structured polyimide surface, which ensures a reliable transfer of the deformations to the sensitive glass fibre core. Inside the segment, a connecting box is placed, in which the optical connectors are stored during concreting. In addition, vibrating wire sensors (VWS) are installed on the inner and outer bending reinforcement, which are used to prove the suitability of the fibre optic system.

(a) (b)

Fig. 4: (a) Installation of the fibre optic sensing cable along the reinforcement bars and (b) test rig for precast tunnel lining segments

3.2 Measurement Results

During the load test, which is described in this paper, the tunnel segment was only loaded vertically and hence, compression (negative strain) on the outer and positive strain on the inner reinforcement can be expected. The vertical load was increased in steps of 25 kN starting from 100 kN. Initially, a pre-load step at 50 kN was performed. During the test, the strain distribution inside the segment was continuously monitored with a sampling frequency of about 0.1 Hz using the OBR controlled by an individual IGMS-developed measurement software.

In
Fig. 5a, the strain profiles along an instrumented bending reinforcement bar in the inner as well as the outer layer at lower load steps up to 175 kN are shown. These results confirm the assumption that negative strains occur on the outer layer of the reinforcement under vertical load. Additional deformations are visible in the centre between ±0.5 m, which can be explained by the bearing point of the cross beam of the test rig. Along the inner reinforcement, increasing strain values can be observed with increasing load. Furthermore, local peaks in the area between ±0.7 m along the strain profiles arise, starting at the load step of 150 kN. These inhomogeneities are related to cracks in the concrete, which start to evolve at a vertical load of 150 kN.

To compare the fibre optic measurement results to these of the vibrating wire sensors, the strain values in the area of the VWS (±0.08 m) can be averaged and their time series can be analysed, see Fig. 5b. Up to a load of 175 kN (dotted line at about 55 min), the measurement data of both technologies fit well to each other, even if small linear deviations can be seen. At the following load steps of 200 kN and higher (60 min and later), the deviations increase significantly, which might be a result of a further opening of the cracks inside the concrete. This suggests that local stress events like cracks cannot be detected by VWS. The fibre optic system is able to monitor the crack opening, although the crack width of a few hundredths of millimetres results in an integrative response of the sensing cable over several centimetres.

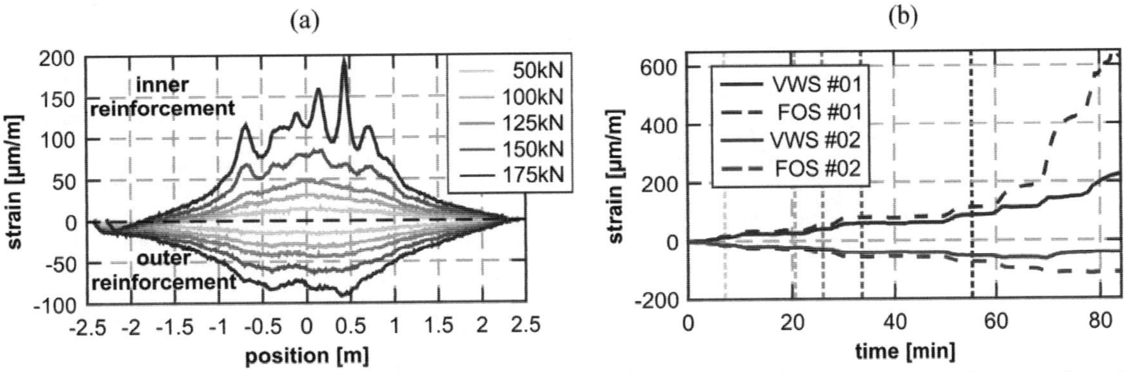

Fig. 5: (a) Measured strain profiles along instrumented inner and outer bending reinforcement bars at selected load steps and (b) comparison between fibre optic measurement system and vibrating wire sensors (dotted lines represent load steps between 50 and 175 kN)

In order to analyse the crack evolution in more detail, further load steps within the test can be considered. Fig. 6b shows the measured strain profiles along an instrumented inner bending reinforcement at different load steps from 200 to 450 kN. Within the central area from -0.7 to 0.7 m, in which inhomogeneities have already become visible at prior load steps, local strain peaks can also be observed at a load of 200 kN. However, the peak heights are significantly increased compared to other strain values along the measured profile, which indicates that a further opening of the cracks occurred. By increasing the test force up to 450 kN, cracks also arise in the area offside the centre. In general, a good correlation between crack positions and the location of the transverse reinforcement bars (Fig. 6a) can be recognized. This implies that cracks along the inner bending reinforcement under vertical load usually appear on positions, where a transverse reinforcement bar is located.

In addition to the in-situ crack monitoring through the fibre optic system, measurements with a camera-based system are performed at selected load steps by the Montanuniversität Leoben. First results in Fig. 6c show the recorded crack pattern at a load step of 300 kN, whereby various cracks in the central area of the tunnel segment are visible. Due to the limited viewpoint of the camera (approximately -0.7 to 1.1 m), cracks at outer positions cannot be recorded. The positions of the captured cracks basically agree with the fibre optic strain peaks, even if differences of some centimetres occur at some locations. However, crack positions, which are very close together (e.g. Pos. 1), may not be separated by the fibre optic system and thus, these cracks appear as one peak in the fibre optic strain profiles. In contrast, the camera system does not deliver internal crack positions, which are not visible on the surface, e.g. Pos. 2.

(a)

(b)

(c)

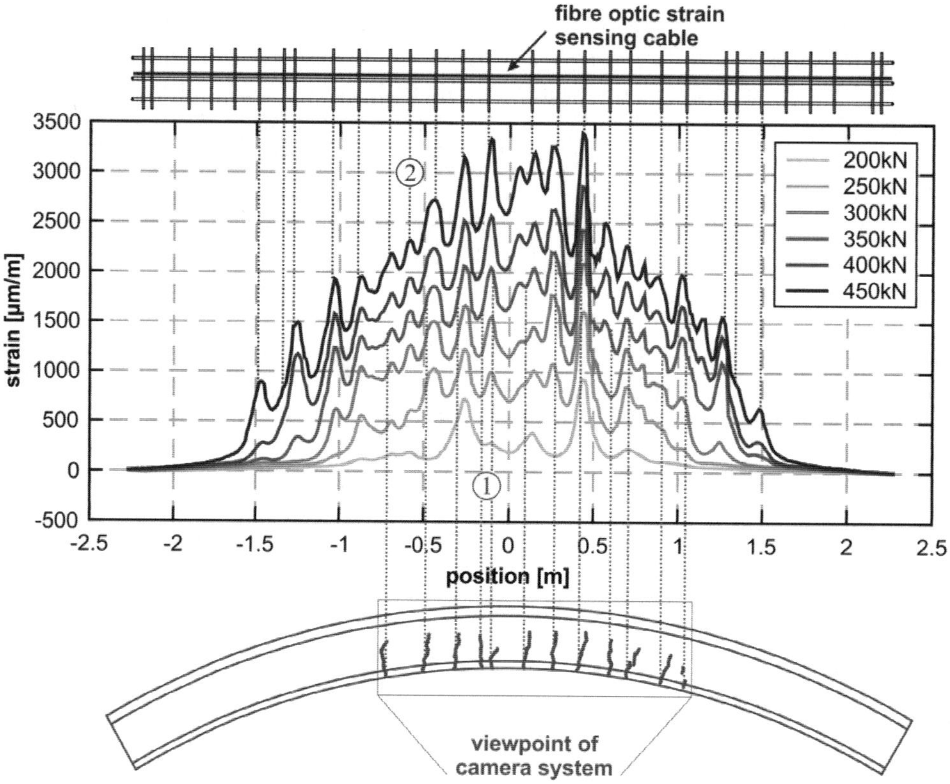

Fig. 6: (a) Schematic representation of instrumented reinforcement, (b) measured strain profiles along inner bending reinforcement at various load steps and (c) crack pattern recorded by camera-based system at 300 KN

4. Monitoring of Shotcrete Tunnel Lining

Due to uncertainties of every geological and geotechnical underground models, geotechnical monitoring and reliable data interpretation are very important in conventional tunnelling. Thereby, systematic and high-frequent monitoring techniques lead to a better understanding of the rock's behaviour, which finally results in an optimisation of the construction process and a safe construction of the tunnel. In geological interference zones, it is state of the art to measure the capacity of the shotcrete using strain gauges and pressure load cells. However, these methods do not give a complete picture of the internal utilization. First attempts to use distributed fibre optic sensors in sprayed concrete were reported by de Battista et al. (2015). As part of a research project in cooperation with the Austrian Federal Railways (ÖBB-Infrastruktur AG) and FMT (Institute of Rock Mechanics and Tunnelling, Graz University of Technology), IGMS developed and installed a fibre optic monitoring system in the shotcrete lining directly at the working face at a construction section of the Semmering Base Tunnel. In the following, the measurement concept is shortly introduced and first results of an autonomous monitoring campaign in spring 2017 are presented.

4.1 Measurement Concept and Sensor Installation

Based on the New Austrian Tunnelling Method (NATM), conventional tunnelling is usually performed in excavation sequences, which enable the rock to support itself (OeGG, 2010). The construction of the tunnel section, in which the fibre optic system was installed, can be split into two essential parts, see Fig. 7a. In a first step, the top-heading area is broken out in sequential parts and the cross section is supported by reinforcement layers and shotcrete. Depending on the geological conditions, this construction method is carried out for several meters. Afterwards, the bench/invert section is excavated and also supported by reinforcement and shotcrete. In order to create a load-bearing ring, the invert is immediately reclosed after the construction.

Fibre optic sensing cables were installed along the reinforcement in peripheral direction of both sections to monitor the load distribution. Due to the harsh environment (e.g. shotcrete) inside the tunnel, cable type V3 was chosen to provide an optimal protection of the sensing fibre. During the curing of the shotcrete, the temperature inside the lining usually varies in a range of 20 to 30 K. For that reason, an additional sensing fibre of type V3 was loosely installed in a cladding tube nearby the strain sensing cable (see Fig. 7b) for temperature compensation. From a connecting box directly at the working face, supply fibres were used to connect the sensing cables to the measurement unit, which was placed about hundred meters behind the instrumented cross section.

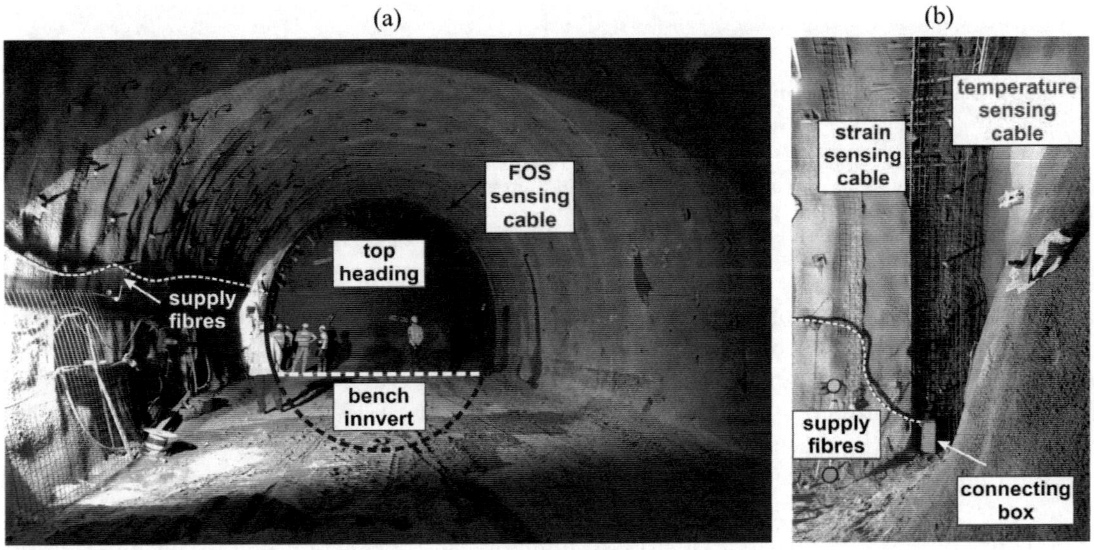

Fig. 7: (a) Overview of tunnel working face and (b) cross section with schematic representation of installed sensing fibres

4.2 Measurement Results

In order to gather complete information about the rock deformation, continuous monitoring of each section was started immediately after applying the shotcrete using the OBR in combination with an own-developed measurement software. The temperature-compensated strain values of the instrumented tunnel profile at selected times are plotted in Fig. 8. It can be seen that first compression zones become visible at the crown and the shoulders (top-left, top-right) shortly after the top-heading construction (Fig. 8a). Compressions in these areas were already known from displacement measurements on the surface at prior cross sections, but could not be located exactly before installation of the distributed fibre optic system. About 48 hours after the top-heading construction (Fig. 8b), the compressions in these selected zones have increased further up to -1100 µm/m. In addition, almost the entire cross section shows negative strain values due to the curing of the shotcrete. The bench/invert was excavated about five days after the top-heading section. Before the refilling of the invert about six hours after construction (Fig. 8c), the strain profiles along the bench/invert section show slight compression with one local inhomogeneity in the bottom area. This might be a result of distortions caused by overlaying reinforcement layers or other sensors (pressure load cells) in this area. In the top-heading section, further compression is visible due to the interacting rock pressure and the curing of the shotcrete. After the invert was reclosed (Fig. 8d), the compression values in the bottom area are significantly higher than these of the sidewall of the bench. In addition, the deformation progress in the top-heading is progressive, but seems to be almost constant along the entire section. This suggests that the installed supporting ring forms a load-bearing system and a uniform distribution of the rock pressure is performed.

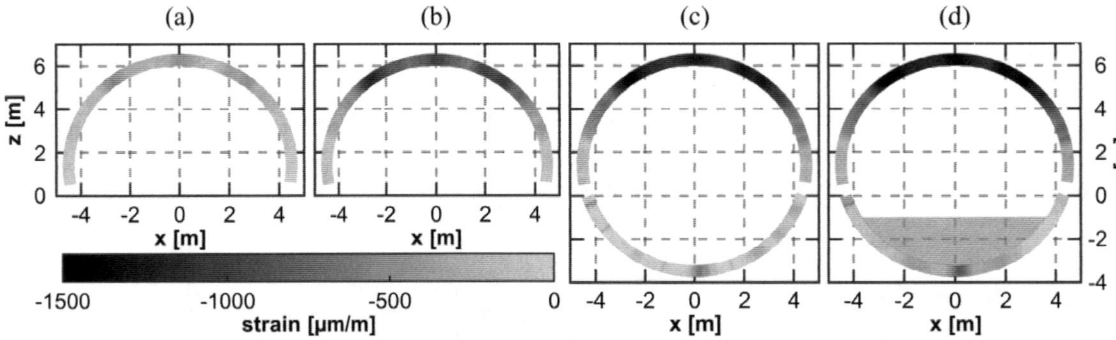

Fig. 8: Distributed strain profiles along top-heading and bench/invert: (a) 6 hours after top-heading construction, (b) 48 hours after top-heading construction, (c) 6 hours after bench/invert construction and (d) 48 hours after bench/invert construction (invert refilled)

The distributed strain profiles measured by the fibre optic system show that various compression zones arise as a result of the interacting rock pressure. In order to analyse this behaviour, five positions along the instrumented cross section are selected (see Fig. 9a) and considered over the entire monitoring campaign of five weeks. Fig. 9b depicts the temporal sequence of the strain values at these locations. By this, it is evident that the positions along the top-heading section (#01-04) show first compressions shortly after the construction. However, the compressions at all positions, expect of #03, remain constant or even decrease after about 18 hours, see period t_1 in Fig. 9b. The further excavation of the tunnel or the curing of the shotcrete might be possible reasons for this progress, but a detailed analysis is still pending. Due to the construction of the bench/invert (period t_2 in Fig. 9b), a similar relaxing of the lining appears at locations along the sidewall (#01-03). The selected position in the invert (#05) shows an immediate reaction to the covering of the shotcrete and the subsequent reclosing. Although the excavation of the entire cross section is finished afterwards, a continuous increase of the compressions can be observed at all selected positions over the remaining period, which is still progressive when the fibre optic interrogation unit was dismantled. For that reason, successive measurement epochs will be carried out in 2017 to assess the further development of the deformations inside the shotcrete lining.

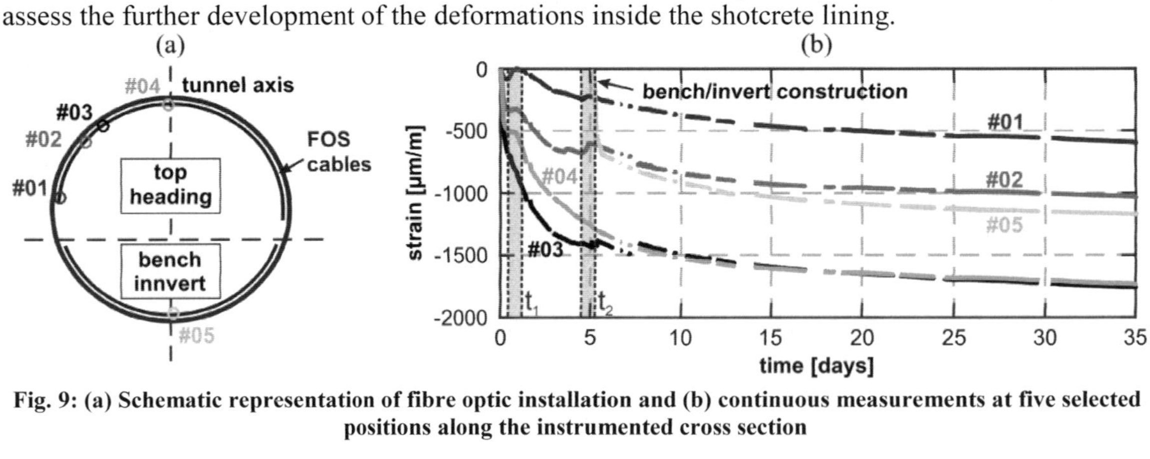

Fig. 9: (a) Schematic representation of fibre optic installation and (b) continuous measurements at five selected positions along the instrumented cross section

5. Conclusions

In this paper, a fibre optic sensing system based on Rayleigh backscattering was introduced, which enables distributed in-situ deformation monitoring with a high spatial resolution in tunnelling applications. For this, robust fibre optic components as well as installation techniques were found to ensure a reliable system within the harsh environments on site. Before the installations, strain calibrations were carried out in the IGMS humidity and temperature-controlled laboratory in order to assess the performance of the fibre optic system. These calibrations show that the derived strain sensitivity coefficient deviates up to 5 % from standard parameters, even though these values are specified for bare fibres.

The developed fibre optic system was utilized in various tunnelling applications and distributed information could be gathered, which cannot be measured using other sensing techniques. During load tests of precast tunnel lining segments at a specifically developed test rig, the in-situ evolution of cracks along the bending reinforcement could be observed and the crack positions were verified by measurements of a camera-based system on the surface. Moreover, an installation in the shotcrete lining directly at the working face in a geological interference zone demonstrates that the fibre optic system is sufficiently robust for tunnelling applications.

Future research will be focussed on the long-term monitoring of the instrumented cross section to assess the further deformation development inside the shotcrete lining. In addition, the fibre optic system is already installed inside a complete ring of tunnel segments, which will be placed into the Koralm tunnel in summer 2017. The results of this operational phase will give new insights into the deformation behaviour of a segment ring and will finally lead to a better understanding of the load distribution within tunnels that are constructed using boring machines.

Acknowledgement

The authors would like to thank the Austrian Federal Railways (ÖBB-Infrastruktur AG) for funding the research project to develop a fibre optic measurement system for tunnelling applications and the opportunity to realize installations within current tunnel projects in Austria. Furthermore, special thanks to the project partners, the Chair of Subsurface Engineering (Montanuniversität Leoben) and the Institute of Rock Mechanics and Tunnelling (Graz, University of Technology), for their support during the load tests and the field installations.

References

Barla, G. 2009, 'Innovative tunneling construction method to cope with squeezing at the Saint Martin La Porte access adit (Lyon-Turin Base Tunnel)', *Proceedings of ISRM Regional Symposium - EUROCK 2009 (keynote lecture),* Cavtat, Croatia, 29-31 October, 10 p.

Battista, N. de & Elshafie, M. & Soga, K. & Williamson, M. & Hazelden, G. & Hsu, Y.S. 2015, 'Strain monitoring using embedded distributed fibre optic sensors in a sprayed concrete tunnel lining during the excavation of cross-passages', *Proceedings of 7th International Conference on Structural Health Monitoring of Intelligent Infrastructure*, Torino, Italy, 1-3 July, 10 p.

Fekete, S. & Diederichs, M. & Lato, M. 2010, 'Geotechnical and operational applications for 3-dimensional laser scanning in drill and blast tunnels', *Tunnelling and Underground Space Technology*, vol. 25, no. 5, pp. 614-628.

Gehwolf, P. & Monsberger, C. & Barwart, S. & Wenighofer, R. & Galler, R. & Lienhart, W. & Harberler-Weber, M. & Moritz, B. & Barwart, C. & Lange, A. 2016, 'Deformation measurements of tunnel segments at a newly developed test rig', *Geomechanics and Tunnelling*, vol. 9, no. 3, pp. 180-187.

Kechavarzi, C. & Soga, K. & Battista, N. de & Pelecanos, L. & Elshafie, M. & Mair, R. 2016, *Distributed Fibre Optic Strain Sensing for Monitoring Civil Infrastructure: A Practical Guide*, ICE Publishing, London, Great Britain.

Kreger, S.T. & Gifford, D.K. & Froggatt, M.E. & Soller, B.J. & Wolfe, M.S. 2006, 'High resolution distributed strain or temperature measurements in single-and multi-mode fiber using swept-wavelength interferometry', *Proceedings of Optical Fiber Sensors 2006*, Cancun, Mexico, 23-27 October, ThE 42, 4 p.

Lienhart, W. & Galler, R. 2016: *Tübbingelement mit Dehnungsmessung*, AT516158 (B1): 28 pages, filed: 10.10.2014, granted: 15.03.2016.

Luna Innovations Inc. 2014a, *Data Sheet: Optical Backscatter Reflectometer (Model OBR 4600)*, Ver. LTOBR4600 REV. 004 02/13/2014, Roanoke, USA, 1 p.

Luna Innovations Inc. 2014b, *User Guide: Optical Backscatter Reflectometer Model OBR 4600*, Ver. User Guide 5, OBR 4600 Software 3.6, Roanoke, USA, 230 p.

Monsberger, C. & Woschitz, H. & Lienhart, W. & Račanský, V. & Haxden, M. 2017, 'Performance assessment of geotechnical structural elements using distributed fiber optic sensing', *Proceedings of SPIE 10168: Smart Structures and NDE*, Portland (OR), USA, 25-29 March, pp. 101680Z1-12.

OeGG, 2010, *NATM - The Austrian Practice of Conventional Tunnelling*, OeGG – Austrian Society for Geomechanics, Salzburg, Austria.

OeGG, 2014, *Handbook – Geotechnical Monitoring in Conventional Tunneling*, OeGG – Austrian Society for Geomechanics, Salzburg, Austria.

Pogats, J. 2016, *Autonomous crack detection on ring joint surfaces of tubing elements*, Master's thesis, Montanuniversität Leoben, Austria

Rabensteiner, K.1996, 'Advanced tunnel surveying and monitoring', *Felsbau*, vol. 14, no. 2, pp. 98-102.

Radončić, N. & Kern, M. & Weissnar, M. & Moritz B. 2015, 'Strain gauges in pre-cast concrete segments: working principle, evaluation and interpretation', *Geomechanics and Tunnelling*, vol. 8, no. 3, pp. 265-272.

Woschitz, H. & Klug, F. & Lienhart, W. 2015, 'Design and calibration of a fiber optic monitoring system for the determination of segment joint movements inside a hydro power dam', *Journal of Lightwave Technology*, vol. 33, no.12, pp. 2652-2657.

Wuilpart, M. 2011, *Rayleigh scattering in optical fibers and applications to distributed measurements* in *Advanced Fiber Optics: Concepts and Technology*, Thévanez, Luc (ed.), EPFL Press, Lausanne, Switzerland.

The 8th International Conference on
Structural Health Monitoring of Intelligent Infrastructure
Brisbane, Australia | 5-8 December 2017

Using Mobile Phones to capture Mode Shapes: A Feasibility Study

D. Hester[1], C. Keenan[1]
[1] *Queen's University Belfast – UK, Email:* d.hester@qub.ac.uk

Abstract

The issue of using the sensing capacity of mobile phones for structural monitoring has received increased attention the past few years. This paper presents a study carried out to examine the feasibility of using a number of mobile phones to capture the mode shapes of a footbridge. The ability of an individual mobile phone to accurately measure acceleration /frequency has been demonstrated by the appearance of a number of different mobile applications (apps) on both android and ios platforms. However, to calculate mode shapes the individual sensors need to be synchronised. For conventional sensing system this is easily achieved as data from all sensors is being recorded by a central data logger. For phones this is not the case, i.e. each phone is logging the data directly to its own memory. Where a unique identifier (e.g. clapper board type event) can be implemented simultaneously at the start of all the phone signals this will allow the signals to be lined up in post processing. However, there is no guarantee that the signals won't drift from one another subsequently. Other issues include, the variation in sensitivity /noise floor of the accelerometers in different phones, and the difficulty some phones have in maintaining a constant scanning frequency. All of the above issues are examined in this paper and it is shown that the drift between individual phones is actually quite small. Moreover it is shown that for simple structures with sufficient levels of vibration, a series of phones can be used to determine an approximate mode shape. However, for lower levels of vibration the MEMS accelerometers in the phones struggle to capture the structure's vibration, i.e. the level of vibration is below the noise floor of the sensor and mode shapes cannot be detected reliably.

1. Introduction

Smart phones have become pervasive in many societies globally and in recent years a small but growing number of researchers have started to look at the possibility of harnessing the sensing capacity of smart phones for various applications, including structural monitoring. For example (Yu et al. 2015) looked at the possibility of using a smart phone as not only a single wireless node but also as a mini-SHM system. Further (Peng et al. 2015) put forward the idea of using data collected by individual smart phone questionnaires to draw a seismic intensity map rapidly after a seismic event. The importance of hoisting operations in new bridge construction is highlighted by (Han et al. 2015), particularly so when hoisting takes place at sea. To assist with these operations they proposal a very clever system where a smart phone is mounted on the girder to be lifted and relevant information from the phones sensors is relayed wirelessly to the hoisting operator.

The study described in thie paper is a further contribution to this new research area. In particular this study looks at the possibility of using acceleration data from multiple smart phones to identify the mode shape of a bridge. Modes shapes, along with frequency and damping, being the foremost modal parameters sought in a modal test, as they are fundamentally important in designing solutions if a structure has a problem with vibration serviceability. (For vibration serviceability, modal mass is also extremely important but this requires forced vibration testing so is outside the scope of this study). It has previously been shown that mobile phones can identify frequency and damping information on timber footbridges (Chang et al. 2014). However, to identify mode shapes requires a number of phones to remain synchronised for the duration of the test, and this is what is studied in this paper. To check that he mode shapes calculated from the phone data are correct, they are s compared to mode shapes from conventional wired accelerometers.

2. Sensing systems used

2.1 Mobile Phones

For this study three iPhones were used to record the acceleration, specifically an iPhone 7, and two iPhone 5S's, (to avoid confusion these are referred to as iPhone 5SA and iPhone 5SB). The iPhone 5s was released in September 2013 and the iPhone 7 was released in September 2016. Figure 1 shows the three phones used in the experiment. Each phone has a triaxial MEMS accelerometer. There are a number of apps available to record acceleration however, the one used in this study was Sensor Kinetics Pro, which is available on iOS and Android platforms. The app allows the user to set the sample rate (up to 100Hz) and once the user hits record the phone will start recording acceleration. Once the test is complete the user hits stop and the data from the phones triaxial accelerometer is saved as a csv file. The csv file is an n x 4 matrix where n is the number of time instants accelerations was recorded at. The first column of the matrix is the time value, and columns 2-4 are the acceleration in the x, y and z directions respectively.

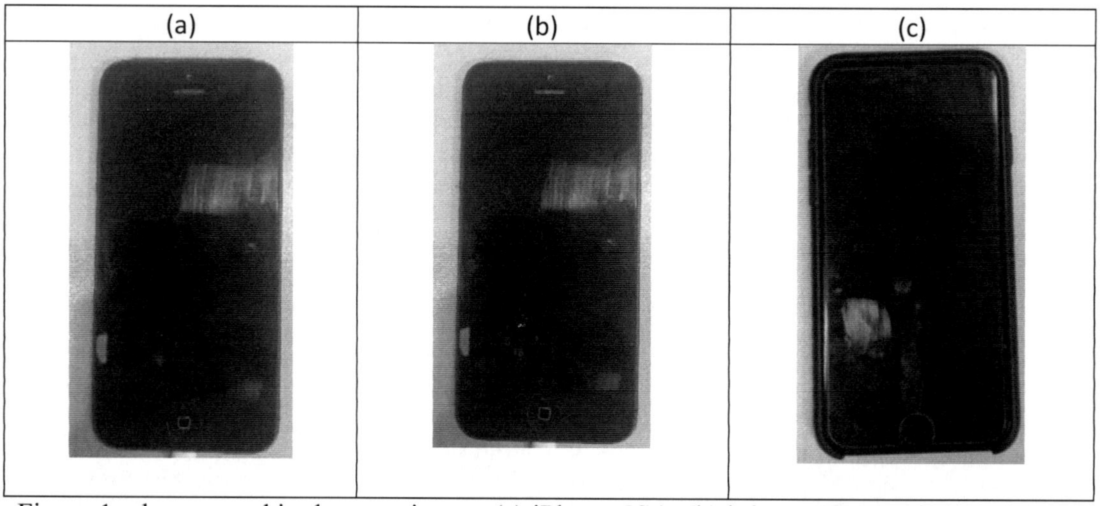

Figure 1, phones used in the experiments (a) iPhone 5SA, (b) iPhone 5SB, (c) iPhone 7.

2.2 Conventional wired acceleration system

In this study the wired accelerometers used were Kistler K-beam MEMS accelerometers, there are ±2g sensors and the product manual quotes the noise density for the sensor as 0.025 mgrms / √Hz. Figure 2 shows the set up used for the wired accelerometers. In the figure the K-beam accelerometers are shown circled and for the purposes of scale they have been left beside the iPhone 5SA and the iPhone 7. The insert in the figure shows a close up view of the K-beam accelerometer (KBA). The signal from the accelerometers was carried to the laptop via a four channel National Instruments (NI) 9234 module (indicated by the dashed rectangle in Figure 2) and was recorded using NI Signal Express.

Figure 2, Conventional wired accelerometers used in the test.

3. Floor test

Before trying to identify mode shapes using data from the mobile phones it was first necessary to examine how well the phones remain synchronised. Unlike the KBAs which all share a common data logger each phone is in effect its own autonomous logger and therefore it is possible that one phone may drift in time relative to the other.

To examine the above, the iPhone 5SA, the iPhone 7, and one KBA were placed on a board as shown in Figure 3. Recording was initiated on all three devices and the board was moved vertically up and down a number of times. In this test the KBA was essentially included as a control. The board was then placed on a quiet laboratory floor for approximately 12 minutes and acceleration recorded. After recording acceleration on the floor the board was again moved up and down and the recording was stopped on all three sensors.

The purpose of manually moving the board up and down at the start and end of the test was to create distinctive common features at the start and end of all three signals. The distinctive features at the start allow the signals to be synchronised in post processing. While the distinctive features at the end allow us to judge if the phones have drifted from one another and if so by how much. Figure 3(a) shows the complete signal recorded by all three sensors after they have been time aligned in post processing. Figure 3(b) shows that at the start to the test the signals are well time aligned. In figure 3(c) it can be seen that after approximately 800 seconds the iPhone 5SA and the iPhone 7 are still synchronised but that the two phones have drifted slightly relative to the KBA. In the context of this test the drift between the KBA and the phones is relatively unimportant (<0.5 seconds in 800 seconds), the very positive finding is that the two phones stayed synchronised over the course of 800 seconds, which means that they can potentially be used to identify mode shapes.

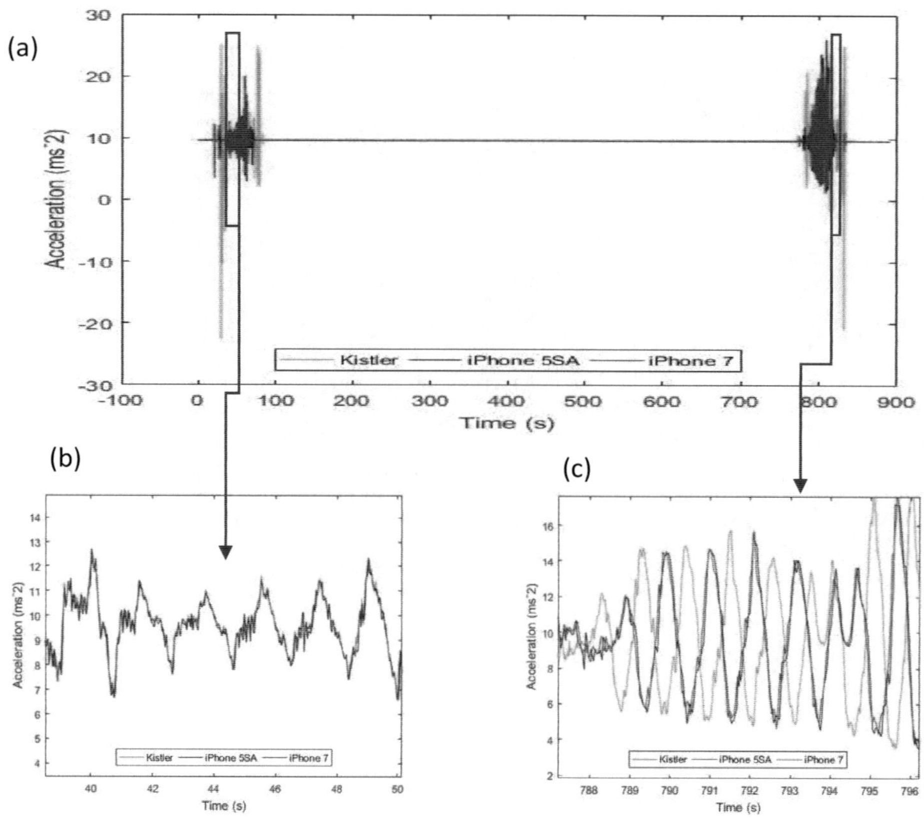

Figure 3, Synchronising phone data in post processing, (a) full time series (b) a zoomed in view of the start of the time series after the signals have been time aligned in post processing (c) zoomed in view at the end of the time series.

4. Bridge test

4.1 Description of bridge

Bridge used in experiment is shown in Figure 4, it is a 28m span timber footbridge with approach ramps on both sides of the bridge. The location of the KBA logging equipment is indicated in Figure 4(b) and a close up view of the equipment is provided later in Figure 6(c).

Figure 4, bridge used in experiment (a) north elevation, (b) south elevation (c) bridge deck

4.2 Data collected using KBAs

To get a good picture of the modal behaviour of the bridge it was decided to use six test points for the KBA modal test, namely approximately quarter span, mid-span and three quarter span on both sides of the bridge. On the day of the test there were four KBA accelerometers available so the data had to be collected in two 'swipes'. Figures 5 (a) and (b) show the arrangement of the KBAs for swipes 1 and 2 respectively. As the bridge is 28m long, quarter span, mid-span and three quarter span are at 7, 14 and 21m from the left hand support respectively. However, due to an error in marking the test points the sensors were actually placed at 9.7m, 14m and 23m from the left hand support. As the main purpose of this test is to compare the phone mode shapes to those from the KBAs the fact that the test points are not exactly at the quarter points is not important.

Figure 6 shows how the phones and the KBAs were set up on site both the phones and the KBAs we were secured to the deck using a suitable mounting/adhesive. The figure gives a sense of the difference in time/effort in setting up both systems on site. The wired KBA system took a couple of hours to set up whereas the phones took just minutes. Originally it was intended to simultaneously record with the phones and the KBAs. However, at the start of the test the battery in the iPhone 5A was found to be dead so we continued with the KBA test and returned the following day to carry out the phone test which is described in the next section, i.e. the data for the phones and the KBAs was collected on two different days

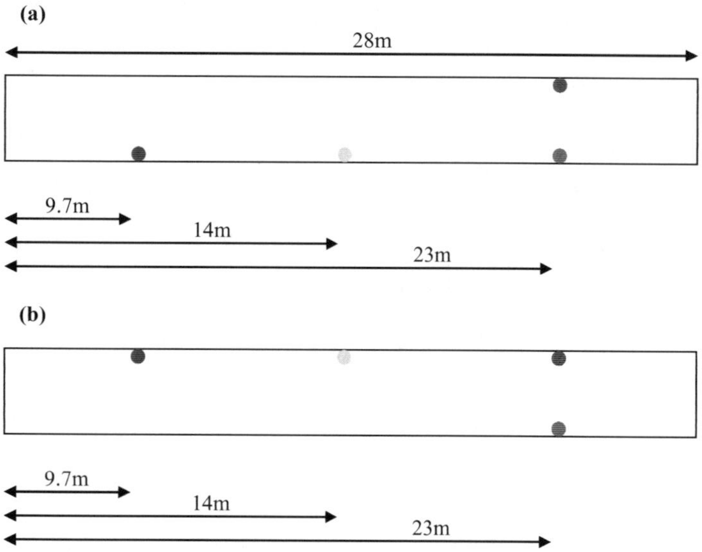

Figure 5, sensor placement for the bridge test (a) swipe 1 layout, (b) swipe 2 layout.

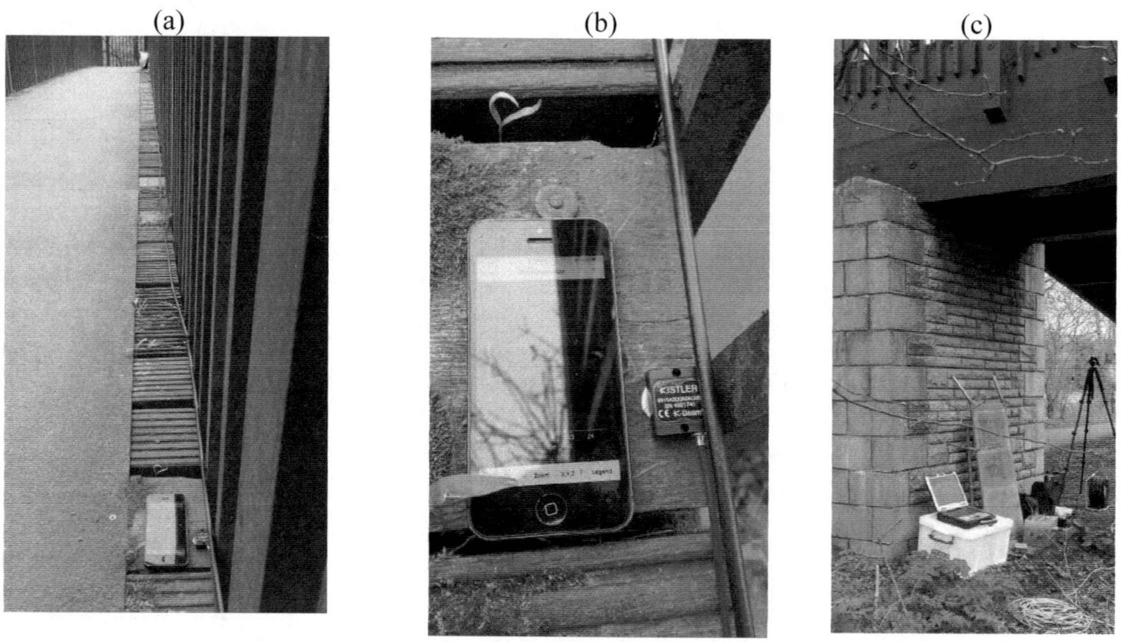

Figure 6, placement of sensors and equipment (a) iPhone 5 and KBA placed on bridge deck, (b) zoomed in view of (a) where both the phone and the KBA are secured to the deck using suitable mounting/adhesive, (c) Logging station for KBA adjacent to west abutment of the bridge.

The bridge is relatively flexible so during the test it was excited by simply walking across it. Figure 7(a) shows the time series recorded during swipe 1 at a scanning frequency of 512 Hz, and the corresponding frequency content is shown in Figure 7(b). The colour of the plots in Figure 7 correspond to the sensor colours indicated in Figure 5, e.g. channel 1 (red sensor) is 9.7m from the left hand support, channel 2 (yellow sensor) is 14m from the left hand support, etc. when the frequency plot in Figure 7(b) is examined it can be seen that there is a clear mode at 2.7 Hz with possibly another mode at approximately 9.5 Hz. Similar plots were observed for the swipe 2 data.

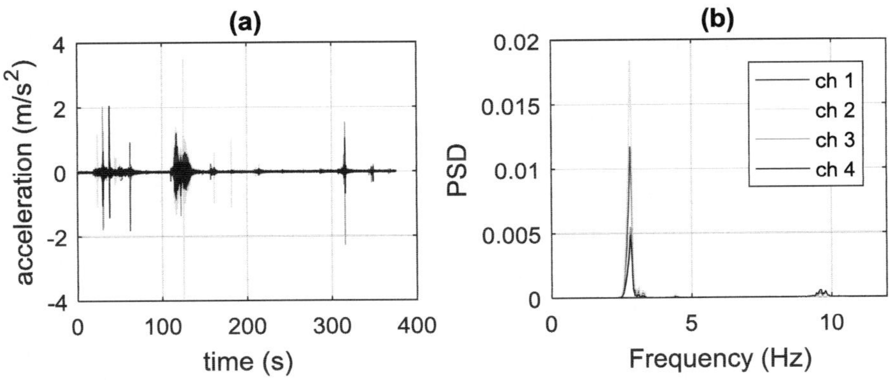

Figure 7, data collected during swipe 1 (a) time series, (b) frequency content for swipe 1

4.3 Data collected with phones

Figure 8(a) shows where the phones were placed on the bridge deck. Unlike with the KBA's where we wanted to capture modal ordinates on both sides of the bridge, with the phones we only instrumented one side of the bridge. This was because the KBA's will give us the global behaviour of the bridge so with the phones it was only necessary to see if they could get one half of the mode shape correct, as this would identify if the phones stayed sufficiently synchronised and if the accelerometers were sensitive enough to capture the vibrations. The maximum scanning frequency the iPhone 5A could operate at was 50 Hz so all three phones we set to scan at 50Hz.

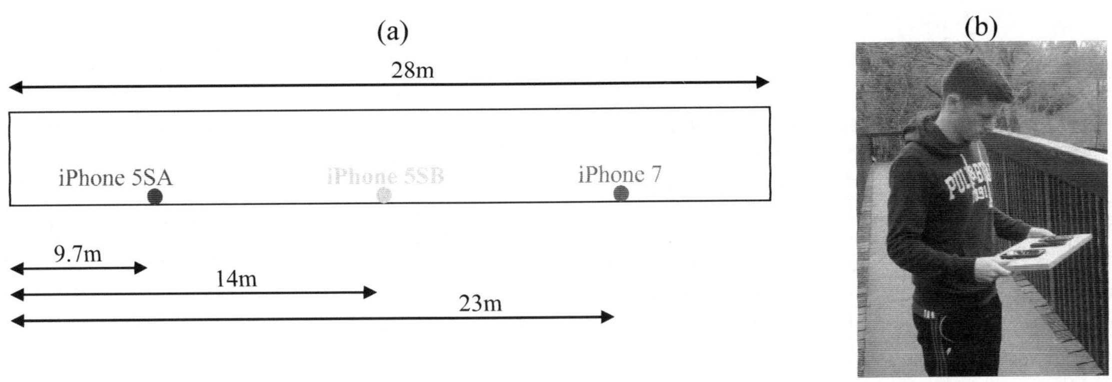

Figure 8, placement of phones on bridge

To allow syncing in post processing the phones were placed on a board at the start of the test and moved up and down to induce common vibrations in all three phones, see Figure 8(b). This process was repeated at the end of the test so it could be checked if the phones remained synchronised during the test or if they started to drift with respect to one another. Figure 9(a) shows the full time series recorded including the synchronisation phase at the start and end of the test, these parts of the signal are indicated using black rectangles in Figure 9(a). Zoomed in views of the synchronisation parts of the signal at the start and the end of the test are shown in Figures 9(b) and (c) respectively. The synchronisation in Figure 9(b) is implement by the analyst by shifting the second and third signals to get them to match with the first signal. However, it can be seen in Figure 9(c) that once the signals have been time alighted at the start they remain time aligned for the duration of the test.

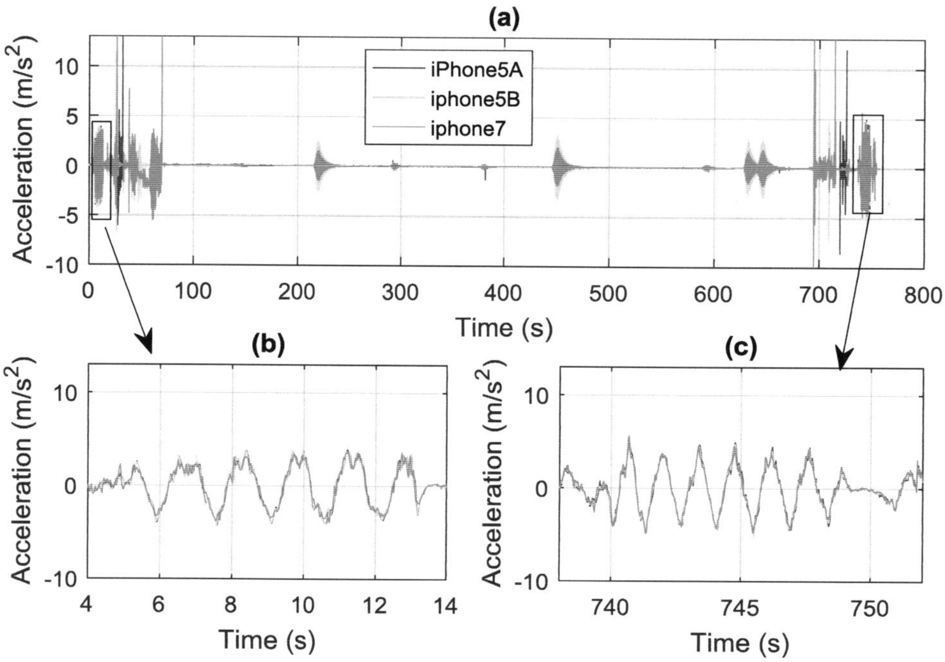

Figure 9, Phone data collected during test (a) full signals, (b) parts of the signal measured while phones were on the board at the start of the test (c) parts of the signal measured while phones were on the board at the end of the test.

After synchronisation the next step was to resample the data. While all three phones were set to scan at 50 Hz they each have independent clocks so for example if there is a data point for the iPhone 5SA at a given instant in time there might not be a corresponding data point in the iPhone 7 or the iPhone 5SB. There will be a data point very close but not necessarily at exactly the same time. To overcome this problem the data from all three phones was resampled (at 50 Hz) so that for a given point in time there was a data point in all three signals. This also overcame that problem that while the phones were scanning close to 50Hz, when the time series from the phones was examined it could be seen that the time data was recorded at, was not in exactly increments of 0.02 seconds for every time step. However, deviations were small. After the data was resampled the parts of the signal at the start and the end that were used to implement and check synchronisation respectively were removed and Figure 10(a) shows the time series that was used for modal identification and Figure 10(b) shows the frequency content of the signals in Figure 10(a). There is a clear peak at 2.7 Hz, however unlike the KBAs (see Figure 7) the phones are showing no peak at 9.5 Hz which indicates the phones are not picking up this mode.

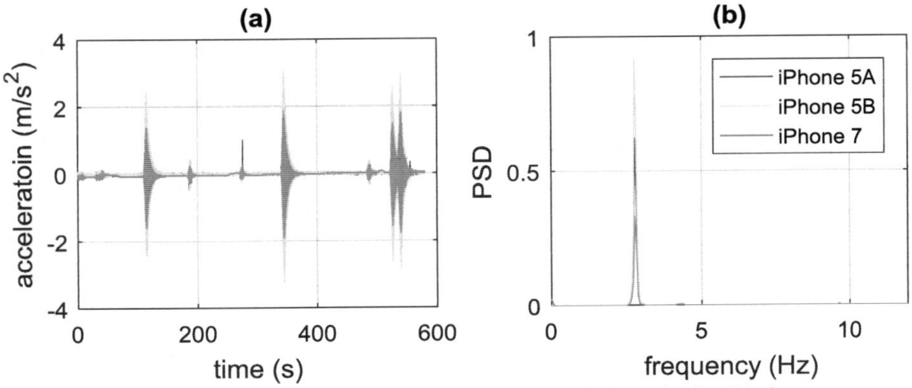

Figure 10, phone data (a) time series data used for modal analysis (b) frequency content of signals in (a)

5. Modal analysis

The modal analysis procedure used to identify the mode shapes will not be described in any detail as it has been well described by others previously. But in summary the method used is the NExT/ERA operational modal analysis procedure (James et al. 1995). This is one of several possible operational modal analysis procedures e.g. (Gentile & Gallino 2008; Van Nimmen et al. 2017). A comprehensive description of the NExT/ERA procedure and the underlying mathmatics is provided in (Brownjohn 2003) and therefore these detials are not reproduced here.

Table 1, Mode shapes identified from KBA and phone data

Mode No	KBA	phones
1	Freq=2.75 Hz 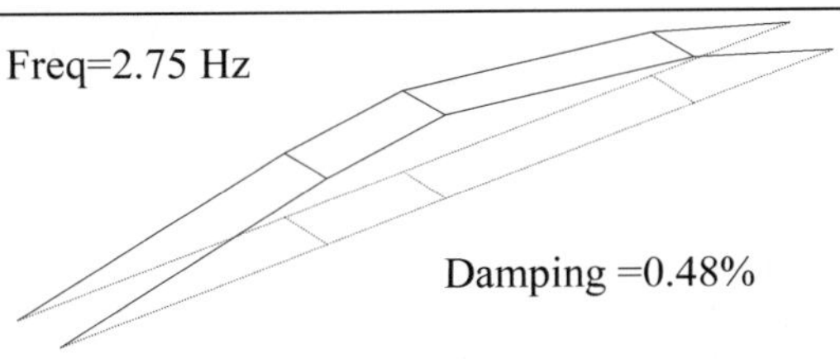 Damping =0.48%	Freq=2.78 Hz 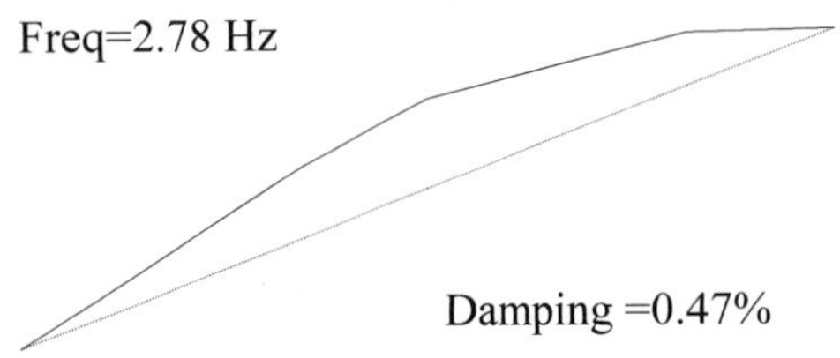 Damping =0.47%
2	Freq=9.5 Hz 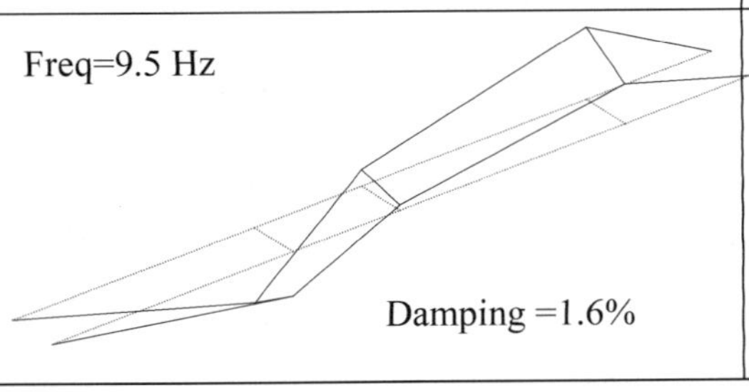 Damping =1.6%	Mode not identifiable in phone data

The results of the modal identification are shown in Table 1. As explained earlier the KBA accelerometers were intended to capture the global modes of the bridge so accelerometers were placed on both side of the bridge. Using this data it was possible to capture the the first and second vertial modes of the bridge. The first mode at 2.75 Hz is a classic first bending mode, and the low damping of 0.48% is consistend with the low damping observed on site. The second mode at 9.5 Hz looks like a combination of second bending and torsion. However, in reality the response of the bridge is dominated by mode 1 response as evidenced by the plots in Figure 7(b). The mode 1 dominance is to be expected as the frequency of the first mode exists within the band of loading frequencys for running or walking humans.

The phones were only placed on one side of the bridge so it was never inteneded to identify the full global mode shape for the bridge. Instead the objective was to see if the phone data could be used to correctly identify the mode shape on one side of the bridge. It was found that for the first mode this is indeed possible, the mode shape is correct and the frequency and damping predicted by the phone data are very similar to those identified in the KBA data. However, it was not possible to identify mode 2 using the phone data. The likely reason for this simply that the amplitude of the vibrations for mode 2 are lower than the noise floor of the phone app/accelerometer, e.g. it can be seen in Figure 10(b) that there is no frequency peak at 9.5 Hz indicating that the phone is simply not picking up these vibrations.

6. Conclusions

This paper shows that for the bridge tested, and the phones used it was possible to identify the first mode of the bridge. However, it should be noted that the bridge tested was quite flexible/lively and that the phones were unable to identify the second mode. From the results it appears the ability of the phone/app to identify mode shapes is limited by the noise in the accelerometer/app, and by the relatively low scanning frequencies that can be achieved, neither of which are unexpected conclusions. However, the phones, in this test at least, appear to remain quite well synchronised for the duration of the test which is encouraging. This paper does not purport to be a comprehensive study on the feasibility of using mobile phones to identify mode shapes. There are various phones and mobile apps for recording accelerations and therefore much works remains to be done to identify which combination of phone/app works best. Moreover it will be important to identify what the exact limitations are, i.e. on what structures will phones not work.

References

Brownjohn, J.M.W., 2003. Ambient vibration studies for system identification of tall buildings. *Earthquake Engineering and Structural Dynamics*, 32(1), pp.71–95.

Chang, W. et al., 2014. Using smartphone to identify dynamic characteristics of timber bridges. In *COST–Timber Bridges Conference 2014*.

Gentile, C. & Gallino, N., 2008. Condition assessment and dynamic system identification of a historic suspension footbridge. *Structural Control and Health Monitoring*, 15(3), pp.369–388. Available at: http://doi.wiley.com/10.1002/stc.251 [Accessed May 30, 2017].

Han, R., Zhao, X. & Yu, Y., 2015. Convenient posture monitoring on girder hoisting of Dalian Xinghaiwan crosssea Bridge based on smart phone. In *Proceedings of the 7th International Conference on Structural Health Monitoring of Intelligent Infrastructure (SHMII'15)*.

James, G.H., Carne, T.. & Lauffer, J.P., 1995. The natural excitation technique (NExT) for modal parameter extraction from operating structures. *International Journal of Analytical and Experimental Modal Analysis*, 10, pp.260–277.

Van Nimmen, K. et al., 2017. Numerical and experimental analysis of the vibration serviceability of the Bears' Cage footbridge. *Structure and Infrastructure Engineering*, 13(3), pp.390–400. Available at: http://dx.doi.org/10.1080/15732479.2016.1160133.

Peng, D. et al., 2015. Research on quick earthquake disaster investigation and evaluation using cloud monitoring method. In *Proceedings of the 7th International Conference on Structural Health Monitoring of Intelligent Infrastructure (SHMII'15)*.

Yu, Y. et al., 2015. Design and initial validation of external sensors board of smart phones for mobile structural health monitoring system. In *Proceedings of the 7th International Conference on Structural Health Monitoring of Intelligent Infrastructure (SHMII'15)*.

The 8th International Conference on
Structural Health Monitoring of Intelligent Infrastructure
Brisbane, Australia | 5-8 December 2017

Investigation of the Influence of Breaking Strain and Temperature on a Binary Crack Sensor

F. Raeisi[1], A. Mufti [2], G. Mustapha[3], D.J. Thomson[4], R. Eden[5]
[1] *University of Manitoba – Canada, Email:* raeisif@myumanitoba.ca [2] *University of Manitoba – Canada,* [3]*Structural Monitoring Technologies - Canada,* [4] *University of Manitoba – Canada,* [5] *Manitoba Infrastructure and Technology – Canada*

Abstract

It is well documented that civil infrastructure, including steel bridges, are aging and are approaching the end of their design service life. One of the challenging problems with aging steel girder bridges is the potential for crack formation and propagation that may result in unusable service conditions. Timely detection of cracks demands frequent manual inspections or the use of Structural Health Monitoring (SHM) systems. SHM has the advantage of providing continuous monitoring. One of the common distributed structural health monitoring systems is based on the use of fibre optic sensors. This technique has been used very successful in the past; however, installation cost and the need for expert interpretation of results make it costly for most owners of steel girder bridges.

In past work we have demonstrated a binary crack sensor that has the potential to be installed on steel girder bridges at a fraction of the cost and can detect the presence of a crack formation with a width of 0.2mm.

The sensor is comprised of a thin copper wire that forms a closed loop circuit and adhesive. As soon as crack forms in the steel girder, the bond between the steel, adhesive and wire will transfer the strains to the adhesive and then to the wire. When the wire reaches its final tensile strain, it will break and the continuity of electrical current will be compromised.

A small scale experimental apparatus was designed to test different material combinations. After selecting the best candidates for the binary sensor, it was tested on a small scale girder in the lab at ambient temperature.

One of the main challenges in developing the binary sensor is to select appropriate materials for both the wire and adhesive. An important characteristic of the wire discussed in this paper is the effect of final tensile strain on the collected test results.

In this work, the effect of pre-straining various types of wire is reported as well as performance versus temperature. In order to prepare for full-scale field tests, the sensor was tested between two extreme temperature of -30ºC and +40ºC in a laboratory environment. The sensor is presently being tested in a harsh, full-scale field environment on a long steel girder bridge. The results will be reported at the conference.

1. Introduction

One of the critical deficiencies in aging steel bridges is crack formation and crack propagation under service load (Ghorbanpoor & Benish 2003). The crack formation may be as a result of material defects, welding defects, etc. and propagates during the service life (Lamtenzan et al.

2000). Detecting cracks at the initial stage of crack formation is critical, so that timely repairs can be used to minimize the probability of significant loss of serviceability (Ellis & Conner 2013), (Chajes et al. 2005). Non-destructive methods such as acoustic emission, eddy current and ultrasound testing is one way of detecting cracks in steel girders. However, these methods are time consuming and access to the point of evaluation is not always possible (Chang et al. 2003), (Chang & Liu 2003). SHM methods provide real time data on the status of structural elements. SHM methods are divided into two categories; discrete monitoring (such as using strain gauges, short gauge or long gauge sensors) and distributed monitoring (such as installing the smart film, fibre optic sensors or broken wire binary sensor)(Raeisi et al. 2017). Distributed monitoring provides comprehensive data since the sensor is installed on the entire length of the girder (Glisic & Inaudi 2011). The broken wire electronic binary sensor is a new cost effective distributed SHM method which detects cracks on steel girders of bridges. The sensor is comprised of an insulated wire bonded to the steel girder using a precisely selected adhesive. As shown in Figure 1, the sensor will be installed over the entire length of a steel girder and as soon as a crack on the girder opens up to 0.2mm, the strains will be transferred through the adhesive to the wire and will cause the wire to break.

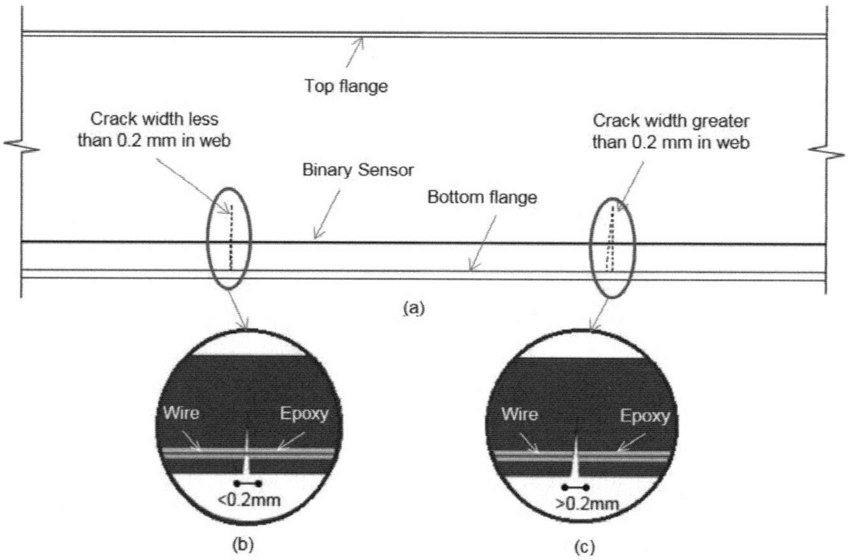

Figure 1 a) Binary sensor will be installed on the entire length of the girder as a distributed monitoring system. b) As soon as a crack forms in the girder, the strains will transfer to the epoxy causing it to fracture. This happens at very small crack openings. c) Over time, the crack will widen and causing the wire to break at 0.2mm crack openings. (Raeisi et al. 2017)

The initial steps of developing the binary sensor was discussed earlier (F.Raeisi et al. 2015), (Raeisi et al. 2017). In this paper, the installation procedure of the binary sensor on a small scale steel girder in the lab will be explained and the effect of changing parameters such as change in the ultimate tensile strain of the wire and change in the temperature will be studied.

2. Introduction to Binary sensor

The binary sensor is a distributed monitoring system comprised of insulated copper wire and epoxy adhesive. The epoxy bonds the wire to steel girder of a bridge. As soon as a crack opens up to 0.2mm width, the strains of the girder transfers to the wire and wire breaks. One of the critical challenges in designing the sensor is to select proper material for both wire and adhesive.

Mechanical properties of materials as well as bonding properties between wire and adhesive will affect the result. In order to choose proper material from the wide range of available options, an experimental apparatus was designed. The apparatus simulates the crack opening on the steel girder, and it is made of two steel plates connected to each other at one end using a flexural hinge (Figure 2). The plates open along the free edge and simulate the crack opening. The selected wire bonds to the plates using the selected adhesive. The complete procedure and results of testing the sensor on the apparatus has been explained in other papers of this author (F.Raeisi et al. 2015), (Raeisi et al. 2017).

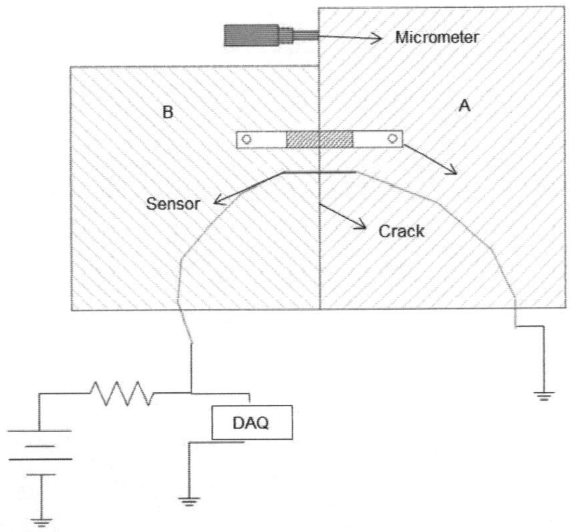

Figure 2 schematic of the apparatus for testing different combination of wire and adhesive. Two steel plates are connected to each other at one end with a flexural hinge and the gap between two plates simulates the crack on a steel girder. Turning the micrometer will cause plate A to move from plate B and widen the crack. The gap opening between two plates will be recorded by the displacement gauge. Two ends of the binary sensor are connected in series with a resistor to an electrical source with the DAQ. As soon as the wire breaks, the resistance will change from a short (Ω=0) to an open ($\Omega=\infty$) (Raeisi et al. 2017)

3. Test on the steel girder in the lab

3.1 Installation procedure

After completing the test on the apparatus, the best candidates (insulated copper wire gauge39 and Loctite Epoxy E-20NS) which had sufficient bonding and reproducible properties were selected for testing on a steel girder in the lab. A W250x67 section steel beam was simply supported over a 3.0 meter span as shown in Figure 3a,b. A crack was created at mid span to test the binary sensor. In order to form the crack, the beam was cut through the bottom flange and was continued to the web. The cut then was filled by re-welding. Using a three point bending system, an actuator applied approximately 60KN force at mid span on top flange and induced a new crack in the welded section. The sensor was installed on the pre-existing crack on girder. A displacement gauge (pi-gauge PI-5-200, Tokyo Sokki Kenkyujo) and two strain gauges were mounted on the other side of the girder to record the crack opening and measuring the strains (Figure 3c,d).

The first step to install the binary sensor is to remove the paint from the surface of the steel girder in order to create the best possible bond between sensor and the surface of the girder. Isopropyl alcohol was used to remove any dirt or contamination from the girder. After cleaning the surface, formwork with 2mm width was mounted on the specified position on the girder and the wire was held temporary at the center of the formwork. Using an applicator, the epoxy was injected into the formwork and covered the wire and bonded it to the steel girder. A blade was used to skim the top of the formwork and remove the excess epoxy. The sensor cured for 24 hours and the formwork was removed after curing time.

The binary sensor is a closed circuit (Figure 4). As soon as the crack opens the wire breaks and the sensor will change to an open circuit. In the short circuit condition a series resistor installed in the circuit causes little or no voltage to be observed. As soon as the wire breaks, the resistance of the circuit will tend to zero and a voltage of several volts is observed across the sensor. The change in the voltage can be easily observed.

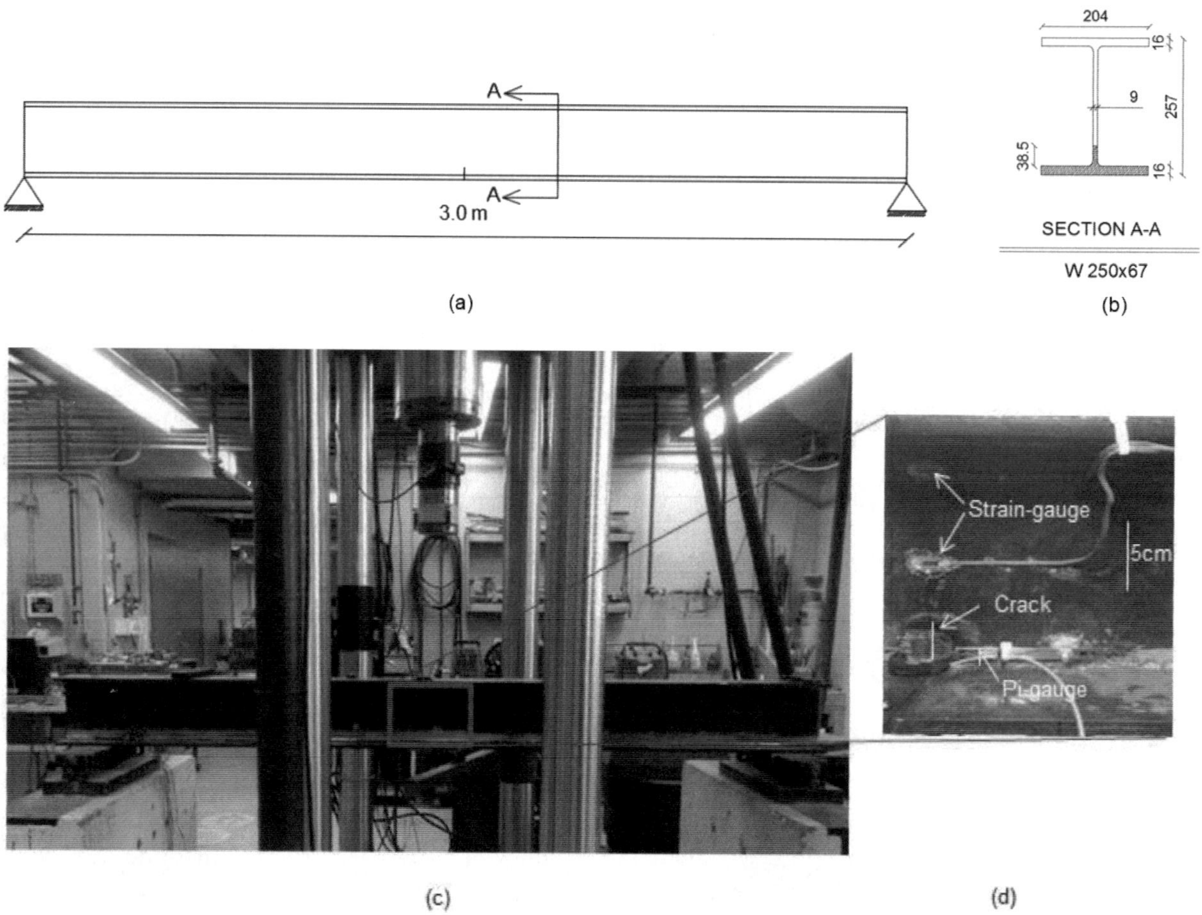

Figure 3 a) 3-meter simply supported steel girder. a crack was created at mid-span. b) Cross-section of the girder, the hatched area demonstrates the pre-existing crack. c) 3-point bending system; the steel girder and its supports under load frame d) a displacement gauge was mounted to record the crack opening (Raeisi et al. 2017)

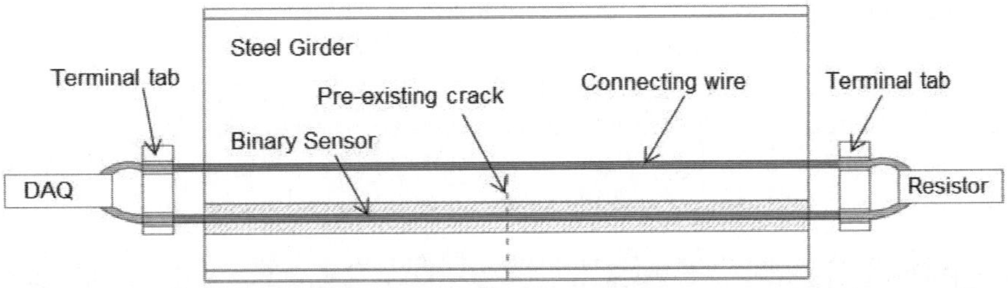

Figure 4 The electrical circuit was formed on the web of a steel girder to monitor the crack opening. Binary sensor, connection wire, DAQ and resistor are the elements of the circuit (Raeisi et al. 2017)

3.2 Test results

After one day of curing the adhesive, the beam was tested. The actuator applied the load gradually. By increasing the load, the steel girder deflected and the pre-existing crack increased in width. The epoxy fractured at less than 0.1mm; and, as the crack opened, the strain in the wire increased and eventually caused the wire to break. When the wire broke a sudden change in the resistance of the circuit occurs. A complete discussion about the test procedure and results can be found in the following reference (Raeisi et al. 2017).

Results with insulated copper wire gauge 39 and Loctite epoxy E20-NS were consistent and they were selected as the final material for testing in the environmental chamber.

4. Test in environmental chamber

Since the binary sensor will be installed on steel girders of bridges, it is important to test the performance of the system under extreme temperature. The same steel girder having a pre-existing crack at mid span was transferred to the environmental chamber. The sensor was installed on the girder at ambient temperature (+22°C) and cured for one day. After curing, the chamber cooled down to -30°C first and a 3-point bending test was completed. After getting the results, the sensor was removed and reinstalled at room temperature. This time the chamber heated up to +40 °C and the sensor was tested at extreme high temperature. Results are shown in Table 1.

Table 1 Test results of Binary sensor on steel girder at ambient temperature, -30°C and +40 °C

Test Number	1	2	3	4	5	6
Crack opening (mm) at Ambient Temperature	0.13	0.13	0.13	0.19	0.14	0.14
Crack opening (mm) at -30 °C	0.26	0.1	0.14	0.11	0.24	0.14
Crack opening (mm) at +40 °C	0.1	0.13	0.17	0.16	0.13	0.16

As it is shown in Table 1, changes in temperature has minimal effect on the results of binary sensor and the sensor can detect cracks with less than 0.2mm opening.

5. Field trial of the binary sensor

The binary sensor was installed on 3.0 meters of a steel girder bridge in Manitoba (Figure 5). Purpose of this demonstration field trial was to demonstrate that the sensor and cellular based data logger can be installed on a bridge girder and will withstand temperatures extremes, bridge vibrations and other environmental conditions commonly observed in the field.

Figure 5 Sensor overlays as seen in SMTs online Analytics software.

The sensor was installed in December and was tested in conditions that reached -30ºC. For the first 3 months system was set to take one sample per day and report once per day and during last 3 months system was set to take a sample every 6 hours and report twice per day.
Some results are shown in Figure 6.

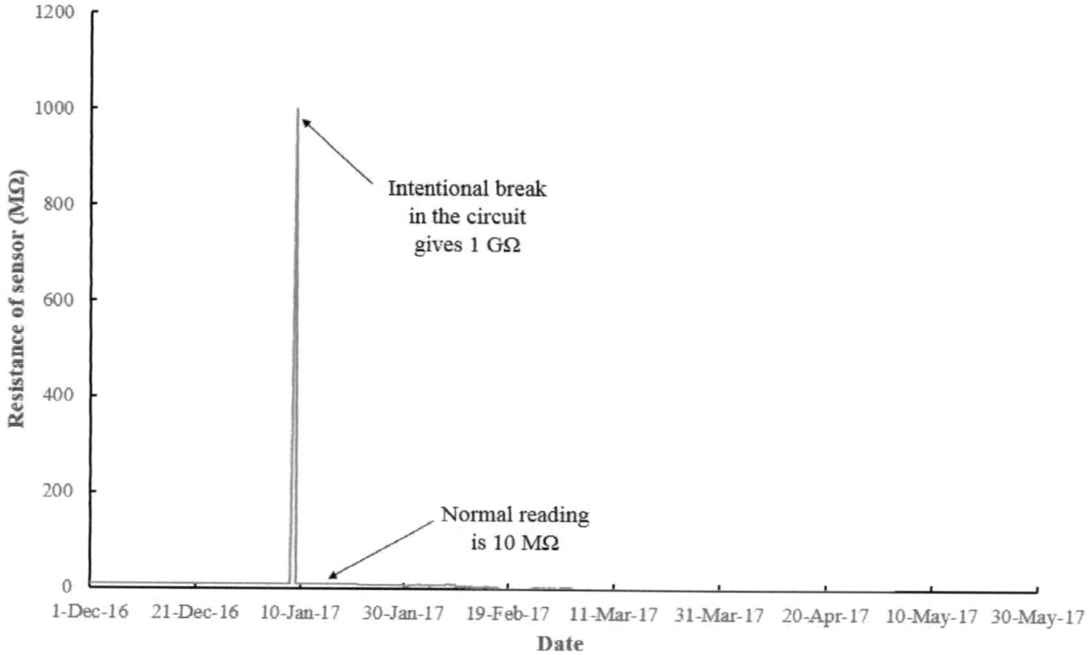

Figure 6. Results of binary sensor installed on a steel girder of a bridge from Dec.2016 to May 2017

6. Pre-straining the wire

As mentioned before, tensile properties of wire will have significant effect on the results. Insulated copper wire is available with different coatings and specifications. In order to select proper wire for the sensor, it is important to study the effect of final tensile strain of the wire on crack opening at which the wire breaks. In this work, magnet wire MW79-C gauge 39 was selected for the study. The MW79-C wire is a ductile material with an ultimate tensile strain of 24%. The wire was installed on the apparatus using Loctite epoxy E20-NS and the test was executed. Results are shown in Table 2.

Table 2 Test results of MW79-C gauge 39 magnet wire in combination with Loctite E-20NS epoxy as the binary sensor on the experimental apparatus. Results are not consistent and are more than desired crack opening of 0.2mm

Test Number	1	2	3	4	5	6	7
Crack width opening(mm)	0.13	0.22	0.27	0.35	0.40	0.58	0.6

As shown in the Table, the crack openings at which the wire breaks are variable and the mean crack opening is more than the desired crack width of 0.2mm. The wide crack opening may be as a result of high tensile strain of the wire and the variation may be due to the kinks on the wire which causes different initial length and finally results in different final tensile strengths. By pre-straining the wire, kinks will be removed from the wire which may lead to more uniform results. The pre-straining will also decrease the ultimate tensile strain of the wire and consequently allow the wire to break at smaller crack openings. In order to investigate the effect of pre-straining of wire on final results, the wire was pre-strained by 5% and 10% and were tested on the apparatus. Results are shown in Table 3.

Table 3 Test results of pre-strained MW79-C gauge 39 magnet wire in combination with Loctite E-20NS epoxy as the binary sensor on the experimental apparatus. Results are consistent and in the range of desired width of 0.2mm

Test Number	1	2	3	4	5
Crack width opening(mm)-5% pre-strained	0.09	0.12	0.13	0.14	0.16
Crack width opening(mm)-10% pre-strained	0.09	0.10	0.12	0.16	0.16

6. Conclusions

A binary sensor for a distributed monitoring system was designed for detecting cracks in steel girders of bridges. Other available methods such as fibre optic sensors are expensive options for most bridges.

A binary sensor constructed of 39 gauge insulated copper wire with epoxy adhesive can detect cracks with less than 0.2mm opening on the steel girder. Tests were conducted at room temperature as well as two extreme temperature (-30 °C and +40°C). Observation from the experiments prove that changes in temperature will have minimal effect on the results and the binary sensor will break at the desired crack width.

Tensile properties of wire and adhesive as well as bonding between these two materials affects the results. In this work, the effect of pre-straining the wire on breaking width were studied. By pre-straining the wire, results were more consistent and was able to detect cracks in the acceptable range.

The binary sensor was tested on a steel girder in the field to test the effect of different parameters such as temperature on its application. The installation procedure of the binary sensor on a large scale girder will be optimized in future. As next step, a Finite Element Model which will help to improve the material selection procedure will be simulated.

References

Chajes, M., Mertz, D. & Quiel, S., 2005. Steel girder fracture on Delaware's I-95 bridge over the Brandywine River. *Proceedings from Structures Congress 2005: Metropolis and Beyond*, pp.1–10.

Chang, P.C., Flatau, A. & Liu, S.C., 2003. Review Paper : Health Monitoring of Civil Infrastructure. *Structural Health Monitoring*, 2(3), pp.257–267.

Chang, P.C. & Liu, S.C., 2003. Recent Research in Nondestructive Evaluation of Civil Infrastructures. *Materials in civil Engineering*, 15(June), pp.298–304.

Ellis, R. & Conner, R., 2013. Investigation and Repair of the Diefenbaker Bridge Fracture. In *Transportation Assosiation of Canada*. Winnipeg.

F.Raeisi et al., 2015. Crack Detection in Steel Girders of Bridges Using a Binary Sensor. In *7th International Conference on Structural Health Monitoring of Intelligent Infrastructure*.

Ghorbanpoor, A. & Benish, N., 2003. *Wisconsin Highway Research Program: Non-Destructive Testing of Wisconsin Highway Bridges*, Milwaukee: Wisconsin DOT.

Glisic, B. & Inaudi, D., 2011. Development of method for in-service crack detection based on distributed fiber optic sensors. *Structural Health Monitoring*.

Lamtenzan, D., Washer, G. & Lozev, M., 2000. *Detection and Sizing of Cracks in Structural Steel Using the Eddy Current Method*, Federal Highway Administration.

Raeisi, F. et al., 2017. Crack Detection in Steel Girders of Bridges Using a Broken Wire Electronic Binary Crack. *Journal of Civil Structural Health Monitoring*, pp.0–23.

The 8th International Conference on
Structural Health Monitoring of Intelligent Infrastructure
Brisbane, Australia | 5-8 December 2017

Smart reinforcement steel bars embedded with MEMS sensors for low-cost structural health monitoring

F. Tondolo [1], A. Cesetti [1], E. Matta [2], A. Quattrone [1], D. Sabia [1].
[1] *Dept. Structural, Geotechnical and Building Eng., Politecnico di Torino – Italy, Email:* francesco.tondolo@polito.it
[2] *Dept. Architecture and Design, Politecnico di Torino – Italy*

Abstract

Structural health monitoring (SHM) is increasingly drawing the attention of researchers and stakeholders engaged in the optimal management of the civil structural and infrastructural asset worldwide. The strain monitoring of reinforced concrete structures is a crucial issue but a limited number of alternative measuring technologies are currently available, mainly including electric strain gages and fibre optic sensors. The difficulty of their installation, the cost of the devices and the uncertain long-term reliability push for alternative low-cost and distributed devices, easier to install and use. With the aim to improve the cost-effectiveness of existing sensing technologies, this paper presents a new concept of smart reinforced concrete steel rebar, achieving strain-sensing capabilities through the incorporation, in the core of the rebar, of an ordinary low-cost embedded MEMS sensor. According to this concept, a cavity is drilled in the rebar and hermetically sealed after being equipped with a pressure and temperature sensor, from whose measurements the cavity volume variations are reconstructed and correlated with the rebar axial deformation. The new sensing system shows significant advantages over current strain-measurement devices, proving cheap, durable, easy to install, robust to mechanical shocks and chemical aggression, and requires simple interrogating equipment. The theoretical framework of the new concept and the results of a preliminary test campaign are here presented.

1. Introduction

Current advancements in the sensor research field are increasingly leading towards the acceptance and implementation of the "smart city" concept: an ultimate "nerves system" intended to monitor the various physical, chemical and biological aspects of manmade surroundings, which is expected to play in the near future a vital role in addressing the economic benefits and ethical need for safe and sustainable structures and infrastructures around the world (Bremer et al. 2017).
One crucial ingredient of this concept will be the employment of appropriate structural health monitoring (SHM) strategies which, combining various sensing technologies with data acquisition and processing capabilities, will provide for a continuous condition assessment of civil engineering assets, increasing their quality and safety levels and reducing their maintenance costs (Spencer et al. 2017). Long-term monitoring of degradation phenomena in aging structures will thus allow for the optimization of preventive maintenance policies and for the early detection of dangerous responses or incipient damages, thus reducing costs associated to repair and down-time. On the other hand, the observation of the structural response after extreme events, such as earthquakes, hurricanes or fires, will allow for a rapid and objective assessment of the structural integrity.
Despite this great potential, SHM is currently not applied yet at a large scale and in a systematic manner to civil structures and infrastructures. This is mostly due to the complexity and cost of

monitoring large-scale constructions, which entails installing and interrogating large numbers of sensors as well as managing huge amounts of data. In this context, the advancement in sensor technologies, the spread of smart materials and structures, and the development of wireless communication platforms are the key opportunities for SHM growth, anticipating a future where every structure will be permanently, pervasively and (nearly-) automatically monitored since its construction (Lynch & Loh 2006).

Currently, various sensor types exist for SHM purposes, including electric strain sensors, fibre optic sensors (FOS), piezoelectric sensors, GNSS-based sensors, accelerometers, inclinometers, acoustic emissions and wave propagation devices. Each of them has a preferable domain of use, as well as genuine challenges when deployed in real-world applications.

Focusing on the SHM of reinforced concrete (r.c.) structures, and with the purpose of improving the cost-effectiveness of existing sensing technologies, this paper presents a new concept of smart reinforcing bar (rebar), achieving strain-sensing capabilities through the incorporation, in its core, of an ordinary low-cost embedded MEMS sensor. According to this concept, a cavity is drilled in the rebar, equipped with a pressure and temperature MEMS sensor and eventually hermetically sealed. The axial deformation in the rebar is finally deduced from the cavity volume variation, on its turn reconstructed from the pressure and temperature variations. The new sensing system shows significant advantages; the specific features and the results of preliminary tests are here reported.

2. The New Measuring System

2.1 Current Strain Sensors for Use in R.C. Structures

Strain monitoring of r.c. structures can be performed by focusing on either the concrete matrix or the reinforcement steel rebars. Because of the intrinsic large scatter in the micro-mechanical properties of concrete and because of the influence of micro-cracking and rheological phenomena at smaller scales, the concrete-oriented approach practically implies using either macro-sensors or small sensors mounted on macro skeletons. In both cases, the average strain along either the macro-sensor or the skeleton length is measured, and a perfect sensor/concrete adhesion is required. As a result, a steel-oriented approach aimed at the strain measurement sounds preferable. It is generally done by a direct measure of strain by means of a local transducer.

In current practice, this local strain transducer is either an electric strain gage glued on the longitudinal lug of the rebar or a FOS glued on the surface of the rebar or incorporated in a longitudinal groove drilled along the bar. In both cases, compensation of temperature effects is needed.

Critical stages for these devices are the sensor installation and the concrete casting. Both strain gages and FOS can be damaged at these stages, or even during in-service operation as a consequence of cracks formation or steel-to-concrete slip. Gluing itself can be compromised by chemical aggression, mechanical action or ageing. In the case of FOS, the protective encapsulation and/or packaging, which is necessary to prevent standard glass fibres from contact with the alkaline environment of the concrete mixture, results in a difference between the strain of the concrete matrix and the strain sensed by the fiber (Quiertant et al. 2012). Moreover, the installation of distributed FOS requires specialized personnel, and the gluing process can be particularly time-demanding (Barrias et al. 2016).

2.2 The Principle of the New Measuring System

The basic idea of the proposed system is to use low-cost existing micro-sensing devices, commonly used for other purposes, as a means to estimate strain in r.c. structural elements, by exploiting in an original way the well-known physical laws of deformable continuum systems.

According to this idea, the rebar axial deformation is measured by an appropriate MEMS sensor which is embedded in a small cavity drilled in its core rather than being applied on the rebar surface. The sensor simultaneously measures both the pressure and the temperature of the fluid contained in the cavity. Any elongation of the rebar will result in a proportional variation of the cavity volume, according to the continuum mechanics, which on its turn will determine pressure and/or temperature variations in the cavity. By measuring the latter variations through the sensor, the volume variation can be computed by applying appropriate thermodynamic state equations to the fluid contained in the cavity, and the axial elongation can be consequently deduced based on a previous calibration.

In greater detail, as long as the fluid is air, the thermodynamic state equation can be approximated by the ideal gas law as follows:

$$pV = nRT \tag{1}$$

where $p = p(t)$ is the absolute pressure in the cavity, $V = V(t)$ is the cavity volume, n is the number of moles of the fluid in the volume, R is the ideal gas constant, $T = T(t)$ is the absolute temperature within the cavity, and t is time.

Denoting as ΔV, Δp and ΔT the variations of the three state variables with respect to their initial values V_0, p_0 and T_0 at $t = 0$, and taking into account that n and R are constant parameters, any volume deformation $\Delta V/V_0$ can be generally computed, at any time t, from the monitored time histories of p and T, as follows:

$$\frac{\Delta V}{V_0} = \left(1+\frac{\Delta T}{T_0}\right) \Big/ \left(1+\frac{\Delta p}{p_0}\right) - 1 \tag{2}$$

In general, the volume deformation $\Delta V/V_0$ of the cavity can be caused either by a variation ΔF of the rebar axial load F (with respect to its initial value F_0) or by a variation ΔT^* of the rebar temperature T^* (with respect to its initial value T_0^*), or by both of them. Assuming for simplicity a linear superposition of the mechanical and thermal effects, this relation can be formalized as follows:

$$\frac{\Delta V}{V_0} = \left(\frac{\Delta V}{V_0}\right)_{\Delta F} + \left(\frac{\Delta V}{V_0}\right)_{\Delta T^*} \tag{3}$$

where $(\Delta V/V_0)_{\Delta F}$ is the volume deformation due to the rebar axial force variations ΔF, and $(\Delta V/V_0)_{\Delta T^*}$ is the volume deformation due to the rebar temperature variations ΔT^*. It is worth noticing that ΔT and ΔT^* may generally not coincide, since the former is the temperature of the fluid in the cavity, while the latter is the average rebar temperature, responsible for the whole rebar thermal deformations.

Complemented by Eq. (2), Eq. (3) plays a fundamental role both at the stage of sensor calibration (including a preliminary thermal calibration and a subsequent mechanical calibration) and at the stage of sensor operation, i.e. during monitoring.

Thermal calibration tests have the scope of applying controlled temperature variations ΔT^* to the mechanically unloaded rebar ($\Delta F = 0$). In these conditions, in facts, the first term of the right-hand side of Eq. (3), i.e. $(\Delta V/V_0)_{\Delta F}$, is null, and Eq. (3) directly provides the experimental relation $f_1(\cdot)$ existing between ΔT^* and $(\Delta V/V_0)_{\Delta T^*}$ as follows:

$$\left(\frac{\Delta V}{V_0}\right)_{\Delta T^*} = \left.\frac{\Delta V}{V_0}\right|_{\Delta F=0} = f_1\left(\Delta T^*\right) \tag{4}$$

where $\Delta V/V_0|_{\Delta F=0}$ is the volume deformations computed according to Eq. (2) on the bases of the measured pressure and temperature variations Δp and ΔT. As long as temperature variations are slow during thermal tests, the rebar temperature ΔT^* can be assumed equal to the cavity temperature ΔT, so that no additional temperature transducers are in this case required for calibration beyond the embedded one.

Mechanical calibration tests are usually conducted in uncontrolled temperature conditions by applying controlled axial force variations ΔF to the rebar. Because temperature variations cannot be avoided (normal fluctuation of temperature in the laboratory), Eq. (3) must be complemented by both Eq. (2) and Eq. (4) in order to provide the experimental relation $f_2(\cdot)$ existing between ΔF and $(\Delta V/V_0)_{\Delta F}$:

$$\left(\frac{\Delta V}{V_0}\right)_{\Delta F} = \frac{\Delta V}{V_0} - f_1\left(\Delta T^*\right) = f_2\left(\Delta F\right) \tag{5}$$

where $\Delta V/V_0$ is computed according to Eq. (2) based on measured Δp and ΔT values, and $f_1(\Delta T^*)$ is computed according to Eq. (4) based on measured ΔT^* values. Notice that, because temperature variations ΔT induced in the cavity by sudden axial force variations need to be transferred to the rebar and eventually dissipated, ΔT^* and ΔT do not generally coincide in this case, and separate measurements would be in principle required (e.g. through adding a further temperature sensor which measure the temperature of the rebar) if a continuous accurate tracking of the axial force time history is desired.

As long as the rebar responds in its linear elastic range and mechanical deformations keep small, a proportional relation is expected between the axial force and the volume deformation, which can be expressed by:

$$\left(\frac{\Delta V}{V_0}\right)_{\Delta F} = f_2\left(\Delta F\right) = \frac{1}{k}\Delta F \tag{6}$$

where k is the mechanical calibration coefficient transforming the measured (corrected) volume variation into the estimated axial load.

Eqs. (4) to (6) express the desired thermal and the mechanical calibration relations.

Even more remarkably, Eqs. (5) and (6) also allows the estimation of the axial force ΔF during sensor operation, which is the main purpose of the monitoring action.

2.3 The Low Cost Sensors

Pressure and temperature measurements in the rebar cavity are provided by common MEMS sensors. A widespread type of MEMS is the barometric sensor, which usually integrates pressure and temperature transducers. Nowadays barometric MEMS are commonly employed in smartphones and smartwatches, IoT devices, wearable and fitness trackers and others, thanks to their small dimensions (within 2x3x1 mm^3). These MEMS are easily available on the market at a cost of about 2 \$. The vacuum technology used for their construction allows to measure the absolute pressure, ranging between 260 hPa and 1260 hPa. This is a valuable characteristic since it establishes an absolute reference for the measuring system over time.

2.4 The System in Details

The measuring system is based on a Patent (Tondolo 2016). In this paper, it is applied on a ribbed steel reinforcing bar in which a cavity is obtained as a transversal drilled hole. The latter determines an internal volume, filled with air, which hosts a hard PCB mounting a barometric pressure sensor LPS25H by STMicroelectronics®. An electrical feedthrough is provided at the entry of the cavity, connecting the hard PCB inside with a soft PCB outside, which on its turn communicates with the acquisition system, exchanging power and data. The cavity is hermetically sealed, and the soft PCB is lodged into a groove machined all along the longitudinal lug of the rebar. Schematics of the system are reported in Figure 1. The cavity determines a reduction of the rebar section and also a critical point for stress increase and concentration. The consequent reduction in resistance and/or ductility may not represent a significant drawback, as long as the exact reproduction of the original rebar mechanical properties is deemed inessential at this stage of development.

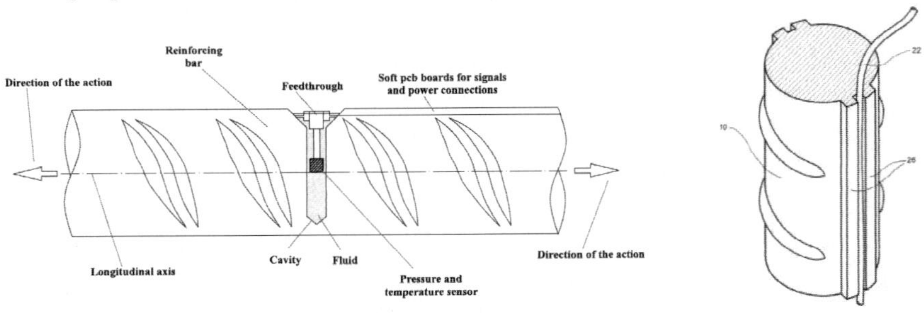

Figure 1. Longitudinal section of the system (left) and axonometric view of the soft PCB hosted in the rebar lug (right).

3. Preliminary Experimental Campaign

3.1 Description of the setup components

The experimental tests focus on a ribbed reinforcing steel bar instrumented with the monitoring system described in the previous paragraphs. Figure 2 shows the system setup during tests.
It consists of a ribbed rebar having 20 mm diameter, equipped with three measuring units spaced 80 mm from each other. A closer look at the ingress of one unit is provided in Figure 2b. A longitudinal groove 2 mm in depth and width is machined into the rebar to host the soft PCBs which connect the pins of the feedthrough to the converter boards (clearly visible at the bottom of Figure 2a). The converter boards acquire the digital pressure and temperature data coming from the embedded sensors and transform them into analogue continuous voltage signals collected by HBM Spider 8 acquisition system. Results from one measuring unit only are considered in this paper.

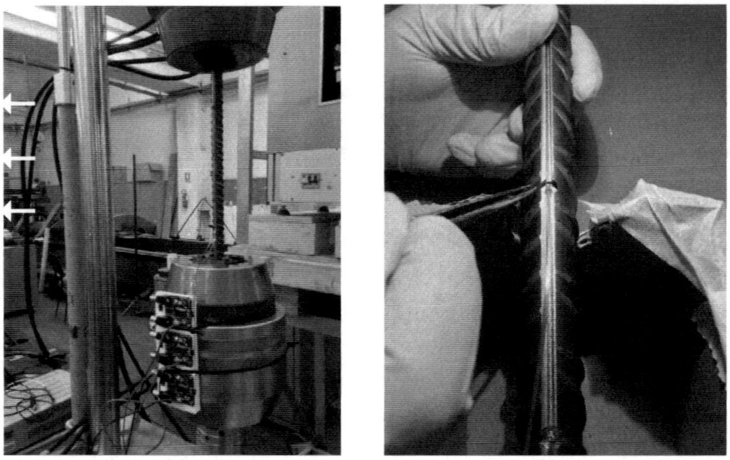

(a) (b)

Figure 2. The experimental setup.

3.2 Temperature calibration test

The temperature calibration test is preliminarily conducted by subjecting the instrumented rebar to temperature variations at constant axial load, in order to get the experimental relation $f_1(\cdot)$ existing between ΔT^* and $(\Delta V/V_0)_{\Delta T^*}$, according to Eq. (4). The rebar is placed into a temperature controlled chamber (Angelantoni Challenge 250) and subjected to slow temperature variations (1°C/6min), approximately spanning 30°C between 14°C and 44°C. Figure 3 shows the controlled chamber (on the left) and a comparison between the experimental curve and its cubic interpolation (on the right). The relation is clearly nonlinear and monotonically decreasing. This counterintuitive result, confirmed by repeated testing of several other measuring units, still deserves an explanation, which will likely come from an accurate finite element thermo-mechanical modelling currently under way. Nevertheless it should be remark that both pressure and temperature measured in the cavity are influenced by the presence of all the components sealed in and their response to the temperature variation. The application of this relation according to Eq. (5) provides, anyway, a substantial correction of environmental temperature trends in axial load tests.

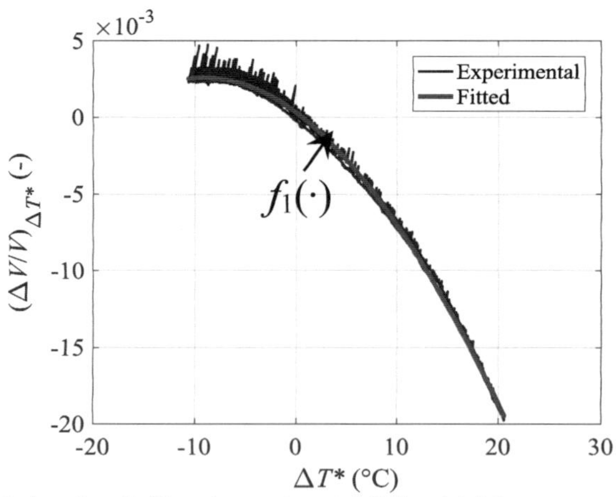

Figure 3. Temperature calibration: controlled chamber (left) and experimental fitting (right).

3.3 Axial load tests

Uniaxial tensile load tests are performed in uncontrolled temperature conditions through a universal testing machine MTS with a maximum capacity of 250 kN. Considering the dimensions of the holes hosting the sensors (4 mm of diameter and 18 mm of depth) and the presence of the longitudinal groove, the rebar transversal section is up to 24% smaller than the nominal section. In order to avoid the insurgence of any plastic strain during the tests due to the effect of the holes, the maximum applied load is limited to 40 kN, assuming a stress intensity factor equal to 3.

At first, a complete loading and unloading cycle is performed through 8 steps of 5 kN each, corresponding to about 20 MPa of tensile stress on the net transversal section of the rebar (axial load test #1). The load variations are applied instantaneously and maintained for about 30 s. According to Eq. (6), this test provides the mechanical calibration coefficient $k = 1.83 \cdot 10^4$ kN. In Figure 4 the time histories are reported for, respectively, the applied force (a), the measured pressure (b) and temperature (c) normalized variations, the corrected volume variations (d), the volume-to-force calibration function (e) and the comparison between the applied and the estimated axial force (f).

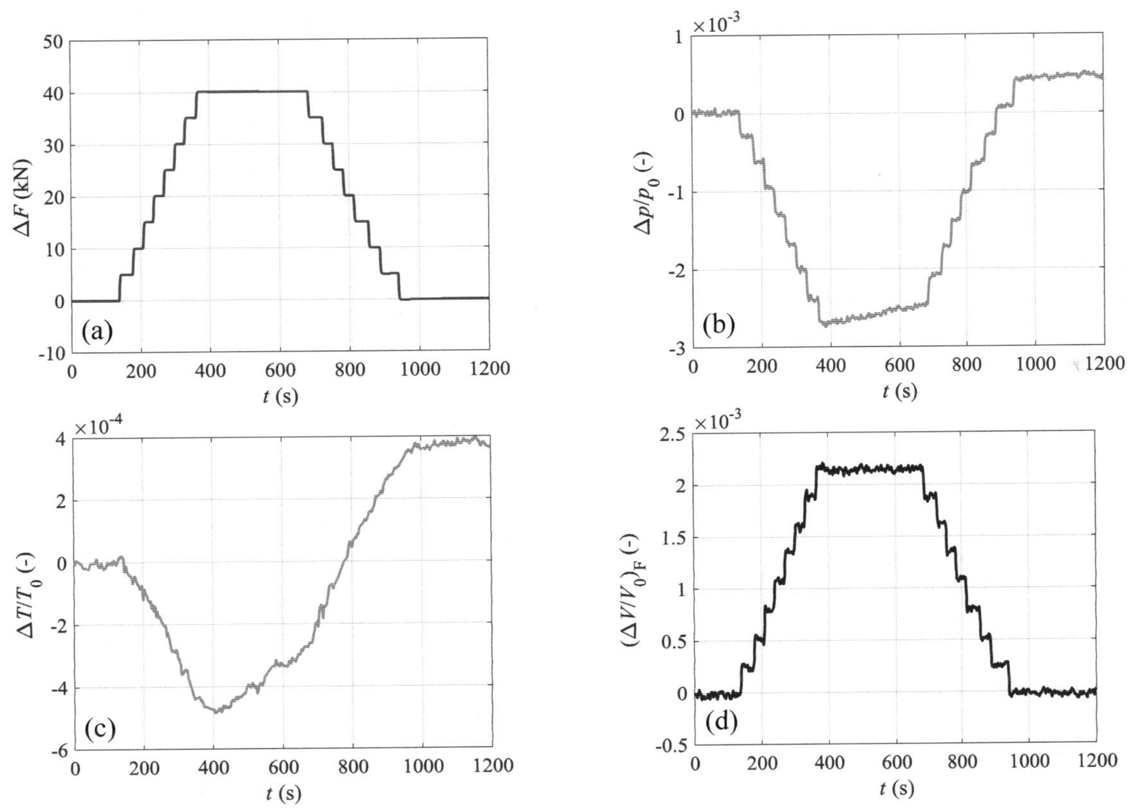

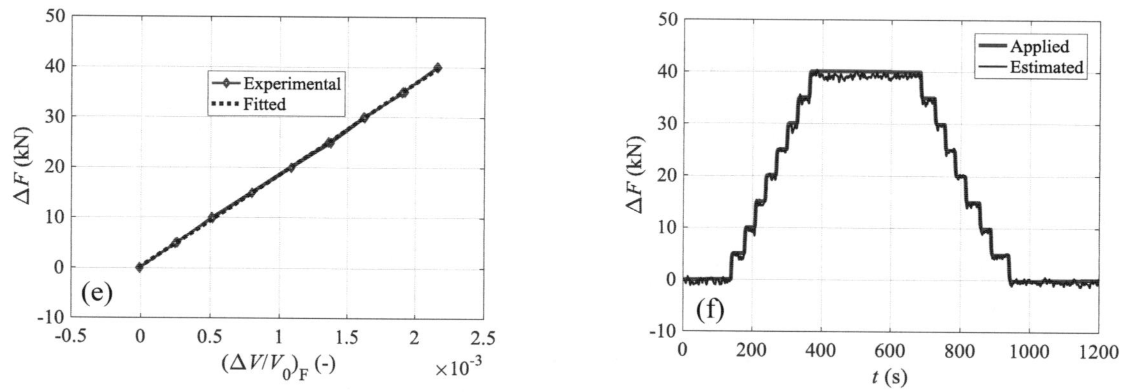

Figure 4. Axial load test #1: (a) applied force; (b) measured pressure; (c) measured temperature; (d) reconstructed force-dependent volume; (e) volume-to-force calibration; (f) applied vs estimated force.

Subsequently, several load tests are conducted by subjecting the same instrumented rebar to a series of alternative force time-histories, in order to verify the accuracy of the axial load estimation in different loading (and temperature) conditions.

In particular, axial load test #2 (Figure 5) presents the same multi-step loading/unloading cycle already applied in axial load test #1; axial load test #3 (Figure 6) presents two consecutive 40 kN one-step loading/unloading cycles; and axial load test #4 (Figure 7) presents a one-step 10 kN loading/unloading cycle superimposed to a ±3.8 kN sinusoidal load test at a frequency of 0.1 Hz.

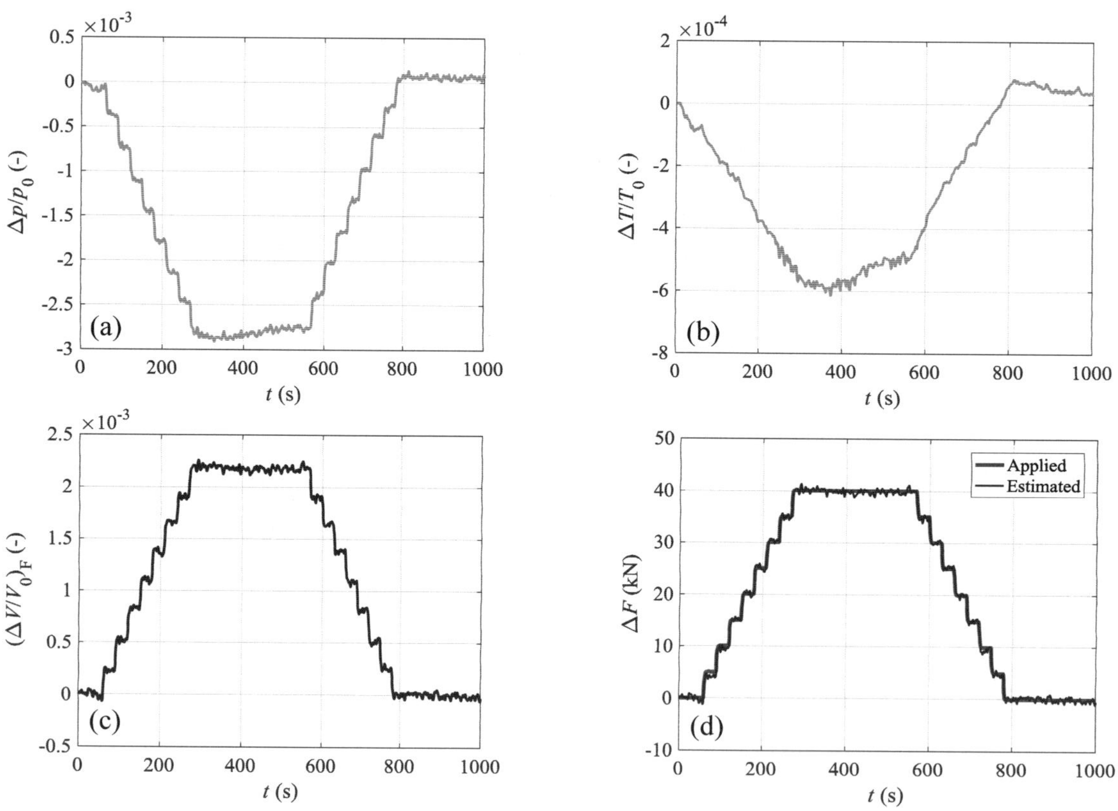

Figure 5. Axial load test #2: (a) measured pressure; (b) measured temperature; (d) reconstructed force-dependent volume; (f) applied vs estimated force. Note: volume-to-force calibration as from test #1.

In all the tests, an accurate estimation of the applied axial load is apparent. The only partial exception is represented by axial load test #4, in which the dynamic character of the input does not allow for a sufficient stabilization of the cavity temperature measurements, implying an imperfect tracking of the imparted load. Even so, the estimated load time-history provides an average value of the sinusoidal amplitude which is quite similar to the actual one. Referring to test #1, in particular, a substantial proportionality and a negligible hysteresis can be appreciated in the volume-to-force relation (Figure 4e).

In all the tests, and particularly in tests #1 and #3, a temperature drift ΔT^* is visible during testing, certainly attributable to environmental effects. A temperature variation ΔT is also measured inside the cavity during the test, as a result of the nearly-adiabatic thermodynamic process occurring inside the cavities during high-rate loading variations (prior to the initiation of the heat conduction process). Presently, not enough experimental information is available to discriminate between the effects of, respectively, ΔT^* and ΔT on the thermal calibration (Eq. 4), in the case when ΔT^* and ΔT do not coincide. For simplicity, Eq. (5) has been here applied by assuming $\Delta T^* = \Delta T$, but further in-depth investigations are necessary on this aspect, possibly requiring the installation of further temperature sensors within (or on the surface of) the rebar. This is however left for future work.

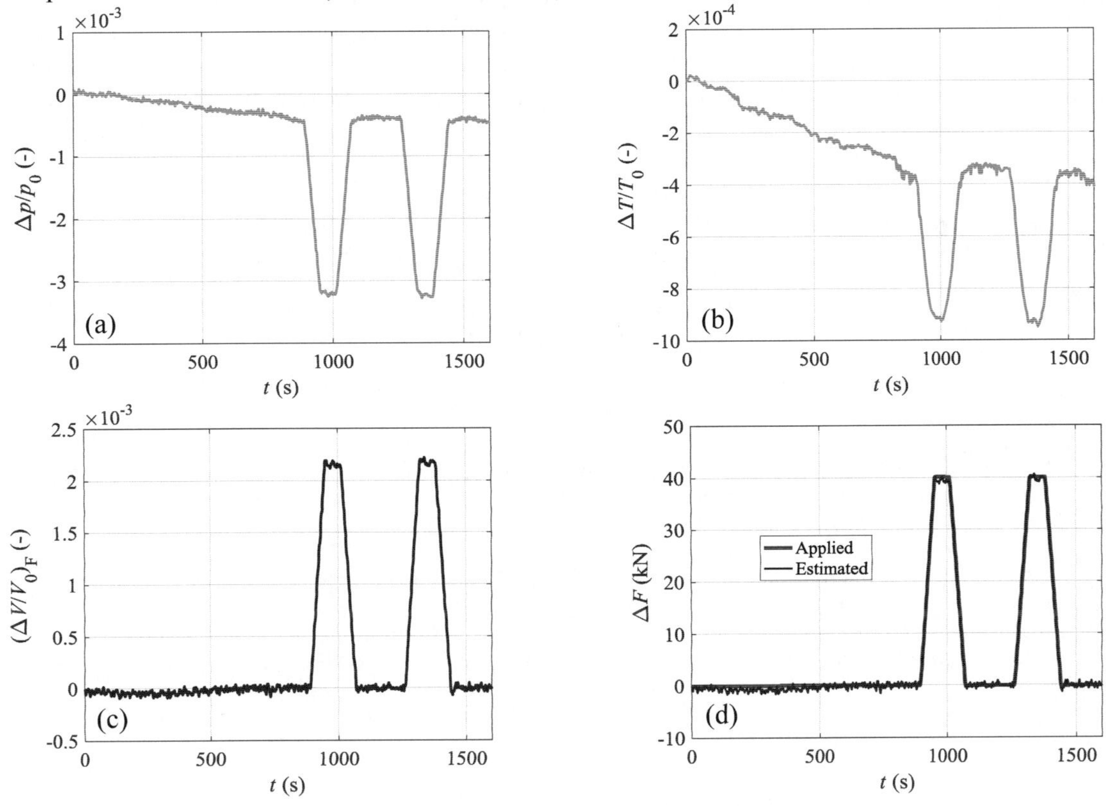

Figure 6. Axial load test #3: (a) measured pressure; (b) measured temperature; (d) reconstructed force-dependent volume; (f) applied vs estimated force. Note: volume-to-force calibration as from test #1.

4. Discussion

The reported test results are extremely encouraging. The estimated axial force, obtained from temperature and pressure measurements through a correction for the environmental temperature drift, closely tracks the applied load time history. The match is slightly less accurate in the case of dynamic tests but still adequate to reliably measure the amplitude of the imparted force.

5. Conclusions

This paper introduces the concept of a new technology for monitoring the axial strain in steel rebars by measuring the pressure and temperature variations occurring in a cavity drilled in the rebar , using commercially available, low cost MEMS sensors.
At first, the basic theoretical principles are explained and used to derive useful calibration and operation formulas, expressing the correlations between the main thermo-mechanical variables.
Then, tests are conducted to ascertain the feasibility and accuracy of the new sensing system. In particular, a thermal calibration test is conducted in temperature controlled conditions under zero axial load, providing the temperature corrections formulas. A mechanical calibration test is then conducted through imparting various axial load cycles in environmentally uncontrolled conditions, providing the required relation between cavity volume variations and axial force. Various mechanical tests are performed to validate the force estimation capability of the device. Results appear very promising, showing a good tracking accuracy in both static and dynamic conditions, and open the way to a new research field in sensor technology, which is here just at its beginning.

References

Bremer, K, Weigand, F, Zheng, Y, Alwis, LS, Helbig, R & Roth, B 2017, 'Structural Health Monitoring Using Textile Reinforcement Structures with Integrated Optical Fiber Sensors', *Sensors*, vol. 17, no. 345, pp. 1-12.

Spencer, BF Jr., Park, J-W, Mechitov, KA, Jo, H & Agha, G, 2017, 'Next generation wireless smart sensors toward sustainable civil infrastructure', *Procedia Engineering*, vol. 171, pp. 5-13.

Lynch, JP & Loh, KJ 2006, 'A summary review of wireless sensors and sensor networks for structural health monitoring', *Shock Vib. Digest*, vol. 38, no. 2, pp. 91-128.

Quiertant, M, Baby, F, Khadour, A, Marchand, P, Rivillon, P, Billo, J, Lapeyrere, R, Toutlemonde, F, Simon, A & Cordier, J, 2012, 'Deformation monitoring of reinforcement bars with a distributed fiber optic sensor for the SHM of reinforced concrete structures', *NDE 2012*, pp. 1-10.

Barrias, A, Casas, JR & Villalba, S 2016, 'A review of distributed optical fiber sensors for civil engineering applications', *Sensors*, vol. 16, no. 748, pp. 1-35.

Tondolo, F 2016, 'Embedded system for measurements of strain/stress in one-dimensional elements', *Patent* n° 102016000118077, Italy.

The 8th International Conference on
Structural Health Monitoring of Intelligent Infrastructure
Brisbane, Australia | 5-8 December 2017

Longitudinal Response of an Existing Cast Iron Tunnel Subjected to Parallel Piggy-back Tunnelling

C.Y. Gue[1], M.M. Alhaddad [1], M.J.Wilcock [1], M.Z.E.B Elshafie [1], R.J. Mair [1]
[1] *University of Cambridge – United Kingdom, Email:* cyg20@cam.ac.uk

Abstract

Tunnelling close beneath an existing tunnel is a common occurrence in an urban environment. Alignment of these tunnels range from a perpendicular undercrossing to perfectly parallel. The latter is much less common, thus much less understood. Conventional assessment methods impose greenfield soil displacements directly on the existing tunnel as a worst case scenario, often leading to high estimates of longitudinal radius of curvature which are beyond the allowable design limit. Subsequently, costly and time consuming mitigation measures would have to be designed and mobilised. The scarcity of high quality case studies of parallel piggy-back tunnelling impedes the continuous effort to understand and improve the efficiency of such impact assessments; this paper presents a detailed case study which focuses on the longitudinal response of an existing 3m diameter, a century old cast iron Royal Mail tunnel that is affected by the construction of an approximately 12m diameter platform tunnel in a parallel piggy-back scenario just 2m directly beneath. Various state-of-the-art instrumentation systems including distributed fibre optic strain sensing, photogrammetry and vibrating wire instrumented bolts and electronic displacement transducers were deployed over a 40m section of the existing Royal Mail tunnel to monitor its response. This enabled the full response of the existing tunnel to be captured in a high level of detail in terms of strain and displacements as the tunnelling face progressed towards and beyond the zone of monitoring for both the pilot and tunnel enlargement phases.

1. Introduction

Conventional assessment of existing tunnels subjected to tunnelling construction below it assumes that the existing tunnel deforms similarly to the soil around it (i.e. assuming greenfield conditions) for both the transverse and longitudinal directions of tunnelling. While this is a simplified approach, it is generally preferred for its simplicity as well as conservatise nature; as there is a lack of field and experimental data, a more accurate impact assessment may not be conducted with confidence.

Attewell et al (1986) developed closed-form solutions for the prediction of a three dimensional tunnelling induced surface settlement profile where the magnitude of settlement at any point within the zone of influence is governed by three factors: relative position from tunnelling face, volume loss and distance from centreline to the point of inflection, *i*; where *i* is a product of the settlement trough parameter '*K*' and depth of tunnel.

Empirical derivation of *K* is taken from the transverse surface settlement profile where it ranges from 0.4 to 0.6 in clays (O'Reilly & New, 1982 and Lake et al, 1992) while Mair and Taylor (1997) suggested that a *K* value of 0.5 is suitable for clays. Generally, *K* is assumed to be the same for both the transverse and longitudinal direction out of convenience and simplicity. For a given tunnel

depth, size and volume loss, a larger K value would result in a wider and shallower settlement profile with larger curvatures (i.e. less differential settlement).
Taking the differential of the derived greenfield vertical settlements, over a section of interest, gives the imposed radius of curvature (ROC). Now assuming that the existing tunnel deforms as a continuous tube; bolt and lining stresses can be assessed based on the imposed ROC by assuming an appropriate location of neutral axis (Euler-Bernoulli beam-theory) for the tunnel cross-section.
The above assessment of ROC holds true for an idealised bending mode deformation where the tunnel deforms as a monolithic structure. On the hand, given the discontinuous nature of cast iron tunnels, the tunnel could deform in a shearing deformation mode as described by Alhaddad et al. (2014), whereby consecutive rings undergo shear displacements or slip relative to adjacent rings in order to accommodate the imposed greenfield settlement. This mode of deformation reduces the forces in the bolts and the extreme fibre strains significantly as the tunnel structure is not acting in flexure.
The applicability of the two extreme idealised deformation modes (bending and shear mode deformations) rely on the pre-existing loads of the circumferential bolts, which are extremely challenging to be determined and likely to exhibit great variability (Wilcock, 2017), as well as stiffening effect from the concrete invert (concrete block at the bottom of the section). Adding to the complexity, flanges of the tunnel linings that were assumed to be rigid in the abovementioned computation of ROC do in fact deform and bend to accommodate bending stresses as discussed later.
Given the number of unknown variables and deformation mechanisms, the bending mode deformation which represents the most onerous conditions is generally favoured, even though the more likely scenario is a combination of both bending and shearing modes. This often leads to very strict ROC limits. Deciding on an appropriate trigger limit would thus rely more on engineering judgement and experience, where validation is carried out through construction monitoring.
Assessing ROC through conventional methods of monitoring such as automated total station surveys (ATS) or manual levelling, given its limited data points and accuracy, is not straight forward. It has also not been clearly defined in industry (Alhaddad, 2017). For example, a similar magnitude of differential vertical settlement among three targets can yield significantly different ROCs depending on the chord length or distance between monitoring points.
There is a pressing need for much more high quality field monitoring data in order to facilitate a better understanding of the underlying deformation mechanisms and hence increase efficiency of future asset impact assessments.
London Crossrail project employs a risk based ROC limit, where the maximum allowable ROC for live tunnels was set at 6km, utility tunnels at 4km and disused tunnels at 2.5km. It will be demonstrated through this case study, that it is possible to exceed these limits without any visible damage to the existing tunnels occurring.

2. Background of Royal Mail tunnel and Crossrail

Royal Mail tunnel (RMT) was constructed in 1917 to convey mail from eastern to western sorting offices in London to reduce delays as a result of traffic congestion on the roads. Serving eight sorting offices along a 10.5km long system, the tunnel was constructed wholly in London Clay. The running tunnels were made up of 7 cast iron segments (including key) to form an external diameter of 2.97m. The entire system was mothballed in 2003 due to high running costs and remains unused today.

The existing RMT is undermined by the construction of an approximately 12m diameter platform tunnel (PTE-E) of the new Crossrail line at London Liverpool Street Station. For a distance of approximately 100m, both tunnels were aligned on top of each other in a parallel piggy-back scenario with only 2m of clear distance separating both tunnels. Construction of PTE-E is carried out in two stages with an initial 6m diameter pilot tunnel which is subsequently enlarged to its full size. An opened face tunnelling construction method was adopted with sprayed concrete lining.
The average axis depth of RMT and PTE-E within the zone of piggy-back alignment was approximately 26.5m and 32.25m below ground level. PTE-E was constructed mainly in London Clay with partial toe into the Lambeth Group. A cross-sectional view within the monitored section along with the soil profile is shown in Figure 1.

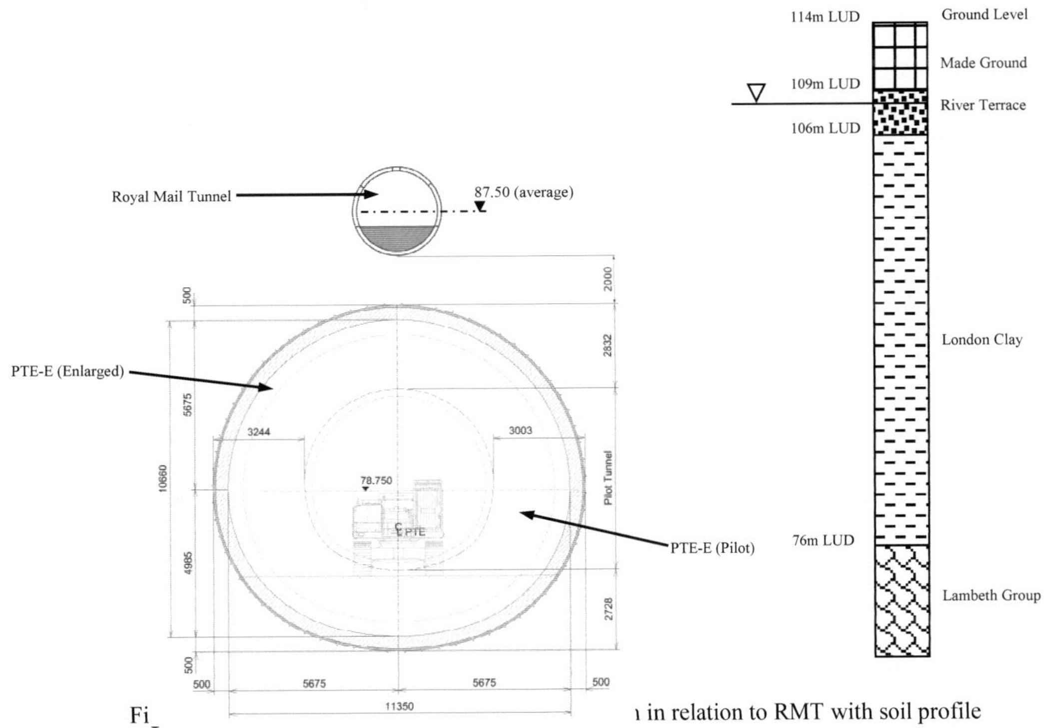

Fi_ ı in relation to RMT with soil profile

3. Construction Progress

Duration of the monitoring spanned over 11 months from 16th April 2013 to 14th of March 2014. Construction of the pilot and tunnel enlargement entered the zone of influence (R2950 to R2870) in early September 2013 and late January 2014 respectively. Construction progress is summarised in Figure 2.

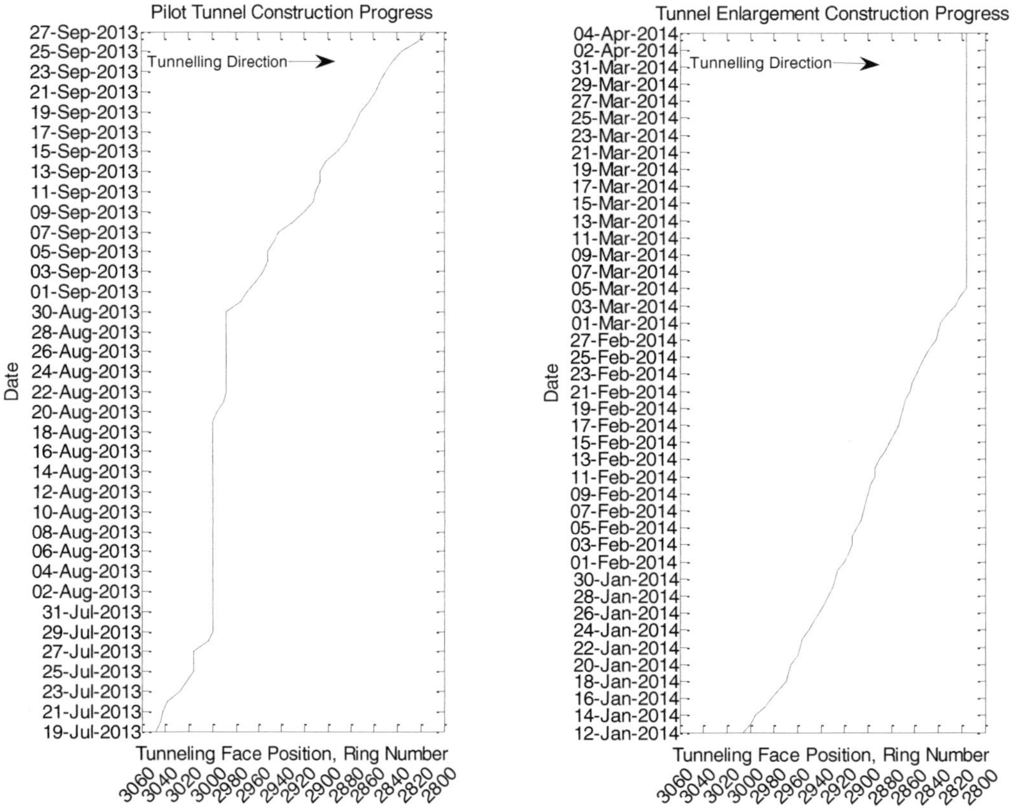

Figure 2 Construction progress of PTE-E for pilot tunnel (left) and tunnel enlargement (Right)

4. Deployed Instrumentations & Layout

Various state-of-the-art instrumentation was deployed to capture the response of RMT during the passage of the tunnelling face including distributed fibre optic strain (DFOS) sensing among others. In addition to conventional automated total station (ATS) and manual levelling systems.

DFOS system based on Brillouin Optical Time Domain Refletometry (BOTDR) relies on the strain dependent frequency shift of backscattered light signals to accurately measure strains along the entire length of the attached fibre optic cable (Mohamad, 2008). The key advantage of this system is its continuous strain sensing ability, which provides a full strain profile along the attached cable length while requiring only power only at the point for the analyser; which can be kilometers away.

A Hitachi tight buffered single core single mode cable was used to measure longitudinal strains along the crown of RMT over a distance of 40m from ring R2970 to R2850. This single continuous cable itself is glued on a series of aluminium L brackets which is in turn bolted on the panels of each ring. An initial prestrain is provided to allow capture of compressive strain. Detailed installation procedure can be found in Gue et al. (2015). Layout of DFOS instrumentation is as shown in Figure 3.

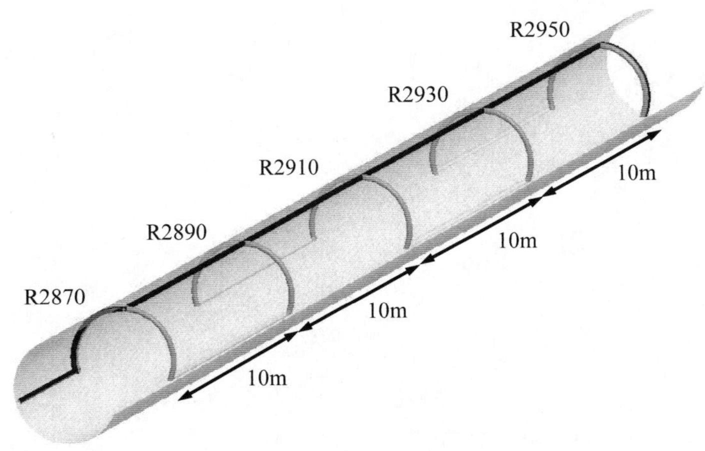

Figure 3 Layout of DFOS in RMT

A series of 120 battery powered, wireless linear potential displacement transducers (LPDT) were set up in pairs (shown in Figure 4) from R2880 to R2940 to measure relative gap movements between consecutive rings, observering both longitudinal and vertical shearing displacements. Temperature data was also captured to compensate the data, however, the monitoring environment proved to be extremely stable, with little temperature variation over time.

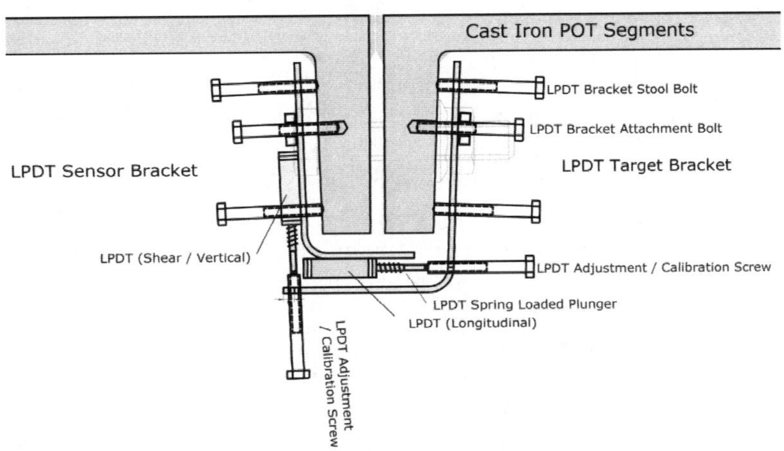

Figure 4 LPDT sensor arrangement

In addition, a new innovative photogrammetry monitoring system (CSattar) which tracks movement of multiple targets from a series of still images taken from off-the-shelf digital single lens reflex (DSLR) camera was deployed to measure both cross sectional convergence as well as longitudinal movements between consecutive rings. The advantages of such system are the low cost and high accuracy. While it is possible to track existing features on the tunnel linings, targets (shown in Figure 5) were included to increse accuracy by increasing contrast.

Figure 5 Layout of various instrumentation systems in Royal Mail Tunnel

5. Global Longitudinal Response of Royal Mail Tunnel

The scattered plots in Figure 6 represent the collective "bow wave" strain profile of multiple points along the fibre optic cable with respect to the distance from the tunnelling face of the pilot tunnel (positive strain denotes tensile strain throughout this paper). The positive horizontal axis represents the normalised distance ahead of the tunnelling face. Greenfield induced strain profiles based on the designed volume loss of 1.25% were superimposed for comparison where the dashed lines were derived from vertical settlements, dotted lines for longitudinal displacements and the solid line which is a combination of both for the full greenfield strain profile.

The theoretical curves are centred at zero for the horizontal axis where a clear separation of tensile strain (ahead of the tunnelling face) and compressive strains (past the tunnelling face) is observed. The measured strain profile indicates that a horizontal shift of the inflection point by approximately 1D ahead of the tunnelling face occurs. It is interesting to note that the bounds of strain is sufficiently captured by the full greenfield profile with minor under prediction for the compressive strains. This under prediction may be caused by the presence of wooden packing between successive rings, providing additional compressibility. Good compatibility between the greenfield prediction and measured strain profile proves that the existing cast iron tunnel behaves flexibly. This is in-line with observations of Standing & Selman (2001). A maximum tensile strain of 0.025% (or ROC of 5.4km) was measured at 2D while maximum compressive strain of -0.03% (ROC of 4.5km) was recorded at approximately -0.5D.

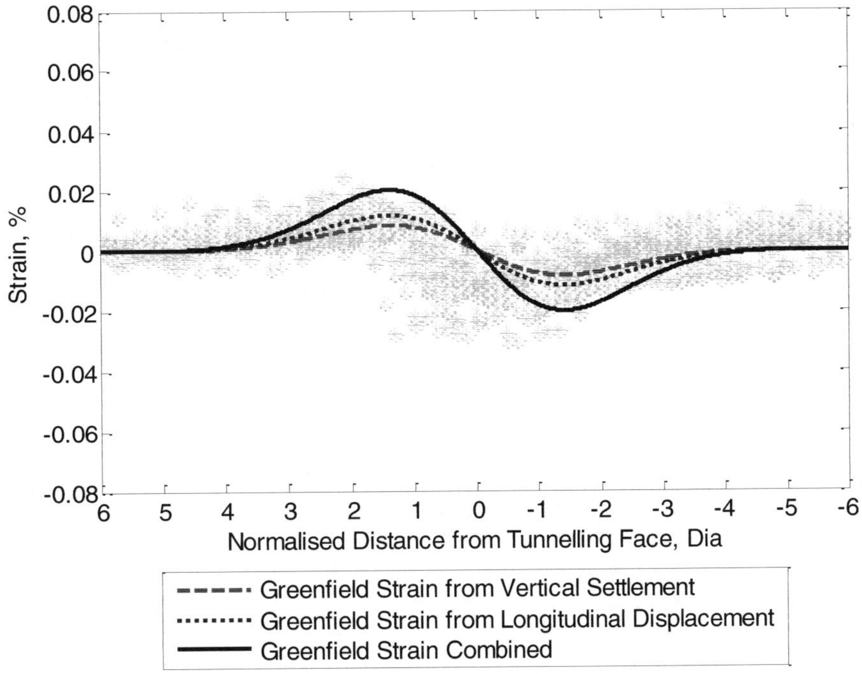

Figure 6 Plot of longitudinal DFOS strain with normalised distance in RMT for pilot tunnel

It is noteworthy, that for the compressibility of the wooden packings to have an affect here, it must be the case that a gap exists between the tunnel rings such that compressibility may occur, hence, under construction it is likely that the tunnel rings were not fully pressed together at the circumferential joints, instead, held apart by a combination of wooden packings and the gapping that must occur as the tunnel lining negotiates curvature (as the circumferential joints are plane and parallel and not tapered, as is the case for most contempoprary concrete tunnel linings).

Similarly, computed curvature values (1/ROC) from CSattar (shown in Figure 7) has good agreement with the observed response from DFOS. Each coloured line represents a fixed location across a group of 5 rings (i.e. 5 ring chord) on the right axis (RiAx). Positive values on the vertical axis denotes sagging curvatures, while ring order number 35 to 44 corresponds to R2903 to R2912 respectively. As the tunnelling face approached the zone of influence on the 10th of September 2013, negative curavtures were observed, signifying hogging behaviour while it rapidly changes to positive curvature or sagging mode when the tunnelling face is directly underneath the point of monitoring. The maximum hogging and sagging ROC recorded were approximately -3km and 3.5km for the phase of the pilot tunnel. Residual curvatures further supports the imprefect recovery observed from DFOS.

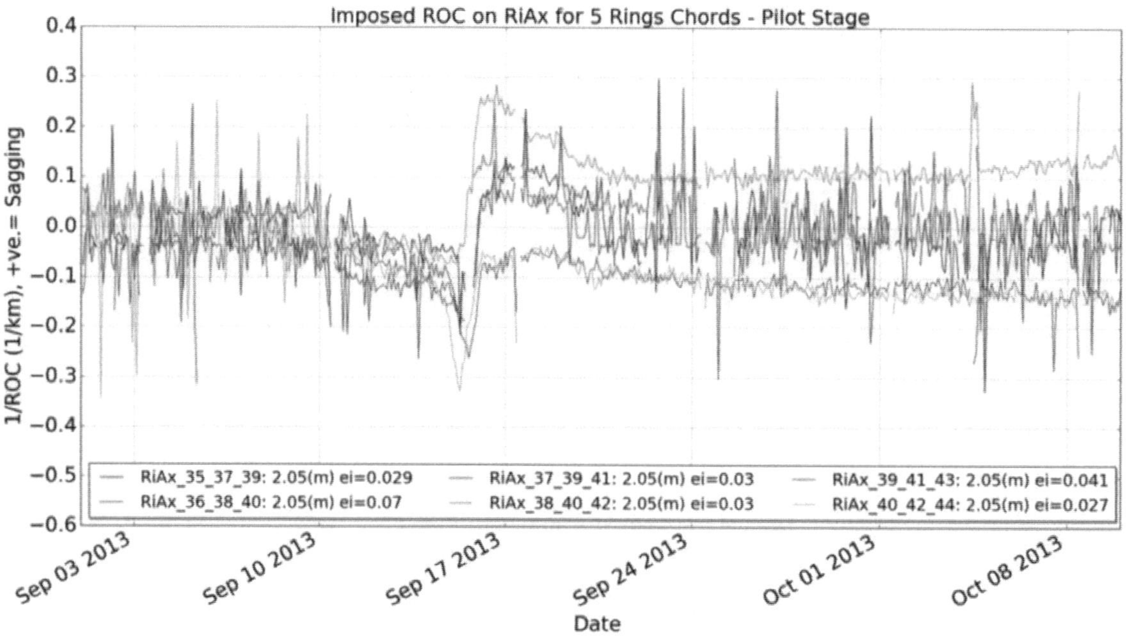

Figure 7 RMT curvature plot from CSattar for pilot tunnel

Longitudinal strains for the full tunnel enlargement (see Figure 8) was approximately doubled from the pilot tunnel with maximum tensile and compressive strain at 0.04% and -0.06% respectively. In contrast with the observations made in the pilot tunnel, a response lag from RMT in reference to the position of tunnelling face was measured. Despite the inflection point agreeing with the theoretical curves at 0, the position of the maximum tensile strain is changed such that it measured at approximately 0.2D. This is due to the presence of the pilot tunnel itself acting as a large dowel ahead of the tunnelling enlargement face which stiffens the response of the surrounding soil. Therefore it requires a much larger soil movement (i.e. closer distance from the point of measurement to the tunnelling face) to induce comparable response from RMT. Magnitude of the tensile strains were also smaller than the full greenfield curve by approximately 30%.

Response lag was not observed once the tunnelling face passes through the point of monitoring since the enlarged tunnel has been constructed (i.e. pilot tunnel no longer exists) and the strain behaviour then has good agreement with the full greenfield strain curve. A remainder of compressive strains was recorded despite recovery from the transitional settlement induced by the new tunnel construction. This, once again, could be caused by the presence of compressive caulking between consecutive rings; hence, a perfect recovery from settlement was not possible.

The overall behaviour thus shows that the tunnel enlargement construction induces a steeper strain gradient (i.e. larger differential settlements) for points ahead of the tunnelling face while it follows greenfield curves past the inflection point.

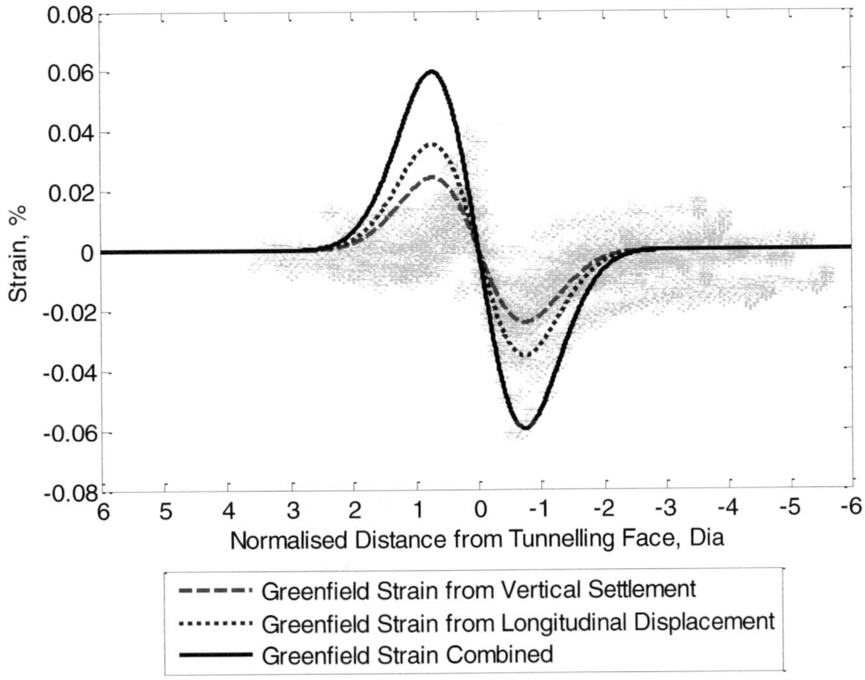

Figure 8 Plot of longitudinal DFOS strain with normalised distance in RMT for tunnel enlargement

Good consistent agreement between the trends of CSattar and DFOS was achieved similarly for the tunnel enlargement (refer to Figure 9). Magnitudes of ROC for the hogging mode were approximately 15%-25% in comparison to the pilot tunnel while the sagging mode observes less change, this is slightly lower from those measured by DFOS. Maximum hogging curvature were consistently measured at RiAx 40_42_44 (R2908-R2912) with an ROC of about -2.3km; a value that exceeds Crossrail standards of -2.5km for unused tunnels. The remaining plots exhibited a ROC within the range of -4km. Maximum sagging ROC of 3.2km was recorded at RiAx 38_40_42 (R2906 to R2910). No damage were observed despite the measured ROC exceeding the prescribed limits which could indicate that partial shearing between rings took place to reduce extreme fibre and bolt strains.

It is important to note that the presented ROC data is based on a 5 ring chord distance while DFOS effectively measures strain between 2 rings. This is the main reason for the slight disparity between DFOS and equivalent strain calculated from measured ROC. Since the existing tunnel is discontinuous, it is not surprising that some rings may settle more than others given the irregularity of bolt numbers (i.e. missing bolts) and unequal bolt forces. This further reinforces the need to have large number of closely spaced monitoring points to reduce excessive averaging which masks the true response.

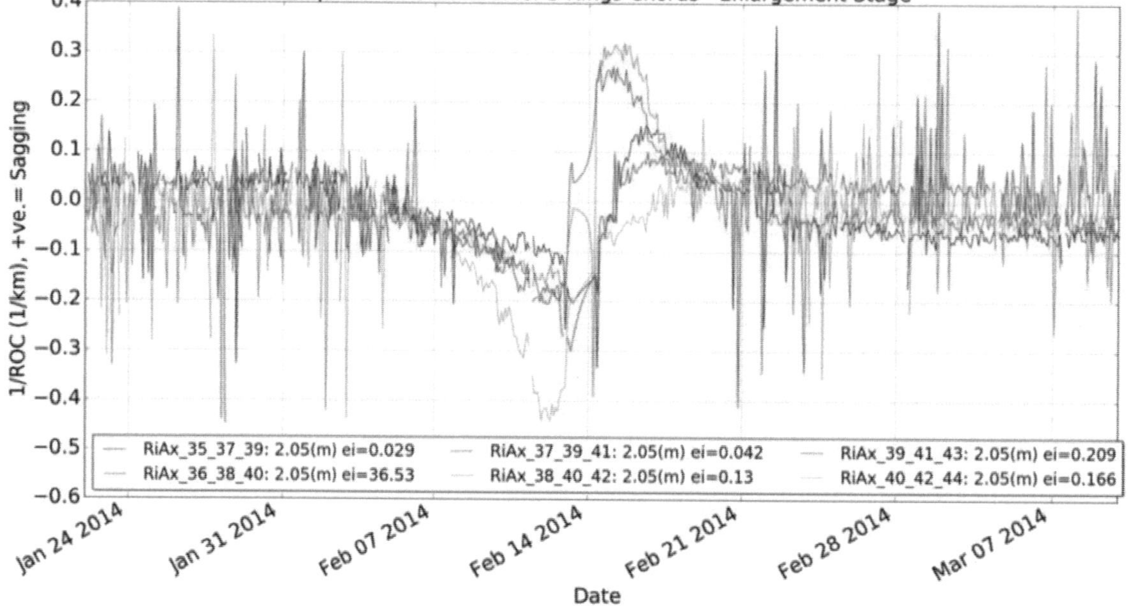

Figure 9 RMT curvature plot from CSattar for tunnel enlargement

For the LPDT measurements, due to the adopted attachment method, and distortions of the circumferential flanges as the tunnel lining was strained, measured LPDT data had to be decoupled in order to isolate longitudinal (bending) and vertical (shearing) displacements. This was done using a 2D plane-strain finite element model to re-produce the likely joint and connection interaction. The shearing displacements are shown in Figure 10 for the tunnel enlargement. The horizontal axis here shows the distance from tunnelling face to the position of shear sensing LPDT, with negitive values denoting the tunnelling approach, and positive values as the tunnelling has progressed past the sensor position. Negative displacement indicates vertical settlement of the leading tunnel ring with respect to the trailing. Data for the pilot tunnel is not shown as the shearing displacements were very small, resulting in poor signal to noise ratio. Further details can be found in Wilcock (2017). Wilcock (2017) concluded that the behaviour of the circumferential flanges is complex and in the case of the RMT exhibit a different interaction mechanism in tensile and compressive modes. Also, the presence of the circumferential bolting is seen as influential in the joint behaviour as the tunnel lining experiences tensile straining; however, distortion of the flanges limit the stresses transferred to the bolts and increase flexibility of the tunnel lining; whilst the intra-ring shearing further allows the tunnel ring to undergo vertical displacement without detriment. The disparity between ROC derived tunnel lining strain and those values recorded by the DFOS system are explained by the presence of intra-ring shearing, here verified by the LPDT sensors.

It is noteworthy in the context of shearing of the tunnel rings, the reporting of tunnel response in terms of ROC is a fallacy; as the flexure (or 'curvature' component) is effectively alliviated by the measure of shearing (horizontal ground movement notwithstanding).

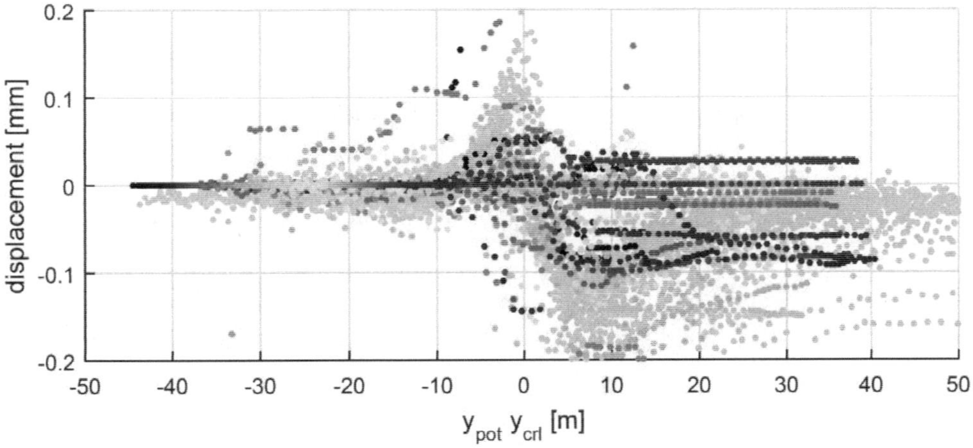

6. Coı

The detailed longitudinal response of an existing cast iron lined tunnel when subjected to a parallel tunnelling directly beneath has been presented in this paper and the following conclusions can be drawn.

- Discontinuous nature of the existing cast iron tunnel lining, including the presence of gaps between the tunnel rings, allows it to respond flexibly in the longitudinal direction that is described by greenfield conditions.
- Behaviour of each individual ring can vary at a local scale due to non-uniformities in the construction (i.e. number of bolts, bolt forces and presence of low stiffness caulking between rings).
- A combination of shearing and bending mode is observed and its effects are transitional. Presence of the pilot tunnel reduces the amount of soil movement ahead of the tunnelling face, which affects the response of the existing tunnel and at the same time reduces the magnitude of strains.
- A tight ROC of less than 2.3km was recorded with no observable damage.

References

Alhaddad, M, 2017, Photogrammetric monitoring of cast-iron tunnels and applicability of empirical methods for damage assessment, PhD. Thesis, University of Cambridge.

Alhaddad, M, Wilcock, M, Gue, CY, Bevan, H, Stent, S, Elshafie, MZEB, Soga, K, Devriendt, M, Wright, P, & Waterfall, P, 2014, 'Multi-suite monitoring of an existing cast iron tunnel subjected to tunnelling-induced ground movements', Proceedings of GeoShanghai International Conference, Shanghai.

Attewell, PB, Yeates, J & Selby, AR, 1986, Soil movements induced by tunnelling and their effects on pipelines and structures. Chapman & Hall, New York.

Gue, CY, Wilcock, M, Alhaddad, MM, Elshafie, MZEB, Soga, K & Mair, RJ ,2015, 'The monitoring of an existing cast iron tunnel with distributed fibre optic sensing (DFOS)', Journal of Civil Structural Health Monitoring, 5, 573-586.

Lake, LM, Rankin, WJ & Hawley, J, 1992, 'Prediction and effects of ground movements caused by tunnelling in soft ground beneath urban areas', CIRIA Project Report 30, Construction Industry Research and Information Association, London.

Mair, RJ & Taylor, RN, 1997, 'Theme lecture: Bored tunnelling in the urban environment.' Proceedings of the 14th International Conference on Soil Mechanics and Foundation Engineering, 2353-2385, A.A. Balkema, Rotterdam.

Mohamad, H, 2008, Distributed optical fibre strain sensing of geotechnical structures, PhD. Thesis, University of Cambridge.

O'Reilly, MP & New, BM, 1982, 'Settlements above tunnels in the united kingdom- their magnitude and prediction' Tunnelling 82, 55-64, The Institution of Mining and Metallurgy, London.

Standing, JR, & Selman, R, 2001, 'The response to tunnelling of existing tunnels at Waterloo and Westminster', Building response to tunnelling: Case studies from construction of the Jubilee Line Extension, London. 2: 509-546, Thomas Telford, London.

Wilcock, MJ, 2016, The behaviour of existing segmental cast iron tunnel linings subject to ground movement from new tunnelling, PhD. Thesis, University of Cambridge

The 8th International Conference on
Structural Health Monitoring of Intelligent Infrastructure
Brisbane, Australia | 5-8 December 2017

Smart Steel Strand Fabricated by the Coaxial Cable Fabry-Perot Interferometer Sensor

Tong Jiao[1], Zhi Zhou[1], Jia Liu[1], Hai Xiao[2]
[1] *Dalian University of Technology – China, Email:* jiaotong@mail.dlut.edu.cn
[2] *Clemson University - USA*

Abstract

To realize full life cycle monitoring of steel strand, coaxial cable Fabry-Perot interferometer (CCFPI) sensors were manufactured by clamping method using SF047 type coaxial cable and were packaged in the glass fiber reinforced plastic (GFRP) to make GFRP-CCFPI smart bar. The mechanical and sensing properties of GFRP-CCFPI smart bars were studied experimentally. Moreover, the core wire of the steel strand was replaced by the smart bar to make CCFPI smart steel strand. The feasibility of using CCFPI smart steel strand to realize structure self-sensing was discussed in experiments.

1. Introduction

In recent years, pre-stressing techniques are widely applied in the construction of critical civil infrastructures such as long span bridges, nuclear stations and high-rise buildings. The stress of strands is of great importance to ensure safety of the pre-stressed structures. It is, thus, very desirable to monitor the stress of strands through the whole life-cycle of structures (Ahlborn 1997). Accordingly, many approaches have been developed to implement the measurement, such as acoustic emission technology, shape memory alloy sensors, acoustoelasticity technology magneto-elastic sensors, vibration-based method, vibrating-wire strain gages, etc. All of these methods with different strengths show good performance in monitoring the stress of strands, however, they struggled to be applied in practical engineering since the installation difficulties and vulnerability in long-term harsh environment. With the development of optical fiber sensors, Zhou et al. integrated fiber Bragg grating (FBG) or Brillouin optical time domain analysis (BOTDA) sensors into fiber-reinforced polymer (FRP) rebar to improve their ruggedness (Zhou 2003). Afterwards, they further developed a FRP-FBG/FRP-BOTDA smart steel strand by replacing the core wire of strand with smart FRP-FBG/FRP-BOTDA bar. The smart steel strand performed good monolithic behavior, self-sensing properties and installation convenience in the process of long-term stress of steel strand monitoring (Zhou 2009 and Lan 2012). Besides, Kim et al. developed a strand by inserting an optical fiber sensor inside a tube installed in the core wire of the strand, they also adopted carbon fiber as the material for the core wire for the sensing strand (Kim 2012 & 2015). The smart steel strands with embedded optical fiber have shown many advantages in practical applications, however, it is difficult to realize whole life cycle monitoring of steel strands due to low measurement range of the optical fiber.
Based on electromagnetic and wave theory, a coaxial cable works similarly as an optical fiber by transmitting an electromagnetic (EM) signal over a long distance (Matick 1968 and Keiser 2003). A typical coaxial cable consists of an inner and outer conductor sandwiched by a tubular insulating layer with a high dielectric constant. Governed by the same EM theory, a coaxial cable and an optical fiber share the common fundamental physics. Inspired by the optical fiber Fabry-Perot interferometer (FPI) sensors, Hai Xiao developed a new coaxial cable Fabry-Perot interferometer (CCFPI) sensor

(Xiao 2014), they proposed the fabrication method and the principle of the CCFPI as well as investigated the sensing mechanism of strain.

From the perspective of practical application, the feasibility of using CCFPI sensor for monitoring the pre-stressed structures was investigated through a development of a CCFPI smart steel strand, obtained by combining the CCFPI sensing unit with the steel strand. Furthermore, the mechanical and the sensing characteristics were analyzed.

2. Sensing mechanism of CCFPI

The schematic of the proposed CCFPI sensor is illustrated in Figure 1. The sensor consists of a pair of reflection points on a flexible coaxial cable. The transmitted electromagnetic wave is partially reflected from the two reflectors and the reflected signals superimposed to form an interference pattern (i.e. an interferogram) that can be observed in the spectrum domain.

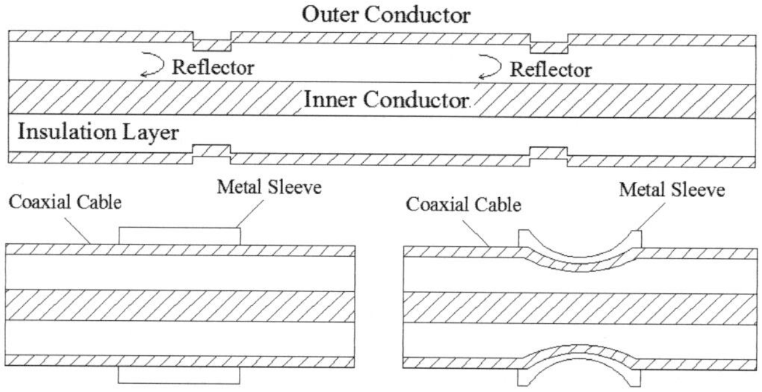

Figure 1 Schematic diagram of CCFPI sensor: (a) CCFPI sensor (b) enlarged reflector fabricated by clamping method

When a strain is applied on the cable, the cavity length and the dielectric constant of the insulating material will change. As a result, a shift in the interferogram can be recorded and used to calculate the applied strain. Each reflection along the cable should be weak enough so that the multiple reflections between the reflectors are negligible. Assuming the amplitude reflection coefficients of the two reflectors are the same, the two reflected waves (U1 and U2) can be written as follows:

$$U_1 = \Gamma(f)e^{-\alpha z}\cos(2\pi ft) \tag{1}$$

$$U_2 = \Gamma(f)e^{-\alpha z}\cos[2\pi f(t-\tau)] \tag{2}$$

$$\tau = \frac{2d\sqrt{\varepsilon_r}}{c} \tag{3}$$

where $\Gamma(f)$ is the amplitude reflection coefficient of the reflector; f is frequency of the EM wave traveling inside the cable; α is the propagation loss coefficient; z denotes the cable axial direction; τ is time delay between the two reflected waves; d is the distance between two reflectors; ε_r is the relative permittivity of the inner dielectric material of the cable; c is the speed of light in vacuum.

The two reflected waves have a time delay and the delay is associated with the distance between the two reflectors and the phase velocity of the wave. The interference signal (U) is the summation of the two reflected waves, which can be written as follows:

$$U = 2\Gamma(f)e^{-\alpha z}\cos(2\pi f\tau)\cos[2\pi f(t+\tau)] \tag{4}$$

According to equation (4), the environmental parameters change (e.g., strain or temperature) will cause a change in time delay. Subsequently, the interference pattern will change. By tracking the shift in the interferogram, the changes of parameters can be found. In real operation, it is easy to follow the shift in the resonant frequency. The N*th* resonant frequency can be deduced from equation (4), which is as follows:

$$f_N = \frac{N}{\tau} = \frac{Nc}{2d\sqrt{\varepsilon_r}} \tag{5}$$

In general, the stretch of the cable will cause an elongation to the cable and a decreasing to the dielectric constant due to the photoelasticity effect. These are the dominant factors that will further influence the interferogram. As a result, the applied strain (ε) can be expressed in terms of changes in distance and relative permittivity of the material ($\Delta\varepsilon_r$) using the following equation (6):

$$\varepsilon = \frac{\Delta d}{d} \; and \; \frac{\Delta\varepsilon_r}{\varepsilon_r} = -P_{eff}\varepsilon \tag{6}$$

where P_{eff} represents the effective Pockels coefficient of the inner dielectric material of the coaxial cable. The N*th* resonant frequency shift (Δf_N) can be deduced from equation (5) and (6) as follows:

$$\Delta f_N = \frac{\partial f_N}{\partial d}\Delta d + \frac{\partial f_N}{\partial \varepsilon_r}\Delta\varepsilon_r \tag{7}$$

The applied strain in terms of N*th* resonant frequency change can be further derived from equation (8), as follows:

$$\varepsilon = \frac{2\Delta f_N}{f_N(P_{eff} - 2)} \tag{8}$$

Equation (8) describes that the applied strain can be calculated by simply measuring the N*th* resonant frequency shift while assuming that the other parameters in the equation are initially known.

3. GFRP-CCFPI smart bar

3.1 Fabrication of the GFRP-CCFPI smart bar

SF047 type coaxial cable with a diameter of 0.89mm was selected to manufacture CCFPI sensor shown in Figure 2, where reflectors were in the red circles. Then the CCFPI was pulled through a heating furnace together with GFRP and epoxy resin to form the GFRP-CCFPI smart bar with a diameter of 5.2mm, as shown in Figure 3.

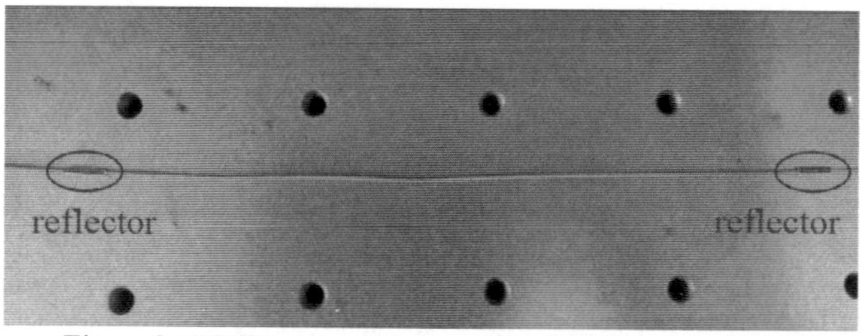

Figure 2. CCFPI sensor fabricated by SF047 coaxial cable

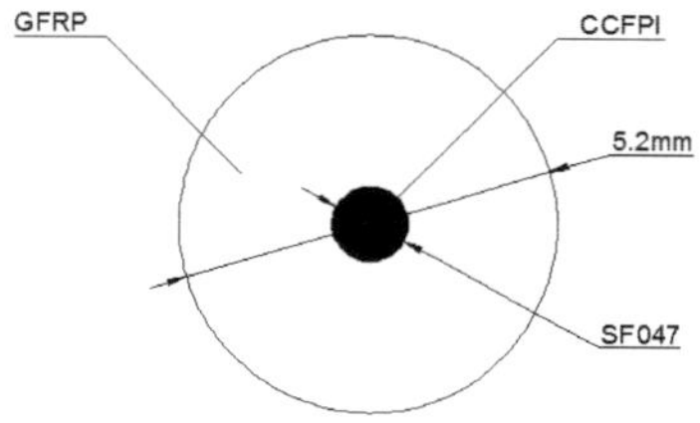

Figure 3. GFRP-CCFPI smart bar

3.2 Mechanical Properties Test

Extrusion anchorages with original length of 80mm were fixed at both ends of the common GFRP bar and the GFRP-CCFPI smart bar with length of 500mm to form tensile specimens. These specimens were stretched by the test device shown in Figure 4 at the rate of 1mm/min uniformly. Stress-strain curve (shown in Figure 5) dropped suddenly when reaching the ultimate strength, which showed brittle failure characteristics obviously. In addition, the common GFRP bar in Figure 6 (a) showed dispersion failure, but the GFRP-CCFPI bar in Figure 6 (b) showed comprehensive characteristics of dispersion failure and splitting failure due to the embedded coaxial cable. Mechanical properties of the common GFRP bar and the GFRP-CCFPI smart bar were listed in Table 1. It is showed that tensile strength of the GFRP-CCFPI smart bar is 81% of that of the common GFRP bar, while elastic modulus of the GFRP-CCFPI smart bar is higher than that of compared the common GFRP bar. That is because the amount of glass fiber in the GFRP-CCFPI smart bar and the common GFRP bar is the same, which makes volume ratio of glass fiber in the GFRP-CCFPI smart bar higher than that of the common GFRP bar.

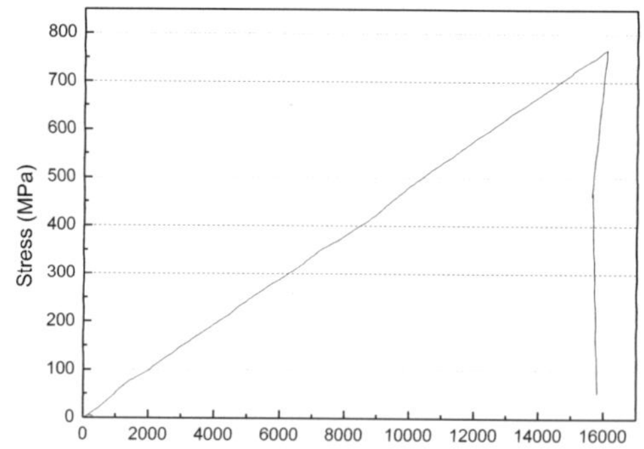

Figure 4. Tension test device of GFRP-CCFPI smart bar

Figure 5. Strain- Stress curve of GFRP-CCFPI smart bar

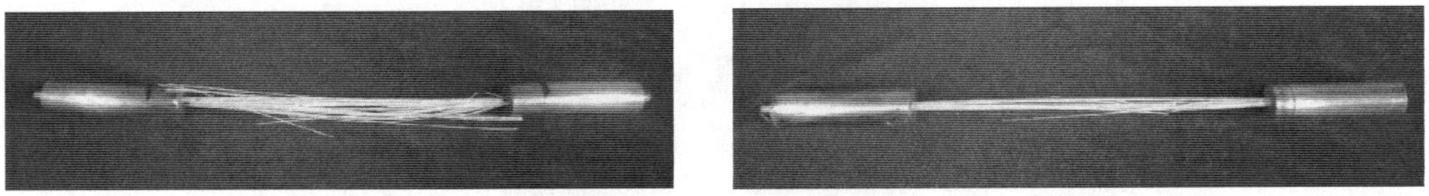

(a) (b)

Figure 6. Failure pattern of tension specimens: (a) Common GFRP bar (b) GFRP-CCFPI smart bar

Table 1. Mechanical properties of the common GFRP bar and the GFRP-CCFPI smart bar

Specimen	Tensile strength (MPa)	Elastic modulus (GPa)	Elongation (%)
Common GFRP	964	46.2	2.04
GFRP-CCFPI smart bar	783	47.8	1.61

3.3 Strain Performances Test

Steel tubes filled with adhesive mixed by epoxy resin, curing agent and iron powder of the same mass and nuts were used as to stretch the GFRP-CCFPI smart bar, as shown in Figure 7. Two dial indicators were used to measure the displacement variation between two nuts. Five strain gauges with a spacing of 200mm were pasted on the smart bar to monitor strain. The smart bar was stretched at a step of 1kN until a significant slip between the smart bar and the anchorages resulted in a failure of load growth. Frequencies of interference spectrum dips measured by vector network analyzer were recorded.

Strain gauges at different position had consistent readings in the whole tensile process, which showed the smart bar was in the uniform tension. The average of five strain gauges was used as the reference value to reduce measurement error. Test results are given in Figure 7 and Figure 8. Strain of GFRP-CCFPI-1 smart bar measured by CCFPI, dial indicators and strain gauges are in good agreement before the tension load reaches 8kN. Strain curve measured by dial indicators turns at 9kN, and then the slope increases, which shows that slip between the smart bar and the bond anchor occurs at 9kN. Strain of GFRP-CCFPI-2 smart bar measured by CCFPI, dial indicators and strain gauges are in good agreement before the tension load reaches 7kN. Strain curve measured by dial indicators turns at 7kN, and then the slope increases, which shows that slip between the smart bar and the bond anchor occurs at 7kN. The test results show that the GFRP-CCFPI smart bar can monitor its strain state with confidence by itself.

Figure 6. Strain sensing performance testing device of GFRP-CCFPI smart bar

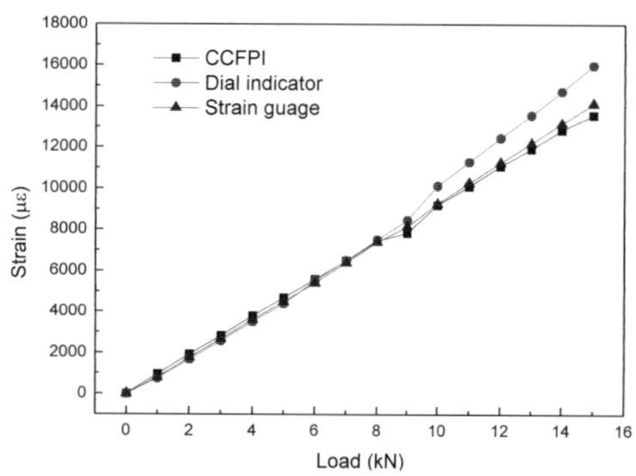

Figure 7. Sensing test results of GFRP-CCFPI-1 smart bar

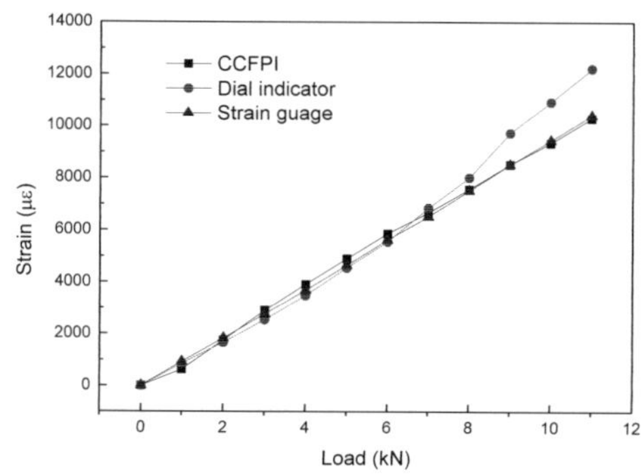

Figure 8. Sensing test results of GFRP-CCFPI-2 smart bar

4. CCFPI Smart Steel Strand

4.1 Configuration of a CCFPI Smart Steel Strand

A smart steel strand defined in this study is a steel member that has sensing capability for structural condition assessment. It consists of a GFRP-CCFPI smart bar of 5mm in diameter and six common steel wires and copper foils around the smart sensor.

The steps to produce the CCFPI smart steel strand are as follows: firstly, copper foils were fixed in a helical form on the smart bar. Then, six steel wires are separated and the sequences of the wires are remembered. Finally, the smart bar with copper foils is winded by the six common steel wires in original sequence. The production process is shown in Figure 9.

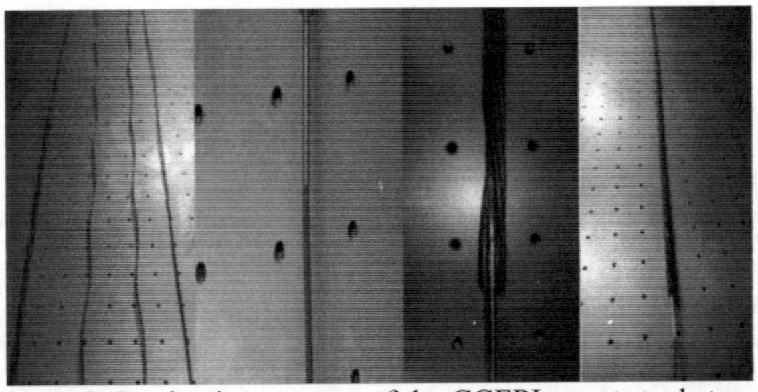
Figure 9. Production process of the CCFPI smart steel strand

4.2 Sensing Principle

If the change of relative dielectric constant of coaxial cable insulation layer is neglected, the relationship between strain and frequency shift can be derived from equation (9), as follows：

$$\varepsilon = \frac{\Delta f_N}{f_N^{'}} \tag{9}$$

where Δf_N is resonant frequency shift of CCFPI; $f_N^{'}$ is the resonant frequency after change. The measured strain of the cable is equal to that of the whole strand due to the cable deforms together with the remaining six steel wires. According to Hook's law, the stress of the strands can be calculated from the strain measured by smart steel strand, as follows:

$$\sigma = E\varepsilon = E\frac{\Delta f_N}{f_N^{'}} \tag{10}$$

where σ is stress of the smart steel strand; E is the elastic module of the smart steel strand; ε is the strain of the smart steel strand.

4.3 Validation Test and Analysis

4.3.1. Mechanical Properties

A common steel strand and a CCFPI smart steel strand were stretched at a rate of 2kN/s uniformly as shown in Figure 10. Length of the two tension specimens is 1m. As shown in Figure 11, the CCFPI smart steel strand (the under one) has the same failure pattern with the common steel strand (the upper one). When reaching the ultimate load, the outer wires are separated from the core wire and all the wires break.

Figure 10. Tension test device of steel strand

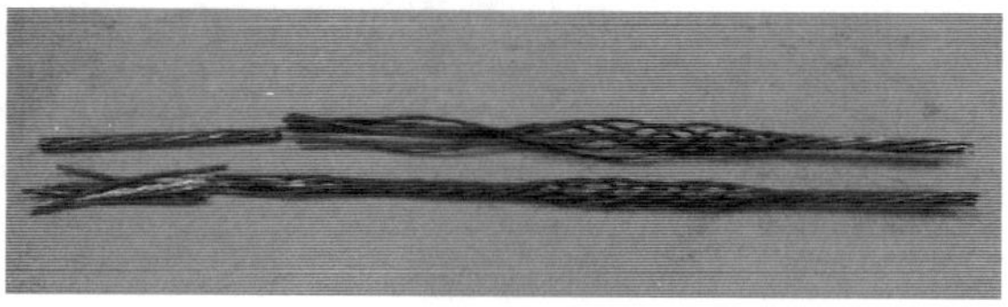

Figure 11. Failure pattern of steel strand

Load- displacement curves of the common steel strand and the CCFPI smart steel strand are shown in Figure 12 and Figure 13. The CCFPI smart steel strand shows the characteristics of ductile failure, but the period from yield to break is short, and there is a significant dip in the curve. That is because six outer wires are separated from the intermediate GFRP-CCFPI smart bar easily due to the difference in elastic modulus when approaching to the ultimate load. Mechanical properties of the common steel strand and the CCFPI smart steel strand are listed in Table 2. Tensile strength and elasticity modulus of the CCFPI smart steel strand are close to those of the common steel strand. So it is possible to take advantage of the CCFPI smart steel strand to replace the common steel strand in part to realize self sensing monitoring of structure.

Table 2. Mechanical properties of the common steel strand and the CCFPI smart steel strand

Specimen	Maximum force (kN)	Yield force (kN)	Tensile strength (MPa)	Elasticity modulus (GPa)
Common steel strand	275	246	1960	188
CCFPI smart steel strand	240	213	1700	168

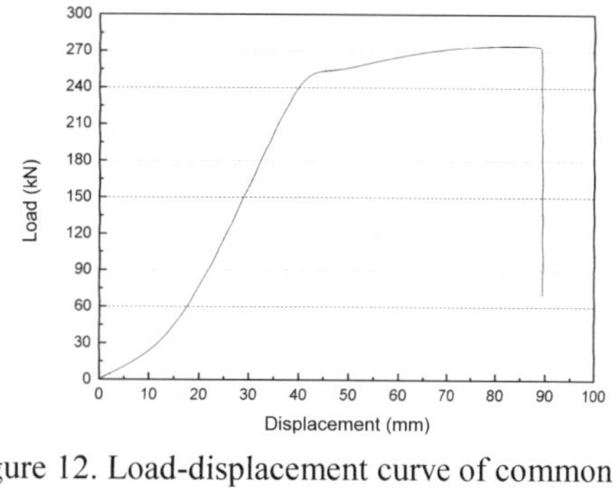

Figure 12. Load-displacement curve of common steel strand

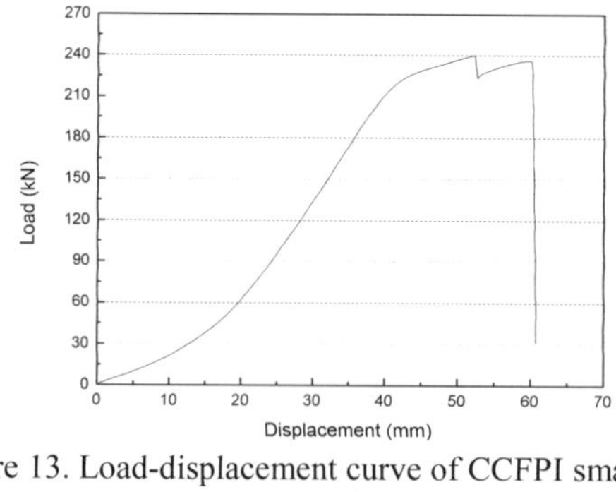

Figure 13. Load-displacement curve of CCFPI smart steel strand

4.3.2. Sensing Properties

The CCFPI smart steel strand with length of 1.3m passed through reaction frame, force transducer and jack and was fixed on both sides of the tensile device shown in Figure 14 with wedge-type anchorages. Three strain gauges with a spacing of 250mm were pasted on the same outer wire of the CCFPI smart steel strand to monitor strain. The smart steel strand was stretched at a step of 10kN until 200kN. Frequencies of interference spectrum dips measured by vector network analyzer were recorded.

Strain gauges at different position had consistent readings in the whole tensile process, which showed the smart steel strand was in the uniform tension. The average of three strain gauges was used as the reference value to reduce measurement error. According to the geometric relationship, the axial strain of the steel strand was derived from the tangential strain of the outer wire. Measurement results of the strain gauge and the CCFPI are compared in Figure 15-17, which show that strain measured by the CCFPI is in good agreement with strain measured by the strain gauge and CCFPI strain curve is of good linearity. Tension monitoring results of three CCFPI smart steel strands are shown in Figure 18, where the horizontal ordinate denotes load measured by force transducer, the longitudinal coordinate denotes tension calculated with strain measured by CCFPI. It is obvious that, the self sensing results of the CCFPI smart steel strand can reflect stress state of itself accurately.

Figure 14. Tension device of CCFPI smart steel strand

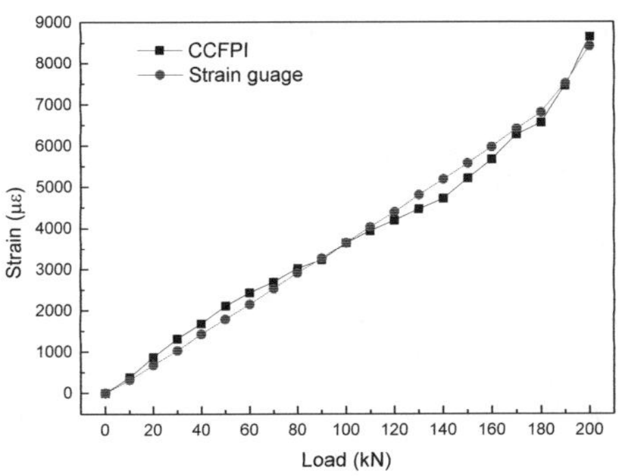

Figure 15. Sensing test results of CCFPI smart steel strand-1

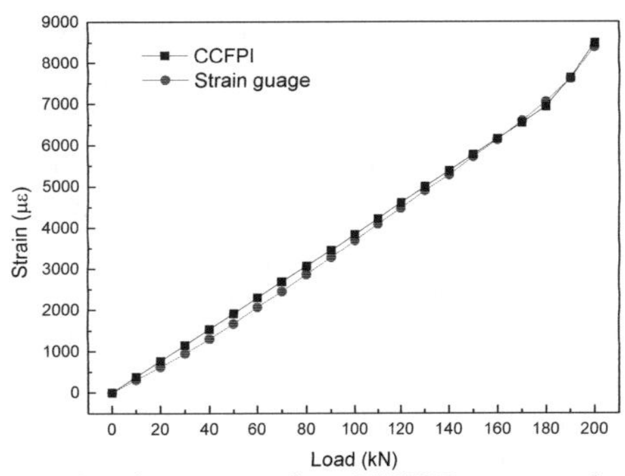

Figure 16. Sensing test results of CCFPI smart steel strand-2

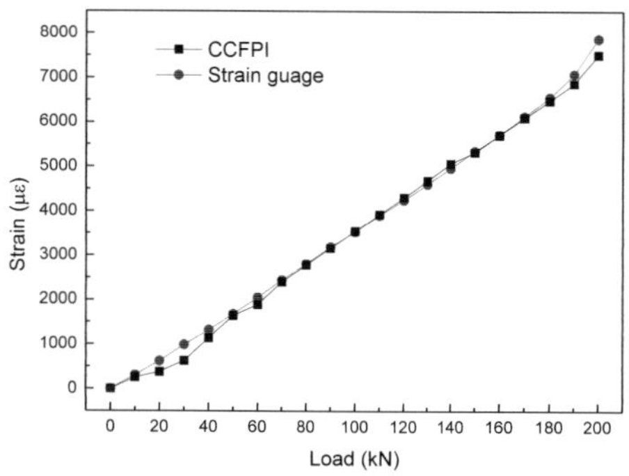

Figure 17. Sensing test results of CCFPI smart steel strand-3

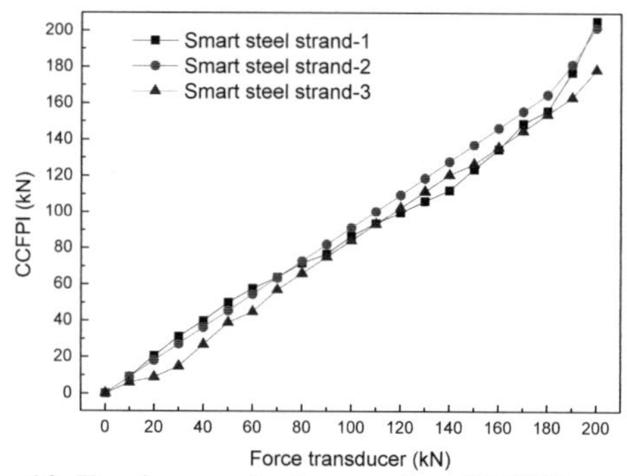

Figure 18. Tension monitoring results of CCFPI smart steel strands

5. Conclusions

CCFPI sensors made by SF047 type coaxial cables were packaged in GFRP to make GFRP-CCFPI smart bars. Tensile strength of the GFRP-CCFPI smart bar is 81% of that of the common GFRP bar, while elastic modulus of the GFRP-CCFPI smart bar is higher than that of the common GFRP bar. The smart bar has stable sensing performance, good linearity and consistent results with strain gauges and dial indicator in strain measurement. Moreover, the core wire of the steel strand was replaced by the smart bar to make CCFPI smart steel strand. The CCFPI smart steel strand has the characteristics of ductile failure. Tensile strength of the CCFPI smart steel strand is 87% of that of the common steel strand, while elastic modulus of the CCFPI smart steel strand is 89% of that of the common steel strand. So it is feasible to use CCFPI smart steel strands to realize structure self-sensing.

References

Ahlborn, TM, Shield, CK, Fench, CW 1997, Full-scale testing of prestressed concrete bridge girders. Exp. Tech. vol. 21, pp. 33-35.

Zhou, Z, Ou, JP 2003, Smart FRP-OFGB Bars and Their Application in Reinforced Concrete Beams. *In Proceedings of the First International Conference on Structural Health Monitoring and Intelligent Structure*, Tokyo, Japan, 13–15 November, pp. 861–866.

Zhou, Z, He, JP, Chen, GD, Ou, JP 2009, A smart steel strand for the evaluation of prestress loss distribution in post-tensioned concrete structures. *J. Intell. Mater. Syst. Struct*. vol. 20, pp. 1901–1912.

Lan, CG, Zhou, Z, Ou, JP 2012, A novel GFRP-FBG smart steel strand and its properties research. *Applied Mechanics and Materials*. vol.105, pp. 1087-1091.

Lan, CG, Zhou, Z, Ou, JP 2012, Full-scale prestress loss monitoring of damaged RC structures using distributed optical fiber sensing technology. *Sensors*, vol. 12, pp. 5380-5394.

Kim, JM, Kim, HW, Park, YH, Yang, IH, Kim, YS 2012, FBG sensors encapsulated into 7-wiresteel strand for tension monitoring of a prestressing tendon. *Adv. Struct. Eng*. vol. 15, pp. 907-917.

Kim, ST, Park, YH, Park, SY, Cho, K, Cho, JR 2015, A sensor-type PC strand with an embedded FBG sensor for monitoring prestress forces. *Sensors*. vol. 15, pp. 1060-1070.

Matick, RE 1968, Transmission line pulse transformers: Theory and applications. *Proceedings of the IEEE*. vol. 56, pp. 47-62.

Keiser G 2003, Optical fiber communications, Wiley Encyclopedia of Telecommunications.

Xiao H, Huang J, Lan X, et al. 2014, Distributed microwave Fabry-Perot interferometer device and method: U.S. Patent Application 14/341,944[P].

The 8th International Conference on
Structural Health Monitoring of Intelligent Infrastructure
Brisbane, Australia | 5-8 December 2017

Applying Phased-based Vision Sensing Method to Modal Feature Identification of Highway Bridge

Z. Y. Wu[1], D. Mo [2], S. Pan [2], R. Christenson[3], and S. Motaref[4]

[1] *Bentley Fellow, Bentley Systems, Incorporated, Watertown, CT, USA, Email:* zheng.wu@bentley.com, [2]*Former Research Intern, Bentley Systems, Incorporated, current Ph.D. Candidate, University of Massachusetts, Amherst, MA, USA.* [3]*Professor, Department of Civil and Environmental Engineering, University of Connecticut, Storrs, CT, USA.* [4]*Assistant Professor in Residence, Department of Civil and Environmental Engineering, University of Connecticut, Storrs, CT, USA.*

Abstract

The current practice for structural health monitoring (SHM) is to place a sensor onto and/or within structure elements to capture the response. Such an approach is effective, but suffers from several drawbacks including (1) a limited number of the available for sensors that can be placed for SHM, (2) difficult, unsafe, or even impossible installation of sensor at some structural components, (3) service interruption e.g. traffic closure, and (4) relatively costly for placing the sensors compared with the vision-based sensors, such as using video cameras as elaborated in this paper. A new approach of using a video camera as the sensor has been developed to remotely capture the motions of the structure. The methods have been formulated and the software tool has been implemented to extract the displacements, velocities and accelerations at the desired points, which can be as many points as needed. The corresponding natural frequencies and modal shapes of the structure are extracted by using the methods and software tools authors developed. The extracted features can be used for assessing structural integrity, safety, serviceability, and operation performance. The approach has been applied to a highway bridge in Connecticut USA, and the results are compared with the conventional sensing methods.

1. Introduction

To monitor the vibration of civil engineering structures, a network of high sensitive sensors, such as accelerometers or strain gages, are commonly used. The sensory data collected can be further processed for obtain the dynamic features such as the natural frequencies and modal shapes (Peeters et al. 2009; Magalhaes et al. 2008 and many others). Although with powerful processing module, data from this type of sensors can yield a greatly accurate result of the structural features, the deployment of such monitoring networks has some limitations. For example, installing the electrical wires to power these sensors, depending on the dimensions of structure, may lead high cost and labor, traffic interruption and difficult installation in many cases. On the other hand, a new monitoring method using high-speed camera and computer vision techniques (Wahbeh et al. 2003; Lee et al. 2006) starts to draw authors' attention, considering its good quality on cost and flexibility. The new monitoring system uses camera to remotely collect images of the target structures. Since dynamic features are hidden in more subtle changes in these images, powerful computer vision algorithms are used to fetch basic physical information first, such as accelerations, velocities or displacements of the interested points. Due to the limitation of optical devices and error introduced in the computer vision algorithm, the data is not as accurate as results sampled by high sensitivity sensors. However, the ease of deployment and low installation cost still encourage researchers to improve their algorithms.

Previous work has implemented many well-known algorithms including digital image correlation (Yoneyama 2007; Iadicola et al. 2012; Kohut et al. 2013; Hoag et al. 2015 and Dworakowski et al. 2016), pattern matching (Busca et al. 2014), and orientation code matching (Fukuda et al 2013; Feng et al 2015; Feng & Feng 2015). However, most conventional methods are heavily dependent on the brightness of the targets and hence the light and the weather may affect the performances. In this paper, an effective approach has been formulated by using the phase-based method (Chen et al. 2016) together with response denoising techniques for structure health monitoring. An integrated framework and software tool have been developed and applied to a highway bridge, where a permanent monitoring system using accelerometers and strain gauges has been installed and used as comparison baseline to validate the vision sensing approach.

2. Methodology

Responses such as the velocities at predetermined points are extracted from a video in three steps. The first step is to convert each frame of the videos to gray scale frame if it is color. Then the complex filter is applied to obtain the local phases of each frame by convolution operation. Finally, by tracing the change of the local phases at the interesting points, the velocities are calculated. When the overall lighting condition does not change in the whole video, the intensity of a small target or an interested point will be constant, even though the point keeps moving. For the $1D$ case, if x represents the position of the interested point, and $I(x,t)$ denotes the intensity of this point at time t, then the intensity of the point forms a contour of constant intensity with respect to time, that is:

$$I(x,t) = c \tag{1}$$

The derivative of the intensity with respect to time is given as

$$\frac{\partial I}{\partial x}\frac{\partial x}{\partial t} + \frac{\partial I}{\partial t} = 0 \tag{2}$$

Therefore, the velocity of the point movement in x direction can be obtained as

$$u_x = \frac{\partial x}{\partial t} = -\left(\frac{\partial I}{\partial x}\right)^{-1}\frac{\partial I}{\partial t} \tag{3}$$

Similarly, the velocity in y direction can be obtained. The extracted response may not be the actual response of a structure. It may contain the motions induced by the camera vibration and other noises caused by external factors. The methods have been proposed for cancelling video camera vibration from the responses and further denoising the responses.

2.1 Video vibration cancellation

Due to many ambient factors, such as wind, traffic etc., video camera is likely vibrating when it is used to record the video of a target structure. The camera vibration will induce the same motion of the recorded video. To decouple the camera motion from the recorded video, the method is proposed for subtracting the camera vibration from the extracted motion responses. The ideal way is to install a sensor on the camera to capture the camera vibration, but not every camera is equipped with such a sensor. We propose to use a reference point in the same video image to represent or approximate the camera motion. The point must be close to video camera and away from the monitoring structure. The motions of such a point are used to represent the motion or vibration of the video camera. To decompose the camera motion from that of the monitoring point on a structure is to subtract the motion of the reference point from the motion of the monitoring point. Let *r* be the reference point and u_{rx} be the velocity in *x* direction for the reference point *r*, the decomposed velocity in *x* direction for monitoring point *p* is:

$$u_{px}^{d} = u_{px} - u_{rx} \tag{4}$$

Similarly, the decomposed velocity in *y* direction for monitoring point *p* is

$$u_{py}^{d} = u_{py} - u_{ry} \tag{5}$$

Decomposed displacement and acceleration for monitoring point *p* can also be obtained by subtracting the corresponding response at the reference point.

2.2 Responses denoising

There are many well-developed methods for denoising videos. In general, the denoising methods are classified in two categories in spatial domain and transform domain. For the simplicity of the application, an average method is proposed for denoising the decomposed and/or extracted responses. For any monitoring point (or pixel in image) *p*, the velocity in each direction is calculated as the average of n_x and/or n_y pixels around pixel *p*, that is:

$$\overline{u_p^d(x,y)} = \left[\sum_{i=1}^{n_x} u_p^d(x,y) + \sum_{j=1}^{n_y} u_p^d(x,y)\right] \Big/ (n_x + n_y) \tag{6}$$

where $u_p^d(x,y)$ is the extracted velocities in x and y directions while n_x and/or n_y are the user-specified number of pixels for the video processing. The average displacements and acceleration can be derived from the averaged velocities.

3 Software Tool Implementation

An integrated software tool VS Extractor (Mo et al. 2015), employing the phase-based method and response denoising and camera motion cancellation, has been developed for processing the videos for SHM. It can extract velocity information of the selected locations in the video images and graphically show the result to the users. In order to provide the best processing performance, the algorithm is implemented in a dynamic linked library in C/C++; the user interface is implemented as a window form application. The workflow of the software tool is shown in Figure 1.

As shown in Figure 1, users start the process by selecting a list of the interested points on the image, and decide the size of the observing area (or averaging area) around the point. This is because the location selected in the image of a video represents a single pixel, it means that the output designates the velocity or displacement for the selected pixel. Considering the uncertainty and inaccuracy of existence of noises in video, it is good practice to calculate the average velocity within the neighborhood of the selected pixel instead of one pixel itself. This method has been implemented by extracting the horizontal and vertical velocities of each pixel in the neighborhood area and then calculating the average velocities as the result for the selected pixel.

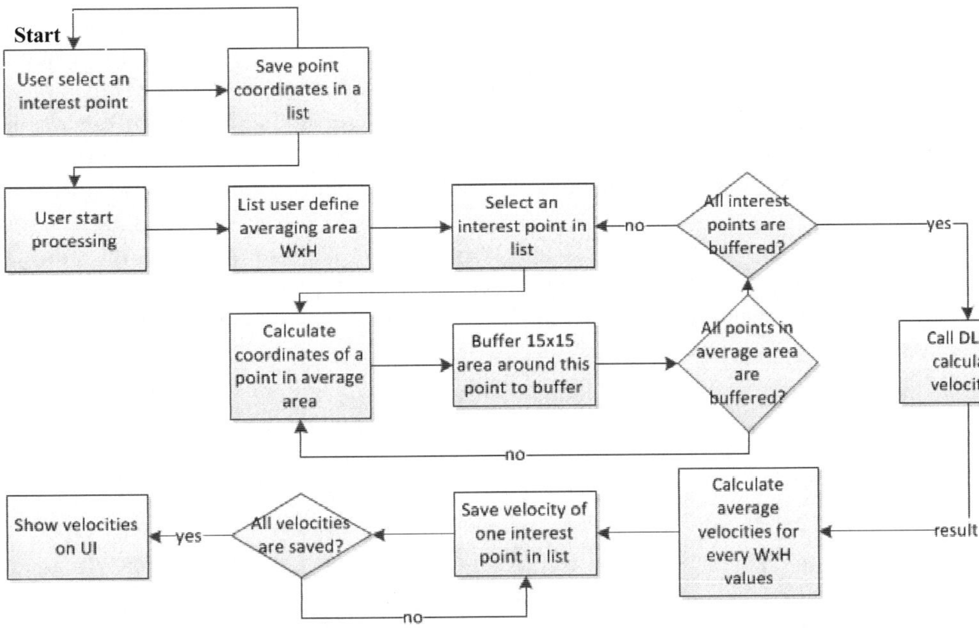

Figure 1 Flowchart of VS Extractor

Another rule of thumb for correctly applying phase-based method is to select the points where good texture contrast presents in the video, such as the edges of the structure components. The results can only be reasonable in the direction perpendicular to the texture e.g. the edges, as the algorithm is based on optical intensity. Slight light intensity changes can affect the performance of the result. Figure 2 shows the interface of the Feature Extractor. The interface can be split into two parts including the left part that allows users to select the video source, control the video playing and choose the interested points, and the right part that displays the features according to the settings and save the features into files.

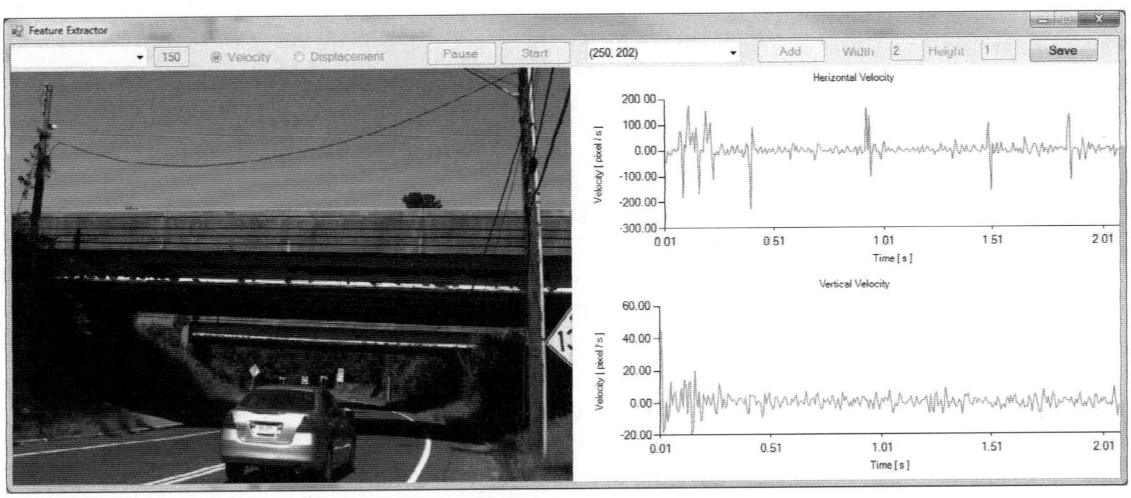

Figure 2 Integrated video analysis tool for applying vision sensor (video camera) to structural health monitoring

After choosing the video source, the first frame will be shown, so that the interested points can be selected easily. The interested points can be selected either by clicking at the positions on the video image directly or by inputting the exact coordinates indirectly. The chosen points will be displayed as red points in the video images. Before the processing, the users are required to input the width and height of averaging area in pixels. It will be default as 1x1 area, which means there will be no averaging applied. After the user select whether velocity or displacement to output, they can click start button to start the process.

During the processing, both horizontal and vertical features of the point (within the predetermined interested points) are shown automatically. After the processing, the feature data will be saved to the files in the format that can be used directly for the DarwinSHM Extractor tool (Wu et al., 2015) for the further analysis.

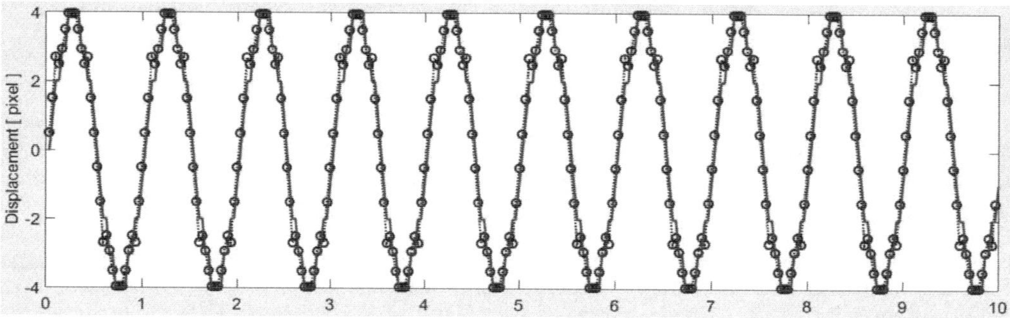

Figure 3 Comparison of Extracted (blue dotted) and True (red solid) Displacement of a Narrow Beam

Figure 3 shows the extracted horizontal displacement from a manufactured video without averaging to validate the implemented algorithm. The manufactured video contains a narrow beam placed along the vertical direction. The narrow beam moves left and right periodically with amplitude of 4 pixels and period of 1 second. It is obvious that the extracted displacement shown in dashed circled line match almost perfectly with the real displacement shown in red solid line. This indicates that the tool extracts the feature correctly.

4 Applications

The developed method has been applied to the real structure, the in-service highway bridge, Bridge No. 03051, in Meriden CT, as shown in Figure 4. The bridge was built in 1964, located on I-91 Northbound in Meriden, Connecticut USA. It is a simple-support single-span eight-girder steel composite bridge of 85 feet long and 55 feet wide. It has less than a 12% skew and 3% longitudinal slope. A wired long-term bridge monitoring system with a total number of 38 sensors has been installed on the Meriden Bridge for research purposes for both structural health monitoring (SHM) and bridge weigh-in-motion (BWIM). The sensors include 18 foil strain gages, 4 piezoelectric strain sensors, 8 piezoelectric accelerometers, 4 capacitance accelerometers (with additional temperature sensing capability), and 4 resistance temperature detectors. Details of the bridge and the monitoring system can be found in Christopher et al (2009, 2011 and 2012).

As illustrated in Figure 4, the test was conducted by setting the video camera on side of the cross road. A Pointgray USB 2 camera was used for capturing the videos of the bridge. It has the maximum resolution of 1280*1024 pixels and the maximum frame rate of 120. The camera was mounted on a tripod, which was set up on the side of the cross road, located at about 150 feet away from the west side of the bridge, and connected with a laptop computer with USB cable. The test

was undertaken in about 2 hours without any impact on traffic. A total of 20 videos have been captured with the frame rate of 60, and average durations of about 5 minutes.
In order to effectively reduce the impact of the video noise, the reference point is selected as shown in Figure 4. In general, the reference point should be selected with good contrast feature that is close to the video camera so that it can represent the motion of the camera itself.

Figure 4 Photo of Meriden bridge with video camera as sensor

Vision Sensor (VS) Extractor is applied to extract velocity and displacement of any selected points from the first frame of the video images. Figure 5 shows an example of the displacements extracted from the video for the selected point. The extracted displacements in both horizontal and vertical directions are plotted in unit of pixels, which can be easily converted in distance unit with the measured physical distance between two points on the image. As shown in Figure 5, the displacements are very small, less than 1 pixel for the one-span highway bridge. Since the displacement is not monitored by the preinstalled monitoring system, it was not possible to validate the displacement measure by the vision sensor. However, the installed accelerometers provided a good baseline for validating the vision sensor.

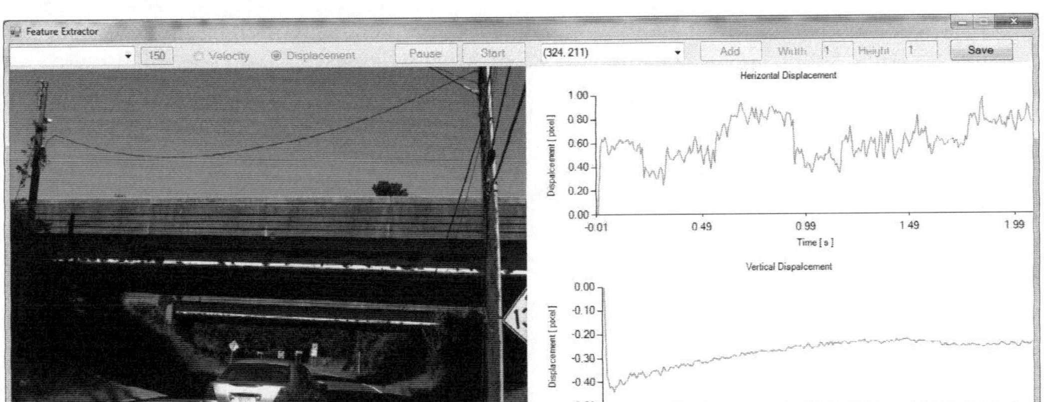

Figure 5 Displacement measured by vision sensor for Meriden bridge

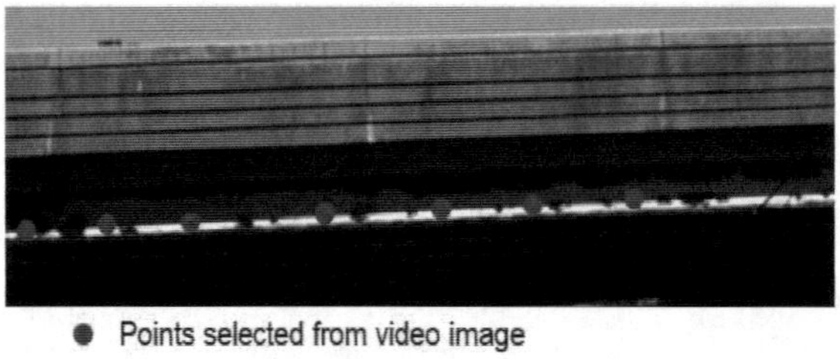

Figure 6 Points selected for extracting velocities from the video of Meriden bridge

To validate the VS Extractor, multiple points, as shown in Figure 6, were selected on the image to extract the velocities as time series data. The velocity of a selected reference point, as shown in Figure 4, was extracted to represent the video camera vibration, which was cancelled from the velocities of the points selected on the bridge by subtracting the reference velocity. Using time domain and frequency domain analysis tool DarwinSHM Extractor, both video-based velocities and accelerometer-based accelerations were used to quickly produce the frequency domain plots. the frequency at the pronounced peaks represent the observed mode of the bridge. Figure 7 compares the frequencies identified by both vision sensor approach and conventional accelerometers. It can be noticed that all the modes identified by the accelerometers are identified by the vision sensor. In addition to those modes, the vision sensor approach also identified two more modes with the frequency 2.3 and 11.3 Hz, respectively. It is believed that these vibration modes do exist but are potentially missed by the limited number of accelerometers installed on the bridge. Although there are 8 accelerometers installed on the bridge, they are allocated in three rows on a plan with three accelerometers aligned up in longitudinal direction, thus on the plan of longitudinal and vertical, the vibrations can only be effectively monitored at three points and in the vertical direction. With the captured video, the velocity was extracted at seven points (as many points as desired can be selected in theory), many more points than those monitored by the accelerometers. Therefore, it is believed that the vision sensor approach using the VS Extractor is effective at capturing not only the accurate dynamic responses as the accelerometers, but also more responses than the limited number of accelerometers used in practical bridge monitoring applications.

The results obtained by VS Extractor for the Meriden bridge case study indicate that the vision sensing via video camera is truly a promising approach for innovating structure health or performance monitoring.

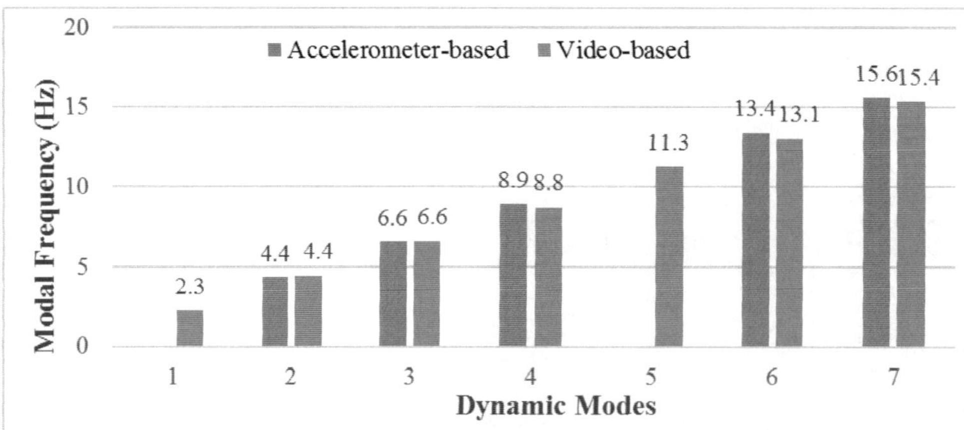

Figure 7 Comparison of frequencies extracted by using accelerometer data and video

5 Conclusions

In this paper, a phased-based method has developed with camera motion cancellation and image denoising. An integrated software tool has been implemented to extract velocity of an interested point from video frames. The developed tool provides user-friendly interface to extract velocity and displacement from selected video source. The approach has been applied to an in-service highway bridge with a conventional bridge monitoring system installed. The vision sensor results are compared with those obtained by conventional sensors on the bridge. The comparison shows that the proposed method and tool can produce the similar results to the accelerometers and may even capture more features than the limited number of accelerometers allows for. This proves that the vision sensor provides a promising alternative for structural health monitoring in practice. Future research will need to focus on the applications of the proposed approach to the large-scale engineering structures.

Acknowledgements

The authors wish to acknowledge the Connecticut Department of Transportation for providing permission to test the phased-based vision sensing method and for providing the bridge monitoring data from the bridge on the Meriden Bridge (Bridge No. 03051), in Meriden, Connecticut USA. The contents of this paper reflect the views of the authors, who are responsible for the facts and accuracy of the data presented herein. The contents do not reflect the official views or policies of the State of Connecticut or the Federal Highway Administration.

References

Busca G., Cigada A., Mazzoleni P., and Zappa E., (2014). "Bridge points by means of a unique vision-based measuring system," Experimental Mechanics, vol. 54, no. 2, pp. 255 – 271.

Chen J. G., Davis A., Wadhwa N., Durand F., Freeman W. T., and Büyüköztürk O., (2016). "Video Camera–Based Vibration Measurement for Civil Infrastructure Applications", *Journal of Infrastructure Systems*, pp. B4016013, DOI: 10.1061/(ASCE)IS.1943-555X.0000348.

Christenson R.E., S.K. Bakulski and A.M. McDonnell (2011), "Establishment of a Dual-Purpose Bridge Health Monitoring and Weigh-In-Motion System for a Steel Girder Bridge", Proceedings of the Transportation Research Board Annual Meeting, Washington DC.

Christenson R. E, S Motaref, AMH McDonnell (2012), "A Dual Purpose Bridge Health Monitoring and Weigh-In-Motion System for a Steel Girder Bridge", 6th International Conference on Weigh-In-Motion (ICWIM 6), pp 397-408.

Christopher, J.W, R.E. Christenson, Anne-Marie H. McDonnell, Alireza Jamalipour (2009).A Non-Intrusive Bridge Weigh-in-Motion System for a Single Span Steel Girder Bridge Using Only Strain Measurements, (pdf 843 kb), Research Report, ConnDOT Report No. CT-2251-3-09-5

Feng M. Q., Fukuda Y., Feng D., and Mizuta M. (2015). "Nontarget vision sensor for remote measurement of bridge dynamic response," J. Bridge Engineering, vol. 20, No. 12.

Fleet D. J. and Jepson A. D., (1990). "Computation of component image velocity from local phase information," Int. J. Comput. Vision, vol. 5, no. 1, pp. 77 – 104.

Fukuda Y., Feng M.Q., Narita Y., Kaneko S. I., and Tanaka T. (2013). "Vision-based displacement sensor for monitoring dynamic response using robust object search algorithm", *IEEE Sensors Journal*, vol. 13, no. 12, pp. 4725-4732, DOI: 10.1109/JSEN.2013.2273309.

Gautama T. and Hulle M. M. V. (2002). "A phase-based approach to the estimation of the optical flow field using spatial filtering," IEEE Trans. Neural Networks, vol. 13, no. 5, pp. 1127 – 1136.

Guizar-Sicairos M., Thurman S. T., and Fienup J. R. (2008). "Efficient subpixel image registration algorithms", *Optics letters*, vol. 33, no. 2, pp. 156-158, DOI: 10.1364/OL.33.000156.

Iadicola M. A., Zobeland R. S., and Ocel J. M. (2012). "Quantitative evaluation of digital image correlation as applied to large scale gusset plate experiments," in Proc., 6th Int. Conf. on Bridge Maintenance, Safety and Management, International Association for Bridge Maintenance And Safety (IABMAS), Stresa, Lake Maggiore, Italy, pp. 1436 – 1443.

Kohut P., Holak K., Uhl T., Ortyl Ł., Owerko T., Kuras P., and Kocierz R. (2013). "Monitoring of a civil structure's state based on noncontact measurements", *Structural Health Monitoring*, vol. 12, no. 5-6, pp. 411-429, DOI: 10.1177/1475921713487397.

Lee J. J. and Shinozuka M., (2006). "A vision-based system for remote sensing of bridge displacement," NDT & E International, vol. 39, no. 5, pp. 425 – 431.

Magalhaes, F., Cunha, A., and Caetano, E., (2008). "Dynamic Monitoring of a Long Span Arch Bridge," Engineering Structures30:3034–3044.

Mo D., Pan S. and Wu Z. Y. (2015). "Development of Phased-based Method and Software Tool for Extracting Dynamic Features from Videos." *Technical Report,* Bentley Systems, Incorporated, Watertown, CT, USA.

Peeters, B., Couvreur, G., Razinkov, O., K¨ undig, C., Van der Auweraer, H., and De Roeck, G., (2009). "Continuous Monitoring of the Øresund Bridge: System and Data Analysis," Structure and Infrastructure Engineering: Maintenance, Management, Life-Cycle Design and Performance 5:395–405.

Wahbeh M., Caffrey J. P., and Masri S. F. (2003). "A vision-based approach for the direct measurement of dispalcements in viberating system," Smart Materials and Structures, vol. 12, no. 5, pp. 785 – 794.

Wu Z. Y., Zhao J., Li J., Cashany M., Zhang J., Christenson R., and Jiang S. (2015). "Improving dynamic feature extraction with frequency domain decomposition for structural health modeling," in *The 3rd Conference on Smart Monitoring, Assessment and Rehabilitation of Structures (SMAR2015)*, Antalya, Turkey, Sep. 2015.

Wu Z. Y, Zhao J., Narayan N., Li J.C., Zhang J. Y. and Cashany M. (2015). "Integrated software tool for dynamic feature extraction." The 7th international conference on SHMII, July 1- 4, 2015, Turin, Italy.

Yoneyama S., Kitagawa A., Iwata S., Tani K., and Kikuta H. (2007). "Bridge deflection measurement using digital image correlation", *Exp Techniques*, vol. 31, no. 1, pp. 34-40, DOI: 10.1111/j.1747-1567.2006.00132

The 8th International Conference on
Structural Health Monitoring of Intelligent Infrastructure
Brisbane, Australia | 5-8 December 2017

Structural Assessment of a Childcare Buildings Rooftop Play Area for Landscaping Works

Max Kudrenko[1], Alan Jeary[2].
[1] *Technical Sales Manager, Mainmark.* [2]*Chief Technology Officer, STRAAM Group.*

Abstract

When an inner Sydney childcare centre wanted to redesign the rooftop area to provide a landscaped play area, they discovered minimal records existed for the extension that created the roof top space. Without drawings or design documentation for the precast plank roof; how was the engineering firm able to design the increased loading to create the mounds and play spaces? The engineering consultants required a system that would determine the current capacity of the roof top. They relied on the highly sensitive STRAAM accelerometers and their extensive experience in analysing the data output to calibrate a reduced node finite element model (RNFEM) of the rooftop structure to make informed design decisions with confidence.

Keywords: Building Assessment, Load Bearing Assessment, Structural Integrity, Monitoring, Testing, Dynamics, Non-linear Damping, FEM Calibration

STRAAM

Mainmark have partnered with STRAAM from New York to introduce the STRAAM structural risk and management system for structures into Australia and New Zealand. The STRAAM system allows for rapid, non-invasive assessment of a structure's current status.

1. Introduction

Figure 1: Roof top view of Child Care Centre

A design engineer had been engaged by the owners of a two storey, trapezoidal shaped, child care centre to plan for a new open-space rooftop garden and children's play area as part of renovation works. The existing roof top was an extension added to the established building 25 years prior. Other than the knowledge that the rooftop extension was constructed with pre-tensioned, hollow core slabs, no other construction records were available. Based on this very limited information, the design engineer was unsure of how to safely arrange the landscaping such that the loading of additional play equipment, soils for garden and a children's sandpit was distributed carefully so that the rooftops safe loading would not be exceeded. To determine this, Mainmark, in conjunction with our partner STRAAM, was then engaged to provide a structural assessment of the building, specifically with a focus on the load bearing capacity of the roof top.

Mainmark was able to record the required data on site over two days. This was completed in a non-disruptive and non-disruptive manner. Shortly after, this data was then analysed, modelled and then reported on. This provided the design engineer the information required to make and informed decision on his design and ultimately resulted in a significant change to the original landscaping design as it was found that the loading of the proposed sandpit above the longest lengths of the buildings rooftop spans would not be within safe limits as identified within AS:1170. Subsequently, the landscaping area was re-designed to be in a safer area in terms of load distrubution. The building was also found to be performing well under likely wind loading events.

2. The Philosophy of Assessment Using Dynamic Measurements With STRAAM

With the developments of electronics and sensor technology, the ability to record very precise measurement properties of monolithic structures has led to a rationale for the management of such facilities with a methodology that enables the augmentation and enhancement of traditional visual inspections (Jeary et al).

Newton's second law of motion is used for the basis of the equations of structures in dynamic motion. This is also applied to a structures individual modes of vibration. The equation to express this sort of motion, under each mode 'r', can be presented as:

$$m_r = \frac{\hat{F}_r}{\hat{x}_r \, 8 \, \zeta_r \, \pi^2 \, f_{r^2}} \quad (1)$$

Where, the displacement is x_r from mode 'r' for and applied for of F_r. The dampening ratio is f_r and the frequency of resonance is F_r.

For application however we are generally most interested in the dampening value and the displacement per unit force value; the reason being is that a dampening value is used as a parameter to give the measurement of energy dissipation on a structure, where the displacement per unit force allows for a comparison with the requirements of design construction. As the modal mass, m_r, does not change in the elastic regime for any single mode, measurements of non-linear dampening characteristics and frequency of resonance provide the necessary information to compare the capacity ratio to the demand. Every structure type has a characteristic vibration modes, which requires that the deflected mode shape that is representative of mode 'r' also needs to be assessed as all of the above parameters apply to individual modes of vibration. For this reason, this method can be applied to many types structures whether they contain stone, steel, timber or concrete.

The methodology requires a force to be applied to the structure. Because a series of structures were tested using induced vibration testing (with a large vibrator system), and these results were correlated against the response of structures from naturally occurring forces, then the non-intrusive methodology has been calibrated in the most precise way possible. Examples of naturally occurring forces include wind, traffic and micro-seismic activity. Because of the precision of the instrumentation used it is no longer necessary to wait for an event such as an earthquake for the state of health of the structure to be assessed. In this way regular or continuous monitoring can also be used to show the gradual ageing process for the structure. Such detailed information allows the management of facilities and the planning of necessary maintenance that is informed by measurement. This type of measurement is capable of discerning information about problem areas of a structure that are not apparent to a visual observation, and therefore offers a significant advantage over a more conventional type of inspection. This type of measurement can also be performed quite quickly structure and is economically efficient. In practice, such as with the child care centre in Sydney, the force can be applied from natural sources such as air movement, traffic or other human agency.

The following describes the response of as single degree of freedom (mode 'r').

$$\boldsymbol{X_r} = \frac{F_r}{8f_r^2\zeta_r M_r \pi^2} \tag{2}$$

Where for mode 'r'
M is the participating mass
f is the frequency of resonance
F is the force
ς is the damping ratio of mode r
Thus the displacement per unit force can be defined as:

$$\frac{X_r}{F_r} = \frac{1}{8\zeta_r f_r^2 M_r \pi^2} \tag{3}$$

- The displacement per unit force is inversely proportional to the damping times the stiffness.
- If the modal mass is unchanging, and the mode shape is unchanging then for a given frequency and damping for a mode 'r' the displacement per unit force will remain unchanged up to the elastic limit.
- In practice there may be small changes in the frequency of resonance (as a function of amplitude – not as a function of time). There will be changes to damping with amplitude. However, we can compensate for such effects by measuring the non-linear damping.

There are several system-level parameters which aid in the conversion from dynamic response (acceleration records) to Dynamic Signature, and identification of damage in a structure during a STRAAM structural assessment:

1. Gross migration of frequency of resonance away from the expected response. In particular, the loss of stiffness in each mode of vibration can be estimated by comparing the differences in measured frequencies of resonance from the expected values. Lower frequencies indicate a loss of stiffness in a particular mode of vibration.
2. Spectral bifurcations whereby a principal mode of vibration shows signs of splitting into several modes of vibration. This indicates that the structure is forming mechanisms and that local resonances are forming.
3. Changes in modal amplitude ratios can be tracked over time to show indications of changes to particular modes of vibration which can indicate that parts of the structure are responding differently from previously. Inferences of damage or strengthening can be made utilizing these changes along with other observations.

4. Energy dissipation as measured through the nonlinear damping parameter. Damping values which are higher than expected are clear indications that energy is being dissipated through cracking, rubbing of structural members or unexpected soil-structure interaction (foundation damage). This measure generally requires a long duration of measurement such as that obtained with continuous monitoring in order to establish the full amplitude dependent damping behaviour.

3. Digital Signal Processing (Baseline Spectral Response Analysis)

The on-site measurements of acceleration are processed using standard and proprietary digital signal processing (DSP) techniques to facilitate the extraction of relevant dynamic properties of the structural system. The main output of the DSP process is the decomposition of the time-series acceleration data into its frequency content (Power Spectral Density). The spectrum is interpreted to identify the modes of vibration (frequencies of resonance, amplitudes and modal ratios) of the structure. This forms the baseline Dynamic Signature and is essentially a fingerprint of how the structure is currently behaving with resonances appearing as isolated peaks in the plots. This baseline Dynamic Signature provides the datum with which future dynamic responses can be compared for assessment of changes to the structure's performance.

4. The Methodology

STRAAM uses a systematic approach, called the STRAAM Protocol, to understand how a structure behaves and for developing the best approach of monitoring the structure for an assessment and then long term monitoring.

The STRAAM Protocol involves several steps necessary to establish the baseline Dynamic Signature of the structure. With regards to the child care center these included:

1. Step 1 (Site Survey, Photos, Drawings, Reports) is to gather field information so that the monitoring plan will achieve the project goals efficiently and the transfer of design drawings and photos. This was accomplished through discussion with the design engineer and the buildings owners.

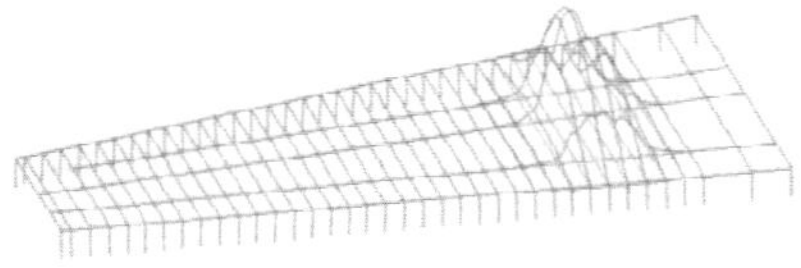

Figure 2: FEM of Child Care Centre rooftop, static deflection with extra 2 kPa at broader end of building

2. Step 2 (Structural Model and Frequency Range Identification) we will use the existing FEM (or as with the case of the child care center, constructed our own) to understand the expected response, mode shapes and frequency ranges for the structure. Additionally, it can help locate areas of the structure which will provide the greatest response for our instruments. This is all integrated into an advanced monitoring plan that is designed to best

achieve the project goals. We performed the FEM on this building to determine the frequency range of each mode of vibration. In this example, to calibrate the FEM a know weight was dropped at a known height with a recorded timestamp to calibrate the model while collecting the data.

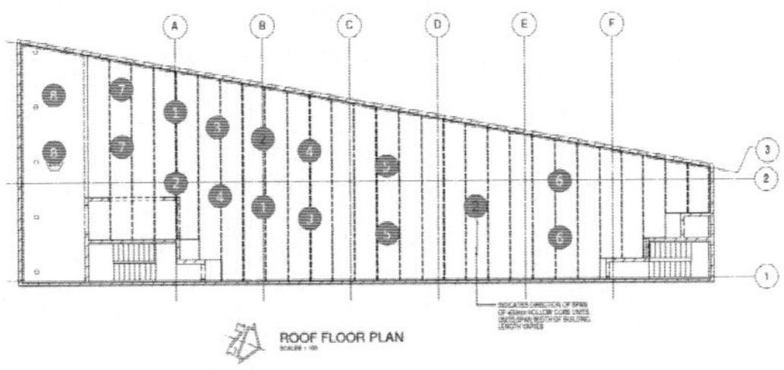

Figure 3: Schematic of Child Care Centre rooftop, accelerometer location (mid span)

3. Step 3 (Data collection/ Signal Conditioning) we proceed to the field to capture response information from the structure in an efficient and effective manner. We performed field measurements in December of 2017. Measurements were taken over a 6 hour period during working hours over two days whilst the child care center remained open and in use. The instruments were placed at strategic points on the roof top (Down each floor of the stair well, incrementally around the edge of the roof, and at multiple lengths across the spans of the hollow core slabs. For this example a total of 33 strategic locations were recorded at. When monitoring, we recorded acceleration (along the X, Y and Z axis) and temperature at a rate of 200 samples per second for the acceleration, and 1 sample per second for the temperature. The sensitivity of the instruments allows STRAAM to capture low amplitude response information to be used for analysis.

We then proceeded with steps 4-7 of the STRAAM Protocol where we process the data into the response spectra and compare the dynamic resonant frequencies in various modes and various conditions. This iterative process completely correlates the measured response with the expected response from the model.

5. Discussion

As a result of the works, a wealth of knowledge of the child care centres current structural state could be reported. Certain results matched intuition, such as:

- The local mode of vibration at the northern portion of the structure matched the frequency equivalent to what would be the expected mode of vibration for the adjacent property, which is approximately twice the height of the child care centre. This suggests that there were some soft connections between the two buildings allowing this mode from the adjacent building to transfer energy. Upon visual inspection these connections were indeed noted, confirming the response.
- Distortions of the roof slab were observed when the modes of whole-body side-sway motion was analysed.
- Energy losses were noted at areas identified with visual cracking on the structure.

Other characteristic were also found, which without this assessment would have been difficult to identify. Examples of these findings included:

- From the requirements of AS3600, the roof slab in its current state had excess capacity. However, because of the unusual geometry, the relationship between the maximum deflection and the applied distributed load was not linear. The capacity limits based off calibrated dampening ratio could then be used with the FEM to determine load bearing capacity.

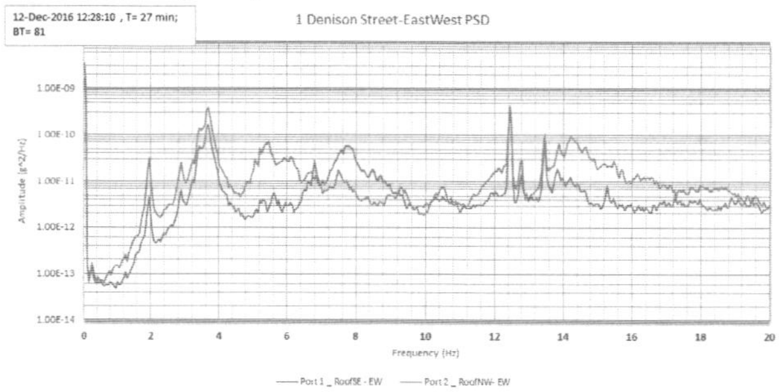

Figure 4: First order East-West Translational Mode

- While the building was still within serviceability requirements for horizontal loading (wind loading), it was found to have a slightly lower capacity then expected for a structure of this type.
- The measurement of the non-linear damping characteristic showed a good correlation with the expected value for a building of this type.

These findings are significant and useful as opposed to STRAAM, traditional methods of assessment often rely largely on:

- The use of strain gauges to only model deflection as a basis for determining bearing capacity.
- Visual inspection which is largely limited to accessibility and cannot match the objectivity and accuracy of measured responses.
- FEM traditionally rely on an assumed dampening ratios, rather than having an actual response to be calibrated against.

On the whole, the building was found to be performing quite well and as expected for a building of its type and construction. The load bearing capacity of the roof top could be checked against loading scenarios to aid with design.

6. Conclusion

Structural risk management assessment and management provide by Mainmark through STRAAM allowed for rapid, real time assessment and then reporting of the child care center. Through measurements of highly sensitive accelerometers, strategically placed a baseline record of the structure can be obtained and built into a finite element model and calibrated against the structures recorded response, greatly increasing the accuracy of the generated model. From the FEM, serviceability and loading conditions can then be modeled and assessed to allow for the concerned parties to have objective information regarding their asset and make informed decisions based on the actual condition of the structure.

As a result of the assessment of the childcare center, in a short amount of time the design engineer was able to make important decisions and necessary changes in order to proceed safely and with confidence with the rooftop landscaping project.

References

Jeary A.P., Winant T, Bunyan J. Some of the measurement of vibration intensity for the assessment of Building Structures. ASCE 8th Foresics Conference. Performance of the Built Environment. Miami Florida. November 2015.

Ellis, B.R., Jeary, A.P. and Severn, R.T.: Integrity monitoring of dams by dynamic tests, *Proceedings of the International Conference on Safety of Dams*, Coimbra, Portugal, p147-152, Apr 1984.

The 8th International Conference on
Structural Health Monitoring of Intelligent Infrastructure
Brisbane, Australia | 5-8 December 2017

Application and Visualization Techniques for Advanced Sensor Networks Case Study: Sensor Installation in Skilled Trades and Technology Centre

G. Mustapha[1], R. Hoemsen[2], R. Spewak[3], K. Knight[4]
[1] *SMT Research Ltd., Canada, Email: gamal@smtresearch.ca,* [2,3,4] *Red River College – Canada* [3] *Red River College, Canada*

Abstract

Structural Health Monitoring of bridges and heritage buildings are essential to ensure modern rehabilitation techniques and materials continue to perform as expected. In addition, SHM systems play a key role to ensure new buildings perform as expected with respect to energy efficiency and durability. This paper focuses on the application of advanced sensor networks and data visualization techniques on a building instrumented as a 'living lab' located in Winnipeg, Canada. The Building Envelope and Technology Access Centre (BETAC) and SMT Research have installed a substantial network of sensors embedded throughout the new Skilled Trades and Technology Centre located on the Red River College campus in Winnipeg, Canada. A total of seven different building envelope types have been instrumented throughout the exterior building envelope as well as the roof and green roof assemblies. The sensor set installed throughout the building consists of heat flux sensors embedded inside specific assemblies along side temperature sensors positioned across the insulation material. In-situ effective R-value can be measured from the installed sensors and compared to controlled tests performed on mock-up test walls in the Dual Environment Test Chamber located in the Centre for Applied Research in Sustainable Infrastructure (CARSI). Field analysis will be performed in real-time during normal building occupancy throughout the temperature extremes inherent to Winnipeg. This paper will demonstrate the methodology and sensor set required to obtain the required quantitative data as well as introduce visualization techniques using augmented reality and virtual tour methods. Augmented reality allows a user to view the building through a Smartphone with live data displayed on the screen and virtual tour methods allowing the user to walk through the BIM model and select sensors to graph and compare.

1. Introduction

This paper covers the methodology and sensor sets used to perform real-time performance analysis of buildings. This includes a unique sensor set for energy performance monitoring as well as a novel approach for data collection and data visualization. The results and analysis will be compiled over a one-year period in order for the building to experience all seasons, the results will be presented at SHMII-9 in 2018. Funded by Manitoba Infrastructure, Red River College is constructing a $60 million, 9300m^2 (100,00ft^2) Skilled Trade and Technology Centre (STTC) that will house laboratory and shop space for up to 1,000 students. The building itself will be used as a 'living lab' as the individual instrumented building components can be tested as per the current method for evaluating energy efficiency as outlined in the BC Building Code, 2012. The building components can be tested in a controlled environment using the facilities of CARSI and then validated in the final constructed STTC building.

The thermal characteristics of materials shall be determined by calculation or by testing in accordance with the applicable product standards listed in the BC Building Code. (BC Building Code, 2012).

The code stipulates that energy efficiency rating for buildings are calculated or simulated. These calculations require inputs such as the building geometry, size, as well as knowledge of specific building components with known R-values. Calculated R-values do not take into account variations in construction, assemblies consisting of various components such as window frames, studs, fasteners, inconsistent insulation, variable air gaps, thermal bridging and deterioration of assembly components over time (Modera, M. P., Sherman, M. H., and Sonderegger, R. C.,1985). The major contribution of this project is to design a system capable of measuring heat flux over a large surface in order to encompass an assembly consisting of different materials. The heat flux in conjunction with the change in temperature across the assembly will give you the in-situ effective R-Value.

The installed sensor suite provides a complete analysis of the roof systems moisture performance as well as the thermal performance of the green roof area. A second set of sensors are installed in specific building components and throughout the different wall systems that comprise the building envelope in order to monitor the moisture and thermal performance of the building envelope.

The impact of this study is to establish a quantitative metric to benchmark the energy efficiency of pre-existing assemblies and materials. This will allow unique materials, assemblies and construction techniques to produce a quantitative measurement that can be recorded and monitored. Energy savings will be achieved by maintaining optimal thermal efficiency over time as the monitored wall structures will essentially be subject to continuous real-time commissioning.

The presentation of the data and visualization will use technology emerging from the Internet of Things (IoT) space as well as the latest software tools used for Augmented and Virtual Reality. These tools will allow anyone to easily inspect wall structures simply by using a Smartphone camera while in the building or remotely by walking through he building in a virtual environment. Currently, the state of the art for sensor visualization is limited to isolated sensor views and static plots distributed over separate views, which are considered inefficient and difficult to use (Clifton Forlines and Kent Wittenburg. Wakame, 2010).

2. Sensor Installation Plan

Moisture Detection Sensor (MDS) were installed in the roof system as well as the building envelope. Heat flux and temperature sensors were installed in strategic locations to measure effective R-value. During construction, other unique assemblies and methods were identified and instrumented as reviewed in this section. Sensors will be compared to external and internal environmental effects, carbon dioxide (CO2), and occupancy sensors along with an external weather station that will be used to correlate cause and effect events in relation to fluctuations with internal sensor telemetry.

2.1 Roof Moisture Detection Grid

Moisture Detection Sensors were installed in 6m (20ft) segments over the entire roof to form a 3m x 3m (10ft x 10ft) grid. The green roof had double the density to form a 1.5m x 1.5m (5ft x 5ft) grid (Hermes, Jamieson, 2015). This sensor grid was installed during roofing so the roof was continuously checked for moisture during and after construction. The roof was segmented in various work packages, each color segment as shown in Figure **1** were installed in one day. Tape coordinates were transferred to a Trimble GPS marking unit that allowed the installers to install the tapes exactly as indicated on the drawing.

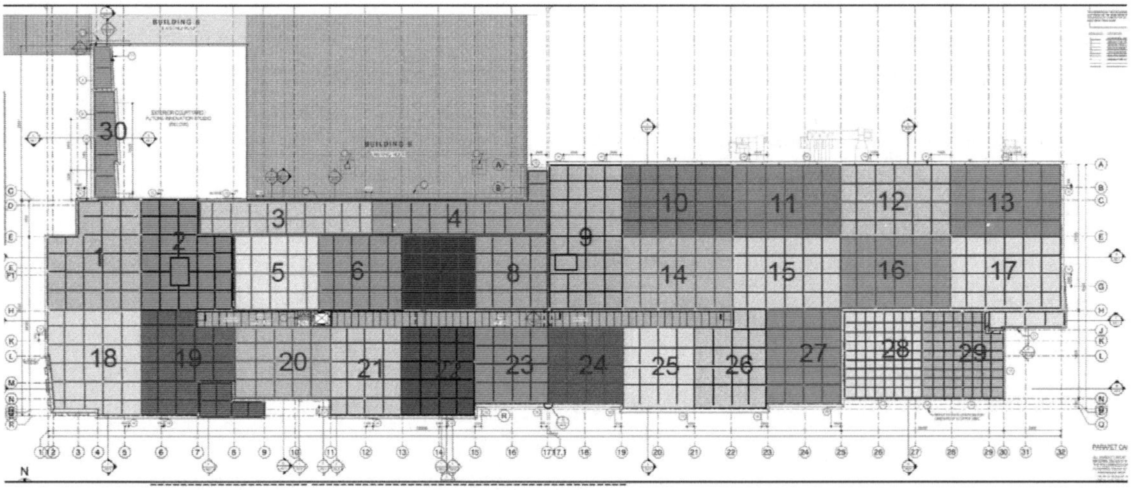

Figure 1. RRC STTC Roof Grid

Figure 2. Sensor Grid Installation

The installation started in the fall of 2016 and continued into the winter. Keeping the roof dry and free of rain and snow were challenging. Figure **2** shows the sensor grid installed with the insulation being installed immediately after.

2.2 Green Roof Performance Monitoring

Moisture Detection Sensors are installed in a grid with double the density as the regular roof. In addition to moisture detection under the membrane the thermal performance of the roof is analyzed using heat flux sensors on the underside of the roof and temperature sensors on the underside and topside of the roof.

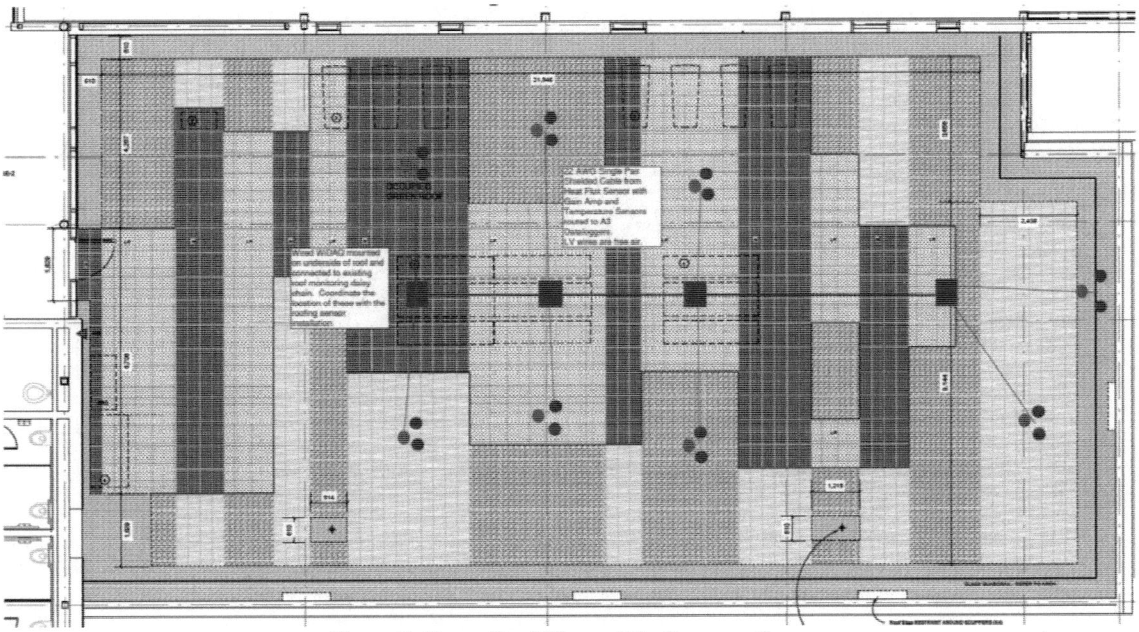

Figure 3. Green Roof Thermal Performance Sensors

The top side of the roof consists of planters in removable trays. Soil moisture in these trays are monitored using wireless data loggers.

2.3 Building Envelope Sensor Installation

The STTC building contains seven different wall types, including variations of masonry, metal siding and curtain walls. The type of sensor and placement is dependent on the wall type and mounting assembly. Moisture Detection Sensor (MDS) tape was placed in areas susceptible to moisture intrusion, Point Moisture Monitoring (PMM) sensors were placed in the plywood mounts for curtain walls and a combination of heat flux and temperature sensors were placed around specific compound assemblies and insulation types to measure thermal efficiency. A typical sensor installation with temperature, heat flux and differential pressure is shown in **Error! Reference source not found.** and Figure **5**. Differential pressure sensors were installed to observe the airflow magnitude and direction through the building envelope assemblies.

Figure 4. Interior Heat Flux Sensor Installation

Figure 5. External Heat Flux, Temperature and Pressure Installation

In-situ R-value is typically measured using a hot-box apparatus in a controlled lab environment and is performed on a single uniform material (Modera, M. P., Sherman, M. H., and Sonderegger, R. C.,1985). This is mainly due to the fact that heat flux sensors are dimensionally small and are unable to encompass a large assembly. Also, the cost of heat flux sensors with high sensitivity is cost prohibitive and the data acquisition is difficult due to the small signal to noise ratio. R-value measurements require heat flux sensors to be mounted on the surface under test with temperature sensors placed across the assembly. Heat flux is calculated as shown in equation (1) (Modera, M. P., Sherman, M. H., and Sonderegger, R. C.,1985).

$$U = \frac{1}{R} = \frac{Q_A}{\Delta T} \tag{1}$$

where: Q_A = Heat flux in W/m^2
T = Temperature difference across assembly under test
R = R-value
U = U-value

RSI (SI) = R-value (U.S.) × 0.1761101838

The ability to perform a heat flux measurement over a large surface area was addressed by designing a large lower cost heat flux sensor used in conjunction with a high gain instrumentation amplifier. Controlled tests were performed on mock-up test walls in the Dual Environment Test Chamber located in the CARSI facility. Field analysis will be performed in real-time during normal building occupancy of the building throughout the temperature extremes inherent to Winnipeg and compared to the results generated in the lab.

2.4 Solar Tube Instrumentation

Solar tubes were installed to route natural light into the building. Potential issues were identified by the owners and building science experts who advised to monitor the perimeter of the solar tube enclosures for accumulated moisture under the roof membrane while monitoring the interior relative humidity and temperature so that the dew point can be calculated.

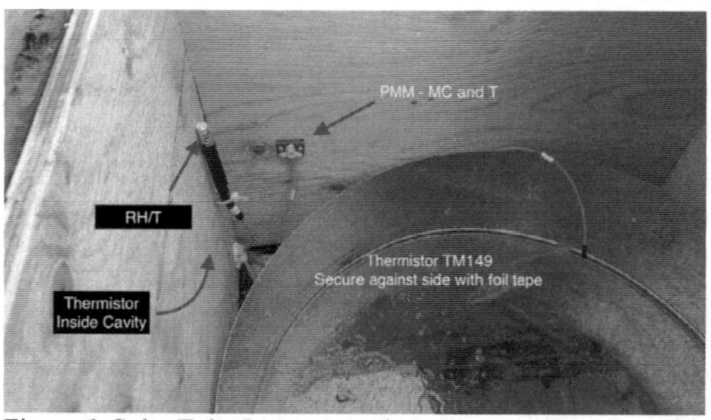

Figure 6. Solar Tube Instrumentation

Figure 7. Trapped Condensation in the Solar Tube

2.5 Parapet Instrumentation

Parapet details are another potential area that owners and building scientists are concerned about. Both curtain wall detail as well as the steel stud detail were recommended for monitoring. Sensors were installed in the curtain wall and steel stud assemblies on all four wall facings in the building. The sensor deployment is shown in Figure **8**.

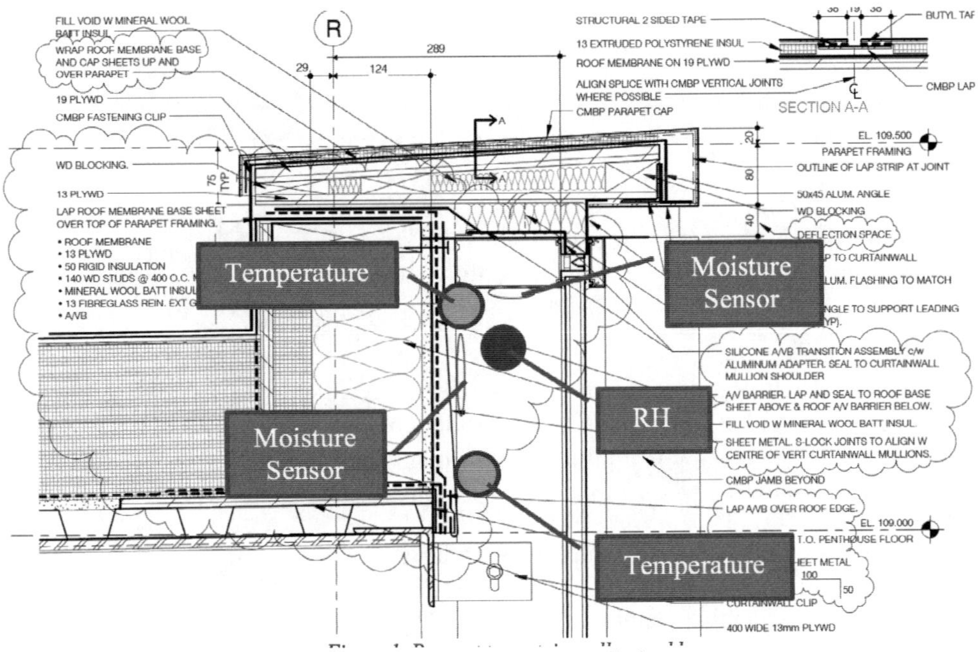

Figure 8. Parapet Sensor Installation

3. Data Acquisition Hardware

The selection of data acquisition hardware and data analysis interfaces are an important aspect of the monitoring system solution. Moisture detection tapes on the roof were routed to 48 channel board multiplexers which were daisy chained to each other using a wired Controller Area Network (CAN) bus (Mustapha, Gamal 2012). Building envelope data acquisition units were connected to the network by joining the daisy chain through 8 channel data acquisition units. The data acquisition units transmit data to a local gateway that synchronizes data with a cloud based server. Sensors in non stationary areas or areas that are difficult to access transmit data wirelessly to a gateway over a unique channel over 802.15.4 in order to not interfere with other internal wireless systems. (Mustapha, Gamal, 2010)

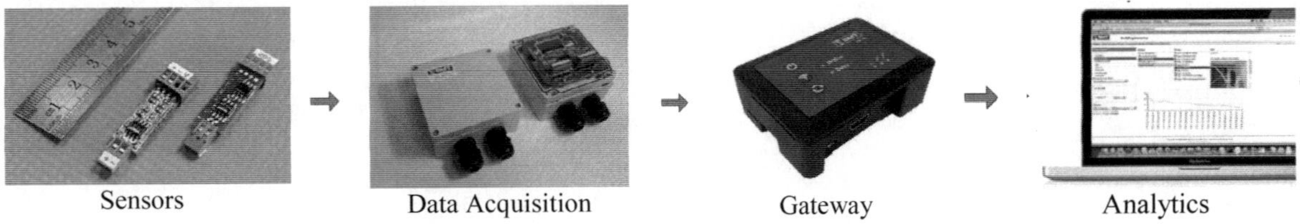

3.1 Sensors

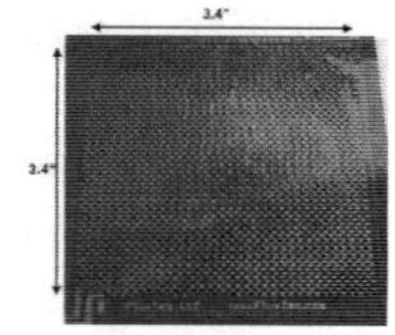

Heat Flux (HF)
A large heat flux array was designed specifically to cover large complex areas so the heat flow of the insulation material and surrounding sensors can be recorded. Heat Flux sensors are connected to instrumentation gain amplifiers prior to being connected to the Data Acquisition unit (Cirenza, Chris, 2017).

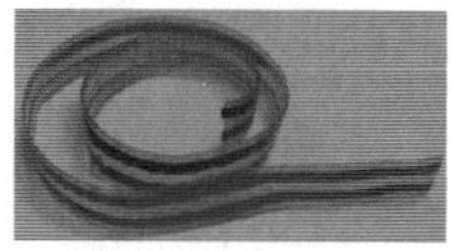

Moisture Detection Sensor (MDS)
Moisture Detection Sensor tapes are adhered under windows and areas susceptible to moisture intrusion. These are the same sensors used in a grid formation on the roof. (Mustapha, Gamal, 2007)

Point Moisture Measurement (PMM) Sensors
PMMs are embedded in the wood frames around the curtain wall assemblies and give a percent wood moisture content reading of the wood. (Mustapha, Gamal, 2007)

Relative Humidity (RH)
RH sensors are used within building envelope cavities, areas such as the solar tubes, and in different air spaces to understand the movement and accumulation of humid air throughout the building (Mustapha, Gamal, 2008).

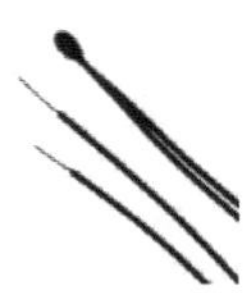

Temperature
Thermistor sensors are used for temperature compensation with Moisture and RH sensors as well as for temperature gradients for heat flux sensors (Mustapha, Gamal, 2008).

Differential Pressure Sensors (DPS)
Differential pressure sensors are used to determine the direction of airflow through the building envelope (Mustapha, Gamal, 2008).

4. Data Visualization Techniques

One of the most important aspects of structural health monitoring is disseminating and understanding the data collected. This involves identification, comparison, and correlation tasks which are performed on vast amounts of spatially embedded sensor data recorded over time (Gennady Andrienko and Natalia Andrienko, 2005). The datasets collected contain challenging features including big data with spatio-temporal attributes. The ability to browse this data using a classical interface is available where sensors can be selected, data can be viewed and sensors can be grouped and graphed together. In addition to this, the ability to overlay sensors on a building drawing allows a user to identify the sensor location and helps in understanding the situational awareness around the sensors.

Two additional data presentation methods will be implemented as part of this project are Augmented Reality and Virtual Reality. These methods will make sensor exploration and analysis more informative and interesting. By creating data analysis tools that are visually accessible and digestible to the bulk of the general population, living labs have become interesting to much more than just building scientists and researchers. Creating interactive tools can spark an interest in education and awareness on sustainable construction and innovative design. These presentation methods also form another branch of research and development for the college to pursue in the future.

4.1 Analytics

Using standard data analysis tools such as sensor overlays and graphing tools, wet areas can be easily identified and data trends pertaining to these areas further analyzed using standard graphing tools. Figure 9 illustrates a typical roof grid sensor overlay, tape segments in red have exceeded a specific threshold. Selecting a sensor graphs the trend as shown in Figure 10.

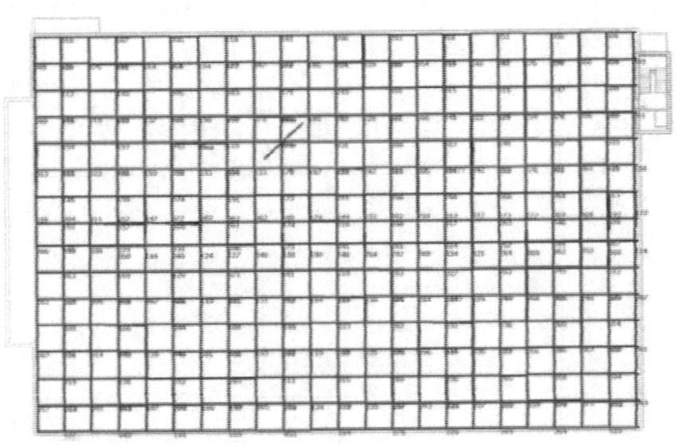

Figure 9. Roof Grid Highlighting Wet Areas

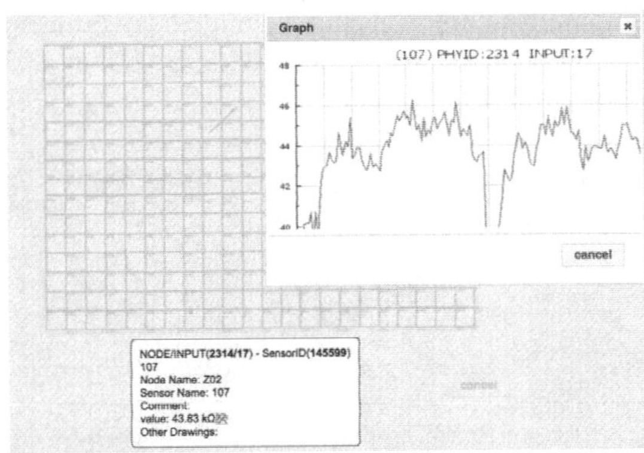

Figure 10. Analysis of Specific Area

4.2 Augmented Reality

Using a custom Smartphone app, data can be extracted from embedded sensors and overlaid over the image in a smartphone display. Once the app recognizes a unique identifier in the camera viewfinder, data pertaining to the sensor location is accessed from the cloud based server and populated on the image, displaying real-time data over the view shown on the camera. The app identified the image as shown in Figure **11** and displayed real time temperature and humidity information from sensors embedded in the wall. This creates a highly interactive environment for educators, students, and visitors of the space. Being able to tangibly associate building assemblies with real-time data with a mobile device also creates an opportunity to very easily investigate the surroundings of the instrumented areas.

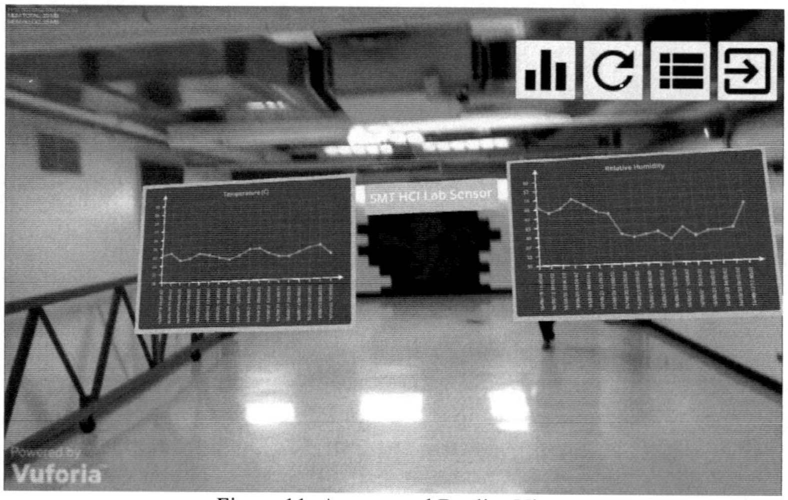

Figure 11. Augmented Reality View

4.3 Virtual Reality

The BIM model for STTC was ported into a gaming software engine, Unity. Sensors were populated throughout the BIM model. Using common gaming controls, users are able to virtually walk through the building and select sensors to view their data. Multiple sensors can be selected, graphed and compared while walking through the building in a virtual environment. Creating a contained environment marrying the real-time sensor data and the BIM model allows for on site and remote exploration of the structure. This allows students and visitors to interact with sensors in restricted and hard to reach areas from a centralized dashboard. For researchers and Building Scientists, this creates an environment to visually investigate both the 3D model, and its corresponding data in a singular analysis ecosystem.

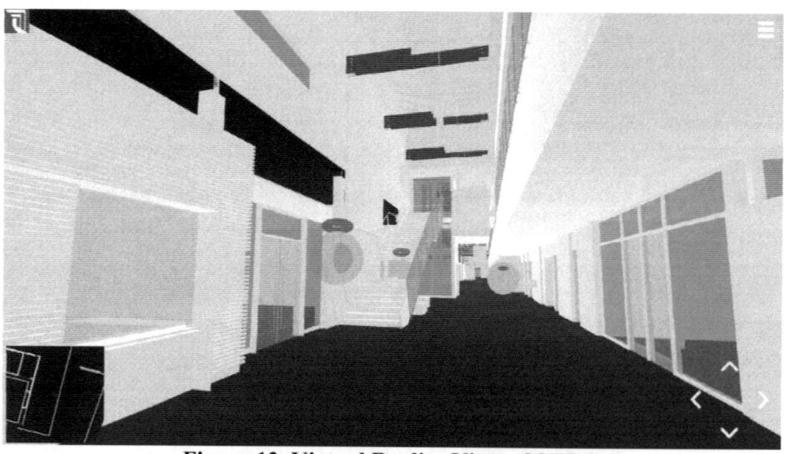
Figure 12. Virtual Reality View of STTC

3. Conclusions

The sensor deployments at the RRC Skilled Trades and Technology Centre proved to be an invaluable tool during construction as it helped quantify the extent of various roof leaks. An understanding of the condensation in the solar tubes during construction has also been an asset as methods to mitigate the stored moisture are being reviewed prior to having the solar tubes freeze during the winter. The data visualization techniques have been tested with building science practitioners. They indicated that the ability to have an overview of the spatio-temporal data will allow for a more in-depth analysis of cause and effect of the building with respect to the data being analyzed. The sensor data will be collected over the period of a year and compared to lab tested assemblies and simulated models to determine if these sensor sets can be used to maintain the status of high performance buildings such as the RRC Skilled Trade and Technology Center. The results will be presented in ISHMII-9 in 2018.

References

BC Building Code, Revision 9. *Housing and Small Buildings* Section 9.36. Energy Efficiency.

Cirenza, Chris, *PHFS-09 Heat Flux Datasheet*, Blacksburg, VA, 2017.

Clifton Forlines and Kent Wittenburg. Wakame: sense making of multi-dimensional spatial-temporal data. In *Proceedings of the International Conference on Advanced Visual Interfaces*, AVI '10, pages 33–40, 2010.

Gennady Andrienko and Natalia Andrienko. Visual exploration of the spatial distribution of temporal behaviors. In *Information Visualisation, 2005. Proceedings. Ninth International Conference on*, pages 799–806. IEEE, 2005.

Hermes, Jamieson, *Membrane Monitoring Technician Manual*, Vancouver, BC, 2015.

Modera, M. P., Sherman, M. H., and Sonderegger, R. C., *Determining the U-Value of a Wall from Field Measurements and of Heat Flux and Surface Temperatures, Building Applications of Heat Flux Transducers*, ASTM STP 885, E. Bales, M. Bomberg, and G. E. Courville, eds., ASTM, 1985, pp. 203–219.

Mustapha, Gamal, *Point Moisture Measurement Datasheet*, Vancouver, BC, 2007.

Mustapha, Gamal, *Moisture Detection Sensor Datasheet*, Vancouver, BC, 2007.

Mustapha, Gamal, *Relative Humidity Sensor Datasheet*, Vancouver, BC, 2008.

Mustapha, Gamal, *Temperature Sensor Datasheet*, Vancouver, BC, 2008.

Mustapha, Gamal, *Differential Pressure Datasheet,* Vancouver, BC, 2008.

Mustapha, Gamal, Controller Area Network CANKey Datasheet, Vancouver, BC, 2012.

Mustapha, Gamal, *I3 Datasheet*, Vancouver, BC, 2010.

The 8th International Conference on
Structural Health Monitoring of Intelligent Infrastructure
Brisbane, Australia | 5-8 December 2017

Development of New Technique in Multi Setup Merging of Vibration Test System Identification

A. Sabamehr[1], F. Mirshafiei[2], A. Bagchi [3].
[1]*Concordia University – Canada, Email:* sabamehrardalan875@gmail.com , [2]*Sensequake Inc. - Canada,* [3]*Concordia University - Canada*

Abstract

In Modal Analysis, multi setup merging is one of the high demand methods when the number of sensors is less than DOFs of structure. The technique is worked based on the selection reference sensors which is fixed during all measurements, and additional sensors are moved or roved in each setup to cover all areas of the structure. Generally, there are three common methods are used for merging multi-setup data including, Post Separate Estimation Re-scaling (PoSER), Post Global Estimation Re-scaling (PoGER) and Pre Global Estimation Rescaling (PreGER). Due to some difficulties in PoSER and PoGER, PreGER have been used recently as an alternative method because the merging process has been conducted before parameter estimation. On the other hand, Random decrement Technique is a time domain procedure which can be used for Operational Modal Analysis test. As OMA can be generated by zero mean Gaussian noise, the noise is removable by applying RDT on the period of time segment with specific trigger which keeps all of frequencies of structure but remove its noise. Therefore, RDT can apply to PreGER to enhance the speed and reduce the data in multi setup merging method. In this paper, Finite Element (FE) model of five-story frame has been tested with multi setup merging. Consequently, the results shows that applying RDT as a pre-process technique to the data can enhance the performance of the system identification process, and confine the number of data in PreGER.

1. Introduction

Vibration tests of civil engineering structures usually require to measure the structure response at a large number of points therefore the mode shapes can be characterized with enough spatial resolution. The limited number of available sensors or acquisition channels imposes the development of several setups with the sensors in different positions. Because it can vary in intensity or frequency content from one setup to the other, it is necessary to keep some sensors in the same positions during the entire test as reference sensors (Manuel 2012).

In this paper, the new technique is developed for multi setup merging modal analysis. The five-story finite element modal has been simulated to verify the result.

2. Background study

The classical strategy which is called Post Separate Estimate Re-scaling (PoSER), is to deal with data collected during ambient vibration tests composed by several setups consists in processing the signals of each setup separately, hence obtaining a set of modal parameters for each setup. The mode shape components identified in each setup are then 'glued' back together using the mode shape components at the reference sections identified in all the performed setups. When a large number of setups is involved, this approach becomes very time consuming because of the number of analyses that have to be performed, one for each setup, and due to the need to pair the results obtained in each individual analysis. This last step is not always evident because in some setups some modes may be missing. This methodology has the disadvantage of requiring a very heavy analysis task when a large number of setups is processed, which, in the case of a parametric method (time domain method), involve the tedious task of interpreting many stabilization diagrams. However, it has the advantage of characterizing the scattering of natural frequency and modal damping ratio estimates during the period of the ambient vibration test. The observed variations are due to the uncertainty of the estimates and also to the influence on modal parameters of environmental or operational factors that may vary during the dynamic test. The Post Separate Estimation Re-scaling (PoSER) has been used in the context of ambient vibration tests of civil engineering structures for a long time (Felber 1993).

Another method is called Post Global Estimation Re-scaling (PoGER) that is the approach to process collected data during several setups with just one analysis. This methodology makes the identification of modal parameters much more straightforward, because only one stabilization diagram has to be analyzed and the mode paring procedure is avoided. The problem is the components at the reference outputs are repeated with the number of setups. Because no previous scaling was applied, the segments are not at the same scale. But, again the mode shape components at the reference outputs can be used to re-scale all the segments to a common level (Manuel 2012).

The Pre Global Estimation Re-scaling (PreGER) approach is innovative method based on a global estimation of modal parameters. In this case, the correlation or the (positive) spectrum matrices, estimated from the signals collected during the developed setups, are scaled before the identification. Consequently, the obtained mode shapes do not need any further post-processing, so the results of the identification are the final modal parameter estimates.

In the case of the frequency domain methods the scaling of the (positive) spectrum matrices is quite simple. Placing the reference outputs in the first lines of the output spectrum matrix, the spectrum matrices of all the setups S_j can be divided in two blocks: $S_{j,ref}$ with the auto and cross-spectra between the reference outputs and $S_{j,mov}$ with the cross-spectra between the other outputs (not reference) and the reference outputs. These three approaches are also systematized in (Parloo 2003).

The Random Decrement Technique (RDT) was proposed by Cole in the end of the sixties and beginning of the seventies within his work at NASA concerning the analysis of the dynamic response of space structures subjected to ambient loads (Cole 1971 and Cole 1973). The concept of RDT is based on response of a system to random input load that is including to three parts: the response to an initial displacement; the response to an initial velocity and the response to the random input loads between the initial state and the time instant t. By averaging a large number of time segments of the response with the same initial condition, the random part of the response will gradually disappear from the average, and what remains is the response of the system to the initial conditions. A usual and good option is to define optimum triggering levels, in the sense of minimization of the variance of the estimate of the RDT functions. In the case of the level crossing triggering condition, this optimum level is $a = \sqrt{2}\,\sigma$ where σ is the standard deviation of the response time history under analysis (Brincker et al 2005) auto $D(\tau)$ and cross $D(\tau)$ RD functions can be mathematically defined by the following expressions:

$$D_{XX}(\tau) = \frac{1}{N} \sum_{i=1}^{N} x(t_i+\tau)|Tx(t_i) \quad (1)$$

$$D_{XY}(\tau) = \frac{1}{N} \sum_{i=1}^{N} y(t_i+\tau)|Tx(t_i) \quad (2)$$

3. Methodology

In this paper, the new technique has been developed in pre-merging technique by use of RDT. Initially, the RDT has been applied to the time domain data, which are collected from each setup to reduce the number of data. It results in reducing time consuming specially when there is large of number of data. Then, the Power Spectrum Density (PSD) has applied to the outcomes to change time domain to frequency domain. After that, using equation 3 has rescaled the PSD in each setup. The rescaled PSD setups have been merged and apply Singular Value Decomposition (SVD) to identify the modal identification.

$$S_{scaled}^{j,mov} = S^{j,mov}. \left(S^{j,ref}\right)^{-1}.\left(\frac{1}{n_s}\sum_{k=1}^{n_s} S^{k,ref}\right) \quad (3)$$

In equation 3, the first part is the roving sensor in setup j, the second part is the reference sensor in setups j and last part is the average of reference sensors in all setups. The detail of methodology is shown in Figure 1.

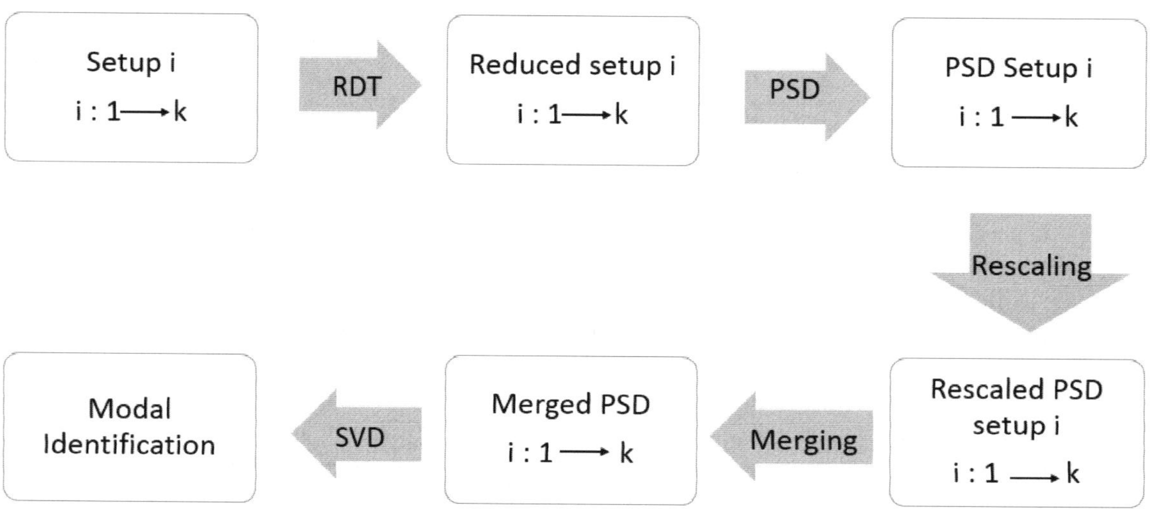

Figure 1. Schematic procedure of RDT – Pre merging multi setup modal analysis

4. Case Study

The FE model of five-story frame has been modelled in SAP2000. The model has been analysed for modal analysis and modal properties have been recorded.

4.1 FE model of Five-story frame

The five-story concrete frame are modelled with 10m length, 5m width and 15m height (3m height in each floor) in SAP2000. The frame is massless and the concentrated mass is located in the centre of each floor.
To simulate the ambient vibration, the White Gaussian Noise (WGN) which is simulated in MATLAB, have been applied to all nodes in each direction. The length of the test is 1000s with the sampling rate 100Hz. Table 1 shows the modal analysis and operational modal analysis with WGN.

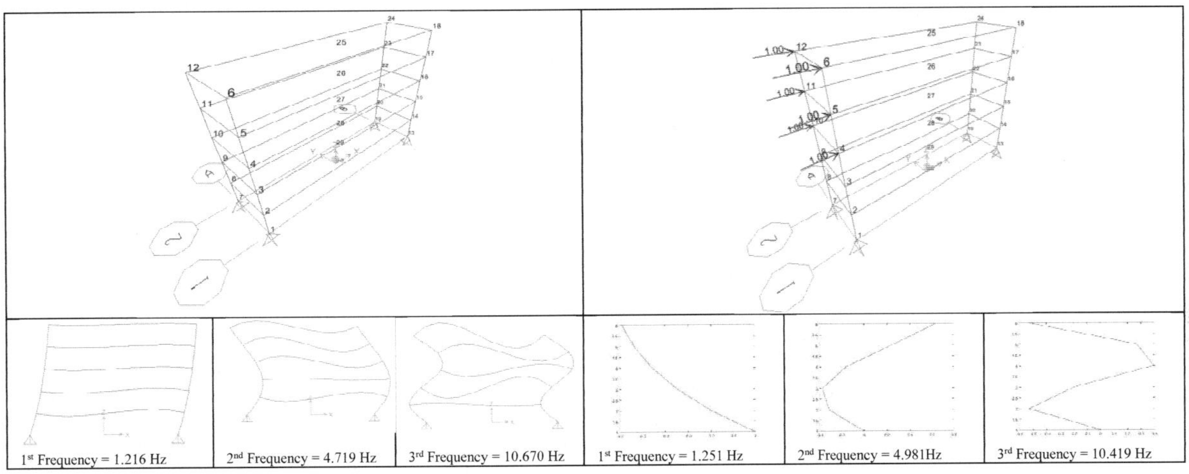

Figure 2. (a) Modal analysis along x direction for first three bending mode (b) Operational modal analysis along x direction by use of Frequency Domain Decomposition (FDD)

Initially, all of data are considered as one setup. The acceleration of each floor has been recorded due to WGN force. The acceleration data are analysed in frequency domain to find the modal properties by use of Frequency Domain Decomposition method (FDD) (Farshchin 2015).
Then, two setups have been generated with 2000 data in each setup (20s length). PreGER method is used to merge the setups and find the modal properties.

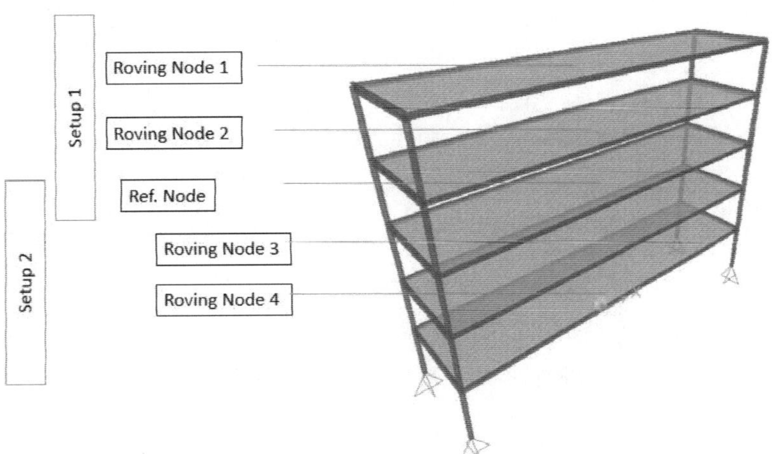

Figure 3. Schematic of multi setup merging in five-story concrete frame

Therefore, we have two setups with two roving and one reference nodes. The details of setups are displayed in Figure 3.

The RDT is applied with 10s time length and the trigger value is standard deviation of data multiply by square root of 2. It is applied to the time domain data collected from sensors. Then, PSD is applied in each channel to convert time domain to frequency domain. To identify the modal properties by use of FDD (frequency domain method), SVD is applied to PSD matrix. After finding the frequencies, the modal properties are compared in three different method such as one setup, multi setups pre merging only and RDT with multi setups pre merging. Table 1 illustrates the modal properties of five-story frame by use of different algorithm in one setup and multi setups.

Table 1. Modal identification by use of pre merging only and RDT with pre merging

One setup with FDD	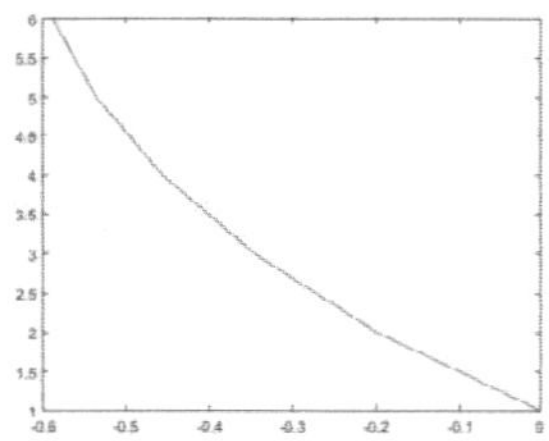 1st Frequency = 1.251 Hz	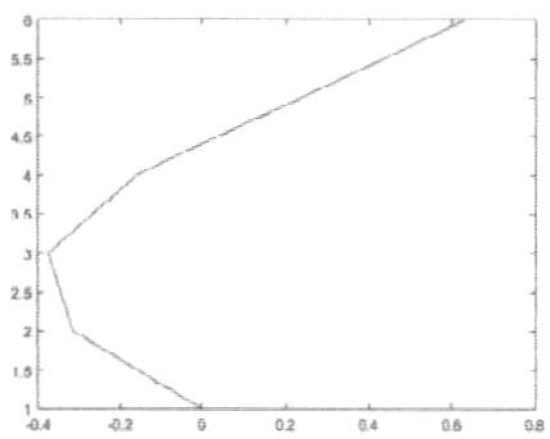 2nd Frequency = 4.981 Hz	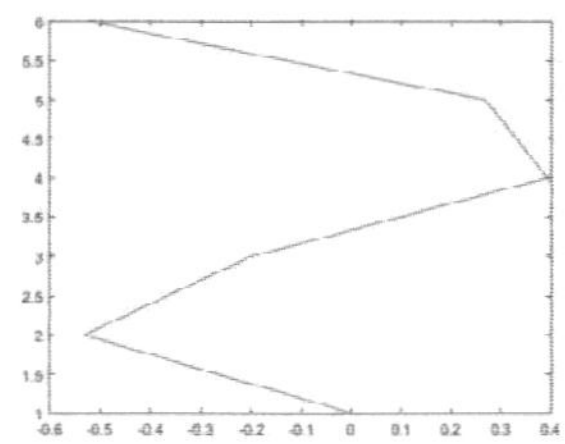 3rd Frequency =10.419 Hz
Multi setups with pre merging	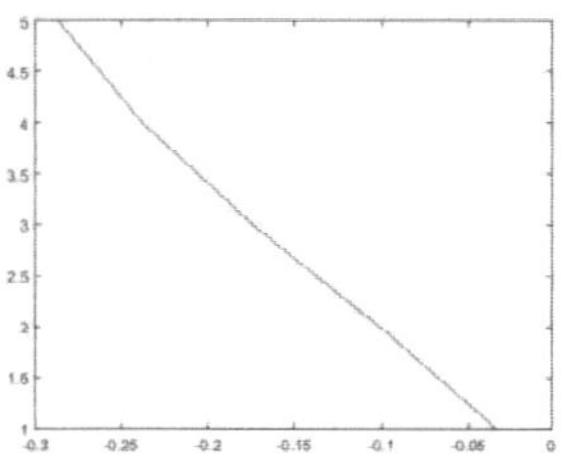 1st Frequency = 1.256 Hz	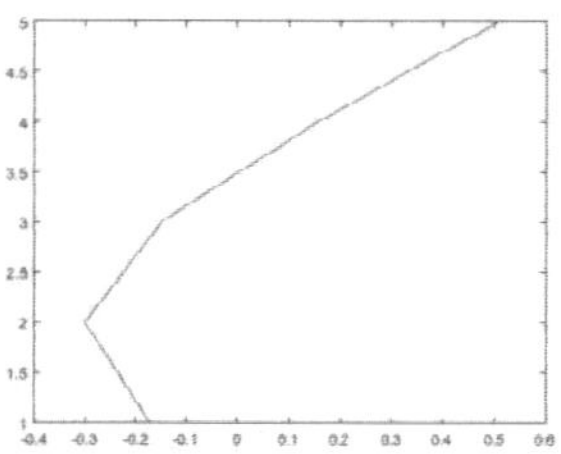 2nd Frequency = 4.83 Hz	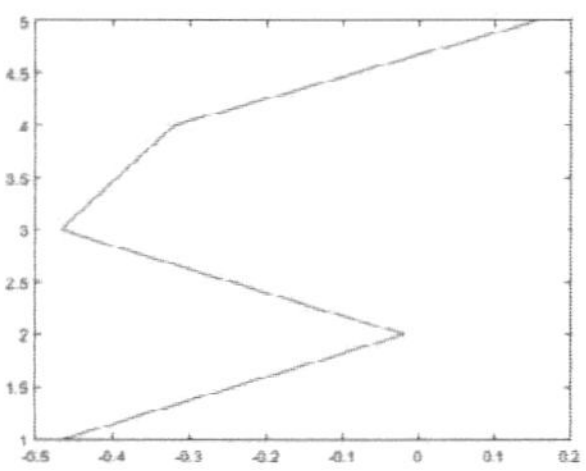 3rd Frequency =10.54 Hz
Multi setups – RDT with pre merging	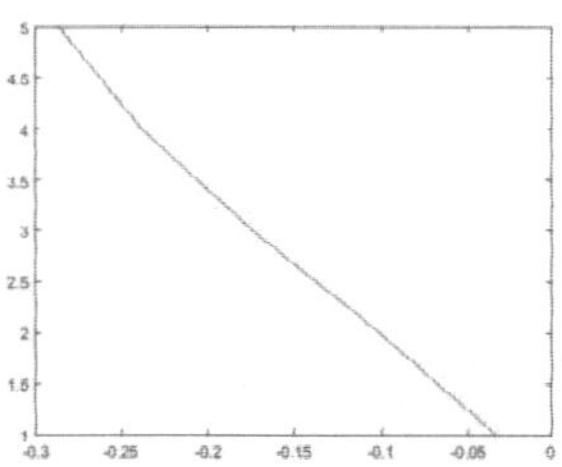 1st Frequency = 1.22 Hz	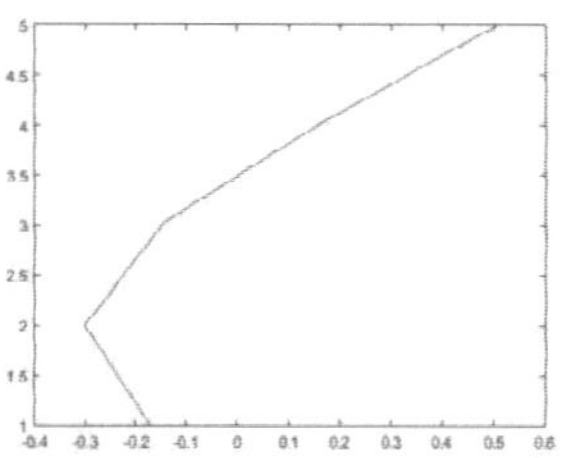 2nd Frequency = 4.76 Hz	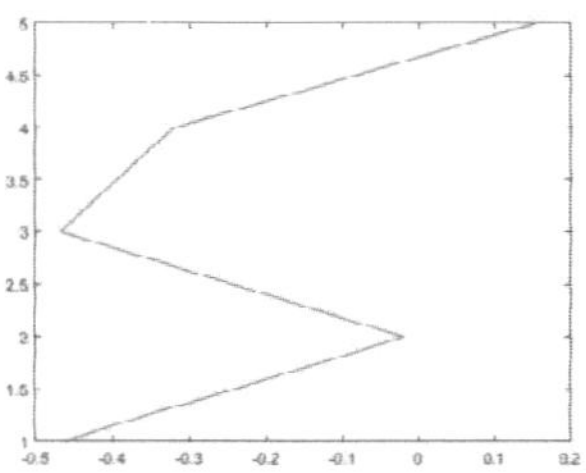 3rd Frequency = 10.54 Hz

Table 2 shows the total modal identification in the frame in different method and their difference with SAP2000 output. It shows that using RDT-PreGER cause to have better result which close to SAP2000 output.

Table 2. Modal identification in Five-story frame by use of different methods

	Mode 1	Difference	Mode 2	Difference	Mode 3	Difference
SAP 2000	1.216 Hz	-------------	4.719 Hz	---------------	10.67 Hz	---------------
FDD	1.251 Hz	-2.88%	4.981 Hz	-5.55%	10.419 Hz	%3.17
PreGER	1.265 Hz	-4.03%	4.83 Hz	-2.35%	10.54 Hz	2.04%
RDT-PreGER	1.22 Hz	-0.33%	4.76 Hz	-0.87%	10.54 Hz	2.04%

5. Conclusion

This study developed the new technique to reduce the number of data and algorithm work faster in merging data. The pre merging technique with RDT cause to reduce the number of data in modal identification. The frequency domain algorithm showed the similar result of pre merging along and pre merging with RDT in five-storey frame. Meanwhile, RDT can remove the noise in ambient vibration test by averaging the time segments in certain trigger value. Among all merging technique, PreGER is the fastest method especially with large number of setups. So, using RDT result in reducing number of data and enhancing the performance of the system identification process. Moreover, the output from RDT-PreGER is closer to the target which is SAP2000.

Acknowledgement

The financial support of the IC-IMPACTS Network (the India-Canada Centre for Innovative Multidisciplinary Partnerships to Accelerate Community Transformation and Sustainability) is gratefully acknowledged.

References

Manuel, F. 2012, "Operational modal analysis for testing and monitoring of bridges and special structures" Doctoral dissertation, University of Porto.

Felber, A. J. 1993, "Development of a Hybrid Bridge Evaluation System" PhD Thesis, University of British Columbia, Vancouver, Canada.

Parloo, E. 2003. Application of Frequency-Domain System Identification Techniques in the Field of Operational Modal Analysis. PhD Thesis, Vrije Universiteit Brussel, Belgium.

Cole Jr, H.A., 1971. "On-line analysis of random vibrations".

Cole Jr, H.A., 1973. "On-line failure detection and damping measurement of aerospace structures by random decrement signatures".

Brincker, R. and Rodrigues, J., 2005. "Application of the random decrement technique in operational modal analysis" In Proc., 1st Int. Operational Modal Analysis Conference (IOMAC).

Farshchin, M. 2015 "Frequency Domain Decomposition" Matlab code for output-only modal analysis

The 8th International Conference on
Structural Health Monitoring of Intelligent Infrastructure
Brisbane, Australia | 5-8 December 2017

Ground Based Interferometric Radar (IBIS-FS) to Delineate a Long Term Plan of SHM for Bridges

M.Bell[1], F.Boscagli [2], E.Shilov [3], M.Twardzik [4].

[1] *IDS Georadar – Australia, Email:* m.bell@idsgeoradar.com, [2] *IDS Georadar - Italy,* [3]*IDS Georadar - Australia,* [4]*IDS Georadar - Australia*

Abstract

With the growing reliability requested to the infrastructure's world, it is necessary to find solutions that allow to monitor and analyse the status and the physical condition of infrastructure facilities, with the highest accuracy possible and with non-invasive monitoring techniques. In particular, bridges and flyovers which have to be continuously open (24/7) to the users and are subject to important stress, must be inspected for their dynamic and static response.
Furthermore, the measuring activities have to be focalized in acquiring trustworthy data to delineate a long-term plan of structural monitoring and maintenance.

The new ground-based radar interferometry technology allows to monitor displacements from a point of sight far from the analysed targets. It provides a continuous mapping of displacements with an accuracy up to 10 μm and a range resolution of 0.5 m.

In this paper, the application of ground-based radar interferometry (IBIS-FS) to evaluate twelve bridges in a railway segment in Australia, is reported. The analysis of these kind of structures is important to assess their dynamic response to the transit of mining train convoy, and consequently to define functional capabilities and levels of acceptable dynamic stress.

The principal survey aims were:

- analyse the status and physical condition of the bridges, comparing the maximum allowed displacement with the real one measured;
- verify the possibility to increase the transportable weight per wagon;
- evaluate the speed amount increasable of the train;
- define a maintenance plan for the whole private railway network, thanks to the collected data.

1. Introduction

In November of 2014, IDS (Ingegneria Dei Sistemi S.p.A.) was commissioned by a mining company to conduct an interferometry survey of bridges on its rail network in Western Australia. The primary objective of the survey was to measure the movement and vibration of twelve bridge structures under load to determine whether they are fit for purpose for current loads and for possible increasing load train.
Moreover, the objective of the survey was to understand if utilizing this technology, as part of standard visual inspection practices, will provide full and ongoing visibility of the condition of

bridge structures on track, allowing more proactive rather than reactive maintenance. Cost benefits will accrue by:

- easy and rapid installation of the instrument compared to standard methods;
- no interruption on the railway traffic during the measurement activities;
- simultaneous multi-point measurement of deflection and vibration;
- reduced risks and improved safety for technicians.

2. Background of Techniques

2.1. Stepped Frequency Continuous Wave Technique (Taylor, 2001)

The principle of this technique is that the radar steps through a range of frequencies from low to high frequencies of the bandwidth at up to 200 times per second (200 Hz). The real and imaginary components of the signal are analysed in the frequency domain. The phase and amplitude information of the signal is retrieved by Inverse Discrete Fourier Transform (IDFT) (Palombo et al., 2011) and an amplitude versus range profile of the radio echoes is constructed for 0.5 m bins forming a one dimensional map called a range profile which shows reflected amplitudes with respect to the range from the radar sensor. Figure 1 demonstrates this concept.

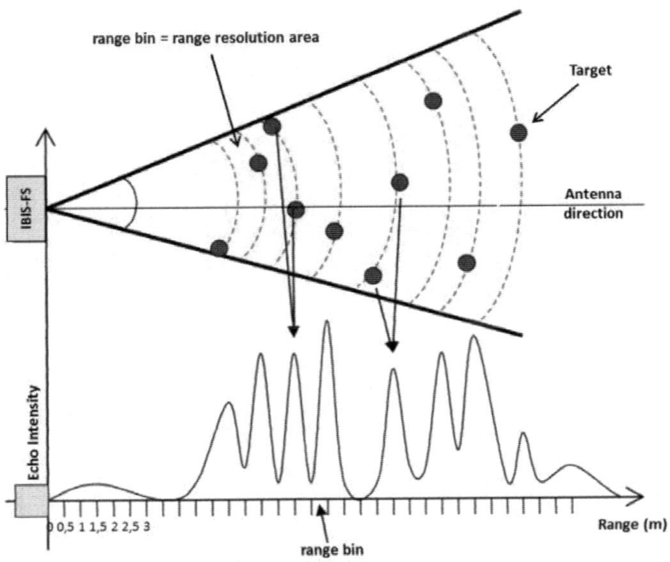

Figure 1. The range profile shows the intensity in dB of the signal reflected back from each object in the transmitted wave path back to the receiver of the radar as a function of each objects distance from the radar.

The sampling frequency controls the maximum frequency measurable by the system with the Nyquist frequency, or maximum detectable frequency, being approximately half of the sampling frequency. The frequency that can be sampled is controlled by the distance from the structure, as for radar waves travelling further to a reflector, the sensor must wait longer to receive the return signal. In practice this means that reflectors at a range of 1 km can only be sampled at 30 Hz, whilst closer reflectors can be sampled at 200 Hz. The sampling rate does not affect the accuracy of measured deflections (up to 0.01 mm) as this is determined by the wavelength of the radar waves.

The IBIS-FS radar has a minimum range resolution of 0.5 meters meaning that only objects with separation greater than this can be resolved as separate objects i.e. they must be at least 1 range bin apart. Objects within the same range bin cannot be resolved as separate objects.

2.2. The Differential Interferometry Technique (Henderson and Lewis, 1998)

Once the measurement data has been assigned to range bins and the peaks selected for the targets of most interest, the deflection of the targets is measured using differential interferometry. The phase of the returned radar waves from each target is measured during consecutive samples with the phase of the return signal calculated using the formula:

$$d_{(p)} \propto \lambda/4\pi \, \Delta\varphi \quad (1)$$

Where dp is the radial displacement, λ is the wave length, and φ the phase change. Figure 2 illustrates the concept.

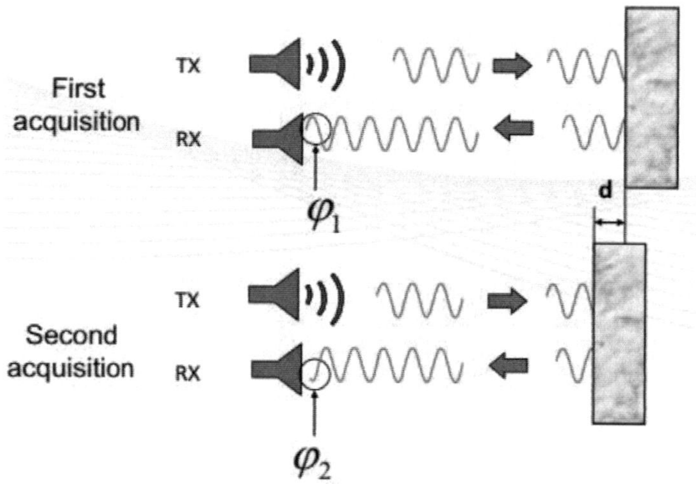

Figure 2. Interferometry concept

The displacement measured is along line of sight so to calculate vertical or horizontal deflection, the geometry of the scenario needs to be calculated. In the example shown in Figure 3, the vertical component of the deflection of the bridge is calculated as:

$$dp = d \sin(\alpha) \quad (2)$$

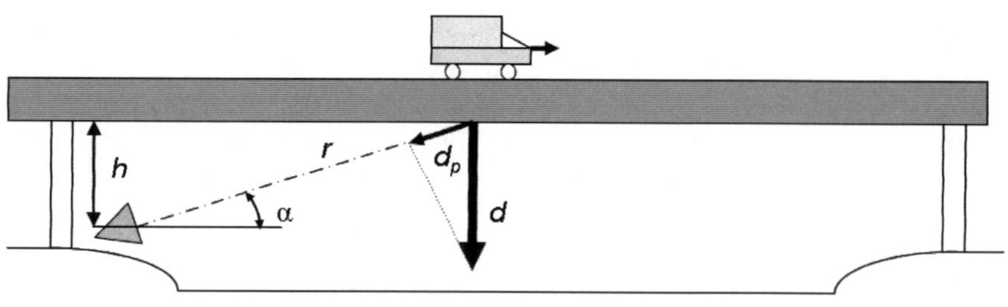

Figure 3. Vertical displacement calculation

In order to calculate deflections (i.e. vertical or horizontal) the alignment of the system to the structure is important. Where the vertical component is required, the alignment must be along the length of the structure and in most cases this would be from beneath the structure. To measure horizontal movement, the alignment would be from the side of the structure. Where both components are required, two systems would need to be synchronized wirelessly.

3. System Description and Specification

The system consists of a sensor, power supply and control unit and has two horn antennas for transmission and reception of the radar signal (see Figure 4).

Figure 4. IBIS-FS system setup

The sensor generates a tune-able sine wave with central frequency of 17.2 GHz and with a bandwidth up to 300 MHz. The radar is classified as a Ku – band, according to standard radar-frequency letter band nomenclature from IEEE Standard 521-1984 (Bernardini et al., 2007).
The sensor is supported on a tripod with a fully rotatable head allowing it to be placed easily at any required angle. Communication between the radar sensor and the control unit is through a standard USB interface. The data logger is a standard laptop running Windows 7 with dedicated software to configure, store, process and view the data.
The power supply is a commercially available 12V battery which can power the system for 8 hours in the field.

4. Survey Methodology and Data Analysis

The bridge surveys were conducted with 2 x IDS IBIS-FS interferometry systems to increase the number of bridge spans that could be monitored over the survey period, with each unit set up under each bridge to monitor the structure. The instrument requires 20 minutes to install and the same time to dismantle. The instrument is specifically designed for outside use and is IP65 standard for all-weather use. A typical system setup from this survey can be seen in Figure 5.

Figure 5. IBIS-FS was set up directly under the bridge to maximize the power of the return signal to the receiver from each range bin

The cross beams shown in Figure 5 were typical and were excellent reflectors of the transmitted radar wave which resulted in excellent data quality for each of the bridges surveyed.
To ensure accurate geometry of the interferometer systems in relation to the bridge, IDS used laser distometers to measure angles and distances from the survey equipment to the underside of the bridge structure.

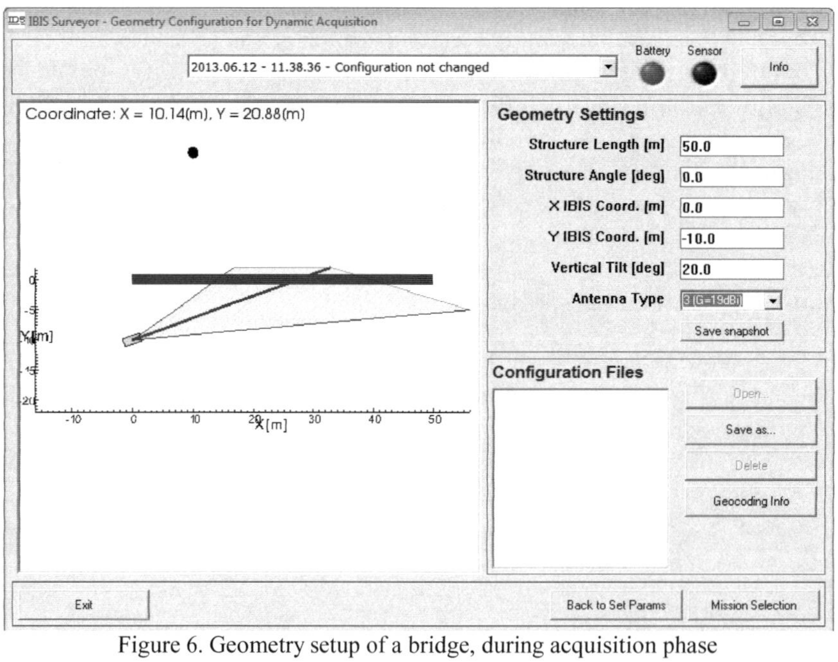

Figure 6. Geometry setup of a bridge, during acquisition phase

Once data acquisition begins, the software showed the range profile (see Figure 7) of the illuminated area. It is important to identify points on the bridge structure that correspond with each of the high peaks within each range profile. These peaks are related to range bins that display an increased amplitude of the return signal and are often associated with good reflective points on bridge structures such as beams, girders and cross beam supports.

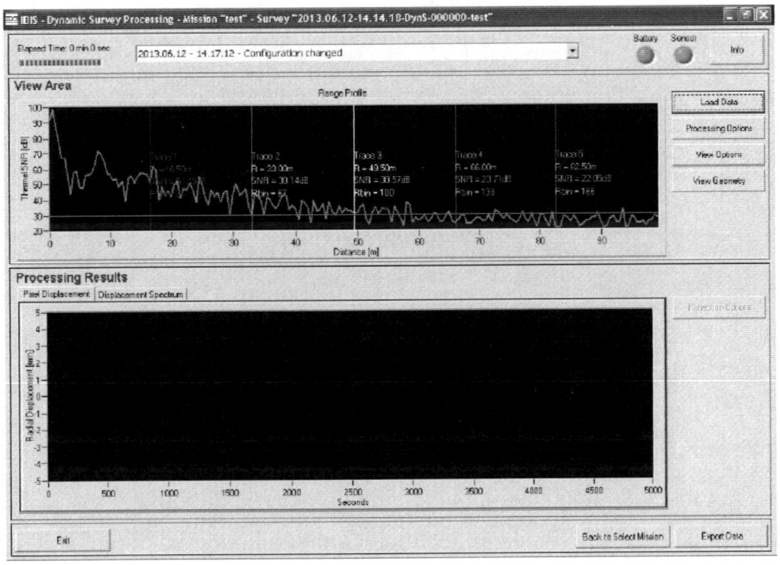

Figure 7. An example of a range profile of the area illuminated by the radar

To identify points on the bridge structure corresponding to peaks of reflected energy within the range profile, the laser distometer is used to determine the distance from the radar head to each of the cross beams under each bridge, which are known to be good reflectors of the transmitted radar wave. From this line of sight distance, the correct peaks associated with each cross beam can be identified, the geometry of the survey environment established (see Figure 8) and the vertical displacement and vibration characteristics of that range bin can be monitored over a specified time frame (see Figure 9 and Figure 10).

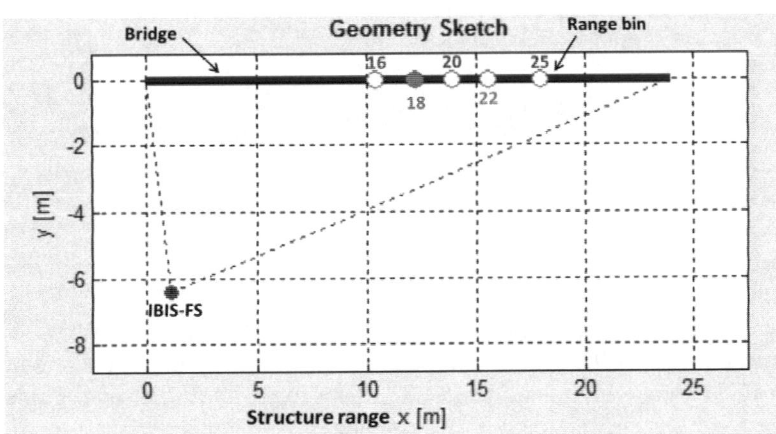

Figure 8. Geometry showing the range bin position that provide maximum power of the reflected radar

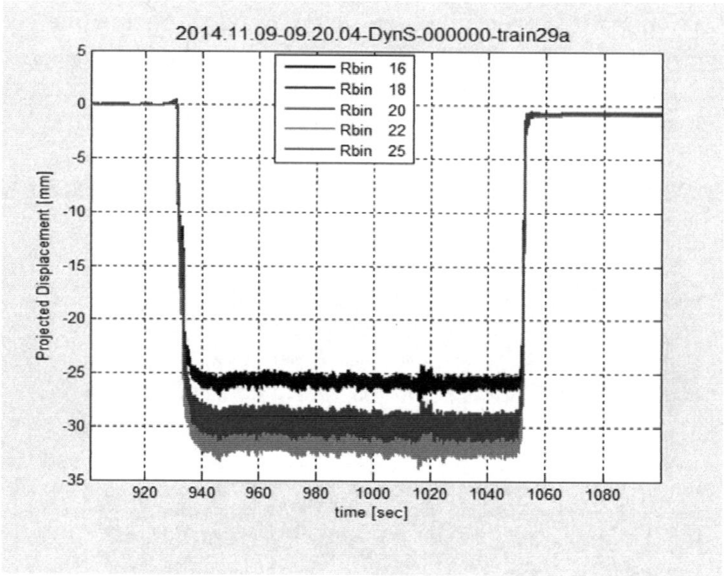

Figure 9. Displacement time graph of various range bins monitored. The time and date of acquisition is also shown on these graphs

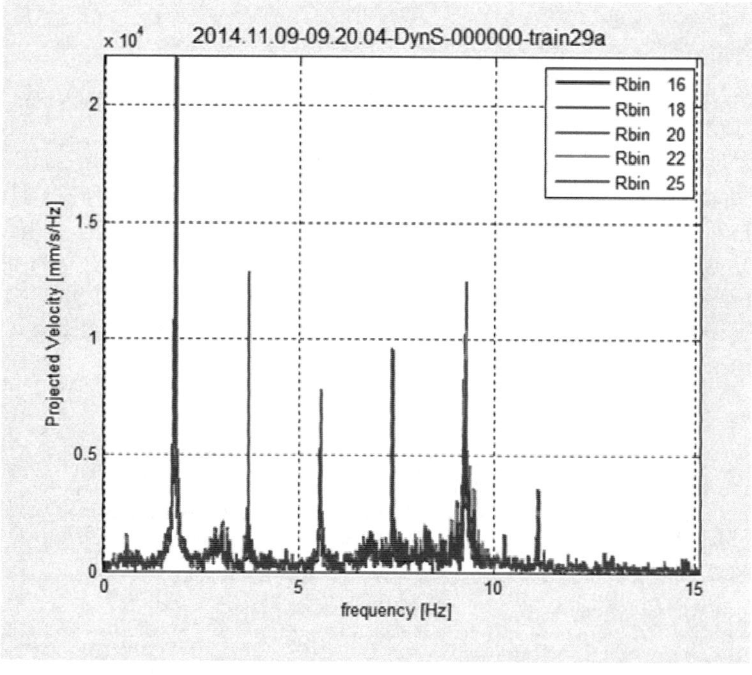

Figure 10. Vibration frequencies of each range bin monitored, with dominant frequency approximately 2 Hz

The vertical deflections over 40 mm and vibrational movement over 3 mm have been highlighted. The threshold for maximum vertical deflection has been guided by the average span of 26,500 mm divided by 640 = 41.41 mm, which is understood to be a maximum tolerance guide. The maximum vibration tolerance has been chosen to be 3 millimeters.

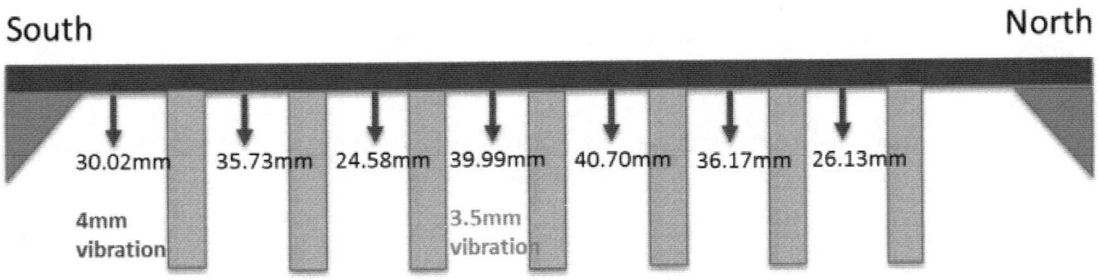

Figure 11. Example of results in one of the surveyed bridge

5. Conclusion

The survey provided detailed information on vertical deflections, vibration movement and vibration frequencies for multiple positions along each bridge structure surveyed, largely demonstrating the capabilities and accuracy of the system.
The data collected have given important information on the possibility to increase the trains load and demonstrate the full applicability of the system into the mining company bridge inspection program.
Furthermore, this field work is only the beginning for an ongoing monitoring and maintenance plan of the mining railway bridges, with the listed activities to be performed:

- periodical monitoring of bridge condition over time;
- regular scanning of areas which exhibit higher vertical deflections or vibration;
- an extended workshop or training program to garner more information from the data collected during the survey;
- continuous improvement of the use of interferometry for bridge monitoring through feedback from the visual inspection and maintenance group.

These information allow the IBIS-FS to be used as a SHM tool (Structural Health Monitoring) to predict damage and, especially, to plan maintenance actions to guarantee and increase the mining productivity.

References

Barnes, R., Lee, T. and Papworth, F., 2011. Interferometric radar for the measurement of structural deflection of concrete structures which in turn was referenced from Taylor 2001

Bernardini, G., Gallino, A., Gentile, C., Ricci, P., 2007. Dynamic monitoring of civil engineering structures by microwave interferometer.

Bernardini, G., De Pasquale, G., Bicci, A., Marra, M., Coppi, F., Ricci, P. and Pieraccini, M., 2007. Microwave interferometer for ambient vibration measurement on civil engineering structures: Principles of the radar technique and laboratory tests.

Gentile, C., 2009. Radar-based measurement of deflections on bridges and large structures: advantages, limitations and possible applications. IV ECCOMAS Thematic Conference on Smart Structures and Materials.

Henderson, F. and Lewis A., 1998, Manual of remote sensing principals and applications of imaging radar, John Wiley and Sons, New York, USA.

Palombo A. et al., 2011, Noninvasive remote sensing techniques for infrastructure diagnostics, Hindawi Publishing Corporation, International Journal of Geophysics Volume 2011, Article ID 204976, 9 pages, doi:10.1155/2011/204976.
Taylor, J., 2001, Ultrawideband Radar Technology, CRC Press

The 8th International Conference on
Structural Health Monitoring of Intelligent Infrastructure
Brisbane, Australia | 5-8 December 2017

Experimental study of temperature effect on WSN tilt node for tunnel structure monitoring

H.W. Huang[1], F. Du[1], D.M. Zhang[1] and Y. Wu[2]

[1] Department of Geotechnical Engineering, Tongji University, China, Email: 09zhang@tongji.edu.cn , [2] Wisen Innovation Ltd, China

Abstract

In this paper, the temperature effects on the output errors of a type of wireless sensor network tilt node (WSN tilt node) integrated by a dual axis MEMS-based inclinometer is studyed by means of temperature experiments. An effective temperature compensation method for the inclination output of the WSN tilt node is proposed for the purpose of decreasing the output temperature drift errors and expanding the application range of the WSN tilt node from a very low to very high temperature without losing the required accuracy and stability. Firstly, a type of WSN tilt node for tunnel and other civil infrastructures monitoring is presented from the point view of hardware composition. The temperature effects on the key sensing element, the MEMS-based inclinometer, are discussed. Then, in order to compensate the output temperature drift errors of the WSN tilt node, the least mean square (LMS) polynomial curve fitting is utilized for building the temperature drift model based on the indoor temperature experiments. Experimental results demonstrate that: (1) the inclination output errors of the WSN tilt nodes vary significantly with temperature variation; (2) to some extent, the temperature drift characteristics of the WSN tilt node differ from the MEMS-based inclinometer after the manufacture; (3) the temperature dependency of the inclination output errors can be fitted by 3rd order polynomial curve over the ambient temperature range from -40°C~80°C and the regression equations can be used to compensate the raw data, and the compensation method proposed is realizable and effective.

1. Introduction

In recent years, Micro-Electro-Mechanical System (MEMS)-based wireless sensor network (WSN) has been rapidly developed for its merit of real-time monitoring and prewarnings on critical infrastructures, such as bridges (Nagayama & Spencer, Jr. 2009) and metro tunnels (Bennett & Soga 2000, Yin & Huang 2015, Wang & Huang 2016). Those infrastructures could be located either in friendly working conditions or in adverse working conditions, such as in the extreme cold region. Among all the adverse factors of working conditions, extreme temperature is one of the most important factors that influence the accuracy of MEMS-based sensors and consequently is one of the largest sources of error in measurement(Xia & Chen 2009). Therefore, in order to expand the application range of the MEMS-based sensors from a very low to very high temperature without losing the required accuracy and stability, temperature compensation should be taken to eliminate the performance degradation of the MEMS-based sensors due to the ambient temperature fluctuations.
To reduce the temperature drift errors of the MEMS-based sensors, the following two processes are needed: (1) Temperature calibration: to develop a temperature model of the sensor errors, i.e., build the relationship between the sensor errors and the sensor temperature; (2) Temperature

compensation: to compensate the temperature drift of the sensor errors according to their temperature during the operation period. Commonly, the MEMS-based sensor errors under different temperature points are tested using temperature chamber (Xia & Chen 2009, Niu & Zhang 2013) . After the temperature calibration, based on the use of Back Propagation (BP) neural networks (Xia & Chen 2009, Fontanella & Accardo 2016) or polynomial fitting (Xia & Chen 2009, Niu & Zhang 2013) to model the relationship between the sensor errors and corresponding temperature (i.e., the temperature model of the MEMS-based sensors). However, these previous researches just focused on the temperature drift and compensation on the MEMS-based sensors. Recently, various wireless sensor nodes prototypes integrated by different MEMS-based sensors and many other electronic components have been developed (Yu & Ou 2009, Ha & Park 2013). For the wireless sensor node, performing the temperature compensation for the measurement output is much more significant for a WSN system to provide reliable and accurate monitoring result.

In this paper, a type of wireless sensor network tilt node (WSN tilt node) for tunnel structure and other civil infrastructures monitoring, integrated by a dual axis silicon capacitive MEMS-based inclinometer, is presented. The electronic components of the WSN tilt node are introduced; and the effects of temperature variation on MEMS-based inclinometer are analyzed. In order to test the temperature drift characteristics of the inclination output, indoor temperature experiments using temperature soak method (Titterton & Weston 2004) have been carried out. Furthermore, the least mean squares (LMS) polynomial curve fitting is applied in constructing the temperature compensation model based on the experiments results; it is then used to compensate the inclination output of the WSN tilt node.

2. Hardware composition of the WSN tilt node

The WSN tilt node studied in this paper, as shown in Fig. 1, is the sensing device in the WSN system. This device is capable of sensing the inclination variation of the structure with the range from -30° to 30° as well as sensing the ambient temperature, processing the measurements and transmitting the data via its internal radio module.

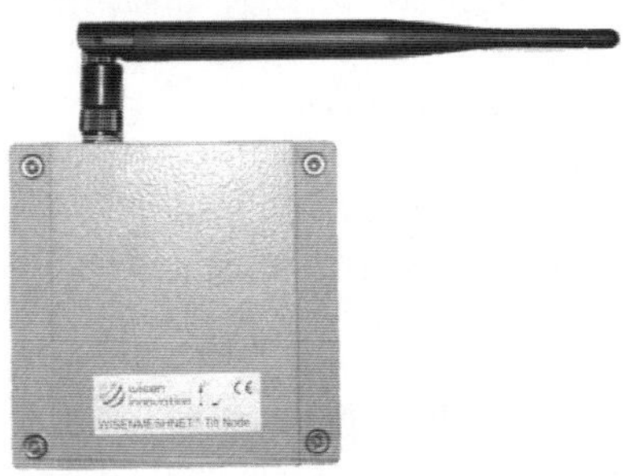

Fig. 1 WSN tilt node prototype

The external component is constructed of aluminum-alloy with the size of 80mm×75mm×57mm, preventing physical damage to the inside electronic components by the harsh environment. The internal modules can be classified into sensing module (Fig. 2) and controling module (Fig. 3), and these two modules are divided into upper and lower layers of printed circuit board (PCB).

As shown in Fig. 2(a), the SCA100T inclinometer is mounted on the lower surface of the lower PCB, measureing the inclination in the two horizontal (X axis and Y axis) direction. The SCA100T inclinometer is a silicon capacitive 3D-MEMS-based dual axis inclinometer that can provide instrumentation grade performance for leveling applications. The inclinometer with high shock durability and time stability is considered suitable for civil engineering and construction application. In addition, a 16 bit analog-digital convertor (ADC), CMOS voltage regulators and switched capacitor voltage converter are placed on this surface of PCB. At the upper surface of lower PCB (shown as Fig. 2(b)), a metal holder is mounted for the installation of a 3.6V lithium polymer battery that is the power supply of the entire WSN tilt node.

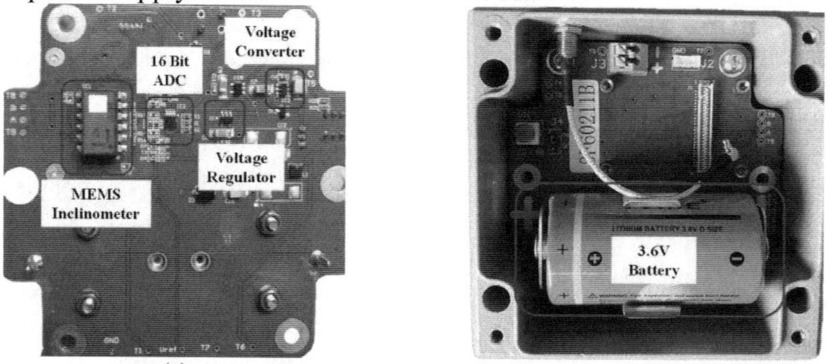

(a) (b)

Fig. 2 Sensing module. (a) Lower surface; (b) Upper surface

Fig. 3 shows the lower surface and upper surface of the controlling module. The controling module mainly consists of a microcontroller with low power 2.4GHz transceiver for IEEE 802.15.4. and a CMOS voltage regulators. The microcontroller, which is responsible for controling the entire WSN tilt node work, receiving the digital signal from the ADC, then processing and transmiting the inclination and temperature measurements using radio, is the core of the whole system. The microcontroller connects a 5dBi antenna with the length of 20cm by a wire.

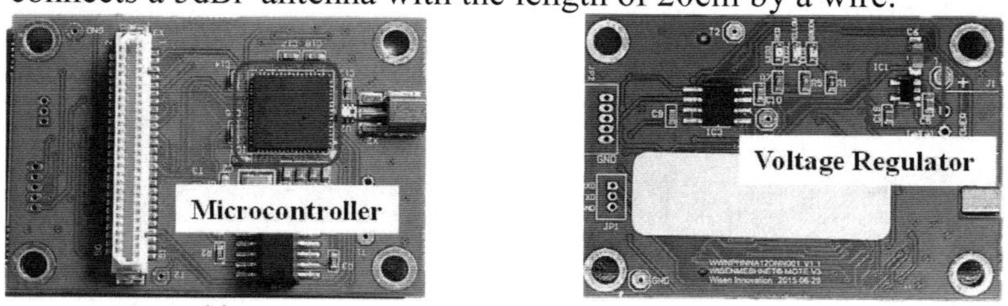

(a) (b)

Fig. 3 Controlling module. (a) Lower surface; (b) Upper surface

In conclusion, the main electronic components inside a WSN tilt node include the MEMS-based inclinometer, AD convertor, microcontroller, CMOS voltage regulators, switched capacitor voltage converter and battery. And the transmitting procedure of the measurement signals inside a WSN tilt node is as follows: At first, the inclination of structure measured by the SCA100T inclinometer output as analog values. Then, the analog values are converted to 16-bit digital values via the 16 bit ADC. The digital signals are transmitted to the microcontroller, the digital signals are modulated to generate the RF transmit signals under the control of the microcontroller. Next, the RF signals are transmitted wirelessly using radio to the gateway by the fully integrated low power radio transceiver for 2.4GHz ISM band.

3. Tepmerature characteristics of the SCA100T inclinometer

As the sensing element, the performances of the SCA100T inclinometer largely determine the performances of the WSN tilt node. The temperature characteristics of the SCA100T inclinometer directly cause the temperature drift errors of the inclination output of the WSN tilt node. Thus, analyzing the temperature characteristics of the SCA100T inclinometer is helpful to understand the temperature dependency of the WSN tilt node output.

The architecture of the MEMS-based sensors is fabricated with silicon, which is a high temperature-sensitive material and its physical characteristics vary greatly with ambient temperature. The elasticity modulus of the silicon will change with the temperature, which will cause the change of the system stiffness so that the resonance frequency of the sensor will be

changed (Tang & Wang 2014). An additional part of the error observed is due to the sensitivity of the packaging and the application specific integrated circuit (ASIC) to temperature. Joo and Choa (Joo & Choa 2007) described the influence of temperature on the package of MEMS-based accelerometer. It generally consists of various materials that have different coefficients of thermal expansion. Therefore, temperature change will induce thermo-mechanical stress that could cause deformation of the MEMS structure, resulting in frequency change or breakage of the structure. Liu et al (Liu & Liu, 2011) presented a model to analyze the influence of temperature on ASIC of MEMS-based accelerometer. Therefore, the measurements of MEMS-based sensors are extremely sensitive to changes in the environmental temperature under the combined effects of these factors.
According to the datasheet of SCA100T inclinometer, typical temperature dependency of offset and sensitivity over the temperature range from -40°C to 85°C is presented in Fig. 4 (Murata Electronics Oy 2017). The mean value and 3σ limits (mean ± 3 times standard deviation) are presented in figure, the 3σ limits represents 99.73% of the SCA100T population.

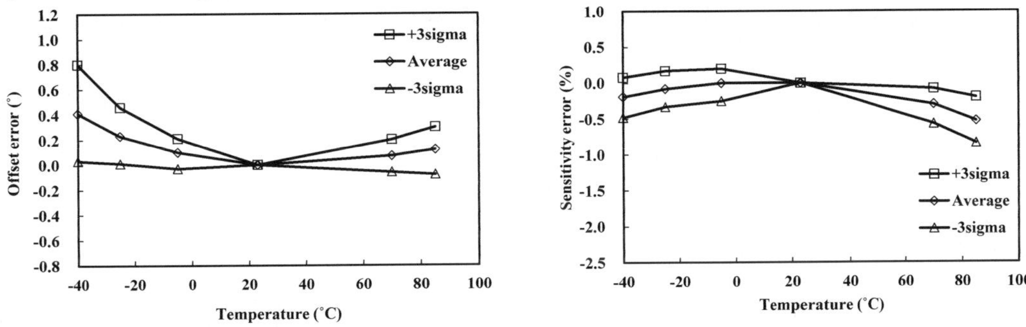

(a) Offset (b) Sensitivity

Fig. 4 Typical temperature dependency of SCA100T inclinometer

In this paper, 3rd order polynomials with LMS theory are used to fit the typical temperature dependency curves of offset and sensitivity. The 3rd polynomial regression equations for offset and sensitive are as following Eq. (1) to Eq. (3) and Eq. (4) to Eq. (6) respectively:

Offset:

$$\text{+3sigma:}\ \ Offcorr = -5.41\times10^{-7}\cdot T^3 + 1.73\times10^{-4}\cdot T^2 - 8.75\times10^{-3}\cdot T + 1.37\times10^{-1} \quad (1)$$

$$\text{Average:}\ \ Offcorr = -3.05\times10^{-7}\cdot T^3 + 8.69\times10^{-5}\cdot T^2 - 4.55\times10^{-3}\cdot T + 6.61\times10^{-2} \quad (2)$$

$$\text{-3sigma:}\ \ Offcorr = -2.42\times10^{-7}\cdot T^3 + 1.31\times10^{-5}\cdot T^2 - 2.15\times10^{-4}\cdot T - 1.46\times10^{-2} \quad (3)$$

Sensitivity:

$$\text{+3sigma:}\ \ Scorr = 6.22\times10^{-7}\cdot T^3 - 6.79\times10^{-5}\cdot T^2 - 2.42\times10^{-3}\cdot T + 1.48\times10^{-1} \quad (4)$$

$$\text{Average:}\ \ Scorr = -2.14\times10^{-7}\cdot T^3 - 7.92\times10^{-5}\cdot T^2 + 2.02\times10^{-3}\cdot T + 8.84\times10^{-3} \quad (5)$$

$$\text{-3sigma:}\ \ Scorr = -8.48\times10^{-7}\cdot T^3 - 9.75\times10^{-5}\cdot T^2 + 5.88\times10^{-3}\cdot T - 1.50\times10^{-1} \quad (6)$$

where Offcorr and Scorr is the 3rd order polynomial fitted to the offset and sensitivity temperature dependency curves. T is the temperature in °C.
For the SCA100T inclinometer, analog output can be transferred to angle (α) using the following Eq. (7) for conversion:

$$\alpha = \arcsin\left(\frac{V_{out} - Offset}{Sensitivity}\right) \quad (7)$$

where Offset is the output of the device at 0° inclination position, Sensitivity is the sensitivity of the device and V_{out} is the analog output of the SCA100T. At 23°C, the nominal offset is 2.5V and the sensitivity is 4V/g.
Combining the Eq. (7) with the Eq. (1) to Eq. (6), the typical temperature dependency curves of inclination output can be obtained, as shown in Fig. 5.

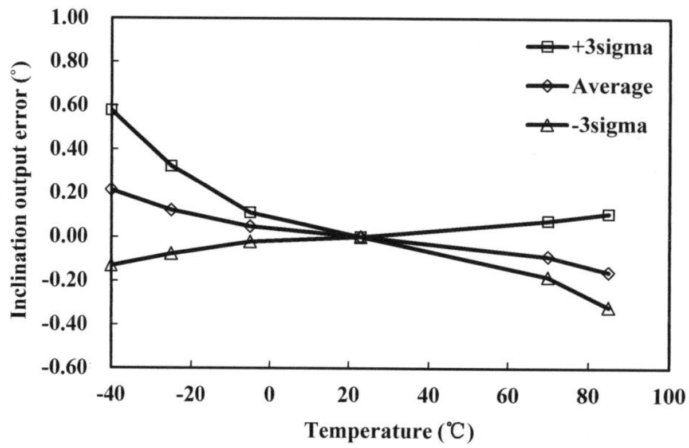

Fig. 5 Typical temperature dependency of SCA100T inclination output

4. Temperature drift characteristic of WSN tilt node and temperature compensation method

In this section, the temperature experiments are presented. The purpose of the temperature experiments is to determine the inclination output error of the WSN tilt node under different temperature point. Based on the experimental results, the temperature compensation method for the WSN tilt node is proposed and demonstrated its effective.

4.1 Temperature experiments

In order to build the relationship between the inclination output error of the WSN tilt node and the operation temperature, indoor temperature experiments using temperature soak method have been carried out. Fig. 6 shows the equipment of the temperature experiment in the laboratory of the Shanghai Institute of Measurement and Testing Technology (SIMT). In the experiments, three WSN tilt nodes with the node ID 2180, 2204 and 2330 were placed at rest in ESPEC EW-0470 temperature chamber, in which temperature variation range is from -70°C to +150°C, with a tolerance of ±1.5°C. The data packets including the inclination and temperature information were transmitted from the WSN tilt nodes to the gateway, then, the gateway transmit the data packets to the server via mobile network. Logging in the local windows server using PC, the inclination data and corresponding temperature data can be obtained in real time.

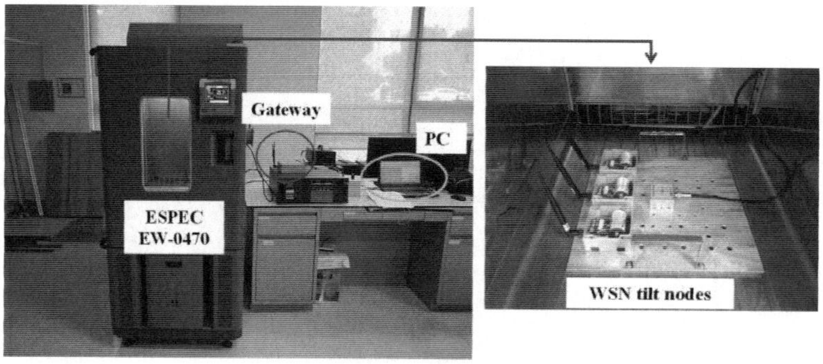

Fig. 6 Temperature experiment equipment

During the course of temperature test, the WSN tilt nodes were statically placed without the lid, the temperature in the temperature chamber was the only environment variable. The temperature value in the temperature chamber increased from -40°C to +80°C selecting the temperature point at -40°C, -30°C, -20°C, -10°C, 5°C, 25°C, 45°C, 65°C and 80°C as the sampling point. The temperature at each sampling point is maintained for at least 60 minutes to ensure that the temperature in the chamber is uniformly distributed and the WSN tilt nodes were fully heated. In this way, any temperature difference between the MEMS-based inclinometer and on-chip temperature sensor of the WSN tilt nodes can be effectively reduced. After that, 40 data packets were collected with the sampling time interval of 8 seconds. According to the temperature output of the WSN tilt nodes, the operation temperature profile of the WSN tilt nodes is shown in Fig. 7.

As shown in Fig. 7, the temperature at each sampling point inside the WSN tilt node was already stabilized after the certain period of constant temperature process. The temperature output is measured by the integrated temperature sensor on the microcontroller chip, and the accuracy of temperature measurement is ±2°C. Hence, the stabilized temperature of the WSN tilt node was not completely identical with the setting temperature of the temperature chamber.

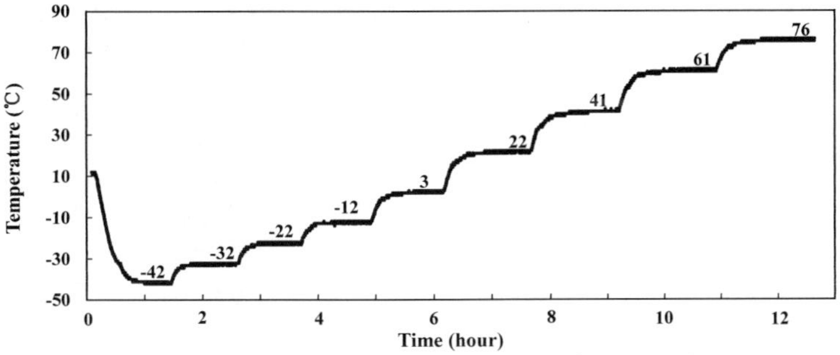

Fig. 7 Temperature profile of WSN tilt node in experiment

4.2 Temperature drift characteristics of WSN tilt node

Take the average value of the 40 inclination output values as the inclination output value of the WSN tilt node at this ambient temperature point. The inclination output at 25°C is set as the zero inclination output. The inclination output error (α_{error}) at each temperature point is calculated using the following Eq. (8):

$$\alpha_{error} = \alpha_T - \alpha_0 \tag{8}$$

where the α_{error} is the inclination output error at temperature T, α_T is the inclination output value at temperature T, α_0 is the inclination output value at 25°C.

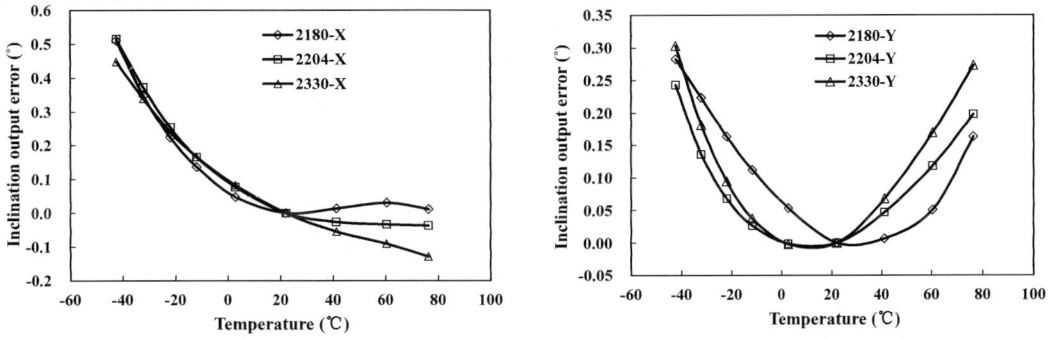

(a) X axis (b) Y axis

Fig. 8 Temperature dependency of inclination output of WSN tilt nodes

Fig. 8 shows the experimental results of temperature dependency of the WSN tilt nodes inclination output in stationary condition for both X axis and Y axis. It is clear that the inclination output errors vary significantly with temperature for both X and Y axis. Both figures show that the temperature drift curve of inclination output slightly varies with different WSN tilt nodes. The temperature drift curves of X axis and Y axis shows different characteristics. For X axis of WSN tilt nodes, the output errors over low temperature range (-40°C~25°C) are obviously greater than the output errors over high temperature range (25°C~80°C). Over the low temperature range, the output errors can be as high as more than 0.45°, while the output errors are less than 0.15° over the high temperature range. However, for Y axis, the differences of output errors over low and high temperature range decrease. To more intuitively illustrate the effect of low and high temperature range for the output errors, the temperature coefficient of output errors (TCO) over different temperature range is calculated. Fig. 9 shows the TCO of different WSN tilt nodes over low and high temperature range. As shown in Fig. 9(a), the TCO of all three WSN tilt nodes over low temperature range is obviously greater than the TCO over high temperature range; therefore, the output errors of X axis are more sensitive to the low temperature. As shown in Fig. 9(b), the TCO over low and temperature range is closed; the output errors of Y axis are sensitive to both the low and high temperature.

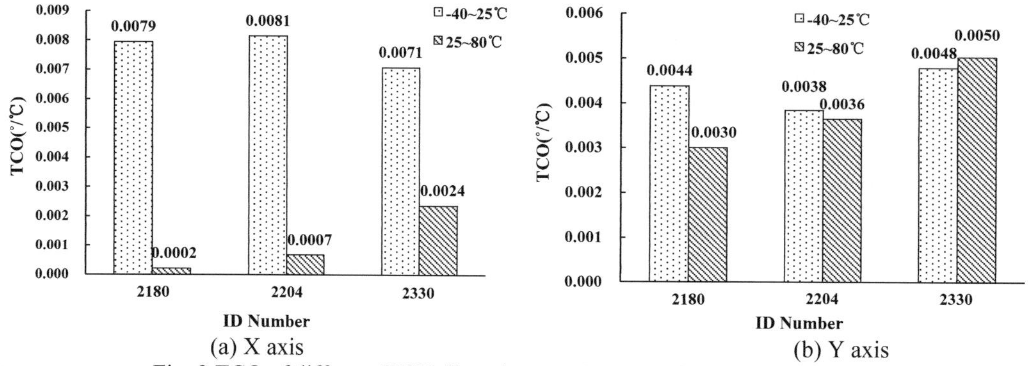

Fig. 9 TCO of different WSN tilt nodes over low and high temperature range

Fig. 10 shows comparison of the temperature drift characteristics between the SCA100T inclinometer and WSN tilt nodes. Over the low temperature range, the inclination output errors of all three WSN tilt nodes for both X axis and Y axis are in the ±3sigma range of the SCA100T inclinometer output errors. However, for Y axis of the WSN tilt nodes, the values of the output errors are more than the values of ±3sigma range of the SCA100T inclinometer over the high temperature range. Thus, comparing with the SCA100T chip, the temperature characteristics of the inclination output of the WSN tilt nodes will change after the manufacture to some extent.

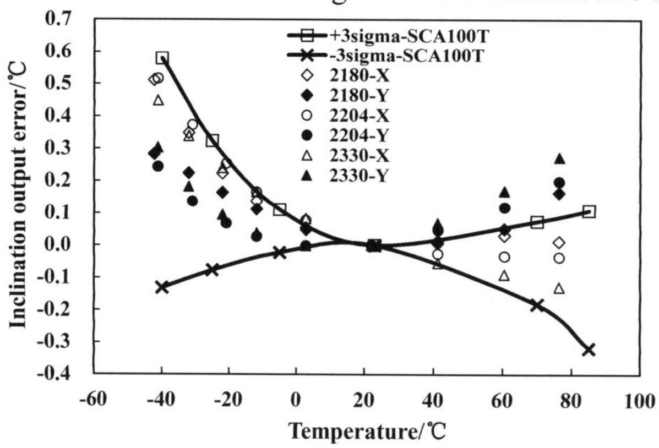

Fig. 10 Temperature dependency of inclination output, SCA100T inclinometer vs. WSN tilt node

4.3 Temperature compensation method for WSN tilt node

According to the analysis of the testing results in Section 4.2, the temperature variation has significant effects on the output accuracy of a WSN tilt node; in addition, the temperature drift characteristics of the WSN tilt node are different with the SCA100T so that the compensation model of the SCA100T can't apply to the WSN tilt node. Therefore, it is necessary to conduct the study on temperature compensation for the WSN tilt node and build the corresponding compensation model. Polynomial fitting has been widely applied, and the LMS curve fitting is reliable to get optimal results. Thus the polynomial fitting method is proposed for simplicity and effectiveness, which can provide satisfying compensating effects.

The main idea of the polynomial fitting compensation is as follows: the correlation between the operating temperature and the output error of the WSN tilt nodes can be found through temperature experiments and its mathematical expression can be obtained through polynomial fitting as the temperature is changed at a certain temperature range. The mathematical function is memorized in the web server background. Thus the corresponding compensation value for each real time temperature tested is calculated and subtracted in the actual inclination output of the WSN tilt node. Ultimately the final compensated inclination output is obtained and shown on the foregrounding of the web server.

In this paper, using 3rd polynomial to describe the approximate function (regression equation) relationship between the inclination output errors and corresponding temperature. One of the WSN tilt node with the node ID 2180 is chosen to illustrate the temperature compensation method. Fig. 11 shows the fitted curve of the inclination output errors for both X axis and Y axis. The goodness of fit R^2 is more than 0.99.

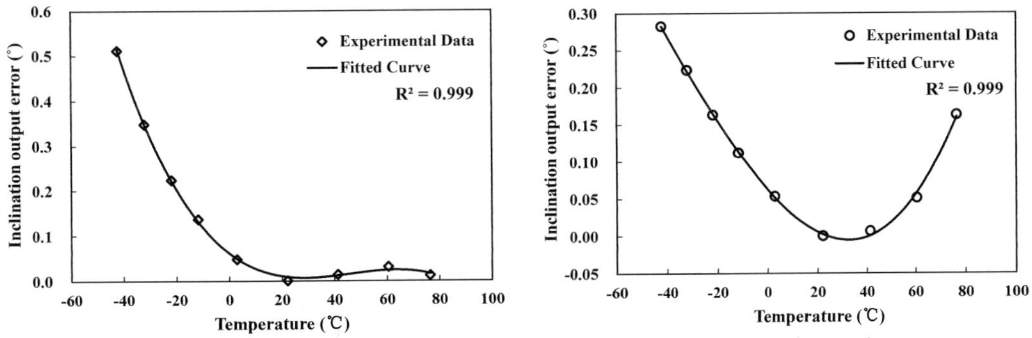

(a) X axis (b) Y axis

Fig. 11 fitted curve of inclination output errors with temperature

After fitting the experimental results with the LMS theory, 3rd polynomial regression equations of inclination output errors for X axis and Y axis are as Eq. (9) and Eq. (10), which can be used for the temperature compensation of the inclination output. Then, by subtracting this regression equation from the raw inclination data, the compensated inclination equation can be obtained as Eq. (11).

$$\alpha_{corr}^{X} = -8.13\times10^{-7}\cdot T^{3} + 1.13\times10^{-4}\cdot T^{2} - 4.47\times10^{-3} + 6.11\times10^{-2} \tag{9}$$

$$\alpha_{corr}^{Y} = 2.98\times10^{-7}\cdot T^{3} + 4.45\times10^{-5}\cdot T^{2} - 3.85\times10^{-3} + 6.24\times10^{-2} \tag{10}$$

$$\alpha_{comp} = \alpha_{T} - \alpha_{corr} \tag{11}$$

where T is the temperature collected by the on-chip temperature sensor inside the WSN tilt node, α_{corr}^{X} and α_{corr}^{Y} is the 3rd polynomial fitted to output errors temperature dependency curve of X axis and Y axis respectively(Fig. 11), α_T is the inclination output before temperature compensation, α_{comp} is the compensated inclination.

The temperature compensation procedure is as follows. For a WSN system, the WSN tilt node measures the inclination of structure and the temperature. The raw data are transmitted from the WSN tilt node to the gateway using radio, then, the gateway transmits the raw data to the local windows server using mobile network. The polynomial fitting equations (Eq. (9) and Eq. (10)) are stored in the web server background and used to calculate the compensation value in terms of the measured temperature, and finally the compensated inclination values are shown on the foregrounding of the web server. To verify the compensation effects, the WSN tilt nodes were placed in the temperature chamber statically and the other group of temperature experiment was carried out using soak method. The experimental process was same with the experiment mentioned before. Fig. 12 shows the inclination output errors of WSN tilt node with node ID 2180 before and after temperature compensation.

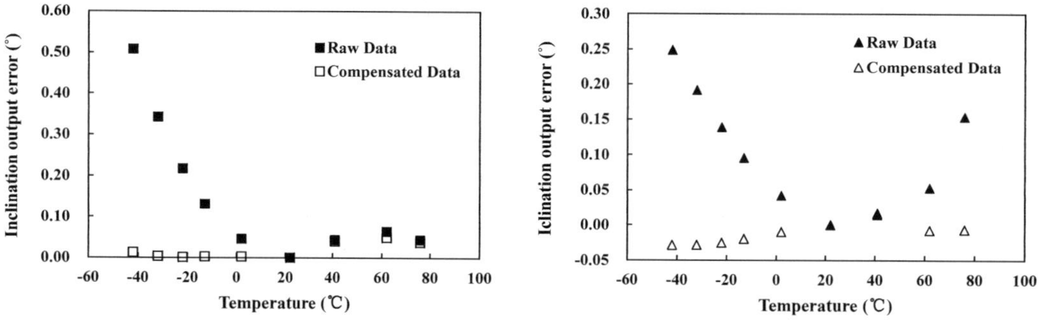

(a) X axis (b) Y axis

Fig. 12 inclination output errors of WSN tilt node before and after compensation

As shown in Fig. 12, the maximum absolute value of output error decreases from 0.508° and 0.249° before compensation to 0.049 (reduced by 90.4%) and 0.029 (reduced by 88.4%) after compensation for X and Y axis respectively over the temperature range from -40°C to 80°C. Another two WSN tilt nodes show the similar results. Therefore, the compensation effect of the proposed temperature compensation method is obvious and effective.

5. Conclusion

In this paper, a type of WSN tilt node for tunnel and other civil infrastructures automatic monitoring is presented. The WSN tilt node is developed based on the MEMS-based SCA100T inclinometer, the other main electronic components consists of the 16-bit AD convertor, microcontroller, CMOS voltage regulators, switched capacitor voltage converter and a 3.6V lithium battery. As the key sensing element of the WSN tilt node, the temperature characteristics of the SCA100T inclinometer are analyzed according to the official datasheet. The temperature drift characteristics of the output errors is obtained by combining the offset temperature dependency, sensitivity temperature dependency and equation of voltage to angle conversion.

Because of the manufacture, the temperature characteristics of the WSN tilt node will be different from the SCA100T inclinometer. In this paper, the temperature drift characteristics of three WSN tilt nodes were tested via indoor temperature experiments using temperature soak method. The experimental results indicate that the inclination output errors vary significantly with temperature for the dual axis. The temperature drift curve of output errors slightly varies with different WSN tilt nodes. And the temperature drift curves of X axis and Y axis show different characteristics. To be specific, the output errors of X axis are more sensitive to the low temperature, and the output errors of Y axis are sensitive to both the low and high temperature.

To improve the accuracy of the WSN tilt node output, LMS polynomial fitting is utilized for building the temperature model the output errors. The temperature drift curves of the output errors can be fitted by 3rd order polynomial curve over the ambient temperature range from -40°C~80°C with great goodness-of-fit, the obtained regression equations are stored in the web server background to be used for temperature compensation. Finally, the results of the temperature compensation tests demonstrate that the proposed compensation method is realizable and effective for the WSN tilt nodes

Acknowledgment

This work was supported by the Transportation Science and Technology Program Research Project of Jilin Province (2015/1/18), Natural Science Foundation Committee program (51278381) and Study on Safety Risk Control Technology and Theory of Super Large Underground Space Construction program (1402XH0014). The support is gratefully acknowledged. The authors want to thank Xiaoyan Huang and Xinwen Xiong for the technical support.

References

Nagayama, T, Spencer Jr, BF & Rice, JA 2009. 'Autonomous decentralized structural health monitoring using smart sensors', *Structural Control & Health Monitoring*, vol. 16, no. 7–8, pp. 842–859.

Bennett, PJ, Soga, K, Wassell, I, Fidler, P, Abe, K, Kobayashi, Y & Vanicek, M 2010. 'Wireless sensor networks for underground railway applications: case studies in prague and london', *Smart Structure and System*, vol. 6, no. 5-6, pp. 619-639.

Yin, JG & Huang, HW 2015. 'Real time monitoring method for the longitudinal settlement of shield tunnel using wireless inclinometer', *2015 Information Technology and Mechatronics Engineering Conference(ITOEC2015)*, pp. 274-280.

Wang, F, Huang, HW, He, B, Wu, Y, Shao, H & Wu, HM 2016. 'Wireless sensing on shield tunnels in Shanghai', *International Conference on Smart Infrastructure and Construction,* pp. 191-196.

Xia, DZ, Chen, SL, Wang, SR & Li, HS 2009. 'Microgyroscope temperature effects and compensation-control methods', *Sensors*, vol. 9, no. 10, pp. 8349-8376.

Niu, XJ, Li, Y, Zhang, HP, Wang, QJ & Ban, YL. 2013. 'Fast thermal calibration of low-grade inertial sensors and inertial measurement units', *Sensors,*vol. 13, no. 9, pp. 12192-12217.

Fontanella, R, Accardo, D, Caricati, E, Cimmino, S & Simone, DD, 2016. 'An extensive analysis for the use of back propagation neural networks to perform the calibration of MEMS gyro bias thermal drift', *Position, Location and Navigation Symposium* pp. 672-680.

Yu, Y, Ou, JP, Zhang, J, Zhang, CW & Li, LY 2009, 'Development of wireless MEMS inclination sensor system for swing monitoring of large-scale hook structures', *IEEE Transactions on Industrial Electronics*, vol. 56, no. 4, pp. 1072-1078.

Ha, DW, Park, HS, Choi, SW & Kim, Y 2013, 'A wireless mems-based inclinometer sensor node for structural health monitoring', *Sensors,* vol. 13, no. 12, pp. 16090-16104.

Titterton, D & Weston, JL 2004 *Strapdown Inertial Navigation Technology, 2nd ed*, American Institute of Aeronautics and Astronomy Press, Reston, VA, USA.

Tang, QJ, Wang, XJ, Yang, QP & Liu, CZ 2014, 'Static temperature analysis and compensation of mems gyroscopes', *International Journal of Metrology & Quality Engineering,*vol. 4 no.3, pp. 209-214.

Joo, JW & Choa, SH 2007, 'Deformation behavior of mems gyroscope sensor package subjected to temperature change' *IEEE Transactions on Components & Packaging Technologies,* vol. 30, no. 2, pp. 346-354.

Liu, MJ, Liu, YF & Dong, JX 2011, 'Temperature characteristic of integrated circuit for capacitive micro-accelerometer'(In Chinese), *Journal of Chinese Inertial Technology*, vol. 19, no. 5, pp. 599-602

Murata Electronics Oy. Data Sheet for the SCA100T dual axis inclinometer series. Available online: http://www.murata.com/~/media/webrenewal/products/sensor/inclinometer/sca100t/sca100t_inclinometer_datasheet_8261800b2_0.ashx?la=en.

The 8th International Conference on
Structural Health Monitoring of Intelligent Infrastructure
Brisbane, Australia | 5-8 December 2017

Embedded wireless sensor systems for long-term SHM and corrosion detection in concrete components

M. Bartholmai[1], S. Johann, C. Strangfeld.
[1] *Bundesanstalt für Materialforschung und -prüfung (BAM) – Germany, Email:* matthias.bartholmai@bam.de

Abstract

State-of-the-art communication standards like RFID and Bluetooth Low Energy enable the development of sensor systems which can be completely embedded into concrete components for long-term SHM and early damage detection. Objective of the project KonSens which is carried out at BAM is the development, implementation, and validation of sensors for measuring of parameters relevant for corrosion, like moisture, pH value, and electrical conductivity, inside steel reinforced concrete components. The primarily addressed application is detection and evaluation of corrosion processes in concrete bridges.

In contrast to cable connected sensors, embedded wireless sensors avoid any pathways for intrusion of moisture and chemicals, e.g., chlorides which could trigger corrosion activity. To allow for long-term, ideally life-time operation, the once embedded sensor systems must work highly energy efficient. One option are passive RFID sensor systems, which work without battery. The energy is transmitted to the system through the electromagnetic field, even to operate sensors. A crucial parameter is the transmission depth in concrete. First experiments with RFID sensors working at frequencies of 13.56 MHz (HF) and 868 MHz (UHF) embedded in concrete specimen resulted positive for transmission depths of up to 13 cm, which is quite promising, considering that corrosion would appear first at the top level of rebars.

A second generation of passive RFID sensor systems has been implemented with improved antenna design. Current experiments using these systems focus on the transmission characteristics in terms of transmission depths and the impact of concrete moisture. Low-energy humidity sensors are used and analysed regarding their capability for measuring the material moisture. Additionally, a relation between transmitted power to the embedded sensor and the moisture content of the concrete specimen caused by energy absorption can be presumed and is under systematic investigation.

1. Introduction

Corrosion of steel rebars as parts of reinforced concrete components is a major problem in the longevity of steel reinforced building structures. The intrusion of moisture and acids, including chloride and CO_2 leads to depassivation of the embedded steel and pitting corrosion. These processes progress under the concrete surface and remain invisible over a long period of time. Often, only when the volume increase of the corroding steel leads to cracks and chipping of the concrete the advanced damage becomes obvious. In this late state of damage process reparations are very expensive and usually require stoppage of operation, e.g., in the case of traffic infrastructure. Hence, this study addresses steel reinforced concrete bridges and comparable structures with high importance for uninterrupted and safe operation. Both, extensive use and harsh environmental conditions and

exposure promote intrusion processes and motivate for technical solutions to gain information about possible corrosion activity.
In recent years, Structural Health Monitoring has gained in importance, since growing age of buildings and infrastructure as well as increasing load requirements demand for reliable surveillance methods. The combination of sensors for condition monitoring and Radio Frequency Identification (RFID)-Technology enables the development of sensor systems, which can be completely embedded into the structure, particularly concrete components. Such wireless, integrated sensors can measure parameters, e.g., corrosion activity or moisture content in situ, without insertion of wires which provide vulnerabilities. At BAM the interdisciplinary project KonSens is carried out, which addresses the field structure integrated sensors. One of the strategies which are followed is the development and implementation of completely embedded sensor systems consisting of energy efficient communication and in situ sensors, and their further application potential (e.g. for precast concrete elements, roadways, wind power plants, and maritime structures).
Completely embedded RFID sensors offer the chance to monitor relevant parameters of structural integrity. Such systems can be integrated in structural components of new buildings during construction or by subsequent integration at crucial positions of existing buildings. For the latter the method of integration must be solved carefully, to minimize the impact on the structure and to assure representative measurements (Lesthaeghe et al. 2013). Actual research concentrates on two main fields of investigation: the measurement of mechanical impact, e.g., strain (Brunner et al. 2016) and the measurement of parameters relevant for corrosion, which is addressed here.
The presented results focus on the implementation of sensor systems consisting of RFID-tag and low-power sensors, which operate passively, i.e., without battery or energy storage and can be embedded completely and permanently in concrete parts. First results could show a relation between transmitted power and embedding depth (Bartholmai et al. 2016) and the suitability of humidity sensors for measuring the material moisture (Strangfeld et al. 2016). Further investigations with an improved system design address the transmission characteristics regarding embedding depth and concrete moisture content. Additionally, the comparison between HF and UHF is examined.

2. System Design and Operation Principle

Long-term embedding in concrete components requires a very robust system design. Besides implementation in passive form without batteries, an accordingly robust encapsulation which resists durably in the aggressive concrete environment is essential. Investigations were carried out, to identify and optimize suitable materials for encapsulation of the systems. Main challenge is the very alkaline environment, which leads to degradation of numerous polymer materials. In contrast to the previously used epoxy resins, thermoplastics like polyethylene, polypropylene and polyvinylchloride (PVC) consisting mainly of simple C_xH_y-structures are very resistant in such conditions. Validation tests carried out in pH 13 solution confirmed this assumption (Johann et al. 2016) and lead to the PVC encapsulation shown in Figures 1 and 2.
The specific antenna geometries for HF and UHF are shape defining for the systems. HF is implemented with a loop antenna with 25 mm diameter, while UHF uses a dipole antenna with length of 101 mm. A porous glass ceramic material (white square surface) is used to cover the capacitive humidity sensor and to allow for moisture transport to the sensor. These systems are additionally equipped with temperature sensors.

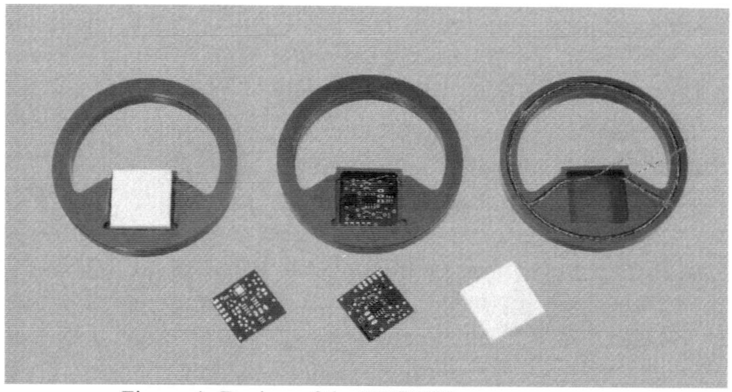

Figure 1: Design of the HF RFID sensor system

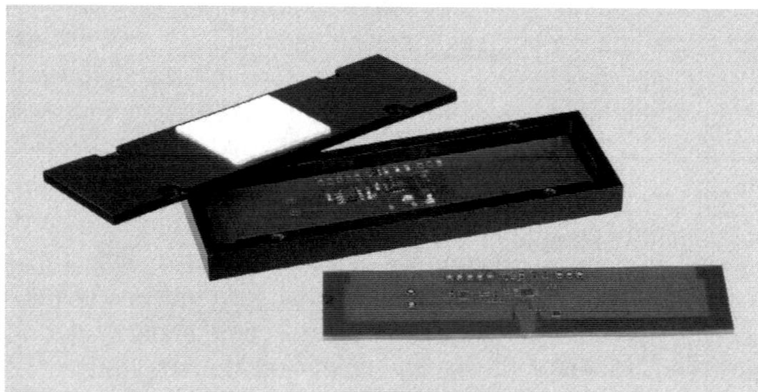

Figure 2: Design of the UHF RFID sensor system

Mode of operation is a passive energy supply which leads to sensor operation, only when an external RFID reader device is applied to the concrete surface close to the embedded sensor system. Then the RFID transponder antenna absorbs energy from the electromagnetic field generated by the reader device. The analog frontend, which is connected to the antenna, transmits the energy through the power management module to operate the external sensors. The signal input is processed by the microchip and communicated back to the reader device through the electromagnetic field via load modulation (HF) or backscattering (UHF).

3. Test Method and Specimen

A measurement routine was established for the systematic investigation of the transmission characteristics (Bartholmai et al. 2016). Its objective is to determine a comparable value, which rates the quality of transmission under specific conditions. The routine delivers the minimal required transmitted power P_{min} to activate a sensor response through RFID operation. Accordingly, a lower value of P_{min} represents a favorable result.
This study is based on measurements with two concrete test specimen. Both specimen have dimensions of 50 x 15 x 10 cm^3. They were assembled by wooden formwork boards and equipped with RFID sensor systems at fixed positions. Subsequently both specimen were filled with the same standard concrete. Figure 3 (left) displays the preparation before concreting of Specimen A with an HF and an UHF sensor system at the same installation depth of 6 cm from the top concrete surface.

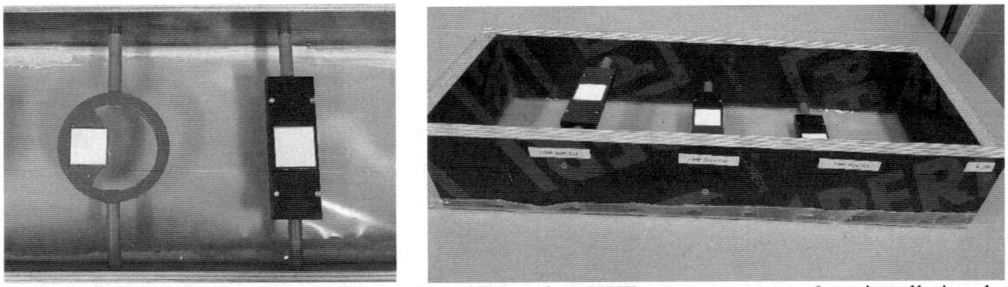

Figure 3. Left: Section of test Specimen A with an HF and an UHF sensor system at 6 cm installation depth. Right: Test Specimen B with three UHF sensor systems at different installation depths.

Figure 3 (right) shows the preparation of Specimen B with three identical UHF RFID sensor systems, which were installed in depths of 3, 6, and 9 cm from the top concrete surface. Measurements were performed periodically, first every day, later with longer time interval. The reader device for operating the measurements was always applied at the same fixed positions at the specimen top surface (Figure 4). Additionally, the sensor results were recorded and the specimen masses were weighed.

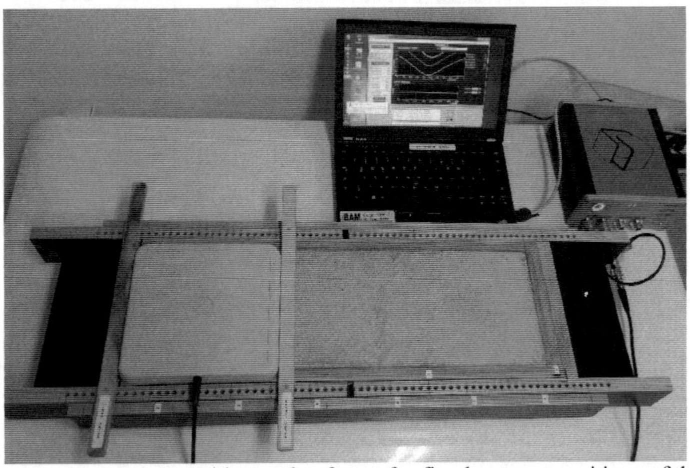

Figure 4: Measurement at a test specimen with wooden frame for fixed antenna positions of the RFID reader device.

4. Results and Discussion

After preparation both specimen were placed in an air-conditioned room with 23 °C and 50 % relative humidity. Figure 5 shows the results of weighted mass and material moisture measured with the UHF RFID sensor system. The mass progression is generated by the drying process of concrete over time and corresponds to the decreasing average moisture content of the specimen. Since the drying proceeds from the surface to the inside of the specimen and gravity has an impact, the moisture distribution is heterogeneous with varying gradients over time. However, the mass curve represents the spatially integral behavior, while the measured relative humidity inside the material results represent the position of the used sensor, in this case 6 cm below the top surface.

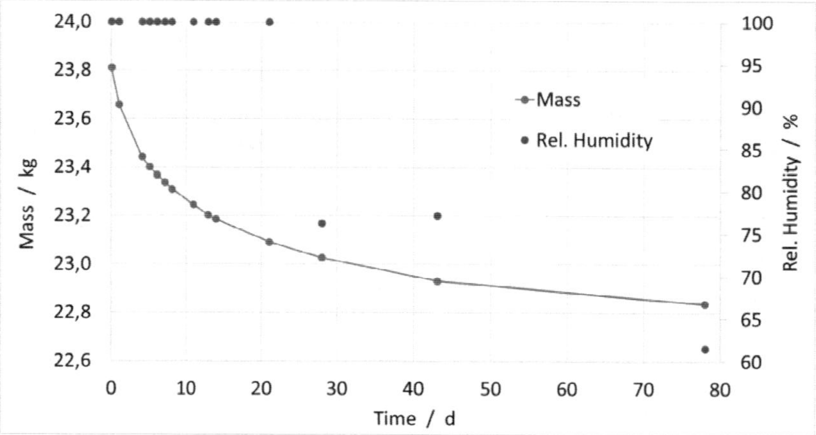

Figure 5: Mass and relative humidity measured at Specimen A.

The sensor results lead to the assumption, that during the first 21 days the sensor is in saturation. The three following measuring points roughly correlate to the mass decrease and give a promising outlook for the further trend of validation tests. The measured relative humidity can be converted into material moisture using the moisture sorption isotherm.

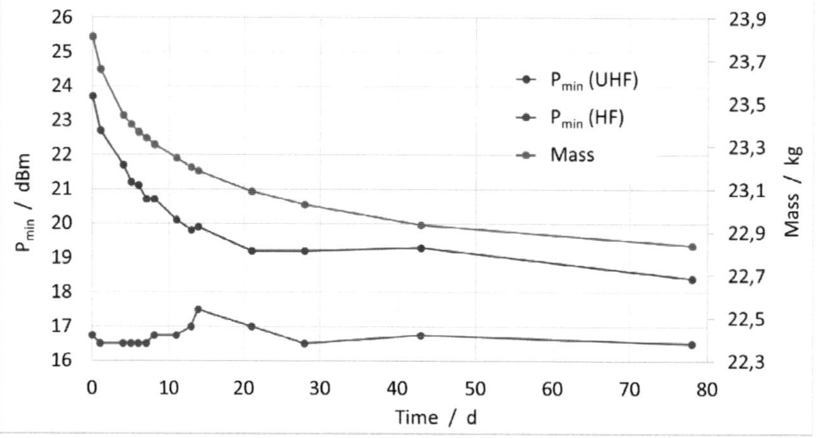

Figure 6: Mass and P_{min} for UHF and HF measured at Specimen A.

Figure 6 displays the mass progression and the progression of P_{min} for the UHF and the HF sensor system. Interestingly, P_{min}(UHF) follows the decrease of mass with time, while P_{min}(HF) remains stable. This behavior becomes even clearer in Figure 7, which shows P_{min} over the specimen mass. At UHF the minimal required power to activate a sensor response relates approximately linearly to the mass and hence to the moisture decrease, while HF is not affected. This behavior can be attributed to the energy absorption by water molecules at UHF, leading to strong signal attenuation and high power requirement at moist conditions. Since HF is rarely affected it allows better transmission through moist concrete. But UHF opens the possibility to use this effect for measuring the moisture content by P_{min}, at least, when other influences are known and respected. This could serve as a redundant or referencing method to the operation of humidity sensors.

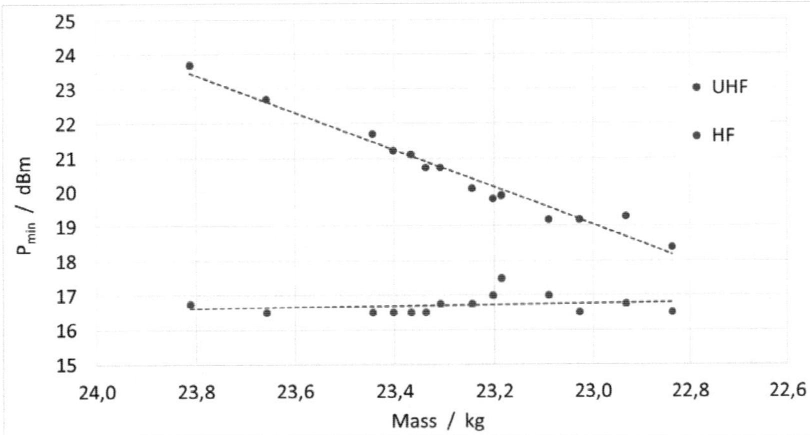

Figure 7: P_{min} for UHF and HF over mass measured at Specimen A.

Similar results are received with Specimen B and shown in Figure 8. P_{min} decreases with decreasing moisture content and accordingly mass. The different sensor installation depths all deliver a roughly linear relation with increasing slope for deeper installation, at least during the recorded time. The sensor system at 9 cm first performs measurements at an already reduced moisture, since the applicable power is limited to 29 dBm (Bartholmai et al. 2016). Compared to installation at 3 and 6 cm the slope is substantially steeper, leading to the assumption that during the recorded time the moisture decrease is particularly high in this local section of the specimen.

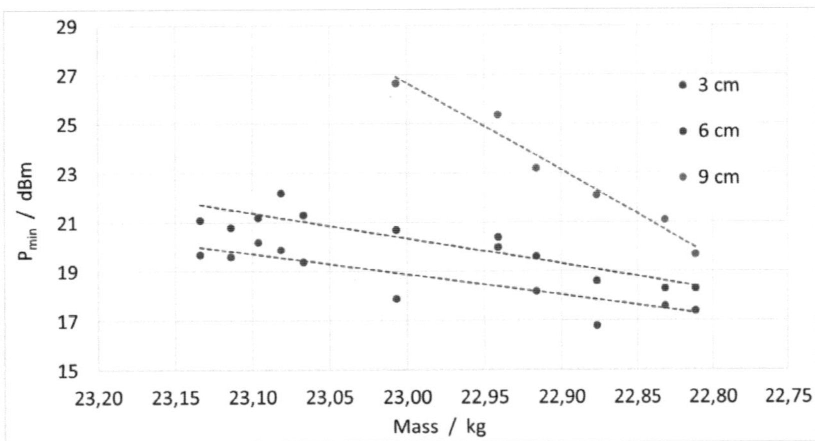

Figure 8: P_{min} for UHF over mass measured at Specimen B at different installation depths.

5. Conclusions and Outlook

The study presents an improved system design of RFID based sensor systems for complete and permanent embedding in concrete as well as results about their transmission characteristics with variation of different influencing parameters. Validation experiments at test specimen demonstrate important relations between minimal required transmission power, concrete moisture and installation depth. Even leading to the idea to use this P_{min} as a reference for the utilized humidity sensors in the case of UHF systems. Additionally, the difference between UHF and HF in signal attenuation based on moisture caused energy absorption is clearly illustrated, indicating the benefits and limits of both systems. The KonSens project is ongoing and addresses the further optimization and characterization of these systems. The presented validation experiments are still running and further tests are planned. A focus will be on the overall long-term behavior as well as on accuracy and reliability of the sensor operation.

References

Lesthaeghe, T, Frishman, S & Holland, SD 2013 'RFID Tags for Detecting Concrete Degradation in Bridge Decks', *Technical report, Institute for Transportation*, Iowa State University, 2013.

Brunner, B, Küng, R, Strässler, D, 2016 ‚Monitoring of Building Constructions with Passive RFID Technology', *Proceedings of the Wireless Congress: Systems & Applications*, Munich, Germany, 9-10 November.

Bartholmai, M, Johann, S, Kammermeier, M, Müller, M & Strangfeld, S 2016. 'Transmission characteristics of RFID sensor systems embedded in concrete', *Proceedings of the IEEE Senors 2016 Conference*, Orlando , Florida, 30 November - 2 December, pp. 1541-1543.

Strangfeld, C, Johann, S, Müller, M & Bartholmai, M 2016. 'Moisture measurements with RFID based sensors in screed and concrete', *Proceedings of the 8th European Workshop on Structural Health Monitoring (EWSHM 8)*, Bilbao, Spain, 5-8 July.

Johann, S, Strangfeld, C, Müller, M & Bartholmai, M 2016, 'RFID sensor systems embedded in concrete – requirements for long-term operation', *Proceedings of the 33rd Danubia-Adria Symposium on Advances in Experimental Mechanics (DAS 33)*, Portoroz, Slowenia, 20-23 September, pp. 68-69.

Bartholmai, M, Johann, S, Kammermeier, M, Müller, M, Strangfeld, S 2016, 'RFID sensor systems embedded in concrete – systematical investigation of the transmission characteristics', *Proceedings of the 8th European Workshop on Structural Health Monitoring (EWSHM 8)*, Bilbao, Spain, 5-8 July.

The 8th International Conference on
Structural Health Monitoring of Intelligent Infrastructure
Brisbane, Australia | 5-8 December 2017

Experimental vibration analysis of the Colosseum in Rome as part of structural health study and preservation effort

G. Bongiovanni [1], G. Buffarini [1], P. Clemente [1], D. Rinaldis [1], F. Saitta [1].
[1] *ENEA – Italy, Email:* paolo.clemente@enea.it

Abstract

The effects of the ambient vibrations on the Colosseum, obtained during the experimental campaign carried out in June 2014 are reported in this paper. Data, which regard the basement and the external northern wall, were analyzed in both time and frequency domain. The vibration amplitudes at different locations were pointed out as well as the dynamic characteristics of the wall. The results obtained are the starting point for future extensive experimental measurements and monitoring. These are needed to plan a suitable preservation and retrofitting interventions.

1. Introduction

Natural events were the cause of the collapse of most of the ancient monuments but one cannot forget that other effects can contribute to the bad conservation. Natural ageing of the materials, associated to the erosion due to wind and rain, as well as to pollutants, and to the effects of the thermal cycles, can determine a continuous deterioration of the exposed surfaces with a significant reduction of the material strength and therefore of the structural capacity. Furthermore, traffic-induced vibrations could increase and accelerate the deterioration processes, so facilitating a significant increase of the vulnerability to static and dynamic actions or even a sudden collapse under seismic actions.

A suitable preservation effort of an historic construction should be based on these considerations and start from an effective knowledge of the characteristics of materials and structure. All these data are fundamental to set up a suitable model for the structural analysis.

It is worth reminding that the structural health analysis of historic structures is particularly hard for several reasons, among these:

- the effective structural size of the various elements (walls, floors, etc.) cannot be evaluated with the needed accuracy;
- the material characteristics are not known with sufficient accuracy;
- structure and materials often exhibit inelastic behaviour;
- the effectiveness of the connections between the horizontal structures and the vertical ones and between the vertical ones is uncertain;
- the depth of the foundation, their geometry and material properties are not known; often the depth is variable as well as the soil characteristics;
- restoration works, which are not always clearly visible, could have changed the original structural behaviour.

Therefore, a mathematical model is quite difficult to implement and the structural health monitoring is necessary to plan suitable preservation or restoration interventions.

The effects of the ambient vibrations on the basement and on the external northern wall of the Colosseum, obtained during the experimental campaign carried out in 2014, are the subject of this paper. The study has the two objectives of analyzing the vibration amplitudes of such vibrations and extracting the dynamic characteristics of the structure as part of preservation effort. Therefore, the results obtained are the starting point for future extensive experimental measurements, needed to plan a suitable preservation plan. The experimental campaign included:

- measurements of electrical tomography, ground penetrating radar and measures in holes,
- active and passive seismic measurements,
- measurement of ambient vibration on the foundation and on the northern wall.

The results of the last aspects are reported in the following.

2. The experimental activities

The experimental vibration analysis was carried out by using 24 velocimetric sensors connected to a Granite acquisition system. In the following the measurements made on the basement and on the northern wall are described separately.

2. The basement

The foundation of the Colosseum is composed by a ring having the same size of the elevation structure and a depth of about 13 *m*, formed by a paving of about 1 m and concrete layers of different characteristics. Four tunnels cross the foundation along the major axis (cryptoporticos E and W) and the minor axis (tunnels N and S). In the S-E quadrant, another tunnel, the so-called passage of Commodus, cross the foundation. Sensors were deployed in two configurations. In the first one (C1b) sensors were deployed at eight locations, three for each of them along the three orthogonal directions (Fig. 1). In the second configuration (C2b), three sensors were kept for each locations 1, 2, 4 and 5 and twelve sensors were deployed in the vertical direction along the internal perimeter of the foundation ring (Fig. 2).

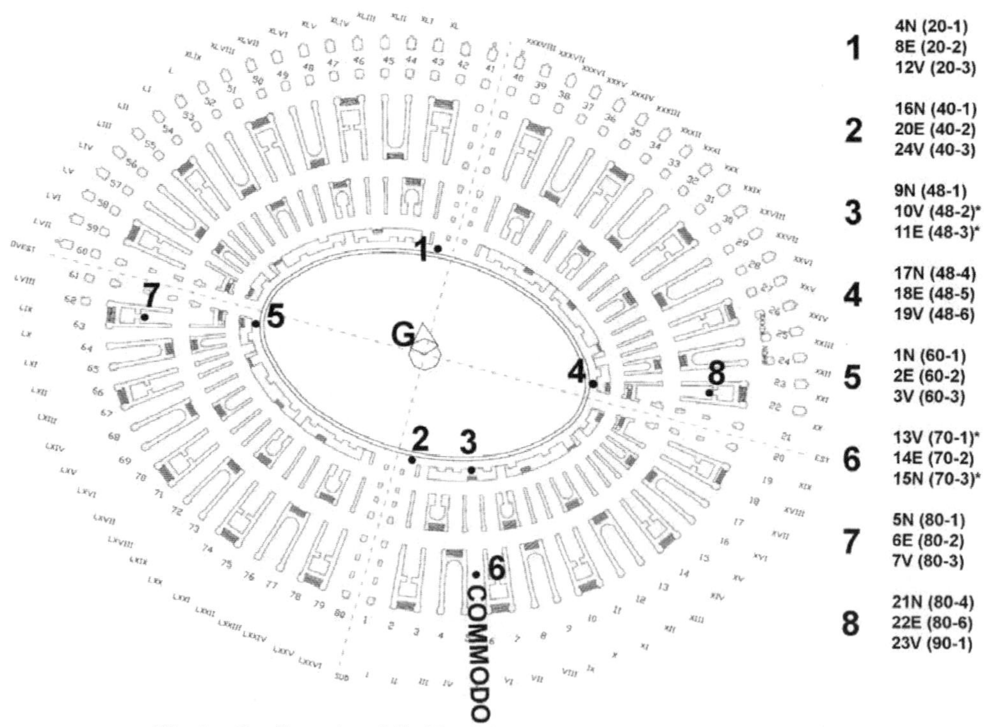

Fig. 1 – Configuration C1b (N = S-N, E = W-E, V = UP)

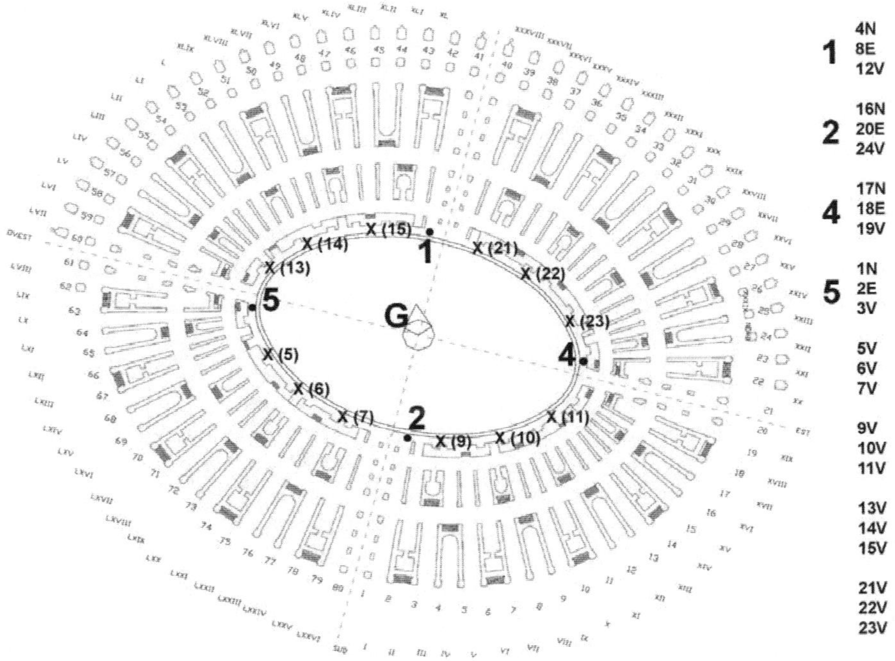

Fig. 2 – Configuration C2b (N = S-N, E = W-E, V = UP)

2.2 Northern wall

Two configurations were considered for the northern wall as well. In configuration C1p, sensors were deployed along four vertical alignments at the arches XXVI, XXXI, XXXVI and XXXVIIII, respectively (Fig. 3). At each of them three sensors were deployed on the ground floor (GF), in the radial, tangential and vertical direction, respectively, one on the first floor (1F) in the radial direction and one on the second floor (2F) in the radial direction. Furthermore, two other sensors were deployed at arches XXXVI and XXXVIIII on the fourth floor (4F) in the radial direction.
In configuration C2p the sensors were kept in the same locations, but sensors on 1F, 2F and 4F were oriented in the tangential directions.
It is worth noting that a building site was operational at the moment of the campaign, which influenced the choice of the sensor locations.

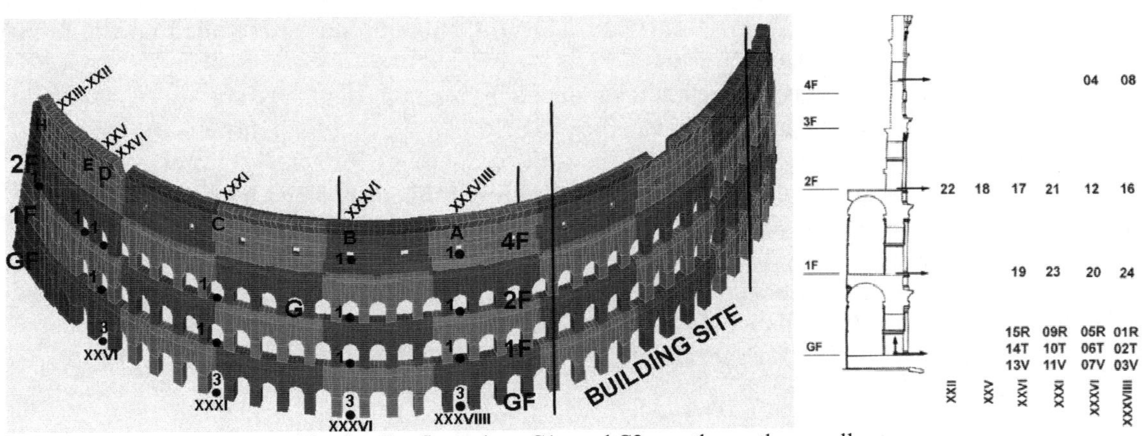

Fig. 3 – Configurations C1p and C2p on the northern wall

3. Data analyses

For each configuration a recording of 30 min was obtained, with a sampling ratio of 200 Hz. The signals were filtered between 0.8 and 20 Hz. Time domain analysis, by calculating the effective value over intervals of 0.2 s, and frequency domain analysis were carried out.

3.1 Basement (Configurations C1b and C2b)

First of all we noticed that:

- for each location, the amplitudes of the three components are almost similar, the UP being the highest and the WE the lowest;
- recordings at the extrados of the foundation, i.e., on the surface level, show amplitudes higher than those in the tunnels, which are at lower levels;

In Fig. 4 the RMS are plotted for all the sensors in configurations C1b and C2b. Figs. 5 and 6 show the PSDs relative to configurations C1b and C2b, respectively.

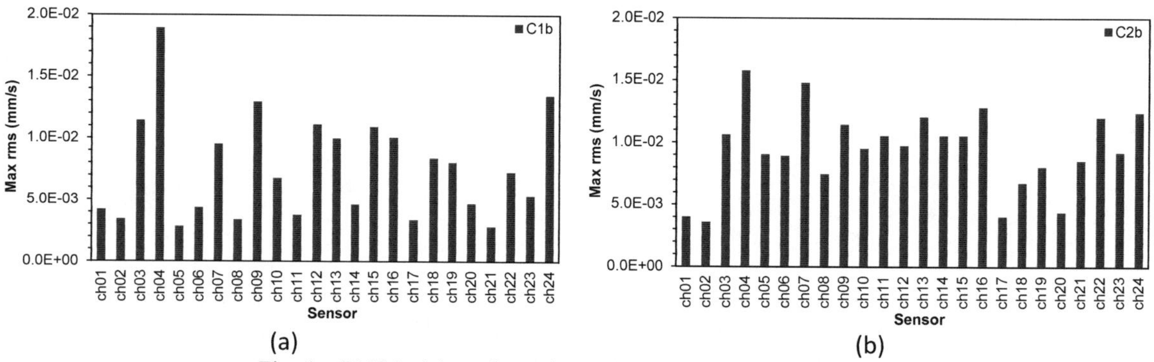

Fig. 4 – RMS in (a) configuration C1b and (b) configuration C2b.

3.2 Northern wall (Configurations C1p and C2p)

The recorded time-histories in the radial direction (C1p) showed as the amplitudes increase with the height as usual. Instead, the three components on the basement have similar amplitudes in each location and also at all the locations. Analogously, the values of RMS (Fig. 7a) at 4F are higher than those at 1F and 2F. It is worth noting that the presence of tourists at 1F could cause higher values of the signals.

With reference to the tangential direction (C2p, Fig. 7b), the amplitudes recorded on the higher floors are similar to those on the basement.

From PSDs the following resonance frequencies can be individualized:

- 1.05 Hz, which interests the portion of the wall close to the Stern's buttress;
- 1.29 Hz, which interests the portion of the walls close to the Valadier's buttress;
- 1.46 Hz, present in a large portion of the walls close to the Valadier's buttress;
- 1.51 Hz, present in a large portion of the walls close to the Stern's buttress;
- 1.61 Hz, which involves the entire wall;
- 1.71 Hz and 1.78 Hz, often not contemporarily present;
- 2.00 Hz and 2.22 Hz, which involve higher modal shapes;
- 2.51 Hz and 2.76 Hz, associated to symmetrical and anti-symmetrical modal shapes, respectively.

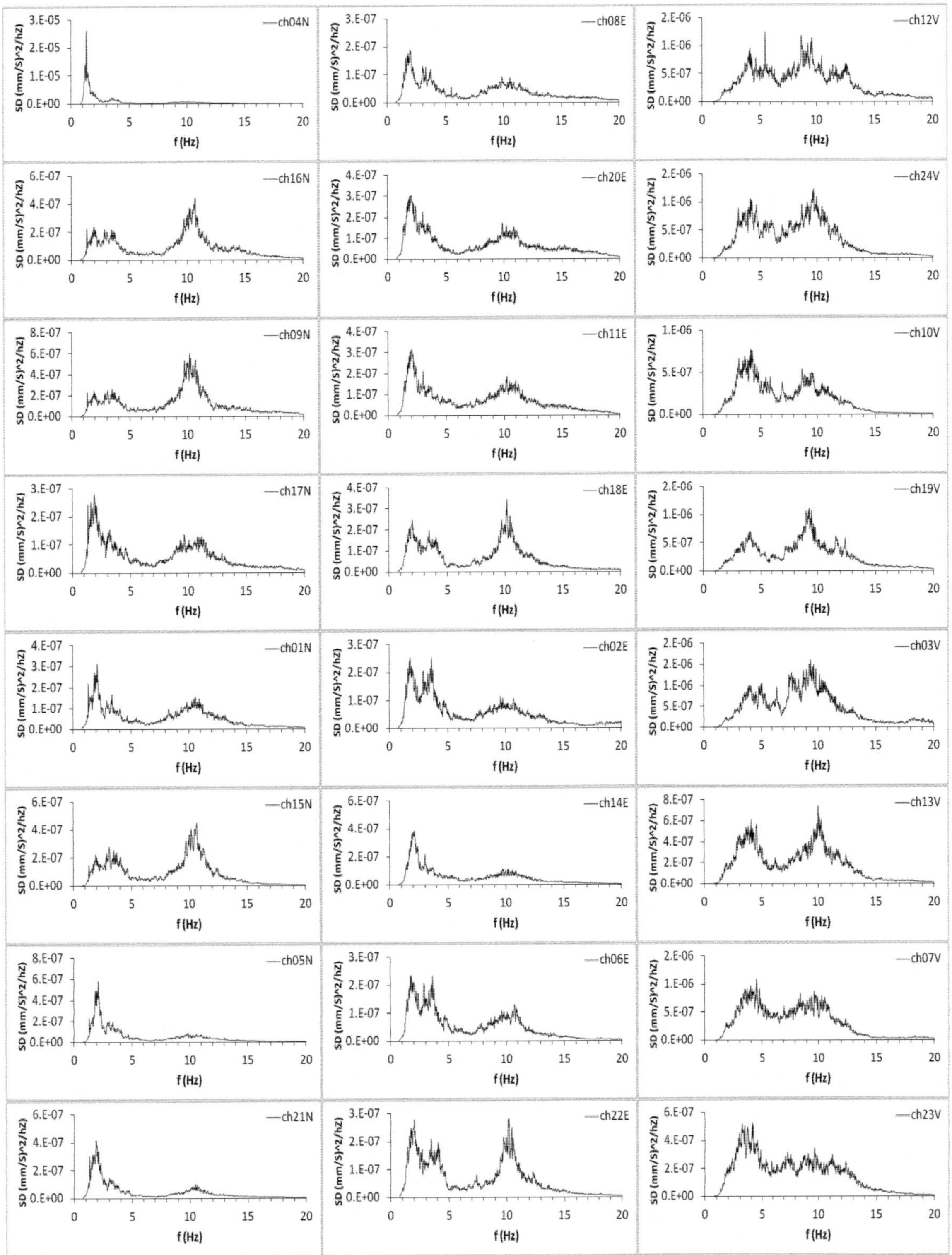

Fig. 5 – PSDs relative to configurations C1b

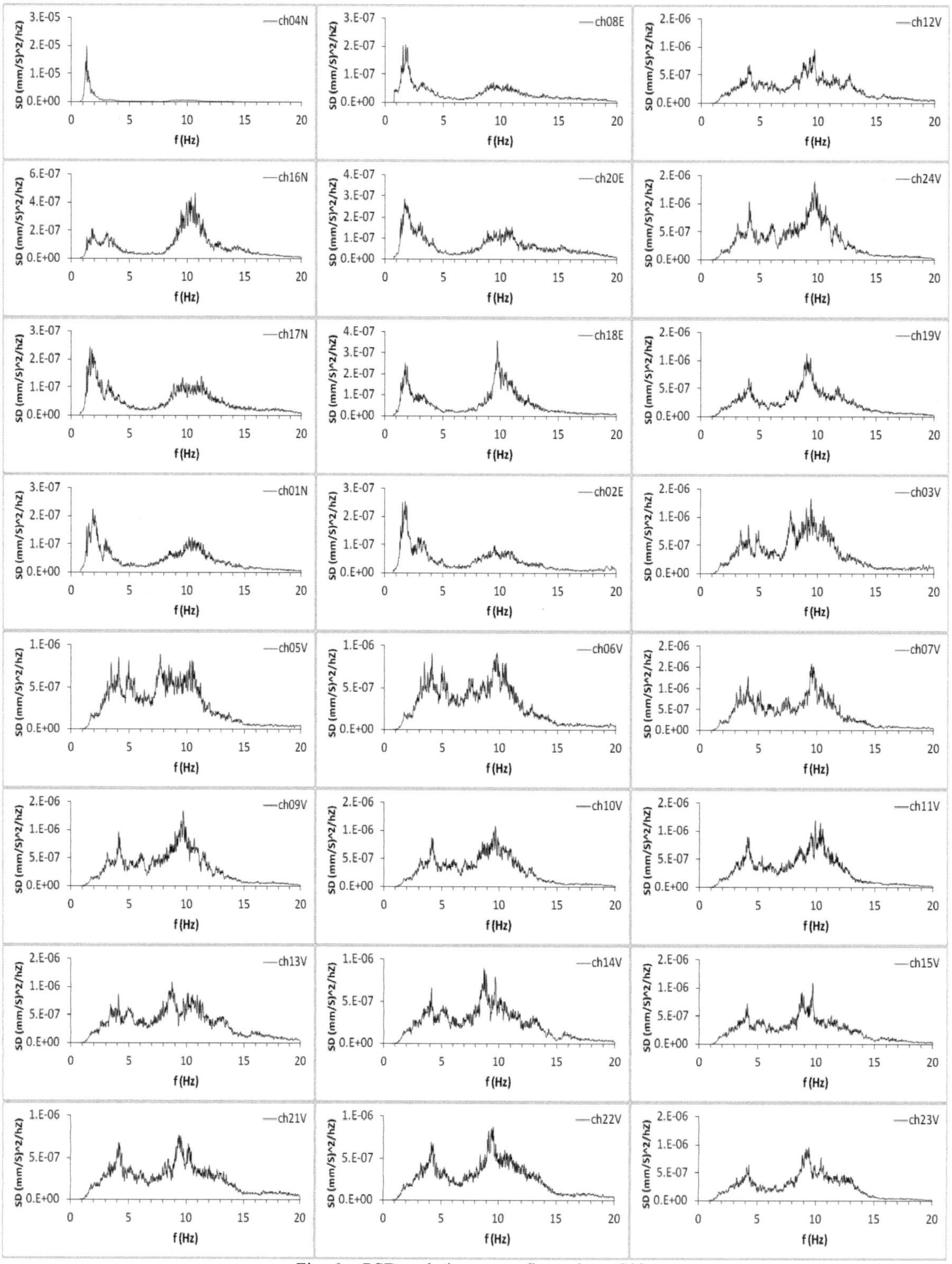

Fig. 6 – PSDs relative to configurations C2b

Figs. 8, 9, 10 and 11 show the PSD relative to the vertical alignments at arches XXVI, XXXI, XXXVI and XXXVIIII, respectively, in configuration C1p. Fig. 12 shows the PSD relative to the vertical alignments at arch XXXI, in configuration C2p.

The cross analysis showed that:

- modal shapes could be associated to the individualized resonance frequencies,
- anyway, the values of phase factors and coherence functions are not always significant; they are often significant between recordings close one to other but becomes not significant for recordings of locations quite far one from the other,
- the previous consideration bring to hypothesize the presence of complex modal shapes.

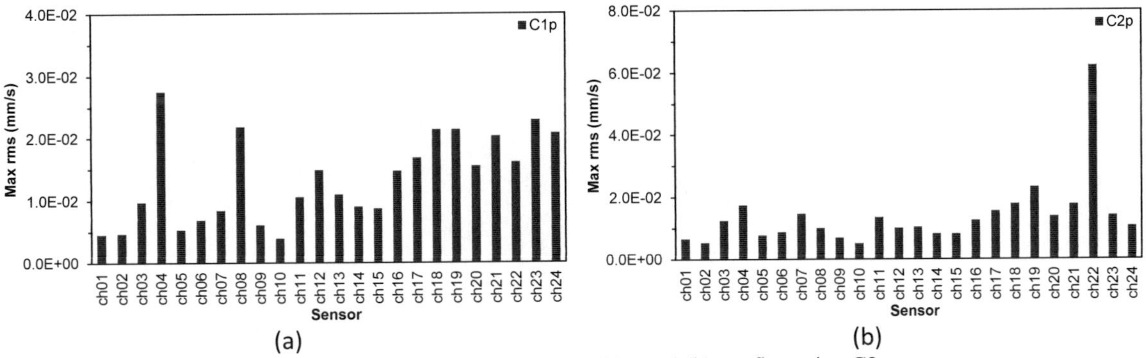

Fig. 7 – RMS of (a) configuration C1p and (b) configuration C2p

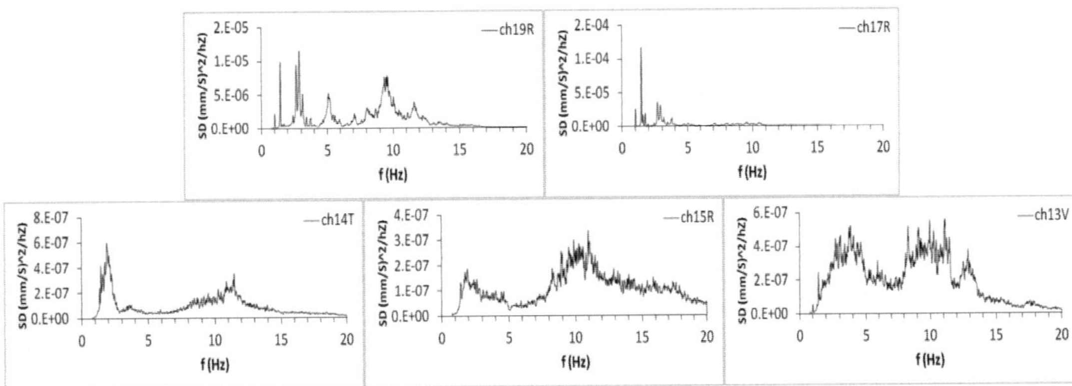

Fig. 8 – PSDs at arch XXVI

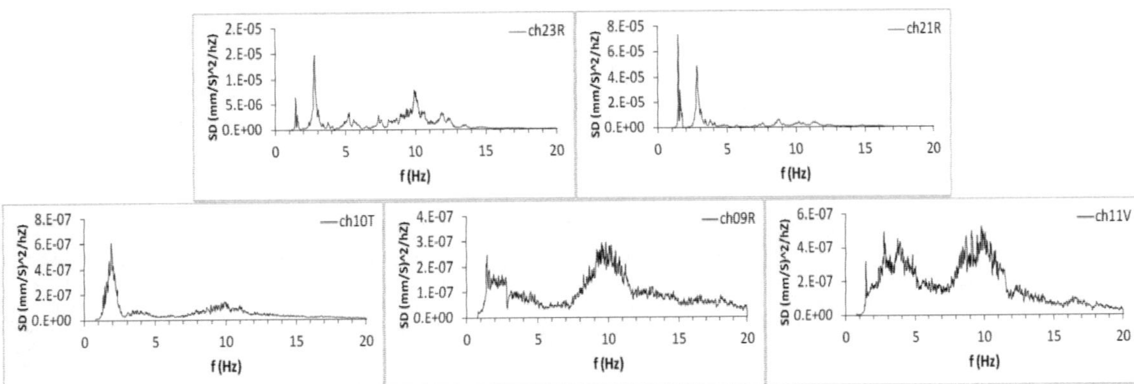

Fig. 9 – PSDs at arch XXXI

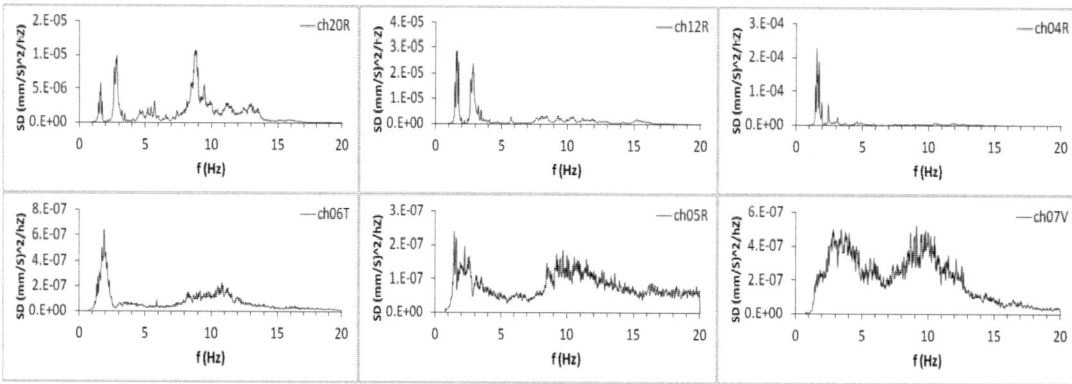

Fig. 10 – PSDs at arch XXXVI

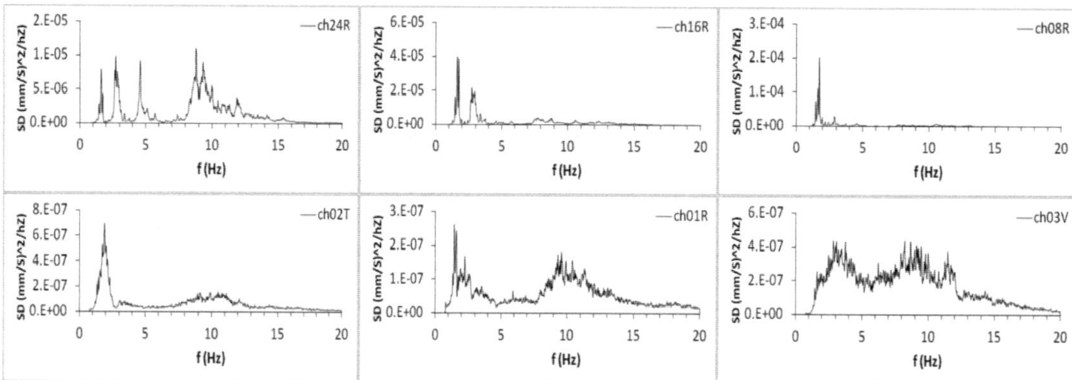

Fig. 11 – PSDs at arch XXXVIIII

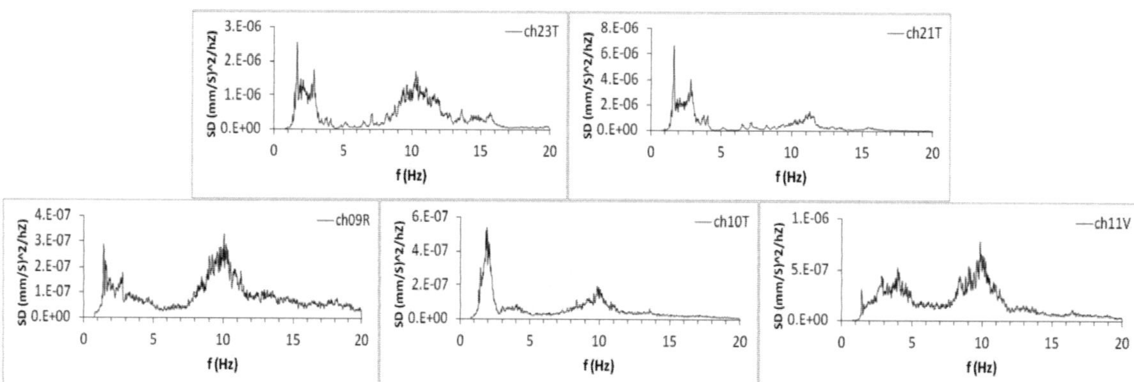

Fig. 12 – PSDs at arch XXXI

3. Conclusions

The ambient vibration analysis on the Colosseum allowed analysing the vibration amplitudes and finding out the dynamic characteristics of the structure, with particular reference to the basement and the northern wall.

The analysis of the recordings on the basement pointed out that the amplitude vibrations were quite similar at all the locations but also the presence of higher peaks at the Valadier's buttress, probably related to the underground, which passes very close to the monument.

Several resonance frequencies were found, most of them are certainly associated to modal shapes of the northern wall or the entire structure. Anyway, the nonlinear behaviour was evident also in presence of ambient vibrations only.
The recorded data will be very useful in setting up a suitable finite element model of the structure, which will allow analysing the structural health status and defining a suitable plan of preservation intervention.

Acknowledgements

The activities here described are part of a research project carried out by ENEA in collaboration with INGV, DICEA Dept. of Sapienza University of Rome, and the *Soprintendenza Speciale per il Colosseo e l'Area archeologica centrale di Roma*. The authors are grateful to M. Azzimato, R. D'imperio, P. Funaro, S. Hailemikael, G. Martini, A. Paciello, S. Paolini, A. Peloso, C. Properzi, A. Screpanti, S. Serafini, V. Verrubbi and A. Zini of ENEA for their precious contribution in the experimental campaign.

References

Bendat JS, Piersol AG 1980, 'Engineering Applications of Correlation and Spectral Analysis', Wiley: New York.

Bergamasco I, Bongiovanni G, Carpani B, Clemente P, Paciello A, Serafini S 2014, 'Ambient vibration analysis for the characterization of soil and coverings of Villa dei Misteri in Pompeii', In Peña F. & Chávez M. (eds) Structural Analysis of Historical Constructions (Proc. SAHC2014, Mexico City, 15-17 Oct), ISBN 04-2014-102011495500-102.

Beste HJ, Clemente P, Conti C, D'Ovidio G, Nakamura Y, Orlando L, Rea R, Rovelli A, Valente G 2015, 'The importance of defining the geometry of foundations and soil layers for dynamic analysis of Colosseum', In Baggio C. et al. (eds), The dynamic interaction of soil and structure (Proc. DISS_13, Roma, 12 Dec 2013), 79-110, DICEAA L'Aquila University, ISBN 978-88-940114-1-8.

Bongiovanni G, Buffarini G, Clemente P, Saitta F 2014, 'Ambient vibrational analysis of Aurelian Column', Proc. 10th U.S. National Conf. on Earth. Eng. (10NCEE, Anchorage, AK, 21-25 July), Paper 1013, EERI, DOI: 10.4231.D35717P1T.

Bongiovanni G, Buffarini G, Clemente P, Saitta F 2014, 'Dynamic characterization of the Trajan's Column', In Peña F. & Chávez M. (eds.) Structural Analysis of Historical Constructions (Proc. SAHC2014, Mexico City, 15-17 Oct), ISBN 04-2014-102011495500-102.

Bongiovanni G, Celebi M, Clemente P 1990, 'The Flaminio Obelisk in Rome: vibrational characteristics as part of preservation efforts', Int. J. of Earth. Eng. and Structural Dynamics, Vol. 19, No. 1, 107-118, John Wiley & Sons.

Caserta A, Clemente P, Conti C, D'Ovidio G, Nakamura Y, Rea R, Rovelli A, Valente G 2015, 'A methodology to analyze ambient vibration in the Colosseum', *In Baggio C. et al. (eds), The dynamic interaction of soil and structure (Proc. DISS_13, Roma, 12 Dec 2013)*, 47-78, DICEAA L'Aquila University, ISBN 978-88-940114-1-8.

Caserta A, Clemente P, Nakamura Y, Rovelli A, Valente G 2014, 'Identification of the mean elasticity modules for soil Colosseum interaction', *In Clemente P. et al. (eds), Dynamic interaction of Soil and Structure (Proc. DISS_12)*, 91-139, DICEAA L'Aquila University, ISBN 978-88-548-3693-8.

Clemente P 1995, 'Traffic-Induced Vibrations on Structures', *IABSE Report Extending the Lifespan of Structures*, Vol. 73.2, 1111-1116, IABSE, Zurich.

Clemente P 2002, *L'analisi dinamica sperimentale nella salvaguardia dei beni culturali, Alcune esperienze dell'ENEA*, ENEA, Roma, ISBN 88-8286-012-4 (in Italian).

Clemente P, Bongiovanni G 1993, 'Ambient Vibration Effects on the Colosseum', *IABSE Report Structural Preservation of the Architectural Heritage*, Vol. 70, 107-114, IABSE, Zurich.

Bongiovanni G, Buffarini G, Clemente P, Rinaldis D, Saitta F 2017, 'Dynamic characteristics of the *Amphitheatrum Flavium* northern wall from traffic-induced vibrations', *Annals of Geophysics, Special issue on Monitoring and seismic characterization of archaeological sites and structures*, INGV, Rome.

Clemente P, Bongiovanni G, Buffarini G 2002, 'Experimental analysis of the seismic behaviour of a cracked masonry structure', *Proc. 12th European Conf. on Earth. Eng. (12ECEE, London 9-13 Sep)*, Paper 104, Elsevier Science Ltd.

Clemente P, Buffarini G 2009, 'Dynamic response of buildings of the cultural heritage', *In Boller C., Chang F.K., Fujino Y. (eds), Enciclopedia of Structural Health Monitoring*, 2243-2252, John Wiley & Sons Ltd, Chichester, UK. ISBN 978-0-470-05822-0.

Clemente P, Rinaldis D 1998, 'Protection of a monumental building against traffic-induced vibrations', *Int. J. Soil Dynamics and Earth. Eng.*, Vol. 17, No. 5, 289-296, Elsevier Science Ltd.

Clemente P, Rinaldis D 2005, 'Design of temporary and permanent arrays to assess dynamic parameters in historical and monumental buildings', *In Ansari F. (ed), Sensing Issues in Civil Structural Health Monitoring*, 107-116, Springer.

Clemente P, Rinaldis D, Bongiovanni G 1994, 'Dynamic characterization of the 'Tempio della Minerva Medica', *Proc., 10th European Conf. on Earth. Eng. (Vienna, 28 Aug – 2 Sept)*, Vol. 2, 981-986, Balkema, Rotterdam.

Clemente P, Rinaldis D, Buffarini G 2007, 'Experimental seismic analysis of a historical building', *Journal of Intelligent Material Systems and Structures*, Vol. 18, No. 8, 777-784, SAGE Publications.

Clemente P, Saitta F, Buffarini G, Platania L 2015, 'Stability and seismic analyses of leaning towers: the case of the minaret in Jam', *The Structural Design of Tall and Special Buildings*, Wiley, Vol. 24, 40-58.

De Stefano A, Clemente P 2009, 'Structural health monitoring of historic buildings', *In Karbhari V.M. and Ansari F. (Eds) Structural Health Monitoring of Civil Infrastructure Systems*, Woodhead Publishing Ltd.

De Stefano A, Matta E, Clemente P 2016, 'Structural health monitoring of historical heritage in Italy: some relevant experiences', *J. of Civil Structural Health Monitoring*, 6(1), 83-106, Springer.

Ewins DJ 1984, *Modal Testing: Theory and Practice*, Research Study Press Ltd.

Pau A, Vestroni F 2008, 'Vibration analysis and dynamic characterization of the Colosseum', *Structural Control and Health Monitoring*, 15: 1105-1121.

The 8th International Conference on
Structural Health Monitoring of Intelligent Infrastructure
Brisbane, Australia | 5-8 December 2017

Comparing algorithms for measurement-system design for bridge load testing

N. Bertola [1], M. Papadopoulou [1], D. Vernay [1], I.F.C. Smith [1,2].

[1] *ETH Zurich, Future Cities Laboratory, Singapore-ETH Centre, 1 CREATE Way, CREATE TOWER, 138602 Singapore, Email:* bertola@arch.ethz.ch, [2] *Swiss Federal Institute of Technology Lausanne (EPFL), Applied Computing and Mechanics Laboratory (IMAC), School of Architecture, Civil and Environmental Engineering (ENAC), Lausanne – Switzerland.*

Abstract

Assessing ageing infrastructure is a critical challenge for civil engineers due to the difficulty of estimating uncertainties in structural models. Error-domain model falsification (EDMF) is a model-based structural identification methodology that uses in-situ measurements to improve knowledge of the real behaviour of structures. In addition to traditional aleatory uncertainties, this methodology accommodates systematic uncertainties induced by sources such as numerical modelling, boundary conditions and model fidelity.

Sensors are usually placed using engineering judgement and experience. However, since sensor-placement is fundamental to the success of structural model updating, a more rational and systematic method is justified. This study presents a measurement-system design methodology to identify the best sensor locations, sensor types and the number of sensors for bridge load testing. A full-scale case study was tested in Singapore.

Three greedy algorithms for optimal sensor-placement using information-entropy-based objective functions are implemented to maximize the information gain of a sensor configuration. As their respective optimal sensor configuration differed, the sensor-placement algorithms are compared using two metrics: the joint-entropy and expected candidate-model-set size. The hierarchical algorithm performs better than sequential algorithms with forward and forward-max strategies in terms of both metrics. In addition, it is shown that the number of sensors used during the case-study could have been reduced from 14 to 6 without any substantial loss of information.

1. Introduction

To avoid the significant economic, social and environmental costs that are associated with replacement of ageing infrastructure, more efficient assessment strategies need to be developed. As bridges have been designed and built in a safe manner, real reserve capacity could be unknown. In such cases, either replacement or costly improvement of some structures is unnecessary.

The aim of model-based structural identification is to use in-situ measurements to improve knowledge of the real behaviour of structures. Several data-interpretation techniques exist to tackle this task, such as residual minimisation and Bayesian updating. Smith (2016) described work concluding that such traditional model calibration methodologies cannot be justified for large structures because they often result in biased identification. To deal robustly with systematic uncertainties, a new approach, called error-domain model falsification (EDMF) was presented by Goulet and Smith (2013a). In this approach, multiple model instances are generated and monitoring data are used only to falsify the model instances that significantly differ from field measurements. The authors highlighted that this approach explicitly considers systematic modeling uncertainties, affecting the sensor placement process.

Regardless of the approach used, a sensor placement strategy is needed to successfully improve knowledge of the structures using measurement data. Papadimitriou (2004) showed that the

information entropy could be used as a sensor placement criterion. A location with a high entropy value of model predictions (Robert-Nicoud et al. 2005a) or offering high entropy reduction of parameter values (Papadimitriou 2004) is considered as a good location. As the number of possible sensor configurations is very large, greedy optimization algorithms, such as sequential algorithms with a forward or backward strategy, were used to reduce computational time (Krause et al. 2008; Kripakaran and Smith 2009). However, the entropy calculation based on sequential optimization strategies disregards the mutual information between sensor locations and can select sensors providing redundant information. Papadopoulou et al. (2014) presented a new hierarchical sensor placement approach using the joint-entropy criterion to take into account mutual information between sensors in the field of wind engineering. Another approach developed by Goulet and Smith (2013b) and then Pasquier et al. (2017) used simulated measurements to deal with this issue probabilistically and predict the usefulness of monitoring for structural identification. These studies used sequential strategies and employed simulated measurements during sensor-placement optimization, whereas the objective function to minimize is the expected candidate-model-set size.
Most of previous sensor placement strategies provide a sensor ranking. However, the evaluation of the performance of a sensor configuration in real systems remains difficult and is a bottleneck for defining the optimal sensor configuration. The present study compares the performance of sensor configurations, in terms of information gain, obtained by algorithms developed by Papadopoulou et al. (2014; 2015) and by a traditional sequential algorithm with a forward strategy. Once the sensor configuration is built by each algorithm, two metrics are used to evaluate the information gain. First, the joint-entropy introduced by Papadopoulou et al. (2014) is calculated. Then a metric is introduced herein based on the work from Goulet and Smith (2013b), Papadopoulou et al. (2016) and Pasquier et al. (2017): the usefulness of measurements to provide probabilistic estimations of the expected number of candidate models obtained with a sensor configuration.
The study is organized as below. Section 2 presents the error-domain model falsification approach. Sections 3 describes the different sensor placement strategies used in the present study and how the sensor configuration will be evaluated. Section 4 presents the full-scale bridge case study and results are discussed in Section 5.

2. System identification through model falsification

Presented by Goulet and Smith (2013a), error-domain model falsification (EDMF) is a structural identification methodology in which multiple model instances are generated from different parameter sets and are falsified if their predictions differ significantly from field measurements. Model instances are instantiations of a model class, in which a combination of primary parameter values $\theta_k=[\theta_1,\theta_2,\ldots,\theta_n]^T$ are assigned in order to generate an initial set of model instances Ω. These primary parameters usually describe unknown geometrical, physical and material characteristics of the structure. Model-instance predictions are compared with field measurements of the structural responses. For each measurement location $i \in \{1,\ldots,n_y\}$, model predictions and measurements are linked to the true behavior using:

$$g_k(x_i,\Theta_k) + U_{i,g_k} = R_i = \hat{y}_i + U_{i,\hat{y}} \quad \forall i \in \{1,\ldots,n_y\} \tag{1}$$

where R_i corresponds to real responses of a structure (unknown) and $\hat{y}_i$ to the measured value at location *i*. Using a finite element analysis (FEA), predictions $g_k(x_i,\Theta_k)$ of the model class G_k are evaluated at location x_i. Θ_k corresponds to adequate values of the parameter vector θ_k. $U_{1,gk}$ and $U_{i,y}$ are modelling and measurement uncertainties respectively. The later uncertainties may be combined following (Robert-Nicoud et al. 2005b).

To select candidate models representing realistic sets of parameter values, the procedure involves falsifying all model instances for which predictions cannot explain measurement data given combined uncertainties within a target reliability of identification ϕ. The set of candidate models obtained after falsification is defined as:

$$\Omega_k'' = \{\theta_k \in \Omega_k | \forall i = 1, \ldots, n_y \; u_{i,low} \leq g_k(x_i, \Theta_k) - \hat{y}_i \leq u_{i,high}\} \quad (2)$$

where the candidate model set (CMS), Ω_k'', is made up of initial model instances, which have not been falsified at one or more measurement locations. $[u_{i,low}, u_{i,high}]$ are the upper and lower threshold bounds representing the shortest intervals including a probability of reliability ϕ^{1/n_y}. They are defined through the probability density function (PDF) of combined uncertainties $f_{U_i}(u_i)$ at each measurement location, using the Šidák correction (Šidák 1967) to maintain a constant level of confidence when many sensors are used to test model instances, such that:

$$\forall i = 1, \ldots, n_y: \phi^{1/n_y} = \int_{u_{i,low}}^{u_{i,high}} f_{U_i}(u_i) du_i \qquad \forall i \in \{1, \ldots, n_y\} \quad (3)$$

All model instances that do not belong to the falsified set, $\Theta_k \in \Omega_k''$, are labeled as candidate models and are assigned an equal probability as expressed in Equation 4. Since so little information is usually available to describe the form of modeling-uncertainty distributions, the EDMF states that every candidate model is equally likely to be the correct model (Goulet and Smith 2013a).

$$\Pr(\Theta_k \in \Omega_k'') = \frac{1}{\int \theta_{k \in \Omega_k''} d\theta_k} \quad (4)$$

Falsified model instances are assigned a probability of 0 as expressed in Equation 5.

$$\Pr(\Theta_k \notin \Omega_k'') = 0 \quad (5)$$

Consequently, Θ_k'' is the vector of random variables describing the parameter values of the candidate model instances given measurement data and its PDF is defined as:

$$f_{\Theta_k''} = \begin{cases} \frac{1}{\int \theta_{k \in \Omega_k''} d\theta_k}, & if\ \Theta_k \in \Omega_k'' \\ 0, & otherwise \end{cases} \quad (6)$$

3. Sensor placement theory

Model-based sensor placement has focused much attention in international research in the last decades. As the number of possible sensor configurations is very high, sensor locations are often selected sequentially, employing greedy sensor placement algorithms (Krause et al. 2008). Such greedy algorithms present the advantage of a shorter computational time compared to stochastic search, such as PGSL (Kripakaran and Smith 2009). After defining an objective function, the strategy can be either forward though adding the best sensor in the network or backward through removing the less useful sensor of the network.

Several objective functions can be used to select the best sensor location. The first approach is based on the information entropy, also called Shannon's entropy, introduced for sensor configuration by Papadimitriou (2004) and Robert-Nicoud et al. (2005a). The information entropy is a measure of information content. For EDMF approach, the information entropy $H(y_i)$ is defined:

$$H(y_i) = -\sum_{j=1}^{N_I} P(y_{i,j}) \log_2 P(y_{i,j}) \quad (7)$$

where y_j is an output variable, such as deflection at a sensor location i, $P(y_{i,j})$ is the probability of the jth interval of a variable's distribution, with $j \in \{1, \ldots, N_I\}$ and N_I is the maximum number of intervals at the ith location; this value depends on the degree of uncertainty. The information entropy at each location j is computed through first counting the number of model instances $m_{i,j}$, for which predictions fall within each interval and then calculating the probability of the interval as $P(y_{i,j}) = m_{i,j}/\sum m_{i,j}$.

In the present study, the first sensor-placement algorithm implemented is a sequential algorithm with a forward strategy, using the information entropy as the objective function to maximize. This sensor-placement algorithm is called the forward algorithm. Another entropy-based sensor-placement algorithm, the forward-max algorithm, was introduced by Papadopoulou et al. (2015) for the assessment of wind around buildings. In the forward-max algorithm, the location with the largest entropy is selected as the first sensor location similar to the forward algorithm. Then, it differs from the forward algorithm as the interval bounds are recalculated for each subset of model predictions of the previously selected locations. For each subset, model-prediction distributions are created at locations that have not been selected. The aim is to find sensor locations that will further separate t hese subsets. The location with the highest information entropy value is selected for all subsets and among all remaining possible locations.

Papadopoulou et al. (2014) introduced a new hierarchical algorithm for sensor placement using the concept of joint-entropy. The joint entropy is an information entropy measure associated with a set of variables. It evaluates the information entropy of a sensor configuration, while assessing the mutual information of the locations. The joint-entropy is less than or equal to the sum of the individual entropies of the variables in the set. For a set of two sensors, it is defined as:

$$H(y_i) = -\sum_{i=1}^{n_s} \sum_{j=1}^{N_I} P(y_{i,j}, y_{i+1,j}) \log_2 P(y_{i,j}, y_{i+1,j}) \quad (8)$$

where $j \in \{1, \ldots, N_I\}$ and N_I is the maximum number of intervals at the i+1 location and $i \in \{1, \ldots, n_s\}$ with n_s the number of potential sensor locations.

The hierarchical algorithm is a sequential algorithm in which model data is organized in a tree structure. At the root is the initial model set, and branches contain subsets of model predictions. Branches from a node represent divisions of the parent model set into smaller groups that can potentially be separated using measurements from the new sensor added to the configuration. This allows calculations of joint entropy of the sensor configurations and therefore avoiding exponential complexity and reducing computational effort. Similar to the forward and forward-max algorithms, the first sensor (i=1) is selected as the one with the largest information entropy value. However, for i >1, the location that maximizes the joint-entropy of the configuration is selected.

Two information-gain metrics were chosen to evaluate the sensor configurations of the three sensor-placement algorithms. First, the joint-entropy will be calculated for each sensor configuration. Although, all three algorithms share the same objective, to maximize the entropy of the sensor configuration, only the hierarchical algorithm evaluates the mutual information during sensor placement. The forward and forward-max algorithms are only able to assess individual-sensor entropy. Therefore, in order to be able to compare the results from the 3 algorithms, the code of the hierarchical algorithm is used to assess the joint-entropy of the configurations build by the forward and forward-max algorithms.

Then, a metric is introduced from Goulet and Smith (2013b) who used simulated measurements to provide probabilistic estimations of the expected number of candidate models obtained with a sensor configuration during sensor placement. Simulated measurements were generated based on the model instances in the model class adding a random value taken from the combined uncertainties. In these studies, sensor locations are selected through calculating their capability to reduce the number of model instances. Sensor locations were evaluated using the 95 % quantile of the expected candidate-model-set size, referred to as expected identifiability. The latter methodology was extended by Pasquier et al. (2017) introducing the expected utility to evaluate the reduction of the candidate-model-set size, corresponding to the quantile 50 %. In both methodologies, sensors are selected sequentially where the location with the smallest expected candidate-model-set size is selected as the best location. However, the procedure was computationally costly, because it required execution of the falsification procedure for a large number of simulated measurements and sensor locations. Therefore, the expected usefulness of measurement will be used only as a metric, similar to Papadopoulou et al. (2016), in order to evaluate the performance of the sensor configurations obtained by the three sensor-placement algorithms previously introduced instead of a sensor-placement objective function as Goulet and Smith (2013b) and Pasquier et al. (2017).

4. Full-scale case study

The full-scale case study is a 32-year-old bridge in Singapore. This pre-stressed concrete bridge is composed of 4 beams carrying three unidirectional traffic lanes over a simply-supported span of 32 meters. The principal characteristics of the bridge are presented in Figure 1.

A static load test, involving 6 trucks (T_i) of approximatively 32 tons located on the deck of the bridge was monitored. The sensor configuration consisted of 2 inclinometers (I_i) on one parapet, 4 deflection targets (P_i) and 8 strain gauges (S_i) distributed on the 4 beams. A laser tracker was positioned on the road below the bridge and used to measure deflections. The sensor configuration was chosen based on engineering judgement.

The following primary parameters were identified as having the most influence on measurements: The Young modulus of site-cast concrete of the deck E_{con}, the Young modulus of the precast concrete of the beams E_{pre}, the Young modulus of the concrete of the barrier E_{bar}, the rotational stiffness of the bearing devices K_{rot} and the vertical stiffness of the bearing devices K_{lon}. Their initial intervals are estimated using engineering heuristics and are presented in Table 1. To reduce model-simplification uncertainties, non-structural elements, such as the asphalt pavement are included in ANSYS-WORKBENCH-17 numerical model. In this 5-parameter space, 1,000 initial model instances were generated using the Latin Hypercube Sampling (LHS).

Upper and lower bounds of model-class uncertainties and measurement uncertainties are presented in Table 2 and uniform distributions are defined between bounds. They are estimated based on engineering judgment, sensor-supplier information and heuristics.

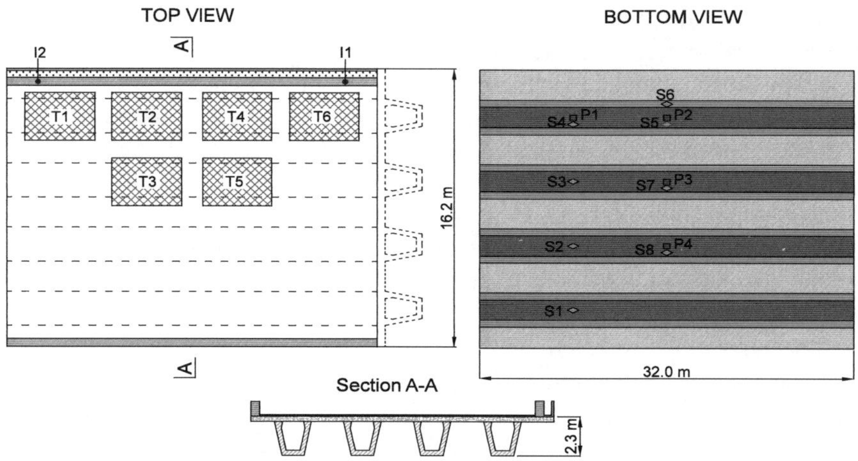

Figure 1: Bridge characteristics within the load-test and sensor configuration.

Table 1: Primary parameters considered and their initial intervals

E_{con} [GPa]	E_{pre} [GPa]	E_{bar} [GPa]	K_{rot} log[Nmm/rad]	K_{lon} log([N/mm])
20-35	25-50	3-40	9-13	8-11

Table 2: Modelling and measurement uncertainties

Uncertainty source	Displacements – (P)		Rotations – (I)		Strains – (S)	
	Min	Max	Min	Max	Min	Max
Model simplifications and FEA (%)	-5	13	-5	13	-5	13
Mesh refinement (%)	-1	1	-1	1	-1	1
Spatial variability (%)	-	-	-	-	-5	5
Additional uncertainty (%)	-1	1	-1	1	-1	1
Sensor accuracy	-0.05 mm	0.05 mm	-1 μrad	1 μrad	-2 με	2 με
Repeatability	-0.15 mm	0.15 mm	-4 μrad	4 μrad	-4 με	4 με
Sensor orientation (%)	-	-	-	-	0	6
Sensor installation (%)	-	-	-5	5	0	5

The three sensor-placement algorithms presented in Section 3 were implemented for 14 possible sensor locations, the actual sensor locations of the test. The selection of sensors is presented in Figure 2 as a histogram with the sensor rank as vertical axis and the sensor identification as horizontal axis. Results of each sensor-placement algorithm are presented with distinctive bars.

The three sensor-placement algorithms select the second inclinometer as the first sensor, due to the use of the same criterion for this sensor selection: the location with the maximum information entropy. After the selection of the first sensor, sensor choices from the sensor-placement algorithms differ. The forward algorithm selects the inclinometers, then deflection measurements and, finally, the strain gauges. After the first inclinometer, the forward-max algorithm selects three deflection targets and the first strain gauge is selected as the fifth sensor. The hierarchical algorithm selects three sensors with different types for the first three sensors.

The forward-max and the hierarchical algorithms select a sensor of each type much earlier than the forward algorithm which may provide distinct information. As the order of sensor selection differs between the three sensor-placement algorithms, a comparison using metrics is needed and will be presented in Section 5.

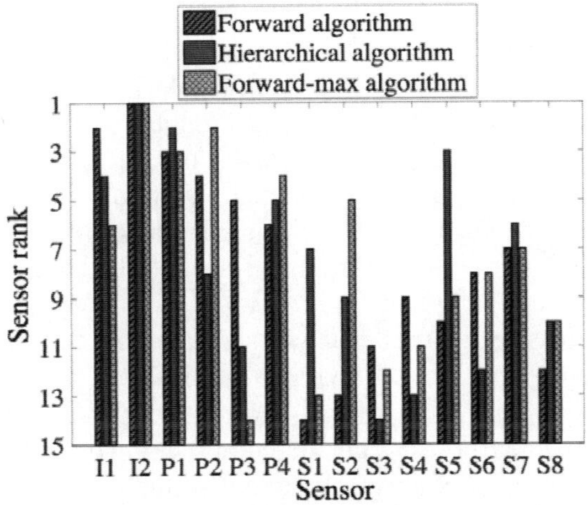

Figure 2: Sensor ranking for three sensor-placement algorithms.

5. Comparison of sensor-placement algorithms

Sensor selections presented in Figure 2 are discussed using two metrics, the joint-entropy and the expected usefulness of measurements, presented in Section 3. Figure 3 presents the joint-entropy as a function of the number of sensors for each sensor-placement algorithm. The joint-entropy evaluates the mutual information between sensors, although for the forward and forward-max algorithms, this metric does not represent the information-entropy objective function used during sensor selection. The sensor-placement algorithms are compared at each step. For all sensor-placement algorithms, the joint-entropy increases with increasing number of sensors. The entropy obtained with the hierarchical algorithm is larger than the forward and forward-max algorithms for the same number of sensors. For the joint-entropy metric, the hierarchical algorithm performs better than the forward and forward-max algorithms. This sensor-placement algorithm discriminates better model instances for the same number of sensors in the sensor configuration.

For the hierarchical algorithm, the joint-entropy gradient decreases with increasing number of sensors. After the 10th sensor, the joint-entropy gradient is small, meaning that little information is gained by adding more sensors to the sensor network.

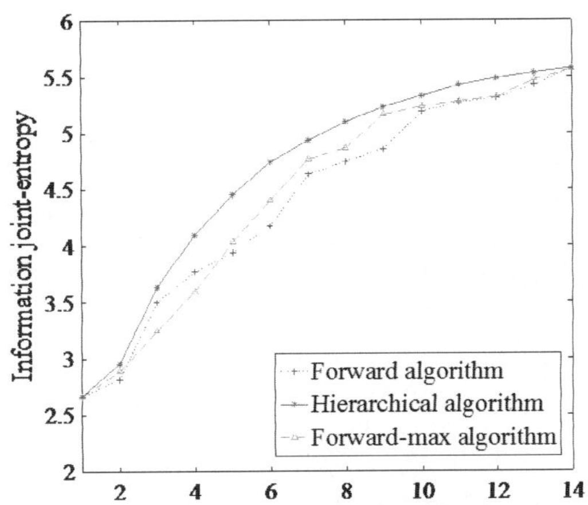

Figure 3: Joint-entropy as function of the number of sensors.

The second metric used in this study is the expected usefulness of measurements. 100,000 simulated measurements are generated from the 1000 model instances in the model class through adding a random value of combined uncertainty from Table 2. Then, the falsification process is performed and, for each simulated measurement, a number of candidate models is obtained. Eventually, the cumulative distribution function (CDF) of the candidate-model-set (CMS) size using the set of simulated measurements is produced. First, results from a defined number of sensors (Fig. 4) will be discussed. Then results will be generalized for the total number of sensors (Fig. 5). Figure 4 presents the CDF for the first 7 selected sensors out of 14 possible sensors obtained by each sensor-placement algorithm.

The hierarchical algorithm provides a smaller CMS size for any cumulative probability than the forward and forward-max algorithms. As the expected number of candidate models is smaller for the same cumulative probability, the hierarchical algorithm performs better than the forward and forward-max algorithms when 7 sensors are considered in the sensor configuration.

For all sensor-placement algorithms, the largest gradient of the cumulative distribution function is observed between 0 and 20 candidate models. As the most likely number of candidate models is at the maximum gradient of the function, the most likely CMS size is between 0 and 20 model instances, independently of the sensor-placement algorithm.

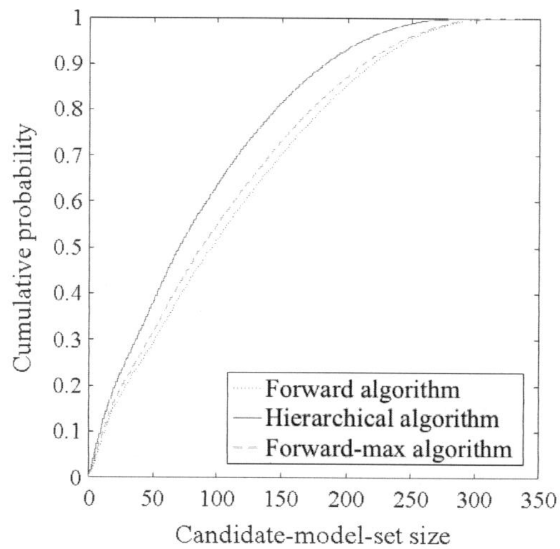

Figure 4: Cumulative probabilities of the candidate-model-set size for a configuration of 7 sensors obtained by each sensor-placement algorithm.

Figure 4 presents results only for the particular case of 7 sensors considered out of 14. This result needs to be extended for every number of sensors considered. Figure 5 presents the quantiles 5, 50 and 95 % of the CDF of the CMS size for the three sensor-placement algorithms as a function of the number of sensors. The observed CMS size using field measurements is presented for comparison. For the results of field measurements, the order of sensor selection is the same as the one obtained with the hierarchical algorithm presented in Figure 2.

For more than 2 sensors, the hierarchical algorithm provides smaller quantiles than the forward-max and forward algorithms. Thus, the hierarchical algorithm performs better than the forward and forward-max algorithms in terms of the expected CMS size.

Considering the hierarchical algorithm results, for all quantiles, an important decrease of CMS size is observed until a configuration of 6 sensors. Then, adding a sensor to the network does not result in a substantial decrease of expected CMS size. Deployment of some sensors could have been avoided without any substantial loss of information. This result is in agreement with results using the joint-entropy metric (Fig.3), where no substantial information gain occurs for more than 10 sensors in the sensor network.

The number of candidate models using field measurements decreases with increasing number of sensors considered in the sensor configuration, following the trend of the expected CMS size of the sensor-placement algorithms. For less than 6 sensors, the CMS size obtained using field-measurements is between the quantiles 5 and 50 %. However, a drop occurs at the 6th sensor selected (S7). From that sensor, the CMS size obtained using field-measurements is slightly larger than the quantile 5 %. Between the 6th and the 14th sensor, the CMS size obtained with field measurements decreases from 10 to 4 model instances. Thus, only a small loss of information would have occurred if only 6 sensors were considered in the network instead of 14.

Considering all sensors, the CMS size observed using field-measurements is only 5 % likely. However, regarding the PDF of the CMS size (not presented herein), it corresponds to the second most likely number of candidate models with an occurrence of approximatively 1,400 over 100,000 simulations. In addition, other factors may explain this result such as a bias in the model, the initial sampling, or one sensor not working properly.

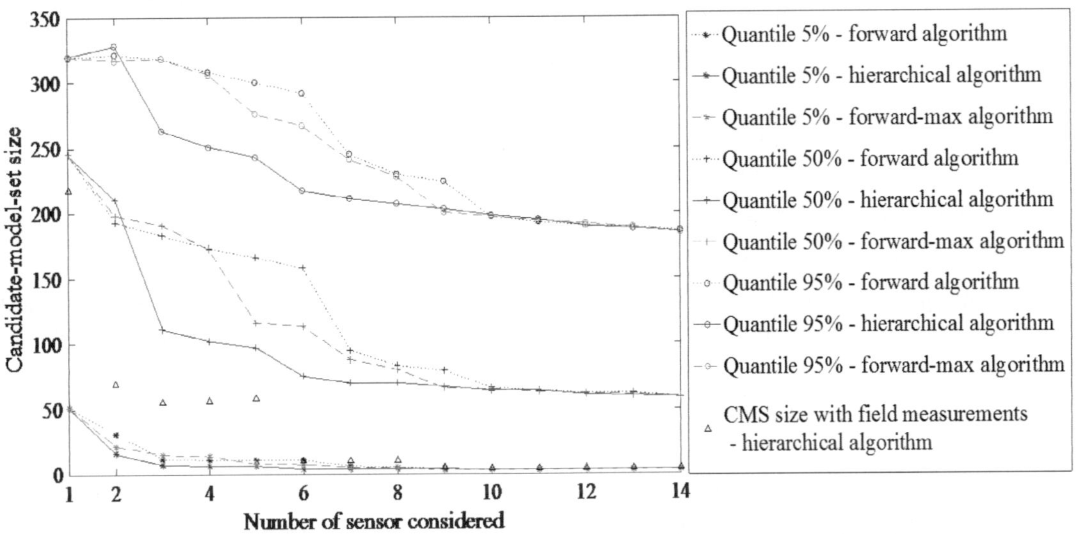

Figure 5: Quantiles 5, 50 and 95 % of the CMS size as function of the number of sensors.

6. Conclusions

A static load test was performed on a full-scale bridge in Singapore. The sensor configuration was designed based on engineering judgement. This configuration was evaluated using three greedy algorithms for sensor system design.

Algorithms were compared using two information gain metrics. For both metrics, the hierarchical algorithm performed better than the forward and forward-max strategies. The hierarchical algorithm is thus recommended as the best sensor-placement algorithm for future studies.

Results related to the candidate-model-set observed with field-measurements showed that a smaller number of sensors could have been deployed with only a limited loss of information. This observation highlights the value of a measurement-system-design methodology.

Two limitations of the present study are to be avoided in future studies. First, the sensor placement was investigated using only 14 locations on the bridge that were chosen based on engineering judgement. However, many other locations are possible and this sensor configuration should be investigated using a wider range of possible sensor locations. Finally, only greedy algorithms were compared to reduce computational cost. A stochastic search algorithm may provide better results in terms of information gain and should be considered for comparison.

Acknowledgements

The research was conducted at the Future Cities Laboratory at the Singapore-ETH Centre, which was established collaboratively between ETH Zurich and Singapore's National Research Foundation (FI 370074016) under its Campus for Research Excellence and Technological Enterprise programme. The authors gratefully acknowledge the support of the Land Transport Authority (LTA) of Singapore for support during load tests in the scope of the bridge case study.

References

Smith, IFC 2016, 'Studies of sensor-data interpretation for asset management of the built environment', *Frontiers in Built Environment*, vol.2, no. 8, pp.1-9.

Goulet, JA & Smith, IFC 2013a, 'Structural identification with systematic errors and unknown uncertainty dependencies', *Computers & Structures*, vol.128, pp. 251–258.

Robert-Nicoud, Y, Raphael, B & Smith, IFC 2005a, 'Configuration of measurement systems using Shannon's entropy function', *Computers & Structures*, vol. 83, no. 8-9, pp. 599-612.

Papadimitriou, C 2004, Optimal sensor placement methodology for parametric identification of structural systems, *Journal of Sound Vibration*, 2004, vol. 278, pp. 923–947.

Kripakaran, P & Smith, IFC 2009, 'Configuring and enhancing measurement systems for damage identification', *Advanced Engineering Informatics*, vol. 23, no. 4, pp. 424-432.

Krause, A, Singh, A & Guestrin, C 2008,'Near-optimal sensor placements in Gaussian processes: theory, efficient algorithms and empirical studies', *Journal of Machine Learning Research*, vol. 9, pp. 235-284.

Goulet, JA & Smith, IFC 2013b, 'Performance-driven measurement system design for structural identification", *Journal of Computing in Civil Engineering*, vol. 27, no. 4, pp. 427-436.

Pasquier, R, Goulet, JA, Smith, IFC 2017, 'Measurement system design for civil infrastructure using expected utility', *Advanced Engineering Informatics*, vol. 32, pp. 40-51.

Papadopoulou, M, Raphael, B, Smith, IFC & Sekhar, C 2014, 'Hierarchical sensor placement using joint entropy and the effect of modeling error', *Entropy*, vol.16, no. 9, pp. 78–101.

Papadopoulou, M, Raphael, B, Smith, IFC & Sekhar, C 2015, 'Optimal sensor placement for time-dependent systems: application to wind studies around buildings', *Journal Computational Civil Engineering*, vol.30, no. 2, p. 04015024.

Papadopoulou, M, Raphael, B, Smith, IFC & Sekhar, C 2016, 'Evaluating predictive performance of sensor configurations in wind studies around buildings', *Advanced Engineering Informatics,* vol. 30, pp. 127–142.

Robert-Nicoud, Y, Raphael, B & Smith, IFC 2005b, 'System identification through model composition and stochastic search', *Journal of Computing in Civil Engineering*, vol. 19, no. 3, pp. 239–247.

Šidák, Z 1967, 'Rectangular confidence regions for the means of multivariate normal distributions', *Journal of the American Statistical Association*, vol. 318, pp. 626–633.

The 8th International Conference on
Structural Health Monitoring of Intelligent Infrastructure
Brisbane, Australia | 5-8 December 2017

The method of optimal sensor placement based on data correlation: Application to cable group of Nanjing No.3 Yangtze River Bridge

S.L. Li[1], J.L. Dong [2], W. Lu [3], H. Li [1], W.C. Xu [4], Y. Jin [4].

[1] Harbin Institute of Technology – China, Email: lishunlong@hit.edu.cn *[2]Zhejiang Scientific Research Institute of Transport - China, [3] Harbin Institute of Technology (Shenzhen) - China, [4] CCCC Highway Consultants CO., Ltd. (HPDI) - China*

Abstract

To evaluate the states of cables effectively and accurately, all cables in Nanjing No.3 Yangtze River Bridge are installed sensors monitoring cables under multiple load. However, a large redundancy is found during the analysis of the monitoring data that many cables show great similarity in the change of cable force, which means the strategy that sensors are arranged on all of cables is not only increases economic investment during the construction of system, but also increases more time to analyze monitoring data in the operational phase. Therefore, the question arises of how to construct a cost-effective health monitoring system by selecting the representative sensor positions considering the similarity existing in cable group, which can obtain maximum amount information of structure by those limited sensors. The tendency of time-varying cable force measured by anchor dynamometer is chosen as research object in this paper, which is acquired by an extraction method based on "B-spline interpolation". The mutual information(MI) based on kernel density estimation, as a general measure of dependencies between variables is applied to detect potential relationship mot only linear in cable group. Then a clustering technology—the bond energy algorithm(BEA) is taken for the optimal placement of cable sensors. Several optimal placements are discussed with different correlation threshold for cable group with 84 stay-cables in the upside of Nanjing No.3 Yangtze River Bridge and verify the effectiveness of this method. Compared with the results that correlation is measured by Pearson Coefficient, the mutual information makes more sense to identify general association existing cable group and to get more reasonable arrangement of sensors.

1. Introduction

Sensor optimization was first studied in the field of dynamic control and system identification of spacecraft (Udwadia 2010, Lim 2012), and then developed to the mode identification and detection monitoring of structural engineering. An experience method of the optimal placement of sensors is the modal kinetic energy method (Chung & Moore 1993), which sensors are arranged in the position where mode of the kinetic energy is larger than other positions; In 1991, Kammer proposed an effective independent algorithm (EI) (Kammer 1990), which can obtain the maximum modal response information by making the sensor points linearly independent for determining modal. Reduction method (Guyan 1965) is also a commonly used in point selection, in this approach the principal coordinate playing a major role in mode reaction are selected as measuring points through successive iteration; Udwadia (Udwadia 2010) considered that methods those based on modal kinetic energy can't get optimal results, and then proposed an optimal algorithm for sensor placement based on Fisher information matrix, which is not only applicable for linear system but also for nonlinear system. Method based on the sensitivity of eigenvector (Dewolf & Zhao 1999) is commonly used in the sensor optimization for structural damage identification, in which degree of freedoms those contain less damage won't be selected to place sensors. In general, there is less research on sensor optimization based on the data correlation, but some studies are researched in wireless sensor networks by using node data correlation to improve the performance of network and compress the redundant data (Gastpar & Vetterli 2003, Pradhan & Kusuma & Ramchandran 2002). An annular spatial correlation model in wireless sensor network is proposed for event monitoring based on that data sampled from nodes with same distance from the event source are highly correlated (Su & Lin 2008). In the file of SHM, the correlation of structural measurement is applied to malfunction diagnosis and optimization of sensors (Lu, Teng, Wen, Zhu & Li 2017). The bond energy algorithm(BEA) is used to optimize the arrangement of sensors, for clustering analysis of correlation matrix of structural response and achieved good results in the large span dome structure (Lu, Wen, Teng, Li & Li 2016).

In this study, considering the redundancy and similarity of cable force found in long-time monitoring data of Nanjing No.3 Yangtze River, a method based on data correlation is employed to optimize the sensor placement. The paper is organized as follows: In section 2, we shortly reviewed the method of optimal sensor placement proposed by Lu W, et al. (Lu, Wen, Teng, Li & Li 2016), and mainly introduced the bond energy algorithm, sensor classification principle, and optimal point selection. The proposed method is applied to cable group in Nanjing No.3 Yangtze River Bridge in section3. The spatial correlativeness of cable group, summarized by long-term monitoring cable force datasets, were first introduced, especially the mutual information is applied to detect the inherent relationship between cable groups. Moreover, reasonable cable force sensor arrangements were optimized by the bond energy algorithm and several optimal schemes under the corresponding correlation threshold verify the effectiveness of the proposed approach. Finally, conclusions and suggestions are given in section 4.

2. Method of optimal sensor placement based on data correlation

2.1 The establishment of the correlation matrix of structural response

Load response matrix with n positions and m load cased can express as $X=[x_1,x_2,...x_n]\in R^{m\times n}$, and it is assumed that γ_{ij} represents the correlation degree of x_i and x_j in different positions, then the correlation matrix R can be expressed as $R=[\gamma_{ij}]_{n\times n}$. Then, an equivalent correlation matrix $D=[d_{ij}]_{n\times n}$ can be transformed from the correlation matrix R by binary processing: if $|\gamma_{ij}|\geq\eta$, $d_{ij}=1$;otherwise, $d_{ij}=0$. Where η indicates the selected correlation threshold. It is clear to see that equivalent correlation matrix D is comprised of only two number 1 and 0, where $d_{ij}=1$ represents the response of position i and j have a strong relation under the constraint of threshold, while $d_{ij}=0$ represents a weak relation.

Thus, the correlation model is established and the strength of correlation in different positions are affected by relevance measure and the selected threshold. Pearson coefficient is usually used to measure the relevance of two variables, also applied to describes the association of structure response in Lu's paper.

2.2 The classification of sensor position by the bond energy algorithm

Cluster association matrix (CA), as shown in Fig. 1, is transformed from equivalent correlation matrix D by the bond energy algorithm(BEA). The purpose of the BEA is to identify and explore natural variable groups and clusters that occur in complex data arrays (Mccormick & Others 1969).

Cluster association matrix (CA) is obtained by permuting the rows and columns of D until the measure of effectiveness (ME) is maximum. The ME for an array $A \in R^{M \times N}$ can be expressed as

$$ME(A) = \frac{1}{2}\sum_{i=1}^{M}\sum_{j=1}^{N} a_{ij} \times \left(a_{i,j+1} + a_{i,j-1} + a_{i+1,j} + a_{i-1,j}\right) \tag{1}$$

The cluster association matrix (CA) is composed of several sub-matrices arranged along the diagonal, the elements in these sub-matrices are equal to 1 and element in other positions are equal to 0. Such block division characteristics uncovers the inner association existing in an array and determining optimal sensor placement. Cable positions can be divided into the following five categories according to the arrangement of the sub-matrix: if there is only one element in a sub-matrix, the corresponding positions belong to ***Type 1*** (position 1 in Fig. 1); if there is only one element in the intersection of two or more sub-matrices, the corresponding sensor positions belong to ***Type 2*** (position 5 in Fig. 1); if there is more than one element in a sub-matrix, and all the elements in matrix do not intersect with other sub-matrices, the corresponding sensor positions belong to ***Type 3*** (position 2 and 3 in Fig. 1); if there is more than one element in the intersection of two or more sub-matrices, the corresponding sensor positions belong to ***Type 4*** (position 3 and 9 in Fig. 1); Other sensor position in the matrix belong to ***Type 5*** (position 4.6.7 and 10 in Fig. 1).

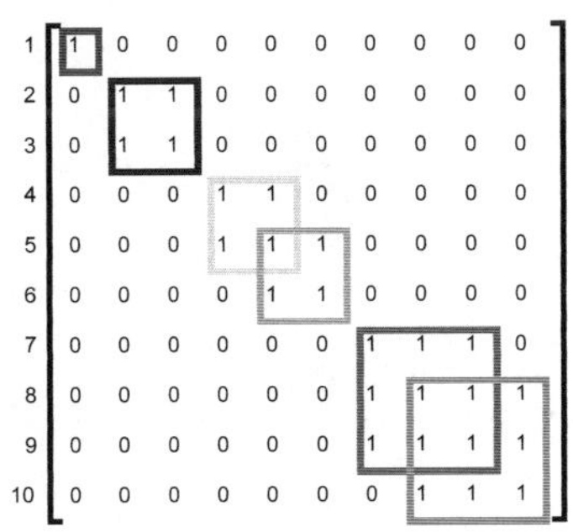

Fig. 1 Clustering Association matrix scheme (CA)

2.3 The selection of optimal monitoring positions

The correlation of positions in the same submatrix are all lager than threshold and correlated with each other more closely than positions in other submatrices. Therefore, one optimal position can be selected as representative form each sub-matrix to reflect the structural response of other positions in the same submatrix. The process of sensor optimal placement is as follows: if the number of ***Type 1*** and ***Type 2*** is m_1 and m_2 ,all positions are chosen as optimal measuring points; If the number of intersection sets contained in ***Type 3*** and ***Type 4*** is m_3 and m_4, respectively, selecting a position from each intersection as the optimal measuring points. Therefore, the total number of optimal placement of sensors is shown in Eq.2.

$$m' = m_1 + m_2 + m_3 + m_4 \tag{2}$$

The selection of optimal measuring points from each intersection in ***Type 3*** and ***Type 4*** is determined by the minimum value of average correlation degree of m' potential positions, the definition of average correlation degree is as follows:

$$\bar{\gamma} = \frac{1}{m'^2}\sum_{i=1}^{m'}\sum_{j=1}^{m'}\gamma_{ij} \tag{3}$$

The positions in ***Type 5*** will always be found with some of positions in ***Type 3*** or ***Type 4*** that belong to the same submatrix, which is already select as optimal position, so there is no need to consider positions in ***Type 5*** for optimal positions.

3. Application to No.3 Yangtze River Bridge

3.1 Pre-processing of monitored cable forces

Nanjing No.3 Yangtze River Bridge is one of the longest cable-stayed bridges in China, comprised of two towers and four cable planes. All 168 cables had anchor load cells installed to monitor the real-time tension. 84 cables on the upriver is taken as optimal object in this paper, marked them as 1~84 as shown in Fig. 2.

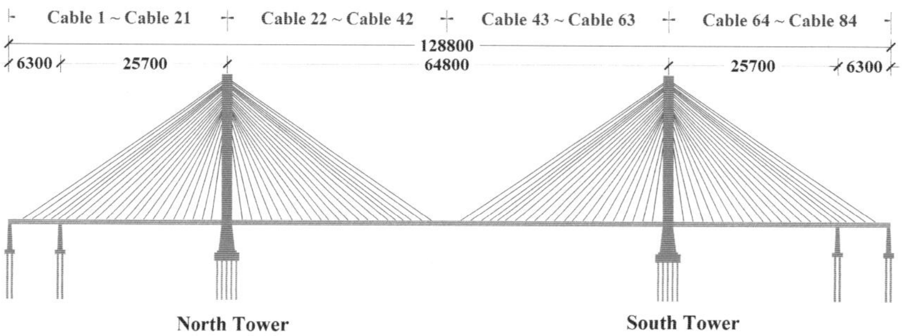

Fig. 2 General description of Nanjing No.3 Yangtze River Bridge (Unit: cm)

The change of cable force in operation phase is the result of multiple loads, and the monitored cable force can be expressed as Eq. (4):

$$T = T_i + T_e + T_v \tag{4}$$

where T, T_i, T_e, and T_v indicate the monitored cable force, cable force caused by dead load, environment load, as well as vehicle load, respectively.

Due to the complicated random distribution of vehicles on the bridge, it is difficult to logically describe the association between the monitored cable forces directly. But environmental load, mainly temperature is basically the same, so cable force except for vehicle load term, called as cable force trend, is employed for optimal sensor placement based in spatial correlation. Firstly, the B-spline interpolation method (Mehrad, Xue & Gu, 2013) is adopted to extract the cable force trend from monitored datasets, the mean value of the cable force trend at each half-hour time segment was adopted, so a dataset with 6120 data points from June 2006 to February 2007 is obtained after removing missing data. The one-day monitored data of one representative cable force can be seen from Fig. 3.

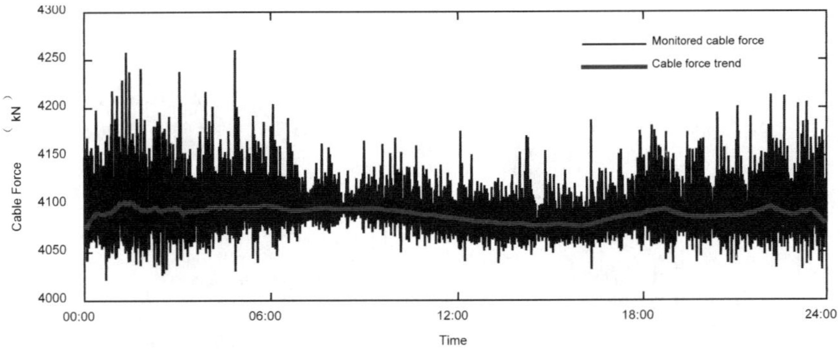

Fig. 3 Extraction of cable force trend from monitored data (00:00-24:00 June 1, 2006)

3.2 Spatial correlation model of cable group

The method mentioned above indicates that the accuracy of the correlation detection between sensor positions greatly affects the rationality of the optimal scheme. While, Pearson coefficient, merely describe linear relationship, is insufficient to reflect the complex association in structural response. Mutual information(MI) measures the uncertainty reduction of variable A given that B has been observed (Tourassi, Frederick, Markey & Jr 2001), so it can be employed as relevance measure. The MI between two discrete random variables A and B is defined in terms of Shannon entropy as:

$$MI(A,B) = H(A) + H(B) - H(A,B) = \sum_i \sum_j p(a_i, b_j) \log \frac{p(a_i, b_j)}{p(a_i)p(b_j)} \geq 0 \tag{5}$$

where $H(\cdot)$ and $p(\cdot)$ indicates the entropy and probability density of variables.

The methods commonly applied for continuous variables to improves the accuracy of mutual information contains histogram based technique (Han & Liang 2011), k nearest neighbor (KNN) based approach (Kraskov, Stögbauer & Grassberger 2004), Bayesian method (Endres & Foldiak 2005) and so on, with high precision, kernel density estimation technique (Moon, Rajagopalan & Lall, 1995, Steuer, Kurths, Daub, Weise & Selbig, 2002) is adopted in this paper. Fig. 4 demonstrate the contrast of MI and Pearson coefficient of 84 cables, it's clear that MI has a better performance in detecting relevance between in cable group especially some cables with nonlinear relationship. Shown in Fig. 4(b), Cable 22 and Cable 63 are two symmetrical shortest cables located in North tower and south tower respectively, a strong correlation should exist in theory, but can't generate any correlation by scatter plot and a low Pearson coefficient (0.697). In contrast, a relatively high MI coefficient 0.877 illustrates the internal association between the two cables. In addition, a stronger contrast appears in Cable 3 and Cable 83, as shown in Fig. 4(c). There, the Pearson coefficient was 0.3007, but MI was 0.7345. An overall view of the contrast of correlation coefficient between MI and Pearson MI for any two cables, MI are generally greater than the Pearson coefficient, as shown in Fig. 4(a), revealing a set of more hidden association not merely linear can be detected by MI and more suitable for establishing the spatial correlation model of cable group.

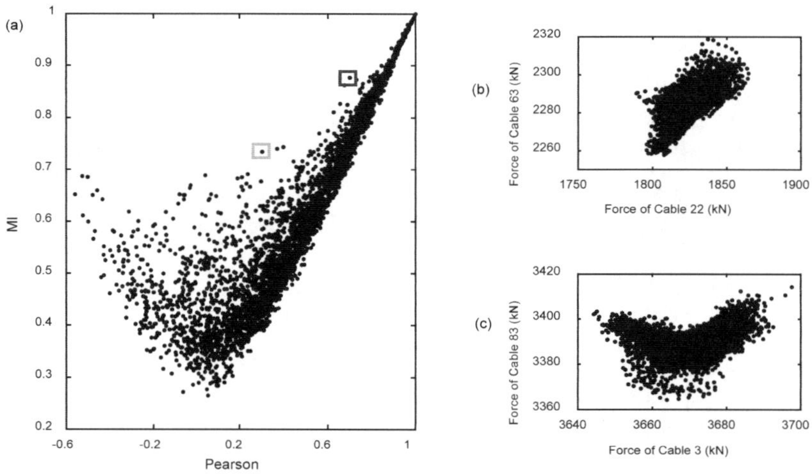

(a) Scatter plot with Pearson and MI

(b) Force of Cable 22 and Cable 63
(c) Force of Cable 03 and Cable 83

Fig. 4 Contrast of correlation coefficient between MI and Pearson

3.3 Optimal sensor placement with threshold $\eta = 0.9$

The optimal method, taking MI as correlation measure is applied to Nanjing No.3 Yangtze River Bridge and the optimal scheme with 84 investigated cables is discussed when threshold $\eta = 0.9$. 50 sub-matrices clustered by BEA are distinguished by boxes of different colours in Fig. 5, explicitly illustrating the correlation in the cable group.

According to the marshalling sequence of positions in cluster association matrix, the corresponding 84 cables can be divided into 5 categories, shown in Table 1. All positions belong to ***Type 1*** and ***Type 2*** (22 and 11 in this study) are chosen optimal measuring points directly, one optimal position should be selected form each intersection in ***Type 3*** and ***Type 4*** (6 and 3 in this study). The average correlation degree of each combination is shown in Fig. 6. The combination with minimum value is consisting of positions $\{64,63,50,84,58,55,41,36,10\}$, coupled with 33 positions in ***Type 1*** and ***Type 2*** constitutes all potential optimal measuring points. But further checking found that positions 33 and 32 are in the same submatrix, position 33 is eliminated in the final scheme to avoid redundancy. Therefore, 48 cables chosen as optimal positions is as follows:

$$Opt_{0.9} = [3,10,12,13,14,15,16,17,18,19,20,23,25,28,29,32,34,36,39,41,43,$$
$$48,50,51,52,55,58,59,60,62,63,64,65,66,67,69,70,72,80,81]$$

Table 1 Classification of cable positions

Category	Cable Position	Total
Type 1	13,14,15,16,17,18,19,20,25,28,29,34,51,52,59,60,62,67,69,70,80,81	22
Type 2	3,12,23,32,33,39,43,48,65,66,72	11
Type 3	{5,10},{21,63},{22,64},{49,50},{53,54,55,56},{82,83,84}	6
Type 4	{36,37},{40,41},{57,58}	3
Type 5	Other positions	30

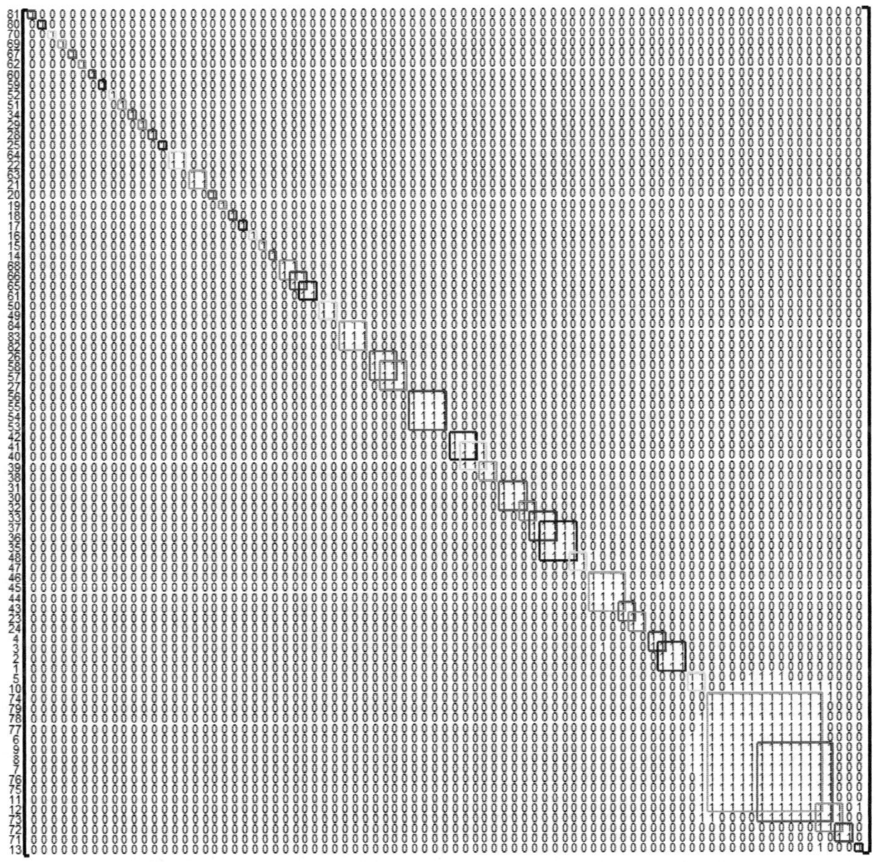

Fig. 5 Clustering correlation matrix of cable group when $\eta = 0.9$

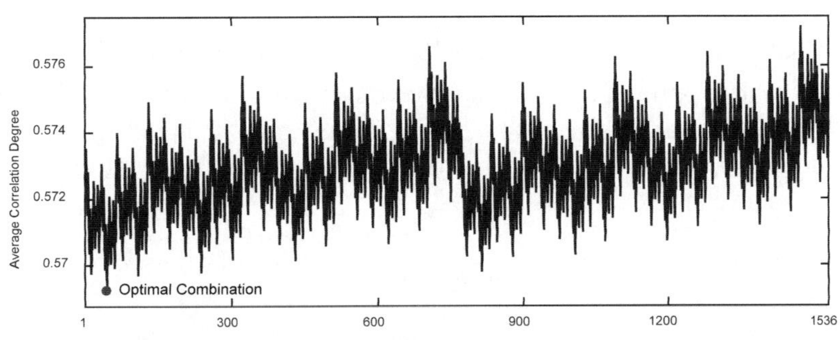

Fig. 6 Average correlation degree of potential optimal position combinations

The final optimal monitoring positions are demonstrated in Fig. 7. It can be seen from Fig. 7 that: (1) Almost half of the investigated cables are selected as optimal positions, which greatly reduced the number of sensors and proved the effectiveness of the method; (2) symmetrical positions like {21,63} and {22,64}, as well as adjacent positions like {53,54,55,56} and {82,83,84} are found in intersections in the ***Type 3*** category for the potential optimal monitoring point, proving that the clustering results based on BEA is consistent with the previous empirical experience; (3) the optimal points near the tower are more dense, indicating that the cable tension of short cables have a large difference from each other, more sensors should be installed to get a comprehensive understanding of the cable force there. For long cables far away from the tower, fewer sensors need to be installed, indicating an obvious correlation between long cables.

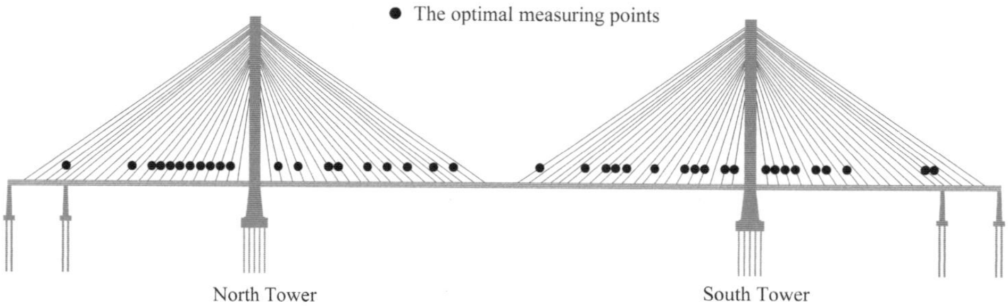

Fig. 7 Optimal sensor placement when $\eta = 0.9$

3.4 Comparison analysis of optimal schemes

Due to limited space, the optimization results with different threshold are given directly. Table 2 summarizes the number of optimal positions with threshold from 0.6 to 0.9, considering MI and Pearson coefficient as relevance measure. It can be seen from Table 2 that the number of optimal monitoring positions in both method, increased with the correlation threshold increase. Moreover, the number of optimal points generated by the MI based method is always less than that of the Pearson based method with the same threshold, which is inseparable from the fact that the mutual information can describe more general relationship between cable group. However, the choice of threshold or number of optimal sensor positions is also significantly influenced by the accuracy, budget, and even the administrative authority.

Table 2 Number of optimal monitoring positions

Threshold	0.9	0.85	0.8	0.75	0.7	0.65	0.6
Mutual information	41	32	26	22	16	15	13
Pearson coefficient	48	35	30	26	22	18	17

26 optimal sensor positions are selected both in MI-based method when $\eta = 0.8$ and Pearson-based method when $\eta = 0.75$. The optimal schemes are demonstrated in Fig. 8, revealing more than 50% of optimal monitoring positions are arranged on the same cable, but the distribution of optimal points of MI-based method is more uniform and reasonable, verifying the superiority of MI in revealing potential association in cable group.

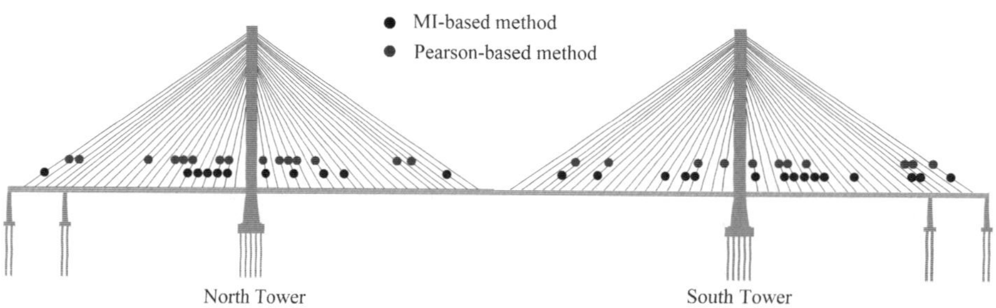

Fig. 8 Optimal sensor placement for two spatial correlation analysis methods

4. Conclusion

Reasonable sensor placement is crucial and essential for building a cost-effective health monitoring system and obtaining the necessary structural response data to precisely evaluate the state of the structure. The method of optimal sensor placement based on data correlation is applied cable group of Nanjing Yangtze River Bridge. In contrast of Pearson coefficient, MI is proved to be a better relevance measure to identify the association deeply, not only linear but also nonlinear relationship in cable group which provides an essential basis for a reasonable optimization. Several schemes of sensor arrangement under different thresholds, using the BEA algorithm, were discussed in this paper. The optimal schemes clearly discovered that the number of optimal sensors calculated by the MI-based method is less than that by Pearson-based method under the same threshold. Further, the optimal sensor points are more intensive among the short cables, while sparse among long cables. The optimal result verified the effectiveness of this proposed method that provides an advisable suggestion that which location should be selected as reasonable monitoring point for a complex structural, which will shed light on the safety assessment and decision making process in the future.

Acknowledgments

The research described in this paper was financially supported by National Natural Science Foundation of China (NSFC Grant No. 51478149, 51678204, 51638007), the 973-Program Project (No. 2013CB036305), the Science and Technology Program of Zhejiang Transportation Department (No. 2017016), and the Special Projects Supported by Scientific Research Institutes of Zhejiang Science Technology Department(No. 2017F30049).

Reference

Udwadia, F. E. 2010, 'Methodology for optimum sensor locations for parameter identification in dynamic systems', *Journal of Engineering Mechanics,* vol. *120, no.* 2, pp. 368-390.

Lim, K. B. 2012, 'Method for optimal actuator and sensor placement for large flexible structures', *Journal of Guidance Control & Dynamics,* vol. *15, no.* 1, pp. 49-57.

Chung, Y. T. & Moore D. 1993,' On-orbit sensor placement and system identification of space station with Limited instrumentations', *Proceedings of the 11th International Modal Analysis Conference, Union College press*, Schenectady, NY, pp:41-46.

Kammer, D. C. 1990, 'Sensor placement for on-orbit modal identification and correlation of large space structures', *Journal of Guidance Control & Dynamics,* vol. *14*, no. 2, pp. 251-259.

Guyan, R. J. 1965, 'Reduction of stiffness and mass matrices', *Aiaa Journal,* vol. *3*, no. 2, pp. 380-380.

Dewolf, J. T., & Zhao, J 1999, 'Sensitivity study for vibrational parameters used in damage detection', *Journal of Structural Engineering,* vol. *125,* no. 4, pp. 410-416.

Gastpar, M. & Vetterli, M. 2003. 'Source-channel communication in sensor networks', *Lecture Notes in Computer Science*, 2634, pp.553-553.

Pradhan, S. S., Kusuma, J. & Ramchandran, K. 2002, 'Distributed compression in a dense microsensor network', *IEEE Signal Processing Magazine,* vol. *19, no.* 2, pp. 51-60.

Su, W. & Lin, Y. 2008, 'An annular spatial correlation model in wireless sensor networks', *Journal of Computer Applications,* vol. 25, no. 6, pp. 1860-1863.

Lu, W., Teng, J., Wen, R., Zhu, J., & Li, C. 2017, 'Malfunction diagnosis of sensors based on correlation of measurements', *Measurement Science and Technology*, vol. 28, no. 2, 024004.

Lu, W., Wen, R., Teng, J., Li, X., & Li, C. 2016, 'Data correlation analysis for optimal sensor placement using a bond energy algorithm', *Measurement,* vol. *91*, pp. 509-518.

Mccormick, W. T., Jr & Others, A. 1969, 'Identification of data structures and relationships by matrix reordering techniques', *Algorithms*, vol. 141.

Mehrad, V., Xue, D. & Gu, P. 2013, 'Prediction of surface reconstruction uncertainties for freeform surface inspection', *Measurement*, vol. 46, no. 8, pp. 2682-2694.

Tourassi, G. D., Frederick, E. D., Markey, M. K. & Jr, F. C. 2001, 'Application of the mutual information criterion for feature selection in computer-aided diagnosis', *Medical physics*, vol. 28, no. 12, pp. 2394.

Han, M. & Liang, Z. P. 2011, 'Correlation analysis of mutual information by probability density estimated from improved averaged-shifted-histogram', *Control Theory & Applications*.

Kraskov, A., Stögbauer, H., & Grassberger, P. 2004, 'Estimating mutual information', *Physical Review E Statistical Nonlinear & Soft Matter Physics,* vol. 69, no. 2, 066138.

Endres, D., & Foldiak, P. 2005, 'Bayesian bin distribution inference and mutual information', *IEEE Transactions on Information Theory,* vol. 51, no. 11, pp. 3766-3779.

Moon, Y. I., Rajagopalan, B. & Lall, U. 1995, 'Estimation of mutual information using kernel density estimators', *Physical Review E Statistical Physics Plasmas Fluids & Related Interdisciplinary Topics*, vol. 52, no. 3, pp. 2318-2321.

Steuer, R., Kurths, J., Daub, C. O., Weise, J. & Selbig, J. 2002, 'The mutual information: Detecting and evaluating dependencies between variables', *Bioinformatics*, vol. 18 Suppl 2, no. suppl_2, pp. S231-240.

The 8th International Conference on
Structural Health Monitoring of Intelligent Infrastructure
Brisbane, Australia | 5-8 December 2017

Development of Bridge Weigh-In-Motion Using Acceleration Response of Concrete Deck Slab

K. Suzuki[1], K. Kawai [2], S. Fukada [3].
[1] *University of Fukui –Japan, Email:* suzuki-k@u-fukui.ac.jp
[2] *University of Fukui - Japan,* [3]*Kanazawa University - Japan*

Abstract

This study develops acceleration-response based Bridge Weigh-In-Motion(BWIM). The acceleration responses of concrete deck slab due to Live-load in the frequency domain is higher than those of beam; therefore, generic accelerometers whose accuracy is not high in low-frequency region, such as less than 1Hz, can be applied for the developed BWIM. Acceleration responses of RC deck fluctuate by the vehicle velocity and weight. In order to subtract the velocity effects, acceleration records are corrected by a velocity dependent coefficient. Then, high-pass filtering is processed to the acceleration data to extract the floor behavior of the deck slab. Acceleration records are converted to deflection data with twice trapezoidal integration process. Drift of the base line and noise effects in the process of integration are eliminated by Kalman filter. Three parameters that control the Kalman filter are determined with the integrated deflection data. The integral calculation is performed for each parameter change, and a parameter combination showing the calculation result that is the closest to the real RC deck's deflection among them is used; that is, RMSE is minimum in the patterns. The calculated slab deflection results show the good agreement with the records of displacement meters. With the use of slab deflection influence lines, BWIM calculation is processed. The results show the maximum error of vehicle weight is about 20% except for one eccentric result.

1. Introduction

Track traffic volume is one of the important indexes for bridge maintenance and management. For example, Load and Resistance Factor Rating (LRFR) uses the ADTT (Average Daily Track Traffic) for the load factor in the legal load rating. Another demand of ADTT is to assess the fatigue environment. Large volume of track traffic is the cause of fatigue damage. The author evaluated fatigue environment for a steel box girder bridge based on the traffic volume of heavy vehicles (Suzuki 2006). Among several methods for obtaining track traffic volume, Bridge Weigh-In-Motion (BWIM) (Moses 1979) is a well-applied method. When a vehicle runs on the bridge, bridge components react with live-load stresses that depend on the load level. The BWIM processes strain data obtained from strain gauges on bridge components. The bridge instrumentation work needs attachment of strain gauges including paint removal if it was a painted structure. That work includes the laborious wiring of connecting leads. In order to aim a labor saving BWIM, another data acquisition BWIM, not using strain gauges, should be developed. The purpose of this study is to develop the BWIM based on acceleration response excited by vehicle passages.

2. Target bridge and bridge instrumentation

The target bridge is a skewed steel plate girder bridge with RC deck slab (Fig1). The bridge length is 36.0m and the width is 9.2m to have 2 lanes. The bridge is simply supported. The framing plan of the bridge and the bridge instrumentation are illustrated in Fig.2.

Fig.1 Target bridge

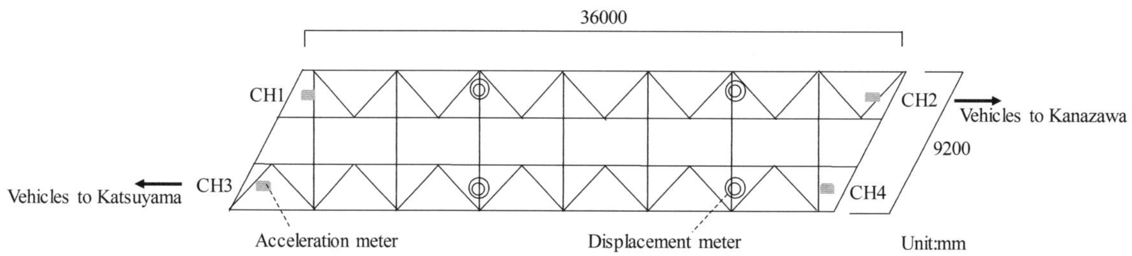

Fig.2 Framing plan and instrumented sensors

4 accelerometers are installed at the bottom surface of the RC deck. The acceleration data are integrated to vertical displacement data for the calculation of BWIM. Since the vertical displacement of the main girder at the span center is large, the data might be suitable for the BWIM calculation if displacement data are used for BWIM. However, the displacement history of the main girder is global behavior of the bridge; that is, the acceleration response less than 1Hz. General accelerometers may have a difficulty for the accurate measurement in the low frequency domain. Hence, acceleration measurement should be done at the component which locally reacts to the live loads even if the displacement response was not large. This is the reason to install the accelerometers at the RC deck. The displacement meters measure the deck deflection which is relative to the girder. These data of displacement meters are referred to verify if the integrated deflection values from acceleration data are valid or not.

For the BWIM process, necessary information is vehicle's entering/leaving time, its velocity, number of vehicle's axles and deformation data of bridge components. The combination of CH1 and CH2 obtain the entering/leaving time of the vehicles for Kanazawa; also, the velocity evaluated by the time difference of the RC deck's live-load response histories at CH1 and CH2. The calculation of vehicle weight uses the data of CH1 or CH4. The data are sampled at 200Hz.

3. Test truck and test run

3.1 Test truck and traveling speed

Fig.3 and Table1 shows the test truck. The truck has three axles, the gross weight of which is 239kN. The truck ran at 40km/h and 60km/h on the lane for Kanazawa. As for the lane for Katsuyama, the speed cases were 40km/h and 47km/h. The reason of 47km/h, not 60km/h, was the slope of the road. The lane for Katsuyama was the up slope, consequently the truck speed could not reach 60km/h. The cases of the test run 6 cases for each lane, in total 26 cases were performed.

Fig.3 Test truck

Table 1

	Axle 1	Axle 2	Axle 3	Gross weight
Weight (kN)	70.4	84.7	84.3	239.4
Axle interval(m)	5.83		1.32	

3.2 Characteristics of the RC deck responses

Fig.4 shows the vertical displacement records of the RC deck of the lane for Katsuayama when the test truck ran. The RC sinks when a vehicle axle passes right above. The displacement history of rear 2 axles is overwrapping and can be seen as one heavy axle. The maximum response is about 0.3mm deflection and the velocity effects, and the dynamic effects can be observed.

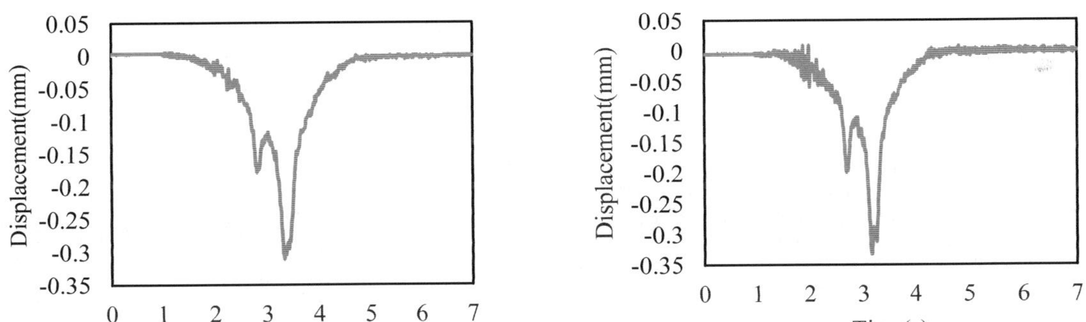

Fig.4 Displacment response of RC deck

Fig.5 is the acceleration responses of RC deck. of the lane for Katsuayama when the test truck ran. The dynamic effects arose more in the case of velocity 47km/h. Since it is not large difference in the displacement response, it is necessary to apply a process to reduce the dynamic effects in the acceleration data. Large acceleration responses occur when the truck passes right above; then, deck vibrations remain until the truck leaves the bridge. The envelope shape of the acceleration history is mostly the same even when the truck speed changes except for the time zone of truck passing immediately above. This fact may support that RC deck's displacement is able to be calculated from acceleration data by reducing the dynamic effects depending on the vehicle velocity.

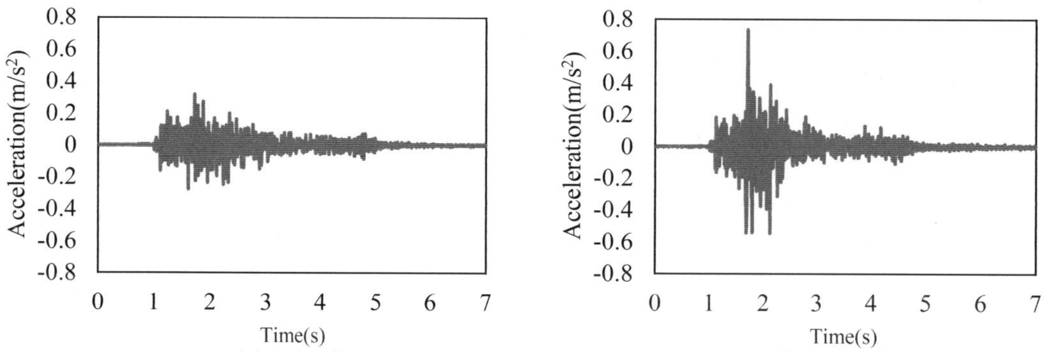

(a) 40km/h (b) 47km/h

Fig.5 Acceleration response of RC deck

4. Calclation from Acceleration response to Displacement response

4.1 Reduction of dynamic effects

In order to reduce the dynamic effects, statistical numerical processing is taken. The acceleration response excited by the higher speed vehicle has the extremely high value very instantaneously. In statistics, such values are far from the median. Accordingly, this study corrects the extreme outlier that is over 3σ from the median, and the outlier is transformed to the average value of neighbors. Fig.6 is the acceleration responses before the reduction process and after it. If the deviation value exceeds three times of the standard deviation, reduction process was applied. Fig.6 shows the high value around 3.2 seconds was corrected.

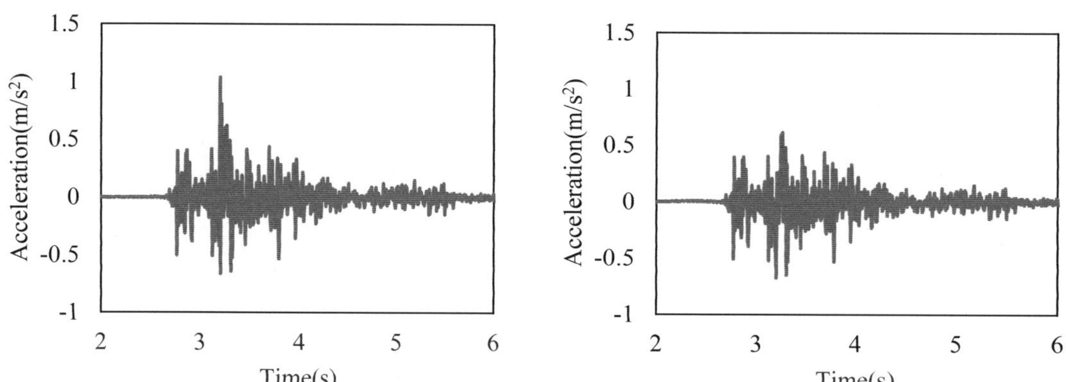

(a) 40km/h (b) 47km/h

Fig.6 Acceleration response of RC deck

4.2 Kalman filter

In the process of twice integration from acceleration to displacement, it is necessary to eliminate the drift of the baseline. For this reason, firstly it is necessary to remove the low frequency component less than 3.0Hz in the acceleration data in order to remove the beam action and extract the floor action. Secondarily, observation errors of devices and noise effects are reduced. For the second step, this study applies the Kalman filter for the integration process.

Eq(1) and Eq(2) are state equation and observation equation.

$$x_t = bx_t^{true} + cw \quad (1)$$

$$y_t = ax_t^{true} + v \quad (2)$$

where, x_t = predicted location at the time of t
y_t = observed data at the time of t
c = Kalman gain
w = error of x_t, w follows normal distribution with variance τ^2, $N(\mu, \tau^2)$
v = error of y_t, v follows normal distribution with variance τ^2, $N(\mu, \tau^2)$

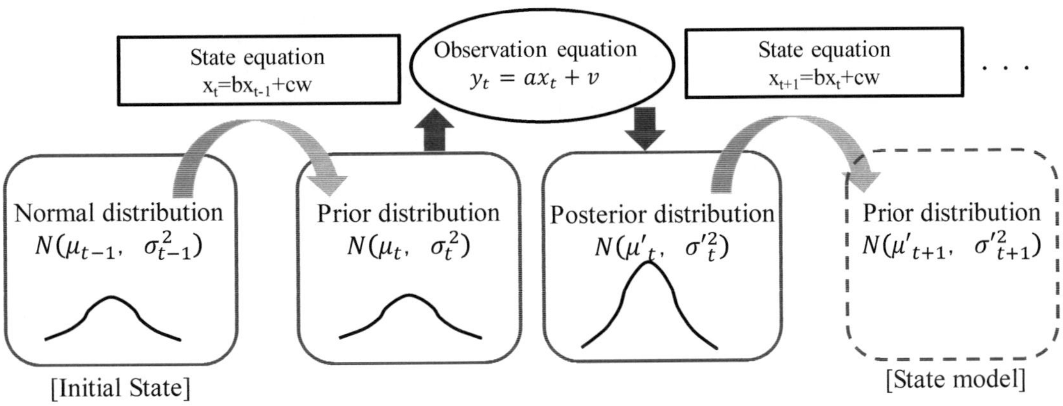

Fig.7 Sequence of calclation

From the initial value, it repeats the sequence of calculations; that is, the condition equation and the observation equation, then state space model is estimated(Fig.7). The prior distribution $N(\mu, \sigma^2)$ is renovated to the posterior distribution $N(\mu', \sigma'^2)$. In this process, the mean value μ is renovated to μ' by the Eq.(3), and also the variance σ^2 is done to σ'^2 by the Eq.(4). In addition, the posterior distribution $N(\mu_{t-1}, \sigma_{t-1}^2)$ is renovated to the new prior distribution $N(\mu_t, \sigma_t^2)$ by the state equation. When the time changes from $t-1$ to t, the mean value μ_{t-1} is renovated to μ_t by Eq.(5) and the variance σ_{t-1}^2 is done to σ_t^2 by Eq.(6). The process needs three control parameters of ρ^2, τ^2, and c.

$$\mu' = \frac{\mu\tau^2 + \sigma^2 y}{\tau^2 + a^2\sigma^2} \quad (3)$$

$$\sigma^2 = \frac{\sigma^2\tau^2}{\tau^2 + a^2\sigma^2} \quad (4)$$

$$\mu_t = b\mu_{t-1} \quad (5)$$

$$\sigma_t^2 = b^2\sigma_{t-1}^2 + c^2\rho^2 \quad (6)$$

4.3 Determination of control parameters

When Kalman filter is applied, the following three parameters are important to control the calculation result, that is, Kalman gain c, noise variance in the state equation ρ^2 and noise variance τ^2 in the observation equation. In this study, we supposed the noise variances ρ^2 and τ^2 were unknown. For the fix of parameters, the root mean square(RMS) was used. Each parameter is independently changed from 0.01 to 1.00. The root mean square is calculated by the measured displacement data and twice integrated displacement from the acceleration data. The parameters were fixed when the RMS minimum.

4.4 Calculation of vertical displacement by twice integration

Fig.8 explain the calculation process of vertical displacement from the measured acceleration data. First, the acceleration data are extracted for the time of vehicle passage. Second, the extracted data are processed with twice trapezoidal integrations. In each integration step, the Kalman filter corrects the data. The three parameters in the Kalman filter is fixed in the whole process.

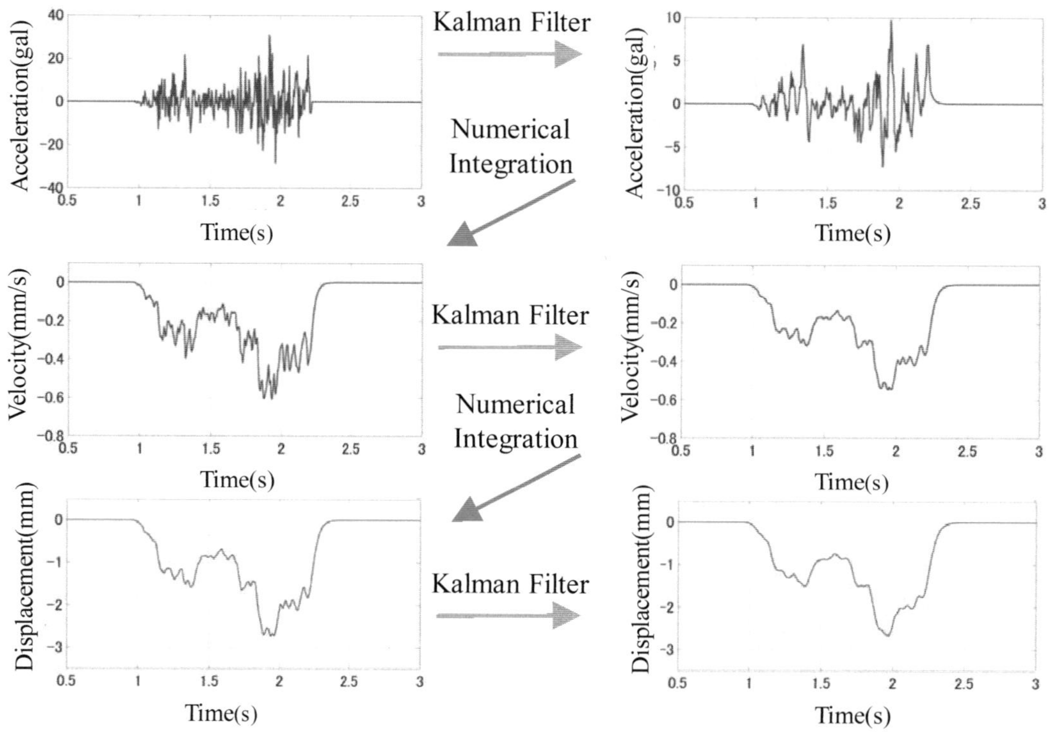

Fig.8 Calclation of vertical displacement(Slab deflection)

4.5 Calculated slab deflection

Fig.9 shows a case of calculated vertical displacement of slab. The calculated results well trace the real measured displacement. A slight difference is observed in the base part of the waveform. Fig.10 is the results of maximum displacement. The horizontal axis means the test number. The calculated results show good agreements with the measured ones. The error is about 10% at the maximum for the both lanes.

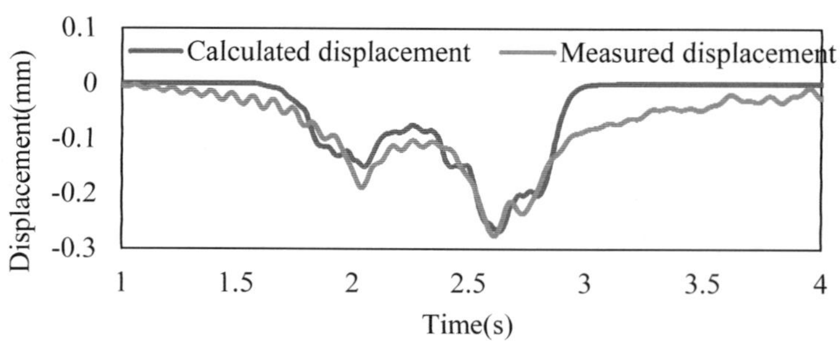

Fig.9 Comparison of displacement histories between the calculated results and measured ones

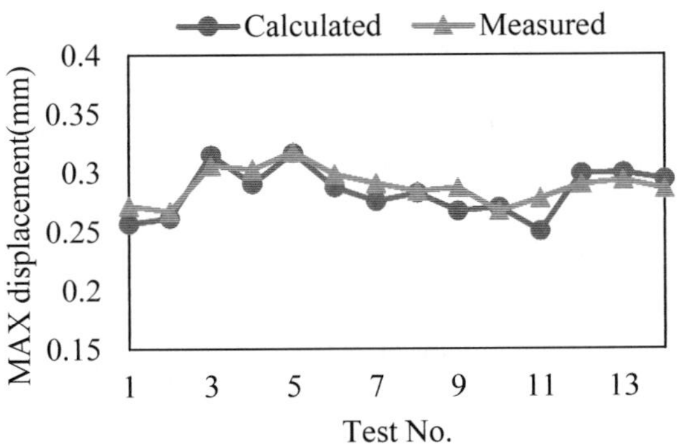

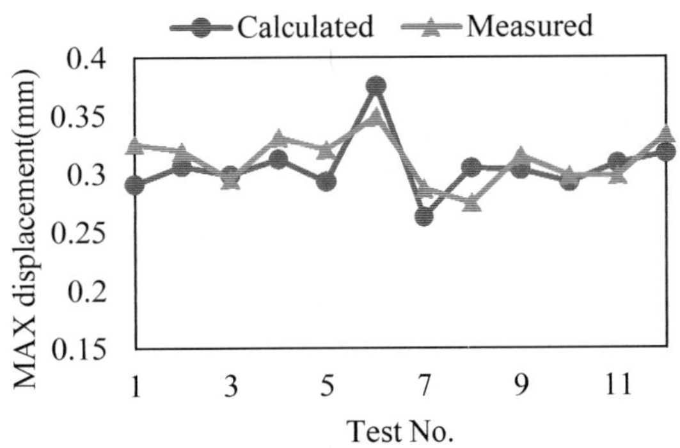

(a) Lane for Kanazawa (b) Lane for Katsuyama

Fig.10 Comparison of the max displacement between the calculated results and measured results

5. Bridge Weigh-In-Motion

5.1 Weight calculation process

Based on the use of slab deflection influence lines, gross weight of the vehicle is calculated. The influence lines are obtained with the process with Fourier transform (Tateishi 1995) by using a vehicle whose axle weight is known. Each influence line for each axle is arranged on the time history of displacement data in accordance with the axle passage time(Fig.11), then the influence lines are multiplied with parameters P, which will be the axle weight, as closest to the real slab deflection by the least square method. Finally, the gross weight of the vehicle is calculated by adding all the axle weight.

5.2 Results of BWIM

Fig.12 displays the results of gross weight for the test truck. The acceleration based BWIM results were close to the real weight of 239kN. Although the test No.8 on the lane for Kanazawa had a large error, the other results are mostly less than 20%. The test of even number is higher velocity cases that were 60km/h for Kanazawa and 47km/h for Katsuyama. Error tends to be larger in the case of 60km/h, hence the dynamic effects might affect the shape of the displacement waveform calculated from the acceleration data.

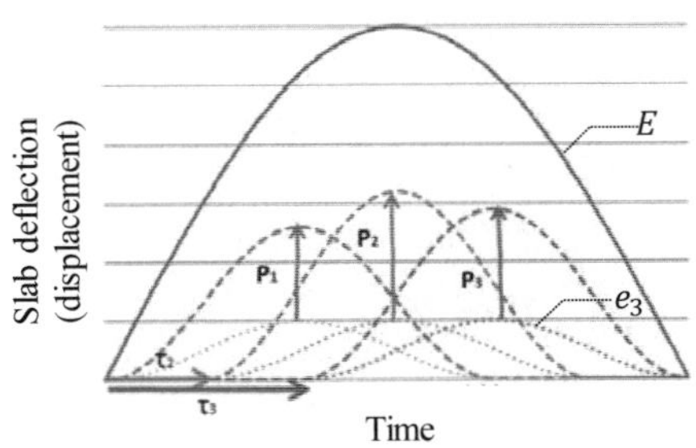

$E \fallingdotseq P_1e_1 + P_2e_2 + P_3e_3$

$G \fallingdotseq P_1 + P_2 + P_3$

where, e_i = Influence line of axle i

P_i = Axle weight of axle i

τ_i = Axle interval time between the 1st axle and the axle i.

E = Real slab deflection

G = Gross weight

Fig.11 Curve fitting process for weight calculation for the case of 3-axle vehicle

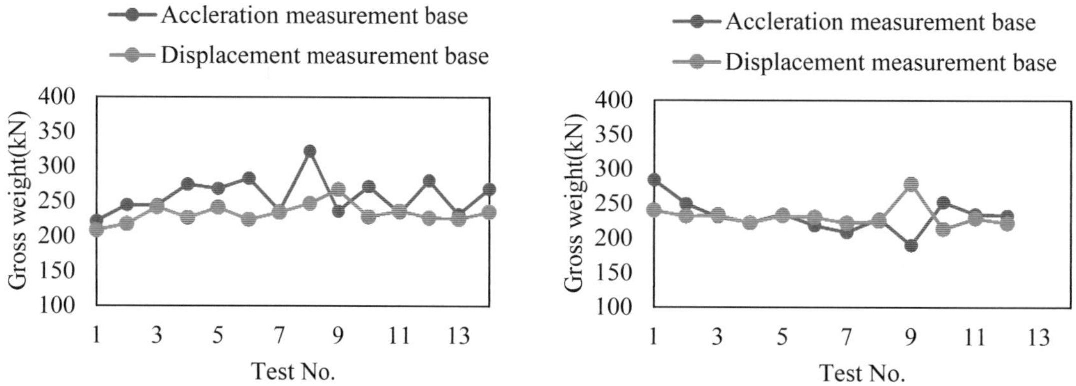

(a) Lane for Kanazawa (b) Lane for Katsuyama

Fig.12 BWIM results

6. Conclusions

This study developed the acceleration-measurement based BWIM. In order to reduce the dynamic effects that occur when vehicle velocity is higher, a statistical numerical processing was performed. For the integration process, Kalman filter was applied to correct the data. The parameters to control Kalman filter was determined by the RMS to fit the acceleration-measurement based displacement to the real one. Finally, BWIM was processed with the use of displacement obtained by the filtered and twice integrated acceleration data. The proposed process is possible to calculate the maximum deflection of RC deck when a vehicle passes with an error about 10%. In addition, it can calculate the vehicle gross weight with BWIM, the error of which is about 20%. However, there might rarely include the eccentric error. The improvement of BWIM accuracy is a future task.

Acknowledgements

This work was supported by JSPS KAKENHI Grant Number 15K18105 and also 16K06463. The authors expresses sincere appreciation to the support.

References

Suzuki K and Yoshikawa S (2013), 'Monitoring of Traffic Axle Loads in Orthotropic Steel Deck Structures', The 13th East Asia-Pacific Conference on Structural Engineering and Construction (EASEC-13) 2013.9, Sapporo, Japan

Moses F (1979), 'Weigh-in-Motion System Using Instrumented Bridges', Transportation Engineering Journal. Proceedings of ASCE. Vol.105 No.TE3. pp.233-239.

Tateishi K, Takenouchi H and Miki C (1995), 'Mechanism for Developing Local Stress at the Connection Details in Steel Bridge Structures', JSCE Journal of Structural Mechanics and Earthquake Engineering. NO.507/I-30 pp.109-119.

The 8th International Conference on
Structural Health Monitoring of Intelligent Infrastructure
Brisbane, Australia | 5-8 December 2017

Efficient Hammering Sound Spectrum Analysis for Concrete Condition Assessment with Distance Metric Learning

Jiaxing. Ye [1], Takumi Kobayashi [1], Masaya Iwata [1], Hiroshi Tsuda [1], Masahiro Murakawa [1]
[1] *The National Institute of Advanced Industrial Science and Technology (AIST) – Japan,*
Email: jiaxing.you@aist.go.jp

Abstract

Hammering inspection, due to its superior efficiency and simplicity, had been extensively applied for concrete structure condition assessment over decades. However, it requires onsite inspector to interpret evaluation results, i.e. judging there exists damage or not by hearing the echoes, therefore individual skill can be influential factor. In this paper, we propose an efficient computerized hammering echo sound analysis system which can automatically discern defective echoes during hammering inspection for concrete structures. Particularly, we formulate the hammering echo (spectrum) analysis through echo-to-echo distance metric learning. In statistical pattern recognition, we mostly rely on conventional Euclidean distances, such as cosine similarity and Mahalanobis distance. Those metrics are established grounded on geometry or statistical hypothesis (Gaussian). For one certain application, e.g. hammering echo spectrum, it is possible to learn a data-driven optimal metric that can significantly improve the performance in pattern classification. In this study, our basic idea is to establish an effective between-echo metric which enables all hammering echoes captured from healthy concrete are tightly gathered in the projected feature space and abnormal echoes are scattered distributed from normal ones. Such (between-class) large margin criterion leads to a convex optimization problem. To validate the metric learning approach for hammering sound analysis, we conducted experiments using several test concrete specimens with various inner defects. Experimental results demonstrated that proposed scheme achieved superior performance for hammering echo analysis for concrete condition assessment. The method can also be generalized to other applications, such as vibration data analysis.

1. Introduction

Aging infrastructure poses the severe challenging issue to human society. For instance, in Japan increasingly more bridges and roads are approaching their design year in service, and the risk of failure arose sharply. In order to ensure the safety of citizens, it is crucial to effectively evaluate status of aging infrastructures. In recent year, many research efforts have been directed to develop efficient and low-cost inspection techniques based on this concern. More advanced and precise non-destructive evaluation approaches had been developed, such as using micro-electromechanical (MEMS) sensors and ultrasonics [Farrar, Charles R., and Keith Worden., 2012]. However, the practical application of those techniques is limited due to the high cost and complex operation. This study focuses on very conventional inspection method of hammering inspection. Although the method is rather simple and condition assessment accuracy is not very high, it has been extensively applied for structural health condition evaluation for decades. The initial literatures describing impact-echo presented from 1970s [Achenbach, 1973]. And later, more studies were carried out in both rhetorical and experimental aspects [Casino et. al, 1986; Sanalone et. al, 1986]. However, there exists one major drawback in current hammering inspection: it relies on inspector's subjective judgement to interpret the imaging data, such as determining the presence of damage. As well as this being costly and time consuming, it leaves the analysis open to human error. Such issue will need to be resolved as concrete condition evaluation continues to play a key role in more structural health assessment tasks. To establish objective criteria for rapid, robust, and reliable hammering investigation, we endeavour to develop computerized hammering inspection system for concrete structures by using signal processing and machine learning techniques. In Fig. 1, we present the high-level flowchart of propose hammering echo pattern analysis system.

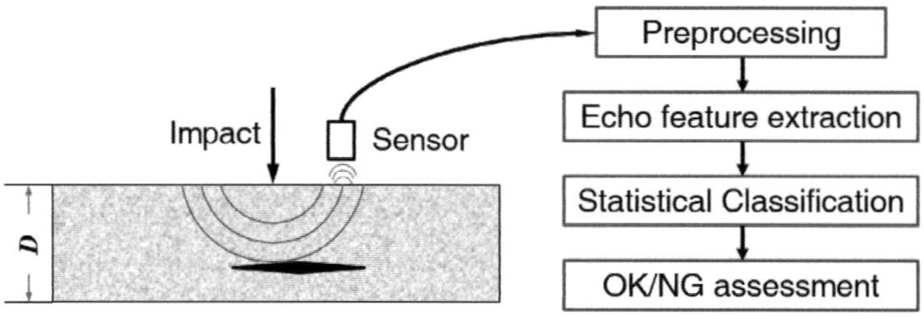

Figure 1: General flowchart of computerized hammering echo analysis system

Distance metric is a key issue in many machine learning applications [Kulis, Brian, 2013]. In this study, we plan to establish optimal distance metric to quantify the similarity between hammering echo sounds, which is anticipated to greatly facilitate healthy/defective pattern classification of echo signal. The commonly used Euclidean distance metric regards that each element of data feature vector contributes equal discriminant information and is independent from others. However, this assumption is usually not satisfied in practical tasks, especially when dealing with data of high dimensionality, where some features may not be tightly related to the topic of interest. As for hammering echo analysis, we convert echo sound to multivariate spectrum vector. It is evident that defective spectral patterns only appear in limited frequency bands and all other frequencies deliver rare information. In this scenario, an efficient distance metric, which could identify important bands and screen out irrelevant feature entries, is preferred. Devising such a distance metric is highly problem-specific and determines the success of the learning algorithm or the computerized hammering echo investigation. To this end, we employ distance metric learning algorithm of large margin nearest neighbour classification (LMNN) [Weinberge et al. 2009]. The fundamental idea is to optimize a metric with the objective that the k-nearest neighbour echoes always belong to the same class (i.e. all collected from healthy concrete structures) while instances from different classes are separated by a large margin. LMNN algorithm had been proved effective for various pattern analysis tasks, such as face recognition and hand-written digits recognition, and it is the first time attempt to apply LMNN for hammering echo sound analysis. In addition, although distance metric learning can be time consuming at training stage, it can run very fast for testing and even enables real-time processing for onsite evaluation. As a result, distance metric learning can be ideal backbone technology for computerized hammering echo analysis for condition assessment of concrete structure.

The contents of this manuscript are organized as follows. In section 2, we deliver a throughout introduction to the proposed hammering echo analysis system using spectral distance metric learning. More concretely, based on echo spectrum feature representation, we first review conventional Euclidean distances (section 2.2), then, we further present the formulation of spectral metric learning algorithm for hammering echo investigation in section 2.3. To validate the proposed method, we conduct experiments on real-world data and exhibit the results in section 3. Subsequently, we conclude this work in Section 4.

2. The proposed approach: learning efficient between-echo metric for hammering sound analysis

This section demonstrates our approach of applying distance metric learning for hammering echo pattern classification. First, we extract spectrum representation from hammering echo signal. Then, a review to conventional Euclidean distance metrics is presented. Subsequently, we show the formulation of distance metric learning based on large margin nearest neighbour objective, which renders multiple merits theoretically.

2.1 Spectral feature extraction for hammering sound

Through decades, time-frequency analysis has been a dominant approach for acoustic signal processing. As for hammering echo analysis, or even more general case of the impact-echo tests, time-frequency analysis had been extensively applied [Zhu, J. Y. & Popovics, J. S., 2007]. Fourier transform (FT) produces the frequency domain spectrum of echo signal, which can facilitate further pattern analysis, and it can be expressed as:

$$x(f) = \int_{-\infty}^{\infty} s(t)e^{-2\pi ift}dt \quad (1)$$

where $s(t)$ is echo waveform to be analyzed and $x(f)$ is the extracted echo spectrum. In the following contents, we will use $x_i \in \mathbb{R}^F$ to denote hammering echo spectrum vector covering F discrete frequency bands. Our pattern classification process is based on echo signal representation of spectrum vector.

2.2 Review of conventional distance metrics

In this section, we present a basic review of distance metrics. From the definition, a mapping $P: X \times X \longrightarrow \mathbb{R}_0^+$ over a vector space X is called a metric if for all vectors $\forall x_i, x_{j,}, x_k \in X$, it satisfies the property:

1. Triangular inequality: $D(x_i, x_j) + D(x_i, x_k) \geq D(x_j, x_k)$
2. Non-negativity: $D(x_i, x_j) \geq 0$
3. Symmetry: $D(x_i, x_j) = D(x_j, x_i)$
4. Distinguishability: $D(x_i, x_j) = 0 \Leftrightarrow x_i = x_j$.

Particularly, if a metric satisfies the first three rules but not the last one, it is called *pseudometric* (Gentle, James E, 2007). The definition of Euclidean distance between two vectors, x_i and x_j, in the p-dimensional space $\mathbb{R}^p$ is defined as

$$d_{Euclidian}(x_i, x_j) = \sqrt{(x_i^1 - x_j^1)^2 + \cdots + (x_i^1 - x_j^1)^2} = \sqrt{(x_i - x_j)'(x_i - x_j)}, \quad (2)$$

and distance between one input vector x_i to the origin is defined as:

$$d_{Euclidian}(x_i, 0) = \sqrt{(x_i^1)^2 + \cdots + (x_i^p)^2} = \sqrt{x_i' x_i}. \quad (3)$$

This Euclidean distance metric implies that all elements of feature vector (i.e. all frequency bins of the echo spectrum vector, x_i) contribute deliver discriminant information to the OK/NG echo pattern classification task. However, in statistics we prefer a distance that for every variable takes its variability into account when determining its distance from the center. Components with high variability, in other words, high uncertainty, should receive less weight compared with those components with lower variability. Such process, which is called Mahalanobis distance metric (Mitchell, Ann FS et al, 1985), can be realized by rescaling the elements of feature vector by the covariance as follows:

$$d_S(x_i, x_j) = \sqrt{(x_i - x_j)' S^{-1} (x_i - x_j)}, \quad (4)$$

and the Mahalanobis distance between input spectrum vector x_i to the origin can be derived by:

$$d_S(x_i, 0) = ||x_i||_S = \sqrt{x_i' S^{-1} x_i}. \quad (5)$$

More concretely, the Mahalanobis distance takes the covariances into consideration, which lead to elliptic decision boundaries (in the 2D case), as opposed to the circular boundary in the Euclidean case. The Euclidean distance can be regarded as a special case of the Mahalanobis distance with equal variances of the variables and zero covariances.

It is obvious that Mahalanobis distance, which is achieved by covariance-based feature rescaling, cannot guarantee optimal classification because it is an unsupervised approach and the label information of training data are not investigated at all for efficient statistical feature extraction. Motivated by the issues, plenty of research efforts had been devoted to the research of learning an appropriate distance metric from labeled examples for superior classification performance [Kulis Brian, 2013].

2.3 Hammering echo distance metric learning formulation

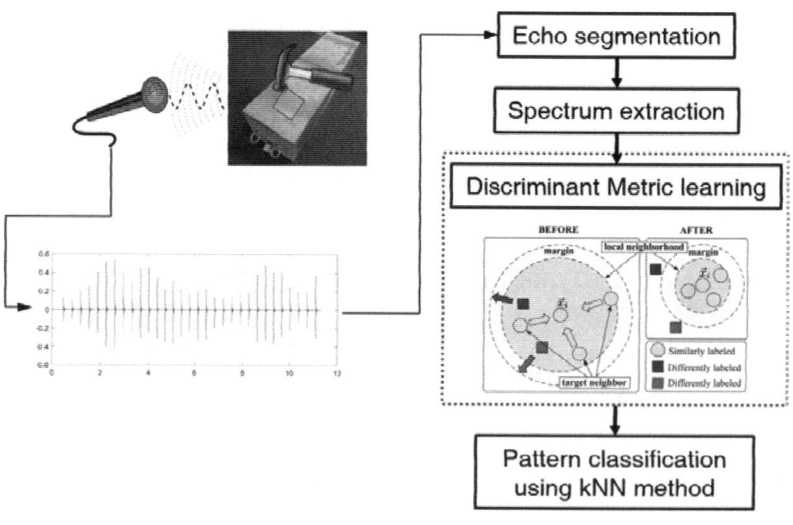

Figure 2: The proposed of computerized hammering echo analysis system based on distance metric learning

In this section, we present our approach of employing large margin nearest neighbour (LMNN) distance metric learning algorithm for hammering echo sound investigation. Intuitively, LMNN is inspired by the k-nearest neighbor classification algorithm (kNN) decision rule that will correctly classify an example if its k-nearest neighbors share the same label (Duda, Richard O. et al. 2012). The LMNN distance metric learning algorithm introduces linear transform to generate robust features for better classification accuracy, which is derived by minimizing a loss function that consists of two components (Weinberger, Kilian Q. et al, 2009). The first component penalizes large distances between examples in the same class that are anticipated to be the k-nearest neighbors, while the second component penalizes small between-sample distances with miss-matched labels. Jointly minimizing these components generates a linear transformation of the input space that can greatly enhance features for classification and weaken influence of irrelevant feature elements as well. We present a chart to demonstrate our hammering echo pattern analysis framework using distance metric learning in Fig. 2.

Let $\{(x_i, y_i)\}_{i=1}^{N}$ denote a training set of N labelled hammering echo spectrum vectors $x_i \in \mathbb{R}^F$ and binary class labels y_i (OK versus NG). Our objective is to learn a linear transformation $L: \mathbb{R}^F \longrightarrow \mathbb{R}^F$, then we are able to compute squared distances as:

$$d_L(x_i, x_j) = \left\|L(x_i - x_j)\right\|_2^2 \tag{6}$$

where L is a positive semi-definite matrix (PSD). It is simple to show that Eq.(6) defines a valid pseudometric. It can be commonly expressed in terms of the square matrix:

$$M = L'L. \tag{7}$$

Matrix M is anticipated to be positive semidefinite since it is formed by a real-valued matrix L. Subsequently, we have following formulation:

$$d_M(x_i, x_j) = (x_i - x_j)' M (x_i - x_j). \tag{8}$$

For the case of setting M to the inverse covariance matrix between x and y, the distance equals to Mahalanobis metric. Here we allow M to be any positive semidefinite matrix and thus Eq. turns out to be generalization of Euclidean metrics.

$$\varepsilon(L) = \textstyle\sum_{ij} \eta_{ij} \left\|L(x_i - x_j)\right\|^2 + c \sum_{ijl} \eta_{ij}(1 - y_{il})[1 + \left\|L(x_i - x_j)\right\|^2 - \|L(x_i - x_l)\|^2]_+. \tag{9}$$

where in the second term $[f]_+ = \max(f, 0)$ denotes the standard hinge loss and $c > 0$ is hyperparameter which can be tuned by cross validation.

Minimize	$\sum_{ij} \eta_{ij}(x_i - x_j)' M (x_i - x_j) + c \sum_{ij} \eta_{ij} (1 - y_{il}) \xi_{ijl}$ **subject to:**
	(1) $(x_i - x_l)' M (x_i - x_l) - (x_i - x_j)' M (x_i - x_j) \geq 1 - \xi_{ijl}$
	(2) $\xi_{ijl} \geq 0$
	(3) $M \succcurlyeq 0$.

The last constraint $M \succcurlyeq 0$ confines that the matrix M is required to be positive semidefinite. It is noteworthy that while this SDP can be solved by standard general-purpose solvers packages, further constraints should be imposed since most of the slack variables ξ_{ijl} never attain positive values. The slack variables ξ_{ijl} are sparse because most labelled inputs are well separated. Furthermore, their resulting pairwise distances do not incur the hinge loss, and we finally obtain very few active constraints. By performing the above explained LMNN algorithm, we can obtain a optimal hammering echo feature representation for the pattern classification task. The real data evaluation is done in the following section 3.

3. Experiments

This part shows the experimental validation of the distance metric learning approach for hammering echo pattern classification. We introduce the contents with 3 subsections.

3.1 Data collection

To evaluation the effectiveness of distance metric learning approach for hammering echo investigation, we prepared three defect concrete blocks with 2cm, 4cm, 6cm depth inner hollows and one solid (healthy) block. Ordinary MIC is used to capture responses of hammering impact over concrete. Hammering echoes was recorded with 44.1kHz sampling rate and 16bit resolution. Particularly, to observe within-class variability of echo sounds, we collected hammering sounds by changing the pose of testing concrete block, the scheme is illustrated in Fig. 3.

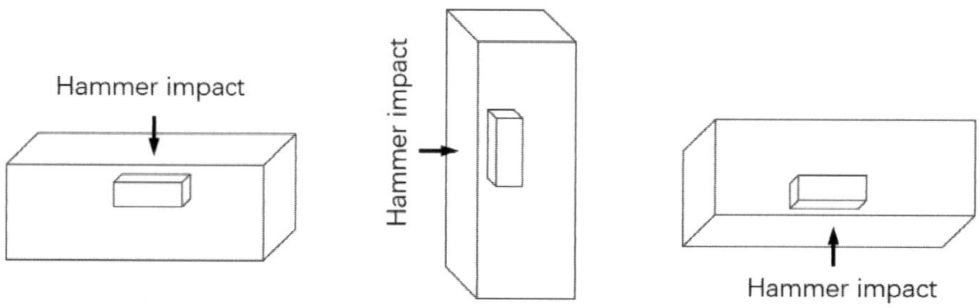

Figure 3. Hammering echo signal capture scheme from test concrete specimen

Tab. 1 showed the number of hammering echoes we collected from both healthy and defective concretes. In addition, random noises had been added during the data collection, including both express way background noise and traffic sound. At condition assessment stage, we adopted 10-folder cross-validation protocol for quantative evaluation.

Table. 1 Hammering echo data collection information

	Healthy	2cm void	4cm void	6cm void
Samp. No. (#)	1165	474	489	558

3.2 Hammering echo pattern visualization

Visualization is critical part of pattern investigation of massive data. In this study, there are mainly four types of hammering echo signal, which are captured from normal concrete specimen and from defective ones with 2cm, 4cm and 6cm inner voids. We first extract all echo spectrum vectors, then show the averaged spectrum for each type in Fig. 4. By comparing Fig. 4 (*a*) ~ (*d*), we can observe that the difference patterns exist only in limit frequency bands. We expect to enhance those defect-induced pattern variations by using distant metric learning algorithm.

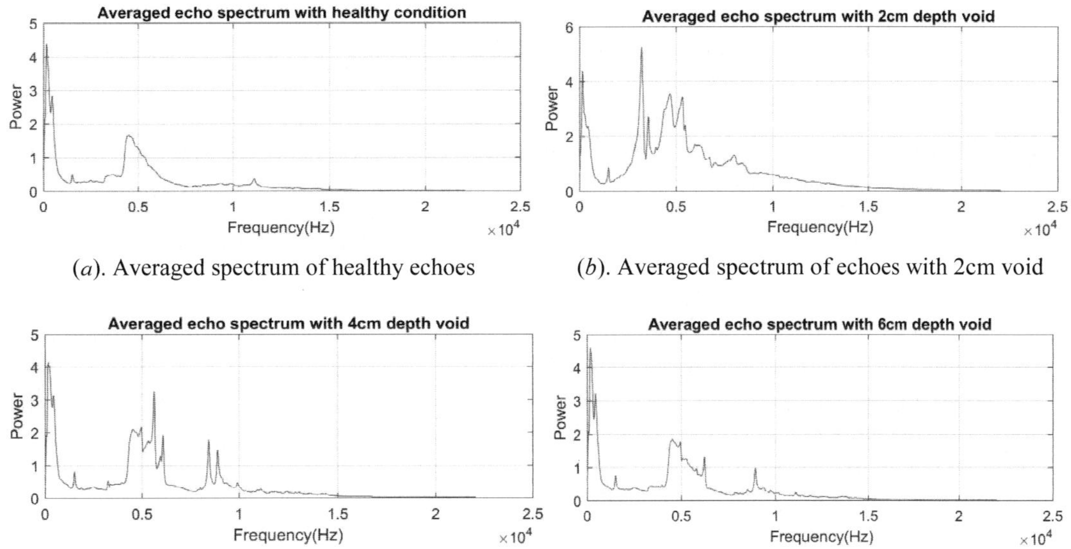

(*a*). Averaged spectrum of healthy echoes

(*b*). Averaged spectrum of echoes with 2cm void

(*c*). Averaged spectrum of echoes with 4cm void

(*d*). Averaged spectrum of echoes with 6cm void

Figure 4. Average spectrum of echoes collected from both healthy and defective concrete blocks

Subsequently, we adopted commonly applied principal component analysis (PCA) method to visualize the massive hammering echo sounds. PCA efficiently reduces the dimension of hammering echo spectrum based on two objectives, which are variance maximization and reconstruction error minimization. The result is presented in Fig. 5.

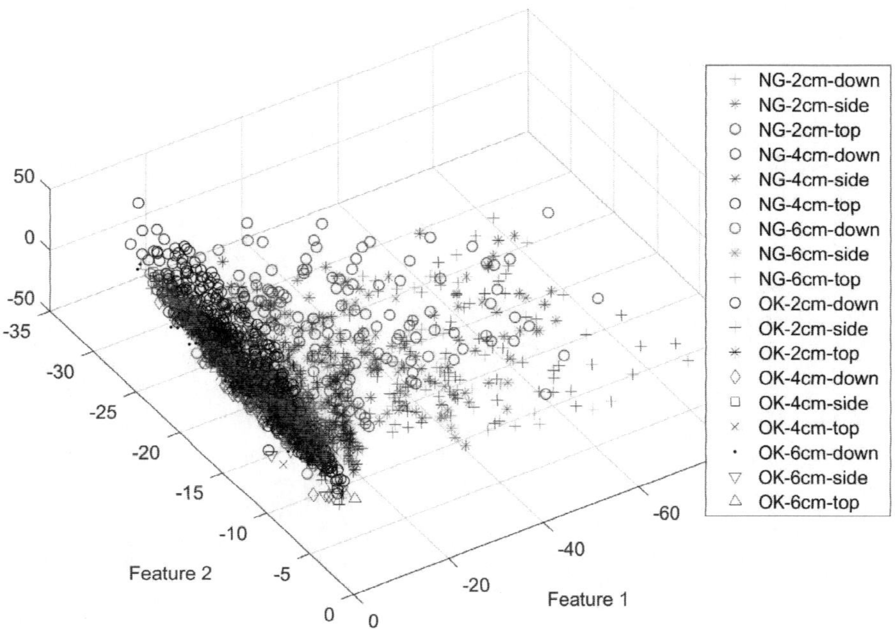

Figure 5. Visualization of all hammering echoes using principal component analysis (PCA) Low-rank projection

According to the data visualization result, we can see defective echoes captured from concrete with 2cm inner void exhibited most distinct patterns in the feature space. Besides, many other echoes overlapped each other, therefore the four-class classification task is rather challenging.

3.3 Condition-based hammering echo classification

To demonstrate the distance metric learning approach for hammering echo investigation, we conducted experimental comparison between other methods. The detailed information of reference methods was summarized in Tab. 2.

Table. 2. Components of hammering echo analysis methods in comparison

	Feature extraction	Classification
M1	Spectrum	kNN method
M2	Mel-scale spectrum	
M3	Spectrum + PCA	
M4	Spectrum + linear discriminant analysis	
M5	Spectrum + LMNN distance metric learning	

We evaluated all five methods using all hammering echoes collection from four types of concrete conditions. Four types of echoes were classified using algorithms, indicating concrete conditions of normal, with 2cm, 4cm and 6cm hollows inside, respectively. The results were shown in Fig. 6. According to results comparison, it is obvious that LMNN achieved superior overall condition-based echo classification accuracy. Nearest neighbor classification using raw echo spectrum and mel-scale spectrum obtained similar precision. The result suggested there is rare benefit of applying mel-scale filter bank. In addition, a sharp drop in classification accuracy can be observed for the case of using principal components of echo spectrum with kNN classifier. The reason is that PCA smears discriminant spectral pattern while the linear projection. Besides, the linear discriminant analysis (LDA) approach didn't produce favorable echo analysis result. The reason is that the distributions of four-class hammering echoes were rather complex and those data cannot be effectively separated by using linear classifier. As a result, LMNN outperformed all other reference methods and it has been proved effective for the task of hammering echo investigation.

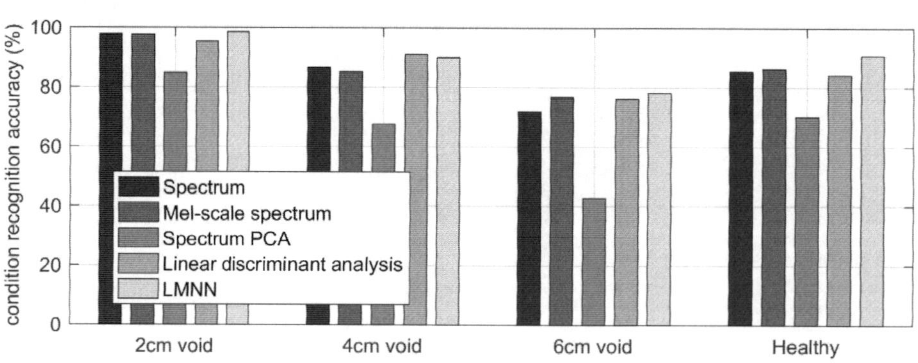

Figure 6. Experimental comparison of echo pattern classification schemes on real echo data

3. Conclusions

In this work, we present novel formulation for hammering echo investigation using signal processing and machine learning algorithms. Unlike conventional impact-echo relying on Euclidean distance metric to assess similarities between echo sound spectrum, we employ distance metric learning approach to extract optimal spectral feature representation via linear projection. The basic idea of distance metric learning is to optimize a metric with the objective that the k-nearest neighbour echoes always belong to the same class (i.e. all collected from healthy concrete structures) while instances from different classes are separated by a large margin, i.e. normal and defective echo sounds. Another advantage of distance metric learning is that it can be performed with high efficiency for the onsite evaluation, e.g. enabling real-time processing. To demonstrate the proposed scheme, we conduct experiments on real hammering echo dataset. An extensive comparison had been drawn between 4 reference methods. The evaluation result demonstrated effectiveness of the propose approach. The current work is still at preliminary stage, more detail settings would be further determined, such as enhancing the scalability and incorporating incremental algorithm design to the current scheme. Those would be our future works.

4. Acknowledgement

This study was partly supported by Cross-ministerial Strategic Innovation Promotion Program (SIP) / Technologies for maintenance, renewal, and management for infrastructure, and supported by the New Energy and Industrial Technology Development Organization (NEDO), Japan.

References

Farrar, Charles R., and Keith Worden. Structural health monitoring: a machine learning perspective. John Wiley & Sons, 2012.

Carino, N. J., Sansalone, M., and Hsu, N. N. 1986. "A point sourcepoint receiver, pulse-echo technique for flaw detection in concrete." ACI J., 83(2), 199–208.

Sansalone M. Impact-echo : the complete story. ACI Struct J. 1997;94:777–86

Zhu, J. Y. & Popovics, J. S. (2007). Imaging concrete structures using air-coupled impactecho. ASCE Journal of Engineering Mechanics, Vol.133, No.6, pp. 628-640

J.D. Achenbach. (1973) "Wave Propagation in Elastic Solids", Elsevier Science

Kulis, Brian. "Metric learning: A survey." Foundations and Trends® in Machine Learning 5.4 (2013): 287-364.

Gentle, James E. Matrix algebra: theory, computations, and applications in statistics. Springer Science & Business Media, 2007.

Mitchell, Ann FS, and Wojtek J. Krzanowski. "The Mahalanobis distance and elliptic distributions." Biometrika (1985): 464-467.

Duda, Richard O., Peter E. Hart, and David G. Stork. Pattern classification. John Wiley & Sons, 2012.

Weinberger, Kilian Q., and Lawrence K. Saul. "Distance metric learning for large margin nearest neighbor classification." Journal of Machine Learning Research, 10. Feb (2009): 207-244.

The 8th International Conference on
Structural Health Monitoring of Intelligent Infrastructure
Brisbane, Australia | 5-8 December 2017

Detection of a void in a concrete bridge pier by a non-destructive technique

M. Abe[1], T. Fujioka [1], Y. Nagata [1].
[1] *CIS., Iwate University – Japan, Email:* abe@cis.iwate-u.ac.jp

Abstract

Post-tensioned prestressed concrete, one means of overcoming concrete's natural weakness in tension, is used widely to produce bridges having a long span. Because of poor workmanship or quality control during construction, however, the ducts sometimes contain prestressing tendons that are not fully filled, leaving voids in the grout where the steel is vulnerable to corrosion. Therefore, we propose a method of detecting such voids using multiple sensors and an impulse hammer. The sensors pick up the vibration wave caused by the impulse hammer. A specimen simulating a bridge pier was made and divided into five regions, each with 0, 1, ..., 4 voids. The two sensor outputs for the region with 0 voids and with 1 void were simulated using the FDTD method. Given a long duration of the time window for analyses, we cannot detect voids because the bridge pier is small, creating many reflections. We can detect a void by analyzing the difference of the initial part of the two simulated outputs because the initial parts were almost identical. Next, the experimentally obtained vibration waves were analyzed. Although we tried to strike the concrete in the same manner, the hammer force output shows small differences among trial, because of which we were unable to detect the void. Therefore, we used a digital filter to convert the experimentally obtained hammer force into the force designed in the FDTD simulation. The same filter was also applied to each sensor output. Results showed that the experimentally obtained sensor output with the filter mentioned above and the simulated output were quite similar. Results also show that the void is detectable by analyzing the difference of two experimentally obtained filtered sensor outputs.

1. Introduction

Prestressed concrete was developed to overcome concrete's natural weakness in tension. The material has characteristics of high-strength concrete in compression and high ductile strength steel for tension. Prestressing tendons (generally of high tensile strength steel cable or rods) are used to provide a clamping load, which produces a compressive stress that balances the tensile stress that the concrete compression member experiences because of a bending load.
The popularity of this form of prestressing for bridge construction increased considerably around the 1950s and 1960s. However, the many difficulties that have been encountered have cast doubt on the long-term durability of such structures.
Because of poor workmanship or quality control during construction, the ducts containing the prestressing tendons are sometimes not fully filled, leaving voids in the grout where the steel is not protected from corrosion. This situation is further exacerbated if water and chloride (from de-icing salts) from a highway are able to penetrate into these voids.

2. Experimental Setup

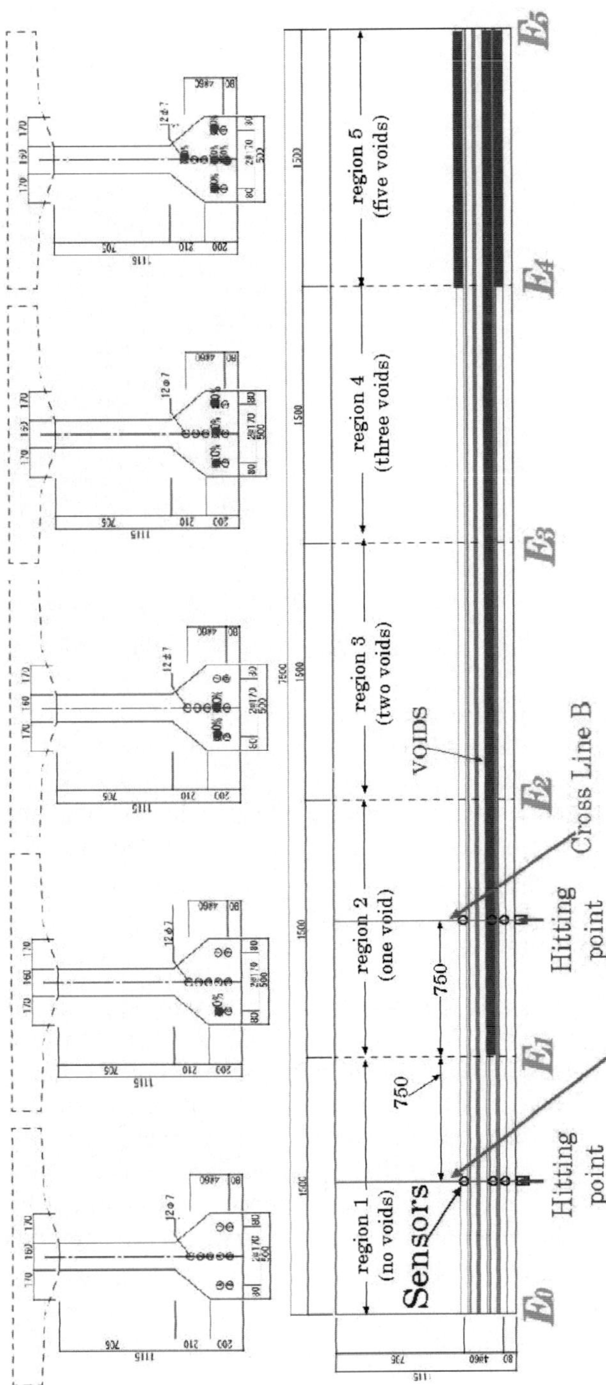

Figure 1 - Side view of the bridge pier as a test specimen.

2.1 Bridge pier

Figures 1 and 2 respectively portray a side view and the sectional view of the bridge pier to be investigated. The nine small circular protrusions in Fig. 2 are tendon edges. Figure 3 depicts the bridge pier setup.

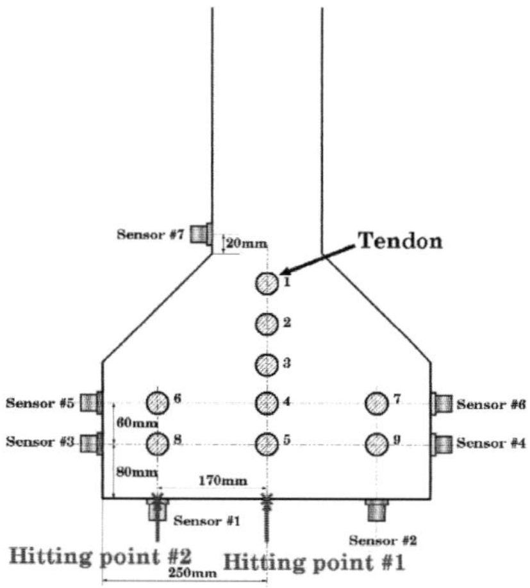

Figure 2 - Sectional view of the bridge pier

Figure 3 - Bridge pier as a test specimen with seven sensors.

The pier comprises five regions as shown in Figs. 1 and 2: region #1 has no voids; region #2 has one void at tendon #6; region #3 has two voids at tendons #4 and #6; region #4 has three voids at tendons #4, #6 and #7; and region #5 has five voids at tendons #4, #5, #6, #7, and #1. As depicted in Fig. 2, seven sensors are attached on cross line A or B, which are at the respective centers of regions 1 and 2. The two hitting points #1 and #2, each 3 cm distant from the line A or B, were struck with an impulse hammer as presented in Fig. 2.

The seven sensors are one-axis vibration accelerometers. The outputs of the sensors and the impulse hammer were stored in a computer after passage through low-pass filters with cut-off frequency of 40 kHz and sampling frequency of 96 kHz.

2.2 Impulse Hammer Outputs

Figure 4 presents the impulse hammer outputs when the bottom surface of the bridge pier was hit five times at region #1, cross line A, where the time of hitting t=0. As portrayed in Fig. 4, the impulse hammer outputs differ, although the worker tried to hit the surface using the same force. Therefore, to investigate details of the sensor output, the hammer output amplitude should be compensated. Figure 5 presents sensor output #5 after compensating only the amplitude of the hammer output. Actually, Fig. 5 shows that a slight difference is observed: it cannot be negligible because it must detect the void by the difference between the sensor output for a region with a defect and that for a region without a defect. That is, not only the amplitude but also the shape of the hammer output itself should be compensated.

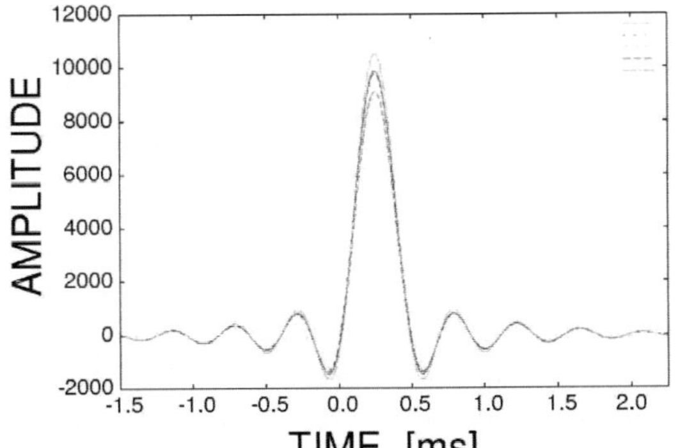

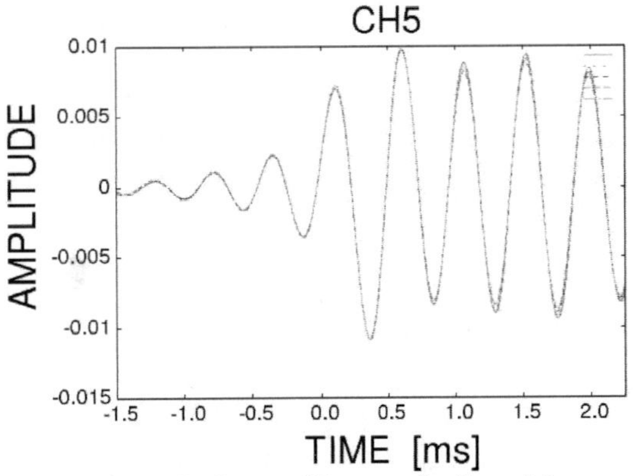

Figure 4 - Hammer force for five trials.

Figure 5 - Sensor #5 output for five trials.

Here, the sensor output starts before the hammer force initiates because of the use of the band-pass filter in the frequency domain for estimating the propagation delay of the direct and/or reflection wave precisely using our method in reference (1), as described later.

3. Finite-Difference Time Domain (FDTD) Method

We developed a nondestructive method for detecting a defect in a caisson used for coastal engineering. During the development process, results show that FDTD method in reference (Fujioka, Nagata, and Abe, 2015) is extremely useful because we can simulate the sensor output waveform by assigning the hammer force and the shape of the concrete structure. Figures 6 and 7 show the assigned hammer force and the sensor #5 output simulated for the specimen shown in Figs. 1 and 2 at region #1, where edge E_m,m=0,1,2,3,4,5 is the absorbing boundary at region #m, i.e., no reflection is generated, meaning that the concrete space extends infinitely.

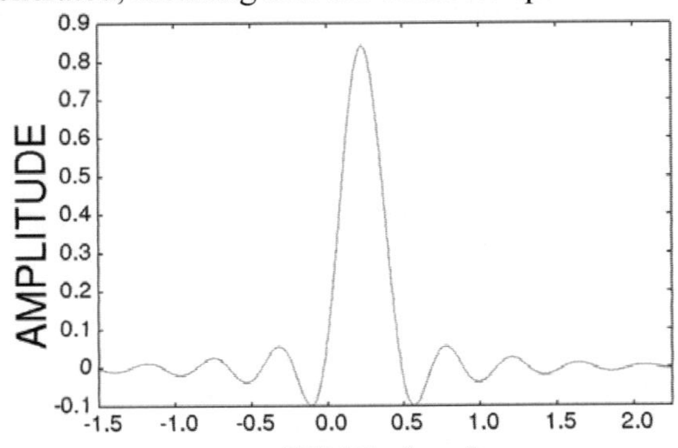

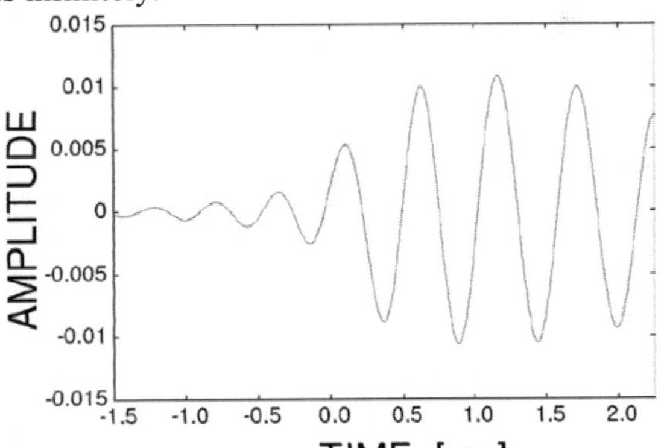

Figure 6 - Assigned hammer force in FDTD simulation.

Figure 7 - Simulated sensor #5 output.

Detailed investigation revealed that Figs. 5 and 7 differ because the velocity assigned in FDTD simulation differed from that of the real specimen.

4. Compensation of the Hammer Force

Experimentally obtained hammer force f_e fluctuates as described in section 2.2. Therefore, we intended to convert $f_e(n)$ into the assigned hammer force f_s used for FDTD simulation. Then, the impulse response $h(n)$ from $f_e(n)$ to $f_s(n)$ is expressed as

$$f_s(n) = f_e(n) * h(n), \tag{1}$$

where n is the time and $*$ is the convolution integral. Here, *h(n)* is calculated using IDFT of $F_s(k)/F_e(k)$, where k stands for the frequency and where $F_e(k)$ and $F_s(k)$ respectively represent the DFT of $f_e(n)$ and $f_s(n)$. Similarly, seven experimentally obtained sensor outputs are compensated using the same impulse response as

$$x'_{pm}(n) = x_{pm}(n) * h(n),\ p = 1,2,\dots,7 \tag{2}$$

where $x_{pm}(n)$ stands for the experimentally obtained *p*-th original sensor output and $x'_{pm}(n)$ signifies the *p*-th compensated sensor output obtained at region *#m*.

Figure 8 portrays the compensated output $x'_{51}(n)$ of sensor #5 for five trials at region #1. Examination of Fig. 8 reveals that the compensated sensor outputs are almost identical.

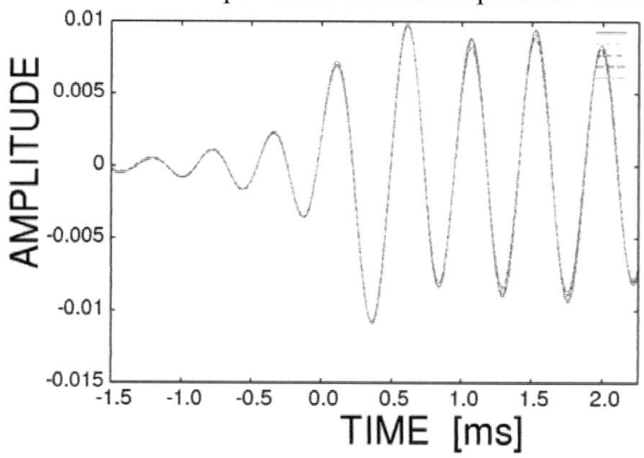

Figure 8 - Compensated output $x'_{51}(n)$ of sensor #5.

5. Difference Between Cases With and Without a Void

Figures 9 and 10 respectively portray the compensated waveforms of sensor #5 at region #1 (no void), region #2 (one void at tendon #6) and their difference for hitting points #1 and #2 (shown in Fig. 2). From the bridge pier shape and the surface wave and the primary wave velocities, the two waves were found to reach sensor #5 at 0.2 ms and 0.07 ms, respectively, for hitting point #1, and at 0.1 ms and 0.04 ms, respectively for hitting point #2. The reflection wave from the void (tendon #6) reaches the sensor #5 at 0.07 ms for hitting point #1, and at 0.04 ms for hitting point #2.

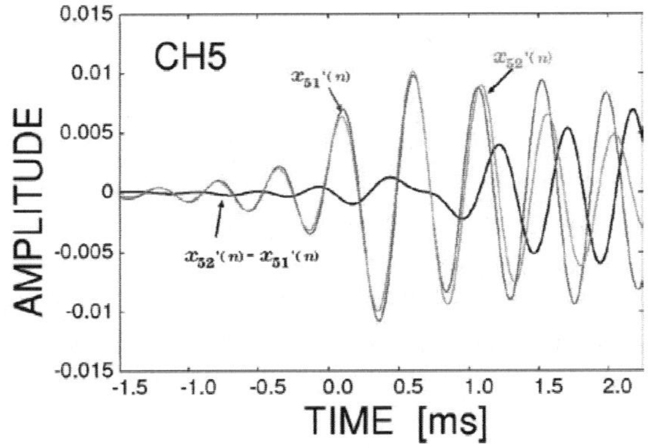

Figure 9 – Compensated waveform for hitting point #1

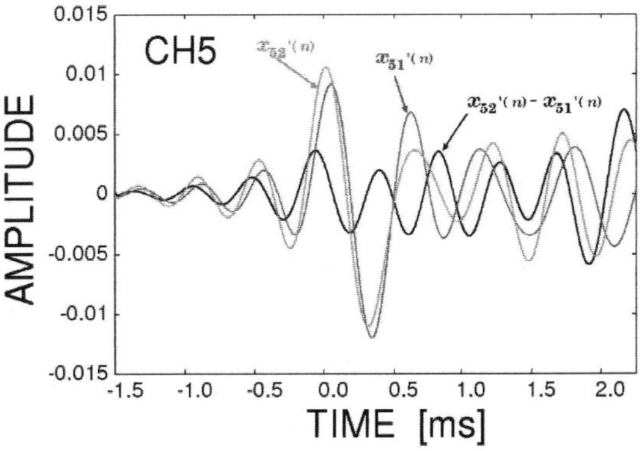

Figure 10 – Compensated waveform for hitting point #2

It is found from Fig. 9 that the difference

$x'_{51}(n)$: compensated sensor #5 output when the sensor is set at region #1 (no void).
$x'_{52}(n)$: compensated sensor #5 output when the sensor is set at region #2 (a void, tendon #6).

between the compensated waveforms of sensor #5 at region #1 (no void) and region #2 (one void at tendon #6) is small around 0 ms – 0.8 ms when the void at tendon #6 is not in the vertically upward direction of the hitting point #1. That is, the reflection from the void is not observed clearly. However, Fig. 10 shows that the difference between the compensated waveforms of sensor #5 at region #1 (no void) and region #2 (one void at tendon #6) is greater than that shown in Fig. 9 around 0 ms – 0.3 ms when the void at tendon #6 is just in the vertically upward direction of hitting point #1. Therefore, the waveform around 0 ms – 0.3 ms is a reflection from the void. The power of the reflection from the void is great when the void is just in the vertically upward direction of the hitting point.
Assuming that region # 1 is healthy, the difference in Fig. 10 represents a reflection from the void. In the next section, a method for estimating the reflection wave propagation delay is explained.

6. Estimation of the Reflection Wave Propagation Delay

We have already reported in reference (Murakami, T, Fujioka, T, Nagata, Y, and Abe, M 2007) that the q-th direct or reflection wave $v_q(n)$ caused by hitting with an impulse hammer is modeled by the convolution between the hammer force $f(n)$ and an impulse response $h_q(n)$ if the bandpass filter bandwidth is sufficiently narrow, as

$$\hat{v}_q(n) = f(n) * h_q(n), \tag{3}$$

and $h_q(n) = \alpha_q \exp(-\beta_q n)\cos\{2\pi\gamma_q(n - \tau_q/\Delta T) + \theta_q\}$, where α_q signifies the amplitude, β_q denotes the decaying factor, γ_q represents the frequency, and τ_q stands for the propagation delay. Also, ΔT is the sampling period, θ_q is the initial phase of the q-th direct or a reflection wave. The five parameters α_q, β_q, γ_q, τ_q, and θ_q are estimated using a genetic algorithm (GA) so that the following estimation error $\boldsymbol{E}$ takes the minimum value:

$$\boldsymbol{E} = \sum_n |e(n)|^2, \tag{4}$$

where

$$e(n) = v(n) - \sum_q \hat{v}_q\,(n) \tag{5}$$

And where $v(n)$ is the vibration wave to be analyzed. $\hat{v}_q$ is the modeled vibration wave. When our estimation method is applied to estimation of the propagation delay of the reflection wave from the void, which represents the residue presented in Fig. 10, the residue $x"_{52}(n) - x"_{51}(n)$ is used as the vibration wave $v(n)$ to be analyzed. Figure 11 presents the results estimated using our method. To minimize the estimation error, results show that two modeled reflection waves $\widehat{v_1}(n)$ and $\widehat{v_2}(n)$ arrive during -0.5 ms and -0.5 ms. When the two vibration waves are assumed, the estimation error $e(n)$ is sufficiently small during 0.0 ms and 0.5 ms. As presented in Fig. 11, the sum of two estimated modeled waveforms #1 and #2 is almost identical to the residue $x"_{52}(n) - x"_{51}(n)$, where each modeled estimated waveform is the convolution between hammer force *f*(n) and the impulse response $h_q(n)$.

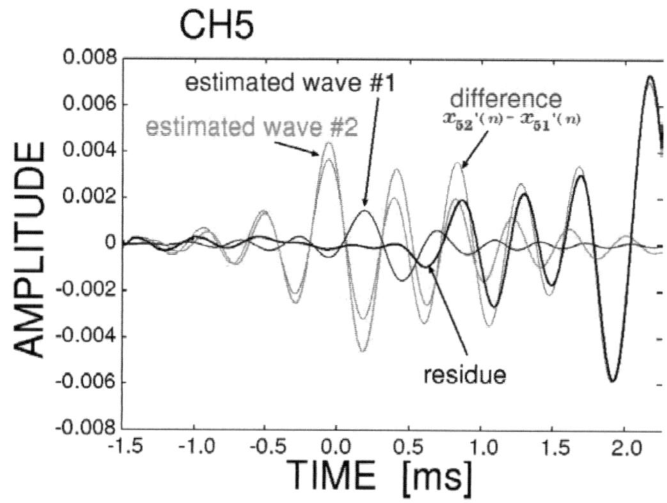

Figure 11 - Residue, two estimated vibration waves, and the estimation error.

Estimated propagation delays $\tau_q, q = 1{,}2$ of the two vibration waves are 0.06 ms and -0.16 ms, respectively, where $\widehat{v_2}(n)$ is the phantom vibration wave caused by the low-pass filter in the frequency domain, as described in the reference (Abe, M, Fujioka, T, and Nagata, Y, 2017), and $\widehat{v_1}(n)$ represents the reflection from the void. Assuming that the primary wave velocity is 4100 m/s, the delay of 0.06 ms is equivalent to propagation distance of 0.23 m, which is almost the same as the theoretically calculated propagation distance 0.22 m <H_2 → void → S_5>, where H_2 is hitting point #2 and S_5 is the #5 sensor.

7. Conclusions

We have developed a method to estimate the propagation delay of a direct or reflected wave caused by an impulse hammer for detecting defects in a caisson. As reported herein, the method is applied to detect voids in the bridge pier. Results of experiments show that our method works well in this application.

References

Fujioka, T, Nagata, Y, and Abe, Y 2015, 'Analysis of impact elastic wave of caisson by using finite-difference time-domain method for nondestructive inspection by percussion method', J. Acoustical Society of Japan Vol. 71, No. 2, pp. 53-64 (in Japanese).

Abe, M, Fujioka, T, Nagata, Y 2017, 'Estimating the propagation delay of a direct or reflected wave in a large concrete caisson by impact with an impulse hammer', J. Acoust. Sci. & Tech. Vol. 38, No. 2, pp. 87-98.

Murakami, T, Fujioka, T, Nagata, Y, and Abe, M 2007, 'Nondestructive technique for estimating crack positions in a concrete structure by subtraction of the surface-wave component', J. Acoust. Sci. & Tech. Vol. 28, No. 5, pp. 310-318.

The 8th International Conference on
Structural Health Monitoring of Intelligent Infrastructure
Brisbane, Australia | 5-8 December 2017

Infrared Thermography Data Analysis Method for Concrete Bridge Condition Evaluation

S. Hiasa [1,2], F.N. Catbas [1].
[1] *University of Central Florida – USA, Email:* hiasa615@Knights.ucf.edu *(1st author)* catbas@ucf.edu *(2nd author)*
[2] *West Nippon Expressway Company Limited (NEXCO-West) - Japan*

Abstract

This paper investigates the effects of the size and shape of delamination on subsurface defect detection by IRT through FE modeling to enhance the reliability and applicability of IRT. It is observed that the temperature difference (ΔT) between a sound area and a delaminated one increases as the size of delamination increases; however, ΔT converges to a certain value when the area approaches 40 cm × 40 cm and a 1 cm thickness. It can be concluded for the shape of delamination that if the aspect ratio for the delamination, which is the ratio of the shorter side length to that of the longer side, is more than 25%, the ΔT of any delamination converges to ΔT of a square-shaped delamination of the same area. If the aspect ratio is 25% or smaller, ΔT is smaller than that of a square-shaped delamination with the same area, and will continue to decrease as the ratio decreases. Based on these findings, this study also investigated the possibility of estimating delamination depth using passive IRT by comparing ΔT from IRT data to ΔT at several depths obtained from FE model simulations. Through an investigation using IRT data from a real bridge deck scanning, this study validated that the estimation method could provide accurate delamination depth by utilizing IRT together with FE modeling.

1. Introduction

Infrared thermography (IRT) is a Non-Destructive Evaluation (NDE) method for detecting existing subsurface deteriorations, including delaminations and voids in concrete, faster than and for less expense traditional sounding tests. IRT is suitable for civil infrastructure inspection since it is a non-contact method and utilizes IR images, which can capture a wide range of concrete structures at one time. Although a fair amount of research has been done on IRT, many of the studies were conducted under different conditions, making it difficult to draw general conclusions. Therefore, there are several uncertainties regarding the accuracy and reliability of IRT for bridge inspection. Past research has been conducted for some of these challenges and uncertainties, such as data collection time, camera specifications, data collection speed based on the size of delamination, in addition to exploring proper methods and conditions for IRT application (Watase *et al.*, 2015; Hiasa, 2016; Hiasa *et al.*, 2016, 2017, Hiasa, Birgul and Catbas, 2016, 2017; Hiasa, Birgul and Necati Catbas, 2017). The size of delamination can be considered to be the most critical factor affecting the detectability of IRT. Some researchers found that the size of delamination affects the detectable depth of the delamination, that is, as the size of delamination increases, the temperature difference between sound and delaminated area also increases (Maierhofer *et al.*, 2005; Yehia *et al.*, 2007). However, the effects of delamination size are not fully known due to insufficient experimental data; especially when looking at the effects of the area, thickness and volume of a delaminated section on detectability of IRT. In existing research, many investigators have utilized artificial defects of different sizes and have calculated different detectable depths for IRT. One of the difficulties of NDE tests is creating a test specimen that can simulate concrete members of bridges without becoming relatively huge.

Therefore, only a small amount of test specimens have been utilized for IRT experiments, which were also performed under similar circumstances, such as passive/active IRT conditions, fixed orientations and limited locations (weather conditions). In order to overcome the limitations and lack of variety, Hiasa, Birgul & F. Necati Catbas (2017) utilized a Finite Element (FE) model simulation to explore sensitive parameters for effective utilization of IRT. Through the FE model simulations, it was found that the most critical factor affecting IRT is the area of delamination. It is also important to note that the thickness affects the temperature differences between the sound and delaminated surface areas. The volume of delamination is not a significant parameter for IRT.

The objective of this study is to investigate the effect of delamination size on the detectability of IRT, in order to enhance the reliability and applicability of IRT for effective bridge inspections. Special consideration was given when exploring the correlation between the size of delamination and thermal contrast of the concrete surface between delaminated and sound areas. Data was gathered to reveal if the effect increases as the size of delamination increases, or to see if it converges to a value at a certain size of delamination. Lastly, the study investigates the possibility of estimating the depth of the delamination by using passive IRT on an in-service bridge (estimating the depths of delamination has been considered one of the biggest limitations of IRT).

2. Method of the study

2.1 Principle of passive infrared thermography

IRT can be classified into two methods; passive and active. Passive IRT is applied under natural circumstances, while active IRT is conducted during thermal excitation. However, active IRT is impractical for bridge inspections due to the difficulties involved in uniformly heating large civil structures over a period of time. The principle of passive IRT is as follows: solar loading heats up the concrete surface. If there is a subsurface defect, air fills that area and acts as a thermal insulator. The air prevents heat from transferring to the concrete beneath the delamination because air has a smaller thermal conductivity (0.0241 W/m·°C) than concrete (1.6 W/m·°C). Consequently, during the daytime, the concrete above the delamination becomes warmer than its surroundings since that area conducts heat faster. Heat travels more quickly from the surface to the delamination than from the surrounding surfaces to the undersurface. Additionally, during the nighttime, that area cools more rapidly than the surroundings. IRT detects subsurface defects by capturing the temperature differences within a concrete surface (Holst, 2000). It is not always possible to detect the delamination of concrete during passive IRT conditions since there are interchange periods between heating and cooling cycles (Hiasa, 2016).

2.2 Finite Element Model developed and validated in a past study

In this study, FE modeling was utilized to investigate the effect of delamination size on thermal contrast of concrete surface between delaminated and sound areas. Previously, a field experiment was conducted on December 19, 2015, at the University of Central Florida using two test specimens with delamination at 1.3 cm (0.5 in.) and 2.5 (1 in.) cm depths from the surface, respectively. Both concrete blocks had the same dimensions of 91.4 cm (3 ft.) × 91.4 cm × 20.3 cm (8 in.) as shown in Fig. 1(a) (Hiasa, 2016). These test specimens were situated outside in a natural environment (passive IRT conditions) as shown in Fig. 1(b). The specimen thickness was designed to simulate a typical bridge deck in the USA. The dimensions of the artificial delamination was 10.2 cm (4 in.) × 10.2 cm × approximately 0.3 cm (1/8 in.). The delamination was simulated using a combination of foamed sheet and cardboard that was wrapped in plastic, as shown in Fig. 1(a). The foamed sheet and cardboard were chosen due to their thermal conductivity properties, which are like those of air. The thermal conductivity of the foamed sheet is 0.024 W/m·°C, and cardboard is naturally porous. In the past study, the same size of FE model was developed by using Heat Transfer Module of COMSOL Multiphysics software. A concrete slab model of the same size as the experiment was established on a larger ground, 6 × 6 × 1 m, and the same size of Styrofoam was installed inside the concrete. In the model, the concrete block was placed on concrete stands (height: 20 cm) to make space for wind flow. The material properties are shown in Table 1.

The boundary conditions for solar radiation, air temperature and convective heat transfer were considered for each side of the concrete block. The primary heat source was solar radiation. The direction of sunlight (the zenith angle and the solar elevation) over the simulation time was automatically computed from the longitude, latitude, time zone, date, and time by the software (COMSOL, 2015), and is as follows: longitude; -81.189103, latitude; 28.591897, time zone (GMT); -5 h. The experiment was located in a test field and the information was taken from Google Maps. The following conditions were defined based on the record of the nearest weather station, approximately 700 m from the test site (WeatherUnderground, 2015) and by means of iterative trials: solar irradiance was 700 W/m^2; heat transfer coefficient was 30 W/m^2·K. The ambient temperature (T_{amb}) was measured throughout the day using thermocouples at the site and then inputted into the software. The external temperature (T_{ext}) (COMSOL, 2015), which is assumed to be temperature of the sky, was calculated as $T_{ext} = T_{amb} - 50$ °C for all exposed surfaces, to account for the effect of radiative cooling in the software (Hiasa, 2016). Fig. 2 depicts the results of the FE model simulation and IRT experiment as a comparison of the temperature differences between the sound and delaminated parts (ΔT) at every instance of time. IRT data was collected every 30 minutes from 7 to 10 AM and every hour thereafter until midnight by an infrared camera, T420 manufactured by FLIR Systems, Inc. The FE model was simulated every 30 minutes. As can be seen in Fig. 2, there are some disparities between each ΔT obtained by IRT and the simulation; however, these disparities are less than ± 0.7 °C, and are considered to fall within an acceptable measurement error range. Since three IR cameras with different specifications were used at the same time, they indicated different ΔT by approximately ± 1 °C at most (Hiasa, 2016). Therefore, it can be concluded that this FE model was properly designed, and ΔT can be simulated by FE modeling.

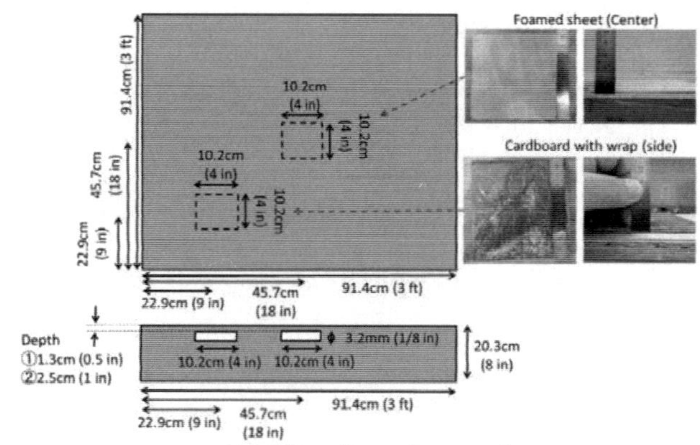

(a) Dimensions of test specimens

(b) Test environment

Fig. 1. Dimensions of concrete test specimen and test environment

Table 1. Material properties of the simulation model

Material Properties	Unit	Concrete	Styrofoam	Ground
Thermal conductivity	W/(m·K)	1.6	0.024	0.6
Heat capacity at constant pressure	J/(kg·K)	880	1130	800
Density	kg/m^3	2300	25	1500
Solar absorptivity		0.6	-	0.94
Surface emissivity		0.88	-	0.76

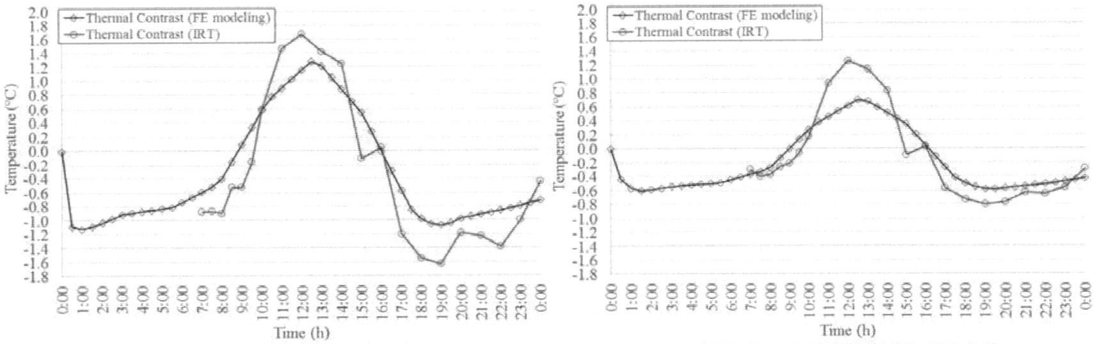

(a) 1.3 cm depth of delamination (b) 2.5 cm depth of delamination

Fig. 2. Temperatures difference from IRT and simulation results

2.3 Model expansion for the present study

In the present study, the same weather conditions of December 19, 2015, and the same material properties for the FE model were used to investigate the effect of delamination size on the detectability of IRT. Adjustments were made to the size of the concrete slab, which was expanded to 300 × 300 × 20.3 cm in this study, as shown in Fig. 3, and to the delamination shapes (in order to analyze effects due to delamination size and shape). The delamination was modeled at the center of the concrete block to avoid the effect of boundary conditions. Regarding the depth of delamination from the surface to the top of the delamination, a depth of mainly 5.1 cm (2 in.) was applied since the top concrete cover for a bridge deck is typically 5.1 cm, and defects often occur around the reinforcing bars. Furthermore, depths of 7.6 cm (3 in.) and 10.2 cm (4 in.) were also simulated for comparative purposes. Table 2 summarizes the size, shape and depth of each delamination model.

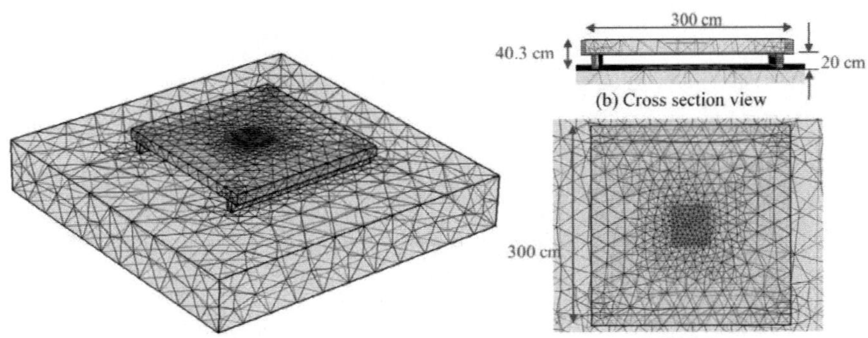

(a) General view (b) Cross section view (c) Plan view

Fig. 3. Expanded FE model for this study

Table 2. Plan size of modeled delamination

Square Depth: 5.1, 7.6 & 10.2 cm		Rectangle Depth: 5.1 cm				Circle Depth: 5.1 cm	
Size (cm)	Area (cm²)	Size (cm)	Area (cm²)	Size (cm)	Area (cm²)	Radius × Thick (cm)	Area (cm²)
10 × 10 × 0.3*	100	5 × 80 × 0.3	400	10 × 90 × 0.3	900	11.3 × 0.3	400
15 × 15 × 0.3	225	8 × 50 × 0.3	400	15 × 60 × 0.3	900	16.9 × 0.3	900
20 × 20 × 0.3*	400	10 × 40 × 0.3	400	20 × 45 × 0.3	900		
30 × 30 × 0.3*	900	16 × 25 × 0.3	400	25 × 36 × 0.3	900		
40 × 40 × 0.3*	1,600						
50 × 50 × 0.3	2,500						
60 × 60 × 0.3	3,600						

* Different thicknesses were modeled at 5.1 cm depth: 0.1, 1, 2, 10, 15 cm

3. Simulation results

3.1 Effect of delamination thickness

In this study, several thicknesses of delamination, 0.1 cm, 0.3 cm, 1 cm, 2 cm, 10 cm and 15 cm, were modeled and analyzed to investigate the effects of delamination thickness. A square shaped delamination with four different areas was used, 10 × 10 cm, 20 × 20 cm, 30 × 30 cm, and 40 × 40 cm. Each delamination was located at a depth of 5.1 cm from the surface. Fig. 4 compares ΔT of each delamination thickness for the respective areas of delamination. It is obvious that the difference of ΔT between 0.1 cm and 15 cm thicknesses increases as the area increases. The maximum difference between two thicknesses for the 10 × 10 cm delamination model is 0.16 °C while it is 0.75 °C for the 40 × 40 cm delamination model. As can be seen from each graph, it is also obvious that 0.1 cm of thickness generates the smallest ΔT, regardless of the area of delamination; consequently, 0.3 cm of thickness yields a smaller ΔT than the other thicker delamination. However, 1 cm, 2 cm, 10 cm and 15 cm thicknesses of delamination generate similar ΔT, there is less than a 0.15 °C difference at most among these four thicknesses at any time and for any area of delamination. Therefore, this result indicates that the effect of thickness increases as the area increases; however, the effect of delamination thickness converges to a certain value of ΔT when the thickness of delamination becomes 1 cm or more.

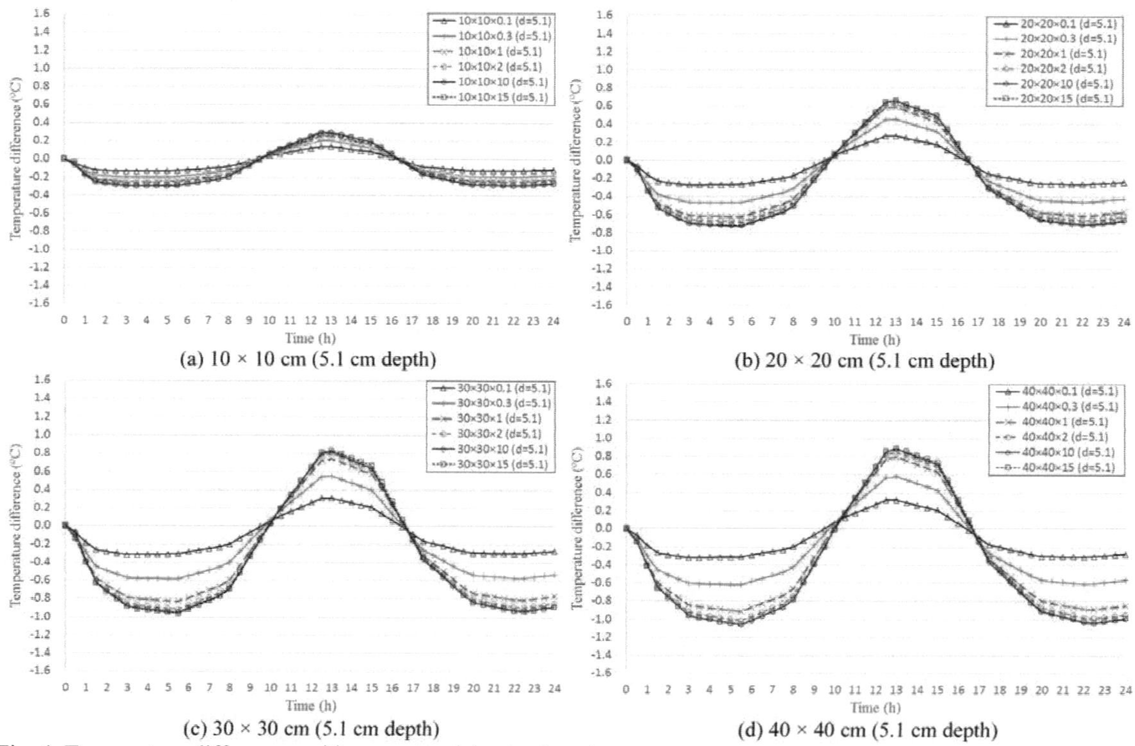

Fig. 4. Temperature differences with respect to delamination sizes

3.2 Effect of delamination area and shape

In this subsection, the effects of delamination area and shape are investigated. Several square shapes of delamination were compared to investigate whether the effect of delamination area also converges to a certain value as it does for delamination thickness. In this study, lengths of a side of a square (L) were modeled as L=10, 15, 20, 30, 40, 50 and 60 cm as shown in Table 2. The thickness for each delamination was set to 0.3 cm for consistency. Fig. 5(a) depicts the simulation result of each delamination model located at 5.1 cm depth. It is obvious that the area of delamination strongly affects ΔT. However, as the delamination area exceeds 30 × 30 cm, the corresponding increase in ΔT is reduced, at most 0.05 °C between 30 × 30 cm and 60 × 60 cm. In fact, when the areas become greater than 40 × 40 cm, the difference between ΔT for each delamination area is less than 0.01 °C. Therefore, it can be concluded that ΔT converges to a certain value when the delamination area is approximately 40 × 40 cm. Another investigation was made for the same areas of delamination, but this time at a depth of 7.6 cm, to see if the same conclusion could be obtained. Fig. 5(b) displays the result. The ΔT of the 7.6 cm depth delamination also converged to a certain value when the delamination area was 40 × 40 cm; thus, it can be concluded that ΔT converges to a certain value when the area approaches 40 × 40 cm, regardless of its depth. It is also important to note that as the delamination depth increases, ΔT becomes smaller.

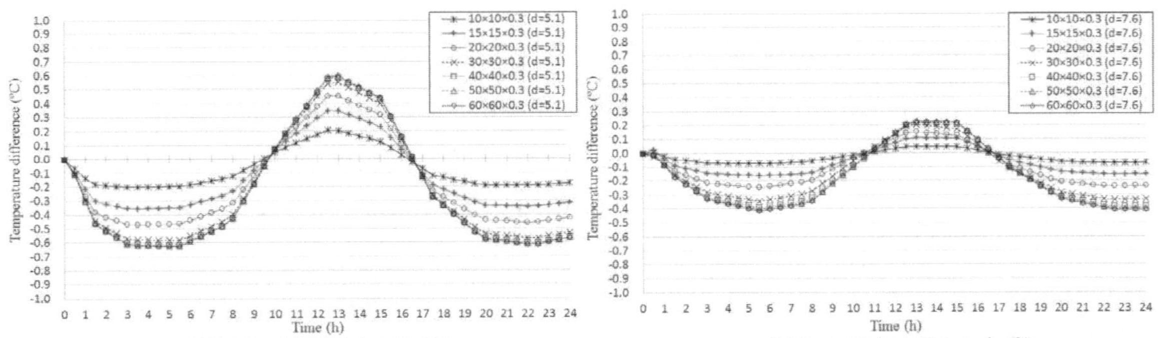

(a) L × L ×0.3 cm (5.1 cm depth) (b) L × L ×0.3 cm (7.6 cm depth)

Fig. 5. Comparison of ΔT with different square shape areas of delamination

The effect of different delamination shapes, such as rectangles and circles, were also explored. Rectangles and circles, each with two fixed areas of delamination (400 cm^2 and 900 cm^2), were modeled and then compared to a square-shaped delamination with the same area. Four different rectangles were modeled; their side lengths and side length ratios are summarized in Table 3. The temperature of two sides of a sound area were compared to investigate how a rectangular-shaped delamination affects the temperature gradients of a concrete surface. Fig. 6 depicts the results of the two cases. "L" denotes the temperature of the concrete that is 10 cm away from the longer side of the rectangle, and "S" indicates the temperature at 10 cm from the shorter side. The maximum difference between the two sides was 0.01 °C in each model throughout the entire day. Therefore, there is no significant difference in ΔT between the longer and the shorter side of the rectangular delamination.

Table 3. Size of rectangular-shaped delamination

Area (cm^2)	400 cm^2				900 cm^2			
Length of sides (cm)	5 × 80	8 × 50	10 × 40	16 × 25	10× 90	15 × 60	20 × 45	25 × 36
Ratio of length (%)	6%	16%	25%	64%	11%	25%	44%	69%

Moreover, when comparing both cases of Fig. 6, the results show that only rectangles which have 25 % or lower ratios of the shorter side length to the longer generated distinctly smaller ΔT than the other delamination, even though they have the same area in each case; 5 × 80 cm, 8 × 50 cm, 10 × 40 cm for 400 cm^2, 10 × 90 cm, 15 × 60 cm for 900 cm^2. It can be assumed that if the aspect ratio (ratio of the length of the shorter side to the longer side of the delamination) is more than 25 %, ΔT of any delamination converges to ΔT of the same area of a square-shaped delamination. Furthermore, if the aspect ratio is less than or equal to 25 %, the ΔT of the rectangular shaped delamination becomes smaller than the ΔT of a square-shaped delamination with the same area, and decreases as the ratio decreases.

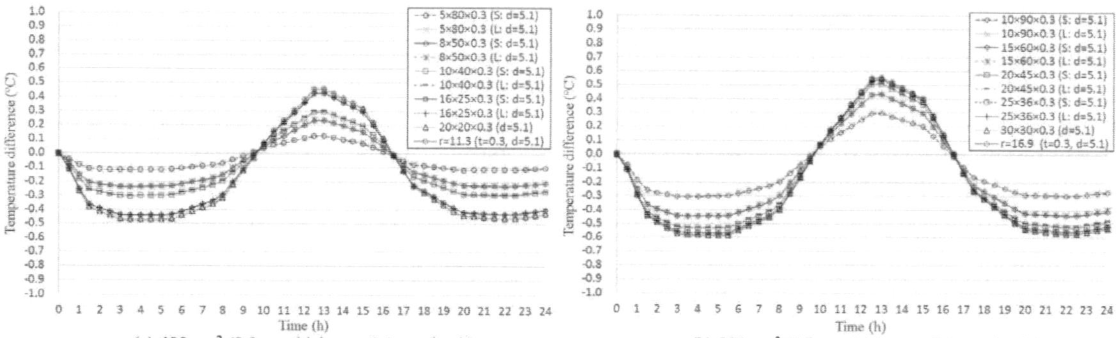

(a) 400 cm² (0.3 cm thickness; 5.1 cm depth) (b) 900 cm² (0.3 cm thickness; 5.1 cm depth)

Fig. 6. Temperature differences depending on the shape of delamination

4. Advanced analysis method: Estimation of delamination depth from IRT data

One of the biggest limitations of IRT is that it cannot provide data about the depth of delamination (Gucunski *et al.*, 2013; Kashif Ur Rehman *et al.*, 2016). However, it has been shown that a strong correlation between the size of delamination and the depth from the surface regarding ΔT exists, as discussed in Section 3. Based on this finding, this study explored whether IRT can provide information about the depth of delamination by using IRT data from an in-service bridge deck. The test bridge is the Haymarket Bridge, which is part of VA Route 15 that passes over Interstate 66 in Haymarket, Virginia. This bridge deck was inspected using several NDE methods by other researchers (Gucunski *et al.*, 2013; Oh *et al.*, 2013). In the field test, 8 core samples were drilled after the data collection of each NDE method, and Fig. 7(a) shows one of the 8 cores drilled at the circle located at the coordinate of F-16 described in Fig. 7(b). Hiasa, Catbas, et al. (2017) scanned the same location with IRT at a normal driving speed without lane closure during nighttime on October 2, 2014, and compared the result to other NDE methods such as impact echo, GPR and chain drag. Fig. 7(b) is the processed IR image (the grid line was assigned at intervals of 60 cm) taken by a cooled type camera (SC5600 manufactured by FLIR Systems, Inc). The indications enclosed by dotted lines in Fig. 7(b) are the potential delamination detected by IRT.

In the present study, the depth of delamination, located in the grids from 15-17 and E-G in Fig. 7(b), was estimated. The actual depth of the delamination is about 6 cm (Oh *et al.*, 2013), and it can be assumed that the surrounding regions are delaminated at the same depth as the core sample as shown in Fig. 7(a). Regarding the size of delamination, the area of the delamination is greater than 40 × 40 cm since one grid line is 60 cm in length in Fig. 7(b) and that delamination occupies about one and a half grids in both the horizontal and vertical directions. Since ΔT converges to a certain value when the dimensions approach 40 × 40 × 1 cm as discussed in Section 3, a 40 × 40 cm square-shaped delamination model was used in this analysis to estimate the depth from the surface. The delamination thickness is an unknown parameter in addition to the depth of delamination. Therefore, two sizes of thicknesses, 0.1 and 1 cm, were modeled in the present study, with ΔT of 40 × 40 × 0.1 cm being the lower limit for ΔT, and ΔT of 40 × 40 ×1 cm being the maximum value. The depth of delamination was estimated by comparing ΔT from IRT data with the minimum and maximum ΔT obtained from FE model simulations at several depths.

For the ΔT of the IRT data, temperatures of delaminated and sound areas were defined by picking three points randomly from the respective parts of IRT data and calculating the averages of the selected points. Fig. 7(c) is the enlarged raw infrared image of the delamination located at coordinates between 15-17 and E-G in Fig. 7(b). The three temperatures written in red are the temperatures of the delaminated area and the other three temperatures written in green are the temperatures of the sound area. From Fig. 7(c), the temperature of the sound area was defined to be 19.02 °C, and the temperature of the delaminated area was 18.69 °C. Hence, ΔT was -0.33 °C.

The same FE model conditions as discussed in Section 2.3 were utilized, and the weather conditions of the simulation were determined by calibrating the deck temperatures of sound areas based on the weather record from the closest weather station (WeatherUnderground, 2014), approximately 15 km away from the bridge. Since the IRT data was collected at 10:30 PM, ΔT of simulation results at 10:30 PM were compared. In the delamination model of 40 × 40 × 0.1 cm, the depths of delamination were changed every 1 cm from 3 cm to 6 cm, and the depths of 40 × 40 ×1 cm model were simulated from 4 cm to 7 cm. Table 4 summarizes ΔT of the IRT data and FE model simulations for each case. In the case of 40 × 40 × 0.1 cm delamination model, ΔT were -0.36 °C when the depth was 5 cm and -0.25 °C when the depth was 6 cm; therefore, the depth of the real delamination can be assumed between 5 and 6 cm if the thickness is 0.1 cm. On the other hand, ΔT of 40 × 40 ×1 cm delamination model was -0.55 °C at 6 cm depth and -0.29 °C at 7 cm depth; thus, the depth of the real delamination can be assumed to be between 6 and 7 cm if the thickness is 1 cm or more. Since the thickness of the delamination is also unknown as mentioned above, the depth of the real delamination can be assumed to be between minimum and maximum depths from the two cases of FE modeling. Therefore, the depth of the delamination indicated by IRT can be estimated to be 5 to 7 cm deep from the surface. This result matches the depth of the core sample that is approximately 6 cm from the surface. Thus, this estimation method worked properly by incorporating IRT with FE modeling.

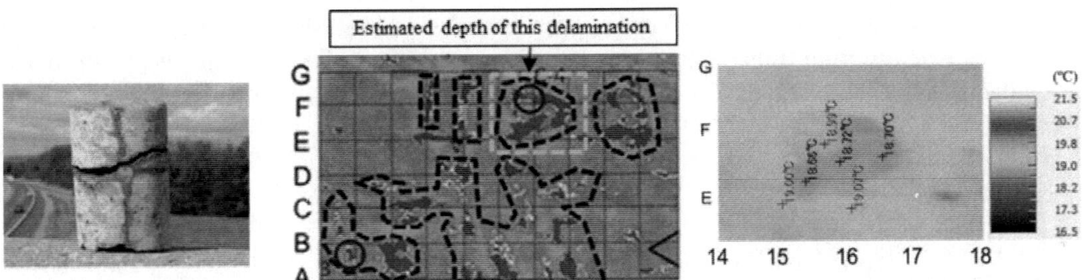

(a) Core sample (F-16)
(b) Processed IR image
(c) Temperatures of delaminated & sound parts

Fig. 7. (a) The core (b), (c) processed and IR images (source: (Gucunski *et al.*, 2013; Hiasa *et al.*, 2017))

Table 4. ΔT of IRT data and FE model simulations for each case

10:30 PM	IRT	Thickness: 0.1 cm				Thickness: 1 cm			
Depth	Unknown	3 cm	4 cm	5 cm	6 cm	4 cm	5 cm	6 cm	7 cm
Del. (°C)	18.69	18.29	18.41	18.53	18.64	17.60	17.97	18.30	18.58
Sound (°C)	19.02	18.90	18.89	18.89	18.89	18.83	18.84	18.85	18.87
ΔT (°C)	-0.33	-0.61	-0.48	-0.36	-0.25	-1.23	-0.87	-0.55	-0.29

5. Discussion

As many researchers suggest, IRT is a suitable approach for the inspection of civil infrastructures since it is a non-contact method and it utilizes IR images, which can instantly portray a wide range of concrete structures at one time. IRT can also scan decks with reasonable accuracy at highway speeds without lane closures, which are usually required to implement traditional methods and other NDE techniques (Hiasa *et al.*, 2017). In addition, IRT can be applied from a distance, which is a big advantage over other NDE methods that require access and contact to the concrete surface. Thus, IRT is able to assess components of a bridge that are too large for other NDE methods, such as the under-side of decks, girders, piers, and railings. Therefore, IRT can be considered as the fastest and easiest NDE methodology in terms of data collection. On the other hand, FHWA (2015) lists the limitations of IRT, such as 1) it is difficult to select the optimal conditions for effective application, 2) obstacles, such as water, stains, or debris, can cause misdetection, 3) unable to provide information about delamination depth, and 4) difficulty of detecting defects located at deeper than 10-13 cm (4-5 in.). As discussed in Section 4, this study presents a possible solution to limitation 3, although more data is needed to validate it.

In addition to this limitation, the authors have clarified other limitations, and explored proper methods and conditions for effective use of IRT. In terms of the optimal conditions, ideal data collection time (nighttime) and proper camera specifications depends on data collection speed (cooled type IR cameras for high-speed scanning) were found in past studies (Hiasa, 2016; Hiasa *et al.*, 2016, 2017). Regarding misdetections introduced by obstacles, those misleading thermal contrasts can be distinguished by comparing IR images with high-definition (HD) images; thus, authors recomend integrated use of IRT and HD imaging for an image-based structural health monitoring at local level (Hiasa *et al.*, 2017). Furthermore, it was also found that debris generate higher thermal contrast than delamination on the concrete surface; thus, it is proposed that temperature differences due to obstacles can be distinguished by analyizing the data (Hiasa, 2016). As for the detectable depth of delamination by IRT, authors found the correlation between size of delamination and detectable depth (Hiasa, Birgul and Catbas, 2017). Kee *et al.*, (2012) were able to detect a 15.2 cm (6 in.) depth of delamination by IRT. Therefore, IRT can detect delamination that is deeper than 10-13 cm if the delamination size is large enough. In addition to these limitations, authors have been developing more objective data interpretation method than just judging the color contrast of IR images by incorporating FE model simulations (Hiasa, Birgul and Catbas, 2016; Hiasa, Birgul and Necati Catbas, 2017). Even though more data is needed to enhance the reliability and usability of IRT for bridge inspection, great improvements in the IRT method have already been achieved through past studies.

6. Conclusions

In this study, the effects of size and shape of delamination on subsurface defect detectability by IRT was explored through FE modeling. The temperature difference between sound and delaminated areas, ΔT, was used as the reference value. Regarding the effect of delamination size, ΔT converges to a certain value when the delamination size is approximately $40 \times 40 \times 1$ cm. As for the effect of delamination shape, it can be concluded that if the aspect ratio (ratio of the length of the shorter side to the longer side of the delamination) is more than 25 %, ΔT of any delamination converges to ΔT of a square-shaped delamination with the same area. If the aspect ratio is 25 % or smaller, ΔT becomes smaller than the ΔT of a square-shaped delamination with the same area, and it continues to decrease as the ratio decreases.

These findings led to an investigation of the possibility in overcoming IRT's depth information limitation. To investigate whether IRT can provide information about the depth of delamination, IRT data from a real bridge deck scanning was utilized. It was assumed that it was possible to estimate the depth of delamination by comparing ΔT from IRT data to ΔT at several depths obtained from FE model simulations. Since ΔT converges to a certain value when the dimensions of the delamination approach that of a $40 \times 40 \times 1$ cm of square, ΔT for two sizes of delamination were used to establish a minimum, ΔT of $40 \times 40 \times 0.1$ cm, and a maximum, ΔT which is ΔT of $40 \times 40 \times 1$ cm. By comparing ΔT from IRT data to ΔT from FE model simulations at several depths, the real delamination's depth from the surface was assumed to be between 5 and 7 cm. Since the actual depth was about 6 cm from the surface, it can be concluded that using IRT and FE modeling together can provide accurate delamination depth information.

References

COMSOL (2015) *Heat Transfer Module User's Guide Version 5.1*.

FHWA (2015) *Federal Highway Administration Research and Technology: Nondestructive Evaluation (NDE) Web Manual*. Available at: https://fhwaapps.fhwa.dot.gov/ndep/DisplayTechnology.aspx?tech_id=14 (Accessed: 25 May 2017).

Gucunski, N., Imani, A., Romero, F., Nazarian, S., Yuan, D., Wiggenhauser, H., Shokouhi, P., Taffe, A. and Kutrubes, D. (2013) *Nondestructive Testing to Identify Concrete Bridge Deck Deterioration*. Washington, D.C.: Transportation Research Board. doi: 10.17226/22771.

Hiasa, S. (2016) *Investigation of Infrared Thermography for Subsurface Damage Detection of Concrete Structures*. Electronic Theses and Dissertations. Paper 5063. <http://stars.library.ucf.edu/etd/5063>. Available at: http://stars.library.ucf.edu/etd/5063.

Hiasa, S., Birgul, R. and Catbas, F. N. (2016) 'Infrared thermography for civil structural assessment: demonstrations with laboratory and field studies', *Journal of Civil Structural Health Monitoring*. Springer Berlin Heidelberg, 6(3), pp. 619–636. doi: 10.1007/s13349-016-0180-9.

Hiasa, S., Birgul, R. and Catbas, F. N. (2017) 'Investigation of effective utilization of infrared thermography (IRT) through advanced finite element modeling', *Construction and Building Materials*, 150, pp. 295–309. doi: 10.1016/j.conbuildmat.2017.05.175.

Hiasa, S., Birgul, R. and Necati Catbas, F. (2017) 'A data processing methodology for infrared thermography images of concrete bridges', *Computers & Structures*, 190, pp. 205–218. doi: 10.1016/j.compstruc.2017.05.011.

Hiasa, S., Catbas, F. N., Matsumoto, M. and Mitani, K. (2016) 'Monitoring Concrete Bridge Decks using Infrared Thermography with High Speed Vehicles', *Structural Monitoring and Maintenance, An International Journal*. Daejeon, Korea: Techno-Press, 3(3), pp. 277–296. doi: http://dx.doi.org/10.12989/smm.2016.3.3.277.

Hiasa, S., Catbas, F. N., Matsumoto, M. and Mitani, K. (2017) 'Considerations and Issues in the Utilization of Infrared Thermography for Concrete Bridge Inspection at Normal Driving Speeds', *Journal of Bridge Engineering*, in press.

Holst, G. C. (2000) *Common sense approach to thermal imaging*. Winter Park, FL, USA: JCD Publishing.

Kashif Ur Rehman, S., Ibrahim, Z., Memon, S. A. and Jameel, M. (2016) 'Nondestructive test methods for concrete bridges: A review', *Construction and Building Materials*. Elsevier Ltd, 107, pp. 58–86. doi: 10.1016/j.conbuildmat.2015.12.011.

Kee, S.-H., Oh, T., Popovics, J. S., Arndt, R. W. and Zhu, J. (2012) 'Nondestructive Bridge Deck Testing with Air-Coupled Impact-Echo and Infrared Thermography', *Journal of Bridge Engineering*, 17(6), pp. 928–939. doi: 10.1061/(ASCE)BE.1943-5592.0000350.

Maierhofer, C., Brink, A., Ro, M. and Wiggenhauser, H. (2005) 'Quantitative impulse-thermography as non-destructive testing method in civil engineering – Experimental results and numerical simulations', 19, pp. 731–737. doi: 10.1016/j.conbuildmat.2005.06.002.

Oh, T., Kee, S., Arndt, R. W., Popovics, J. S., Asce, M. and Zhu, J. (2013) 'Comparison of NDT Methods for Assessment of a Concrete Bridge Deck', *Journal of Engineering Mechanics*, 139(March), pp. 305–314. doi: 10.1061/(ASCE)EM.1943-7889.0000441.

Watase, A., Birgul, R., Hiasa, S., Matsumoto, M., Mitani, K. and Catbas, F. N. (2015) 'Practical identification of favorable time windows for infrared thermography for concrete bridge evaluation', *Construction and Building Materials*. Elsevier Ltd, 101, pp. 1016–1030. doi: 10.1016/j.conbuildmat.2015.10.156.

WeatherUnderground (2014) *Hourly Weather History & Observations*. Available at: https://www.wunderground.com/history/airport/KHEF/2014/10/2/DailyHistory.html?req_city=Haymarket&req_state=VA&req_statename=&reqdb.zip=20168&reqdb.magic=1&reqdb.wmo=99999&MR=1.

WeatherUnderground (2015) *Weather History for Orlando, FL [KFLORLAN179]*. Available at: http://www.wunderground.com/personal-weather-station/dashboard?ID=KFLORLAN179#history/tdata/s20151219/e20151219/mdaily.

Yehia, S., Abudayyeh, O., Nabulsi, S. and Abdelqader, I. (2007) 'Detection of Common Defects in Concrete Bridge Decks Using Nondestructive Evaluation Techniques', *Journal of Bridge Engineering*, 12(April), pp. 215–225. doi: 10.1061/(ASCE)1084-0702(2007)12:2(215).

The 8th International Conference on
Structural Health Monitoring of Intelligent Infrastructure
Brisbane, Australia | 5-8 December 2017

Structural Damage Identification Using Millimeter Wave Imaging and Image Processing

A. Noori Hoshyar[1], S. Kharkovsky [2], B. Samali [3]
[1] *Centre for Infrastructure Engineering, SCEM, Western Sydney University, Australia*
Email: Azadeh.H@westernsydney.edu.au

Abstract

In the past decades, structural health monitoring (SHM) has received wide attention in preventing the sudden failure of structural components by identifying the damages in its early stages. Basically, to ensure the safety and reducing the serviceabili Otsu's thresholding ty of civil infrastructure it is important to inspect and assess the physical and functional condition of structures. Currently, manual inspection is the main form of assessing the conditions to ensure structure still meets the safety requirements. However, there are still several accidents that are reported as a result of insufficient inspection and conditional assessment of structures. In order to prevent further incidents, it is necessary to continuously inspect and assess the condition of structures with appropriate techniques. This is why the development and application of efficient non-destructive testing and computer vision methods for infrastructure health monitoring are in demand. This paper present the developed smart damage detection system for a local infrastructure health monitoring which complement the image-based damage detection methods in hazardous scenarios. The system is based on a relatively simple millimeter wave continuous wave reflectometer with an open-ended waveguide antenna for the purpose of automatic imaging of flaws such as cracks in a steel plate at different standoff distances, monitoring the crack to avoid further deformation through a sequence of inspections at intervals. However, in some cases the original images based on measured data do not provide desired information, leading to developing the image processing algorithm to enhance the imaging result. The proposed algorithm is based on Otsu's thresholding method and Prewitt approximation in six directions in an image which has been created from the measured data to facilitate the structural damage identification. The proposed algorithm can successfully enhance the quality of images and visualize the crack in a steel plate under dielectric coating at different standoff distances.

1. Introduction

Structural health monitoring (SHM) is defined as the detection strategy of structural defects, Wu et al. (2009). In this respect, a damage observation has become a crucial concern in different structural tests in civil engineering. Damages in the specific case of metal structures are often concerning the presence of cracks. Since crack damages can arise to secondary undesirable affects like deformations, the localization of cracks in early stages are important factors for the maintenance, safety and durability of metal structures. The patterns and the density of crack distributions disclose remarkable information to observe for experts, such as various failure modes and degradation which are useful for preventing further deterioration. Therefore, detection of such cracks prior to its visibility is of extreme interest for metal structures.

There are different methods available for crack detection and measurement, such as visual inspection, image processing, smart film technique, distributed optical fibre and optical backscattered reflectometer. However, the conventional methods for monitoring the cracks in the metal encounter some limitations. The visual inspections are not able to detect the metal cracks location properly. In other words, the location of the appeared crack is not known priori. The main issue is not only the location but also the crack qualifying in the timely manner. In addition, the instability in connecting components experimental devices or any part of the structure can threat the human safety. Therefore, distinguishing the damage over the whole measurement is difficult and inaccurate. However, the other limitation of this technique is that the, the inspection efficacy can be easily influenced by the expert decision and will results in large measurement errors, Dasog et al. (1993). In light of these important issues, the need of automatic damage detection system to help experts for precision and safe investigation is of extreme necessity and offers the cost-effective substitute way of inspection, Yan et al. (2002). It involves obtaining the crack patterns via signals and images through cameras, sensors and computer programs incorporating with laboratory monitoring systems, allowing image processing techniques, Nazmul et al. (2008).

In addition, an enormous range of imaging devices and the novel image acquisition techniques has been developed in different applications. Accordingly, various number of image processing techniques are being investigated to handle imagery data. These methods are adopted as modern image processing approaches such as partial differential equation methods, Fourier analysis, Markov random fields and etc., Chan et al. (2005), or statistical and mathematical based techniques, Forsyth et al. (2002), for the sake of image interpretation. Therefore, in automatic detection systems, the selection of appropriate imaging techniques is of high priority. Nondestructive testing (NDT) are recognized as tools that render a more comprehensive picture of an inspection problem. Mariapaola et al. (2015) proposed the application of NDT imaging techniques as an effective tool for visual inspection. These techniques can be applied to map inhomogeneity and to recognize the highest risk damage areas in different structures. The different NDT imaging techniques are photogrammetry, IR-thermography, X-ray radiography, microwave and ToF-stress wave tomography. In this regards, the use of microwave imaging techniques for the purpose of non-destructive testing (NDT) in civil infrastructure has been raised widely, Andrés et al. (2015), Feng et al. (2013), Adam et al. (2015). Microwave imaging is the technique applied to sense a given object by means of examining microwave pulses. This technique offers comprehensive descriptions for the imaging systems. In applying imaging techniques, there are a number of features such as noncontact, one-sided, low-power, real time, in-field and etc., Kharkovsky et al. (2007), Kharkovsky et al. (2009), which makes the techniques very attractive for on-line inspections. In common microwave imaging inspections, the reflection from the test specimen is recorded by the RF detector and used for detection of damage inside the specimen, Ahanian et al. (2015). Due to the great effect of noise while scanning, this paper employs the noise reduction technique to reduce this effect.

By recording signals and generating images through Matlab software and application of signal processing and image processing, the images are enhanced and accomplished by segmenting the damage areas from the healthy state. The segmentation algorithms are generally extended based on similarity and can be categorized as; thresholding, Cheriet et al. (1998), template matching, Warfield et al. (2000), region growing, Mat-Isa et al. (2002), edge detection, Paik et al. (1991), and clustering, Yang et al. (2008). These techniques have been widely applied in various areas of statistics and recently found crucial applications in civil engineering, Gang et al. (2014), Yuan-Sen et al. (2015), Chun et al. (2013). Valença et al. (2013) proposed an automatic monitoring system using photogrammetry and image processing techniques. In their system, the photogrammetric data is employed to determine the crack areas. A high-pass filter and Otsu's method followed by mathematical morphology are the main image processing techniques have been used to identify the

cracks. The major drawback of their research was incorrect detection of cracks along the test. In the research work of Hutchinson et al. (2006), the combination of canny edge detection algorithm and the wavelet transform is presented to evaluate the concrete damage. In another research, Oh et al. (2009) designed a crack detection system to extract cracks in the bridge substructures by employing the canny detection algorithm. Their algorithms reached the good image segmentation results. Qader et al. (2003) performed a comparison study on crack detection in the bridge surface by employing the wavelet transform, Fourier transform, Sobel filter, and Canny filter. Cotič et al. (2013) proposed the fusion of GPR and thermographic phase data and improving the visualization of defects in concrete using the fuzzy c-mean algorithm.
Although there are various researches in the area of defect detection using image analysis in different materials, such as concrete, Wang et al. (2010), it is still less research on image analysis of metal. Different ND imaging techniques can be employed to identify the structural damages. However, the exploring of computer vision and image-based algorithms in such structures has been still rarely adopted. Furthermore, there has been less research on monitoring the inception or expansion of damages in the structure. This research investigates one of the common structural damage in structural engineering as fatigue cracks on metal. Although this consideration is narrow in comparison with other possible structural damages, the proposed algorithms can be generalized to include them in respect to the proper imaging.

According to the importance of defect detection and also the available literature, the task of this paper is formulated as microwave imaging system problem. Compared with the results in 2009 in this paper, we perform systematic investigation on the problem which is followed by image proceeding approaches. The methodology of this research relies on the modern computer vision techniques instead of traditional photogrammetric category. This set of solution algorithms are proposed in this paper is called as "Damage Image Analysis System" to facilitate the structural damage identification. Thereby, an early damage assessment that may be followed by a rapid disaster response is identified. It is cost effective, while increasing the accuracy and throughput of the applications. The rest of this paper is structured as follows: We shortly go over the measurement approach and specimens in Section II. In Section III, the proposed image processing algorithms for damage detection are explained in detail. The results are demonstrated in Section IV, and Section V is assigned to conclusion.

2. Measurement Approach and Specimen

In this research, an open-ended flange mounted rectangular waveguide probe was used as an open-ended waveguide antenna (OEWA). The OEWAs are robust, reliable, efficient and relatively simple antennas for crack detection, Zoughi (2000), Kharkovsky and Zoughi (2007), Zoughi and Kharkovsky, (2008), Kharkovsky et al. (2009). Measurement system consisted of a computerized imaging system including a millimeter-wave reflectometer with a single frequency oscillator, isolator, magic tee, detector, variable short, the OEWA, and a scanning mechanism providing a raster scan of the reflectometer over the sample under test. All measurements were conducted at 90 GHz at different distances between the OEWA and the surface of specimen (referred to as standoff distance). The OEWA irradiated the specimen under test and picked up the signal reflected from the specimen. This reflected signal was combined with the reference signal in the magic tee, the combined data was sent to the detector and the computer recorded its output voltage for every data point over a plane for each specimen. The recorded data are analysed through signal processing and the images of the specimen were generated through proposed image processing algorithms. Figure 1 shows the measurement setup of this study.

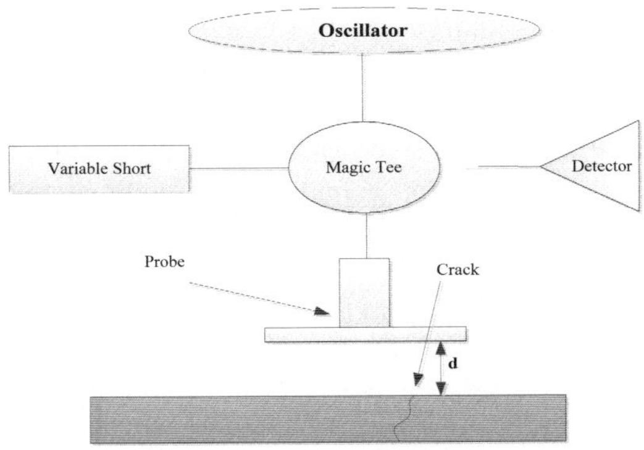

Figure 1. Measurement Setup

The specimen is a section of a 12-mm-thick A-36 steel plate with a few through fatigue cracks created during cyclical loading. A W-band open-ended flangemountedn rectangular waveguide probe with an aperture dimension of 2.54 mm x 1.27 mm and with a square (19 mm x 19 mm) flange was used in this investigation. Two scanned areas with dimensions of 50 mm by 70 mm and 30 mm by 30 mm include cracks obtained at a standoff distance of d~0 mm, d~0.8 mm,d~1.7 mm.

3. The Proposed Image Processing Approach

The proposed algorithm is to monitor the structural damage in civil infrastructure especially in metal. The basis of this research relies on the fact that structural components provide pixels which are visible by the high-resolution 'sensors' that can demonstrate the topographical appearance characteristics of structural integrity. The changes in structural integrity can be recognized by variation of image intensities in different intervals.

The proposed algorithms are able to supervise the suspected critical cracks using the spot measurement. Since the produced images from the antenna are not clear and have low resolution, the new image processing approach has been investigated and run upon the same conditions in Matlab to achieve the enhanced images and segmenting the area of defect. Since in image processing, there is no single best approach, the algorithms should be dictated by the problems at hand, by the methods being used, and by the results required.

To begin with, upon the antenna measurements of metal specimens, the information of returned signals from the metal target is gathered as the raw data. The recorded signals include the magnitude and phase at different frequencies. These data is processed to achieve the two-dimensional imagery through Matlab software. The measurement of sweeping frequency to produce the two-dimensional imagery has been obtained by the angles from -10 to +10. The phase and magnitude at each frequency are multiplied by a 'kaiser' window function to minimize the FFT affect over a noninteger numbers. The windowed frequency data is transformed to the time domain by employing the inverse Fourier transform. The win factor is adjusted to 3.5 per frequency band which can best demonstrate the results. The frequency range between 80 and 90 KHZz is appropriate for monitoring the metal crack detection system. The corresponding magnitude and phase in a time domain has been obtained by the process to perform enhancement.

The antenna time domain measurement result is subsampled as an image in Matlab. Since the purpose of this process is to obtain the high level variations between intensities of the image to detect the edges of crack, it would be optimal to convert the image into greyscale in which the minimum intensity valuc is cqual to 0 and the maximum intensity value is equal to 255. Since LAB is one of the useful color models which represent every color through three components of luminance, red/green and blue/yellow, it could be beneficial to transform the RGB to LAB using XYZ as an intermediate color space. The luminance would present the greyscale image of the crack. Figure 2 shows the tristimulus curves of LAB color space which has been obtained by data tables of CIE 1964 Supplementary Standard Colorimetric System, Wyszecki et al. (1982).

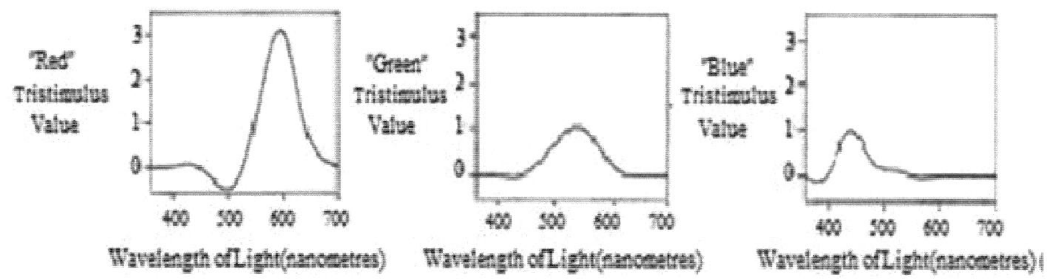

Figure 2. Tristimulus Curves.

Secondly, the averaging filter and contrast enhancement techniques are used to precede the enhancement of raw images. Enhancement in this case consists sharpening the image, while preserving as much as its gray tonality as possible. Here, a rectangular averaging filter of size 3×3 is considered.

Since average filter is a separable filter, it can be implemented by dividing into two smaller filters. In this paper, firstly, the 3×1 average filters is applied to the raw image and then the 3×1 average filter proceeds the result. This process will result in less computational time. The new gray value of the corresponding pixel is calculated by taking the average of all nine values in the 3×1 and 3×1 masks. This process is followed by the grayscale transformation function to adjust the intensity. Since our images are of the class 'uint8', the function multiplies the values by 255 to specify the actual values in an image.

To achieve the segmented crack from the background, the rest of this process is followed by two different methods. It calculates the gradient magnitude by computing the gradients in six directions in an image with the prewitt filter for edge detection. The edges of crack are returned at hose points where the gradient of image is maximum. Therefore, the edges are acheived using the prewitt approximation to the derivative. The detail algorithm is described below. The detail algorithm is described below. P is a crack image; P(x, y) is the intensity of a pixel (x, y).

(1) Assuming the profile of the image as f(x)

(2) Computing the derivative as f'(x)

$$\frac{dy}{dx} = \lim_{h \to 0} \left(\frac{f(x+h) - f(x-h)}{h} \right) \tag{1}$$

Where h has the value of '1' as the smallest possible value can occur in an image, therefor the discrete version of the derivative expression is

$$f(x+1) - f(x-1) \tag{2}$$

(3) Leaving the scaling factor, the horizontal, vertical, and diagonals filters are

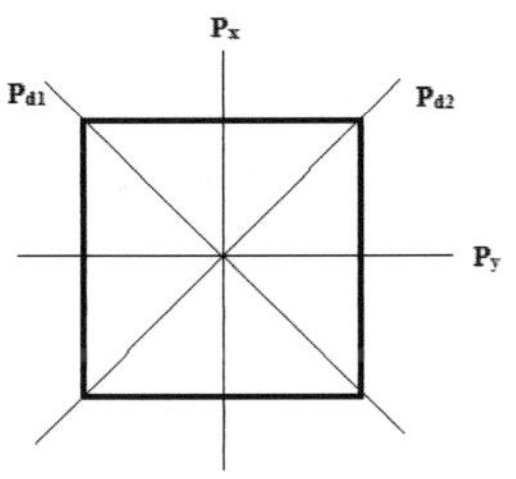

(3)

- Horizontal: (prewitt filter)

$$P_x = \begin{pmatrix} -1 & 0 & 1 \\ -1 & 0 & 1 \\ -1 & 0 & 1 \end{pmatrix} \tag{4}$$

- Vertical:

$$Py = Px' \tag{5}$$

- Diagonal (Left to right):

$$P_{d1} = \begin{pmatrix} 0 & 1 & 1 \\ -1 & 0 & 1 \\ -1 & -1 & 0 \end{pmatrix} \tag{6}$$

- Diagonal (Right to Left):

$$P_{d2} = \begin{pmatrix} -1 & -1 & 0 \\ -1 & 0 & 1 \\ 0 & 1 & 1 \end{pmatrix} \tag{7}$$

These filters will find Horizontal, Vertical, Diagonal (Left to right) and Diagonal (Right to Left) edges in an image while smoothing the results.

4) Assuming px, py, p_{d1} and p_{d2} as the gray values produced by applying above filters to an image, the gradient magnitude is computed by

$$\sqrt{p_x^2 + p_y^2 + p_{d1}^2 + p_{d2}^2} \tag{8}$$

5) Determining the threshold value of image by Otsu's thresholding method

6) Obtaining the binary image containing edges only applying the threshold value

We introduced a new approach that allows experts to accurately identify defects that are significantly associated with any subset or linear combination of cracks. The proposed method can greatly simplify health monitoring in structures and can be used to identify the cracks significantly associated with deterioration. Using imaging technology, we demonstrate that our method attains accurate measures of detection over a range of relevant scenarios.

4. Results

In this section, the results have been investigated and analysed. The images are 90 GHz images of cracks in a steel plate at different standoff distances with an open-ended waveguide probe. The resulted images have been enhanced and segmented in different distances. Figures 3 show the signals have been recorded through the open-ended waveguide antenna under 50*70 scanning obtained at three standoff distances of 0 mm, 0.8 mm and 1.7 mm.

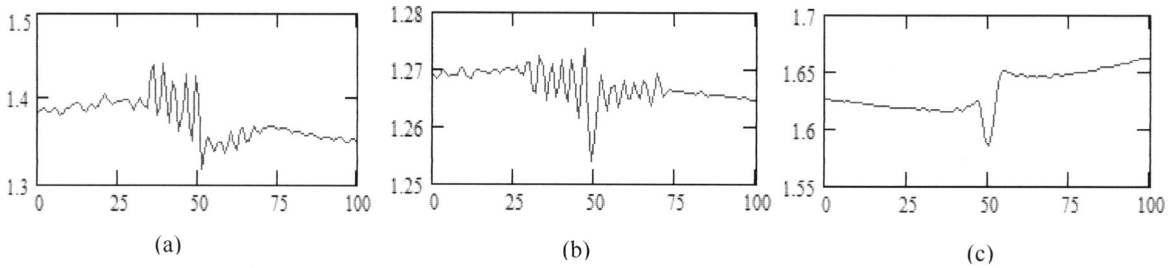

Figure 3. Signals recorded under 50×70 scanning obtained at three standoff distances: (a) 0.2 mm, (b) 0.8 mm and (c) 1.7 mm.

Figure 4 and figure 5 show the generated images of crack in Matlab for two scanned areas of 50×70 and 30×30 with standoff distances of d~0.2, d~0.8 and d~1.7, respectively. As it is clear from figure 3 and figure 4, the original images are blurring, and limited information can be gained from images of crack.

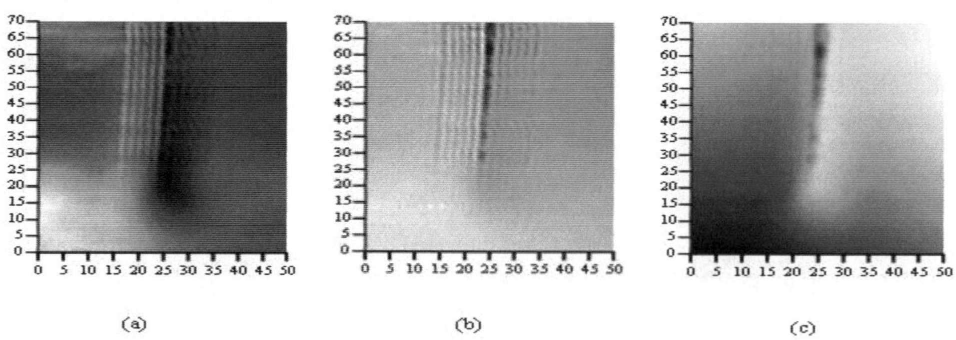

Figure 4. Original images of the portion of covered crack under 50×70 scanning obtained at three standoff distances: (a) 0.2 mm, (b) 0.8 mm and (c) 1.7 mm.

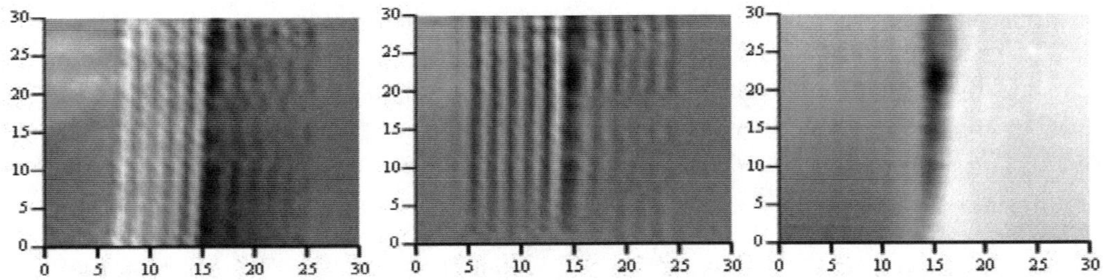

(a) (b) (c)

Figure 5. Original images of the portion of covered crack under 30×30 scanning obtained at three standoff distances: (a) 0.2 mm, (b) 0.8 mm and (c) 1.7 mm.

Figure 6 shows the results of image enhancement process. The raw images have been enhanced to provide more observable information.

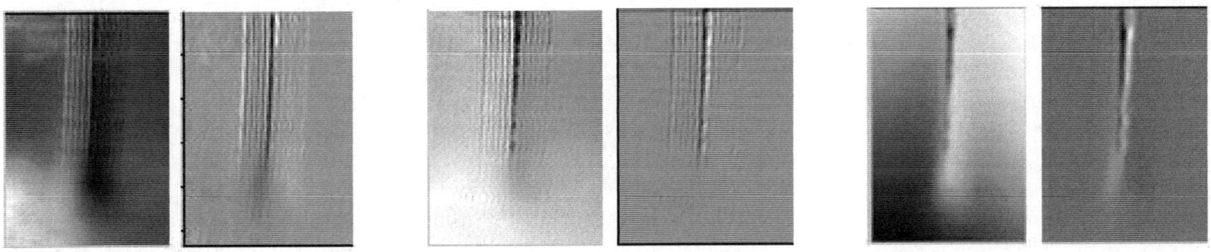

(a) (b) (c)

Figure 6. a) Left. Original image with d~1.7, a) Right. Enhanced image with d~1.7, b) Left. Original image with d~0.8, b) Right. Enhanced image with d~0.8 c) Left. Original image with d~0.2, c) Right. Enhanced image with d~0.2 mm

Figure 7 show the images achieved by employing the proposed algorithm by calculating the gradient magnitude in six direction of x, y, d1, d2 followed by Otsu's thresholding method.

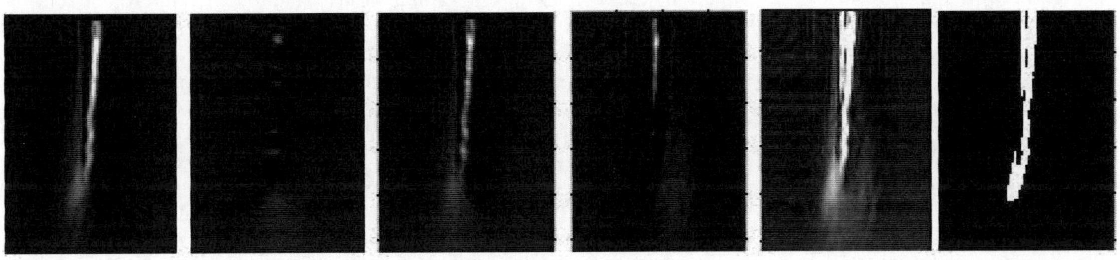

(a) (b) (c) (d) (e) (f)

Figure 7. a) Gradient in x direction b) Gradient in y direction c) Gradient in d1 direction, d) Gradient in d2 direction, e) Gradient Magnitude f) Otsu's thresholding.

Imaging results confirm that the developed algorithm can identify the crack edges while eliminate the noises on the target region. It is shown that the algorithm provides the image with pure texture and strong edges.

5. Conclusions

The structuring of automatic damage detection system in civil structures have been proposed and developed. The microwave imaging technique has been used to present the discontinuities inside the structure. The algorithm employs signal processing and image processing techniques to manipulate, enhance and segment the crack images in structures.

In this paper, the experiment study is performed to identify crack in a steel plate at different standoff distances with an open-ended waveguide probe. The imaging results of experiment study show that our proposed algorithm could effectively identify and manifest the cracks in infrastructure. The imaging results are compared with the original images for validation purpose.

The purpose of this study is to minimize the need of manual inspections for civil infrastructures. It seeks to complement edge-based damage detection methods in hazardous scenarios.

References

Adam, B, Aled, J & Sam, K 2015,' Microwave processing of cement and concrete materials – towards an industrial reality?', *Cement and Concrete Research*, vol. 68, pp.112–123.

Ahanian, S.H, Sadeghi, n & Moini, R 2015,' An array waveguide probe for detection, location and sizing of surface cracks in metals', *NDT&E International*, vol. 70, pp. 38–40.

Andrés, F, Castro, M, Valcuende & B, Vidal 2015,'Using microwave near-field reflection measurements as a non-destructive test to determine water penetration depth of concrete', *NDT&E International*, vol. 75, pp. 26–32.

Cheriet, M, Said, J. N & Suen C. Y 1998, 'A recursive thresholding technique for image segmentation', *IEEE Transactions on Image Processing*, vol. 7, no. 6, pp. 918-921.

Chun, L, Chao-Sheng, T, Bin, Sh & Wen-Bin. S 2013,' Automatic quantification of crack patterns by image processing', *Computers & Geosciences*, vol. 57, pp. 77-80.
Cotič, P, Niederleithinger, E, Bosiljkov, V & Jagličič, Z 2013, 'NDT Data Fusion for the Enhancement of Defect Visualization in Concrete', *Key Engineering Materials*, vol. 569-570, pp. 175-182.

Dasog, G.S & Shashidhara, G.B 1993, 'Dimension and volume of cracks in a Vertisol under different crop covers', *Soil Science,* vol. 156, no. 6 , pp. 424–428.

Feng, M.Q, Roqueta, G & Jofre, L 2013,' 22 – Non-destructive evaluation (NDE) of composites: microwave techniques', *Non-Destructive Evaluation (NDE) of Polymer Matrix Composites*, pp. 574-616.

Gang, Li a, Shuanhai, He. B , Yongfeng, Ju & Kai, Du 2014,' Long-distance precision inspection method for bridge cracks with image processing', *Automation in Construction*, vol. 41, pp. 83–95.

Hutchinson, T.C & Chen, Z.Q 2006, 'Improved image analysis for evaluating concrete damage', *J. Comput. Civ. Eng*, vol. 20, no. 3, pp. 210–216.

Kharkovsky, S & Zoughi, R 2007,'Microwave and millimetre wave nonsdestructive testing and evaluation: Overview and Recent Advances ', *IEEE Instrumentation & Measurement Magazine*, vol. 10, no. 2, pp. 26-38.

Kharkovsky S, Ghasr M.T. and Zoughi R., 2009, "Near-field millimeter wave imaging of exposed and covered fatigue cracks," *IEEE Transactions on Instrumentation and Measurement*, 58(7): 2367-2370.

Mariapaola, R, Jakub, S & Steffen, F 2015, ' Application of imaging techniques for detection of defects, damage and decay in timber structures on-site', *Construction and Building Materials*.

Mat-Isa, N. A, Mashor, M. Y & Othman, N. H 2002, 'Automatic seed based region growing for pap smear image segmentation', *International Conference on Biomedical Engineering*, Kuala Lumpur, Malaysia.
Nazmul, I & Matsumoto, T 2008, 'High resolution COD image analysis for health monitoring of reinforced concrete structures through inverse analysis', *Int. J. Solids Struct.*, vol. 45, pp. 159–74.

Oh, J.K, Jang, G & Oh, S 2009, 'Bridge inspection robot system with machine vision Autom', *Construction,* vol. 18, no. 7, pp. 929–941.

Paik, J. K, Park, Y. C & Park, S. W 1991, 'An edge detection approach to digital image stabilization based on tri-state adaptive linear neurons', *IEEE Transactions on Consumer Electronics*, vol. 37, no. 3, pp. 521-530.

Qader, A, Abudayyeh, O & Kelly, M 2003, 'Analysis of Edge-Detection Techniques for Crack Identification in Bridges', *J. Comput. Civ. Eng.*, vol. 17, no. 4, pp. 255–263.

Valença, J, Dias-da-Costa, D, Júlio, E, Araújo, H & Costa, H 2013, 'Automatic crack monitoring using photogrammetry and image processing', *Measurement*, vol. 46, no. 1, pp. 433-441.

Wang, P & Huang, H 2010, 'Comparison analysis on present image-based crack detection methods in concrete structures', *3rd International Congress on Image and Signal Processing (CISP)*, vol. 5, pp. 2530 –2533.

Warfield, S. K, Michael, K, Jolesz, F. A, & Ron, K 2000, 'Adaptive, template moderated, spatially varying statistical classification', *Medical Image Analysis*, vol. 4, no. 1, pp. 43-55.

Wu, C, Kuo, W, Yang, Y, Hwang, S, Elwood, K, Loh, C & Moehle, J 2009, 'Collapse of a nonductile concrete frame: shaking table tests', *Earthq Eng Struct Dynam*, vol. 38, no.2, pp. 205–24.

Wyszecki, G & Stiles, W 1982, 'Color Science: Concepts and Methods, Quantitative Data and Formulae'. *New York: John Wiley and Sons.*

Yan, A, Wu, K & Zhang, X 2002, 'A quantitative study on the surface crack pattern of concrete with high content of steel fiber', *Cement and Concrete Research*, vol. 32, no. 9, pp. 1371–1375.

Yang, X, Weidong, Z, Yufei, C & Xin, F 2008, 'Image segmentation with a fuzzy clustering algorithm based on Ant-Tree', *Signal Processing*, vol. 88, no. 10, pp. 2453-2462.

Yuan-Sen, Y, Chung-Ming, Y & Chang-Wei, H 2015,' Thin crack observation in a reinforced concrete bridge pier test using image processing and analysis', *Advances in Engineering Software*, vol. 83, pp. 99-108.

Zoughi, R., 2000, "Microwave non-destructive testing and evaluation", Kluwer Academic Publishers, Dordrecht, Netherlands.

Zoughi R. and Kharkovsky S., 2008,"Microwave and millimeter wave sensors for crack detection," *Fatigue & Fracture of Engineering Materials & Structures*, 31(8): 695-713.

The 8th International Conference on
Structural Health Monitoring of Intelligent Infrastructure
Brisbane, Australia | 5-8 December 2017

Evaluation of the Coefficient of Thermal Expansion for a Temperature Driven Method of Structural Health Monitoring

J. Reilly[1], M. Yarnold [2], B. Murphy [2], B. Glisic [1].
[1] *Princeton University – USA, Email:* jpr2@princeton.edu, [2] *Tennessee Technological University - USA*

Abstract

Structural Health Monitoring (SHM) seeks to characterize the performance and health of a structure based on recorded sensor data and analytic techniques. In most cases, SHM entails building some relationship between mechanical loads and monitored parameters. Temperature is normally considered noise in this analysis, and many techniques seek to filter out any thermal effects during analysis. Daily thermal loads on a structure can cause comparable strains to daily traffic loads, prompting temperature's leading role in Temperature Driven (TD) – SHM. Temperature can be easily measured on structure using a variety of commercially available temperature sensors, an advantage when compared to the difficulty in measuring daily traffic loads. TD-SHM relates input temperatures readings to output strain and displacement measurements in order to create a set of damage sensitive, three-dimensional signatures for the structure. An initial stage of the formulation of TD-SHM is the evaluation of the Coefficient of Thermal Expansion (CTE) of the structure. In concrete structures the CTE can often vary over time, which complicates its evaluation. This paper presents a method for evaluation of the CTE through examination of the relationship between changing temperature and strain at different locations on the structure, and relative displacements at support. These CTE evaluations serve as a basis for understanding the thermal behaviour of the structure, and for later development of TD-SHM. The method is validated on data collected from the Streicker Bridge, on campus at Princeton University.

1. Introduction

Temperature changes play a large role in the day-to-day behaviour of any bridge, but are correspondingly underrepresented in contemporary Structural Health Monitoring (SHM) methods. In many cases daily temperature changes on a bridge produce strain comparable to daily traffic loadings (Yarnold 2013). Temperature Driven - Structural Health Monitoring (TD-SHM) considers temperature as the driving factor in structural monitoring. TD-SHM relates an input temperature on the structure to an output strain and displacement at different locations on the structure. These three parameters will form a three dimensional signature that is sensitive to damage in the structure.
The Coefficient of Thermal Expansion (CTE) commands this relationship between temperature, strain, and displacement. While many materials have a stationary CTE throughout their lifetime, concrete can have a range of CTE values which can be difficult to assess in situ. Many design codes specify a standard for different types of concrete, typically near 11 or 12 $\mu\varepsilon/^{o}C$, though one study found values between 7 and 8 $\mu\varepsilon/^{o}C$ to be more accurate (Ndon 1995). Another study found that the CTE of early age concrete can change by up to 60% with changes in relative humidity, while being unaffected by the age of the concrete (Lim 2015). This paper explores a method for the evaluation of the CTE of built concrete structures using long term temperature and strain measurements. This method relies on assumptions of near uniform temperature on the structure and on the negligence of

creep and shrinkage effects on the material, and addresses ways to maximize the likelihood of achieving conditions as close to these assumptions as possible. These thermal gradients and rheological strains can obscure the relationship between temperature and strain. This process minimizes these effects to show a clear and definable relationship between temperature and strain and a reasonable evaluation of the CTE.

These methods are formulated using data taken from a real structure, the Streicker Bridge on campus at Princeton University. This deck-stiffened arch pedestrian bridge has over 100 fibre optic sensors embedded in to concrete deck measuring temperature and strain, complemented by displacement sensors at the abutment of the southeast leg. Figure 1 shows an elevation view of the Streicker Bridge and locations of primary sensors.

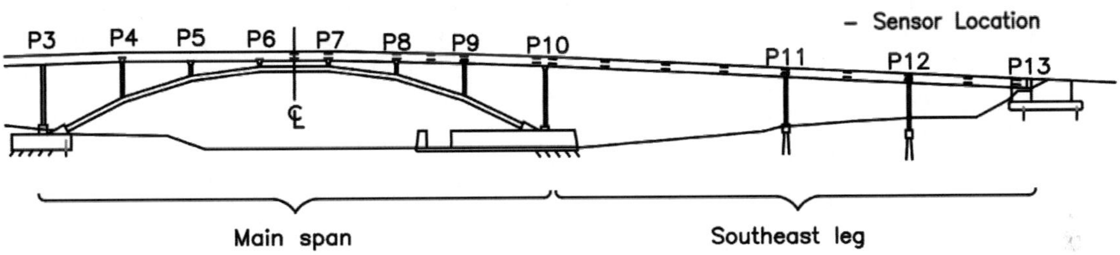

Figure 1. Streicker Bridge Elevation View

This paper primarily uses the parallel Fibre Bragg Grating (FBG) sensors shown in Figure 1. Further details on the Streicker monitoring system can be found in previous work. (Sigurdardottir 2015).

2. Coefficient of Thermal Expansion

Long term strain sensors embedded in concrete experience the total strain on the material, summarized in Equation 1 (Glisic & Inuadi 2007).

$$\varepsilon_{Tot} = \varepsilon_{Mech} + \varepsilon_{Thermal} + \varepsilon_{Creep} + \varepsilon_{Shrink} \quad (1)$$

The total strain, ε_{Tot}, is the combination of the mechanical, ε_{Mech}, thermal, $\varepsilon_{Thermal}$, creep, ε_{Creep}, and shrinkage strains, $\varepsilon_{Shrinkage}$. The evaluation of the CTE relies on two main assumptions in order to isolate the thermal strain. First, over short enough time periods, rheological strains will be negligible. Removing creep and shrinkage strains results in Equation 2.

$$\varepsilon_{Tot} = \varepsilon_{Mech} + \varepsilon_{Thermal} \quad (2)$$

Next, if the structure is determinate (unrestrained) in the direction of the measured strain, then thermal changes will not induce a mechanical response. Mechanical strains from loading on the structure can be filtered out by analysis of the measurements and monitoring schedule. If the structure is free to expand, then the total strain can be further broken down in Equation 3.

$$\varepsilon_{Tot} = \varepsilon_{Thermal} = \alpha \Delta T \quad (3)$$

Equation 3 relates the thermal and total strain to changes in temperature under linear free expansion, where α is the CTE and ΔT is a change in temperature. With these assumptions, the CTE of the material is the rate of change of the total strain on the sensor with temperature, and can be evaluated as the slope of the best fit line between the temperature and total strain. In many cases, however, one or both of these assumptions are invalid. For example, in early age concrete the shrinkage of the first month will be significant, affecting the intercept of the best fit line of temperature and total strain. If the structure is not completely free to expand, as almost all structures are to some degree, there will be some mechanical strain response to the thermal strain, affecting the rate of change of total strain with temperature, directly impacting the CTE evaluation. Also, while uniform temperature changes do not induce a mechanical response in determinate structures, non-linear thermal gradients can cause stresses even in determinate structure. All of these effects need to be minimized in order to achieve an accurate evaluation of the CTE.

2.1 Streicker Bridge Example

This process for evaluating the CTE of a built concrete structure was tested on data taken from the Streicker Bridge. Figure 2, below, shows data taken in 2010 from one sensor located at the top of the section at Pier 12.

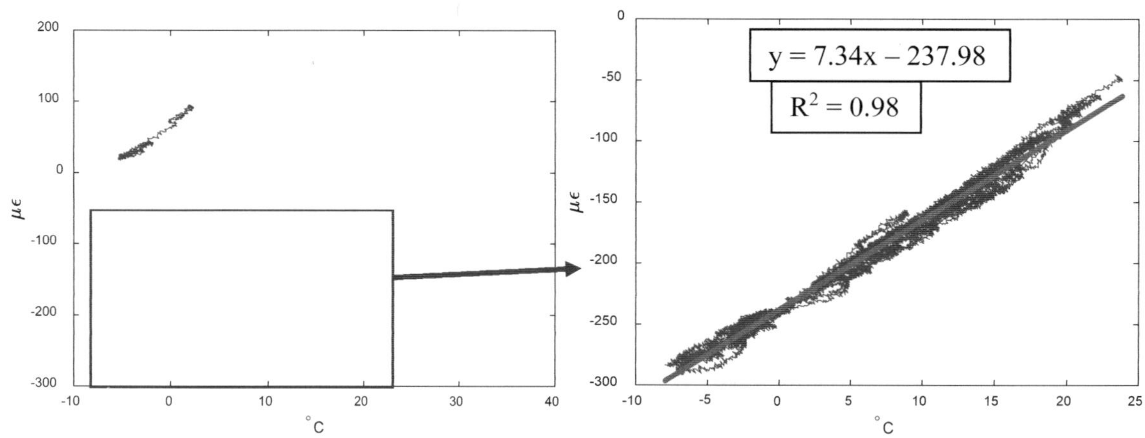

Figure 2. A) Strain vs Temperature, P12up 2010 B) Strain vs Temperature, P12up Fall 2010

Figure 2.A shows temperature and strain data recorded for the entire year of 2010. The data can be divided into three distinct time periods of monitoring: January, summer, and fall (clockwise from top left). The gaps in data are from periods where the bridge was not monitored. While there is a strong relationship between temperature and strain in Figure 2.A, it can be hard to quantify. The concrete in the Streicker Bridge was poured in the fall of 2009, months before the data shown above. There is a non-negligible amount of creep and shrinkage occurring in the time period. Figure 2.B shows just the fall of 2010 for the same sensor, showing a much stronger fit, using a linear regression (York 2004), evaluating the CTE as 7.3 με/°C for that location of the bridge. The fall has a good fit because of the relatively small thermal gradients the structure endures. Figure 3 shows the same sensor for the summer of 2010.

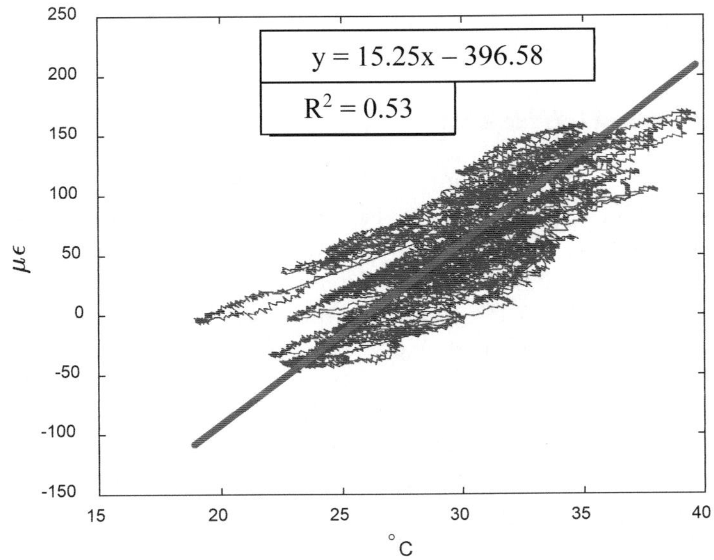

Figure 3 Temperature vs Strain for P12up, Summer 2010

Figure 3 shows a poor linear fit between temperature and strain, as well as a vastly different CTE evaluation (slope of best fit line) than for the fall time period. The summer has much larger thermal gradients than the fall, affecting this rate of change between temperature and strain, as well as a non-negligible amount of rheological strain, affecting the intercept of the best fit line. While the main assumptions are violated in the summer time period, this paper presents methods for mitigating the impact of these violations.

In order to mitigate the effects of creep and shrinkage, the summer time period can be broken up into nine separate weeks of data, shown in Figure 3.

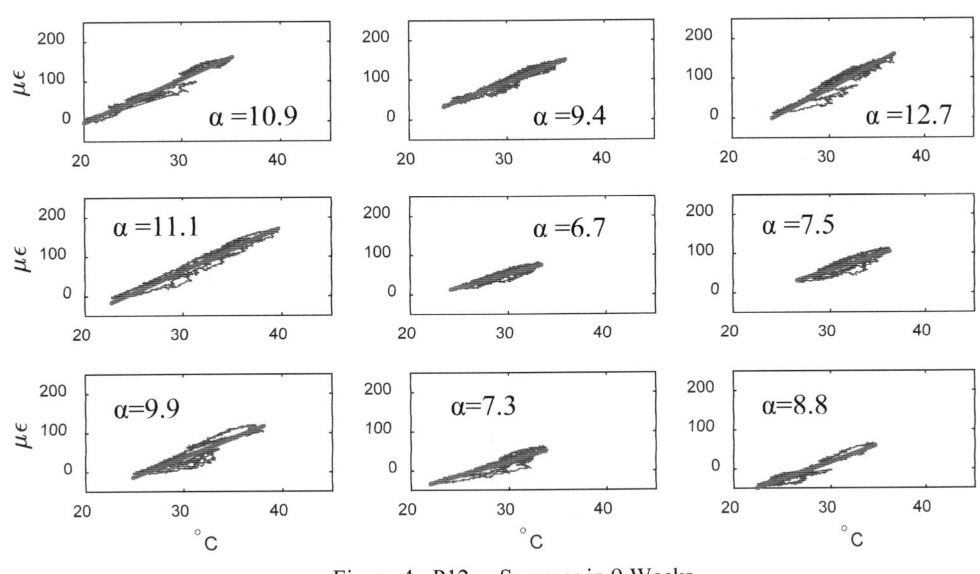

Figure 4. P12up Summer in 9 Weeks

The creep or shrinkage over any single week is negligible, allowing for stronger linear correlation between the temperate and strain changes. Despite minimizes the rheological strains, there is still a dispersion of results caused by thermal gradients in the structure.

3. Mitigating Effects of Thermal Gradients

Thermal gradients can be difficult to measure on a structure, and are often non-linear in nature (Emanuel 1978). These nonlinear gradients can cause non-linear strain distributions, as well as deplanations of the cross section. Two methods are used to mitigate the adverse effects of these thermal gradients by identifying time periods of minimal thermal gradient (Reilly 2016).
First, the 95% Range Method (95 MR) identifies time periods when 95% of all temperature measurements on the structure are within some chosen maximum range. This method ensures that almost all of the temperature sensors are reading similar values, with the allowance for some outliers. The value of the maximum range is chosen based on finding the largest possible amount of time periods, while not compromising the minimization of the thermal gradients in the structure.
Next, the Maximum Local Gradient (MLG) method seeks to directly minimize the local, vertical thermal gradients in the structure. These vertical thermal gradients cause many of the unwanted structural effects. The MLG method identifies time periods where each location with a pair of sensors has a local gradient under some chosen maximum gradient. Equation 4 defines the local thermal gradients.

$$\left|\frac{T_{top}-T_{bot}}{h}\right| < MLG \qquad (4)$$

T_{Top} and T_{Bot} refer to the temperatures at the top and bottom of the section, while h refers to the distance between the sensors. The MLG method is not disrupted by different temperatures in different locations on the bridge, only that each vertical pair of sensors has an acceptable local gradient. Both of these methods can vary the bounds during identification, allowing for the method to be tailored to each specific application.

4. Evaluating the CTE

Figure 5 shows the process for evaluating the CTE of built concrete structures. The data is broken up into small time periods to minimize the influence of creep and shrinkage strains. Then, each data set is filtered for time periods of minimal thermal gradients. A weighted average of the slope of the best fit lines of each of these time periods of minimal thermal gradient produces the final CTE evaluation.

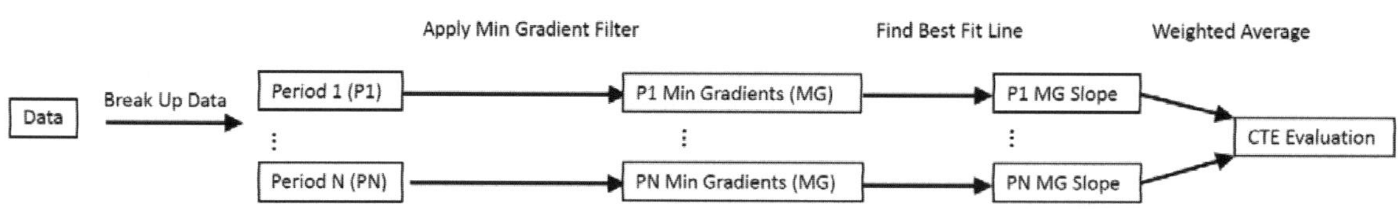

Figure 5. CTE Evaluation Process

This process seeks to maintain the strain based assumptions made previously, in order to produce the most accurate estimation of the CTE as possible. Table 1 shows a summary of results for several sensors on the Streicker Bridge.

Table 1. CTE Evaluations

	Fall Fit	Summer Fit	Broken S Fit	95MR 3.5	MLG 6
P8up	9.6	4.0	4.4	10.0	12.2
P8down	9.0	21.0	22.9	11.8	11.6
P9up	10.1	7.4	6.2	9.7	10.9
P9down	9.8	16.3	14.7	12.2	12.0
P10up	8.9	8.6	6.4	8.8	10.2
P10down	9.0	10.7	8.2	8.7	8.4
P12up	7.3	15.2	9.4	10.3	11.5
P12down	9.4	45.9	20.1	11.6	11.8

Table 1 shows the CTE evaluations for eight sensors in a variety of cases. Four locations are shown, each with a top and bottom sensors. The values shown correspond to the slope of the linear best fit line of the temperature and strain plots for each specific case. The fall time period is assumed to provide accurate evaluations at each location on the bridge. The summer time period provides poor evaluations because of the occurrences of creep, shrinkage, and thermal gradients. Breaking the summer up into nine separate weeks of data improves the fits somewhat (Broken S Fit), but the real improvements come with the minimal gradient filters. The column '95 MR 3.5' shows the results from the process outlined in Figure 5 using the 95MR method with a boundary of 3.5 °C, while 'MLG 6' use a boundary of 6 °C/m with the MLG method. To judge how well each method evaluates the CTE, the values for the top and bottom sensors are compared. The paired sensors should be similar, or as similar as in the fall. The actual values of the CTE should be close to the fall values, though not necessarily the same.

The minimum gradient filters show a clear improvement over the simple summer fit, though produce different evaluations. While in some cases the two methods produce nearly the same value, P8up, for example, sees a difference of over 20%. The MLG produces more consistent evaluations between top and bottom sensors, though not as close to the fall values of the 95MR method. These methods produce different results because of their differences in gradient filtering technique.

The values shown in Table 1 are approximations of the CTE of different locations of the Streicker Bridge. While these values may be near the true, material CTE of concrete, these approximations find a more global CTE. These evaluations include the thermal behavior of the reinforcing steel, the pre-stressing steel, and the relative restraint imposed on the sections by the columns or other sections of the bridge. It is unlikely that this process for CTE evaluation will yield the same results as laboratory testing using ASTM procedures, but these evaluations still give valuable insight into the thermal behavior of the bridge.

5. Conclusions

This paper presents a method to the evaluation of the coefficient of thermal expansion of built concrete structures. This method relies on an idealized linear relationship between temperature and strain, formed by two main assumptions on the negligence of creep and shrinkage strains and the absence of thermal gradients in the structure. By sectioning the data and applying a minimal gradient filter, these assumptions are upheld in order to produce accurate evaluations of the CTE. These techniques were validated on data taken from the Streicker Bridge on campus at Princeton University. This process for evaluation of the CTE will aid in the formulation of a temperature driven method for SHM. Future work will examine the CTE at the centroid of the cross section in order to minimize bending effects from the structural system of the bridge.

Acknowledgements

This research has been supported by National Science Foundation Grant CMMI- 1434455. Any opinions, findings, and conclusions or recommendations expressed in this material are those of the authors and do not necessarily reflect the views of the National Science Foundation.

This research was aided by previous work completed by Dorotea Sigurdardottir and Yao Yao, and by theoretical discussion and advice from Xi Li, Hiba Abdel-Jaber, and Kaitlyn Kliewer.

References

Yarnold, M 2013, *Temperature-Based Structural-Identification and Health Monitoring for Long-Span Bridges*, Drexel University, Philadelphia.

Ndon, U & Bergeson, K 1995, 'Thermal Expansion of concretes: case study in Iowa', *Journal of Materials in Civil Engineering,* vol. 7, no. 4, pp. 246-251.

Lim, J, Moon, K, Falchetto, A, & Jeong, J 2015, 'Testing and Modelling of Hygro-thermal Expansion Properties of Concrete', *KSCE Journal of Civil Engineering,* vol. 20, no. 2, pp. 709-717.

Sigurdardottir, D. H & Glisic, B 2015, 'On-site validation of fiber-optic methods for structural health monitoring: Streicker Bridge', *Journal of Civil Structural Health Monitoring,* vol. 5, no. 4, pp. 529-549.

Glisic, B & Inuadi, D 2007, *Fibre Optic Methods for Structural Health Monitoring,* John Wiley & Sons Ltd, Wiltshire.

York, D, Evensen, N, Martinez, M & Delgado, J 2004, 'Unified equations for the slope, intercept, and standard errors of the best straight line', American Journal of Physics, vol. 72, no. 3, pp. 367-275.

Emanuel, J 1978, 'Temperature Distributions in Composite Bridges', Journal of the Structural Division, vol. 104, no. 1, pp. 65-78.

Reilly, J, Abdel-Jaber, H, Yarnold, M & Glisic, B 2016, 'Identification of steady-state uniform temperature distributions to facilitate a temperature driven method of Structural Health Monitoring', *Proceedings of SPIE Smart Structures and Materials + Non-destructive Evaluation and Health Monitoring (SPIE)*, Las Vegas, USA, 1 April.

The 8th International Conference on
Structural Health Monitoring of Intelligent Infrastructure
Brisbane, Australia | 5-8 December 2017

Comparative Performance Assessment of Response-Only Modal Identification Algorithms for a Skewed Concrete Highway Bridge

Abdou K. Ndong[1], Abdollah Bagheri[3], Osman E. Ozbulut[2], Devin K. Harris[2]

[1] Department of Civil and Environmental Engineering, University of Virginia, USA, Email: akn7se@virginia.edu, *[2] Department of Civil and Environmental Engineering, University of Virginia, USA, [3] Department of Mechanical Engineering, University of Maryland-College Park, MD, USA*

Abstract

Modal identification methods that utilize only measured response are attractive options for large-scale civil engineering applications. This paper compares performance of four operational modal analysis techniques in identifying the modal parameters of a skewed reinforced concrete slab bridge derived from ambient excitations. The approaches examined include frequency domain decomposition (FDD) method, enhanced frequency domain decomposition (EFDD) method, stochastic subspace identification (SSI) method, and recently proposed variational mode decomposition (VMD) method. During the field experiment, the vibration response of the selected bridge are recorded at sixteen measurement points using wireless accelerometers. Both the ambient excitation and impact excitation tests are conducted. The collected data is analyzed by four output-only modal analysis techniques to extract natural frequencies, damping ratios and mode shapes and the results are compared. The differences in modal parameter estimation by using different excitations are explored. Results from the examination highlighted the strength of the VMD method, which decomposes an input signal into a number of distinct sub-signals (modes) where each mode has a center frequency and collectively reproduce the input signal. The method is entirely non-recursive and the modes are extracted concurrently.

1. Introduction

As the core of dynamic analysis and finite element model updating, modal parameter identification plays an important role in structural health monitoring. For civil engineering structures that are usually large and complex, operational modal analysis (OMA) is an important technique used to estimate the modal parameters. OMA techniques extract modal characteristics such as natural frequencies, mode shapes and damping ratios without requiring the measurement of input forces. The modal parameter estimation in OMA is performed either in time domain using the correlation functions among output responses or in frequency domain using the output power spectral density functions. A variety of dynamic excitation source including vehicles, wind, impact and shaker loading has been considered for output-only modal identification of bridge structures. However, there is a lack of consensus on the most appropriate excitation method for bridge testing (Alwash et al. 2009).

This paper discusses modal parameter estimation of a skewed reinforced concrete bridge using different output-only system identification techniques and different excitation types. First, a novel method based on variational mode decomposition (VMD) for modal parameter estimation as well as three other well-known OMA techniques are briefly described. The techniques considered in this study include the frequency domain decomposition (FDD) method, enhanced frequency domain decomposition (EFDD), and data-driven stochastic subspace identification (SSI) method. Frequency Domain Decomposition (FDD) Technique, Enhanced Frequency Domain Decomposition (EFDD) Technique and Stochastic Subspace Identification (SSI) are the most well-known and used methods among the OMA techniques. FDD uses Complex Mode Indicator Function approach for processing the measured responses and then determine the natural frequencies and mode shapes of the structure with the Peak Picking Technique which is relatively easy to implement and use as it requires fewer complex calculations. However, the SSI method is a time domain method and it is more complex and time consuming compared to FDD and EFDD. Then, the type and characteristics of an in-service bridge selected for field testing is described. The instrumentation of the bridge and vibration tests that include two different excitation types are described. Results from the tests are presented and discussed.

2. Output-only Modal Parameter Estimation Methods

Many output-only system identification techniques for the modal parameter estimation have been developed and examined. In this section, a new modal estimation method based on VMD is first described. Then, three commonly used operational modal analysis techniques, two frequency domain methods and one-time domain method, are briefly discussed.

2.1. Variational Mode Decomposition (VMD)

VMD method is a novel methodology for adaptive and quasi-orthogonal signal decomposition (Dragomiretskiy and Zosso 2014), which can decompose the signal into various modes or intrinsic mode functions (IMFs) using calculus of variation. In order to obtain a set of IMFs, the recursive sifting method is used in empirical mode decomposition (EMD), which does not allow for backward error correction. On the contrast, VMD solves the variation problem efficiency using a classical optimization methodology called alternate direction method of multipliers (Boyd et al. 2011). VMD can be essentially interpreted as signal decomposition with a wavelet-packet transform filter bank structure, and VMD is an adaptive decomposition method with more refinement time-frequency divisions than EMD (Wang et al. 2015). Therefore, researching the signal characteristic extraction based on VMD has important and practical value in the modal identification field.

The VMD decomposes an acceleration signal $A(t)$ into a set of sub-signals (modes), $A_k(t)$, $k=1, 2, ..., K,$ which have compact bandwidths in spectral domain. Each sub-signal is assumed to be compacted around a corresponding center frequency ω_k. Therefore, each sub-signal denotes response of structure to a particular mode and the center frequencies ω_k represent the natural frequencies of structure. The damping ratio for each mode can then be determined using the corresponding modal response and logarithmic decrement method. Using the modal response of structure at different sensor locations, the mode shapes can also be identified. In particular, the values of the modal response at a local maximum or minimum for each sensor location are used to form the mode shape vector.

2.2. Frequency Domain Decomposition (FDD)

FDD technique is based on decomposing the power spectral density function matrix using the Singular Value Decomposition. It has been shown that this decomposes the spectral response into a set of single degree of freedom systems (SDOF), each corresponding to one individual mode. However, the accuracy of modal frequency estimates is limited by the frequency resolution of the spectral density estimate and damping cannot be estimated. A detailed description of the method can be found in (Brincker et al. 2000).

2.3. Enhanced Frequency Domain Decomposition (EFDD)

EFDD method (Brincker et al. 2001), which usually provides improved estimates of modal parameters, transforms the SDOF power density function back into the time domain. The natural frequencies are then obtained by calculating number of zero-crossings as a function of time and the damping ratio is estimated from the logarithmic envelope of the corresponding SDOF correlation function using the logarithmic decrement method.

2.4. Stochastic Subspace Identification (SSI)

SSI techniques are the most well-known and used time domain methods and perform the modal parameter estimation over raw measured time series. SSI takes the advantage of state space modeling, which reduces the second order problem into two first order problems. In general, SSI looks for a set of parameters that will minimize the deviation between the predicted system response of the model and measured system response. SSI techniques can be classified into two groups: (i) covariance-driven SSI and (ii) data-driven SSI techniques (Van Overschee and De Moor 1996). Data-driven SSI does not require pre-processing of the data compared to covariance-driven SSI that is a two-stage method. In this study, data-driven SSI using principal component algorithm is adopted to process the acquired data.

3. Bridge Description and Test Instrumentation

The selected bridge for the field testing is a concrete slab bridge, named as War Branch and built in 1976. The two-span simply-supported bridge is located in Rockingham County, Virginia, USA. Each span of the bridge has a total length of 9.8 m. The thickness of the concrete slab is 0.53 m. The deck slab has voids with a diameter of 0.3 m and spacing of 0.45 m in the direction of traffic. The bridge shown is skewed at an angle of 45 degrees to the main road. The forced and ambient vibration testing of one of the spans are conducted in this study.

The dynamic bridge assessment procedure involves the attachment of nine accelerometers underneath of the span of the bridge at 16 measurement points in two set-ups as shown in Figure 1. Note that two common sensors are used as reference in each set-up. The uniaxial accelerometers with a measuring range of ±5g are used. The response of the bridge to ambient excitations or an impact excitation generated by a modal impact hammer is measured. Ambient vibrations are generated by the passing traffic, wind and walking people and recorded for a total of 15 minutes. For impact excitation, a large sledge impulse hammer with a force capacity of 22.2 kN is used. Three impact locations are chosen and excited by the impact hammer for five repetitions and for a duration of 15 seconds. All data is collected with a sampling frequency of 500 Hz.

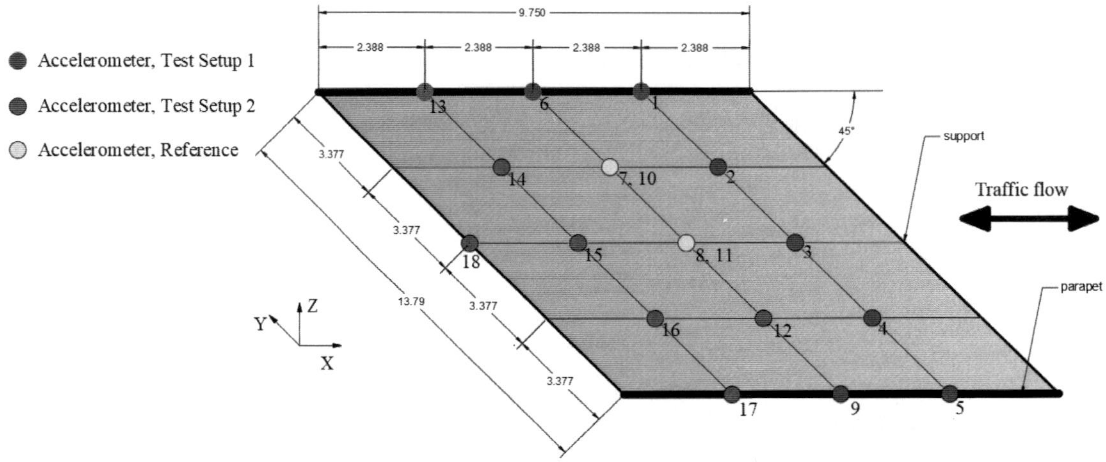

Figure 1. Sensor instrumentation plan for one of the bridge span with two setups (all dimensions are in meters)

4. Results

The modal parameters of the reinforced concrete slab bridge described above are estimated using both the ambient vibration data and impact excitation data using four output-only system identification methods (VMD, FDD, EFDD and SSI). Typical measured acceleration time histories recorded by sensor 8 during ambient excitation and impact hammer excitation tests are shown in Figure 2. The results for the VMD method are obtained using MATLAB and the results for other system identification techniques are obtained from the ARTeMIS Modal Pro software.

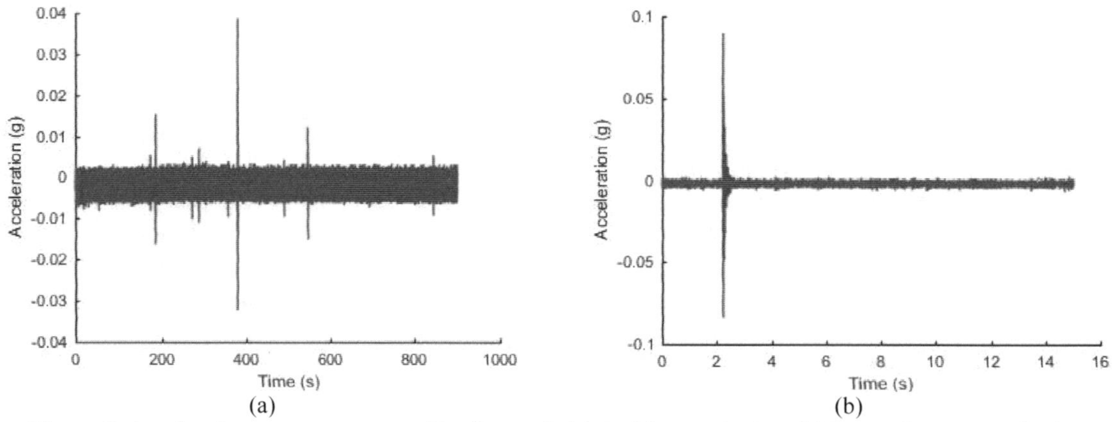

Figure 2. Acceleration response measured by Sensor 8: (a) Ambient excitation, (b) Impact hammer excitation

4.1. VMD Method

To obtain the modal parameters using the VMD methodology, the measured vibration signal is first decomposed and then the bridge parameters are extracted from the envelop spectrum of each mode component. Figure 3 illustrates a typical decomposed acceleration response for the first three modes for the signal shown earlier in Figure 2. Each decomposed signal represents a modal response with a certain frequency. It can be seen that the time-domain responses of higher modes have higher vibration frequencies. In addition, the response for each mode decays with time due to the damping. The phase angle plots of the modal responses, which becomes a linear line for a signal with single frequency content, are evaluated. The slope of each phase angle plot gives the corresponding natural frequency. To estimate the damping ratio, a linear function is fit to the decaying amplitude of the modal responses using a linear least-square regression. The slope of the fitted linear function is used to compute the damping ratio. This procedure is repeated for the data obtained from each sensor of two set-ups. Table 1 and 2 provide the first three natural frequencies and corresponding damping ratios of the tested bridge for ambient vibration testing and impact hammer test, respectively. Note that for the impact hammer test, the modal properties cannot be extracted from the data obtained from all sensors. The averaged values of first three modal frequencies obtained from the VMD algorithm using ambient vibration data are 25.61, 31.63, and 39.14 Hz, respectively, while the same frequencies are determined as 26.07, 30.59, and 38.71 Hz using the impact hammer data.

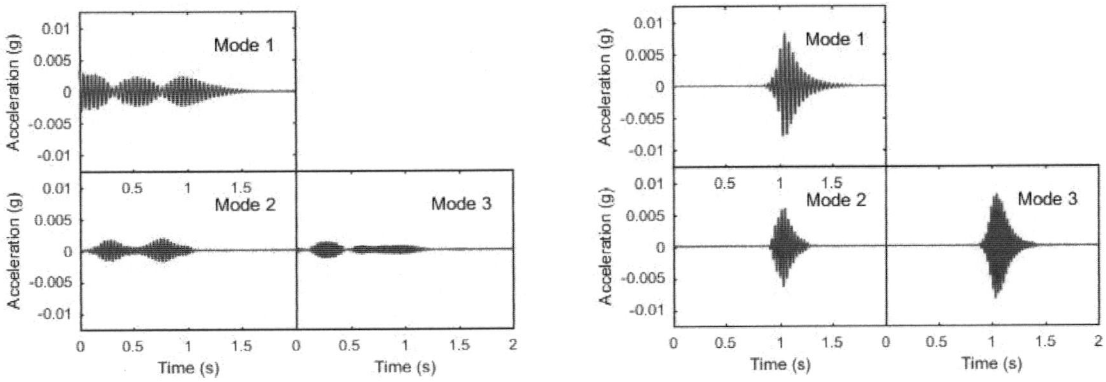

(a) (b)

Figure 3. Decomposed signals for: (a) Ambient excitation; (b) Impact excitation

Table 1. Natural frequencies and damping ratios obtained from ambient excitations using VMD

		Modal Frequency (Hz)			**Damping Ratio**		
	Sensor	**1**	**2**	**3**	**1**	**2**	**3**
First set-up	1	25.52	29.49	41.39	0.045	0.017	0.022
	2	25.67	32.59	41.32	0.029	0.017	0.021
	3	26.17	31.57	39.14	0.012	0.034	0.032
	4	24.90	28.19	34.51	0.028	0.034	0.052
	5	25.65	29.61	41.36	0.038	0.006	0.020
	6	25.79	28.55	37.73	0.009	0.060	0.019
	7	25.13	31.18	41.25	0.023	0.023	0.022
	8	25.68	28.92	37.36	0.015	0.033	0.047
	9	24.57	33.18	39.42	0.019	0.032	0.020
Second set-up	10	25.72	32.42	38.78	0.028	0.033	0.004
	11	26.06	35.80	41.59	0.064	0.007	0.004
	12	26.50	32.20	34.98	0.013	0.000	0.003
	13	25.48	33.01	38.01	0.028	0.017	0.031
	14	25.34	32.13	40.05	0.013	0.023	0.013
	15	25.64	33.44	41.63	0.047	0.005	0.009
	16	25.32	31.24	38.95	0.014	0.016	0.007
	17	26.66	33.12	35.60	0.014	0.004	0.030
	18	25.24	32.70	41.43	0.004	0.020	0.029
	Ave.	**25.61**	**31.63**	**39.14**	**0.025**	**0.021**	**0.021**

Table 2. Natural frequencies and damping ratios obtained from impact excitations using VMD

		Modal Frequency (Hz)			Damping ratio		
	Sensor	1	2	3	1	2	3
First set-up	1	25.45	32.66	40.29	0.041	0.052	0.037
	2	25.75	31.74	38.76	0.039	0.052	0.046
	3	26.57	-	40.29	0.039	-	0.040
	4	26.55	30.49	39.37	0.044	0.044	0.033
	5	23.17	26.63	33.87	0.060	0.034	0.052
	6	26.19	-	39.18	0.041	-	0.045
	7	29.53	-	40.40	0.042	-	0.043
Second set-up	8	23.73	27.01	33.86	0.047	0.031	0.052
	9	26.35	-	39.59	0.042	-	0.021
	10	27.32	31.30	39.65	0.032	0.036	0.024
	11	25.68	31.99	36.41	0.043	0.038	0.061
	12	26.30	33.57	38.43	0.038	0.048	0.028
	13	25.20	29.94	40.21	0.034	0.019	0.038
	14	27.12	-	41.69	0.062		0.058
	Ave.	**26.07**	**30.59**	**38.71**	**0.043**	**0.039**	**0.041**

The mode shape vectors for the first three modes are derived from the corresponding modal responses. The mode shapes along the longitudinal and transverse axis of the bridge are illustrated in Figures 4 and 5, respectively.

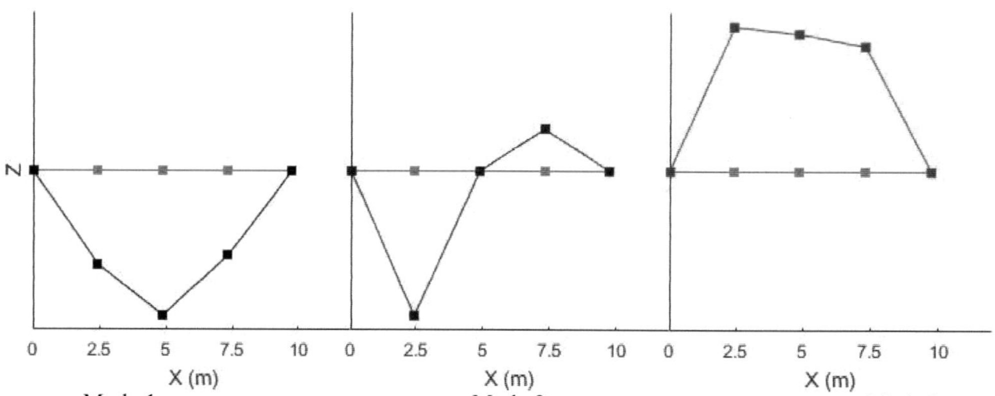

Mode 1 Mode 2 Mode 3

Figure 4. Mode shapes along the middle longitudinal axis of bridge and obtained using VMD

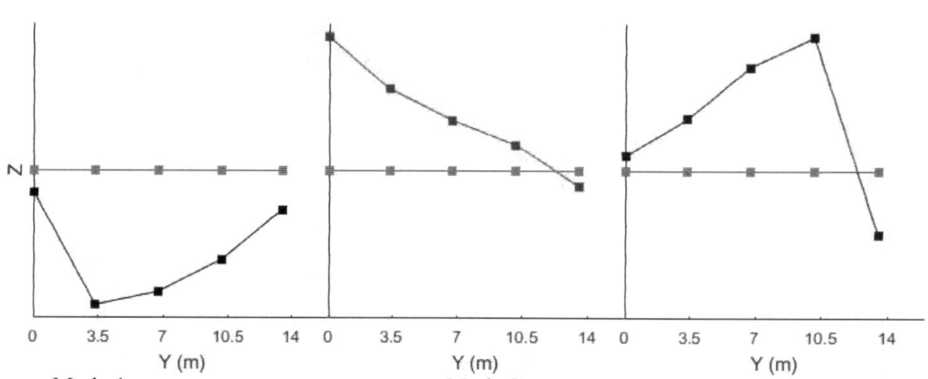

Mode 1 Mode 2 Mode 3

Figure 5. Mode shapes along the middle transverse axis of bridge and obtained using VMD

4.2. FDD, EFDD, and SSI Methods

To provide a comparison for the modal parameters obtained from the VMD method, the vibration data is also analyzed using three other commonly used OMA techniques. Data obtained from both ambient excitations and impact are analyzed using each method. The singular values of the spectral density function matrix for the ambient vibration data to estimate the modal parameters using FDD are shown in Figure 6(a), while the stabilization diagram illustrating the stable and unstable poles for the SSI is shown in Figure 6(b). The EFDD method is also employed to obtain both natural frequencies and damping ratios for the first three modes. Table 3 and 4 summarize the findings of all methods during ambient and impact tests for the modal parameters of the studied bridge.

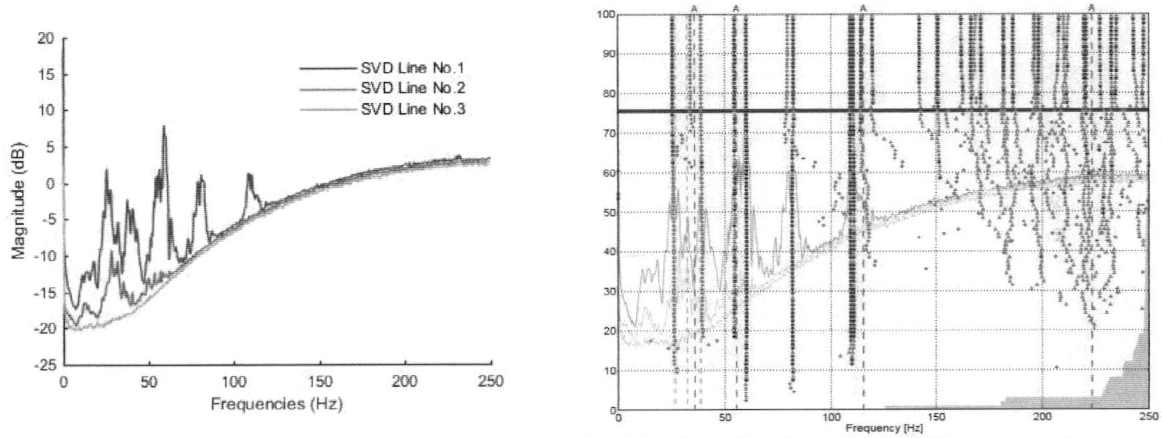

Figure 6. (a) Singular values for FDD method, (b) Stabilization diagram for SSI method

The modal shapes along the longitudinal and transverse axis of the bridge are also extracted from the ambient vibration data using each method and illustrated in Figures 7 to 12.

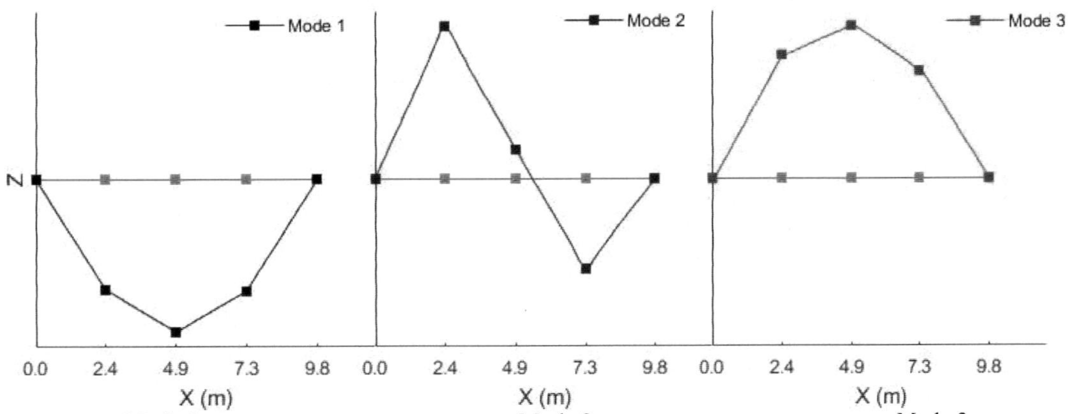

Mode 1 Mode 2 Mode 3

Figure 7. Mode shapes along the middle longitudinal axis of bridge and obtained using FDD

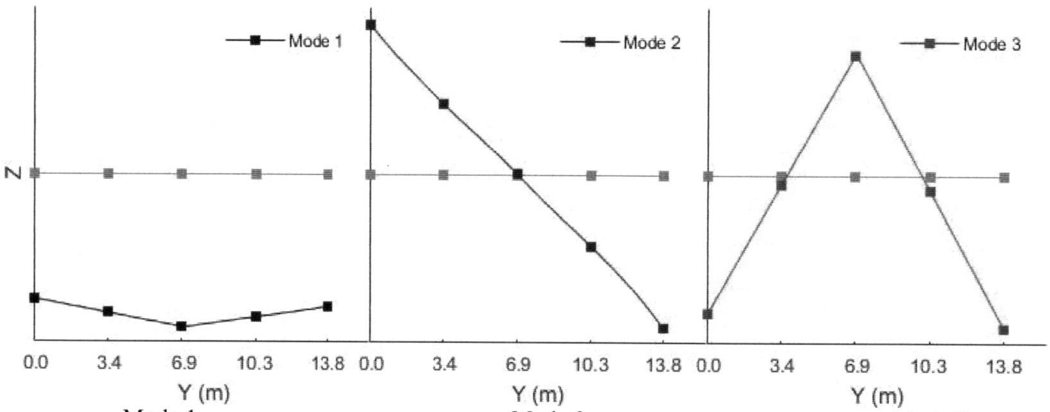

Mode 1 Mode 2 Mode 3

Figure 8. Mode shapes along the middle transverse axis of bridge and obtained using FDD

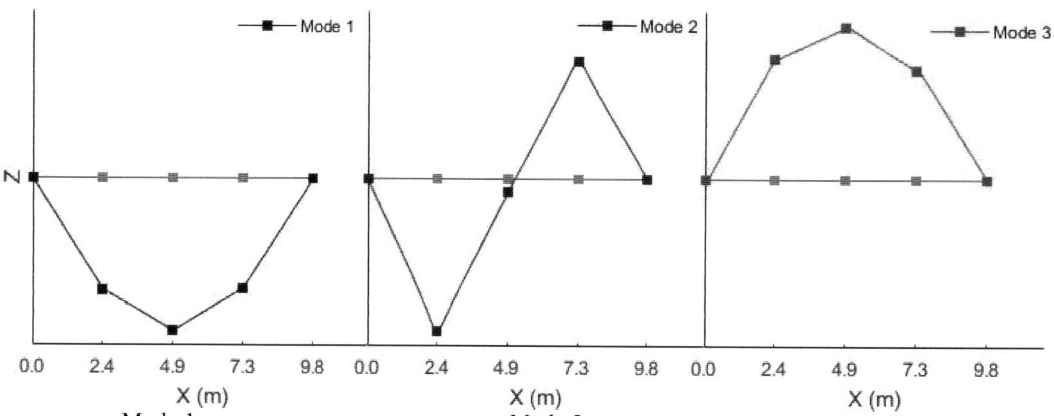

Mode 1 Mode 2 Mode 3

Figure 9. Mode shapes along the middle longitudinal axis of bridge and obtained using EFDD

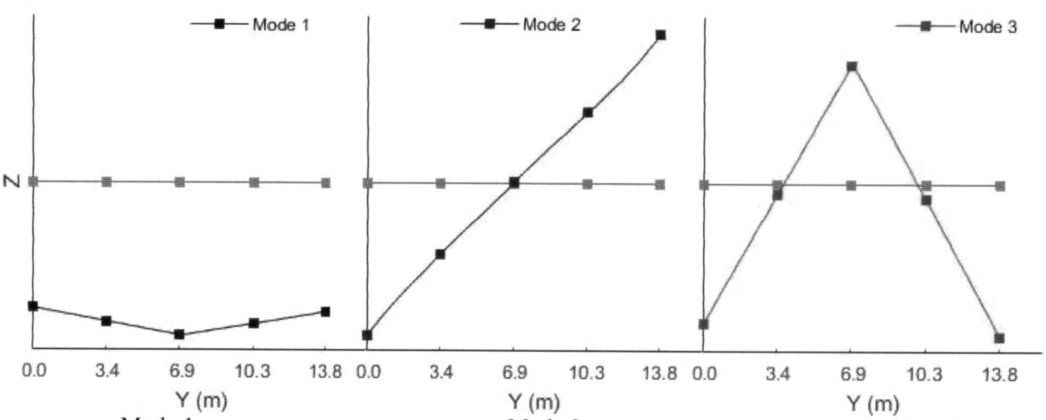

Mode 1 Mode 2 Mode 3

Figure 10. Mode shapes along the middle transverse axis of bridge and obtained using EFDD

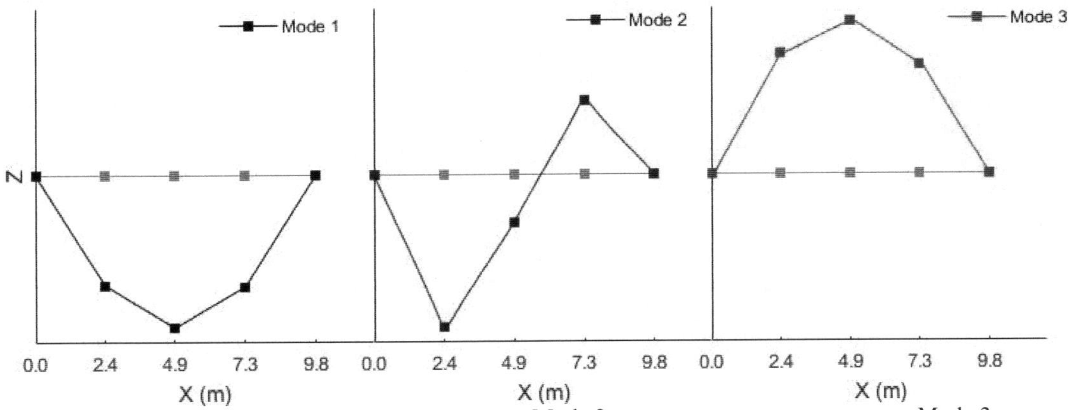

Mode 1 Mode 2 Mode 3

Figure 11. Mode shapes along the middle longitudinal axis of bridge and obtained using SSI

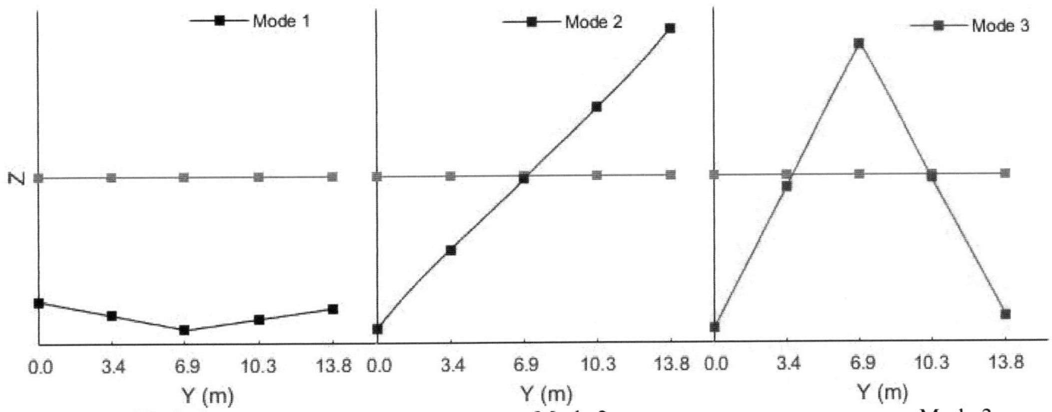

Mode 1 Mode 2 Mode 3

Figure 12. Mode shapes along the middle transverse axis of bridge and obtained using SSI

5. Discussions

Because two setups over the bridge are implemented in ambient and impact hammer testing, the obtained modal parameters provide a global estimation. The natural frequencies of the bridge are in the range of 25.37 to 39.59 Hz. As can be seen from Tables 3 and 4, the frequency differences among different identification techniques are generally small. In particular, reliable natural frequency values could be extracted from either the frequency domain methods or the time domain methods and small differences amongst the different modal parameter identification technique exists. It can be also seen that the new VMD method is able to yield mutually consistent natural frequency estimations from either ambient and impact excitations.

When the results of ambient vibration and impact excitation tests are compared, it is seen that all methods provide a similar estimate for natural frequencies in both cases. In particular, the deviations in the first natural frequency between the results obtained using different excitations are 1.80%, 3.82%, 1.92% and 3.71%. However, it is clear that the differences in damping estimation between the tests with two excitation types (ambient and impact) are significant. Small vibration amplitudes observed during ambient excitation caused larger uncertainties and scatter in damping estimations for all methods. In addition, the SSI method does not provide a realistic damping estimate for the second mode for the ambient vibration test. It is also noted that data-driven SSI consumes considerably more computational resources and needs relatively long calculation time to obtain results compared to other methods studied here.

Compared to frequency domain output-only system identification methods, which assume infinite modes of vibration, the VMD method identifies modal frequencies and mode shapes more accurately as it decomposes an acceleration signal into a given number of vibration modes which is physically meaningful.

Table 3. Modal frequencies and damping ratios obtained from ambient excitations using different OMA methods

	Modal Frequency (Hz)				Damping ratio %			
Mode	FDD	EFDD	SSI	VMD	FDD	EFDD	SSI	VMD
1	25.63	26.61	25.37	25.61	-	4.48	4.68	2.46
2	31.98	32.22	33.65	31.63	-	1.64	25.67	2.12
3	37.84	38.41	38.68	39.14	-	2.69	5.58	2.14

Table 4. Modal frequencies and damping ratios obtained from impact excitations using different OMA methods

	Modal Frequency (Hz)				Damping ratio %			
Mode	FDD	EFDD	SSI	VMD	FDD	EFDD	SSI	VMD
1	26.61	26.10	26.31	26.07	-	4.17	5.97	4.31
2	32.72	32.33	31.97	30.59	-	4.16	6.78	3.93
3	39.06	39.59	39.40	38.72	-	2.71	4.85	4.11

6. Conclusions

This paper presents the results of a comparative study of various output-only identification techniques for evaluating bridge modal properties from measured vibration data. A new OMA technique based on variational mode decomposition is discussed first. Then, vibration tests on a skewed reinforced concrete bridge are described. Both impact excitation tests and ambient vibration tests are conducted. The results obtained from VMD are compared with well-known other OMA techniques ranging from basic frequency domain identification techniques to the advanced time domain statistics subspace identification technique. The first three natural frequencies of the bridge are determined using different methods. Natural frequencies of the bridge estimated by four techniques are found to be in good agreement while the damping ratios and mode shapes have more scatters. It is also noted that the natural frequencies extracted from ambient vibration response are similar to the corresponding values determined by impact excitation response.

References

Alwash, M., Sparling, B. F., & Wegner, L. D. (2009). Influence of excitation on dynamic system identification for a multi-span reinforced concrete bridge. Advances in Civil Engineering, 2009.

Boyd S, Parikh N, Chu E, et al, *Distributed optimization and statistical learning via the alternating direction method of multipliers*, Foundations and Trends in Machine Learning, 2011, 3(1): 1-122.

Brincker, Rune, Lingmi Zhang, and P. Andersen. "Modal identification from ambient responses using frequency domain decomposition." *Proc. of the 18*'International Modal Analysis Conference (IMAC), San Antonio, Texas*. 2000.

Brincker, Rune, C. Ventura, and Palle Andersen. "Damping estimation by frequency domain decomposition." *19th International Modal Analysis Conference*. 2001.

Dragomiretskiy K, Zosso D, *Variational mode decomposition, IEEE Transactions on Signal Processing*, 2014, 62(3): 531-544.

Van Overschee, P., and De Moor, B. 1996, Subspace Identification for Linear Systems: Theory-Implementation-Applications, Kluwer Academic Publishers, Dordrecht, The Netherlands.

Wang Y, Markert R, Xiang J, et al, *Research on variational mode decomposition and its application in detecting rub-impact fault of the rotor system*, Mechanical Systems and Signal Processing, 2015, 60: 243-251.

The 8th International Conference on
Structural Health Monitoring of Intelligent Infrastructure
Brisbane, Australia | 5-8 December 2017

Efficient and Large Scale Monitoring of Retaining Walls along Highways using a Mobile Mapping System

W. Lienhart [1], S.Kalenjuk [1], C. Ehrhart [1].
[1] *Institute of Engineering Geodesy and Measurement Systems, Graz University of Technology, Austria*
Email: werner.lienhart@tugraz.at

Abstract

The safety of highways in mountainous areas is strongly dependent on the stability of the retaining walls, which support either the highway itself or the slopes adjacent to it. Failure of such a retaining wall can lead to death of highway users, to repair works and thus to the closure of highway lanes.
An early detection of negative changes of these retaining walls is therefore of high importance. Due to the large number of retaining walls in alpine areas, it is impossible to equip every retaining wall with a monitoring system. The usual approach is to monitor only the walls, which are known to have problems. This leaves most of the walls uninspected. Thus, an efficient method is needed which is capable to monitor all retaining walls along a highway.
In this paper, we present a new approach based on a mobile mapping system (MMS). We use a measurement platform consisting of two laser scanners, an inertial measurement unit (IMU), a differential GNSS sensor and multiple cameras. This platform is mounted on a standard commercial car. Whilst the car travels with up to 100 km/h along the highway, the data of the multi sensor system is continuously recorded with high frequency. For instance, each laser scanner records 1 million points per second. As a result, georeferenced high resolution point clouds of all retaining walls along the highway are obtained. We further analyse the point clouds to derive safety critical parameters like tilt changes of the retaining walls. We demonstrate that it is possible to determine the tilt of retaining walls with an accuracy of better than 0.1°. Furthermore we can reliable detect the movement of individual blocks of the retaining wall and identify areas where material fell off.

1. Introduction

Retaining walls are frequently used to construct highways in mountainous regions to support either the highway itself or the slopes adjacent to it. Failure of these retaining walls can lead to the death of highway users and the repair works may cause delays due to closures of highway lanes. For instance, in 2012 a truck driver was killed on the Austrian Highway Brennerautobahn (A13) because of a collapse of a retaining wall (Die Presse, 2012).
An early detection of negative changes of these retaining walls is therefore of high importance. With respect to geotechnical structures, the Eurocode 7 (EN 1997-1, 2004) distinguishes between serviceability limit state (SLS) and the ultimate limit state (ULS). The SLS defines the limit of the functionality of the structure whereas the safety of the users and the stability of the structure are in danger if the ULS is exceeded. Tilt changes of the structure are an important parameter to define the current state of the structure. The maximum allowed tilt changes for retaining walls are between 0.057° and 0.382°for the SLS and 0.764° for the ULS. These tilt changes are usually monitored

with manual or automatic total stations (TS) or tilt sensors (T) as depicted in Fig. 1. Total stations measure horizontal (Hz) and vertical angels (V) as well as distances (D) to reflective targets (retroreflective foils or glass prisms). These measurements can be used to calculate 3D positions of the targets on the retaining wall. Usually, several targets are placed at different heights in one cross section. Therefore, the tilt of the retaining wall (α) can be derived. On the other hand, the tilt of the wall can directly be measured with a tilt sensor.
However, it has to be noted that in both approaches the behaviour of the structure is only monitored at a few single points. Damages (d) which occur between the measurement positions remain undetected, see Fig. 1.

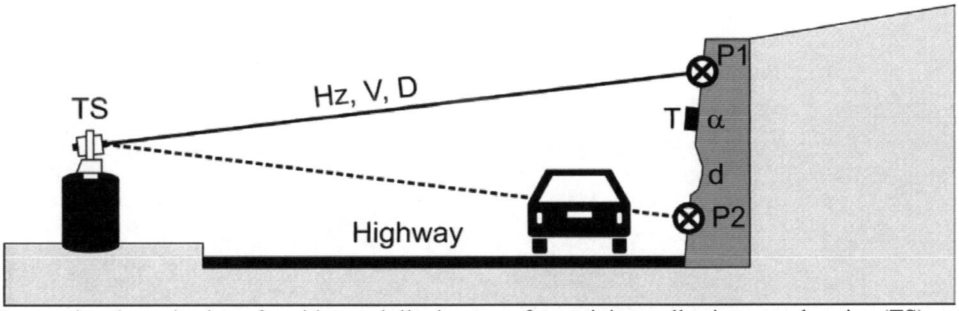

Fig. 1 Conventional monitoring of position and tilt changes of a retaining wall using a total station (TS), reflective targets (P1 and P2) and a tilt sensor (T). Undetectable damage (d)

Furthermore, the conventional approaches are costly and require access to the retaining walls. Because of this and due to the large number of retaining walls in alpine areas, it is impossible to equip every retaining wall with a monitoring system. The usual approach is to monitor only the walls, which are known to have problems. This leaves most of the walls uninspected. Thus, an efficient method is needed which is capable to monitor all retaining walls along a highway.

2. Proposed Approach using a Mobile Mapping System

In order to overcome the aforementioned problems, we propose a new approach based on a mobile mapping system (MMS) which delivers high resolution point clouds of the retaining walls. The whole system should not require physical access to the retaining walls and shall not influence the traffic flow. For a working system the position and orientation of the vehicle has to be known continuously and the measurements to the wall have to be performed with high frequency and high precision.

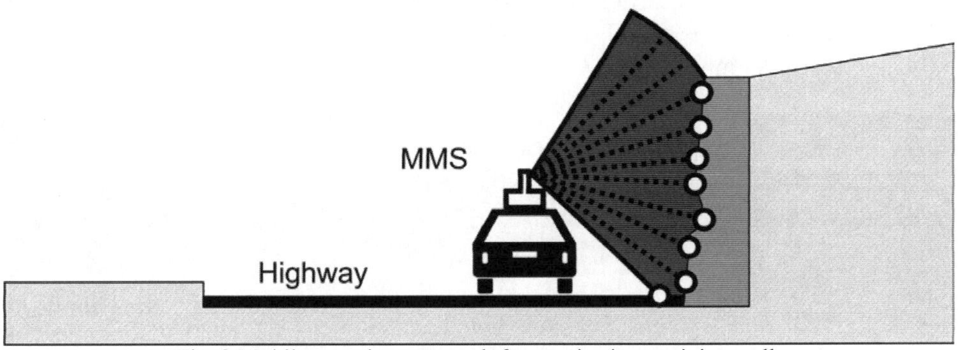

Fig. 2 Mobile mapping approach for monitoring retaining walls

2.1 Hardware

Based on the requirements, our used MMS consists of a geodetic GNSS antenna and a receiver, an inertial measurement unit (IMU), an odometer, several cameras and two laser scanners, see Fig. 3. All sensors are calibrated with respect to each other. The whole platform is mounted on a standard commercial car. The position and orientation of the vehicle are determined using the GNSS in combination with the IMU and the odometer. The distances and angles from the platform to the wall are recorded with the two profile laser scanners. To acquire high density point clouds also at high speeds, each laser scanner measures 1 million points per second. The cameras are used to provide true colours for the point clouds and to validate the results of the automated defect classification.

Fig. 3 Mobile mapping system for monitoring retaining walls

3. Field Study

To evaluate the proposed approach and our developed data processing algorithms, several different retaining walls were measured with the MMS along the Austrian highways. The retaining walls were passed several times with different velocities (60, 80 and 100km/h) in order to verify the repeatability of the results. In this article we discuss the results of two gravity walls (Fig. 4 left and centre) and three anchored walls (Fig. 4 right and Fig. 5). The maximum height of the gravity walls is 8m and the maximum length 255m.

Fig. 4 Gravity walls (left and centre) and one of the anchored walls (right) used in the field study

The anchored walls are higher and also longer (maximum height >23 m and length >300m). The wall A6 is also equipped with a conventional monitoring system. As can be seen in Fig. 5, glass prisms (P) are placed in two different heights in different profiles.

Fig. 5 Anchored walls used in the field study, Left: Retaining wall A6 with total station targets at heights 1 and 2 in profiles a and b. Right: Dissolved anchor wall A9

3.1 Data Acquisition and Data Preparation

Whilst the car travelled along the highway, the data of the MMS was continuously recorded with high frequency. Based on the raw data of the individual sensors we developed following methodology to determine tilt changes and to detect deficiencies.

In a first processing step, high accurate GNSS positions are determined using the recorded GNSS data together with reference station data from the nation-wide permanent reference GNSS network APOS (Austrian Positioning Service). Together, with the odometer and IMU data, the position and orientation of the platform and thus the trajectory are known at any time

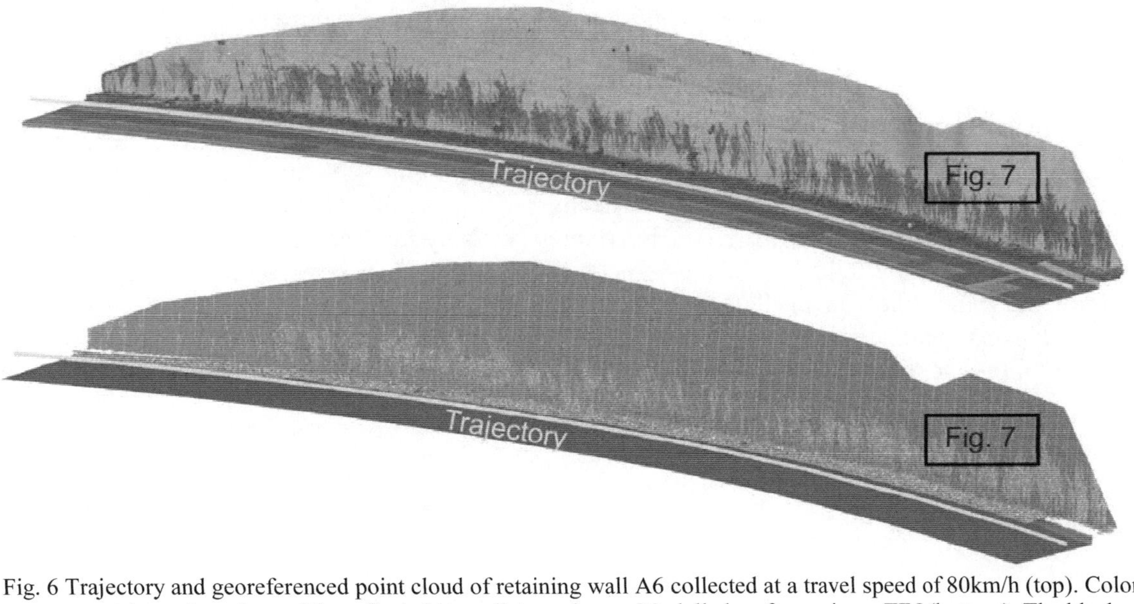

Fig. 6 Trajectory and georeferenced point cloud of retaining wall A6 collected at a travel speed of 80km/h (top). Colors represent intensity values of the reflected laser distance beam. Modelled surface using a TIN (bottom). The black rectangle indicates the area which is shown in more detail in Fig. 7

Fig. 7 Detail of point cloud (top) and modelled surface (bottom) of retaining wall A6

This information is now combined with the data of the two profile laser scanners which results in a high resolution georeferenced point cloud of the whole retaining wall Fig. 6 top and Fig. 7 top. Next, the surface of the retaining wall is reconstructed e.g. as a TIN, Fig. 6 bottom and Fig. 7 bottom. We generate vertical profiles of the retaining walls by intersecting the surface model with planes orthogonal to the trajectory of the vehicle. Fig. 8 left demonstrates the principle. Displayed are the intersecting lines for profiles with 5m separation. For our detailed analysis a profile was generated every 5cm along the trajectory (Fig. 8 right). Thus, more than 6000 profiles were obtained for the more than 300m long retaining wall A6.

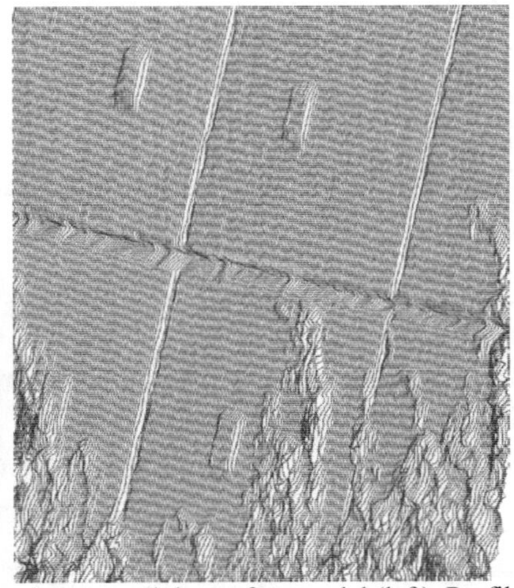

Fig. 8 Intersection lines between vertical planes every 5m along the trajectory and the surface model (left). Profiles with 5cm spacing for detailed analysis (right)

3.2 Detailed Analysis and Interpretation

Robust adjustment methods are finally used to extract the retaining wall from the vertical profiles and to subsequently determine the tilt of the retaining walls without outliers. It has to be noted that the threshold values for outlier determination have to be adapted to the roughness of the retaining walls. If the threshold values are too small, too many valid measurements will be excluded whereas if the threshold values are too large, other elements like vegetation will be used for the tilt calculation and incorrect tilt values will be reported.

In order to verify the reliability of the MMS based tilt determination, the same retaining wall A6 was measured several times. Two measurement runs were made with 60km/h, two with 80km/h and one with 100km/h. For each measurement run, the tilt of the wall is computed at every vertical profile. Fig. 9 displays the distribution of the tilt differences $\Delta\alpha = \alpha_{t2} - \alpha_{t1}$ of two measurement runs made at t1 and t2. Individual outliers were already excluded in the analysis.

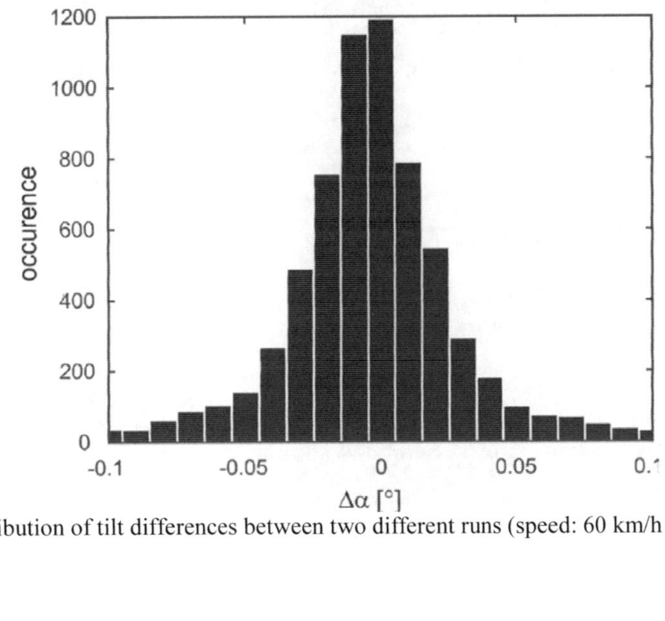

Fig. 9 Distribution of tilt differences between two different runs (speed: 60 km/h and 80 km/h)

Furthermore, the empirical standard deviation $s_{\Delta\alpha}$ of the tilt differences can be calculated to quantify the achievable precision of our proposed approach. It has to be noted that the precision is also dependent on the roughness of the structure. An overview of the results of all five walls is given in Table 1.
The question arises if these precisions are sufficient to detect tilt changes relevant for the SLS and ULS. In deformation analysis, a measured difference is only than stated as significant if the calculated test statistic is higher than the value of the corresponding distribution at a given significance level, usually 5%. Table 1 quotes the minimum size of a tilt change to be detected reliably. In case of the first four retaining walls, it can be seen that tilt changes of less than 0.1° can be detected reliably with the developed approach. Problems only occur with the dissolved anchored wall. For this special construction tilt changes may also not be the relevant parameter to be observed.

Table 1 Standard deviation of tilt change measurements and detectable tilt change at a significance level of 5%

Retaining wall	Type	$s_{\Delta\alpha}$ [°]	Minimal detectable tilt change [°]
A3	Gravity wall with smooth surface	±0.007	0.013
S1	Gravity wall with rough surface	±0.047	0.093
A4	Anchored wall with smooth surface but heavy vegetation	±0.007	0.014
A6	Anchored wall with smooth surface	±0.026	0.050
A9	Dissolved anchored wall	±0.074	0.145

Finally we take a closer look at the outliers which are automatically flagged prior to the tilt calculation. Outliers are caused for instance by vegetation. On the other hand, outliers i.e. deviations from the general shape of the profile, may also give valuable information about structural deficiencies. In first test, we applied artificial deficiencies by placing obstacles on the surface of the retaining wall. After the first successful test, the analysis was extended to all measured retaining walls. An interesting result is displayed in Figure 10.

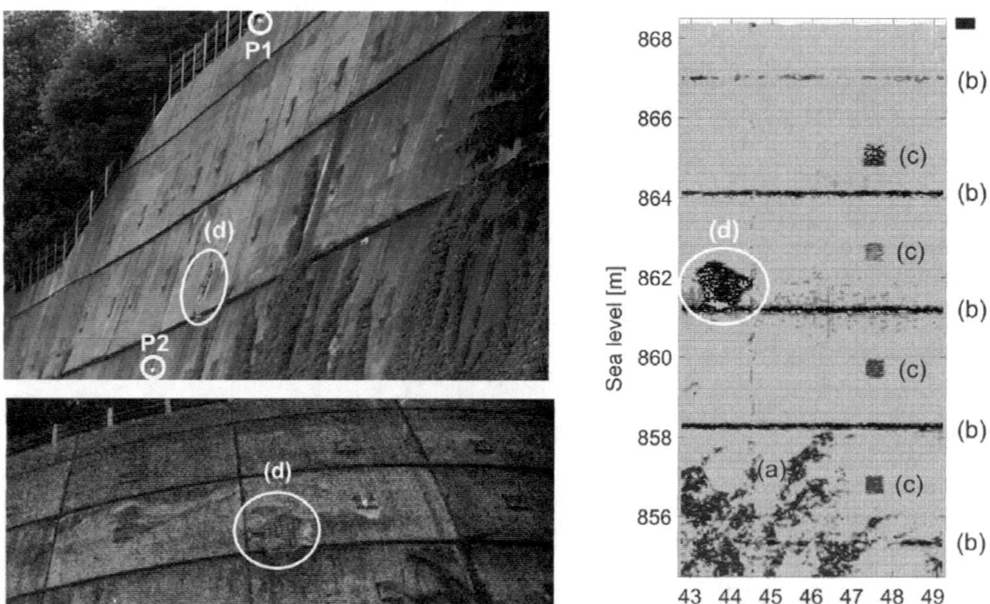

Fig. 10: Position of prism targets P1 and P2 and damage (d) in retaining wall (left), color coded significant deviations from fitted lines with categories (a) vegetation, (b) horizontal joints, (c) anchor heads and (d) damage

Different types of outliers are highlighted in Fig. 10 right. We automatically group the outliers by using the DBSCAN algorithm (Ester et al., 1996) and classify them thereafter. Group (a) belongs to vegetation, group (b) are the horizontal joints and group (c) are the anchor heads. Interesting is group (d). This is a deficiency where material has fallen off and the rebar of the reinforced concrete is visible as can be seen in Fig. 10 left. Interestingly, this retaining wall is also monitored with conventional methods and two target positions are located in the vicinity of the damaged area. However, as stated at the beginning of this paper, conventional methods cannot detect local damages and do not deliver a complete picture of the retaining wall.

4. Conclusions and Outlook

The proposed approach of collecting high resolution georeferenced point clouds proved to be suitable to derive critical deformations like tilt changes and deficiencies like pop outs fully automatically of all retaining walls along the motorway. A detailed analysis confirmed that tilt changes of better than 0.1° can be determined if outliers are automatically removed in the analysis. The data collection can be performed at high travel speeds and thus does not adversely affect the traffic flow. We are repeating the measurement campaign this year and are planning to assess the absolute accuracy by comparing position and tilt changes derived from total station measurements with the results from the mobile mapping system.

5. Acknowledgements

We want to thank the Austrian Research Fund (FFG), the Autobahnen- und Schnellstraßen-Finanzierungs-Aktiengesellschaft (ASFINAG), the Austrian Federal Railways (ÖBB) and the Vereinigung Österr. Bohr-, Brunnenbau- und Spezialtiefbauunternehmungen (VÖBU) for supporting this project

References

Die Presse 2012, 'Brennerautobahn: Laster unter Betonmauer begraben', *newspaper article*, 20.03.2012

EN 1997-1:2004, 'Eurocode 7: Geotechnical design - Part 1: General rules', The European Union Per Regulation 305/2011, Directive 98/34/EC, Directive 2004/18/EC.

Ester Martin, Kriegel Hans-Peter, Sander Jörg, Xu Xiaowei 1996, 'A density-based algorithm for discovering clusters in large spatial databases with noise', *Spatial,Text & Multimedia*. KDD-96, pp. 226–231.

The 8th International Conference on
Structural Health Monitoring of Intelligent Infrastructure
Brisbane, Australia | 5-8 December 2017

Signal stationarization in damping estimation for bridges from traffic loadings

S. Kim[1], H.K. Kim[1]
[1] *Seoul National University – South Korea, Email:* hokyungk@snu.ac.kr

Abstract

This study focuses on the enhancement in damping estimation of cable-stayed bridges from traffic loadings by a signal stationarization. Since the response of bridge in a sensing position is represented by an envelope-like function as traffic loading moves over the deck, the operational monitoring data can be nonstationary when the number of lanes of the bridge is limited one or two for one direction with small volume of traffic. In this case, the estimation of damping from operational modal analysis becomes more challenging. An amplitude-modulating (AM) function was proposed for minimizing this sort of nonstationarity. The AM function was obtained from temporal root-mean-square function of measured response. An approximated stationary signal was extracted from dividing the original signal by the AM function. This process reduced nonstationarity in traffic-induced response significantly. Using these stabilized damping ratio, a correlation analysis was performed with the representative environmental factors.

1. Introduction

The many of research demonstrates the temperature dependency of the natural frequency of the civil infrastructures. However, the effect of environmental conditions on damping ratio is still not clearly demonstrated since other factors also contribute to the variation of damping ratios, and the error bound of operational modal analysis (OMA) based damping estimation is relatively high. Although some researchers have investigated the relationship between the estimated modal damping ratio and corresponding ambient temperatures, no obvious pattern of variation was observed (Xia et al., 2006; Moser and Moaveni, 2011).

This study aims to reduce the uncertainties in OMA-based damping estimation by a regularization process based on proper algorithm parameter selection and signal stationarization method. The effect of proposed regularization process was verified in Natural Excitation Technique-Eigensystem Realization Algorithm (NExT-ERA) by numerical simulation and field monitoring data, respectively. Finally, the parameter dependency of OMA-based estimated damping ratio from field monitoring data was analyzed by a correlation analysis between the distribution of damping ratio and the representative environmental factors.

2. Regularization process based on parameter selection and signal stationarization

In order to reduce the algorithmic error in OMA-based damping estimation, the regularization process based on the proper parameter selection and the signal stationarization was proposed.

2.1. Parameter-based regularization in NExT-ERA

NExT-ERA is the output-only system identification method to estimate a damping ratio from a measured acceleration. NExT estimates an impulse response function without any input information by calculating the auto/cross correlation function of the measurements. This correlation function is obtained from an inverse Fast Fourier transform (FFT) of the power spectral density function of the measurements. For that, a Hanning window with a length of 2^{15} is applied with an overlapping of 50%. With these window size, overlapping and record lengths of 3,600 seconds, the frequency resolution is 0.0031Hz, and the number of average is 21 which are sufficient value for accurate identification (Kim et al., 2016).

In addition, OMA based damping estimation is very sensitive to a system order in ERA. To obtain a converged value, a system order is selected based on a concept of stabilization diagram and its median value. A series of modal analysis is performed for all range of the system order from 1 to 50, and the estimated modal properties which satisfies the criteria (frequency range, EMAC and physically reliable damping ratio) is recorded. Then, the median value of the identified damping ratios is selected as a representative value.

2.2. Signal stationarization

One of the reasons for uncertainty in OMA based damping estimation is nonstationary in loading. NExT procedure is based on the assumption of stationary white noise excitation. In reality, however, the nonstationary loadings such as earthquake, extreme wind or traffic loading are mainly applied to civil structures, and it causes nonstationary responses. Thus, this strong violation of excitation condition for the basic assumption of classical OMA can be a reason for a poorly damping estimation.

Chiang and Lin (2008) proposed that the approximate stationary acceleration can be extracted from a nonstationary ambient acceleration which can be modeled as a product of white noise and an amplitude-modulating (AM) function. Then, to transform the nonstationary response into stationary process, AM function is calculated as the temporal root-mean-square (RMS) function of the measured acceleration. Finally, the approximate stationary acceleration is acquired to divide the sample record by envelop of AM function.

3. Analysis of simulated data

A numerical simulation of ambient vibration test was performed based on 2-DOF model. The mass and stiffness matrices of Magalhães et al. (2010) were used to the simulation with the damping matrix was viscous as shown in Table 1. The corresponding modal properties were also listed in Table 2.

Table 1 Structural properties of a 2-DOF model

Stiffness (kN/m)	Damping (kN-s/m)	Mass (ton)
$\begin{bmatrix} 100 & 0 \\ 0 & 100 \end{bmatrix}$	$\begin{bmatrix} 0.3275 & -0.0725 \\ -0.0725 & 0.3275 \end{bmatrix}$	$\begin{bmatrix} 1.25 & 0.25 \\ 0.25 & 1.25 \end{bmatrix}$

Table 2 Modal properties of a 2-DOF model

Parameters	1st mode	2nd mode
Natural frequency	1.30Hz	1.60Hz
Damping ratio	0.52%	2.00%
Mode shape	[1 1]	[1 -1]

3.1. Load condition

The stationary loading, utilized amplitude-modulating function and nonstationary loading are displayed in Figure 1, respectively.

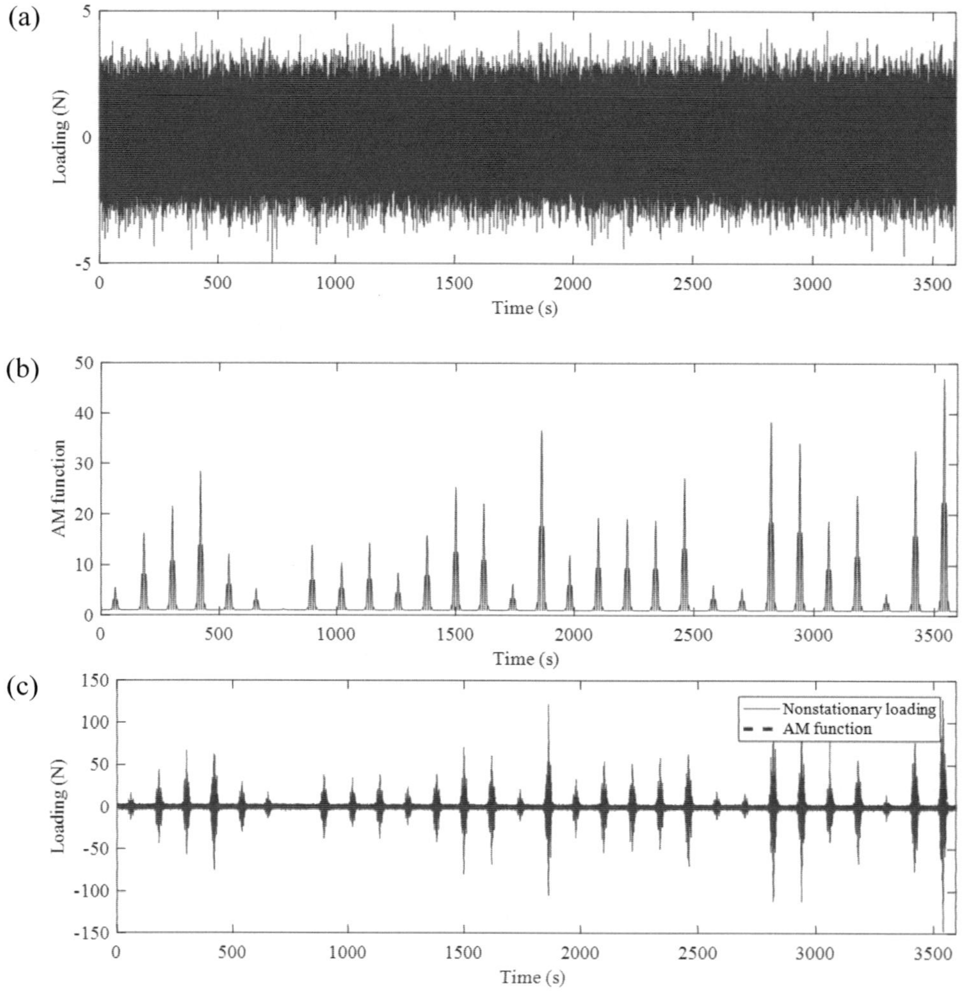

Figure 1. Time histories of (a) stationary loading, (b) AM function of Gaussian pulse and (c) nonstationary loading generated by multiplying of (a) and (b)

For the stationary excitation, a total of 100 random white noise loading was applied to both DOFs to simulate ambient vibrations, respectively. The structural acceleration response was evaluated by the dynamic analysis of discrete-time state-space model with a sampling frequency of 100 Hz and a duration of 3,600 seconds.

As previously mentioned in Section 2.2, a nonstationary loading can be modeled as a product of white noise and an AM function. For the comparative study, the same random process introduced in a previous paragraph was utilized with the AM function was produced by a Gaussian pulse to reproduce a recursive pulse loading as shown in Figure 2 (b). In order to check the effect of nonstationarity on the accuracy, the number of pulses was varied from 1 to 10.

3.2 Estimated damping ratio

Figure 2 presents the distribution of estimated first modal damping ratio for the 100 simulations from a simulated stationary/nonstationary/stationarized accelerations (no. of pulses = 3).

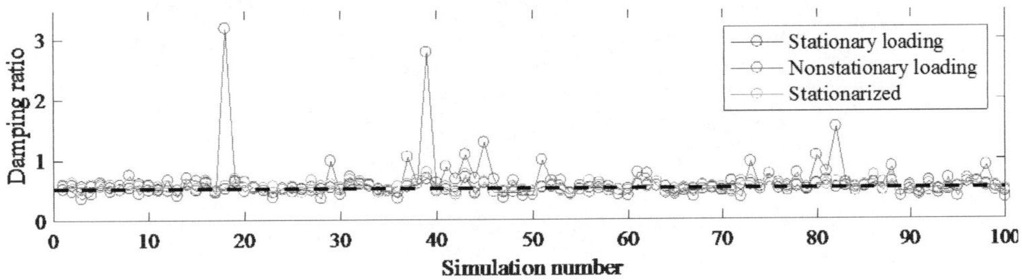

Figure 2. The distribution of estimated first modal damping ratio

The mean values of damping ratios for each case were almost similar to the input value of 0.52%. However, in terms of a scattering, the estimated damping ratios under nonstationary excitation showed over/underestimation compared to the other cases. Due to this scattering, the standard deviation of estimated damping ratio under nonstationary excitation was 0.19, relatively high compared to that of the other cases of 0.06%. After a signal stationarization, over/underestimated damping ratios under nonstationary excitation were recovered to the values under stationary excitation, and it leads to reduce a scattering in terms of decreased standard deviation.

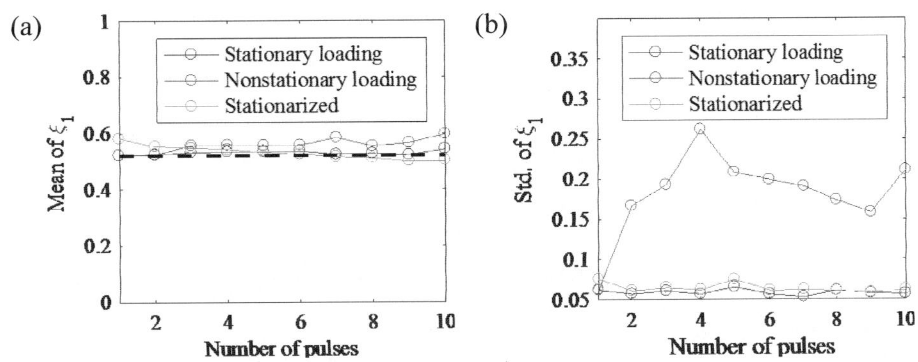

Figure 3. (a) Mean value and (b) standard deviation of estimated damping ratio according to the number of pulses in AM function

Figure 3 illustrates the relationship between the standard deviation of estimated damping ratio and the number of pulses for cases of stationary/nonstationary/stationarized acceleration, respectively. As the number of pulses (which can be considered as the nonstationarity in the loading) increases, this scattering in estimated damping ratios in nonstationary case also increased. It is clear that the stationarization contributed to reduce this scattering due to nonstationary loading.

4. Application: field monitoring data

4.1 Investigated bridge

The examined bridges consist of two parallel cable-stayed bridges as shown in Figure 4. The Bridge 2, which is the newer bridge, was opened in 2005 due to the increased traffic volume (Seo et al., 2013). The 1-hour acceleration data recorded at the center of the main span for 3 days (from October 15 to 17, 2012) was utilized for OMA (Kim et al., 2013).

Figure 4. Investiagted bridge

4.2 Damping estimation according to the regularization process

Figure 5 illustrates the distribution of estimated damping ratio of the investigated bridge during 3 days according to the regularization. As can be seen, the proposed regularization process was effective to reduce a scattering in estimated damping ratio. The coefficient of variance (COV) was dramatically reduced from 95.07% to 29.73% by applying regularization process.

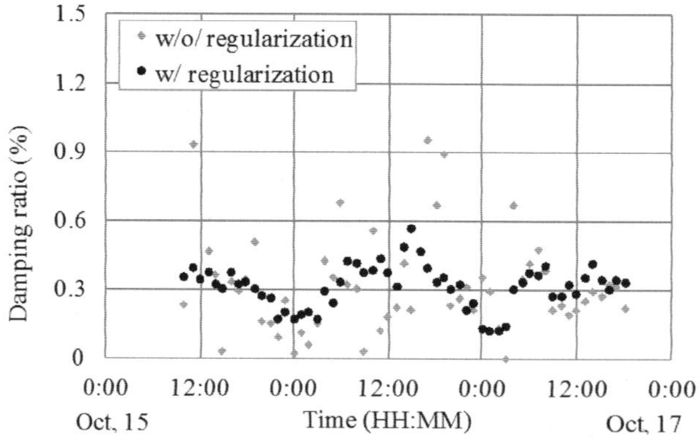

Figure 5. Estimated damping ratio during 3 days according to the application of regularization

5. Environmental effect on the damping ratio variation

5.1 Aerodynamic effect

The aerodynamic damping of the first vertical mode due to wind effect can be estimated with experimentally obtained flutter derivatives for the bridge sections based on the single-mode flutter analysis theory. In a lock-in range, however, vortex shedding occurs at the bridge deck, and it causes a complex shedding force which cannot be evaluated in a flutter derivatives accurately. Therefore, only the data set of which mean wind velocity was out of lock-in range was utilized in OMA, and the other cases, within the lock-in range, were discarded.

5.2. Correlation analysis

As can be seen in Figure 5, the damping ratio rose during from morning till night, and declined from the night to dawn. This trend was similar to that of considering environmental factors. Actually, the time evolution of the standardized variables (zero-mean and unit-variance) shared similar time-varying pattern as shown in Figure 6.

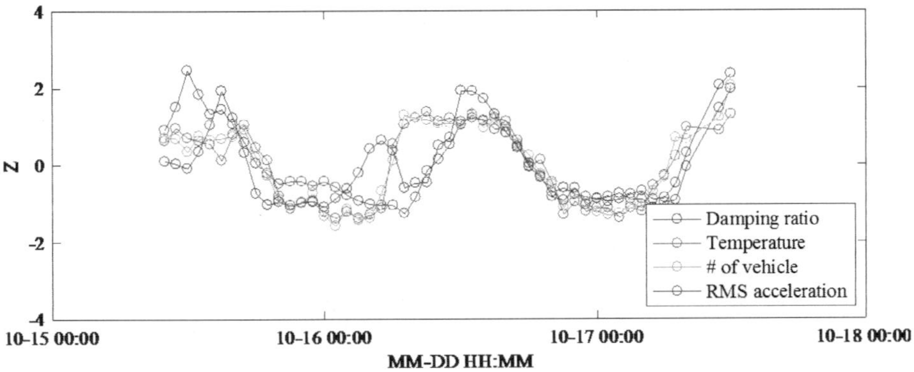

Figure 6. The distribution of standardized variables

Based on this distribution, these three environmental parameters were selected for a correlation analysis with the estimated damping ratio. A correlation analysis was performed with the objective of identifying the most dominant environmental factor on the variation of estimated damping ratio. Table 3 presents the correlation coefficient between the estimated damping ratio and environmental factors of a temperature, the volume of traffic or the amplitude of acceleration with/without regularization. As can be seen, before the regularization, the correlation coefficient was quiet low for all of environmental factors due to the large analytic error. By applying the regularization process, the large error bound was significantly reduced and the correlation appeared clearly. Among them, the most dominant parameter on the variation of damping estimation is the number of vehicles and corresponding RMS amplitude.

Table 3. Correlation coefficient between damping ratio and environmental factors with/without regularization

Damp.	Temperature	# of vehicle	RMS acceleration
w/o/ regularization	0.273	0.294	0.205
w/ regularization	0.447	0.805	0.761

5. Conclusions

This research investigated the parameter dependency of OMA-based estimated damping ratio. In order to remove the uncertainty level of damping, a regularization process based on parameter selection and signal stationarization was implemented which resulted in more accurate damping estimation compared to the direct application of classic OMA. The bias and variance error in estimates were decreased, as well as the amplitude-dependency of damping ratio appeared more clearly compared to the result without a regularization process. By applying the regularization, the correlation between damping ratio and environmental factors appeared more clearly. A correlation analysis found that the number of vehicle and corresponding RMS acceleration was the most dominant environmental factor on the variation of damping ratio.

Acknowledgements

This work was supported by the National Research Foundation of Korea (NRF) grant funded by the Korean government (MSIP) (2017R1A2B4008973) and partially supported by grants (17SCIPB119960-02) from the Ministry of Land, Infrastructure and Transport of the Korean Government.

References

Chiang, D. Y., & Lin, C. S. (2008). Identification of modal parameters from nonstationary ambient vibration data using correlation technique. *AIAA journal*, *46*(11), 2752-2759.

Kim, S. J., Kim, H. K., Calmer, R., Park, J., Kim, G. S., & Lee, D. K. (2013). Operational field monitoring of interactive vortex-induced vibrations between two parallel cable-stayed bridges. *Journal of wind engineering and industrial aerodynamics*, *123*, 143-154.

Kim, S., Park, J., & Kim, H. K. (2016). Damping Identification and Serviceability Assessment of a Cable-Stayed Bridge Based on Operational Monitoring Data. *Journal of Bridge Engineering*, 04016123.

Magalhães, F., Cunha, Á., Caetano, E., & Brincker, R. (2010). Damping estimation using free decays and ambient vibration tests. *Mechanical Systems and Signal Processing*, *24*(5), 1274-1290.

Moser, P., & Moaveni, B. (2011). Environmental effects on the identified natural frequencies of the Dowling Hall Footbridge. *Mechanical Systems and Signal Processing*, *25*(7), 2336-2357.

Seo, J. W., Kim, H. K., Park, J., Kim, K. T., & Kim, G. N. (2013). Interference effect on vortex-induced vibration in a parallel twin cable-stayed bridge. *Journal of wind engineering and industrial aerodynamics*, *116*, 7-20.

Xia, Y., Hao, H., Zanardo, G., & Deeks, A. (2006). Long term vibration monitoring of an RC slab: temperature and humidity effect. *Engineering Structures*, *28*(3), 441-452.

The 8th International Conference on
Structural Health Monitoring of Intelligent Infrastructure
Brisbane, Australia | 5-8 December 2017

Structural Health Monitoring of Elastic Concrete Based on the Acoustic Emission Technique

Jie Xu[1], Guang Yang[1], Qinghua Han[1].

[1] *School of Civil Engineering, Tianjin University /Key Laboratory of Coast Civil Structure and Safety of Ministry of Education, Tianjin University, Tianjin 300072, China, jxu@tju.edu.cn*

Abstract

The application of the acoustic emission (AE) technique to monitoring the fracture process-es and crack characterization in elastic concrete beams were investigated in this paper. Thir-ty-six elastic concrete three-point bend beams were tested by using AE technique. Four dif-ferent rubber contents, 0%, 5%, 10% and 15% were taken into consideration. Test results show that the AE energies of elastic concrete beams improve significantly with the increas-ing rubber content. The present experimental study may be useful for the quantitative measurement of fracture energy of elastic concrete and also it may be useful to relate AE energy to fracture energy of elastic concrete.

1. Introduction

In recent decade, the application of AE has received more attention as a potential non-intrusive and real time monitoring technique in engineering structures (Carpinteri et al. 2012; Han et al. 2015). Elastic concrete, also called crumb rubber concrete or recycled tire rubber-filled concrete, is a new type of concrete using crumb rubber as a composition of cement concrete materi-als. According to previous studies, the deformability of concrete increases significantly after adding rubber crumb in it. Furthermore, elastic concrete has ductile failure and bet-ter crack resistance. Besides, its acoustical properties, wearing resistance, aging re-sistance and erosion resistance are all better than ordinary concrete (Han et al. 2015; Xing et al. 2016). However, there is few research on the application of monitoring the elastic concrete by acoustic emission.

2. Test set-up and Results

The experimental investigations are carried out on a series of three-point bend elastic concrete specimens of the same geometry with different (notch/depth) ratios of 0.15, 0.30, and 0.50. All specimens were cast in a specially made steel moulds conforming to RILEM recommendations. Three specimens of each type (total 36 specimens) were tested for measuring the data during fracture process and concurrently AE parameters were also measured. A schematic diagram of the three-point bend test specimen was shown in Figure 1.

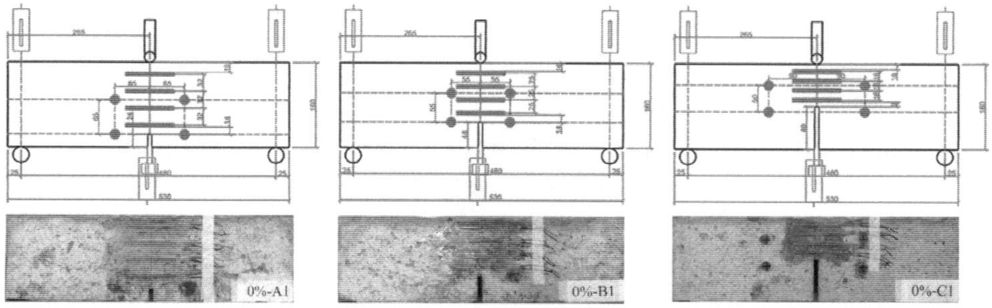

Figure 1. Schematic diagram and photos of the three-point bend test specimens.

Figure 2 shows the typical recorded plot of cumulative AE energy and load with time. It is observed that before peak load very little AE activity has taken place. On the onset of crack formation occurring at or near the peak load, the release of AE energy started. Hence, the initial behavior of the specimens with different contents of rubber particles is similar but varies as the fracture progresses. The rate of AE increases till the failure of specimen. It can be seen from Figure 2 that the load-time curve for high content of rubber particles is smoother than the load-time curve for low content. The ellipses in Figure 2 refer to the sudden change of cumulative AE energy. The elastic concrete will have more than one time of the sudden change. This means that elastic concrete correspond to more than one main crack. Rubber particles can help concrete disperse into two or more cracks. We can conclude that elastic concrete has advantages in ductility.

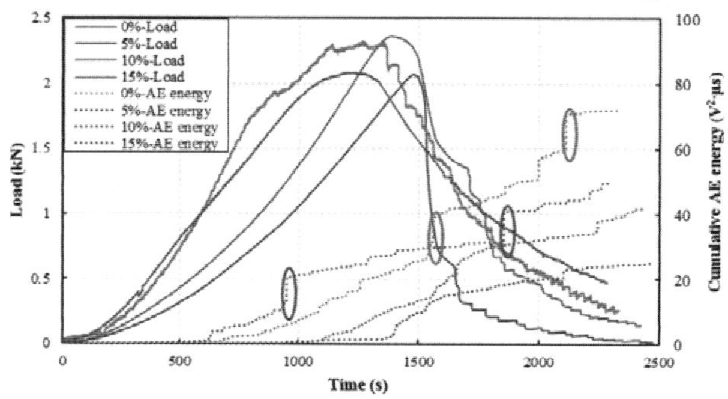

Figure 2. Variation of load and AE energy with time (different contents of rubber particles).

3. Conclusions

The result of this research shows that the AE is a useful technique for monitoring the mechanical properties of the elastic concrete beam. For high content of rubber particles, the top of load-time curve is much flatter than the load-time curve for low content. In the AE energy curves, two or more sudden changes can be seen. This phenomenon cor-responds to more than one crack. Rubber particles can help concrete disperse into two or more cracks. It has advantages in ductility. The present experimental study may be useful for the quantitative measurement of fracture energy of elastic concrete and also it may be useful to relate AE energy to fracture energy of elastic concrete. An equation has been proposed to obtain fracture energy of elastic concrete by knowing acoustic emission energy.

Acknowledgements

The authors would like to thank National Science Foundation of China (No.51408408 & No.51525803), Tianjin Natural Science Foundation (16JCQNJC07400) and Science and technology project of MOHURD (2016K3004) for providing the fund to carry out this experimental works.

References

Carpinteri, J. Xu, G. Lacidogna, A. Manuello: "Reliable onset time determination and source loca-tion of acoustic emissions in concrete structures", Cem. Con. Comp., 2012; 34: 529-537.

Q. Han, J. Xu, A. Carpinteri, G. Lacidogna: "Localization of acoustic emission sources in structural health monitoring of a masonry bridge", Struct. Control. Hlth., 2015; 22(2): 314–329.

Qinghua Han, Yihong Wang, Jie Xu, Ying Xing. Static behavior of stud shear connectors in elastic concrete–steel composite beams, J. Constr. Steel. Res., 2015; 113(6):115-126.

Ying Xing, Qinghua Han, Jie Xu, Qi Guo, YihongWang. Experimental and numeri-cal study on static behavior of elastic concrete-steel composite beams, J. Constr. Steel. Res., 2016; 123:79–92.

The 8th International Conference on
Structural Health Monitoring of Intelligent Infrastructure
Brisbane, Australia | 5-8 December 2017

A Study on Accuracy Evaluation of GPS Vertical Monitoring Outcomes on Main Span of Can Tho Bridge, Vietnam

H. L. T. Ho[1], H. V. Le [1], L. T. Nguyen [1].
[1] *University of Transport and Communications – Vietnam, Email:* lanhuongho0507@gmail.com

Abstract

Can Tho bridge is a cable-stayed structure with two wire planes constructed in Vietnam, which has the longest span in South East Asia. A structural health monitoring system (SHM) was installed on the bridge right after it had been put into operation. The monitoring outcomes are used to keep an eye on the bridge's conditions, review the design, work out the maintenance and repair schedule and provide warnings in case of hazards.

GPS is the major component in the SHM. However, use of GPS altitude has not been taken into consideration in the high-level technical measuring standards currently in use since Vietnam has not possessed the standard Geoid compatible to the territory, able to be put into massive practice throughout the country. Also, the gravity points are scarce, therefore, GPS altitude when shifted to ordinary elevation often provides inconsistent outcomes in different areas.

For GPS monitoring spots successively installed on the bridge, the works'coordinates monitored in three directions at different times are valuable outcomes, which can also enable us to know of the bridge's oscillation by time as well as by three directions.

This article presents a study on the method of assessing the data accuracy of vertical GPS continuous monitoring based on the actual figures statistically obtained of Can Tho bridge span center, the results showed that (1) when the selected model is appropriate, the statistic method can be used to assess the accuracy of monitorig data; (2) the GPS monitoring data of Can Tho bridge along to vertical direction have high accuracy which is, hereby, can be used to assess the bridge's conditions.

Keyworks: SHM, GPS, cable-stayed bridge, regression analysis

1. Introduction

Can Tho bridge links Vinh Long and Can Tho provinces and is one of the two most important bridges under national-level inspection. The bridge lies on the sole route from Ho Chi Minh city to the provinces in Western Vietnam. In 2010, after its inauguration, Can Tho bridge became the cable-stayed structure with two wire planes, which has the longest main span in South East Asia (Figure 1).

The bridge is 26m wide, the main bridge itself is 2,750m long, the tower pillar is 164.8m high, the tower is 134.7m high from the floor, the clearance is 39m, there are 216 stayed cables and the bridge is designed as per AASHTO LRFD standards. Particularly, on the 550 metre main span, its 210 metre center section is structured in the form of fabricated steel box girder and fitted up with concrete boxes cast in situ at both bridge heads.

Figure 1. Can Tho Bridge

The SHM was installed on Can Tho bridge right after it came into operation. The system was designed and installed by Japanese contractors, and it is composed of GPS antenna, wind metering sensor, observation camera, structural temperature metering sensor, air temperature metering sensor, rainfall metering sensor, tensimeter, acceleration meter, tension meter for stayed cables, and the like (NTT Data corp. 2010). The data collected is sent via optic cables to the data archive and conversion unit and the bridge management head office. Based on the real-time measurements by the SHM, the bridge's conditions are analysed, assessed and reported every quarter.

In Can Tho bridge's SHM, GPS antennae function as the key component of the system, utilised for monitoring displacement in three directions (Le et al 2015). These displacement values are then used to work out the structure's own oscillation frequency as well as its cross and vertical displacement (Kaloop et al 2009&2011), therefore, GPS is fitted at specific positions, such as the tower top, the bridge center, the one-fourth spot on the bridge, etc. Can Tho bridge has 2 GPS bases and 9 rovers installed on the main bridge (Figure 2).

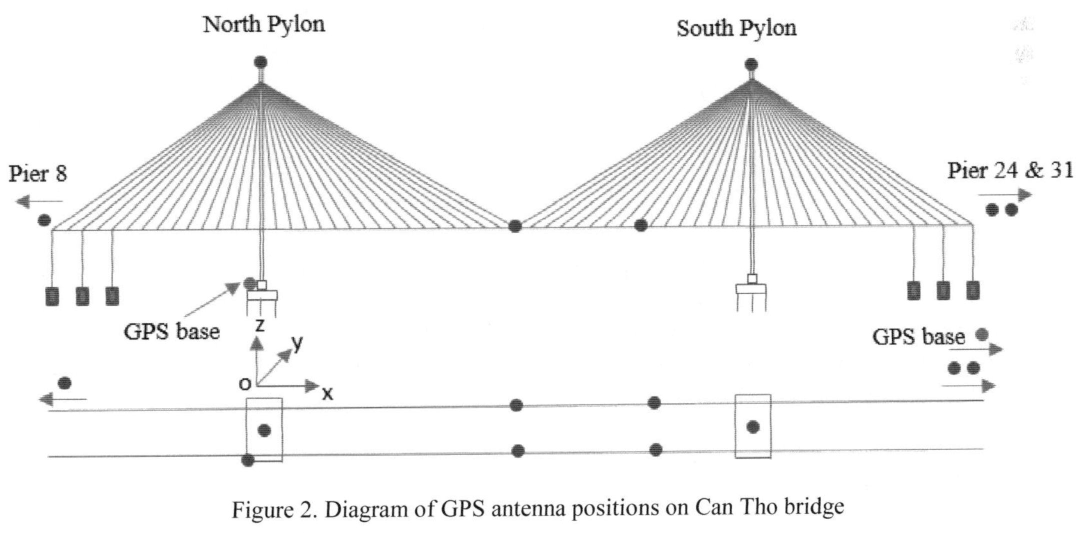

Figure 2. Diagram of GPS antenna positions on Can Tho bridge

The GPS continuous monitoring makes use of Real Time Kinematic (RTK). Two antennae are fixed at the management office and near abutment A1. These spots are grouted by concrete on drilling piles 125m deep at certain coordinates and altitude (the bases). The immobile antennae are installed at specific spots of the bridge (the rovers) (Figure 2). The GPS data is collected in real-time mode at frequency 2HZ. According to the catalogue, the horizontal accuracy is

±(10mm+1ppm) and the vertical accuracy is ±(20mm+1ppm). The GPS data is processed by Leica GNSS Spider (NTT Data corp. 2010), resulting in the coordinates and altitude of all monitoring points in the bridge's coordinate system. All GPS values are numerically archived in the computer every single minute, every ten minutes, every hour and every day and demonstrated on the screen in the form of a time and displacement relationship chart.

2. Developing a measuring value model and assessing accuracy of GPS altitude data

When GPS is applied in measuring, the value obtained is the geodetic altitude. If the plumbing abnormality is ignored (which is small in the Mekong delta, Vietnam), the relationship between the geodetic altitude and the ordinary elevation is presented in the fomular as follows (Li et al 2001):

$$H_{t_M} = H_{td_M} + \xi \qquad (1)$$

H_{t_M} : Ordinary elevation at monitoring spot M

H_{td_M} : Geodetic altitude

ξ : Altitude anomaly

In reality, the values of GPS monitoring in direction Z are the monitoring altitude values of a spot at different times, which are utilised to calculate the oscillation amplitude in vertical direction, meaning the substraction of altitude values, as a result, the altitude anomalies of a spot at two different times are equal.

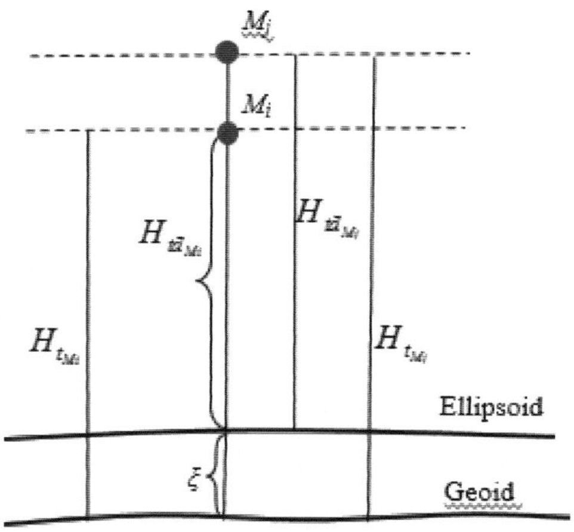

Figure 3. Altitude of a monitoring spot at times i and j

According to Fig. 3, when M is supposedly the spot for GPS antenna installation on the bridge, the positions of which at times *i* and *j* (the cross deviation when anomaly is examined in this case is found very little) are:

For spot M at time *i*: $\xi = H_{t_{M_i}} - H_{td_{M_i}}$

For spot M at time *j*: $\xi = H_{t_{M_j}} - H_{td_{M_j}}$

$$\Rightarrow H_{td_{M_j}} - H_{td_{M_i}} = H_{t_{M_j}} - H_{t_{M_i}} \Rightarrow h_{td} = h_t \qquad (2)$$

From equation (2), it is obvious that the geodetic altitude substraction equals the ordinary elevation substraction when monitoring spots on the bridge are regarded, which means the altitude anomaly does not affect the outcomes in use for oscillation monitoring on cable-stayed bridges via GPS.

According to a research (Ho & Le 2016) on Can Tho bridge, the vertical displacement of the span center is mainly caused by the temperature with the correlation coefficient between the temperature and the displacement exceeding 0.8. Thus, when developing a model from the data collected by GPS vertically measuring, we can possibly utilise the single variable linear model with the temperature as the major variable whereas the other factors are considered as random causes of the displacement.

The model for monitoring values at time i is demonstrated below (Rice 1995, Rousseeuw et al 1987):

$$Y_i = a + bX_i + \varepsilon_i \tag{3}$$

In which:
Y_i is the data on continuous displacement in vertical direction of spot GPS-RTK at time i;
X_i is the variable of the single variable model; the cause of displacement is the temperature at time i; a, b is the regression ratio;
ε_i is the residual between the values obtained and the model;

The estimated values (approximation) is calculated by the linear regression model:

$$\overline{Y}_i = a + bX_i \tag{4}$$

The deviation between (3) and (4) is the surplus between the model and the data determined by the formula:

$$\varepsilon_i = Y_i - \overline{Y}_i = Y_i - a - bX_i \tag{5}$$

ε_i is the random error of the estimation model, which is more accurate to the reality when the sum of squares of all deviations is smaller, thus, via the least square method, the coefficients a and b can be sought in such a way to ensure that the sum of squares of error ε_i is the smallest.

$$\sum_{i=1}^{n} \varepsilon_i^2 = (Y_i - a - bX_i)^2 = \min \tag{6}$$

To solve an extreme problem with the function of two variables, we have:

$$b = \frac{\sum_{i=1}^{n}(X_i - \overline{X})(Y_i - \overline{Y})}{\sum_{i=1}^{n}(X_i - \overline{X})^2} = \frac{\sum_{i=1}^{n} Y_i X_i - n\overline{X}\overline{Y}}{\sum_{i=1}^{n} X_i^2 - n(\overline{X})^2} = \sum_{i=1}^{n} \frac{x_i y_i}{x_i^2} \tag{7}$$

$$a = \overline{Y} - b\overline{X} \tag{8}$$

In which: $\overline{X}$, $\overline{Y}$, x_i, y_i is calculated in turns by the formula below:

$$\overline{X} = \frac{\sum_{i=1}^{n} X_i}{n}; \qquad \overline{Y} = \frac{\sum_{i=1}^{n} Y_i}{n} \tag{9}$$

$$x_i = X_i - \overline{X}; \qquad y_i = Y_i - \overline{Y} \tag{10}$$

When an appropriate model has been set up, all conditions of the function have been examined, then the error of the vertical displacement measuring values can be computed by formula (11)

$$m = \sqrt{\frac{[\varepsilon\varepsilon]}{n-1}} \tag{11}$$

Apart from error quantification of deviation ε_i, the regression model could also be considered as appropriate if deviation ε_i abides by the Gaussian distribution law (Shumway 2010, Weisberg 2005). To examine Gaussian distribution of deviation ε_i, it is possible to make use of the normal distribution chart and such statistical inspecting methods as Chi-square or Lilliefors test.

3. Experimental computing with the vertical measurement data collected of Can Tho bridge span centre

The data for experimental computing is Can Tho bridge's average time series of 10 minutes quoted in January 2016, including the vertical GPS data of the span centre and the corresponding atmospheric temperature data. To examine the feasibility of the assessing method, the data in use is collected from long-term monitoring, taking two cases with corresponding data length into account; 2 days (288 data) and 4 days (576 data) incessantly.

The charts illustrating the data used for computing are given in Fig. 4 with the blue line inside demonstrating GPS coordinate data; the charts underneath depicting the corresponding temperature; the red dash line in each chart separating the data of each day. It is shown in Fig. 4 that when the temperature rises, the girder's altitude decreases, i.e. there is a negative correlation between the atmospheric temperature data and the GPS data.

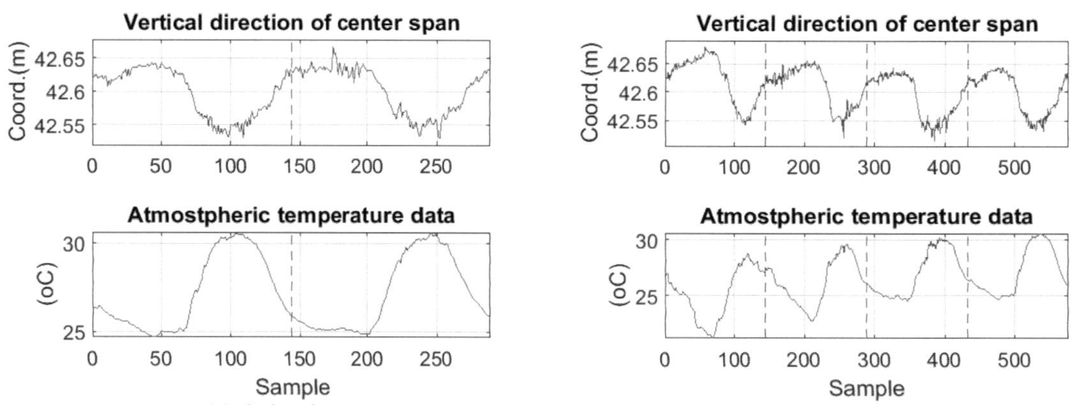

(a) 2-day data (b) 4-day data

Figure 4. Plots of GPS data and corresponding temperature data

The chart presenting the correlation and the coefficient r between the vertical GPS coordinate data and the atmospheric temperature data is depicted in Fig. 5. The correlation chart shows that the correlation between the atmospheric temperature and the vertical displacement data of the span centre is greatly negative (coefficient: ~ -1). Thus, the single variable linear regression model with the atmospheric temperature as the variable is agreeable to the vertical coordinate data of Can Tho bridge span centre.

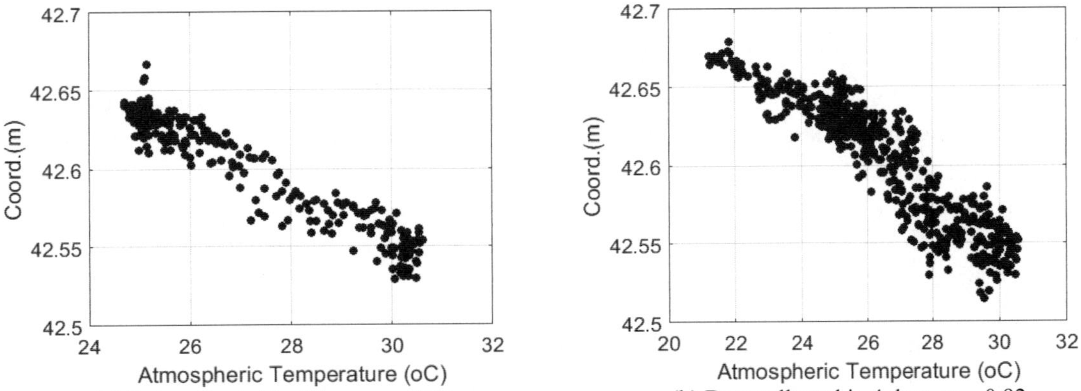

(a) Data collected in 2 days, r = -0.96 (b) Data collected in 4 days, r = -0.92

Figure 5. Temperature and displacement correlation scatter

The single variable linear regression model by formula (3) has been applied to develop a regression model for the vertical coordinate data of Can Tho bridge span centre. The single variable linear regression analysis comes out as follows:

- The regression model for 2-day data: $z_i = 43.0469 - 0.0164{*}t_i$
- The regression model for 4-day data: $z_i = 43.0216 - 0.0158{*}t_i$

In which, z_i is the regression value of vertical coordinates by the atmospheric temperature variable t_i. The outcomes from the regression model show that when the atmospheric temperature varies 1^0C, the vertical coordinate value changes -16.4mm for the data in 2 successive days, and -15.8mm for the data in 4 successive days.

The charts present the linear regression outcomes, the model's determining coefficient R^2 and 95% of confidence interval in Fig. 6. It can be noted that the determining coefficient R^2 is very high (0.92 & 0.84), which implies that the single variable linear regression model by temperature is appropriate to the real-time displacement mode of Can Tho bridge span centre.

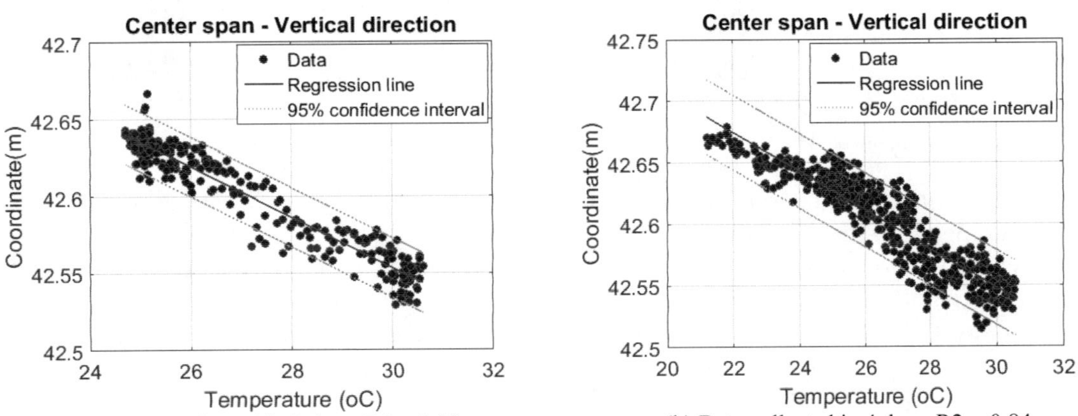

(a) Data collected in 2 days, R2 = 0.92 (b) Data collected in 4 days, R2 = 0.84

Figure 6. Plots of regression analysis and 95% confidence interval

Deviation ε_i between the regression model and the real-time data is calculated by formula (5) and depicted in Fig. 7. The central error of deviation ε_i is calculated by formula (11):

$$m = \sqrt{\frac{[\varepsilon\varepsilon]}{n-1}}$$

The results of deviation ε_i error are sequentially ± 0.0098m and ± 0.0154m for both cases of the data collected in 2 days and 4 days. These outcomes show that GPS bridge monitoring provides relatively high accuracy. The chart that examines deviation ε_i distribution also reveals that deviation ε_i abides by the normal distribution law, presented in Fig. 8. However, to draw an accurate conclusion on deviation ε_i distribution, it is essential to apply such standards of statistical inspection as Chi-square or Lilliefors test.

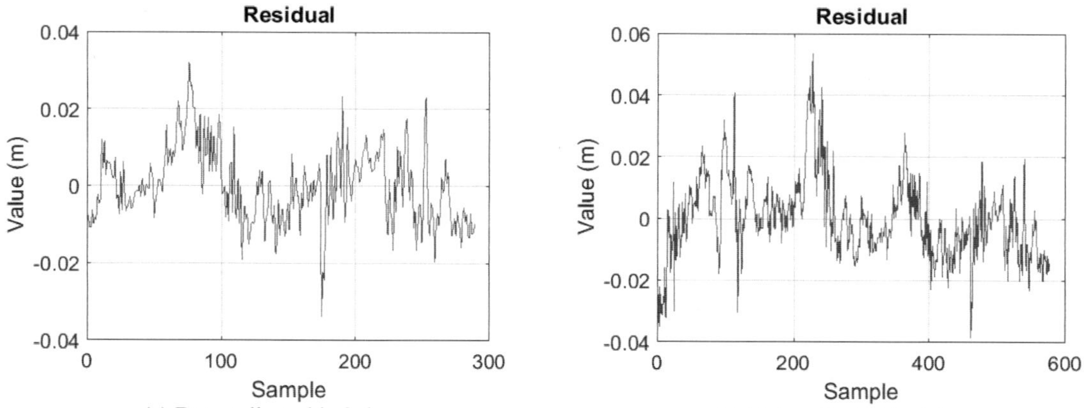

(a) Data collected in 2 days (b) Data collected in 4 days

Figure 7. Plots of residual εi

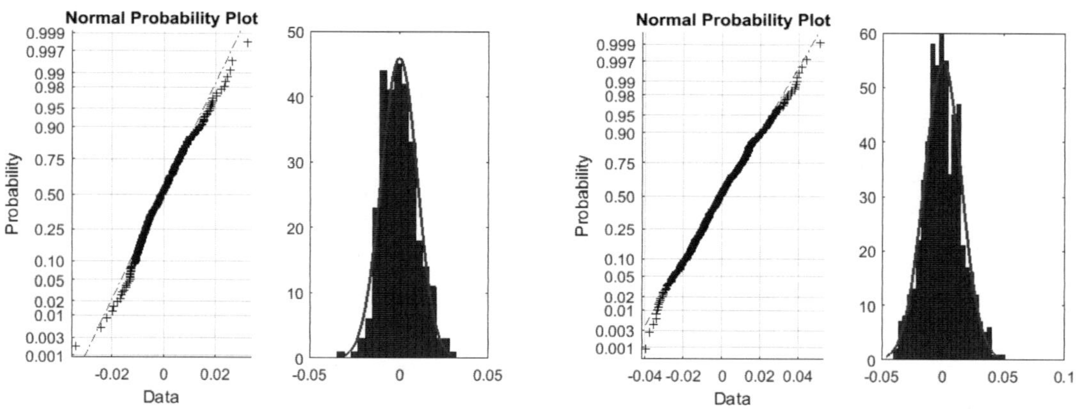

(a) Data collected in 2 days (b) Data collected in 4 days

Figure 8. Charts examining deviation εi normal distribution

Deviation ε_i normal distribution is examined by Chi-square test with the assumption H_0 that "Deviation ε_i is distributed by the normal distribution law". The results of assumption h and probability coefficient p examined for each case show that the assumption is acceptable, which means the deviation between the regression model and the real-time data abides by the normal distribution law, or deviation ε_i is the random error. The results are entered in Table 1 below.

Table 1.Results of examined deviation ε standard dislocation

2-day case		4-day case	
h	P	h	p
0	0.0558	0	0.1982

4. Conclusions

Can Tho bridge plays a very important role in the national transport network linking the Southern economic hub to West Vietnam. Therefore, it is absolutely crucial to ensure the safety for the bridge as well as the accurate analysis of the bridge's conditions by real-time monitoring data. The analysis and assessment of GPS data collected of Can Tho bridge shows that the GPS data is relatively consistent, highly reliable, quite accurate and relevant to the theory-based estimation analysis and agrees to the bridge's characteristics.

Use of the altitude measuring outcomes in computing oscillation and displacement from GPS data in the SHM is not affected by altitude anomaly error (oscillation computing takes only amplitude and oscillation times per second into account)

Statistical methods can be used to assess the GPS measurement accuracy, however, the chosen model should suit the measurement data, and all assumptions should be well examined to opt for the right error computing fomular, relevant to error distribution features.

References

M. R. Kaloop, H. Li 2009. 'Monitoring of bridge deformation using GPS technique', *KSCE Journal of Civil Engineering*, vol. *13, no* 6, p. 423-431.

M. R.Kaloop, H. Li 2011. 'Sensitivity and analysis GPS signals based bridge damage using GPS observations and wavelet transform', *Measurement*, vol. *44, no.* 5, p. 927-937.

H. V. Le, M. Nishio 2015, 'Time-series analysis of GPS monitoring data from a long-span bridge considering the global deformation due to air temperature changes', *Journal of Civil Structural Health Monitoring*, vol. 5, no. 4, pp. 415-425.

H. V. Le, M. Nishio, H. Yamada, H. Katsuchi 2015, 'Statistical condition assessment of a cable-stayed bridge using GPS monitoring data'. *The 7th International Conference on Structural Health Monitoring of Intelligent Infrastructure-SHMII 2015.*

X. Li, H. J. Go¨tze 2001, 'Tutorial Ellipsoid, geoid, gravity, geodesy, and geophysics. *Geophysics*, vol.66, no6, p. 1660–1668.

H. L. T. Ho, H. V. Le 2016,'Study on correlation between wind, temperature data and displacement data of some locations on the Can Tho bridge', *Transport and Communications Science Journal*, vol 54, pp. 39-45.

H. L. T. Ho, H. V. Le 2016,' Establisment of prediction model for displacement of the girder at mid – span cross-section of Can Tho bridge using the monitoring data and regression analysis', *Transport and Communications Science Journal*, vol.55,pp. 3-7

R. A. Rice 1995, *Mathematical Statistic and data Analysis*. Book, California, USA.

P. J. Rousseeuw, A.M. Leroy 1987, *Robust Regression and Outlier Detection*. Book, Wiley-Interscience.

R. H. Shumway, D. S. Stoffer 2010, 'Time series analysis and its applications: with R examples'. *Springer.*

S. Weisberg 2005, *Applied Linear Regression*. Book, third edition. Wiley-Interscience.
NTT Data Corp. 2010, Design of structural health monitoring system for Can Tho bridge construction project

The 8th International Conference on
Structural Health Monitoring of Intelligent Infrastructure
Brisbane, Australia | 5-8 December 2017

Tunnel inspection with high-performance devices. Performance Indicators.

F. Sánchez-Domínguez[1], J.A. Ramos-García[2], F. Mata-Aroco[2], E. Calvo-Haro[2]
[1]*Euroconsult NT-Spain, Email:* fsanchezd@euroconsult.es, [2]*Euroconsult NT-Spain*

ABSTRACT

The management and maintenance of currently in service tunnels are required of specific lining inspection techniques with very high performance scanning systems. Besides, most of the infrastructures maintenance contracts (including tunnels) are managed through Key Performance Indicators. In order to assess the different tunnel state indicators, it is necessary the use of objective inspection tools (not only visual inspection performed by surveyors) to measure and control the evolution of deteriorations.
This paper presents different automated techniques to perform this objective evaluation. Some of them are already fully implemented in the infrastructure maintenance market (such as LIDAR), nevertheless, it also includes the latest inspection techniques, giving the possibility of gathering this extra information of the deterioration state of tunnels in all types of lining. Therefore, the results obtained in currently tunnel visual inspections performed by specialist surveyors can be complemented with the combination of the latest techniques, mainly based on imaging and laser. The automated inspection offers a better knowledge of the state of the tunnel due to its objectivity, avoiding subjectivity problems associated to the traditional visual inspection. With all this information, the overall status of the tunnel can be evaluated to define both service levels and status indicators properly.

1. Tunnelinspection: background

Over the last years, the number and length of different type of tunnels (road, railway, and utilities) have increased significantly in the developed countries. Furthermore, some of the oldest infrastructures have had an important increase of passengers, traffic, occupancy, and operation time. However, the available maintenance time periods has been reduced. In this context, it is necessary to implement new techniques to inspect the condition of infrastructures (which are critical for cities mobility), allowing a preventive maintenance of the structural stability without affecting the traffic. The maintenance inspections allows to analyze the preservation and deterioration condition of tunnels on a routine basis in terms of cracks, breaks, significant bulges of the lining affecting the clearance, damp areas, etc. We must keep in mind that all these deterioration signs are indicative of possible structural problems, hence the importance of these inspections.
Scheduled routine inspections are key to detect safety issues and to prevent structural, geotechnical or infrastructure/operation issues or failure. Moreover, if these inspections are conducted with high-performance equipment, which allows an objective and automated analysis of the real state of the infrastructure, it is possible to accurately assess the key performance indicators from a preventive maintenance perspective and to consider the most appropriate investments for a safe tunnel operation.

2. Available techniques

Detailed inspections of tunnels, conducted with non-destructive methods (radar, impact-echo, ultrasonic technologies…) present the drawback of being slow, entailing complex tests and demanding many material resources. On the other hand, visual inspections of the tunnel conducted

by specialized staff do not ensure the most comprehensive and swift information, as they are subjective and time-consuming. In order to solve both issues, different new inspection systems have been under development during the last years in order to get fast and, most importantly, accurate inspections of the tunnel condition focused in non-destructive and detailed visual techniques.
The currently used techniques in the market for high-performance inspection in new-built tunnels and, particularly, in currently-built tunnels are the following:

- 3D linear laser cameras
- LIDAR
- Thermography
- Cameras with near infrared illumination

Among all these techniques, the 3D linear laser cameras have been the last one to enter in the market and they are currently in constant improvement. These cameras are changing the methodology of damage inspection and management for different infrastructures (pavements, railways, tunnels) due to their high productivity, performance, and their ability to incorporate automatic algorithms to the acquired images analysis. The inspection systems incorporate these 3D cameras combined with the thermography and LIDAR information in order to ensure a more comprehensive inspection. In addition to this, techniques such as near infrared cameras are capable of offering the user a 360° field of view inside the tunnel allowing the surveyors to conduct an in-situ inspection.
These techniques are summarized below, where we will see the 3D cameras in closer detail, as these are new systems in the domain of tunnel inspection.

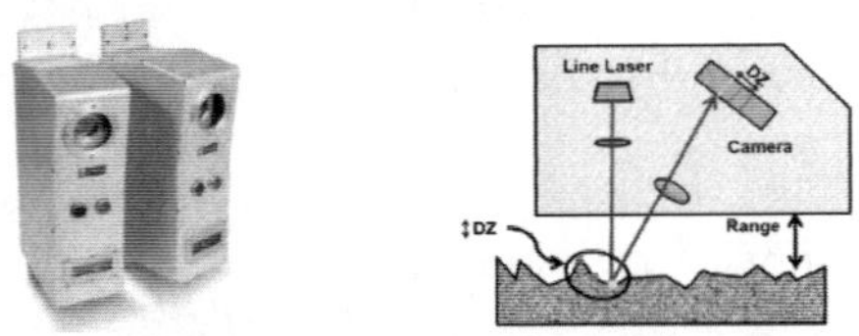

Figure 1 - 3D cameras technology

2.1. 3D laser cameras

The 3D laser cameras apply the triangulation principle to extract 3D information from the sensors, as shown in Figure 1. A pattern of known light (a line in this case) is projected from the laser onto the object to be inspected. The line is recorded by an area-scan camera placed at a fixed distance with an oblique angle relative to the projected light. The relative position (rotation + translation) between camera and laser is obtained by means of a precise calibration process.
The 3D sensor cameras used in tunnel applications (Figure 2) are the LCMS camera-laser units originally developed for road pavement surface inspection tasks by INO (National Optics Institute, Canada), and marketed by Pavemetrics. These sensors have been adapted and integrated for tunnel inspection within the *Tunnelings* system [1].

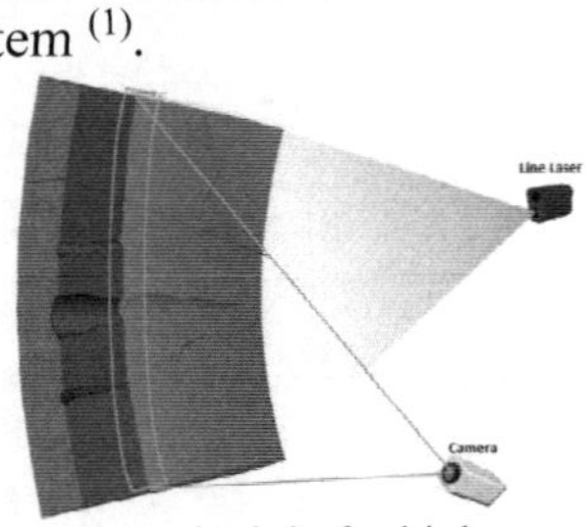

Figure 2 - Triangulation measurement principle for high-speed acquisition of 3D information

The 3D laser cameras are triggered by a Distance Measurement Instrument (DMI) to acquire profiles of 3D points, creating images by means of profiles aggregation.
Each camera of the *Tunnelings* system captures an array of 2352 points per profile in a width of approx. 2 m with three-dimensional information. The profiles are acquired every time the system receives a signal from the Distance Measurement Instrument (DMI) of the vehicle. The DMI measures the displacement and distance in the driving direction of the survey with millimeter resolution. This provides, for example, for every 1000 consecutive profiles merged, a 2352 x 1000 mm resolution image each camera saved in a common file.
With this technology, it can be obtained the luminance images (Figure 1-a), that any camera could offer, and at the same time the distortions of the laser line, which can be represented on a range image that shows (Figure 1-b) the distances between sensor and surface. Using both images, it is possible to carry out a later 3D reconstruction (Figure 1-c) and develop numerical algorithms of damage analysis. In below image are included the three images described.

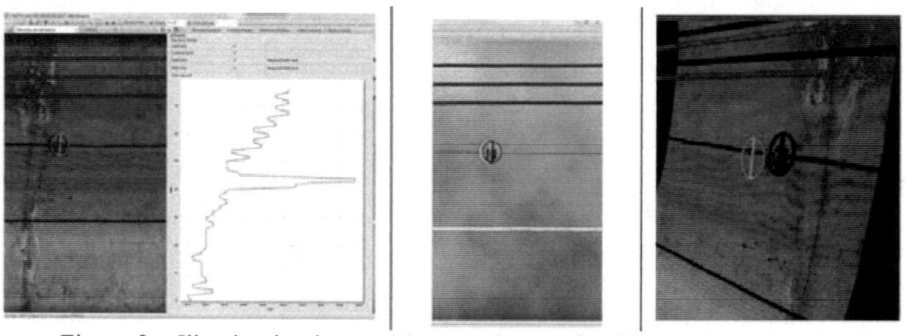

Figure 3 – Illumination image (a), range image (b), 3D reconstruction (c)

The high resolution image allows to take surveys on it as well as and to program deterioration analysis algorithms. In the figure 4 is shown a 3D high resolution image.

Figure 4 – High resolution 3D reconstruction

2.2. LIDAR

The LIDAR is a technique used currently in the mobile mapping and topography for the 3D reconstruction in a wide range of infrastructures. This technology measures the distance to the object to be analyzed through an emitted pulsed laser, measuring different physical principles of the reflected light from the object.
Depending on the time used, the LIDAR system is based on the physical principle of *'Time Of Flight (TOF)'* or *'Phase-Shift'* to take the measurements. The principle of *'Time Of Flight (TOF)'* measures the time that a punctual light beam takes to be reflected on the surface, while the principle of *'Phase-Shift'* measures the difference of the emitted wavelength phase. The utilization of one principle or other affects in the precision of the survey and its operation in dynamic conditions.

The measurement range of LIDAR systems usually covers distances from approx. 0.5 m up to (or greater than) 80 m, with a measurement accuracy greater than 5 mm in depth even in the most accurate systems. Considering that 360° are covered, the transverse resolution depends on the distance, as the system performs a circular scanning and registers data at certain angles. The resolution then varies in proportion to the distance covered by the laser. Below is included an image with two LIDAR sensor installed on the tunnel inspection vehicle to scan 190º x 2.

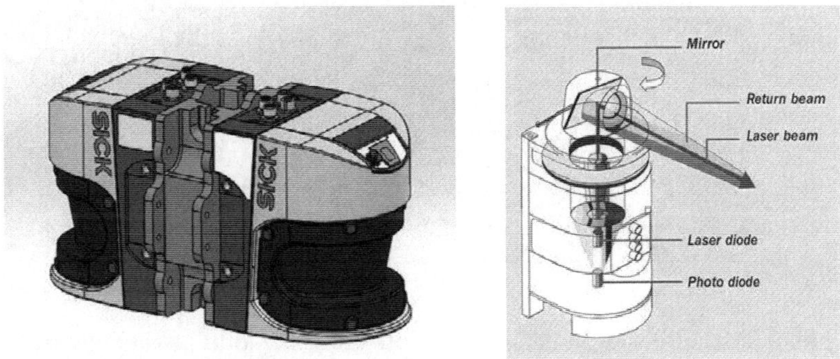

Figure 5 - Image of 2 LIDAR sensor mounted in parallel and graphic description of the working principle

The LIDAR system is really useful to generate a 'map' of the inspection setting, but it is difficult to apply it for defect analysis purposes due to the acquisition speed, its resolution and accuracy in comparison with 3D vision cameras. However, it offers the advantage of being able to detect obstacles at great distances, such as the slopes in portals, tunnel headwalls and clearance (figure 6).

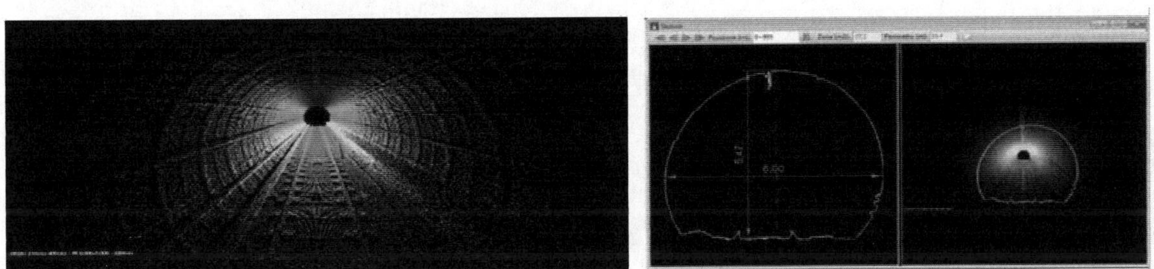

Figure 6 - Image of LIDAR sensor within a tunnel and obtaining cross sections clearance

2.3. Thermography

Infrared thermography allows the acquisition of lining high resolution surface temperatures while measurements are taken with the laser sensors (3D cameras and LIDAR). The measurement of the surface temperature contributes to a better analysis of the dampness present at the lining. It also allows for surveying hidden damages, for instance leakages, delamination concrete hollows or cavities that might be present in the materials composing the support or within its backfill. The thermography is registered in synchrony with the rest of the sensors, allowing the combined usage of these techniques for the detection of eventual damages.

Thanks to the thermography technique and though their post-processing, it is possible to show temperature images with their pixels in different colors in concordance to the temperature interval selected, and with a resolution of a tenth of a degree. The thermography cameras installed in the tunnel inspection vehicle described in this document have the following features:

- Temperature range: 0-250 °C
- Spectral range: 7.5 to 13 μm

- Detector: 120 Hz
- Pixel size: 4 cm
- Accuracy in absolute temperature measurement: ±2 °C or ±2 %

Below is included an image (Figure 7) acquired with this type of sensors in an underground tunnel.

Figure 7 – Thermographic images synchronized with the rest of the sensors

The camera has a software that makes it possible to make punctual measurements on the acquired image, to escalate the temperature gradient and to generate temperature histograms so as to identify elements of with a different temperature in the tunnel.
With the infrared image it is also possible to inspect all the electrical equipment installed in the tunnel, including cables, connections or splices that are, for example, either hidden or located in exposed areas of the tracks. In this case, the goal is to detect potential problems that could result in a future failure or even risk of fire.

2.4. Cameras with visible infrared illumination

This type of cameras (Figure 8) make possible the acquisition of images in unlit environments from which a global vision of the tunnel is required. The equipment includes a set of infrared illuminators and cameras able to acquire high-resolution images in completely dark environments.

Figure 8 - Near infrared cameras

The cameras have a CMOS sensor of e2v which features a high quality and sensitivity at the near-infrared. Besides, these cameras also include a global shutter that allows for an accurate moving snapshots and reproduction of sharp contrasts. The figure 9 shows an example of a complete darkness tunnel images captured with these cameras.

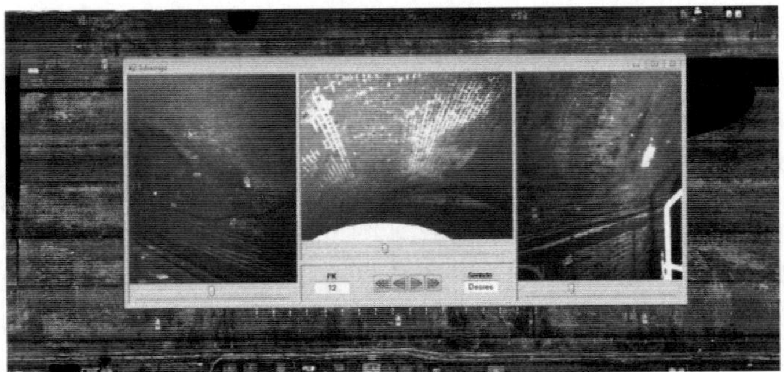

Figure 9 – Images of a very dark tunnel with visible infrared cameras.

Though the combination of the techniques explained in this paper, it is possible to inspect from large-diameter to small-diameter tunnels with a reasonable measurement speed and ensuring maximum resolution in each sensor.
The following image shows how the scanning system has been installed on the inspection vehicle to be used in different types of infrastructure and in tunnels of different diameters with different support materials.

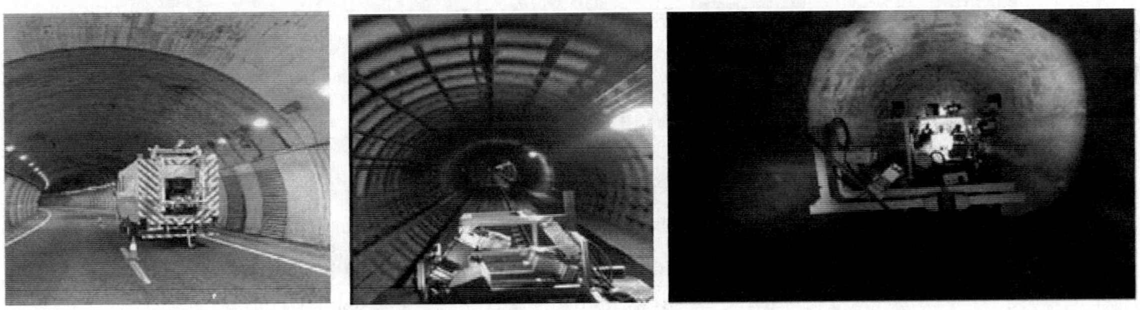

Figure 10 – Different installations for a road, railway and water tunnel

Even if all the described techniques are combined for the analysis of the structural condition of the tunnel, the analysis of damages is mainly based on 3D cameras due to their high resolution and their capacity to reconstruct three-dimensional images that may be automatically processed for detection. The analysis of the results will vary according to the type of support of the tunnel. Obviously, the type of visible deterioration in a brick tunnel or a masonry tunnel will not be the same as those observed in a segment tunnel or a metal frame tunnel.
In general, we can identify four main types of tunnel support [(2)] if the classification criterion is that of homogeneity in the visible results after an automatic analysis.

- Metal frame tunnels.
- Concrete tunnels with precast segments.
- In situ concrete tunnels and shotcrete tunnels.
- Brick and masonry tunnels.

These four groups evidently present defects that could be present in all of them –for instance, dampness and cracking. As for the cracking, it might be classified according to its location (at the keystone or at the arches), or its development and orientation (longitudinal, transverse, oblique). The severity of cracks may be later considered in terms of extent, width and depth. Other frequent defects in all these different tunnels –although with their own specificities depending on the mechanical behavior of the lining material– are spalling or birdsmouthing, bulging or clearance deformations [(3)].

Following it is shown some cracking and spalling images in different linings.

Figure 11 – Longitudinal in a brick lining tunnel

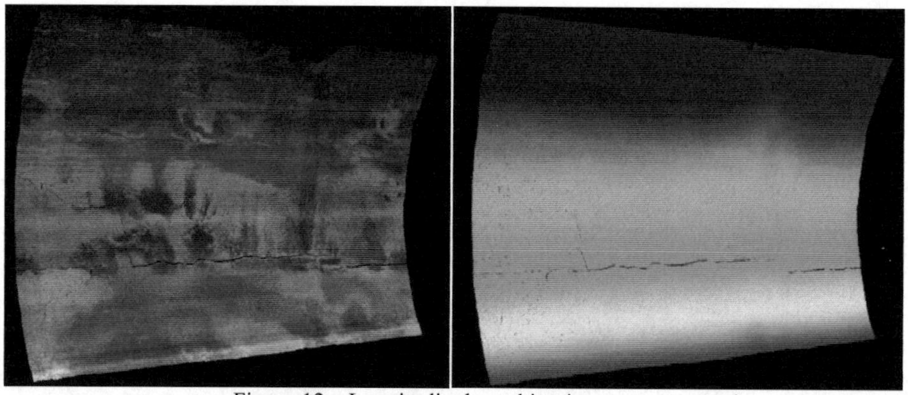

Figure 12 – Longitudinal cracking in concrete tunnel

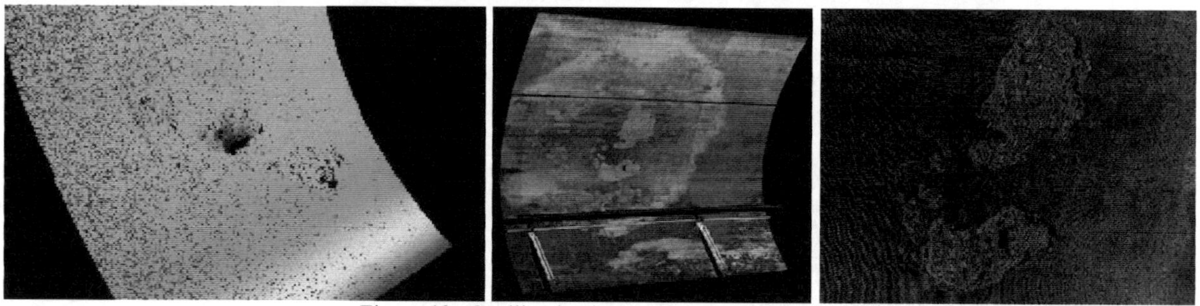

Figure 13 - Spalling in situ concrete tunnel

In addition to these defects which are common to all these linings, there are other incidents that are particular to metal frame tunnels, to concrete tunnels with precast segments, metal structures and to brick and masonry tunnels.

2.4.1. Metal Frame Tunnels

For the metal frame tunnels, attention must be paid to the areas presenting corrosion and surface calcareous deposits, either at the joints or the center of the metallic segments, and to the condition of the bolts joining such segments (if they are missing, deflected or the nuts are missing). Some examples of deterioration in metal frame tunnels are shown below. They are analyzed according to the image and the reconstruction obtained thanks to the 3D cameras.

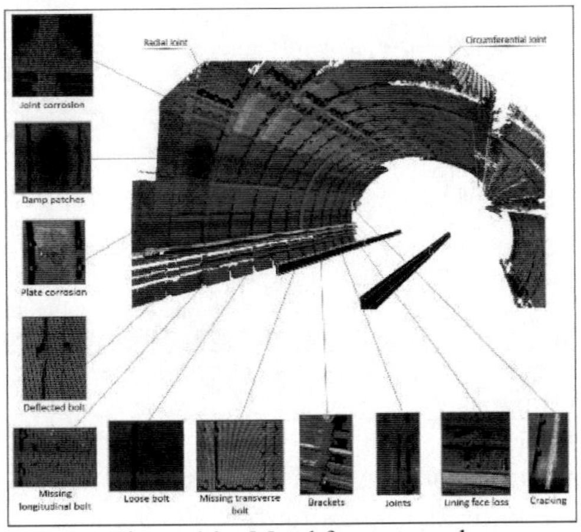

Figure 14 – Metal frame tunnels

2.4.2. Concrete Segment Tunnels

In concrete segment tunnels, the monitoring of the faulting and the opening between segments may be as accurate (and less) than 1 mm thanks to the 3D cameras system [(4)]. Thus, it is possible to measure the faulting and the opening along the whole joints and compare it with a second inspection to check whether any changes have taken place.
In the figure 15 are shown examples of the measurement of such faulting and their representation along the tunnel.

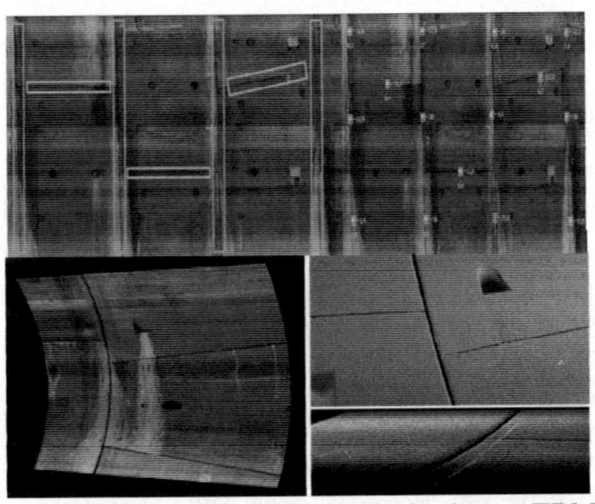
Figure 15 – Faulting and opening analysis along a TBM tunnel

2.4.3. Brick Tunnels

In brick and masonry tunnels, the loss or fall of bricks and the deterioration of joints are treated specifically. To do so, the algorithms calculate the presence of joints with mortar loss and with a depth greater than a certain threshold; then they are also used to calculate the affected volume and to identify which elements have fallen off the support. Below are included some examples of detection of brick loss and mortar loss using 3D cameras. As can be noted, with this information it is possible to obtain measures and generate sections of deterioration.

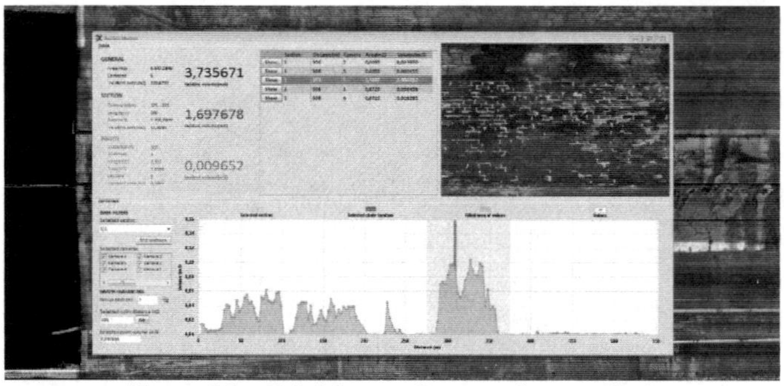

Figure 16 – Brick loss and mortar loss analysis

3. Calculation of indicators

Regarding the cracking, it is possible to calculate its extent and severity per rings or per sections of fixed distance (5 m, 25 m, etc.), with identification of their location in the transverse section of the tunnel (keystone, shoulders or arches), and to determine whether the cracking is longitudinal, transverse or oblique. As for the dampness, the presence of water on its surface is analyzed to obtain the percentage of area affected. Regarding to breaks and chipping, it is possible to calculate the volume of material affected.

All this information, together with more specific data in cases such as those of brick tunnels, segment tunnels or metal frame tunnels, is statistically processed to obtain adapted key performance indicators allowing the global condition of the tunnel assess (5).

4. Monitoring

The use of a high-performance system for the registration of deterioration and defects, with automatic analysis and storage of every single element in a database, offers a higher interesting asset, as it makes it possible to compare different inspections. This comparison is necessary for the monitoring and supervision of the evolution of the tunnel.

If the system is capable of storing an automatically detect cracks (with all its characteristics such as location, width and depth) and to compare it with successive inspections conducted under the same conditions, then it will be able to evaluate the evolution of defects and determinate how the deterioration is progressing to prevent potential risks. The figure 17 shows an example of a comparison of 2 surveys.

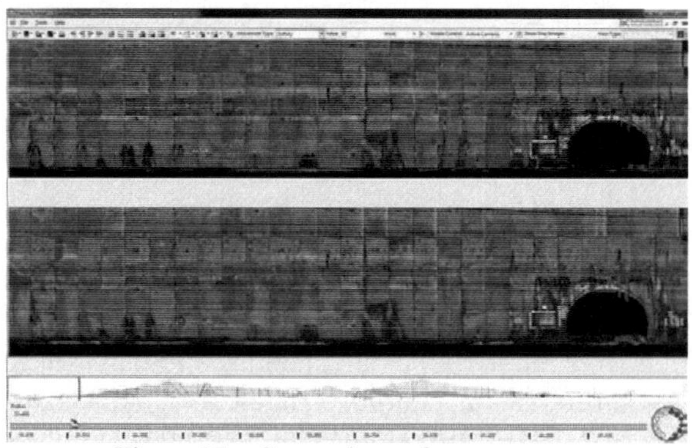

Figure 17 – Comparison of two surveys

From a geometrical perspective, it is possible to generate a relative mapping of the deformations (6) that provides information about section changes and, at the same time, reveals any bulging or deformation which could alter the service clearance of the tunnel. An example deformation mapping is shown in the figure 18. It is intended to indicate the relative distance from the support to the center of the tunnel so large-scale deformations of the tunnel section can be observed. On the other hand, it is also possible to compare the position of the segments inside the tunnel so as to determine whether millimeter movements are taking place.

Figure 18 – Deformation mapping in a metallic tunnel

5. Conclusions

High-performance systems incorporating sensors of 3D, LIDAR and thermography cameras allow for the calculation of objective key performance indicators that make it possible to assess the condition of any type of lining in different types of tunnel.

Thanks to them we are able to calculate the severity and the extent of the most important damages and to prioritize maintenance activities. To this end, we would need to categorize the intervention areas, from those points which are in good condition and those with minor defects which would be simply subject to monitoring, to those areas where repair works need to be done without delay due to stability risks.

References

Sánchez- Domínguez, García- Ramos & et al 2017, "Pavemetrics LCMS system". Retrieved from http://www.pavemetrics.com/.

Sánchez- Domínguez, García- Ramos & et al 2015 "High-performance Automated Structural Inspections in Tunnels". *XXV Road World Congress, World Road Association, PIARC*, Seoul, South Korea.

Sánchez- Domínguez, García- Ramos & et al 2015 "Automated Structural Evaluation of Tunnels in Service". *ITA WTC 2015 Congress and 41st General Assembly, Lacroma Valamar Congress Center*, Dubrovnik, Croatia.

Sánchez- Domínguez, Duval A. & et al 2014 "Use of 3D Scanning Technology for Automated Inspection of Tunnels". *Proceedings of the World Tunnel Congress (WTC) 2014*, Foz do Iguaçu – Brazil.

Gavilán, Sanchez- Domínguez, & et al 2013 "Mobile inspection system for high-resolution assessment of tunnels". *The 6th International Conference on Structural Health Monitoring of Intelligent Infrastructure (SHMII-6)*, Hong Kong – China.

Sánchez- Domínguez, García- Ramos & et al 2016 "Euroconsult Tunnelings" http://www.euroconsult.es/pdf/tunnelingspliegos-baja.pdf

The 8th International Conference on
Structural Health Monitoring of Intelligent Infrastructure
Brisbane, Australia | 5-8 December 2017

Spherical Imaging and Virtual Environments for Structural Health Monitoring

R. Napolitano[1], A. Blyth[1], B. Glisic[1]
[1] *Princeton University – USA, Email:* rkn2@princeton.edu

Abstract

Thorough documentation is pivotal in the life cycle management and performance monitoring of infrastructure. As technology continues to progress, 3D models and virtual environments are becoming increasingly popular for designing or documenting a structure. These models and environments can be optimized for life cycle management if they are paired with current information about a structure including past monitoring results and possible interventions. The work presented here reflects a methodology and digital workflow, which uses 360 degree spherical imaging as well as novel tools for virtually experiencing infrastructure and associated data. This project was tested on a real-world case study, Streicker Bridge in Princeton, NJ, USA. Streicker Bridge is a pedestrian footbridge spanning 350 feet, which opened in 2010. This bridge has been monitored since construction began in 2008 using fiber-optic strain gauges. The bridge environment was captured utilizing the Ricoh Theta spherical imaging camera and processed utilizing the Kolor Panotour virtual environment software. The virtual environment of Streicker Bridge was made interactive by adding responsive points of interest; these points enable certain types of damage to be highlighted and explored for future maintenance and life cycle management. Exploration of the damage in the virtual environment include additional 2D images which accentuate aspects of structural damage, hyperlinks to papers associated with this work, as well as additional forms of metadata. The results of this research will be recommendations for how the different groups working together on the life cycle management and performance of the bridge can document their site and meaningfully convey data to the appropriate audiences.

1. Introduction

Structural Health Monitoring is a process that aims to provide in-time information about how a building, bridge, or other structure is functioning under certain loading conditions. In previous literature, SHM has been equated to being the nervous system of a structure as its main purpose is to detect unusual structural behavior in near-real time (Glisic and Inaudi 2007). There are five main activities in the SHM process; these activities are outlined in Figure 1.

Monitoring strategy	*Installation of SHM system*	*Maintenance of SHM system*	*Data management*	*Closing activities*
Monitoring aim	Installation of sensors	Providing for electrical supply	Collecting data (reading of sensors)	Interruption of monitoring
Selection of monitored parameters	Installation of accessories (boxes, cables...)	Providing for com-munication lines (wired or wireless)	Storage of data (local or remote)	Dismantling of monitoring system
Selection of monitoring systems	Installation of reading units	Implementation of maintenance plans for hardware	Providing for access to data	Storage of monitoring components
Design of sensor network	Installation of software	Repairs and replacements	Visualization	
Schedule of monitoring	Interfacing with users		Export of data	
Data explo-itation plan			Interpretation and data analysis	
Costs			The use of data	

Figure 1: Main activities of SHM process (Glisic and Inaudi, 2007)

(1) Develop a monitoring strategy with an aim, preliminary design, as well as a preliminary budget

(2) Install the sensors along with any necessary reading units, software, interfaces, and accessories

(3) Maintain the system by ensuring electrical connections, maintenance plans, etc.

(4) Develop a data management plan for the collecting and storing of data.

(5) Close activities (if the project is not continuous)

This paper expands on current literature on the activities related to data management (Glisic and Inaudi 2007, Koo et al 2011, McNeill 2009, Zhou et al 2006). As can be seen in Fig. 1, data management encompasses collecting, storing, providing access to, visualizing, exporting, interpreting, and using data. This paper in particular focuses on creating and implementing a novel visualization workflow for SHM and validating this method on a case study.

2. Existing methods for visualization

Currently, various forms of 2D and 3D visualization exist for SHM. The location of physical sensors and graphs of resulting datasets are often expressed in 2D. These are sometimes expressed with corresponding still images of the structure. Many SHM companies supply their own 2D visualization software (Colombo and Bittencort 2016). Individually these data forms are useful for communication. However they are difficult to integrate with one another or meta-data such as information on the functioning status of sensors. Current methods also lack the ability to adequately describe a 3D sensor network, for example a sensing sheet wrapped around a complex element. 3D systems such as Building Information Models (BIM) are alternative, effective tools for communication. They are useful to show metric details about the inner workings of a structure and its sensors. This level of detail is often unnecessary for the general maintenance and monitoring of a structure. In these instances where project groups will not take full advantage of 3D models' capabilities, they are inefficient with regard to time and cost of their construction.

However due to the 3D nature of a structure, 2D visualization can be inadequate for presenting SHM data. The danger of miscommunication because of a lack of clarity becomes more acute when a 3D sensor network is represented with 2D documentation. No existing methods provide an efficient visualization system for a structure that does not require a detailed 3D model. There is a clear gap in SHM visualization systems that demonstrates a need for visualization methods for projects that have 3D networks but will not necessitate a 3D model.

2. Virtual Tours and Informational Modeling

Virtual tours (VT) are commonly used for real estate, tourism, and educational purposes to give broad audiences an understanding of a 3D space. These tours are created using spherical panoramas that are stitched together to recreate the physicality of a structure. As demonstrated in Figure 2A spherical panoramas are able to display an entire view of a structure, as one would experience it in reality. In comparison, planar panoramas (2B) do not capture the floor and ceiling of a structure, only the 360° view perpendicular to the camera lens (Napolitano et al 2017). The entire view provided by a spherical panorama is crucial for an accurate spatial understanding. It becomes especially important if the SHM system is implanted in the floor or ceiling of a structure.

A

B

Figure 2: Comparison of panorama types: a) planar panorama, b) spherical panorama.

Combining virtual reality (VR) and informational modeling (IM) could be an effective way to convey SHM data, where informational modeling is the inclusion of sensor charts or datasets within the virtual environment. A VT/IM model is a plausible solution to fill the gap in current visualization methods. This preliminary project develops and tests a VT/IM method for SHM on Streicker Bridge.

3. Tools and software

This project uses 36 spherical panorama images taken with a RICOH THETA S camera, 12 megapixels, to reconstruct a pedestrian bridge (Ricoh 2017). These images provide comprehensive views from the top and bottom of the bridge. The RICOH THETA S camera was chosen because unlike other cameras, it automatically converts images into panoramas and eases the process of creating the virtual tour. For the sake of clarity, the tripod used to support the camera was removed and the images were lightened using Photoshop. This was especially important for images taken on top of the bridge deck where the tripod sometimes overlapped with the location of embedded sensors. By removing the tripod, sensor locations are more discernible.
The Kolor Panotour Pro software program was used to stitch together a virtual environment with a customized interface in which users can access PDFs, images, text, databases, etc. through "hotspots." "Hotspots" are on-click conditionals embedded in the environment that perform a predefined action. "Hotspots" give the user access to the associated datasets and navigation (Kolor 2017).

4. Case study

Streicker Bridge is a pedestrian bridge on Princeton's campus. Its construction began the fall of 2008 and opened in the summer of 2010. The bridge is constructed as a main span deck-stiffened arch with four continuous girder approach ways, which will be referred to as "legs" throughout this paper. The deck and legs are reinforced post-tensioned concrete; the arch and supporting columns are weathering steel. Figure 3A depicts a view of the bridge's main span from the road below. The two eastern legs of the bridge as seen from the main span are depicted in Figure 3B (Janberg 2017).

A

B

Figure 3A. View of Streicker Bridge mainspan from the road below. Figure 5B. View of eastern legs from on top of the bridge deck.

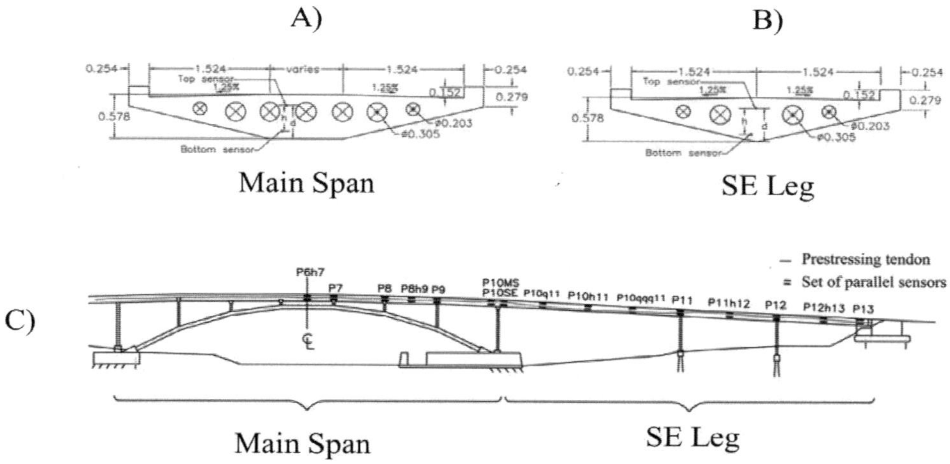

Figure 4: Streicker Bridge sensor locations: A) main-span cross-section, B) SE leg cross-section, and C) elevation of the main span and the SE leg.

The sensor network embedded in the bridge is 3D. The mainspan deck and southeast leg are outfitted with strain, displacement, and temperature gauges and a sensing sheet. Figure 4 shows the placement of these sensors (Abdel-Jaber 2014, Abdel-Jaber 2017). Figure 4A is a section cut of the mainspan that demonstrates the typical parallel upper and lower sensors in the deck. Similarly, the section cut in Figure 4B shows the top and bottom sensors in the southeast leg. Figure 4C shows the relative location of individual sensors in an elevation view. The VT/IM system produced for this project tests the integration of these sensors and associated data into their physical locations to enhance understanding of the sensing network.

5. Results

The product of this work is a VT/IM that showcases how the 3D sensing network of Streicker Bridge can be implemented and understood easily in a virtual environment. Figure 5 demonstrates the customizable interface provided, which allows the user to easily navigate the bridge in three ways. (1) Adjacent panoramas are stitched together allowing the user to virtually walk through the space by clicking on navigational "hotspots." (2) The dropdown menu in the upper right corner of the screen provides navigation to dissociated panoramas. (3) The Google Maps plug-in on the left of the screen allows selection of a panorama based on the coordinates of where the image was captured (Napolitano et al 2017).

Figure 5: VTIM interface: (1) list of panoramas (top right), (2) an geographic map (left), and (3) "hotspots" inside the environment link to adjacent panoramas (bottom left).

Additional "hotspots" are programmed to display the sensor network data. A graphic image indicates the type of sensor at that location, its functional status, and identifying information (i.e. "P9 Down E"). A legend is accessible through a toggle button on the lower right corner of the screen. Clicking on the graphic "hotspot" opens either an image or website database based on the type of information stored for the particular sensor or project type. Figure 6 shows the "hotspot" and demonstrates what a user sees after clicking on the "hotspot" (Napolitano et al 2017).

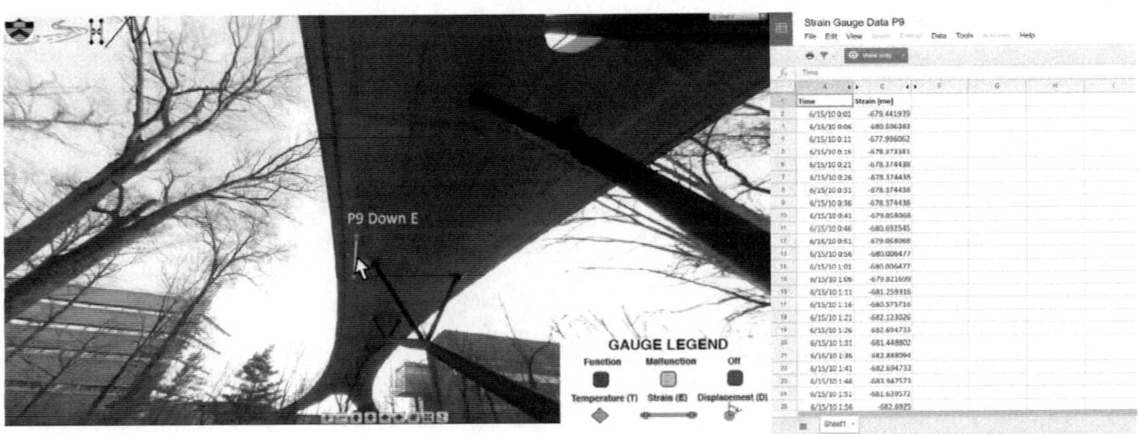

Figure 6: Screen capture of the VT/IM user interface showing sensor "hotspots" and a strain gauge dataset stored online that launches with activation of strain gauge "hotspot."

6. Conclusions

The VT/IM model created for Streicker Bridge illustrates how this technology can be implemented to enhance communication between different professional groups working on a SHM project. During life cycle modeling it is imperative that the data being collected is well organized and easily accessible to prevent confusion or failure of the project. The simplistic 3D environment proposed here provides an economical alternative to time consuming and costly existing 3D documentation methods. Its intuitive nature allows the user a better understanding of the data than 2D documentation methods allow. The methods presented in this paper can be implemented on 3D SHM projects to address the gap in technology that currently exists for documenting 3D systems that do not have a need for a full 3D model.

Acknowledgments

The authors would like to acknowledge the members of the SHM lab at Princeton for their assistance with this work. Additionally we would like to acknowledge the School of Engineering and Applied Sciences at Princeton.

References

Glisic, Branko & Inaudi, Daniele 2007, *Fibre Optic Methods for Structural Health Monitoring,* John Wiley & Sons, Chichester, West Sussex, England.

Koo, Ki, Battista, Nicky & Brownjohn, James, 2011, 'SHM Data Management System Using MySQL Database with MATLAB and Web Interfaces.' , *Proceedings of the 8th International Conference on Structural Health Monitoring of Intelligent Infrastructure (SHMII5)*, Cancun, Mexico, 11-15 December.

McNeill, D. K. 2009, "Data management and signal processing for structural health monitoring of civil infrastructure systems.' CRC Press, Boca Raton, FL.

Zhou, HF, Chan, TK, Wang, JY, and Ni,YQ 2006, 'A structural health monitoring data management system for instrumented cable-supported bridges.' *Proceedings of the Asia-Pacific Workshop on Structural Health Monitoring*, Yokohama, Japan, 4-6 December.

Colombo, A.B. and T.N. Bittencort, 2016, 'Development of a web interface for structural health monitoring data visualization and structural performance assessment,' In *Maintenance, Monitoring, Safety, Risk and Resilience of Bridges and Bridge Networks,* edited by Túlio Nogueira Bittencourt, Dan M. Frangopol, and André T. Beck, 227, Foz Do Iguaçu, Brazil.

Napolitano, Rebecca, Blyth, Anna, Glisic B., 2017, 'Virtual Environments for Structural Health Monitoring.' 11th International Workflow on Structural Health Monitoring (IWSHM), 12-14 September, Stanford, USA.

Theta, 2017, 'RICOH THETA S Camera.' Web. 09 May 2017.

Kolor, 2017, 'Panotour - Virtual Tour Creation Software.'. Web. 09 May 2017.

Janberg, Nicolas, 2016, 'Structurae: Image-ID:258296.' Structurae. Web. 09 May 2017.

Abdel-Jaber, H. 2014. 'Structural Health Monitoring for the Evaluation of Prestressing Forces and Pre-Release Cracks.' Master's thesis, Princeton University.

Abdel-Jaber, H. 2017. 'Comprehensive Strain-based Methods for Monitoring Prestressed Concrete Beam-like Elements.' PhD thesis, Princeton University.

The 8th International Conference on
Structural Health Monitoring of Intelligent Infrastructure
Brisbane, Australia | 5-8 December 2017

Short-term Wind Speed Forecasting for Traffic Control during Strong Wind on Bridges

W.H Jung[1], S.J Kim[2], J.Y Park[3], H.K Kim[4]
[1] *Seoul National University – South Korea* [2] *Institute of Construction and Environmental Engineering – South Korea,*
[3] *Seoul National University – South Kore,* [4] *Seoul National University – South Korea Email:* hokyungk@snu.ac.kr

Abstract

During storng wind, bridge management system decides to limit vehicle speed or passage of vehicles in order to reduce the risk of vehicle accident when field-monitored wind speed data exceeds specific a threshold. But this management system, based on real-time data, can not respond immediately to the occurrence of strong winds and cope with ever-changing wind speed. In order to ensure safety of vehicles and efficient operation of the bridge during strong winds, wind speed forecasting technique is required. In this study, the applicability of short-term wind speed forecasting to bridge management system in order to predict strong wind is investigated. Strong wind during typhoon season is main target to predict. Outlier Robust Extreme Learning Machine (ORELM), which is machine learning algorithm, and Wavelet Packet Decomposition (WPD), which is signal decomposition algorithm, are combined and applied during forcasting process. Case study is performed to verify forcasting algorithm, using field-monitored wind speed data which are measured at two locations: Incheon airport weather station and bridge site in Busan.

1. Introduction

Since vehicles on a bridge are vulnerable to strong wind, the probability of wind-induced accident, such as overturning or sideslip, increases during strong wind. In order to reduce wind-induced risk of the vehicle, several studies have been conducted on calculation of critical wind speed which is risky wind speed to vehicle and assessment of the number of days for traffic control (Baker, 2015; Kim et al., 2016). An actual bridge management system decides to limit vehicle speed or passage of vehicles when field-monitored wind speed data exceeds a threshold. Generally, in South Korea, traffic control is initiated when 10-minute average of field-monitored wind speed exceeds 25 m/s. But this management system, based on real-time data, can not respond immediately to the occurrence of strong winds and cope with ever-changing wind speed. Especially, it is difficult to decide when to start and end the traffic control.

In order to ensure safety of vehicles and efficient operation of the bridge during strong winds, wind speed forecasting technique is required. Wind speed forecasting techniques have been developed in the area of wind power generation. However, they focused on general wind speed around 15m/s (Liu et al., 2015; Wang et al., 2015; Meng et al., 2016), the accuracy of predicting strong winds should be verified.

In this study, Outlier Robust Extreme Learning Machine (ORELM), which is machine learning algorithm, and Wavelet Packet Decomposition (WPD), which is signal decomposition algorithm, are combined as wind speed forecasting algorithm. The accuracy of predicting strong winds using wind speed forecasting algorithm is evaluated and applicability of the algorithm to bridge management system is checked. In particular, strong winds during typhoons are targeted to be predicted. As a method to improve prediction accuracy of strong winds during typhoon, a training data for machine learning algorithm, which contain wind speed data at the time of previous typhoon occurrence, applied. Case study is performed to verify forcasting algorithm, using field-monitored wind speed data which are measured at two locations: Incheon airport weather station and bridge site in Busan. Data type of wind data is maximum 10-minute averaged wind speed for each hour.

2. Methodology of Wind Speed Forecasting

As a machine learning algorithm, Extreme Learning Machine (ELM) was proposed by Huang et al.(2006) and was known to have shorter computation time and higher prediction accuracy than other algorithms such as Support vector machine, back propagation neural network and Radial basis function neural network (Huang et al., 2006; Cancelliere et al., 2013; Sun and Liu, 2016). ELM was modified by Zhang and Luo(2015) to enhance robustness against outliers and ORELM was proposed. ORELM showed more accurate prediction performance than ELM when prediction step become larger(Peng et al., 2017).
Generally, wind speed data is nonsationary and nonlinear, so it is difficult to expect high prediction accuracy when wind speed data is directly applied to machine learning algorithm and wind speed prediction is performed. In order to reduce nonstationarity and nonlinearity of wind speed data, signal decomposition algorithms are utilized. The prediction accuracy can be improved by applying decomposed wind speed data to prediction. Liu et al.(2015) combined 4 different decomposition algorithm such as Empirical Mode Decomposition (EMD), Fast Ensemble Empirical Mode Decomposition (FEEMD), Wavelet Decomposition (WD) and Wavelet Packet Decomposition (WPD) with ELM, and he found that the prediction accuracy of short-term wind speed was the highest when ELM was combined with WPD. In this study, WPD is used as signal decomposition algorithm and ORELM is used as machine learning algorithm.

2.1 Wavelet Packet Decomposition (WPD)

In Wavelet analysis, a signal is divided into two subbands using analysis filters (Laine and Fan, 1993; Gokhale and Khanduja, 2010). Analysis filters are composed of low pass filter, which extracts low frequency subband named approximation of the signal, and high pass filter, which extracts high frequency subband named detail of the signal. WPD decomposes both of subbands into approximation and detail of subbands at each level of decomposition. Figure 1 shows that level 3 decomposition with WPD.

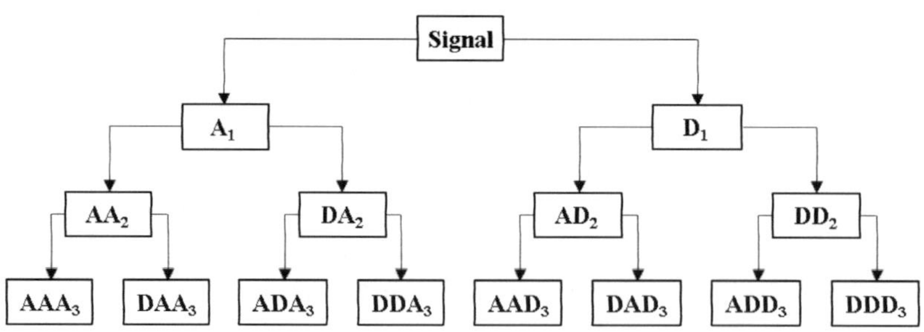

Figure 1. Level 3 decomposition using WPD

2.2 Outlier Robust Extreme Learning Machine (ORELM)

ORELM is machine learning algorithm for Single-hidden Layer Feedforward Neural Networks (SLFNs). Figure 2 shows the structure of neural network in ORELM. It randomly generates input weights $\boldsymbol{\omega}_i$ and biases b_i, which connect input layer and hidden layer, then optimizes output weights $\boldsymbol{\beta}_i$, which connects hidden layer and output layer.

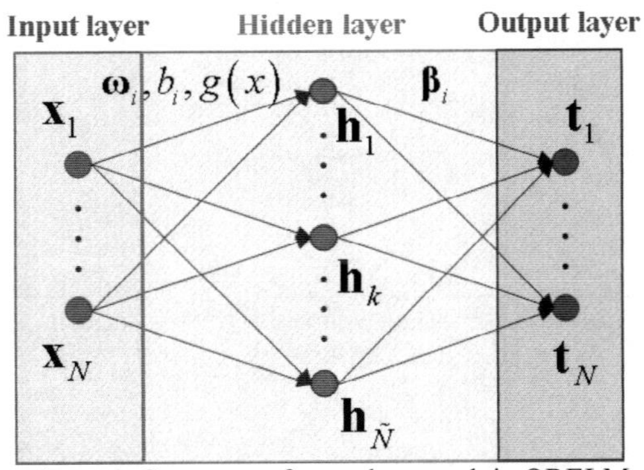

Figure 2. Structure of neural network in ORELM

For data set containing N samples $\{(\mathbf{x}_i, \mathbf{t}_i)\}_{i=1}^{N}$, where $\mathbf{x}_i = [x_{i1}, x_{i2}, \cdots, x_{in}]^T \in \mathbb{R}^n$ and $\mathbf{t}_i = [t_{i1}, t_{i2}, \cdots, t_{im}]^T \in \mathbb{R}^m$ refer to input vector and output vector respectively.
ORELM with $\tilde{N}$ hidden nodes and activation function $g(x)$ can be modeled mathematically as

$$\sum_{i=1}^{\tilde{N}} \beta_i g(\boldsymbol{\omega}_i \cdot \mathbf{x}_j + b_i) = \mathbf{t}_j,\ j = 1, 2, 3, \cdots, N. \tag{1}$$

The above equation can be written in matrix form as

$$\mathbf{H}\boldsymbol{\beta} = \mathbf{T} \tag{2}$$

where $\mathbf{H} = \begin{bmatrix} g(\boldsymbol{\omega}_1 \cdot \mathbf{x}_1 + b_1) & \cdots & g(\boldsymbol{\omega}_{\tilde{N}} \cdot \mathbf{x}_1 + b_{\tilde{N}}) \\ \vdots & \cdots & \vdots \\ g(\boldsymbol{\omega}_1 \cdot \mathbf{x}_N + b_1) & \cdots & g(\boldsymbol{\omega}_{\tilde{N}} \cdot \mathbf{x}_N + b_{\tilde{N}}) \end{bmatrix}_{N \times \tilde{N}}$, $\boldsymbol{\beta} = \begin{bmatrix} \beta_1^T \\ \vdots \\ \beta_{\tilde{N}}^T \end{bmatrix}_{\tilde{N} \times m}$, $\mathbf{T} = \begin{bmatrix} \mathbf{t}_1^T \\ \vdots \\ \mathbf{t}_N^T \end{bmatrix}_{N \times m}$.

ORELM optimizes output weights $\boldsymbol{\beta}$ corresponding training data with following optimization problem as

$$\min\left(\|\boldsymbol{\varepsilon}\|_1 + \frac{1}{C}\|\boldsymbol{\beta}\|_2^2\right) \text{ subject to } \mathbf{t} - \mathbf{H}\boldsymbol{\beta} = \boldsymbol{\varepsilon} \tag{3}$$

where $\|\ \|_1$ and $\|\ \|_2^2$ refer to ℓ_1-norm and ℓ_2-norm respectively.

2.3 Short-term wind speed forecasting with WPD-ORELM

The process of predicting short-term wind speed with WPD-ORELM is shown on Figure 3. In this study, Wind speed data is decomposed at level 6 decomposition which means that there are 64 decomposed data. Each decomposed data is applied to ORELM and trains neural networks. In machine learning algorithm, number of input data and hidden neurons are set to 6 and 50 respectively. Number of output data, which means forecasting time or step, is set to 1, 3, 6 and 12.

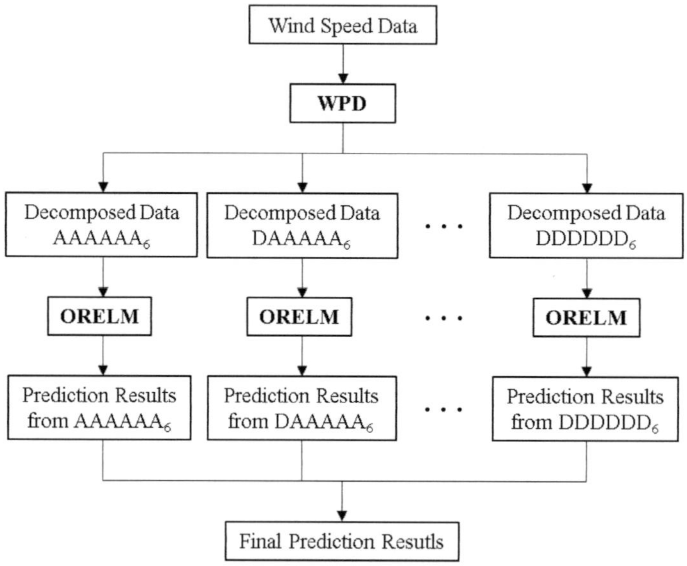

Figure 3. The process of wind speed forecasting with WPD-ORELM

The prediction accuracy is quantified with two criteria, which are mean absolute error (MAE) and mean absolute percent error (MAPE). They are calculated with following equations as

$$\text{MAE} = \frac{1}{N}\sum_{i=1}^{N}\left|t_i - \hat{t}_i\right| \tag{6}$$

$$\text{MAPE} = \frac{1}{N}\sum_{i=1}^{N}\left|\frac{t_i - \hat{t}_i}{t_i}\right| \times 100 \tag{7}$$

where N is total number of prediction points, t_i is measured wind speed and $\hat{t}_i$ is forecasted wind speed.

3. Case Study

Wind speed data measured in two different location, which are Incheon international airport weather station and bridge site in Busan, are used. At each location, wind speed data, which has 24 samples, during typhoon is set to wind speed to be predicted before 1, 3, 6 and 12 hours, or steps. All forecasting trials are executed 30 times independently. In this study, two different types of training data, which has 744 samples, or hourly wind speed data during 31 days, are applied to construct neural networks. Training data (a) is general wind speed data around 6m/s, and the other, training data (b) is wind speed data, which contains wind speed at the time of typhoon that occurred before the forecast origin.

3.1 Case 1 : Incheon airport weather station

The wind speed on August 28, 2012 when strong wind was generated by typhoon Volaven is targeted to predict. The test data are presented on Figure 4 and the last 24 samples on the graph are prediction targets. Two different types of training data are used and they are shown on Figure 5. Training data (a) is general wind speed data of which average is 6.6 m/s and standard deviation is 2.7 m/s. Training data (b) is wind speed data which contains wind speed generated by typhoon Muifa on August, 2011 and their average is 7.4 m/s and standard deviation is 3.8 m/s.

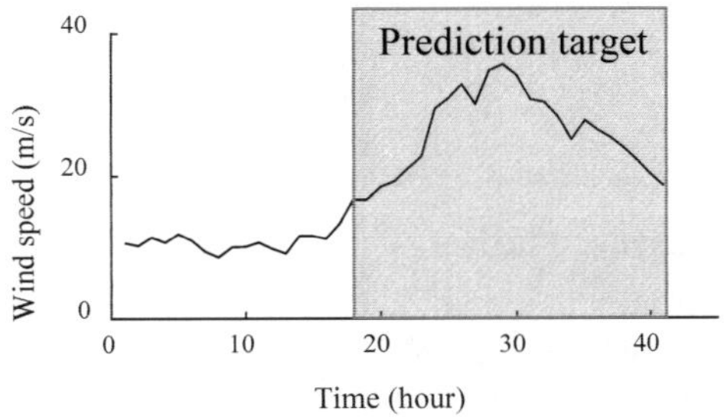

Figure 4 Test data from Incheon airport weather station

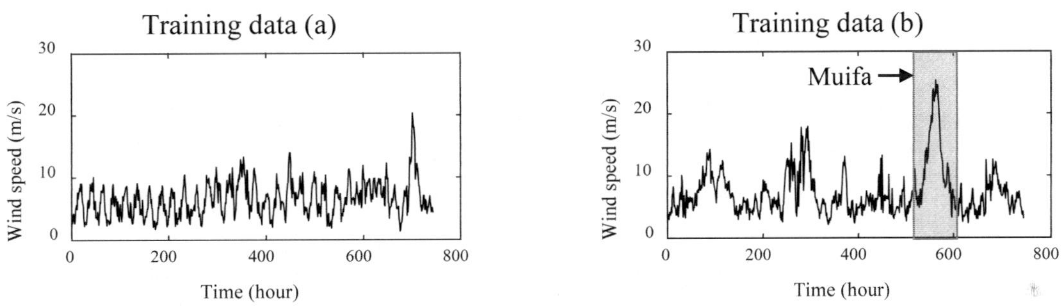

Figure 5. Two types of training data from Incheon airport weather station

3.2 Case 2 : Bridge site in Busan

The wind speed on September 17, 2012 when strong wind was generated by typhoon Sanba is targeted to predict. The test data are presented on Figure 6 and the last 24 samples on the graph are prediction targets. Two different types of training data are used and they are shown on Figure 7. Training data (a) is general wind speed data of which average is 5.0 m/s and standard deviation is 2.6 m/s. Training data (b) is wind speed data which contains wind speed generated by typhoon Volaven on August, 2012 and their average is 6.5 m/s and standard deviation is 3.9 m/s.

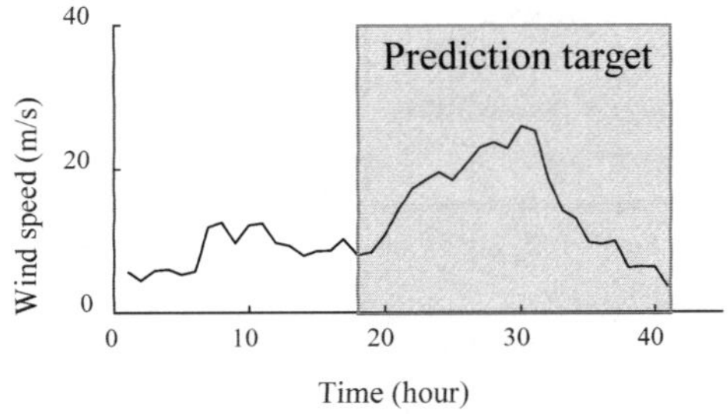

Figure 6 Test data from bridge site in Busan

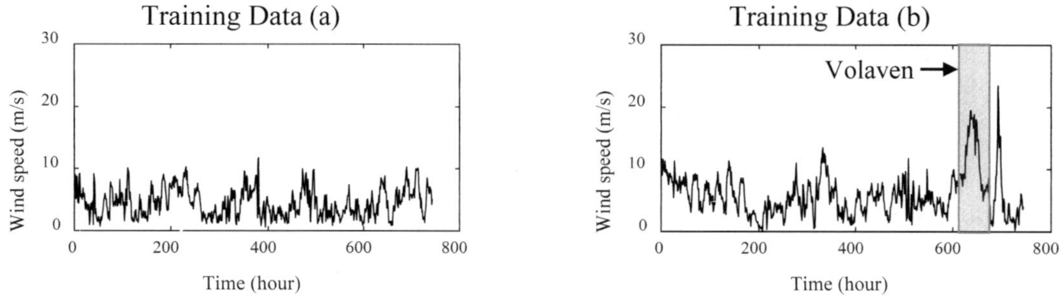

Figure 7. Two types of training data from bridge site in Busan

4. Forecasting results

4.1 Case 1 : Incheon airport weather station

Each forecasting results are presented on Figure 8 and forecasting errors are summarized in Table 1. In each cases, the range of predicted wind speed during 30 times executions was shaded on the graphs and red circle represents average of predicted wind speed regarded as final prediction.

	Training Data (a)	Training Data (b)
1 Step	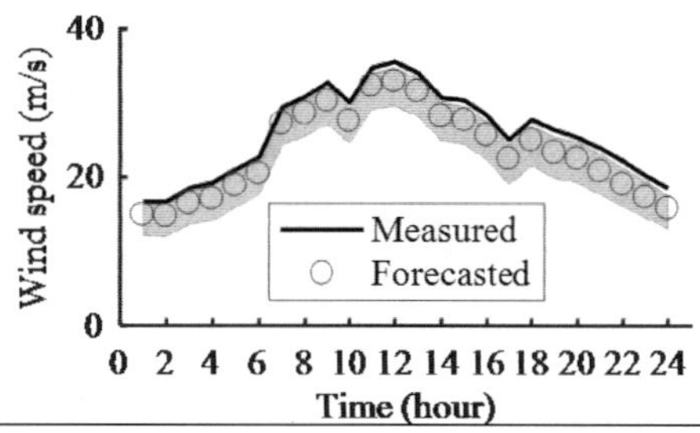	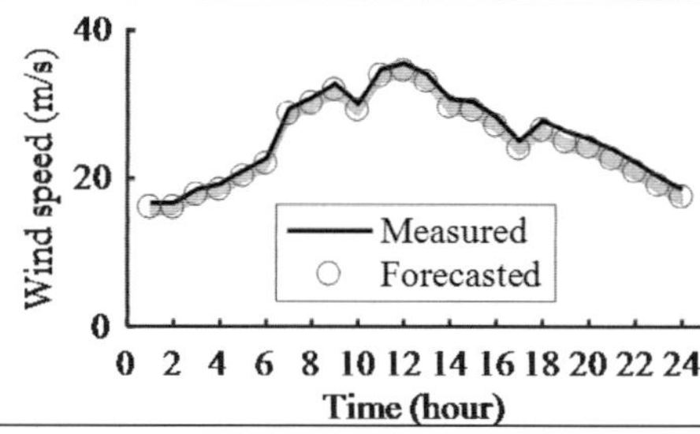
3 Step	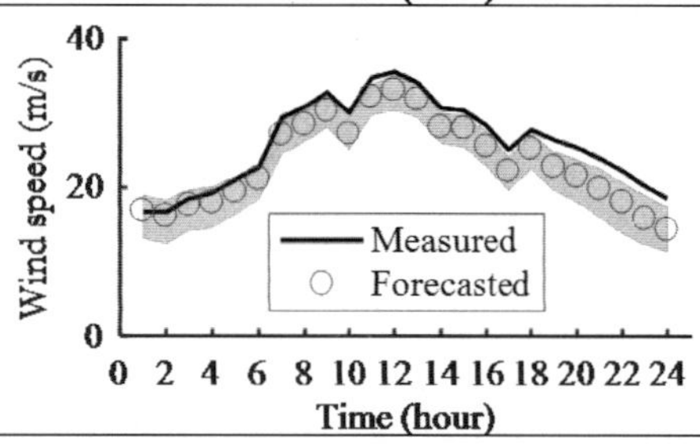	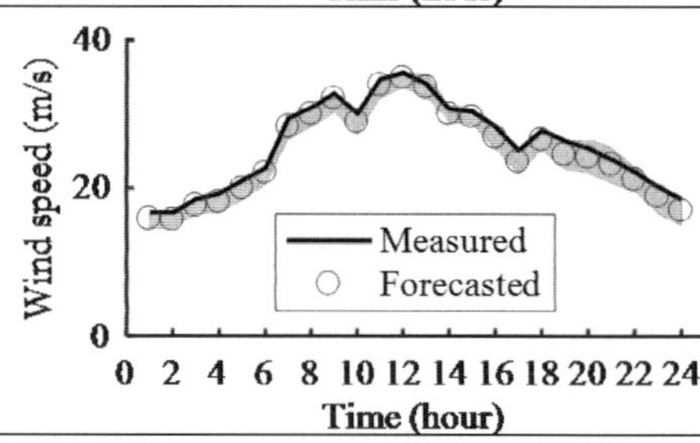
6 Step	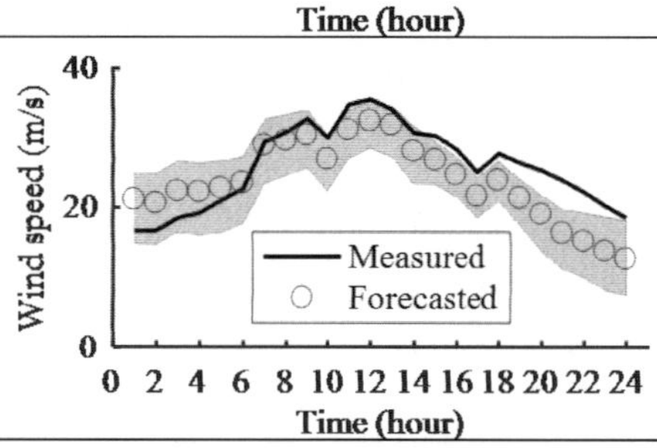	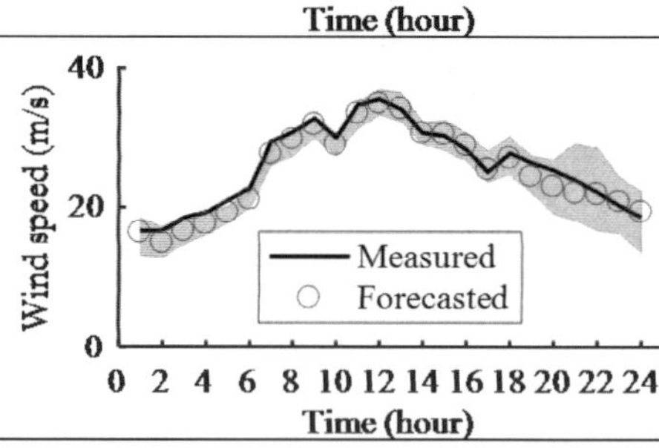

12 Step

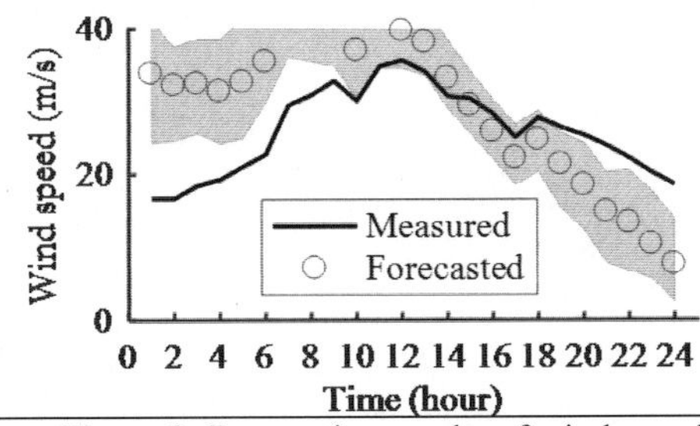

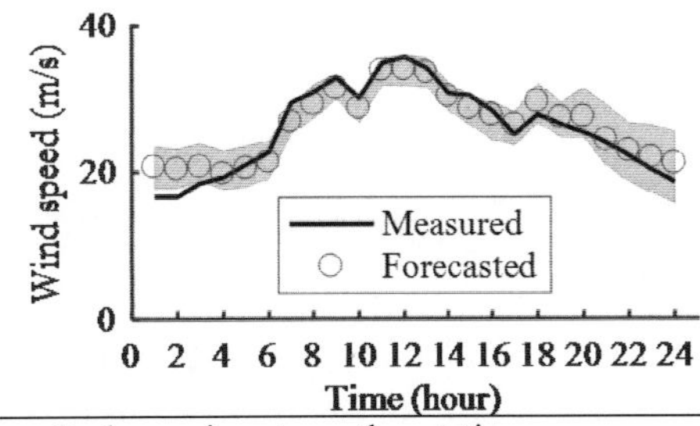

Figure 8. Forecasting results of wind speed data from Incheon airport weather station

Table 1. Forecasting errors of wind speed data from Incheon airport weather station

	Training Data (a)		Training Data (b)	
	MAE (m/s)	MAPE (%)	MAE (m/s)	MAPE (%)
1 Step	2.54	10.2	1.05	4.2
3 Step	2.55	10.0	1.09	4.5
6 Step	3.74	15.9	1.08	4.5
12 Step	8.26	36.9	1.54	6.7

As shown on the graphs, the larger prediction step is applied, the greater forecasting errors occurs. Comparing the forecasting accuracy of training data (a) and training data (b) shows that the accuracy is 2.5 to 5.5 times higher when training data (b) is used. In the case of using training data (b), wind speed is predicted within 5% error up to 6 steps.

4.2 Case 2 : Bridge site in Busan

Each forecasting results are presented on Figure 9 and forecasting errors are summarized in Table 2. In each cases, the range of predicted wind speed during 30 times executions was shaded on the graphs and red circle represents average of predicted wind speed regarded as final prediction.

	Training Data (a)	Training Data (b)
1 Step	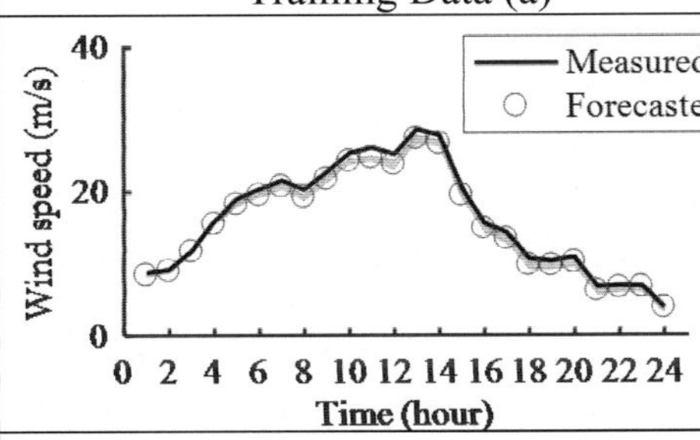	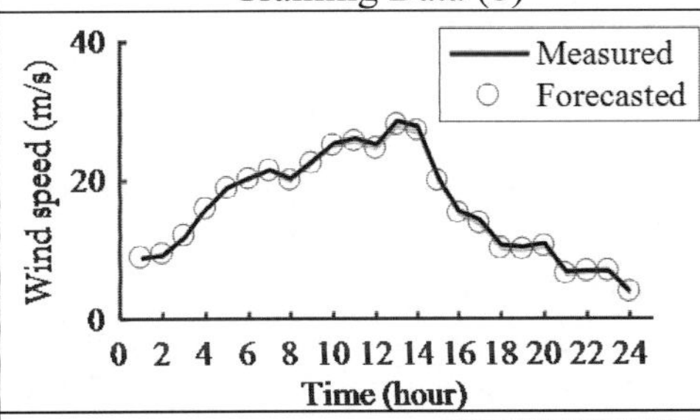
3 Step	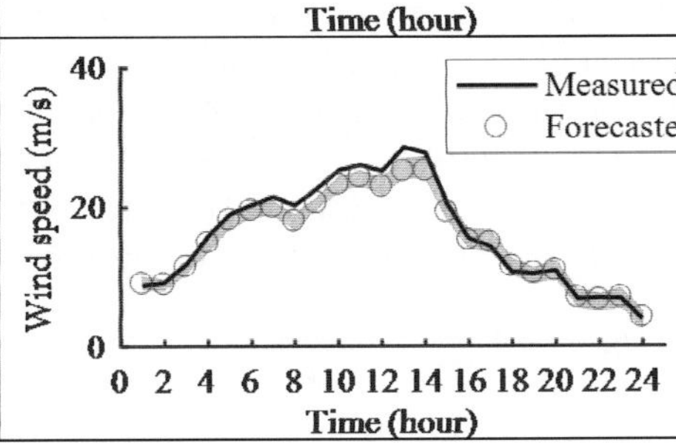	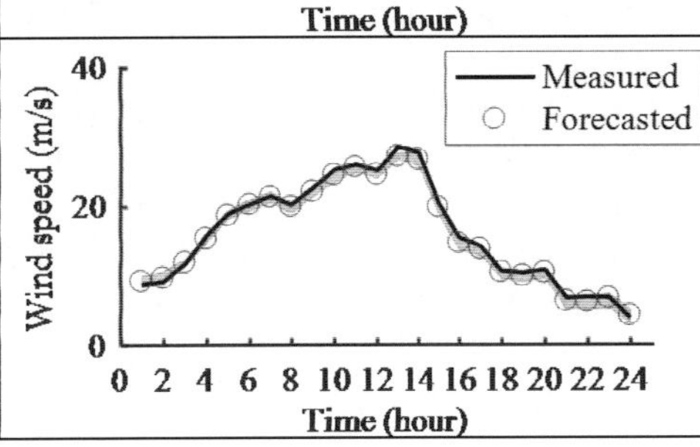

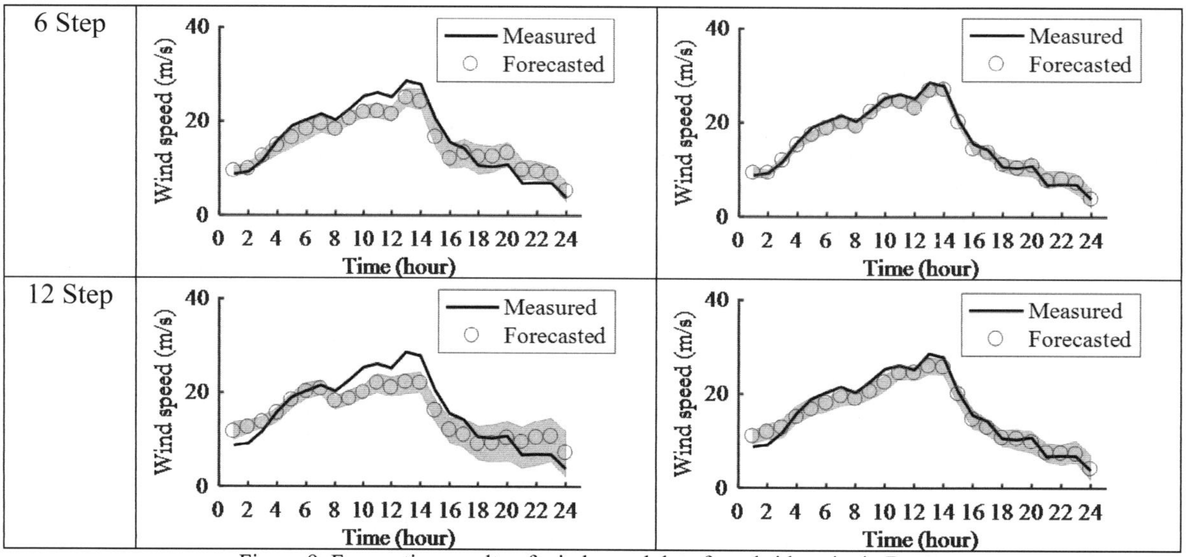

Figure 9. Forecasting results of wind speed data from bridge site in Busan

Table 2. Forecasting errors of wind speed data from bridge site in Busan

	Training Data (a)		Training Data (b)	
	MAE (m/s)	MAPE (%)	MAE (m/s)	MAPE (%)
1 Step	0.70	4.8	0.28	2.0
3 Step	0.96	5.5	0.47	3.8
6 Step	2.09	16.5	0.72	5.0
12 Step	2.64	23.1	1.23	9.0

As shown on the graphs, the larger prediction step is applied, the greater forecasting errors occurs. Comparing the forecasting accuracy of training data (a) and training data (b) shows that the accuracy is 2.0 to 3.0 times higher when training data (b) is used. In the case of using training data (b), wind speed is predicted within 5% error up to 6 steps.

4. Conclusions

As a way to improve the bridge management system during strong wind, applicability of short-term wind speed forecasting technique is investigated in this study. In order to verify prediction accuracy to strong wind, short-term wind speed forecasting is applied to predict strong wind during typhoon before 1, 3, 6 and 12 steps. As the prediction step increases from 1 to 12, the prediction error increases by about 3 times. In the case of training with wind speed data of past typhoon season, the accuracy of prediction is 3 times higher than that of general wind speed training data. In addition, when using wind speed data of past typhoon season as training data, wind speed can be effectively predicted within 5% error up to 6 steps. It can be said that short-term wind speed forecasting can be applied to bridge management system in order to predict the occurrence of strong wind during typhoon. In this study only focused on strong wind during typhoon. Further research will cope with strong wind during monsoon season.

Acknowledgements

This research was supported by a grant (17SCIP-B119964-02-000000) from the Ministry of Land, Transport and Maritime Affairs of the Korean government through the Korea Bridge Design & Engineering Research Center of Seoul National University. The first author of this study was supported by the BK21 PLUS research program of the National Research Foundation of Korea. Also, Busan Infrastructure Corporation provided field-monitored wind data for this research.

References

Baker, C 2015, 'Risk analysis of pedestrian and vehicle safety in windy environments', *Journal of Wind Engineering and Industrial Aerodynamics*, vol. 147, pp. 283-290.

Cancelliere, R, Gosso, A & Grosso, A 2013, 'Neural networks for wind power generation forecasting: A case study', *10th IEEE International Conference*, Evry, France, 10-12 April, pp. 666-671.

Gokhale, MY & Khanduja, DK 2010, 'Time domain signal analysis using wavelet packet decomposition approach', *Int'l J. of Communications, Network and System Sciences*, vol. 3, no. 3, pp. 321-329.

Huang, GB, Zhu, QY & Siew, CK 2006, 'Extreme learning machine: theory and applications', *Neurocomputing*, vol. 70, no. 1, pp. 489-501.

Kim, SJ, Yoo, CH & Kim, HK 2016, 'Vulnerability assessment for the hazards of crosswinds when vehicles cross a bridge deck', *Journal of Wind Engineering and Industrial Aerodynamics*, vol. 156, pp. 62-71.

Laine, A & Fan, J 1993, 'Texture classification by wavelet packet signatures', *IEEE Transactions on pattern analysis and machine intelligence*, vol.15, no. 11, pp. 1186-1191.

Liu, H, Tian, HQ & Li, YF 2015, 'Four wind speed multi-step forecasting models using extreme learning machines and signal decomposing algorithms', *Energy Conversion and Management*, vol. 100, pp. 16-22.

Meng, A, Ge, J, Yin, H, & Chen, S 2016, 'Wind speed forecasting based on wavelet packet decomposition and artificial neural networks trained by crisscross optimization algorithm', *Energy Conversion and Management*, vol. 114, pp. 75-88.

Peng, X, Zheng, W, Zhang, D, Liu, Y, Lu, D & Lin, L 2017, 'A novel probabilistic wind speed forecasting based on combination of the adaptive ensemble of on-line sequential ORELM and TVMCF', *Energy Conversion and Management*, vol. 138, pp. 587-602.

Sun, W & Liu, M 2016, 'Wind speed forecasting using FEEMD echo state networks with RELM in Hebei, China', *Energy Conversion and Management*, vol. 114, pp. 197-208.

Wang, J, Hu, J, Ma, K & Zhang, Y 2015, 'A self-adaptive hybrid approach for wind speed forecasting', *Renewable Energy*, vol. 78, pp. 374-385.

Zhang, K & Luo, M 2015, 'Outlier-robust extreme learning machine for regression problems', *Neurocomputing*, vol. 151, pp. 1519-1527.

The 8th International Conference on
Structural Health Monitoring of Intelligent
Infrastructure Brisbane, Australia | 5-8 December 2017

Validating Strain Gauge Placement Methods for Structural Health Monitoring of Large Cable Support Bridge

Z. Y. Wu[1], K. Zhou[2], H. W. Shenton III[3] and M. J. Chajes[3]

[1] *Bentley Fellow, Bentley Systems, Incorporated, Watertown, CT, USA, Email:* zheng.wu@bentley.com, [2]*Formerly Research Intern, Bentley Systems, Incorporated, Watertown, CT, USA,* [3]*Professor, University of Delaware, DE, USA*

Abstract

The field measurements that provide engineers with adequate useful information are required for effectively monitoring and evaluating structural performance. Strain information generally shows powerful capacity for detecting the local structural property variations. Thus strain gauges are commonly used for monitoring and testing structural performance. Placing strain gauges is the key step towards a cost-effective monitoring or testing program. This paper presents the applications of strain gauge placement optimization for a large cable supported bridge to validate the methods that are developed for maximizing the strain contributions to the structural performance assessment. The sensor placement solutions are optimized using a genetic algorithm-based optimization solver, leading to a number of features that enrich the application flexibility. A case study has been carried out to demonstrate the effectiveness of the developed tool for real world applications.

1. Introduction

Structural health monitoring has gained growing attention from the engineering community to improve the reliability and sustainability of civil infrastructures, and thus minimize the economic or human life loss of unpredicted structural failure. In recent decades, along with the advancement of finite element modeling and sensing technologies, vibration measurement-based structural performance monitoring and assessment have been increasingly undertaken in practical activities (Liu et al. 2009; Sun et al. 2013). The degradation of structural performance oftentimes is caused by the material, geometry defects/damages within the structure, which can be reflected by the measured vibration responses. To adequately and accurately implement performance evaluation, the data acquisition system needs to be set up appropriately for the purpose of collecting useful information. The general data acquisition system consists of many types of sensors, in which the accelerometer is the most popularly used sensor among them. Accelerations measured by accelerometers and further converted to modal information are widely adopted for structural performance evaluation and health monitoring (Lynch et al. 2006; Sun et al. 2015; Kaloop et al. 2016). On the other hand, strain measurement is the essential information used to monitor the performance of the structure with local property deterioration (Kesavan et al. 2008; Wu and Xu 2014; Zhang et al. 2015). However, even using high sensitive strain measurements, structural performance may still be assessed ineffectively if the strain gauges installed cannot adequately provide the useful strain information related to local property change. In engineering practice, the validity of measured strains is solely ensured by strain gauge placement proposed by the professional engineers with empirical experience and judgement. Such decision of strain gauge placement sometimes lacks of the support from rigid engineering discipline and thus possibly becomes irrational. Hence, it is imperative to formulate physics-oriented metrics to facilitate strain gauge placement (Zhou and Wu 2017). In this paper, an integrated approach has been applied to a large cable supported bridge to validate the methods and software tool DarwinSHM Pretest (Zhou et al. 2017) for sensor placement.

2. Methodology

The detailed methodology formulations were given by Zhou and Wu (2017) for conducting the strain gauge placement. As mentioned, there are two different approaches, i.e., strain difference coverage (SDC) and strain contribution coverage (SCC), which attempt to evaluate the strain gauge sensing performance. Specifically, SCC is implemented by analogy with the quantification of mode shape rank contribution in the effective independence method (Li et al. 2007). As compared with SCC, SDC is conceptually straightforward as it directly computes the strain response change due to the possible damages, and the strain gauges are placed to capture as many damages as possible so that the structural performance can be sufficiently evaluated by the strain data accordingly. The Darwin Optimization Framework (Wu et al. 2012) is applied to optimize the sensor placement solution based upon the metrics defined in the proposed approaches. The different optimization objectives are defined for the two proposed methods, the stain gauge placement optimization model is generalized as:

- Search for N_s strain gauge locations, $\mathbf{S} = \{S_1 \ S_2 \ ... \ S_{N_s}\}$
- Maximize one of coverage capacity objectives (Zhou and Wu 2017)
- Subject to entire solution space (*end cross-sections of all members*)

The strain gauge location candidates are discrete decision variables sampled from solution space. The optimization problem is solved in two stages, as illustrated in Fig. 1, including (1) database creation for solution evaluation; (2) optimization via Darwin Optimization Framework, i.e., a generic optimization software tool built upon the genetic algorithms (Wu and Simpson 2001). It enables parallel optimization on a local many-core machine and HPC cluster of computers, and solves for single and multi-objective optimization problems with/without constraints.

The general implementation workflow is illustrated in Fig. 1, which is achieved by three components including *OSPEngine*, *OSPFitness*, *OSPSolverUI*. The data exchange between the components is supported by well-defined Application Programming Interfaces (APIs). The roles and functionalities of those components are elaborated as follows.

2.1 OSPEngine

This component undertakes database generation based on above two methods, as can be seen in Fig. 1. During database generation process, *OSPEngine* resorts to some dependencies, including openSTAAD [34] that is an object library with rich APIs for creating new analysis scenarios, performing the finite element analysis and retrieving the analysis results. Since current version of openSTAAD is not able to directly output strain response, we use the simple beam bending theory to approximately estimate the strain of the member cross-section based on the internal force, bending moment of cross-section retrieved by OpenSTAAD (Bentley 2012). The tensile/compressive strain at certain point of cross-section in one member can be calculated by using the internal force and moment of cross-section, respectively.

It is previously mentioned that the strain gauges are to be installed at the end cross-sections of members without specifying the point locations, since the strain gauges are assumed to measure the maximum strains throughout entire end cross-sections. This is realized by evenly subdividing the one cross-section into different pieces with different distances y. By calculating and comparing their strains, the maximum value of this cross-section can be found. The integrated methods are implemented by iteratively calibrating the FE model with introduced damage, the FE solver is called via openSTAAD API in a loop as shown in Fig. 2.

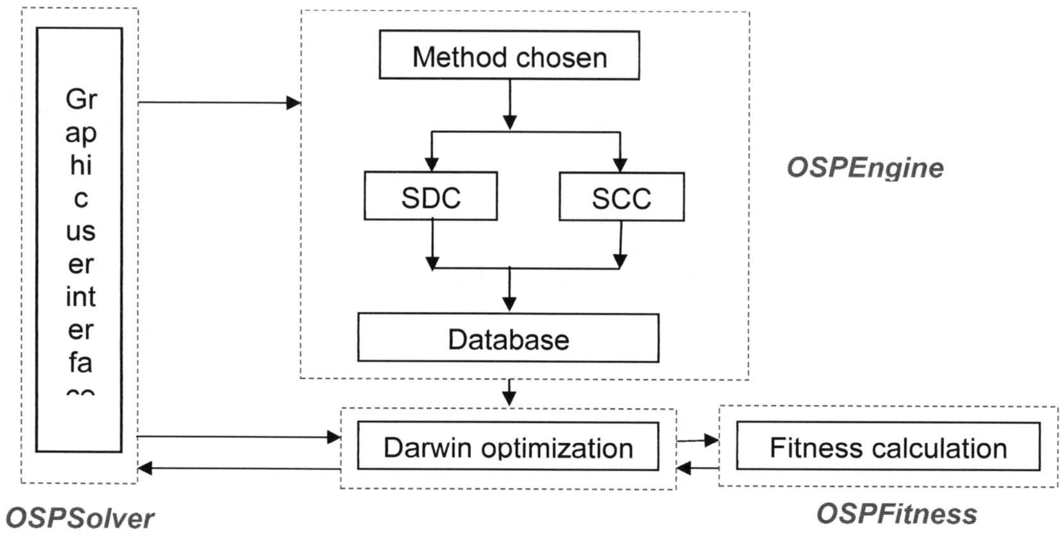

Figure 1 DarwinSHM Pretest architecture overview.

2.2 Darwin Optimization Framework

This component is to perform the solution search for strain gauge placement. Each solution candidate will be generated and passed to *OSPFitness* for objective evaluation as follows.

2.3 OSPFitness

The responsibility of this component is to prepare the optimization problems in the format as required for Darwin optimization application. Specifically, the decision variables must be specified in the predefined format, e.g. the variable name, lower and upper bounds together with the solution increment between the bounds. Each of the possible solutions is generated by and then passed from Darwin Optimization Framework to this component to calculate one of optimization objectives. The calculated objective value is returned to Darwin Optimization Framework for enhancing the solution search.

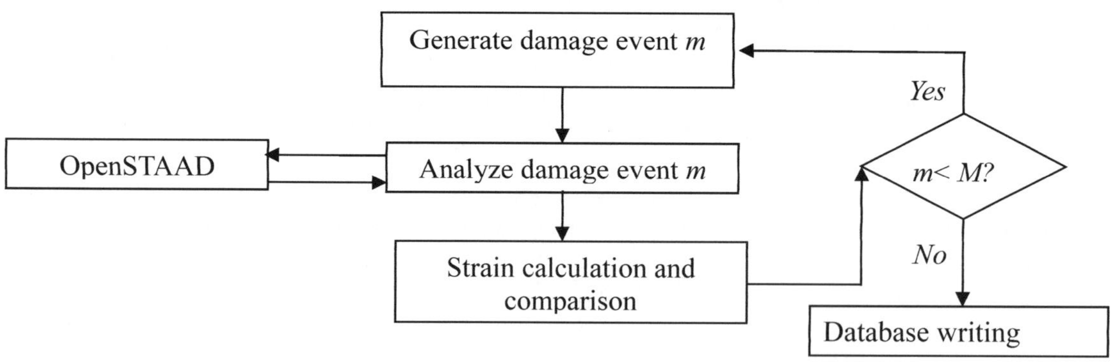

Figure 2 Database creation for strain gauge placement

2.4 OSPSolverUI

This is the Graphical User Interfaces (GUIs) that allow users to specify parameters and review the results analyzed. All GUIs have been integrated into DarwinSHM Pretest, as shown in Fig. 3.

Figure 3 Method selection and data operation user interface.

The tool is developed as the wizard that guides a user to conduct the strain gauge placement in two primary steps.

1) Database generation. These two methods are built upon the similar underlying idea, resulting in the common user-specified parameters. These parameters include: (1) file path of FE engine solver; (2) file path of given FE model; (3) file path of database to be generated. (4) sensor accuracy threshold to distinguish the measured strain change caused by damage, as indicated in Eq. (3) and Eq. (10); (5) the number of damage events which are analyzed iteratively for comparison; (6) the range of stiffness reduction coefficients; and (7) the maximum and minimum numbers of damaged elements in each event.

2) Solution optimization and evaluation. Once all parameters for database generation are set up, one database containing the comparison results of evaluated tensile/compressive strains will be created. It indeed offers the versatility of strain gauge placement study, including (1) optimizing the placement of a given number of strain gauges without any pre-determined strain gauges, (2) optimizing the additional number of strain gauges in combination with pre-determined ones in an incremental manner; (3) evaluating a pre-determined strain gauge

placement solution (without optimization), and also (4) exploring other features to facilitate and customize the placement optimization analysis.

4. Case Study

The tool has been applied to the strain gauge placement of the Indian River Inlet Bridge (IRIB) in Delaware. A permanent monitoring system, with more than 140 sensors, was installed on the IRIB during construction (Shenton et al. 2010, 2014 2015; Chajes et al. 2015). A 3D finite element model has been built with 470 nodes and 767 members (Shenton et al 2013; Marquez et al. 2013). In this case, we define two-side truck moving load. Since 40 strain gauges has been installed during the construction. The placement of the sensor, shown in Fig.4(b), was conducted by experience. The tool is applied to validate the methods.

Taking SDC method as an example, the database for evaluating strain gauge placement is generated with the parameters defined as follows.

1) The stiffness reduction coefficient is specified in the range of 0 and 0.5. For the illustration purpose, Young's Modulus is used to represent the reduction of element/member stiffness.
2) The number of possible damaged elements/members is defined from 5 to 10, and all the elements/members, including boundary and interior elements/members in the FE model, is likely subject to the stiffness reduction.
3) The strain difference threshold is defined as 8×10^{-4}.

With the parameters defined above, 1,000 damage scenarios were then randomly generated and iteratively analyzed under the two-way moving load. For each damage case, the strain difference between the damaged and healthy scenarios is calculated for both ends of an element and compared with the threshold to designate the sensitivity of a damage event at a possible strain gauge location. If the strain difference is less than the threshold at one location, the location is deemed not sensitive to the damage scenario. Therefore, the strain gauge place at such a location will not be able to detect the damage scenario. In other word, the damage scenario is not covered by a strain gauge placed at such a non-sensitive location. Thus, the strain difference is employed to assess the damage capture capability of each possible strain gauge location. After iterative FE evaluation and comparison, the evaluation database is eventually generated for evaluating the strain gauge placement.

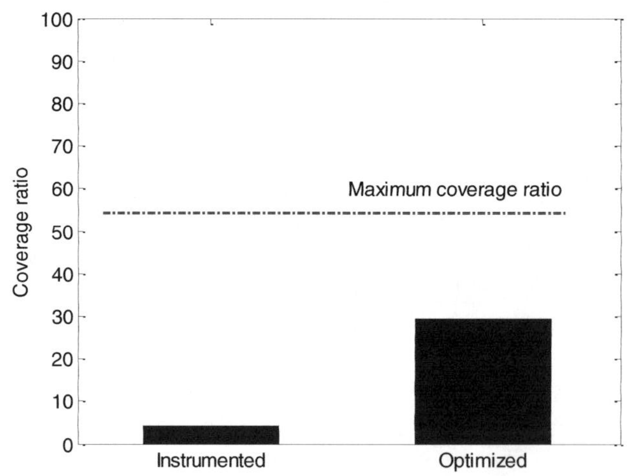

(a) Performance comparison of IRIB strain gauge placement

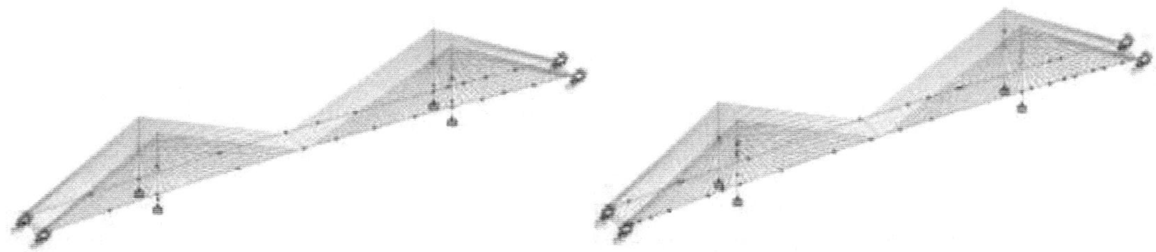

(b) Instrumented sensor locations (c) Optimized sensor locations

Figure 4 Results of strain gauge placement for IRIB

By applying the sensor placement tool using the created database and the defined optimization objective, a 40 gauge optimized layout was obtained as shown in Fig. 4(c). Red dot indicates the strain gauge, which is located at the end cross-section of element/member. The coverage ratio improvement by two placement solutions can be seen in Fig. 4(a). The significant improvement in coverage capacity has been achieved by our tool based on this benchmark validation analysis.

We have further investigated the sensing performance with respect to different numbers of the sensors. The performance curve, as illustrated in Fig. 5 indicates that 180-strain gauges results in the most cost-effective placement, since the very small gap yielded with respect to maximum coverage ratio. Meanwhile, the associated placement with the members and the affiliated end cross-section distribution is given in Fig.6.

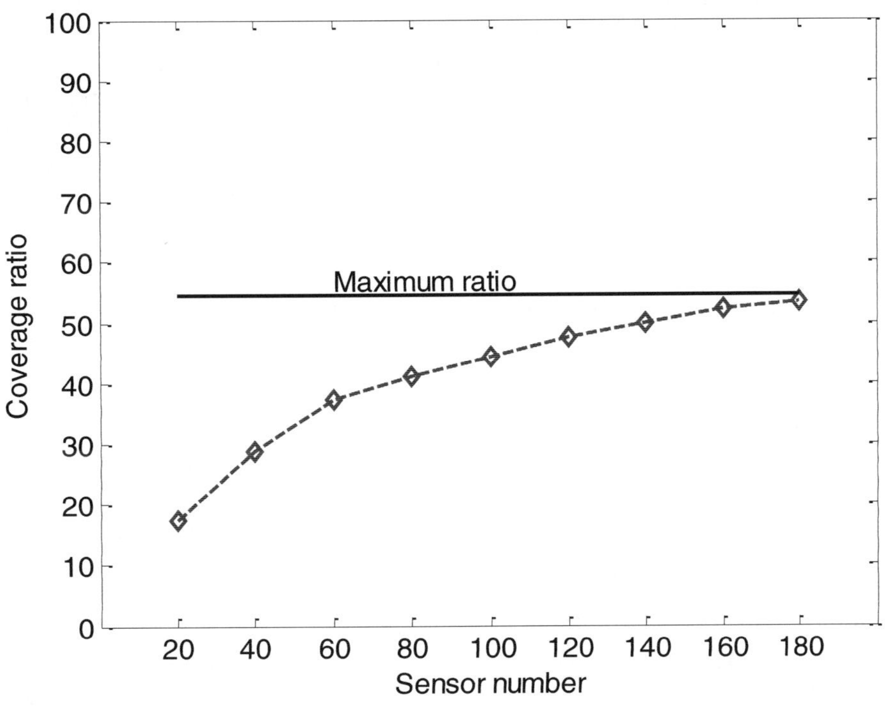

Figure 5 IRIB strain gauge placement by strain difference coverage (SDC) method

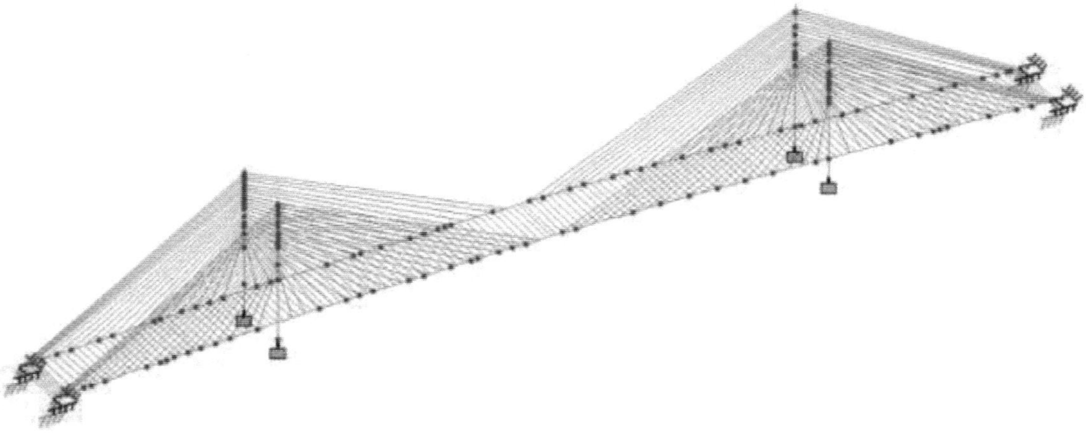

Figure 6 Optimized placement of 180 strain gauges for IRIB

5. Conclusions

This paper demonstrates the application of an integrated approach and software tool for optimizing strain gauge placement. Two strain gauge placement methods are integrated with a generic optimization solver. A user-friendly user interface facilitates the optimization process by enhancing the application practice in terms of engineering perspectives. As demonstrated with the case study, the proposed methods are effective at identifying the solutions of good quality for strain gauge placement. However, like many modeling methods and tools, this tool still has practical limitations since the real uncertainties, e.g., ambient randomness and unpredicted load transfer, etc., are not taken into account. Therefore, engineering judgement oftentimes is also required and important to achieve the sensible results for practical applications, and the final solution must be determined by the sound engineering judgement while the methods and the well-integrated tool are capable of significantly facilitating the decision-making process of strain gauge placement.

References

Bentley Systems. (2012). *OpenSTAAD V8i(SELECTseries 4), Users' Manual*. Yorda Linda, CA, USA.

Bernini, R., Fraldi, M., Minardo, A., Minutolo, V., Carannante, F., Nunziante, L. and Zeni, L. (2005). "Damage detection in bending beams through Brillouin distributed optic-fibre sensor." *Bridge Structures,* 1(3), 355-363.

Chajes, M.J., Shenton, H.W., Al-Khateeb, H.T., Wenczel, G.R., and Ramanna, N. (2015). "Structural Health Monitoring of the Indian River Inlet Bridge: Results from Controlled Load Tests Conducted over the First Two Years of Service." *Proceedings of the 7th International Conference on Structural Health Monitoring of Intelligent Infrastructure*, Torino, Italy.

Kaloop, M. R., Hu, J. W., Sayed, M. A. and Seong, J. (2016). "Structural Performance Assessment Based on Statistical and Wavelet Analysis of Acceleration Measurements of a Building during an Earthquake." *Shock and Vibration,* 2016.

Kesavan, A., John, S. and Herszberg, I. (2008). "Strain-based structural health monitoring of complex composite structures." *Structural Health Monitoring,* 7(3), 203-213.

Li, D. S., Li, H. N. and Fritzen, C. P. (2007). "The connection between effective independence and modal kinetic energy methods for sensor placement." *Journal of Sound and Vibration,* 305(4-5), 945-955.

Liu, M., Frangopol, D. M. and Kim, S. (2009). "Bridge system performance assessment from structural health monitoring: A case study." *Journal of Structural Engineering,* 135(6), 733-742.

Lynch, J. P., Wang, Y., Loh, K. J., Yi, J. H. and Yun, C. B. (2006). "Performance monitoring of the Geumdang Bridge using a dense network of high-resolution wireless sensors." *Smart Materials and Structures,* 15(6), 1561-1575.

Marquez, P., Chajes, M.J., Shenton III, H.W., Al-Khateeb, H., Wenczel, and Cardinal, J. (2013). "Structural Health Monitoring: Establishing the Baseline Performance of Delaware's Indian River Inlet Bridge." *Proceedings of the 6th International Conference on Structural Health Monitoring of Intelligent Infrastructure*, Hong Kong.

Shenton III, H.W., Carson, P.D., Chajes, M.J., O'Shea, D., Kursinsky, C., (2010). "Delaware's Indian River Inlet Bridge: a comprehensive plan for long-term structural monitoring," Transportation Research Record: Journal of the Transportation Research Board, No. 2201, 148-153.

Shenton III, H.W., Chajes, M.J., Wenczel, and Davidson, K. (2013). "Lessons Learned in the Construction and Early Experiences with the Indian River Inlet Bridge SHM System." *Proceedings of the 6th International Conference on Structural Health Monitoring of Intelligent Infrastructure*, Hong Kong.

Shenton III, H.W., Chajes, M.J., Wenczel, G., Ramanna, N., Al-Khateeb, H., Davidson, K., and Marquez, P. (2014). "Structural Health Monitoring of Delaware's Indian River Inlet Bridge: Year One Update." *Proceedings of the 2014 Structures Congress*, Boston, Massachusetts.

Shenton, H.W., Fernandez, M, Ramanna, N, Chajes, M.J., and Al-Khateeb, H.T. (2015). "Structural Health Monitoring of a Cable-Stayed Bridge: Using Tiltmeter Data to Determine Edge Girder Deflections." *Proceedings of the 7th International Conference on Structural Health Monitoring of Intelligent Infrastructure*, Torino, Italy.

Sun, H. and Buyukozturk, Oral. (2015). "Optimal sensor placement in structural health monitoring using discrete optimization." *Smart Materials and Structures,* 24(12), 125034.

Sun, H., Waisman, H. and Betti, Raimondo. (2013). "Nondestructive identification of multiple flaws using XFEM and a topologically adapting artificial bee colony algorithm." *Numerical Methods in Engineering*, 95(10), 871-900.

Wu, Z. Y. and Simpson, A. R. (2001). "Competent genetic-evolutionary optimization of water distribution systems." *Journal of Computing in Civil Engineering,* 15(2), 89-101.

Wu, Z. Y. and Xu, G. Q. (2014). "A Novel Formulation for Finite Element Model-based Damage Detection" International Conference on Bridge Maintenance, Safety and Management, July, 2014, Shanghai, China.

Wu, Z. Y., Wang, Q., Butala, S., Mi, T. and Song, Y. (2012). Darwin optimization framework user mannual, http://communities.bentley.com/communities/other_communities/bentley_applied_research/w/bentley_applied_research__wiki/software-prototypes.

Zhang, J., Guo, S. L., Wu, Z. S. and Zhang, Q. Q. (2015). "Structural identification and damage detection through long-gauge strain measurements." *Engineering Structures,* 99, 173-183.

Zhou, K. and Wu, Z. Y. (2017). “Strain gauge placement optimization for structural performance assessment”, Engineering Structures 141(15):184-197.

Zhou, K., Wu, Z. Y., Yi, X. H., Zhu, D. P., Narayan, R. and Zhao, J. (2017) Generic Framework of Sensor Placement Optimization for Structural Health Modeling”, ASCE Journal of Computing in Civil Engineering 31(4):04017018

The 8th International Conference on
Structural Health Monitoring of Intelligent Infrastructure
Brisbane, Australia | 5-8 December 2017

Uncertainty in determining the deformed shape of beams using the conjugate beam method

C. Kavanaugh[1], B. Glisic [2]

[1] Princeton University, Civil and Environmental Engineering Department, 41 Olden Street, Princeton, NJ, 08540, USA. Email: ckavanaugh9@gmail.com *[2] Princeton University, Civil and Environmental Engineering Department, 41 Olden Street, Princeton, NJ, 08540, USA.*

Abstract

The deformed shape of a structure can be defined as the curvature of the centroid line of the structure after deformation. This curvature can be used to indicate abnormal behavior in structures – particularly those with beam-like configurations. The deformed shape is an extremely useful tool in the field of structural health monitoring, as it can help with evaluating the serviceability and overall stability of a structure. However, monitoring the deformed shape over long periods of time has proven difficult to achieve with necessary levels of accuracy. The deformed shape can be analyzed using both direct and indirect monitoring techniques. Direct monitoring techniques are sensitive to environmental factors, and therefore accurate deformation measurement using such methods is difficult to achieve. Double integration of a structure's curvature, a method of indirect monitoring, is also limited in its scope, as it involves manually correcting the deformed shape after integration, and is difficult to use if sensors are not evenly spaced on the structure. Other indirect methods also require detailed knowledge of boundary conditions in order to determine the deformed shape. However, alternative methods exist for indirectly determining the deformed shape of the structure. The aim of this paper is to analyze the deformed shape of the beam via the conjugate beam method, which applies the measured curvature as a load on the structure in order to determine its deformed shape. This study presents an analysis of the uncertainties and errors associated with the conjugate beam method, as well as an evaluation of the number of sensors needed when using this method in order to obtain the desired degree of accuracy. The error in this method depends on the type and positioning of loading on the structure, the magnitude of the curvature of the structure, and the number of sensors that the structure is equipped with. This method is also tested on data from a laboratory specimen and a real structure in order to validate the method as a way of accurately determining the deformed shape.

1. Introduction

Structural Health Monitoring (SHM) uses sensors placed on or around structures in order to monitor behavior and performance over the lifetime of the structure. These sensors can detect damage and unusual structural behavior via a variety of different parameters. The deformed shape of the structure is one of the most useful parameters for analyzing stability and serviceability, particularly in beam-like configurations such as continuous girders. These unusual behaviors in strain and curvature can be directly correlated to the deformed shape of the structure. Therefore, the deformed shape plays an important role in design-criteria for structures that behave like beams (ACI, 2008 and AIS, 2005). The deformed shape is therefore of particular interest when evaluating the long-term performance of structures.

Current methods determine the deformed shape via direct measurements or indirect derivation (Sigurdardottir et al., 2017). Unfortunately, direct measurements are vulnerable to environmental effects, such as temperature and humidity, and contain challenges for accurate implementation, and indirect measurements require knowledge about the structure's support conditions (Sigurdardottir et al., 2017).

Sheng, et al. offers another method for determining the deformation of a beam using the observed curvature of the structure and the conjugate beam method, which was developed by H. Muller-Breslau in 1865 (Sheng et al., 2010 and Hibbeler, 2012). This method requires dividing the structure into cells. Each of these cells are equipped with a set of parallel strain sensors. The curvature can be measured using the strain measurements from the top and bottom sensors at each location. This curvature is then used as the load on the conjugate beam (Sheng et al., 2010). The slope of the real beam is numerically equivalent to the shear in the conjugate beam, and the displacement of a point on the real beam is numerically equivalent to the bending moment of the same point on the conjugate beam (Hibbeler, 2012). In Sheng, et al., this method is used in conjunction with observed curvature values in order to determine the deformation of the beam at each sensor location (Sheng et al., 2010). A linearly interpolated curve is assumed between each sensor (Sheng et al., 2010). The method provides good correlation to laboratory results (Sheng et al., 2010). However, there is little discussion of the errors associated with this method.

This aim of this research is to examine the errors associated with the conjugate beam method. Preliminary results are shown in this paper.

2. Determination of Deformed Shape via the Conjugate Beam Method and Associated Errors

Understanding uncertainties associated with specific methods of analysis is crucial in interpreting results related to structural health monitoring, and helps to determine the proper sensor number and configuration necessary for achieving the desired level of accuracy in results (Sigurdardottir et al., 2017). As such, it is necessary to perform a rigorous analysis of the errors associated with the method of data analysis – in this case the conjugate beam method. The uncertainty of error associated with the conjugate beam method depends primarily on the beam length and geometry, sensor number, loading type and magnitude, and support configurations on the beam. Below is a description of how the error limits for this method were developed. For simplification of calculations, it was assumed that the sensors were spaced equidistantly across the beam.

2.1 Beam Theory – Theoretical Method for Beam Deformation

The deformation of a beam at a specific location is related to the curvature of the beam and the internal moment in the beam at that location. By taking the double integral of the curvature, the deformation of the beam can be determined (Hibbeler, 2012). Similarly, the theoretical deformation of the beam can be found by taking the double integral of the theoretical moment in the beam, and then dividing by the Young's Modulus, E, and second moment of inertia, I, of the beam (Hibbeler, 2012). In this way, an equation for the theoretical deflection of the beam can be determined:

$$v(x) = -\iint \frac{M(x)}{EI} dxdx + C_1 x + C_0 = \iint K(x) dxdx + C_1 x + C_0 \tag{1}$$

Where K is the curvature and the values of the constants C1 and C2 can be determined using the boundary conditions of the beam.

2.2 Conjugate Beam Method

In the conjugate beam method, the curvature of a real beam is applied as loading on a conjugate beam. Under these loading conditions, the slope of the real beam is equal to the shear in the conjugate beam, and the bending moment in the conjugate beam is equal to the deformation in the real beam (Hibbeler, 2012).
The deformation in the real beam can be determined by finding the bending moment diagram due to the curvature loading on the conjugate beam, if support conditions are known. Equation 2 describes the deformation of the real beam, assuming that the sensors are evenly spaced. (Equation 2):

$$v(x_i) = \frac{il}{n}V_A + M_A - \frac{l^2}{n^2}\sum_{j=1}^{i}\left[\left(i - j + \frac{1}{2}\right)K_{j-1} + \left(i - j + \frac{1}{3}\right)\left(K_j - K_{j-1}\right)\right] \quad (2)$$

Where i is the index of the sensor at which the deformation is being determined, l is the length of the beam, n is the number of sensors on the beam, VA is the vertical reaction of the leftmost support on the beam span, MA is the moment reaction at the leftmost support, and K is the curvature at a given sensor location.
In this study, theoretical error limits were derived for a simply supported beam, with MA = 0. The vertical reaction, in this case can be calculated using the observed curvature as shown below:

$$V_A = \frac{l}{2*n}\sum_{j=1}^{n}\left(K_j + K_{j-1}\right) - \frac{l}{n^2}\sum_{j=1}^{n}\left[\left(\frac{i}{2} - \frac{1}{3}\right)K_{j-1} + \left(\frac{i}{2} - \frac{1}{6}\right)K_j\right] \quad (3)$$

2.3 Error in the Conjugate Beam Method

There are two sources of error that must be considered when determining the uncertainty of this method. The first is the error associated with the measurements of the monitoring system, which can be described using the standard deviation in measurements of the sensors. Equation 4 shows the uncertainty in the curvature, using the propagation formula described in Rabinovich, 2005 (Rabinovich, 2005 and Sigurdardottir et al., 2017):

$$\delta K = \frac{\delta\varepsilon_b}{h} + \frac{\delta\varepsilon_t}{h} + \frac{|\varepsilon_b - \varepsilon_t|}{h^2}\delta h = \frac{2}{h}\delta\varepsilon + \frac{|K|}{h} \quad (4)$$

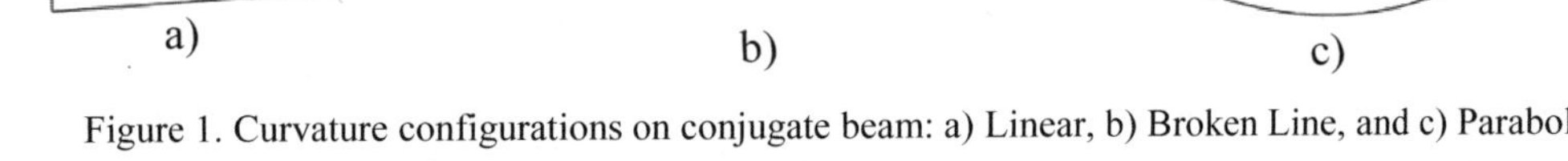

a) b) c)

Figure 1. Curvature configurations on conjugate beam: a) Linear, b) Broken Line, and c) Parabolic.

Table 1 – Error limits of conjugate beam method with linear interpolation on a simply supported beam.

Curvature Type	Error
Linear	0
Broken-line	$\delta v(x_i) = -\frac{1}{8}K(x_i)\frac{l^2}{n^2}$
Parabolic	$\delta v(x_i) = -\frac{1}{12}K(x_i)\frac{l^2}{n^2}$

The second error when using the conjugate beam method is due to the interpolation of the curvature between each sensor measurement. The value of the error depends upon the support conditions of the beam, the magnitude and configuration of the loading on the beam, and the number and positioning of sensors on the beam. The curvature that describes the deformation of the beam depends upon the type of loading that is applied to the beam, the length of the beam, and the geometric and material properties of the beam. The three main curvature distributions considered in this study are linear, broken-line, and parabolic, as shown in Figure 1. The difference between the deformations predicted by Equations 1 and 2 was then found, and is numerically equal to the error due to the linear interpolation. The error was found to depend upon the magnitude of the observed curvature, the length of the beam, l, the number of sensors, n, and a constant C, which depends upon the specific curvature type and load placement (Equation 5):

$$\delta v = CK\frac{l^2}{n^2} \tag{5}$$

Table 1 summarizes the error for the three curvature types evaluated for a simply supported beam. The error limit for the broken-line curvature was found for the loading configuration where there is a concentrated load halfway between sensors (resulting in the maximum error). The parabolic curvature distribution error limits were found for a uniformly distributed load across the whole span of the beam.

3. Laboratory Validation

In order to validate the equations and uncertainties derived above, the conjugate beam method was tested on deflection data collected from an aluminum beam in the lab (Figure 2). Data and knowledge about the laboratory setup for this section was taken from (Sigurdardottir et al., 2017). The beam had a cross-section of 254 x 9.3mm, a modulus of elasticity of 72000 N/mm2, and was equipped with five fiber Bragg-grating (FBG) strain sensors, which were installed on top of the beam with equal spacing. The total length of the beam was 1710 mm, and the sensors were placed from 383 mm to 1193 mm, with a spacing of 201 mm. Because no parallel topology was used in this setup, it was assumed that the strain in the top sensor was equal to the strain in the bottom sensor, so that the curvature could be calculated. For this study, the beam was tested as a simply supported beam, and a point load and uniform load were applied at varying locations along the length of the beam. The uncertainty of the sensors' mechanical strain measurement was about 0.01-0.05 mm. Four LVDT displacement sensors were also placed on the beam to act as controls, and to provide the necessary data to transform the deflection of the beam into the deformed shape (Figure 2) (Sigurdardottir et al., 2017).

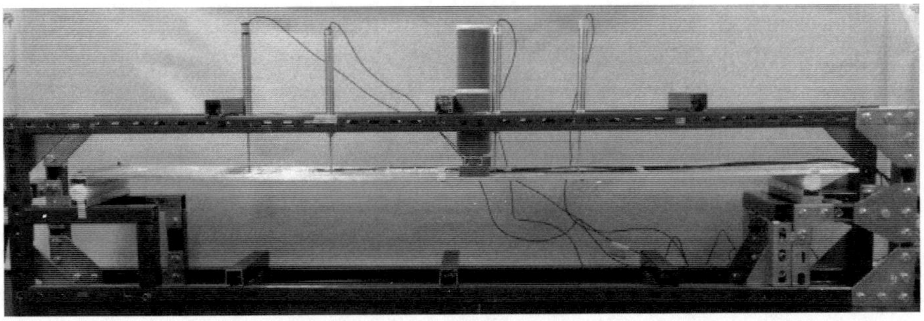

Figure 2. Laboratory setup with sensors for simply supported beam (Photo Credit: Sigurdardottir et al., 2017).

A point load of 82.9 N was applied to the beam when testing the uncertainty equations derived for a broken-line curvature (Sigurdardottir et al., 2017). This load was applied at various locations on the beam (Figure 3). One advantage of the conjugate beam method is that the deformed shape can be found directly if the boundary conditions are known. In this case, there were no sensors at the supports, but the curvature at the supports was assumed to be zero in order to calculate the deformed shape of the beam. For direct comparison to the equations shown in Table 1, which assume equal spacing of sensors, the deflections calculated via the conjugate beam method were transformed using measurements from the LVDT meters (Figure 3) (Sigurdardottir et al., 2017). The uncertainty due to the load being placed between sensors is relatively small, and falls within the error limits shown in Table 1, indicating that there is excellent agreement between the deformation predicted by the conjugate beam method and the deformation predicted by beam theory for the case of broken-line curvature.

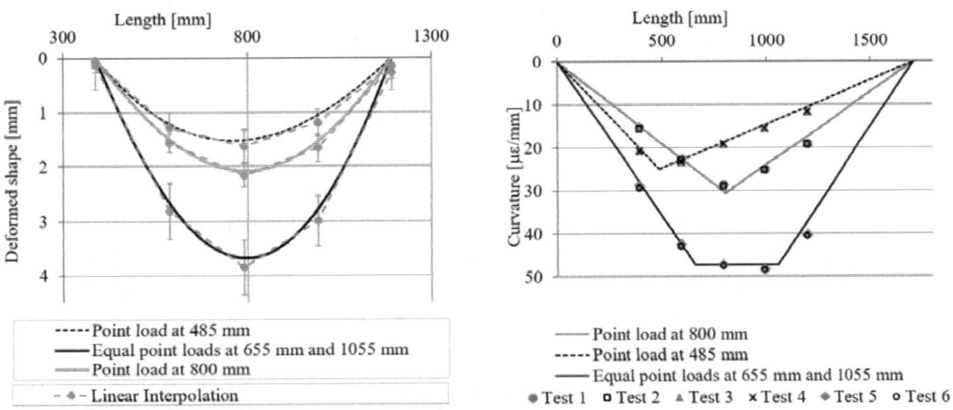

a) Displacement b) Curvature

Figure 3. a) Corrected theoretical and predicted displacement for point load at various locations on the beam and b) theoretical and measured curvature values for each loading configuration.

4. On-Site Validation

Streicker Bridge is a pedestrian footbridge that is equipped with long-gage fiber-optic strain sensors (Figure 4). These sensors are installed in a parallel typology. The data used in this study was collected from the sensors on the southeast leg, which can be considered as a continuous curved post-tensioned concrete girder. Data and knowledge about the sensor configuration for this section was taken from Sigurdardottir, et. al., 2015 and Sigurdardottir and Glisic, 2015 (Sigurdardottir et al., 2017 and Sigurdardottir and Glisic, 2015). The southeast leg is also equipped with a BOTDA distributed sensor, which was not used in this paper. There are three spans along the southeast leg, each supported by columns, P10, P11, P12, and P13, where P13 is a rubber bearing at the end of the bridge (Sigurdardottir and Glisic, 2015).

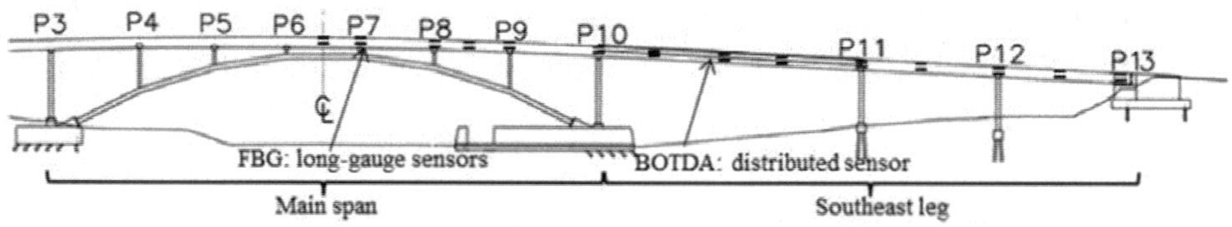

Figure 4. Sensor distribution on Streicker Bridge [7].

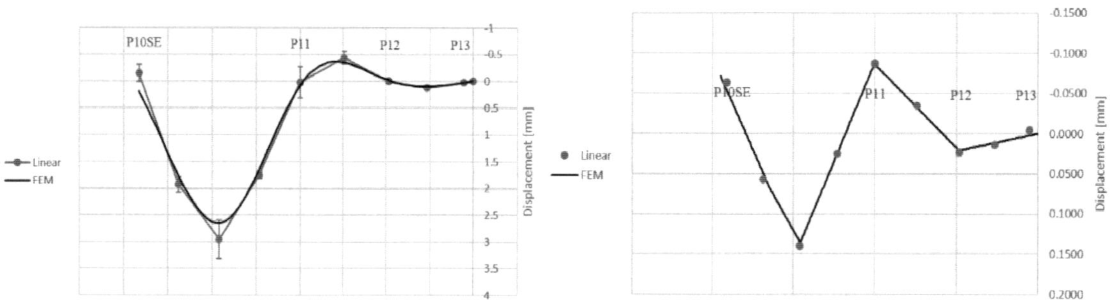

a. Displacement

b. Curvature

Figure 5. a) FEM and conjugate beam method predicted displacements for truck test at middle of longest span on SE leg of Streicker Bridge and b) Measured curvature and curvature predicted by SHM for truck test.

Tests were performed on the bridge in April 2014. In these tests, a four-metric-ton truck was parked at approximately the middle of the longest span, between P10 and P11. The observed curvature obtained from sensor measurements is shown in Figure 5b. Direct measurements of vertical displacement for direct comparison were not available, so the results of the approximation were compared to curvatures and deformations obtained using a validated Finite Element Model (FEM) (Sigurdardottir et al., 2017). It should be noted that the main span of the bridge is also equipped with FBG strain sensors. However, the main span has a greater stiffness than the southeast leg, and therefore has a smaller displacements in response to loading. The displacements measured on the main span due to loading from the four-metric-ton truck, which is the maximum allowable load on the bridge, were smaller than the noise observed by the sensors. Therefore, the main span is not used as a case study for this paper.

In order to predict the deformed shape of the structure, it was assumed that the sensors were equidistant, and that the curvature at the end of the bridge was zero. The deformed resulting from the analysis is shown in Figure 5a. Results show that the deformed shape falls prediominantly within the acceptable error limits predicted by the equations in Table 1 for broken-line curvature. This indicates that there is good correlation between the deformed shape predicted by this method and the deformation predicted by the FEM model. Howver, it should be noted that the deformed shape falls outside of the error limits for location P10SE. This discrepancy is currently under further evaluation, but could be a result of assumptions regarding error limits boundary conditions. Furthermore, the vertical displacement predicted by the FEM model at P10SE corresponds to a shortening of the column at this location. Thus, the tests done on Streicker Bridge validate in a good part the conjugate beam method for the broken-line curvature loading configuration.

5. Conclusion

This study assesses the use of the conjugate beam method as a method for long-term monitoring of the deformed shape of a structure. This deformed shape can be transformed into the actual vertical displacement of the structure if rotation, displacement, and elongation of supports are known.
The novelty of this preliminary study is that it offers an analysis of the errors associated with using a linear interpolation between curvature measurements in conjunction with the conjugate beam method. The theoretical error limits for three curvature configurations – linear, broken-line, and parabolic – were derived, as well as generalized equations for finding the deformed shape of the structure.
Laboratory tests were conducted in order to validate the method. For the case of a broken-line curvature distribution, the deformed shape predicted by the conjugate beam method closely matched that predicted by beam theory, and the error associated with the method was within the derived theoretical error limits.
On-site evaluation of the method was conducted using data taken from Streicker Bridge. For the concentrated load, which resulted in a broken-line curvature distribution, the deformed shape predicted by the conjugate beam method was approximately the same as that predicted by FEM. Errors in the deformed shape were within the derived error limits. However, discrepancy was noted at one sensor location, and it is subject of the further research.

References

ACI, 2008, *Building Code Requirements for Structural Concrete (ACI 318-08) and Commentary. An ACI Standard.* American Concrete Institute. Farmington Hills, MI.

AIS, 2005, *Steel Construction Manual (13th ed).* American Institute of Steel Construction. Chicago, IL.

Sigurdardottir, D. H., J. Stearns, & B. Glisic, 2017. 'Uncertainty in determination of the deformed shape of beams using strain measurements', *Smart materials and Structures*.

Sheng, S., W. Zhishen, Y. Caiqian, W. Chunfeng, T. Yongsheng, & W. Gang, 2010. 'An Improved Conjugated Beam for Deformation Monitoring with a Distributed Sensitive Fiber Optic Sensor', *Structural Health Monitoring*, 9(4): 361-378.

Hibbeler, R. C, 2012. *Structural Analysis: Eighth Edition*. Pearson Prentice Hall. Upper Saddle River, NJ.

Rabinovich, S. G, 2005. *Measurement errors and uncertainties, theory, and practice (3rd ed)*. Springer. Berlin.

D. H. Sigurdardottir & B. Glisic, 2015. 'On-Site Validation of Fiber-Optic Methods for Structural Health Monitoring: Streicker Bridge', *Journal of Civil Structural Health Monitoring*, 5(4):529-549.

The 8th International Conference on
Structural Health Monitoring of Intelligent Infrastructure
Brisbane, Australia | 5-8 December 2017

Low cost vision-based systems using smartphones for measuring deformation in structures for condition monitoring and asset management

R. Kromanis[1] and A. Al-Habaibeh[2].
[1] *Nottingham Trent University, Civil Engineering Department – UK,*
Email: rolands.kromanis@ntu.ac.uk [2]*Nottingham Trent University, Product Design Department – UK.*

Abstract

Asset management is an important area in civil engineering. Monitoring deformations and movements of structures is an essential tool for the asset management and condition monitoring. Currently, a range of technologies such as contact (e.g. strain gauges and displacement sensors) and non-contact sensors (e.g. lasers, radars and cameras) is available for monitoring structural response. However, there is a need for low cost systems, which could be used for real life situations and laboratory environments (for teaching and research). This paper presents a novel approach of using smartphones to monitor movements in structures. The low-cost and easy to use hardware (i.e. smartphone) is readily available for the majority of people. The idea is that a smartphone camera is used to monitor specific markers (points of interest) on a structure. Movements of the markers can be obtained and calibrated for accurate movements of the structure. This paper presents results obtained from both lab structures and a full-scale bridge. Results show that the technology is a simple, low-cost and provides indicative and at the same time accurate information about structural response.

1. Introduction

Continuously aging civil infrastructure requires prudent, timely and accurate assessment. Asset maintenance costs are rising, especially for strategically and economically important assets such as bridges (Aktan et al., 1996). These assets have to provide an uninterrupted traffic flow, which plays a vital role in the modern society (Brownjohn, 2007). A large proportion of bridges have reached the end of their design life. Their maintenance costs are raising and their soundness is frequently questioned. The collapse of the I35W Bridge (Minneapolis, USA) in 2007 took lives of 13 people and injured 145 and daily economic loss was estimated to raise up to $220,000 (Wald & Davey, 2008). The closure of the Forth Road Bridge (Edinburgh, UK) in late 2015 left a significant impact on economy in Scotland considering that a closure of a carriageway for one day can cost up to £650,000 (Hannan, 2015). Bridge collapses and closures could be if damages were detected early, which can be achieved with the aid of structural health monitoring (SHM) systems.
Currently, a range of SHM technologies such as contact sensors (e.g. strain gauges and linear variable differential transformer) and non-contact sensors (e.g. lasers and cameras) are available for condition monitoring of bridges. Installation of sensors can be a complicated process due to concerns related to access requirements. If contact sensors are chosen, an access to desired sensor locations is required. Access challenges can be avoided when considering non-contact measurement systems. Non-contact (or remote) measurement systems allow capturing structural changes without coming in a physical contact with the structure. A wide range of proprietary technologies such as spectral analysis, digital image correlation and radars (Vaghefi et al., 2012) are commonly deployed for the condition assessment of bridges and laboratory structures. However, there is a need for low-cost systems with potential applications in both real-life and lab (teaching and research) environments.
A number of researchers have investigated the advantages of vision-based strategies for monitoring structures (Ho, Lee, Park, & Lee, 2012; Lee, Fukuda, Shinozuka, Cho, & Yun, 2007; Lee & Shinozuka, 2006; Zaurin & Necati Catbas, 2010). Vision-based monitoring systems typically consist

of cameras, which capture digital images of a structure (or part of it), and software, which analyses images using image processing techniques. Methods, which can be employed to quantify displacements in bridges using image processing, have been validated on lab and full-scale structures (Lee et al., 2007), however low-cost (or freeware) tools for deformation monitoring are not commonly made available.

Vision-based systems can also indicate the location, number and types of vehicles on a bridge. This information can be coupled with measurements of structural response for damage detection (Zaurin & Catbas, 2010). Vision-based techniques can also be used to capture effects of ambient conditions, and, in particular, those due to temperature variations (Kromanis, 2015). At present, vision-based technologies are mainly deployed for short-term monitoring. Applications to long-term monitoring of bridges are still under development. Vision-based technologies hold a vast potential and may constitute a holistic monitoring system that can track vehicular traffic (Coifman, Beymer, McLauchlan, & Malik, 1998; Koller, Weber, & Malik, 1994; Liu & You, 2007), thermal loads (Kromanis, 2015) and human activities (Zhao & Chellappa, 2003), and also measure structural response.

This research investigates the capability of a low-cost vision-based system using a standard smartphone camera to measure structural deformations. A smartphone (Lenovo A806) is employed to capture images (and record videos) of structures under loadings. Collected visual information is then analysed using Matlab image processing toolbox (Mathworks, 2017). The performance of the proposed system has been investigated on laboratory structures and a full-scale bridge. Results show that structural deformations can be measured at required accuracies using the proposed system.

2. Methodology

All structures deform when subjected to loadings. Deformations of simple structures such as beams can be calculated using well-known engineering formulae. For example, a vertical deflection (δ) at any point of the length (l) of a cantilever beam, which is subjected to a point load (P) at its free end, can be calculated using the following equation:

$$\delta = \frac{Px^2}{6EI}(3l - x) \qquad (1)$$

where x is the distance from the support at which δ is calculated, E is Young's modulus of the material and I is the second moment of area of the beam. For a rectangular beam with a known breadth (b) and height (h) I is calculated as follows:

$$I = \frac{bh^3}{12} \qquad (2)$$

Eq. 1 shows that the deflection of a beam is directly related to P, E and I values. When considering a cantilever beam (with known E and I values) subjected to a point load at the free end, its deflection (δ) should be directly related to the applied load (P). However, if current δ is larger than previously observed δ when the beam is subject to the same P, a change in E and/or I values has occurred. For the above example, it is unlikely that E value would change leading to a conclusion that I value has changed, and the beam is damaged. In previous studies damage is introduced by making a cut in the beam (Farrar & Jauregui, 1998; Kim, Park, & Lee, 2007). When considering a rectangular beam, a cut in the top or bottom part of the section reduces the height of the beam, thus reducing I value.

A schematic representation of a vision-based system for measuring deformations such as vertical deflections in structures is provided in Figure 1. A cantilever beam is considered as a structure under loadings. The beam has artificial or natural markers (or points of interest) on its surface. Their relative location is tracked by the software, and structural deformations are estimated. A smartphone camera is used to capture images while the beam is under loadings. The images are processed with algorithms analysing the movement of markers. Deformations of the beam are computed for each image. The load-response knowledge may then be used to the assess condition of the beam, for example using Eq. 1. For more complex structures such as bridges, the proposed vision-based system could be employed to obtain measurements that are used for structural identification and condition assessment.

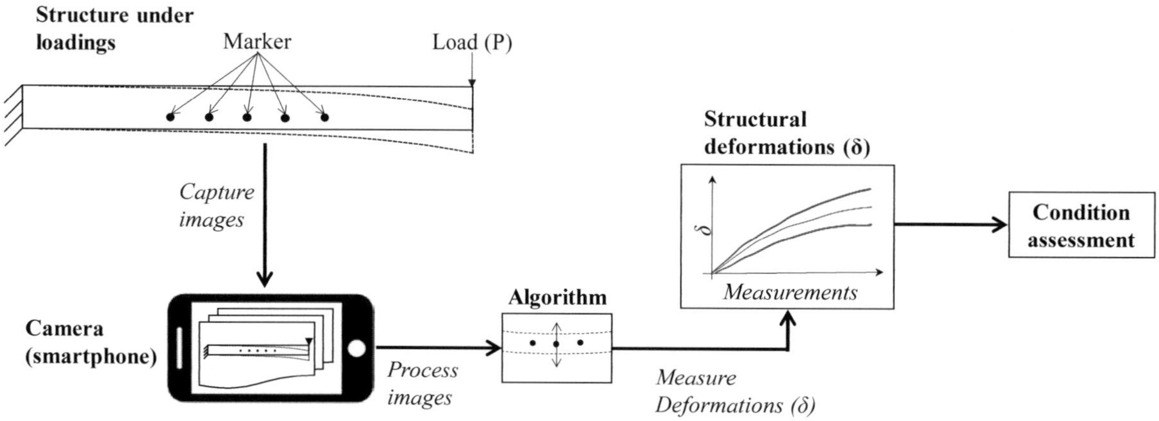

Figure 1 A schematic representation of condition assessment of a structure under loadings using a low cost vision-based measurement collection system with a smartphone camera.

3. Approach and technology

Vision-based technologies and image processing techniques have now been developed to an extent when they can enable fast and robust extraction of data from images. For example, video streams of traffic from a bridge can be combined with displacement measurements collected with contact sensors to create influence lines, which then serve as input features into damage detection methodologies (Zaurin & Catbas, 2010). For the successful application of the proposed vision-based system, algorithms, which focus solely on movements of selected markers, have to be developed. Processes involved when selecting markers on structures and analysing their movements are shown in the flow chart in Figure 2.

Once images of a structure under loadings are captured and saved/stored in a selected directory, information of structural movements can be obtained using image processing tools. Firstly, the reference image is selected. A desirable number of areas containing markers is set. A rectangle is used to draw a boundary identifying an areas with a marker. These are then considered as separate images. Image adjustment parameters such as sharpness, intensity and number of pixels are altered to optimize the quality of the images with markers. Once the above processes are complete, an iterative process is initiated, in which images (starting from the first or selected image) are analysed in a consecutive order as follows:

a) parameters for each image containing a marker are adjusted using adjustment parameters set for the reference images;
b) background of the image is removed (only the marker is left in the image);
c) the location of marker is calculated inputting a known measure;
d) the pixel information is converted to a standardized measurement unit such as a millimetre;
e) movement of the marker is saved for the ith image.

The above process is repeated for all images of the structure under loadings. The movements of the selected markers are appended in a matrix forming time series of structural movements.

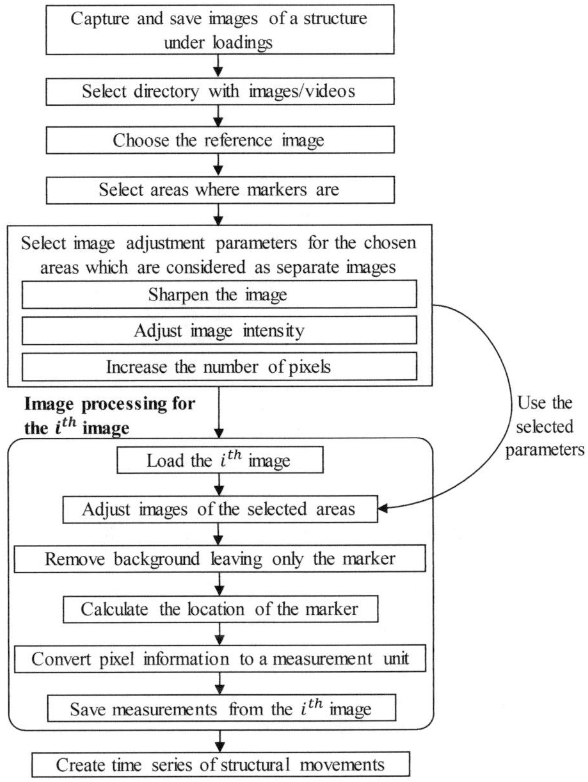

Figure 2 A flowchart of implemented processes in the proposed vision-based system.

3.1 Materials and patterns

Normally, each material has a unique texture and surface patterns. Some structural material such as aluminium and steel have uniform surface texture and may have no natural markers or patterns that can be used to track local deformations. However, the main focus of this study is on the development of the vision-base system for deformation monitoring using smartphones. For reasons of brevity, two structural materials (concrete and timber shown in Figure 3) with natural and artificial markers are discussed. Artificial markers are dots drawn with a marking pen. Natural markers are air voids in the concrete beam (Figure 3 (left)) and knots in the timber beam (Figure 3 (right)). Movements of markers can be obtained when estimating the location of markers in each image.

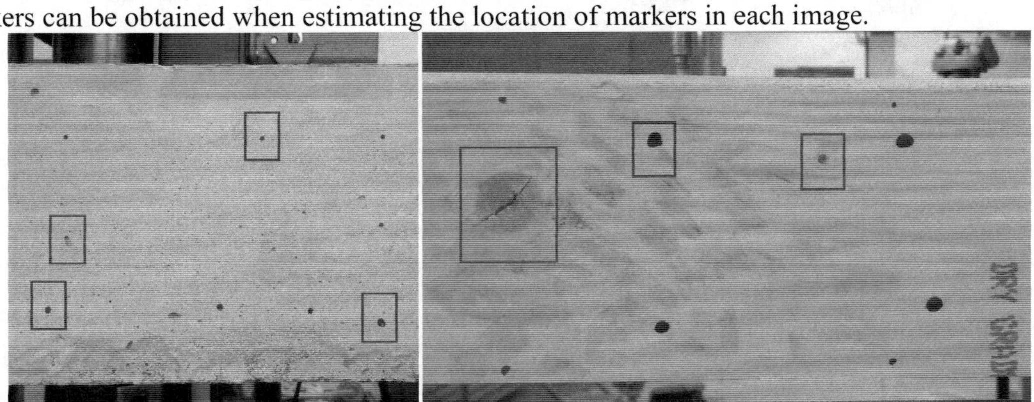

Figure 3 Concrete (left) and timber (right) beams with natural and artificial (created with a marking pen) markers or points of interest. Artificial and natural markers are shown in red and blue rectangles, respectively.

3.2 Application of image processing techniques

The general concept of digital image processing is to analyse an information content available in an image. For example, an image of a circle, which represents a marker on the beam, is captured using four resolution settings and shown in Figure 4. Figure shows that the quality of images depends on the set resolution of the camera. Higher resolution images offer more details than lower resolution images. Figure 4 shows images of the circle captured with low resolution cameras (a) 12 by 16 (12×16) pixels (px), (b) 12×16 px, (c) 24×32 px and with a high resolution camera (d) 768×1024 px.

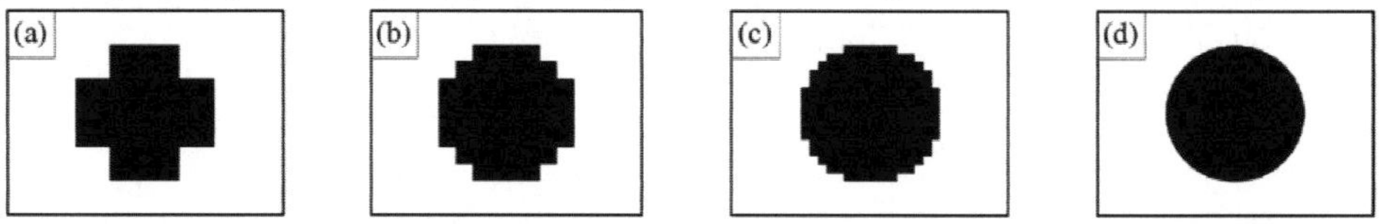

Figure 4 Images of a circle. The aspect ratio of the images is 4:3. The number of pixels in the images is as follow (a) 6×8 px, (b) 12×16 px, (c) 24×32 px and (d) 768×1024 px.

The movement of an object, say a circle shown in Figure 4, can be recognized when comparing the present image with the previous image. When multiple images are taken, while the object of interest moves, time series of object movements can be created. The resolution of an image plays an important role when estimating the movement of an object within the image. Pixel values are converted to standardised measurement units such as millimetres. The size of a pixel in millimetres (u_{px}) can be expressed using Eq. 3, in which a known distance (u) is divided by a number of pixels (N_p).

$$u_{px} = \frac{u}{N_p} \tag{3}$$

As an example, consider images (b1) and (c1) from Figure 5. The size of the frame of the image with a circle is 30 by 40 mm. From this information, the pixel size in vertical and horizontal directions can be calculated. The size of one pixel in the vertical and horizontal directions in image (b1) and image (c1) is $30/12 = 2.5\ mm$ and $30/24 = 1.25\ mm$, respectively. In images (b2) and (c2) the object has moved 1.25 mm down and to the right. However, this movement can only be detected in image (c2), which has higher resolution than image (b2). In images (b3) and (c3) the object has moved 2.5 mm down and to the right. This movement is recognized in both images.

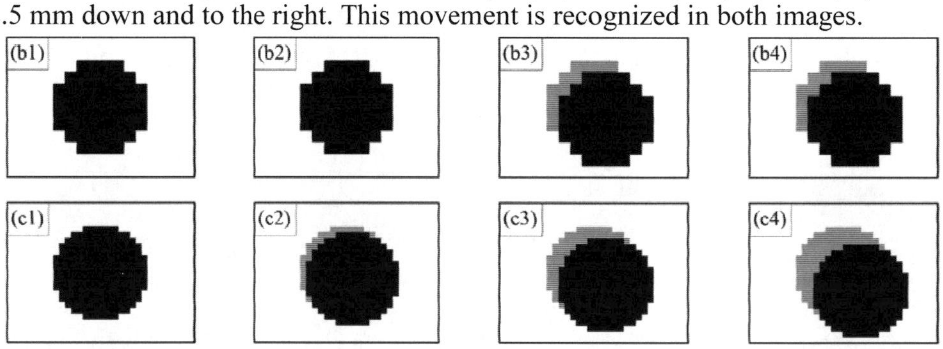

Figure 5 Movement of a circle in 12×16 px (b1, b2, b3, b4 (top)) and 24×32 px (c1, c2, c3, c4 (bottom)) images.

The algorithmic steps employed in image processing in this study is shown in Figure 6. Initially an area with a marker (point of interest) is selected in the reference photo. In this example the reference photo is chosen to be a concrete beam shown in Figure 3 (left). The selected area with the marker (Figure 6 (a)) is taken from the bottom left part of the beam. A rectangle is drawn around the area containing a marker. The area should be large enough to cover the range of possible movements of the marker. The colour image of the area is converted to a grayscale image (Figure 6 (b)), therefore reducing the size of the information contained in the image and time required to analyse the image.

Next, the sharpness and contrast of the grayscale image is adjusted (Figure 6 (c)). Then the adjusted image is converted to a binary (black and white) image. In these steps, an operator has to decide which parameters have to be altered to obtain the required quality of the image. In order to increase measurement precision, the number of pixels in the binary image are increased (Figure 6 (d)). For the given example, the scale of the image is increased four times. In the next step, connected pixels (components) are found. In Figure 6 (d) these are black dots, which are air voids in the concrete and the marker. Only the largest of the recognized components (the marker) is retained for further image analysis (Figure 6 (e)). Now, when only the marker is left in the image, its relative location can be calculated. The previous location of the marker can be compared with the present location of the marker, and its movement can be computed (Figure 6 (f))

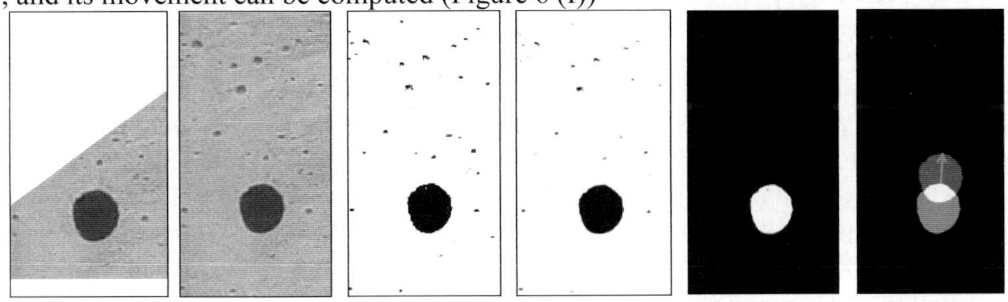

Figure 6 Steps involved in the image processing of an area with a marker. The selected area with a marker (a) is converted to a greyscale image (b) and then to a binary image (c). The number of pixels are increased (d), unnecessary markers are removed (e) and the movement of the marker can be tracked (f).

Datasets of each ith binary image, which are extracted from photos taken of the structure under loadings, are analysed as matrices.

$$\mathbf{U} = \begin{bmatrix} y_{11} & \cdots & y_{1n} \\ \vdots & \ddots & \vdots \\ y_{p1} & \cdots & y_{pn} \end{bmatrix} \quad (4)$$

y is a binary value (0 or 1) for a pixel in p by n matrix $\mathbf{U}$, where p and n is a row and column, respectively. When calculating the vertical location of the marker (δ_i) for each ith image, only relevant pixel information of the top part of the marker is considered. The following algorithm is use:

Input

- A binary matrix **U** of the selected area with the marker for each image.

Output

- A time series of vertical movements of the marker (δ_M) for the selected monitoring period.

Algorithm:

For each image i_j

For $m = 1$ to n in $\mathbf{U}$

$N_a =$ the number of the first row p with a nonzero value, $N_a \neq 0$

if a nonzero value is not found, proceed to the next column and do not put a value in N_a

Next column n in $\mathbf{U}$

$k =$ the number of elements in N, $k \leq n$

$\delta = \frac{1}{k}\left(\sum_{o=1}^{k} N_o\right)$

$\delta_M(i) = \delta$

Next i_j

4. Results

Three examples are selected to demonstrate the application of the proposed vision-based system for measurement collection using smartphones. In this study Lenovo A806 was used to capture images and videos. Initially the system is evaluated on two laboratory set-ups (i) a simply supported concrete beam and (ii) cantilever timber beam. Beams are subjected to both mechanically and manually applied loadings. Then a 20X optical lens is attached to the smartphone to evaluate if the proposed system can be employed to measure displacements of a footbridge exposed to forced vibrations.

4.1 Laboratory structures

An illustration of the proposed vision-based system using a smartphone for deformation monitoring of a laboratory structure is shown in Figure 7. Markers are drawn on the face of the structure. A smartphone is fixed on a tripod at a preferred distance from the lab structure. Setting the correct distance, at which photos (or video) are taken, has a direct impact on the resolution of images and calculated structural deformations. Understanding of structures is required to provisionally estimate expected movement of the structure. Then a suitable distance where a smartphone should be placed can be chosen. The collected images are analysed using the proposed image processing approach and algorithms, and time series of estimated structural movements are created.

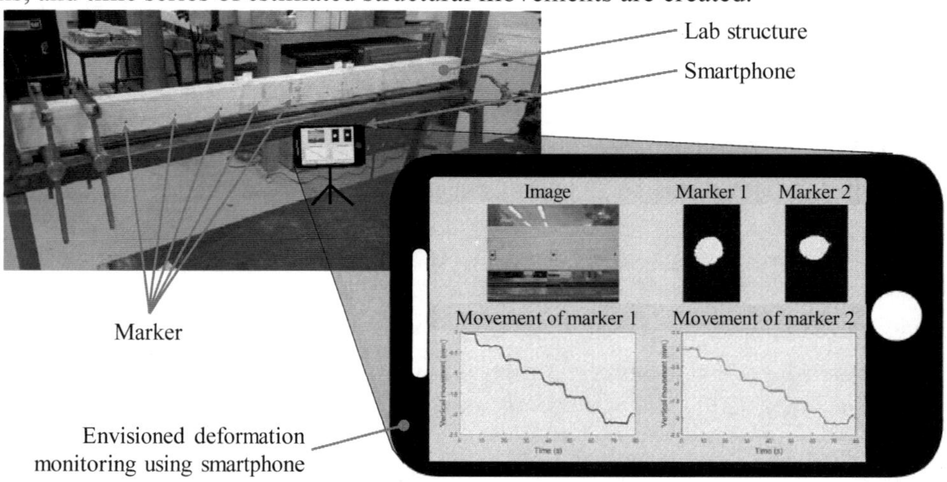

Figure 7 Envisioned set-up of the vision-based system using a smartphone for deformation monitoring of a beam in the structure's laboratory at Nottingham Trent University.

<u>Concrete beam test</u>

A simply supported concrete beam with the dimensions of 80 mm width, 130 mm height and 1200 mm length is subjected to three point flexural loadings. In the experimental set-up the supports are moving upward, therefore, vertical movements of the beam close to the right support are monitored. Markers are drawn on the surface of the beam. A smartphone is set 400 mm away from the beam. The camera is focused on the beam, and a 3120×4208 px image is taken once in two seconds. Two controlled loading scenarios are considered. The first loading scenario: at 140 s load is increased from 0 to 10 kN and kept for 50 s, then the load is increased to 14 kN and kept for 50 s before it is reduced to pre-load, which is around 4 to 6 kN. The second loading scenario: at 405 s load is increased from pre-load to 11 kN, kept for 20 s, increased to 12 kN, kept for 25 s, increased to 15 kN and kept for 25 s before it is removed (reduced to 0 kN). Four areas with markers are selected to analyse structural movements of the beam. The binary images of these areas are shown in Figure 8 (right). Time series of computed displacements of the markers are plotted in Figure 9. Displacement time series of the markers closely replicate the applied loading scenarios.

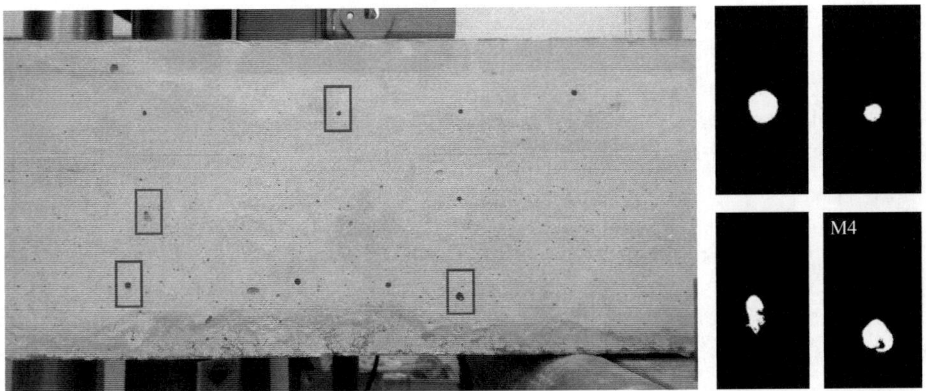

Figure 8 A part of the concrete beam under monitoring (left) and binary images of selected areas with markers (right). Red and blue rectangles on the photo of the concrete beam represent the selected areas with artificial and natural markers (Mi, where i is the number of a marker, $i = 1, 2, 3, 4$), respectively.

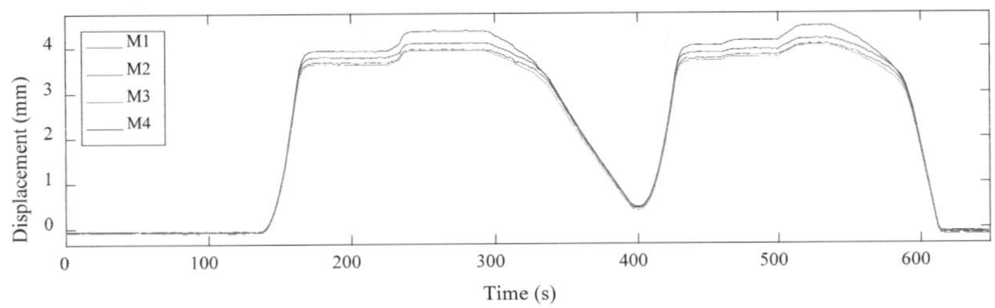

Figure 9 Time series of vertical movements of the selected markers on the concrete beam.

<u>Timber cantilever test</u>

The second laboratory set-up evaluates the performance of the proposed system on a timber cantilever beam with 120 mm height, 45 mm with and 1000 mm length. A hanger with a plate is attached to its free end to accommodate applied loadings. A smartphone is placed 400 mm away from the beam to capture deformations of the free end of the beam (see Figure 10 (right)) and focused. A 3072×4096 px image is taken once in three seconds. The load is applied manually using the following load steps: 0N, 50 N, 100 N, 200 N, 300 N and 400 N. The load is removed as follows: 400 N, 300 N, 200 N, 100 N, 50 N and 0 N. This process describes one loading cycle. After the loading cycle has been repeated twice, a damage, in a form of a 27 mm cut in the top side of the beam close to its fixed end, is created. And the loading cycle is repeated.

Figure 10 A part of the timber cantilever beam under monitoring (left). Red rectangles represent selected areas of artificial markers (Mi, where i is the number of a marker, $i = 1, 2$), which are analysed (right) using the proposed image processing approach.

Two areas with markers are selected (see Figure 10 (left)). Their binary images are shown in Figure 10 (right). Figure 11 shows time series of computed movements of the markers, which represents beam response to applied loadings. The load steps are clearly discernible in the plot. As anticipated, displacements of M2 are larger than those of M1, at similar loads. M2 is located further from the fixed end of the beam, therefore, according to Eq. 1, the vertical deflection at this location is expected to be the highest. Displacement time series are not as smooth as those computed for the concrete beam (see Figure 9). This can be related to the nature of applied loadings. The beam oscillates slightly when weights (in a form of steel plates) are applied. Noisy measurements between 750 s and 800 s, period when structure is being damaged, can be observed. After the onset of damaged, the magnitude of vertical deformations is larger than previously observed deformations (when structure was healthy). This indicates alteration in the cross section area of the beam (see Eq. 1 and Eq. 2) – damage.

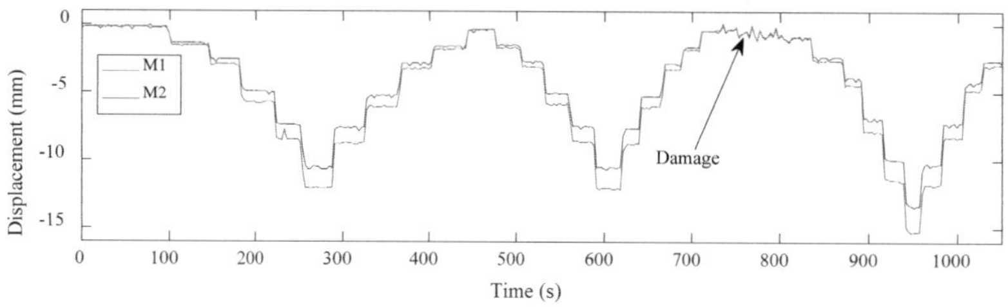

Figure 11 Time series of vertical movements of the selected markers on the timber beam.

4.2 Full-scale structure

The performance of the vision-based system is also evaluated on a full-scale bridge. The Wilford Bridge (see Figure 12) crosses the River Trent and is both a footbridge and aqueduct bridge. The bridge has a single 69 m long span. The smartphone is placed on a small tripod approximately 40 m away from the mid-span of the bridge, at which it is focused on. The bridge is excited by a group of people jumping at its mid-span. Videos are recorded while the bridge is exposed to forced vibrations. The frame size and collection rate of the videos are set to 1088×1920 px and 30 frames per second, respectively. While bridge response is collected, strong wind blows affecting the quality and collection of videos. While analysing some videos measurement noise is evident. For reasons of brevity, this is not shown in this study. A part of time series of bridge movements is shown in Figure 13. The first modal frequency can be extracted from time series. The modal frequencies is found to be 1.688 Hz. This is very close to the model frequency obtained using Global Navigation Satellite System derived displacements by Psimoulis (2016), which is 1.7 Hz

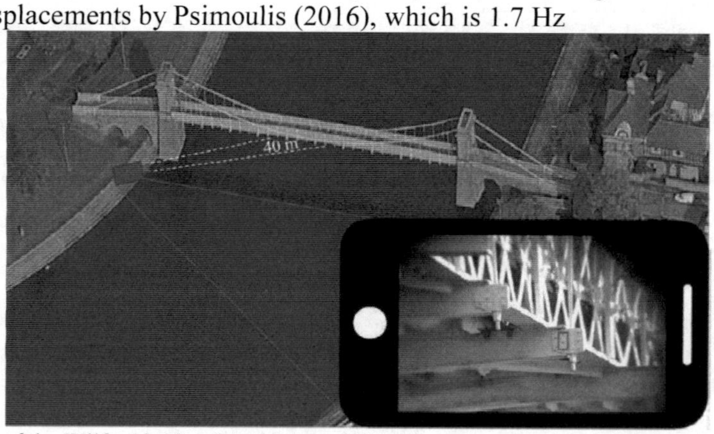

Figure 12 Monitoring of the Wilford bridge (Nottingham, UK) using the proposed vision-based system. Blue rectangle (in the image shown in the smartphone illustration) is the selected area with a natural marker (a head of the bolt).

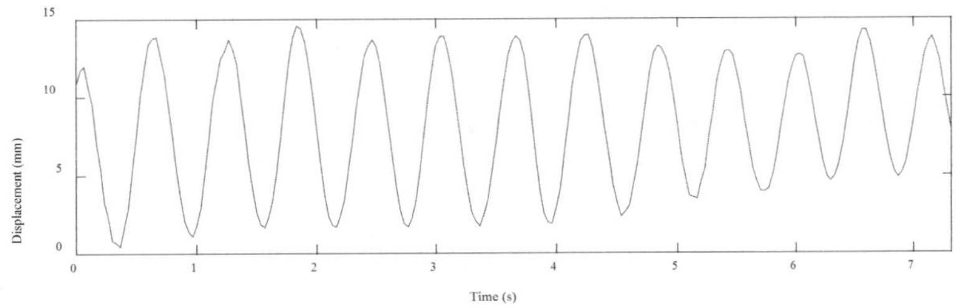

Figure 13 Time series of vertical movements measured at the mid-span of the bridge.

5. Conclusions

In structural health monitoring context the focus is on monitoring and analysing structural deformations. This study proposes a low cost vision-based system using smartphones for monitoring deformations. The premise is that the collected images or videos of the structure under loadings can be analysed to obtain structural deformations. The image processing approach is initiated with choosing a reference photo. Then areas with identified (or artificially created) markers (points of interest) on the structure are selected. A set of algorithms is employed to analyse the movement of these markers in subsequent images. This study draws the following conclusions:

- Indicative information of structural deformations can be obtained from collected images. Measurements obtained from the laboratory tests are within expected accuracies required for teaching and research. The accuracy is predominantly governed by the focus of the camera (image sharpness), distance from the structure under loadings, image resolution and the stability of the camera.
- Deformation monitoring of full-scale bridges could be made possible when adding an optical zoom lens to a smartphone. Results from the full-scale experiment show that sufficiently accurate deformations can be obtained to derive the first modal frequency of the bridge.

Low cost vision-based systems deploying smartphones for deformation monitoring have potential application in laboratory experiments. These systems could reduce costs of data acquisition systems and labour. The performance of proposed system has to be calibrated and compared with professional cameras and proprietary sensing systems (both contact and non-contact). It is anticipated that in near future smartphones will become faster and more powerful in their vision-based capacities than they are now. Smartphones will allow taking high quality photos at high frequencies, therefore broadening SHM possibilities.

Acknowledgment
The authors would like to thank Dr Panagiotis Psimoulis and Dr Lukasz Bonenberg from the University of Nottingham for helping with measurement collection on the Wilford Bridge.

References

Aktan, B. A. E., David, N. F., Vikram, L. B., Helmicki, A. J., Hunt, V. J., & Shelley, S. J. (1996). Condition assessment for bridge management. *Journal of Infrastructure Systems*, *2*(3), 108–117.

Brownjohn, J. M. W. (2007). Structural health monitoring of civil infrastructure. *Philosophical Transactions of the Royal Society a-Mathematical Physical and Engineering Sciences*, 589–622. https://doi.org/DOI 10.1098/rsta.2006.1925

Coifman, B., Beymer, D., McLauchlan, P., & Malik, J. (1998). A real-time computer vision system for vehicle tracking and traffic surveillance. *Transportation Research Part C: Emerging Technologies*, *6*, 271–288. Retrieved from http://www.sciencedirect.com/science/article/pii/S0968090X98000199

Farrar, C. R., & Jauregui, D. A. (1998). Comparative study of damage identification algorithms applied to a bridge: I. Experiment. *Smart Materials and Structures*, *7*(5), 704.

Hannan, M. (2015). Forth Road Bridge closure will have huge impact on Scottish economy. Retrieved January 26, 2016, from http://www.thenational.scot/news/forth-road-bridge-closure-will-have-huge-impact-on-scottish-economy.10831

Ho, H. N., Lee, J. H., Park, Y. S., & Lee, J. J. (2012). A synchronized multipoint vision-based system for displacement measurement of civil infrastructures. *The Scientific World Journal*. https://doi.org/10.1100/2012/519146

Kim, J. T., Park, J. H., & Lee, B. J. (2007). Vibration-based damage monitoring in model plate-girder bridges under uncertain temperature conditions. *Engineering Structures*, *29*(7), 1354–1365. https://doi.org/10.1016/j.engstruct.2006.07.024

Koller, D., Weber, J., & Malik, J. (1994). *Robust multiple car tracking with occlusion reasoning*. Springer Berlin Heidelberg. Retrieved from http://link.springer.com/chapter/10.1007/3-540-57956-7_22

Kromanis, R. (2015). *Structural Performance Evaluation of Bridges : Characterizing and Integrating Thermal Response*. University of Exeter.

Lee, J. J., Fukuda, Y., Shinozuka, M., Cho, S., & Yun, C. (2007). Development and application of a vision-based displacement measurement system for structural health monitoring of civil structures. *Smart Structures and Systems*, *3*(3), 373–384.

Lee, J. J., & Shinozuka, M. (2006). A vision-based system for remote sensing of bridge displacement. *NDT & E International*, *39*(5), 425–431. https://doi.org/10.1016/j.ndteint.2005.12.003

Liu, Z., & You, Z. (2007). A Real-time Vision-based Vehicle Tracking and Traffic Surveillance. In *Eighth ACIS International Conference on Software Engineering, Artificial Intelligence, Networking, and Parallel/Distributed Computing (SNPD 2007)* (pp. 174–179). Ieee. https://doi.org/10.1109/SNPD.2007.56

Mathworks. (2017). MATLAB image processing toolbox.

Psimoulis, P. A., Peppa, I., Bonenberg, L., Ince, S., & Meng, X. (2016). Combination of GPS and RTS measurements for the monitoring of semi-static and dynamic motion of pedestrian bridge . In *In: 3rd Joint International Symposium on Deformation Monitoring (JISDM), 30 March - 1 April 2016, Vienna, Austria*.

Vaghefi, K., Oats, R. C., Harris, D. K., Ahlborn, T. (Tess) M., Brooks, C. N., Endsley, K. A., … Dobson, R. (2012). Evaluation of commercially available remote sensors for highway bridge condition assessment. *Journal of Bridge Engineering*, *17*(6), 886–895.

Wald, M. L., & Davey, M. (2008, January 16). States Advised to Check for a Bridge Design Flaw. *New York Times*, p. A17. Retrieved from http://www.nytimes.com/2008/01/16/washington/16bridge.html?_r=0

Zaurin, R., & Catbas, F. N. (2010). Integration of computer imaging and sensor data for structural health monitoring of bridges. *Smart Materials and Structures*, *19*(1), 15019. https://doi.org/10.1088/0964-1726/19/1/015019

Zaurin, R., & Necati Catbas, F. (2010). Structural health monitoring using video stream, influence lines, and statistical analysis. *Structural Health Monitoring*, *10*(3), 309–332. https://doi.org/10.1177/1475921710373290

Zhao, W., & Chellappa, R. (2003). Face recognition: A literature survey. *Acm Computing Surveys (ACSU)*, *35*(4), 399–458. Retrieved from http://dl.acm.org/citation.cfm?id=954342

T rnational Conference on
Structural Health Monitoring of Intelligent Infrastructure
Brisbane, Australia | 5-8 December 2017

Damage detection of structures based on relative variation of wavelet total energy and experimental study

Y.Q. Xiang[1], Y.K. Jia[2]
[1]*Zhejiang University – China, Email:*xiangyiq@zju.edu.cn [2]*Zhejiang University – China*

Abstract

The damage of bridge would intensify with the increase of service life, so it confronted the bridge engineer with searching for the appropriate damage detection method. After applying discrete wavelet transform to dynamic response signal at the measuring point, wavelet total energy could be calculated. Setting the intact structure as the reference state, the detection method based on the relative variation of wavelet total energy was proposed. To verify the effect of proposed damage detection method in practical application, six 6m simple-supported beams were analyzed in lab. Each beam was divided into 10 parts, and nine acceleration sensors were arranged at division points except boundary points. Dynamic impact load was imposed by hammer, and damage was simulated by 3mm width notch with different height. The experiment results show that the proposed damage detection method can accurately determine the existence and location of the damage and it is sensitive to damage. What's more, it also shows robustness to mother wavelet type, data length and impact location. If the impact place is away from the damage place, it can be better to detect the damage place.

1. Introduction

By the end of 2016, about 800,000 highway bridges and 200,000 railway bridges have been built in China. Meanwhile, a large number of existing bridges are under interaction of complex environment and multi-hazard load.Damage of these structures will be inevitable with increase of time. Damage will seriously affect the durability and security of the bridge structures.So, developing themonitoring, inspection and maintenance technology of the bridge infrastructures in service were of great importance forall countries in recent years.

Structural damage is essentially changes of local stiffness and mass, which leads tochanges of structural dynamic properties. According to the method usedin the monitoring data, the damage detection methods can be divided into static (Abdo 2012), dynamic detection method and Hybrid method. Damage detection methods based on frequency (Bandara, Chan & Thambiratnam2014), model shape(Lieven&Ewins1988, Pandey, Biswas & Samman 1991), flexibility matrix(Hosseinzadeh Amiri & Razzaghi et al 2016) and wavelet transform(Rucka & Wilde 2006) belong tothe main dynamic detection method. But there exists some limits for the above-mentioneddetection methods.

For frequency-based detection method, significant damage only cause small changes of natural frequency of the structures, especially for large structures (Chandrashekhar & Ganguli 2009). Modal Curvature (Jing & Xiang 2008), Modal Assurance Criterion (MAC), and Enhanced

Coordinate Modal Assurance Criterion (ECOMAC) are representatives of modal shape-based methods. These mode-based detection methods need stable modal shape, and the detection results are affected by incompleteness and error of the testdata (Jing & Xiang 2008). Methods based on flexibility matrix involve complex stiffness correction, parameter conversion and equation solving, and the detection reliability is low. For traditional wavelet transform-based detection methods, the damage indexes are usually affected by wavelet type, decomposition scale and noise.

The wavelet energy is an important character of wavelet transform.It is sensitive to small changes in the signal. Traditional energy-based detection method usually adopt wavelet packet energy (Han,Ren&Sun 2005), but band division in wavelet packet analysis is too meticulous, which is time-consuming relatively. **Discrete wavelet transform**(DWT) can also give adaptive time-frequency resolution or space-scale resolution, and can reduce the redundancy and computation amount.

The aim of this article is to propose the damage detection method based on **relative variation of wavelet total energy(RVWTE**). The wavelet coefficients at different scales are gathered after dealing with the signals at every measured point of structure by using DWT. Then the wavelet total energy of the measured points is calculated by summing up all the wavelet energy at each scale.In order to verify validity and reliability of the proposed method, taking the typical simply-supported steel beam as example, setting the intact structure as the baseline state, analyzing and test dynamic signal and response in different damage cases, calculating the correspondingdamage index to judge whether the structure damage occur or not and what is damage degree.It is of great significance to the damage detection and assessment of similar bridge structures.

2 Basic theory and method

2.1 DWTand definition of damage index

Corresponding to the special mother wavelet, the continuous wavelet transform of a set of signals in time domain is defined as below (Kim &Melhem 2004):

$$(W_{\psi}f)(a,b)=\frac{1}{\sqrt{a}}\int_{-\infty}^{+\infty}f(t)\psi^{*}(\frac{t-b}{a})dt \tag{1}$$

Where a and$b$ are scale and translation parameters, respectively and ψ^{*} is complex conjugate ofψ .

$(W_{\psi}f)(a,b)=<f(t),\psi_{a,b}(t)>$, if$a=2^{-j}$,$b=2^{-j}k$, $j,k\in Z$, DWT can be writtenas following:

$$(DW_{\psi}f)(j,k)=<f(t),\psi_{j,k}(t)> \tag{2}$$

Where$\psi_{j,k}(t)=2^{j/2}\psi(2^{j}t-k)$, $j,k\in Z$.

There is one-dimensionDWTtoolbox in MATLAB. Beforeapplying DWT to the signals, wavelet decomposition has tobe donefirst.At this moment,the wavelet coefficients havecontained in the decomposition results, i.e.

$$[C,\ L] = \text{wavedec}(X,\ N,\ \text{'wname'}) \tag{3}$$

Where $\boldsymbol{X}$ is signals for analysis and N is the wavelet decomposition scale. 'wname' is the mother wavelet type. $\boldsymbol{C}$ is the wavelet decomposition vector and $\boldsymbol{L}$ is the label vector. Take the wavelet decomposition in N scales as example, it is shown in Fig.1.

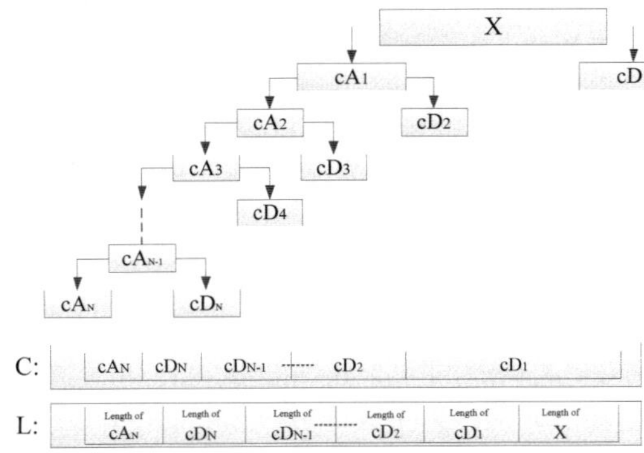

Fig.1 Structure of wavelet decomposition at N scales

As is depicted in Fig.1, the wavelet decomposition vector is written as follows (Gun, Soon & Joan et al 2011):

$$C=[C_1,C_2,C_3,\cdots,C_{N+1}]=[cA_N,cD_N,cD_{N-1},\cdots,cD_1] \tag{4}$$

Where A_N is the highest scale approximation, $D_i(i=N,N-1,\cdots,1)$ is the details from all scales. If the length of vibration signal sample is l, A_N has $l/2^N$ coefficients of wavelet, and D_i has $l/2^i$ coefficients of wavelet.

$\boldsymbol{C}$ is a column vector after DWT of the signal, and the wavelet coefficients should be reconstructed.Each column in $\boldsymbol{C}$ after reconstruction corresponds to the approximation and detail coefficients at a certain scale.

The wavelet energy of each sub-band $C_i(i=N+1,N,\cdots,1)$ is the sum of the squares of allthe coefficients for that sub-band, as given in Eq.5(Rosso, Martin & Figliola et al 2006).

$$E_j=\sum_k |C_j(k)|^2 \tag{5}$$

The wavelet total energy is the sum of all the wavelet energy at each sun-band, as given in Eq.6 (Rosso, Martin & Figliola et al 2006).

$$E_{tot} = \sum_{k} E_j = \sum_{j} \sum_{k} \left| C_j(k) \right|^2 \quad (6)$$

In actual, when arranging rational measuring points on the specified structure, different dynamic response signals, such as acceleration, displacement or velocity signal etc., can be acquired through different sensors and data acquisition system. If damage exists, dynamic response signal at damaged place will be more disordered, and the coefficients will change significantly, which lead to the increasing of wavelet energy. So $E_{tot,dam}$ at the damaged place will increase most than otherplaces without damage. In order to detect the existence of damage and locate the damaged place, the damage index *DSI*is proposed based on **RVWTE**.

$$DSI = (E_{tot,dam} - E_{tot,undam}) / E_{tot,undam} \quad (7)$$

Where $E_{tot,dam}$ and $E_{tot,\mathrm{un}dam}$ are the wavelet total energy before and after damage at each measuring point. *if* $DSI < 0$, we make $DSI = 0$.

The rule for detecting the damaged place is that DSI_{dam} at damaged place is larger than DSI of approach points on both sides.

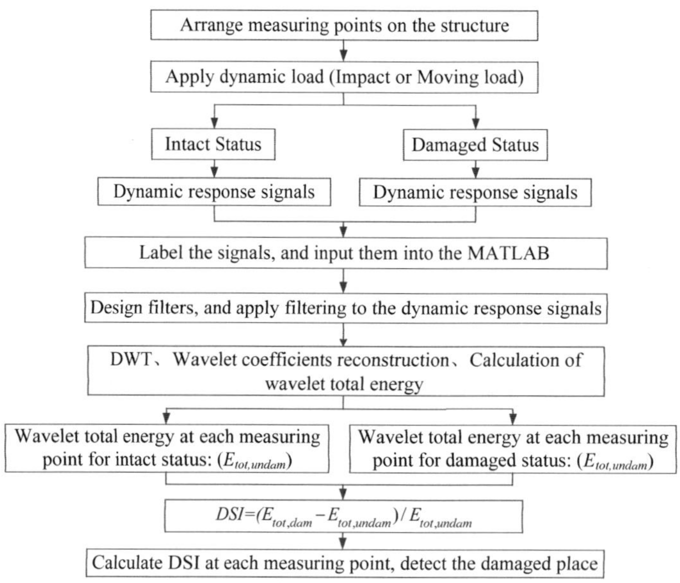

Fig.2 Damage detection procedure based on**RVWTE**

2.2 Steps for damage detection based on RVWTE

The damage detection program is designed by MATLAB software, and the detailed flow chartis shown in Fig.2.

3 Experimental model andtest

3.1 Experiment beam model

In order to verify validity and reliability of the proposed method, the typical simply-supported steel beam was chosen as analytical and test example, as shown in Fig. 3. The beam is steel I-section beam. Length of the beam is 6 m. The material is steel grade Q-345.The density, young's modulus and Poisson's ratio of the steel material are 7850 kg/m^3,2.06×10^{11}Pa and0.3 respectively.

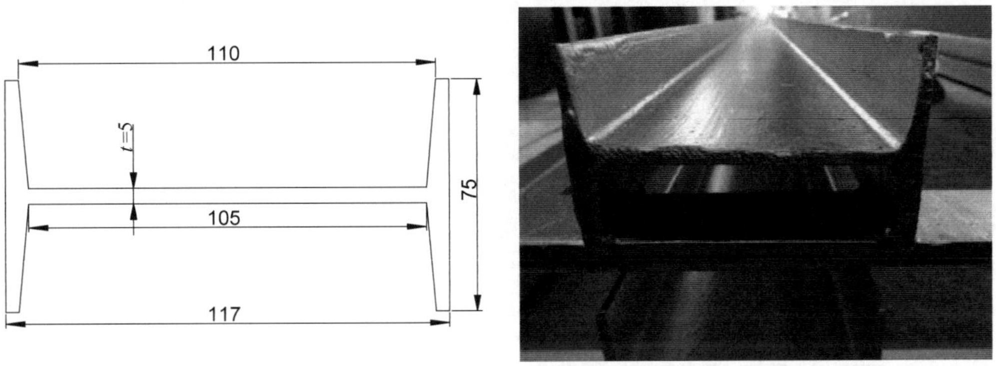

Fig. 3 Cross section size of experiment beam (Unit： mm)

The beam is divided into 10 equal segments, and 9 accelerometers are arranged on each divide except for the two boundary points.The labels and sensitivities of the accelerometers are shown in Tab. 1. Boundary condition of the experiment beam is simple-supported.

Tab. 1 Measuring points number, labels and sensitivities of accelerometers

Measuring points Label	N1	N2	N3	N4	N5	N6	N7	N8	N9
Label of accelerometers	1063	1066	1080	1065	322	1064	1077	1067	1076
Sensitivity (pc/ m·s^{-2})	321.3	331.4	326	321	360	311.2	329	324.5	338.3

3.2 Test instrument and data acquisition

The experiment beam usednine KD 1300 piezo-electric accelerometers. The hardware used to record the signals was TST5912 dynamic data acquisition system. Data processing was carried out byMATLAB. The beam was excited by LC-02A impact hammer with the sensitivity of 3.86pc/N.

3.3 Damage sceneries

Damage at the measuring point is simulated by a 3 mm width notch, and different depths correspond to different damage levels. The definition of damage level is as follows:

$$\alpha_n = H_n^d / H \tag{8}$$

Where H_n^d and H are depth of beamsectionin damaged and intact status, respectively.

In actual test, six beams with same cross section are designed tosimulatedepthsofbeam in the damagelocation in the different damage levels, in whichone is intact beam, and the other beams have different damage conditions. Five beams with different damage cases are listed in table2, which includesthree single damage cases (SD1 ~ SD3) and two double damages (DD1 ~ DD2). Labels and coordinates of the measuring points are shown in Fig. 4. In Fig. 4,*x* is the distance between measuring point and left supporting point of experiment beam.

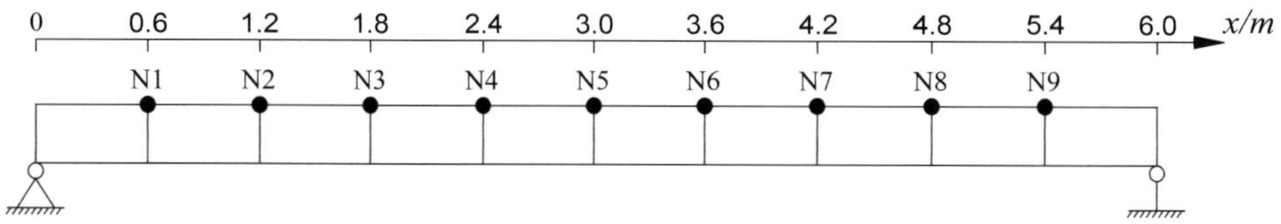

Fig. 4 Layoutand coordinates of measuring points

Tab. 2DamageSceneries of the experiment beams

Damage Sceneries		damage place	damage level
Single Damage	SD1	N5	40%
	SD2	N7	30%
	SD3	N4	20%~30% by 5%
Double Damage	DD1	N2、N5	40%
	DD2	N3、N7	30%

3.4 Load condition

Different location of the impact load will lead to different excitation modes, and four load locations on the experiment beams are set at the centralpoints of first four segments, as shown in Fig. 5.

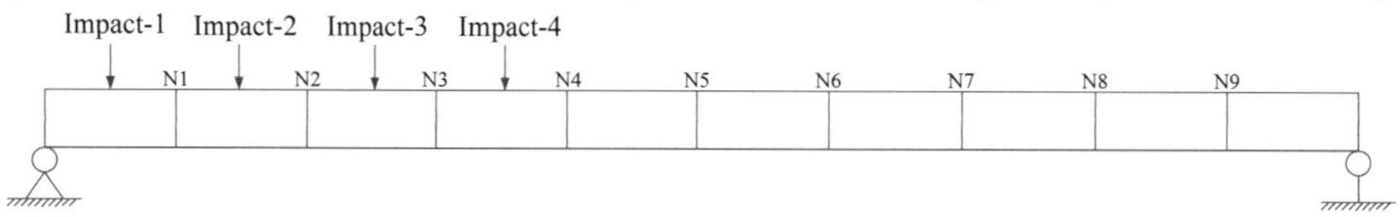

Fig.5 Segments and Different impact place in the experiment beams

4. Results analysisandverification

4.1 Data pre-processing

The sampling frequency is 1 kHz and 500 Hz for impact load and moving load, respectively. The force applied by impact hammer is discrepant each time, so the impact load shall be normalized. The normalization is as follows:F_1 and $F_2$are two different impact loads, and F_{norm} is the normalized impact load. The normalization of $F_1$and F_2 is like the following:

$$\delta_1 = F_1 / F_{norm}, \quad \delta_2 = F_1 / F_{norm} \tag{9}$$

The corresponding reduction of acceleration signal is like:

$$a_1^{norm} = \delta_1 \cdot a_1 , \quad a_2^{norm} = \delta_2 \cdot a_2 \tag{10}$$

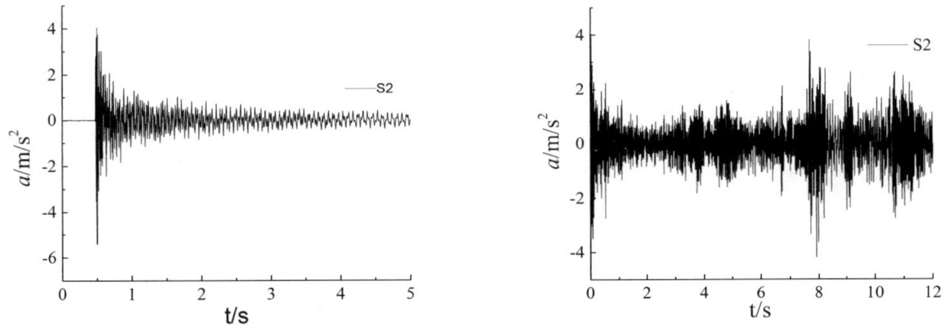

(a) Impact 1 (b) Moving load (v=0.3m/s)

Fig.6 Accelerationresponsesat N2 under Impact 1 and moving load (v=0.3m/s) before damage

Typical measured acceleration of intact beam under impact 1 and moving loadare shown in Fig.6.After applying FFT to the acceleration at each measuring point, natural frequency of the beam can be obtained. Taking SD1 for example, the changeof natural frequency of the experiment beamSD1 before and after damage is listedin Table3. The f_{int} and f_d strand for the frequency at intact and damaged status, respectively.

Tab. 3 Variety of natural frequency of beam before and after damage

Order	f_{int} /Hz	f_d /Hz	$(f_d - f_{int})/ f_{int}\times 100$ （%）
1.00	3.512	3.510	-0.341
2.00	14.101	14.099	-0.071
3.00	29.878	29.877	-0.070
4.00	53.871	53.864	-0.050
5.00	88.908	88.837	-0.080

From thetable 3, natural frequency of the experiment beam SD1decrease after damage, but the changeis very little. In additionto the environmental and noise effect, in fact, the frequency method cannot detect and identify damage of beam SD1.

4.2 Damage detection results by RVWTE

The filter is Elliptic. For impact load, the cutting frequency of the filter is 100 Hz to obtain the first four vibration mode and remove the effect of high frequency components on signals. Daubechies wavelets are orthogonal wavelets with limited compact support, and have strong time-frequency localization. So db1 is chosen for mother wavelet with scale 8 for DWT of the signals.

4.2.1 Detection result for SD1 condition

As shown in Fig.7, DSI at each measuring point is not uniform, implying the existence of damage.Besides, DSI at N5 isobviouslylarger than other place, which means that damage occursin aboutN5 of beam. The detection result is same as the pre-set damage location. Thus, it is found thatthe proposed method can detect damage near the mid-span of the steel beam in SD1 condition.

4.2.2 Detection result for SD2condition

As shown in Fig.8, DSI at each measuring point is also not uniform, implying the existence of damage.Here, DSI at N7 isdistinctlylarger than other place, which means that damage occursin aboutN7of beam. The detection result is almost same as the pre-set damage location. Thus, it is found that the proposed method can detect N7 damage of thesimply-supported steel beam inSD2 condition.

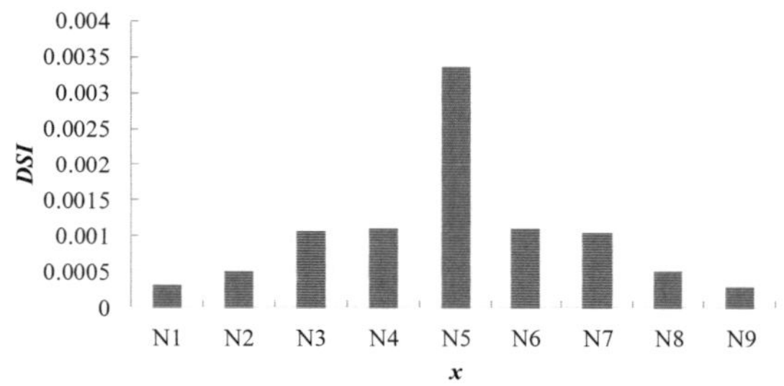

Fig. 7 Detection result for SD1 condition under impact 1

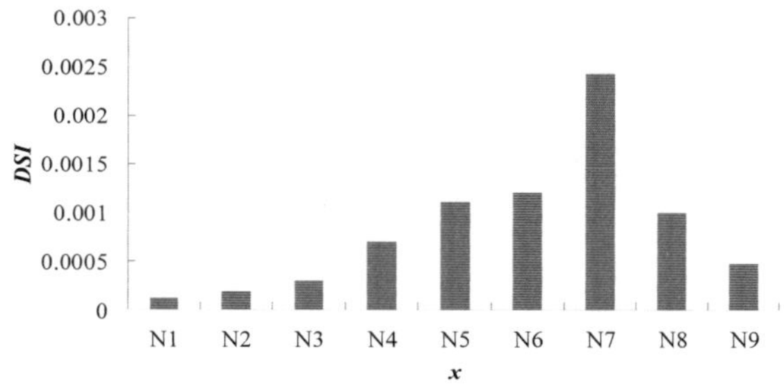

Fig. 8 Detection result for SD2 condition under impact 1

4.2.3 Detection result for DD1 condition

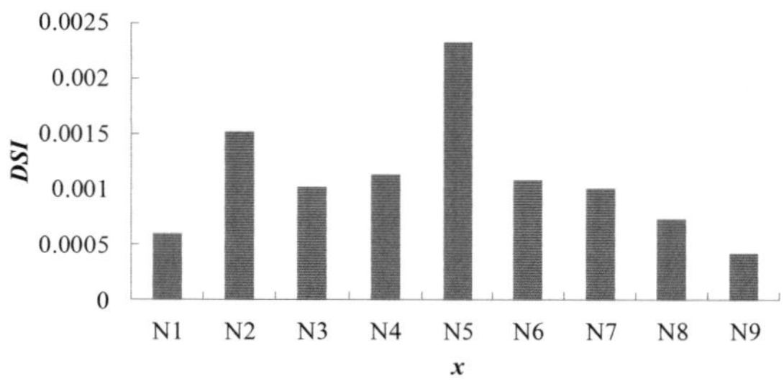

Fig. 9 Detection result for DD1 condition under impact 1

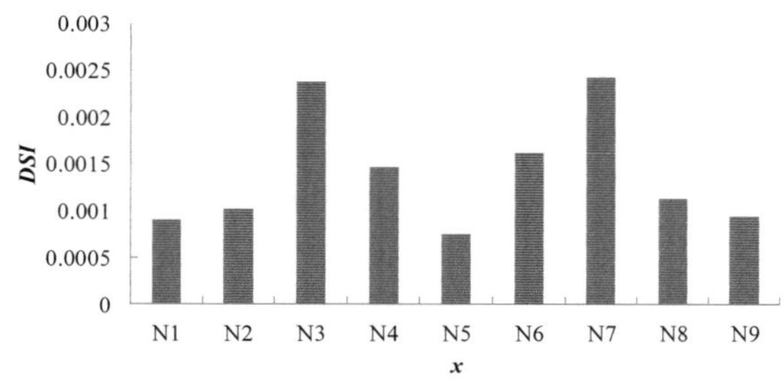

Fig. 10 Detection result for DD2 condition under impact 1

As shown in Fig. 9, DSI at each measuring point is not uniform, implying the existence of damage.It is noticed that DSI at N2 and N5 are larger than onesat both sides of these two nodes, which means that near N2 and N5 pointsof the beam the damage may occur. The detection result is same as the pre-set damage location.

4.2.3 Detection result for DD2condition

As shown in Fig. 10, DSI at each measuring pointis also not uniform, implying the existence of damage. It is noticed that DSI at N3 and N7 are larger than onesat both sides of these two nodes, which means that damage exists in the beamnearN3 and N7 points. The detection result matches with the pre-set damage location. The above results showed the proposed method can predict the double damage cases of the simply-supported steel beam, such as DD1 or DD2 condition.

4.3 Analysis ofdamage sensitivity

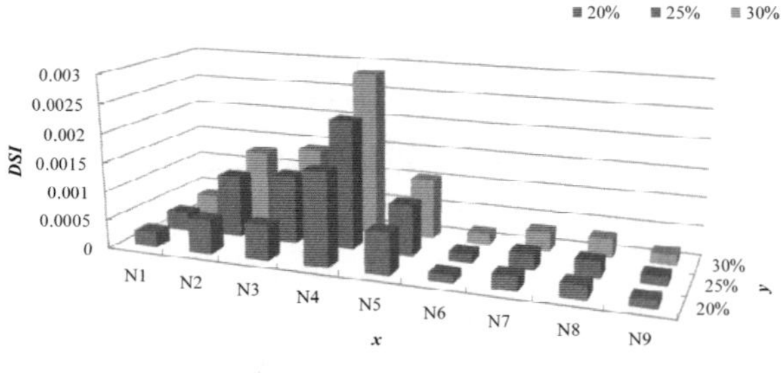

Fig. 11 Damage detection results for SD3 under different level of damage

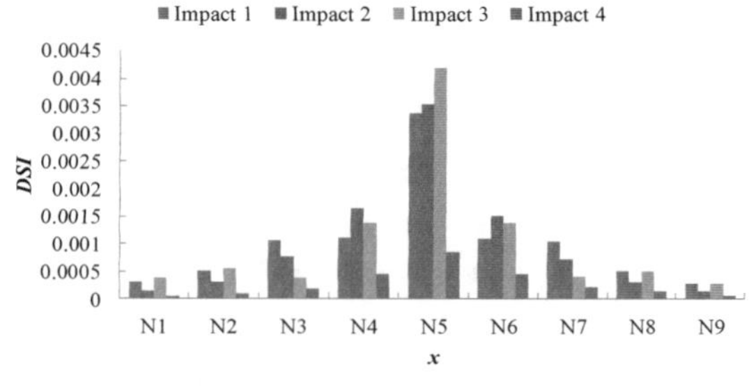

Fig. 12 Detection result for SD1 in case of different loading position

In order to verify thesensitivity of the damage index, different levels of damage are applied to Node N4 (SD3). The acceleration responses under impact 1 are chosenfor analysis. Fig. 11 gave damage detection results for SD3 under different levels of damage.

As shown in Fig. 11, the detection method based on RVWTE is able to detect damage with the different damage level.The damage index DSI increased with the increasing of damage level. So the proposed damage index is sensitivity to single damage. But there still wasfurther irnprovement for damage level detection because of the measuring error and external disturbances.

5 Factor analysis on damage detection

5.1 Effect of load location

Due to different load location, the excited model shape may be a little differenteven for the same kind of excitation type. Four cases are set to analyzeeffect of load location on detection results. Taking SD1 condition for example, Fig. 12 showed the detection result for different load locations.

As shown in Fig. 12, DSI at N5 is always larger than one at both sides of N5when impact load is exerted on segment 1- segment4 of the beam, which means that the damage exists nearN5 ofthe beam. The proposed method still can detect where damagearewhen impact load location moves, but the detection result has some difference. DSI under impact 4 (when impact load act on segment 4 of the beam)is lower than other DSI under other impact load, and the detection result at impact 3 is the best one. Reason may beas follows:

The wavelet total energy E_{tot} at measuring point near the load point is generally big, so the DSI is relative lower under the same $\Delta E_{tot}=E_{tot,dam}-E_{tot,\mathrm{un}dam}$ according to Eq. (7). DSI in whichload point isclose todamage place is less than DSI in which the load locationis away from damage place. So in real application,you can better detect the damage place by adjusting the damage location.

5.2 Effect of mother wavelet type

In order to analyze effect of mother wavelet type on damage detection results in Lab, five different types of wavelet are chosen for **DWT** of acceleration signals under impact 1 for SD1 condition. The resultsare shown in Fig. 13. From Fig. 13, the damage place can still be detected bythose five different wavelets, and the DSI nearthe damaged place of the beam changes little. Effect of wavelet on the proposed method is negligible. Thefilter of mother wavelet can be avoided in real application. The proposed damage index has some robustness to mother wavelet type.

5.3 Effect of data length

Data length is 5120 in this paper for impact load. Choosing the appropriate data length is very important to fast and efficient damage detection. Taking SD1 damage detection for example, five acceleration data length, such as 1024, 2048, 3072, 4096 and 5120, are set for analysis. The resultsareshown in Fig. 14.

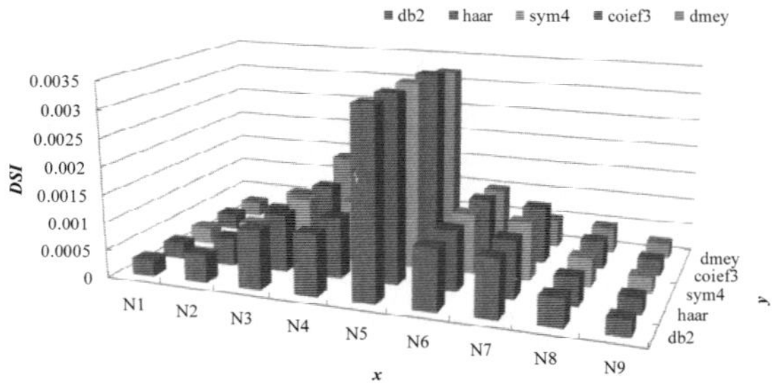

Fig. 13 Detection result for SD1 damage condition using different kinds of wavelet

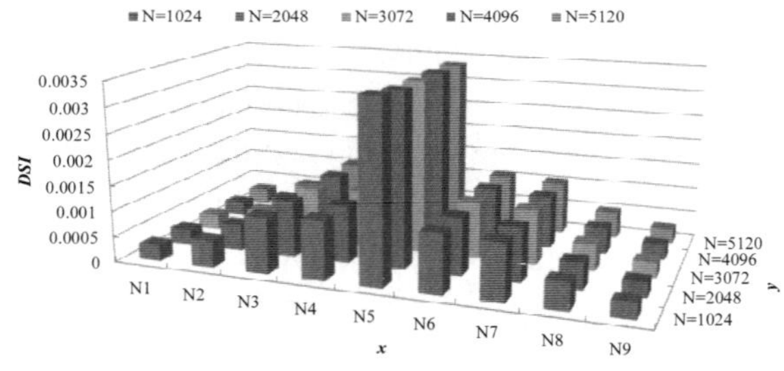

Fig. 14 Detection result by analyzing different length of acceleration data

From Fig. 14, it is found that when accelerationresponse adopts different lengths of signals, varying trend of DSI at each measuring points isbasically same.So, the proposed method almost can detect the damage place for different data length.But the data length only has some difference to wavelet energy.As shown in Fig. 15, when data length is larger than 3072, the wavelet total energy at N1 (E_0^1) tends to be stable. But DSI (Taking DSI at N5 as example) change little for different data length. In real application, it has good flexibility when choosing the right length of signals.

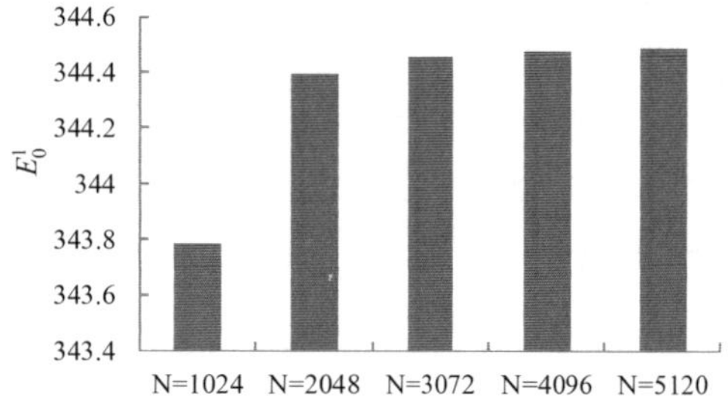

a) E_0^1 of the beamfor different data length

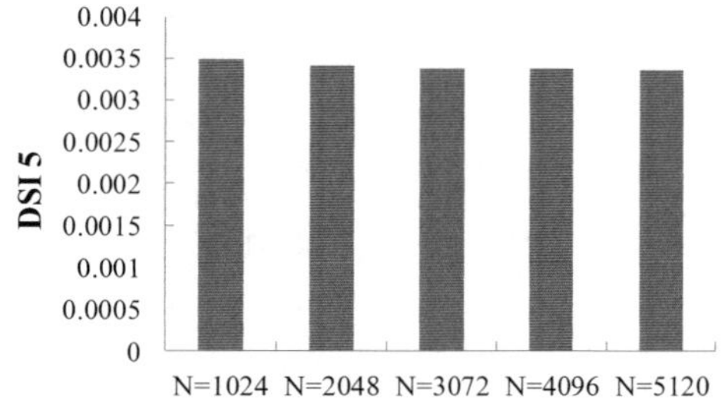

b) DSI in theN5of the beam fordifferent data length

Fig.15 E_0^1 using different lengths of data and DSI at damage place N5

6 Conclusions

After applying **DWT** to dynamic acceleration signals, wavelet total energy at the measuring points could be calculated. The damage method based on **RVWTE** was proposed, Six 6 m-long beams were made for experimentto verify effectiveness of the proposed method. 3 single damages and 2 double damages sceneries were set, and impact load is the excitation. Sensitivity and effect factors of damage index were analyzed, and conclusions were drawn as follows:

1) The damage detection method based on RVWTE can detect the existence and location of single and double damages in the beams accurately.

2) The proposed damage index is sensitive to damage level. But in real application, the detection result is not obvious for damage less than 20% due to measuring error and external disturbances.This is to be further improved.

3) The proposed damage index shows some robustness to mother wavelet type and data length. For

different impact location, the proposed method can accurately detect the damage place, but the result shows a little difference. If the impact place is away from the damage place, it can be better to detect the damage place.

Acknowledgments
This work is financially support by National Natural Science Function of China (51541810) and the Cyrus Tang foundation of China.

References:

Abdo ABM 2012, 'Parametric study of using only static response in structural damage detection', *Engineering Structures*, vol. 34, no. 1, pp. 124-131.

Bandara RP, Chan TH & Thambiratnam D P 2014, 'Structural damage detection method using frequency response functions', *Structural Health Monitoring*, vol. 13, no. 4, pp. 418-429.

Lieven NAJ & EwinsDJ.'Spatial correlation of mode shapes: the coordinate modal assurance criterion (COMAC)', *Proceedings of the Sixth International Modal Analysis Conference*, Florida, America, pp. 690-695.

Pandey AK, Biswas M & Samman MM 1991, 'Damage detection from changes in curvature mode shapes', *Journal of Sound and Vibration*, vol. 145, no. 2, pp. 321-332.

Hosseinzadeh AZ, Amiri GG, Razzaghi SAS, et al2016, 'Structural damage detection using sparse sensors installation by optimization procedure based on the modal flexibility matrix', *Journal of Sound and Vibration*, vol. 381, pp. 65-82.

Rucka M & Wilde K 2006, 'Application of continuous wavelet transform in vibration based damage detection method for beams and plates', *Journal of Sound and Vibration*, vol. 397, no. 3-5, pp. 536-550.

Chandrashekhar M & Ganguli R 2009, 'Structural damage detection using modal curvature and fuzzy logic', *Structural HealthMonitoring*, vol. 8, no. 4, pp. 267-282.

Jing LJ & Xiang YQ 2007, 'Damage identification for the main girders of large-span cable-stayed bridge based on modal analysis technique', *China railway Science*, vol. 28, no. 6, pp. 54-59. (In Chinese)

Jing LJ & Xiang YQ 2007, 'Damage identification of main girders for long span cable-stayed bridge based on flexibility matrix', *Journal of Zhejiang University (Engineering Science),* vol. 42, no. 1, pp. 164-169. (In Chinese)

Han JG, RenWX & Sun ZS 2005, 'Wavelet packet based damage identification of beam structures', *International Journal of Solids & Structures*, vol. 42, no. 26, pp. 6610-6627.

Kim H & Melhem H 2004, 'Damage detection of structures by wavelet analysis', *Engineering Structures*, vol. 26, no. 3, pp. 347-362.

Gun JY, Soon GL, Joan C, et al 2011, 'Decentralized damage detection using wavelet signal analysis embedded on wireless smart sensors', *Engineering Structures*, vol. 33, no. 7, pp. 2162-2172.

Rosso OA, Martin MT, Figliola A, et al 2006, 'EEG analysis using wavelet-based information tools', *Journal of Neuroscience Methods*, vol. 153, no. 2, pp. 163-182.

The 8th International Conference on
Structural Health Monitoring of Intelligent Infrastructure
Brisbane, Australia | 5-8 December 2017

Development of a vision based SHM for bridge damage identification

D. Lydon[1], S.E. Taylor[1] D.Robinson[1], M.Lydon[1] D. Hester[1]
[1]School of Natural and Built Environment
Queens University Belfast
BT9 5AG
[1]*Email:* dlydon01@qub.ac.uk

Abstract

Structural Health Monitoring (SHM) techniques can provide vital information on the performance and capacity of new and existing structures. Generally sensors are embedded or surface mounted to assess the integrity of the structure and provide a means of damage detection or failure prediction. However, existing systems are limited by the need for contact with the structure and demand onerous set up procedures. This paper details the field and laboratory testing of a low-cost camera based SHM system. The system has been developed to provide bridge deflection data. A post-processing method has been developed within MATLAB to allow for damage and associated losses in stiffness to be successfully located on the structure, enhancing the damage identification process. The displacement identification algorithms have been tested and validated through a series of laboratory trials. On completion of the laboratory calibration this vision based system was subsequently tested using live traffic on an existing 19m span beam and slab bridge in Northern Ireland. The findings of the field trial confirm the ability of this system to accurately relate measured bridge deflections. The developed algorithms transform the measured camera or video images into a form that is highly damage-sensitive/change-sensitive for bridge assessment within the context of Structural Identification with input and output characterization. This research will significantly advance current sensor-based structural health monitoring with computer-vision techniques, leading to practical applications for damage detection of complex structures with a novel approach.

1. Introduction

A software algorithm was developed for monitoring the deflection of points on a civil structure as part of a project involving Queens University Belfast, University College Dublin and the University of Central Florida. Before filed testing could take place, it was important to test the performance of the algorithm in the laboratory under varying light conditions and monitoring distances. It was also decided to determine the change in accuracy between using natural features of the beam vs an applied target.
A scale model of a bridge was simply supported and loaded using an accurately calibrated 600kN capacity hydraulic actuator in the laboratory and an apparatus was attached to the test rig in order to apply load to the beam at two points simultaneously as shown in Figure 1.

Figure 1 Test Rig in the Laboratory

A pattern was applied to the beam under the points that load would be applied at in order to determine the accuracy of the target based feature extraction. Three Linear Variable Differential Transducers (LVDTs) were also placed in series at these points on the beam, they were spaced 1168mm apart and set to a scanning rate of 2Hz. A Nikon D810 camera was set up and triggered using Digicam Control.

2. Algorithm development

The steps involved in an algorithm for calculation of displacement of civil structures is the following:

1) Camera Calibration

Camera Calibration is a method of determing the intrinsic and extrinsic paramenters of the camera used to record the structure in order to both remove lens distrortion effects and to provide a scaling factor for the conversion from pixel units to engineering units. The process used in the system was developed by (Bouguet 2015). There are two types of lens distroirton present in a standard camera, radial distortion and tangential distortion. Radial distortion occurs when light rays bend more near the edges of a lens than they do at its optical center. Tangential Distorton occurs when the lens and the image plane are not parallel.

The system obtains the scaling factor by determining the length of a checkerboard square in pixels (found by corner detection between the black and white squares using the Harris method(Harris & Stephens 1988)) and then performing a division operation with this number and the known length of the square in mm.

2) Feature Extraction

The second step is to obtain a set of features on the structure that can be reliably tracked, in a process known as registration. The process selected for use in the algorithm was Binary Robust Invariant Scalable Keypoints(BRISK)(Leutenegger et al. 2011). The keypoint detection used by BRISK is an extension of AGAST, which is in itself an extension of FAST(Rosten & Drummond 2005). The keypoints are invariant of scale and are estimated in continous scale-space. The keypoints are composed of binary string, simple brightness comparison tests. Neighbourhood sampling is performed in deterministic, equally spaced concentric circles to detect features.Several long distance sample point comparisons to determine oreientation are performed and the points are returned in order to be tracked by the next stage of the algorithm.

3) Feature Tracking

The algorithm tracks the features returned by BRISK using a Kanda-Lucas-Tomasi(KLT)(Tomasi & Kanade 1991) object tracking method. The algorithm works by generating an image pyramid, where each levels is down-sampled by a factor of two in width and height. The system begins tracking each point in the lowest resolution level, and continues tracking until all levels have been checked. This method is applicable to the recordings gathered because there will not be a large displacement between frames as the videos were captured at a rate of 60fps. The method uses a threshold value for the amount a feature can be expected to move between frames to eliminate points that cannot be reliably tracked, reducing the error in the display of results. The positions of the points in each frames are stored in a matrix in Matlab, and the Euclidean distance between the points in subsequent frames is calculated and plotted with respect to time in order to obtain a plot of displacement of points in a video.

3. Laboratory work

3.1. Details of Concrete Beam used in Test

The concrete beam used in the test was a 1:5 scale model of a bridge platform used in a previous testing scenario. The specifications of the bridge are as follows:

- Cross Sectional Area: 30.238 x 10^3mm^2
- Depth: 250mm
- Width: 230mm
- Neutral Axis: 137.09mm
- E-Value: 26.6 x 10^3N/mm^2
- I-value: 4.05 x 10^9 mm^3

The dimensions of the model bridge are shown in Figure 2.

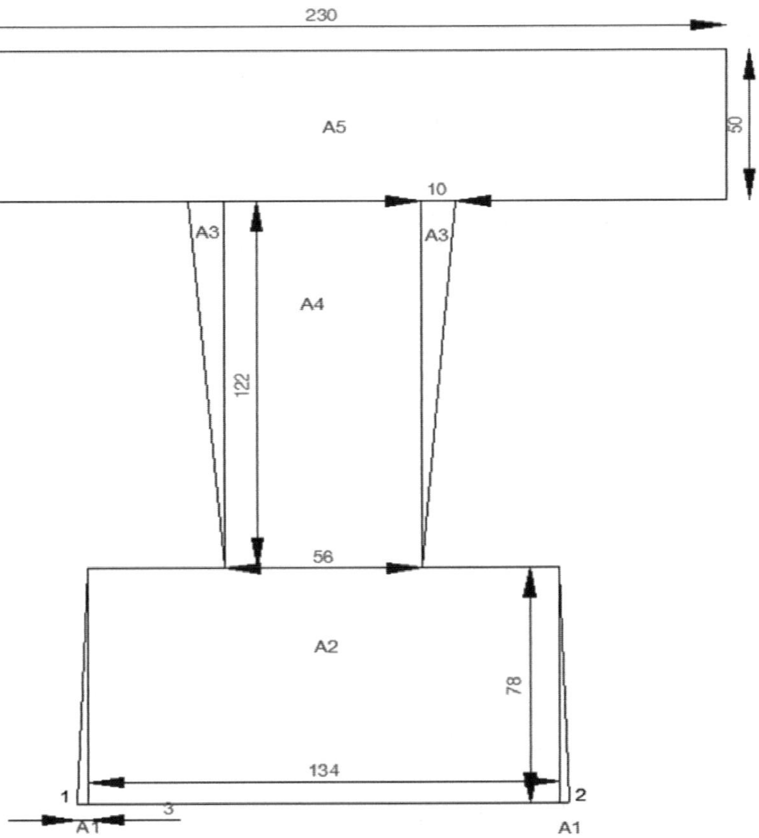

Figure 2 Dimensions of Concrete Beam(mm)

3.2. Testing plan

A series of loading cycles were carried out on the beam, the details have been presented in Table 1.

Table 1 Test Plan for Concrete Beam Series

Test Number	Monitoring Distance(m)	Lighting Conditions
1	2	Artificial
2	2	Moving
3	2	Natural
4	4	Artificial
5	4	Moving
6	4	Natural
7	8	Artificial
8	8	Moving
9	8	Natural

A load of 10kN was applied and deflection was monitored by LVDT. The lighting was altered by the use of artificial lights and the monitoring distance was changed by moving the cameras further from the test specimen. The results from the 2M (Figure 3) and 8M (Figure 4) tests are shown below.

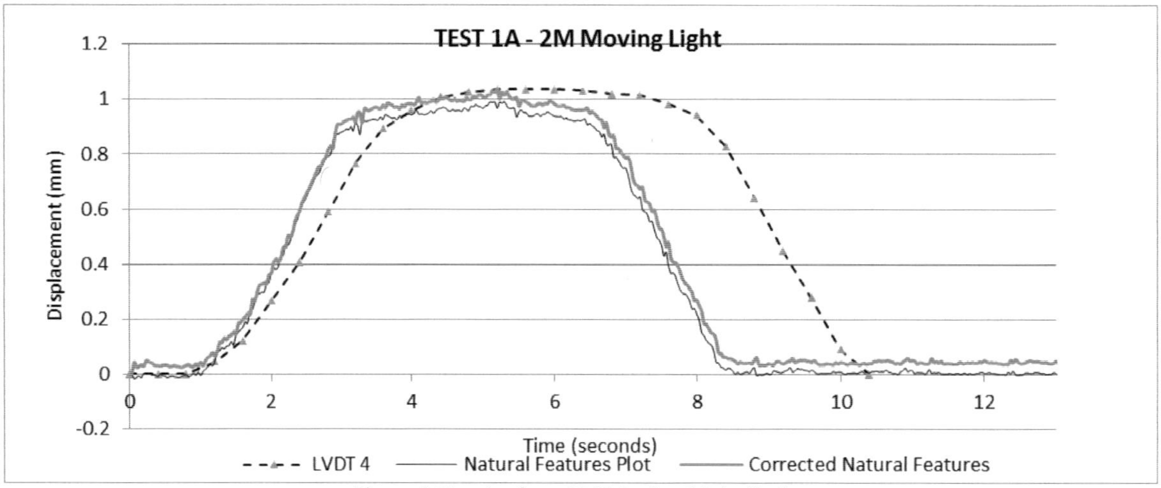

Figure 3 Results from 2M Moving Light Series

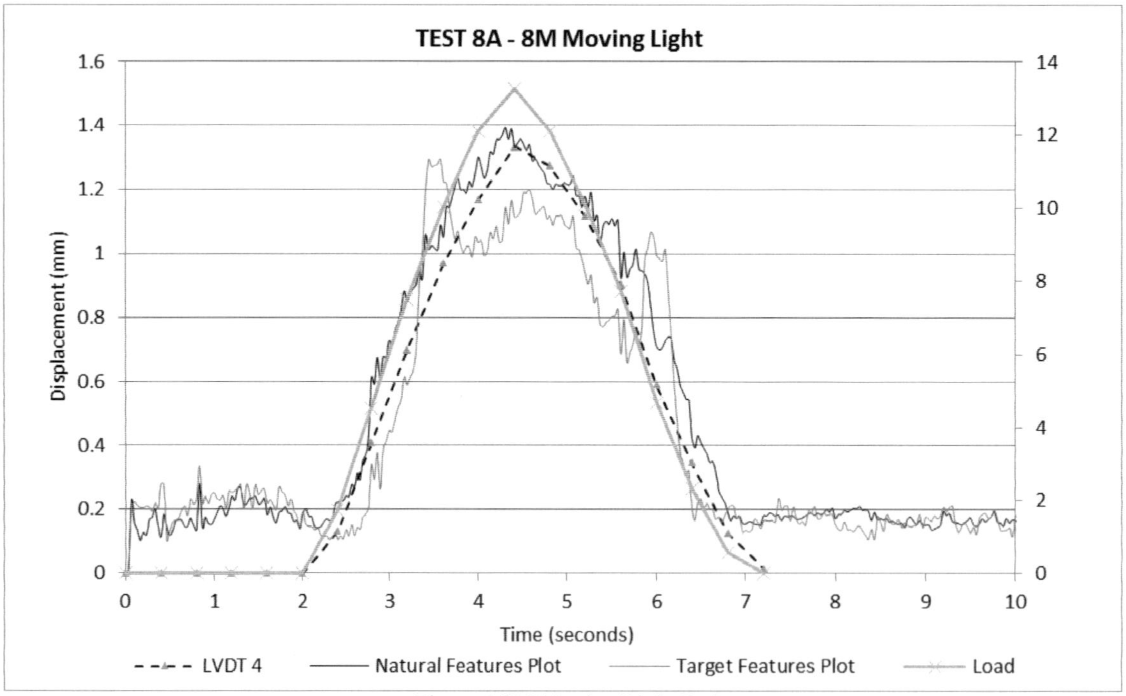

Figure 4 8M Moving Light Series

The maximum deflection as determined by the algorithm was compared to that of the LVDT and an error percentage generated for each test series as is shown in Table 2.

Table 2 Results from Concrete Beam Series

Distance(m)	Lighting	Average Error Percentage
2	Artificial	3%
2	Natural	2%
2	Varying	2%
4	Artificial	2%
4	Natural	2%
4	Varying	2%
8	Artificial	3%
8	Natural	4%
8	Varying	7%

The performance of the algorithm was judged to be acceptabel, and a set of field trials were scheduled.

4. Field work
4.1. Verners Bridge

A bridge that was marked for repair for Transport NI was made available for testing in order to validate the performance of the algorithm in field conditions.. A series of passes by a calibrated weight truck over this bridge were recorded and displacement was calculated – the results from a series of passes are shown in Figure 5.

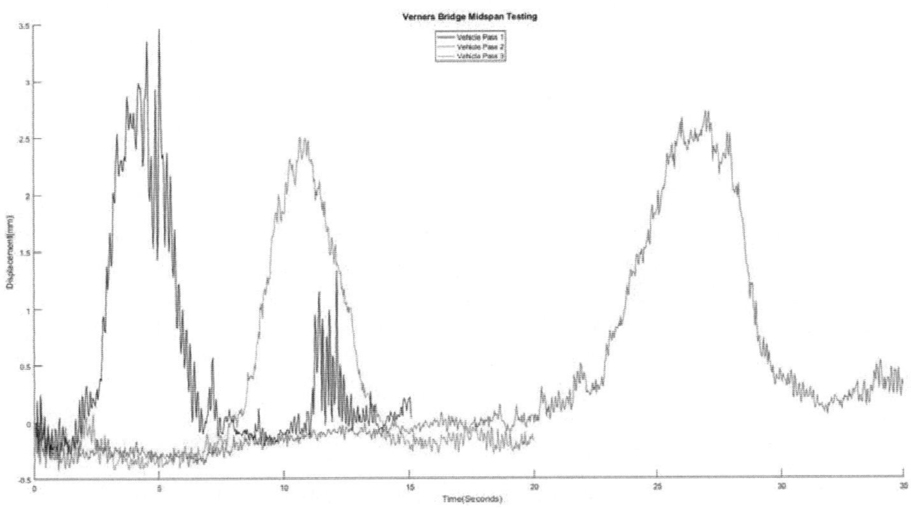

Figure 5 Results from Midspan Monitoring on Verners Bridge

Due to an error with calibration, the results from the accelerometers used for ground truth verification were not available; nevertheless there is a close correlation between readings of the camera solution for subsequent passes on the same location. The difficulty with accelerometer setup provides a further validation of the validity of vision systems as a method, with no issues occurring with setup or monitoring with the camera equipment.

4.2. Majors bridge

Majors Bridge is a masonry arch bridge which carries two lanes of traffic on the A34 Lisnagole road north of Lisnaskea. The bridge has a cumulative span of 7.41m, Cover to arch ring 0.11m (approx. 0.2m measured on site recently) and a deck width 9.0m.
A few approximate measurements are presented below and the east elevation can be seen in Figure 6.

- Voussoir depth at keystone 0.77m.
- Soffit to top of parapet wall 1.87m.
- Parapet wall to road surface 0.9m.
- Parapet wall thickness 0.48m.

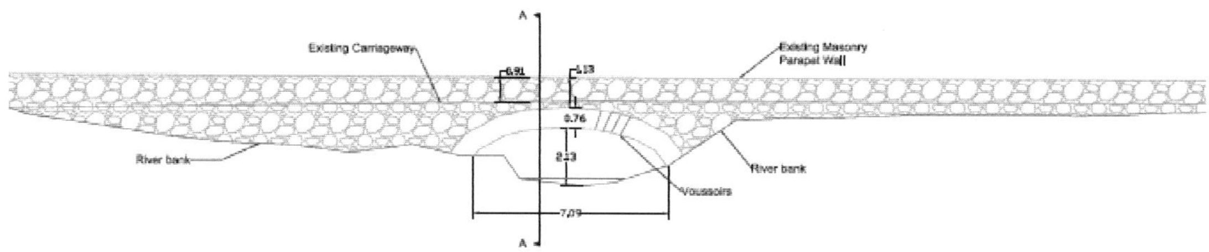

East Elevation
Figure 6 Site layout elevation at Majors Bridge

Concerns about the bridge condition were raised subsequent to a traditional bridge inspection carried out by Transport NI. Excessive movement was reported under normal traffic loading, in order to assess this, an initial site investigation was carried out to determine the suitability of the proposed system.
Access was gained to the site through a housing development on the south east of structure, this area provided a safe location of parking and setting up of equipment. During this site visit some short term monitoring was carried out using the camera based system. It was also intended to use a dial gauge system to provide validation of the results but this was not possible during this visit due to the very uneven ground conditions.

4.5. Monitoring setup

The camera was set up at a distance of 2.9m from the spandrel wall on the south east side of the structure. Monitoring was carried out under live traffic to determine if the system could provide information on the bridge behavior. Targets were attached to the underside of the bridge using a non-destructive heavy duty double sided tape. The target locations are shown in Figure 6 and this allowed for accurate calculation of the movement based on the known geometry of the target.

Figure 7 Targets attached to bridge soffit

4.5.1. Trial data collection

The results shown in Figure 7 show the measured response of the bridge structure under traffic loading. The actual weight of the vehicles is unknown but images of the vehicles have been embedded on the graph to give an indication of vehicle type. In this case a maximum deflection of ~1.2mm was measured at the mid span.

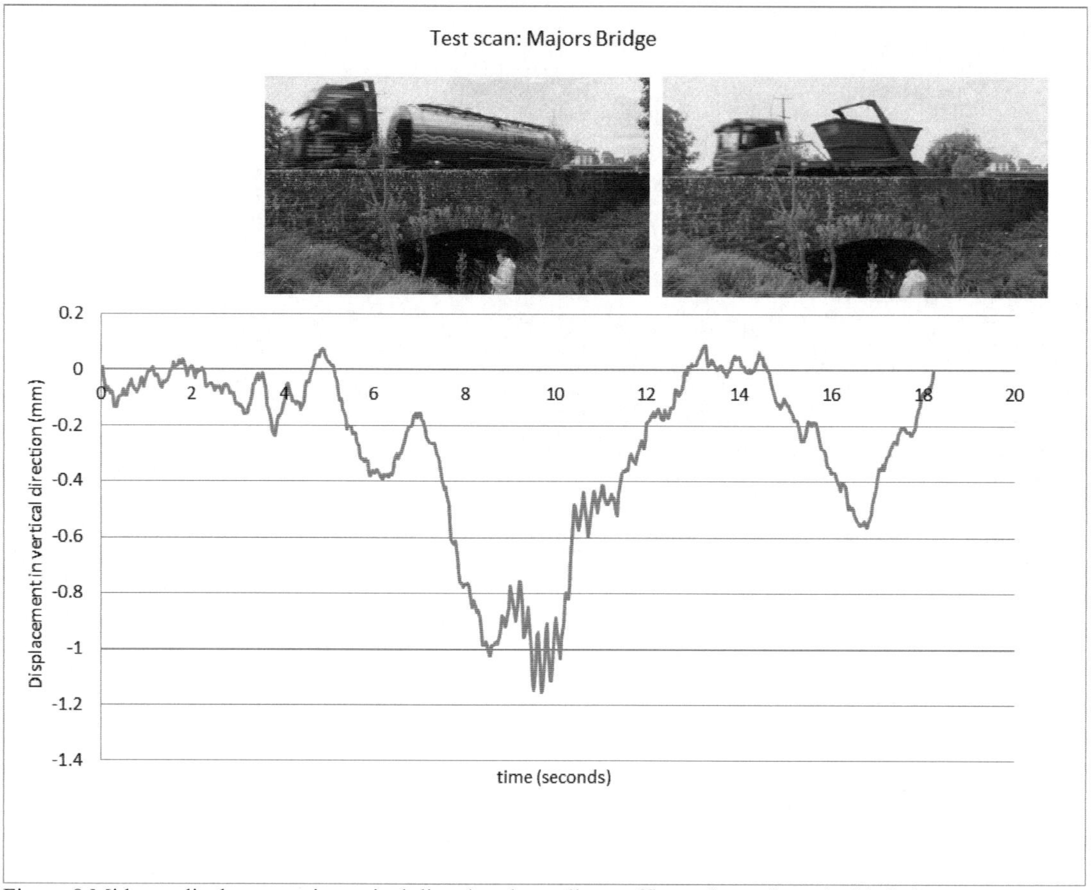

Figure 8 Midspan displacement in vertical direction due to live traffic as shown in embedded picture.

The purpose of this initial visit was to determine the suitability of the site layout for camera based monitoring. The trial monitoring carried out indicates this method of assessment is suitable for this site. However, to provide a more critical analysis of this bridge a more exhaustive monitoring program would be required along with knowledge of the specific traffic loading.

5. Conclusions

The work shown here has proven that reliable deflection readings can be obtained from a low cost camera system that can be easily deployed in the field. The next step in the development of the system is to incorporate the deflection readings gained into a system for damage detection based on varying displacement either from calibrated weight vehicles, or through use of a long term system where the weights of trucks passing over a bridge are averaged over time.

References

Bouguet, J.-Y., 2015. Camera Calibration Toolbox for Matlab. Http://Www.Vision.Caltech.Edu/Bouguetj/Calib_Doc/Index.Html#Ref. Available at: http://www.vision.caltech.edu/bouguetj/calib_doc/.

Harris, C. & Stephens, M., 1988. A combined corner and edge detector. Proceedings of the 4th Alvey Vision Conference, pp.147–151. Available at: http://citeseerx.ist.psu.edu/viewdoc/download?doi=10.1.1.434.4816&rep=rep1&type=pdf.

Leutenegger, S., Chli, M. & Siegwart, R.Y., 2011. BRISK: Binary Robust invariant scalable keypoints. In 2011 International Conference on Computer Vision. IEEE, pp. 2548–2555. Available at: http://ieeexplore.ieee.org/document/6126542/ [Accessed November 24, 2016].

Rosten, E. & Drummond, T., 2005. Fusing Points and Lines for High Performance Tracking. Proceedings of the IEEE International Conference on Computer Vision, 2(October), pp.1508–1511.

Tomasi, C. & Kanade, T., 1991. Detection and Tracking of Point Features. Carnegie Mellon University Technical Report, pp.91–132. Available at: http://www.lira.dist.unige.it/teaching/SINA/slides-current/tomasi-kanade-techreport-1991.pdf.

The 8th International Conference on
Structural Health Monitoring of Intelligent Infrastructure
Brisbane, Australia | 5-8 December 2017

Prognosis of Wire Fracture through Adaptive Wavelet Analysis of Acoustic Signals

H. Qu[1], T. Li[1], J. A. Cain[1], G. Chen[1]
[1] *Missouri University of Science and Technology – USA, Email:* gchen@mst.edu

Abstract

In this study, acoustic emission (AE) features were characterized with adaptive wavelet analysis and used to predict and detect wire fracture in a seven-wire strand that was instrumented with a pair of AE sensors at two ends and tested up to 89 kN. The cross section of one wire was locally reduced up to 90% in 10% increment at center. The AE parameters (hits, energy, and counts) changed little up to 80% reduction in cross section of the partially cut wire, and suddenly jumped at the fracture (under 73 kN) of the notched wire with 90% reduction in cross section. The acoustic signals of inter-wire slippage and fracture initiation are significantly shorter in time duration than the signal of fracture. The frequency band of the fracture signal is significantly broader than that of either the fracture-induced echo or artificial tapping noises. The time duration of artificial tapping noises is substantially longer than that of either fracture or fracture-induced echo. These distinct time-frequency characteristics allow an early detection and localization of wire fracture following the proposed procedure.

1. Introduction

Steel cables have been widely used in cable bridges for decades. These bridges provide critical paths in transportation. However, bridges deteriorate with aging and environmental effects, and can be damaged due to incidents. Early detection of deterioration and damage is most helpful in alerting engineer-in-charge with incipient catastrophic issues to prevent a complete collapse of bridges and thus ensure public safety.

Visual inspection as current practice in bridge maintenance can be subjective, difficult, expensive, dangerous, and disruptive. As a nondestructive detection method, AE has been applied into bridge cable monitoring due to its simplicity in sensor attachment and low attenuation (Rizzo & di Scalea 2001, Li & Ou 2008, Jin et al. 2008, Betti et al. 2016). Most of these researchers analyzed AE waveform in time domain for the detection and identification of wire fracture (Paulson & Cullington 1998, Drummond et al. 2007), impact source (Sun & Qian 2011), and inter-wire friction (Zejli et al. 2006). AE waveforms were also analyzed in frequency domain with the change in Fourier spectra (Oskouei et al. 2009). To preserve key time-varying features in frequency domain, short-time Fourier transform (STFT) and wavelet-based techniques were applied to verify modeling techniques (Jin et al. 2008), accuracy improvement of source localization algorithms (Hamstad et al. 2002), damage assessment of deteriorated RC structures (Yoon et al. 2000), and denoising effect of filtering techniques (Grosse et al. 2004). Most, if not all, of these time-frequency analyses were applied to the AE signals of fiber reinforced polymer composite materials (Qi 2000, Lu et al. 2011), and glass fiber reinforced polymer bridge cables (Xu et al. 2016). Little has been done to understand the full potential of time-frequency analysis of AE signals for the extraction of pre-fracture features in damage prediction of steel cables and distinguish pre-fracture and fracture signals from fracture-induced echoes and noises so that false prediction and detection cases of wire fracture are prevented.

In this study, a seven-wire steel strand was axially loaded and unloaded in displacement control, each with the cross section of one wire reduced up to 90% in 10% increment and instrumented with one AE sensor at each end of the strand. AE parameters (hits, energy, and counts) were used to detect and locate wire fracture. Pre-fracture signals were identified and differentiated from fracture signals, fracture-induced echo responses, and artificial tapping noises for the early detection of wire fracture in bridge cables by analyzing them with the recently proposed adaptive wavelet analysis (Qu et al. 2016) and comparing their features in time-frequency domain.

2. AE Test Setup of Strand Specimen

Each stay cable of the I-70 Mississippi River Bridge consists of a number of steel strands, each with seven twisted high strength wires. In this study, a 6.4-meter-long strand from the bridge construction was axially loaded with one outer wire notched at center to simulate a damage-induced reduction of cross section area. The properties and manufacture specifications are included in Table 1, and the lay length is 19.05 cm.

Table 1. Specifications and properties of a high strength steel strand (270 ksi)

Strand Parameter	Value
Nominal diameter	0.6 in. (~1.52 cm)
Minimum breaking strength	58,600 lbf (~260.7 kN)
Minimum yield point at 1% extension	52,740 lbf (~234.6 kN)
Nominal modulus of elasticity	28.5×10^6 psi (~196.5×10^3 MPa)
Nominal cross sectional area	0.217 in^2 (~1.4 cm^2)
Nominal weight	740 lbs/1,000 ft (~110 kg/100 m)

Figure 1 shows the overall test setup of a strand of seven twisted wires under axial loading. At each end of the strand, composite beams were used as anchorage support, and they are anchored by four Dywidag rods to prevent the red beam from moving and rotating during tests. Once the anchorage support is in place, a strand was passed through the gap between two channels of the red beam, a load-distribution steel platen, a hydraulic jack, another steel platen, and a three-piece anchor wedge. At the other end of the strand, the anchorage support is identical except that the hydraulic jack was replaced by a load cell.

Figure 1. Test setup of a seven-wire strand

One narrowband AE sensor (Model R1.5I) with a resonant frequency of 14 kHz and another wideband AE sensor (Model F15I-AST) with a nearly constant frequency response between 100 kHz and 400 kHz were deployed at each end of the test strand for comparison. The two AE sensors were attached in orthogonal directions with silicon compound on the surface of the three-piece anchor wedge. AE signals were recorded from a 32-channel Micro-II system. The time window length was set to 8 ms to ensure that the dominant features of pre-fracture, fracture, and fracture-induced echo signals can be captured properly. Each strand was tested up to 20 kips (~ 89 kN) for three times at each 10% increment of area (Figure 2). Finally, with 90% of the cross section area reduced, the strand was tested to failure. The cut locations of the two strands (600 cm clear span length) are 300 cm and 550 cm from their left anchor points, respectively.

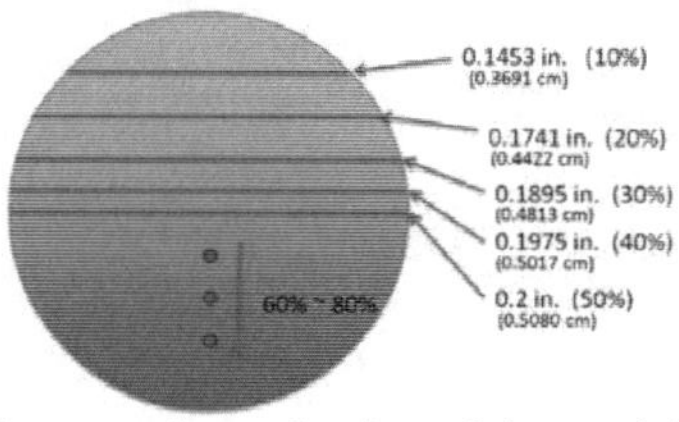

Figure 2. Cross section reduction of the partially cut wire

Axial loads were applied up to 20 kips (~ 89.0 kN) in about 3 min. In addition to external loading, four types of artificial noises were generated by dropping four different types of foreign objects on the test strand at the height of 1 ft (~0.3 m). The four objects were a steel bar, a marker, a steel washer, and a penny as shown in Figure 3.

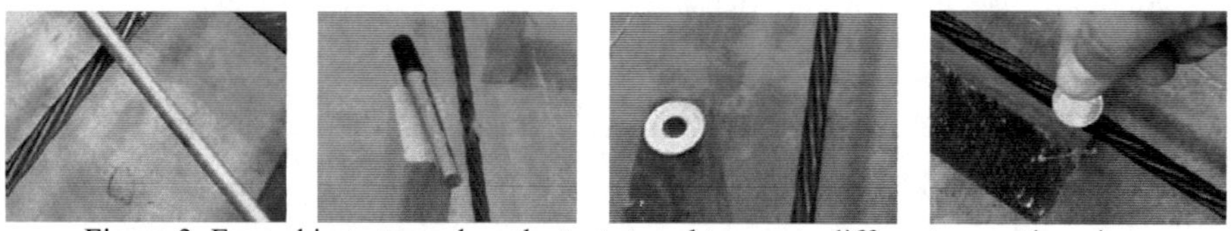

Figure 3. Four objects tapped on the test strand to create different acoustic noises

3 Fracture Detection and Localization

When the reduced cross section area of the cut wire was up to 80%, the recorded AE signals changed little, indicating no acoustic event. When the cross section area of the cut wire was reduced by 90%, acoustic events were detected until the cut wire was broken at a load of approximately 73.4 kN. Figure 4 presents the typical AE parameters accumulated over time at the stage of 90% reduced cross section area, which were recorded from the strand with the left R1.5I sensor. The hits, energy and counts of AE signals suddenly increase, indicating fracture of the cut wire at 163.2 s.

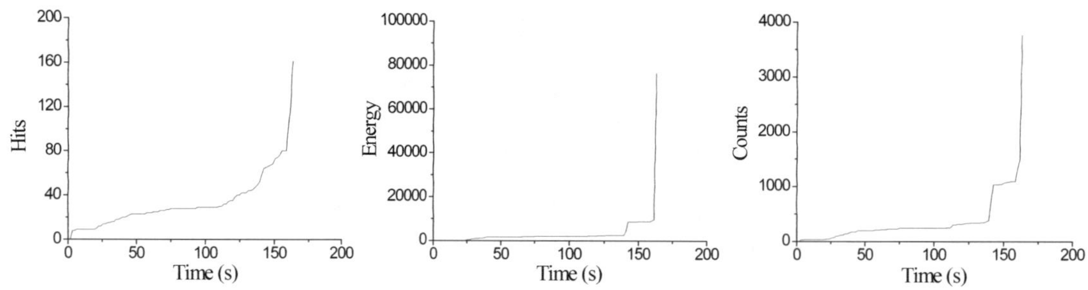

(a) Cumulative hits of AE events (b) Cumulative energy of AE event (c) Cumulative counts of AE events

Figure 4. Accumulated AE parameters from the left R1.5I sensor

The source of a fracture event can be determined from occurrence time instants of the signals recorded by at least two AE sensors. Figure 5 presents the number of events from various locations based on the measurements of two AE sensors (Model F15I-AST and R1.5I). All occurrence time instants of the captured acoustic events are marked with reference to the moment of fracture. The low frequency R1.5I sensors recorded more events than the high frequency F15I-AST sensors.

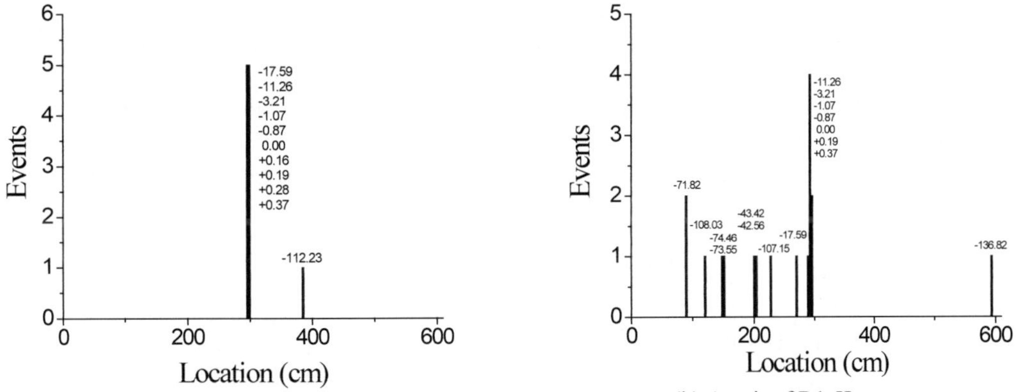

(a) A pair of F15I-AST sensors (b) A pair of R1.5I sensors

Figure 5. Number of events from various locations (30 dB threshold)

To quantify the accuracy of location detection from both types of the sensors, the identified sources of fracture events are compared with their ground truth data. The notch was physically located at 300 cm from the left support while the highest AE events from both pairs of sensors occurred from 299 cm. The error in location prediction is approximately 0.5%, which are both small and demonstrate the feasibility of fracture localization using the AE.

The accumulated parameters of the AE signals as shown in Figure 4 all indicate the onset of fracture at 145.61 s. The accumulated hits as shown in Figure 4(a) indicates that slippage likely occurs between the wires in contact at the early stage of tests setup. The significant hits, energy and counts at 50.97 s and their rapid increase at 145.61 s appear good indicators of the fracture event, which are likely associated with the effect of fracture initiation.

To understand the sequence of various events, the occurrence time instants of acoustic events captured by a pair of F15I-AST sensors as shown in Figure 5(a) is presented in Table 2. It also includes the relative occurrence time with respect to the fracture event, which are the same as those included in figures 5(a). The first event corresponded to an inter-wire slippage away from the mid-span, which occurred at 112.23 s prior to the fracture. The next five events represented the fracture initiation activities starting at 17.59 s prior to the fracture. Finally, four fracture-induced echo responses were observed up to 0.37 s after the fracture event.

Table 2. Occurrence time of various events from the F15I-AST sensor pair on the strand

Event Category	Occurrence Time (s)	Relative Time (s)
Inter-wire Slippage	50.97	-112.23
Fracture Initiation	145.61	-17.59
Fracture Initiation	151.94	-11.26
Fracture Initiation	159.99	-3.21
Fracture Initiation	162.13	-1.07
Fracture Initiation	162.33	-0.87
Fracture	163.20	0.00
Fracture-induced Echo	163.36	+0.16
Fracture-induced Echo	163.39	+0.19
Fracture-induced Echo	163.48	+0.28
Fracture-induced Echo	163.57	+0.37

4 Adaptive Wavelet Analysis of AE Signals for Prognosis

In practical applications, steel wires fracture as their cross sections are reduced due to corrosion. Steel corrosion is a slow process during which the maximum stress in wires under normal operation loads changes little. As a result, the fracture initiation of a steel wire could happen long before the wire is broken and usually scatter over a period of time with only one or two acoustic events at a time, which is different from a short-time loading test with a continuously increased tension force. It is thus critical to predict any incipient wire fracture event so that immediate actions can be taken to prevent steel strands from fracture in application.

The adaptive wavelet analysis (AWA) differs from the continuous wavelet transform (CWT) in making the mother wavelet parameters change over time for various parts of a signal. In the case of Morlet wavelet as indicated in Eq. (1), the center frequency ω_c is a function of time:

$$\psi(t) = \frac{1}{\sqrt{2\pi}} e^{-\frac{t^2}{2}} e^{i\omega_c(t)t} \tag{1}$$

Therefore, the AWA can be mathematically expressed into:

$$AWA\{x(t)\}(a,b) = \int_{-\infty}^{\infty} x(t)\,\overline{\psi_{a,b}(t)}dt = \int_{-\infty}^{\infty} x(t)\frac{1}{\sqrt{2\pi}} e^{-\frac{(t-b)^2/a(t)^2}{2}} e^{-i\omega_c(t)(t-b)/a(t)}dt \tag{2}$$

where $\overline{\psi_{a,b}(t)}$ represents the complex conjugate of the wavelet function ψ with scaling and shifting factors of a and b, respectively. As the center frequency changes from $\omega_c(t_1)$ to $\omega_c(t_2)$, $a(t)$ changes from $a(t_1)$ to $a(t_2)$accordingly to keep the same frequency range of interest. In other words, $a(t_1)/\,a(t_2)=\omega_c(t_1)/\,\omega_c(t_2)$ for each frequency value.

For wire fracture prediction, the fracture signal as indicated by the strongest AE event in Figure 5 is not available and the pre-fracture signals, such as inter-wire slippage and fracture initiation, must thus be uniquely identified over time as precursors of the wire fracture. Due to low amplitude of the pre-fracture signals, noises are characterized in order to remove their effect in the early detection of wire fracture. In addition, the echo effect of surrounding objects and strand boundary conditions, associated with wire fracture, is comparable in amplitude with the pre-fracture signals and thus characterized in order to prevent a false positive identification of the following fracture event.

Due to its computational efficiency and effectiveness both in frequency and time identification, the AWA is applied into the inter-wire slippage, fracture initiation, fracture, fracture-induced echo, and four types of artificial tapping noises. As shown later, the inter-wire slippage and fracture initiation signals stop within 0.002 s while others tend to last much longer. Therefore, high frequency resolution is selected for the first 0.002 s signal and high time resolution is used beyond 0.002 s to preserve time information. As such, a time-dependent center frequency $\omega_c(t) = 2\pi f_c(t)$ is selected in the AWA as follows:

$$f_c(t) = \begin{cases} 11 - 5000t & t < 0.002\ s \\ 1 & t > 0.002\ s \end{cases} \text{Hz} \tag{3}$$

The hit definition time (HDT) was set to 600 μs, which triggered a sudden termination of the two types of pre-fracture signals as shown in figures 6 and 7. Moreover, because of high sampling rate of the AE signal, the edge effect of wavelet transform is negligible. Figure 6 shows the inter-wire slippage signals and their time-frequency scalograms of the strand recorded from a pair of R15I-AST sensors at -112.23 s and a pair of R1.5I sensors at -74.45 s. These events occurred much earlier than the fracture event at different locations as sparsely distributed in Figure 5. These short events are rich in low frequency components. The inter-wire slippage signals from the two types of AE sensors are consistent since their dominant frequency is lower than the characteristic frequency of the AE sensors.

Without loss of any generality, two pairs of fracture initiation signals were randomly selected from a large number of data sets recorded. Figure 7 shows the fracture initiation signals and their time-frequency scalograms. These signals are short in time, and their energy is distributed over a wide range of frequency. All the recorded signals are consistent and repeatable in application. The randomly selected two pairs of signals respectively occurred at -1.07 s (from the F15I-AST sensors) and -0.87 s (from the R1.5I sensors).

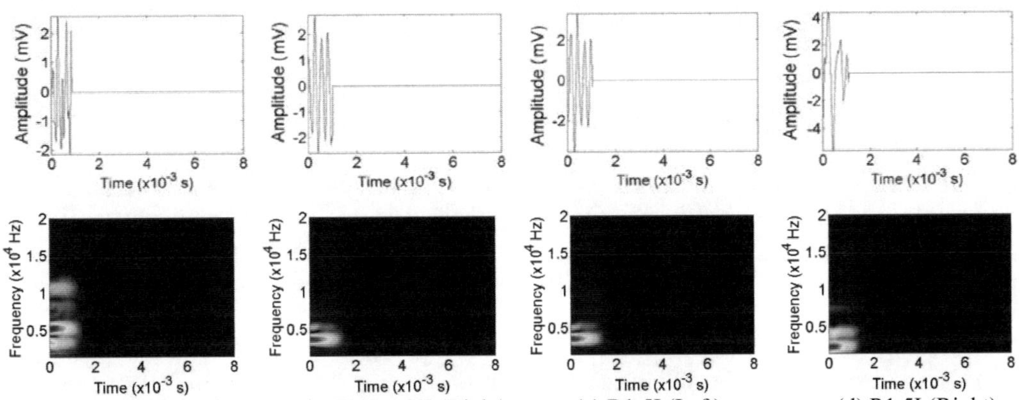

(a) F15I-AST (Left) (b) F15I-AST (Right) (c) R1.5I (Left) (d) R1.5I (Right)

Figure 6. Inter-wire slippage signals

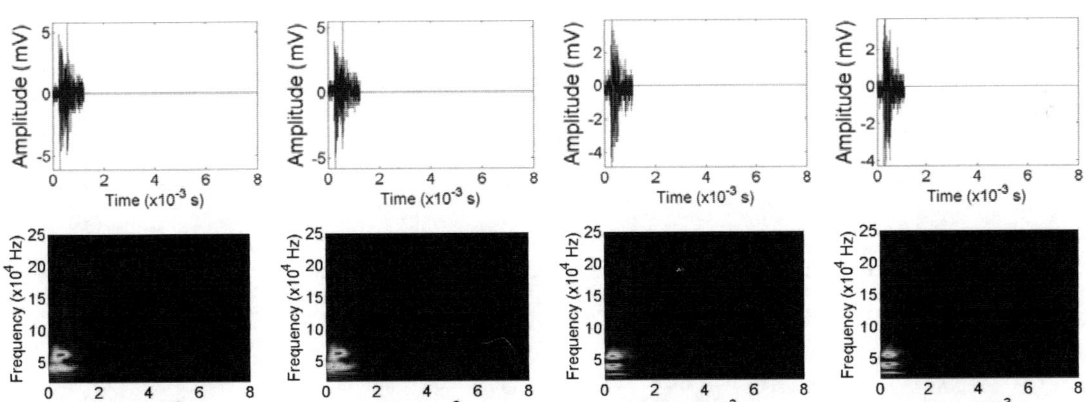

(a) F15I-AST (Left) (b) F15I-AST (Right) (c) R1.5I (Left) (d) R1.5I (Right)

Figure 7. Fracture initiation signals

Unlike the multiple inter-wire slippage and fracture initiation signals, only a single event recorded from two pairs of sensors was associated with the wire fracture as shown in Figure 8. Due to an excessive amount of energy released from the fracture, all the time functions last long over time. Such energy release may also reach beyond the measurement range of the sensors (10 V), causing signal saturation in time domain. This leads to a clear separation of fracture event from others by solely considering the signal amplitude in time domain.

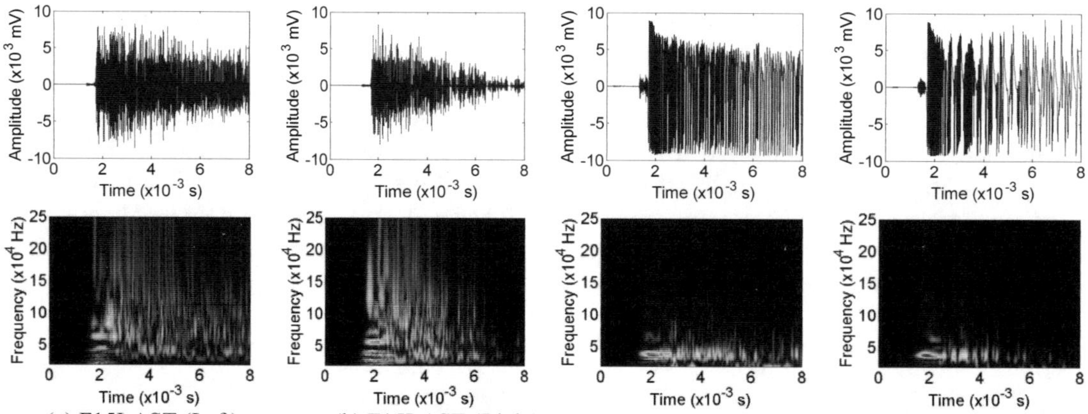

(a) F15I-AST (Left) (b) F15I-AST (Right) (c) R1.5I (Left) (d) R1.5I (Right)

Figure 8. Fracture signals from the 1st strand

Two pairs of fracture-induced echo responses captured by the two types of AE sensors were also randomly selected. Figure 9 shows the echo responses and their time-frequency scalograms at +0.16 s (F15I-AST sensors) and +0.19 s (R1.5I sensors). All the acoustic signals attenuate over time, likely because the echo responses result from multiple wave reflections at surrounding objects and strand boundaries.

Traffic alarms, rain drops, and accidental impact by foreign objects such as sand particles and small gravels can potentially affect the identification results of wire fracture from AE signals. Traffic-induced noises may be filtered out with guard sensors as they are generated from outside the sensing range. However, the rain drops on steel cables and the accidental impact of foreign objects can be difficult to remove. Artificial tapping noises as indicated in Figure 3 were thus generated during the tension test. Figure 10 presents the AE signals of artificial tappings with four types of objects when the first strand was subjected to a load of 89 kN. As the characteristics of the tapping signals recorded by two types of AE sensors are similar, only one signal from the left F15I-AST sensor of an intact strand is presented for each type of tapping noise. The noises decay rapidly over time since tapping represents an impact loading.

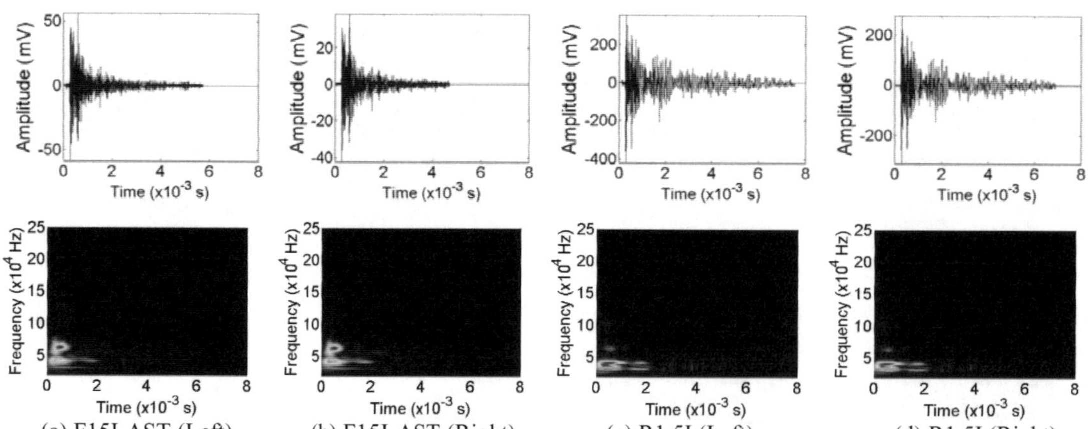

(a) F15I-AST (Left) (b) F15I-AST (Right) (c) R1.5I (Left) (d) R1.5I (Right)

Figure 9. Fracture-induced echo signals from the 1st strand

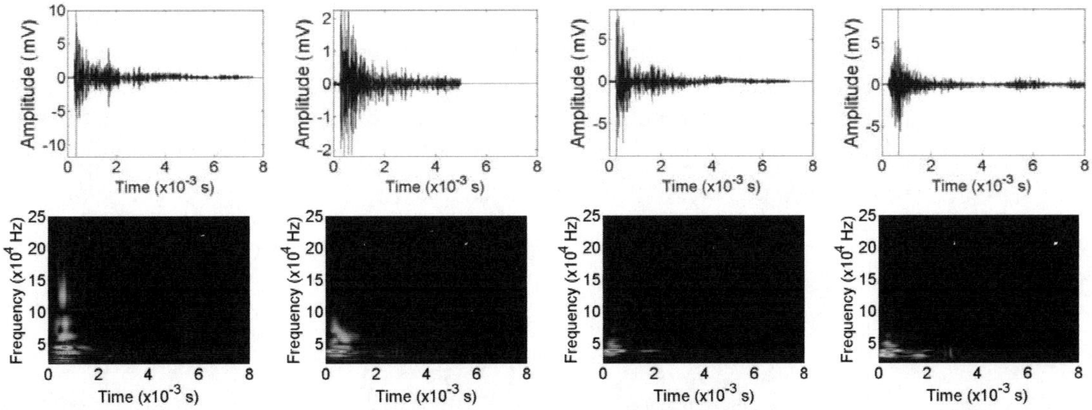

(a) Steel bar (b) Marker (c) Steel washer (d) Penny

Figure 10. Artificial tapping noises

Figures 6-10 indicate that the amplitudes of the recorded AE signals are up to 10 mV due to inter-wire slippage, fracture initiation and artificial tapping, up to 50 mV due to fracture-induced echo, and up to 10 V due to wire fracture. Thus, distinguishing the inter-wire slippage and fracture initiation signals from the tapping noises is critically important for the early detection of wire fracture in steel strands and cables. The inter-wire slippage and fracture initiation signals are both short in time, but the latter possesses wide frequency bandwidth. The fracture-induced echo responses and noises attenuate significantly over time, and they all have long time duration.

Based on the distinct features of various signals, the fracture initiation events can be identified from all acoustic events recorded from the strand. Figure 11 summarizes the identified fracture initiation events. Both pairs of sensors detect the fracture initiation events at virtually the same area. The false positive identification for the fracture initiation events occurs only from the R1.5I sensors. This is likely because the fracture initiation and inter-wire slippage signals are similar in many aspects and, more importantly, their frequencies fall into two sides of the resonant frequency (14 kHz) of R1.5I sensors so that the AWA can be amplified locally around the resonant frequency.

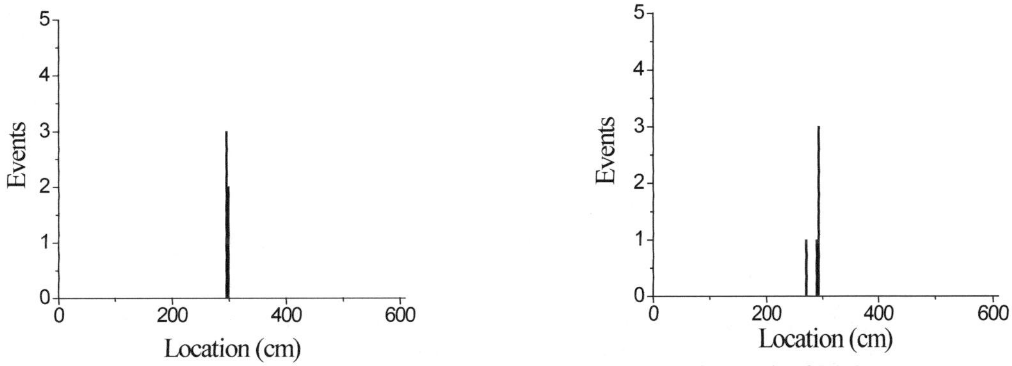

(a) A pair of F15I-AST sensors (b) A pair of R1.5I sensors

Figure 11. Number of fracture-initiation events

5. Conclusions

Prior to wire fracture, acoustic events of inter-wire slippage and fracture initiation were observed from a steel strand of seven wires. In comparison with the fracture initiation signals, the inter-wire slippage signals have substantially lower frequency components and similar time duration.

Wire fracture can be detected with any of hit, energy, and count parameters calculated from a recorded acoustic emission signal. It can be located from a distribution of events over the length of a test strand, which are registered by at least two acoustic emission sensors. The frequency bandwidth of wire fracture signals is much wider than that of fracture-induced echo responses or artificial tapping noises. Both the fracture signals and the noises are long in time duration.
Acoustic signals of fracture initiation possess unique time-frequency characteristics such as wide frequency bandwidth and short time duration. These characteristics can be used as fracture warning in application. For wire fracture prediction, the high frequency acoustic sensors result in less number of false positive identification

References

Rizzo, P & di Scalea, FL 2001, 'Acoustic emission monitoring of CFRP cables for cable-stayed bridges' *6th Annual International Symposium on NDE for Health Monitoring and Diagnostics*, International Society for Optics and Photonics.

Li, D & Ou, J 2008, 'Acoustic emission monitoring and critical failure identification of bridge cable damage', *The 15th International Symposium on: Smart Structures and Materials & Nondestructive Evaluation and Health Monitoring*, International Society for Optics and Photonics.

Jin, T, Sun, Z & Sun, LM 2008, 'Acoustic emission monitoring of stayed cables based on wavelet analysis', *The 15th International Symposium on: Smart Structures and Materials & Nondestructive Evaluation and Health Monitoring*, International Society for Optics and Photonics.

Betti, R, Sloane, MJD, Khazem, D & Gatti, C 2016, 'Monitoring the structural health of main cables of suspension bridges' *Journal of Civil Structural Health Monitoring*, vol.6, no.3, pp. 355-363.

Paulson, PO 1998, 'Continuous acoustic monitoring of suspension bridges and cable stays', *Non-Destructive Evaluation Techniques for Aging Infrastructure & Manufacturing*, International Society for Optics and Photonics.

Drummond, G, Watson, JF & Acarnley, PP 2007, 'Acoustic emission from wire ropes during proof load and fatigue testing' *NDT & E International*, vol. 40, no.1, pp. 94-101.

Woodward, RJ 1989, 'Detecting fractures in steel cables', *Wire Ind*, vol. 56, no. 18, pp. 401-406.

Sun, L & Qian, J 2011, 'Experimental study on wire breakage detection by acoustic emission' *Frontiers of Architecture and Civil Engineering in China*, vol. 5, no.4, pp. 503-509.

Zejli, H, Laksimi, A, Tessier, C, Gaillet, L & Benmedakhene, S 2006, 'Detection of the broken wires in the cables' hidden parts (anchorings) by acoustic emission' *Advanced Materials Research*, Vol. 13, pp. 345-350.

Oskouei, AR, Ahmadi, M & Hajikhani, M 2009, 'Wavelet-based acoustic emission characterization of damage mechanism in composite materials under mode I delamination at different interfaces', *Express Polym Lett*, vol.3, no.12, pp. 804-813.

Hamstad, MA, O'GALLAGHER, A & Gary, J 2002, 'A wavelet transform applied to acoustic emission' J. Acoust. Emiss, vol. 20, pp. 39-61.

Yoon, DJ, Weiss, WJ & Shah, SP 2000, 'Assessing damage in corroded reinforced concrete using acoustic emission', *Journal of engineering mechanics*, vol. 126, no. 3, pp. 273-283.

Grosse, CU, Finck, F, Kurz, JH, & Reinhardt, HW 2004, 'Improvements of AE technique using wavelet algorithms, coherence functions and automatic data analysis', *Construction and building Materials*, vol. 18. no.3, pp. 203-213.

Qi, G 2000, 'Wavelet-based AE characterization of composite materials', *NDT & E International*, vol. 33, no. 3, pp. 133-144.

Lu, C, Ding, P, & Chen, Z 2011, 'Time-frequency analysis of acoustic emission signals generated by tension damage in CFRP', *Procedia Engineering*, vol. 23, pp. 210-215.

Xu, C, Ai, S, Xie, J, Chen, G & Li, C 2016, 'Fatigue Failure Process Characterization for Carbon Fiber Sheet Reinforced Steel Rods Using the Acoustic Emission Technique', *Journal of Nondestructive Evaluation*, vol. 35, no. 2, pp. 1-13.

Qu, H, Li, M, Chen, G, Sneed, LH & Anderson, NL 2016, 'Multi-bandwidth wavelet transform of impact echo signals for delamination detection of thin concrete slabs', *Journal of Civil Structural Health Monitoring*, vol. 6, no. 4, pp. 649-664.

The 8th International Conference on
Structural Health Monitoring of Intelligent Infrastructure
Brisbane, Australia | 5-8 December 2017

Corrosion detection of steel reinforcements of reinforced concrete structures using distributed long gauge sensors

Nariman Fouad[1], Mohamed A. Saifeldeen[1], Huang Huang[2] and Zhishen Wu[3]

[1] *Aswan University - Egypt,* [2] Southeast University – China, [3] *Ibaraki University- Hitachi, Japan E-mail;* zhishen.wu.prof@vc.ibaraki.ac.jp

Abstract

This paper presents a new approach in corrosion monitoring system of steel rebar in reinforced concrete (RC) flexural structures based on distributed long-gauge sensors. Three different approaches, continuous-strain ratio (CSR), distributed-strain ratio (DSR), and section fiber model (SFM) were proposed to evaluate corrosion levels using strain measurements. Two main cases will be considered in this study. In the first case, the corrosion occurs inside concrete beam before cracks and the second case is after cracking. Distributed packaged long gauge carbon fiber line (PCFL) sensors with self- temperature compensation mounted on steel reinforcement and on compression fibers of corroded beam were used to monitor the corrosion under a sustained service load. Two RC beams with the same size and reinforcements were used, the corrosion started after the load was sustained. For the first beam, the load was sustained at 15 % of the ultimate design load to represent the non-cracked case and the other beam was loaded by 30 % of the ultimate design load to represent the cracked case. A comparison between the efficiency of the three different approaches in the two cases under considerations was presented. The results showed that, in case of the cracked beam, the most suitable approach was DSR approach; however the most suitable one for the non-cracked beam was the CSR approach. Finally it can revealed that, the corrosion locations and levels can be detected effectively by using distributed PCFL sensors based on a suitable approach according to the state of the structure under study.

Keywords: Corrosion monitoring, Reinforced concrete, Carbon sensor, distributed sensors, corrosion identification

1. Introduction

Corrosion is considered to initiate when the chloride concentration around the reinforcement reaches a threshold to cause dissolution of the protective film. When the corrosion of steel bars develop significantly, it not only affects the structural serviceability by cracking, or even spalling of the concrete cover, but also has an impact on the structural safety by decreasing the load-bearing capacity of reinforcement concrete members,(Maslehuddin et.al. 1990, Pantazopoulou & Papoulia 2001, and Yu et.al. 2015) . Corrosion monitoring, especially at the early stages of steel rebar corrosion, is one of the important measures for the maintenance of an RC structure, and it can reveal the onset and the progress of steel corrosion damage inside structures.
The assessment of infrastructures based on visual inspections is limited and has varying degrees of effectiveness. Additionally, visual inspections cannot identify deficiencies hidden within a structure. Thus, major deterioration issues may only become apparent once the structure is already in need of rehabilitation. A better management system to optimize maintenance strategies is required to

prolong the service life of structures, ensure safety, and reduce costs. One potential strategy to improve structure management is to use sensor technologies, such as long gauge sensors, for regular measurement and analysis of important structural and environmental parameters under operating conditions to identify problems at an early stage, (Davis et.al. 2016, ElBatanouny et.al. 2012, and Villalba & Casas 2013).

In recent decades, a lot of researches studied the long gauge sensors for macro-strain measurements. Recently, a new technique for measuring strain levels with carbon fiber line (CFL) sensor was proposed. Moreover, a post-tensioning method was proposed to stretch the slack fibers and control the initial creep of the CFL sensor to enhance the output signal of the CFL sensor [7-10].(Huang & Wu 2012, Huang et.al. 2012 ,Saifeldeen et.al. 2016, Saifeldeen et.al. 2016)

A reliable inspection method is required at an early stage before the functionality of the RC structure is seriously damaged due to steel corrosion. The inspection should be done without damaging the RC structures, both new and old ones. Most of non-distractive testing (NDT) methods faced a lot of limitations such as, they are too local so they are not suitable for large scale structures, cannot provide quantitative information about embedded steel corrosion, and cannot provide continuous corrosion monitoring. In the present paper, a new approach in corrosion monitoring system of steel rebar in RC beams based on distributed PCFL sensors mounted on steel bars and compression fibers of corroded beam under a sustained service load. Two main cases will be study in this paper. The first case, when the corrosion occurs inside concrete beam before cracks. The second case, when the corrosion occurs with slight cracks, this case was studied before by the authors' group (Fouad et.al. 2017). The distributed measured strains utilized to evaluate the deterioration of the structures health with the advance of corrosion time. The comparison between the two cases under consideration and the ability of each approach to calculate the corrosion levels in each case were discussed. Finally, the corrosion degrees calculated using each approach was compared with the weight loss of external model.

2. Packaged CFL sensors

Recently, Fouad et.al (Fouad et.al. 2016) proposed a packaging system for CFL sensors using a basalt fiber reinforced polymer (BFRP) tube to protect the sensors for enhancing their sustainability in long-term monitoring as shown in Figure 1. The proposed packaging system consists of two CFL sensors—of which one functions as a working sensor and the other is used for temperature compensation. Both these sensors consist of two main parts each—the two fixed ends and the sensing part, which is located between the two ends. For the working sensor, the sensing part is kept free inside a manually fabricated BFRP tube with a 2-mm inner diameter, and the two fixed ends are fixed directly to the BFRP tube with epoxy resin. However, only one end of the compensation sensor is fixed to the BFRP tube. Then, both the BFRP tubes are bonded together with epoxy resin to realize a PCFL sensor.

The principle of a CFL sensor is to establish a relationship between the electrical resistance change ratios (ΔR/R) and strains (ε). Carbon fiber is a low conductivity semiconductor material in which its electrical resistance can be expressed as:

$$R = \frac{\rho * L}{A} \tag{1}$$

where R, ρ, L and A are, respectively the initial electrical resistance, the electrical resistivity, the gauge length and the cross-section area of CFL sensor. A Wheatstone bridge measuring circuit was used to measure the change in electrical resistance of the sensor. The relationship between ΔR/R and strain is as follows:

$$\varepsilon = \left(\frac{\Delta R}{R}\right) / \mathrm{GF} \tag{2}$$

where GF is the gauge factor

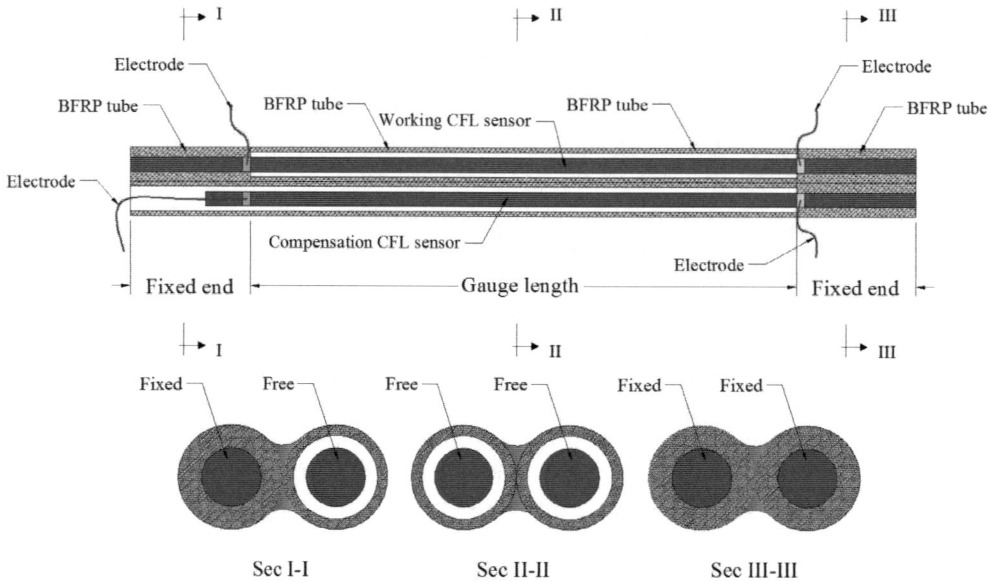

Fig. 1 Schematic of proposed PCFL sensor

3. Methodologies of monitoring corrosion using strain measurements

3.1.CSR based corrosion identification technique

In case of the localized corrosion, all the distributed strains along the steel reinforcements will increase with the advance of corrosion resulting from the increments of the beam curvature but by various values according to the percent of damages occurred in the cross section of the beam. For a certain cross section of the beam the measured strains in the intact, cracked, and corroded states can be formulated as:

$$\varepsilon_o = \frac{M * y_o}{EI_o} \qquad \varepsilon_{co} = \frac{M * y_{co}}{EI_{co}} \tag{3}$$

where M is the applied moment, and $\varepsilon, EI, and\ y$, are the flexural strain , flexural stiffness, and height of neutral axis at a certain position and the subscript (o), and (co) represent the intact, and corroded states respectively. Under a constant applied moment the corrosion levels in a certain location Ψ based on CSR approach at a certain time can be expressed as:

$$\Psi = \frac{EI_o - EI_{co}}{EI_o} * 100 = \frac{(\varepsilon_{co}) - (\varepsilon_o)}{(\varepsilon_{co})} * 100 \tag{4}$$

3.2.DSR based corrosion identification technique

This technique is suggested by comparing the strains of two different sensors at two different locations, one at the target location and another as a reference sensor to eliminate the effect of the small curvature variations. The reference sensor will be selected at a location with less probability of cracks and corrosion occurrence that usually located near the support.

For the intact state, the strain at any section solely depends on the magnitude of the moment as $(EI_i = EI_r)$ in which the ratio between reference strain (ε_r) and target strain (ε_i) can be expressed by using Eq.4 as:

$$\left(\frac{\varepsilon_i}{\varepsilon_r}\right)_o = \frac{M_i.y}{EI_i} \Big/ \frac{M_r.y}{EI_r} = a_1 \quad where\ a_1 = \frac{M_i}{M_r} \tag{5}$$

where, EI_i and EI_r are the flexural stiffness of the cross section at the target and reference locations respectively and the subscript (o) represents the intact case. The slope of this state (a_1) represents the ratio of moment.

For the corroded state, it is supposed that the ratio between target and reference strains to remain constant under the same load until the corrosion starts in certain locations causing addition reduction in stiffness $(EI_i^{co} = \beta_i^{co} EI_i)$ .where, subscript (co) represents the corroded state and β_i^{co} is the damage index which represent the ratio between flexural stiffness of intact and corroded conditions in which $(0 \leq \beta_i^{co} \leq 1)$.

The regression line shifts with an increase slope of (a_2) as follows:

$$\left(\frac{\varepsilon_i}{\varepsilon_r}\right)_{co} = \frac{M_i.y}{EI_i^{co}} \Big/ \frac{M_r.y}{EI_r} = a_2 \qquad where\ a_2 = a_1\left(\frac{1}{\beta_i^{co}}\right) \tag{6}$$

Finally, the corrosion level Ψ can be determined from this formula:

$$\Psi = \left(\left(\frac{\varepsilon_i}{\varepsilon_r}\right)_{co} \Big/ \left(\frac{\varepsilon_i}{\varepsilon_r}\right)_o - 1\right) * 100 = \left(\frac{a_2}{a_1} - 1\right) * 100 \tag{7}$$

In case of absence of corrosion the slope will not changes where ($a_2 = a_1$) then $\Psi = 0$

3.3.SFM based corrosion identification technique

To calculate the actual decrease in cross sectional area of steel reinforcement from the measured strains, nonlinear analysis based on section fiber modelling method are carried out regarding the class bending theory using the measured tensile strains of steel and compressive strains of the upper surface of concrete beam.

Concrete in compression is modeled taking into consideration the fracture energy of concrete which gives stress in terms of strains as:

$$f = f_c\left(\frac{2\varepsilon}{\varepsilon_o} - \left(\frac{\varepsilon}{\varepsilon_o}\right)^2\right) \tag{8}$$

where $\varepsilon_o = \frac{2f_c}{E}$, and E is initial modulus of concrete.

Concrete in tension before cracking is supposed to be linear elastic$(\sigma = E.\varepsilon)$. In this study the concrete compressive was measured experimentally it was found to be f_c=32 N/mm^2 and tensile strength $f_t = f_c/10$, and the yield stress of steel reinforcement was f_y= 375 N/mm^2. Reinforcing steels are treated as a linear elastic-perfectly plastic material in which $\sigma_s = E \text{ x } \varepsilon_s$ where σ_s and ε_s are the stress and strain of steel rebar.

Based on the assumption that plane sections remain plane and normal to the longitudinal axis, the nonlinear behaviour of RC beam disregarding shear and bond-slip can derive from the constitutive relations of concrete and reinforcing steel fibers into which each section is divided, the summation of the internal forces can be determined as follows:

$$\sum \mathrm{N} = \sum_{i=1}^{n} \sigma_i \mathrm{A}_i + \sigma_s \mathrm{A}_s + \acute{\sigma_s} \acute{\mathrm{A}_s} \tag{9}$$

Take $\sum \mathrm{N} = 0$ for equilibrium and Calculate A_s with the progress of corrosion Finally, the corrosion damage Ψ represented by the decrease in area of steel as follows:

$$\Psi = \left(\frac{A_{s_o} - A_{s_c}}{A_{s_o}} \right) * 100 \tag{10}$$

The previous three approaches were discussed in details in (Fouad et.al. 2017) in case of corrosion occurrence after cracks.

4. Experimental setup

In this study, a comparison between the three mentioned approaches CSR, DSR, and SFM in the two cases will be conducted. In the non-cracked case, the prepared RC beam was loaded by using a hydraulic machine to a load of 10 KN (15% of the ultimate design load) with keeping this load constant. In the cracked case the beam was loaded by 20 kN (30% of the ultimate design load), then the load maintain constant. Subsequently, the erosion of one tension steel bar starts using accelerated corrosion process using salt solution mentioned in (Fouad et.al. 2017). To insure the sustainability of corrosion progress, the salt solution was changed every about two hours with carefully cleaning the corroded part by a small soft brush. The sustained load and corrosion process continued for 16 hours.

Two RC beams of 150 x 200 x 1800 mm with two deformed bars with a diameter of 12.7 mm as tensile reinforcements, and 10 stirrups with a mild steel of 10 mm diameter as shear reinforcements were used in this study, see figure 2. The corrosion will be applied onto one of the main steel reinforcements at the designed location through a small opening with a size of (50 mm width x 100 mm length x 40 mm depth) introduced on the bottom surface of the RC beam. The PCFL sensors were prepared as mentioned before with a gauge length of 250 mm, thereafter they fixed from its all length using the epoxy resin on the designed locations and divided into two groups as follows;

(1) SC group: this group consists of six sets of PCFL sensors which were mounted on the corroded tension bar.

(2) CT group: only four sets of PCFL sensors were installed on the compression zone of the RC beam at a height of 160 mm from the bottom surface.

The signals of all sensors mounted on the inner steel bar and top zones of the RC beam were recorded simultaneously by the data measurement systems The change of room temperature during the corrosion test was recorded and it was founded to be varied within 3 ^{0}C.

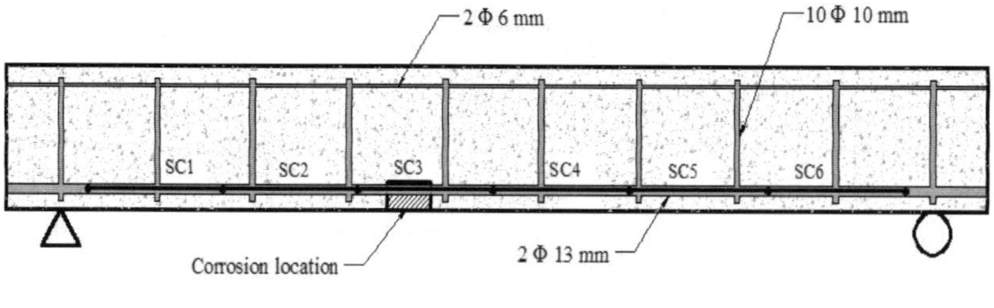

Fig. 2 Beam reinforcements and sensors placement

5. Results and discussions

5.1.Based on CSR approach

The corrosion levels values were calculated using CSR approach for each section from Eq.4. Figures 3 (a), and (b) gives the calculated corrosion levels along the beam by using the fitted values of the distributed strains of (SC) group for the two cases.
From these figures it can be noted that, by using CSR approach the location of induced corrosion can be detected effectively in the two cases. However, small values were founded in the other locations which can be considered due to the small changes in strains due to the curvature variations especially in the cracked case. Based on that, this approach is effective in localizing the corrosion with some errors in the corrosion levels values.

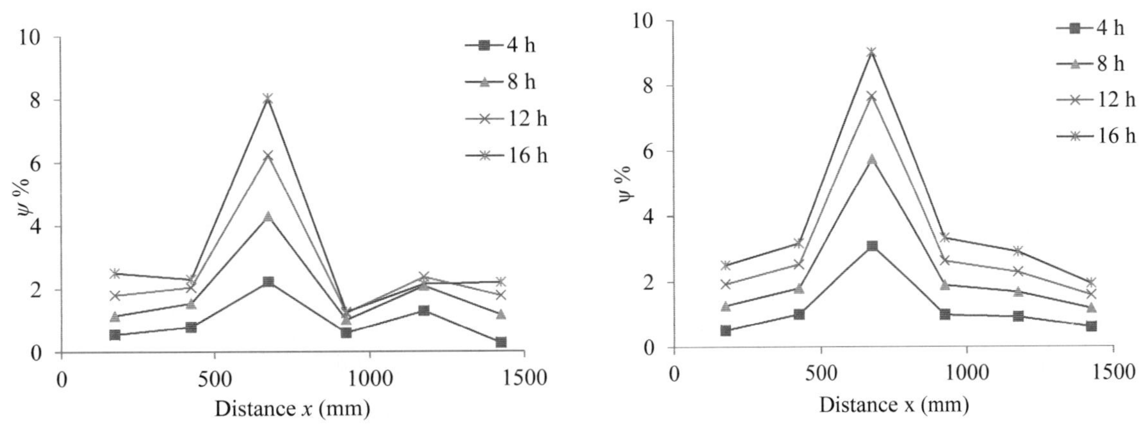

Fig. 3 Corrosion levels along the length of the beam based on CSR for (a) non-cracked beam, and (b) cracked beam

5.2.Based on DSR approach

In this experiment the reference sensor was chosen to be the end sensor near the right support to be far from the corrosion location. The corrosion levels were calculated by using Eq.7 are plotted in Figure 4 (a) and (b). It can be observed from the calculated corrosion levels using DSR method that, all the locations changes randomly with the progress of corrosion and no clear information about corrosion location by using this method in the non-cracked case. The small noises in the measured strains of the target or reference sensors result in large errors in the corrosion levels calculations and lead to non-trustable information about the corrosion locations. However for the cracked stage it can clearly observe that, the corrosion damages locations can be found obviously, and by using this method the effect of curvature variations can be eliminated and it can be considered that the calculated values related mainly to the decrease in area steel.

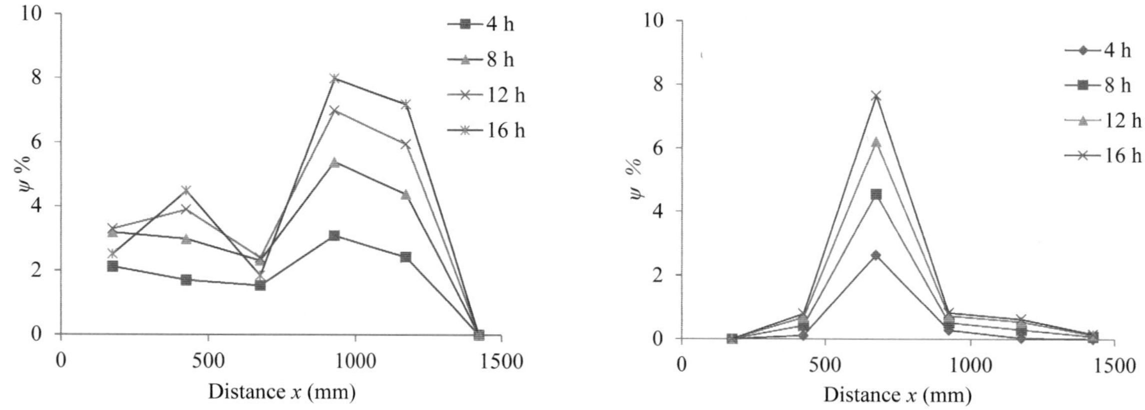

Fig. 4 Corrosion levels along the length of the beam based on DSR for (a) non-cracked beam, and (b) cracked beam

5.3.Based on SFM approach

Based on the equilibrium of the internal forces of each cross section of the beam with the advance of corrosion, the actual A_s in each cross section can be calculated during the corrosion progress as well as the corresponding corrosion levels using Eq.9 and 10 as shown in Figures 5 (a) and (b). It can be clearly identified from these figures that, the A_s values calculated for all sections of the beam are not reliable and not represents the area loss due to corrosion effects in the non-cracked case which caused by the noises in the measured strains of both CT and SC groups.
In case of the cracked stage, the A_s values and the corresponding corrosion levels calculated for all sections of the beam are almost constant rather than the location of corrosion. Regarding to the negative calculated values, it is resulting from the errors in the measured compression strains with the advance of corrosion in this stage. Eventually it can be confirmed that, by applying this method the effect curvature can be separated from effect of decrease in area steel in a certain cross section.

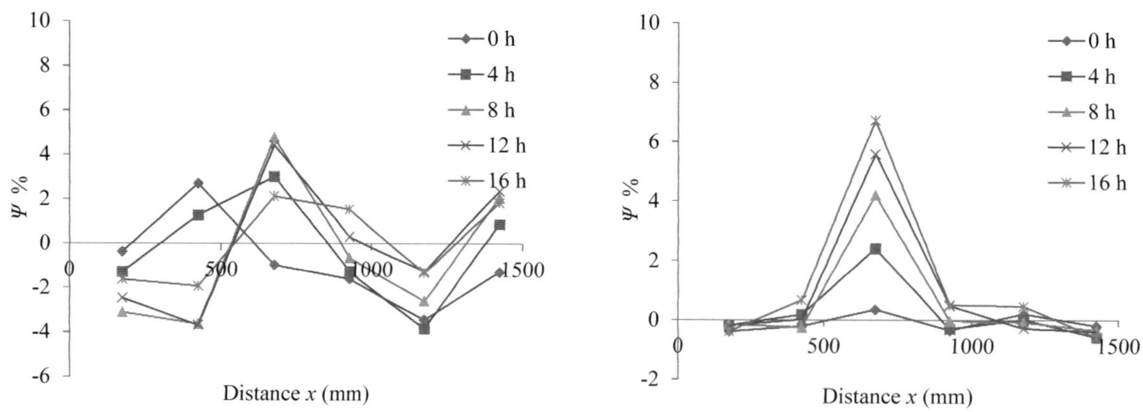

Fig. 5 Corrosion levels along the length of the beam based on SFM for (a) non-cracked beam, and (b) cracked beam

5.4. Total comparison between three different approaches

The calculated corrosion degrees based on CSR, DSR, and SFM were compared with the weight loss of external model every for 4 hours of corrosion time as shown in Figures 6 (a) and (b). In case of the cracked case, it was found that, there is a good agreement between the measured corrosion levels evaluated based on DSR, SFM, and actual weight loss of corrosion model. The small variations between the model and the calculated corrosion degrees based on strain measurements related to the efficiency of corrosion and the difficulty of effective cleaning of the corroded part during the corrosion process. The calculated values based on CSR method have divergent values than the others which have been expected as a result of curvature variations with the advance of localized corrosion, however only the CSR method was applicable in the non-cracked stage due to the noise in the measured strains in this stage.

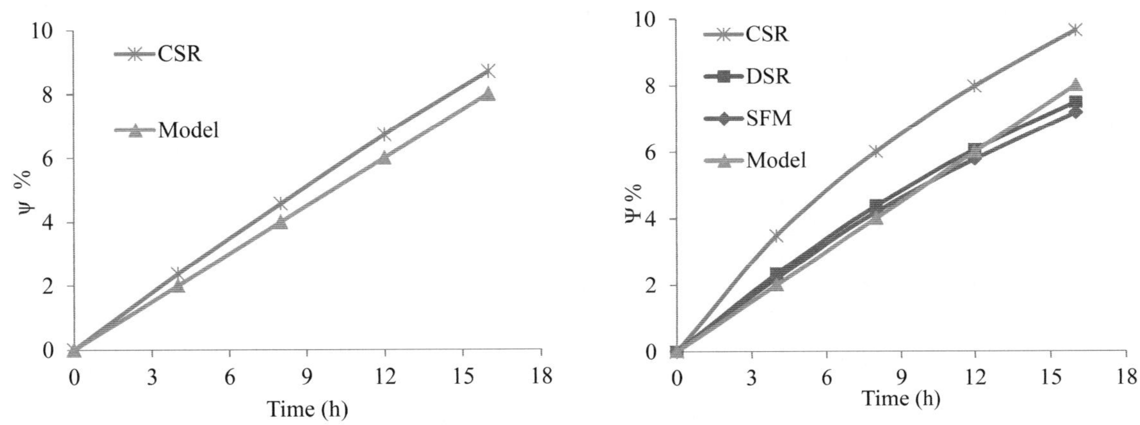

Fig. 6 comparison between corrosion levels calculated from the three approaches with the weight loss of external model for (a) non-cracked beam, and (b) cracked beam

Conclusions

This paper presents a new technique of steel corrosion monitoring of flexural RC structures using distributed macro-strains of long-gauge carbon fiber sensors. Three different approaches were proposed to detect the actual corrosion degrees. Based on corrosion monitoring experimental results in the non-cracked and cracked cases, the following conclusion can be addressed:

1. In case of corrosion occurrence inside concrete before cracks, the first approach CSR will be most suitable to evaluate the corrosion degrees because the changes on steel strains in a certain location will be mainly related to the decrease in bar diameter. However, the use of DSR and SFM will not be effective and not trustable to identify and localize the corrosion because of the effect of noise on the small measured strains in this stage.
2. In case of early corrosion levels after cracks initiation, the DSR and SFM will be more suitable than CSR because at the CSR affected by the curvature variations
3. By using distributed PCFL sensors mounted onto steel reinforcements of the RC structures, the corrosion damages locations and values can be effectively found at either the cracked or the non-cracked stages.

Finally, it can be considered that using the distributed long-gauge PCFL sensors is the promising technique for the continuous corrosion monitoring of steel bars of the large scale structures with a high accuracy.

References

Maslehuddin, M., Allam, I.A., Al-Sulaimani, G.J., Al-Mana, A. and Abduljauwad, S.N., 1990. 'Effect of rusting of reinforcing steel on its mechanical properties and bond with concrete.' *Materials Journal*, 87(5), pp.496-502..

Pantazopoulou, S.J., and Papoulia, K.D. 2001 'Modeling cover-cracking due to reinforcement corrosion in RC structures.' *Journal of Engineering Mechanics*127.4 (2001): 342-351.

Yu, L., François, R., Dang, V.H., L'Hostis, V. and Gagné, R., 2015. 'Distribution of corrosion and pitting factor of steel in corroded RC beams.' *Construction and Building Materials*, 95, pp.384-392.

Davis, M., Hoult, N.A. and Scott, A., 2016. 'Distributed strain sensing to determine the impact of corrosion on bond performance in reinforced concrete.' *Construction and Building Materials*, 114, pp.481-491.

ElBatanouny, M.K., Mangual, J., Vélez, W., Ziehl, P.H., Matta, F. and González, M., 2012. 'Monitoring corrosion in prestressed concrete beams using acoustic emission technique.' *In SPIE Smart Structures and Materials+ Nondestructive Evaluation and Health Monitoring*, pp. 834718-834718. International Society for Optics and Photonics.

Villalba, S. and Casas, J.R., 2013. Application of optical fiber distributed sensing to health monitoring of concrete structures. *Mechanical Systems and Signal Processing*, 39(1), pp.441-451.

Huang, H., and Wu, Z.S. 2012. 'Static and dynamic measurement of low-level strains with carbon fibers.' *Sensors and Actuators A* 183: 140–147.

Huang, H., Yang, C., and Wu, Z.S. 2012. 'Electrical sensing properties of carbon fiber reinforced plastic strips for detecting low-level strains.' *Smart Materials and Structures* 21(3): 035013.

Saifeldeen, M.A., Fouad, N., Huang, H. and Wu, Z., 2016. Advancement of long-gauge carbon fiber line sensors for strain measurements in structures. *Journal of Intelligent Material Systems and Structures*, p.1045389X16665974..

Saifeldeen, M. A., Fouad, N., Huang, H., & Wu, Z. S. 2016. 'Stabilization of electrical sensing properties of carbon fiber sensors using pre-tensioning approach.' *Smart Materials and Structures*, 26(1), 015012.

Fouad, N., Saifeldeen, M. A., Huang, H., & Wu, Z. 2017. 'Corrosion monitoring of flexural reinforced concrete members under service loads using distributed long-gauge carbon fiber sensors.' *Structural Health Monitoring*, 1475921717698973.

Fouad, N., Saifeldeen, M. A., Huang, H., & Wu, Z. S. 2016. 'Early corrosion monitoring of reinforcing steel bars by using long-gauge carbon fiber sensors.' *Journal of Civil Structural Health Monitoring*, 6(4), 691-701.

The 8th International Conference on
Structural Health Monitoring of Intelligent Infrastructure
Brisbane, Australia | 5-8 December 2017

Structural Health Monitoring in the Sydney Harbour Bridge Using Spectral Moments

M. Makki Alamdari[1] & N. Khoa[2] & T. Rakotoarivelo[3], H. Kalhori[4], J. Li[5].

[1]*School of Civil and Environmental Engineering, University of New South Wales,* m.makkialamdari@unsw.edu.au,
[2-4] *Data 61|CSIRO– Australia, Email:* Khoa.Nguyen@data61.csiro.au, Thierry.Rakotoarivelo@data61.csiro.au, Hamed.Kalhori@data61.csiro.au.
[5] Curtin University– *Australia, Email:* Junli@curtin.edu.au.

Abstract

The motivation behind this paper is to develop a spectral-based damage identification scheme using output only acceleration responses. The presented method is in the context of non-model-based damage identification methods and does not require any representative numerical/analytical model of the structure. The method utilizes spectral moments of the response as damage sensitive feature. Spectral moments directly retrieve information from the power spectral density of the response. Unlike the modal data that only provide information at a limited number of eigen-frequencies, spectral moments capture information from the entire spectra, hence they can distinguish any subtle differences between a normal and distorted signal. The feasibility of the approach in damage identification was validated using real data from the Sydney Harbour Bridge. There are approximately 800 jack arches over a total distance of 1.2 km need to be continuously monitored. For this study, two instrumented jack arches were considered. These joints are located on the eastern side of the bridge underneath the bus lane near the north pylon. One of these two joints had a known crack in 2012, along the front face propagating toward the surface of the deck, while the other joint was intact. This damage was repaired in 2013. Acceleration data were collected from tri-axial accelerometers mounted on the base of each joint before and after repair. The presented spectral-based method along with the hypothesis testing involving the *KS*-test were applied to obtain a decision on whether or not the structure is damaged. Spectral moments with different orders were also investigated. It was demonstrated that the proposed spectrum-driven feature can reliably distinguish between the healthy and damaged joints which is of great importance for the asset owner. The presented results illustrated high potential of this approach to identify damage in real-life structures.

Keywords: Damage, Power Spectral Density, Spectral Moment, *KS*-test, Jack Arch.

1. Introduction

The fact that civil structures are getting older and they are becoming more and more vulnerable to natural and human-induced damage presents a challenge to the engineering community about how to monitor the integrity of these structures in a systematic way. Structural Health Monitoring (SHM) is a cutting-edge technology that develops and implements damage identification strategies for an existing structure to monitor its serviceability and safety (Farrar, 2007).

Basically, SHM is a multidisciplinary approach which combines a variety of technologies and techniques and it involves three major steps including: data acquisition, feature extraction and statistical analysis (Makki Alamdari, 2015a). A large amount of research has been conducted in SHM to extract useful features out of the vibration responses which can be categorized into time domain features or frequency domain features. In the first category, statistical models and identification techniques are often utilized. Several autoregressive-based (AR) feature extraction approaches have been adopted in the previous works (Mujica, 2008). In the second category, modal-based or spectral-based features are employed for SHM (Makki Alamdari, 2015b).

Basically, informative features about damage can be extracted in the frequency domain in systems under stochastic loadings by using the Power Spectral Density (PSD) of the output response. Application of PSD has been widely reported in the literature for damage identification (Liberatore, 2004, Bayissa, 2007). Despite the employment of PSD in SHM has been studied in the previous works, in this study it has been reformulated according to the concept of spectral moments. In addition, this work aims to investigate the feasibility of the approach in a large-scale SHM application. The spectral moments retrieve information directly from the PSD and they provide some important information about the statistical properties of a random stationary signal (Antoni, 2006); e.g. the variance of a random process is the zero-order spectral moment of that observation. Although spectral moments have non-dimensional nature, they provide adequate characterization of a system by a straightforward computation (Alamdari, 2017). They capture information from the entire spectra and hence they can distinguish any subtle differences between a normal and distorted signal. Their explicit dependence on the frequency content of the response and provision of information from the entire frequency range are some appealing features with spectral moments that makes them suitable for SHM applications.

Statistical tests have been used to make decision in damage detection. In (Nair, 2006), the authors proposed a time series algorithm for damage detection and localization. An autoregressive moving average (ARMA) time series was modelled from vibration signals and the first three coefficients were used to extract a damage sensitive feature. Then two-sample *t*-test was applied on the feature to distinguish between before and after damage states. The technique was validated using data from the numerical simulations and lab experiments of a four-story structure. The authors in (Hae Young Noh, 2009) applied the same approach using acceleration and strain data obtained from a three story structure at the National Center for Research on Earthquake Engineering (NCREE) in Taiwan. The results suggested that acceleration data are useful for damage detection while strain data can be used for damage localization. In (Shahidi, 2015), two-sample *t*-test was utilized to test the change in two different damage-sensitive features obtained from the regression coefficients of linear regression models. However, in order to use *t*-test, it is required that two data samples should come from normal distributions with equal variances and are obtained independently. Kolmogorov–Smirnov (*KS*) test is a non-parametric test without any assumption about data distribution. In (Liu Mei, 2016), the authors used two-sample *KS*-test to distinguish data sets of time series model residuals for damaged and undamaged conditions.

This paper aims to present the results of a large scale SHM on the Sydney Harbour Bridge (SHB), one of the Australia's iconic structures. A spectrum-driven feature using the spectral moments of the output acceleration response is employed as indicative feature to identify the existing crack in structural components of the bridge. The damage decision is made by examining the statistical difference in the values of the moments obtained from the observed samples from the healthy and damaged states. *KS*-test is adopted to determine whether the samples from a potentially damaged jack arch are different enough from the samples from an undamaged jack arch. The main advantages of the proposed approach include using output-only data, describing the characteristics of the structure by a robust and sensitive indicative feature and employing statistical hypothesis testing to decide about the health state of the jack arches in a rigorous manner with explicit statistical confidence level.

2. Feature extraction

For a system under Gaussian stochastic process, the response process can be characterized by correlation function in the time domain as,

$$R_{\mathbf{xx}}(\tau) = E[\mathbf{x}(t)\mathbf{x}(t+\tau)] \tag{1}$$

where $\mathbf{x}$ is the measured time response and $E[*]$ operator is the probabilistic expected value. τ refers to the lag operator. The autocorrelation function, $R_{\mathbf{xx}}(\tau)$ defines how a signal is correlated with itself, with a time separation τ.

The frequency characterization of such random stationary process can be done using PSD function which is calculated by taking the Fourier transform of the autocorrelation function as,

$$S_{\mathbf{xx}}(\omega) = \int_{-\infty}^{+\infty} R_{\mathbf{xx}} e^{-iw\tau} d\tau \tag{2}$$

where $S_{\mathbf{xx}}(\omega)$ is the PSD of the response at frequency ω.

For a given PSD, the nth-order spectral moment for location i, can be expressed by,

$$\lambda_{\mathbf{x}_i}^{n} = \int_{-\infty}^{+\infty} |\omega|^n S_{\mathbf{xx}_i}(\omega) d\omega \tag{3}$$

where $S_{\mathbf{xx}_i}$ is the PSD of the response at sensor location i, and frequency ω. n is the order of the spectral moment. As illustrated by Equation (3), spectral moments retrieve information directly from PSD without any further manipulation. From this equation, it can be inferred that spectral moment is the summation of the weighted spectral information. Any distortion in a system such as damage affects specific frequencies. By summing up the individual contribution of each frequency, spectral moments provide indication of change in the system. The higher order moments give a weight to higher frequency components however, according to the previous research, very high order moments, more than 3, are not informative in general (Di Paola, 1986).

The nth order spectral moment for a discretized signal is obtained by,

$$\lambda_{\mathbf{x}_i}^{n} = \frac{2}{N^{n+1}} \sum_{0}^{\lfloor N/2 \rfloor} S_{\mathbf{xx}_i}(j)\left(\frac{j}{\Delta t}\right)^n \qquad j \in [1{:}N/2] \tag{4}$$

where $S_{\mathbf{xx}_i}(j)$ is the discrete N-point PSD and Δt is the sampling period.

In this study, the spectral moments of the structural response obtained from a large population of the samples, at a current state and a benchmark state are statistically compared to identify any change in the structure. In the next section, real data from the Sydney Harbour Bridge are adopted to identify an existing damage in one of the structural components using the proposed spectrum-driven feature.

3. Field test investigation

The Sydney Harbour Bridge (SHB) was opened in 1932. Traffic lane 7 is a dedicated bus and taxi lane on the eastern side of the bridge. Lane 7 consists of an asphalt road surface on a concrete deck supported by concrete and steel jack arches. There are approximately 800 jack arches over a total distance of 1.2 km, see Figure 1. The jack arches are very difficult to access and are typically inspected at two yearly intervals according to standard visual inspection practices. At the request of the asset manager, Data 61 designed and developed a monitoring system including sensing, data analysis and user applications to continuously monitor each of the approximately 800 structural jack arch components and to make recommendations about action required to be taken by asset manager/ bridge inspector (Runcie, 2015).

Figure 1. Illustration of a series of jack arches underneath the bus lane.

3.2. Data acquisition

Each of the 800 jack arches has been instrumented with a three low-cost tri-axial MEMS accelerometers. An '*Event*' is loosely defined as a time period during which a '*significant*' physical action is applied on the joint, such as a motor vehicle driving across that joint. More specifically, an event is started for a joint when one of its sensors (sensors 1 to 3) reports a significant difference from the rest position, i.e. the difference is greater than a predetermined threshold value.

In an ideal scenario, sensor 2 at rest (i.e. when no event is in progress) should return an acceleration $A:(0,0,-g)$, where g is the gravity. For sensors 1 and 3 due to its arbitrary orientation once glued on the joint, the gravity component will be distributed along its 3 axes: $A:(gx, gy, gz)$. Moreover due to hardware biases, the real acceleration measured by a deployed sensor will be $A:(gx+Ex, gy+Ey, gz+Ez)$, with (Ex,Ey,Ez) being the bias vector.
Average acceleration at rest, Ar for a sensor is defined as the average of A over a given number of samples M (currently M = 200) when there are no event occurring. Ar is defined as $Ar:(Arx, Ary, Arz)$.
During the search for an event, a sensor records data at a given sampling time i as an instantaneous acceleration: $Ai:(Aix, Aiy, Aiz)$. For that given sample i, one scalar difference metric is defined between the instantaneous and the rest accelerations as: magnitude of the differences: $V= |Ai - Ar|$. If this value V is greater than the event detection threshold (20mg) then we assume that an event has started. This default threshold may be different for different nodes. When an event has been detected, a fixed number of samples N around that event are collected.

3.4. Case study

For this case study, two instrumented jack arches are considered labeled joint 1 and joint 2. These joints are located close to the north pylon and north main span. Joints 1 and 2 are separated by an expansion joint as schematically illustrated in Figure 2. In 2012, a known crack existed on joint 2 while joint 1 was in good condition (In February 2013, damaged joint 2 was repaired).
Analysed data were collected from tri-axial accelerometers mounted on the base of each joint, sensor 2. Each tri-axial accelerometer measures acceleration responses in the direction of passing traffic (x), in the direction of gravity (z) and in the lateral direction (y). These three directions have been illustrated in Figure 2.
Before conducting any repair on damaged joint 2, the event data were collected during October 2012 (a total number of 150 events were recorded.). Data acquisition setting for event data collection in 2012 was based on recording event data for a period of 1.6 seconds at a sample rate of 375 Hz, resulting 600 samples.
In February 2013, damaged joint 2 was repaired and another set of event data was collected (a total number of 150 events were recorded.). The data after repair were collected with a frequency sampling of 250 Hz for duration of 2 seconds resulting 500 samples.

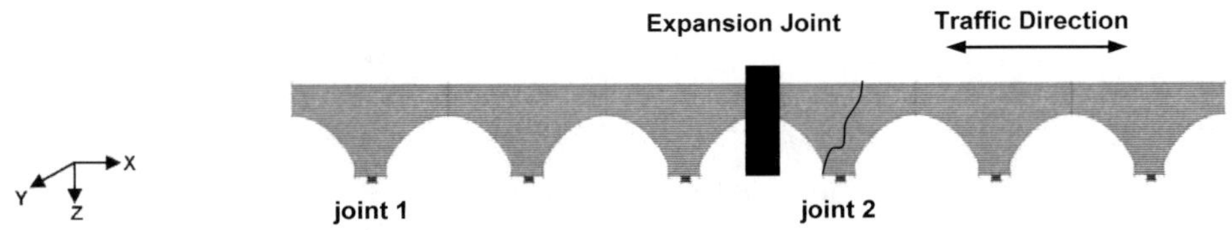

Figure 2. Schematic of the evaluated joints: joint 1 is a healthy joint and joint 2 is a damaged joint.

A comparison was made between the normalized acceleration responses in *x, y* and *z* directions using collected event data from the healthy joint (joint 1) and the damaged joint (joint 2). Comparison was made both before and after repair and the results are, respectively, presented in Figure 3 and Figure 5.

From the time history of events shown in Figure 3, dominant peaks are observed in the acceleration responses of the damaged joint 2 in *x, y* and *z* directions whereas the peaks are less pronounced in the healthy joint 1. The reason for these pronounced peaks in the time history is that before conducting the repair on damaged joint 2, a temporary steel plate was placed on top of the damaged area which introduced uneven road profile. Hence, as a consequence of passing traffic noticeable energy was injected into the structure by hitting the plate.

It should be noted that not all the events have the same vibration pattern as shown in Figure 3. Figure 4 summarizes typical vibration patterns observed in *z* direction from different events in damaged joint 2. As seen some events have multiple separated peaks due to the fact that it corresponds to more than one vehicle and some events have multiple closely-spaced peaks due to the fact that the number of axles in some vehicles is higher, Figure 4 (1-b). Also in some cases, the peaks are not pronounced because the event is triggered by passing traffics in other lanes, Figure 4 (c).

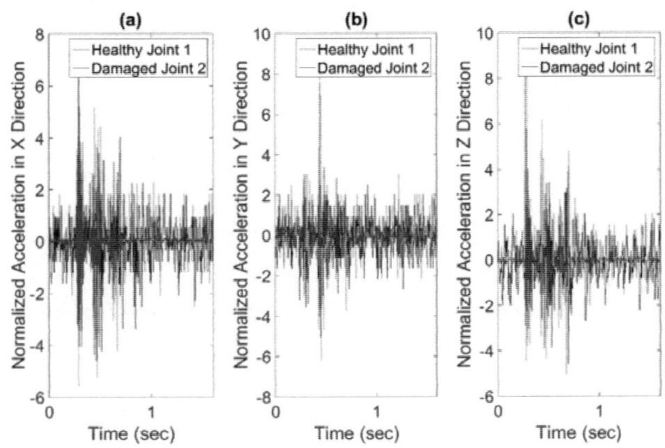

Figure 3. Illustration of acceleration responses before repair obtained from the healthy joint 1 and damaged joint 2 (a) in *x* direction, (b) in *y* direction and (c) in *z* direction.

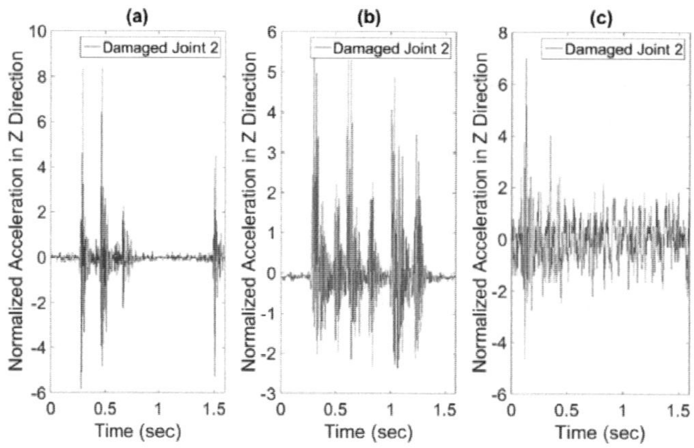

Figure 4. Illustration of acceleration response in z direction in various events in damaged joint 2.

After repair, the acceleration responses obtained from the repaired joint 2 and the healthy joint 1 were again compared in x, y and z directions and the results are presented in Figure 5. As illustrated, no dominant peaks, such as before, are observed in the vibration response.

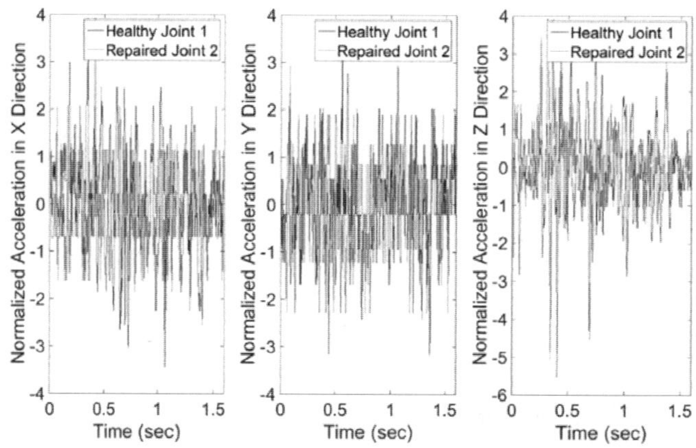

Figure 5. Illustration of acceleration response after repair obtained from the healthy joint 1 and repaired joint 2 (a) in x direction, (b) in y direction and (c) in z direction.

4. Damage identification results

For each joint, a total number of 300 events (150 events before repair and 150 events after repair) are considered to calculate the zeroth, first and the second order spectral moments according to Section 2.

Measured time data in x, y and z directions for each event were separately normalized with respect to the mean and the standard deviation as,

$$\bar{\mathbf{x}}(t) = \frac{\mathbf{x}(t) - \mu}{\sigma} \tag{5}$$

where μ, σ, $\mathbf{x}(t)$ and $\bar{\mathbf{x}}(t)$ are, respectively, the mean value, the standard deviation, the original time signal and the normalized time signal. The normalized responses in x, y and z directions are adopted to calculate the time history of response magnitude for each event.

Damage identification procedure starts by generating the PSD of response (magnitude) for each event. Since the data acquisition set up (frequency sampling and number of samples in each event) were not the same for events before and after repair, special considerations were made to make sure that the constructed PSDs, before and after repair, had the same frequency resolution and the same frequency range. To obtain this, only 1.2 second of each event was considered to ensure that the frequency resolution is the same for events before and after repair. Additionally, in order to get the same frequency range, the same number of spectral lines was considered.

The zeroth, first and second order spectral moments are calculated for the response magnitude and are, respectively, labelled as λ_m^0, λ_m^1 and λ_m^2 where subscript 'm' refers to the response magnitude. A summary of statistics is visualized with a boxplot in Figure 6. The horizontal axis in these figures shows the joint ID (1 or 2) and the vertical axis illustrates the spectral moment.

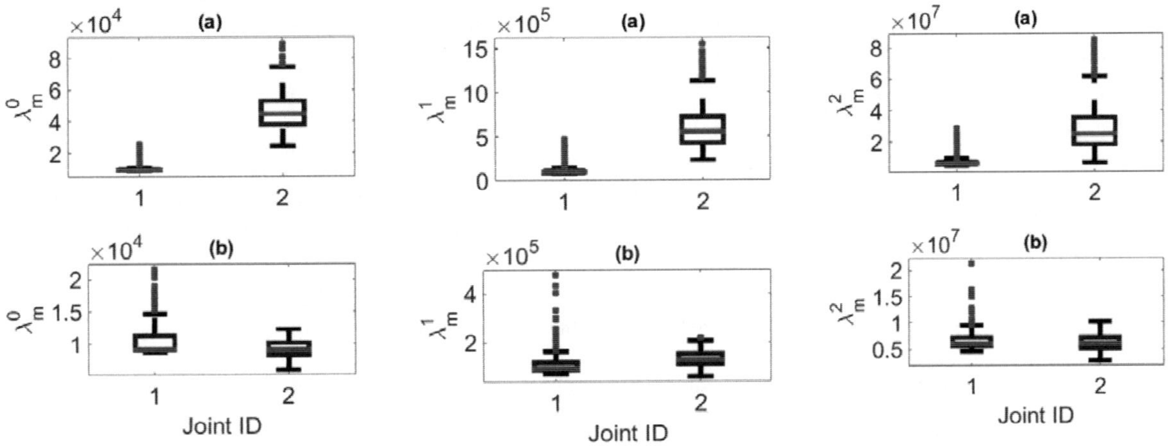

Figure 6. Comparison between the zeroth, first and second order spectral moments; (a) before repair and (b) after repair.

From the obtained boxplots it is quite obvious that there are variations in the spectral moments obtained from different events. This variation can be attributed to different traffic patterns on the bridge e.g. vehicles with different speed, weight and number of axles would result in different patterns of acceleration time responses and consequently different spectral moments. This change also can be attributed to environmental condition e.g. temperature changes within various events. The maximum variability in Figure 6 corresponds to the damaged joint 2 before repair (Figure 6 (a)) which demonstrates the fact that there is a noticeable difference between acceleration responses obtained from different events. This result coincides with the findings presented in Figure 4. As discussed earlier, the source of this variation is highly relevant to the temporary steel plate located on top of the damaged zone.

Figures 6 also illustrates that before repair the median of the spectral moments at damaged joint 2 is considerably different from the healthy joint 1 no matter which order of the spectral moment is adopted, see Figure 6 (a). This clearly conveys the message that structural behavior of these two joints are different. Interestingly, after repair no noticeable difference between the median of the moments between the healthy joint 1 and the repaired joint 2 is observed. This implies that the structural behaviors of these two joints are similar.

4.1. Statistical hypothesis testing

Statistical hypothesis testing is widely adopted to make decisions about the structural state in real-life applications. It enables explicit statistical confidence levels attached to the decisions. Most importantly, statistical hypothesis testing provides a useful framework to distinguish the healthy and damaged states of a structure in a rigorous and standardized manner based on available data.

In the present work, the goal is to detect the occurrence of damage in the jack arches by determining whether samples from a potentially damaged jack arch are different enough from samples from an undamaged jack arch. Therefore, statistical hypothesis testing is adopted to determine the statistical significance in the differences of two populations e.g. the pre- and post-repair.

The two-sample *KS*-test is a nonparametric hypothesis test that evaluates the difference between the empirical cumulative density functions (ECDF) of the two distributions to determine if two distributions differ or not. There is no any assumption about the distribution normality required as in the *t*-test. Also *KS*-test is sensitive to differences in both location and shape of the ECDFs of two samples. The two samples are obtained from the magnitudes of spectral moments extracted from vibration data before and after damage.

Let consider spectral moment λ be the characteristic feature that specifies the system (structure). Damage detection decision is made by testing for statistically significant differences between the features of the baseline (healthy) and current (healthy or damaged) states of the structure. The hypothesis testing is derived as:

$H_0 : SM_{\lambda,b} - SM_{\lambda,c} = 0$ Null hypothesis- Healthy structure

$H_1 : SM_{\lambda,b} - SM_{\lambda,c} \neq 0$ Alternative hypothesis- Damaged structure (6)

where H_0 and H_1 are the null and the alternative hypotheses describing the healthy and the damaged states, respectively. SM is the sample set of spectral moment magnitudes and subscripts b and c are, respectively, indicating the benchmark and the current states of the structure.

In this work, the significance level is set to $\alpha = 0.05$. If the p-value is smaller than or equal to α, H_1 is accepted otherwise H_0 is accepted. *KS*-test also returns a k-value representing a distance between the ECDFs of the two samples.

In our case study since we do not have any dataset for damaged joint 2 before damage occurrence, therefore, we use the collected samples after repair for benchmark state and the samples collected before repair is considered for current unknown state needs to be determined. Table 1 presents the results of the *KS*-test for joints 1 and 2.

Table 1. Results of damage decision using the Kolmogorov-Smirnov test

	damage decision			k-value			p-value		
Feature	λ_m^0	λ_m^1	λ_m^2	λ_m^0	λ_m^1	λ_m^2	λ_m^0	λ_m^1	λ_m^2
Joint 1	H1	H1	H1	1	0.9667	0.9267	≈ 0	≈ 0	≈ 0
Joint 2	H0	H0	H0	0.1800	0.1600	0.1267	0.1575	0.2680	0.5553

The results in Table 1 show that *KS*-test successfully captured the damage in joint 1 by rejecting the hypothesis in all value of spectral moment orders with p-values were nearly zero. On the other hand, it accepted the hypothesis in all values of spectral moment orders in case of joint 2. This shows the potential of using this hypothesis test for damage detection.

In order to validate the effectiveness of the proposed approach in real life implementation, the size of the samples required for testing of the current state of the structure is evaluated. Figure 7 shows the k-value of the *KS*-test for different sizes of test sample for the spectral moment order 2, given the same benchmark sample. The results suggest that a test sample of 50 events is enough for this hypothesis testing. In the case of the SHB data where we collected traffic events with a 2-minute sleep between events, it took about 100 minutes to have enough data for the hypothesis testing. Even though it is not a real-time monitoring, it is still reasonable.

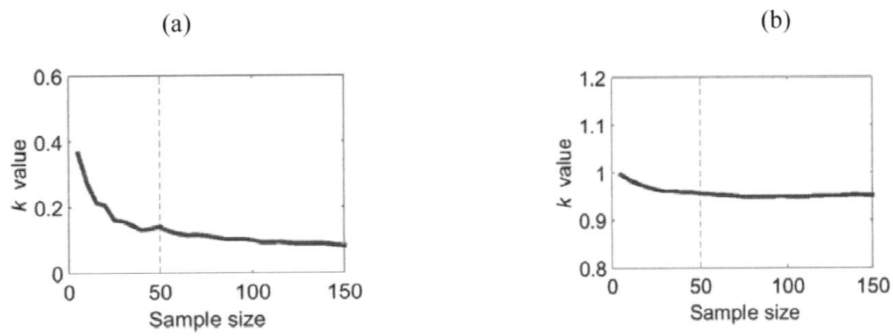

Figure 7. Illustration on effect of the test population size on the k-value, (a) joint 1, (b) joint 2.

5. Conclusion

Developing non-model-based damage characterization schemes using output only responses is in high demand by SHM community since this approach does not require service interruption and expensive devices to excite the structures. To this aim, this paper presented a damage identification method based on the concept of spectral moments. A response parameter known as spectral moment was implemented to characterize damage. Then *KS*-test was used to detect the occurrence of damage in a structure by determining whether a sample collected at the current state of a structure is different enough from a sample collected in a benchmark healthy state of the same structure. The method was successfully applied on concrete jack arches which are major structural components in the Sydney Harbour Bridge. It was demonstrated that *KS*-test can reliably differentiate between the healthy joint (after repair) and damaged joint (before repair). Besides the promising results, the method is easy to implement as damage sensitive feature is a scalar parameter. Moreover, the proposed feature contains broadband frequency information from the entire frequency range thus it can distinguish any subtle differences between a normal and distorted signal. Given its low computation cost and promising results, this approach could provide interesting benefits for real life SHM applications.

6. Acknowledgement

The authors wish to thank the Road and Maritime Services (RMS) in New South Wales, Australia for provision of the support and testing facilities for this research work. NICTA is funded by the Australian Government through the Department of Communications and the Australian Research Council through the ICT Centre of Excellence Program. CSIRO's Digital Productivity business unit and NICTA have joined forces to create digital powerhouse Data61.

References

Antoni, J., 2006. The spectral kurtosis: a useful tool for characterising non-stationary signals. Mechanical Systems and Signal Processing, 20(2), pp.282-307.

Bayissa, W.L. and Haritos, N., 2007. Structural damage identification in plates using spectral strain energy analysis. Journal of Sound and Vibration, 307(1), pp.226-249.

Di Paola, M. and Muscolino, G., 1986. On the convergent parts of high order spectral moments of stationary structural responses. Journal of sound and vibration, 110(2), pp.233-245.

Farrar, C.R. and Worden, K., 2007. An introduction to structural health monitoring. Philosophical Transactions of the Royal Society of London A: Mathematical, Physical and Engineering Sciences, 365(1851), pp.303-315.

Liberatore, S. and Carman, G.P., 2004. Power spectral density analysis for damage identification and location. Journal of Sound and Vibration, 274(3), pp.761-776.

Alamdari, M.M., Rakotoarivelo, T. and Khoa, N.L.D., 2017. A spectral-based clustering for structural health monitoring of the Sydney Harbour Bridge. Mechanical Systems and Signal Processing, 87, pp.384-400.

Makki Alamdari, M., 2015a. Vibration-based structural health monitoring (Doctoral dissertation).

Makki Alamdari, M., Samali, B., Li, J., Kalhori, H. and Mustapha, S., 2015b. Spectral-based damage identification in structures under ambient vibration. Journal of Computing in Civil Engineering, 30(4), p.04015062.

Mei, L., Mita, A. and Zhou, J., 2016. An improved substructural damage detection approach of shear structure based on ARMAX model residual. Structural Control and Health Monitoring, 23(2), pp.218-236.

Mujica, L.E., Vehí, J., Ruiz, M., Verleysen, M., Staszewski, W. and Worden, K., 2008. Multivariate statistics process control for dimensionality reduction in structural assessment. Mechanical Systems and Signal Processing, 22(1), pp.155-171.

Nair, K.K., Kiremidjian, A.S. and Law, K.H., 2006. Time series-based damage detection and localization algorithm with application to the ASCE benchmark structure. Journal of Sound and Vibration, 291(1), pp.349-368.

Rincie, P., An Update on a Large Scale SHM Deployment on Sydney's Harbour Bridge and Associated Research Activities, 2015.

Shahidi, S.G., Nigro, M.B., Pakzad, S.N. and Pan, Y., 2015. Structural damage detection and localisation using multivariate regression models and two-sample control statistics. Structure and Infrastructure Engineering, 11(10), pp.1277-1293.

The 8th International Conference on
Structural Health Monitoring of Intelligent Infrastructure
Brisbane, Australia | 5-8 December 2017

Spectrogram Analysis of Local Excitation Method (LEM) for SHM Improvement of Damage Detection

T. Oshima[1], Y. Miyamori [1], T. Yamazaki [1], S. Mikami [1], S. Yasue [2], Y. Hashidume [1]
[1] *Kitami Institute of Technology – Japan, Email:* oshima@kitami-it.ac.jp
[2] *Kitamirai Giken Co., Ltd. - Japan*

Abstract

A method of damage identification for civil infrastructures based on local excitation method (LEM) is dealt with in this paper. An algorithm through Local Excitation Method (LEM) (Beskhyroun, Oshima, Mikami et al. 2003, 2004, 2005, 2006, 2007, 2011, 2013) analyzing the change of PSD (Power Spectrum Density) on received acceleration signal of structure at each location, is used. Spectrogram analysis is applied to improve damage detection through LEM. Spectrogram analysis over time history of received acceleration signal helps the detailed analysis of PSD changes which may be the effect of local and small damages.

1. Introduction

The identification of structural damage is an essential goal of health monitoring for civil infrastructures. (Doebling, et al. 1996, Farrar, et al. 1994, 1996, Sampaio, et al. 1999, Peeters, et al. 2001, Xue, et al. 2006) Damage of a structure reflects on a change of modal parameters, such as natural frequencies, mode shapes and modal damping. A method of damage identification for civil infrastructures based on Local Excitation Method (LEM) is dealt with in this paper. The method is described theoretically and applied to the experimental data obtained from a steel bridge. Several damage scenarios were considered to the test structure. Experimental results show that the proposed approach may be successfully implemented to detect damage and to locate regions where damage occurred. In this experiment the artificial damage is introduced by adding some weights on steel beam. Reason why we took this case is that we needed a small structural condition change to be detected by the proposed LEM. The algorithm through LEM by analyzing the change of PSD on received acceleration of structure at each location, is used. The algorithm is used to detect damage, locate its position and monitor the increase of damage using only the measured data without the need for any modal identification or numerical models. Spectrogram analysis is used to improve damage detection through LEM. LEM uses sweep wave function with continuously increasing frequency for input wave on structure by piezoelectric actuator. Running spectrum (Spectrogram) analysis over the time history of received acceleration helps detail analysis of PSD changes which reflects on the improvement of local and small damage detection. Those results will be shown in this paper.

2. Damage Identification Algorithm

Many techniques have been proposed in the area of non-destructive damage detection using changes in modal parameters. In many structures, however, only few modes are available in the analysis, so that it may decrease the accuracy of detecting and localizing damage using these techniques. In order to overcome the problem of the limited number of identified modal parameters,

PSD changes are investigated on the various accelerometer readings. It is considered at all frequencies in the measurement range and not just at the modal frequencies. The results will be compared before and after damage using the proposed method.
PSD is determined from the acceleration time histories without the need to measure the excitation forces.

Let $G_i\,(f)$ denotes the PSD magnitude measured at channel number i at frequency value f. The absolute difference in PSD magnitude before and after damage can then be defined as

$$D_i = \frac{|G_i(f) - G^*{}_i(f)\,|}{1+|G_i(f)\,|} \tag{1}$$

where $G_i\,(f)$ and $G_i^*\,(f)$ represent PSD magnitude for the undamaged and damaged structures, respectively. In the denominator of Eq. (1) 1 is added for analytical reason not to set zero just in case. The excitation forces used for the undamaged and damaged structure must have the same amplitude, location and waveform in order to ensure that the changes in PSD data are mainly due to damage and not due to the change in excitation force characteristics.
When the change in PSD is measured at different frequencies on the measurement range from f_1 to f_m , a matrix [**D**] can be formulated as follows

$$\mathbf{D} = \begin{bmatrix} D_1(f_1) & D_2(f_1) & \cdots\cdots & D_n(f_1) \\ D_1(f_2) & D_2(f_2) & \cdots\cdots & D_n(f_2) \\ . & . & \cdots\cdots & . \\ . & . & \cdots\cdots & . \\ . & . & \cdots\cdots & . \\ D_1(f_m) & D_2(f_m) & \cdots\cdots & D_n(f_m) \end{bmatrix} \tag{2}$$

where n represents the number of measuring points. In matrix [**D**], every row represents the changes in PSD at different measuring channels but at the same frequency value. The summation of PSD changes over different frequencies can be used as the indicator of damage occurrence and the increase in damage. In other words, the first damage indicator is calculated from the sum of columns of matrix [**D**] as total change (**TC**)

$$\mathbf{SM} = \left\{ \sum_f D_1(f) \;\; \sum_f D_2(f) \;\; . \;\; . \;\; . \;\; \sum_f D_n(f) \right\} \tag{3}$$

Then total changes **SM** in PSD is calculated from the sum of columns of matrix [D] on each node.

However, it was found that total change **SM** is a weak indicator of damage localization. A statistical decision making procedure is employed to determine the location of damage. The first step in this procedure is the selection of the maximum change in PSD at each frequency value (the maximum value in each row of matrix [**D**]) and dividing all other changes in PSD measured at other nodes. For example in matrix [**D**], if $D_3\,(f_1\,)$ is the maximum value in the first row, then this value will be used as denominator $D_3\,(f_1\,)$ and all other values in this row are divided as shown in Eq.(4). The same process is applied to the different rows in matrix [**D**] to formulate the matrix of maximum changes of PSD at different frequencies, [C]

$$C = \begin{bmatrix} C_1(f_1)=\dfrac{D_1(f_1)}{D_3(f_1)} & C_2(f_1)=\dfrac{D_2(f_1)}{D_3(f_1)} & \cdots & C_n(f_1)=\dfrac{D_n(f_1)}{D_3(f_1)} \\ C_1(f_2)=\dfrac{D_1(f_2)}{D_2(f_2)} & C_2(f_2)=\dfrac{D_2(f_2)}{D_2(f_2)} & \cdots & C_n(f_2)=\dfrac{D_n(f_2)}{D_2(f_2)} \\ \vdots & \vdots & & \vdots \\ \vdots & \vdots & & \vdots \\ C_1(f_m)=\dfrac{D_1(f_m)}{D_5(f_m)} & C_2(f_m)=\dfrac{D_2(f_m)}{D_5(f_m)} & \cdots & C_n(f_m)=\dfrac{D_n(f_m)}{D_5(f_m)} \end{bmatrix} \quad (4)$$

In order to monitor over all frequencies for damage detection at any channels, a new matrix [**C**] is formulated. This procedure can normalize the big differences over frequency ranges at each measuring points. For example in the matrix [**C**], a value of 1 is used corresponding to the locations of $C_3(f_1)$, $C_2(f_2)$ and so on, as shown in Eq.(4)

$$\mathbf{SC} = \left\{ \sum_f C_1(f) \quad \sum_f C_2(f) \quad . \ . \ . \quad \sum_f C_n(f) \right\} \quad (5)$$

Then **SC** is also the indicator of damage occurrence and it is called damage detection (**DD**). The damage localization indicator **D I 0** is defined as the scalar product of {***SM***} and {***SC***} as shown in the following expression;

$$\mathbf{DI0} = \{SM(1) \times SC(1) \quad SM(2) \times SC(2) \ \ldots \ SM(n) \times SC(n)\} \quad (6)$$

3. Experimental Setup and Equipment Arrangement for Damage Identification of Steel Railway Bridge

3.1 Experimental Setup

Fig. 1 Railway Bridge Photo used for Measurement
(Assembled steel I beam, length 6.34m Railway bridge)

Bridge which was constructed in the year of 1909, has elements of 2 main girders (assembled I beams by rivets) and 3 diaphragms at center and at two ends of girder. Horizontal laterals are connected at the upper flange of girder by rivets as shown in Fig. 1. In experiment bridge is supported on two wooden blocks at the end with fixation. Detail dimension of bridge is 6.450m (length), 1.944m (width) and 0.805m (height) as shown in Fig. 1 . The dimensions of web plate and upper and lower flange plates are 758 × 10, 267 × 22 and 267 × 25 (mm), respectively. Cross sectional dimension(L type angle) of vertical stiffeners is 80 × 80 × 10(mm).

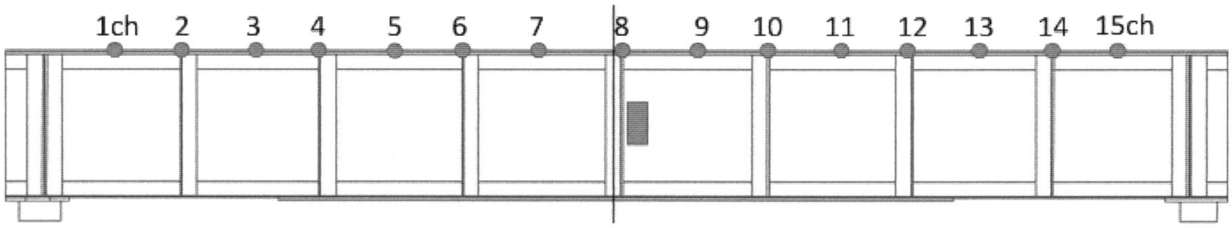

Fig. 2 Location of Accelerometers (from 1ch to 15ch) and Actuator (at center)

Fig. 3 Accelerometers on flange plate and actuator setting

3.2 Sensor Arrangement

Sensors are 15 accelerometers and located in horizontal direction as shown in Fig. 2 and Fig. 3. Actuator is set on the center of web plate at the middle of girder around at ch8 in horizontal direction as shown in Fig. 3. Then out of plane vibration of web plate is generated by this setup.

Excitation: Piezoelectric actuator is permanently fixed at the center point of the web plate of girder. Excitation force amplitude is 50 (N). Excitation force direction: horizontal, Excitation wave form: Sine sweep wave, Excitation frequency: 1~800Hz, Excitation time: 20 sec, Recording time: 25 sec, and Sampling rate: 10000 Hz.

[1] Piezoelectric actuator: A wave function generator is used to adjust the required excitation frequency range, the time in which the actuator reaches the maximum frequency, and the excitation wave form. Then, the adjusted wave is transferred to the power supplier which in turn provides the actuator with the adjusted power. The actuator is fixed to the structure by a magnetic holder. This magnetic holder is suitable for fixing the actuator to steel structures; A steel spring is fixed upon the actuator to control the excitation force amplitude. The produced force amplitude is constant but the excitation frequency gradually increases over time until it reaches its maximum value after the designated time. It should also be noted that because the actuator is pressed but not glued to the test structure, the actuator provides pressure force only. In this case, the excitation wave form is a sweep function.
[2] Piezoelectric accelerometer: The accelerometer has a nominal sensitivity of 10 mv/g, a specified frequency range of 5~4000Hz, and an amplitude range of 200 m/s^2.

4. Damage Detection

4.1 Modelled Damage Details

Fig. 4 Artificial Condition Change by Cramp and Small Steel Plates for Case 1(left) and Case 2(right)

In order to simulate the damage detection artificial ciondition changes are made by adding small weight as shown in Fig. 4 at the location ch5. Small steel plates are used to add small weight on the flange plate. For Case 1 one small steel plate is fixed by clamp and total additional weight is 9.8kg including the weight of clamp (0.83% as additional weight, weight of one girder is 1182.4kg) and for Case 2 two steel plates are fixed and total weight is 13.9kg(1.17 % as additional weight). Location of the weight is the opposite side of accelerometer ch5.

4.2 Damage Identification for Case 1 and Case 2

The results for Damage Indicator **DI0** by using Eq. (6) are shown in Fig. 5(a) for Case 1 and Fig. 5(b) for Case 2. In Eq. (1) differences of PSD between intact condition and damage condition (weight added) are calculated over every channels and frequencies. Then following through Eq. (2) to Eq. (6) we get the results of Damage Indicator as **DI0.** We can see that in Fig. 5(a) and (b) damage locations are well detected and it is needed more detailed analysis on the results.

In this paper the damage is modelled by adding small weight but in our previous paper (Mikami , et al. 2007) the result of crack detection are shown by using same procedure of damage detection.

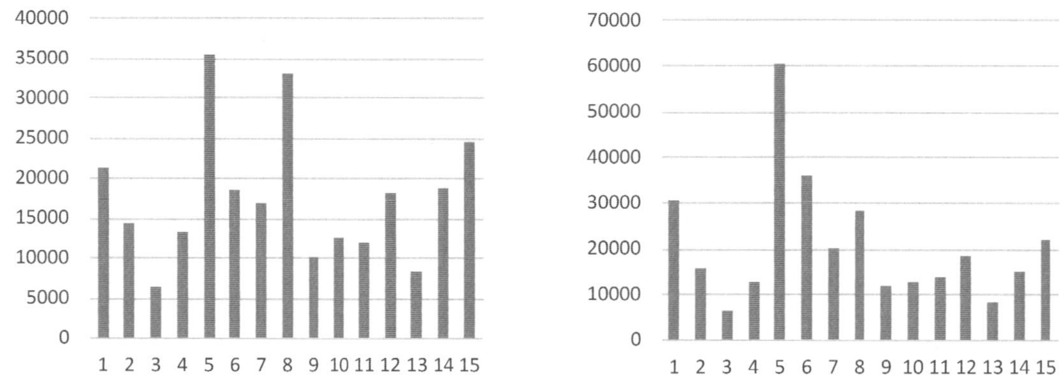

(a) Case 1 (b) Case 2

Fig. 5 Damage Indicator DI0 for Case 1 and Case 2 (Scale: Unit)

5. Spectrogram Analysis

Spectrogram analysis is a time running spectrum analysis which works on Matlab. The result of spectrogram analysis is given in the space of time (horizontal) and frequency (vertical) axes. Hamming window is used to calculate Fast Fourier Transform (FFT) of received signal. 2048 digital data length is used as a window size and overlap its half length of 1024 data to move over all received signal up to 25sec from the beginning of recording. The result of FFT amplitude (absolute of real part) is shown in logarithmic scale as shown in Fig. 6.

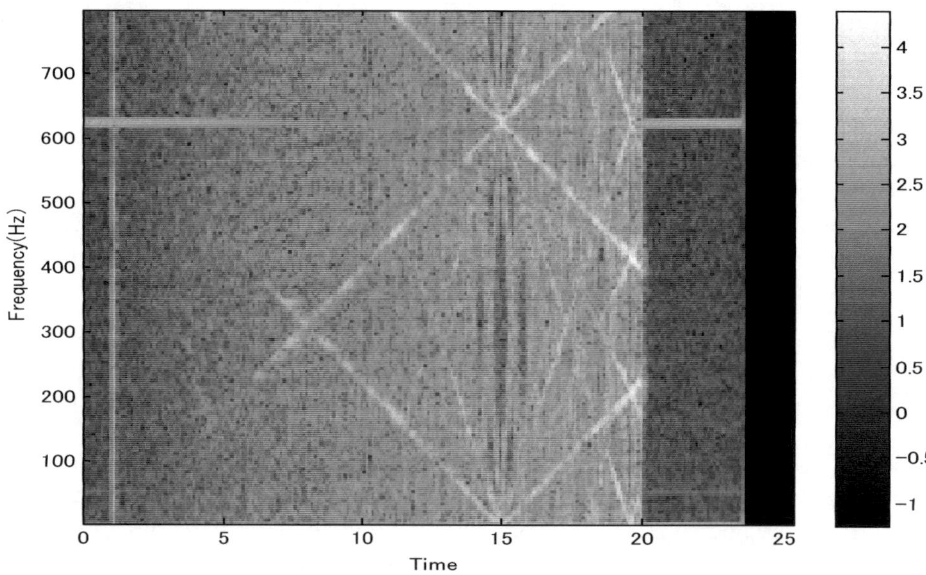

Fig. 6 Result of Spectrogram Analysis at ch 5 in the Case of Intact Condition

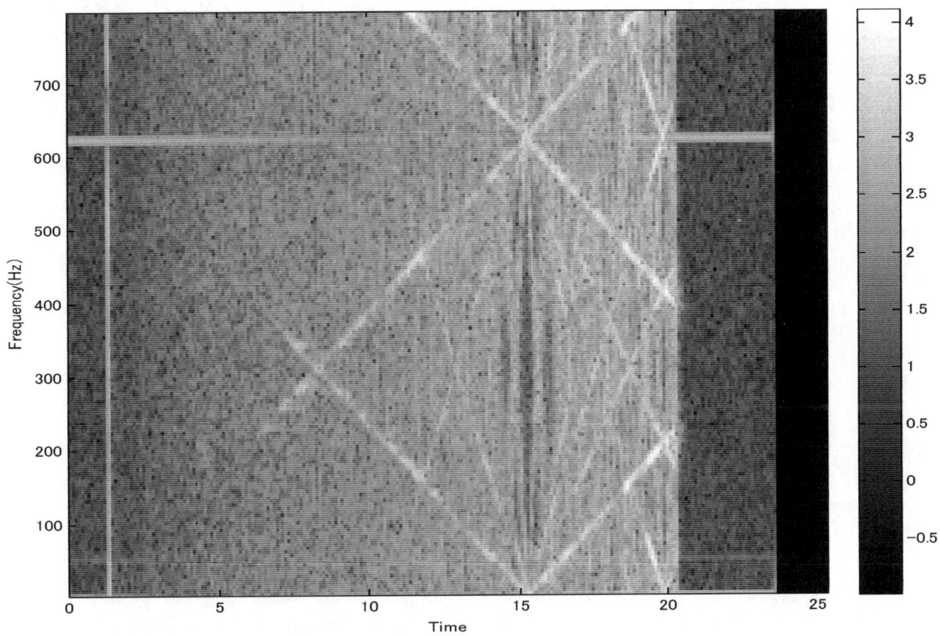

Fig. 7 Result of Spectrogram Analysis at ch 5 in Case 2

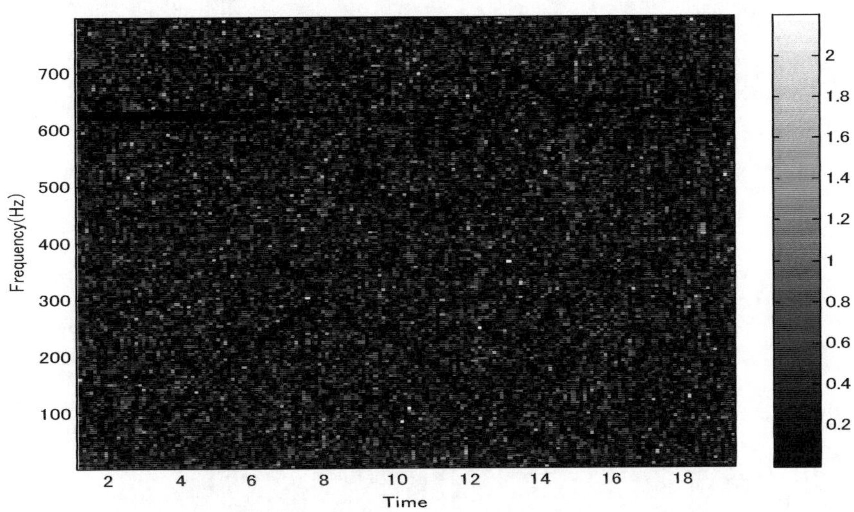

Fig. 8 Result after the subtraction from intact condition at ch 5 in Case 2

5.1 Result of Spectrogram Analysis for Case 2

Fig. 6 shows the result of spectrogram analysis of the received signal at ch5 in the case of intact condition (without additional weight). Range of frequency in FFT analysis is taken up to 800 Hz. As the anxiety of aliasing on the digitized data is considered, FFT analysis is done up to 2000 Hz and low pass filter is also used to check the result. Around at 620 Hz in Fig.6 there is a continuous constant amplitude from the beginning (white horizontal line), so that it is the noise of measurement system. And in the beginning of horizontal time axis (around at 1.2sec) vertical straight line is seen. This is also the noise of actuator input which is started after the recording system is started. Excitation is ended in around 20 sec and recording of signal is ended at 25 sec. Timing of excitation start is adjusted in this figure on Matlab. This result of adjusting start time is used later to subtract the result of FFT amplitude between before and after the damage to get the effect of damage only. In Fig. 7 the result of spectrogram analysis on the data of Case 2 at ch 5 is shown. And then the result of subtraction of Case 2 data from intact condition is shown in Fig. 8. Around at 620 Hz black line is seen, so that in the frequency domain subtraction between intact and damage states is done well. Small white dots (higher values) are distributed over the space. So that we see that the effect of damage is distributed over the space of both frequency and time domains.

5.2 Analysis of Frequency Contribution

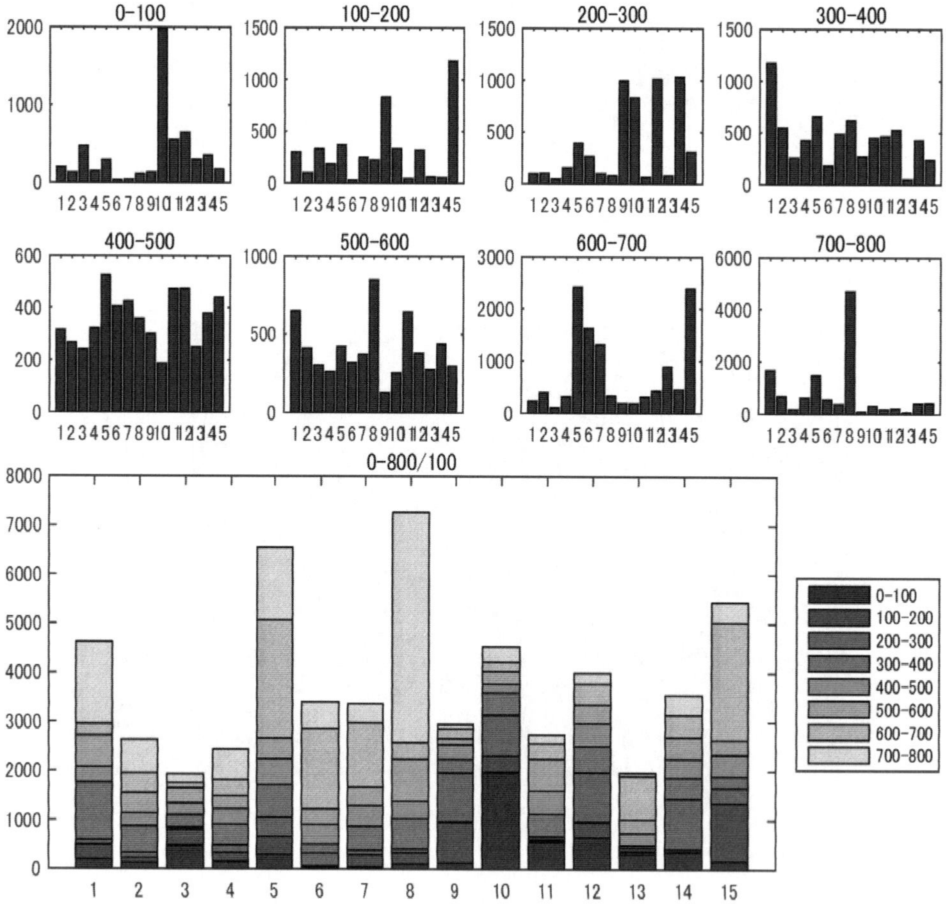

Fig. 9 Analysis of Frequency Contribution in Case 1 (Scale : Unit)

In previous article it is shown that frequency domain analysis could give clearer information on damage detection. So that in the procedure to calculate the damage indicator **DI0** followed from Eq. (1) to Eq.(6) , frequency contributibution analysis can be done. By dividing into several frequency ranges in matrix elements the contribution of each frequency ranges on the damage indicator is given. The result for the Case 1 is shown in Fig. 9 and for Case 2 in Fig. 10, respectively. Vertival axis is again for the value of damage indicator given by Eq. (6) at every 15 channels.

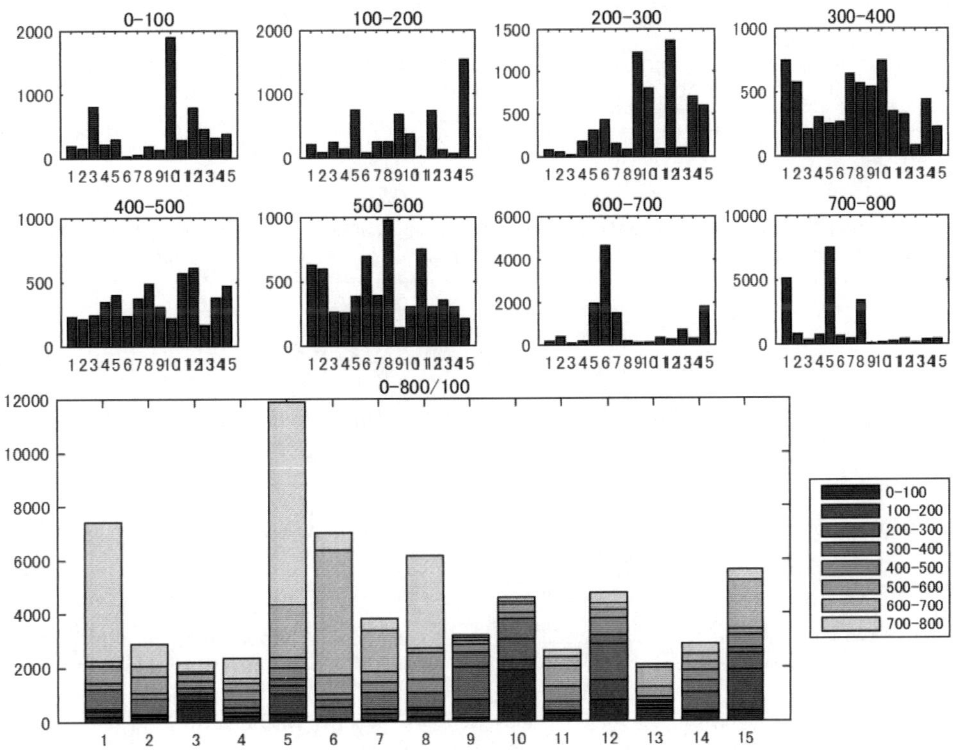

Fig. 10 Analysis of Frequency Contribution in Case 2 (Scale : Unit)

6. Conclusion and Remarks

By using spectrogram analysis the actual distribution of effect of damage over the space of both frequency and time domains is obtained. And the distribution of frequency contribution on damage indicator is also obtained. The effect of small damage to the vibration of structure is seen in the higher frequencies. So that in the damage detection through Local Excitation Method (LEM) analytical algorithm needs to focus more on the effect of higher frequency responses.

In order to apply to long term SHM using the prpposed LEM, the measurement before and after the structural condition change like damage ocurrence, needs to be done within a certain environmental condition. From our previous work the temperature difference during the mesurements before and after a damage needs to be less than 10 ℃ to get good result.

References

Oshima T., Yamazaki T., Onishi K. and Mikami S., 2002, Study on damage evaluation of joint in steel member by using local vibration excitation, (In Japanese), Journal of Applied Mechanics , JSCE, Vol.5, pp.837-846 .

Beskhyroun S., Oshima T., Mikami S., Yamazaki T., 2003, Damage detection and localization on structural connections using vibration based damage identification methods, Journal of Applied Mechanics, JSCE, Vol.6, pp. 1055-1064 .

Beskhyroun S., Mikami S., Oshima T. and Yamazaki T., 2004, Modified damage identification algorithm based on vibration measurements, Journal of Applied Mechanics, JSCE, Vol.7, pp. 97-107 .

Beskhyroun S., Oshima T., Mikami S., Tsubota Y., 2005, Structural damage identification algorithm base on change in power spectral density , Journal of Applied Mechanics, JSCE, Vol.8, pp. 73-84 .

Beskhyroun S., Mikami S., Oshima T., 2006, Structural damage detection and localization using changes in phase angle, Journal of Structural Engineering, JSCE, Vol.52A, pp. 659-670 .

Beskhyroun S., Mikami S., Oshima T. , 2006, Nondestructive damage detection scheme for steel bridges, Journal of Applied Mechanics, JSCE, Vol.9, pp. 63-74 .

Mikami S., Beskhyroun S., Miyamori Y., Oshima T., 2007, Application of a vibration-based damage detection algorithm on a benchmark structure , Structural Health Monitoring of Intelligent Infrastructure, Vol.3 , CD ROM(No. 55) .

Mikami S., Beskhyroun S., Yamazaki T., Oshima T., 2007, Damage detection in concrete structure using tunable piezoelectric actuators, Journal of Applied Mechanics, JSCE, Vol.10, pp.77-88 .

Mikami S., Beskhyroun S., Oshima T. , 2011, Wavelet packet based damage detection in beam-like structure without baseline modal parameters, Journal of Structure and Infrastructure Engineering, Maintenance, Management, Life-Cycle Design and Performance, Vol.7, pp. 211-227 .

Kumar R. P., Oshima T., Yamazaki T., Mikami S., Miyamori Y., 2012, Detection and localization of small damages in a real bridge by local excitation using piezoelectric actuators, Journal of Civil Structural Health Monitoring, Vol.2, No. 2, pp. 97-108 .

Beskhyroun S., Oshima T., Mikami S., Miyamori Y., 2013, Assessment of vibration-based damage identification techniques using local excitation source , Journal of Civil Structural Health Monitoring, Vol.3, No. 3, pp. 207-223 .

Doebling S. W., C. R. Farrar, M. B. Prime, and D. W. Shevitz, 1996, Damage Identification and Health Monitoring of Structural and Mechanical Systems from Changes in their Vibration Characteristics, A Literature Review, Los Alamos National Laboratory Report, LA-13070- MS .

Farrar C. R. et al., 1994, Dynamic Characterization and Damage Detection in the I-40 Bridge Over the Rio Grande, A Literature Review, Los Alamos National Laboratory Report, LA-12767- MS .

Farrar C. R. and D. A. Jauregui, 1996, Damage Detection Algorithms Applied to Experimental and Numerical Model Data from the I-40 Bridge, Los Alamos National Laboratory Report, LA-12979-MS .

Sampaio R. P. C., Maia N. M. M. and Silva J. M. M., 1999, Damage detection using the frequency-response-function curvature method, Journal of Sound and Vibration, 226(5), pp. 1029-1042 .

Peeters B., Maeck J. and De Roeck G., 2001, Vibration-based damage detection in civil engineering: excitation sources and temperature effects, Smart Materials and Structures, 10, pp.518-527 .

Xue S. T. , N. Fujitani, Z.B. Wei and H.S. Tang, 2006, Comparison of variations of natural frequencies for wooden structural models with and without damages part 2, shaking table based results, Proceedings of the international conference on Structural Health Monitoring and Intelligent Infrastructure, China, pp 1101-1104 .

The 8th International Conference on
Structural Health Monitoring of Intelligent Infrastructure
Brisbane, Australia | 5-8 December 2017

Dynamic curvature based monitoring in a highway overpass

K. Kliewer[1], B. Glisic[2]

[1] *Princeton University, Department of Civil and Environmental Engineering US, Email:* kkliewer@princeton.edu
[2]*Princeton University, Department of Civil and Environmental Engineering US, Email:* bglisic@princeton.edu

Abstract

With the growing challenge of aging infrastructure and the increasing cost for replacement and repair, structural health monitoring (SHM) offers an approach to address these challenges. It has been found in the literature that curvature and strain based methods may offer a more reliable approach to dynamic SHM compared to other methods such as acceleration and frequency based approaches. This research focuses on the application of a curvature based damage detection method, the normalized curvature ratio (NCR), to an in-service highway overpass. This method is widely applicable to beam like structures because it permits a structure to remain in-service, utilizes the service loads for monitoring, and is adaptable to a variety of sensor arrangements. Additionally, this research will focus on the use of long-gage fiber Bragg grating (FBG) sensors as they offer numerous benefits compared to other sensors currently available, such as low cost, multiplexing capabilities and the ability for both static and dynamic monitoring. Fiber optic sensors also allow for the instrumentation of large areas of a structure with long-gages sensors which helps enable global monitoring of the structure. Previous research applying this curvature based method in both small-scale laboratory testing and applied this method to a girder on an inservice highway overpass, demonstrating the feasibility of this method as a potential damage sensitive feature. This resaerch will focus on the applications of this method to girder 2 of the in-service highway overpass. Long term dynamic strain measurements from vibrations due to traffic loading on the structure have been measured through a series of FBG strain sensors instrumented on the structure. This research shows encouraging results and the potential for the NCR to be used as a simplistic metric for damage detection using FBG strain sensors.

1. Introduction

As the average age of structure in the US increase, there is a growing need for effective structural health monitoring (SHM) methods to help address the challenge of aging infrastructure and optimize repair costs. This research presents a dynamic structural health monitoring (SHM) method for beam like structures, the normalized curvature ratio (NCR). This is a curvature based damage detection parameter that utilizes a series of long-gage fibre Bragg grating (FBG) strain sensors. A major benefit of using strain sensors for curvature based analysis methods is that the curvature can be determined directly from the strain measurements with minimal uncertainty in the calculation as opposed to other methods such as those based on accelerations which will introduce a higher uncertainty in the curvature calculations. Additionally, it has been found in the literature that curvature and strain based analysis may be a more reliable means for structural monitoring as they show more sensitivity to damage (Pandy et al. 1991). FBG sensors also offer many benefits such as long gage sensor possibilities as well as static and dynamic monitoring abilities, they are durable and lightweight, immune to electro-magnetic interference and offer multiplexing capabilities

(Glisic & Inaudi 2008). Additionally, this research will focus on the use of long-gage sensors as opposed to point sensors as they are not influenced by local inhomogeneity of monitored material and increase the chance of detecting damage due to the larger coverage.
Previous research on the normalized curvature ratio has been performed in small scale laboratory tests (Kliewer & Glisic 2016). Additionally, the NCR was applied previously applied to girder 5 of the existing highway overpass that is the focus of this paper (Kliewer & Glisic 2017). Based on the success of this application to girder 5 of the structure, the NCR will be applied to girder 2 of this in-service highway overpass. An overview of the highway overpass, the monitoring system and the NCR method will be presented along with the results of the application to girder 2 of the structure.

2. Overview of Highway Overpass

Figure 1. Monitored highway overpass (Sigurdardottir & Glisic 2013)

The highway overpass that will be the focus of this study is located in the United States. The overpass was part of international bridge study (IBS) within the long-term bridge performance (LTBP) program by the federal highway administration (FHWA). The overpass is a simple composite structure consisting of multiple spans with built-up steel girders of varying heights with a concrete deck. This structural design is very common in the United States and thus provides the opportunity to apply these methods to a typical structure. This research will only focus on one span of the structure, as shown in figure 2a. This span is skewed such that each girder in the span has a different length. There is a total of 8 built-up steel girders on this span with perpendicular bracing trusses, shown in figure 2b, connecting the webs of the girders and there is additional lateral bracing connecting the exterior girders.

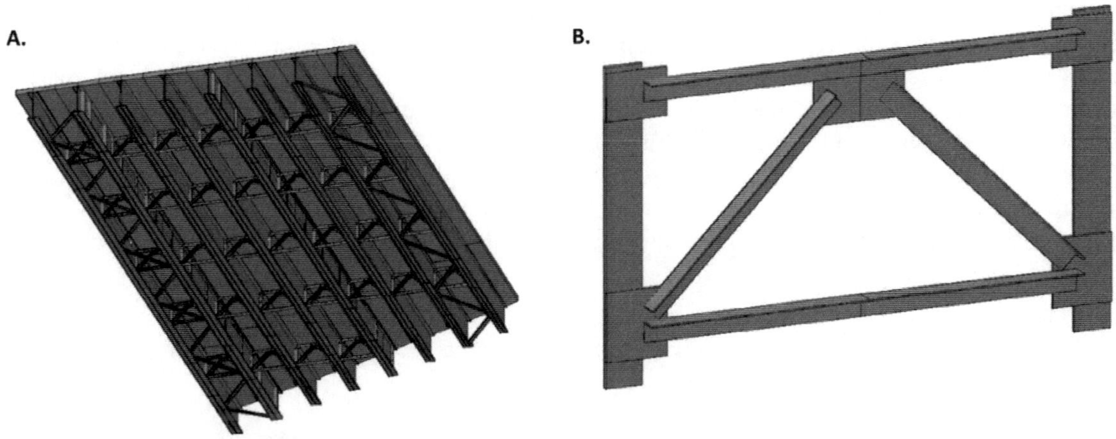

Figure 2: Structural details of the underside of the highway overpass with the western girder 1 on the left and the eastern girder 8 on the right (a) and detail of the perpendicular truss between girders (b) (Sigurdarddottir, Hwang, & Glisic 2012).

3. Monitoring System

The southbound span of the structure is instrumented with sensors on girder 2 and girder 5. Both girders are instrumented with fibre Bragg grating strain sensors at three locations on the girder, the two quarter spans and the mid spans. The length of the girders, the sensor gage-lengths and spacing of the FBG sensor locations are shown in figure 3. Sensors are installed in parallel topology with a strain and temperature sensor installed on the top flange and installed at the bottom flange. For each girder, there are a total of 6 strain and 6 temperature sensors installed.

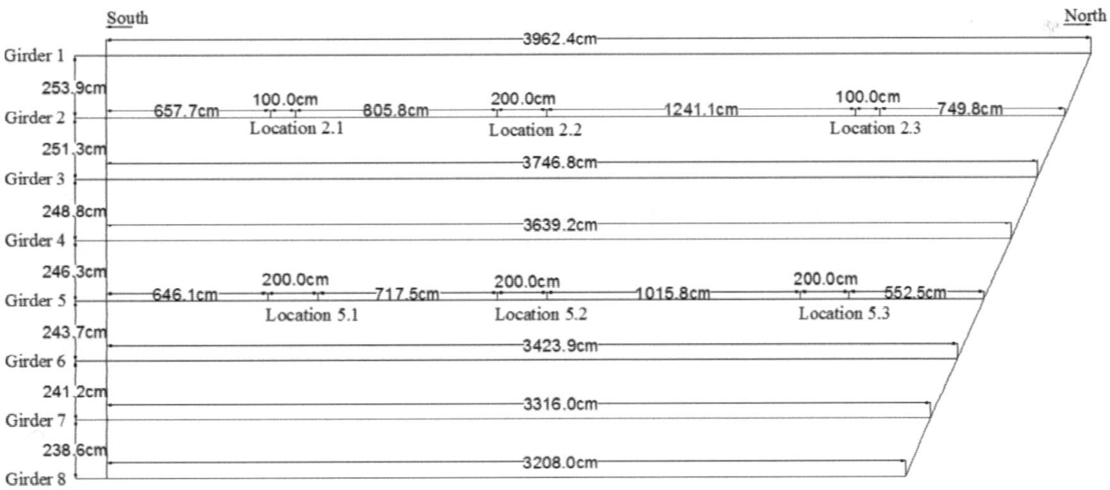

Figure 3: Sensor layout on structure (Flanigan 2014).

The sensors were installed on the structure in 2011 as part of the international bride study. Since the installment of the monitoring system on the structure, the Princeton SHM lab has maintained periodic data collection site visits to the structure which occur several times throughout the year and have been ongoing for almost six years. During the data collection site visits, the structure remains in-service during data collection and the measurements consist of the strain response of the structure caused by the service loads from the traffic loading. The strain response of the structure is recorded for approximately 1 hour for each site visit and based on the reading unit used, the data is recorded with a sampling rate of 250 Hz. As mentioned previously, these methods have previously been applied to girder 5 and the focus of this research will be applying similar methods to girder 2 of the structure. An example of a strain response from a heavy weight vehicle passing over the structure is shown in figure 4. The figure shows several peak strain responses on the structure followed by periods of free vibration which are the result of heavy weight vehicles passing over the structure. It is these periods of free vibration that follow the large peaks that will be used in this paper for the analysis.

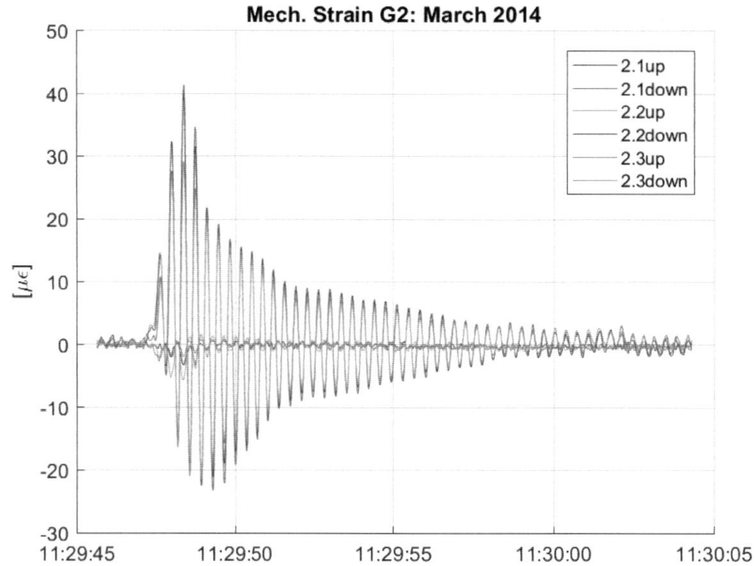

Figure 4: Girder 2 mechanical strain response, March 2014

The monitoring of girder 2 presents several challenges that were not present for the analysis of girder 5. Due to the diagonal bracing located between girder 1 and girder 2, the local DOT has noted fatigue cracking occurring on girder 2 of the structure. These cracks have been repaired using crack arrest holes, however limited information is currently available regarding these repairs. Additionally, because of the location of girder 2 relative to the traffic lanes on the overpass, girder 2 experiences significantly lower magnitude loads relative to girder 5.

4. Normalized Curvature Ratio

4.1 Experimental NCR

In order to compare the states of the structure, the normalized curvature ratio (NCR) was developed. Using the strain response from the sensors on the structure, the curvature at the location of the sensors on the bridge can be determined. This is done for each of the sensor locations and for each time step. As mentioned previously, the FBG sensors are installed along the top and bottom flange of the steel girder and are separated by a known distance, h. The curvature is related to the strain at each location by the following equation where, where κ, ε_t and ε_b are the curvature, the strain in the sensor at the top flange and the strain in the sensor at the bottom flange here below:

$$\kappa = \frac{\varepsilon_t - \varepsilon_b}{h} \quad (1)$$

To determine the NCR, the period of free vibration such as those shown in figure 4 are extracted from the strain measurements obtained during site visits. The curvature is determined using equation 2 and the peak curvature values are used. A ratio of these peak values at different sensor locations is then taken. Provided there is no change in the structure, this ratio should remain consistent. Any change in the structure should be reflected by a change in the curvature ratios observed. This method allows for a comparison of the different states of the structure without the development of a complex model of the structure. Additionally, the NCR is independent of the magnitude of the load applied to the structure.

4.2 Theoretical NCR

In addition to obtaining the experimental values of the NCR, the theoretical values of the NCR can be obtained for the structure. This provides a helpful means of comparison and validation. Because the beam deflection is equal to the double integral of the curvature, an equation for the curvature mode shape of the structure, $\kappa(x)$, can be obtained bed on the theoretical general modal shape for a beam, $\phi(x)$. The equation for the displacement mode shape, $\phi(x)$, and the derived general equation for curvature, $\kappa(x)$, are shown below where the b values are coefficients determined thorough boundary conditions or curve fitting and is the eigenvalue parameter for the associated mode. The equation can be used for any known or assumed boundary conditions of the structure. For the overpass used in this study, each girder is simplified as a simply supported beam.

$$\phi(x) = b_1 sin\, k_n x + b_2 cos\, k_n x + b_3 sinh\, k_n x + b_4 cosh\, k_n x \quad (2)$$

$$\kappa(x) = -b_1 k_n^2 sin\, k_n x - b_2 k_n^2 cos\, k_n x + b_3 k_n^2 sinh\, k_n x + b_4 k_n^2 cosh\, k_n x \quad (3)$$

Using these equations, the theoretical NCR for each combination of sensor locations can be determined.

5. Results

Dynamic events from a total of 25 site visits from June 2011 to December 2016 were used in the analysis of girder 2. The analysis utilized strain time histories with periods of free vibration and a high enough magnitude of its peak strain values. There were between one and nine dynamic strain events that occurred and were recorded for each site visit that were used in this analysis. Using the free vibration time histories from the strain response of the FBG sensors, the curvature values at each sensor location were determined using equation 1. Lower magnitudes of strain values contribute much higher values of uncertainty in the results, so a cut-off of 5 microstrain was used in this analysis. Because girder 2 is located to the edge of the structure where there is a lower volume of traffic and a lower traffic speed, the peak magnitudes of the strain response is lower than the strain response from girder 5. The normalized curvature ratios and associated uncertainties were determined for each sensor location on girder 2, which can be seen in figure 5. Additionally, the theoretical NCR values were determined for girder 2. These values are listed in table 1 and are also shown in figure 5 as a comparison with the experimental data results.

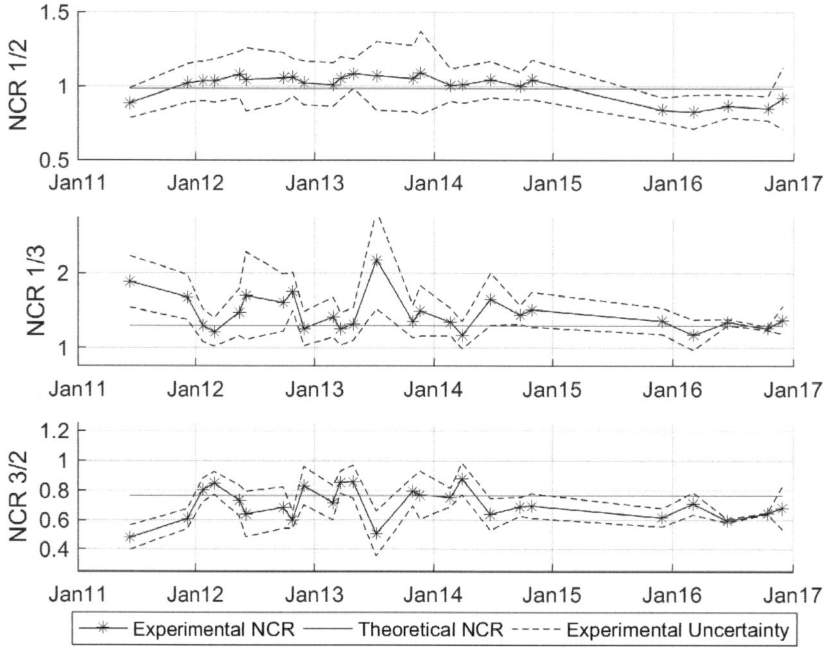

Figure 5: Experimental NCR values for girder 2 of structure

Table 1: Girder 2 theoretical NCR values

NCR 1/2	0.98
NCR 1/3	1.29
NCR 3/2	0.76

The NCRs indicate there is no a significant change from the beginning of the monitoring for this sensor system indicating there is no significant changes in the structural performance that could affect NCR. When comparing the experimental NCR values with the theoretical value, there are times where the theoretical NCR falls outside of the experimental values uncertainty bounds. The agreement between the theoretical and experimental value is not as good for girder 2 compared to what was observed for girder 5 (Kliewer & Glisic 2017). One potential reason is that the simply supported beam model used to determine the theoretical NCR values may not fully represent the behaviour of girder 2. This may be partially explained by the additional bracing that occurs with girder 1. For NCR 1/2 (sensor location 1 and location 2), and for NCR 3/2 (sensor location 3 and location 2), the theoretical value frequently falls outside of the experimental values in the more recent data measurement sessions. These results are consistent with previous research on this girder that has observed unusual behaviour of the structure in the more recent data measurement sessions (Kavanaugh 2017).

6. Conclusions

This research investigated the use of the normalized curvature ratio to analyse an existing, in use bridge located in the United States. The work uses the long term dynamic strain response of the structure caused by heavy weight traffic loading on the structure. Additionally, because this research relied on the free vibration response of the structure from the traffic, the structure was able to remain in-service during testing, causing no costly closures or disruptions to the bridge traffic that can be necessary for other SHM and NDE testing methods. The measurements were obtained from a system of long-gage FBG strain sensors that have been installed on the structure since 2011. Using the strain response of the structure, the curvature values were determined and used for the analysis. The method was previously successfully applied to girder 5 of the structure and the focus of this research was the application of the NCR method to girder 2 of the structure. Both the theoretical and experimental NCR values were determined and analysed. While no significant change in the structural behaviour was determined, the results indicate the possibility of unusual structural behaviour. Current research is exploring the unusual behaviour observed.

Acknowledgements

This material is based upon work supported by NSF GRFP Grant No. 1148900, NSF CMMI-1362723, and USDOT-RITA DTRT12-G-UTC16. Any opinions, findings, and conclusions or recommendations expressed herein are those of the authors and do not necessarily reflect the views of the funding agencies. The authors would like to thank the following individuals for their assistance with this research: Corrie Kavanaugh, Dorotea Sigurdardottir, Dennis Smith, and Joe Vocaturo. The project on the US202/NJ23 highway overpass in Wayne has been realized with the important support, great help and kind collaboration of several professionals and companies. We would like to thank SMARTEC SA, Switzerland; Drexel University, in particular Professor Emin Aktan, Professor Frank Moon (now at Rutgers University), and graduate student Jeff Weidner (now Assistant Professor at University of Texas, El Paso); New Jersey Department of Transportation (NJDOT), and in particular Nat Kasbekar and Eddy Germain; Long-Term Bridge Performance (LTBP) Program of Federal Highway Administration; PB Americas, Inc., Lawrenceville, NJ, in particular Mr. Michael S Morales, LTBP Site Coordinator; Rutgers University, in particular Professors Ali Maher and Nenad Gucunski; All IBS partners; and Kevin the lift operator. We would also like to thank Yao Yao who helped with the sensor installation.

References

Pandey, A., Biswas, M., & Samman, M. 1991, 'Damage detection from changes in curvature mode shapes', *Journal of sound and vibration*. vol. 145, no. 2, pp. 321-332.

Glisic, B. & Inaudi, D. 2008, *Fibre optic methods for structural health monitoring*, John Wiley & Sons.

Kliewer, K. and Glisic, B. 2016, 'A curvature based approach in dynamic monitoring using long-gage Fiber optic Sensors', *Proceedings of the 8th European Workshop on Structural Health Monitoring (EWSHM)*, Bilbao, Spain.

Kliewer, K. and Glisic, B. 2017, 'Application of the normalized curvature ratio to an in-service structure', *Proceedings of SPIE Smart Structures and Materials+ Nondestructive Evaluation and Health Monitoring*, Portland, Oregon.

Sigurdardottir, D. and Glisic, B. 2013, 'Neutral axis as damage sensitive feature', *Smart Materials and Structures,* vol. 22, no. 7, pp. 075030.

Sigurdardottir, D.H., Hwang, T., Glisic, B. 2012, 'In-service Assessment of the US202/NJ23 Bridge Health Condition using Dynamic Long-Gauge Fiber-Optic Sensors', *Proceedings of the 5th European Conference on Structural Control*, Paper No. 088, Genoa, Italy.

Flanigan, K. 2014, 'Determination of deformed shape from dynamic long-gage strain measurement', Senior thesis. Princeton University.

Kavanaugh, C. 2017, 'Evaluation of Long-Term Behavior of Neutral Axis and its Performance in Detecting Structural Damage', Senior thesis. Princeton University.

The 8th International Conference on
Structural Health Monitoring of Intelligent Infrastructure
Brisbane, Australia | 5-8 December 2017

Design of SHM system for My Thuan cable-stayed bridge - disadvantages and proposed changes based on acoustic emission

L.M. Chinh[1], S. Grzegorz [2],
[1] *Thuy Loi University – Vietnam, Email:*chinhlm@tlu.edu.vn, [2] *Kielce University of Technology - Poland,*

Abstract

The My Thuan Bridge, first cable-stayed bridge in Vietnam, had begun construction in 1997 and was put into operation in 2000. The bridge, which takes over the Mekong River, connects Cái Bè District of Tiền Giang Province with Vĩnh Long City of Vĩnh Long Province in Vietnam. The project was developed by a joint venture between the governments of Australia and Vietnam. The bridge, which cost A$91 million, was the largest overseas assistance project undertaken by the Australian government. The bridge was designed under Australian Ausroads – 92 standard and complies with Vietnamese 22TCN 18-79 standard. Since construction, there was no SHM system designed for the bridge. Therefore, in 2015 the Directorate for Roads of Viet Nam (DRVN) decided to implement SHM system for My Thuan bridge. However, the design of SHM system at that time had a lot disadvantages. This paper describes those disadvantages and proposes some changes. The new system is based on the acoustic emission (AE) technology, which utilizes the destructive process identification method (RPD). The application of AE technique allows monitoring and tracking of the processes that occur over the entire volume of a structure, which is independent from strain and stress and dependent mainly on the integrity of the structure that is of utmost importance for the service life and safe operation of the bridge.
Keywords: SHMS, cable stayed bridge, monitoring, damage, cracks, accoustic emission.

1. Introduction

My Thuan (Mỹ Thuận) Bridge, the first cable-stayed bridge in Vietnam, had begun construction in 1997 and was put into operation in 2000. The Bridge joins the banks of the Tien Giang and makes up a segment of the 1A Highway, the main traffic route across Vietnam. The bridge, which was built between July 1997 and May 2000, connects Cai Be District of Tien Giang Province on the south bank with Vinh Long City of Tan Ngai District on the north bank (Figure 1).
The Mỹ Thuận cable-stayed bridge with two suspension planes was designed to comply with Australian AUSROADS-92 bridge design standard, subjected to H30 – XB80 live load test and compared to Vietnamese standard 22 TCN 18-79. The following assumptions were taken into account in the design:

- Category II arterial road (design speed 80km/h, 100 "kN" axle load);
- Bridge width: 23.66 "m" (4 lanes for automobiles, 2 lanes for motorbikes);
- Inclination: 5% along the bridge and 2% across;
- Clearance for ships passing under the bridge: BxH = 50 x 200 "m", at the lowest water level in the river from the sea level;
- Designed to withstand earthquakes magnitude 7 on the Richter scale;
- Design lifespan: 100 years.

Figure 1. General view of My Thuan Bridge.

1.1. Technical design of My Thuan Bridge

The total length of the bridge is 1535.2 “m” with viaducts and feeder roads. The main bridge has three spans with dimensions 150+350+150 “m” supported by two main pylons (Figure 2).

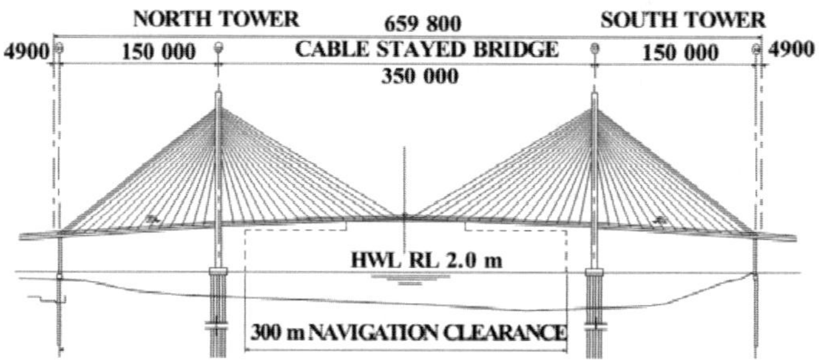

Figure 2. The main parameters of My Thuan Bridge

The main spans are made of prestressed concrete class 50 (fc’ = 50 “MPa”) and have two main girders internally attached by a set of floor beams. The bridge was constructed with the use of the cantilever method. The height of the main spans is 1760 “mm” with the width ranging from 1200 to 1400 “mm”. The deck slab is 250 “mm” thick. The end spans are suspended on 4 x32 stay cables and each stay is composed of 22 - 29 strands 15.2 “mm” in diameter, made from seven single/independent wires, with blue HPDE sheathing. A system of dampers was installed on each side of the cable stays to control vibration. The two pylons made of reinforced concrete class 50 have a splayed H-frame and the height of 123.5 “m” (from the foundation base) and 84.43 “m” (from the deck). Each pylon is supported by 16 bored piles, with 2.5 “m” diameter, driven into the riverbed to the depth of -90 “m” (northern pylon) and -100 “m” (southern pylon). The upper level of the foundation base is +1m from the sea level. The stay cables are in semi-harp configuration, with two suspension planes 18.6 “m” apart, arranged symmetrically along the axis of the bridge vertical alignment. The smallest inclination angle of the end stay cables is 31.031o, whereas the largest angle of inclination of internal stay cables near the pylons is 77.39o. The end stay cables are anchored in the main girders 5.0 “m” from the axis of the tie-down pier in the direction of access viaducts. The total number of stay cables, 128, is divided into 8 groups, with 16 stay cables in each group. The largest number of strands (68) was used in the end cables, while the smallest number (22) of strands was used in internal cables. The strands are 15.2 “mm” in diameter and made from seven steel wires. The cables and anchorages were supplied by Freyssinet. Each stay cable has three complementary protective barriers against external factors formed by galvanization and an HDPE sheath extruded onto the strand. The interstices were filled with petroleum wax. In this way, the stays meet the requirements of the BS 5896 British standard. The anti-vibration system (Damping Guide Deviators - DGD’s) was supplied by Freyssinet through the general contractor, BHE. The deviators are located on each side of the cable stays to damp excessive vibration and wind-induced resonance vibration in the stay cables. Tie-down piers with dimensions 1500 x 3500 "mm" in cross section, made of B40 prestressed concrete are located on both sides of the main bridge to restrain displacement of the pylons. The piers, without heads, are rigidly connected with the main spans. Each pier is founded on two 2.5 "m" diameter bored piles extending to the depth of -60 "m" (on the north side) and -74 "m" and -84 "m" on the south side.

1.2. Current condition of the bridge

After 15 years in operation the My Thuan bridge is generally in good condition. Every 5 years a thorough review of the entire bridge is conducted to assess the state of the bridge construction. The biggest problem of the bridge is the large number of cracks that appear on the bridge, especially at the cable to plate fixing area, on the connection between the crossbeam and deck slab, on the pylon near the crossbeam and on the deck slab (Figure 3).

There were also some measurment instruments installed on the brigde, except for the weather stations, such as humidity level and wind speed and direction gauges, installed in the center of the main span and on one of the pylons; the water level monitoring camera and traffic monitoring camera. No monitoring system (SHMS) was initiated for the Mỹ Thuận Bridge during the design and construction stages. The current condition of these devices is unknown.

Figure 3. A lot of cracks found on the bridge: a) the cable to plate fixing area, b) connection between the crossbeam and deck slab c) on the pylon near the crossbeam, d) on thedeck slab

2. The SHM system of My Thuan Bridge

2.1. General guidelines of the SHM system for My Thuan Bridge.

The SHM system for My Thuan Bridge main functions is monitoring the influence of weather and climate on the object and determining the actual state of the object during exploitation. Measurement and monitoring equipment must be modern and durable; existing equipment (if still useful) must be included in the newly built SHM system (sensors, data collection system, and data acquisition system, connection between these components, control and reporting software). The SHM system should consist of the following three components (L. M. Chinh 2014):

- Structural observation system;
- Meteorological monitoring system;
- Image monitoring system.

There are two main target function groups of the My Thuan Bridge SHM system:

- Maintenance and rehabilitation works, which:

- Provide data to detect and evaluate deformation as well as degradation of structures,
- Provide data to develop appropriate and timely maintenance and maintenance plans.

- Management of the safety of communication, which:
  - Uses the monitoring data for bridge traffic management to ensure vehicles safety on the bridge.

Therefore, the design of the SHM system for My Thuan Bridge, that enable it to work properly and optimally, is essential. The SHM system should be designed based on the technical and design features of My Thuan Bridge, as well as the need to identify the main physical quantities of Measurement and monitoring function. The selection of monitoring equipment for this system should consider the following factors (L. M. Chinh 2014):

- The design features, structure of the bridge, natural surroundings,
- Components to be monitored, quantities to be monitored,
- Economic efficiency,
- Accuracy and durability of the sensors.

2.2. Options of SHM system for My Thuan Bridge

To monitor the entire behavior of My Thuan bridge structure, the best solution is to install monitoring equipment on all structural parts of the bridge. Due to the symmetrical design of the bridge, the monitoring system could be located on a half of the bridge, which is still sufficient and reduces the overall equipment cost. However, the SHM system shall still collect all necessary information and data to control bridge during exploitation stage. There are two options of SHM system proposed for My Thuan Bridge construction (Table 1).

Table 1. Comparing 2 options of SHM system for My Thuan Bridge.

Item/Sensors	Quantity	OP1	OP2	Location
GPS	02	Yes	-	
Anemometer	01	Yes	Yes	On the top of the desk
Thermometer and huminidy	01	Yes	Yes	On the top of the desk
Precipitation gauge (Rainfall)	01	Yes	Yes	On the top of the desk
Biaxial tiltmeter	03	Yes	Yes	On the top, middle and the bottom of pylon P12
Scour sensor	01	Yes	-	-
Seismometer	01	Yes	-	-
Temperature inside the cable	01	Yes	-	-
Peak sensor	02	Yes	-	-
IP Camera	06	Yes	Yes	On the pylon
Cable accelerometer	04	Yes	Yes	4 longest cables
Displacement joint of main girders	02	Yes	Yes	Between main bridge with estacade
Stress and strain gauge	20	Yes	Yes	In 5 cross-sections of the main girders (4 sensor for one cross section)
Data logger	01	Yes	Yes	
TOTAL	**46**	**46**	**31**	

Based on table above, option 2 is proposed for the following reasons:

- The placement of sensors should ensure that the monitoring data is collected in the most appropriate way. Therefore, the design of SHMS is necessary to obtain data on the deformation of the structure under the influence of strong winds;
- To manage and fully monitor structure of the bridge, the accelerators and tilt gauges are needed;
- To undertake and monitor traffic effectively, wind sensors and rain gauges are also necessary;
- The number of sensors must adequately report reliable data for warnings to be made in the correct and appropriate manner.

2.3. Design of SHMS of My Thuan Bridge

In 2015, based on the guidelines for the SHM system of My Thuan Bridge, the SHM system was invented by ITST (Institute Of Transport Science And Technology) and accepted by DRVN (Directorate for Roads of Vietnam). Detailed of sensors locations are presented in the table below (Table 2).

Table 2. Types and locations of SHM system sensors designed by ITST for My Thuan Bridge

No	Notation	Item	Description	Location (Figure 4 & 5)	Unit	Quantity
I. 1. Sensors						
01	DN	Pylon declination sensor	Monitors the declination angle of the Pylon	On top of Pylon P12, Horizontal direction	Set	02
02	CV-KCG	Expansion joint displacement sensor	Monitors changes in dimension of the expansion joint	Abutment A1 in the North and Abutment A2 in the South	Set	02
03	DD-DC	Main girder vibration sensor	Monitors the one-direction vibration of the main girder	Middle span of the floor in main bridge	Set	01
05	GIO	Wind speed and direction sensor	Wind monitoring by ultrasonic waves	Middle span of the floor in main bridge	Set	01
06	DD	Stayed-cable vibration sensor	Monitors vibration of longest stayed cable	4 crossed end stayed cables pylon P12 and P13	Set	04
07	US	Main girder stress and deformation sensor	Measures stress and deformation of main girder	Girder edge under the floor	Set	20
08	NĐ-DV	Stayed cable inside temperature sensor	Monitors temperature changes	Middle span of the floor in main bridge	Set	01
09	NĐ-ĐA	Air temperature and relative moisture sensor	Monitors the ambience temperature and humidity	On the bridge desk at the middle of the bridge	Set	01
10	LM	Rainfall sensor	Rain monitoring	On the bridge desk at the middle of the bridge	Set	01
11	CAM	IP Camera	IP Camera with high resolution	At the pylon P12, bridge entrance and exit	Set	02
I.2 – Datalogers						
01	MR	Multifunction dataloger	Data processing US, LM	Inside technical box	Set	01
02	TR	Temperature dataloger	Data processing NĐ-DV		Set	01
03	ACC	Acceleration sensor	Data processing DĐ		Set	01

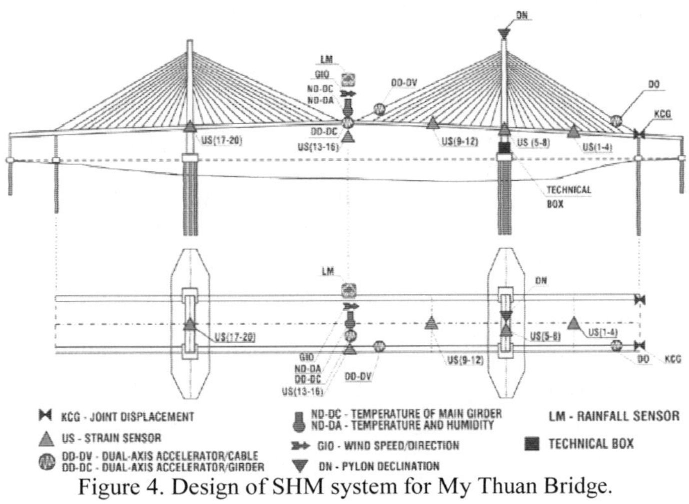

Figure 4. Design of SHM system for My Thuan Bridge.

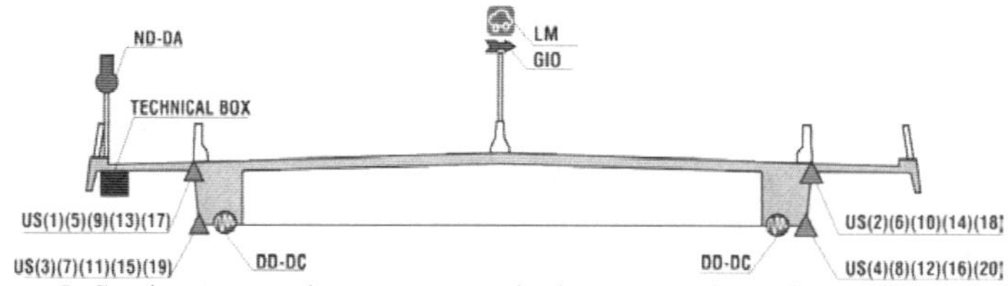

Figure 5. Strain gauges placement at typical cross section of My Thuan Bridge.

2.4. Disadvantages of the My Thuan bridge SHM system

There are three main disadvantages in the designed SHM system for My Thuan Bridge:

- The bridge symmetry was used to design SHM system and the system is practically installed on one half of the My Thuan Bridge, which does not objectively express the overall behaviour of the bridge, especially with the non-symmetrical loads;
- The concrete structure of the bridge was experiencing a lot of cracks during operation phase, therefore, it should have taken into account the possibility of monitoring the behaviour of these cracks in the future;
- There is no bridge condition at the first phase of operation – phase "zero". Therefore, to determine exactly the positions of the sensors for the entire SHM system, a FEM model should be developed to analyse the behaviour of the bridge.

3. Hybrid SHM system with Accoustic Emission technology

My Thuan Bridge is a large bridge made of reinforced concrete. Over time, the main beam and the pylons and anchorage sites started to experience large shrinkages and creeping changes, which are the characteristic phenomena for concrete and reinforced concrete structures with a large cubature. The aforementioned construction elements have suffered some displacement and deformation during their service time. Therefore, measuring and controlling the stresses, deformation, displacement, and temperature changes of the bridge components during operation is very essential. In addition, the bridge is also heavily influenced by weather and climate, especially from strong wind and tropical storms.

The size and current conditions of the My Thuan Bridge (with a lot of cracks) require that the following factors have to be accounted for while developing an objective for the new SHM system (L. M. Chinh & G. Swit, 2015):

- the need to locate damage over the entire volume of the structure or at least in its major parts,
- difficulties in defining real load acting on the structure, far divergent from the results of the load tests. For example, dynamic loads on a bridge, resulting from convergent travel of several lorries with loads and at the speeds far exceeding permissible values, that induce excessive loads on its individual elements,
- difficulties in defining interaction between various defects; a detailed spatial location of these defects is necessary, as well as the knowledge about the loads,
- difficulties in defining the degree to which the detected defect affects the capacity for load of the entire structure, that is, determining whether a given defect is active (developing) or passive (static). In many cases, serious-looking cracks are just remnants of initial deformations which have resulted in the redistribution of stresses leading to fault propagation in a different region,
- defining real operating conditions of the structure, such as temperature, rainfall, wind load, that have a noticeable effect on the development of deterioration. The diagnostic and monitoring system has to take into account all these conditions to be able to provide objective and unequivocal information.

The system is based on the measurements of acoustic emission generated by active (growing) destructive processes. By tracking their dynamics, it is possible to assess their influence on the condition of the structure and potential risks.
Identifying destructive processes with crack widths allows assessing their effect on the safety of structures by monitoring and observation of the structure's current condition and predicting service life changes in particular in building structures. The risk assessment criteria and codes are summarised in Table 3. Active monitoring and processes registration are possible owing to the proper distribution of acoustic sensors so as to enable to explore the entire volume of structure under test (G. Świt 2009).

Table 3. Characteristic classes of signals, assigned symbols and Codes

Colour/ symbol	●	■	■	▲	■	●
Class	No. 1	No. 2	No. 3	No. 4	No. 5	No. 6
Risk code	5	4	3	2	1	0

where:

No. 1. Crack formation in the cement paste,
No. 2. Crack formation at the paste/aggregate boundary,
No. 3. Crack formation in the tensile zone,
No. 4. Crack propagation,
No. 5. Loss of adhesion near the crack,
No. 6. Compression/Lateral buckling of struts/rebar/ compression crushing of concrete/bar rupture.

3.1. Application of FEM method to structural analysis

In 2016, a bridge deformation was measured with the support of Freyssinet under the bridge review program. At the same time, FEM model was developed to analyse the bridge behaviour. The bridge review program measurement results were just slightly different from the FEM model values (N.H. Thuan et all 2016). The FEM method is used to model the deformation behavior of the bridge structure. In particular, the two main girders and horizontal girders are modeled by two node linear element. The bridge deck is modeled by four node plate elements. The H-beam tower is modeled with a two-node linear element and the connection between the bridge girder and the bridge tower at the crossbar position was also simulated to ensure the freedom relative to the fixed and movable pillars of the two vertical beams. The cable systems are modelled using cable elements, which can be used to calculate wire tension and nonlinear geometry analysis (Figure 6).

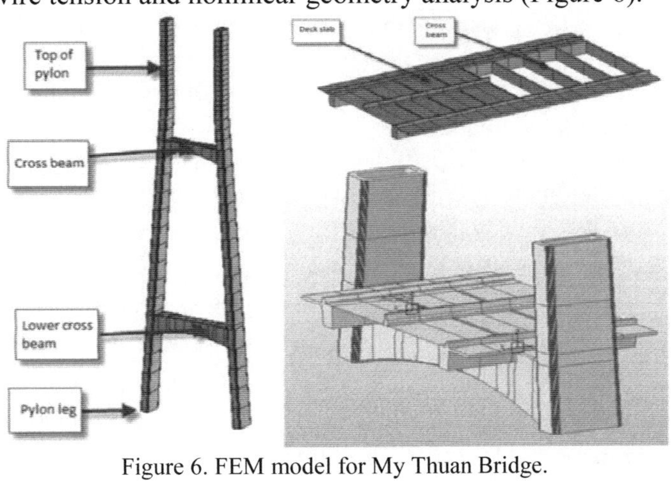

Figure 6. FEM model for My Thuan Bridge.

The results show that the analysis model and experimental results are quite appropriate. The basic mode of vibration identification and its characteristics have relatively small differences between the experiment results and the experimental model values (Figure 7).

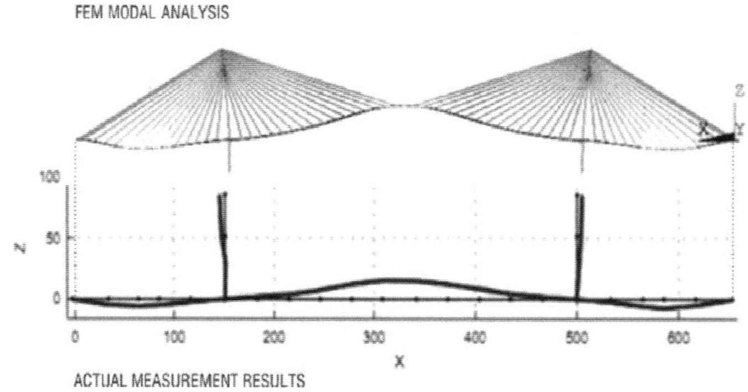

Figure 7. Comparison of deformation of My Thuan Bridge between FEM model analysis and measurement during bridge review.

However, the results of the analysis also show that there is a difference between the experimental and analytical models. In the analysis model, an unknown vibration mode was found in the experimental results. This difference can be explained by the fact that the parameters in the model cannot be simulated in the real-life behavior of the bridge (Table 3).

Table. 3 Comparison between FEM model analysis values and measurement result (Công ty TNHH GTVT, 2016).

Vertical oscillation frequency of My Thuan Bridge (Hz)			
Mode	Measurement result	FEM model analysis	Δ %
Mode 1 - 2	0,299	0,289	3,2%
Mode 2 - 3	0,387	0,368	4,9%
Mode 3 - 8	0,588	0,598	-1,8%
Mode 4 - 10	0,653	0,659	-0,8%
Mode 5 - 12	0,708	0,723	-2,1%

Based on these results, we can conclude, that using the FEM model for bridge behavior analysis is useful and should be included in the development of the SHM system for My Thuan Bridge and it could be a powerful tool for analyzing future bridge behavior.

3.2. My Thuan bridge SHM system with acoustic emission solution

For monitoring entire My Thuan Bridge, there are two components that need to be added to the current system (L. M. Chinh & G. Swit 2016):

Component 1 – using accelerometer to measure the dynamic characteristic of the whole span under excitation induced by wind (monsoonzone and strong winds), rain and service load. The amount of "vibration", amplitudes, etc. can also be used for fatigue analysis. The other objective of this component is condition and forces diagnosis for stay cables with an auxiliary system activated when acceleration recording commences in the span measurement system. The main tasks of this component are (Figure 8):

- Measurement of mid-span vertical (z) and transverse (y) accelerations (two-axis accelerometers – 100Hz) and development of data record thresholds in terms of wind speed, rainfall or acceleration leap caused by heavy vehicles (this initiator should be developed after a 12 – 24 month observations and subsequent comparison with the assumed values and with the results from the computational model). There are 03 measurement points with 02 sensors on each.

- Measurement of stay cable accelerations (three-axis accelerometers), with a high measurement frequency of min. 500Hz, at the mid-length (seems hardly possible). An "extended" data acquisition apparatus, a recording equipment, a "good" and tried algorithms for calculating and evaluating the results and a trial measurement are required to define basic values and forms of natural vibrations of the stay cables (referenced to the computational model). The measurement points are selected on at least 8 stay cables (three directions at each point).

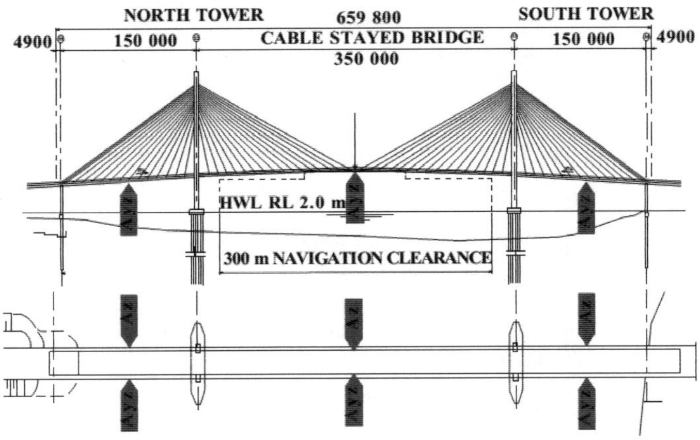

Figure 8. Placement of accelerometer to measurement dynamic characteristic of My Thuan Bridge

Component 2 – using the acoustic sensor to measure the AE signals, generated in the spans and pylons by excitations induced by wind (monsoon zone and strong winds), rainfall and service loads, and to diagnose the structural condition and long-term changes occurring in the structure of the spans and pylons. The main task of this component will be a short-term and long-term measurement of the AE signals generated by destructive processes. Sensors mounted on several points of the structure, selected based on statistical analyses results, which indicate the most critical points from the strategic point of view. There shall be 9 to 16 measurement points in the optimal configuration. The measurement points should be in the centers of spans, in the span above pylons and in each pylon leg (Figure 8).

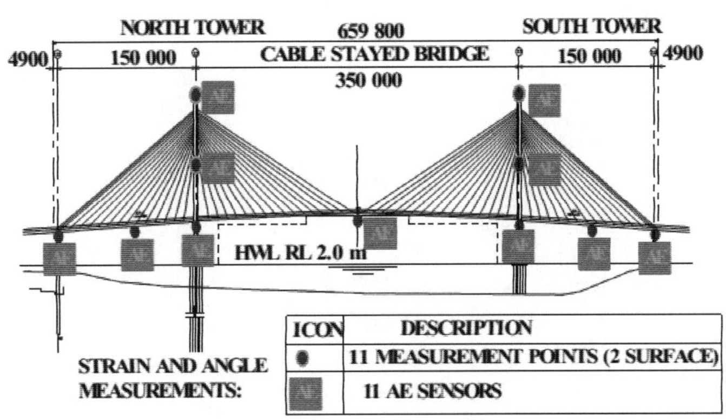

Figure 8. Placement of AE sensor to cracks monitoring of My Thuan Bridge

4. Conclusions

Based on the SHM system designed by ITST for My Thuan Bridge, the author noted some shortages and disadvantages of the system. With the experience of concrete structure and large bridges, some changes to the current system have been proposed. The proposed system of global evaluation and monitoring of concrete structures, based on the measurement of acoustic emission accompanying destructive processes, covers the entire volume of the element under test or its selected part and allows locating and identifying active destructive processes and their dynamics under real service loads. The data collected can be the basis for the determination of the structural condition of the construction and risk potential. Also, the data indicate the areas where either point methods can be used for AE source location or other methods of damage identification can be applied.

References

L. M. Chinh, 2014. BRIMOS – SHM system for cable-stayed Can Tho Bridge, *Inżynieria I Budownictwo – ISSN 0021-0315*. Poland. No 7/2014:.397-400.

L. M. Chinh, 2014. Long term structural health monitoring system for cable stayed bridge in Vietnam. *Tạp chí Khoa học kỹ thuật thủy lợi và Môi trường - ISSN: 1859-3941*. No. 44 (3/2014).

L. M. Chinh, G. Swit, 2015. Application of the acoustic emission method of identification and location of destructive processes to the monitoring of the technical state of pre-stressed concrete bridges. *Hội nghị khoa học Công nghệ Giao thông vận tải lần thứ III*, năm 2015.

G. Ś wit, " Diagnostic of Presterssed Concrete Structures by Means of Acoustic Emission," *8th ICRM' S 2009, Proceeding of IEEE, Chengdu, China*, pp. 958-962, 2009.

Công ty TNHH GTVT, 2016. Công ty Cổ phần Tư vấn Đầu tư và Xây dựng công trình giao thông, Công ty TNHH Fressinet Việt Nam (2016), *Báo cáo kết quả kiểm định cầu dây văng Mỹ Thuận (Report of My Thuan Bridge cables test results)*.

N.H. Thuan, B.T. Thanh, N.V. Minh. 2016. Đo đạc thực nghiệm và mô phỏng số nhận dạng dao động kết cấu nhịp cầu dây văng Mỹ Thuận. *Tạp chí Giao thông Vận Tải – ISSN 2354-0818*. 9/2016.

L. M. Chinh. G. Swit. 2016. A Prototype System for Acoustic Emission-Based Structural Health Monitoring of Mỹ Thuận Bridge. *Prognostics & System Health Management Conference—Chengdu (PHM-2016 Chengdu, China)*.

The 8th International Conference on
Structural Health Monitoring of Intelligent Infrastructure
Brisbane, Australia | 5-8 December 2017

Dynamic assessment of a multi-directional damage index for bridge structures

G. Lomiento[1], N. Bonessio[2]
[1] *California State Polytechnic University, Pomona, Civil Engineering Department – US, Email:* glomiento@cpp.edu
[2] *University of California, Irvine, Department of Mechanical and Aerospace Eng. – US, Email:* bonessio@uci.edu

Abstract

The experimental assessment of a novel multi-directional damage index for bridge elements is presented. The index relies on the combination of multiple vibration modes in different directions. The multi-mode combination automatically selects most relevant dynamic features, avoiding the need for any pre-screening. Dynamic tests on a bridge model are presented, with the aim of assessing the accuracy of the multi-directional damage indicator. Comparative results are presented for the proposed and a traditional damage index approach.

1. Introduction

The use of vibration-based damage identification methods for civil structures has been extensively investigated in the past three decades (Carden and Fanning 2004). Among all the currently used methods, Damage Index (DI) methods proved to be particularly suitable for damage identification of flexible structures (Fan and Qiao 2011). DI methods have been numerically and experimentally investigated on a variety of structures, including bridges and slabs. These methods rely on localized variation in strain or curvature between an undamaged and a possibly damaged structure for the identification, localization, and severity assessment of the damage as a stiffness reduction (Kim and Stubbs 2002). Strain and curvature are extracted from normalized vibration modes, and require a preliminary system identification phase to extract relevant dynamic characteristics of the structure. The selection of the relevant vibration modes for the damage identification is crucial for the accuracy of the severity assessment of the damage. When damages are multi-directional in nature, i.e. have different effects on the multi-directional stiffness properties of a structural member, the interpretation of the results and the estimate of the damage severity are particularly difficult.
Flexibility methods are preferable to DI methods for damages with multi-directional effects, as they can identify changes in a multiplicity of stiffness terms of the global stiffness matrix of a system. The inclusion of all possible stiffness components increases the size of the identification problem, which may result in computational problems, and inevitable inaccuracies. DI methods, on the other hand, are computational effective and can provide accurate estimate of the damage, but are aimed at identifying stiffness variations in predetermined directions. A modal strain energy approach has been proposed for damage identification in 2D elements, but has a limited capability in terms of damage severity estimate (Cornwell et al. 1999). Most of the existing studies on DI methods focus on mono-directional effects of damages, such as the reduction of the flexural stiffness of beams and columns in a predetermined direction. When 3D vibration modes are included in the analysis, these methods lack of consistency in results. If a single damage has significantly different effects in multiple directions, the severity assessment provided by these methods depends on the main direction of inflection of the selected vibration mode. The use of multiple vibration modes with different directions of inflection results into a multiplicity of severity estimates for each single damage. The consequent redundancy of data leads to discrepancy in results, as shown in (Barroso

and Rodriguez 2004) for a 3D structural frame, for which damage severity indicators, false positives, and missing detection errors differ significantly for modes in two different directions of vibration.

This study investigates the possibility of combining multiple 3D vibration modes to identify unitarily multi-directional effects of the damages. The proposed Multi-mode Multi-directional Damage Index (MMDI) method relies on the definition of a combination rule for individual modes results. The combination factor is finally translated into a directionality vector, which is indicative of the prevalent direction of the damage effect. The method has been investigated on flexural structural members. An experimental test has been carried out to demonstrate the applicability of the method to a portal frame, representing a downscaled model of a bridge structure.

2. Damage Indices

The proposed MMDI method uses the Damage Index definition provided in (Bonessio et al. 2012) and (Benzoni et al. 2013) but can be extended to any similarly defined indices based on variations of modal strain. The selected DI method was originally developed to localize and quantify damages in structural bridge components, as well as in seismic response modification devices such as isolators and energy dissipators. The method was compared to similar DI methods in terms of accuracy by means of numerical simulation of vibrations occurring along predominant direction of deflections.

For an individual i-th mode, the localization parameter that compares the modal strain energy variation of the j-th sub-element with respect to a k-th sub-element is evaluated as:

$$\beta_{i,jk} = \frac{\int_{L_j} \varepsilon^{*2} dl}{\int_{L_j} \varepsilon^{2} dl} \cdot \frac{\int_{L_k} \varepsilon^{2} dl}{\int_{L_k} \varepsilon^{*2} dl} \qquad (1)$$

where L is the sub-element length, and ε , ε^* are the generic modal strain terms (i.e. modal curvature for flexural elements), in the undamaged and damaged condition, respectively. A value $\beta_{i,jk}$ >1 indicates damage in the i-th sub-element, while $\beta_{i,jk}$ <1 indicates damage in the k-th sub-element. A normalized damage localization term is further defined for the j-th element as:

$$\beta_{i,j} = \frac{\beta_{i,jk}}{\beta_{i,k_{min}}} \geq 1 \qquad (2)$$

where $\beta_{i,k_{\min}}$ is the minimum value of the localization parameter $\beta_{i,jk}$ of Eq. (1).

A statistical significance criterion is used to identify the locations of damage through the definition of the damage localization index:

$$Z_{i,j} = \frac{\beta_{i,j} - \bar{\beta}_i}{\sigma_{i,\beta}} \qquad (3)$$

where the parameters $\bar{\beta}_i$ and $\sigma_{i,\beta}$ are the mean and the standard deviation of $\beta_{i,j}$, respectively. The j-th sub-element of a structure is considered damaged if $Z_{i,j}$ exceeds a pre-determined confidence level. In this study, the damaged condition is recognized by $Z_{i,j} > 1.28$, corresponding to a one-tailed test at 90% confidence level.

The severity index $\alpha_{i,j}$ represents the fractional change in stiffness of the j-th element:

$$\alpha_{i,j} = \left(\beta_{i,j}\right)^{-0.5} - 1 \qquad \alpha_{i,j} \geq -1 \tag{4}$$

For each vibration mode, the fractional change in stiffness is associated with type and direction of the strain measure used in Eq. 1. For flexural elements, for instance, the severity index quantifies the flexural stiffness reduction in the direction in which the modal curvature is gathered.

In this study, a formulation is added in order to extend the method to multi-directional damage estimates from multiple 3D vibration modes. The method requires an arbitrary subdivision of the structure into sub-elements not necessarily corresponding to physical components. The damage localization and severity estimation is based on the variations caused by damages on the modal strain values of the sub-elements.

3. MMDI Method

The fractional change in stiffness defined in Eq. 4 is associated with type and direction of the strain measure of Eq. 1. When multiple vibration modes are used, a multiplicity of damage localization and severity indices are defined, each of them referring to damage effects in the direction of the modal deflection.

The proposed MMDI method combines mono-directional damage indices from individual modes into global indices that are representative of the multi-directional nature of the damage:

$$\beta_j = \sum_{i=1}^{n} \gamma_i \beta_{i,j} \, . \tag{5}$$

The combination factor is based on the average deviation q_i of the $\beta_{i,j}$ values from the unit value:

$$\gamma_i = \frac{\left(1/q_i\right)^2}{\sum_{i=1}^{n}\left(1/q_i\right)^2} \; , \; \sum_{i=1}^{n} \gamma_i = 1 \tag{6}$$

in which n is the number of modes of vibration included in the analysis. The γ_i factor measures the relevance of the i-th mode for the damage detection. The MMDI localization index Z_j and severity index α_j are determined by substituting $\beta_{i,j}$ with β_j into Eqs. (3) and (4).

The γ_i factor is then used to identify the prevalent directionality of the multi-mode damage severity estimation, through the directionality vector

$$\mathbf{v_j} = \sum_{i=1}^{n} \sqrt{\gamma_i}\, \mathbf{v_{i,j}} \tag{7}$$

where $\mathbf{v_{i,j}}$ is the Euclidean vector associated with the direction of the i-th mode stiffness reduction. Eq. 7 is based on the assumption that the multi-directional stiffness reduction has an elliptical shape. The multi-mode formulation relies entirely on results from individual modes, and can be applied to any DI method providing localization and severity indices.

4. Experimental investigation

A downscaled bridge model was used as a testbed for the validation of the method and subjected to dynamic tests with white noise excitations. A total of 19 shake table tests were performed in the undamaged and five damaged conditions, as described in the following.

4.1 Bridge model

The one-span steel bridge model was conceived to allow removing parts on columns and deck elements and to accommodate viscous dampers and friction isolators. The model is composed by two identical columns and one single-span deck structure, realized by two longitudinal beam elements connected by transversal secondary beams. The column section is made of four flanged vertical portions, each one constituted by a central hollow rectangular section HSS8x4x1/4 and four peripheral C4x7.25 sections, and a cap beam for the connection with the deck. The total column height is 82", while the width of the cap beam is 51". The deck is made by two steel boxes, realized with ½" thick plates, connected by two tri-parted longitudinal W6x15 beams, with additional bottom and top ¼" plates to augment flanges thickness. The deck structure is 126" length by 64" wide. A total of 44 structural parts (16 in each column and 12 in the deck) are removable to simulate local damages in the structural bridge components. The use of removable parts to simulated damages is justified by the experimental program objective of not exceeding the elastic range in all the structural components, while at the same time reducing their stiffness in a controlled way. After damages are simulated, the previous undamaged condition can be restored by placing back the removed parts and the model can be used for further tests. Additional mass is provided through steel plates located over the deck structure (3,600 lb) to rise the vibration period of the system. Selected schematic drawings of the columns and the bridge assembly are presented in **Fig. 1** and **Fig. 2**. In **Fig. 3** and **Fig. 4**, a plan view and removable parts from the deck and the columns are presented, respectively. The additional steel plates are visible on top of the shake table installation of the bridge model shown in **Fig. 5**.

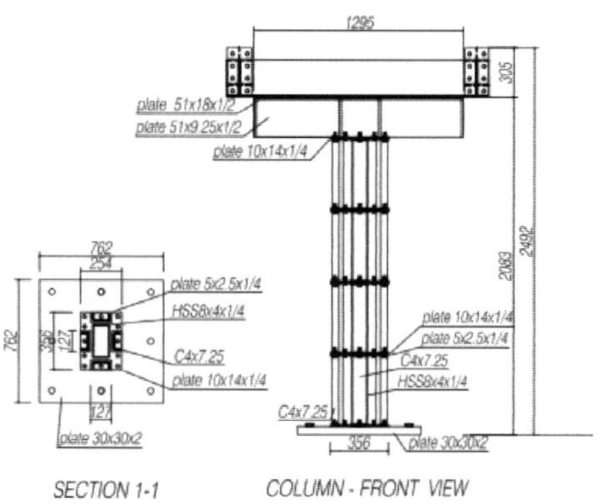

Fig. 1. Bridge model column: front view and section.

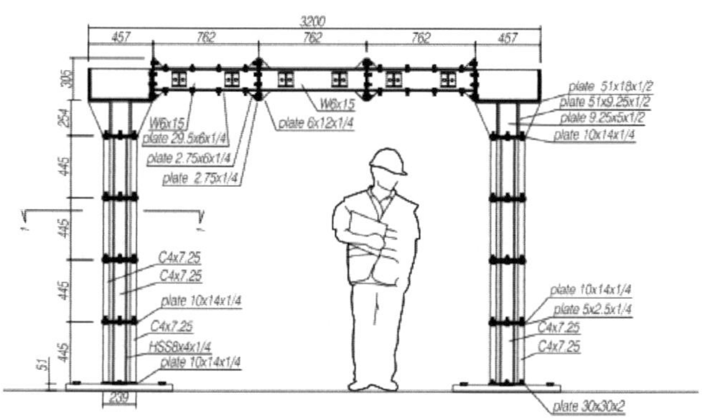

Fig. 2. Bridge model assembly.

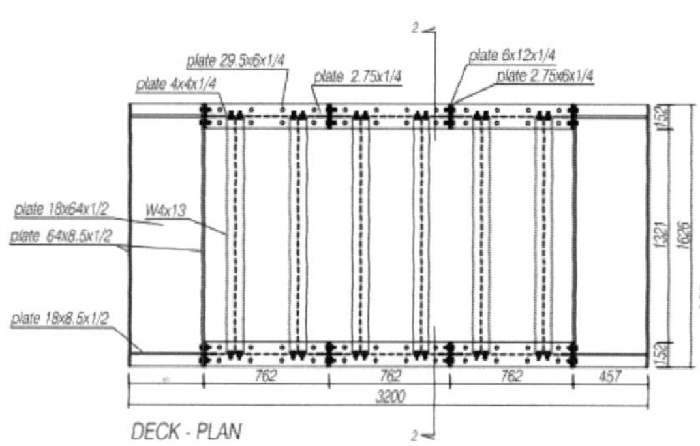

Fig. 3. Bridge deck model. Plan view.

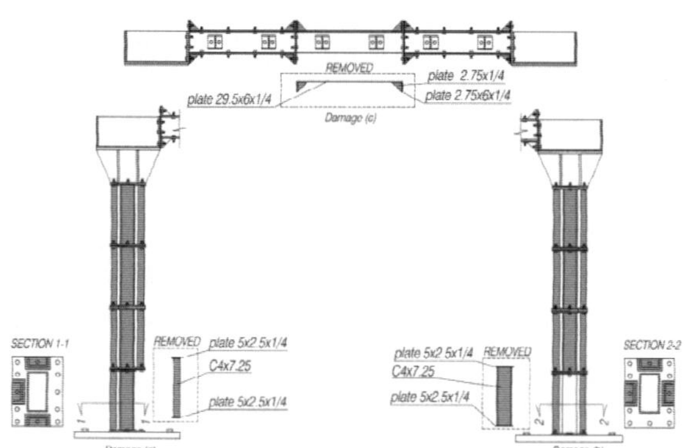

Fig. 4. Bridge model removable parts.

4.2 Test setup

The present investigation deals with the bridge configuration without isolators and dissipating devices, for which the deck is directly connected to the columns through bolted connections able to transfer flexural moments. The bridge model was subjected to shake table tests at the Caltrans Seismic Response Modification Devices test facility at the University of California, San Diego. The test installation consisted in (1) the connection of the base-plates of the bridge columns to the shake table, with the orientation shown in Fig. 5, (2) the connection of the deck structure (with clamped steel plates for additional mass) to the column cap-beams, in order to simulate a moment-resisting connection, and (3) the instrumentation with displacement transducers and wired accelerometers.
In details, the sensor system was composed by:

- 5 displacement transducers (**Fig. 6**),
- 9 tri-ax wired accelerometers, labeled #1to9 (cyan colored in **Fig. 7**), and
- 8 mono-ax wired accelerometers (4 in Y dir + 2 in X dir + 2 in Z dir), labeled #10to17 (green colored in **Fig. 7**).

Additional feedback was provided by the 6 mono-ax accelerometers of the shake table. Prior the shaking tests execution, a Finite Element Model (FEM) was implemented to analyze dynamic characteristics of the bridge model and calibrate the excitation input.

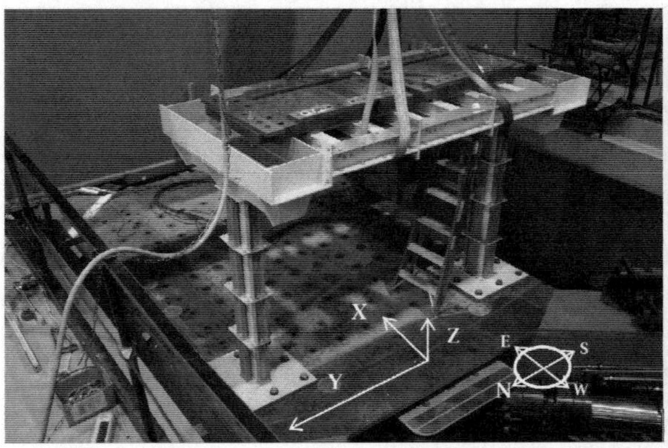

Fig. 5. Shake-table bridge model setup.

Fig. 6. Location and direction of the displacement transducers installed on the bridge.

Fig. 7. Location and direction of the accelerometers installed on the bridge model.

The model has been tested in the "undamaged" and in 5 "damaged" conditions (obtained by removal of one or more removable parts). A total of 19 dynamic tests were performed with three types of white noise input excitations with flat (constant) power spectral density over a fixed frequency range obtained through a random generation algorithm specifically developed in Matlab®. This type of input was chosen as representative of ambient vibrations that can be recorded on a road bridge structure during normal operation. Each input is represented by an acceleration time history which was imposed in displacement control to the shake table. Type A and B inputs, with a flat (constant) power spectral density over the frequency range 1-10Hz, were used for the horizontal motion of the table. Input A was used in all the mono-directional tests, while input B was used as lateral horizontal input in multi-directional tests when input A was used in longitudinal direction. A type C input was used as vertical input with a higher frequency content ranging from 5 to 15Hz. The peak acceleration imposed in vertical direction is 0.230g, which is approximately 50% higher than in the horizontal directions. A summary of main input functions characteristics is presented in **Table 1**.

Table 1. Type A (horizontal) input motion characteristics and expected bridge response values.

INPUT A					
D_{max} base	0.180	in	=	4.6	mm
V_{max} base	2.264	in/s	=	57.5	mm/s
A_{max} base	0.154	g			
RMS displacement	0.063	in	=	1.6	mm

INPUT B					
D_{max} base	0.177	in	=	4.5	mm
V_{max} base	2.504	in/s	=	63.6	mm/s
A_{max} base	0.146	g			
RMS displacement	0.035	in	=	0.9	mm

INPUT C					
D_{max} base	0.024	in	=	0.6	mm
V_{max} base	1.206	in/s	=	30.6	mm/s
A_{max} base	0.230	g			
RMS displacement	0.009	in	=	0.2	mm

4.3 Test protocols and simulated damages

Tests 1-13 were performed on the undamaged bridge model, in order to identify the baseline dynamic characteristics of the structure. Three levels of input intensity were used, specifically 10%, 25%, and 50% of the reference input. The accelerations associated with the 25% input intensity represent commonly encountered values due to ambient vibration of common bridges in service, and were selected for the tests on the damaged scenarios. Tests 14-15 refer to damaged configurations identified with labels from #1 to #5, as shown in **Table 2**. The damage in the North Column was simulated by removal of the bottom C section on the South side of the column, and affects primarily the stiffness of the column for oscillations in longitudinal direction (see **Fig. 5** for the orientation of the bridge model). The bottom C section on the West side of the column was removed to simulate a lateral damage in the South Column. The middle upper plate of the West side beam was removed to simulate damage in the deck (see **Fig. 4**).

5. System Identification

For the identification of the modal characteristics, the output-only Subspace Stochastic Identification Covariance (SSI-Cov) (Peeters and De Roeck 2001) method was used, since tests were performed to simulate ambient-like vibrations with unknown input. Common tolerances values of 1%, 5%, and 2% for frequency, damping ratio, and the Modal Assurance Criterion on the mode shape, respectively, were used to identify stable poles from a 200 order model. Sample stabilization diagrams for an X-dir and a 3D test are shown in **Figs. 8** and **9**. The envelope of the

Power Spectral Density functions evaluated for all the records is added to the plot to highlight the presence of peaks for frequency values associated with physical poles. In the plot, poles that respect all the stability conditions are represented with blue markers, while red markers identify poles that are stable in terms of at least one among frequency, damping ratio, and mode shape. The identification of a torsional mode in the X-dir test is indicative of accidental eccentricity of the supported mass.

Table 2. Summary of the dynamic tests on the damaged configurations

	Input Type-Intensity[1,2]			
Test #	**Dir-X**	**Dir-Y**	**Dir-Z**	**Simulated damage[3]**
14	-	A-25%	-	Damage #1
15	A-25%	B-25%	C-25%	Damage #1
16	A-25%	B-25%	C-25%	Damage #2
17	A-25%	B-25%	C-25%	Damage #3
18	A-25%	-	-	Damage #4
19	-	-	C-25%	Damage #5

Notes

[1] Input type A = reference input for horizontal direction (frequency band 1-10Hz).
Input type B = input for Y direction in 3D shaking (frequency band 1-10Hz).
Input type C = reference input for Z direction (frequency band 5-20Hz).

[2] Input intensity is expressed as percentage of a reference input, for which the bridge model reaches the estimated ultimate capacity.

[3] Damage #1 = North Column.
Damage #2 = North Column + South Column.
Damage #3 = North Column + South Column + Deck.
Damage #4 = South Column + Deck.
Damage #5 = Deck.

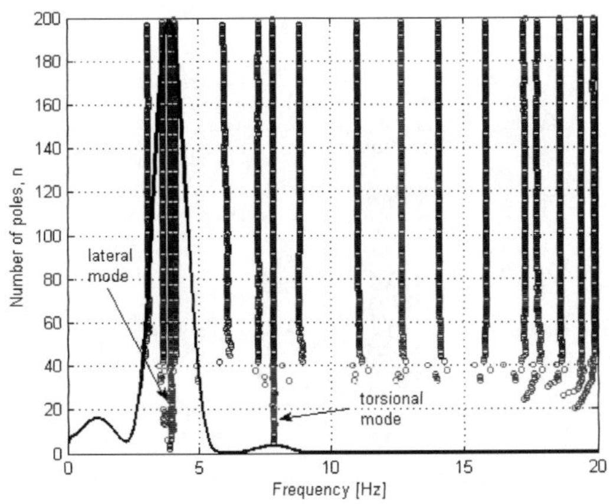

Fig. 8. Stabilization chart for X-dir test

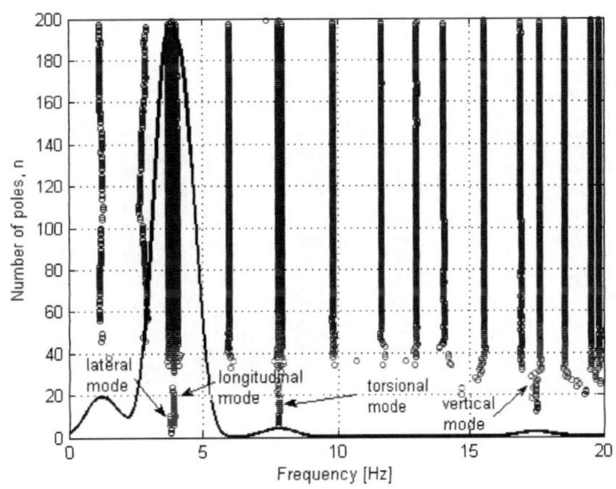

Fig. 9. Stabilization chart for X-dir test

6. Damage identification

Effects of damages were visible in terms of reduction in vibration frequency, as summarized in **Table 3**, while a clear trend in terms of damping was not visible. The mode shapes identified by means of the SSI-Cov procedure were used to assess the damage severity of the damages from individual modes, and through the proposed MMDI method.

Table 3. Average value ± standard deviation of frequencies and damping ratios.

		Vibration mode			
		1. Lateral (X dir)	2. Longitudinal (Y dir)	3. Torsional (Z axis)	4. Vertical (Z dir)
undamaged	frequency (Hz)	3.87 ± 0.12	4.10 ± 0.09	7.95 ± 0.16	17.88 ± 0.80
	damping ratio (%)	1.48 ± 0.90	0.43 ± 0.26	0.84 ± 0.87	0.80 ± 0.51
damaged	frequency (Hz)	3.59 ± 0.01	3.96 ± 0.08	7.49 ± 0.05	16.95 ± 0.06
	damping ratio (%)	0.68 ± 0.29	2.22 ± 1.51	0.44 ± 0.33	0.28 ± 0.13

Sample DI from individual modes are presented in **Fig. 10** for the damage in the North Column for test #15. This specific damage consists of a longitudinal damage in the bottom portion of the North Column (see **Fig. 4**). Results are presented in terms of $\beta_{i,j}$ index. From the plots, a significant difference in results is evidenced by difference in the index distribution along the column. The torsional mode (mode 3) identified highest levels of damage. The lateral and vertical modes proved to be poorly correlated to the actual damage of the column, which mostly affected the flexibility in longitudinal direction. The poor correlation is evidenced by values of the index $\beta_{i,j}$<1. Similar scatter of the results were observed for the damages on South-Column (predominant lateral vibration mode), and for the damages on the deck beams (predominant vertical vibration mode). The values of the combination factor γ_i are reported for each of the mode. The longitudinal mode 2 is associated with a γ_i≈0.72, indicating that the mode is most correlated to the damage. The torsional mode 3 is characterized by γ_i≈0.15, demonstrating its high correlation to the damage for this test. In both tests, lateral and vertical modes are characterized by small γ_i values. The Multi-Mode localization index shown in the figure is the weighted average of the indices from the individual modes (Eq. 5). The localization term β, localization index Z and quantification index α from test #15 are presented in **Fig. 11** for the Multi-Mode method.

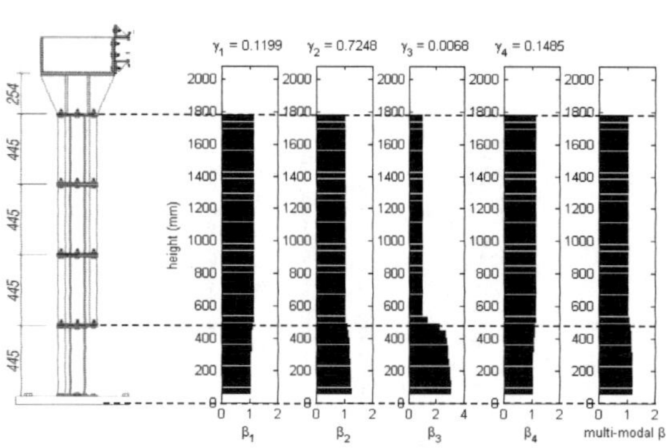

Fig. 10. Damage assessment in the North-Column – test #15

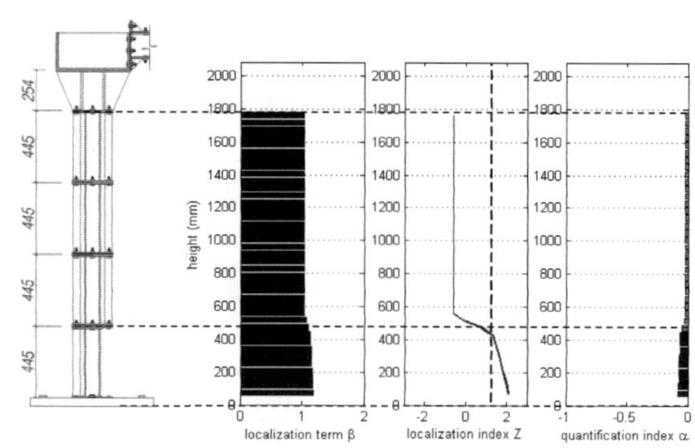

Fig. 11. Damage assessment in the North-Column – test #15

The Multi-Mode damage indices are correlated to a specific direction by the vector defined in Eq. 7. A polar representation of the Multi-Mode severity indices is presented in **Fig. 12**. The identified damages from the tests are compared with the theoretical reduction of stiffness induced into the different structural elements. The dashed line represents the undamaged condition, characterized by full stiffness in any direction, while the black line identifies the expected stiffness reduction. The results obtained with MMDI methods appear in significant agreement with the polar representation of the damage effects. The main advantages of the MMDI approach over the individual-mode approach consist of (i) providing a single damage severity estimate for each test, i.e. for each set of accelerometric records, thereby reducing the redundancy of data available from multiple modes, and (ii) identifying the damage in terms of its multi-directional effects.

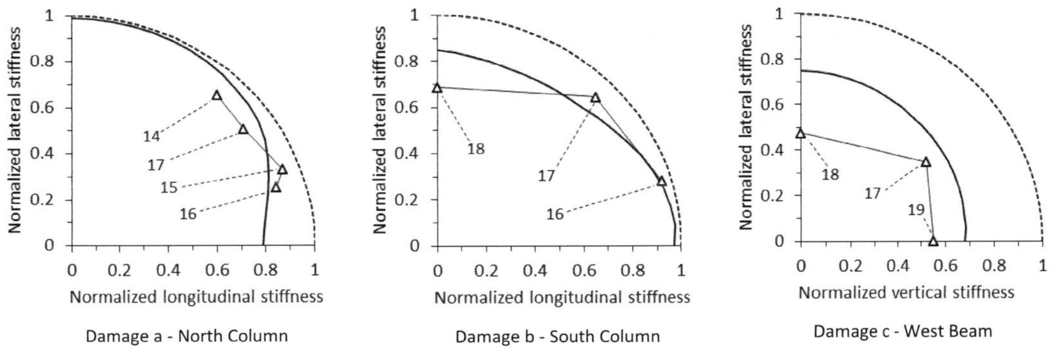

Fig. 12. Polar representation of the normalized flexural stiffness reductions from the multi-mode assessment

The overall accuracy of the MMDI method appears satisfactory. Recognition of the multi-directional effects of damages allowed reduction of severity estimate errors for two out of three simulated damages, with the exception of the damages in the beam. For this damage the highest accuracy in the damage severity assessment is obtained by using individual vertical vibration modes. The MMDI severity estimate of the beam damage is however comparable with the most accurate results obtained by using individual vertical modes. The localization accuracy of the beam damage is also improved in comparison with results from individual lateral and torsional modes. In this sense, the MMDI formulation appears effective in identifying modes mostly affected by damages, and condensing dispersed damage information from individual modes into global damage measures

3. Conclusions

A new method for identification of multi-directional damage effects from a multiplicity of vibration modes is presented. The method can be paired with any Damage Index method based on variations in modal strain between and undamaged and a damaged structure. The improved accuracy of the method derives on the use of properly defined combination factors in order to identify most relevant modes associated to a given damage scenario. The severity assessment of the damage is also improved by associating a directionality to the severity estimate.

The proposed method has been experimentally validated on a portal frame, representing a downscaled bridge model subjected to dynamic testing. Damages were simulated in columns and beam elements of the structure and identified based on fundamental lateral, longitudinal, torsional and vertical vibration modes. As expected, the accuracy of the damage assessment based on individual modes appeared significantly correlated to the selection of the mode shapes. The proposed method proved able to provide an accurate estimate of the damages without any preselection of the vibration modes. The present study supports the feasibility of using the method for similar structures, mainly composed by flexural members. After this preliminary investigation, the method will be used on full-scale monitored bridges for further validation.

References

Carden, EP, Fanning, P 2004, 'Vibration based condition monitoring: a review' *Structural Health Monitoring*; vol. 3, no. 4, pp. 355-377.

Fan, W, Qiao, P 2011, 'Vibration-based Damage Identification Methods: A Review and Comparative Study' *Structural Health Monitoring*; vol. 10, no. 1, pp. 83-111.

Kim, JT, Stubbs, N 2002, "Improved damage identification method based on modal information" *Journal of Sound and Vibration*, vol. 252, no. 2, pp. 223–238.

Cornwell, P, Doebling, SW, Farrar, CR 1999, 'Application of the strain energy damage detection method to plate-like structures' *Journal of Sound and Vibration*, vol. 224, no. 2, pp. 359-374.

Barroso, L, Rodriguez, R 2004, 'Damage Detection Utilizing the Damage Index Method to a Benchmark Structure' *J. Eng. Mech., ASCE*, vol. 130:2, no. 142, pp. 142-151.

Bonessio, N, Lomiento, G, Benzoni, G 2012 'Damage identification procedure for seismically isolated bridges' *Structural Control and Health Monitoring*; vol. 19, no. 5, pp. 565-578.

Benzoni, G, Bonessio, N, Lomiento, G 2013 *Structural health Monitoring of Bridges with Seismic Response Modification Devices. Report no. SSRP-13/02* Department of Structural Engineering UCSD, San Diego.

Peeters, B, De Roeck, G 2001 'Stochastic system identification for operational modal analysis: a review' *Journal of Dynamic Systems, Measurement, and Control,* vol. 123, no. 4, pp. 659-667.

The 8th International Conference on
Structural Health Monitoring of Intelligent Infrastructure
Brisbane, Australia | 5-8 December 2017

A Baseline-free Modal Strain Energy Method with Bayesian Data Fusion Technique for Damage Localization

U. Baneen[1]
[1] *University of Engineering & Technology, Lahore – Pakistan, Email:* u_baneen@uet.edu.pk

Abstract

Modal Strain Energy Method (MSEM) as compared to other vibration parameters such as mode shapes and modal curvatures, proved to be an effective tool to localise damage in structures. The conventional MSEM requires the information of a baseline model or data from healthy structure in order to identify structural damage. For most of the existing structures the data from their pristine state are not available, additionally, a numerical model representing the baseline model has to be accurate to generate acceptable results. In a previous published work, a baseline-free approach with MSEM was successfully presented to improve the accuracy of damage detection in beam-type structures. The effectiveness of the technique is however, lowered due to the presence of noise in the measurement data. In this paper, a Bayesian data fusion technique is employed to exploit the benefits of the baseline-free approach of MSEM by reducing the noise while highlighting the useful information in terms of damage localization. Various damage scenarios on a beam structure are considered which includes single and multiple damage with smaller to higher severity. Results from the experimental testing on the beams are presented by employing Bayesian data fusion technique and the effectiveness of the technique is highlighted by comparing the results with the previous published work.

1. Introduction

Structural health monitoring is a continuous evolving tool to help ensure the safety of the structures and their critical parts (Mangalgiri & Harinarayana 2017, Janapati et al. 2016). The global dynamic techniques for structural damage identification involve acquiring the dynamic response of the structure which could be in the form of natural frequencies and/or modal displacements. The change in the vibration response could be taken as a confirmation of the presence of structural damage. This step is the first and most important in the structural damage identification process which should be accomplished earliest in order to avoid any catastrophic event and to prevent any huge cost for structural repair.

Several studies have revealed that structural dynamic parameters such as natural frequencies, mode shapes, modal damping and curvature mode shapes, can be used as damage indicator in the structure (Yin, Jiang & Yuen 2017, Chang & Kim 2016, dos Santos et al. 2016). As compared to natural frequencies, mode shapes are better indicator of damage, for the reason that natural frequencies although easier to measure, show slight change even for severe damage (Gillich & Praisach 2014, Zhao et al. 2016, Mishra, Nigam & Dixit 2016). Nonetheless, mode shape based methods are also prone to some limitations. Primarily, mode shapes are easily affected by environmental noises, including ambient loads or inconsistent sensor positions, thus making it insensitive to small damage even with high density mode shape measurements. A novel idea of using the root-mean-square property of response signals to generate power mode shapes was proposed (Fang & Perera 2009). By considering numerical and experimental cases, it was shown that under random excitation the proposed indices have the potential to detect damage in real structures at an early stage. In order to improve the damage detection process, methods based on recurrence plots are compared with classical modal analysis techniques (Iwaniec & Kurowski 2017). It was experimentally verified that these methods are much more responsive to variations in dynamic properties due to damage, as compared to methods based on classical modal analysis.

An alternative to using mode shapes is its second derivative such as modal curvatures (Jaisawal & Pande 2015, Cao et al. 2014, Dessi & Camerlengo 2015). Damage either in the form of crack or slot causes a reduction in flexural stiffness which tends to increase the curvature of the mode shapes in the vicinity of the damage. As mentioned previously, the extracted mode shapes are always contaminated with measurement noise. By taking second difference of mode shapes to acquire curvatures, the noise is also amplified which reduces the accuracy of the curvature based methods. Modal strain energy method (MSEM) can be considered a special case of modal curvature based method, in particular to beam- or plate-like structures. The MSEM utilises a change in fractional modal strain energy for damage detection. The damage index from modal strain energy indicates the degradation of the flexural rigidity of the structure due to damage (Radzienski, Krawczuk & Palacz 2011, Li 2010). As compared to curvature based methods, MSEM has better stability against measurement noise. The method requires a baseline model or data from healthy structure so that damage can be detected by comparing the variations in strain energy from both healthy and damaged structure. This necessity to compare current/measured data to the data from healthy structure/baseline data presents several problems. In most cases, when the data from healthy structure is not available, a baseline numerical model is acquired which has to correspond as well as possible to the real structure (Hemez 2004). However, the uncertainties present in the model along with the errors associated with noise in the measurement data, limit the usefulness of this method.

In previous published work, the author presented a baseline-free approach to improve the accuracy of MSEM (Baneen & Kausar 2017). Various damage scenarios were considered including single and multiple damage with severity as small as 25 percent and as large as 50 percent thickness reduction. The proposed approach showed promising results from both numerical simulation and experimental testing. For each damage scenario, accurate detection was achieved from numerical simulation. While considering the experimental data, the proposed approach was found to generate results with comparatively fewer noisy peaks in the damage index however, damage was not detected in any of the damage scenario. In this paper, in continuation to the previous published work a Bayesian data fusion technique is implemented on the damage indices generated from the proposed baseline-free approach of MSEM. The aim of using Bayesian data fusion technique is to localise both single and multiple damage by suppressing the noise.

2. The Modified MSEM

This method is applicable to beam- and plate-like structures. The details of conventional MSEM and the proposed modified MSEM can be consulted from the previous published work. Similar terms will be used here to define the damage index from modified MSEM. The modified damage index δ_{ij} for the *j*th element in the *i*th mode can be calculated as

$$\delta_{ij} = \left[\left(\frac{\int_{L_j}^{L_{j+1}} [\kappa_i^d(x)]^2 dx}{\int_0^L [\kappa_i^d(x)]^2 dx} \frac{\int_0^L [\kappa_i^f(x)]^2 dx}{\int_{L_j}^{L_{j+1}} [\kappa_i^f(x)]^2 dx} \right) - 1 \right] \times \left| \frac{\kappa_i^f(x)}{\left(\kappa_i^f(x)\right)_{max}} \right| \tag{1}$$

In Eq. 1, κ_i^d represents ith modal curvature of potentially damaged structure and κ_i^f represents the fitted curvature for the ith mode of a hypothetically healthy beam. Both κ_i^f and κ_i^d can be obtained as

$$\kappa_i^f(x) = \sum_{j=0}^{3} \rho_j x^i \tag{2}$$

$$\kappa_i^d(x) = \left(u_i(x+h) - 2u_i(x) + u_i(x-h)\right) / h^2 \tag{3}$$

The coefficients ρ_j in Eq. 2 can be obtained by fitting the measured curvature κ_i^d for ith mode using curvature elements $\kappa_i^d(x-2)$, $\kappa_i^d(x-1)$, $\kappa_i^d(x+1)$, and $\kappa_i^d(x+2)$ (Ratcliffe 2000). The curvature κ_i^d can be acquired from ith modal displacement u_i by using central difference approximation. In Eq. 3, $u_i(x)$ is the ith mode shape obtained at a distance x along the beam length and h is the uniform sampling interval between two measurement points.

3. Bayesian Data Fusion Technique

In a previous published work, the damage indices obtained from modified MSEM showed less noise as compared to the conventional MSEM technique. However, it could not detect the damage in any of the damage scenario considered. These damage indices from each mode provide positive localization if most of their peaks are consistent at the location of the damage. With the aim of reducing the effect of noise a Bayesian fusion process is applied. This technique has been developed and assessed for structural damage identification in quite a few published cases, which can be consulted for more details (Baneen & Guivant 2015 & 2013, Baneen et al. 2012). The likelihood functions representing damage index of each mode are initially defined. For that, the product of likelihoods is performed signifying the fusion of multiple sources of information to construct a probability density function, as described in Eq. 4.

$$p(x) \propto p^0(x) \times \prod_{i=1}^{n} L_i(x) \tag{4}$$

In Eq. 4, the factors $L_i(x)$ correspond to the individual likelihoods acquired from damage index of each mode. The factor $p^0(x)$ is the *a priori* belief regarding the locations of damage. As per previous investigation, the damage would cause greatest change in modal parameters if its location is within 20% of either end of the beam for the first four bending modes (Baneen & Kausar 2017). Thus based on this information, the probability density function is generated by using beta distribution, '*Beta*(α,β)' as *a priori* with shape parameters α = 2 and β = 2. A triangular kernel (or any unimodal kernel) of suitable width is used in convolving the damage index functions so that the likelihoods can be obtained as expressed in Eq. 5,

$$L_i(x) = \int_{v=v_1}^{v_2} k(v)\cdot M_i(x-v)\cdot dv \quad (5)$$

In Eq. 5, $M_i(x)$ is damage index function of each ith mode and $k(v)$ denotes a kernel function.

In order to avoid the inconsistency due to non-modelled critical errors or outliers in the likelihoods, a pruning process is used. The pruning process considers K of the N available individual likelihoods by selecting the K higher values of the N likelihoods assessed at the locations of interest as

$$p(x_m) = p^0(x_m).\prod_{k=1}^{K} L_{i(k)}(x_m), \quad (6)$$
$$L_{i(k)}(x_m) \geq L_j(x_m), \forall j \,/\, j \neq i(r), 1 \leq r \leq K$$

4. Experimental Results

Data from same experimental set up published previously, is used in this paper (Baneen & Kausar 2017). The experimental modal analysis with the set up shown in Fig. 1, was carried out to obtain mode shapes.

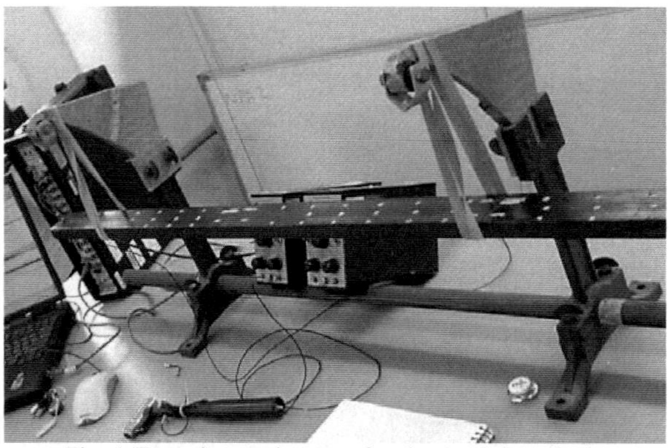

Fig. 1 Experimental set up for the modal analysis

A mild steel beam (*Dimensions*, 1000×50×20 mm; *Young's Modulus, E* = 200 GPa; *Poisson's ratio, v* = 0.3; *density, ρ* = 7850 kg.m^{-3}) was supported in a way to simulate the free-free boundary conditions. For a normalised length of the beam, the damage in the form of crack/slot was created near mid-span at 0.54 from the left-end of the beam. The severity of the damage was increased from 25% (5 mm deep) to 50% (10 mm deep) of the depth of cross section so that the comparison of the increase in severity of the damage could be assessed. For multiple damage case, another crack/slot was introduced at 0.19 from the left-end of the beam.

The displacement mode shapes and hence the modal curvatures, of the beam with different damage scenarios were obtained at 21 points along the beam length and damage indices from the improved MSEM were generated by using Eq. 1. As concluded in the previous work, the damage indices from improved MSEM for each damage scenario as shown in Figs. 2-4, are much smoother, clearer and comparatively less clouded with noisy peaks. However, except first mode, no other mode information is accurate for single damage case. Similarly, for multiple damage case although there are true peaks at damage location in few modes but these are accompanied with some false peaks. The location of damage, *d*, is indicated by means of an arrow in the figures.

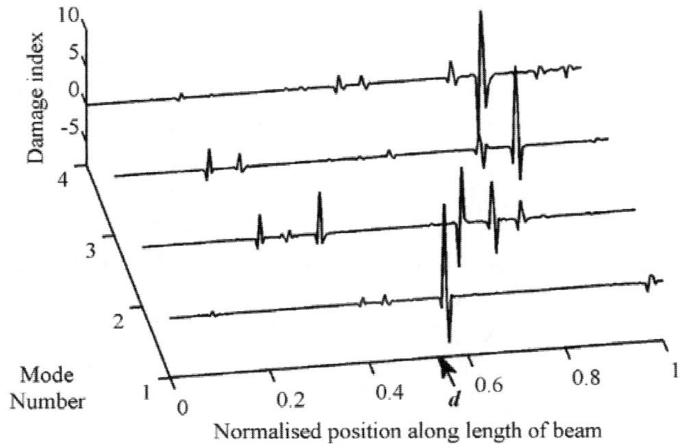

Fig. 2 Damage indices from first four bending modes for 5 mm deep slot, using improved MSEM

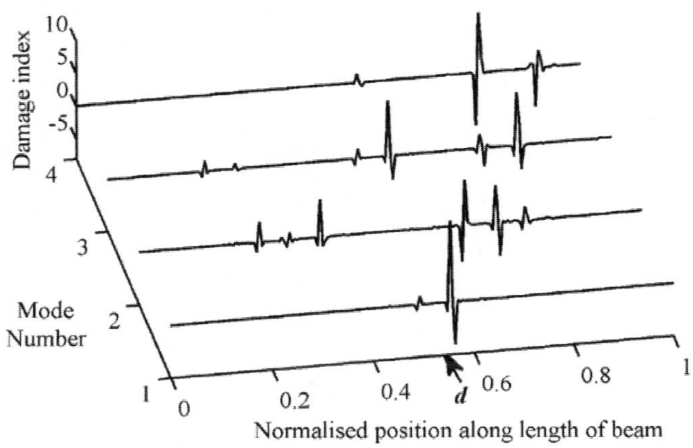

Fig. 3 Damage indices from first four bending modes for 10 mm deep slot, using improved MSEM

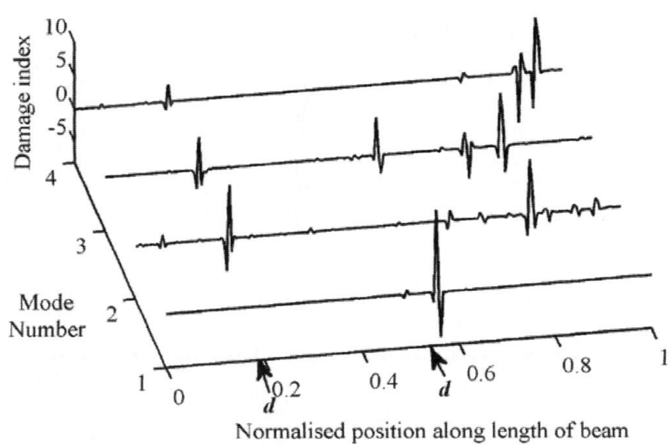

Fig. 4 Damage indices from first four bending modes for 10 mm deep multiple slots, using improved MSEM

The Bayesian data fusion technique is then employed on the damage indices to suppress the false information and highlight useful information in the form of estimated damage indices. The damage indices from the MSEM have both negative and positive peaks, which do not respect the non-negativeness condition for qualifying as likelihood functions. Convolving these damage indices with a positive kernel creates the possibility of concealing the true peaks or making false peaks more prominent. In order to overcome this issue with the MSEM, the damage indices are scaled for satisfying the non-negativeness condition and the likelihoods are generated as

$$L_i(x) = \int_{v=v_1}^{v_2} k(v) \cdot |M_i(x-v)| \cdot dv \tag{7}$$

It should also be stated that the normalised absolute values of the damage indices are used in Eq. 7. The likelihoods acquired from Eq. 7 are then used to obtain the estimated damage indices. The results shown in Figs. 5-7, are acquired by considering maximum likelihoods K for $N = 4$. In order to check the accuracy, the results for each damage scenario are obtained by increasing the number of samples or in other words, reducing the sampling interval by a factor of 2.5, 5 and 10 (shown in Fig. 5-7 as S/2.5, S/5 and S/10, where S is the sampling interval).

With the help of Bayesian data fusion technique, the damage is successfully localised for both single damage cases that is 5 mm and 10 mm deep as can be seen in Fig. 5 and 6, respectively. In these cases, the results from all sample sizes are almost similar, indicating a single high dominant peak at the damage location.

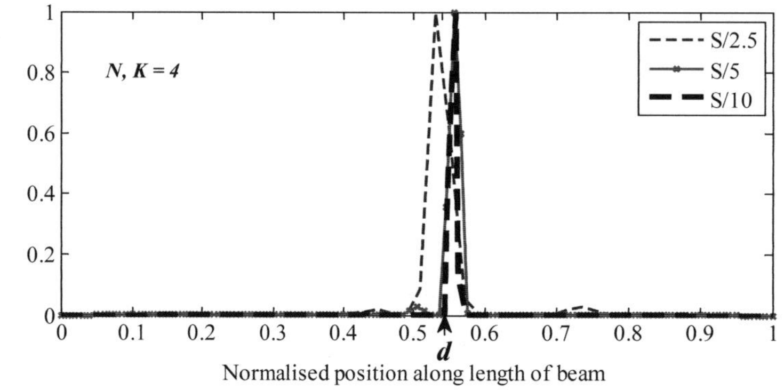

Fig. 5 Estimated damage indices using Bayesian data fusion method – 5 mm deep slot (S: *sampling interval*)

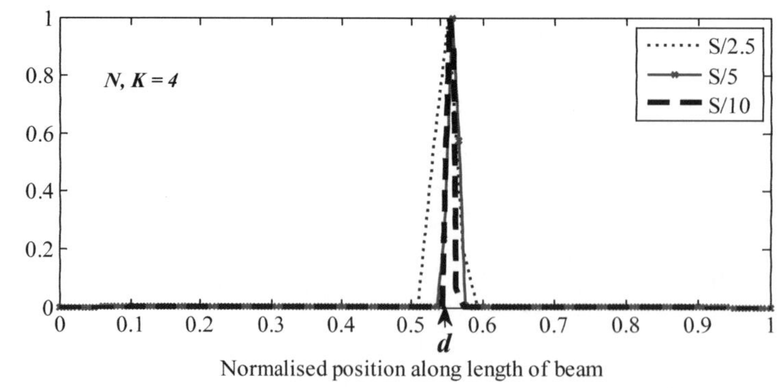

Fig. 6 Estimated damage indices using Bayesian data fusion method – 10 mm deep slot (S: *sampling interval*)

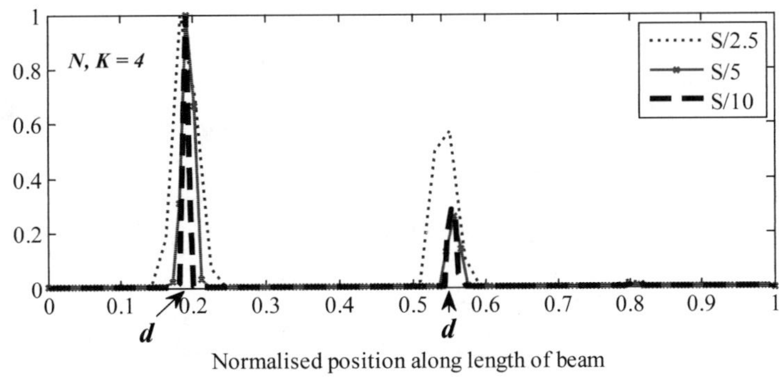

Fig. 7 Estimated damage indices using Bayesian data fusion method – 10 mm deep multiple slots (S: *sampling interval*)

For multiple damage case, both left-end and mid-span damage are clearly detected for all sample sizes as can be seen in Fig. 7. In case of multiple damage, the severity of both slots was equal to 50% thickness reduction and both are successfully localised. Now in order to check whether the developed method works for small size damage too, another case of multiple damage is considered with severity of 2.5% (0.5 mm deep) thickness reduction. For a normalised length of the beam, the locations of both slots are 0.43 and 0.8 from left-end of the beam. Initially conventional MSEM is applied and damage indices are generated as can be seen in Fig. 8. It shows so many false peaks in each mode. Then results are generated by using improved MSEM as shown in Fig. 9. Clearly, the results are comparatively less noisy. However, except the damage indices of the first two modes which indicate the second slot only, none of the others provide any confident result.

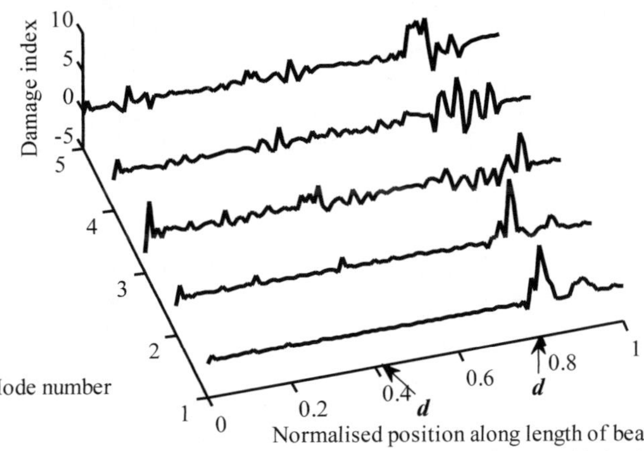

Fig. 8 Damage indices from first four bending modes for 0.5 mm deep multiple slots, using MSEM

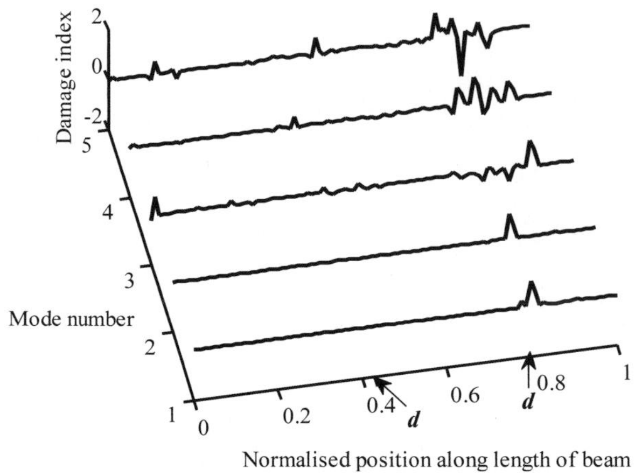

Fig. 9 Damage indices from first four bending modes for 0.5 mm deep multiple slots, using modified MSEM

Now Bayesian data fusion technique is applied to generate estimated damage indices. As this case involves very small damage, hence estimated damage indices are obtained for each K of $N = 4$ likelihoods. As can be seen in Fig. 10, the second slot is clearly visible, however the first slot is not detected for $K = 4$. A small peak for first slot is beginning to appear for $K = 3$, which becomes of equal magnitude in case of $K = 2$.

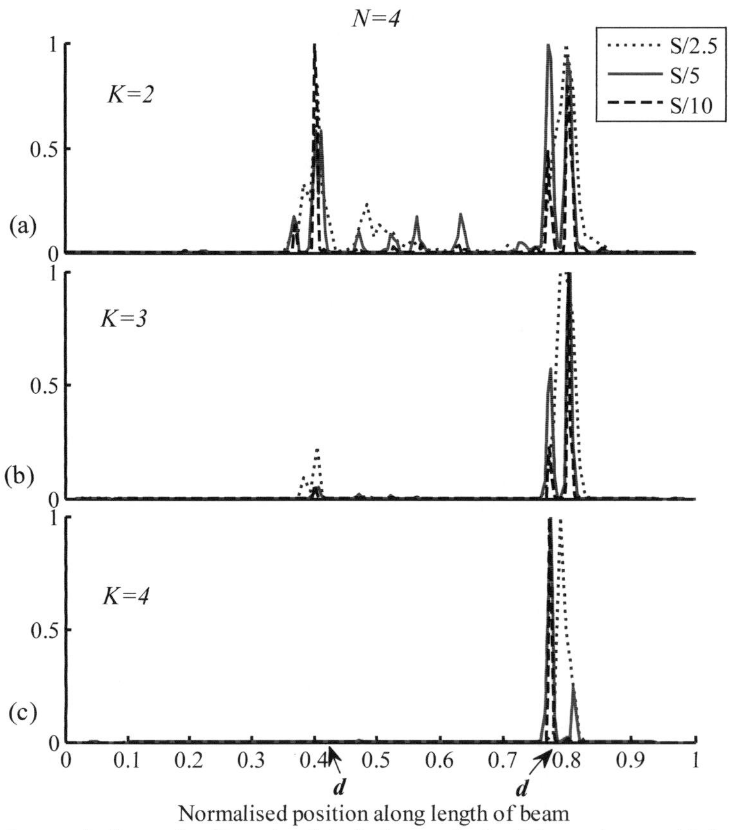

Fig. 10 Estimated damage indices using Bayesian data fusion method – 0.5 mm deep multiple slots (S: *sampling interval*)

5. Conclusions

In this paper, a Bayesian data fusion technique is employed to localise the damage in beam-like structure. In continuation to a previous published work, the damage indices which were generated with less noise, are used here with Bayesian data fusion technique. The different damage scenarios involving the single and multiple damage cases with severity as small as 25% and as severe as 50%, are considered here. After the application of Bayesian data fusion, damage was successfully localised in both single and multiple damage cases. In order to assess the method further, a multiple damage case with very small severity such as 2.5% has also been considered. The developed technique successfully localise one damage while the other damage location was also observable by reducing the likelihoods.

References

Mangalgiri, PD & Harinarayana, K 2017, *Structural health monitoring*, In Aerospace Materials and Material Technologies pp. 449-477, Springer, Singapore.

Janapati, V, kopsaftopoulos, F, Li, F, Lee, SJ & Chang, FK 2016, 'Damage detection sensitivity characterization of acousto-ultrasound-based structural health monitoring techniques', *Structural Health Monitoring*, vol. 15, no. 2, pp. 143-161.

Yin, T, Jiang, QH & Yuen, KV 2017, 'Vibration-based damage detection for structural connections using incomplete modal data by Bayesian approach and model reduction technique', *Engineering Structures,* vol. 132, pp. 260-277.

Chang, KC & Kim, CW 2016, 'Modal-parameter identification and vibration-based damage detection of a damaged steel truss bridge', *Engineering Structures*, vol. 122, pp. 156-173.

dos Santos, FLM, Peeters, B, Van der Auweraer, H, Goes, LCS & Desmet, W 2016, 'Vibration-based damage detection for a composite helicopter main rotor blade', *Case Studies in Mechanical Systems and Signal Processing*, vol. 3, pp. 22-27.

Gillich, GR & Praisach, ZI 2014, 'Modal identification and damage detection in beam-like structures using the power spectrum and time-frequency analysis', *Signal Processing*, vol. 96, pp. 29-44

Zhao, B, Xu, Z, Kan, X, Zhong, J & Guo, T 2016, 'Structural damage detection by using single natural frequency and the corresponding mode shape', *Shock and Vibration*, vol. 2016, Article ID 8194549.

Mishra, R, Nigam, M & Dixit, A 2016, 'Experimental study on structural damage detection based on column like structure using natural frequencies: review', *International Journal of Engineering Development and Research*, vol. 4, no. 2, pp. 1378-1380.

Fang, SE & Perera, R 2009, 'Power mode shapes for early damage detection in linear structures', *Journal of Sound and Vibration*, vol. 324, no. 1, pp. 40-56.

Iwaniec, J & Kurowski, P 2017, 'Experimental verification of selected methods sensitivity to damage size and location', *Journal of Vibration and Control*, vol. 23, no. 7, pp. 1133-1151.

Jaisawal, NG & Pande, DW 2015, 'Sensitizing the mode shapes of beam towards damage detection using curvature and wavelet transform', *International Journal of Scientific & Technology Research*, vol. 4, no. 4, pp. 266-272.

Cao, M, Radzienski, M, Xu, W & Ostachowicz, W 2014, 'Identification of multiple damage in beams based on robust curvature mode shapes', *Mechanical Systems and Signal Processing*, vol. 46, no. 2, pp. 468-480.

Dessi, D & Camerlengo, G 2015, 'Damage identification techniques via modal curvature analysis: overview and comparison', *Mechanical Systems and Signal Processing*, vol. 52, pp. 181-205.

Radzienski, M, Krawczuk, M & Palacz, M 2011, 'Improvement of damage detection methods based on experimental modal parameters', *Mechanical Systems and Signal Processing*, vol. 25, pp. 2169-2190.

Li, YY 2010, 'Hypersensitivity of strain-based indicators for structural damage identification: a review', *Mechanical Systems and Signal Processing*, vol. 24, pp. 653-664.

Hemez, FM 2004, 'Uncertainty quantification and the verification and validation of computational models', *Damage Prognosis for Aerospace, Civil and Mechanical Systems*, pp. 201-220, John Wiley & Sons Ltd, Chichester, England.

Baneen, U & Kausar, Z 2017, 'A baseline-free modal strain energy method for damage localisation', *International Journal of Civil Engineering*, pp. 1-12, doi:10.1007/s40999-017-0149-z.

Ratcliffe, CP 2000, 'A frequency and curvature based experimental method for locating damage in structures', *Journal of Vibration and Acoustics*, vol. 122, no. 3, pp. 324-329.

Baneen, U & Guivant, JE 2015, 'A 2D Bayesian approach for damage detection in plate-type structures', *Insight (Northampton): non-destructive testing and condition monitoring*, vol. 57, no. 3, pp. 144-152.

Baneen, U & Guivant, JE 2013, 'Damage detection in fiber-reinforced composite beams by using a Bayesian fusion method', *Journal of Vibration and Acoustics: Transactions of the ASME*, vol. 135, no. 6, pp. 061008-061008-11.

Baneen, U, Kinkaid, NM, Guivant, JE & Hersberg, I 2012, 'Vibration based damage detection of a beam-type structure using noise suppression method', *Journal of Sound and Vibration*, vol. 331, no. 8, pp. 1777-1788.

The 8th International Conference on
Structural Health Monitoring of Intelligent Infrastructure
Brisbane, Australia | 5-8 December 2017

Experimental Investigation on the Feasibility of Drive-by Pavement Roughness and Bridge Vibration Monitoring

S. Hasegawa [1], P.J. McGetrick [2], C.W. Kim [1], K.C. Chang [1], S. Nakajima [1].
[1] *Kyoto University – Japan, Email:* kim.chulwoo.5u@kyoto-u.ac.jp, [2] *Queens University Belfast – United Kingdom*

Abstract

This study investigates the feasibility of utilizing vehicle accelerations both for bridge condition monitoring and for pavement roughness monitoring, so called drive-by monitoring. This approach copes with disadvantages caused by deployment and maintenance of the vibration sensors on bridges that is generally costly and time-consuming. Another aim of this study is to utilize accelerations of the vehicle for pavement roughness monitoring instead of utilizing data from an expensive sensor system. In this study, laboratory experiments are carried out to investigate feasibility of identifying pavement roughness from the acceleration of the same vehicle used in the drive-by monitoring. This study also discusses the feasibility of bridge monitoring by extracting bridge frequencies from vehicle acceleration. Observations from the laboratory experiment showed the feasibility of detecting the damage of bridge. The roadway surface profile paved on the experimental bridge was also successfully identified although the accuracy was affected by the vehicle speed.

1. Introduction

The demand for monitoring and repairing roads and bridges has increased in recent years. Hence structural integrity inspection and pavement roughness monitoring are two major inspection tasks in terms of preventative bridge maintenance. Visual inspection methods for highway bridges are still widely used, while the road surface profiler is also generally adopted for pavement roughness monitoring. However, it is true that there are great needs to reduce time consuming and labour-intensive inspections inherent to the conventional approaches.

In order to solve this problem, the vibration-based bridge monitoring, so called bridge health monitoring, has been considered to enhance the efficiency of structural integrity inspections (Chupanit. & Phromsorn 2012). In this method, the inspection results are the output of the sensors and do not greatly depend on inspectors' experiences. However, this method of using sensors attached directly on the bridge has also disadvantages such as long sensor installing time and high cost of sensors. Hence a vehicle monitoring system deployed with sensors and a measuring system has been developed to improve the inspection speed. In this way, sensors are installed in an inspection vehicle and the inspection vehicle runs on the bridge or road. When the inspection vehicle is passing, the bridge vibration can transfer to the inspection vehicle and then to sensors in the inspection vehicle. In such a way, it is possible to extract the information of bridges or roads by using the inspection vehicle. This indirect method (Lin & Yang 2005), so called the drive-by bridge inspection (Kim et al. 2014), enables a reduction in the time and cost of monitoring because it is enough to set a few sensors in the inspection vehicle instead of installing on the bridge, and inspection can be conducted only by the inspection vehicle travelling on bridges. If high accuracy of this indirect method is verified, it can be an improved and balanced method of monitoring.

Table 1 Nomenclature

$\mathbf{C_v}$	Damping matrix of vehicle model
$C_{s,i}$	Damping constant of axle *i*
$\mathbf{d}$	Measured acceleration vector
D_i	Distance from axle *i* to centre of gravity
Err	Sum of Least square error
$\mathbf{f_v}$	Dynamic interaction force vector of vehicle
$F_{t,i}$	Dynamic wheel contact force under wheel *i*
$\mathbf{g}$	Unknown force vector of vehicle
$\mathbf{I}$	Identity matrix
$\mathbf{K_v}$	Stiffness matrix of vehicle model
$K_{s,i}$	Spring constant of axle *i*
$\mathbf{M_v}$	Mass matrix of vehicle model
$\mathbf{Q}$	Selection matrix
$w_{v,i}$	The total displacement at under wheel *i*
$\mathbf{X}$	Zero-th order state variable vector
$\hat{\mathbf{X}}$	First order state variable vector
$\mathbf{y_v}$	Displacements vector of vehicle
λ	Regularization parameter

The drive-by inspection method has been studied in recent years focusing on extraction of bridge frequency (Yang et al. 2004), damage detection (Bu et al. 2006), road profile identification (McGetrick et al. 2013), and so on. There are a few researches on the drive-by inspection by mathematical verification or simulation. However, experimental verification (Yang & Chang 2009) has not been conducted widely yet. In this study, a scaled model experiment is conducted in the laboratory in order to confirm the feasibility of the drive-by inspection as a monitoring method for bridges and roads.

2. Road Profile Identification

2.1 Algorithm of road profile identification

In the algorithm adopted here (McGetrick et al. 2013), a 2 degree-of-freedom (DOF) half-car model is used, with the 2 DOF corresponding to vehicle body bounce (vertical translation) and pitch rotation . The equation of motion of the 2 DOF half-car model is shown in Eq. (1). Symbols used hereafter are explained in **Table 1**, where the variable of *i* represents the number of axle (*i*=1 means front axle and *i*=2 means rear axle).

$$\mathbf{M_v\ddot{y}_v} + \mathbf{C_v\dot{y}_v} + \mathbf{K_v y_v} = \mathbf{f_v} \quad (1)$$

Here, $\mathbf{f_v}$ contains dynamic vehicle forces corresponding to vertical translations and pitch rotations. This equation of motion can be converted to Eq. (2) by using state-space formulation.

$$\dot{\mathbf{X}} = \mathbf{AX} + \overline{\mathbf{f_v}} \quad (2)$$

Here, $\mathbf{X}$, $\mathbf{A}$ and $\overline{\mathbf{f_v}}$ are defined as Eq. (3)

$$\mathbf{X} = \begin{Bmatrix} \mathbf{y_v} \\ \dot{\mathbf{y}}_\mathbf{v} \end{Bmatrix}, \quad \mathbf{A} = \begin{bmatrix} \mathbf{0} & \mathbf{I} \\ -\mathbf{M_v}^{-1}\mathbf{K_v} & -\mathbf{M_v}^{-1}\mathbf{C_v} \end{bmatrix}, \quad \overline{\mathbf{f_v}} = \begin{bmatrix} \mathbf{0} \\ \mathbf{M_v f_v} \end{bmatrix} \quad (3)$$

Moreover, Eq. (3) can be rewritten to Eq. (4) by using the standard exponential matrix representation. This is referred as a zeroth order system (Busby & Trujillo 1997).

$$\mathbf{X}_{j+1} = \mathbf{M}\mathbf{X}_j + \mathbf{G}\mathbf{g}_j \quad ; \; j = 1, \; \dots, \; N \tag{4}$$

Here, $\mathbf{M} = e^{\mathbf{A}h}$ where h is the time step. $\mathbf{g}_j$ is a vector of vehicle forces, which is target of identification, and represented as Eq. (5).

$$\mathbf{g}_j = \{-F_{t,1} - F_{t,2}\, , -D_1 F_{t,1} + D_2 F_{t,2}\}^T \tag{5}$$

where, $F_{t,i} = K_{s,i} w_{v,i} + C_{s,i} \dot{w}_{v,i}$ (i=1 and 2 for the front and rear axles respectively).
Matrix **G** is expressed as shown in Eq. (6).

$$\mathbf{G} = \left(\mathbf{A}^{-1}(\mathbf{M} - \mathbf{I})\right) \begin{bmatrix} \mathbf{0} \\ \mathbf{M_v}^{-1}\mathbf{I} \end{bmatrix} \tag{6}$$

In respect of regularization terms, a first order system which regularizes the derivative of the unknown forces can give a more precise and smooth solution compared to a zeroth order system. Hence Eq. (4) is converted to Eq. (7) by defining the new state variable vector, $\widehat{\mathbf{X}}_j = [\mathbf{X}_j \quad \mathbf{g}_j]^{\mathrm{T}}$.

$$\widehat{\mathbf{X}}_{j+1} = \begin{bmatrix} \mathbf{M} & \mathbf{G} \\ \mathbf{0} & \mathbf{I} \end{bmatrix} \widehat{\mathbf{X}}_j + \begin{bmatrix} \mathbf{0} \\ \mathbf{I} \end{bmatrix} \dot{\mathbf{g}}_j \;\; ; \;\; j = 1, \; \dots, \; N \tag{7}$$

If the measured acceleration vector is defined as $\mathbf{d}_j$, the relationship between $\mathbf{d}_j$ and $\widehat{\mathbf{X}}_j$ can be written as $\mathbf{d}_j = \mathbf{Q}\widehat{\mathbf{X}}_j$ by using the selection matrix **Q**. The least square minimization problem is defined by utilizing Tikhonov regularization (Tikhonov & Arsenin 1977). The error sum is formulated as shown in Eq. (8).

$$Err(\widehat{\mathbf{X}}_k, \mathbf{r}_k) = \sum_{k=1}^{m} (\mathbf{d}_k - \mathbf{Q}\widehat{\mathbf{X}}_k, \mathbf{I}(\mathbf{d}_k - \mathbf{Q}\widehat{\mathbf{X}}_k)) + (\dot{\mathbf{g}}_k, \lambda\, \mathbf{I}\dot{\mathbf{g}}_k) \tag{8}$$

For two example vectors of **x** and **y**, $(\mathbf{x}, \mathbf{y})$ means the vector product as $(\mathbf{x}_k, \mathbf{y}_k) = \sum_{j=1}^{N} \mathbf{x}_{k,j}\, \mathbf{y}_{k,j}$ for each k. The reason of introducing the Tikhonov first regularization is that this inverse problem of road profile identification is an ill-posed problem. In other words, the values of $\widehat{\mathbf{X}}_j$ are too sensitive to the values of $\mathbf{d}_j$ and slight changes in values of $\mathbf{d}_j$ can make a big difference in the output of road profile values.
The optimal value of λ can be determined by L-curve method (Hansen 1992). In this method, firstly the array of candidates of λ is defined and E_{norm} and F_{norm}, shown in Eq. (9) and Eq. (10) respectively, are calculated for every values of λ. Then all pairs of E_{norm} and F_{norm} are plotted in the log-log graph and they are connected to make one curve line. Finally, the value of λ at the point of the largest positive slope in the curve is chosen as the optimum value.

$$E_{norm} = \sqrt{\sum_{k=1}^{m} (\mathbf{d}_k - \mathbf{Q}\widehat{\mathbf{X}}_k, \mathbf{I}(\mathbf{d}_k - \mathbf{Q}\widehat{\mathbf{X}}_k))} \tag{9}$$

$$F_{norm} = \sqrt{\sum_{k=1}^{m} (\dot{\mathbf{g}}_k, \dot{\mathbf{g}}_k)} \tag{10}$$

By the L-curve method, the optimum values of $\widehat{\mathbf{X}}_j$ and $\dot{\mathbf{g}}_j$ are determined, and $\mathbf{g}_j$ is also identified as a part of $\widehat{\mathbf{X}}_j$. Finally, the displacements under the wheel i are calculated by solving Eq. (11) using the Runge-Kutta method.

$$\mathbf{g} = \begin{Bmatrix} \sum_{i=1}^{2} (K_{s,i} w_{v,i} + C_{s,i} \dot{w}_{v,i}) \\ -\sum_{i=1}^{2} (-1)^i D_i (K_{s,i} w_{v,i} + C_{s,i} \dot{w}_{v,i}) \end{Bmatrix} \quad (11)$$

2.2 Experiment model of bridge

The scaled bridge model used in the experiment is shown in **Fig. 1**. The main observation span (I girder span) is 5.4 m as shown in the **Fig. 2**. The test vehicle runs on a pair of rails (right and left) paved with road profiles on their surfaces. These road profiles were modelled by scaling the real road profile on a bridge with span length of 40.4 m (Kim et al. 2005). The model road profiles were scaled at 1/10 of the original profiles. The properties of the scaled bridge model are shown in **Table 2** and the natural frequencies and damping ratio of the scaled model bridge were identified via free vibration tests.

Fig. 1 Picture of scaled model bridge

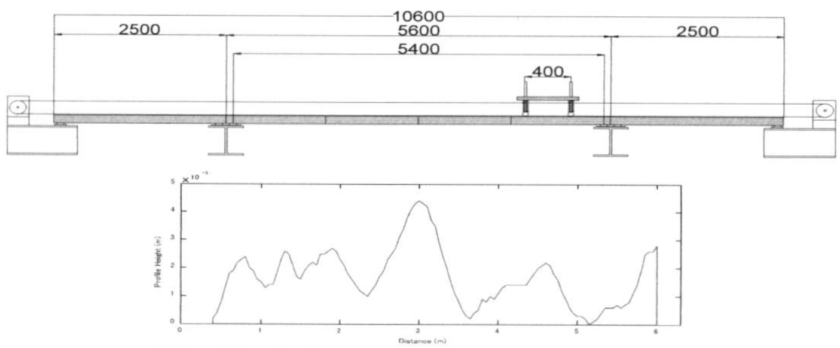

Fig. 2 Scaled model bridge and road profiles

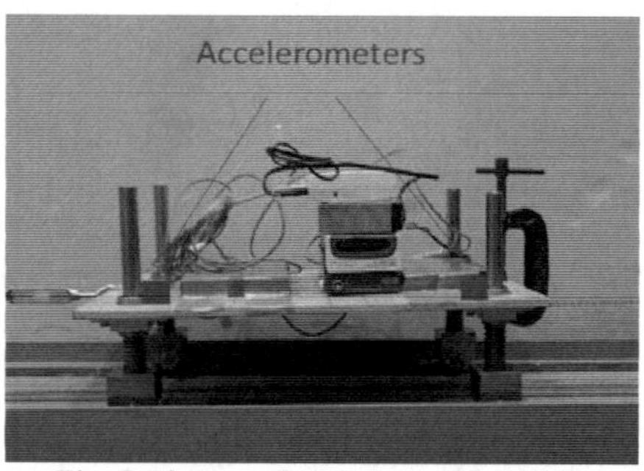

Fig. 3 Picture of scaled model vehicle

Table 2 Model bridge properties

Span length	5.4(m)
Young's Module	2.1×10^{11}(N/m^2)
Moment of inertia of Area	11.36×10^{-7}(m^4)
1st theoretical natural frequency	3.47 (Hz)
2nd theoretical natural frequency	13.86 (Hz)
Damping ratio	0.011

Table 3 Model vehicle properties

Mass (kg)	Total	19.67
	Front	7.39
	Rear	12.28
Spring constant (N/m)	Front	3697
	Rear	6141
Natural frequency (Hz)	Front	3.56
	Rear	3.56
	Bounce	3.56
Suspension damping ratio	Front	0.056
	Rear	0.052

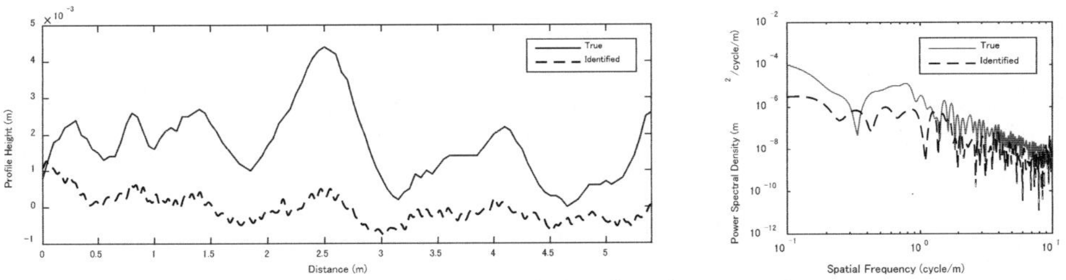

(a) Speed S1 = 0.55 m/s (continued next page)

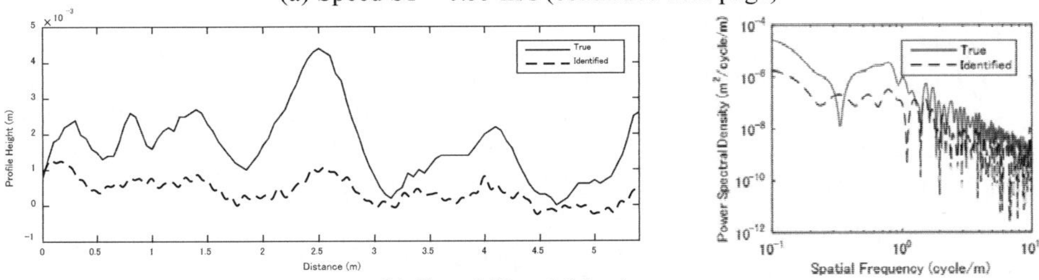

(b) Speed S2 = 1.05 m/s

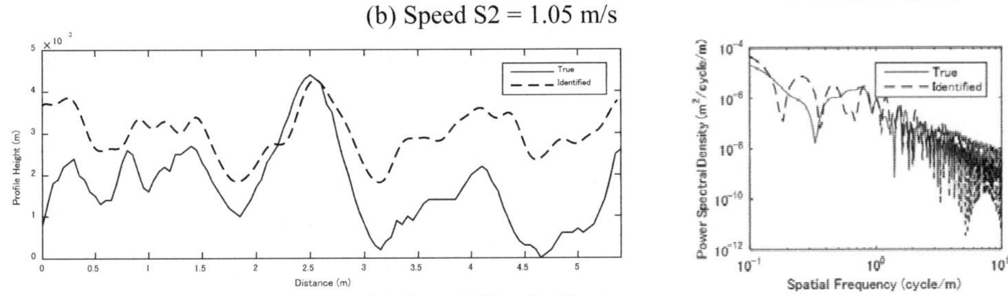

(c) Speed S3 = 2.45 m/s

Fig. 4 Experiment results of identified road profiles and PSD

2.3 Experiment model of vehicle

The photo of the scaled vehicle model is shown in **Fig. 3**. The model vehicle is composed of steel plate body, axles, springs, data logger (DC-104R model by Tokyo Sokki Kenkyujo Co.), wireless router, mass, two accelerometers, battery and vice. The accelerometers were attached respectively on the front and rear axles as shown in **Fig. 3**. The sampling frequency of measuring the accelerations is 100 Hz. The properties of vehicle model are shown in **Table 3** and the damping ratios were identified by the free vibration test. The model vehicle is driven by an external motor connected with a wire. The times of entry and exit of the model vehicle were detected by the optical sensors installed at both ends of 5.4 m the observation span. Three vehicle speeds were tested; S1=0.55 m/s, S2=1.05 m/s, and S3=2.45 m/s.

2.4 Experiment results

An experiment of road profile identification was conducted. The experiments were carried out using the scaled model vehicle and bridge outlined in section 2.3. The comparisons of true and identified road profiles for each speed are shown in **Fig. 4**. Comparisons of the power spectral densities of true and identified road profiles for each speed are also shown in **Fig. 4**. Roughness profile was well identified under the speed S3 in both of simulation and experiment. However, for other vehicle speed conditions differences between the true and identified profiles were observed.

Observations demonstrated that among three speed conditions higher the vehicle speed is higher the accuracy of profile identification becomes. Similar trends were observed in PSD, i.e. the accuracy of identification is better at high vehicle speed than low vehicle speed. It also should be noted that the road profile identified by the algorithm contains the dynamic and static displacements of the bridge. It means that there is a potential to extract some bridge information, i.e. frequency or bridge displacements.

3. Damage detection

3.1 Theoretical background

In this section, the theoretical background for damage detection is outlined, while more details are available in the work by Kim et al. (2016). Two vehicles were modeled as sprung masses $m_{v,1}$ and $m_{v,2}$, each supported by a spring of spring constant $k_{v,1}$ and $k_{v,2}$, as shown in **Fig. 5**. They were assumed to move at a constant speed v. The road surface roughness is defined as:

$$r(x) = \sum_i d_i \cos(n_i vt + \theta_i) \tag{12}$$

where $n_{s,i}$ is the ith spatial frequency, d_i the amplitude, and θ_i the random phase angle. Neglecting the damping effects of vehicles and bridge for simplicity, the equations of motion for the two vehicles and bridge can be written as

$$m_v\ddot{y}_{v1}(t) + k_v(y_{v1}(t) - y_b(x_1, t) - r(x_1)) = 0 \tag{13}$$

$$m_v\ddot{y}_{v2}(t) + k_v(y_{v2}(t) - y_b(x_2, t) - r(x_2)) = 0 \tag{14}$$

$$m_b\ddot{y}_b + EIy_b = f_{c,1}(t)\delta(x - x_1) + f_{c,2}(t)\delta(x - x_2) \tag{15}$$

where y_{v1}, y_{v2} and y_b are the displacement response of the front vehicle, rear vehicle and bridge, $f_{c,1}(t)$ and $f_{c,2}(t)$ the contact forces for the front vehicle and rear vehicle respectively. Solving these differential equations by the mode-superposition method, one can get the displacement responses. Differentiating the displacements twice, one can further get the acceleration for a specific vehicle as

$$\begin{aligned}\ddot{y}_v(t) = \sum_{n=1}^{\infty}\{&A_{1,n} \cos((n-1)l(t)) - A_{2,n} \cos((n+1)l(t)) + A_{3,n} \cos \omega_v t \\ &- A_{4,n} \cos(\omega_n t - nl(t)) + A_{5,n} \cos(\omega_n t + nl(t))\} \\ &+ \sum_i d_i \frac{\omega_v^2}{\omega_v^2 - (n_i v)^2}\{-(n_i v)^2 \cos(n_i vt + \theta_i) \\ &+ \omega_v^2 \cos\theta_i \cos\omega_v t + n_i v\omega_v \sin\theta_i \sin\omega_v t\}\end{aligned} \tag{16}$$

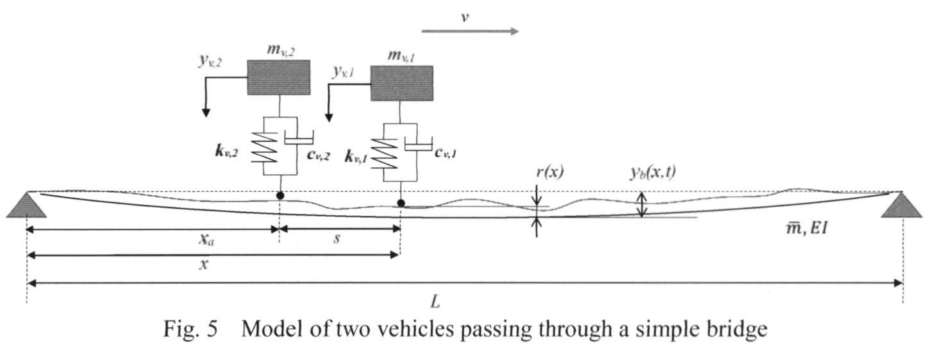

Fig. 5 Model of two vehicles passing through a simple bridge

Herein the details of coefficients $A_{1,n}$, $A_{2,n}$, $A_{3,n}$, $A_{4,n}$ and $A_{5,n}$ are omitted but available in Yang et al. (2012). It is shown in Eq. (16) that the bridge-related responses (of frequency $\omega_n t \pm nl(t)$, where $l(t)=\pi vt/L$) is carried in the vehicle responses, which motivates the development of the drive-by method. To remove or reduce the effect of the roughness-induced vibrations, it was proposed to subtract their Fourier spectra from each other, in acknowledging that theoretically the roughness-induced vibration components in the vehicles' responses are the same when passing the same contact point on the bridge (see Eqs. (17) and (18) below for roughness-related terms).

$$\ddot{R}_{v1,r}(t) = \sum_i d_i \frac{\omega_{v,1}^2}{\omega_{v,1}^2 - (n_i v)^2} \{-(n_i v)^2 \cos(n_i vt + \theta_i)\} \tag{17}$$

$$\ddot{R}_{v2,r}(t) = \sum_i d_i \frac{\omega_{v,2}^2}{\omega_{v,2}^2 - (n_i v)^2} \{-(n_i v)^2 \cos(n_i vt + \theta_a)\} \tag{18}$$

Here θ_a is defined as $\theta_a = n_i s + \theta_i$. It has been shown that the vehicle responses carry bridge-related vibrations. When the bridge is damaged, its dynamic characteristics may be changed, and such a changing characteristic could be carried in and detected from the vehicle responses. This technique for detecting bridge damage from the vehicle responses were tested in the following laboratory experiment.

Fig. 6 Scaled model tractor and trailers

Table 4 Model trailer properties

Mass (kg)	Total	6.70
	Front	3.35
	Rear	3.35
Spring constant (N/m)	Front	719
	Rear	719
Natural frequency (Hz)	Front	2.49
	Rear	2.49
	Bounce	2.49
Suspension damping ratio	Front	0.057
	Rear	0.059

Table 5 Damage scenarios

Scenario	Mass (kg)	1st natural frequency (Hz)	Damping ratio
INT (Intact)	0	3.64	0.011
DMG1 (Damage1)	8.3	3.5	0.009
DMG2 (Damage2)	12.8	3.47	0.008
DMG3 (Damage3)	17.2	3.42	0.007

3.2 Experiment model

In the experiment for damage detection by the moving vehicle, train of vehicles composed of one tractor and two trailers as shown in **Fig. 6** was adopted. The tractor is same as the vehicle outlined in section 2. The two trailers have same dimensions and properties. **Table 4** shows properties of the trailers. The tractor and trailers are connected in order of tractor, trailer 1 and trailer 2. The accelerometers were installed on the front axle and rear axle in the same way as shown in **Fig. 3**. The acceleration measured at "trailer *i*" is called as "acc *i*", where the "acc *i*" indicates the mean values of the accelerations measured at the front and rear axles of "trailer *i*".

There are two purposes to use not only the tractor but also trailers for the experiment. The tractor serves as an actuator and trailers can serves as moving sensors receiving the bridge vibrations excited by the tractor. For the drive-by bridge inspection, if the target bridge is well excited it is expected that much more information on bridge vibrations will be included in the vibrations measured at the trailers. The reason for utilizing two trailers is to eliminate influence of roadway roughness on measured accelerations. To eliminate the influence of roadway roughness in the measured accelerations of trailers, first the responses of two trailers at the same contact point are synchronized, and then the acceleration Fourier spectrum of one trailer is subtracted from the other as discussed theoretically in section 3.1. This enables the highlighting of bridge vibration components in the measured accelerations. The scaled model bridge and road profile model is same with ones explained in section 2 and its property has been given in **Table 3**.

3.3 Experiment scenarios

In the damage experiment, the bridge damage was imitated by adding masses as additional mass leads to decrease of natural frequency of the bridge. The additional mass was placed at the middle of the observation span of the bridge. Three different damage severities were considered as shown in as **Table 5**, where INT indicates intact case without additional mass with bridge. For the damage scenarios DMG1, DMG2 and DMG3, different additional masses were considered as shown in **Table 5**. Changes in the natural frequency and damping ratio of the bridge due to additional mass are also shown in **Table 5**. The speed of the vehicles was S2=1.05 m/s.

The damage may change the spectral distribution pattern of the test vehicle acceleration responses, which could serve as a damage sensitive feature. In this study Mahalanobis distance (MD) (Taguchi & Jugulum 2000) was used to quantify the change in pattern.

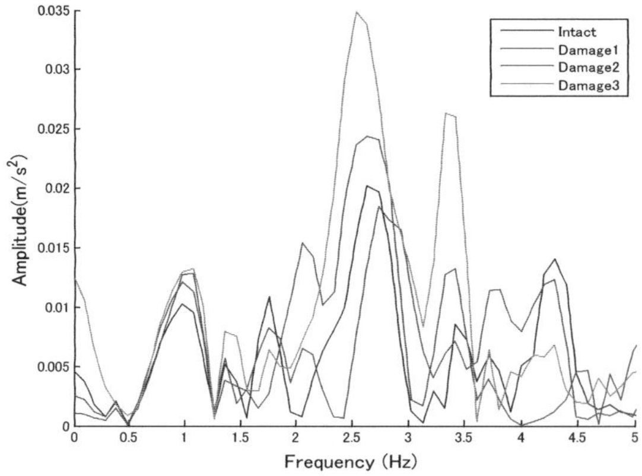

Fig. 7 Comparison of the damage scenarios FFT[acc1]-FFT[acc2]

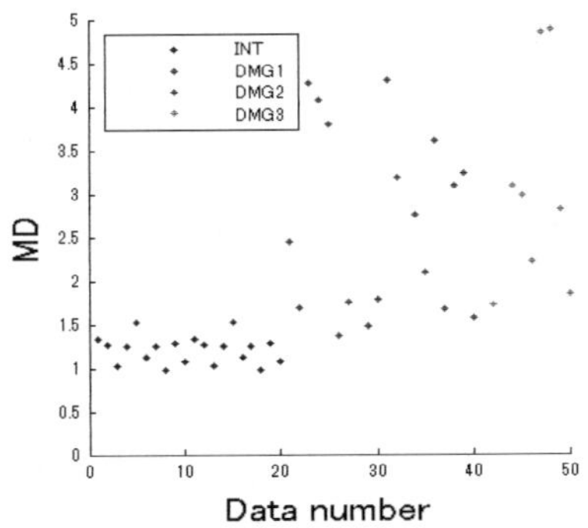

(a) MDs of FFT[acc1]-FFT[acc2] (3.0- 3.5 Hz)

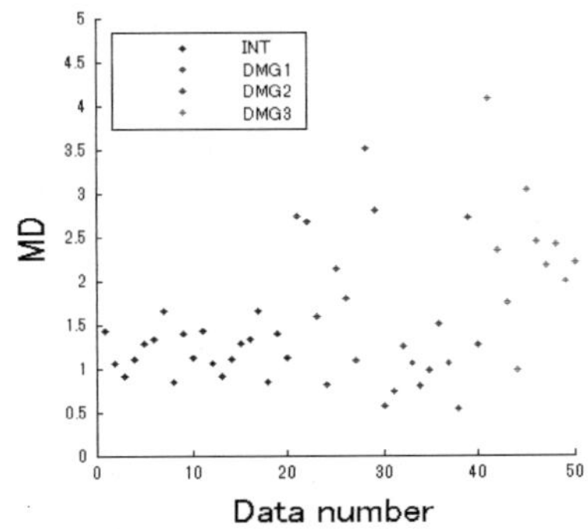

(b) MDs of FFT[acc1]-FFT[acc2] (3.5- 4.0 Hz)

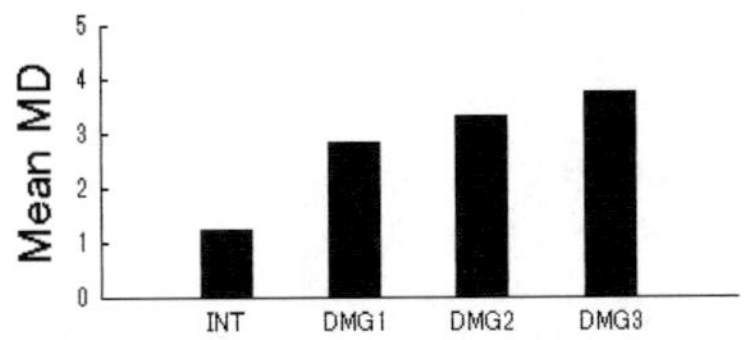

(c) Mean MDs of FFT[acc1]-FFT[acc2] (3.0- 3.5 Hz)

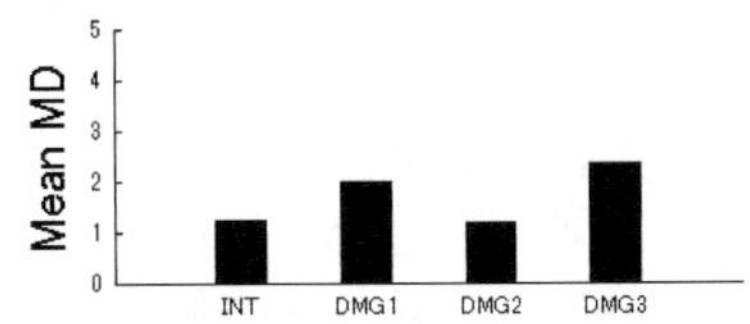

(d) Mean MDs of FFT[acc1]-FFT[acc2] (3.5- 4.0 Hz)

Fig.8 Damage detection results by MD

The measured accelerations are converted to the Fourier spectrum and MD of areas of Fourier spectra between frequencies is calculated in each frequency band at intervals of 0.5 Hz. The MD values of each damage scenario are compared with MD values of INT scenario. If MD values are deviated from those of INT scenario it means changes in integrity of the observed bridge.

3.4 Experiment results

Fig. 7 shows the residual spectra, FFT[Acc1]−FFT[Acc2], obtained by subtracting one spectrum from the other one. In **Fig. 7**, the changes of spectra due to the damage level changes are observed, especially in the band of 2.5-4.0 Hz, where the bridge natural frequency lies.

Fig. 8 shows the MDs estimated from laboratory experiments for FFT[Acc1]-FFT[Acc2], taking the amplitudes in the frequency range of 3.0 Hz to 3.5 Hz and 3.5 Hz to 4.0 Hz as variables respectively. These ranges were selected because the bridge natural frequencies in the four scenarios were around 3.5 Hz. 20 runs of the test for Intact case and 10 runs for Damage1, Damage2 and Damage3 were conducted. It is observed that MDs of FFT[Acc1]-FFT[Acc2] in 3.0 Hz to 3.5 Hz showed noticeable changes between the test scenarios. It is also obvious that the MD value increases as more mass added, hence the damage detection by utilizing MD has a potential to identify not only the existence of damage but also damage levels.

4. Conclusions

In this research, the feasibility of the drive-by inspection was investigated experimentally. Observations from the study are summarized as follows.

(1) It was possible to identify the road roughness profile utilizing acceleration responses of the vehicle.

(2) The accuracy of roadway road roughness identification was better at high vehicle speed than low vehicle speed, comparing three different vehicle speeds.

(3) It was possible to detect the changes in natural frequency of bridge with an acceptable range of accuracy by running the test vehicle linked with two trailers.
(4) Taking the observations from the intact case as reference and comparing MDs, MDs of areas of Fourier spectra between frequencies increased as damage level increased, and demonstrated feasibility of the fault detection by the drive-by monitoring.

The next step for this study is to increase the accuracy of the roadway roughness identification algorithm and carry out a field experiment on a real bridge with real vehicles.

References

Bu, Q.J., Law, S.S. & Zhu, Q.X., 2006, 'Innovative bridge condition assessment from dynamic response of a passing vehicle', *J. of Engineering Mechanics*, vol.132, no.12, pp.1372-1379.

Busby, H.R. & Trujillo, D.M., 1997, 'Optimal regularization of an inverse dynamics problem', *Computers and Structures*, vol.63, no.2, pp.243-248.

Chupanit, P. & Phromsorn, C., 2012, 'The importance of bridge health monitoring', *Int'l Scholarly and Scientific Resarch and Innovation*, vol.6, no.6, pp.389-392.

Hansen, P.C., 1992, 'Analysis of discrete ill-posed problems by means of the L-curve', *SIAM Review*, vol.34, no.4, pp.561-580.

Kim, C.W., Kawatani, M. & Kim, K., 2005, 'Three dimentional dynamic analysis for bridge-vehicle interaction with road way roughness', *Computers and structures*, vol.83, pp.1627-1645.

Kim, C.W., Inoue, S., Sugiura, K., McGetrick, P.J. & Kawatani, M., 2016, 'Extracting bridge frequencies from dynamic responses of two passing vehicles', *Insights and Innovations in Structural Engineering, Mechanics and Computation*, vol.6, pp.1858-1864

Kim, C.W., Isemoto, R., McGetrick, P.J., Kawatani, M., & OBrien, E.J., 2014, 'Drive-by bridge inspection from three different approaches', *Smart Structures and Systems*, vol.13, no.5, pp.775-796.

Lin, C.W. & Yang, Y.B., 2005, 'Use of a passing vehicle to scan the fundamental bridge frequencies: An experimental verification', *Engineering Structures*, vol.27, no.13, pp.1865-1878.

McGetrick, P.J., Kim, C.W., González, A. & OBrien, E.J., 2013, 'Dynamic axle force and road profile identification using a moving vehicle', *Int'l J. of Architecture, Engineering and Construction*, vol.2, no.1, pp.1-16.

Taguchi, G. & Jugulum, R., 2000, 'New trends in multivariate diagnosis', *Indian J. of Statistics*, vol. 62(B), pp.233-248.

Tikhonov, A.N. & Arsenin, V.Y., 1977, *Solutions of Ill-posed Problems*, New York: Wiley.

Yang, Y.B., Li, Y. & Chang, K.C., 2012, 'Effect of road surface roughness on the response of a moving vehicle for identification of bridge frequencies', *Interaction and Multiscale Mechanics*, vol.5, iss.4, pp.347-368.

Yang, Y.B., Lin, C.W. & Yau, J.D., 2004, 'Extracting bridge frequencies from the dynamic response of a pssing vehicle', *J. of Sound and Vibration*, vol.272, iss.3-5, pp.471-493.

Yang, Y.B. & Chang, K.C., 2009, 'Extraction of bridge frequencies from the dynamic response of a passing vehicle enhanced by the EMD technique', *J. of Sound and Vibration,* vol.322, pp.718-739.

The 8th International Conference on
Structural Health Monitoring of Intelligent Infrastructure
Brisbane, Australia | 5-8 December 2017

A damage index method for detection of localized corrosion in reinforcement bars using scattering of longitudinal guided waves

Rajeshwara C. S. [1,2,3], Sauvik Banerjee [2], Ye Lu [3]

*[1]IITB-Monash Research Academy, Indian Institute of Technology Bombay, Powai, Mumbai 400076- India, Email:*rcsriramadasu@gmail.com, *[2] Department of Civil Engineering, Indian Institute of Technology Bombay, Powai, Mumbai 400076 - India, [3] Department of Civil Engineering, Monash University, Clayton, VIC 3800 - Australia*

Abstract

Corrosion of steel is a major cause of failure in reinforced concrete structures. Recently, the focus of research has been to identify different kinds of damage that could occur simultaneously in reinforced concrete. In the recent literature, it is reported that the presence of damage such as corrosion and debond in reinforced concrete could lead to a phenomenon called zero effect state due to their complementary effect on wave characteristics. In this paper, two different damage scenarios in reinforcement are considered. It is demonstrated that the guided wave mode L(0, 1) can be used to assess the level of pitting corrosion in the reinforcement bars. A novel damage index is developed using the reflection and transmission signatures of wave modes from the front and rear edges of the corroded region to assess the corrosion of steel rebars. The damage index at various points on the rebar is plotted, which gives the information on the location, intensity, and axial extent of the corrosion. The numerical modelling is carried out using Abaqus/CAE and results for various levels of corrosion are validated through laboratory experiments. The study is further extended to determine the minimum quantifiable corrosion size for the current experimental specimen.

1. Introduction

Steel bars are extensively used as reinforcement and as tendons in concrete structures. Corrosion of steel is identified as a major cause of failure in these structures. Guided waves have found their use in monitoring of mechanical and aerospace structures (Lu et al. 2007; Lu et al. 2014; Sakamoto et al. 2013; Sikdar et al. 2015) and are currently being tested for their applicability in monitoring corrosion of rebars in reinforced concrete (Sharma & Mukherjee 2010; Miller et al. 2013; Lu et al. 2013; Mustapha et al. 2014). It is widely believed that guided waves have the advantage over conventional chemical and mechanical methods due to their long range of inspection.

Guided wave features such as amplitude, time of flight are widely used to identify corrosion in the reinforced concrete (Wu & Chang 2006; Miller et al. 2013). Time of flight obtained by using cross-correlation technique is used to detect the existence of uniform corrosion in a rebar by few researchers. The difference in time of flight of an undamaged rebar and damaged rebar has been used to confirm that the TOF varies with the diameter of the rebar (Amjad et al. 2015). Few researchers have tried to use principal components as damage indices however such methods are limited in their capability to identify the location of damage (Lu et al. 2013).

In recent studies, it is reported that corrosion process leads to a reduction in diameter and debond. These two features were reported to exhibit a complementary effect on wave characteristics (Ervin & Reis 2008). It is also reported that the presence of two different types of damage can lead to

concealment of damage (Rajeshwara et al. 2017). Hence, it is necessary to develop methods which can distinguish different damages and feasibly identify concealed damage.
Studies indicate that excessive acidic environment often leads to pitting type of corrosion in which the corrosion is limited to a small length of the reinforcement bar (Cao & Cheung 2014; Masc 2015). Pitting corrosion is recognised to be more dangerous than uniform corrosion, as it reduces the cross section of reinforcement to a point where rebar fails to carry the load and can lead to a catastrophic failure(Ahmad 2003). The methods developed for uniform corrosion such as time of flight can no longer be applied for pitting corrosion as the time of flight and arrival times are not affected much. There is an imminent need to develop a new method for detecting the pitting type of corrosion in rebars.
In this study, an attempt is made to identify parameters that can be used to detect pitting corrosion in rebars. Guided wave characteristics for various states of corrosion in reinforcement bars are examined. A damage index method developed by utilising the scattered wave modes is presented.

2. Wave propagation in reinforcement bar with damage

Wave motion is particle movement coupled with the transfer of energy from one particle to subsequent particle. The concept of wave propagation has been extensively used to study the mechanical behaviour of matter. Rayleigh and Lamb are recognised for the discovery of guided waves and Lamb waves respectively (Rayleigh 1885; Lamb 1917). In this section, wave propagation in reinforcement bar is presented and various scattering modes generated due to wave interaction with damage are explained.
The wave propagation in a cylindrical waveguide is governed by equation (1)

$$(\lambda + \mu)\nabla\nabla \cdot \mathbf{U} + \mu\nabla^2\mathbf{U} + \rho f = \rho\ddot{\mathbf{U}} \tag{1}$$

where **U** is the displacement vector, f is the body force vector, ρ is the density, and λ and μ are Lame constants. The solution suggested by Gazis to the above governing equation is acclaimed to hold good for longitudinal waves (Gazis 1959). An alternate solution proposed by Sun et al., can be used to estimate longitudinal as well as torsional modes (Sun et al. 2005). The displacement vector **U** can be expressed as in equation (2) using Helmholtz decomposition.

$$U = \nabla\phi + \nabla \times \mathrm{H} \quad and \quad \nabla \bullet \mathrm{H} = 0 \tag{2}$$

The general solution presented by Sun et al. is given by equation (3)

$$\phi = f(r)e^{im\theta}e^{i(kz-\omega t)} \qquad H_r = h_r(r)e^{im\theta}e^{i(kz-\omega t)}$$
$$H_\theta = h_\theta(r)e^{im\theta}e^{i(kz-\omega t)} \qquad H_z = h_z(r)e^{im\theta}e^{i(kz-\omega t)} \tag{3}$$

Three traction free boundary conditions on the outer surface and gauge invariance condition for an infinitely long reinforcement bar lead to an eigenvalue problem [C][A]=0, where **C** is a 4×4 matrix. Theoretical dispersion curves can be plotted using the eigenvalues of above equation (Rose 2014). The presence of damage and changes in the interface conditions render a complicated form to the above set of analytical equations. Hence, numerical solutions are employed to solve such problems.

The existence of corrosion in reinforcement leads to contamination of the received signal due to scattering at the damage edges. The change in geometric properties of the reinforcement necessitates a change in group velocity in damaged area. Hence, a change in the time of flight can be observed in that area. However, this change in time of flight cannot result in an appreciable change in the time of flight of the initial transmitted signal of the entire structure.

Depending upon the location and relative size of the damage, with respect to the reinforcement bar, an infinite number of modes are possible. The prevalence of any particular mode depends on the incident wave energy and reflection coefficients. A few basic transmission and reflection modes that occur due to damage front and rear edges are shown in Figure-1. A part of the initial signal is unaffected by the damage and arrives at the sensor at time T. The times of flight for various portions of the rebar are indicated by t1, t2 and t3. Arrival times for various damage scattered modes is listed in the Table-1.

Table 1 scattered wave modes and their arrival time at the receiver.

Mode No.	Time of arrival at sensor	Mode No.	Time of arrival at sensor
1	T	6	T+2*t1+2*t2
2	3T	7	T+2*t2+2*t3
3	T+2*t1	8	T+4*t2+2*t3
4	T+2*t2	9	T+4*t3
5	T+2*t3	10	T+4*t1

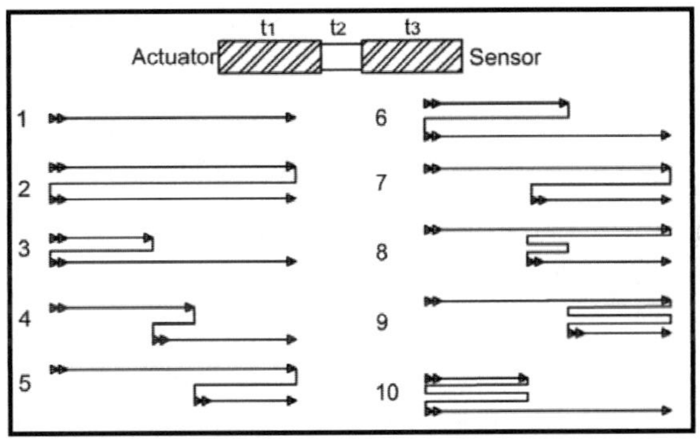

Figure 1 Reflection and transmission pattern in damaged rebar.

When damage is located at the centre or when the actuator and sensor are arranged equidistant from the damage, 3 & 5 arrive at the same time and 6, 7, 9 & 10 arrive at same time. Thus, in the received signal these two sets form two distinct peaks which have been verified by numerical simulations and experiments. The difference between the time of arrival of these two peaks can be used in the calculation of the length of damage.

3. Numerical Modelling

A numerical model of rebar with simply supported boundary condition is developed in Abacus/CAE 6.14, a commercial finite element analysis software. The length and diameter of rebar are taken as 600 mm and 20 mm respectively. Material properties of the rebar are mentioned in the Table-2. Mesh size of 2 mm and time increment of 1×10^{-7} sec are used to satisfy Courant-Freidrich-Levy condition.

Table 2 Material properties of rebar.

Material	Density (kg/m^3)	Poisson ratio	Young's Modulus (GPa)
Steel	7850	0.30	200

It has been widely reported in the literature that the presence of ribs has a negligible effect on the wave characteristics for a high ratio of wavelength to dimensions of ribs (Beard 2002; Miller et al. 2002; Ervin et al. 2006; Ervin et al. 2009). In this study, a simplified solid rebar model is used instead of ribbed rebar for simulation as typical rib dimension for 20 mm rebar is in the range of

1.25 mm to 1.35 mm. For the frequency of testing used in the current study, which is 100 kHz, the wavelength to rib dimension is estimated to be greater than 2. Hence, the ribs have limited effect on wave characteristics.

A five cycle hanning pulse at 100 kHz is applied at one end of the rebar to generate longitudinal waves. A rebar without damage is used as a baseline. The baseline signal shows a clear transmission signature at time 'T' and boundary echo after time '3T'. Corrosion is simulated in rebar by reducing the diameter of selected portion with all the other modelling parameters unchanged. A typical model with 100 mm wide corrosion located at the centre is shown in Figure-2(a). Dispersion curves for a rebar with a diameter of 20 mm are plotted using the group velocity that is obtained from numerical results. These dispersion curves were verified with experiments on bare rebar by changing the excitation frequency. The results are shown in Figure-2(b).

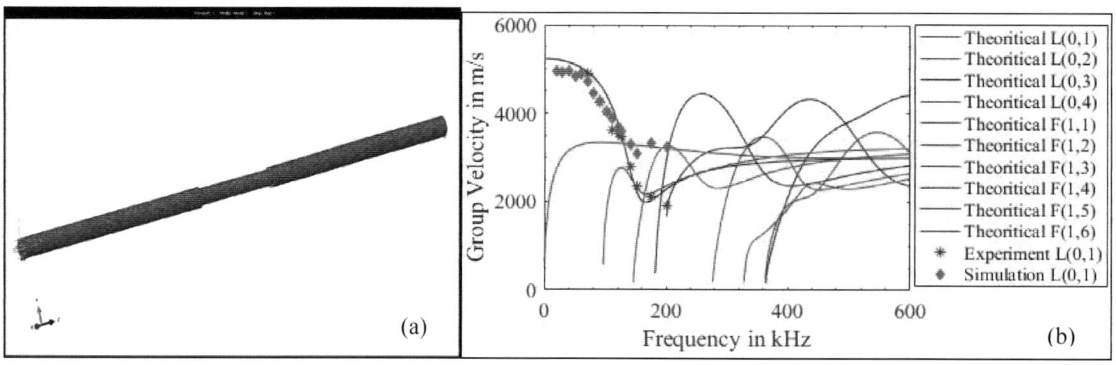

Figure 2 (a) FEA Model of reinforcement with corrosion and, (b) dispersion curves for a 20 mm rebar.

The response signal obtained from a damaged model is compared with baseline signal. From Figure-3 we can observe that the damage signature is generated due to reflection from the damage edges. Energy envelopes using a Hilbert transform are plotted in Figure-4 to analyse the signal. Two distinct peaks are observed in a corroded specimen, due to a combination of wave reflections at the front and rear edges of damage. This reflection pattern can be used as a metric to identify corrosion.

4. Experimental Validation

Simulation results were verified by conducting experiments on various samples of rebars shown in Figure-5(a). Corrosion in rebars is simulated by artificially scrapping the material from rebars. Typical drawing of a baseline rebar and a damaged rebar with pitting corrosion is shown in Figure-5(b). Contact transducers, of diameter 12.5mm, supplied by Olympus were used in experiments to generate and capture the longitudinal waves. The input signal was generated with a Tektronix arbitrary function generator, AFG3102, and the response is collected with a digital oscilloscope, DPO1034B. A sampling frequency of 10 MHz was used for recording the response signal. The received signal is averaged over 512 cycles, using the inbuilt function available in the oscilloscope, to reduce the effect of random noise.

The simulation and experiment results were compared in Figure-6 and Figure-7. The direct transmitted signal and boundary reflected signal can be recognised clearly in the Figures. The arrival time of the L(0,1) mode from experiments (125.2 μs) is close to that of arrival time from simulation (129.5 μs). The group velocity (4040 m/s) of the first wave packet is close to theoretical group velocity (4430 m/s) of L(0,1) mode obtained through dispersion curves which are calculated using DISPERSE software for a plain rebar.

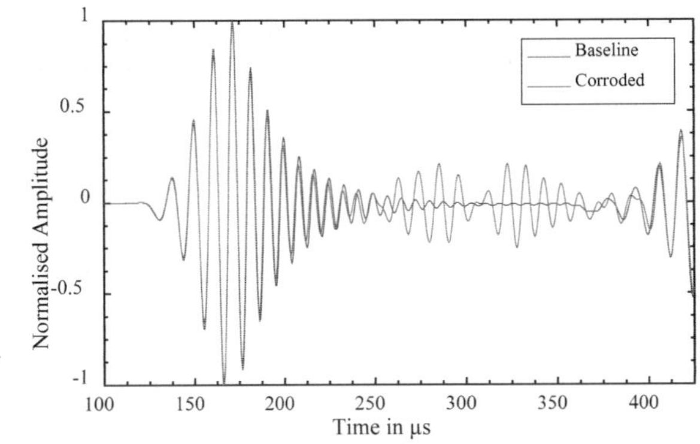

Figure 3 Comparison of signal from a corroded rebar with baseline signal using FEA.

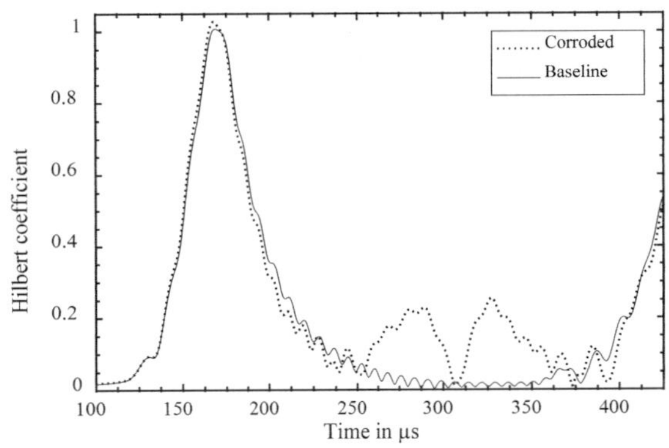

Figure 4 Comparison of energy envelopes of signal from corroded rebar with baseline using FEA.

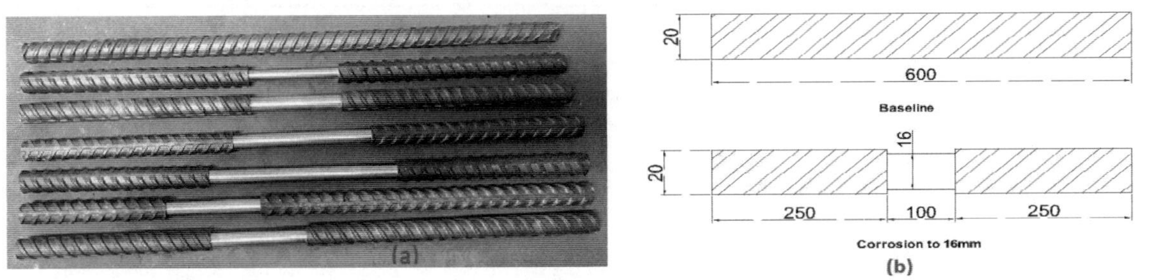

Figure 5 (a) Rebar samples used in experiment and, (b) typical reinforcement bar used as baseline and a reinforcement bar with corrosion.

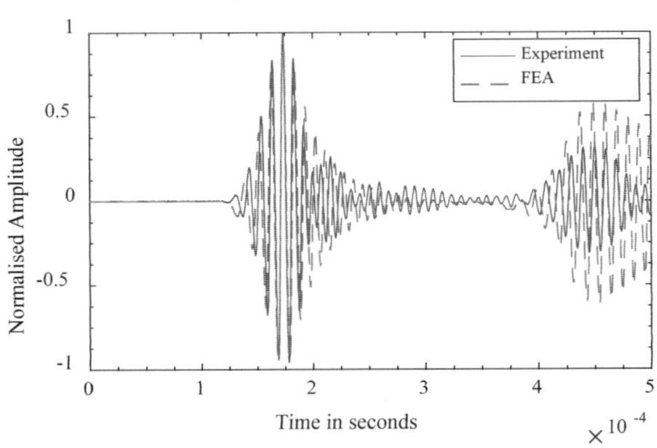

Figure 6 Comparison of baseline signal from experiment and FEA.

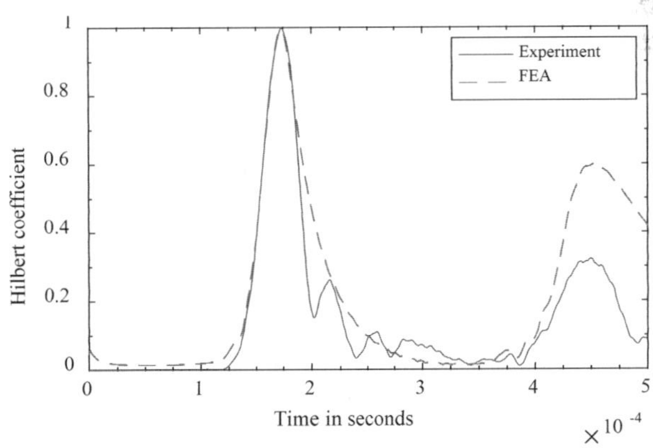

Figure 7 Comparison of energy envelope of baseline signal from experiment and FEA.

5. Results and Discussion

The reflection pattern generated due to damage edge is used to develop a damage index. In this method, the rebar is discretised into a few nodes and damage is assumed to exist at each of the nodes. The damage index for each of these nodes is assigned with the value of strength of reflected signal. Figure-8 shows a typical discretization of rebar in which the rebar is divided into 10 elements with 11 nodes. If there is a change in properties at node 2, a part of the signal will be reflected after time 't1' else there will be no reflection. For an undamaged rebar, the damage index will be typically equal to zero. Basically, the damage index captures the edge reflected signal. The level of corrosion in the rebar is determined by the damage index value and the length of corrosion is determined by the distance between the peaks in damage index plot.

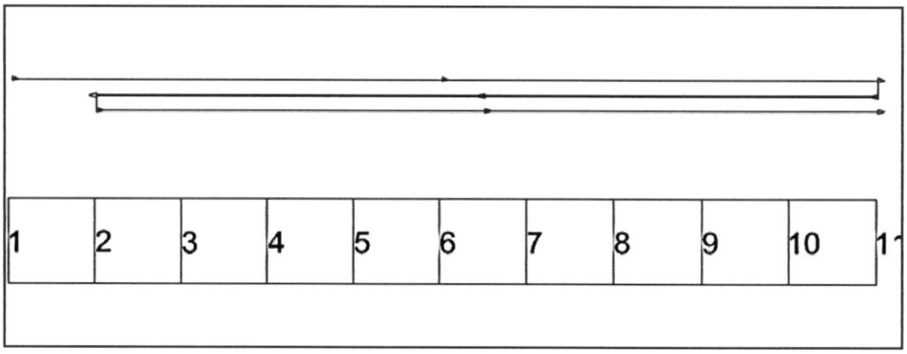

Figure 8 Plot showing discretization in a typical rebar for damage index calculation and path of signal in case of damage at node 2.

Damage index DI is defined by the following equation

$$DI(x) = \sqrt{\int_{t_1}^{t_2} (S - B)^2 \, dt} \tag{4}$$

Where $t_1 = (L + 2x)/V$

$t_2 = t_1 + \Delta$

V is the velocity of a specific mode

L is length of rebar and x is distance of node from sensor

Δ is the bandwidth of the input signal (50 μs for 100 kHz)

S is the signal from the specimen with damage.

B is the baseline signal.

To quantify the effect of pitting corrosion in a reinforcement bar, rebar models with various levels of corrosion are tested. Two different states of corrosion in a rebar is achieved in the following manner. First, the length and location of corrosion are kept constant and intensity of corrosion is varied by changing the diameter of rebar. Second, the diameter and location are kept constant and length of the corroded portion is varied.

5.1 Rebar with varying corrosion intensity

The presence of corrosion in a rebar causes scattering of wave energy by damage boundaries. The scattered waves can be used to identify the presence of corrosion and quantify the corrosion through various stages. The effect of an increase in corrosion intensity on damage index is examined by modelling various rebar samples. Figure-9 shows the damage index, calculated for FEA results, for rebars with a corrosion damage of 100 mm with a change in corrosion intensity. The diameter of the rebar in the corroded portion is decreased and damage index is plotted at discrete points on the rebar. From the damage index shown in figure-9, it can be observed that as the corrosion increases the damage index increases. Using these damage index plot, the location of the corrosion can be precisely located and the spread of the damage can be estimated. The length of damage estimated from the damage index is listed in Table-3. The area under the damage index can serve as a metric for the extent of corrosion.

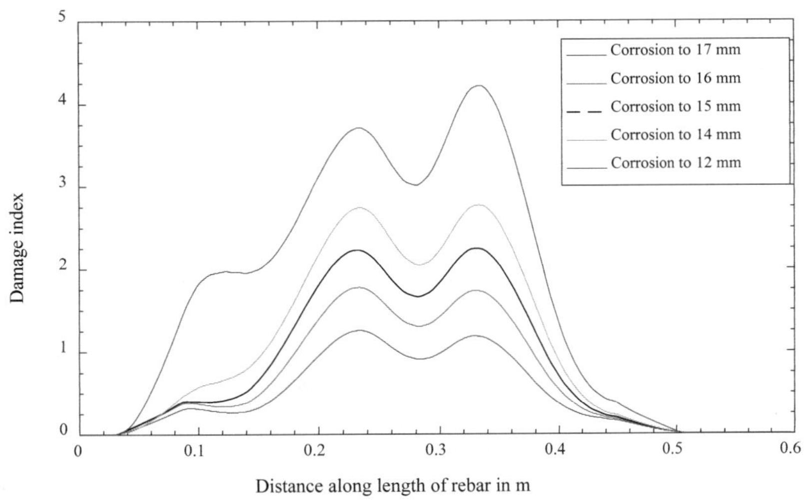

Figure 9 Damage index of steel bars corroded in middle 100 mm for various extent of corrosion.

Table 3 Estimation of damage length using damage index plot for a 100 mm corrosion damage in rebar.

Corroded diameter (mm)	First peak (mm)	Second Peak (mm)	Damage size (mm)	Error in length estimation (%)
17	234	330	96	4%
16	234	330	96	4%
15	231	333	102	2%
14	234	333	99	1%
12	231	333	102	2%

Simulation results of few of the samples are verified with experiments. Figure-10 shows the received signal from rebars with different intensity of corrosion. The corrosion is located at the centre of the rebar and the length of the corroded portion is maintained as 100 mm. The distinct peaks due to the scattering by the damage edges can be observed in experiment results similar to FEA results. The estimates of length and location are tabulated in Table-4. The experimental results demonstrate that the damage location and extent can be estimated with reasonable accuracy from the damage index shown in Figure-11.

Table 4 Corrosion length estimate from damage index plots using FEA.

Corroded rebar diameter (mm)	First peak (mm)	Second Peak (mm)	Damage size (mm)	Error in length estimation (%)
16	246	342	96	4%
12	255	345	90	10%

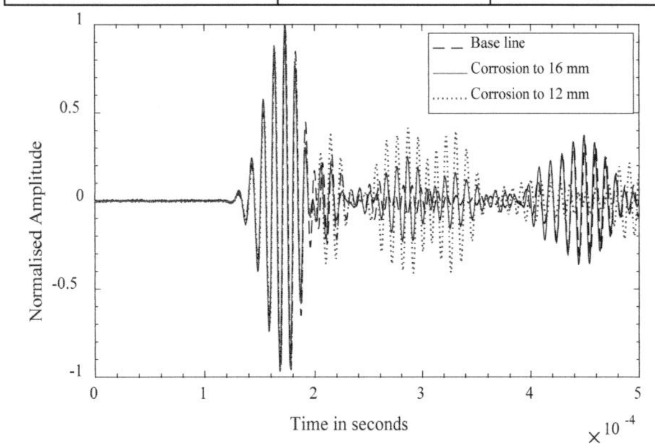

Figure 10 Comparison of signal from rebars with various intensity of corrosion.

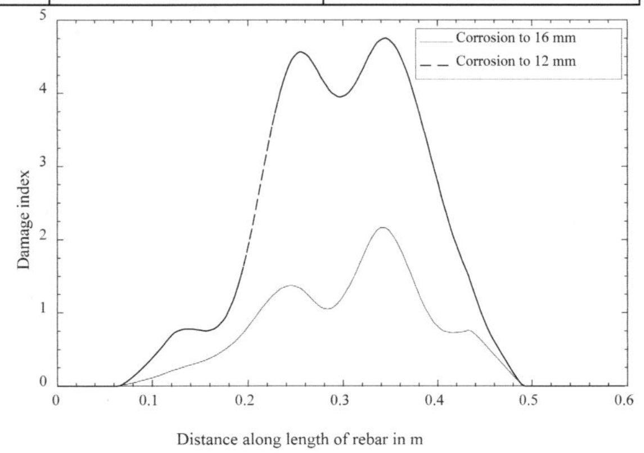

Figure 11 Damage index plots for rebars with various levels of corrosion.

5.2 *Rebar with varying axial extent of corrosion*

The study is extended to quantify the damage with various axial extents. Rebars with varying damage were modelled in FEA and damage index method was used to estimate the axial length of corrosion. Figure-12 shows damage index of various damage scenario using FEA. The damage index for corrosion lengths 100 mm, 150 mm and, 200 mm depict that damage can be monitored when the axial extent of pitting corrosion in rebars gets increased. These results are confirmed with experiments various rebar samples. The damage index obtained from experiments is shown in Figure-13. The length estimated for various size of damage is tabulated in Table-5.

The presence of distinct peaks in the received signal can only be observed for the extent of corrosion exceeding a cut-off damage length. Several rebars models with varying damage length were analysed using FEA and the cut-off damage size is found to be 70mm for the current case. Hence, the present method can be used to estimate the length with reasonable accuracy for damage sizes above this cut-off length. Using the current method for damage size less than cut-off length results in a error, as observed for the pitting corrosion of length 50 mm.

Table 5 Estimate of corrosion length using damage index plot for damage located equidistant from transducers.

Actual corrosion length (mm)	Damage size (simulation) (mm)	Error (Simulation) (%)	Damage size (experiment) (mm)	Error (Experiment) (%)
50	102	100%	-	-
100	99	4%	96	4%
150	129	14%	132	12%
200	178	11%	192	4%

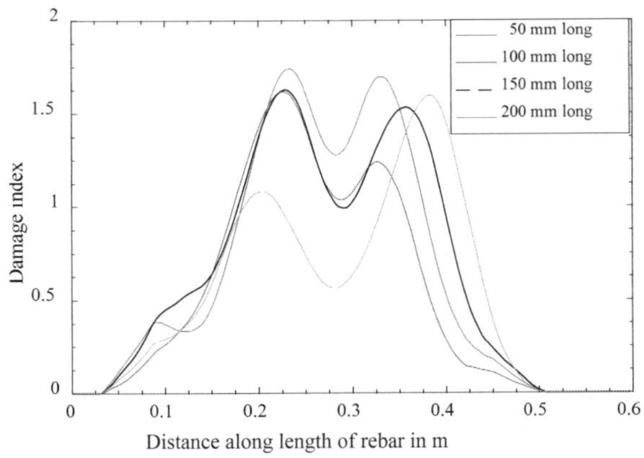

Figure 12 Damage index for various length of damage located at the centre using FEA.

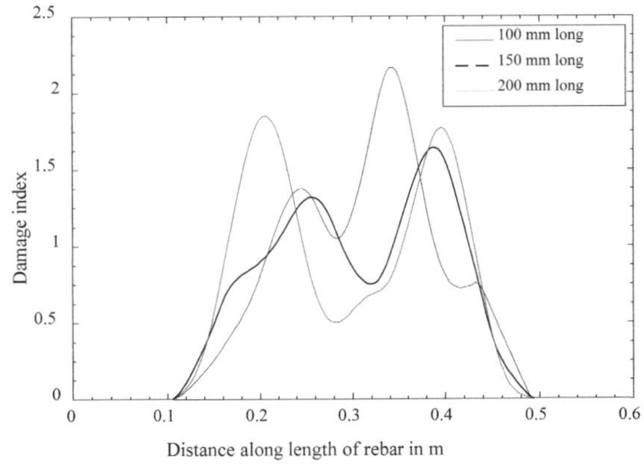

Figure 13 Damage index for various length of damage located at the centre using experiment.

6. Conclusion

The applicability of guided waves for detection of corrosion in reinforcement steel bars is explored in this paper. A careful examination of the received signal has shown that damage scattered signal can be used to locate and assess the condition of corrosion. A damage index method is proposed to quantify the pitting corrosion in reinforcement bars using the scattered signal. The proposed damage index can indicate the location and extent of corrosion in reinforcement bars. It is planned to develop a comprehensive method for a damage size less than the cut-off length.

References

Ahmad, S., 2003. Reinforcement corrosion in concrete structures, its monitoring and service life prediction–a review. *Cement and Concrete Composites*, 25(4), pp.459–471.

Amjad, U., Yadav, S.K. & Kundu, T., 2015. Detection and quantification of diameter reduction due to corrosion in reinforcing steel bars. *Structural Health Monitoring*, 14(5), pp.532–543.

Beard, M.D., 2002. Guided wave inspection of embedded cylindrical structures. *Imperial College, Department of Mechanical Engineering*, Doctoral dissertation.

Cao, C. & Cheung, M.M.S., 2014. Non-uniform rust expansion for chloride-induced pitting corrosion in RC structures. *Construction and Building Materials*, 51, pp.75–81.

Ervin, B.L. et al., 2006. Estimation of general corrosion damage to steel reinforced mortar using frequency sweeps of guided mechanical waves. *Insight - Non-Destructive Testing and Condition Monitoring*, 48(11), pp.682–692.

Ervin, B.L. et al., 2009. Monitoring corrosion of rebar embedded in mortar using high-frequency guided ultrasonic waves. *Journal of Engineering Mechanics*, 135(1), pp.9–19.

Ervin, B.L. & Reis, H., 2008. Longitudinal guided waves for monitoring corrosion in reinforced mortar. *Measurement Science and Technology*, 19(5), p.55702.

Gazis, D.C., 1959. Three-dimensional investigation of the propagation of waves in hollow circular cylinders. I. Analytical foundation. *The Journal of the Acoustical Society of America*, 31(5), pp.568–573.

Lamb, H., 1917. On waves in an elastic plate. *Proceedings of the Royal Society of London. Series A, Containing papers of a mathematical and physical character*, pp.114–128.

Lu, Y. et al., 2013. Guided waves for damage detection in rebar-reinforced concrete beams.

Construction and Building Materials, 47, pp.370–378.
Lu, Y. et al., 2007. Quantitative evaluation of crack orientation in aluminium plates based on Lamb waves. *Smart Materials and Structures*, 16, pp.1907–1914.
Lu, Y. et al., 2014. The application of guided waves for debonding identification in FRP reinforced civil structures.
Masc, R.R., 2015. Concrete deterioration detection using distributed sensors. *Strructures and Buildings*, 168(SB2), pp.118–126.
Miller, T., Hauser, C.J. & Kundu, T., 2002. Nondestructive inspection of corrosion and delamination at the concrete-steel reinforcement interface. *Nondestructive Evaluation*, pp.121–128.
Miller, T.H. et al., 2013. A new guided wave–based technique for corrosion monitoring in reinforced concrete. *Structural Health Monitoring: An International Journal*, 12(1), pp.35–47.
Mustapha, S. et al., 2014. Damage detection in rebar-reinforced concrete beams based on time reversal of guided waves. *Structural Health Monitoring*, 13(4), pp.347–358.
Rajeshwara, C.S., Banerjee, S. & Lu, Y., 2017. Identification of zero effect state in corroded RCC structures using guided waves and embedded piezoelectric wafer transducers (PWT). *Procedia Engineering*, 188, pp.209–216.
Rayleigh, Lord, 1885. On waves propagating along the plane surface of an elastic solid. *Proc. London Math. Soc.*, 7(1), pp.4–11.
Rose, J.L., 2014. *Ultrasonic waves in solid media*, Cambridge university press.
Sakamoto, W.K. et al., 2013. Polymer matrix-based piezoelectric composite for structural health monitoring. In *2013 Joint IEEE International Symposium on Applications of Ferroelectric and Workshop on Piezoresponse Force Microscopy, ISAF/PFM 2013*. pp. 295–297.
Sharma, S. & Mukherjee, A., 2010. Longitudinal guided waves for monitoring chloride corrosion in reinforcing bars in concrete. *Structural Health Monitoring*, 9(6), pp.555–567.
Sikdar, S., Banerjee, S. & Ashish, G., 2015. Ultrasonic guided wave propagation and disbond identification in a honeycomb composite sandwich structure using bonded piezoelectric wafer transducers. *Journal of Intelligent Material Systems and Structures*, pp.1–13.
Sun, Z., Zhang, L. & Rose, J.L., 2005. Flexural torsional guided wave mechanics and focusing in pipe. *Journal of Pressure Vessel Technology*, 127(4), p.471.
Wu, F. & Chang, F.-K., 2006. Debond detection using embedded piezoelectric elements in reinforced concrete structures-part I: experiment. *Structural Health Monitoring*, 5(1), pp.5–15.

The 8th International Conference on
Structural Health Monitoring of Intelligent Infrastructure
Brisbane, Australia | 5-8 December 2017

The use of mid-span acceleration measurement of bridges under in-service loads as a parameter for use in a Structural Safety Evaluation (SSE) system

J. C. Moore[1], S.S. Mahini[2], R Glencross-Grant[3].
[1] *University of New England – Australia, Email: jmoore30@une.edu.au,* [2]*Member of ISHMII, Queensland University of Technology, Australia,* [3]*University of New England - Australia*

Abstract

Bridge vibration frequencies and Frequency Response Functions (FRF) produced by impact loading have been reported by several authors, but these responses have not been used to predict the probability of failure, whereas mid-span deflection signals have. It has also been reported that the change in such frequencies caused by structural deterioration may be too low to detect early deterioration. In order to create low cost Structural Health Monitoring (SHM) systems for use on small to medium sized bridges on low-trafficked local roads, it is important to define suitable parameters that can be economically measured in order to determine the onset of structural deterioration and monitor the Structural Health of bridges. Methods relating mid-span deflection to acceleration signals have been reported in both theoretical and experimental contexts, but there is still no well-defined relationship with a probability of failure. The current availability of low cost accelerometers will enable the construction of improved SHM systems and further work is justified to establish a reliable relationship. In this paper, experimental data from simultaneous mid-span deflection and acceleration measurements are examined for the cases of a contemporary concrete bridge and an aged timber beam bridge. The outcome is the identification of a peak acceleration parameter that can be used in a SSE system. This parameter can be used to identify the onset of structural deterioration and enable changes in the Modulus of Elasticity and second moment of area to be monitored.

1. Introduction

Bridge vibrations induced by moving vehicles can cause a reduction in bridge long-term serviceability and an increase in the level of maintenance required (Sammartino et al., 1999; Shi and Cai, 2008; Lilley and Winslade, 2014). In the early 19th century it was not clear to engineers if there would be enough time for a rapidly moving load to deform a structure (Frýba, 2013; Timoshenko, 1953, p. 173). This was no doubt as a result of low vehicle velocities at the time. However, as will be reported in this paper, sufficiently fast acting sensing systems are contemporarily available to measure the deflections that are produced by a faster moving vehicular load. What is still not commercially available though is a low cost Structural Health Monitoring (SHM) (Worden et al., 2007) system that can be used on small to medium span bridges on local roads. Such SHM systems are needed to monitor contemporary concrete bridges, as well as aged timber beam bridges; timber beam bridges that were constructed according to designs created mid-19th century (Dare, 1903) and that are still extant (Moore et al., 2013; Samali et al., 2011), albeit extensively maintained during their service life.
One of the earlier interpretations of SHM was to use sensing systems to identify a level of structural damage (Worden et al., 2007). However, it can be more important to identify a level of degradation so that maintenance can be planned and a level of Structural Safety Evaluation (SSE) identified so

that a structure can be kept operating in a safe manner, as predetermined by a service specification (Chan et al., 2011). Chan et al. (2011) reported how SSE could be applied to the operation of a large suspension bridge - the Tsing Ma Bridge in Hong Kong. It has been suggested by Inaudi (2011) that the overall cost of implementing SHM can be as high as 30% of the overall structural cost, whereas the necessary instrumentation might only cost about 3%. Hence for a bridge of value about $500 000 (for example, Hortons Creek Bridge, on the Armidale to Grafton Road, New South Wales (NSW) as shown in Figure 1), the required total instrumentation cost might be expected to be c. $15 000 or less. A similar case could be made for Munsies Bridge (Gostwyck, Uralla, NSW as shown in Figure 2), which was used as a test case (see following Methodology). The savings (cited 30%) are obtained by replacing components at the end of life and before they fail; not by removing them prematurely, nor by waiting for catastrophic failure. Damaged components should be identified by a change in the effective SSE. The difficulty is to limit the total cost of the sensing system to a value below $15 000. Therefore, the aim in this paper was to determine if a mid-span acceleration parameter for a bridge girder, such as identified by Equation 2 (see following Background Review), can be used in a low cost SSE/SHM system for small to medium size bridges.

Figure 1: A single span timber beam road bridge-Hortons Creek Bridge, Armidale-Grafton Road, NSW (Source: Moore, 2009)

Figure 2: Easternmost span of the test timber bridge (Case 2) shown together with test vehicle and measurement tripods (Source: Moore, 2009)

2. Background Review

Detailed theoretical numerical analyses have been reported in relation to the examination of the dynamic response of bridges as a result of moving vehicular loads (Lalthlamuana and Talukdar, 2014; Frýba, 1996; Frýba, 2013). In particular Frýba (1996, Table 4.1) and Bishop and Johnson (1960) report the characteristics of natural vibrations of single span beams. In addition, experimental test loadings have also been reported (Feltrin et al., 2010), but there are still aspects of the dynamic response that are not fully explained, some aspects of which are discussed in this paper. Samali et al. (2007) reported that a Frequency Response Function (FRF) could be identified for a timber beam bridge, but did not identify how this response could be related to a probability of failure. In more recent work, Samali et al. (2011) reported how an FRF could be related to the location of damage in a new experimental structure but Hamm et al. (2016) also reported similar results for a timber footbridge, and concluded with the question "Can the damage be identified early enough, or is the influence of the damage on the Eigen frequency too low to be detected in time?" (Hamm et al., 2016, Conclusions).

The concept of utilising measured vibration frequencies to identify structural integrity has long been recognised, but it has also been reported that "the computations that would be required for large engineering structures could be prohibitive" (Salawu, 1997, p721). Salawu (1997, p721) also reported that "natural frequency changes alone may not be sufficient for a unique identification of the location of structural damage. This is because cracks associated with similar crack lengths but at two different locations may cause the same amount of frequency change". While computing capability has significantly improved since 1997, maybe several thousand times (Moore, 2006), extensive in-service calculations require power and this resource is often limited in remote locations. The asset manager of a bridge is interested in knowing when a structure will fail and a parameter measured with an SHM system must not incur high system costs, neither in terms of the required computing power, nor in the equipment required (e.g. the number of sensors).

The measurement of the frequency response of a bridge can be achieved with the use of an accelerometer and there are two low cost types readily available. One is described as a micro-electro-mechanical system (MEMS) architecture accelerometer and the other is a seismic Geophone (Neitzel et al., 2007; Neitzel et al., 2012). The MEMS type consists of a miniature vibrating beam housed in an electronic integrated circuit, and the Geophone consists of a moving magnetic mass within a coil. An introduction to the accelerometers that are currently available is provided by Neitzel et al. (2007). As identified by Neitzel et al. (2012, p8) a MEMS accelerometer is "a suitable sensor for modal analysis of structures" and that in their experiments "Natural frequencies, damping coefficients as well as derived displacements could be determined with high accuracy". Because of the extremely small size of the active sensor in the MEMS type it has a natural resonant frequency of about 5000 Hz whereas the much larger Geophone type has a natural resonant frequency of about 5 Hz. Since the frequencies of interest in bridge structures are in the range 1 to 300 Hz (Hamm et al., 2016; Samali et al., 2007; Feltrin et al., 2010) the MEMS device is the more appropriate sensor to use.

Feltrin et al. (2010) provided an excellent report of the measurement of the mid-span deflection and acceleration of a wooden trough bridge, of span 19.6 m, loaded with a 40-tonne test vehicle moving at both 15 $km.h^{-1}$ and 60 $km.h^{-1}$. This bridge, across the Wigger River at Egolzwil, Switzerland was built in 2005. The measured mid-span displacements were as shown in Figure 3 with a peak deflection of 18 mm. The test vehicle was reported to have entered the span at a speed of 60 $km.h^{-1}$ at the abscissa time of 2.0 seconds and departed at about 3.5 seconds (refer Figure 3). The measured mid-span accelerations were as shown in Figure 4. This Figure contained two curves: the first was the measured dynamic acceleration which was clearly seen; the second dotted curve was reported as 'the quasi-static curve' (Feltrin et al., 2010), which consisted of only the frequency components of the signal below about 2 Hz. This latter curve was not clearly shown and was obscured behind the first curve. These authors also reported examining the first differential of the mid-span deflection, that is the velocity, but did not report the second differential of the mid-span deflection, that is the mid-span acceleration nor compare it with the actual measured acceleration.

There was significant signal activity prior to the 'two second' point and the first, the major maximum, peak of -1.1 $m.s^{-2}$ was reached at a time of 1.6 second which occurred before the vehicle entered the span (refer Figure 3). However, the reason for this activity was not reported. The positive peak amplitude of the quasi-static maxima of 0.11 $m.s^{-2}$ occurred at about 2.5 seconds in the middle of the deflection signal peak, at which time the peak acceleration was about 0.8 $m.s^{-2}$. Although it was not reported by Feltrin et al. (2010), it can be inferred from Figure 4 that, in the period 3 to 6 seconds, the Q factor of the transient was about 27 and the damping factor 0.018 (Giangrandi, 2014; Bishop and Johnson, 1960).

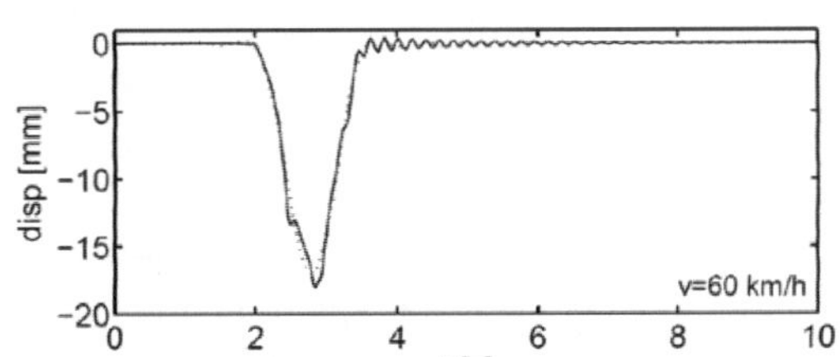

Figure 3: Displacements due to crossing of the 40-tonne test truck at 60 $km.h^{-1}$ measured with

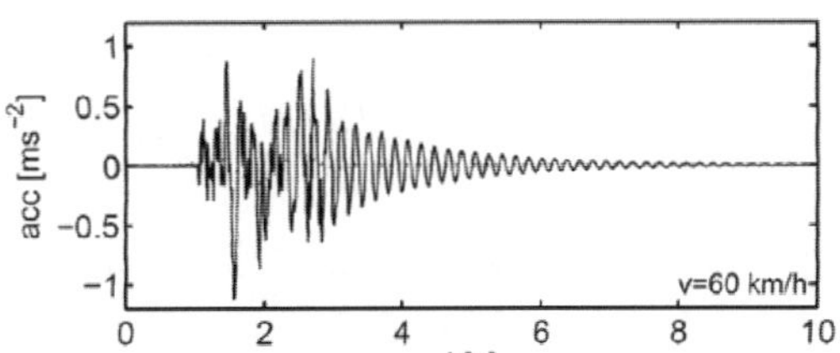

Figure 4: Mid-span accelerations due to crossing of the 40-tonne test truck at 60 $km.h^{-1}$; source Feltrin et al. (2010, Figure 12)

LVDT sensors; source Feltrin et al. (2010, Figure 9)

Feltrin et al. (2010) also reported an approximate equation, Equation 1, which could be used to relate the quasi-static acceleration to the vehicle speed. However, it does not relate the acceleration to the bridge material parameters; the Modulus of Elasticity (*E*) and the second moment of area (*I*). Similarly, it was not reported how it could be directly used in an SHM/SSE system.

$$a_{max} \approx 2 \times d_{max} \times \left(\pi \frac{v_{truck}}{(L+2w)}\right)^2 \qquad (1)$$

Moore and Mahini (2015) reported an alternative analytic closed form solution relating the bridge material parameters and the test vehicle speed and mass with the measured mid-span acceleration as shown in Equation 2. If, therefore, the vehicle velocity and the peak acceleration are measured and the span is known, then the product *EI* can be calculated. Such a measurement could be used in an SHM/SSE system to detect any changes in *EI* indicative of a change in Condition State such as from State 1 to State 2 (RTA, 2007). This signal, a_{vp}, appears (see following results) to be similar to the quasi-static signal reported by Feltrin et al. (2010), and is not the dominant signal typically recorded with an accelerometer (refer Figure 3). However, one advantage of this approach is that the value of *EI* is inversely proportional to the measured parameter, a_{vp}, whereas if the measured parameter is the frequency of oscillation then *EI* is proportional to the square root of the measured quantity (Bishop and Johnson, 1960; Frýba, 1996).

$$a_{vp} = \frac{P v_{truck}{}^2 L}{4 E I} \qquad (2)$$

Arraigada and Partl (2006) evaluated the use of accelerometers for measuring the deflection of pavements under in-service traffic. They found that the "amplification of low frequency errors of the recorded acceleration leads to unacceptable deflection baseline drifts". To overcome these errors and those caused by double integration they developed an algorithm employing a polynomial derived from the root mean square (rms) value of the acceleration signal that was then subtracted from the double integrated signal to provide a deflection signal. It is unclear how such an algorithm could be used in an SHM/SSE system because of the difficulty of integration, plus the concern that the signal to be integrated is a small fraction of the total acceleration (refer Figure 4, dotted line).

After an extensive semi-analytical analysis Lalthlamuana and Talukdar (2014) reported results where the mid-span girder acceleration appeared to be proportional to the velocity of the test and not to the square of that velocity. However, they only reported that "The mean peak central deflection is found to increase with the vehicle speed" (Lalthlamuana and Talukdar, 2014, Section 4.2.1). Their model bridge was a T-beam Reinforced Concrete (RC) bridge with a span of about 17 m. Their acceleration signal (Lalthlamuana and Talukdar, 2014, Figure 8) was not of a form similar to either that shown in Figure 3, nor that described by Equations 1 and 2. Therefore, the reason for this disparity needs to be determined.

3 Methodology

Two cases were examined, Case 1 was Toms Gully Bridge and Case 2 Munsies Bridge. Toms Gully Bridge (see Figures 5 and 7), a three span prestressed concrete and steel beam bridge constructed c. 2005 and located on the Boorolong Road, 20 km west of Armidale, New South Wales (NSW). Munsies Bridge (see Figures 2 and 8), a six span composite timber and concrete bridge was commissioned in 1938. The superstructure of this bridge was timber, supported on concrete piles and located at Gostwyck, Uralla, NSW across Salisbury Waters.

The mid-span movements of the girders were measured directly using Laser Distance Sensors, type Micro-Epsilon optoNCDT 1402-100, and acceleration sensors type ADXL345 (Earl, 2016; Analog

Devices, 2009). The distance sensors were mounted on surveying tripods with their top surfaces in the range 100 mm to 120 mm below the girder mid-span and the accelerometers, mounted in protective plastic housings, were rigidly attached directly to the underside of the girder mid-span (see Figure 6). The resolution of the laser distance sensors was 0.01 mm and the resolution of the accelerometers was 0.04 $m.s^{-2}$. The distance sensors had a dynamic measurement range of 50 to 150 mm and the accelerometer signal range was nominally 0 to 16 G (0 to 150 $m.s^{-2}$).

The monitoring of these spans was part of a short-term project undertaken by Civil Engineering students at the University of New England. The particular test bridge spans were chosen because of their low height above the ground line, which enabled the girder mid-spans to be accessed without the need for supplementary scaffolding. A TMR-200 logging system (Tokyo Measuring Instruments Laboratory Co. Ltd. (TML), 2013) was used to record the sensor signals with the period between samples set to 5 milliseconds (a sampling frequency of 200 Hz). A small portable generator was used to provide mains voltage (240 Vac), electric power to the measurement system. The span was loaded by driving a test vehicle, of mass about 2.2 tonnes at velocities from 10 $km.h^{-1}$ to 100 $km.h^{-1}$. This mass included the kerb mass of the vehicle (Ford Motor Company, 2016) together with the mass of the driver and the fuel load. The on-site experiment took about two hours including equipment installation, recording measurements and equipment removal.

Figure 5: Concrete test Bridge (Case 1)

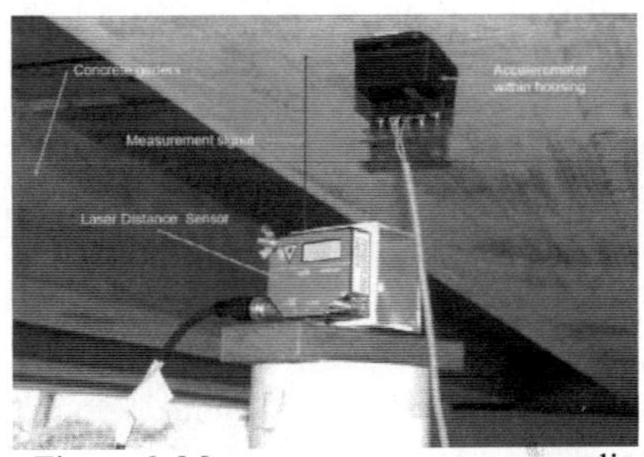

Figure 6: Measurement sensors as applied to Case 1 concrete bridge girders

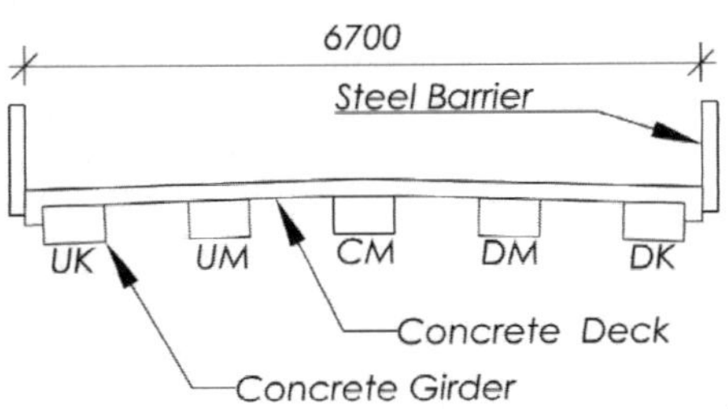

Figure 7: Cross-section schematic of test Concrete Bridge

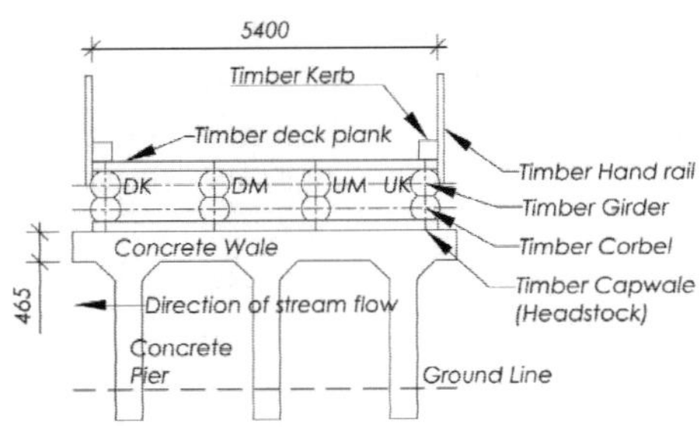

Figure 8: Cross section schematic of test Timber Bridge, span 6

3.1 Case 1: Concrete bridge (Toms Gully)

In this case the westernmost span was measured. Although there were five girders supporting the concrete deck the mid-spans of only four were instrumented (see Figure 6). These were, the Upstream Kerb (UK), the Upstream Main (UM), the Central Main (CM) and the Downstream Main (DM) girders (see schematic Figure 7). The Downstream Kerb (DK) girder was not instrumented because of limited sensors available at the time. The expected deflections were in the range 0.1 to 10 mm and the expected accelerations were 0.1 to 10 $m.s^{-2}$, well within the sensor dynamic ranges.

3.2 Case 2: Timber Bridge (Munsies Bridge)

Case 2 involved the measurement of the four girders of the easternmost span (see Figures 2 and 8). The accelerometers were screwed to the underside of the girder at the mid-span and monitored as for Case 1. Because of the alignment of the bridge approach roads the maximum test vehicle velocity was limited to 80 $km.h^{-1}$.

4 Results

4.1 Case 1: Concrete bridge (Toms Gully)

Because this bridge was a two lane bridge and comprised five girders, the test vehicle was driven predominantly above the upstream main (UM) girder when travelling west to east. In this direction, the vehicle moved from the road pavement onto the measured span and then onto the middle span. The peak deflection signals were less than one millimetre and the peak accelerations less than 0.5 $m.s^{-2}$; both within the dynamic range of the sensors. The limited space in this paper precludes providing a full set of data and only graphs directly related to the aim are provided. In one example set of measurements of the four girders for an east to west traverse at 100 $km.h^{-1}$ the deflection signal was less than about ± 0.1 mm before the test vehicle arrived at the bridge. Under loading, the peak deflection of the UM girder was about 0.55 mm (see Figure 9), with the other girder deflections being 0.3 mm or less - a nominal effective bridge stiffness of about 36 $kN.mm^{-1}$. The peak acceleration for the upstream main girder was a maximum at about 0.2 $m.s^{-2}$ (see Figure 10). The peak accelerations for the other girders were of a similar magnitude despite the deflections for those girders being less than that for the UM girder. The acceleration signal was less than ± 0.1 $m.s^{-2}$ prior to the arrival of the vehicle (see Figure 10) with a dominant oscillatory character both during the vehicle traverse and for more than one second after the vehicle had left the measured span. In the two second period before the test vehicle arrived at the measured span it oscillated at 60 Hz at low amplitude.
In the period after three seconds (abscissa value, Figure 10), as the test vehicle moved from loading the measured span to loading the adjacent span, the frequency of the signal changed. During the loading period the frequency of the transient was about 20 Hz with a damping factor of about 0.011. After the vehicle left the span, the frequency was about 9 Hz with a damping factor of about 0.013. In Figures 11 and 12 are shown the spectra of the deflection and acceleration signals. The dominant frequencies were 0.68 Hz (caused by vehicle loading) and resonances at about 9 Hz, 20 Hz and 60 Hz. In Figure 13 is shown the peak mid-span girder deflections after the signals were filtered with a low pass 0 to 5 Hz filter. These data included signals for the four measured girders and for both passes. Figure 14 shows the acceleration signal (refer Figure 10) during the loading period after filtering with a 2 Hz (as per Feltrin et al. (2010)) low pass filter (red curve), and the acceleration (blue curve) computed by numerically doubly differentiating the deflection signal (refer also Figure 9). The point that would be determined by use of Equation 2 is the peak of the blue curve coincident with the peak deflection. This value was not calculated for the concrete bridge because E and I were not known with any accuracy. The peak magnitude of the filtered acceleration signal was about 12% of the measured acceleration signal.

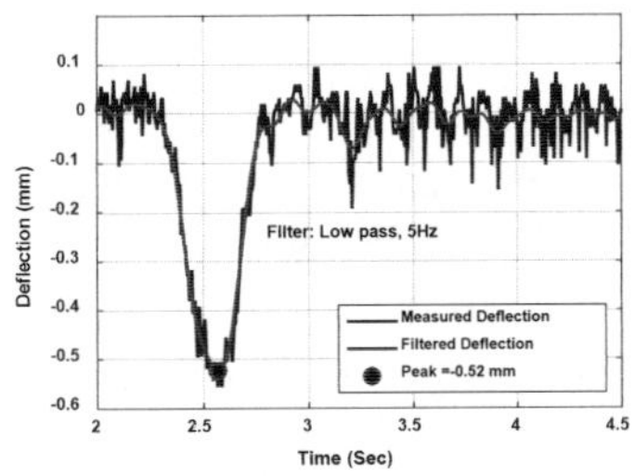

Figure 9: The deflection signal for the UM concrete girder

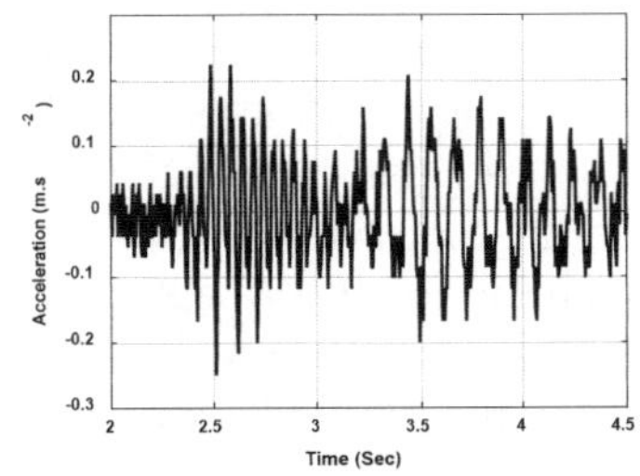

Figure 10: The acceleration signal for the UM concrete girder

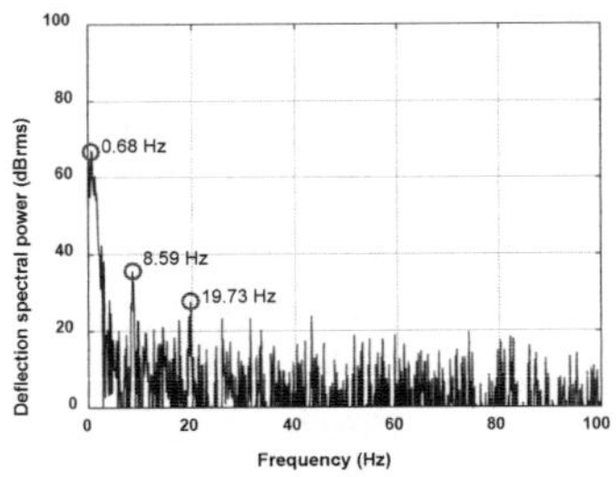

Figure 11: Spectrum of the deflection signal for the DM concrete girder, traverse speed 100 km.h^{-1}

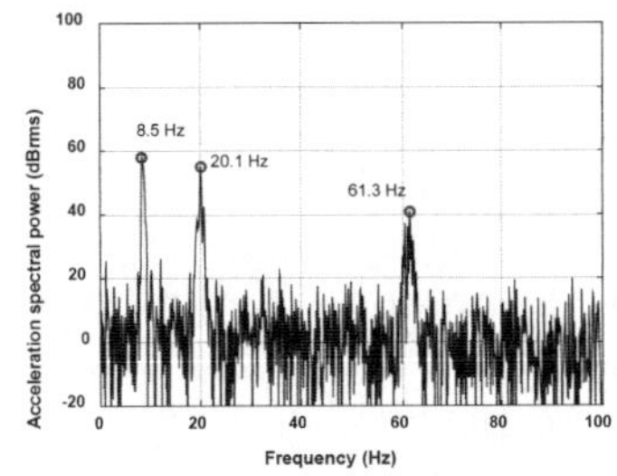

Figure 12: Spectrum of the acceleration signal for the DM concrete girder, traverse speed 100 km.h^{-1}

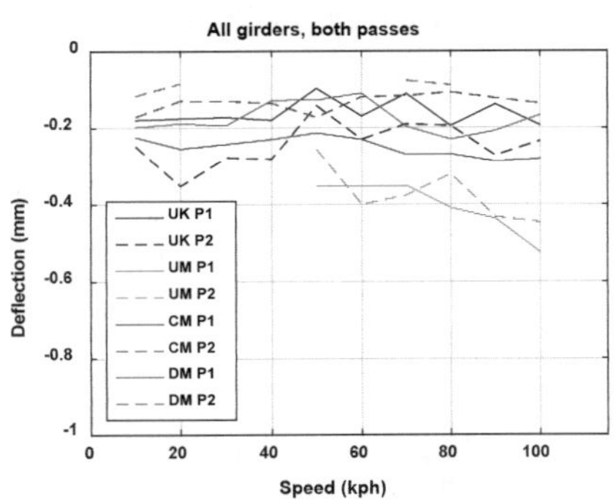

Figure 13: Filtered peak girder deflections-v-traverse speed for the concrete girders (Low Pass filter cut off ~ 5Hz)

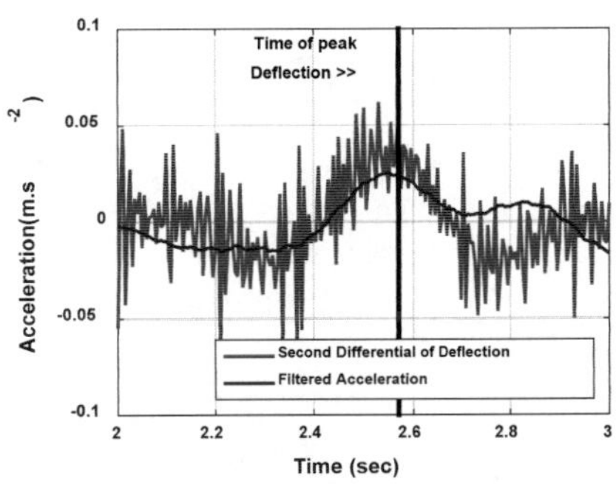

Figure 14: Second differential of the deflection and acceleration below 2 Hz; DM concrete girder at a traverse speed of 100 km.h^{-1}

4.2 Case 2: Timber Bridge (Munsies Bridge)

The test vehicle was driven along the centreline of the timber bridge with the wheels nominally directly above each of the respective main girders (DM and UM). In a manner similar to that for the concrete bridge, this pass involved moving from the road pavement onto the most easterly span, which was the measured span and then on to an adjacent unmeasured span. The peak deflection of the main girders was about 4.0 mm (see Figure 15 for the case of the DM girder), whereas the peak kerb girder deflection was about 2.5 mm; the nominal effective bridge stiffness being about 4.5 kN.mm^{-1} - eight times less than the concrete bridge. The peak of the signal magnitude envelope of the main girder accelerometers were about 8 m.s^{-2} (see Figure 16) and the kerb girder signals were about 3 m.s^{-2}. A downward deflection was caused by the test vehicle loading in the period 3.8 to 4.4 second and an upward deflection in the period from 4.4 second to about 5 second as the vehicle loaded the adjacent span. There was significant oscillatory behaviour recorded in the deflection signal both during the traverse and for more than four seconds after the vehicle had departed the measured span. The dominant frequency in the deflection signal after the test vehicle had departed was about 7 Hz with a damping factor of about 0.005 and a peak amplitude of about 0.3 mm. The related amplitude of the acceleration signal at this frequency was about 0.43 m.s^{-2}; a value computed from the deflection amplitude times the square of the angular frequency. The amplitude of this acceleration signal was small, at about 9%, compared to the other acceleration components caused by natural vibration. The oscillatory behaviour of the DM girder acceleration signal both during the traverse and immediately afterwards was as shown in Figure 16. The dominant frequency was about 20 Hz and the damping factor was about 0.04, a factor which was about three times the damping factor for the concrete case. The spectrums for the deflection and acceleration signals were complex and a simple FRF was not readily identified. Common frequencies occurred at about 8 Hz and 15 Hz in both the deflection and acceleration spectrums, but there were 10 or more additional confounding signals below 20 Hz. The peak deflections, after filtering with a 5 Hz low pass filter, are shown in Figure 17. The main girders (UM and DM) deflected about 3 mm and the kerb girders about 1.5 to 2 mm, these deflections being substantially independent of speed. Because of prior experimentation (Moore et al., 2013) the effective girder parameters were known, which enabled the calculation of the mid-span acceleration according to Equation 2. This value, 0.28 m.s^{-2}, about 4% of the measured peak acceleration, was plotted together with the acceleration signal filtered with a 2 Hz low pass filter and the second differential of the deflection, as shown in Figure 18.

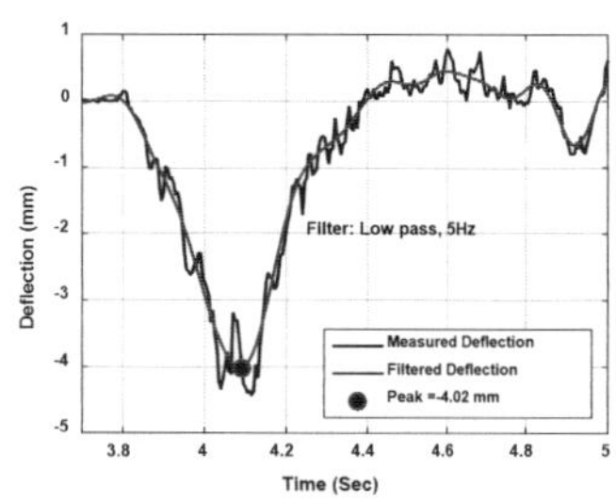

Figure 15: The deflection signal for the DM timber girder, traverse velocity 80 km.h^{-1}

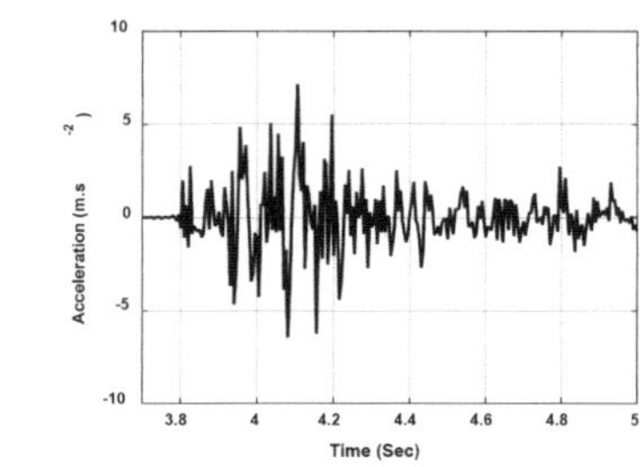

Figure 16: Acceleration signal for the DM timber girder, traverse velocity 80 km.h^{-1}

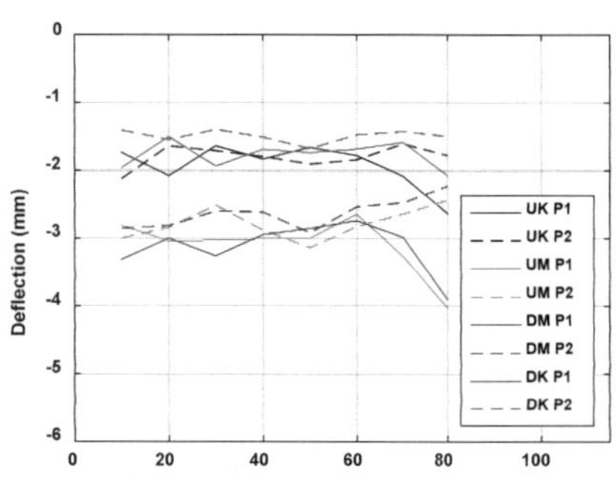

Figure 17: Filtered Peak deflection-v-speed for the timber test bridge

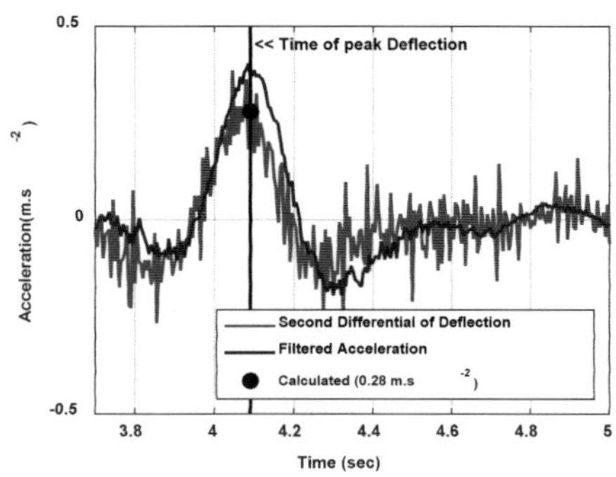

Figure 18: Calculated acceleration (See Eqn 2) compared with the acceleration signal below 2 Hz, and the second differential of the deflection; DM timber girder, traverse speed 80 km.h^{-1}

5 Discussion

The original aim of the work that produced the data reported here was to examine whether low cost accelerometers could be used effectively to examine the Structural Health of bridge girders. The outcome was exceptionally positive, but the most surprising outcome, was that it should be possible

to predict dynamic impact by measuring the transient behaviour of bridge girders with low cost accelerometers. As can be seen in Figures 13 and 17, the mid-span girder deflections seem to be independent of the traverse speed, which was not surprising. What was surprising though was the amplitude of the purely oscillatory behaviour and that the girders in both test cases were effectively undamped with damping factors less than 0.01. This has implications for the life time of such structures and needs to be examined in much more detail than was possible in these experiments.
The aim of this evaluation was to determine whether an SSE signal, such as defined by Equation 2, could be determined using accelerometers and again the outcome seemed to be very positive. As can be seen in Figure 18, application of Equation 2 appeared to produce a signal that was verified experimentally, but only after, the excessive signals caused by natural vibration, were removed by low pass filtering (see Figure 18, blue curve). The peak of this filtered acceleration curve could serve as an SSE signal to be used to identify changes in *EI*.
The peak amplitude of the vibrations in the acceleration signal were about 20 times that directly caused by the vehicular load deflecting the girders. The quasi-static signal characterised by Equation 1 also appeared to be similar to that characterised by Equation 2. A further surprising result was that the measured frequencies, typified in Figure 10, appeared to be anomalous in that they were high when loaded and low when not loaded. This result also needs to be repeated and examined in further experiments before appropriate comment can be made.

6 Conclusions

It was concluded firstly, that Equation 2 was valid and defined the peak of the acceleration signal when that signal was filtered to remove the natural frequencies of vibration. Secondly the signal calculated using Equation 2 was small, less than 15% of the measured peak accelerations, compared to the large signal caused by the natural oscillatory behaviour of the structures. Thirdly, the two test bridges, and probably all bridges (see for example Figure 4), are vibrating at unnecessarily high magnitudes because the damping of the vibrations on the test bridges was so low that there was no effective damping, neither in the timber bridge nor the concrete bridge. Fourthly, while the timber test bridge vibrated at a wide range of frequencies, the concrete test bridge vibrated at very specific frequencies, and fifthly the low frequency peak mid-span deflections seemed to be independent of vehicle velocity.
These conclusions are predicated on the two test bridges examined, but are believed valid because both the concrete bridge and the timber bridge exhibited somewhat similar behaviour. However, these results should justify further research on a sufficiently large number of bridges to provide statistically valid interpretations.

Nomenclature

a_{max}	Maximum acceleration at mid-span (mm.s^{-2})
d_{max}	Maximum displacement (mm)
v_{truck}	The truck speed (m.s^{-1})
L	The bridge span (m)
w	The distance between the front and rear axle of the truck (m)
a_{vp}	Peak vertical acceleration at mid-span (m.s^{-2})
P	The weight of the test vehicle (kN)
E	The MoE of the bridge span (m)
I	The second moment of area of the bridge span (m)

References

Analog Devices. 2009. *ADXL 345 Three Axis Digital Accelerometer* [Online]. Available: www.analog.com [Accessed September 8 2016].

Arraigada, M. & Partl, M. 2006. Calculation of displacements of measured accelerations, analysis of two accelerometers and application in road engineering. *Swiss Transport Research Conference.* Monte Verità, Ascona, Switzerland.

Bishop, R.E.D. & Johnson, D.C. 1960. *The Mechanics of Vibration.*

Chan, T.H., Wong, K., Li, Z. & Ni, Y. 2011. Structural Health Monitoring for Long-Span Bridges - Hong Kong experience and Continuing onto Australia. *In:* CHAN, T. & THAMBIRATNAM, D. (eds.) *Structural Health Monitoring in Australia.* New York: Nova Science Publishers.

Dare, H.H. 1903. Recent Road-Bridge Practice in New South Wales. *Institution of Civil Engineers.* London.

Earl, B. 2016. *ADXL345 Digital Accelerometer User's guide* [Online]. Adafruit Industries. Available: https://learn.adafruit.com/adxl345-digital-accelerometer [Accessed 8 November 2016].

Feltrin, G., Steiger, R., Gsell, D., Gulzow, A. & Wilson, W. 2010. Serviceability assessment of a wooden trough bridge by static and dynamic tests. *World Congress of Timber Engineering (WCTE).* Riva Del Garda, Italy.

Ford Motor Company. 2016. *Ford Ranger 22 Aug 2016 eBrochure* [Online]. Available: http://www.ford.com.au/ [Accessed 1 September 2016].

Frýba, L. 1996. *Dynamics of Railway Bridges,* London, Thomas Telford Services Ltd.

Frýba, L. 2013. *Vibration of solids and structures under moving loads,* e-book, Springer Netherlands.

Giangrandi, I. 2014. *Measuring the Q-factor of a resonator with the ring-down method* [Online]. Switzerland. [Accessed 3 November 2016].

Hamm, P., Schänzlin, J., Francke, W. & Scheuble, S. Monitoring of Timber Bridges by Repeatedly Measuring the Natural Frequency. Proceedings of the World Conference on Timber Engineering (WCTE 2016), August 22-25 2016 Vienna, Austria.

Inaudi, D. 2011. Cost-Benefits Analysis in SHM Projects. *5th International Conference on Structural Health Monitoring of Intelligent Infrastructure (SHMII-5).* Cancun, Mexico.

Lalthlamuana, R. & Talukdar, S. 2014. Dynamic response of bridge subjected to eccentrically moving flexible vehicle: a semianalytical approach. *Shock and Vibration,* 2014.

Lilley, D. & Winslade, J. 2014. Traffic-Generated Vibration of Highway Bridges. *ASEC 2014 Conference Structural Engineering in Australasia - World Standard.* Auckland, NZ.

Moore, G.E. 2006. Cramming more components onto integrated circuits, Reprinted from Electronics, volume 38, number 8, April 19, 1965, pp. 114 ff. *IEEE Solid-State Circuits Newsletter,* 3, 33-35.

Moore, J.C. & Mahini, S. 2015. Measuring Dynamic Impact on Aged Timber Bridges; Some Experimental Options. *PLSE 2015 - Second International Conference on Performance-based and Life-cycle Structural Engineering* Brisbane, Australia.

Moore, J.C., Mahini, S.S., Glencross-Grant, R. & Patterson, R.A. 2013. Structural health monitoring of older timber bridge girders using laser-based techniques. *Australian Journal of Structural Engineering,* 14, 27-42.

Neitzel, F., Niemeier, W., Weisbrich, S. & Lehmann, M. Investigation of low-cost accelerometer, terrestrial laser scanner and ground-based radar interferometer for vibration monitoring of bridges. Proceedings of the 6th European Workshop on Structural Health Monitoring (EWSHM 2012), 2012. 542-551.

Neitzel, F., Schwanebeck, T. & Schwarz, W. 2007. Zur Genauigkeit von Schwingwegmessungen mit Hilfe von Beschleunigungs- und Geschwindigkeitssensoren. *Allgemeine Vermessungs-Nachrichten (AVN).*

RTA 2007. Bridge Inspection Procedure Manual. Second ed.: Roads and Traffic Authority of New South Wales.

Salawu, O. 1997. Detection of structural damage through changes in frequency: a review. *Engineering structures,* 19, 718-723.

Samali, B., Li, J. & Crews, K. 2007. Load Rating of impaired Bridges Using a Dynamic Method. *Electronic Journal of Structural engineering,* Special Issue, 66 - 75.

Samali, B., Li, J., Dackermann, U. & Choi, F. 2011. Vibration-Based Damage Detection for Timber Structures in Australia. *In:* CHAN, T. & THAMBIRATNAM, D. (eds.) *Structural Health Monitoring in Australia.*

Sammartino, G., Amirouche, F., Chen, Z.L. & Wang, M.L. 1999. Effects of heavy vehicle speeds on the structure and vibration of bridges. *American Society of Mechanical Engineers, Design Engineering Division (Publication) DE,* 101, 139-147.

Shi, X. & Cai, C. 2008. Suppression of vehicle-induced bridge vibration using tuned mass damper. *Journal of Vibration and Control.*

Timoshenko, S.P. 1953. *History of Strength of Materials,* New York, USA, Dover Publications, Inc.

Tokyo Measuring Instruments Laboratory Co. Ltd. (TML) 2013. Multi-recorder TMR-200 with TMR 231 Voltage recording unit.

Worden, K., Farrar, C., Manson, G. & Park, G. 2007. The fundamental axioms of structural health monitoring. *Proceedings of the Royal Society,* 463, 1639-1664.

The 8th International Conference on
Structural Health Monitoring of Intelligent Infrastructure
Brisbane, Australia | 5-8 December 2017

Damage Identification using Temperature-Driven SHM

M. Yarnold[1], B. Murphy[2], J. Reilly[3], and B. Glisic[3]
[1] *Texas A&M University - USA, Email: myarnold23@gmail.com,* [2] *Tennessee Technological University – USA,*
[3] *Princeton University - USA*

Abstract

Temperature-driven structural health monitoring (SHM) of bridge structures has progressed over the past several years due to its advantages over other conventional forms of monitoring. The concept for temperature-driven evaluation utilizes thermal inputs as excitations to the structural system. Therefore, data acquisition of these excitations, along with the corresponding responses (e.g. strains or displacements), allows for identification of unique signatures (or baselines) of the structure. Some of the logistical advantages of temperature-driven SHM are the large measurement signal-to-noise ratios, minimal data storage and time synchronization requirements, and relatively inexpensive equipment. However, research has indicated the biggest advantage is the sensitivity of temperature-driven signatures to realistic structural changes, potentially as a result of damage. Recently, a long-span cantilever truss bridge has been instrumented with a temperature-driven monitoring system that includes localized temperature, strain, and displacement measurements. A numerical study of the structure is presented, which evaluates the sensitivity of the SHM system to realistic damage scenarios.

1. Introduction

Structural Health Monitoring (SHM) systems for long-span bridges have traditionally been vibration-based. This typically includes the measurement of acceleration at discrete points along the structure. From these measurements, a dynamic signature is tracked, which is based in some form on the natural frequencies and/or mode shapes of the structure. These systems have shown benefits; however, they suffer several widely recognized limitations (Brownjohn et al. 2011; Catbas et al. 2007). One of the biggest limitations is that vibration-based signatures have minimal sensitivity to structural damage. Normal environmental conditions can change these signatures in similar ways making it difficult to identify damage (De Roeck 2003). Another problem is that these systems require significant data storage and data processing. Other drawbacks are the unknown nature of the inputs, which are assumed as wide-banded white noise. They are also predicated on modal theory assumptions (e.g., linearity, stationarity, etc.) and have small signal-to-noise ratios in situations of local damage.

To address these challenges, researchers have studied other forms of static systems. One particular type that has shown promise is temperature-driven systems (Cao et al. 2011; Gaoxin and Youliang 2015; Kromanis and Kripakaran 2016; Kromanis et al. 2015; Reilly 2016; Warren and Dubbs 2017; Xia et al. 2017; Yarnold and Moon 2015). These systems measure the environmental changes with a focus on temperature change. The resulting responses of the structure due to these inputs are also recorded. The appreciable response generated from thermal "loads" makes this methodology attractive. Initial studies have also shown the input-output relationship to be sensitive to changes in the structure (Yarnold and Moon 2015).

The study presented in this paper provides an overview of a temperature-driven SHM system installed along a long-span cantilever truss bridge. Then, a numerical study is presented which evaluates the sensitivity of the temperaure-driven system to realistic damage scenarios. The objective is to establish the potential structural changes that can and cannot be identified with the implemented temperature-driven system.

2. Hurricane Bridge SHM System

The Hurricane Bridge is the long-span cantilever truss bridge monitored as part of this study. The structure is located in Tennessee, USA, and is owned by the Tennessee Department of Transportation. Overall, the length is roughly 545 m (1787 ft), which includes a four-span Warren deck truss (Figure 1).

Figure 1: Hurricane Bridge

The middle of the bridge includes two suspended sections. These sections are simply supported between a pin and hanger connection and the center pier (Figure 2). The main truss members are built-up steel sections, which are common for the age of contruction (1949). The floor system supporting the roadway is a conventional stringer-floorbeam system with reinforced concrete deck. A significant rehabilitation project was performed in 2011. The main modifications were widening the deck, strengthening select truss members, and installing a "catch system" to increase the redundancy of the pin and hanger details.

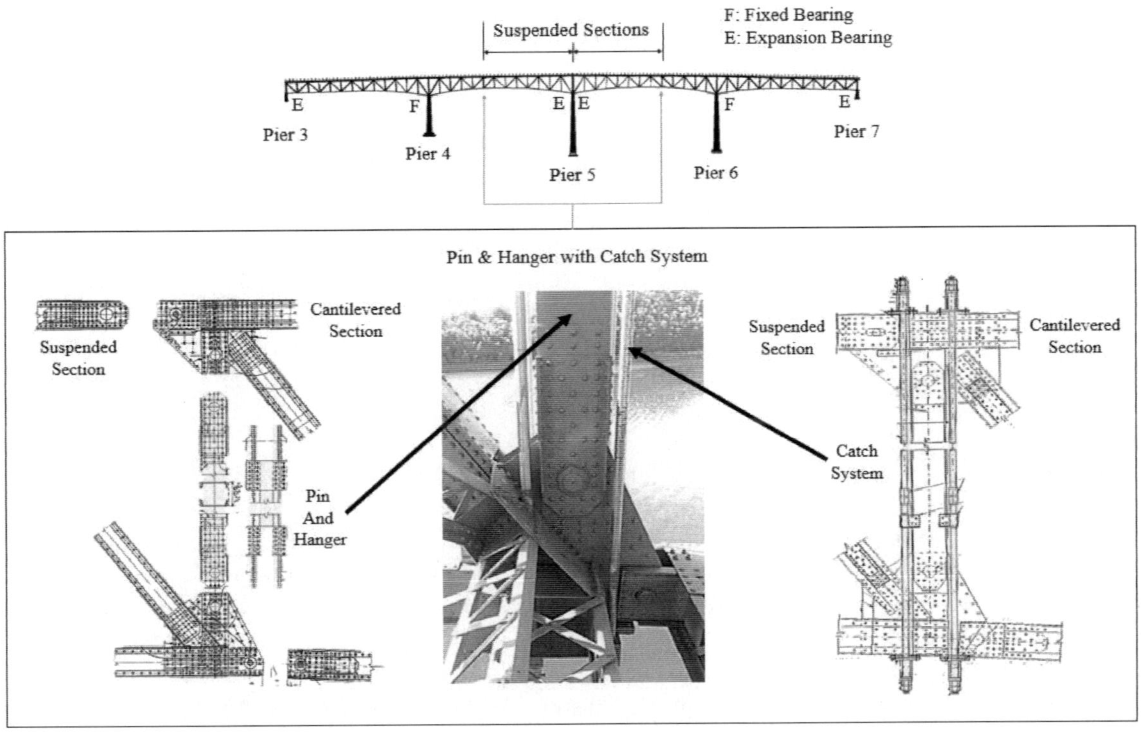

Figure 2: Hurricane Bridge Overview: Pin & Hanger and Catch System Detail

In 2016, a long-term SHM system was implemented along the Hurricane Bridge. The temperature-driven concept was utilized for design of the system (Yarnold et al. 2012). This concept utilizes the continual thermal "loads" as the excitation to the structural system. The corresponding responses (e.g. strains and displacements) are also recorded to obtain unique temperature-driven signatures of the structure. An overview of the Hurricane Bride SHM system design and installation are provided below.

2.1 Design

A 3D finite element (FE) model was developed prior to implementation of the SHM system. The FE model includes the primary superstructure and substructure components of the bridge. This model was then utilized for running sensitivity studies to best select measurement locations for the long-term SHM system. Further details regarding the FE model and sensitivity studies used for the design can be found in Murphy and Yarnold (2017).

The final SHM system design is shown in Figure 3. Due to the symmetric nature of the bridge only half of the main truss spans were instrumented. To capture member strains, four truss cross-sections were selected. This includes two strain gages per truss member on the upper chord, diagonal, and lower chord for both the upstream (North) and downstream (South) sides (12 strain gages per truss section). Eight strain gages were also installed along the pin and hanger system. This resulted in 56 strain gages, which also include thermistors for temperature measurements. In addition, eight displacement gages (with thermistors) were included at two expansion bearing and joint locations.

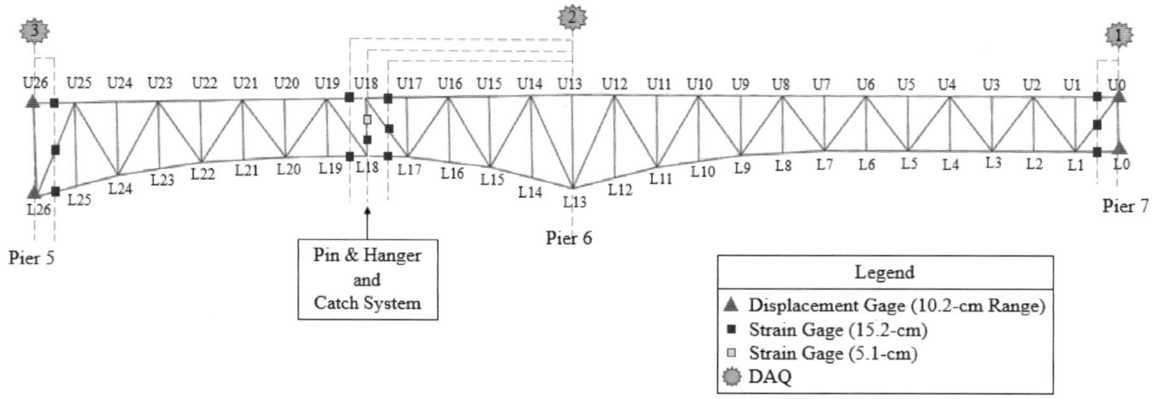

Figure 3: Final SHM System Design

2.2 Installation

The installation of the SHM system was performed in roughly one week with an average of four people on-site. Vibrating wire sensors were selected for the study to provide stable long-term measurements. These sensors were divided into three groups. Each group of sensors were cable-connected to a local data acquisition system (DAQ). The local DAQ's collect the data and, then through wireless communication, transmit to a ground station installed at the base of Pier 7. This allows for easy future access to the system if required. All DAQ's (including the ground station) are powered with solar panels. To balance the measurement resolution desired with power and data storage demands, a sampling rate of once every five minutes was selected for all sensors (56 strain, 8 displacement, and 64 temperature). Further details regarding the SHM system installation can also be found in Murphy and Yarnold (2017).

3. Sensitivity Study

When exposed to temperature variations, structures expand/contract accordingly. Under ideal conditions, the structure behaves as designed (e.g. expansion bearings are able to move with little to no impedence, all truss members are intact and fully functioning, etc.). With these conditions, the behavior of the movement mechanisms of the Hurricane Bridge (shown in Figure 4) would have zero rigidity and move freely. However, in reality, bridge behavior is rarely so simple. Most likely, the expansion bearings are exhibiting some degree of impedance due to the normal degradation of being a nearly 70-year-old structure. To represent this, "realistic" conditions were assumed as approximately 5% rigidity and were applied to the translational movement of the bearings and the rotational movement at both pins of the pin and hanger connection. The structure was simulated by a finite element software using these conditions, producing the primary baseline model used for the sensitivity study.

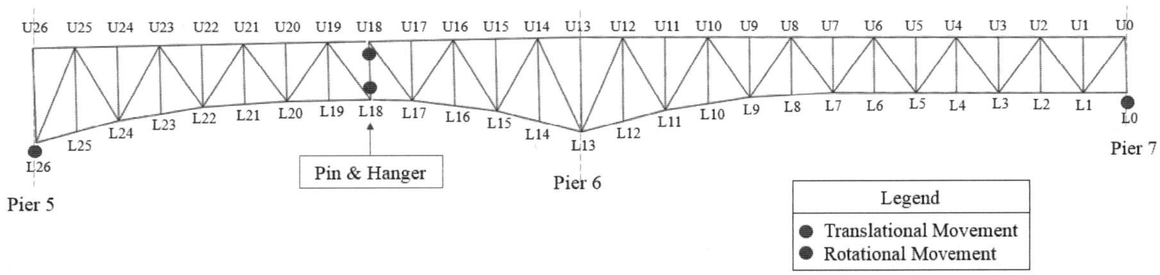

Figure 4: Locations of Movement Mechanisms

Damage identification is directly related to a structure's sensitivity to the measurements taken. For this project, six damage scenarios were chosen to determine the ability of the installed SHM system to identify damage. The first damage scenario (DS1) was chosen to be a frozen bearing near the end of the bridge at Pier 7. To simulate this, the behavior of the northern expansion bearing at Pier 7 was assigned as fully rigid (100% rigidity), which prevented any relative translational movement (between the truss and pier) at that location. Similarly, the second damage scenario (DS2) was chosen to be a frozen northern bearing near the middle of the bridge at Pier 5.

The remaining scenarios were chosen to investigate the structure's sensitivity to fracture of a tension member. Clearly, fracture of a member directly instrumented would be highly sensitive, but the sensitivity of the surrounding members was unknown. These scenarios were simulated by simply removing the member from the model. The third damage scenario (DS3) involved the fracture of a bottom chord member of the suspended section adjacent to the instrumented truss bay (Beam L19-L20 North). The next member (Beam L20-L21 North) was also investigated (DS4). The final two damage scenarios involved the fracture of a top chord member of the cantilevered section. DS5 investigated the sensitivity of the cantilever member adjacent to the instrumented truss bay (Beam U16-U17, North). The final damage scenario (DS6) pertained to the fracture of member U15-U16 North.

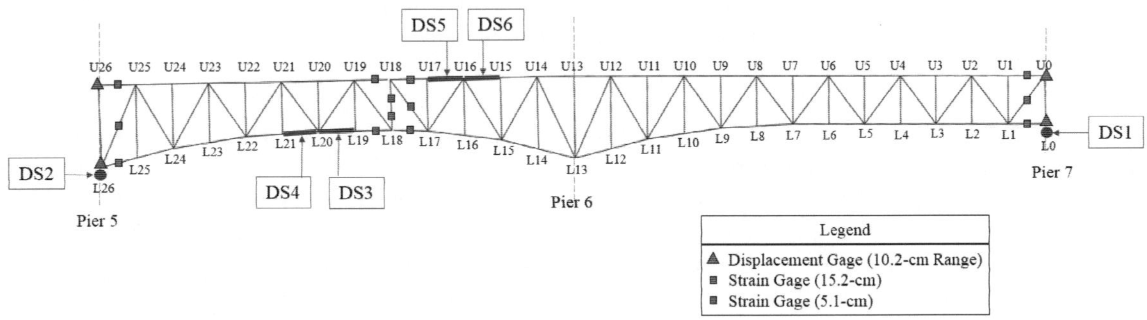

Figure 5: Damage Scenario Locations

The conditions from the damage scenarios were individually applied to the baseline model developed earlier and then simulated with an annual temperature change of 60°C (108°F) to determine the temperature-driven SHM system's ability to identify the damage conditions. Figure 6 shows the restrained (or mechanical) strain sensitivity of each damage scenario compared to the baseline for several of the most sensitive members. The distance from the scenario result and the baseline is indicative of the member's sensitivity and thus its potential to identify damage of that nature. A result in close proximity to the baseline indicates the member is less sensitive to that particular damage scenario. Conversely, a result far from the baseline indicates high sensitivity, and the damage scenario can potentially be identified by the instrumentation of that member.

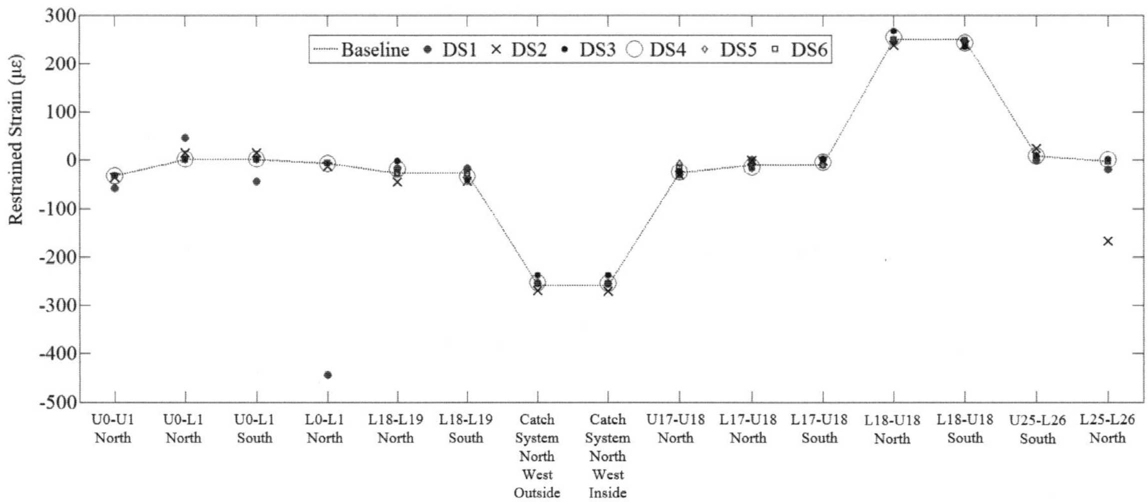

Figure 6: Member Sensitivity to Damage Scenarios

The results from Figure 6 are shown numerically in Table 1 below. The highlighted results were found to be the most sensitve members for each damage scenario. Based on the simulated results, damage corresponding to a frozen bearing (DS1 and DS2) can be identified by the installed SHM system. The damage scenario DS1 resulted in multiple members with significant response, especially beam L0-L1 North which experienced a 439 με of compression from -6 με to -445 με after the temperature change. The upper chord (U0-U1 North) and both diagonals (U0-L1 North and South) also experienced appreciable responses of greater than 25 με. Though not as significant in magnitude, DS2 also produced adequate response for damage identification. The most significant response from this scenario is seen in beam L25-L26 North with a compression of 165 με from -2 με to -167 με after temperature change. The effects of this damage can be seen in several other members with responses around 15 με.

Damage detection of a fractured member was possible as long as the fracture occurred within one truss bay of the instrumented member, as shown by the results of DS3 and DS5. The results also indicate that the detection of damage beyond one bay of the instrumented member (similar to DS4 and DS6) is more difficult, and less likely, with the current SHM system. These scenarios produced member responses of less than 10 με for the most sensitive members. Responses of that magnitude make damage identification difficult because they are approaching the limitations of the sensing equipment (±1 με) and can be masked by other factors or conditions (small signal-to-noise ratio).

Table 1: Sensitivity to Damage Scenarios: Restrained Strain

Bridge Member	Simulated Restrained Strain, $\mu\varepsilon$ (Absolute Difference from Baseline)						
	Baseline	DS1	DS2	DS3	DS4	DS5	DS6
U0-U1 North	-32	-58 (26)	-36 (4)	-33 (1)	-32 (0)	-32 (0)	-32 (0)
U0-L1 North	2	47 (43)	15 (13)	5 (3)	3 (1)	2 (0)	2 (0)
U0-L1 South	2	-43 (45)	16 (14)	5 (3)	3 (1)	2 (0)	2 (0)
L0-L1 North	-6	-445 (439)	-13 (7)	-6 (0)	-6 (0)	-5 (1)	-5 (1)
L18-L19 North	-27	-16 (11)	-44 (17)	-1 (26)	-18 (9)	-26 (1)	-27 (0)
L18-L19 South	-27	-16 (11)	-43 (16)	-40 (13)	-33 (6)	-28 (1)	-27 (0)
Catch System North-West-Outside	-258	-254 (4)	-269 (11)	-237 (21)	-253 (5)	-252 (6)	-255 (3)
Catch System North-West-Inside	-259	-255 (4)	-270 (11)	-238 (21)	-254 (5)	-253 (6)	-256 (3)
U17-U18 North	-25	-30 (5)	-31 (6)	-22 (3)	-24 (1)	-6 (19)	-16 (9)
L17-U18 North	-10	2 (12)	0 (10)	-16 (6)	-14 (4)	-16 (6)	-13 (3)
L17-U18 South	-10	2 (12)	-1 (9)	3 (13)	-4 (6)	-10 (0)	-9 (1)
L18-U18 North	249	246 (3)	238 (11)	268 (19)	254 (5)	250 (1)	250 (1)
L18-U18 South	249	246 (3)	239 (10)	235 (14)	244 (5)	248 (1)	248 (1)
U25-L26 South	10	0 (10)	24 (14)	10 (0)	9 (1)	10 (0)	9 (1)
L25-L26 North	-2	-19 (17)	-167 (165)	3 (5)	1 (3)	-1 (1)	-2 (0)

The installed SHM system also showed potential for damage identification of bearing behavior via displacement measurements, as shown in Table 2. DS1 and DS2 produced significant displacement responses regardless of which bearing was damaged. However, damage from a fractured member could not be determined accurately due to the minimal responses obtained. No response from a fracture damage scenario was greater than 0.2 cm.

Table 2: Sensitivity to Damage Scenarios: Displacement

Sensor Location	Simulated Displacement, cm (Absolute Difference from Baseline)						
	Baseline	DS1	DS2	DS3	DS4	DS5	DS6
L0 North	7.9	0.6 (7.3)	10.3 (2.4)	8.0 (0.1)	7.9 (0.0)	7.9 (0.0)	7.9 (0.0)
L0 South	7.9	1.3 (6.6)	10.3 (2.4)	8.0 (0.1)	7.9 (0.0)	7.9 (0.0)	7.9 (0.0)
L26 North	-6.7	-13.1 (6.4)	-3.6 (3.1)	-6.5 (0.2)	-6.6 (0.1)	-6.7 (0.0)	-6.7 (0.0)
L26 South	-6.7	-13.2 (6.5)	-3.9 (2.8)	-6.5 (0.2)	-6.6 (0.1)	-6.7 (0.0)	-6.7 (0.0)

To further show the sensitivity of a temperature-driven SHM system, Figure 7 illustrates the 3D relationship between restrained strain, displacement, and temperature variation. This plot was numerically generated using the FE model at L0-L1 North under DS1 conditions. Graphically it can be seen the variation in strains and displacements with temperature change. The temperature-driven SHM methodology utilizes this sensitivity to maxize the potential for identification of structural damage.

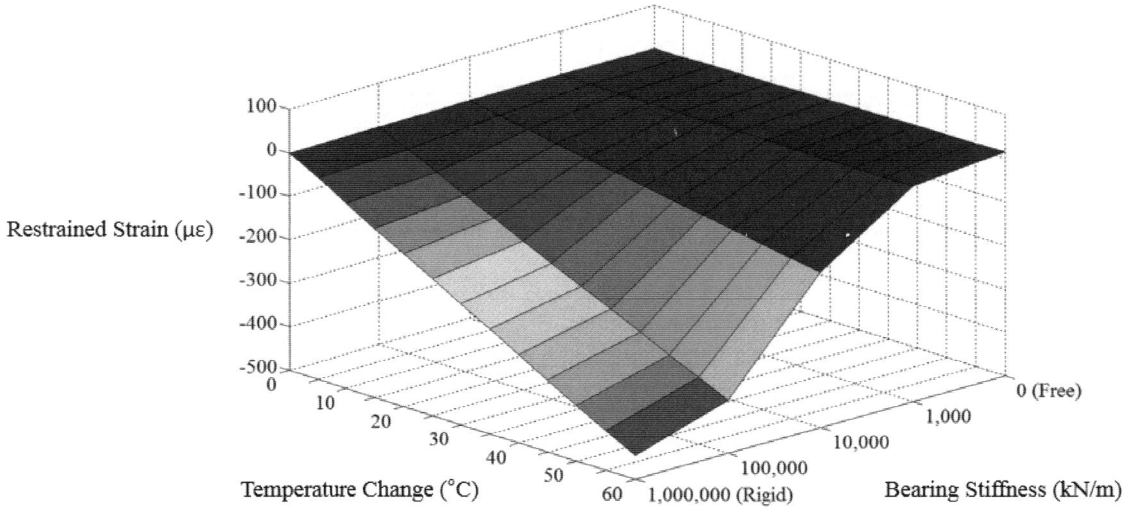

Figure 7: Lower Chord (L0-L1) North Sensitivity to Temperature and Pier 7 North Bearing Stiffness

4. Conclusions and Future Work

The potential for temperature-driven SHM systems to identify structural damage was investigated through evaluation of a recently instrumented long-span cantilever truss bridge. The approach of the study was to numerically simulate realistic damage scenarios. Then, the sensitivity of a temperature-driven baseline for the instrumented locations was recorded due to the fact that the higher the sensitivity the greater likelihood for identification. A total of six damage scenarios were simulated that included breakdown of two different expansion bearings and fracture of four tension members. The numerical study indicated that bearing failure could be identified with the temperature-driven approach. However, fracture of a member could only be identified if the same bay, or an adjacent bay, was instrumented. A fracture occurring further from the sensors would likely not be identified. Overall, the study indicates promise for temperature-driven SHM systems to identify structural damage. The strength of the methodology has shown to be for evaluation of movement systems such as bearings, joints, and connections. Future research will include further damage scenarios analysis of this and other structures. In addition, simulation of vibration-based signatures will be generated so the sensitivity of the natural frequences and mode shapes can be compared.

References

Brownjohn, J. W., Stefano, A., Xu, Y.-L., Wenzel, H., and Aktan, A. E. (2011). "Vibration-based monitoring of civil infrastructure: challenges and successes." *J Civil Struct Health Monit*, 1(3-4), 79-95.

Cao, Y., Yim, J., Zhao, Y., and Wang, M. L. (2011). "Temperature effects on cable stayed bridge using health monitoring system: a case study." *Structural Health Monitoring*, 10(5), 523-537.

Catbas, F. N., Ciloglu, S. K., Hasancebi, O., Grimmelsman, K., and Aktan, A. E. (2007). "Limitations in Structural Identification of Large Constructed Structures." *Journal of Structural Engineering*, 133(8), 1051-1066.

De Roeck, G. (2003). "The state-of-the-art of damage detection by vibration monitoring: the SIMCES experience." *Journal of Structural Control*, 10(2), 127-134.

Gaoxin, W., and Youliang, D. (2015). "Research on monitoring temperature difference from cross sections of steel truss arch girder of Dashengguan Yangtze Bridge." *International Journal of Steel Structures*, 15(3), 647-660.

Kromanis, R., and Kripakaran, P. (2016). "SHM of bridges: characterising thermal response and detecting anomaly events using a temperature-based measurement interpretation approach." *J Civil Struct Health Monit*, 6(2), 237-254.

Kromanis, R., Kripakaran, P., and Harvey, B. (2015). "Long-term structural health monitoring of the Cleddau bridge: evaluation of quasi-static temperature effects on bearing movements." *Structure and Infrastructure Engineering*, 1-14.

Murphy, B. R., and Yarnold, M. T. (2017). "Temperature-Driven Assessment of a Cantilever Truss Bridge." *Structures Congress*.

Reilly, J., Abdel-Jaber, H., Yarnold, M., Glisic, B. "Identification of Steady-State Uniform Temperature Distributions to Facilitate a Temperature Driven Method of Structural Health Monitoring." *Proc., Proceedings SPIE*, 980521-980521-980527.

Warren, M. E., and Dubbs, N. C. (2017). "Periodic Temperature-Based Bearing Assessment of a Steel Multigirder Bridge." *Transportation Research Record: Journal of the Transportation Research Board*, 2642, 26-34.

Xia, Q., Cheng, Y., Zhang, J., and Zhu, F. (2017). "In-Service Condition Assessment of a Long-Span Suspension Bridge Using Temperature-Induced Strain Data." *Journal of Bridge Engineering*, 22(3).

Yarnold, M. T., and Moon, F. L. (2015). "Temperature-Based Structural Health Monitoring Baseline for Long-Span Bridges." *Engineering Structures*, 86(0), 157-167.

Yarnold, M. T., and Moon, F. L. (2015). "Temperature-based structural health monitoring baseline for long-span bridges." *Eng. Struct.*, 86, 157.

Yarnold, M. T., Moon, F. L., Dubbs, N. C., and Aktan, A. E. (2012). "Evaluation of a Long-Span Steel Tied Arch Bridge using Temperature-Based Structural Identification." *International Association for Bridge Management and Safety*Stresa, Italy.

The 8th International Conference on
Structural Health Monitoring of Intelligent Infrastructure
Brisbane, Australia | 5-8 December 2017

Sparse Regularization for Structural Damage Identification

R.R. Hou[1], Y. Xia [1].
[1] *Department of Civil and Environmental Engineering, The Hong Kong Polytechnic University – Hong Kong, China,*
Email: rongrong.hou@connect.polyu.hk

Abstract

The vibration-based damage detection which utilizes measured vibration characteristics to identify structural damage is an inverse problem. In general, the number of potential damage locations is greater than that of available measurements that will result in an underdetermined problem. The conventional vibration-based damage detection methods employ the Tikhonov regularization to deal with this underdetermined problem. The Tikhonov regularization tends to provide over smooth solutions, which causes the identified damage distributed to many structural elements. However, this result does not match the practical situation. Compared with the total elements of the entire structure, structural damage typically occurs at a small number of locations only. In this study, the l_1 regularization technique is employed to exploit the sparsity of the structural damage. The sensitivity-based model updating is adopted using frequency and mode shape changes for structural damage detection. Two numerical examples including a cantilever beam and a planar truss are utilized to demonstrate the effectiveness of the proposed sparse damage detection method. The results showed that the proposed sparse damage detection method is effective in identifying single and multiple damages for different types of structures even under the condition of small frequency and mode shape changes. Moreover, the robustness of the proposed method to noise is investigated through simulation studies.

1. Introduction

Structural damage detection has become an important research topic and a considerable amount of research work has been carried out in this area in the past decades. Due to its efficiency in practice and simplicity of implementation, the vibration-based damage detection has attracted considerable attention (Doebling *et al.* 1996, Sohn *et al.* 2003). The vibration-based damage detection methods are based on the change of the global vibration characteristics of the structure, such as frequencies (Salawu 1997, Kim & Stubbs 2003), mode shapes (Shi *et al.* 2000, Xia *et al.* 2002), and their variants (Pandey *et al.* 1991, Pandey & Biswas 1994), to identify damage.
Finite element (FE) model updating is a widely used technique that uses vibration properties to locate and quantify structural damage. Natural frequencies are often more accurate, reliable and simpler to be measured than other vibration properties, but they are not very sensitive to local damage especially for large scale structures. Mode shapes could provide both local and spatial information of a structure and thus more sensitive to local damage. However, it is difficult to measure high-order mode shapes accurately. Other variants, such as mode shape curvature and flexibility, also have several pros and cons (Carden & Fanning 2004). The major difficulty is associated with the process of deriving these variants from mode shapes which is prone to noise amplification and will produce some unavoidable errors. Among previous studies, frequencies and mode shapes are the most widely used indexes for damage detection.

Civil structures generally contain a large number of components, especially for large-scale structures. However, for the sensitivity-based model updating, the number of available vibration measurements is typically less than that of potential damage locations resulting in an underdetermined problem. In addition, the sensitivity-based model updating for damage detection is usually ill-posed that a small perturbation in the input data (for example, measurement noise) would lead to enormous changes in the solution. Therefore, conventional vibration-based damage detection methods commonly employ Tikhonov regularization (or l_2 regularization) to deal with this problem (Ahmadian *et al.* 1998, Weber *et al.* 2009, Li & Law 2010). The Tikhonov regularization tends to provide over smooth solutions, which causes the identified damage distributed to many structural elements. However, this does not match the practical situation in which damage usually occurs in a few elements or sections only especially at the early stage.

Structural damage which occurs at a small number of locations is sparse in spatial domain. Therefore, the identification of the damaged elements can be treated as a sparse recovery problem. The sparsity of a vector is defined that only a small number of elements are nonzero and the rest are zero or very close to zero. According to the sparse recovery theory, the sparse vector can be recovered using a small number of measurement data only. The sparse recovery theory has received widely used in solving inverse problems which possess sparsity in time domain or spatial domain (Theodoridis *et al.* 2012). However, the sparse recovery theory starts to attract interest and begins to be applied in structural health monitoring (SHM) until very recently (Candès 2006, Baraniuk 2011). Wang and Hao (2010) discussed several applications of compressive sensing (CS) in structural engineering. These potential applications included improvement of sampling, optimization of model updating and proposition of innovative damage indicators. Bao *et al.* (2010) applied the CS technique to vibration data compression. Compared to the traditional data compression methods, CS enabled directly compressing the measurement data. Morever, Yang and Nagarajaiah have conducted a series of research about applying the sparsity and low rank algorithms to structural identification (Yang & Nagarajaiah 2013, 2015) and SHM problems (Yang & Nagarajaiah 2014).

For damage detection, Zhou *et al.* (2013) proposed a damage identification method based on the substructural sensitivity analysis and the sparse constraints regularization. Hernandez (2014) presented a l_1-based sensitivity approach to localize and quantify isolated structural damage using incomplete modal information. Zhou *et al.* (2015) developed a new l_1 regularization approach to detect structural damage using the first few frequency data. Experimental results showed that the proposed method is advantageous over the traditional damage detection approach based on the l_2 regularization. Zhang *et al.* (2017) presented a new algorithm which combines the extended Kalman filter and l_1-norm regularization to identify structural damage using free vibration responses. The numerical and experimental studies indicated that the l_1-norm regularization term could able to suppress the interference of measurement noise and thus improve the identification accuracy.

While natural frequencies are usually more accurate and reliable than other vibration properties. they are not sensitive to structural damage. Ambient and environmental effects may cause significant variations (5%-10%) in nature frequencies which are difficult to distinguish from the damage-induced frequency changes. Compared to modal frequencies, mode shapes could provide local and spatial information of a structure and thus more sensitivity to local damage. In this study, the sensitivity-based model updating is adopted using frequency and mode shape changes for structural damage detection. The l_1 regularization technique is employed to exploit the sparsity of the structural damage in order to obtain more accurate and reliable damage identification result. The effectiveness of the proposed method is demonstrated through applications to a numerical cantilever beam and a numerical planar truss.

2. Sparse damage detection using frequencies and mode shapes

The sensitivity-based model updating minimizes the discrepancy between analytical predictions and experimental data in order to find changes in structural parameters. The Tikhonov regularization is commonly introduced to stabilize this inverse problem

(1)

where $R(\{p\})$ and $\{R^E\}$ are the analytical modal parameters and the measured ones, respectively; $\{p\}$ ($-1 \le p_i \le 0$) is the stiffness reduction factor (SRF); and $\beta > 0$ is the regularization parameter. The residual norm $\|R(\{p\}) - \{R^E\}\|_2^2$ is the data-fitting term measuring the fidelity of the updated parameters to the measurement data, the solution norm $\|p\|_2^2$ is the regularization term quantifying the fidelity to the prior information of the solution. $p_i = 0$ indicates that the *i*th element is intact; while $p_i = -1$ means the element is completely damaged. Therefore, $\{p\}$ is a sparse vector which has only several non-zero items at the damaged locations. The values of SRF indicate both damage location and damage severity.
As mentioned previously, the structural damage is sparse in spatial domain and thus the l_1-regularization instead of the Tikhonov regularization is introduced to the model updating to enforce sparsity in the solution

(2)

In this study, both natural frequencies and mode shapes are utilized in model updating for damage detection. Based on Equation (2), the optimization problem is further rewritten as

(3)

where $\{\lambda_i\}$ and $\{\phi_{ji}\}$ are the *i*th eigenvalues and the *i*th mode shape at the *j*th point, respectively, *m* is the number of measured eigenvalues, *np* is the number of measurement points, *n* is the element number in the FE model, and subscripts "*A*" and "*E*" represent the items obtained from the analytical FE model and experiment, respectively. Eigenvalues are used instead of frequencies for the convenience of the sensitivity calculation. In addition, in order to make the three terms in Equation (3) comparable, the eigenvalue residual, eigenvector residual and the regularization term are divided by the length of the vectors (i.e., *m*, $m \times np$ and *n*, respectively).

3. A Cantilever Beam

3.1. Model Description

A cantilever beam as shown in Figure 1 is first utilized as a numerical example for preliminary study. The total length of the beam is 900 mm and the size of the cross section is $50.75 \times 6.0\,\text{mm}^2$. The mass density and Young's modulus are estimated as $7.67 \times 10^3\,\text{kg/m}^3$ and $2.0 \times 10^{11}\,\text{N/m}^2$, respectively. The beam is modelled by 90 Euler-Bernoulli beam elements each 10 mm long with 91 nodes. The damage is simulated by the reduction of Young's modulus of the element. Two damage scenarios as listed in Table 1 are considered. One has single damage and the other contains two damaged elements.

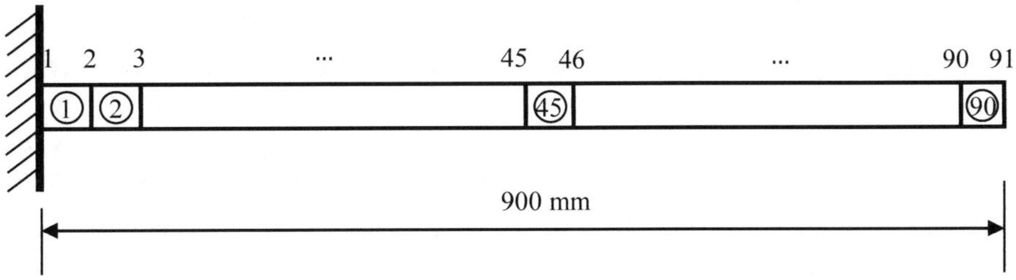

Figure 1. Geometric configuration of a cantilever beam (not to scale)

Table 1. Damage scenarios of the beam

Scenario	Element no.	Damage severity
1	1	40%
2	1	50%
	45	50%

In order to obtain the mode shape of the beam, the vertical accelerations are assumed to be measured every 100mm counting from the clamped end (i.e., nodes 11, 21, 31, 41, 51, 61, 71, 81 and 91). The first six nature frequencies and mode shapes are employed for damage identification. In consideration of the measurement noise effect, random noise with the normal distribution is directly added to modal parameters. For practical ambient vibration test, it is suggested that natural frequencies may contain 1% noise and mode shapes may contain 8%-10% noise (Mottershead & Friswell 1993, Allemang 2003). For damage scenario 1, 1% and 10% random noise were introduced into the natural frequencies and mode shapes. To examine the robustness of the proposed sparse damage detection method to noise, two noise levels are considered for damage scenario 2. For noise level 1, 1% and 10% random noise were introduced into the natural frequencies and mode shapes, respectively; for noise level 2, 2% and 20% random noise were added to the natural frequencies and mode shapes, respectively. The natural frequencies and modal assurance criterion (MAC) of the undamaged and damaged structure are listed in Table 2. Since random noise is directly added to modal parameters, a few natural frequencies in the damaged states are higher than the undamaged ones (Table 2).

Table 2. Frequencies and MAC of the beam in the undamaged and damaged states

Mode no.	Undamaged	Damage Scenario 1		Damage Scenario 2 (Noise level 1)		Damage Scenario 2 (Noise level 2)	
	Freq. (Hz)	Freq. (Hz)	MAC	Freq. (Hz)	MAC	Freq. (Hz)	MAC
1	5.88	5.77 (−1.84)	99.94	5.71 (−2.82)	99.95	5.67 (−3.56)	99.87
2	36.81	35.98 (−2.27)	99.72	35.55 (−3.42)	99.49	36.52 (−0.80)	98.90
3	103.01	103.08 (+0.06)	99.34	102.11 (−0.87)	99.47	99.85 (−3.07)	98.99
4	201.73	197.35 (−2.17)	99.55	195.60 (−3.04)	98.47	196.69 (−2.50)	99.20
5	333.22	335.45 (+0.67)	99.40	331.59 (−0.49)	99.07	332.56 (−0.20)	98.28
6	497.30	490.15 (−1.44)	99.05	482.18 (−3.04)	99.28	470.44 (−5.40)	97.71
Average of change (%)		−1.16	99.50	−2.27	99.45	−2.59	98.82

Values in parentheses are the frequency change ratios (%) between the damaged and undamaged states

3.2. Damage Detection using Sparse Regularization

For the l_1-regularized model updating problem in Equation (3), the regularization parameter β controls the trade-off between the data fidelity and the sparsity of the solution. The optimal regularization parameter should keep the residual norm and solution norm small at the same time. In this study, the regularization parameter β is determined utilizing the curves of the residual norm and solution norm versus the regularization parameter. Firstly, the objective function is solved for different β ranging from 0.005~1.0 with the increment of $\Delta\beta$=0.005 which is proved to have sufficient resolution through numerical studies. Then, the associated residual norm and solution norm are calculated for each β.

For damage scenario 1, the resulting curves of the residual norm and solution norm are displayed in Figure 2. With the increase of β, the residual norm increases and the solution norm decreases. When $\beta \geq 0.25$, the solution norm is zero, indicating the solution of the l1-regularized problem is zero, i.e., no any damage is detected. Therefore, the regularization parameter β should be selected less than 0.25 in order to obtain a reasonable solution. When β is in the range of 0.065 to 0.25, the values of the residual norm and solution norm are almost unchanged with the increase of β. Therefore, the inflection point β=0.065 is selected as the optimal regularization parameter which keeps these two norms small at the same time. The corresponding damage identification result that is represented by SRFs is shown in Figure 3. The damage location and severity are determined accurately and no false identification occurs. Figure 4 illustrates the identified SRFs for damage scenario 2. For noise level 1, the damaged elements 1 and 45 are identified correctly while there are small errors with the damage severities. For noise level 2, the damage locations can be determined, but the identified damage severities are less than the true values (i.e. $SRF_{1,45}$= 50%). The identification result is still acceptable even under a high noise level.

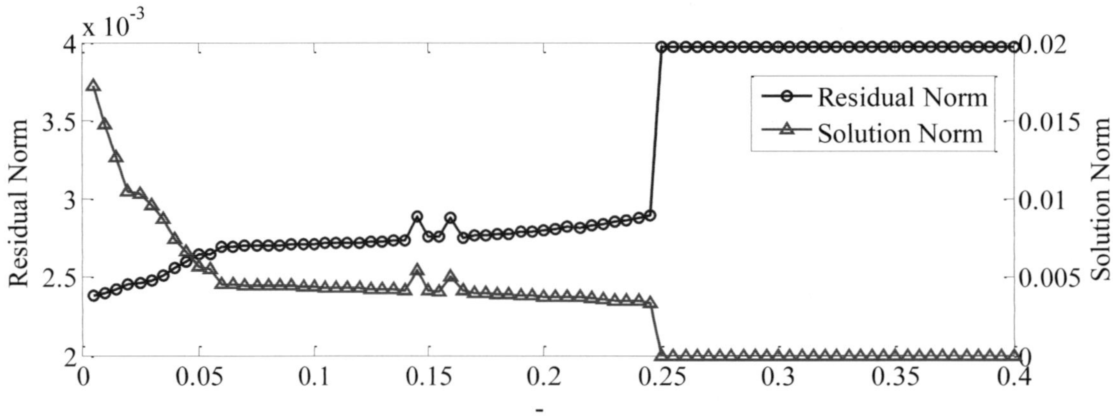

Figure 2. Residual norm and solution norm for different β for damage scenario 1

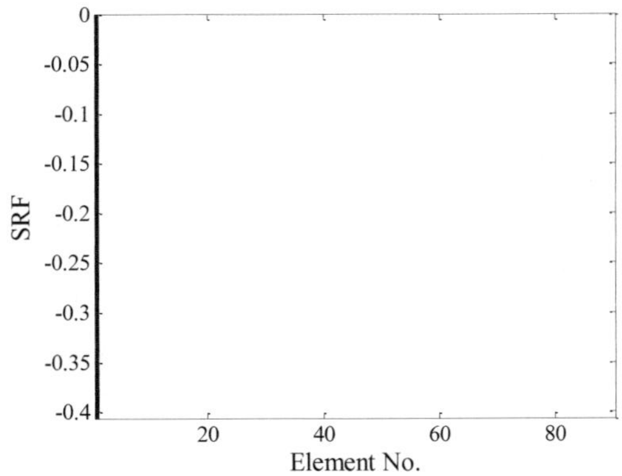

Figure 3. Damage identification result for damage scenario 1 (β = 0.065)

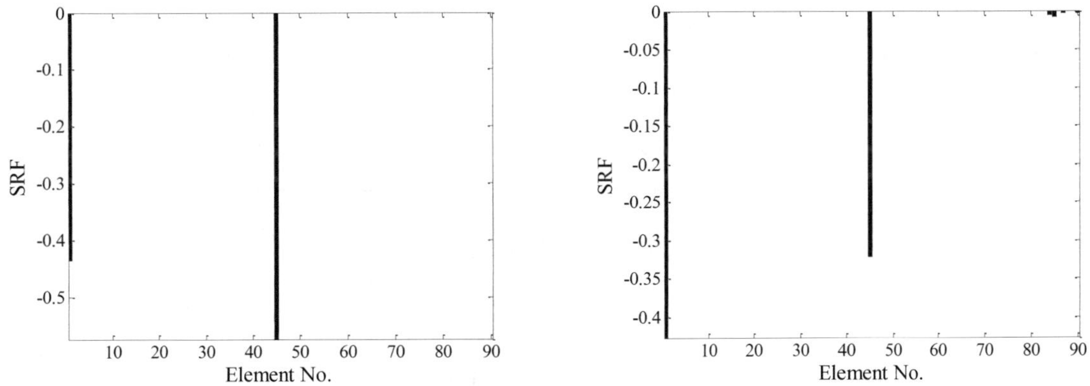

(a) Noise level 1 (β = 0.05)

(b) Noise level 2 (β = 0.055)

Figure 4. Damage identification result for damage scenario 2

4. A Planar Truss

4.1. Model Description

A 6-bay planar truss structure shown in Figure 5 is used as the second numerical example. It is simply supported and composed of 31 bar elements with 25 DOFs. The length of the horizontal and vertical bars is 1.50 m and the diagonal bar is 2.12 m long. The section dimension of each bar is $0.05\text{m} \times 0.05\text{m}$. The mass density and Young's modulus are estimated as 2.77×10^3 kg/m^3 and 7.0×10^{10} N/m^2, respectively. In this study, two elements are considered to be damaged and the damage is simulated by the reduction of Young's modulus of the element while mass remaining unchanged. The horizontal element 3 is damaged with 40%, i.e., SRF_3=−40% and the damage severity of the diagonal element 24 is SRF_{24}=-50%.

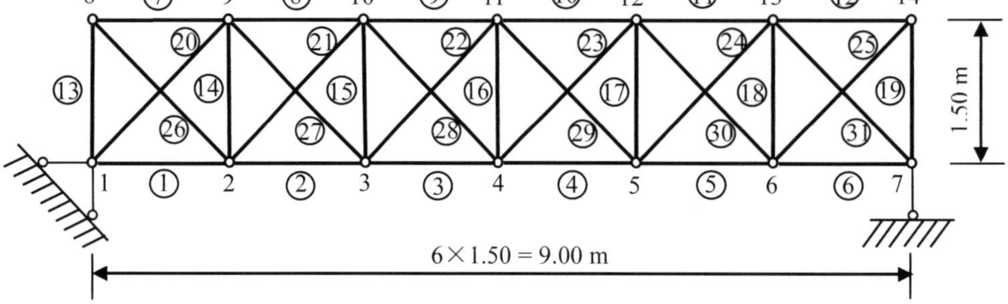

Figure 5. Geometric configuration of the truss structure

The accelerations of all 25 DOFs are assumed to be available in order to obtain the complete mode shape. The first six nature frequencies and mode shapes are employed for damage identification. As introduced above, 1% and 10% random noise were introduced into the natural frequencies and mode shapes, respectively. The natural frequencies and MAC of the undamaged and damaged structure are listed in Table 3.

Table 3. Frequencies and MAC of the truss in the undamaged and damaged states

Mode no.	Undamaged state	Damaged state	
	Freq. (Hz)	Freq. (Hz)	MAC
1	36.92	34.66 (–6.12)	99.97
2	77.11	75.99 (–1.46)	99.84
3	135.59	136.44 (+0.62)	99.99
4	226.36	216.55 (–4.33)	95.79
5	253.52	252.25 (–0.50)	98.94
6	364.15	345.91 (–5.01)	92.82
Average of changes (%)		–2.80	97.89

Values in parentheses are the frequency change ratios (%) between the damaged and undamaged states

4.2. Damage Detection using Sparse Regularization

The optimal regularization parameter is determined as β=0.045 using the previous strategy. Figure 6 shows the damage identification result for the truss structure. The damage locations are determined correctly and the identified SRFs are very close to the true values. There is no false identification for other undamaged elements.

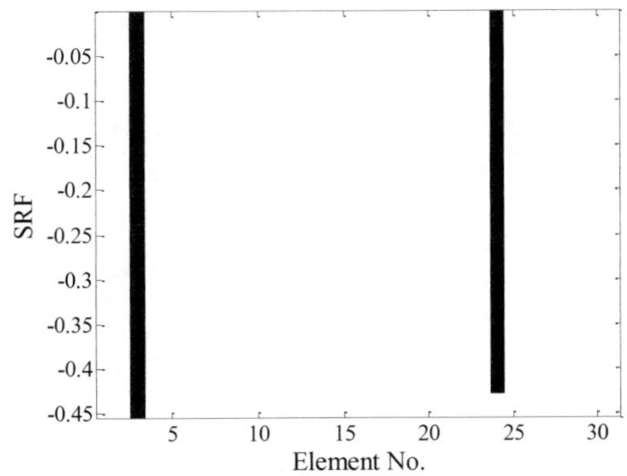

Figure 6. Damage identification result for the truss structure (β = 0.045)

5. Conclusions

This study developed a new sensitivity-based model updating method for damage identification using frequency and mode shape changes. In consideration of the sparsity of the structural damage, the l_1 regularization technique is employed to locate the sparse damage and quantify damage severity using a relatively small number of measurements. The advantages of using frequency changes and mode shape changes for damage detection are that frequencies are usually more accurate and reliable than other vibration properties and mode shapes are more sensitive to local damage. Two numerical examples demonstrate that the proposed damage detection method is effective in identifying single and multiple damages even under the condition of small frequency and mode shape changes. Moreover, numerical studies show that the proposed method has good robustness to noise. The damage identification result is still acceptable when the vibration data is under severe noise contamination.

References

Doebling, SW, Farrar, CR, Prime, MB & Shevitz, DW 1996, 'Damage identification and health monitoring of structural and mechanical systems from changes in their vibration characteristics: a literature review', *Los Alamos National Laboratory Report*.

Sohn, H, Farrar, CR, Hemez, FM, Shunk, DD, Stinemates, DW, Nadler, BR & Czarnecki, JJ 2003, 'A review of structural health monitoring literature: 1996–2001', *Los Alamos National Laboratory Report*.

Salawu, OS 1997, 'Detection of structural damage through changes in frequency: a review', *Engineering Structures*, vol. 19, no. 9, pp. 718-723.

Kim, JT & Stubbs, N 2003, 'Crack detection in beam-type structures using frequency data', *Journal of Sound and Vibration*, vol. 259, no. 1, pp. 145-160.

Shi, Z, Law, S, & Zhang, L 2000, 'Damage localization by directly using incomplete mode shapes', *Journal of Engineering Mechanics*, vol. 126, no. 6, pp. 656-660.

Xia, Y, Hao, H, Deeks, AJ & Zhu, X 2008, 'Condition assessment of shear connectors in slab-girder bridges via vibration measurements', *Journal of Bridge Engineering*, vol. 13, no. 1, pp. 43-54.

Pandey, A, Biswas, M & Samman, M 1991, 'Damage detection from changes in curvature mode shapes', *Journal of Sound and Vibration*, vol. 145, no. 2, pp. 321-332.

Pandey, A & Biswas, M 1994, 'Damage detection in structures using changes in flexibility', *Journal of Sound and Vibration*, vol. 169, no. 1, pp. 3-17.

Carden, EP & Fanning, P 2004, 'Vibration based condition monitoring: a review', *Structural health monitoring*, vol. 3, no. 4, pp. 355-377.

Ahmadian, H, Mottershead, J & Friswell, M 1998, 'Regularisation methods for finite element model updating', *Mechanical Systems and Signal Processing*, vol. 12, no. 1, pp. 47-64.

Weber, B, Paultre, P & Proulx, J 2009, 'Consistent regularization of nonlinear model updating for damage identification', *Mechanical Systems and Signal Processing*, vol. 23, no. 6, pp. 1965-1985.

Li, X & Law, S 2010, 'Adaptive Tikhonov regularization for damage detection based on nonlinear model updating', *Mechanical Systems and Signal Processing*, vol. 24, no. 6, pp. 1646-1664.

Theodoridis, S, Kopsinis, Y & Slavakis, K 2012, 'Sparsity-aware learning and compressed sensing: An overview', November, [online] Available: http://arxiv.org/abs/1211.5231.

Candès, EJ 2006, 'Compressive sampling', *Proceedings of the International Congress of Mathematicians*, Madrid, Spain, pp. 1433-1452.

Baraniuk, RG 2011, 'More is less: signal processing and the data deluge', *Science*, vol. 331, no. 6018, pp. 717-719.

Wang, Y & Hao, H 2010, 'An introduction to compressive sensing and its potential applications in structural engineering', *The 11th International Symposium on Structural Engineering*, Guangzhou, China, 18–20 December, pp. 1089–1094.

Bao, Y, Beck, JL & Li, H 2010, 'Compressive sampling for accelerometer signals in structural health monitoring', *Structural Health Monitoring*, vol. 10, pp. 235-246.

Yang, Y & Nagarajaiah, S 2013, 'Output-only modal identification with limited sensors using sparse component analysis', *Journal of Sound and Vibration*, vol. 332, no. 19, pp. 4741-4765.

Yang, Y & Nagarajaiah, S 2015, 'Output-only modal identification by compressed sensing: Non-uniform low-rate random sampling', *Mechanical Systems and Signal Processing*, vol. 56-57, pp. 15-34.

Yang, Y & Nagarajaiah, S 2014, 'Structural damage identification via a combination of blind feature extraction and sparse representation classification', *Mechanical Systems and Signal Processing*, vol. 45, no. 1, pp. 1-23.

Zhou, S, Bao, Y & Li, H 2013, 'Structural damage identification based on substructure sensitivity and l_1 sparse regularization', *Civil Structural Health Monitoring Workshop (CSHM-4)*, San Diego, California USA, 12–15 March.

Hernandez, EM 2014, 'Identification of isolated structural damage from incomplete spectrum information using l_1-norm minimization', *Mechanical Systems and Signal Processing*, vol. 46, no. 1, pp. 59-69.

Zhou, XQ, Xia, Y & Weng, S 2015, 'L_1 regularization approach to structural damage detection using frequency data', *Structural Health Monitoring*, vol. 4, no. 6, pp. 571-582.

Zhang, C, Huang, JZ, Song, GQ & Chen, L 2017, 'Structural damage identification by extended Kalman filter with l_1-norm regularization scheme', *Structural Control and Health Monitoring*.

Mottershead, J & Friswell, M 1993, 'Model updating in structural dynamics: a survey', *Journal of Sound and Vibration*, vol. 167, no. 2, pp. 347-375.

Allemang, RJ 2003, 'The modal assurance criterion–twenty years of use and abuse', *Sound and Vibration*, vol. 37, no. 8, pp. 14-23.

The 8th International Conference on
Structural Health Monitoring of Intelligent Infrastructure
Brisbane, Australia | 5-8 December 2017

Improved eigensystem realization algorithm to distinguish spurious modes

C.X. Qu[1], T.H. Yi[2], H.N. Li[2]
[1] *Dalian University of Technology – China, Email:* quchunxu@dlut.edu.cn[2] *Dalian University of Technology – China,*

Abstract

During damage identification or assessment of bridges, the modal parameters, such as frequency, damping and modal shape are also utilized, which shows the importance to identify the modal parameters precisely. Eigensystem realization algorithm (ERA) is one of the most popular identification methods. However,the noises in practical environment can influent the effectiveness of ERA, which can introduce spurious modes. This paper proposes an improved ERA to distinguish the spurious modes. Any two Hankel matrices with one time step shift used in ERA have been proved. The singular value matix is obtained by singular value decomposition of Hankel matrix, whose rank is concerned by truncated the different singular values. The noises can make the singular values hard to be truncated. Through moving the start point of time recording to compose two Hankel matrices can help to observe the change of singular values. If the singular vaues are changed with the moving time recording, the identified modes are spurious. To determine the rank initially, the stabilization diagram is employed, which can give the interested modes first that may contain spurious. Then the moving data diagram can be built to show the singular value changing, which picks the exact modes. Finally, the proposed method is verified by a numerical example. The results demonstrate that the improved ERA can distinguish the spurious modes.

1. Introduction

The modal identification for civil engineering has attracted many researchers, which can be used for damage identification, response evaluation and so on. For civil engineering, the excitation is unknown and cannot be measured. The operational modal analysis (OMA) is preferred for engineers, which only use the measured data. There are many different methods to be developed and tested, such as ARMA models (Autoregressive Moving Average) (Akaike, 1969, Vu et al., 2011), NExT (The natural excitation technique) (James et al., 1995, Caicedo, 2011), ITD method (The Ibrahim Time Domain Technique) (Ibrahim & Pappa, 1982, Olsen & Brincker, 2013), random decrement method (RDT) (Ibraham, 2001), stochastic subspace identification method (SSI) (Van Overschee & De Moor, 2012), eigensystem realization algorithm (ERA) (Juang & Pappa, 1985, Qin et al., 2001), Blind modal identification (Nagarajaiah & Yang, 2015), the identification based on Bayesian method (Au et al., 2013, Yan & Katafygiotis, 2015).

Due to the complicate environment of practical engineering, the measurement always contains noise, which can introduce spurious modes during modal identification process. To reduce the spurious mode effect, the singular value decomposition (SVD) technique is always used to estimate system order in identification process because the property of SVD can filter noises to some extent

(Mevel et al., 2002). Magalhaes mentioned stabilization diagram is the most popular approach to separate the physical and spurious modes (Magalhaes et al., 2009). P. LI chose the close number of block-rows and block-columns of Hankel matrix in ERA to separate signal and noise matrices (Li et al., 2011). F.S.V analyzed the sensitivity of poles and provided estimates for the pole error in ERA (Bazan, 2004). Y. Zhang et al. gave the component energy index and used stabilization diagram to indicate spurious modes (Zhang et al., 2005). Cara presented expectation maximization algorithm to identify modes with the parameters estimated by SSI, which is useful to discard the spurious modes (Cara et al., 2012). Marchesiello investigated on the effect of spurious poles on nonlinear dynamics (Marchesiello et al., 2016). However, two problems still have not been solved. First, the noises can affect all the singular values such that the real singular value is hard to be extracted. Second, due to the effect of noises, it is hard to find the stable modes from stabilization diagram clearly.

This paper presents an improved stabilization diagram method in ERA to distinguish the spurious modes, which adds an additional process after stabilization diagram to determine which mode is spurious. The proposed process is achieved by changing the choice of hankel matrices. According to the formulation about the similar state matrix calculation, the noise effect on SVD and structural eigenvalues are illustrated. Finally, the proposed method is investigated by a practical engineering with structural health monitoring (SHM) system.

2. ERA and Stabilization Diagram

2.1 Eigensystem Realization Algorithm

In order to present clearly the spurious mode identification problem, first ERA (Juang & Pappa, 1985) and the stabilization diagram (Peeters & Roeck, 2001) are reviewed briefly. ERA is performed under the condition of impulse input. The formulation of the free pulse response is utilized to deduce the state matrix.

Considering an n degree-of-freedom (DOF) system, the dynamics in discrete time domain is:

$$\begin{aligned}\mathbf{x}_{k+1} &= \mathbf{A}\mathbf{x}_k + \mathbf{B}\mathbf{u}_k \\ \mathbf{y}_k &= \mathbf{C}\mathbf{x}_k + \mathbf{D}\mathbf{u}_k\end{aligned} \tag{1}$$

where k means k-th time step; $\mathbf{A}, \mathbf{B}, \mathbf{C}, \mathbf{D}$ are state matrix, input matrix, output matrix and feed-through matrix in discrete time domain, respectively; $\mathbf{x}_k, \mathbf{y}_k, \mathbf{u}_k$ are state vector, output measurement vector and input vector, respectively. For the impulse input, which means $\mathbf{u}_0 = 1$ and $\mathbf{u}_k = 0, (k \neq 0)$, the measurement $\mathbf{y}_k$ can be used to generate Hankel matrix $\mathbf{H}(k-1)$ as follows:

$$\mathbf{H}(k-1) = \begin{pmatrix} \mathbf{C} \\ \mathbf{CA} \\ \cdots \\ \mathbf{CA}^{p-1} \end{pmatrix} \mathbf{A}^{k-1} \begin{pmatrix} \mathbf{B} & \mathbf{AB} & \ldots & \mathbf{A}^{k+q-1}\mathbf{B} \end{pmatrix} = \mathbf{O}_p \mathbf{A}^{k-1} \mathbf{C}_q \tag{2}$$

where $\mathbf{O}_p, \mathbf{C}_q$ are the observability and controllability matrices, respectively. $\mathbf{H}(k-1)$ and $\mathbf{H}(k-1)$ can be decomposed by SVD as follows:

$$\begin{aligned}\mathbf{H}(k-1) &= \mathbf{P}\mathbf{A}^{k-1}\mathbf{Q}\\ \mathbf{H}(k) &= \mathbf{P}\mathbf{A}^{k}\mathbf{Q}\end{aligned} \tag{3}$$

The SVD for Eq.(3) is expressed as,

$$\mathbf{H}(k-1) = \mathbf{P}\mathbf{T}\mathbf{T}^{-1}\mathbf{A}^{k-1}\mathbf{R}^{-1}\mathbf{R}\mathbf{Q} = \mathbf{U}\mathbf{\Gamma}^{2}\mathbf{V}^{\mathrm{T}} \tag{4}$$

where

$$\mathbf{U} = \mathbf{P}\mathbf{T},\ \mathbf{V} = \mathbf{Q}^{\mathrm{T}}\mathbf{R}^{\mathrm{T}},\ \mathbf{\Gamma}^{2} = \mathbf{T}^{-1}\mathbf{A}^{k-1}\mathbf{R}^{-1} \tag{5}$$

$$\begin{aligned}\mathbf{U}^{\mathrm{T}}\mathbf{U} &= (\mathbf{P}\mathbf{T})^{\mathrm{T}}\mathbf{P}\mathbf{T} = \mathbf{T}^{\mathrm{T}}\mathbf{P}^{\mathrm{T}}\mathbf{P}\mathbf{T} = 1\\ \mathbf{V}^{\mathrm{T}}\mathbf{V} &= \mathbf{R}\mathbf{Q}(\mathbf{R}\mathbf{Q})^{\mathrm{T}} = \mathbf{R}\mathbf{Q}\mathbf{Q}^{\mathrm{T}}\mathbf{R}^{\mathrm{T}} = 1\end{aligned} \tag{6}$$

Due to Eq.(5) and Eq.(6), the following can be obtain,

$$\mathbf{\Gamma}^{-1}\mathbf{U}^{\mathrm{T}}\mathbf{H}(k)\mathbf{V}\mathbf{\Gamma}^{-1} = \mathbf{\Gamma}^{-1}\mathbf{T}^{-1}\mathbf{A}^{k-1}\mathbf{R}^{-1}\mathbf{R}\mathbf{A}\mathbf{R}^{-1}\mathbf{\Gamma}^{-1} = \mathbf{\Gamma}\mathbf{R}\mathbf{A}\mathbf{R}^{-1}\mathbf{\Gamma}^{-1} = \hat{\mathbf{A}} \tag{7}$$

where $\hat{\mathbf{A}}$ is also minimization realization state matrix which has the same eigenvalue of $\mathbf{A}$. Therefore, it is obvious that the SVD of $\mathbf{H}(k-1)$ can be used into $\mathbf{H}(k)$ to get the minimization realization state matrix.

2.2 Improved stabilization diagram

It is obvious that the property of eigenvalue changing with different *k* can be utilized to distinguish whether the eigenvalue is affected by noise or not. If the eigenvalue is not changed a lot along *k* rising, it means that the noise does not affect the eigenvalue too much. While the eigenvalue is varying obviously, it means that the noise affects the eigenvalue a lot. Then *k* is still rising, and the eigenvalue should be stable, which means that the eigenvalue has been all polluted by noise.
Another important problem is the truncated order of $\mathbf{\Gamma}^{2}$. If the concerned order is chosen as n_t, then $\mathbf{U}$, $\mathbf{\Gamma}$, $\mathbf{V}$ in Eq.(4) are truncated as,

$$\begin{aligned}\mathbf{U}_{n_t} &= \mathbf{U}_{k-1}(:,1:n_t)\\ \mathbf{V}_{n_t} &= \mathbf{V}_{k-1}(:,1:n_t)\\ \mathbf{\Gamma}_{n_t} &= \mathbf{\Gamma}_{k-1}(1:n_t,1:n_t)\end{aligned} \tag{8}$$

Then the order n_t is fixed and the power k is changed, which is used to check whether the eigenvalue of $\hat{\mathbf{A}}_k$ is stable or not. So the improved stabilization diagram is formed by the frequency horizontal axis and the power k vertical axis with fixed order k. The truncated order is determined by the component energy index (CEI) proposed by reference (Zhang et al., 2005).The procedure can be summarized as follows:

Step 1: Determine the truncated order n_t by CEI.

Step 2: Fixed n_t, set k=1, initialize the upper limit for k.

Step 3: Form $\mathbf{H}(k-1)$ and $\mathbf{H}(k)$, do the SVD of $\mathbf{H}(k-1)$, and obtain $\mathbf{U}_{n_t}$, $\mathbf{V}_{n_t}$, $\mathbf{\Gamma}_{n_t}$ in Eq.(8).

Step 4: Obtain the state matrix as shown in Eq.(7).
Step 5: Calculate the frequencies, damping ratios, and modal shapes
Step 6: Plot the stable point in the improved stabilization diagram with power k and frequencies.
Step 7: If the power k does not reach the upper limit, go back to Step 3; If it does, go to Step 8.
Step 8: Find the vertical straight lines formed by the stable points, which are physical mode; Others are spurious.

3. Numerical example

In order to prove the propose method, an numerical example is investigated. To illustrate the spurious mode problem clearly, the example is revised from the reference (Wang et al., 2009) simulation example to be 8 DOF in-plane lumped-mass model. The mass and stiffness for each floor are the mass and stiffness of the first 8 floor in the reference (Wang et al., 2009), whose parameters are list in Table. 1.

Table. 1 Model parameters

Floor No.	1	2~5	6~8
Mass ($\times 10^6$ kg)	1.126	1.100	1.100
Stiffness ($\times 10^6$ N/m)	862.07	862.07	554.17
Damping	Rayleigh damping with first two damping ratios are 5%		

The example structure is excited by an impulse at the beginning time. The sensors are supposed to collect the relative displacements for all the floors. The measurement noises are assumed to be 20% of the variance for the structural free pulse response. The truncated order can be determined by CEI as shown in Fig. 1.

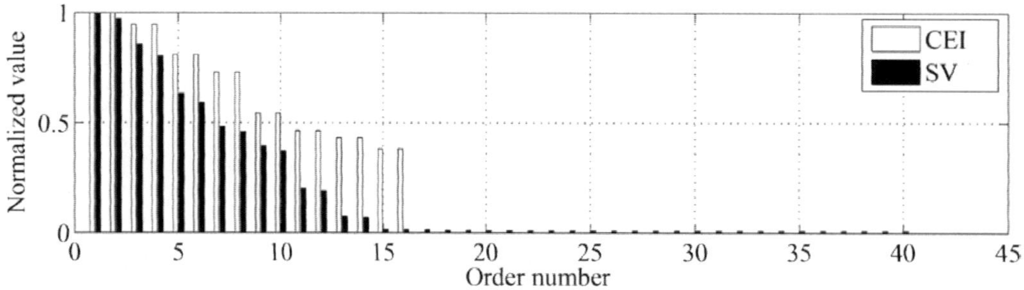

Fig. 1. Normalized singular values and CEIs

where SV means singular value. Therefore, the truncated order n_t is 16. Following the steps in the previous section, the improved stabilization diagram is executed as shown in Fig. 1.

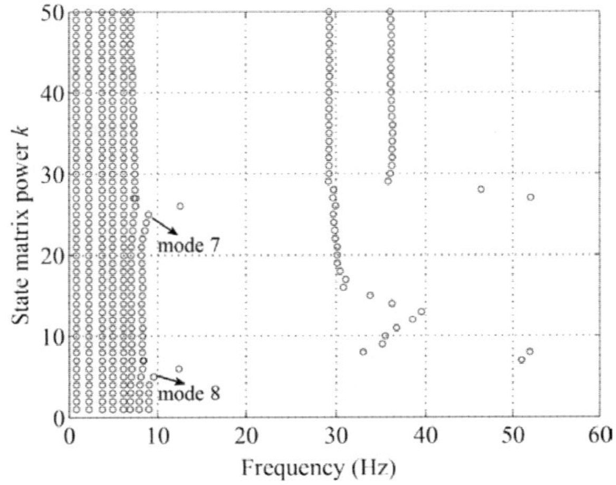

Fig. 2. Improve stabilization diagram for $n_t = 16$.

From Fig. 2, it is founded that mode 7 is affected when k is larger, while mode 8 is almost polluted at the beginning. Along k is rising, the points are not stable, which means that the noises introduce spurious modes. To display the spurious modes, the mode comparisons are shown in Fig. 3.

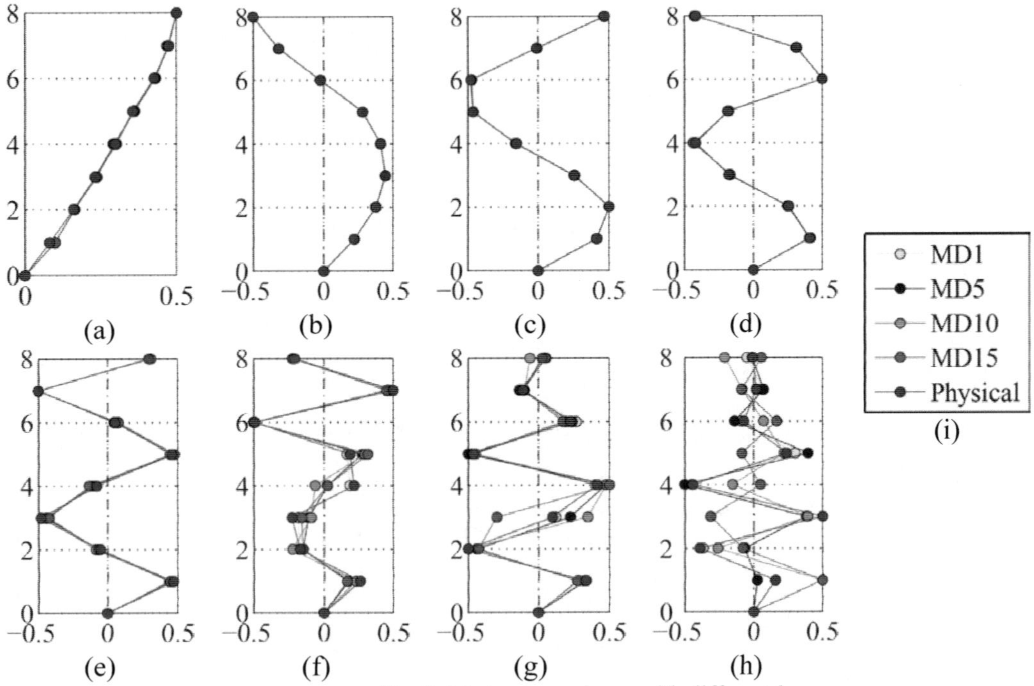

Fig. 3. Mode comparisons with different k.

3. Conclusions

This paper presents an improved stabilization diagram in the ERA to distinguish spurious modes. The different choice of two Hankel matrices with one time step shift is used in the ERA. The truncated order n_t is determined by component energy index. The improved stabilization diagram can be performed by determining truncated order n_t first and plotting the diagram with different power k to distinguish the spurious modes. The proposed method is finally investigated by a numerical example. The results show that the proposed method can distinguish the physical modes and spurious modes clearly.

References

Akaike, H 1969, 'Fitting autoregressive models for prediction', *Annals of the institute of statistical mathematics,* vol. 21, no. 1, pp. 243-247.

Vu, VH, Thomas, M, Lakis, AA & Marcouiller, L 2011, 'Operational modal analysis by updating autoregressive model', *Mechanical Systems and Signal Processing,* vol. 25, no. 3, pp. Pages 1028–1044.

James, GH, Carne, TG & Lauffer, JP 1995, 'The natural excitation technique (NExT) for modal parameter extraction from operating structures', *Modal Analysis-the International Journal of Analytical and Experimental Modal Analysis,* vol. 10, no. 4, pp. 260-277.

Caicedo, JM 2011, 'Practical guidelines for the natural excitation technique (NExT) and the eigensystem realization algorithm (ERA) for modal identification using ambient vibration', *Experimental Techniques,* vol. 35, no. 4, pp. 52-58.

Ibrahim, SR & Pappa, RS 1982, 'Large modal survey testing using the Ibrahim time domain identification technique', *Journal of Spacecraft and Rockets,* vol. 19, no. 5, pp. 459-465.

Olsen, P & Brincker, R (2013) Using random response input in Ibrahim Time Domain. IN ALLEMANG, R., CLERCK, J. D., NIEZRECKI, C. & WICKS, A. (Eds.) *Special Topics in Structural Dynamics.* New York, Springer.

Ibraham, SR 2001, 'Efficient random decrement computation for identification of ambient responses', *Proceedings of the 19th IMAC,* Orlando, FL, 2001.

Van Overschee, P & De Moor, BL (2012) *Subspace identification for linear systems: theory-implementation-applications,* Springer Science & Business Media

Juang, JN & Pappa, RS 1985, 'An eigensystem realization algorithm for modal parameter identification and model reduction', *Journal of guidance, control, and dynamics,* vol. 8, no. 5, pp. 620-627.

Qin, Q, Li, HB, Qian, LZ & Lau, C-K 2001, 'Modal identification of Tsing Ma bridge by using improved eigensystem realization algorithm', *Journal of sound and vibration,* vol. 247, no. 2, pp. 325-341.

Nagarajaiah, S & Yang, Y 2015, 'Blind modal identification of output-only non-proportionally-damped structures by time-frequency complex independent component analysis', *Smart Structures and Systems,* vol. 15, no. 1, pp. 81-97.

Au, SK, Zhang, FL & Ni, YC 2013, 'Bayesian operational modal analysis: theory, computation, practice', *Computers & Structures,* vol. 126, no., pp. 3-14.

Yan, WJ & Katafygiotis, LS 2015, 'A novel Bayesian approach to structural model updating accommodating statistical modal information from multiple setups', *Structural Safety,* vol. 52, no., pp. 260-271.

Mevel, L, Benveniste, A, Basseville, ME & Goursat, M 2002, 'Blind subspace-based eigenstructure identification under nonstationary excitation using moving sensors', *IEEE Transactions on Signal Processing,* vol. 50, no. 1, pp. 41-48.

Magalhaes, F, Cunha, A & Caetano, E 2009, 'Online automatic identification of the modal parameters of a long span arch bridge', *Mechanical Systems and Signal Processing,* vol. 23, no. 2, pp. 316-329.

Li, P, Hu, S & Li, H 2011, 'Noise issues of modal identification using eigensystem realization algorithm', *Procedia Engineering,* vol. 14, no., pp. 1681-1689.

Bazan, FSV 2004, 'Eigensystem realization algorithm (ERA): reformulation and system pole perturbation analysis', *Journal of sound and vibration,* vol. 274, no. 1, pp. 433-444.

Zhang, Y, Zhang, Z, Xu, X & Hua, H 2005, 'Modal parameter identification using response data only', *Journal of Sound and Vibration,* vol. 282, no. 1, pp. 367-380.

Cara, FJ, Carpio, J, Juan, J & Alarcon, E 2012, 'An approach to operational modal analysis using the expectation maximization algorithm', *Mechanical Systems and Signal Processing,* vol. 31, no., pp. 109-129.

Marchesiello, S, Fasana, A & Garibaldi, L 2016, 'Modal contributions and effects of spurious poles in nonlinear subspace identification', *Mechanical Systems and Signal Processing,* vol. 74, no., pp. 111-132.

Peeters, B & Roeck, GD 2001, 'Stochastic system identification for operational modal analysis: a review', *Journal of Dynamic Systems, Measurement, and Control,* vol. 123, no. 4, pp. 659-667.

Wang, Y, Lynch, JP & Law, KH 2009, 'Decentralized H_∞ controller design for large-scale civil structures', *Earthquake Engineering & Structural Dynamics,* vol. 38, no. 3, pp. 377-401.

The 8th International Conference on
Structural Health Monitoring of Intelligent Infrastructure
Brisbane, Australia | 5-8 December 2017

Modal identification of a flexible footbridge using output-only methods

M. Domaneschi[1], G.P. Cimellaro[2], K. Kliewer[3], B. Glišić[3].
[1] *Politecnico di Torino, Department of Structural, Geotechnical and Building Engineering IT,*
Email: marco.domaneschi@polito.it [2] *Politecnico di Torino, Department of Structural, Geotechnical and Building Engineering IT,* [3] *Princeton University, Department of Civil and Environmental Engineering US*

Abstract

This work researches the modal frequencies identification of a footbridge structure by output only technique. The structural system of the footbridge herein considered is a continuous curved girder, i.e. belongs to a frequently used one all around the world in urbanized areas. It is located on the Princeton University campus in the US and serves as an on-site laboratory for short- and long-term research and educational purposes.

The investigation utilizes measurements collected by long-gauge fiber optic strain sensors installed along the main-span and south-east leg of the structure. The power spectral density of the distributed long-gauge dynamic strain response in terms of curvature is used for detecting the modal frequencies of the footbridge. Pedestrian loading represents the external excitation. However, detailed information on the load intensity, frequency and distribution are lacking. This aspect represents the challenge of the present research with the aim to obtain as much information as possible about the modal characteristics of the footbridge from very limited dynamic information (only on the structural response) and without knowing the input conditions. Previously published research serves as a guide for evaluating the effectiveness of the results and the effectiveness of the employed methodology.

This preliminary step paves the path to further research, i.e., (i) to the implementation of a refined finite element model of the entire structure for dynamic analysis, and (ii) to the development of a new wireless sensor network, by using as a reference the results from the existing embedded fiber optic sensors.

1. Introduction

Information on the structural condition and safety can be gained through the analysis of the structural parameters collected by monitoring systems and processed by suitable methods. This activity is usually termed as structural health monitoring (SHM) and can be divided in four main phases, from the essential *assessment of presence of damage* to the most complicated *prognosis* of the structural state and remaining life. *Localization* and *quantification* of damage are the intermediate phases.

SHM is also connected to the innovative view of resilience for civil structures and infrastructure that is attracting the attention of several specialists all over the world, particularly in the bridge engineering field. The time required to recover the original performance of structure after a damaging event, is the crucial resilience aspect to be investigated. The capacity to assess damage presence, localization and intensity is obviously essential for developing resilient structures and communities.

The exclusive use of structural response variables for the system parameters identification is a recently proposed technique and the methodologies that belong to it are usually termed *output-only*.

They are particularly attractive owing to the wide range of applicability: e.g. to the cases when it is of paramount importance that the structure remain in-service and operating without disrupting the normal activities (e.g. transportation network) (Domaneschi et al, 2017; Sigurdardottir & Glišić, 2015).

Despite the number of SHM methodologies and the extensive literature with several applications, designers and owners relatively rarely employ SHM on real structures. In addition, the most effective SHM methodologies are usually restricted to specific conditions (type of structure, loading conditions, etc.). Furthermore, education in SHM, while expanding, is still limited.

This work is focused on the assessment of the modal characteristics of the Streicker Bridge on the Princeton University campus. An output-only method is applied to the response data in terms of longitudinal strains and curvature. The monitoring system consists in a fiber optic sensor network embedded in the footbridge concrete elements. The results are compared to previous outcomes from literature.

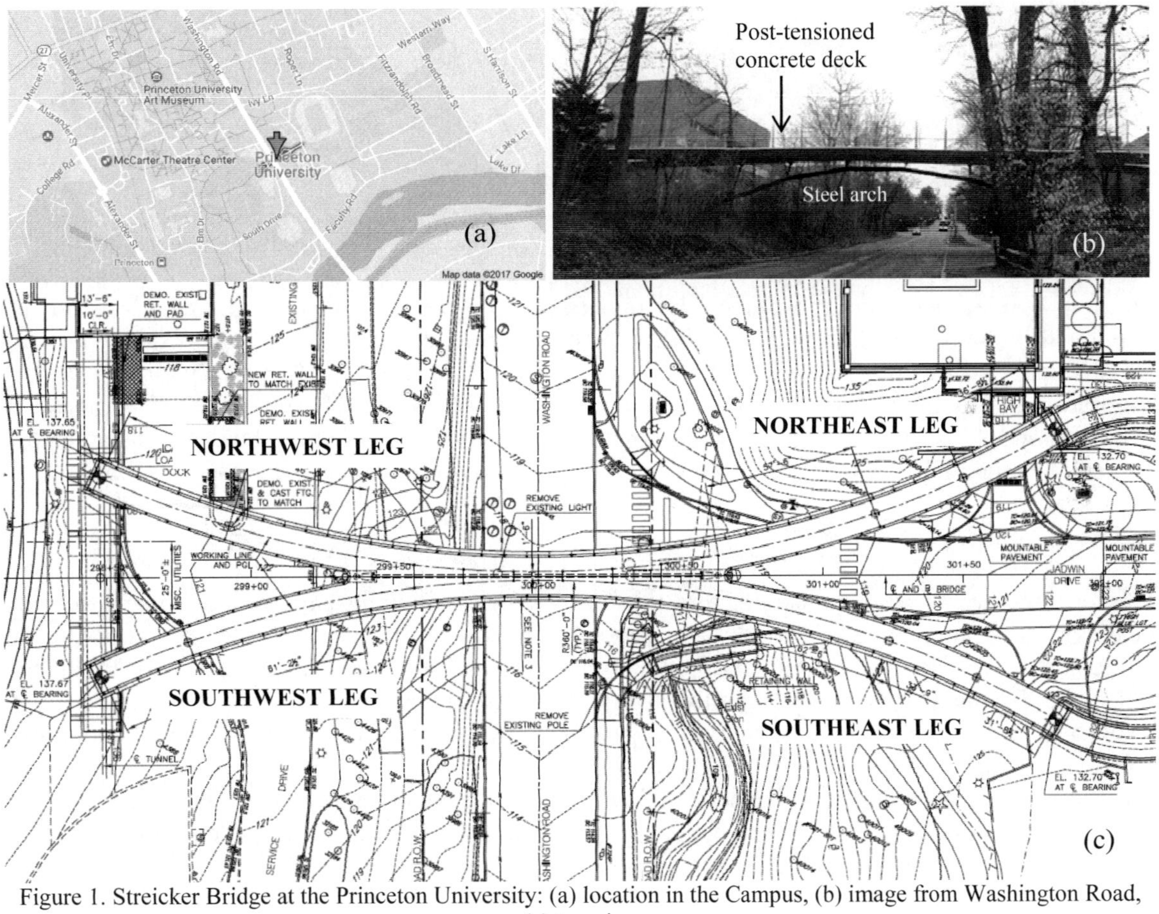

Figure 1. Streicker Bridge at the Princeton University: (a) location in the Campus, (b) image from Washington Road, (c) top view.

2. Streicker Bridge

Streicker Bridge is a 104-m long footbridge completed in 2010 on the Princeton University Campus overpassing Washington Road (Fig. 1a,b). The top view (Fig. 1c) of the bridge highlights a central main span and four lateral approaching legs. A steel arch supports the main span over Washington Road, while the deck is a post-tensioned high-performance concrete girder. There are also four post-tensioned high performance concrete lateral legs on the bridge which are supported by steel "Y" columns (Fig. 2). More details about the bridge can be found in (Sigurdardottir & Glišić 2015).

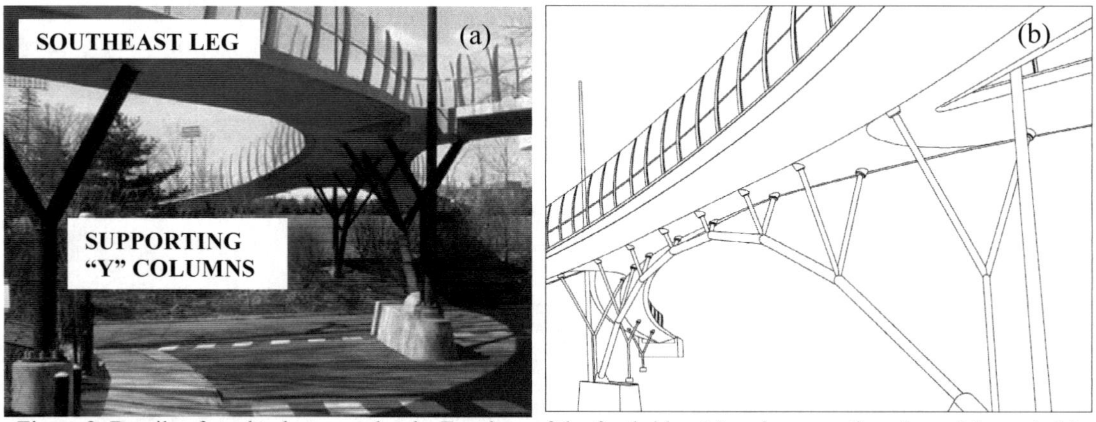

Figure 2. Details of steel columns under the East legs of the footbridge (a) and perspective view of the arch (b).

The legs connect the main span to the lateral bearings on the ground. The original shape in the horizontal plane provides horizontal stability to the footbridge despite the intrinsic slenderness of the steel supporting columns. Vertical stability is provided by the arch (Figure 2b) in the central main span and by the supporting columns under the legs. The width of the cross section increases from the middle of the main span to the connections with the legs which have a constant cross section. The maximum deck depth is 0.6 m and the cross-section diameter of the arch is 0.32 m.

3. The monitoring system

Previous research has demonstrated the capabilities of using long-gage strain sensors for the detection of local damage and for the assessment of the dynamic characteristics of a structure. In particular, fiber-optic sensors can show high sensitivity and measure beyond kilohertz frequency range. Furthermore, they can be employed in aggressive environments where they have proven stable and resistant to external perturbations such as chemical agents and environmental conditions (corrosion, electromagnetic interference, etc.).

Streicker Bridge is equipped with a series of fiber-optic strain and temperature sensors that provide the structure with an effective monitoring system. The bridge and the embedded monitoring system is frequently used for research and teaching on SHM. Because the bridge is rather symmetric in plan, only half of the main span and the longest leg (southeast) are equipped with the monitoring system.

Fiber-optic sensors were embedded in the concrete bridge deck during construction, providing monitoring of the structural response from the birth of the structure. Two types of fiber optic sensors were instrumented on the structure, discrete fiber Bragg-grating (FBG) long-gauge strain and temperature sensors and distributed Brillouin optical time domain analysis (BOTDA). For the purposes of this study, only the FBG sensors are used due to their ability to measure dynamic strain. Long gauge length of sensors is not sensitive to the local inhomogeneity that introduce discontinuities in the local mechanical properties of concrete.

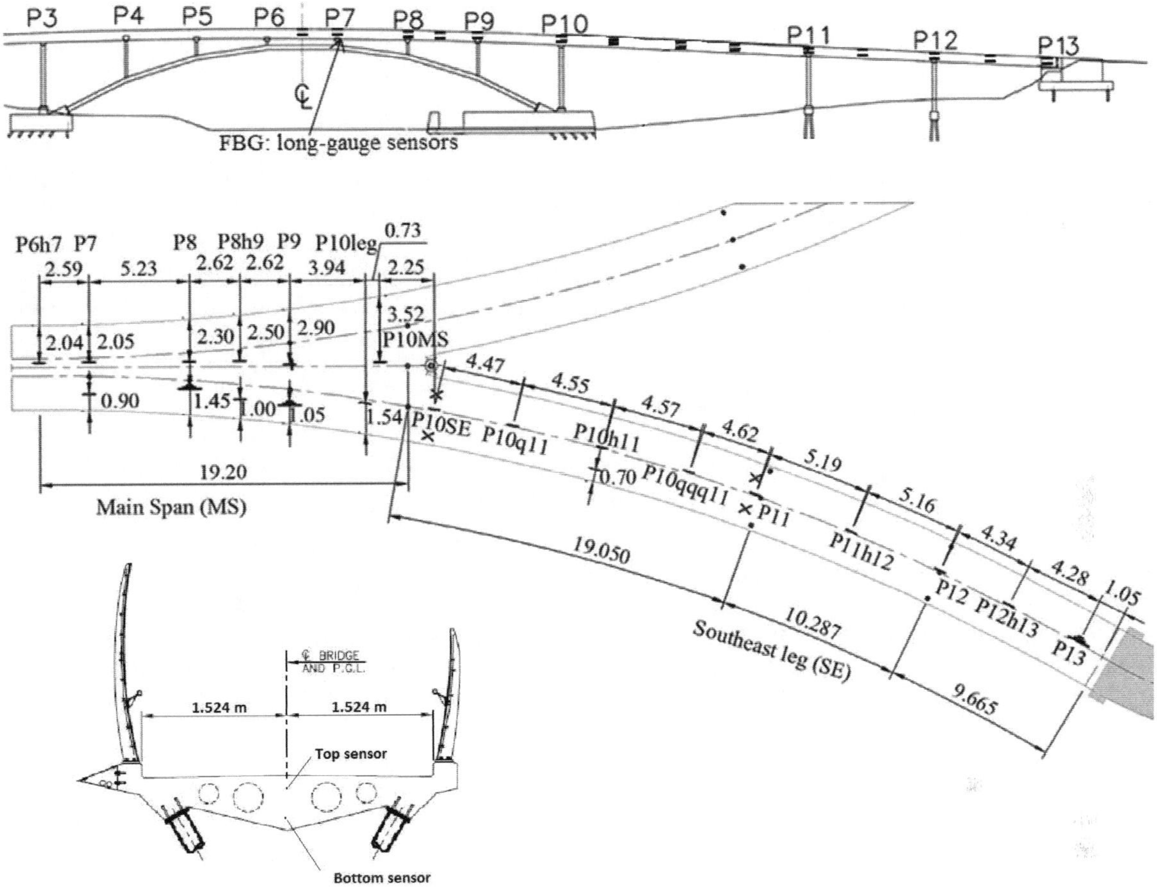

Figure 3. Sensors location on the deck and within the cross section (modified from Sigurdardottir and Glisic, 2015).

The monitored parameter in the bridge girder is the longitudinal strain and the derived parameter is the curvature (Domaneschi et al. 2017) that is determined from a couple of fiber optic sensors at the top and at the bottom of the cross section. During the dynamic tests on the structure, the strain measurements were recorded at a sampling rate of 250 Hz with an expected error in frequency of ± 0.06 Hz. Additional details on the specification of the monitoring system can be found in Sigurdardottir & Glišić (2015).

Figure 3 depicts the distribution of the discrete sensors on the bridge deck. The long-gauge FBG sensors are installed in different topologies. Only the parallel topology is considered in this work and it consists of two discrete sensors parallel to the longitudinal axis of the deck, approximately aligned with the vertical axis of symmetry, one sensor at the top and one at the bottom of the cross section. Parallel sensors capture strain due to normal force and bending moment that create axial strain in the cross section. Bending moment in particular creates curvature in the vertical plane. No significant shear or torsion forces are expected. Curvature χ can be calculated using available strain data at the top and the bottom of the cross section ($\varepsilon_{top}, \varepsilon_{bottom}$) as follows:

$$\chi = (\varepsilon_{bottom} - \varepsilon_{top})/h$$

(1)

where h is the vertical distance between the parallel sensors.

Every long-gauge FBG sensor is equipped with an FBG temperature sensor for temperature compensation and measurements. In (Sigurdardottir & Glišić 2015) additional details can be found for a more comprehensive description of the monitoring systems.
The main span of the footbridge is divided into seven segments between columns P3–P10 (see Figure 3). Parallel sensors are at columns P7, P8, P9, and P10 and at two locations mid-span between P6 and P7 (P6h7) and between P8 and P9 (P8h9). The southeast leg (three spans) is fixed to the deck of the main span at one end and simply supported at the other end, with sensors installed at columns P10-P12, at abutment P13, in the middle of each span (P10h11, P11h12 and P12h13) and in the quarters of the longest span (P10q11 and P10qqq11). The gauge length for the parallel sensors is set to 0.60 m, based on principles described in Glišić (2011).

4. Dynamic characteristics from literature

The footbridge has been extensively tested in several experimental campaigns. Furthermore, finite element (FE) simulations have been also performed with simple plane beam models. Therefore, literature data is available for comparison. In particular, the natural frequencies of the structure determined from the experimental tests were 3.11 and 3.72 Hz. On the other hand, the natural frequencies determined from the FEM model were equal to 3.21 and 3.77 Hz (Sigurdardottir & Glišić 2015). Vibration monitoring using a wireless MEMS accelerometer board is also reported in literature (Sabato 2015) and the modal frequencies are consistent with previous researches (3.08 Hz and 3.75 Hz). Table 1 summarizes the results from literature.

Table 1. Dynamic parameters from literature

	f_1 *[Hz]*	f_2 *[Hz]*
Experimental Tests (Sigurdardottir & Glišić 2015)	3.11	3.72
FEM (Sigurdardottir & Glišić 2015)	3.21	3.77
Experimental Tests (Sabato 2015)	3.08	3,75

All the reported results have been collected by employing group of pedestrians running or jumping at specific frequencies or randomly. The present study investigates the dynamic characteristics of the footbridge without knowing the characteristics of the input within an SHM output-only approach.

5. Results

The data currently available for analysis were collected in 2014. For the recorded sets of signals the input is unknown. The strains time histories from the parallel sensors are computed for obtaining the dynamic curvature (Equation 1). The curvature is computed for extracting modal properties of the footbridge. It has been verified that curvature processing allows halving the number of vectors to process without losing information from the strain data.
The experimental curvature data was processed in the frequency domain by computing the power spectral density (PSD) of the bridge response signals through the periodogram function. The PSD of a stationary random process is, by definition, a real symmetric function mathematically related to

the autocorrelation sequence by the discrete-time Fourier transform. The periodogram is an estimate of the PSD of a wide-sense stationary (weakly-stationary) random process, such as a finite-length segments of a signal.

Tests from 2014 were performed on the southeast leg of the bridge. The loading conditions are unknown, but they consisted of pedestrians that were jumping, running randomly over the bridge or marching at synchronized but unknown frequencies.
Figure 4 depicts the processed results from the set termed as *DynamicPeaks1* in the database in both the time and the frequency domain. Unfortunately some issues with the reading of sensors at P7 and P6h7 occur and data for those sensors were not available.
Figures 5 to 10 show the results in the frequency domain for all of the remaining data sets from the 2014 tests. Figure 5 also depicts an overview of the time history records for *DynamicPeaks1* data set. The sensors between P10-P12 are characterized by the highest intensity responses.

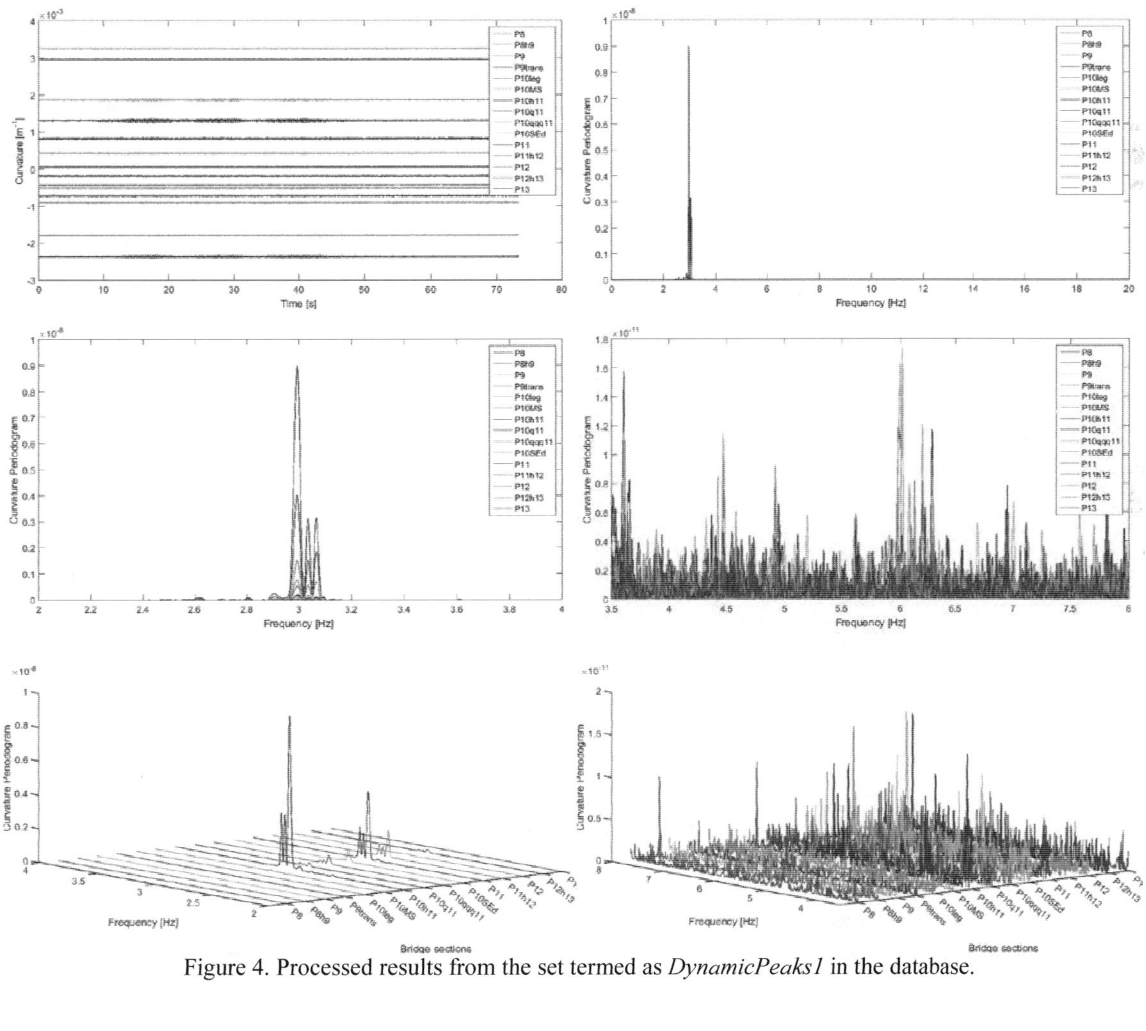

Figure 4. Processed results from the set termed as *DynamicPeaks1* in the database.

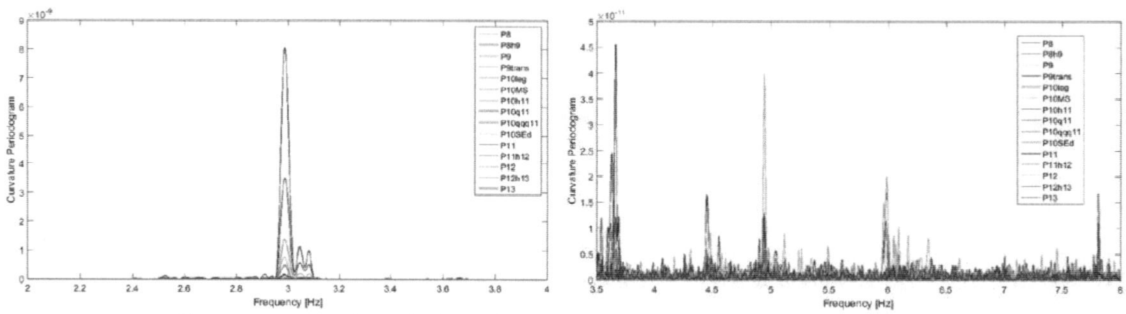

Figure 5. Processed results from the set termed as *DynamicPeaks2* in the database.

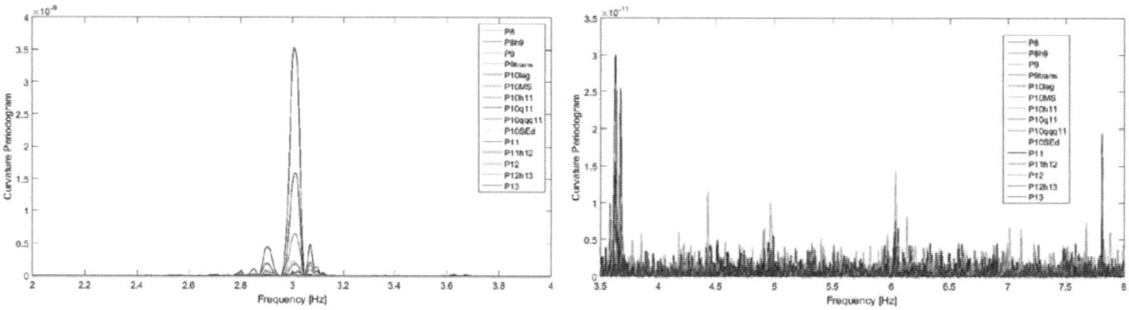

Figure 6. Processed results from the set termed as *DynamicPeaks3* in the database.

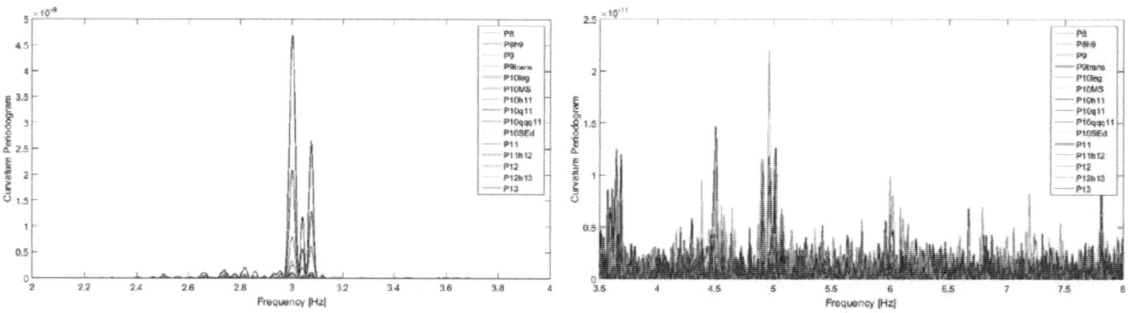

Figure 7. Processed results from the set termed as *DynamicPeaks4* in the database.

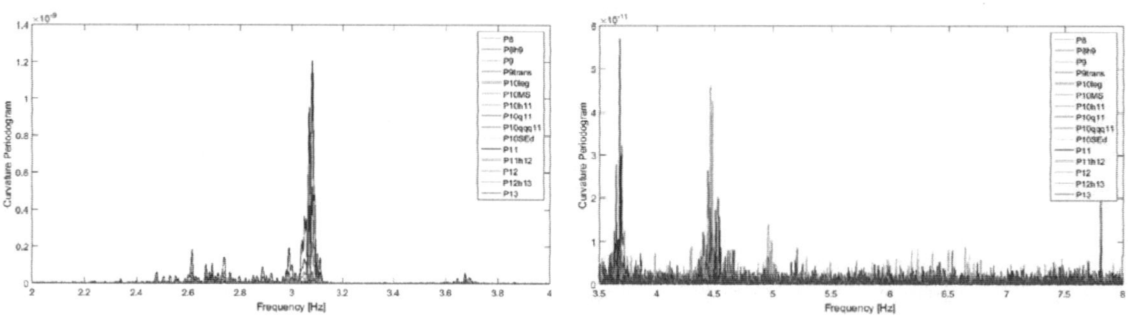

Figure 8. Processed results from the set termed as *RunningPeaks* in the database.

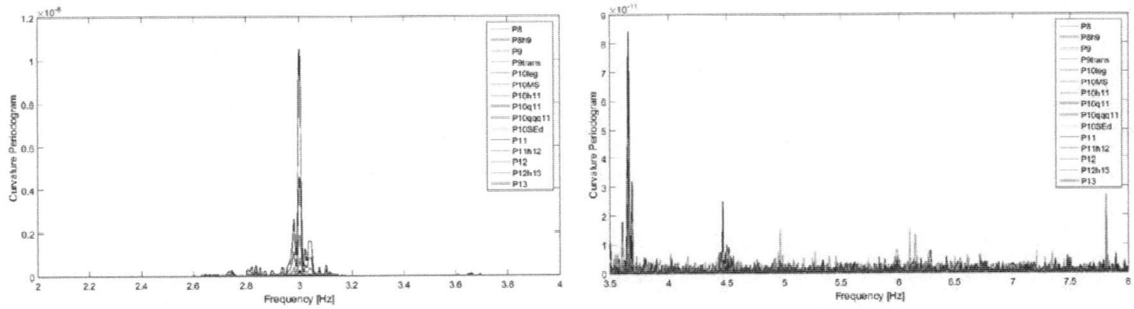

Figure 9. Processed results from the set termed as *DynamicPeaks41* in the database.

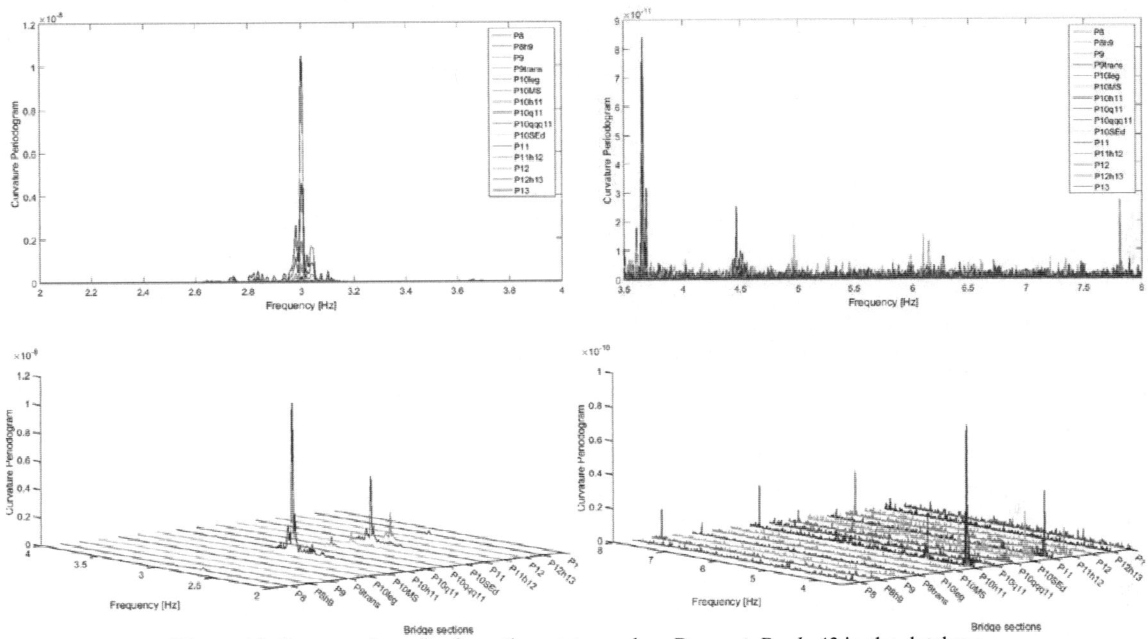

Figure 10. Processed results from the set termed as *DynamicPeaks42* in the database.

Table 2. Natural frequencies.

	f_1 *[Hz]*	f_2 *[Hz]*	f_3 *[Hz]*	f_4 *[Hz]*	f_5 *[Hz]*	f_6 *[Hz]*
DynamicPeaks1	*2.94*	*3.61*	*4.47*	*4.92*	*6.01*	*7.81*
DynamicPeaks2	*2.98*	*3.66*	*4.45*	*4.94*	*5.99*	*7.81*
DynamicPeaks3	*3.00*	*3.63*	*4.43*	*4.96*	*6.03*	*7.81*
DynamicPeaks4	*3.00*	*3.64*	*4.50*	*4.96*	*5.99*	*7.81*
DynamicPeaks41	*3.00*	*3.66*	*4.46*	*4.96*	*6.09*	*7.80*
DynamicPeaks42	*2.99*	*3.68*	*4.42*	*4.94*	-	*7.81*
RunningPeaks	*3.08*	*3.67*	*4.46*	*4.96*	-	*7.81*

From these results, the first amplification peak in the frequency domain is determined to be approximately 3 Hz for all of the experimental tests. This is a lower value with respect to those from literature in Table 1. Also, the second peak has lower values (approximately 3.6 Hz). The available results in the range [3.5 Hz, 8 Hz] highlight a third peak around 4.4 Hz, a fourth at 4.9, a fifth one at 6 Hz and last one at 7.8 Hz. Table 2 summarizes the results of the analyses for each recorded signal.

Due to the location and topology of the embedded sensors, the recognized natural frequencies are reasonably associated to vertical flexural natural modes. Furthermore, comparable highlighted intensities in all data sets and the redundancy in the frequency values allow considering these outcomes accurate with satisfactory confidence.

The damping ratio of the first mode associated to 3 Hz natural frequency can be estimated by the applying the logarithmic decay formula at the sensor P10h11 and it is evaluated as 0.57%. This value, not reported by other research works in literature on the same footbridge, is compatible with others of similar slender structures (Shahabpoor et al., 2017; Van Nimmen et al, 2017).

6. Conclusions

This work investigates the modal characteristics of the existing footbridge on the Princeton University campus in Princeton US, the Streicker Bridge. The investigation uses measurements collected by long-gauge fiber optic strain sensors embedded along the entire structure. The power spectral density of the distributed long-gauge dynamic curvature is used. The damping ratio associated to the first identified mode is also computed.

Pedestrian loading represents the external excitation. However, detailed information on the load intensity, frequency and distribution are lacking and the research can be ascribed within the *output-only* procedures. Comparison with previous works from literature mostly validate the proposed procedure, as slightly lower frequency values have were noted.

The present investigation represents the first step in further research for consolidating the results and deepening the knowledge on the footbridge characteristics.

Acknowledgements

This material is based upon work supported by NSF GRFP Grant No. 1148900, NSF CMMI-1362723, and USDOT-RITA DTRT12-G-UTC16. Any opinions, findings, and conclusions or recommendations expressed herein are those of the authors and do not necessarily reflect the views of the funding agencies. Many thanks to HNTB, SMARTEC, Micron Optics, and faculty, staff, and students of Princeton University.

References

Domaneschi, M, Sigurdardottir, D, Glišić, B, 2017, 'Damage detection on output-only monitoring of dynamic curvature in composite decks', *Structural Monitoring and Maintenance*, Vol. 4, No. 1, pp. 1-15

Sabato, A, 2015, 'Pedestrian bridge vibration monitoring using a wireless MEMS accelerometer board', *IEEE 19th International Conference on Computer Supported Cooperative Work in Design (CSCWD)*, Calabria I.

Sigurdardottir, D, Glišić, B, 2015, 'On-site validation of fiber-optic methods for structural health monitoring: Streicker Bridge', *J Civil Struct Health Monit*, Vol. 5, pp. 529–549.

Glišić, B, 2011, 'Influence of gauge length to accuracy of long-gauge sensors employed in monitoring of prismatic beams', *Measurement Science and Technology*, Vol. 22, No. 3, art. no. 035206 (13pp).

Shahabpoor, E, Pavic, A, Racic V, Zivanovic, S, 2017, 'Effect of group walking traffic on dynamic properties of pedestrian structures', *Journal of Sound and Vibration*, Vol. 387(2017), pp. 207–225.

Van Nimmen, K, Lombaert, G, De Roeck, G, Van den Broeck, P, 2017, 'The impact of vertical human-structure interaction on the response of footbridges to pedestrian excitation', *Journal of Sound and Vibration*, Vol. 402(2017), pp. 104–121.

The 8th International Conference on
Structural Health Monitoring of Intelligent Infrastructure
Brisbane, Australia | 5-8 December 2017

Assessment of Flexural Stiffness and Load Carrying Capacity Using Substructural System

S. Jamali[1], KY. Koo[2], THT. Chan[3], A. Nguyen[4], DP.Thambiratnam[3]
[1] *PhD Candidate, Queensland University of Technology – Australia,*
[2]*Lecturer, University of Exeter- United Kingdom,*
[3]*Professor, Queensland University of Technology – Australia, Email: tommy.chan@qut.edu.au*
[4] *Research Fellow, Queensland University of Technology – Australia*

Abstract

The evaluation philosophy of an existing structure must differ from the design philosophy of a new structure. Many parameters are related to the structural properties and among them, stiffness and boundary conditions are governing factors affecting the resistance of the structure to external loading. However, due to the aging and deterioration of structures, effects of boundary conditions change over time and it becomes intricate to understand their current condition for structural assessment. This study was set out to assess the feasibility of estimating flexural stiffness in beam-like structures using substructural bending rigidity identification (SBI) method when no information about 'as-is' fixity of boundary conditions is available. The results of numerical and experimental analyses showed that the SBI method accurately estimates the flexural stiffness of various beam profiles having unknown boundary conditions. It was found that using estimated stiffness, analytical model can be tuned with updated support fixities and structural and modal characteristics, which are then used for detection of the critical member, worst load effect and evaluation of serviceable load carrying capacity. Findings from this study provided evidence that SBI can be efficiently implemented as a short-term monitoring tool with easy testing setup for in-operation structural integrity assessment of beam-type elements.

Keywords: stiffness, beam, boundary condition, substructural system, load rating, load carrying capacity

1. Background

Studies over the past two decades have provided important information on using structural health monitoring (SHM) techniques for condition assessment of structural elements. By measuring the dynamic response of a structure such as a bridge under known/unknown loading, one can obtain vibrational parameters after analysing the data and correlating it to structural parameters, since any change in vibrational behaviour of a structure leads to a corresponding change in structural parameters (Chan & Thambiratnam 2011; Jamali *et al.* 2017a). Load carrying capacity (LCC) of structures such as bridge girders decreases over time due to various factors such as environmental condition, over-loading and material deterioration. On the other hand, increase in vehicular loading requires substantially higher LCC to accommodate change in loading pattern; which directly affects safety and maintenance costs. In Australia and many other countries, bridge standards provide specifications for bridge load rating when safe bearing capacity of bridge is uncertain (Jamali *et al.* 2017b, c). For structural integrity assessment, conventional non-destructive testing such as infrared thermography, acoustic emission, and visual inspection are conducted to determine integrity index of a structure. Nevertheless, these tests are merely extensions of design standards and can provide limited information about structural safety, which leads to conservative results. Therefore, current structural conditions must be considered for evaluation of existing structures. Reviewed studies clearly indicate the importance of stiffness as a measure of structural integrity index, akin to LCC. To evaluate stiffness of beam-like structures as a whole system, measurement of global parameters such as deflections, rotations and natural frequencies are involved (Hoffmann *et al.* 2007, Lakshmanan *et al.* 2008, Gao *et al.* 2015). Previous investigations have found the stiffness of bridges by simplifying the bridge as a single degree-of-freedom system using Euler beam theory (Chowdhury 1999; Higgs & Tongue 1999; Chen *et al.* 2002; Islam *et al.* 2014). Similarly, Subramaniam (1994), Samali *et al.* (2007), Moore *et al.* (2013) and Brashaw *et al.* (2009) attempted to evaluate the current stiffness of timber bridges by assuming 'as-designed' boundary conditions. One major issue in previous studies that has not been dealt with is lack of information on support conditions. It is well-established that the behaviour of a structure varies with its aging; for instance, roller supports might change to provide partial fixity over time, which affects the load path of the structure by restraining rotational deflection and hence giving different load distribution and capacity. Therefore fixity of boundary conditions at structural joints such as abutments are prone to change which must be correctly reflected in structural condition assessment. This paper presents a new approach for flexural rigidity assessment of beam-type elements using 'conquer and divide' strategy. Substructural bending rigidity identification (SBI) approach divides the beam into an internal substructural system that correlates acceleration input at interfacial location of substructure to that centrally output through an optimization scheme. Significant advantages of using SBI is that no prior knowledge of boundary conditions is needed and it uses an efficient sensory system (5 sensors), which makes it simple to setup and deliver. Due to page constraints, further details on the mathematical model development of SBI can be found in (Jamali *et al.* 2017c). In this paper, implications of SBI for various beam-like structures are assessed, and then relationship between estimated stiffness and LCC is explored. In the next section, the rationale behind SBI's theory is evaluated which is followed by concluding remarks.

2. Verification study

The complete process of SBI is shown below and it is deatiled in (Jamali *et al.* 2017c). SBI layout is prepared based on the length of the beam to capture the bending defomration udner flexural mode. Using the first bending mode, the vertical and rotational accelrations are recorded and using nonlinear curve square fitting, reposne of interfacial accelration is minimized with respect to the central repsone. Minizmed objective fucntion (ξ) of optizmiation is used for estimation of bending stiffness using Classical beam theory. Follwoing sections illustrate the application of SBI for stiffness esitmation and calibrtion of analytical model.

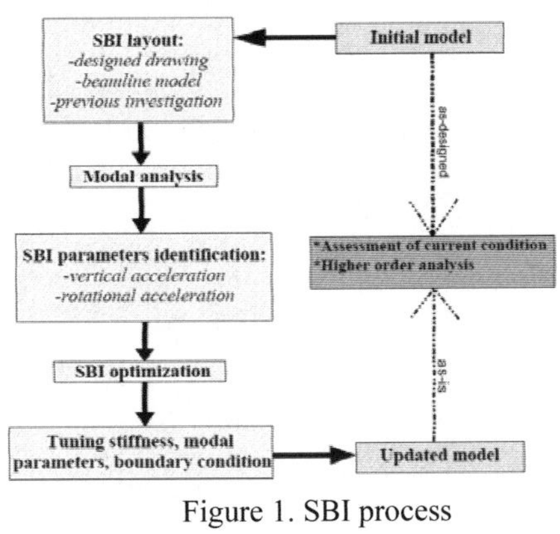

Figure 1. SBI process

2.1 Contuinous Beam

A continuous beam with four equal spans of 8m each is divided into a series of substructural systems meshed at 0.1m as shown in Figure 2. Band limited white noise is applied at span two to excite the fundamental mode of vibration of the rectangular beams simulated in CSiBridge®. In traditional approach, L_{eff} is used for effective length of span as support-to-support scaping. In SBI, L_{sub} approximates the fixity of boundary conditions using internal substructural length chosen arbitrarily and independent of boundaries. As an example of SBI input for span one, accelerations at joints 2&3 are simulated to obtain vertical acceleration $\langle \left[\frac{joint(2)_{acceleration} + joint\ (3)_{acceleration}}{2} \right] \rangle$ and rotational acceleration at left end interfacial$\langle \left[\frac{joint(2)_{acceleration} - joint\ (3)_{acceleration}}{distance\ between\ two\ joints/sensors} \right] \rangle$, while joints 19&20 are used for right end interfacial. The response of interfacial ends is then minimized with regard to mid-span vertical acceleration response at joint 11. In the same way, response of each span is recorded separately for SBI optimization.

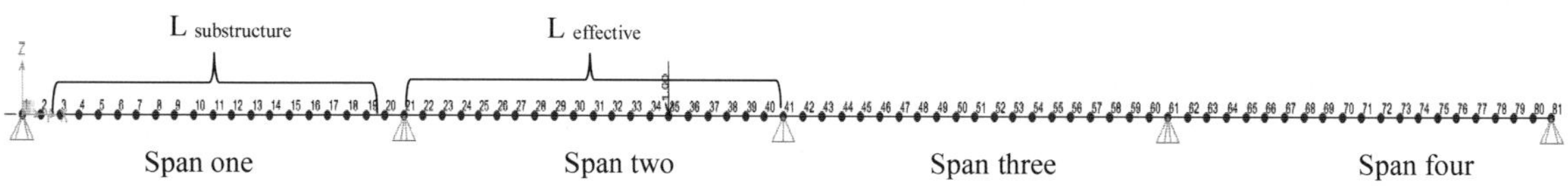

Figure 2. Elevation of continuous beams

After taking Fast Fourier Transform (FFT) of signals, it was noted that due to the existence of nodal points; some higher modes are not captured as seen in power spectral density of span three (see Figure 3-a). However, only first bending mode is needed for SBI, which was accurately detected by all substructural spans. In addition, spans located further away from excitation source had more noise component in comparison to span two, implying that excitation source needs to be sufficiently high to excite the early modes of the structure. This view is supported by (Memory *et al.* 1995) who concluded that response of a structure to a dynamic load mobilises a higher modulus of elasticity in comparison with that to a static load.

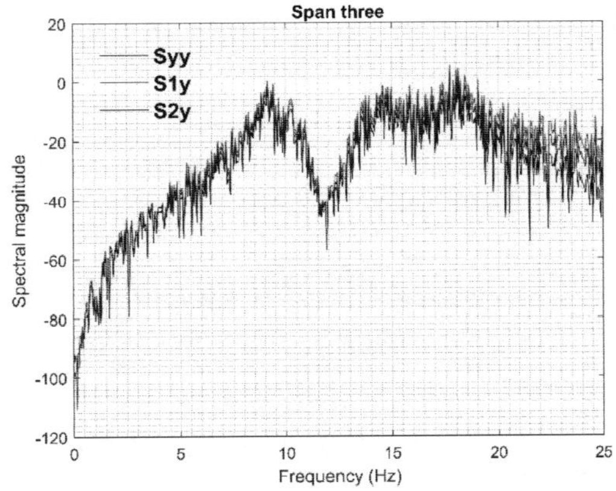

(a)

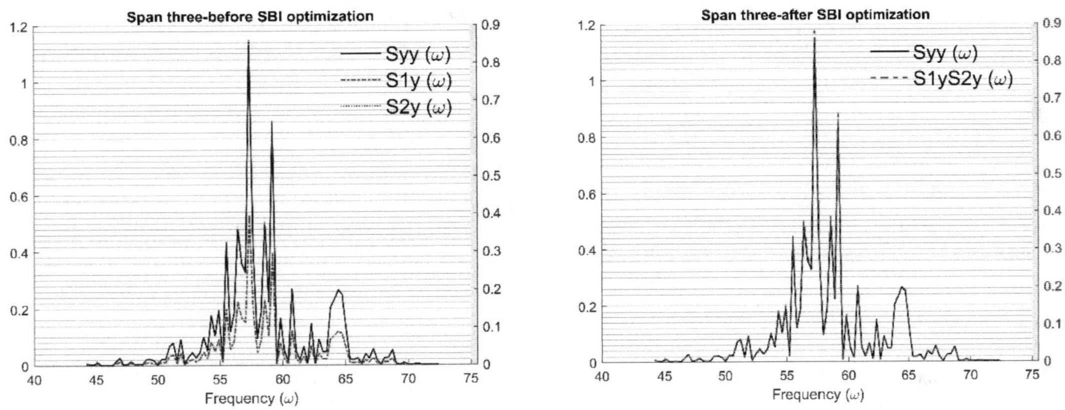

(b)

Figure 3. (a) Spectral densities of substructural system at span three (b) Optimization of bending frequency

Having the first flexural mode at 9.13 Hz, a range 7 to 11.5 Hz was selected as bending region for optimization using nonlinear curve-square fitting method. Figure 3(b) compares the optimization of substructural system input-output response at span three. Clear trend of minimization in bending peak can be observed after implementing SBI; this confirms that optimization has globally converged with very low residuals. From the correlational analysis and using SBI formula (Jamali *et al.* 2017c):

$$EI_{SBI} = \frac{\rho A \omega_1^2 L_{sub}{}^4}{16\xi^4} = \frac{2.4028\times10^3\times0.08\times\left(9.1\times2\times\pi\right)^2\times7.7^4}{16\times1.5124^4} = 2.638\times10^7\,N.m^2 \qquad (1)$$

Where A is the cross sectional area, ρ is mass density and ω is angular frequency that is $\omega = 2\pi f_{bending}$. Estimated SBI for span three differs 0.49% to intact EI (2.652×10^7 N.m^2), which gives the fixity of support to be in pinned condition when substituted into the Classical beam theory as shown in equation (2). Stiffness of the other spans approximated to be: span 1 = 2.644×10^7 (0.28%), span 2 = 2.6651×10^7 (0.02%), span 4 = 2.653×10^7 (0.6%); which denotes that the adequate intensity of excitation needs to be distributed across each span for better stimulation of modal modes and estimation of stiffness. Thus, random moving excitation would be a good alternative which is investigated in the next case study.

$$f = \frac{\lambda^2}{2\pi L_{eff}^{\;2}}\sqrt{\frac{EI_{SBI}}{m}} \rightarrow 9.13 = \frac{\lambda^2}{2\pi \times 8^2}\sqrt{\frac{2.601\times 10^7}{192.2}} \rightarrow \lambda = 3.159 \quad (2)$$

Using updated stiffness and boundary conditions and applying load-displacement relationship for linear system with allowable deflection, bending capacity would be:

$$K = \frac{48EI}{L_{eff}^{\;3}} = 2438437.5(N/m)\times \Delta_{\max}(0.032m) = 78.03\,\text{kN} \quad (3)$$

This means that the span remains within permissible deflection as long as the applied load does not exceed this value. Allowable deflection can be code-defined value or specified threshold measured from previous investigations. Taking into account the rigidity of span in load resistance and nonlinear behaviour, a variation of ±10% from obtained value is allowed as safety margin since structure reacts to load differently as opposed to theoretical assumptions (Nowak & Collin 2012).

2.2 Solid Concrete Girder Bridge

To assess the performance of SBI for bridge model updating and load rating, a 20m reinforced concrete girder bridge is investigated. Layout of bridge is as illustrated in Figure 4, which consists of two traffic lanes with parapet on each side.

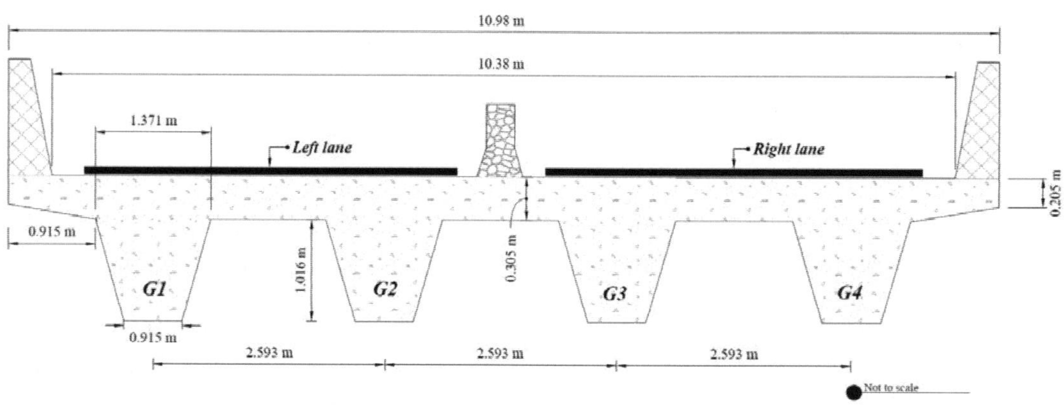

Figure 4. Cross sectional view of the bridge

In order to mimic the testing condition, various truck loadings as per AS 5100.2 (Australian Standards 2004) were applied to represent the ambient traffic. To prevent interruption of traffic flow and easier accessibility, measurement was recorded at the base of left exterior girder (G1). For the purpose of measurements, interior girders may be considered as well. Nonetheless, due to the passage of vehicles that induce additional mass on deck, estimation of stiffness would be less accurate; so measurement after passage of vehicles is preferred for interior girders. Another important aspect of SBI is L_{sub}, which is the length of internal substructure. Determination of L_{sub} is mainly dependent on the total support-to-support length of structure (L_{eff}), and preferably it is suggested to keep L_{sub} long enough so that bending behaviour of beam is closely represented. With regard to the concrete girder, L_{sub}

was taken as 18.8m as internal substructure length. Figure 5 provides the spectral densities of bending peak at interfacial (*S1y*/*S2y*) and central (*Syy*) locations. This isolated bending peak is used as bending region for optimization.

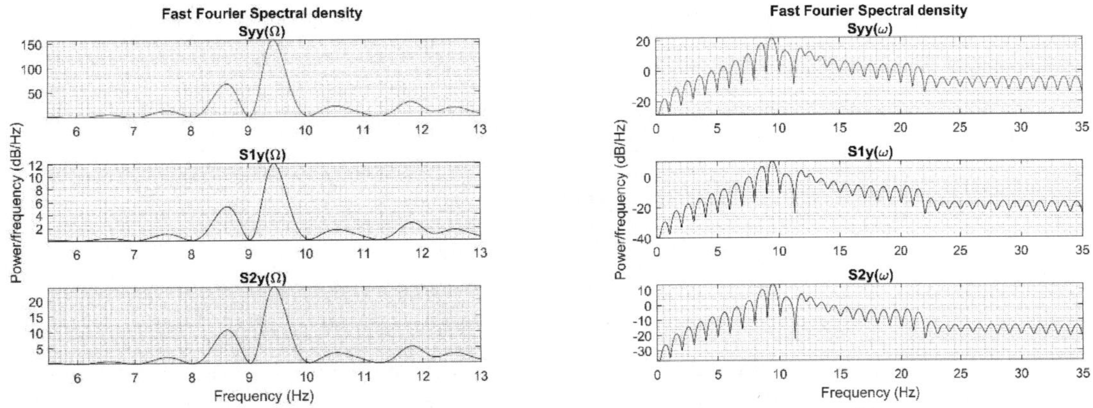

(a) (b)

Figure 5. Power spectral density of G1, (a) Hertz scale, (b) Log scale

Output of optimization resulted in *EI*= 2.912 $\times 10^{10}$ N.m^2, which is very close to intact *EI* (2.9304$\times 10^{10}$ N.m^2). Having estimated the stiffness, analytical model could be updated to match the modal and structural parameters obtained by SBI. After that, the deck is load rated as per specification of TMR guidelines (Queensland Department of Transport and Main Roads 2013) for three travel cases as illustrated in Figure 6 to find out the most adverse load effect. For the load assessment of bridge components, two assessment vehicles are included for ease of calculation, viz. 45.5t semi-trailer as reference vehicle and 44t as coexisting vehicle with traveling speed of 45 km/hr as follows:

- Case 1: Reference vehicle is travelling along bridge centreline without any other vehicles.
- Case 2: Assessment vehicle is positioned adjacent to the kerb and second assessment vehicle is positioned adjacent to the carriageway centreline
- Case 3: Reference vehicle is travelling on the lane with coexisting vehicle in the other lane

Results of multi-step static analysis revealed that largest load effects for shear force and bending moments occur at G4 and G2 respectively in case two, when reference vehicle is positioned along the kerb and other vehicle is adjacent to the centerline of deck. What follows is an illustrative description of shear assessment of G4 for live load case (SF_{LL}) (Jamali *et al.* 2017c):

$$SF_{LL} = \frac{\phi SF_{ULS} - \gamma_{DL} SF_{DL} - \gamma_{SDL} SF_{SDL}}{\sum_{i=1}^{n} \gamma_{LL} SF_{LL}(1+DLA) * \mathrm{AVF}} \quad (4)$$

where

- Ultimate shear capacity of girder (ϕSF_{LL}) = 1480 kN
- Dead load shear force (SF_{DL}) = 469 kN
- Superimposed dead load shear force (SFS_{DL}) = 88 kN
- Live load shear force from reference vehicle (SF_{LL}) = 267 kN
- Live load shear force from accompanying vehicle (SF_{LL}) = 42 kN $\times$accompanying vehicle factor (AVF)

▪ Dynamic load allowance (DLA) = 0.4

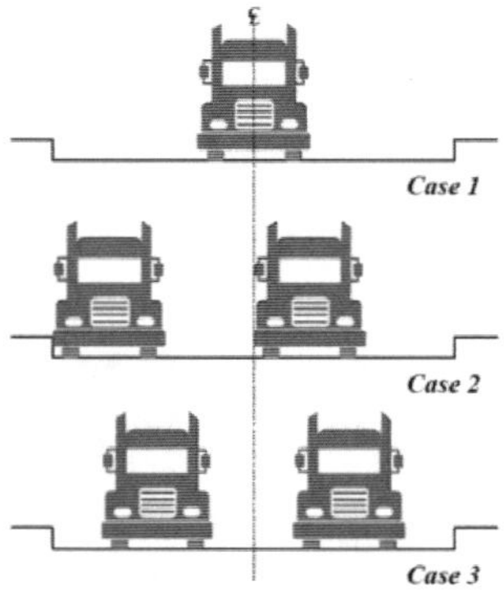

Figure 6. Lateral position of assessment vehicles for different cases

Using appropriate load factors, the shear force ratio is found to be 0.881. This highlights that G4 has no reserved capacity for this travel loading case and suitable management plan such as load limitation or retrofitting is required for functional performance of the girder. It is also worth noting that similar approach can be carried out for assessment of other load effects, for example when bending is critical at mid-span.

2.3 Aluminium Beams

As pointed out in the introduction to this paper, SBI can efficiently estimate the stiffness of beam-like structures with unknown boundary conditions. In an experimental study, two square hollow section (SHS) beams made of aluminum were tested to better understand the effect of L_{sub} and damage on SBI output. For the sake of stable testing, beams were seated on plank at support locations and held by clamping without creating any specific boundary condition as shown in Figure 7.

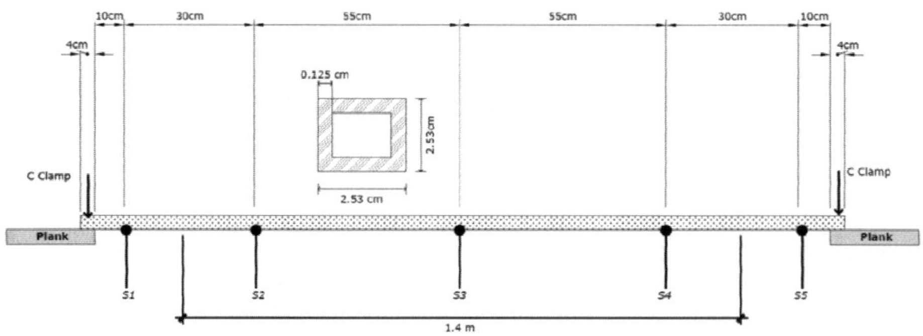

(a)

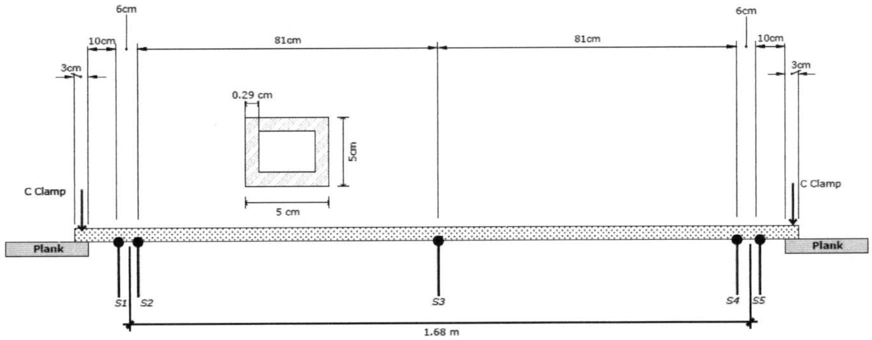

(b)

Figure 7. Testing setup for SHS beams, (a) Small SHS, (b) Large SHS

For both beams, excitation was provided by random hammer tapping along the length of the beam for logging acceleration data at sensor locations (S1 to S5). Figure 8 shows the optimization inputs for small SHS as well as the spectral densities found by Welch method for each beam.

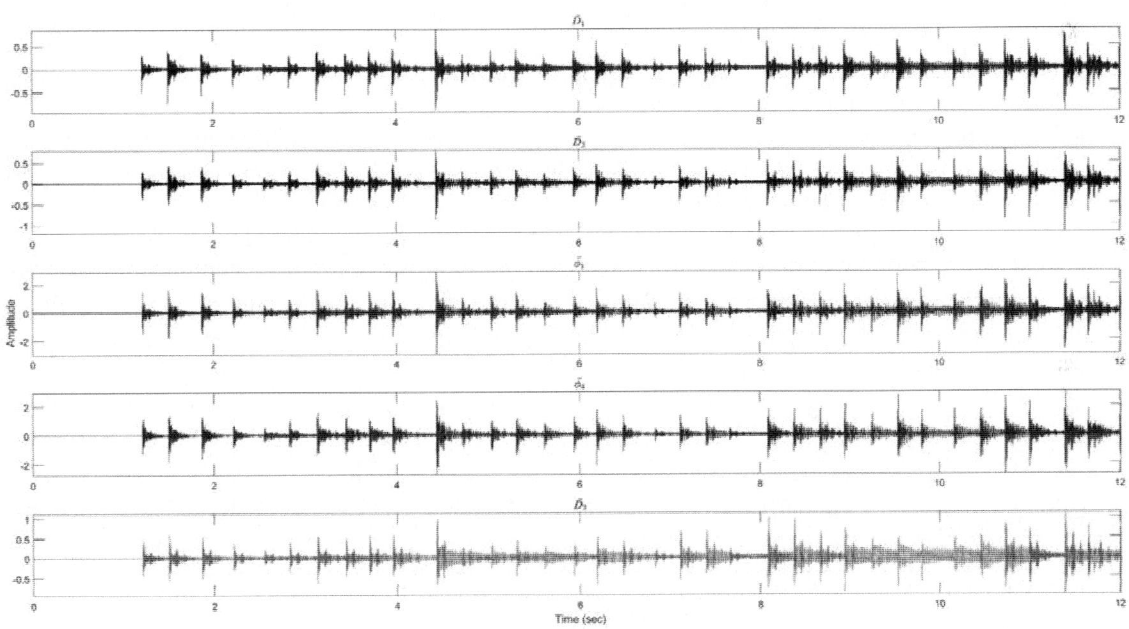

(a)

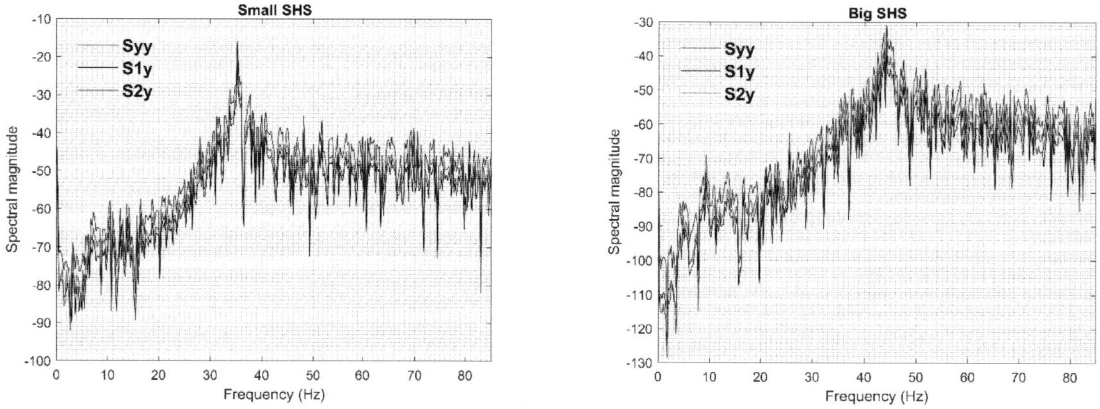

(b)

Figure 8. (a) Input-output response of sub-structural length for small SHS, (b) Spectral density of first bending mode

For both beams, Young's modulus is 70×10^9 N/m^2 while those estimated for small SHS was 64.5 $\times10^9$ N/m^2 and for large SHS was 69.8 $\times10^9$ N/m^2. This discrepancy could be attributed to the spacing of interfacial sensors for obtaining rotational acceleration. Lack of accuracy on approximation of rotational acceleration was due to the distance between S1-S2 and S4-S5, while for large SHS; this distance was only 6cm. Moreover, considering the relationship of L_{sub} in SBI equation (with power of four), short sensor spacing for end conditions leads to longer Lsub, thus having more length of beam covered to represent its bending behavior. Using calculated stiffness and equation (2), fixity of supports are found to be pinned-fixed condition which assisted in updating the analytical model. By adopting load-deflection relationship, bending capacity within linear range for large SHS is 1.6 kN. Large SHS was then further tested as a damaged beam by placing solid mass midway between S2-S3. It can be seen from the data in Table 1 that for a range of damages with different severities, SBI effectively reflected the degradation of beam by estimating the reduction trend for bending frequency and flexural stiffness. It appears that the change in global frequency affects the ξ value in Eq. (1) and in turn *EI*. Since frequency is not a very good indicator of damage (level 1 and level 2 damages), trivial changes in *EI* can be seen due to small damage severity. Yet, severe damage will have significant impact on stiffness which can be confirmed by quick visual inspection in case of real structures. In addition, the merit of SBI is that it is dependent on global behaviour of the structure, and as long as the global characteristics change, the optimized *EI* will reflect these changes as well.

Table 1. Effect of damage on big SHS

Damage state	Bending mode (Hz)	EI (N.m^2)	ξ (average)
No damage	44	14196	1.4028
0.5 kg mass	41.2	13144	1.3833
1 kg mass	37	11600	1.3525
1.5 kg mass	35.3	10919	1.3412

3. Conclusion

Variation in loading conditions and surrounding environment change the properties of a structure, in which such changes must be considered for structural assessment. Stiffness and boundary conditions are important indices of structural integrity that affect the LCC. Underestimating the LCC of structures such as bridges will cause load-posting and traffic limitation that undoubtedly affect the road network functionality. Towards this end, this paper developed and applied a novel method for estimating the flexural stiffness and the boundary conditions of beam-like structures. The SBI method was assessed for different beam configurations made up of diverse materials and support conditions. The results of this study indicate that SBI can be successfully carried out for continuous beams and bridge girders under impact, random and moving load excitations. Furthermore, it was identified that using the estimated stiffness, modal and structural properties of a structure along with updated boundary conditions can be used for calibration of finite element model for further assessment. Experimental investigations also confirmed that SBI accurately reflects the effect of damage in the beam-like elements which could be applied to the load-displacement relationship to approximate the linear LCC within permissible deflection range. Owing to the latest developments in sensory equipment, for future works it is recommended that effect of using gyroscope be assessed to determine the rotational acceleration. Above findings suggest that SBI can be considered as a cost-effective method for in-service assessment of beam-type structures on periodic basis.

Acknowledgments

The first author is thankful for the research funds provided by the Queensland University of Technology. The authors have no conflict of interests regarding the publication of this paper.

References

Chan, THT & Thambiratnam, DP 2011, *Structural Health Monitoring in Australia*, Nova Science Publishers, New York.

Jamali, S, Chan, THT, Thambiratnam, DP, Pritchard, R, & Nguyen, A 2017a, 'Pre-test finite element modelling of box girder overpass-application for bridge condition assessment', *Proceedings of Australian Structural Engineering Conference (ASEC),* Brisbane, Asutralia, 23-25 November, pp. 1-8.

Jamali, S, Chan, THT, Thambiratnam, DP, & Nguyen, A 2017b, 'Comparative study of grillage analogy and finite element method for bridge heavy load assessment', *Proceedings of Austroads Bridge Conference*, Melbourne, Australia, 3-6 April, pp. 1-10.

Jamali, S, Koo, K-Y, Chan, THT, Nguyen, A & Thambiratnam, DP 2017c, 'Stiffness-based capacity estimation for beam-like structures using substrucutral bending rigidity identification method, (Submitted to a journal for possible publication).

Hoffmann, S, Wendner, R, Strauss, A, & Bergmeister, K 2007, 'Comparison of stiffness identification methods for reinforced concrete structures', *Proceedings of International Workshop on Structural Health Monitoring: Quantification, Validation, and Implementation,* Stanford, California, pp. 354-363.

Lakshmanan, N, Raghuprasad, B., Muthumani, K., Gopalakrishnan, N, & Basu, D 2008, 'Identification of reinforced concrete beam-like structures subjected to distributed damage from experimental static measurements', *Computers and Concrete,* vol. 5, no. 1, pp. 37-60.

Gao, Q-f, Wang, Z-l, Lv, X-h, Zhang, W, & Yin, L-h 2015, 'Formulas for natural frequencies of girder bridge and continuous rigid frame bridge', *The IES Journal Part A: Civil & Structural Engineering,* vol. 8, no. 3, pp. 155-164.

Chowdhury, MR 1999, 'Rapid beam load capacity assessment using dynamic measurements', *Proceedings of SPIE, The International Society For Optical Engineering*, Orlando, Flordia, pp. 973-979.

Higgs, J, & Tongue, D 1999, 'Dynamic Bridge Testing Systems (DBTS) for the evaluation of defects and load carrying capacity of in-services bridges', *Proceedings of International Conference on Current and Future Trends in Bridge Design Construction and Maintenance,* Singapore, 4-5 October, pp. 1-21.

Chen, S, Siswobusono, P, Delatte, N, & Stephens, B 2002, *Feasibility Study on Dynamic Bridge Load Rating (Report No. 01221),* University Transportation Center for Alabama, Alabama.

Islam, A KM A, Li, F, Hamid, H, & Jaroo, A 2014. *Bridge Condition Assessment and Load Rating using Dynamic Response (Report No. 134695)*, Ohio Department of Transportation, Ohio.

Subramaniam, H. 1994, 'A Non-destructive Method for Determining the Adequacy of In-service Corbelled Timber Girder Bridges', *Proceedings of 17th ARRB Conference,* Gold Coast, Austrlaia, pp. 103-119.

Samali, B, Li, J, Crews, K, & Al-Dawod, M 2007, 'Load Rating of impaired bridges using a dynamic method', *Electronic Journal of Structural Engineering*, vol. special issue, pp. 66-75.

Moore, J. C., Mahini, S., Glencross-Grant, R., & Patterson, R. 2013, 'Monitoring the Structural Integrity of Timber Bridges under Dynamic Traffic Loads using a High Speed Camera', *Proceedings of 6th International Conference on Structural Health Monitoring of Intelligent Infrastructure*, Hong Kong, 9-11 December.
Brashaw, BK, Vatalaro, RJ, Wang, X, Verreaux, M, & Sarvela, K 2009, *Development of Flexural Vibration Inspection Techniques to Rapidly Assess the Structural Health of Rural Bridge Systems (Report No. CTS 08-22),* University of Minnesota, Minneapolis.

Memory, T, Thambiratnam, D, & Brameld, G 1995, 'Free vibration analysis of bridges', *Engineering Structures,* vol. 17, no. 1, pp. 705-713.

Nowak, AS & Collins, KR 2012. *Reliability of structures*. McGraw Hill, New York.

Australian Standards 2004. *Bridge Design- Part 2: Design Loads (AS 5100.2)*, Sydney, Australia.

Queensland Department of Transport and Main Roads 2013, *Tier 1 Heavy Vehicle Bridge Assessment Criteria*, TMR, Queesnland.

AUTHOR INDEX

AUTHOR INDEX

AUTHOR INDEX

AUTHOR INDEX

AUTHOR INDEX

International Society for Structural Health Monitoring of Intelligent Infrastructure
University of Manitoba
Agricultural & Civil Engineering Bldg.
A250 - 96 Dafor Road
Winnipeg, Manitoba R3T 2N2
Canada

ISBN 978-1-5108-6457-3

CURRAN ASSOCIATES INC.
proceedings.com

8th International Conference on Structural Health Monitoring of Intelligent Infrastructure (SHMII 2017)

Structural Health Monitoring in Real-World Application

Brisbane, Australia
5-8 December 2017

Volume 2 of 3

Editors:

Tommy Chan **Saeed Mahini**

ISBN: 978-1-5108-6457-3

Printed from e-media with permission by:

Curran Associates, Inc.
57 Morehouse Lane
Red Hook, NY 12571

Some format issues inherent in the e-media version may also appear in this print version.

Copyright© (2017) by International Society for Structural Health Monitoring of Intelligent Infrastructure
All rights reserved.

Printed by Curran Associates, Inc. (2018)

For permission requests, please contact International Society for Structural Health Monitoring of Intelligent Infrastructure at the address below.

International Society for Structural Health Monitoring of Intelligent Infrastructure
University of Manitoba
Agricultural & Civil Engineering Bldg.
A250 - 96 Dafor Road
Winnipeg, Manitoba R3T 2N2
Canada

Fax: 204-474-7519

central@ishmii.org

Additional copies of this publication are available from:

Curran Associates, Inc.
57 Morehouse Lane
Red Hook, NY 12571 USA
Phone: 845-758-0400
Fax: 845-758-2633
Email: curran@proceedings.com
Web: www.proceedings.com

TABLE OF CONTENTS

VOLUME 1

KEYNOTE LECTURES

REGULAR SESSIONS

RS1: SMART AND OTHER ADVANCED SENSORS

RS5: DAMAGE IDENTIFICATION/LOCALIZATION AND EVALUATION/ASSESSMENT OF BRIDGES AND BUILDING STRUCTURES

SUB-TOPIC 1: DETECTION AND LOCATION OF STRUCTURAL DAMAGE (NOTE: DAMAGE INDEX OR LOCATION)

SUB-TOPIC 2: STRUCTURAL PARAMETERS AND DAMAGE IDENTIFICATION (NOTE: PARAMETER IDENTIFICATION)

VOLUME 2

SUB-TOPIC 3: STRUCTURAL CONDITION ASSESSMENT OF BRIDGES

SUB-TOPIC 4: CONDITION ASSESSMENT OF BUILDING STRUCTURES

RS6: MODEL UPDATING, SAFETY EVALUATION AND RELIABILITY FORECAST

RS7: DAMAGE CONTROL, REPAIR AND STRENGTHENING

RS8: LIFE-CYCLE PERFORMANCE-BASED DESIGN AND PERFORMANCE MONITORING

RS10: SMART MATERIALS, STRUCTURES AND REHABILITATION

RS11: GLOBAL POSITIONING SYSTEM (GPS) AND RELATED SYSTEMS

RS12: REMOTE MONITORING SYSTEMS

RS13: INTEGRATED SYSTEMS AND IMPLEMENTATIONS OF SHM

RS15: STANDARDIZATION OF SHM SYSTEMS

RS16: CRITICAL ISSUES FOR SHM SUCH AS COST EFFECTIVENESS OF SHM SYSTEMS

RS17: RESILIENCE OF INFRASTRUCTURE SYSTEMS

MINI-SYMPOSIA

MS1: RECENT SHM ADVANCES IN AUSTRALIA

VOLUME 3

MS2: DATA-DRIVEN STRUCTURAL DAMAGE IDENTIFICATION AND PERFORMANCE ASSESSMENT

MS3: TEMPERATURE BEHAVIOUR OF STRUCTURES

SPECIAL SESSIONS

SS1: RECENT ADVANCES IN RAILWAY SYSTEM MONITORING IN CHINA

SS2: CASE STUDIES AND NEW METHODS FOR MODAL-BASED SHM OF CIVIL STRUCTURES

SS3: OPTICAL BASED MONITORING TECHNIQUES FOR INFRASTRUCTURAL MAINTENANCE AND SAFETY

The 8th International Conference on
Structural Health Monitoring of Intelligent Infrastructure
Brisbane, Australia | 5-8 December 2017

An Energy-based Damping Evaluation for the Local Damage Detection of an Existing Steel Truss Bridge

S. Mustafa[1], Y. Matsumoto [2], H. Yamaguchi [3].
[1] Assistant Professor and Head of the Department of Civil Engineering, JIS College of Engineering – India, Email: samim@alumni.iitg.ernet.in [2, 3] Professor, Saitama University - Japan

Abstract

An energy-based damping evaluation (EBDE) was introduced in this paper for the vibration-based structural health monitoring (SHM) of an existing steel truss bridge. The damage detection by the proposed EBDE was carried out by estimating the contribution of modal damping ratios from different structural elements utilizing experimentally identified modal damping ratios, and estimated modal strain and modal potential energies from an updated finite element model (FE-model) of the structure under consideration. The advantage of using damping as a damage indicator is that the damping change in global modes affected by the local damage can be identified with a small number of sensors. In a previous study, vibration measurements were conducted with five sensors at an existing Warren steel truss bridge with damage in a diagonal member before and after an emergency repair work. A substantial increase in the modal damping ratio of a global mode coupled with the vibration of a damaged diagonal member was observed, although no significant change in the natural frequency was observed for this mode due to local damage. The present study utilized the EBDE to identify the cause of the modal damping increase by observing the change in the contribution from different structural elements on the modal damping ratios.

1. Introduction

In the current standard bridge management strategy in Japan, primary periodic inspections rely on visual inspection, although various nondestructive test methods have been used in second step detailed inspections (Fujino & Siringoringo 2008). The current scheduled visual inspections are often time consuming, costly and require the components to be readily accessible. Furthermore, the results of visual inspection are subjective, and vary with the knowledge and experience of the inspectors. To address these issues, damage detection based on physical behaviors through measurements of deformations, stresses, vibrations are required.

There are some serious issues with the practical application of vibration-based SHM to detect local damage. The most important concern for its practical application is the sensitivity issue of modal parameters to local damage. There are many vibration-based damage detection approaches have been proposed in the literature and an extensive review of those can be found in Doebling et al. (1998) and Fan & Qiao (2011). Most of these studies for damage detection have been mainly based on change in modal frequencies and/or change in mode shape related indices. However, the problem related to low sensitivity of damage features due to local damages remains a concern for the practical application of these methodologies. On the other hand, it has been recognized that the damping is more sensitive to the local damages especially due to crack or some internal changes in the structural property (Modena et al. 1999; Curadelli et al. 2008; Yoshioka et al. 2010; Frizzarin et al. 2010). This is the main motivation of considering damping as a damage indicator for vibration-based SHM of steel truss bridges in this study. Another issue with the practical application of vibration-based SHM is the availability of limited number of sensors due to the budget constrain. In case of large structures such as steel truss bridges, a number of sensors may be required to detect local damage which would lead to trade-off between monitoring cost and ability of damage detection. Therefore, it would be worthy to develop a method to detect damage with limited number of sensors.

In this paper, a practical and effective vibration-based SHM was introduced to overcome the aforementioned difficulties which used an EBDE for the detection of local damage. The advantage of using damping is that the damping change in global mode, affected by the local damage, can be identified with a small number of sensors (Yamaguchi et al. 2015). The EBDE was then carried out to estimate the damping parameters of the test structure utilizing experimentally identified modal damping ratios. As an application, the damage detection by the proposed framework was carried out by observing the change in the contribution of modal damping ratios from different structural elements.

2. Energy-based Damping Model for Test Bridge

The bridge studied was a warren type steel truss bridge constructed in 1965 and it consists of five main spans, each having length of 70.77 meters. All the spans are simply supported. The tension diagonal members are a H-section, and the compression members are made of a box section with elliptical holes to reduce steel weight in the web of tension diagonal members. The bridge has a composite deck of steel plate girders and reinforced concrete slab. A photographic view of the truss bridge is shown in Fig. 1.

Fig. 1. View of the truss bridge

Using energy-based definition, the damping of structures in terms of equivalent viscous damping ratio can be defined as the ratio of the dissipated energy per cycle, D_n to the maximum potential energy in a cycle, U_n. Hence, the nth modal damping ratio, ξ_n can be expressed as:

$$\xi_n \equiv \frac{D_n}{4\pi U_n} \tag{1}$$

The nth modal dissipating energy, D_n can be expressed as the summation of dissipating energies due to different sources, such as damping capacity of materials, friction at structural connections, and energy dissipation at supports. The studied steel truss bridge was considered to be composed of five sub-structures as shown in Fig. 2: diagonal members (D), girders (G), upper chord members (UC), top lateral bracings (TLB), bottom lateral bracings (BLB), and supports (S). After calculating D_n, the nth modal damping ratio for the steel truss bridge can be evaluated using Eq. (1) as

$$\xi_n = \frac{2\pi\eta_d V_{d,n}}{4\pi U_n} + \frac{2\pi\eta_g V_{g,n}}{4\pi U_n} + \frac{2\pi\eta_{uc} V_{uc,n}}{4\pi U_n} + \frac{2\pi\eta_{tlb} V_{tlb,n}}{4\pi U_n} + \frac{2\pi\eta_{blb} V_{blb,n}}{4\pi U_n} + \frac{8A_{s,n}\mu_s R}{4\pi U_n} \tag{2}$$

where η_d, η_g, η_{uc}, η_{tlb}, η_{blb} are the equivalent loss factors of diagonal members, girders, upper chord members, top lateral bracings and bottom lateral bracings, respectively. In the present study, it was assumed that the equivalent loss factors accounted for the material damping as well as the damping at structural connection for simplicity. $V_{d,n}$, $V_{g,n}$, $V_{uc,n}$, $V_{tlb,n}$ and $V_{blb,n}$ are nth modal strain energies of sub-structures described above. $A_{s,n}$, μ_s and R are the nth modal amplitude at the movable support, dynamic friction coefficient and support's reaction due to vertical load respectively. The unknown loss factors and friction coefficient can be evaluated by using experimentally identified modal damping ratios and corresponding energy ratios obtained from the updated FE-model (Dammika et al. 2014; Mustafa & Matsumoto 2017).

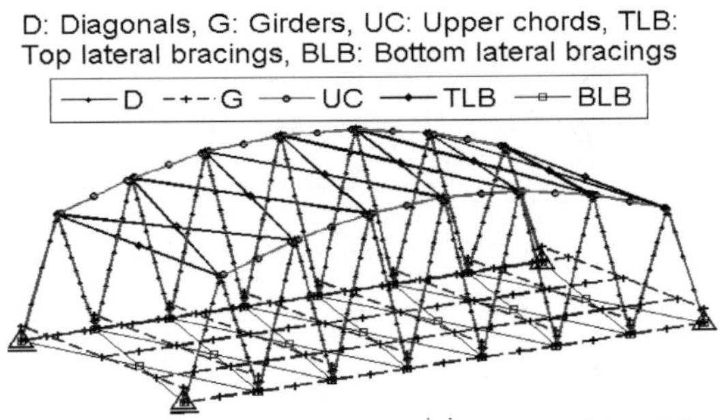

Fig. 2. Three dimensional FE-model with sub-structures considered for damping analysis

Using the same principle, the contribution of each internal element to the modal damping ratio, $\xi_{l,n}$ can also be obtained by using the strain energy of that element and corresponding loss factor

$$\xi_{l,n} = \frac{2\pi\eta_l V_{l,n}}{4\pi U_n} \tag{3}$$

where η_l and $V_{l,n}$ are the equivalent loss factor and nth modal strain energy of lth element. The equivalent loss factor of the lth element, η_l can be assumed to be the same as the loss factor of that sub-structure to which it belongs. By looking at these contributions from all the elements, one can have a clear understanding about those elements which have dominant contribution to damping for a particular mode which can be useful information from the SHM point of view.

3. Energy-based Damping Evaluation

3.1 Vibration Measurements and System Identification

In a separate study (Yamaguchi et al. 2015), the vibration acceleration measurement was conducted by a car running test using two sensor layouts at the first span of the bridge which was not having any damage. In measurement Case 1 shown in Fig. 3(a), a total of thirteen vertical sensors were used to identify the global vibration modes. A total of five data sets of 10-min measurements were collected at a sampling rate of 100 samples per second. An example of full length recorded data measured at U5 location is shown in Fig. 4(a). For the identification of diagonal modes, another two sets of measurements were carried out using the sensor setup in measurement Case 2 shown in Fig. 3(b). Hence, a total of seven sets of measurements were recorded. Free vibration (FV) responses were extracted from full length recorded data by checking the traffic free time on the bridge from the video of traffic recorded during measurements. Fig. 4(b) shows one such FV response extracted from full length recorded data as shown in Fig. 4(a). Seven global modes were identified from the measurement Case 1 and four diagonal modes were identified from the measurement Case 2.

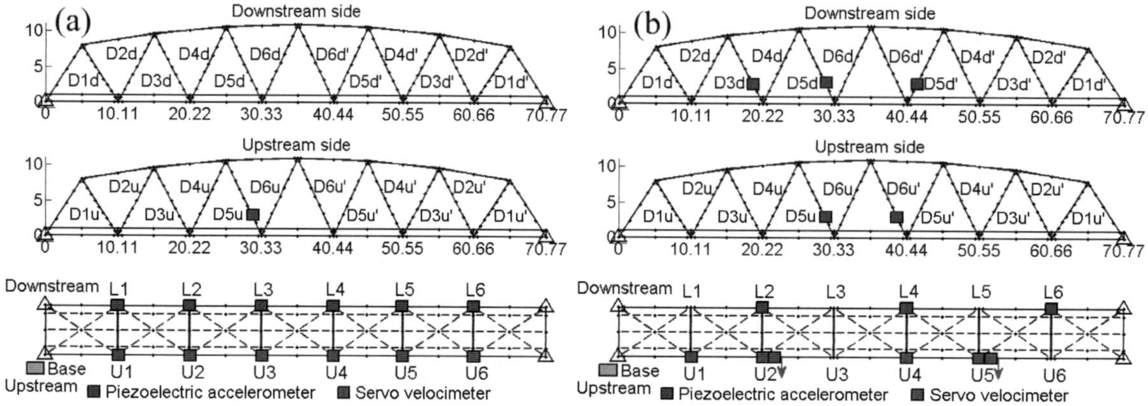

Fig. 3. Sensor layout at the first span; (a) Measurement case 1, (b) Measurement case 2

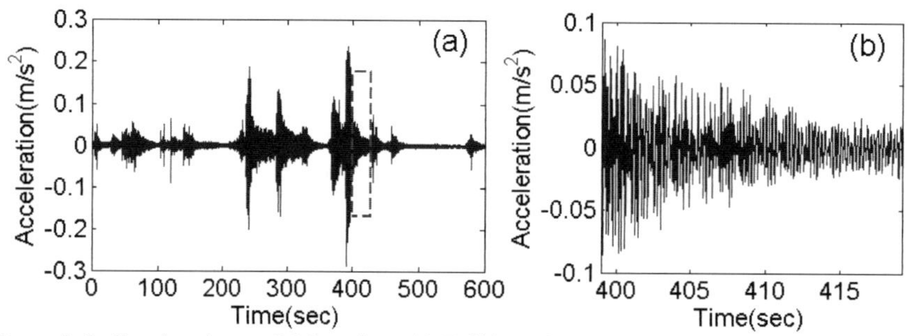

Fig. 4. Recorded vibration data at L2 location; (a) Full length recorded data, (b) Extracted FV response

Fig. 5 shows the modal identification results for damping and corresponding natural frequency for all the FV records by taking the average of the identified values obtained from SSI and ERA. There were slight fluctuations of the identified modal damping ratios as can be seen from Fig. 5. However, these fluctuations in the identified modal damping ratios were not significantly large and this kind of fluctuations is typical for experiments conducted to determine damping, mainly because of the variation in amplitude of vibration owing to different traffic conditions (Chen et al. 2016).

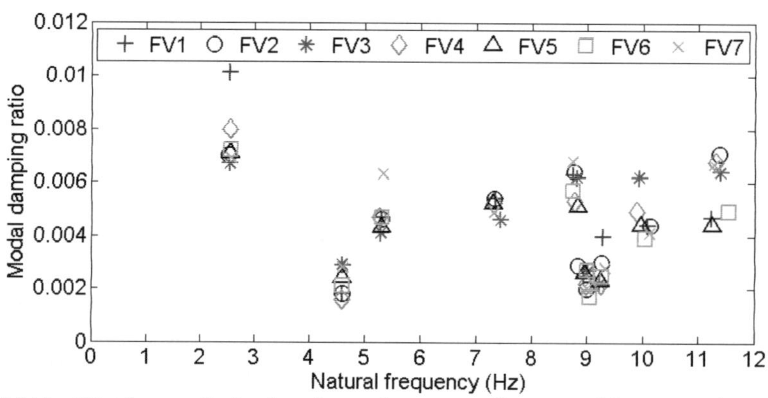

Fig. 5. Modal identification results for damping and corresponding natural frequency for all the FV records

3.2 Identification of Loss Factors and Friction Coefficients

The strain energy of each element was calculated using updated mode shapes and updated stiffness parameters and the corresponding subsystem stiffness matrices. Then the strain energy of each sub-structure was estimated by adding the strain energies of elements belonging to that sub-structure. Equivalent loss factors and friction coefficients were then identified following the steps described in Table 1. At Step 1, the equivalent loss factor of diagonal members was estimated by considering diagonal dominant mode on the assumption that the energy dissipated owing to the diagonal members only. The mean value of the estimated loss factors corresponding to different FV records was taken as the equivalent loss factor of diagonal members and used in the next step for the determination of other unknown damping parameters. At Step 2, the equivalent loss factors of girders, upper chord members, top lateral bracings and bottom lateral bracings and friction coefficient of supports were estimated by considering six global modes. Table 2 shows the estimated equivalent loss factors for all the FV records along with the average estimated values. Different damping parameters were estimated for different FV records owing to the fluctuation in the identified modal damping ratios. However, the reasonability of the estimated damping parameters can be judged by the fact that all estimated values are larger than the loss factor of steel which ranges from 0.0001-0.0006 (Zoghi 2014), suggesting the assumed energy dissipation in the sub-structures.

Table 1. Steps for estimating loss factors and friction coefficient

Analysis step	Considered mode	FV record	Estimated parameter	Analysis conditions
Step 1	3rd diagonal (Dominant diagonal mode)	FV 1 to 6	η_d	Assumed that energy dissipated only from diagonal members
Step 2	1: 1st bending 2: 2nd bending 3: 3rd bending 4: 2nd torsional 5: 4th bending 6: 3rd torsional	FV 1 to 7	η_{uc}, η_g, η_{tlb}, η_{blb}, μ_s	Average value of η_d from Step 1 was used NNLS was employed to get equivalent loss factors

Table 2. Estimated equivalent loss factors and friction coefficient

Analysis step	Para-meter	FV1	FV2	FV3	FV4	FV5	FV6	FV7	Avg.
Step 1	η_d	0.0085	0.0068	0.0073	0.0100	0.0068	0.0073	-	0.0078
	η_{uc}	0.0206	0.0245	0.0000	0.0277	0.0183	0.0136	0.0226	0.0212
	η_g	0.0085	0.0077	0.0121	0.0090	0.0088	0.0074	0.0075	0.0087
Step 2	η_{tlb}	0.0000	0.1532	0.0914	0.1041	0.0000	0.0346	0.0982	0.0963
	η_{blb}	0.0000	0.0196	0.0021	0.0000	0.0027	0.0211	0.0692	0.0229
	μ_s	0.7688	0.0025	0.7753	0.0288	0.2545	0.4476	0.0000	0.3796

3.3 Evaluation of Analytical Modal Damping Ratios

After getting the equivalent loss factors, the analytical modal damping ratios were evaluated using Eq. (2) by substituting the loss factors and corresponding energy ratios. Fig. 6 shows the comparison between analytical modal damping ratios estimated using average loss factors and friction coefficient shown in Table 2 and corresponding experimental damping by taking average over all the FV records. It can be seen that analytically evaluated modal damping ratios agree well with the experimentally identified ones.

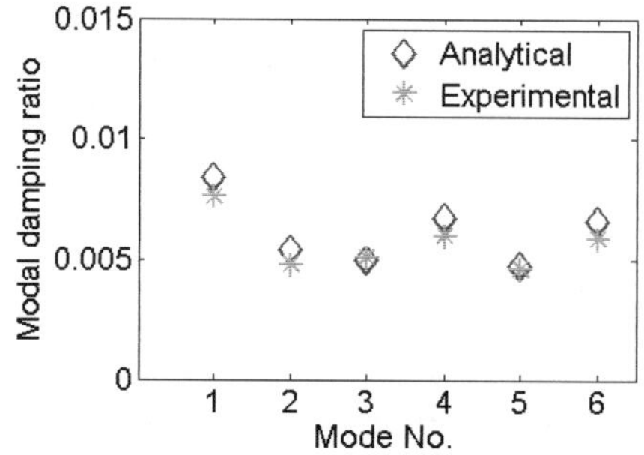

Fig. 6. Comparison of average experimental modal damping ratios and corresponding analytical modal damping ratios

The contribution of each sub-structure to the modal damping ratios was then estimated by using corresponding loss factor and energy ratio. Fig. 7 shows the contribution of damping from each sub-structure to the modal damping ratios estimated analytically with the average damping parameters shown in Table 2. From Fig. 7(a), it can be seen that there is a significant damping contribution from support for the 1st bending mode as it is expected due to the slip of the movable support in longitudinal direction. There is also dominant damping contribution from upper chord members due to bending of these members in this mode. For other bending modes, there are significant damping contribution from upper chords, girders and diagonal members, however, the damping contribution from supports is much less than the 1st bending mode. In Fig. 7(b), the significant contribution of modal damping from top and bottom lateral bracings for the 1st torsional mode express the torsional effect in that mode. Similar trends can also be found in the 2nd and 3rd torsional modes as can be seen in Figs. 7(e) and 7(i). In the diagonal mode, the dominant damping contribution from diagonal members can be observed from Fig. 7(f).

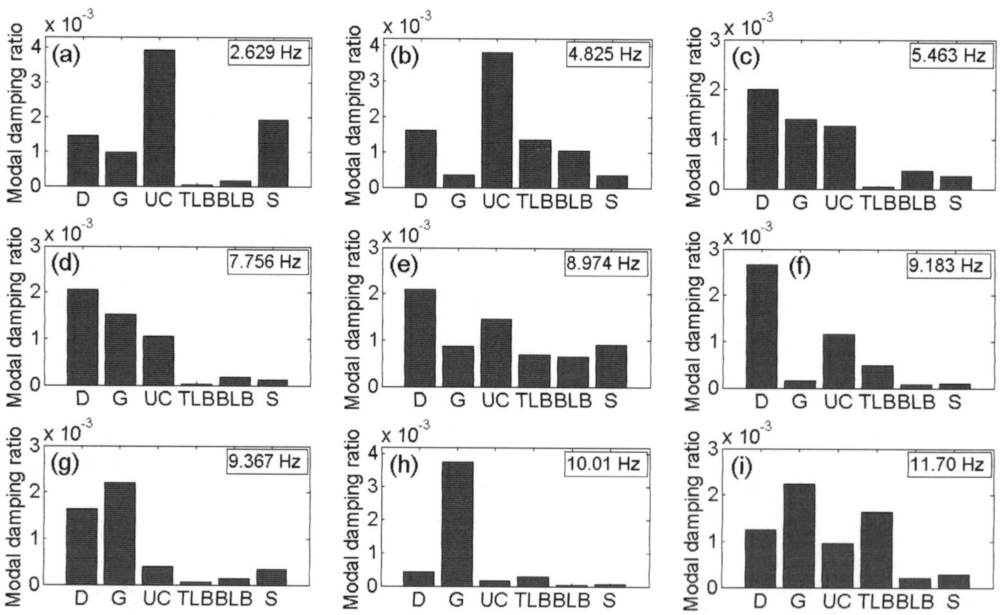

Fig. 7. Contribution of each damping source to modal damping ratios for: (a) 1st bending, (b) 1st torsional, (c) 2nd bending, (d) 3rd bending, (e) 2nd torsional, (f) 3rd diagonal, (g) 4th bending, (h) 5th bending and (i) 3rd torsional mode

3.3 Contribution of Modal Damping from Each Element

Using the same energy-based damping definition, the contribution of modal damping ratios from each element was estimated by using corresponding loss factor and modal energy ratio (MER) in Eq. (3). In this paper, the MER was defined as the ratio of modal strain energy of a structural component to the modal potential energy of the whole structure. The assumption here was that all the elements in a sub-structure had the same loss factor which was equal to the loss factor of that sub-structure. Fig. 13 shows the elements contributing largely to the modal damping ratio for the 1st bending mode, 1st torsional mode and 3rd diagonal mode respectively. The dissipation of energy occurs mainly from the upper chord members due to the bending in vertical direction for the 1st bending mode as can be seen from Fig. 8(a). In case of the 1st torsional mode, the elements from upper chords, top and bottom lateral bracings and diagonals have dominant contribution to damping as can be seen from Fig. 8(b). The diagonal elements which have dominant damping contribution can be observed in Fig. 8(c). From similar plots for other modes, one should be able to identify the elements contributing largely to the energy dissipation for a particular mode.

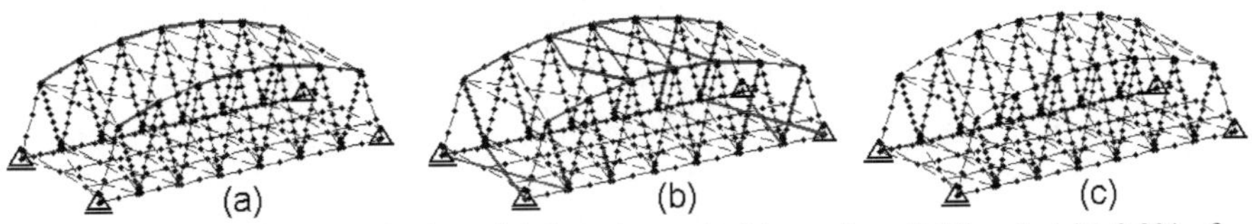

Fig. 8. Elements having significant analytical modal damping ratio (shown in solid lines): (a) >0.3% of modal damping for 1st bending, (b) >0.1% of modal damping for 1st torsional, and (c) >1% of modal damping for 3rd diagonal mode

4. Application to SHM

4.1 Problem Description

A large crack at the lower end part of D5u diagonal member was observed in the fourth span of the same bridge as shown in Fig. 9(a). As an emergency measure, reinforcing plates were installed inside and outside of the flange and web. Fig. 9(b) shows the damaged D5u member and its reinforced condition. Field vibration measurements of this span were carried out before and immediately after reinforcement using five accelerometers only as shown in Fig. 9(a). Table 3 shows the modal identification results by ERA for frequencies and damping that were averaged over identified parameters corresponding to nine sets of recorded data for before and after reinforcement conditions of D5u diagonal member. It was observed that the changes in natural frequencies of the global modes due to damage are much smaller than the change in modal damping ratios. However, the change in modal damping ratio of 3rd bending mode was much higher than the changes in other modes. The details can be found in Yamaguchi et al. (2015).

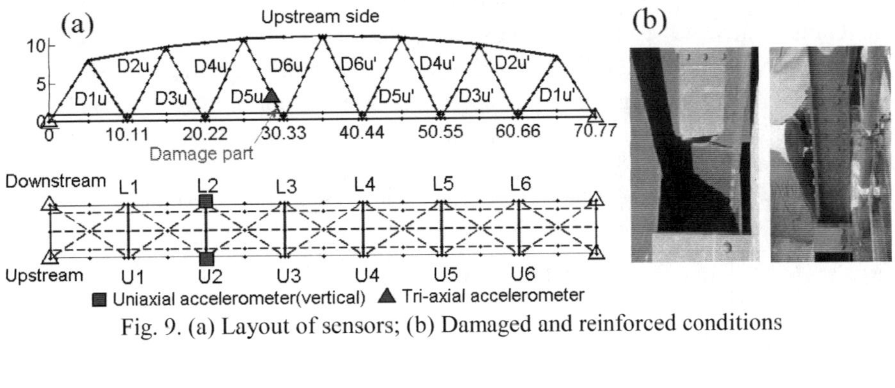

Fig. 9. (a) Layout of sensors; (b) Damaged and reinforced conditions

Table 3. Modal identification results of frequencies and damping for the fourth span

Identified modes	Natural frequency (Hz)					Modal damping ratio					
	Experimental			FE-model		Experimental			Analytical		
	Before	After	Rate	Before	After	Before	After	Rate	Before	After	Rate
1st bending	2.577	2.604	-1%	2.615	2.629	0.0093	0.0069	35%	0.0092	0.0070	31%
2nd bending	5.254	5.313	-1%	5.394	5.463	0.0072	0.0061	18%	0.0078	0.0061	28%
3rd bending	7.143	7.295	-2%	7.604	7.756	0.0106	0.0060	77%	0.0097	0.0059	64%
Diagonal	7.135	9.783	-27%	7.164	9.184	0.0055	0.0039	41%	0.0061	0.0046	33%

In this study, an analytical explanation for this substantial change in experimentally identified modal damping ratio for 3rd bending mode was given and possible reasons were discussed. In order to simulate the damage in the FE-model, the sectional properties of all the elements of D5u diagonal member were reduced to half. The frequencies of the FE-model showed a reasonable agreement with the corresponding experimental identification, as observed in Table 3. It can be observed that the frequency of the diagonal mode came very close to the frequency of the 3rd bending mode due to damage. Because of this closeness in frequency, dynamic coupling of D5u diagonal member was observed in the 3rd bending mode as can be seen in Fig. 10(b).

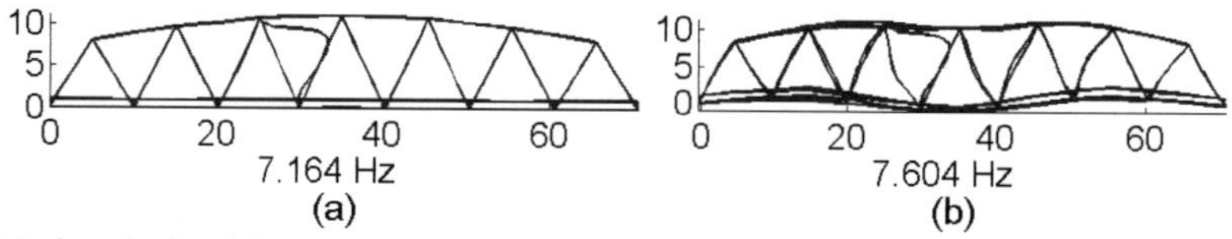

Fig. 10. Modes obtained from damaged FE-model of the fourth span; (a) Diagonal, (b) 3rd bending

4.2 Identification of Loss Factors for Damaged Span

To understand possible causes of the significant change in experimentally identified modal damping ratios for global modes due to damage in D5u diagonal member, the analytical modal damping ratios were evaluated for this span. Due to the limitation of number of identified modes shown in Table 3, only the loss factors of D, G and UC were estimated using the experimentally identified modal damping ratios at the fourth span. The loss factors of TLB and BLB and the dynamic friction coefficient at support were assumed to be the same as those of the first span for which experimental data with 17 locations were available. At first, the analytical modal damping ratios corresponding to the after reinforcement (AR) condition were evaluated. For before reinforcement (BR) condition, the equivalent loss factor of damaged D5u member (η_{d5}) was estimated by considering D5u diagonal in-plane mode as shown in Fig. 2(a). In this step, the equivalent loss factor for other diagonal members was taken as 0.0115 which corresponds to that in AR condition. Table 4 shows the identified damping parameters for the fourth span for AR and BR conditions. From Table 3, it can also be seen that the analytically evaluated modal damping ratios agree well with the corresponding experimental ones while there was increase in loss factor of girders along with the increase in loss factor of damaged diagonal member for BR condition.

Table 4. Damping parameters corresponding to AR and BR conditions

Damping parameters	η_{d5}	η_d	η_g	η_{uc}	η_{tlb}	η_{blb}	μ_s
AR	0.0115	0.0115	0.0130	0.0078	0.0963	0.0229	0.3796

4.3 Investigation on Change in Loss Factors and Damping

The increase in the loss factor of girders might have attributed to the fact that the same values of loss factors and friction coefficient of the first span were used in the fourth span for top and bottom lateral bracings and supports, although it was observed that the modal damping ratios of the fourth span are different from the first span even for AR condition. Although the contributions of damping from these sub-structures were smaller individually, collectively their contribution can be significant and could affect the results as well.

Possible reasons for the increase in the loss factor of girder could include the amplitude dependence of the loss factor. To investigate more on this, the initial modal amplitudes of vibration at measurement locations obtained from ERA were listed in Table 5. From this table, it can be seen that the initial modal amplitudes of girders for 1st and 2nd bending modes are much higher in BR condition compared to AR condition which might have caused the increase in the modal damping for these two modes. For the 3rd bending mode, a significantly large amplitude of D5u member was observed in BR condition. If the loss factor of D5u is amplitude-dependent, then the loss factor of D5u member for the 3rd bending mode in BR condition would be much higher compared to that for the damaged diagonal mode and this increase in the loss factor might have caused the significant increase in the modal damping ratio for the 3rd bending mode. Considering the above hypothesis, the loss factor of damaged D5u member for the 3rd bending mode was evaluated as 0.1047, keeping the loss factors of all other sub-structures same as the AR condition, which was clearly much higher than the corresponding value of 0.0172 for the damaged diagonal mode. Here, the consideration of amplitude-dependent loss factor could be reasonable as the damage discussed in this paper was due to cracks that could induce significant increase in frictional damping which is expected to be dependent on the amplitude of vibration. Hence, the analytical model for damping using energy-based approach could be modified considering amplitude-dependent loss factors to express the substantial change in the modal damping ratio of coupled global mode more reasonably in case of damage in diagonal member of steel truss bridges, which could be the future works in this field.

Table 5. Initial modal amplitudes at measurement locations for AR and BR conditions

Mode	Amp. at AR condition (m/s^2)			Amp. at BR condition (m/s^2)		
	Amp. at U2	Amp. at L2	Amp. at D5u	Amp. at U2	Amp. at L2	Amp. at D5u
1st bending	0.0058	0.0062	0.0001	0.0255	0.0270	0.0050
2nd bending	0.0126	0.0128	0.0033	0.0261	0.0246	0.0015
3rd bending	0.0053	0.0043	0.0021	0.0053	0.0083	0.1488
Diagonal	0.0008	0.0014	0.0342	0.0024	0.0011	0.0583

5. Conclusions

In this paper, an EBDE was introduced for detecting damage at local level by using minimum number of sensors. The main conclusions of this paper can be summarized as:

1. The modal damping ratios can be analytically evaluated by using energy-based damping definition and a method to know the contribution of modal damping ratios from different structural elements of a FE-model was introduced.
2. The results of EBDE implied that substantial change found experimentally in the damping for the global mode coupled with diagonal mode was caused by significant increase in the equivalent loss factor of damaged diagonal member.

3. The proposed approach may be able to detect damage at local level by investigating analytically the contribution from structural elements to the damping of global mode coupled with vibration of local member in steel truss bridges.

References

Fujino, Y & Siringoringo, DM 2008, 'Structural health monitoring of bridges in Japan: an overview of the current trend.' *Proc. of the 4th International Conference on FRP composites in Civil Engineering*, Zurich.

Doebling, SW, Farrar, CR & Prime, MB 1998, 'A summary review of vibration-based damage identification methods.' *Journal of Shock Vibration. Digest,* vol. 30, no. 2, pp. 91–105.

Fan, W & Qiao, P 2011, 'Vibration-based damage identification methods: a review and comparative study.' *Structural. Health Monitoring,* vol. 10, pp. 83–129.

Modena, C, Sonda, D & Zonta, D 1999, 'Damage localization in reinforced concrete structures by using damping measurements, Damage assessment of structures.' *Proc. international conference on damage assessment of structures, DAMAS 99*, pp. 132–141.

Curadelli, RO, Riera, JD, Ambrosini, D & Amani, MG 2008, 'Damage detection by means of structural damping identification.' *Engineering Structure,* vol. 30, no. 12, pp. 3497-3504.

Yoshioka, T, Yamaguchi, H & Matsumoto, Y 2010, 'Structural health monitoring of steel truss bridges based on modal damping changes in local and global modes.' *Proc. of 5th World Conference on Structural Control and Monitoring, International Association for Structural Control and Monitoring*: Los Angeles, vol. 167.

Frizzarin, M, Feng, MQ, Franchetti, P, Soyoz, S & Modena, C 2010, 'Damage detection based on damping analysis of ambient vibration data.' *Structural Control and Health Monitoring,* vol. 17, pp. 368–385.

Yamaguchi, H, Matsumoto, Y & Yoshioka, T 2015, 'Effects of local structural damage in a steel truss bridge on internal dynamic coupling and modal damping.' *Smart Structure and Systems,* vol. 15, no. 3, pp. 523-41.

Dammika, AJ, Kawarai, K, Yamaguchi, H, Matsumoto, Y & Yoshioka, T 2014, 'Analytical Damping Evaluation Complementary to Experimental Structural Health Monitoring of Bridges.' *Journal of Bridge Engineering,* vol. 20, no. 7.

Mustafa, S & Matsumoto, Y 2016, 'Bayesian model updating and its limitations for detecting local damage of an existing truss bridge.' *Journal of Bridge Engineering*, vol. 22, no. 7, pp. 04017019-1-14.

Chen, GW, Beskhyroun, S & Omenzetter, P 2016, 'Experimental investigation into amplitude-dependent modal properties of an eleven-span motorway bridge.' *Engineering Structure,* vol. 107, pp. 80-100.

Zoghi, M 2014, 'The international handbook of FRP composites in civil engineering.' *CRC Press.*

The 8th International Conference on
Structural Health Monitoring of Intelligent Infrastructure
Brisbane, Australia | 5-8 December 2017

Effect of Ballast State on Dynamic Parameters of a Multispan Ballasted Prestressed Concrete Railway Bridge

Y. Zhang[1], Y. Miyamori[1], T. Oshima[1], Y. Shirakawa[1], S. Mikami[1], T. Saito[1]
[1] *Kitami Institute of Technology - Japan, Email:* miyamoya@mail.kitami-it.ac.jp

Abstract

In this paper a phenomenon that the structural dynamic parameters of a multispan ballasted prestressed concrete railway bridge vary significantly according to the different ballast states is presented. In high latitude regions, the ballasts on the railway bridges would be frozen in winter annually owning to the repeated melting and freezing of snow when the temperature fluctuates near 0 °C. Based on the bridge dynamic experiment results and visual checks of ballast states in different seasons, a hypothesis was proposed that the state of ballasts could affect the structural stiffness of the bridge directly, and it was also assumed that a reasonable explanation of the seasonal variation of structural dynamic parameters exists. In order to prove the rationality of the above assumption, a high resolution finite element model was performed which was strictly associated with the real structure. The ballasts in frozen state were simulated by increasing the stiffness of the solid elements of the ballasts and the stiffness was assumed empirically considering to the high difficulty of field stiffness measurement of the frozen ballasts. The finite element model analysis result of the structural natural frequencies was in good agreement with the filed dynamic experiment result. Thus, the rationality of the hypothesis that the frozen ballasts could increase the structural stiffness and vary the dynamic parameters of the bridge was well evaluated. Meanwhile, the irregularity of damping ratio variations from the experimental results was also discussed. The nonuniform distribution of the ice and snow may be considered an acceptable explanation of this phenomenon.

1. Introduction

Ballasted prestressed concrete railway bridge is a familiar structure type which was widely used in the 20th century, and now these existed ballasted prestressed concrete railway bridges have been constructed for more than several decades. As railway bridges are the structures which are subjected to high intensity moving loads, the structural health statuses of these ballasted railway bridges have been drawn much attention.

With the development of structural health monitoring (SHM) technology, the vibration-based SHM method is expected to be a promising damage detection method (Boller & Chang 2009) owning to its high efficiency and high potential to detect the invisible structural damage which is difficult to be found by normal visual inspections. The core concept of the vibration-based structural damage detection method is measuring the dynamic behavior (natural frequencies, damping ratios, mode shapes) of the global or local structure to evaluate the health status of the structure (Farahani & Penumadu 2016) (Dilena &Morassi 2011) (Chang & Kim 2016).

However, in cold regions the low temperature could cause very severe natural phenomena and complex changes of structural material properties. This would affect the dynamic characteristics of the structure. For ballasted railway bridges, the material properties of ballasts would have more complicated changes owning to the emerging of ice. The effect should be studied to clarify how much it would affect the dynamic parameters of the bridge to prevent the overestimate of the structural health state.

As this study is very detailed in the SHM field, this kind of research is still insufficient. The preference (Gonzales & Ülker-Kaustell 2013) proposed the hypothesis that the change in natural frequencies of a ballasted railway bridge was due to changes in the stiffness parameters of some materials with the onset of frost. In the reference (Zabel & Brehm 2010), the ballasted railway bridge vibration experiment results showed that the natural frequencies varied up to approximately 30% between summer and winter. This trend was also confirmed in the dynamic parameter measurement for road bridges. The preference (Yang & Dutta 2007) reported variations in the natural frequencies of the first transversal modes of a road bridge of several spans due to seasonal variations. The frost in the ground was found to be a likely cause. However, the magnitude of frequency increase for the railway bridge was much higher than the road bridges.

Based on the above background, a study about the effect of ballast states on the dynamic characteristics of the ballasted prestressed concrete railway bridge is presented by conducting field vibration experiments and employing finite element model analysis. The process of the study was shown in Fig 1.

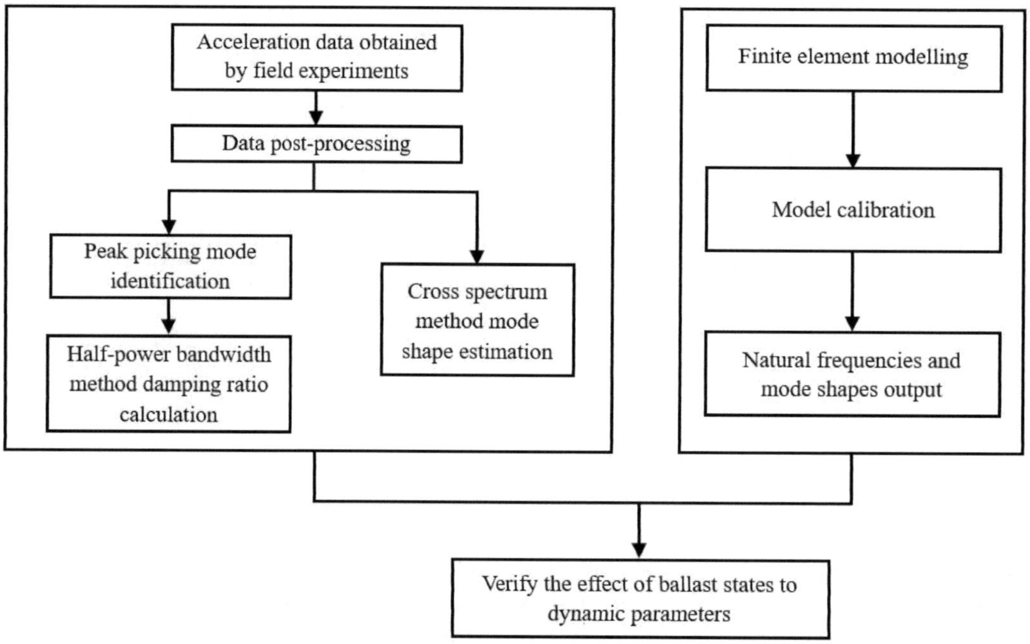

Fig. 1 The process of this study

2. Bridge description

The Mukagawa Bridge is a 5-span ballasted prestressed concrete simply supported beam bridge which is located in Kitami City, Hokkaido, Japan. The construction of the bridge was completed in 1976. The bridge connects the Kitami City and Ikeda County. A lateral view of the 3 spans of the bridge is shown in Fig. 2 and a general drawing is shown in Fig. 3.

The section views of the bridge are shown in Fig. 4. On the two sides of each span, the section view was designed as Fig. 4 (a) with wide girders. And on the middle of each span, the section view was designed as Fig 4 (b) with narrow girders. The girder section changed at the left and right quarter points. There are three cross beams on all the quarter points. Handrails were also constructed on the edges of the walkways.

Fig. 2 Three spans of the bridge

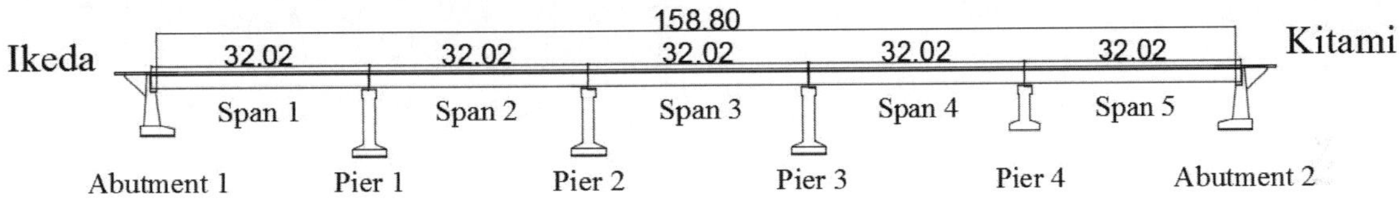

Fig. 3 General drawing of the bridge (Unit: m)

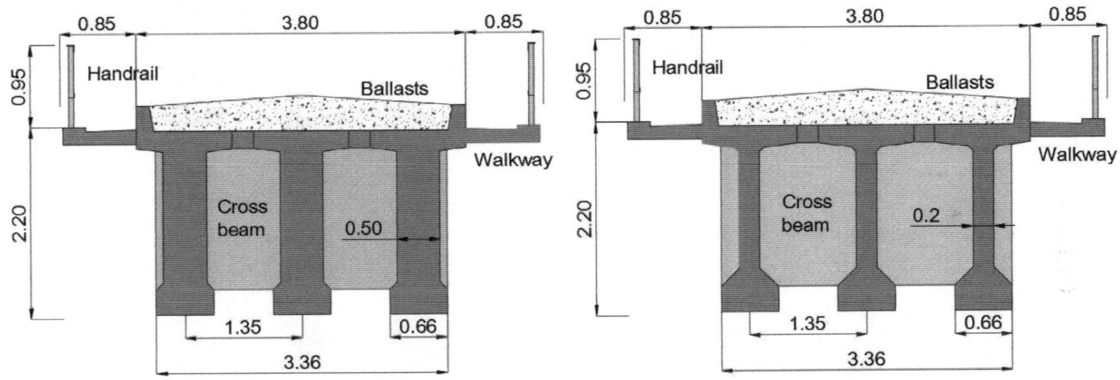

(a) section of two sides of girders (b) section of middle of girders

Fig. 4 Section views of the bridge (Unit: m)

The decks are supported at the bottom of the two sides of each girder by line bearing supports, as shown in Fig. 5. The distance between the two sides of the deck is 32.02 m and the distance between the two side supports is 31.3 m.

The ballasts are distributed on the deck, as shown in Fig. 6. As the bridge is already out of service, the sleepers and tracks have been destructed for several years.

3. Bridge dynamic experiments

3.1 Experiment Introduction

In order to investigate the dynamic parameters of the bridge, a series of vibration experiments were carried out from the autumn of 2015. The time table of the experiments is shown in Table 1. In each experiment, there were several tests which were conducted span by span. All the tests were concentrated on only one span. At the beginning of this study on the No.1 experiment, 10 sensors were symmetrically installed in two rows in one span. For efficiency considerations, from the No.2 experiment 6 sensors were installed on only one span, and the sensor distribution map was shown in Fig. 7. The reduction in the number of sensors does not affect the accuracy of the experiment result.

Fig. 5 Line bearing support

Fig. 6 Distributed ballasts on the deck

Table 1 Dates of the experiments

Experiment number	Date
No. 1	11. 16th. 2015
No. 2	2. 11th. 2016
No. 3	8. 18th. 2016
No. 4	10. 18th. 2016
No. 5	11. 29th. 2016

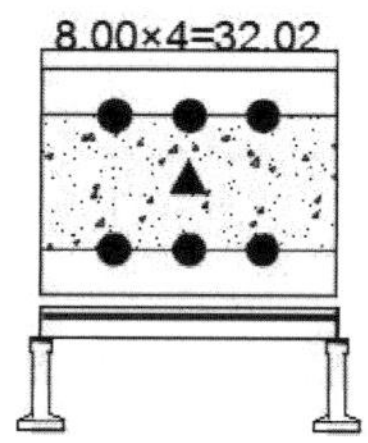

Fig. 7 Sensor distribution positions (unite: m)

The excitation method in this series of experiments was human jumping and landing simultaneously on the deck as an impulse excitation. On center and quarter points symmetrical and unsymmetrical excitation were applied on every span to obtain the 1st and 2nd bending and torsional modes. As the higher order modes were not as accurate as the first four modes, higher order mode were not discussed in this study.

The updated smart wireless acceleration sensors were used as the main measuring equipment in this series of experiments. Based on the original smart sensor (Rice & Mechitov 2010), waterproof function was added by fixing the sensor boards in a plastic box. Meanwhile, the power supply system was also optimized and the battery life was greatly enhanced. The measuring setting of smart sensors was 280 Hz with 60 s sampling period. On 5 s and 35 s the excitations were applied.

3.2 Modal Parameters Identification

The acceleration data were collected by wireless communication to the computer. The frequency domain power spectrum was obtained by fast Fourier transform. As only simple impulse excitations were applied, the dynamic behavior of the bridge was very regular. Thus, the natural frequencies were identified by peak picking method in the frequency domain power spectrum. The peak picking method is based on the fact that the frequency response functions reach an extreme value approximately at the natural frequency. The damping ratios were calculated by half-power bandwidth method. And the mode shapes were estimated by cross-spectrum method.

3.3 Experiment Result

The identified natural frequencies and damping ratios of every span in all the experiments are shown in Fig. 8 and 9. And the mode shapes of Span 1 in the No. 2 experiment are shown as an example of the experimental mode shapes in Fig. 10.

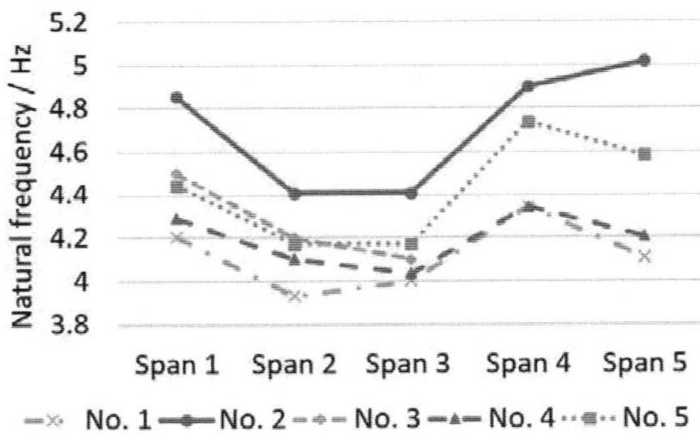

(a) Mode 1

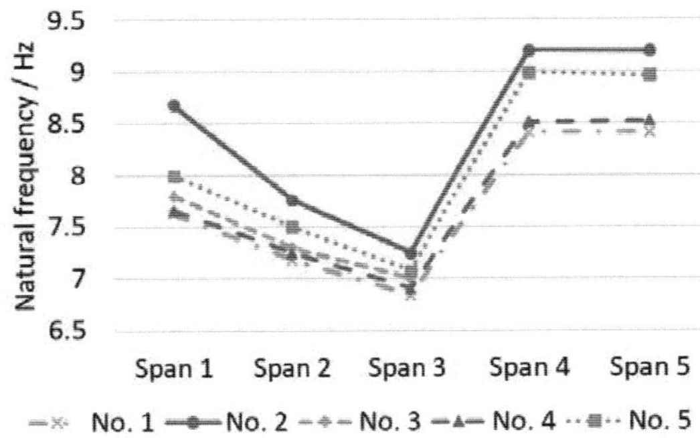

(b) Mode 2

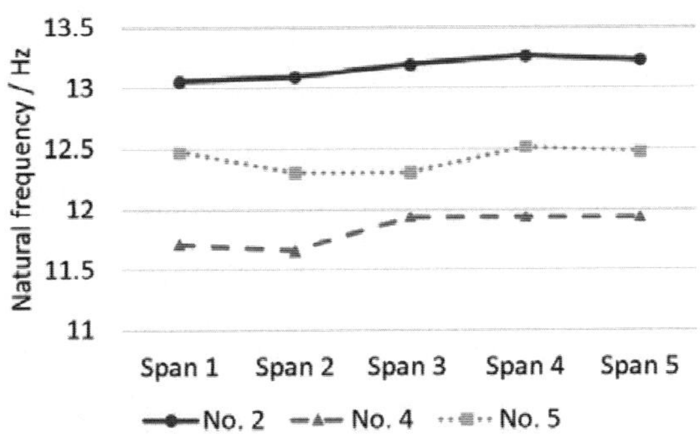

(c) Mode 3

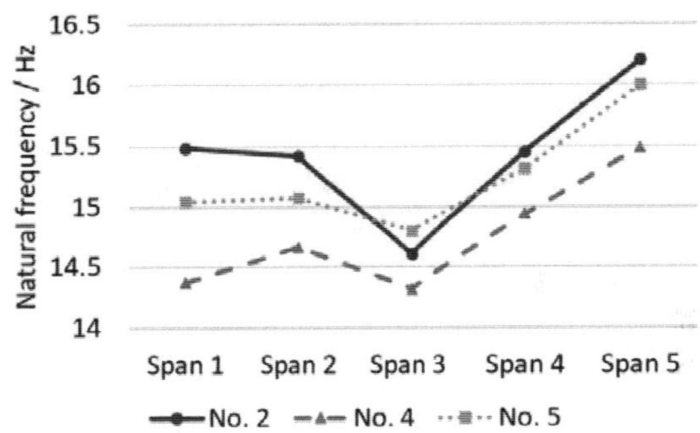

(d) Mode 4

Fig. 8 Natural frequencies

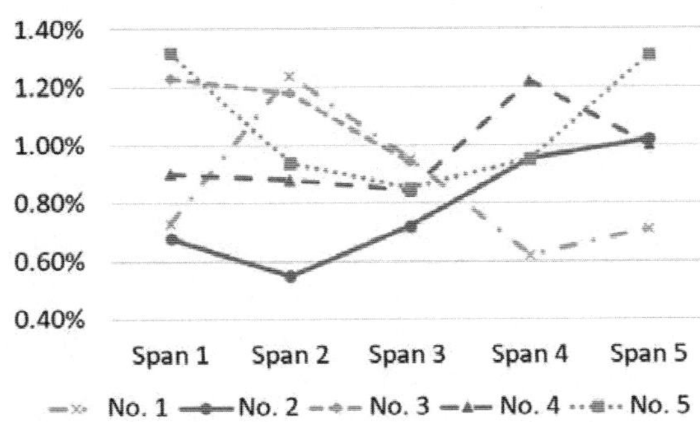

(a) Mode 1

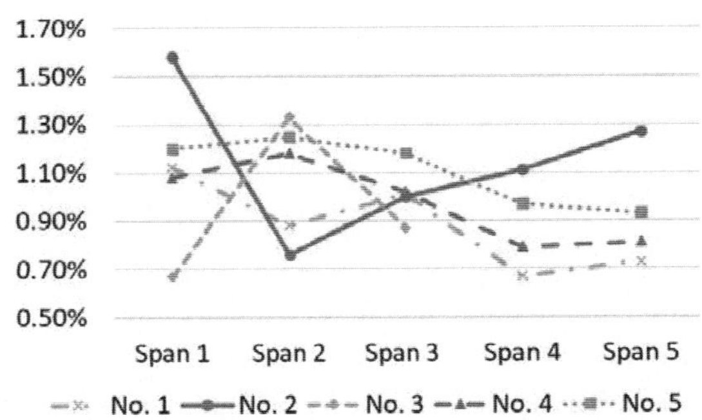

(b) Mode 2

Fig. 9 Damping ratios

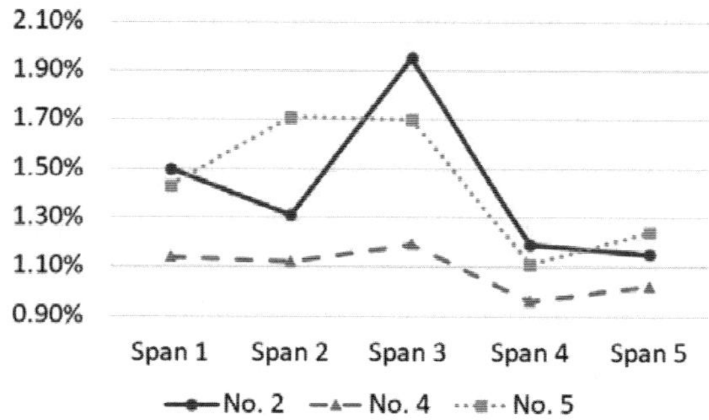

(c) Mode 3

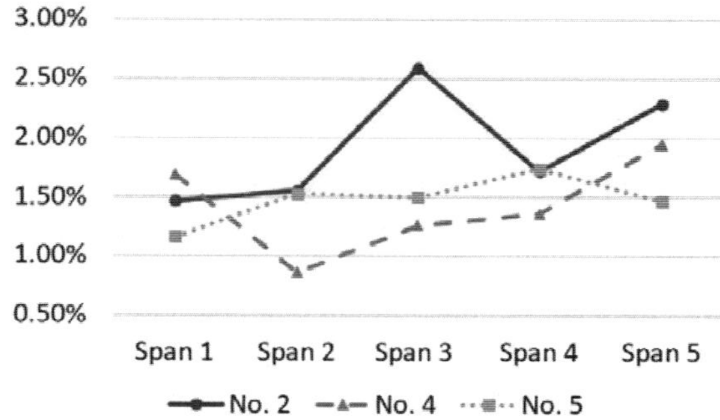

(d) Mode 4 (e)

Fig. 9 Damping ratios (continue)

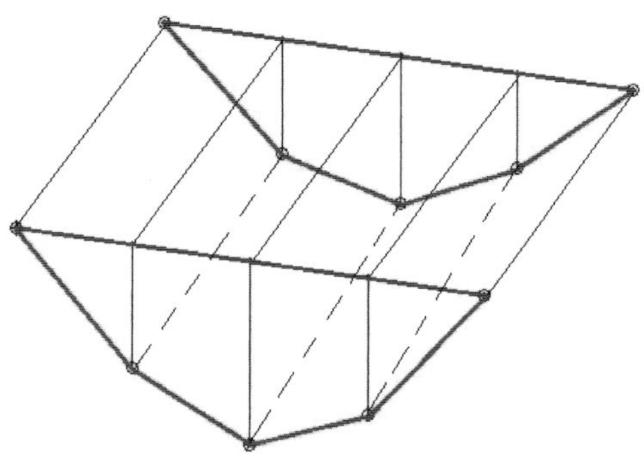
(a) Mode 1

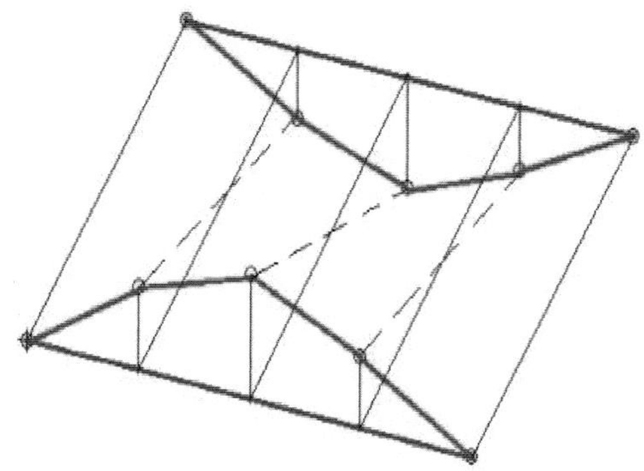
(b) Mode 2

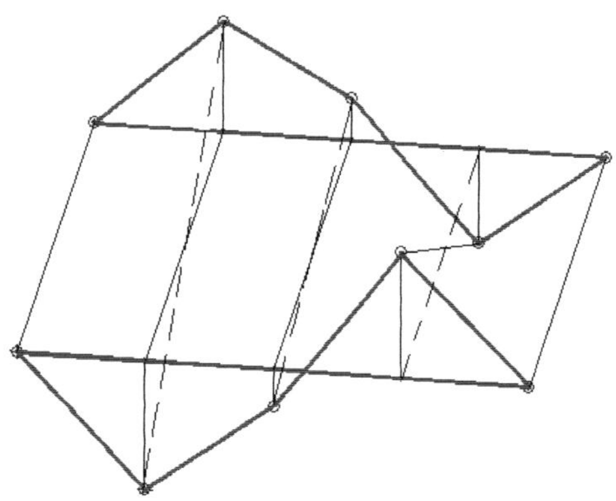
(c) Mode 3

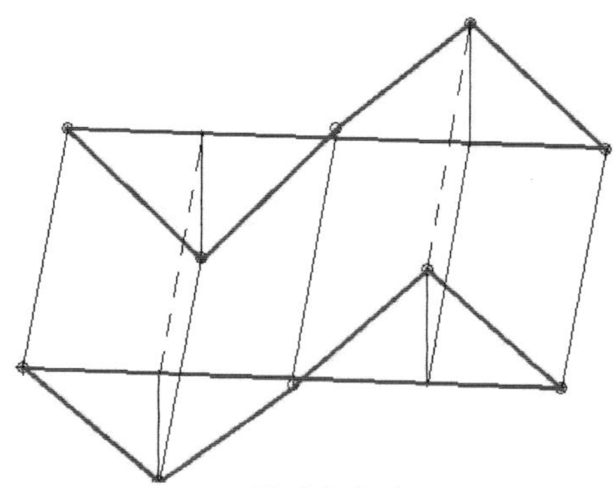
(d) Mode 4

Fig. 10 Experimental mode shapes

In this study four modes were discussed. From low to higher order, the four modes were the 1st bending mode, the 1st torsional mode, the 2nd torsional mode, and the 2nd bending mode. As human jumping and landing was the only excitation method in these experiments, only vertical vibration modes were efficiently excited. Thus, no horizontal vibration mode was identified.

For the results of natural frequencies in Fig. 8, a clear phenomenon was presented that the natural frequencies in the No. 2 experiment were much higher than in other experiments even though some data was lost due to some technical problem. In order to find the reasonable explanation for this phenomenon, a series of investigations were carried out. Normally, the difference of concrete stiffness which was caused by temperature variation was considered as the main reason of this phenomenon

(Xia & Chen 2012). However, the temperature gap between the No. 2 and 5 experiments was only 2 °C. The 2 °C temperature difference could not cause so obvious natural frequency differences. Thus this assumption was excluded. The frozen supports could be another possible reason to cause this phenomenon (Alampalli 1998). Freezing of accumulated moisture at the supports may cause partial fixity at the ends of the embedded girders, and resulted in the apparent increase in natural frequencies. However, there was no ice or snow on the supports by visual inspection so that the mechanical structure of the supports didn't change, leading to the same the constrain condition of the bridge. This is also not the acceptable explanation of this phenomenon. In this case, great attention was paid to the frozen ballasts. In this series of experiments, three typical ballast states were obtained, dry, frozen, and wet, as shown in Fig. 11.

(a) Dry (b) Frozen (c) Wet

Fig. 11 Three ballast states

Firstly, the dry ballast state [Fig. 11 (a)] was presented in the No. 1 experiment. There was no rainfall during the week before the experiment so the water content of the ballasts could be very low. So in the No. 1 experiment, the dry ballasts performed independently and the bonding effect in the ballasts was almost nonexistent. Secondly, the frozen ballast state [Fig. 11 (b)] appeared in the No. 2 experiment. The ballasts were totally frozen. The ice filled the gaps between the ballasts and froze the ballasts together to form a new composite material block. As the special climatic conditions in winter of Hokkaido, the ballasts had gone through a special temperature variation process. When the daily temperature fluctuated around 0 °C, the snow on the ballasts would melt into water and seep into the space within the ballasts. After the temperature dropped to 0 °C, the seeped water would freeze. After several days on which the daily temperature fluctuated around 0 °C, the ice which filled in the space within the ballasts would be accumulated more and more. When the ice content reached a relatively high level, the material properties of the frozen ballasts would change obviously. It could be seen as a totally different composite material. In this hypothesis, the ice acts as filler, and the ballasts act as aggregate. Thirdly, the wet ballast state [Fig. 11 (c)] was surveyed in the No. 5 experiment. In this case, the ballasts were a little wet, but not frozen. This ballast state could be seen as the intermediate state between the dry and frozen ballast states. The increase of water content of the ballasts would affect the material properties by reducing the friction and increasing the stickiness. However, this property change was still too mild comparing to the freezing.

For damping ratios as shown in Fig 9, no clear regularity could be found within these data. The damping ratios did not show the obvious distribution as natural frequencies which performed a much higher value in winter than other seasons. The damping ratios just varied unproportionally with the season altering. The particularity of damping ratio variations could be explained by the complexity of the ballasts. The difference of ballast shapes leads to the different spaces and interaction mechanisms in ballasts. And this would directly affect the water content, fiction, and stickiness of the ballasts. Furthermore, with the onset of more complicated ice material in winter, the ununiform distribution of frozen ballasts would be much severe. This could be an acceptable explaination of this complex damping ratio variation phenomenon.

4. Numerical modelling

4.1 Model Description

The three-dimensional finite element model of the Mukagawa Bridge was established by Midas Civil software. A general view of the numerical model of one span bridge was shown in Fig. 12 and a local view of the section conversion joint was shown in Fig. 13. In order to reduce the difference between the actual structure and the numerical model, all the structural elements including girders, cross beams, walkways, ballasts and deck were modelled with solid elements. In total the structure was divided into 10956 solid elements with 16654 nodes.

The geometric and mechanical parameters of the numerical model of the single span bridge such as elastic modulus, density and Poisson's ratio were shown in Table 2. As low magnitude human jumping and landing excitation method was applied in this series of experiments, the vibration amplitude of the bridge kept in very low level. Thus, the contact mechanism between the ballasts and the deck was considered as fixed. The inhomogeneous distribution of the ballasts was not considered in the model. All the ballasts were modelled in same properties.

Owning to the high complexity of restraint mechanism of the line bearing supports, the bridge behaves that on some translation and rotation directions it allows slight displacements. Because of the inexistence of the nonlinear information of the supports, the supports were modelled as typical simply supported case in the beginning modelling phase, which means the bridge had pinned supports at one end and roller supports at the other end. However, the typical simply supported boundary condition leaded to a very high natural frequency difference comparing to the experimental results. Even the error rates of the 1st bending and torsional modes were up to 75%. Thus, the model of the supports was optimized by trial and error method. Finally, the supports are modelled by employing a combination of direction fixed constrains and spring element constrains. The z direction of the supports of the left and right girders of the pinned side are modelled by spring elements. The stiffness of z direction spring elements was 60 kN/mm^2.

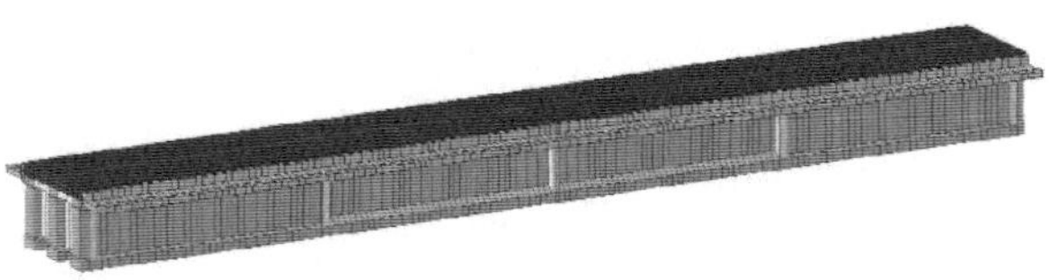

Fig. 12 General view of the one span model

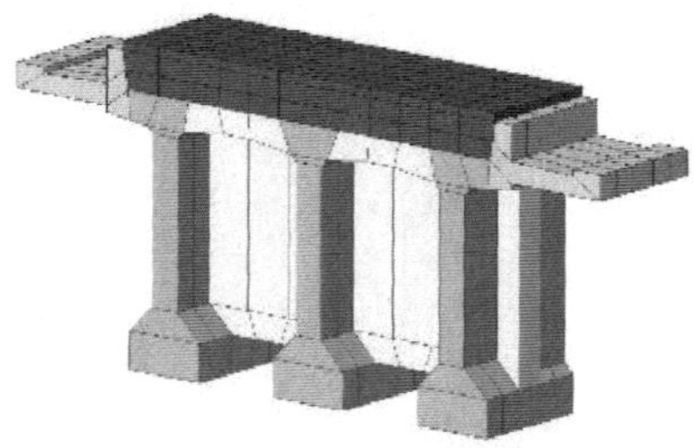

Fig. 13 Section conversion joint

Table 2 Material properties of the model

	Elastic modulus (kN/mm^2)	Density (kN/m^3)	Poisson's ratio
Girder concrete	34.00	25.00	0.1667
Cross beam concrete	26.00	24.50	0.1667
Walkway concrete	26.00	24.50	0.1667
Ballasts	16.00	20.00	0.25
Frozen walkway	30.00	24.50	0.1667
Frozen ballasts	30.00	20.00	0.25

The frozen ballasts on the numerical model were taken into account by increase the stiffness of the solid elements of ballasts. As the high difficulty of measuring the stiffness of the frozen ballasts by field test, the stiffness of the frozen ballasts on the numerical model was assigned to 16 kN/mm^2 initially and empirically. And after calibration, the stiffness of the ballasts on the frozen ballasted model was adjusted to 30.00 kN/mm^2. Considering the existence of the 3 cm ice on the walkway in the No. 2 experiment, the elastic modulus of walkway was increased to 30 kN/mm^2.

4.2 Analytical Result

The analytical mode shapes were shown in Fig. 14. Four following modes were discussed: the 1st bending mode, the 1st torsional mode, the 2nd torsional mode, and the 2nd bending mode. The natural frequencies of the fundamental model, frozen ballasted model, and bridge vibration experiments were shown in Table 3 to make a comparison.

From the comparison of analytical results and the vibration experimental results in Table 3, the differences between the analytical results and the vibration experimental results of mode 1 and 2 were 0% and less than 5% in mode 4. On mode 3, the differences between the experimental and analytical result is about 20% in both models. By adjusting the support simulation model, it could not get a very good correlation between the experimental and analytical natural frequencies on mode 3. The natural frequency of mode 3 is not sensitive to the support condition and the optimization of natural frequency of mode 3 is still an unsolved issue which would be investigated in further study. However, the increase of ballast stiffness still caused obvious natural frequency increase of mode 3 which is the same variation trend as other modes. Therefore, the rationality of the assumption could be proved that in winter the frozen ballasts would increase the structural stiffness and this would directly lead to the increase of the structural natural frequencies.

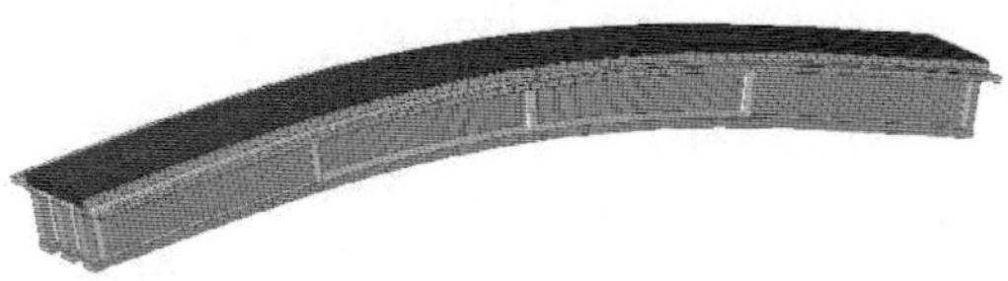

(a) Mode 1

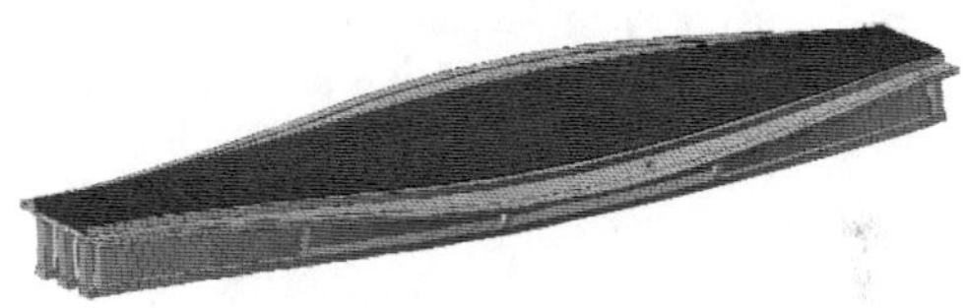

(b) Mode 2

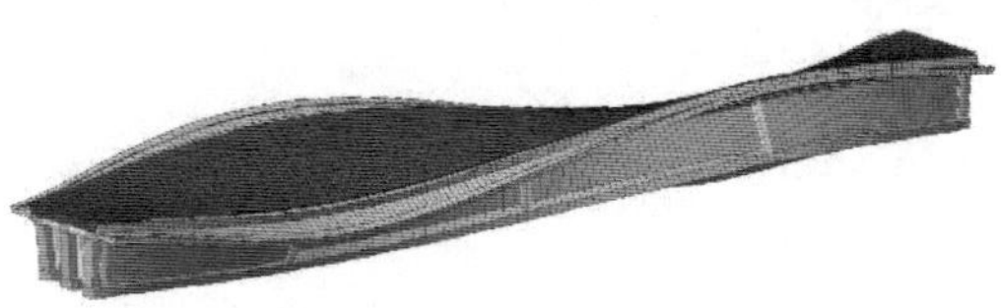

(c) Mode 3

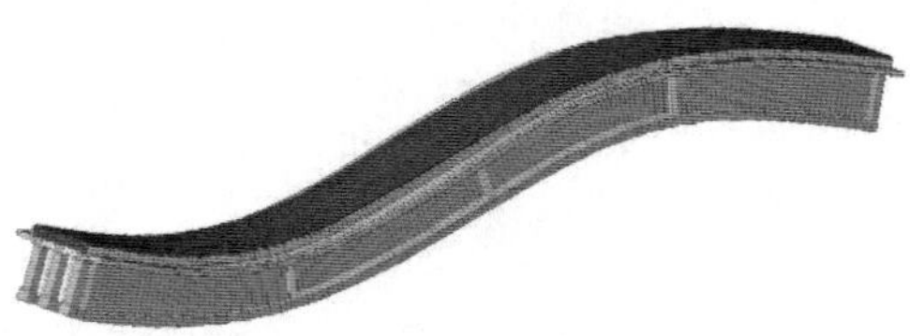

(d) Mode 4

Fig. 14 Mode shapes of the numerical model

Table 3 Natural frequencies of numerical models and vibration experiments

Mode	Span 2 on the No. 4 experiment			Span 2 on the No. 2 experiment		
	Vibration experiment / Hz	Fundamental model / Hz	Difference	Vibration experiment / Hz	Frozen ballasted model / Hz	Difference
Mode 1	4.1	4.1	0%	4.4	4.4	0%
Mode 2	7.2	7.2	0%	7.8	7.8	0%
Mode 3	11.7	14.4	23.1%	13.1	15.4	17.6%
Mode 4	14.7	14.1	-4.1%	15.4	15.0	-2.6%

5. Conclusion

In this paper, a research about the effect of ballast state on dynamic parameters of a multispan ballasted prestressed concrete railway bridge has been performed. Based on the results of bridge vibration experiments in different seasons and the analytical results of a finite element model, two conclusions were drawn as following.

Firstly, it was confirmed that the environment variations have significantly effects on natural frequencies and damping ratios. For natural frequencies, we got a clear phenomenon that the natural frequencies of every mode in all the experiments shifted synchronously with the season altering. For damping ratios, the damping ratios varied without any obvious regularity.

Secondly, for this kind of ballasted railway bridges, the ballast state can directly affect the material property of the ballasts, and this will directly lead to the change of structural natural frequencies. Comparing to the temperature effect on the natural frequencies, the ballast state has more significant effect on the natural frequencies.

As the structural natural frequencies are the most important fundamental information to estimate the health condition of the structure by applying the vibration-based SHM method, the effect of ballast state on natural frequencies could not be neglected when conducting a SHM project on a ballasted railway bridge. How to offset this effect on natural frequencies to avoid the overestimation of the structure health condition is already on the schedule of the study. And the study of the modal damping ratio variations due to the ballast states is also in progress with the purpose of getting a better comprehension of the dynamic characteristics of the ballasted railway bridge.

6. Acknowledgements

This research was supported by Japan Society for the Promotion of Science (JSPS) Grant-in-Aid for Scientific Research(C) Grant Number 15K06176. The contributions of Mr. Takanori Kadota, Dr. Tomoyuki Yamazaki and the Kitami City Government were greatly acknowledged.

References

Boller, C, Chang, F, & Fujino, Y 2009, *Encyclopedia of Structural Health Monitoring*, Wiley, West Sussex.

Farahani, RV & Penumadu, D 2016, 'Damage identification of a full-scale five-girder bridge using time-series analysis of vibration data', *Engineering Structures*, 115, pp. 129-139.

Dilena, M & Morassi, A 2011, 'Dynamic identification of a reinforced concrete damaged bridge', *Mechanical Systems and Signal Processing*, 25(8), pp. 2990-3009.

Chang, KC & Kim, CW 2016, 'Modal-parameter identification and vibration-based damage detection of a damaged steel truss bridge', *Engineering Structures*, 122, pp. 156-173.

Gonzales, I & Ülker-Kaustell, M 2013, 'Seasonal effects on the stiffness properties of a ballasted railway bridge', *Engineering structures*, 57, pp. 63-72.

Zabel, V & Brehm, M 2010, 'The influence of temperature varying material parameters on the dynamic behavior of short span railway bridges', *Proc. International Conference on Noise and Vibration Engineering (ISMA),* Leuven, Belgium, 20-22 September, Vol. 74.

Rice, JA, & Mechitov, K 2010, 'Flexible smart sensor framework for autonomous structural health monitoring', *Smart structures and Systems*, 6(5-6), pp. 423-438.

Yang, ZJ & Dutta, U 2007, 'Seasonal frost effects on the soil–foundation–structure interaction system', *Journal of Cold Regions Engineering*, 21(4), pp. 108-120.

Xia, Y & Chen, B 2012, 'Temperature effect on vibration properties of civil structures: a literature review and case studies', *Journal of civil structural health monitoring*, 2(1), pp. 29-46.

Alampalli, S 1998, 'Influence of in-service environment on modal parameters', *Proceedings-SPIE The International Society for Optical Engineering*, Colorado, American, 10-13 May, Vol. 1, pp. 111-116.

The 8th International Conference on
Structural Health Monitoring of Intelligent Infrastructure
Brisbane, Australia | 5-8 December 2017

Early Damage Detection using Multivariate Data-driven Approaches – Application to Experimental Data from a Cable-stayed Bridge

E. Sousa Tomé [1], M. Pimentel [2], J. Figueiras [3]

[1] *CONSTRUCT-LABEST, Faculty of Engineering (FEUP), University of Porto – Portugal, Email:* emanuel.tome@fe.up.pt, [2] *CONSTRUCT-LABEST, Faculty of Engineering (FEUP), University of Porto – Portugal,* [3]*CONSTRUCT-LABEST, Faculty of Engineering (FEUP), University of Porto – Portugal*

Abstract

The implementation of automatic and real-time data processing algorithms in order to reduce the big amount of data down to a human and useful scale is often pointed out as a key step for increasing the value of Structural Health Monitoring (SHM). One of the reasons usually pointed out for the low usability of the structural monitoring data is the fact that damage events are usually masked by the environmental and/or operational effects. Indeed, the robustness and accuracy of the damage (or novelty) detection methods depend on how successfully the changes in the structural response due to damage can be discerned from the normal environmental and operational effects. The process of removing the environmental and/or operational effects from the structural response is usually termed as data normalisation.

In this context, the present work describes the adopted methodologies for data normalisation and novelty detection implemented in the Corgo Bridge, a cable-stayed bridge located in northern Portugal recently opened to traffic and wherein a long-term monitoring system has been installed. Data normalisation is accomplished by means of application, alone and combined, of two well-established multivariate statistical tools: multiple linear regression analysis and principal component analysis. The Hotelling T^2 control chart is used to track the existence of abnormal values. The performance of the chosen data normalisation methods are evaluated and compared. The first year of data is used to establish the multivariate models and the remaining data is used to validate the fitted models and the ability for novelty/damage detection. Since the bridge is new and sound, damage scenarios are numerically simulated.

1. Introduction

Bridges, as part of the infrastructure system, are key elements in modern societies. The structural inefficiency of these structures have important, and sometimes irreversible, negative consequences to the society and to the economy.

Notwithstanding the large efforts placed on the development of Structural Health Monitoring (SHM) technologies, the SHM systems often include a large number of sensors and generate big amounts of data which in many circumstances is of low usability by the bridge owners/operators (Zhang and Zhou, 2007, Webb et al., 2014, Brownjohn, 2007, Ko et al., 2009, Omenzetter and Brownjohn, 2006). The implementation of automatic and real-time data processing algorithms in order to reduce the big amounts of data to a human and useful scale is often pointed out as a key step for increasing the value of SHM. In particular, algorithms enabling the robust detection of small structural damages and thereby supporting decision-making on asset management are still a challenge to be addressed.

In general, the robustness and accuracy of the anomaly detection algorithms depend on how successfully the changes in the structural response due to damage can be discerned from the normal environmental and operational effects. In the scope of SHM, the process of removing the environmental and operational effects from the structural response is usually termed as data normalisation. Depending on the type of structure, this process can be complex and constitutes one of the main reasons for the slow transition of SHM from academia to real practical applications.
In this context, two well-established multivariate statistical tools, multiple linear regression analysis (MLR) and principal component analysis (PCA), are applied, alone and combined, in order to suppress the environmental and operational effects from the sensor readings. These multivariate statistical methods were chosen since they are relatively simple, intuitive and can be easily implemented for real-time damage detection in the existing SHM system. Moreover, their ability for early damage detection has already been demonstrated (Magalhães et al., 2012, Comanducci et al., 2016) when modal frequencies are used as damage sensitive features. Since the selected case study is a new bridge recently opened to traffic, the assumption that the bridge is currently being observed in an undamaged state and that its response to thermal loads is linear is justifiable. The Hotelling T^2 control chart is used to track the existence of abnormal values. The first year of data is used to establish the multivariate models and the remaining data is used to validate the fitted models and the ability for novelty/damage detection. Since the bridge is new and sound, damage scenarios in the stay cables are numerically simulated.

2. Data normalisation and novelty detection: theoretical background

2.1 Multiple linear regression (MLR)

The linear regression analysis is probably the simplest multivariate statistical tool to relate the observed environmental and/or operational factors with the observed structural responses and/or features. This statistical tool can be used to predict one or more responses (dependent variables) from a collection of predictors (usually called as predictor, regressor or independent variables) and to assess the influence of the predictors on the dependent variables (Johnson and Wichern, 2013). The multilinear regression model is expressed by (Johnson and Wichern, 2013):

$$\mathbf{Y} = \mathbf{Z}\boldsymbol{\beta} + \boldsymbol{\varepsilon}_{MLR} \tag{1}$$

where $\mathbf{Y}$ is a n-by-m matrix of the dependent variables, being n the number of observations and m the number of dependent variables, $\mathbf{Z}$ is a n-by-(r+1) matrix with the corresponding n values of r selected predictor variables, $\boldsymbol{\beta}$ is a (r+1)-by-m matrix with the parameters to be determined that weight the contribution of each predictor variable, and $\boldsymbol{\varepsilon}_{MLR}$ is a n-by-m matrix with the random error of the MLR model. The estimates of the model parameters ($\hat{\boldsymbol{\beta}}$) are usually obtained through the least squares method, being given by:

$$\hat{\boldsymbol{\beta}} = (\mathbf{Z}^T\mathbf{Z})^{-1}\mathbf{Z}^T\mathbf{Y} \tag{2}$$

Therefore, the obtained model can be used to obtain predictions of the dependent variables ($\hat{\mathbf{Y}}$), being the error of the forecasts the difference between $\mathbf{Y}$ and $\hat{\mathbf{Y}}$. Within the scope of SHM, the multilinear regression model should ideally be constructed using data of an entire year (Magalhães et al., 2012).

2.2 Principal component analysis (PCA)

Principal Component Analysis (PCA) can be defined as a statistical tool for explaining the variance-covariance structure of a set of variables through linear combinations of these variables (Johnson and Wichern, 2013) and it has been extensively used in the scope of SHM, namely for the recognition of patterns in the data, data cleansing and data compression (Cavadas et al., 2013, Figueiredo et al., 2009). The fact that PCA is a latent-variable method, that is, a method that is able to remove the environmental effects using only the structural responses, without any knowledge about the input, is usually pointed out as its most attractive feature (Santos et al., 2013, Yan et al., 2005, Bellino et al., 2010).

Considering an *n-by-m* matrix $\mathbf{Y}$ with the original variables, where m is the number of sensors or/and features and n is the number of observations in time, a transformation to another set of m variables $\mathbf{Z}$, the principal component scores, can be made by:

$$\mathbf{Z} = \mathbf{Y} \cdot \mathbf{T} \tag{3}$$

where $\mathbf{T}$ is the transformation (or loading) matrix, an m-by-m orthonormal matrix that applies a rotation to the original coordinate system (Magalhães et al., 2012). The covariance matrix of the original variables in the training period, $\mathbf{\Sigma}$, is related to the covariance matrix of the principal component scores, $\mathbf{\Lambda}$, by the following equation,

$$\mathbf{\Sigma} = \mathbf{T} \cdot \mathbf{\Lambda} \cdot \mathbf{T}^T \tag{4}$$

being the $\mathbf{T}$ and $\mathbf{\Lambda}$ matrixes obtained by the singular value decomposition of the covariance matrix $\mathbf{\Sigma}$ of the original variables. The columns of $\mathbf{T}$ are the singular vectors and the diagonal matrix $\mathbf{\Lambda}$ contains the singular values of the matrix $\mathbf{\Sigma}$ in descending order. The singular values stored in $\mathbf{\Lambda}$ are the variances of the components of $\mathbf{Z}$. Moreover, the matrix $\mathbf{\Lambda}$ can be split into a matrix with the first p singular values and in a matrix with the remaining m-p singular values, which are not relevant to explain the variability of $\mathbf{Y}$.

After the choice of p, the first p components of the matrix $\mathbf{Z}$ can be calculated using equation (3) and a transformation matrix $\hat{\mathbf{T}}$ built with the first p columns of $\mathbf{T}$. Those first p components can be re-mapped to the original space (Magalhães et al., 2012, Yan et al., 2005) using the following equation:

$$\hat{\mathbf{Y}} = \hat{\mathbf{Z}} \cdot \hat{\mathbf{T}}^T = \mathbf{Y} \cdot \hat{\mathbf{T}} \cdot \hat{\mathbf{T}}^T \tag{5}$$

being then the residual error matrix estimated from:

$$\boldsymbol{\varepsilon}_{PCA} = \mathbf{Y} - \hat{\mathbf{Y}} \tag{6}$$

The residual error matrix $\boldsymbol{\varepsilon}_{PCA}$ is expected to be insensitive to the effects modelled by the PCA and can be used to detect damage. However, it should be borne in mind that this approach assumes that the environmental factors have a linear or a weakly non-linear effect on the structural responses (Yan et al., 2005). Once again, to account for a large range of variation, with the aim of removing environmental effects from the monitored structural responses, the covariance matrix to which the singular value decomposition is applied should be estimated from an entire year of data.

Finally, this tool can be used alone to remove the environmental factors on the structure response (Yan et al., 2005) or it can be used to remove the environmental factors not explained by the regression model (Magalhães et al., 2012). On the scope of this work, the obtained results using the two approaches are presented and conclusions about their performance are put forward.

2.3 Hotteling T^2 control chart

After removing the environmental factors from the measured structural response, a control chart can be used to track the existence of abnormal values, which can be related to the presence of damage. The so called control limits define the accepted process variability. If an observation exceeds those control limits, the observation is said to be an out-of-control observation. In context of SHM, this out-of-control observation may be associated with the presence of damage in the structure.

In order to have only one control-chart instead of having one for each variable/sensor, the multivariate Hotelling T^2 control chart was adopted, being the T^2-statistic calculated from the following expression:

$$T^2 = r(\bar{\mathbf{x}} - \bar{\bar{\mathbf{x}}})^T \mathbf{S}^{-1}(\bar{\mathbf{x}} - \bar{\bar{\mathbf{x}}}) \quad (7)$$

where r is the number of observations considered (window size), $\bar{\mathbf{x}}$ is the average of the observations inside the window, $\bar{\bar{\mathbf{x}}}$ is the process average when it is in control and $\mathbf{S}$ is the process covariance matrix, also estimated in the reference period. The lower control limit is zero and the upper control limit (UCL) is obtained from:

$$UCL = \frac{m(s+1)(r-1)}{s \cdot r - s - m + 1} F_{m,\, s \cdot r - s - m + 1}(\alpha) \quad (8)$$

where $F_{m,\, s \cdot r - s - m + 1}(\alpha)$ is the α percentage point of the F distribution with m and $s \cdot r - s - m + 1$ degrees of freedom, being m the number of sensors and/or features and s the number of subgroups (or windows) collected during the reference period. This statistic, as all the multivariate control charts, has the advantage of condensing all sensors responses into a scalar indicator, working in the context of SHM as a damage indicator. However, after an abnormal behaviour is flagged, it is not possible to know which sensor(s) are responsible for that behaviour.

3. Case study: Corgo Bridge

3.1 Description of the bridge

The Corgo Bridge (Figure 1) is a 2790m long prestressed concrete box-girder bridge located in northern Portugal. The bridge belongs to the motorway A4 connecting the industrialised coastal area to Northeast border with Spain. The bridge is divided into three continuous sub-viaducts: the West Sub-Viaduct (WSV), the Central Sub-Viaduct (CSV) and the East Sub-Viaduct (ESV) with lengths of, 855m, 768m and 1167m, respectively. The WSV and the ESV are continuous frame bridges with the majority of the spans being 60m long and were constructed using movable scaffolding systems. The CSV is a cable-stayed bridge with a 300m long central span balanced by 126m long adjacent spans and two continuous end spans with 48m and 60m on each side (see Figure 2). The suspension system consists of one single central plane with four symmetric semi-fans of 22 stay-cables. The central span and the adjacent spans were built using the balanced cantilever method, while the continuous end spans were constructed using the same movable scaffolding systems of the WSV and ESV.

The deck of the CSV is 28m wide and is constituted by a 3.5m high box-girder with overhangs supported by prefabricated concrete struts spaced at regular intervals of 3m (see Figure 1 (b)). The pylons receiving the forces from the stay cables are monolithically connected to the deck and have a total height of about 193m. Below the deck, the pylons have an octagonal-shaped box cross-section with a variable transversal dimension along the height. The piers, with heights varying between 18m and 113m, have a constant octagonal-shaped box cross-section. The deck is supported on the piers through pot bearings which are either fixed in both longitudinal and transversal directions or only transversely fixed. Further information about the Corgo Bridge can be found in Barata (2012).

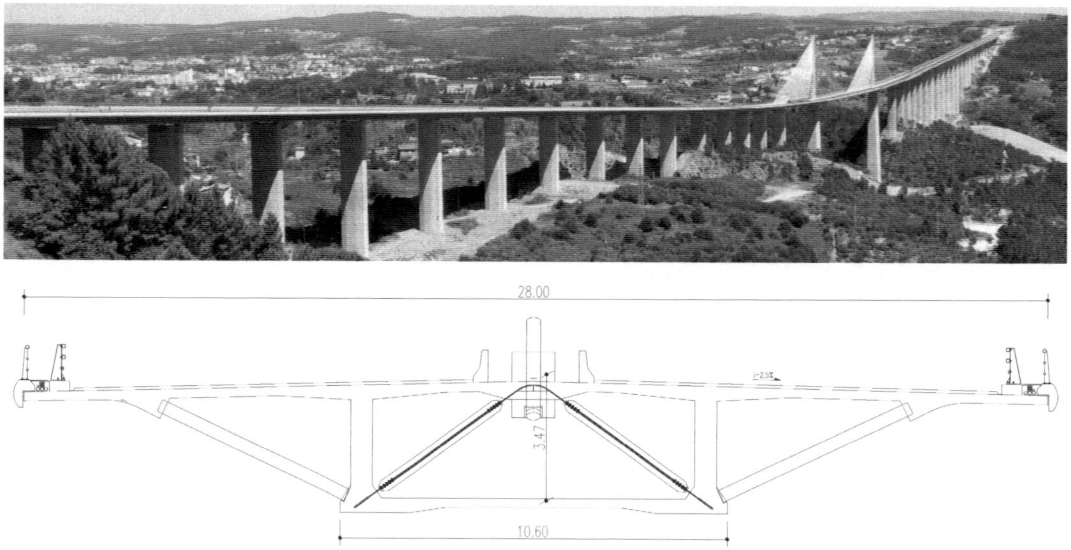

Figure 1 – Corgo Bridge: (a) general view, (b) cross-section of the CSV.

3.2 Monitoring system

A comprehensive structural monitoring system was implemented in the Corgo Bridge, with a particular focus on the CSV. The system comprises the measurement of bearing displacements, vertical deflections, rotations, cable forces, average concrete strains, strains in the steel diagonals, ambient and concrete temperatures, ambient and concrete humidity, corrosion potential and rebar corrosion rate. The measurements are performed by fibre-optic sensors (span deflections, steel, concrete strains and concrete temperatures of the CSV) and electric sensors (relative bearing displacements, rotations, cable forces, concrete strains of the ESV, temperature, humidity, corrosion potential and corrosion rate) (Félix et al., 2012). An elevation of the CSV is presented Figure 2 with the locations of the subset of sensors used in these work for application of the novelty detection algorithms.

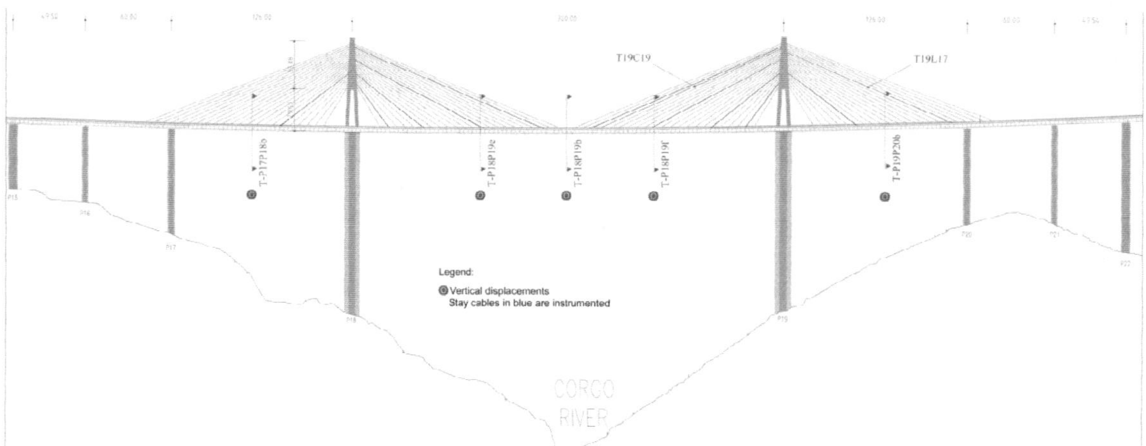

Figure 2 – Side elevation of the Central Sub-Viaduct with the location of the monitored sections.

4. Results and discussion

The present section describes the followed procedures in order to remove (or at least minimise) the environmental effects on the selected sensors of the structural monitoring system. In order to account for the yearly variation of the environmental action, the first year of data was used to establish the multivariate models. The remainder data were used to validate the fitted models, evaluate the quality of the forecasts of the models and evaluate the ability for novelty detection.

The stay-cables are the most crucial elements of the cable-stayed bridges since the cable forces are the best indicators of the internal force distribution as well as the alignments of the cable-supported structures (Ren et al., 2008). Therefore, with the aim of detection damage in the suspension system of the bridge, five vertical displacements of the girder and ten cable forces were chosen from the monitoring system of the bridge for inclusion in the MLR and PCA models.

In order to test the ability of the multivariate techniques presented in Section 2 for removing the environmental effects and detecting damage, several damage scenarios involving the reduction of the stiffness of eight stay-cables was analysed. For the sake of clarity, only the scenarios corresponding to a stiffness reduction of 1.5%, 2.5% and 5% in stay-cables T19C19 and T19L17 (see Figure 2 for stay-cables localisation) are presented. The first stay-cable (T19C19) illustrates a case where the damage is easier to detect since it is located in the central span near to instrumented sections. The second one (T19L17) illustrates a case where the damage detection is more difficult since it is located in the lateral span, far from the instrumented sections and close to pier P20.

For the construction of the MLR models, three independent variables were chosen: stay-cable temperature ($T_{stay-cable}$) ; average of the concrete temperature sensors of a girder cross-section ($T_{u,sensors}^{Girder}$) ; average of the concrete temperature sensors of a cross-section of a pier ($T_{u,sensors}^{Pier}$) . $T_{stay-cable}$, $T_{u,sensors}^{Girder}$ and $T_{u,sensors}^{Pier}$ are expected to be, respectively, representative of the temperature of the stay cables, of the uniform temperature component of the girder and of the uniform temperature component of the pylon and piers. Indeed, the majority of the sensor responses are mainly function of the uniform temperature component of the different structural elements of the bridge (girder, stay-cables and piers/pylons), being the contribution of this temperature component higher than 90% for the majority of the sensors (Tomé et al., 2016).

The PCA models were constructed both using the residuals of the MLR models and the raw data. One of the key steps of the PCA is the choice of the number of p components to retain. Although several criteria, many of them ad hoc, have been proposed for the choice of the p components to retain (Jolliffe, 2002, Peres-Neto et al., 2005, Jackson, 1993), still there is not a definitive answer (Johnson and Wichern, 2013). However, in the context of SHM "the selection of an appropriate dimension p is not so critical as it appears" (Yan et al., 2005) since what is being looked for is a change in the hyperplane defined by the p principal components adopted from the reference state to the current state. Therefore, stable results can be obtained using different p principal components (Yan et al., 2005). For these reasons, in the present work the number of retained principal components was chosen in order to minimise the number of false positives and maximise the ability for damage detection.

The MLR and PCA models were adjusted for hourly and daily average data. However, only the results obtained using the daily average are presented in this work since considerable better results were obtained. This happens essentially due to two reasons. First, the averaging process allows the reduction of some of the autocorrelation that is characteristic of time series, making it easier the task of construct models with uncorrelated residuals. Secondly, the process of daily averaging the data can be seen as a low-pass filter, removing the sudden changes that are of difficult modelling and reduce the effect of sudden changes induced by traffic.

One could state that important information can be lost and the ability for detection of a change in the structural response of the bridge is reduced when the daily average values are used. However, for the present case study, when small damages are introduced in the bridge, the daily variations are very similar to the healthy state, and the differences between the undamaged and damage structural response are smaller or of the same order of magnitude of the noise level of the sensors. In other words, for the sensitivity of the used sensors, the stiffness of the bridge is the same before and after small damages as those considered in this work.

In order to compare the adopted data normalisation methods (MLR, MLR-PCA and PCA), the percentage of positives for the undamaged state and three levels of damage in stay-cables T19C19 and T19L17 are presented in Table 1. It should be referred that the percentage of false positives presented were obtained for a UCL calculated using equation (8) where $\alpha = 1\%$. According to the data presented in Table 1, the number of false positives (type I error) is generally only lower than 1% when a considerable number of principal components are used. Considering that damage is clearly flagged only when at least 50% of the points after the introduction of damage are above the UCL, only a stiffness reduction of 5% in stay-cable T19C19 can be clearly flagged (using for example the PCA model with 5 principal components (PCs) and a window size of one day for the computation of the T^2 control chart – see Figure 3 (a)).

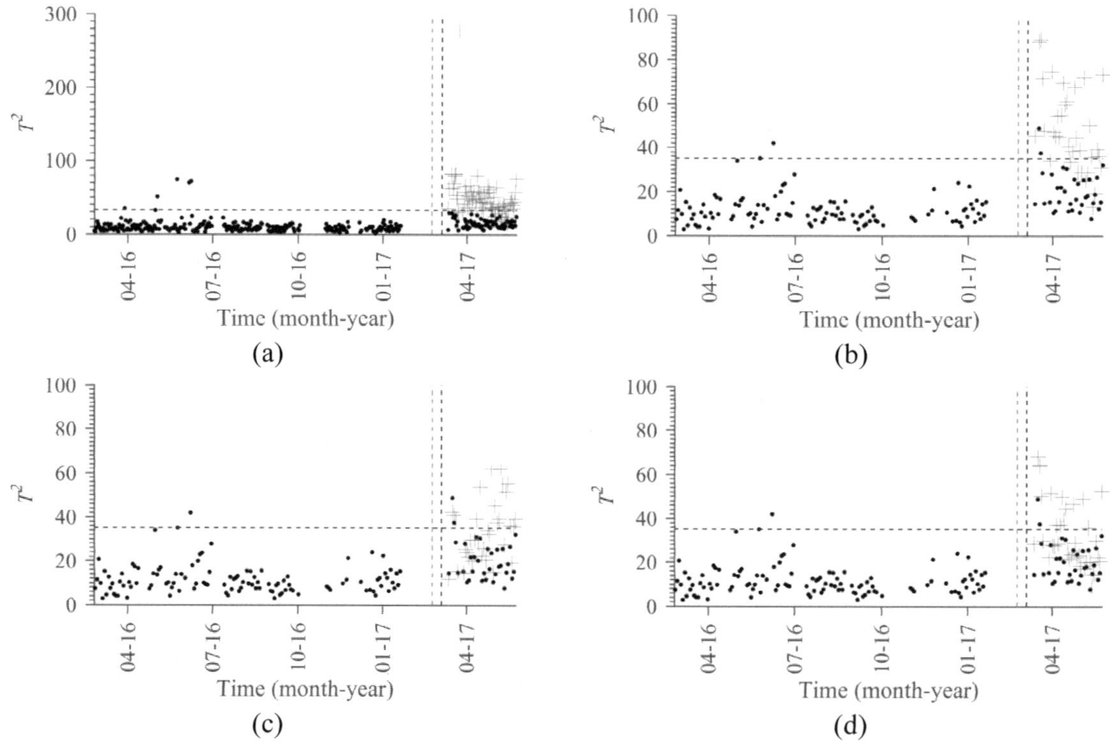

Figure 3 – Hotelling T^2 control charts for: (a) stay-cable T19C19 (-5%) – PCA model with 5 PCs and a window of one day; (b) T19C19 (-2.5%) – MLR-PCA model with 6 PCs and a window size of two days; (c) T19L17 (-5%) – MLR-PCA model with 6 PCs and a window size of two days; (d) T19C19 (-1.5%) – MLR-PCA model with 6 PCs and a window size of two days. The green dashed vertical line corresponds to the end of the reference period and the blue vertical line corresponds to the time instant of introduction of damage.

However, since the residuals of the fitted models usually have some level of autocorrelation, a higher number of false positives is expected. Therefore, if a percentage of false positives of 5% is

Table 1 – Percentage of positives after the introduction of damage.

Model	No. Comps	Undamaged			T19C19 (-1.5%)			T19L17 (-1.5%)			T19C19 (-2.5%)			T19L17 (-2.5%)			T19C19 (-5%)			T19L17 (-5%)		
		Window size (d)			Window size (d)			Window size (d)			Window size (d)			Window size (d)			Window size (d)			Window size (d)		
		1	2	3	1	2	3	1	2	3	1	2	3	1	2	3	1	2	3	1	2	3
Raw	-	10%	35%	71%	20%	65%	81%	13%	35%	76%	34.3%	82.4%	95.2%	11.4%	38.2%	95.2%	95.7%	100.0%	100.0%	20.0%	79.4%	100.0%
MLR	-	19%	76%	100%	49%	100%	100%	27%	76%	100%	90.0%	100.0%	100.0%	25.7%	85.3%	100.0%	100.0%	100.0%	100.0%	48.6%	100.0%	100.0%
MLR-PCA	1	14%	53%	86%	40%	100%	100%	14%	62%	95%	82.9%	100.0%	100.0%	17.1%	61.8%	100.0%	100.0%	100.0%	100.0%	35.7%	94.1%	100.0%
	2	11%	38%	76%	37%	97%	100%	13%	47%	95%	78.6%	100.0%	100.0%	17.1%	55.9%	95.2%	100.0%	100.0%	100.0%	32.9%	85.3%	100.0%
	3	10%	26%	71%	31%	91%	100%	11%	41%	90%	70.0%	100.0%	100.0%	17.1%	52.9%	95.2%	100.0%	100.0%	100.0%	32.9%	82.4%	100.0%
	4	10%	24%	67%	26%	88%	100%	10%	35%	90%	64.3%	100.0%	100.0%	12.9%	44.1%	95.2%	100.0%	100.0%	100.0%	32.9%	82.4%	100.0%
	5	7%	18%	62%	19%	62%	81%	10%	29%	71%	44.3%	94.1%	100.0%	10.0%	41.2%	76.2%	100.0%	100.0%	100.0%	27.1%	70.6%	100.0%
	6	3%	6%	33%	9%	38%	67%	3%	6%	33%	30.0%	73.5%	95.2%	1.4%	17.6%	42.9%	95.7%	100.0%	100.0%	11.4%	47.1%	90.5%
	7	3%	6%	19%	9%	32%	57%	3%	6%	29%	25.7%	64.7%	90.5%	1.4%	17.6%	42.9%	90.0%	100.0%	100.0%	10.0%	44.1%	85.7%
	8	1%	3%	10%	1%	9%	19%	1%	3%	14%	1.4%	14.7%	57.1%	1.4%	5.9%	23.8%	31.4%	85.3%	95.2%	5.7%	41.2%	76.2%
	9	1%	0%	10%	1%	6%	10%	1%	0%	10%	1.4%	8.8%	33.3%	1.4%	0.0%	14.3%	15.7%	58.8%	85.7%	1.4%	8.8%	38.1%
	10	1%	0%	0%	1%	0%	5%	1%	0%	0%	1.4%	5.9%	14.3%	1.4%	0.0%	4.8%	8.6%	35.3%	71.4%	1.4%	5.9%	38.1%
	11	1%	0%	0%	1%	0%	5%	1%	0%	0%	1.4%	2.9%	4.8%	1.4%	0.0%	4.8%	1.4%	11.8%	47.6%	1.4%	0.0%	33.3%
	12	0%	0%	0%	0%	0%	5%	0%	0%	0%	0.0%	0.0%	4.8%	0.0%	0.0%	4.8%	0.0%	8.8%	19.0%	0.0%	0.0%	28.6%
	13	0%	0%	0%	0%	0%	5%	0%	0%	0%	0.0%	0.0%	4.8%	0.0%	0.0%	0.0%	0.0%	5.9%	9.5%	0.0%	0.0%	4.8%
	14	0%	0%	0%	0%	0%	5%	0%	0%	0%	0.0%	0.0%	4.8%	0.0%	0.0%	0.0%	0.0%	0.0%	4.8%	0.0%	0.0%	4.8%
PCA	1	11%	35%	67%	21%	62%	86%	11%	35%	76%	35.7%	85.3%	95.2%	14.3%	41.2%	90.5%	95.7%	100.0%	100.0%	18.6%	79.4%	100.0%
	2	10%	32%	67%	19%	59%	86%	11%	29%	76%	34.3%	79.4%	95.2%	12.9%	35.3%	90.5%	94.3%	100.0%	100.0%	17.1%	76.5%	100.0%
	3	7%	21%	67%	13%	47%	86%	7%	21%	76%	27.1%	76.5%	95.2%	8.6%	29.4%	85.7%	92.9%	100.0%	100.0%	15.7%	73.5%	100.0%
	4	6%	15%	52%	13%	41%	81%	7%	15%	57%	24.3%	73.5%	95.2%	8.6%	17.6%	71.4%	91.4%	100.0%	100.0%	14.3%	50.0%	100.0%
	5	1%	9%	43%	11%	35%	71%	1%	12%	38%	21.4%	67.6%	85.7%	2.9%	14.7%	47.6%	85.7%	97.1%	100.0%	10.0%	44.1%	95.2%
	6	1%	9%	29%	10%	32%	71%	1%	12%	24%	18.6%	64.7%	85.7%	2.9%	14.7%	42.9%	77.1%	97.1%	100.0%	10.0%	38.2%	95.2%
	7	1%	6%	19%	7%	29%	62%	1%	9%	24%	17.1%	50.0%	81.0%	1.4%	11.8%	33.3%	71.4%	94.1%	100.0%	5.7%	32.4%	90.5%
	8	1%	6%	10%	1%	12%	33%	1%	6%	10%	1.4%	14.7%	61.9%	1.4%	5.9%	19.0%	11.4%	50.0%	76.2%	1.4%	8.8%	47.6%
	9	1.4%	2.9%	9.5%	1.4%	11.8%	23.8%	1.4%	5.9%	9.5%	1.4%	14.7%	61.9%	1.4%	5.9%	19.0%	10.0%	50.0%	76.2%	1.4%	8.8%	38.1%
	10	1.4%	2.9%	9.5%	1.4%	5.9%	19.0%	1.4%	2.9%	9.5%	1.4%	11.8%	52.4%	1.4%	2.9%	9.5%	8.6%	47.1%	71.4%	1.4%	5.9%	33.3%
	11	0.0%	0.0%	4.8%	0.0%	2.9%	9.5%	0.0%	0.0%	4.8%	0.0%	8.8%	9.5%	0.0%	0.0%	4.8%	0.0%	14.7%	57.1%	0.0%	0.0%	4.8%
	12	0.0%	0.0%	4.8%	0.0%	2.9%	9.5%	0.0%	0.0%	4.8%	0.0%	2.9%	9.5%	0.0%	0.0%	4.8%	0.0%	14.7%	42.9%	0.0%	0.0%	4.8%
	13	0.0%	0.0%	4.8%	0.0%	2.9%	4.8%	0.0%	0.0%	4.8%	0.0%	2.9%	9.5%	0.0%	0.0%	4.8%	0.0%	11.8%	33.3%	0.0%	0.0%	4.8%
	14	0.0%	0.0%	4.8%	0.0%	2.9%	4.8%	0.0%	0.0%	4.8%	0.0%	2.9%	4.8%	0.0%	0.0%	4.8%	0.0%	8.8%	9.5%	0.0%	0.0%	4.8%

accepted, the models marked in green in the columns of Table 1 that refer to undamaged state are considered effective in removing the environmental and operational effects from the data. Even

though, only a stiffness reduction of 5% in stay-cable T19C19 can be flagged (see for example the two scenarios marked in blue for MLR-PCA and PCA). However, if a percentage of false positives of 6% is accepted, the damage scenario where the stiffness of the stay-cable T19C19 is reduced in 2.5% is now also clearly flagged (see Figure 3 (b)) and the stiffness reduction of 5% in stay-cable T19L17 and of 1.5% in stay-cable T19C19 are now near to be clearly flagged (see Figure 3 (c) and (d)).
From the analysis of Table 1 it can be concluded that the bigger the window size used for the computation of the T^2-statistic, the higher is sensitivity to damage. However, it is also more prone to a higher number of false positives. Finally, the low ability of the MLR model to normalise de data is mainly because the MLR models are only able to remove actions that were measured. For instance, the MLR models were incapable of removing the linear trend due to the concrete rheological effects (see Figure 4 where the residuals for the midspan vertical displacement DV-T-P18P19b are presented).

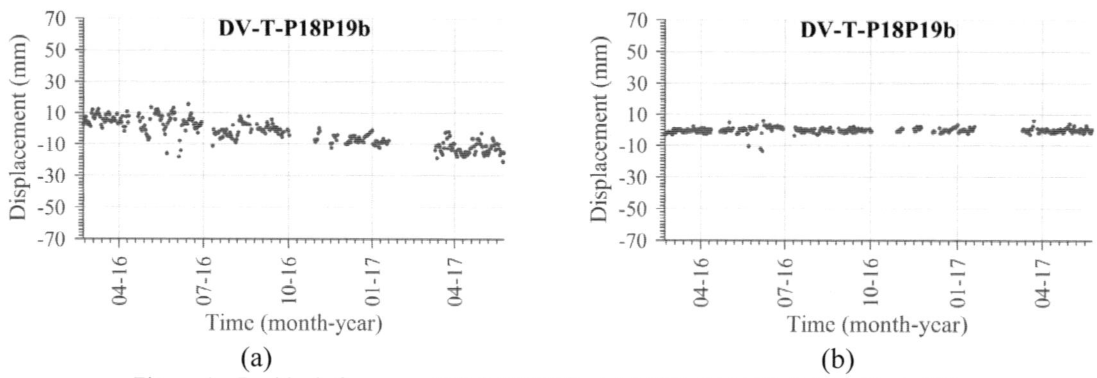

Figure 4 – Residuals for sensor DV-T-P18P19b: (a) MLR, (b) MLR-PCA with six PCs.

5. Conclusions

The ability to detect small damages in the suspension system of the Corgo Bridge using data-driven techniques applied to the structural response of the bridge under environmental and operational effects was evaluated in this work. The considered subset of sensors was adopted with the aim of damage detection in the suspension system of the bridge. The ability to detect small damages in the suspension system of the Corgo Bridge using the installed structural monitoring system and the adopted techniques was demonstrated. It was concluded that the ability for damage detection was significantly improved when daily average values were used and that MLR-PCA models have a slightly better ability for damage detection than the PCA models. Moreover, it was shown that the MLR models did not perform well since they are unable to remove effects that are not measured as the concrete rheological effects.

Acknowledgements

The authors would like to acknowledge to the bridge owner, AutoEstradas XXI, in the persons of engineers Nuno Costa and Filipe Mendes, by the availability of the data from the meteorological station of the Corgo Bridge and by the interest in the carried out studies. The authors also acknowledge to CAETXXI and to Sociedade de Contrustruções Soares da Costa, S. A., in the persons of engineers Davide Borges and Luís Afonso, for their interest and collaboration in all the works developed in the Corgo Bridge by CONSTRUCT-LABEST. This work was financially supported by: Project POCI-01-0145-FEDER-007457 – CONSTRUCT – Institute of R&D in

Structures and Construction funded by FEDER funds through COMPETE2020 – Programa Operacional Competitividade e Internacionalização (POCI) – and by national funds through FCT - Fundação para a Ciência e a Tecnologia. The support by the FCT through the Ph.D. grant SFRH/BD/91536/2012 attributed to the first author is also greatly acknowledged.

References

Barata, V. 2012. Viaduto do Corgo da A.E. Transmontana. *In:* Faria, R., Sarmento, A. M., Pimentel, M., Ferraz, M., Nunes, S. & Lourenço, M. (eds.) *Encontro Nacional Betão Estrutural 2012.* Porto: FEUP/GPBE.

Bellino, A., Fasana, A., Garibaldi, L. & Marchesiello, S. 2010. 'PCA-based detection of damage in time-varying systems'. *Mechanical Systems and Signal Processing,* vol. 24, no. 7, pp. 2250-2260.

Brownjohn, J. M. 2007. 'Structural health monitoring of civil infrastructure'. *Philos Trans A Math Phys Eng Sci,* vol. 365, no. 1851, pp. 589-622.

Cavadas, F., Smith, I. F. C. & Figueiras, J. 2013. 'Damage detection using data-driven methods applied to moving-load responses'. *Mechanical Systems and Signal Processing,* vol. 39, no. 1-2, pp. 409-425.

Comanducci, G., Magalhaes, F., Ubertini, F. & Cunha, A. 2016. 'On vibration-based damage detection by multivariate statistical techniques: Application to a long-span arch bridge'. *Structural Health Monitoring,* vol., no., pp.

Félix, C., Rodrigues, C., Faria, R. d., Figueiras, J., Afonso, L. & Barata, V. 2012. Conceção e implementação do sistema de monitorização estrutural do Viaduto do Corgo. *In:* Faria, R., Sarmento, A. M., Pimentel, M., Ferraz, M., Nunes, S. & Lourenço, M. (eds.) *Encontro Nacional Betão Estrutural 2012.* Porto: FEUP/GPBE.

Figueiredo, E., Park, G., Figueiras, J., Farrar, C. & Worden, K. 2009. Structural Health Monitoring Algorithm Comparisons Using Standard Data Sets. Los Alamos National Laboratory.

Jackson, D. A. 1993. 'Stopping Rules in Principal Components Analysis: A Comparison of Heuristical and Statistical Approaches'. *Ecology,* vol. 74, no. 8, pp. 2204-2214.

Johnson, R. A. & Wichern, D. W. 2013. *Applied Multivariate Statistical Analysis,* Harlow, Pearson.

Jolliffe, I. T. 2002. *Principal Component Analysis*, Springer.

Ko, J. M., Ni, Y. Q., Zhou, H. F., Wang, J. Y. & Zhou, X. T. 2009. 'Investigation concerning structural health monitoring of an instrumented cable-stayed bridge'. *Structure and Infrastructure Engineering,* vol. 5, no. 6, pp. 497-513.

Magalhães, F., Cunha, A. & Caetano, E. 2012. 'Vibration based structural health monitoring of an arch bridge: From automated OMA to damage detection'. *Mechanical Systems and Signal Processing,* vol. 28, no., pp. 212-228.

Omenzetter, P. & Brownjohn, J. M. W. 2006. 'Application of time series analysis for bridge monitoring'. *Smart Materials and Structures,* vol. 15, no. 1, pp. 129-138.

Peres-Neto, P. R., Jackson, D. A. & Somers, K. M. 2005. 'How many principal components? stopping rules for determining the number of non-trivial axes revisited'. *Computational Statistics & Data Analysis,* vol. 49, no. 4, pp. 974-997.

Ren, W.-X., Liu, H.-L. & Chen, G. 2008. 'Determination of cable tensions based on frequency differences'. *Engineering computations,* vol. 25, no. 2, pp. 172-189.

Santos, J. P., Crémona, C., Orcesi, A. D. & Silveira, P. 2013. 'Multivariate statistical analysis for early damage detection'. *Engineering Structures,* vol. 56, no., pp. 273-285.

Tomé, E. S., Pimentel, M. & Figueiras, J. 2016. Análise e simulação da resposta estrutural de um obra de arte atirantada sob ação térmica. *Encontro Nacional Betão Estrutural 2016.* Coimbra: FCTUC.

Webb, G. T., Vardanega, P. J. & Middleton, C. R. 2014. 'Categories of SHM Deployments: Technologies and Capabilities'. *Journal of Bridge Engineering,* vol. 20, no. 11, pp. 04014118.

Yan, A. M., Kerschen, G., De Boe, P. & Golinval, J. C. 2005. 'Structural damage diagnosis under varying environmental conditions—Part I: A linear analysis'. *Mechanical Systems and Signal Processing,* vol. 19, no. 4, pp. 847-864.

Zhang, Q. & Zhou, Y. 2007. 'Investigation of the applicability of current bridge health monitoring technology'. *Structure and Infrastructure Engineering,* vol. 3, no. 2, pp. 159-168.

The 8th International Conference on
Structural Health Monitoring of Intelligent Infrastructure
Brisbane, Australia | 5-8 December 2017

Using SHM Data to Determine Continuous Load Ratings for a Cable-Stayed Bridge

Hadi T. Al-Khateeb[1], Michael J. Chajes[2], and Harry W. Shenton III[3]
[1]*Former Research Associate, University of Delaware, USA,* [2]*Professor, University of Delaware, USA, Email: chajes@udel.edu* [3]*Professor, University of Delaware, USA*

Abstract

A bridge's load rating is perhaps the single most important measure of its overall structural condition. During design, bridges are evaluated using analytical models that are subjected to code prescribed design loads. The modelling results are in turn used to compute the initial load ratings. Once a bridge is in service, visual inspections are typically used to determine if changes to the bridge condition has occurred. If changes in condition are identified, the analytical models are adjusted and the bridge rating is re-computed. Not only is this process most often based on qualitative inspection results, for complex bridges, the analytical models are complex, time consuming to develop and use, and provide only approximate results. Due to these limitations, there is a need for more efficient methods for determining accurate load ratings for complex bridges throughout their service life.
To meet this need, a method has been developed that uses data from Structural Health Monitoring (SHM) systems in combination with conventional load rating equations, to compute continuous rating factors throughout the life of a bridge. The method can yield accurate ratings since it utilizes SHM data which accounts for actual in-service response.
A comprehensive SHM system was installed during the construction of Delaware's 533-m cable-stayed Indian River Inlet Bridge (IRIB). In this paper, data collected from the IRIB's SHM is used to illustrate the proposed method for computing continuous rating factors. These rating factors are being used by the Delaware Department of Transportation (DelDOT) to both assess the condition of the IRIB and to manage this important structure.

1. Introduction

Transportation agencies evaluate bridge safety through visual inspection and the computation of bridge load ratings. Rating factors for a bridge do not change over time unless damage or section loss occurs. The accuracy of current evaluation methods depends on the accuracy of visual inspection and the subsequent evaluation and characterization of damage/section loss, as well as the understanding of actual bridge loading and response. The advent of robust SHM systems presents the opportunity for more accurate evaluation and rating techniques. With this in mind, this paper shows how data from a continuous SHM system can be used in conventional rating equations to calculate bridge rating factors continuously over time. The new approach uses SHM data which includes live load effects, thermal loads, prestressing losses and other time-dependent effects to yield more accurate rating factors.

2. Indian River Inlet Bridge

Delaware's Indian River Inlet Bridge was completed in 2012. It was designed and built under a design/build contract by Skanska/AECOM. The structure carries four lanes of US 1 traffic over the Indian River Inlet, a tidal inlet that connects the Indian River Bay to the Atlantic Ocean. Being located on the Atlantic Ocean, the bridge is exposed to severe salt spray and high winds due to hurricanes. The 533-m long stay-cable structure has a main span of 290 m and twin back spans of 122 m. The bridge is 32.3-m wide with two lanes and a shoulder in each direction. A 3.7-m wide pedestrian path is located on the east side of the bridge, shifting the centerline of the roadway toward the west edge girder. The pylon locations were designed to be outside of the navigable waterway in order to avoid scour issues that plagued the previous bridge. Figure 1 provides a plan view of the bridge.

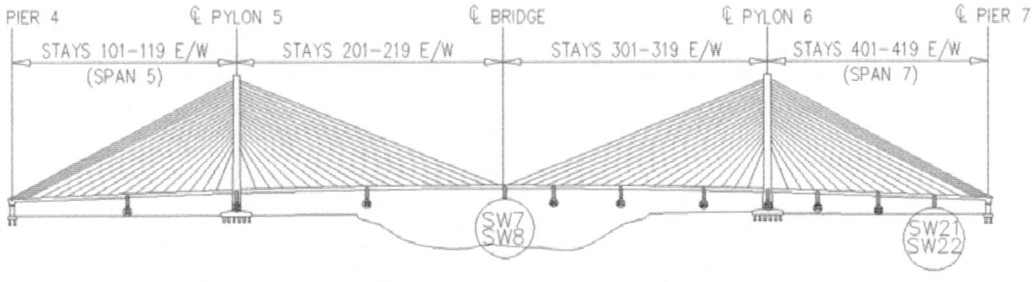

Figure 1. Plan view of the Indian River Inlet Bridge

The edge girders resemble oblique rectangles with dimensions of 1.8-m tall by 1.5-m wide. They are reinforced with standard grade rebar and have concrete strengths averaging 56.5 MPa. Transverse girders support the deck between the edge girders and consist of a combination of pre-cast and cast-in-place beams. The transverse girders are spaced approximately every 3.66 m along the bridge. A 216-mm cast-in-place deck includes 41.3 mm's of latex modified concrete as a wearing surface. The edge girder, transverse girders, and deck were all cast together making the superstructure act monolithically. Transverse post-tensioning running through the transverse girders and anchored in the edge girders also help to accomplish a monolithic state. Post tensioning is also present in some of the back spans of the edge girder.

Each pylon anchors 38 stays, 19 in each direction. The stays are anchored at approximately every 7.3 m to the underside of the cast-in-place edge girders and are composed of 1,862 MPa, 15.7-mm diameter, seven wire strands. The strands are individually sheathed in grease and plastic and surrounded by a helical high-density polyethylene (HDPE) pipe for corrosion protection. The sizes of the stays vary from 19 to 61 strands.

The pylons are cast-in-place concrete boxes. At the foundation, each pylon is composed of two slightly tapered concrete boxes that transition to one concrete box at the road deck level. In the stay cable anchorage portion of the pylon the dimension is 3.66 m by 3.35 m and the cables are anchored to steel boxes contained in the pylons. The pylon foundation is supported by 42 to 49, 0.91-meter square prestressed concrete piles.

3. Structural Health Monitoring System

Early in the planning process for the bridge, the Center for Innovative Bridge Engineering (CiBRe) at the University of Delaware (UD) proposed installing a permanent SHM system on the bridge to the owner, DelDOT. To better manage and maintain such an important piece of infrastructure, both CiBRe and DelDOT believed that it would be very beneficial to have a system in place that could provide quantitative data about actual bridge behavior.

CiBRe researchers designed and oversaw the installation of a comprehensive fiber-optic SHM system that runs the entire length of the bridge. The 100 plus sensors used to monitor various parameters include 72 strain sensors, 9 tilt meters, 3 displacement sensors, 27 accelerometers, and 2 anemometers (see Figure 2). Most sensors are embedded in the concrete members, such as the pylon, edge girder, and road deck. The data will be collected over the life of the bridge and used to determine how the structure has behaved over time and provide early warning signs of potential structural problems. The system will also capture data for specific events such as heavy weight vehicles and extreme weather events. The data collection rate can be varied from 10 to 250 Hz as required.

The SHM system has been in operation since the bridge was opened. The system collects data from all sensors at 15 minute intervals and at higher sample rates during high wind events and when heavy vehicles cross the bridge.

Figure 2. Sensor layout on Indian River Inlet Bridge

4. Using SHM Data for Continuous Load Rating

In the US, the Federal Highway Administration (FHWA) requires a Load and Resistance Factor Rating (LRFR) factor be computed for all in-service bridges. A rating factor greater than 1.0 is required for all applicable limit states. Modern bridges are designed according to the Load and Resistance Factor Design (LRFD) specifications, where the HL-93 live load is used as the design live load. The Manual for Bridge Evaluation (MBE) states that bridges that have been designed and checked with the HL-93 loading may not require load rating calculations until changes (section loss, damage, etc.) occur for bridge components and reduce its rating below the design load level. The conventional rating equation used can be seen in Eq.1

$$RF = \frac{C-(\gamma_{DC})(DC)-(\gamma_{DW})(DW)\pm(\gamma_P)(P)}{(\gamma_{LL})(LL+IM)} \tag{1}$$

where RF is the conventional rating factor; C is the capacity, (γ_{DC}) (DC) is the factored dead load effects, (γ_{DW}) (DW) is the factored wearing surface effects, (γ_P) (P) is the factored permanent loads effects, and (γ_{LL})(LL+IM) is the factored live load effects with dynamic load allowance.

The American Association of State Highway and Transportation Officials (AASHTO) code requires designers to design for thermal loads. Nevertheless, thermal effects are ignored when designing and rating conventional type bridges (short and medium bridges), because of the small effects comparing to the major loads. However, for long-span and segmental bridges, thermal effects can be much more significant. In fact, the AASHTO Manual for Condition Evaluation of Bridges (MCEB), states that thermal effects should not be considered in calculating load ratings “except” for long span bridges and concrete arches.

In order to use the monitored data correctly, understanding the components of the SHM data is essential. Assuming a prismatic simply supported beam with top and bottom sensors, Figure 3 shows the theoretical strain distribution along the section.

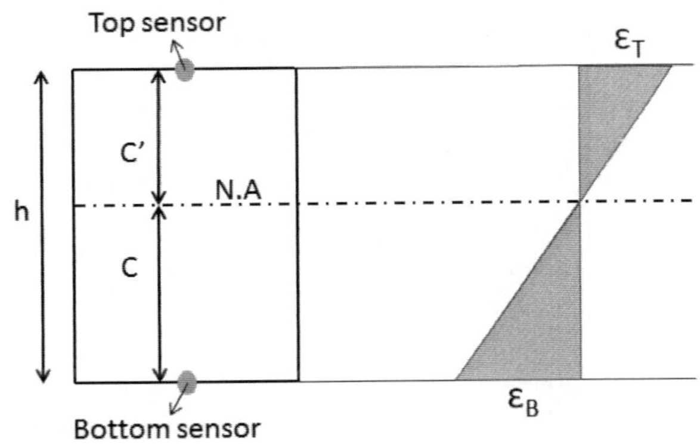

Figure 3. Strain distribution along the cross section for the superposition case

The measured strain at the top and bottom sensors can be represented by Eq. 2 and 3.

$$\varepsilon_T = -\frac{Mc\prime}{EI} + \frac{P}{EA} + \alpha\Delta T_T \tag{2}$$

$$\varepsilon_B = +\frac{Mc}{EI} + \frac{P}{EA} + \alpha\Delta T_B \tag{3}$$

Eqs. 2 and 3 represent the total strain, which is composed of mechanical strain and thermal strain due to temperature change and live load effects. In general, the strain induced by temperature can be broken into two components, strain in a force free status and strain caused by stresses induced by temperature in boundary restrained conditions (Li et al. 2010). Therefore, the total strain that the senor reads has two components; one of them is related to free expansion ($\alpha*\Delta T$) and the other is the mechanical strain due to the stresses induced by temperature and live loads.
In order to estimate rating factors based on SHM data, structural forces and stresses are calculated using the monitored strain. The total strain measured by a strain sensor includes all types of loads affecting the bridge (e.g. thermal loads, live loads, wind loads, prestressing losses, and others).
If the strain and the temperature change are recorded at both the top and bottom locations of the section, the moment associated with structural health monitoring (M_{SHM}) can be calculated using SHM data as shown in Eq. 4, and the axial load associated with structural health monitoring (P_{SHM}) can be calculated using SHM data as shown in Eq. 5.

$$M_{SHM} = \frac{EI}{h}[(\varepsilon_B - \varepsilon_T) - \alpha(\Delta T_B - \Delta T_T)] \tag{4}$$

$$P_{SHM} = \frac{EA}{h}[c'(\varepsilon_B - \alpha\Delta T_B) + c(\varepsilon_T - \alpha\Delta T_T)] \tag{5}$$

Once the structural forces (M_{SHM} and P_{SHM}) are known, the associated SHM stresses can be calculated. As the strain collected by the sensor has two components, strain free status ($\alpha*\Delta T$) and the strain caused by the stresses induced by any loads in the presence of boundary conditions, the strain caused by all loads (thermal, live, etc.) can be obtained by subtracting the strain free status from the total measured strain. Eqs. 6 and 7 show the stresses obtained by SHM data.

$$E\varepsilon_T - E\alpha\Delta T_T = -\frac{M*c\prime}{I} + \frac{P}{A} \tag{6}$$

$$E\varepsilon_B - E\alpha\Delta T_B = +\frac{M*c}{I} + \frac{P}{A} \tag{7}$$

The structural forces and stresses acquired by SHM data can be directly used in combination with the section capacity and dead load effects to obtain rating factors. These forces and stresses represent actual load effects on the bridge at the sensor location.
Determining the actual in-service loads effecting a bridge has long been of interest to bridge owners. By having two strain sensors, one at the top and one at the bottom of a representative section, as shown, obtaining these in-service structural forces and stresses are possible. When doing this using SHM data, the frequency of the collected data will determine the types of the loads captured in the collected strain. For example, vehicular and wind loads change at a much higher frequency than do thermal load. Therefore, if one wants to capture vehicular and wind loads, high frequency data must be recorded.
In this research, high frequency data was collected at 25 Hz. The data was then processed to separate the trend signals (with thermal effects) from the live load signals (caused by vehicles). Figure 4 shows the recorded signals and the separated signals. The total collected signal is the blue line, the red signal represents the trend signal, and the green signal (obtained by subtracting the red signal from the blue signal), represents the live load vehicle data.

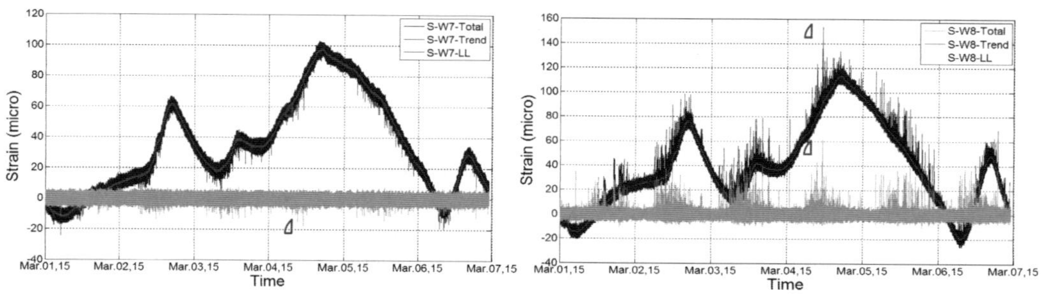

Figure 4. Separating live load signal from trend signal

Once the signals are separated, member forces can be computed using Eqs. 4 and 5. An example of the rating factor found using the SHM terms for Strength1-Flexural is given by Eq. 8.

$$RF_{SHM} = \frac{CM - M_{DL} - M_{DL-SHM-LF}}{M_{LL-SHM-HF}} \tag{8}$$

where RF_{SHM} is the rating factor found using SHM data; CM is the design moment capacity, M_{DL} is the dead load design moment, $M_{DL-SHM-LF}$ captures the actual change in dead load moment due to time dependent effects (thermal changes, creep, and shrinkage) measured by low frequency SHM data, and $M_{LL-SHM-HF}$ captures the actual live load moments measured by high frequency SHM data.

5. Continuous Load Rating Factor Results

To illustrate the results of computing continuous RF's using SHM data (Eq. 8), we will consider data collected during two, three-month time periods that are one year apart. The first time period will be September to November of 2015, and the second time period will be September to November of 2016. The strain data is from the west edge girder, sensors SW21 and SW22, which are located on the top and bottom of the girder (see Fig. 1). This is the structural element and location along that element that governs the conventional rating of the bridge (that RF is 1.17). As mentioned, the high frequency SHM data is recorded at 25 Hz, and peak strains are captured for all heavy vehicle passes (see the scatter plot showing peak strains in Fig's. 5 and 6, as well as associated histograms of the peaks strains in Fig's. 5 and 6). Only peak strains above a defined threshold are kept. These strains are then used to compute the RF_{SHM} as described previously. A scatter plot of the computed RF_{SHM} are shown for the two three-month time periods in Fig's 7 and 8.

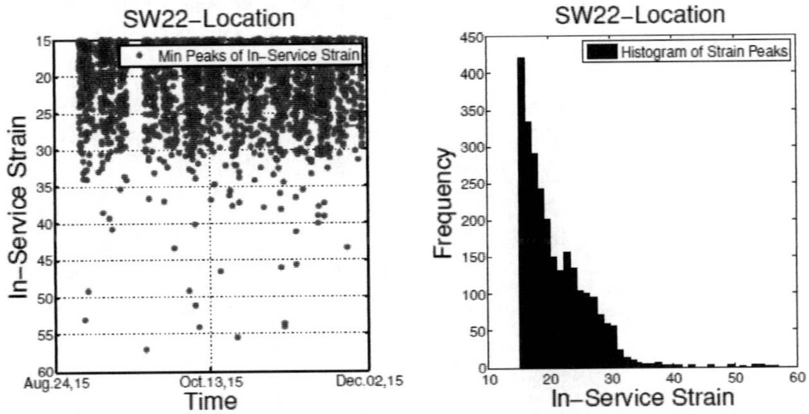

Figure 5. Peak strains recorded at SW22 from September through November of 2015

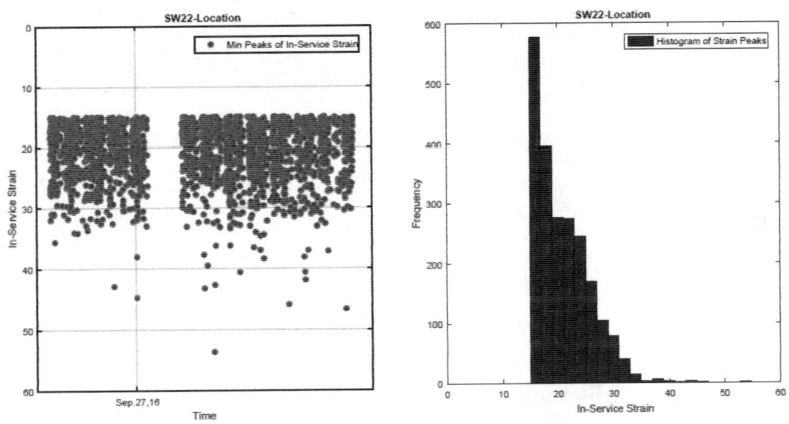

Figure 6. Peak strains recorded at SW22 from September through November of 2016

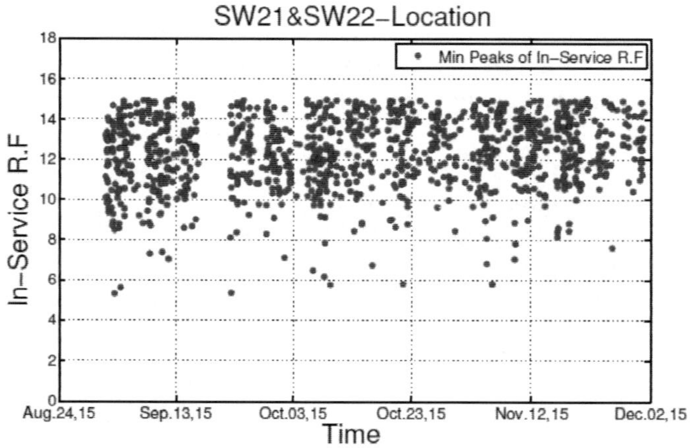

Figure 7. Continuous rating factors at SW21/22 from September through November of 2015

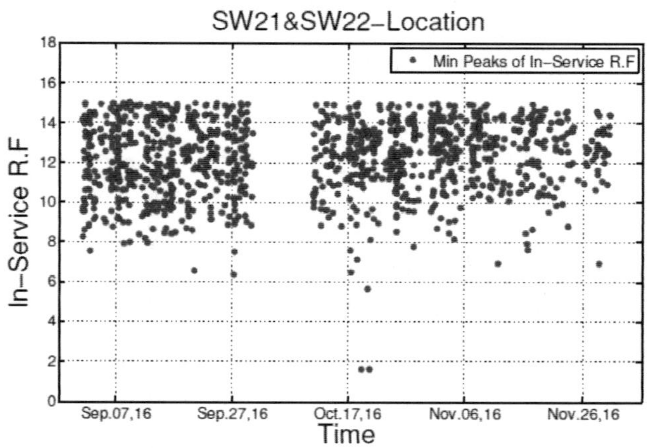

Figure 8. Continuous rating factors at SW21/22 from September through November of 2016

Looking at the recorded peak live load strains, several observations can be made.

1. The cut off for peak strains was 15 $\mu\varepsilon$.
2. There was one time during each three-month window when the system was down.
3. The large majority of the strains recorded below 35 $\mu\varepsilon$.
4. The largest strains recorded were between 55 and 60 $\mu\varepsilon$.
5. The data during the two three-month windows is fairly similar.

Looking at the RF_{SHM} results, several observations can be made.

1. The vast majority of the rating factors are above 8.0. Note that this is in comparison to the conventionally computed rating factor that is based on design live loads and corresponding design moments of 1.17.
2. Only two rating factors from large peak strain events are below 4.0.
3. The large strain above 50 $\mu\varepsilon$ that occurred around October 20th of 2016 led to a rating factor slightly below 2.0.
4. The large strains above 50 that occurred during 2015 (there were seven of them), did not produce rating factors lower than 4.0.

6. Conclusions

Based on the design of the IRIB, and using conventional LRFR methodology, a rating factor of 1.17 was determined. This rating is based on strength and service limit states considering material properties and response based on the design parameters and on the design loads. Using data from a fiber-optic SHM system, continuous rating factors were determined. Results of these rating factors over two three-month periods indicate that the actual rating factor for the bridge incorporating actual live loads and time dependent effects (thermal, creep, and shrinkage) are generally above 8.0 and in all but two instances above 4.0. The fact that all large strains do not result in equally low ratings is due to the significant effects of temperature on the resulting strains and ensuing ratings. The direct inclusion of these thermal effects is one great benefit of using in-situ structural response.

The results indicate that continuous rating factors found using SHM data is a promising way of evaluating the health and performance of a bridge. Over time, it will be interesting to see the trends of the continuous rating factors. How do they vary annually? Do they maintain a similar pattern from year to year? Do they get smaller over time? Do large loads, perhaps from permit vehicles, show up as extreme values? Do any values approaching 1.0 occur?

Rating factors derived from SHM data present a very promising approach for assisting bridge owners in the complex task of managing large bridges like the IRIB. The research team is continuing to work closely with DelDOT to integrate this approach into their existing bridge management process.

Acknowledgments

We would like to thank the DelDOT and the Federal Highway Administration for funding the work and would also like to acknowledge the efforts of Doug Robb (formerly DelDOT), Craig Stevens and Jason Arndt (DelDOT), Jim Zammataro (formerly Cleveland Electric Labs), and Keith Chandler of Cleveland Electric Labs (formerly Chandler Monitoring Systems). Finally, we would like to acknowledge the many University of Delaware students who have worked on this project including Jack Cardinal, Patrick Carson, Kent Davidson, and Pablo Marquez.

References

Al-Khateeb, H.T. (2016). "Bridge Evaluation Utilizing Structural Health Monitoring Data." Dissertation, University of Delaware.

Chajes, M.J., Shenton, H.W., Al-Khateeb, H.T., Wenczel, G.R., and Ramanna, N. (2015). "Structural Health Monitoring of the Indian River Inlet Bridge: Results from Controlled Load Tests Conducted over the First Two Years of Service." Proceedings of the 7th International Conference on Structural Health Monitoring of Intelligent Infrastructure, Torino, Italy.

Shenton, H.W., Fernandez, M, Ramanna, N, Chajes, M.J., and Al-Khateeb, H.T. (2015). "Structural Health Monitoring of a Cable-Stayed Bridge: Using Tiltmeter Data to Determine Edge Girder Deflections." Proceedings of the 7th International Conference on Structural Health Monitoring of Intelligent Infrastructure, Torino, Italy.

Shenton III, H.W., Chajes, M.J., Wenczel, G., Ramanna, N., Al-Khateeb, H., Davidson, K., and Marquez, P. (2014). "Structural Health Monitoring of Delaware's Indian River Inlet Bridge: Year One Update." Proceedings of the 2014 Structures Congress, Boston, Massachusetts.

Marquez, P., Chajes, M.J., Shenton III, H.W., Al-Khateeb, H., Wenczel, and Cardinal, J. (2013). "Structural Health Monitoring: Establishing the Baseline Performance of Delaware's Indian River Inlet Bridge."

Proceedings of the 6th International Conference on Structural Health Monitoring of Intelligent Infrastructure, Hong Kong.

Shenton III, H.W., Chajes, M.J., Wenczel, and Davidson, K. (2013). "Lessons Learned in the Construction and Early Experiences with the Indian River Inlet Bridge SHM System." Proceedings of the 6th International Conference on Structural Health Monitoring of Intelligent Infrastructure, Hong Kong.

Shenton III, H.W., Carson, P.D., Chajes, M.J., O'Shea, D., Kursinsky, C., (2010). "Delaware's Indian River Inlet Bridge: a comprehensive plan for long-term structural monitoring," Transportation Research Record: Journal of the Transportation Research Board, No. 2201, 148-153.

AASHTO, 2011. "Manual for Bridge Evaluation (2nd Edition)," Washington, D.C.: American Association of State Highway and Transportation Officials (AASHTO).

Marquez, P., 2013. LOAD RATING OF THE INDIAN RIVER INLET BRIDGE USING A STRUCTURAL HEALTH MONITORING SYSTEM. University of Delaware.

The 8th International Conference on
Structural Health Monitoring of Intelligent Infrastructure
Brisbane, Australia | 5-8 December 2017

Satellite-Based Monitoring of a Highway Bridge in Canada - Challenges, Solutions and Value

D. Cusson [1], K. Trischuk [1], D. Hébert [2], G. Hewus [3] M. Gara [4], P. Ghuman [4]

[1] *National Research Council Canada – Canada, Email: daniel.cusson@nrc.gc.ca,* [2] *Transport Canada – Canada,* [3] *Federal Bridge Corporation Limited - Canada,* [4] *3v Geomatics - Canada*

Abstract

An innovative space-borne bridge displacement monitoring technology was demonstrated on a major highway bridge in Cornwall (Canada) over a two-year period. Two major challenges had to be overcome and solutions were identified and implemented. The first challenge related to the proper selection and analysis of satellite imagery to optimize the number and signal quality of point targets on the bridge for adequate movement detection and analysis. The second main challenge was the comparison and validation of independent sets of data having different measurement methods, reference baselines, and distributions in space and time.

The key findings are as follows: (i) The regression analysis of Interferometric SAR (InSAR) data from the satellite produced the best coherence when fitted against three independent variables, being height, temperature, and time; (ii) The bridge railings appeared to be excellent SAR reflectors with their sharp edges and regular spacing along the bridge, showing a return of over 3,000 clear InSAR point targets to the satellite; (iii) The SAR displacement thermal sensitivity data was found to compare well to numerical modelling thermal data when computed along the longitudinal axis of the bridge. Results show great promise and value in applying satellite-based technology for the remote monitoring of highway bridges to optimize preventive maintenance management, extend structural lifespan, minimize traffic disruptions due to late repairs, and ensure structural integrity after extreme events such as major earthquakes and flooding.

1. Introduction

Highway bridges are critical to Canada's road transportation network and are in urgent need of rehabilitation, owing to ageing, limited capital investment, and insufficient inspection and maintenance. Bridge owners in North America have been struggling for decades with how to cost-effectively rehabilitate their large portfolios of highway bridges (U.S. DOT/FHWA/FTA 2007).

As a complement to current inspection and maintenance protocols, the judicious use of structural health monitoring (SHM) of highway bridges can contribute to addressing many of these challenges and slowing down the ageing of such critical structures. Preliminary studies (Marinkovic *et al.* 2007; Cusson *et al.* 2012) have shown that remote monitoring with radar satellites can provide ground measurements with sub-centimeter accuracy. The RADARSAT-2 Satellite, for instance, is capable of monitoring large structures by detecting and measuring movements on hundreds point targets (PTs) on a single structure, while hundreds of these structures can be monitored and analysed in a single scene when acquired over major urban centres (Cusson *et al.* 2012).

The main objectives of this demonstration project were (i) to deploy a bridge displacement monitoring technology based on satellite imagery (Satellite Aperture Radar, or SAR) and (ii) to validate it against numerical bridge model predictions of thermal displacements, and measurements from displacement\temperature sensors installed on a highway bridge.

Figures 1-3 present the new 335 m long, 4-span, low-level North Channel Bridge (being the subject of this study) and the old high-level North Channel Bridge (being demolished at the time), crossing the St-Lawrence River between City of Cornwall (North) to Cornwall Island (South), located in Ontario, Canada. The new bridge was completed and opened to traffic in January 2014, while the demolition of the old steel bridge superstructure started in Sept. 2014 and ended in Nov. 2015.

Fig. 1: Drone view of the new low-level bridge and the old high-level bridge in Cornwall, ON

Fig. 2: View of bridges looking South (2015-09-15)

Fig. 3: View of bridges looking South-West (2015-10-21)

2. Acquisition Plan and InSAR Analysis of Satellite Imagery

Among the different modes of radar images that can be acquired with RADARSAT-2, the Spotlight mode is best-suited for bridge monitoring applications, which provides highest accuracy (but smallest coverage ~18 km x 8 km). Anticipating possible image viewing issues given the complex configuration and timing of the construction/demolition work, four stacks of images from the Canadian Space Agency were requested, each with a repeat cycle of 24 days. For each ascending pass and each descending pass of the right-looking satellite, two stacks of images with low (26°) and high (47°) incidence angles were selected. Only one stack was to be retained for final analysis.

In an initial analysis, challenges related to the proper selection of satellite imagery were addressed to optimize of the number and stability of PTs for bridge movement detection. These challenges included:

- Shadow effects – due to the relative positioning of the bridges (Fig. 1), with the upper structure creating SAR shadows on the image of the lower structure, reducing the number of PTs (Ferretti *et al.* 2007);
- Layover effects – also due to the relative positioning of the bridges and the nature of SAR data collection, where the upper bridge can appear on the opposite side of the lower bridge depending on the viewing angle and the relative slant ranges of the bridges from the satellite, thus requiring additional processing of PTs to separate valid mixed targets (Ferretti *et al.* 2007);
- Target stability – due to construction activity on the bridges (e.g. cranes on new bridge, Fig. 2; and the gradual dismantling of the old bridge's steel superstructure, Fig. 3), reducing the number of valid consistently stable point targets for the final analysis;
- Multi-bounce effects – occurring when microwaves bounce off more than one physical object or entity (e.g. body of water) before returning to the SAR sensor, requiring additional processing or contributing to signal noise when mixed physical targets are not themselves stable;
- Height fit – as precise localization of point targets on the bridge requires accurate 3D positioning, which in turn requires derivation of an accurate estimate of each point target's height relative to a known digital elevation model (DEM). This can be a challenging task when there is a lot of variance in the data due to noise or unmodelled strong thermal effects.
- Number of scenes – where 12 to 15 valid scenes in a given stack are typically required to obtain sufficient measurement accuracy from an InSAR analysis of RADARSAT-2 satellite data.

The preliminary analysis revealed that, for the case at hand, the *sla74-asc* stack (ascending pass with a 26.5° incidence angle) was the most suitable stack for the subsequent detailed analysis.

For the full analysis, SAR data (radar wave phase information) were processed with a multi-model linear regression analysis to extract different sets of data from the *sla74-asc* imagery for over 3000 point targets registered on the new bridge deck. Key parameters included:

- Height (m) – fitted against perpendicular baseline (stereo sensitivity) data (Ferretti et al. 2007);
- Displacement thermal sensitivity (cm/°C) – fitted against measured temperatures at the bridge;
- Displacement time rate (cm/year) – fitted against time from the first image acquisition;
- Coherence (ranging from 0 to 1) – confirming the quality of fit and the reliability of processed displacement data, where only PTs with coherence ≥ 0.8 should be trusted for InSAR analysis (Ferretti *et al.* 2007). In general, the choice of coherence cut-off is dataset- and sensor-specific and must be judicially chosen. The coherence is computed following a multi-model regression fit, yielding a statistic on both the quality of fit as well as the noise level of the point target.

A Python script was used to extract, de-noise, and visualize selected post-processed data by displaying displacement thermal sensitivity collected over the deck in 3D plot, as shown in Fig. 4 (X axis=West-East distance (reversed); Y axis=North-South distance; Z axis=Height). The figure reveals very useful details:

- The displacement thermal sensitivity shows gradual variation along the bridge (more later);
- The height and steady slope of the deck along the length of the bridge (1.2° over 335 meters) are very well captured by the point cloud;
- The two steel railings of the bridge returned a large amount of clear signals to the satellite. The sharp angles of the railings appear to be excellent natural corner reflectors. Another advantage of railings for InSAR is that snow and ice do not accumulate easily on them due to their exposure to wind and sun (since accumulating snow and ice on the deck may affect SAR measurements);
- Four groups of outliers leaning towards East appear to match exactly the locations of the four piers of the old bridge (East-West reversed). Despite being on another bridge, these points are very useful to confirm the exact positioning of the cloud of points of interest, since it is otherwise difficult to determine precisely where the new bridge starts and ends in a continuous point cloud over the road. Regular Google optical images do not have high-enough resolution.

Figure 5 (left) presents the resulting displacement time rate data superposed onto a Google Earth optical image. With all points being either yellow or green, one can conclude that very small movements have occurred over the relatively short monitoring period. Figure 5 (right) illustrates the coherence of the displacement data over a Google Earth image, confirming the high quality and reliability of the processed data, with most points being dark blue (well above 0.8) – a coherence of 0.8 or higher on directional data is comparable to a standard deviation of less than 0.45 radians.

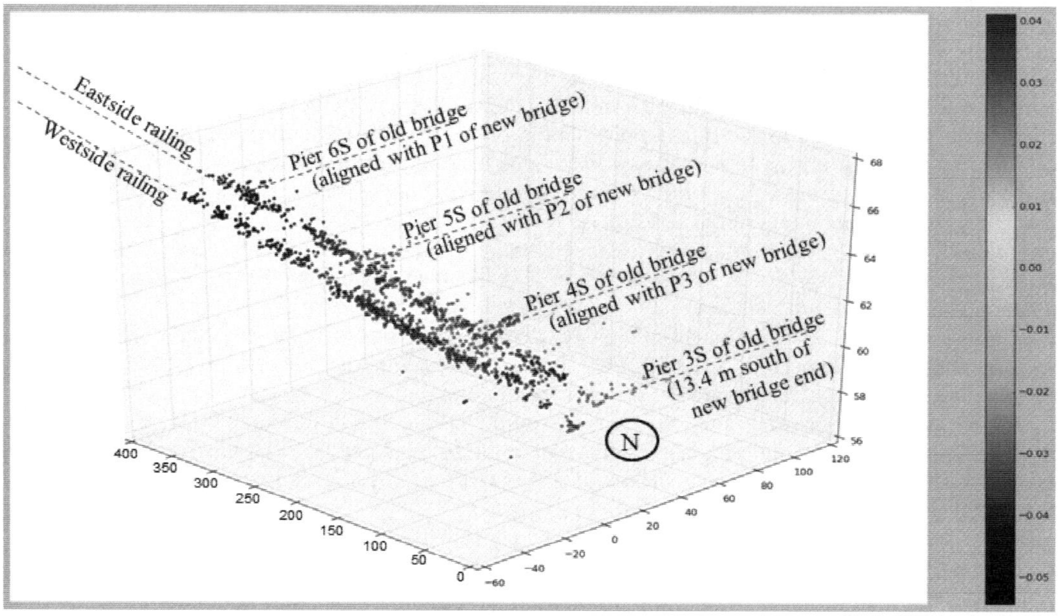

Fig. 4: 3D plot (m) of InSAR displacement thermal sensitivity data (cm/°C)

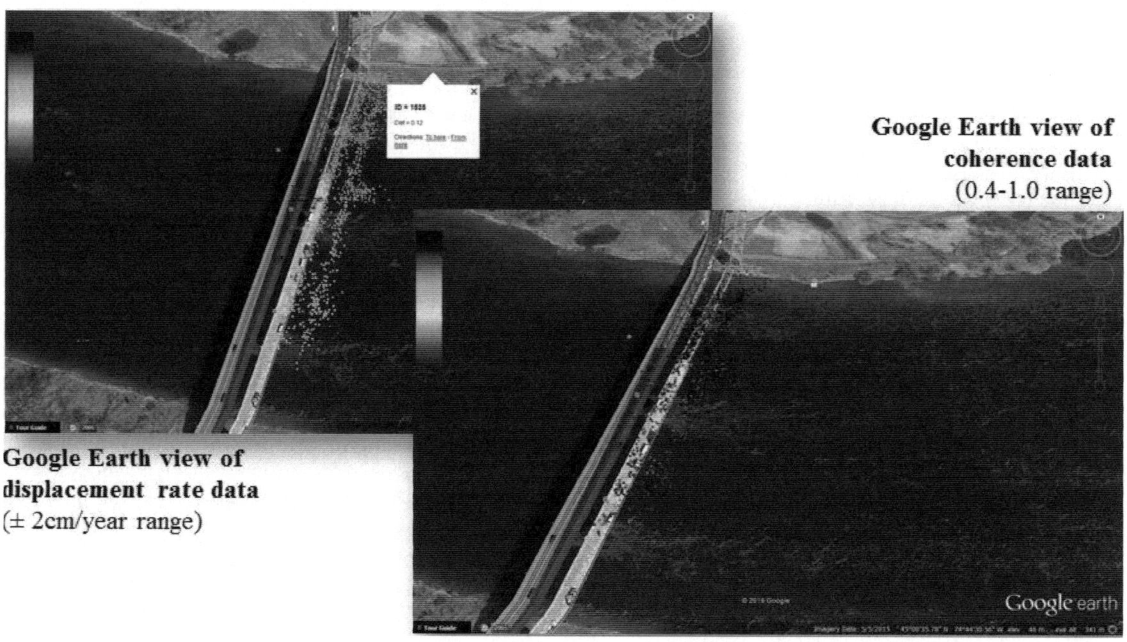

Fig. 5: Processed displacement time rate data (left) and corresponding coherence data (right)

3. Sensor-Based Bridge Monitoring

To validate the proposed satellite-based monitoring technology on the North Channel Bridge, the measurements from RADARSAT-2 need to be compared to similar measurements from other known ground-based technologies, including (i) temperature and displacement sensors installed on the bridge, and (ii) a finite-element numerical model of the bridge calibrated on sensors measurements from a truck load test.

3.1 Measurement of Ambient Conditions

Ambient temperature plays a crucial role in this study, as it is required in the proposed InSAR analysis of the satellite data, since our adopted approach is to fit the satellite displacement data with temperature data in order to remove the thermal movement and focus on mechanical movements of the bridge that could be due to settlement, problematic joints or bearing, excessive loading, etc.

A first set of temperature data was obtained from temperature sensors installed inside the twin steel box girders of the bridge, as shown in Figure 6. Being in direct contact with the steel girder, these sensors are expected to provide values similar to ambient air temperatures, due to the high thermal conductivity of steel. A second set of temperature data was obtained from the nearby Massena Intl. Airport's weather station (NY State, USA; 23 km away), as it was the closest station that could provide hourly values of air temperature. Using bridge temperature data over airport data to fit SAR displacement data resulted in a small improvement in measured coherence (not shown).

Figure 7 illustrates the correspondence between the bridge sensor temperatures and the Massena airport temperatures at different times, where each data point corresponded to the time of a satellite pass over the bridge. The strong correlation between the two sets of temperature data and the small improvement in observed coherence suggest that if sensor measurements are not available at the site being monitored, local airport temperatures are suitable for the proposed InSAR analysis approach.

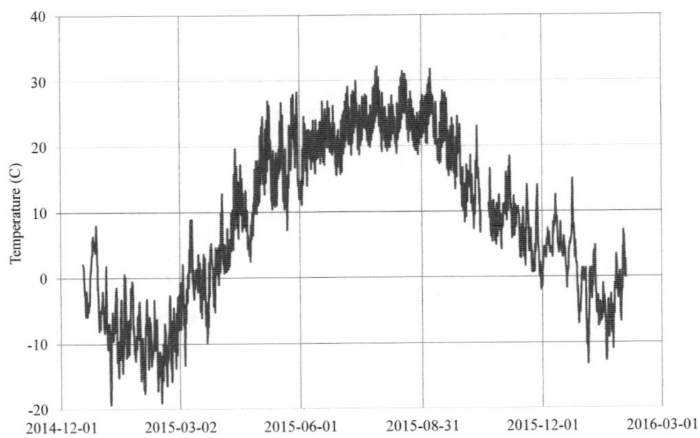

Fig. 6: Temperature measured inside steel box girder of bridge

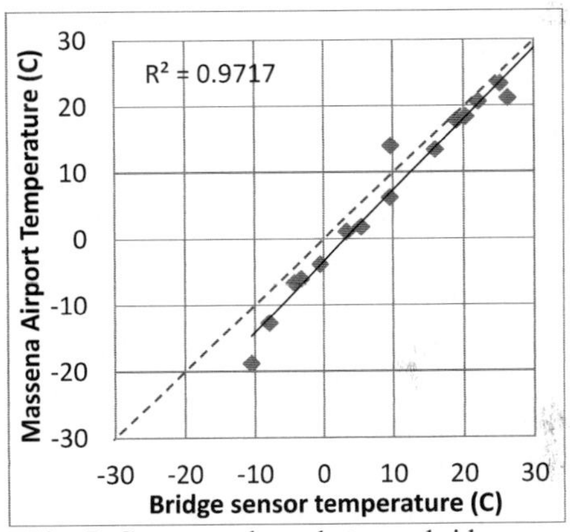

Fig. 7: Correspondence between bridge sensor temperature and local airport temperature

3.2 Measurement of Live Loads

Two truck load tests were conducted on February 2, 2016 during the evening (low traffic period), including a first test on the southbound (SB) lane and a second test on the northbound (NB) lane, with the 2nd last truck axle (Figure 8) aligned with mid-span position on Span 4 (first from North).

Wireless fiber-optic displacement sensors at the bottom of the twin steel box girders were used to measure the longitudinal deformations at the quarter-span and mid-span locations of Span 4. Figure 9 presents the deformations measured during the northbound lane test. It can be seen that the load test produced a horizontal elongation of 0.087 mm when the truck came into position, and an identical horizontal shortening when the truck left the bridge. Measurements from the two load tests were used to calibrate the finite-element model of the bridge by simulating identical load tests, comparing with the measurements, and adjusting Young's modulus of the reinforced concrete deck.

Fig 8: Truck used for load test on bridge (SB lane)

Truck load = 640 kN
Axle spacings = 4.3, 1.5, 6.5, 2.5, 1.5, 1.5 m

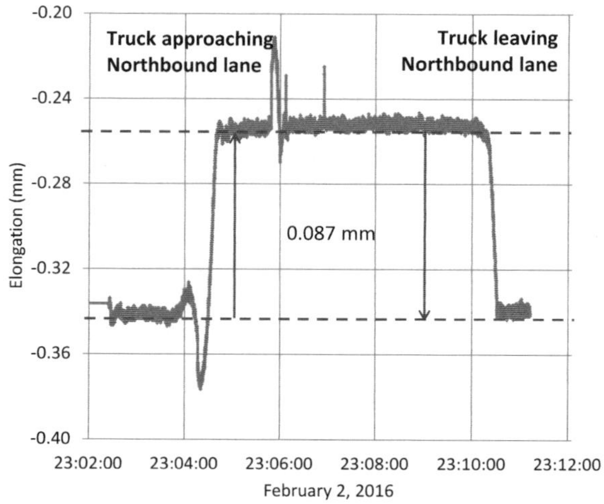

Fig 9: Load test data from mid-span sensor in East girder

4. Bridge Numerical Modelling

Since radar satellite sensing provides line-of-sight displacement measurements without distinguishing thermal effects from mechanical effects, or between the X, Y, and Z components of displacement, another tool is required to help determine the orthogonal components of mechanical effects on the bridge that could lead to structural issues if left undetected. This tool could be numerical modelling to determine the thermal movements of the bridge.

The commercial TNO-Diana finite element software package was used to build a detailed 3D numerical model of the North Channel Bridge in Cornwall, Ontario (steel-reinforced concrete deck on twin steel box girders). The construction drawings of the bridge and its construction specifications were used to obtain detailed information on bridge geometry, boundary conditions, and structural material properties. For a bridge under service loads, a structural static analysis and isotropic linear material models were deemed more than adequate for the determination of bridge displacements as a function of service loads and temperature effects.

Figure 10 illustrates the bridge model with the superstructure, three piers and two abutment walls (incl. 10,000 nodes, 11,154 elements, 6 materials, and 61 properties). The piers were modelled from their fixed connection to the superstructure down to their fixed connection to the river bed rock. The load-bearing portion of each abutment wall was modelled assuming uniaxial bearing connection (i.e. free movement in longitudinal direction, and blocked in transverse and vertical directions).

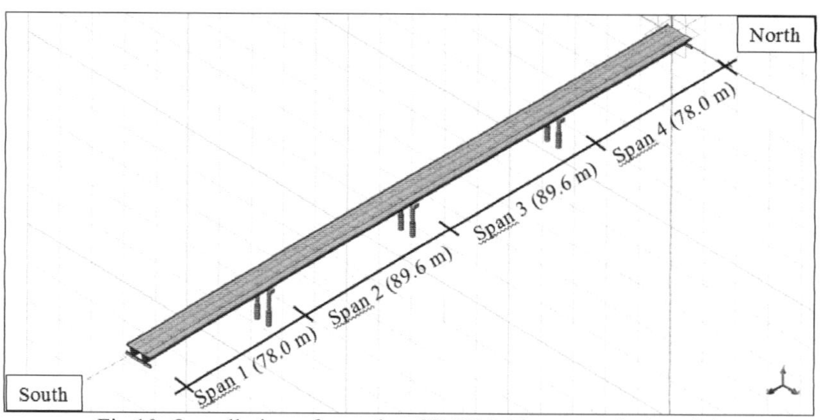

Fig 10: Overall view of complete 3D FE bridge numerical model

4.1 Numerical Model Calibration with Load Test Results

The truck load test conducted on the bridge on February 2, 2016 was simulated using a structural linear analysis of the bridge model for the 2 truck passes (SB and NB lanes). Figure 11 indicates the locations of the applied axle loads with the vertical black arrows, and those of the 4 wireless sensors on the steel box girders. After small adjustments to Young's modulus of the deck's reinforced concrete (set to 35 GPa), the longitudinal deformations of bottom layer of steel girders at mid-span and quarter-span compared well with sensor measurements. For example, at mid-span on SB lane (1st test), the predicted vs. measured values were 0.10 mm vs. 0.10 mm in the West girder, 0.08 mm vs. 0.07 mm in the East girder; and at mid-span on NB lane (2nd test), the predicted vs. measured values were 0.09 mm vs. 0.07 mm in the West girder, and 0.11 mm vs. 0.09 mm in the East girder.

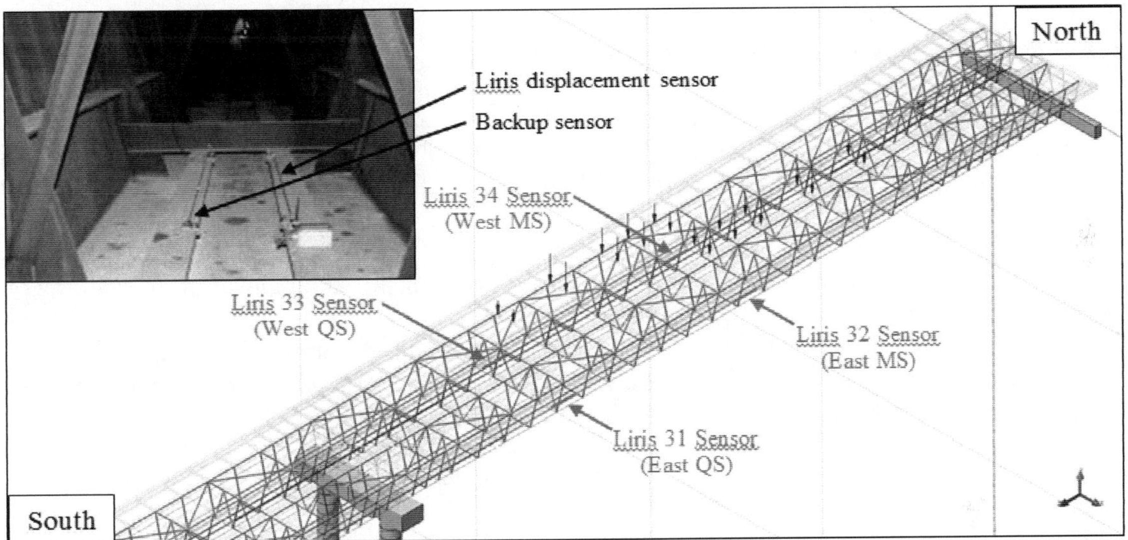

Fig 11: Detailed view of FE model showing locations of sensors (green arrows) and truck axle loads (black arrows)

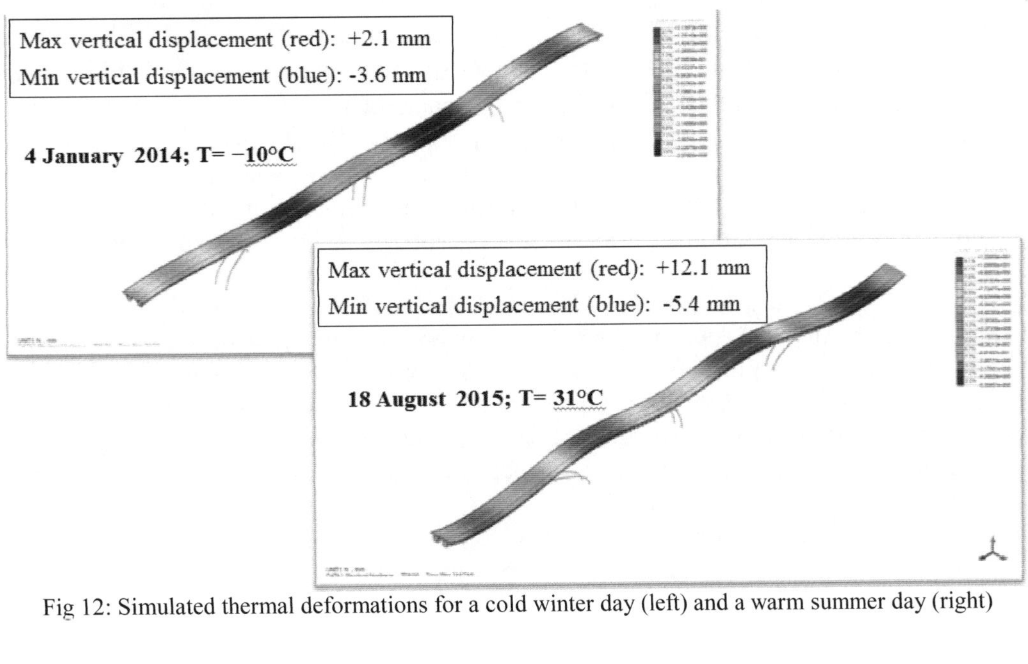

Fig 12: Simulated thermal deformations for a cold winter day (left) and a warm summer day (right)

4.2 Simulation of Thermal Deformation

Measured bridge temperatures (Figure 6) were used as input data in a structural nonlinear analysis of the bridge model, with time steps ranging from 0 day (2014-12-17) to 412 days (2016-02-02).

For example, Figure 12 (left) above presents the thermal deformation (vertical displacement) of the bridge computed for January 14, 2014 corresponding to the satellite pass with the coldest temperature of the year. One can see the deck contraction towards the middle pier, producing inward bending of the two other piers, and bending of the concrete deck. Figure 12 (right) presents the thermal deformation (vertical displacement) of the bridge computed for August 18, 2015, corresponding to the satellite pass with the warmest temperature of the year. One can observe the elongation of the bridge deck away from the middle pier, producing an outward bending of the two other piers, and the bending of the concrete deck.

5. Validation of Satellite-Based Monitoring Technology

Movement obtained from InSAR analysis of satellite data represents the line-of-sight (LoS) movement, which is the change in the distance between the satellite and the point target being monitored on the ground (herein on a bridge). Since the satellite incidence angle is not 0° (from the vertical), detected movement includes horizontal and vertical components of unknown proportion to each other. For example, a positive LoS displacement (i.e. movement away from satellite) may be the result of a downward vertical displacement, a horizontal displacement away from satellite, or an unknown combination of both.

To allow comparison of SAR data with FEM data, it was decided to convert the FEM displacement data (knowing exactly the horizontal and vertical components) into LoS displacement (given the satellite viewing geometry) with the following equation:

$$\Delta LoS = -\Delta V\,(\cos\alpha\cos\gamma - \sin\alpha\sin\beta\sin\gamma) - \Delta H\,(\sin\alpha\sin\beta\cos\gamma - \cos\alpha\sin\gamma) \qquad (1)$$

where ΔLoS is the line-of-sight displacement; ΔV is the vertical displacement; ΔH is the horizontal displacement; α is the incidence angle of the satellite line-of-sight direction (26.5° for SLA74-asc); β is the angle between the azimuth of satellite track on earth and the azimuth of bridge (26° between the SLA74-asc satellite track and the North Channel Bridge); and γ is the inclination angle of the bridge deck from North to South (1.2° for the North Channel Bridge).

An interesting special case for the above equation is when the bridge longitudinal direction is parallel to the satellite path (i.e. $\beta = 0°$), in which case Eq. 1 would reduce to $\Delta LoS = -\Delta V \cos\alpha$. As a direct consequence, the sensibility of InSAR LoS measurements to any horizontal movement will be negligible. On the other hand, if β was 90°, the sensitivity to horizontal movement would be at its highest level.

Using Equation 1 above, the SAR displacement temperature sensitivity data can be compared directly to FEM thermal data when FEM data is expressed in terms of LoS displacement as a function of change in temperature from a reference temperature (set here as T = 0°C). To that effect, these data were plotted in Figure 13 for 9 different positions on the bridge deck, being over vertical supports and mid-span locations. Two key observations can be made from Figure 13:

- A strong correlation exists between LoS displacement and temperature at any given location on the bridge;
- The slope of the linear regression analysis for each curve is a parameter that can be directly compared to the SAR line-of-sight displacement temperature sensitivity.

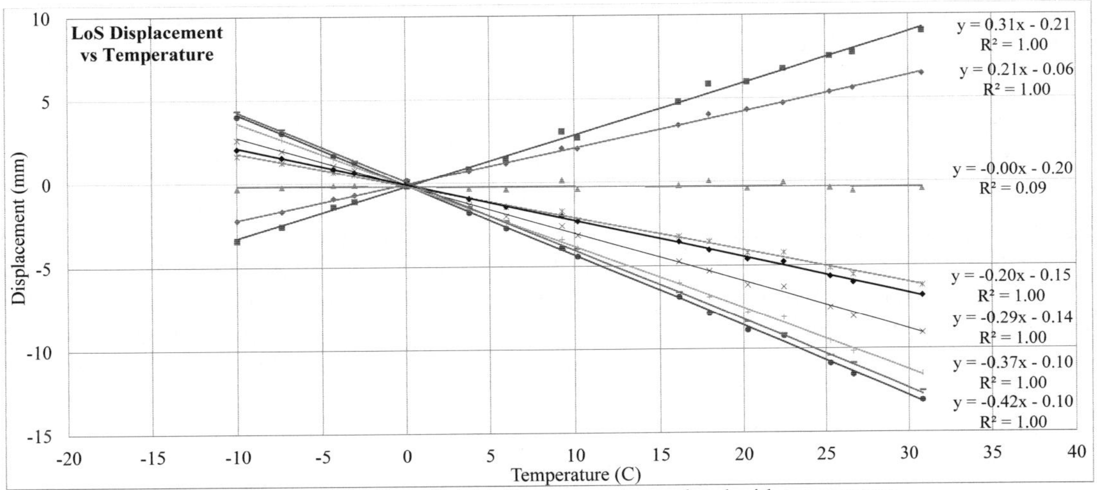

Fig. 13: FEM line-of-sight displacement correlated with temperature

The slopes of the linear regressions presented above were then expressed as the FEM thermal sensitivity along the longitudinal distance on the bridge, and illustrated in Figure 14. It is interesting to see how the thermal sensitivity varies along the bridge. Figure 14 also compares the two sets of independent SAR and FEM data, where a good fit can be observed. Note that there were not enough valid SAR point targets near the south end of the bridge to generate points for x > 275 meters. Longer monitoring, adding more valid PTs to the InSAR analysis, would eliminate this issue.

It is noted that the 2 peaks at approximately x = 125 m and 225 m appear more accentuated for the FEM data than for the SAR data. It was therefore suspected that the support conditions at the three piers could play a role and affect the thermal sensitivity of the FEM displacement data.

To investigate this further, the bridge model was modified to change the support conditions, i.e. to release the longitudinal movement of the connections of Piers 1 and 3 to the deck structure. In other words, unidirectional bearings were simulated in Piers 1 and 3, as opposed to the as-designed fixed connection. As shown in Figure 15, the two peaks appear to be flatter than with the previous support condition, and the correlation with the SAR data seems to have changed slightly. More monitoring and inspection data are required before concluding on the status of the support conditions.

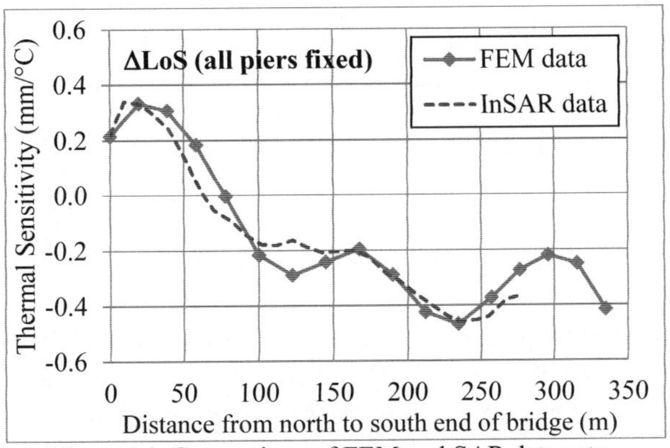

Figure 14: Comparison of FEM and SAR data sets for thermal sensitivity of line-of-sight displacement (for actual bridge support condition: all piers fixed)

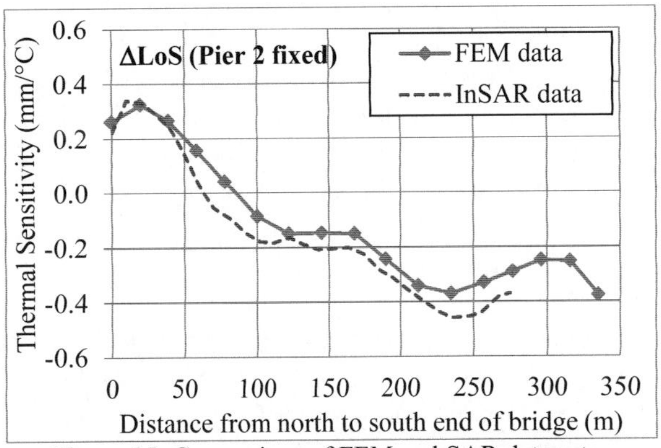

Figure 15: Comparison of FEM and SAR data sets for thermal sensitivity of line-of-sight displacement (for fictitious bridge support condition: only P2 fixed)

Nevertheless, an important point is made from the above discussion and illustrated data in Figures 14 and 15 that the SAR/FEM comparison of the LoS displacement thermal sensitivity data can be used to identify the following bridge issues:

- frozen or damaged bearings
- pier flexibility issues, or
- numerical modelling issue!

As mentioned above, our adopted approach is to use multi-model regression analysis to fit SAR satellite data against height and temperature in order to properly position point targets on the bridge, identify valid data points, and remove thermal movement in order to obtain the remaining residual dataset of displacement time rate, which represents the mechanical movement of the bridge that could be due to unexpected settlement, problematic joints or bearing, excessive loading, etc.

6. Conclusions

The key findings of this study are as follows:

- Regression analysis of SAR data from the satellite produced the best coherence when fitted against three independent variables: height, temperature, and time. For example, when temperature was excluded from the fit, the data coherence reduced significantly.
- The bridge railings appeared to be excellent SAR reflectors with their sharp edges and regular spacing along the bridge, returning over 3,000 clear signals to the satellite;
- The SAR displacement thermal sensitivity was found to compare well to FEM thermal data when computed along the longitudinal axis of the bridge. Such data was found promising, for example, to detect possible issues with problematic bearing conditions.
- Results show great promise and value in applying satellite-based technology for the remote monitoring of highway bridges to optimize preventive maintenance management, extend structural lifespan, minimize traffic disruptions due to late repairs, and ensure structural integrity after extreme events such as major earthquakes and flooding.

References

Cusson D., Ghuman P., Gara M., McCardle A. Remote Monitoring of Bridges from Space, 54th Annual Brazilian Congress on Concrete, IBRACON, Oct. 2012, 25 p.

Ferretti A., Monti-Guarnieri A., Prati C., Rocca F., Massonnet D. InSAR Principles: Guidelines for SAR Interferometry Processing and Interpretation, ESA Publication TM-19, Feb. 2007, 48 p.

Marinkovic P., Ketelaar G., Van Leijen F., Hanssen R. InSAR quality control - Analysis of five years of corner reflector time series, 5th Intl. Workshop on ERS/Envisat SAR Interferometry, `FRINGE07', Frascati, Italy, Nov 26-30, 2007.

U.S. DOT, FHWA, FTA. The 2006 Status of the Nation's Highway Bridges and Transit: Conditions & Performance, Report to Congress, 2007.

Acknowledgments

Direct financial contribution to this project was provided by Transport Canada. Technical assistance and logistics were provided by National Research Council Canada, Federal Bridge Corporation Ltd., 3v Geomatics, Osmos Canada, Seaway Intl. Bridge Corporation, Morrison Hershfield, and Canadian Space Agency.

The 8th International Conference on
Structural Health Monitoring of Intelligent Infrastructure
Brisbane, Australia | 5-8 December 2017

Fast and automatic intelligent detection of shield tunnel lining defects

Y. Xue[1], Y. Li [2], HW. Huang[2].
[1] *Department of Civil Engineering, Tongji University – China, Email:* yadongxue@126.com [2] *Department of Civil Engineering, Tongji University – China*

Abstract

Tunnel lining defects are an important indicator reflecting the safety status of shield tunnel. This paper presents an automatic intelligent classification and detection methodology for subway tunnel lining defects. The methodology is inspired by the state-of-the-art computer science—deep learning. A new Fully Convolutional Network (FCN) model for image classification is proposed. Defects information from CCD cameras was used to train this model. GoogLeNet and VGG are used to compare and evaluate our model. The best-set accuracy of our model for classification is over 95% at a test-time speed of 48ms per image. For defects detection, feature images of tunnel lining defects from large scale image were computed by FCN and then detected with the tool of Region Proposal Network (RPN) and Position-sensitive RoI pooling. Three indices (detection rate, detection accuracy and detection efficiency) were used to evaluate this approach synthetically. The result proves that the proposed approach is very fast and efficient for automatic intelligent classification and detection of tunnel lining defects.

1. Introduction

Subways are an important means of urban transportation. However, owing to poor geological conditions, lack of timely maintenance and damage incurred by humans, subway tunnels very often suffer from surface and internal structural damage with increasing operation time. The most prominent defects in urban subway shield tunnels are cracks and leakages. These defects reduce the strength of the concrete, dampen the tunnel lining, negatively affect the structure safety, and shorten the tunnel lifetime; consequently, these defects significantly increase safety-related risks of operating urban rail traffic. Thus, in the case of massive construction of urban subway, it is very important to develop methods that would allow to detect, evaluate, and treats defects associated with tunnel lining.

The traditional inspection of tunnel lining structure is manual inspection, recording and taking pictures, then analysis, evaluating and giving results indoor. This method suffers from low efficiency, poor accuracy, and cannot meet the demands of massive tunnel defects detection. While the rise of machine vision has changed such situation. For example, automatic crack detection system based on linear array CCD camera has been carried out to detect tiny crack (Yu et al., 2007). Besides, COMS industrial camera is also used in machine vision for defect detection (Zhang et al., 2014). Faced with a large amount of image data, the current problem is how to process it efficiently, quickly and intelligently.

Large number of image processing techniques (IPTs) have been used to detect defects in civil engineering. The edge-detection algorithms (Fast Harr transform (FHT), Fast Fourier transform (FFT), Sobel filtering, and Canny filtering) were compared and FHT was defined to be the best one for detecting cracks in bridges (Abdel-Qader et al., 2003). Some IPTs based on mathematical morphology were also proposed (Landstrom and Thurley, 2012). These methods could be divided into three steps including contrast enhancement, mathematical morphological processing and information extraction by filter. And some other theory-based methods, like artificial neural network (Jin and Zhou, 2014), support vector machine (Qu et al., 2010), Adaboost (Cord and Chambon, 2012), K-nearest neighbor algorithm (Lei and Zuo, 2009) were also used in civil engineering and demonstrated good performance. However, the disadvantages of these traditional methods are low speed and poor robustness in the case of complex background. With the rise of deep learning, the state-of-the-art method could change this situation.

Deep learning has been the most advanced topic in the field of Computer Science in recent years. Although some of its methods had been carried out in the 80s and 90s of the last century (LeCun et al., 1998; Hopfield, 1982), it is in 2006 that deep learning came to reality from ideal. (Hinton and Salakhutdinov, 2006). Generally, deep learning is a method which uses a cascade of many layers of nonlinear processing units for feature extraction and transformation (LeCun et al., 2015). The application in the field of image processing can be divided into two parts: image classification and object detection (including image segmentation). Image classification problem in image processing is that of determining whether or not the image data contains some specific object or feature. The object detection is to identify and locate specific objects in a complete image. To solve this kind of problem, Convolutional Neural Network (CNN) was widely used. AlexNet (Krizhevsky et al., 2012), VGG (Simonyan and Zisserman, 2014), GoogLeNet (Szegedy et al., 2015), ResNet (He et al., 2015) were proposed and stood out in LISVRC series. In object detection, recent state-of-the-art deep learning method, such as r-cnn (Girshick et al., 2014), SSD (Liu et al., 2016), R-FCN (Li et al., 2016), made some good performance in object detection tasks like pascal VOC Challenge (Everingham et al., 2015). And there have been some applications in civil engineering. Deep CNN was used to detect road crack (Zhang et al., 2016) and a classifier with a CNN model for detecting concrete cracks from images was also built, which obtains an accurate result in the dataset (Cha et al., 2017).

The successful history of applying CNNs to image processing has inspired us to use deep learning methods for detection of tunnel lining defects. The objective of the present study is to develop a method for quick, efficient, and simultaneous detection of defects on large-scale complex backgrounds. To address this problem, we built two datasets, one for classification of tunnel lining features, the other for tunnel lining defects detection. We used the deep learning approach with these datasets, to obtain a method for detection of cracks and leakages. The main advantages of this method are its high computation speed, accuracy, and suitability for massive defect detection tasks.

2. Datasets

In this paper, we used an advanced Movable Tunnel Inspection (MTI) system to acquire tunnel lining surface images. The MTI system consisted of six CCD cameras and twelve light-emitting diodes (LEDs) as sources of light, which could scan over 13 meters of a tunnel's surface. Several detection tasks (Figure 1) were conducted in the Shanghai subway No.1, 2, 4, 7, 8, and these data constituted the dataset on which we developed and validated our model.

We extracted samples image with size of 256×256 pixels from origin image. Data for the image classification consists of 9520 images, manually labeled with the presence of one of five object categories including leakage, crack, segment joint, pipeline and lining. Each image contained only one ground truth label. We divided them into two part, one was training set, and the other was test set. The number of each categories for training and test shows in Table 1.

Figure 1 MTI-100 in detection task

Table 1

Datasets for image classification

	label	training set	Test set	Total
leakage	1	1200	400	1600
crack	2	1680	560	2240
segment joint	3	1500	500	2000
pipeline	4	1350	450	1800
lining	5	1410	470	1880
		7140	2380	9520

Data for the object detection also came from images by MTI. We used whole image for each camera with the size of 3000×3724 pixels. The dataset contained a total of 4139 annotated images. We labeled three classes including crack, leakage. In each image, we annotated in the ground truth with class and bounding box. To annotate the bounding box, we used a rectangular box that contains the coordinates of the upper left corner and the lower right corner. In this dataset, image could contain more than one ground truth label. Figure 2 shows some of the samples we labeled. We divided the images into two parts, 3000 for training-val set and 1139 for test set.

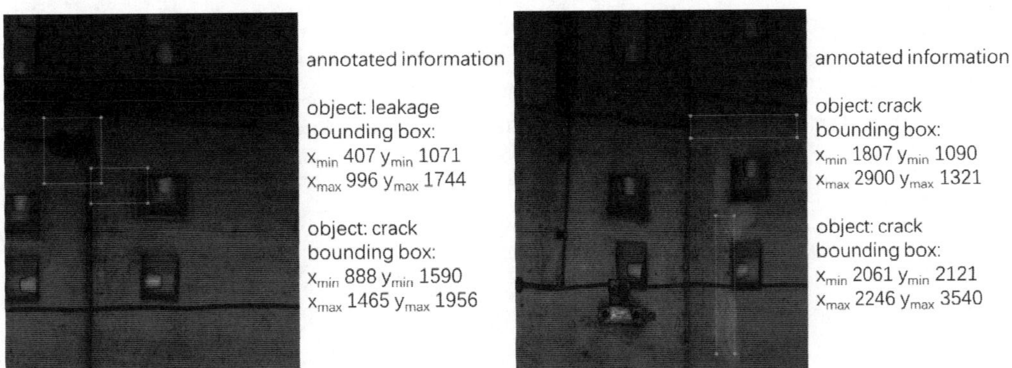

Figure 2. Samples in dataset for object detection

3 Methodology

Fully Convolutional Network (FCN) is a kind of CNN with only the last layer fully-connected, which is removed and replaced in object detection. In this section, we designed a FCN model for the backbone architecture. Based on that, we consulted the main idea of R-FCN (Li et al., 2016) to realize defect detection in large-scale image with region proposal network and Position-sensitive RoI pooling. Figure 3 shows the whole process of the method.

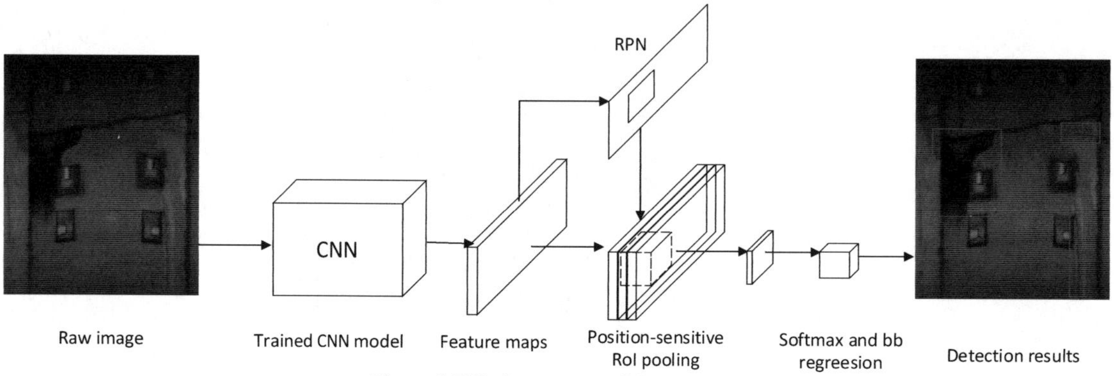

Figure 3 Whole process of the method

3.1 Backbone architecture

The backbone architecture is a CNN model for image classification. Compared with the origin neural network, the peculiarity of CNN is that there are many hidden layers. Convolution layer and pooling layer are most common ones.

Convolution layer and pooling layer

Convolution layer origins from the receptive field from the research of the visual nerve cells in cats (Hubel and Wiesel, 1968). Different from the connection mode of origin neural network, the neurons in convolution layer are sparsely connected. Because the image has the property of locality and invariance, the number of parameters can be greatly reduced by this kind of connection. In a convolution layer, there are n kernels with the size of $l\times l$ convoluting with the feature map with the size of $w\times h\times b$, and sliding over the stride of s. Thus the output with the size of $(h-l)/(s+1)\times(w-l)/(s+1)\times n$ is obtained. Each convolution layers can be understood as the extraction of the feature image.

The pooling layer is also called sub-sampling layer, which is usually set after the convolution layer. The calculation is similar with convolution layer. While the activation function doesn't work here. Two ways, mean pooling and max pooling, are used in pooling layer. Pooling layer can reduce the parameters of the model, learn the spatial signature of the image, achieve the invariance of displacement, scaling and distortion, and decrease the occurrence of over fitting.

Our model

The FCN model inspired by GoogLeNet and VGG. As we all know, the advantages of GoogLeNet are inception module and fully convolutional, while VGG's is that small convolution kernels are used in the best-set model. Thus, we improved inception module and 7×7 convolutional layer in the first layer in GoogLeNet. For inception module, two 3×3 convolutional kernels were used to replace the 5×5 one (shows in figure 4). As for the 7×7 convolutional layer, we used four 3×3 convolutional layers to replace. Table 2 shows the overall architecture schema of our model.

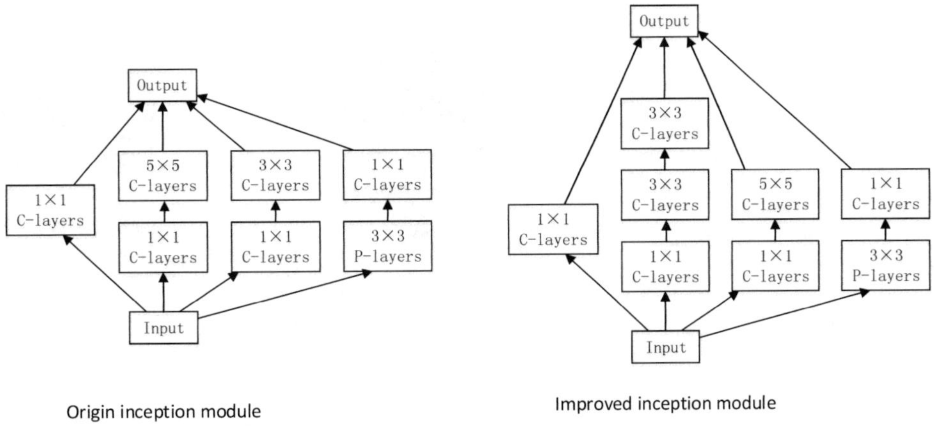

Figure 4. Origin inception module and improved inception module. C means convolutional and P means pooling.

Table 2 Overall architecture schema of our model

type	size/stride	output size	depth
convolution	3×3/1	224×224×32	1
convolution	3×3/1	224×224×32	1
max pooling	3×3/2	112×112×32	0
convolution	3×3/1	112×112×64	1
convolution	3×3/1	112×112×64	1
max pooling	3×3/2	56×56×64	0
convolution	3×3/1	56×56×198	1
max pooling	3×3/2	28×28×198	0
inception		28×28×256	3
inception		28×28×256	3
max pooling	3×3/2	14×14×480	0
inception		14×14×512	3
inception		14×14×512	3
inception		14×14×528	3
inception		14×14×832	3
max pooling	3×3/2	7×7×832	0
inception		7×7×832	3
inception		7×7×1024	3
avg pooling	7×7/1	1×1×1024	0
dropout		1×1×1024	0
linear		1×1×1000	1
softmax		1×1×4	0

3.2 Region Proposal Network

Region proposal is a region that has possibility to be the region of object. In the large-scale image, to generate region proposal is an important step to detect the defects. Traditional methods for extracting them are Selective Search (SS) (Uijlings et al., 2013), Edge Boxes (EB) (Zitnick and Dollár, 2014) and so on. But those methods can't set in the end to end deep learning model and they are time-cost. Region proposal network (RPN) was proposed in 2015 (Ren et al., 2015), it can take an image of any size as input and outputs a set of rectangular object proposals, the procedure is showed as figure 5.

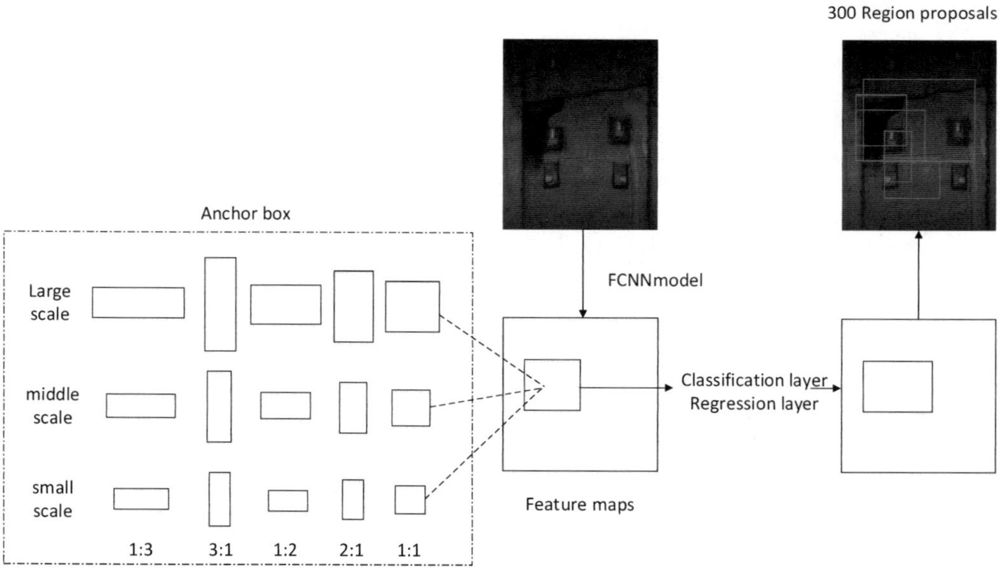

Figure 5 Region proposal network

Anchor box is set as objection regions' template, and it will slide over the feature map in the FCN model. There are two parameters of anchor box, aspect ratio and scale. In this paper, we set three scale of the anchor box. Faster R-CNN used anchor box with the ratio of 1:2, 1:1, 2:1 (Ren et al., 2015). While because the shape of the crack is usually very long, we added aspect ratio with 1:3 and 3:1. Two function layers (classification layer and regression layer) are used to preliminarily demonstrate if the region of sliding window belongs to the object and regress the bounding box of the object. The region with ground truth of the object would be computed in RoI-wise subnet in 3.3.

3.3 Position-sensitive RoI pooling

FCN has always been very deep to realize translation invariance in image classification. While in object detection, what we want is translation variance. A state-of-art method to overcome this dilemma is Position-sensitive RoI pooling (Li et al., 2016). Based on the feature map CNN model extracted, convolution layer with n kernels is used to compute the position sensitive feature map. We spread the feature map to k2 with the depth of l by means of location information. Than a ROI pooling layer (Girshick, 2015) (a single degree pyramid pooling (He et al., 2015)) is used to divide each w×h RoI rectangle into $k \times k$ bins by a regular grid with size of $w/k \times h/k$. An array of $k \times k \times n$ is obtained. Because there are two corners of each region, so for bounding box regression, there would be $4n$ convolution kernels and obtain an array of $k \times k \times 4n$.In this paper, in order to get an accurate result, we set the size of score map 7, and the category is 2. So there are 147 convolution kernels for softmax and 588 kernels for bounding box regression.

4 experiments

We realized the method in the deep learning framework Caffe (Jia et al., 2015). On the one hand, based on the datasets in 2.2, we trained and tested FCN model to evaluate the model we designed. One the other hand, combining RPN and Position-sensitive RoI pooling, the tunnel lining defects detection methodology was set up to detect defects in the image of 3000×3724 pixels.

4.1 experiments on feature map classification

We performed experiments on Datasets for image classification, we trained the models on the training set, and evaluated on test set. In the last layer in CNN, we used a softmax layer to compute the probability of each category.

A comparison was carried out between the top 1 model in ISVRC2012 (AlexNet) and in ISVRC2014 (GoogLeNet) and our model. We used Nesterov´s Optimal Gradient Method to train the model with the minibatch size of 40 in training and 20 in test. We used a small initial learning rate of 0.009 and reduce to 0.0009 in 20 epoch and to 0.00009 in 40 epoch.

Figure 6 shows the results of accuracy with each epoch. For AlexNet, the max accuracy was 91.47% at the 57th epoch at a test time of 46ms; for GoogLeNet, it was 94.58% at the 49th epoch at a test time of 50ms; and for our model, it was 95.84% at the 58th epoch at a test time of 48ms.

We constructed confusion matrices to analyze the number of correct classifications for each image category. The rows and columns of these confusion matrices represent the labels of different categories. The *i*-th value in the *j*-th column indicates the number of times the image with the ground truth label *i* was classified by the model into *j*-th category. The diagonal elements in these confusion matrices report correct classifications.

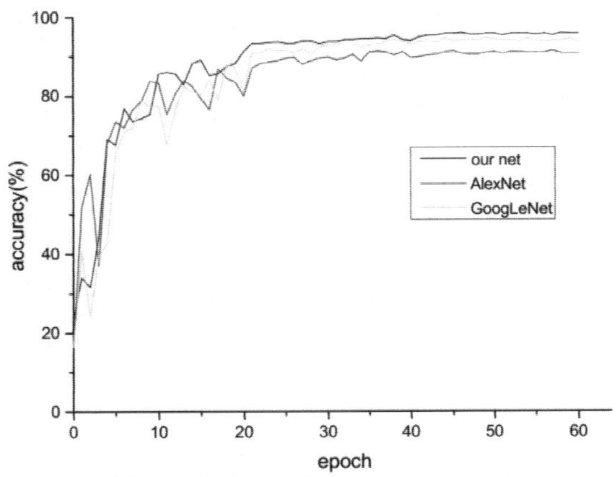

Figure 6 Curve of accuracy vs epoch

Table 3 Confusion matrix of test set of AlexNet

	1	2	3	4	5	
1	352	18	11	3	16	88.00%
2	13	505	6	3	33	90.18%
3	10	28	443	13	6	88.60%
4	4	6	6	432	2	96.00%
5	1	18	2	4	444	94.88%

Table 4 Confusion matrix of test set of GoogLeNet

	1	2	3	4	5	
1	389	4	1	4	2	97.25%
2	9	541	7	1	2	99.61%
3	13	8	459	13	7	91.80%
4	15	7	6	419	3	93.11%
5	1	21	3	2	443	94.26%

Table 5 Confusion matrix of test set of our model

	1	2	3	4	5	
1	391	5	1	3	0	97.75%
2	4	555	1	0	0	99.11%
3	7	14	464	12	3	92.80%
4	6	6	12	425	1	94.44%
5	7	14	3	0	446	94.89%

We compared our model with another model that performed exceptionally in ISVRC2014—VGG16 model. Although we extensively varied the hyper-parameters and some other parameters in VGG16 model, this model did not exhibit convergence when applied to our dataset. In the training phase, the performance of VGG was very slow. This can be attributed to a large number of parameters in VGG16, which makes it difficult to train this model. For users and researchers who do not have much experience with parametric tuning of CNNs, the complexity of training can become an issue. Thus, for engineering applications our model appears to be more suitable.

4.2 Experiments on defect detection

Next, we evaluated our net on datasets for object detection. With a pre-computed FCN model, we set the learning rate as 0.001 for first 10k iterations and 0.0001 for next 20k iterations. We used approximate joint training to train an end to end model with SGD.

We evaluated our results of detection by three indicators:

- Detection rate: detection rate refers to ratio of the number of actual results detected to the number of defects we labelled in the test set.
- Detection accuracy: detection accuracy refers to average value probability of all the detected regions to a certain category.
- Detection efficiency: detection efficiency refers to the speed of per image in the detection.

We tested the data in test set with a total of 1139 images. There were 1867 defects in the data, the detection rate was 92.1%, the detection accuracy was 83.4% and detection efficiency was 0. 267s per image. Figure 8 shows some detection results

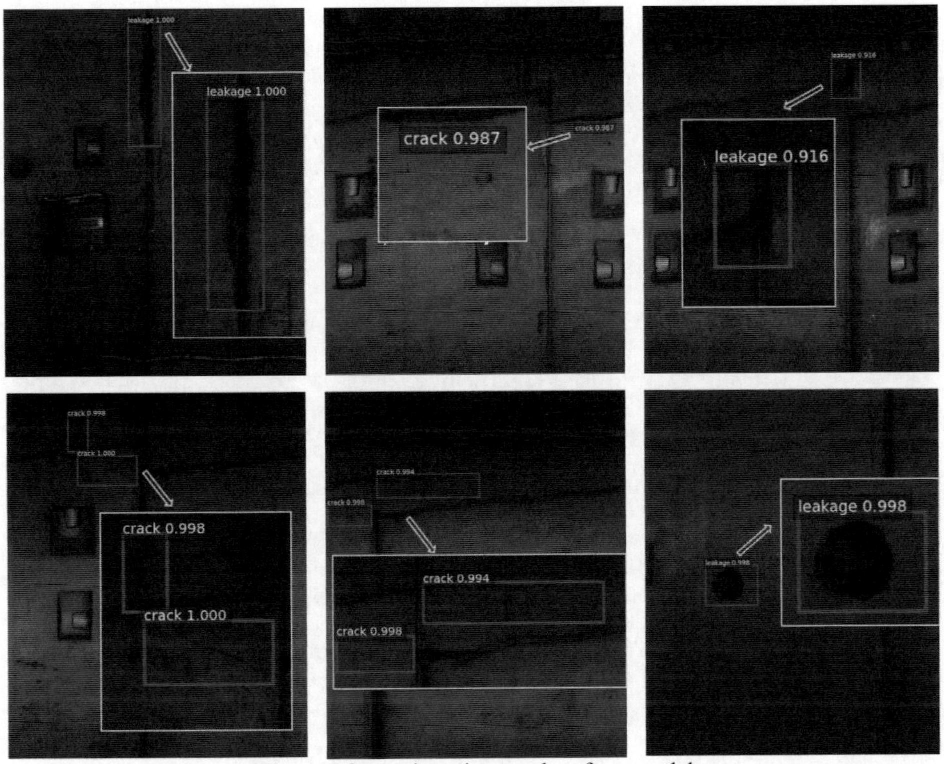

Figure 7 Some detection results of our model

5 Conclusion and future work

A deep-learning method for shield tunnel lining defects fast-detection via fully convolutional network is proposed in this paper. Based on an improved FCN model, the feature is extorted with high the accuracy (95.84%), and this model is easy to train and apply. Furthermore, we use RPN and Position-sensitive RoI pooling to set up the tunnel lining defects detection method. The performance of the method was evaluated on the large-scale images of 3000×3724 pixels. Results shows the method is very fast, accurate and efficient. The methodology could be combined with the machine vision inspection to realize defects detection in the fully operational subway tunnel. The machine vision inspection is served to gather pictures of tunnel lining in the un-operation hours, and the methodology can deal with the pictures synchronously.Still, the proposed method has some limitations. Although our dataset contained thousands of images, this is still not quite sufficient for deep learning. Thus, we plan to conduct more MTI-based defect detections, to generate more labelling data. Further, we plan to develop deep learning-based methods for extracting defect-related features.

References

Abdel-Qader, I, Abudayyeh, O, & Kelly, ME 2003, 'Analysis of edge-detection techniques for crack identification in bridges', *Journal of Computing in Civil Engineering*, vol. 17, no. 2, pp. 255-263.

Cha, YJ, Choi, W, & Buyukozturk, O 2017, 'Deep learning-based crack damage detection using convolutional neural network', *Computer-Aided Civil and Infrastructure Engineering*, vol. 32, no. 3, pp. 2013-2014.

Cord, A, & Chambon, S, 2012, 'Automatic road defect detection by textural pattern recognition based on AdaBoost', *Computer-Aided Civil and Infrastructure Engineering*, vol. 27, no. 4, pp. 244-259.

Everingham, M, Eslami, SA, VanGool, L, Williams, CK, Winn, J, & Zisserman, A 2015, 'the pascal visual object classes challenge: A retrospective', *International Journal of Computer Vision*, vol. 111, no. 1, pp. 98-136.

Girshick, R, Donahue, J, Darrell, T, & Malik, J 2014, 'Rich feature hierarchies for accurate object detection and semantic segmentation', *The 27th IEEE Conference on Computer Vision and Pattern Recognition (CVPR) (CVPR 2014)*, Columbus, OH, USA, 24-27 June, pp. 580-587.

Girshick, R. 2015, 'Fast r-cnn', *The 28th IEEE Conference on Computer Vision and Pattern Recognition (CVPR) (CVPR 2015)*, Boston, MA, USA, 7-12 June, pp. 1440-1448.

Hubel, DH, & Wiesel, TN 1968, 'Receptive fields and functional architecture of monkey striate cortex', *The Journal of physiology*, vol. 195, no. 1, pp. 215-243.

Hopfield, JJ 1982, 'Neural networks and physical systems with emergent collective computational abilities', *Proceedings of the national academy of sciences*, vol. 79, no. 8, pp. 2554-2558.

Hinton, GE, & Salakhutdinov, RR 2006, 'Reducing the dimensionality of data with neural networks', *Science*, vol. 313, no. 5786, pp. 504-507.

He, K, Zhang, X, Ren, S, & Sun, J 2015, 'Spatial pyramid pooling in deep convolutional networks for visual recognition', *IEEE transactions on pattern analysis and machine intelligence*, vol. 37, no. 9, pp. 1904-1916.

He, K, Zhang, X, Ren, S, & Sun, J 2016, 'Deep residual learning for image recognition', *The 29th IEEE Conference on Computer Vision and Pattern Recognition (CVPR) (CVPR 2016)*, Las Vegas, Nevada, USA, 26 June-1 July, pp. 770-778.

Jin, T, & Zhou, ZY 2014, 'Leakage Detection Method for Piping Network Based on BP Neural Network', *Applied Mechanics and Materials* Vol. 470, pp. 738-742.

Jia, Y, Shelhamer, E, Donahue, J, Karayev, S, Long, J, Girshick, R, & Darrell, T, 2014, 'Caffe: Convolutional architecture for fast feature embedding', *Proceedings of the 22nd ACM international conference on Multimedia (ACM Multimedia 2014),* Orlando, Florida, USA, 3-7 November, pp. 675-678.

Krizhevsky, A, Sutskever, I, & Hinton, GE 2012, 'Imagenet classification with deep convolutional neural networks', *Advances in Neural Information Processing Systems 25 (NIPS 2012)*, Stateline, Nevada, USA, 3-8 December, pp. 1097-1105.

LeCun, Y, Bottou, L, Bengio, Y, & Haffner, P 1998, 'Gradient-based learning applied to document recognition', *Proceedings of the IEEE*, vol. 86, no. 11, pp. 2278-2324.

Lei, Y, & Zuo, MJ 2009, 'Gear crack level identification based on weighted K nearest neighbor classification algorithm', *Mechanical Systems and Signal Processing*, vol. 23, no. 5, pp. 1535-1547.

Landstrom, A, & Thurley, MJ 2012, 'Morphology-based crack detection for steel slabs', *IEEE Journal of selected topics in signal processing*, vol. 6, no. 7, pp. 866-875.

LeCun, Y, Bengio, Y, & Hinton, G 2015, 'Deep learning', *Nature*, vol. 521, no. 7553, pp. 436-444.

Liu, W, Anguelov, D, Erhan, D, Szegedy, C, Reed, S, Fu, CY, & Berg, AC 2016, 'SSD: Single shot multibox detector. *The 14th European Conference on Computer Vision (ECCV 2016)*, Amsterdam, The Netherlands, pp. 21-37.

Li, Y, He, K, & Sun, J 2016, 'R-fcn: Object detection via region-based fully convolutional networks', *Advances in Neural Information Processing Systems 29 (NIPS 2016),* Barcelona, Spain, 5-10 December, pp. 379-387.

Qu, Z, Feng, H, Zeng, Z, Zhuge, J, & Jin, S 2010, 'A SVM-based pipeline leakage detection and pre-warning system', *Measurement*, vol. 43, no 4, pp. 513-519.

Ren, S, He, K, Girshick, R, & Sun, J 2015, 'Faster r-cnn: Towards real-time object detection with region proposal networks', *Advances in Neural Information Processing Systems 28 (NIPS 2015)* Montreal, Quebec, Canada, 7-12 December, pp. 91-99.

Simonyan, K., & Zisserman, A. (2014). Very deep convolutional networks for large-scale image recognition. arXiv preprint arXiv:1409.1556.

Szegedy, C, Liu, W, Jia, Y, Sermanet, P, Reed, S, Anguelov, D, ... & Rabinovich, A 2015, 'Going deeper with convolutions', *The 28th IEEE Conference on Computer Vision and Pattern Recognition (CVPR) (CVPR 2015)*, Boston, MA, USA, 7-12 June, pp. 1-9.

Uijlings, JR, Van De Sande, KE, Gevers, T, & Smeulders, AW 2013, 'Selective search for object recognition', *International journal of computer vision*, vol. 104, no 2, pp. 154-171.

Yu, SN. Jang, JH, & Han, CS 2007, 'Auto inspection system using a mobile robot for detecting concrete cracks in a tunnel', *Automation in Construction*, vol. 16, no. 3, pp. 255-261.

Zitnick, CL, & Dollar, P 2014, 'Edge boxes: Locating object proposals from edges', *The 12th European Conference on Computer Vision (ECCV 2014),* Zurich, 6-12 September, pp. 391-405.

Zhang, W, Zhang, Z, Qi, D, & Liu, Y 2014, 'Automatic crack detection and classification method for subway tunnel safety monitoring', *Sensors*, vol. 14, no. 10, pp. 19307-19328.

Zhang, L, Yang, F, Zhang, YD, & Zhu, YJ 2016, 'Road crack detection using deep convolutional neural network', *2016 International Conference on Image Processing (ICIP 2016)*, Phoenix, Arizona, USA, 25-28 September, pp. 3708-3712.

The 8th International Conference on
Structural Health Monitoring of Intelligent Infrastructure
Brisbane, Australia | 5-8 December 2017

Image Recognition of Shield Tunnel Cracks by Deep Learning

Q.T. Li[1], Y.D. Xue[1], H.W. Huang[1]
[1] *Department of Geotechnical Engineering, Tongji University– China, Email:* huanghw@tongji.edu.cn

Abstract

Along with sharp increases in requirements of urban metro tunnel maintenance, metro shield tunnel structure defects, especially cracks, need to be inspected rapidly and accurately. In recent years using deep learning (DL) for structure defects inspection of shield tunnel is a thriving trend around the world. However, the results of image recognition cannot meet the requirements of practical engineering. Considering the influence on cracks from interference of metro shield tunnel lining surface, this paper presents a study on DL-based image recognition of metro shield tunnel cracks using fully convolutional networks. The number of crack images required for training and validation are augmented to 185400 by cropping and sliding window operation and random rotation. The final pixel resolutions of all crack images used are 600 × 600. In order to solve the problems that segmented crack regions are discontinuous and some small noises are existed, we take a morphological image processing as post-processing step to connect discontinuous cracks and remove noises. The length and the width of crack are calculated according to half of crack region's perimeter and diameter of the biggest incircle, respectively. Compared against frequently-used traditional methods, region growing algorithm and adaptive thresholding algorithm, experimental evaluation is conducted based on segmented results, inference time and error rates by four typical cracks including transverse crack, vertical crack, diagnal crack and wavy crack. The study shows that metro shield tunnel cracks can be rapidly and accurately recognized through the proposed method which will support a delighted help for structure health monitoring and maintenances for tunnel and underground works.

1. Introduction

Shield tunnel is a typical civil infrastructure for urban transportation all over the world. In recent years, more and more shield tunnels, especial metro tunnels, are used in China which undergo rapid economic development. However, due to fast construction, limited project schedule, lack of sufficient work force, as well as complicated segment assembling and environmental disturbance around metro tunnels, shield tunnels occasionally suffer from defects such as cracks, water leakage etc. Among these defects, cracks are most commonly observed (Liu et al., 2013). For example, a metro tunnel interval in Shanghai had serious and dense defects because of an unexpected surcharge 7m in height by construction activities nearby. Owing to the lack of timely inspection of these defects, safety of the lining structure and its serviceability were seriously threatened (Shao et al., 2016). Therefore rapid and accurate defects inspection for metro tunnel is urgently needed to improve maintenance level and safety operation practices.

In past years, inspection of tunnel lining cracks had been implemented manually via experienced workers, which is subjective, low efficient, high costly (Asakura & Kojima, 2003, Yuan et al., 2013). Nowadays in order to deal with these shortcomings of manual inspection, structure health monitoring (SHM) techniques of metro tunnels, based on computer vision (CV), is a powerful trend which can overcome workers' subjective and provide high quality inspections by collecting several types of data (Montero et al., 2015). For instances, in light of existing problems of conventional image recognition methods for tunnels cracks, an automatic crack recognition method based on local grid characteristics in images was presented (Wang et al., 2012). For tunnel lining inspection, Gavilan et al. (2013) presented a novel system consisting of a CV-based software which facilitates inspection of cracks and relative displacement between assembled segments. In order to implement metro tunnel health monitoring, Zhang et al. (2014) presented a distance histogram-based shape descriptor that effectively differentiate the spatial shape between cracks and other irrelevant objects in feature extraction process.

Along with significant advancement in the field of data acquisition equipment and computational hardware, such as Graphics Processing Unit (GPU), a novel powerful machine learning technique named deep learning (DL) has been proposed (LeCun et al., 2015). The key advantage of DL is that multiple layers of features are not designed by human engineers but via a general-purpose learning procedure. DL was selected as one of 10 breakthrough technologies by MIT technology review in 2013[1]. High-level features were extracted by convolutional neural networks (CNN) from 17 low-level features which are grey value, edge, frequency, entropy, texture, histogram of oriented gradients (HOG) etc. And multi-layer perceptron was chosen as a detector for crack identification achieving very fast prediction (Makantasis et al., 2015, Protopapadakis & Doulamis, 2015). Zhang et al. (2016) proposed a supervised deep CNN for pavement crack detection by demonstrating that learned image features with deep learning framework showed superior crack recognition capability when compared with features extracted by traditional hand-crafted methods. Cha et al. (2017) proposed a DL-based algorithm using a deep architecture of CNN for inspecting concrete cracks, which showed better performances compared with traditional Canny and Sobel edge detection methods.

In this study, we propose a novel algorithm to recognize concrete cracks from metro tunnel lining surface images based on DL. Metro shield tunnels have many assembled segment joints and electric wires on tunnel lining surface, which pose great challenges for image recognition of cracks. The main objective of this study is to build a robust classifier that is less influenced by noises caused by lighting, shadows, cement grout blots, segment joints, electric wires etc. The content of the study is described as follows. Section 2 presents characteristics of cracks and interference of metro shield tunnel lining surface. Section 3 introduces the architecture of the proposed algorithm and explains detailed methodologies. Section 4 demonstrates experimental evaluation by comparing to traditional frequently-used methods. Section 5 concludes this article.

2. Cracks and interference of metro shield tunnel lining surface

Metro shield tunnel is a tubular underground structure which is one dimensional in weak light environment. The surface of tunnel lining has a certain curvature and contains some interference on image recognition which are segment joints, cables, pipes, LED lighting, bolt holes et al. See Figure 1 for detailed information about metro shield tunnel lining surface. Due to the interference, crack image recognition of metro shield tunnel would face four big challenges as below:

- Lots of segment joints: shield tunnel is made of segments via Tunnel Boring Machine (TBM). If mileage is 1 km, the length of segment joints will be 20 km correspondingly for metro tunnels .
- High randomicity of crack patterns: crack regions are of arbitrary size and shape. In other words, cracks in a class will be highly different.
- Numerous noises: some noises such as cement mortar, surface scratches, manual marks, water blots are inevitably left on tunnle lining surface.

[1] https://www.technologyreview.com/s/513696/deep-learning/

- Uneven light condition: tunnel lining surface is in weak light environment and captured images are easily influenced by shadows.

Figure 1. Metro shield tunnel lining surface

There are some images simultaneously containing cracks and interference, as shown in Figure 2. Cracks are smaller compared to inferference, which obviously increase the difficulty of crack identification. For instances, segment joints are similar to cracks to some extent because they are all long and narrow. It is hard to locate a wirelike crack in low light condition rather than in high light condition. When using some global thresholding algorithms, like Otsu's method, cables and cracks will be all segmented because of their close grey values, making it hard to distinguish cables and cracks easily by computers. Moreover, the edges of bolt hole are also similar to cracks, which disturbs the successful identification of cracks.

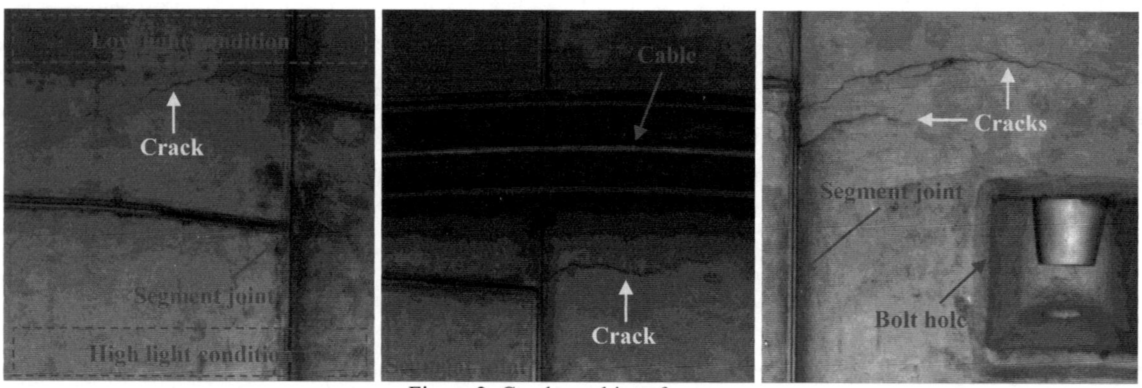

Figure 2. Cracks and interference

3. Proposed method

With crack images of metro shield tunnel lining, the objective of the study is to compute crack's semanteme and location by using global information to resolve what is crack while using local information to resolve where is crack. The definiton of a crack in our study is that it should be identifiable in images via naked human eyes. In this section, we build an architecture of fully convolutional networks (FCN) that combines semantic information from deep, coarse layer and appearance information from shallow, fine layer to produce accurate and detailed segmentation for metro shield tunnel cracks. All of the computation in this article are implemented on a desktop computer with one GPU (CPU: Intel core i7-4790@3.60Hz, 8 core processors, RAM: 16GB, GPU: NVIDIA GTX-1080, 8G and system type is 64-bit). CAFFE, a state-of-the-art DL framework (Jia et al., 2014), is used to perform this study

3.1 Data preparation

Crack images of metro shield tunnel lining are taken from metro tunnel inspection equipment MTI-100 which is developed by Tongji University and Shanghai TRISK Co., Ltd. (Huang et al., 2017). The total number of captured images containing cracks is 405 with 3,000 × 22,344 pixel resolutions. The 405 captured images are cropped into smaller images of 1000 × 1000 pixel resolutions for convenient image processing. Then we use sliding window operation and random rotation to generate images of 600 × 600 pixel resolutions for building crack image dataset from those cropped images, which is a frequently-used method used to augment image numbers. After augmenting, the final crack image dataset has 185,400 images of 600 × 600 pixel resolutions. The ratio between image numbers of training and validation is 4:1 (Cha et al., 2017) which result in 148,320 training images and 37,080 validation images, respectively. For training, there is a one-to-one correspondence between raw images and ground truth which is carefully and manually labled by experts. Figure 3 shows some raw images and corresponding ground truths in our crack image dataset.

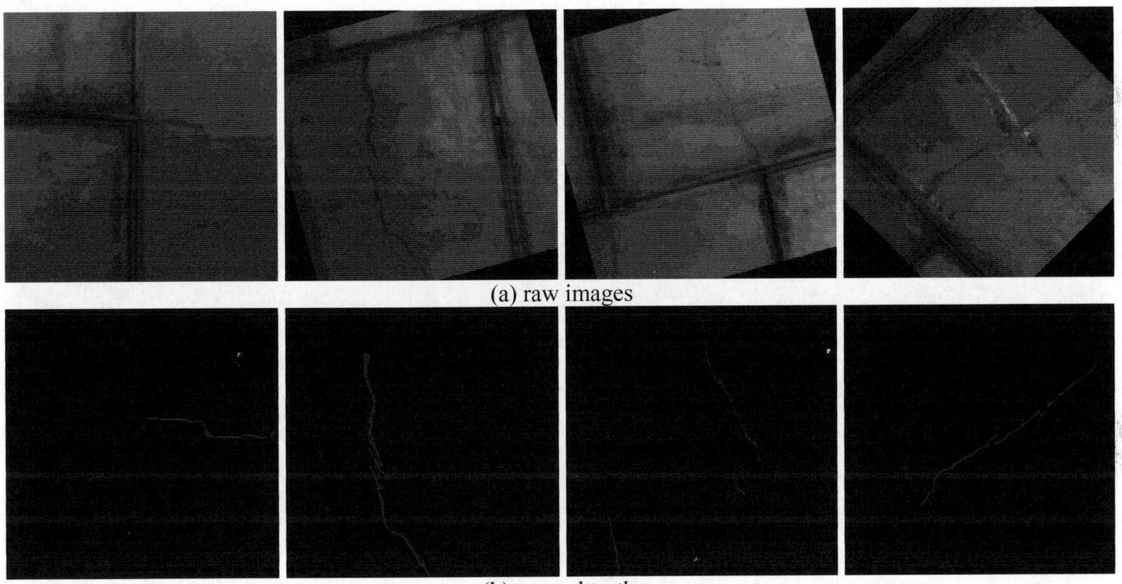

(a) raw images

(b) ground truths

Figure 3. Some raw images and corresponding ground truths in our dataset

3.2 FCN architecture

FCN, one specific CNN, is proposed by Long et al. (2015) to conduct semantic segmentation of images. Compared to CNN, FCN can take input of arbitrary size and produce correspondingly-sized output with efficient inference and learning, for the reason that classification nets are converted into fully convolutional nets used to achieve pixelwise predictions. Basic components of FCN are convolution, pooling, activation functions etc. Hierachies of deep features are fused to define a nonlinear local-to-global representation across several layers.

Figure 4 presents flowchart of our FCN architecture. Firstly, raw images and corresponding ground truths are input to the networks. The first layer is the raw image with pixel size 600 × 600, and 3 channels. After several convolution and pooling operations, a lot of abstract feature maps are generated. The number of feature maps from different layers are typed by transverse figures in Figure 4. Backward convolution is a special operation to upsampling feature maps in order to output corresponding-sized predicted maps in the end. Loss function's value will be computed by

$$E = \frac{1}{m}\sum_{k=1}^{m} E_k \tag{1}$$

$$E_k = \frac{1}{2}\sum_{j=1}^{l} (\hat{y}_j^k - y_j^k)^2 \tag{2}$$

where m is the number of training images, E_k is mean squared error for a certain training image, $\hat{y}_j^k$ is pixel value of predicted map and y_j^k is pixel value of ground truth. The goal of training is to reduce loss function by Stochastic Gradient Descent (SGD) algorithm. When the loss value is converged after a number of iterations which consist of forward inferences and backward learnings, the trained model is acquired. Learning curve of our training is shown by Figure 5. After about 40,000 iterations, loss function is converged.

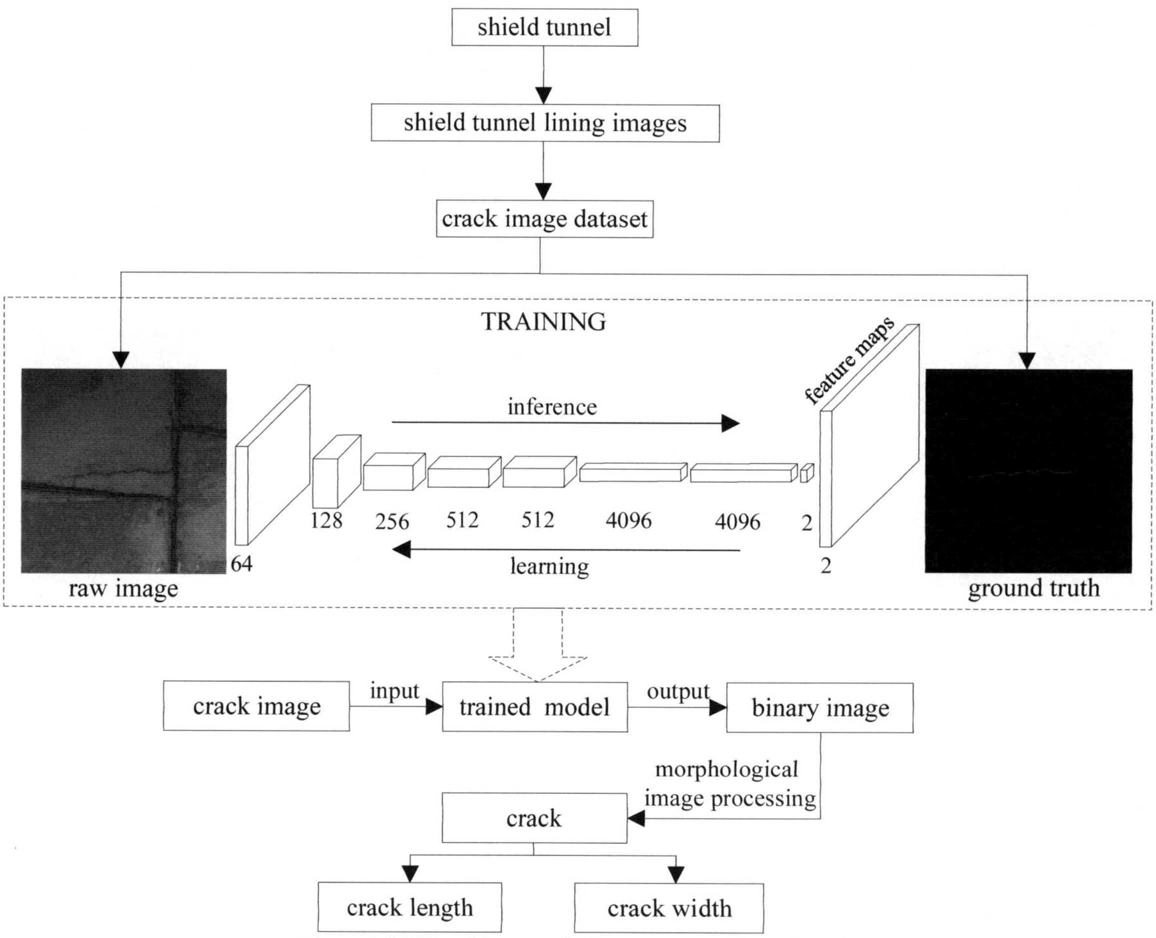

Figure 4. FCN architecture for image recognition of crack

Next step, crack images are fed into trained model and binary images will be obtained subsequently. In the binary images, pixels of grey value 0 are background and pixels of grey value 1 belong to cracks. However, the segmented crack regions are discontinuous and some small noises are identified as cracks unfortunately. In order to solve this problem, we take a post-processing step which is morphological image processing to connect discontinuous cracks and remove noises. Figure 7 shows the post-processing of binary images.

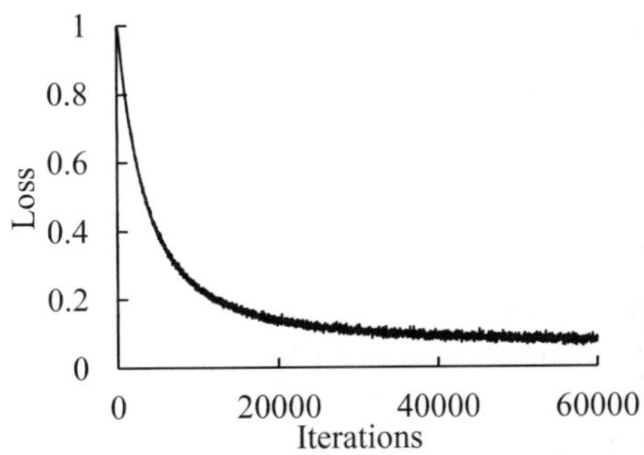

Figure 5. Learning curve

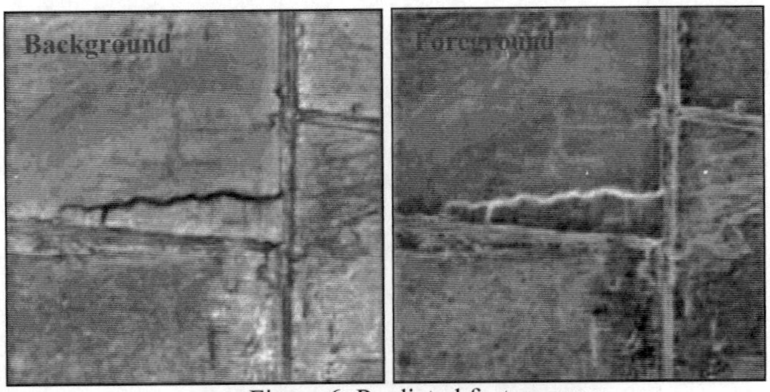

Figure 6. Predicted feature maps

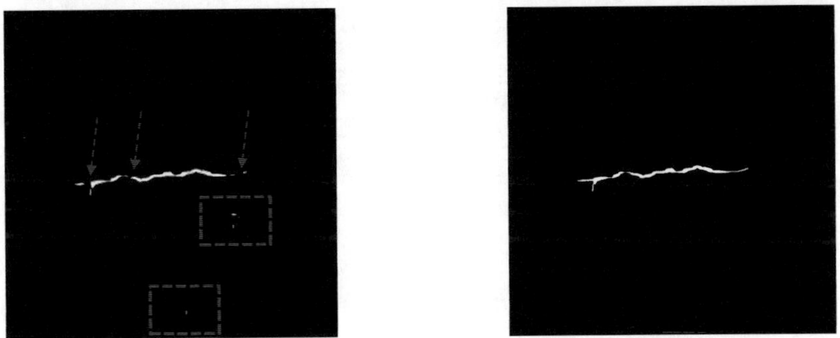

(a)segmented crack regions (b)final result

Figure 7. Postprocessing of binary images

The length and width of crack are calculated after final segmented results are gotten. We consider half of crack region's perimeter as crack length and diameter of the biggest incircle as crack width. Taking the final result of Figure 7 for example, crack length and width are 579 pixels and 7 pixels, respectively.

4. Experimental evaluation

The performance of proposed method is compared against frequently-used traditional methods which are Region Growing Algorithm (RGA) and Adaptive Thresholding Algorithm (ATA). RGA is a simple region-based image segmentation method that examines neighboring pixels of initial seed points and determines whether pixel neighbors should be added to the region (Kamdi & Krishna, 2012). ATA is used to separate needed foreground, i.e. image objects, from background based on the difference in pixels' grey value of each region (Zhang et al., 2017). Dou et al., (2011) employed a similar global thresholding operation named Otsu's method to segment regions of water leakage from infrared radiation images for tunnels. Experimental evaluation is conducted based on segmented results, inference time and error rates of four typical cracks which are transverse crack, vertical crack, diagnal crack and wavy crack. The four typical crack images all have 1000 × 1000 pixels in particular.

4.1 Segmented results

Segmented results of different algorithms are shown in Figure 8. We can see that the proposed method provides clearer crack information than other methods compared to ground truths. DL-based approach is not susceptible to interference of metro shield tunnel lining surface. RGA hardly identify whether pixels belong to crack or not. Crack region of RGA is limited and smaller than proposed method. Inversely, ATA recognize bigger crack region including segment joints, bolt hole etc. Segment joints are similar to crack's image feature for the reason that grey values of them are approximately the same. Inevitably, ATA has produced many noises interfering in accurate segmentation of crack regions.

4.2 Inference time

Inference time of different algorithms is detailedly shown in Table 1. It is known that inference time of the proposed algorithm is 0.20 seconds per image on average while RGA and ATA are 3.17 seconds and 0.35 seconds, respectively. In other words, Inference time of the proposed algorithm is about 15.85 times shorter than RGA and 1.75 times shorter than ATA. If there are a mass of crack images, total inference time will be a crucial factor to be considered. See Figure 9 for bar chart of inference time.

Table 1. Inference time of different algorithms (seconds)

Crack types	Proposed	RGA	ATA
Transverse crack	0.20	3.35	0.34
Vertical crack	0.20	2.93	0.35
Diagonal crack	0.20	3.86	0.34
Wavy crack	0.20	2.54	0.35
Mean value	0.20	3.17	0.35

4.3 Error rates

In this study error rate is defined as the ratio between the number of mislabeled pixels and the number of total pixels in a image by

$$Error\ rate = \frac{N_{false}}{N_{total}} \times 1000‰ \tag{3}$$

where N_{false} is the number of mislabeled pixels, N_{total} is the number of total pixels in a image. In light of our crack images, N_{total} is always 1,000,000. Table 2 shows error rates of different algorithms. Error rate of our proposed algorithm on average is 2.68 ‰ while RGA and ATA are 4.04 ‰ and 84.57 ‰, respectively. In other words, Error rate of the proposed algorithm is about 1.51 times smaller than RGA and 31.56 times smaller than ATA. Bar chart of error time can be seen from Figure 10.

Table 2. Error rates of different algorithms

Crack types	N_{false} (pixels)			N_{total} (pixels)	*Error rate* (‰)		
	Proposed	RGA	ATA		proposed	RGA	ATA
Transverse crack	2156	2583	75243	1000000	2.16	2.58	75.24
Vertical crack	1077	1326	80661		1.08	1.33	80.66
Diagonal crack	3701	5686	70374		3.70	5.69	70.37
Wavy crack	3798	6554	111986		3.80	6.55	111.99
Mean value					2.68	4.04	84.57

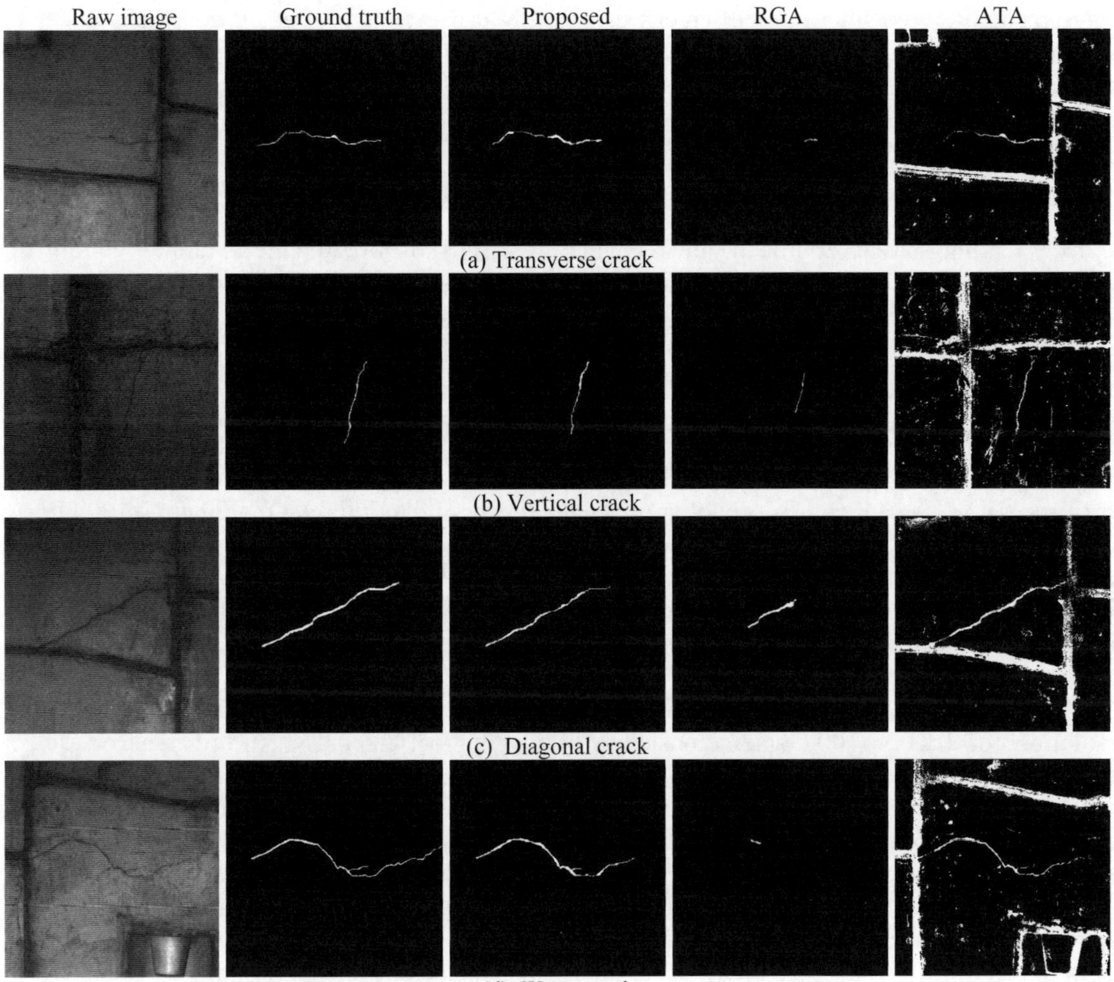

(a) Transverse crack

(b) Vertical crack

(c) Diagonal crack

(d) Wavy crack

Figure 8. Segmented results comparison of different algorithms

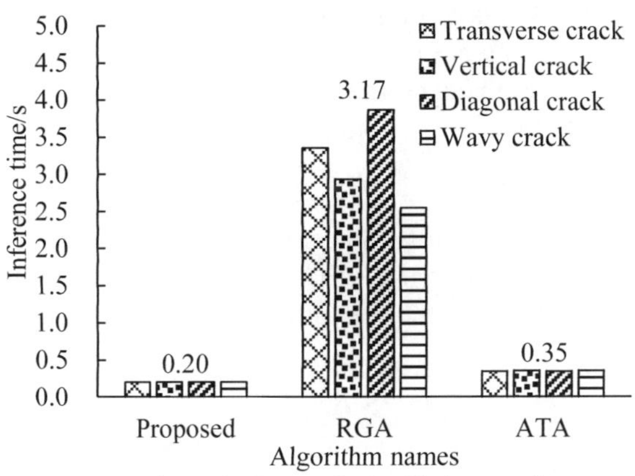

Figure 9. Comparison of inference time

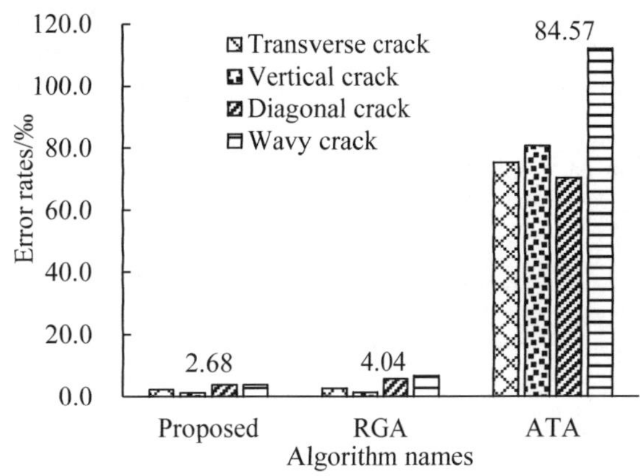

Figure 10. Comparison of error rates

In these comparative studies by four typical cracks, the proposed method shows better performance based on segmented results, inference time, error rates than those of RGA and ATA. DL-based

Image recognition of shield tunnel cracks using FCN can effectively recognize cracks under the influences of interference such as segment joints, bolt holes and etc., and shows excellent robustness in overcoming terrible light condition. Metro shield tunnel cracks are rapidly and accurately identified by our method.

5. Conclusions

A deep learning-based approach for recognizing cracks of metro shield tunnel rapidly and accurately is proposed in this paper using fully convolutional networks (FCN). Image acquisition of metro shield tunnel cracks is performed by metro tunnel inspection equipment MTI-100 which is developed by Tongji University and Shanghai TRISK Co., Ltd. A total of 405 images containing cracks are chosen by naked human eyes. The number of crack images required for training and validation are augmented to 185,400 by cropping and sliding window operation and random rotation. The final pixel resolutions of all crack images used are 600 × 600 for the reason that small images are easy to be processed. After about 40,000 iterations, FCN model is converged. We take a post-processing step which is morphological image processing to improve the final results and have achieved satisfactory binary images from raw crack images. In binary images, cracks are the regions that contain the pixels of grey value 1 and the other pixels, i.e. pixels of grey value 0, are non-cracks. Crack positions are easily acquired in image coordinate. The length and width of crack is calculated according to half of crack region's perimeter and diameter of the biggest incircle, respectively.

The performance of proposed method is experimentally evaluated by 4 large images with resolutions of 1000 × 1000 pixels consisting of four typical cracks which are transverse crack, vertical crack, diagnal crack and wavy crack. Compared against two frequently-used traditional methods, region growing algorithm (RGA) and adaptive thresholding algorithm (ATA), great superiority of proposed method is shown based on segmented results, inference time and error rates. In terms of segmented results, the proposed method provides clearer crack information than other methods compared to ground truths. It means that the proposed method is not susceptible to interference of metro shield tunnel lining surface. Inference time of the proposed algorithm is about 15.85 times shorter than RGA and 1.75 times shorter than ATA. Error rate of the proposed algorithm is about 1.51 times smaller than RGA and 31.56 times smaller than ATA. All in all, deep learning-based approach with our proposed method can be used to rapidly and accurately recognize cracks for structure health monitoring and maintenances of metro shield tunnels.

In the future, we will build a more high-efficiency architecture to simultaneously implement semantic segmentation of multiple defects like cracks, water leakage, spalling etc.

Acknowledgements

This study is supported by the Natural Science Foundation Committee Program of China (Grant No. 51538009 , Grant No. 51778474), the Fundamental Research Funds for the Central Universities of China (Grant No. 0200219129) as well as the Science and the Technology Commission of Shanghai Municipality (Grant No. 16DZ1200402).

References

Asakura, T. & Kojima, Y. 2003, 'Tunnel maintenance in Japan', *Tunnelling & Underground Space Technology Incorporating Trenchless Technology Research,* vol. 18, no. 2–3, pp. 161-169.

Cha, Y. J., Choi, W. & Buyukozturk, O. 2017, 'Deep learning-based crack damage detection using convolutional neural networks', *Computer - aided Civil & Infrastructure Engineering,* vol. 32, no. 5, pp. 361-378.

Dou, H. T., Huang, H. W. & Xue, Y. D. 2011, 'Model test on infrared radiation feature of tunnel seepage and image processing', *Chinese Journal of Rock Mechanics & Engineering,* vol. 30, no. s2, pp. 3386-3391.
Gavilan, M., Sanchez, F., Ramos, J. A. & Marcos, O. 2013, 'Mobile inspection system for high-resolution assessment of tunnels'. *Proceedings of the 6th International Conference on Structural Health Monitoring of Intelligent Infrastructure (SHMII6)*, Hong Kong, China, 9-11 December.

Huang, H. W., Sun, Y., Xue, Y. D. & Wang, F. 2017, 'Inspection equipment study for subway tunnel defects by grey-scale image processing', *Advanced Engineering Informatics,* vol. 32, no., pp. 188–201.

Jia, Y.Q., Shelhamer, E., Donahue, J., Karayev, S., Long & Jonathan 2014, 'Caffe: Convolutional architecture for fast feature embedding', *Eprint Arxiv,* vol. no., pp. 675-678.

Kamdi, S. & Krishna, R. K. 2012, 'Image segmentation and region Growing algorithm', *International Journal of Computer Technology & Electronics Engineering,* vol. 1, no. 2, pp. 103-107.

LeCun, Y., Bengio, Y. & Hinton, G. 2015, 'Deep learning', *Nature,* vol. 521, no. 7553, pp. 436-444.

Liu, C., Lai, J. X. & Liu, Q. 2013, 'A review of research situation on shield tunnel diseases', *Applied Mechanics & Materials,* vol. 438-439, no., pp. 943-948.

Long, J., Shelhamer, E. & Darrell, T. 2015, 'Fully convolutional networks for semantic segmentation'. *Proceedings of IEEE Conference on Computer Vision and Pattern Recognition*, Boston, MA, USA, 7-12 June, pp. 3431-3440.

Makantasis, K., Protopapadakis, E., Doulamis, A. & Doulamis, N. 2015, 'Deep convolutional neural networks for efficient vision based tunnel inspection'. *Proceedings of IEEE International Conference on Intelligent Computer Communication and Processing*, Cluj-Napoca, Romania, 3-5 September, pp. 335-342.

Montero, R., Victores, J. G., Martínez, S., Jardón, A. & Balaguer, C. 2015, 'Past, present and future of robotic tunnel inspection', *Automation in Construction,* vol. 59, no., pp. 99-112.

Protopapadakis, E. & Doulamis, N. 2015, 'Image based approaches for tunnels' defects recognition via robotic inspectors'. *Proceedings of International Symposium on Visual Computing*, Las Vegas, Nevada, USA, 14-16 December, pp. 706-716.

Shao, H., Huang, H. W., Zhang, D. M. & Wang, R. L. 2016, 'Case study on repair work for excessively deformed shield tunnel under accidental surface surcharge in soft clay', *Chinese Journal of Geotechnical Engineering,* vol. 38, no. 6, pp. 1036-1043.

Wang, P., Huang, H. & Xue, Y. 2012, 'Automatic recognition of cracks in tunnel lining based on characteristics of local grids in images', *Chinese Journal of Rock Mechanics & Engineering,* vol. 31, no. 5, pp. 991-999.

Yuan, Y., Jiang, X. & Liu, X. 2013, 'Predictive maintenance of shield tunnels', *Tunnelling & Underground Space Technology Incorporating Trenchless Technology Research,* vol. 38, no. 3, pp. 69-86.

Zhang, D. J., Li, Q. Q., Chen, Y., Cao, M., He, L. & Zhang, B. L. 2017, 'An efficient and reliable coarse-to-fine approach for asphalt pavement crack detection', *Image and Vision Computing,* vol. 57, no., pp. 130-146.

Zhang, L., Yang, F., Zhang, Y. D. & Zhu, Y. J. 2016, 'Road crack detection using deep convolutional neural network'. *Proceedings of IEEE International Conference on Image Processing*, Phoenix, Arizona, USA, 25-28 September, pp. 3708-3712.

Zhang, W.Y., Zhang, Z.J., Qi, D. & Liu, Y. 2014, 'Automatic crack detection and classification method for subway tunnel safety monitoring', *Sensors,* vol. 14, no. 10, pp. 19307-19328.

The 8th International Conference on
Structural Health Monitoring of Intelligent Infrastructure
Brisbane, Australia | 5-8 December 2017

Ground Penetrating Radar as a Monitoring Tool: Mapping Internal Features with Attribute Analysis

I. Morris[1], H. Abdel-Jaber[2], B. Glisic[2].
[1] *Princeton University – USA, Email:* imorris@princeton.edu, [2] *Princeton University –USA.*

Abstract

In reinforced concrete structures, including those instrumented with embedded sensor networks, severe structural problems often begin as minor issues or maintenance oversights. Unusual behaviors like corrosion or cracking can be detected with embedded sensors as deviation from a reference state of the structure, but verifying the actual cause of the deviation, i.e. localization and characterization, is often limited to targeted nondestructive investigations. Ground penetrating radar (GPR), for example, has been used to investigate defects in structures and can be a powerful tool to compare the physically constructed details of the structure to construction documents. In this work, a post-tensioned concrete pedestrian bridge instrumented with embedded fiber optic sensors was investigated (1) to locate and map continuous internal features of the bridge deck and (2) quantify the differences between the concrete used in each of two construction phases. To this end, a survey of the main span and one of four approaches of Streicker Bridge was conducted using a 2.6 GHz GPR antenna. Analysis of the data using standard processing and filtering techniques was complemented by attribute analysis for improved mapping of features, focusing on the post-tension tendons, embedded air voids, and connections between elements. These results indicate the value of combining nondestructive investigation with SHM sensing approaches for verification of anomalies and understanding of the as-built condition of a structure, as well as providing details for continued SHM of the structure.

1. Introduction

Streicker Bridge crosses Washington Road on the campus of Princeton University, connecting two busy areas of the campus (figure 1). In plan, the bridge is formed by two intersecting arches which are mirrored about a central axis to form a main span with four approach legs. The main span (104 m) is a deck-stiffened arch supported by y-shaped columns; the legs are similarly supported continuous girders (figures 1b and 1c). The steel elements are made of weathering steel and the deck is made of post-tensioned high performance concrete with a specified 28-day strength of 6000 psi.

The bridge was constructed in two main phases because of a delay in the construction schedule. The main span and three approaching legs were constructed in August 2009, and the final phase including the southeast leg was constructed two months later in October 2009. Compressive core tests on the concrete used in each phase indicate that, although the same properties were specified in each mix, different batches and curing conditions lead to different final properties of the cast-in-place concrete. At 28 days, the first phase of construction has a compressive strength of 48.5 MPa (7040 psi), while the second phase has a compressive strength of 59.7 MPa (8660 psi).

Figure 1. Streicker Bridge: (a) near-plan view: main span and four approaching legs, (b) main span: deck-stiffened arch, (c) southeast leg: curved continuous girder, and (d) springing of main span.

During construction, Streicker Bridge was partially instrumented with a structural health monitoring (SHM) system. The system consists of parallel sets of long-gauge fibre optic strain and temperature sensors. Only half of the main span and the southeast leg are instrumented due to the symmetry of the bridge and the research oriented objectives (figure 2). The monitoring system has presented several results, including detection and characterization of cracks (Hubbel & Glisic 2013) (Abdel-Jaber & Glisic 2016), assessment of prestressing forces (Abdel-Jaber & Glisic 2014), quantification of the thermal expansion coefficient (Reilly et al. 2017), and other strain-based parameters (Sigurdardottir & Glisic 2015).

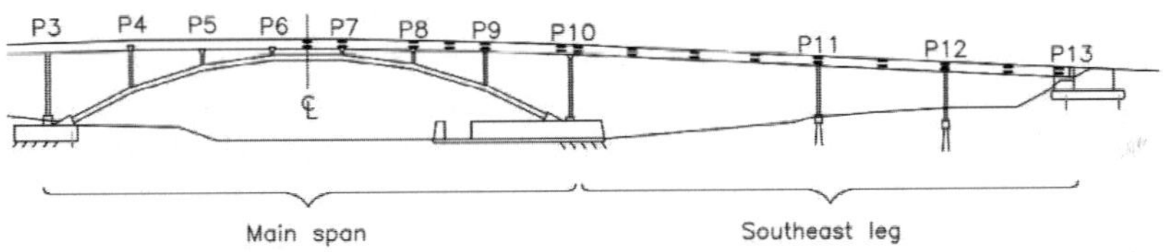

Figure 2. Elevation view of locations of parallel sensors along the main span and southeast leg of Streicker Bridge.

1.2 Ground Penetrating Radar

As a non-destructive technique, ground penetrating radar (GPR) offers an excellent option for inspecting subsurface features in globally homogeneous materials like concrete. GPR is a technique by which electromagnetic pulse signals are transmitted and recorded. The reflections of these signals identitfy interfaces and features by change in the electromagnetic properties of different media (relative dielectric constant, electrical conductivity, and magnetic permeability). Civil infrastructure applications of GPR include mapping and locating utilities (Boniger & Tronicke 2011), generating the layout of rebar in concrete (Sensors & Software 2015), and general concrete inspection for various features and anomalies (Annan 2003).

Most modern GPR antennas used for infrastructure inspection have a fixed transmission and receiving antenna. These antennas are classified by the centre frequency of the signal pulses they emit. High frequency antennas (900 MHz -2.6 GHz) have smaller wavelengths and are able to resolve small targets (i.e., rebar, delamination) but have limited penetration depth (up to 1 m with a 900 MHz in ideal conditions). Lower frequency antennas (as low as 30 MHz) are used for deeper exploration of geological strata, while middle frequency ranges (270-900 MHz) are used for a variety of mid-range applications, from utility locating to archaeological studies (Boniger & Tronicke 2011) (Zhao et al. 2013) (Conyers 2016). These antennas collect data in the form of transects (also called radargrams, scans, profiles, or lines). One transect represents one path over which the antenna has been moved and recorded signals. Each recorded signal is called a trace, which has a defined spatial location and records the arrival time (in nanoseconds) and amplitude of incident EM signals. The time of significant arrivals can be used in conjunction with an estimate or measurement of the group velocity of the waves in the material to convert the time axis to the depth axis (Annan 2003). For a more detailed discussion of the principles and methods of GPR surveys, the interested reader is referred to the many excellent books on the subject (e.g., Annan 2003, Jol 2009).

Different aspects of the recorded GPR data have been used for a variety of applications in life-cycle management, infrastructure monitoring, and other civil SHM applications. Specifically, GPR has been used in damage identification in different ages and mixes of concrete (Yehia et al. 2014), robust estimates of pavement thickness (Barnes et al. 2008) (Pedret-Rodes et al. 2015) (Olhoeft 2000), and identification of voids and other anomalies in civil structures (White et al. 2014) (Smith & Olhoeft 2000). In addition, a collection of quantitative measures of GPR data, called attribute analysis, have been successfully used to differentiate plastic, concrete, and metal (dos Santos et al. 2014) (Dogan & Turhan-Sayan 2016), identify contamination by centroid frequency shifts, and interpret subtle features in challenging data sets (Zhao et al. 2013) (Burton et al. 2004) (Allroggen & Tronicke 2016). Attribute analysis includes amplitude and frequency based attributes like energy (Dogan & Turhan-Sayan 2016), image processing methods (Zhao et al. 2013), and other quantitative measures (dos Santos et al. 2014) (Allroggen & Tronicke 2016). Attribute analysis has recently been used on the data presented in this survey to explore the slight difference between the concrete in each construction phase using GPR attributes (Morris et al. 2017).

In this work, Streicker Bridge was surveyed with a high frequency (high resolution) GPR antenna. The aims of the survey are to characterize the reflections of different internal features and obtain a reference state for the bridge deck's internal arrangement of features. The objectives are to identify the reflections patterns that characterize the internal air voids, reinforcing, and post-tension strands and investigate if amplitude attributes can quantify the different features present in each scan. The study exploring the differences between the materials is presented in (Morris et al. 2017).

3. Methods

3.1 GPR survey

The GPR survey was conducted in November, 2016 in dry and sunny weather, without recent precipitation. A 2.6 GHz GSSI StructureScan HR system was used to collect parallel transects along the SE leg, main span, and NE leg of Streicker Bridge. The transects used in this work were collected at a transect spacing of 15 cm and are between 8-9 m in length (figure 3). Also apparent is the variable depth of the cross section, which varies up to about 55 cm. The location of each scan relative to the internal features, such as air voids, is shown for a section of the deck at mid-span between columns (figure 4).

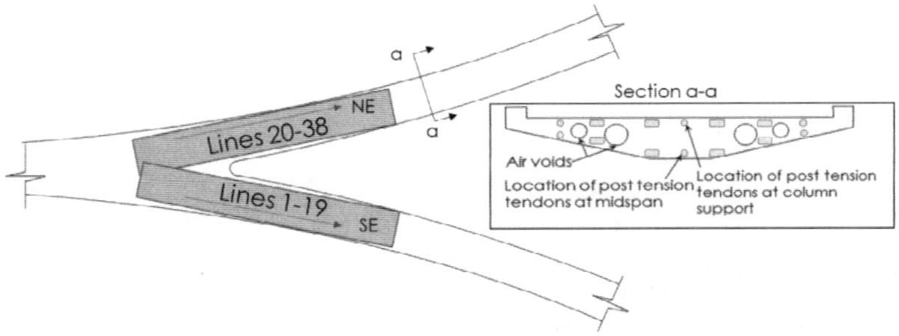

Figure 3: Location of Cross section of Streicker bridge deck, showing variable position of post-tension tendons and location of air voids. Only scans of the southeast leg are included in this work. The rebar cage is omitted from the section drawing for clarity (Morris et al. 2017).

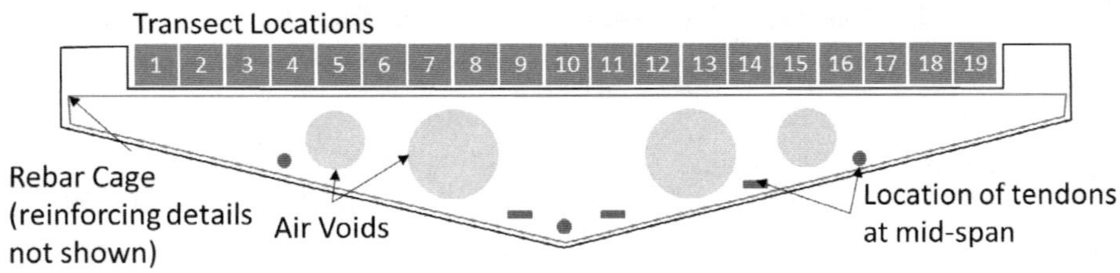

Figure 4: Section taken between columns (where the post-tension tendons are at their lowest point) showing location of GPR scans relative to internal features. The reinforcing in the section is not included for clarity; only the main rebar cage location is shown. Transects run longitudinally down the SE leg of Streicker Bridge.

3.2 Data Processing

Minimal processing is necessary to identify the features in each transect. The same process was applied to all scans. First, the time-zero was adjusted to remove the concrete-air interface above the surface of the deck. Second, the data was centred at zero and low frequency noise was removed by dewowing and removing the DC component of the data. Finally, AGC (automatic gain control) with five points was applied to increase the visibility of features deeper in the deck, especially the post-tension tendons and lower surface of the deck. Though AGC does alter the natural values of the amplitudes in the data, this does not change the relative values of each kind of reflection. Furthermore, it is applied to all transects used in this analysis so the effect is uniform. From geometric hyperbola analysis, the average velocity of the signal is found to be 0.11m/ns. This value is used to verify the depth of features and convert the time axis to depth axis. The processed data can be visually interpreted to identify the locations of the air voids, post-tension tendons, and lower surface of the deck (figure 5). The hyperbolic reflections along the upper edge of the deck are caused by the rebar cage. These were not migrated (collapsed) in this study to assess the ability of the attribute analysis in a more robust manner.

3.3 Analysis and Mapping

The post-tension tendons, air voids, and location of the connection can be quantitatively compared in the processed data using attribute analysis. To supplement and quantify the differences, basic amplitude based attribute analysis was performed. Two reflection attributes and three propagation attributes were computed for each trace in transects in the southeast leg. The reflection attributes included maximum amplitude and maximum absolute amplitude. The propagation attributes computed were the area of each trace, area of the absolute value of each trace, and total energy of the trace (Dogan & Turhan-Sayan 2016). The total energy is computed by integrating the squared amplitude over time (Dogan & Turhan-Sayan 2016). These are all amplitude based attributes, so the relative strengths of the internal feature reflections are well represented in the analysis.

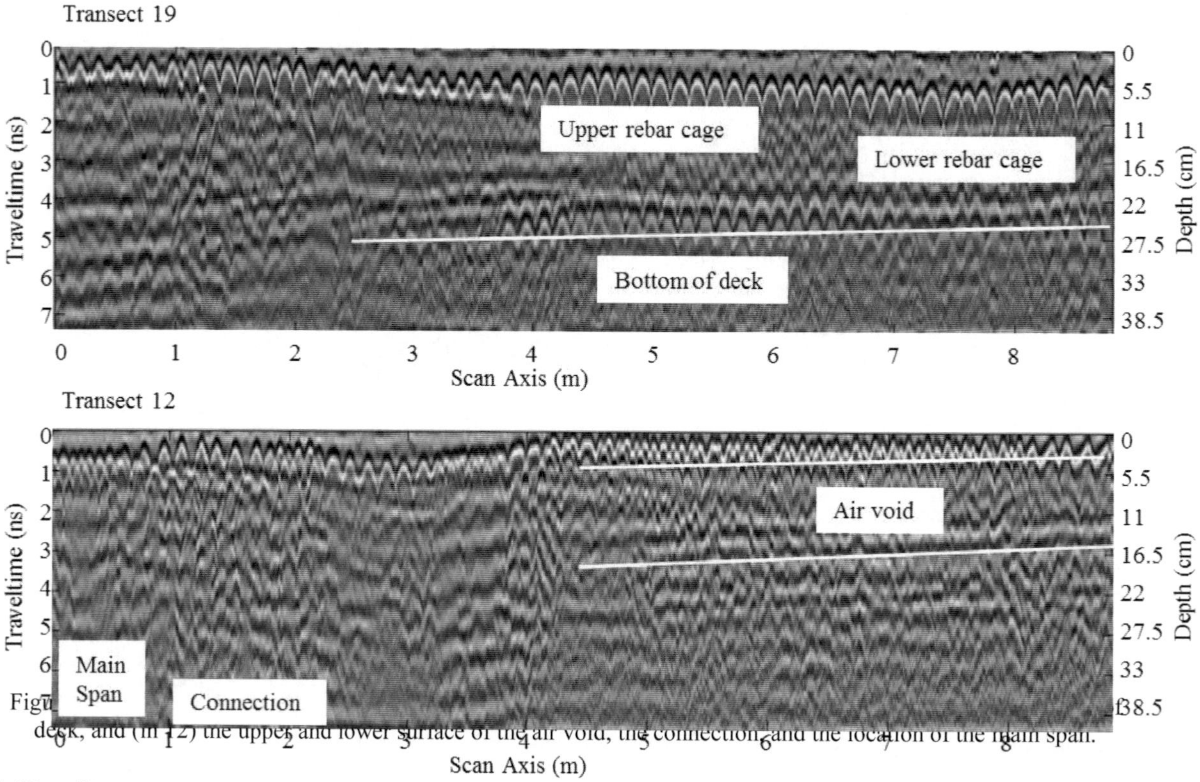

Figu… deck, and (in 12) the upper and lower surface of the air void, the connection, and the location of the main span.

4. Results

4.1 Feature Identification

The air voids are highly visible in the processed data (figure 5). This is because the air-concrete interface represents a sharp contrast between materials, specifically the relative dielectric constant (the concrete in Streicker bridge has a dielectric of about 8.5, whereas the dielectric constant of air is defined as 1). The air voids become present about 4 meters into the scan; this is corroborated by the construction drawings of the bridge. The air voids are characterized qualitatively as "positive" or M-wavelets (Annan 2003). They are also recognizable by the location of the rebar cage in the deck. The air voids were not secured to the bottom of the deck; when the concrete was poured, their buoyancy pushed up the upper surface of the rebar cage. This is why the rebar cage appears to sag between the connection and the air voids located in the main span and in the southeast leg (figure 5). This location for the air voids is also provided by the attribute analysis.
Qualitatively, the bottom surface of the deck is also recognizable in the processed transects as another air-concrete interface, for those transects less than 40cm deep (figure 6). This horizon occurs at is a different time for each transect of the deck because of the variable depth of the cross section (figure 4). The reflections caused by the post-tension tendons are less obvious in most of the transects, especially at later times or deeper in the section. This is due to a combination of factors. First, the tendons are much smaller than the air voids and are therefore more difficult to resolve in deeper areas. Second, the tendons are grouted into place, causing a series of material interfaces rather than the "sharp" interface of the air voids. However, the characteristic draped

profile of the tendons over the column locations aids in their identification (figure 6). Finally, for the same reason that the tendons are more difficult to see deeper in the section, the details of the southeast leg connection to the main span are still unclear. The most obvious detail is the overlapping/interlocking of the upper rebar cage to ensure continuous structural behaviour. The vertical interface between the two concretes, which was likely grouted, may be present (figure 6).

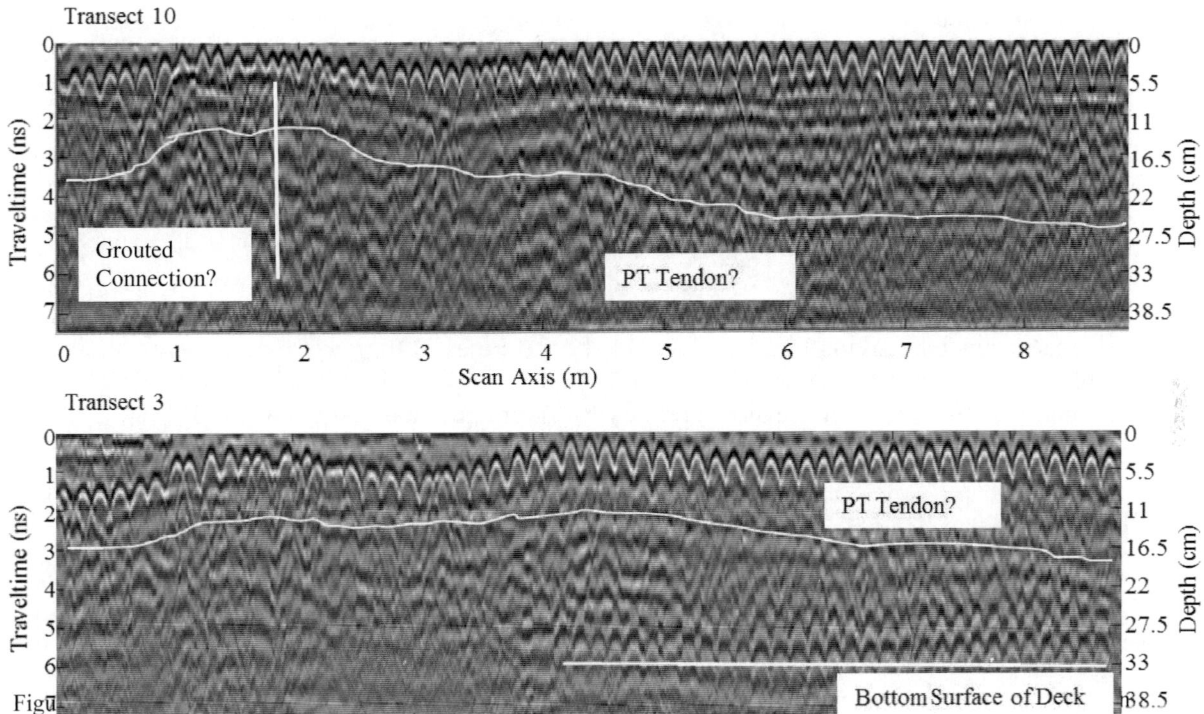

Figu... the latter. The post-tension tendons can be seen near the top of section between 1-3 m and deeper in the section towards 8m. The depth of the deck at transect 10 is at its maximum, so it is not visible.

4.2 Attribute Analysis

Attribute analysis of transects containing primarily reflections of air voids, post-tension tendons, or neither was performed to quantitatively compare the GPR reflections of each feature. For all five attributes computed (amplitude based reflection and propagation attributes), the transects containing air voids had the highest values (Table 1). The values of each attribute are derived directly from the amplitudes of the traces. Since the air-concrete interface has the highest contrast, and both the top and bottom of the air void cause a reflection, this result confirms the use of amplitude attributes to classify reflection strength. Conversely, the transects containing post-tension tendons or no features are separated by as little as 1% (average maximum amplitude of trace) and as much as 15% (average area of trace). These attributes are therefore a simple and direct way to identify the locations of the air voids in this structure by their strong reflections, or other composite reinforced structures with a variety of constituent materials. This method is more straightforward than visual interpretation and is not impeded by the significant rebar reflections.

Table I: Average trace attributes presented for a sample transect containing either no internal feature, a post-tension tendon, or and air void. The units are arbitrary.

	Maximum	Absolute	Total	Area	Absolute

		Maximum	Energy		Area
No feature	1.54 e5	2.29 e5	1.00 e12	-1.68 e6	1.04 e7
Air Void	2.07 e5	2.75 e5	1.59 e12	-1.31 e6	1.23 e7
Tendon	1.56 e5	2.30 e5	9.49 e11	-1.98 e6	1.06 e7

5. Conclusions

GPR is an excellent tool for structural investigations, providing useful information about everything from contaminant spread to crack localization. In this work, a high frequency GPR survey of a pedestrian bridge was conducted to investigate the internal features of Streicker Bridge, including air voids and post tension tendons. GPR has proven to be a useful and complimentary method of validation and investigation for SHM monitoring. This work provides excellent comparison with the construction drawings of the bridge based on visual interpretation of the reflection patterns from different interfaces. Additionally, the reflections interfaces can be quantified using amplitude attributes to quantitatively compare the reflections present in the data from post-tension tendons, air voids, the rebar cage, or even the lower surface of the deck in some cases. The air-concrete interface reflects the GPR signal more strongly than the interface at the post-tension tendons, giving higher values for the attributes in transects where there are air voids present. This information can be used to validate visual interpretation, identify anomalies in construction, and provide the details necessary for future decision making and SHM of a structure. The details of the connection between the two construction phases was impeded by the reflection hyperbolas from the rebar cage blocking the reflections of the features below. Migrating the hyperbolas would likely provide some additional insight into the details of the connection between the southeast leg and the main span. Specifically, this survey can be used to monitor corrosion risks by tracking the depth of concrete cover protecting the rebar, developing more accurate structural models, and provide a basis for tracking the continued behaviour of the connection between the two construction phases of Streicker Bridge.

References

Hubbell, D & Glisic, B 2013, 'Detection and Characterization of Early-Age Thermal Cracks in High-Performance Concrete,' *ACI Materials Journal,* vol. 110, no. 3.

Abdel-Jaber, H. & Glisic, B 2015, "Monitoring of pre-release cracks in prestressed concrete using fiber optic sensors," *Proceedings of SPIE 9437, Structural Health Monitoring and Inspection of Advanced Materials, Aerospace, and Civil Infrastructure,* San Diego, USA, p. 943720.

Abdel-Jaber H & Glisic, B 2014, "A method for the on-site determination of prestressing forces using long-gauge fiber optic strain sensors," *Smart Materials and Structures*, vol. 23, no. 7, p. 75004.

Reilly, J, Abdel-Jaber, H, Yarnold,M. & Glisic, B 2017, "Evaluating the coefficient of thermal expansion using time periods of minimal thermal gradient for a temperature driven structural health monitoring," *Proceedings of SPIE 10169, Nondestructive Characterization and Monitoring of Advanced Materials, Aerospace, and Civil Infrastructure,* Portland, USA, p. 1016929.

Sigurdardottir, D.H. & Glisic, B. 2015, "On-site validation of fiber-optic methods for structural health monitoring: Streicker Bridge," *Journal of Civil Structural Health Monitoring,* vol. 5, no. 4, pp. 529–549.

Böniger U & Tronicke, J 2011, "Subsurface utility extraction and characterization: Combining GPR symmetry and polarization attributes," *IEEE Transactions on Geoscience and Remote Sensing*, vol. 50, no. 3, pp. 736–746.

Sensors & Software, 2015, *Concrete Scanning with GPR Guidebook*, Sensors & Software, Mississauga.

Annan, AP 2003, *Ground Penetrating Radar Principles, Procedures, and Applications,* Sensors & Software , Mississauga.

Zhao, W, Forte, E, Pipan, M, & Tian, G 2013, "Ground Penetrating Radar (GPR) attribute analysis for archaeological prospection," *Journal of Applied Geophysics*, vol. 97, pp. 107–117.

Conyers, LB 2016, "Ground-penetrating radar mapping using multiple processing and interpretation methods," *Remote Sensing*, vol. 8,

Jol, H 2009, *Ground Penetrating Radar Theory and Applications*. Amsterdam: Elsevier.

Barnes, C, Trottier, J-F & Forgeron, D 2008, "Improved concrete bridge deck evaluation using GPR by accounting for signal depth-amplitude effects," *NDT&E International.*, vol. 41, pp. 427–433.

Pedret Rodes, J, Perez-Gracia, V, & Martinez-Reguero, A 2015, "Evaluation of the GPR frequency spectra in asphalt pavement assessment," *Construction and Building Materials*, vol. 96, pp. 181–188.

Olhoeft, GR 2000, "Maximizing the information return from ground penetrating radar," *Journal of Applied Geophysics*, vol. 43, pp. 175–187.

White, J, Hurlebaus, S, Shokouhi, P, Wittwer A & Wimsatt, A 2014, "Noncontact techniques for monitoring of tunnel linings," *Structural Monitoring and Maintenance,* vol. 1, no. 2, pp.197-211.

Smith, SS, Olhoeft, GR 2000, "Automatic processing and modeling of GPR data for pavement," *Proceedings of the Eighth International Conference on Ground Penetrating Radar,* Gold Coast, Australia, vol. 4048, pp.188-193.

dos Santos, VRN, Al-Nuaimy, W, Porsani, JL, Hirata, NST, & Alzubi, HS 2014, "Spectral analysis of ground penetrating radar signals in concrete, metallic and plastic targets," *Journal of Applied Geophysics,* vol. 100, pp. 32–43.

Doğan, M. & Turhan-Sayan, G 2016, "Preprocessing of A-scan GPR data based on energy features," in *Detection and Sensing of Mines, Explosive Objects, and Obscured Targets XXI, SPIE*, **city, country**, vol. 9823, p. 98231E1-98231E8.

Burton, BL, Olhoeft, GL, & Powers, MH 2004, "Frequency spectral analysis of GPR data over a crude oil spill," *Proceedings of the Tenth International Conference on Ground Penetrating Radar, 21-24 June, 2004*, Delft, the Netherlands.

Allroggen N, & Tronicke, J 2016, "Attribute-based analysis of time-lapse ground-penetrating radar data," *Geophysics*, vol. 81, no. 1.

Morris, IM, Abdel-Jaber, H & Glisic, B 2017, "Ground penetrating radar for material characterization in structural health monitoring applications" *Proceedings of 11th International Workshop on Structural Health Monitoring, IWSHM,* Palo Alto, CA, (in review).

The 8th International Conference on
Structural Health Monitoring of Intelligent Infrastructure
Brisbane, Australia | 5-8 December 2017

Hidden Defects In Bridges – their identification, management and prevention

T.Pape[1], P. Burnton[2], J. Collins[3], A. Ghose[4], J. Webb[5], D. Ashurst[3], C. Christodoulou[5], P.Sparkes[5]
[1] *AECOM - Australia, Email:* torill.pape@aecom.com [2] *Arup – Australia, Email:* *peter.burnton@arup.com* ,
[3]*Arup – United Kingdom,* [4]*Waterman– United Kingdom,* [5]*AECOM – United Kingdom*

Abstract

In recent decades, significant effort has been made to improve asset management processes of bridges, of which a core requirement includes the collation and review of component condition information. This is typically carried out through visual inspections, where a competent inspector will sight visible bridge components and make an assessment on their condition (usually referencing guidelines provided in standard inspection manuals). This information is central to enabling asset owners to determine the impacts of condition on the performance of the bridge, which in turn informs maintenance and capital works programs, as well as risk management strategies to maintain the ongoing safe serviceability of the structure. However, the success of the process is, among other items, very sensitive to the quality of the inspections. If an inspection does not include all components because, for example, they are hidden and not easily examinable, the adverse effects can be significant or in the worst cases catastrophic.

Recent catastrophic and unexpected failures due to hidden defects (such as Stewarton rail bridge, Ayrshire, De la Concorde bridge in Canada and more recently the Italian concrete bridge near Lecco) prompted members of the Bridge Owner's Forum in the UK to develop a guide that will recommend good practice for inspecting, identifying and maintaining bridges with hidden components. Commissioned by CIRIA and jointly facilitated by Arup and AECOM, this work is complete and the CIRIA Guide C764 was published in June 2017. The current paper provides a summary of findings from the guide, with highlights on aspects of defects in hidden components, proposed risk assessment techniques for management, and recommendations for the development of more targeted inspection and maintenance regimes (that may include health monitoring techniques) to critical areas. Some Australian case studies and applications are also discussed.

1 Introduction

Bridge owners, operators and engineers including designers, assessors and maintainers all play key roles in bridge management. A core component of a bridge management programme is on-site examinations by a competent bridge inspector. These inspections are central to recording bridge condition and are used by bridge owners to ensure defects are identified and rectified within reasonable timeframes as well as identifying larger maintenance works. When coupled with a refined numerical assessment of capacity, the risk posed by a bridge and / or individual components to threatening safety can be better defined. Generally, this works well in Australia: there are few serious failures. However, amongst other items the success of the process is sensitive to the quality of the access available during the inspection. If an inspection does not include all elements since, for example they are hidden and not easily examinable, the adverse effects can be significant or in the worst cases catastrophic.

High profile cases in recent years such as at Stewarton, de la Concorde, Hammersmith and Boston Manor have bought the importance of hidden defects in bridge components to the forefront of bridge owners' focus. Prompted by these and other cases, the Bridge Owner Forum (BOF) identified the need to define good practice when inspecting, identifying and maintaining hidden bridge components. BOF approached the publisher CIRIA to produce a new guidance document covering these areas. In

December 2014, CIRIA procured the joint services of AECOM and Arup to write this report, recently published as CIRIA Guide C764 "Hidden Defects in Bridges – Guidance for Detection and Management" (Collins et al, 2017). Guidance is given to:

- Define the component that may contain a defect. Common defect types are described.
- Inspection guidance on
- Identification and maintenance
- Design guidance.

This paper provides a brief summary of guidance provided in C764. Whilst this guide was written for a UK and Ireland audience, this paper has been expanded to provide an Australian context to this issue. The paper discusses what a hidden defect is, some examples where hidden defects have been instrumental in catastrophic structural failures, an overview of current practices in identifying and managing such defects and components, and then outlining potential solutions in how to manage bridges with hidden components.

2 What is a Hidden Defect?

In a bridge, a hidden component is an element that would not usually be visually inspected as part of a bridge inspection i.e. to within touching distance using normal inspection techniques such as visual techniques and hammer tapping. The CIRIA report goes further than physically obscured bridge areas by stating that shortcomings in inspection quality may lead to a component being to extent hidden. There may be very good reasons for this. For example, an element may be inaccessible at the time of inspection e.g. water levels preventing access to a bridge pier. Whilst bridge inspectors generally do an excellent job, unfortunately poor practices such as inaccurate recording and low levels of detail are regularly found.

It is also important to consider untypical hidden defects and untypical hidden components. This may include unusual or bespoke detailing, accidental inclusion of temporary works, or any form or structure (intentional or otherwise) that may lead to a detrimental condition (such as water retention), and thus a compromise in the structural integrity of a component or the overall structure itself. A hidden component can also be a one that no one knows exists, but may cause a defect that is detected by non-visual means.

The CIRIA guidance goes into considerable detail on all the commonly encountered bridge types in the UK and Ireland, and their possible hidden defects. Areas covered are summarised in Table 1. Guidance is given to:

- Define the component that may contain a defect. Common defect types are described.
- Inspection guidance
- Identification and maintenance
- Design guidance

Table 1 Summary of bridge types / elements and typical hidden areas and defects that may be contained therein

Bridge type / element	Hidden components	Defects
Iron / steel bridges	Components under ballast / surfacing Protected steel surfaces Insides of boxes Members behind non-structural elements Abutting structures Deck / buckle plates, jack arches Suspension bridge main cables & hangers Cable stays Components deemed too difficult to inspect / missed	Metallurgical Section loss from corrosion Latent weld defects Fatigue cracking Brittle fracture Bolting and rivets

Bridge type / element	Hidden components	Defects
Concrete bridges	Within the concrete body Reinforcement Prestressing wires / strands and anchorages Voided and cellular structures Half-joints Obscured surfaces Concrete hinges Temporary Works	Alkali-silica reaction Sulphate attack Presence of chlorides Carbonisation Cracking Reinforcement / prestressing corrosion Missing / inadequately fixed reinforcement / prestressing
Masonry arch bridges	Arch rings Spandrels Masonry piers	Hollowness Mortar loss Masonry deterioration Cracking
Timber	Timber interior Connections	Biological attack Splitting
Bearings and expansion joints	Poor access Uninspectable details Inspection at the "wrong" time	Metallurgical and corrosion Bearing seizure Sliding surface deterioration Elastomeric deterioration Bearing installation errors Other bearing defects Expansion joint hidden defects
Substructure	Piles Abutment walls Retaining walls Internal voids/chambers Cladding fixtures over components Poor/no access (e.g. piles in water) Un-inspectable details	Corrosion Cracking Crushing Chemical Attack Scour
Hidden durability components	Drainage, waterproofing, paint, galvanising, sealants, cladding	
Hidden safety components	Parapets, surfacing, access / walkways / gantries	
Other bridge components	Moveable bridges	
Ancillary components	Services, bays and ducts	

3 Examples of Failures due to Hidden Defects

In collating the case studies for the CIRIA report and on presentation to various audiences, it has become apparent that this document may have an important secondary function. As a collection of interesting and often difficult works to existing bridges, the case studies provide useful background reading for bridge professionals and those wishing to understand the daily challenges faced by the industry. A short summary of five case studies is given below, two of which are Australian examples, and the rest are those that became central to forming the new guidance of the CIRIA report.

3.1 Examples from United Kingdom

3.1.1 *Stewarton Bridge, Ayrshire, UK*

At 06:12 on 27th January 2009 in Stewarton, Ayrshire, a train carrying a mixed consignment of gas oil, diesel and kerosene derailed as it crossed a half through steel underbridge (Figure 1). Fuel was leaked into the local environment. There were no fatalities or injuries. Whilst multiple factors contributed to the failure, the underlying cause was failure of a defective hidden component (RAIB, 2010). The ensuing investigation demonstrated a shear failure to two of the three main girders. The web plate thickness had reduced due to heavy corrosion in areas that were buried under ballast. In some locations, complete loss of section had occurred.

Prior to the collapse, concerns had been raised on corrosion to visible areas of the bridge. It was also noted that at defined locations there were areas under ballast that may be prone to severe corrosion and lamination leading to loss of section. Following this urgent defect report, the bridge owner inspected the main girder to assess the issues raised, suggest actions and recommendations. The full extent of the corrosion was not identified in this inspection and no recommendations were made to repair the defect.

(a) Overview of collapse (b) heavy corrosion to the buried web

Figure 1: Stewarton Rail Bridge

3.1.2 *Hammersmith Flyover, London*

The M4 Hammersmith flyover had an inspection of hidden post-tensioning cables undertaken in 2011. To the surprise of the inspectors, the condition of the cables was so poor that an emergency closure and strengthening works were undertaken on one of London's busiest roads (Figure 2).

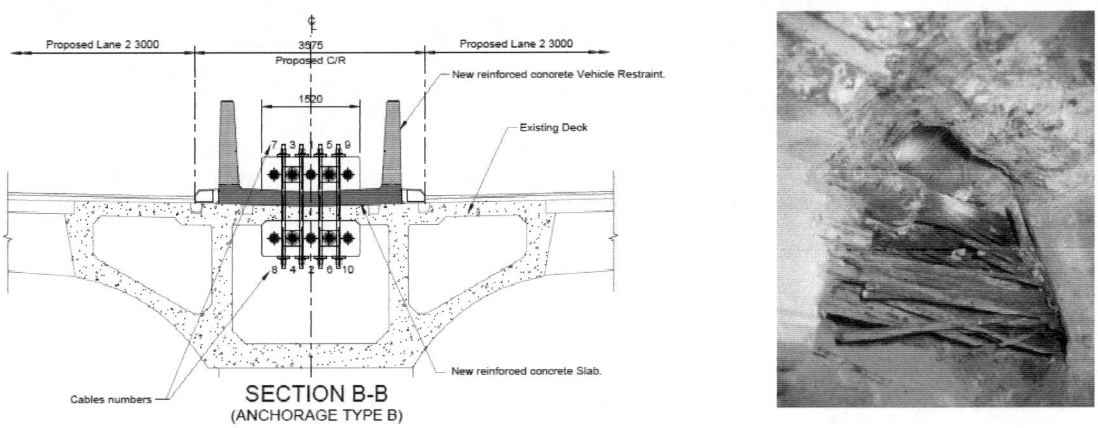

(a) temporary strengthening works by external post-tensioning (b) corroded tendons

Figure 2: Hammersmith Flyover bridge.

A feature of the flyover (constructed 1963) was that in every one of the 16 spans, temporary works that should have been removed had not been, and had been a major cause of the deterioration of many of the tendons in the viaduct. The temporary works were mainly void fillers or void seals used with shuttering, in materials that would absorb and hold water for the 50 years and more since 1963, thus leading to the accelerated corrosion. However, being temporary works they had not been recorded in as-built records, and many were hidden by being cast into concrete and grout. Only when the emerging visibility of corrosion became visible through the concrete did the identification of the tendon corrosion occur. Repair costs were in the order of over £100m.

3.1.3 *Boston Manor, West London*

The 1 km Boston Manor viaduct is in West London (Wearne, 2014) (Nicholls, 2015) carrying the dual carriageway of the M4 motorway (Figure 3). Opened in 1965, its girders and trusses have over 600 electroslag butt welds. Electroslag welding is rarely encountered in bridgeworks: Boston Manor is thought to be the only electroslag welded bridge on the UK trunk road network. Following concerns over the method's susceptibility to fatigue and brittle fracture, the method was superseded in the late 1970's.

As late as 2011's principal inspection, no cracks breaking through the paint system had been found. Aware of the historic issues of electroslag welding, the Highways Agency (bridge owner) commissioned a numerical assessment of the welds. No issues were found. To verify the assessment, non-destructive testing (NDT) and material samples were taken. The rapid development of events was remarkable:

- Mid March 2012, start of NDT inspection.
- 22 March 2012, first crack in electroslag weld identified.
- 23 March 2012, further cracks found and 17.5t emergency weight restriction imposed.
- 28 March 2012, increased incidence of cracks found, 7.5t emergency weight restriction imposed.

In total, 111 cracks were found in electro-slag welds. It was stated that the cracks would require remediating within a maximum of 250 weeks to avoid critical failures. A summer embargo on structural works was approaching around the London 2012 Olympic and Paralympic Games, during which all weight restrictions required removing. Work to remove cracks by grinding / burring started immediately.

(a) Overview of Viaduct (b) crack in electroslag weld

Figure 3: Boston Manor Viaduct.

The deepest and longest crack on the bridge was identified in a truss section of the bridge. During the final part of its grinding on 30 June 2012, two weeks before the Olympic embargo and lifting of the weight restriction, a small inclusion in the weld material was uncovered. Hourly monitoring showed the crack to be propagating under fatigue cycles. Further attempts to remove the crack by burring lead to continued propagation: the crack was being "chased". Just seven days before the embargo, the length of the crack and its propagation rate posed such risks to the safety of traffic that the road was closed to all traffic: brittle fracture was a real possibility. The crack was finally removed by burring under monitoring by thermal image detection. The bridge reopened 13 July 2012 to 7.5 tonne traffic.

3.2 Examples from Australia

3.2.1 *Sorell Causeway Bridge, Tasmania*

The Sorell Causeway Bridge was located west of Hobart in Tasmania's south in an aggressive marine environment (Pape and Melchers, 2013). It was a reinforced and post-tensioned concrete bridge of 34 spans (totalling 1.08 km), and it was the longest bridge of its kind when completed in 1955. Evidence of corrosion initiation in the piles, headstocks and the girders were identified as early as 1975, with cracking, spalling and rust staining in some components. The structure continued to deteriorate, and concerns regarding its integrity peaked in the mid to late 1990's, with numerous inspections, investigations and testing initiated in an attempt to quantify the performance of the structure. Results were inconclusive and based on an assessment on network and public safety risks as well as the route criticality, the bridge was eventually decommissioned and replaced in 2002.

A detailed post-demolition examination of several girders identified that the corrosion was much more significant than first thought (Figure 4), which was reflected in the poor performance and catastrophic failure of the girders during load testing. Severe pitting in the post-tensioning strands had occurred in areas that were not previously identified or offered no external visual evidence of deterioration. Conversely, zones of intact steel free of corrosion was observed in other areas despite highly elevated chloride levels. Non-destructive testing was useful in some instances, but careful consideration of results and trends was required to provide a more accurate representation of the corrosion deterioration.

(a) Overview of Cracking following post-tensioning ducts

(b) overview of corrosion in wires

Figure 4: Sorell Causeway Bridge.

3.2.2 *Ipswich Bus Station*

The bus station is partly set below ground level with secant pile walls in an arched form retaining the fill and roads as well as supporting the structural deck above the bus station. An unreinforced block wall acted as an architectural facing in front of the piled retaining wall. Inspectors noted that the block wall was often damp in one length of the wall. This observation was nearly a decade after the station had opened. Further investigations revealed that soil had been washed into the space between the piles and blockwall by groundwater. This in turn had caused a stormwater pipe to rupture and increase the water flow and hence rate at which soil was lost below the road. The block wall was acting as a retaining structure for which it was not designed, its curved shape providing the strength to prevent collapse. A large hole had formed below the adjacent road – the source of the material now in the void between pile and blockwork. Two potentially catastrophic hidden defects were identified; overloading and potential collapse of the block wall and formation of a large sinkhole in the road above (see Figure 5).

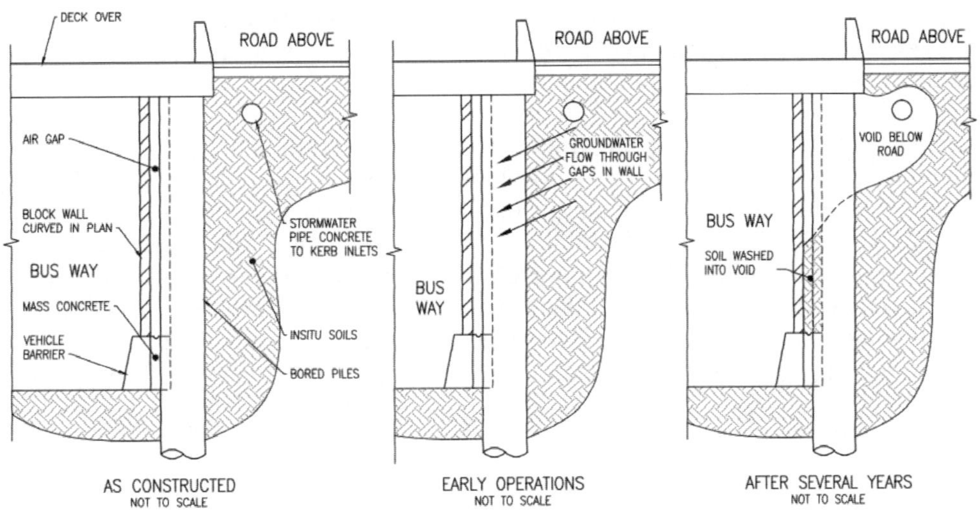

Figure 5: Formation of void between piles and blockwork at the Ipswich Bus Station.

Similar hidden sinkholes have been known to form in several situations including below the curved spill through surface protection at bridge abutments, at joints in stormwater culverts, and behind retaining walls, including bridge abutments, adjacent to rivers or stormwater channels. Cladding panels are also known to complicate inspections and conceal deterioration of the structure behind the cladding

4 Existing current method of managing hidden defects

Responsibility for the management of bridge infrastructure rests with many different organisations and authorities worldwide. The management of their infrastructure is typically governed by a series of documents, guidelines, standards, specifications or technical notes. In some cases where organisations are small, have limited resources or are responsible for a small number of bridges in their network, these are often managed implicitly or on an individual basis. The following sections provide an overview of current bridge asset management practice in the UK and Australia, particularly in light of where components are hidden.

4.1 UK

Roads

Typical bridge management practice for trunk roads is defined in the Design Manual for Roads and Bridges (e.g. BD 63 (Highways Agency, 2007)). For local roads, the document "Well Managed Highway Infrastructure – A Code of Practice" (UKRLG, 2016) is preferred. For the purposes of identification and mitigation of hidden defects, the approaches outlined in the two documents is similar. Bridge condition is recorded from a regime of inspections including general, principal, and special inspections. Inspections provide key information for all maintenance interventions, including decisions for monitoring or testing.

These guidance document currently do not explicitly detail a regime for uncovering and inspecting hidden components for general or principal inspections. Typically, a hidden component would only be uncovered as part of a special inspection if either: concern had arisen from a principal inspection as to the condition of a hidden element (rust staining, evidence of local failure etc.); a regime of special inspections and detailed assessment is instigated, typically following a collapse or concern of a common detail (Highways Agency, 2004, 2009a and 2015). Examples of this latter case include half-joints (due to the de la Concorde collapse), deck hinges and post-tensioned bridges (based on conditional issues identified in a number of older post-tensioned bridges in the 1980's (Woodward, 2001)).

It is remarked that highway bridge management's techniques are reactive in nature: a visible possible symptom is reserved or known issue identified prior to commencing an active search for defects.

Rail

The owner and maintainer of bridges on the British heavy rail network, Network Rail, operates inspections similar to the highways' authorities' inspections, conducting routine and detailed examinations of varying frequency and using a risk-based approach. Under this regime, it would not be usual practice to expose hidden components. Following the collapse of the Stewarton Bridge in 2009, Network Rail reviewed their inspection approach to incorporate hidden critical elements (HCE). This interesting, proactive work is detailed in the CIRIA guide and has resulted in intrusive inspection to approximately 4000 bridge spans. This has led to a programme of works, including bridge replacement and strengthening (Figure 6). HCE examination has now entered the regular cycle of bridge inspections.

Figure 6: Extensive corrosion in ballast retention plates

(Whilst ballast retention plates had been provided at Southend High Street, they were hidden. During a HCE inspection lids were removed exposing extensive corrosion. Here, maintenance painting appears to have been carried out only to the top of ballast plate level. This bridge was replaced.)

4.2 Australia

In Australia, a range of asset management documentation relating to bridges exists. From a high level perspective, Austroads provides several documents that presents collated information and best-practice summaries of bridge management practices in Australia. These guidance documents include:

- Guidelines for Bridge Management – Structure Information (AP-R252-04)
- Guide to Asset Management (currently being updated)
- Guide to Bridge Technology (currently being updated)

All state road authorities have formal documentation that provide high-level asset management guidance for their infrastructure. In relation to bridges, various documents of varying levels of sophistication and depth exist to provide more specific guidance. Relating to the inspection of bridges, again most state road authorities have developed their own bridge or structure inspection manuals which contain advice on the inspection, condition rating and maintenance requirements of different types of bridges and materials. These generally contain similar guidance and steps to those documented in UK bridge management manuals. Minimal guidance has been published for rail authorities.

However it should be noted that only a handful of manuals specifically mention hidden components, and a lesser number to provide guidance for their management. With the exception of the Queensland Department of Transport and Main Roads (TMR), there are no publically available guidelines that provide risk management strategies for bridge managers/owners to identify, inspect and manage structure-critical components that are hidden or inaccessible.

In their "Level 3 Inspection Criteria for Potentially Structurally Deficient Bridges" (TMR, 2016), TMR identifies the "possibility of hidden distress" as a factor to be considered when assessing a bridge for an immediate risk to public safety. A specific example is provided in relation to a lightly reinforced component as being potentially structurally deficient in shear, where "shear cracking may not be visible on the surface" [p10]. On inspection reports, the inspector must make mention of any areas not able to be inspected due to their obscurement or inaccessibility.

VicRoads provides some references to hidden deterioration in their Structure Inspection Manual for some specific applications (VicRoads, 2014). For example, deep I-beams that are in contact with footway soil backfill may have the "potential for severe corrosion to the hidden surface" [p77]. Pot bearings are also mentioned because they "may be hidden by stainless steel skirts" [p88]. VicRoads also makes provisions for the identification of components that are hidden in their inspection reporting .

Transport for NSW through the Assets Standards Authority has released a Technical Note TN 072 (TfNSW, 2014) that more specifically identifies hidden structures, used by both road and rail authorities. They define hidden structures as "structures or components of a structure that are obscured in such a way that they are not readily visible" and state that "the examination of hidden structures is important to assure ongoing structural integrity" [p1]. It provides a list of examples of hidden structures which are listed below, and also provides a list of pedestrian subway structures that fall into this category.

- superstructure beams hidden by false ceilings
- columns hidden by architectural coverings
- bridge piers, parapets or abutment walls obscured by advertising panels

The guideline provides recommendations for the inspection of hidden components, what key defects the inspector should be aware of and reporting recommendations and responsibilities. There is also mention of incorporating the maintenance requirements of a hidden structure in a Technical Maintenance Plan, however exact guidance is not provided.

Advice provided in Structures Examination Engineering Manual (RailCorp, 2009) relates to the identification of defects that may be indicators of hidden components deterioration, such as "stain trails indicating hidden corrosion or working members".

From a code perspective, Part 8 of the newly published Australian Standard AS 5100 – Bridge Design is the only section that mentions the use techniques to inspect hidden areas in certain applications, such as timber bridge drilling or Ground Penetrating Radar for void detection. However guidance is minimal and provides no broader recommendations.

In summary, there is currently minimal published guidance for bridge asset owners or managers to assist in identifying and developing strategies to manage hidden components and their defects. Risk management strategies dealing with items such as half-joints, hinges, post-tensioning cables or other special or low-redundancy hidden components are generally dealt with on a unique bridge or issue-specific case. Consideration should be given to this area, particularly relating to structural capacity.

5 Proposed method of managing hidden defects for existing structures

The CIRIA report provides detailed guidance, largely based on the approach taken by Network Rail post-Stewarton, on how a bridge owner / operator / maintainer can manage the risk posed by hidden defects to their portfolio of bridges. It is representative of a proactive approach and highlights three key stages, summarised in Figure 7.

5.1 Understanding risk features

It is anticipated that the vast majority of bridge owners already have a bridge management system from which the task of identifying at-risk structures can commence. This is likely to include considerations towards structure redundancy, ductility environmental conditions, bridge loading history and the impact of condition on the structural capacity of the bridge.

Each bridge portfolio managed will be distinct, though there may be common features that can streamline management. Following identification of all bridges, hidden components should be grouped. In general terms, the criticality of these components can be ranked to assist in prioritising works.

A key factor in understanding risks of this nature is to collate and review as much historic information about a structure as possible. This includes past inspection reports, drawings, maintenance and refurbishment activities and any other information relating to the management or operation of the bridge.

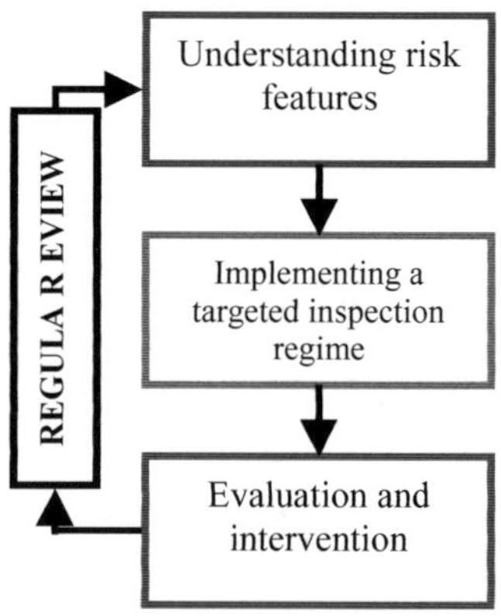

Figure 7: Simplified flow chart of suggested process for managing hidden defects

5.2 Implementing a targeted inspection regime

Structure-by-structure, hidden components are identified by desk study. The inspector and asset manager should remain inquisitive, and ask not only what can be seen, but also what remains unseen or hidden. An appropriate inspection strategy can then be carried out, which may require intrusive investigations. This stage may identify further hidden components.

5.3 Evaluation and intervention

As in any bridge management regime, high quality inspections are only of value if they are accurately interpreted and remedial works enacted upon. This stage of bridge management is generally well recorded in existing guidance. Accurate interpretation may include consideration of how the identified condition of the hidden element effects bridge capacity: numerical assessment may be required. Then, if unacceptable risk remains, remedial works would follow that may include:

- Monitoring
- Load restrictions
- Further numerical assessment
- Component replacement
- Alternative load path provision
- Strengthening
- Bridge replacement
- Updating of emergency plans

5.4 Regular review

Risk posed by a hidden defect may change over time due to:

- Condition degradation
- Load changes
- Changes in load path
- Developments in structural understanding
- Cultural change in risk perception

It is hence vital that once the hidden defects programme is set up, it is regularly reviewed. This should include incorporation of intrusive works to hidden areas as part of the established inspection regime.

5.5 Design

The above discussion outlines strategies for existing bridges. For new structures risks associated with hidden components and their defects can be mitigated at the design phase. This requires the designer to review the overall structure for areas where components may become hidden or difficult to inspect, understand the criticality of the hidden components, the processes that may result in defects at these hidden components and the opportunity to eliminate these hidden components. The review should also consider moisture traps or other mechanisms that may increase deterioration of a component.

The ultimate aim is to eliminate such components or issues, however this is not always be possible, for example piles will be in the ground and reinforcement embedded in concrete. Therefore, the designer should aim to provide a considered approach to mitigation of this risk. Other options is to allow for long-term monitoring strategies to be installed to either monitor the bridge (such as structural health monitoring, see Section 6 for further discussion) or to provide on-going mechanical support to eliminate the risk (e.g. dehumidification in post-tensioning applications or cathodic protection applications).

Beyond the specific guidance given in this section, the designer of a new bridge looking to avoid critical defects in hidden deck components is directed to Hendy and Iles (2015) and Soubry (2001)

6 The Future – Structural Health Monitoring Systems

Other engineering disciplines have demonstrated the technical and commercial advantages bought about by extensive monitoring and data analysis afforded by technological advances in database systems. Rolls-Royce's engine health management is an excellent case study (Waters, 2009). It is felt that the nascent application on bridges as structural health monitoring systems (SHMS) has much potential in the detection of hidden defects. This is true both for installation on existing bridges (Anderson, 2012) and, as a preventative action, on new bridges. Stonecutters Bridge, Hong Kong, was perhaps the first new bridge to have an exhaustive SHMS installed (Figure 8).

A number of infrastructure authorities and research agencies have also explored the use of SHM in monitoring existing structures for performance in Australia (e.g. Aflatooni et al, 2013, Aflatooni et al, 2015, Chan & Thambiratnam 2011, Nguyen et al, 2015), but these have typically been for shorter term applications. Ongoing practical research in this area, particularly with the aim of identifying and monitoring components with hidden defects is recommended for Australian applications.

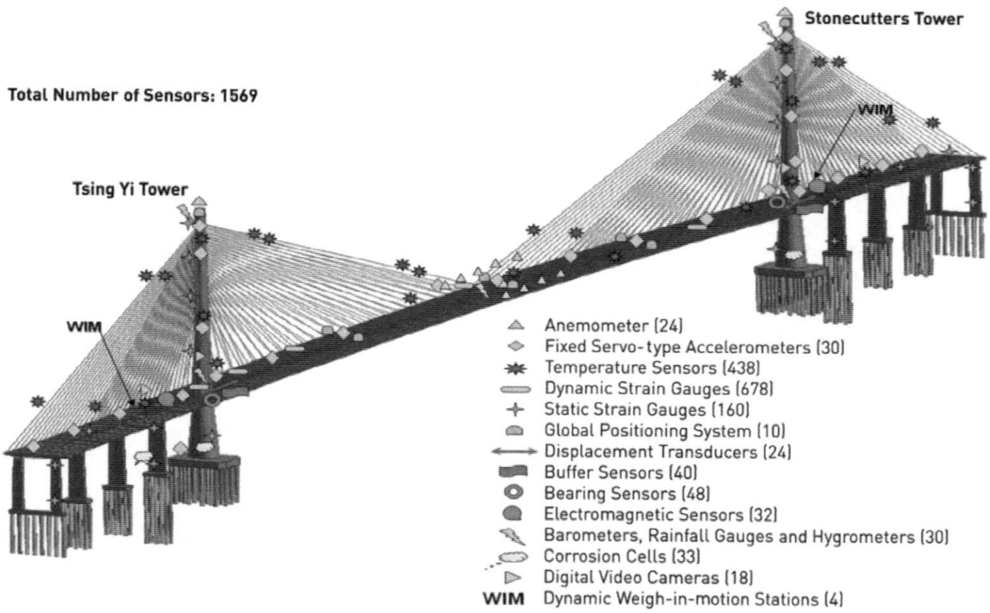

Figure 8: SHMS on Stonecutters Bridge, Hong Kong.
Image courtesy of Highways Department, Hong Kong SAR

7 Conclusions

Defective bridges may put the travelling public, those working on the bridge and third parties at risk. Where a defect is not known and if applicable, appropriate monitoring is not undertaken, the associated risk cannot be managed or mitigated appropriately. In severe cases this can lead to severe network disruptions and at worst, fatalities, as observed for the collapse of the Stewarton rail Bridge in Scotland in 2009. A brief review of current bridge management publications in the UK and Australia highlights the lack of guidance on managing bridges with hidden components.

Based on these failures and lack of guidance, CIRIA published C764 "Hidden Defects in Bridges – Guidance for Detection and Management", which provides considerable detail on all the common bridge types and their possible hidden defects. Guidance is given to:

- Define the component that may contain a defect. Common defect types are described.
- Inspection guidance
- Identification and maintenance
- Design guidance

Similarly in Australia, such manuals and guidance have not been developed to account for uncovering, inspecting and managing hidden components and subsequent defects. With aging bridge stock ever more heavily used, effective management of hidden bridge components is going to increase in importance unless extensive actions are taken. Bridge owners, managers and inspectors are urged to consider this most important subject in their day-to-day work.

Further Information

The CIRIA report was published in June 2017. The report authors would welcome any further discussion on this paper or the report. Please contact Torill Pape (torill.pape@aecom.com) or Peter Burnton (peter.burnton@arup.com).

Acknowledgements

C764 was funded by Freyssinet, Welsh Government, ICE, Transport Scotland, Highways England, Transport Infrastructure Ireland, Network Rail, TransportNI, Environment Agency, Ramboll, Mouchel, Strainstall, Arup and AECOM. Report contents have been shaped by the project steering group including members from the sponsorship organisations. The Sorell Causeway case study is based on research carried out on behalf of the Department of State Growth, Tasmania (formerly the Department of Infrastructure, Energy & Resources), whose funding is gratefully acknowledged.

References

Aflatooni, M., Chan, T.H., Thambiratnam, D.P. and Thilakarathna, I., 2013. Synthetic rating system for railway bridge management. Journal of Civil Structural Health Monitoring, 3(2), pp.81-91.

Aflatooni, M., Chan, T.H. and Thambiratnam, D.P., 2015. Condition monitoring and rating of bridge components in a rail or road network by using SHM systems within SRP. Structural Monitoring and Maintenance, 2(3), pp.199-211.

Anderson, M, Gale, J, St Leger, J, 2012, "The role of structural monitoring in the management of bridges", Structural Faults + Repair 2012 proceedings, Edinburgh

Austroads, 2004, "Guidelines for Bridge Management: Structure Information", Sydney (AP-252/04).

Austroads, (in publication), "Guide to Asset Management", Sydney

Austroads, 2009, "Guide to Bridge Technology", Sydney

Chan, T.H. and Thambiratnam, D.P., 2011. Structural health monitoring in Australia. Nova Science Publishers.

Collins, J, Ashurst, D, Webb, J, Sparkes, P, Ghose, A, 2017, "Hidden defects in bridges - Guidance for detection and management", CIRIA, London

Hendy, C.R. and Iles, D.C. (eds), 2015, Guidance notes on best practice in steel bridge construction, P185/5, sixth issue, Steel Bridge Group, The Steel Construction Institute, Ascot, UK (ISBN: 978-1-85942-196-3)

Highways Agency, 2007, "Inspection of highway structures", Design manual for roads and bridges. BD 63/07, The Stationary Office, London

Highways Agency, 2012, Interim Advice Note 171/12, Risk Based Principal Inspection Intervals, IAN 171/12 , Highways Agency, London

Highways Agency, 2015, "Management of post-tensioned concrete bridges" , Design manual for roads and bridges. BD 54/15, The Stationary Office, London

Nicholls T, Armstrong C, El-Belbol S, Harrison J, Makinson N, Smith Da, 2015, "Repairs to electro-slab weld defects on M4 Boston Manor viaduct", Proc ICE Bridge Engineering Institution of Civil Engineers, London, Vol 168(3), pp259-272

Nguyen, V.V., Dackermann, U., Li, J., Alamdari, M.M., Mustapha, S., Runcie, P. and Ye, L., 2015, "Damage Identification of a Concrete Arch Beam Based on Frequency Response Functions and Artificial Neural Networks", Electronic Journal Of Structural Engineering, Vol 14, Issue 1

Pape TM, Melchers RE, 2013, Performance of 45-year-old corroded prestressed concrete beams. Proceedings of the Institution of Civil Engineers-Structures and Buildings. Nov; 166(10):547-59.

Queensland Department of Transport and Main Roads (TMR), 2016, Structures Inspection Manual Part 3 - Appendix F: Guidelines for the Management of Sub-Standard and Defective Bridges

RailCorp, 2009, Structures Engineering Manual – Structures Examination, TMC 301, v2.0

RAIB, 2010, Derailment of a freight train near Stewarton, Ayrshire, Report 02/2010, Rail Accident Investigation Branch, Derby, UK

Soubry, M, 2001, Bridge detailing guide, C543, CIRIA, London, UK (ISBN: 978-0-86017-543-8) www.ciria.org

Transport for NSW (TfNSW), 2014, Examination of Hidden Structures, Technical Note TN 072:2014

UKRLG, 2016, Well-Managed Highway Infrastructure - A Code of Practice, The Stationary Office, London

VicRoads, 2014, Road Structures Inspection Manual Part 1 and 2

Waters, N, 2009, "Engine Health Management" Ingenia Issue 39 June 2009, Royal Academy of Engineering, London

Wearne C, Armstrong C, Nichols T, Harrison J, 201, “Collaboration in an emergency: repairing London’s M4 Boston Manor viaduct”, Proc. ICE Civil Engineering, Institution of Civil Engineers, London, Vol 167(1), pp33-39

Woodward, R, 2001 , “Durability of post-tensioned tendons on road bridges in the UK”, Durability of post-tensioning tendons, Workshop 15-16 November 2001, Ghent (Belgium), Technical Report Bulletin 15, fib Lausanne, Switzerland and IABSE Zurich, Switzerland

The 8th International Conference on
Structural Health Monitoring of Intelligent Infrastructure
Brisbane, Australia | 5-8 December 2017

Tilt and seismic response monitoring of a rehabilitated building

D. Murià-Vila[1] J. Camargo[1], G. Rodríguez[1], B. D. Aldama[1].
[1] *Institute of Engineering, Universidad Nacional Autónoma de México (UNAM) – México,*
Email: dmv@pumas.ii.unam.mx

Abstract

The main results of the response of a building instrumented with accelerometers during 18 seismic events, and the analysis of the displacement and verticality obtained with a GPS system between February 2011 and March 2017 are presented. The building consists of a tower of 22 stories and three low-rise structures with a basement in common. From the beginning the tower presented differential settlement and tilting. To control these problems there has been four underpinning and a restructuring of the tower. The aim of the article is to show the structural performance of the building and determine whether rehabilitation has helped to control the loss of verticality.

1. Introduction

This work focuses on a building which is one of the several structures that were damaged during the Michoacán earthquake in 1985 and it was later retrofitted. The building consists of a tower of 22 stories and three structures of two stories founded on soft soil (Figure 1). The four structures have a common basement. The foundation of the low-rise structure was constructed over a 4 m embedded box and the tower over 7 m embedded box partially supported on 156 piles. From the beginning of its construction in 1964, the tower presented differential settlement and tilting and uplifting of the low-rise structures. In order to control these problems, there have been four underpinning in different years and the reinforcement of the tower structure between 2008 and 2009. The last restoration consisted in stiffening the tower with steel diagonals in the T component and the several coupling beams strengthened at theirs ends with a U-type wrapping of one layer of carbon fibre-reinforced polymer strips (L component). At the base of the tower on the south side, expanded in 1987, 30 control piles were added and reinforced concrete walls in transversal component were built between the basement of the tower and the low-rise structure of the north side. Geotechnical studies show that regional settlements in the study area are 7 cm per year (López Acosta and Ávila 2007, Rodríguez *et al.* 2009).

In order to monitor the structural response of the building, a seismic instrumentation was installed and is operational since December 2008; as well as a high-resolution GPS system to measure the relative displacements between the tower roof and the base. The main results of the seismic response of a building instrumented with accelerometers, and the analysis of the displacement and verticality obtained with a GPS system between February 2011 and March 2017 are presented. The aim of the paper is to show the structural performance of the building during seismic events, and determine whether rehabilitation has helped to control the loss of verticality to keep the tower of the building in operative conditions.

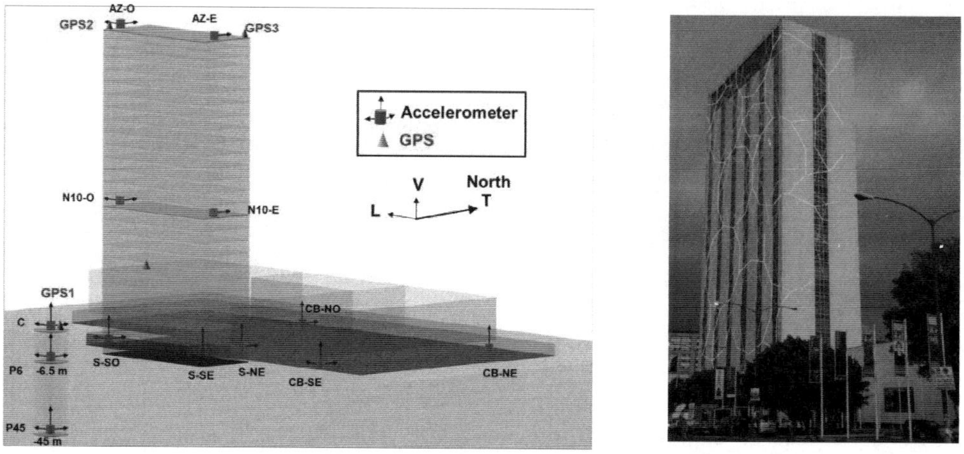

Figure 1. The building and its instrumentation

2. Structural health monitoring system

The building's seismic instrumentation consists of a network of 28 high-resolution uniaxial servo-accelerometers located at different points in the building and surrounding terrain, as shown in Figure 1. The resolution of the monitoring system is suitable for recording both ambient vibration and high intensity earthquakes. With the accelerometer network, 18 seismic events have been recorded and analyzed between February 2011 and March 2017, and some ambient vibration tests have also been recorded. The characteristics of the seismic events recorded, peak ground acceleration (PGA) and the Arias intensity (I_{Arias}) of the L and T components of the soil (Arias 1970) are shown in Table 1. The most intense earthquakes that have been registered are the ones of December 10, 2011 (event 11-3), June 16, 2013 (event 13-2), March 20, 2012 (event 12-1), April 18, 2014 (event 14-1) and May 8, 2014 (event 14-2). For events 11-3, 13-1 and 13-2 small cracks were observed in the finishing of some masonry walls. With event 12-1, the most intense recorded to date, the mechanism of one of the elevators was damaged and cracks and partial fall of the finishing of several concrete and masonry walls were observed. The maximum accelerations were 176.3 and 31.8 cm/s² on the roof and surrounding ground respectively. For event 14-1, with an intensity similar to event 12-1, an increase in cracks of some masonry walls from events 11-3, 13-1 and 13-2 was observed. In event 14-2, no more cracks were detected other than those observed after event 14-1, it should be mentioned that the cracks of this event had not been repair. The other remaining events did not cause visible damage.

The seismic events were analyzed with a structural health monitoring system "AlertaE" to estimate the possible global damage of the building (Murià-Vila *et al.* 2010). The system is based on five indicators. The first two are ground motion parameters PGA and I_{Arias}. The other three are structural demand parameters: seismic coefficient (Cs), interstory drift (SD), and variation of fundamental horizontal translation frequencies (VF) identified in the coda segment of the seismic response. For each indicator, a scale of three levels of possible damage is proposed: possible light damage (Level 1), possible intermediate damage (Level 2) and possible severe damage (Level 3). The three levels of damage are defined for each indicator by setting thresholds, which are a function of the building characteristics (structure and non-structural elements), site location characteristics, and the structural condition when monitoring begins. The description of light, intermediate and severe damage is presented in (Murià-Vila *et al.* 2010). The threshold values used for the building are shown in Table 2. The structural condition of the building is established by weighting the five indicators, based on four colors, using the following logic rules: Green, when VF, SD, PGA, I_{Arias} and Cs belong to Level 1. Yellow, when VF or SD, or two of the remaining intensity indicators match Level 2. Orange (building inspection) may be determined when VF or SD is Level 3 and at least one indicator PGA, I_{Arias} or Cs is Level 2, or VF is Level 2 and one of the other indicators is Level 3. Orange state, indicates an urgent need to inspect the building to exclude a red-state case. Red, is established when the inspection suggests evacuating the building.

Table 1. Characteristics of seismic events recorded and structural response in the building

Event	Date	M_W	Ep. Dist. in km	I_{Arias}, in cm/s	PGA in cm/s^2	A_{max}, in cm/s^2 L	A_{max}, in cm/s^2 T	D_{max}, in cm L	D_{max}, in cm T	$D(GPS)_{max}$, in cm L	$D(GPS)_{max}$, in cm T	f in Hz L	f in Hz T
11-1	25-feb-11	6	430	0.02	0.9	4.8	6.7	<1.0	<1.0	NA	NA	0.67	0.59
11-2	07-abr-11	6.7	539	0.08	1.6	13.3	15.0	<1.0	1.1	NA	NA	0.64	0.58
11-3	10-dic-11	6.5	214	2.85	17.2	57.7	78.3	3.6	5.6	4.3	5.8	0.62	0.57
11-4	16-dic-11	4.6	364	<0.01	0.4	2.0	2.0	<1.0	<1.0	8.5	3.0	0.66	0.58
12-1	20-mar-12	7.4	358	15.04	31.8	176.3	163.0	16.4	14.0	17.2	15.9	0.56	0.55
12-2	02-abr-12	6	356	0.41	7.9	27.2	35.3	2.4	3.1	2.9	3.0	0.56	0.55
12-3	11-abr-12	6.4	475	0.42	5.0	32.5	27.9	3.1	2.6	3.1	2.9	0.57	0.55
12-4	15-nov-12	6.1	225	0.52	6.1	15.2	33.4	1.3	2.5	1.4	2.8	0.58	0.56
13-1	21-abr-13	5.8	392	0.57	7.8	42.7	59.4	3.7	5.5	3.3	6.4	0.57	0.55
13-2	16-jun-13	5.8	162	3.44	20.6	73.4	108.1	6.2	10.1	7.2	12.0	0.57	0.54
13-3	21-ago-13	6	305	0.95	8.9	39.5	54.0	3.2	4.6	3.7	3.9	0.54	0.54
14-0	09-mar-14	5.8	408	0.05	1.3	8.4	16.7	<1.0	1.5	1.1	3.2	0.57	0.55
14-1	18-abr-14	7.2	352	12.86	30.0	132.2	192.9	13.0	17.5	13.2	20.4	0.53	0.53
14-2	08-may-14	6.4	336	8.50	29.4	99.0	168.8	9.3	18.8	9.1	19.3	0.56	0.53
14-3	10-may-14	6.1	345	0.92	8.5	42.3	46.7	4.1	4.7	4.4	5.3	0.56	0.54
14-4	07-jul-14	6.9	844	0.17	2.1	15.7	18.2	1.6	1.7	1.9	1.8	0.55	0.54
14-5	29-jul-14	6.4	392	0.16	3.4	16.6	24.3	1.5	2.0	1.9	2.2	0.55	0.55
15-1	20-mar-15	5.4	168	0.09	3.8	7.8	13.6	<1.0	1.1	<1.0	1.3	0.59	0.56

L- Longitudinal component, T- Transversal component

Table 2. Thresholds of the indicators for RC buildings of medium height on Mexico City soft soil

Level	PGA in cm/s^2	I_{Arias} in cm/s	Seismic coefficient (Cs)	Interstory drift (SD) in %	Variation of frequency (VF) in %
1	< 15	< 5	< 0.03	< 0.12	< 7
2	15 - 130	5 - 140	0.03 - 0.16	0.12 - 0.60	7 - 25
3	> 130	> 140	> 0.16	> 0.60	>25

In May 2009, a GPS system composed of a reference station (GPS1) was installed in the base and two antennas (GPS2 and GPS3) were installed on the roof of the tower to monitor the verticality. The GPS1 is located on the roof of the low-rise structures, for security reasons it was not possible to install it in the field station. It should be mentioned that measurements made by locating GPS1 at the field station resulted in similar relative shifts. Since November 2011, the GPS system has been operating in parallel with the recording system of accelerometric signals, in order to directly measure the relative displacements of the roof with respect to the reference station. With the GPS system, the displacements on the tower roof can be directly obtained, including movements due to seismic events. To follow the verticality, the GPS data obtained during one hour of recording are averaged.

3. Analysis of the structural response records

The acceleration and the maximum displacements of seismic records roof (D_{max}) are included in Table 1, in all events the displacements calculated with the double integration of the acceleration and measured with the GPS system ($D(GPS)_{max}$) are very similar. The maximum relative accelerations and displacements in the L component were 176.3 cm/s^2 and 16.4 cm during event 12-1, and in the T component were 192.9 cm/s^2 and 18.8 cm, during events 14-1 and 14-2, respectively. Figure 2 compared the roof displacements in the NE corner of the building obtained from GPS3 and those calculated with the accelerograms during events 11-3, 12-1, 13-1, 13-2, 14-1 and 14-2. These displacement histories show that GPS data performed well, except for events 12-1, 14-1, and 14-2. The displacements obtained with the GPS system are available from event 11-3. For event 12-1, there are no GPS2 records of the SW roof corner since it was temporarily out of service. Additionally, GPS3 on the NE corner of the roof had problems with the satellite signals which caused an alteration of the amplitude and the loss of certain segments of the displacement history. Despite this, the maximum displacement amplitudes were obtained in the L component and very close to the maximum value in the T component. For events 14-1 and 14 -2 the GPS data from both corners of the roof was recorded but also presented a similar problem and some data was lost. Nonetheless, the most significant amplitudes were registered.

The response spectra derived from all seismic events recorded are show in Figure 3. The most significant spectral ordinates that are observed correspond to the most intense events that have been recorded (12-1, 14-1 and 14-2). The highest dominant ground periods estimated from the horizontal components were determined in a range between 1.3 and 2.0 s.

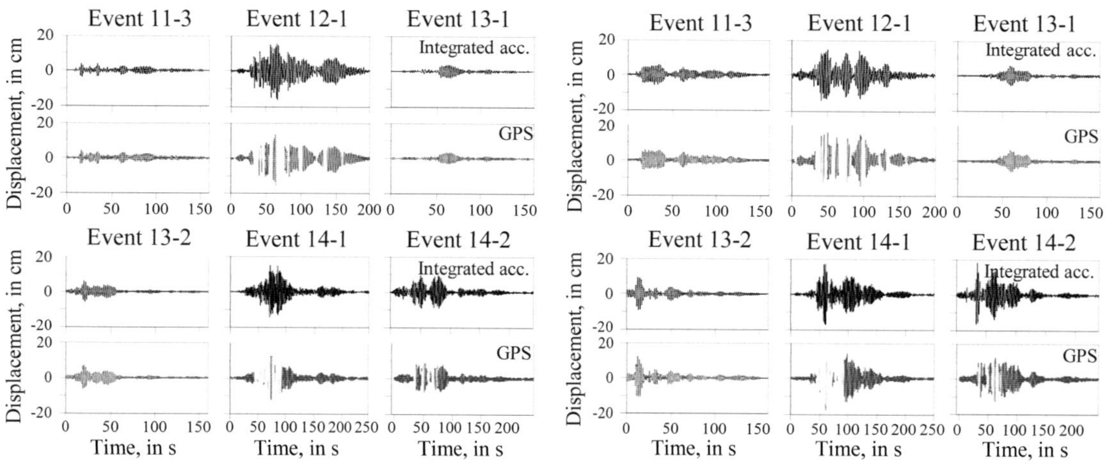

a) L Component b) T Component

Figure 2. Comparison of time-histories of the NE corner roof displacements measured with GPS3 and double-integrated acceleration

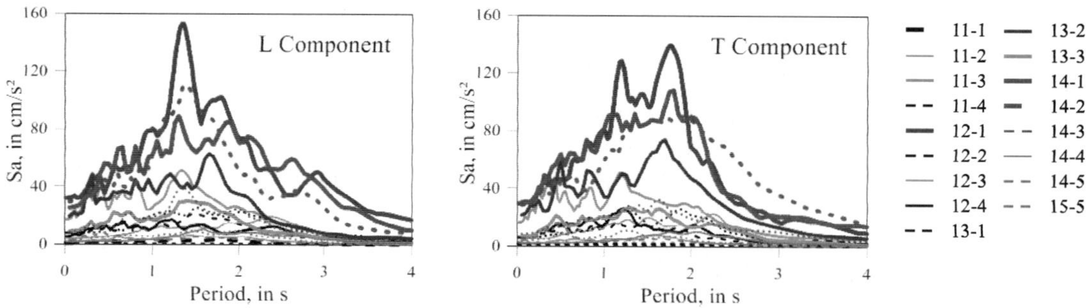

Figure 3. Response spectra calculated with the records of the field-surface site

The spectral analysis of the ambient vibration records obtained before and after the last rehabilitation showed an increase in fundamental vibration frequencies of the structure after restoration. The increase for the L, T and torsion components were 65, 26 and 45%, respectively.

On the other hand, the fundamental frequencies determined by the spectral analysis of the records of the first event (11-1, low seismic intensity) are practically the same as those obtained from the ambient vibration tests performed after the rehabilitation, with differences of less than one percent. In order to determinate the fundamental vibration frequencies of the structure, a spectral analysis of all the recorded seismic events was conducted. Figure 4 shows the calculated Fourier spectral ratios between roof and ground of components L and T, in the frequency band where the fundamental vibration frequencies of the tower of the building were detected. It is appreciated that the forms of the curves and the intervals of maximum amplitudes vary from one event to another, indicating that the fundamental frequencies changes. On the other hand, for some events appeared more than one significant ordinate in a small interval of time. This suggests that the system exhibits nonlinear behavior, in particular with the higher intensity events. Note the event 12-1, as it is where a reduction of 16.4 and 6.0% of the fundamental frequencies of L and T components occurs with respect to the first event 11-1 recorded after its rehabilitation, which implies a reduction of stiffness in the building by 35 and 12%, respectively. In this event the corresponding average SD (interstory drift) were 0.20 and 0.14%. The variation of the structural parameters of the system can be attributed mainly to rearrangements of the non-structural elements and also rehabilitated structural elements. The fundamental frequencies identified in the L and T components are included in Table 1. These frequencies or periods of vibration of the building are within the range of the predominant periods of the site (Figure 4).

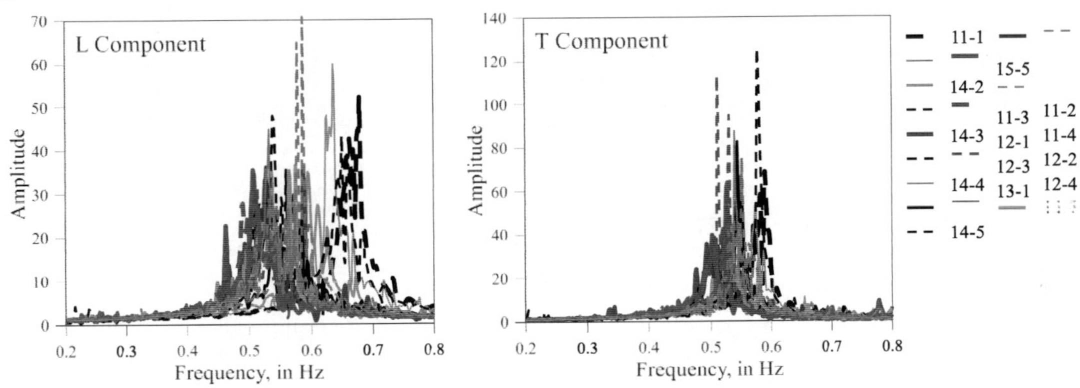

Figure 4. Fourier spectral ratios between the motion at roof and at field-surface site

3.1 Structural condition estimated with AlertaE system

Several seismic events have been recorded between February 2011 and March 2017. From these, 18 have been analyzed based on the AlertaE System implemented in the building (Murià-Vila *et al.*, 2013). Results are presented in Table 3. When the AlertaE system evaluated the seismic records, it determined that events 11-3, 12-1, 13-2, 14-1 and 14-2 reach the Yellow state. For these events, the building was inspected and the damage previously described was observed. The ground motion indicators (PGA and I_{Arias}) were at maximum at the event 12-1. In the case of event 11-3 the warning system detects Yellow state because the VF and PGA indicators exceed the Level 1 damage thresholds (Table 2). Subsequent inspection showed evidence of cracks in masonry walls, but the width of the cracks was small so it is considered as slight damage (Green state). In event 11-4, the VF indicator virtually recovers its initial value, which confirms the level of slight damage of the previous case, and the fact that the initial value was not fully recovered is evidence of deterioration due to seismic events consecutive despite its small intensity (Murià-Vila *et al.*, 2001). Event 12-1 was classified as Yellow state and almost all indicators were Level 2. The inspection confirmed that the level of damage was intermediate. From this event, the VF indicator of component L has not recovered the initial reference value. Therefore, the level of damage remains intermediate for subsequent events. Although for the events 12-2, 12-3, 12-4 and 13-1 the AlertaE system assigns Green state to the other indicators because it is of small intensity. Events 13-2, 14-1 and 14-2 again reach the Yellow state, but the remaining events of small intensity evidence that VF is practically the same as event 12-1.

Table 3. Characteristics of seismic events and estimated structural condition of the building

Event	$I_{Arias,}$ in cm/s	Cs		SD in %		VF in %		Damage level from event 11-1
			L	T	L	T	L	
11-1 (V)	0.02	<0.01	<0.01	<0.01	<0.01	0	0	Light
11-2 (V)	0.08	<0.01	<0.01	<0.01	0.01	4	2	Light
11-3 (A)	2.85	0.01	0.01	0.04	0.05	7	4	Intermediate
11-4 (V)	<0.01	<0.01	<0.01	<0.01	<0.01	2	1	Light
12-1 (A)	15.04	0.05	0.03	0.20	0.14	16	6	Intermediate
12-2 (V)	0.41	<0.01	<0.01	0.03	0.03	16	6	Intermediate
12-3 (V)	0.42	<0.01	<0.01	0.04	0.03	15	7	Intermediate
12-4 (V)	0.52	<0.01	<0.01	0.02	0.02	13	4	Intermediate
13-1 (V)	0.57	0.01	0.01	0.05	0.06	16	6	Intermediate
13-2 (A)	3.44	0.02	0.02	0.07	0.09	16	8	Intermediate
13-3 (V)	0.95	<0.01	<0.01	0.04	0.04	19	8	Intermediate
14-0 (V)	0.05	<0.01	<0.01	<0.01	0.02	15	6	Intermediate
14-1 (A)	12.86	0.03	0.03	0.17	0.18	21	9	Intermediate
14-2 (A)	8.50	0.02	0.04	0.13	0.19	16	9	Intermediate
14-3(V)	0.92	0.01	<0.01	0.05	0.05	14	7	Intermediate
14-4(V)	0.17	<0.01	<0.01	0.02	0.02	17	7	Intermediate
14-5(V)	0.16	<0.01	<0.01	0.02	0.02	18	7	Intermediate
15-1(V)	0.09	<0.01	<0.01	<0.01	<0.01	12	4	Intermediate

L- Longitudinal component, T- Transversal component

3.2 Parametric analysis of the event 12-1

It is appreciated that the dynamic properties of the building are sensitive to the intensity of the movement of the ground. In order to identify the variation of the frequencies and fractions of damping in the most intense events 12-1 a parametric analyzes of the records were made. The analysis with parametric techniques adopted was based on the modal minimization method proposed by Beck and Jennings (1980). For the identification of the structural parameters, a computer program was used (Camargo and Murià-Vila, 2011), in which multiple inputs and outputs signals were established based on the seismic records. By means of this technique, sequential analyzes are performed for short duration time windows, allowing the study of the variation of the structural parameters from window to window. The Event 12-1 was divided into 24 windows of 6 to 15 s, in each one the frequencies, damping and the contribution in the response of the significant modes were identified, with which it is possible to reproduce the measured acceleration response in the building.

Figure 5 shows the variation of the vibration frequencies and contribution of the identified modes together with the comparison of the measured and calculated signals. In the T component the fundamental mode frequency (1T mode) was 0.54 Hz which dropped to 0.49 Hz in the intense phase of the earthquake and returned almost to its initial value in the final phase. In the L component there was a greater decrease of the fundamental frequency (mode 1L), from 0.61 to 0.50 Hz from the initial window to the intense phase and a final value of 0.55 Hz, which indicates that the initial value was not recovered. These results agree with the frequency variation reported by the AlertaE system. For the torsion, the frequencies varied between 0.78 and 1.06 Hz and there were no differences between the values of the initial and final windows. It is noteworthy that in the torsion as well as in the second modes there are greater uncertainties in the estimation. In the T component, the variation of the frequency in the final phase was 2.6% with respect to the initial value, compared with the estimated value at the beginning of the monitoring the difference is 9.8%. In the L component, these variations were 10.3 and 18.6%, respectively. In most windows in the L and T components the contribution of the respective fundamental modes (Figure 5), is greater than 80% and the total contribution is greater than 90%. In the signal associated with torsion, the contribution of the first mode in this component (mode 1R) ranged from 11 to 68%, while the total contribution fluctuated between 30 and 79%.

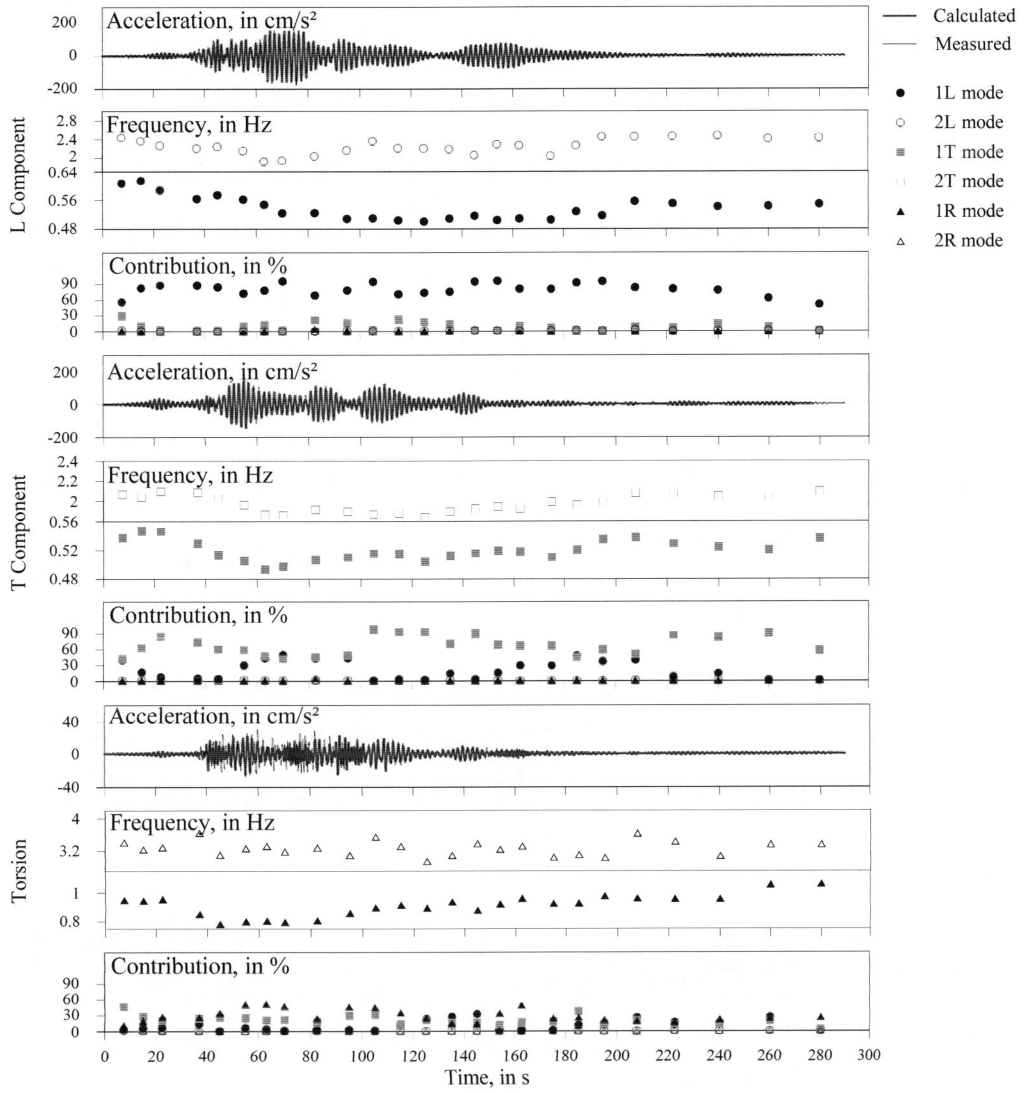

Figure 5. Identified and estimated frequencies for the building during event 12-1

The critical damping fraction for modes 1L and 1T was estimated in the range of 0.02 and 0.051 and a correlation was observed between damping and the intensity of movement, while all other modes had values between 0.015 and 0.10 without an evident relationship with the amplitudes of the movement.

3.3 Analysis of GPS System Records

From October 30, 2011 to date there is a continuous record of GPS data which allows to calculate the tower´s tilt. To estimate the displacements an initial reference position for the three GPS must be established. The initial position of GPS1, GPS2 and GPS3 antennas was determined as average value of the coordinates recorded during 24 h with a sampling frequency of 10 Hz (Fig. 1). Once stablished the initial reference coordinates of the three antennas the tower´s tilt was calculated. The horizontal displacements due to the tower´s tilting calculated after the initial continuous monitoring in the roof where the GPS2 and GPS3 antennas are located were 0.271 and 0.310 m in L component respectively, and 1.017 and 1.020 m in T component.

In addition to the orthogonal coordinates, the GPS system provides the number of satellites and the quality of the data. Only the data that have a possible deviation in the measurement less than 0.025 m and when the system was able to make corrections of position in real time is considered in the analysis. Figure 6 shows the tilt of the tower in the L and T components of GPS2 (SW corner) and GPS3 (NE corner) in the period from November 2011 to March 2017. The precipitation and temperature data from a weather station near the building (PEMBU, 2016) is also presented. The L component shows a downward tendency at an annual rate of about 1 cm. In the T component a very small tendency to decrease tilt was observed until 2015, and from 2016 onwards there is an apparent stabilization. In this T component there is a periodic fluctuation that can be correlated with the rainy and drought seasons as well as with the temperature.

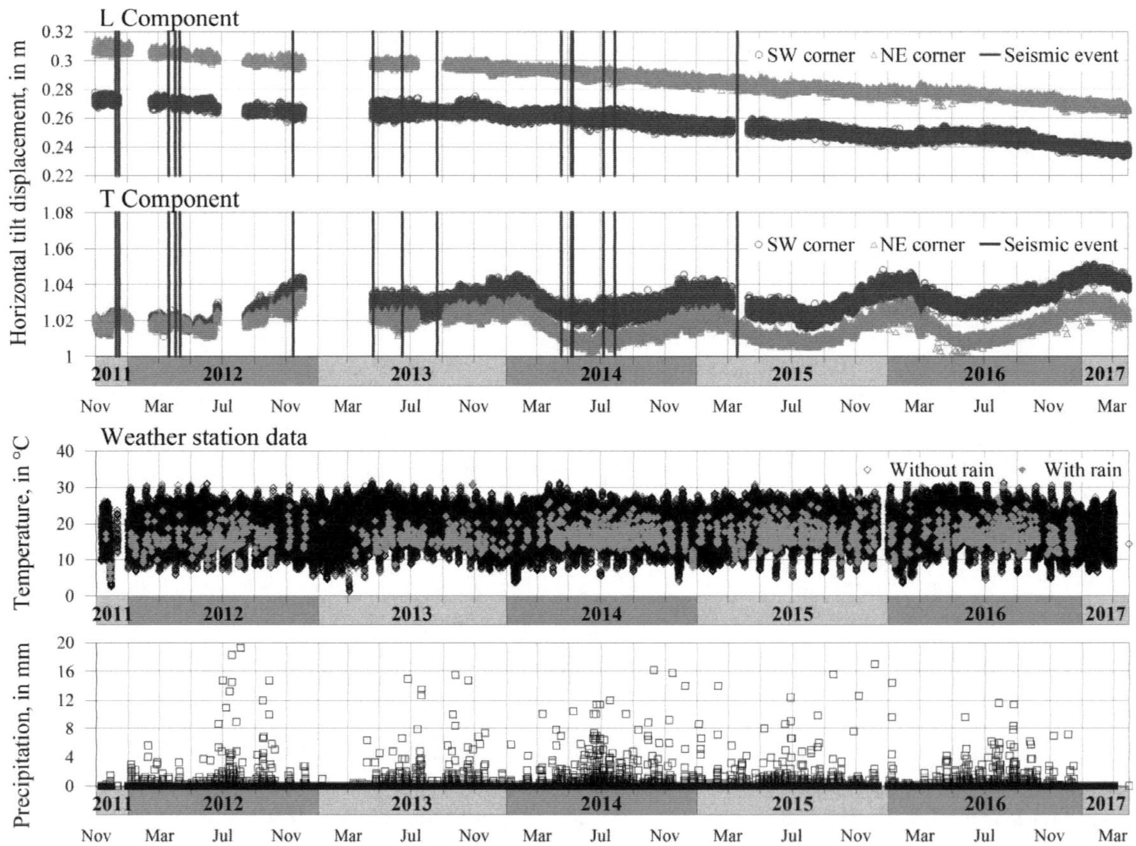

Figure 6. Hourly variation of the tower´s tilt between November 2011 and Mars 2017

Figure 7 shows the evolution of the horizontal resultant displacement obtained with the orthogonal displacements of the components T and L in the SW and NE corners of the tower between July 1965 and March 2017, data previous the GPS system was obtained from topographic surveys. In this graphics is also indicated the most significant seismic events that have occurred in the period, as well as the installation and period of operation of a pumping and water filtration system. It can be seen in the figure that between July 1965 and October 1985, the tower tilted up to 0.96 m. After the earthquakes of 1985 the tilt increase of 3 to 4 cm and with the earthquake of April 25 of 1989 increased approximately 0.3 cm. In 1991 a system of 21 pumping wells was installed for the extraction of water in certain zones and 13 infiltration boreholes for the injection of water in other zones in order to control the verticality and settlements of the building. This system operated until August of 2005; during this period, the tower´s tilt decreased from 1.04 m to 0.97 m (López Acosta and Ávila 2007). From August of 2005 to October 2009 the tilting increased to 1.06 m. From the last rehabilitation in the tower, from October 2009 to March 2017 the tilt maintained their value between 1.04 and 1.08 m. The effects of earthquakes after 1989 are not seen to have influenced the tower's tilt. The analysis of the recent values suggests that the last underpinning and reinforcement of the structure have contributed in the control of the verticality of the tower.

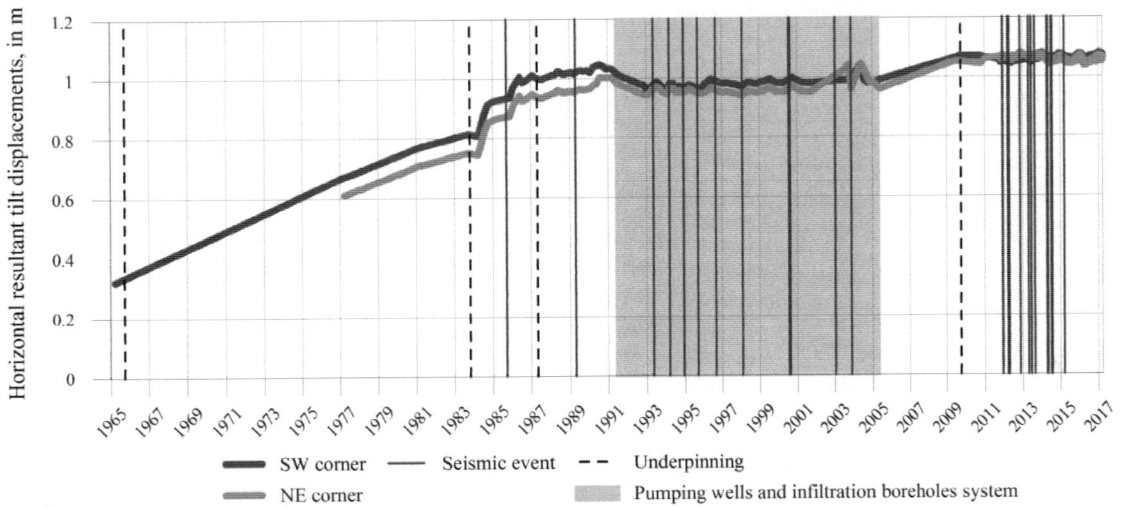

Figure 7. Variation of the tower´s tilt between July 1965 and March 2017

4. Conclusions

The building has suffered some damage, especially in the earthquake of March 20, 2012 (event 12-1), which is the first major earthquake to occur after rehabilitation. Their effects caused the fundamental vibration frequencies to decrease by 16.4 and 6.0% in the L and T components, respectively. The earthquakes of April 18 and May 8, 2014 (events 14-1 and 14-2), although of a similar intensity to event 12-1, did not imply an appreciable additional change in the values of the vibration frequencies, since their values decreased by less than 1%. The variation of the structural parameters is attributed to different sources of non-linearity in the structure, mainly to rearrangements of the non-structural elements and rehabilitated structural elements.

The GPS displacements show a good performance of this measuring system, except for the three most intense seismic events. These events presented problems with the satellite signals that caused the alteration of the amplitude and the loss of some segments of the displacement history. For this reason, the problems presented in the GPS system during the events of greater intensity should be studied in detail to obtain a possible solution. Regarding the tower´s tilt of the building after the last rehabilitation, it shows very small variation and the tendency suggests even a small recovery of verticality which correlates with the seasonal changes of precipitation and temperature. According to these measurements, the tilt values are practically stable in T component and show a slight recovery in L component. It is expected that the monitoring of the verticality in the coming years will confirm the stability of the tower's tilt. For the safety of the structure, it is important to continue monitoring its verticality, the GPS system has proved to be very efficient for this monitoring process, and complements well with the periodic topographic measurements. Regarding the follow-up of the tower's tilt of the building, the suggestion by Çelebi (2013) of the use of GPS systems to contribute to the monitoring of the structural health of buildings is valid in this case.

Acknowledgments

The study was carried out with the financing provided by the Government of the Mexico City and the National Autonomous University of Mexico. The authors wish to thank the support of the UNAM. We thank L.A. Aguilar, M. Ayala, P.H. Galván, M.A. Macías, D. Almora, L. Alcántara, J.M. Velasco, C. Pérez, A.L. Ruiz, R. Delgado, A. Mora, K. Pérez, H. Sandoval, M. Torres, G. Castro, R. Vázquez, I. Molina and L. Ramírez for the operation and maintenance of the instrumentation of the building, as well as for the collaboration in the pre-processing or analysis of the registers. The authors also thank the valuable comments and recommendations of S. Alcocer, S. Loera and R. Meli.

References

Arias A. (1970). “A Measure of Earthquake Intensity”, Seismic Design for Nuclear Power Plants, Editor R. J. Hansen, MIT Press.

Beck J.L. and Jennings P.C. (1980), “Structural identification using lineal models and earthquake records”. Earthquake Engineering and Structural Dynamics, vol. 8, pp. 145-160.

Camargo J. and Murià-Vila D. (2011): Implementation of genetic algorithms in a system identification method for instrumented buildings. 5th International Conference on Structural Health Monitoring of Intelligent Infrastructure, (SHMII-5), Cancún, México.

Çelebi M. (2013). Chapter 2: Seismic Monitoring of Structures and New Developments, Earthquakes and Health Monitoring of Civil Structures, M. Garevski, ed., Springer Environmental Science and Engineering, pp. 37-83.

López Acosta N.P. and Ávila J.A. (2007). Anexo 6.13 del Libro Blanco del Centro Cultural Universitario Tlatelolco (CCUT): Proyecto de rehabilitación de la estructura y cimentación de la Torre del CCUT, Instituto de Ingeniería, UNAM.

Murià Vila D., Aldama B.D. and Loera S. (2010). “Structural warning for instrumented buildings”, Proceedings of 14th European Conference on Earthquake Engineering, Ohrid, Macedonia, 29 Auguste-3 September, pp.1-8.

Murià Vila D., Camargo J., Aldama B.D., Rodríguez G., Aguilar L.A. and Ayala M. (2013), "Structural health monitoring of an instrumented building in Mexico with accelerometers and GPS sensors", 6th International Conference on Structural Health Monitoring of Intelligent Infrastructure, The Hong Kong Polytechnic University, Hong Kong, 9-11 December, pp. 1 9.

Murià Vila D., Rodríguez G., Zapata A. and Toro A.M. (2001). "Seismic response of a twice-retrofitted building", ISET Journal of Earthquake Technology, vol. 38, n° 2-4, pp. 67-92.

PEMBU (2016). “Programa de Estaciones Meteorológicas del Bachillerato Universitario de la Universidad Nacional Autónoma de México”, http://www.ruoa.unam.mx/pembu/index.html

Rodríguez J.F., López Acosta N.P. and Auvinet G. (2009). “3D Numerical modeling of the behavior of large buildings founded on Mexico City soft clays”, Proceedings of the 2nd International Workshop on Geotechnics of soft soils-Focus on ground improvement. Glasgow, U.K., September 3, 2008, pp. 87-94.

The 8th International Conference on
Structural Health Monitoring of Intelligent Infrastructure
Brisbane, Australia | 5-8 December 2017

Verification of Seismic Response Estimation Method Based on Limited Number of Sensors by Using a Shaking Table Test Data of 6-Story RC Wall Frame Building

T. Morii[1], T. Sagawa [2], K. Sugimoto [2], K. Okada [2], M. Shiraishi [2].

[1] *Institute of Technology, Shimizu Corporation - Japan, Email:* t.morii@shimz.co.jp
[2] *Institute of Technology, Shimizu Corporation - Japan*

Abstract

Several methods have been proposed to estimate seismic response of each floor of a building by utilizing sensors installed on a limited number of stories. Most of these methods are techniques based on the modal analysis, and have already attained to a practical application level. However, applicability of those methods has not been examined in detail as for nonlinear deformation of the building.

In this paper, we will examine the applicability of our estimation method according to the modal shape calculated on the basic of the design model of building by using experimental data of a shaking table test of the 1/3 scaled 6-story RC building executed in January 2015 at *E-Defense* (Sugimoto, K., et al, 2017). RC building specimen was consisted of two moment-resisting, and one was frame structure in longitudinal direction, another was frame with structural walls in transversal direction. Response states of the whole building were evaluated on each direction respectively by using 3 points of sensors. From the viewpoint of structural health monitoring, we focus on the maximum story drift angle of buildings.

Finally, we can draw our conclusions as follows. Estimation results from 3 points of sensors (1, 3 and RFL) correspond approximately to the observed results up to 0.015 radians in the frame direction. However, in the transverse direction with shear walls, even if the story drift angle is less than 0.01 radians, estimation results depart from observed data. We think that the change of modal shape in specimen due to several cracks in shear walls decreased estimation precision. Therefore we think the estimation method should be introduced mechanism like a mode update etc, in order to improve the estimation method in the nonlinear region.

1. Introduction

Recently, demand has increased to evaluate the structural safety and function recovery of buildings shortly after they are subjected to an earthquake. The seismic response of each floor of a building can be easily measured if sensors are installed on every floor. High performance sensors are generally expensive, so it is difficult to install sensors on every floor of a building, especially a high-rise building.

Several methods have been proposed to estimate the seismic response of each floor of a building by utilizing sensors installed on a limited number of stories (Suzuki, Y. et al., 2008). Most of these methods are techniques based on modal analysis, and have already reached a practical application level. The main aim of these methods is to estimate structural health determination and damage-identification on each floor by using a limited number of high-precision conventional servo-type accelerometers. Some papers have been published on the estimation method based on model updating (Saito, T., 2013), others have been published on the evaluation method by using less sensors, for example only 2 sensor points (Okada, K. et al, 2016). However, the applicability of these methods has not been examined in detail for nonlinear deformation of buildings.

In this paper, we will examine the applicability and effects of our estimation method (Morii, T., et al., 2017) according to the mode shape calculated on the basis of the design model of a building by using experimental data of the large shaking table test of the 1/3-scale 6-story RC building executed in January 2015 at *E-Defense* (Sugimoto, K., et al, 2017). Observed ground motions in 1995 Kobe earthquake were used as input for the large shaking table test, and the amplitude levels of the input motions were repeatedly increased until the model specimen collapsed. We installed 25 servo-type accelerometers in the model specimen on each floor. Response states of the whole building were evaluated by using a few sensor points. From the viewpoint of structural health monitoring, we focus on the maximum story drift angle of buildings. We analyzed the estimated results from limited point accelerometers by comparing them with observed data from all sensors in addition to other data such as displacement meters.

2. Estimation Method of Whole Building Seisemic Response

Fig.1 shows the general outline of an estimation method of whole building seismic response based on a limited number of sensors and mode shape. In this paper, we assume that the mass and the stiffness of each story in the building are determined from the structural design model. The distributions of the mass and stiffness are represented as $\boldsymbol{m}=[m_1 \cdots m_{n_f}]^T$ and $\boldsymbol{k}=[k_1 \cdots k_{n_f}]^T$, respectively, where n_f is the number of stories.

The participation matrix $\boldsymbol{\Gamma}=[{}_1\gamma \cdots {}_{n_f}\gamma]^T$ is calculated from the mass matrix $\boldsymbol{M}$ and stiffness matrix $\boldsymbol{K}$ by solving an eigenvalue program, where ${}_j\gamma$ is a column vector whose element is a participation factor of jth mode. Supposing the first n_m modes dominate the seismic response of a building, the response absolute acceleration $\boldsymbol{a_{ps}}$ at each sensor is approximated from

$$\boldsymbol{a}_{ps} = \boldsymbol{D}_m \cdot u + {}_m\boldsymbol{\Gamma}_s \cdot {}_m\boldsymbol{q} \tag{1}$$

where $\boldsymbol{D}_m=[1 \cdots 1]^T \in \mathbb{R}^{n_s}$, the ground acceleration u, ${}_m\boldsymbol{\Gamma}_s$ is a submatrix of $\boldsymbol{\Gamma}$ where the rows corresponding to the floors with sensors and the corresponding to the first n_m modes are extracted. ${}_m\boldsymbol{q}=[{}_1q \cdots {}_{n_m}q]^T$ is the modal response relative acceleration vector for the 1st to n_m modes. In this time, the number of modes n_m equates with the number of sensors n_s installed on the floors of the building except the 1st floor. Then the estimate of the modal response is obtained from

$${}_p\boldsymbol{q} = {}_m\boldsymbol{\Gamma}_s^{-}(\boldsymbol{a}_o - \boldsymbol{D}_m \cdot u) \tag{2}$$

where ${}_m\boldsymbol{\Gamma}_s^{-}$ is the inverse matrix of ${}_m\boldsymbol{\Gamma}_s$, $\boldsymbol{a_o}$ is observed in the data of the attached accelerometers. The response absolute acceleration $\boldsymbol{a_p}$ on each story is now estimated by

$$\boldsymbol{a}_p = \boldsymbol{D} \cdot u + {}_m\boldsymbol{\Gamma} \cdot {}_p\boldsymbol{q} \tag{3}$$

where $\boldsymbol{D}=[1 \cdots 1]^T \in \mathbb{R}^{n_f}$, ${}_m\boldsymbol{\Gamma}$ is a submatrix of $\boldsymbol{\Gamma}$ where the columns corresponding to the first n_m modes are extracted. Then, the calculations of story drift displacement $\boldsymbol{d_p}$ are obtained by taking differences between displacements of adjacent floors that are calculated from $\boldsymbol{a_p}$ by numerical integration. Next, the story drift angle is defined by dividing the story drift displacement $\boldsymbol{d_p}$ and the height of the story. In addition, the shear force Q_i on each floor is calculated from following mass m_k and estimated acceleration $a_{p,k}$.

$$Q_i = \sum_{k=i+1}^{n_f} m_k \cdot a_{p,k} \tag{4}$$

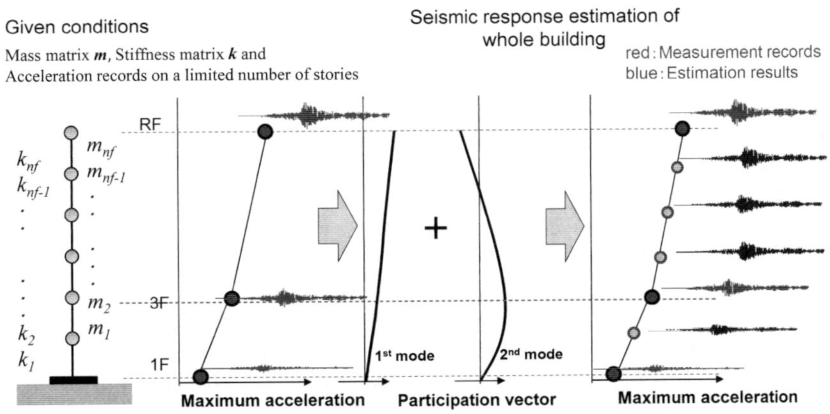

Fig.1 – General outline of estimation method of whole building based on a limited number of sensors and mode shape

3. Shaking Table Test of 6-story RC Wall Frame Building

The large shaking table test of a 1/3-scale 6-story RC wall frame building model was executed in December 2015 at *E-Defense* (Sugimoto, K., et al, 2017).

3.1 Outline of Shaking Table Test

Fig.2 shows the 1/3-scale 6-story RC wall frame building. The specimen was designed 1/3 scaled 6-story RC wall frame building, the plan is 3.6×5.4m, the height is 5.4m and the weight is 3193kN. The specimen was consisted of two moment-resisting, and one was frame structure in longitudinal direction, another was frame with structural walls in transversal direction. And the model had nonstructural wall with gap along the column side and bottom, which was hanging from the girder of upper floor, in X1, X4 frames on 2nd to 6th story.

Employed input motions for the shaking table test were an observed ground motion as illustrated in Fig.3. The input motion was observed record at Japan Meteorological Agency Kobe in 1995 Kobe earthquake, called JMA Kobe wave. These waves were input in the three(x,y,z) direction of the specimen, Figure 3 shows only horizontal 2 waves. The input motions were adjusted in time by scaling factor 1/3.

The amplitude levels of the input seismic motions were repeatedly increased until the model specimen collapsed as listed in Table 1. The random wave with the same amplitude level was employed after a seismic motion in order to understand the vibration characteristics of the model specimen.

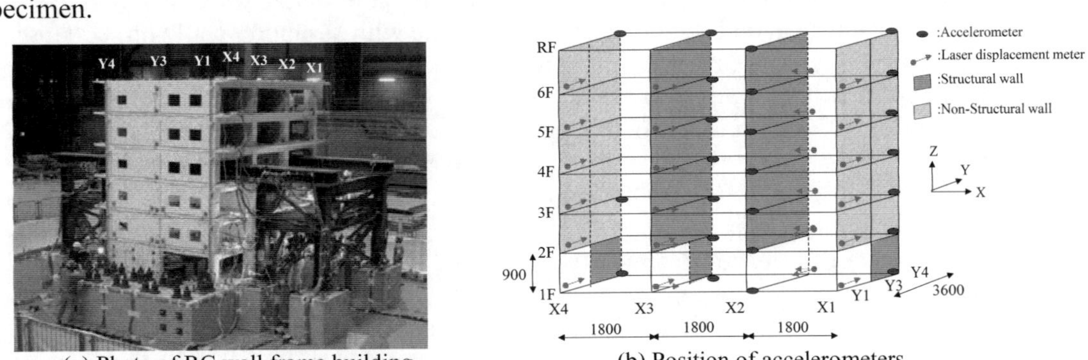

(a) Photo of RC wall frame building (b) Position of accelerometers

Fig.2 – 1/3 6-story RC wall frame building and structural health monitoring system

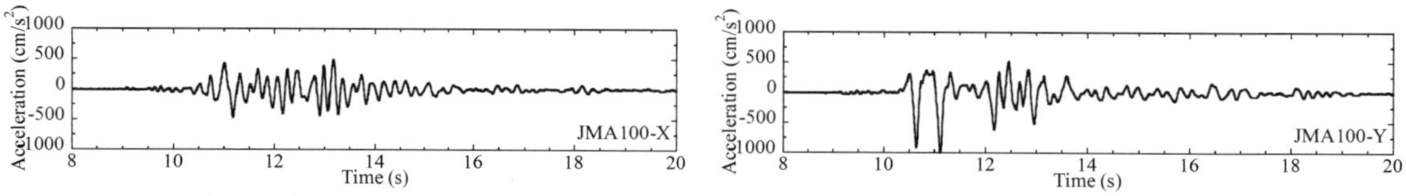

(a) Longitudinal direction (b) Transversal direction

Fig.3 – Input motions for shaking table test in RC wall frame building model (JMAKobe 100%)

3.2 Structural Health Monitoring System

A structural health monitoring system generally consists of servo-type accelerometers, a data recording system and an analysis system based on a personal computer. We installed 25 servo-type accelerometers in the model specimen on each floor to verify the estimation method as depicted in Fig.2. The accelerometers placed at X2Y4 points, X3Y1 points and X4Y4 points on each floor, and X1Y4 points on 1, 3 floors and the roof floor. The accelerometers were fixed on concrete slab by using bolts.

We dealt with the experimental data between case JMA010 and case JMA140-1 as listed in Table 1. In this model specimen, the acceleration data converted to displacement by numerical integration after elements of less than 0.3Hz were cut in the frequency domain. To understand the dynamic behavior of the model specimen, others sensors such as displacement meters were set up in the 6-story RC wall frame building specimen. We analyzed the results estimated from limited point accelerometers by comparing them with observed data from all sensors in addition to experimental data such as displacement meters.

3.3 Damage Condition and Vibration Characteristics of Specimen

(a) Maximum drift angle and damage condition of specimen

Input waves for shaking table test and damage condition of the specimen are listed in Table 1. We select several input motions such as cases JMA55-1, 70-1, 100, 120 and 140-1, and describe the maximum story drift angle R_{max} and damage status of the specimen as follows.

In case JMA55-1, the maximum story drift angle R_{max} is 1/146 radians in longitudinal direction, 1/765 radians in transversal direction. Some column main bars were beginning to yield. In case JMA070-1, some girders and structural walls were yielding, R_{max} is 1/100 radians in longitudinal direction, 1/457 radians in transversal direction. As for case JMA100, R_{max} is about 1/100 radians in both directions, structural walls of X1 frame in the 1st story were crushing concrete. Then, in case JMA120, R_{max} reaches to about 1/76 radians in longitudinal direction. In transversal direction, R_{max} is 1/29 radians and the response base shear coefficient was recorded to be maximum value 1.18. After that, seismic motions hadn't input in longitudinal direction, because of the aims of experiments is to understand behaviors of frame with structural wall until collapse.

In case JMA140-1, response base shear coefficient decrease and R_{max} increase 1/11 radian in transversal direction, structural walls of 1st and 2nd floors were failed in shear manner. Finally, input motion changed to JR120 in order to collapse the specimen, the specimen collapsed from 1st to 3rd story, leans against support frame. The joint bars at the gap of nonstructural wall were fractured on 2nd and 3rd floors of X1 and X4 frames.

(b) Vibration characteristics of specimen

The vibration characteristics of the specimen were investigated from the results of random wave excitation. Fig.5 shows the variations of the natural period of the specimen in the 1st to 2nd modes. And then, Mode shape of 1st natural period in both directions is shown in Fig.6.

Natural frequencies of the specimen were identified by using a nonlinear least squares method (Nakagawa, T., et al., 1982) from the results of the spectral ratios of the roof floor (RFL) to the 1st

Table1 – Results of shaking table test in RC wall frame model

Case name of input wave※1	Input Acceleration [cm/s²] (Multiplying factor)		Maximum story drift angle [rad] (Story)		Damage
	Longitudinal (X) diection	Transversal (Y) diection	X Dir. (Y1 frame)	Y Dir. (X2 frame)	
00-r※2	45 (-)	45 (-)	-	-	Elastic deformation (No damage)
JMA10	49 (0.1)	76 (0.1)	1/2495 (3F)	1/9589 (1F)	Ditto
JMA40	205 (0.4)	414 (0.4)	1/262 (2F)	1/1496 (1F)	Cracking
JMA55-1	268 (0.55)	584 (0.55)	1/146 (2F)	1/765 (1F)	Yielding of column main bar
JMA70-1	351 (0.7)	671 (0.7)	1/100 (2F)	1/457 (1F)	Yielding of girder and wall
JMA55-2	245 (0.55)	469 (0.55)	1/141 (2F)	1/602 (1F)	Ditto
JMA70-2	293 (0.7)	599 (0.7)	1/120 (2F)	1/422 (1F)	Ditto
JMA100	491 (1.0)	989 (1.0)	1/101 (2F)	1/99 (1F)	Fracture of joint bar at nonstructural wall concrete
JMA55-3	279 (0.55)	442 (0.55)	1/162 (2F)	1/169 (1F)	—
JMA120	515 (1.2)	1111 (1.2)	1/76 (2F)	1/29 (1F)	Maximum shear capacity
JMA140-1	—	1293 (1.4)	—	1/11 (1F)	Damage of nonstructural and structural walls were proceeding failure at 1st and 2nd story
JMA140-2	—	1280 (1.4)	—	1/14 (1F)	
JR120	—	918 (1.2)	—	1/7 (1F)	

※ 1 Case name of random wave is represented with the " -r " after the case name of input wave.

※ 2 Random wave excitation in virgin state of specimen.

※ 3 Afeter JMA140-1, shakeing table test is performed only in transversal direction.

floor. The spectral ratios of the roof floor (RFL) to the 1st floor were computed from observed acceleration, which were averaged by four sensors on the same floor, based on the method of fast Fourier transform (FFT). For examples, Fig.6 shows the spectral ratios of the specimen during case 00-r, JMA70-1-r, JMA100-r and JMA120-r.

In the initial state (case 00-r), the 1st natural period is 0.23s in longitudinal direction consisted only frame, and that one is 0.1s in transversal direction added structural wall. By comparison of mode shape of 1st natural period in the initial state, observed results approximately matches to that of design model in both directions, as shown in Fig7.

In the longitudinal direction, 1st natural period gently increase depended on the experiential maximum drift angle and cracking of column. From this figure, the natural periods in the 2nd mode seem not to vary more than 1st mode. The participation vector of the 1st mode in longitudinal direction is also varied gradually, depicted in Fig.5. However, in the transversal direction added structural wall, the variations of natural period is made difference. The natural period of the 1st mode hardly change until case JMA070-1-r, the 1st natural period alters greatly when the structural walls on 1th story are fractured in case JMA100-r. Then, the 1st natural periods increased as proceeding to failure of structural walls, that one become about 0.4s during case JMA140-1-r, so it is about 4 times that of the initial state. Finally, in case JMA140-2-r, 1st natural period is shorten a little due to settle the gap of nonstructural wall and column.

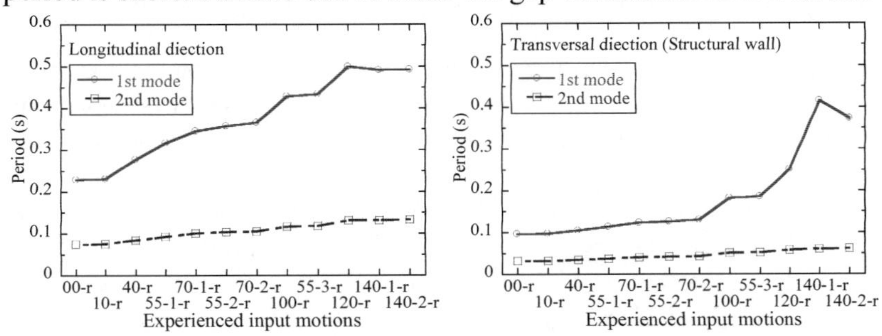

(a) Longitudinal direction (Frame) (b) Transversal direction (Shear wall)

Fig.4 – Vibration characteristics of RC wall frame model

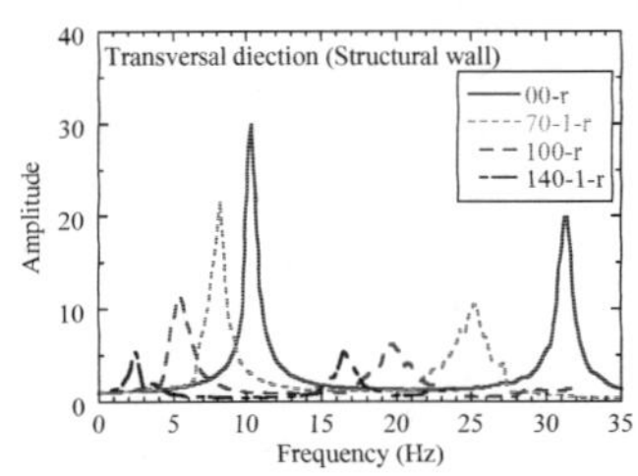

Fig.6 –Example of spectral ratios (Transversal direction)

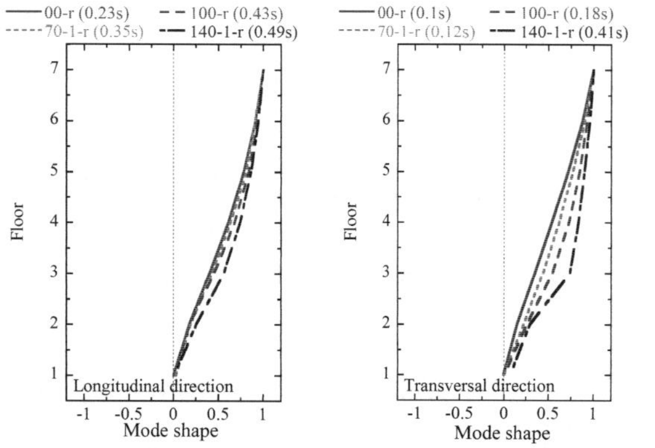

(a) Longitudinal direction (b) Transversal direction

Fig.5 –Observed mode shape of the specimen model

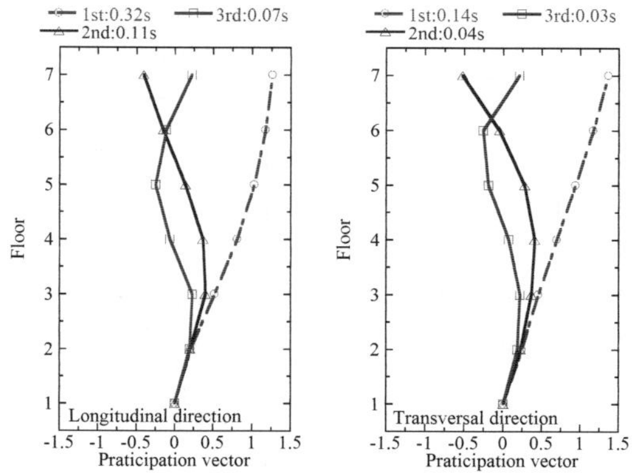

(a) Longitudinal direction (b) Transversal direction

Fig.7 –Participation vectors of a design model

4. Application for Shaking Table Test of 6-story RC Wall Frame Building

The proposed our estimation method based on the mode shape and a limited number of sensors were applied to the large shaking table test of the 1/3-scale 6-story RC wall frame building.

4.1 Eigenvalue Analysis Model for Response Estimation

Table 2 shows the information of the structural design model of the model specimen, and Fig.7 shows the participation vector of the specimen as the result of eigenvalue analysis. The structural design model is determined from the result of push-over analysis by using a frame model of the specimen. The natural periods and the participation vector shape of the structural design model are similer to the observed results as illustrated in Fig.7. We evaluated the seismic response of the model specimen from the results of eigenvalue analysis based on this structural design model as follows.

Table2 – Information of structural design model

Story	Height (mm)	Weight (kN)	Intial stiffness(kN/mm)	
			Longitudinal (X) diection	Transverse (Y) diection
6	900	306	182.6	730.8
5	900	305	220.0	1151.4
4	900	305	210.0	1491.0
3	900	305	191.9	1842.5
2	900	305	200.5	2264.9
1	900	305	309.7	2078.9

4.2 Estimation results

The seismic responses of the 1/3-scale 6-story RC wall frame building were estimated by using 3 sensors, which were installed on 1, 3, and the roof floor. The position of these sensors was made considering the mode shape of specimen. The number of sensors on the upper floor is 2, the estimation method considered up to the 2th mode.

Fig.8 shows the example of evaluated response results in transversal direction in case JMA70-1. Time history data have been extracted between 18s and 30s including principal shock from the response wave. The story drift angle was about 1/500 radians, the deformation level of the specimen stay linear region. From these figures, the estimated results of both the acceleration and the story drift angle accurately correspond to the observed data in the time history. Furthermore, the relationship of story drift angle and shear force can be estimated the observed results in 1st story.

The distributions of maximum story drift angle in transversal direction are illustrated in Fig.9, and these one in longitudinal direction are illustrated in Fig.10. In these figures, red lines denote the estimation values from a limited number of sensors, blue lines denote observation values from accelerometers on every story and green lines denote experimental values from displacement meters. In longitudinal direction, although the specimen reaches to a nonlinear region, which is 0.015 radians in story drift angle (JMA120) with some damage such as the yielding and fracture of part of the column, the results of the estimation are almost identical with the observed values and experimental values on the entire building model in all response indices. We think that our estimation method can evaluate the observed value and experimental value accurately because the natural periods and participation vector of the specimen are not much changed until case JMA120, as shown in Fig.5. On the other hands, in transversal direction, estimated maximum story drift angle were departed from observed and experimental data in case JMA100. In case JMA100, shear fractures of structural wall occurred on 1st story in the specimen, and participation vector of the specimen are changed from initial state. Therefore, we think that accuracy of estimation method in nonlinear region is depended on a change of participation vector in the specimen.

As mentioned above, in longitudinal direction, the results are estimated from only 3 sensor points (1, 3 and RFL) that correspond approximately to the observed results and experimental results until the specimen is 0.015 radians. However, in transversal direction, the estimated results are not corresponded approximately to the observed results in 0.01 radians.

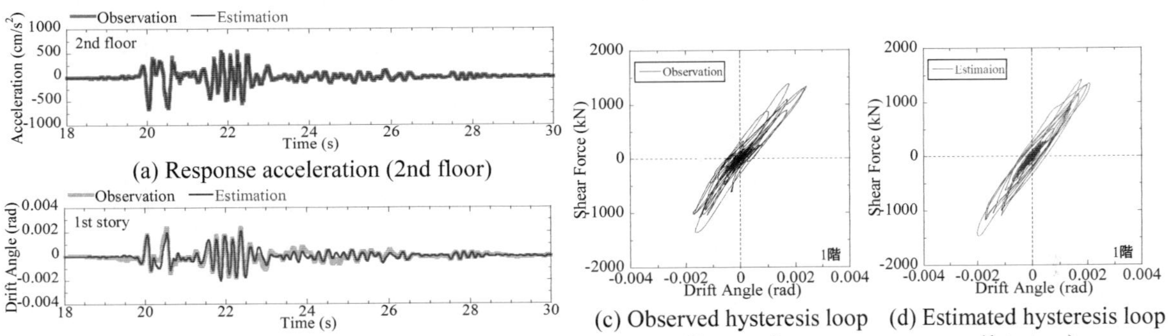

(a) Response acceleration (2nd floor)

(b) Story drift angle (1st story)

(c) Observed hysteresis loop (1st story)

(d) Estimated hysteresis loop (1st story)

Fig.8 – Comparison of response data in transvers direction (JMA70-1)

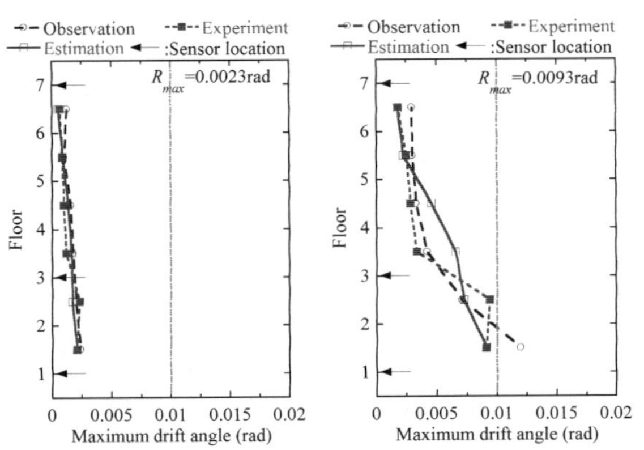

(a) JMA70-1

(b) JMA100

Fig.9 – Maximum drift angle in transvers direction

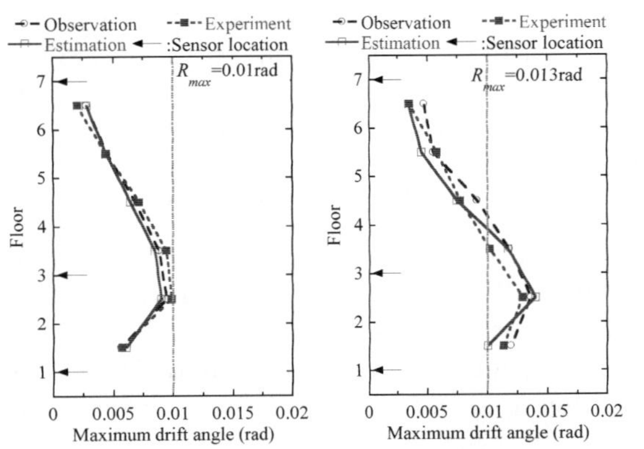

(a) JMA70-1

(b) JMA120

Fig.10 – Maximum drift angle in longitudinal direction

5. Effects of Model Updating Mehod on Estimation Results

We think that the change of modal shape in specimen due to fracture of shear walls decreased estimation precision. Therefore, we tried to improve the estimation method by introduced mechanism like a mode update in the nonlinear region. In nonlinear region, we confirmed the stiffness in some stories decrease from the relationships of shear force and story drift angle.

5.1 Model updating method based on equivalent stiffness

The outline of the improved estimation method of whole building seismic response based on a limited number of sensors and mode shape is shown in Fig.11. We try to introduce mechanism like a mode update in original estimation method. In improved estimation method, the initial stiffness replaced on the equivalent stiffness on each floor, if estimated maximum story drift angle is over a yielding story drift angle. A yielding story drift angle is set according to the results of push-over analysis the design model. The equivalent stiffness is calculated from relationships of shear force and story drift computed by the initial stiffness in model. The standard point for calculation of the equivalent stiffness is set in the estimated maximum story drift angle.

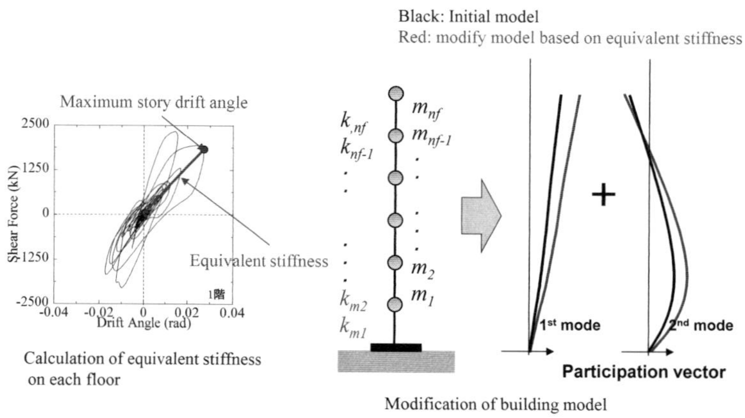

Fig.11 – General outline of model updating method based on equivalent stiffness

5.2 Effects of model updating method on estimation results in nonlinear region

In order to confirm the effects of model updating on estimation results, the estimated error *Er* and fit index *Bf* were introduced. From the results of the analysis, the ratio of the estimated response *Pr* to the observed response *Or* was defined as the fit index *Bf* on each story and each floor. Then, the estimated error *Er* is determined as follows:

$$Er = \sum_{i=1}^{n_f} \sqrt{(1 - Bf_i)^2} \; / \; n_f \tag{5}$$

where n_f is the number of stories. The fit index *Bf* indicates that the accuracy degree is high as the value of *Bf* is closer to 1. Then the estimated error *Er* expresses that the accuracy degree becomes high as the value of *Er* becomes small. The fit index *Bf* and estimated error are calculated for each response index such as maximum response acceleration and maximum story drift angle. In this section, we dealt with the observed values from all sensors as true values, and compared them with the estimated values.

Finally, we examined the effects of model updating on estimation results in nonlinear region. The experimental data between case JMA70-1 and case JMA140-1 were investigated. Fig.12 shows the influence of the model updating method on the applicability of a nonlinear response level.

We focus on the story drift angle, Er_R indicated the estimated error of maximum story drift angle, and Bf_{Rmax} indicated fit index of the maximum story drift angle of the specimen R_{max}. This figure shows the relationship between the maximum story drift angle of the specimen R_{max} and the estimated error Er_R or fit index Bf_{Rmax}. The estimated error Er_R and fit index Bf_{Rmax} of an original method change depending on the deformation level of the specimen. From these figures, the effects of model updating method aren't confirmed much until about 0.01 radians. However, the results of the model updating method are improved estimation accuracy in strongly nonlinear region, which is maximum story drift angle is over 0.03 radians.

Fig.13 shows distribution of maximum story drift angle and hysteresis loop on 1st story, by comparison of deformation level. From these figures, the effects of model updating method were confirmed in strongly nonlinear region more than 0.03 radians. From the views points of structural health monitoring of buildings, we think estimation accuracy in weakly nonlinear region around from 0.01 to 0.03 radians is most important, and need to improve the proposed method.

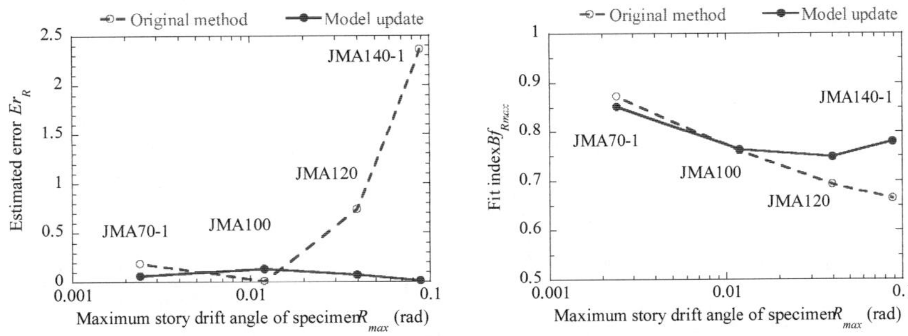

(a) Estimated error Er_R (b) Fit index Bf_{Rmax}

Fig.12 – Estimated error Er_R and fit index Bf_{Rmax} from JMA70-1 to JMA140-1

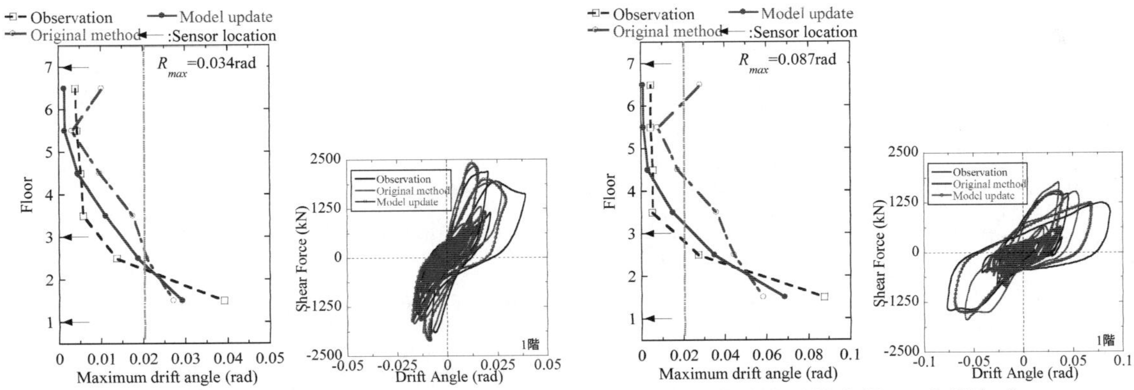

(a) JMA120 (R_{max}= 0.034rad) (b) JMA140-1 (R_{max}= 0.087rad)

Fig.13 – The effects of model updating method by comparison of deformation level

6. Conclusions

In this paper, we will examine the applicability of our estimation method according to the modal shape calculated on the basic of the design model of building by using experimental data of a shaking table test of the 1/3 scaled 6-story RC building executed in January 2015 at *E-Defense* (Sugimoto, K., et al, 2017). RC building specimen was consisted of two moment-resisting, and one was frame structure in longitudinal direction, another was frame with structural walls in transversal direction. Response states of the whole building were evaluated on each direction respectively by using 3 points of sensors. From the viewpoint of structural health monitoring, we focus on the maximum story drift angle of buildings. In this paper, we can draw our conclusions as follows.

1) Estimation results from 3 points of sensors (1, 3 and RFL) correspond approximately to the observed results up to 0.015 radians in the frame direction. However, in the transverse direction with shear walls, even if the story drift angle is less than 0.01 radians, estimation results depart from observed data.
2) The change of modal shape in specimen due to fractures in shear walls decreased estimation precision in nonlinear region. Therefore we proposed a model updating method based on the equivalent stiffness calculated from relationship story drift angle and shear force on each story.
3) The Effects of model updating method based on the equivalent stiffness were appeared in strongly nonlinear region more than 0.03 radians. From the views points of structural health monitoring of buildings, we think estimation accuracy in weakly nonlinear region around from 0.01 to 0.03 radians is most important, and need to improve the proposed method.

Acknowlegements

This study is partially supported by the sub-project II (Project Director: Prof. Masayoshi Nakashima, Kyoto University, Japan) of the Special Project for Reducing Vulnerability for Urban Mega Earthquake Disasters of the Japanese Ministry of Education, Culture, Sports, Science and Technology (MEXT). The shaking table test of a 6-story RC wall frame building is performed by the sub-working group of RC structures (WG leader: Prof. Minehiro Nishiyama , Kyoto University, Japan, WG manager: Mr. Hideo Katsumata, Obayashi Corporation, Japan). Many useful advices from the sub-working group of structural health monitoring of superstructures (WG leader: Prof. Jun Tobita, Nagoya University, Japan, WG manager: Dr. Michihito Shiraishi, Shimizu Corporation, Japan) is also appreciated.

References

Suzuki,Y. Et al., 2008 ,'Real-time building damage estimation system base on obsereved building response', *In proceedings of the 14th World Conference on Earthquake Engineering*, No.0439.

Saito, T., 2013,'Bayesian model updating for evaluation of building seismic response and probabilistic damage estimation', *Proceedings of the 11th International Conference on Structural Safety and Reliability*.

Okada, K., Kataoka, S., 2016,'Overall Building Response Estimation Method on Records from Two Seismographs', *Journal of Japan Association for Earthquake Engineering* Vol. 16, No.3, 94-113 (in Japanese).

Nakagawa, T., and Koyanagi, Y., 1982, 'Experimental data analysis by the least squares method - program SALS', University of Tokyo Press (in Japanese).

Sugimoto,K. et al., 2017, 'Shaking table test of 6-Story wall frame building to investigate collapse process of RC buildings, 16th world conferenc on earthquake', Puper No.0615.

Morii,T. et al., 2017, 'Estimation of whole building seismic response using Limited number of sensors –Verification of large shaking table test of 18-story stell high rise building-', 16th world conferenc on earthquake, Puper No.4088.

The 8th International Conference on
Structural Health Monitoring of Intelligent Infrastructure
Brisbane, Australia | 5-8 December 2017

Identification of System Parameter and Aerodynamic Force from Wind Tunnel Test Data

Y. Hui[1], S.S. Law [2]
[1] *Key Laboratory for Wind and Bridge Engineering of Hunan Province, College of Civil Engineering, Hunan University. E-mail: alihui@hnu.edu.cn*
[2] *School of Civil Engineering, Chongqing University.*

Abstract

System identification has been extensively studied to identify unknown parameters and excitations of a multi-degrees-of-freedom system. It is, however, found that if a single-degree-of-freedom (SDoF) system is under forced vibration and neither the system parameters nor the excitations are known, the identification is not possible with any existing system identification method. This paper solves this problem by assuming a stationary excitation with its energy smoothly varying close to the natural frequency of the system. An identification strategy is proposed for the SDoF system based on the Extended Kalman Filter with Unknown Input (EKF-UI) technique. Numerical example is presented to validate the proposed method on the excitation and system identification. This method is then applied to identify the aerodynamic force and system parameters of a high-rise building model in wind tunnel test. The vortex shedding induced excitation in the cross-wind direction can be clearly noted in the power spectrum of the identified wind load.

1. Introduction

System identification and damage detection techniques are popularly researched in the last few decades. In most of the existing methods, such as methods of least-squares estimation (LSE) and the extended Kalman filter (Loh, et al., 2000; Yang and Lin, 2004), the external excitations should be available for the identification. Although some techniques do not require exact values of the input, yet they require a stationary Gaussian white noise input, such as the covariance of covariance matrix technique (Law, et al., 2012; Hui, et al., 2017), random decrement technique (Huang and Yeh, 1999), and stochastic subspace identification technique (Yu and Ren, 2005). These methods are not applicable in cases with non-white process input (Spanos and Zeldin, 1998; Li and Chang, 2012). An analytical recursive solution for the LSE with unknown inputs (LSE-UI) has been proposed recently (Yang, et al., 2007a). Extended Kalman filter method has also been developed further to identify system parameters and unknown inputs (Yang, et al., 2007b; Pan, et al., 2016). These methods can be used to identify both system parameters and non-white excitations.
These methods, however, have been examined for MDoFs systems only where the number of unknown inputs is less than the number of DoFs. The problem of identification of physical parameters and unknown input of a system when the number of unknown input equals to the number of DoFs has not been analytically studied. This paper addresses this problem with the EKF-UI technique applied to a SDoF system, and a strategy is proposed for the identification with numerical validation. The aerodynamic force, stiffness and damping of a high-rise building model in wind tunnel test are further identified with the proposed approach illustrating problems and limitations of this application.

2. Background

2.1 Extended Kalman Filter with Unknown Input (EKF-UI)

The equation of motion of a m-DOFs structural system can be expressed as

$$\mathbf{M}\ddot{\mathbf{x}}(t)+\mathbf{C}\dot{\mathbf{x}}(t)+\mathbf{K}\mathbf{x}(t)=\mathbf{L}\mathbf{F}(t) \tag{1}$$

where **M**, **C** and **K** are $m\times m$ mass, damping and stiffness matrices of the structure. Rayleigh damping is assumed for the structure. $\mathbf{F}(t)=[f_1(t), f_2(t),...,f_q(t)]'$ is a vector of forces acting at q DoFs of the structure, **L** is a $m\times q$ mapping matrix associated with $\mathbf{F}(t)$. The elements of matrix $\boldsymbol{F}(t)$ can be either null or normally distributed random numbers.
Assuming there are n unknown parameters in the system such as the elemental stiffness and damping, the extended state vector with dimension $2m+n$ can be expressed as

$$\mathbf{Z}(t)=[\dot{\mathbf{x}}(t)^{\mathrm{T}},\mathbf{x}(t)^{\mathrm{T}},\boldsymbol{\theta}^{\mathrm{T}}]^{\mathrm{T}} \tag{2}$$

in which $\theta=[\theta_1,\theta_2,...,\theta_n]^{\mathrm{T}}$ is the n-unknown vector. When written in the state space format,

$$\dot{\mathbf{Z}}(t)=\mathrm{g}(\mathbf{Z},\mathbf{F},t)+\mathbf{w}(t) \tag{3}$$

where **F** is the unknown excitation and $\mathbf{w}(t)$ is the system noise vector with zero mean and covariance matrix $\mathbf{Q}(t)$. The system parameters are assumed non-varying with time with $\dot{\theta}=0$.
The observation vector of output can then be expressed in discrete form as

$$\mathbf{Y}_k=\mathrm{h}(\mathbf{Z}_k,\mathbf{F}_k,k)+\mathbf{v}_k \tag{4}$$

where $\mathbf{Y}_k$ is the l-dimension observation vector and v_k is the measurement noise vector with zero-mean and covariance matrix **R**.
Function g(•) can be obtained after linearization of Eq.(1) as

$$\mathrm{g}(\mathbf{Z},\mathbf{F},t)=\mathbf{A}\times\mathbf{Z}(t)+\mathbf{B}\times\mathbf{F}(t)=\begin{bmatrix}\mathbf{A}_1 & \mathbf{0}_{2m\times n}\\ \mathbf{0}_{n\times 2m} & \mathbf{0}_{n\times n}\end{bmatrix}\times\mathbf{Z}(t)+\begin{bmatrix}\mathbf{B}_1\\ \mathbf{0}_{n\times q}\end{bmatrix}\times\mathbf{F}(t) \tag{5}$$

where $$\mathbf{A}_1=\begin{bmatrix}\mathbf{0}_{m\times m} & \mathbf{M}\\ \mathbf{M} & \mathbf{C}\end{bmatrix}^{-1}\begin{bmatrix}\mathbf{M} & \mathbf{0}_{m\times m}\\ \mathbf{0}_{m\times m} & -\mathbf{K}\end{bmatrix},$$

and $$\mathbf{B}_1=\begin{bmatrix}\mathbf{0}_{m\times m} & \mathbf{M}\\ \mathbf{M} & \mathbf{C}\end{bmatrix}^{-1}\begin{bmatrix}\mathbf{0}_{m\times m}\\ \mathbf{I}_{m\times m}\end{bmatrix}\mathbf{L}=\begin{bmatrix}-\mathbf{M}^{-1}\mathbf{C}\mathbf{M}^{-1} & \mathbf{M}^{-1}\\ \mathbf{M}^{-1} & \mathbf{0}\end{bmatrix}\begin{bmatrix}\mathbf{0}_{m\times m}\\ \mathbf{I}_{m\times m}\end{bmatrix}\mathbf{L}=\begin{bmatrix}\mathbf{M}^{-1}\mathbf{L}\\ \mathbf{0}_{m\times m}\end{bmatrix}$$

If acceleration is included in the observation vector, then

$$\mathrm{h}(\mathbf{Z}_k,\mathbf{F}_k,k+1)=\mathbf{D}\times\mathbf{Z}_k+\mathbf{E}\times\mathbf{F}_k \tag{6}$$

where $\mathbf{D}=\mathbf{N}\left[-\mathbf{M}^{-1}\mathbf{C},-\mathbf{M}^{-1}\mathbf{K},\mathbf{0}_{m\times n}\right]$, $\mathbf{E}=\mathbf{M}^{-1}\mathbf{L}$, and **N** is a $l\times(2m+n)$ transition matrix indicating the location of measured responses.
If the unknown parameters θ and unknown excitation $\mathbf{F}(t)$ are required to be identified, the state vectors $\mathbf{Z}_k$ and $\mathbf{F}_k$ are needed. Let $\hat{\mathbf{Z}}_k$ and $\hat{\mathbf{F}}_k$ be the estimates of $\mathbf{Z}_k$ and $\mathbf{F}_k$. Then $\mathbf{g}_k$ and $\mathbf{h}_{k+1}$ can be expressed in terms of $\hat{\mathbf{Z}}_k$, $\hat{\mathbf{Z}}_{k+1|k}$ and $\hat{\mathbf{F}}_k$ by linearizing Eqs. (3) and (4) as

$$\mathbf{g}_k\approx\hat{\mathbf{g}}_k+\mathbf{G}_{k|k}(\mathbf{Z}_k-\hat{\mathbf{Z}}_k)+\mathbf{B}_{k|k}(\mathbf{F}_k-\hat{\mathbf{F}}_{k|k}) \tag{7}$$

$$\mathbf{h}_{k+1}\approx\hat{\mathbf{h}}_{k+1|k}+\mathbf{H}_{k+1|k}(\mathbf{Z}_{k+1}-\hat{\mathbf{Z}}_{k+1|k})+\mathbf{D}_{k+1|k}(\mathbf{F}_{k+1}-\hat{\mathbf{F}}_{k|k}) \tag{8}$$

where $\mathbf{g}_k=\mathbf{g}(\mathbf{Z}_k,\mathbf{F}_k,k\Delta t)$, $\mathbf{h}_{k+1}=\mathbf{h}(\mathbf{Z}_{k+1},\mathbf{F}_{k+1},k+1)$, $\hat{\mathbf{g}}_k=\mathbf{g}(\hat{\mathbf{Z}}_{k|k},\hat{\mathbf{F}}_k,k\Delta t)$, $\hat{\mathbf{h}}_{k+1|k}=\mathbf{h}(\hat{\mathbf{Z}}_{k+1|k},\hat{\mathbf{F}}_{k|k},k+1)$, and matrices $\mathbf{G}_{k|k}$, $\mathbf{B}_{k|k}$, $\mathbf{H}_{k+1|k}$, and $\mathbf{D}_{k+1|k}$ are Jacobian matrices.

With Eqs.(7) and (8) and the observations (y_1, y_2, ..., y_{k+1}), the estimates of $\mathbf{Z}_{k+1}$ and $\mathbf{F}_{k+1}$, $\hat{\mathbf{Z}}_{k+1|k+1}$ and $\hat{\mathbf{F}}_{k+1|k+1}$ can be calculated through minimizing the objective function J_{k+1} expressed as

$$J_{k+1} = \sum_{i=1}^{k+1} \mathbf{\Delta}_i^{\mathrm{T}} \mathbf{R}_i^{-1} \mathbf{\Delta}_i = \overline{\mathbf{\Delta}}_{k+1}^{\mathrm{T}} \mathbf{W}_{k+1}^{-1} \overline{\mathbf{\Delta}}_{k+1}, \qquad (11)$$

where $\mathbf{R}_i$ is the covariance matrix of measurement noise at time $t=i\Delta t$, and $\mathbf{\Delta}_i$ is the error between the observation $\mathbf{y}_i$ and the linearized output $\mathbf{h}(\mathbf{Z}_i, \mathbf{F}_i, i)$ with

$$\mathbf{\Delta}_i = \mathbf{y}_i - \mathbf{h}(\mathbf{Z}_i, \mathbf{F}_i, i), \quad \overline{\mathbf{\Delta}}_{k+1} = [\mathbf{\Delta}_1^{\mathrm{T}}, \mathbf{\Delta}_2^{\mathrm{T}}, ..., \mathbf{\Delta}_{k+1}^{\mathrm{T}}]^{\mathrm{T}}, \quad \mathbf{W}_{k+1}^{-1} = \begin{bmatrix} \mathbf{R}_1^{-1} & 0 & 0 & 0 \\ 0 & \mathbf{R}_2^{-1} & 0 & 0 \\ 0 & 0 & \ddots & \vdots \\ 0 & 0 & \cdots & \mathbf{R}_{k+1}^{-1} \end{bmatrix}$$

and vectors $\hat{\mathbf{Z}}_{k+1|k+1}$ and $\hat{\mathbf{F}}_{k+1|k+1}$ can be obtained from Yang, et al. (2007b).

In the case where displacement $\mathbf{x}(t)$ and acceleration $\ddot{\mathbf{x}}(t)$ are measured, the expression of function $\mathbf{h}(\mathbf{Z}_k, \mathbf{F}_k, k)$ is different from the case when only $\ddot{\mathbf{x}}(t)$ is measured. According to the discrete form of state-space function of the system,

$$\begin{bmatrix} \dot{\mathbf{x}}_{k+1} \\ \mathbf{x}_{k+1} \end{bmatrix} = (\mathbf{I}_{m\times m} + \mathbf{A}_1 \Delta t) \begin{bmatrix} \dot{\mathbf{x}}_k \\ \mathbf{x}_k \end{bmatrix} + \Delta t \mathbf{B}_1 \mathbf{F}_k = (\mathbf{I}_{m\times m} + \mathbf{A}_1 \Delta t) \begin{bmatrix} \dot{\mathbf{x}}_k \\ \mathbf{x}_k \end{bmatrix} + \Delta t \begin{bmatrix} \mathbf{M}^{-1}\mathbf{L} \\ \mathbf{0}_{m\times m} \end{bmatrix} \mathbf{F}_k \qquad (12)$$

One can see that the lower $m\times m$ elements of $\mathbf{B}_1$ are all zeros, indicating that the force at the kth time instant $\mathbf{F}_k$ does not have any effect on the displacement response of the system $\mathbf{x}_{k+1}$ at the (k+1)th time instant such that

$$\begin{bmatrix} \dot{\mathbf{x}}_{k+2} \\ \mathbf{x}_{k+2} \end{bmatrix} = (\mathbf{I}_{m\times m} + \mathbf{A}_1 \Delta t)^2 \begin{bmatrix} \dot{\mathbf{x}}_k \\ \mathbf{x}_k \end{bmatrix} + \Delta t (\mathbf{I}_{m\times m} + \mathbf{A}_1 \Delta t) \begin{bmatrix} \mathbf{M}^{-1}\mathbf{L} \\ \mathbf{0}_{m\times m} \end{bmatrix} \mathbf{F}_k \qquad (13)$$

Let $\mathbf{U} = (\mathbf{I}_{m\times m} + \mathbf{A}_1 \Delta t)^2$ and $\mathbf{V} = \Delta t (\mathbf{I}_{m\times m} + \mathbf{A}_1 \Delta t) \begin{bmatrix} \mathbf{M}^{-1}\mathbf{L} \\ \mathbf{0}_{m\times m} \end{bmatrix}$, we have

$$\begin{bmatrix} \dot{\mathbf{x}}_{k+2} \\ \mathbf{x}_{k+2} \end{bmatrix} = \mathbf{U} \begin{bmatrix} \dot{\mathbf{x}}_k \\ \mathbf{x}_k \end{bmatrix} + \mathbf{V}\mathbf{F}_k = \begin{bmatrix} \mathbf{U}_1 \\ \mathbf{U}_2 \end{bmatrix} \begin{bmatrix} \dot{\mathbf{x}}_k \\ \mathbf{x}_k \end{bmatrix} + \begin{bmatrix} \mathbf{V}_1 \\ \mathbf{V}_2 \end{bmatrix} \mathbf{F}_k, \qquad (14)$$

$$\mathbf{x}_{k+2} = \mathbf{U}_2 \times \begin{bmatrix} \dot{\mathbf{x}}_k \\ \mathbf{x}_k \end{bmatrix} + \mathbf{V}_2 \times \mathbf{F}_k. \qquad (15)$$

Matrix $\mathbf{U}_3$ is partitioned as $\mathbf{U}_3 = \mathbf{N}[\mathbf{U}_2 \quad \mathbf{0}_{m\times n}]$ to match the dimension of the state vector $\mathbf{Z}(t)$ and the locations of sensors. The relation between the observation vector and the excitations can finally be expressed as

$$\mathbf{Y}_k = \begin{bmatrix} \mathbf{x}_{k+2} \\ \ddot{\mathbf{x}}_k \end{bmatrix} = \begin{bmatrix} \mathbf{U}_3 \\ \mathbf{D} \end{bmatrix} \times \mathbf{Z}_k + \begin{bmatrix} \mathbf{V}_2 \\ \mathbf{E} \end{bmatrix} \times \mathbf{F}_k + \mathbf{v}_k. \qquad (16)$$

The error of identified excitation can further be examined based on the following equation

$$\text{Error of Identification} = \frac{\left\| \mathrm{F}_{\mathrm{id}}(\mathrm{t}) - \mathrm{F}_{\mathrm{real}}(\mathrm{t}) \right\|}{\left\| \mathrm{F}_{\mathrm{real}}(\mathrm{t}) \right\|} \times 100\% \qquad (17)$$

where, $\|\bullet\|$ is the second order norm of the vector, $\mathrm{F}_{\mathrm{id}}(\mathrm{t})$ is the vector of identified excitation and $\mathrm{F}_{\mathrm{real}}(\mathrm{t})$ is the vector of real excitation.

3 Proposed Strategy for identification of excitation and system parameter of a SDoF System

3.1 Identification of System Stiffness

3.1.1 Basic Theory

The displacement response of a SDoF system can be expressed in frequency domain as

$$X(i\varpi) = H(i\varpi)F(i\varpi) = \frac{1}{k}\left[\frac{1}{(1-\beta^2)+i(2\xi\beta)}\right]\cdot F(i\varpi), \tag{18}$$

where H is the response function, F is the Fourier transform of the applied force f, k is the stiffness, ϖ is frequency of the applied force, $\beta = \varpi/\omega$ is the ratio of the excitation frequency to the undamped natural frequency of the system, and ξ is the damping ratio.
Assuming two SDoF systems with unit mass whose natural frequency and stiffness are ω_1, k_1 and ω_2, k_2, respectively. Subscript 1 denotes the real system parameters and subscript 2 denotes an arbitrary system. The two systems have the responses $x_1(t)$ and $x_2(t)$ under different input $f_1(t)$ and $f_2(t)$. $X_1(i\omega)$ and $X_2(i\omega)$ are the Fourier transform of the displacement. Let $X_1(i\omega_1)$ and $X_2(i\omega_1)$ be the responses of the two systems with the excitation frequency $\varpi = \omega_1$. According to Eq. (18), let $X_1(i\omega_1) = X_2(i\omega_1)$ and we have

$$H_1(i\omega_1)F_1(i\omega_1) = H_2(i\omega_1)F_2(i\omega_1), \tag{19}$$

$$\frac{1}{k_1}\left[\frac{1}{(1-\beta_1^2)+i(2\xi_1\beta_1)}\right]\cdot F_1(i\omega_1) = \frac{1}{k_2}\left[\frac{1}{(1-\beta_2^2)+i(2\xi_2\beta_2)}\right]\cdot F_2(i\omega_1) \tag{20}$$

where $\beta_1 = \omega_1/\omega_1 = 1.0$, $\beta_2 = \omega_1/\omega_2$. The ratio between the amplitude of real load $F_1(i\omega_1)$ and identified excitation $F_2(i\omega_1)$ can be expressed as

$$R = \left|\frac{F_1(i\varpi_1)}{F_2(i\varpi_1)}\right| = \left|\frac{k_1}{k_2}\left[\frac{(1-\beta_1^2)+i(2\xi_1\beta_1)}{(1-\beta_2^2)+i(2\xi_2\beta_2)}\right]\right| = \frac{k_1}{k_2}\frac{2\xi_1}{\left|(1-\beta_2^2)+i(2\xi_2\beta_2)\right|}, \tag{21}$$

with $k_1/k_2 = m\omega_1^2/m\omega_2^2 = \beta_2^2$. Eq. (21) can further be simplified as

$$R = \frac{2\xi_1\beta_2^2}{\left|(1-\beta_2^2)+i(2\xi_2\beta_2)\right|}, \tag{22}$$

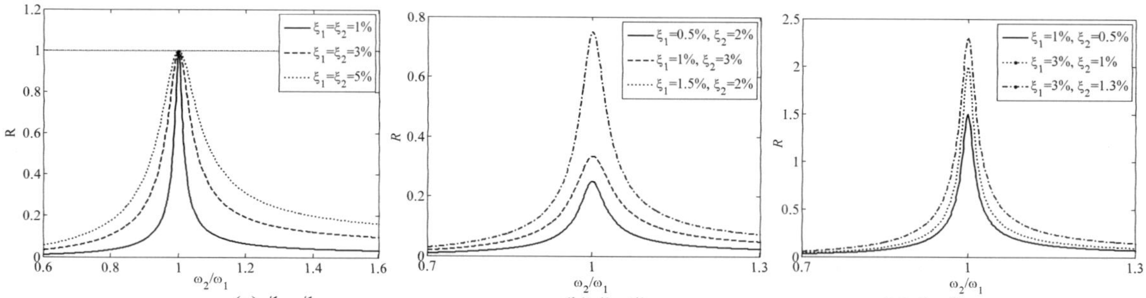

(a) $\xi_1=\xi_2$ (b) $\xi_1<\xi_2$ (c) $\xi_1>\xi_2$

Fig. 1 Ratio between amplitudes of excitations varies with ω_2/ω_1

Fig. 1(a) shows the variation of R with the change of ω_2/ω_1 for $\xi_1=\xi_2$. It can be clearly seen that $R<1.0$ for all the cases, except when $\omega_2/\omega_1=1.0$ where $R=1.0$. Figs. 1(b) and 1(c) give the values of R when $\xi_1<\xi_2$, and $\xi_1>\xi_2$ respectively. It is noted that two different systems can have exactly the same responses but with different sets of system parameters and excitations. Fig. 1 shows that when a system has its natural frequency different from that of the real system, the corresponding excitation at the natural frequency would be larger than that when the two systems have matching natural frequency, i.e. ω_2/ω_1=1.0. This property is noted independent of the damping ratios in the two systems.

3.1.2 Identification Procedure

The above feature of the two systems is then adopted as a tool of identification. For a SDoF system vibrating in the steady state, the corresponding excitation should have a minimum energy input at the resonant frequency. This behavior is independent of the damping values of the system.

The procedure to identify the stiffness of a SDoF system can be realized in the following steps:

(1) Specify several stiffnesses $\mathbf{K}_i$ (i=1,2,...,n) of the target system around the true stiffness estimated earlier, then identify the applied excitations by adopting each K_i and the assumed set of damping values.

(2) Calculate the power spectrum of the identified excitation corresponding to different system stiffness values, and obtain value of the highest spectral peak close to the prior judged resonant frequency.

(3) Plot all the spectral values versus the corresponding assigned stiffnesses to get a "U" shape curve. The true stiffness **K** of the target system should, therefore, associate with the minimum value of the curve.

3.2 Identification of System Damping and Excitation

3.2.1 Basic Theory

Again, let $X_1(i\bar{\omega}_1)$ and $X_2(i\bar{\omega}_2)$ be the displacement responses of the two systems with the same stiffness (natural frequency is $2\pi\omega_1$) but with different damping ratios of ξ_1(true system damping ratio) and ξ_2 (identified damping ratio) respectively. Let $X_1(i\bar{\omega}) = X_2(i\bar{\omega})$. According to Eq. (18), we have

$$H_1(i\bar{\omega})F_1(i\bar{\omega}) = H_2(i\bar{\omega})F_2(i\bar{\omega}), \tag{23}$$

$$\frac{1}{k}\left[\frac{1}{(1-\beta^2)+i(2\xi_1\beta)}\right]\cdot F_1(i\bar{\omega}) = \frac{1}{k}\left[\frac{1}{(1-\beta^2)+i(2\xi_2\beta)}\right]\cdot F_2(i\bar{\omega}), \tag{24}$$

where $\beta = \bar{\omega}/\omega_1$. The ratio R between the amplitudes of the two excitations $F_1(i\bar{\omega})$ and $F_2(i\bar{\omega})$, can be expressed as

$$R = \left|\frac{F_2(i\bar{\omega})}{F_1(i\bar{\omega})}\right| = \left|\frac{(1-\beta^2)+i(2\xi_2\beta)}{(1-\beta^2)+i(2\xi_1\beta)}\right| \tag{25}$$

The change of R with β is plotted in Fig. 2. For two systems with different damping ratios having exactly the same responses, it is noted that the magnitudes of excitation of the two systems are almost identical in the whole frequency domain except in the vicinity of the resonant frequency. If the identified damping is further away from the true system damping, the peak ($\xi_2>\xi_1$) or valley ($\xi_2<\xi_1$) close to the resonant frequency of the identified excitation will become higher. This means that if damping is not correctly identified, it will lead to result with the identified excitation fluctuating sharply close to the resonant frequency. In the present study, it has been assumed that the applied excitation is smooth around the resonant frequency of the target system. Thus, the identified damping ratio with the EKF-UI technique should associate with a relatively smooth spectrum of excitation close to the resonant frequency of the structure.

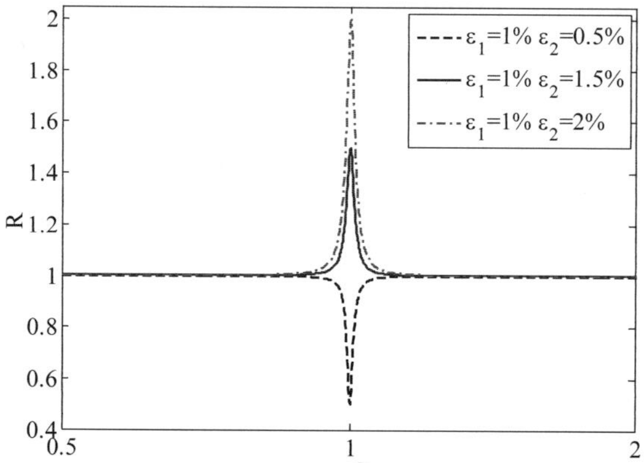

Fig. 2 Ratio between amplitudes of excitations

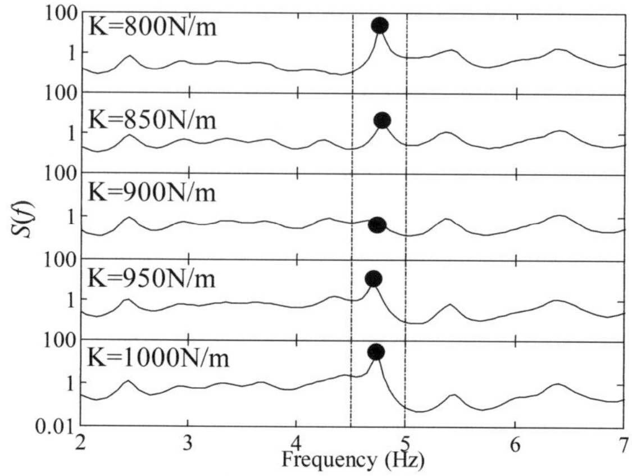

Fig. 3 Power spectra

3.2.1 Identification Procedure

The above feature of the two systems is adopted as a tool of identification. The damping ratio of the system can be estimated within a narrow range with the following procedure:

(1) Specify several damping $\mathbf{C}_i$ (i=1,2,..,n) of the target system within a selected frequency range encompassing the natural frequency of structure. Then identify the applied excitation for each C_i using the EKF-UI technique.

(2) Calculate the power spectrum of the identified excitations, and plot the spectrum within the selected range. The correct estimate of damping is selected as the one associated with a more smooth power spectrum of excitation close to the system natural frequency. The corresponding set of excitation is the identified excitation.

It should be noted that the true excitation theoretically has a smooth curve around the resonant frequency as shown in Fig. 2 and it associates with the correct system damping, yet it is almost impossible to find the exact damping values and the true excitation in practice. This is because the power spectrum of the real excitation fluctuates within the selected frequency range in practice. Therefore, the estimated damping will be in the form of a range within which the power spectrum of the corresponding excitations still has some fluctuations around the system natural frequency.

4. Numerical Example

For a SDoF system with mass M=1.0kg, K=875N/m, C=0.8Ns/m, the natural frequency and damping ratio of the system are 4.71Hz and 1.35%, respectively. A 200s long Gaussian white noise excitation is applied to the system, and displacement and acceleration are recorded simultaneously at a sampling frequency of 200Hz. The noise-to-signal ratio is assumed to be 1.0%. K, C, and F(t) are assumed to be unknown, but it is known that the stiffness K of the system is within the range of 800N/m to 1000N/m, which means the natural frequency is within the range of 4.5Hz to 5.0Hz. Damping ratio is known in the range of 0.5% to 2.0%. Therefore, the system parameters are estimated firstly by assuming K equal to 800N/m, 850N/m, 900N/m, 950N/m and 1000N/m in turn. Initial damping ratio is assumed as 1.0%. The power spectra of the identified excitations are shown in Fig. 3. The identified damping and the highest spectral peak values in the selected frequency range are shown in Table 1.

Table 1 Identified spectra maxima corresponding to different stiffnesses

Stiffness **K**(N/m)	800	850	900	950	1000
Identified damping **C**(Ns/m)	0.79	0.84	0.86	0.80	0.82
Maximum spectral value of force within [4.5Hz, 5.0Hz]	22.0	2.4	0.84	7.9	25.9

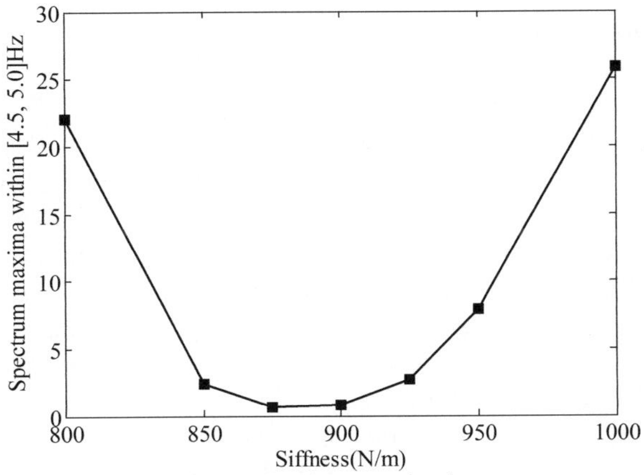

Fig. 4 - Maximum spectral value

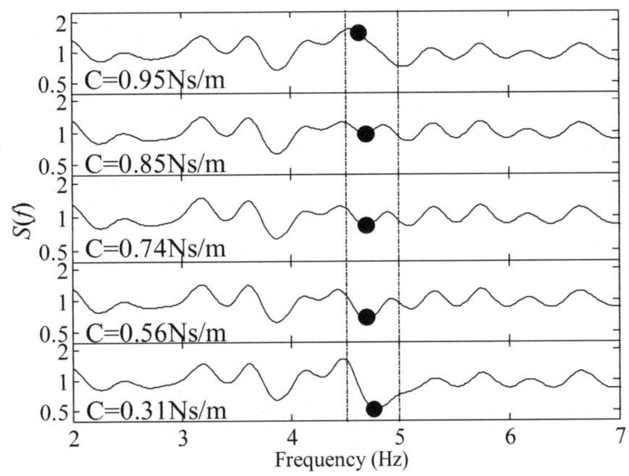

Fig. 5 Power spectra of excitation

It is found from Table 1 that the maximum value of the force spectrum within [4.5Hz, 5.0Hz] has the smallest values when K∈(850N/m, 900N/m). To find the true stiffness, two values of 875N/m and 925N/m are additionally selected within this range for the identification. The maximum values of the spectra within [4.5Hz, 5.0Hz] of all the examined cases are plotted against their corresponding stiffness values in Fig. 4. It is clear that the real stiffness should be around 875N/m. Therefore, the stiffness is taken as 875N/m. It should be noted that the accuracy of the identified stiffness is subjected to a fine grid of stiffness selected for the study.

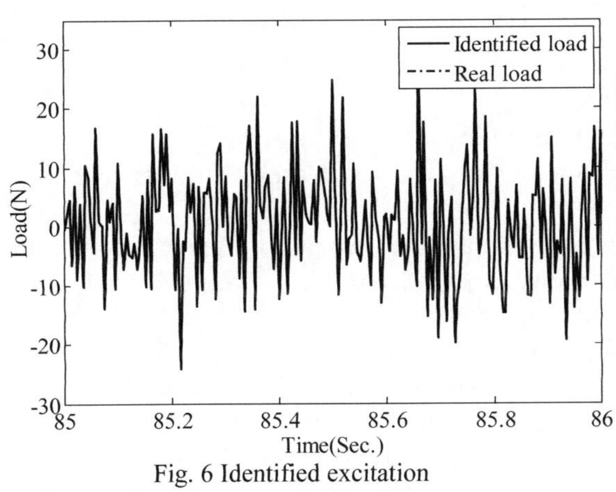

Fig. 6 Identified excitation

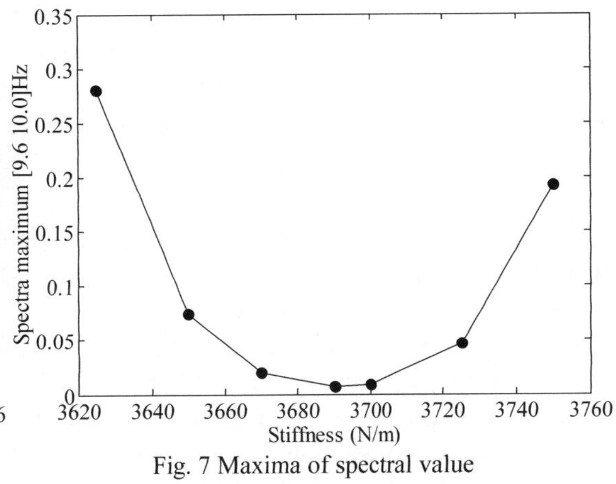

Fig. 7 Maxima of spectral value

In the next step to identify the damping ratio, the initial error covariance matrix of extended state vector is set as $\mathbf{P}_{Z,0|0} = diag\{0.1, 0.1, 0.01\}$. The power spectrum of the identified excitations corresponding to some selected damping values are shown in Fig. 5.

It can be clearly seen that, with the increase of damping from 0.31Ns/m to 0.95Ns/m, the peak valuc within the selected range gradually increases. Since it has been assumed *a prior* that the energy of the applied excitations should be smoothly varying around the resonant frequency of the system, damping values associate with sudden jump in the power spectrum of the excitation will be discarded. Therefore, based on results shown in Fig. 5, the damping of the target system should be approximately in the range of [0.6, 0.9]Ns/m. Fig. 6 compared the identified and real loadings when C= 0.75Ns/m (other value within [0.6, 0.9]Ns/m give similar results). It is noted that the difference between the identified and real loadings is very small and negligible.

5. Wind Load and System Parameter Identification for an Aero-elastic Model

There is a consensus in wind engineering that aerodynamic load and wind-induced vibration could not be measured simultaneously even in well controlled wind tunnel experiment. This is because if an aero-elastic structural model is adopted for wind induced responses measurement (say displacement, acceleration, et al.), the loads recorded by the High Frequency Force Balance underneath the base of model will include both the aerodynamic force and motion-induced base constraining force. The aerodynamic force cannot be obtained directly via this measurement.

SDoF system is the most commonly used model in wind tunnel experiment for high-rise building, as the higher modes can hardly be excited by the incident wind especially when the wind speed is lower than the design wind speed. The aerodynamic force, wind induced aerodynamic stiffness and damping of such a model are identified simultaneously by adopting the proposed method.

The aero-elastic model of a solar tower measures 10.8cm diameter and 1.22m high. The natural frequency, damping ratio of the first mode and weight are 9.75Hz (K=3750N/m), 0.7%, and 1.0kg respectively. The wind speed is measured at 1.0m high above the floor. The displacement and acceleration in the cross-wind direction were simultaneously measured at the top of building with a sampling frequency of 200Hz. The measured data are all low-pass filtered to reduce the influence of high frequency components.

It is known that Vortex Induced Vibration (VIV) in the structure in the cross-wind direction has the vortex shedding frequency matching the structural resonant frequency. This results in resonance of the system and the occurrence of strong negative aerodynamic damping which may greatly reduce the total damping ratio of the coupling wind and structural system. Preliminary check with the system shows that the VIV occurs at around 6.5 m/s. The proposed method and the procedures described in section 3 are adopted to identify the system stiffness and damping of the system respectively when the wind speed is 6.5 m/s.

The initial error covariance matrix of the extended state vector is set as $\mathbf{P}_{Z,0|0} = diag\{0.1, 0.1, 10^{-6}\}$.

Firstly, the wind load is identified with different assumed values of K with the numerical value of C assumed as 0.1Ns/m.

The highest spectral peak values of the identified excitation near the resonant frequency from using different K values are plot in Fig. 7. The smallest value is found corresponding to K=3690N/m. Therefore, the stiffness of the system is taken as 3690N/m.

By setting K=3690N/m, different initial guesses of damping within the range of [0.1, 0.5] Ns/m are used to obtain different wind load estimates. The power spectra of some of these wind load estimates are examined in Fig. 8. The amplified spectra close to the resonant frequency for C=0.1Ns/m and 0.2 Ns/m are similar and have a smoother distribution than the other ones.

The final identified wind load time series is shown in Fig. 9(a), and the corresponding identified damping is 0.155Ns/m. Fig. 9(b) is the power spectrum of the identified wind load. A spectral peak is clearly noted matching that in the power spectrum of the recorded acceleration. This indicates that the tower undergoes VIV at this incident wind speed. This means the proposed strategy can identify excitation that is not a white process. There is another peak at the first harmonic of the resonant frequency, and this is typical of response resulting from a periodic impulsive excitation. It may be concluded that the proposed strategy can successfully identify excitation and parameters of a SDoF simultaneously.

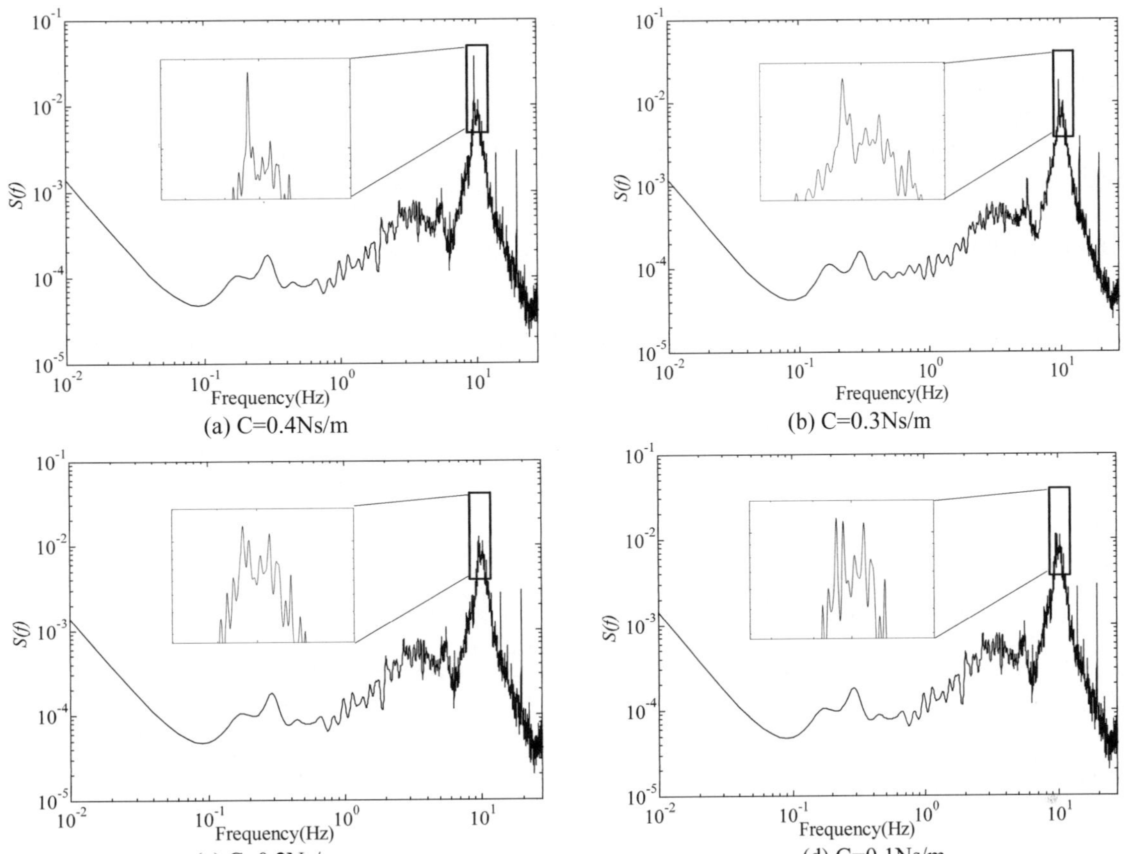

(a) C=0.4Ns/m

(b) C=0.3Ns/m

(c) C=0.2Ns/m

(d) C=0.1Ns/m

Fig. 8 Power spectrum of wind load identified based on different damping

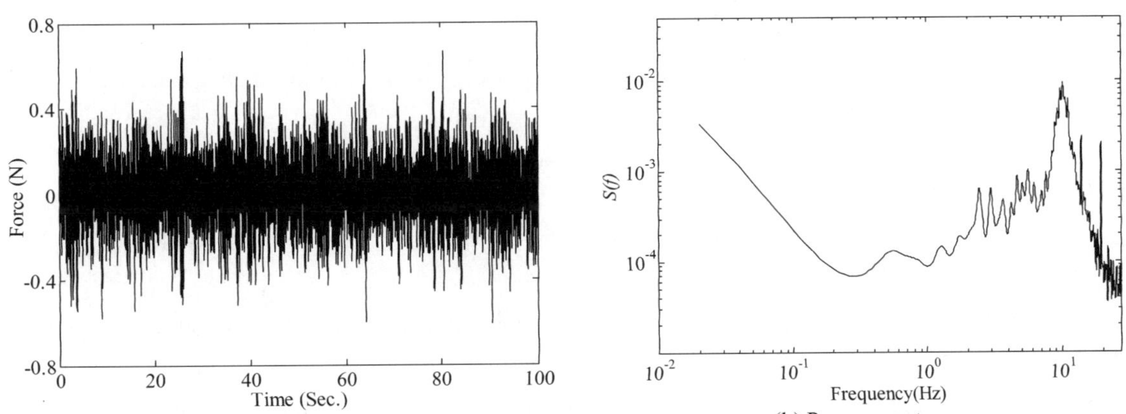

(a) Time series

(b) Power spectrum

Fig. 9 Identified wind load

6. Conclusions

An identification strategy based on the EKF-UI technique is proposed to identify the unknown excitation and system parameters of a SDoF system based on output only. The applied excitation is assumed stationary random with the energy smoothly distributed in a frequency range close to the natural frequency of the target system. The proposed strategy can successfully identify the wind load acting on a high-rise building as well as the system parameters of the wind-structure coupling system. The negative aerodynamic damping and stiffness can then be identified from wind tunnel test data.

Acknowledgements

This study is supported by the National Natural Science Foundation of China (51408207), and the Key laboratory of Bridge Wind Resistant and New Technology of Hunan province (15K028). Support from the 111 project (B13002) is also acknowledged.

References

Loh, CH, Lin, CY & Huang, CC. 2000. 'Time domain identification of frames under earthquake loadings.' *J. Eng. Mech.* (ASCE), vol. 126, pp. 693–703.

Yang, JN & Lin, S. 2004. 'On-line identification of nonlinear hysteretic structures using an adaptive tracking technique.' *Int. J. NonLinear Mech.*, vol. 39, pp. 1481–1491.

Law, SS, Lin, JF & Li, XY. 2012. 'Structural Condition Assessment from White Noise Excitation and Covariance of Covariance Matrix.' *AIAA J.*, vol. 50, pp. 1503–1512.

Hui, Y, Law, SS & Ku, CJ. 2017. 'Structural damage detection based on covariance of covariance matrix with general white noise excitation.' *J. Sound Vib.* vol. 389, pp. 168-182.

Huang, CS & Yeh, CH. 1999. 'Some Properties of Random dec Signatures.' *Mech. Syst. Signal. Pr.*, vol. 13, pp. 491–507.

Yu, DJ, & Ren, WX. 2005. 'EMD-based stochastic subspace identification of structures from operational vibration measurements.' *Eng. Struct.*, vol. 27, pp. 1741–1751.

Spanos, PD & Zeldin, BA. 1998. 'Generalized Random Decrement Method For Analysis Of Vibration Data.' *J. Vib. Acoust.*, vol. 120, pp. 806–813.

Li, Z, & Chang, C. 2012. 'Tracking of Structural Dynamic Characteristics Using Recursive Stochastic Subspace Identification and Instrumental Variable Technique.' *J. Eng. Mech.* (ASCE), vol. 138, pp. 591–600.

Yang, JN, Pan, S, & Lin, S. 2007a. 'Least-Squares Estimation with Unknown Excitations for Damage Identification of Structures.' *J. Eng. Mech.* (ASCE), vol. 133, pp. 12–21.

Yang, JN, Pan, S, & Huang, H. 2007b. 'An adaptive extended Kalman filter for structural damage identifications II: unknown inputs.' *Struct. Control Health Monit.*, vol. 14, pp. 497–521.

Pan, S, Xiao, D, Xing, S, Law, SS, Du, P, & Li, Y. 2016. 'A general extended Kalman filter for simultaneous estimation of system and unknown inputs.' *Eng. Struct.*, vol. 109, pp. 85–98.

The 8th International Conference on
Structural Health Monitoring of Intelligent Infrastructure
Brisbane, Australia | 5-8 December 2017

Operational Modal Analysis Using Time-Frequency Stochastic System Identification

C.M. Chang[1], S.K. Huang [2]

[1] *Assistant Professor, Department of Civil Engineering, National Taiwan University, Taiwan, Email:* changcm@ntu.edu.tw [2] *Assistant Researcher, National Center for Research on Earthquake Engineering, Taiwan*

Abstract

Structural health monitoring assesses structural integrity by observing vibration responses of structures. Operational modal analysis is one of structural health monitoring approaches aiming at extracting the dynamic characteristics of structures from the vibration responses when these structures are under their operating conditions. These characteristics include natural frequencies, damping ratios, and mode shapes. Deviations in these characteristics represent the changes in structural properties and also imply possible damage to structures. In this study, a new operational modal analysis is proposed using stochastic subspace system identification based on multivariate time-frequency distributions. These time-frequency distributions are computed from the short-time Fourier transform, and a time-frequency matrix is subsequently obtained by stacking these distributions with respect to time. As the derivation in the data-driven stochastic subspace system identification, the future time-frequency matrix is projected onto the past time-frequency matrix. By exploiting the singular value decomposition, the system and measurement matrices of a stochastic state-space representation are derived. Consequently, the dynamic characteristics of a structure are extracted. As compared to the time-domain stochastic subspace system identification, the proposed method utilizes the past and future matrices with a lower dimension in projection. A spectral magnitude envelope can be applied to the time-frequency matrix to highlight the major frequency components as well as to eliminate the components with less influence. To demonstrate the effectiveness of the proposed method, both simulation and experimental data are employed to derive dynamic characteristics of structures. The proposed method is also compared with other operational modal analysis methods such as the stochastic subspace identification and frequency domain decomposition. As seen in the result, the proposed method is superior to the conventional methods, in particular of the dynamic characteristics at higher modes.

1. Introduction

Visual inspection is currently the most available practice in monitoring the safety of civil structures. High costs and unreliability limit the use of visual inspection to infrequency occurrences. Structural health monitoring (SHM) of civil infrastructure in real time can reduce inspection and repair costs, as well as downtime, all while providing increased public safety. The principles of SHM and its application will aid not only in the inspection of existing infrastructure, but also facility enhanced monitoring of future construction projects as well. Moreover, after natural disasters, it is imperative that emergency facilities and evacuation routes, including bridges and highways, be assessed for safety in a timely manner. Continuous and automated structural health monitoring can facilitate these goals.

Operational modal analysis is a SHM method that extracts the dynamic characteristics of structures from vibrational responses, such as natural frequencies, damping ratios, and mode shapes. Deviations in these dynamic characteristics indicate the changes in structural properties. Farrar and James (1997) extracted the modal properties of a bridge by means of system identification. Brownjohn (2003) identified tall buildings for exploring the natural frequencies and mode shapes. Lynch *et al.* (2006) employed wireless sensors to perform operational modal analysis on a bridge. Siringoringo and Fujino (2008) developed a systematic system identification method and applied this method on a long-span bridge. Weng *et al.* (2008) facilitated a wireless sensor network to carry out a structural health monitoring campaign on a cable-stayed bridge and explored the dynamic characteristics of the bridge deck and cables. Jang *et al.* (2010) demonstrated structural health monitoring using a dense array of wireless sensors on a bridge. The archived data were employed to discover the dynamic characteristics of this bridge. Operational modal analysis has become a commonly-used strategy for long-term monitoring of structures in order to assure their serviceability and safety.

Stochastic system identification is often exploited in operational modal analysis because of unavailable input excitation. This sort of system identification methods translate measured responses of structures into a mathematical model in a stochastic sense and subsequently extract the modal properties from this model. In time domain, Van Overschee and De Moor (1991) proposed the stochastic subspace system identification method which utilized the matrix orthogonal projection to derive a stochastic state-space representation from the extended observability and controllability matrices. Then, the modal properties were extracted from the stochastic system and measurement matrices. Peeters and De Roeck (1999) further extended this method by introducing reference outputs for the derivation of the extended observability and controllability matrices, resulting in reduced computational loading. Moreover, the number of states requires engineering judgements, and this type of methods may yield nonstructural modes to be identified. Thus, Peeters and De Roeck (2001) proposed the stabilization diagram for subspace system identification to improve the identification quality. By searching a number of identified modes, the stable ones are viewed as the significant modes of structures. However, these modified subspace system identification methods highly depends on users' experience and may not always result in high-quality outcomes.

To lower the chance of deriving the nonstructural modes, many researchers have integrated the signal processing methods into stochastic system identification. One of the examples is to apply the singular spectrum analysis (Broomhead *et al.* 1987) to measured outputs before performing stochastic system identification. This method retains the most influential content and eliminates noise and uncertainties in signals. Loh and Liu (2013) applied the singular spectrum analysis in a system identification problem of a bridge, and the energy of the contaminated modes were successfully reduced in the identification process. Chang and Loh (2015) proposed an improved stochastic subspace system identification method by considering the singular spectrum analysis and carried out this method to extract the modal properties of a bridge. To some extent, this method is helpful for avoiding the nonstructural modes to be identified. With this time series analysis, the stochastic subspace system identification, however, still needs to search a large amount of identified modes and to generate a stabilization diagram for assuring structural modes.

Alternatively, the stochastic subspace system identification can be performed in frequency domain. McKelvey *et al.* (1996) proposed a subspace-based multivariable system identification method using transfer functions. An input-output system identification method was realized by mapping the subspace system identification into a multiple-input and multiple-output transfer function matrix. Cauberghe (2004) further extended this method into operational modal analysis. This method formed a Hankel matrix by the frequency responses, which were calculated using the discrete Fourier transform to the time-domain responses. By utilizing the stabilization diagram, the modal properties of a structure can be obtained. Similarly, this method requires prior knowledge to render high-quality results. To identify a specific mode found in the spectral density curve, success is not guaranteed. In addition, another operational modal analysis in frequency domain is the frequency domain decomposition to a power spectral density matrix (Brincker *et al.* 2000). Through the peak-picking technique (Bendat and Piersol 1993), the structural modes can be picked from the eigen-domain spectral density function. Then, both natural frequencies and mode shapes were identified. In this method, the quality of identified natural frequencies highly replied on the number of spectral lines used. Insufficient spectra lines may result in distorted identified modal properties. Improvements can be made to these methods in order to deliver high-quality identification results.

In this study, a new stochastic subspace system identification method is introduced using multivariate time-frequency distributions. First, a multivariate time-frequency distribution matrix is formed using the discrete Fourier transform, and this matrix has the outputs stacked in order along the timewise row, resulting in a time-frequency Hankel matrix. The columns of this time-frequency matrix represents the transformed frequency responses at a specific frequency over time. As described in Van Overschee and De Moor (1991), this matrix is divided into the "past" and "future" matrices. By projecting the future matrix onto the past one, an observability matrix can be obtained from this subspace. The modal properties are finally extracted from this observability matrix. As compared to the time-domain subspace system identification, the size of the time-frequency Hankel matrix is reduced to a half because of the symmetry properties. The proposed method can window the time-frequency data within a specific frequency range and then perform the matrix orthogonal projection. Thus, this method extracts the modal properties in a single mode at a time. A numerical example of lumped-mass buildings with different heights is provided in this study to demonstrate all these features. In addition, the proposed method is also applied in a system identification problem of an eight-story building using shake table testing. This method successfully identifies the natural frequencies and mode shapes of this building with high quality and reduced computational time.

2. Time-Frequency Stochastic Subspace System Identification

The proposed method consists of four steps: 1) forming a time-frequency Hankel matrix, 2) matrix orthogonal projection, 3) complex-domain state-space realization, and 4) mode extraction. The Hankel matrix is formed as time and frequency in rows and columns, respectively, while the outputs are stacked in rows at a time. The matrix orthogonal projection is employed to obtain the subspace that contains the complex-domain extended observability matrix. Then, the system and measurement matrices are derived by specifying a number of complex states. The modal properties including the natural frequencies, damping ratios, and mode shapes are consequently calculated based on the system and measurements matrices.

Consider a discrete-time stochastic state-space model given by

$$\begin{aligned} &\mathbf{x}[t+1]=\mathbf{A}_d\mathbf{x}[t]+\mathbf{w}[t] \\ &\mathbf{y}[t]=\mathbf{C}_d\mathbf{x}[t]+\mathbf{v}[t] \\ &\mathbf{x}\in\mathbb{R}^{n_s\times 1},\ \mathbf{y}\in\mathbb{R}^{n\times 1},\ \mathbf{w}\in\mathbb{R}^{n_s\times 1},\ \mathbf{v}\in\mathbb{R}^{n\times 1} \end{aligned} \tag{1}$$

where $\mathbf{x}$ and $\mathbf{y}$ are the state and output measurement vector at the time step t; $\mathbf{A}_d$ and $\mathbf{C}_d$ are the system and measurement matrices in the stochastic state-space representation; $\mathbf{w}$ and $\mathbf{v}$ are the input and output Gaussian noise with a zero mean; and n_s and n are the number of states and outputs, respectively. The time span is defined as $t\in[1,N]$ where N is the total number of samples. Thus, the frequency domain state vector $\mathbf{X}[k]\in\mathbb{C}^{n_s\times 1}$ can be presented by a recursive equation (McKelvey and Viberg 2001), given by

$$W_{N_{de}}^{k}\mathbf{X}[k]=\mathbf{A}_{\mathrm{d}}\mathbf{X}[k]+\left(\mathbf{I}-W_{N_{de}}^{k}\mathbf{A}_{\mathrm{d}}^{N_{de}-1}\right)\mathbf{K}\mathbf{V}[k]$$
$$\mathbf{Y}[k]=\mathbf{C}_{\mathrm{d}}\mathbf{X}[k] \tag{2}$$

where $W_{N_{de}}=e^{\frac{j2\pi}{N_{de}}}$; $j=\sqrt{-1}$; k represents the k-th spectral lines; N_{de} represents the window length, which is equal to the number of discrete Fourier transform points; $\mathbf{V}[k]\in\mathbb{C}^{n\times 1}$ is the frequency-domain noise measurement noise, and $\mathbf{K}$ is the Kalman gain (Van Overschee and De Moor 1991). For N samples of $\mathbf{y}$ in Eq. (1), a Hankel matrix in time domain can be formed by

$$\bar{\mathbf{H}}=\left[\frac{\bar{\mathbf{Y}}_{\mathrm{p}}[1:N_{de}+l-1]}{\bar{\mathbf{Y}}_{\mathrm{f}}[l+1:N]}\right]=\begin{bmatrix}\mathbf{y}[1] & \mathbf{y}[2] & \cdots & \mathbf{y}[N_{de}]\\ \mathbf{y}[2] & \mathbf{y}[3] & \cdots & \mathbf{y}[N_{de}+1]\\ \vdots & \vdots & \ddots & \vdots\\ \mathbf{y}[l] & \mathbf{y}[l+1] & \cdots & \mathbf{y}[N_{de}+l-1]\\ \mathbf{y}[l+1] & \mathbf{y}[l+2] & \cdots & \mathbf{y}[N_{de}+l]\\ \mathbf{y}[l+2] & \mathbf{y}[l+3] & \cdots & \mathbf{y}[N_{de}+l+1]\\ \vdots & \vdots & \ddots & \vdots\\ \mathbf{y}[N-N_{de}+1] & \mathbf{y}[N-N_{de}+2] & \cdots & \mathbf{y}[N]\end{bmatrix} \tag{3}$$

where $l=(N-N_{de})/2$; "p" and "f" denote the "past" and "future" output Hankel matrix. Take the discrete Fourier transform to each row in $\bar{\mathbf{H}}$, and Eq. (3) becomes

$$\mathbf{H}=\left[\frac{\mathbf{Y}_{\mathrm{p}}}{\mathbf{Y}_{\mathrm{f}}}\right]=\begin{bmatrix}\mathbf{Y}[0] & \mathbf{Y}[1] & \cdots & \mathbf{Y}[N_{de}]\\ W_{N_{de}}^{-1}\mathbf{Y}[0] & W_{N_{de}}^{-1}\mathbf{Y}[1] & \cdots & W_{N_{de}}^{-1}\mathbf{Y}[N_{de}]\\ \vdots & \vdots & \ddots & \vdots\\ W_{N_{de}}^{-(l-1)}\mathbf{Y}[0] & W_{N_{de}}^{-(l-1)}\mathbf{Y}[1] & \cdots & W_{N_{de}}^{-(l-1)}\mathbf{Y}[N_{de}]\\ W_{N_{de}}^{-l}\mathbf{Y}[0] & W_{N_{de}}^{-l}\mathbf{Y}[1] & \cdots & W_{N_{de}}^{-l}\mathbf{Y}[N_{de}]\\ W_{N_{de}}^{-(l+1)}\mathbf{Y}[0] & W_{N_{de}}^{-(l+1)}\mathbf{Y}[1] & \cdots & W_{N_{de}}^{-(l+1)}\mathbf{Y}[N_{de}]\\ \vdots & \vdots & \ddots & \vdots\\ W_{N_{de}}^{-(N-N_{de})}\mathbf{Y}[0] & W_{N_{de}}^{-(N-N_{de})}\mathbf{Y}[1] & \cdots & W_{N_{de}}^{-(N-N_{de})}\mathbf{Y}[N_{de}]\end{bmatrix} \tag{4}$$

When $\mathbf{Y}_{\mathrm{f}}$ is projected on to $\mathbf{Y}_{\mathrm{p}}$, denoted by $\mathbf{Y}_{\mathrm{f}}/\mathbf{Y}_{\mathrm{p}}$, then the subspace $\mathbf{O}$ can be obtained by

$$\mathbf{Y}_{\mathrm{f}}/\mathbf{Y}_{\mathrm{p}}=\mathbf{O}=\begin{bmatrix}\mathbf{C}_{\mathrm{d}}\\ \mathbf{C}_{\mathrm{d}}\mathbf{A}_{\mathrm{d}}\\ \vdots\\ \mathbf{C}_{\mathrm{d}}\mathbf{A}_{\mathrm{d}}^{l-1}\end{bmatrix}\begin{bmatrix}\mathbf{X}[0] & \mathbf{X}[1] & \cdots & \mathbf{X}[N_{de}]\end{bmatrix}=W_{N_{de}}^{l}\mathbf{Y}_{\mathrm{f}}\mathbf{Y}_{\mathrm{p}}^{H}\left(\mathbf{Y}_{\mathrm{p}}\mathbf{Y}_{\mathrm{p}}^{H}\right)^{-1}\mathbf{Y}_{\mathrm{p}} \tag{5}$$

where $\mathbf{O}$ is the extended observability matrix, and $\mathbf{\Gamma}$ is the extended controllability matrix. $W_{N_{de}}^{l}$ in Eq. (5) represents the shifted phase, which doesn't affect the identification results. By utilizing the singular value decomposition, the matrices $\mathbf{A}_{\mathrm{d}}$ and $\mathbf{C}_{\mathrm{d}}$ can be derived from

$$\mathbf{O}=\mathbf{U}_{\mathrm{O,m}}\mathbf{\Lambda}_{\mathrm{O,m}}\mathbf{V}_{\mathrm{O,m}}^{T}+\mathbf{U}_{\mathrm{O,0}}\mathbf{\Lambda}_{\mathrm{O,0}}\mathbf{V}_{\mathrm{O,0}}^{T} \tag{6}$$

and

$$\mathbf{\Gamma}=\mathbf{U}_{\mathrm{O,m}}=\begin{bmatrix}\mathbf{C}_{\mathrm{d}}\\ \mathbf{C}_{\mathrm{d}}\mathbf{A}_{\mathrm{d}}\\ \mathbf{C}_{\mathrm{d}}\mathbf{A}_{\mathrm{d}}^{2}\\ \vdots\\ \mathbf{C}_{\mathrm{d}}\mathbf{A}_{\mathrm{d}}^{l-1}\end{bmatrix},\ \mathbf{A}_{\mathrm{d}}=\begin{bmatrix}\mathbf{C}_{\mathrm{d}}\\ \mathbf{C}_{\mathrm{d}}\mathbf{A}_{\mathrm{d}}\\ \mathbf{C}_{\mathrm{d}}\mathbf{A}_{\mathrm{d}}^{2}\\ \vdots\\ \mathbf{C}_{\mathrm{d}}\mathbf{A}_{\mathrm{d}}^{l-2}\end{bmatrix}^{\dagger}\begin{bmatrix}\mathbf{C}_{\mathrm{d}}\mathbf{A}_{\mathrm{d}}\\ \mathbf{C}_{\mathrm{d}}\mathbf{A}_{\mathrm{d}}^{2}\\ \mathbf{C}_{\mathrm{d}}\mathbf{A}_{\mathrm{d}}^{3}\\ \vdots\\ \mathbf{C}_{\mathrm{d}}\mathbf{A}_{\mathrm{d}}^{l-1}\end{bmatrix} \tag{7}$$

where the subscripts "m" and "0" denote the main and null spaces of $\mathbf{O}$; $\mathbf{\Lambda}$ is a diagonal matrix containing the singular values in a descending order; $\mathbf{\Gamma}$ represents the extended observability matrix and has a column rank n_s; $(\ )^\dagger$ denotes the pseudo inverse. $\mathbf{A}_d$ is derived from Eq. (7), and $\mathbf{C}_d$ is obtained from the first block matrix in $\mathbf{\Gamma}$. Finally, the natural frequencies, damping ratios, and mode shapes can be extracted from $\mathbf{A}_d$ and $\mathbf{C}_d$ by the eigen analysis. Thus, all these modal properties in the *q-th* mode are represented by

$$\mathbf{A}_{\mathrm{d}}\boldsymbol{\eta} = \lambda_{\mathrm{d}}\boldsymbol{\eta}$$
$$\frac{1}{\Delta t}\ln\left(\lambda_{\mathrm{d},q}, \lambda_{\mathrm{d},q}^{*}\right) = -\xi_q\omega_{\mathrm{n},q} \pm j\omega_{\mathrm{n},q}\sqrt{1-\xi_q^2},\ q = 1 \sim n_s \quad (8)$$
$$\boldsymbol{\varphi} = \mathbf{C}_{\mathrm{d}}\boldsymbol{\eta}$$

where λ_{d} is one of the eigenvalues in $\mathbf{A}_d$; λ_{d}^{*} is the complex conjugated λ_{d}; $\boldsymbol{\eta}$ is the eigenvector with respect to the state vector; ω_n and ξ are the natural frequency and damping ratio; Δt is the sample time; and $\boldsymbol{\varphi}$ is the mode shapes with respect to the outputs. By calculating all eigenvalues and eigenvectors from $\mathbf{A}_d$, the modal parameters can be extracted.

3. Numerical Example

In this study, a 10-story lumped-mass buildings is chosen as an example to demonstrate performance of the proposed identification method. The natural frequencies of these three buildings are listed in Table 1. The damping ratio at each mode is 0.01. All simulation data are sampled in 1/200 seconds.

Table 1. Natural frequencies of the numerical building.

Mode unit: Hz	1	2	3	4	5	6	7	8	9	10
10-story building	0.95	2.81	4.62	6.32	7.89	9.27	10.45	11.40	12.08	12.51

The proposed system identification method is used in these examples. The measurement duration is 102.4 seconds, which is equivalent to 20,480 samples. The output response measured is absolute floor accelerations. The number of delays used to form the Hankel matrix in Eq. (3) is 2,048. Because the symmetry in the discrete Fourier transform is considered, the resulting dimension of **H** in Eq. (4) is (18,342×n) × 1025. Figures 1 exhibits the identified mode shapes as compared to the true ones. To understand the quality of identified mode shapes, the mode assurance criteria is employed and given by

$$\mathrm{MAC}(q) = \frac{\boldsymbol{\varphi}_{true}^{T}\boldsymbol{\varphi}_{id}}{\|\boldsymbol{\varphi}_{true}\|\|\boldsymbol{\varphi}_{id}\|} \quad (9)$$

$\boldsymbol{\varphi}_{true}$ and $\boldsymbol{\varphi}_{id}$ denote the true and identified mode shapes. For the this building, the last modes cannot be successfully identified because the natural frequencies are too close to each other. The identified mode shapes at these modes are coupled, and they are not shown in Figure 4. The mode assurance criteria values for this case are 0.999, 0.994, 0.976, 0.992, 0.988, 0.995, 0.988, and 0.915 from the 1st mode to 8th mode.

4. Experimental Validation

An 8-story building is excited by a white noise signal using shake table testing. As shown in Figure 2, each story is 1.1 m high and supported by four uniform steel columns. The moment of inertia of these I-shaped columns is 3.69×10^{-6} m^4. In testing, additional mass blocks are added to floors, and the resulting mass including the floor plate at each floor is 760 kg. Accelerometers are attached to the side of floor plates and measures floor accelerations along the strong axis. The peak ground acceleration of input white noise excitation is around 0.05g. The duration is 80 seconds, and the sampling period is 0.005 seconds.

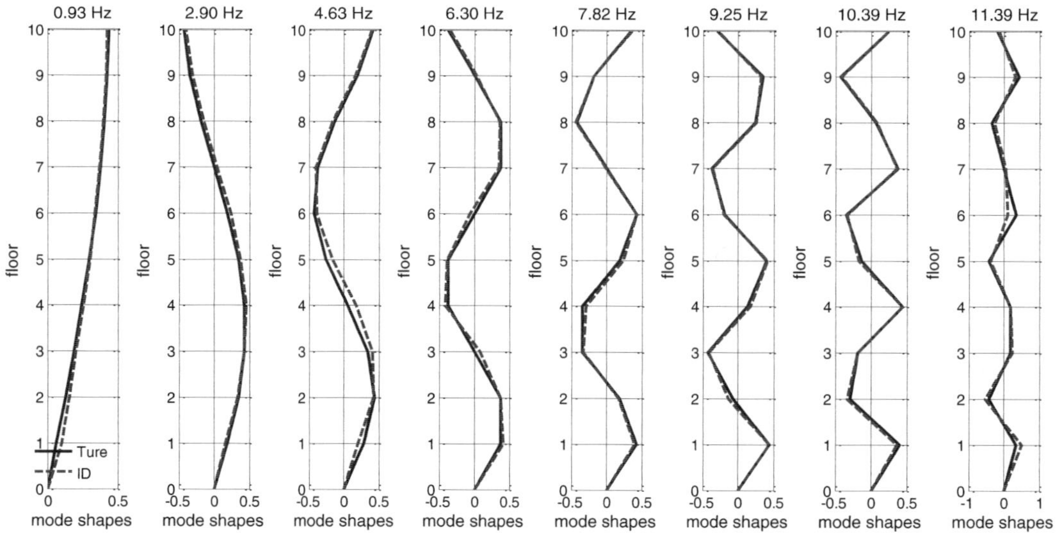

Figure 1. Identified mode shapes of 10-story building.

(a) (b)

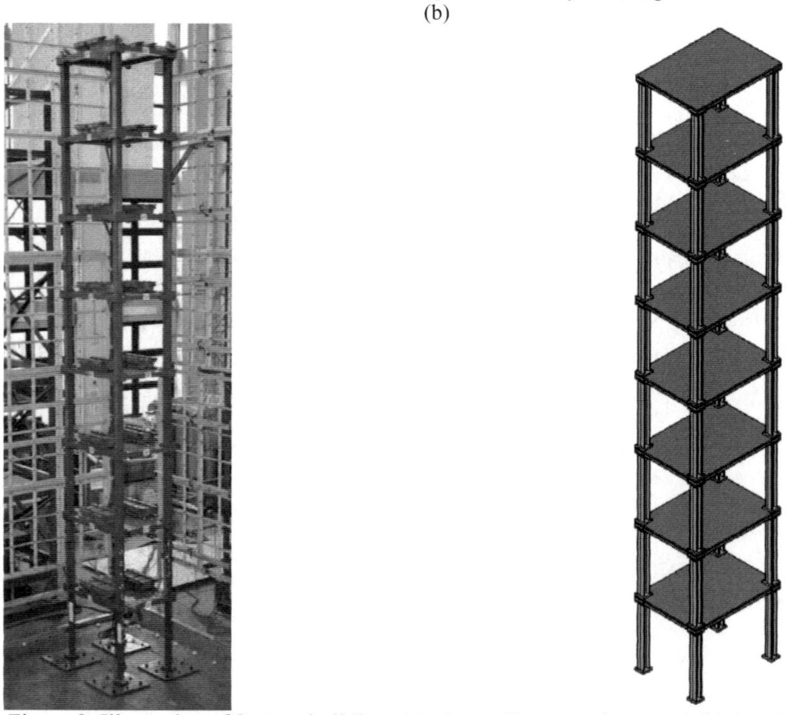
Figure 2. Illustration of 8-story building: (a) photo of test specimen and (b) drawing.

All floor acceleration data are used in system identification. These data are preprocessed to eliminate the signal offsets. The idol portion of the data is removed, and the resulting duration is 62.5 seconds. Typically, resampling the data to a lower sampling rate is employed before performing system identification, but the sample rate remain unchanged (i.e., 200 Hz) in analysis. The number of fast Fourier transform points used is 2048, and the resulting dimension of **H** in Eq. (4) is 83,632 × 1025. The number of picked natural frequencies from the frequency domain decomposition method is 7. Because a low-pass filter with a 50-Hz cutoff frequency is embedded in the data acquisition system, the frequency amplitudes are much lower above 45 Hz. Figure 3 illustrates the power spectral density generated from the frequency domain decomposition method and its residual spectral density in eigen domain. The natural frequencies obtained from the frequency domain decomposition method are 1.86, 6.74, 12.99, 19.43, 25.88, 32.13, and 36.23 Hz.

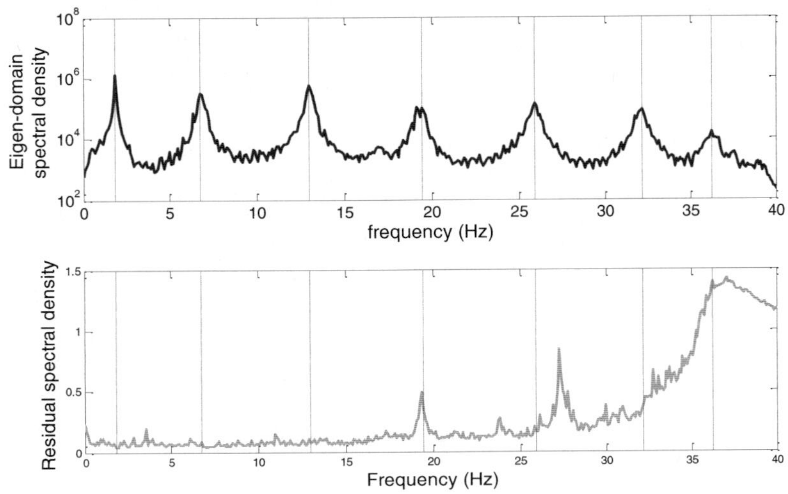

Figure 3. Peak-picking results from the power spectral density computed by the frequency-domain decomposition method.

As found in the results, the identified natural frequencies are 1.87, 6.80, 13.02, 19.45, 25.87, 32.13, and 36.25 Hz. The difference between the picked and identified natural frequencies from the 1st mode to 7th mode are 1.01%, 0.86%, 0.23%, 0.06%, 0.04%, 0.01%, and 0.06%. Moreover, Figure 4 displays the identified mode shapes using the proposed system identification method. The identified mode shapes are also compared to the results using the method in Chang and Loh (2015). As can be seen, the proposed method is able to identify all visible modes, while the other method can only obtain six modes. The mode shapes obtained from both methods look similar, and the mode assurance criteria values between these two methods are 0.995, 0.999, 0.999, 0.993, 0.993, 0.997, and 0.994. Therefore, the proposed method outperforms the method used in Chang and Loh (2015).

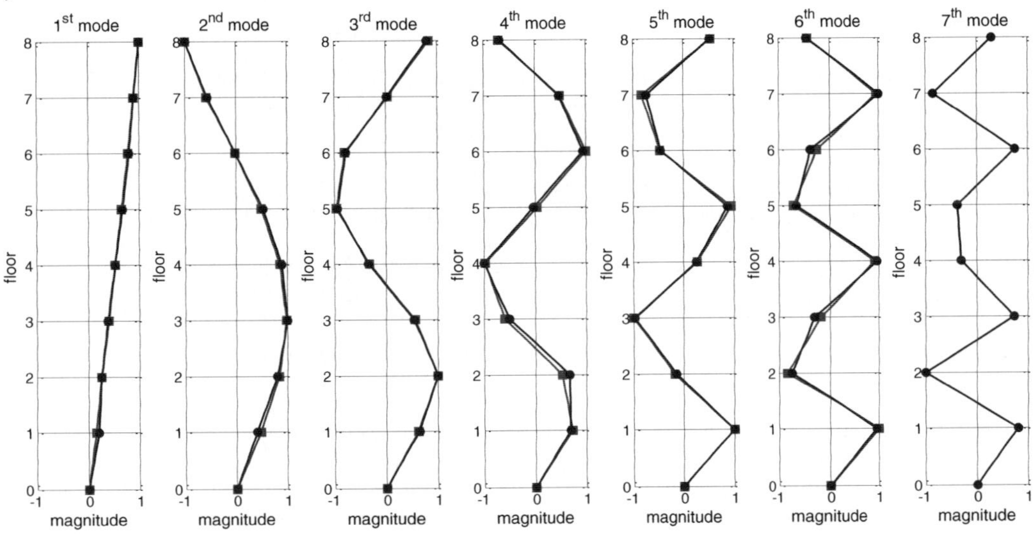

Figure 4. Identified mode shapes of experimental 8-story building.

4. Conclusions

This study proposed a new stochastic subspace system identification method based on time-frequency distributions of measured responses. The discrete Fourier transform was employed to generate the time-frequency distributions. These time-frequency distributions were subsequently stacked as a Hankel matrix. By utilizing the matrix orthogonal projection, a subspace containing the extended observability matrix was obtained. Then, the modal properties of a structure were extracted from this extended observability matrix. Because the discrete Fourier transform yielded a symmetric spectrum, the dimension of the Hankel matrix was significantly reduced. Moreover, the proposed method allowed the identification process performing mode by mode that lowered the computational loading. Three numerical examples demonstrated the effectiveness of the proposed method. Meanwhile, this method was also applied to experimental data. As a result, this method successfully extracted the modal properties of the experimental building.

Acknowledgements

This research is supported by Ministry of Science and Technology in Taiwan under Grant No. MOST 106-3011-F-009-003.

References

Farrar, C.R. and James, G.H. 1997, "System identification from ambient vibration measurements on a bridge", *Journal of Sound and Vibration*, vol. 205, no. 1, pp. 1-18.

Brownjohn, J.M.W. 2003, "Ambient vibration studies for system identification of tall buildings", *Earthquake Engineering and Structural Dynamics*, vol. 32, pp. 71-95.

Lynch, J.P., Wang, Y., Loh, K.J, Yi, J.-H., and Yun, C.-B. 2006, "Performance monitoring of the Geumdang Bridge using a dense network of high-resolution wireless sensors", *Smart Materials and Structures*, vol. 15, pp. 1561-1575.

Siringoringo, D.M. and Fujino, Y. 2008, "System identification of a suspension bridge from ambient vibration response", *Engineering Structures*, vol. 30, no. 2, pp. 462-477.

Weng, J-H., Loh, C-H., Lynch, J.P., Lu, K.-C., Lin, P.-Y., and Wang, Y. 2008, "Output-only modal identification of a cable-stayed bridge using wireless monitoring systems", *Engineering Structures*, vol. 30, no. 7, pp. 1820-1830.

Jang, S.A., Jo, H., Cho, S., Mechitov, K.A., Rice, J.A., Sim, S.J., Jung, H.J., Yun, C.B., Spencer, B.F., Jr., and Agha, G. 2010, "Structural health monitoring of a cable-stayed bridge using smart sensor technology: deployment and evaluation", *Smart Structures and Systems*, vol. 6, no. 5-6, pp. 439-459.

Van Overschee, P. and De Moor, B. 1991, "Subspace algorithms for the stochastic identification problem", *Proceedings of the 30th IEEE Conference on Decision and Control*, Brighton, UK, pp. 1321-1326.

Peeters, B. and De Roeck, G. 1999, "Reference-based stochastic subspace identification for output-only modal analysis", *Mechanical Systems and Signal Processing*, vol. 13, no. 6, pp. 855-878.

Peeters, B. and De Roeck, G. 2001, "Stochastic system identification for operational modal analysis: a review", *Journal of Dynamic Systems, Measurement, and Control*, vol. 123, no. 4, pp. 659-667.

Broomhead, D.S., Jones, R., and King, G.P. 1987, "Singular system analysis with application to dynamical systems", *In E.R. Pike & L.A. Lugaito (Eds.) Chaos, noise, and fractals*, pp. 15-27.

Loh, C.H. and Liu, Y.C. 2013, "Application of recursice SSA as data pre-processing filter for stochastic subspace identification", *Smart Structures and Systems*, vol. 11, no. 1, pp. 19-34.

Chang, C.M. and Loh, C.H. 2015, "Improved stochastic subspace system identification for structural health monitoring", *Journal of Physics: Conference Series (DAMAS)*, vol. 628, 012010.

McKelvey, T., Akcay, H., and Ljung, L. 1996, "Subspace-based multivariable system identification from frequency response data", *IEEE Transactions on Automatic Control*, vol. 41, no. 7, pp. 960-979.

Cauberghe, B. 2004, "Applied frequency-domain system identification in the field of experimental and operational modal analysis", *Praca doktorska*, VUB, Brussel.

Brincker, R. Zhang, L. and Andersen, P. 2000, "Modal identification from ambient response using frequency domain decomposition", *Proceedings of the International Modal Analysis Conference (IMAC)*, San Antonio, TX, USA, pp. 625-60.

Bebdat, J. and Piersol, A.G. 1993, "Engineering applications of correlation and spectral analysis", John Wiley & Sons, NJ, USA.

Aguero, J.C., Yuz, J.I., Goodwin, G.C., and Delgado, R.A. 2010, "On the equivalence of time and frequency domain maximum likelihood estimation", *Automatica*, vol. 46, no. 2, pp. 260-270.

McKelvey, T., Atala, H., and Parada, D.O.B. 2008, "A subspace method for frequency selective identification of stochastic systems", *Proceedings of the 17th World Congress the International Federation of Automatic Control*, Seoul, Korea, pp. 8846-8851.

McKelvey, T. and Viberg, M. 2001, "A robust frequency domain subspace algorithm for multi-component harmonic retrieval", *Proceedings of 35th Asilomar Conference on Signals, Systems and Computers*, Pacific Grove, CA, USA, pp. 1288-1292.

The 8th International Conference on
Structural Health Monitoring of Intelligent Infrastructure
Brisbane, Australia | 5-8 December 2017

Estimation of Hydration Stresses in a Full Scale PSC Railway Bridge

P. Nellisseri[1], S. Chikermane[1], P. Banerji [2]
[1] IIT Roorkee – India, Email: pbanerji.iitb@gmail.com [2] IIT Bombay - India,

Abstract

The setting and hardening process in concrete is always associated with a heat development and related temperature variation in the domain due to the heat of hydration. Several models exist in literature which model this phenomena. Most of these models indicate a non-linear distribution of tensile stresses which may crack the girder as they may exceed the strength of green concrete. These models very often give different results based on the assumptions and changing boundary conditions. In this paper various such models are evaluated and calibrated with the existing data acquired during the setting process for two girders. The numerical methods presented, characterize the hydration temperatures and stress/strain evolution in the concrete in the first days post concreting. Since these are field results, they inherently encompass the environmental interaction and the different phases of concrete.
The data corresponds to a full scale health monitoring program initiated by Indian railways on a bridge designated for mixed traffic involving both freight and metropolitan passenger transport. In this program two girders – a service girder and an identical control girder were comprehensively instrumented using fibre optic embedded sensors. While calibrations of heat models have been done extensively in the laboratory, a full scale test of this nature serves to give realistic estimates of behaviour of PSC girders in actual site conditions.

1. Introduction

In any large volume of concrete having dimensions sizable enough, there is always a possibility of thermal stresses being generated due to hydration effects. Mass concrete is any volume of concrete with dimensions large enough to require measures to be taken to overcome heat generation from cement hydration, by reducing the volume change in order to minimize cracking as defined in 'Cement and Concrete Terminology'(1).
The heat generation within a concrete block creates a thermal gradient – which if the concrete cannot move freely generates associated thermal stresses. If these thermal stresses induced are tensile in nature – and exceed the tensile strength of early concrete – cracking of the elements can result. Large concrete elements constrained by formworks generally are susceptible to such actions. Many studies have been done with regard to this aspect of concrete behavior and have been comprehensively listed by Azenha (2).
In this paper a detailed thermal analysis has been done of a Pre-stressed concrete girder based upon the information given about the concrete used. The thermal effects have been modeled and studied using finite element software and instrumentation has been done on the girder to estimate the strains and the temperatures at different locations. These have then been compared to the strength of concrete used and possibility of cracking has been estimated.

2. Description of Bridge

The girder forms a part of a railway track designed predominantly for suburban train traffic as an extension of the vast suburban rail network of Mumbai city. Although the estimated traffic is predominantly suburban, the girder has been designed to allow freight to pass over it in the event of such movement being warranted. Since the load of freight traffic far surpasses the suburban counterpart, the freight loading is hence critical for girders of this nature.

The bridge – under completion is shown in Figure 1 and the cross sections of the girder at different locations are shown in figure 2.

Figure 1 – overview of the bridge

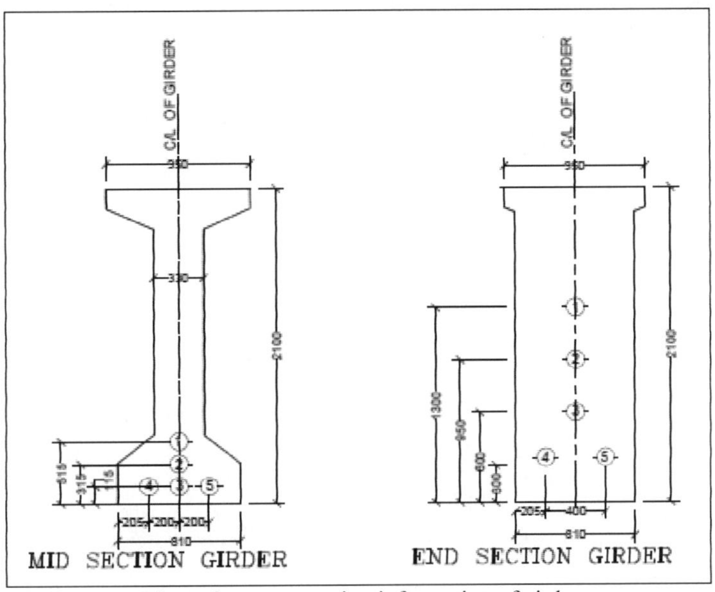

Figure 2 – cross section information of girder

2.1. Instrumentation Details

The girder was instrumented using fiber optic strain and temperature sensors. The sensors were installed prior to concreting on the rebars and were embedded in the concrete thus giving the data from actual onset of concrete hardening. The instrumentation scheme is shown in Figure 3 and Figure 4.

Girder 1 Detail

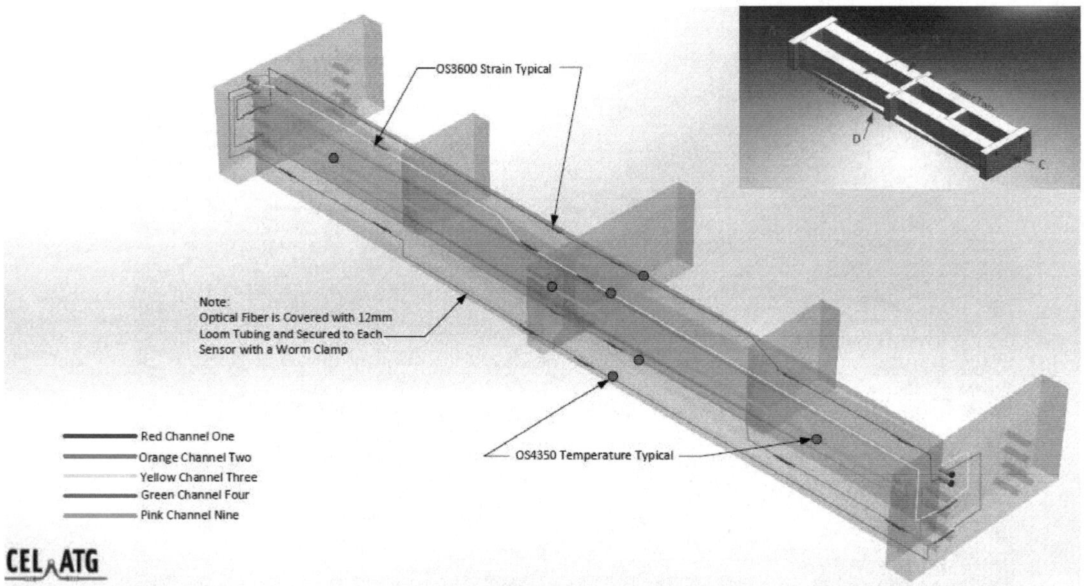

Figure 3 – instrumentation of girder – description and placement of sensor arrays

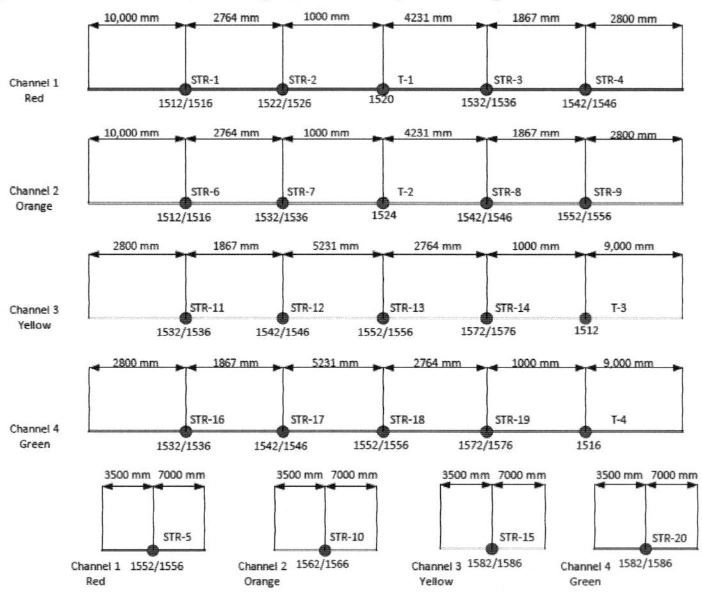

Figure 4 – sensor information for each channel in Figure 3

The data acquisition was done using a Micron Optic SM125 Interrogator. In the present paper, only the data from the girder for the first few hours after concreting is being considered. The data has been acquired at a sample frequency of 0.5 Hz or one data point every 2 seconds.

3. Thermal analysis of concrete

One of the primary objectives of analyzing thermal effects is to determine the temperature field in the medium resulting from the boundary conditions and the heat generations due to the hydration reactions. Once this distribution is known, heat flow by conduction on any point in the medium can be identified by Fourier's law.

Consider a control volume having dimensions $\Delta x, \Delta y \text{ and } \Delta z$, and a temperature field of $T(x, y, z)$. The general heat transfer equation analyzes the flow of heat through this control volume. Let c be the specific heat of the material, k the thermal conductivity, ρ the specific mass and $\frac{dT}{dx}$ the thermal gradient in the x direction, then

$$k\Delta x\Delta y\Delta z\left(\frac{\partial^2 T}{\partial x^2}+\frac{\partial^2 T}{\partial y^2}+\frac{\partial^2 T}{\partial z^2}\right) = c\rho\Delta x\Delta y\Delta z\frac{\partial T}{\partial t} \tag{1}$$

Or

$$k\nabla^2 \mathrm{T} = \mathrm{c}\rho\frac{\partial \mathrm{T}}{\partial \mathrm{t}} \tag{2}$$

If the body generates heat in the form of $\dot{q}$ then there is an additional term in the equation incorporating this and the generalised equation for heat transfer then becomes

$$k\nabla^2 \mathrm{T} + \dot{\mathrm{q}} = \mathrm{c}\rho\frac{\partial \mathrm{T}}{\partial \mathrm{t}} \tag{3}$$

This equation fundamentally states that at any point in the medium, the net rate of energy transfer by conduction to the interior of a unit volume plus the volumetric rate of thermal energy generation must be equal to the rate of variation of the thermal energy accumulated inside this volume.

For the heat of hydration estimates a model has been used which has been developed by Jian Lu et al (3) for the three gorges dam in China. This model states that

$$\dot{q} = q_i(1 - e^{-\alpha t^\beta}) \tag{4}$$

Where the constants α, β, q_i are dependent upon the cement and concrete composition.

Thermal properties used to determine concrete temperature additionally, are

1. Thermal conductivity
2. Density
3. Specific heat
4. Convective heat transfer coefficients.

These properties depend on age of the concrete and temperature, cement type, w/c, aggregate type and unit weight. It may be difficult to determine these values because the variations of these properties are large and the test methods for acquiring them are also very complex. For the analysis done here, the mid-range estimations of all of these are used as per ACI code (4). The thermal properties used in the analysis are shown in Table 1

Table 1 – thermal properties used in analysis

Items	Unit	Value
Conductivity	kcal/m·h/°C	2.3
Density	Kg/m^3	2400
Thermal expansion	/°C	1×10^{-6}
Poisson's ratio		0.18
Placement temperature	°C	28
Convection (Steel form)	$kcal/m^2 \cdot h \cdot °C$	12
Analysis time	hr	120
Specific heat	kcal/kg/°C	0.23

The element used for modelling is the PLANE 70 3D element in ANSYS. This is an eight noded brick element and can be used as either an hexahedral or tetrahedral formulation. Since the geometry cannot be meshed using a hexahedral element, a tetrahedral element is used and mesh sensitivity performed on the maximum nodal temperature till a convergence of 0.01 degrees centigrade is achieved. The solid geometry model is shown in Figure 5.
The model is first run for the thermal analysis based upon the formulation developed above and then the results are imported into the stress analysis in a subsequent run. The thermal analysis is done as a transient non-linear analysis and the stress analysis as a linear static analysis. The results are then compared with the response quantities measured on site and conclusions drawn.

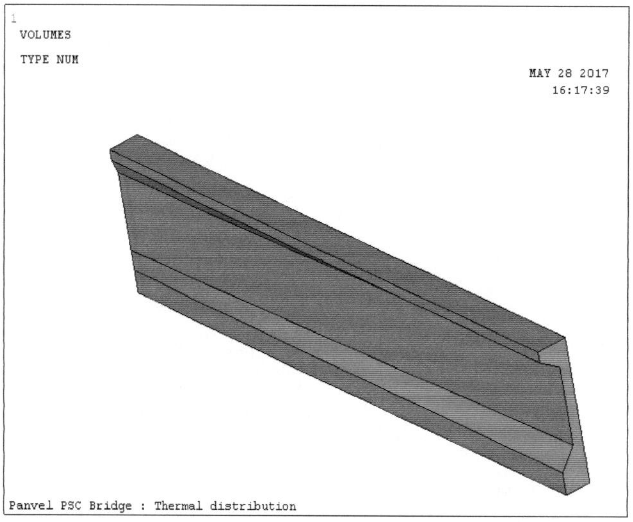

Figure 5 – solid model of quarter section of bridge

The bridge has only been modelled as a quarter section – incorporating symmetry boundary conditions to reduce the modelling efforts. The final model has 50,794 elements and 11,344 nodes. The meshed model is shown in Figure 6.

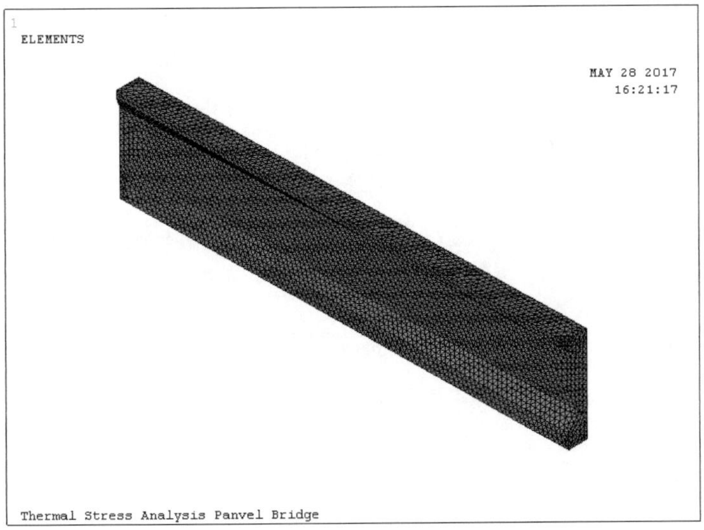

Figure 6 – meshed model of bridge girder

4. Results

The graphs in Figures 7 and 8 show the correspondence between the expected values of Temperatures and strains from the numerical model and those recorded at site. In both of these, it can be seen that there are significantly higher temperatures recorded at site as compared to those expected. This would result in higher values of thermal stresses and increase the risk of cracking of the structure.

Location 1 of Temperature data is at the mid-point close to the end of the girder and location 2 is at the midpoint at centre span. Since the thickness of the web is much larger at the end span, it is natural that there would be a higher value of temperature at this location. This trend is borne out both in the numerical data and the site recordings.

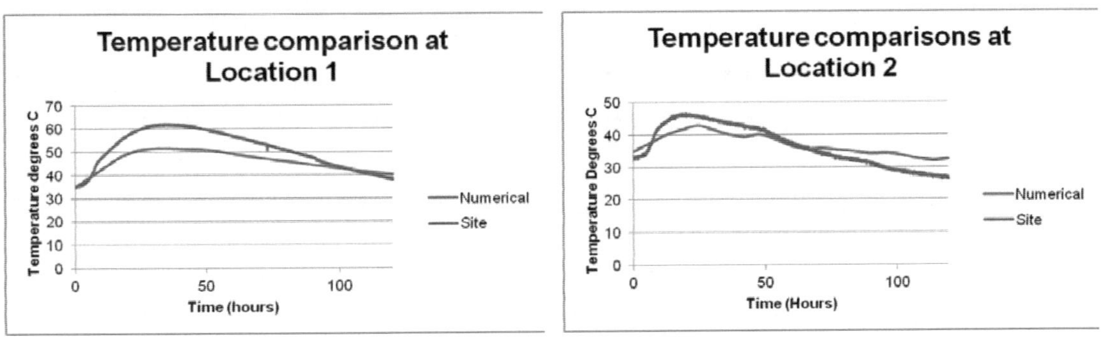

Figure 7 – Temperature comparisons at different locations

The numerical model indicates a value of tensile stress for the model under the conditions of hydration and curing. The maximum values of this tensile stress are about 0.6 MPa, which indicates that the M50 concrete used for the bridge would be sufficient for the tensile stresses generated – even considering the reduced strength at early age. The stress contours for this are shown in Figure 9 and the time history of the tensile stress are indicated in figure 10.

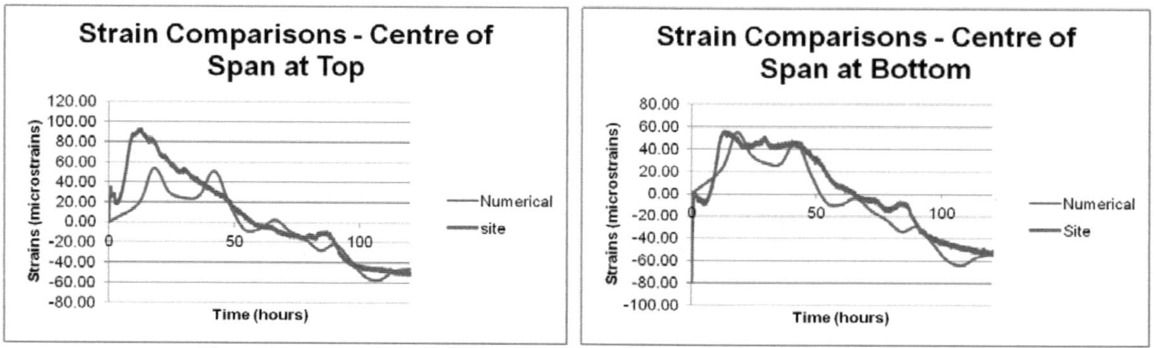

Figure 8 – comparisons of strains between numerical model and site recording

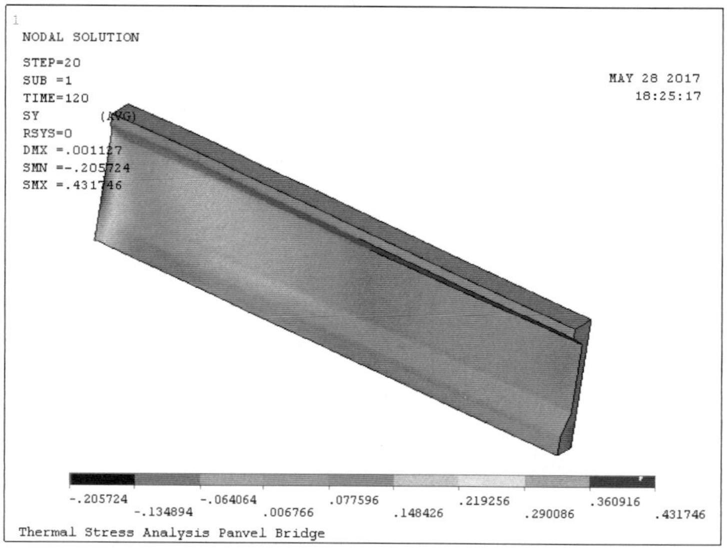

Figure 9 – stress contours for thermal stresses - +ve indicates tension

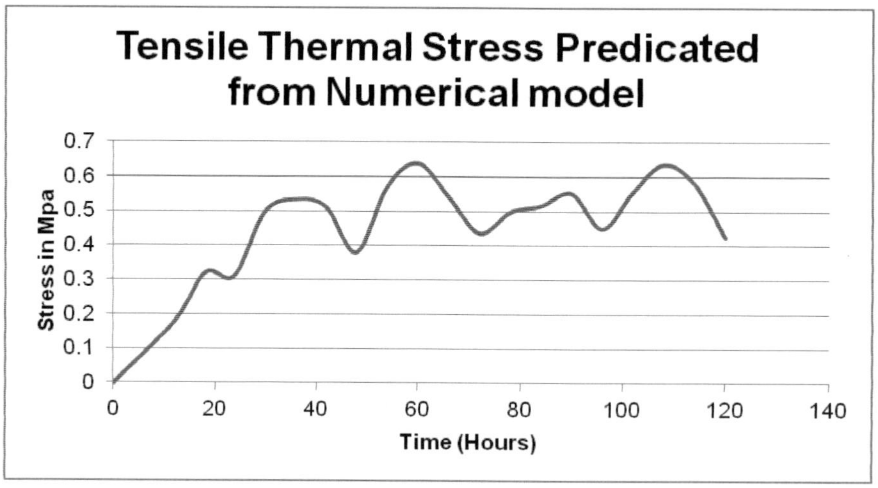

Figure 10 – Estimate of Thermal Tensile stress from Numerical model

3. Conclusions

In conclusion, the site recordings of temperature show a higher value than those expected from numerical models generated from mid-range estimations of thermal properties. However, the long term – post 120 hour values of both the numerical models and site recordings give very good correlations. The thermal stress values are reasonable – about 0.6 MPa as the estimate of tensile strength for M50 concrete is 4.9 MPa from the Indian codes(5). So this value represents about 15% of the full age strength for this particular concrete, which should be acceptable.

However, in the view of the increased temperatures and hence the possibilities of increased stresses generated, the thermal stresses may creep into the region where cracking could be a concern for this structure. Hence it is recommended that efforts should be made to reduce the heat of hydration. This could be achieved through infusion of fly ash, reducing the temperature of the mixing water or by various means.

Hence the methodology adopted of instrumentation and modelling can give significant benefits to real life projects by indicating areas of concerns.

References

'Cement and Concrete Terminology', 2005, American Concrete Institute , ACI 318 Farmington Hills

Azenha, M.A.D, 2009, 'Numerical simulation of the structural behaviour of concrete since its early ages', *Dr. Thesis,* Engineering Department, University of Porto, Porto

Jian Liu, Liping Qiao, Pujian Li & Chun-Fang Dai, 2010, 'Laboratory Test and Numerical Simulation of Time-Dependent Thermo mechanical behaviour of the Three-Gorges Dam', *Journal of Materials in Civil Engineering* 22(2), pp. 111-123.

ACI (2005). Building Code Requirements for Structural (ACI 318-05) Concrete and Commentary (ACI 318R-05). ACI Committee 318

IS-456(2000). Indian standard for Plain and Reinforced Concrete

The 8th International Conference on
Structural Health Monitoring of Intelligent Infrastructure
Brisbane, Australia | 5-8 December 2017

Full Probabilistic Model for Traffic Loads on Bridges Based on Weigh-In-Motion Data

J. Kim [1], J. Song [2].
[1] *Department of Civil and Environmental Engineering, Seoul National University – South Korea, Email: wlghks4599@snu.ac.kr,* [2] *Department of Civil and Environmental Engineering, Seoul National University – South Korea*

Abstract

When designing a new bridge or evaluating the condition of a bridge in operation, it is essential to accurately estimate the traffic load effects on the bridge to assure safety and serviceability. Although many research efforts have been made, the developed traffic load models mostly rely on assumptions or engineering judgment, and thus cannot fully consider effects of various uncertain characteristics of vehicles and traffic flows on traffic loads. To overcome these limitations, this study aims to develop a full probabilistic model of bridge traffic loads based on Weigh-In-Motion (WIM) data collected from three sites of South Korea. Detailed statistical studies are performed for each of the random variables introduced to describe important characteristics of traffic flow and vehicles. In particular, statistical models of vehicles characteristics such as axle weight and axle spacing are first developed using empirical distribution models. During the development process, the statistical correlation between random variables is fully considered. Next, high-precision statistical models are developed to describe the traffic flow characteristics such as headway, heavy vehicle ratio, and traffic volume. Using these developed statistical models, WIM data are generated by Monte Carlo simulation (MCS) for a time period of interest, and traffic load effects are evaluated using the generated WIM data and the influence lines of the bridge. The MCS results in Gumbel probability papers show good agreements with those calculated by actual WIM data, which validates the developed models successfully. In the future research, the developed full probabilistic model will be used to create a generic traffic load model that can provide the traffic load distribution for a given site based on local values of critical parameters.

1. Introduction

In general, traffic load effects on a bridge are site-specific because the characteristics of vehicles passing over the bridge and the traffic conditions highly depend on the bridge location. In many design codes of the bridge, however, traffic load models are based on a single truck load and lane load which cannot fully consider site-specific conditions of the bridge. These load models are slightly different in design codes but most models rely on conservative assumptions and engineering judgment. This limitation might result in inaccurate evaluation of traffic loads, which may increase costs for construction and maintenance of bridge or likelihoods of unexpected failures.

Recent developments of Weigh-In-Motion (WIM) technologies made it possible to obtain accurate measurements of various data related to traffic loads on a bridge. These sets of WIM data usually include vehicles' velocity, axle loads, gross vehicle weight, vehicle length and passing time, etc. WIM technologies have been used mainly to determine whether vehicles on a road or a bridge violate the rules regarding weights, size, etc. Many governments or owners of bridges or roads have installed WIM systems at many highways and bridges and a large amount of WIM data has been collected.

Therefore, researchers have paid attentions to estimating the traffic load effects on bridges using WIM data in their research efforts for cost effective design or accurate evaluation of a bridge in operation.

There are two main approaches available for estimating the traffic load effects on a bridge. The first approach is fitting statistical distributions to the calculated load effects computed by influence line analysis employing collected WIM data. These distributions are then used to directly predict the load effects during the service life of a bridge through an extrapolation process, e.g. using a normal probability paper (Nowak 1993, 1994). The second approach is to run Monte Carlo simulation (MCS). In this MCS-based approach, they first establish the statistical distributions representing various random variables related to traffic conditions based on measured WIM data. Then, they generate WIM data based on the established statistical distributions after choosing the distribution parameter values to reflect given traffic conditions of the bridge of interest. Then, the traffic load effects are computed using the generated WIM data, and a selected type of distribution model is fitted to the calculated traffic load effects. Finally, important characteristics of the traffic load effects are investigated through extrapolation process toward the service life of the bridge (O'Connor and O'Brien 2005, Caprani et al. 2008, Ruan et al. 2012, and Enright and O'Brien 2013).

In these previous research efforts using WIM data, it has been observed that a governing state of traffic flow depends on the span length of a bridge. For a bridge with a "short and medium span length" (15~200m), the free flow state is more important because the weight of one or two heavy trucks dominates the traffic load effects. On the other hand, for a bridge with a "long span length," the congestion flow state is considered important because the number of vehicles passing over the bridge dominates the traffic load effects (Flint and Jacob 1996).

In this study, following the MCS-based approach described above, probabilistic models of traffic load effects on a bridge are developed based on WIM data collected from three highway sites in South Korea. This study does not just simply follow the aforementioned MCS-based procedure, but considers detailed characteristics of vehicles and traffic to the full extent. By establishing detailed statistical models, several important characteristics of traffic flow are identified and modeled appropriately. Using the developed statistical models, artificial WIM data set is generated by Monte Carlo simulation for a given bridge service life. Traffic load effects are calculated by influence line analysis of the bridge employing the generated WIM data. Through comparison between these results and those by the original measured WIM data, the developed probabilistic model is validated.

2. Description of WIM Data Used in This Study

This study uses WIM data collected from three highway sites in South Korea: Gimcheon, Sunsan and Waegwan for eleven months in the year of 2013, as detailed in Table 1. Each measurement location has a primary lane where heavy-weight vehicles such as heavy trucks and trailers mainly pass while the other lanes are called secondary lane or fast lane where mainly small cars and buses pass. This study focuses on the primary lanes of each site, which creates maximum traffic load effects on a bridge in general.

Table 1. Details of WIM data used in this study

Location	Gimcheon	Sunsan	Waegwan
Number of measured lanes	3	2	4
Annual average daily traffic	17,885 veh/day	29,484 veh/day	53,595 veh/day
Annual average daily traffic (truck) in primary lane	2,405 veh/day	4,307 veh/day	3,483 veh/day
Measurement period	Feb. 2013~Dec. 2013	Feb. 2013~Dec. 2013	Feb. 2013~Dec. 2013

The measured WIM data need to be pre-processed in order to eliminate the unreliable data points which might result in inaccurate estimation of traffic load effects. For the pre-processing process, this study follows the guidelines provided by Sivakumar et al. (2011). However, the characteristics of measured WIM data and WIM systems are different from country to country so the guidelines were modified based on the author's judgment.

3. Statistical Models Developed Based on WIM Data

After data cleaning procedure, WIM data are used to establish the statistical models which will represent traffic and vehicle conditions of the bridge in generating WIM data by MCS. In this study, statistical models are developed separately for vehicle characteristics and traffic flow characteristics. In the following subsections, details of the two categories of statistical models are provided.

3.1 Statistical Models Developed for Vehicle Characteristics

The WIM measurement systems used in this study automatically classify the passing vehicles into 12 types following the rules provided by the Ministry of Land, Infrastructure, and Transport (MOLIT) of South Korea. These rules are established in terms of the number of axles and maximum axle spacing location. The vehicle type classifications are shown in Figure 1 (MLTM 2012).

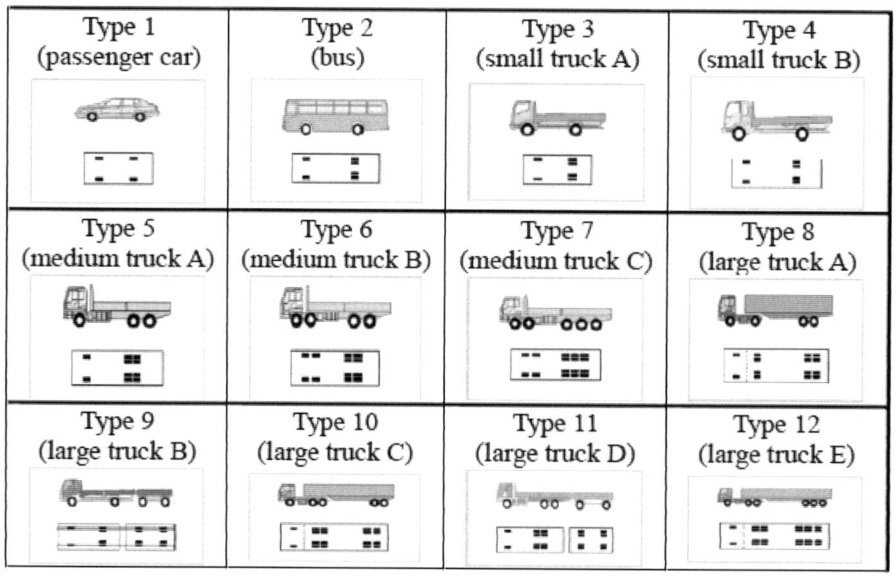

Figure 1. Vehicle type classifications by WIM systems in Korea (MLTM 2012)

Random variables are introduced to describe the following vehicle characteristics: GVW (Gross vehicle weight, i.e. the sum of the axle weights), axle weights, total axle spacing and axle spacing. Kennedy et al. (1992) assumed that these variables follow normal distributions. Grave et al. (2000) used Weibull distribution. Multimodal normal distributions were used for GVW by Cooper (1995) and O'Brien et al. (2006). The histograms of these variables generally show multiple peaks and do not follow a typical statistical distribution model. Therefore, this study uses empirical distribution models as alternatives to typical statistical distribution models so that each vehicle characteristic is described accurately. Furthermore, correlation relationships between these random variables are identified and modeled to generate WIM data in MCS with correlation fully considered. Crespo-

Minguillón et al. (1997) and Srinivas et al. (2006) considered correlation relationship between axle loads using copula functions. However, these studies used rank correlation coefficients which cannot describe linear correlation relationship in MCS.

To overcome the limitation, in this study, the linear correlation between random variables is considered by constructing Nataf distribution models (Liu and Der Kiureghian 1986, Der Kiureghian 2005). Specifically, a Nataf distribution model describes the relationship between the correlation coefficient between the original random variables, ρ_{ij} and that for the corresponding standard normal random variables, ρ'_{ij} by an integral equation

$$\rho_{ij} = \int_{-\infty}^{\infty}\int_{-\infty}^{\infty}\left(\frac{x_i-\mu_i}{\sigma_i}\right)\left(\frac{x_j-\mu_j}{\sigma_j}\right)\varphi_2(z_i, z_j; \rho'_{ij})dz_i dz_j \qquad (1)$$

where x_i and x_j are the original random variables, μ and σ denote their means and standard deviations, and z_i and z_j are the standard normal variables for which the Nataf distribution model is constructed. For given correlation coefficients (ρ_{ij}), means and standard deviations of the original models (estimated by WIM data), correlation coefficients of z_i and z_j, denoted by ρ'_{ij}, can be obtained by solving Eq. (1) iteratively. For MCS, the standard normal random variables $z_i, i = 1, \ldots, N$ are first generated based on the correlation coefficient matrix $\mathbf{R}' = [\rho'_{ij}]$, then transformed to the original random variables by $x_i = F_{X_i}^{-1}[\Phi(z_i)]$ where $F_{X_i}(\cdot)$ is the empirical cumulative distribution function (CDF) of the original random variable and $\Phi(\cdot)$ is the standard normal CDF. For some vehicle types, the correlation coefficient matrix $\mathbf{R}' = [\rho'_{ij}]$ may not satisfy the positive semi-definite condition. In such cases, approximate values satisfying the positive semi-definite condition are used.

3.2 Statistical Models Developed for Traffic Flow Characteristics

Most of the MCS-based approaches developed for estimating the traffic load effects on a bridge describe traffic flow characteristics in terms of traffic volume, headway, the heavy vehicle ratio, and velocity. These four variables have significant statistical dependence between each other, which should be incorporated into modeling traffic flow characteristics based on measurement data together with theories of transportation engineering.

As traffic volume varies significantly over time, we need to consider the hourly variation of traffic volume as well. Therefore, traffic volumes are obtained from measured WIM data in the unit of veh/5 min at every given hour and these are fitted to empirical distribution models. The unit of veh/5min is used instead of the widely-used unit veh/1hour to describe congestion situation more accurately. In transportation engineering, Greenshield's model is widely used to describe the relationship between velocity and traffic volume (Adolf 1990). This relatively simple macroscopic traffic flow model is not perfect, but considered fairly accurate and useful.

Figure 2(a) shows two traffic states identified by Greenshield's model: "free flow" state and "congestion state." The model shows that, even if the traffic flow has the same traffic volume, the velocity of a vehicle can be considerably different. This phenomenon is observed in the WIM data measurement from the three sites as shown in Figure 2(b)-(d), in which traffic volume on the x-axis is the number of vehicles passing for 5-minute intervals, and the velocity on the y-axis is the mean of the velocity of passing vehicles for 5-minute intervals. It is also seen that, at the primary lane at Gimcheon, congestion state does not occur while the primary lanes at Sunsan and Waegwan experience congestion state.

It is noted that the congestion state does not occur every day. For example, the congestion state occurs on only 39 days during the measurement period at Sunsan, and on 5 days at Waegwan. To consider this, the probability of the daily occurrence of congestion state is introduced into the traffic flow characteristics model. If MCS indicates congestion state for a given day, the average velocity per 5-minute interval is randomly generated from the conditional empirical distribution of the velocity given the occurrence of a congestion state. If congestion state does not occurs for a given day, the conditional distribution given no congestion is used instead. The empirical distribution is fitted to velocity data grouped in terms of traffic volumes (with 10veh/5min intervals). To simulate variability in the velocity of each vehicle, randomly generated Gaussian noise is added to the mean velocity.

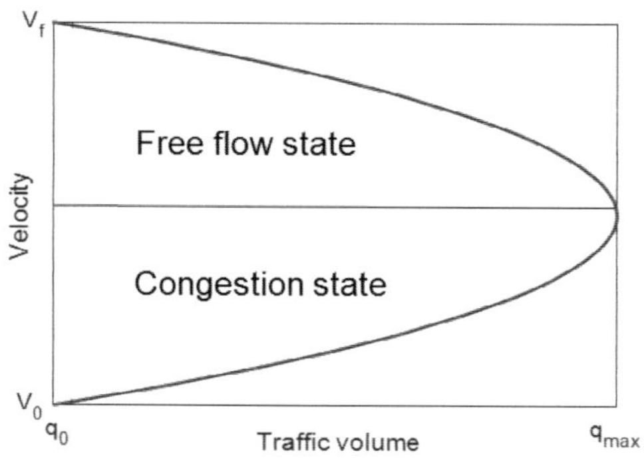

Figure 2. (a) Traffic volume-velocity relationship in Greenshield's model

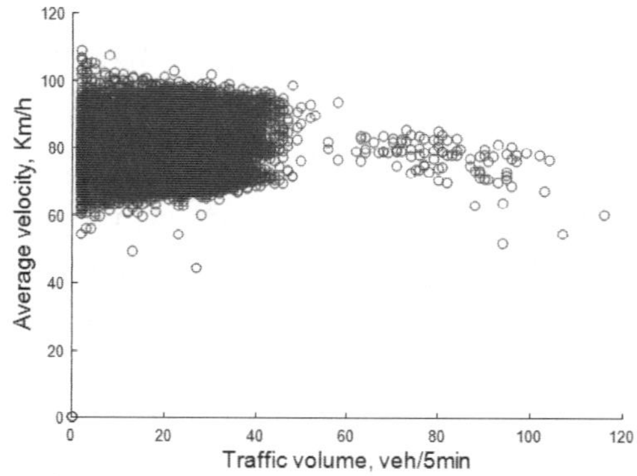

Figure 2. (b) Traffic volume-Velocity relationship at primary lane of Gimcheon

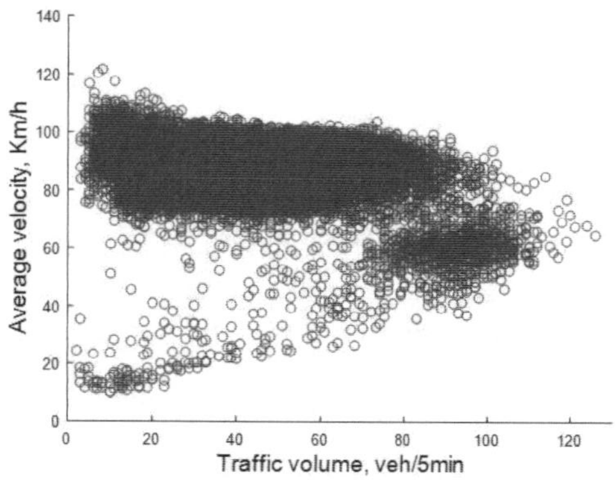

Figure 2. (c) Traffic volume-Velocity relationship at primary lane of Sunsan

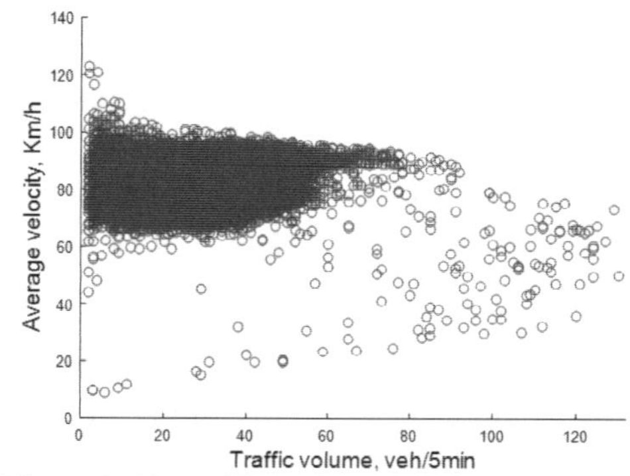

Figure 2. (d) Traffic volume-Velocity relationship at primary lane of Waegwan

The twelve vehicle types ratio is assumed to follow normal distribution fitted to the measured data with monthly variations considered. This study also investigates the heavy vehicle ratio, which affects the traffic load effects significantly. The vehicles belonging to Type 5 to 12 in Figure 1 are considered "heavy vehicles" in this study to cover medium and large trucks. Data analysis in this study confirms that it is important to consider the variation of the heavy vehicle ratio which might induce the maximum traffic load effects on bridges.

The heavy vehicle ratio is estimated for each hour and each day of the week as shown in Figure 3 and 4. Figure 3 shows that the heavy vehicle ratio on a weekday during the night or dawn is higher than that during the daytime at all primary lane of three sites. This is because truck drivers tend to avoid times showing heavy traffic in order to transport the freight more quickly. Additionally, the heavy vehicle ratio between the time period from Saturday 9 pm to Sunday 4 am decreases, and then increases again on Sunday night at the primary lane at Sunsan, as shown in Figure 4. The other two sites have the same tendency as shown in Figure 4 because many truck drivers take a rest on holidays.

Therefore, if variation of the heavy vehicle ratio by hour and day of the week is not considered in generating WIM data, the traffic load effects on a bridge might be underestimated. Therefore, it is crucial to consider the time variation of heavy vehicle ratio.

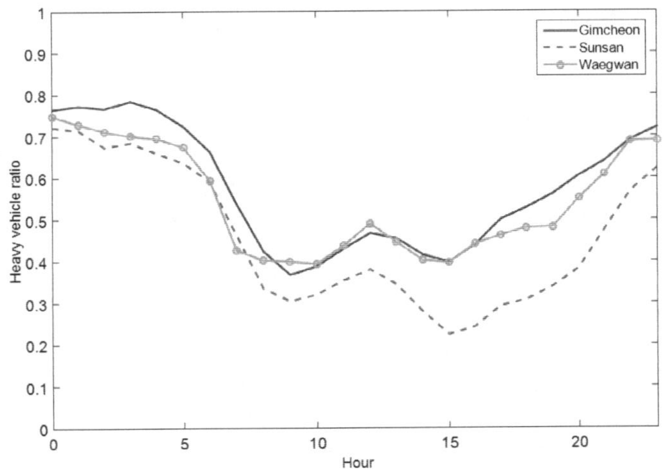

Figure 3. Heavy vehicle ratio in Weekday at three sites

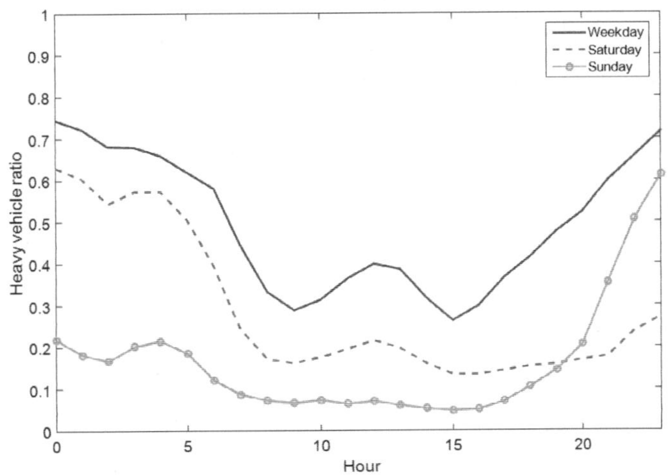

Figure 4. Heavy vehicle ratio of primary lane at Sunsan

"Headway", an important characteristic in traffic flow description, is defined as the distance or time between two successively passing vehicles. The headway shows a large uncertainty propagated from influence of traffic volume, driving habits, safety distance, interactions between multiple vehicles, etc. Many studies have been carried out to establish headway models separately for different flow states depending on traffic volumes. The congestion and queue state cause short headways while the free flow state results in relatively long headways. In existing studies, for congestion state, headway is described by a deterministic model developed based on the safety distance rules and measurement data. On the other hand, for free flow states on short-medium-span bridges, the arrivals of vehicles are assumed to follow a Poisson process, and thus the corresponding interval times are assumed to follow an exponential distribution. However, MCS using exponential distribution may generate inter-arrival times close to zero, which is impossible in real traffic situation. To address this issue, Cowan

(1975) used the shifted exponential distribution model to incorporate the minimum inter-arrival time between vehicles into the probabilistic model.

O'Brien and Caprani (2005) and Enright and O'Brien (2013) used three piecewise quadratic curves fitted to empirical cumulative distributions of headways up to 4 seconds of headways. For headways greater than 4 seconds, a shifted exponential distribution was used. In particular, to consider drivers' behavior more accurately, in these research efforts, the headway is defined as the distance between the rear axle of the leading vehicle and the front axle of the following vehicle, so-called inter-axle gap instead of the distance between the front axle of the leading vehicle and the front axle of the following vehicle. This is because the inter-axle gap is not dependent on the overall vehicle length.

In this study, we use an approach similar to those by O'Brien and Caprani (2005) and Enright and O'Brien (2013) for developing the headway model, i.e. inter-axle gap is used as the measure of headway. For headways smaller than 4 seconds, an empirical distribution is fitted to headway data grouped in terms of traffic volumes (with 10veh/5min intervals). The headways over 4 seconds are modeled by a shifted exponential distribution based on Poisson process assumption.

4. Validation of Probabilistic Model: Daily Maximum Traffic Load Effects

WIM data is generated by MCS using the developed statistical models described in detail in the preceding sections. Traffic load effects are then calculated by use of moment influence line of a simple beam bridge. It is assumed that overall traffic volumes will not change significantly for the remaining service life of the bridge, and the velocity of each vehicle remains constant while a vehicle passes over the bridge. The calculated load effects are compared with those by actual WIM data in a Gumbel probability paper to validate the developed probabilistic models. In existing studies, the Gumbel probability paper has been used to extrapolate the load effects or find the suitable statistical distribution for calculated load effects in block maximum method. If the plotted data follow the exact Gumbel distribution, the data will be plotted in a linear line. If the plotted data show significant nonlinearity in the paper, the data set is considered to follow a different type of distribution.

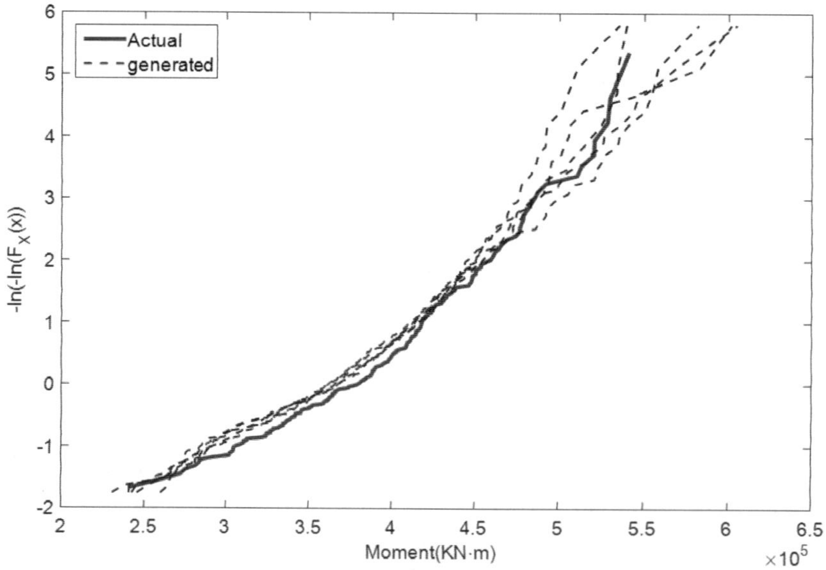

Figure 5. Gumbel probability paper of daily maximum moment at mid-span, Gimcheon

Traffic load effects are calculated using 5 sets of WIM data generated by MCS for one year period and the daily maximum load effects are recorded and plotted in the Gumbel probability paper in Figures 5, 6 and 7. The results are compared with those by the actual WIM data set for validation. Larger variability is observed at the upper ends of the curves because of the low likelihoods of extreme events. Although there is a slight bias, the MCS results of daily maximum moment show good overall agreement with those by actual WIM data at all three sites. These results confirm validity of the full probabilistic model developed in this study in estimating the traffic load effects on a bridge.

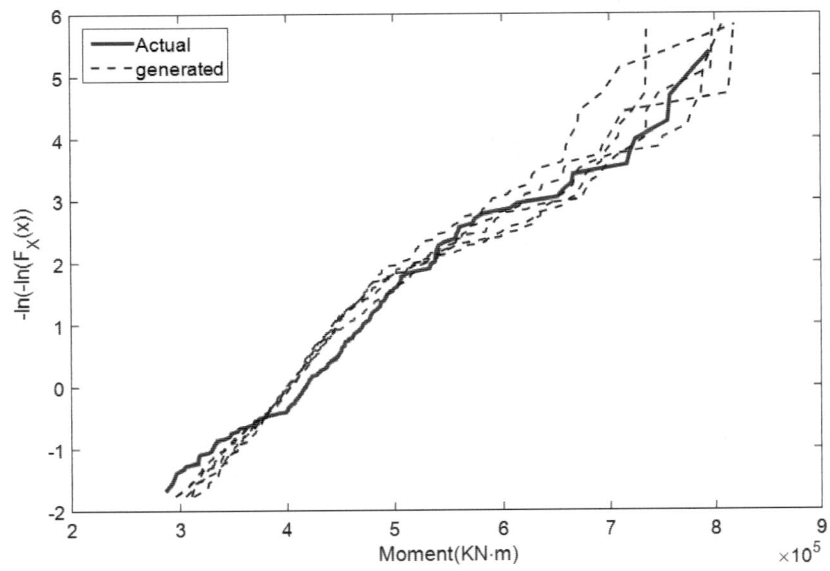

Figure 6. Gumbel probability paper of daily maximum moment at mid-span, Sunsan

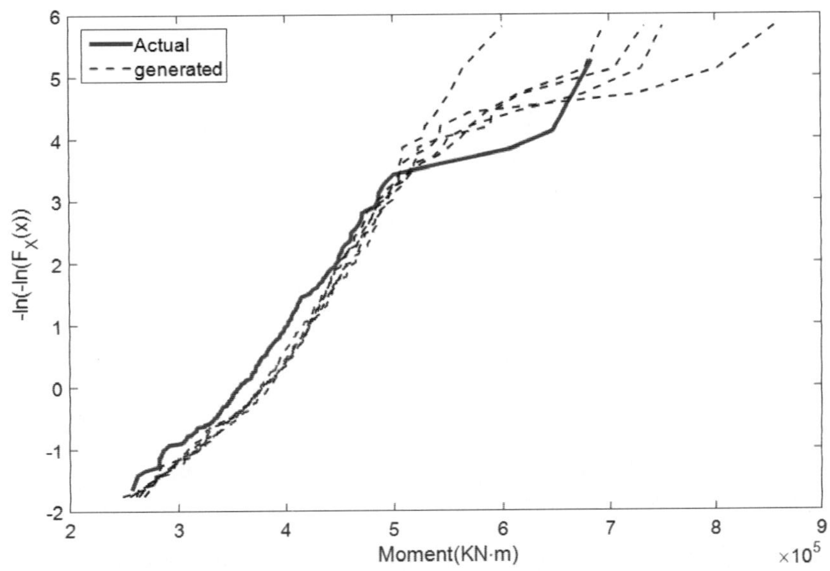

Figure 7. Gumbel probability paper of daily maximum moment at mid-span, Waegwan

5. Conclusions

This paper presents a full probabilistic model for traffic load effects on bridges, which is developed based on WIM data collected from three sites of Korea. To facilitate accurate Monte Carlo simulation of traffic load effects, detailed statistical studies are performed for each of various characteristics of traffic flow and vehicles and integrated.

In the vehicle characteristics model, four random variables representing the characteristics of each vehicle, i.e., gross vehicle weight, axle weights, total axle spacing and axle spacing are modeled by empirical distribution models fitted to the WIM data. In particular, correlation between the random variables is fully considered using a Nataf model. In the traffic flow characteristics model, traffic volumes are obtained in the unit of veh/5min to describe congestion state. The relationship between velocity and traffic volume in Greenshield's model is identified by the measured WIM data. This velocity-traffic volume relationship and the likelihood of congestion occurrence are incorporated into the model development process. The headway distribution models are constructed for free flow and congestion state separately.

Using these developed statistical models, WIM data are generated at all three sites and the corresponding traffic load effects are calculated with moment influence line of the simple beam bridge. To validate the developed model, the daily maximum moments are recorded from the generated traffic load effects and compared with those by actual WIM data in Gumbel probability papers. The plots show good agreements between load effects by generated and actual WIM data and validate the developed probabilistic model in terms of probabilistic distribution of maximum traffic load effects on bridges.

In the future, a thorough investigation will be performed to find appropriate statistical distribution models of local traffic load effects through MCS with longer time periods. Moreover, the developed models will allow us to develop a generic traffic load model that can provide the probabilistic distribution for a local site, which is obtained by using local values of critical parameters representing characteristic of traffic flow and passing vehicles.

Acknowledgement

The authors would like to gratefully acknowledge the support by the research project, "Development of Life-cycle Engineering Techniques and Construction Methods for Global Competitiveness Upgrades of Cable Bridges" of the Ministry of Land, Infrastructure and Transport (MOLIT) of the Korean Government (Grant No. 17SCIP-B119964-02).

References

Nowak, A.S., 1993, 'Live load model for highway bridges', *Structural Safety*, vol. 13, pp. 53-66.

Nowak, A.S., 1994, 'Load model for bridge design code', *Canadian Journal of Civil Engineering*, vol. 21, pp. 36-49.

O'Connor, A., & O'Brien, E.J., 2005, 'Traffic load modelling and factors influencing the accuracy of predicted extremes', *Canadian Journal of Civil Engineering*, vol. 32, no. 1, pp. 270–278.

Caprani, C. C., OBrien, E. J., & McLachlan, G. J., 2008, 'Characteristic traffic load effects from a mixture of loading events on short to medium span bridges', *Structural Safety*, vol. 30, no. 5, pp. 394-404.

Ruan, X., Zhou, X., & Guo, J. 2012, 'Extreme Value Extrapolation for Bridge Vehicle Load Effect Based on Synthetic Vehicle Flow', *Journal of Tongji University (Natural Science),* vol. 40, no. 10, pp. 1458-1462.

Enright, B., & O'Brien, E.J., 2013, 'Monte Carlo simulation of extreme traffic loading on short and medium span bridges', *Structure and Infrastructure Engineering*, vol. 9, no. 12, pp. 1267-1282.

Flint, A.R., & Jacob, B., 1996, 'Extreme traffic loads on road bridges and target values of their effects for code calibration', *Proceedings of IABSE Colloquium*, Delft, Netherlands, pp. 469-478.

Sivakumar, B., Ghosn, M., & Moses, F., 2011, *Protocols for collecting and using traffic data in bridge design (Vol. 683*), Transportation Research Board, Washington D.C.

MLTM, 2012, 'Korea Bridge Design Code (Limit state design), Ministry of Land, Transportation and Maritime Affairs in Korea.

Kennedy, D. L., Gagnon, D. P., Allen, D. E., & MacGregor, J. G. ,1992. 'Canadian highway bridge evaluation: load and resistance factors', *Canadian Journal of Civil Engineering*, vol. 19, no. 6, pp. 992-1006.

Grave, S.A.J., O'Brien, E.J., & O'Connor, A.J., 2000, 'The determination of site-specific imposed traffic loadings on existing bridges', *Bridge Management 4,* London, U.K., pp. 442–449.

Cooper, D.I., 1995, 'The determination of highway bridge design loading in the United Kingdom from traffic measurements', *First European Conference on Weigh-in-Motion of Road Vehicles*, Zurich, Switzerland., pp. 413-421.

O'Brien, E.J., Caprani, C.C., & O'Connell, G.J., 2006, 'Bridge assessment loading: A comparison of West and Central/East Europe', *Bridge Structures*, vol. 2, no. 1, pp. 25–33.

Crespo-Minguillón, C., & Casas, J.R., 1997, 'A comprehensive traffic load model for bridge safety checking', *Structural Safety*, vol. 19, pp. 339–359.Srinivas, S., Menon, D., & Meher Prasad, A., 2006, 'Multivariate simulation and multimodal dependence modeling of vehicle axle weights with copulas', *Journal of transportation engineering*, vol. 132, no. 12, pp. 945-955.

Liu, P.L., & Der Kiureghian, A., 1986. 'Multivariate distribution models with prescribed marginal and covariances', *Probabilistic Engineering Mechanics*, vol. 1(2), pp. 105-112.

Der Kiureghian, A., 2005, 'First- and second-order reliability methods', In: Nikolaidis E, Ghiocel DM, Singhal S, editors. *Engineering design reliability handbook.* Boca Raton (FL): CRC Press [Chapter 14].

Adolf, D.M., 1990, 'Fundamentals of Traffic Flow', *Prentice - Hall*, Inc. Englewood Cliff New Jersey 07632, second edition.

Cowan, R.J., 1975, 'Useful headway models', *Transportation Research*, vol. 9, no. 6, pp. 371-375.

O'Brien, E.J., & Caprani, C.C., 2005, 'Headway modelling for traffic load assessment of short-to medium-span bridges', *The Structural Engineer*, vol. 83, pp. 33–36.

The 8th International Conference on
Structural Health Monitoring of Intelligent Infrastructure
Brisbane, Australia | 5-8 December 2017

Probabilistic Risk Assessment of Vehicle Safety on a Bridge Using Field-monitored Data

S.J Kim[1], H.K Kim [2]
[1] *Institute of Construction and Environmental Engineering – South Korea,*
[2] *Seoul National University – South Korea Email:* hokyungk@snu.ac.kr

Abstract

New approaches to evaluate vulnerablity of bridge for strong wind with respect to vehicle safety are suggested and analysed. Expected time that exceeding critical wind speed is estimated as risk index through probabilistic approach. Field-monitored wind data is used together with long-term wind data obtained from nearby weather station. 20-year long-term wind data collected from nearby weather station and one year field-monitored wind data are utilized for case study. Three different data processing methods for estimating probability distribution model are attempted: Method 1) correction of wind speed with consideration of surface roughness and height, Method 2) measured-correlate-predict method, Method 3) applying relationship between daily mean wind speed – daily wind distribution. In short, new approaches give much lower risk value than existing one which means that original method evaluate risk quite conservatively.

1. Introduction

Strong gust wind can increase risk of vehicle and induce overturning or sideslip accident, especially when the vehicle is running on a wind-exposed structure. In order to prevent such accidents, wind screen is installed, which can reduce wind speed effectively. However, wind screen construction can induce aerodynamic problem of bridge girder and high construction cost, traffic control is usually applied to ensure vehicle safety. Yet, if traffic control by strong winds is too frequent, traffic control can also lead to high economic losses, so proper decision-making is needed to choose the most reasonable method.

In South Korea, the number of days for traffic control N_E is usually for decision-making which is an expected frequency of day that exceed critical wind speed in one year (Kwon, 2005; Kim et al, 2016). If this risk index exceed a standard value, 6 days, it is recommened to construct windscreen at interest area. However, 6 days is too uncertain to make reasonable and economical decision-making. In order to alleviate this uncertainty, Kim(2011) attempted a decision-making process for wind screen construction by applying a benefit-cost ratio analysis. He calculated benefit and cost for situations that with or without windscreen and compared each of them. He calculated accident cost based on the expected number of accidents during the year which is calculated from the number of days for traffic control. However, according to this method, it is neccessary to assume that the wind speed is constant throughout the day, and it is not realistic because most wind speed values are lower than daily maximum wind speed. In order to more accurately calculate the cost, a new index of lower units, such as seconds or hours, should be applied.

This study aims to investigate new probabilistic assessment methods for evaluating vulnerability of a bridge for strong wind with respect to vehicle safety. A new risk index, T_E, is adopted, which is the estimated time over a certain wind speed per year. Field-monitored wind data is used to estimate probability distribution model in conjunction with long-term wind data obtained from a nearby weather station. This study investigated and compare three diffrent data processing methods: method 1) correction of wind speed with consideration of surface roughness and height, method 2) measured-correlate-predict(MCP) method, method 3) applying relationship between daily mean wind speed – daily wind distribution. A long-span bridge in Busan City were selected for case study. 20-year long-term wind speed data collected from a nearby weather station and 1-year field-monitored wind data are utilized. The data types of long-term wind data are houly data and daily mean data and are formed in 10-minute averaged wind speed.

2. Methodology

A key point to evaluate expected time T_E is to estimate the probability distribution function at the girder height of the target bridges from long-term data and field-monitored data. Then, based on the estimated probability function, the excess probability of the specific wind speed is calculated and T_E is estimated by using appropriate coefficient. Three data processing methods for estimating probability distribution model are considered as follow.

2.1 Method 1. Correction of wind speed with consideration of roughness and height

First method uses a correction factor that converts wind speed from weather station to target location, taking into account the surface roughness and height. The correction factor is calculated according to KSCE(2006) as follows:

$$\begin{aligned} C &= \left(\frac{z_{G1}}{z_1}\right)^{\alpha 1} \times \left(\frac{z_2}{z_{G2}}\right)^{\alpha 2}, \quad z_2 \geq z_b \\ &= \left(\frac{z_{G1}}{z_1}\right)^{\alpha 1} \times \left(\frac{z_b}{z_{G2}}\right)^{\alpha 2}, \quad z_2 < z_b \end{aligned} \tag{1}$$

where subscript 1 refers to the weather station and subscript 2 refers to the bridge site; α is the exponent that governs the shape of the wind profile; z is the height of the measurement point; z_G is the gradient height of the wind profile, and z_b is the minimum height.
After converting wind speed data by multiplying the correction factor, probabilistic distribution function is estimated. In this study, Generalized Extreme Value (GEV) distribution model is adopted which is a model combining three typical extreme value distribution models: Gumbel, Fréchet and Weibull distributions. The equation of the cumulative distribution function for the GEV distribution model is expressed as follows:

$$F(X < x; \mu, \sigma, \xi) = \exp\left\{ -\left[1 + \xi \frac{x-\mu}{\sigma} \right]^{-1/\xi} \right\} \tag{2}$$

where σ is the scale parameter, μ is the location parameter, and ξ is the shape parameter. Finally, using this cumulative distribution function, T_E can be obtained by following equation.

$$T_E = 365(day) \times 24(hour) \times 60(\min) \times F(U > V_C) \tag{3}$$

where V_C is critical wind speed and F(U> V_C) is probability that wind speed U exceeds V_C.

2.2 Method 2. Measured-Correlated-Predict (MCP) method

According to correlation analysis, the correlation coefficient between the wind data of weather station and the field-monitored data is 0.70, which means a clear correlation. Therefore, the Measured-Correlated-Predict (MCP) method, which is one of the correction methods for the long-term wind data considering the wind characteristics of the target area, is adopted. Wind property of target area can be estimated using field-monitored data. Then by analysing a relationship between wind data at the weather station and field-monitored wind data, long-term wind data is converted to interested location. Two analysing method for estimating relationship between two data are considered as follows.

2.2.1 Method 2-1. Linear regression

This method considers mean and standard deviation value as most important characteristics of wind properties. Therefore, it provides a simple linear equation to convert wind speed distribution with respect to mean and standard deviation. The equation is established as follows:

$$U_T = \frac{\sigma_T}{\sigma_W} U_W + m_T - \frac{\sigma_T}{\sigma_W} m_W \quad (4)$$

where U, σ, m are wind speed, standard deviation of wind speed and mean of wind speed, respectively. Subscript W refers to the weather station and subscript T refers to the bridge site. When calculating σ_W and m_W, only the long-term data measured in the same period as field-monitored data are used because the nature of the wind changes with time. After converting the wind data, expected time T_E is obtained through probabilistic approach in the same manner as in Method 1.

2.2.2. Method 2-2. Parameter scale

Second MCP method calculate a ratio of coefficients of probability distribution functions of both data set rather than using mean and standard deviation value. First, each distribution function for each data set of weather station and bridge site are estimated, and parameters of each function are extracted, in this case three parameters for each. Then, ratios of each parameter are calculated. As same as linear regression, only data of weather station measured in same period with that of bridge site can be used when the ratios are calculated. At last, by multiplying these ratios to parameters of probability function, which is estimated from whole long-term wind data, final probability distribution function of bridge site can be obtained. It is very simple process but this have advantages that it can reflect the properties of distribution function of bridge site.

2.3 Method 3. Applying relationship between daily mean wind speed – daily wind distribution

This method utilize daily wind data rather than daily maximum data. First, probability distribution function of daily mean wind speed at bridge site is obtained in same manner with method 2-2. Then using this probability function, expected time T_E is easily obtained as follows:

$$T_E = 365 \times 24 \times 60 \times \int_0^\infty p(\bar{U}) \cdot p(U > V_C \mid \bar{U}) d\bar{U} \quad (5)$$

where $\bar{U}$ is daily mean wind speed and $P(\bar{U})$ is probability function of daily mean wind speed. $p(U > V_C \mid \bar{U})$ is conditional probability that wind exceeds critical wind speed when daily wind speed is given.

In order to calculate the conditional probability $p(U > V_C \mid \bar{U})$, relationship between daily mean wind speed and daily wind distribution which can be obtained from field-monitored data. First, all field-monitored data is divided with one-day interval, and then daily mean wind speed and parameters of distribution function are extracted for each day. By performing correlation analysis, linear equation is obtained between daily mean wind speed and each distribution parameters. Figure 1 shows overall sequence of process of estimation of relationship. Selecting appropriate probability model is important when estimating each daily wind distribution function. Weibull function is adopted in this study because it shows good performance for fitting accuracy.

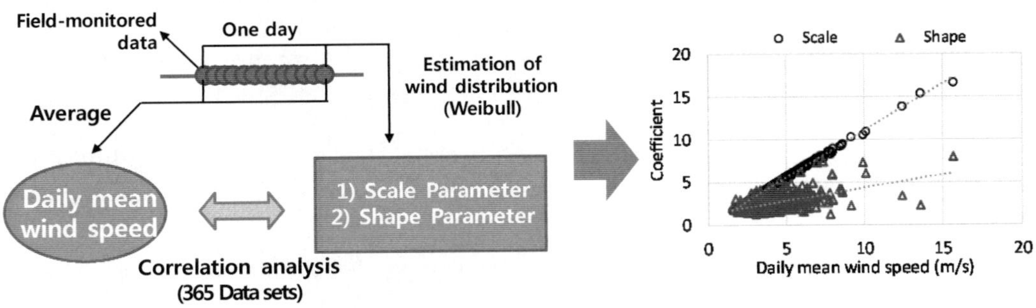

Figure 1. Estimation of relationship between daily mean wind speed and daily wind distribution

As a result, correlation coefficient is obtained for each parameters: 0.9991 for scale parameter and 0.5367 for shape parameter. Below equations are established which shows relationship between daily mean wind speed and two parameters.

$$\text{Scale parameter } \sigma = 1.09\bar{U} + 0.09 \tag{6}$$

$$\text{Shape parameter } \xi = 0.30\bar{U} + 1.31 \tag{7}$$

However, shape parameter has low correlation coefficient about 0.5367. This can occur quite error and it should be improved in future.

Finally, using equation (6) and (7) wind distribution function can be estimated for arbitrary daily mean wind speed and can calculate conditional probability $p(U > V_C \mid \bar{U})$.

3. Results and Discussions

Expected time T_E obtained from above-mentioned methods are compared each other and the result from previous research (Kim et al. 2016). According to original approach, using MCP method and GEV distribution function, N_E is calculated as 0.17 day, which can be converted to 259 minutes by multiplying 24 hours and 60 minutes. Ratio between T_E and N_E for each methods are calculated. The result is shown in Table 1.

Table 1. Expected exceeding times T_E

	Method 1	Method 2		Method 3
		Method 2-1	Method 2-2	
Expected exceeding time T_E	1.03	1.97	28.3	2.71
Ratio with original method	0.4%	7.0%	10.9%	1.1%

According to Table 1, results of new approaches are only 0.4%~11% compared with original approach. This means that new methods evaluate exceeding frequency much lower than original one. The reason of this result is, as mentioned above, that most of wind speed during a day is lower than daily maximum wind speed. Therefore, according to this result, it can be said that using the number of traffic control as a risk index can give too conservative result when somebody perform economic analysis.

In addition, parameter study is performed to investigate how sensitive of vulnerability is according to critical wind speed. Variation of expected exceeding time T_E according to critical wind speed are plotted in Figure 2.

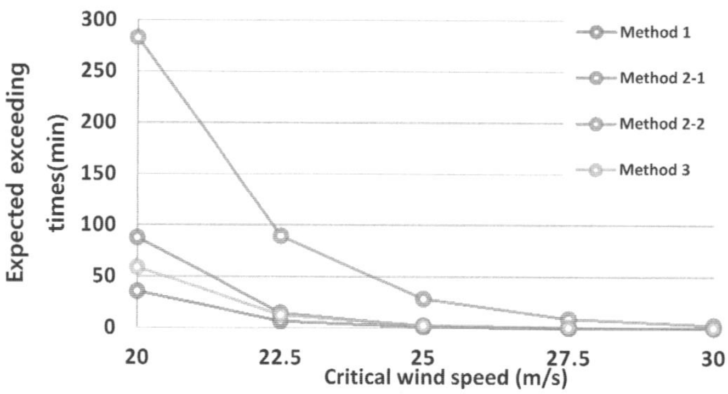

Figure 2. Variation of T_E according to critical wind speed

According to Figure 2, exceeding frequency decrease exponentially when critical wind speed increases. About 3~7 times increment could be observed when critical wind speed increase 2.5m/s. This result means that choosing limit wind speed for traffic control can be an essential factor for bridge operation during strong wind. Decreasing critical wind speed can be one of the effective method for securing vehicle safety.

4. Conclusions

New probabilistic assessment methods to evaluate vulnerability of a bridge for strong wind with respect to vehicle safety are investigated in this study. As a new risk index, expected time T_E that exceed specific wind speed is calculated. Field-monitored wind data and long-term wind data are utilized and three different data processing methods to estimeate probability distribution model are considered. Based on the assessment results, two concluding remarks are obtained. (1) Results of new approaches are only 0.4%~11% compared with previous approach. It can be said that using the number of traffic control as a risk index can give too conservative result when somebody perform economic analysis. (2) Frequency of exceeding critical wind speed decrease exponentially when critical wind speed increases. It can be also said that decreasing critical wind speed can be one of the effective method for securing vehicle safety. This research is just a beginning step, so all proposed methods are needed to be improved and verified with respect to probabilistic treatment for accurate assessment.

Acknowledgements

This study and presentation were supported by grants(17SCIP-B119964-02) from the Ministry of Land, Infrastructure and Transport of the Korean Government and also supported by a National Research Foundation of Korea (NRF) grant funded by the Korean government (MSIT) (NRF-2017R1A2B4008973). The authors are thankful to Busan Infrastructure Corporation for sharing long-term wind data obtained from the Gwangan bridge.

References

Kim, DH, Kwon, SD, Lee, IK & Jo, BW 2011. 'Design criteria of wind barriers for traffic. Part 2: decision making process'. *Wind and Structures*, Vol. 14, pp.71-80.

Kim, SJ, Yoo, CH & Kim, H K 2016. 'Vulnerability assessment for the hazards of crosswinds when vehicles cross a bridge deck.' *Journal of Wind Engineering and Industrial Aerodynamics*, Vol 156, pp. 62-71.

Korean Society of Civil Engineers 2006. *Design Guidelines for Steel Cable-Supported Bridges*, Korea.

Kwon, SD & Jeong UY 2005, 'An overall wind shielding program for enhancing driving stability', *Journal of Korean Society of Steel Construction*, Vol. 17, no. 3, pp-263-270

The 8th International Conference on
Structural Health Monitoring of Intelligent Infrastructure
Brisbane, Australia | 5-8 December 2017

Structural model updating for reserve capacity estimation in concrete bridges

M. Proverbio[1,2], D.G. Vernay[1], I.F.C. Smith[2]

[1] *ETH Zurich, Future Cities Laboratory, Singapore-ETH Centre, CREATE, Singapore, Email: marco.proverbio@epfl.ch* [2]*Applied Computing and Mechanics Laboratory (IMAC), School of Architecture, Civil and Environmental Engineering (ENAC), Swiss Federal Institute of Technology (EPFL), Lausanne, Switzerland*

Abstract

Updating of physics-based model parameters involves identifying suitable model classes and values for parameters that influence real behavior. Error domain model falsification (EDMF) is a robust population methodology for structural identification and helps identify sets of plausible models by comparing their predictions with measurements at sensor locations. EDMF is easy to understand for practicing engineers and can provide robust parameter identification without assumptions on values of uncertainty correlations between measurement locations. This study focuses on the application of EDMF to assess the reserve capacity of concrete bridges for serviceability limit states (SLS). A case study, the structural identification of a reinforced-concrete bridge in Singapore, illustrates the methodology developed for the estimation of reserve capacity. In addition, a cross-validation method is employed to verify the robustness of the approach through comparing measurement values with predictions at sensor locations that are not used for model updating. Finally, traffic-load specifications of design-stage codes (British Code - 1978) and assessment codes (Eurocodes) are compared. This study demonstrates that, in spite of traffic-load specifications having been increased, the bridge still has SLS reserve capacity. Moreover, advanced methods of analysis and assessment help quantify reserve capacity that remains hidden when conservative design-like approaches are performed.

Keywords: structural model updating, reserve capacity, error-domain model falsification, serviceability limit states, concrete bridges.

1. Introduction

Design-stage behaviour models need first and foremost to be safe – more than sophisticated and precise. Once a structure is built the estimation of reserve capacity for asset management should involve more enhanced models than those used at the design stage. These models need to be capable of representing aspects such as as-built geometry, in-situ boundary conditions, deterioration and real material behaviour. Since design and existing structure evaluation are tasks that are performed with different goals in different contexts, assessment of structures has been formalized specifically in some codes (FIB 2013). Semi-probabilistic and full-probabilistic verifications are defined according to the amount of statistical information available on probability distributions of actions and effects (Brühwiler et al. 2012).

Many authors (Tapan et al. 2008, Frangopol et al. 2014) evaluated the strength of deteriorated existing structures by adding into the analysis deterioration mechanisms and their consequences such as reduction of steel diameter, loss of concrete cover and reduction of stiffness due to cracking. In (Ghasemi and Nowak 2016) weight-in-motion data are employed to improve the assessment of live loads, which lead to updated mean and standard deviation of deflection for serviceability limit state

(SLS) verifications. In the USA, information of bridge conditions from inspection is incorporated into the safety factor for the assessment of load rating of existing structures (Gao 2013).
Structural identification methods are used to improve knowledge of structural behaviour. Model-free methodologies based on vibration monitoring have been employed by many researchers to infer the status of existing structures. Although these techniques may be appropriate for damage detection, there is weak support for reserve capacity estimation and they are of no use when alternative scenarios (repair, retrofit, improvement) have to be evaluated (Smith 2016).
When structures are modelled, for instance using finite elements (FE), structural identification is often referred to as model updating. In model updating, measurements are carried out and used to improve the accuracy of model predictions. However, while measurements provide additional information for assessment of structures, structural identification involves many assumptions and sources of uncertainties. Error-domain model falsification (EDMF) is a robust approach to structural model updating that can provide parameter identification without assumptions on levels of uncertainty correlations. EDMF was employed (Goulet and Smith 2013) to evaluate the reserve capacity of a bridge in Switzerland with respect to its critical limit state of deflection. Pasquier et al. (2014) developed an approach based on EDMF to evaluate the fatigue reserve capacity.
This paper focuses on the evaluation of reserve capacity for the serviceability limit state (SLS) for concrete bridges. In the next section EDMF is introduced. Then, in Section 3, EDMF is applied to a full-scale case study and the reserve capacity of the bridge for the SLS is assessed.

2. Structural identification using EDMF

Error-domain model falsification (EDMF) – initially proposed in (Goulet and Smith 2012) – helps identify plausible physics-based models using information provided by measurement data. Plausible models are defined by parameter values that are assigned to a model class. Each model class has a unique parameterization that includes characteristics such as material properties, geometry, boundary conditions and actions.
Let n_y be the number of measurement locations. For each location $i \in \{1, \dots, n_y\}$, R_i denotes the real responses of a structure (unknown in practice) and y_i corresponds to the measured value at location i. Predictions $g_i(\boldsymbol{\theta})$ are evaluated at location i through assigning a vector of parameter values $\boldsymbol{\theta}$ to the selected model class. Evaluation of predictions is usually carried out using finite-element analyses. Since model-prediction uncertainty $U_{i,g}$ and measurement uncertainty $U_{i,y}$ are unavoidable, model predictions and measurements are linked to the real behavior using the following equation:

$$g_i(\theta) + U_{i,g} = R_i = y_i + U_{i,y} \tag{1}$$

Rearranging the terms and combining the two sources of uncertainty $U_{i,g}$ and $U_{i,y}$ in a unique source U_c:

$$g_i(\boldsymbol{\theta}) - y_i = U_{i,c} \tag{2}$$

where the combined uncertainty U_c is the difference between uncertainties U_y and U_g. The left-hand side of Equation (2) is the difference between a model prediction and a measurement at location i, which is often called the residual $r_i = g_i(\boldsymbol{\theta}) - y_i$.
The error in measurements U_y is usually estimated by conducting multiple series of tests under site conditions. Manufacturer specifications are often optimistic lower bounds on the as-installed uncertainty. The error in the model class U_g is estimated using values taken from the literature, stochastic methods (to estimate uncertainties of parameters that are not included in the model class

parameterization), engineering judgment and local knowledge. In practical situations, uncertainties associated with the model class are usually much larger than measurement uncertainties.

EDMF selects plausible behavior models by falsifying those for which residuals exceeds thresholds that are defined in the uncertainty domain (i.e. the error domain). First, an initial model set (IMS) is generated through assigning parameter values to the model class using sampling techniques. Then, threshold bounds are defined at each sensor location, according to a given confidence level – in many engineering problems a value of 0.95 is employed.

Models for which residuals are within threshold bounds at each sensor location are included in the candidate model set (CMS). Models for which residuals exceed these bounds, at one or more sensor locations, are falsified – and therefore, rejected. When a candidate model set is identified, prediction tasks can be performed employing the CMS to predict at unmeasured locations and, for example, to assess the reserve capacity of the structure.

3. Case study

The flyover, which consists of four prestressed concrete beams, has a single span of 32 m and is supported at each end by 4 bearing devices (Figure 1a-c). The beams support and are connected to a reinforced concrete slab that is 22 cm thick. Figure 1d presents the components of the FE model. This model includes also non-structural elements such as the asphalt pavement and precast concrete barriers. These elements, which are usually included as permanent loads in design-stage models, are considered in order to reduce model-simplification uncertainties.

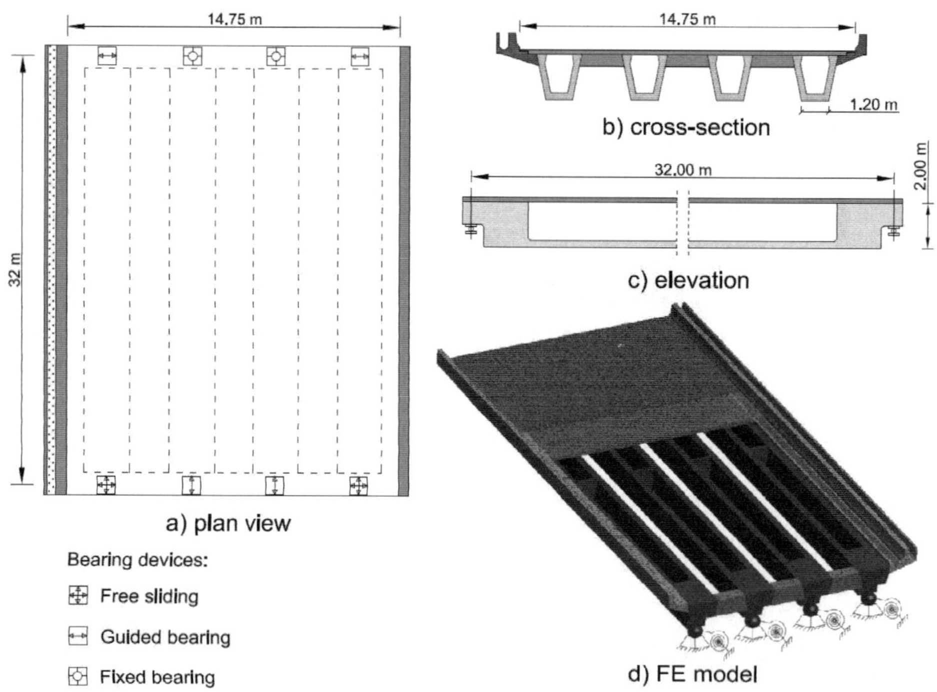

Figure 1: Plan view a), cross-section b), elevation c) and FE model d) of the flyover.

A static load test involving 6 trucks, each with a gross weight of 33 tons, is performed. The measurement system consists of a laser tracker, 8 strain gauges (S) and 2 inclinometers (I). In order to increase the accuracy of deflection measurements, 4 prisms (P) are placed on the bottom faces of the main beams. Truck configuration and sensor locations are depicted in Figure 2.

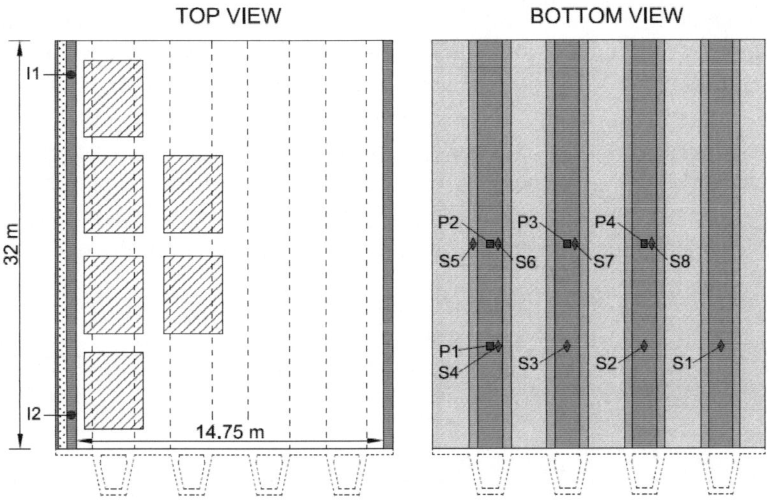

Figure 2: Truck configuration and sensor placement.

3.1 Uncertainty sources

The initial model set is generated through sampling the five-dimensional parameter space defined by the Young's modulus of cast-in-place concrete, the Young's modulus of precast concrete, the Young's modulus of barrier concrete, the rotational and the vertical stiffness of the bearing devices. The initial interval for each parameter, defined according to engineering judgment, is reported in Table 1.

Table 1: Parameter initial intervals

Parameters	**Lower bound**	**Upper bound**
θ_1 - Young's modulus of cast-in-place concrete	20 GPa	35 GPa
θ_2 - Young's modulus of precast concrete	25 GPa	50 GPa
θ_3 - Young's modulus of barrier concrete	3 GPa	40 GPa
θ_4 - Rotational stiffness of bearing devices	9 log(Nmm/rad)	13 log(Nmm/rad)
θ_5 - Vertical stiffness of bearing devices	8 log(N/mm)	11 log(N/mm)

Table 2 shows the bounds of the uniform probability distributions that describe modelling and measurement uncertainty sources. The bounds are expressed as a percentage of the mean value of model predictions for modelling uncertainties and as a percentage of the measured value or in absolute terms for measurement uncertainties. The uncertainty associated with model simplifications and the FE method is estimated using engineering judgement and considering the model class features. The FE model is constructed using solid elements with perfectly connected interfaces between adjacent layers. Therefore, this model class is likely to overestimate the real stiffness of the bridge. Moreover, this source takes into account that only the most sensitive parameter uncertainties are selected for identification. Bounds of uncertainty associated with mesh refinement and bounds for additional uncertainty are taken from (Goulet et al. 2010). The source of uncertainty associated with spatial variability originates from strain sensors measuring behavior that is affected by the variability

of material properties. Uncertainty sources estimated in this paper are similar to those employed in previous studies that focus on bridges (Goulet et al. 2010; Pasquier and Smith 2016).
Measurement uncertainties are estimated by comparing multiple measurements under site conditions and considering the type of sensors that are employed. Sensor accuracies – which represent the lowest source of measurement uncertainty – are based on manufacturer specifications. The measurement repeatability was assessed by taking multiple measurements under the same load case. Since the noise was mainly due to the vehicle traffic across a lane that was kept open during the test, a uniform distribution was used with bounds corresponding to the maximal and minimal values recorded. For strain gauges, an uncertainty also arises from the imperfect alignment of gauges with respect to the bridge longitudinal axis. Finally, additional noise associated with sensor installation have been considered for inclinometers and strain gauges using engineering judgment.

Table 2: Modelling and measurement uncertainty sources

Uncertainty source	**Displacements – (P)**		**Rotations – (I)**		**Strains – (S)**	
	Min	**Max**	**Min**	**Max**	**Min**	**Max**
Model simplifications and FE method (%)	-5	13	-5	13	-5	13
Mesh refinement (%)	-1	1	-1	1	-1	1
Spatial variability (%)	-	-	-	-	-5	5
Additional uncertainty (%)	-1	1	-1	1	-1	1
Sensor accuracy	-0.05 mm	0.05 mm	-1 μrad	1 μrad	-2 με	2 με
Repeatability	-0.15 mm	0.15 mm	-4 μrad	4 μrad	-4 με	4 με
Sensor orientation (%)	-	-	-	-	0	6
Sensor installation (%)	-	-	-5	5	0	5

3.2 Candidate model set (CMS)

In order to perform model falsification, for each measurement location, a combined uncertainty is computed and threshold bounds are determined for a confidence level fixed at 95%.
Figure 3 shows the results of structural identification. The vertical axes represent parameter values (θ_i) and predictions at sensor locations, which are referenced in accordance with Figure 2. Each grey line represents an instance of the initial model set along with the predictions of this instance at each sensor location. Models for which residuals are within threshold bounds at each sensor location are included in the CMS (red lines).
The reduction of parameter ranges after falsification is related to the measurement system adopted. Some parameters such as the vertical stiffness (θ_5) and the Young's modulus of precast concrete (θ_2) are well identified, while for other parameters such as the Young's modulus of site-cast concrete (θ_1), falsification does not reduce the initial interval. The performance of identification in reducing the initial parameter uncertainties depends on factors such as the initial choice of parameters, the sampling technique and the sensor configuration.

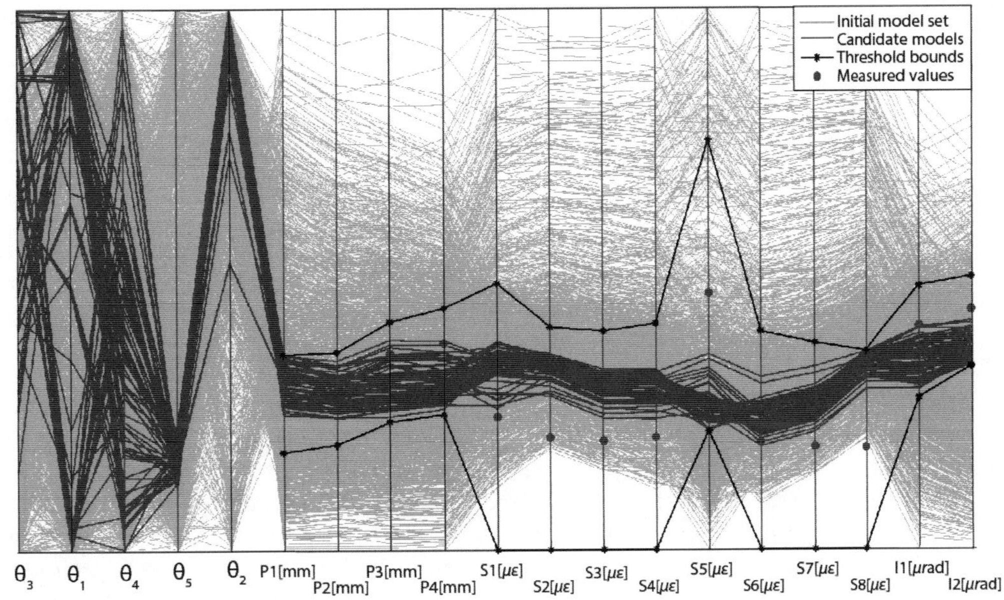

Figure 3: Parallel axis plot of structural identification using EDMF. The vertical axes represent parameter values ($\boldsymbol{\theta_1}$ *to* $\boldsymbol{\theta_5}$) and predictions at sensor locations for prisms ($\boldsymbol{P_1}$ *to* $\boldsymbol{P_4}$), strain gauges ($\boldsymbol{S_1}$ *to* $\boldsymbol{S_7}$) and inclinometers ($\boldsymbol{I_1}, \boldsymbol{I_2}$).

3.3 Cross-validation of model predictions

Strong methods for result validation are required to assess the performance of model updating techniques. In this study, a cross-validation is carried out using candidate models that have been identified to predict at locations that were not used for model updating.

Figure 4 shows an example of cross-validation. Prediction Q at locations i are obtained by combining model uncertainties with predictions of candidate models $g(\boldsymbol{\theta}^*)$ according to Equation (3).

$$Q_i = g_i(\boldsymbol{\theta}^*) + U_{i,g} \tag{3}$$

The CMS prediction distributions Q are shown in Figure 4 (background).

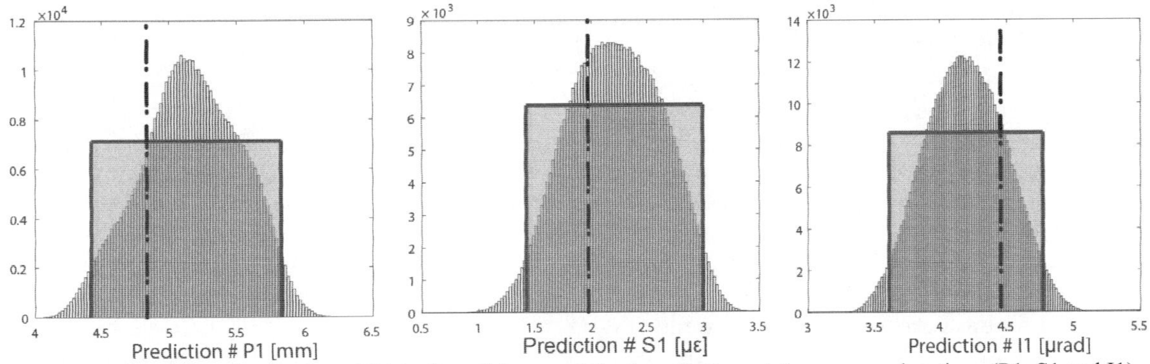

Figure 4: Probability distributions (PDs) of candidate-model-set predictions at three sensor locations (P1, S1 and I1) that have not been used for falsification. The vertical dash-dotted lines represent the measured values.

Since EDMF involves the hypothesis that all candidate models are equivalent likely, 95% confidence intervals are computed using those distributions. Uniform probability distributions (PDs) are assumed between bounds as there is little information to justify a more precise assumption (Figure 4, foreground). The PDs at three locations (P1, S1 and I1), which have not been used for falsification, are plotted. For each sensor, the prediction bounds include the measured value (vertical dash-dotted lines). In the remainder of this paper the CMS is employed to assess the reserve capacity (RC) of the bridge.

3.4 Serviceability limit state (SLS) design requirements

The analysed flyover was built in Singapore in 1985 following the requirements of the current code at the time – the British Codes (BS 5400-2 1978 and BS 5400-4 1984). Nowadays Singapore adopts the Eurocodes for bridge design (EN 1991-2 and 1992-2) that have been extended with the National Annexes.

Considering the serviceability limit states (SLS), the structural requirements are similar in these two codes. Under the load configuration for the SLS, the codes require to limit i) the stress level (compression in concrete and tension in rebars and tendons), ii) the crack width (according to the exposure classes in the EN and the bridge categories in the BS) and iii) the maximum deflection. In this case study, the crack width limitation was found to be the critical requirement.

For this prestressed bridge, considering the exposure class in EN and the bridge category in the BS, the control of crack width is satisfied by ensuring that no tensile stress occurs in the concrete around bonded tendons.

Although the requirements for crack control in the two codes are similar, the design traffic loads have been updated during the last 30 years. Increased average weight and number of vehicles has led to an increase of design traffic loads. Figure 5 shows the traffic-load design configurations for the serviceability limit states (SLS) according to the current Eurocodes (EN) and the British Code 1978 (BS).

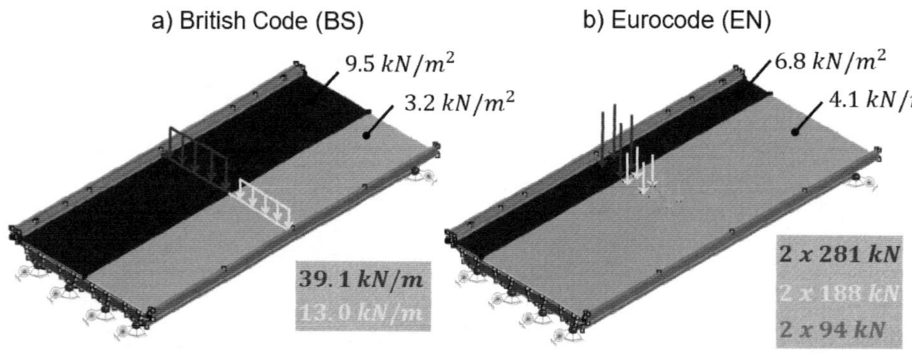

Figure 5: Traffic-load design values and configurations according to the British Code (BS 5400-2 1978) and the current Eurocode (EN 1991-2) for serviceability limit states.

Despite the different subdivisions of the bridge carriageway into notional traffic lanes, both codes consider a combination of uniformly distributed loads and concentrated loads, referred to as knife-edge loads in the BS or tandem loads in the EN.

The BS requires a higher uniformly distributed load than the EN. In the BS, the first two notional lanes are subjected to the heaviest distributed load (blue footprint in Figure 5a), while the EN assigns the heaviest load only to the first notional lane (blue footprint in Figure 5b). Considering the half left-

hand side of the bridge carriageway (7.37 m width), the distributed load required by the EN (average of 5.2 kN/m^2) is 45% lower than the load prescribed by the BS (9.5 kN/m^2).
On the contrary, the EN concentrated tandem loads have a greater impact than the knife-edge loads of the BS. Considering the half left-hand side mid-span section of the bridge, the EN tandem load ($\approx$ 1,023 kN) is more than three times the magnitude of the BS knife-edge load ($\approx$ 288 kN).
Since BS and EN prescribe different traffic-load configurations, a direct comparison is not straightforward. In order to assess which traffic load is the most critical, the half left-hand side of the bridge (i.e. subjected to the heaviest loads) can be analyzed as a simply supported beam (32m long and 7.37m width). Since the flexural behavior governs the SLS, the maximum values of bending moment can be compared. Under the EN traffic load, the mid-span bending moment is 16% higher than the moment obtained considering the BS traffic load. Therefore, the load configuration required by the current EN is the most severe.

3.5 Reserve capacity for SLS

The reserve capacity (RC) of bridges can be defined as the extra carrying capacity for the critical limit state. For example, when truck loading dominates, the reserve capacity is the ratio of as-built maximum truck load to as-designed maximum truck load.

$$\text{RC} = \frac{\textit{as-built max traffic load } (\gamma_Q)}{\textit{design traffic load } (\gamma_Q)} \tag{4}$$

The as-built maximum truck load should be determined after measurements and site observations through enhanced models and applying all relevant safety factors (γ_Q). The as-designed truck load is taken from current code requirements.
Sometimes simplified design-stage models, instead of enhanced models that have been updated using measurements, are preferred to evaluate the as-built maximum traffic load because of their simplicity. Although conservative simplified models are acceptable at design stages, they are not capable of representing the real behaviour of the structure and often lead to underestimation of the RC. The simplified model shown is Figure 6, where only the most loaded simply-supported beam is considered and effective loads are estimated according to the area of influence of the beam, is used to assess the reserve capacity of the flyover (Figure 7).

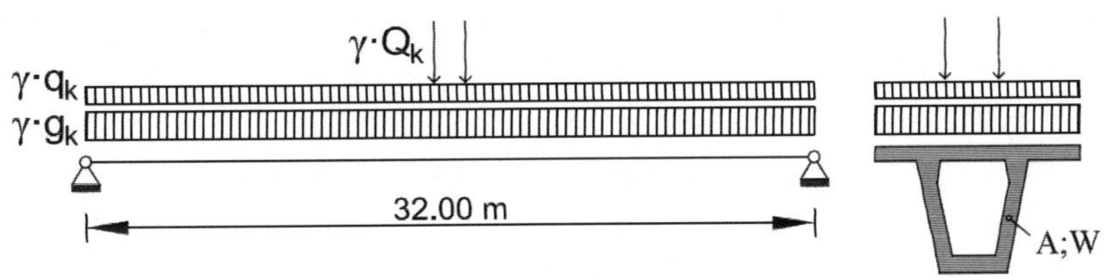

Figure 6: Simplified design model of the most loaded beam.

Under the assumption of perfectly elastic behaviour, the maximum values of stress in bending σ_{max} are estimated according to Equation 5, combining the contribution of the axial stress $N(p)$ – due to the prestress – with the contribution of the flexural stress $M(q,p)$ – due to dead and live loads and equivalent prestress loads.

$$\sigma_{max} = \frac{N(p)}{A} \pm \frac{M(q,p)}{W} \tag{5}$$

In Equation 5, the cross-section properties of the beam are the area A and the section modulus W. Figure 7 shows the assessment of SLS RC using the simplified model described above. While under the traffic load of the BS 1978, the bridge shows a SLS reserve capacity of 1.15, when the current EN load configuration is considered in the simplified analysis, the bridge is not able to comply with current requirements (RC<1).

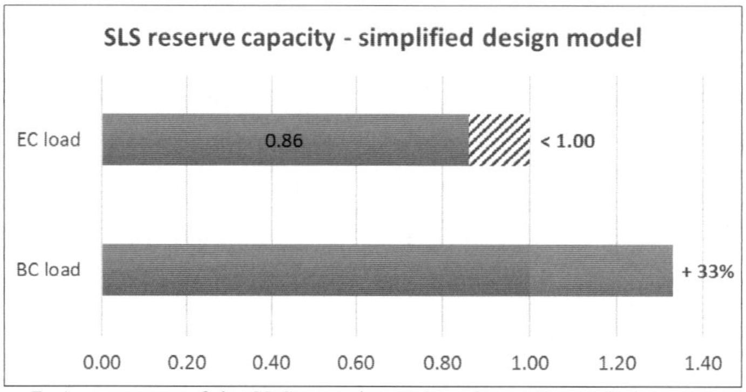

Figure 7: Assessment of the SLS capacity (RC) using a simplified design model.

Figure 8 shows the assessment of SLS RC using the most critical model that has been identified through EDMF and applying partial safety factors for traffic loads (γ_Q) to both EN and BS loads. Alternatively, a full probabilistic approach may be considered. In such a case, the RC probability distribution is calculated considering predictions of all candidate models including the model uncertainty (see Equation 3). Then, the RC is assessed for a given probability of exceeding the critical limit state. Although a full-probabilistic approach is applicable, the partial-safety-factor approach is preferred in this study, since it allows a direct comparison with design code requirements. Moreover, a full probabilistic approach would require the traffic-load probability distributions as input, and the direct computation of the *probability of failure*.

Under the traffic load of the BS, the bridge shows a SLS RC of 1.41. However, using an advanced model and updated parameter values, the bridge shows a RC of 1.11 even when the current – most severe – EN load configuration is considered. Therefore, the bridge is able to comply with current SLS requirements because uncertainties have been reduced.

In this study the identification of the critical limit state for the structure is still missing, since only the SLS behaviour is investigated. Future work, involving non-linear simulations, will focus on the assessment of reserve capacity for the ultimate limit state.

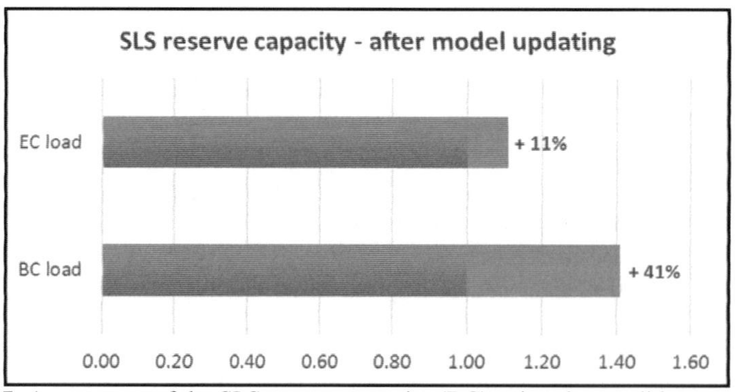

Figure 7: Assessment of the SLS reserve capacity (RC) using the candidate model set.

4. Conclusions

This study, involving a full-scale bridge, demonstrates that EDMF is a simple and robust methodology to perform structural identification. In addition, most of existing bridges, have reserve capacity beyond the safety factors provided by structural codes. In this study, for example, when the candidate model set is used to estimate the reserve capacity, the bridge is able to comply with SLS requirements of current codes. Finally, design-stage simplified models provide weak support to assess the reserve capacity of existing bridges.

Acknowledgements

The research was conducted at the Future Cities Laboratory at the Singapore-ETH Centre, which was established collaboratively between ETH Zurich and Singapore's National Research Foundation (FI 370074016) under its Campus for Research Excellence and Technological Enterprise programme. The authors gratefully acknowledge the support of the Land Transport Authority (LTA) of Singapore for support during load tests in the scope of the case study.

References

Fédération Internationale du Béton, Fib, 2010. Model Code for concrete structures 2010.

Brühwiler, E., Vogel, T., Lang, T. and Lüchinger, P., 2012. Swiss standards for existing structures. Structural Engineering International, 22(2), pp.275-280.

Gao, L., 2013. Load Rating Highway Bridges in the United States: The State of Practice. Structural Engineering International, 23(3), pp.327-331.

Tapan, M. and Aboutaha, R.S., 2008. Strength evaluation of deteriorated RC bridge columns. Journal of Bridge Engineering, 13(3), pp.226-236.

Frangopol, D. and Tsompanakis, Y. eds., 2014. Maintenance and safety of aging infrastructure. CRC press.

Smith, I.F.C., 2016. Studies of Sensor Data interpretation for Asset Management of the Built environment. Frontiers in Built Environment, 2, p.8.

Ghasemi, S.H. and Nowak, A.S., 2016. Reliability analysis for serviceability limit state of bridges concerning deflection criteria. Structural Engineering International, 26(2), pp.168-175.

Goulet, J.A. and Smith, I.F.C., 2013. Structural identification with systematic errors and unknown uncertainty dependencies. Computers & structures, 128, pp.251-258.

Pasquier, R., Goulet, J.A., Acevedo, C. and Smith, I.F.C., 2014. Improving fatigue evaluations of structures using in-service behavior measurement data. Journal of Bridge Engineering, 19(11), p.04014045.

Goulet, J.A. and Smith, I.F.C., 2012. Performance-driven measurement system design for structural identification. Journal of Computing in Civil Engineering, 27(4), pp.427-436.

Goulet, J.A., Kripakaran, P. and Smith, I.F.C., 2010. Multimodel structural performance monitoring. Journal of Structural Engineering, 136(10), pp.1309-1318.

Pasquier, R. and Smith, I.F.C., 2016. Iterative structural identification framework for evaluation of existing structures. Engineering Structures, 106, pp.179-194.

The 8th International Conference on
Structural Health Monitoring of Intelligent Infrastructure
Brisbane, Australia | 5-8 December 2017

Automated Bridge Model Calibration Using SHM Controlled Load Test Data

H. Al-Khateeb[1], C. Natalicchio[1], H. Shenton[2], M. Chajes[2], Z. Wu[3], and W. Elhaddad[4]

[1]Research Associate, Dept. of Civil and Environmental Engineering, University of Delaware, Newark, DE, USA [2]Professor, Dept. of Civil and Environmental Engineering, University of Delaware, Newark, DE 19716, USA [3]Bentley Fellow, Bentley Systems, Incorporated, Watertown, CT, 06795, USA [4]Software Engineer, Bentley Systems, Incorporated, Anaheim, CA, 92806, USA

Abstract

Frameworks for structural health monitoring very often rely on calibrating a computer model of the structure (i.e., system identification). The calibrated model is then used to evaluate the performance of the structure, identify changes or damage to the structure, or predict the response to other types of excitation. Historically, calibration has been done manually, using single, unique, input datasets, a process that is labor and time intensive. With current SHM systems that can provide near continuous data at high rates, there is the potential to update models at near similar rates. To do this, however, requires an automated approach that will take the most recent data, and perform a calibration with little user intervention. Tools for executing model calibration in an automated sense are just now being developed and tested. These tools offer the potential to bring SHM from a mostly research/academic pursuit, to state-of-the-practice; i.e., to put it in the hands of practicing engineers and owners.

In this study the Indian River Inlet Bridge (IRIB), located in Delaware, USA, was used as a test case to verify a methodology for automated calibration of a complex bridge model using SHM data. The calibration tool is prototype software being developed by Bentley Systems, Inc., for use with SHM systems. Controlled load tests of the IRIB have been conducted at different intervals. Strain data from key locations on the bridge were used as input. The study found that calibration of a complex, long span bridge model using pseudo-static strain data poses several challenges, but offers the potential for greater accuracy, when compared to, for example, calibration based on only ambient vibration measurements. The paper presents the methodology and results of the calibration using a 6-parameter model and an 18-parameter model.

1. Introduction

The Charles W. Cullen Bridge, also known as the “Indian River Inlet Bridge” (IRIB), is a 1,750 foot cable-stayed bridge located in Delaware, on the east coast of the United States. The bridge carries State Route 1 over the Indian River Inlet, a strong tidal channel that connects the Indian River Bay to the Atlantic Ocean. The current bridge replaced a steel girder structure that had to be replaced because of severe scour around piers that were located in the channel. Traffic opened to the bridge in the spring of 2012.

In the conceptual phase of this project researchers from the University of Delaware (UD), Center for Innovatie Bridge Engineering (CIBRe), proposed the idea of installing a permanent Structural Health Monitoring (SHM) system on the bridge to the owner, the Delaware Department of Transportation (DelDOT). With DelDOT’s support, and the support of the Federal Highway Administration, a comprehensive SHM system was designed and installed on the bridge during construction (which began in 2009). The bridge and SHM system have been reported on in other ISHMII conferences (Shenton, et al, 2011: Shenton, et al, 2013, Shenton, et al, 2015). From among others, there are 44 strain gauges located in the longitudinal edge girders of the bridge, the data from which was used in this study. The SHM system and strain sensors will be described in more detail later.

Controlled live load tests have been conducted on the IRIB on five different dates. The first of those was in 2012 just before the bridge was opened to traffic, and the most recent being in May 2016. The controlled live load tests were performed using trucks of known weights and axle spacing. During the tests, the loaded trucks made passes over the bridge while strain data was collected. The live load tests will be described in more detail later.

Today, signature bridges, such as the IRIB, are designed using high fidelity computer models. Once the bridge is built the model can be calibrating using data collected from a controlled load test or a vibration study of the structure. Response data can come from sensors that are permenantly mounted to the structure, such as in the case of the IRIB, or from sensors that are temporarily mounted to the bridge. The benefit of the permenant SHM system over a temporary deployment is that a test can be conducted on very short notice, with minimal effort, and therefore usually more frequently than might be the case when relying on a temporary sensor setup. The benefits to the owner of having a calibrated model are: (1) more accurate load ratings of the bridge, (2) accurate permit vehicle evaluation, (3) to support visual inspection by identifing changes or damage to the structure, and (4) to use in the immediate aftermath of extreme event to assess the condition of the structure. Calibrations today, however, are typically done manually, using a "brute force" approach, and are labor and time intensive.

To take advantage of the near real-time, continuous data that SHM systems can provide today, an automated calibration method is needed that will streamline and simplify the process. Only then will SHM and model calibration move from a mostly research/academic pursuit, to state-of-the-practice, where these modern tools will be in the hands of practicing engineers and owners.

Bentley Systems, Inc. is developing a protoype software tool called the "Darwin SHM Calibrator", that works with their Darwin Optimization interface to calibrate a STAAD structural model based on SHM/test data from a structure (ref.). The CIBRe at UD is collaborating with Bentley Systems, Inc to test the prototype calibration tool with a complex STAAD.Pro model of the IRIB, using the strain data collected during the controlled load tests. Calibration using strain data can be directly linked to the elastic modulus of the concrete in the main load carrying edge girders as the modulus is a property of concrete and effects the response of the structure. Calibration utilizing strain data has the potential to provide more accurate results than other forms of SHM data, such as vibration or displacement measurments.

Calibrations performed using the prototype Darwin SHM Calibrator with the IRIB STAAD model and data obtained during the controlled load tests are presented in this paper. The calibrated compressive strengths results are compared to known compressive strengths of the IRIB that were measured from cyclinder break tests of the cast concrete.

2. Descripton of the Bridge

The Indian River Inlet Bridge is a cable stayed design with a total length of 533m (1,750 ft). The main span of the bridge is 290m (950 ft) in length and both backspans are 122m (400 ft). The bridge consists of four pylon towers with 152 stays supporting the three spans of the bridge. Steel was precluded by DelDOT as a material due to the bridge's proximity to the Atlantic Ocean and the persence of chlorides. The total width of the bridge is 32.3m (106 ft 2 in) It carries two travel lanes and a shoulder in each direction. A 3.7m (12 ft) pedestrain walkway located on the east side of the bridge shifts traffic lanes towards the west side of the bridge. The longitudinal edge girders are roughly rectangular in shape and are 1.83m by 1.52m wide (6 ft by 5 ft).

The bridge was constructed using a combination of precast and cast-in-place reinforced concrete. The portion of the bridge over land was constructed on falsework; the edge girders in this region were cast in roughly 22m (72 ft) increments. The portion of the bridge over the water was constructed using a form traveler, which cast the edge girders, transverse beams, and deck together in 24 foot sections. Construction spanned nearly two years from the start to completion, therefore portions of the bridge were cast more than a year and a half apart. Data collected from cyclinder breaks at 56-days show a maximum, average, and minimum compressive concrete strength of 98, 57, and 38 Mpa (14.2, 8.2, and 5.5 ksi), respectively. One would therefore expect to have similar variability in the concrete modulus along the length of the edge girders. More detail on the bridge design and construction can be found in Nelson (2011).

3. Description of SHM System

A permanent SHM system was installed on the bridge during construction. The SHM system has a total of 120 sesnors distributed over the bridge, which include, but are not limited to: accelerometers, stain sensors, temperature sensors, and tiltmeters. Strain sensors make up the bulk of the sensors on the bridge and are the focus of this paper. The fiber-optic strain gauges are embedded in the concrete, anchored to the longitudinal reinforcing steel. Strains are measured at 11 different locations along the edge girders. Each location has 4 sensors, the top and bottom of the cross-section of both the east and west girders. This results in a total of 44 unique strain readings. The 11 sensor locations are shown on both the east and west edge girder in Figure 1.

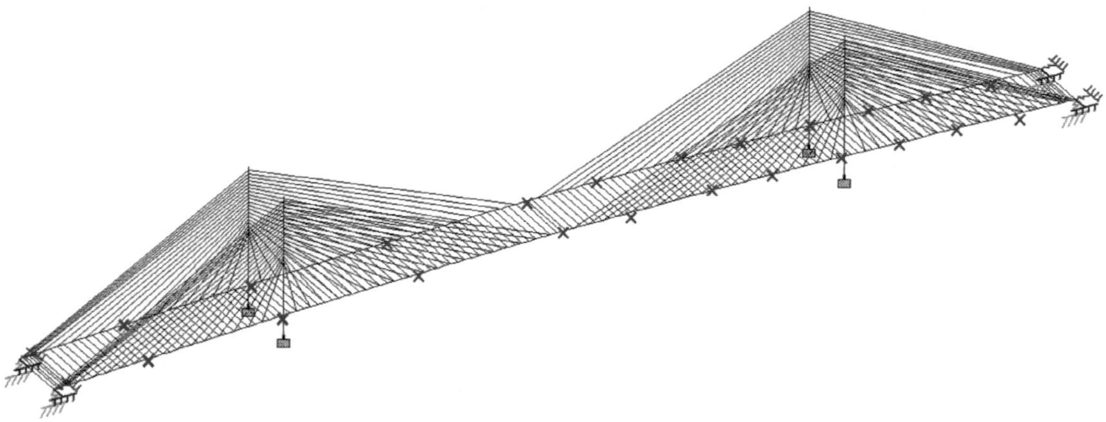

Figure 1 Edge Girder Strain Sensor Locations on the Indian River Inlet Bridge

4. Description of Load Tests

Five controlled load tests have been conducted on the IRIB. The timing of these were prior to opening, at 6 months, 1 year, 2 years, and 4 years after it was opened to traffic. Load tests were performed using fully-loaded 10-wheel dump trucks. DelDOT weighed the trucks off-site and the weights were verified on-site with portable truck scales; the axle spacings of each truck was recorded. The average weight of each truck was 278 kN (62.4 kips). The bridge was closed during the load tests to assure that the trucks were the only vehicles on the bridge at the time. Multiple passes were performed with different speeds, number of trucks, and truck configurations, based on the purpose of each pass. The pass using four trucks side-by-side was used in this study: one truck located in each lane of the bridge, all traveling southbound. This pass was used for the study as it produced some of the greatest strain responses in the bridge. The trucks traveled between 8 and 16 kph (5 and 10 miles per hour) over the entire length of the bridge to minimize any dynamic response. This pass was considered quasi-static due to the speed of the trucks and elimination of dynamic effects. The strain measurements from the load tests are then able to be compared to strain that would develop due to a concentrated load of equal weight. The data used in this study was from the load test performed on Novemeber 28, 2012, which was six months after the bridge was opened to traffic. The strain data was recorded at a rate of 125 samples-per-second during the controlled load tests.

5. Description of STAAD Model

The STAAD.Pro model of the IRIB, shown in Figure 1, consists of 767 elements. The east and west edge girders are modeled using 320 beam elements. Nine load cases were created in the STAAD model to represent 9 locations of a concentrated load produced by the four side-by-side trucks used in the pass. Each truck was represented by a single 278 kN (62.4 kip) point load. The 9 load case locations are at the points directly over the strain sensor locations, shown in Figure 2.

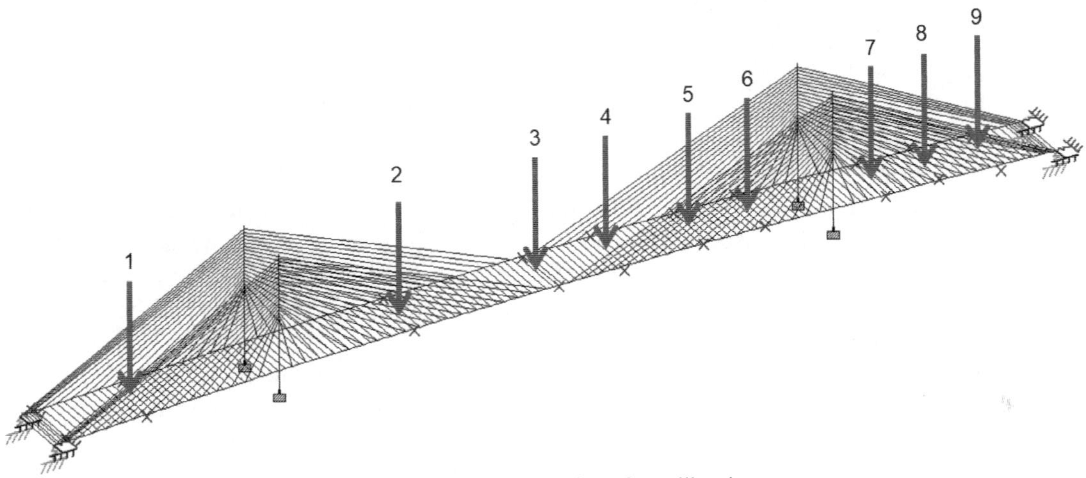

Figure 2: Load case locations for calibration

Each arrow in this image represents 4 point loads in the model, to account for the four trucks used in the pass. The transverse positions of the point loads correspond to the center of the four travel lanes of the bridge.

6. Calibration Tool and Approach

Darwin SHM Calibration tool (Wu et al 2013; 2011) uses SHM/test data and a computer model of the structure to perform an automated calibration. At the core of the software is the Darwin Optimization Tool (Wu et al. 2011). The calibration tool utilizes a genetic algorithm as the optimization method to calibrate the model with the user input data to optimize the user defined parameters. The software has the capability to use multiple types of SHM data, e.g., static or dynamic, from different types of sensors, e.g., displacement, strain, and acceleration, to optimize different parameters, e.g., elastic modulus, density, dimensions, and spring supports of the model. The tool has been tested on a building (Zhao and Wu 2013) and a small bridge (Wu et al. 2013) using measured dynamic characteristics, i.e., frequencies and mode shapes, of the structure. The purpose of this study was to use quasi-static data from a controled load test of a large, signature bridge, to test the calibration tool. The use of strain data from controlled load tests for model calibration has the potential for greater accuracy than other forms of data.

The objective of the optimized calibration is to search for a set of model parameters that minimizes the difference between the real structure and the model. In this study, the optimized parameters were limited to the elastic modulus of the edge girder beam elements. These are the main longitudinal load carrying members of the bridge and are also the members that control the load rating of the bridge. The objective function can be expressed in general as:

$$E_T = E_{static} + E_{freq} + E_{mode} \quad (1)$$

where E_{static} , E_{freq} and E_{mode} are the static response-based, frequency-based and mode shape-based error function, respectively. Here, only the static strain reponses are used. The calibration tool offers several different error functions: the specific error function adopted in this work is the root-mean-square of the absolute difference in strain, which can be expressed as

$$E_{static} = \sqrt{\frac{1}{n}\sum_{i=1}^{n} \omega_i \left|\varepsilon_i^A - \varepsilon_i^O\right|^2} \quad (2)$$

in which ω_i is a weighting parameter, and ε_i^A and ε_i^O are the model predicted and observed (SHM data) strain in the edge girder at location *i*.

In this investigation the calibration parameter used was the Elastic Modulus (E), also known asYoung's Modulus, of the east and west longitudinal edge girders. Young's Modulus is a property of concrete that is related to the strength and stiffness of the material, and therefore effects the strain response of the bridge. The calibration tool uses the STAAD model of the structure, user input member cross-section properties, target strain measurments, and beam element groupings (parameters) to optimize the elastic modulus of the model. The target strain measurements are the strain measurements gathered from the controlled load tests, for the 9 load positions shown in Figure 2. The calibration works to create a model that accurately represents the response of the bridge based on the strain data by optimizing the modulus.

The specific parameters given to the calibration tool correspond to elements in the STAAD model and make up groupings of beam elements for which the elastic modulus is to be optimized. Parameters consisted of the 320 beam elements that the edge girders are composed of in the STAAD model. In this paper results are shown of calibrations using 6 parameters and 18 parameters. Figure 3 a and b show the 6-parameter and 18-parameter groupings of the east and west edge girders, respectively.

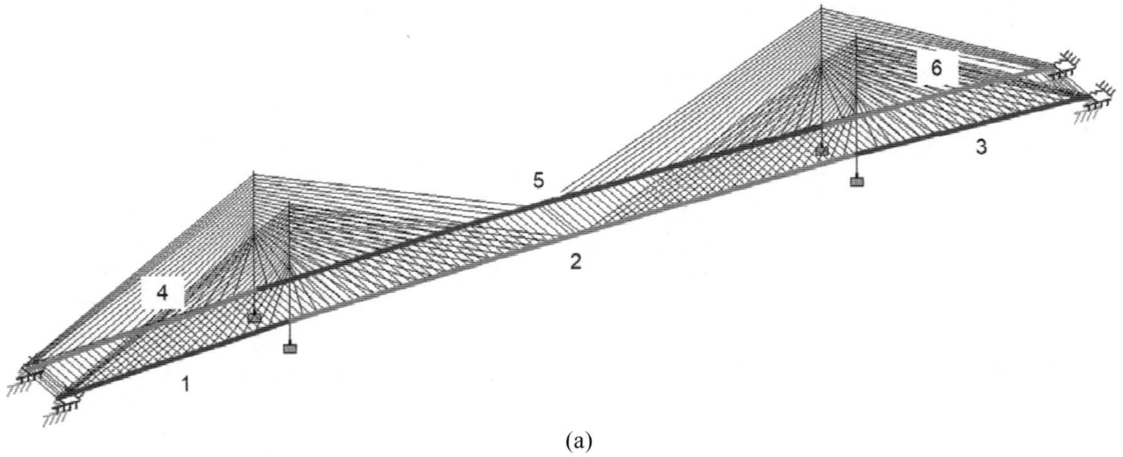

(a)

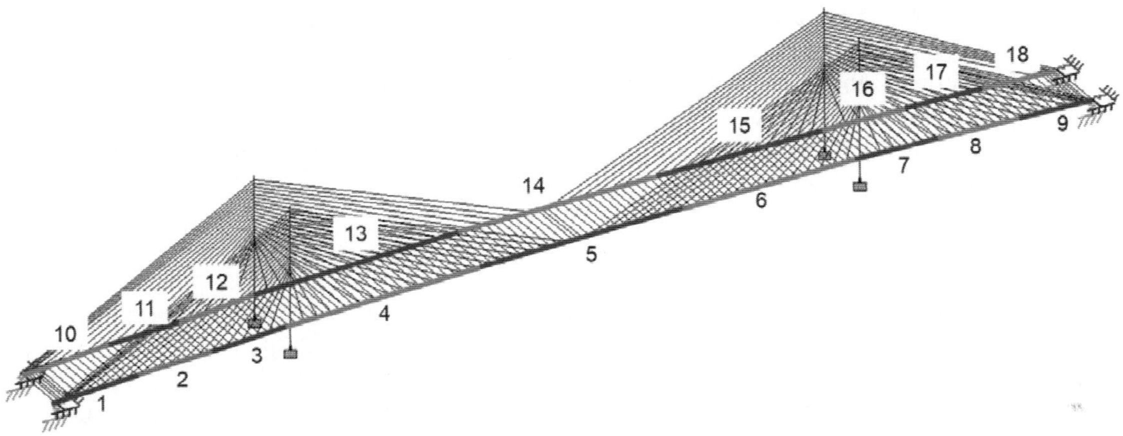

(b)

Figure 3: Parameter Groupings (a) 6 Parameters (b) 18 Parameters

In the 6-parameter model, the groups in the near and far spans (parameters 1, 3, 4, and 6) each consist of 36 beam elements and the groups in the mid-span (parameters 2 and 5) each consist of 88 beam elements. The six parameters are each split into thirds to create the 18-parameter groupings. The modulus is constant within a single parameter group. Therefore, in the 6-parameter case the calibration optimizes 6 moduli, while in the 18-parameter case it optimizes 18 unique moduli. By using one modulii to represent a large portion of the bridge the results are limited because in reality the concrete of the edge girders was poured using numerous concrete mixes which will all have varying moduli values. On the other hand, by increasing the number of parameters the calibration tool must work through a larger number of possible solutions that may effect the final solution.
In order for the calibration tool to run there must be more strain readings input by the user than parameter groups. In the two cases discussed here there were the 9 load cases used, which results in a total of 62 independent target strain values. During each load case there are areas of the edge girders which develop low strain values, these values are considered negligible due to sensitivity of the sensors and were not used in the calibration. The 62 indpendent strain values used were all above 2.0 microstrain.

The target strain values input in the calibration tool were determined by identifying the time that the four trucks were located at the nine sensor locations shown in Figure 2. The strain measurments were then determined for all sensors for the 9 load cases.
The user defined parameter range and increment determines the upper and lower limits, and resolution, of the parameters. It also has a large effect on the run time as these factors set the parameter space and number of possible solutions for the genetic algorithm. The parameter range for concrete moduli was defined as 22.2 to 55.6 Gpa (3224 ksi to 8060 ksi), which corresponds to a compressive strength range of 22 to 138 Mpa (3.20 ksi to 20 ksi). This is well above and below the known compressive strengths determined from the 56-day cyclinder breaks which produced values ranging from 38 to 98 Mpa (5.5 to 14.2 ksi), as reported earlier. Properties of concrete are not explictly known since concrete is dependent on it's environment and continues to strengthen as it ages, unlike properties of other materials such as steel. The load test used in this study was performed over two years after portions of the edge girders were cast. It is therefore conceivable to see slightly higher moduli values than those found from the 56-day cylinder breaks due to the increase in compressive strength of concrete over time.
In the calibration tool, optimization results are based on the "fitness value" which is given in equation (1.2). The calibration tool continues to optimize the moduli to minimize the fitness value. The same target strain values were used in both cases presented here, therefore, the fitness value can be compared without being normalized.
As mentioned previously, the optimization is completed using a genetic algorithm, which is the basis for the Darwin Optimization Tool.

7. Results

Presented in Figure 4 are the calibrated elastic modulus values for the 6-parameter calibration. The x axis in the plot corresponds to the parameter number that was shown on Figure 3(a) and the y axis is elastic modulus values in ksi. The dashed lines on the figure show the maximum, average, and minimum moduli that were recorded from the 56-day cyclinder breaks. The fitness value calculated using the root-mean-square-error function was 8.64 while the initial model, which was based on a uniform modulus value of 27.8 GPa (4030 ksi (strength of 5 psi) had a fitness value of 19.41.

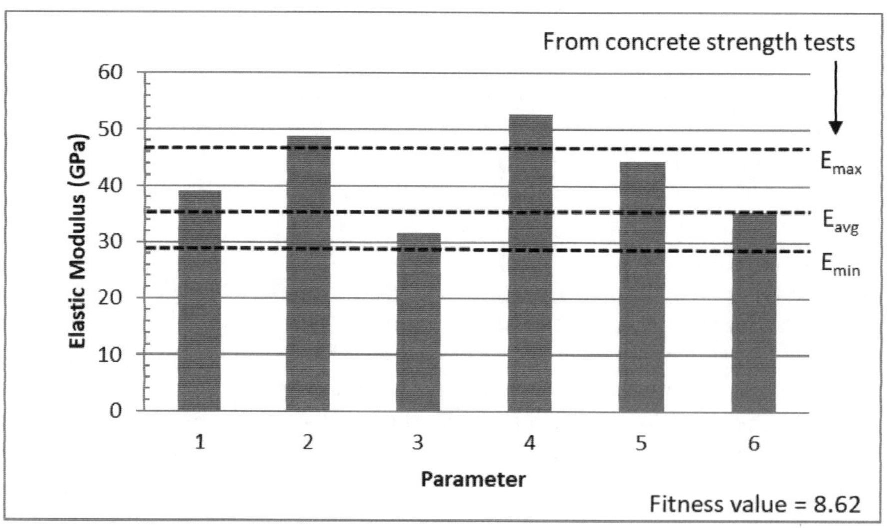

Figure 4 Calibrated elastic modulus using 6 parameters

One can see from Figure 4 that four out of six of the calibrated modulii fell within the range of measured modulii based on the concrete strength. Furthermore, the average calibrated modulii is equal to 42 GPa (6105 ksi), which is within 18.3% of the average measured value.

Shown in Figure 5 are the calibrated moduli for the 18-parameter calibration. The fitness value for the calibration was 4.54. As one might expect the 18-parameter calibration produced a model that more accurately represents the bridge response than the 6-parameter calibration, demonstrated by the lower fitness value. By allowing the tool to optimize smaller sections of beam elements the calibration was able to match the input target strain values more closely. Using the 6-parameter calibration, the optimizer was limited since full span lengths were considered to have a constant modulus value, which is not realistic based on the construction sequence of the edge girders. In both cases the calibrated models are an improved representation of the bridge compared to the initial model which had a fitness value of 19.41 when both edge girders have a uniform modulus of 27.8 GPa (4030 ksi).

One can see from Figure 5 that half of the calibrated moduli fall within the range of the measured moduii, and half do not. Only one value falls below the measured minimum; eight are greater than the maximum, with the largest calibrate moduli being 18.7% greater than the maximum measured modulii.

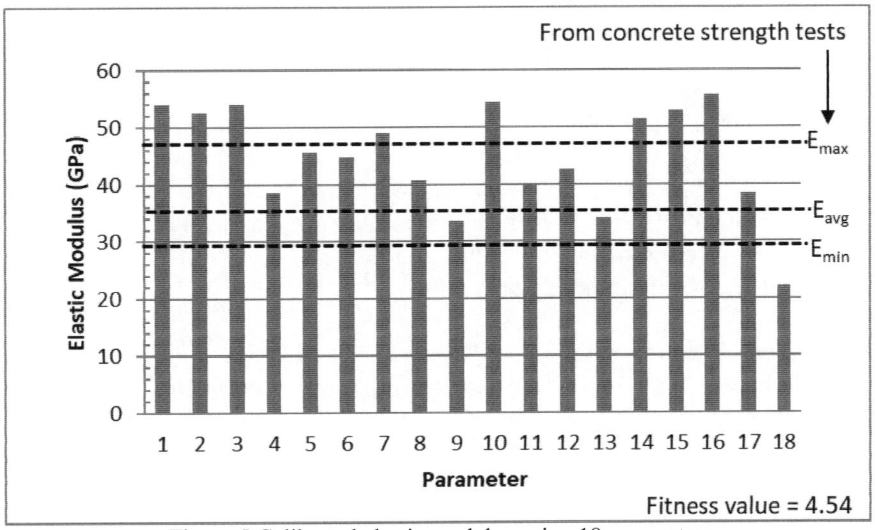

Figure 5 Calibrated elastic modulus using 18 parameters

8. Conclusions

Results have been presented to demonstrate the use of an automated model calibration tool to calibrate a structural model based on SHM data. The calibration tool used was a prototype tool developed by Bentley Systems, Inc. that works with the Darwin Optimization tool. The IRIB model and strain data from controlled load tests performed on the bridge 6 months after the bridge was opened to traffic were used. The parameter that was optimized in the calibrations was the elastic modulus of the longitudinal edge girders.

Calibrations were performed using 6-parameter and 18-parameter cases. When comparing the final fitness values of these calibrations, the 18-parameter model produced results that more accurately represent the response of the bridge based on the target strain data. Overall, both models significantly improved the initial model and produced accurate models of the bridge. It is important to consider the portions of the edge girders that are expected to have the same modulus value when creating the parameters, which would be determined from the construction sequence. These calibrations used arbitrary parameter groupings and did not consider the construction sequence of the bridge.

The software tool provides a streamlined, automated process for model calibration. The tool is not in commercial production at this time, but shows tremendous promise as a modern engineering tool that can be added to the bridge engineer's "toolbox", that takes advantage of the data provided by an SHM system.

References

Nelson, E. T. (2011) "Indian River Inlet Bridge – Surviving the Storms," Aspire the Concrete Bridge Magazine, Precast/Prestressed Institute, Chicago, IL, Winter 2012

Shenton III, H.W., Carson, P.D., Wenczel, G., and Chajes, M.J. (2011) "Structural health monitoring of the Indian River Inlet Bridge," Proceedings of the 5th International Conference on Structural Health Monitoring of Intelligent Infrastructure (SHMII-5) 11-15 December 2011, Cancún, México

Shenton, H.W., Chajes, M.J., Wenczel, G., Davidson, K. (2013), "Lessons Learned in Construction and Early Experience with the Indian River Inlet Bridge SHM System," Proceedings of the 6th International Conference on Structural Health Monitoring of Intelligent Infrastructure (SHMII-6) 9-11 December 2013, Hong Kong, PRC

Shenton, H., Fernandex, M., Ramanna, N., Chajes, M., Al-Khateeb, H. (2015), "Structural Health Monitoring of a Cable-Stayed Bridge: Using Tiltmeter Data to Determine Edge Girder Deflections," Proceedings of the 6th International Conference on Structural Health Monitoring on Intelligent Infrastructure (SHMII-7) 1-3 July 2015, Torino, Italy

Wu, Z. Y., Yi, X., Zhao, J., Jeary, A, Winant, T., Love, R. and Bhide S. (2013). " Optimized Calibration of Finite Element Model for Verrazano Narrows Bridge Span." *The 6th International Conference on Structural Health Monitoring of Intelligent Infrastructure*, Dec. 9 - 11, 2013, Hong Kong, China.

Wu, Z. Y., Wang, Q. and Butala S. (2011). "Generalized Framework for high Performance Infrastructure System Optimization." *International Conference on Computing and Control for the Water Industry*, Sept. 5-7, 2011, Exeter, UK.

Zhao, J. and Wu, Z. Y. (2013) "Finite Element Model Calibration of UCLA Factor Building by Darwin Parallel Optimization." The 6th International Conference on Structural Health Monitoring of Intelligent Infrastructure, December 9-11, 2013, Hong Kong, China.

The 8th International Conference on
Structural Health Monitoring of Intelligent Infrastructure
Brisbane, Australia | 5-8 December 2017

Improving prediction capability of finite element models of bridges using static and dynamic data

W.-J. Cao[1,2], D. Vernay[2], C.G. Koh[1] and I.F.C. Smith[2,3]

[1] *Department of Civil and Environmental Engineering, National University of Singapore - Singapore, Email: caowenjun@u.nus.edu* [2] *ETH Zurich, Future Cities Laboratory, Singapore-ETH Centre, CREATE- Singapore* [3] *Applied Computing and Mechanics Laboratory (IMAC), School of Architecture, Civil and Environmental Engineering (ENAC), Swiss Federal Institute of Technology (EPFL), Lausanne - Switzerland*

Abstract

Although finite-element models are commonly used in designing civil infrastructure, they may fail to represent the true operational behaviour. Measurements from monitoring and inspection can provide valuable information to improve the prediction capability of models. A significant amount of research has focused on system identification using either dynamic or static measurements separately. However, few research includes the systematic nature of many sources of uncertainties.
In this paper, the methodology of error-domain model falsification recently proposed by Goulet and Smith is adopted with emphasis on the combination of both dynamic and static measurements. This methodology is most useful when uncertainties are systematic such as those originating from epistemic sources. Two case studies are presented to demonstrate this methodology. The first involves a simply supported beam in which the static and dynamic responses can be derived analytically. In the second case, a field test on an in-service prestressed concrete highway bridge in Singapore was conducted. The values of deflection and inclination as well as natural frequencies were measured. The results of model falsification using both static and dynamic measurements show higher falsification capacity compared with using only dynamic measurements. It is also shown that updating models using a model-falsification approach is more robust than using a traditional implementation of Bayesian inference.

1. Introduction

To increase the knowledge of real behavior of complex structure, structural identification (SI) was introduced decades ago (Aktan et al. 1997). Predictions of initial finite element models and measured responses are often different from each other due to the lack of knowledge of the structure.
A significant amount of research has focused on SI of full-scale civil engineering structures (Brownjohn 2007). Classical methods aim to find the optimal parameter values by adjusting them so that the finite-element-analysis results best match the corresponding measurements (Araujo et al. 2011). However, this approach ignores modeling uncertainties and measurement uncertainties which leads to error compensation at sensor locations during the optimal process.
Probabilistic FE model updating using Bayesian inference scheme was introduced by (Katafygiotis and Beck 1998). Instead of looking for an optimal solution, this approach estimates the level of confidence of model updating results. Many applications on bridges can be found in the literature (Behmanesh et al. 2015). However, these probabilistic studies do not include modelling uncertainties, which can significantly affect the model updating results.
Error-domain model falsification was proposed by Goulet and Smith (Goulet and Smith 2013) to overcome the limitations of other approaches. Based on a multi-model approach, this method

generates possible model instances from an initial model space defined by engineering judgement, then falsifies wrong model instances using measurements. Model instances are falsified if the difference between finite-element-analysis results and measurements falls outside the threshold bounds computed by combining modelling and measurement uncertainties.
Vibration monitoring data is commonly used in structural identification of bridges because of its lower cost and because there is no requirement to interrupt traffic. It reflects the structural stiffness, the mass and the damping. Another merit of vibration data is that the accuracy requirements of accelerometers and data acquisition systems are not as high as for strain gauges, deflection prisms and inclinometers.
The mass distribution of the bridge and the stiffness of materials are regular unknown parameters of the identification process. However, depending solely on dynamic properties (natural frequencies and mode shapes) is not enough because mass and stiffness parameters can be changed by the same ratio while still giving the same eigen-solutions. On the other hand, the deflections, strains and inclinations are only related to the stiffness. Thus, static and dynamic measurements have the potential to complement each other. This paper contains a study of such potential using a case study of a highway bridge in Singapore.

2. Methods

2.1 Error-domain model falsification

The error-domain model falsification methodology was proposed by Goulet and Smith. The fundamental equations are shown in Eqs. (1) and (2). As-built structures often have high uncertainties originating from many sources. Some are due to the unknown parameter values, while others are due to the modeling errors. Measurements can be used to better understand the behavior of existing structures. However, errors are always present in measurement data.
For any comparison point i, the sum between a prediction $\mathbf{g}_i(\boldsymbol{\theta}^*)$ from a model class using the right parameter values $\boldsymbol{\theta}^*=[\theta_1^*,\theta_2^*,\cdots,\theta_n^*]^T$ and the modeling error associated with the model class ($\epsilon_{model,i}^*$) is equal to the "true" value Q_i. On the other hand, the "true" value Q_i is the sum of measurement y_i and measurement error $(\epsilon_{measure,i}^*)$.

$$\mathbf{g}_i(\boldsymbol{\theta}^*)+\epsilon_{model,i}^*=Q_i=y_i+\epsilon_{measure,i}^* \quad (1)$$

$$\mathbf{g}_i(\boldsymbol{\theta}^*)-y_i=\epsilon_{measure,i}^*-\epsilon_{model,i}^*=\epsilon \quad (2)$$

In Eq. (2), the difference between prediction $\mathbf{g}_i(\boldsymbol{\theta}^*)$ and measurement y_i is a combination of modeling error and measurement error, which are both random in nature.
Model instances are falsified if the differences between their predictions and the measurement data fall outside the threshold bounds defined using the combined uncertainty $U_{c,i}$, shown in Eq. (3) and (4). Let n_m be the number of comparison points and ϕ be the target probability. $F_{U_{c,i}}^{-1}(x): x \in [0,1]$ represents the inverse cumulative distribution function of the combined uncertainty.

$$T_{low,i}=F_{U_{c,i}}^{-1}\left(\frac{1}{2}\left(1-\phi^{1/n_m}\right)\right) \quad (3)$$

$$T_{high,i}=F_{U_{c,i}}^{-1}\left(1-\frac{1}{2}\left(1-\phi^{1/n_m}\right)\right) \quad (4)$$

2.2 Bayesian approach

Bayesian approach is based on Bayesian conditional probability. The prior probability of physical parameters $P(\boldsymbol{\theta})$ is updated using a likelihood function $P(\mathbf{y}|\boldsymbol{\theta})$ and measured data **y.** The posterior probability $P(\boldsymbol{\theta}|\mathbf{y})$ is obtained using the normalization constant $P(\mathbf{y})$.

$$P(\boldsymbol{\theta}|\mathbf{y}) = \frac{P(\mathbf{y}|\boldsymbol{\theta})P(\boldsymbol{\theta})}{P(\mathbf{y})} \tag{5}$$

3. Theoretical case study

An idealized simply supported beam is studied to demonstrate the improved model-updating results when dynamic tests are used in addition to static tests. Error-domain model falsification and the Bayesian approach are compared.

This beam has a length L=8.92m, a Young's modulus E^*=7x10^{4}MPa, a moment of inertia I^*=7.0125xe^{-4}m^4, a mass distribution m^*=229.14kg/m, and is subject to a uniform force q=2245.6N/m. For this beam, parameters to be identified are Young's modulus, moment of inertia and mass distribution. The choice of parameter ranges is based on engineering judgement and should contain the true values of parameters. Four scenarios are evaluated. In scenario I, the parameter ranges contain the true parameter values. In scenario II, the ranges of Young's modulus and moment of inertia do not contain the true parameter values. This can happen in practice when engineering judgement is false. In scenarios III and IV, the ranges of all three parameters do not contain the true parameter values. In scenarios I, II and III, only dynamic measurements are used. In scenario IV, both dynamic and static measurements are adopted. The specific parameter ranges are shown in Table 1. 1000 model instances are generated as every parameter has 10 possible values.

Table 1 Ranges of the parameters in scenarios

	E(MPa)	I(m^4)	m(kg/m)
Scenario I	[6x10^4,8x10^4]	[6x10^{-4},8x10^{-4}]	[200,250]
Scenario II	[8x10^4,9x10^4]	[8x10^{-4},9x10^{-4}]	[200,250]
Scenario III &IV	[8x10^4,9x10^4]	[8x10^{-4},9x10^{-4}]	[350,400]

The first three bending natural frequencies and the deflections at $L/6$ $L/4$, $L/2$ are calculated according to Eqs. (6) and (7). Six simulated measurements are obtained by adding a random value of measurement error to the true values as expressed in Eqs. (8) and (9).

$$\omega = \left(\frac{n\pi}{L}\right)^2 \sqrt{\frac{EI}{m}} \quad n = 1\sim3 \tag{6}$$

$$y_{1/6} = \frac{205qL^4}{31104EI}, y_{1/4} = \frac{57qL^4}{6144EI}, y_{1/2} = \frac{5qL^4}{384EI} \tag{7}$$

$$\omega_{sm} = \omega_{true} + \epsilon^d_{meas} \tag{8}$$

$$y_{sm} = y_{true} + \epsilon^s_{meas} \tag{9}$$

For dynamic measurements, it is assumed that the measurement error and modeling error follow the normal distribution $\epsilon^d_{meas} \sim N(0,0.5)$ and $\epsilon^d_{model} \sim N(0,0.5)$. For static measurements, it is assumed that the measurement and modeling errors follow the normal distribution $\epsilon^s_{meas} \sim N(0,0.0002)$ and $\epsilon^s_{model} \sim N(0,0.0002)$.

Scenario I: In this scenario, correct assumptions are made about E and I, which indicate that the true values fall within the respective ranges. Both error-domain model falsification and Bayesian updating lead to correct identification. Candidate models using each method are shown in Figure 1 and Figure 2.

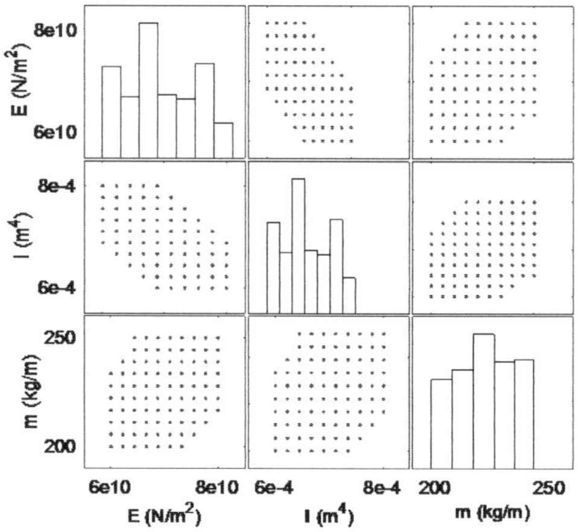

Figure 1: Scenario I (EDMF): Pairwise comparison of parameters found in the candidate model set

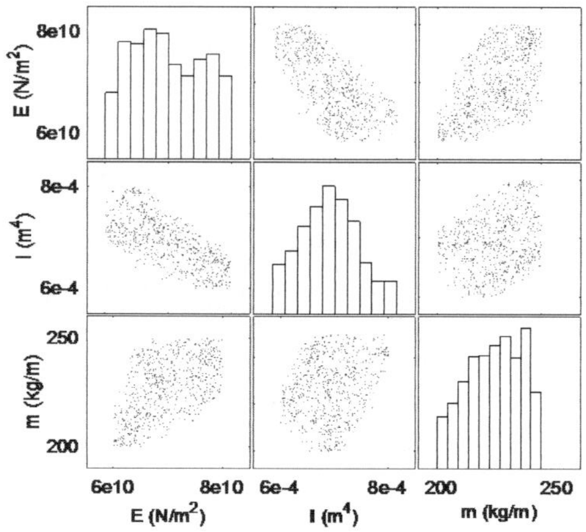

Figure 2: Scenario I (Bayes): samples from 2-D marginal posterior distribution of parameters

Scenario II: In this scenario, wrong assumptions are made about the values of E and I. In error-domain model falsification, all model instances are falsified, which is reasonable considering the initial ranges of parameters are wrongly defined. However, the Bayesian approach fails to detect this problem, shown in Figure 3.

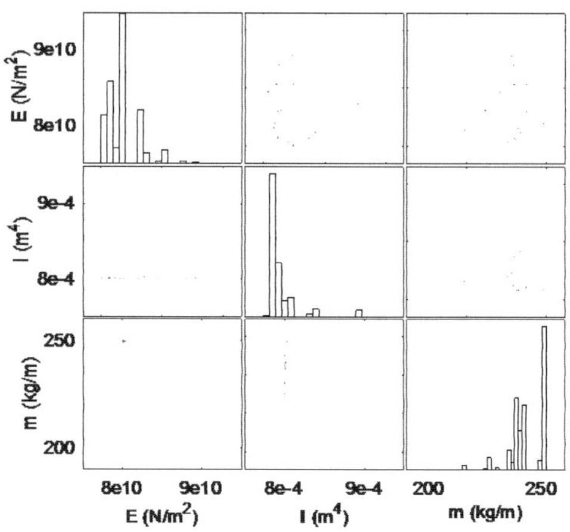

Figure 3: Scenario II (Bayes): samples from 2-D marginal posterior distribution of parameters

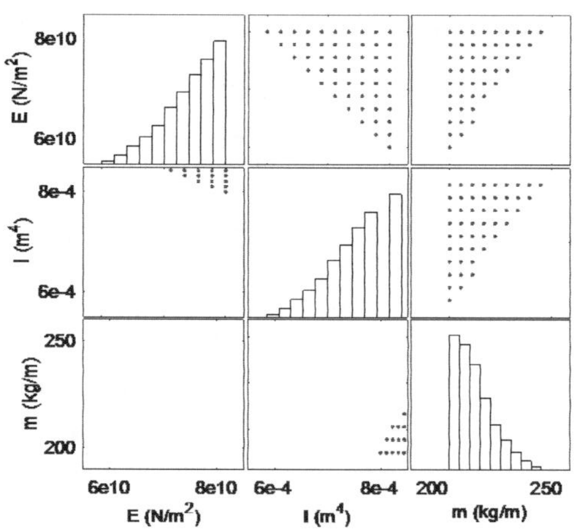

Figure 4: Scenario III (EDMF): Pairwise comparison of parameters found in the candidate model set

Scenario III: In this scenario, the initial ranges of parameters are all wrong. Hence, there should not be any candidate model. However, this time EDMF retained wrong model instances (shown in Figure 4). This is considered type-II error in statistics. In practice, the best way to reduce this error may be to include other measurements.

Scenario IV: In this scenario, by using both dynamic and static measurements, all model instances are successfully falsified. Bayesian approach failed to detect this problem (shown in Figure 5).

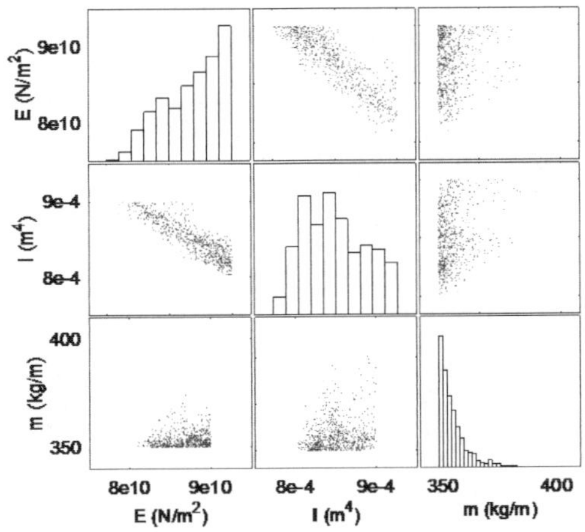

Figure 5: Scenario IV (Bayes): Pairwise comparison of parameters found in the candidate model set

The advantage of EDMF over the Bayesian inference approach is demonstrated through scenario II, in which EDMF successfully detects that all models are wrong as opposed to the Bayesian inference approach. However, this advantage of EDMF may not always exist, especially when falsification is performed only with dynamic data. By including static measurements, there is a higher chance that EDMF successfully falsifies all wrong models.

4. Full-scale case study

4.1 Description of the bridge

The full-scale case study, as shown in Figure 6, is a simply supported bridge in Singapore with a length of 32.8 m, a width of 16 m, and 3 traffic lanes. The superstructure consists of 4 main girders (with a depth of 2.0 m). The girders are precast while the deck and the diaphragms are cast on site.

Figure 6 Photos of the bridge (Image Credit: FCL-CCI)

Nine measurements including four natural frequencies, four deflections and one inclination are used in this study. The deflections and inclinations are obtained through a static load test involving 6 trucks, each with a gross weight of 33 tons. The deflections are measured at different location under different girders. And the inclination is measured at the end of the bridge. The measured values are shown in Table 2.

Table 2 Dynamic and static measurements

	Vertical bending 1 (f1)	Torsion 1 (f2)	Lateral bending 1 (f3)	Vertical bending 2 (f4)
Observed frequency	4.00Hz	6.11Hz	10.02Hz	14.8Hz
	1/4L under Girder 1 (P1)	1/2L under Girder 1 (P2)	1/2L under Girder 2 (P3)	1/2L under Girder 3 (P4)

Deflection	4.832 mm	6.063 mm	6.101 mm	4.552 mm
	End of the Girder (I1)			
Inclination	4.4603×10^{-4} rad			

4.2 Error domain model falsification

Nine parameters are chosen in this case, and the ranges of all parameters are defined based on engineering experience and information obtained from drawings. These ranges are shown in Table 3. Note that all girders are assumed to have the same values for parameters.

Table 3 Lower bound and upper bound of each parameter

Parameter	Description	Lower bound	Upper bound
E_barrier (MPa)	Young's modulus of barrier	3000	40,000
E_precast (MPa)	Young's modulus of precast concrete	20,000	35,000
E_cast-in-situ (MPa)	Young's modulus of cast-in-situ concrete	25,000	50,000
D_cast-in-situ (kg/m3)	Density of cast-in-situ concrete	1800	3000
D_precast (kg/m^3)	Density of precast concrete	1800	3000
Log_T (N-mm/rad)	log of bending stiffness of bearing	9	13
Log_V (N/m)	log of vertical stiffness of bearing	8	11
Log_L (N/m)	log of longitudinal stiffness of bearing	9	11

It is computationally expensive to execute finite element analysis for many thousand times using traditional grid-based sampling or Latin hypercube sampling. The underlying relationships between system parameters and responses can be computed using advanced interpolating functions. Surrogate models are adopted under this circumstance. In this case study, Gaussian process is used (Rasmussen and Williams 2004). First, a set of 1000 model instances generated by Hammersley sampling technique is calculated by finite element analysis. Then, surrogate models are built based on Gaussian process. Finally, 6^8 (=1,679,616) model instances are calculated using the surrogate models. Besides measurement errors and modeling errors, additional errors due to the application of surrogate models are introduced.

Three scenarios are designed. The first one considers only dynamic measurements. The second one considers only static measurements. Finally, the last one considers both dynamic and static measurements. Combining both measurements provides the smallest number of candidate models (Table 4). The updated parameter ranges in different scenarios are presented in Table 5. EDMF is able to narrow down the ranges of E_cast-in-situ to 40%, E_precast to 40%, D_precast to 60%, Log_T to 20%, Log_V to 20%. The prediction ranges of candidate models in scenario III are represented in Table 6.

Table 4 Error domain model falsification in different scenarios

Scenario	Model instances	Falsified models	Candidate models
I: Using only dynamic measurements	6^8=1,679,616	1,679,465	151
II: Using only static measurements	6^8=1,679,616	1,531,918	147,698
III: Using both static and dynamic measurements	6^8=1,679,616	1,679,557	59

Table 5: Parameter ranges of candidate models in different scenarios

Scenario		E_barrier (Mpa)	E_ cast-in-situ (Mpa)	E_precast (Mpa)	D_ cast-in-situ (kg/m^3)	D_precast (kg/m^3)	Log_T (Nmm/rad)	Log_V (N/m)	Log_L (N/m)
I	min	3000	20,000	25,000	1800	1800	9	8.6	9
	max	40,000	35,000	50,000	3000	3000	13	11	11
II	min	10,400	20,000	25,000	1800	1800	12.2	9.2	9.4
	max	40,000	26,000	35,000	3000	3000	13	11	11
III	min	10,400	20,000	25,000	1800	2280	12.2	9.2	9.4
	max	40,000	26,000	35,000	3000	3000	13	9.8	11

Table 6: Prediction ranges of candidate models in scenario III

Scenario		P1 (mm)	P2 (mm)	P3 (mm)	P4 (mm)	f1 (Hz)	f2 (Hz)	f3 (Hz)	f4 (Hz)	I1 (rad)
III	min	4.314	5.537	5.430	3.925	3.987	6.105	10.677	13.496	0.00036313
	max	5.017	6.472	6.338	4.607	4.344	6.352	10.898	13.89	0.00049527

Conclusions

It is demonstrated that updating finite element models using a model falsification approach is more robust than using traditional implementations of Bayesian updating.
Surrogate models based on Gaussian processes have been used to infer the underlying relationships between system parameters and responses in order to reduce computation times associated with finite element simulations.
Furthermore, error-domain model falsification has been adapted to combine information provided by both dynamic and static measurements. The results of model falsification using both measurements show higher falsification capability compared with using only static or dynamic ones.

Acknowledgements

The research was conducted at the Future Cities Laboratory at the Singapore-ETH Center, which was established collaboratively between ETH Zurich and Singapore's National Research Foundation (FI 370074016) under its Campus for Research Excellence and Technological Enterprise programme. The authors would like to gratefully acknowledge the support of the Land Transport Authority of Singapore (LTA) to perform the case study.

References

Aktan, A.E. & Farhey, D.N. 1997, 'Structural identification for condition assessment: experimental arts', *Journal of Structural Engineering*, 123(12), pp.1674-1684.

Brownjohn, J.M. 2007, 'Structural health monitoring of civil infrastructure', *Philosophical Transactions of the Royal Society of London A: Mathematical, Physical and Engineering Sciences*, 365(1851), pp.589-622.

Araujo, I.G., Maldonado, E. & Cho, G.C. 2011, 'Ambient vibration testing and updating of the finite element model of a simply supported beam bridge', *Frontiers of Architecture and Civil Engineering in China*, 5(3), pp.344-354.

Beck, J.L. & Katafygiotis, L.S. 1998, 'Updating models and their uncertainties. I: Bayesian statistical framework', *Journal of Engineering Mechanics*, 124(4), pp.455-461.

Behmanesh, I., Moaveni, B., Lombaert, G. & Papadimitriou, C. 2015, 'Hierarchical Bayesian model updating for structural identification', *Mechanical Systems and Signal Processing*, 64, pp.360-376.

Goulet, J.A. & Smith, I.F. 2013, 'Structural identification with systematic errors and unknown uncertainty dependencies', *Computers & structures*, 128, pp.251-258.

Rasmussen, C.E. and Williams, C.K. 2006, *Gaussian processes for machine learning (Vol. 1)*, Cambridge: MIT press.

The 8th International Conference on Structural Health Monitoring of Intelligent Infrastructure Brisbane, Australia | 5-8 December 2017

Bayesian Structural Identification of a Suspension Bridge Using Temperature and Traffic Loading

A. Jesus[1], M.R. Salami[1], R. Westgate[2], K. Koo[3], J. Brownjohn[3], I. Laory[1]

[1] *Civil Research Group, School of Engineering, University of Warwick, CV4 7AL – United Kingdom, Email:* andrehjesus@openmailbox.org, i.laory@warwick.ac.uk [2] *Atkins Ltd – United Kingdom* [3] *College of Engineering, Mathematics and Physical Sciences, University of Exeter, EX4 4QF – United Kingdom*

Abstract

Large civil infrastructure such as long suspension bridges are critical connecting elements, whose behaviour is difficult to discern and should be constantly monitored for safety. Critical factors relevant for this purpose usually are non-directly measurable, having therefore to be inferred, based on monitored data. This structural identification is usually highly susceptible to uncertainties, having numerous methodologies been developed for this purpose. This paper presents an enhanced modular Bayesian approach for structural identification, able to comprehensively quantify uncertainties in data interpreted by a physics-based model. Gaussian processes are used as emulators of the model response surface and existent modelling discrepancies. The Metropolis Hastings algorithm was used to expand the original methodology for multiple parameters inference. This method was applied to calibrate a finite element model of Tamar suspension bridge using long-term monitoring data. Measurements of temperature, traffic, mid-span displacement and natural frequencies in a one year span were used for identification of the prestress in the main/side cables of the bridge and stiffness of its thermal expansion gap. Results indicate a relative increase in the force of the suspension cables, while the expansion gap stiffness remains considerably low. This work is believed to pave a new way for identification of critical properties in large-scale infrastructures.

1. Introduction

Large civil infrastructures such as suspension bridges represent a capital investment from stakeholders, justifying a dedicated long-term monitoring system to ensure its serviceability. Designing, deploying and maintaining such a system and interpreting its data are typical challenges of Structural Health Monitoring (SHM).

Data interpretation can be accomplished by use of a physics-based model, which firstly has to be calibrated, validated and verified against experimental data. Unfortunately calibration parameters are often unknown, can only be estimated through expensive in-situ tests and can be masked by other uncertainties. It is relevant to learn about these parameters and their uncertainties, since they often represent critical physical attributes, e.g. soil stiffness on a foundation layer, prestress level of a tendon cable, etc. The study of the correlation between experimental/simulated data and identification of these parameters is known as structural identification (st-id).

For this purpose, it is necessary to investigate the sensitivity between environmental/operational factors, measured responses and the parameters/properties. For example, Cross et al. (2013) showed that traffic, temperature and wind are the most influential factors of Tamar bridge dynamic response, by decreasing order of relevance. Similarly, Westgate & Brownjohn (2011) manually examined the influence of stiffness parameters (stiffening structure, suspension cables) in the modal frequencies of a detailed Tamar bridge Finite Element (FE) model. Finally, a comparison between simulated and measured data should in principle calibrate the model and learn what values those properties/parameters should attain. This can be seen in subsequent studies of Tamar bridge FE model considering traffic effects from Westgate et al. (2015b), and radiation induced temperature variations by Westgate et al. (2015a). Unfortunately this approach did not provide an optimal solution, and the authors perspective in Westgate et al. (2015b) of the issue is that: "*In fact, Tamar Bridge has so far presented a challenging case study for model calibration, lending support to the view that no single model provides a perfect representation of a structure when matched to provided experimental data*". In other words, all the assumptions, simplifications and inadequacies of the model interfere severely with identification of its parameters. The final conclusion is that, to use a model for data interpretation it is crucial that all of its uncertainties are properly quantified, and in particular from the two most prevalent sources, parametric and model discrepancy.

Several methodologies exist for such a purpose, such as fuzzy numbers Erdogan et al. (2014), Markov processes Lam et al. (2015), amongst others. A relevant example is the model falsification method from Goulet & Smith (2013), which was used to identify Tamar bridge parameters by Laory et al. (2013). Another significant area of research for uncertainty quantification are Bayesian methodologies, which have been extensively used for st-id, as seen in Beck & Yuen (2004) or Behmanesh & Moaveni (2016). However, in these works model discrepancy was generally considered under strong assumptions, and a considerable computational effort was often required.

This work includes several contributions: st-id of Tamar bridge with a modular Bayesian approach, based on simulations of a FE model and available monitoring data during a one year time-frame; for the first time, two of Tamar bridge most relevant environmental/operational effects, temperature and traffic, are considered and simulated simultaneously. An equally significant aspect of this contribution is the enhancement of the methodology, by use of the Metropolis Hastings algorithm to infer multiple calibration parameters. To the best of our knowledge this is the first implementation of this formulation, which allows for multiple responses and multiple calibration parameters simultaneously. In Section 2 a summary description of the methodology and of the developments that have been added can be found; Section 3 describes Tamar bridge SHM system and the detailed FE model under consideration; Section 4 presents the st-id results and their discussion; and finally Section 5 highlights the major conclusions of the present work.

2. Enhanced Modular Bayesian Approach

The modular Bayesian approach (MBA) formulation was initially developed for a single response case by Kennedy & O'Hagan (2001a). Arendt et al. proved that unless under some specific conditions, the single response case will fail to identify the true structural parameters, and proposed instead a multiple response approach, i.e. with multiple measurements, which allows for a more informative model. For a more comprehensive description the reader should check Arendt et al. (2012). In a previous work by Jesus et al. (2017) this methodology was applied in practice, on a reduced scale model of a bridge.

2.1. General Workflow of the MBA

The modular Bayesian approach aims to solve a fundamental equation of model calibration, written as

$$Y^e(X^e) = Y^m(X^e, \theta^*) + \delta(X^e) + \varepsilon \tag{1}$$

where: Y^e are observations, dependent on design variables X^e; Y^m are simulations of the model, dependent of the design variables and a vector of structural parameters θ^*; $\delta(X^e)$ is a discrepancy function that translates the difference between the model and the true process and ε is an observation error, that is assumed to follow a Gaussian distribution $\mathcal{N}(O,\Lambda)$. Essentially Eq. (1) represents a data-model, given by a calibrated status of a physics-based model when $\theta = \theta^*$, a discrepancy function and an experimental error.

To solve this equation while considering all of the involved uncertainties, the process is broken in four modules, hence the name modular Bayesian approach. For a flowchart of the algorithm see Fig. 1. In module 1 and 2 the computer model and the discrepancy function are fitted by statistical models, known as multiple response Gaussian processes (mrGp), whose parameters (hyperparameters ϕ) have to be estimated. These hyperparameters characterize the mrGps and account for an approximation of its associated uncertainties. Afterwards, these statistical models are used to set up a global data model that explains both simulations and observations for a set of given structural parameters θ. This is done in the third module of the approach, where the posterior distribution of these parameters is estimated by Bayesian inference. The fourth and final module predicts the observed process, by updating the mrGps previously determined, with the posterior information of θ.

The next two subsections present developments over the above described formulation, namely in module 2 and module 3 of the MBA.

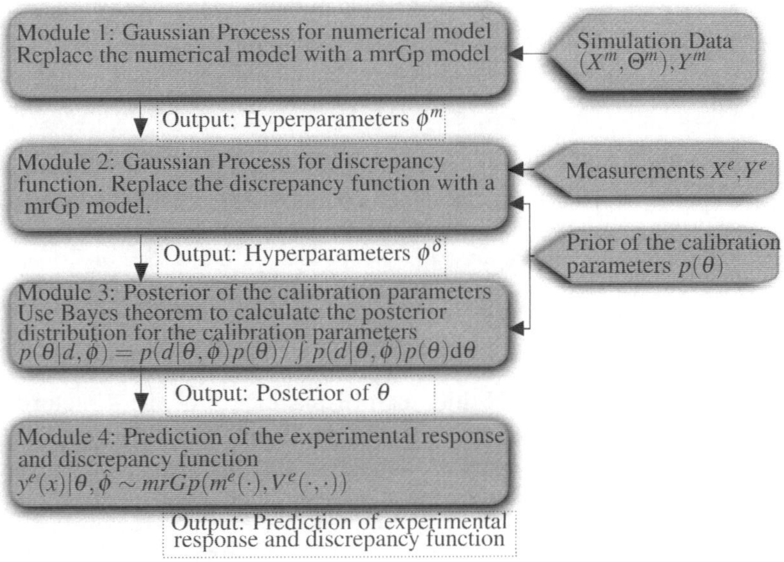

Figure 1: Flowchart of the modular Bayesian algorithm

2.2. Numerical Integration of the Observation MrGp - Module 2

One development over the original formulation has been implemented in the second module of the MBA, where the discrepancy function δ is approximated by a mrGp. The problem at this stage is that both, the structural parameters θ and the discrepancy function are still unknown. To surpass this issue, this metamodel of the observations is integrated with respect to the prior distribution of the parameters $p(\theta)$: This integration was accomplished through closed form solutions, specifically for Gaussian and uniform priors. The full integral expressions can be seen in Arendt et al. (2012), Eq. (C3) and in Kennedy & O'Hagan (2001b), Eq. (7). In this work a MATLAB module for numerically integrating this solution was developed, based on a package for multi-dimensional Markov Chain integration named Gail, from Choi et al. (2015).

2.3. Markov Chain Monte Carlo Sampling of Posterior Distribution - Module 3

On module 3 Bayesian inference is used for determining the posterior distribution of the structural parameters θ, according to Bayes' theorem.

$$p(\theta|d,\hat{\phi}) = \frac{p(d|\theta,\hat{\phi})p(\theta)}{\int p(d|\theta,\hat{\phi})p(\theta)\mathrm{d}\theta} \tag{2}$$

where $p(\theta|d,\hat{\phi})$, $p(d|\theta,\hat{\phi})$ and $p(\theta)$ are the posterior, likelihood and prior probability density functions (PDF), respectively, $\hat{\phi}$ are the hyperparameters of the mrGps estimated in the previous modules and d is the full data-set of simulations plus observations.
The integral in the denominator is the marginalized posterior distribution for the data. Previous authors have determined this quantity for low order examples with Gauss-Legendre quadrature.
To expand the modular Bayesian approach for multiple calibration parameters, it is necessary to calculate this integral using Markov Chain Monte Carlo methods. For this purpose it has been decided to use the Metropolis Hastings algorithm. This choice implies that a proposal distribution, as similar to the target distribution as possible, has to be specified. In this work a multivariate normal distribution was chosen. The computational effort was reduced by considering multiple Markov chains.

3. Tamar Bridge

In this section Tamar bridge's SHM system and the experimental data under consideration are detailed. Additionally a description of the FE model developed to study the infrastructure behavior, and its assumptions/simplifications are also shown.

3.1. Structural Health Monitoring System

A diagram of some sensors of Tamar's bridge SHM system are displayed in Fig. 2. For the purpose of this work the natural frequencies and mid-span displacement of the structure, for a given cable temperature and traffic loading, are the main quantities of interest. This data was observed in the

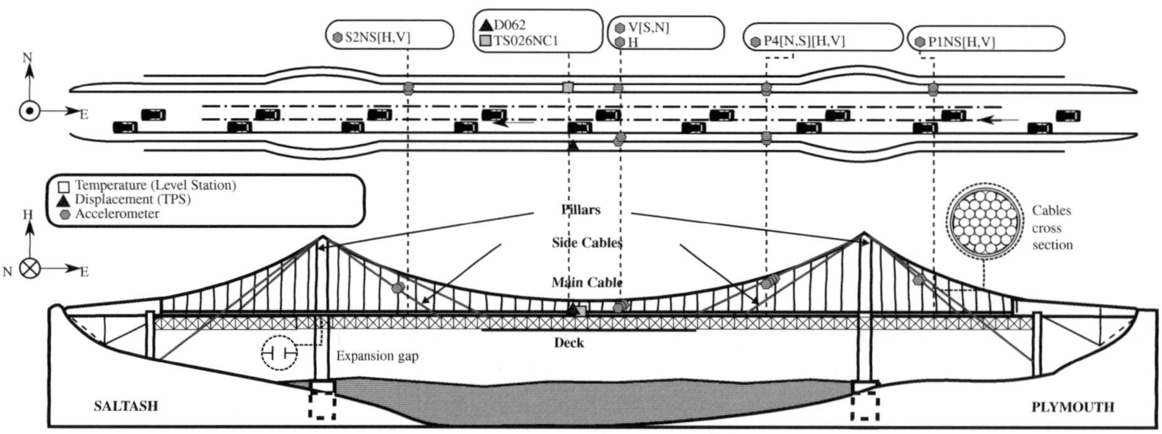

Figure 2: Diagram of Tamar bridge SHM system – location of: accelerometers from which natural frequencies/mode shapes are estimated; temperature; and displacement reflector.

time period betweeen May of 2009 to March of 2010, since synchronized temperature/traffic/modal data was richer during this period. Furthermore it is believed that a year timescale would be a good reference period for calibration/validation of the FE model.
Firstly, natural frequencies of the structure were determined from a Stochastic Subspace Identification (SSI) technique, based on the readings of the accelerometers shown in Fig. 2. At each half-hour, 10 minute acceleration recordings are post-processed to determine the natural frequencies and mode shapes of the infrastructure. Nomenclature of these mode shapes is the first vertical symmetric (VS1), first lateral symmetric (LS1a) and first vertical asymmetric (VA1). See Section 4 of Cross et al. (2013) for a more extensive description.
Secondly, displacements at the middle span of the bridge in three directions, vertical, East and North were also considered as responses for calibration/validation. These were obtained from a Total Positioning System (TPS) and a set of reflectors located at specific points of the bridge.
Thirdly, traffic monitoring is available only from hourly toll gates vehicle counts from the Plymouth side. These raw measurements have been interpolated to provide an half-hourly count. For more details on assumptions of this operation and a description of the monitored traffic consult Westgate et al. (2015b) paper, in particular Table 1, which shall be used in this work to determine a total mass due to traffic.
Finally, temperature readings from the cable shown in Fig. 2 have been down sampled from a 10 s to an half hour count. Intervals in the whole data-set that were out-of-match were removed, and although the modular Bayesian approach is able to correctly identify outliers these were also removed. After applying these changes only 2419 data points remained, and can be seen in Fig. 3 and 4 plots.
By examining Figs. 3(a) and 3(b) it can be seen that some summer months have been left out, which explains the isolated islands in the region of higher temperature. Still though, this data-set approximately covers uniformly the entire temperature-traffic space. Analysis of Fig. 4 shows the relationship

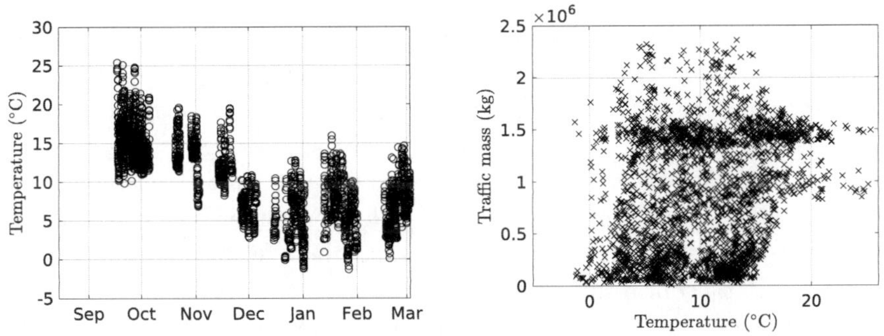

(a) Cable temperature – daily and seasonal variations
(b) Traffic temperature measurement space

Figure 3: Post-processed data – May of 2009 to March of 2010 time period - temperature and traffic

of the temperature/traffic with the natural frequencies/mid-span displacement, which tends to be linear. The only exception is the traffic/displacement relation. Hence a linear correlation function was assumed for the mrGp that fits the discrepancy function and the model.
In the next Section considerations for the FE model of Tamar bridge shall be presented.

3.2. Finite Element Model

A detailed FE model of Tamar bridge was developed by Westgate & Brownjohn (2011) using ANSYS Parametric Design Language (APDL) source code. It consists of around 45000 elements, from

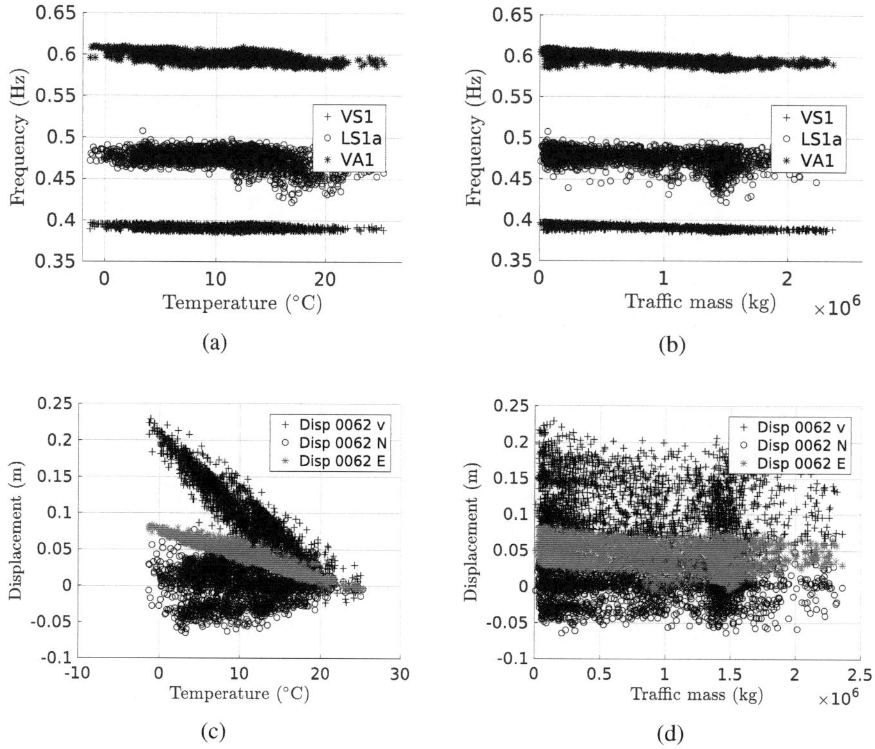

Figure 4: Post-processed data – May of 2009 to March of 2010 time period - natural frequencies and mid-span displacements

which: expansion joints were modeled with spring elements; truss members with fixed-rotation beams; deck/towers with shells and the cables and hangers with uni-axial, tension only elements.
Considering firstly the temperature effect, a comprehensive transient thermal analysis from Westgate et al. (2015a) showed that the actual thermal behaviour of Tamar bridge is very complex, but some relationships can be established between different temperatures for the cables, lower truss and deck of the bridge, relative to a given air temperature (sensor location can be seen in Fig. 2). In this work it has been assumed: a static steady state thermal analysis, i.e. thermal equilibrium on all elements; and the relations between materials/air from Westgate work.
At the same time, the traffic load is assumed as a set of distributed mass points, evenly spread longitudinally across the bridge deck, and asymmetrically in the lateral direction, as depicted in Fig. 5. Each point has half the mass of a standard vehicle $m = 1660$ kg. It must be noted however that the

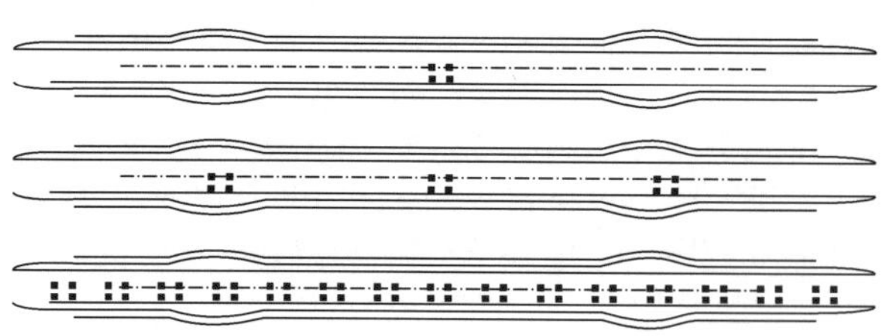

Figure 5: Traffic modeling of Tamar bridge FE model (top view of deck) for 2, 6 and 30 vehicles

actual amount of on-bridge traffic mass is not the same as the total-mass reported from the toll gates. According to Westgate et al. (2015b), the average number of vehicles on the bridge in the tolled direction is 1/43rd of the total half-hourly count, being this relation used to simulate the traffic effect. Having with the proper assumptions and simplifications aimed to reproduce the same experimental setup as described in the previous Section, it is now possible to address the calibration of the FE model. The structural parameters of interest are: the prestress in the main and side cables of the bridge, and the stiffness of the springs at the thermal expansion gap of the deck, Fig. 2 for reference. Simulations of the FE model have been performed in a Latin hypercube space, where uniformly random values of temperature, traffic mass and structural parameters were generated and the corresponding natural frequencies computed. A flowchart of this process is shown in Fig. 6. On a first

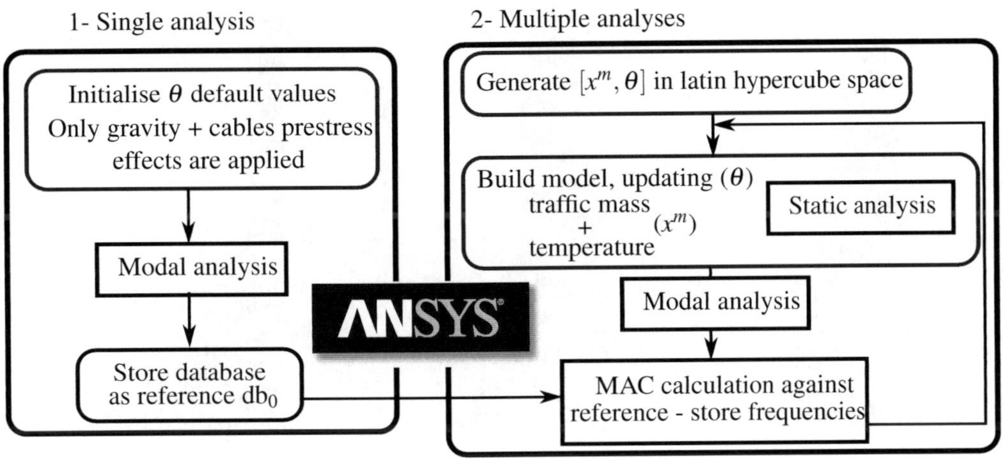

Figure 6: Simulation flowchart

phase, the natural frequencies were simulated and stored without any loading applied and with default values of structural parameters. On a second phase, for each combination of loading/parameters the frequencies of each run were classified and stored, using the modal assurance criterion (MAC), with a fit of at least 80 %.

Table 1 presents a summary of all the input data given to the modular Bayesian approach, and the structural parameters θ that have to be inferred. Having defined the experimental setup and the FE

Table 1: Data-set for Tamar bridge.

	Description
θ	prestress in the main/side cables and stiffness of modeled springs at deck expansion gap
X^e	readings of temperature at bridge cable and total mass due to traffic
Y^e	natural frequencies determined by SSI and measured mid-span displacement
$[X^m, \Theta]$	combination set of temperature/traffic/θ parameters on a latin hypercube space with: temperature [-3; 25] °C; traffic load [0; 2.5×10^6] kg; main cables [410; 2700]$\mu\varepsilon$ and side-span cables [500; 3700]$\mu\varepsilon$ initial strain; and deck stiffness [1×10^{-4}; 10×10^1] kN/mm
Y^m	simulations from the FE model for $[X^m, \Theta]$

model of Tamar bridge, the next Section will present results of the inference of the structural parameters and provide predictions of the metamodel of Eq. (1).

4. Structural Parameters Inference and Discrepancy Surfaces Prediction

In this section inference of the structural parameters under consideration is presented.

The metropolis hastings algorithm has been tuned to sample 2×10^4 samples from the likelihood

PDF, having reached an acceptance ratio of 39%. The resulting samples are shown in Fig. 7 and its moments in Table 2. The inference of these parameters accounts for temperature/traffic variability, noise/residual variations and modeling discrepancies. Due to the comprehensiveness of the methodology these parameters should, in theory, represent their true physical counterparts.

It is reassuring to notice that the PDF of the expansion gap stiffness is shifted towards low values

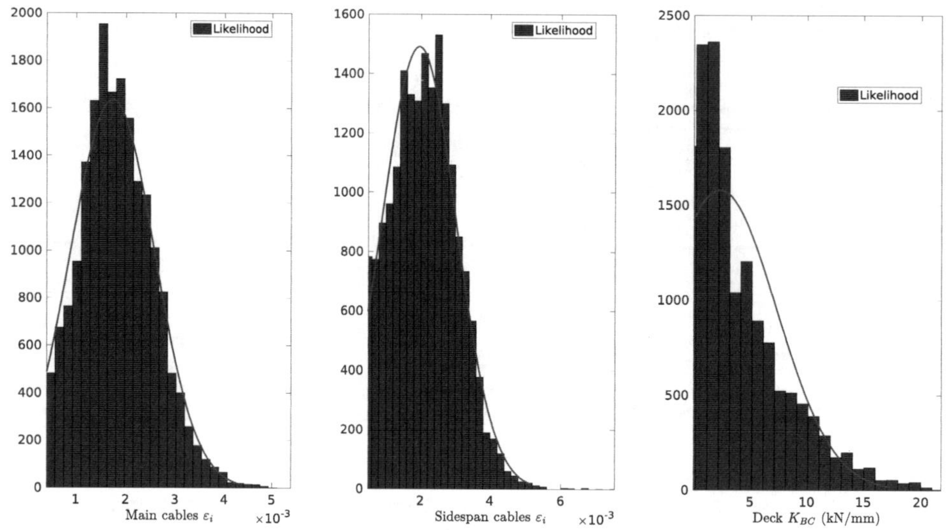

Figure 7: Inference of structural parameters: main and side-span cables initial prestress (left and center) and stiffness of deck expansion gap (right)

within the sampled space, otherwise this could indicate an obstruction of the gap moving parts.

On the other hand an increase in the prestress of the tendom cables could indicate internal damage,

Table 2: Means and standard deviations of marginalized posterior distribution

	Main cables ε_i	Side cables ε_i	K_{BC} deck
	$\mu\varepsilon$		kN/mm
Mean :	1723	1900	2.331
Std :	800	1060	4.631

such as broken wires, corrosion, cracks and wear. The bridge's suspension cables are made from 31 locked coil wire ropes (as shown in Fig. 2), being each 60mm in diameter, and the overall diameter of the main cable is about 380 mm, resulting in a cable area of 882.36 cm^2. Assuming the Young Modulus as 155 GPa it is possible to estimate the cable forces and compare them against two previous identifications reported in Westgate & Brownjohn (2011), as shown in Table 3. These values indicate a 13 % increase of cable force in a time frame of eight years.

Notice that in Laory et al. (2013) the stiffness of the expansion gap is also similarly skewed towards the lower region of the parameter space, but it had much higher values, arguably realistic (order of up to 1×10^7 kN/mm). Current efforts are targeted at reducing the variance of the likelihood in Fig. 7.

Finally, to demonstrate the ability of the modular Bayesian approach to quantify model discrepancies, Fig. 8 shows the discrepancy mean surfaces for the second natural frequency and the vertical mid-span displacement. A surface closer to zero values has less discrepancy and indicates a better

Table 3: Identification of suspension cables forces and comparison against previous reported identifications

	Forces (kN)				
Year	Main susp. cable		Side susp. cable		Method
	E[θ]	$\sigma[\theta]$	E[θ]	$\sigma[\theta]$	
1961	20296		20597		iterative shape finding
2001	20296		20597		iterative shape finding
2009	23564	10941	25985	14450	modular Bayesian approach

performance of the model, relatively to values further from zero. Note that these mean surfaces are sandwiched between a prediction interval, which is now shown here for the sake of simplicity. For example, the LS1a frequency displays a combined effect with worst performance of the model near lower temperature/traffic values. On the other hand the vertical displacement in the middle span of the bridge is more strongly influenced by the temperature alone, with better predictability near extreme temperatures and performing worst near the 21°C range.

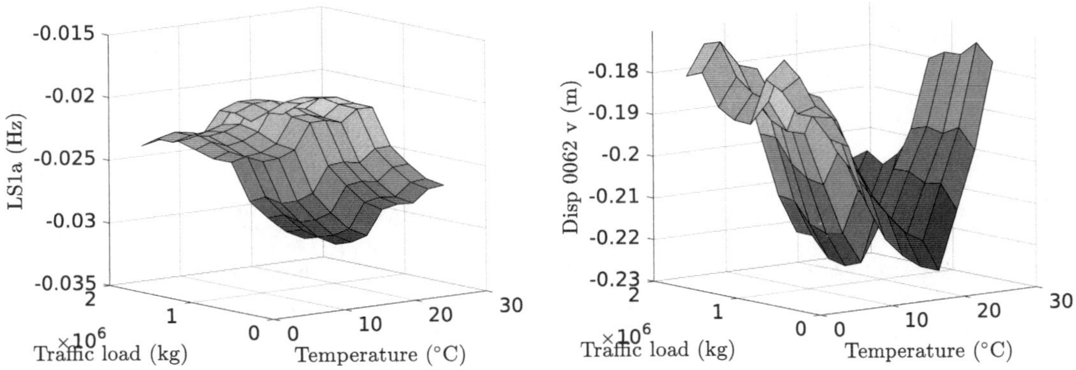

Figure 8: Discrepancy surfaces between FE model and measurements for second natural frequency and vertical displacement on the bridge's mid-span

5. Conclusions

In this work a modular Bayesian approach has been applied for structural identification of Tamar bridge's suspension cables prestress and the stiffness of its deck expansion gap. For the first time traffic and temperature were simulated simultaneously and used to consider environmental and operational conditions during the identification process. The methodology: makes use of multiple response Gaussian processes to aid in uncertainty quantification; it is able to highlight deficiencies of the model; and has been expanded to identify multiple structural parameters simultaneously.

Results were compared against work from previous authors, and they indicate a growth in the cables force and a reassuring low value of stiffness in the expansion gap. It is also possible to obtain probabilistic predictions and quantify the usefulness of the model, so that it can be more effectively updated, and/or used for damage detection purposes.

Acknowledgments

This work was supported by the Engineering and Physical Sciences Research Council (EPSRC) reference number EP/N509796.

References

Cross, E., Koo, K., Brownjohn, J. & Worden, K. 2013. “Long-Term Monitoring and Data Analysis of the Tamar Bridge”. en. In: *Mechanical Systems and Signal Processing* 35.1-2. ISSN: 08883270.

Westgate, R. J. & Brownjohn, J. M. W. 2011. “Development of a Tamar Bridge Finite Element Model”. In: *Dynamics of Bridges, Volume 5*. New York, NY: Springer New York. ISBN: 978-1-4419-9824-8 978-1-4419-9825-5.

Westgate, R., Koo, K.-Y. & Brownjohn, J. 2015b. “Effect of Vehicular Loading on Suspension Bridge Dynamic Properties”. en. In: *Structure and Infrastructure Engineering* 11.2. ISSN: 1573-2479, 1744-8980.

— . 2015a. “Effect of Solar Radiation on Suspension Bridge Performance”. en. In: *Journal of Bridge Engineering* 20.5. ISSN: 1084-0702, 1943-5592.

Erdogan, Y. S., Gul, M., Catbas, F. N. & Bakir, P. G. 2014. “Investigation of Uncertainty Changes in Model Outputs for Finite-Element Model Updating Using Structural Health Monitoring Data”. en. In: *Journal of Structural Engineering* 140.11. ISSN: 0733-9445, 1943-541X.

Lam, H.-F., Yang, J. & Au, S.-K. 2015. “Bayesian Model Updating of a Coupled-Slab System Using Field Test Data Utilizing an Enhanced Markov Chain Monte Carlo Simulation Algorithm”. en. In: *Engineering Structures* 102. ISSN: 01410296.

Goulet, J.-A. & Smith, I. F. 2013. “Structural Identification with Systematic Errors and Unknown Uncertainty Dependencies”. en. In: *Computers & Structures* 128. ISSN: 00457949.

Laory, I., Westgate, R. J., Brownjohn, J. M. W. & Smith, I. F. C. 2013. “Temperature Variations as Loads Cases for Structural Identification”. In: *Proceeding of the of 6th International Conference on Structural Health Monitoring of Intelligent Infrastructure (SHMII), Hong-Kong, 2013*. Hong Kong.

Beck, J. & Yuen, K. 2004. “Model Selection Using Response Measurements: Bayesian Probabilistic Approach”. In: *Journal of Engineering Mechanics* 130.2.

Behmanesh, I. & Moaveni, B. 2016. “Accounting for Environmental Variability, Modeling Errors, and Parameter Estimation Uncertainties in Structural Identification”. en. In: *Journal of Sound and Vibration* 374. ISSN: 0022460X.

Kennedy, M. C. & O’Hagan, A. 2001a. “Bayesian Calibration of Computer Models”. In: *Journal of the Royal Statistical Society: Series B (Statistical Methodology)* 63.3. ISSN: 1467-9868.

Jesus, A., Brommer, P., Zhu, Y. & Laory, I. 2017. “Comprehensive Bayesian Structural Identification Using Temperature Variation”. In: *Engineering Structures* 141. ISSN: 0141-0296.

Arendt, P. D., Apley, D. W., Chen, W., Lamb, D. & Gorsich, D. 2012. “Improving Identifiability in Model Calibration Using Multiple Responses”. en. In: *Journal of Mechanical Design* 134.10. ISSN: 10500472.

Kennedy, M. C. & O’Hagan, A. 2001b. *Supplementary Details on Bayesian Calibration of Computer Models*. Tech. rep.

Choi, S.-C. T., Ding, Y., Hickernell, F. J., Jiang, L., Rugama, L. A. J., Tong, X., Zhang, Y. & Zhou, X. 2015. “GAIL - Guaranteed Automatic Integration Library in MATLAB: Documentation for Version 2.1”. In: *Computer Science* abs/1503.06544.

The 8th International Conference on
Structural Health Monitoring of Intelligent Infrastructure
Brisbane, Australia | 5-8 December 2017

Structural Reliability Analysis of Existing Steel Plate-girder Bridges using Posterior Distributions of Model Parameters estimated by Dynamic Monitoring Data

M. Nishio[1], R. Kuroda [2]
[1] *Department of Civil Engineering, Yokohama National University – Japan, Email:* nishio@ynu.ac.jp
[2] *Department of Civil Engineering, Yokohama National University - Japan*

Abstract

This paper presents applicability of the posterior distributions of finite element (FE) model parameters to the structural reliability analysis of existing bridges. The target here was a steel plate-girder bridge that showed corrosions in its girders and bearings. A nominal FE model of the target bridge was first constructed, and statistics of its resonant frequencies were acquired by multiple dynamic measurements for use as the observation data in the Bayesian inference. The prior distributions of the model parameters that represented prior uncertainties related to the structural conditions were determined by using the inspection data. Then, the sensitivity analysis had a key role in the determination of parameters to be applied to the Bayesian inference that must had contributions to the measurement data, i.e., resonant frequencies. The posterior distributions of spring stiffness at the ends of steel girders, which were given to represent the performance of corroded bearings, and stiffness of the concrete slab were estimated by adopting Markov Chain Monte Carlo method (MCMC). The uncertainties of those model parameters could then be reduced in the posterior distributions, and they also showed consistencies with actual conditions recognized in the inspection. The posterior distributions could be used to calculate the structural reliability index of the target bridge. The probability density functions of maximum von Mises stress on the steel girders under the design live-load were derived by Monte Carlo FE model calculations using both the prior and posterior distributions for comparison. In the results, the reliability index β against the criteria of steel yield stress in the case of using the posterior distributions became higher than that in the case of using the prior distributions. It was concluded that the structural reliability of the existing bridge could be assessed by applying the posterior distributions that were the updated uncertainties by measurement.

1. Introduction

In the operation of existing bridges with large population, some kind of decision making is often required based on expected structural capacities; e.g., the priority setting of maintenance and repair to the several bridges, in which damages or deteriorations are recognized. This study aims to make those decision makings conducted based on the quantitative comparison of the structural capacities to the required performance that can be derived by the structural reliability calculations (Thoft-Cristensen, P. & Baker, K.J. 1982) using the numerical model of target structure with considering parameter uncertainties under the actual structural conditions.

This paper presents the applicability of the Bayesian posterior distributions of the finite element (FE) model parameters to the structural reliability analysis of the existing bridge (Kuroda & Nishio 2016). The posterior distributions of model parameters are the updated uncertainties by the measurement data acquired under the actual structural condition. They can be applied to the calculation of the structural reliability index β that indicates the capacity to the limit state that is set as the required performance in the structure design.

There are some previous studies that tried to assess the conditions of bridges with some kind of damages based on the structural reliability analysis. Especially for steel bridges, for instance, Kayser, et al 1989 predicted the performance decrement according to the structural reliability index β calculated by the numerical analysis with applying the degradation model due to the corrosion at the ends of steel plate-girders. Okasha, et al 2011 evaluated the decreased load bearing capacity of a steel plate-girder bridge by the reliability index β calculated using the numerical model that reflected estimated corrosion thinning occurred in the plate-girders by the strain measurement. They also assessed the deterioration over time based on the structural reliability. However, there has been little previous study that consider how to deal with the uncertainties of structural properties related to the observed damages/deterioration in the discussion of structural reliability of existing bridges.

The significant point of this paper is that we apply the Bayesian posterior distributions of model parameters, which are the updated probabilities using the measurement data, directly to the calculation of the structural reliability index β to assess the structural capacity of the existing bridge with considering uncertainties on the structural conditions. The target bridge is an existing steel plate-girder bridge, in which the corrosions were observed in its girders and bearings. We firstly conduct the dynamic measurement to use resonant frequencies for the observation data in the Bayesian inference. The FE model of the target bridge is constructed, and the uncertain parameters and their prior distributions are set regarding the inspection sheet that mansion about the corrosion. After the posterior distributions were then estimated, the structural reliability index β for the maximum Mises stress under the design live load against the steel yield stress is calculated by Monte Carlo (MC) calculation. By comparing the two reliability indexes β that are calculated from the "prior" and the "posterior" distributions of FE model parameters, we verified the effectiveness of reducing the uncertainties relevant to the actual structural conditions by using structural monitoring data for assessing structural capacity of existing bridge at the time.

2. Target Bridge and Dynamic Data Acquisition

The target bridge in this study was an existing steel plate-girder bridge that consisted of six I-beam girders (G1-G6) and the RC slab, with the span of 11.7m and the width of 7.2m, as shown in Fig.1. The bridge was completed in 1963, and some significant deterioration; cracks in the RC slab and corrosions at the ends of girders and the bearings, had been found in the inspection.

Here, the dynamic measurement test was conducted to extract resonant frequencies for use as the observation data in the estimation of posterior distributions of the FE model parameters. The multiple measurements of acceleration were acquired under the normal traffic. Figure 2 shows the locations of six accelerometers (PCB 393B04) to capture the first and second bending and torsional modes. Each data acquisition was taken for five-minutes continuously, with the sampling frequency of 2kHz, and 19 data sets were acquired in total.

The resonant modes were identified from the cross-power spectrum density (CPSD) of acquired data that was derived by the Welch's method taking seven-time averages with 150,000 FFT length, 50% overlap, and the Hanning windowing. The peak frequencies were extracted by the curve-fitting procedure, which was locally applying to the frequency range around each of three peak frequencies clearly recognized in the CPSDs. The mode shapes were also identified by the phase plots. Table 1 shows the statistics of extracted resonant frequencies of the 19 data sets. It was also recognized here that the 1st mode was the first bending, the 2nd was the first torsion, and the 3rd mode was also the bending mode in the lateral direction.

3. Estimation of Posterior Distributions of FE Model Parameters

An FE model of the target bridge was constructed, and the model parameteres that must be considered relatively large uncertainties due to the damages/detoriorations were picked up. The prior distributions of those parameters were set by also taking advantages of the inspection data, and the global sensitivity analysis took a key role to understand which model parameters could be treated in the estimation of posterior distributions. The posterior distributions of model parameters, which were the parameters of the RC slab, the boundary conditions of each steel girder, were then estimated by the Markov Chain Monte Carlo method (MCMC).

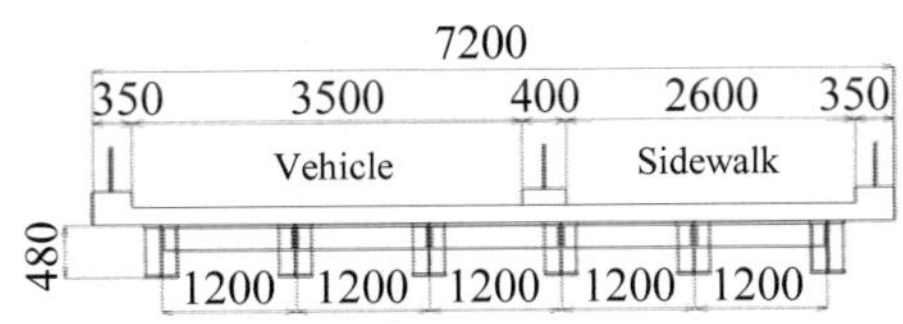

Fig. 1: Cross-section of target bridge (unit: mm)

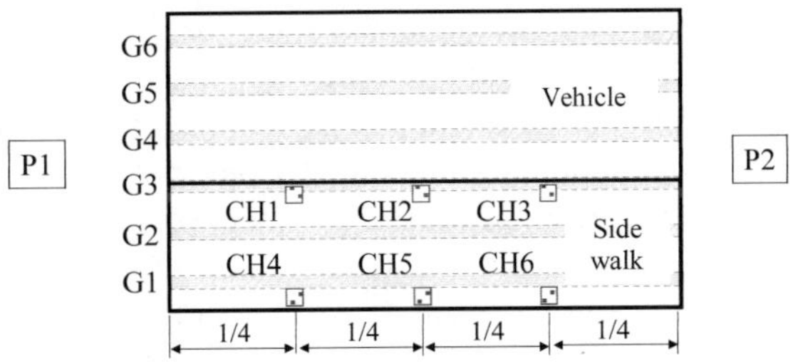

Fig. 2: Layout of accelerometers in the data acquisition

Table 1: Statistics of extracted 1st to 3rd modes resonant frequencies

Mode	Mean (Hz)	Standard deviation (Hz)	Coefficient of variance (%)
1st	9.8	0.148	1.51
2nd	11.7	0.124	1.06
3rd	18.4	0.143	0.78

3.1 Finite Element Model of Target Bridge

The nominal FE model was constructed by using one of general-purpose FEA software, Abaqus 6.14. All structural members including the RC slab were described by the shell element, and the nominal values of material properties and the boundary conditions for the simply-supported beam were assigned. The model verification was taken to confirm the element type and the mesh configuration by taking into account both the convergence of purpose outputs, resonant frequencies, here, and the calculation cost. The four-node reduced integration first-order shell element was then adopted, and the number of element became 124,108 in total. The mode shapes and resonant frequencies of the 1st to 3rd modes, which are identified in the measurement, are shown in Fig.3.

(a) 1st (8.78Hz) (b) 2nd (11.0Hz) (c) 3rd (18.1Hz)

Fig. 3: Resonant modes calculated in the constructed FE model

3.2 Uncertainties of Model Parameters

In the structural engineering field, the parameter uncertainties are often considered against the material properties and the size/dimension of structural members. In those cases, some data of statistic, which are indicated in the previous works or the official data-sheets of material properties, can be used. On the other hand, the uncertainties of properties of existing structures include those that are come from the effect of aging and damages/deterioration. In the case of this study, the uncertainties of model parameters associated with the corrosions at the ends of girders and bearings, and the cracks in the RC slab were considered significantly.

Figure 4 shows the pictures of recognized corrosions and cracks taken in the inspection work. From these pictures, we could recognize the detail members and locations of those deteriorated parts.

In the FE model, the effect of corrosion on the steel girder is considered as the thinning of plate thickness. It could be recognized from the pictures, such as Fig.4(a) and (b), that most of those corrosions occurred on the bottom of lower flanges and lower half of web members, near the ends of girders, as shown in Fig.5 (a). The crack on the RC slab in Fig.4(c) can be treated as the reduction of bending stiffness of the shell element of the slab member. To represent the effect of corrosion at the steel bearings, an imaginary linear spring in the longitudinal direction was given to each of bearings, as shown in Fig.5 (b). By assigning a large value as the spring stiffness, the movability of a steel bearing can be reduced. In the observation of pictures, e.g., Fig.4(e) and (f), the severe corrosion especially occurred especially at the inside girders (G2-G5).

The uncertainties of model parameters were then given by setting the prior distributions of related model parameters, which is summarized in Table 2. Here, the prior distribution is the uniform distribution with lower and upper limits. The range between lower and upper limits represents the degree of uncertainty.

(a) Corrosion on lower flange at G3

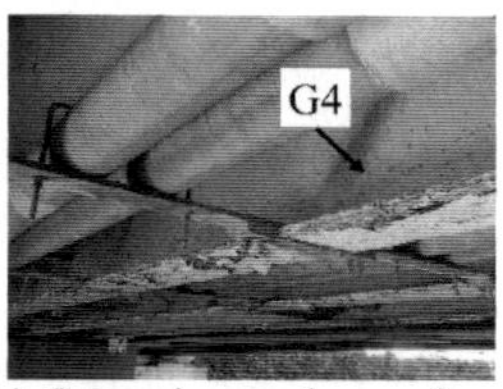

(b) Corrosion on lower flange and web at G4

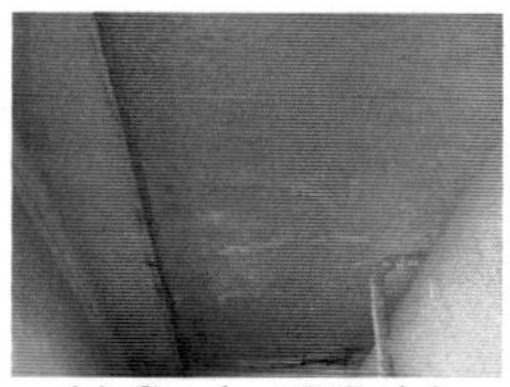
(c) Crack at RC slab

(e) G1 movable side bearing

(f) Corrosion at G2 movable bearing

Fig. 4: Pictures of corrosions and cracks taken in the inspection work

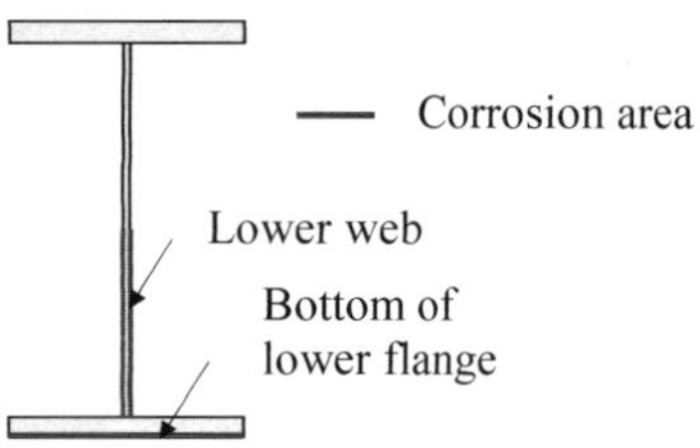

(a) Locations where thickness is lost from corrosion

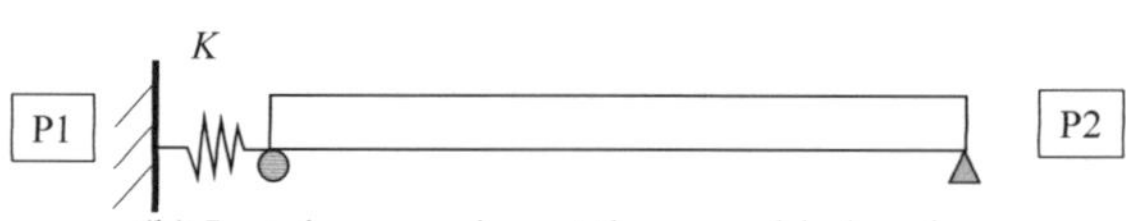

(b) Imaginary spring at the movable bearing

Fig. 5: Effects of corrosion in the FE model

In the description of corrosions at the steel girders in the FE model, we used the statistics of thickness reduction due to corrosion in actual bridges in Japan that were obtained in some previous studies (e.g., Kaita, et al, 2005). Here, it was mentioned that, in the corrosion of steel plate-girders, the amount of thickness reduction at the lower flanges was 1.1-3.2 times larger than that at the webs. Therefore, we dealt with only the thickness reduction at the lower flange by assigning the value of 50% of thickness reduction at the flange to the thickness reduction at the web. In the distribution of thickness reduction in the previous studies (e.g., Kaita, et al, 2005), it was also reported that 2mm was the maximum order of thickness reduction in the steel plate-girders of actual bridge due to corrosion. We then applied uncertainties of lower flange thickness reduction by the uniform distribution with the range of 0mm to 2mm. The parameters are Tb_1, …, Tb_6 for girders G1, …, G6, respectively. For the spring stiffness K_1, …, K_6 at the bearings, it was previously verified that, if the spring stiffness is in the range smaller than $K=1\times10^6$ N/m, the bearing behaved the same as the simply-supported boundary condition without restriction; if the spring stiffness of larger than $K=1\times10^{10}$ N/m is applied, the bearing behaved as the completely fixed-end boundary condition. Therefore, the uncertainty range of K_1, …, K_6 was determined to 1×10^6 N/m in lower limit to 1×10^{10} N/m in upper limit.

The bending stiffness reduction of the RC slab due to cracks was represented by the reduction of Young's modulus of the shell element of the slab member E_S in the FE model. The uncertainty was assigned by the uniform distribution with the range of 0% (no stiffness reduction) to −10%.

In addition, it was considered that the uncertainty of the density of the slab shell element D_S highly contributed to the characteristics of resonant modes. The uncertainty of this parameter was relatively high because it consisted of not only the uncertainty of RC member but also those of properties of pavement, weigh or bridge railing, and so on. We thus apply the uniform distribution with the range of ±10% against the nominal value.

Table 2: Uncertainty parameters and prior uniform distribution range

Members	Parameters (unit)		Lower	Upper
G1	Lower flange thickness (mm)	Tb_1	23	25
	Bearing spring constant (N/m)	K_1	10^4	10^{13}
G2	Web bottom side thickness (mm)	Tb_2	8	9
	Lower flange thickness (mm)		23	25
	Bearing spring constant (N/m)	K_2	10^4	10^{13}
G3	Web bottom side thickness (mm)	Tb_3	8	9
	Lower flange thickness (mm)		23	25
	Bearing spring constant (N/m)	K_3	10^4	10^{13}
G4	Web bottom side thickness (mm)	Tb_4	8	9
	Lower flange thickness (mm)		23	25
	Bearing spring constant (N/m)	K_4	10^4	10^{13}
G5	Web bottom side thickness (mm)	Tb_5	8	9
	Lower flange thickness (mm)		23	25
	Bearing spring constant (N/m)	K_5	10^4	10^{13}
G6	Web bottom side thickness (mm)	Tb_6	8	9
	Lower flange thickness (mm)		23	25
	Bearing spring constant (N/m)	K_6	10^4	10^{13}
RC slab	Young's modulus (GPa)	Ec	22. 5	25. 0
	Density (kg/m^3)	Dc	2160	2640

3.3 Global Sensitivity Analysis

The global sensitivity analysis (GSA) can investigate the contribution of each parameter uncertainties to the required structural response. By removing the parameters that show low contributions to the comparative feature in the estimation, the ill-posed of inference can be prevented. The GSA here was realized by the analysis of variance (ANOVA) based on the full-factorial experimental design of the prior distributions in Table 2. This ANOVA determines the R^2 statistic that can indicate the contribution of variation (uncertainty) of each influential factor (model parameter) to the target output responses (resonant frequencies). The R^2 statistic can be calculated by the following formula:

$$R^2 = 1 - \frac{\sum_{j=1}^{m}\sum_{i=1}^{n}\left(y_{ij} - \bar{y}_j\right)^2}{\sum_{i=1}^{n}\sum_{j=1}^{m}\left(y_{ij} - \bar{y}\right)^2}, \tag{1}$$

where y_{ij} is output under ith (i=1-n) influential factor in level j (j=1-m). To reduce the calculation cost, the total 14 uncertain model parameters in Table 2 were divided into two groups (Group 1: Lower flange thickness, Young's modulus of deck, deck density; Group 2: Bearing spring constant, deck density), and generated 2^8 = 256 samples for Group 1, and 2^7 = 128 samples for Group 2. The contributions of model parameter uncertainties to the resonant frequencies can then be compared relatively from the R^2 statistics that are shown in Fig. 6. Here, it can be seen that density of slab element Dc shows very high contribution. Compared to the D_C, the spring stiffness K_1, ..., K_6 showed relatively high contributions; therefore, those seven parameters were determined to be applied to the estimation of posterior distributions.

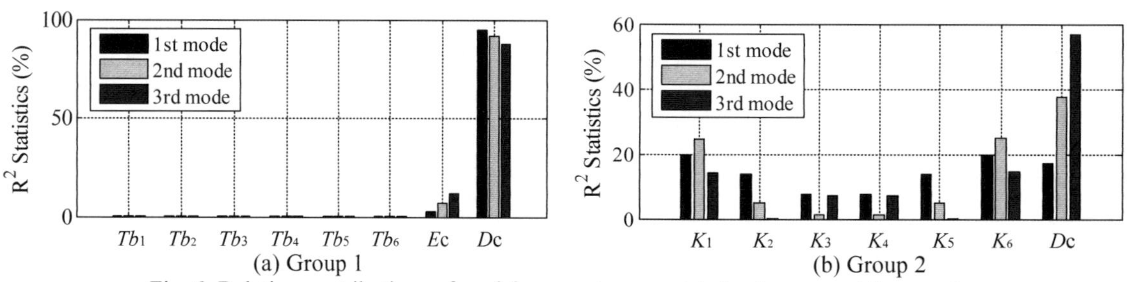

Fig. 6: Relative contributions of modal parameter uncertainties to resonant frequencies

3.4 Estimated Posterior Distributions of Model Parameters

The Markov chain Monte Carlo (MCMC) methods was adopted in deriving the posterior distributions based on the Bayesian inference in this study. Here, we also applied the surrogate model based MCMC calculation (Higdon, et al., 2008), that made possible to derive the posterior distributions of model parameters by using the resonant frequencies as the comparative feature, which was the different parameter from the objective estimation parameters, FE model parameters. This made it possible to make MCMC calculation lower cost by avoiding the high-cost FE model calculation at each MC sampling step.

Here, the measurement output y is expressed as in the following equation:

$$y = \eta\left(\hat{\boldsymbol{\theta}}\right) + \delta + \varepsilon, \tag{2}$$

where $\eta(\hat{\boldsymbol{\theta}})$ is true numerical output with true input model parameter $\hat{\boldsymbol{\theta}}$, δ is disturbance due to numerical modeling error, and ε is measurement error expressed by white noise. To estimate the posterior distribution of $\boldsymbol{\theta}$ here (Higdon, et al., 2008), the likelihood function is not set to only the measurement data, but the defined data matrix $\mathcal{D} = \left(\mathbf{y}^T, \boldsymbol{\eta}^T\right)$, where $\mathbf{y}$ is the comparative features extracted from the measurement data, i.e., the resonant frequencies in this study, and $\boldsymbol{\eta}$ is the numerical data for the surrogate model. This is the matrix of the calculated comparative features under the prior distributions of the FE model parameters by generating 200 samples in the Latin hypercube sampling (LHS). By assuming the Gaussian process to $\eta(\boldsymbol{\theta})$ and δ, a hierarchical Bayesian equation is then expressed as in the following equation:

$$\begin{aligned}&\pi\left(\boldsymbol{\theta}, \mu, \lambda_\eta, \rho_\eta, \lambda_\delta, \rho_\delta, | \mathcal{D}\right) \\ &\quad \propto L\left(\mathcal{D} | \boldsymbol{\theta}, \mu, \lambda_\eta, \rho_\eta, \lambda_\delta, \rho_\delta, \Sigma_y\right) \times \pi(\boldsymbol{\theta}) \times \pi(\mu) \times \pi\left(\lambda_\eta\right) \times \pi\left(\boldsymbol{\rho}_\eta\right) \times \pi\left(\lambda_\delta\right) \times \pi\left(\boldsymbol{\rho}_\delta\right),\end{aligned} \tag{3}$$

where μ_η and μ_δ (=0) are means, ρ_η and ρ_δ are correlation coefficients, which describe a covariance function, and λ_η, and λ_δ penalty factors of numerical output η and modeling error δ, respectively. The left side of Eq. (3) is the intended posterior distribution, and the right side is in the form of multiplying likelihood function $L(.)$, which consists of measurement data, against the objective estimation parameter $\boldsymbol{\theta}$ and the prior distribution $\pi(.)$ of $\boldsymbol{\theta}$, μ, $\lambda_{rp}\rho_\eta$, λ_δ, and ρ_δ.
Figure 7 shows the posterior distributions of target model parameters; spring stiffness K_1, …, K_6 and density of slab shell element D_C. They were the histograms of adopted 50,000 samples with 5000 step burn-in in the Metropolis-Hastings algorithm, which was one of the MCMC calculations. Notice that the range of x-axis in each of six histograms in Fig.7 is equal to the range of each prior uniform distribution. There is actually less uncertainty reduction in slab density D_C; on the other hand, the updating of probability distributions can be recognized in the posterior distributions of bearing spring stiffness K_1, …, K_6. By seeing the results of K_1 and K_6, which represent the bearing performances of two outside girders G1 and G6, the probability becomes relatively low in the range of stiffness value larger than 1×10^9 N/m. This indicates that the possibilities of movability reduction in the bearings of girders G1 and G6 are low. On the other hand, the probabilities are biased toward the stiffness value of 1×10^{13} N/m in the posteriors of K_2 to K_5 that are the inside girders. This indicates that there is possibility that the bearings of girders G2 to G5 become close to immovable due to the corrosion.

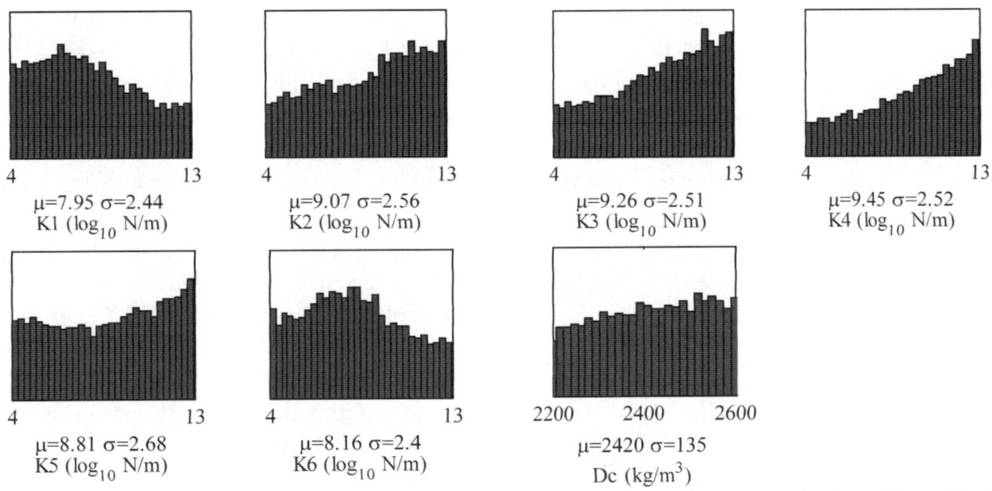

Fig. 7: Posterior distributions of model parameters (the ranges of x-axis are the ranges of prior uniform distributions)

4. Structural Reliability Analysis using the Posterior Distributions

The live load capacity of the target bridge was assessed by the structural reliability analysis. Here, the estimated posterior distributions of model parameters were used to consider actual structural conditions. The effectiveness of using the posterior distributions was verified by comparing structural reliability index β derived from the "prior" and "posterior" distributions.

4.1 Calculation of Structural Reliability Index β

The structural reliability analysis is the procedure to assess the structural capacity to the limit state under the variabilities of load and resistance statistically (Thoft-Cristensen, P. & Baker, K.J. 1982). The structural reliability index β is the statistic measure to indicate the capacity quantitatively. The simplest method to calculate the β is the first-order second-moment method (FOSM). In the FOSM, the resistance R and the load action S are assumed to be mutually independent, and if both are governed by the normal or log-normal distributions, the mean and standard deviation of the performance function $Z=R-S$ can be derived by:

$$\mu_Z = \mu_R - \mu_S \tag{4}$$

$$\sigma_Z^2 = \sigma_R^2 + \sigma_S^2 , \tag{5}$$

where μ_R and μ_S are the normal or lognormal mean values, and σ_R and σ_S are the standard deviations of R and S, respectively. When μ_Z and σ_Z are obtained for the performance function Z, the reliability index β can be derived as follows:

$$\beta = \frac{\mu_Z}{\sigma_Z} . \tag{6}$$

4.2 Structural Reliability Index β for the Target Bridge

The performance of bridge against the live load can be assessed by applying the desining static live load, which is indicated in *Japan specifications for highway bridges*, to the numerical model, as shown in Fig.8. When the live load was applied to the FE model of the target bridge with varying the spring stiffness at the bearings, the Mises stress distribution and the its maximum value near the end of the steel girder was changed. Therefore, the load action S in this study was the maximum Mises stress within the area near the end of steel girders under the static live load. Here, the variability of S, i.e., μ_S and σ_S, were due to the uncertainties of model parameters. The distribution of load action S could be obtained by the Monte Carlo (MC) calculation. The resistance R here was the yield stress of the steel, and its statistics for μ_R and σ_R for Eq. (4) and (5) were referenced values in the statistical data. Note that the distribution of the yield stress of SS400 steel material was the lognormal distribution.

The uncertainty parameters considered here for the load action S were the bearing spring stiffness, K_1, …, K_6 and the slab element density Dc; the seven parameters in total. Here, we set two cases for verification, Case I and Case II. Case II is the case of using "prior" distributions; 500 samples were generated from the prior uniform distributions of those seven parameters by the Latin Hypercube sampling (LHS). In Case II, which is the case of using "posterior" distributions, the last 500 samples were picked up from the MCMC 50,000 samples of the posterior distributions. In each case, the FE model calculation to extract maximum Mises stress in the steel girders under the designing live load was taken in each sampling.

To apply the FOSM for the reliability index β, the distribution of Mises stress must also show the normal or lognormal distributions. It was confirmed that the distributions in both Case I and II satisfied the lognormal distributions by the Kolmogorov-Smirnov test with the p-values under the significance level of 1%.

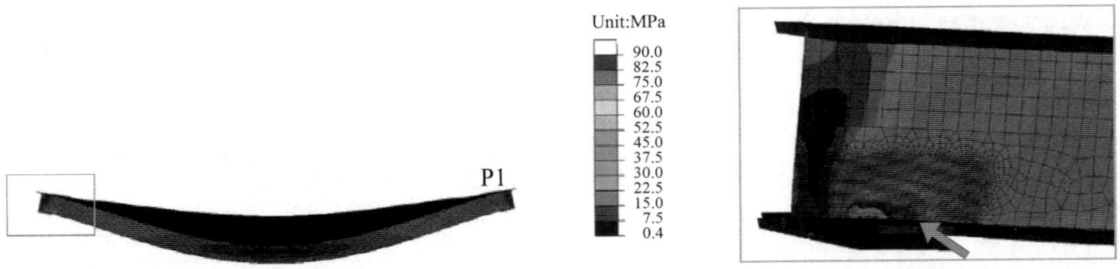

(a) Deformation (b) Maximum Mises stress near the end of girder

Fig. 8: Mises stress distribution of the target bridge under static live load

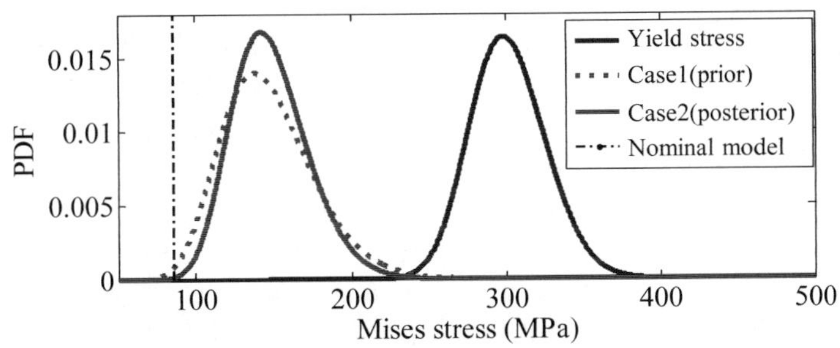

Fig. 9: Relationship of probability density functions in lognormal of the steel Mises stress and the yield stress

The lognormal distributions of Mises stress in Case I (prior) and Case II (posterior), that is the load action S, and the lognormal distribution of steel yield stress, the resistance R, are shown in Fig.9. The logarithmic mean, and logarithmic standard deviation of the Mises stress in Case I are μ_{S1} =147 MPa and σ_{S1} =29.2 MPa, respectively. In Case II, μ_{S2}=148 MPa and σ_{S2} =24.3 MPa. The difference of the mean between two cases was only 1MPa; however, the standard deviation in Case 2 was reduced around 15% against that in Case1. Notice that this was exactly the contribution of uncertainty reduction by the Bayesian inference.

The structural reliability index β against the steel yield stress was then calculated by applying the logarithmic means and standard deviations to the FOSM in each case. The results were then β=3.44 in Case I and β=3.94 in Case II. First, in both cases, there were enough capacity to the limit state; the b became larger than β=2.5, which indicate that the probability of collapse is smaller than 1% in the structural reliability.

The most significant point was that the β=3.94 in Case II became around 14% higher than β=3.44 in Case I. This means that the structural safety of the target bridges was not overestimated in the dangerous direction by using the posterior distributions, which were updated uncertainties by the measurement. The effect of uncertainty reduction could increase the reliability index β.

Thus, in this study, it was possible to show the validity of using model parameter posterior distribution when conducting reliability assessment of an existing bridge in which uncertainty in the actual state of a bridge is reduced by using measurement data. The result here could show the effectiveness of using the posterior distributions estimated by the Bayesian inference with in-situ measurement for assessing the structural capacity of existing structures.

5. Conclutions

In this study, the applicability of the posterior distributions of finite element (FE) model parameters to the structural reliability analysis of an existing steel plate girder bridge was verified. The results are summarized as below:

- The posterior distributions of model parameters related to the bridge bearing performance could be estimated by the resonant frequencies. The movability of corroded bearings could be assessed from the obtained posterior distributions of spring stiffness.
- By reducing the uncertainties of model parameters, the standard deviation of maximum Mises stress under the live load was reduced. This can be said that the uncertainty was reduced by the measurement.
- The capacity of maximum Mises stress in the steel girder to the limit state of the steel yield stress was assessed by deriving the structural reliability index β. The index β was compared between the calculated values in two cases, one was the case of using "prior" distributions of model parameters, the other was the case of using the ones of "posterior". The β value in the case of using "posterior" became higher than that in the case of "prior". This was because the standard deviation of maximum Mises stress could be reduced by the uncertainty reduction.

The structural reliability index β derived by the posterior distributions of model parameters, that is the statistical measure related to the sutural condition, is expected to be applicable to the decision makings in the bridge maintenance and operations. In our future work, more effective data acquisitions in the estimation of posterior distributions to reduce the model parameter uncertainties related to specific damages in existing bridges are considered. Furthermore, the applicability of this statistical measure to the decision makings in the practical maintenance and operational works will be verified in the collaboration with bridge owners.

References

Higdon, D., Gattiker, J., Williams, B., Rightley, M.: Computer Model Calibration Using High-Dimensional Output, *Journal of the American Statistical Association*, Vol. 103, pp. 570-583, 2008.

Kaita, T. et al., An experimental study on remaining bending strength of corroded plate girder, *JSCE Journal of Structural Engineering*, Vol. 51A, pp. 139-148, 2005. (In Japanese)

Kayser, J. R. and Nowak, A. S., Reliability of corroded steel girder bridges, *Structural Safety*, 6(1), pp.53-63, 1989.

Kuroda, R., Nishio, M., Reliability assessment of an existing steel girder bridge using posterior distributions of model parameters, *Journal of JSCE ser. A1*, Vol.72, No.3, pp.380-392, 2016. (In Japanese)

Okasha, N. M., Frangopol, D. M. and Orcesi. A. D., Automated finite element updating using strain data for the lifetime reliability assessment of bridges, *Reliability Engineering and System Safety*, 99, pp139-150, 2011.

Thoft-Cristensen, P., and Baker, K.J., Structural Reliability Theory and Its Application, Springer, 1982 (1985 in Japanese).

The 8th International Conference on
Structural Health Monitoring of Intelligent Infrastructure
Brisbane, Australia | 5-8 December 2017

Modal Parameters Variation of PC beams and its influence on Failure Probability

O.S. Luna Vera [1], C.W. Kim [1], Y. Oshima [2].
[1] *Department of Civil & Earth Resources Engineering, Kyoto University – Japan,*
Email: kim.chulwoo.5u@kyotou.ac.jp, [2]*Center of Advanced Engineering Structural Assessment and Research, Public Works Research Institute - Japan*

Abstract

This study aims to examine how changes in dynamic properties of a prestressed concrete (PC) beam (i.e. mode shapes, frequencies and damping) influence on their probability of failure. This study makes use of collected experimental information from the PC beam under static loading tests. In the static loading tests, the PC beam was put under several loading and unloading levels until failure, and an artificial breakage was also introduced to the PC tendon. Forced vibration tests were carried out along the static loading test to assess changes in the dynamic properties of the PC beams with different health conditions. The reliability of the PC beams is calculated by using a well-known approach such as first-order reliability method (FORM), and then the probability of failure in different crack scenarios is examined. The frequency response function (FRF) can be easily obtained by transfer functions estimations, since an impact hammer was used in the dynamic tests, and modal parameters were identified utilizing the FRF. Finally, the study investigates the influence of the variation of modal parameters on the probability of failure.

1. Introduction

The use of structural health monitoring for performance description could be a challenging approach for making a decision in maintenance of structures. Natural frequencies have shown being sensitive features related to strength reduction when evaluating concrete bridges with 50 years of life (Quattrone, et al., 2012). Damping ratio can be also a sensitive damage indicator for reinforced and prestressed concrete structures. Dammika, et al. (2015) showed that higher values of damping ratio in aged bridges can link to deterioration. Regardless, more evidence is needed to prove the link between SHM and structural resistance. Inconsistencies between experimental testing and theoretical models come up because different reasons that need to be handled (Udwadia, 2005). Damage in structures is interpreted as a decay of its mechanical properties, or in other words, a decrease in stiffness. However, the tendency of natural frequencies to the development of damage shows an abnormal increase/decrease from one configuration to another, even experimentally (Dilena & Morassi, 2011).

This study focuses on the use of two different techniques to evaluate performance of a PC-beam, such as First Order Reliability Method and Frequency Response Functions. The first method is to assess the reliability in different scenarios and the second one to monitor the modal parameters. In order to have a better understanding of the relationship between techniques, experimental data was used to update a FEM model.

2. The PC-beam Benchmark Study

This study utilizes experimental data collected in a previous research (Luna, et al., 2017) where a PC beam was tested in the facilities of Kyoto University as shown in Figure 1. The type of data gathered are displacements from static tests, load-unload in different levels, and accelerations from vibrations tests, impact hammer tests before and after each static test. The location of the displacement transducers and the piezoelectric servo-accelerometers can be seen in Figure 2.

2.2 Damage Scenarios

Eight scenarios (E1 to E8) are extracted from the experimental data, from which the first four scenarios, no superficial cracks are introduced but internal grouting cracks are implied due to the tendons' cut and static load tests of 150kN. The last tendon was severed between E2 and E3 (see Figure 3).

Figure 1. PC-beam tested at Kyoto University, Japan

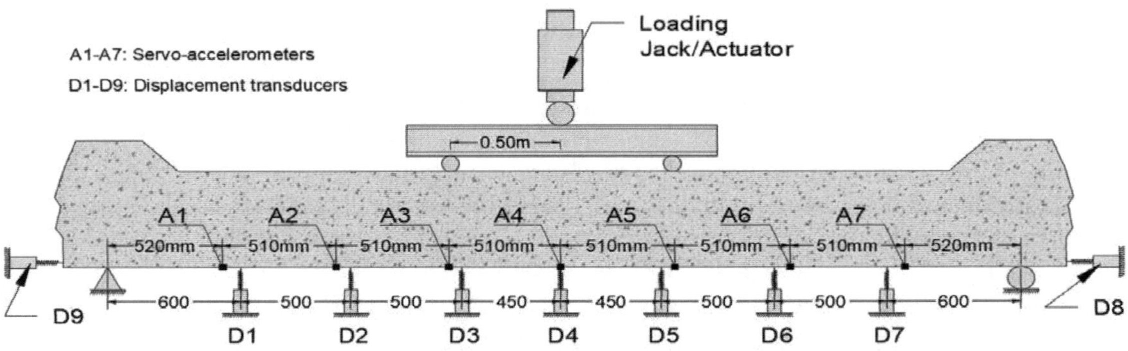

Figure 2. Sensors location

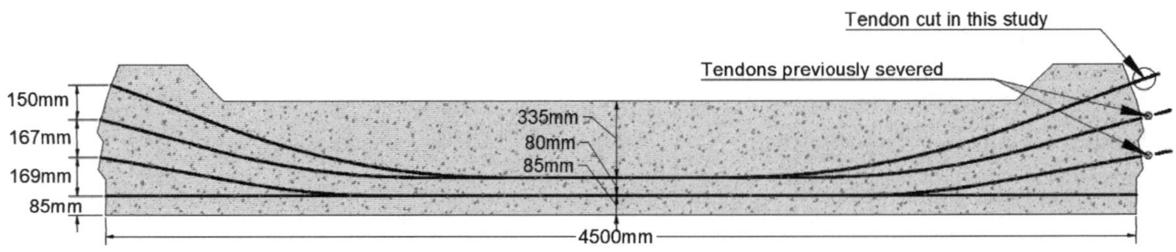

Figure 3.Tendons' profile and cut location

Then, the last four scenarios introduce the propagation of superficial cracks, that were measured carefully, because static load-unload increments (200kN, 250kN, 300kN, 350kN). Table 2 shows the summary of the damage scenarios along with the performance calculated by means of energy.

Table 1. Cracks length and performance

Damage Scenario	Cracks length (Front face)	Cracks length (Back face)	Cracks length (Average)	Energy dissipated (ΔW)	Work applied (W)	Performance $Z = 1 - \frac{\Delta W}{W}$
E1	0	0	0	15.95 Joules	119.61 Joules	86.67%
E2	0	0	0	8.67 Joules	115.14 Joules	92.47%
E3	0	0	0	10.78 Joules	117.74 Joules	90.85%
E4	0	0	0	9.77 Joules	118.29 Joules	91.74%
E5	666.49 mm	657.96 mm	662.23 mm	67.27 Joules	330.96 Joules	79.67%
E6	2121.05 mm	2030.60 mm	2075.83 mm	191.80 Joules	837.92 Joules	77.11%
E7	3602.75 mm	3695.59 mm	3649.17 mm	574.08 Joules	1948.13 Joules	70.53%
E8	5359.94 mm	4877.01 mm	5118.48 mm	2841.17 Joules	4817.31 Joules	41.02%

2.2 Experimental modal data collected

The experimental modal parameters were obtained from the frequency response functions, where the frequencies come from the peaks and the damping ratio is in function of the bandwidth frequency of 3dB from the maximum value of the peak, so called damping factor.

Table 2. Experimental modal parameters collected

Damage Scenario	1st frequency Hz (mean/std)	2nd frequency Hz (mean/std)	3th frequency Hz (mean/std)	1st damp. ratio % (mean/std)	2nd damp. ratio % (mean/std)	3th damp. ratio % (mean/std)
E1	50.00/0.00	136.69/0.56	421.86/0.71	0.70/0.02	1.91/0.14	1.01/0.11
E2	50.00/0.00	138.09/0.40	419.68/0.70	0.75/0.03	1.61/0.12	0.88/0.20
E3	50.00/0.00	137.76/0.56	422.69/5.59	0.81/0.04	1.69/0.14	1.20/0.30
E4	49.73/0.25	136.77/0.64	422.55/6.02	0.99/0.05	1.73/0.12	1.14/0.33
E5	49.50/0.00	135.90/0.31	423.09/5.75	0.92/0.06	1.71/0.74	1.15/0.32
E6	49.50/0.00	131.13/0.71	423.23/3.93	0.85/0.06	2.67/0.19	1.04/0.37
E7	49.50/0.00	132.52/1.18	420.89/6.31	0.76/0.04	2.55/0.38	1.05/0.31
E8	48.89/0.21	130.41/1.76	410.72/0.77	0.97/0.07	2.98/0.31	1.21/0.17

2.3 Finite Element Model of the PC-beam

The PC beam is modelled with 16 one-dimensional beam elements that also allows axial forces in its centroid. Node and element numbering are shown in Figure 4. The static load is described as P and distributed equally in node 6 and 12. The tendons' tension is described as a horizontal force T in all nodes plus a moment ΔM due to the tendons' profile. ΔM is in function of the tendons' resultant force T and the eccentricity to the centroid of the cross-section located in each node.

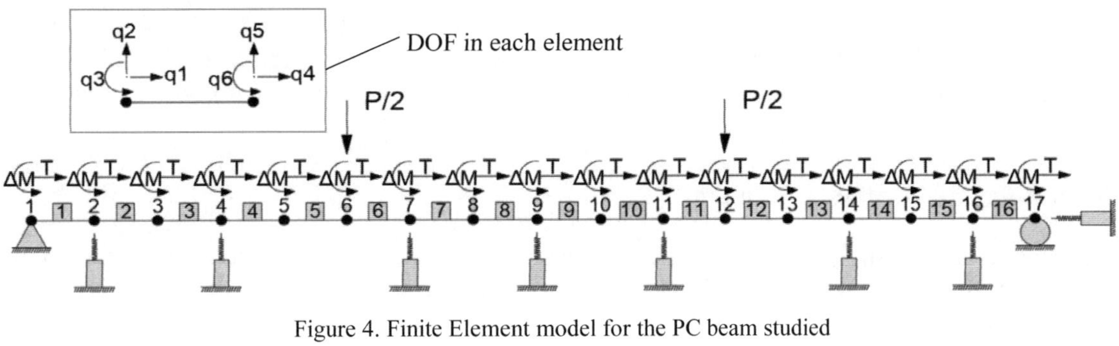

Figure 4. Finite Element model for the PC beam studied

Boundary conditions apply to node 1 as pinned and as roller in node 17. Nodes 2, 4, 7, 9, 11, 14, 16, 17 is where the displacement transducers were deployed in the experiment. The local stiffness matrix for each element is:

$$K = \begin{bmatrix} \frac{AE}{L} & 0 & 0 & -\frac{AE}{L} & 0 & 0 \\ 0 & \frac{12EI}{L^3} & \frac{6EI}{L^2} & 0 & -\frac{12EI}{L^3} & \frac{6EI}{L^2} \\ 0 & \frac{6EI}{L^2} & \frac{4EI}{L} & 0 & -\frac{6EI}{L^2} & \frac{2EI}{L} \\ -\frac{AE}{L} & 0 & 0 & \frac{AE}{L} & 0 & 0 \\ 0 & -\frac{12EI}{L^3} & -\frac{6EI}{L^2} & 0 & \frac{12EI}{L^3} & -\frac{6EI}{L^2} \\ 0 & \frac{6EI}{L^2} & \frac{2EI}{L} & 0 & -\frac{6EI}{L^2} & \frac{4EI}{L} \end{bmatrix} \quad (1)$$

According to the previous model, the displacement equations that correspond to the location of the seven vertical transducers, described in Figure 2 (D1-D7), are $(\delta_1 \cdots \delta_7)$ and also the equation δ_H correspond to the horizontal transducer located at the right (D8).

$$\begin{Bmatrix} \delta_1 \\ \vdots \\ \delta_7 \end{Bmatrix} = \frac{P \cdot L^3}{E \cdot I} \begin{Bmatrix} \frac{4599}{551368} \\ \frac{46849}{3308208} \\ \frac{119009}{6616416} \\ \frac{63271}{3308208} \\ \frac{119009}{6616416} \\ \frac{46849}{3308208} \\ \frac{4599}{551368} \end{Bmatrix} + \frac{T \cdot h \cdot L^2}{E \cdot I} \begin{Bmatrix} \frac{386628}{215378125} \\ \frac{476223}{215378125} \\ \frac{28017}{21012500} \\ 0 \\ -\frac{28017}{21012500} \\ -\frac{476223}{215378125} \\ -\frac{386628}{215378125} \end{Bmatrix} \quad (2)$$

$$\delta_H = \frac{17 \cdot T \cdot L}{2 \cdot A \cdot E} \quad (3)$$

Therein P is the load, L is the clear span length (4100mm), I is the moment of Inertia, E is young's modulus or elastic modulus, T is the total tendons' force, h is the height of the beam (500mm) and A is the Area of the cross-section ($b*h$: the base is 250mm).

3. Reliability Analysis of the PC Beam

The method for reliability analysis of the beam is the well-known First Order Reliability Method (FORM). This method is based on the calculation of safety index β (also called reliability index) and from there, the probability of failure is $P_f = \Phi(-\beta)$, where $\Phi(.)$ is the standard normal cumulative distributed function.

3.1 Variables Description as Random

Before the reliability analysis, it is necessary to describe the variables as random. From which, one way to represent the change in stiffness along the experiment stages is by sampling the elastic modulus using Eq. (2) with the data collected experimentally (load and displacement in different locations). Load and displacements from the range 0kN to 150kN were used to approximate the Young's modulus. The range is in the linear response so that nonlinearities can be avoided in the analysis. With this information, histograms were constructed for every scenario. Figure 5 shows the histograms for the E1 and E8 scenarios.

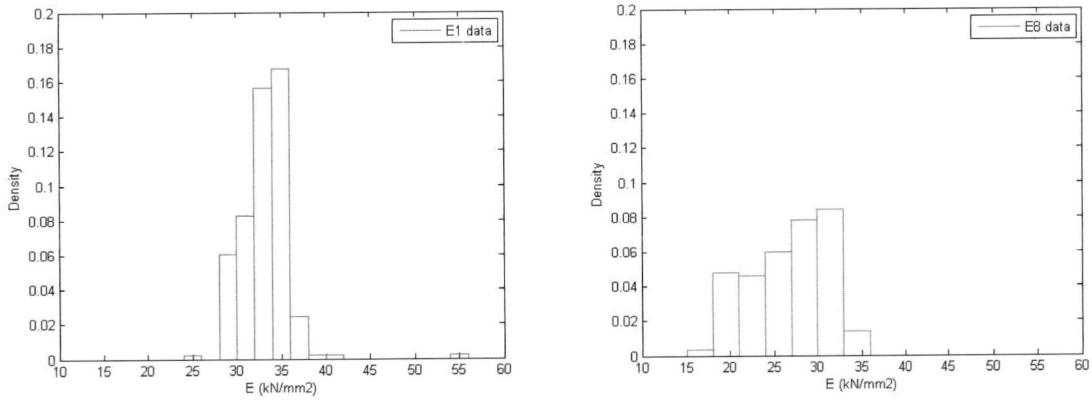

Figure 5. Histogram of Elastic modulus in scenarios E1 and E8

Once obtained all histograms, a Gaussian fit (normal distribution) is done for every scenario (see Figure 6) and parameters such as mean, standard deviation and coefficient of variation, can be found in Table 3.

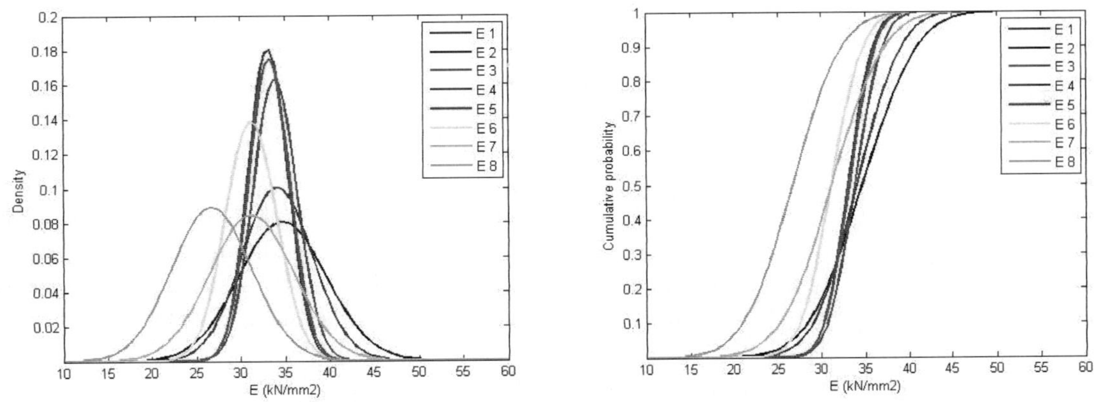

Figure 6. Gaussian fit of the Elastic modulus in different scenarios

Table 3. Summary of elastic modulus as random variable in each scenario

Damage Scenario	Elastic Modulus		
	Mean kN/mm^2	Standard deviation	Coefficient of variation
E1	33.10	2.21	0.067
E2	34.69	4.95	0.143
E3	33.88	2.45	0.072
E4	34.10	3.97	0.116
E5	33.24	2.28	0.069
E6	31.17	2.88	0.092
E7	31.23	4.72	0.151
E8	26.69	4.49	0.168

Then, the rest of the variables have a small variance compared to the measured value like Length and Inertia with a coefficient of variation (cov) of 0.004 and the static load P is 0.01. The geometrical values come from several manual measurements along the whole girder, and the cov of the load comes from the variance between the target loads in the control room of the jack actuator and the data logger values recorded.

3.2 Limit State Function Considered

Building codes based on limit states functions, such as International Building Codes (ICB), specify that the typical deflection limit for bending in a simple supported beam under live load is:

$$\frac{L}{360} > \delta \tag{4}$$

where L stands for the clear span of the beam and the failure domain can be define as:

$$\frac{L}{360} - \delta < 0 \tag{5}$$

Since the explicit equation for the mid span displacement was obtained from the finite element model in chapter 2.3, the limit state function for the specific case of the beam under study is:

$$\frac{L}{360} - \frac{63271 \cdot P \cdot L^3}{3308208 \cdot E \cdot I} = 0 \tag{6}$$

4. Results and Discussions

The reliability analysis in each damage scenario was conducted by considering the change of the elastic modulus only (a technique to represent the change in stiffness). The probability of failure values obtained in the range below 150kN was very small and difficult to analyze. Another way to monitor the reliability change is by observing the development of the safety (or reliability) index instead (Figure 7). The safety index gives an erratic trend in the first scenarios, but when the cracks started and propagated to more parts of the beam, the safety index trend stabilizes and becomes a decreasing tendency. Probability of failure values between 150kN and 350kN assume a linear change in the young modulus, cause only experimental data below 150kN (load-displacement) was used to avoid nonlinear analysis. A nonlinear analysis will bring the issue of a distribution of the young modulus with at least two peaks, which will be a future work.

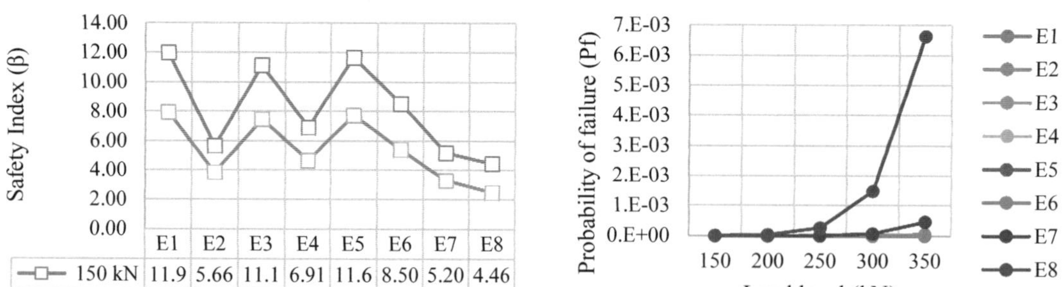

	E1	E2	E3	E4	E5	E6	E7	E8
150 kN	11.9	5.66	11.1	6.91	11.6	8.50	5.20	4.46
350 kN	7.91	3.86	7.46	4.67	7.73	5.40	3.32	2.48

Figure 7.Reliability Index and Probability of failure development

The analysis of the modal parameters, on the other hand, are too extensive to summarize it in one unique value. Parameters, such as frequency, have no alteration when damage like tendon breakage occurs (Döhler, et al., 2014) (Limongelli, et al., 2016). The study of the relationship between static stiffness and modal stiffness of structures indicates that the first mode dominates in the summation of all modal stiffness. This is because greater number of modes considered, the smaller the difference between the total modal displacement and the static modal displacement of a structure (Wahyuni & Ji, 2010). The reason for the average of the damping ratios studied, and also frequencies (3 first natural bending modes), is to compare them to the safety indexes through the development of all damage scenarios (see Figure 8, Figure 9 and Figure 10).

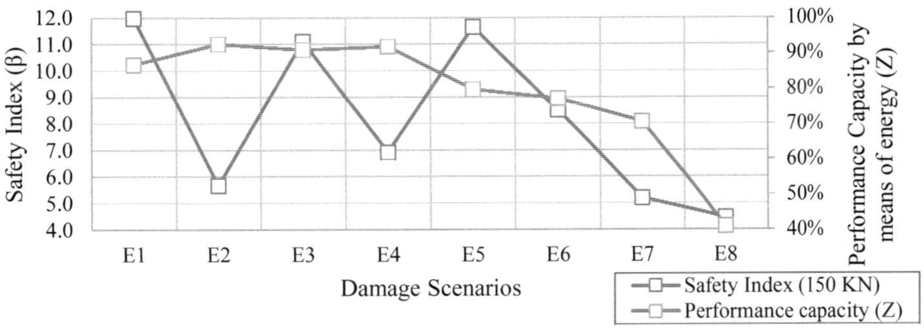

Figure 8. Comparison of safety index with the performance capacity by means of energy (Z)

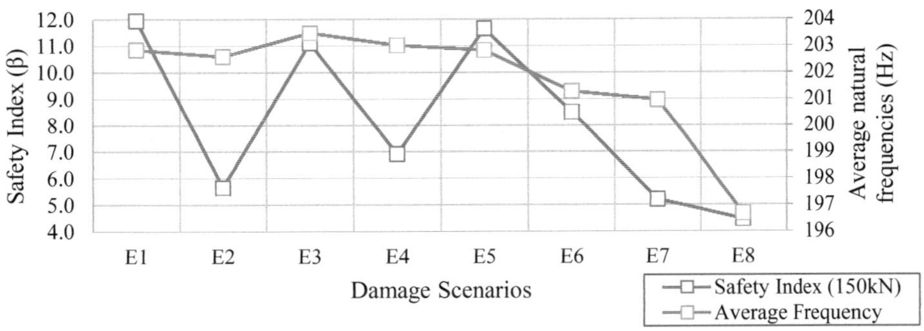

Figure 9. Comparison of safety index with the average from the first 3 bending frequencies

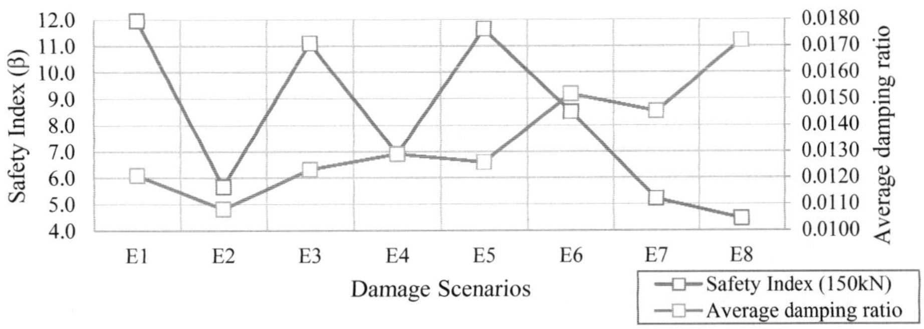

Figure 10. Comparison of safety index with the average from the first 3 bending damp. Ratio

5. Conclusions

- The safety index has an inconsistent trend in the safe domain, but opposite when big damage occurs such as large cracks.
- The summation of frequencies and damping ratios seems to stabilize a trend correlated with change in performance.
- The influence of modal parameters in the probability of failure was not clear due to the assumption of a linear system response. Nonlinearities needs to be introduce in order to obtain more realistic correlations.

References

Dammika, A. J. et al., 2015. *An investigation on modal damping ratio as an indicator of invisible damage in PC bridges.* London, Taylor & Francis Group, pp. 404-410.

Dilena, M. & Morassi, A., 2011. Dynamic testing of a damaged bridge. *Mechanical Systems and Signal Processing,* pp. 1485-1507.

Döhler, M., Hille, F., Mevel, L. & Rücker, W., 2014. Structural health monitoring with statistical methods during progressive damage test of S101 Bridge. *Engineering Structures,* Volume 69, pp. 183-193.

Limongelli, M. P. et al., 2016. Damage detection in a post tensioned concrete beam - Experimental investigation. *Engineering Structures,* pp. 15-25.

Luna, O., Kim, C. W. & Oshima, Y., 2017. *Performance variation influence on experimental modal parameters of a prestressed concrete beam. Proc. of Applied Mechanics Conference,* JSCE, Kyoto, Japan.

Quattrone, A. et al., 2012. *Vibration tests on dismounted bridge beams and effects of deterioration.* s.l., s.n.
Udwadia, F. E., 2005. Structural Identification and Damage Detection from Noisy Modal Data. *Journal of Aerospace Engineering,* 18(3), pp. 179-187.

Wahyuni, E. & Ji, T., 2010. Relationship between Static Stiffness and Modal Stiffness of Structures. *IPTEK, The Journal for Technology and Science,* 21(2).

The 8th International Conference on
Structural Health Monitoring of Intelligent Infrastructure
Brisbane, Australia | 5-8 December 2017

Probabilistic Diagnosis and Prognosis of Steel Plate Corrosion by Particle Filter Assimilation of Force-Displacement Observations

S. Yi [1], J. Song [2].
[1] *Department of Civil and Environmental Engineering, Seoul National University – South Korea, Email:* yisangri@snu.ac.kr [2] *Department of Civil and Environmental Engineering, Seoul National University – South Korea*

Abstract

Accurate diagnosis and prognosis of deterioration are essential to ensure operational safety and effective maintenance of structures. Bayesian-filter-based assimilation of measurement data and phyisical model facilitates diagnosis and prognosis of the deterioration propagation, as well as calibration of parameters in the deterioration evolution models. Among existing Bayesian filter methodologies, particle filter is considered an effective tool especially for nonlinear models subjected to non-Gaussian uncertainties. This paper presents a novel application of the particle filter approach in which mechanical, i.e. force-displacement observations are assimilated to infer corrosion progresses. For this purpose, the spatial distribution of deterioration outcome, e.g. thickness of a steel plate, is described by a random field model whose parameters are updated by Markov Chain Monte Carlo (MCMC) sampling given measurement data on loads and displacements. Using a particle filter, various static, time-varying and stochastic model parameters related to deterioration phenomenon are inferred for the purpose of diagnosis and prognosis. The results of numerical examples of a steel plate subjected to pitting corrosion show good performance in prediction of effective thickness pattern and identification of model parameters under various senarios. The proposed approach is expected to promote optimal condition-based maintenance stratagies.

1. Introduction

Reliable prediction of structural deterioration, such as corrosion and fatigue, is crucial for establishing safe and effective maintenance strategies, but is known to be challenging due to its complicated and irregular nature. Monitoring and inspection data in operational condition are often utilized to supplement this unpredictable attribute, along with a deterioration prognosis model, but this approach may suffer due to the use of an imperfect physical prediction model, noisy measurements, and often infeasible direct measurement during non-destructive tests (NDT).
The structural prognosis methods can be classified into two types – full-data-driven methods and physics-based methods. It is noted that physics-based methods show superior performance in deterioration prediction and do not require massive sets of training data unlike data-driven method (An *et al.* 2014). Among physics-based methods, Bayesian filters such as Kalman filter, extended Kalman filter, and particle filter gained popularity because they can handle successive data input effectively and calibrate system model parameters efficiently. In particular, Kalman filter and extended Kalman filter, which are the analytical solutions of the linear Gaussian Bayesian filter and a linearly approximated solution of nonlinear Bayesian filter, respectively, are applied extensively to track fatigue and corrosion propagation of structural elements (Ray & Tangirala 1996, Papadimitriou *et al.* 2011). On the other hand, particle filter, which is a simulation-based Bayesian filter approach,

took subsequent but remarkable spotlight in recent years due to its versatility and drastic improvement of computational performance (Cadini & Avram 2009, Lee *et al.* 2015).

Particle filter allows us to incorporate various indirect nonlinear measurements so as to infer physical deterioration state. However, mechanical responses such as displacement and strain, which can be often acquired conveniently, has been limitedly introduced as *diagnostic variates* in Bayesian filter frame work, i.e. changes in such types of responses are used to diagnose deterioration. For instance, Sbarufatti *et al.* (2016) used measurements of mechanical property to estimate deterioration state via particle filter. They used artificial neural networks (ANNs) surrogate model to relate strain observations with the underlying deterioration state, i.e. fatigue crack length, which was trained by means of off-line finite element analysis (FEM). However, the level of damage that can be estimated by the framework is restricted to that occurs at a predefined location. On the other hand, Neerukatti *et al.* (2016) proposed fatigue crack localization approach via particle filter, but NDT signal was directly utilized, and was limited to the damage at a single location.

This paper proposes an integrated particle filter framework which utilizes the on-line observation of mechanical inspections as corrosion diagnostic variates, and retrieves full-domain diagnosis and prognosis of spatial distribution of deterioration while embracing the damage growth rate and spatial propagations. In order to infer physical corrosion state via mechanical load-displacement data, a Bayesian physical parameter identification method is also proposed using stochastic spectral representation of deterioration variate and Markov Chain Monte Carlo (MCMC) sampling scheme. In order to fully utilize the ability of particle filter approach, various types of parameters are calibrated, such as the time-varying model parameters and statistical parameters, as well as the static model parameters. Recently, some researchers treated parameter identification (Lee *et al.* 2015, Sbarufatti *et al.* 2016, Neerukatti *et al.* 2016) as a part of particle filter research, but to the best of authors' knowledge, no previous works have examined such extensive range of parameters altogether.

In the following section, a brief introduction on particle filter is presented. Subsequent two sections provide detailed models that are employed in the proposed particle filter approach. Specifically, Section 3 proposes a model that bridges the gap between the mechanical observations from inspection and the corrosion states of interest while Section 4 introduces the stochastic corrosion evolution model. Numerical examples are provided in Section 5 based on synthetic measurements, including the verification of the measurement model in Section 3.

2. Particle Filter Approch for Measurement-Model Assimilation

2.1 Bayesian Filter

When periodic structural health monitoring inspection results are available, the Bayesian filter can be used for short-term prediction of deterioration propagation dynamics for each inspection interval, and for calibration of deterioration parameters. This section briefly introduces basic requirements and procedure of Bayesian filter. Bayesian filter generally consist of recursive two-step procedures – *propagation* of state estimation based on the system model, followed by the observation-based estimation *updating* based on measurement model (Simon 2006). The two-step procedure can be described as

$$\boldsymbol{x}_{k+1} = f(\boldsymbol{x}_k, \boldsymbol{\theta}_k, w_k) \tag{1}$$

$$\boldsymbol{y}_k = h(\boldsymbol{x}_k, v_k) \tag{2}$$

where k is time index, $\boldsymbol{x}_k$ is the vector of state of interest, $\boldsymbol{\theta}_k$ is vector of model parameters and $\boldsymbol{y}_k$ are the measured quantities. The functions $f(\cdot)$ and $h(\cdot)$ respectively denote the *system* model that describes the linear or nonlinear propagation of deterioration state, and the *measurement* model that relates deterioration state and a measured quantity. The variables $\{w_k\}$ and $\{v_k\}$ represents the uncertainty in the system and measurement equations, respectively, and they are generally assumed to be statically independent.

The propagation step (step 1) can be described probabilistically via the transition probability density function (PDF) of state, i.e. $p(\boldsymbol{x}_{k+1}|\boldsymbol{x}_{0:k},\boldsymbol{\theta}_{0:k})$ or $p(\boldsymbol{x}_{k+1}|\boldsymbol{x}_k,\boldsymbol{\theta}_k)$ under the Markov assumption. The updating step (step 2), assuming that k-th observations $\boldsymbol{y}_k$ are acquired at the time step k determines the optimal estimation of states and parameters by combining the estimation from system model and additional measurement information via Bayes' theorem, i.e.

$$p(\boldsymbol{x}_k,\boldsymbol{\theta}_k|\boldsymbol{y}_k) = cL(\boldsymbol{y}_k|\boldsymbol{x}_k,\boldsymbol{\theta}_k)p(\boldsymbol{x}_k,\boldsymbol{\theta}_k) \tag{3}$$

where $L(\boldsymbol{y}_k|\boldsymbol{x}_k,\boldsymbol{\theta}_k)$ denotes the likelihood of measurement given current estimations, $p(\boldsymbol{x}_k,\boldsymbol{\theta}_k)$ is prior estimation given in terms of PDF, and c is normalization constant. The left-hand side term $p(\boldsymbol{x}_k,\boldsymbol{\theta}_k|\boldsymbol{y}_k)$ represents the posterior PDF given noisy observation at time instant k. Note that Eq. (3) is also based on first order Markov assumption (Simon 2006).

2.2 Particle Filter

Particle filter is one of the Bayesian filters, which is also termed as sequential Monte Carlo method. In this method, the PDFs in the algorithm are approximated by the histograms of realizations, and the propagations are described by dynamic simulation of each realizations. This realized set of random state and random parameters are called *particles*. State transition and likelihood are calculated by direct simulations in Eq. (1) and (2). Therefore, particle filter enables us to engage nonlinear models and non-Gaussian distributions without additional efforts. There are several approaches to perform Bayesian updating described in Eq. (3) via particle filter, and this paper uses one of the classical approaches, termed as sequential importance resampling (SIR) method. Additionally, the technique proposed by Liu & West (2001) is employed to handle *particle depletion* problem. The technical details can be referred to An *et al.* (2013).

2.3 Particle Filter Approach for Structural Deterioration Prediction

Since particle filter can deal with complex and highly nonlinear measurement models, it can be usefully adopted for the deterioration prediction by exploiting various types of SHM data. This research applies the approach to physical index of corrosion in which the effective thickness is not directly observable whereas displacement measurements obtained from periodic field load test are available. The acquired datasets have displacement measurements of predefined load setting with the noise error included. While describing the likelihood of the displacement measurement, as shown in Eq. (3), the progress of corrosion can be indirectly inferred and the information is transmitted through the particle filter framework for the prediction of future deterioration state. Furthermore, to infer the corrosion pattern of the whole surface of the member using discrete and sparsely located displacement measurement, an additional scheme that can spatially interpolate the corrosion progress surface is required to the measurement model. The subsequent section introduces a detailed description of the measurement model used in particle filter.

3 Stochastic System Parameter Estimation Given Structural Responses

In this paper, periodic structural inspections by static loading tests is concerned. The measurement state (structural response) is related to deterioration state (corrosion depth) by equilibrium equation of the structural system. In general system identification (SI) methods, the structural parameters are inferred based on available structural responses through optimization procedures employing repeated structural analysis. Instead of optimization, this research uses Bayes' theorem to compute the conditional probability of corrosion depths given structural responses.

3.1 Spectral Representation of Random Field

The spatial distribution of effective thickness, i.e. deterioration state, is described by a random field model. To parameterize the random field into a set of independent random variables, a spectral representation method, namely Karhunen–Loève expansion (KL-Expansion), is employed. This section briefly introduces KL-expansion focusing on the implementation while detailed derivation and explanation can be found from the literature (Ghanem & Spanos 2003).
Firstly, the covariance function of a random field is decomposed by

$$C(\boldsymbol{x}_1, \boldsymbol{x}_2) = \sum_{m=0}^{\infty} \lambda_m \varphi_m(\boldsymbol{x}_1)\varphi_m(\boldsymbol{x}_2) \tag{4}$$

where $\boldsymbol{x}_1$ and $\boldsymbol{x}_2$ are locations of interest, the scalar parameters $\{\lambda_i\}$ and functions $\{\varphi_i(\cdot)\}$ are obtained by solving Fredholm equation of second kind (Ghanem & Spanos 2003) :

$$\int_{\mathrm{D}} C(\boldsymbol{x}_1, \boldsymbol{x}_2)\varphi_m(\boldsymbol{x}_2)dx_2 = \lambda_m \varphi_m(\boldsymbol{x}_1) \tag{5}$$

Note that Eq. (5) is an eigenvalue problem in which λ_i is positive and finite eigenvalue and $\varphi_i(\cdot)$ is eigenfunction with the orthogonality property. Then, the random field with covariance in Eq. (4) can be parameterized in terms of infinite number of random variables (Ghanem & Spanos 2003), i.e.

$$d^{ex}(\boldsymbol{x}) = \mu_d(\boldsymbol{x}) + \sum_{m=1}^{\infty} \sqrt{\lambda_m}\varphi_m(\boldsymbol{x})\xi_m \tag{6}$$

where $\mu_d(\boldsymbol{x})$ is the mean function of the random field, i.e. a deterministic function of spatial coordinates, and $\{\xi_m\}$ are independent KL random variables that determine the stochastic property of the random field. When the original random field is Gauissian, $\{\xi_m\}$ are independent standard normal random variables. $\varphi_m(\cdot)$ and $\sqrt{\lambda_m}$ in Eq. (6) are deterministic basis function and scaling factor of random variables. Finally, using eigenfunctions with significant scaling factors only, the random field approximated in terms of a finite number of standard normal random variables, i.e.

$$d(\boldsymbol{x}) = \mu_d(\boldsymbol{x}) + \sum_{m=1}^{M} \sqrt{\lambda_m}\varphi_m(\boldsymbol{x})\xi_m \tag{7}$$

3.2 Bayesian Inference of KL Random Variables

The PDFs of KL random variables are updated given the displacement measurements. With the measurements at N_D locations $\boldsymbol{D}^{obs} = \{D_n^{obs}\}$, $n = 1, \dots, N_D$, the posterior probability of KL-random variables are,

$$p(\boldsymbol{\xi}|\boldsymbol{D}^{obs}) = c_{\xi} p(\boldsymbol{D}^{obs}|\boldsymbol{\xi})p(\boldsymbol{\xi}) \tag{8}$$

where $\boldsymbol{D}^{obs}$ denotes the measurements of displacement, $p(\boldsymbol{\xi})$ is the prior PDF, c_{ξ} denotes normalizing constant, and $L(\boldsymbol{D}^{obs}|\boldsymbol{\xi})$ is the likelihood function that is computed by structural analysis and predefined measurement error with PDF $f_v(\cdot)$, i.e.

$$L(\boldsymbol{D}^{obs}|\boldsymbol{\xi}) = f_v(\boldsymbol{D}^{FEM}(\boldsymbol{\xi}) = \boldsymbol{D}^{obs}) = \prod_{n=1}^{N_D} f_v(D_n^{FEM}(\boldsymbol{\xi}) = D_n^{obs}) \tag{9}$$

in which $\boldsymbol{D}^{FEM}(\boldsymbol{\xi})$ is the response calculated by finite element analysis. However, Eq. (8) cannot be analytically derived due to complexity in the likelihood function and the unknown normalizing constant. One of effective sampling methods used for such Bayesian updating is Markov Chain Monte Carlo (MCMC) method. In the method, the posterior PDF is directly approximated by the samples, without requiring the knowledge of normalizing constant c_{ξ}. Using MCMC, the posterior PDF of KL random variables in Eq. (7) is estimated by the samples $\{\xi_m^j\}$ and corresponding estimate on KL random variables is obtained as

$$\hat{\xi}_m = \int_{-\infty}^{\infty} p(\xi_m|\boldsymbol{D}^{obs})\,\xi_m d\xi_m \approx \frac{1}{N_\xi}\sum_{j=1}^{N_\xi} \xi_m^j \tag{10}$$

By substituting $\{\hat{\xi}_m\}$ into Eq. (7), one can obtain the expectation of random field $\hat{d}^{obs}(x)$ as deterministic function in the continuous space. The whole procedure in this section is to obtain function $h(\cdot)$ in Eq. (2), where y_k represent the measured displacments and $\boldsymbol{x}_k$ are effective thickness.

4 Pitting Corrosion Growth Model

By taking advantage of the acquired spatial distribution of effective thickness, the corrosion propagation can be evaluated for each of densely selected spatial points. The stochastic corrosion evolution model introduced in Wang *et al.* (2016) is adopted for this work. The model is formulated based on the assumptions that **a)** corrosion damage is irreversible, **b)** corrosion damage is initially dynamic but will reach stationary eventually, and **c)** corrosion initiation time and corrosion rate are finite. Considering the stochastic nature of corrosion, the logarithm of corrosion rate is described as a Brownian bridge process:

$$\ln(V(t)) = \Lambda(t) + W(t) - \frac{t - t_o}{t_f - t_o} W(t_f) \tag{11a}$$

$$W(t) \sim N(0, \sigma_o^2 t) \tag{11b}$$

where $W(t)$ is a generalized Wiener process with parameter σ_o, and drift term $\Lambda(t)$ is defined as

$$\Lambda(t) = \ln(V_o) + (\alpha - 1)\ln\left(1 + \frac{t - t_o}{\tau}\right) \tag{12}$$

The parameters t_o and t_f respectively denote corrosion initiation time and repassivation time, i.e. time when corrosion gets stationary. V_o is the initial corrosion rate and $\alpha < 1$ and τ are positive constants. In this study, corrosion model parameters $\boldsymbol{\theta} = \{\alpha, \tau, t_o, V_o, \sigma_o\}$ are estimated by particle filter based on measurements of structural responses. Corrosion depth are obtained by integrating random corrosion rate $V(t)$.

$$D(t) = \int_0^t V(s)\,ds \tag{13}$$

Finally, the effective thickness at the location can be obtained by subtracting corrosion depth from initial depth, i.e. $d(t) = d_o - D(t)$.

5 Numerical Examples

5.1 Problem Definitions and Verification of Measurement model

A steel plate element of dimensions 2.5×2.5×0.1 m subject to the pitting corrosion, is considered as a numerical example of the proposed method. As a reference corrosion time history to be estimated, a synthetic corrosion depth pattern is generated according to the growth model in Eq. (11), with randomly realized noise. In order to introduce spatial dependence of the reference corrosion depth, as evidenced in real-world phenomena, we impose spatial correlation during the Gaussian sampling of the Weiner process as described in Eq. (11b). With this presumed exact state configuration at a certain time instant, the corresponding exact displacement values of static loading test are calculated by finite element analysis, and the synthetic measurement observation sets are acquired by adding randomly generated measurement noise to the reference displacement.
Figure 1 illustrates the integrated particle filter framework proposed in this paper. The measurement model introduced in section 4 is used to convert displacement measurement into continuous function of effective thickness. The 5-year-period static loading test is assumed as inspection action following the configuration of experiment illustrated in Figure 2. The total applied load equally distributed at the top edge is 50 MN. To characterize the basis function of spatial pattern of corrosion, the auto-correlation function is defined as

$$C(\boldsymbol{x}_1, \boldsymbol{x}_2, t) = \sigma(t)^2 \exp\left(-\frac{\sqrt{|\boldsymbol{x}_1 - \boldsymbol{x}_2|^2}}{L_d}\right) \tag{14}$$

where $\sigma(t)$ is the standard deviation of spatial pattern of the effective thickness and is assumed to be proportional to the square-root of the operational time, and $L_d = 0.5\text{m}$ is predefined correlation length.

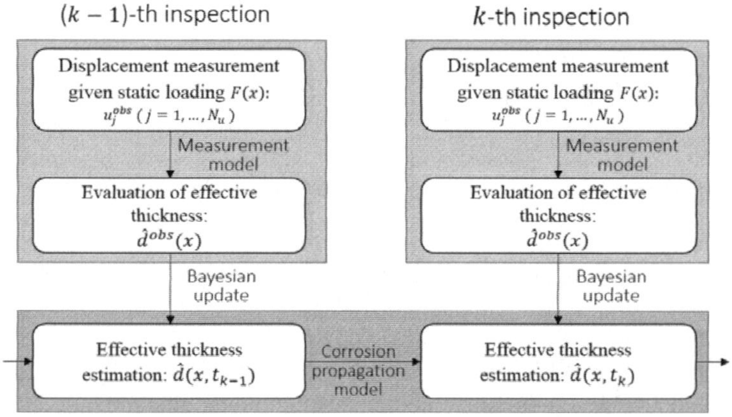

Figure 1 Procedure of particle filter framework for diagnosis and prognosis of corrosion in steel plate

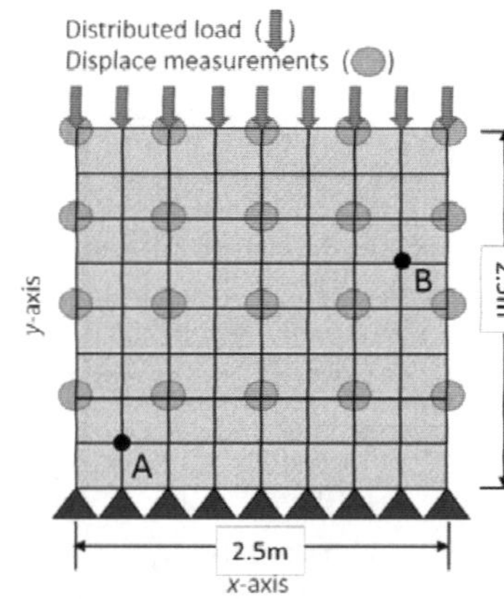

Figure 2 Schematic of periodical static loading test

Without the loss of generality, it is assumed that repeated measurements are conducted once the sensors are installed so that the white noises are eliminated, and therefore, measurement errors are considered to be trivial (relative error of 0.5%). Figure 3 shows comparison between the synthetic reference (left) and evaluated effective thickness from displacement measurements (right). MCMC is conducted to estimate the means of KL-variables of the random field. It is noted that the proposed measurement model well reconstructs the reference state of effective thickness based on measurements of structural responses.

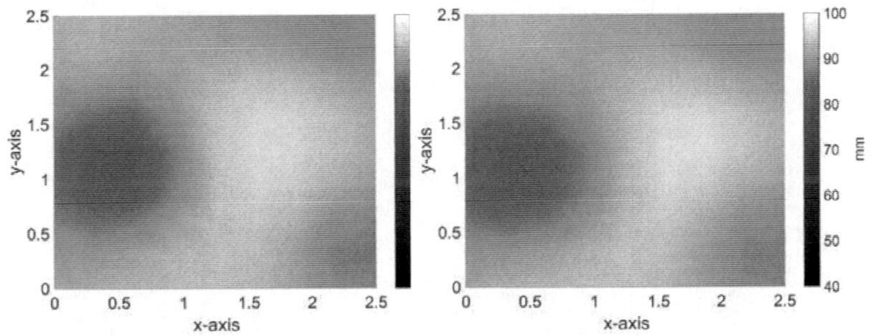

(a) Reference state ($t = 22$yr) (b) MCMC-based evaluation ($t = 22$yr)

Figure 3 Result of MCMC-based effective thickness evaluation

For the corrosion growth model, five parameters $\boldsymbol{\theta} = \{\alpha, \tau, t_o, V_o, \sigma_o\}$ in Eqs. (11) and (12) are regarded as unknown constant or dynamic parameters. Initial knowledge of σ_o is assumed to follow a lognormal PDF and others follow normal PDFs. Note that σ_o is not a direct model parameter but the statistical parameter of the PDF of a random variable $W(t)$ in Eq. (11b).

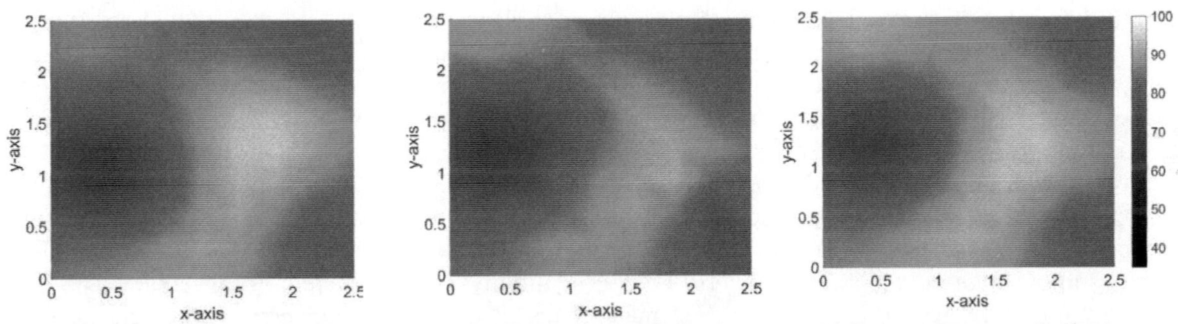

(a) Reference state (year 50)

(b) Estimated pattern of effective thickness (initially over-predicted)

(c) Estimated pattern of effective thickness (initially under-predicted)

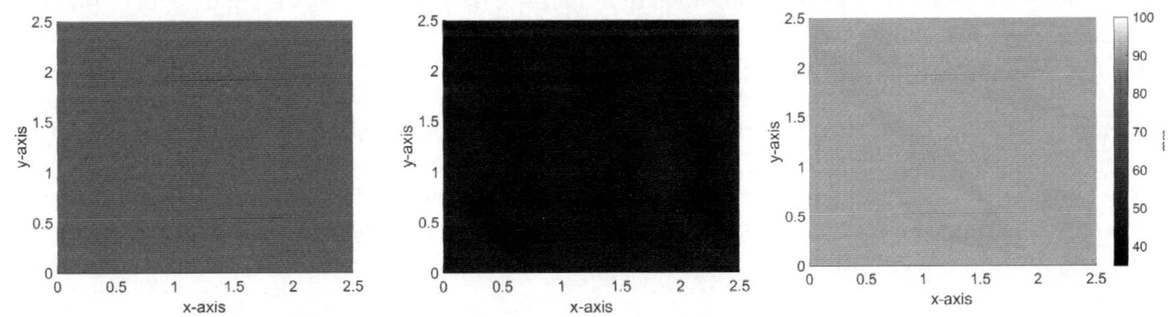

(d) Initial prediction of effective thickness of year 50 (with exact initial guess of parameters)

(e) Initial prediction of effective thickness (over-predicted)

(f) Initial prediction of effective thickness (under-predicted)

Figure 4 Prediction of effective thickness via particle filter (year 50)

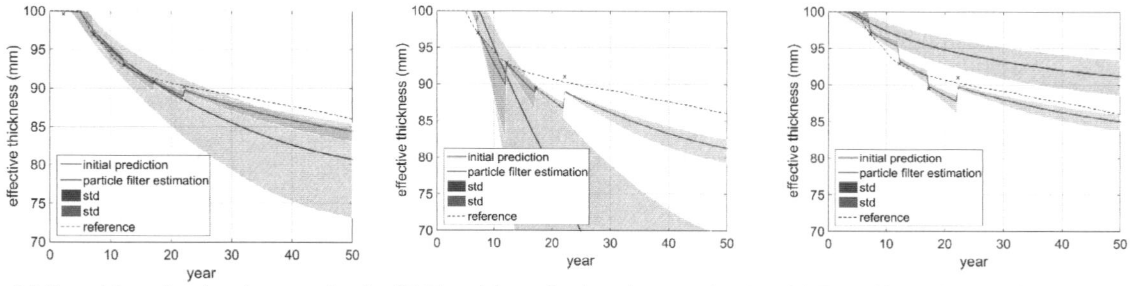

(a) Transition of estimation at point A (exact initial guess of parameters) (b) Transition of estimation at point A (initially over-predicted) (c) Transition of estimation at point A (initially under-predicted)

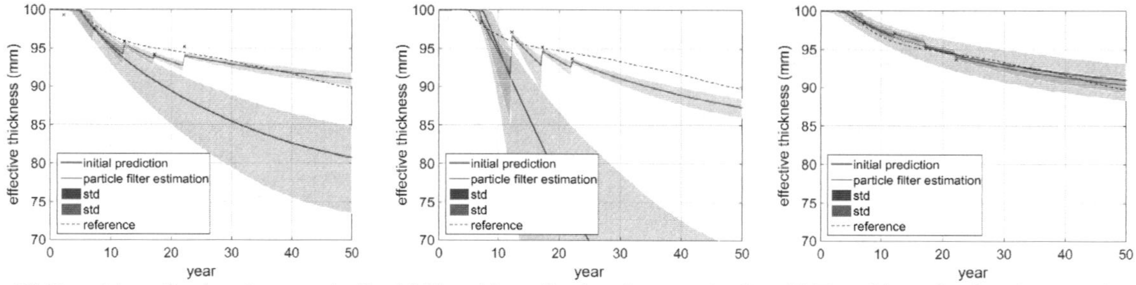

(d) Transition of estimation at point B (exact initial guess of parameters) (e) Transition of estimation at point B (initially over-predicted) (f) Transition of estimation at point B (initially under-predicted)

Figure 5 Prediction and adjustment of effective thickness via particle filter at specific locations A and B

5.3 Numerical Results and Discussions

Figure 4 shows the predicted pattern of corrosion after 50 years, for which actual model parameters in Eq. (11) are $\{\alpha, \tau, t_o, V_o, \sigma_o\} = \{0.15, 4, 5, 1.2, 0.2\}$ and are homogeneous along the plate surface. We imposed 10~40% errors on initial knowledge of each parameter. Total five measurement sets until 22-year are assimilated with stochastic propagation model. Figure 4(a) illustrates the reference deterioration pattern to be predicted. With the exact initial guess of model parameters, mean of the error, defined as absolute residual normalized by mean corrosion depth, is 2.1%, and the maximum error appears to be 7.7% near the central area. These discrepancy arises from the propagation uncertainty. Figures 4(b) and (c) illustrates the predicted deterioration state, assuming incomplete initial knowledge on model parameters. For both initially over-predicted (mean error 3.0% and maximum error 8.6%) and under-predicted (mean error 2.0% and maximum error 5.5%) deterioration rate cases, the proposed particle filter leads to the satisfactory prediction for future states. The initial prediction of effective thickness is provided in figures 4(d)-(f) to highlight the robustness of proposed method. Note that initial guess of future deterioration is homogeneous since initial assumptions on the parameters were homogeneous. Detailed time history of estimations in point A and B in the Figure 2 is displayed in Figure 5. It is shown that even in the case with initial *exact* information on model parameters are attained, the uncertainty in corrosion evolution leads to improper prediction, but proposed approach shows ability to enforce the estimations back to the track. It is noted that the speed of corrosion is also automatically adjusted by calibrating parameters, so that discrepancy rate gets lower as more updatings are conducted. On-line estimation and updating of model parameters are presented in Figure 6, uniformly selected 81 points are shown together with the reference values.

Since locations show different tendencies of corrosion transition, their parameters converge to different values. However, on average parameters seem to approach their reference values.

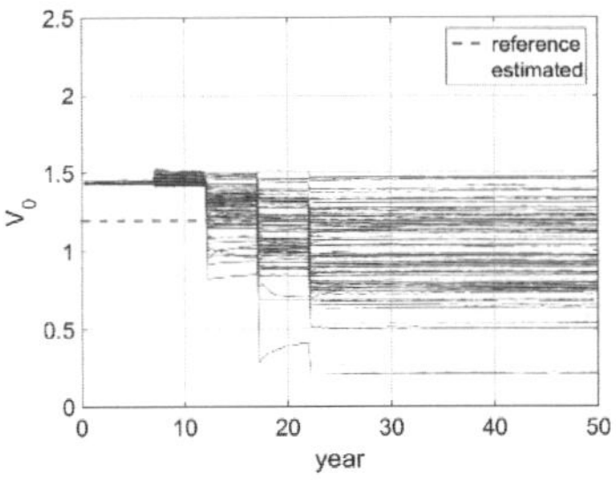

(a) parameter V_o

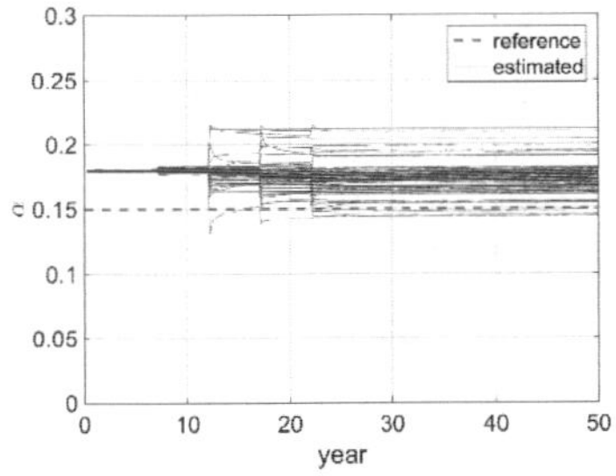

(b) parameter α

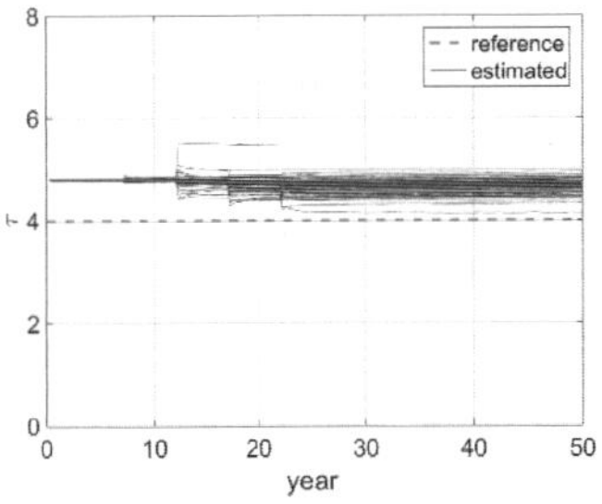

(c) parameter τ

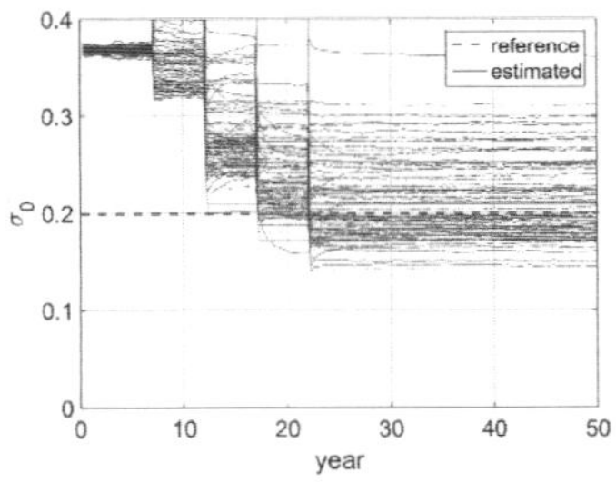

(d) parameter σ_o

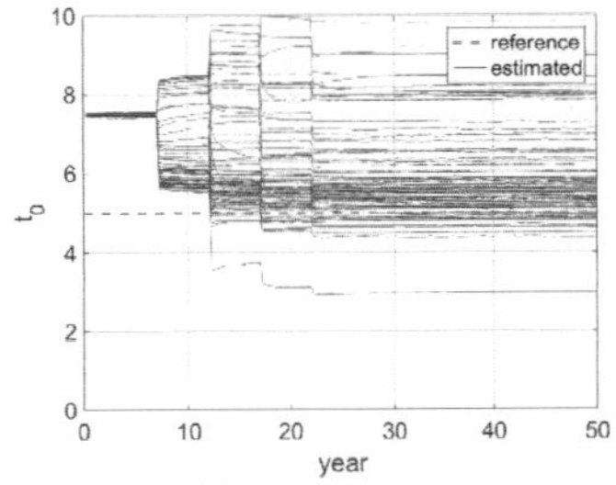

(e) parameter t_o

Figure 6 Transient of parameter estimation from 5-year-iteval measurement data (selected 81 nodes)

The next example considers a case in which the parameter V_o, which is assumed to be constant in measurement model in Eq. (12), is gradually increasing along the operational year. This is an example of situations in which the mathematical form of the model cannot provide a correct description on real-world phenomena. Figure 7 confirms that, even in this case, deterioration pattern can be predicted well while the time-varying model parameters are tracked down successfully.

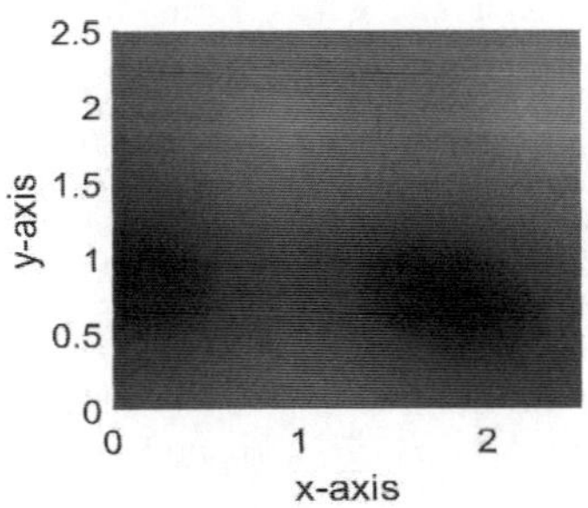

(a) Reference state

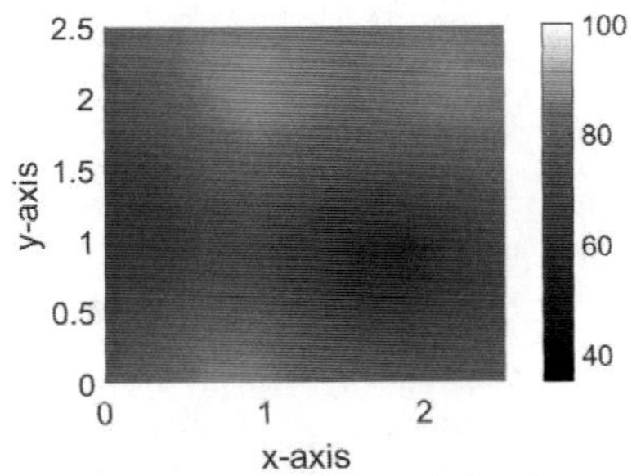

(b) Estimated pattern of effective thickness

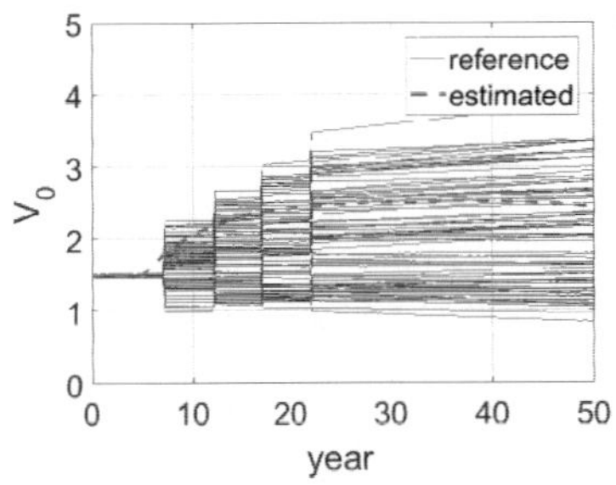

(c) Estimation of parameter V_o at selected 81 nodes

Figure 7 Estimation of effective thickness via particle filter with the time-dependent parameter

6 Conclusions

This paper presents an integrated particle filter frame work for diagnosis and prognosis of the spatial corrosion pattern of steel plate element based on noisy and indirect mechanical inspection measurement. To relate inspection data with whole-surface corrosion pattern, Bayesian theory based structural parameter estimation scheme is successfully verified and adopted. The numerical result confirmed that the particle filter estimation well tracks actual corrosion state as well as uncertain process, and shows good performance in prediction stage. Even when an initial estimation is

incorrect, the particle filter shows ability to calibrate its parameters to estimate and track the actual parameter.
This framework should be further substantiated by involving industrially applicable models. Furthermore, it can be combined with other corrosion progress models and other loading/measurement configurations. Steel plate element in this paper can be modified to predict deterioration of pipeline or other marine structures that is vulnerable to the corrosion immersion. Moreover, this method can be further augmented to continuously track the deterioration under the situation of unexpected and rapid performance degradation, such as natural disasters. The improved prediction of stochastic deterioration state can be readily combined with condition-based maintenance strategy.

Acknowledgements

The authors would like to gratefully acknowledge the support by the research project, "Development of Life-cycle Engineering Techniques and Construction Methods for Global Competitiveness Upgrades of Cable Bridges" of the Ministry of Land, Infrastructure and Transport (MOLIT) of the Korean Government (Grant No. 16SCIP-B119960-01).

References

An, D, Kim, NH & Choi, JH 2014, 'Practical options for selecting data-driven or physics-based prognostics algorithms with reviews', *Reliability Engineering & System Safety*, Vol. 133, 223-236.

Ray, A & Tangirala, S 1996, 'Stochastic modeling of fatigue crack dynamics for on-line failure prognostics', *IEEE Transactions on Control Systems Technology*, vol. 4, no. 4, pp. 443-451.

Papadimitriou, C, Fritzen, CP, Kraemer, P, & Ntotsios, E 2011, 'Fatigue predictions in entire body of metallic structures from a limited number of vibration sensors using Kalman filtering', *Structural Control and Health Monitoring*, vol. 18, no. 5, pp. 554-573.

Cadini, F, Zio, E, & Avram, D 2009, 'Monte Carlo-based filtering for fatigue crack growth estimation', *Probabilistic Engineering Mechanics*, vol. 24, no.3, pp. 367-373.

Lee, SJ, Zi, G, Mun, S, Kong, JS, & Choi, JH 2015, 'Probabilistic prognosis of fatigue crack growth for asphalt concretes', *Engineering Fracture Mechanics*, vol. 141, pp. 212-229.

Sbarufatti, C, Corbetta, M, Manes, A, & Giglio, M 2016, 'Sequential Monte-Carlo sampling based on a committee of artificial neural networks for posterior state estimation and residual lifetime prediction', *International Journal of Fatigue*, vol. 83, pp. 10-23.

Neerukatti, RK, Hensberry, K, Kovvali, N, & Chattopadhyay, A 2016, 'A novel probabilistic approach for damage localization and prognosis including temperature compensation', *Journal of Intelligent Material Systems and Structures*, vol. 27, no. 5, pp. 592-607.

Simon, D 2006, *Optimal state estimation: Kalman, H infinity, and nonlinear approaches*, John Wiley & Sons, New Jersey.

Liu, J & West, M 2001, *Combined parameter and state estimation in simulation-based filtering.* In Sequential Monte Carlo methods in practice (pp. 197-223), Springer, New York.

An, D, Choi, JH, & Kim, NH 2013, 'Prognostics 101: A tutorial for particle filter-based prognostics algorithm using Matlab', *Reliability Engineering & System Safety*, vol. 115, pp. 161-169.

Ghanem, RG & Spanos, PD 2003, 'Stochastic finite elements: a spectral approach', Courier Corporation, New York.
Wang, H, Yajima, A, Liang, RY, & Castaneda, H 2016, 'Reliability-based temporal and spatial maintenance strategy for integrity management of corroded underground pipelines', *Structure and Infrastructure Engineering*, vol. 12, no. 10, pp. 1281-1294.

The 8th International Conference on
Structural Health Monitoring of Intelligent Infrastructure
Brisbane, Australia | 5-8 December 2017

A Finite Element Model Updating Method Considering Changes of Boundary Conditions during Construction

Xixin Gao [1], Yong-feng Luo[1], Lei Wang [2], Junhua Feng [1].
[1] *Department of Structural Engineering, Tongji University, Shanghai 20092-China.*
Email: 300gaoxixin@tongji.edu.cn [2] *Tongen Civil Engineering Technology CO., Ltd., Shanghai, China*

Abstract

To ensure the safety of a large-span and complicated structures during construction, modern construction process monitoring techniques adopting the finite element (FE) model updating method are widely used in the servo-control of construction processes. The model updating technique is indispensable to the control of construction processes. During construction, boundary conditions of a structure may change substantially for several times, which can significantly influence the distribution of the member stresses and deformations of the structure. Therefore, boundary condition will be one of critical parameters for FE model updating during construction. In this paper, based on the step-by-step modeling (SSM) method for analyzing the mechanical behavior of a structure during construction, an updating method of structural FE model using static measurements considering changes of boundary conditions is proposed. The proposed method is used to update the FE model of Zhejiang Lu Bridge, in Shanghai, during the moving process. The results show that the proposed method is effective and practical.

1. Introduction

To ensure the safety of large-span and complicated structures during construction, modern construction process monitoring techniques adopting the finite element (FE) model updating method are widely used in the servo-control of construction processes (Chacon & Zorrilla 2015). In such processes, construction mechanical analysis is required and interruptible calculation for the FE model updating is requisite, which can be realized using the step-by-step modeling (SSM) method (Liu & Guo 2008). Then current finite element model updating method is not perfect and suitable for accurate construction analysis of complicated structures. Therefore, it is essential to investigate or improve the FE model updating method during construction.

Many scholars have made significant progress in the FE model updating method recently. A lot of research achievements on the modal updating method using dynamic measurements have been published (Esfandiari et al. 2009, 2010; Ren & Chen 2010; Shadan et al. 2016). Because the accuracy and reliability of measured stress and displacement are generally higher, static monitoring is widely applied to construction monitoring of large-span and complicated structures, and the research have been focused on FE model updating methods using static measurements in recent years (Sanayei et al. 1997, 2012). However, there has been no practical and perfect method for updating the FE model of a structure during construction till now.

During construction, boundary conditions of a structure may change substantially for several times. The change of boundary conditions can significantly influence the distribution of the member stresses and deformations of the structure during construction. Therefore, boundary condition is one of critical parameters for FE model updating during construction (Mottershead et al. 1996; Sanayei et al. 1999). In 2015, Zhejiang Lu Bridge on the Suzhou River, in Shanghai, was repaired. It was first toke down wholly and moved to a temporary shop for overhaul. After repair, it was re-installed to the original position over the river. In this construction process, changes of boundary conditions of the bridge occurred several times. In this paper, based on the step-by-step modeling (SSM) method, a new updating method of structural FE model using static measurements during construction， considering changes of boundary conditions, is proposed. The proposed method is adopted to update the FE model of Zhejiang Lu Bridge during the moving process. The accuracy and efficiency of the method are verified.

2. The step-by-step modeling (ssm) method

An overall construction procedure is generally divided into several stages according to the construction scheme, and a FE model can be established and calculated in each construction stage. Based on the general finite element method, the element birth and death method (ANSYS Inc. 2004) and the SSM method (Liu & Guo 2008), which can consider the nonlinearity of a structure, are the two main tools to analyze the mechanical behavior of the structure during construction. The SSM method is more suitable for construction continuous calculation because calculation interrupt for model updating can be realized for construction mechanical analysis in this method.

In the SSM method, an FE model is established in each construction stage with the sequential addition or removal of members according to the construction process, and the structural stiffness matrix is gradually formulated. In this case, modeling and solving occur in turns.

Based on the construction scheme, a construction procedure is generally divided into N construction stages. Because a steel structure is usually in the elastic state during construction, material nonlinearity is ignored in the construction simulation analysis of steel structures, and only geometric nonlinearity is considered. Geometric nonlinear problems are usually solved using incremental theory (Belytschko 2000). In this method, in the ith construction stage, the load $\Delta \boldsymbol{F}_i$ applied on the structure can be divided into several steps. In the jth load step, the load increment $\Delta \boldsymbol{F}_i^j$ and the corresponding displacement increment $\Delta \boldsymbol{u}_i^j$ are obtained. Therefore, the structural incremental equilibrium equations in this stage can be represented as

$$ {}^{t}\boldsymbol{K}_i^j \Delta \boldsymbol{u}_i^j = \Delta \boldsymbol{F}_i^j \tag{1} $$

$$ {}^{t}\boldsymbol{K}_i^j = {}^{0}\boldsymbol{K}_i^j + {}^{L}\boldsymbol{K}_i^j + {}^{\sigma}\boldsymbol{K}_i^j \tag{2} $$

where ${}^{t}\boldsymbol{K}_i^j$ represents the tangent stiffness matrix at the jth load step in the ith construction stage, consisting of the elastic stiffness matrix ${}^{0}\boldsymbol{K}_i^j$, the displacement stiffness matrix ${}^{L}\boldsymbol{K}_i^j$ and the stress stiffness matrix ${}^{\sigma}\boldsymbol{K}_i^j$.

The Newton-Raphson iteration algorithm (Belytschko, 2000) is adopted to solve Eq. (1), and the nodal displacement vector at the ending moment in the ith stage $\Delta \boldsymbol{u}_i$ and the stress increment $\Delta \boldsymbol{\sigma}_i$ can be obtained. Finally, the nodal displacement vector $\boldsymbol{u}_i$ and the stress vector $\boldsymbol{\sigma}_i$ at the ending moment of the ith stage are obtained from Eq. (3) and Eq. (4).

$$ \boldsymbol{u}_i = \boldsymbol{u}_{i-1} + \Delta \boldsymbol{u}_i \tag{3} $$

$$ \boldsymbol{\sigma}_i = \boldsymbol{\sigma}_{i-1} + \Delta \boldsymbol{\sigma}_i \tag{4} $$

Furthermore, the structural tangent stiffness matrix is obtained and stored to modify the structural initial stiffness matrix in the $i+1^{\text{th}}$ construction stage.

According to the predetermined construction sequence, the mechanical behavior analysis of a steel structure in each construction stage can be conducted according to the above processes. Eventually, the mechanical behavior analysis of the steel structure during construction is completed.

3. The FE MODEL updating method considering changes of boundary condition during construction

A system of structural monitoring during construction is composed of the front-end monitoring system and the back-end monitoring system, as shown in Figure 1. The model updating module, in which some critical parameters in a FE model are reasonably updated according to monitoring data from the front-end monitoring system, belongs to the back-end monitoring system.

During a process of model updating in a construction stage, monitoring data are introduced in the original structural model, and critical parameters are updated. Afterwards, the mechanical behavior of the structure in this stage is analyzed using the updated model. Finally, the structural model in the next stage is established and analyzed according to the updated model in the current stage.

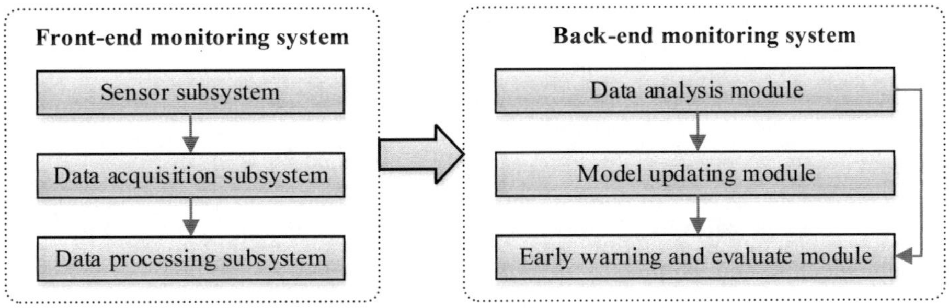

Figure.1 The system of structural monitoring during construction

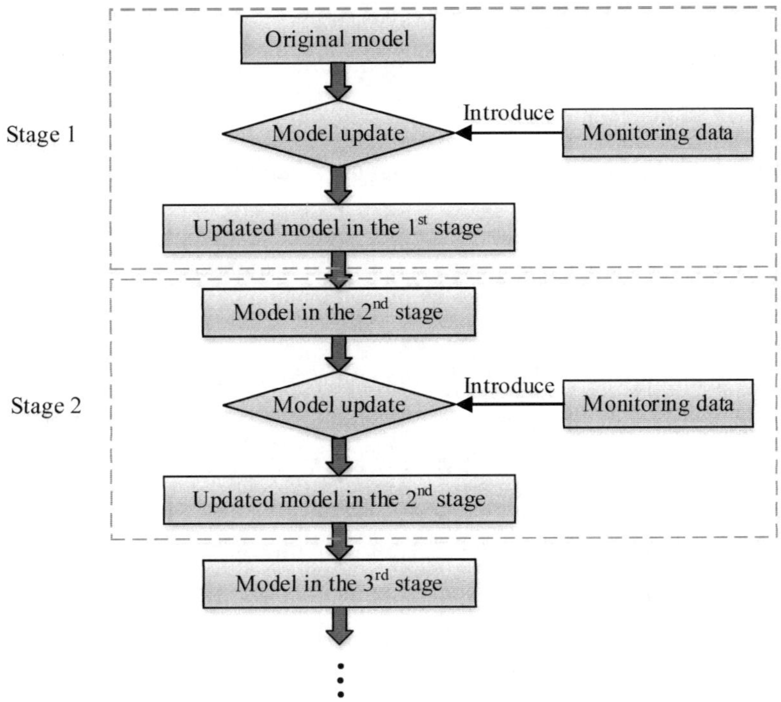

Figure.2 The process of model updating during construction

In a FE model, support conditions are generally simplified to hinged supports or fixed supports, which cannot consider the actual stiffness of the supports. To simulate the support conditions more accurately, elastic supports are usually used in the mechanical behavior analysis of a structure during construction. However, the stiffness of the elastic supports is indeterminate, because the range of the support joint region is uncertain and the support joint construction is complicated. The uncertainty of the stiffness of the elastic supports substantially influences the distribution of the member stresses and deformations. Therefore, the stiffness of the elastic supports is a critical parameter of the FE model updating during construction.

3.1. Realization of elastic support in the FE model

Suppose the stiffness of the support, corresponding with the component of the ith nodal displacement, u_i, is k_i, and the number of DOFs on elastic supports is m. Therefore, the stiffness matrix including elastic supports $\boldsymbol{K}_d$ is expressed as shown in Eq. (5)

$$\boldsymbol{K}_d = \begin{bmatrix} k_1 & & & & \\ & \ddots & & & \\ & & k_i & & \\ & & & \ddots & \\ & & & & k_m \end{bmatrix} \tag{5}$$

Considering the boundary conditions, the basic force-displacement relationship of the FE model can be expressed as block matrix

$$\begin{bmatrix} \boldsymbol{K}_{aa} + \boldsymbol{K}_d & \boldsymbol{K}_{ab} \\ \boldsymbol{K}_{ba} & \boldsymbol{K}_{bb} \end{bmatrix} \begin{Bmatrix} \boldsymbol{u}_a \\ \boldsymbol{u}_b \end{Bmatrix} = \begin{Bmatrix} \boldsymbol{F}_a \\ \boldsymbol{F}_b \end{Bmatrix} \tag{6}$$

where $\boldsymbol{u}_a$ and $\boldsymbol{u}_b$ are boundary nodal displacement vector and non-boundary nodal displacement vector respectively, and $\boldsymbol{u}_a$ is known and $\boldsymbol{u}_b$ is unknown; $\boldsymbol{F}_a$ and $\boldsymbol{F}_b$ are the unknown and known nodal loads vectors respectively.

3.2. The FE model updating method considering changes of boundary conditions during construction

At the beginning of the ith construction stage, the structural incremental equilibrium equation can be represented as

$$\boldsymbol{K}_i^0 \Delta \boldsymbol{u}_i = \Delta \boldsymbol{F}_i \tag{7}$$

where $\Delta \boldsymbol{F}_i$ is the load increment applied on the structure, and $\Delta \boldsymbol{u}_i$ is the corresponding displacement increment; the initial structural stiffness matrix $\boldsymbol{K}_i^0$ can be expressed as block matrices in Eq. (8)

$$\boldsymbol{K}_i^0 = \begin{bmatrix} \boldsymbol{K}_{aa} + \boldsymbol{K}_d & \boldsymbol{K}_{ab} \\ \boldsymbol{K}_{ba} & \boldsymbol{K}_{bb} \end{bmatrix} \tag{8}$$

Solve the incremental equilibrium equation of the structure by Newton-Raphson method, and the structural tangent stiffness matrix $\boldsymbol{K}_i$ is obtained as shown in Eq. (9)

$$\boldsymbol{K}_i = \begin{bmatrix} \boldsymbol{K}'_{aa} + \boldsymbol{K}_d & \boldsymbol{K}'_{ab} \\ \boldsymbol{K}'_{ba} & \boldsymbol{K}'_{bb} \end{bmatrix} \tag{9}$$

Establish the objective function of model updating

$$F(X) = (FEM(X) - TEST)^2 \tag{10}$$

where $F(X)$ is an objective function which reflects the difference between the static responses obtained by calculating the FE model of the structure and the measured responses; X is a FE model updating parameter set; *TEST* is the measured responses of static monitoring and *F*(*X*) is the static responses obtained by calculating the FE model of the structure corresponding to *TEST*.

Many FE model updating problems are multi-objective optimization problems, the objective function of which can expressed as the vector form

$$F(X) = (F(X)_1^*, F(X)_2^*, \ldots, F(X)_m^*)^{\mathrm{T}} \tag{11}$$

where $F(X)_i^*$ is the ith sub-objective function.

Therefore, the mathematical model of the FE model updating method using static measurement can be shown as Eq. (12)

$$\begin{cases} \min F(X) = (F(X)_1^*, F(X)_2^*, \ldots, F(X)_m^*)^{\mathrm{T}} \\ \text{s.t. } X_l \le X \le X_u \end{cases} \tag{12}$$

where X_l and X_u are the lower and upper limit of the FE model updating parameters.

In general, indirect algorithm, such as linear weighting method and min-max optimization, can be used to solve the multi-objective optimization problem. In this paper, the linear weighting method is adopted to solve the objective optimization problem, and the aforementioned mathematical model is translated to the following form

$$\begin{cases} \min U(X) = \sum_{i=1}^{n} \lambda_i \cdot F(X)_i^* \\ \text{s.t. } \mathrm{X}_l \le X \le X_u \end{cases} \tag{13}$$

where λ_i is the weight of the ith sub-objective function $F(X)_i^*$.

In this paper, only the stiffness of elastic supports is the parameter of model updating during construction. Therefore, the stiffness on each direction of the support has equal weight, and the weight of the each direction on the ith support is represented as

$$\lambda_i = \frac{1}{m} (i = 1, 2, \ldots, m) \tag{14}$$

Afterwards, appropriate optimization algorithms are selected to solve he single objective optimization problem. After solving the optimization problem, the optimal solutions to the stiffness of supports can be obtained. Moreover, the optimal solutions are plugged into the structural incremental equilibrium equation. Solve the incremental equilibrium equation of the structure by Newton-Raphson method, and the stresses and displacements in the ith construction stage are obtained. Finally, save the structure stiffness matrix at the final instant of the ith construction stage for the model updating of the i+1th construction stage.

3.3. Analysis process of the FE model updating method considering changes of boundary conditions during construction

The flow chart of the FE model updating method considering changes of boundary conditions is proposed in this paper and shown in Figure.3.

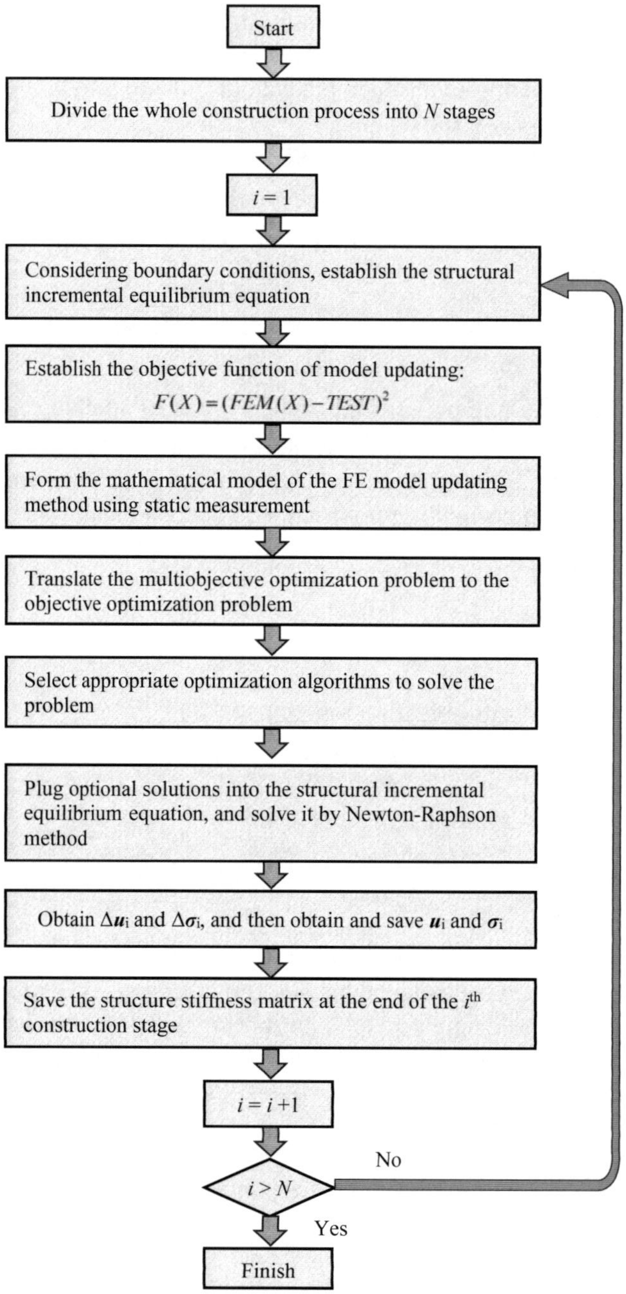

Figure.3 Flow chart of the FE model updating method

4. Analysis of engineering project

4.1. Project profile

Zhejiang Lu Bridge is located on the Suzhou River in Shanghai, which was constructed in 1908. The bridge has a fish-bellied steel truss structure system with the span of 59.74m, which was overhauled in 2015 in order to prolong the service life. The distance between two fish-bellied trusses is 9.03m, and the cantilever outward from each truss for sidewalk is 4.96m. Because the bridge cannot be overhauled on the river smoothly, the entire bridge had to be moved to a temporary shop on the north shore for repair, and back to the original position on the river after overhauled. In this paper, the FE model updating of the bridge during the process of moving to the original positon on the river was adopted for construction analysis (Figure.4).

The members of the FE model, as shown in Figure.5, are simulated by spatial beam elements, and each node of an element has three translational DOFs and three rotational DOFs. A uniformly distributed load of $q = 1.7\text{kN}\cdot\text{m}^{-2}$ which is caused by concrete bridge deck pavement and a liner load of $p = 1.38\text{kN}\cdot\text{m}^{-1}$ which is caused by the handrails are considered. The structural weight is also considered.

Figure.4 Zhejiang Lu Bridge transporting to the original position on the river

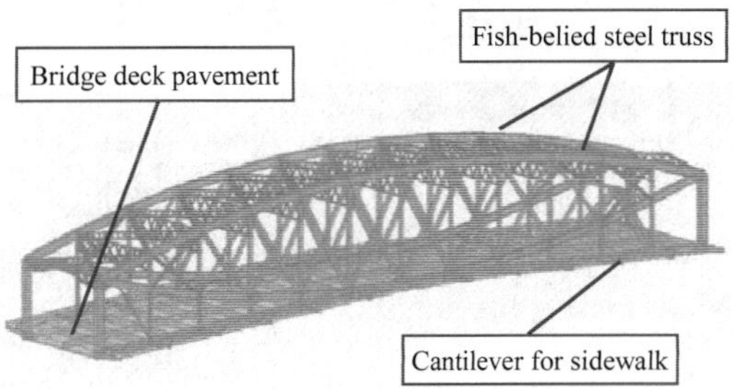

Figure.5 The FE model of the bridge

4.2. Construction scheme

The overall moving process to the original position on the river was divided into 6 stages, and the positions and numbers of the supports in all stages are shown in Figure.6. In this process, boundary conditions changed among stages, and the FE models in any stages had different boundary conditions. As shown in Figure.7, the slings on crawler crane, hinged supports on the caisson and on the buoyancy tank served as the supports in different stages respectively. It can be seen that the structure of actual supports are complex, and the stiffness of the supports was uncertain. Therefore, the stiffness of the supports was the critical parameter of the FE model updating during the transporting process.

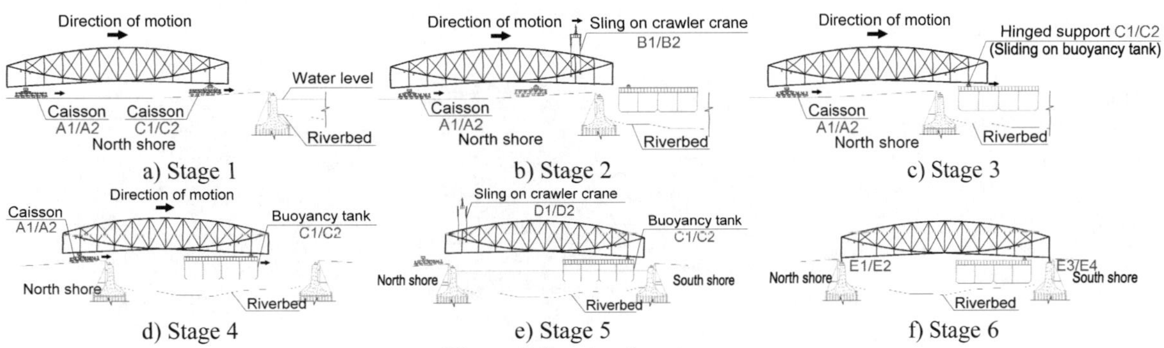

Figure.6 Construction stages

a) Slings on crawler crane b) Hinged support on the caisson c) Hinged support on buoyancy tank

Figure.7 Actual supports

4.3. Monitoring tactics

During the moving process of the bridge, changes of boundary conditions occurred in each stage, which lead to the redistribution of the member stresses and deformations of the bridge. So it is essential to monitor the stresses and displacements of the bridge during the moving process.

In this section, only stresses are used to construct the optimization object function. The monitoring points for stresses were arranged on (Luo 2014): 1) the members on which the values of stresses are the largest in all stages; 2) the members on which the values of stress variation are large; 3) the members on the significant positions such as members near supports. The layout of the stress monitoring points is shown in Figure.8. Vibrational chord strain gauges were used to measure the strain of members, and some pictures about the stress monitoring points on site are shown in Figure.9.

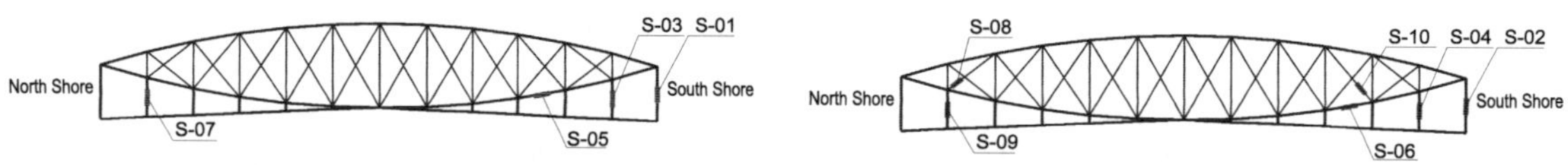

a) The truss on the east site b) The truss on the west site

Figure.8 Layouts of stress monitoring points

a) On the chord member b) On the column c) On the end column

Figure.9 Monitoring points of stress

4.4. FE model updating

The FE model updating method, proposed in this paper, using static measurements considering changes of boundary conditions was used to update the model of the Zhejiang Lu Bridge. In the original FE model of the bridge, the stiffness of all supports are infinite. The first step method was used for optimization. Table.1 and Table.2 shows the theoretical and measured values of key monitoring points before and after updating the model.

Table.1 Stress values of key monitoring points before updating models

		Moving stages											
		1		2		3		4		5		6	
		Value/ Mpa	Error/%	Value/ Mpa	Error/%	Value/ Mpa	Error/%	Value/ Mpa	Error/%	Value/ Mpa	Error/%	Value/ Mpa	Error/%
S-03	Measured	-54.8	107.72	19.97	34.40	-40.33	182.25	-51.07	122.89	-52.4	117.23	14.68	10.76
	Theoretical	-113.83		13.1		-113.83		-113.83		-113.83		13.1	
S-06	Measured	37.47	44.09	18.09	93.48	32.08	34.69	33.02	36.55	33.93	38.26	44.92	6.97
	Theoretical	20.95		1.18		20.95		20.95		20.95		48.05	
S-10	Measured	-12.51	63.87	-37.62	39.82	-16.98	20.73	-12.08	69.70	-20.1	1.99	-2.08	95.19
	Theoretical	-20.5		-52.6		-20.5		-20.5		-20.5		-0.1	

Table.2 Stress values of key monitoring points after updating models

		Moving stages											
		1		2		3		4		5		6	
		Value/ Mpa	Error/%	Value/ Mpa	Error/%	Value/ Mpa	Error/%	Value/ Mpa	Error/%	Value/ Mpa	Error/%	Value/ Mpa	Error/%
S-03	Measured	-54.8	33.21	19.97	8.26	-40.33	66.43	-51.07	37.91	-52.4	45.86	14.68	3.95
	Updated	-73		18.32		-67.12		-70.43		-76.43		14.1	
S-06	Measured	37.47	27.94	18.09	56.44	32.08	22.23	33.02	33.07	33.93	34.87	44.92	4.85
	Updated	27		7.88		24.95		22.1		22.1		47.1	
S-10	Measured	-12.51	46.28	-37.62	26.79	-16.98	12.19	-12.08	47.93	-20.1	1.49	-2.08	84.13
	Updated	-18.3		-47.7		-19.05		-17.87		-20.4		-0.33	

As shown in Table.1 and Table.2, several conclusions can be drawn:

1) The relative errors between the theoretical and measured values on the monitoring point S-03 are large in the 1st, 3rd, 4th and 5th moving stage. After the FE model updating, the errors are smaller than the original results. In particular, the maximum variation of the relative error from 182.25% to 66.43% occurred in the 3rd stage. The reason is that S-03 is on the member nearby the support C1 in these stage. The updating for the stiffness of supports obviously influences the updating values of S-03.

2) The effects of the FE model updating on the monitoring point S-06 and S-10 are not obvious. The maximum variation of the relative error from 93.48% to 56.44% occurred on the point S-06 in the 2nd stage. The reason is that these points are far from the supports. Moreover, only one parameter with the stiffness of supports, which has little influence on the updating results of these points, is used to update the model in this project.

5. Conclusions

Based on the SSM method, the FE model updating method using static measurements, considering changes of boundary conditions, was proposed in this paper. The proposed method was used to update the FE model of Zhejiang Lu Bridge, in Shanghai, during the moving process, which verified the effectiveness and validity of the proposed method. The FE model updating results show that the scope of the relative errors on some monitoring points are relatively large. The main reason is that other critical model updating parameters such as section size and joint stiffness are ignored in this paper.

The future research will mainly focus on the FE model updating method with multiple parameters, considering the aforementioned parameters, to reduce the relative errors of the static responses.

References

Ahmadian H, Gladwell GML & Ismail F 1997, 'Parameter selection strategies in finite element model updating', *Journal of Vibration and Acoustics-Transactions of the Asme*, vol.119, no.1, pp.37-45.

ANSYS Inc. 2004, 'Release 10.0 documentation for ANSYS', Canonsburg: ANSYS Inc.

Belytschko, Liu T, Moran WK, et al. 2000, 'Nonlinear Finite Element Analysis for Continua and Structures'

Chacon R & Zorrilla R 2015, 'Structural Health Monitoring in Incrementally Launched Steel Bridges: Patch Loading Phenomena Modeling', *Automation in Construction*, vol. 58, pp.60-73.

Esfandiari A, Bakhtiari-Nejad F, Rahai A, et al. 2009, 'Structural model updating using frequency response function and quasi-linear sensitivity equation', *Journal of Sound and Vibration*, vol.326, no.3-5, pp.557-573.

Liu XW and Guo YL 2008, 'Construction mechanics analytical procedures for steel structures in view of the geometric nonlinearity', *Journal of Xi'an University of Architecture &Technology (Natural Science Edition)*, vol.40, no.2, pp.161-169.

Luo YF, Ye ZW, Chen XM, et al. 2014, 'Research on key parameter selection and monitoring point arrangement in construction monitoring of spatial steel structures' *Journal of Building Structures*,vol.35, pp.108-115.

Mottershead JE, Friswell MI, Ng GHT, et al. 1996, 'Geometric parameters for finite element model updating of joints and constraints', *Mechanical Systems and Signal Processing*, vol.10, no.2, pp.171-182.

Ren WX & Chen HB 2010, 'Finite element model updating in structural dynamics by using the response surface method', *Engineering Structures*, vol.32, no.8, pp.2455-2465.

Shadan F, Khoshnoudian F & Esfandiari A 2016, 'A frequency response-based structural damage identification using model updating method', *Structural Control & Health Monitoring*, vol.23, no.2, pp.286-302.

Sanayei M, Imbaro GR, McClain JAS, et al. 1997, 'Structural model updating using experimental static measurements', *Journal of Structural Engineering-Asce*, vol.123, no.6, pp.792-798.

Sanayei M, McClain JAS, Wadia-Fascetti S, et al. 1999, 'Parameter estimation incorporating modal data and boundary conditions', *Journal of Structural Engineering-Asce*. vol.125, no.9, pp.1048-1055.

Sanayei M, Phelps JE, Sipple JD, et al. 2012, 'Instrumentation, Nondestructive Testing, and Finite-Element Model Updating for Bridge Evaluation Using Strain Measurements', *Journal of Bridge Engineering*. vol.17, no.1, pp.130-138.

The 8th International Conference on
Structural Health Monitoring of Intelligent Infrastructure
Brisbane, Australia | 5-8 December 2017

Resilience assessment of high damping rubber bearings in "beyond-design" conditions

M. Domaneschi[1], L. Martinelli [2], G.P. Cimellaro [3]
[1] Politecnico di Torino, Department of Structural, Geotechnical and Building Engineering, Turin IT, Email: marco.domaneschi@polito.it [2] Politecnico di Milano, Department of Civil and Environmental Engineering, Milan IT,[3] Politecnico di Torino, Department of Structural, Geotechnical and Building Engineering IT

Abstract

Passive isolation systems are an established solution for the design of civil engineering structures that are required to provide superior performances in the case of a seismic event. Although their application to the seismic protection of special structures as nuclear buildings is currently limited, isolation systems are likely to become more widespread in the design of strategic infrastructures and facilities. In this work numerical investigations on the ultimate limit state conditions of filled high damping rubber bearings under cyclic shear loading are presented, focusing on the influence of the axial load with respect to the device.

1. Introduction

In the recent past, passive isolation systems have been largely used as a valuable earthquake-resistant strategy in the design of civil engineering structures, particularly with reference to the large class of ordinary buildings. Indeed, passive, anti-seismic, systems have already been used to protect more than 23000 structures such as bridges and buildings, both existing and of new construction, in more than 30 countries. However, their introduction for the seismic protection of strategic structures, such as nuclear power plants, is not as widespread as their use for buildings (Basu et al 2014).
Nevertheless, seismic isolation systems are likely to become an established highly reliable solution in the design of special structures and facilities for which superior performances are needed. In particular, reliability and functionality after a seismic event are of utmost importance (De Grandis et al 2009, Perotti et al 2013).
The anticipated better performance of isolated structures consists in reducing the values of the seismic acceleration transmitted to the superstructure. This attenuation is obtained at the price of larger relative displacements (Domaneschi 2010).
Where isolation devices are installed they are likely to become the most critical components in a seismic assessment. Thus, all factors affecting the devices performance have to be carefully considered. A reliable and close to reality evaluation of the constitutive behaviour, up to the device limit state, need to be fully addressed (Perotti et al 2013).
Failure of isolators is obviously related to "beyond design" loading conditions. They consist in extreme seismic events which lead to amplified multi-directional loading paths. Such extreme circumstances cannot be excluded during the structure's life and their study requires reliable constitutive models of the devices (Perotti et al 2013).
Finite element (FE) modelling of seismic isolation devices and the evaluation of their performance for moderate, to large, shear strain values has been successfully accomplished. However, for very

large values of the shear strain cyclic constitutive models are however still lacking (Domaneschi et al 2015).
This research is devoted to the implementation and validation of special hyper-elastic-dissipative material models for the phenomenological interpretation of rubber bearings' response for very large shear strains. Since a variation of the vertical load carried by the devices has to be expected (Warn and Whittaker 2008), the isolators models are tested in the case of different values of the axial force. Finally, the influence of the device shape is also highlighted.
This research can give a contribution to the assessment of isolators reliability in "beyond design" conditions. Therefore, it is connected to the view of resilience for civil structures and infrastructures (Cimellaro et al. 2010).

2. Physical and numerical models of the isolator device

In the class of isolator devices, a basic distinction is based on shape, materials and dimensions. The most commonly adopted solution in the world is represented by the steel-laminated rubber bearings, especially high damping rubber bearings (HDRB) and the lead rubber bearings (LRB) for which the largest experience has been gathered and detailed literature is provided (see e.g. Basu et al. 2014). A schematic geometry of an elastomeric isolator is depicted in Fig. 1 with a physical example.

Figure 1. Physical example of the isolator device.

The relevant role fulfilled by seismic isolation devices in seismic engineering is associated to their reliability in ensuring large deformations in the horizontal plane and providing high stiffness in the vertical direction. These effects are obtained through the low compressibility of rubber and the introduction of thin reinforcing steel plates. Therefore, steel-laminated rubber bearings represent the most commonly adopted solution in the world (Basu et al 2014).
The failure of the isolators in the case of very strong seismic events is related to extreme multi-directional loading paths. To study the device behaviour under these circumstances, FE models employing the overlapping discretization are a promising approach. This technique is based on the concept that the material behaviour can be split into a number of parallel fractions, each one with conventional mechanical properties and hence available constitutive laws (hyper-elastic, elastic-plastic and viscoelastic). The global behaviour is achieved by a suitable superposition of simple constitutive models and adequate material parameters. In this way, robust FE models for HDRB devices can be developed (Milani & Milani 20112, Gracia et al 2010), on the basis of overlapped constitutive laws and fitting of standard laboratory test data.
The translation of this concept in FE analyses requires the generation of a numerical model with several identical overlaid meshes, sharing the same nodes, each one presenting a different constitutive law. For these devices, an adequate mathematical model should not only reproduce the basic elastic behaviour of the rubber, but also account for dissipative properties within a large deformation range. Damping factors for a HDRB ranges from 10% to 15%, while shear modulus (G) lies in the 0.8-1.4 MPa range (Perotti et al 2013).

Table 1. Properties of the first HDRB device modelled.

Diameter	1350 mm
Total rubber thickness	256 mm
Thickness of each rubber layer	16 mm
Number of steel layers	15
Shape factor	19.7
Rubber lateral cover thickness	25 mm
Rubber shear modulus	1.4 MPa
Horizontal stiffness (100%)	7.82 kN/mm
Vertical stiffness	5792 kN/mm

Table 2. Properties of the device 601 in (Nagarajaiah & Ferrell 1999).

Diameter	10 in
Total rubber thickness	2.00 in
Thickness of each rubber layer	0.25 in
Number of rubber layers	8
Rubber shear modulus	1.4 MPa

Table 3. Properties of device 302 in (Nagarajaiah & Ferrell 1999).

Diameter	5 in
Total rubber thickness	2.00 in
Thickness of each rubber layer	0.25 in
Number of rubber layers	8
Rubber shear modulus	1.4 MPa

Furthermore, the extreme value of the axial load can be achieved at the maximum relative horizontal displacement that is usually related to the dimension of the device. To study this effect, extreme possible geometries of the isolation device have been selected in the FE simulations.

Three FE models of circular HDRBs (Fig. 2) have been developed within the MARC® FE code, according to the geometrical properties listed in Tables 1-3. The characteristics of the first HDRB device closely reproduce the prototypes designed and tested for the ELSY reactor within SILER project (2013). The remaining FE models in Fig. 2e and Fig 2f reproduce the geometries of bearings 601 and 302 in (Nagarajaiah & Ferrell 1999).

The FE mesh for the first (SILER) device has been prepared considering symmetry of geometry and boundary conditions. In the end, it consists in about 35000 isoparametric hexahedral solid elements and 25000 nodes. According to the overlapping simulation approach, rubber layers are composed by two overlapping meshes. The first one accounts for the hyper-elastic behaviour while the other one the elastic-plastic behaviour.

The models from (Nagarajaiah & Ferrell 1999) have a mesh consistent with the first one. It is obtained by a geometric scaling operation without changing the number of elements and nodes.

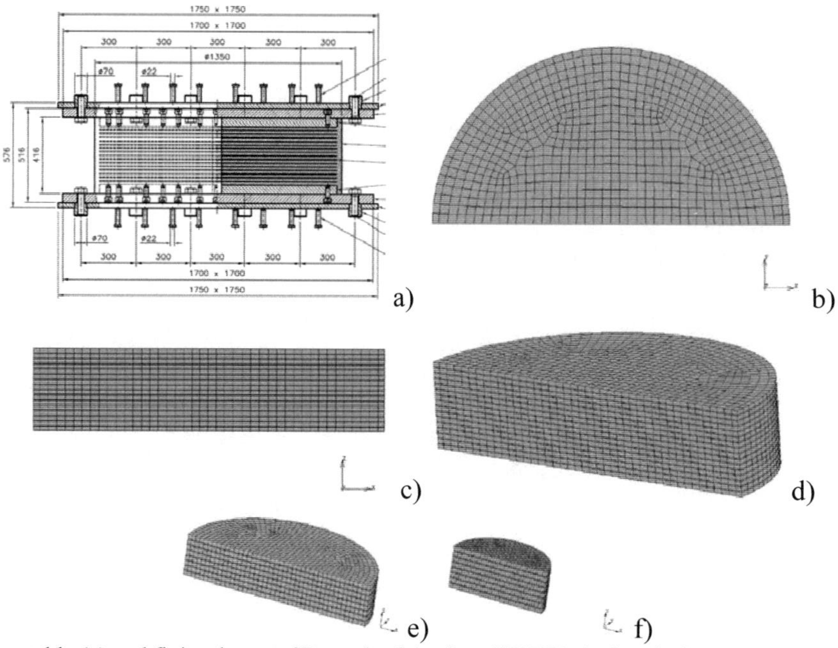

Figure 2. Design table (a) and finite element 3D mesh of the first (SILER) device (b-d). FE mesh of the devices 601 and 302 (e-f) respectively in (Nagarajaiah & Ferrell 1999).

For all FE models, the hyper-elastic behaviour is simulated according to a three-constants Mooney-Rivlin model (Yeoh model). The material parameters are assumed to be equal to those derived in (Gracia et al 2010). They have been identified through laboratory tests on HDRBs under cyclic relative displacements and constant axial force up to the specimen failure. Their corresponding values are listed in Table 4 and are also in agreement with the those independently identified in (Bianchi et al 2011a, 2011b) for similar conditions.

Table 4. Value of the constants in the Mooney-Rivlin material model.

C10	4.989×10^{-1}
C20	3.47×10^{-2}
C30	2.28×10^{-3}

The elastic-plastic behaviour assigned to the second mesh is derived from the von Mises' theory. It occupies the same volume of the elastomeric material and coincides with the previous first mesh with the Mooney-Rivlin constitutive law. The preliminary fitting of the experimental outcomes have been processed assuming a perfectly elastic-plastic behaviour defined by values in Table 5.

Table 5. Material parameters for the elastic-plastic material model.

Young modulus E (MPa)	11.28
Poisson's ratio ν	0.00
Yielding stress f_y (MPa)	0.49

Finally, a non-linear elastic perfectly plastic material (Young modulus 210000 MPa, Poisson's ratio 0.3, yielding 300 MPa) has been assigned to the steel layers.
The boundary conditions on the FE model consist in fixed displacements at the device base. Uniaxial horizontal displacement (100% of rubber thickness) at the upper isolator surface is imposed with constant vertical load. MARC® type 7 3D, eight-node, first-order, isoparametric

elements have been used. Fig. 3 depicts the comparison between laboratory data and FE model simulation.

3. Numerical results

The loading conditions herein presented aim to analyse the influence of the axial force on the hysteresis cycles. The material parameters have been identified on the base of an experimental uniaxial horizontal test. Then, different vertical loads have been applied on the finite element model of the device. Furthermore, the limit condition at failure (delamination mode), as reproduced in laboratory tests, has been simulated by FE analyses in order to identify the corresponding stress state. This development are based on numerical analysis on FE models of the device carried out within the MARC® FE code.

Initial validation of the adopted material values has been performed with reference to data from the physical tests of the isolator (SILER 2013). The boundary conditions in the FE model consist in fixed displacements at the device base, imposed uniaxial horizontal displacement (100% of rubber thickness) at the upper isolator surface, constant vertical load at the top. Fig. 3 depicts the comparison between the FE model simulation and laboratory data.

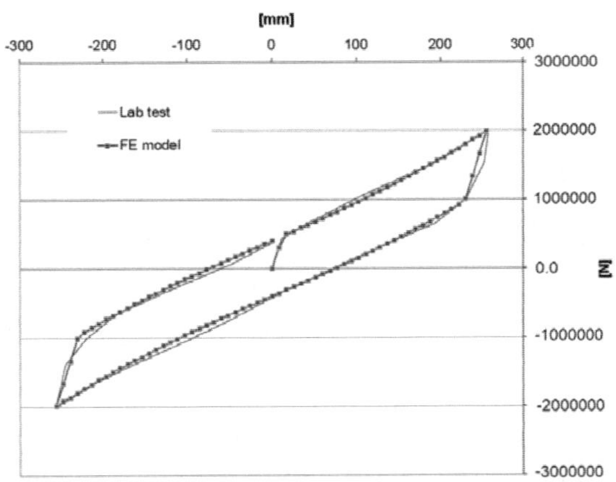

Figure 3. Comparison between laboratory tests hysteresis cycle (100% - 256 mm - horizontal uniaxial displacement) at constant axial load (8075 kN, halved due to symmetry) and FE simulation. SILER isolator.

The validated FE model has been exploited for evaluating the influence of axial load variations on the performance of the device. The analyses take into account mechanical and geometric non-linearities in order to study possible beyond design conditions.

At first, the same loading path of Fig. 3 has been reproduced with double (16500 kN) and triple axial force (24800 kN). Furthermore, zero axial force is also investigated. As it can be inferred from Fig. 4a there is no influence of this static parameters at the adopted shear deformation values.

The same axial forces have been used to study the device with monotonically increasing horizontal displacements. Figures 4b and 4c report the loading paths in terms of force–displacements at 300% and 600% of shear strain, respectively.

All the responses appear equivalent, although for the horizontal displacements at 600% shear strain minor differences among the curves can be detected. For the same 600% condition the analysis has not been completed due to excessive deformation in some finite elements.

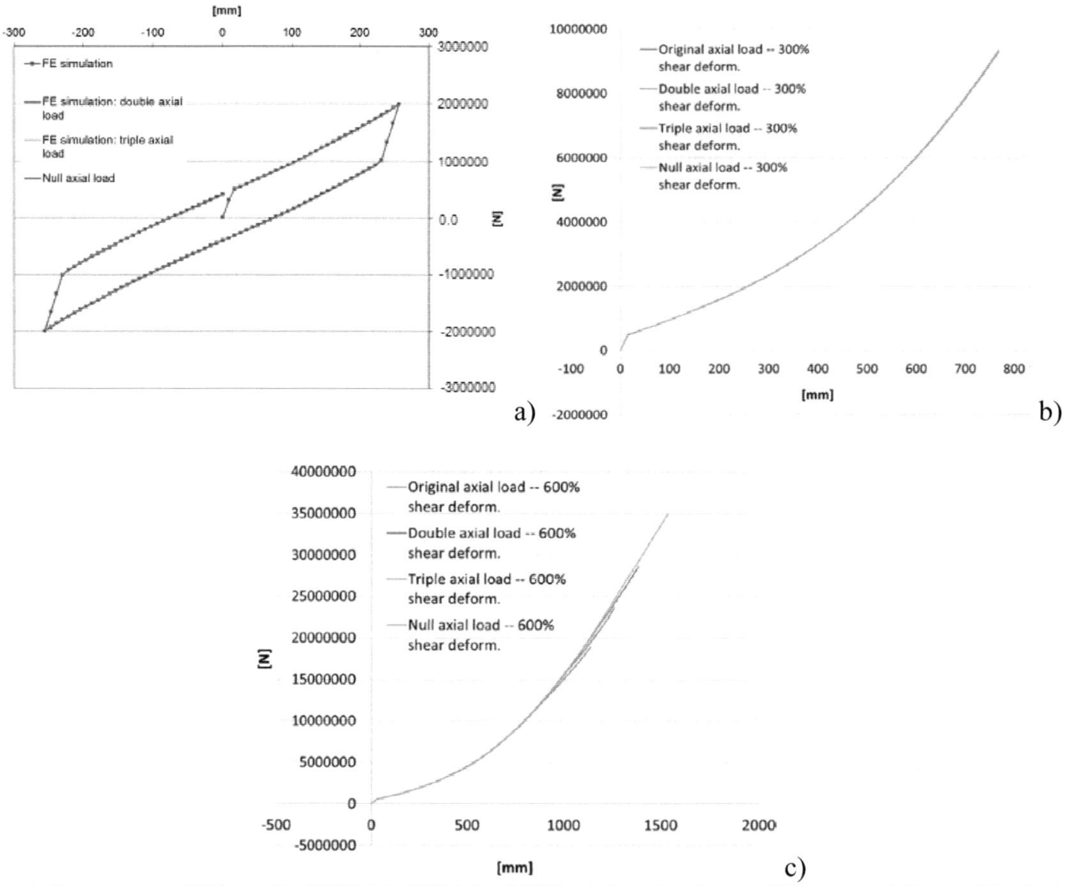

Figure 4. Comparison of FE results: 100% (a), 200% (b), 300% (c) shear loading at different axial forces. Siler isolator.

Laboratory experimentations focused on the evaluation of the ultimate limit state for delamination have been performed. In particular, a monotonic test up to failure has been performed according to the EN15129:2009 (2009) norm. The maximum allowable vertical load and the horizontal displacement at 300% of shear deformation are the ultimate conditions. This same testing condition has been simulated also numerically with the developed isolator model. The resulting stress state at failure has been depicted in Figures 5a-5c. Figure 5a and 5b report the maximum stresses at failure, while Fig. 5c the minimum ones.

The overlaid FE technique for the elastomeric layers does not allow to easily obtain the resultant stress-field in the rubber volume, due the superposition of two different constitutive models. The stress value at the steel-rubber delamination limit state has to be computed as the stress at the surface of the steel layers.

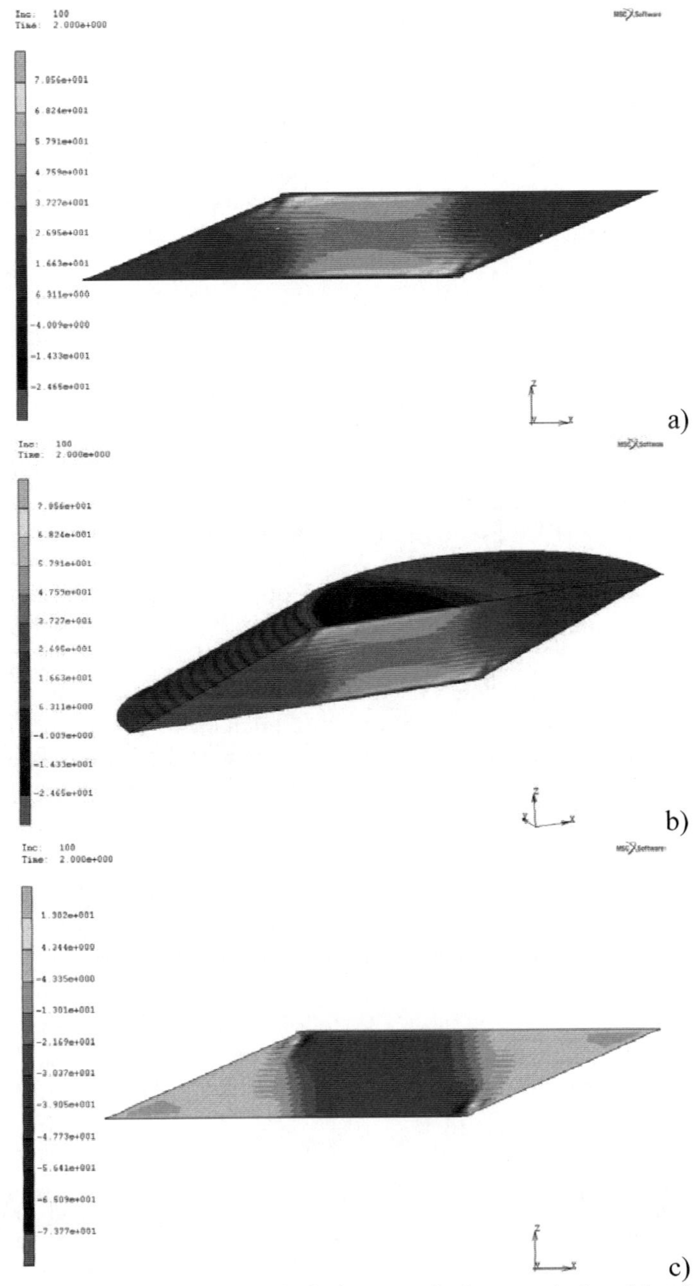

Figure 5. Maximum (a,b) and minimum (c) principal stresses in the numerical model at failure (300% shear deformation, and maximum axial load 24800kN). SILER isolator.

4. Scaled devices

Coming to the isolator models 601 and 302, Figures 6 and 7 report the curves in terms of shear force and horizontal relative displacement (Nagarajaiah & Ferrell 1999).

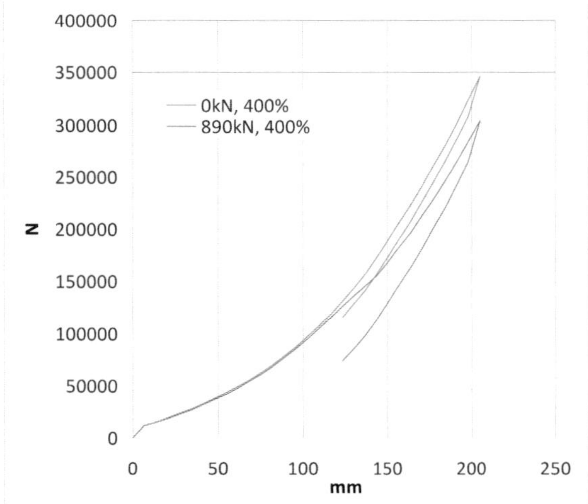

Figure 6. Comparison of FE results: shear loading at different axial forces for device 601 at 400% shear strain.

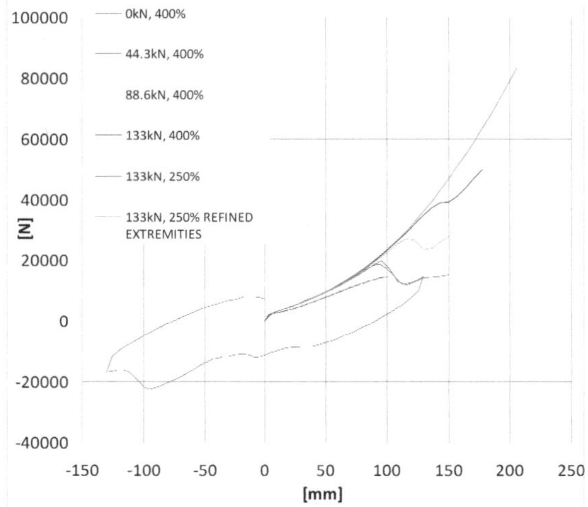

Figure 7. Comparison of FE results: shear loading at different axial forces for device 302 at 400% and 250% shear strain.

Several values of the axial load have been adopted to provide data for comparison with published results. Furthermore, the simulations allows to assess the influence of both shape and axial load on the device response.

Differently from the SILER device, that was really thin and large, models 601 and 302 show an increasing influence of axial force with the increasing slenderness of the device. Particularly, the most slender device (isolator 302) shows an unstable branch with a critical load that reduces at the increase of the axial load. This is consistent with results coming from experiments on devices of similar shape.

The distribution of the internal stresses is also of interest in order to avoid local material or bonding failures. The internal stresses in device 601 are depicted in Figures 8 and 9, while those of device 302 are in Figures 10-13. The limit value of 79 MPa for principal maximum stresses, as highlighted from the SILER isolator FE analysis, is used for establishing the ultimate limit state conditions.

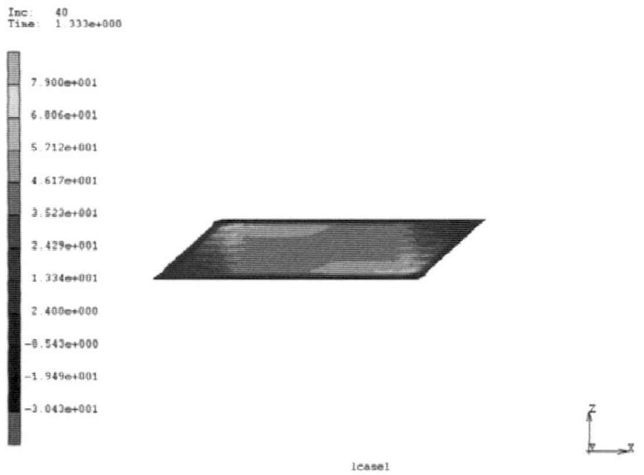

Figure 8. Principal stresses in the numerical model of device 601: axial load 890kN and 140% shear deformation. The maximum principal stresses reach the ultimate value of about 79 MPa.

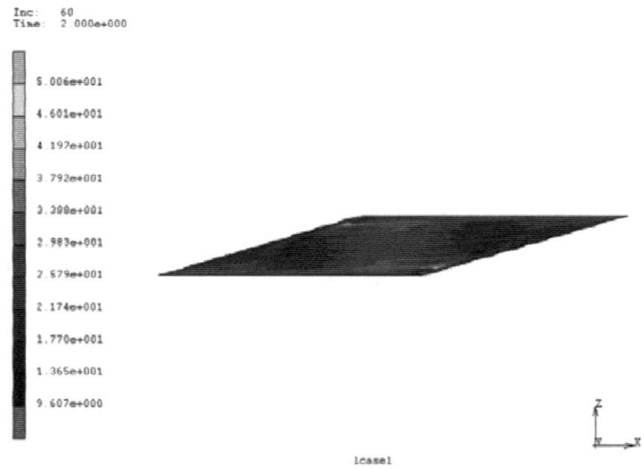

Figure 9. Principal stresses in the numerical model of device 601: zero axial load and 400% shear deformation. The maximum principal stresses stay below the ultimate value of about 79 MPa.

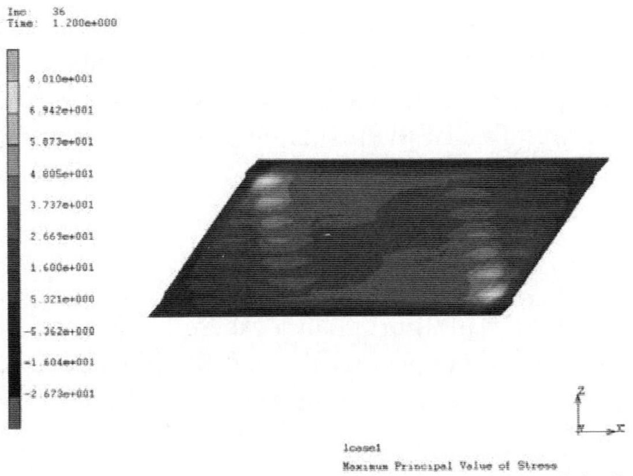

Figure 10. Principal stresses in the numerical model of device 302: axial load 133kN and 85% shear deformation. The maximum principal stresses reach the ultimate value of about 79 MPa.

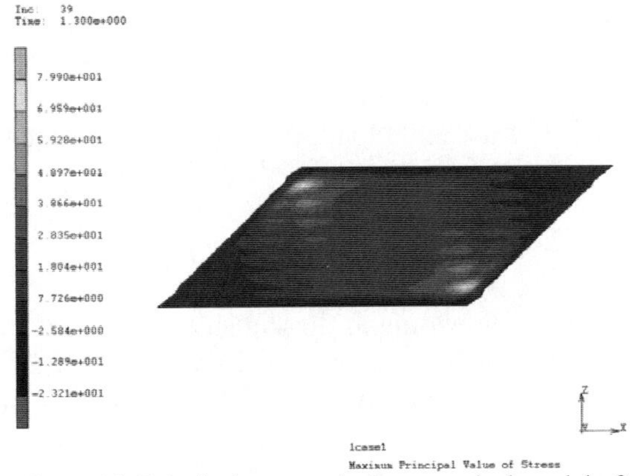

Figure 11. Principal stresses in the numerical model of device 302: axial load 88kN and 130% shear deformation. The maximum principal stresses reach the ultimate value of about 79 MPa.

As these figures show, the experimentally defined limit value of about 79 MPa, that corresponded to debonding in SILER tests of real devices, was reached under a decreasing shear deformation at the increase of the axial load. This is an interesting and significant finding for the design and the implementation of such devices in special structures.

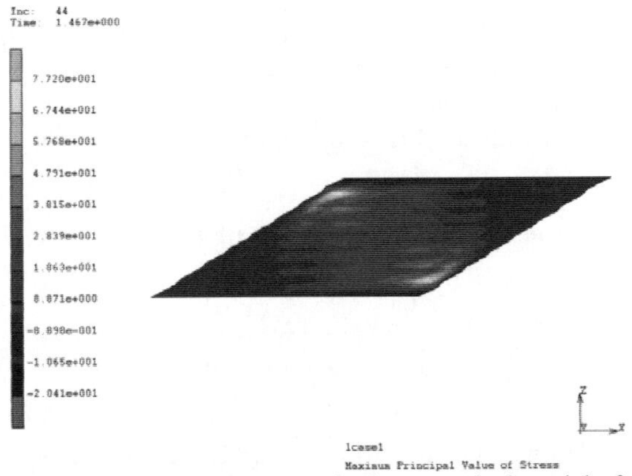

Figure 12. Principal stresses in the numerical model of device 302: axial load 44kN and 200% shear deformation. The maximum principal stresses reach the ultimate value of about 79 MPa.

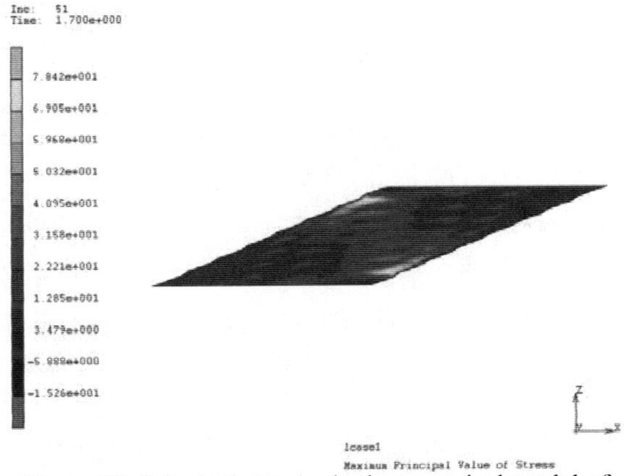

Figure 13. Principal stresses in the numerical model of device 302: zero axial load and 290% shear deformation. The maximum principal stresses reach the ultimate value of about 79 MPa.

5. Conclusions

Scope of the paper is to study the limit state and the "beyond design" response of rubber isolators having different aspect ratios.

If thin and large geometries for the devices are considered, the numerical simulations do not show essential differences in the response at increasing horizontal shear strain and axial loads. On the contrary, increasing slenderness of the devices, the influence of axial force essentially affects the isolator behaviour. Particularly, unstable branches in the shear force-displacement characteristics are shown with a critical load that reduces at the increase of the axial load. This is consistent with results coming from laboratory experiments on devices of similar shape.

The limit stress state condition of the device for local material or bonding failures has been also assessed (79 MPa). This has been fixed by comparing the results of laboratory experimentations at failure and the internal stresses from the numerical analyses. The results pointed out as at the increase of the axial load, the experimentally derived limit value of the principal stresses is reached for a smaller value of the shear deformation.

Acknowledgements

The research leading to these results has received funding from the European Research Council under the Grant Agreement n° ERC_IDEal reSCUE_637842 of the project IDEAL RESCUE—Integrated DEsign and control of Sustainable CommUnities during Emergencies.

References

Basu, B et al. 2014, 'An EACS joint perspective. Recent studies in civil structural control across Europe', *Structural Control & Health Monitoring*, vol. 21 no. 12, pp. 1414–1436.

Bianchi, G, Corradi dell'Acqua, L, Domaneschi, M, Mantegazza, DC & Perotti, F 2011, 'HDRB isolating devices for nuclear power plants: FE modelling and damage/failure characterization', *Structural Engineering World Congress (SEWC2011)*, Cernobbio (CO), Italy.

Cimellaro, GP, Reinhorn, AM & Bruneau, M 2010, 'Framework for analytical quantification of disaster resilience', *Engineering Structures*, vol. 32, pp. 3639–3649.

De Grandis, S, Domaneschi, M & Perotti, F 2009, 'A numerical procedure for computing the fragility of NPP components under random seismic excitation', *Nuclear Engineering and Design*, vol. 239, no.11, pp. 2491-2499.

M. Domaneschi (2010), "Feasible control solutions of the ASCE benchmark cable-stayed bridge", Structural Control and Health Monitoring, 17(6):675-693.

Domaneschi, M, Martinelli, L, & Perotti, F 2012, 'Seismic Rotational Components on Isolated Structures', *Proc. of 15th World Conference on Earthquake Engineering*, Lisbon, Portugal.

Domaneschi, M, Martinelli, L, Perotti, F & Tomasin, M 2015, 'Assessing the performance of a high damping rubber bearing in beyond-design conditions', *Proc. of 10th International Workshop on Structural Health Monitoring, IWSHM 2015*, vol. 2, pp. 1073-1080.

Gracia, LA, Liarte, E, Pelegay, JL & Calvo, B 2010, 'Finite element simulation of the hysteretic behaviour of an industrial rubber. Application to design of rubber components', *Finite Elements in Analysis and Design*, vol. 46, pp. 357-368.

Milani, G & Milani, F 2012, 'Stretch–Stress Behavior of Elastomeric Seismic Isolators with Different Rubber Materials: Numerical Insight', *Journal of Engineering Mechanics*, vol. 138, pp. 416-429.

Nagarajaiah, S & Ferrell, K 1999, 'Stability of elastomeric seismic isolation bearings', *Journal of Structural Engineering*, vol. 125, no. 9, Paper No. 15573.

Perotti, F, Domaneschi, M & De Grandis, S 2013, 'The numerical computation of seismic fragility of base-isolated NPP buildings', *Nuclear Engineering and Design*, vol. 262, no. 189–200.

SILER, European Commission 7th EURATOM Framework Program 2007-2013.

Standard CSN EN 15129 2009. Anti-seismic devices.

Warn, G & Whittaker, A 2008, 'Vertical Earthquake Loads on Seismic Isolation Systems in Bridges', *Journal of Structural Engineering*, vol. 134, no. 11, pp. 1696–1704.

The 8th International Conference on
Structural Health Monitoring of Intelligent Infrastructure
Brisbane, Australia | 5-8 December 2017

Monitoring of innovative, temporary traffic bridge using surface mount fibre-optic strain gauges

F. Fischli [1], E. de Bruin [2], M. van Bezooijen [2], M. Iten [1].

[1] *Marmota Engineering AG – Switzerland, Email:* frank.fischli@marmota.com [2] *Brem Funderingsexpertise BV - Netherlands*

Abstract

The monitoring of road related structures becomes more and more important. This is not only a safety aspect due to the more intense and heavier traffic or the aging structures. It can also be an economic aspect when it comes to new road pricing models or new materials being tested that have a longer lifetime or are more cost-efficient. Furthermore, during the design period of a bridge or other structure, assumptions are made about the intensity of the traffic during its lifecycle. Gathering information about the effective loads that are acting on the structure can also affect the remaining lifetime of a structure, both in a positive or negative way.

Bridges are an ideal structure for implementation of structural health monitoring systems: the sensors can be installed from underneath without disturbing the traffic and the behaviour under load is normally well known.

This paper focusses on a practical application of structural health monitoring on a temporary road bridge. The bridge was designed using a novel material made from fibre-reinforced polymers (FRP) which is promising in terms of costs, production, weight, strength and installation procedure.

The instrumentation consists of several pairs of short-length, surface mount fibre-optic strain gauges (FBG) installed perpendicular to each other at the bottom side of the bridge deck. It was originally designed to be able to understand and proof the structural behaviour of the temporary bridge. The monitoring system however could also be used to gain a lot of information about the traffic passing on the bridge. Parameters such as number of axes per vehicle and driving speed could be obtained from the measured data. Additional information such as a line of influence or the load per axis could also be gained after a calibration of the system.

The structural health monitoring system will be described regarding its practical implementation, its capability and limitations, as well as the results and conclusions that could be obtained out of it.

1. Introduction

Many post world-war bridges in Holland are due to be rehabilitated nowadays. As those bridges are usually part of important traffic routes, it is not possible to completely close them for traffic over the period of the rehabilitation works. It is usually a requirement to keep a few lanes open, e.g. for public transport, emergency vehicles or lighter traffic such as pedestrians or cyclists. As a consequence, auxiliary constructions are necessary to keep this part of the traffic running.

These auxiliary structures are often complex and expensive to build, especially for heavy traffic. Therefore, new design or construction approaches which allow for lighter constructions become more and more interesting since they are often promising in terms of cost and construction efficiency.

This paper describes the instrumentation of a temporary road bridge for public transport buses using fibre-optic strain gauges. The intention was to attach the temporary bridge deck laterally to the old existing bridge in such a way that no new foundations were required. Therefore, a lightweight structure was preferable which could be achieved by building the additional lane out of fibre-reinforced polymers (FRP). The purpose of the instrumentation was to be able to verify the design assumptions, as this was one of the first applications of this novel type of material. The strain gauges were placed on the bottom side of the bridge deck which was easily accessible from a scaffold constructed under the bridge.

The structural health monitoring of bridges using fibre optic sensors has been adopted for a long time and many research and commercial projects are reported e.g. in (Maaskant et al. 1994), (Inaudi et al. 1996) or (Glisic & Inaudi 2010). Different technologies and applications are discussed and presented in (Casas & Cruz 2003). The use of such monitoring techniques for testing the application of new materials in early stage has been described e.g. in (Brönnimann et al. 1998) for carbon-fibre-reinforced polymer wires. Advanced applications of bridge monitoring installations such as weight-in-motion systems are reported e.g. in (Lydon et al. 2017).

Installations of fibre-optic sensing systems for monitoring of infrastructure in many areas of civil engineering including their applicability and limitations are reported e.g. in (Iten et al. 2016) or (Fischli et al. 2013).

2. Measurement technology

A fibre-optic strain measurement system was chosen for monitoring of the temporary bridge deck. Fibre-optic sensors can be multiplexed which is a great advantage in terms of sensor routing and number of connecting cables. Furthermore, the sensor itself is small and lightweight, intrinsically safe and passive.

Among the different fibre-optic technologies, Fibre Bragg grating (FBG) sensors were chosen to suit best for the required monitoring needs as the technology is well matured and the sensors are easy to install (Morey et al. 1990). The system shall be able to measure continuously and automatically, at high sampling speeds and able to accommodate several sensors per measurement channel.

Two types of sensors were installed in the present project: strain sensing and temperature sensing FBGs (c.f. Figure 1). The strain sensing FBGs are aimed to measure strain or indirectly displacements between two fixations. Temperature sensing FBGs are used to compensate thermal effects that are induced in the strain sensing FBGs. The strain sensing FBG is sensitive to strain as well as temperature changes whereas the temperature sensing FBG is packaged strain free and thus only sensitive to temperature changes.

The strain sensors have a gauge length of 250 mm and can be mounted to surfaces using the anchors at each end of the actual sensing distance. The strain sensor has a typical accuracy of 7.5 microstrain ($\mu\varepsilon$) and a precision of 5 microstrain. The temperature sensors have a length of 210 mm and are packaged in a robust steel tube. The typical temperature accuracy is listed as < 1 °C and the precision as 0.2 °C.

All sensors were connected to a state-of-the-art measurement unit on 4 individual optical channels and were logged continuously. The maximum sampling frequency of the unit was set to 100 Hz during the actual measurements which allowed for detection of strain changes induced by fast vehicles.

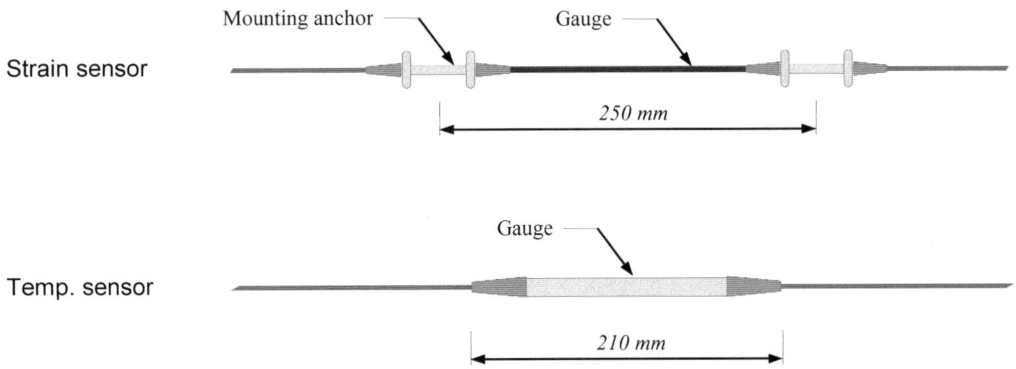

Figure 1: Schematic drawing of the two sensor types

3. Monitoring project

The steel-concrete composite arch bridge, built in 1951, is typical of its type for post-world-war constructions. It has a total length of 138 meters and accommodates 2 lanes for vehicles and one cycling lane across the river "IJssel" outside of the Village Doesburg in The Netherlands (Figure 2). The bridge needed to be generally renovated (bead blasting and painting of existing steel beams) and the existing steel supporting beams had to be reinforced in 2016. As the concrete bridge deck had to be replaced as well, the bridge needed to be closed for general traffic for most of the duration of the renovation works.
The bridge is a very important connection in the local infrastructure around Doesburg and feasible alternative routes do not exist. Therefore, one lane always had to be kept open for vehicles of the emergency services and for public transportation busses. The solution was to remove the cycling lane on the outside of the bridge and replacing it with an innovative high strength FRP deck. The temporary deck was installed on the existing load-bearing structure for the former cycling lane, without any need for additional reinforcement to the steel structure (Figure 2).

Figure 2: Bridge to be monitored with scaffold for instrumentation (left); construction of temporary FRP deck (right)

The bridge deck was divided into panels of 8.10 m length in longitudinal direction by the existing, transversal load bearing beams of the former cycling lane. The sensors were attached on the bottom side of the auxiliary bridge deck in 4 cross-sections over two panels with the aim of measuring the bending moments (c.f. Figure 3). The sections were chosen in such a way that different bending moments in longitudinal directions could be captured. Two transversal locations were chosen in each cross-section. One in the centre of the lane where little load is applied by the vehicles and one in the axis of the tyres of the passing vehicles where a maximum load is applied from directly above (Figure 4). A pair of sensors was placed in each location, one sensor measuring the strain in longitudinal direction (sensors no. 3 and 5) and one sensor measuring the transversal direction

(sensors no. 2 and 4). The detailed layout of the sensors installed on the lower side of the temporary bridge deck is shown in Figure 4.
Additional sensors were installed on the existing transversal supporting beams (sensor no. 1 in Figure 4) to get a complete picture of the bearing behaviour of the auxiliary bridge. All strain sensors were paired with a temperature sensor for compensation of thermally induced effects which was installed in close proximity.

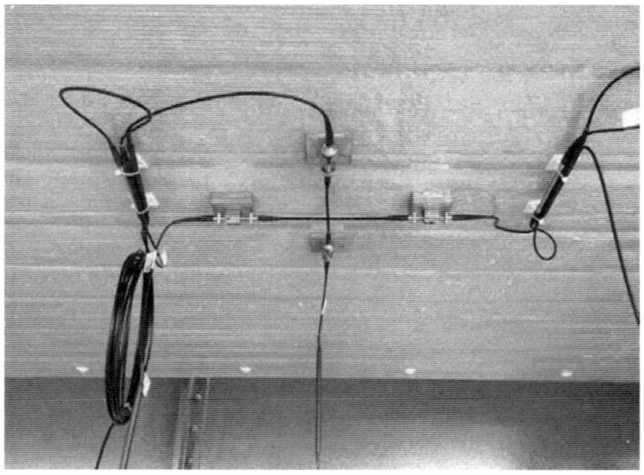

Figure 3: Instrumented sections in one panel along the bridge deck (left) and a pair of mounted strain sensors with paired temperature sensors (right)

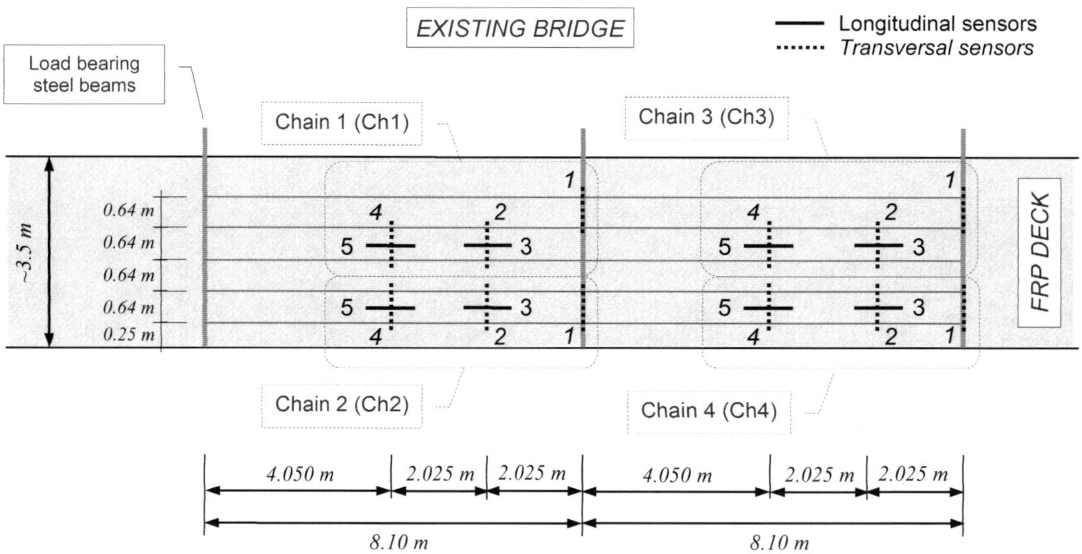

Figure 4: Schematic layout of the installed sensor chains on the bottom side of the temporary FRP bridge deck and on the load bearing steel beams

4. Results/measurements and discussion

The following diagrams show strain measurements for a typical dataset for different groups of sensors over time. The measurement data was collected during the passing of a public service bus.

The speed of the passing vehicles was limited to 24 km/h. The presented strain measurements are all compensated for thermal fibre-optic effects using the assigned, closest temperature compensation FBG. Each sensor chain is represented by a different colour.

4.1. Longitudinal strain sensors

The longitudinal sensors allow for measuring the influence of the load on the strain at the sensor position in the direction of the passing vehicles. Figure 5 shows a sample data set for the two sensor chains on the outer side of the bridge deck (Ch. 2 and Ch. 4) in longitudinal direction including the sensors on the supporting beam. The longitudinal sensors no. 3 and 5, mounted on the bottom side of the bridge deck experience tension when the load is positioned exactly above the sensor. The transversal sensors on the supporting beam (no. 1) on the other hand goes into compression.

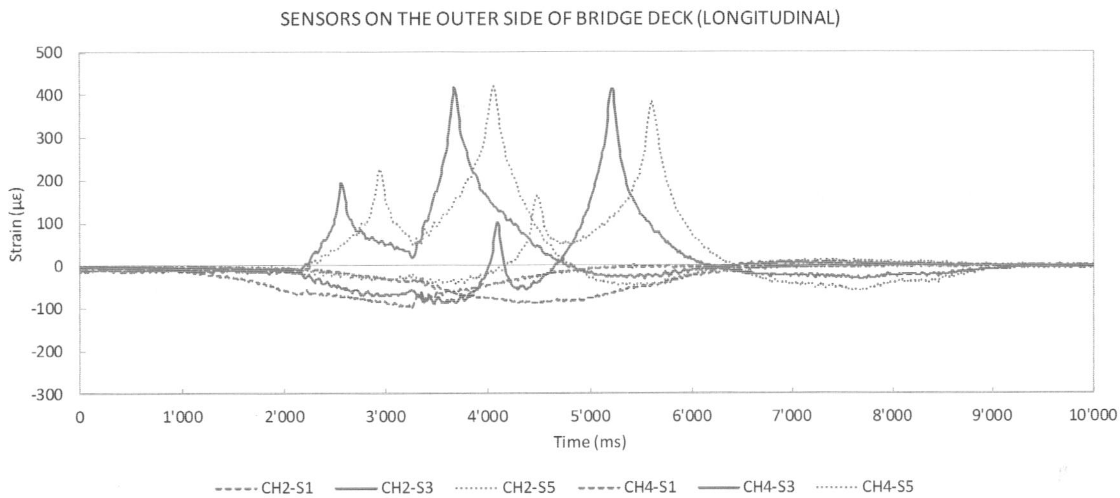

Figure 5: Strain measurements of the longitudinal sensors and the sensor on the transversal supporting beam on the outer side of the bridge deck

The strain measurements for the longitudinal sensors and the corresponding sensor on the supporting beam in the centre of the bridge (Ch. 1 and Ch. 3) are shown in Figure 6. These sensors show the same tension/compression behaviour as for the sensors on the outer side of the deck, though with different amplitudes.

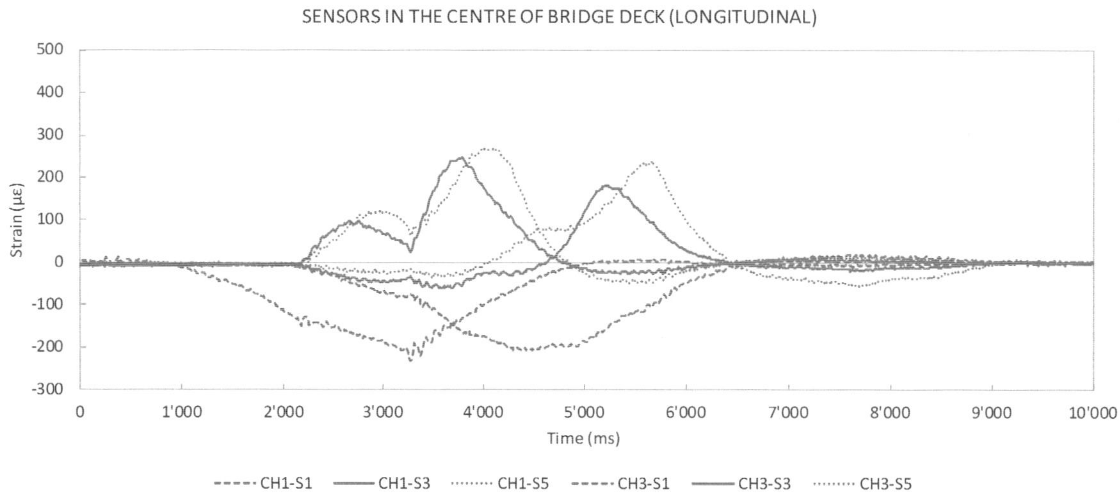

Figure 6: Strain measurements of the longitudinal sensors and the sensor on the transversal supporting beam on the side of the bridge deck

4.2. Transversal strain sensors

The strain measurements for the transversal sensors in the two sections are shown in Figure 7 and Figure 8. The sensors on the outer side of the bridge deck show a similar behaviour as the longitudinal sensors, i.e. sensors on the bridge deck experience tension whereas the sensors on the supporting beams go in compression.

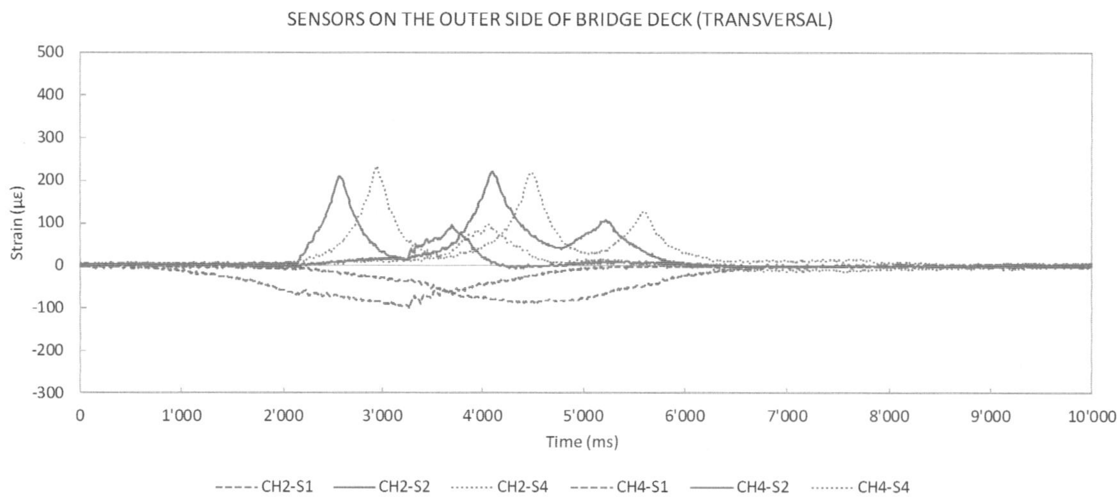

Figure 7: Strain measurements of the transversal sensors and the sensor on the transversal supporting beam on the outer side of the bridge deck

The transversal sensors in the centre of the bridge deck show a different behaviour than the other three sets of sensors. The sensors mounted on the bridge deck also experience compression, but the peaks are not that clearly distinguishable and the amplitude is much smaller.

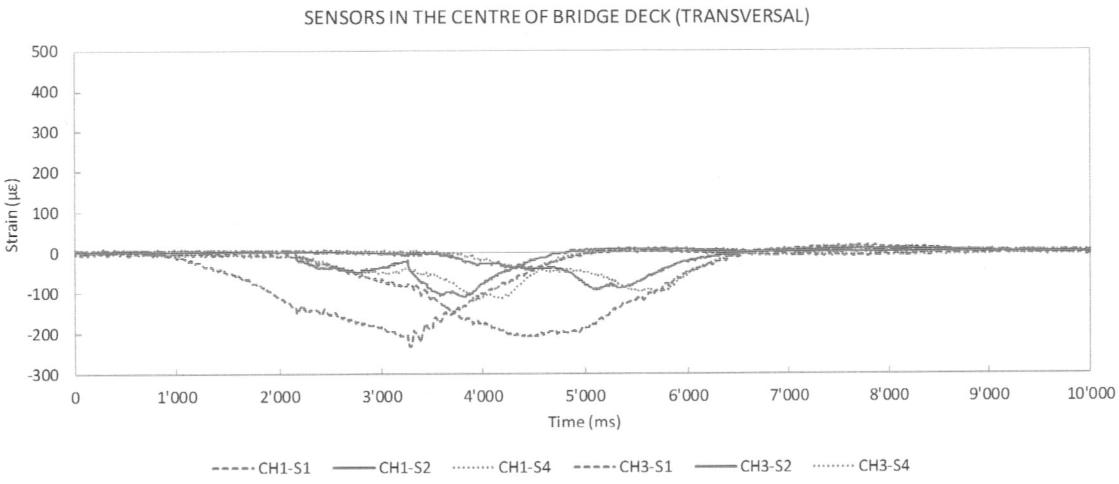

Figure 8: Strain measurements of the transversal sensors and the sensor on the transversal supporting beam in the centre of the bridge deck

4.3. General discussion of the strain measurements

The peaks on the outer side of the bridge (directly below the applied load) have a spikey shape whereas the sensors in the centre picture have a rounded, wider peak form. This makes sense from a structural point of view since the sensor in centre experience an indirect and therefore less distinct loading. These peaks allow for immediate determination of the direction of movement and also the number of axes or groups of axes respectively. The assessment of number and arrangement of the axes however might not always be straightforward and unique.

It can be observed that for the longitudinal strain sensors, the second peak generally has the higher amplitude than the first one. For the transversal sensors on the outer side this is inverse, i.e. the first peak has a higher amplitude than the second one. A feasible explanation for this effect could not be found so far and is subject to further investigations. It might be a depending on the type of vehicle.

4.4. Velocity of the vehicle

The calculation of the speed of a vehicle is straightforward using a strain measurement over time and knowing the distance between the strain gauges in longitudinal direction. The passing of a vehicle at the location of a strain gauge is indicated by strain peak (c.f. Figure 9). The distance between two sensors of the same number in different panels is 8.1 m. Taking for instance the two peaks at 3680 and 5220 ms, an average velocity of 18.94 km/h can be calculated.

The time resolution of this monitoring system sampling at 100 Hz is 10 milliseconds (ms). Thus, the resolution and accuracy of the average speed measurements are improved with increasing distance between the two sensors. The speed resolution for distance of 8.1 m and a velocity of approximately 20 km/h is around 0.1 km/h whereas it's only 0.5 km/h for a distance of 2.03 m at the same velocity.

The determination of the exact peak time, therefore, needs to be done carefully since the calculation of the velocity is relatively sensitive to the exact time measurement. This especially applies to shorter longitudinal distances between the measurement sensors.

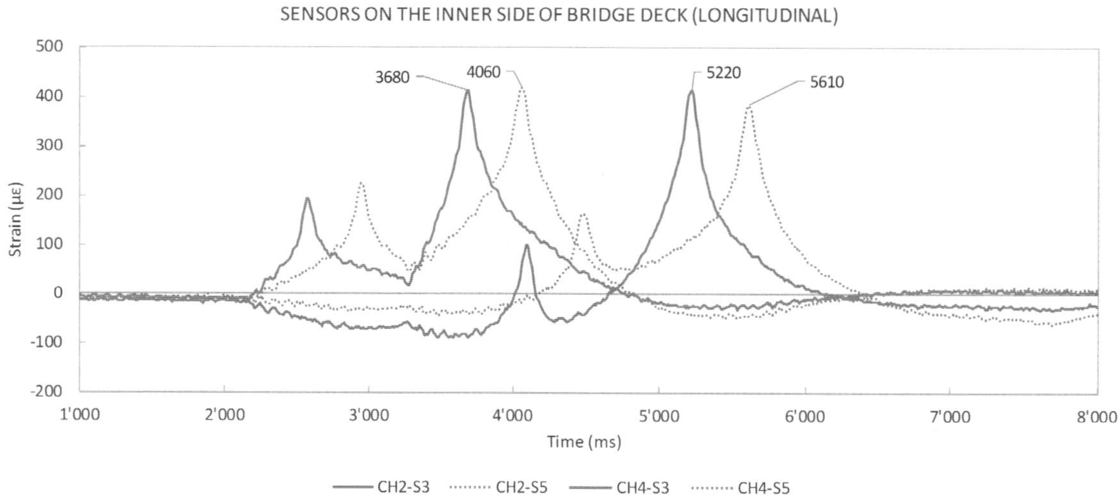

Figure 9: Strain measurements of the longitudinal sensors with indicated peaks of vehicle passing over a sensor location for calculation of velocity

4.5. Assessment of structural behaviour

A typical strain measurement over time presumably allows for determination of the point in time when the load of the vehicle passes from one panel to the next one (c.f. vertical dash-dotted lines in Figure 10). Such measurements basically allow for construction of a line of influence for the strains in a determined location for a moving load on the bridge deck. This is a very powerful tool for engineers when verifying the load bearing behaviour of new construction materials or for assessing structural health of existing structures.

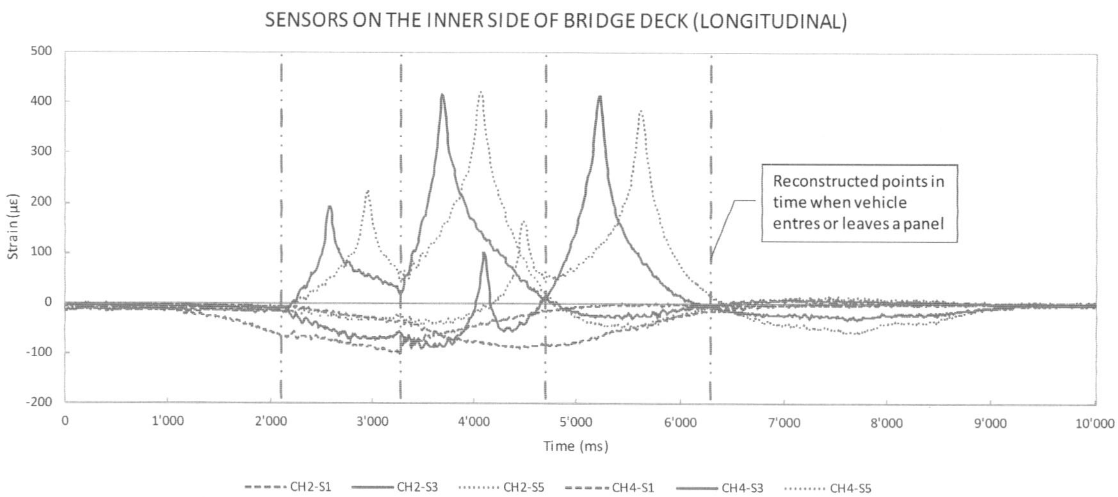

Figure 10: Strain measurements of the longitudinal sensors with reconstructed points in time when an axis enters or leaves a panel of the bridge deck

5. Conclusions

The use of fibre-optic sensors turned out to be a very convenient solution for monitoring of the auxiliary bridge deck in the current project. The sensors are lightweight and easy to install from a scaffold underneath the bridge. The cabling of the sensors is straightforward and reduced to a minimum since many sensors can be multiplexed on one channel. The readout unit can be placed in a secure and dry location away from the sensors as, in contrary to other technologies, the signal can easily be transported over long distances with minimal loss. The data acquisition is done on a proprietary software and can be customised for different acquisition scenarios. As the sensors are mounted to the surface of the deck, the sensors can be removed after the required monitoring period and can be re-used for further monitoring projects.

The monitoring system was capable of measuring the bending strains in the bridge that are induced by the load of the vehicles. These measurements helped to verify the assumed bearing behaviour of the new innovative bridge deck which in turn gave a certain safeguarding for the structural engineers in charge. Unfortunately, the available information about the bearing behaviour or the expected strains out of the design were limited and thus it was not possible to carry out a direct comparison between the expected and measured strains.

Instead, it could be demonstrated that with the installed measurement system, additional information such as direction and speed of a passing vehicle can be gained with little additional effort. The determination of the number of axes or the type of vehicle might also be possible but requires some experience and learning of the system. The implementation of a weight-in-motion system based on the described sensors could be a next logical step providing the system can be calibrated accordingly.

The potential applications of fibre-optic sensors for bridge monitoring are numerous. New materials or construction methods can be monitored in an efficient way which allows for optimisation of such structures. The same method can be used for structural health monitoring or assessments of load bearing on existing structures. Lines of influence, which is a very powerful tool for structural engineers, can be constructed with relatively little effort.

A fair investment for installation of such a monitoring system can potentially decrease the maintenance effort of a bridge considerably since the actual remedial actions can be assessed and planned more refined and based on the real condition of the structure determined by load measurements.

Acknowledgement

The authors greatly acknowledge the Dutch Ministry of Infrastructure and the Environment Rijkswaterstaat and Civiele Technieken deBoer BV for their help and assistance during the monitoring project.

References

Brönnimann, R, Nellen, PM & Sennhauser, U 1998, 'Application and reliability of a fiber optical surveillance system for a stay cable bridge', *Smart Materials and Structures*, Volume 7, Number 2.

Casas, JR & Cruz PJS, 2003, 'Fiber Optic Sensors for Bridge Monitoring', *Journal of Bridge Engineering*, vol. 8, no. 6, pp. 362-373.

Fischli, F, Iten, M, Carrera, A & Puzrin, A 2013, 'Laboratory study of fibre-optic strain sensor fixation for monitoring tunnel linings', *Proceedings of SMAR 2013 the 2nd Conference on Smart Monitoring, Assessment and Rehabilitation of Civil Structures: [9-11 September 2013, Istanbul TR], ITU, Instanbul Technical University, Istanbul, Turkey.*

Glisic, B, Inaudi, D 2010, 'Distributed fiber optic sensing and integrity monitoring', *Transportation Research Record (TRR), Journal of Transportation Research Board of National Academies (TRB)* No. 2150 "Infrastructure Maintenance and Preservation", pp. 96-102.

Inaudi, D, Vurpillot, S & Casanova, N 1996, 'Bridge monitoring by interferometric deformation sensors', *Proc. SPIE 2895, Fiber Optic Sensors V, 34 (September 24, 1996),* doi:10.1117/12.252146.

Iten, M, Fischli, F & Puzrin, A 2016, 'Geotechnical monitoring of infrastructure using distributed fibre-optic sensing: project examples, results and limitations', *CSHM-6 Belfast, Ireland*, M.2.

Lydon, M, Taylor, S, Doherty, C, Robinson, D, OBrien, EJ & Žnidarič, A 2017, 'Bridge weigh-in-motion using fibre optic sensors', *Proceedings of ICE - Bridge Engineering*. DOI: 10.1680/jbren.15.00033

Maaskant, R, Alavie, AT, Measures, RM, Ohn, MM, Karr, SE, Glennie, DJ, Wade, C, Tadros, G & Rizkalla, S 1994, 'Fiber optic Bragg grating sensor network installed in a concrete road bridge', *Proc. SPIE 2191, Smart Structures and Materials 1994: Smart Sensing, Processing, and Instrumentation, 457 (May 1, 1994),* doi:10.1117/12.173977.

W. W. Morey, WW, Meltz, G and Glenn, WH 1990, 'Fiber Optic Bragg Grating Sensors', *Proc. SPIE 1169*, Fiber Optic and Laser Sensors VII, 98, doi:10.1117/12.963022;
ember, pp. 72-94.

The 8th International Conference on
Structural Health Monitoring of Intelligent Infrastructure
Brisbane, Australia | 5-8 December 2017

Experiences with continuous monitoring of deformation and modal properties of an arch dam

P. Moyo[1], P. Bukenya [2]
[1] University of Cape Town – South Africa, Email: pilate.moyo@uct.ac.za [2] University of Cape Town – South Africa

Abstract

Abstract: Dam safety is concerned with ensuring that the structural integrity of a dam is maintained throughout its service life. This requires a sound understanding of the loading environment, structural behaviour, material deterioration mechanisms and structural performance of a dam at any given time. The most common approaches for dam safety evaluation are visual inspections, intermittent instrumented monitoring and in some cases finite element modeling. This paper reports on the integration of continuous ambient vibration monitoring and static monitoring of a concrete arch dam. By developing relationships between the dam's response (natural frequencies and dam wall deformations) and the operating loading (reservoir level, temperature) features of operational behavior of the dam have been revealed. The results of the monitoring system show the importance of continuous monitoring of critical parameters in arch dams.

1. Introduction

Dam safety is concerned with ensuring that the structural integrity of a dam is maintained throughout its service life. This requires a sound understanding of the loading environment, structural behaviour, material deterioration mechanisms and structural performance of a dam at any given time. The most common approaches for dam safety evaluation are visual inspections, instrumented monitoring and finite element modelling. The work presented in this paper follows from previous work carried by the authors (Moyo & Oosthuizen, 2013) which focused on demonstrating the application of ambient vibration measurements for structural health monitoring and finite element model validation and updating of concrete arch dams. A unique feature of this paper is the integration of continuous ambient vibration monitoring (AVM) and static monitoring of a concrete arch dam. By developing relationships between the dam's response (natural frequencies and dam wall deformations) and the operating loading (reservoir level, temperature) the onset of anomalous behaviour can be detected and the future response of a dam can be predicted. This paper reports on the continuous monitoring of Roode Else Berg Dam, a double curve concrete arch dam located in Western Cape, South Africa.

2. Description of subject dam

Roode Elsberg Dam (Figure 1) is located in the Western Cape about 130km North of Cape Town near the town of Worcester. The dam mainly supplies water for irrigation of vineyards and limited domestic use via a 7 kilometre tunnel to an adjacent valley.

Figure 1: Roode Elsberg Dam

Roode Elsberg Dam is a double curvature concrete arch dam with a centrally located spillway. The dam has a gross capacity of 8.21 million m^3. The height to lowest foundation point is 72m and the length of the crest is 274m. Two galleries have been provided, one following foundation level and a horizontal instrumentation gallery. Figure 2 shows a typical cross-section through the dam and the relative positions of the ISETH Sliding micrometers installed in boreholes during the 1980s to monitor the dam and foundation.

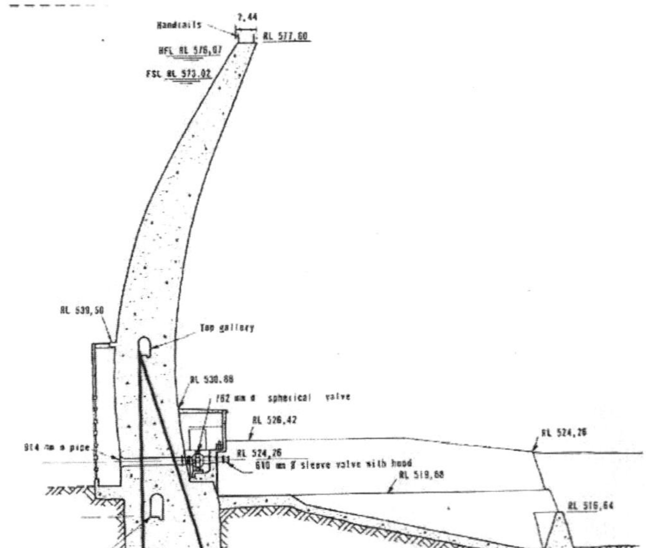

Figure 2: Typical cross-section of Roode Elsberg Dam

3. Roode Else Berg Dam Continuous Monitoring Systems

3.1 Introduction

A suite of static monitoring instrumentation has been in operation since 1984. They include, clinometers and pendulum clinometer, geodetic survey targets, crack gauges as well as Trivecs. The latest additions in 2010 and 2013 respectively are; a continuously monitored GPS system at four survey beacons and an ambient vibration measurement (AVM) system. In addition, there is a weather station to monitor the ambient environmental conditions at the dam as well as a system to measure water temperatures.

3.2 Description of the Ambient Vibration Monitoring System

The ambient vibration monitoring system consists of three force balanced of tri-axial accelerometers (Figure 3) with a dynamic range of more than 125 dB at ± 2 g full scale within a frequency range of 0.1Hz to 50 Hz. The accelerometers are connected to a 24bit digitizer which communicates with the industrial computer. The data acquisition system (industrial computer and the digitizer) is placed on the dam wall inside the upper gallery of the dam to avoid any environmental effects (Figure 4.1). The data is sampled at 50Hz and stored at hourly intervals.

Figure 3: Accelerometers on the dam crest and data acquisition system in the upper dam gallery

3.3 Description of the Global Navigation Satellite System

Displacements of Roode Elsberg have been monitored by Department of Water and Sanitation (DWS) using geodetic survey methods twice a year since 1984. In 2010 the DWS implemented an automatic deformation monitoring system on the dam using Global Navigation Satellite Systems (GNSS) as shown in Figure 4. The system comprises of the following:

i. Four GNSS receivers, each powered by solar energy and two 12 V batteries, two secondary sensors on the crest of the dam structure, one on each side of the spillway and two base station sensors up and downstream of the dam wall;

ii. Poynting Wi-Fi network streaming all the receiver data to a server in a central control room; and

The two secondary GNSS sensors on the crest of the dam wall (P203 on left flank and P206 on right flank) are used to calculate real-time kinematic positions of the points and to detect any sudden movement at the crest of the dam, as well as the base points. The two base GNSS sensors up and downstream of the dam wall are used as fixed stations. All the satellite data is accessible to DWS staff via a web interface.

Figure 5: GPS network on Roode Elsberg dam

Two base stations (Figure 5) were set up on the existing Geodetic Deformation network stations of the dam, and run continuously. These base stations are used as reference stations for the relative positioning in the post-processing mode. The base stations are occupied permanently in order to achieve high accuracy in the estimation baseline lengths between base stations.

3.4 Measurement of Environmental and Operational Factors

Water levels on Roode Elsberg dam are measured by simple elevation gauges i.e. level staff gauges. The fluctuation levels of the Roode Elsberg reservoir have been monitored on a daily basis since 1974. For this paper, only water levels recorded from the 25th November 2013 are considered as this ties with the installation of the dynamic monitoring system. A weather station (Figure 6) was installed on 25th November 2013, on the dam crest to measure ambient temperatures and there are thermometers that are embedded into the dam wall to measure reservoir water temperatures at different water levels.

Figure 6: Weather station installed on the dam crest near the spillway

4. Results and discussions

This paper is focussed on the continuous monitoring systems installed on the dam, namely, the GPS system, the AVM system and the temperature monitoring system. At least one year and half of reliable monitoring data is now available from the continuous monitoring system at Roode Elsberg dam. This provides an opportunity to understand the normal condition of the dam structure i.e., how the structure responds to normal variations in environmental and operational conditions. The results are divided into two main sections namely, summary of loading (water level and temperature) observations, summary of response observations and their association with loading.

4.1 Reservior Level

Figure 7 shows the variation of the reservoir level over the period November 2013- June 2015. The reservoir level is characterised by a long period of full capacity (June 2014 – December 2014). It is also noted that the dam level went below 25m in 2015 where as it remained above 40m in 2014.

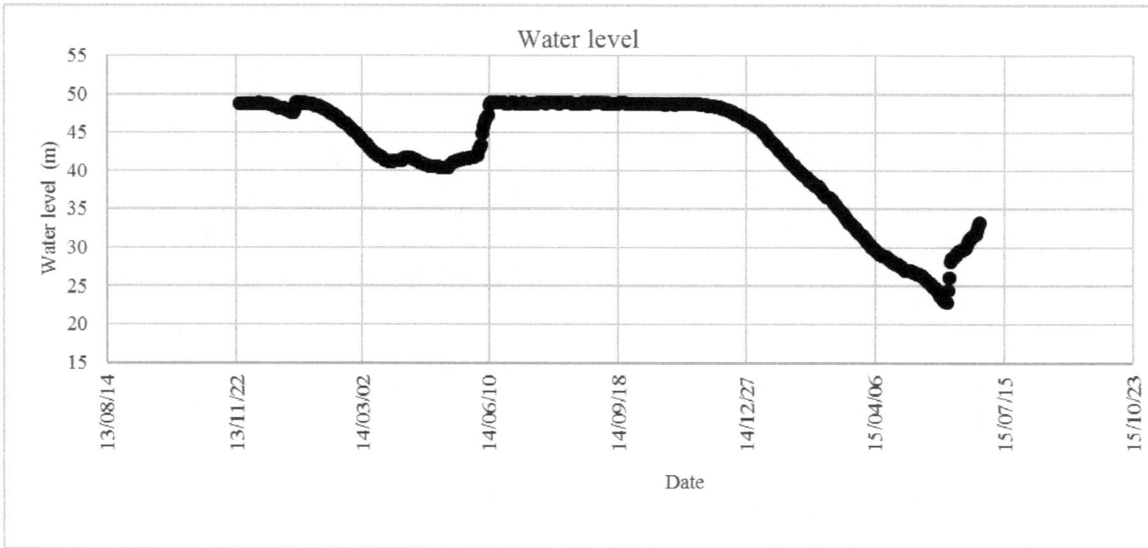

Figure 7: Water level

4.2 Ambient Vibration Monitoring Results

The dynamic response (natural frequencies) were extracted from acceleration time series recorded by the dynamic monitoring system. Data driven stochastic subspace identification (SSI-data) method was used to estimate the natural frequencies of the dam. SSI-data method is one of the time based output-only modal analysis methods that utilizes stabilization diagrams to estimate the modal parameters (Zhang et al., 2005). This technique has been shown to perform better than other output only techniques such as frequency domain techniques (Bukenya, 2012). Figure 8 shows typical singular values of spectral densities of the recorded accelerations. Figure 9 shows a typical stabilization diagram used in the estimation of the natural frequencies using the SSI-data method.

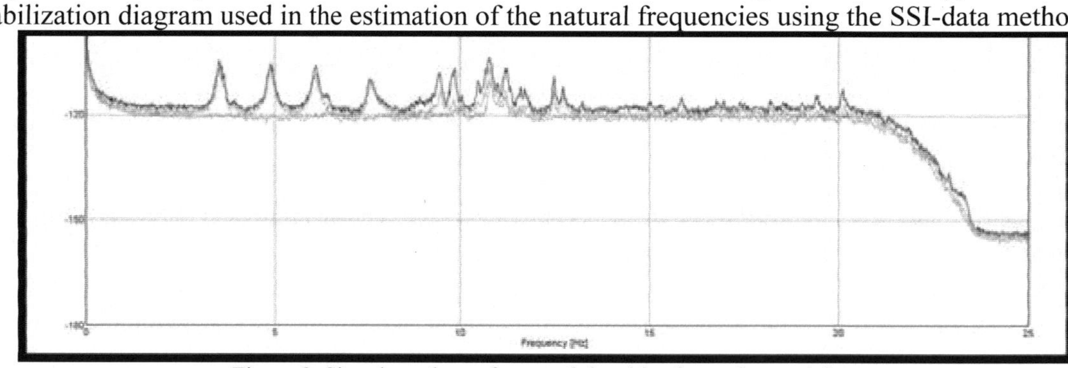

Figure 8: Singular values of spectral densities from observed data

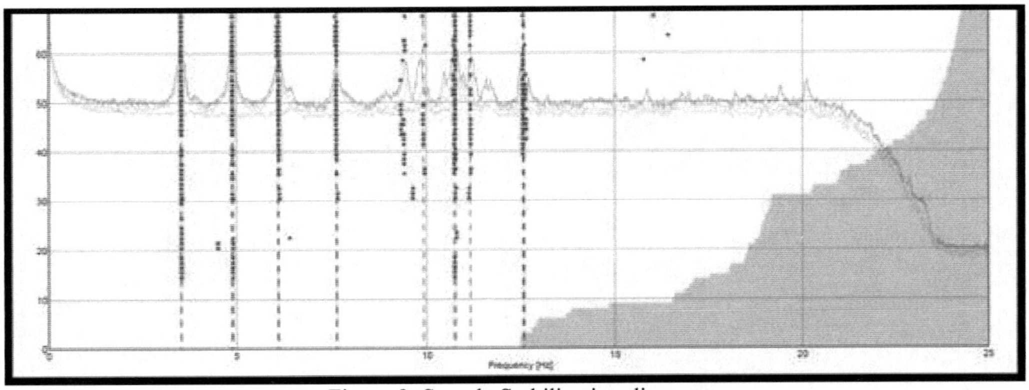

Figure 9: Sample Stabilization diagram

The time variation of the first two natural frequencies during the 19 month period is shown in Figures 10 and 11. The first two natural frequencies show strong relationship between dynamic behaviour and the reservoir level (Figure 12 and Figure 13). The influence of temperature variation on dynamic properties of the dam is not obvious from a comparison of temperature data to natural frequency data.

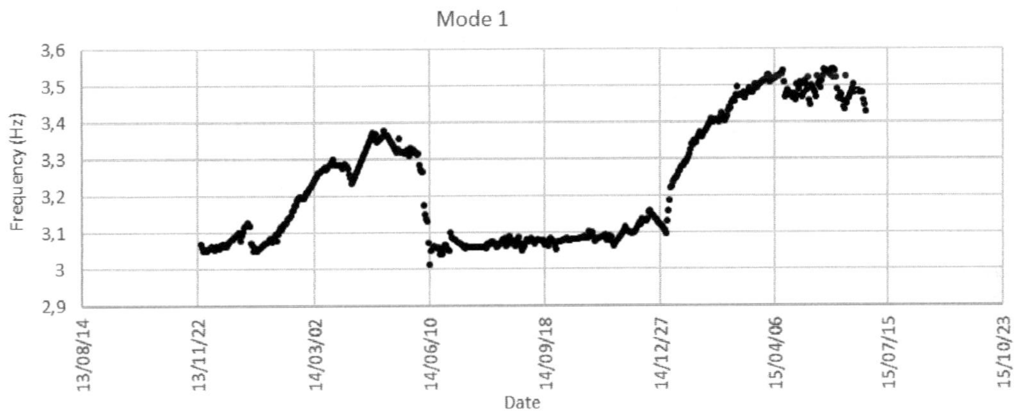

Figure 10: Variation of mode 1 with time

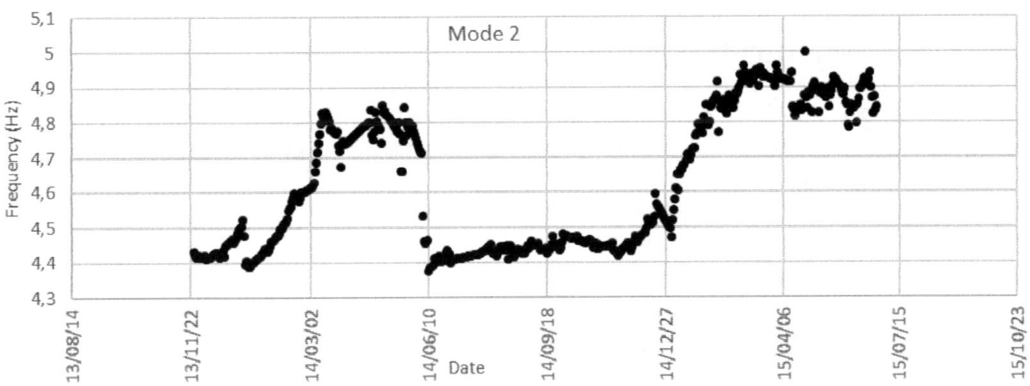

Figure11: Variation of mode 2 with time

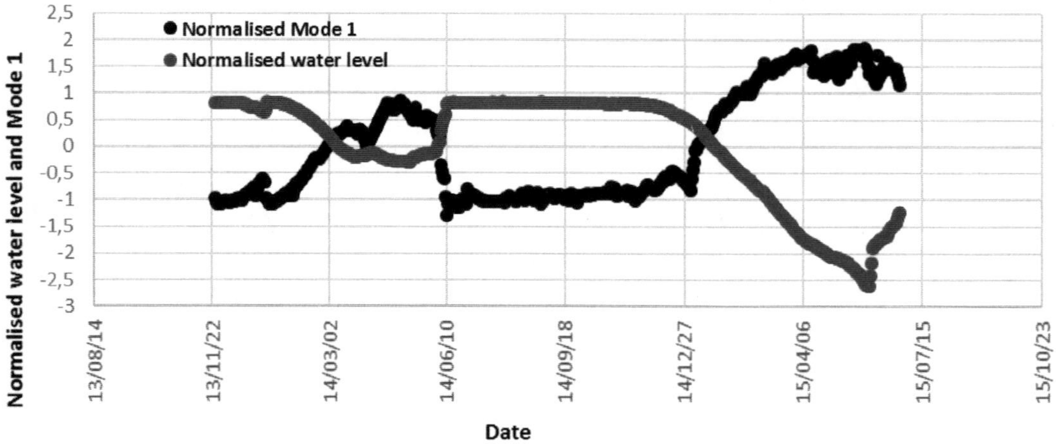

Figure 12: Variation of mode 1 and water level with time

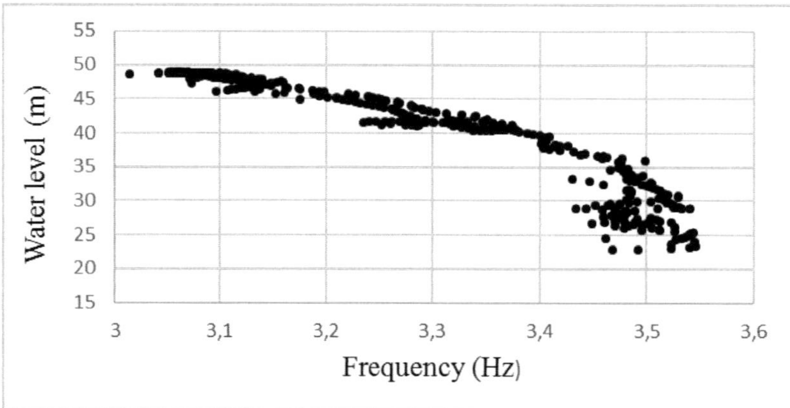

Figure 13: Water level versus mode 1

4.2 Results of GPS Monitoring

The data available for the present study is for the period January 2014 – November 2014. The data is not scaled to absolute displacements and thus presented as normalised values. The data was normalised using the mean and variance. Figures 14 shows the variation of normalised displacements over a period of 10 months, i.e. from January 2014 to end of November 2014. The data shows a strong dependency of displacements on temperature variations (Figure 15).

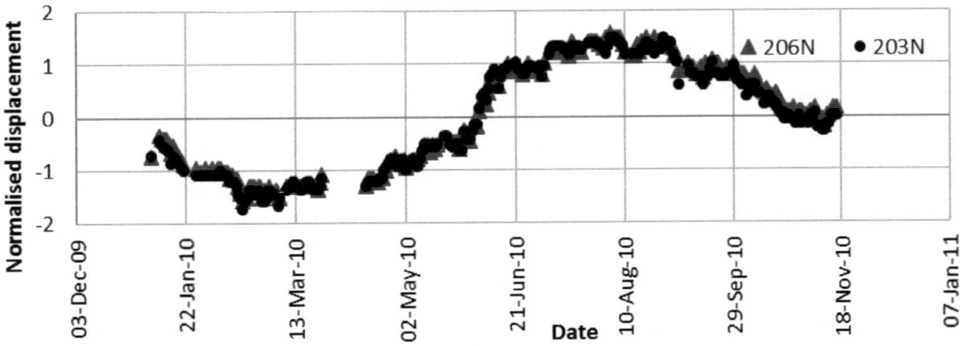

Figure 14 Normalised displacements in the radial direction

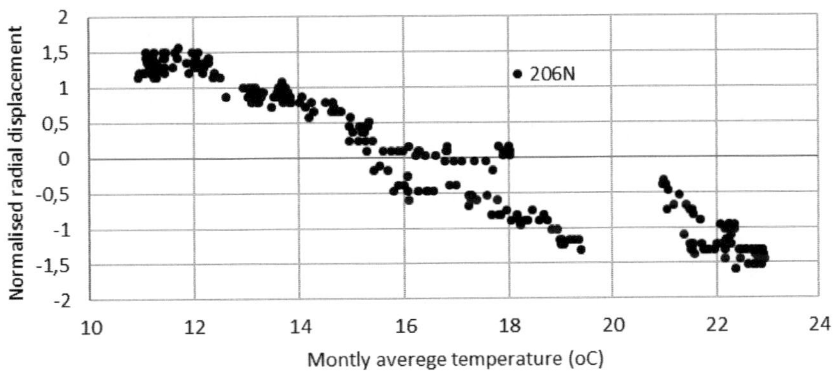

Figure 15: Normalised radial displacement versus monthly average air temperature

4.2 Concluding Remarks

Dams in operation are subjected to hydrodynamic loading by the impounded water body as well as environmental loading such as temperature and wind loading. An objective of the study was to develop a better understanding of the relationship between the loading environment and the response of the dam through continuous monitoring. The following key conclusions are made from the continuous monitoring data;

i) The dynamic behaviour of the dam is strongly dependent on the reservoir level. The natural frequencies of the dam decrease with water level. The dam's dynamic behaviour is weakly influenced by temperature variation.
ii) The seasonal deformation of the dam is strongly dependent on temperature variation. While the seasonal variation of deformations is largely driven by temperature variation, the reservoir level does influence the deformation of the dam especially as the dam fills rapidly.

5. Acknowledgements

This work is funded by the Water Research Commission (WRC) South Africa whose support is gratefully acknowledged. The authors would like to thank Department of Water and Sanitation for allowing us carry out this research on their dam. The Concrete Material and Structural Integrity

0Research Unit (CoMSIRU) is highly recognized for the support provided to the first author in this research.

References

Bukenya, P., Moyo, P. & Oosthuizen, C. 2012. Modal parameter estimation from ambient vibration measurements of a dam using stochastic subspace identification methods. *Proceedings International Conference on Concrete Repair, Rehabilitation and Retrofitting.* 3-5 September 2012, Cape Town, South Africa. CRC Press (Taylor & Francis), pp. 540-545.

Moyo, P., & Oosthuizen, C. (2013). *Structural Health Monitoring of Arch Dams Using Dynamic and Static Measurements*. WRC report No. K5/2025/1/13, ISBN 978-1-4312-0453-3.

Zhang LM, Brinker R, Andersen P (2005*). An overview of operational of modal analysis: major developments and issues*. In: Proceeding of 1st International Operational Modal Analysis Conference, Copenhagen

The 8th International Conference on
Structural Health Monitoring of Intelligent Infrastructure
Brisbane, Australia | 5-8 December 2017

Assistant Decision-making System for Structural Health Monitoring in Three Gorges Project

Chen Bo[1, 2, 3], Shi Zhongwen[1, 2, 3], Wu Zhongru[1, 2, 3]
[1] *College of Water Conservancy and Hydropower Engineering, Hohai University, Nanjing, China, Email:chenbohhu@126.com,* [2] *State Key Laboratory of Hydrology-Water Resources and Hydraulic Engineering, Hohai University, Nanjing, China,* [3] *National Engineering Research Centre of Water Resources Efficient Utilization and Engineering Safety, Hohai University, Nanjing, China*

Abstract

In view of the abnormal or unsafe factors in the health monitoring and analysis of the Three Gorges Dam, an expert system is built, from the overall safety of the project, to provide assistant decision-making suggestions for the safe operation of engineering buildings. The system mainly consists of first-hand database, expert consultation service module, assistant decision-making module and data warehouse. The main function is designed as follows: (1) When the cause of structural abnormality of the Three Gorges project has been identified by the dam health monitoring and analysis system, assistant decision-making suggestions can be provided by visiting the knowledge base and conducting knowledge reasoning analysis. (2) When the cause of structural abnormality cannot be found by the dam health monitoring and analysis system, assistant decision-making suggestions can be provided by conducting expert consultation about diagnosing difficult structural problems, during which the expert group conducts the safety evaluation after inquiring the data and analyzing the calculation. In order to support the expert consulting diagnosis, an association intelligent query processing is proposed in this system to provide information including engineering design and construction archives, health monitoring data, analysis and evaluation results, structural positive forward and inverse analysis results and expert knowledge base, etc. Moreover, positive forward, inverse and mixed inference machine is designed to select, understand and explain the relevant knowledge, to form the preliminary assistant decision-making suggestions. Lastly, the membership degree of the expert judgment weight is conformed to comprehensively consider the authority, specialties and engineering familiarity of consulting experts, to establish a multi-level security evaluation model.

1. Introduction

During the construction of the dam and its ancillary buildings, a large number of safety monitoring instruments are embedded, to monitor the reservoir water level, temperature, rainfall and other environment variables, as well as the deformation, seepage, stress, strain and other effect results. The measured data of these instruments imply the trend or abnormality of the dam safety behaviour (Chen 2016). Therefore, comprehensive evaluation of dam health and emergency decision of dam risks can be conducted by the experienced experts group, through the monitoring model of these information (Wu 2000, Wu 2003). The knowledge and practical experience of experts are one of the key factors in the above process. With the development of intelligent technologies, the structural health decision can be completed in the case of expert offline, by establishing an intelligent expert system, with storing engineering information and expert knowledge in the knowledge base and building inference rules by intelligent technologies of simulating expert decision-making ideas (Su 2000, Yue 2001).

In this paper, in view of the abnormality or unsafe factors in the health monitoring and analysis of the Three Gorges Dam, an expert system is built, from the overall safety of the project, to provide assistant decision-making suggestions for the safe operation of engineering buildings. In the expert system, the safety comprehensive evaluation and assistant decision-making suggestions of the Three Gorges Project can be achieved, through qualitative or quantitative analysis and interpretation of the building safety information.

2. Function, structure and process of the system

The assistant decision-making system for structural health monitoring in Three Gorges Project mainly consists of first-hand database, expert consultation service module, assistant decision-making module and data warehouse. The main function is designed as follows: (1) When the cause of structural abnormality of the Three Gorges project has been identified by the dam health monitoring and analysis system, assistant decision-making suggestions can be provided by visiting the knowledge base and conducting knowledge reasoning analysis. (2) When the cause of structural abnormality cannot be found by the dam health monitoring and analysis system, assistant decision-making suggestions can be provided by conducting expert consultation about diagnosing difficult structural problems, during which the expert group conducts the safety evaluation after inquiring the data and analyzing the calculation. The structure diagram and the workflow chart are shown in Fig.1 and Fig.2.

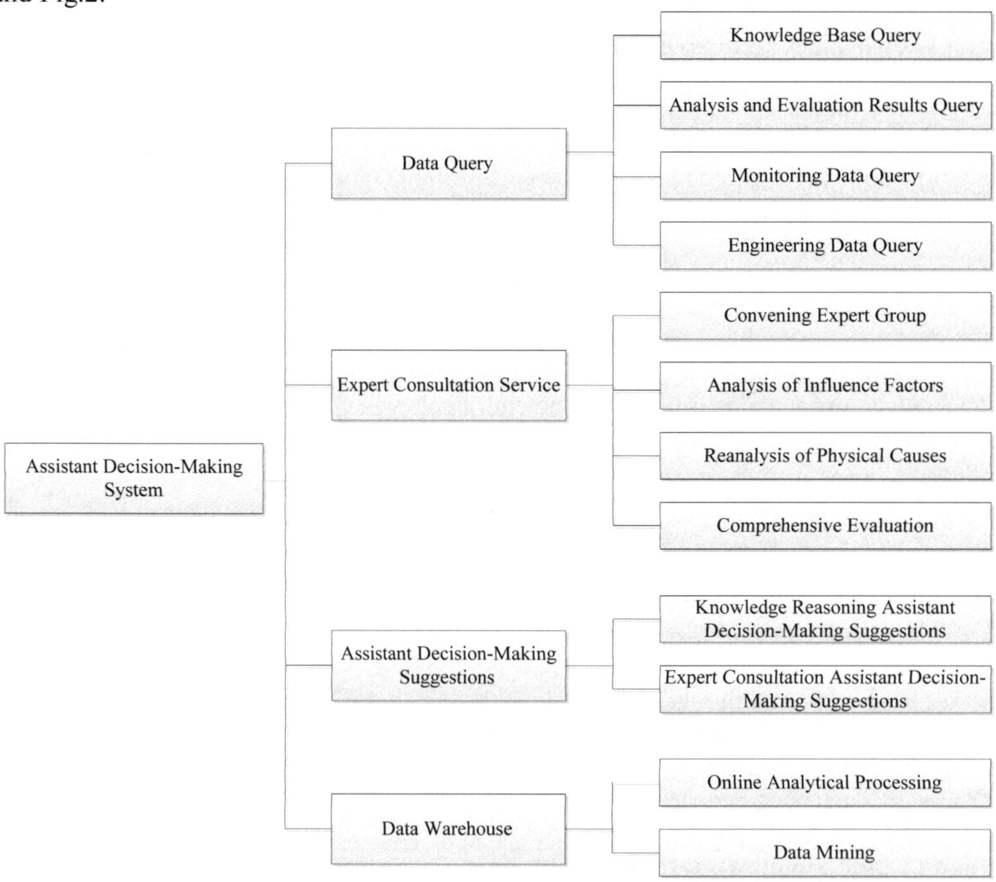

Fig.1 the structure diagram of assistant decision-making subsystem

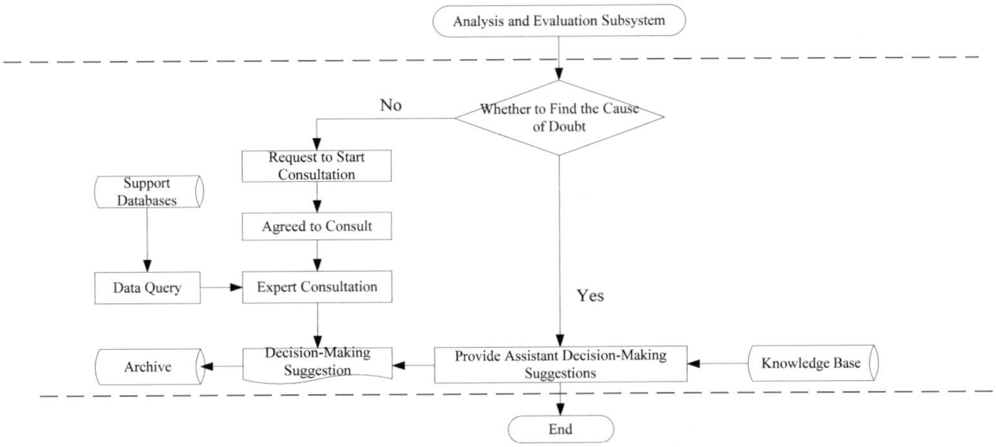

Fig.2 the workflow chart of assistant decision-making system

3. Design of association intelligent searching for data query module

The experts need enough first-hand data to evaluate the safety of the building comprehensively and give a decision-making advice. These data include the relevant monitoring variables and monitoring data of the corresponding environment variables, the evaluation results of analysis and evaluation subsystem, structural forward and inverse analysis results, the relative design and construction data in the parts where abnormal danger or incurable diseases occur. This system designs an association intelligent search processing function, which provides information support for expert consultation diagnosing. After the prediction and reasoning of the user's query plan, intention and interest, this system conducts intelligent search and provides the user with an effective response. According to whether the query attribute is clear, the system would conduct direct query or indirect query respectively. That is, when the query attribute is clearly expressed in the query, the query system will give a clear answer directly; when the query attribute is not clearly expressed in the query, the system would conduct an indirect query through the context and related knowledge.

4. Design of expert consultation function

When the analysis and evaluation subsystem cannot find the cause of the structural abnormality of the structure, it is necessary to conduct an expert consultation. At first, relevant experts are gathered to set up an expert group. Then, in the advanced network consultation support environment, the experts comprehensively evaluate the Three Gorges project, through inquiring relevant information about structural abnormality of the building, analyzing the influencing factors of the problem and then reanalyzing the physical causes. The workflow chart of expert consultation service subsystem is shown in Fig.3.

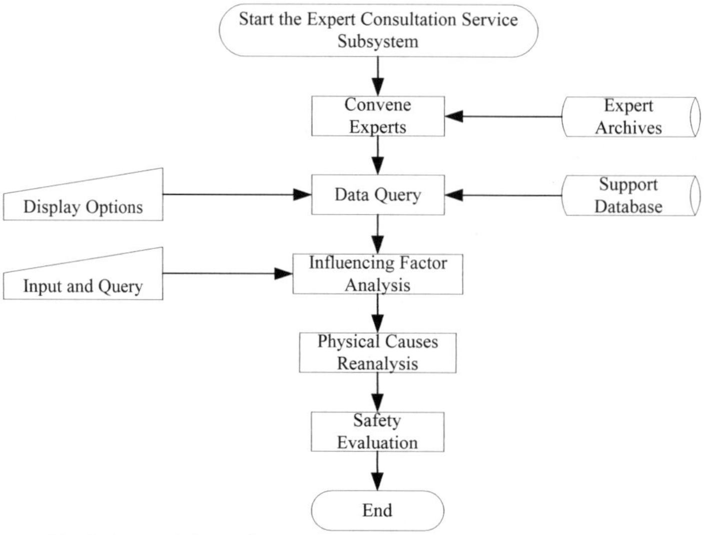

Fig.3 the workflow chart expert consultation service subsystem

When convening the expert group, the weight of expert evaluation is determined by considering the authority, specialty and engineering familiarity of the experts. A multi-level safety evaluation model is established to consider the expert's suggestion and his weight. In this process, the factors, which affect the incurable diseases, are variety and contain certain degrees of randomness. Therefore, it is necessary to construct an evaluation model to analyze the significance of the influence doubt and the factors. Upon the main factors put forward by the experts, the system re-analyze the physical causes online or offline through the inverse analysis of the calculation model and the calculation parameters, and then analyzes the deformation, stress and stability coefficient produced by the main factors which affect the safety, according to the revised calculation model and calculation parameters. A three-level safety evaluation model of Three Gorges project is constructed here, using the multi-level threshold fuzzy evaluation theory, considering the system complexity of multi-level and multi-objective in the health evaluation of Three Gorges project structures.

The system creates the online consultation support environment to provide experts with an advanced technology platform by using modern network technology and multimedia technology, and the high-speed network as the transmission medium. When the Three Gorges project encounters a major danger, it is possible to organize experts who are familiar with the project immediately, conduct consultation diagnosing in the local area network of main office, and then provide safety decision-making advices, so as to eliminate the potential safety hazard of the Three Gorges project buildings quickly and efficiently.

5. Design of knowledge reasoning for assistant decision-making suggestions

The assistant decision-making suggestion subsystem is working, when the physical causes of the structural abnormality has been analyzed and the first or second level alarm is proposed. In this subsystem, the knowledge in the knowledge base is used to conduct reasoning judgment to propose the preliminary assistant decision-making suggestions on the structural abnormality. If the physical causes of the structural abnormality cannot be analyzed by this subsystem, then the expert consultation subsystem will startup directly, to propose assistant decision-making suggestions for the relevant departments through the comprehensive expert diagnosis.

5.1 Design of system knowledge reasoning

In view of the knowledge network of the system and the safety features of the Three Gorges project, a variety of reasoning and analysis mechanisms are adopted to deal with the expert knowledge and experience in the knowledge base, for orderly, convenient and quick reasoning analysis.

(1)Forward reasoning

Starting from the facts of the problem, the system searches the matching knowledge in the knowledge set. When there is a conflict, a matching knowledge is chosen in the conflict set. Sometimes a new knowledge is matched, so that the reasoning of the problem continues until a certain state. If the conclusion contains the facts that have been generated, then the conclusion is solvable. If the knowledge in the conflict solution is unavailable, expert consultation is used to fix it. The expression above can be expressed as equation 1:

$$\begin{array}{l} p \rightarrow \begin{cases} \Omega(i) \mid p = F(p) \\ \Omega(j) \mid_{j\rightarrow p, j=1,2,\ldots p} = F(p) \end{cases} \\ F(p) \rightarrow \begin{cases} F(m) \rightarrow F(n) \rightarrow \ldots\ldots \\ \phi \end{cases} \end{array} \quad (1)$$

where p is the fact; $\Omega(i)$ is the knowledge set; $F(p)$ is the knowledge that matches p; $\Omega(j)\mid_{j\rightarrow p, j=1,2,\ldots p}$ is the conflict knowledge set; $F(m)$ is the new knowledge drawn from $F(p)$; $F(n)$ is the new knowledge drawn from $F(m)$; ϕ is an empty set.

(2) Backward reasoning

The conclusion of a known or hypothetical problem is introduced into a generated database as the top node of network, and then the matching solution is conducted in the knowledge set. If there is, the solution is solvable. If it does not match, then a new knowledge is reselected in the knowledge set; if it matches, there is a solution. When all the knowledge does not match, then the expert comprehensive diagnosis is carried out. If there is an optional knowledge that cannot be affirmed or denied, then put it into the database.

(3) Mixed reasoning

Mixed reasoning combines the advantages of forward reasoning and backward reasoning, and avoids their own imperfections. Some complex problems can be solved by mixed reasoning, including forward-backward mixed reasoning and backward-forward mixed reasoning. Forward-backward mixed reasoning refers to selecting a goal first by forward reasoning, that is, selecting the target from the partial results deduced from the known facts, and then solving the target by backward reasoning. In the process of reasoning which is inaccurate or difficult to determine, sometimes the reliability of the forward reasoning is not high, and these conclusions are used as the target to conduct backward reasoning, so as to improve the reasoning efficiency and the reliability of the conclusion. Backward-forward mixed reasoning refers to assuming a goal or a conclusion first and then carrying out backward reasoning to get some information about the problem, and using this information to conduct forward reasoning to find possible solutions.

5.2 Expert consultation decision-making

At present, the commonly used assistant decision-making suggestions, include controlling the operating water level, reducing uplift pressure or ground water level, and engineering treatment measures, etc. However, for the large-scale water conservancy project such as the Three Gorges, it is difficult to make decisions on whether to reduce the water level of the reservoir or not, which involves a wide range of factors and, if without adequate consideration, will produce adverse effects on flood control and power generation benefits. The engineering treatment measures can overcome the problems caused by low reservoir water level, but it takes a long time to achieve the purpose of danger eliminating. The expert consultation decision-making module ensures the safety of the building in Three Gorges project upon the safety decision-making suggestions with a comprehensively using of the above emergency measures, according to the type, scope, severity of structural abnormalities and the supplementary knowledge reasoning proposed by expert inquiry, analysis, calculation, and review, repair and assistant knowledge deduction.

(1)Reducing uplift pressure or ground water level suggestions

When the uplift pressure or ground water level is abnormal or reaches the alarm index, the drainage is firstly made to reduce the uplift pressure or the ground water level below the alarm index.

(2)Reducing operating water level suggestions

When the alarm index or level of the analysis and evaluation subsystem reaches the first or second level, it should be controlled below the third level by reducing the operating water level, meanwhile monitoring the abnormality or the deformation of dangerous parts. Through the results of analysing the strength, stability and deformation under the load conditions of abnormality or danger in the database generated by expert query or through the strength, stability and deformation results after the back analysis (when the database has no results), the following analysis is made:

$$\begin{aligned} &\Omega(H,T,P,...)\rightarrow\begin{Bmatrix}\Omega(\delta)\\ \Omega(\sigma)\\ \Omega(\kappa)\end{Bmatrix} \\ &\Omega_0(H,T,P,...)\rightarrow\begin{Bmatrix}\Omega_0(\delta)\\ \Omega_0(\sigma)\\ \Omega_0(\kappa)\end{Bmatrix}\end{aligned} \tag{2}$$

$\sigma \le [\sigma]$ and $k \ge [K]$ are used as constraints to find the corresponding load set:

$$\Omega([\sigma],[\kappa])\rightarrow\Omega_m(H_m,T_m,P_m,...) \tag{3}$$

where $\Omega(H,T,P,...)$ is the load set; H is water level; T is temperature; P is uplift pressure; $\Omega(\delta)$, $\Omega(\sigma)$ and $\Omega(\kappa)$ are the deformation, stress and stability safety coefficient of $\Omega(H,T,P,...)$; $\Omega_0(H,T,P,...)$ is the load corresponding to the abnormality or danger; $\Omega_0(\delta)$, $\Omega_0(\sigma)$ and $\Omega_0(\kappa)$ are the deformation, stress and stability safety coefficient of $\Omega_0(H,T,P,...)$; σ and κ are calculated stress and stability safety coefficient; $\Omega([\sigma],[\kappa])$ is a set of allowable stress and stability coefficient; $\Omega_m(H_m,T_m,P_m,...)$ is a load control set that satisfies the allowable stress and stability safety coefficient; H_m is operating controlling water level; T_m is controlling temperature load; P_m is controlling uplift pressure load.

(1)Engineering treatment measures

In order to give full play to the benefits of the project, it is often necessary to adopt engineering measures to deal with the abnormality or danger immediately without reducing the water level or after adopting urgent measures to reduce the water level. As the engineering measures are related to the safety and investment of the project, it is necessary to propose the preliminary engineering measures at first and to analyze the safety and economy of the project to formulate the scheme of engineering measures. In order to find the optimal scheme, the primary scheme is evaluated by the discrete target method. The formula is as follows:

$$R = \sum_{i=1}^{n} P_{mi} \beta_i \tag{4}$$

where R is the optimization scheme of engineering measures; P_{mi} is the remark or rating given by the ith expert on engineering measures; β_i is the comprehensive judging right of the ith expert.

6.Design of data warehouse for assistant decision-making system

Analysis based on the source database often has some limitations, for example, extracting data leads to "spider web" phenomenon, and data are often difficult to translate into directly useful information, which affects the efficiency of the system. The Three Gorges dam project is large in scale and has tens of thousands of monitoring instruments and huge amounts of safety monitoring data. Therefore, high-performance database software is needed to meet the need of decision-making support.

Data warehouse technology is a kind of information integration technology. The data warehouse obtains the data from a plurality of information sources, and the data are stored in the internal database of the data warehouse after processed. Through providing the users with the access tools, the data warehouse provides a unified, coordinated and integrated information environment to support the enterprise's overall decision-making process and the in-depth comprehensive analysis of enterprise management. A data warehouse contains seven parts: data sources, data extraction, conversion and loading tools, data modeling tools, core warehousing, data warehouse target databases, front-end data access and analysis tools, and data warehouse management tools.

The data warehouse needs to have corresponding tools to extract data from the data source. Before entering the data warehouse, the data warehouse must be tested to eliminate errors that may be hidden in the data. In order to meet the needs of decision-making support and in-depth analysis, the data are arranged, processed and reorganized by the data extraction and conversion tools and then loaded into a database of one or more data warehouses. The data warehouse system provides a complete set of data access and analysis tools and in-depth data analysis functions, namely data mining functions such as data comparison, trend analysis, pattern recognition, etc. OLAP is used as a verification analysis tool, and data mining is used as a mining analysis tool for data warehouse design to provide the basis for analysis and decision-making.

The following is to design the decision-making system with data warehouse as the core, and online analysis and data mining tools as the methods.

(1)Data warehouse support for decision-making

The data warehouse, OLAP, data mining and model base, method base and knowledge base are combined to form an advanced comprehensive decision-making support system. The data warehouse can realize the integration and storage of the corresponding data of decision-making theme; OLAP realizes multidimensional data analysis; data mining finds the implicit knowledge and laws in the data warehouse (including the database); the model base and method base achieves the comprehensive assistant decision-making of a large number of models (software package and program); the knowledge base uses knowledge reasoning for qualitative analysis. The structure of the comprehensive decision-making support system is shown in Fig.4.

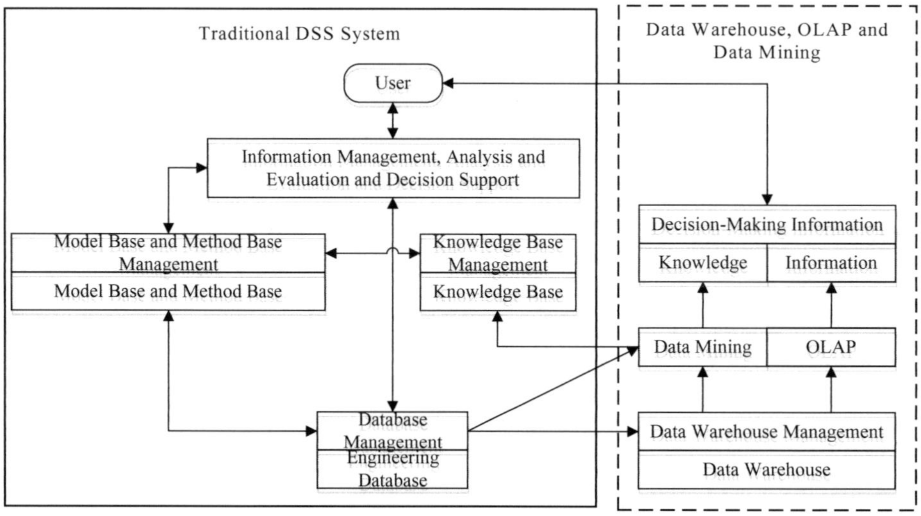

Fig.4 The structure of the comprehensive decision-making support system

The architecture of the comprehensive decision-making support system consists of three levels: the first level is the engineering database system and the model method base system, which is the basis of decision-making support, and provides assistant decision-making information of quantitative analysis for the safety monitoring problem; the second level is the data warehouse which extracts and synthesizes data from the engineering database, and OLAP which analyzes on the basis of the data warehouse and gets comprehensive information; the third level is the combination of data mining and knowledge base. Data mining is applied to data warehouse and database, and the knowledge and the law are extracted which can be directly used in decision-making. At the same time, the knowledge and the law extracted are integrated into the knowledge base and through knowledge reasoning, the qualitative analysis assistant decision-making is achieved.

(2)Safety monitoring data warehouse design

The safety monitoring data warehouse of the Three Gorges project is established according to the structure of the safety monitoring data warehouse system shown in Fig.5.

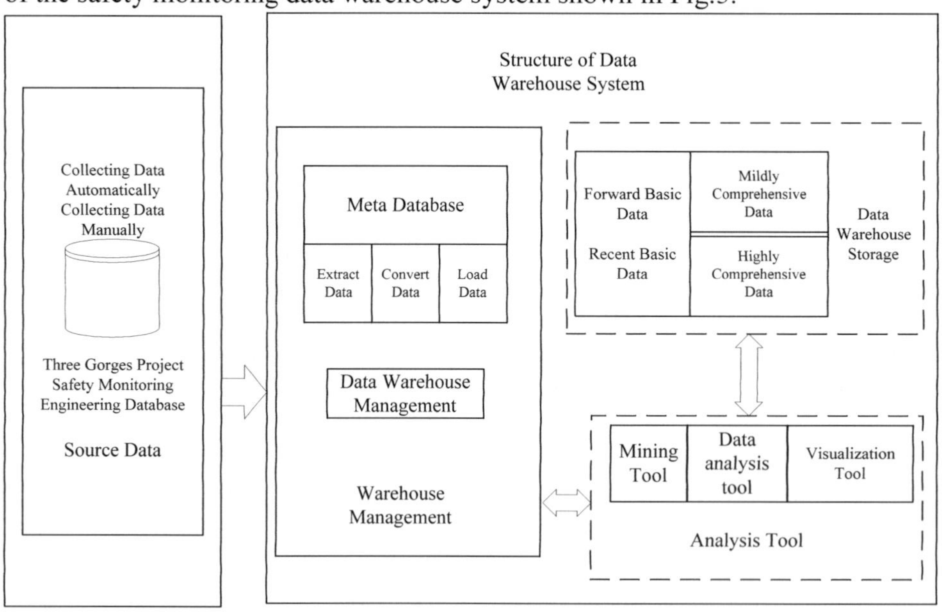

Fig.5 Logical structure of data warehouse system

Source data is automatically collected or manually collected in the safety monitoring system stored in the engineering database.

Warehouse management is the combined process of data extraction, conversion, loading and defining, which makes the system automatically extract data from data sources at different levels and then move the data to the data warehouse, and give maintenance. Metadata is used to store data model, define data structure and manage data safety, maintenance, back-up and recovery.

Analysis tools include data mining tools, data analysis models and model bases, visualization tools and so on. Among them, data mining tools are used to discover new knowledge, rules and new data relationship; data analysis tools are used to form the current basic data layer from the source data according to the required subject of decision-making, and then form a mildly and highly integrated decision-making data layer according to the requirements of the comprehensive decision-making.

Acknowledgement

This work was supported by the research funds from the National Natural Science Foundation of China (grant nos 51139001, 51379068, 51279052, and 51179066), Jiangsu Natural Science Foundation (grant no. BK20140039), Jiangsu Basic Research Program (grant no. BK20160872) and the Priority Academic Program Development of Jiangsu Higher Education Institutions (grant no. YS11001).

References

Chen Bo, Wu Zhongru, Liang Jiachen, et al. 2016, "Time-Varying Identification Model for Crack Monitoring Data from Concrete Dams Based on Support Vector Regression and the Bayesian Framework", Mathematical Problems in Engineering, vol. 2017, Article ID 5450297, 11 pages, https://doi.org/10.1155/2017/5450297.

Wu Zhongru, Gu Chongshi, Hu Qunge, et al. An Investigation on the Dam Safety Comprehensive Appraisal Expert System [J]. International Journal Hydroelectric Energy, 2000,18(2):1-5.

Wu Zhongru. Safety Monitoring Theory and Its Application of Hydraulic Structures[M]. Beijing, China: High Education Press, 2003.

Su Huaizhi, Gu Chongshi, Wu Zhongru. Application of Artificial Intelligent Theory to Dam Safety Monitoring[J]. Dam Observation and Geotechnical Tests, 2000, 24(3):7-9.

Yue Jianping, Lei Weigang. The Mixed Knowledge Bank based on Formulas and Artificial Neural Network for Dam Safety Monitoring[J]. Shuili Xuebao, 2001,5: 54-57.

The 8th International Conference on
Structural Health Monitoring of Intelligent Infrastructure
Brisbane, Australia | 5-8 December 2017

Experimental Study on Optimum Proportion of Carbon Fiber Reinforced Cement Intelligent Mortar

S. Yan[1], S.Y. Zhang[1], C.F. Jin[2]. [1.] *School of Civil Engineering, Shenyang Jianzhu University, Shenyang Liaoning, 110168 - China, email:* syan1962@163.com [2.] *Liaoning Provincial College of Communications, Shenyang Liaoning, 110122 - China.*

Abstract

Carbon fiber-reinforced cement mortar (CFRM) which mixes in ordinary cement mortar with chopped carbon fibers can improve its strength and electrical conductivity. CFRM has the special properties of piezoresistivity which is closely associated with the dispersion of carbon fiber in the preparation of CFRM. In this paper, working performance of the mixture was tested by adding MC, HPMC and HEC with different content of dispersant, respectively. CFRM samples with seven different carbon fiber contents were monotonously loaded to obtain the optimum carbon fiber content. The CFRM with maximized sensitivity for the piezoresistivity was selected as optimum one. The optimum CFRM was subjected to monotonic loading and cyclic loading, and the stress operating range with linear characteristics and the piezoresistivity model was both obtained. The experimental results show that the fluidity of the CFRM decreases and the setting time increases with the increase of dispersant content. When the content of dispersant is 0.2% of the mass of cementing material, CFRM can achieve good fluidity and setting time. The piezoresistivity sensitivity of CFRM was the maximum valve when the content of carbon fiber was 0.6% of the mass of cementing material.

1. Introduction

In recent years, earthquakes, floods, storms and other natural disasters have caused varying degrees of impact on these infrastructures. As a result, the performance of the infrastructures needs to be regularly tested, finding the infrastructures security hidden danger in time, and early warning of possible damage (Housner 1997). Therefore, the evaluation of the safety performance has become an inevitable requirement for future infrastructures, and it is becoming an important field of the development. Smart material based damage identification (DI) technology can play an important role in the field (Bahar 2008).
CFRM is a kind of smart material which is produced with cement mortar as matrix material, in which add chopped carbon fiber as conductive component and other additives, so that the cement mortar or concrete has become a pressure sensitive, temperature sensitive (Seebeck effect) of

intelligent materials and electromagnetic shielding properties, referred to as CFRM (Chung 2000). Compared with ordinary cement based materials, the tensile strength, compressive strength and flexural properties of the composites are greatly improved by adding carbon fiber. It is showed that the short cut carbon fiber as conductive component dispersed in the cement mortar can not only reduce the resistivity of CFRM, but also regularly result in a change in the resistivity with the change of stress state, namely stress sensitive properties (Zhu 2007, Fu 1996). According to functional and structural characteristics of materials, it is usually embedded in the structure as an on-line DI sensor or directly as structural material. The pressure sensitivity can be used to the stress state and damage degree detection, realizing in fields of the online real-time structural health monitoring (SHM) and damage assessment of bridge, dam, houses and other structures, or in monitoring highway traffic (Wang 2013). Therefore, it is of great importance to study the piezoelectric properties of CFRM, such as the electrical and mechanical properties of CFRM, and the quantitative relationship between stress and resistivity.

The principle of pressure sensitivity of carbon fiber reinforced cement-based composites (CFRCC) mainly depends on its conductivity. CFRCC usually exists four kinds of conductive path forms: a path formed by a connected fiber network; a conductive form of a dispersed non-interconnected fiber network; a passage formed by a connected water network; a conductive form through a cement matrix (Peng 2014). In dry condition, the latter two contributions to the conductivity are quite limited, and the conduction is mainly through the carbon fiber conductive network to achieve. On the one hand, under the action of stress, the gap of the CFRM and the hydrated colloidal particles, which is the potential barrier, has increased or decreased. It results in the change of their ability to hinder the electron transition (Wang 2005, Li 2015, Zhao 2011). Under the compressive stress, the porosity and colloidal particles become smaller, the potential barrier is smaller and the probability of electron transition becomes larger, and the resistivity of the CFRM is smaller. On the contrary, the tensile stress at the barrier becomes larger, the probability of electron transition becomes smaller, and the resistivity of the CFRM becomes larger (Wang 2007, Wen 2006b, Wen 2005). On the other hand, the pressure of carbon fiber plug phenomenon occurs, when the carbon fibers in the cement mortar occurred during insertion and removal, the number of overlap between the carbon fibers are changed, the original local has formed or connected conductive network changed dense or loose, changing the conductivity of carbon fiber cement mortar seepage the conductive mode, displaying changes of resistivity of mortar (Wen 2001, Cao 2001, Zhou 2011).

With the development of CFRM, the influence of the preparation process of CFRM, the dispersion of fiber and the length of fiber on the strength and conductivity of CFRM are relatively mature (Wen 2006a, Wen 2006c). The description of the type and dosage of dispersant on the properties of the mixture of carbon fiber reinforced cement mortar are unclear. In the range of the carbon fiber content not affecting other properties of carbon fiber reinforced cement mortar, the sensitivity of carbon fiber content on the adhesion of influence is less reported. The goal of this paper is to establish a model which can describe the relationship between electrical performance and mechanical one for CFRM. Toward this, an experimental investigation is performed. The CFRM with three different dispersants was added to study the mortar fluidity, carbon fiber dispersibility and coagulation time. The results can help to evaluate the optimal proportion of concrete work stress through the research on the influence of pressure sensitivity, carbon fiber cement mortar with different carbon fiber dosage, and obtain the linear stress working range under the optimal dosage, provide reference for its application in practical engineering.

2. CFRM Fluidity Based Experiment

2.1 Test Materials and Setup

Raw materials of specimens in the test are described as follows: 42.5 (brand) ordinary Portland cement, 5mm PAN short carbon fiber (the main performance parameters are shown in Table 1), dispersant (HPMC, MC and HEC), water reducing agent, defoamer (Tributyl phosphate), and silica fume. The mix ratio of the CFRM is shown in Table 2.

Table 1. Carbon fiber properties

Fiber diameter /μm	Tensile strength /GPa	Tensile modulus /GPa	Carbon content	Elongation	Density /g/cm^3	Volume resistance /Ω·cm
9	5.04	232	≥95%	2.10%	1.8	1.5×10^{-3}

Table 2. The mix ratio of CFRM

Water	Cement	Sand	Water reducer	Defoaming agent	Silica fume	Carbon fiber
0.5	1	1.5	0.80%	0.20%	15%	0.60%

The setup used in the test is JJ-5 cement mortar mixer and fluidity tester, shown in **Figure** 1.

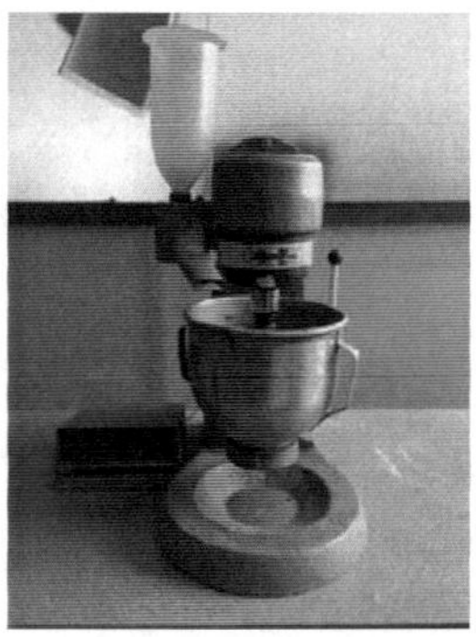

Figure 1. The experimental setup: the cement fluidity tester (left) and mortar mixer (right).

2.2 Test Method

In the experiment, carbon fibers were dispersed into cement mortar by wet mixing method (Dong 2016). The producing process was described as follows. Firstly, the half of the total amount of water was used to disperse the dispersant. After the dispersant was fully dissolved, the carbon fibers were added to the solution. Then the defoaming agent was added after fully mixing. The cement, sand and silica fume were mixed into the container. Finally, the dispersed carbon fibers

were added to the vessel and stirred for 3 minutes, and a water reducing agent was added.
The test design was referred to the code of determination of fluidity of cement mortar (GB/T 2419-2005) of China. By using the jump table, the fluidity of the cement mortar under a certain vibration condition was measured according to the expansion range. In this study, the fluidity test with three dispersants of HPMC, MC and HEC were performed by the amount of admixture of 0%, 0.2%, 0.4%, and 0.6%, respectively.

2.3 Test Results and Analysis

The fluidity of CFRM under different dispersants and its measured content in the fluidity test was shown in **Figure** 2. It can be seen that the fluidity degree of CFRM respectively mixed with HPMC, MC and HEC of three dispersants shows a decreasing trend with the increase of dispersant dosage.

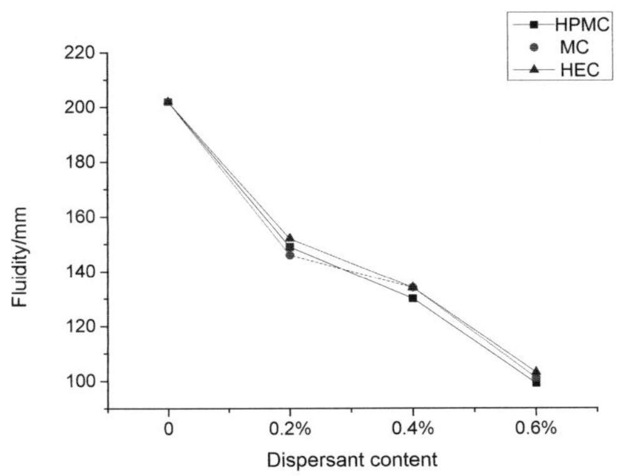

Figure 2. Effect of different dispersants and contents on the fluidity of CFRM.

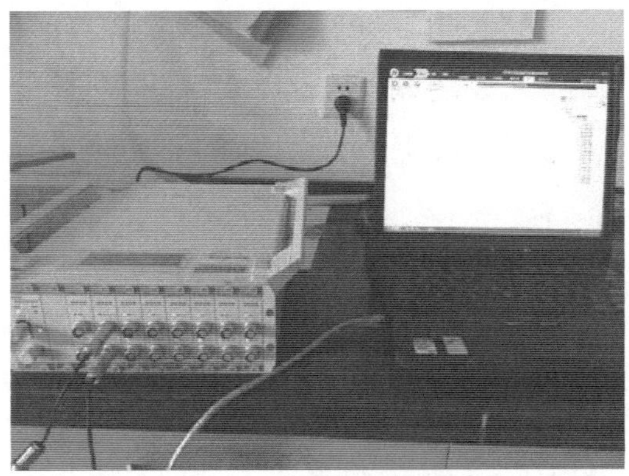

Figure 3. The CFRM dispersion based ER Experimental setup.

When the dispersant was dissolved in water, a gel solution with a certain viscosity can be formed, so that the carbon fibers can be dispersed into a single filament without agglomeration. However, when the amount of dispersants is too high, the viscosity of the solution will be too large, which is not good for the dispersion of carbon fibers, leading to the reduced fluidity of the mixture, which is not good for the vibration molding of the CFRM block.

3. CFRM Dispersion Based Electrical Resistance Experiment

3.1 Test Materials and Setup

The goal of the experiment is to investigate the influence law of the CFRM dispersion properties on the electrical resistance (ER) by using different types and amounts of the dispersants. Toward this, the used material and ratio in the test are consistent with the above-mentioned fluidity test.
The setup used in the test is a JJ-5 cement mortar mixer, a ZT96 cement mortar vibration compacting table, a standard curing box, a Donghua DH5922 dynamic signal acquisition system, a DC power supply, and a 500Ω standard resistance. The setup of the test is shown in **Figure** 3.

3.2 Test Principle

In this study, the carbon fiber ER method was used to evaluate the carbon fiber dispersion properties (Han 2016). Under the unchanging condition of raw materials, curing process and test methods, the dispersion properties of carbon fibers in cement-based composite materials can be indirectly evaluated by testing electrical resistivity for different components and content of carbon fibers, and by calculating the ER variation coefficient. The specimen ER depends on the number of carbon fibers to form conductive paths, so it is controlled by the carbon fiber distribution in the worst section. In the case of the certain carbon fiber content, the better dispersion of carbon fibers will bring the better electrical conductivity, and the smaller coefficient of resistivity variation. Formula (1) and (2) are described as follows:

$$CV_R = \frac{S(R)}{\bar{R}} \times 100\% \tag{1}$$

$$S(R) = \left[\frac{1}{n-1}\sum_{i=1}^{n}(R_i - \bar{R})^2\right]^{\frac{1}{2}} \tag{2}$$

where n—the sample number (n=6).

$\bar{R}$—the mean value of the tested ER.

$S(R)$—the standard deviation of the tested ER.

CV_R— the electrical resistivity coefficient of variation (ECVR, CV_R) of the tested ER.

When R_i=, $S(R)$=0 and CV_R=0, the carbon fibers can be completely dispersed. The bigger CV_R will bring the more uneven dispersion of carbon fibers.

3.3 Test Methods

The wet mixing method was used to disperse carbon fibers in the experiment. The specific process is shown in **Figure** 4. All specimens usually have the six standard steps to produce before the test. The control time is very important and keeps each step the same time for the different specimens.

The mix ratio in the test was consistent with that of the above fluidity test. The amounts of HPMC, MC and HEC were 0, 0.2% and 0.4%, respectively. The specimen size of 40mm×40mm×160mm is designed. The total of 7 groups of specimens was produced, in which one group is not mixed with dispersants, and others are mixed with the different amount of dispersants. Each group had three test specimens. The 21 blocks were firstly tested for the anti-bending strength, and the 42 blocks were then tested for the compression strength.

The schematic of the test system is shown in **Figure** 5. A 5V DC power supply and a series circuit of standard ER and the tested specimen were used in the test. R_0 is a standard resistor of 500Ω and R_x is the measured resistance. The voltages across the two resistors were respectively measured by the Donghua DH5922 dynamic signal acquisition system. The following Equation (3) can be applied to calculate the ER of the tested block by using the Ohm's law.

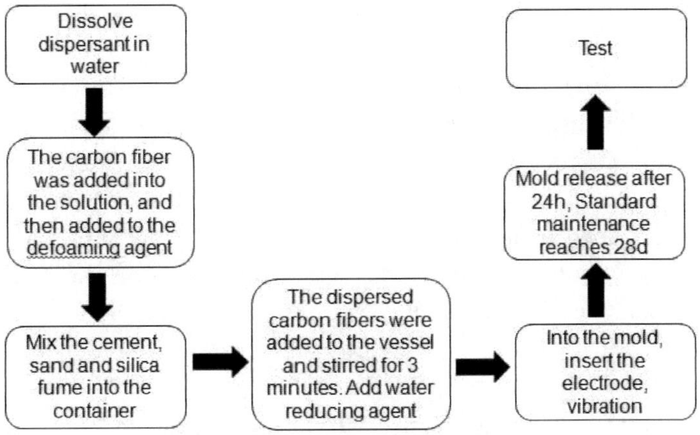

Figure 4. The CFRM preparation process.

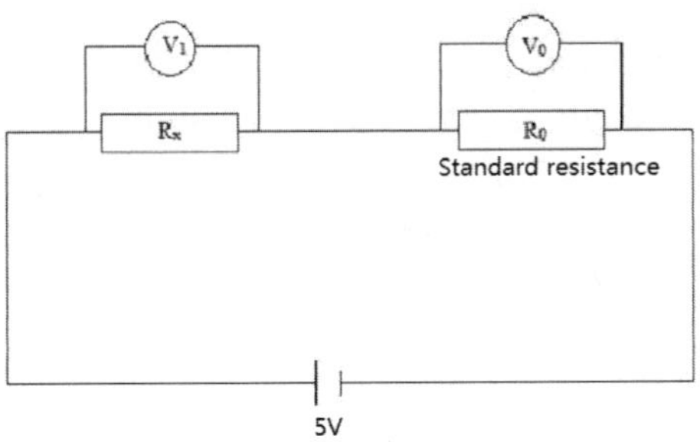

Figure 5. The schematic of the carbon fiber dispersion test system.

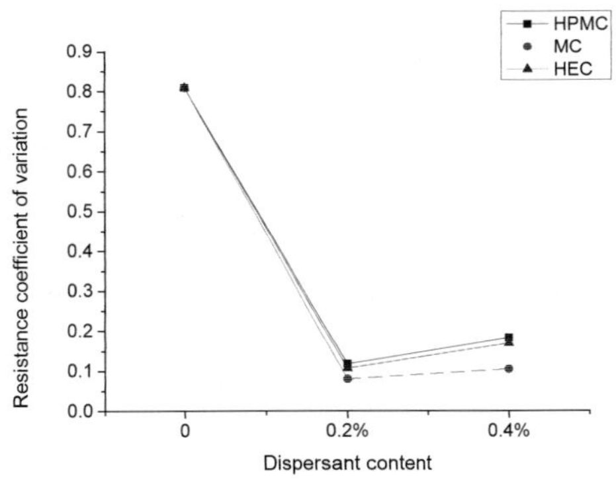

Figure 6. The relationship between CVR and dispersants.

$$R_x = \frac{V_1}{V_0} R_0 \tag{3}$$

where V_1 and V_0 are the volts of the tested block and the standard resistance, respectively.

The relationship between the three dispersants obtained by the hardening resistance test method and their contents and the CV_R is shown in **Figure** 6.

It can be seen from the figure that the carbon fiber dispersion in the cement-based mortar can be greatly improved by the addition of dispersants. When the dispersant dosage is 0.2% of the mass of the cementitious material, the CV_R of the specimen corresponding to the MC type is the smallest. The CV_R of the MC and HEC types increased slightly with the increase of the content of HPMC, while the CV_R decreased slightly with the increase of the dosage. The carbon fiber amounts for HPMC, MC and HEC corresponding to the lowest CV_R have optimal values of 0.2%, respectively, which can be explained by the carbon fiber dispersion mechanism.

The dispersant is dissolved in water to form a solution with a certain viscosity, and the surface of the carbon fiber is formed with a thin layer of lubricating oil after the adsorption of the dispersant solution, so that the friction of the fiber is reduced, and the agglomeration phenomenon is not formed. The addition of dispersant can reduce the surface tension of liquid. The bubble has a certain role in promoting the dispersion of the fiber, even if the fibers are separated from each other, thus preventing the fiber from contacting, preventing the fiber from sticking and forming, and the fiber is uniformly dispersed in the solution. However, when the dispersant dosage is too low, it is not conducive for carbon fiber dispersants to be fully coated by the solution. When the dispersant dosage is too high, the solution viscosity is too large, impacting on the dispersant liquidity, and making the carbon fibers bonded together, which is not conducive to the carbon fiber full dispersion.

4. Effect of Carbon Fiber Content on Piezoresistivity of CFRM

4.1 Piezoresistivity Mechanism

Domestic and foreign scholars on the CFRM piezoresistivity mechanism of a wide range of studies, based on the conductivity mechanism studies have shown that there are roughly the following models: (1) Carbon fiber insertion and extraction effect model; (2) Conductive channel model; (3) Tunneling model. The conductivity and piezoresistivity of CFRM are mainly determined by the degree and content of carbon fiber lap. Under normal circumstances, the above-mentioned conductive model exists at the same time, but by the carbon fiber content and the degree of overlap in some way the main way.

4.2 Test Ratio

The materials used in the test were consistent with the above test, and the test pieces were made in a size of 40 mm×40 mm×160 mm. The test compositions were as shown in Table 3.

Table 3. Test ratio

Number	Water /g	Cement /g	Sand/g	Water reducer/g	Dispersant /g	Defoaming agent/g	Silica fume/g	Carbon fibre/g
a	330	561	990	13.2	1.32	1.32	99	2.64
b	330	561	990	13.2	1.32	1.32	99	3.96
c	330	561	990	13.2	1.32	1.32	99	5.28
d	330	561	990	13.2	1.32	1.32	99	6.60

4.3 Experiment Method

The content of carbon fiber was 0.4%, 0.6%, 0.8%, 1.0% of the mass of the cementitious material, the equipment used in the test and the type and amount of the third part dispersant were as same as the previous test.

4.4 Test Results and Analysis

The electromechanical relationship of each test block is shown in Figure 7, each of which consists of three pressure-sensitive test blocks. Where the *x*-axis is represented by σ/σ_{max} and the *y*-axis resistance change rate $\Delta R/R_0=(R-R_0)/R_0$ is used to achieve dimensionless variable. It can be seen that the curve exhibits negative piezoresistivity, that is, the change rate of resistivity decreases with the increase of stress, and the resistance change rate is minimized when the stress reaches the maximum value.

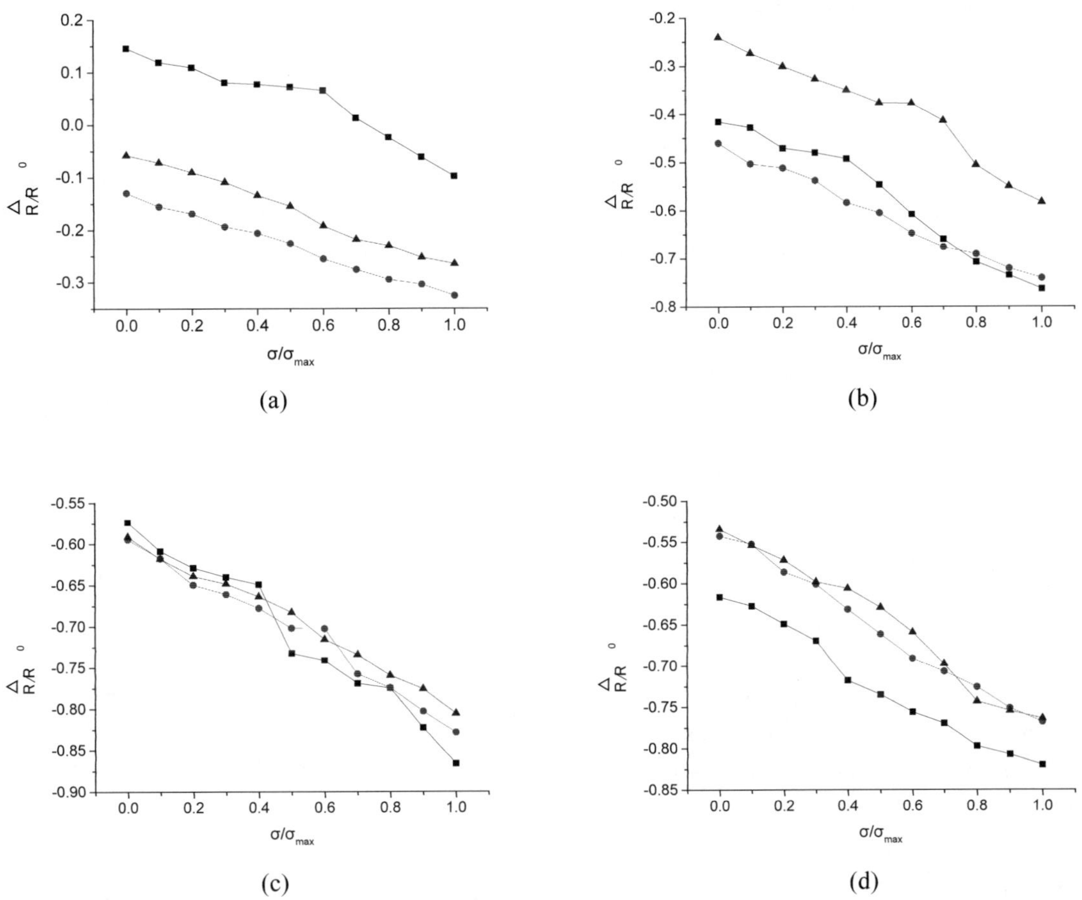

(a) (b) (c) (d)

Figure 7. The piezoresistivity curve of carbon fiber content of: (a) 0.4%, (b) 0.6%, (c) 0.8% and (d) 1.0%.

It is found that the three stages of the curve are linearly fitted, that is, the form of $y=kx+b$, and the fitting formula is shown in Table 4. Among them, the slope k of the fitting formula is defined as the sensitivity, and the physical meaning is the magnitude of the resistivity change rate under the unit force. The b is the case where x=0 in the fitted formula, and its physical meaning is the initial resistivity change value of the specimen to be tested. The value can be calculated from the initial resistance measured by each test block.

Table 4. The average fitting formula of each group

Number	fitting formula
a	y=-0.215x+b
b	y=-0.331x+b
c	y=-0.239x+b
d	y=-0.235x+b

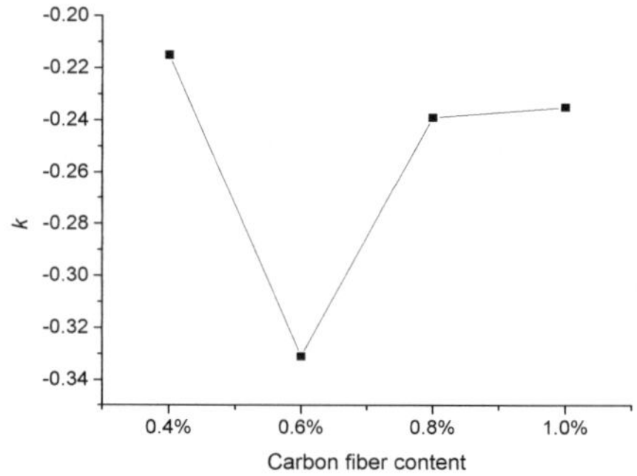

Figure 8. Comparison of piezoresistivity sensitivity values of each group.

When the sensitivity $k<0$, k smaller, indicating the higher the sensitivity of the pressure sensitivity curve. The pressure sensitivity of each group is shown in Fig. It can be seen from the figure, when the carbon fiber content of 0.6%, the sensitivity of the best sensitivity.

5. Conclusions

In this paper, the optimal dosage of dispersant and carbon fiber in CFRM was obtained by the experimental method. The main conclusions are as follows:

(1) HPMC, MC, HEC three dispersants with the increase in the amount of CFRM mixture will reduce the flow, and thus affect the production of CFRM vibrator. Therefore, the content of these three dispersants is not too large;

(2)The incorporation of HPMC, MC and HEC dispersants can improve the dispersibility of carbon fiber in cement paste. When the content of the three dispersants is 0.2%, the dispersion of carbon fiber is the best.

(3)The amount of carbon fiber has a reasonable range, below or above this range, will reduce the performance of CFRM. From the CFRM piezoresistivity sensitivity analysis, CFRM test block has the best sensitivity to piezoresistivity when the carbon fiber content is 0.6% in the case of the test piece according to the above test size and proportion.

Acknowledgments

This work was financially supported in part by the Natural Science Foundation of Liaoning Province of China (2015020595, 20141092), and the Science Foundation of Liaoning Education Department of China (LJZ2016029).

References

Bahar, D & Salih, Y 2008, 'Thermoelectric behavior of carbon fiber reinforced lightweight concrete with mineral admixtures', *Journal of New Carbon Materials*, vol. 23, no. 1, pp. 21-24.

Cao, JY & Chung, DDL 2001, 'Carbon fiber reinforced cement mortar improved by using acrylic dispersion as an admixture', *Journal of Cement and Concrete Research*, vol.31, no.11, pp.1633-1637.

Chung, DDL 2000, 'Cement reinforced with short carbon fibers: a multifunctional material', *Journal of Composites Part B*, Vol.31, pp.511-526.

Dong, X 2016, 'Study on mechanical properties of carbon fiber reinforced cement mortar'. *Journal of Sichuan Cement*, Vol.12, no.8, pp.62-63. (in Chinese)

Fu, XL & Chung, DDL 1996, 'Self-monitoring of fatigue in damage in Carbon fiber reinforced cement', *Journal of Cement and Concrete Research*, vol.26, no.1, pp.15-20.

Han, BG, Zhang, LQ & Zhang, CY 2016, 'Reinforcement effect and mechanism of carbon fibers to mechanical and electrically conductive properties of cement-based materials', *Journal of Construction and Building Materials*, Vol.125, pp.479-489.

Housner, GW, et al 1997, 'Structural control: past, present, and future', *ASCE, Journal of Engineering Mechanics*, vol.123, no.9, pp.897-971.

Li, BL, Peng, L & Wang, XY 2015, 'Study on the mechanical properties of carbon fiber reinforced cement mortar', *Journal of Sichuan Building Materials*, vol.41, no.5, pp.31-32. (in Chinese)

Peng, HL, Gao, PW & Wang, HQ 2014, 'Study on mechanical and electrical properties of cement mortar with different functional materials', *Journal of Silicate Bulletin*, vol.33, no.1, pp. 208-210. (in Chinese)

Wang, S & Chung, DDL 2007, 'Negative piezoresistivity in continuous carbon fiber epoxy-matrix composite', *Journal of Mater Sci*, vol.42, pp.4987-4995.

Wang, YL & Zhao, XH 2005, 'Positive and negative pressure sensitivities of carbon fiber-reinforced cement-matrix composites and their mechanism', *Journal of Acta Materiae Compositae Sincia*, vol.22, no.4, pp.40-44. (in Chinese)

Wang, YL, Zhao, XH, Zhu, DL & Du, JH 2013, 'Sensitivity and mechanism analysis of carbon fiber cement based composites', *Journal of Silicate Bulletin*, vol.32, no.9, pp.1817-1819. (in Chinese)

Wen, S & Chung, DDL 2001, 'Uniaxial compression in carbon fiber reinforced cement, sensed by electrical resistivity measurement in longitudinal and transverse directions', *Journal of Cement and Concrete Research*, vol.31, pp.297-301.

Wen, S & Chung, DDL 2005, 'Pitch-matrix composites for electrical, electromagnetic and strain-sensing applications', *Journal of Mater Sci*, vol.40, pp.3897-3903.

Wen, S & Chung, DDL 2006, 'Self-sensing of flexural damage and strain in carbon fiber reinforced cement and effect of embedded steel reinforcing bars', *Journal of Carbon*, vol.44, no.8, pp.1496-1502.

Wen, S & Chung, DDL 2006, 'Model of piezoresistivity in carbon fiber cement', *Journal of Cement and Concrete Research*, vol.36, pp.1879-1885.

Wen, S & Chung, DDL 2007, 'Piezoresistivity-based strain sensing in carbon fiber-reinforced cement', *Journal of ACI Materials*, vol.104, pp.171-179.

Zhao, XH, Li, GB & Wang, YL 2011. 'Piezoresistivity of carbon fiber reinforced cement-matrix composites', *Journal of Acta Materiae Compositae Sincia*, vol.28, no.5, pp.215-217. (in Chinese)

Zhou, L & Liu, HT 2011, 'Calculation method of flexural bearing capacity of carbon fiber reinforced concrete beam', *Journal of Applied Mechanics and Materials*, vol.71-78, pp.5080-5083.

Zhu, SR & Chung, DDL 2007, 'Theory of piezoresistivity for strain sensing in carbon fiber reinforced cement under flexure', *Journal of Mater Sci*, vol.42, pp.6222-6233.

The 8th International Conference on
Structural Health Monitoring of Intelligent Infrastructure
Brisbane, Australia | 5-8 December 2017

Fatigue assessment in steel bridges: Integrating field measurements and numerical modelling to compute hot spot stresses

J. Kwad[1,2], P. Kripakaran[1]
[1]University of Exeter, College of Engineering, Mathematics and Physical Sciences, Vibration Engineering Section, North Park Road, EX4 4QF, +44 (0)1392 72 6421, Exeter, United Kingdom, Email: Jeek201@Exeter.ac.uk, *[2]Civil Engineering Department, University of Anbar, Ramadi, Anbar, Iraq*

Abstract

The hot spot stress approach is often employed to evaluate the fatigue strength of welded structures in cases where the nominal stress is difficult to estimate reliably because of geometric and/or loading complexities. Its predictions are mostly conservative since the nominal stresses are estimated for code-specified loadings that are much larger than the loads experienced by a bridge. This paper presents a novel methodology to evaluate the hot spot stress at a weld detail in a fatigue critical connection using strain data measured further away from areas of significant stress concentration with a finite element model (FEM) of the connection. Strains measured at various points around the physical connection are used to compute the forces and moments applied at the connection. These forces are then applied to a numerical model of the connection to predict the stresses at the detail. The methodology is investigated with a full-scale case study: The Bascule Bridge in Exeter (UK). The stress time-history produced using the proposed approach with in-situ strain measurements is compared to that generated using the modified hot spot stress approach. Results show that the predicted stress time-history is more accurate and reliable, and hence promises to be more appropriate for fatigue evaluation.

1. Introduction

Metallic bridges of all ages, including those that are approaching or have exceeded their design life and those that are relatively new having been built over the last few decades, are vulnerable to fatigue damage. Fatigue damage is progressive damage created as a consequence of cyclic loading and is a concern mainly at locations of high stresses such as connections in civil structures. In this paper, we specifically focus on the welded details in metallic bridges although the presented ideas are applicable to any metallic structure. Fatigue risk is particularly higher for older bridges due to a combination of much greater traffic volumes in recent years and significant age-related deterioration (Zhou, 2006). In the UK, there are 6000 metallic bridges, which are over 50 years old, and fatigue is highlighted as the major risk factor to their bridge safety (Adasooriya & Siriwardana, 2014). Thus, the assessment of the remaining fatigue life of steel bridges is of great importance both nationally and internationally for making decisions regarding rehabilitation and replacement of existing bridges.
Significant amount of research has gone toward evaluating effectively and accurately the remaining fatigue life of metallic bridges. However, most of the current approaches for fatigue life assessment focus on the use of fatigue design rules (Zhou & Chan 2007), which has two key limitations. First, these approaches are based on the nominal stress in the desired section, without taking account of the local stress concentration effects in a specific weld detail (Park & Kim 2014). Moreover, in the

case of large structures with complex details, an accurate estimation of the nominal load effects in these details are often difficult to obtain (Tveiten et al. 2007; Ye et al. 2012). Using the nominal stress in such cases can yield unrealistic results. Second, these approaches, since they exaggerate greatly the actual applied live loads and in turn the stress cycles, often result in overly conservative estimates of remaining fatigue life. In some cases, the approaches even gives a remaining fatigue life that is negative, implying that the bridge should have failed by now while in reality the structure is in-service and without any visible evidence of fatigue cracking (Kwon et al. 2012). The unnecessary retrofits or repairs necessitated by such conservative assessment procedures can significantly increase bridge maintenance costs.

Consequently, an alternative method proposed in several standards such as Eurocodes is a hot spot stress method. It enables assessing the fatigue strength of welded structures in cases where a clear definition for the nominal stress cannot be made because of geometric and/or loading complexities (Aygül et al. 2012; Cho et al. 2010; Liu et al. 2014). Hot spot stress is calculated commonly through finite element analysis (FEA) of a numerical model of the relevant welded connection for standard fatigue design loads. The rules and recommendations for computation of the hot spot stress including the detailed process of experiments and modelling are provided by The International Institute of Welding (IIW) (Hobbacher 2009). A major advantage of using the hot spot stress method is that stress concentrations in the locality of the weld toe are taken into consideration in the fatigue stress evaluation. A second advantage is that because the fatigue life depends on the weld type rather than the constructional details as for the nominal stress method, fewer number of S-N curves is required for fatigue life evaluation. However, the approach still results in overly conservative estimates of remaining fatigue life since it uses design daily traffic loads which are much larger than real traffic on bridges (Aygül et al. 2012).

With recent advances in sensing technology, field strain measurements have been increasingly employed in assessing the remaining fatigue life of steel bridges. These are found to be more reliable than assessments based on stress predictions purely from numerical simulations (Schumacher & Nussbaumer 2006). Strain sensors are usually installed further away from fatigue-critical locations to obtain the nominal strains and stresses (Ni et al. 2010). The nominal stresses are multiplied by the Stress Concentration Factor (SCF) to take into account the stress concentration and dimension effects. This approach is now widely recognised and known as a modified hot spot stress method in many codes and standards such as The International Institute of Welding (IIW) 2000 and (EN 1993-1-9 2005). However a recent comprehensive review by (Saini et al., 2016) examining the SCFs of welded joints of metallic structures concludes that accurately computing the SCFs, which determine directly the reliability of the modified hot-spot stress method, is difficult due to its dependence on loading conditions which are generally unknown.

Few researchers have attempted to evaluate hot spot stresses at welds using in-situ measured data. (Schumacher 2003, Schumacher & Nussbaumer 2006) showed that the measured hot spot stresses at locations of fatigue cracking were lower than the hot spot stresses calculated using current design codes. However, measuring directly the hot spot stress is a challenging task for many structures due to site limitations. Pasquier et al (2016) proposed a methodology that enables evaluating hot spot stresses using finite element models of the bridge that were identified based on measurements from static load tests. The internal stresses, namely axial, shear and bending stresses in the bridge elements, that were predicted by the models for traffic loads detected using a weigh in-motion (WIM) system were used to compute hot spot stresses through stress amplification with SCFs. This methodology however overestimates hot spot stresses due to the known conservativeness in SCF values (Schumacher & Nussbaumer 2006). Moreover, since the stresses were not measured in-situ but predicted through models of the bridge, modelling uncertainties such as due to environmental conditions (e.g. temperature) and local conditions such as actual speed and location of vehicle, and road profile are not taken into account.

This paper proposes a novel methodology for fatigue life assessment that is based on computing the in-situ hot-spot stresses experienced by plate weld details. It integrates field measurements of strains taken from around the weld detail with a numerical model of the component having the weld detail to evaluate the in-situ hot-spot stresses. Hence unlike the modified hot spot stress method, knowledge of the SCF is not required for fatigue life assessment but its effect is instead explicitly accounted for within the modelling process. Also, as in-situ strains are measured around the detail, dynamic effects and variations due to ambient conditions are accounted for in the methodology. The novel methodology is illustrated in this paper using measurements from a full-scale bridge.

2. Methodology

The methodology for fatigue life assessment proposed in this paper is illustrated in Figure 1 and is broadly as follows. The first step is to create a numerical model of the full-scale steel bridge under investigation to identify the fatigue-critical bridge connections. Then we develop a local model of the identified connections. The local models are linear elastic models that comprise all the connection components (e.g. angles, or weld) and small portions of the structural elements that are part of the connection such as the beams that make up a main girder-stringer connection. Simultaneously we develop and implement a field measurement plan, for the fatigue-critical connections. This will include installing strain gauges on the structural elements relevant to the connection in a manner that will enable evaluating the time-history of internal forces that are transferred via the connection. These force time histories are then applied to the local model to evaluate the time histories of relevant hot-spot stresses. The hot-spot stress time-histories are then used for fatigue life assessment using standard approaches such as using the Miner's Rule.
The individual steps in the methodology are now explained in detail. To create the local model, we use the International Institute of Welding (IIW) recommendations (Niemi et al. 2006) to define the mesh technique, type of critical hot spot points and stress extrapolation techniques. In the IIW recommendations, three types of hot spot stress points *a*, *b* and *c* are classified under welded details according to the location of hot spot and weld form. *a* and *c* are for the weld toe on base plate where welds attaching a stiffener plate and *b* corresponds to the weld toe at base plate edge. For types *a* and *c*, linear extrapolations should be used and stress evaluation points are on nodal points $0.4t$ and $1.0t$ away from the hot spot weld toe, where t is the thickness of base plate. For the hot spot stress of type *b*, stress evaluation points should be coincident with the nodal points at 4mm, 8mm and 12mm distances from the weld toe, and quadratic extrapolation should be used. The structural elements that intersect at the connection are not modelled fully but only a short length as sufficient to ensure stress concentration effects from the connection are absent at the boundary.
The field measurement plan needs to consider several factors including strain gauge locations and sampling rates for data collection and processing. Measuring strain histories for this purpose requires a modern data acquisition system with large storage capacity and a high sampling rate (~50 Hz). Strain gauges are installed in a manner that will allow evaluating the internal forces. To this end, install the sensors at locations corresponding to the boundary of the local model. For each beam element, install sufficient strain gauges to enable computing the three internal forces - bending moment, axial force and shear force. We assume that the element experiences negligible torsion or bending about the minor axis. Typically this will mean installing strain gauges at the top and bottom of the cross-section and a rosette near the neutral axis. Together these gauges will enable computing the exact location of the neutral axis, and subsequently the internal forces through simple Euler-Bernoulli beam bending assumptions.
The internal forces are computed for each measurement time-step are applied to the local model of the connection. Since the local model is a linear-elastic model with stresses linearly related to the apply loads, this process of evaluating the model for the internal forces can be further simplified.

We compute load scale factors (LSF), which are essentially the stress at the hot-spot stress location for an internal force of magnitude unity. The internal forces evaluated from in-situ strain measurements are multiplied by the corresponding LSFs to evaluate the relevant hot-spot stresses. Once the stress time-histories are available, we calculate the stress histograms, representing the number of cycles versus the corresponding distribution of stress ranges, by using a cycle counting algorithm, e.g. the rainflow algorithm (ASTM E1049 - 85 2005). Then suitable S-N curves that describe the fatigue resistance of the detail are adopted from the Eurocode 3 (EN 1993-1-9 2005). The fatigue damage factor is evaluated using a linear damage accumulation model, as proposed by Miner (Miner M 1945), according to which the damage factor, *D*, is given by Equation (1),

$$D = \frac{n_1}{N_1} + \frac{n_2}{N_2} + \frac{n_3}{N_3} + \ldots = \sum_{i=1}^{k} \frac{n_i}{N_i} \leq 1.0 \tag{1}$$

where n_i is the number of applied load cycles for a given stress range and N_i denotes the number of resisting load cycles for a given stress range. Fatigue failure is reached for D = 1. The damage factor is used to arrive at an estimate of the fatigue life of the component, i.e. the time remaining before the component or structure could be expected to fail by fatigue based on the current loading conditions.

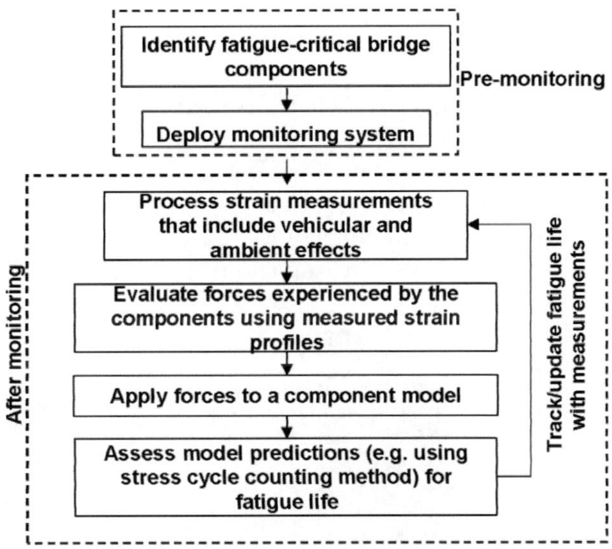

Figure 1: Proposed Methodology

3. Case study - The Bascule Bridge

The Bascule Bridge, which is 17.28 m in total length, is a metallic bridge in Exeter UK that carries road traffic. A global finite element model for the Bascule Bridge was originally developed by (Jalil et al. 2017). More than 28,496 nodes and 26,891 shell elements are included in the Bascule Bridge FE model. The global model has already been shown to give reliable stress results for both static and dynamic loading (Jalil et al. 2017). Details of the model are not presented here for reasons of brevity. Fatigue studies (Jalil et al. 2017) using design traffic loads based on National Annex to BS EN 1991-2-3 (2003) identified that the most vulnerable detail with respect to fatigue are the fillet welds in the main girder-cross beam connections. In these connections, a fillet weld is used to attach a vertical stiffener to the web of the main girder. Each vertical stiffener is also connected to the cross-beam via a series of bolts.

3.1 Local hot spot analysis for critical member

In this section, a local model of the connection is presented. It consists of an assembly of a main-girder, cross beam and vertical web stiffener. The welded connection is assumed to be fully rigid within the FE model. This assumption is acceptable as the large number of heavy bolts and weld in this connection are likely to make the connection behave in a rigid manner (Zamiri Akhlaghi, 2009).

This model, shown in Figure 2 (a), was made based on the IIW recommendations (Hobbacher, 2009) for analysis with a coarse mesh composed of shell elements without modelling the actual weld geometry. Three-dimensional 8-noded shell elements were used to model the cross beams, main girders and vertical stiffener. Figures 2 (b) and (c) shows the two types of boundary conditions used for the two ends of the main-girder. In Figure 2 (b), all translational and rotational degrees of freedom of the node at the neutral axis of the main beam either side of the cross-beam is assumed to be fixed based on an almost zero strain assumption at the neutral axis. In Figure 2 (c), all the nodes on the boundary of the main beam either side of the cross-beam are assumed to be fixed based on the Garg et al (1982) assumption. While both are different from the real in-situ boundary conditions, these are still reasonable as the relative vertical displacement at the two ends of the main beam is likely to be minimal. However, as the rotations are difficult to gauge, we attempt the two boundary conditions.

The ANSYS® parametric design language (APDL) was used to build the model of the connection. To comply with the IIW recommendations, the element size at the type '*a*' hot spot region was kept as 8mmx8mm (i.e. $0.4t$x$0.4t$ with the base plate thickness t = 20mm) and 20mmx20mm (i.e. $t*t$) in the elements adjacent to those elements. The local model mesh consists of 29,400 elements with a total of 31,659 nodes. Typical analyses running times are approximately 1 minute on a PC with 8GB of RAM. Materials were assumed linear elastic and isotropic (E=205 GPa and ν=0.3).

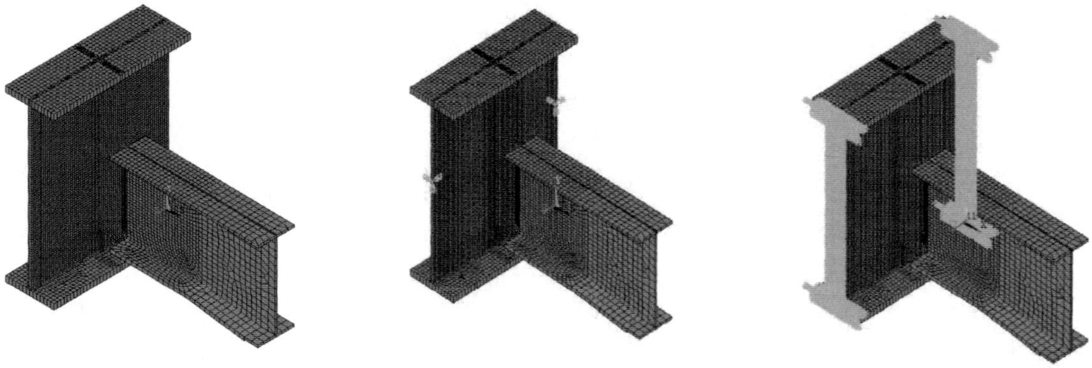

(a) (b) (c)

Figure 2: (a) Shell element model without the welds modelled, (b) Boundary applied on the model (fixed two nodes at natural axis) and (c) Boundary applied on the model (fixed all nodes).

Unit internal forces are applied on the end of the cross beam at 1040 mm distance from the connection in order to calculate LSFs at the hot spot reference points of the connection. The results are illustrated in Table 1.

Table 1. Loading scale factor of the Bascule Bridge typical connections

Boundary condition scenario	Unit load	Stress factor at 10 mm from weld toe (MPa)	Stress factor at 20mm from weld toe (MPa)
Scenario 1	1 kNm	0.064985	0.065395
	1 kN	1.148000	0.471500
	1 kN	0.102500	0.164000
Scenario 2	1 kNm	0.031980	0.012505
	1 kN	2.740850	1.732250
	1 kN	1.250500	0.779000

3.2 Verification of the methodology

A monitoring system consisting of 7 strain gauges, data logger and power source was installed in locations of interest near the bridge. K- and J-type strain gauges were instrumented in the real connection following the IIW recommendations as shown in Figure 3. Strains were recorded from the real live-loads response at 2000 Hz sample rate for a period of 6 hours. 2000 Hz is a lot higher sampling rate than required but this was used as the instrumentation offered this capability. The purpose of the monitoring was to directly measure the strains caused by the site specific traffic, in various locations of the strain gauges. Noise in the measurements is eliminated from measured strains using a low-pass filter. Following the methodology described previously, the internal forces are computed for measurement each time-step. The forces are used in conjunction with the finite element models discussed in section 3.1 and Matlab to determine the stress time history at the weld toe of the connection.

A comparison has been made between measured and calculated stresses time history at hot spot points i.e the weld toe using two type of boundary conditions. The stress from multiplied LSFs from the FE analyses and sectional forces from measured strain were calculated at the same points at which the strain gauges were placed. This comparison was also used as a means of verifying the FE model. The results show a good agreement between experimental stresses and stresses from numerical model under real life loads for scenario one of the boundary conditions. The error is less than 2% as illustrated in Figure 4 and 5. Hence, the local FE model seems to be appropriate to simulate the local stress values induced by the real traffic.

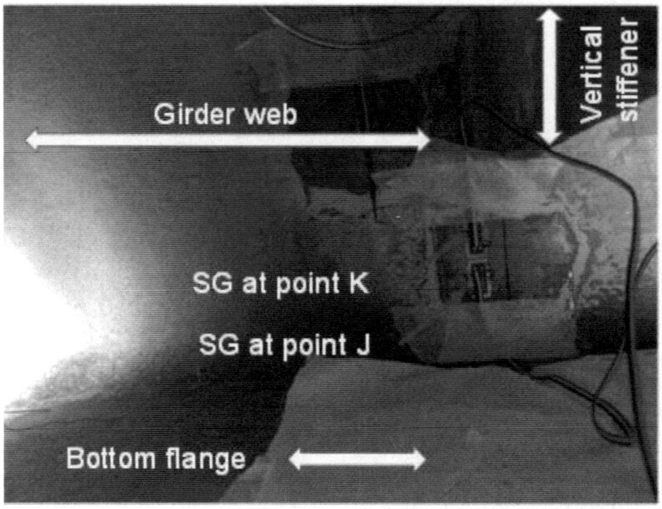

Figure 3: Weldable strain gauges installed to stiffener- web gap region

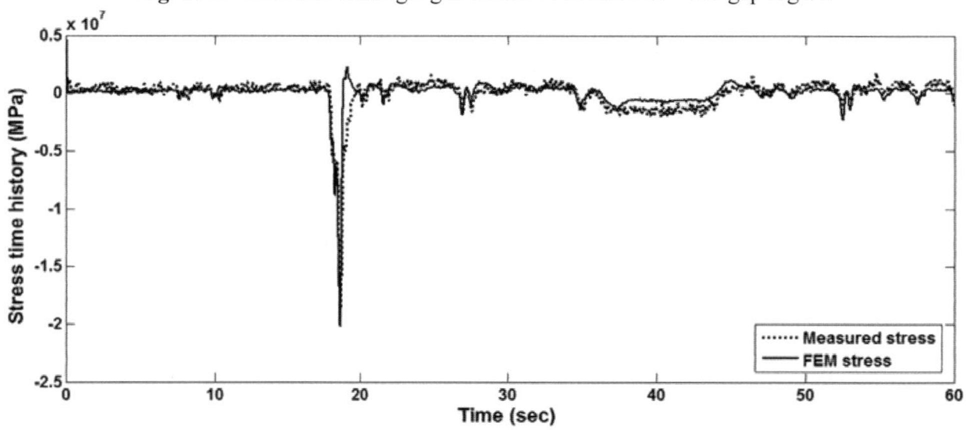

Figure 4: Comparison stress time history (60 sec) between FEM and measured stress at the position of the strain gauge K for scenario one of boundary conditions scenarios.

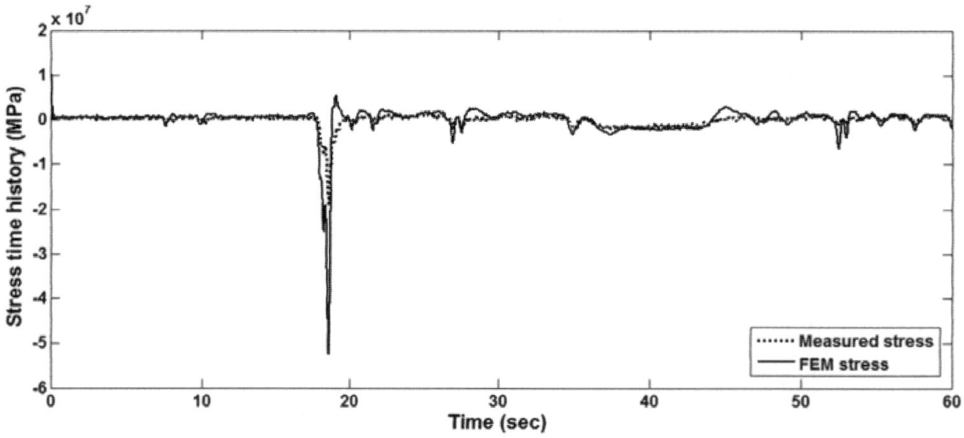

Figure 5: Comparison stress time history (60 sec) between FEM and measured stress at the position of the strain gauge

K for scenario two of boundary conditions scenarios.

4. Validation the proposed methodology

In this section, a comparison has been made between measured and calculated hot spot stresses at hot spot points i.e the weld toe using scenario one. For this purpose, firstly, SG-E, SG-F, SG-G, SG-H, SG-J and SG-K strain gauges were instrumented as illustrated in Table 2. Strains after post processing using low pass filter to eliminate the noise effects were multiplied by modulus of elasticity of steel (E=205 GPa) to get the stress response. In this study, the hot spot stress point of the most critical fatigue damage location is classified as a type '*a*'. For type '*a*', linear extrapolations is used and stress evaluation points are on points 8 mm and 20 mm away from the hot spot weld toe, as shown in Figure 3. Therefore, strain measurements from SG-J and SG-K strain gauges are used to obtain the hot spot stress value. On the other hand, based on proposed methodology, sectional forces from SG-E, SG-F, SG-G and SG-H measured strains are calculated and multiplied by LSFs calculated from FE model to evaluate hot spot stress at weld toe. This comparison is also used as a means of verifying the proposed methodology as illustrated in Table 3.

Table 2. Summary of the locations of strain gauges and thermocouple installed along the longitudinal main girder and cross beam.

Girder having strain gauge	Sensors	Position relative to main girder	Description
Cross-beam	SG-E and SG-F strain gauges	1040 mm west	Upper and underside of the flange
Cross-beam	SG-,SG-G,SG-H, and SG-I strain gauges	1040 mm west	Mid web
East main girder	SG-J and SG-K strain gauges	0 mm	the fillet weld between the vertical stiffener welded to the web of the girder

Table 3. Hot spot stress of the Bascule Bridge typical connections

Connection type	Stress at 0.4t (=8mm) (MPa)	Stress at 1t (=20 mm) (MPa)	Hot spot stress σ_{hs} (MPa) based strain measurements	Hot spot stress σ_{hs} (MPa) based proposed methodology
Vertical stiffener to the girder web	20.31	8.38	28.30	27.38

The results reveal that a good agreement between hot spot stress based proposed methodology and hot spot stress from measured strain under real life load with less than 3.2% error. Consequently, the proposed methodology successfully evaluates the hot spot stress at the weld toe, which seems to be appropriate to simulate the local stress values induced by the real traffic.

5. Conclusions

This paper presents a novel fatigue assessment procedure for metallic bridges that is based on measured strain time-histories collected while the bridge is in service in contrast to currently available techniques that employ artificial strain histories predicted for code-specified loads. The developed procedure is illustrated for a welded connection that is part of a main beam-cross girder junction in a full-scale steel bridge. Conclusions from this research are as follows.

1. The strains predicted by the local model of the connection upon using internal forces computed from measured strains is within 2% of the true values. This demonstrates that the real stresses at fatigue critical details can be evaluated using the proposed methodology.
2. The hot spot stresses evaluated from the proposed methodology also matches the in-situ stresses at the hot-spot locations on the connection to within 3.2%.

Future research will investigate the boundary conditions that need to be applied for the local model to more accurately represent the physical connection. The project will also investigate ideas to extrapolate fatigue life results generated for one instrumented connection to other un-instrumented connections through using global response measurements.

Acknowledgements

The first author would like to acknowledge the financial support of The Higher Committee for Education Development in Iraq (HCED IRAQ) scholarship reference D11000218. The authors would also like to acknowledge the Vibration Engineering Section (VES) at the University of Exeter for providing the NI equipment used in this investigation and the on-site support given by VES research team. The authors would like to thank the Bridges and Structures team at Devon County Council for sharing data and providing access to the Bascule Bridge.

References

ASTM E1049 - 85, 2005. Standard Practices for Cycle Counting in Fatigue Analysis 1. *ASTM International, West Conshohocken, PA.*, 85(Reapproved 2011), pp.1–10.

Aygul, M., 2012. Fatigue Analysis of Welded Structures Using the Finite Element Method. , pp.1–56.

Aygül, M., Al-Emrani, M. & Urushadze, S., 2012. Modelling and fatigue life assessment of orthotropic bridge deck details using FEM. *International Journal of Fatigue*, 40, pp.129–142.

BS NA EN 1991-2 2003 (English) UK National Annex to Eurocode 1. Actions on structures. Traffic loads on bridges.
Cho, T. et al., 2010. Fatigue Reliability Analysis for the Crack Propagation Compared with LRFD Specification. , 10(1), pp.35–49.

EN 1993-1-9, 2005. *Eurocode 3: Design of steel structures - Part 1-9: Fatigue Fatigue strength of steel structures, European Committee for Standardization, Brussels.*

Fricke, W., 2003. Fatigue analysis of welded joints : state of development. *Marine Structures*, 16, pp.185–200.

Garg, V. K., Chu, K. H. and Wiriyachai, A. 1982. "Fatigue Life of Critical Members in a Railway Truss Bridge." *Earthquake Engineering and Structural Dynamics*, 10, p.779-795.

Hobbacher, A.F., 2009. The new IIW recommendations for fatigue assessment of welded joints and components – A comprehensive code recently updated. *International Journal of Fatigue*, 31(1), pp.50–58.

Kwad, J. Alencar, G. Correia, J. Jesus, A. Calçada, R. Kripakaran, P., 2017. Fatigue assessment of an existing steel bridge by finite element modelling and field measurements. Unpublished conference paper at: 6th International Conference on Fracture Fatigue and Wear, 26-27 July 2017, Porto, Portugal.

Kwon, K. et al., 2012. Probabilistic Fatigue Life Estimation of Steel Bridges by Using a Bilinear S - N Approach. *Journal of bridge engineering (ASCE)*, 17(1), pp.58–70.

Liu, R. et al., 2014. Hot spot stress analysis on rib-deck welded joint in orthotropic steel decks. *Journal of Constructional Steel Research*, 97, pp.1–9. Available at: http://dx.doi.org/10.1016/j.jcsr.2014.01.012.

Miner M, 1945. Miner Cumulative Damage in Fatigue.pdf. *Journal of Applied Mechanics*, 12(3), pp.A159–A164.

Ni, Y.Q., Ye, X.W. & Ko, J.M., 2010. Monitoring-Based Fatigue Reliability Assessment of Steel Bridges : Analytical Model and Application. *Journal of Structural Engineering*, 136(12), pp.1563–1573.

Niemi E, Fricke W, Maddox S J, 2006. Fatigue Analysis of Welded Components: Designer's Guide to the Structural Hot-spot Stress Approach [M] .*Woodhead Pub*.

Park, J.Y. & Kim, H., 2014. Fatigue Life Assessment for a Composite Box Girder Bridge. , 14(4), pp.843–853.

Pasquier, R., D'Angelo, L., Goulet, J.-A., Acevedo, C., Nussbaumer, A., and Smith, I. F. C. (2016). Measurement, data interpretation and uncertainty propagation for fatigue assessments of structures. *Journal of Structural Engineering*. doi:10.1061/(ASCE) BE.1943-5592.0000861.

Saini, D.S., Karmakar, D. & Ray-Chaudhuri, S., 2016. A review of stress concentration factors in tubular and non-tubular joints for design of offshore installations. *Journal of Ocean Engineering and Science*, 1(3), pp.186–202.

Schumacher, A, 2003. Fatigue behaviour of welded circular hollow section joints in bridges. PhD thesis EPFL n°2727, Swiss Federal Institute of Technology (EPFL), Lausanne.

Schumacher, A. & Nussbaumer, A., 2006. Experimental study on the fatigue behaviour of welded tubular K-joints for bridges. *Engineering Structures*, 28(5), pp.745–755.

Tveiten, B.W.T., M, X.W. & V, S.B., 2007. Fatigue Assessment of Aluminum Ship Details by Hot-Spot Stress Approach. *ABS Technical Papers*, 135(November), pp.13–17.

Yang, M. et al., 2016. Fatigue behavior and strength evaluation of vertical stiffener welded joint in orthotropic steel decks. *Engineering Failure Analysis*, 70, pp.222–236. Available at: http://dx.doi.org/10.1016/j.engfailanal.2016.05.001.

Ye, X.W., Ni, Y.Q. & Ko, J.M., 2012. Experimental evaluation of stress concentration factor of welded steel bridge T-joints. *Journal of Constructional Steel Research*, 70, pp.78–85.

Zamiri Akhlaghi, F. 2009. Fatigue life assessment of welded bridge details using structural hot spot stress method, a numerical and experimental case study. Master's thesis. Chalmers University of technology.

Zhou, T.Q. & Chan, T.H.T., 2007. Hot spot stress analysis of fatigue for Tsing Ma Bridge critical members under traffic using finite element method. *Progresses in Fracture and Strength of Materials and Structures, 1-4*, 353–358(3), pp.925–928. Available at: <Go to

Zhao XL, Packer JA, 2000. IIW (International Institute of Welding). Fatigue design procedure for welded hollow section joints. Doc. XIII- 1804-99, XV-1035-99.

The 8th International Conference on
Structural Health Monitoring of Intelligent Infrastructure
Brisbane, Australia | 5-8 December 2017

Establishing Fatigue Life Evaluation Method for Deteriorated Steel Elements

Lee-Sak An[1], Yeun Chul Park [2], Ho-Kyung Kim[3].
[1] *Department of Civil and Environmental Engineering, Seoul National University – Korea, Email:peacefulan@snu.ac.kr,*
[2] *BK21PLUS, Department of Civil and Environmental Engineering, Seoul National University - Korea,* [3] *Department of Civil and Environmental Engineering, Seoul National University – Korea*

Abstract

This study establishes a method of a reliability based fatigue life evaluation for deteriorated steel elements to revise safety inspection guides reasonably. Two significant inspection items, corrosion and crack, are considered to evaluate remaining fatigue lives of steel elements. Three fatigue life evaluation scenarios based on inspection and a fatigue life evaluation method reflecting conditions of corrosion and fatigue crack are developed. Fatigue limit state functions for each scenario are derived by utilizing linear elastic fracture mechanics and stress-strain approach. An equation of a stress concentration factor for a semispherical corrosion pit is derived to determine critical pit size which could initiate fatigue crack. An algorithm of a fatigue life evaluation based on inspection results is suggested. A case study is conducted to verify effectiveness of proposed method. Measured data obtained from field monitoring is considered for the case study.

1. Introduction

Bridge evaluation is important for maintenance of deteriorated steel bridge. Initial step for the evaluation is inspection of each member. Based on the inspection results, condition assessment of the bridge could be conducted and then future maintenance plan could be set. There are several bridge evaluation manuals and guidelines available from U.S., Europe and other countries. However, most of them recommends quantitative evaluation rather than qualitative evaluation. For example, grading steel members in AASHTO Bridge Element Inspection Guide Manual depends on the judgement of inspector (AASHTO 2010). Ontario Structure Inspection Manual provides quantitative criteria for corrosion, but does not suggest criteria for crack (Ontario Ministry of transportation 2000). Korea Infrastructure Safety & Technology Corporation (KISTEC 2012) categorizes conditions of steel elements into five (A to E) levels to evaluate deteriorated steel bridges (Table 1). Grading steel member is conducted by measuring crack length, uncoated and corroded area or section loss due to corrosion. However, ranges of the damages were somehow determined without any rational judgement.

This study conducted preliminary study for reliability-based remaining fatigue life evaluation for deteriorated steel bridge member. Reliability index for a steel bridge member which included corrosion or fatigue crack was calculated. Realistic deterioration scenario was established to simulate deterioration of steel bridge member.

Table 1. Categories of Damaged Steel Members by KISTEC (2012)

Levels	Damage Types	
	Corrosion	**Crack**
A	- None	- None
B	- Uncoated area < 10% - Corroded area < 2%	- Local crack in secondary member
C	- Uncoated area > 10% - Corroded area between 2% and 10%	- Large crack in secondary member - Local crack in primary member
D	- Corroded area > 10% - Section loss due to corrosion < 10%	- Large crack in primary member
E	- Section loss due to corrosion > 10%	- Large crack in primary member (> 20% of its depth)

2. Deterioration scenarios of steel members

Two types of damages, which are corrosion and fatigue crack, are considered. During inspection, three cases of damages could be found. Those are finding fatigue crack, corrosion or both. Since this study only focuses on fatigue assessment, corrosion failure is not considered. Three deterioration scenarios are established as per the inspection results and presented in Table 2. Details of each scenario are:

- Scenario I: This scenario demonstrates the case that both fatigue crack and pitting corrosion are found at same location. In general, corrosion pit growth rate decreases as it grows and crack propagation rate increases as it propagates. Although it is unknown that which one occurred earlier, but it is expected that the member will be failed due to fatigue crack.
- Scenario II: This scenario demonstrates the cases when only fatigue crack is found.
- Scenario III: This scenario demonstrates the case when fatigue crack initiated from corrosion pit. Corrosion pit develops stress concentration and micro-cracks. Thus, fatigue crack could be formed from the corrosion pit.

Table 2. Fatigue Life Evaluation Scenarios Reflecting Inspection Results

Scenarios	Types of Defect		Expected Time Line
	Corrosion	**Crack**	
I	O	O	Beginning of service → Corrosion nucleation → Transition from pitting corrosion to fatigue crack initiation → Inspection → Failure OR Beginning of service → Crack initiation → Corrosion nucleation → Inspection → Failure
II	X	O	Beginning of service → Fatigue crack initiation → Inspection → Failure
III	O	X	Corrosion nucleation → Inspection → Transition from pitting corrosion to fatigue crack initiation

3. Establishing fatigue limit state

As per the scenarios, both corrosion and fatigue crack propagation should be considered for defining fatigue limit state. The parameter which commonly represent both fatigue crack propagation and corrosion is time. Therefore, fatigue limit state function is defined in terms of time as shown in Eq. (1)

$$g = t_f - t \tag{1}$$

where, t_f and t denote expected life of steel member and years of operation, respectively.

3.1 Scenario I and Scenario II

For both Scenarios I and II, fatigue limit state is defined for crack propagation. Although corrosion is found in Scenario I, it is anticipated that fatigue crack propagation may lead to failure irrespect to existence of corrosion. Thus, it is expected that the steel member will be failed due to fatigue crack. Linear elastic fracture mechanics (LEFM) using Paris' Law [Eq. (2)] is widely accepted for simulating fatigue crack growth (Paris & Erdogan 1963).

$$\frac{da}{dN} = C(\Delta K)^m \tag{2}$$

where, a and N indicate fatigue crack length and the number of cycles, respectively. C and m are material coefficients. ΔK is range of stress intensity factor at the crack tip. It can be calculated as follows.

$$\Delta K = 2.24 \cdot S_{re} \sqrt{\frac{a}{\pi}} \cdot F \tag{3}$$

$$F = 1 + 1.2\left(\frac{a}{T} - 0.5\right) \tag{4}$$

where, S_{re}, F, T denote equivalent stress range, correction factor and plate thickness, respectively. If initial crack length (a_0) and critical crack length (a_c) are known, expected number of cycles to failure (N_f) can be calculated by integrating Eq. (2) as follows.

$$N_f = \int_{a_0}^{a_c} \frac{da}{C(\Delta K)^m} \tag{5}$$

Since the fatigue limit state is defined in terms of time, the number of cycles to failure could be converted into time by dividing average daily stress cycles ($ADSC$). Finally, fatigue limit state function for Scenario I and II are presented as follows.

$$g = t_f - t = t_{cp} - t = \int_{a_0}^{a_c} \frac{da}{C \cdot \Delta K^m \cdot 365 \cdot ADSC} - t \tag{6}$$

3.2 Scenario III

This scenario demonstrates the case only corrosion was found from inspection, but fatigue crack is initiated from corrosion pit. Finding fatigue crack from corrosion pit is not usual and has often been reported (Kondo 1989). Corrosion pit develops stress concentration and micro-cracks around the pit. Thus, fatigue crack could be formed from the corrosion pit under repeated loading. Therefore, pit depth threshold (d_{th}), which could develop fatigue crack initiation, should be found. Once fatigue crack initiated, it is expected that fatigue crack leads to failure

Expected life of steel member, t_f, in Eq. (2) should include expected time to form corrosion nucleation (t_0), required time for growing corrosion pit to pit depth threshold (t_{cg}) and required time to failure by fatigue crack propagation as shown in Table 2. Fatigue limit state function for Scenario III is defined as follows.

$$g = t_f - t = t_0 + t_{cg} + t_{c,cp} - t \tag{7}$$

The first part of this scenario (t_0) is not well understood yet, because it depends on factors such as electrolytes, loads, materials (Shi & Mahadevan 2001). Therefore, the first part, t_0, is assumed as a random variable.

Pit depth threshold is determined by estimating fatigue threshold proposed by Roy and Fisher (2005). It was developed based on the stress-strain approach suggested by Smith et al. (1970). Roy and Fisher's study successfully estimated maximum stress range which may not initiate fatigue crack under constant stress range for UIT treated welded cover plate connection. The result of this study was adopted by Park (2015) and applied to estimate fatigue threshold for various tube-to-transverse plate welded connection. In order to find pit depth threshold, the critical condition which may initiate fatigue crack is defined as per stress-strain approach (Eq. (8)).

$$\sqrt{(\sigma_{max}\sigma_a)} = \sqrt{(\sigma_a + \sigma_0)\sigma_a} = \frac{F_u}{2} \tag{8}$$

where, σ_{max}, σ_a, σ_0 and F_u denote maximum stress, stress amplitude, mean stress and ultimate strength of steel, respectively. All stresses include stress concentration due to corrosion pit. Thus, stress concentration factor developed by corrosion pit should be applied. Cerit et al. (2009) proposed a stress concentration factor for corrosion pit, but it only considered very small corrosion pit, which the depth is smaller than 1 mm. However, 1 mm pit depth might be too small for this study. Thus, stress concentration factor for hemispheric shape pit is developed by a parametric study. Parametric finite element analysis is conducted for various pit depths. Result of the parametric study is shown in Figure 1. Irrespective to pit depth, stress concentration of hemispheric shape pit is proportional to pit depth to plate thickness ratio. A formula for stress concentration factor was derived and presented in Eq. (9).

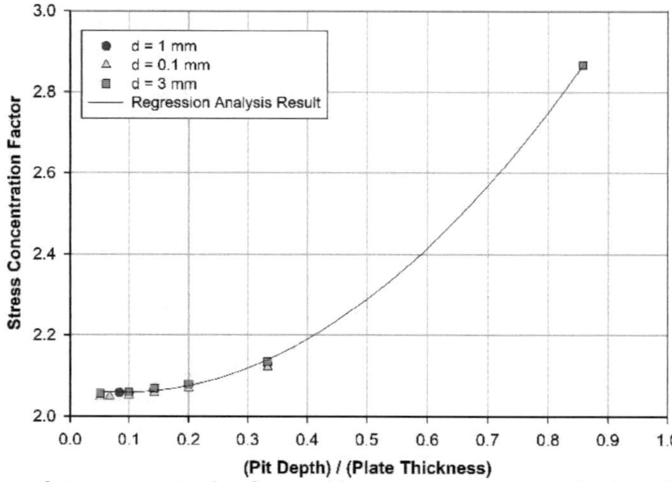

Figure 1. Variation of stress concentration factor with respect to penetration depth to plate thickness ratio

$$\mathrm{SCF} = 2.055 - 0.0249\left(\frac{d}{T}\right) + 0.7279\left(\frac{d}{T}\right)^2 \tag{9}$$

where, d and T are pit depth and plate thickness, respectively. Finally, pit depth threshold could be obtained by solving Eq. (10).

$$1 + \frac{\left\{2.055 - 0.0249\left(\frac{d}{T}\right) + 0.7279\left(\frac{d}{T}\right)^2\right\} - 1}{1 + \frac{a}{d}} = \sqrt{\frac{F_u^2}{4(S_a + S_0)S_a}} \tag{10}$$

where, S_a and S_0 are nominal stress amplitude and mean stress, respectively.
Komp (1987) proposed a corrosion growth model, which corrosion growth can be determined against time.

$$t_{c,cg} = \left(\frac{d_{th}}{p}\right)^{\frac{1}{q}} \tag{11}$$

where, p and q are environmental parameters.
For fatigue crack growth, d_{th} can be considered as an initial crack length. Finally, fatigue limit state function for Scenario III is determined as follows.

$$g = t_f - t = t_0 + t_{cg} + t_{c,cp} - t = t_0 + \left(\frac{d_{th}}{p}\right)^{\frac{1}{q}} + \int_{d_{th}}^{a_c} \frac{da}{C \cdot \Delta K^m \cdot 365 \cdot ADSC} - t \tag{12}$$

3.3 Fatigue evaluation algorithm

An algorithm of a fatigue evaluation method which reflects the scenarios and the limit state functions by utilizing inspection results is presented in Figure 2. The red, green, and blue lines indicate scenario I, scenario II, and scenario III respectively.

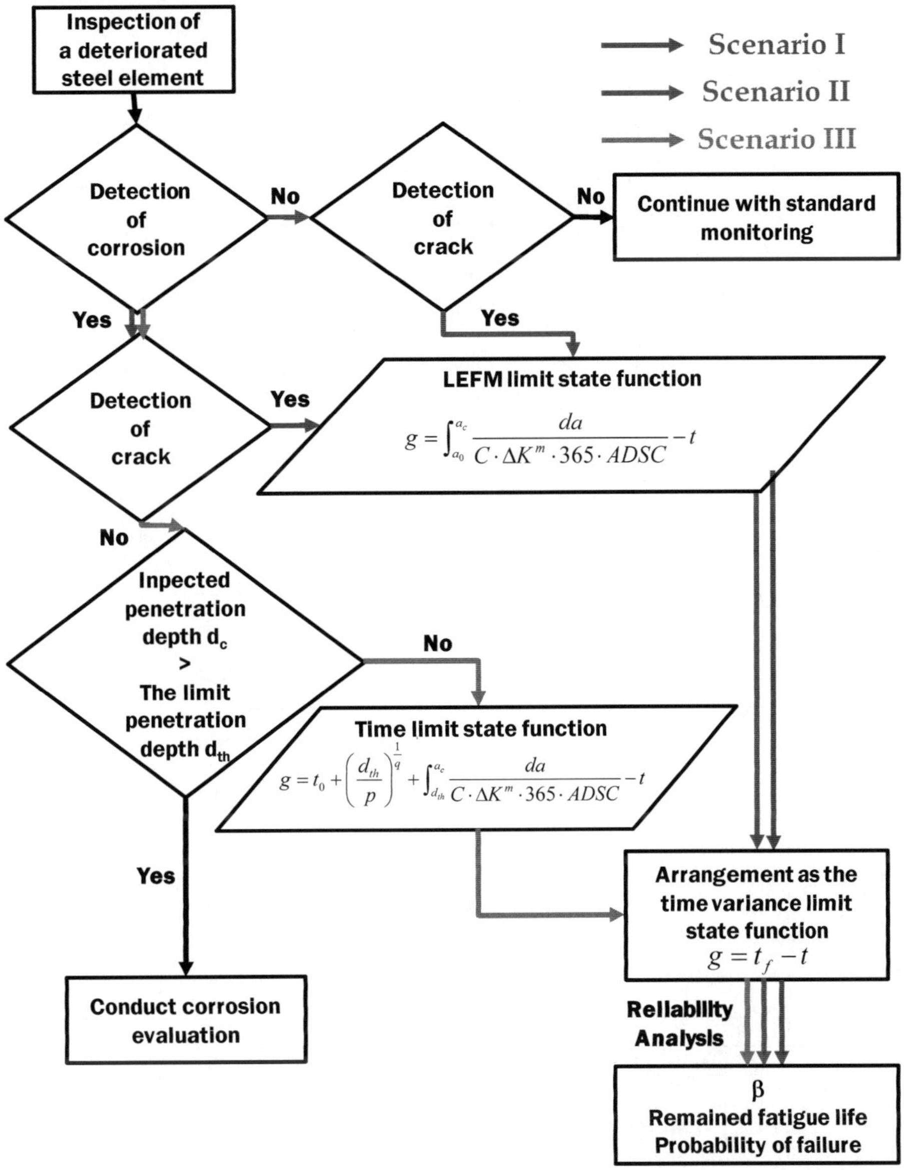

Figure 2. Fatigue evaluation algorithm reflecting inspection results

4. Case study

A case study is conducted for existing 80 year-old bridge by applying the fatigue evaluation method suggested in this study. Actual measurement of stress range for fatigue critical detail is used for reliability analysis (Figure 3). The fatigue critical detail was fillet-welded cover plate welded connection at the bottom of main girder. AASHTO fatigue detail category of cover plate detail is E′, which is the lowest among other welded details in the bridge. It was assumed that initial crack was semi-circular edge crack occurred at the middle.

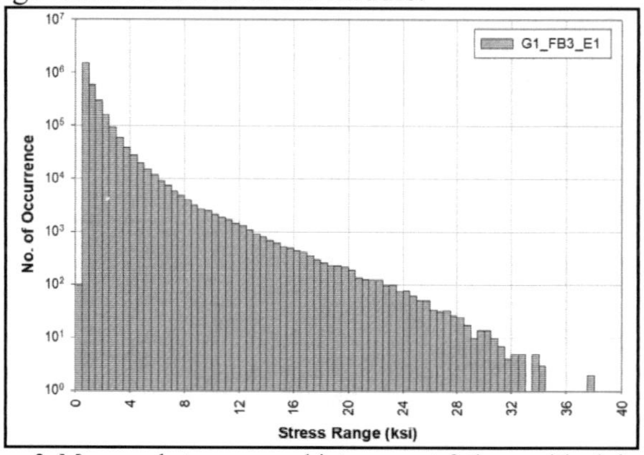

Figure 3. Measured stress range histogram at fatigue critical detail

First-Order Reliability Method (FORM) is utilized for the reliability analysis. Parameters that are used for the reliability analysis are presented in Table 3.

Table 3. Statistics of Parameters

Parameter	Mean	Coefficient of Variance	Distribution type	Reference
T (mm)	51	-	Deterministic	Girder flange thickness
F_u (MPa)	400	-	Deterministic	Ultimate strength of steel
S_0 (MPa)	145	-	Deterministic	Actual measurement
S_a (MPa)	30	-	Deterministic	Actual measurement
a_c (mm)	51	-	Deterministic	Girder flange thickness
m	3	-	Deterministic	Rolfe and Barsom (1977)
ADSC (cycles/day)	3323	0.3550	Lognormal	Actual measurement
t_0 (day)	1500	0.01/0.5/0.95	Lognormal	Shi and Mahadevan (2001)
C (MPa$\sqrt{mm}$MPa· mm$^{1/2}$)	7.12 x 10^{-9}	0.63	Lognormal	Fisher (1984)
p (10^{-6} m)	80.2	0.42	Lognormal	Albrecht and Naeemi (1984)
q	0.593	0.4	Lognormal	

Result of fatigue reliability analysis for Scenarios I and II is shown in Figure 4. Reliability indices decrease from approximately 5.0 to 2.0 for the operation time (t) from 10 to 100 years. The trend of the reliability indices shows a sharper decrease at the early years of operation than the later years of operation. As per Moses et al. (1987), target reliability index for existing bridge evaluation is 2.0. It is expected to drop below the target reliability index after about 95 years, which confirms safety of the analysed detail.

Results of fatigue reliability analysis for scenario III is shown in Figure 5. Time to corrosion nucleation (t_0) is considered for reliability analysis. It was assumed as a random variable due to a lack of information about the phenomenon. Even though coefficient of variance (COV) is changed from 0.01 to 0.95, reliability indices do not show significant discrepancies. It indicates that the time to corrosion nucleation does not have significant influence on fatigue failure of scenario III.

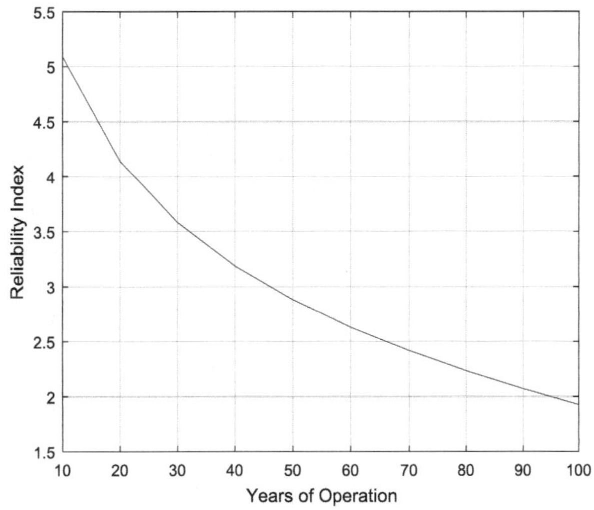

Figure 4. Fatigue reliability analysis for Scenario I and Scenario II

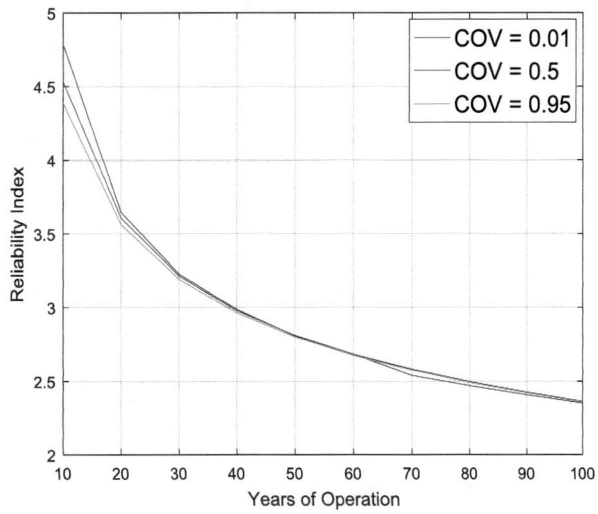

Figure 5. Fatigue reliability analysis results for Scenario III

5. Conclusions

As a preliminary study for developing reasonable guideline for evaluating deteriorated steel bridge member, fatigue reliability assessment procedures were studied. Since fatigue crack propagation corrosion are the two major damages of deteriorated steel members, fatigue limit state functions which reflects corrosion and crack propagations simultaneously are developed. Stress-strain approach is used to determine a limit penetration depth of a pitting corrosion. Case studies were conducted to verify effectiveness of proposed evaluation procedure.

Acknowledgements

This research was supported by a grant (17SCIP-B119964-02-000000) from the Ministry of Land, Transport and Maritime Affairs of the Korean government through the Korea Bridge Design & Engineering Research Center of Seoul National University.

References

AASHTO 2010, *Bridge Element Inspection Guide Manual*, American Association of State Highway and Transportation Officials, Washington, D.C.

Ontario Ministry of transportation 2000, *Ontario Structure Inspection Manual(OSIM)*, Ministry of transportation, Ontario.

KISTEC 2011, *Safety Inspection and Precision Safety Diagnosis Detailed Guide(Bridge)*, Korea Infrastructure Safety and Technology Corporation, Goyang-city.

Paris, P. C., Gomez, M. P. & Anderson, W. E. 1961, 'A rational analytic theory of fatigue', *The Trend in Engineering*, 13, pp. 9-14.

Paris, P. C. & Erdogan, F. 1963, 'A critical analysis of crack propagation laws', *Journal of Basic Engineering,* vol. 85, no. 4, pp. 528-533.

Kondo, Y. 1989, 'Prediction of fatigue crack initiation life based on pit growth', *Corrosion*, vol. 45, no. 1, pp. 7-11.

Rolfe, S. T. & Barsom, J. M. 1977, *Fracture and fatigue control in structures: applications of fracture mechanics*, ASTM International.

Shi, P. & Mahadevan, S. 2003, 'Corrosion fatigue and multiple site damage reliability analysis', *International Journal of Fatigue*, vol. 25, pp. 457-469.

Roy, S. & Fisher, J. W. 2005, 'Enhancing fatigue strength by ultrasonic impact treatment', *International Journal of Steel Structures*, vol. 5 no. 3, pp. 241-252.

Smith, K.N., Watson, P. & Topper, T.H. 1970, 'A stress-strain function for the fatigue of metals', *Journal of Materials*, vol. 5, no. 4, pp. 767-778.

Park, Y. C. 2015, 'Developing fatigue resistant tube-to-transverse plate welded connections for highway sigh, signal, and luminaire support structures', Ph.D. dissertation, Lehigh University, Bethlehem.

Cerit, M., Genel, K. & Eksi, S. 2009, 'Numerical investigation on stress concentration of corrosion pit', *Engineering Failure Analysis*, vol. 16, pp. 2467-2472.

Komp, M. E. 1987, 'Atmospheric corrosion ratings of weathering steels-Calculations and significance', *Material Performance*, vol. 26, no. 7, pp. 42-44.

Albrecht, P., & Naeemi, A. H. 1984, *NCHRP report 272: Performance of weathering steel in bridges*, Transportation Research Board of the National Academics, Washington, D.C.

Moses, F., Schilling, C. G. & Raju, K. S. 1987, *NCHROP report 299: Fatigue evaluation procedures for steel bridges*, Transportation Research Board of the National Academics, Washington, D.C.

Guo, T. & Chen, Y.-W. 2013, 'Fatigue reliability analysis of steel bridge details based on field-monitored data and linear elastic fracture mechanics', *Structure and Infrastructure Engineering*, vol. 9, pp. 496-505.

Jiang, X.-C. & Staehle, R. 1997, 'On the activation energy in the chemical-mechanical correlation model', *Corrosion*, vol. 53, pp. 869-879

The 8th International Conference on
Structural Health Monitoring of Intelligent Infrastructure
Brisbane, Australia | 5-8 December 2017

Experimental Qualification of a Monitoring System in an underground tunnel for future French Radioactive Waste repository Cells

R. Farhoud[1], S. Delepine-Lesoille[2], N. Gilardi[2], J. Bertrand[2], G. Hermand[2].
[1] *Andra, rd 960, 55290 Bure, France – France,* radwan.farhoud@andra.fr, [2] *Andra, 1-7 rue Jean Monnet, 92298 Chatenay-Malabry, France - France*

Abstract

Monitoring system's qualification for the future French underground radioactive waste repository is carried out in different steps. Tests are performed in laboratory to validate measurement's concept, and in situ tests to validate technique's adaptability and installation procedures in constrained environment. In the Andra's underground research laboratory (URL), 500m deep, a concrete tunnel standing for an intermediate level waste repository cell, has been realised and highly instrumented. Implemented sensing techniques were largely used in civil engineering, like Vibration Wire Extensometers (VWE), or newly developed for this application, like Optical Fiber Sensors (OFS). We tested respectively Brillouin and Raman scatterings to obtain respectively strains and temperature distributed measurements. For strain sensing, 2 different types of OFS cables were evaluated: Flexible, easy to install versus a reinforced cable, more difficult to splice and to manipulate but expected to be more robust.
Several sections of the tunnel liner were instrumented in order to compare installation techniques or measuring methodology. For instance, different optical fiber strain sensing cables were installed in the poured concrete. Sensing cables were connected, with either splices or connectors, to a telecom cables paired with remote measuring devices. Both connection techniques has to be evaluated in order to choose the most appropriate (robust) technique, regarding construction constraints, durability and sensing performance. Similarly, we tested two different VWE techniques, with either one coil or two coils; the second case works at lower voltage which could improve their lifetime.
As a first result, measurements provided the two OFS are in good agreement. We observe differences of sensitivities to thermal effects within the two OFS strain sensing cables. Strain provided by the two VWE sensor types also fit well. The 2 tested technologies, innovative and traditional sensors, placed at the same orientation in the cross section provide similar measurements.

1. Introduction

Andra is the French radioactive waste management agency. It is in charge all French radioactive wastes. For Intermediate level (IL) and high level (HL) long-lived radioactive wastes, a future industrial underground disposal, named Cigéo is the reference solution.
Cigéo's design consists of horizontal tunnels, made of poured concrete for IL wastes and by steel tubes for HL wastes, excavated in a claystone at 500m under the surface. The monitoring of Cigéo is a big challenge because sensors placed in the repository cells will have to operate in harsh conditions (temperatures around 90°C, Gamma ray, corrosion, etc.) for unusual time scale (several decades) without possible maintenance after implementation.
To face these difficulties, a specific qualification program (Buschaert S. & al. - 2012) has been elaborated in order to qualify a monitoring system for the target applications

(Buschaert S. & al. - 2013): monitoring in several radioactive environments (according to the waste type), several locations or phase of construction. For that, experiments were led in the Andra's underground research laboratory (URL), in the framework of full-scale demonstrators, to validate sensors or systems qualified under controlled conditions, in a similar conditions of Cigéo (host rock, construction constraints, safety requirements), but without radiations.
For each sensing technology, this monitoring program is divided in 4 steps, inspired from the Qualification guide FD CEN/TR 14748 dedicated to "Non-destructive testing" methods:

1- A state of the art in sensing is performed for each parameter of interest. It is meant to select the most promising technology. It requires depth understanding of sensors' functioning and construction (material, metrology, uncertainties, life-time, etc.);
2- Tests are performed in laboratories under controlled conditions;
3- Qualification and long term tests of complete monitoring systems:
   a. In construction sites with surface tests on real structures
   b. In the URL, i.e. in the target environment (host formation, concrete), in the framework of demonstrators
4- Hardening of technologies to radiation, hydrogen, corrosion, etc.

Metrological aspects are taken into account in each qualification step.
In the EU project Modern, we provided an extensive review of optical fiber sensors as a function of the spatial resolution, the optical principle implemented inside the system, the sensing parameter etc. (MODERN European project, 2013). However, their applications to monitor structure made of clay were not addressed.The conference proceedings will be published in an electronic format only, while all accepted abstracts will also be printed in the abstract book. The full paper in Microsoft Word file should be written in compliance with these instructions and submitted in Word format via the online submission system.
Authors are required to carefully proofread material before submitting.
Papers must be prepared in correct English (preferably Australian format). SI-units should be used.
Paper length should not exceed 10 pages.
Papers should describe the background of the subject, the authors' work, including the methodologies used, discussion on the importance of the work and concluding remarks. Technical terms should be explained unless they may be considered to be known to the broader scientific community. Terms, acronyms and abbreviations should be mentioned at their first appearance.

2. Demonstrator description

Andra's major research facility is the Underground Research Laboratory sited at Bure in the Meuse district, located in the north-Est of France (see Figure 1). It is dedicated to the study of the feasibility of a reversible geological disposal of high and intermediate level long-lived radioactive waste in a Callovo-Oxfordian claystone. The lab itself contains no radioactive waste, and never will. Instead, the underground facility is focusing on testing the rock and prototype waste-containment strategies.
One of the test programs is dedicated to evaluate different concrete liner types for the construction of the nuclear cell dedicated to the intermediate level type waste. It consists of a gallery (tunnel) excavated into a clay rock, according to the minor horizontal stress field, by a tunnelling machine named MAP (Punctual Attack Machine) over 83m length. This gallery has a large diameter drilling about 5.40 m covered by shotcrete liner in 3 steps (3, then 7 and finally 11cm) and poured concrete liner in 3 steps also (basement-vault, then slab and finally vault) over 30cm thick. Shotcrete was placed by pass of 1.2m, after excavation and poured concrete liner by passes of 3.6m long.

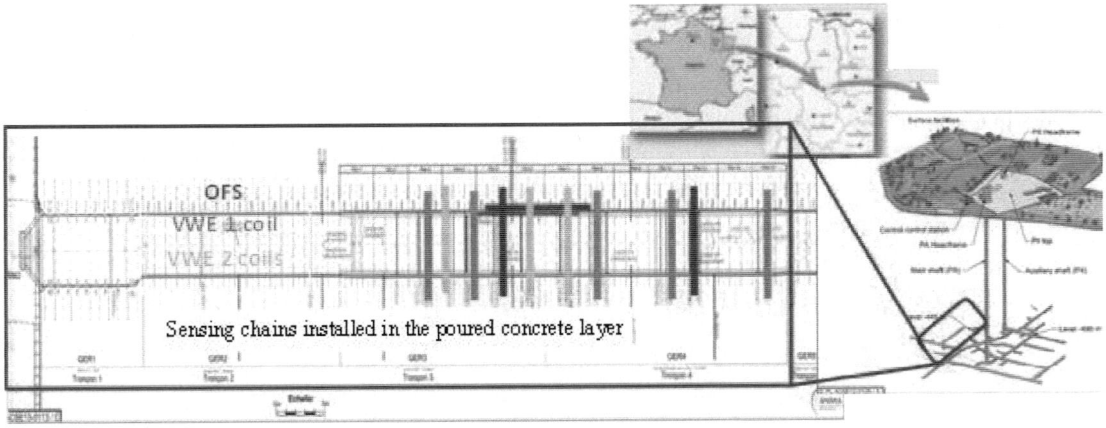

Figura 1: Location of Andra's URL in France, surface (top) and underground installation (bottom – right). A schematic of the instrumented gallery, with VWE (blue and yellow) and OFS technologies (red) is provided in the bottom – left part.

The elapsed time between shotcrete and poured concrete liner grouting, about 9 months, aims to allow clay rock convergences before its limitation with the poured concrete liner. This liner is a C60 concrete type (60 MPa at 28 days), without metallic reinforcement. This choice is meant to avoid corrosion, that could appear in Cigéo, due to interaction between concrete reinforcement and radioactive wastes.

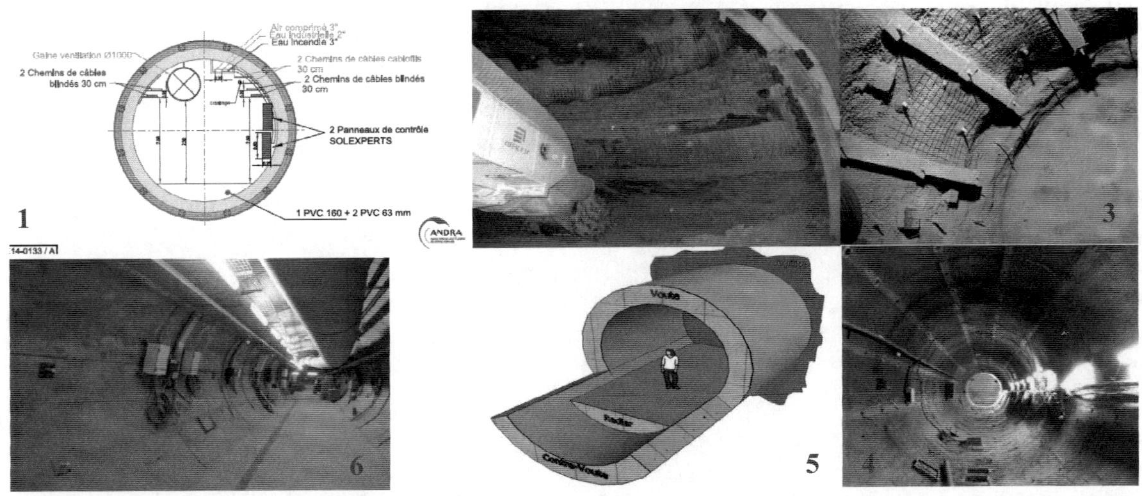

Figura 2 Gallery's construction phases with final concept (1), excvation machine (2), compressive beams & bolts (3), the entire gallery after excavation (4), concrete liner pouring phases (5) and the gallery after all works (6).

In the Andra URL demonstrator, several sensors were installed in order to improve the rock and concrete liners behaviour understandings, in addition to the sensors' qualification program.

3. Sensors implementation methods

3.1 Installation's methods adopted to avoid sensors' damage

The sensors were installed in both the shotcrete and the poured concrete layer. Implementation was prepared in order to maximise survival rate of sensors and avoid failures due to construction works

or phases. Feedback obtained on previous demonstrator (Farhoud et al., 2014) showed that sensors are principally damaged during construction phases and when location of sensors or cables are unknown or note clearly identified.
In view of measurement interpretation, we paid attention to sensor location inside the structure. As soon as sensors were installed in or above the shotcrete, a 3D scan was performed to identify sensors, cables and boreholes. Topographers provide accurate sensors locations (see Figure 3).

THE GRAPHIC ON THIS PAGE WAS NOT AVAILABLE ON THE E-MEDIA VERSION OF THIS PAPER.

Figura 3: Localization of installed sensors and cables in the gallery (right) provided by a 3D scan which is the input for installation of sensors, identified in direction by color and localization by topographer (left)

3.2 Improvement on sensors' installation procedures

We tested respectively Brillouin and Raman scatterings to obtain respectively strain and temperature distributed measurements. For strain sensing, 2 different types of OFS cables were evaluated: Flexible, easy to install (called AF) versus a reinforced cable (noted V3), more difficult to splice and to manipulate but expected to be more robust. Distributed temperature measurements are obtained through Raman scattering into multimode fibers. The flexible cable offers several optical fibers, one being multimode. What is more, we implemented a second section. Optical lines (from the 2 sensing cable of each type) were carefully localised in the structure with heating events created while acquisition was going on.
OFS cables can be installed in the tunnel with high operational constraints due to co-activity: several companies were working in a single and reduced area at the same moment.
Sensing cables lengths were longer than target sensing areas; connexion with telecommunication fibers were placed far from the working areas and protected into junction boxes.

THE GRAPHIC ON THIS PAGE WAS NOT AVAILABLE ON THE E-MEDIA VERSION OF THIS PAPER.

Figura 4: Picture of junction boxes implemented next to the instrumented section, either for spliced optical fiber sensing cables (left) or connected OFS (right) to a communication cable.

To obtain accurate location of the sensing cable inside the concrete, we used fiberglass bars that provide the arc shape and can be tightly attached to the rock. OFS cables were then attached with simple scotch, to reduce crush (Farhoud et al. 2014). In the first section, OFS cable was attached to the fiberglass by a scotch every 0.5m; cables moved when concrete was poured in the basement part of the tunnel. For the second section, scotch distance was reduced down to 0.25m and Rilsan every 0.5m. This second solution appeared enough robust to concrete pouring, construction phases and workers interventions.
In addition to the fixation test, connexion between fibers, along a sensing line between the sensitive part of the cable and the sensing instrument, were tested by two methods. First method consists of splicing fibers from the instrument to the sensing location. It's the best way to connects together, with low losses, low optical reflexion and long durability. The second method, fast and flexible in construction conditions, consists of installing OFS cables equipped with connectors. In this case, connexion is only a plug of 2 connectors. It induces optical loss in the optical line. However, this method (connectors) enables to install sensors very quickly (which significantly reduces overall cost), to avoid splicing in dusty environment, to enable reconfiguration in case of failure. Pictures of junction boxes are illustrated in figure 4.

4. Results obtained

Prior to results, it's worthness to describe distributed sensing principal based on OFS for strains and temperatures measurements. The term distributed sensor designates the case in which the optical fiber itself becomes a sensor. It is thus no longer necessary to implement anticipated sensor positions since measurements are being performed all along the optical fiber connected to the reading device (as well as within the extension cables).
As detailed in (A. Rogers, 1999) and in (Lantick et al., 2009) truly distributed measurements rely on a sensing process, based on Rayleigh, Raman or Brillouin scatterings in optical fibres. An additional work, performed by JM. Henault (JM. Henault et al., 2010), provided specific methods developed and used to evaluate the whole sensing chain (optoelectronic instruments and sensing cable).

4.1 Comparison between fiber connection techniques (splice vs connector)

Comparison between both methods is illustrated in figure 5. After several months of measurements, both connected and spliced fibers provide similar strain measurements. Contrary to expectations, strains measured by cables linked by connectors, were more accurate that those liked by splices, especially for flexible cable. Indeed, the measurements dispersion observed is about 1μm/m for cables with connectors, while it's about 3μm/m for the second case.

THE GRAPHIC ON THIS PAGE
WAS NOT AVAILABLE ON
THE E-MEDIA VERSION OF
THIS PAPER.

Figura 5: Relative strains measured by Brillouin scattering into 2 different OFS cables, reinforced (Blue and green) and flexible (red and black), at the same location (Middle of the eight, see the star). The influence of the connection method is illustrated: spliced cables (blue and red) versus connectors (black and green). Raman temperature measurement at the same location is the orange curve (y axis on the right). Strain raw data from Brillouin scattering are provided with dotted lines. Solid lines are strain compensated from thermal effects.

2. Paper format

Authors are recommended to use this template to prepare their papers. This template is an A4 format with 2 cm margins left, right; 3.5 cm margin top and 2 cm margin bottom.
All text paragraphs should be typed in single space. Double spacing should only be used before and after headings as shown in this example. No spaces should be placed between paragraphs.

4.2 Thermal effect's compensation

The raw measurements obtained in figure 5 from Brillouin scattering requires compensation from thermal effects. Indeed, the Brillouin frequency is sensitive both to strain and temperature variations:

$$\Delta \nu_B = C_\epsilon \Delta\epsilon + C_T \Delta T \tag{1}$$

where $\Delta\epsilon$ and ΔT are variations of strain and temperature respectively, and $C\epsilon$ and CT are the Brillouin sensitivities, measured in laboratory on the selected sensing cable.
As temperature can vary significantly inside the concrete ring, it is necessary to compensate its effects to focus on strain. Mohammad et al. (Mohammed et al. 2014) showed that for OFS in concrete, differential dilatation should be considered:

$$\Delta \nu_B = C_\epsilon \Delta\epsilon + (C_\alpha + C_\epsilon \alpha_{concrete})\Delta T \tag{2}$$

where $C\alpha$ is the Brillouin thermal coefficient of the naked optical fibre, and αconcrete is the thermal expansion coefficient of the host concrete.
We used this last formula to compensate the effect of temperature on Brillouin frequency variation, with the coefficients:

- $\alpha_{concrete}$: 12 µm/m/°C (equal to the reference VWE wire thermal expansion coefficient)
- $C\epsilon$: 0.0465 MHz/µm/m (from laboratory measurements) (I. Planes et al., 2016)
- $C\alpha$: 0.96 MHz/°C (from Kurashima et. al, 1990).

In Figure 6 we can see for the reinforced OFS cable that strains in the inner part of the concrete ring are more sensitive to temperature's changes than in the extrados part, as expected. We also noticed that strain in the inner and outer parts of the concrete tunnel (considered in the middle part of the cross section) are very similar, which indicates that the concrete ring may not yet be loaded by the surrounding rock.

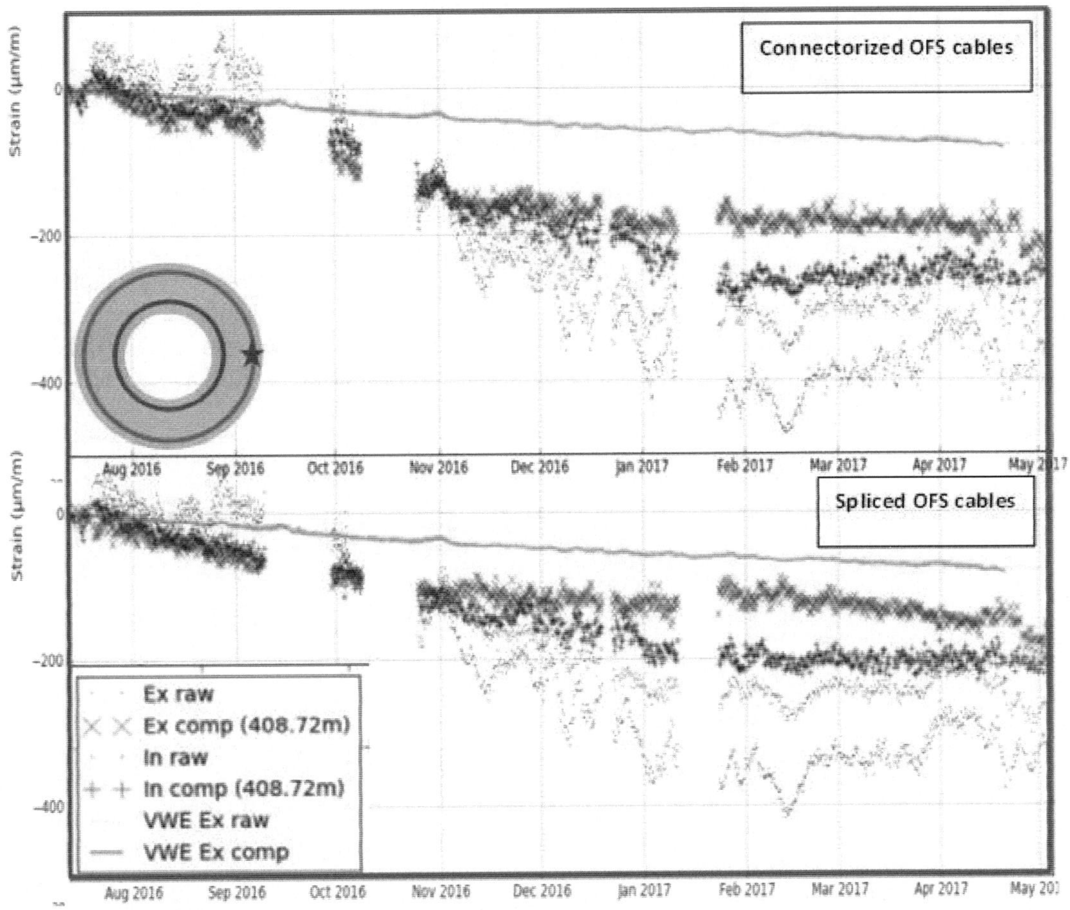

Figura 6: Relative strains measured by VWE(red) and compared with strains measured by reinforced OFS in the same location of an instrumented section of the tunnel, in extrados part (green) and intrados part (blue). Doted lines represent raw measurements, when + and x are compensated measurements to thermal effects. Temperatures are measured by Raman scattering technique into OFS cables at same location.

2.1 Page Numbering, Header, Footer

4.3 Comparison of data provided by VWE in several sections with two different techniques

In addition to distributed strain measurements obtained provided by optical fibers in 2 sections, strain measurements were acquired by VWE (Vibrating Wire Extensometer) based on (i) one coil with pulse interrogation technique in 5 sections or (ii) 2 coils based on maintained interrogation mode in 3 sections. Figure 7 illustrates VWE sensors based on tow different technologies, with functionning principal and photo of installtion in the same section in the tunnel.

Figura 7: VWE, double coils with interrogation in maintained mode (left), simple coil with pulse interrogation mode (right) and photo of both sensor types installed in the demonstrator (middle) in the same location

In Figure 8 presents measured relative strain. Initial measurement was taken 14 days after concrete pouring in order to avoid thermal effects. We can observe that over one year, strains provided by the two WVE technologies are in good accordance. Different behaviour during the first months is attributed to thermal effects on the measurements and can be removed by adequate correction of raw data.

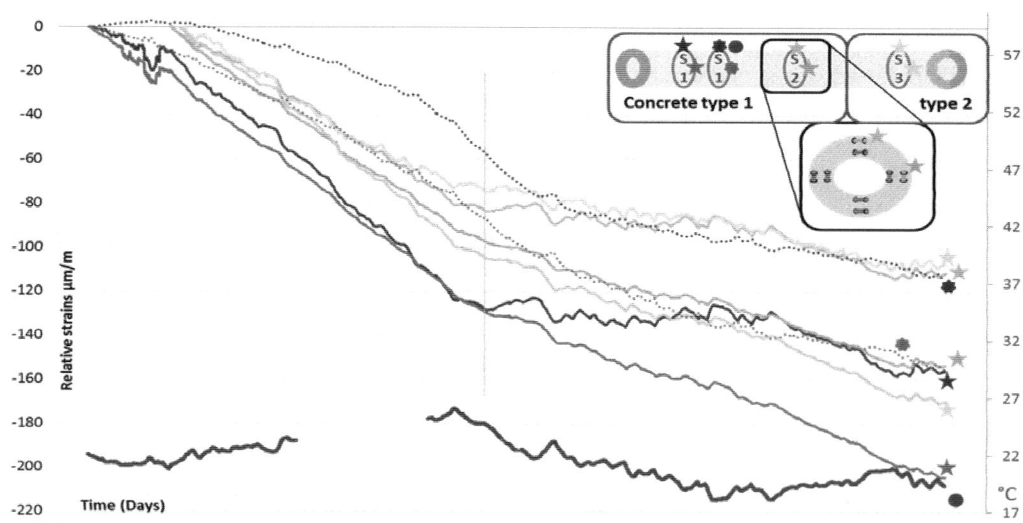

Figura 8: Relative strains and temperature measured by VWE with 2 different technologies, 1 coil (dotted curves) and 2 coils (solid curves), in the same locations, in 3 concrete sections at extrados part in the side and top part of the concrete tunnel

4.4 Survival rate of sensors

In order to assess robustness of sensors intended to be use for the monitoring of the underground radioactive waste disposal, the following table illustrates survival rate after more than a year.
Table 1 illustrates the success of sensors' installation in the poured concrete liner. We reached 100% survival rate.
Sensors installed in the shotcrete faced several failure because of the important strains applied. We proved feasibility of shotcrete instrumentation but implementation requires improvement in the future.

Table 1: survival rate of sensors after more than a year

Sensor technolgy	Sensors by section	Localisation			Total sensors	Failed sensors	% Failed	Remarks
		Clay	Shot-crete	Poured concrete				
VWE 1 coil	24		■	■	120	-	-	Sensor not proposed for monitoring of disposal
VWE 2 coils	18			■	54	0	0%	-
OFS (ε + T°)	1	■			2	2	100%	Possible high bending on OFS cable
OFS (ε + T°)	2		■		6	3	50%	Strains out of rang, Temperatures OK
OFS (ε + T°)	2			■	2	0	0%	Longitudinal strains
OFS (ε + T°)	12			■	24	0	0%	-
OFS (ε + T°)	12			■	12	0	0%	Strains out of rang

5. Conclusions

Qualification of sensing chain for the monitoring of a concrete tunnel, as a radioactive waste disposal is a big challenge. Andra faced this challenge, of installing sensors to operate in harsh conditions without maintenance for unusual long period, by a R&D program with demonstrators performed in representative environment. The demonstrator instrumented in Andra's URL proved that VWE sensors, equipped with 2 coils and interrogated in maintained interrogation mode, provide measurements in good accordance with other techniques. The first technique is VWE sensors with 1 coil. The second one is a distributed optical fiber sensing technique based on Brillouin Optical Time-Domain Analysis (BOTDA). We proved that connecting OFS with connectors (versus splicing) allowed fast operation, more compatible with construction site constraints. In the future, we will assess whether connected and spliced cables show similar durability over time. We presently evaluate the method to suppress thermal influence on strain measurements.
Data provided by OFS and VWE, in the poured concrete layer, proved to be similar after about a year.

6. Acknowledgments

A lot of thanks to Marmota, GTC, Solexperts and Geo-Instrumentation companies for efficient sensors installations.

References

S. Buschaert, S. Lesoille, J. Bertrand, S. Mayer, P. Landais, 2012. "Developing the tools for geologic repository monitoring - Andra's monitoring R&D program" - WM2012 Conference, February 26-March 1, 2012, Phoenix, Arizona, USA

S. Buschaert, S. Delepine-Lesoille, J. Bertrand, JP. Dubois, R. Farhoud, G. Hermand, 2013. "Monitoring strategy of Cigéo reversible disposal" - MoDeRn International Conference and Workshop - Luxembourg – March 19-21, 2013

MODERN European project (2013). State of Art Report on Monitoring Technology DELIVERABLE (D-N°: 2.2.2), 16/05/2013

R. Farhoud, N.Gilardi, S. Delepine-Lesoille, G. Hermand and J. Zghondi, 2014, "Tunnel monitoring system qualification for an underground radioactive waste disposal", ICCEA (International Conference on Advanced Technology Innovation), June 2017, Koh Samui, Thailande, 2017

A. Rogers, 1999. "Distributed optical-fiber sensing", Measurement Science Technology, pp. 75-99, 1999.

V. Lanticq, E. Bourgeois, P. Magnien, L. Dieleman, G. Vinceslas, A. Sang and S. Delepine-Lesoille, 2009. "Soil-embedded optical fiber sensing cable interrogated by Brillouin optical time-domain reflectometry (B-OTDR) and optical frequency-domain reflectometry (OFDR) for embedded cavity detection and sinkhole warning system." Measurement Science and Technology, Vol. 20, No. 3, 2009.

JM. Henault, G. Moreau, S. Blairon, J. Salin, JR. Courivaud, F. Taillade, E. Merliot, JP. Dubois, J. Bertrand, S. Buschaert, S. Mayer and S. Delepine-Lesoille, 2010. "Truly Distributed Optical Fiber Sensors for Structural Health Monitoring: From the Telecommunication Optical Fiber Drawling Tower to Water Leakage Detection in Dikes and Concrete Structure Strain Monitoring," Advances in Civil Engineering, vol. 2010, Article ID 930796, 13 pages, 2010. doi:10.1155/2010/930796

H. Mohamad, K. Soga, and B. Amatya, 2014, "Thermal Strain Sensing of Concrete Piles Using Brillouin Optical Time Domain Reflectometry", Geotechnical Testing Journal, vol. 37, pp 333-346.

I. Planes, S. Girarda, A. Boukentera, E. Marina, S. Delepine-Lesoille, R. Farhoud, J. Zghondi, F. Fischli and Y. Ouerdane, "Distributed Strain Monitoring in Tunnel Shotcrete (with and without yieldable concrete wedges) by fiber-based TW-COTDR technique", APOS2016 (Asia-Pacific Optical Sensors Conference), ISBN: 978-0-9600380-5-3, paper F1A.4, 2016, 11–14 October 2016, China.

T. Kurashima, T. Horiguchi and M. Tateda, 1990, "Thermal Effects on the Brillouin Frequency Shift in Jacketed Optical Silica Fibres," Appl. Opt., Vol. 29, pp. 2219–2222.

The 8th International Conference on Structural Health Monitoring of Intelligent Infrastructure Brisbane, Australia | 5-8 December 2017

Identifiability based sensor configuration analysis for Bayesian structural identification

A. Jesus[1], Y. Zhu[1], K. Koo[2], J. Brownjohn[2], I. Laory[1]

[1] *Civil Research Group, School of Engineering, University of Warwick, CV4 7AL – United Kingdom, Email:* andrehjesus@openmailbox.org, i.laory@warwick.ac.uk [2] *College of Engineering, Mathematics and Physical Sciences, University of Exeter, EX4 4QF – United Kingdom*

Abstract

When performing structural health monitoring, the design of an optimal sensor configuration aims at improving the effectiveness of structural identification (st-id) and anomaly detection while reducing costs/redundant data. Since this activity takes place when only simulated data is available, model uncertainties have a great impact on the results. This paper studies the application of a hybrid Bayesian methodology for optimal sensor configuration improvement of st-id. It exhibits a comprehensive formulation for uncertainty quantification, aided with multiple response Gaussian processes, and a reduced computational effort relatively to previous works. Information of the optimal sensor position, amount and diversity of data that should be collected, before the actual monitoring phase, can be obtained. An aluminium bridge scale model instrumented with strain gauges/thermocouples, and supported at one end with springs, whose stiffness have to be identified, is used as a case-study. The methodology is applied to a finite element model of the actual structure, and the optimal location of strain gauges is obtained. Validation is carried out by comparison of these results against a subsequent st-id of the bridge supports stiffness, when measurements are made available. Results show that: the used performance criterion is a reliable index of how accurate the identification is (identifiability); the computational effort of the method can be three times less than previous works; and even with poorly informative responses it is possible to design a reliable and cost-effective sensor configuration.

1. Introduction

Due to the occurrence of natural hazards and progressive aging of existent structures there is a raising concern over serviceability and maintenance of critical infrastructures. To this end, the structural health monitoring (SHM) technology is currently under active research, to prove its feasibility over traditional approaches. It relies on the design of a monitoring system, its deployment, data collection/interpretation and finally structural risk assessment and prognostics.
In this way, the design of an optimal sensor configuration usually takes place when only simulation data, e.g. with aid of the finite element (FE) method, is available. In turn this leads to the problem that the uncertainties of the model will affect the results. Most of the research community agree in that, the two most prevalent sources of model uncertainties are modeling discrepancies and parametric uncertainty. These are associated with the inaccuracy in modeling of the physical system Castro-Triguero et al. 2013 and the differences between the physical attributes and associated model parameters, respectively. Papadimitriou et al. 2000 has an extensive research on design of suitable sensor configurations in the presence of parametric uncertainty, minimizing the so called information entropy. On the other hand modeling discrepancies did not receive as much attention, having been initially assumed as a

zero-mean uncorrelated Gaussian by a number of authors Katafygiotis et al. 1998. With the aim of lessening this strong assumption, other authors have reformulated it to account for existent correlations, although still considering a zero-mean function, as can be seen in Simoen et al. 2013 or Vincenzi & Simonini 2017. This indicates an existent interest in considering both of these uncertainties, with as much detail as possible.

In the light of these developments we center our attention in a hybrid modular Bayesian approach (MBA) developed by Kennedy & O'Hagan 2001, which has been extended for computation of the optimum sensor configuration, as a preposterior analysis by Arendt et al. 2016. The uncertainties are quantified with the aid of a robust framework, based on multiple response Gaussian processes. In a previous work by Jesus et al. 2017, the practical feasibility of this methodology for st-id has been shown, in an aluminium truss bridge subject to thermal expansion. Main advantages of these methodologies include their overall comprehensiveness towards the two sources of uncertainty (parametric and modeling discrepancy) and the low computational effort when performing st-id. Disadvantages include a strong assumption on normality, non consideration of the uncertainties 'second order' effects, and a lack of identifiability Yuen 2010, which can however be minimized, as seen in recent research from Arendt et al. 2012a.

Therefore in this work we aim to: develop an alternative approach to the optimal sensor configuration proposed by these authors, the Fast Preposterior Analysis (FPA) with a reduced computational effort; present a broader view of the performance index of this methodology and how it can be used to predict/improve the identifiability during st-id; and demonstrate its feasibility for practical SHM applications, by revisiting the thermally expanded aluminum bridge case-study. Validation is carried out by comparing the results against a st-id combinatorial analysis, when monitoring data is made available.

2. Enhanced modular Bayesian approach

The modular Bayesian approach (MBA) formulation was initially developed for a single measured response case by Kennedy & O'Hagan 2001. Arendt et al. 2012b demonstrated that unless under some specific conditions, this approach will fail to identify the true value of structural parameters, and proposed instead a multiple response case, which allows for a more informative data model. The question of how to select the responses that maximize identifiability shall be detailed later on in Section 3. In this Section a short summary and relevant details of the implementation are presented. The MBA aims to solve a fundamental equation of model calibration, which can be written as

$$Y^e(X^e) = Y^m(X^e, \theta^*) + \delta(X^e) + \varepsilon \tag{1}$$

where: Y^e are observations, dependent on design variables X^e; Y^m are simulations of the model, dependent of the design variables and of a vector of structural parameters θ^*; $\delta(X^e)$ is a discrepancy function that translates the difference between the model and the true process and ε is an observation error, that is assumed to follow a Gaussian distribution $\mathscr{N}(O, \Lambda)$.

To solve this equation while considering all of the involved uncertainties, the process is broken into four modules, hence the name *modular* Bayesian approach. For a flowchart of the algorithm see Fig. 1 (cf. Fig. 1 of Jesus et al. 2017 for a detailed version). In module 1 and 2 the computer model and the discrepancy function are fitted by statistical models, known as multiple response Gaussian processes (mrGp), whose parameters (hyperparameters ϕ) have to be estimated. These hyperparameters characterize the mrGps and account for an approximation of their associated uncertainties. Afterwards the mrGps are used to set up a global data model that explains both simulations and observations for a set of given structural parameters θ. This is done in the third module of the approach, where the

posterior distribution of these parameters is estimated by Bayesian inference. The fourth and final module predicts the observed process, by updating the mrGps previously determined, with the posterior information of θ.
In the first three modules of the MBA the mrGps are approximated to the available data by maximization of their log-likelihood function. To increase the numerical stability of these operations, we avoid the inversion of a covariance matrix in the last term of the log-likelihood function. This is possible since this matrix is symmetric and positive-definite, being implicit that it can be written in the Cholesky factorized form.

3. Fast Preposterior Analysis

3.1. Original Preposterior Analysis

Having presented the formulation of the multiple response MBA, it is now possible to investigate the question of which responses should be chosen in order to improve the identifiability of the method. The original preposterior analysis is effectively very similar in nature to the MBA. According to Arendt et al. inference of the structural parameters θ of Eq. 1 is most accurate, when its posterior distribution, determined in module 3 of the MBA, is as narrow as possible. It is an objective of this work to assess the validity of this claim. Also it should be noted that it is not possible to infer these structural parameters before having any monitoring data available. Likewise it is not possible to establish the discrepancy function $\delta(X)$ before actually having any monitored responses. The only possible approach is to: establish prior knowledge on any unknown quantities (hyperparameters of mrGp, structural parameters); to generate realizations of these quantities in a Monte Carlo sampling space; and to average across results of the structural parameters posterior. The inferred covariance matrix of the posterior probability density function (PDF) is thereafter called a preposterior covariance, in the sense that it is based on prior knowledge. A final note is that if there is any prior information, or any heuristics relatively to the most relevant measured responses they can be incorporated in any of these unknown variables.

3.2. Numerical Approach

Our strategy relies on minimizing the computations required during the Monte Carlo loop of the original preposterior algorithm. A flowchart of our approach is depicted in Fig. 1. It is divided into two phases, which are then subdivided into three steps. The first phase can be thought of as a preliminary run, with all the sensors active, hence all available q responses are being simulated. The discrepancy function δ mrGp hyperparameters, $\phi_j^{(i)}$ with $j = 1 \dots q$, are saved during this phase, at each iteration i of the Monte Carlo (MC) loop. The second phase is a combinatorial or optimization run, similar to what has been presented by Jiang et al. 2015, with the particularity that the required hyperparameters of δ do not have to be calculated. Instead the corresponding hyperparameters $\phi_k^{(i)}$ for each combination k are selected from what has been stored in the first phase.
The choice between using a combinatorial analysis with $2q-1$ runs or an optimization routine depends heavily on the number of sensors under consideration. As q grows the computational cost of the combinatorial analysis becomes prohibitive, and optimization algorithms such as genetic algorithms are better suited to obtain the optimal sensor configuration.
These strategies combined with the factorization described in Section 2 drastically reduce the computational effort required for each MC loop.
In the next sections application of this algorithm to a case-study of an aluminum bridge shall be presented to illustrate its performance.

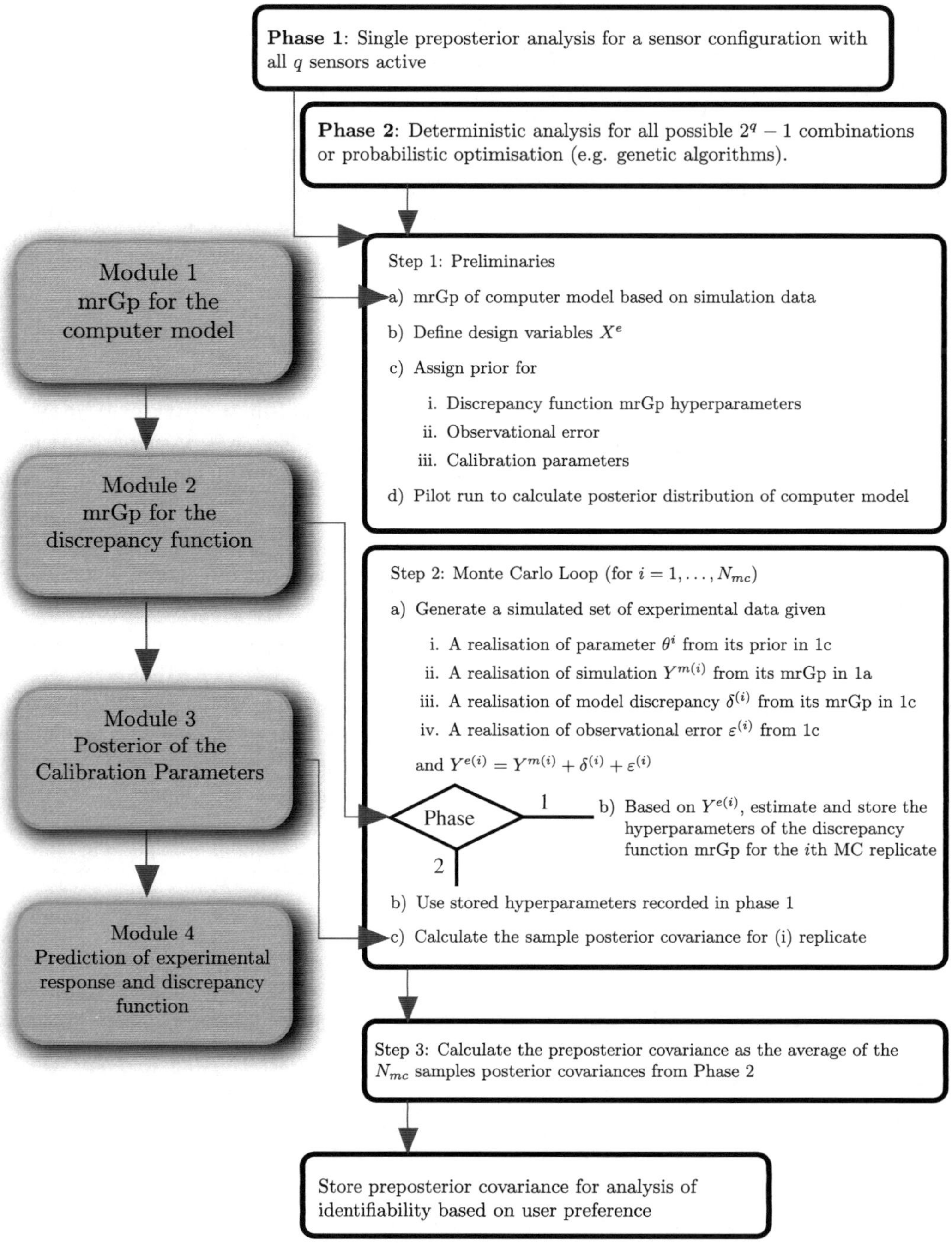

Figure 1: Flowcharts of: modular Bayesian approach on the left and Fast Preposterior Analysis on the right

4. Aluminum Bridge Subjected to Thermal Loading

A reduced-scale laboratory aluminum bridge inspired by the New Joban Line Arakawa (Japan) railway bridge, was built at the Warwick Civil Engineering Laboratory and subjected to thermal loading due to infrared heaters. Typical daily ambient temperature in the laboratory ranged from approximately 18°C up to 21°C. A numerical model of the structure was also developed to study the phenomena. The stiffness of a pair of springs located at one of the ends of the bridge will be considered as a structural parameter to be inferred. This inference by the MBA is not the primary goal of this work, but it shall be used as a validation tool to the optimal sensor configuration determined by the FPA. Nine readings/simulations of strain are used as input to the MBA and FPA methodologies, and shall be detailed in this Section as well. Section 5 details results of application of the MBA and FPA.

4.1. Experiment

The truss structure in Fig. 2(a) is simply supported at its ends, being constrained on one of them by two linear springs Fig. 2(b) (supplier reference value is $K = 552.26$ N/mm).

Geometrical information and the measurement setup of the bridge are displayed in Fig. 3. Material

(a) (b)

Figure 2: Aluminum bridge subjected to thermal loading (a) and detail - springs constraint (top view) (b).

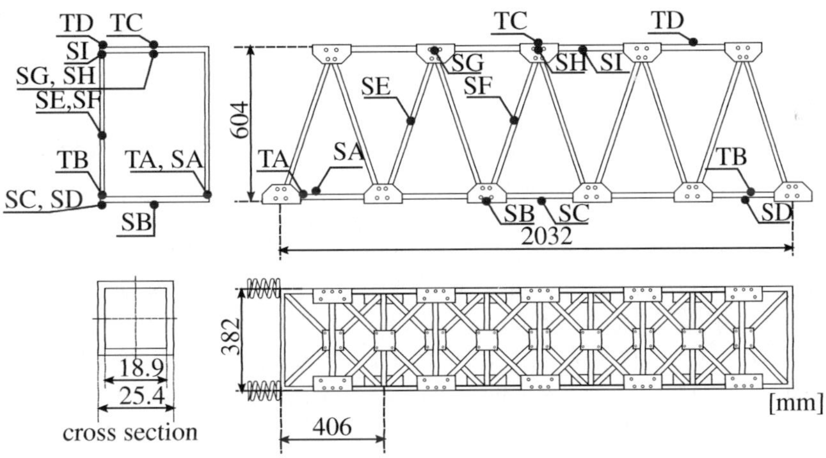

Figure 3: Aluminum bridge measurements diagram (dimensions in mm) - thermocouples TA-TD and strain gauges SA-SI. Placement of labels above/below bar reflects the sensor position.

properties of the bridge are: aluminum alloys of grading 1050AH14 and 6082T6 for the gusset plates and for the box section beams, respectively. Stainless steel M6 8.8 bolts connect the multiple members of the bridge. Eleven strain measurements and four thermocouple readings were made during the experiment, at a sampling rate of 1 Hz. A proportional-integral-derivative (PID) controller on a Labview routine was used to control the infrared heaters.

Temperature and strain readings during the main experiment, which took approximately half an hour, are displayed in Figs. 4(a) and 4(b). The reason why the strain at the top-side of the bars is in compression, is that all the strain gauges have been placed on the bottom side of the bars. Therefore, despite the global bending of the structure that leaves the top bars in traction there is a localized bending at the top bars, which is measured as a compression on their bottom side. The temperature-strain relation is not linear because the measurements are performed on the surface of a squared hollow section, which cools down faster than the internal cross section and will, therefore, still have some residual thermal deformation when cooling down.

Although the strain gauges are thermally compensated for aluminum, the data from the strain gauges

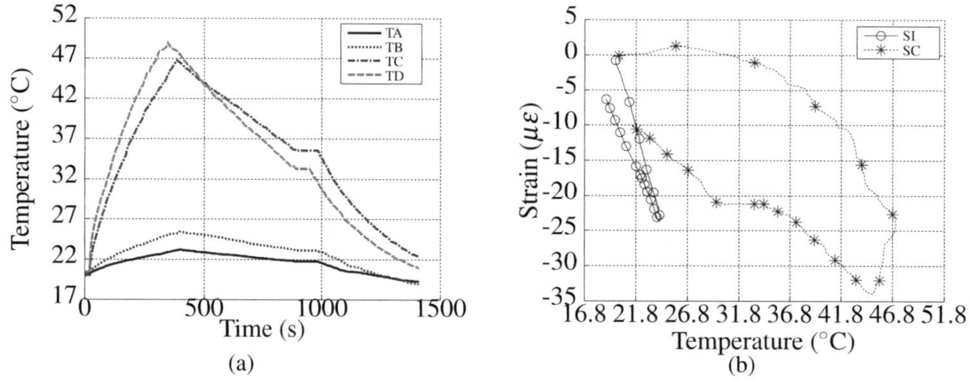

Figure 4: Aluminum bridge heating/cooling cycle - Temperature readings (a) and strain measurements (b), see Fig. 3 for reference.

has been post-processed to remove the remnant thermal output effect, which is related to the natural thermal expansion of the gauge.

4.2. Finite Element Model

In this Section a FE model of the scale aluminum bridge is presented. The model was coded with APDL (ANSYS parametric design language).

Beam elements with rotational stiffness were used to represent the bars of the bridge. The material model is isotropic, linear-elastic with Young modulus $E = 70 \times 10^6$ kPa, Poisson coefficient $\nu = 0.35$ and coefficient of thermal expansion $\alpha = 23.1 \times 10^{-6}{}^\circ C^{-1}$, as standard aluminum. Reference temperature was set as $T_0 = 20$°C. A uniform distribution of temperature was applied through all the bar elements, and is based on the mean of the thermocouples measurements TA and TB at the top of the bridge seen in Fig. 4(a). Essentially nine linearly spaced points of thermal loading from 20.19°C to 45.49°C were simulated as a quasi-static analysis, with different values of the spring stiffness. Each analysis took approximately 0.128 seconds.

Fig. 5 shows the strain output of the bridge model for this loading condition. It is considerably easier to model the infrastructure behavior with a uniform temperature gradient on all of its elements, but obviously this is a model discrepancy, since in the laboratory experiment the top of the bridge is much hotter than the bottom side, as seen in Fig. 4(a).

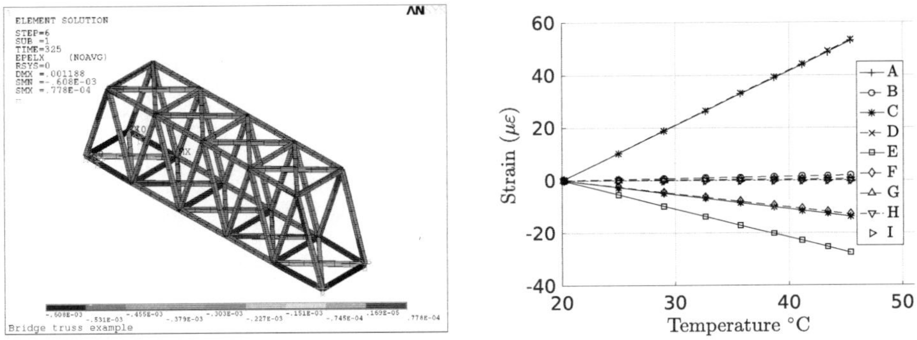

(a) (b)

Figure 5: a) FE model with linear springs at support, maximum temperature of heating cycle. b) Strain/temperature variation at stiffness $K = 300$ N/mm

5. Results and Discussion of the Fast Preposterior Analysis and its Validation

5.1. Considerations of the Analyzes and computational effort

Having established the responses from the experiment and model presented in Section 4, non-informative uniform priors have been assigned to the: structural parameter $\theta = [300 \quad 900]$ MPa; the discrepancy function mrGp hyperparameters $\omega_\delta = [0.01 \quad 10]$, $\Sigma_\delta^2 = [1 \times 10^{-8} \quad 1 \times 10^1]$, $\beta_\delta = [-2 \quad 2]$; and the observation error $\lambda = [0 \quad 1 \times 10^3]$. The functions in the mean and covariance of the mrGps are set as, a polynomial regression $H(\bullet) = 1$ and a linear correlation function. Finally, 36 and 14 temperature sites in the range of Fig. 4(a) were used for measurements and simulations sites. The computational effort of the whole combinatorial analysis was approximately six hours on a single-processor, 3.6 GHz, i7 machine. According to our estimate Arendt et al. 2016 approach would take 15.3 hours, which is roughly three times what the FPA took.
Finally, the MBA was also run for all possible combinations, i.e. $2q - 1 = 511$ runs, being the measurements from the experiment effectively used to identify the properties of the spring stiffness PDF (mean value $E[\theta]$ and its posterior variance). The relative bias $\varepsilon = \frac{|K - E[\theta]|}{K}$ was also calculated. The same data size, K prior structure and regression/correlation functions as the FPA were considered.

5.2. Optimal sensor configuration based on the Fast Preposterior Analysis

This Section presents the results of the application of the FPA and the MBA methodologies to the aluminum bridge case study. Fig. 6 displays the relative bias ε against the posterior variance relationship, that were achieved by the MBA combinatorial analysis. As can be seen there is a clear tradeoff between these two quantities, which is a common scenario when using an estimator such as Bayesian inference. This result counters the claim from Arendt et al. that, a decrease in the variability of the posterior PDF improves the identifiability of the solution. This is a limitation of the MBA and we believe that this will tend to be the norm, rather than what Arendt and Jiang et al. claimed. The question then is, if it is possible to quantify this tradeoff with the Fast Preposterior Analysis.
Subsequently, the sensor configurations analyzed by the FPA were grouped according to their number of sensors, and Fig. 7(a) shows the minimum and maximum values of preposterior covariance for each of these groups. In general both the maximum and minimum values tend to decrease with an increasing number of sensors. At this point it is useful to compare, how the percentual bias that these sensor configurations attain (calculated with the MBA) behaves at these maximum/minimum values

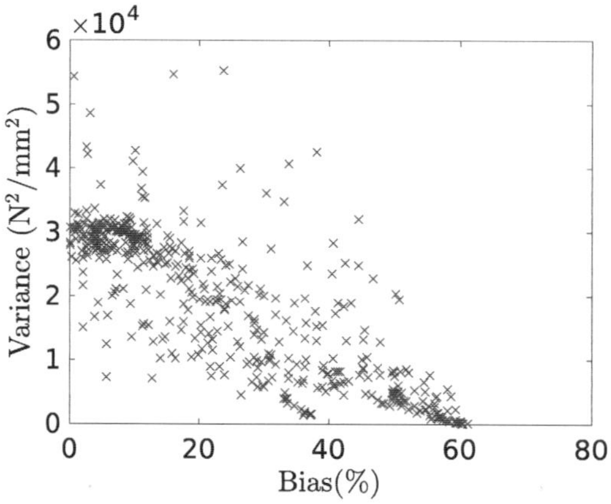

Figure 6: Bias-variance tradeoff observed for all sensor configurations (crosses) of the MBA combinatorial analysis

of $\hat{\Sigma}^2$, which can be seen in Fig. 7(b). It is important to mention that these figures display a relationship between two outputs of the two algorithms for the same sensor configurations, and should be analyzed collectively.

The difference between the maximum and minimum values of $\hat{\Sigma}^2$ is larger in the middle region (3, 4

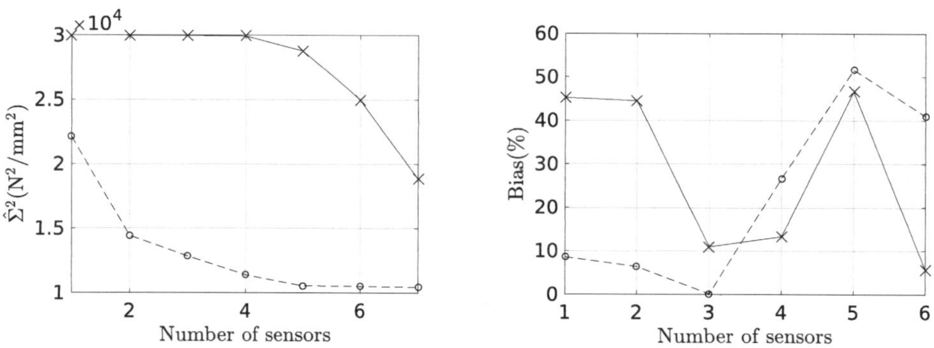

(a) (b)

Figure 7: a) Minimum/maximum values of the preposterior covariance obtained by the FPA (dashed line with circles and solid line with crosses, respectively) b) Percentual bias obtained from the MBA for the sensor configurations displayed in Fig. 7(a)

and 5 sensors), and an inverse relation (shorter difference) exists relatively to their equivalent bias in Fig. 7(b). More importantly, it is noticeable that in this central region the bias increases as the number of sensors increase. Therefore both the minimum and maximum preposterior covariances should be examined carefully to locate this central region.

Examining the sensor configurations with minimum preposterior covariance in this region, Fig. 8 highlights their placement on the actual structure and moments of the posterior PDF inferred by the MBA, respectively. These configurations have measurements at the top and bottom of the bridge and almost exclusively at its center (sensor A is the exception). As the number of sensors increases the remaining bottom/top locations (sensor B and G) are also being selected.

Figure 9 displays the prior and posterior PDFs of the spring stiffness inference. One possible way to decrease the overall posterior variance would be to consider differently natured responses, such as

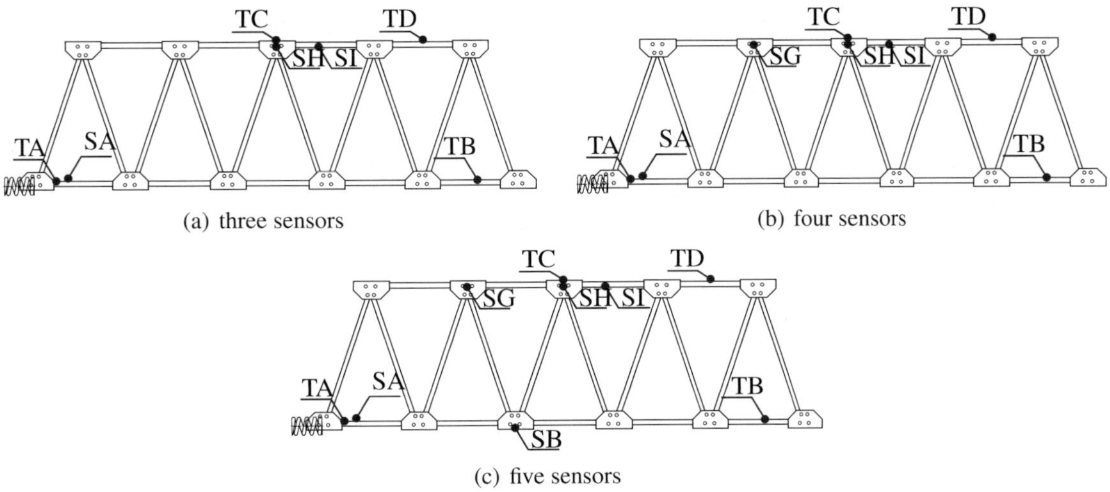

Figure 8: Diagrams of best sensor configurations obtained by the FPA. Thermocouples TA-TD and strain gauges SA-SI.

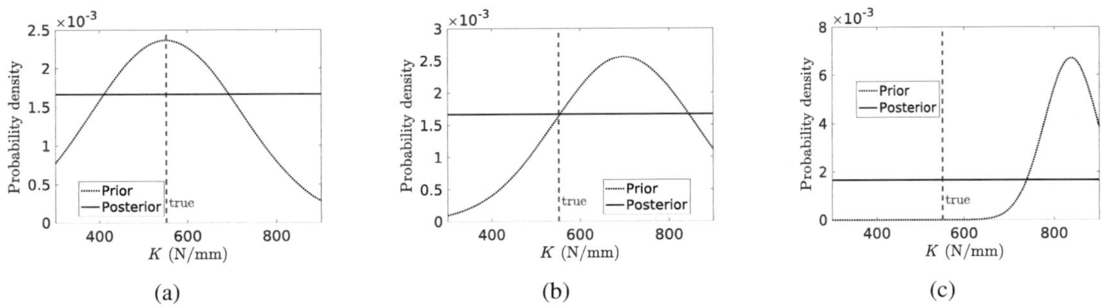

Figure 9: Posterior PDF of spring stiffness determined by the MBA, for minimum preposterior covariance values returned by the FPA, with a) three sensors b) four sensors and c) five sensors.

displacement, load in addition to strain. In other words to increase the informativeness supplied by the structural system.

It is important to mention that these results were obtained on the basis of non-informative prior knowledge, which would imply that it is an over-conservative solution. Despite this, from a total number of nine sensors an optimal solution with five sensors would reduce costs to almost 50 %.

6. Conclusions

In this paper we have developed a Fast Preposterior Analysis that quantifies the identifiability of a modular Bayesian approach, in order to reduce the uncertainties that occur during inference of structural parameters. No monitoring data is required for this analysis. This methodology selects an optimal sensor configuration based on available prior knowledge, and uses the minimum, complemented by the maximum, preposterior covariance as a performance criterion. Results of its application to an aluminium bridge scale model have been validated by a combinatorial analysis of the MBA. Main conclusions are:

- it has been shown that a tradeoff between identifiability and precision exists and cannot be quantified by the minimum preposterior covariance alone; the FPA quantifies this tradeoff by using the maximum and minimum preposterior covariance as performance criteria;

- this methodology can obtain a set of sensor configurations for practical structural identification in small-scale case-studies;
- Numerical strategies such as: the Cholesky factor decomposition; and avoiding estimation of hyperparameters in the FPA Monte Carlo loop, considerably reduce the computational effort which was required by the original formulation.

Acknowledgments

This work was supported by the Engineering and Physical Sciences Research Council (EPSRC) reference number EP/N509796.

References

Castro-Triguero, R., Murugan, S., Gallego, R. & Friswell, M. I. 2013. "Robustness of Optimal Sensor Placement under Parametric Uncertainty". en. In: *Mechanical Systems and Signal Processing* 41.1-2. ISSN: 08883270.

Papadimitriou, C., Beck, J. L. & Au, S.-K. 2000. "Entropy-Based Optimal Sensor Location for Structural Model Updating". In: *Journal of Vibration and Control* 6.5.

Katafygiotis, L. S., Papadimitriou, C. & Lam, H.-F. 1998. "A Probabilistic Approach to Structural Model Updating". en. In: *Soil Dynamics and Earthquake Engineering* 17.7-8. ISSN: 02677261.

Simoen, E., Papadimitriou, C. & Lombaert, G. 2013. "On Prediction Error Correlation in Bayesian Model Updating". en. In: *Journal of Sound and Vibration* 332.18. ISSN: 0022460X.

Vincenzi, L. & Simonini, L. 2017. "Influence of Model Errors in Optimal Sensor Placement". en. In: *Journal of Sound and Vibration* 389. ISSN: 0022460X.

Kennedy, M. C. & O'Hagan, A. 2001. "Bayesian Calibration of Computer Models". In: *Journal of the Royal Statistical Society: Series B (Statistical Methodology)* 63.3. ISSN: 1467-9868.

Arendt, P. D., Apley, D. W. & Chen, W. 2016. "A Preposterior Analysis to Predict Identifiability in the Experimental Calibration of Computer Models". en. In: *IIE Transactions* 48.1. ISSN: 0740-817X, 1545-8830.

Jesus, A., Brommer, P., Zhu, Y. & Laory, I. 2017. "Comprehensive Bayesian Structural Identification Using Temperature Variation". In: *Engineering Structures* 141. ISSN: 0141-0296.

Yuen, K.-V. 2010. "Bayesian Methods for Structural Dynamics and Civil Engineering". In: *Bayesian Methods for Structural Dynamics and Civil Engineering*. John Wiley & Sons, Ltd. ISBN: 978-0-470-82456-6.

Arendt, P. D., Apley, D. W., Chen, W., Lamb, D. & Gorsich, D. 2012a. "Improving Identifiability in Model Calibration Using Multiple Responses". en. In: *Journal of Mechanical Design* 134.10. ISSN: 10500472.

Arendt, P. D., Apley, D. W. & Chen, W. 2012b. "Quantification of Model Uncertainty: Calibration, Model Discrepancy, and Identifiability". In: *Journal of Mechanical Design* 134.10. ISSN: 1050-0472.

Jiang, Z., Apley, D. & Chen, W. 2015. "Surrogate Preposterior Analyses for Predicting and Enhancing Identifiability in Model Calibration". In: *International Journal for Uncertainty Quantification* 5.4.

The 8th International Conference on
Structural Health Monitoring of Intelligent Infrastructure
Brisbane, Australia | 5-8 December 2017

Pre-Heated Epoxy CFRP to Concrete Joints for Reinforced Concrete Structures

J. D. Tan[1], S.S. Mahini[2], K. Hoad[3]
[1]*Honours Student, School of Civil Engineering and Surveying, University of Southern Queensland, QLD, Australia Corresponding author; Email: u1097616@umail.usq.edu.au*

[2]*Senior Lecturer, Discipline of Civil and Environmental Engineering, University of New England, NSW, Australia*
[3]*Graduate Student, Discipline of Civil and Environmental Engineering, University of New England, NSW, Australia*

Abstract

Previous research into the bonding of Carbon Fibre Reinforced Polymer (CFRP) to concrete to strengthen and maintain existing concrete structures has resulted in a problem where the bond fails under loading. This has been a limiting factor to the amount CFRP is used to strengthen and maintain existing structures. Furthermore, research into increasing the bond strength is necessary so that this underestimated material can be applied to its full potential. Research conducted by Hoad (2014) concluded that an anchorage configuration with 5 anchors in a 'T' pattern could increase the load-slip ratio. A number of tests were conducted at the University of New England Engineering testing laboratory in Armidale NSW, Australia in 2016. The result of such testing was then compared with that of Hoad (2014) as well as previous published research results by Smith (2010) and Zhang and Smith (2011). The tests included using a different installation technique for an existing configuration to increase the ultimate load that could be applied in the Universal Testing Machine (UTM). The results were presented in a load-slip graph and concluded that the technique of preheating the epoxy to 50°c for 1 hr before application, would improve the exothermic chemical reaction and hence improve the bond strength in the joint. The application of preheated epoxy increased the bond strength such that the CFRP plate failed before the bond failed; meaning the capacity of the joint is much higher.

1. Introduction

Many early built structures are facing deterioration, degradation and sometimes need to support a greater design load. An increase is required in the effort to extend the life expectancy of existing structures because the cost to replace outweighs the cost to maintain. However, this means studies have to be undertaken to improve the technology we use to do such activities. Hence a forerunner for this significant problem is to use Carbon Fibre Reinforced Polymer (CFRP) externally bonded to concrete.

Masuelli's (2013) introductory report into Fibre Reinforced Polymers (CFRP) defined CFRP as a composite material of a polymer matrix reinforced with fibres such as carbon. CFRP composites are non-corrosive, lightweight, low density, exhibit high specific strength, high specific stiffness, are easily constructed and can be designed to meet specific design requirements (Masuelli, 2013).With carbon being light weight and very strong it makes it an adequate material to reinforce the polymer. From past research into fibre reinforced polymer to concrete, it is understood that using Carbon as the reinforced fibre for the polymer will increase the load capacity shown by the area under the load vs slip curve, hence meaning the bond strength is increased.

Past Externally bonded CFRP composites are a mechanical approach to strengthen and repair reinforced concrete structures. Proven to be an effective applicant to multiple materials it is increasingly becoming vastly popular as it provides strength, flexibility, low cost, and light weight characteristics. However, premature debonding of the materials has put a halt on the rapid growth of this innovative technology (Hoad, 2014).

Studies have provided research in relation to CFPR to concrete, anchorage devices associated with such methodology and an analysis of the current status of the problems faced in industry today. This study will compare a variety of peer reviewed journal articles based on the bond between CFRP and concrete. Following, a detailed overview of the report 'Optimisation of Carbon FRP-to-concrete anchorage configurations' will be conducted to firmly advance the experimental research data collected during the research work. Results taken from past research as well as new data collected for this paper will be summarised, showing a comparison of the maximum yielding loads and the slip values that correspond with the failure at those points.

2. Review of Past Studies

Smith (2010) reviewed research into the strengthening of concrete and other construction materials with externally fitted CFRP composites. With the aim to enhance the understanding of the interfacial behaviour, whilst improving the bond strength between the CFRP composite and the construction material. This was done by adding a single CFRP anchor doweled into the concrete, fanned at the surface, and connected via the use of the epoxy to the CFRP plate. Graphing a controlled specimen (CN) (no anchors) against a specimen with a single anchor (I1) showed that there is a clear distinguishable difference between the two. With both specimens having a similar initial debonding point, but the specimen with the anchor yielding at almost double the maximum load and 10 times the slip capacity. It was discussed that the CFRP to concrete will generally experience interfacial or debonding failure in the concrete joints. Conclusions suggested that further study and research will be undertaken at The University of Hong Kong on the application of CFRP to concrete (Smith, 2010).

Mechanical anchorage can improve the strain capacity of externally bonded CFRP composite which are used to strengthen concrete structures. Lacking research of CFRP in isolation or in groups is however limiting their design. Zhang and Smith (2011) researched a series of tests and reported the effects of having none, one or 3 anchors (I3) in the concrete structure. Tests between rigid and flexible anchors were also conducted. The optimal placement of the anchors allowed for a threefold increase in the strength of the CFRP anchored joints versus the unanchored control joints average. The slip capacity was generally increased by fivefold. As a result of the test observations and measurements, it is recommended that the flexible anchors are the anchor of choice (Zhang & Smith, 2011).

CFRP – strengthened reinforced concrete structures are likely to experience significant temperature variations. Such variation can have significant impact on the bond performance of CFRP-to-concrete interfaces. Gao, Teng and Dai (2012) presented an analytical solution for the full-deformation process of CFRP-to-concrete bonded joints under combined thermal and mechanical loading. Using a bilinear bond-slip model and closed form expressions a solution was determined. The legitimacy of the solution is demonstrated and proven through comparisons with both experimental results and finite-element predictions. The results are presented to show the effect of thermal loading on the interfacial shear stress and slip distributions. Results illustrated that a temperature rise (summer conditions) will increase the ultimate load, whereas a decrease in temperature (winter conditions) will decrease the ultimate load (Gao, Teng, & Dai, 2012).

The influence of CFRP plate thickness, width and elastic modulus when applied to CFRP to concrete bonds are considered to be important parameters in strengthening the bond between the CFRP and the concrete. Hence in 2013 Zhang and Smith aimed to advance the knowledge of the influence that the above parameters would have on improving the bond between the CFRP and concrete. This was done by conducting a number of tests, both controlled and with anchors. From the results it was determined that an increase in plate width will considerably enhance the joint capacity, whilst the influence of the anchor begins to decrease. Thickness and elastic modulus were shown to be less influential (Zhang & Smith, 2013).

Kalfat et al. (2013) researched into the concept of metallic anchorage devices to strengthen the bond between the CFRP and the Concrete. The study concluded that using this method was the most effective way of externally bonding the CFRP, however undesirable issues arise including the difficulty to install and metal being relatively easily corroded.

Hoad (2014) researched the 'optimum' configuration of anchors in order to improve the debonding nature between the CFRP and the reinforced concrete.

The anchor that was tested was made from CFRP and commonly called the spike anchor. Three anchorage configurations were analysed during this study. Following the research that Kalfat et al. did in 2013, but CFRP was used rather than metal plates. The load-slip and strain results obtained determined that the T5-1 configuration was optimum (Hoad, 2014).

Mertog˘lu, Anil and Durucan (2016) researched the bond slip behaviours and strain distributions at the CFRP and reinforced concrete interface. The main variables within the research were width, length of the bond and the number of CFRP anchors. Results represented that using anchorage has large effects on the ultimate load capacity, strain distribution and energy dissipation capacity. Also increasing the bond strength with an increasing number of anchors.

From the research obtained there is a clear gap in the knowledge and understanding of FPR-to-concrete. Therefore it is important that the data is correlated together on a single graph to graphically show the difference between each anchor method. It has been determined that using multiple anchors will give the highest yielding strength. This study will use a new installation technique of preheating the epoxy before application on the anchored specimen which could also prove to be influential on improving the bond strength.

3. Experimental Details

When conducting research into the bonding nature between CFRP and concrete, generally there are five steps that the preparation can be broken into. An easier understanding and clearer results can be formed from breaking down the preparation stage into different steps.

The 50 MPa concrete was ordered and delivered with a 10% leniency regarding the AS1012.2.
The concrete was then poured into moulds measuring 200 mm x 200 mm x 400 mm.
A sample taken from the batch where tests to determining the slump, flexural, splitting and Unconfined Compressive Strength (UCS) were performed. Following the AS1012.8.1 Clause 9.2.1 the specimens were left to cure in ambient conditions for 28 days.

Once the concrete samples cured, the process of marking out the precise positions for the CFRP sheets and CFRP anchors can be completed. The dimensions for the T5 layout are as follows:

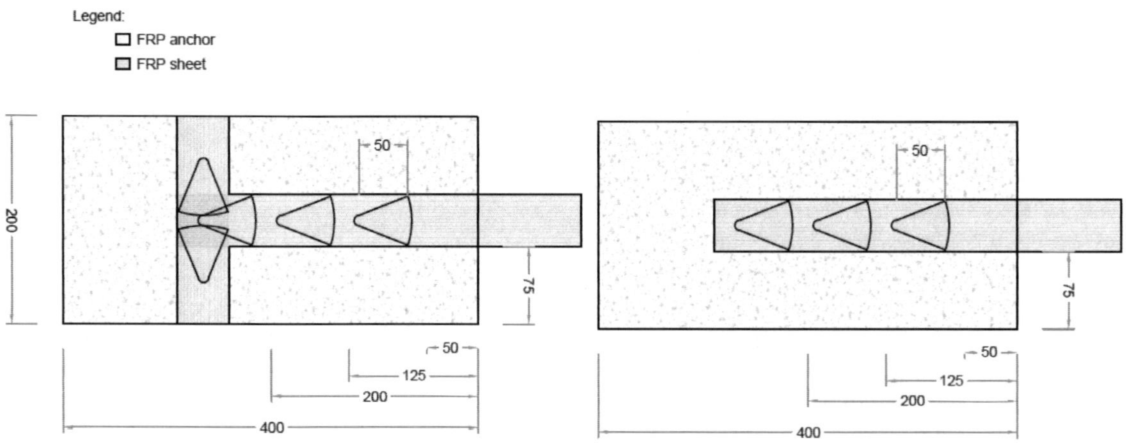

Figure 1: (a) T5 Anchorage Configuration (Plan View). (b) I3 Anchorage Configuration (Plan View).

The 14mm holes for the anchors were set out according to figure 1(a) and drilled ensuring that the masonry drill piece was marked at 40 mm as a means to prevent the hole being too deep. Following this, the surface of the sample was scabbled to improve the cohesion between the concrete surface and the epoxy. This was done using a hammer along with a cold chisel; alternatively a 100 mm grinder could have been used. Dust and particles of broken concrete was removed with an air jet, finishing the cleaning process with a wipe down with acetone to remove any unwanted oils from the surface.

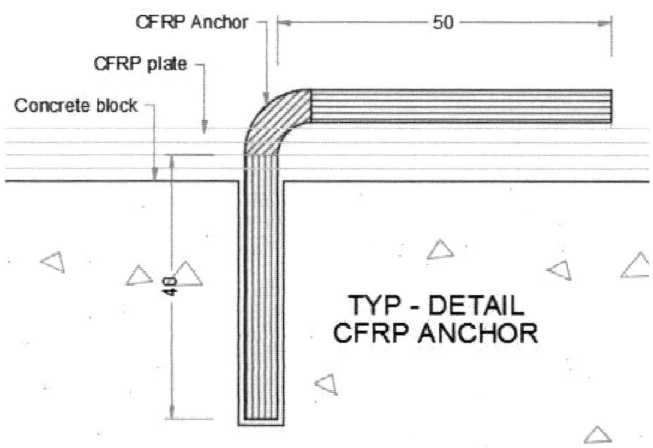

Figure 2: Dimensions of a typical CFRP anchor (This Study).

The next process was to prepare the CFRP anchors. This was done by cutting 90 mm long carbon fibre sheets comprising of 50 mm length in the fan and 40 mm embedded portion. The two part epoxy Sikadur-330 was heated for 1hr at 50°c. Compounds were mixed to form the resin to be applied to only 20 mm of the end of the carbon sheet previously cut. Firmly rolling the sheet and compressing the fibres into a smooth finish, ensuring that the remained of the anchor sheet remained free from the epoxy. Finally the anchors were placed in a dust free environment for one day to cure.

For each sample, there were 3 layers of the CFRP sheets. Each sample had two types of sheet; the longer ones (vertical part of the T) which overlapped each layer of the shorter sheet (horizontal part of the T).

The next part was to apply the sheets and anchors to the concrete blocks. Sikadur-330 epoxy was heated for 1 hr at 50°c and mixed. The dowels were set into the concrete and the rest of the void/hole was filled with the epoxy. The fibres in the sheet at the location of the dowels were separated to allow the dowel to stick through as per Figure 2. The sheets were applied as per above, more resin was applied at each layer. The dowels were spread into a fan and folded down flat with the sheet as seen in Figure 3. Finally more epoxy was applied to the surface to achieve saturation.

The specimens were then covered in cling wrap and a flat heavy weight was sat on top to assist with holding the fan flat. The weight was removed after 24 hrs and curing took place for 28 days.
Once cured, strain gauges were applied to the specimen at the locations set out in Figure 4. Specimens were set up into the Universal Testing Machine to test.

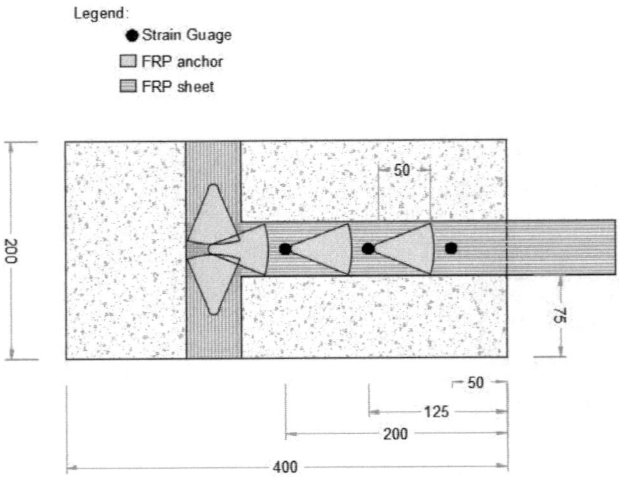

Figure 3: Location of strain gauges.

Figure 4: Specimen in testing apparatus.

4. Experimental Results

The results contained herein are presented for firstly the comparison between the I3 configurations tested. Then a comparison between different prepared T5 anchor configurations as shown in Figure 1 against a T5 prepared using the preparation method as discussed.
The comparison of the experimental results will be collated into Table 1 and will show the maximum load, slip, strain and the type of failure.

(i) Smith (2010)

The joints that were tested were the control joint (no anchors) and the single anchor joint. It was evident that the addition of an anchor increased the maximum load on the specimen as well as an increase in capacity. In the table below, the maximum load and slip are given for each type of joint that was tested.
During the test of the single anchor joint, the debonding begun at the loaded end at a similar load to that of the controlled specimen. The anchor was then essentially engaged as the strain moved through it and the cracking became prevalent. The stiffness in the joint reduces as the length of the crack increased. Then there was a drop in the load, from which it began to pick up again as a cause of friction in the joint. Once the specimen had failed the load decreased at a constant rate, until the remaining CFRP anchor was fully ruptured. The slip capacity of the single anchored joint extended to 10 times that of the control joint.

(ii) Zhang and Smith(2011)

The joints that were tested were a number of different configurations which included control joints, single anchor joints, double anchored joints, and triple anchored joints. From the tests it showed that comparatively the 3FA75 (or I3 Configuration) had the highest average maximum load and the maximum slip capacity. In the table below, the maximum load and slip are given for each type of joint that was tested.
During the test the third anchor away from the loaded end was found to be close to ineffective, however it may add to the overall strength of the joint. It was also noted that the debonding began at the end closest to the load, from which once the strain passed through the first anchor and into the second anchor the debonding shown my cracking became more prevalent.

(iii) Hoad (2014)

The joints will be used in comparison are the T5 prepared joints. This methods follows closely to a method studied previously by (Kalfat et al, 2013) but uses an anchored CFRP plate rather than the steel plate. In the table below the maximum load and slip are given for the T5 prepared anchored CFRP joint that was tested.

From the test conducted by Hoad, it is noticeable that the T5 configuration yielded the highest load, and was considered to have the highest slip capacity of the specimens tested. The results show a stage of debonding at about 20-25 kN of load, but still reaches such a higher maximum load. This suggests that the CFRP debonds slightly before the strain gets to the first anchor, and once it passes the anchor it is able to continue rising until complete debonding. It is noted that different climates and curing times could have been a reason as to why the results of the T5 configuration were lower than the result specimens previously tested by Zhang and Smith.

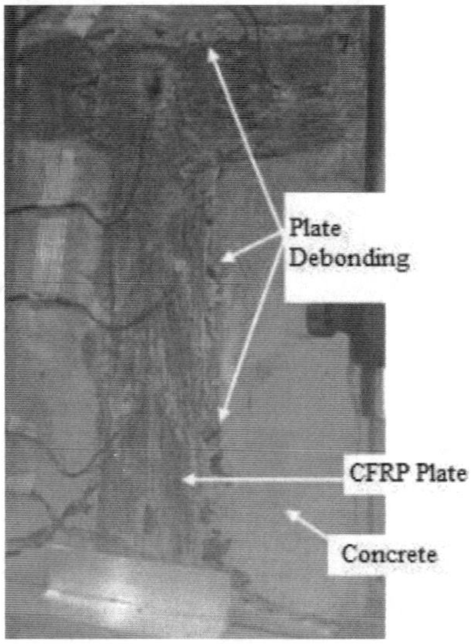

Figure 5: T5-1 post-test showing failure.

(iv) Tan (2016)

For this paper, new preparation techniques were used to improve the already discussed T5 configuration set out by Hoad (2014). From the use of preheating the epoxy before mixing at 50°c for 1 hr before applying the maximum yielding load and the slip correlated to that point was tested. In the table below the results for the above are shown. From the test conducted during this study, it is noticeable that the heating of the epoxy ultimately causes a rupture of the CFRP plate.

Figure 6: T5-3 post-test showing failure.

The above Figure 6 shows the failure of the method conducted for this study. Showing that the plate before the first anchor was mostly ruptured but the plate past the first anchor remained intact and still fully bonded.

Table 1: Comparison on experimental results

Author	Specimen Code Identification		Load Pu (kN)	Slip S (mm)	Max Strain (με)	Failure Mode
Hoad (2014)	Five tee-anchored series	T5-1	39.8	3.5	23406	c
This Study (pre-heated epoxy)		T5-2	43.0	2.7	-	a
		T5-3	46.1	3.4	8786	a
Smith (2010)	Control CN	Joint	17.5	0.9	-	
	Single anchor I1	series	30	2.1	-	
Zhang & Smith (2011)	Three anchor I3	series	43.8	1.75	11750	b

Failure Modes:

(a) Plate mostly of fully ruptured, but not fully debonded.

(b) Plate mostly or fully ruptured post debonding.

(c) Plate fully or mostly debonded

5. Analytical Comparison

Figure 7 presents the comparison of the maximum loads which is typically momentarily prior to the failure of the joint. The comparison shows that the load resulted from the preheated epoxy on the T5 anchor configuration (T5-2 & T5-3) are considerably higher than the standard prepared T5 configuration (T5-1). This suggests that within the same climate and following the same procedure (apart from the preheating of the epoxy) the use of preheated epoxy significantly increases the maximum load that can be applied. Figure 7 also shows that one specimens (T5-2) with the preheated epoxy has similar results (0.8 kN difference) to that of the I3 presented by Zhang and Smith (2011), whereas the other specimen (T5-3) is higher than the I3 by 2.3 kN.

The performance of the joints under testing can be conveniently summarised by the load-slip response and hence will be displayed in a load vs slip graph in this section. As shown in figure 8 the area under the graph can also determine the capacity of the joint. The comparative study shows that the control joint having the smallest curve, followed by the single anchor set out buy Smith in 2010. Following, again the graph shows that the two T5 specimens set out in this study are considerably higher than the T5 set out by Hoad and Mahini in 2014. This suggests that using the preheated epoxy will increase the load-slip response of the joint under increasing load. For the I3 configuration (Zhang & Smith, 2011) once the load yielded, the joint did not resist any further load meaning that it was a complete debond failure. Whereas the specimens tested in this study had a statistical mean higher then and were able to resist small amounts of load after the failure meaning that the capacity is higher. Also important is how the specimen reacted to the strain throughout the test. Figure 9 shows the comparison of the T5-3 conducted in this study, the T5-1 conducted by Hoad and Mahini in 2014, and the I3 configuration conducted by Zhang and Smith in 2011. The comparison shows that the I3 was able to distribute the strain throughout the specimen more efficiently then the T5 configuration. It also shows that the T5-3 was carrying the majority of the strain in the second strain gauge (50mm from the loaded end). This indicates that the bond past the anchor was that so strong that it was almost completely unaffected by the load applied.

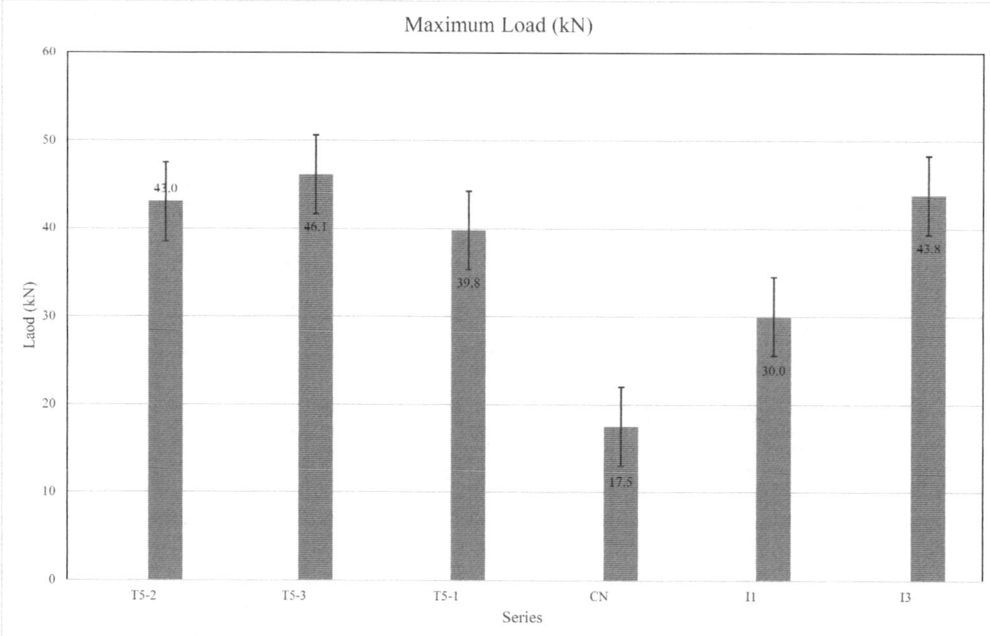

Figure 7: Maximum Load graph showing a comparison of results.

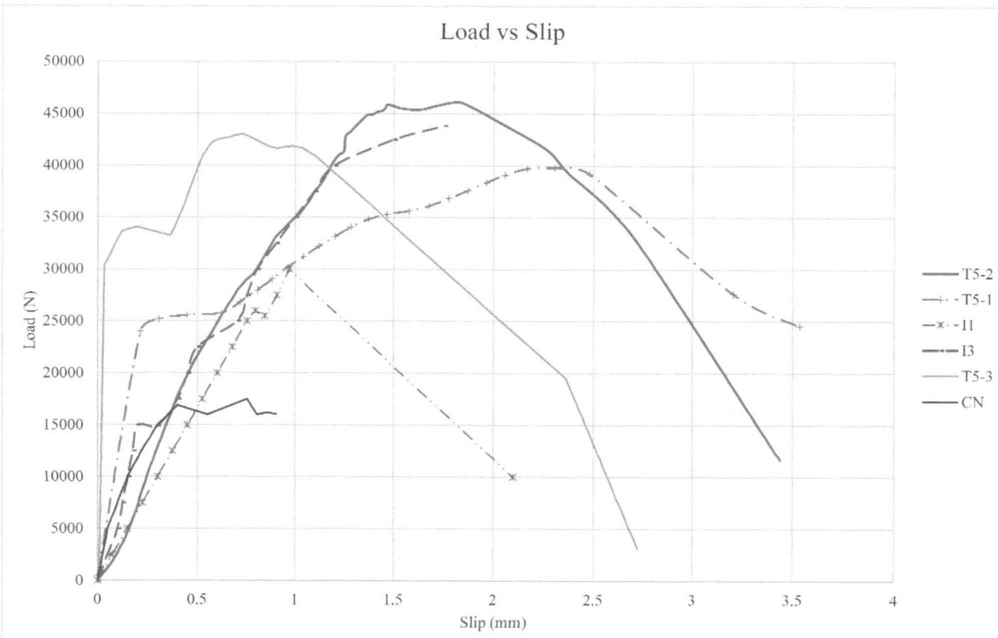

Figure 8: Load-slip graph showing a comparison of results.

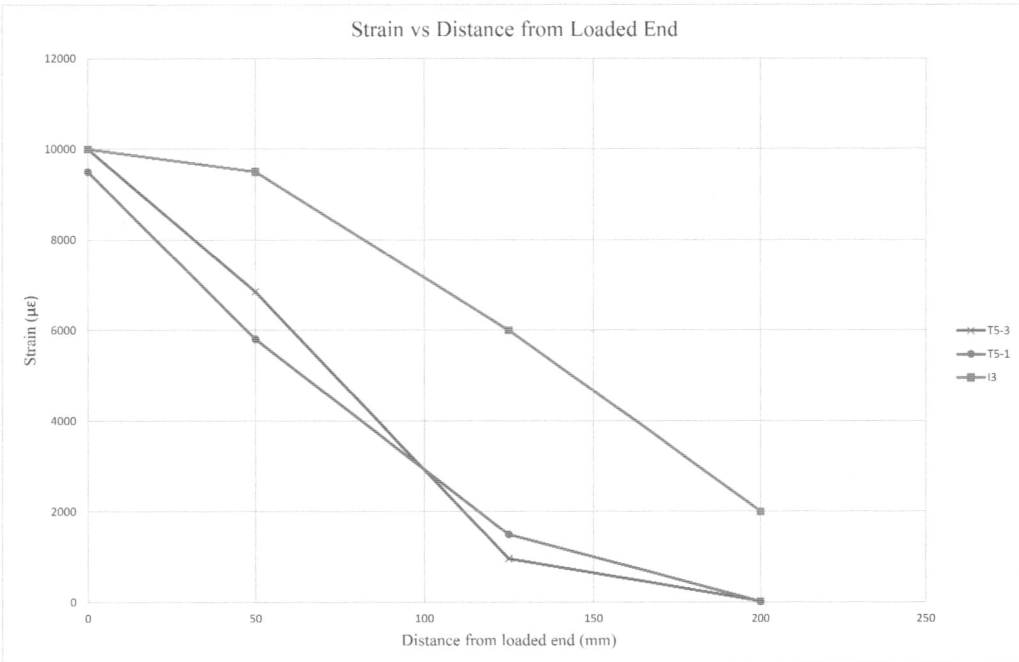

Figure 9: Comparison of specimen showing Strain vs Distance from applied showing.

6. Discussion

In an effort to advance the current knowledge of CFRP to concrete, tests were conducted to maximise the bond between the CFRP and the concrete. Different preparation techniques were used and the results were presented in a load-slip graph against previously published results in this field of expertise. The area under the curve of the graph indicated the capacity of the bond. Results show that preheating the epoxy before application would increase the bond strength to a point where the CFRP would rupture rather than a debonding failure. Initially errors and differences between the results set out in previously published papers and the results set out in this study could have arisen from using different climates and geographical locations. This may have affected the specimens curing time and conditions. However with these results presented, the uses of CFRP to concrete to maintain and strengthen existing concrete structure can be wider developed. This could mean that the cost of rebuilding a concrete structure could be avoided simply by using CFRP to maintain and strengthen the existing structures. In a Local Government context this could increase the funds available to asset owners and used in other areas, whilst increasing the quality and longevity of infrastructure. Further research would include testing for the optimal temperature of the preheated epoxy, also using an additional layer of CFRP.

7. Conclusion

The result of a series of tests which consisted of 5 different preparation methods, ranging from a control joint (no anchors) to a T5 joint (5 anchors with a preheated epoxy), were presented in this paper. It has already been proven that adding CFRP anchors into the concrete will enhance the bond strength and reduce the debonding failure. In this study a new preparation method was summarised which was the T5 anchor configuration set out by Hoad and Mahini in 2014. To advance this, test were done to show the effect that preheating the epoxy would have on the bond strength. In previous studies the CFRP plates were commonly debonding from the concrete, which as shown is a debonding failure. However, the method of preheating the epoxy has proven to increase the bond strength to a point such that the CFRP fails and ruptures rather than the bond. The load-slip responses have been graphically displayed against a number of already published results. Following, the maximum strain has been compared with the already published results. The comparison showing that the preheating of the epoxy increases the ultimate load by 5.2%.

Acknowledgements

The funding providing by the University of New England to perform the experiment and laboratory work is gratefully appreciated and acknowledged. The author would like to thank the project Supervisor, Dr Saeed Mahini, for providing detailed knowledge about the topic, and the help provided throughout the weekly meetings with him. The author acknowledges the research conducted by Katelyn Hoad and Saeed Mahini in 2014. Furthermore the authors of past research papers and data produced are acknowledged. The Author also acknowledges his peer Mr Aaron Blake and Technical Officer Mr Tony McKinnon for the help they provided with setting up the laboratory work.

References

Masuelli, M. A., 2013. *Introduction of Fibre-Reinforced Polymers - Polymers and Composites: Concepts, Properties and Processes,* San Luis, Argentina: InTech.

Hoad, K. L., 2014. *Optimisation of Carbon FRP-to-concrete anchorage configurations,* Bachelor thesis submitted to the Discipline of Civil and Environmental Engineering, University of New England, Armidale, Australia.

Smith, S. T., 2010. Strengthening of Concrete, Metallic and Timber Construction Materials with FRP Composites. *The 5th International Conference on FRP Composites in Civil Engineering* .

Zhang, H. W. & Smith, S. T., 2011. FRP-to-concrete joint assemblages anchored with multiple FRP anchors. *Composite Structures.*

Gao, W. Y., Teng, J. G. & Dai, J.-G., 2012. Effect of Temperature Variation on the Full-Range Behaviour of FRP-to-Concrete Bonded Joints. *Journal of Composites for Construction.*

Zhang, H. & Smith, S. T., 2013. Fibre-reinforced polymer (FRP)-to-concrete joints anchored with FRP anchors:testand experimental trends. *NRC Research Press.*

Kalfat, R., Al-Mahaidi, R. & Smith, S. T., 2013. Anchorage Devices used to improve the performance of reinforced concrete beams retrofitted with FRP strips.. *Journal of composites for construction,* pp. 221-227.

Mertoglu, C., Anil, O. & Durucan, C., 2016. Bond slip behavior of anchored CFRP strips on concrete surfaces. *Construction and Building Materials.*

Standard Australia, 1986-2000. Methods of Testing Concrete.

The 8th International Conference on
Structural Health Monitoring of Intelligent Infrastructure
Brisbane, Australia | 5-8 December 2017

Improvement of Accelerated Weathering Resistance of Epoxy Resin Based on Hydrogenated Diglycidyl Ether of Bisphenol A (HDGEBA) Using Multi-Walled Carbon Nanotubes (MWCNTS) to Protect Timber Bridges

S. A. Awad [1, 2], C. M. Fellows[1] and S. S. Mahini[3]

[1] *Chemistry, The University of New England, Armidale NSW 2351, Australia,*

[2] *College of Education for Pure Science, University of Anbar, Al-Anbar, Iraq*

[3] *Civil and Environmental Engineering, The University of New England, Armidale NSW 2351, Australia.*

Email: sawad3@myune.edu.au.

Abstract

There are several strategies to reduce environmental degradation risks of outdoor timber bridges. Among these, one of the most important is coating with epoxy resin to protect timber from UV irradiation and moisture. However, most epoxy products used for these purposes were developed for indoor applications, and their long-term stability in outdoor environments are not well understood. Modifying structural epoxy systems and developing monitoring technique to reinforce and increase the longevity of timber are therefore necessary. This study aims to compare and evaluate the effects of durability under accelerated weathering conditions (UV radiation and moisture) of an epoxy resin based on hydrogenated diglycidyl ether of bisphenol A (HDGEBA-TMDA) and a similar resin incorporating 0.5% Multi-Walled Carbon Nanotubes (MWCNTs). All specimens were exposed alternately for two weeks of UVA radiation of maximum intensity at 340 nm and two weeks of 100% humidity at 50±5°C for total periods of 1, 2, 3, 4 and 6 months. Before and after exposure to artificial accelerated weathering, the spesimens were investigated using mechanical testing. Preliminary results showed a significant improvement in mechanical properties and more resistance to different accelerated weathering times when 0.5% MWCNT filler is added to HDGEBA-TMDA adhesive epoxy.

Keywords: Epoxy adhesives, MWCNTs, Mechanical properties, Accelerated weathering.

1. Introduction

Demand for advanced composite materials with high-performance properties has been increasing to meet new requirements or replace standard materials such as metals. Epoxy composites have excellent strength, stiffness, toughness, hardness, and heat distortion temperature compared to the properties of metals. Recently, epoxy adhesive composites have been studied for broad applications including construction, and aerospace automotive and aerospace industries (Gabr 2010). Epoxy resins are widely used as polymer matrices for advanced composite materials because epoxy composites made of materials outstanding corrosion resistance, electric insulation, thermal stability, and durability (Mahfuz 2004; Kawaguchi 2003). Bridges supported on timber are common in rural regions of Australia, though the superstructure is typically constructed using steel and reinforced concrete. These timber bridges have been exposed to heavier and faster traffic in modern times, while at the same time they have undergone weathering and require rehabilitation (Andrawes, 2017). This is primarily due to timber degradation, and as a result, many bridge timbers are in poor decayed condition when they are exposed to several environmental factors such UV irradiation, humidity, high temperature. Hence there is a need to retrofit these bridges either through adding conventional natural or chemical materials, particularly in the early stages of degradation (Balendra, Wilson, & Gad, 2010).

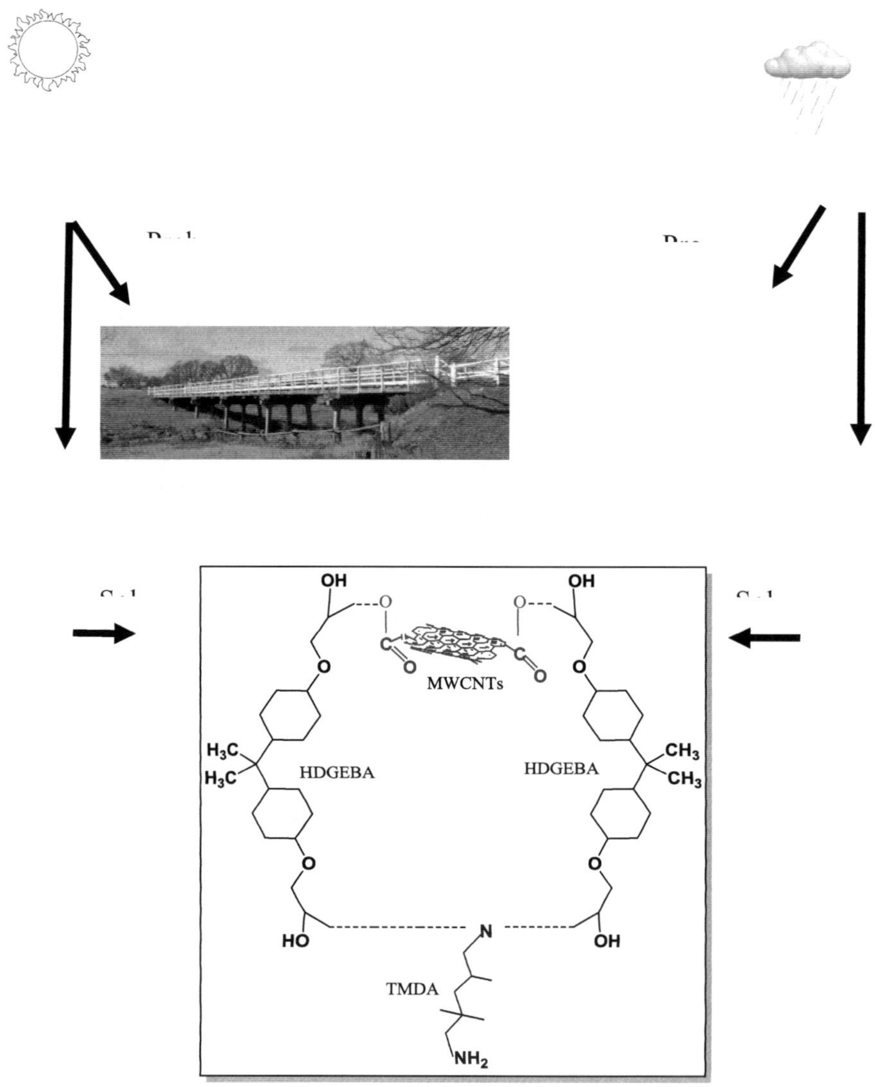

Figure 1: The problem of continued exposure to accelerated weathering (UV irradiation and moisture)

Although epoxy resin has excellent properties, it is weaker and more brittle than many other resins when it is exposed to long-term weathering. One way of reducing this problem is to add fillers that can strengthen epoxy resins as well as stabilising them against degradation. Multi-walled carbon nanotubes (MWCNTs) are used in this work to compensate the weak and brittle properties of the epoxy resin and to improve its resistance to degradation. Among the above fillers, MWCNTs are one of the most prospective fillers for high strength epoxy composites due to their ultrahigh tensile strength, low density, high aspect ratio, and high thermal and electrical conductivity (Coleman 2006; Gojny 2016; Kathi 2008). Moreover, MWCNTs have a broad surface area to volume ratio, which improves adhesion with the epoxy adhesive matrix, enhancing the properties of composites. In recent years, many researchers have reported that incorporation of multi-walled CNTs that were concluded (MWCNTs) with an epoxy matrix improved the mechanical and thermal properties (Guadagno 2011; Kim 2006). MWCNTs are considered to be an ideal reinforcing additive for epoxy molding composites with high performance (Zhou 2007; Chen 2009).
In this research, neat epoxy and epoxy adhesive-0.5%MWCNTs composites were exposed to different accelerated weathering time, and the properties before and after exposure were evaluated by mechanical testing.

2. Experimental Program
2.1 Materials

Epalloy 5000 epoxy adhesive (HDGEBA), an epoxy monomer, was provided from CVC Thermoset Specialties (, Brenntag Australia Pty Ltd. Victoria, Australia), and Sikadur 330 B (2, 2, 4-trimethylene-1, 6-hexadiamine, TMDA) resin as a curing agent was purchased from Sikadur Australia, NSW, Australia. Multi-Walled Carbon Nanotubes (MWCNT) were supplied from ANT Applied Nano Technologies, Australia).

2.2 Preparation of HDGEBA-TMDA-0.5%MWCNT composites

The MWCNT filler was dispersed in epoxy resin (HDGEBA) using high-intensity sonication over 2 h to achieve a uniform dispersion. Then, the curing agent TMDA was added to the HDGEBA-0.5% MWCNT mixture at room temperature at a weight ratio of 4:1 HDGEBA:TMDA. The epoxy composites were heated to 50°C for 10 h to reduce bubble formation and create a more homogeneous mixture of epoxy adhesive and filler.

2.3 UV irradiation and Moisture Exposure

Both epoxies were treated using an accelerated weathering chamber, equipped with fluorescent UV lamps to generate UVA radiation of maximum intensity at 340 nm. The specimens were exposed to UV irradiation for one week, then turned over and exposed for a further week after which they were exposed to 100% humidity at elevated temperature (50±5°C) for two weeks. Specimens were tested after this procedure. This process was then repeated, to give total accelerated weathering times of 1, 2, 3, 4 and 6 months for each specimen.

2.4 Tensile Strength Tests

Tensile strength testing was performed by a Universal Testing Machine (UTM) (Industrial Series DX 300KN, Instron Ltd., Bayswater VIC). The specimens were tested according to ASTM D638 using the dog-bone shape of all specimens with dimensions 50 mm length, 20 mm width, and 2 mm thickness. The testing speed was 10 mm/min. Specimens were positioned vertically between the grips of the testing machine, and stress-strain curves were plotted during the test. Five specimens were tested for each set of conditions.

Figure 2: The tensile testing rig

3. Results and Discussion

3.1. Mechanical properties

3.1.1. Tensile characteristics

Measurements of the engineering properties can refer to the environmental degradation resistance (Woo et al. 2008). In tensile tests, the stress-strain curves of HDGEBA epoxy resins and HDGEBA-0.5% MWCNT composites were captured. Figures 3 and 4 provide a comparison of tensile stress-strain results of epoxy resin with 0.5% amounts of MWCNT as a reinforcing agent before and after exposure to different accelerated weathering times. All specimens were overdrawn immediately after the tensile stress reached the maximum point. The mechanical strength is primarily dependent on interfacial bonding between the filler and the epoxy. Therefore, the first behavior of the resistance of the MWCNT-epoxy composites was caused by MWCNT-epoxy interfacial bonding, and the excessive addition of MWCNTs to the epoxy causes reduced tensile strength performance. The epoxy adhesive with 0.5% MWCNT showed (Fig. 4) the highest tensile strength of 28.6 MPa, compared with neat epoxy 14.3 MPa before exposure to accelerated weathering. The epoxy adhesive composites founded to be more resistant during exposure further accelerated weathering (six months) which was about 21.2 MPa, while the neat epoxy decreased to 9.4 MPa.

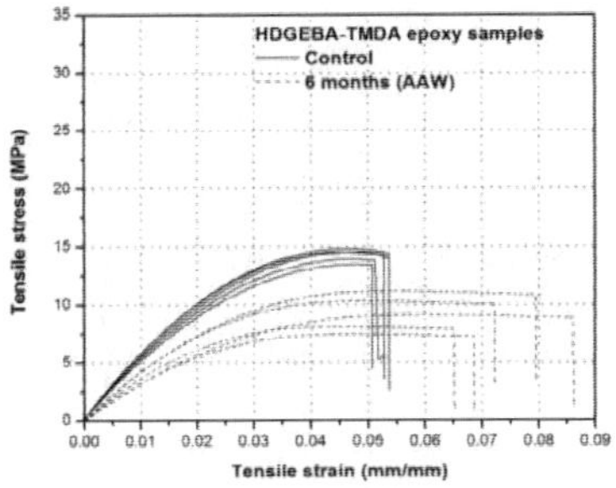

Figure 3: Five replicate Stress-Strain Curves of HDGEBA epoxy specimens before accelerated weathering (solid black line) and five replicate curves after 6 months accelerated weathering (AAW) (dashed line)

The test results of strain at break (Fig. 5) showed that neat epoxy adhesives have the highest strain before exposure (5.4%) and after exposure of six months (7.5%), compared with epoxy adhesives containing 0.5% MWCNT composites that had a more constant elongation before (2.5%) and after exposure to accelerated weathering (3.6%) (Fig. 6). The results proved that the incorporation of 0.5% MWCNT in the epoxy adhesive composites inhibited the outdoor environmental conditions that caused by accelerated weathering conditions (UV irradiation and moisture).

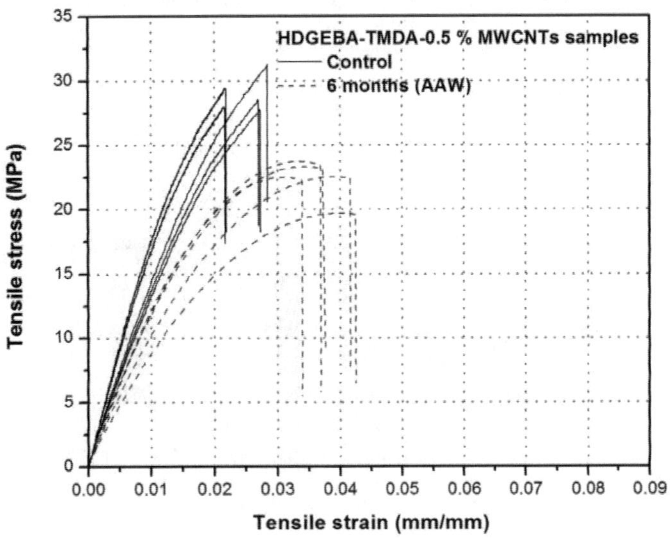

Figure 4: Five replicated Stress-Strain Curves of HDGEBA -0.5% MWCNT specimens before accelerated weathering before accelerated weathering (solid black line) and five replicate curves after 6 months accelerated weathering (AAW) (dashed line).

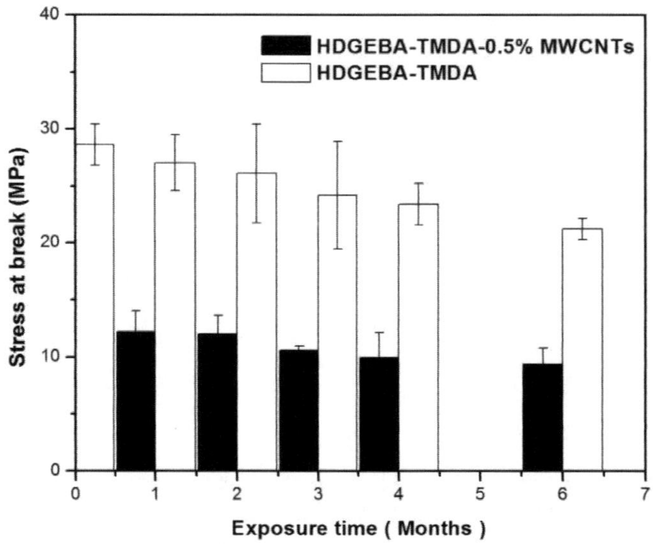

Figure 5: Stress at break data of HDGEBA-TMDA epoxy and HDGEBA-TMDA-0.5%MWCNTs specimens after 6 months accelerated weathering

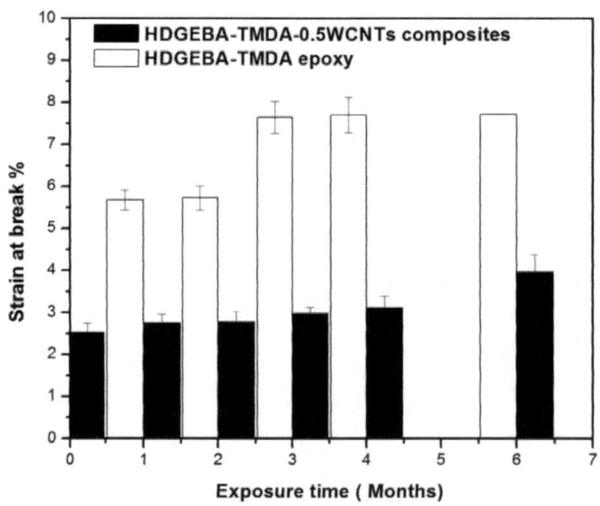

Figure 6: Stress at break data of HDGEBA-TMDA epoxy and HDGEBA-TMDA-0.5%MWCNTs specimens after 6 months accelerated weathering

4. Conclusion

This project developed experimental solutions to reduce environmental degradation of timber bridges by repairing or coating epoxy to protect timber bridges from UV irradiation and moisture. Tensile tests of the weathered epoxy and epoxy–filler composite specimens were carried out by using a UTM according to ASTM D 638 in order to compare with weathered samples. The purpose of the tests was to determine the relationship between tensile strength and increasing exposure of specimens for different accelerated weathering times by UV irradiation and moisture conditions. In these tests, the mechanical properties of 0.5% MWCNTs reinforced epoxy adhesive composites after exposure and the results were compared with specimens before exposure. The incorporation of MWCNTs into epoxy resins (HDGEBA-TMDA) appeared to be more beneficial as composites and have better resistance to the loss in tensile strength before and after exposure to accelerated weathering compared with the epoxy systems without fillers. The measurements of the mechanical properties of HDGEBA-TMDA-0.5%MWCNTs composites also showed that the maximum tensile strength increased significantly and its elongation as well as durability improved before exposure compared with epoxy adhesives (HDGEBA-TMDA).

References:

Andrawes, B & Kim, K.H.E 2017, 'Retrofitting of bridge abutment timber piles using fiber-reinforced polymer composites,' *Journal of Structural Integrity and Maintenance,* vol.2, no. 1, pp. 29-38.

Awad, S.A., Fellows, C.M., Mahini, S. S 2016, 'Rehabilitation of timber-adhesive system by applying smart chemical fillers against environmental degradation', *Australian Network of Structural Health Monitoring*, Newsletter, Vol. 9, pp. 14-7.

Balendra, T., Wilson, J. L., & Gad, E. F 2010, 'Review of Condition Assessment and Retrofitting Techniques for Timber Bridge Assets in Australia,' *Advances in Structural Engineering,* vol.13, no. 1, pp. 171-180.

Chen, Z.-K., Yang, J.-P., Ni, Q.-Q., Fu, S.-Y. & Huang, Y.-G. 2009, 'Reinforcement of epoxy resins with multi-walled carbon nanotubes for enhancing cryogenic mechanical properties,' *Polymer*, vol. 50, no. 19, pp. 4753-9.

Coleman, J.N., Khan, U. & Gun'ko, Y.K. 2006, 'Mechanical Reinforcement of Polymers Using Carbon Nanotubes,' *Advanced Materials*, vol. 18, no. 6, pp. 689-706.

Gabr, M.H., Elrahman, M.A., Okubo, K. & Fujii, T. 2010, 'Effect of microfibrillated cellulose on mechanical properties of plain-woven CFRP reinforced epoxy', *Composite Structures*, vol. 92, no. 9, pp. 1999-2006.

Genedy, M., Daghash, S., Soliman, E. & Taha, M. 2015, 'Improving Fatigue Performance of GFRP Composite Using Carbon Nanotubes', *Fibers*, vol. 3, no. 1, p. 13.

Gojny, F.H., Wichmann, M.H.G., Köpke, U., Fiedler, B. & Schulte, K. 2004, 'Carbon nanotube-reinforced epoxy-composites: enhanced stiffness and fracture toughness at low nanotube content', *Composites Science and Technology*, vol. 64, no. 15, pp. 2363-71.

Guadagno, L., De Vivo, B., Di Bartolomeo, A., Lamberti, P., Sorrentino, A., Tucci, V., Vertuccio, L. & Vittoria, V. 2011, 'Effect of functionalization on the thermo-mechanical and electrical behavior of multi-wall carbon nanotube/epoxy composites', *Carbon*, vol. 49, no. 6, pp. 1919-30.

Kathi, J. & Rhee, K.Y. 2008, 'Surface modification of multi-walled carbon nanotubes using 3-aminopropyltriethoxysilane', *Journal of Materials Science*, vol. 43, no. 1, pp. 33-7.

Kawaguchi, T. & Pearson, R.A. 2003, 'The effect of particle–matrix adhesion on the mechanical behavior of glass filled epoxies. Part 2. A study on fracture toughness', *Polymer*, vol. 44, no. 15, pp. 4239-47.

Kim, J.A., Seong, D.G., Kang, T.J. & Youn, J.R. 2006, 'Effects of surface modification on rheological and mechanical properties of CNT/epoxy composites', *Carbon*, vol. 44, no. 10, pp. 1898-905.

Mahfuz, H., Adnan, A., Rangari, V.K., Jeelani, S. & Jang, B.Z. 2004, 'Carbon nanoparticles/whiskers reinforced composites and their tensile response', *Composites Part A: Applied Science and Manufacturing*, vol. 35, no. 5, pp. 519-27.

Mashhadzadeh, A. H, Fereidoon, A. & Ghorbanzadeh Ahangari, M., 'Surface modification of carbon nanotubes using 3-aminopropyltriethoxysilane to improve mechanical properties of nanocomposite based polymer matrix: Experimental and Density functional theory study', *Applied Surface Science*.

Moore, J.C., Mahini, S., Glencross-Grant, M. R., Patterson, R 2013, 'Structural Health Monitoring of Older Timber Bridge Girders Using Laser-Based Techniques' *Australian Journal of Structural Engineering,* vol.14, no.1, pp. 27-42.

Woo, R.S.C., Zhu, H., Leung, C.K.Y. & Kim, J.-K. 2008, 'Environmental degradation of epoxy-organoclay nanocomposites due to UV exposure: Part II residual mechanical properties,' *Composites Science and Technology*, vol. 68, no. 9, pp. 2149-55.

Zhou, Y., Pervin, F., Lewis, L. & Jeelani, S. 2007, 'Experimental study on the thermal and mechanical properties of multi-walled carbon nanotube-reinforced epoxy', *Materials Science and Engineering: A*, vol. 452–453, pp. 657-64.

The 8th International Conference on
Structural Health Monitoring of Intelligent Infrastructure
Brisbane, Australia | 5-8 December 2017

Effect of FRP Length on Damage and Seismic Performance of Retrofitted Ordinary RC Frames

S. Jafari[1], S.S. Mahini [2].
[1] Master Student, Department of Civil Engineering, Shiraz University of Techlogy, Iran, Email: S.Jafari@sutech.ac.ir
[2] Senior Lecturer, Discipline of Civil and Environmental Engineering, University of New England - Australia

Abstract

This paper reports an investigation into the effectiveness of different lengths of FRPs in enhancing seismic performance of ordinary RC retrofitted at joints by FRP sheets. The flexural stiffness of FRP-retrofitted joints of the frame is first determined using nonlinear Finite Element (FE) analysis. The stiffness of the retrofitted joints is then implemented into the FE model of the frame and nonlinear static (pushover) analysis is carried out using OpenSees. The seismic performance level of a retrofitted frame with different lengths of FRPs is then compared with those of the original frame and also with available past studies. Two types of retrofitting methods; web and flange bonding; are applied and studied. The results show that the performance level of the web-bonding RC frame at the target displacement is significantly enhanced in comparison with the original frame. In addition, the frames retrofitted with longer lengths of FRPs have better performances, whereas the flange-bonding is not very effective. However, the load carrying capacity of the flange-bonded frame is enhanced in a higher rate at the target displacement compared to the web-bonding. It is also found that the web-bonding can upgrade a RC ordinary moment resisting frame (OMRF) to an intermediate level and even to a higher ductility level when is applied in a length of 400-800 mm, while by the flange-bonding, this improvement does not occur even though with a longer length of FRP sheets. Finally, the global damage indexes of the frames at the performance point (PP) and at the life safety (LS) are determined from the capacity curves. This evaluation demonstrates that the damage degree of the frame is reduced after retrofitting by FRPs.

1. Introduction

The use of FRP sheets to create composite concrete structures has increased in recent years. This is particularly true for seismic upgrading of existing reinforced concrete (RC) structures, many of which were designed for gravity loads only or according to the older codes of practice. The seismic retrofit strategy for an RC frame may include strengthening deficient members such as beams, columns and beam-column joints to increase strength, stiffness and/or ductility and to enhance the overall seismic performance of the frame by increasing lateral strength, reducing drift and/or increasing ductility. In the last decade, a vast amount of research has been carried out on the FRP retrofitting of RC frame members, such as beams (Niroomandi et al. 2010) and columns (Promis et al. 2009), all indicating the effectiveness of the retrofit technique and some highlighting certain problems associated with the technique including brittle debonding failure and the methods to overcome these problems (Mosallam & Banerjee, 2007).

Beam-column joints are also crucial components of a frame both in terms of structural stability and its seismic performance. Li et al. (2002) reported that the designed hybrid FRP reinforcement greatly improves the stiffness and load carrying capacity of the joint; it also delays the crack initiation at the joint through FRP confinement. Mahini and Ronagh (2010) developed web-bonded FRP sheets for strengthening of exterior beam-column joints in order to evaluate the possibility of relocating the plastic hinge away from the column face. Their experimental studies showed that the FRP repairing/retrofitting system can restore/upgrade the integrity of the joint, keeping/upgrading its strength, stiffness and ductility, as well as, shifting the plastic hinge from the column facing further into the beam, in such a way that the joint remains elastic (Mahini and Ronagh, 2011). The practicality and effectiveness of using web-bonded FRPs on plastic hinge relocation has also been reported by Smith and Shrestha (2006).

Despite the large amount of research conducted on the behaviour of different RC frame components retrofitted with FRP, limited research has been carried out on the behaviour of FRP retrofitted RC frames. In an experimental study, Balsamo et al. (Balsamo et al 2005) indicated that the repaired frame had a large displacement capacity without exhibiting any loss of strength, while providing almost the same energy dissipation of the original frame. In another experimental study Di Ludovico et al. (2008) showed the effectiveness of the FRP retrofitted configuration in improving the global performance of the structure in terms of ductility and energy dissipation capacity and enhancing its load capacity by 50%. In a numerical study, Zou et al. (2007) indicated that FRP confinement increases the strength of columns but has little effect on their stiffness, which is an important advantage in seismic retrofit, as larger stiffness lead to higher seismic forces. Niroomandi et al. (2010) presented a study on the effectiveness of FRP sheets (500 mm) retrofitting the joints in enhancing the seismic performance level and the seismic behaviour factor (R) of ordinary RC frames. In a research by Zarandi et al. (2015), it was determined that the flange-bonded scheme is practical and can be applied to actual 3D frames as it takes into account the presence of the cross beams and the integrated floor slab, whereas for the purpose of relocating a plastic hinge, the web-bonded scheme can only be effectively applied to 2D frames. Mahini (2015) evaluated performance of retrofitted a RC MRFs using steel-braced as well as the FRP-web bonded systems and determined global damage index of the frames at the performance point (and at the life safety (LS)). This evaluation demonstrated that the damage degree of the frame is reduced after retrofitting particularly by FRPs. In addition to that, Hadigheh et al. (2014 and 2016) indicated that FRP retrofitting is more effective than steel braces for low- and medium-rise OMRFs.

One challenge for civil engineers is to assess the amount of damage to a structure caused by moderate-to-strong earthquakes (Aghagholizadeh & Massumi 2016). A seismic damage model actually reflecting strength and stiffness degradation under cyclic loading is the basis of structural safety margin evaluation and post-earthquake restoration (Zeng et al. 2017). In terms of the damage evaluation, Haoxiang et al. (2013) presented a global damage index based on multiple linear force-deformation curves in pushover analysis (i.e. performance-based seismic design) in order to evaluate the integrated damage of RC structures. Huanjun et al. (2015) investigated the damage behaviour of 4 twelve-story frame buildings with lightweight concrete floors and glass-fiber reinforced beams and 1 reference traditional frame using a software of inelastic damage analysis of structures (IDARC) under three earthquake time-history excitations i.e. very severe, severe, and moderate intensity.

This paper assesses the performance of an existing plain RC OMRF and also FRP-retrofitted frames using web/flange-bonding. For this purpose, an eight storey RC building which was retrofitted by Maheri and Akbari (2003) using steel bracing systems is retrofitted with the CFRP web/flange-bonded techniques in different lengths. The retrofitted joints stiffness in the form of moment–rotation relation is determined in detailed FE modelling of the composite joints conducted by other researchers and then the results is utilized to carry out nonlinear pushover analysis of a selected full frame. The seismic performances (via force– displacement capacity curve) of the FRP-retrofitted frames are then compared with the corresponding plain frame. Finally, the global damage indexes from capacity curve of plain/retrofitted frames at the capacity (performance) point as well as LS are obtained and the results are compared using performance based-design criteria.

2. Selectes frames

In a number of experimental and numerical studies, Maheri & Ghaffarzadeh (2008) investigated the effectiveness of using directly connected internal steel bracing to retrofit RC frames. For comparative purposes, a three-bayed, eight-story RC frame investigated by Maheri and Akbari (2003) is selected and retrofitted by CFRP laminates at joints in the present study. In designing the moment resisting frame, the design dead and live loads were assumed to be 27.50 kN/m and 17.50 kN/m, respectively. The compressive strength, f_c' and tensile strength, f_t of concrete are taken as 27.46 MPa and 3.668 MPa, respectively. In addition, the elastic modulus of the concrete E_c is taken as 24.63 GPa and the yield stress of steel reinforcement was assumed to be 412 MPa. Design base shears, V are determined for a Peak Ground Acceleration (PGA) of 0.3g. In estimating the equivalent static earthquake load, the weight of the system is taken as the dead load plus 20% of live load, based on the Iranian seismic code (Iranian seismic code 2800, 1998). The moment resisting frame is designed based on *'weak-beam strong column'* principle using ACI-95 Code (ACI-95, 1989). Dimensions and flexural reinforcements of the designed beam and column sections are shown in Fig. 1. In this figure, ρ_t ,ρ_s and ρ_s' are the total steel ratio of columns and the tensile and compressive steel ratios of the beams, respectively. All members and joints reinforcements were designed to achieve the desirable strength and ductility. This frame is used in the present study with the web-and-flange-bonded, FRPs. In designing the retrofitted joints, the RC frame details are kept the same as those used in reference. The thickness of the FRP sheets with 400-800 mm length for the joints are then determined on the basis of relocating the plastic hinge away from the column face to the cut-off point of the FRP sheet in the beam. The FRP sheets length is first chosen to be 500 mm, based on the Pauley and Priestly (1992) design approach for obtaining the desirable plastic hinge relocation (Niroomandi et al. 2010). The properties used for the FRP material are the same as those given in Table 1.

The required FRP thickness for the large beam-column joints of the first and second floor is calculated to be 4.95 mm, and decreased to a minimum of 1.65mm as the height of the story level increases and the size of the joints decreased. It is evident that the demand on FRP web-bonding to change the behaviour of a large joint and to relocate the plastic hinge away from the column face is much larger than the web-bonded FRP demand for smaller joints. Schematic illustrations of typical exterior/ interior joints and analytical modelling of FRP retrofitted frame/joints (web or flange bonding), together with the calculated FRP sheet thickness for different joints are shown in Figs 1 and 2.

3. OpenSees modelling

In this study, RC frames and the FRP sheets are modeled by Open System for Earthquake Engineering Simulation (OpenSees). The OpenSees is a software framework for simulating the seismic response of structural and geotechnical systems. OpenSees has advanced capabilities for modelling and analysing the nonlinear response of systems using a wide-range of material models, elements, and solution algorithms (PEER, 2016). The Tcl scripting language is chosen to support the OpenSees commands, which are used to define the problem geometry, loading, formulation and solution. These commands are one-line commands which have specific tasks (Mazzoni. et al., 2009).

Table 1. Mechanical properties of CFRP sheets* used for modelling (Niroomandi et al. 2010).

<table>
<tr><td rowspan="3">Modulus of elasticity (MPa)</td><td>In fibers direction</td><td>$E_x = 240,000$</td><td rowspan="3">Compressive strength (MPa)</td><td>In fibers direction</td><td colspan="2">$\sigma_x = 80$</td></tr>
<tr><td rowspan="2">Perpendicular to fibers direction</td><td>$E_y = 18,581$</td><td rowspan="2">Perpendicular to fibers direction</td><td colspan="2">$\sigma_y = 80$</td></tr>
<tr><td>$E_z = 18,581$</td><td colspan="2">$\sigma_z = 80$</td></tr>
<tr><td rowspan="3">Tensile strength (MPa)</td><td>In fibers direction</td><td>$\sigma'_x = 3,900$</td><td rowspan="3">Shear modulus (MPa)</td><td>$G_{xy} = 12,576$</td><td rowspan="3">Poisson's ratio</td><td>$\upsilon_{xy} = 0.2$</td></tr>
<tr><td rowspan="2">Perpendicular to fibers direction</td><td>$\sigma'_y = 53.7$</td><td>$G_{xz} = 12,576$</td><td>$\upsilon_{xz} = 0.2$</td></tr>
<tr><td>$\sigma'_z = 53.7$</td><td>$G_{yz} = 7,147$</td><td>$\upsilon_{zy} = 0.3$</td></tr>
</table>

* FRP sheet has 1 mm cured thickness

3.1. Materials

Concrete, steel and FRP sheets are modelled through Uniaxial Material Commends. Uniaxial Material concrete01 commend is used to construct a uniaxial Kent-Scott-Park concrete material object with degraded linear unloading/reloading stiffness according to the work of Karsan-Jirsa and no tensile strength. Steel bars are modelled by Uniaxial Material Steel01 which is used is to construct uniaxial bilinear steel material object with kinematic hardening and optional isotropic hardening described by a non-linear evolution equation. Elastic commend is used to model FRP sheets on web of specific beams. This commend is used to construct an elastic uniaxial material object. Through Minimax material commends, the stress-strain behaviour of the FRP material is defined. If however the strain ever falls below or above certain threshold values, the other material is assumed to have failed. From that point on, values of 0.0 are returned for the tangent and stress (PEER, 2016).

3.2. Elements

Beams and columns in the original frame are modelled through nonlinear Beam Column elements. This commend is used to construct a non-linear beams or columns element objects, which is based on the non-iterative or iterative force formulation, and considers the spread of plasticity along the elements. This command allows determining the locations and the sections of elements (PEER, 2016). To model the elements in the FRP-retrofitted frame force Beam Column element commend is used. This command is used to construct a force Beam Column element object, which is based on the non-iterative (or iterative) flexibility formulation. The locations and weights of the element integration points are based on so-called plastic hinge integration, which allows specifying plastic hinge lengths and sections at the element ends. GeaomTransf P-Delta is used to construct the P-Delta Coordinate Transformation (PDeltaCrdTransf) object, which performs a linear geometric transformation of beam stiffness and resisting force from the basic system to the global coordinate system, considering second-order P-Delta effects (PEER, 2016).

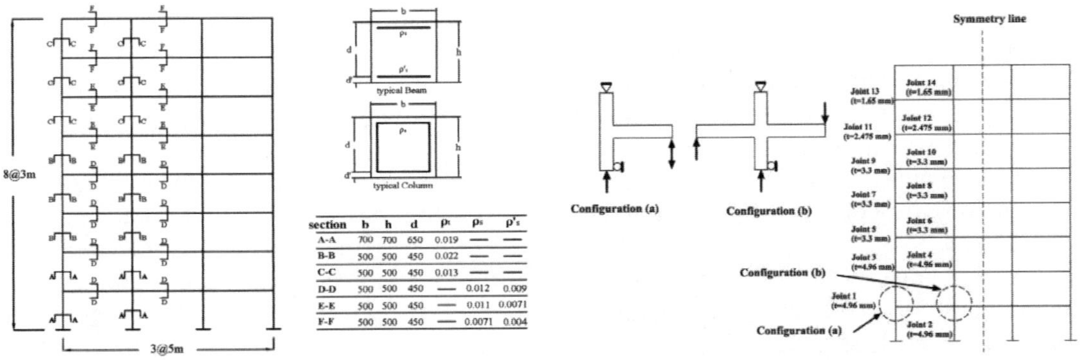

section	b	h	d	ρ_t	ρ_s	ρ'_s
A-A	700	700	650	0.019	—	—
B-B	500	500	450	0.022	—	—
C-C	500	500	450	0.013	—	—
D-D	500	500	450	—	0.012	0.009
E-E	500	500	450	—	0.011	0.0071
F-F	500	500	450	—	0.0071	0.004

Fig. 1. Selected momenrt resisting frame (Maheri & Akbari, 2003) and schematic illustration of the exterior and interior joint and FRP sheet thickness at each joint as used in the nonlinear FE analysis (Niroomandi et al. 2010).

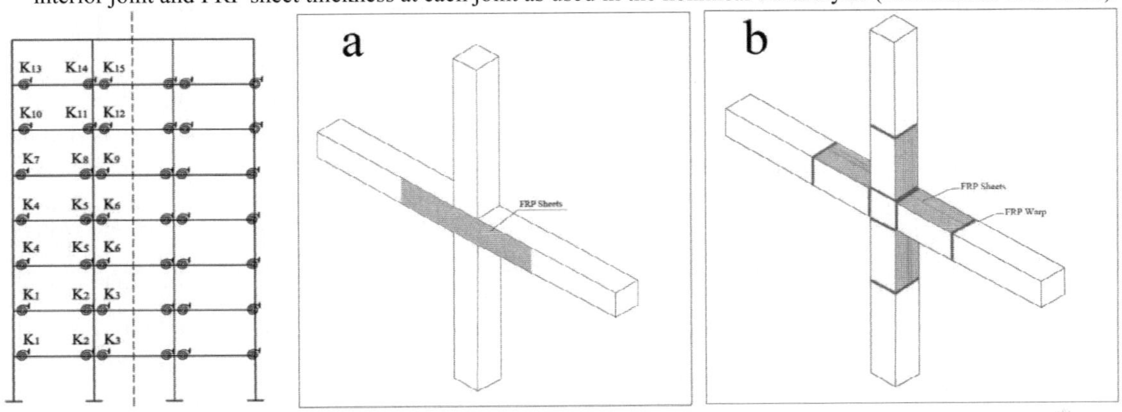

Fig. 2. Analytical modelling of the web/flange-bonded CFRP frames and typical interior RC joint retrofitted by web-bonding (a) flange-bonding (b).

3.1. Recorders

In order to generate base shear-roof displacement diagrams, two recorders with reaction and displacement argument are used. These recorders can record all reactions and displacements in all degree of freedoms and directions of any node of the model during the analysis (PEER, 2016). Recorder with plastic rotation argument is defined to record the rotations of the end of elements.

4. Nonlinear static (pushover) analysis of the frames

In the nonlinear static pushover procedure, monotonically increasing forces are applied to a nonlinear mathematical model of the structure until the displacement of the control node exceeds the target displacement. The NEHRP guidelines (FEMA 356, 1997) indicate that, for a specific earthquake, the building should have enough capacity to withstand a specified roof displacement. This is called the target displacement and is defined as an estimate of the likely building roof displacement in the design earthquake. The guidelines give an indication of how to estimate the target displacement, using the following expression:

$$\Delta_t = C_0 C_1 C_2 C_3 S_a \frac{T_e^2}{4\pi} g \tag{1}$$

where, the modification factors C_0, C_1, C_2 and C_3, that expresses the spectral displacement and expected maximum elastic displacement at the roof level, are the expected maximum inelastic

displacements to displacements calculated for linear elastic response, the effects of stiffness degradation, strength deterioration, and pinching on the maximum displacement response and the increased displacements due to dynamic second order effects respectively. Also, T_e, represents the effective fundamental period of the building in the direction under consideration, calculated using the secant stiffness at a base shear force equal to 60% of the yield force; and S_a is the response spectrum acceleration at the effective fundamental period and damping ratio of the building.
In order to verify models created in OpenSees, pushover analysis is carried out on a frame, which has been studied by Niroomandi et al. (2010). For this purpose, a constant gravity load, equal to the total dead load plus 20% of the live load based on the Iranian earthquake code (Iranian seismic code 2800, 1998), is applied to each frame. An inverted triangular distribution over the height is used as the lateral load pattern. Force-deformation criteria for plastic hinging is defined based on FEMA 356 (1997) patterns. The base shear versus roof displacement curves for the original and retrofitted frames are shown in Fig. 3. This figure indicates that the results of OpenSees modelling and SAP2000 outcomes (Niroomandi et al., 2010) have an acceptable similarity. The base shear versus roof displacement curves for the all frames retrofitted with different length of FRP sheets from 400 mm to 800 are shown in Fig. 4. As it is shown, the web-bonded FRP strengthening of the joints resulted in an increase in the lateral load carrying capacity of the original RC frame. The load carrying capacity of the frame with flange-bonding is enhanced in a higher rate at the target displacement compared to the web-bonding. It can also be seen that, unlike the original frame, the FRP retrofitted frame has a large displacement capacity without exhibiting any loss of strength. For both retrofitting schemes (web and flange bonded frames), it is clear that for longer FRP sheets, the lateral load carrying capacity and the stiffness of the retrofitted frame is increased. This is more effective in web-bonding and with longer FRP length (400mm in compare with 700-800 mm).

5. Performance and damage assessment of the retrofitted frames

5.1. Performance assessment

Basically, the seismic performance level of a structure describes its state of damage on a capacity spectrum curve (Niroomandi et al. 2010). The state of damage of the whole structure may be represented by an index. This damage index can be used as an indicator to describe the reserve capacity of existing structures and the state of the lateral load-carrying capacity of the building. Based on the fundamental parameters, various local and global damage indices have been proposed to evaluate RC structures. Structural ductility, storey drift, element and connection rotation, energy dissipation, and fatigue of the structure are the parameters that have been considered for damage assessment (Haoxiang et al. 2013). Based on the Iranian Code, 5% damped response spectrum with earthquake acceleration base design of 25% of the gravity acceleration value is used to obtain the performance point of each frame. In this study, the exact solution of the ATC-40 document is used in order to calculate the PP of all systems (ATC 40, 1999). Figure 5 shows the Acceleration Displacement Response Spectra (ADRS) curve and the performance level of all frames.
According to this figure, the performance point has a spectral displacement of 31.11cm. The performance points of the web-bonded frames from 400 to 800 mm also have spectral displacements of 34.18, 26.48, 23.07, 20.50 and 19.65cm, respectively. Such a reduction in the spectral displacement for the FRP retrofitted frames indicates that the inelastic lateral load resistance has been enhanced through web-bonded FRP retrofitting. Also, the shift in the performance point of the frame due to joint retrofitting has resulted in an increase in the spectral acceleration value from 0.211g to 0.317g. This indicates that there has been an increase in the seismic load capacity for the FRP retrofitted frames. For flange-bonded frames, the performance

and life safety points have been reduced in compare to the original frame, whereas the amount of these parameters almost remain stable for different length of FRP.

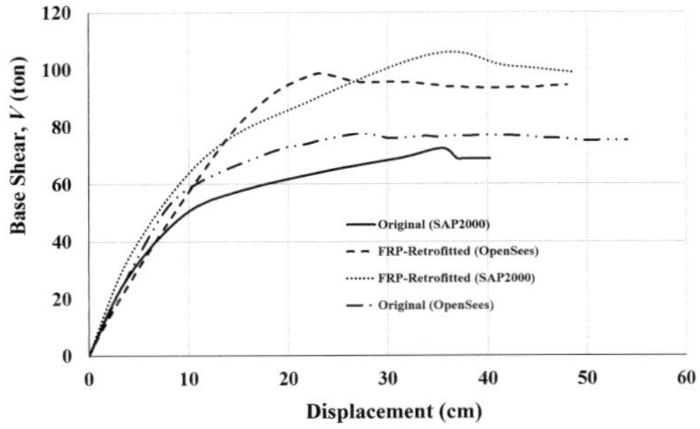

Fig. 3. Base-shear roof-displacement curves of original and FRP-Retrofitted frames.

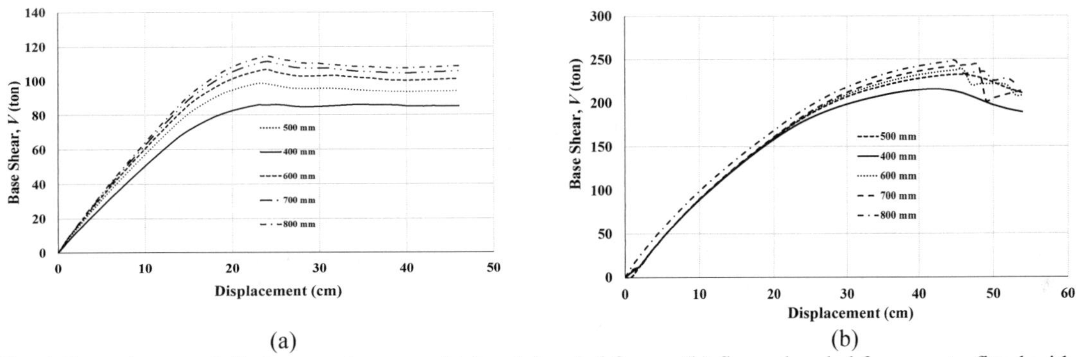

Fig. 4. Base-shear roof-displacement curves of (a) web-bonded frames (b) flange-bonded frames retrofitted with different length of FRP sheets from 400 to 800 mm.

5.2. Damage assessment based on capacity curve

The general global damage index which has a range of 0 to 1 is calculated based on base shear-top floor deflection curve obtained from only one pushover analysis. It is suitable to obtain the damage index of any displacement such as the PP and LS points. The general damage index for capacity curve can be obtained by Eq. (2).

$$D_C = 1 - \frac{\sum_{i=0}^{n-1} k_i (u_{i+1} - u_i) k_n (u_c - u_n)}{k_0 u_C} \tag{2}$$

where k_0 is the initial stiffness at the elastic stage and u_c is the current displacement. The typical skeleton curve or capacity curve of the RC frame can be presented with equivalence multiple linear (see Figure 6(a)) according to Eq. (2).

The global damage indexes are calculated based on Eq. (2) at the PP and the LS points. The global damage indices of all frames are shown in Table 3. Damage assessment is established according to the damage degree description in Table 3 and the damage degree and the limit values as illustrated in Figure 6(b) (Mahini, 2015).

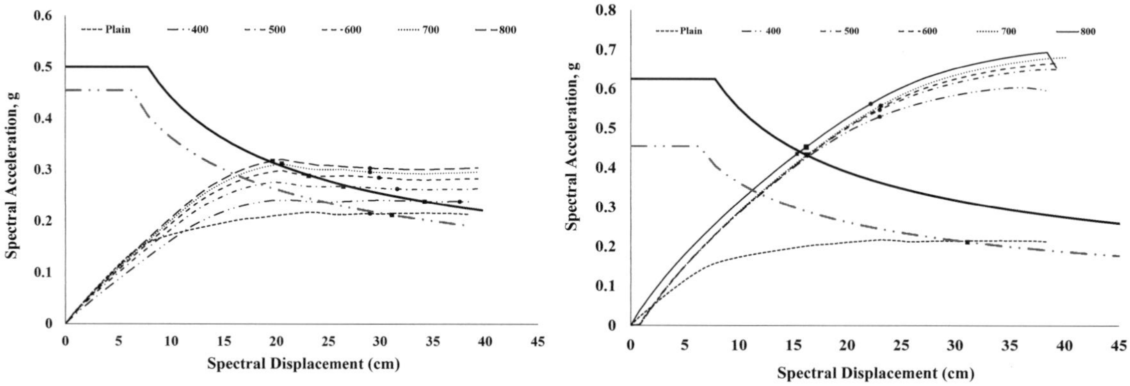

(a) (b)

Fig. 5. ADRS curve and the performance level of (a) each flange-bonded frame; (b) web-bonded frame (● Shows the life safety point and ■ Shows the performance point)

According to Table 3, the plain frame is in severe damage at the adjusted performance point whereas; the FRP-retrofitted frames are at moderate degrees of damage. On the other hand, at life safety, web-bonded frame with 400 mm length of FRP indicates a severe degree of damage compared with other FRP-retrofitted frames with longer FRP sheets which experiences moderate damage at LS. However, the moderate degree of damage of the plain frame is not acceptable as the LS point is reached before the PP, which is not satisfactory based on the ATC-40 standard. Further, the global damage indexes of all flange-bonded frames are more satisfactory. In Figures 7 (a) and (b) this trend is noticeable. All retrofitting systems have reduced the degree of damage of the plain frames at the performance point (PP). This improvement is more clearly indicated with respect to the life safety (LS) demand where the frame retrofitted with longer FRP sheets which have the lowest degree of damage among the retrofitted frames. It is noticeable that the plain frame has been rejected in terms of performance based assessment criteria.

6. Conclusions

It is evident that the weak stories (strong beam-weak column is more obvious) in the plain RC OMRFs can result collapse of the building in a strong earthquake. Therefore, retrofitting may become necessary for these frames in earthquake prone regions. The results of this investigation show that the performance level of the web-bonded RC frame at the target displacement is significantly enhanced in comparison with the original frame. In addition, the frames retrofitted with longer FRP sheets have better performances, whereas the flange-bonding is not promising. However, the load carrying capacity of the frame with flange-bonding is more effectively enhanced at the target displacement compared with the web-bonding. The damage index determination is an effective method to assess the degree of damage of existing retrofitted RC structures subjected to seismic action. For this purpose, a global damage index is presented based on multiple linear force-deformation curves obtained from pushover analysis. Based on the results, it is concluded that the performance level of FRP retrofitted RC frame for LS is satisfactory expect for the web-bonded frame with 400 mm length of FRP. This parameter however is satisfactory for all flange-bonded frames. Moreover, the frames retrofitted with longer sheets of FRP have the lower degree of damage index in web-bonded frames, whereas the degree of damage remains stable in all flange-bonded frames. The damage frames retrofitted with 500 mm in length of FRP or more is moderate at both PP and LS for web-bonded frames. Consequently, the web-bonded and flange-bonded CFRP system (when carefully designed) could upgrade the performance of the existing OMRFs. It can also improve the damage degree of the frame from severe to moderate.

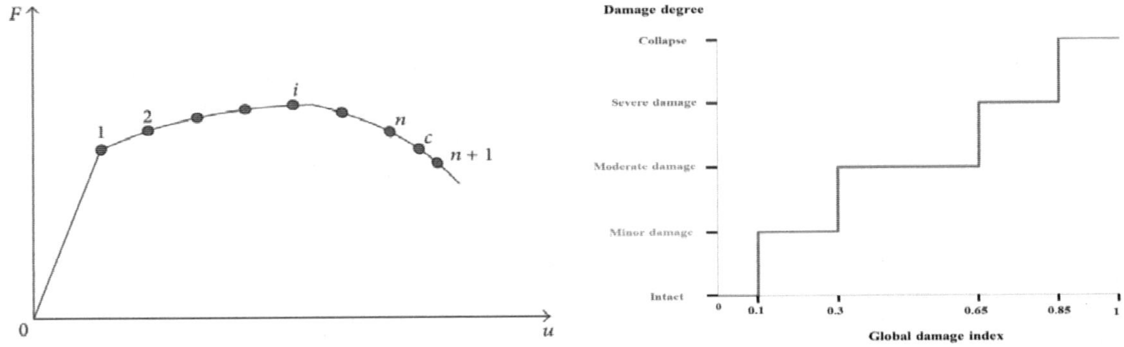

(a) (b)

Fig. 6. Multiple linear force-deformation curve (a) (Haoxiang et al. 2013), (b) Damage degrees and the limit value for different damage indices (Mahini, 2015).

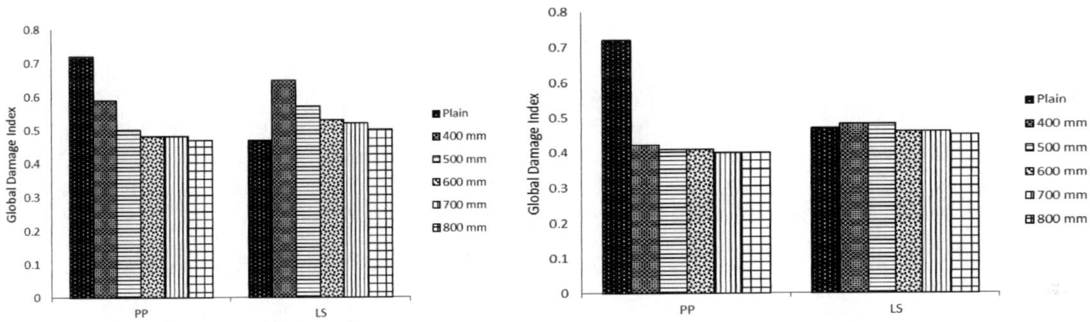

(a) (b)

Fig. 7. Global damage index of the plain and retrofitted frames at PP and LS (a) web-bonded frames, (b) flange-bonded frames).

Table 3. Global Damage Index (D_c) of all systems.

Frame	D_C @ PP	Damage Degree @ PP	D_C @ LS	Damage Degree @ LS
Plain	0.72	Severe	0.47	Moderate
Web-bonded frames				
400 mm	0.59	Moderate	0.65	Severe
500 mm	0.50	Moderate	0.57	Moderate
600 mm	0.48	Moderate	0.53	Moderate
700 mm	0.48	Moderate	0.52	Moderate
800 mm	0.47	Moderate	0.50	Moderate
Flange-bonded frames				
400 mm	0.42	Moderate	0.48	Moderate
500 mm	0.41	Moderate	0.48	Moderate
600 mm	0.41	Moderate	0.46	Moderate
700 mm	0.40	Moderate	0.46	Moderate
800 mm	0.40	Moderate	0.45	Moderate

References

Niroomandi A, Maheri A, Maheri Mahmud A, Mahini SS 2010, 'Seismic performance of ordinary RC frames retrofitted at joints by FRP sheets', *Journal of Engineering Structures*, vol. 10, pp.1016.

Promis G, Ferrier E, Hamelin P 2009, 'Effect of external FRP retrofitting on reinforced concrete short columns for seismic strengthening', *Journal of Compos Structures*, vol. 88, pp.367-379.
Mosallam AS & Banerjee S 2007, 'Shear enhancement of reinforced concrete beams strengthened with FRP composite laminates', *J Compos Part B* vol. 38, pp. 781-793.
Li J, Samali B, Ye L, Bakoss S 2002, 'Behaviour of concrete beam-column connections reinforced with hybrid FRP sheet', *Journal of Compos Structures*, vol. 57, pp.357-365.
Mahini SS & Ronagh, HR 2010, 'Strength and Ductility of FRP Web-Bonded RC Beams for the Assessment of Retrofitted Beam-Column Joints', *Journal of Composite Structures*, Elsevier, vol.92(6), pp.1325-1332.
Mahini SS and Ronagh, HR, 2011, Web-Bonded FRPs for Relocation of Plastic Hinges away from the Column Face in Exterior RC Joints, *Composite Structures*, Elsevier, 93 (2011), pp. 460–2472.
Smith ST & Shrestha R 2006, 'A review of FRP strengthened RC beam column connections', *Proceeding of the third international conference on FRP composites in civil engineering*, pp. 661-664.
Balsamo A, Colombo A, Manfredi G, Negro P, Prota A 2005, 'Seismic behaviour of a full scale RC frame repaired using CFRP laminates', *Journal of Engineering Structures*, vol.27, pp.769-780.
Di Ludovico M, Prota A, Manfredi G, Cosenza E 2008, Seismic strengthening of an under-designed RC structure with FRP', *Journal of Earthquake Engineering Structures Dynamic*, vol. 37, pp.141-162.
Zou XK, Teng JG, De Lorenzis L, Xia SH 2007, 'Optimal performance-based design of FRP jackets for seismic retrofit of reinforced concrete frames', *J Compos Part B*, vol.38, pp.584-597.
Zarandi S & Maheri MR 2015, 'Seismic performance of RC frames retrofitted by FRP at joints using a flange-bonded sheme', *Iranian Journal of Science and Technology*, vol. 39, no. C1, pp. 103-123.
Mahini, SS 2015, 'Damage and Seismic Performance Assessment of FRP-Retrofitted Multi-Storey RC Buildings', *Electronic Journal of Structural Engineering*, vol. 14(1).
Hadigheh, S. A., Mahini, S.S. and Maheri, M.R. 2014, 'Seismic Behaviour of FRP-Retrofitted Reinforced Concrete Frames', *Journal of Earthquake Engineering*, vol. 18, pp. 1171-1197.
Hadigheh S.A., Mahini, S. S., Setunge S. & S.A. Mahin 2016 Rehabilitation of the Buildings Subjected to Earthquakes: Resilience and Performance Parameters, Earthquake and Structures, An international Journal, vil. 11, no. 6.
Aghagholizadeh M & Massumi A 2016, 'A new method to assess damage to RCMRFs from period elongation and Park'Ang damage index using IDA', Int J Adv Struct Eng, vol. 8, pp. 243-252.
Zeng L, Xiao Y, Chen Y, Jin S, Xie W, Li X 2017, 'Seismic Damage Evaluation of Concrete-Encased Steel Frame-Reinforced Concrete Core Tube Buildings Based on Dynamic Characteristics', *Journal of Applied Science*, vol. 7, pp. 314.
Haoxiang H., Maolin C. and Yongwei L 2013, 'Earthquake Damage Assessment for RC Structures Based on Fuzzy Sets', Hindawi Publishing Corporation, *Mathematical Problems in Engineering*, Article ID 25486525, pp. 1-11.
Huanjun Jiang, Bo Fu, Xilin Lu and Linzhi Chen 2015, 'Seismic Damage Assessment of RC Members by a Modified Park-Ang Model', *Advances in Structural Engineering*, vol. 18, no. 3.
Maheri MR & Akbari R 2003, 'Seismic behaviour factor, R, for steel X-braced and knee-braced RC buildings', *Journal of Engineering Structures*, vol. 25, pp.1505-1513.
Maheri MR & Ghaffarzadeh H 2008, 'Connection over-strength in steel-braced RC frames', *Journal of Engineering Structures*, vol. 30(7), pp.1938-1948.
Iranian code of practice for seismic resistance design of buildings, 1999, Standard No. 2800. 2nd. Ed.
ACI Committee 318 1989, *Building code requirements for reinforced concrete (ACI 318-95) and commentary (ACI 318R-95),* American Concrete Institute, Detroit (Michigan).
Pauley T & Priestley MJN 1992, 'Seismic design of reinforced concrete and masonry buildings', John Wiley & Sons.
PEER 2016, *Opens system for earthquake engineering simulation (OpenSees),* Version 2.5.0, Pacific Earthquake Eng. Research Center, Univ. of California, Berkeley, http://opensees.berkeley.edu/.
Mazzoni S. McKenna F. Scott MH. Fenves L 2009, O*pensees Command Language Manual*, Pacific Earthquake Eng. Research Center, Univ. of California, Berkely, http://opensees.berkeley.edu/.
American Society of Civil Engineering (ASCE) 2000, *Pre-standard and commentary for the seismic rehabilitation of buildings*, Prepared for the Federal Emergency Management Agency, FEMA 356.
Federal Emergency Management Agency (FEMA) 1997, *NEHRP provisions for the seismic rehabilitation of buildings,* Report, FEMA 273 and 274, Washington (DC).
ATC 40 1996, *Seismic Evaluation and Retrofit of Concrete Buildings,* ATC-40, Applied Technology Council, Redwood City.

The 8th International Conference on
Structural Health Monitoring of Intelligent Infrastructure
Brisbane, Australia | 5-8 December 2017

Non-Destructive Evaluation Techniques for Identification of Interfacial Cracking in FRP-Concrete Connections

M.M. Sarker, S.A. Hadigheh*, D. Dias-da-Costa
*The University of Sydney, School of Civil Engineering – Australia (*Corresponding Author Email: ali.hadigheh@sydney.edu.au)*

Abstract

One of major concerns in FRP-strengthened concrete structures is crack propagation within the interface region. When a reinforced concrete beam is subjected to external forces, high tensile and normal stresses develop and contribute to the separation of the FRP plate from substrate. This is a common failure pattern for externally bonded (EB) connections. Fracture of interface in FRP-concrete joints can occur in different ways, namely fracture inside the composite (FRP delamination), FRP-adhesive separation (cohesion failure), or within the concrete substrate (debonding).

Currently there are several techniques that can be used to monitor the process of strain localisation and failure mechanisms. For example, photogrammetry and image processing have been successfully used to obtain full-field measurements of displacements and strains in concrete members. However, the precision is typically not enough to assess the small differential deformations which are expected to develop along the FRP. The use of active infrared thermography can provide more data and pinpoint sources of potential damages and delamination inside the member and in the immediate vicinity of the interface.

This paper examines five different non-destructive evaluation techniques to identify localised damage to FRP-concrete composite connections and further evaluates the interfacial properties considering different fracture mechanisms.

1. Introduction

Fibre-reinforced polymer (FRP) materials are increasingly considered as a viable flexural strengthening method for existing structures, structures that need to resist high tensile forces and structural rehabilitation. FRPs are light-weight, corrosive resistant with high strength, stiffness and toughness, and highly tailorable to satisfy strength and serviceability requirements.

FRP strengthening has shown promising outcomes in terms of improving the strength of elements in the post-cracking stage (ACI 440.2, 2008). As such, the bond behaviour of FRP-to-concrete joints is of interest for structural health monitoring and characterisation of fracture process in FRP-strengthened members. In these joints, failure modes occur mainly through FRP delamination, failure in the concrete substrate, or adhesive failure (Figure 1). Crack development in the adhesive layer often goes undetected as the strengthened surface hides it from visual inspection. Methods of non-destructive evaluation can assist in detecting the development of such failure mechanisms.

The objective of this paper is to comparatively evaluate the different monitoring techniques presently used in research to study the effects of damage on the FRP-concrete interfacial connection. The methods of detection explored in different non-destructive techniques, include: digital image correlation, infrared thermography, ultrasonic, x-ray tomography, and acoustic emission. Focus is given to high accuracy detection and visualisation of defects and crack propagation, which is of crucial importance to the understanding of failure mechanism in FRP-concrete joints. The fracture mechanisms of joints and the detection methods are herein reviewed, with main advantages and limitations outlined and the potential for interfacial crack monitoring evaluated. This paper serves as the first part of a review of existing research which provides a benchmark to define advanced monitoring schemes for an experimental programme on FRP-strengthened concrete structures.

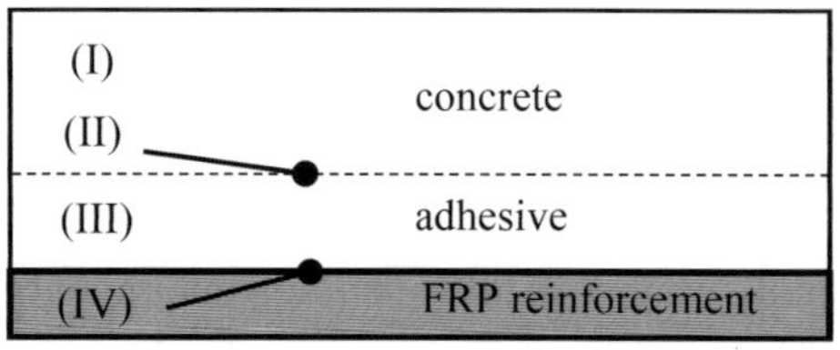

Figure 1: Bond failures of FRP-concrete interfaces: (I) failure in the concrete, (II) failure in the interface between the concrete and adhesive layer (adhesion failure), (III) adhesive failure (cohesion failure), (IV) failure between fibres and adhesive layer.

2. Monitoring techniques

The fracture mechanisms of FRP-concrete connections require the evaluation of microscopic crack development and propagation which can lead to critical cracking in FRP-concrete bond. Monitoring these failure mechanisms requires cost-effective and reliable damage assessment techniques to enable the visualisation of the progression of delamination and cracking in the interface. The following section provides a brief description on each non-destructive technique.
The assessment of strain over an entire area of interest instead of a single point allows for a comprehensive evaluation of the bonded joint. The effective strain relates to design considerations which predict the contribution of FRP to the shear strength and flexural capacity of strengthened elements (ACI 440.2, 2008). High strain rates in FRP-concrete joint can cause microcrack development. Crack propagates and strain localisation finally leads to the failure of the connection. ACI 440.2 (2008) guidelines suggest that monitoring techniques are required to measure a minimum delamination size. For example, in the case of wet layup installation, defects smaller than 1,300 mm^2 are acceptable provided the delaminated area is less than 5% and there are no more than 10 such defects per each square metre.

2.1. Digital image correlation

Digital image correlation (DIC) refers to a form of non-contact monitoring which relies on registering and identifying surface points and characteristics. Through an image processing algorithm, the analysis of grey-scale variations is made on samples prepared with high-contrast speckle patterns. Subsets of the total region of interest (ROI) are overlapped with each subset containing a unique light intensity for the system to recognise pattern changes at different stages. The subset in the reference (initial) stage of image registration is correlated with the changes in the subset pattern over the ROI using advanced algorithm methods. Such changes are then related to displacement and strain changes. The measurements can be made over the entire line-of-sight of the camera continuously in what constitutes a full field measurement technique (Carr et al., 2016).

In 3D measurements, a triangulation procedure is utilised to measure the relative displacement of objects and develop a coordinate system using stereovision methods (Helfrick et al., 2011). For general photogrammetry methods, circular optical targets are mounted on the surface and the centres of the targets are found through an ellipse or circle finding algorithm. This contrast with DIC in the sense that it does not rely on random speckles to track displacements. A sensor captures multiple images in separate time stamps and the surface geometry is monitored for deformation. An example of the experimental schematic and stereoscopic setup of DIC is expressed in Figure 2.

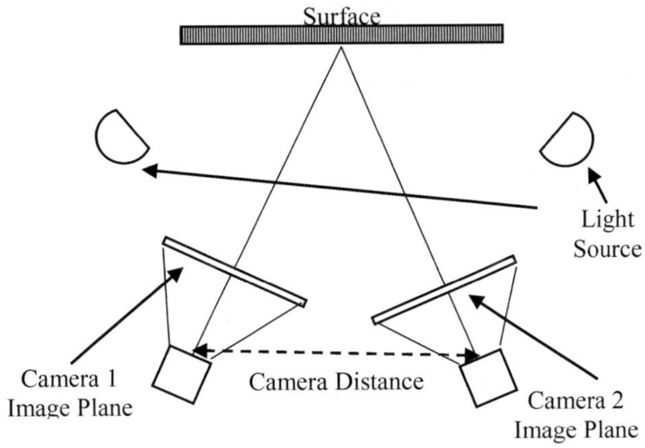

Figure 2: 3D stereoscopic representation of DIC technique using triangulation.

Sabau et al. (2017) presented a non-contact measurement approach based on DIC to experimentally analyse the bond behaviour between fibre-reinforced cementitious matrix composites and concrete. The experimental methodology involved both traditional contact methods of measurement (i.e. LVDTs and strain gauges) and non-contact image correlation systems, including point-tracking system and digital image correlation techniques. DIC was used to evaluate the displacement and strain fields on the surface of the matrix. The speckled pattern was created by painting the surface of the ROI in white, and then spraying it with a black airbrush to create the highly-contrasted pattern. A subset size of 10 pixels, at a step of 10 pixels, was used that provided a measuring accuracy of 0.037 pixels. In terms of displacements, the accuracy was found to be 0.006 mm, and in terms of strains the accuracy was 1000 $\mu\varepsilon$.

D'Antino et al. (2016) utilised DIC to measure the strain field on the surface of the bonded composites. Displacements and strain measurements were obtained for different subsets in 5-pixel step size (~0.81 mm). Square subset with edge sizes between 21 and 60 pixels were used to calculate the longitudinal strain on the matrix surface of specimens in order to assess the influence of subset dimensions. Results obtained with subsets and steps with dimensions less than 1 mm which are affected by the inhomogeneity of the material. The use of subsets with greater dimensions allows for reducing the noise due to the matrix granularity while precision is reduced.

Pohoryles et al. (2017) addressed the use of non-destructive evaluation (NDE) techniques for monitoring strain, displacements, cracking and debonding of FRPs and FRCMs. The study utilised DIC in strain and displacement monitoring and was shown to be comparatively more accurate than traditional contact methods of measurement and acoustic emissions (AE). The experimental methodology involved a correlation algorithm developed by Khennouf et al. (2010). The post-processing of DIC images was done through a photogrammetric image measurement application Vision Measurement Software (VMS), *VMSCapture* (Robson and Shortis, 2005). Correlation is achieved through least-squares method (LSM) with subset deformation. Pattern combinations are then matched through a spline interpolation within the grayscale images. For strain and displacement calculations, the subset size was 31 × 31 pixels with a 9-pixel step. To compare the strain measurements, the ratio between DIC and strain-gauge strains is calculated. The accuracy of the sensor is determined by mean value of the ratio over all recorded data points, along with the standard deviation. The average ratio of strains for DIC-strain gauges was 1.04, whilst the average standard deviation (SD) was 0.30. This indicates that there is more variance in the data set for DIC readings. However, it was shown that DIC provided highly accurate strain readings which were independent of surface preparation and range of strain values, showing reliable average strains for measurements above and below 1000 με. In one sample reinforced with FRP, the graph strain vs. time graph for the beam showed that the strain assessed using DIC and strain-gauges correlated well, eventually failing at 4000 με. The study found that DIC cannot measure deformations below the surface of the FRP, and therefore strain gauges and LVDTs are required to measure debonding. However, the results showed that there is strong correlation between DIC and LVDT readings, with an average ratio of 1.04, and an average SD of 0.028 – much lower than that of strain-measurements. This suggests that there are lower measurement errors in vertical deflections, and DIC is a reliable tool for considering the influence of deflections on crack development in FRP-concrete interfaces.

2.2. Infrared thermography

The use of infrared thermography (IRT) is frequently applied in the inspection of debonding and delamination in FRP-concrete connections. Thermographic methods can be characterised as being either based on passive or active systems.

Passive thermography is applied when the temperature gradient is supplied by the external environment, i.e. the surrounding ambient heat transfer and sunlight exposure. Due to the size of concrete structures, structural elements typically have low thermal conductivity. As such, the day-time temperatures of the ambient surroundings are not synchronised with day-time changes in concrete. Passive thermographic methods are commonly used for bridge decks where the direct influence of solar heating provides the additional thermal gradient – improving the detection capabilities of subsurface defects.

Active systems based on pulsed thermography can be used for detection of debonding between FRP-strengthened structures and underlying concrete substrate (Ghosh and Karbhari, 2011). Active thermography applies heat directly to the surface of a region of interest in the FRP-concrete connection. The application is usually done over a short period and depends on the type of heat source, the predicted depth of the defect, and the thermal diffusivity of the material. As the heat source is applied and subsequently removed, the thermal decay at the surface is monitored over time as the temperature diffuses into the material.

Lai et al. (2010) studied the characterisation of flaws and deterioration of interfacial bond strength of externally bonded concrete elements reinforced with carbon fibre-reinforced polymers (CFRP). The two types of examined defects addressed were flaws associated with the poor application of the FRP and delamination in the adhesive layer. The aggressive environment which replicated rapid degradation conditions was simulated using a water bath with temperatures ranging from 25-60°C. Results showed that the use of infrared thermography gave up to 88% accuracy when compared to actual flaw sizes determined through image correlation analysis. Flaw sizes varied across each temperature gradient, though showed high values for sample at 25°C where the average was 282 mm^2. Comparatively, the average defect area detected at 40°C and 60°C was 173 mm^2 and 184 mm^2, respectively. Delamination was measured after 50 weeks, and showed areas ranging from 272 mm^2 for a sample immersed at 60°C to 1725 mm^2 for a sample at 25°C. The difficulty lied in defining delamination boundaries using IR thermography, and finding consistent results when compared to image correlation methods. It was suggested that the high-water temperatures affected the bonds in the epoxy resin creating moisture fatigue and ultimately debonding.

Tashan and Al-Mahaidi (2014) recorded the thermal responses of CFRP-concrete bonds to detect and characterise defects in the bond zone between single- and multi-layers. Pulse thermography

(PTT) infrared thermography with pulse intervals of 1 s, 3 s, and 5 s were used as thermal loading for all specimens and halogen lamps were positioned at different distances from the investigated surfaces. Thermal signals and thermal contrasts gave the indication of the thermal responses of PTT-IRT. The indicators were calculated by considering a region where discontinuities were not present as a reference point, and compared to a pixel-by-pixel computation of the time-dependent evolution of thermal signals and contrasts. Thus, defect detection becomes much clearer to the user. The results of the study showed that by increasing the distance between the heat source and the investigated surface, the ratio of the thermal signal of a defect under a single CFRP layer to a defect under a double layer increases. By increasing the heat source distance to 1.2 m, it was shown that the ratio between single and multiple layers rises to 400%. The maximum thermal signal detection under a single CFRP layer is double that of the two CFRP layers when the heat source is positioned at 50 cm. In terms of delamination, it was found that the results became unreliable due to the noise level in the PTT-IRT readings. Increasing the number of CFRP layers means that the contrast in the identification of delamination will produce unacceptable noise levels and produce irrelevant values, making it an unreliable tool to measure delamination in a FRP-concrete interface.

2.3. Acousto-ultrasonic

Non-destructive analysis of interfaces in polymer composite joint using acousto-ultrasonic testing (AU) generally has a reduced ability to detect small cracks. The method employs small pulses of ultrasound passing through a composite material interface, and later examined to detect deformities. The presence of deformities such as delamination and debonding are expressed through changes in acoustic energy of the pulses. Limitations of the method arise from the use of low frequencies (50-500 kHz) and long-duration signals (100 microseconds), in which case it is difficult to distinguish between delamination or debonding in multilayer applications. Furthermore, at normal incidence, the penetration of the waves is restricted to FRP <7 mm. This limitation is overcome by introducing rolling sensors with an incident angle different than 0° (Ekenel et al., 2005).
The use of AU in non-destructive evaluation of delamination has shown promise in research. Ekenel et al. (2005) demonstrated the technology in the monitoring of CFRP laminates on bridges. Delamination was induced over a region of interest (305 mm × 305 mm) such that it would not affect the structural integrity of the bridge. The use of AU showed that delamination can be detected at the interface between the FRP and concrete substrate, without the direct influence of interior reinforcement, coarse aggregate or surface orientation affecting the detection. To characteristically verify the accuracy of the testing, a traditional method of bonding failure inspection (tap test) was utilised. The experimental verification was repeated with three different operators for each section, however no significant variations was reported. Comparable to other methods of non-destructive evaluation, the method is seen to be user-friendly, and labour intensity is minimal, though it requires specialised operator.

2.4. X-ray tomography

The use of x-ray tomography employs a method of inspecting and characterising microcracks in polymer composites due to mechanical and thermal damage. High resolution x-ray microtomography has gained popularity for its use in the visualisation of structural defects at a spatial resolution of approximately 50-100 μm (Stock, 1999). As a non-destructive testing technique, x-ray tomography can monitor the development of cracks from internal damages. This evolves from detecting voids, delamination and microcracking propagation. Fundamentally, past research into the area of polymer matrix composite analysis has been limited to the characterisation of damage crack propagation between fibre and epoxy laminates (Dunkers et al., 2002).

Schlling et al. (2005) undertook a study to assess the capabilities and limitations of micro-CT the characterisation of delamination and microcracking in FRP–matrix composite laminates. The results indicated that the critical experimental parameter is the magnification. As such, the sample size is limited to approximately 1.5 mm, a point at which the crack opening (0.5–1 μm) is ~20% of the reconstructed pixel width (~4 μm). The use of dye penetrant agents developed contrast images of the damage reconstruction for enhanced radiography in larger specimens. It was found that the three-dimensional configuration of cracks and microcracks were characterised with the use of a dye. However, the damage being assessed needs to have adequate connectivity to allow for the penetration of the dye. As a result, debonding failures are difficult to detect if microcracking existed within the bonded joints, which would potentially lead to large interfacial crack propagation. It would be preferable to conduct experiments without the use of dye penetrants; however, the clarity of the crack would have to be considered. Detection through the dye allows for the contrast to identify the crack, even when it is <5% of the pixel width. Microcracks were detected in similar tests within the study using dye for specimens 1.4 mm × 15 mm in cross-section, and showed that the spatial resolution decreases due to lower magnification. At lower magnification level, microcracks were not adequately identified. A magnification of 75×, which corresponds to a reconstructed pixel size of ~4 μm, was required to identify those. Hence the method could potentially become a viable non-destructive evaluative technique if magnification of crack propagation can be viewed without the use of a dye penetrant.

2.5. Acoustic emission

The use of acoustic emission (AE) testing requires stress waves emitted by cracks in material under loading. Flaws in the composite interface as a result of thermal or mechanical deformation are detected by identifying these waves. An increase in AE detection reflects the development of microcracking and subsequently the localisation and propagation of macrocracks. One of the limitations of this method is that flaws can only be detected as the material is undergoing damage. Van Tittelboom et al. (2012) found that the crack development can be detected before structural failure occurs. However, the research relied on the self-healing of concrete to regain stiffness during evaluation. Focus should be placed on the identification of cracking and interfacial debonding without relying on the strengthening through crack repair.

Aggelis et al. (2013) developed an experimental study to monitor the behaviour of glass fibre-reinforced polymer (GFRP) cross-ply laminates under tensile loading. Stress waveform analysis and AE response due to sustained loading were examined at each loading step. As such sensitivity changes, due to increasing loading, effectively examine failure modes as the fracture mechanism changes from cracking in the matrix to delamination. The study addresses the issue of early detection of failure process. Active failure mechanisms were reported as a result of damage accumulation however acoustic emission does not place actual insight into the nature of the mechanism, whether it is matrix cracking, fibre rupture or delamination. Thus, development of this process means incremental loading steps need to be employed to assess each stage in the fracture process, as opposed to characteristic failure due to degradation of modulus or matrix failure.

3. Comparison of techniques

Empirical measurements using strain-gauges, and LVDTs require wiring and preparation which may induce unexpected mass or stiffness changes in loading response. Though the measured data is easily transferable, it is only available at the location of the transducers. Since transducers are not subject to rigid body motions, sensitivity varies at each transducer position. Although measurements and data processing are in real-time and relatively fast, it is limited in measuring displacement and strain.

The use of photogrammetric methods has proved to be effective in visualising the entire region of interest as well as high resolution identification of defects. DIC provides a high level of accuracy for complex spatial deformations, which is appropriate for flexible and rigid body structures. Full field DIC measurement allows for strain to be measured across the entire surface of ROI, and the strains for initial bond failure and post-failure strain. The measurement of effective fibre strain is an important evaluative parameter for monitoring of interfacial crack development and failure mechanisms in FRP-concrete connections. Therefore, DIC has the advantage of monitoring the effective strain development in the bond between FRP and concrete, and it requires less cumbersome and laborious instrumentation setups.

Due to the anisotropic nature of heating, there is the possibility of variation in the detection capabilities of infrared thermography. In order to mitigate concerns of reliability, active thermographic methods require the efficient, uniform distribution of thermal radiation, such as the one that can be achieved through quartz halogen bulbs (Milovanovic and Pecur, 2016). Passive thermographic methods, whilst non-contact, non-invasive and require limited equipment, have strict guidelines and controls during detection. For defects and delamination to be successfully detected, diurnal and nocturnal temperatures need to vary upwards of 10°C for a successful measurement. This is recommended for open surface inspections, such as a bridge decks. However, the area of interest cannot be too large as subsurface delamination would be hard to detect due to the thermal variation associated with the surface emissivity (Janku et al., 2016). One of the considerations for the application of thermal energy on an interfacial bond surface is the uniform input of heat. The area of interest of the composite should diffuse temperature in a relatively linear manner. To visualise detection for bond defects, Tashan and Al-Mahaidi (2014) recommended that the input heat flux and the pulse length need to be greater than 500 W/m^2 and 1 s, respectively. The primary determination of defects and interfacial debonding comes from the spatial resolution and atmospheric characteristics of the site. Limitations of the method arise when the inspected surface requires cleaning to reach a constant thermal signature. The inconsistent diffusion and non-uniform application of heat results in a noisy image that can mask defects. For consideration of civil structures strengthened with FRP, the efficiency in inspection requires guidelines for thermography (Washer et al., 2010).

The implementation of x-ray tomography allows for the 3D characterisation of defects in the FRP-concrete interface, including its nature and shape. Ultrasonic methods can determine the depth and location of internal flaws, however it is limited in its use when analysing defects near the equipment due to multiple reflections and scatterings in composites. Dye penetrants required to fully visualise failure mechanisms, such as delamination, and the cost of labour and operation of the machinery often make this method laborious.

Comparatively, acoustic emission methods are widely used in field tests and experiments due to its highly sensitive nature. The method relies on the progression of failure modes in the FRP-concrete interface for adequate analysis. As such, for multilayered specimens, defect size measurements and locations are difficult to achieve. In comparison to DIC and visual analysis, acoustic emission methods give more depth in the fracture mechanism analysis as it can identify and classify cracks below the surface. A summary of the critical advantages and limitations of each method, as well as the potential applications considered from the literature are presented in Table 1.

Table 1: Summary of ND Techniques.

Technique	Advantages	Limitations	Applications/ Considerations	Precision
DIC	Rapid, simple, inexpensive	Sub-surface flaws difficult to detect, should be used in conjunction with other methods	Surface cracking, delamination	Microstrain (<1000 με) and displacements at 0.006 mm (Sabau et al., 2017)
Infrared Thermography	Rapid area coverage, remote sensing capabilities	Anisotropy of heat may give contradicting information; surface contamination varies detection	Fatigue damage, crack development	Delamination between 150 – 2000 mm^2 (Lai et al., 2010)
X-ray Tomography	3D visualisation reveals nature and shape of subsurface defects	Damage caused by fatigue and delamination hard to be detected without penetrating dye, operation costs due to equipment	Microcracking and crack propagation	Microscopic Crack Propagation ~50-100 μm (Schlling et al., 2005)
Ultrasonic Testing	Depth and location of crack development can be determined	Sensitivity of technique makes it difficult to detect failure near equipment	Depths of crack/defects, delamination	< 7 mm (Ekenel et al., 2005)
Acoustic Emission	Suitable for field experimentation, high sensitivity	Suitable for fracture mechanism nearing failure, crack depths difficult to measure, sensitive to noise, difficult to measure multi-layered specimens.	Cracks, delamination, fibre rupture	Highly accurate qualitative analysis (Aggelis et al., 2013)

4. Conclusion

Cracks and fracture mechanisms resulting from debonding of FRP-concrete joints are precursors to the deterioration of integrity, serviceability and health of structural elements. The most common failure modes in FRP-strengthened structures result from delamination in the FRP or debonding between the FRP and concrete substrate. This paper has examined five non-destructive evaluation techniques which can monitor the progression of crack development and propagation in the interface of FRP-concrete connections. The methodology for each technique and use in previous research has been examined, and critical review of the strength and limitations of each technique was presented. With evolving technology, the measurement of interfacial defects and crack development can be monitored with highly precise quantitative and qualitative evaluation techniques. Each technique, or a combination of techniques, can be used to monitor the progression of microstrain deformations and microcracking. As strain and displacement are important design considerations, it becomes vital to define advanced monitoring schemes for an experimental programme to capture these parameters. Identification of delamination and localised damage to bonded joints requires reliable inspection and detection tools, necessary to prevent failure mechanisms, and monitor the health of structural elements strengthened with FRP materials.

References

ACI 440.2 2008, 'Guide for the Design and Construction of Externally Bonded FRP Systems for Strengthening Concrete Structures – ACI 440.2 R-08'.

Aggelis, DG, Barkoula, NM & Matikas, TE 2013, 'Acoustic Emission as a Tool for Damage Identification and Characterization in Glass Reinforced Cross Ply Laminates', *Applied Composite Materials*, vol. 20, no. 4, pp. 489-503.

Carr, J, Baqersad, J, Niezrecki, C & Avitabile 2016, 'Full-Field Dynamic Strain on Wind Turbine Blade Using Digital Image Correlation Techniques and Limited Sets of Measured Data from Photogrammetric Targets', *Experimental Techniques*, vol. 40, no. 2, pp. 819-831.

D'Antino, T, Sneed, LH, Carloni, C & Pellegrino, C 2016, 'Effect of the inherent eccentricity in single-lap direct-shear tests of PBO FRCM-concrete joints', *Composite Structures*, vol. 142, pp. 117-129.

Dunkers JP, Sanders DP, Hunston DL, Everett MJ & Green WH 2002, 'Comparison of optical coherence tomography, X-ray computed tomography, and confocal microscopy results from an impact damaged epoxy/E-glass composite', *The Journal of Adhesion*, vol. 78, no. 2, pp. 129-154.

Ekenel, M, Galati, N, Myers, J, Nanni, A & Godinez, V 2005 'Acousto-Ultrasonic Technology for Nondestructive Evaluation of Concrete Bridge Members Strengthened by Carbon Fiber–Reinforced Polymer', *Transportation Research Record: Journal of the Transportation Research Board*, vol. 1928, pp. 245-251.

Ghosh, KK, Karbhari, VM 2011 'Use of infrared thermography for quantitative non-destructive evaluation in FRP strengthened bridge systems', *Materials and Structures*, vol. 44, no. 1, pp. 169-185.

Helfrick, MN, Niezrecki, C, Avitabile, P & Schmidt, T 2011, '3D Digital Image Correlation Methods for Full-Field Vibration Measurement', *Mechanical Systems and Signal Processing*, vol. 25, no. 3, pp. 917-927.

Janku, M, Brezina, I & Grosek, J 2016 'Use of Infrared Thermography to Detect Defects on Concrete Bridges' *Procedia Engineering*, vol. 190, pp. 62-69.

Khennouf, D, Dulieu-Barton, JM, Chambers, AR, Lennard, FJ & Eastop, DD 2010, 'Assessing the feasibility of monitoring strain in historical tapestries using digital image correlation', *Strain*, vol. 46, no. 1, pp. 19-32.

Lai, WL, Kou, SC, Poon, CS, Tsang, WF & Lai, CC 2010 'Characterization of the deterioration of externally bonded CFRP-concrete composites using quantitative infrared thermography', *Cement and Concrete Composites*, vol. 32, no. 9, pp. 740-726.

Milovanovic, B & Pecur, IB 2016 'Review of Active IR Thermography for Detection and Characterization of Defects in Reinforced Concrete', *Journal of Imaging*, vol. 2, no. 11.

Pohoryles, DA, Melo, J, Rossetto, T, Fabian, M, McCague, C, Stavrianaki, K & Lishman, B 2017, 'Use of DIC and AE for Monitoring Effective Strain and Debonding in FRP and FRCM-Retrofitted RC Beams', *Journal Of Composites for Construction*, vol. 21, no. 1, pp.

Robson, S, & Shortis, MR 2005, 'Vision measurement software (VMS)', http://discovery.ucl.ac.uk/43219/.

Sabau, C, Gonzalez-Libreros, JH, Sneed, LH, Sas, G, Pellegrino, C & Taljsten, B 2017, 'Use of image correlation system to study the bond behavior of FRCM-concrete joints', *Materials and Structures*, vol. 50, no. 172.

Schilling, PJ, Karedla, BR, Tatiparthi, AK, Verges, MA & Herrington, PD 2005, 'X-ray computed microtomography of internal damage in fiber reinforced polymer matrix composites', *Composites Science and Technology*, vol. 65, no. 14, pp. 2071–2078.

Stock SR 1999. 'X-ray microtomography of materials', *International Materials Review*, vol. 44, no. 4, pp. 141-164.

Tashan, J & Al-Mahaidi, R 2014, 'Bond defect detection using PTT IRT in concrete structures strengthened with different CFRP systems', *Composite Structures*, vol. 111, pp. 13-19.

Van Tittelboom, K, De Belie, N, Lehmann, F, & Grosse, CU 2012, 'Acoustic emission analysis for the quantification of autonomous crack healing in concrete', *Construction and Building Materials*, vol. 28, no. 1, pp. 333–341.

Washer, G, Fenwick, R & Bolleni, N 2010, 'Effects of solar loading on infrared imaging of subsurface features in concrete', *Journal of Bridge Engineering*, vol. 15, no. 4, pp. 384-390.

The 8th International Conference on
Structural Health Monitoring of Intelligent Infrastructure
Brisbane, Australia | 5-8 December 2017

Research to Improve the Bond between CFRP and Mild Steel for Strengthening Steel Structures

A. Blake[1], S. Mahini, M.ASCE[2], K. B. Tan[3]

[1]*Undegraduate Student, Discipline of Civil Engineering, University of Southern Queensland, QLD, Australia*
Corresponding author; Email: u1095809@umail.usq.edu.au
[2]*Senior Lecturer, Discipline of Civil and Environmental Engineering, University of New England, NSW, Australia*
[3]*Graduate Student, Discipline of Civil and Environmental Engineering, University of New England, NSW, Australia*

Abstract
Currently there is a need to develop innovative methods to strengthen or rehabilitate steel structures that are deteriorating and are required to withhold a certain load demand. CFRP is currently being used as a strengthening technique to strengthen these structures. This paper has achieved its aim by providing an insight and an overall summary of current peer reviewed journal articles that have investigated the bond between Carbon Fibre Reinforced Polymer (CFRP) and metal surfaces. This paper has also enhanced the knowledge on improving the bond strength between CFRP and mild steel by developing a new technique of heating the epoxy before application. The epoxy was heated before application and after application for 80 minutes at 50°C. This technique improves procedures taken to rehabilitate and strengthen steel structures. Future research topics have been identified as; testing different types of epoxy with different temperatures, investigating the delamination failure mode and the CFRP wrap technique.

1. Introduction

This paper provides an insight and overall summary of current, peer reviewed, journal articles that have investigated the bond between CFRP and metal surfaces including the methodology that each investigation used. The data and information collected from a past study at the University of New England (Tan,2015) has been investigated to further develop a technique. This paper aims to develop the load-slip relationship for CFRP to steel bonding and the different failure modes by developing new techniques that will enhance the strength of steel structures by using CFRP and concludes the existing gap in the knowledge base lies.

There is a need to develop new methods to strengthen and rehabilitate steel structures. This is because of the large number of steel structures that are deteriorating and still required to withstand previously determined or increased loads (Zhao & Zang, 2007).

CFRP is currently being used as a strengthening technique to strengthen these deteriorating structures. Throughout the 21st century many researchers have studied CFRP and steel bonding. Most past studies have concerned this technique of providing an insight on increasing the bonding strength.

CFRP is “a composite material made of a polymer matrix reinforced with fibres” (Alberto Masuelli, 2013) with these fibres made of carbon. The CFRP is bonded to the steel by an adhesive epoxy applied between the two metal surfaces.

The current problem with this technique is the bond between the CFRP and the steel substance is failing. Typically three types of failure occur. These are; failure between two surfaces (adhesive failure), failure within the adhesive (cohesion failure) and a combination of both adhesive and cohesive failure.
Clearly research is needed to investigate how the bond can be strengthened to improve the load carrying capacity and strength of these steel structures.

2. Literature Review

Hollaway and Cadei (2002) produced a paper that reviewed problems that are encountered when bonding polymer composites to metallic structures while examining possible methods to overcome these problems. The paper showed that metallic structures pose a different and more difficult set of problems compared to reinforced concrete structures (Hollaway & Cadei, 2002).
The paper stated that the weakest component of the polymer composite is the adhesive, whilst outlining that the durability of the polymer composite was an area needing to be assessed. This was supported with a review of some of the practical applications that have been used, and noting that only a limited number of papers that have been published based on laboratory testing.
Following this study Zhao and Zhang (2007) provided an innovative review of areas that have only received limited investigation and those areas of research that have been rapidly developing. These areas are; the bond between CFRP and steel, strengthening of steel hollow section members, and fatigue crack propagation in the CFRP applied to steel structures (Zhao & Zang, 2007).
The bond strength and test method was reviewed by focusing on the de-bonding modes that occur, while analysing the bond-slip relationship of CRFP and steel.
Zhao and Zhang (2007) identified three areas for future research. These were; the stability of CFRP strengthened steel structures, fatigue crack propagation modelling and the bond-slip relationship.
Smith (2010) produced a review and research data on FRP to steel on past literature. This article states the three test parameters; surface preparation, metal type and bond length, need to be investigated and highlights the fact that there has been little research into these parameters. To date research has been focused on developing the bond strength and bond stress- slip models (Smith, 2010).
In particular, testing was conducted with both mild and stainless steel while using surface techniques such as; sanding, grinding and needling. It was believed that the epoxy should be applied to the metal straight after surface preparation occurred in order to prevent oxidisation of the metal surface. Overall, Smith (2010) stated that there "appeared to be a correlation between rougher topology of the metal's surface and reduced bond strength."
Teng et al. (2010) prepared a paper researching and examining results from experimental work on steel surface treatment and adhesive properties. The test specimens were prepared using three techniques; solvent cleaned, hand grinded and grit blasted. The grit blasting was most successful in removing the steel's weak surface layer. Three different sizes of grit blasting were conducted all of which resulted in the similar surface characteristics.
The article concluded that surface preparation techniques continuously varied the bond strength. When using grit blasting with appropriate adhesive, adhesion failure between the epoxy and the steel can be prevented (Teng, Fernando, Yu, & Zhao, 2010). Overall Teng et al. (2010) believed there was a possibility that a standard surface preparation technique could be developed.
Kim et al. (2010) describes and explains a number of different tests on the shear strength and behaviour of CFRP applied to metal joints. The testing investigated two types of steel (mild and stainless steel) and multiple surface preparation methods. Every testing sample was loaded with a displacement control to allow for the failure process to be monitored and to identify the type of failures.

Kim et al. (2010) used four surface preparation methods; no treatment, sanding, grinding the surface and needling. This time from the beginning of preparation to the application of the CFRP was restricted to a maximum of three hours.
It was concluded that the most effective surface preparation was grinding for mild steel and sanding for stainless steel. The time taken preparing mild steel should to be monitored and minimised in order to limit oxidation and thereby enhance the bond (Kim et al., 2010)
Narmashiri et al. (2012) conducted research investigating the effect of applying different bond lengths for CFRP strips for strengthening steel. The investigation included parameters such as; load bearing capacity, failure mode, strain on CFRP, strain on adhesive, strain on steel beam, lateral deformation, and vertical deflection. 3D modelling and a nonlinear analysis was used to undergo the testing. Narmashiri et al. (2012) concluded that using different bond lengths does affect the structural behaviour of strengthened steel.
Mosawe et al. (2015) stated that the bond between the steel members and CFRP is the main issue needing to be considered when rehabilitating steel structures. The paper documents and analyses experimental and numerical effects on CFRP properties and the bond between CFRP and steel using quasi-static loads. Mosawe et al. (2015) concluded that CFRP is very sensitive to investigate the properties in the bond between CFRP and steel, and finally determined that the ultra-high CFRP modulus with low tensile strength significantly effects the bond.
Tan (2015) completed a research project investigating and testing different surface techniques used for strengthening the bond between CFRP to mild steel. The experimental work Tan (2015) carried out comprised altering the preparation of the metal surface by using an electric sander and grinder with different grit sizes.
From the data produced Tan (2015) produced graphs and calculations for the load-slip distribution, strain distribution, and the ultimate tensile and shear stress. For each sample the failure mode was identified and recorded. Although these graphs and calculations were provided Tan (2015) did not provide any further conclusions other than stating which preparation had the highest tensile strength. All the literature has advancing knowledge concerning the bond strength between CFRP and steel. In terms of increasing bond strength surface preparation is deemed to be the major factor followed by metal type and the bond length. Currently surface preparation using a small grit size is the most effective method of increasing the bond strength.

3. Methodology

3.1 Details of test specimen set-up

The design including relevant dimensions of the single-lap CFRP to metal joint assembly tested is shown in Figure 1. Assembling the specimens was straightforward and reasonably quick to prepare.

Every test specimen was restrained and stiffened to ensure the mild steel plates did not twist or bend while applying the tensile load through the use of an extra steel plate.
All specimens used Sikadur -30 Normal Epoxy to create the bond between the CFRP and mild steel components. All specimens were carefully constructed to maximise the precision and accuracy and, to minimise any potential errors.
Tan (2015) applied the epoxy in winter ambient conditions, in the laboratory for New England Region New South Wales, Australia. This was reasonably cold with the epoxy only cured for seven days before testing was carried out.

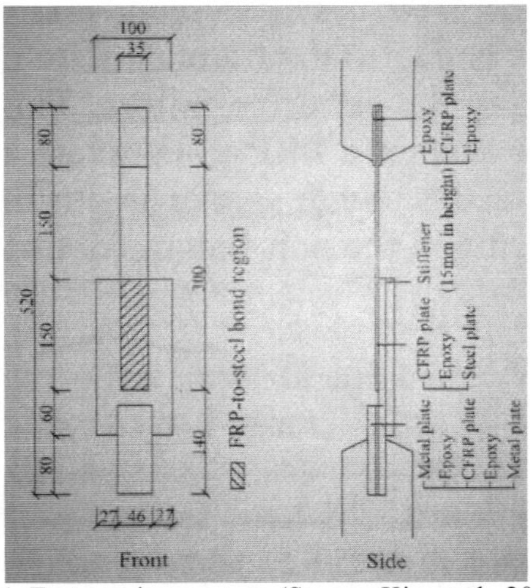

Figure 1: Test specimen set-up (Source: Kim et al., 2010)

3.2 Surface Preparation

Kim et al. (2010) prepared the surface of their specimens by using four different techniques; i) no treatment, ii) sanding with electric sand using 220 grit sandpaper, iii) grinding with a grinding wheel and, iv) needling the metal surface with a needle scaler. Prior to the application of the CFRP each surface was cleaned by using a wiping solvent. All surface techniques were prepared for 3 minutes and application of the CFRP was restricted to a maximum of three hours for each specimen. The three hour limitation was in order to prevent and minimise oxidisation of the top layer on the mild steel (Kim, Smith, & Young, 2010).

Tan (2015) prepared the surface of their specimens by using three techniques; i) electric sander with 80 grit sandpaper, ii) electric sander with 220 grit sandpaper and, iii) electric grinder with 40 grit wheel. Figure 2 shows the surface when prepared by an electric grinder.

The surface was also cleaned with a solvent, to remove weak layers, dust, grease and any other potential contaminants. This was accomplished by the use of a wipe or brush, ensuring the solvent evaporates quickly and solvent was never re-used for any purpose. Tan (2015) used acetone as their solvent to complete the cleaning of the surfaces and also used the time limit that Kim et al. (2010) used for application of the epoxy and CFRP.

3.3 Application

Two specimens were created following the method that Tan (2015) used with the electric grinder and a 40 grit disc. The method was altered by heating the epoxy before application for 80 minutes at 50°C. Once the epoxy was applied, the specimens were placed in oven again for 80 minutes at 50°C. Strain gauges were placed along the bond length of the CFRP plate to be able to capture the axial strains and to determine the interfacial shear stresses. The strain gauges were installed on the centre line of the CFRP and were attached to the connector pad by soldering the wires between them. Figure 3a shows the final arrangement of two specimens prior to setup in the Universal Testing Machine (UTM) for testing.

The strain gauges were connected to a data logger system and the strain data was collected through the National Instrument (Tan, 2015). This is a common method of testing the specimens. Kim et al. (2010) used a very similar method. Figure 3b shows the setup that was used in the UTM.

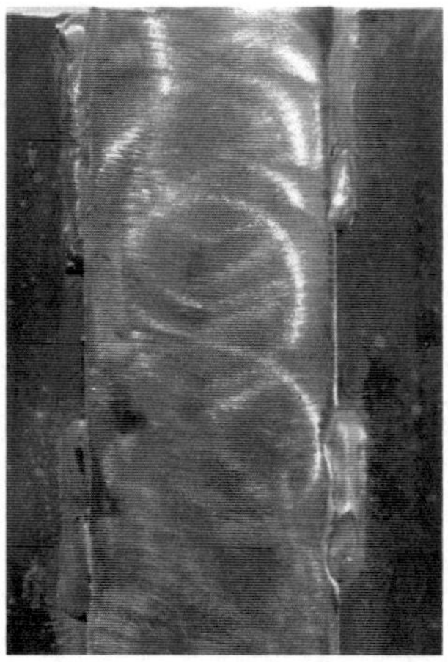
Figure 2: Grinded surface preparation (Blake, 2016)

Figure 3a: Final specimen arrangement (Blake, 2016)

Figure 3b: UTM Specimen set-up (Blake, 2016)

As seen in Figures 3a and 3b the setup is reasonably simple. However it takes some time because of the delicate size of the strain gauges.

4. Results

Tables 1-3 shows the test results from multiple tests. The results were used to provide an analysis of the behaviour of each specimen. These results include a brief summary of the mild steel samples and the following details; surface preparation type, adhesive thickness, failure loads (Pu) and failure mode.

Kim et al. (2010) tested three mild steel (MS) specimens for both no treatment (NT) and grinding the metal surface with a disc (GR). The results are presented in Table 1. The following results in Table 2, were obtained by Tan (2015). Besides those shown in Table 2 Tan (2015) investigated other samples. However paper focuses on the grinding technique as grinding has been one of the most effective techniques.

Two specimens have been constructed and tested by altering the temperature (T) of the epoxy, the results are shown below in Table 3.

Table 1: Test Results, (Kim et al. 2010)

Specimen	Adhesive Thickness (mm)	P_u (kN)	Failure Mode
MS-NT-1	1.5	28.0	Mode 1
MS-NT-2	1.4	22.9	Mode 1
MS-NT-3	1.3	30.4	Mode 1
MS-GR-1	1.2	33.0	Mode 2B
MS-GR-2	1.5	34.5	Mode 2B
MS-GR-3	1.3	33.8	Mode 2B

Table 2: Test Results,(Tan, 2015)

Specimen	Adhesive Thickness (mm)	P_u (kN)	Failure Mode
MS-GR-1	1.35	32	Mode 2B
MS-GR-2	1.37	26	Mode 2B

Table 3: Test Results, (Blake, 2016)

Specimen	Adhesive Thickness (mm)	P_u (kN)	Failure Mode
MS-GR-T-1	1.99	32	Mode 3
MS-GR-T-2	2.13	36	Mode 3

Mode 1 was adhesion end failure in the oxidisation layer, which is shown in Figure 4a.

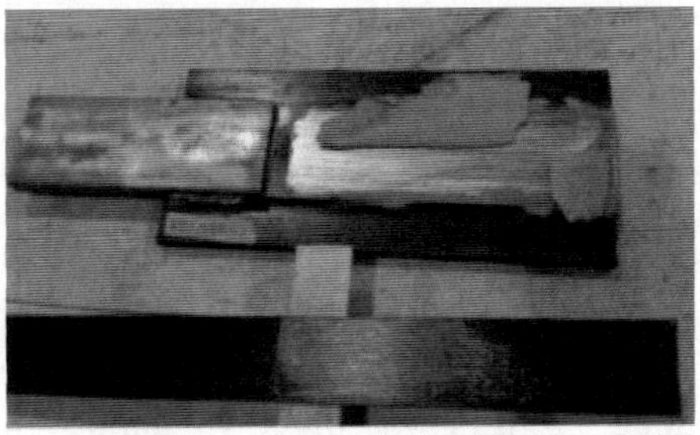

Figure 4a: Mode 1 Failure- Adhesion (Tan, 2015)

Mode 2B was thin-layer cohesive failure in CFRP to adhesive interface with less than 10% of the adhesive remaining on the steel face (Kim, Smith, & Young, 2010). This is shown in Figure 4b.

Figure 4b: Mode 2B Failure- Cohesion (Tan, 2015)

Mode 3 is failure in the CFRP by delaminating which is shown in Figure 4c.

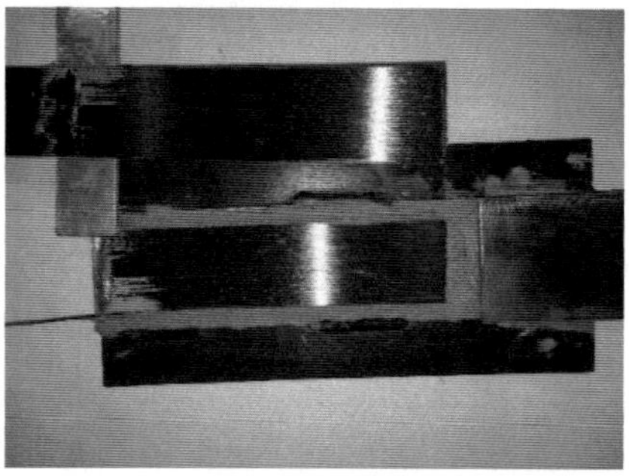

Figure 4c: Mode 3 Failure- Delamination (Blake, 2016)

Tan (2015) provided the data sets that were recorded when testing were conducted. These data sets included slip, extension, load, time, and strain measurements for each strain gauge. The data sets were also collected from the two test samples that were carried out in this study with temperature variation. This allowed the load-slip model, maximum load (P_u) and the strain-distance model to be developed. Extra data from Kim & Smith (2010) was interpreted from the models that were provided in the corresponding papers.

5. Graphical Analysis

While all specimens failed, the failure was different for each sample. The failure modes were different for NT, GR and GR-T. Kim et al. (2010) explained that all three NT specimens failed by adhesion end failure in the oxidised layer based either in the oxidised layer of metal interface or in a combination between the epoxy and the oxidised layer. Both Kim et al. (2010) and Tan et al. (2015) stated that all GR specimens tested failed by thin layer cohesive failure between the CFRP and the epoxy interface. This included the majority of the epoxy being left on the CFRP with some splitting in the carbon fibre plate occurring parallel to the fibres. Therefore, with either mode debonding has occurred between the CFRP and the mild steel substrate. Samples of GR-T failed by delamination failure in the CFRP. This means that the bond between the CFRP and the mild steel substrate has failed. An improvement in the bond strength was made by heating the epoxy. After processing the data collected Figure 5a, b and c have been generated.

Figure 5b shows the typical Load-Slip relationship for each different specimen tested. This graph has been created from data from other graphs and raw data collected from test samples. Examining the graph it can be seen that the GR temperature, altered epoxy samples can withstand the greatest load and have the highest capacity evidenced by the each graph's peak and the area under the curve optimum load-slip relationship involves a reasonable amount of slip with the highest load before failure.

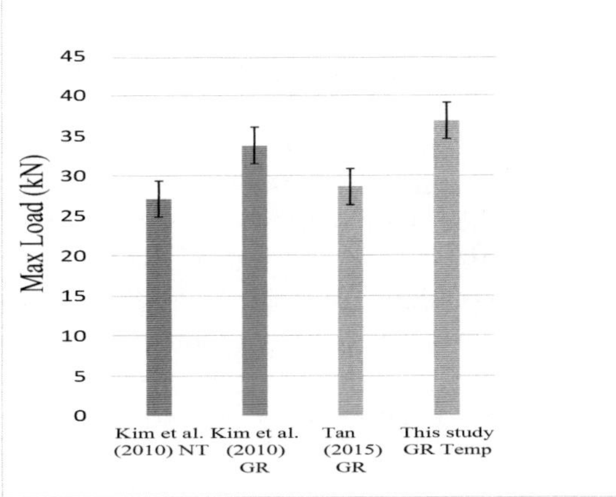

Figure 5a: Maximum Load Achieved

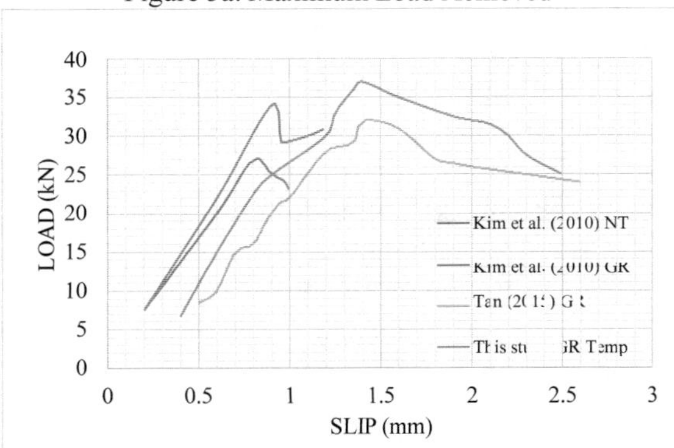

Figure 5b: Load Slip Relationship

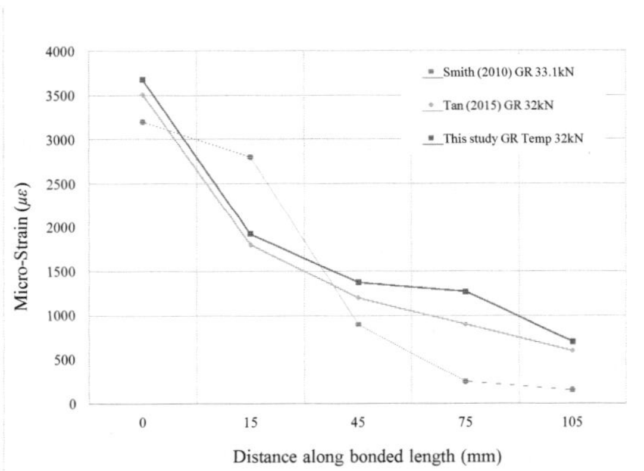

Figure 5c: Strain Distance Relationship

Figure 5c presents the strain-distance relationship, which shows that the strain decreases as the distance increases from the loaded end for all samples. This means that the strain is being transferred through the bond length with the maximum strain being registered at the load end. The area under the line represents the energy absorption of each sample.

It is apparent from an interpretation of the graphs that the sample GR-Temp is similar to the other samples in distributing the strain but is absorbing more energy. This is shown by a comparison of the area under the curves.

6. Discussion

From the analysis of the results the differences in the preparation methods is shown in the forgoing graphs. The reasoning for the trends in the test results is discussed in this section followed by topics requiring further research. When preparing the surface of the mild steel the oxidised layer must be fully removed this is evidenced by examining the failure modes and the strength of the bond of the non-treated specimen compared to other specimens shown on Figures 5a and 5b. This supports the Kim et al. (2010) statement that detailed the importance of minimising the time between preparing the surface and installation of the epoxy, in order to reduce oxidisation in the top layer of the mild steel. This explains the result that was gained from the NT specimen which could be a small factor in other samples where the oxidised layer has not been removed completely during the preparation stage.

The samples of Tan (2015) were tested under ambient conditions similar to those conditions under which the GR-T samples were tested. Unfortunately the ambient conditions for the Kim et al. (2010) test are not known which would mean the curing of the epoxy most probably would have occurred at a different rate. If this was the case it may explains why the Kim et al. (2010) GR samples are yielding a higher capacity and maximum load than the samples tested by Tan (2015).

The ambient conditions may be heating the epoxy to a temperature similar to that which occurred artificially for the GR-T samples. If this was the case then it would seem that there could be a nominal time of the year to strengthening a deteriorating steel structure using the CFRP procedure.

It can be determined from the graphs produced that the techniques being developed are improving the bond between CFRP and mild steel although it is believed that there is still the potential for further improvement. Areas for improvement can be reducing errors ambient conditions are considered. Error that occurred in the GR-T samples that should be mentioned are the increase in epoxy thickness and the jaws of the UTM slipping on the CFRP.

After examining past research and the laboratory work that has been conducted further research that could be undertaken involves investigating methods of strengthening the carbon fibre plate to avoid delamination failure. In addition in this case further tests should include determining the optimum temperature in order for the epoxy to gain the maximum bond strength. As Tan (2015) briefly mentioned a CFRP wrap could be installed to strengthen the joint. This CFRP wrap technique could also be tested in order to find an optimum configuration that will assist in failures in the CFRP to steel joints. Lastly, further research that could be undertaken investigating different types of epoxy possibly resulting in a stronger bond.

7. Conclusion

Overall this paper has comprises a state of the art review of the techniques that have been developed to increase the bond between CFRP and mild steel substrates. Summaries of past research which have been undertaken and published have been provided to indicate the current state of knowledge concerning CFRP steel bonds. The typical methodology that has been used throughout the world is explained in detail to assist readers and/or future researchers understand how the investigations have been carried out. Through the research and investigations that have been carried out it has been confirmed that surface preparation does increase the strength of the bond between CFRP and mild steel. The load-slip and strain-distance relationship models and graphs provided confirm that preparation using an electric grinder is one of the most effective surface preparation techniques to strengthen the bond. The description of the debonding modes indicate some of the issues occurring with this technique of CFRP strengthening and which need further research.

This paper has also confirmed that for two samples surface preparation involving grinding combined with heating the epoxy before application is beneficial. Both these samples showed the bond strength between the CFRP and mild steel substrate was increased prior to the CFRP delaminating. It is therefore concluded that further comprehensive investigations should be undertaken regarding heating the epoxy and confirming the appropriate temperature and ambient conditions under which the strengthening should be undertaken. It is believed that this grinding/heating technique could contribute significantly to the strength of mild steel CFRP bonds for rehabilitating and strengthening steel structures.

Acknowledgments

The first author gratefully acknowledges the support and data results attained from Mr Kim Bock (Kelvyn) Tan. Also, he would like to acknowledge the guidance and supervision of Dr Saeed Mahini and technical support of Mr Tony McKinnon, the discipline technician. Further the authors also would like to acknowledge the University of New England for the funding, the materials and the equipment provided to undergo the laboratory work that was required. The authors would also like to acknowledge the unit coordinator, Mr Rex Glencross-Grant for making it possible for this study to be undertaken. Lastly the first author would like to thank a fellow student Mr Jacob Tan for assisting and providing support with laboratory work and technical assistance.

References

Al- Mosawe, A., Al- Mahaidi, R., & Zhao, X.-L. (2015). Effect of CFRP properties, on the bond characteristics between steel and CFRP laminate under quasi-static loading. *Construction and Building Materials*, 489.

Alberto Masuelli, M. (2013). Introduction of Fibre-Reinforced Polymers- Polymers and Composites: Concrepts, Properties and Processes. *InTech*, 1-38.

Blake, A. (2016). *Advancing Research on Improving the Bond between CFRP and Mild Steel for Strengthening Steel Structures,* Bachelor thesis submitted to the Discipline of Civil and Environmental Engineering, University of New England, Armidale, Australia.

Fernando, D., Teng, J. G., Yu, T., & Zhao, X. L. (2012). Debonding Failures in CFRP-Strengthened Steel Structures.
Hokkaido University Japan.

Hollaway, L. C., & Cadei, J. (2002). Progress in the technique of upgrading metallic structures with advanced polymer composites. *Progress in Structural Engineering and Materials*, 131-148.

Kim, S. J., Smith, S. T., & Young, B. (2010). Effect of Surface Preperation on the Strength of FRP-to-Mild Steel and FRP-Stainless Steel Joints. *The 5th International Conference in Civil Engineering*, 869-872.

Narmashiri, K., Zamin Jumaat, M., & Ramli Sulong, N. (2012). Strengthening of Steel I-Beams Using CFRP Strips: An Investigation on CFRP Bond Length. *Advances in Structural Engineering, 15*(12), 2191-2204.

Schnerch, D., Dawood, M., Rizkalla, S., Summer, E., & Stanford, K. (2006). Bond behavior of CFRP strengthened steel structures. *Advances in Structural Engineering 9(6),* 1-29.

Silvestre, N., Young, B., & Camotim, D. (2008). Non-linear behaviour and load-carrying capacity of CFRP- strengthened lipped channel steel columns. *Engineering Structures*, 2613-2630.

Smith S.T. (2011) Strengthening of Concrete, Metallic and Timber Construction Materials with FRP Composites. In: Ye L., Feng P., Yue Q. (eds) Advances in FRP Composites in Civil Engineering. Springer, Berlin, Heidelberg

Tan, K. B. (2015). *An Experimental Investigation into Potential for De-bonding of FIbre Reinforced Polymer (FRP) on Metal Substrate when Used as a Strengthening Technique,* Bachelor thesis submitted to the Discipline of Civil and Environmental Engineering, University of New England, Armidale, Australia.

Teng, J. G., Fernando, D., Yu, T., & Zhao, X. L. (2010). Treatment of Steel Surfaces for Effective Adhesive Bonding.
The 5th International Conference on FRP Composites in Civil Engineering (pp. 865-868). Beijing China: University of Hong Kong.

Xia, S. H., & Teng, J. G. (2005). Behaviour of FRP-TO- Steel Bonded Joints. *Proceedings of the International Symposium on Bond Behaviour of FRP in structures*, 411-418.

Yu, T., Fernando, D., Teng, J., & Zhao, X. (2011). Experimental study on CFRP-to-steel bonded interfaces.
Engineering Part B Composites, 43(5), 2279–2289.

Zhao, X.-L., & Zang, L. (2007). State-of-the-art-revew on FRP strengthened steel structures. *Engineering Structures*, 1808-1823.

The 8th International Conference on
Structural Health Monitoring of Intelligent Infrastructure
Brisbane, Australia | 5-8 December 2017

Full-scale measurements of the physical response of WTT building with the use of Real-Time Kinematic Global Navigation Satellite System

M. Cwik[1], M.A. Gizejowski [2].
[1] *Warsaw University of Technology – Poland, Email:* michal.cwik@op.pl [2] *Warsaw University of Technology – Poland*

Abstract

This paper presents full-scale measurements of 208 m high Warsaw Trade Tower (WTT) building using Real-Time Kinematic Global Navigation Satellite System (RTK-GNSS) devices. The aim was to evaluate physical response of the WTT building caused by measured wind action. The scope of the research included continuous displacement measurements the top of the WTT building and simultaneous measurements of wind speed and direction as well as the processing of obtained data.
The paper provides the description of a measuring system based on three GNSS receivers and a weather station. General operating scheme as well as more detailed issues including synchronising time between GNSS devices and weather station, problems with real-time data acquisition and wireless transferring processes and its influence on measurements are described. Estimation of the accuracy of the RTK-GNSS measurements, determination of the neutral position of the top of the WTT building during negligible environmental actions and observed displacement during strong winds, as well as measured wind speed and direction, are presented.
Full-scale investigations presented in this paper forms a part of the wider research project the primary aim of which was to investigate a possibility of evaluating average wind pressure distribution over the building's façade, corresponding to conditions observed during full-scale measurements of building's displacements. As a result the assessment of structural health of high-rise buildings could be monitored by modelling and evaluating its response to wind action and comparing theoretical to physical displacements of structures.

1. Introduction

The design working life of the high-rise buildings usually ranges between 50 to 100 years, sometimes even more. At that period a structure should remain safe, serviceable and durable without major repair being necessary. Apart of the standard maintenance activities, it is necessary to undertake periodic inspections of the structure. Such inspections should provide information on the condition of the building. Due to the possibility of fatigue, corrosion as well as the deterioration caused by other phenomena, comprehensive inspections require access to all components of a structure. For a number of reasons, such inspections would be expensive, time-consuming and sometimes introduce significant technical challenges. Consequently, these problems may result in adjusting or even limiting the scope of inspections to suit the needs of specific buildings.
Having regard that it is almost impossible to access and evaluate every part of a structure, Cwik et al. (2013) proposed solution, referring to high-rise buildings, which principle is not to examine the structure itself but rather to examine and analyse its response to a known loading. A comparison of the theoretical and measured responses would then constitute the indicator of a structural condition. Authors expect to obtain a fair degree of similarity in theoretical and physical displacements of high-rise buildings which are in a good structural condition, and clear differences for buildings with structural deficiencies. Proposed method is based on comparison of displacements of the top of the

high-rise building caused by wind action. Such an approach requires evaluating wind pressure distribution over the building's façade, which would be used for determining theoretical displacement of the building's structure. The scope of the wide research project includes full-scale measurements, wind-tunnel tests and CFD simulations of the WTT building. Review of the research project was given by Cwik (2014).
This paper presents comprehensive description of full-scale measurements of the WTT building comprising a instrumentation, system integration, communication, data acquisition, accuracy estimation, determination neutral position of the top of the WTT building during negligible environmental actions and displacement during strong winds, as well as wind speed and direction.

2. Aim and scope of the research

The aim of full-scale measurements was to evaluate physical response of the WTT building caused by measured wind action. The scope of this part of the research included displacement measurements of the building's roof and simultaneous measurements of wind speed, direction and other meteorological quantities. The results will represent the physical response of the structure subjected to specific wind loads. After installing instrumentation and conducting tests, appropriate measurements started in January 2012. Including maintenance breaks and technical break-downs, data collection period lasted for 6 months.

3. Warsaw Trade Tower building and surroundings

The WTT building (Fig. 1a), formerly Daewoo Center, was constructed during 1997-1999. It is 208 m high (184 m excluding the spire). In 2012 during full-scale measurements it was the third tallest building in Poland and since 2016 it is fourth. It is used mainly for commercial offices purpose. Building structure is composed of a covered exterior steel framed tube lateral system and an interior reinforced-concrete core. It is 52 × 27 m in plan. The shape of the building reminds semi-cylinder with flat western façade and with cuboids intersecting the curved rest of the façade.
The building is located in not well developed northwest part of the city centre (Figs 1b, c). According to field measurements of the height of surrounding buildings and terrain elevation, in an area including in a circle with a radius of 500 m and centre at the location of the WTT building, surroundings are 18.4% built-up with buildings exceeding 3 m and 10.5% with buildings exceeding 15 m. The average height of the roughness elements is of 4 m. Field measurements showed also that analysed area has a constant elevation and is representative for other surrounding areas.
The WTT building is directly exposed to wind action with no considerable neighbouring structures.

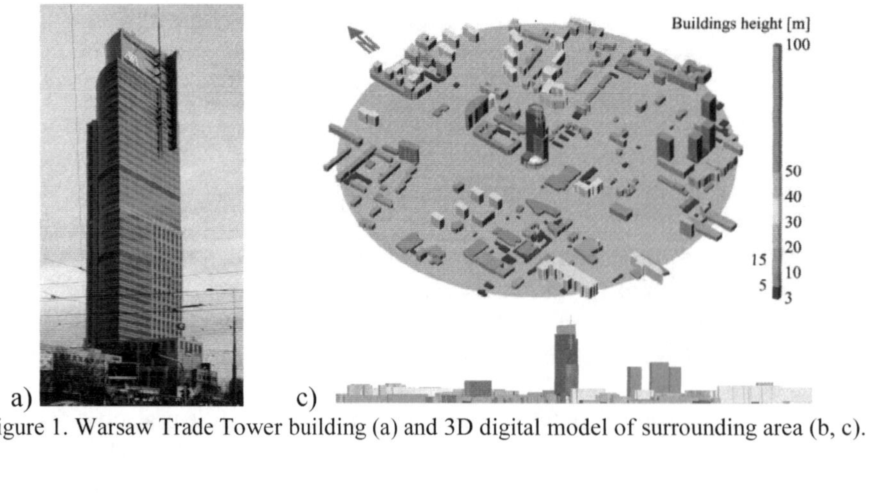

Figure 1. Warsaw Trade Tower building (a) and 3D digital model of surrounding area (b, c).

4. Instrumentation

The measuring system was based on RTK-GNSS devices manufactured by Leica Geosystems and a weather station manufactured by Vaisala. The central unit of the system was a Twinhead Durabook tablet operating under Windows XP. The tablet was equipped with external GSM/UMTS modem and antenna.
Measuring devices were installed on the roofs of two buildings: the WTT building at height of 184 m and the Faculty of Civil Engineering of Warsaw University of Technology (FCEWUT) building at height of 27 m. The buildings are spaced 2805 m to each other.
Two GNSS receivers acting as rover stations (GMX902 with external antenna AS10 and GMX901 with integrated antenna), the weather station (WXT520), tablet (T7M), GSM/UMTS modem with external antenna and auxiliary equipment were installed on the roof of the WTT building (Figs 2, 3a, b, c). The third GNSS receiver acting as a reference station (GMX902 with external antenna AS10) with auxiliary equipment was located on the roof of FCEWUT building (Fig. 3d).

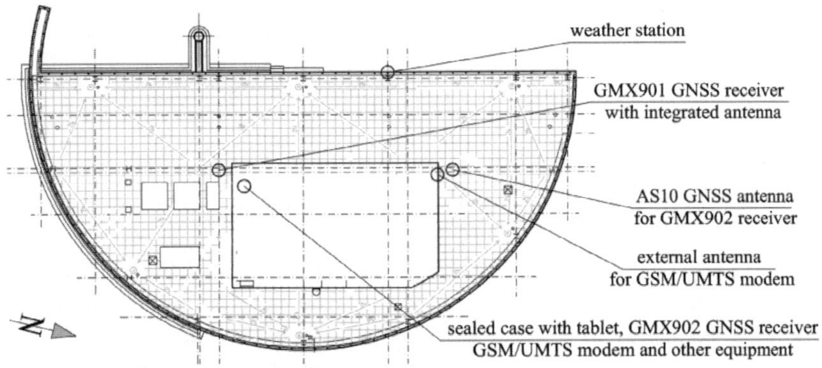

Figure 2. Location of the measuring devices on the roof of the WTT building.

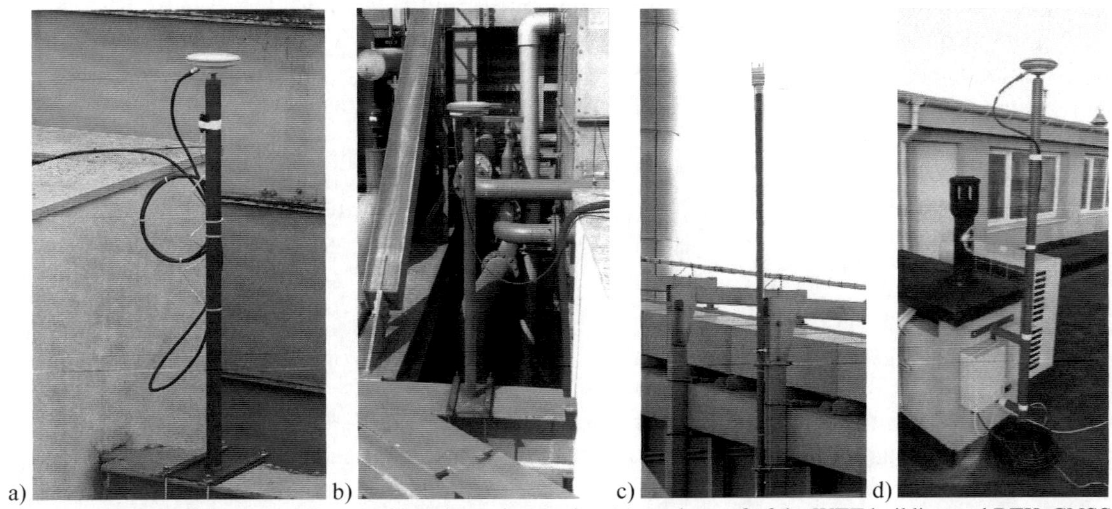

Figure 3. RTK-GNSS rover stations (a, b) and weather station (c) on the roof of the WTT building and RTK-GNSS reference station (e) on the roof of FCEWUT building.

GMX902 receivers supports GPS and GLONASS satellites at L1 and L2 frequency while GMX901 supports only GPS satellites at L1 frequency. WXT520 weather station measures wind speed and direction, precipitation, atmospheric pressure, temperature and relative humidity.

5. System integration, communication and data acquisition

The rover stations and the tablet were connected by copper cables provided by the manufacturer of the devices and USB cables. Due to the considerable distance it was decided to use the RS-232 communication interface to connect station the weather station and the tablet. Wireless GSM/UMTS transmission was used to provide data transmission between the tablet and Internet. The reference station was connected directly to local network and Internet by fibre cable. General operating scheme of the measuring system is shown in Figure 4.

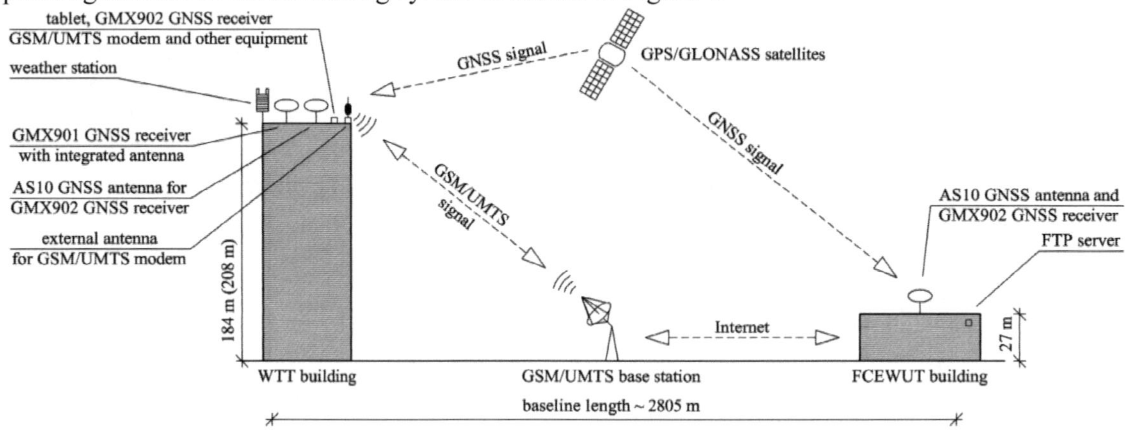

Figure 4. General operating scheme of the full-scale measuring system.

Sampling frequency of the rover stations was of 20 Hz for GMX902 and 1 Hz for GMX901. The reference station GMX902 was operating at a frequency of 10 Hz. Weather station measurements were conducted with a frequency of 1 Hz. All data from the measuring devices were transmitted in real-time to the tablet. Leica GNSS Spider, Vaisala Configuration Tool and MeteoDataLogger programs running on the tablet were responsible for managing the devices and for data acquisition, management and storage. Displacement measurements, wind speed and direction as well as other meteorological quantities after pre-processing were transmitted to a FTP server where they were stored and make available for further processing. Figures 5 and 6 present detailed operating schemes of the weather station and the RTK-GNSS measuring systems.

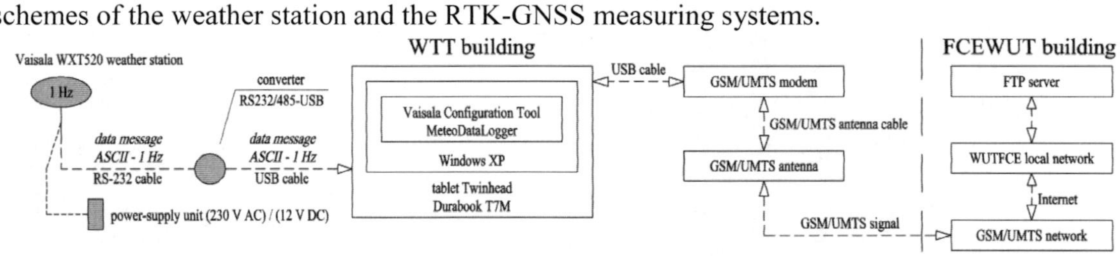

Figure 5. Detailed operating scheme of the weather station measuring system.

Mean daily data volume transmission over Internet was of ~3 GB. During the tests GSM/UMTS signal strength and quality on the roof of the WTT building was found to be insufficient to provide necessary transfer rate. Due to other media transmitters located on the roof and possible signal interference, building maintenance services did not allow to mount active GSM/UMTS antenna equipped with signal amplifier. At the same time, there was no technical possibility to use cable connections between the tablet and local network and Internet. Therefore it was decided to use external passive antenna. Nevertheless, the strength and quality of GSM/UMTS signal was still very variable and sometimes not sufficient. This in turn had a direct and significant impact on the quality of RTK-GNSS measurements. Temporary signal constraints caused interruption in data transmission from the reference station which resulted in insufficient accuracy of measurement of the rover stations.

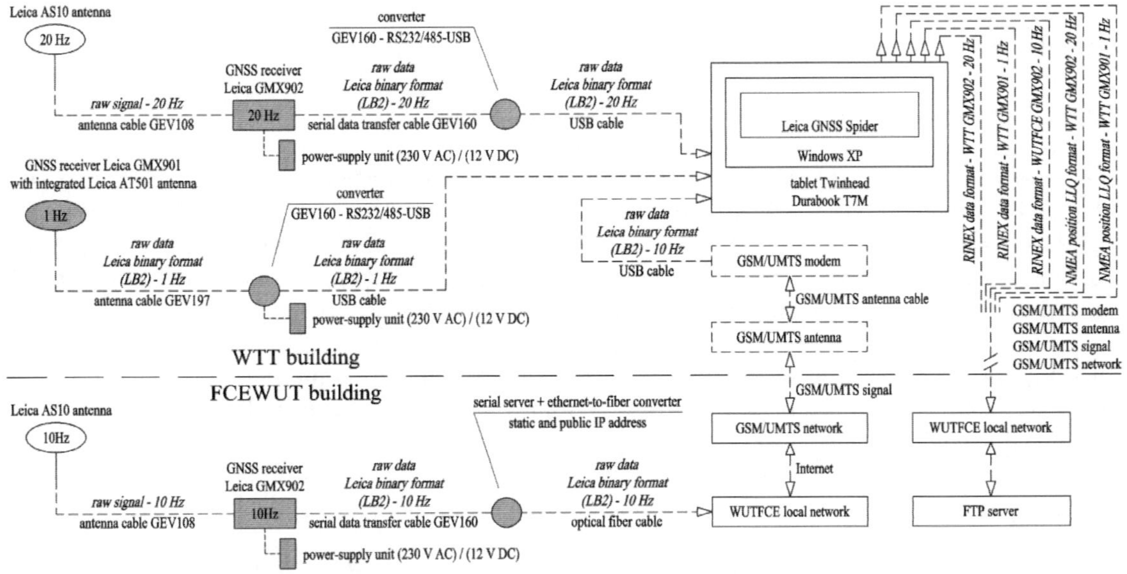

Figure 6. Detailed operating scheme of the RTK-GNSS measuring system.

6. Synchronizing time between GNSS devices and weather station

During the tests preceding appropriate measurements the lack of time correlation between RTK-GNSS and weather station measurements was found. The divergence between timestamps of both types of measurements was variable over time and ranged from tens of seconds to several minutes. Due to the fact that accurate time synchronization was of particular importance, determination of reasons of observed issue and applying effective solutions was necessary.

The RTK-GNSS measuring system labelled timestamps using "GPS time", while the weather station used Windows XP system time which is synchronized with Coordinated Universal Time (UTC) via Network Time Protocol (NTP). GPS time started counting from 00:00 5-6 January 1980. At that time it was synchronized with UTC. Since then in 1981, 82, 83, 85, 88, 90, 91, 92, 93, 94, 96, 97, 99, 2006, 09, 12, 15 and 16, the International Earth Rotation and Reference Systems Service (IERS) added leap second to UTC in order to synchronize UTC to solar time (IERS 2017). Leap second corrections are not applied for GPS time (National Coordination Office for Space-Based Positioning, Navigation and Timing 2012). As a result, at the time when the measurements were conducted, the timestamps of weather station measurements were ahead of RTK-GNNS measurements by 15 seconds. Solution applied in the research was adding an additional 15 seconds to the RTK-GNSS timestamps.

The second reason for the lack of time synchronization was the significant delay of the system time on the tablet between successive time synchronizations with NTP servers. Observed divergence increased over time and reached several minutes. Default weekly synchronization frequency was found to be insufficient. Possible reason of this problem was the tablet's CMOS battery. It powers Real Time Clock which runs continuously, whether the computer is on or off. Low battery power, battery failure or even air temperature could affected system time and caused mentioned delays. An experiment was conducted to estimate the maximal interval between successive connections to NTP servers after which divergence between timestamps of RTK-GNSS and weather station measurements would not exceed 1 second. It was found that this period was 15 minutes and this frequency of synchronization with NTP servers was applied for further measurements.

Implemented solutions allowed to reduce divergence between timestamps of both types of measurements from several minutes to acceptable less than 1 second.

7. Data assessment and processing

Initially, the time synchronization of weather station measurements with RTK-GNSS measurements was performed. Subsequently, RTK-GNSS measurements with phase ambiguities not fixed, characterized by accuracy of less than 1 m, were removed from further analysis. One of the reason for the lack of ambiguities fixed were temporary signal constraints. They resulted in lack of data from the reference station essential for accurate measurements of the rover stations. The second factor affecting the accuracy of the measurements was the visibility and constellation of the satellites. Also, RTK-GNSS measurements from the periods were meteorological measurements were not obtained or they were corrupted were removed. This was mainly due to maintenance breaks and technical break-downs of the weather station. After described above data filtration of measurements from 6 months of observation the total time length of acceptable measurements was approximately 1012 hours i.e. just over 42 days.

8. Accuracy estimation

The accuracy of the RTK-GNSS measurements was estimated with a use of three methods: part per million (PPM) values of GNSS receivers, Coordinate Quality (CQ) indicators introduced by devices manufacturer and measurements of the neutral position of the top of the WTT building during negligible environmental actions, in particular minimal wind speed.

8. 1 Part per million

Most manufacturers of GNSS receivers provides PPM as one of the characteristics of the receiver specifying its accuracy. Usually it is expressed in millimetres as the sum of a fixed value and a quotient of a constant coefficient and a baseline length divided by one million. It should be noted that calculated accuracy does not take into account specific measurement conditions such as current satellites constellation and visibility, atmospheric condition, multi-path effect, etc. However, the calculated value, to some extent, indicates the level of accuracy to be taken into account together with other accuracy measurement methods. The accuracy values calculated using this method for GNSS receivers used in this research are given in Table 1.

Table 1. Accuracy of GMX901 and GMX902 receivers according to manufacturer information.

Receiver	Horizontal ppm	Vertical ppm	Horizontal accuracy [mm]	Vertical accuracy [mm]
GMX901	5 mm + 2 ppm	10 mm + 2 ppm	11	16
GMX902	10 mm + 1 ppm	20 mm + 1 ppm	13	23

Baseline length of 2805 m was used in calculations. The PPM values were taken from the data provided by the device manufacturer (Leica 2007, 2010). Calculated PPM values indicate that horizontal accuracy is of ~12 mm and vertical accuracy is twice lower.

8. 2 Coordinate Quality indicator

The second method used for estimation of the accuracy of RTK-GNSS measurements was based on the Coordinate Quality indicator proposed by the devices manufacturer. CQ is calculated and assigned to each position measurement. According to the manufacturer's information (Leica 2009), the receiver's position differs from the correct position with a probability of 2/3 by no more than the CQ value. Indicator is calculated with the use of root mean square of range measurement errors and the Dilusion Of Precision (DOP) matrix. It should be noted that CQ value refers to the accuracy of the point position over time taking into account at same time horizontal and vertical components. This explains much higher CQ values from the standard deviation calculated for the individual component positions and the accuracy estimated by other methods. For the purposes of this paper the CQ indicators will be given together with measurements.

8. 3 Neutral position during negligible environmental actions

Determination of the neutral position of the top of WTT building during negligible environmental actions was used as the third method of estimating the accuracy of RTK-GNSS measurements. Minimum wind speed was the essential criterion for selecting measurement sequences for analysis. Additionally, in order to avoid the effects of solar radiation which may cause one side of the building's facade to be heated and therefore the displacement of the building, measurements only from periods between sunset and sunrise were used. Preliminary selection of measurement sequences showed the inability to use the same measurement sequences for both rover receivers. The reason was the lack of measurements after performing data filtration described in Section 7. Therefore, it was decided to separately estimate neutral position for GMX901 and GMX902 receiver. Three independent 60-minute measurement sequences for both receivers were included in the analysis. Additionally two 5-minute extracted sub-sequences characterized by the highest accuracy for the GMX902 receiver, were also chosen for analysis. Most of the measurements were made with the visibility of 6-8 satellites. Mean wind speed ranged from 0,5 to 3,1 m/s, wind direction was variable. In terms of determining the neutral position and estimating the accuracy of RTK-GNSS measurements it was assumed that location of the top of the WTT building during selected periods was constant and equal to estimated during this analysis.
Firstly, taking into account in particular measurements characterized by the lowest CQ and standard deviation values, the neutral positions of both rover stations were estimated. The neutral position coordinates are shown in Table 2.

Table 2. Estimated coordinates of neutral positions of rover stations.

Receiver		Coordinate [m]	CQ indicator [mm]	standard deviation [mm]
GMX901	Grid easting	-1186,148	45-46	6-10
	Grid northing	5787220,705		13-14
	Altitude	318,735		36-46
GMX902	Grid easting	-1189,907	16-20	2-10
	Grid northing	5787236,900		2-8
	Altitude	318,740		7-18

Next, the accuracy of RTK-GNSS measurements was estimated. It was found that the accuracy of GMX901 receiver of horizontal position measurements is of ±20 mm and ±45 mm for vertical position measurements. The accuracy of GMX902 receiver is of ±10 mm for horizontal position measurements and ±30 mm for vertical. Accuracy values were estimated with a confidence level of about 90%. At the same time, it was found that for both types of receivers approximately half of the CQ indicator value corresponds to the accuracy of the horizontal position measurement with a confidence level of 90%.

8. 4 Summary

Table 3 presents the accuracy of the RTK-GNSS measurements estimated with a use of three methods described above. Presented values refer to visibility of 6-8 satellites and level of confidence assumed for further research of 90%.

Table 3. Estimated accuracy of rover stations with 90% level of confidence.

Receiver	Horizontal accuracy [mm]	Vertical accuracy [mm]	CQ indicator [mm]
GMX901	±20	±45	40
GMX902	±10	±30	20

9. Physical response during strong winds

Finally, analysis of the displacement of the top of the WTT building during strong winds was carried out. From a set of 6-month observations, 30-second sequence from 11.03.2012 and time from 6:36:00 to 6:36:30 am, characterized by the highest measured wind speed, was chosen for investigation. Figure 7 presents time history of wind speed and direction measurements for selected sequence together with preceding and following 30-second periods. During measurements, rain with intensity of 2.5 cm/h was observed. Temperature was of 1.9°C and humidity of 86%.

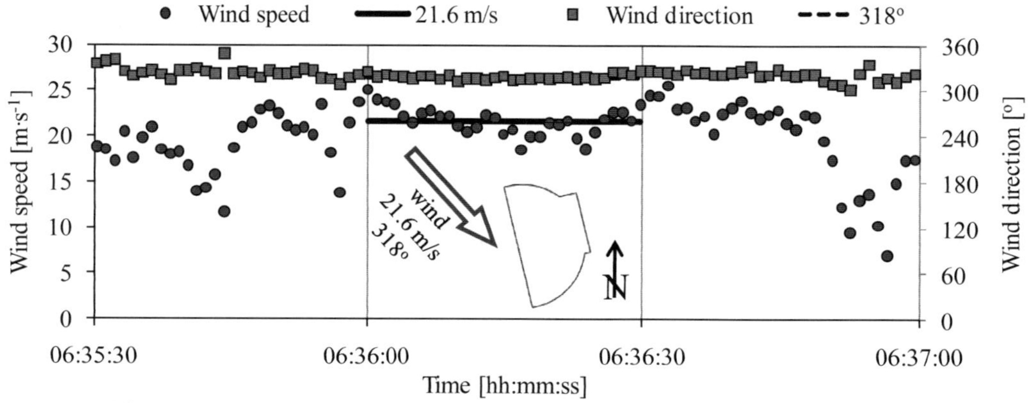

Figure 7. Time history of wind speed and direction for selected 30-second sequence and preceding and following 30-second periods.

After performing data filtration described in Section 7 the lack of GMX901 receiver measurements for selected sequence was found. Consequently, only measurements taken by GMX902 receiver were used for further investigation. Measurements were made with the visibility of 15 satellites (GPS and GLONASS). The number of positions determined during selected sequence was of 516. Theoretical number of measurements in 30 seconds period at 20 Hz sampling is 600. The differences between the theoretical and the obtained numbers are natural and related to the quality of the satellite signals and the specificity of the RTK-GNSS receiving and processing system. Results of the measurements are shown in Figure 8. Table 4 presents 30-second averaged results of full-scale measurements. Zero coordinates constitute estimated neutral position the top of WTT building. Due to the downtown character of the surrounding area and the early hours of the measurements, i.e. 38 minutes after sunrise, the possible displacement of the top of the building due to non-uniform thermal load of the façade caused by solar radiation was considered as negligible.
According to the analysis presented in Section 8 and taking into account good visibility of satellites the accuracy of the horizontal measurements is of ±10 mm, while the vertical is ±30 mm with a confidence level not less than 90%.

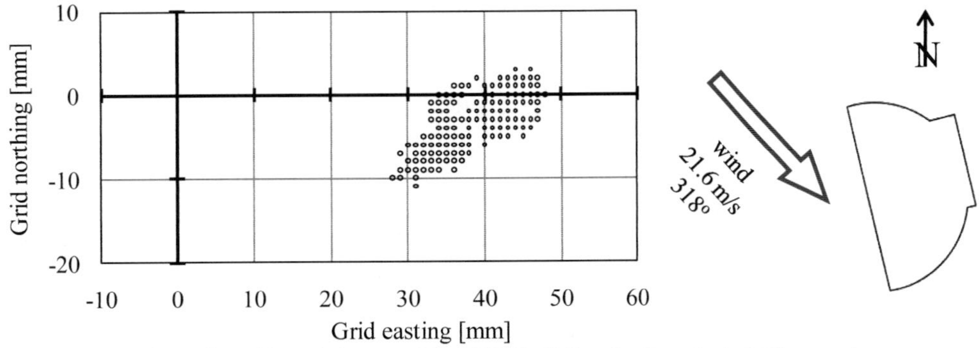

Figure 8. Horizontal position of the top of the WTT building for the selected 30-second sequence.

Table 4. Average 30-seconds values of full-scale measurements during strong winds.

Wind speed [m/s]	Wind direction [°]	Displacement [mm]			
		Grid easting	Grid northing	Altitude	CQ indicator
21.6	318	39	-3	2	39

The results obtained for other sequences with lower wind speed values, are characterized by a lower response of the building. Significant difference in displacement in the direction of the main axis of the building was observed. In addition, it was found that as wind direction raises, displacements in the south direction increases relatively to the east direction.

10. Conclusions

The paper provides detailed description of full-scale measurements of the high-rise WTT building with a use of RTK-GNSS technology and the weather station. Brief outline of the WTT building and surroundings area is given. Instrumentation and system integration as well as data acquisition, management, transmission and storage processes are described. Three methods used for estimation of the accuracy of the RTK-GNSS measurements is described. Determination of the neutral position of the top of the WTT building during negligible environmental actions and displacement measurements during strong winds, as well as measured wind speed and direction, are presented.
Technical problems found during the tests and solutions applied are presented. The problem that was not completely solved was the insufficient GSM/UMTS signal strength and quality on the roof of the WTT building, which caused interruption in data transmission from the reference station resulted in insufficient accuracy of measurement of the rover stations. External passive GSM/UMTS antenna only slightly improved signal. Due to the fact that most roofs of tall buildings are used for the installation of media transmitters, in similar RTK-GNSS measurements of a high-rise building one should consider using more reliably communication with Internet such a cable connection. Another problem found was the lack of time correlation between RTK-GNSS and weather station measurements. The divergence was variable over time and ranged from tens of seconds to several minutes. Two identified sources of the problem and solution applied are presented.
Observed displacement of the top of the WTT building was of 39 mm. Due to the downtown character of the surroundings and early hours of the measurements, impact of solar radiation on displacement was considered as negligible. During measurements heavy rain was observed.
The accuracy of the RTK-GNSS horizontal measurements was of ±10 mm. Measurements were made with the visibility of 15 satellites. Relative level of the accuracy of 25% should be considered as unsatisfactory. Further work is required to obtain more reliable results. This could be possible by

improving the RTK-GNSS measuring system as well as continuing full-scale measurements in anticipation of higher wind speeds and consequently more significant displacements.
Obtained results proved the possibility of evaluating physical response of the high-rise WTT building caused by measured wind action. Measured wind speed and direction were therefore used in wind-tunnel tests and CFD simulations for evaluating pressure distribution over the building's façade under conditions corresponding to those observed during full-scale measurements. Evaluated average wind pressure distribution could be used to evaluate theoretical response of the structure subjected to wind loads, which correspond to those observed in full-scale measurements with a use of finite element method (FEM) calculations. A comparison of the theoretical and measured responses would then constitute the indicator of structural condition. It is expected to obtain a fair degree of similarity for buildings in a good structural condition, and evident discrepancy for buildings with structural deficiencies.

Acknowledgement

This paper is a review of the first co-author's part of doctoral dissertation defended at the Warsaw University of Technology (Cwik 2013). The author would like to express his greatest appreciation to the dissertation supervisors – the second co-author of the paper Prof. Marian Gizejowski and Dr. Adam Goliger of the Council for Scientific and Industrial Research in Pretoria, RSA for providing an overall PhD study guidance. The gratitude goes also to the Warsaw University of Technology, Polish Ministry of Science and Higher Education and the National Research Foundation for the financial support, e.g. made available through statutory subsidies, dean's grants and the project MONIT grant „Monitoring of Technical State of Construction and Evaluation of its Lifespan".

References

Cwik, M., Gizejowski, M. A., and Goliger, A. M. 2013, 'Wind-tunnel investigations of pressure distribution over high-rise buildings'. *Research and Applications in Structural Engineering, Mechanics & Computation: Proceedings of the Fifth International Conference on Structural Engineering, Mechanics & Computation (ed. A. Zingoni),* Cape Town, South Africa, 2-4 September, pp. 235-236 [full text on CD, 645-650].

Cwik, M. 2014, 'Modelling of wind action on high-rise buildings – Warsaw Trade Tower case study'. *Proceedings of the XIII Conference of the Italian Association for Wind Engineering – In-Vento 2014,* Genova, Italy, 22-25 June, [full text on CD].

International Earth Rotation and Reference Systems Service 2017, Bulletins C 53, 9 January 2017. Available from: http://www.iers.org [25 May 2017].

National Coordination Office for Space-Based Positioning, Navigation and Timing, Official U.S. 2012, Government information about the Global Positioning System (GPS) and related topics. Timing Session of the Civil GPS Service Interface Committee (CGSIC), 17 September 2012. Available from: http://www.gps.gov [25 May 2017].

Leica 2007, *Leica GMX901 User Manual. Version 1.0*, Leica Geosystems AG, Heerbrugg, Switzerland.

Leica 2010, *Leica GMX902 GG/GMX902 GNSS User Manual. Version 1.3*, Leica Geosystems AG, Heerbrugg, Switzerland.

Leica 2009, *Leica SmartNet UK & Ireland Network RTK User Guide*, Leica Geosystems Ltd. (accessed October 2011).

Cwik, M 213, *Modelling of wind action on high-rise buildings – Warsaw Trade Tower building case study*, PhD thesis (in Polish), Publishing House of Warsaw University of Technology, Warsaw University of Technology.

The 8th International Conference on
Structural Health Monitoring of Intelligent Infrastructure
Brisbane, Australia | 5-8 December 2017

Outlier detection of ARIMA model coefficients estimated from GPS thermal deformation data for assessing structural condition of a cable-stayed bridge

H. V.Le[1], M. Nishio[2]
[1] *University of Transport and Communication – Vietnam, Email:* hienlv84.uct@gmail.com
[2] *Yokohama National University - Japan*

Abstract

Air temperature is one of the most significant environmental effects that affect the responses of civil structures, including the long-span bridges. The target structure in this study was a cable-stayed bridge located in Vietnam. In the long-term GPS monitoring here, the periodic global thermal deformation, which was caused by the daily air temperature changes, was clearly recognized. This paper especially presents the numerical study to show our strategy of taking advantage of this periodic global deformation to assess the structural condition. A finite element model of a general cable-stayed bridge was constructed to calculate global thermal deformations by inputting periodic air temperature histories. The numerical data; i.e., calculated displacements with measurement noise, were obtained under different structural conditions with changes of the boundary conditions and the tensions of cables. The Autoregressive Integrated Moving Average (ARIMA) model was applied to the calculated thermal deformations to use the coefficients of estimated models as the features. The Mahalanobis distance of the feature vector, which was composed of AR and MA coefficients, in each structural condition was then compared. In the results, it was shown that the global structural changes; both the boundary condition and the cable tension changes, could be recognized statistically by the outlier detection of Mahalanobis distance of the proposed feature vector. In addition, the Mahalanobis distance distribution derived from the actual long-term GPS monitoring data acquired in the target cable-stayed bridge was also presented. The distances were distributed in the same range of value under the structural condition, in which significant structural changes have not been reported. It was concluded that the global thermal deformation data that could be acquired by the GPS could become one of effective responses to be used for assessing especially the global structural changes.

1 Introduction

Structural Health Monitoring (SHM) system is necessary to install for assessing the conditions of long-span bridges, which are special constructions with complicated structures and numerously affected by environmental factors during their operations. Study on monitoring data analysis can realize the structural condition changes due to damages or boundary condition changes to guide the managers find out methods for maintenance, retrofit or replacement.

Outlier analysis is a fundamental theory in statistic which in order to detect the abnormal data from acquired data set, then abnormal event occurred on structure can be detected. Several studies tried to apply outlier analysis for damage detection of structures, i.e. Worden et al, 2000, this study used outlier analysis to find out damage in both simulation and experimental cases. Despite some assumptions were used in calculation, this study showed that outlier analysis based on Mahalanobis distance could realize damage of structures. Furthermore, Nishio et al 2011 applied Mahalanobis distance outlier detection to statistical model validation to compare several extracted features from experimental system, then the results showed that this method was useful for statistical model validation.

The traditional methods based on the changes of frequency or mode shape using for damage detection of structural monitoring or bridge structural monitoring that were used in several researches. One of example is Bandara et al, 2014 who applied the frequency response functions for structural damage detection of three-story bookshell structure in experiment, and this method showed reliable results in experimental data. Otherwise, Ni et al 2015 proposed the multi-novelty indices formulated by monitoring-derived modal parameters for structural damage alarming and localization of cable-supported bridge. The proposed method was also validated with two cable-supported bridge that were the Tsing Ma bridge and the Tsing Kau bridge. These results showed that the multi-novelty indices method could identify the damage region of the target bridges.
In term of time-series analysis application considering the effects of environmental factors, some studies applied time-series models such as Autoregressive (AR) or Autoregressive with Exogenous input (ARX) etc. to investigate the change in extracted features in order to detect outlier or damage of structures. Sohn et al 2000 and Sohn et al 2001 applied the AR time-series model for dynamic signals to extract AR coefficients, then signals coming from damage or undamage system could be realized by examining the change of AR coefficients. Furthermore, Omenzetter & Brownjohn 2006 applied Autoregressive Integrated Moving Average (ARIMA) model for strain data of bridge recorded under normal operational and environmental conditions. The extracted coefficients were then examined their changes in trend to detect structural condition change or damage.
Considering the effect of atmospheric temperatures to the global deformation of cable-stayed bridge, the current study focusses on outlier detection of ARIMA model coefficients estimated from GPS thermal deformation for assessing structural condition. This paper especially presents the numerical study to show our strategy of taking advantage of this periodic global deformation to assess the structural condition. A finite element model of a general cable-stayed bridge was constructed to calculate global thermal deformations by inputting periodic air temperature histories. The numerical data; i.e., calculated displacements with measurement noise, were obtained under different structural conditions with changes of the boundary conditions and the tensions of cables. The ARIMA time-series model was applied to the calculated thermal deformations to use the coefficients of estimated models as the features. The Mahalanobis distance of the feature vector, which was composed of AR and MA coefficients, in each structural condition was then compared. In addition, the Mahalanobis distance distribution derived from the actual long-term GPS monitoring data acquired in the target cable-stayed bridge was also presented.

2 Global deformation mode of the cable-stayed bridge due to periodic temperature variation

The target bridge in this study is a cable-stayed bridge in Vietnam which is the longest cable-stayed bridge in the South East Asia. The properties of bridge are 2,750m in total length, 26m of girder width, 171m of tower height, and 500m length of the centre span which is made by concrete box-girder, but a part of 210m length at centre made by steel box-girder to increase the loading capacity. By using the elastic rubber bearings to connect towers and girders, the girders are free to slide longitudinal at the limitation of bearings.
The installed SHM system includes many kinds of sensors such as anemometers, accelerometers, temperature sensors, and GPS. The GPS system including nine sensors has been installed to capture the displacement of bridge in the interested locations such as on the top of towers, at the middle and quarter of main span, and some piers. The locations of GPS rover sensor are figured out by red circle in the Fig. 1. Furthermore, the adopted GPS equipment are products of Leica co., ltd with GMX 902 model having precisions based on real time kinematic technique are (±10mm ± 1ppm)for horizontal plane and (±20 mm ± 1ppm) for the vertical direction, respectively. GPS signal at each rover station was acquired in 20Hz, and then they were taken average in ten-minutes. Moreover, to observe the changing atmospheric temperature, one temperature sensor was installed at the middle

center span on the deck of bridge (the green circle on Fig. 1). The time interval of temperature data acquisition was set as same as the GPS data.

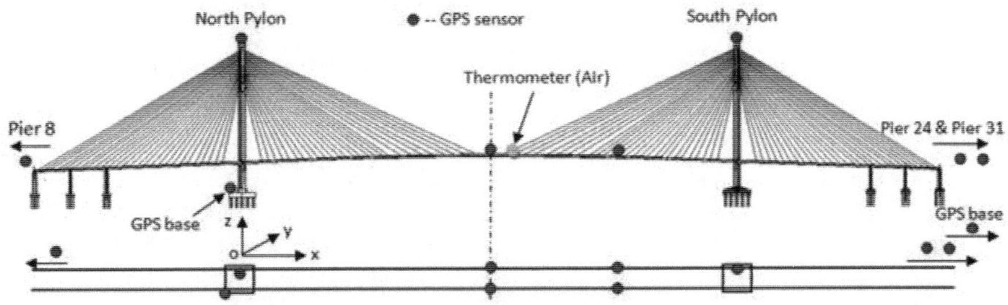

Figure 1 GPS sensors on the Can Tho bridge (Le & Nishio 2015)

In our previous study (Le & Nishio 2015), the ten-minutes-average GPS time-series data were used to analyze the correlation with the atmostpheric temperature to study how the effects of temperature to the displacements of the interested positions of bridge. The results of correlation analysis showed that the correlation between tower points and air temperature changes along to the longitudinal direction of bridge are recognized in both quite high positive and negative values (<0.6), it means that the towers deform with same behaviour on the inward side when the air temperature change. Otherwise, the correlation coefficients between the temperature and the GPS data at the girder (center and quarter spans) showed the results that they were much close to -1 according to vertical direction.This means the girders show clearly global deflections as the air temperature increases. In addition, the correlation coefficients were very low coefficients along to longitudinal and lateral directions except the one in longitudinal direction of the quarter point.This can be explained by the target bridge is almost symmetric while the quarter span point is the non-symmetric point.Moreover, the frequency spectra was derived to analyze the dominated periods of GPS and air temperature time-series data. The results showed that the correlation between GPS data and temperature data are mostly dominated by the one-day periodic behaviour (24 hours-cycle).The global deformation modes of the target bridge due to daily periodic temperature changes was them be identified as shown in Fig. 2.

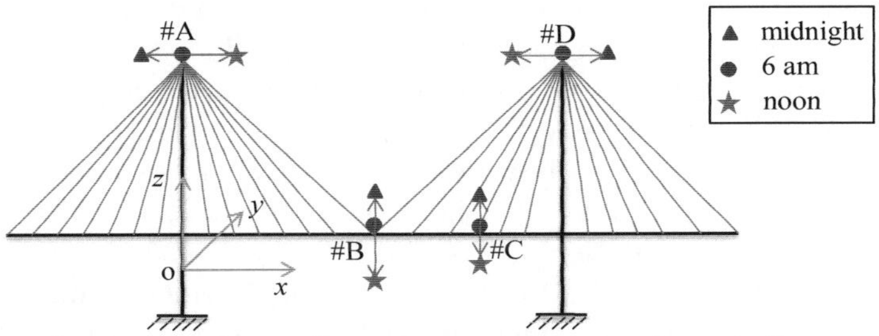

Figure 2 In-plane global deformation modes due to the day-cycle periodic temperature change(Le & Nishio 2015)

3 Numerical study on structural condition assessment by outlier detection of ARIMA model coefficients

3.1 Numerical model description

A numerical model of a general cable-stayed was constructed for verification here. The total span of the bridge was 420m with 220m main span, the width of the girder was 15.6m, and the two towers with the height of 90m. The general layout of the model is shown in Fig.3, and the properties of structural elements are also shown in Table 1.

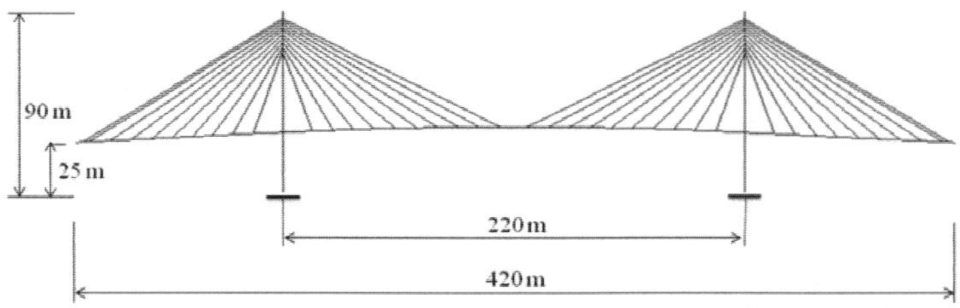

Figure 3 Constructed FE model of a general cable-stayed bridge

Table 1 Material and sectional properties of the numerical model

Material ID	Name	Modulus of Elasticity (kN/m^2)	Poisson's ratio	Weight Density (kN/m^3)	Area (m^2)	I_{xx} (m^4)	I_{yy} (m^4)	I_{zz} (m^4)
1	Cable	1.9613×10^8	0.3	77.09	0.0052	0.0	0.0	0.0
2	Girder	1.9995×10^8	0.3	77.09	0.3092	0.007	0.1577	4.7620
3	Pylon	2.78×10^7	0.2	23.56	9.2	19.51	25.567	8.123
4	CBeam-Girder	1.9995×10^8	0.3	77.09	0.0499	0.0031	0.0447	0.1331
5	CBeam-Pylon	2.78×10^7	0.2	23.56	7.2	15.79	14.472	7.992

The fixed-end boundary conditions were applied to the bases of the two towers and the bases of two anchorages base are fixed condition; and using rigid links to connect between main girders and bearings.

3.2 Time-series data acquisition

In this step, the ten-minute average temperature was generated by one day length of sinusoidal type which was used as input factor to the numerical model. The coordinates of all nodes and cable tensions were updated step by step in each case of temperature changes based on the nonlinearity analysis. The displacements of all model nodes were then acquired in both cases: normal condition and structural condition changes. The scenario of structural condition changes were that: the typical bearing properties of boundary conditions were changed to nearly fix conditions at two towers and one of anchorage; and the tension of three selected cables were reduced into 5%, 10%, 25%, and 50% for case 1, and 10% for other cases. Description of the case studies of the structural condition changes and the selected points are shown in Fig.4 and summarized in Table 2. Here, there were four selected points at the tower locations (#Pt1 and #Pt4), the main girder locations (#Pt2 and #Pt3) for analysis. In addition, the outcome time-series displacement data were then added the random white noise which consulted the range of white noise from the real target bridge. The ratio of error and displacement range along to horizontal plane (x- and y-direction) is around 25% while this ratio along to vertical direction is 12%.

4 Application of ARIMA time-series model for structural condition assessment

Generally, time-series analysis is appropriate for studying the monitoring time-series data, especially considering the effects of environmental factors (Sohn et al 2000, Omenzetter et al 2006).

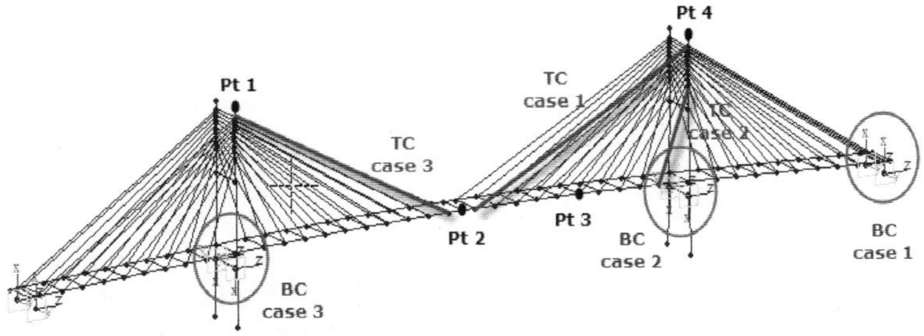

Figure 4. Layout of structural condition change cases and analysis points

Table 2 Case studies of structural condition changes

Case study of boundary condition change (BC)	Bearing property change	Notice of location	**Case study of cable tension changes (TC)**	Portion of tension reduction	Notice of location
BC case 1	SDz: $*10^4$ SRy: $*10^5$	Anchorage	TC case 1	5% 10% 25% 50%	Tension change in case 1
BC case 2	SDz: $*10^4$ SRy: $*10^5$	Right tower	TC case 2	10%	Tension change in case 2
BC case 3	SDz: $*10^4$ SRy: $*10^5$	Left tower	TC case 3	10%	Tension change in case 3
BC case 4	SDz: $*10^{20}$ SRy: $*10^{20}$	Right tower			
BC case 5	SDz: $*10^{20}$ SRy: $*10^{20}$	All towers			

According to the results of correlation analysis, the correlation between GPS data and temperature are mostly dominated by the 24-hour periodic behaviour. Thus, the time-series data here were divided to day by day, and the Autoregressive Integrated Moving Average (ARIMA) model was then applied in this study for both actual long-term GPS data and numerical data to extract features.

4.1 Description of ARIMA time-series model and sequence of application steps

The ARIMA time-series model is usually used to describe the non-stationany process which has a general form as ARIMA(p,d,q) defined by (Le & Nishio 2015):

$$\left(1-\phi_1 B-\phi_2 B^2-\cdots-\phi_p B^p\right)\left(1-B\right)^d y_t = c+\left(1-\psi_1 B-\psi_2 B^2-\cdots-\psi_q B^q\right)\varepsilon_t \, , \tag{1}$$

where p, d and q are the orders of model; y_t is the t-th component of the target time-series vector **y**, andε_t is the white noise error term. ϕ_i (i =1-p) and ψ_j (j =1-q) are coefficients of the autoregressive term and the moving average term, respectively, and B is the lag-operator, which is defined as:

$$B^k y_t = y_{t-k} \, ; \tag{2}$$

Therefore, $(1\text{-}B)^d$ in Eq.(1) indicates the d-th order difference of the original time-series. By taking the difference, the trend in the time-series can be removed, and the non-stationary process can be transferred to the stationary process.

The sequence of application steps are proposed as:

(1) Checking and transform the time-series data to the stationarity by taking difference and applying autocorrelation and partial autocorrelation functions (ACF & PACF);
(2) Definition of the appropriate orders by applying ACF, PACF and Bayesian Information Criterion (BIC);

(3) Application of defined ARIMA model for the day by day time-series data to extract the AR-MA coefficients.

4.2 Application of ARIMA model for numerical data

For validation of the ARIMA model application in the GPS time-series which the ARIMA(1,1,1) was defined, thus here this model was then applied for numerical data. The length of numerical data here was taken to be 20 days that was divided to day by day. According the GPS global deformation behaviour acquired from the real bridge, some significant directions of considering points were selected to be analysed that were the *x*-direction (longitudinal) of the left and right tower points (#Pt1 and #Pt4), the quarter span point (#Pt3), and the *z*-direction (vertical) of the center and quarter center points (#Pt2 and #Pt3).

Figure 5 shows one example results of one-day ARIMA model performance which the fitting ARIMA (1,1,1) model to time-series data in the Fig. 5a, and the standardized distribution of residual in the Fig.5b. Statistically, the ARIMA (1,1,1) is the appropriate model if the fitting residual should be white noise. Therefore, the Ljung-Box Q-test was used to check whether the residual error distribution showed the white noise process with normal distribution. The null hypothesis here was; "the residual is the white noise". In the results, the statistical possibility *p*-values at lags (3, 6, 9) were (0.1503, 0.2767, 0.0808) that are all larger than the 5% significant level, and the test statistic values were calculated at (2.0695, 5.1049, 12.6617), which are all smaller than the critical values at 5% significant level (3.8415, 9.4877, 14.0671). It means that the null hypothesis was not rejected. The residuals are thus identified as the white noise process.

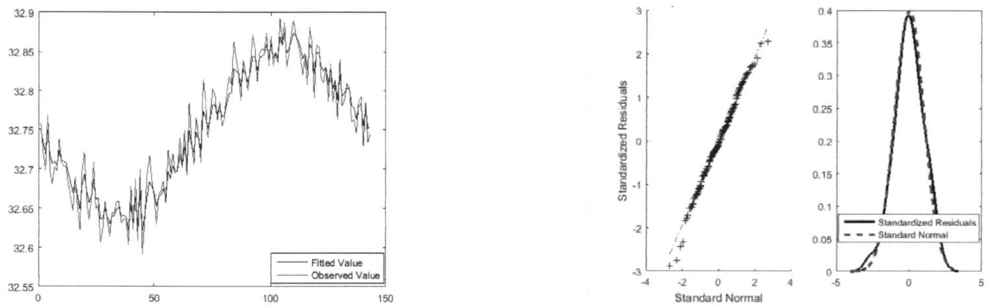

(a) Overlay of estimated and numerical series (b) Distribution of the residual error

Figure 5 Estimated results of one ARIMA (1,1,1) model

4.3 Plotting of extracted features

The extracted AR-MA coefficients of ARIMA model application were then plot on figures in both cases of normal condition and structural condition changes. Figure 6 shows the overlaid plot of coefficients comparing two cases: BC 2 and BC 4 which the bearing properties at right tower was changed at different values. Here, the blue star symbol denotes the coefficients in normal condition case, and the red circle symbol shows the coefficients of low value change while green circle symbol shows the high value change case. Figure 7 plots the overlaid coefficients of cable tension change in case 1 in three different changing value of tension reduction (-5%, -25%, -50%). The legend of symbols shows on the Figure 7a.

The Figure 6 & 7 represent for all cases, and they show the same situation that even the trend in coefficient distributions seem to change when the structural condition change, this trend is not clear to recognize in all plots. Therefore, the Mahalanobis distance method needs to be applied for AR-MA coefficients to test its ability to recognize the structural condition changes.

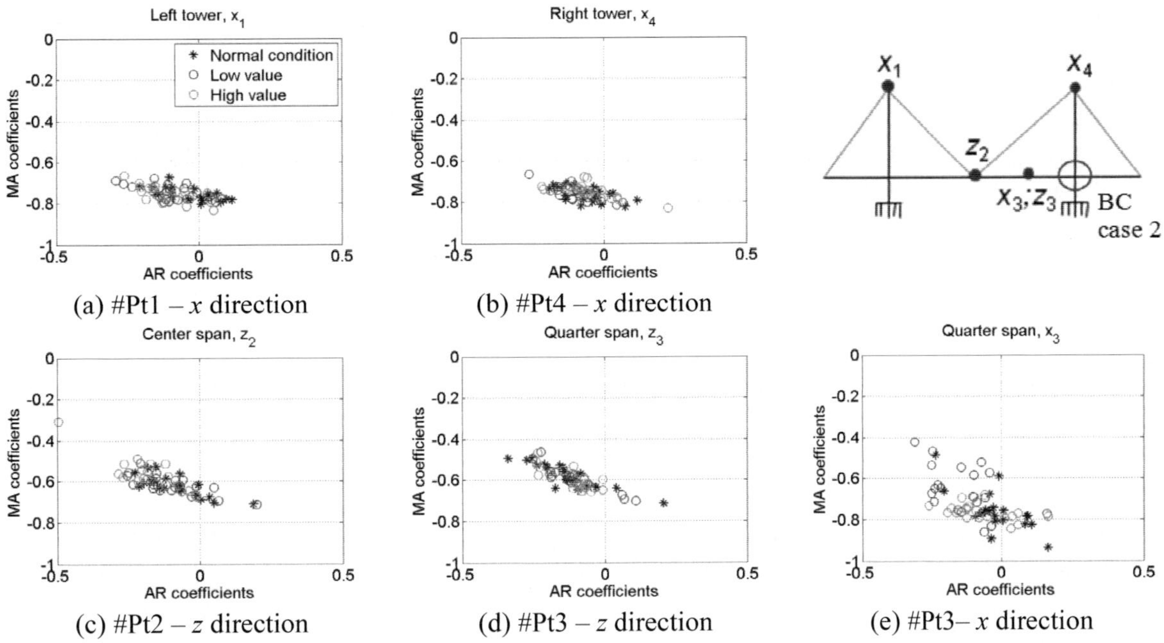

(a) #Pt1 – x direction (b) #Pt4 – x direction

(c) #Pt2 – z direction (d) #Pt3 – z direction (e) #Pt3– x direction

Figure 6 Overlaid plot of AR-MA coefficients in the cases of BC case 2 and BC case 4

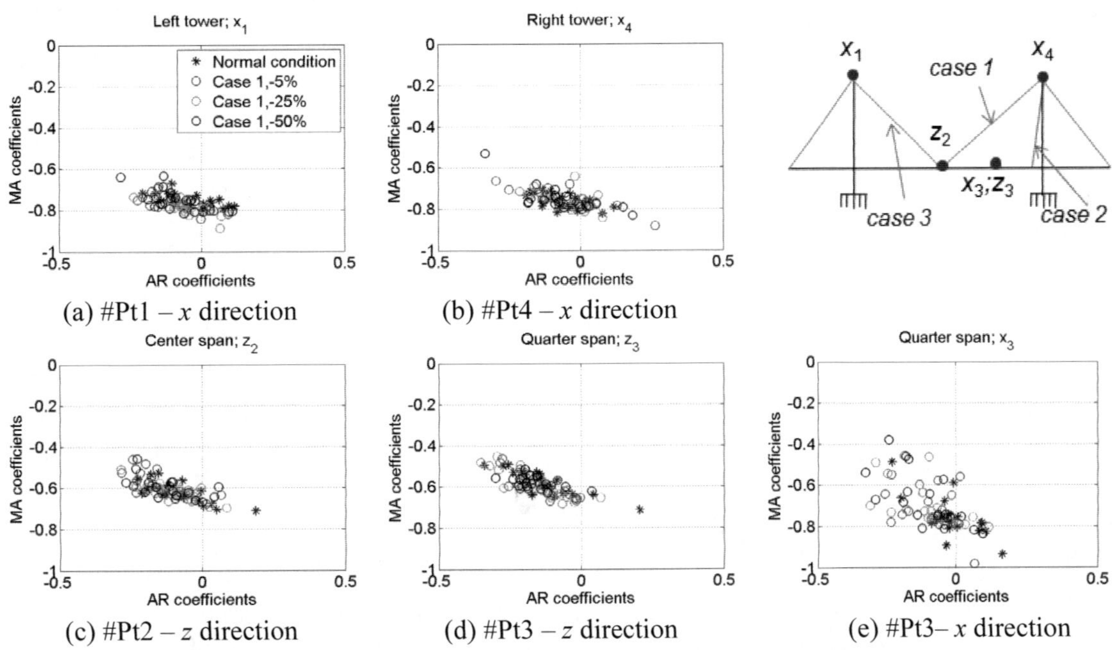

(a) #Pt1 – x direction (b) #Pt4 – x direction

(c) #Pt2 – z direction (d) #Pt3 – z direction (e) #Pt3– x direction

Figure 7 Overlaid AR-MA coefficient plots of the tension change in case 1

5 Mahalanobis distance application for structural condition assessment

5.1 General Mahalanobis distance method

The Mahalanobis distance method had proposed for comparing the multivariate feature vectors and the algorithm of statistical structural damage detection (Nishio et al 2011). The theory of this method is to normalize distance from the mean. Thus, the general method was applied for this study that was done in two steps: (1) calculate the distances of base coefficients in normal condition using as baseline; (2) using the estimated covariance matrix and the mean vector of baseline to calculate the distances of estimated coefficients in cases of structural condition changes, as following equations:

$$MD_j^B = \sqrt{(\mathbf{x}_j^B - \overline{\mathbf{x}}^B)^T \mathbf{S}_B^{-1} (\mathbf{x}_j^B - \overline{\mathbf{x}}^B)} \tag{3}$$

$$MD_k^E = \sqrt{(\mathbf{x}_k^E - \overline{\mathbf{x}}^B)^T \mathbf{S}_B^{-1} (\mathbf{x}_k^E - \overline{\mathbf{x}}^B)} \tag{4}$$

In whichthe subscript B denotes the base data sets; $\overline{\mathbf{x}}^{\mathrm{B}}$ is the mean vector of obtained feature vectors x_j of the baseline; $\mathbf{S}_{\mathrm{B}}$ is the estimated covariance matrix of the baseline.

5.2 Application of Mahalanobis distance for numerical data

The algorithm of Mahalanobis distance mentioned above was applied for all AR-MA coefficients which the coefficients of normal condition case were used as baseline. The Mahalanobis distance of each structural condition change case was then calculated by using the estimated mean and covariance matrix of baseline. The Mahalanobis distance of baseline and estimation were then figured out to compare for outlier detection. Figure 8 & 9 are the representative figures in the changes of boundary condition and cable tension reduction, respectively. The upper and lower bounds in all figures were made by the range of baseline distances.

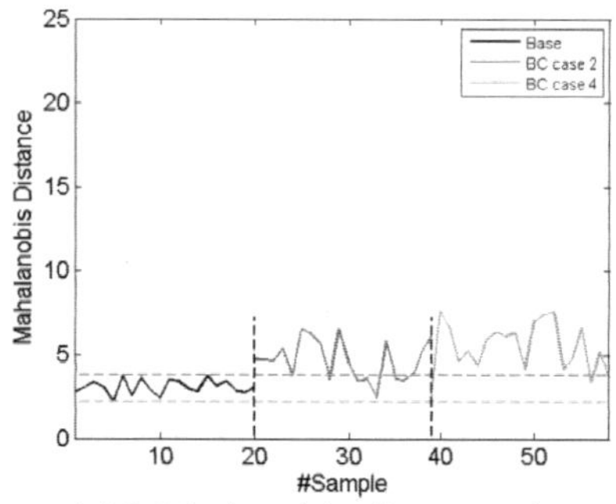

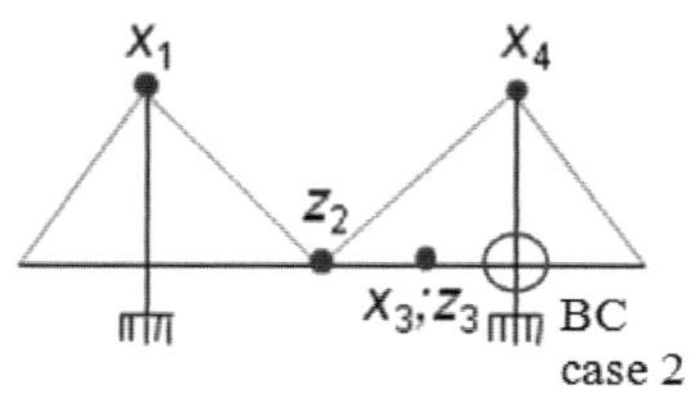

(a) Mahalanobis distance plots (b) Locations of boundary condition changes

Figure 8 Comparing Mahalanobis distance results between BC 2 and BC 4

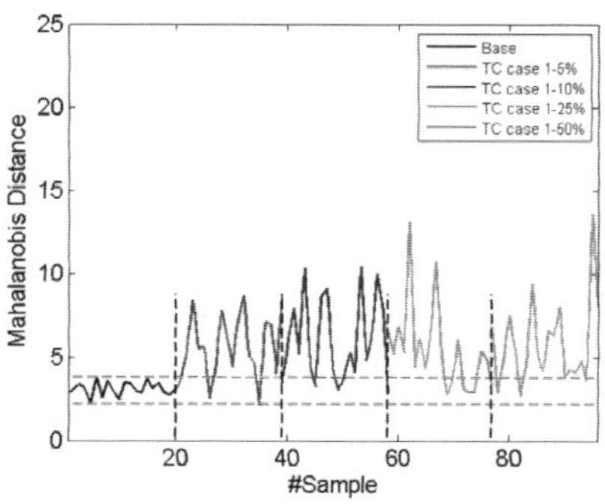

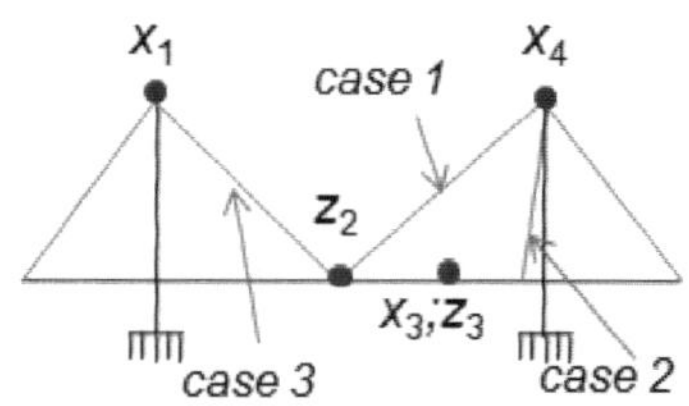

(a) Mahalanobis distance plots (b) Locations of cable tension changes

Figure 9 Comparing Mahalanobis distance results of cable tension change in case 1 (TC1)

It can be seen that the estimated distances are much larger than the baseline in all cases of structural condition changes. Statistically, there were more than 70% of the estimated distances that were out of baseline distance range. Therefore, the outlier distances can be easily recognized based on plotting distances (see Fig. 8a & 9a), this also means that the structural condition changes could be recognized by applying the Mahalanobis distance method for multivariate AR-MA coefficients.

5.3 Application of Mahalanobis distance method for the real GPS data

The algorithm of Mahalanobis distance method was then applied for the AR-MA coefficients that extracted from almost one year GPS time-series data of the target bridge. To investigate the effects of baseline size to the outlier features, three cases of baseline were considered to study that were one-month, three-month, and six-month baseline coefficients, respectively. The estimated mean and covariance matrix of baselines were also used for calculating the estimated Mahalanobis distance. The Mahalanobis distance results are figured out in Fig. 10 where the dark line shows the baseline distance, the red dash line shows the estimated distances, and the green dash lines show the baseline range. It can be seen that when the size of baseline coefficient increases the outlier distances by baseline range decrease. Statistically, there was 86.6% of estimated distances that were out of baseline range in case of one-month baseline while this portion were 23.9% and 7.2% for the cases of three-month, and six-month baseline, respectively. This means that the size of baseline coefficient influences on the estimated distance. In this target bridge, it was confirmed that there was no significant change in the structural condition by the inspection. Therefore, the baseline distribution from the six-month data is appropriate to construct the outlier detection algorithm of the structural condition assessment. Moreover, it was also considered that the baseline distribution should be carefully defined in relation to the actual structural condition firstly, and it would be better to take as large number of base data sets as possible.

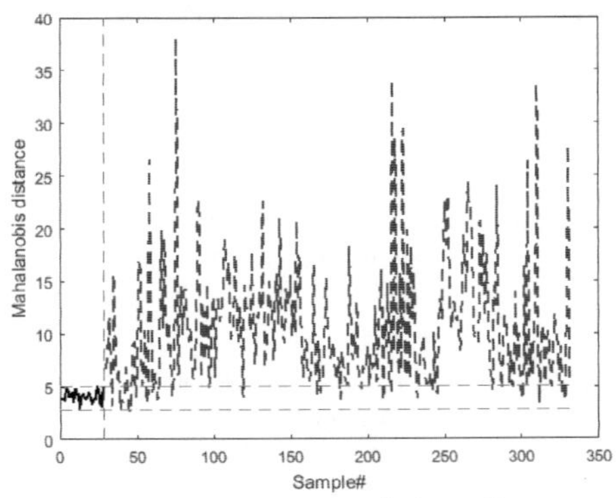

(a) One-month baseline

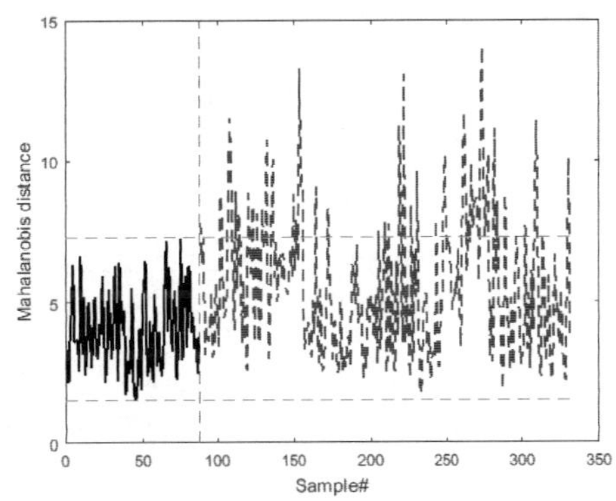

(b) Three-month baseline

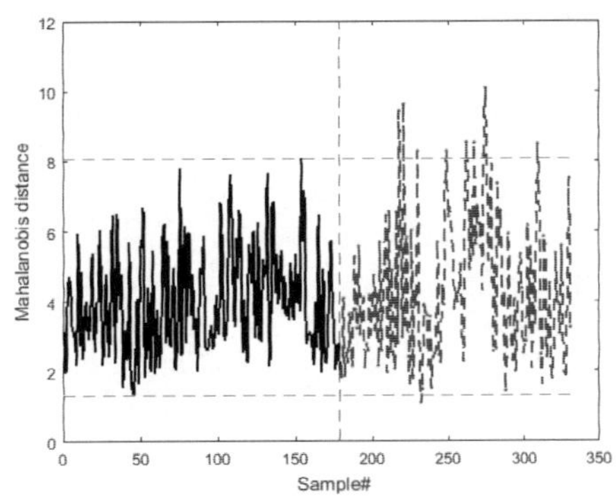

(c) Six-month baseline

Figure 10 Mahalanobis distance application for actual GPS data with different baselines

6 Conclusions

This study summarizes the GPS thermal deformation acquired from the real target cable-stayed bridge, and especially presents the strategy of structural condition assessment by taking advantage of periodic global deformation applying for numerical study. The ARIMA time-series model was taken to apply for novelty numerical time-series data to extract AR-MA coefficients as features. The distributions of these coefficients conclude that even there is a trend in change of coefficient distribution when structural condition change, this trend is not clear to recognize. However, the Mahalanobis distance results applying for numerical AR-MA coefficients conclude that structural condition change behaviours could be easily recognized by the outlier distances in their distributions. Application of Mahalanobis distance for the real target bridge also proved the capability of its application for structural condition assessment. This study also shows that the global thermal deformation data that acquired by the GPS could become one of effective responses to be used for assessing the global structural changes.

References

K. Worden, G. Manson2000, 'Damage detection using outlier analysis', *Journal of Sound and Vibration*, vol. 229, no. 3, pp. 647-667

M. Nishio, F. Hemez, K. Worden, G. Park, N. Takeda, C. Farrar 2011,'Feature extraction for structural dynamics model validation', *Linking Models and Experiments*, vol. 2, pp. 153-163

R. P. Bandara, T. H. T. Chan, D. P. Thambiratnam 2014, 'Structural damage detection method using frequency response functions', *Special issue Article of Structural Health Monitoring*, vol. 13, no. 4, pp. 418-429.

Y. Q. Ni, J. Wang, T. H. T. Chan 2015, 'Structural damage alarming and localization of cable-supported bridges using multi-novelty indices: a feasibility study', *Structural Engineering and Mechanics*, vol. 54, no. 2, pp. 337-362.

H. Sohn, J. A. Czarnecki, C. Farrar 2000, 'Structural Health Monitoring using statistical process control', *Journal of Structural Engineering*, vol. 126, no. 11, pp. 1356-1363.

H. Sohn, C. R. Farrar, N. Hunter, K. Worden 2001, *Applying the LANL statistical pattern recognition paradigm for structural health monitoring to data from a surface-effect fast patrol boat*, Los Alamos National Laboratory, Los Alamos.

P. Omenzetter, J. M. W. Brownjohn 2006, 'Application of time series analysis for bridge monitoring', *Smart Material Structure*, vol. 15, no. 1, pp. 129-138.

H. V. Le, M. Nishio 2015, 'Time-series analysis of GPS monitoring data from a long-span bridge considering the global deformation due to air temperature changes', *Journal of Civil Structural Health Monitoring*, vol. 5, no. 4, pp. 415-425.

The 8th International Conference on
Structural Health Monitoring of Intelligent Infrastructure
Brisbane, Australia | 5-8 December 2017

Remote Monitoring of Steel Radar Towers in the Canadian Arctic

V. Limaye[1], F. Horvath [1], C. Davies [2], T. Rudowicz [3].

[1] *SHM Canada – Canada, Email:* Vidya.limaye@shmcanada.com [2] *Cleland Jardine Engineering Limited – Canada,* [3]*Ontario – Canada*

Abstract

The North Warning System comprises a series of radar stations located in the Canadian Arctic. Each rotating radar antenna is supported by a structural steel tower and enclosed within a geodesic dome. Typically, these radar towers are subjected to unusual environmental and operating conditions – such as large temperature changes, ice accretion and high winds as well as significant eccentric loads rotating several times a minute. Some of the towers have experienced cracking of the support structure, as well as premature wear of operational components.

The purpose of this project was to evaluate the response of each structure to real-life conditions at three such radar sites and to compare the results with related finite element (FE) modelling of the structures performed earlier. At present, study on one radar tower has been completed. Its scope included visual inspections, finite element modelling, and validation of the model through measurement of strain, wind, and temperature using an automated Structural Health Monitoring (SHM) system at each site.

The challenges faced by the team working on site included arctic weather conditions, the presence of strong electromagnetic fields surrounding high-energy radar equipment, and remote location. The data acquired on site were transmitted to a base station in Halifax at regular intervals over a period of five months where they were reviewed for quality assurance and provided to the analysis team. As the site is unmanned for the majority of the year, a highly reliable system had to be employed that did not require human intervention.

This paper discusses the key design aspects and features of the SHM system installed, challenges of field installation, data transmission through satellite link, and ensuring functionality of the system over the five-month data collection period. Comparison between model data and measured data showed good agreement except for locations where there were structural anomalies.

1. Introduction

The North Warning System comprises a series of radar stations located across the Canadian Arctic. Each radar antenna is supported by a 15 m high structural steel tower and enclosed within a geodesic dome. Typically, these radar towers are situated on high-altitude plateaus and are subjected to unusual environmental and operating conditions – such as large temperature changes (daily and seasonal), ice accretion and high winds, as well as significant eccentric loads oscillating several times a minute. Some of the towers were experiencing cracking of the support structure, as well as premature wear of operational components. In most circumstances it is possible to easily replace a cracked structural component; however, a critical component failure at a remote radar station could cause significant damage to the installation, and have serious security and financial implications

1.1. Purpose of Monitoring

Cleland Jardine Engineering Ltd (CJE) had previously conducted finite element (FE) analyses of the three different radar tower designs used in the North Warning System, and concluded that the dominant load effect on areas of the structures experiencing fatigue was due to rotation of the radar.

The FE models also indicated that environmental conditions (i.e. wind and temperature variations) did not cause significant stress variations at these critical locations (contrary to previous reports by others, which claimed that wind caused significant stresses).

During site investigations, it was also noted (at one site in particular) that several connections had gapping between steel mating surfaces – and it was hypothesized that in combination with the procedures used to level the radar, this could lead to significant out-of-flatness at bearing surfaces – potentially causing the premature bearing-wear that had been reported in this tower design.

The use of strain gauges was initially proposed to test the hypothesis that gapping in connections and installation sequences relating to tensioning of these connections caused out-of-flatness conditions at bearings. During client discussions, the scope quickly grew to a multi-month monitoring project with 32 strain gauges, temperature sensors and an anemometer at each site. The main goals of the project were to:

- Verify the conclusions of structural FE models – that (a) the rotating antenna loading caused high cyclical stress ranges in the structure beneath it, and (b) that wind conditions and temperature variations did not cause high stress variations in these parts of the structure;
- Test the accuracy of the finite element analysis model – to determine how closely the model predicted stress-ranges reflected real-life response to varying load conditions;
- Test the hypothesis that the installation methods in combination with gapping of connection surfaces was distorting bearing mounting surfaces;
- Look for unexpected anomalies in structural response – i.e. stress variations that were not predicted by the analyses conducted to date

2. System Design and Installation

SHM Canada Consulting Limited was selected to design, install, and operate structural health monitoring systems for the three radar towers under study.

2.1. Challenges Faced

The following challenges encountered in the implementation of the monitoring program were successfully addressed:

2.1.1. Remoteness of Sites

The sites, located in the Canadian Arctic Region, are isolated stand-alone facilities. Each site is accessible from client's forward base only by chartered flights using small aircrafts or helicopters. Hence, sound logistics was key to the success of the project. The SHM system was assembled prior to shipping to confirm completeness, then disassembled and shipped to the forward base. Similarly, all tools and accessories required for the installation to the last screw and clip were packed and shipped in advance, as running to the local hardware store was not an option.

The remoteness of the site also required that the designed system was reliable enough to function unmanned for prolonged periods of time and be managed remotely.

2.1.2. High EMF Environment

At the onset of the project it was understood that high EMF environment existed within the work area and the SHM system was designed to address those concerns.

Signals produced by the strain gauges are in the millivolt range and are easily influenced by high EMF environment. Those effects were mitigated by:

- Installing a signal amplifier with bridge completion circuitry adjacent to each strain gauge, and,
- Using a triple shielded signal cable between the amplifier and the data logger.

2.1.3. Rigid Work Schedule

Each radar station has limited living facilities which are shared by a number of work crews in rotation who fly in to perform their tasks in a certain number of days, and depart. The plane that would carry the crew out also brings in the next group of workers. Hence, a realistic installation schedule was developed and agreed with the client to ensure satisfactory installation, testing and commissioning of the system without affecting other planned activities.

2.1.4. Harsh Weather Conditions

Harsh weather conditions are a major factor in carrying out any fieldwork in the arctic region. The window of opportunity available even at the best times of the year is narrow.
Accessing the site can be difficult due to sudden changes in the weather locally or regionally. It affects the work at three levels.
Personnel: Ensuring health and safety of personnel is critical. Hence, the team members were equipped with appropriate arctic gear to ensure their survival even under adverse conditions.
Equipment: The equipment must be able to withstand extremely low temperatures and consistently high winds for prolonged periods of time. The data loggers were installed in one heated enclosure to maintain relatively stable operating conditions. Ultrasonic anemometers with built-in de-icing system, specially designed for sub-zero temperatures were selected. The support systems for anemometer and satellite antenna were designed to withstand high winds and allow for quick installation.
Weather Related Delays: The weather conditions in the region are unpredictable and are a primary factor influencing the work schedule. Adverse weather conditions can prevent the installation team from reaching the radar installation, or leave them stranded on the site for days at a stretch.

2.2. System Design and Installation

There were three parameters to be monitored at each radar site with minor differences in total number of sensors and their location. Hence, a common system configuration was adopted for all three sites. General system configuration, challenges faced, and system installation are discussed in this section.

2.2.1. General System Configuration

The SHM system was designed with the following parameters:

- Strain gauge locations were selected so that acquired data could be compared to data gained from the finite element model
- Weldable strain gauges were used to avoid problems that bondable strain gauges may cause in widely varying temperatures
- Signal amplifiers were positioned close to the strain gauges to minimize EMI effects of the radar transmission. The amplifiers deliver an output signal of 2.5 V ±1 V. Triple shielded cables were used to transmit the amplified signal to the data logger. This technique was able to yield very low noise levels in the acquired data.
- A special program was created for the data logger to achieve proper synchronization of data acquisition to radar antenna rotation
- As the number of strain gauges was exceeding the input capacity of one data logger, multiple loggers were used that communicated with each other for proper synchronization
- Data transmission was accomplished using satellite communication due to lack of wired or cellular connections

The system comprised the following major components:

- Sensors
  - Weldable strain gauges with signal amplifier & bridge completion module (32 to 34)
  - Temperature sensors (2)
  - Ultrasonic high-precision anemometer (1)
- Data Collection System
  - Data loggers with 1GB data storage module (3)
- Data Transmission System
  - Satellite communication modem (1, on-site)
  - Satellite antenna (1, on-site)
  - High-speed internet modem (1, at base station in Halifax, NS)

The block diagram of the SHM system is shown in **Figure 1**. The strain gauges are placed in a standard Wheatstone bridge configuration. The bridge completion resistors are integrated into the amplifier unit.

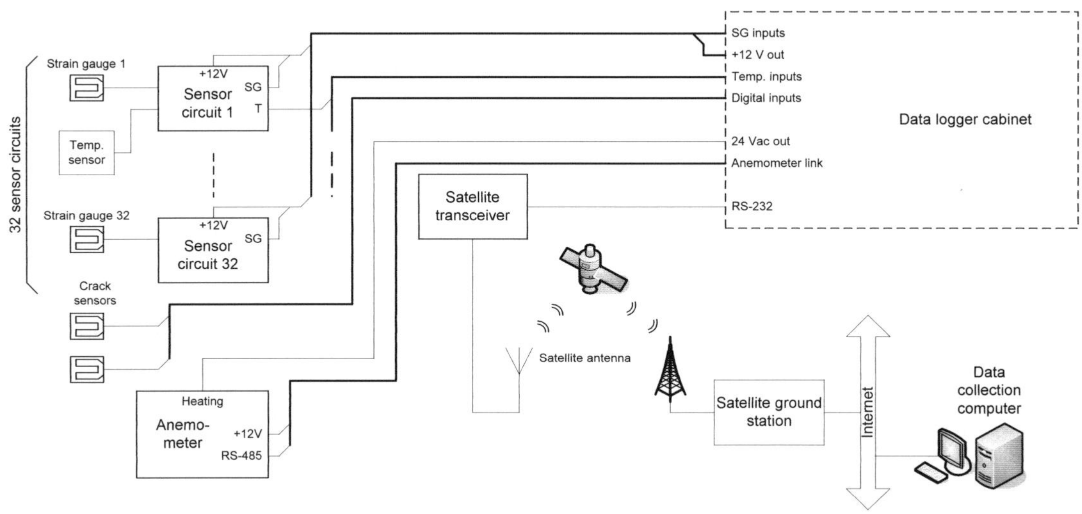

Figure 1. Block diagram of the SHM system

2.2.2. Installation of the SHM system

The installation was carried out by a three-person team over 5 days and included installation of the weldable strain gauges, signal amplifier units, data loggers, anemometer, satellite antenna and cabling, followed by the testing and commissioning of the system. The data loggers were installed in a heated stainless steel cabinet. Installation of the anemometer on the external structure was carried out by a specially trained rope access technician. Due to space limitations, only representative photographs showing key features of the system are shown in Figures 2, 3, and 4.

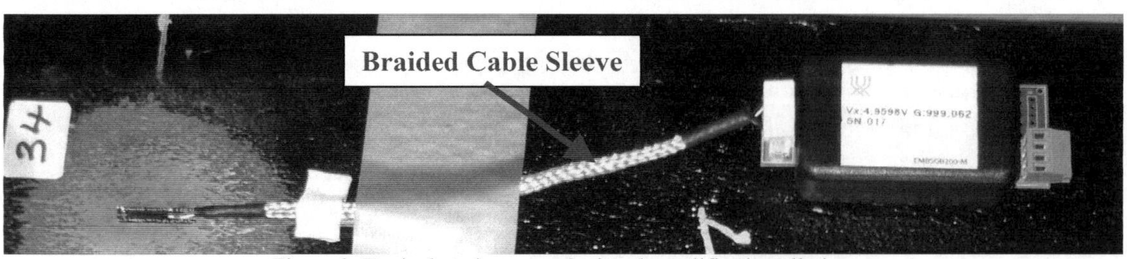

Figure 2. Typical strain gauge & signal amplifier installation

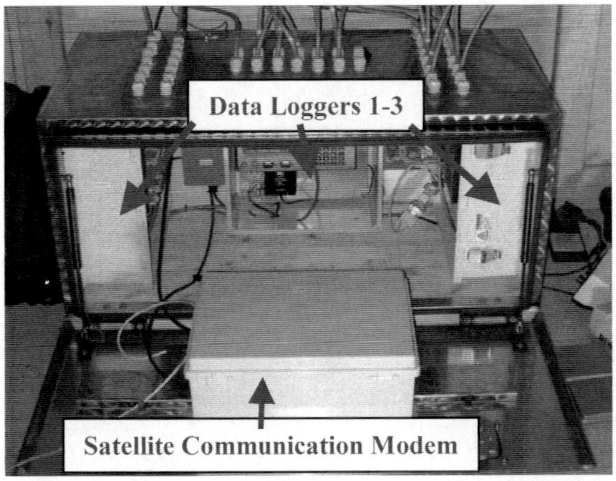

Figure 3. Data logger cabinet

Figure 4. Anemometer Installation

2.2.3. Data Collection and Recording

The specified rotation period for the radars is 12 seconds. To accurately measure changes in strain it was decided to acquire and record a strain data set at each 3 degrees of rotation, meaning 120 strain readings would be recorded in one rotation of the radar. This yields 100 ms sampling rate. The 3° angular resolution gives enough granularity of data to recognize anomalies in the behaviour of the structure while it does not generate excessive amount of data (see discussion in 2.2.4).
To improve data quality, strain values from five consecutive data sets (n = 5) were recorded and averaged for each 3° position of the antenna (Figure 5).

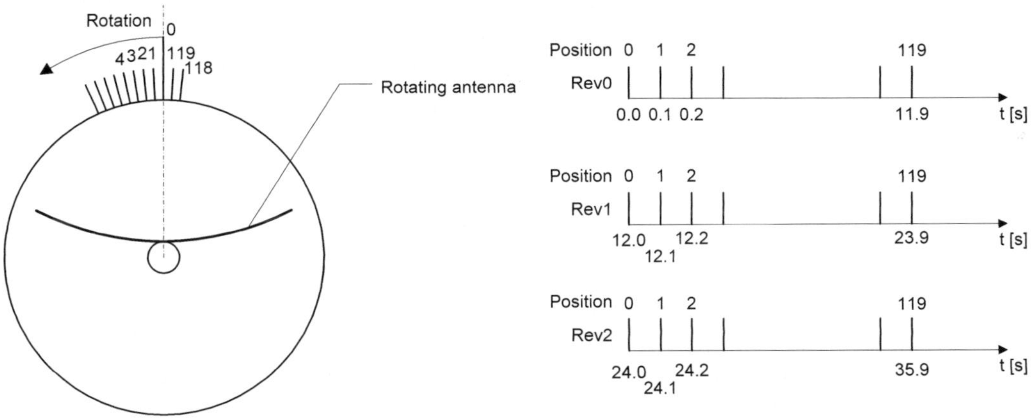

$$\text{Average}[0] = \frac{\text{Rev0}[0]+\text{Rev1}[0]+\text{Rev2}[0]+\ldots+\text{Rev}(n-1)[0]}{n}$$

$$\text{Average}[1] = \frac{\text{Rev0}[1]+\text{Rev1}[1]+\text{Rev2}[1]+\ldots+\text{Rev}(n-1)[1]}{n}$$

$$\text{Average}[2] = \frac{\text{Rev0}[2]+\text{Rev1}[2]+\text{Rev2}[2]+\ldots+\text{Rev}(n-1)[2]}{n}$$

$$\vdots$$

$$\text{Average}[119] = \frac{\text{Rev0}[119]+\text{Rev1}[119]+\text{Rev2}[119]+\ldots+\text{Rev}(n-1)[119]}{n}$$

Rev = one revolution of the antenna.
Averaging over n revolutions.
The indexes correspond to the numbered angular positions.

Figure 5. Raw data averaging protocol

This procedure was repeated on an hourly basis. One data set (record) included strain, ambient temperature, wind velocity and direction, data logger temperature, and data logger battery voltage. Each data record was time-stamped by the data logger.
The data were locally stored on the data logger and regularly transmitted to the base station in Halifax. To account for the possibility of interruption in data transmission, 1 GB storage capacity was provided in each data logger for additional data storage.

2.2.4. Data Transmission

Transmitting large amounts of data through wired and cellular data communication is relatively inexpensive. However, in the Canadian Arctic Region, there is no Internet or cellular communication available with the exception of a few local communities. After exploring other communication options, a satellite based data transmission option was found to be the most practical. Data transmission via satellites is an expensive option that has to be managed judiciously to transfer desired information at reasonable cost. This means restricting the amount of data that should be transmitted.
The communication costs are incurred on a per-second basis and start when the data logger system connects to a satellite orbiting overhead and end when the system signs off. Some percentage of the total satellite time billed is expended in engaging and disengaging with the satellite and is non-productive, i.e. no data are transmitted during that period. So, data transmission effected at closer intervals would result in greater amount of unproductive time as compared to larger amount of data transmitted at greater intervals. Ideally, the data could be transmitted on a daily basis if uninterrupted satellite contact could be maintained for longer periods of time. However, a significant number of satellites orbiting over the polar region tend to turn their communication mode off to save on power. Hence the opportunities to find larger communication windows are limited. There is also a risk of losing connection during longer transmission periods. Following several trials, it was determined that data transmission effected every 6 or 8 hours provided the most cost-effective solution.

3. Results

Data monitoring was conducted between October 2013 and March 2014. Data were collected and initially evaluated by SHM Canada for completeness and any anomalies, and then input to a Microsoft Excel file and sent to CJE weekly for further processing and evaluation.
Over this 5-month period the data collected indicated that the exterior temperature ranged from +20°C to -32°C and recorded wind speeds reached as high as 112 km/h. In this timeframe the antenna was temporarily shut down for maintenance and antenna bearing replacement (an event that took several days, involving several iterations of levelling and adjusting connections).
Upon commissioning the system, it was immediately apparent that the 12 second cycle of the rotating antenna was a dominant feature in every strain gauge data set. It was determined that for each strain gauge sensor data record there were two primary attributes – the *average stress per cycle* and *stress range per cycle*, and these 2 attributes formed the basis for detailed data evaluation.

3.1. Effects of Antenna Rotation

As mentioned above, the 12-second cycle of the rotating antenna was the dominant feature of all strain gauge data sets – typically producing a sinusoidal curve when stress was plotted vs. time (Figure 6). Data from a few sensors also showed significant noise – particularly from those located close to the motor (Figure 7). Due to the averaging methods employed in data sampling, it was not possible to investigate these higher frequency effects further.

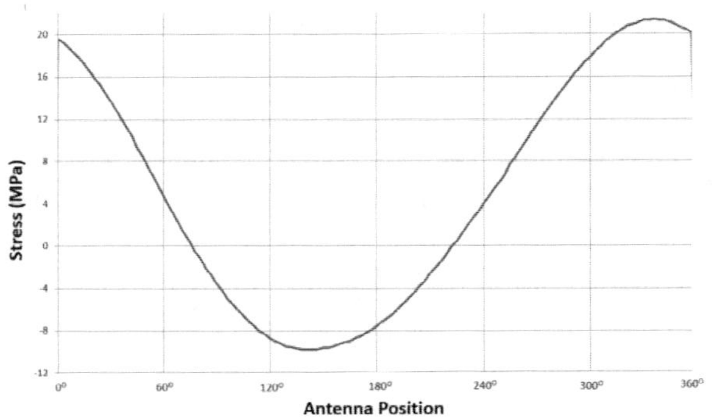

Figure 6. Typical stress cycle

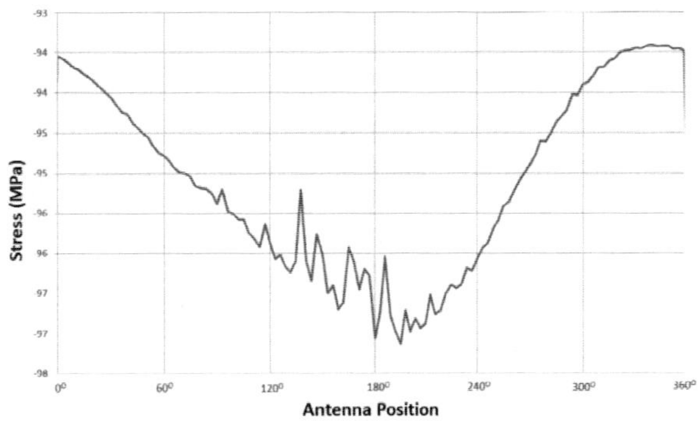

Figure 7. Cycle from sensor located near motor

3.1.1. Validation of Finite Element Model & Anomalies

Stress ranges calculated at each of the 32 strain gauge locations by the FE model for a typical cycle of antenna rotation were compared with stress ranges measured by the strain gauges (stress range was generally consistent over the 5 months, although some variations were seen as will be discussed below).

In general, the measured and calculated values were in very close agreement. The difference between calculated and measured stress ranges at 14 of the 32 gauges was within 1.0 MPa, 27 of the 32 gauges were within 2.5 MPa, and 3 of the remaining 5 gauges were within 4.0 MPa. This was thought to be very close, particularly considering (a) the accuracy of the gauges (+/- 0.4 MPa), (b) the simplifications inherent in the assumptions of any linear FE model, and (c) that the lowest fatigue stress threshold for welded connections (i.e. level of interest) was 31 MPa.

The difference between calculated and measured stress ranges due to antenna rotation at the remaining two gauges was within 10 MPa (for example at one, calculated stress range was 22.8 MPa, whereas measured stress range was 30.8 MPa before the antenna bearing was replaced, and 34.0 MPa after the bearing replacement).

These two gauges were located in the same area of the structure (above and below a support point – Figure 8). The higher than expected stress ranges were indicative that something unexpected was occurring at this corner (the structure was flexing and stressing more than expected), and this subsequently became the focus of further study. The SHM project was therefore fundamental to identifying that this area of the structure that as experiencing unexpectedly high cyclical stresses, so that action could be taken before damage (e.g. fatigue cracking) occurred.

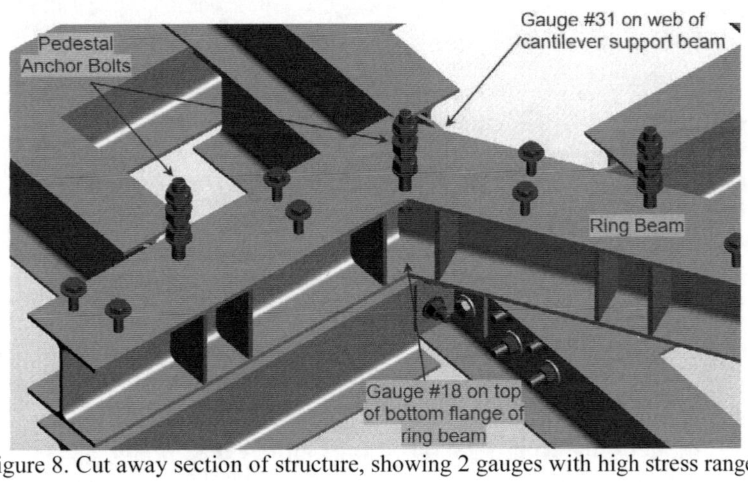

Figure 8. Cut away section of structure, showing 2 gauges with high stress ranges

3.2. Effects of Wind

Both wind speed and direction data (3-minute moving average) were transmitted every 10 minutes. As the strain gauge sampling frequency was not the same (average of five 12 second cycles once per hour), this hampered the data comparison. It was also not possible to determine whether the wind speed recorded was a steady speed, or a gust. Wind sampling protocol may be revised for future studies. Over the 5-month period, there were many windy periods – with multiple days exceeding 40 km/h, 5 days when the wind speed exceeded 80 km/h, up to 112 km/h (Figure 9).

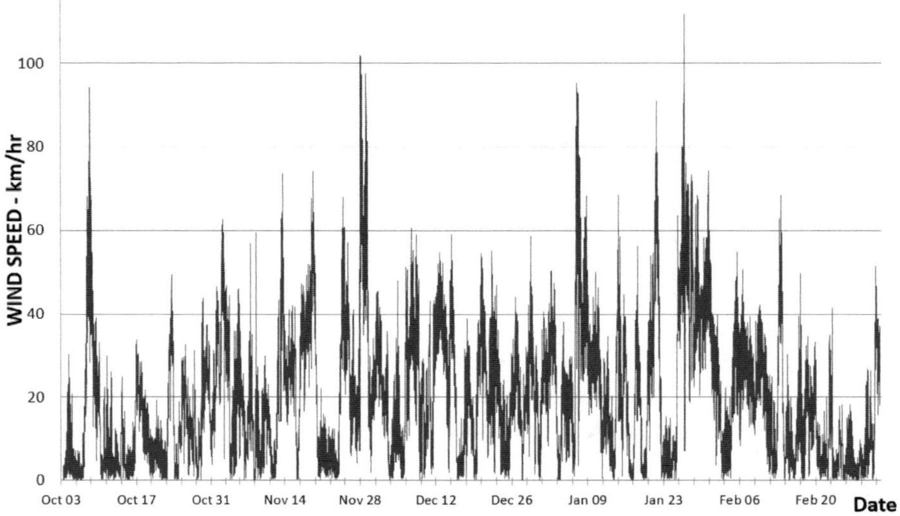

Figure 9. Recorded wind speed

To correct for potential effects of different wind directions, we focussed on individual wind events (each perhaps a day or two long) – looking at conditions before (when it was calm), during the windy period, and again after (when calm conditions had returned). Wind direction was found to be fairly consistent throughout each event (e.g. Figure 10).

October 8, 2013

Time of Day	Average Wind Speed	Average Wind Direction
00:00 - 06:00	51 km/h	208°
06:00 - 12:00	57 km/h	189°
12:00 - 18:00	50 km/h	193°
18:00 - 24:00	37 km/h	235°

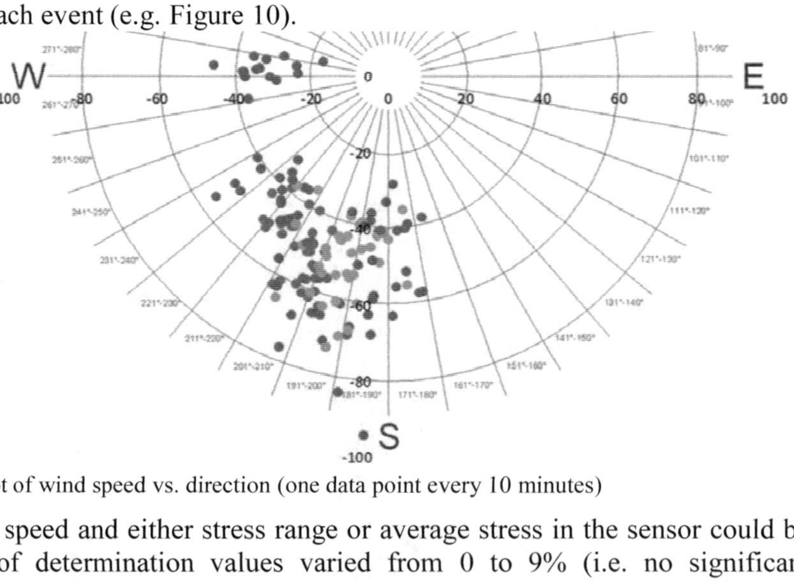

Figure 10. Plot of wind speed vs. direction (one data point every 10 minutes)

No correlation between wind speed and either stress range or average stress in the sensor could be identified – R^2 coefficient of determination values varied from 0 to 9% (i.e. no significant correlation). This is consistent with the conclusions of previous studies by the author and the predictions of the FE model – that the wind has a negligible effect on stresses in the radar base and the structure immediately below it.

3.3. Effects of Temperature

Temperature sensors were installed both on the exterior of the structure, and within the unheated radome enclosing the antenna. The two temperature readings were very strongly correlated – with a coefficient of determination (R^2) of 95%. Correlation of average stress (per cycle) with temperature varied – with R^2 values varying from 3% to 92% depending on sensor location on the structure (example in Figure 11 below). Unsurprisingly, there was no significant correlation of stress range (per cycle) to temperature.

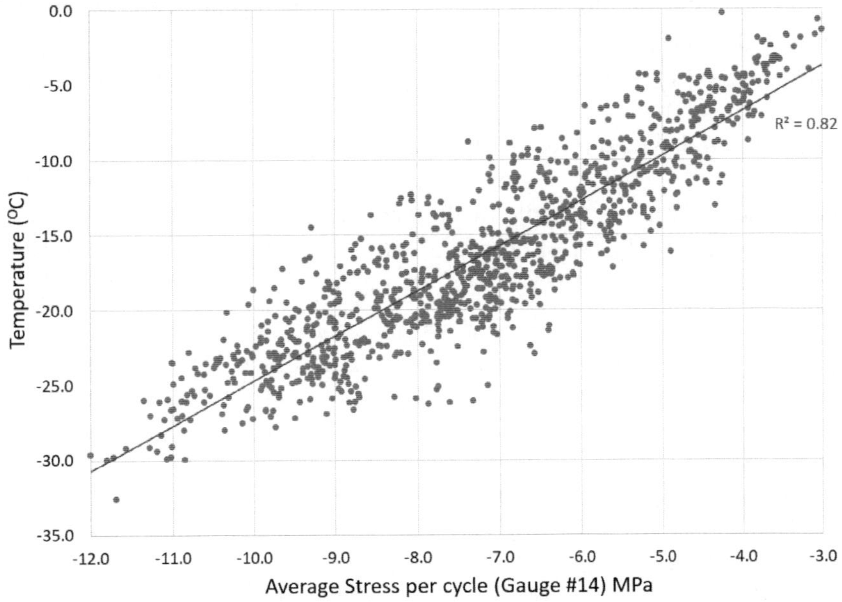

Figure 11. Typical correlation of average stress per cycle vs. temperature

The stress variation attributable to temperature change was typically low – for a 20°C temperature change the average stress variation was 4.7 MPa, with only 2 gauges indicating variations exceeding 8 MPa (14 & 16 MPa) under this change. Considering the low frequency of temperature changes (particularly in comparison to antenna rotation), thermal variations and average stress per cycle were not considered to be significant factors in the consideration of fatigue.

The correlation of temperature-related stress variations predicted by the FE model vs. measured was generally poor – with R^2 correlation values varying from 3% to 88%. This could be because the FE model was overly simplistic is assuming that all elements experienced that same temperature variation simultaneously, or because the temperature sensors did not measure the temperature at the strain sensors. However, as measured stress ranges were typically lower than those predicted by the model and neither the FE model nor the SHM data indicated that temperature effects were relevant to the key goals of the project, this was not investigated further.

3.4. Effects of Maintenance

Mid-way through the monitoring period the antenna bearing was replaced – an operation that takes several days and requires the connection between the antenna and the support structure to be adjusted and re-levelled.

Throughout this period many of the gauges indicated large stress variations – particularly as the connections were being adjusted. Some gauges indicated short-term stress changes up to 200 MPa, and 3 gauges indicated an increase in cyclical stress range after the adjustments exceeding 2 MPa.

Overall, the magnitude of the stress variations was surprising. The SHM program clearly validated the initial hypothesis and demonstrated that the installation methodologies and maintenance procedures were distorting the structure and, in turn, the bearing mounting surfaces. This is likely to be a major contributing factor to premature wear.

3.5. Lessons learned

Analysis of measured data (at strain gauge locations) demonstrated that:

- Stress ranges were generally consistent throughout the project (with a few locations being affected by maintenance procedures).
- Changes in average stress in most gauges were strongly correlated with exterior temperature changes.
- Wind was found to have no measurable effect on either stress range or average stress.
- The finite element model is a good representation of the structure and tower/antenna interface, particularly to predictable loading conditions (e.g. rotating equipment).
- The FE model was unreliable in predicting stress variations due to temperature effects, but stress variations were typically less than calculated.

Significance of results - From the owner/operators' point of view, the project demonstrated that:

- Wind and temperature variations do not have a significant effect on the structure directly beneath the radar antenna.
- Rotation of the radar antenna is the dominant load condition causing cyclical stresses in the structure below.
- Cyclical stresses due to antenna rotation can generally be predicted by the FE model, but structural anomalies exist that can amplify the stress ranges. The strain monitoring program clearly identified one area of the structure that was not fully supported and was experiencing higher than anticipated stress variations that could be addressed before leading to premature deterioration or fatigue cracking.
- The existence of gaps between steel connection surfaces presents significant potential for distortion of the structure and the antenna – causing large variations in both average stress and stress ranges – and likely a major contributing factor to premature bearing wear.
- Sampling rate of wind data was too low to correlate wind gusts with strain data. This must be improved in future designs.

4. Conclusions

The custom-designed SHM system proved to be reliable and accurate enough to deliver data useful for the study, without any human intervention at the remote site. The custom-designed and written program was able to handle transmission interruptions that are inherent to satellite communications. Program parameters were remotely adjustable from Halifax. Satellite communications costs were within acceptable limits for the whole 5-month period.

In conjunction with the finite element modelling of the tower structure, the structural health monitoring project was very useful in furthering the understanding of how this complex structure behaves and responds to varying load conditions. Fortunately, over the 5-month monitoring period the structure experienced a variety of different conditions – including winds up to 112 km/h, temperature variations over 50°C, and regular maintenance of the antenna.

The results demonstrate that SHM data were instrumental in identifying weaknesses and anomalies in the structural system. The monitoring program also provided a basis to validate the analytical model through comparison of the measured strains with the values derived from the analysis.

The 8th International Conference on
Structural Health Monitoring of Intelligent Infrastructure
Brisbane, Australia | 5-8 December 2017

Deflection Measurement of Concrete Bridge by Moiré Imaging Technique

S. Ri [1], H.Tsuda [1], T. Narita [2], K. Yamada [2], Y. Hayasaka [3], M. Kobayashi [3].
[1] *National Institute of Advanced Industrial Science and Technology – Japan, Email:* ri-shien@aist.go.jp
[2] *East Nippon Expressway Company Ltd. - Japan,* [3]*NEXCO-Engineering Tohoku Company Ltd. - Japan*

Abstract

In this study, we applied the sampling Moiré method for measuring the deflection of the concrete bridges using a digital camera and grating markers. The applicability and effectiveness of the present method were verified through a field experiment of a new constructed prestressed reinforced concrete (PRC) bridge. In the field experiment, one or two 20 tons vehicles run at a speed of 60 km/h over the 35 meters long PRC bridge. The deflection measured by the convetional ring-type displacement sensor and our Moiré imaging technique was compared. The displacement measured from our imaging method was very close to that measured with the convetional ring-type displacement sensor. Both methods measured a deflection of 0.6 mm at the center of the bridge when a 20 tons truck passed over the central location. This optical displacement measurement method is useful for structural health monitoring of various infrastructures such as long bridges and high buildings.

1. Introduction

Measuring the static and dynamic displacement of in-service structures is an important issue for the purposes of design validation, performance monitoring, and safety assessment. For instance, a long-term continuous monitoring of structures permits constant safety oversight and an immediate warning can be issued when unusual displacements is detected. The Global Positioning System (GPS) (Xu *et al.* 2002) and laser Doppler vibrometer (LDV) (Nassift, Gindy & Davis 2005) have successfully been applied for monitoring the structural health of large-scale structures. GPS can provide real-time displacement measurements with a millimeter-level accuracy. LDV can provide accurate displacement measurement within its applicable range. These instruments, however, are basically point measurement systems and are quite costly for distribution measurement.

Since about 40% (3,700 km) of the total length of highway bridges in Japan (9,000 km) has been in-service for over 30 years, the deflection measurement technique for concrete bridges with low-cost and high accuracy is required. Most of the existing structural health monitoring (SHM) systems are using contact-type displacement sensors and/or accelerometers. These systems are highly expensive to operate, mainly due to cumbersome and complex installation of sensors and their data acquisition systems. As for deflection measurement from conventional ring displacement sensors, it was required several days for constructing scaffolding and equipping piano wires and the sensors. In addition, a strain-to-displacement calibration procedure is needed before measurement because the sensor measure not displacement but strain. One other main bottleneck lies in that multi-point measurement is difficult to be performed since the cost and labor are directly proportional to the number of the sensors.

Image/optical measurement techniques have been recently paid much attention from viewpoints of low-cost and multi-point sensing ability. Wahbeh et al. (Wahbeh *et al.* 2003) presented a vision-base approach for measuring displacements of vibrating systems. In this method, a high-resolution digital camera was used to recorded two high-resolution red lights (LED) mounted on the objects to be measured and a nonlinear Gaussian regression curve fit was utilized to determine the center of the red spot of the LEDs. Lee and Shinozuka (Lee & Shinozuka 2006) also presented a vision-based system for remote displacement sensing of bridges by image processing. The used targets had four white spots with known geometry and black background and the centers of four white spots were determined from the black and white image. In these image processing techniques, however, the accuracy of detection of the central position of spots is limited. Digital image correlation (DIC) technique is another kind of image processing technique for measuring displacement distributions (Sutton *et al.* 1986). DIC technique using random patterns or natural patterns can obtain the deformation of surface by comparison of the digital images before and after deformation. Recently the DIC technique was applied to measure the deflection of a bridge in service (Yoneyama *et al.* 2007). In the application of civil engineering, D. Feng *et al.* have proposed vision-based structural health monitoring called vision sensor. They applied the vision sensors to field tests of the Manhattan ridge whose maximum span length was 415 meters during train passing (Feng & Feng 2017). They also measured the response of a 16.75 meters plate girder railway bridge to passing freight trains (Feng & Feng 2015). I. Choi *et al.* have proposed a target-less vision based displacement sensor (TVDS) through image convex hull optimization (Choi et al. 2016). T. Schumacher *et al.* have proposed a monitoring methodology based sensor (VVS) that utilizes Eulerian specification by the intensity analysis. They performed a field test of a bridge in Oregon (Schumacher & Shariati 2013). K. Walker *et al.* have applied the Eulerian-based virtual visual sensors (VVS) based on the intensity monitoring to health monitoring of a 21 meters long USDA Forest Service pedestrian wood-made bridge in western Oregon (Walker et al. 2018).

Recently, a fast, simple, and robust phase measurement technique called the sampling Moiré method (Ri, Fujigaki & Morimoto 2010) has been developed for accurate small-displacement distribution measurements. In this method, a grating pattern with a known pitch attached on an object to be monitored is captured with a digital camera. Then, the phase distribution of Moiré fringes is analyzed from the grating image through image processing such as down-sampling and intensity-interpolation. The displacement distribution can be evaluated from the phase difference of the Moiré fringe before and after deformation. In this Moiré imaging technique, required preparation is only attaching grating markers on the target and setting a digital camera to record the grating image. From the simple preparation, this technique allows low cost but accurate displacement at multi-point simultaneously. Whereas the DIC method works with a random pattern, the sampling Moiré method does just the opposite, reducing processing complexity by requiring the structure to incorporate a regular high-constrast grating pattern.

Despite the aforementioned advances, study of phase-based imaging measurement technique is still at an early stage. As an effort to explore the potentials of the sampling Moiré method for low-cost SHM, this study aimed to demonstrate the usefulness of the optical displacement method. Field experiments of a newly constructed PRC bridge were conducted to confirm the remote, dynamic and multi-point measurement capability of the image-based measurement technique using a single digital camera.

2. Principle of Displacement Measurement

The sampling Moiré method is an effective phase analysis technique that relies on the phase distribution of the Moire fringe from a single image by utilizing the Moire effect. In this study, we applied the sampling Moire method to measure the deflection of a new bridge.

2.1 Down-sampling and intensity interpolation for generating Moiré fringes

The basic principle of the sampling Moire method based on the down-sampling and intensity interpolation image processing is illustrated in Fig. 1. First, when a regular grating marker with pitch p (unit: mm) is captured with a CCD camera as shown in Fig. 1(a), the recorded intensity of the grating [Fig. 1(b)] is presented in Eq. (1).

$$I(i,j)=I_a\cos(2\pi i/P+\varphi_0)+I_b=I_a\cos[\varphi(i,j)]+I_b \tag{1}$$

where I_a and I_b represent the amplitude of the grating intensity and background intensity. P is the captured grating pitch as pixel unit in the digital image, and $\varphi(i,j)$ is the phase value of the grating.

In the sampling Moiré method, multiple phase-shifted Moiré fringes can be obtained by down-sampling the recorded grating image. In Fig. 1(b), as example, the sampling pitch is 4-pixel. When every four pixels from the first sampling point is picked up, a Moiré fringe with pitch $P\times T/|T-P|$ can be obtained. If the second, third, and fourth sampling points are selected as the sampling start point, multiple phase-shifted Moiré fringes can be obtained, as shown in Fig. 1(c). This process corresponds to the phase shifting. Next, by performing an image interpolation using neighbouring sampled intensity data, smooth phase-shifted Moiré fringes can be recovered. The k-th phase-shifted Moiré fringe image, shown in Fig. 1(d), can be expressed as follows:

$$I_m(i,j;k)=I_a\cos\left[2\pi\left(\frac{1}{P}-\frac{1}{T}\right)i+\varphi_0\right]+I_b=I_a\cos\left[\varphi_m(i,j)+2\pi\frac{k}{T}\right]+I_b \tag{2}$$

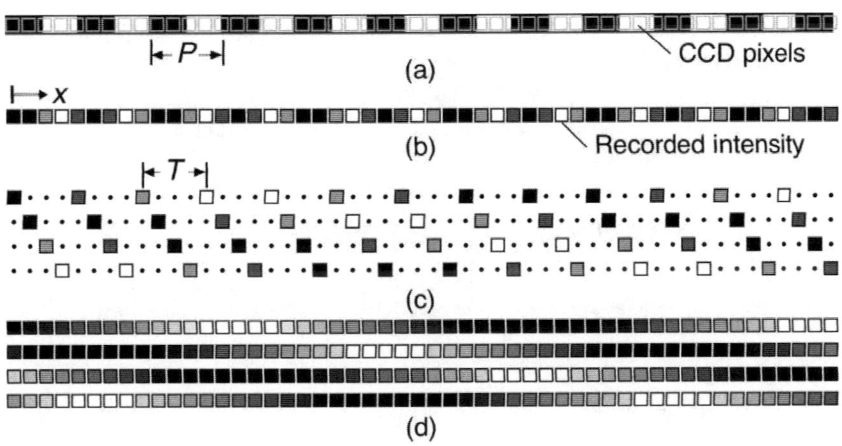

Fig. 1 Principle and image processing procedure of the sampling Moiré method

2.2 Phase analysis by phase-shifting and discrete Fourier transform

Using the multiple phase-shifted Moiré fringes, the phase value of the Moiré fringe can be calculated by phase-shifting method using discrete Fourier transform (DFT) algorithm, as presented in Eq. (3).

$$\varphi_m(i,j) = -\tan\frac{\sum_{k=0}^{T-1} I_k(i,j;k)\sin(2\pi\frac{k}{T})}{\sum_{k=0}^{T-1} I_k(i,j;k)\cos(2\pi\frac{k}{T})} \tag{3}$$

2.3 Displacement calculation from phase difference of Moiré fringe

The phase distribution of the Moiré fringe after deformation can be obtained in the same manner. The displacement can be obtained from the *phase difference* of the Moiré fringes $\Delta\varphi_m(i,j) = \varphi_m'(i,j) - \varphi_m(i,j)$ obtained from the grating image before and after deformation. The displacement is directly proportional to the phase difference of the Moiré fringe as follows.

$$u(i,j) = -\frac{p}{2\pi}\Delta\varphi_m(i,j) \tag{4}$$

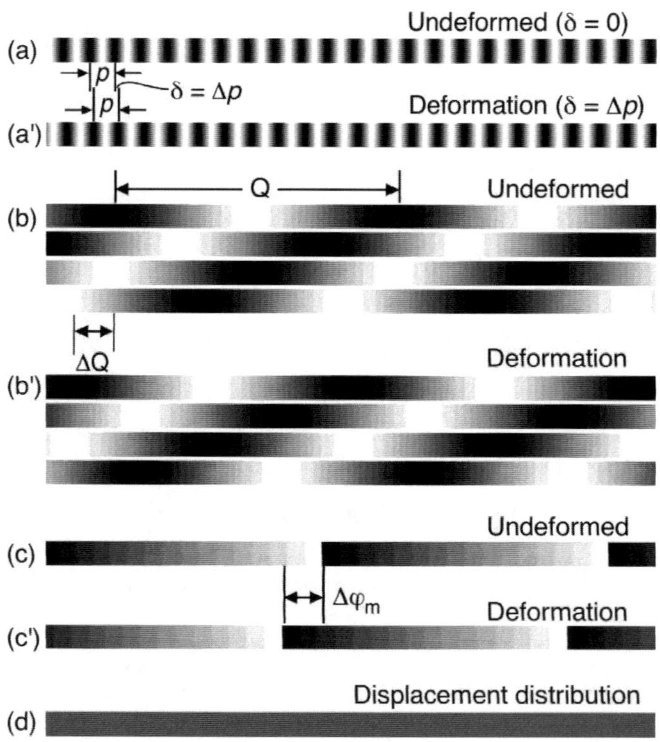

Fig. 2 Displacement measurement by the sampling Moiré method

Figure 2 shows an example of the displacement distribution measurement based on the sampling Moiré method by computer simulation. From our early studies, the measurement accuracy of the sampling Moiré method can be rearched to 1/500 of the grating pitch. Becuase the displacement is measured from the phase information instead of intensity information in the sampling Moiré method, it much robust to light change during measurment. For the field experiment, currently, we can only measurement accurately the displacement within 50 m distance between the camera and the structures. This imaging method is weak to thick fog and heat haze for a bad weather. To remove the effect on fog and heat haze, it is our future work.

3. Field Experiment For Displacement Measurement of a New Bridge

We conducted this field experiment to compare the displacement measurement accuracy of a conventional contact-type displacement sensor and the image-based displacement measurement using the sampling Moiré technique.

3.1 Bridge and Loading Conditions

Figure 3 shows schematics of a prestressed reinforced concrete (PRC) bridge on which vehicle running tests were conducted. The total length of the bridge is 138.5 meters and the width is 10 meters. In this tests, we measured the deflection of the 35.1 meters long section between A1 and P1, as illustrated in left side of Fig. 3. Both dynamic and static loading experiments were carried out. In the dynamic test, a 22 tons water truck passed through the bridge at a speed of 60km/h, as shown in Fig. 3(a). In the static test, a 22 tons truck or two trucks of 42 tons of total weight were used for loading. The deflection between the section A1-P1 was measured when these trucks were stopped at the 1/4 point (quarter-span) or the 1/2 point (mid-span) of the section A1-P1, or the 1/2 point of the section P1-P2 (the next mid-span), as shown in Fig. 3(b).

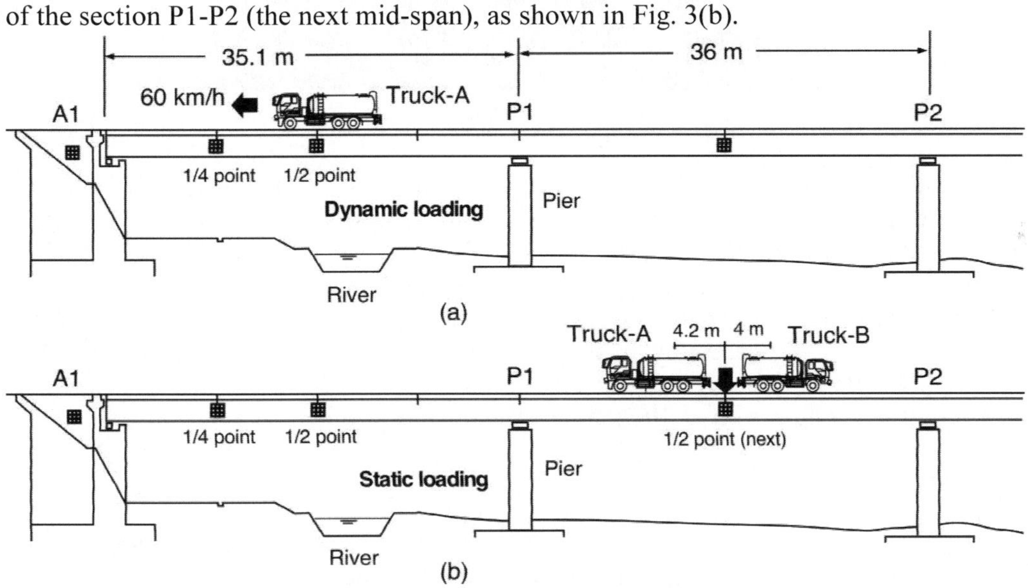

Fig. 3 Schematics of the tested bridge and loading conditions at (a) dynamic and (b) static tests

3.2 Experimental Setup

Figure 4 shows a photograph of the field experiment for measuring deflection of a new highway PRC bridge. Before measurement, we attached 11 Moiré markers with a 100 mm grating pitch on the bridge surface. As denoted the parts of b and c in Fig. 4, two ring-type displacement sensors (Tokyo Sokki Kenkyujo; OU-30) were installed at the 1/4 and 1/2 points of the section A1-P1.

Fig. 4 Photograph of field experiment for measuring deflection of a new highway PRC bridge: a. Moiré marker; b. ring-type displacement sensor; c. wiring of the sensor

Fig. 5 Photograph of the test water trucks.

Figure 5 shows a photograph of the trucks used for loading. The size and weight of the truck-A are 12 m (L)←3 m (W) ←3.8 m (H) and 22 ton, respectively. The size and weight of the truck-B are 9 m (L)←2.5 m (W) ←3.7 m (H) and 20 ton, respectively. To record the grating image in static measurement and a movie in dynamic measurement, a cinema DSRL camera (Canon, EOS-1D C; 4096←2016 pixel @ 24fps) attached with a prime lens (Canon, CN-E24mm T1.5 LF; focus length, f = 24 mm) was used. The parameter of image recording are: ISO = 100, Aperture = 4, shutter speed = 1/60 s. The distance between the camera to the bridge was about 26 m. In the displacement analysis, the grating pitch was set as 100 mm and the down-sampling pitch was 19-pxiel.

3.3 Experimental Results

Figure 6 shows the measured displacement vector in y-direction at 11 points under the dynamic deflection measurement. When the truck passed near the center of the bridge, a bow-shape deflection distribution was observed. The experimental results of dynamic deflection measurement at 1/4 point and 1/2 point in time series are plotted in Figs. 7(a) and 7(b), respectively. The black and red lines indicated the results of measured displacement from conventional sensor and our method. The displacement results obtained by the sampling Moiré method corresponded well with the result obtained from the conventional contact-type displacement sensor. Figure 8 shows the experimental results of the static deflection measurement of the bridge at different locations, i.e., 1/4 point, 1/2 point and the next 1/2 point of the total span of the bridge. When the trucks placed at the next 1/2 point, opposite deflection was measured. Because the truck-A and truck-B have a 12-m-long and 9-m-long, the structure deformation behave is not simply linear according to the one or two distributed loading.

Fig. 6 Measured displacement vector in y-direction under the dynamic deflection measurement

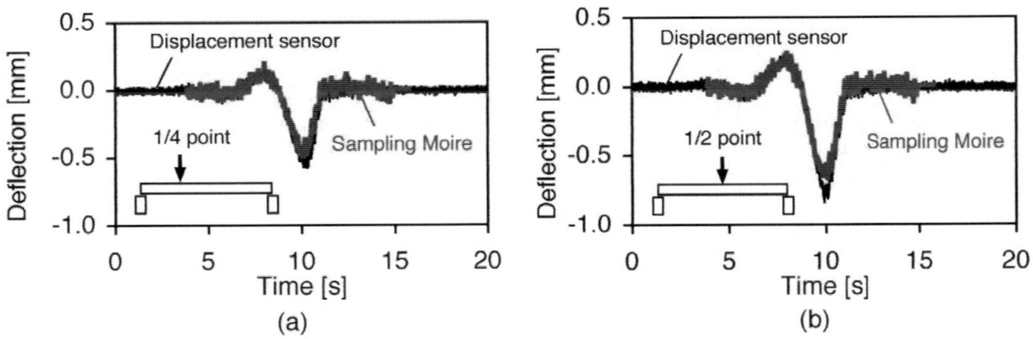

Fig. 7 Experimental results of dynamic deflection measurement in time series when a 22-ton truck passing through the bridge at the speed of 60 km/h

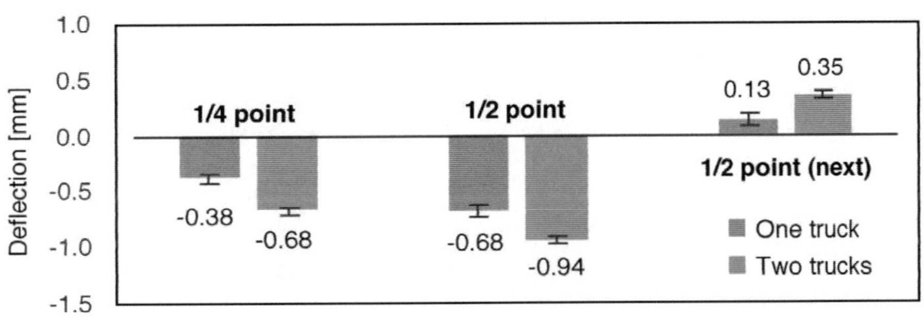

Fig. 8 Experimental results of static deflection measurement at 1/2 point when the trucks stopped at the location of 1/4 point, 1/2 point and the next 1/2 point.

4. Conclusions

Study of the phase-based imaging displacement measurement method for low-cost SHM of civil engineering structures such as long bridge or high building is still at an early stage. While existing studies mostly focus on the measurement accuracy evaluation, this study validated the usefulness of the imaging method for practical application by using a digital camera and an advanced Moiré methodology. The remote and multi-point measurement capability of the sampling Moiré method was confirmed through field tests of a newly constructed PRC bridge with a camera set 26 m away from the bridge. Field experiments demonstrated that both static and dynamic displacement at multiple points could be simultaneously and accurately measured using a single digital camera and grating markers.

References

Choi, I., Kim, J & Kim, D 2016, 'A target-less vision-based displacement sensor based on image convex hull optimization for measuring the dynamic response of building structures', *Sensors*, vol. 16, 2085 (17pp).

Feng, D & Feng, MQ 2015, 'Model updating of railway bridge using in situ dynamic displacement measurement under trainloads', *J. Bridge Eng.*, vol. 20, no. 12, 04015019 (12pp).

Feng, D & Feng, MQ 2017, 'Experimental validation of cost-effective vision-based structural health monitoring', *Mechanical Systems and Signal Processing*, vol. 88, pp. 199-211.

Lee, JJ & Shinozuka, M 2006, 'A vision-based system for remote sensing of bridge displacement', *NDT & E Int.*, vol. 39, pp. 425-432.

Nassift, HH., Gindy, M & Davis, J 2005, 'Comparison of laser Doppler vibrometer with contact sensors for monitoring bridge deflection and vibration', *NDT & E Int.*, vol. 38, pp. 213-218.

Ri, S., Fujigaki, M & Morimoto, Y 2010, 'Sampling Moiré method for accurate small deformation distribution measurement', *Experimental Mechanics*, vol. 50, no. 4, pp. 501-508.

Schumacher, T & Shariati A 2013, 'Monitoring of structures and mechanical systems using virtual visual sensors for video analysis: fundamental concept and proof of feasibility', *Sensors*, vol. 13, pp. 16551-16564.

Sutton, MA., Cheng, M., Peters, WH., Chao, YJ & McNeil, SR 1986, 'Application of an optimized digital correlation method to planar deformation analysis', *Image and Vision Computing*, vol. 4, no. 3, pp. 143-150.

Wahbeh, AM., Caffrey, JP & Masri SF 2003, 'A vision-based approach for the direct measurement of displacements in vibrating systems', *Smart Materials and Structures*, vol. 12, pp. 785-794.

Walker, K., Miller, T.H., Gupta, R., Shariati, A & Schumacher, T 2018, 'Development of virtual visual sensor applications for wood structural health monitoring', *Journal of Testing and Evaluation*, vol.46, no. 1, to be published.

Xu, L., Guo, JJ & Jiang JJ 2002, 'Time-frequency analysis of a suspension bridge based on GPS', *J Sound Vib.*, vol. 254, no. 1, pp. 105-116.

Yoneyama, S., Kitagawa, A., Iwara, S., Tani, K & Kikuta, H 2007, 'Bridge deflection measurement using digital image correlation', *Experimental Technique*, vol. 31, no. 1, pp. 34-40.

The 8th International Conference on
Structural Health Monitoring of Intelligent Infrastructure
Brisbane, Australia | 5-8 December 2017

Long-Term Structural Health Monitoring of Suramadu Bridge

M.R. Salami[1], A. Jesus[2], A.P. Pamuncak[3], I. Zarkasi[4], B. Suhendro[5], B. Budiono[6], P. Suprobo[7], I. Laory[8]

[1] Research Fellow, University of Warwick – United Kingdom, Email: m.salami@warwick.ac.uk,
PhD student, University of Warwick – *United Kingdom[2], Indonesian Ministry of Public Works and Housing – Indonesia[3], Indonesian Ministry of Public Works and Housing – Indonesia[4], Gadjah Mada University – Indonesia[5], Bandung Institute of Technology – Indonesia[6], Sepuluh Nopember Institute of Technology – Indonesia[7], Assistant Professor, University of Warwick – United Kingdom[8]*

Abstract

Infrastructure systems deteriorate while in service due to routine use and exposure to extreme conditions (e.g. corrosive environment and dynamic loads). Many countries tackle with the serviceability of their infrastructure and failing to raise sufficient funding to repair and replace them. In order to optimize the maintenance, long-term reliability and life-cycle cost of infrastructure systems, Structural Health Monitoring (SHM) has received significant public attention in recent years. This paper reports on SHM and data measurement of the Suramadu Bridge, in Indonesia. Data from accelerometer, anemometer, thermocouple, tiltmeter, weight-in-motion (WIM) and electro-magnetic (EM) sensors were recorded since 2012 to 2015, considered in this paper which provide an extensive database for long-term monitoring of the bridge. In addition, a fully detailed finite element (FE) model was developed using SAP2000 and its dynamic behaviour compared with modal analysis. This information collected from the monitoring system in parallel with the FE model can be used for anomaly detection, performance assessment and maintenance planning of the bridge.

1. Introduction

In recent years, sensors have increasingly been used to collect data concerning the actual loading, environmental conditions and responses of civil infrastructure (Jang et al. 2010; Cho et al. 2010; Zhu et al. 2010; Nguyen et al. 2013). The main underlying idea behind SHM approaches is that the data obtained through monitoring systems can be used to reduce the uncertainties associated with the deterioration and loading processes (i.e. operational and environmental conditions) of structures. This can result in more accurate structural performance assessments and predictions, paving the way towards a better allocation of limited funds for maintenance purposes and to a timely detection of abnormal behaviors, thus preventing failures (Brownjohn 2007).

Irrespective of the SHM approach taken, some common challenges exist. One of the most important is the changing environmental (temperature, humidity, wind, etc.) and operational conditions (live loads, etc.) to which structures are subjected. These variations have an impact on the measured signals, making data interpretation a complex task. In effect, failure to quantify the input-output relationship between environmental and operational conditions, on the one hand, and structural responses (e.g. strains, displacements) or derived quantities (e.g. natural frequencies), on the other, can make it impossible to distinguish structural changes of interest due to damage/deterioration from changes induced by operational and environmental conditions (Sohn, 2007). Therefore it is warranted to investigate the effect of environmental condition on SHM. This paper is a part of an extensive research on SHM of Suramadu bridge in Indonesia. The objective of this paper is to present the measured data, which will be used in future studies regarded to system identification, anomaly detection and performance assessment of the bridge. This paper consists of a brief description of the Suramadu bridge, its monitoring system, presentation of measured and post-processed data and a detailed FE model.

2. Bridge Description

Suramadu Bridge is a cable-stayed bridge which is part of a 5.4km long connection between Madura Island and Surabaya on the island of Java in Indonesia. Constructed over six years (2003–2009). For future reference in this work, X axis represents Surabaya-Madura (south-north) direction, Y-axis represents Gresik-Pasuruan (west-east), and Z axis represents vertical direction. There are 3 parts of this bridge. The first one is fly-over on both side, covering 1458m from Surabaya and 1818m from Madura. The second part is the connectors (approach bridge) of 627m each. The last part is a central bridge which consists of two 192m side-extensions and a 434m main span. Figure 1 shows the schematic representation of the central Suramadu Bridge alongside with the location of sensors used in this paper. The bridge comprises two symmetrical cable-stayed cantilever sections. The deck is 30m wide and comprises two longitudinal steel girders, approximately 2.8m deep, stringers and floor beams and a reinforced concrete slab with 250mm in depth. The cable plane is aligned with the central line of the main girders. There are 70 stay cables in each cable plane and 140 cables in the whole bridge. The longitudinal spacing of cables is set to be 12m for standard girders segments and for 2 segments under both towers is set to be 18.5m. In the pylon tower, the distance between the intersecting points of the central lines of stay cables with the central line of the pylons tower are all 2.2m.

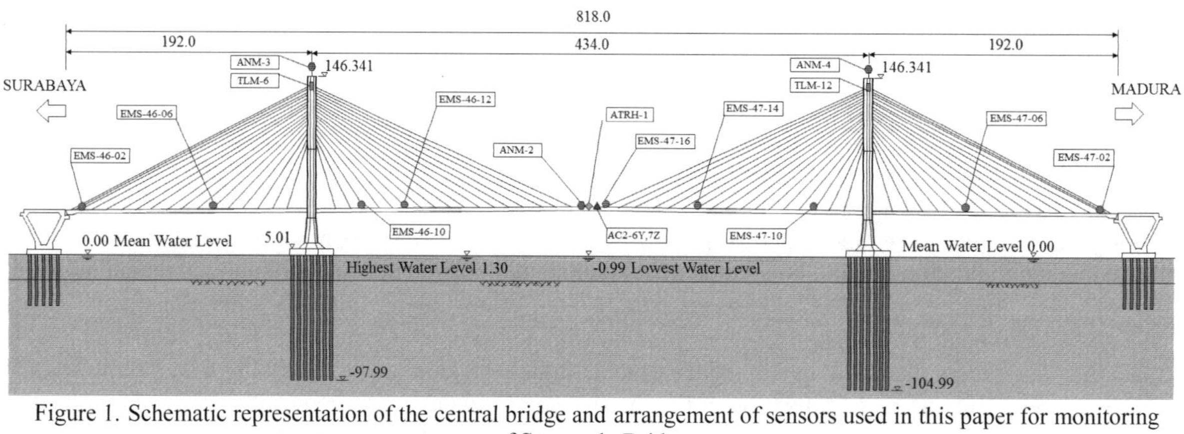

Figure 1. Schematic representation of the central bridge and arrangement of sensors used in this paper for monitoring of Suramadu Bridge.

To improve the earthquake-resistance performance of the main bridge, 4 sets of longitudinal viscous dampers are located on lower pylon cross beams under the main deck. In the transverse direction (across the bridge deck), the positive devices adopt rubber blocks with an area of 120x100 cm and thickness of 30 cm to suppress the bridge deck movement in this direction. The longitudinal slope from one end to the mid-span is 1%, and the vertical circular curve with the R=18000m is set at the mid-span, whose tangent length is 180 m and vector height is 0.9 m. This vertical curve can enhance the comfort feeling when driving and also brings aesthetic appearance of the bridge.

Extention joints and vertical bearing are located at both ends to allow the longitudinal elongation of the bridge due to change in temperature.
Cast-in-situ group piles foundation with a large pile diameter is used in Suramadu main bridge. The diameter of those piles are 2.4m and they are vertically bored. The pile's foundation is set to be quincunx in plan, and the minimum distance between piles from center to center is 6m. The external diameter of steel pile casing is 2.7m and wall thickness is 20mm. After finishing the construction, the steel pile casing become part of the permanent structure, however, the contribution of the pile casings to the bearing capacity of piles was not taken into account in the calculation. The pile cap is an octagonal cube with a thickness of 6m.

3. Bridge Monitoring System

Suramadu bridge is an essential part of transportation for eastern Java in Indonesia. After the construction, the Indonesian Ministry of Public Works and Housing (IMPWH) invested in monitoring and maintenance of the bridge due to the fact that resiliency and serviceability of the bridge were essential for the economy of the region. After the construction of the bridge, an extensive set of sensors were installed in 2009 and 2011. Anemometers were implemented in order to measure the magnitude and direction of wind speed. The acquired data can be used for traffic management, where a gate will be closed when the wind speed is over the threshold (40km/h for motorcycle lanes and 60km/h for vehicle lanes).
Monitoring the temperature and humidity is important to record environmental impact. Air temperature and relative humidity (ATRH) sensors were installed to measure ambient temperature and humidity in pylon and girder. Tiltmeters were deployed in order to monitor the angular rotation of the pylons due to the load applied to the bridge. In Suramadu bridge, two uniaxial tiltmeters are combined to form bi-axial tiltmeter. Accelerometers are important device in SHM to measure the vibration occurred in the bridge. Bi-axial accelerometers are formed by combining 2 uniaxial accelerometers. GPS are shown recently to be a valuable device in SHM (Schaal & Larocca 2009). In Suramadu bridge GPS are used to monitor the bridge movement, two references are installed outside the bridge while 14 rovers are installed in the bridge.
Weigh-in-motion (WIM) sensors provide valuable information about the non-stationary traffic loads such as total number of vehicles and their weight on each axle, which can be used for fatigue analysis (Guo et al. 2012). Other sensors such as strain sensors, displacement transducer, corrosion sensors (for steel bar, concrete and girder) cathodic protection monitoring and rain sensors were installed in Suramadu bridge, which provides an extensive database for monitoring and research about the performance of the bridge. Table 1 summarizes the monitoring system, the quantity of each device and sampling frequency. Figure 3 presents few samples of sensors installed on the bridge. It is worth mentioning that maintenance and regular inspection of sensors in long-term monitoring is a challenge in SHM. An inspection was carried out in 2014 which provides significant information about the functionality of the sensors. Quality and accuracy of data directly depend on inspection and maintenance of the sensors and is essential for long-term monitoring of any infrastructure.

Table 1. List of functioning sensors in Suramadu Bridge.

Sensor	Installed Device	Quantity	Sampling Rate (Hz)
Tri/Bi-Axial Anemometer	Young 81000/85000	2/2	4
ATRH	Campbell CS215-L	8	0.025
Bi Axial Tiltmeter	Geokon 6350	16	0.025
Single/Bi/Tri Axial Accelerometer	Silicon 2220-002/2220-002/2460-002	6/11/1	5
Seismic Accelerometer	PCB Piezotronic 393B31	2	5
EM Sensor	IIS EM Sensor PowerStress EM-II-xxx	32	0.00027
GPS	Lieca GR1230X/GR10/GR1200	10/4/2	1
WIM Sensor	Kistler	4	-
Displacement Transducer	Microepsilon WDS 2000 P96	8	5
Temperature Sensor (FBG)	BGK-FBG-4700S	16	1
FBG Strain Sensor	BGK-FBG-4150/4000/4000T/4150	42/78/50/4	1
Steel Bar Corrosion	FBG-FBG-4150	8	1
Cathodic protection monitoring (CPM)	IIS B-0L-001 CeBox Flux	81	0.00004
Rain Sensor	Texas Instrument TE525	2	-

(a) (b) (c) (d)

Figure 2. Sensors installed in Suramadu Bridge (a) Anemometer, (b) Tiltmeter, (c) GPS and (d) EM sensor

4. Measured Data

Location of sensors and the broad number of sensors provide an extensive information about the different stage of the bridge. Processing the data, it was found that temperature variation in the site is not remarkable. The average temperature for 2014 is 29.1°C with a standard deviation of 1.2°C and the average humidity is about 70.5% with a deviation of 4.6%. In Suramadu Bridge, maximum system temperature is taken to be 40°C and minimum system temperature is taken to be 15°C.

Although the variation of temperature is not considerable, it has a significant impact on structural performance. Figure 3(a) shows the cable forces at the right side of the bridge connected to the right pylon (EMS-47 2,6,10). It can be seen that with an increase in air temperature the cable force (EMS) decreases. Cyclic change in cable forces for 4 days is shown in Figure 3(b). This cyclic behavior can cause fatigue in cables in the lifetime of the bridge and is important to consider in the long-term performance of the bridge. Environmental condition (temperature, humidity and wind) has a considerable impact on the performance of the bridge. It is important to separate the environmental impact from traffic loads and vibration measurement from accelerometer recorded in order to use for system identification/damage detection (Ko et al. 2002; Teughels & Roeck 2004; Ren et al. 2005; Weng et al. 2008).

Wind load is a dynamic force has a considerable impact on the long slender cable-stayed bridge. The basic wind velocity in design was taken as 27 m/s. Figure 4 (c and d) show the wind speed and cable forces connected to the left pylon (EMS-46 2,6,10). The figure provides a correlation between anemometer located in the mid span (ANM-2) and the one located at top of the left pylon (ANM-3). It can be seen that compared to temperature the wind speed has less impact on cable forces. Wind speed is changing with height as it can be seen in the x-axis of the plots, that the wind speed at top of the pylon is 3 times of the wind speed at deck level.

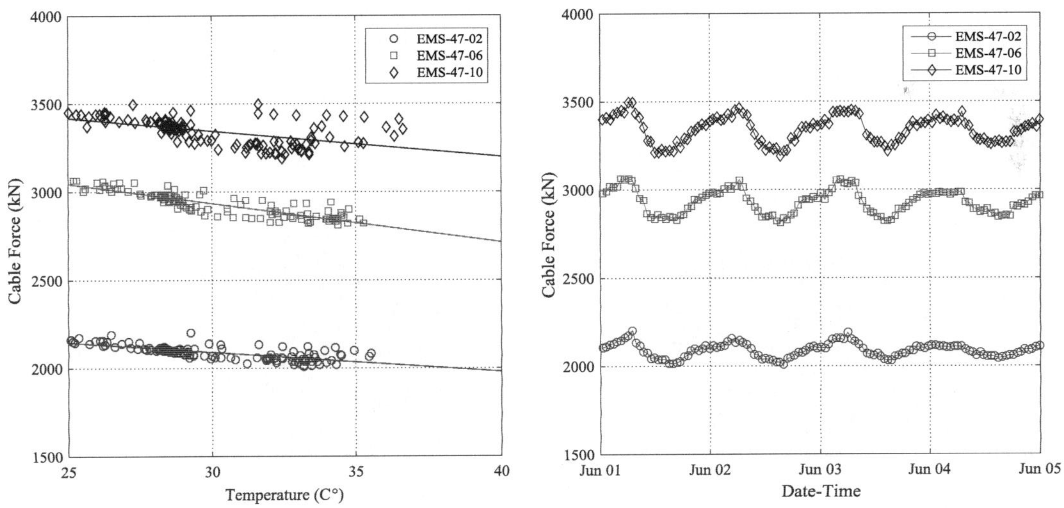

(a) (b)

Figure 3. (a) Cable force-temperature relationship for the Madura-side of the central bridge

Pylons are instrumented with tiltmeter which provide constant and reliable data information about the pylons since they are located inside the pylons they were less affected by harsh environment (rain, sun, wind, etc.). Figure 5(a) shows the tilt at top of the Surabaya-side pylon (TLM-6) for both directions (X is longitudinal and Y is across the bridge). It can be seen that the pylon is sensitive in the longitudinal direction to change of temperature. This is because of the variation in cable forces as a result of elongation of cables due to change of temperature as discussed previously. Figure 5(b) shows the tilt versus wind speed at the mid-span. Similar to temperature variation, the pylon is sensitive in the longitudinal direction.

Figure 6(a) shows the total number of vehicle passing the bridge for 24 hours at 27th of April 2014. WIM data can be used in traffic control and traffic loads estimation for future design of the bridges in the area. Maximum traffic occurs from 10 am to 6 pm with an average of 450 cars passing per hour. Figure 6(b) shows the hourly traffic mass for the same time. This data can be used to estimate traffic mass in FE model and generate a realistic simulation of the bridge.

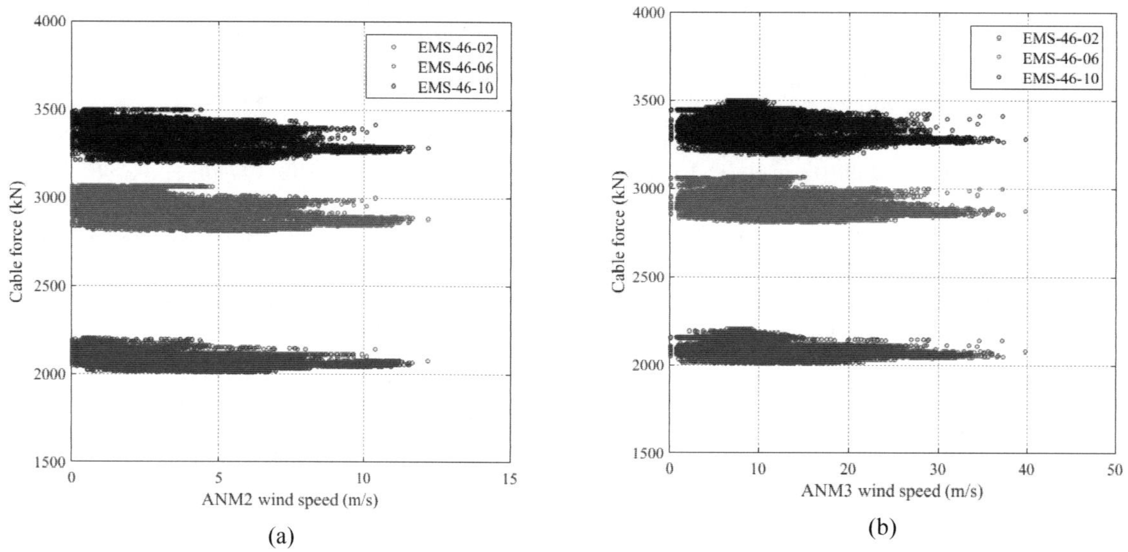

Figure 4. Cable forces versus wind speed located at mid-span and top of the pylon.

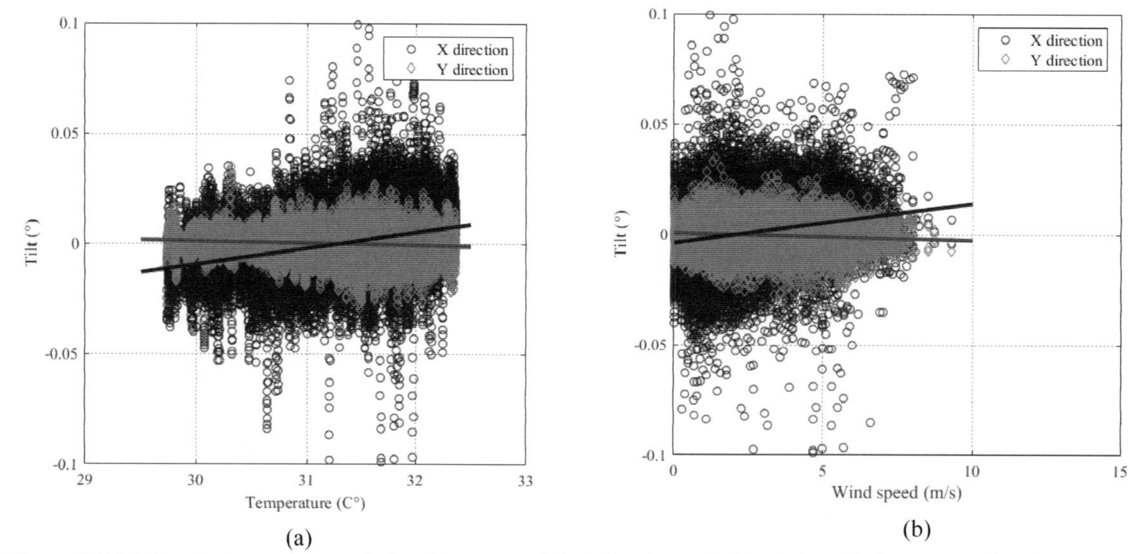

Figure 5. (a) Tiltmeter-temperature relationship at top of the left pylon and (b) relationship between pylon tilt with wind speed (ANM2)

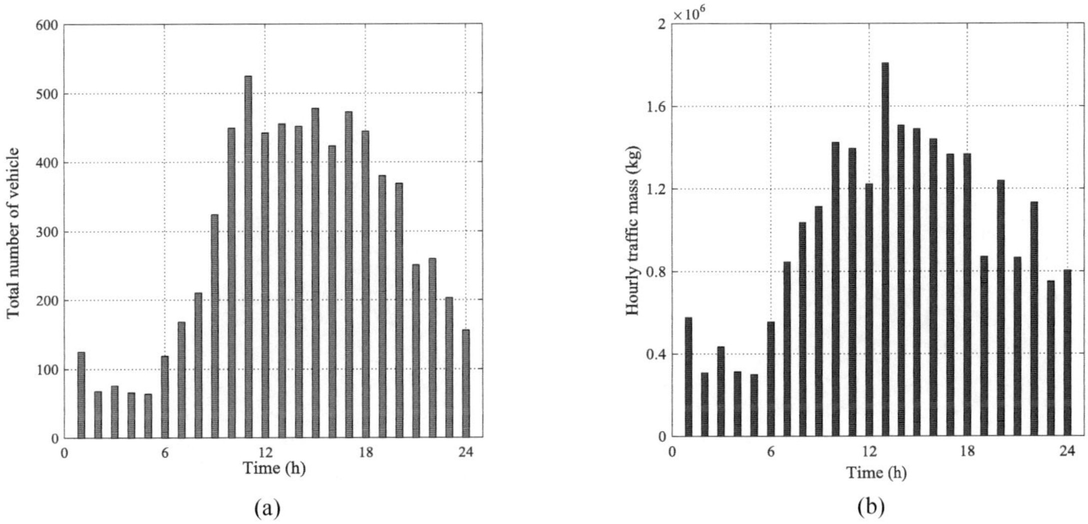

Figure 6. (a) Total number of cars passing the bridge for 24 hours at 27th of April 2014 and (b) hourly traffic mass

5. Finite Element Modelling

A finite element (FE) model of the Suramadu Bridge is constructed for validation of the analysis results of the measured data based on detailed drawings and design documents. A commercial structural analysis software, SAP2000 (Inc, 2000), is used. The bridge's main box girder is modeled by 426 frame elements with 3 different sectional properties. Additional masses are appended to the girder to represent the pavement, guardrails, water supply pipes, curbs and diaphragms (105 kN/m). Each of two pylons is modeled by 185 frame elements with 5 different sectional properties. The pylons footings are on the 6 m foundation over the piles and thus modeled as fixed boundary conditions. The cables are modeled by elements that consider the nonlinear effect caused by self-weight of cables with resulting tension forces and sag. Link elements with elastic stiffness are used for expansion joints on both sides of the bridge. The dampers were modeled using springs. Figure 7 shows the extruded representation of the FE model. A preliminary validation of the FE model is achieved by comparing the computed modal properties with those extracted from acceleration responses measured in 2012. The natural frequencies of the vertical and lateral modes are obtained from modal analysis and the results are shown in Figure 8. First six mode shapes evaluated from the FE model, including lateral, vertical and torsional modes are shown in Figure 9. The first few peak frequencies (i.e. 0.1978, 0.2021, 0.2511, 0.3234 and 0.3561 Hz) are in very good agreements with the FE analysis results which shows the general validity of the FE model. The higher modal frequencies were not captured due to the low sampling frequency of the accelerometer. Using a higher sampling frequency data and updating of the FE model may increase its accuracy for comprehensive SHM of the bridge which is the topic of future research.

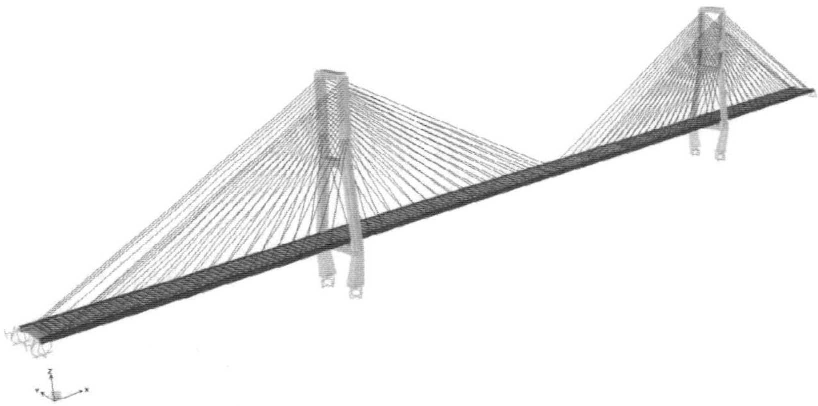

Figure 7. FE model of Suramadu Bridge

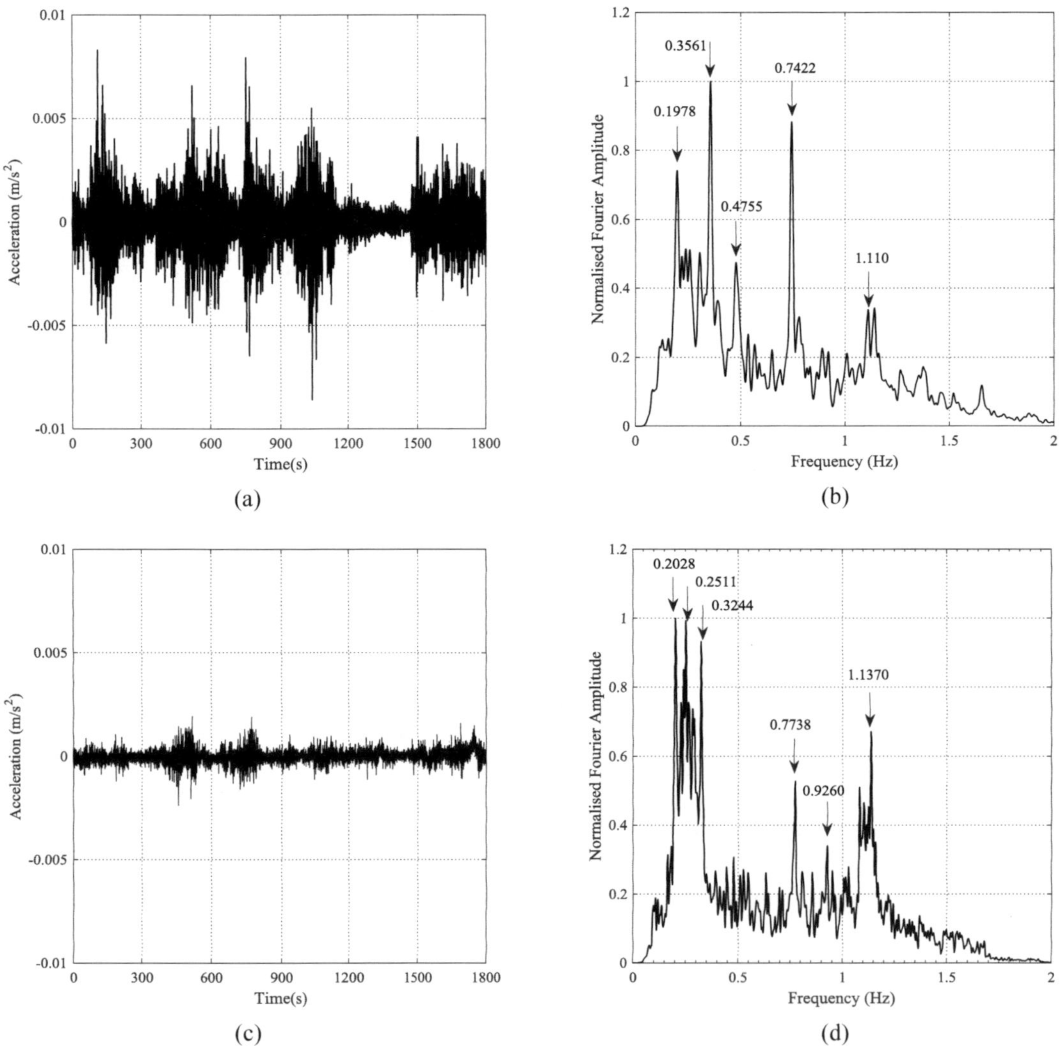

Figure 8. Power spectral density (PSD) of lateral (a) and vertical (b) acceleration record collected in August 2012 at mid-span

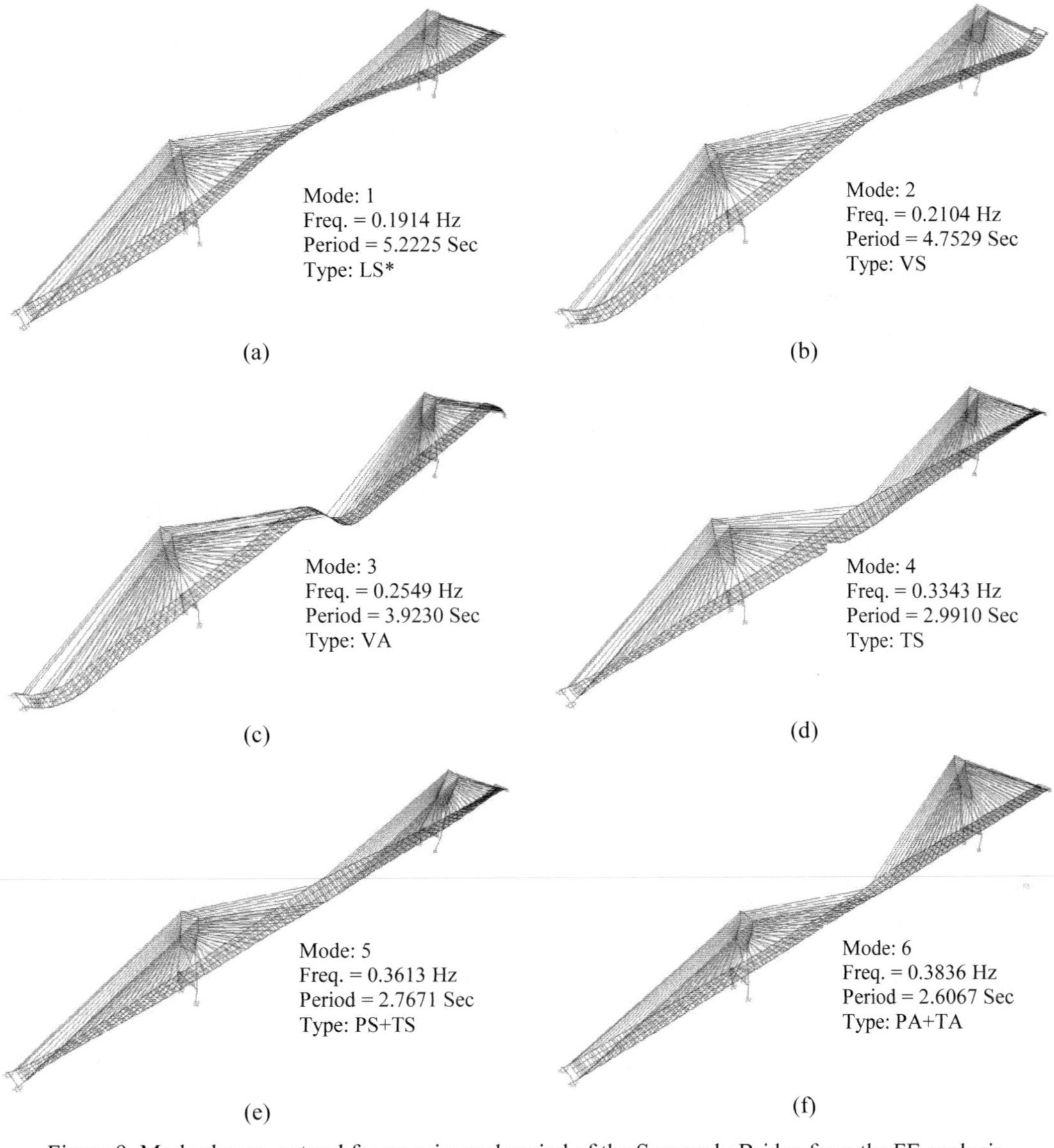

Figure 9. Mode shapes, natural frequencies and period of the Suramadu Bridge from the FE analysis
*V, vertical; L, lateral; T, torsional; P, pylon; S, symmetric; A, antisymmetric.

5. Conclusions

This paper presents an overview of the SHM system of Suramadu Bridge in Indonesia. Sensors have been installed after bridge construction and provide an extensive amount of data for structural performance analysis and anomaly detection. Data recorded since 2012 to 2015 are used in this paper. The main objective of this study was to evaluate the influence of environmental variations such as temperature, humidity and wind loads on structural behaviour. It was found that temperature variation has a considerable impact on the structural responses and should be considered in future structural analysis, damage detection and identification.

Acknowledgement

This study is funded by the British Council (Grant ID: 217544274). The writers would also like to acknowledge the supports from the Indonesian Ministry of Public Works and Housing (IMPWH) and the University of Warwick, UK. Any opinions, findings and conclusions or recommendations expressed in this paper are those of the authors and do not necessarily reflect the views of the IMPWH.

References

Jang S, Jo H, Cho S, Mechitov K, Rice J, Sim SH, Jung HJ, Yun CB, Spencer BF, & Agha G 2010, 'Structural health monitoring of a cable-stayed bridge using smart sensor technology: deployment and evaluation', *Smart Structures and Systems*, vol. 6, no. 5–6, pp. 439–459.

Cho S, Park J, Jung H, Yun C, Jang S, Jo H, Jr BFS, & Nagayama T 2010, 'Structural health monitoring of a cable-stayed bridge using acceleration data via wireless smart sensor network', *Smart Structures and Systems*, vol. 6, no. 5, p. 87786.

Zhu J, Ye G, Xiang Y, & Chen W 2010, 'Dynamic behavior of cable-stayed beam with localized damage', *Journal of Vibration Control*, vol. 17, no. 7, pp. 1080–1089.

Nguyen K, Kim J, & Shin SW 2013, 'Long-term Monitoring of the Hwamyung Cable-Stayed Bridge using Wireless Sensor Network', *Proceedings of the 2013 World Congress on Advances in Structural Engineering and Mechanics,* Jeju, Korea, September, pp. 2745–2758.

Brownjohn JMW 2007, 'Structural health monitoring of civil infrastructure', *Philosophical Transactions of the royal Society A-Mathematical Physical and Engineering Science*, vol. 365, no. 1851, pp. 589–622.

Schaal RE & Larocca AP 2009, 'Measuring Dynamic Oscillations of a Small Span Cable-Stayed Footbridge: Case Study Using L1 GPS Receivers', *Journal of Surveying Engineering*, vol. 135, no. pp. 33–37.

Guo T, Frangopol DM, & Chen Y 2012, 'Fatigue reliability assessment of steel bridge details integrating weigh-in-motion data and probabilistic finite element analysis', *Computers and Structures*, vol. 112–113, pp. 245–257,.

Ko JM, Sun ZG, & Ni YQ 2002, 'Multi-stage identification scheme for detecting damage in cable-stayed Kap Shui Mun Bridge', *Engineering Structure*, vol. 24, no. 7, pp. 857–868.

Teughels A, & Roeck GD 2004, 'Structural damage identification of the highway bridge Z24 by FE model updating', *Journal of Sound Vibration*, vol. 278, no. 3, pp. 589–610.

Ren, WX, Peng XL, & Lin, YQ 2005, 'Experimental and analytical studies on dynamic characteristics of a large span cable-stayed bridge', *Engineering Structure*, vol. 27, no. 4, pp. 535–548.

Weng JH, Loh CH, Lynch JP, Lu KC, Lin PY, & Wang Y, 2008, 'Output-only modal identification of a cable-stayed bridge using wireless monitoring systems', *Engineering Structure*, vol. 30, no. 7, pp. 1820–1830.

Computers & Structures Inc. SAP2000-Integrated finite element analysis and design of structures. Berkeley, CA: Computers & Structures Inc; 2000.

The 8th International Conference on
Structural Health Monitoring of Intelligent Infrastructure
Brisbane, Australia | 5-8 December 2017

Project VIMTO: a new system for the vibration and impact monitoring of tram operations

P.R.A. Fidler [1], S. Hartley [2], J.P. Talbot [2]
[1] *University of Cambridge, UK, Email:* prafl@cam.ac.uk, [2] *University of Cambridge, UK*

Abstract

Disturbance to building occupants caused by tram-generated ground-borne noise and vibration presents a significant barrier to the expansion of tram networks in our cities. Furthermore, such disturbance is often an indicator of deteriorating track infrastructure. Monitoring and understanding ground-borne noise and vibration is therefore a key priority for tram operators. Project VIMTO (Vibration and Impact Monitoring of Tram Operations) is concerned with developing a new system for monitoring vibration and impact of trams, whereby the vehicles themselves are used as the primary monitoring instrument. Low-cost vehicle-mounted instrumentation is being used to record axle-box vibration signatures, along with positioning data, to 'map' a tram network in terms of its propensity to generate vibration. Such mapping aims to offer near real-time continuous monitoring, enabling the formulation of more efficient, optimised maintenance strategies. This paper describes the current development system and presents some initial results from trials on the Midland Metro, UK.

1. Introduction

The surface roughness of railway track increases during service. Along with other forms of deterioration, such as settlement at rail joints, this leads to an increase in the level of vibration and wheel-rail impacts generated as a vehicle runs along the track. From its origin at the wheel-rail interface, vibration propagates through the track structure, through the ground and into buildings, where it may then cause disturbance as perceptible vibration and/or re-radiated noise (Lang 1988, Howarth & Griffin 1991, Talbot 2013). A particular concern can be the development of corrugation (Figure 1), a short-wavelength irregularity (typically 20 mm to 200 mm) that forms along the rail head and leads to particularly severe, tonal noise due to the periodic wheel-rail interaction (Grassie 2009).

Figure 1: Examples of rail corrugation on mainline track
(Images: RailMeasurement Ltd. Used with permission)

There are over 290 km of modern tramway in the UK alone, the majority of which run through densely populated urban areas. Proximity to residential properties and sensitive buildings such as concert halls and hospitals means that ground-borne vibration is a major concern of tram operators, who must work continuously at track maintenance to keep noise and vibration within acceptable levels. Currently, this work relies on regular, manual track inspections to identify areas of deterioration before they become a significant source of vibration. Such inspections are time consuming and costly. There is a clear need for a more efficient monitoring method.

Vehicle-based infrastructure monitoring is not a new concept. In particular, indirect measurement of railhead irregularities using accelerometers mounted on the axle-boxes of bogies was first trialled on mainline track in the mid-to-late 1980s (Grassie, 1989 & 1996). This initial work has led to the development of at least one commercially available system (RailMeasurement Ltd) which has since been tested specifically for managing wheel/rail rolling noise with the French railway operator SNCF (Bogini et al. 2011). However, this previous work has been concerned with specialist trials on mainline track, and has so far been limited to conditions when vehicle speeds and track behaviour remain similar to those under which the system is calibrated. Despite the apparent research successes, there is little or no evidence of such monitoring being used regularly in service with either mainline trains or trams. Project VIMTO aims to develop a low-cost vehicle-based system for track monitoring that may be permanently installed on service trams running on an operational network. The focus is on monitoring the track in terms of its propensity to generate ground-borne noise and vibration. As such, the ultimate objective is to develop a system that tracks changes in the underlying roughness of the railhead, in the form of the roughness spectrum as defined in standards such as EN 15610:2009 (European Committee for Standardization, 2009).

2. The Midland Metro tram system

The Midland Metro service (Figure 2) first opened in 1999, operating on both on-street and dedicated rail sections over 20 km of track between Wolverhampton St George's and Birmingham Snow Hill stations.

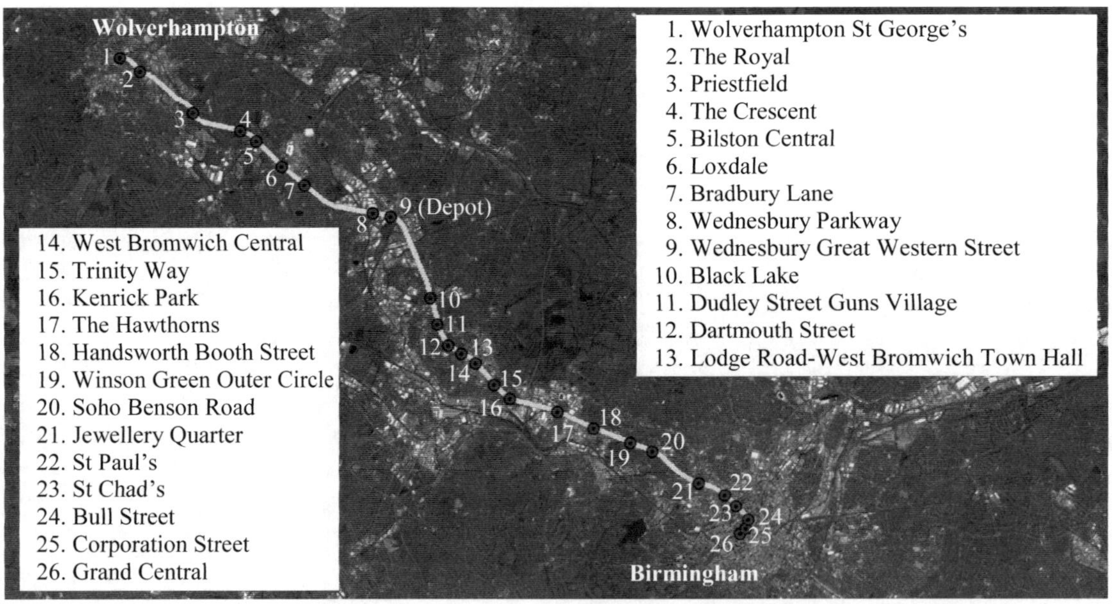

Figure 2: The Midland Metro tram network operates between Wolverhampton and Birmingham, UK (Image: Google, © 2017 Infoterra Ltd & Bluesky)

In 2016 the system was extended further into Birmingham city centre, with a new stop at St Chad's replacing the original Snow Hill terminus, and new stops added at Bull Street, Corporation Street and Grand Central Station. Further extensions to the network are planned.

Since September 2014, the system has operated a fleet of CAF Urbos 3 trams (Figure 3), providing a service of 10 trams per hour in each direction. The trams are 33 metres long and consist of five modules with four articulations, three of the modules being mounted on bogies and the other two modules being suspended. The maximum operating speed of the trams is 70 km/h.

Figure 3: A CAF Urbos 3 Tram at the Bull Street tram stop
(Image: Jennifer Schooling. Used with Permission)

3. Monitoring system overview

The aim of Project VIMTO is to develop a low-cost vehicle-based system for track monitoring that may be permanently installed on service trams running on an operational network. The system described here is the current development system. It uses a total of five accelerometers, mounted at the ends of the axle bridges – the steel beams mounted off the primary suspension that support the axle boxes of each wheelset (Figure 4). These offered the most practical locations at which to mount the accelerometers whilst being as close as possible to the wheel-rail interface, in order to minimise the mechanical filtering effect of the various axle/suspension components between the rail head and the measurements.

As indicated in Figure 5, four accelerometers (P1-P4) are located on one of the power bogies, with a fifth (T1) located on the central trailer bogie. This arrangement includes a degree of redundancy that is expected to be reduced as part of the development work. In principle, just two accelerometers should be sufficient; one for each rail. However, a significant amount of the development work involves assessing the quality of the accelerometer signals and the degree of variability between measurement locations. Although the power and trailer bogies are based on a common bogie frame, the influence of differences in unsprung mass, and the effect of the motors in the power bogies has yet to be determined.

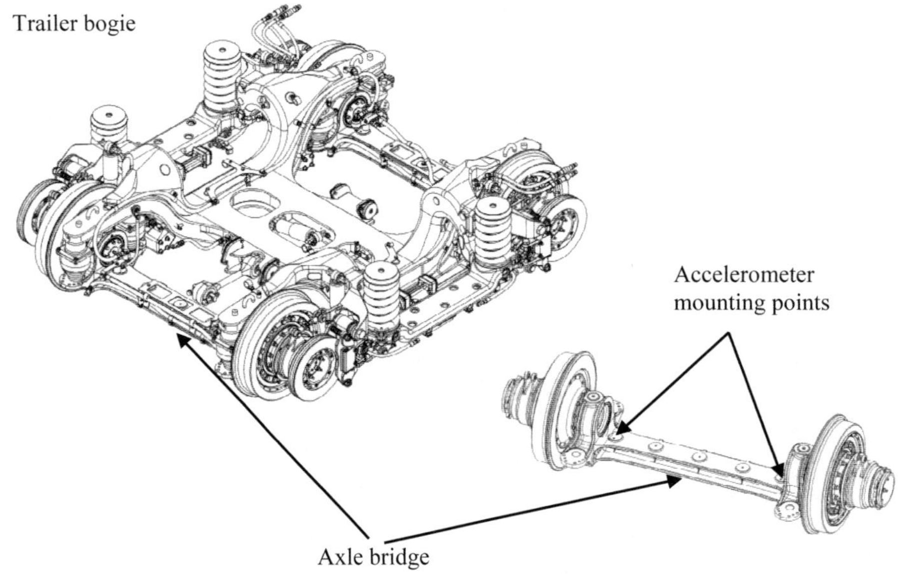

Figure 4: Accelerometer locations on the axle bridge
(Images: CAF. Used with permission)

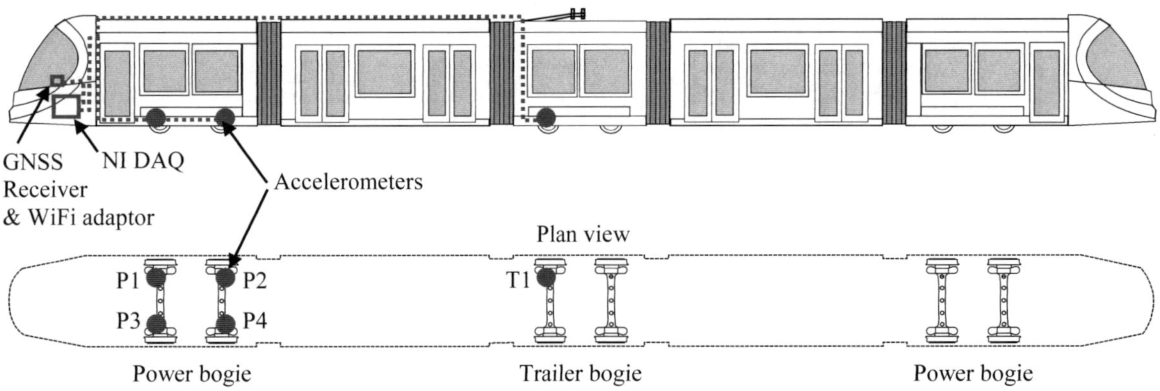

Figure 5: Diagram of system installed in the tram, showing the approximate locations of the cabling and accelerometers, P1-P4 (power bogie) and T1 (trailer bogie)

3.1 Hardware

The data acquisition (DAQ) hardware consists of: a NI CDAQ-9132 data acquisition unit, with a 128 GB SD card; two NI 9234 analogue-to-digital (ADC) modules and one NI 9401 8-channel digital I/O module; all from National Instruments. All five accelerometers are Dytran 3055D4 single-axis ICP-type piezoelectric devices, with a sensitivity of 50 mV/g and a dynamic range of 100 g. These were selected based on previous experience of axle-box vibration measurements (Degrande et al. 2006, Lees & Talbot 2014), which typically indicate peak vibration levels in the region of 50 g. The accelerometers are stud-mounted in tapped holes in the axle beams, using a proprietary thread-locking compound. Electrical connections, for the combined ICP power and accelerometer signal, are made to the ADC modules with standard RG 174 coaxial cable.

In addition to the accelerometers, the system uses a GNSS receiver to determine tram location and speed, which is connected to the data acquisition unit via USB. An additional CAF speed sensor, of the same type used on the tram to provide speed information to the driver, was also originally installed on one wheel of the central trailer bogie, and connected to the digital I/O module using 4-core cable.
To facilitate data transfer to and from the system, a Netgear WNCE2001 Ethernet to WiFi adapter was installed to allow connection to the Internet within the tram depot. WiFi Internet connectivity was provided by Centro (Transport for West Midlands).

3.2 Installation

During an initial visit to the tram depot, a cabinet in one of the two driver's cabins had been identified as a suitable location for the data acquisition unit. Installation of the monitoring system on the tram took place over a number of visits to the depot between November 2015 and July 2016. The cable installation took place on 5th November 2015, when four coaxial cables for accelerometers P1-P4 were routed to the power bogie behind the driver's cabin. A further three coaxial cables and the 4-core cable for the CAF speed sensor were routed over the roof of the tram, past the two intervening articulations and then down to the trailer bogie. All cabling was routed within fire-resistant flexible trunking (Figure 6) and is hidden from view, whether under the tram, within cable runs inside the tram cowlings or along the roof.

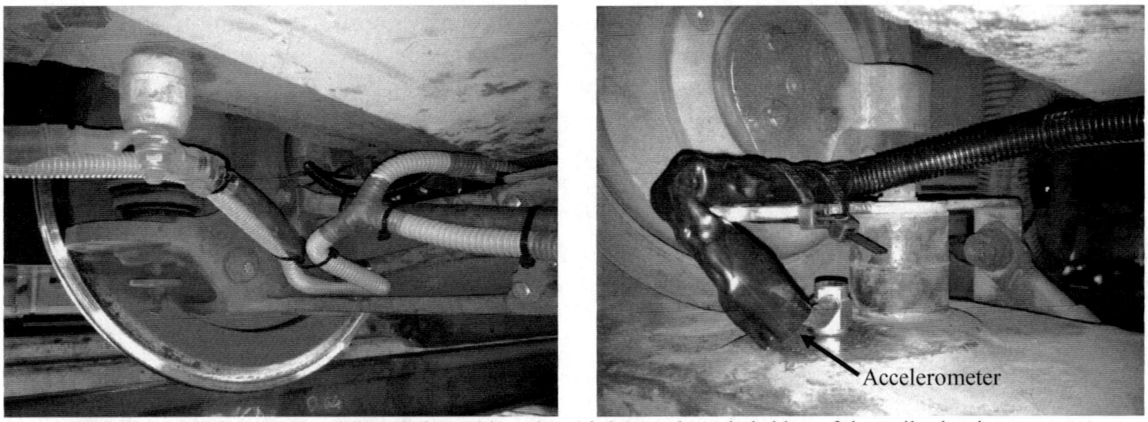

Figure 6: Accelerometer cabling (left) and location (right) on the axle bridge of the trailer bogie

The data acquisition unit and accelerometers were connected on 17th February 2016, along with the GPS receiver and the WiFi adapter. These were placed above the cabinet housing the data acquisition unit but hidden from view on the left-hand side of the driver's cabin. Calibration tests were performed on the accelerometers on 4th May 2016.

3.3 Hardware issues

The cable runs proved challenging, and a number of suspected cable breakages during installation necessitated a reduction in the number of sensors finally commissioned; two of the three accelerometers planned for the trailer bogie, together with the CAF speed sensor, had to be omitted. Furthermore, although data could initially be logged from the GPS receiver, there was no accurate location data – only status data indicating that no GPS fix was available. The Olimex GPS unit was disconnected and replaced by a Navilock NL-8003P GNSS receiver on 10th August 2016, which has since proven to be reliable.

The Netgear WiFi initially failed in October 2016. The data acquisition unit continued to log data on the SD card until 11th November 2016, when the card became full. An initial attempt to reconfigure the WiFi adapter in December 2016 appeared to have been successful but the adapter failed again in January 2017 and was subsequently replaced with a TP-Link TL-WR802N device on 27th April 2017. So far, this new WiFi adapter appears to be working well.

A further lack of robustness concerns the threaded cable connectors to the accelerometers, which have demonstrated a tendency to work loose under the high-vibration levels seen on the bogies. No thread-locking compound was originally applied to these relatively delicate connectors, to allow easy disconnection in the event of a bogie change on the tram.

These hardware issues have provided valuable lessons as part of the commissioning of the development system, and will undoubtedly inform a more robust design of future installations.

3.4 Data acquisition software

The NI CDAQ-9132 data acquisition unit runs a real-time Linux distribution. A custom LabView program, compiled into a real-time Linux executable, is used to acquire and store data from the sensors. As a development system, raw data is acquired and saved permanently: no data reduction or analysis is performed on the unit, and no data is presented to any user interface. The program has a single real-time thread that records data from the five working accelerometers and from the GNSS receiver, and a non-real-time thread that saves data to log files. Acceleration data is stored in National Instruments' TDMS file format, while GNSS location data is stored in text files. A single clock is used to time-stamp each entry in both files, thereby providing a common time base.

Data are recorded from 5am to 11pm every day. Acceleration data are recorded at 1652 Hz, whereas the GNSS data are recorded at 10 Hz. These sampling frequencies were chosen to achieve the necessary compromise between spatial and frequency resolution, and without producing unmanageable quantities of data. The tram typically records 264 MB per hour, equating to 4.8 GB of data per day. Data are transferred from the tram to a server in Cambridge via Rsync over SSH. This can currently only be done while the tram is in the depot, typically overnight. Whilst this currently limits any real-time analysis or presentation that may be done, it is possible that improved communications in future may make it possible to stream acceleration data in real time.

4. Initial results and analysis

The GNSS location data along the route are usually good, but suffer from errors in areas where the tram goes through a cutting, under a bridge, or near tall buildings (Figure 7). As expected, no location data are available for the section of track between Wednesbury Great Western Street and Black Lake where the track goes through a tunnel. The GNSS location is also not accurate enough to determine which of the two running tracks the tram is using. However, although there are points at various locations along the route at which a tram could be switched to the opposite track, this only actually happens at the ends of the route, with the tram normally driving on the left-hand track with respect to the direction of travel.

Data files are currently processed using some bespoke Python scripts. These first check the location data log files to check whether there is any significant movement of the tram (on some days it remains parked in the depot) and, if there is, the acceleration data and location data are combined into a single file for ease of processing.

For the purpose of initial analysis, the route from Wolverhampton St George's station to Grand Central station in Birmingham has been divided into 50 m sections (except for the tunnel and two under-bridge sections where the sections are longer). The root mean square (RMS) acceleration level of each of the accelerometer signals is then calculated for each 50 m section of track.

Figure 7: GNSS data from the suburban area near Wednesbury Parkway (left) and in Birmingham city centre at the corner of Bull Street and Corporation Street (right). (Images: Google)

To account for inaccuracies in the GNSS data, the tram is determined to be within a track segment if the GNSS position is within a polygon that extends 50 m along the track and approximately 35 m either side. The scripts are written to allow a customised polygon to be used where necessary, such as for sections of the track near tall buildings, under bridges or within tunnels. Journeys from Wolverhampton to Birmingham are treated separately from those from Birmingham to Wolverhampton. For simplicity, all data presented here are from journeys in the Wolverhampton-to-Birmingham direction (data acquired in the reverse direction are similar).

Figure 8 shows the RMS acceleration levels for the entire route between Wolverhampton and Birmingham on a selection of different days between August 2016 and April 2017.

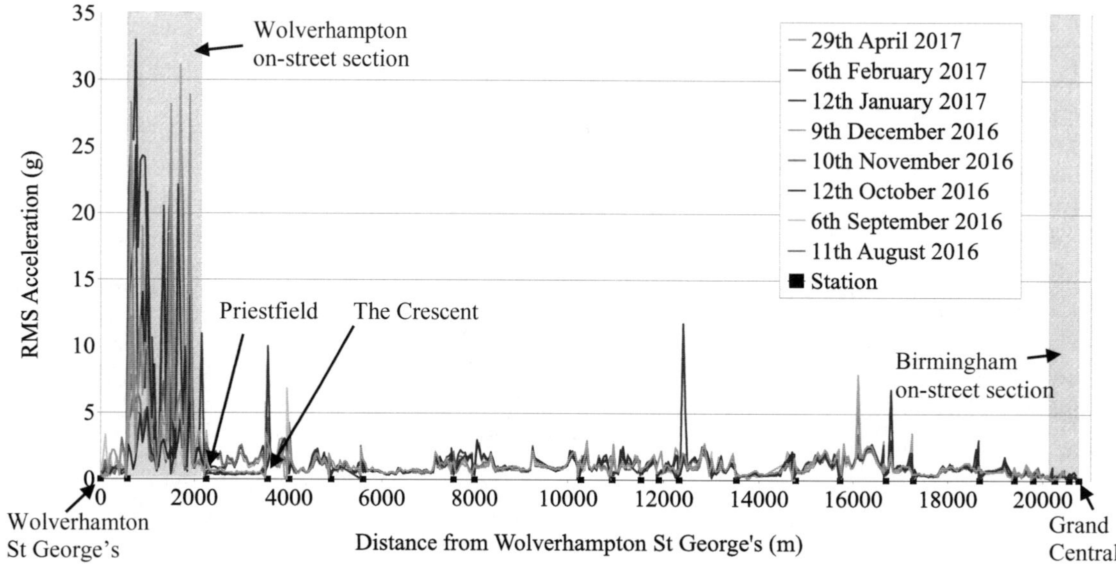

Figure 8: Graph showing RMS acceleration levels from accelerometer P1, recorded on a selection of different days for each 50 m section of the entire route between Wolverhampton and Birmingham.

The data for the track sections near Wolverhampton show considerably higher RMS accelerations than elsewhere. This is attributed to the poorer track quality between The Royal and Priestfield tram stops (600 m to 1950 m), which is an on-street section where trams must brake and accelerate more frequently than on the dedicated rail sections, as they operate in a mixed-traffic environment with other road users.

For most of the remainder of the route, from The Crescent through to Grand Central (3500 m to 21750 m), the RMS accelerations recorded from journeys on different days generally show similar values. The overall consistency and repeatability between journeys, each made several weeks apart, provides a degree of confidence that the system is responding to actual features present on the track, and is not unduly influenced by factors such as changes in weather or the driving style of individual tram drivers. Although occasional peaks are evident in the data, these are usually at or near tram stops, where breaking and acceleration of the tram is required. Another peak (at 16100 m) corresponds to a set of points.

One observation from Figure 8 is that the section between the Priestfield and The Crescent stops (2250 m to 3500 m) appears to show a marked *improvement* (lower RMS accelerations) between November and December 2016. This is surprising as, without intervention, track would normally be expected to wear and deteriorate over time, and we are currently unaware of any maintenance having been undertaken over this section. Figure 9 shows an enlarged view of the data, highlighting the section between Priestfield and The Crescent; Figure 10 shows both a sample of the raw acceleration data acquired on this section and also the improvement in RMS acceleration over time, as seen on accelerometers P1 and P2. At the time of writing, there is no clear explanation for the apparent improvement.

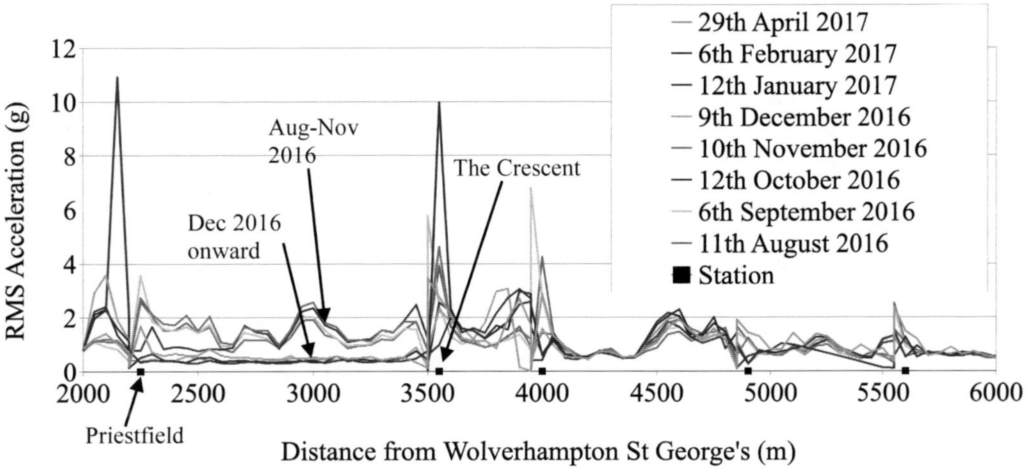

Figure 9: Graph showing the RMS accelerations recorded by P1 near the section between Priestfield and The Crescent

One key area of concern for the tram operator is the section of track at the corner of Bull Street and Corporation Street, where the tram must make a 90° turn. This places high demands on the track, and makes it a particularly sensitive section in terms of both track wear and noise generation (rail lubrication is currently being trialled by the operator over this section to reduce noise generation). The RMS acceleration levels recorded in this area are currently relatively low (< 1 g), as they are for all of the newly laid section between St Chad's and Grand Central, and the average velocity of the tram is only 15 km/h (which also contributes to low acceleration levels). It nevertheless remains an area of interest.

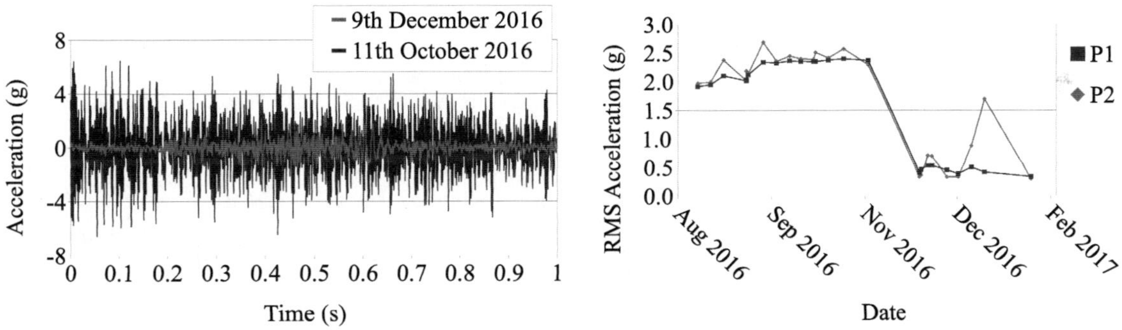

Figure 10: Plots showing the apparent improvement on the track section between Priestfield and The Crescent, showing (left) a 1 s sample of raw acceleration data from P1, and (right) the improvement as recorded by P1and P2

5. Future work

This paper has reported on the initial development work of VIMTO. The project remains work in progress and there is considerably more research to be done, firstly in terms of understanding the quality of the data and, secondly, in terms of developing the necessary signal processing and supporting models to relate the measured accelerations to the desired rail roughness spectra. There is currently no allowance made for the effect of tram speed on the magnitude and frequency content of the measured accelerations. Better treatment of the data is also required at tram stops, as currently no attempt is made to omit acceleration data recorded while the tram is stationary.

One important piece of future work is to use a direct method of measuring surface roughness along a section of the track in order to validate and properly calibrate the data from the VIMTO system.

To achieve the goal of a near real-time monitoring system, a web-based 'dashboard' or reporting system is required. This will highlight those sections of track where the inferred rail roughness has changed over time, and hence highlight locations of potential wear.

6. Conclusions

Despite initial difficulties commissioning the system, axle-box acceleration data has now been collected from a tram over a period of several months. Initial analysis of the data indicates broadly consistent RMS acceleration levels for each section of track over this time. Sections of track that have exhibited differences require further investigation.

With more detailed analysis and subsequent development work, it is hoped to achieve the ultimate objective of developing a system that tracks changes in the underlying rail roughness spectrum, thereby offering near real-time continuous monitoring to enable the formulation of more efficient, optimised maintenance strategies.

Acknowledgements

This work was funded by the EPSRC Impact Acceleration Account (Grant No. EP/K503757/1) and by the Cambridge Centre for Smart Infrastructure and Construction (EP/I019308/1). The work was facilitated by Centro, the West Midlands Integrated Authority, now known as Transport for West Midlands, and by National Express. The assistance provided by Colin Robey, Sophie Allison, Khera Lakhwinder, Lee Ashmore, Tony Ross, Des Coulson and the staff at the Wednesbury Depot is gratefully acknowledged. Data supporting this paper is available from the University of Cambridge repository at: https://doi.org/10.17863/CAM.9665.

References

Lang, J 1988, 'Ground-borne vibrations caused by trams, and control measures', *Journal of Sound & Vibration*, vol. 120, no. 2, pp. 407-412.

Howarth, HVC & Griffin, MJ 1991, 'The annoyance caused by simultaneous noise and vibration from railways', *Journal of the Acoustical Society of America*, vol. 89, no. 5, pp. 2317-2323.

Talbot, JP 2013, 'Lift-over crossings as a solution to tram-generated ground-borne vibration and re-radiated noise', *Proceedings of the Institute of Mechanical Engineers Part F – J. Rail & Rapid Transit*, vol. 228, no. 8, pp. 878–886.

Grassie, SL 2009, 'Rail corrugation: characteristics, causes and treatments', Proceedings of the Institution of Mechanical Engineers, Part F: Journal of Rail and Rapid Transit. vol. 223 issue: 6. pp 581-596. DOI: 10.1243/09544097JRRT264

Grassie, SL 1989, 'Corrugation on Australian National: cause, measurement and rectification', 4th International Heavy Haul Railway Conference, Brisbane, Australia, pp. 188-192.

Grassie, SL 1996, 'Measurement of railhead longitudinal profiles: a comparison of different techniques', Wear, vol. 191, pp. 245-251.

RailMeasurement Ltd. Corrugation measurement or "noise mapping" at line speed: HSRCA, High Speed Rail Corrugation Analyser factsheet, Cambridge. http://railmeasurement.com/hsrca/ (accessed 22nd May 2017).

Bongini, E, Grassie, SL & Saxon, MJ 2011, 'Noise mapping' of a railway network: validation and use of a system based on measurement of axlebox vibration in T. Maeda et al. (Eds.), *Noise and Vibration Mitigation for Rail Transport Systems*, NNFM 118, Springer-Verlag.

European Committee for Standardization 2009, EN 15610:2009 'Railway applications - Noise emission - Rail roughness measurement related to rolling noise generation'.

Degrande, G, Schevenelsa, M, Chatterjeeet, P et al. 2006, 'Vibrations due to a test train at variable speeds in a deep bored tunnel embedded in London clay', *Journal of Sound & Vibration*, vol. 293, pp. 626–644.

Lees, A & Talbot, JP 2014, 'Mapping of ground-borne vibrations created by trams', Fourth-year undergraduate project, Cambridge University Engineering Department.

The 8th International Conference on
Structural Health Monitoring of Intelligent Infrastructure
Brisbane, Australia | 5-8 December 2017

Desktop Experiment Rigs as Teaching Aids for Structural Dynamics

N. Haritos
Strucomp P/L – Australia, Email: strucomp@bigpond.com

Abstract

This paper presents examples of where experiment kits for TechnoLab's desktop Pixi-Frame based system for statics and mechanics can be easily adopted to perform studies of the dynamic properties of the specimens under test in these kits. The examples presented principally feature the use of video processing photogrammetric software, Tracker, for performing the required analysis.

1. Introduction

The author has accumulated extensive experience in the teaching of statics, mechanics of solids, mechanics of structures and structural dynamics to undergraduate and post-graduate students and the delivery of structural dynamics workshops to practicing engineers.

Over the past three years, he has devoted a great deal of his energy and resources to the development of desktop experiment rigs aimed at facilitating the teaching of a range of engineering topics, in the above areas of study, to students and professional engineers.

Of particular interest to those wishing to explore material relevant to vibration based SHM techniques, are the experiment rigs he has developed for the study of single and multi-degree of freedom dynamic systems.

In this paper, the author describes innovations he has introduced in the development of these test rigs, and how they can be used to demonstrate features such as:

- The properties of simple single degree of freedom structural systems, (natural frequency, damping and relationship to mass and stiffness),
- The properties of simple multi-degree of freedom structural systems, (modal frequencies, modal damping and relationship to mass and stiffness matrices),
- Base excitation of single and multi-degree of freedom structural systems, (Swept Sine Wave, modelled earthquake and random excitation),
- The influence of changes in stiffness and damping on the dynamic properties of single and multi-degree of freedom structural systems, (relevant to SHM and damage detection techniques)

Examples are offered of the practical use of these systems in workshops and demonstrations.

2. TechnoLab™ Mechanics and Dynamics Teaching Platform Concept

The TechnoLab™ mechanics and dynamics teaching platform was principally developed by the author to address the perceived need of Tertiary Institutions for affordable, effective "hands-on" testing equipment to support engineering undergraduate and post-graduate courses in subjects relevant to the suite of experiments produced for this purpose in this platform.

TechnoLab™ features a number of innovative features in experiments performed in its test frames that distinguish this system from experiments produced by manufacturer/suppliers of engineering laboratory teaching equipment that are well-established in the international marketplace.

2.1 Principal features of the TechnoLab™ platform

Principal features of the TechnoLab™ platform include:

- The platform makes prolific use of a purpose-built Pixi Frame™, with an attachable Window Frame™, for installing the majority of the experiment test rigs developed for the system.
- The Pixi-Window Frame™ combination, as shown in Figure 1, is designed as a desktop system that can be easily placed on desks/tables in tutorial or other teaching/classroom spaces, negating the requirement for special laboratory facilities thus freeing these up for research/other pursuits.

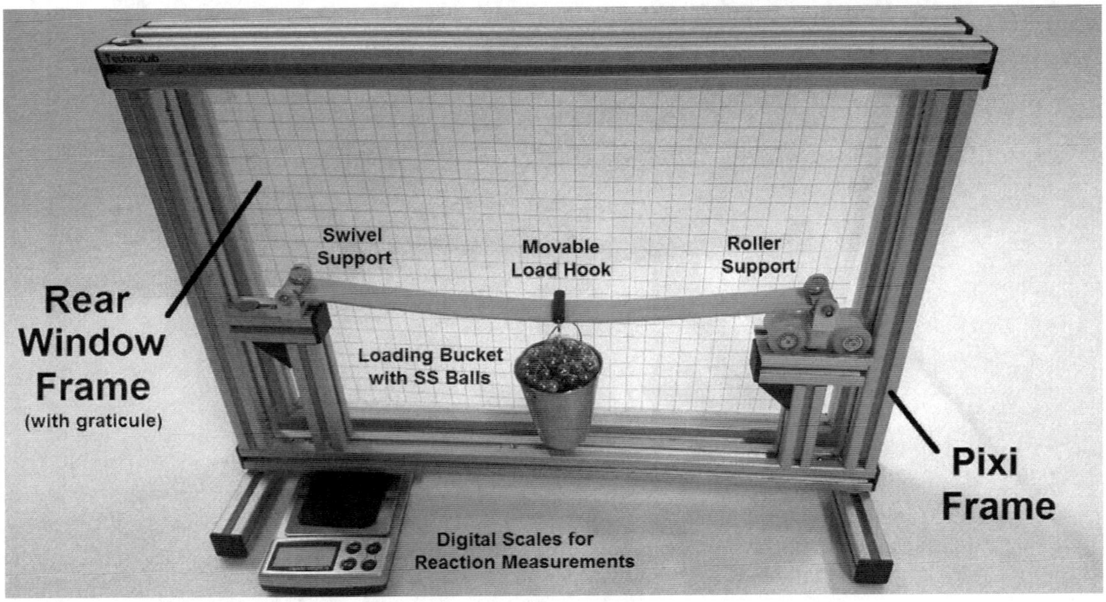

Figure 1: Pixi-Window Frame™ Combo Depicting a Simply Supported Beam with Central Point Load Experiment

- The size of the Pixi-Window Frame™ combo is approx. A3 and is supplied in a Polypropylene satchel or Aluminium carrying case, which, together with ancillary items needed for the experiment test setups of interest to the particular class, can be easily stored away in cupboard space or in one or more "medium to larger sized" suitcases - as many as would be needed for the number of frame setups used in the class concerned – typically 12 units in a classroom setup for students engaged "hands-on" in pairs for the bundled set of experiments for that particular class.
- Deflections/deformations of test specimens in the purpose designed TechnoLab™ experiment test rigs are clearly visible to the naked eye so that rather basic methods for their measurement, such as "tracing" on the removable transparent graticule sheet on the Window Frame™ or the use of rulers or digital calipers (supplied with each Pixi Frame), can easily be used to take measurements of the test deflections/deformations to normally acceptable levels of accuracy.
- Where improved accuracy in deformation or displacement measurements than can be afforded by the naked eye is required, (such as in TechnoLab's "Forces in a Statically Determinate Truss" experiment), then use is made of a simplified form of photogrammetry to perform the measurements required, (see Figure 2).
- For dynamic experiments, TechnoLab™ measures the acceleration response using logger-units with on-board accelerometers, smart phones and/or a prototype purpose-built tiny Bluetooth accelerometer, currently under development. Tracker (2017) video processing software is used for determining the time history response of user-nominated features in the videoed experiments.

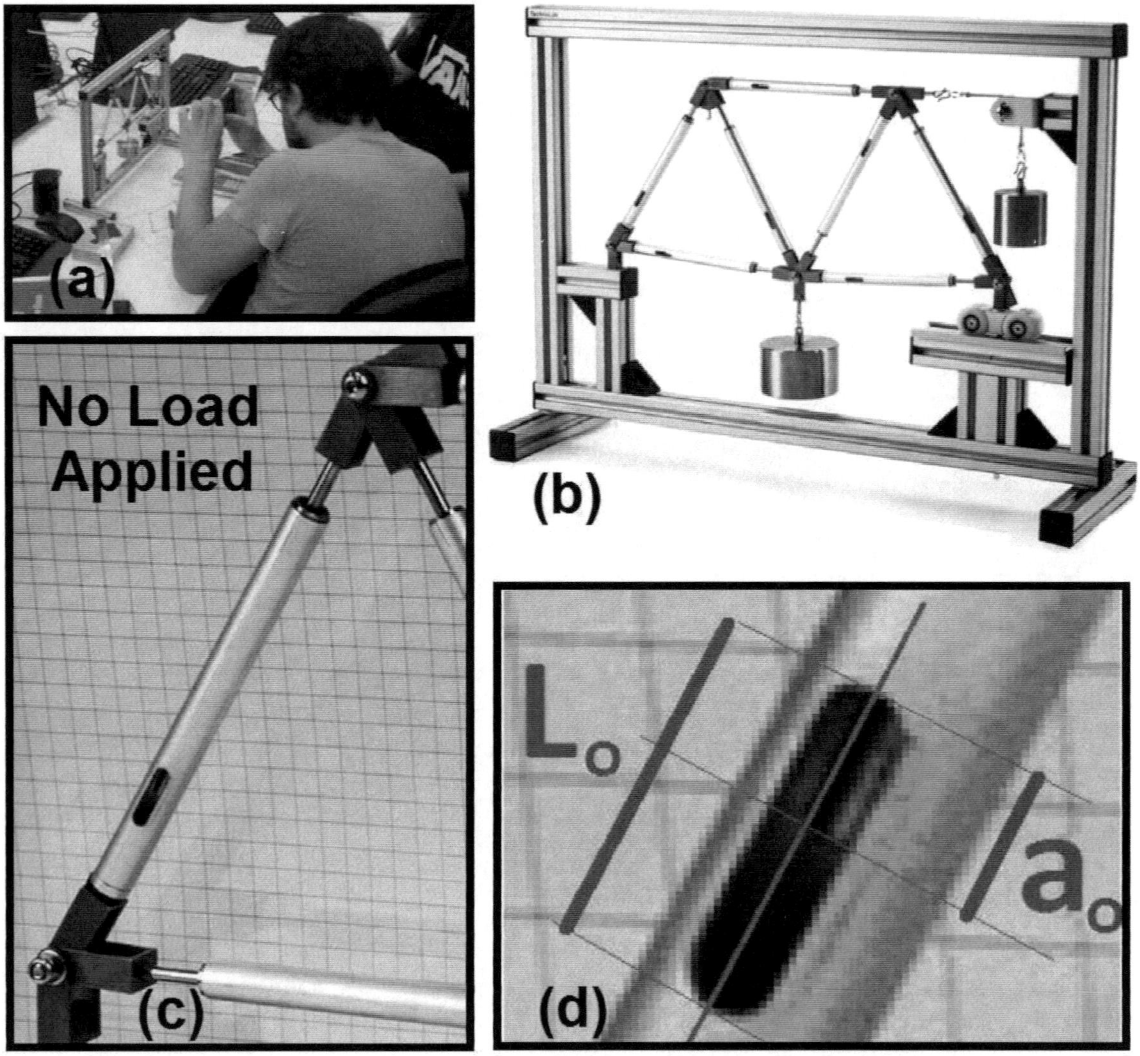

Figure 2: TechnoLab's "Forces in a Seven Member Planar Truss" Experiment
(a) Using a phone camera to take a "close-up" photograph of pointer within slot for Member#1
(b) Seven Member Planar Truss with Two Nodes Loaded (note perceptible nodal displacements)
(c) Digital photograph depicting the left hand side support node with slot of Member#1 clearly visible
(d) Magnified digital image of 13.0mm long slot with internal pointer clearly visible to identify "a_o" and "L_o" $\rightarrow a_o = 13.0 \times L_o$

3. TechnoLab's Structural Dynamics Experiment Kits

The TechnoLab™ kits that support the dynamics series of experiments mostly involve minor enhancements to those kits established for Statics or Mechanics tests in the series, and in a few cases, purpose-designed experiment rigs for the particular experiment at hand, (eg its ground motion simulator).

Basic to a fundamental understanding of structural dynamics is the recognition that both the stiffness properties and the distribution of mass in a structure determine its intrinsic dynamic characteristics of natural frequencies (modal frequencies) and associated mode shapes of vibration.

This predicates that stiffness properties of a structure would need be well investigated, as would the mass distribution in a structure, in order to understand its dynamic (modal) properties.
TechnoLab™ experiment kits, primarily designed for statics and mechanics, can be geared towards establishing stiffness/flexibility properties of a structure, which at the simplest level of application can lead to a Single Degree Of Freedom (SDOF) model of its dynamic characteristics, and in the more general case, Multi Degree Of Freedom (MDOF) modelling of these characteristics.

3.1 Close-coiled helical spring stiffness - enhancements for dynamics

The TechnoLab™ statics kit for the experimental determination of the stiffness of close-coiled helical springs consists of setups that allow the compression force vs spring contraction characteristics of a single spring or two springs in series or in parallel to be observed. Figure 3 depicts a typical setup for investigation of the stiffness of a pair of springs in series. A full turn of the thumb nut compresses the spring assemblage by 1mm, so that spring assembly contraction can be determined in terms of the whole number and proportion of turns of the thumb screw and the force in the springs directly obtained from reading the reaction force on the digital scales.

Figure 3: TechnoLab's "Stiffness of a Pair of Springs in Series" – Compression Experiment

Adjustments to this spring setup allows for hanging weights to be progressively added to an attachment at the end of the spring assembly to alternatively obtain the tensile force vs spring elongation characteristics of the spring configuration under test. The spring force for this arrangement is the weight force applied to the attachment. Spring elongation (displacement) is obtained from spring end position points noted on the Window Frame, and/or using a simplified photogrammetric technique not too dissimilar to that outlined in Figure 2.
It is in this version of the determination of spring stiffness that students observe oscillation of the spring-mass assembly to occur – a "gentle" introduction to a SDOF dynamic system, (see Figure 4). In terms of their Statics investigation for spring stiffness, "k", they are required to suppress these oscillations in order to identify the spring elongation for the weight force applied, (associated with the mass of the load attachment). Astute tutors may point out to these students that manually positioning the oscillating spring/mass assembly to the half-way position of the up/down oscillations would virtually eliminate these oscillations and allow easy observation of spring elongation. In addition, the tutors could also point out that, as additional incremental mass is added to the load attachment, the rate of oscillations (their frequency) progressively reduces.

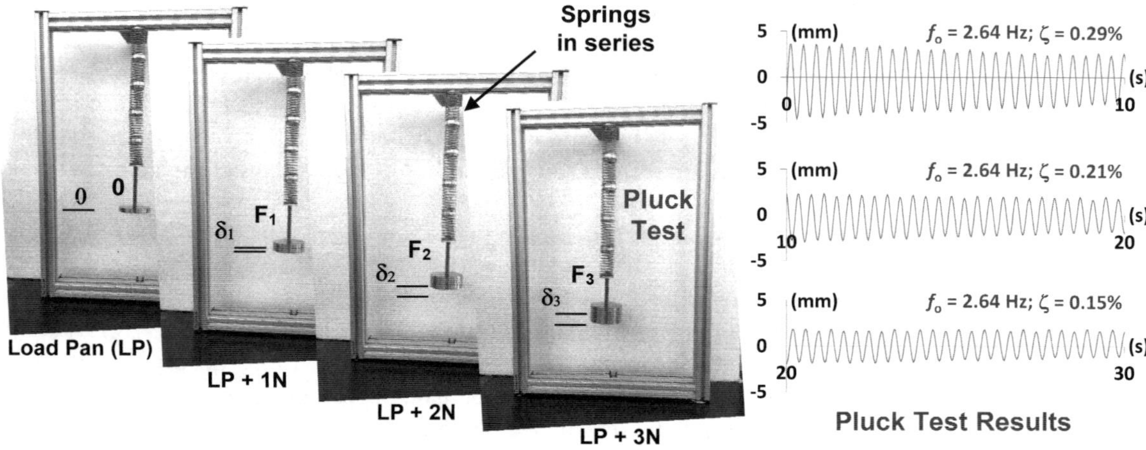

Figure 4: TechnoLab's "Stiffness of a Pair of Springs in Series" (Tension Experiment) and Pluck Test Investigation

Naturally, this simple spring/mass assembly experiment configuration for statics lends itself, after some minor adjustment, towards a more formal investigation of the dynamic properties of a SDOF dynamics system, and this is in fact what is done in the TechnoLab dynamics series of experiments. Additional masses and/or additional springs in series are assembled to provide a matrix of stiffness "*k*" and end mass "*m*" combinations for students to explore the well-known theoretical relationship for frequency of oscillations, "f_o":

$$f_o = \frac{1}{T} = 2\pi\sqrt{\frac{k}{m}} \tag{1}$$

Simple cycle counting or Tracker video processing software (Diffraction Ltd Design, 2017) can be used to obtain accurate estimation of the natural period, "*T*" in equation (1). Excel based software supplied with this TechnoLab experiment can be used on Tracker data to also evaluate amplitude decay properties (damping) of the spring/mass assembly under test, (see example in Figure 4).

3.2 Simple Flexural Systems - enhancements for dynamics

The TechnoLab product series also supports a number of simple flexural experiments in statics:

- simply supported beam with point loads (point masses) or distributed load ("continuous" mass)
- simply supported beam with overhang (point/distributed loads)
- cantilever beam with point loads (point masses) or distributed load ("continuous" mass)
- two-span continuous beam (point/distributed loads)
- propped cantilever with point loads (point masses) or distributed load ("continuous" mass)

using "ruler-like" rectangular cross-sections in various materials, including steel, aluminium, timber, polycarbonate, carbon fibre amongst others.

These setups can be easily configured to assess associated dynamic characteristics of these flexural systems in a manner similar to that used for spring/mass assemblages.

A simple "pluck" test can be performed on the setup concerned, (Figure 5 is relevant here) and a procedure similar to that used in §2.1 for spring/mass assemblages can be adopted to evaluate natural period, "*T*", (and therefore natural frequency "f_o"), and critical damping ratio "ζ_o" (when Tracker software is used to record the response), for the primary mode of vibration. TechnoLab experiments designed for statics can thus also cost-effectively be used to study structural vibration.

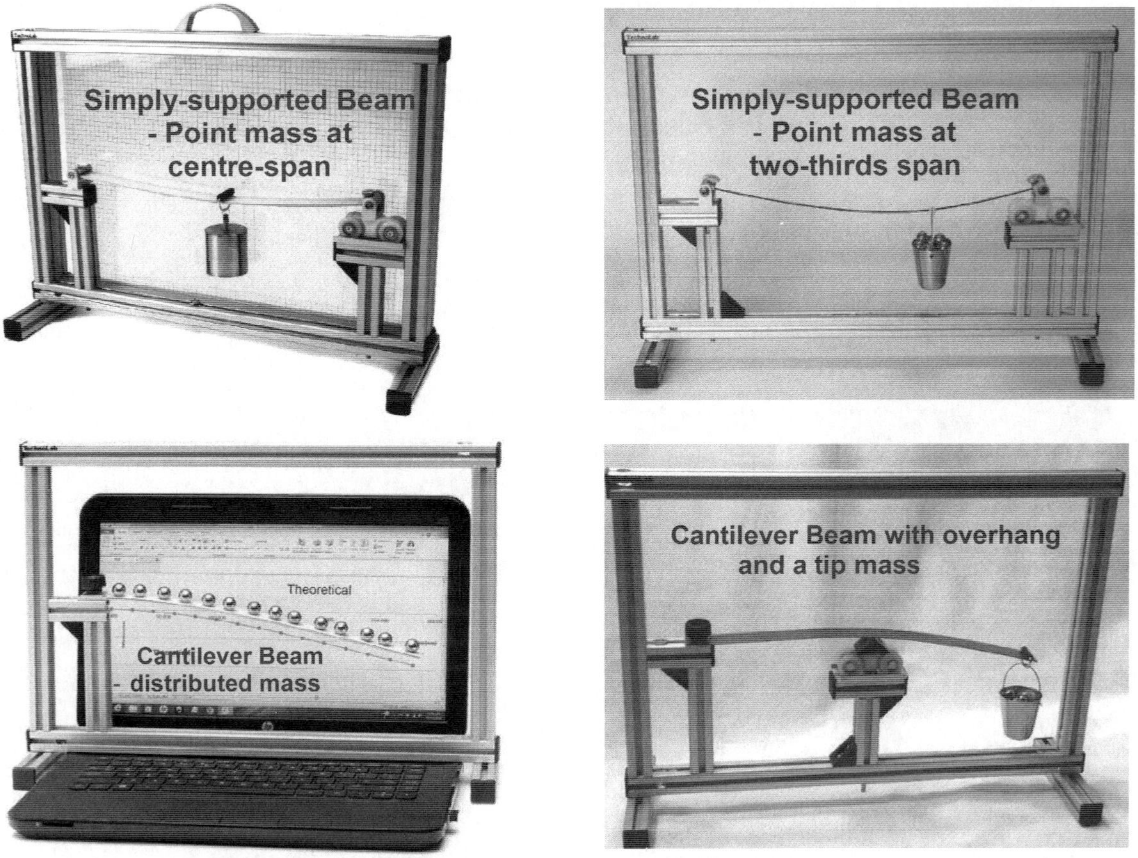

Figure 5: TechnoLab's Simple Flexure – Dynamics Experiments

Figure 6 depicts the experimental results of a pluck test on a cantilever with distributed mass. The experimentally obtained natural frequency and mode shape of the first mode of vibration are compared by students with their theoretical predictions and found to be in good agreement. Damping values are observed to be quite small and show amplitude dependent.

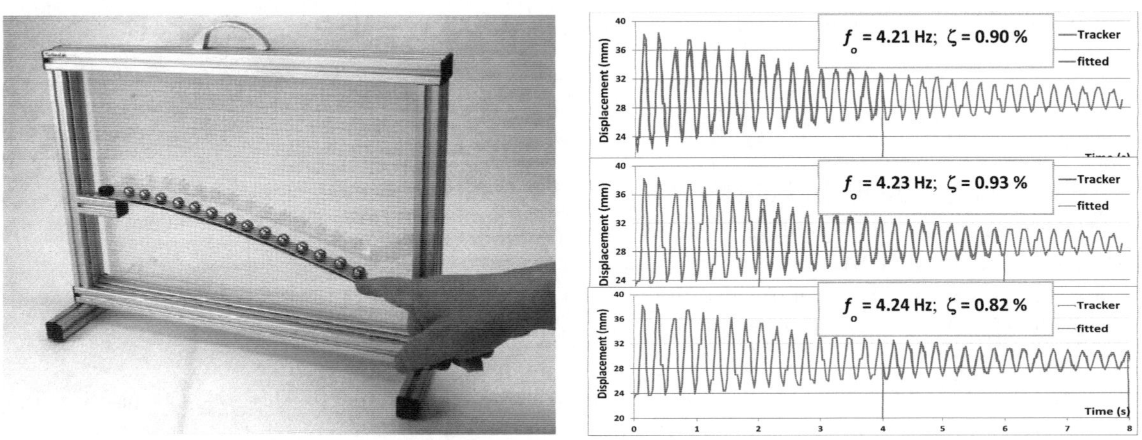

Figure 6: Results from a Pluck Test

3.3 Two Storey Sway Frame System – Dynamic Properties

Figure 7 depicts a TechnoLab model two-storey sway frame, with nominally identical column members (of uniform rectangular cross-section) throughout. The "Mass" on the top storey is largely associated with an iPhone5 that is connected mid-span on the stiff beam. Two steel strips with a mass equivalent to the iPhone5 are bolted onto the stiff beam of the first storey.

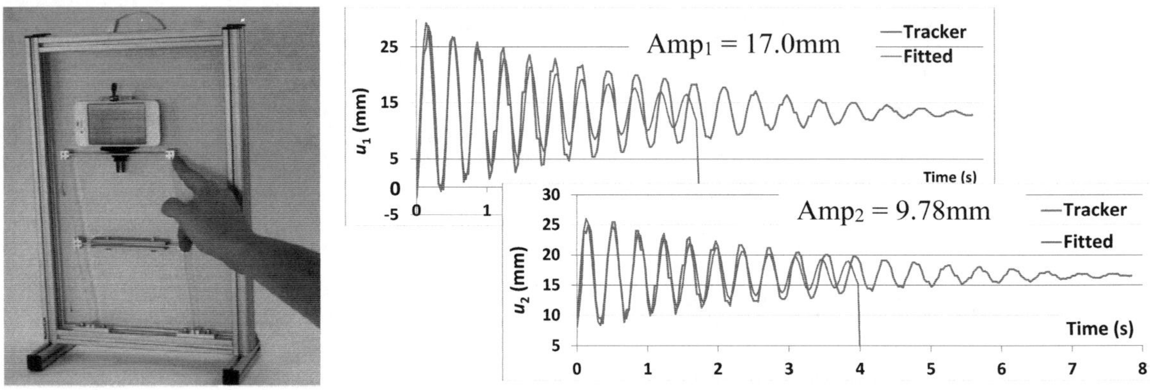

Figure 7: Two-storey Sway Frame – "Tracker" Verification of "Vibration" App Results

Modal Analysis: For a sway frame, the storey stiffness, "k_s",is given by:

$$k_s = 2 \times \frac{12EI}{\ell^3} = \frac{24EI}{\ell^3} \qquad (2)$$

where "EI" is the flexural rigidity (product of Young's modulus "E" of the column material and second moment of area of its cross-section, "I"), and "ℓ" is the column length.
If "u_1" is the sway displacement at the top storey and "u_2" the sway displacement of the first storey, then the stiffness and mass matrices, [K] and [M], respectively, in this two-storey sway frame, become:

$$[K] = k_s\begin{bmatrix} 1 & -1 \\ -1 & 2 \end{bmatrix}; \quad [M] = m\begin{bmatrix} 1 & 0 \\ 0 & 1 \end{bmatrix} \qquad (3)$$

where m is the mass of storey#1 and #2.
A traditional eigenvalue analysis on these matrices yields:

$$f_1 = 0.098\sqrt{\frac{k_s}{m}}, \quad f_2 = 0.258\sqrt{\frac{k_s}{m}}; \qquad [\Phi] = \begin{bmatrix} 1.0 & 1.0 \\ 0.618 & -1.618 \end{bmatrix} \qquad (4)$$

where "f_1" and "f_2" are the modal frequencies in Hertz and corresponding mode shape vectors are associated with successive columns of the eigenvector matrix, [Φ].
Experimental Model Stiffness Matrix: The stiffness matrix for the experimental model $[K]_m$ can be determined using a simple experimental procedure. This procedure requires the evaluation of the Flexibility Matrix of the experimental model, as an intermediate objective, the steps of which are listed as follows:

- Measure the displacement in the top storey and in the bottom (first) storey, $u_{1,1m}$ and $u_{2,1m}$ respectively, due to a horizontal force P_1 applied to the top storey.
- Evaluate entries for the first column of the model flexibility matrix, $[F]_m$: $f_{1,1m} = u_{1,1m} / P_1$ and $f_{2,1m} = u_{2,1m} / P_1$

- Measure the displacement in the top storey and in the bottom (first) storey, $u_{1,2m}$ and $u_{2,2m}$ respectively, due to a horizontal force P_2 applied to the bottom (first) storey.
- Evaluate entries for the second column of the model flexibility matrix, $[F]_m$: $f_{1,2m} = u_{1,2m} / P_2$ and $f_{2,2} = u_{2,2m} / P_2$
- Invert the Flexibility matrix $[F]_m$ to obtain the model stiffness matrix $[K]_m$, so that:

$$[K]_m = [F]_m^{-1} = \begin{bmatrix} f_{1,1} & f_{1,2} \\ f_{2,1} & f_{2,2} \end{bmatrix}_m^{-1} = \frac{1}{f_{1,1} \times f_{2,2} - f_{1,2} \times f_{2,1}} \begin{bmatrix} f_{2,2} & -f_{1,2} \\ -f_{2,1} & f_{1,1} \end{bmatrix}_m \quad (5)$$

Special Result for Perfect Sway Frame. For a perfect experimental frame (all columns identical), it would be observed that: $f_{1,1m} = 2 f_{2,1m}$ and $f_{1,2m} = f_{2,2m} = f_{2,1m} = 1/k_s$. Consequently,

$$[K]_m = \frac{1}{f_{1,1} \times f_{2,2} - f_{1,2} \times f_{2,1}} \begin{bmatrix} f_{2,2} & -f_{1,2} \\ -f_{2,1} & f_{1,1} \end{bmatrix}_m = \frac{1}{\frac{2}{k_s} \times \frac{1}{k_s} - \frac{1}{k_s} \times \frac{1}{k_s}} \begin{bmatrix} \frac{1}{k_s} & -\frac{1}{k_s} \\ -\frac{1}{k_s} & \frac{2}{k_s} \end{bmatrix}_m = k_s \begin{bmatrix} 1 & -1 \\ -1 & 2 \end{bmatrix} \quad (6)$$

which is in agreement with the theoretical form for a "perfect" sway frame in Equation(3).
Tracker data captured from the video recording of both the top and bottom storey motion of the frame when subjected to "Pluck Tests" with different initial shapes before release can be used to identify both modal frequencies and corresponding mode shapes. Similarly, acceleration response records of these "Pluck Tests", captured using the iPhone App "Vibration" (Diffraction Ltd Design, 2017), even though the measurement may be conducted at only one point, can also be used to not only identify the modal frequencies, but also their corresponding mode shapes. Figure 7 depicts Tracker analysis results for a "traditional" Pluck Test with initial displaced shape of $u_2 = u_1/2$, the static displaced shape for a force-induced horizontal displacement of the top storey in order to verify results obtained independently from the "Vibration" App – a TechnoLab student experiment exercise for this experiment setup. The modal amplitude ratio of the first story to the top storey is $9.78/17.0 \simeq 0.58$ which is slightly below the result for a "perfect" uniform sway frame of 0.62.

4. TechnoLab's Linear Actuator – Shaker Systems

TechnoLab has developed a number of Shaker Systems to act as ground motion simulators (eg for observing simulated earthquake response of structural models) and/or for identification of mode shapes and natural frequencies using Swept Sine Wave (SSW) ground motion. In their simplest and smallest form they are best suited to demonstrating basic structural dynamics concepts in a very visually graphic way – eg for the concept of resonance; distinction between forced and free vibration, etc. The example use of a model with a footprint of 110mm by 60mm is considered first. Figure 8 depicts the Tracker produced response traces for the point mass horizontal motion to a SSW ground motion input (frequency of Sine linearly varying from 0.5 to 16.5 Hz over 64 s) for a sequence of three simple lightweight cantilevers (lengths 240, 215 and 185 mm respectively), with the same point mass attached at their tip. The traces are shown alongside "double exposure" images produced by superimposing four consecutive video frames of each cantilever to provide a visual indication of the mode shape. The resonant behaviour is clearly observed and the resonant frequency, f_o, easily identified from the time associated with observed peak resonant response, t_m, as: $f_o = 0.5 + t_m/4$ Hz, and noted within the time traces. The SDOF model prediction for each cantilever is presented in brackets within the cantilever images. A small rotation observed at the cantilever support suggests the models are a little more flexible than "true" cantilevers. This would largely account for values in observed resonant frequencies being lower than theoretical values.

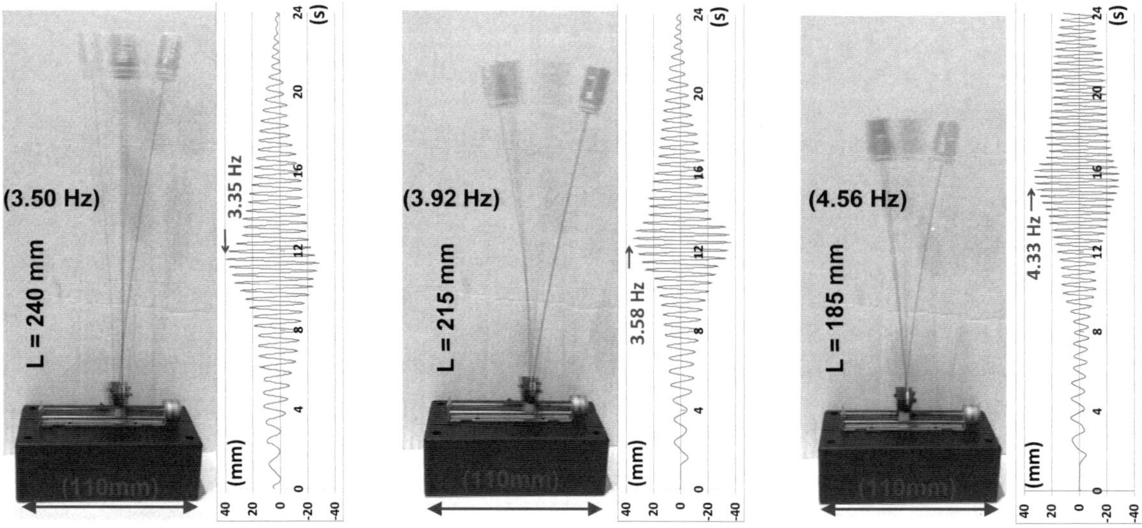

Figure 8: Double Exposure Consecutive Images at Resonance of Three Cantilevers subjected to SSW Ground Motion

The more advanced shaker systems from TechnoLab, in addition, can also be used to perform more formal experimental setups and investigations such as for: single and multi-degree-of-freedom dynamic systems; the influence of damping on response; Response Spectra for earthquakes, Flexural-Torsional coupling in idealised physical models of low-rise building structures, etc.
Figure 9 depicts the first two modes of a 3-storey building frame model obtained from testing using SSW ground motion, set to vary linearly from 0.5 Hz to 16.5 Hz over 64 s, from the touch panel.

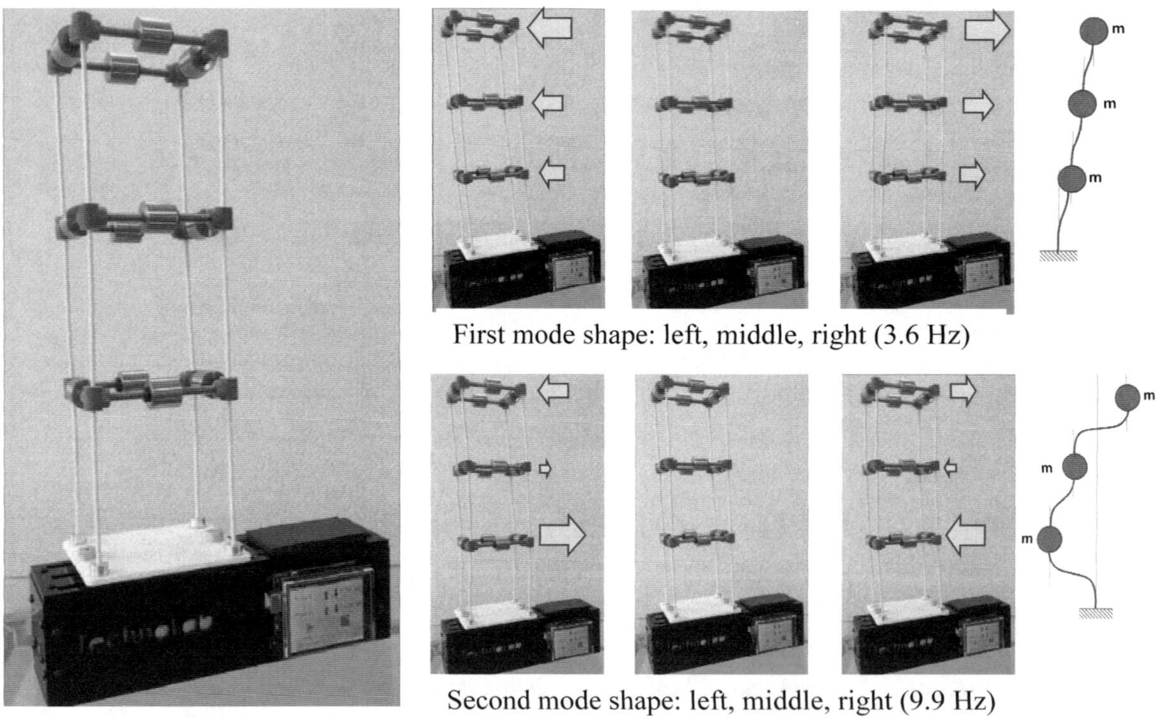

Figure 9: Ground Motion Simulator with Touch Panel Selection for Preset Earthquake Records and SSW Control

The Tracker records of the relative displacement to ground of the top corner of each of the three storeys are depicted in Figure 10 below. An iPhone6 was used to produce the video (30 frames per second) investigated by Tracker. The relative movement records for the 2nd and 3rd storeys have been displaced by 30 and 70 mm respectively to visually separate details of these records from the 1st storey response, as otherwise these records would be superimposed on each other.

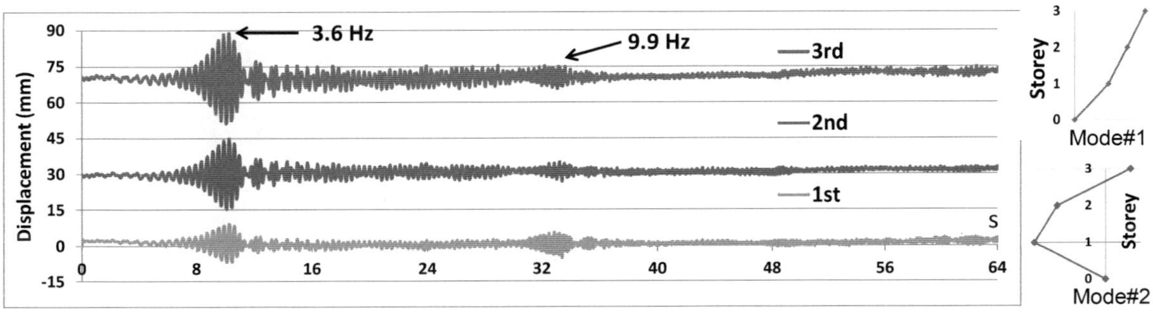

Figure 10: Relative Displacement Response of 3 Storey Model Building to SSW (0.5 to 16.5 Hz in 64s) Ground Motion

The resonant "build up" of displacement through the resonant frequencies of the two identified dynamic vibration modes and the relative magnitudes of the displacement at the three storeys of the associated mode shapes can be gauged from inspection of the peak to peak displacements at the two resonant frequencies. The peak response shapes at the resonant frequencies are reproduced in normalised form alongside the response record traces in Figure 10. These response shapes are seen to match very well with the theoretical mode shapes for a "uniform" three-storey sway frame building depicted in Figure 9.

4 Concluding Remarks

Experiment kits for TechnoLab's desktop Pixi-Frame based system for statics and mechanics can in most cases be easily formulated to perform studies of the dynamic properties of the specimens under test in those experiments. Selected examples of such applications have been presented in this paper principally featuring the use of video processing photogrammetric software, Tracker, for performing the required analysis. In addition, an example has also been presented for the dynamic property investigation of a two-storey sway frame where the mass of an iPhone was integrated within the physical model and where the App Vibration on the iPhone was used to capture acceleration response data for performing the dynamic analysis in conjunction with Tracker.

Two of the models from the TechnoLab's desktop ground motion simulator series and their use in classroom demonstrations and for performing structural dynamic property investigations of model building structures have also been demonstrated in this paper.

A key outcome from the early use of these desktop dynamic investigation kits and the small scale experiments these can only accommodate is the immediate need for a lightweight wireless tri-axial accelerometer and data acquisition software for accessing streamed data simultaneously and contemporaneously from a set of such accelerometers for conducting modal studies on 3-dimensional structural forms at laboratory scale. TechnoLab is currently developing such a system.

References

Diffraction Limited Design LLC 2017, 'Vibration', https://itunes.apple.com/au/app/vibration/id301097580?mt=8

Tracker 2017, Video Analysis Software Version 4.9.8 http://www.opensourcephysics.org/items/detail.cfm?ID=7365

The 8th International Conference on
Structural Health Monitoring of Intelligent Infrastructure
Brisbane, Australia | 5-8 December 2017

Structural Health Monitoring of PSC Girder Using Embedded Fibre Optic Sensors

P. Banerji[1], S. Chikermane[2], P. Nellisseri [2]
[1] *IIT Bombay – India, Email:* pbanerji.iitb@gmail.com [2] *IIT Roorkee - India*

Abstract

Prestressed concrete girders form a significant component of the bridge stock of Indian railways. The long term performance of these bridges is a cause of concern to the bridge authorities and several of these bridges have shown varying degrees of deterioration over the time of their use. The problem is further exacerbated by the immense problem of dismantling and disposal of these girders if they are found to be inadequate. While several tests have been conducted in laboratory conditions, a full scale real life testing on a bridge intended for mixed traffic of freight and metropolitan transport is relatively rare.

With a view to a better understanding of the behaviour of these girders, one prestressed girder has been comprehensively instrumented with fibre optic sensors that were embedded in the bridge during construction. Data has been acquired for this girder over the different stages of its construction including curing, prestressing over different stages, erection and finally a comprehensive load test on the girder.

The entire procedure has been described in this paper, along with the data acquired over different stages, and conclusions have been drawn with comparisons with design stresses estimated for the girder.

1. Introduction

Indian Railways has traditionally used both steel girders and PSC (Pre-stressed Concrete) girders for their bridges. While PSC girders are largely cost efficient, they have been posing regular problems for the railways through issues such as loss of pre-stress, cracking of girders etc. The authors had also worked on a distressed girder for Indian Railways in the past (Banerji et al 2013). With the view of better understanding the long term behaviour of PSC girders, Indian Railways commissioned a project that involved embedding optical sensors in a concrete girder during the construction process. These sensors would then give the complete load history of the bridge from its casting to commissioning and during its design life.

This paper describes the procedures followed, and gives some of the data acquired during the various phases of construction and attempts to correlate these results with the design expected values.

2. Description of Bridge

The bridge is a viaduct that has been built to augment the suburban rail network of Mumbai. Although the traffic intended on this bridge is predominantly suburban passenger trains, the bridge has been designed for an eventuality of freight loads also using the same network. Since these loads are significantly larger than the suburban loads, these predominate the bridge design. The bridge is shown in Figure 1.

Figure 1 – view of the bridge under construction

The grade of concrete used in the precast girders is of M50 – generating a cube strength of 50 MPa at 28 days and the concrete used in the cast-in-situ components such as the deck slab is of grade M35 generating a 28 day cube strength of 35 MPa.

The volume of the pre-cast girder is 30 m^3 and the volume of the deck slab is 40 m^3, giving a total volume of concrete of 100 m^3 (2 girders + deck slab). The entire concreting of the pre-cast girders is done in a single pour and so too is the deck slab – hence reducing the number of cold joints in the structure.

In order to inhibit the effects of corrosion, all the reinforcement bars used for the girder are epoxy coated and there is a provision on site to epoxy paint all the cut edges to ensure there is no direct metal open surface. Similarly the binding wire used to tie the reinforcements is also PVC coated to prevent and direct exposure of metal.

3. Instrumentation and Data Collection

A detailed description of fibre optic sensors has been done by many researchers (e.g. Scott et al 2015) and these sensors were preferred for the particular project due to their robustness and the ability of giving noise free data, which is the particular advantage of fiber optic sensors. The technology used by the sensors was a Fiber Bragg Grating that is burned directly on the optical fiber. The strain sensors also have a companion temperature sensor along them. While the strain sensors are directly temperature compensated, additional compensating is done using the companion sensor. In all two girders were instrumented with a total of 40 optical strain sensors and 14 additional temperature stand-alone sensors.

3.1. Instrumentation Scheme

The data interrogator used was a Micron Optics Sm125 instrument. This is a sixteen channel instrument with each channel having the capacity of measuring frequencies between 1510 and 1590 nano meters. The fundamental principle used in fiber optic technology as applied here is that there is a shift in wavelength caused due to strain in the Bragg grating burned on the sensor. This shift is calibrated to estimate the level of strain the sensor is being subject to. The strain sensitivity being in the order of pico-meters, it is possible to intelligently design sensors so that several sensors can be multiplexed on a single channel. This was done in this project to ensure about 5 sensors on each channel, the rest of the channels being left open for redundancy. The multiplexed sensors hence form what is known as sensor arrays.

It was decided to have an instrumentation at end-spans, quarter spans and mid-spans. The sensors applied at mid-spans and quarter-spans were at the top and bottom of the section to estimate the bending strains coming on the girder. This was done because this is the most predominant strain in these types of systems.

At the end span, the diaphragms were instrumented to estimate the level of strains coming at these locations. Diaphragm design is among the most complicated part of PSC girder designs. The instrumentation has been done in this manner to actually estimate whether the diaphragms are actually being subject to the loads that they have traditionally been estimated for.

The instrumentation scheme has been shown here in Figures 2-3.

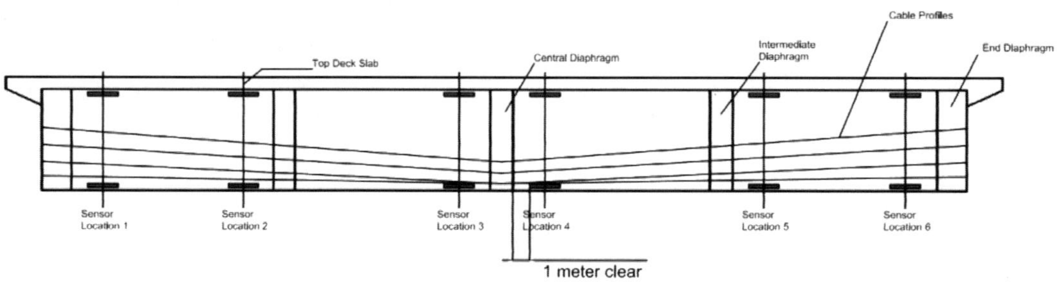

Figure 2 – location of sensors on girders

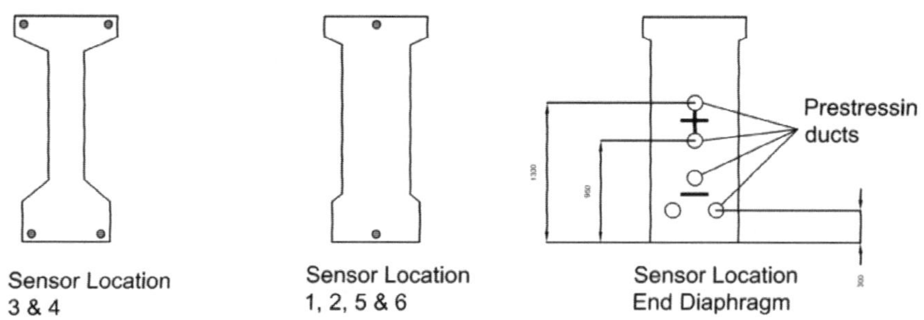

Figure 3 – sensors at instrumentation locations shown in Figure 2

3.2. Load Cases

All events classify as a load case. The entire sequence of events from casting the girder to final commissioning have been recorded. Since the actual commissioning is still pending, a full scale load test was conducted on the girder. This load case incorporated 125% of the design load and maintained it in place for 24 hours before unloading. The entire sequence of load cases presented here are:

a. First stage prestressing – where the three ducts in the centroidal axis of the girder are prestressed as shown in Figure 3
b. Hoisting the girder and putting it in place on its supports – to see if there are any erection stresses generated
c. Casting the diaphragms and the slab – majority of the dead load is now in place
d. Second stage prestressing – can only be done after casting the slab as these ducts are not centroidal on the girder and hence if prestressed earlier will cause large torsional stresses in the system
e. Full load test and removing the load to see if the initial stresses are regained

In addition to these the first few hours of the girders would generate some hydration stresses. A different paper has been submitted focusing on these specifically. Since the hydration stresses – if less than the modulus of rupture are not very significant otherwise, these are not being dealt with here. It is assumed in this paper that these stresses are lower than the rupture modulus and are hence insignificant.

4. Response of the girder for different load cases

Due to the volume of the data, only the results for the centre span are being presented here. This is because these are the most critical results for the design of the girder as the maximum stresses during service operations would occur at these locations.

4.1. First Stage Prestressing

The results due to the first stage prestressing are shown in Figure 4. The first stage prestressing comprised prestressing of three cables in the centroidal axis of the girder. The sequence of prestressing was from top to bottom, where the top cables were prestressed first, the middle ones next and the bottom ones last. The profile of the cables are shown in Figure 2.

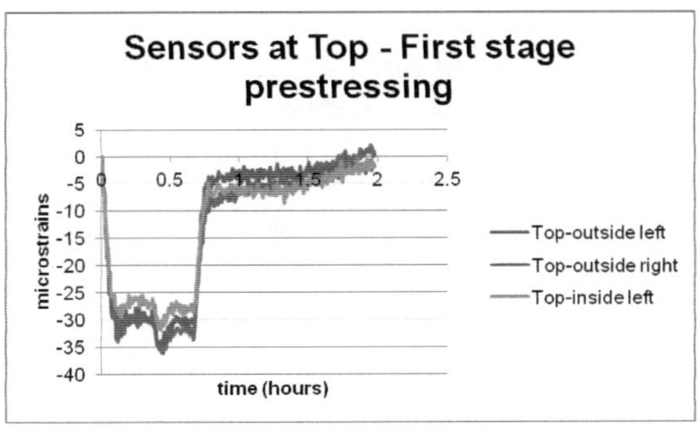

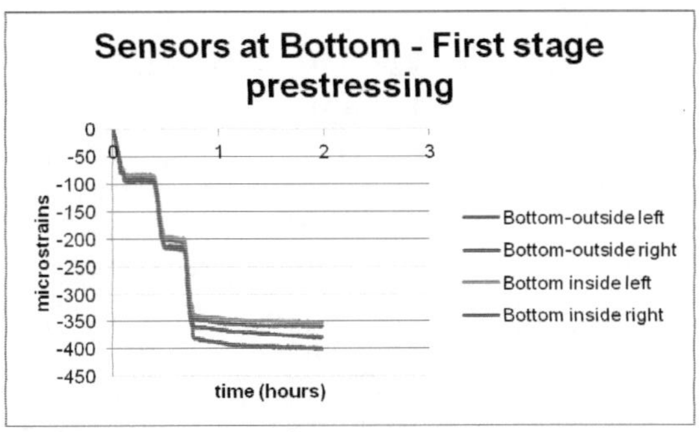

Figure 4 – response due to first stage prestressing

The locations of the sensors are defined as the final placement on the bridge and looking from the instrumentation hut. Inside is the face that is towards the other girder and outside looks towards the hut. Left and right are also accordingly defined. Further, in this paper, for brevity these will be referred as for example TIL (Top-inside-left) etc.

4.2. Lifting the girder and hoisting it

This is to identify any erection stresses coming in the girder. The lifting was done in two stages, first the girder was lifted off the shuttering plate and placed on the ground on blocks and then it was hoisted up to be placed on top of the pier cap. All the data has been spliced on the ends of each other to give a composite picture of the entire lifting process. The entire process can be seen in Figure 5.

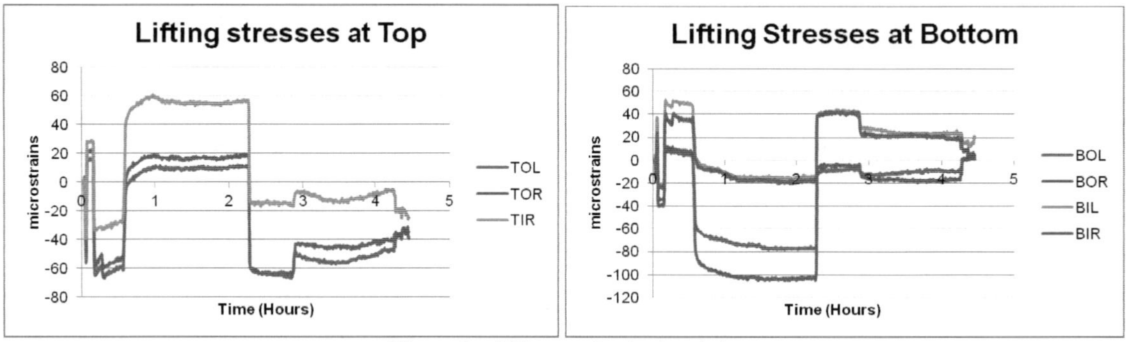

Figure 5 – lifting strains in the girders

Very interestingly in the above figure it is seen that although there are significant stresses induced in the girders due to lifting, the final stress levels are back to almost zero levels. All load cases are calibrated to ensure that starting is at zero so the actual picture of the effect of the load case can be directly seen. Actually, it is expected that the dead load strains of the girder would occur during the lifting off the shuttering plate. This effect is actually not seen here. It is hence postulated that during the prestressing operation, due to an eccentric load being applied on the girder, the girder actually "lifts" off the shuttering plate and the effect of the dead load strains are incorporated in the prestressing strains seen.

4.3. Casting of diaphragms and slabs

The results for casting of the slabs and Diaphragms are given in Figure 6.

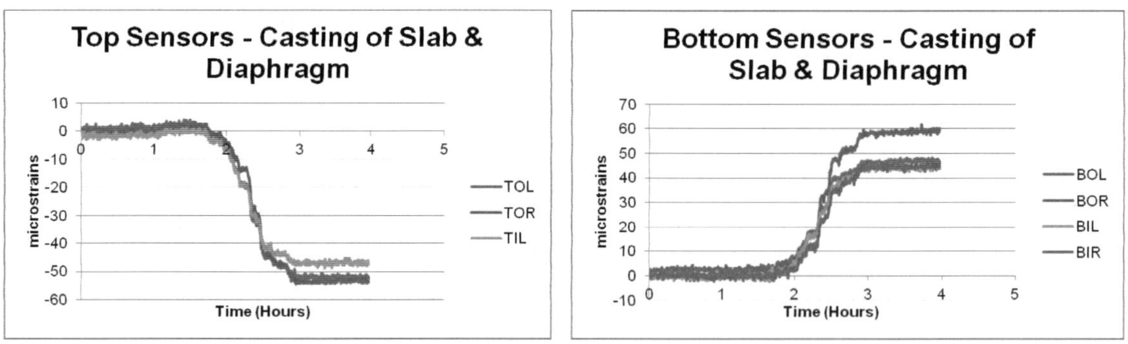

Figure 6 – Response at top and bottom during casting of slabs and diaphragms

4.4. Second Level Prestressing

After casting of the slab, the second level prestressing is undertaken. This prestressing is primarily to deal with the live load and since it incorporates tendons which are not on the girders centroidal axis, it can only be done after the slab has been cast. The response results are given in Figure 7

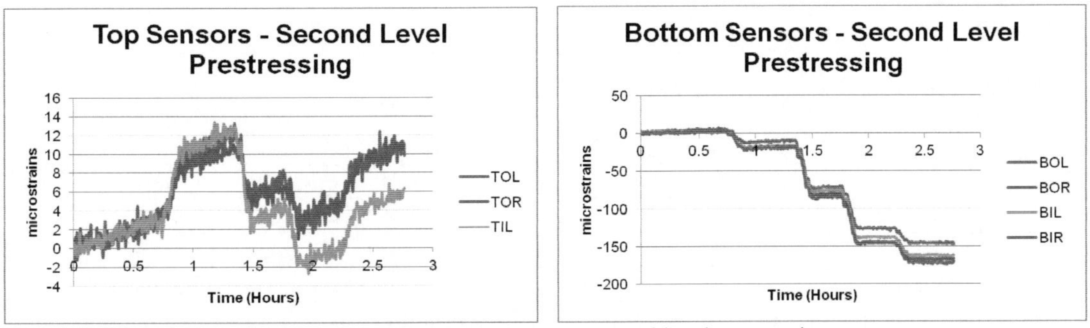

Figure 7 – response due to second level prestressing

Interestingly it can be seen here that while the top sensors are virtually immune to prestressing and indicate negligible response, the bottom sensors are showing 4 peaks. Two of these are due to prestressing of the instrumented girder, and the other two are due to the prestressing of the companion girder. Hence, we can see two major jumps and two relatively smaller jumps.

4.5. Load Test

The final load applied was a load test, which was conducted using large concrete blocks. The total load was computed as per the design assumptions, and a load was applied which was 125% of this load. The total load applied was of 55 blocks of concrete of size 1.5 meters cube. Each of these blocks weigh about 8 Tons resulting in a total applied weight of 440 Tons. This load was maintained for 24 hours before being removed. The results are shown in Figure 8.

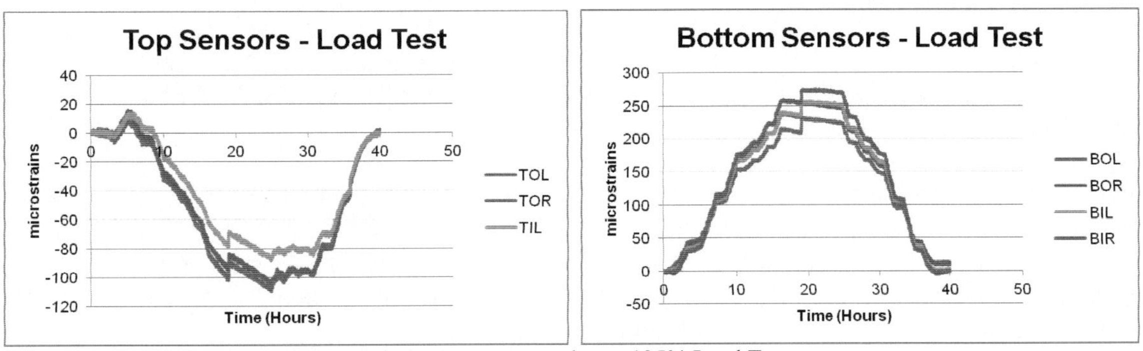

Figure 8 – response due to 125% Load Test

Again interestingly it can be seen that all sensors achieve pre-load values at the end of the load test, indicating that the structure is elastic for this level of response.

5. Analysis of Results

For each sensor, the strain value is shown and the maximum and minimum strains achieved are noted. This will give the envelop of responses seen by each of the sensors for all the load cases enumerated above.

5.1. Listing for Sensors

The listing for all the sensors is given in Table 1.

Table 1 – strain levels for all sensors

		TOL	TOR	TIL	BOL	BOR	BIL	BIR
1st prestress	Max val	2	0	0	0	0	0	0
	Min Val	-36	-35	-32	-360	-380	-354	-400
	Final Val	0	-2	-2	-359	-378	-353	-399
Lifting	Max val	17	23	61	15	15	53	43
	Min Val	-66	-65	-39	-79	-105	-21	-25
	Final Val	-34	-39	-24	1	4	20	5
Casting slab	Max val	4	3	1	48	62	46	47
	Min Val	-54	-55	-49	-2	0	-3	-2
	Final Val	-52	-54	-47	47	60	44	45
2nd Prestress	Max val	13	12	13	7	3	4	3
	Min Val	-1	-1	-3	-147	-169	-163	-174
	Final Val	10	11	6	-147	-167	-162	-170
Load Test	Max val	10	16	14	233	255	258	275
	Min Val	-110	-105	-88	-1	-1	-2	-6
	Final Val	3	-2	6	12	10	2	-1

At the end of the 1st prestress for example the value of strain at BOL is -359. This then becomes the reference for the next load case. So the maximum value in lifting would now be -359 + 15 or -344. Similarly the minimum value will be -359 – 79 or -432 and the final value will become -358. This process is carried out for all the sensors and the results are given in Table 2. This is the resultant net value considering sequence of loadings. Since the net value for 1st prestress is the same, it is not considered further.

Upon analysis of the data given in the table, the minimum value is found to be -483 microstrains and it corresponds to sensor BOR during lifting, and the maximum value is found to be 105 microstrains and it corresponds to sensor BIR during the load test. Taking a value of modulus of elasticity of concrete as $5000\sqrt{f_{ck}}$ where f_{ck} is the cube strength of concrete (IS456 2000), it works out to 35.355 GPa for concrete of grade M50 as defined earlier.
The corresponding stress values for the earlier computed strains hence work out as -17.07 MPa (Compression) and 3.71 MPa in Tension. The permissible values for concrete in compression as per the Indian code are 16 MPa and in Tension the modulus of rupture is 4.95 MPa.
Ideally the maximum tension induced in concrete needs to be safely within this value. Hence the compressive value is actually exceeded and, while the tensile value is not directly exceeded, it is too close for comfort. These values are actually recorded values and there is a high possibility that these values may be exceeded for other girders as the uncertainty of concrete is reasonably large.

Table 2 – resultant net value for all sensors

		TOL	TOR	TIL	BOL	BOR	BIL	BIR
Lifting	Max val	17	21	59	-344	-363	-300	-356
	Min Val	-66	-67	-41	-438	-483	-374	-424
	Final Val	-34	-41	-26	-358	-374	-333	-394
Casting slab	Max val	-30	-36	-23	49	66	66	52
	Min Val	-88	-94	-73	-1	4	17	3
	Final Val	-86	-93	-71	48	64	64	50
2nd Prestress	Max val	-39	-42	-34	54	63	48	48
	Min Val	-53	-55	-50	-100	-109	-119	-129
	Final Val	-42	-43	-41	-100	-107	-118	-125
Load Test	Max val	20	27	20	86	88	96	105
	Min Val	-100	-94	-82	-148	-168	-164	-176
	Final Val	13	9	12	-135	-157	-160	-171

6. Conclusions

Several interesting insights have come up during the instrumentation and subsequent analysis of the girder. Firstly, the concept that the dead load of the girder is activated along with the prestressing is not a common engineering assumption. This, however, does not vitiate any design principle but is an interesting insight.

Erection stresses are rarely considered by the designer, primarily as they are not easy to quantify and there is no standard depicting them. In the data analysis it is seen, however, that not only are they fairly high but also form one of the critical load cases. Since there is no upper bound that these may be given, it is strongly recommended that some form of representation of these needs to be made in design of PSC girder systems.

The sensors have been seen to behave quite accurately and with a good repetitive action (similar locations giving similar patterns and values) over 10 months of embedding. The exercise, thus, seems to be successful in generating data that can be quite valuable.

A detailed analysis is underway which would answer the questions as to why there are differences between the assumptions and actual responses. PSC girders are typically designed as no tension systems and the presence of a tensile stress of a magnitude so close to the modulus of rupture is not warranted. In addition to this, there would be a long term loss of prestress due to creep and relaxation – in which case, the tensile stress would definitely increase.

There seems to be, thus, several areas where the design of PSC girders as practiced may need to be augmented to generate a more reliable and robust structural system.

References

Banerji, P., Chikermane, S., Scott, R.H., Grattan, K.T.V., Basheer, M., Sun, T. and Surre, F. (2013). "*Commissioning and evaluation of a Fibre Optic Sensor system for bridge monitoring*". IEEE Sensors Journal 13(7) 2555-62

Scott, R.H., Banerji, P., Chikermane, S., Surre, F., Sun, T. & Grattan, KTV, 2015, 'Assessment of fibre optic strain gauges for field use in India', *Int. J. Sustainable Materials and Structural Systems,* Vol. 2, Nos. ½

IS-456(2000). Indian standard for Plain and Reinforced Concrete

The 8th International Conference on
Structural Health Monitoring of Intelligent Infrastructure
Brisbane, Australia | 5-8 December 2017

Monitoring the axial displacement of a high-rise building under construction using embedded distributed fibre optic sensors

N. de Battista[1,2], N. Cheal [3], R. Harvey [4], C. Kechavarzi [1,2].
[1] *Centre for Smart Infrastructure and Construction, Cambridge University – UK, Email:* n.debattista1@gmail.com
[2] *Epsimon Ltd - UK,* [3]*Multiplex Construction Europe Ltd - UK,* [4]*WSP Group - UK*

Abstract

The floor-to-floor axial shortening of vertical load-bearing elements is an important factor in the design and construction of high-rise buildings. Contractors need to allow for the expected final compression of columns and walls due to superimposed load, concrete creep and shrinkage, particularly when installing finishes and partitions in lower floors, while the building has not yet been completed. An added complication arises from the differential shortening between elements of different stiffness.
This axial shortening is predicted by designers using empirical models, in advance of construction. However, in practice, the shortening at every level cannot be measured continuously using traditional surveying measurement techniques during construction. Therefore, a monitoring system using distributed fibre optic sensors (DFOS) measuring strain and temperature, is being installed during the construction of Principal Tower, a 50-storey reinforced concrete building in London. DFOS sensors are being embedded inside two columns and two walls as the construction progresses. Using the strain and temperature data acquired from this system, the axial deformation relative to the ground level can be calculated along the whole height of the completed elements, at any time during the construction. Thus, the engineers and contractors are able to verify their predictions and adjust their assumptions if necessary.
A selection of the data acquired during the construction of the first 17 levels of the building is presented. These data have shown that the amount of shortening experienced by a member is influenced by the member's stiffness and size. The monitoring data have also revealed that thermal movement has a significant effect on the overall axial displacement of the building.

1. Introduction

The topic of axial deformation of tall buildings has been investigated for several decades. It is well known that, as a building is being constructed, the additional dead load from a newly constructed level, results in elastic shortening of the vertical load-bearing elements beneath it. In addition, these elements also deform axially due to thermal expansion and contraction, material creep and, in the case of concrete structures, shrinkage.
The net effect is time-variable and difficult to predict, as it depends on the interaction of several parameters related to the structural design, member geometry, material properties and environmental factors. Thus each individual load-bearing element will deform differently, leading to differential shortening between adjacent columns, and between columns and core walls. This can give rise to additional forces in the connecting floor slabs and beams (Pan et al. 1993), which need to be accounted for in the structural design. Differential shortening can also damage non-structural

elements such as finishes, external cladding, mechanical services and partition walls, if these are not installed with sufficient tolerance and inbuilt flexibility (Fintel et al. 1986).

The shortening of each vertical load-bearing element is estimated in advance of construction, in order to guide the pre-setting of the column and wall heights at each level. This ensures that, upon completion of the construction, the building's floors are as close as possible to the intended levels. Several methods have been proposed to assist engineers in predicting the elastic and inelastic shortening of individual structural members (Pan et al. 1993; Fintel et al. 1986; Fintel & Khan 1969), and to compensate for the expected differential shortening (Park 2003; Park et al. 2013). However, there is a great degree of uncertainty in these calculations. In addition, these predictions only account for the permanent shortening of the members and do not take into account transient thermal effects.

In order to verify the shortening predictions during construction, it is common for contractors to use surveying techniques to measure the actual shortening of individual members at a selection of levels. This usually consists of manual measurements using laser-based instruments, typically carried out once every few weeks or even months. Being one-off measurements, there can be little control, if any, over the time of day and the environmental conditions under which these measurements are taken. Therefore the transient thermal effects cannot be accounted for, making it difficult to do a meaningful comparison between the measurements and the predicted deformation: essentially one is comparing the predicted shortening due to load, creep and shrinkage, with the actual shortening due to all actions including thermal expansion or contraction at the time of measurement.

An alternative, or addition, to manual measurements is to have an embedded displacement monitoring system, such as the ones described by Kim & Cho (2005) using vibrating wire strain gauges (VWSGs), by Choi et al. (2013) using VWSG-based wireless sensor nodes, and by Glisic et al. (2013) using fibre optic long-gauge strain sensors (SOFO). Such embedded systems have thus far used point sensors, and therefore were limited to measuring displacement at just one or a few levels. This limits their usefulness as they cannot provide a measure of the local member shortening at un-instrumented levels, or of the global shortening of the whole building.

The limitations of current techniques for measuring the actual shortening of tall buildings motivated the development of a monitoring system using distributed fibre optic sensors (DFOS). The fibre optic sensing cables are embedded in reinforced concrete elements during construction, while, at the same time, a fibre optic analyser takes automated measurements of temperature and strain from the embedded cables. This system is being deployed for the first time at Principal Tower, a 50-storey building currently under construction in central London. This paper describes the monitoring system and reports on a selection of the data acquired from the first 17 levels of the building.

2. Principal Tower

Principal Tower (Figure **2.1** left) is a residential building currently under construction in central London. It is designed by world-renowned architectural firm Foster + Partners, and is scheduled for completion in mid-2018. The overall project contractor is Multiplex and the tower is being constructed by Careys. When complete, the 163 m-tall building will consist of 50 storeys above ground, overlying two basement storeys. Construction of the tower started in July 2016 and, at the time of writing (May 2017), the first 25 levels had been built, with the facade installation and internal fit-out ongoing in the lower seven levels (Figure **2.1** right).

Figure 2.1. Principal Tower: illustration showing the design of the tower (left), and the tower currently under construction (right).

2.1 Structural system and prediction of axial shortening

The structural system, designed by WSP, consists of a central reinforced concrete core, 200-225 mm-thick post-tensioned concrete floor slabs and 12 columns around the buildings' square floor plan perimeter. The ground level (level 0) has a 5 m-high foyer and the rest of the floors have a clear height of approximately 2.9 m. A section of the tower footprint, including four columns and part of the core, spans over a new railway tunnel excavated directly beneath the ground floor. This required a complex load transfer structure to be incorporated into the ground floor slab. The columns that are supported directly by this structure, rest on vibration-isolating rubber bearings.
During the design stage, WSP undertook a construction stage analysis in order to predict the axial shortening of the core walls and each individual column. The calculations account for the geometry and reinforcement area of the elements, as well as the time-dependent concrete properties and superimposed load as the building is being constructed. These predictions are being used during the construction of the tower to pre-set the concrete casting levels of the columns and walls.

2.2 Construction of Principal Tower

The basement levels were constructed using a top-down method, allowing the tower to be constructed simultaneously. The tower structure is being built using an automated jumpform, which uses an array of 20 tonne-capacity jacks to lift the rig up off the cast concrete, typically 2 days after it has been poured.
For the first nine levels, only the core was erected using the jumpform, while the columns were cast using traditional falsework and shuttering. At the tenth level (level 9), the rig was adapted and enlarged to encompass the whole floor footprint, such that the core and all the columns then started being constructed using the jumpform system. The slab construction follows a maximum of three floors lower. This was the first building in the UK to use such a technique, allowing the construction to proceed at a rate of one floor per week.

3. Distributed fibre optic monitoring system for measuring axial displacement

The monitoring system developed for measuring axial displacement is based on long-distance distributed fibre optic sensing (DFOS) using the Brillouin scattering technique. This technique is explained in detail by Kechavarzi et al. (2016) and by Motil et al. (2016). In a nutshell, a fibre optic spectrum analyser launches light pulses into a single-mode optical fibre and measures the peak Brillouin frequency of the backscattered light signal, ν_B, coming from everywhere along the fibre. The signal is discretised at regular distance intervals along the fibre, which is calculated from the return time of the light. ν_B is directly related to the strain and temperature of the optical fibre at the location where the backscatter occurred. Therefore, by comparing ν_B to the peak Brillouin frequency of an earlier baseline measurement, ν_{B0}, one can derive the change in strain and / or temperature that occurred at each measurement point along the fibre.

3.1 Fibre optic sensor system

At Principal Place, two adjacent columns (referred to as C8 and C9) and two locations in the core walls (referred to as W1 and W2) were chosen to be monitored (Figure **3.1** left). All four elements are supported directly by the basement foundations and not by the load transfer structure.
Column C8 has a rectangular profile of 2.1 x 0.35 m and terminates at level 42. Column C9 has a rectangular profile of 1.25 x 1.25 m at levels 0 and 1, an L-shaped profile of 1.95 x 0.36 m and 1.265 x 0.47 m from levels 2 to 23, a rectangular profile varying between 1.27 x 0.61 m and 2.09 x 0.56 m from levels 24 to 42 and a circular profile of 0.525 m diameter from levels 43 to 49. Walls W1 and W2 start as 0.6 m thick at level 0, changing to 0.5 m thick at level 19 and 0.4 m thick at level 36. W1 terminates at level 42 while W2 extends all the way to level 49.
Each location is being instrumented with a pair of fibre optic cables: a 4-core single-mode fibre cable with a tightly bonded nylon outer jacket and a cross-section of 5.2 × 1.3 mm, which measures strain and temperature (Fujikura JBT-03813, referred to as the strain cable), and a 4-core single-mode fibre loose-tube cable with a thermoplastic outer jacket and a 6 mm diameter, which measures only temperature (Excel 205-300, referred to as the temperature cable). The cables are cable-tied to the reinforcement of the instrumented elements during steel fixing, and subsequently embedded in the concrete (Figure **3.1** right). The sensor cables start from level 0 (ground floor) in the case of C8, W1 and W2, and from level 1 in the case of C9. The cables were prepared with sufficient length to extend all the way up to the top of the elements and wound on cable reels that are housed in the jumpform rig. This enables the rig operatives to un-reel the cables as the construction progresses upwards.
Before the columns and walls started being concreted at the ground floor, the sensing fibres in the cables were connected together with a combination of fusion splices and mechanical optical fibre connectors to form two closed-loop circuits, one for columns C8 and C9 and another for walls W1

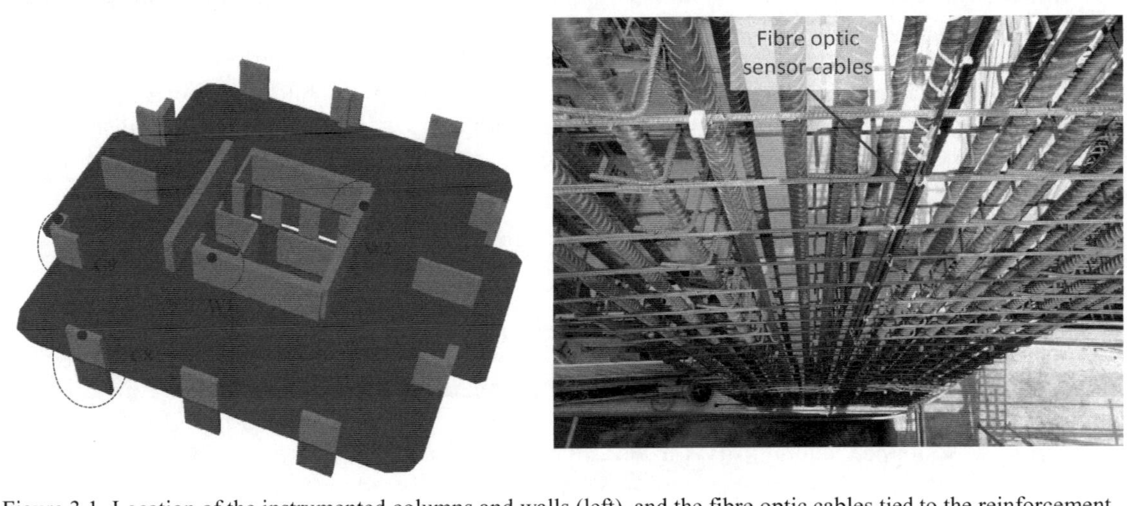

Figure 3.1. Location of the instrumented columns and walls (left), and the fibre optic cables tied to the reinforcement prior to concreting (right).

and W2. Each circuit was plugged into a channel of a two-channel Brillouin optical time domain analysis (BOTDA) spectrum analyser (Omnisens DITEST STA-R), which is housed permanently in a secure monitoring room inside a building close to the tower. Having the columns and walls monitored from separate optical fibre circuits meant that, if the instrumentation in one of the elements developed a fault or was damaged, monitoring on its respective circuit could be stopped while it was being repaired, without affecting the monitoring of the other circuit. More details about the sensor system are provided by de Battista et al. (2017).

3.2 Monitoring

Automated monitoring started on 3rd September 2016, nine days before the lowest part of the core was concreted. Since then, measurements from each optical fibre circuit have been taken at least twice every hour. The monitoring was stopped temporarily on a number of occasions, either to carry out repairs on the sensing cables following accidental damage, or due to unexpected software problems which were resolved in the early months of the monitoring.

The measurements provide the peak Brillouin frequency at 0.2 m intervals along the embedded sensing fibres, with a spatial resolution of 1 m. Thus the sensor system provides around 3000 measurements points along the four instrumented locations. Each measurement is the averaged result of 3000 individual readings and takes about 8 minutes per channel to complete. The data files are then stored in the analyser's internal memory and transferred automatically to an off-site secure server over an Internet connection.

3.3 Data processing

The Brillouin frequency data from the individual measurement files are post-processed to derive the change in temperature and total strain at 0.2 m intervals along the constructed levels of the concrete elements, with respect to a baseline measurement. Each level has its own baseline measurement, which, by default, is taken as the first measurement recorded just after midnight immediately following the concreting of the element. When data are not available at the time of concreting, the baseline is taken as the first available measurement recorded just after midnight.

The thermal strain of the concrete is then estimated from the temperature data, using a thermal expansion coefficient of 10 $\mu\varepsilon$/°C. It is assumed that the measured temperature is representative of the whole cross-section, and that the columns / walls are free to move unrestrained. By removing the estimated thermal strain from the total strain, a measure of "mechanical strain" is also obtained.

Finally, the total and mechanical axial displacement at each measurement point along the elements are estimated by integrating the total and mechanical strain, respectively. The displacement is calculated for each level individually, relative to the bottom of that level. The levels are then combined to obtain the cumulative displacement along the whole height of the instrumented elements, relative to a datum at the lowermost measurement points: the bottom of level 0 for C8, W1 and W2, and the bottom of level 1 for C9.

4. Results and discussion

The total axial displacement derived from the monitoring system during the construction of the first 17 levels of the building, are shown in Figure **4.1** for columns C8 and C9, and in Figure **4.2** for walls W1 and W2. These plots show the displacement averaged over a 1 m section (5 measurement points) at mid-height of each level. There are a number of periods with missing data when the monitoring was stopped temporarily for various reasons, as explained in Section 3.2. Since each level has its own initial baseline measurement, the displacement data following the resumption of monitoring were unaffected by the period of missing data.

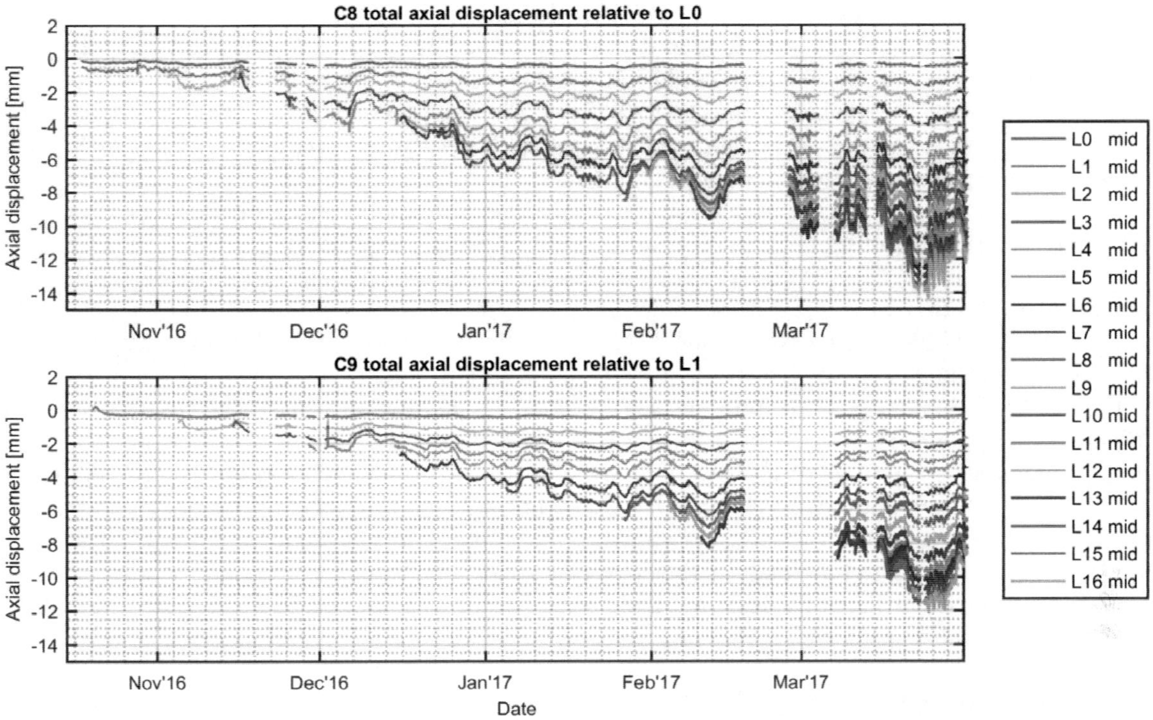

Figure 4.1. Axial displacement of columns C8 (top) and C9 (bottom) estimated from strain measurements taken between mid-October 2016 and end of March 2017. Negative displacement indicates shortening.

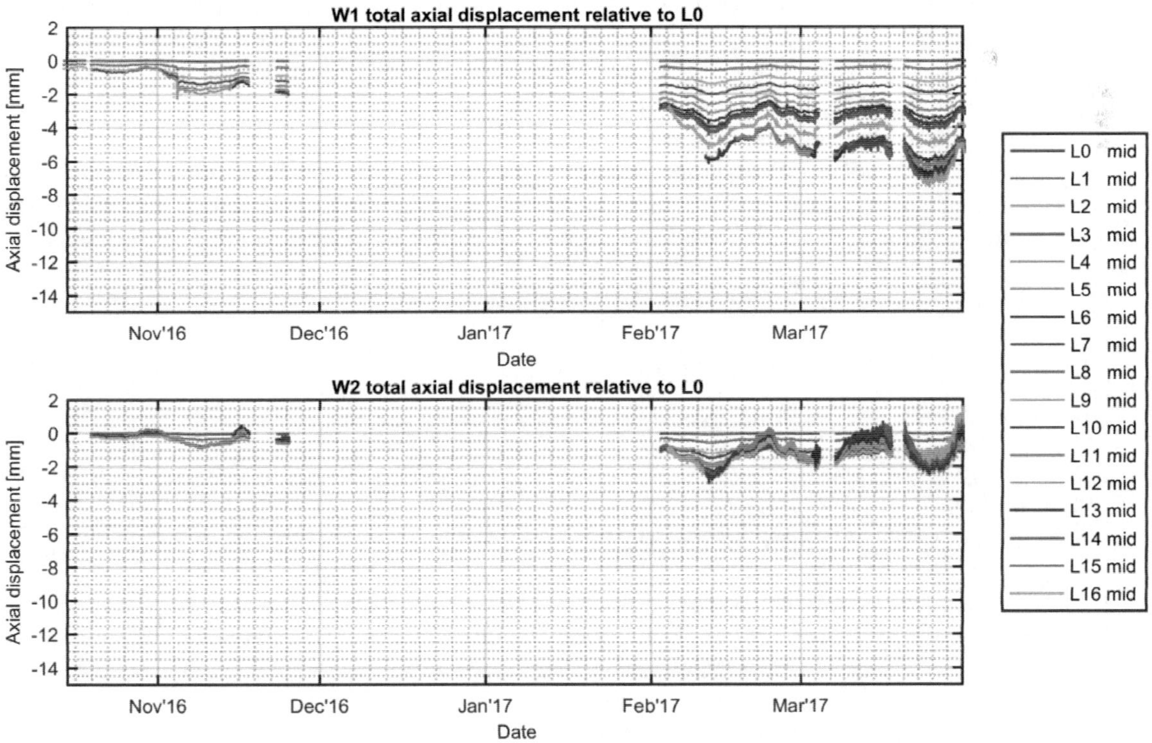

Figure 4.2. Axial displacement of columns W1 (top) and W2 (bottom) estimated from strain measurements taken between mid-October 2016 and end of March 2017. Negative displacement indicates shortening.

In the case of the walls, there was a long break in the monitoring from 24th November 2016 to 3rd February 2017. This was due to accidental damage sustained in one of the sensing cables when the core was up to level 7. The damage could not be repaired before early February 2017, since the jumpform rig was being adapted to incorporate the whole building perimeter, from December 2016 to January 2017. During this period, only one additional level was constructed on the core (level 8), for which the baseline was taken 58 days after concreting, when the monitoring was resumed.

4.1. Differential shortening

As expected, the four instrumented elements have deformed at different rates. Column C8, which has approximately half the cross-sectional area and a similar amount of steel reinforcement to C9, has shortened around 15 – 25% more than C9 (Figure **4.1**), despite supporting approximately 12% less dead load. Similarly, both columns are shortening significantly more than the core walls (Figure **4.2**), as the walls have a larger cross-sectional area and are stiffer.

While the displacement data of the two columns and wall W1 indicate a general shortening trend over time, the deformation pattern of wall W2 (Figure **4.2** bottom) is rather unexpected. W2 has shortened much less than W1, despite having the same thickness and a similar amount of reinforcement. The data show that, at times, W2 has even elongated, particularly during warmer periods. It is possible that this section of the wall, which is only 3.6 m long and bounded by perpendicular walls at each end, is more restrained. This requires further investigation.

4.2. Thermal effects

The data show that the axial shortening of all four elements did not increase uniformly with time, contrary to what is assumed in the prediction calculations. There is a strong thermal effect that gives rise to occasional fluctuations in displacement. One example is the decrease in shortening that happened during a warmer period in the second week of January 2016, when uncharacteristically high daytime maximum temperatures of around 10°C were recorded around London. Conversely, there was an increase in shortening during a particularly cold period in mid-February 2017, when the maximum recorded temperature was around 2°C.

An inspection of the mechanical axial displacement estimated from the data, shows that thermal effects can account for around half of the total deformation during temperature extremes. An example of this is shown in Figure **4.3** for column C9. This can have unexpected consequences

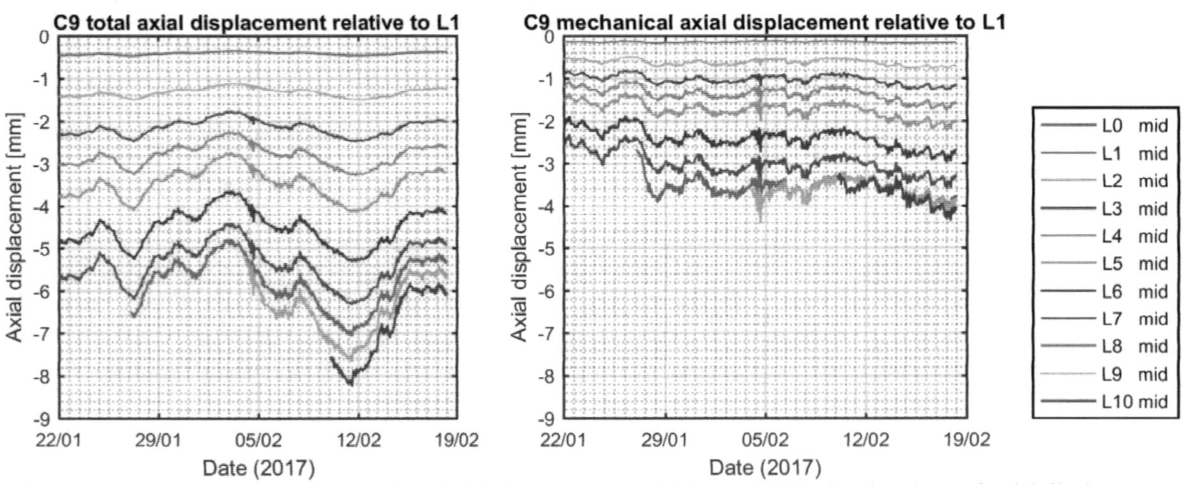

Figure 4.3. Data extract from a four week period during January and February 2017, showing the total axial displacement (left) and the mechanical axial displacement, i.e., excluding thermal effects (right), of column C9.

when using shortening estimates to pre-set column heights, particularly when concreting during hot or cold periods, since these estimates do not normally account for such transient thermal effects.

4.3. Effect of abrupt loading

Since the monitoring system is recording data continuously, it is possibly to capture events which cannot be observed with periodic or occasional measurements. Figure **4.4** shows one such example, where, at around 6pm on 4th November 2016, the cumulative shortening over six levels at wall W1 more than doubled suddenly, going from -1.1 mm to -2.3 mm between one reading and the next. Upon inspection of the construction logs, it became apparent that this one-off event happened when the jumpform rig was locked onto the core, after it had been raised from level 5 to level 6. At this point, the core concrete at level 5 was only one day old. It is likely that the transfer of the load from the jacks onto the locking pins was unusually abrupt close to W1, thus causing the sudden local shortening. Although the shortening started decreasing immediately after this event, about 0.5 mm cumulative residual deformation remained in the wall over the six levels.

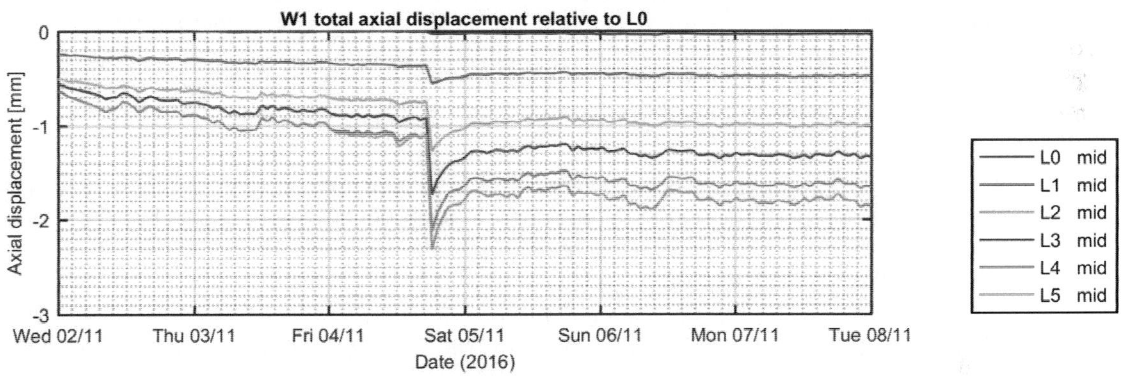

Figure 4.4. Data extract from six days in November 2016, showing a sudden increase in the shortening (negative displacement) of wall W1.

5. Conclusion

Axial shortening is an important aspect of the design and construction of tall buildings. The combined effect of elastic and inelastic shortening of individual load-bearing columns and walls needs to be accounted for in advance, during construction. However, predictions of the time-dependent shortening of concrete elements are notoriously inaccurate as they try to account for the interaction between several parameters; it is not uncommon for these predictions to be quoted with a margin of error in the region of 60%.

Therefore, it is imperative that construction does not rely on shortening predictions alone, but that contractors also measure the actual shortening as construction progresses, so that the predictions can be adjusted accordingly if necessary. This is commonly done by means of manual measurements using surveying techniques, which are typically taken once every few weeks or months, and only at a small selection of levels. These measurements are therefore of limited use.

In this paper, a novel monitoring system for continuously monitoring the axial displacement along the whole height of a reinforced concrete tall building, as it is being constructed, has been presented. The system uses distributed fibre optic sensor (DFOS) cables, which are embedded in the concrete at selected columns and walls, as the construction progresses. At the same time, a fibre optic Brillouin spectrum analyser measures the strain and temperature along the DFOS cables. The

data are post-processed to obtain the axial displacement profile along the whole height of the instrumented elements, at any point in time.
This system is being installed at Principal Tower to monitor two columns and two core walls. The data that have been acquired so far, provided the displacement time histories of the instrumented elements with unprecedented detail. These have shown how an element's shortening is affected by its profile and stiffness, with smaller and less stiff elements shortening more.
The continuous data have also shown that transient thermal effects can play a significant role in axial shortening, at times accounting for as much as 50% of the total deformation. This is a particularly important observation, in light of the fact that shortening predictions prior to construction do not take into account such thermal effects. Another advantage of having a continuous monitoring system is that it is possible to observe occasional and unexpected events. This has been illustrated with an example where the monitoring system picked up a sudden increase in shortening in one of the walls, most likely arising from an abrupt load transfer of the jumpform rig from the jacks onto the locking pins.
Even from the small selection of data that has been presented in this paper, it can be seen that a DFOS-based continuous axial shortening monitoring system can be very beneficial to contractors and design engineers during the construction of tall buildings. It is envisaged that systems similar to the one deployed at Principal Tower will be used in other tall buildings in the near future.

Acknowledgements

The authors would like to acknowledge the contributions made by (listed in alphabetical order) Giuseppe Narcisio, Jason Shardelow, Paul Fidler and Peter Knott (CSIC); Diego Padilla-Philipps (WSP Group); Mason Brown (Multiplex Construction); and the Careys site team of engineers and operatives, in particular Costel Niculcea (Careys).
This research has been made possible through funding under EPSRC grant EP/N021614/1 and Innovate UK grant 920035, as well as funding by WSP, Multiplex Construction and Careys.

The data presented in this paper can be obtained from https://doi.org/10.17863/CAM.10423

References

Choi, S.W. et al., 2013. Field monitoring of column shortenings in a high-rise building during construction. *Sensors*, 13(11), pp.14321–14338.

de Battista, N., Harvey, R. & Cheal, N., 2017. Distributed fibre optic sensor system to measure the progressive axial shortening of a high-rise building during construction. In *Proceedings of the 39th IABSE Symposium – Engineering the Future*. Vancouver, BC, Canada.

Fintel, M., Ghosh, S.K. & Iyengar, H., 1986. *Column shortening in tall buildings - Prediction and compensation*, Skokie, IL.

Fintel, M. & Khan, F.R., 1969. Effects of column creep and shrinkage in tall structures - Prediction of inelastic column shortening. *ACI Journal Proceedings*, 66(12).

Glisic, B. et al., 2013. Ten-year monitoring of high-rise building columns using long-gauge fiber optic sensors. *Smart Materials and Structures*, 22(5), p.55030.

Kechavarzi, C. et al., 2016. *Distributed fibre optic strain sensing for monitoring civil infrastructure*, London, UK: ICE Publishing.

Kim, H. & Cho, S., 2005. Column shortening of concrete cores and composite columns in a tall building. *The Structural Design of Tall and Special Buildings*, 14(2), pp.175–190.

Motil, A., Bergman, A. & Tur, M., 2016. State of the art of Brillouin fiber-optic distributed sensing. *Optics & Laser Technology*, 78, pp.81–103.

Pan, L.B., Liu, P.C. & Bakoss, S.L., 1993. Long- term shortening of concrete columns in tall buildings. *Journal of Structural Engineering*, 119(7), pp.2258–2262.

Park, H.S., 2003. Optimal compensation of differential column shortening in high-rise buildings. *The Structural Design of Tall and Special Buildings*, 12(1), pp.49–66.

Park, S.W., Choi, S.W. & Park, H.S., 2013. Moving average correction method for compensation of differential column shortenings in high-rise buildings. *The Structural Design of Tall and Special Buildings*, 22(9), pp.718–728.

The 8th International Conference on
Structural Health Monitoring of Intelligent Infrastructure
Brisbane, Australia | 5-8 December 2017

Technical Feasibility Study of Condition Monitoring and Safety Warning System of Generic Bridge Structure

Yufeng Zhang [1,2,3,4,5], Jiayi Peng [1,2,3,4,5,6], Jinhua Yang [7], Song Xu [1,2,3,4,5]

[1] Jiangsu Transportation Institute – China, Email: kolya@jsti.com [2] State Key Laboratory of Safety and Health for In-service Long Span Bridges – China, [3]Jiangsu Highway Bridge Engineering Technology Research Center - China, [4]Key Laboratory of Large-span Bridge Health Inspection & Diagnosis Technology of Transport,Ministry of Transport,PRC - China, [5]Jiangsu Health Monitoring Data Center of Long Span Bridges - China, [6] Hohai University - China, [7] Nanjing University - China

Abstract

The construction of CMSW system (which is short for the condition monitoring and safety warning system) of bridges contributes to finding and fixing safety problems of bridges in time. Building a mature CMSW system is of great economic and social benefits. Small working scope, scattered system and difficult later-stage maintenance are common problems for current CMSW systems. Aiming at these difficulties, this paper proposes a series of technical solutions. The girder deviation, stress of key parts, concrete crack, structural deformation, cable force and icing condition on bridge deck are main fields in the system. Applying prepositive A/D converter, simplifying the data acquisition system with existed transmission network of road system and enhancing the importance of discontinuous monitoring and qualitative analysis make it possible to solve above problems. The technical feasibility of CMSW technique is also discussed in this paper,.

1. Introduction

As a key component of road network in China, bridge construction involves huge investment and strategic importance. Bridge network is the main driving force of regional economic development and serves as the function of connection and promotion in economic development. It promotes economic exchanges and cooperation among economic circles, develops regional economic division and complement, improves regional investment climate, accelerates regional urbanization and globalization and promotes the human resources flow and industrial structure upgrading.
Nowadays, China's economy is witnessing a rapid development which raises a higher demand on road transportation. According to the latest statistics, the number of bridges in China is about 805,300[1], which is after United States (the number is 607,380) [2,3] in the world. The number, dimension and technical content of the bridge constructions in China all represent top levels among the world. 200,000 new bridges are about to be built in China by the end of 2020. However, the safety situation of bridges is more and more serious. Problems are found Among the large number of aging bridges. Some bridges have potential problems during the design and construction stage. Some bridges have suffered vehicle overloading frequently and are seriously damaged. The design load of some bridges cannot meet the current requirement. Due to natural disasters, some bridges have already been damaged.. For those disabled and dangerous bridges, if their injuries cannot be detected and repaired in time, sudden damage and collapse could happen and cause serious loss of life and property which brings about bad social influence. According to the latest statistics in China,

among 805,300 bridges, the number of disabled bridges is close to 100,000 and among which one fifth are dangerous. With the aging of bridges and rapid increase of traffic especially the heavy trucks, the number of dangerous bridges is still increasing in spite of repair. In recent years, the collapse of bridges happen frequently. For example, the collapse of Chongqing Qijiang Rainbow Bridge caused 49 deaths. He'nan Luan Chuan Bridge collapsed and caused 28 deaths. In July 2016, four bridges collapsed successively during nine days. They are Jiangsu Yancheng 328 Provincial Road Tongyu River Bridge (11th July), Fujian Wuyishan Gongguan Bridge (14th July), approach bridge of Hangzhou Qianjiang No.3 Bridge (15th July) and Beijing Baoshansi Baihe Bridge (19th July). Those bridge accidents caused great economic loss and indelible pain, which exerts bad influence to our society.

Based on the past experience and lesson of bridge accidents at home and abroad, the design and construction defects often showed up completely after the bridge serves 10 to 20 years. Meanwhile, it is possible that the accumulated structural damage and fatigue during long-term operation can endanger safety of bridges. However, the damage is not detectable which causes collapse. Compared to the long-span bridges, the funding, labour and recognition on general bridges are limited. Numerous generic bridges often suffer heavy loads and cannot be repaired timely. Because of the large quantity of disabled bridges and labour shortages of administrative department, it is difficult to acquire condition of bridges, especially the girder deviation (caused by earthquake, washout, centrifugal force of vehicles and temperature effect), serious crack of topside structure (caused by overloading, disrepair). For these damages, it is usually difficult to approach their site and to detect timely by manual inspection. As a result, sudden collapse could happen during operation stage of the bridge and cause great loss. Consequently, the development of simple monitoring system aiming at generic bridges is of great importance for the safety of bridge structures and transportation as well as decreasing the life and property loss As long as the possible failure is detected by the monitoring system, emergency measures can be taken like early alert and traffic control.

2. Research Status

Traditionally, the evaluation of bridge structure is conducted through manual visual inspection or measurement of portable devices. The manual inspection of bridge can be categorized into regular inspection, periodic inspection and special inspection. But manual inspection is limited in practical application for some reasons. Firstly, manual inspection needs a lot of labours and resources. Secondly, there are many blind spots during inspection. Thirdly, it is hard to analyse damage by manual inspection because of strong subjectivity. Lastly, manual inspection can influence daily traffic and is inefficient. Due to the aforementioned drawbacks and, traditional inspection is not sufficient to make timely forecast of safety condition of bridges [4,5].

With the rapid development of modern sensor technology, computer and communication technology as well as signal processing and analysis technology, structural health monitoring technology and condition evaluation technology of bridge have been greatly concerned by engineers and researchers all over the world in recent years. The structural health monitoring analysis of bridge is conducted through the measurement of data which can reflect the environmental excitation and structural response. Real-time monitoring and evaluation of bridge condition contribute to the performance of safety forecast when the structural emergency occurs. Also, the monitoring data can provide statistic support for later repair[6]. Different from the traditional inspection methods (including numerous non-destructive testing technology), safety forecast system of bridges focuses on diagnosing conditional and environmental factors that can lead to structural damages as well as evaluating symptom and trend of structural performance. Which helps provide schemes for timely maintenance and repair. As a result, the structural health monitoring and safety evaluation system has a transformative impact on bridge health system.

The upgrade of sensor and automatic remote monitoring technology in structural health monitoring makes the safety forecast system more practical which has been applied on many bridges all over the world. At the aspect of engineering application, the structural health monitoring and safety forecast system of bridge is developing towards simplicity, economy and reliability. And the system has been put into practice on Jiangyin Yangtze River Highway Bridge, Nanjing Yangtze River No.2 Bridge, No.3 Bridge, Runyang Yangtze River Highway Bridge, Sutong Yangtze River Highway Bridge and so on.

However, it should also be noted that due to the complexity and importance of long-span bridges, the system on the long-span bridges needs to focus on the overall safety, workability, reliability and some other important structural conditional and environmental factors demands a quantitative standard which is complex and costly[7]. However, for those generic bridges on the highway, relatively simple structure and low cost lead to obvious main factors that affect their safety. For example, for the bridges in mountainal area and quickwater region, main attention should be paid to deflection of infrastructures due to landslide and scour, since the deflection of infrastructures can lead to the decrease of security, stability and anti-seismic capacity. For the bridges in earthquake-prone area and typhoon-prone, serious deflection of superstructures should be prevented. So, the falling beam is the most important issue. For the bridges which have served for a long time and suffered serious overloading, main problem could be the crack of concrete. As a result, the structural health monitoring sytems of these bridges should develop towards simplicity, economy and reliability. Main attention should be paid to safety and forecast function.

Considering that there are some potential risks on generic bridges, the construction of CMSW system is helpful to detect the safety problems and solve them timely, which is of great economic and social benefits. Previously, people paid more attention to structural health monitoring technology of long-span bridges. Its application on generic bridge is limited because of funding, labour and some other factors. In fact, on the aspect of workability, the application and research of the structural health monitoring system have obtained some achievement. Compared with long-span bridgs, the monitoring item, complexity of construction of system, data processing and safety forecast of normal bridgs are relatively simple. The existed communication net of the road (especially highway) is also helpful to transmit monitoring data. Based on the existed technical condition, a CMSW system featured by reliability, effectiveness, simplicity and economy can be constructed and fit for generic highway bridges.

3. Major Technical Scheme

3.1 Principle of System Construction

CMSW system of the normal highway bridge focuses on monitoring the main indexes in the operational period and has the function of safety forecast. Attention should be paid to the main indexes, for instance, the stress variation of key components, the cracking and later-development of concrete, large structural deformation and displacement of expansion joint and bearing. The significant part of the system lies on safety forecast. Since generic highway bridges have characteristics of large scope,, relatively simple structure and limitation of funding and labour, the system should be featured by effectiveness, reliability, economy and flexibility.

1) Effectiveness: the monitoring data and forecast signal should have relatively high accuracy to avoid meaningless repetitive check and reduce the cost of emergency treatment.

2) Reliability: the system should be featured by low fault rate and relatively high durability in order to decrease maintenance cost.
3) Economy: the system should cost less and have a higher performance to fit for the numerous demands of generic highway bridges.
4) Flexibility: the sensor technology, data acquisition and transmission method should be flexible. Multiple plans should be available for selection and the system should be combined with manual inspection conveniently, so as to fit for different bridges which have the specific features.

3.2 Solution

3.2.1 Monitoring items:

According to the features of generic bridges, the current proposed seven technical approaches of condition monitoring items includes vehicle loads (overload management), girder deviation, structural stress of key parts, concrete crack, structural deformation, pull rod (hanger rod) cable tension and bridge deck icing monitoring:

(1) Vehicle Loads (Overload Treatment)

Necessity: The frequent vehicle overload can damage bridge seriously. Each year, 70% of numorous road traffic accidents are related to the vehicle overload in the country.

Concept: Vehicle Loads (Dynamic Weighing) System is a relatively mature technology, the relatively independent overload management system is structed by "High Speed Weighing Subsystem" + "Overlimited Information releasing and guidance Subsystem" + "Intelligent Vehicle License Plate Recognization Subsystem" + "Video Surveillance Subsystem". The current technical problem is how to improve measurement accuracy of dynamic weighing instrument.

Methods: dynamic weighing instrument (Curved plate, quartz, piezoelectric and so on), License Plate Capture System and Electronic Display Screen.

(2) Girder Deviation

Necessity: In real projects, the accident of girder unseating happened frequently caused by the earthquake or overload. However, in most cases, these force effects cause the girder vertical or lateral deviation instead of the girder unseating. Fortunately, the most girder deviations are not caused by only one large event, but are accumulated day by day. For instance, the girder deviation of ramp bridge can be increased by braking force and temperature. The increased girder deviations greatly reduced the ability of resistance overturning of bridges. The deviated girder is a potential danger of the bridge collapse under heavy load and eccentric vehicle load. Therefore, it is necessary to monitoring the girder deviation of the bridge which is located at a bad geological area, a bad hydrological area, a earthquake-prone area and a small curvature radius ramp bridge.

Concept: The scenario is to monitor girder deviation with displacement sensors or displacement overrun triggers. When the girder deviation increases over the threshold, the surveillance system will send alarm signal and activate the video recorder (The purpose of activating the video recorder is to facilitate the manager to inspect the girder's situation and to make a judgement in order to eliminate the false alarm). If the alarm can be confirmed, the manager can send the notification to the warning board, together with the traffic manage department to control traffic and start the emergency repairing plan to ensure the traffic safety.

Methods: Displacement sensor, wire alignment transducer, displacement trigger device, and so on.

(3) Stress monitoring in key parts of structure:

Necessity: Structural stress is not only the base of fatigue analysis of steel structure in long-span bridges, but also the base of component safety evaluation of generic bridges. Considering that the bridge structure is usually a high-order statically indeterminate structure, the internal force will transfer to other components and redistribute when the damage of bridge happens. Therefore, the stress monitoring in key parts of structure is helpful and necessary to identify the structural damage of the bridge.

Concept: Stress monitoring is a traditional monitoring item. The current method is that according to the theoretical analysis to confirm stress monitoring sites in key parts of structures with stress sensors (for example, vibrational strain sensors, resistive strain sensors and fiber bragg grating strain sensors). It is also possible to monitor the regional stress with long gauge fiber bragg grating strain sensors and then to identify structural damage by strain modal analysis.

Methods: Traditional point strain sensors (vibrational strain sensors, resistive strain sensors and fiber bragg grating strain sensors) and long gauge fiber bragg grating strain sensors.

(4) Distributed monitoring of girder crack:

Necessity: Crack is a common problem of concrete bridges, but also a direct reflection of the bridge self-safe state. Therefore, it is a significant part of routine state inspection (monitoring) in highway bridges. Observation closely should be assisted with bridge inspection vehicles and other large auxiliary equipment since the crack maybe often happens at the bottom of the beam. It will impact the normal traffic seriously. It is often difficult to find the crack in the daily inspection. Unfortunately, it is difficult to accurately predict the location of the crack after it occurs. So, the traditional 'point' type of detection equipment cannot monitor the crack information if there are some distances between the crack site and the sensor site. And laying out a large number of 'point' sensors costs too much, it also does not match the demand of crack monitoring. Therefore, the distributed crack monitoring technology is necessary to solve above problems.

Concept: Distributed sensor equipments (for instance, distributed fiber bragg grating sensor, long-gauge strain monitoring sensor, flexible conductive coating, crack monitoring smart grid) can be layed regions which are easy to crack. The detection of the bridge can be executed monthly, and the crack over 0.1mm can be detected promptly.

Methods: distributed fiber bragg grating sensor, long-gauge strain monitoring sensor, flexible conductive coating, crack monitoring smart grid and other distributed crack monitoring technology.

(5) Structural deformation monitoring:

Necessity: The structural deformation of the bridge is an important parameter for the direct evaluation of the bridge state, which is a necessary part of bridge monitoring.

Concept: Structural deformation monitoring is a traditional monitoring item. GPS and closed tube are the usual methods for monitoring structural deformation of long-span bridges. However, Under the condition of current technology, the first problem is that GPS is costly (A single installation of GPS usually costs more than one hundred thousand yuan, and in order to keep the high accuracy, the layout of base station is also necessary. So, at least two GPS should be installed on one bridge). Another problem is that the measurement accuracy of the vertical displacement can only reach about 1 cm. it is difficult to satisfy the accuracy requirements of the deformation measurement of the generic bridge. Closed tube measurement has serious delay and often needs maintenance. It is hard to satisfy the stable and maintenance-free requirement of monitoring system. Therefore, combined with the crack monitoring technology, using distributed fiber bragg grating strain sensor to measure the structural deformation by the cross section angle transformation. In our previous research and engineering experiment test, it has been proved that this method can be used to measure the structural displacement of the generic bridge, the accuracy of which can reach 0.2 mm, under the assumption of the density of two long gauge sensors per meter.

Methods: Long gauge fiber bragg grating sensor.

(6) Cable force monitoring

Necessity: Cable is one of the most significant components of the bridge. Cable force is an important parameter in the evaluation of the state of the cable and the bridge, which must be monitored. For the normal cable supported bridge on highway, as cable-stayed bridge, extradosed bridge, self-anchored suspension bridge, half-through and through arch bridge, pull rod (hanger rod) cable force is also the necessary item to be monitored.
Concept: For the relative long cable (or rod), acceleration sensor is recommended to monitor the cable force using the theory of string vibration. For the short cable (or rod), magnetic flux sensor is recommended to monitor the cable force.
Methods: acceleration sensor and magnetic flux sensor.

(7) Bridge deck icing condition monitoring:

Necessity: In winter, since the temperature of the bridge deck usually lower than that of the road pavement, the bridge deck usually get frozen before the road pavement. Cars often skid on icy deck under relative high speed and are difficult to be controlled. According to the statistics, this is one of the main reasons result in serious traffic accidents in winter.Accident happens in such situation may cause collapse problem on bridges. The best way to avoid this kind of accident is to prevent the road frozen, and to inform the people of driving slowly and carefully. However, the number of the stuffs of highway management department is limited, as well as the number of the bridges is huge. It is difficult to master the information about the icy deck and broadcast in public in time. It is also impossible to wipe off all ice in time. Therefore, if we establish the deck icing conditon monitoring system of the highway bridge, we can inform the drivers of the ice alarm to let them drive slowly and carefully. This kind of traffic accidents could decrease greatly.
Concept: The temperature and humidity on bridge deck can be monitored. Once the temperature of the deck reaches 0℃ and the humidity is high, the warning indicator could be triggered. And then, the information about icy deck could be sent to variable message sign, mobile phones, and vehicle navigation system in order to prevent the accident.
Methods: non-contact temperature sensing techniques and image and discriminant techniques for the presence of water.

3.2.2 Method of system construction

Comparing with health monitoring system of the long-span bridge, the CMSW system of highway generic bridge has the following characteristics in the system design in Table 1, which must be fully considered:

Table 1 Characteristics and countermeasures of CMSW system of the generic bridge

N.O.	environmental characteristics of implementation	countermeasure
1	**The system size of the single bridge is smaller:** compared with the health monitoring system of long-span bridge, the CMSW system of highway generic bridge is smaller.	**Simplify the monitoring item:** highlight the safety indexes (such as stress of the key parts, crack, fallen beam, large deflection, etc.). additionally, appropriate attention can be paid to the usability index of bridge which is the requirement of the management, such as winter deck icing warning; **Consider the intermittent monitoring system firstly:** in order to reduce the cost, the system can be made as simple as possible. Consider the intermittent monitoring firstly, which should be more closely integrated with the manual inspection. **Emphasis the qualitative analysis and judgment of the evaluation result:** since the monitoring indicator is less and simple, data is limited. Various indexes depends on the correlation analysis cannot be applied. Therefore, compared with the structural health monitoring system of the large bridge, the qualitative analysis and judgment will be emphasized in generic bridge.
2	**Systems are scattered strongly:** the monitoring equipment is relatively less in each CMSW system of highway generic bridge. However, the whole system includes many sub-systems in every bridge. Therefore, the sensors and acquisition system of the generic bridge monitoring system have a higher dispersion than the large bridge monitoring system which has the same volume as that of the former.	**Set A/D converter preposition as long as possible:** due to the high dispersion of data collector, the transmission distance of analog signals is limited, so A/D converter is required to be set preposition as soon as possible. **Make full use of the existing transmission network of highway system:** make full use of the existing transmission network of the road (especially highway) to transfer monitoring data. The sensor signal which has been A/D converted should be transferred to the nearby transmission network of the road by the wireless digital radio or fiber.
3	**Debugging and maintenance are difficult:** Due to the dispersion and the limited condition of the bridge, it is more difficult to debug and maintain the CMSW system of the highway generic bridge compared to that of the large bridge.	**Simplify the data acquisition system as long as possible:** the more equipment the system has, the more faults it makes. Therefore, it is necessary to simplify the monitoring system to ensure stability under unattended operation situation.

Based on above characteristics, the CMSW system of highway generic bridge will give priority to adopt the sensor signal into a special single chip microcomputer system. After the A/D conversion (when the sensors are few, connecting sensors directly to the A/D conversion module can be considered), the data should be transferred to the nearest highway electromechanical engineering monitoring system. After being collected to the area station (the server can be set up in the highway management company or local traffic management department), the data should be transferred to professional research and implementation institution of bridge health monitoring through the IP line communication (in order to ensure the data security and confidentiality) before being analysed by professional engineers. The topological graph of the system is shown in Figure 1.

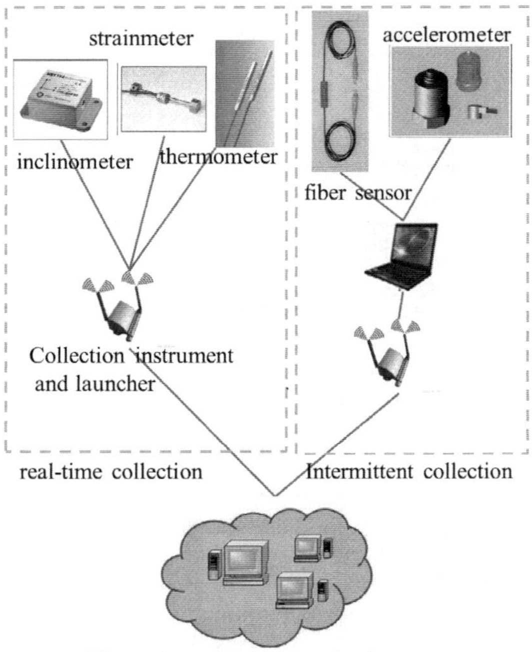

Figure 1 topological graph of system

Considering the decentralization of the system, the cost of data acquisition and transmission subsystem will be the main part of the whole system cost, even more than the cost of sensor subsystem. Therefore, in order to further simplify the system to reduce the investment, intermittent monitoring approach could be considered as well which means connecting the sensor to the signal collection box without setting data acquisition equipment in each bridge monitoring system. Engineers collect data for a short time (24 hours) monthly by connecting the data acquisition system to the collection box.

4. Conclusions

In conclusion, aiming at the CMSW system of generic bridge featured by small scope, strong dispersibility and difficult maintenance, this paper proposes a series of approaches including the testing of vehicle load, girder deviation, stress of key parts, concrete crack, structural deformation, cable force and bridge deck icing problem, prepositive A/D converter, simplifying the data acquisition system, making full use of existing transmission net of the road system, giving priority to intermittent monitoring and more concerning qualitative analysis and judgment in the outcome assessment. With these measurements and analysis methods, the CMSW system can be established to help detect and solve safety problems of bridges. From all above, the establishment of the CMSW system has a lot of economic and social benefits.

References

Ministry of Transport of the People's Republic of China. Statistical bulletin of transportation industry development in 2016[M].

Number of bridges in the US that need repair? More than 55,000[M]. CNN Newsource Sales, Inc., 2017 Feb 15.

Report: Over 65,000 U.S. bridges in need of repair. https://www.usatoday.com/story/news/nation/2013/09/15/bridges-need-repair/2816881/

Messina A et al. Structural Damage Detection by a Sensitivity and Statistical - based method. Journal of Sound and Vibration , 1998 ,216 (5) : 791 - 818.

Ku W , Storer R H , Georgakis C . Disturbance detection an d isolation by dynamic principal component analysis[J]. Chemometries And Intelligent Laboratory Systems, 1995, 30(1): 179～196.

Charles R. Farrar, Keith Worden. An Introduction to Structural Health Monitoring[M]. Springer Vienna , 2010 , 365 (1851) :303-315

JM Ko , YQ Ni. Technology developments in structural health monitoring of large-scale bridges[J]. Steel Construction , 2008 , 27 (12) :1715-1725

The 8th International Conference on
Structural Health Monitoring of Intelligent Infrastructure
Brisbane, Australia | 5-8 December 2017

Design of Intelligent Decision Support System for Hydraulic Structural Safety

Zhongru Wu[1,2,3], Jiachen Liang[1,2,3], Bo Chen[1,2,3], Chongshi Gu[1,2,3], Bangbin Wu[1,2,3]

[1] *College of Water Conservancy and Hydropower Engineering, Hohai University, Nanjing, China, Email:liangjjjc@126.com,* [2] *State Key Laboratory of Hydrology-Water Resources and Hydraulic Engineering, Hohai University, Nanjing, China,* [3] *National Engineering Research Centre of Water Resources Efficient Utilization and Engineering Safety, Hohai University, Nanjing, China*

Abstract With the development of systems engineering and intelligent methods, a hydraulic structural safety intelligent decision support system (HSSIDSS) is established. The system is built upon three subsystems, including information management subsystem, analysis & evaluation subsystem, intelligent support decision-making subsystem. By the cooperation of the three subsystems, the system achieves various monitoring functions such as data processing, abnormality recognition, physical genesis analysis, graded early-warning, expert diagnosis and support decision-making. The whole system is supported by five bases, including project database, method base, knowledge base, model base and graphic base. With Artificial Neural Networks (ANN) and other machine leaning algorithms integrated in the support bases, some intelligent monitoring models are established to deal with massive monitoring data and non-linearity. Data mining methods, such as on-line analytical processing (OLAP), clustering algorithms and decision-theoretic rough sets (DTRS), are also presented to diagnose structural health adaptively. An on-line consulting service platform, which provides queryable and visual data, is built for experts to carry out comprehensive evaluation and make decisions upon the results of prototype observations and system analysis. The real-time monitoring and scientific decision-making for structural safety can be achieved by adopting these methods in the system.

1. Introduction

To take advantage of hydropower and water resources scientifically and orderly, many reservoirs have been constructed in Yangtze river, Yellow river and the Yalong river in P. R. China, such as the Three Gorges Dam with 185m height, the Laxiwa dam with 250m height and the Jinping I with 305m height. These projects are constructed in complex geology environment and they are operated under huge water pressure, alternative temperature and accidentally seismic, etc. Hidden troubles are developed in these hydraulic structures over time. Providing the necessary guarantees of health for the dams is important to the national economy and the people's livelihood.

To monitor health, a group of monitoring instruments are arranged in the hydraulic structures, whose number is sometimes up to hundreds or thousands. The large amount of prototype observations, which are accumulated by these instruments, provide information to guide construction, verify designation and evaluate structural health. Currently, the mathematical models and finite element method (FEM) are applied in structural health calculation. Decision support systems (DSS) are also used for dam health monitoring effectively. For example, a classical type of DSS named DAMSAFE is able to summon experts to share knowledge and information in dam safety cooperative management (Comerfordete, 1993). Dam monitoring system developed by Elecricite de France(EDF) in early stage managed measurement data of more than 150 dams.

However, in these systems, analytic ability of the complex problem remains to be strengthened. With the development of systems engineering and artificial intelligent methods, more and more intelligent models are applied in structural health monitoring. Chen (Chen, 2017) builds a time-varying identication model for crack monitoring data using support vector regression (SVR) and the Bayesian evidence framework (BEF). The model can provide predicted results that are more accurately nonlinearity fitted and is suitable for use in evaluating the behavior of cracks in dams. Salazar (Salazar, 2015) introduces random forests in his predicting model and gets high accuracy in displacement and seepage prediction of an arch dam. Su (Su, 2011) established an early-warning system of dam health (EWSDH) based on wavelet frames and rough sets theory to diagnose the unexpected dam behavior adaptively. However, current researches of DSS always focus on model base and knowledge base, but ignore the importance of group experts (Chen, 2009). Therefore, in consideration of the advantages of intelligent method in solving massive data and experts in dealing with unstructured problems, HSSIDSS integrated intelligent monitoring models and expert consulting module to provide scientific support for structural health monitoring.

2. System global structure and workflow

2.1 Global structure of HSSIDSS

HSSIDSS is an important part of water project. This large and complex system, which aims to monitor the health state of the dams and other hydraulic structures, consists of multiple monitoring items, numerous monitoring instruments and several computer networks. Various kinds of monitoring data is real-timely and orderly collected and managed in the system. The monitoring data reflects the behaviour change of the dam body, its foundation, slopes and grottos. By fast processing the monitoring data online, making qualitative or quantitative analysis accurately, and intelligent judging & evaluating the analysis results, the system can real-timely analyze and monitor the safety state of hydraulic structure. The system mainly performs the following 3 core functions:

- Processing and managing prototype monitoring measurement of various kinds and other data relevant to safety decision-making, including structural design, construction information, analytical methods, expert knowledge and experience.
- Assisting in analysing corresponding geneses in response to abnormal or unsafe signs of the structures, evaluating the structural safety state, giving graded warning according to operation characteristics of the structure.
- Providing advice of safety operation, bringing the structure into full play under the condition of guaranteeing the overall structural safety.

In a sum, the major functions of the system can be defined into the following three: information management, analysis & evaluation and intelligent decision-making. The three corresponding subsystems build up the 2nd layer of the structure. The support base cluster, which is the 3rd layer of the structure, includes project database, method base, knowledge base, model base and graphic base, offering quantitative data, expert knowledge, analysis software and visual image for the three subsystems. In this way, the support base cluster can be called the total information repository for the three subsystems. The decision support function is based upon expert diagnosis and system analysis in normal conditions or advice given by authorities in the case of emergency.

From the above, the system is a hierarchical structure of four levels including total control layer, subsystems layer, support bases layer, and bottom layer, as shown in Figure 1.

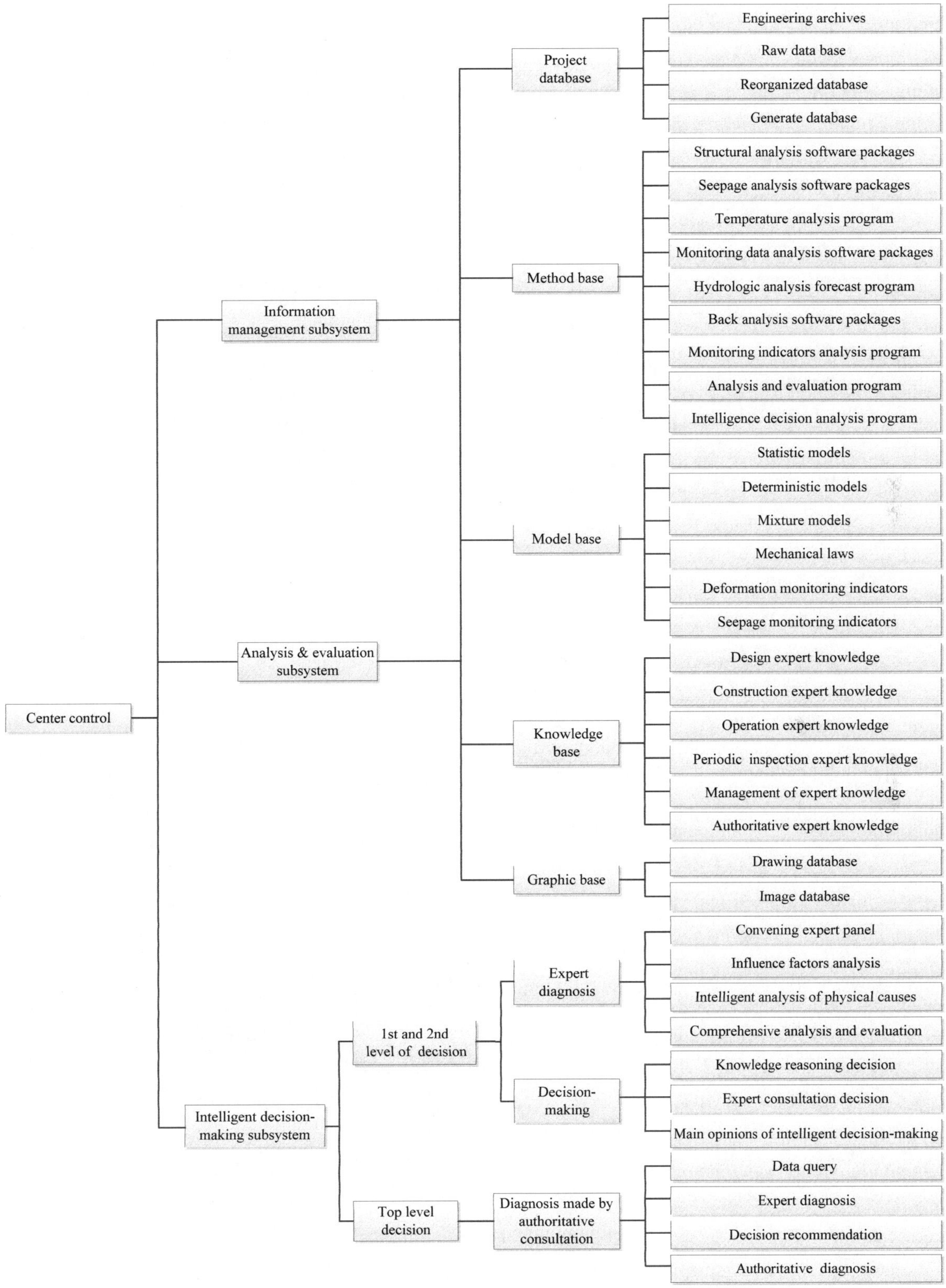

Fig.1 Overall structural diagram of HSSIDSS

2.2 Workflow of HSSIDSS

Corresponding to the three major functions of the system, the workflow of the system is composed of three major steps, including information management, analysis & evaluation and intelligent decision-making, as shown in Figure 2. For elaboration on each step please refer to section 3.

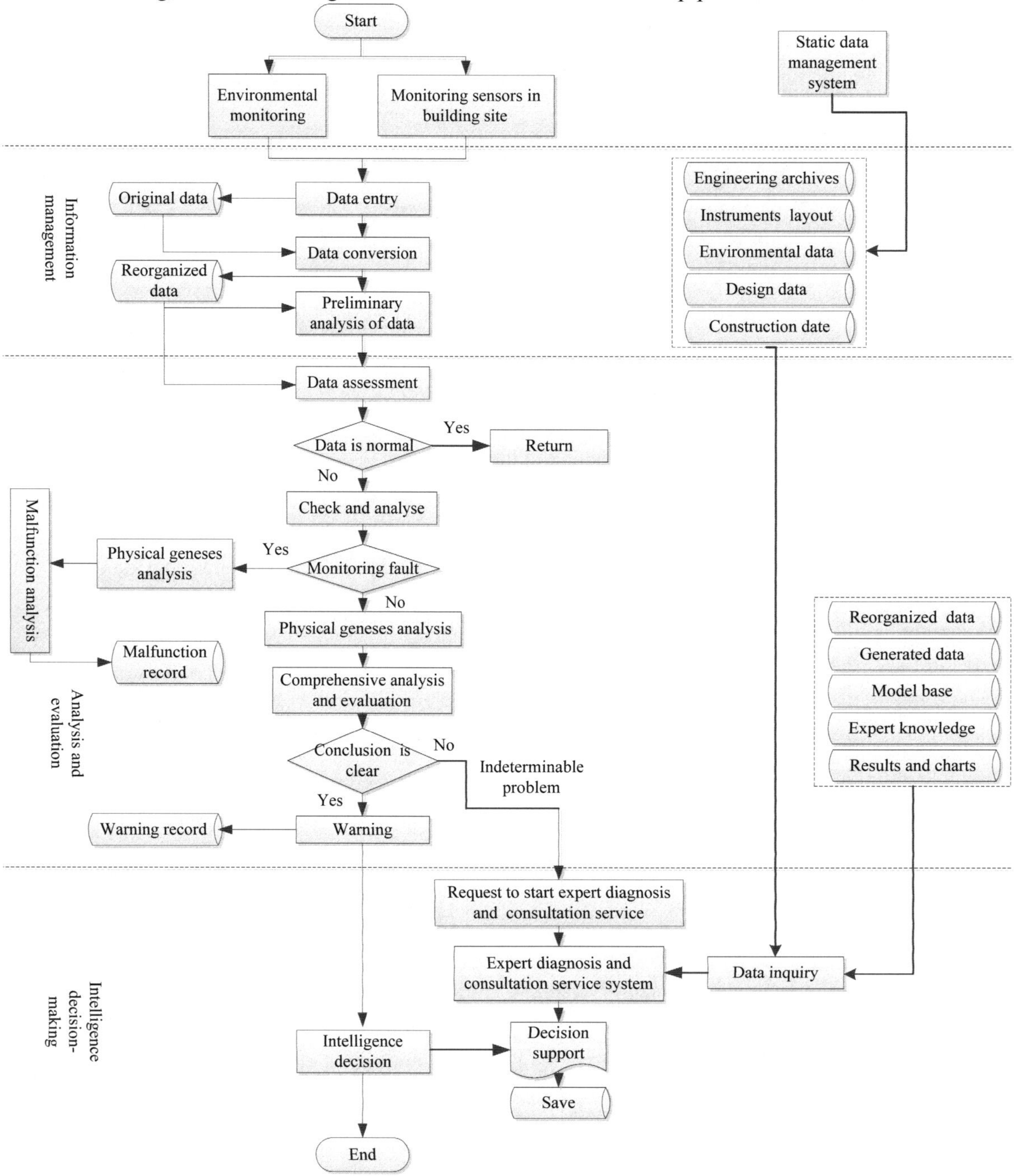

Fig.2 Workflow chart for HSSIDSS

3. Main functions and workflow of subsystems

3.1 The subsystem of information management

Major functions of information management subsystem are as follows: (1) Managing various kinds of data acquired from monitoring data collecting system and other sources effectively and providing instructions for data collection of engineering safety monitoring system. (2) Recording data into original database and processing the data (storing, transforming, preliminary analysis, checking original monitoring data, correcting or eliminating invalid data and finding doubtful data) to ensure that the monitoring data entered into the system can reflect the safety state of engineering structure in a timely and accurate manner. (3) Providing preliminary analysis results of monitoring data for users in forms of charts regularly or irregularly. (4) Maintenance of support database cluster. The monitoring data would be deleted or imported as needed to ensure efficient operation of the system. The subsystem of information management consist of 6 modules, including information input module, information conversion module, initial analysis module, document management module, information output module and database maintenance module. For structural diagram and workflow chart of the information management subsystem please refer to Figure 3 and Figure 4 respectively.

3.2 The subsystem of analysis & evaluation

The analysis & evaluation subsystem, which analyzes and evaluates structure safety based on monitoring data and knowledge base, performs the following major functions: (1) Data evaluation. The subsystem uses various kinds of evaluation criteria to determine whether the monitoring data is normal to locate doubtful and abnormal parts of the structure and their problems. (2) Preliminary check. By methods of correlation analysis of similar monitoring points, the subsystem quickly determines whether the doubt or abnormity is caused by monitoring errors or by the structural changes in itself. (3) Monitoring fault inspection. As for abnormal data caused by monitoring errors, the subsystem inspects data records and monitoring system to correct error measurements or provide guide to repair malfunction of the monitoring system. (4) Analysis on physical geneses. For abnormal data caused by structural changes, the subsystem analyzes the external factors (environment) first, then geneses of abnormal change of seepage, deformation and stability of the structure foundation on the state of the whole structure. Thereby physical geneses of abnormity and its effects are identified. (5) Comprehensive evaluation & analysis and warning. The subsystem summarizes the results of physical cause analysis, generates comprehensive evaluation on safety state of the structure and raising warnings for abnormal structural changes. For structural diagram and workflow chart of the evaluation & analysis subsystem, please refer to Figure 5 and Figure 6 respectively.

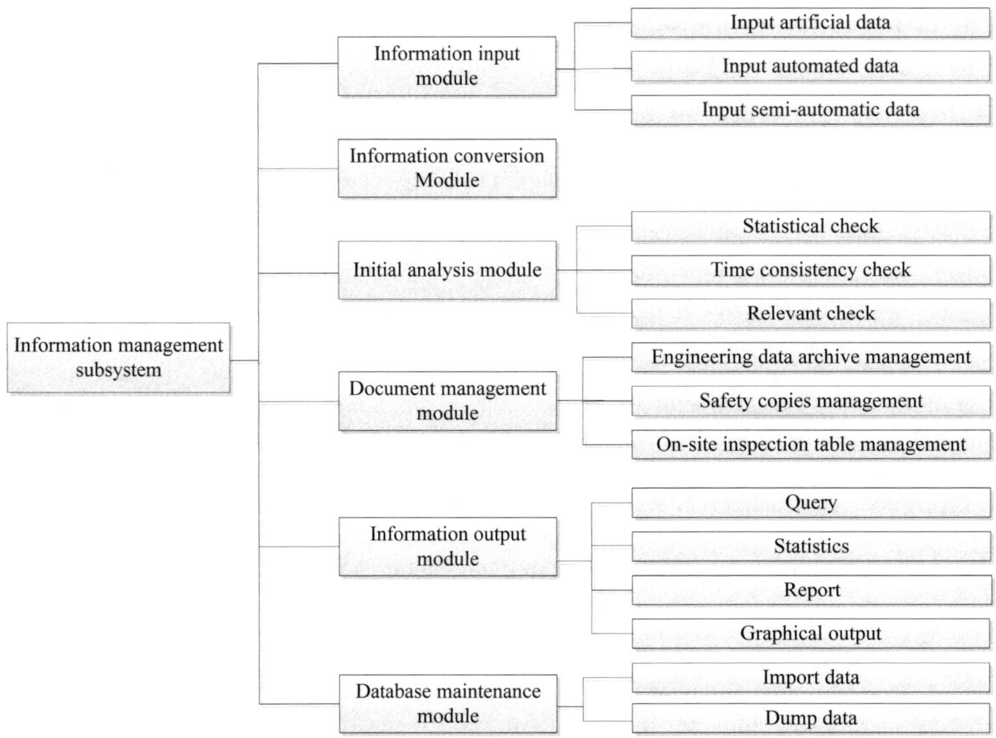

Fig.3 Structural diagram of information management subsystem information

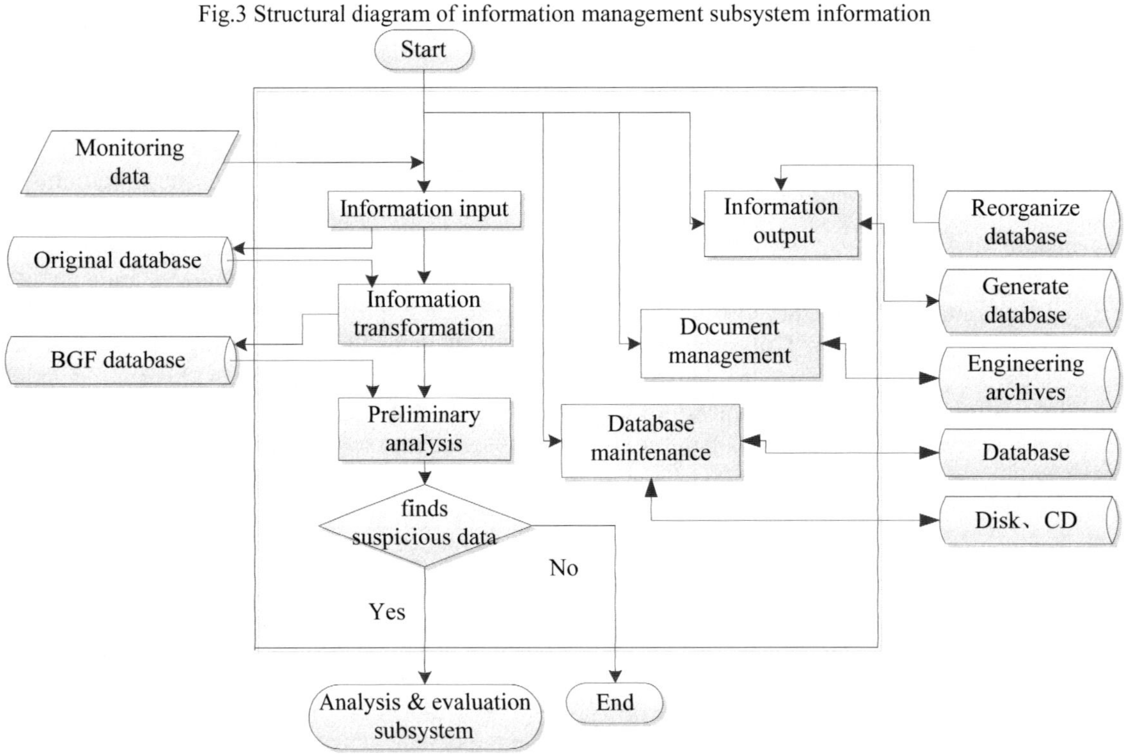

Fig. 4 Workflow chart of information management subsystem

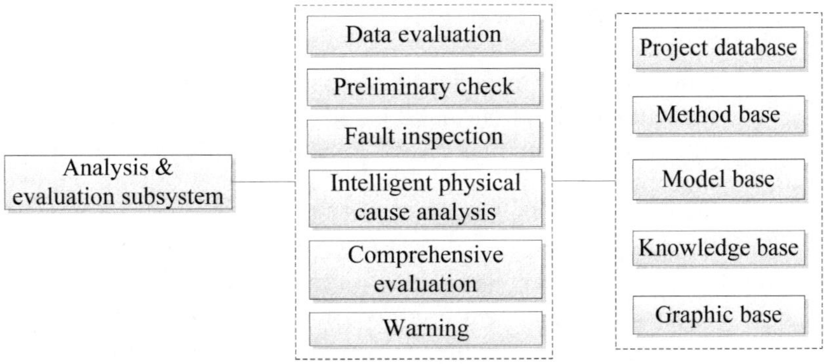

Fig.5 Structural diagram of analysis & evaluation subsystem

The working process is as shown in figure.6

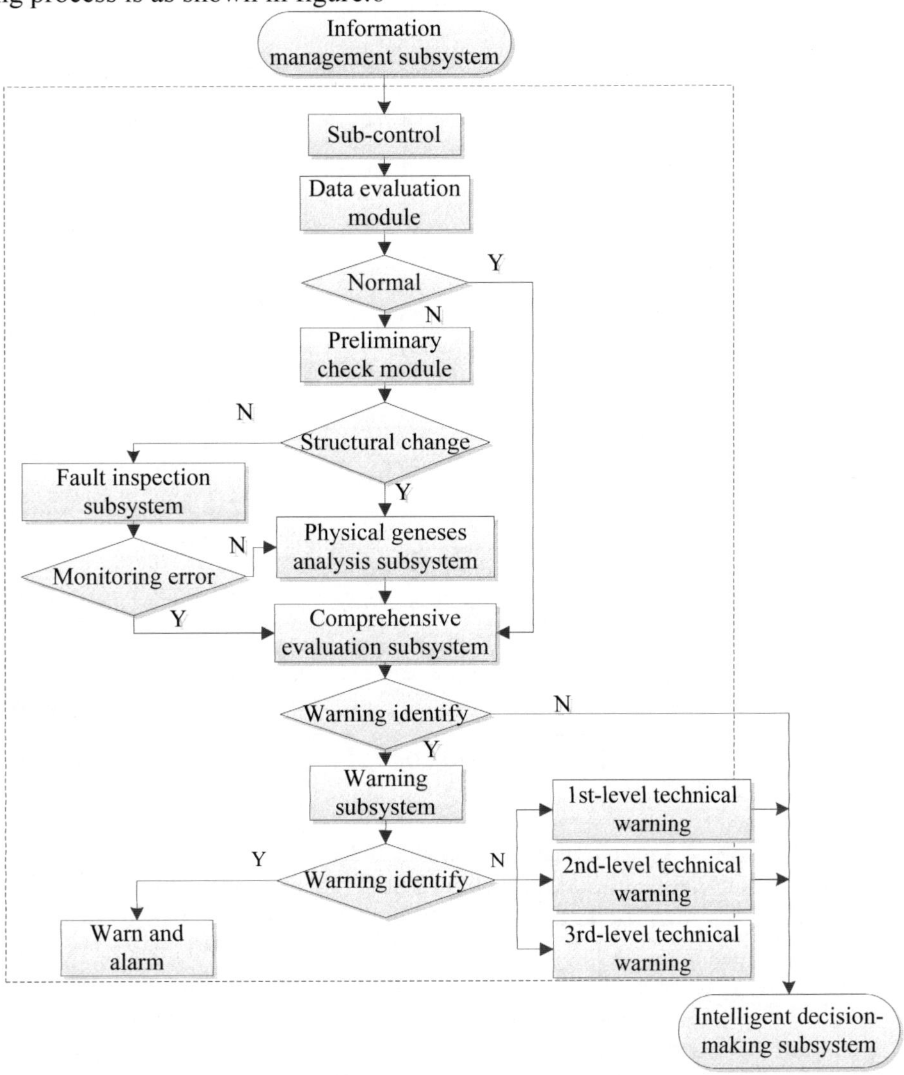

Fig.6 Workflow chart of analysis & evaluation subsystem

3.3 The subsystem of intelligent decision-making

Dealing with structural abnormity, unsafe factors and special conditions identified by analysis & evaluation subsystem, the intelligent decision-making subsystem gives intelligent decisions for safe operation of the overall water project. The major functions of the subsystem include: (1) Identifying geneses of structural abnormity by making reasoning analysis based on knowledge base. (2) Generating expert decisions based on calculation analysis and safety analysis. Experts could refer to data and information in database while online expert consultation is carried out. (3) Carrying out online remote control for decision implementation via mobile applications.

This subsystem not only considers the overall safety of water conservancy project, but also the relative independence of its each substructure. The subsystem performs four main functions, including data query, expert diagnosis, intelligent decision-making and expert consultation which can be encrypted. The structural diagram is shown in Figure 7.

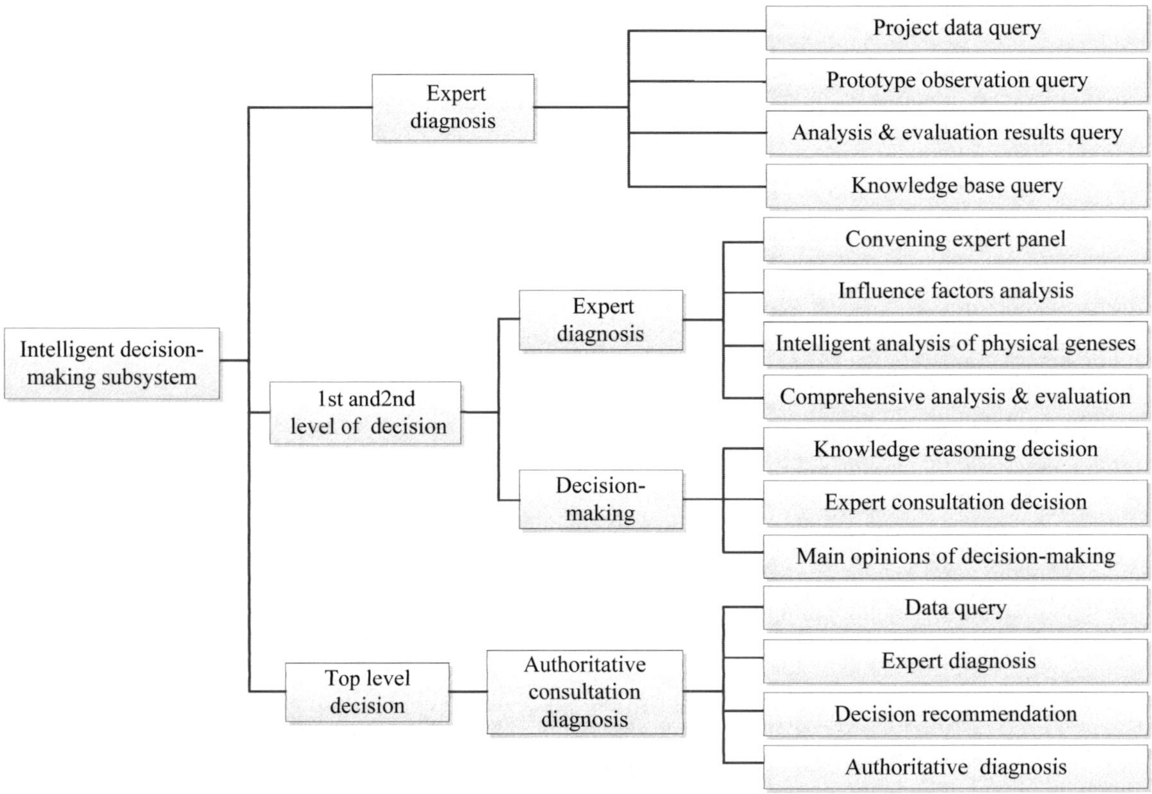

Fig. 7 Structural diagram of intelligent decision subsystem

The workflow of intelligent decision-making subsystem (Figure 8) is as follows: (1) When analysis & evaluation subsystem has identified geneses of structural abnormity, intelligent decision-making can be done by reasoning analysis based on knowledge base. (2) When analysis & evaluation subsystem fails to identify geneses of structural abnormity, experts will be summoned to make data queries and calculation analysis. Intelligent decision-making will be based on the experts' opinions on safety evaluation. (3) In the case of special circumstances or emergency, expert consultation is provided. Construction structures can be damaged or undermined under special conditions such as strong earthquakes, catastrophic floods, geographical disasters and wars. Experts consultation will be organized immediately based on relevant data using encrypted subsystem. In this way the optimal decision is made timely based on experience from authorities with weight analysis methods.

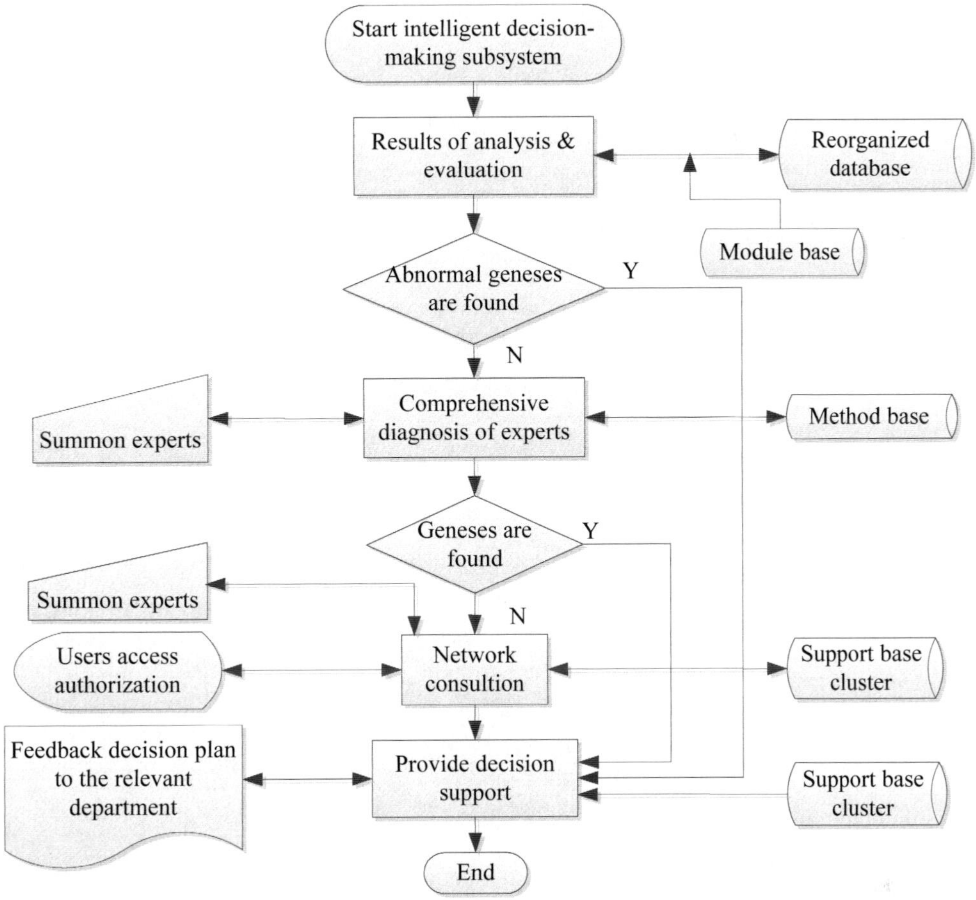

Fig.8 Workflow chart of intelligent decision-making subsystem

3.4 Support databases

The three subsystems share information collected from structural safety monitoring system, including project archives, environmental variables and effect quantities. Various kinds of graphs reflecting structural changes are processed and different calculation methods and processing procedures are applied to realize different functions. It requires many kinds of calculation methods, models and knowledge, so the support base cluster is established to offer comprehensive and reliable support for information management, analysis & evaluation and intelligent decision-making. The support base cluster covers five bottom layer bases, including project database, method base, model base, knowledge base and graphic base. The structural diagram of support database is shown in Figure 9.

4 Intelligent processing of HSSIDSS

4.1 Intelligent processing of monitoring models

(1) Intelligent modeling based on neural network

Artificial Neural Networks (ANN) models with the characteristics of self-organization, self-adaptability, associative power, self-learning ability and strong non-linear mapping ability, are appropriate to solve the complicated nonlinear problems encountered in engineering. The BP neural

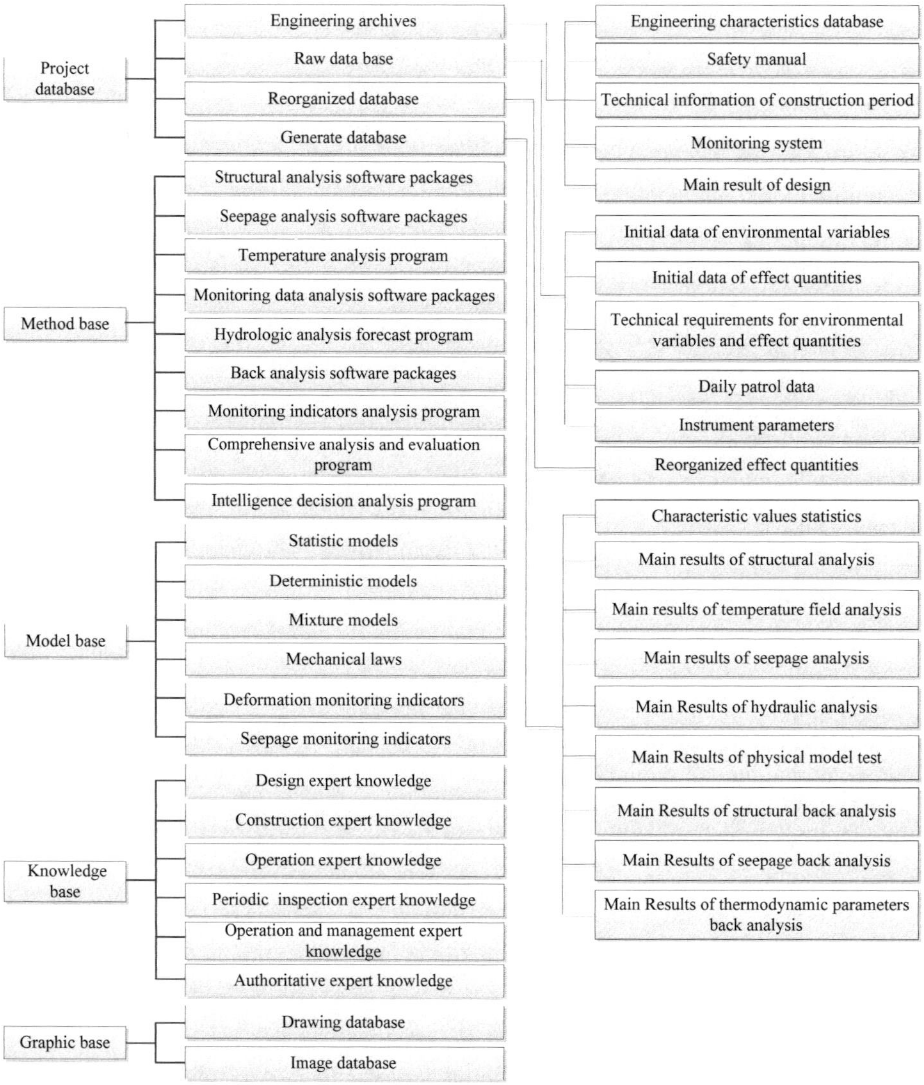

Fig.9 Structural diagram of support databases

network model, which is the most common used kind of ANN models in civil engineering, is integrated in HSSIDSS. The network architecture is the following (Figure 10):

①Factors in input layer are：

• Water level factors, which reflect the hydrostatic load of reservoir water, for example, H, H^2, H^3.

• Temperature factors, which reflect temperature effect changing with seasonal change, for example, $\sin\frac{2\pi t}{365}$, $\cos\frac{2\pi t}{365}$, $\sin\frac{4\pi t}{365}$, $\cos\frac{4\pi t}{365}$.

• Time-effect factors, which reflect irreversible change of structural behaviour over time, for example, θ and $\ln\theta$.

• Special factors, for example, precipitation factors and hysteresis factors in seepage monitoring models.

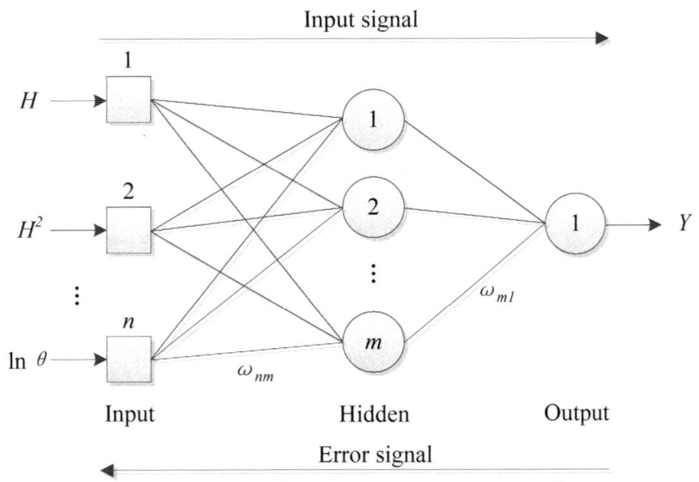

Fig.10 Structural diagram of BP neural network model

② One hidden layer is chosen. Due to the large computation amount of back-propagation algorithm, parameters (number of hidden neurons m and the connection weight ω) are randomly initialized, and updated by genetic algorithm. Learning rate parameter is optimized by heuristic methods (Jacobs, 1988).

③ The output is the effect quantity Y, so there is one neuron in output layer.

There are many other useful intelligent models for structure monitoring analysis such as least squares support vector machine (LSSVM), K-nearest neighbours (KNN) and boosted regression trees (BRT). They are mentioned more briefly than ANN models in this section because of lower popularity. But they are integrated in HSSIDSS as a supplement and comparison to ANN model in important points.

(2) Adaptive variable weight combination model

At present, each model has its limitation. There is no model suitable for every situation (Salazar, 2015). Therefore, it is necessary to combine different models according to the specific task.

In view of the fact that a single model can only simulate a particular time period (the construction period, the initial reservoir filling period or the operation period) or a specific monitoring program (displacement, seepage, stress, etc.), the system selects different models adaptively, gives them appropriate weights, and finally combines them. Therefore, the adaptive variable weight combination model based on least square method (LSM) is established.

Assume that there are m kinds of available models to solve the problem F. And then equation $x \in X \subset R^n \xrightarrow{y_i} y \in Y \subset R$ is used to represent the number i model. The combination model is described as the linear combination of different models:

$$\begin{cases} y(x) = \sum_{i=1}^{m} \omega_i y_i(x) \\ \sum_{i=1}^{m} \omega_i = 1 \end{cases} \qquad (i = 1, 2 \cdots m) \tag{1}$$

Where $\omega_i(x)(i = 1, 2 \cdots m)$ is the weight of model $y_i(x)$.

Assume that $\omega_i(t)$ is a continuous function of time t. According to the Weierstrass approximation theory, any continuous function can be approximated by an appropriate number of polynomial functions. Therefore, $\omega_i(t)$ is expressed as the sum of p polynomials.

$$\omega_i(t){=}a_{i0} + a_{i1}t + \cdots + a_{ip}t^p \qquad (i = 1,2\cdots m) \tag{2}$$

Where a_i is obtained by LSM and then the weight $\omega_i(t)$ is obtained.

4.2 Intelligent processing of decision support

In addition to statistical method, visualization technology and query technology, different tools and technologies, such as On-Line Analytical Processing(OLAP), clustering method and decision-theoretic rough sets (DTRS) are integrated in HESIDSS for data mining and expert decision-making.

(1) OLAP and hierarchical cluster analysis

The advantages of combined application of OLAP and cluster analysis in monitoring data analysis are as follows:

• It overcomes the shortcoming of the processing and analysis of a single observation point but the whole health state of structure from the perspective of the all items and different parts.

• The frequently active area of the structure can be obtained directly from the results of cluster analysis, which makes it possible to focus on the key area of the structure in operation.

• It can provide useful information for experts to explore the deformation and seepage mechanism of the structure, and then evaluate whether it is in accordance with the mechanics laws and establish the monitoring index of typical section.

The OLAP model is mainly defined by three concepts: "measurement value", "dimension" and "hierarchy". In HSSIDSS, the “measurement value” is the dynamic quantitative data, which is the prototype observation data. The "dimension" is the static analysis axis, which represents the perspective of analysis. For example, the dimension of a model includes the environment variable DIM, the inspection DIM, the effect quantity DIM, time DIM, and space DIM(figure 11). The "Hierarchy" refers to the aggregation of data according to specific characteristics in some dimensions. For a dynamic hierarchical OLAP dimension, we need to use hierarchical clustering algorithm to construct a hierarchical structure.

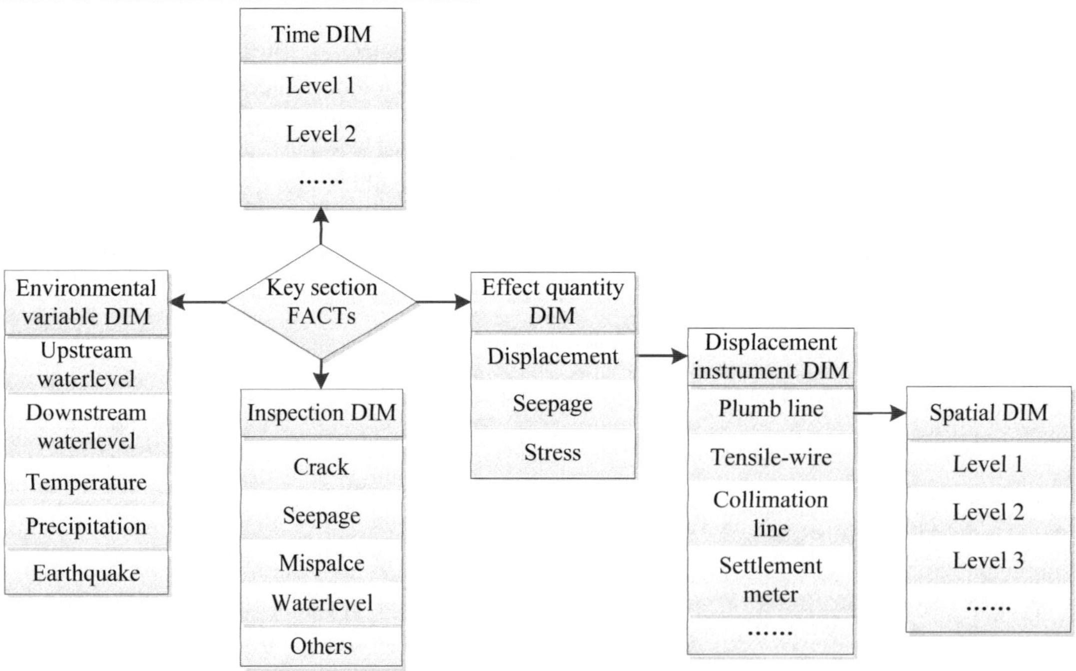

Fig. 11 A star model of OLAP cube

(2)A diagnosis model based on decision-theoretic rough sets (DTRS)
Rough set theory, which doesn't require prior knowledge and can deal with the uncertain problems, has been widely used in expert system, machine learning and pattern recognition areas.
The diagnostic model in HSSIDSS is based on DTRS proposed by Yao (Yao, 2009; Yao, 2008). Based on the Bayesian theory, the diagnostic model compares the risk of various classification decisions, and finds optimal classification of the minimum risk loss. In this way, objects are divided into positive, negative or boundary region. The diagnosis model includes the following steps:
(1) Environmental factors and other possible physical geneses are analysed according to the changes of structural behavior. The most influential factors are chosen by attribute reduction method to obtain condition attributes. In addition, based on risk minimized attribute reduction and hierarchical analysis, safety evaluation index are simplified to obtain decision attributes. Then the decision information table can be created by combining condition attributes and decision attributes.
(2) The health state of hydraulic structure was diagnosis by multi-classification attribute decision method. Usually, the structural safety evaluation is a "multi-classification" problem including five categories: "normal", "basic normal", "light abnormal", "severe abnormal", "malignant abnormal". By two-phase method, the multi-classification problem can be transformed into several two-classification problems which can be solved by Three-way Decision-theoretic. The architecture of the diagnostic model is shown in figure 12.

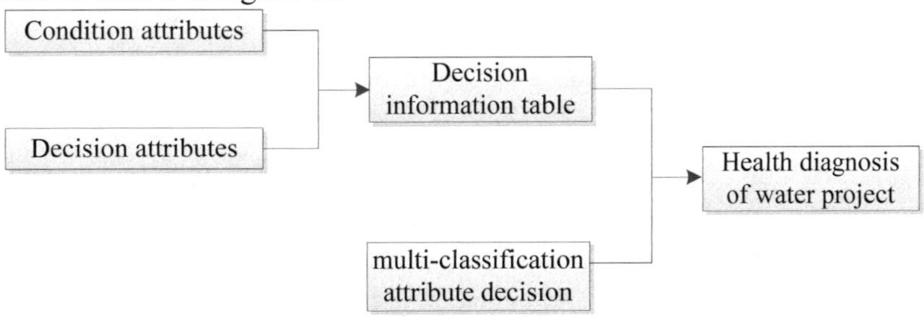

Fig.12 architecture of the diagnostic model based on DTRS

5 Conclusions

(1) HSSIDSS is a thorough monitoring system that achieves many monitoring functions such as data processing, anomaly recognition, physical genesis analysis, warning, expert diagnosis and decision-making proposal.
(2) Hydraulic structure behaviour shows strong non-linear and dynamic characteristics. Intelligent processing focuses on the combination of machine learning models and data mining methods. By absorbing the intelligent algorithms into the system, the ability of comprehensive health evaluation and handling complex problems with massive data is increased.
(3) Expert experience is still critical for interpreting the results of observed values and decision-making. In HSSIDSS, more useful information and a more convenient consulting service platform is provided for experts, which can improve the speed and precision of expert judgment and decision making, especially when special disasters come.

Acknowledgement
This work was supported by the research funds from the National Natural Science Foundation of China (grant nos 51139001, 51379068, 51279052, and 51179066), Jiangsu Natural Science Foundation (grant no. BK20140039), Jiangsu Basic Research Program (grant no. BK20160872) and the Priority Academic Program Development of Jiangsu Higher Education Institutions (grant no. YS11001).

References

Comerfordete JB, 1993, 'The Role of AI technology in the management of dam safety: The DAMSAFE System', *Dam Engineering,* vol.3, issue. 4, pp. 265-275.

Chen B, Wu ZR, Liang JC, *et al.* 2017, 'Time-Varying Identification Model for Crack Monitoring Data from Concrete Dams Based on Support Vector Regression and the Bayesian Framework', *Mathematical Problems in Engineering* , vol. 2017, Article ID 5450297, 11 pages.

Fernando Salazar, Miguel Ángel Toledo, Eugenio Oñate *et al.* 2015, 'An empirical comparison of machine learning techniques for dam behavior modelling', *Structural Safety,* vol. 5, no. 1, pp. 1-27.

Su HZ, Wen ZP, Wu ZR, 2011, 'Study on an Intelligent Inference Engine in Early-Warning System of Dam Health'. *Water Resourced Management,* vol. 25, no. 6, pp. 1545-1563.

Chen X, Wang ZQ, 2009, 'Overview of Theory and Methods of Decision Support Systems', *Control and Decision*, vol. 21, no.9, pp. 961-968

Jacobs RA, 1988, 'Increased rates of convergence through learning rate adaptation', *Neural Networks*, vol. 1, no. 4, pp: 295-307.

Fernando Salazar, Rafael Moran, Miguel A. Toledo *et al.* 2015, 'Data-Based Models for the Prediction of Dam Behavior: A Review and Some Methodological Considerations', *Archives of Computational Methods in Engineering*, vol. 24, no. 1, pp. 1-21.

Yao YY, 2009, 'Three-way decision: An interpretation of rules in rough set theory', *Lecture Notes in Artificial Intelligence 5589: Proceedings of the 4th International Conference on Rough Sets and Knowledge Technology (RSKT 2009)*, Gold Coast, Australia, pp. 642–649.

Yao YY, Zhao Y, 2008, 'Attribute reduction in decision-theoretic rough set models', *Information Sciences*, vol. 178, no. 17, pp. 3356–3373.

The 8th International Conference on
Structural Health Monitoring of Intelligent Infrastructure
Brisbane, Australia | 5-8 December 2017

Structural Health Monitoring Developments in TRUSS Marie Skłodowska-Curie Innovative Training Network

A. González[1], F. Huseynov[2], B. Heitner[3], M. Vagnoli[4], J.J. Moughty[5], A. Barrias[5], D. Martinez[1], S. Chen[1], E. OBrien[1], D. Laefer[1], J.R. Casas[5], R. Remenyte-Prescott[4], T. Yalamas[3], J. Brownjohn[2].
[1]*University College Dublin – Ireland, Email:* arturo.gonzalez@ucd.ie, [2]*Full Scale Dynamics Ltd – UK,* [3]*Phimeca Engineering – France,* [4]*University of Nottingham – UK,* [5]*Universitat Politecnica Catalunya – Spain*

Abstract

This paper reports on recent contributions by the Marie Skłodowska-Curie Innovative Training Network titled TRUSS (Training in Reducing Uncertainty of Structural Safety) to the field of structural safety in rail and road bridges (http://trussitn.eu). In TRUSS, uncertainty in bridge safety is addressed via cost efficient structural performance monitoring and fault diagnostics methods including: (1) the use of the rotation response due to the traffic traversing a bridge and weigh-in-motion concepts as damage indicator, (2) the combination of design parameters in probabilistic context for geometrical and material properties, traffic data and assumption on level of deterioration to evaluate bridge safety (via Bayesian updating and a damage indicator based on real time measurement), (3) the application of a fuzzy classification technique via feature selection extracted using empirical mode decomposition to detect failure, and (4) the testing of alternative vibration based damage sensitive features other than modal parameters. Progress is also made in improving modern technologies based on optical fiber distributed sensing, and sensors mounted on instrumented terrestrial and on aerial vehicles, in order to gather more accurate and efficient info about the structure. More specifically, the following aspects are covered: (a) the spatial resolution and strain accuracy obtained with optical distributed fiber when applied to concrete elements as well as the ideal adhesive, and the potential for detecting crack or abnormal deflections without failure or debonding, (b) the possibility of using the high-resolution measurement capabilities of the traffic speed deflectometer for bridge monitoring purposes and, (c) the acquisition of bridge details and defects via unmanned aerial vehicles.

1. Introduction

Buildings, energy and transport infrastructure are subject to live loads (environmental, traffic, etc.) that tend to be highly variable. Therefore, the structural strength of this infrastructure relies on properties of engineering materials that also vary with time and stress path. There is a statistical uncertainty associated to loads and strength that requires structural safety calculations to be performed within a probabilistic framework (i.e., the probability of load effects exceeding structural strength for the life of the structure). The large degree of uncertainty in the design stage is covered by overestimating loading conditions and underestimating structural strength through the use of conservative safety factors. However, while in-service, over-conservatism can lead to unnecessary replacement or strengthening of existing infrastructure. Indeed the greatest benefit of assessing an existing structure may result from taking measurements that will reduce much of the uncertainty contained in the structural and load models used during design. For this reason, the TRUSS project places emphasis upon the rapid evolution of Structural Health Monitoring (SHM) technology and the urgent need for the next generation of experts, trained in these new techniques to assess structural safety accurately. A wide range of activities is encompassed under five Work Packages (WPs) from January-2015 to December-2018 (Table 1).

Table 1. Work Packages in TRUSS

WP	Description
1	Management (http://trussitn.eu/about-truss/structure/): It deals with the management, governance, recruitment and mobility of the programme.
2	Dissemination and outreach (http://trussitn.eu/dissemination/): It disseminates the novel outputs produced by TRUSS structured research program via presentations at national workshops, world's leading international conferences, publications in peer-reviewed journals of high-impact factor, reports made available free online and outreach activities.
3	Structured training (http://trussitn.eu/training/): It governs the provision of appropriate training modules for the recruited fellows, to ensure that the quality of training meets best practice and that the training needs of the fellows, as planned in their Project Career Development Plan, are being adequately addressed.
4	Research on buildings, energy and marine infrastructure (http://trussitn.eu/research/buildings-energy-marine-infrastructure/): It aims to reduce uncertainty via advanced material, load & structural modelling and monitoring systems. More specifically addressing uncertainties in the response of not-well known materials, in the variation of the non-linear dynamic response of sliding structures under water, in the safety of large wind turbine towers and in the remaining life of assets in extreme and harsh marine conditions.
5	Research on rail and road infrastructure (http://trussitn.eu/research/rail-and-road-infrastructure/): It aims to reduce uncertainty, improve structural assessments and management of land transport infrastructure via development of new technologies that will allow more efficient data collection, and new algorithms that will process the data collected from the structure to estimate its safety more accurately than current approaches.

The following sections describe TRUSS research on bridges within WP5. This WP counts with expertise in traffic load and vehicle-bridge interaction modelling, uncertainty modelling, reliability analysis, field measurements and SHM from University College Dublin and Arup (Ireland), University of Nottingham, Full Scale Dynamics Ltd, AECOM, Microlise Ltd and TRL (UK), Universitat Politecnica Catalunya, COMSA and COTCA (Spain), Phimeca Engineering and Université de Nantes (France) and Greenwood Engineering (Denmark). SHM for bridges can rely on a high number of sensors as early warning system. For example, 323 networked sensors were embedded in the new bridge that replaced the bridge collapsing in Minneapolis in 2007. While this complex SHM arrangement could be adopted for selected new infrastructure based on its strategic importance, clearly the cost of extending such an approach to the entire bridge stock is prohibitive. Therefore, TRUSS further develops SHM algorithms and technologies to improve safety in a cost-efficient manner that will allow monitoring more bridges more frequently. Section 2 discusses four novel SHM algorithms based on displacement, rotation or acceleration measurements that can be gathered from traditional sensors directly attached to discrete locations of the bridge. Section 3 reviews advances by TRUSS in three promising modern SHM technologies using Distributed Optical Fiber Sensing (DOFS), using sensors mounted on a vehicle driving over the bridge (i.e., drive-by SHM), and using sensors carried by an Unmanned Aerial Vehicle (UAV) flying around the bridge (i.e., fly-by SHM). Finally, Section 4 provides conclusions.

2. SHM Algorithms Based on Traditional Sensors

2.1 Damage Detection Method Using Rotation Based Bridge-WIM

TRUSS is investigating a new damage detection method based on monitoring rotation at supports. Figure 1(a) shows a 20 m long 1-D bridge model where a hypothetical inclinometer is placed at the left support to record rotation while traversed by a two-axle vehicle. Figure 1(b) depicts the rotation-time history obtained from the hypothetical inclinometer for healthy and damaged cases. Damage is modelled as 10%, 30% and 50% stiffness reductions over 1 m length around midspan. The graph shows how rotations become larger the more severe the damage.

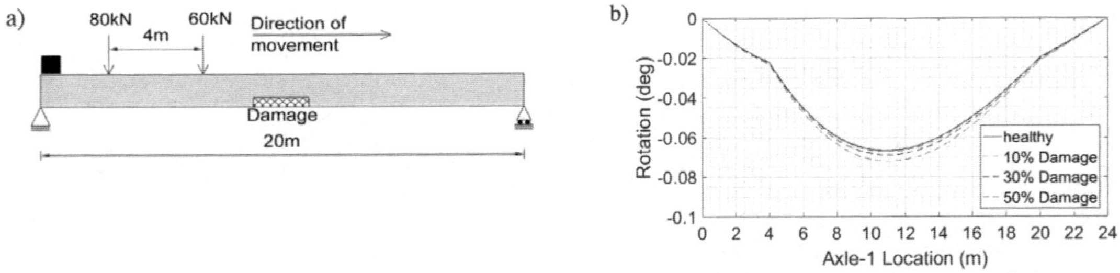

Figure 1. a) 1-D bridge model, b) Rotation time history obtained at left support under two-axle vehicle loading

Developing a successful SHM system without knowing the applied load is challenging. Real traffic data obtained at one Weigh-In-Motion (WIM) site in the U.S under the Federal Highway Administration's Long-Term Pavement Performance program is employed (Leahy 2013). Figure 2(a) shows the general trend of the histogram of Gross Vehicle Weight (GVW) for 5-axle trucks, which is a predominant vehicle at the site of interest, over 4 years. There is a clear consistency in the statistical distribution, i.e., two peaks every year that represent the sets of loaded and unloaded trucks. Having identified this trend, a damage detection method is proposed based on the statistical repeatability in traffic data. In essence, any shift in the known statistical distribution will indicate damage. Figure 2(b) shows the monthly average tandem weights of loaded 5-axle trucks in tandem-1 and tandem-2 space. The dashed blue line defines the 95% confidence area. If there was any structural damage, then estimated tandem weights would fall outside of the confidence area.

Figure 2. a) GVW histogram of 5-axle trucks at WIM site, b) Average monthly tandem weights in 5-axle trucks

The proposed method is numerically tested by simulating 30% damage from July 2011 at three different locations: 1/4, 1/2 and 3/4 span. Then, a rotation-based Bridge-WIM system is used to estimate weights. Figure 2 shows how for all three cases, the estimated load felt outside of the confidence area denoting the presence of damage.

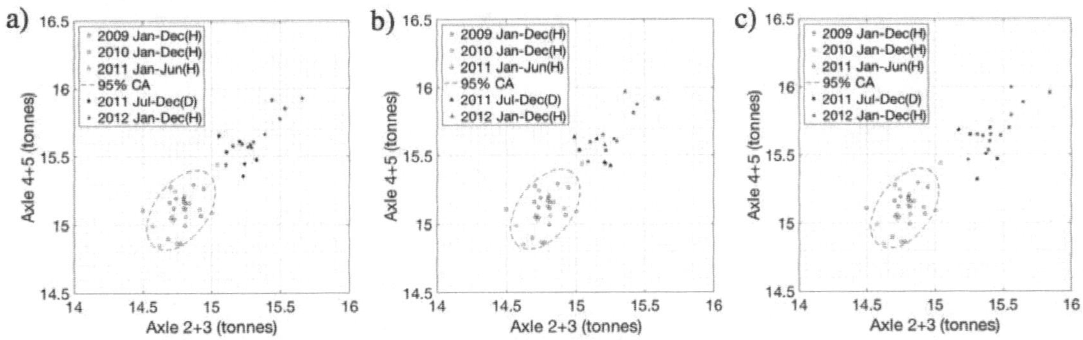

Figure 3. Average tandem weights of loaded 5-axle trucks predicted by rotation-based B-WIM system for damaged and healthy cases. Damage is simulated at a) quarter span b) mid-span c) three quarter span

2.2 Probabilistic Bridge Safety Model

TRUSS demonstrates how SHM data can be incorporated into a bridge safety model and how to quantify the effectiveness of SHM systems. For example, corrosion is the most important deterioration process in the life of reinforced concrete structures, leading to extensive cracking, spalling and above all, reduced capacity. Several models exist in the literature to model the process of corrosion depending on the environment, geometry and materials of the bridge (Marsh & Frangopol 2007, Décatoire 2015, Val et al 1998, Suo & Stewart 2009). Given that these models are strongly based on assumptions and on limited number of experimental results (mostly applying accelerated corrosion), engineers should apply them with caution. The most proficient way to improve an estimation on corrosion would be to gather real-time direct measurements, however, as reinforcement is embedded in the concrete, this is not a simple task. But corrosion leads to reinforcement loss affecting not only the capacity, but also the stiffness and as a result, the response of the bridge. If changes in the response were sufficiently large, then they can be detected with SHM. What is understood by 'large changes' may vary depending on the attributes of the chosen SHM system including no. sensors, sensitivities to changes in the response, accuracies and robustness of the sensors, etc. Here, TRUSS develops a probabilistic bridge safety model using Bayesian updating to update the information of reinforcement loss due to corrosion based on commonly applied corrosion models. The method has been tested using different damage indicators based on single location deflection, strain and rotation measurements during ambient traffic conditions. The bridge has been modelled as a reinforced concrete slab and corrosion has been assumed to be uniform along the length of the bridge. In the near future, random field modelling will be applied to account for the spatial correlation of corrosion. A damage indicator built from image-based deflection measurements (Ribeiro et al 2014, Ojio et al 2016) is plotted against the level of reinforcement loss in Figure 4(a). It can be seen that the mean line in the graph is monotonic increasing, therefore, the damage indicator is robust. However the graph reveals a high level of variability coming from the sensor accuracy and the variability of ambient traffic. Similar graphs can be used to compare different damage indicators on the basis of the slope of the mean value and its variance. Figure 4(b) shows an example of the distribution of the prior estimate on reinforcement area loss and possible posterior distributions after updating using 10, 20 or 50 measurements. The quality of the updating can be evaluated in different ways while focusing on the improvement compared to rely solely on the prior estimate. I.e., the improvement (decrease) in the standard deviation either the 95% confidence interval or the improvement in the mean value (how much closer it is to the actual value) can be good indicators of the effectiveness of the updating. The effect of updating on reliability index can also be evaluated to compare the performance of several damage indicators.

a) b)

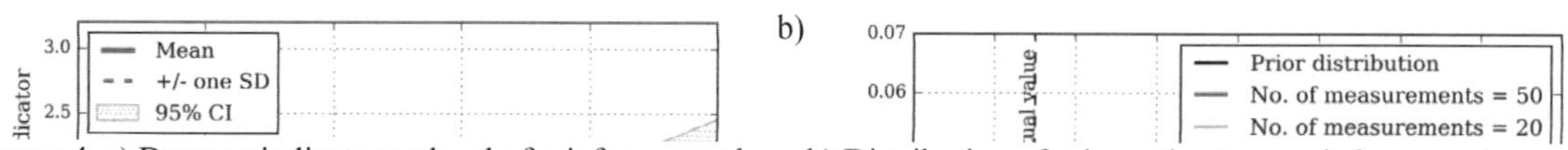

Figure 4. a) Damage indicator vs level of reinforcement loss, b) Distribution of prior estimate on reinforcement area loss and possible posterior distributions

2.3 Fault diagnostics based on Machine Learning Method

Bridges are among the most critical assets of the railway network, due to their long-life and strategic position on the railway network (Vagnoli et al 2017). Indeed, due to a limited number of alternative routes, if a railway bridge experiences a failure, the availability of the whole railway network is affected (Chen & Duan 1999). Hence, in order to improve the safety and availability of the railway network, TRUSS proposes a data-driven condition monitoring and fault detection method, which is able to automatically assess the health state of the bridge, by analysing the data provided by sensors installed on the bridge. In a case study, the method has been applied to a steel truss bridge, where a truss member was cut at different times and with different magnitudes. 'Failure 1' represents the bridge with a vertical truss element cut by half, and 'Failure 2' is obtained by cutting the vertical element completely. The data provided by the sensors are divided by considering three time windows based on the actual position of the vehicle: i) a position when no vehicles are travelling over the bridge, and the bridge response is influenced only by environmental factors such as wind; ii) a forced response, when a vehicle is passing over the bridge; iii) a free vibration response, when the vehicle leaves the bridge, but continues to vibrate due to the influence of the vehicle. For each time-window, 10 statistical features of the data are assessed per second. For example, the mean value and standard deviation of the response are computed at every second, and these features are used as input to an Empirical Mode Decomposition (EMD) analysis to extract the trend of each feature over time (Figure 5).

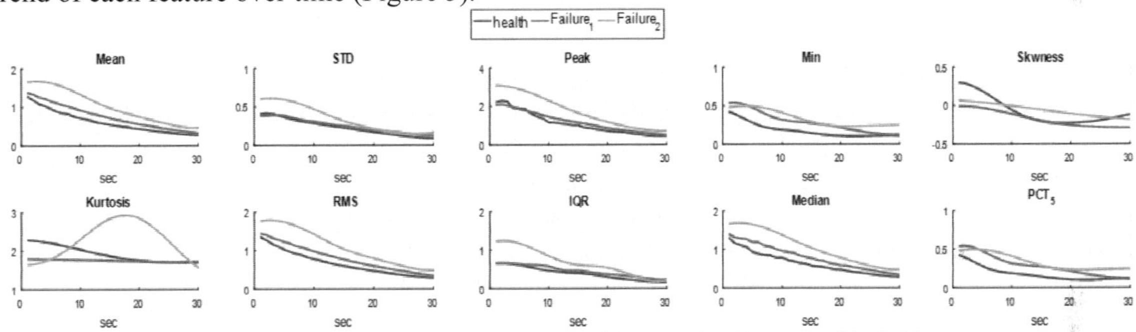

Figure 5. EMD residuals of the features of the three health states of the bridge

A set of behaviours is then randomly sampled with the aim of training a machine learning algorithm, that will automatically assess the bridge condition when a new unknown behaviour is provided by the sensors. Whenever a new unknown measurement becomes available, the trend of the computed statistical features of the unknown behaviour will be used as an input to the machine learning algorithm in order to assess the health state of the bridge. I.e., while the bottom chart in Figure 6 shows the real health state of the bridge for three scenarios, the top chart shows the probability of a health state for the different measurements of each scenario according to the machine learning algorithm. It should be noted that the 1st unknown pattern (to the left of the figure) is classified as behaviour of a healthy bridge mainly, whereas, the 2nd pattern is classified as 'Failure 1' with almost 100% probability, and finally, the 3rd pattern is classified as 'Failure 2'.

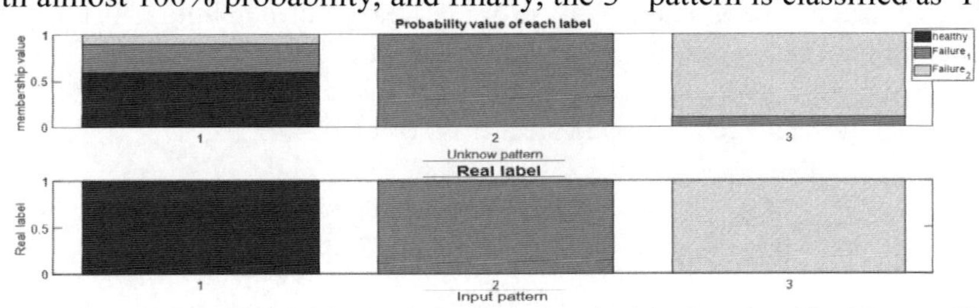

Figure 6. Estimated health state of the bridge by the proposed method (top), and real health state of bridge (bottom)

2.4 Damage Sensitivity of Untested Vibration-based Parameters

Traditional damage detection methods that monitor variations in modal-based parameters have been shown to be supported by a well-established theoretical base, but their real-world application has suffered from various pitfalls such as their susceptibility to environmental and operational conditions and assumption of linear structural behaviour and signal stationarity, as highlighted by Moughty & Casas (2017a). TRUSS investigates the damage sensitivity and real-world application of numerous previously untested vibration-based parameters and benchmark their performance against traditional modal-based damage sensitive features. Initial work began by identifying numerous suitable vibration parameters whose damage sensitivity was assessed under ambient conditions using the progressive damage test data from a post-tensioned three-span bridge (Moughty & Casas 2017b). This assessment demonstrated considerable damage sensitivity for energy-based vibration parameters, which can be attributed to increased kinetic energy induced by a reduction in structural potential energy due to damage. Moreover, when a frequency-based component is used as a denominator to the energy parameter, the resulting damage sensitivity increases. Examples of such parameters are the Destructive Potential Factor (Araya & Saragoni 1984) and Vibration Intensity, as per Equation (1).

$$I = \frac{Energy}{freq} = \mathrm{a}_0{}^2/f \tag{1}$$

More recently, the primary focus of research has centred on the application of EMD on non-stationary signals to attain Hilbert-Huang Transformation (HHT) (Huang et al 1998) spectrums for the purpose of early-stage local damage identification in bridges subjected to operational effects. The premise of this investigation stems from the fact that bridges are designed to behave linear-elastically under operational design loads, which implies that healthy bridges subjected to normal operational conditions should produce a nonlinear, but stationary vibration response. However, the presence of damage will induce a nonlinear-inelastic structural behaviour in the vicinity of the damage location, resulting in signal non-stationarity that should be identifiable in the HHT spectrum. This thesis was applied to measured acceleration data from a progressive damage test conducted on a steel truss bridge subjected to single-vehicle induced excitation (Kim et al 2014). Initial results presented in Moughty & Casas (2017c) note that initial application of EMD failed to sufficiently decompose the recorded signals into their individual modes due to the presence of varying levels of noise per sensor location and damage state. This led to the application of a superior method of signal decomposition known as Ensemble EMD (Wu & Huang 2004) coupled with a robust polynomial spline smoothing technique (Garcia 2010) that uses a discrete cosine transformation to discard small high frequency components so that an estimate of signal noise variance can be obtained. The advantage of separating signal from noise in this manner is that one need not have prior knowledge of the structure's modal characteristics, as would be the case if the typical band-pass filter method were employed. Additionally, as the HHT spectrum presents instantaneous frequency and amplitude components simultaneously, this allows for their combination to produce instantaneous Vibration Intensity, which, as mentioned, is one of the originally assessed vibration parameters that exhibited excellent damage sensitivity and resolution under ambient conditions, but normally requires the use of a Fourier-based transform that assumes linear-stationarity. Figure 7 presents the performance of instantaneous Vibration Intensity for varying damage states of the aforementioned steel truss bridge.

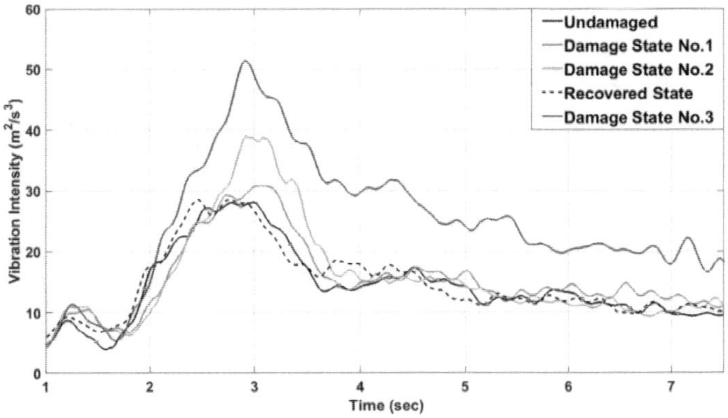

Figure 7. Instantaneous Vibration Intensity at sensor 4 for various condition states of a steel truss bridge

3. Modern SHM Technologies

3.1 Distributed Optical Fiber Sensing

The use of fiber optics sensors in SHM applications in infrastructures is something that has been studied and practiced with encouraging results. However, when dealing with bridges, the necessary number of discrete sensors required to perform a global and general monitoring action, can increase rapidly together with the necessary number of connection cables and acquisition systems, which exacerbate the complexity of the entire monitoring system (Barrias et al 2017). For this reason, TRUSS investigates the applicability of DOFS, which take advantage of their distributed and spatial continuity reading capability for measuring strain and temperature data with considerable resolution and high accuracy over extensive length stretches, using down to one single sensor (Barrias et al 2016). Some challenges arise with the bonding procedure of these sensors, especially when dealing with concrete structures due to the roughness and heterogeneity of the aggregates (Villalba & Casas 2012). Hence, the goals of TRUSS are related to the analysis of the possible spatial resolution and strain measurement accuracy achievable with this technology as well as to the assessment of its effectiveness for damage detection without failure or debonding of the optical fiber sensor. In order to address the later, a study of the most suitable adhesives for the bonding of the sensor to the structural material and the different techniques of attachment of the sensor itself have been carried out. Two small single reinforced concrete beams have been instrumented with a 5 m DOFS each and then undergone a three-point load test. The innovative aspect of this experiment is the use of DOFS attached directly on the reinforcement of the structural element without the need of previously mechanizing the rebar and with a very minimal coating assuring the optimal strain transfer scenario between the two elements. Furthermore, the same sensor is used on the outside surface of the concrete after its hardening allowing for a direct comparison between the strain being measured at the rebar and concrete surface. Additionally, two different adhesives have been used to bond the sensor to the rebar – cy-anoacrylate and epoxy for beam 1 and 2 respectively. Figure 8 reveals how this segment of the fiber was able to measure the evolution of developed strains with the increase of load for the entire length of the reinforcement bars. Moreover, it is possible to verify a good agreement between DOFS and data measured from strain gauges, validating the applicability of this bonding technique.

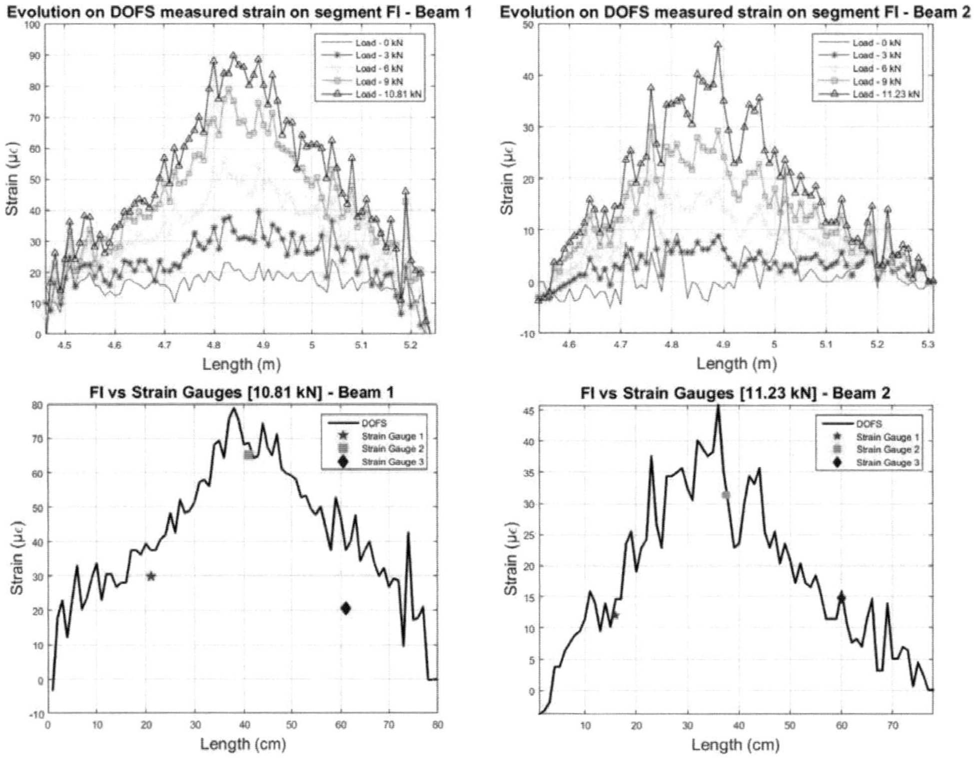

Figure 8. Strains measured by the DOFS on the segment embedded on the concrete for Beam 1 (up-left) and Beam 2 (up-right) and comparison with Strain Gauges for Beam 1 (down-left) and Beam 2 (down-right)

3.2 Drive-by SHM System

Direct instrumentation of bridges is commonly used in SHM. Unfortunately, this method is not economically feasible to be applied to all bridges. Using drive-by monitoring, the bridge condition is assessed using the response measured on a passing vehicle, which removes the need for installations or power source at the bridge. Drive-by monitoring is gaining in popularity in recent years, although only a limited number of practical applications have been reported (Malekjfarian et al 2015, Yang et al 2004, Yan & Chen 2016). TRUSS aims to exploit the high resolution capabilities of the Traffic Speed Deflectometer (TSD) for drive-by SHM purposes.

The principle applied by TRUSS to drive-by lies in the relationship of curvature with bending moment and stiffness. If a local increase in curvature occurs at a damaged location, this peak will be related to the stiffness loss. Curvature can be calculated from the second derivative of the displacement function, i.e., considering deflections at three bridge locations closely and equally spaced for an instant fixed in time. The curvature measured by a moving vehicle is an Instantaneous Curvature (IC) represented in Figure 9(a) for damaged and non-damaged situations. The TSD does not measure deflections, but the relative velocity between the vehicle and the bridge. Relative measurements can be also used for calculating the Rate of Instantaneous Curvature (RIC), which removes the influence of the vehicle as per Equation (2).

$$\mathrm{RIC}(x,t) = \frac{\dot{q}(x-\Delta x,t)-2\dot{q}(x,t)+\dot{q}(x+\Delta x,t)}{\Delta x^2} \tag{2}$$

where $\dot{q}$ is the relative velocity and Δx is the spacing between two consecutive TSD sensors. Six sensors are employed to calculate four RICs. Then, the Moving Average Difference (MAD) defined by Equation (3) is used as damage indicator to reduce the influece of noise and road profile.

$$\mathrm{MAD}(x) = \frac{\frac{1}{Z}\sum_{n=-\frac{(Z-1)\times L}{2f}}^{\frac{(Z-1)\times L}{2f}} \mathrm{RIC}^{\mathrm{damaged}}_{\mathrm{average}}(x+n) - \frac{1}{Z}\sum_{n=-\frac{(Z-1)\times L}{2f}}^{\frac{(Z-1)\times L}{2f}} \mathrm{RIC}^{\mathrm{healthy}}_{\mathrm{average}}(x+n)}{\min\left(\mathrm{RIC}^{\mathrm{healthy}}_{\mathrm{average}}\right)} \times 100\ (\%) \tag{3}$$

where f, L and Z are sampling frequency, bridge length and number of points respectively. Figure 9(b) shows the MAD applied to the same damage situation as in Figure 8(a). Despite the noisy conditions and the road profile, MAD shows to be a consistent damage indicator.

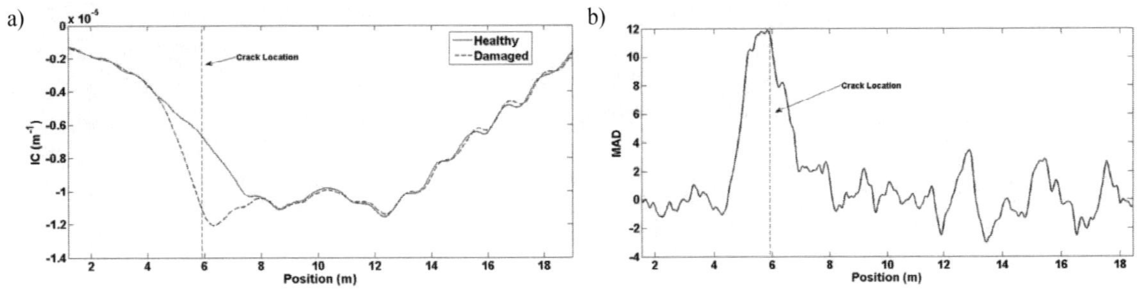

Figure 9. a) Instantaneous Curvature before & after damage @ 5.9m, b) Moving Average Difference for crack @ 5.9m

3.3 Fly-by SHM System

Chen et al (2016) review how the technology of civilian UAVs has improved significantly in recent years. In the meantime, the price of mature commercial UAVs is dropping constantly. Nowadays, with a budget about €2000, it is easy to find a functional UAV equipped with 3 axis gimbal, high-resolution action camera, obstacle detection and collision avoidance system. Those changes prompted UAVs becoming a competitive alternative for bridge visual inspection. Comparing to other visual inspection methods, UAVs is safer and more efficient than in-person inspections (Chan et al 2015), and much cheaper and more portable than bridge inspection tracks with a robotic arm (Giberti et al 2014) or a terrestrial laser scanner (Truong-Hong & Laefer 2015). In previous exploratory research, Kim et al (2015) used UAVs for images based inspection. The collected images were used for crack detection via their image processing algorithms. However, the detection was based on thousands of scattered images. Even though damages were identified in single images, to locate the crack in a real structure is still challenging.

To further improve the efficient and applicability of UAV based inspection, TRUSS has adopted 3D reconstruction technology. An easy way is using a stereo camera, such as the Kinect, for depth information acquisition and reconstruction. But the mono camera system is more compact and common in UAVs. Thus, TRUSS has combined structure from motion (SFM) method (Ullman 1979) and regular mono camera for 3D reconstruction. By detecting key points in each image taken from multiple viewpoints, the geometric relationship between images can be calculated and used for depth reconstruction. In practice, it has been noted that the quality of the SFM result is highly related to the data acquisition strategies, such as the capturing distance, camera angle and image overlapping rate. A laboratory experiment has indicated that the camera angle plays a decisive role compared to other factors and that combining different image sets can significantly improve the data quality (Chen et al 2017). By the end of the experiment, data accuracy was brought to millimeter level comparable to the laser scanned reference. Improving the data quality using 3D point cloud facilitates its use for damage detection. By analyzing the information, the damages in different structure surfaces, such as concrete (Figure 10(a)), masonry (Figure 10(b)) or pavement (Figure 10(c)) were succesfully detected. In addition, those damages or any other extracted features can be highlighted in a 3D model for further inspection or documentation (Figure 10(d)). Finally, low price commercial UAVs have shown to possess the capability to acquire high-quality images that can be used for 3D documentation and surface damage detection. Future works will be focused on using different sensors, such as an infrared camera or multispectral camera for hidden damage detection.

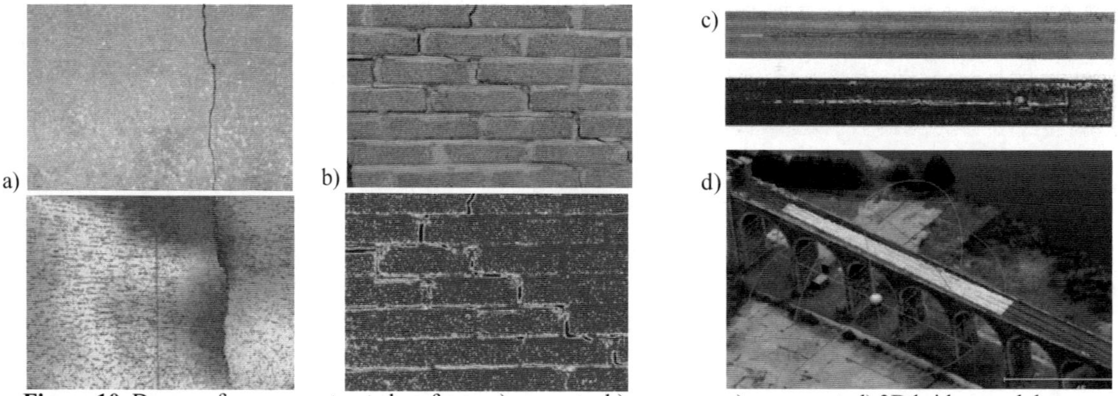

Figure 10. Damage from reconstructed surfaces: a) concrete, b) masonry, c) pavement, d) 3D bridge model

4. Conclusions

Given the age of some existing infrastructure, often deteriorating at faster rates than anticipated due to extreme environmental conditions and increasing demand in traffic loads, continuous monitoring of bridges is becoming increasingly important. This paper has presented developments of cost-effective algorithms and modern SHM sensing technologies for damage detection and structural assessment. TRUSS research is currently ongoing and updates about the achievements of the project can be followed on http://trussitn.eu. Ultimately, the project aims to optimize the available resources allowing to monitor a larger bridge stock, and to make infrastructure networks last longer and be safer.

Acknowledgements

The TRUSS ITN project is funded by the European Union's Horizon 2020 research and innovation programme under the Marie Skłodowska-Curie grant agreement No. 642453.

References

Araya, R & Saragoni, R 1984, 'Earthquake accelerogram destructiveness potential factor', *Proceedings of the 8th World Conference of Earthquake Engineering*, San Francisco, USA, vol. 2, pp. 835–841.

Barrias, A, Casas, JR & Villalba, S 2016, 'A review of distributed optical fiber sensors for civil engineering applications', *Sensors*, vol. 16, no. 5, pp. 748.

Barrias, A, Casas, JR & Rodriguez, G 2017, 'UPC – BarcelonaTech experience on the use of Rayleigh based distributed optical fiber sensors for SHM of concrete structures', *Proceedings of the 27th European Safety and Reliability Conference (ESREL 2017)*, Portoroz, Slovenia, 18-22 June.

Chan, B, Guan, H, Jo, J & Blumenstein, M 2015. 'Towards UAV-Based bridge inspection systems: a review and an application perspective', *Structural Monitoring and Maintenance*, vol. 2, no. 3, pp. 283–300.

Chen, WF & Duan, L 1999, *Bridge Engineering Handbook*, CRC press, Boca Raton, FL, United States.

Chen, S, Laefer, D. & Mangina, E 2016, 'State of technology review of civilian UAVs', *Recent Patents in Engineering*, pp.1–15.

Chen, S, Laefer, DF, Byrne, J & Natanzi AS 2017, 'The effect of angles and distance on image-based, three-dimensional re-constructions", *Proceedings of the 27th European Safety and Reliability Conference (ESREL 2017)*, Portoroz, Slovenia, 18-22 June.

Décatoire, R 2015, *Optimisation of inspection plans for structures submitted to a stochastic degradation context: application on RC structures*, PhD Thesis. University of Nantes, Faculty of Science and Technology, France.

Garcia, R 2010, 'Robust smoothing of gridded data in one and higher dimensions with missing values', *Computational Statistical Data Analysis*, vol. 54, pp. 1167–1178.

Giberti, H, Zanoni, A, Mauri, M & Gammino, M 2014, 'Preliminary study on automated concrete bridge inspection', *Proceedings of the ASME 2014 12th Biennial Conference on Engineering Systems Design and Analysis*, Copenhagen, Denmark.

Huang, NE, Shen, Z, Long, SR, Wu, MC, Shih, HH, Zheng, Q, Yen, N-C, Tung, CC & Liu, HH 1998, 'The empirical mode decomposition and the Hilbert spectrum for nonlinear and non-stationary time series analysis', *Proceedings Royal Society London*, Series A, vol. 454, pp. 903–993.

Kim, CW, Chang, KC, Kitauchi, S, McGetrick, PJ, Hashimoto, K & Sugiura, K 2014, 'Changes in modal parameters of a steel truss bridge due to artificial damage', *Proceedings of the 11th International Conference on Structural Safety and Reliability, (ICOSSAR)*, New York, pp. 3725-3732.

Kim, J-W, Kim, S-B, Park, J-C & Nam, J-W 2015, 'Development of crack detection system with unmanned aerial vehicles and digital image processing', *Proceedings of Advances in Structure Engineering and Mechanics*, Incheon, Korea.

Leahy, C 2013, *Using weigh-in-motion data to predict extreme traffic loading on bridges*, PhD thesis, College of Engineering and Architecture, University College Dublin, Ireland.

Malekjafarian, A, McGetrick, PJ & OBrien, EJ 2015, 'A review of indirect bridge monitoring using passing vehicles', *Shock and Vibration*, vol. 2015, pp. 16.

Marsh, PS & Frangopol, DM 2007, 'Reinforced concrete bridge deck reliability model incorporating temporal and spatial variations of probabilistic corrosion rate sensor data', *Reliability Engineering & System Safety*, vol. 93, pp. 394-409.

Moughty, JJ & Casas, JR 2017a, 'A state of the art review of modal based damage detection in bridges: development, challenges and solutions', *Applied Sciences*, vol. 7, no. 5, 510.

Moughty, JJ & Casas, JR 2017b, 'Performance assessment of vibration parameters as damage indicators for bridge structures under ambient excitation,' *Proceedings of the 10th International Conference on Structural Dynamics (EURODYN 2017)*, Rome, Italy, September 10-13.

Moughty, JJ & Casas, JR 2017c, 'Evaluation of the Hilbert Huang Transformation of transient signals for bridge condition assessment', *Proceedings of the 27th European Safety and Reliability Conference (ESREL 2017)*, Portoroz, Slovenia, June 18-22.

Ojio, T, Carey, C, OBrien, EJ, Doherty, C & Taylor, S 2016, 'Contactless bridge Weigh-in-Motion', *Journal of Bridge Engineering*, vol. 21, no. 7, 04016032.

Ribeiro D, Calçada R, Ferreira J & Martins, T 2014, 'Non-contact measurement of the dynamic displacement of railway bridges using advanced video-based system', *Engineering Structures*, vol. 75, pp. 164-180.

Suo, Q & Stewart, MG 2009, 'Corrosion cracking prediction updating of deteriorating RC structures using inspection information', *Reliability Engineering and System Safety*, vol. 94, pp. 1340-1348.

Truong-Hong, L & Laefer, DF 2015, 'Documentation of bridges by terrestrial laser scanner', *Proceedings of IABSE conference 2015*, Geneva, Switzerland.

Ullman, S 1979, 'The interpretation of structure from motion', *Proceedings of the Royal Society of London. Series B, Biological Sciences*, vol. 203, no. 1153, pp. 405-426.

Vagnoli, M, Remenyte-Prescott, R & Andrews, J 2017, 'A fuzzy-based Bayesian Belief Network approach for railway bridge condition monitoring and fault detection', *Proceedings of the 27th European Safety and Reliability Conference (ESREL 2017)*, Portoroz, Slovenia, 18-22 June.

Val, DV, Stewart, MG & Melchers, E 1998, 'Effect of reinforcement corrosion on reliability of highway bridges', *Engineering Structures*, vol. 20, no. 11, pp. 1010-1019.

Villalba, S & Casas, JR 2012, 'Application of optical fiber distributed sensing to health monitoring of concrete structures', *Mechanical Systems and Signal Processing*, vol. 39, no. 1, pp. 441–451.

Wu, Z & Huang, NE 2004, 'A study of the characteristics of white noise using the empirical mode decomposition method', *Proceedings Royal Society A*, vol. 460, pp. 1597–1611.

Yang, Y-B, Lin, C & Yau, J 2004, 'Extracting bridge frequencies from the dynamic response of a passing vehicle', *Journal of Sound and Vibration*, vol. 272, no. 3, pp. 471-493.

Yang, Y-B & Chen, WF 2016, 'Extraction of bridge frequencies from a moving test vehicle by stochastic subspace identification', *Journal of Bridge Engineering*, vol. 21, no. 3.

The 8th International Conference on
Structural Health Monitoring of Intelligent Infrastructure
Brisbane, Australia | 5-8 December 2017

Flood Resilience of Masonry Bridges: Analysing Structural Performance under Hydrodynamic Forces and Scour

P. Kripakaran[1], L. Walter[1], R. Kahraman[1], M. Riella[1], M.E. Ebrahimi[1], G. Tabor[1], S. Djordjevic[1]
[1] *University of Exeter – United Kingdom, Email:* p.kripakaran@exeter.ac.uk

Abstract

Masonry arch bridges are critical elements of the transport infrastructure in the UK and abroad. Ensuring their structural performance during extreme weather events including floods is essential for a resilient transport network as bridge closure due to failure or damage during or in the aftermath of floods hampers severely rescue and recovery efforts. Current procedures for assessing flood-induced risks to masonry arch bridges, which determine the interventions taken in preparation to future flood events, are fundamentally unreliable since (i) the actual forces experienced by structures during floods are yet to be reliably evaluated, and (ii) our scientific understanding of the fluid-structure interaction is inadequate. This paper presents results from an ongoing research project that aims to address both these knowledge gaps. First an approach to predict hydrodynamic forces and scour is developed using computational fluid dynamics models that have been validated using measured data from flume experiments. Next the predicted force distributions are applied systematically to numerical models of bridges to study their behaviour under flood conditions. This paper illustrates the developed approach using a numerical model of a full-scale bridge as a case study. Preliminary results from analysis are given in this paper. Results support identifying potential regions of structural damage and also suggest opportunities for measuring structures under flood conditions to enhance post-flood assessment of these structures.

1. Introduction

The structural performance of bridges during floods is a critical factor in the resilience of transport networks. The closure of a small number of bridges can drastically increase travel times and even completely isolate communities, thereby severely hampering rescue operations and increasing greatly the time for post-flood recovery. For instance, the floods in Cumbria, UK in November 2009 (Rowson & Wynne 2009), when 314mm of rain fell in 24 hours, caused 29 bridges to collapse. Furthermore, over 20 bridges, mostly masonry bridges, remained closed for many months after the flood due to damage resulting in huge additional travelling distances. More recently during the winter of 2015/16 , repeated storm events that included storms Desmond, Eva and Frank damaged hundreds of bridge in Cumbria with some villages such as Braithwaite even completely cut off from the transport network.

Masonry arches, due to their special geometry that typically includes short span and low clearance, are particularly vulnerable to flooding. These structures constitute 40% of the UK bridge stock (Bridle & Sims 2009) and are essential components of transport infrastructure in the UK and other countries (McKibbins et al. 2006). The majority of these structures are hundreds of years old with many designated to be of great importance to cultural and engineering heritage. These were also typically designed to withstand flood events much less severe than those happening today (Mountain & Benn 2005). Given that climate change is expected to increase both the frequency and severity of future flood events (Kirby et al. 2015), methods for assessing reliably the structural performance of these bridges under flooding become of even greater importance.

Previous research on the topic of the fluid-bridge structure interaction during flooding has mostly focused on the topic of scour around bridge piers. Little attention has been given to scour around single or multi-span masonry arch bridges. This is particularly important considering the significant number of failures of single span structures on small watercourses due to abutment scour (Benn 2013). Moreover even the studies on piers (Pagliara & Carnacina 2011; Pagliara & Carnacina 2013) are limited to geometries and arrangements that are not representative of masonry bridge piers. Few studies have investigated the impact of innundation on masonry arch bridges. Hulet et al. (Hulet et al. 2006) studied the static load-carrying capacity of partially or fully submerged masonry arch bridges. They found that the load-carrying capacity of a bridge can be reduced by a factor of 1.6-1.8 in such conditions. However no study has examined what happens to masonry arch bridges during the turbulent flow conditions in real life.

There is therefore limited guidance available to practitioners when it comes to assessing masonry bridges during flooding. For example, UK Highways Agency bridge design guidance (Highways Agency 2012; Highways Agency 1994) acknowledges that water flowing past a bridge pier will exert a hydrodynamic force on the pier. The force is split into two components – one in the direction of flow known as drag, and the other normal to the flow referred to as lift. The guidance also notes that if the water level reaches above the soffit level of a bridge, flowing water will exert an upward force on the bridge deck. However the equations given in the guidance are based on simplified physics models that ignore the highly turbulent flow conditions or the actual free surface profile of the flow in real life. Existing guidance also fail to adequately account for the effects of debris accumulation on flow and thereby the uplift and lateral forces on the structure. Uplift and lateral forces can pose a major threat to the integrity of masonry bridges which rely on a pre-compression due to self-weight to carry loads. Moreover these structures being much less flexible than modern bridges can be taken to collapse by even small movements.

This paper presents results from a research project aimed at filling the current knowledge gap in our understanding of the fluid-structure interaction for masonry arch bridges during floods. It briefly describes how the flow conditions around these bridges are simulated for the first time using computational fluid dynamics (CFD) models. Then the predicted pressures and forces are applied to a numerical model of a full-scale masonry arch bridge to evaluate its performance during flooding. As the project is ongoing, only preliminary results are included in the paper. The results show that the CFD predictions can facilitate assessment of these structures for flooding risks.

2. Methodology

The research presented in this paper is part of a larger project aimed at investigating the effects of debris blockage on flow pattern, scour and hydrodynamic loads at masonry bridges under flood conditions. Debris accumulation at bridges is known to worsen scour and hydrodynamic forces by constricting flow but the underpinning science is not adequately understood to enable mitigating the risks from debris blockage. This project aims to fill this knowledge gap. This project has three phases:

(i) Laboratory experiments using a flume,

(ii) Simulation using computational fluid dynamics (CFD) models, and

(iii) Development of guidance for practitioners.

The goal of the first phase is to test reference scenarios of flow, debris and structure combinations to provide data for the development of numerical models and their validation in the second phase. Scenarios at full-scale with real-life hydraulic conditions will then be simulated via developed CFD codes, and results from both phases will be used to develop an assessment guidance in the third phase of the project. Project outcomes will eventually be built into the existing UK guidance for assessment of bridges under hydraulic action, and is expected to enable improved management of bridges at risk to flooding. A detailed review of the project background and methodology is presented in Ebrahimi et al. (Ebrahimi et al. 2016). This paper only briefly discussess the experiments and the creation of the CFD models. Instead it focuses on the results obtained from CFD models for flow around innundated masonry bridges, and how the predictions are applied to numerical models of a single bridge.

2.1 Flume experiments

The experiments are carried out in a recirculating flume 0.6m-wide, 0.65m-deep and with a 10m-long working section. The flume is equipped with a centrifugal pump capable of recirculating solid particles up to 3 mm in size, which enables it also simulate sediment transport crucial for studying scour. Initial experiments used a model of a masonry bridge pier. This is shown in Figure 1. The scaled single pier features two triangular cutwaters which are a typical feature of masonry bridges in the UK. Each cutwater is an isosceles triangle with base angle 45°. A pier width of 50 mm was chosen. Pier stream-wise length is 156 mm resulting in a width-to-length ratio of 0.32 that matches the average aspect ratio estimated for several masonry bridge piers mainly in Devon, UK.

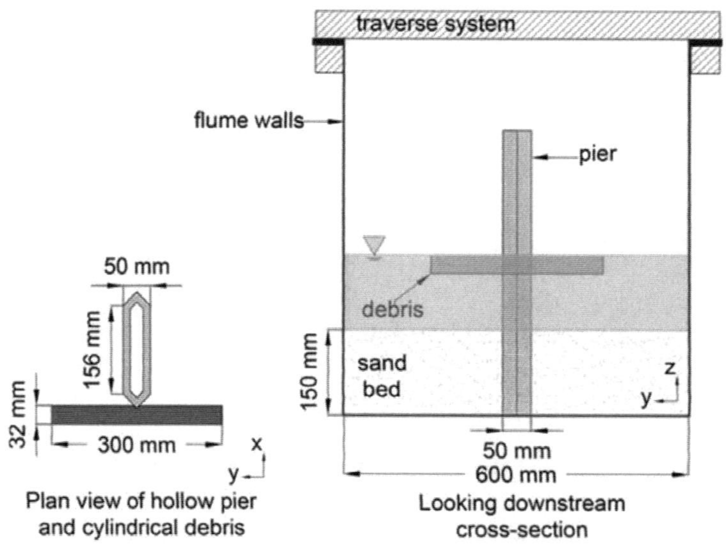

Figure 1: Schematics of experiment. Left: Plan view of pier (x shows flow direction); Right: Cross-section of flume with pier.

A number of hydraulic and flow parameters were measured during the experiments which were conducted for various combinations of flow depths and discharge rates. The flow rate was measured using an electromagnetic flowmeter (resolution ±0.1 l/s). Approach flow depth was quantified using a digital point gauge (resolution 0.01 mm) mounted sufficiently upstream of the pier to read undisturbed free surface elevation. Velocity measurements were carried out with the aid of a Vectrino Profiler ADV (accuracy 0.5% of measured velocity), installed on a 3-axis traverse system supported on the flume. Velocity measurements were performed on a grid of 20 predefined (x, y, z) coordinates. Velocity data collection duration at each point was between 3-20 minutes depending on the degree of turbulence and velocity fluctuations at that point. Final scour geometry was measured after stopping the flow carefully to avoid any disturbance to the scour topography. Measurements were carried out using the distance measurement feature of ADV probe (accuracy of 0.5 mm) by moving the probe over a grid of ~ 240 predefined (x, y) points on one side of the pier.

We are also currently setting up to run hydraulic experiments with a model of a single span masonry arch bridge. A schematic of the model is shown in Figure 2. The model is a 1:10 scaled version of Canns Mill Bridge in Devon, UK. The abutment widths are however reduced to minimize the ratio of structure width to that of flow. The ratio of the bridge span length (466 mm) to width is kept equal to 0.73 which is the estimated average for several masonry bridges mainly in Devon, UK. The pressures and the forces applied on the model will be measured live in the laboratory. Figure 2 shows the pressure taps in the model used for measuring hydrodynamic pressures. The uplift forces will be measured using load cells (not shown in the figure) that keep the model in place. The laboratory measurements for both bridge piers and the masonry arch model are used for calibrating CFD models.

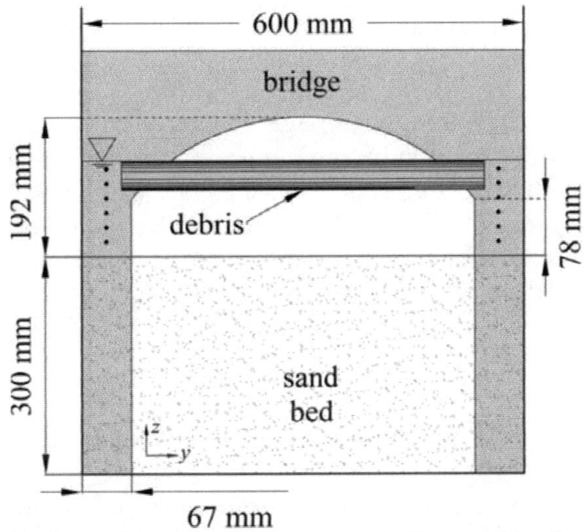

Figure 2. Cross-sectional schematic of a single span arch. Streamwise length is 340 mm. Pressure measurement taps shown by dots on the abutments.

2.2 Computational modelling

Computational Fluid Dynamics (CFD) is utilised to model effectively the experimental conditions found within the flume. The purpose is to validate CFD models using laboratory measurements and then to use CFD to simulate for useful full-scale scenarios. The Unsteady Reynolds Averaged Navier-Stokes (URANS) equations governing fluid flow are solved numerically using the open-source CFD toolbox, OpenFOAM.

Two sets of simulations have been conducted: single phase and multi phase simulations. The single phase simulations exclude air and model the surface as a flat surface using a slip boundary. The multiphase simulations account for both water and air phases with an interface separating them. The main motivation is to investigate the improvement in simulation predictions in relation to increase in the computational time in moving from a single to a multi phase simulation.

The fluid domain is a complete replica of the experimental flume's geometry. This ensured continuity between both experiment and numerical modelling and facilitated direct comparisons. A fully structured mesh consisting of 8 million and 15 million cells, respectively, is used throughout both simulations. The average stream-wise velocities are sampled over the whole depth of the flow and are compared with those results recorded by the ADV. Results from CFD simulations are observed to be comparable to laboratory measurements. These comparisons are shown in a separate paper by the authors (Ebrahimi et al. 2017) and are not shown here for reasons for brevity.

In this paper, the calibrated CFD models are then used to simulate flow corresponding to a scenario when the water level reaches up to the parapets of a masonry arch bridge. The pressures applied on the arch under these conditions are captured from the CFD simulation. These pressures are subsequently applied to a finite element model of the masonry arch bridge to evaluate its structural behaviour. For simplicity, it is assumed that the channel is perfectly straight and of rectangular cross-section, with the vertical channel walls perfectly flush with the bridge ends. This is not representative of real-life conditions. Rivers are aligned at an angle to the bridge referred to as an angle of attack, which average around 30° but can reach up to 90° in worst cases. The river gradient and boundaries are also often non-zero and arbitrary respectively.

2.3 Numerical modelling of masonry arch bridges

The masonry arch is simple in concept but its structural behaviour can be complex. It relies on the arch ring being in compression as masonry has negligible tensile strength i.e. the arch does not hold when in tension. Masonry structures heavily rely on gravity i.e. the weight of the masonry and the backfill, to pre-load the arch ring in compression. The integrity of the arch is maintained as long as any tension in the arch ring due to additional load (e.g. traffic axle loads, upward hydrodynamic loads) is less than the tensile strength of the mortar or the ring elements. This paper investigates the interplay between stresses in the arch ring and flood hydrodynamic loads.

2.3.1 Model parameters

A linear, elastic numerical model of the Canns Mill Bridge (see Figure 2) at real-scale is created. The arch ring and the spandrel walls are modelled using shell elements. The effect of the backfill, which transfers load from the top and between the spandrel walls, is simulated using crosswalls.

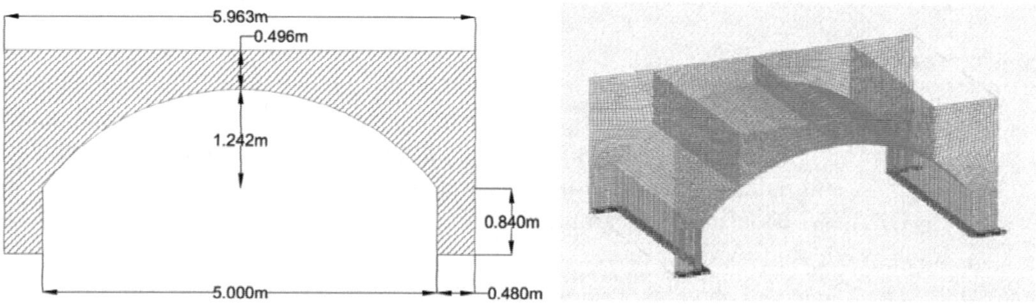

Figure 3. Dimensions of the bridge model (left) and an image of the numerical model of the bridge in Oasys GSA (right)

There are several critical parameters that affect the behaviour of masonry arches under loading. The effects of the following two key parameters are investigated in detail in this research.

- *Backfill*: Davey (Davey 1953) was one of the first to suggest that the fill had any real structural contribution to the arch. He conducted a series of tests on real bridges, which in some cases found that the bridge was two and a half times stiffer, in the presence of the fill compared to its absence. It has been known for at least a century that the soil backfill material placed around the barrel of a masonry arch bridge can have a significant beneficial effect on its ultimate load-carrying capacity. The study by Callaway et al. (Callaway et al. 2012) concluded that the backfill is known for transmitting and distributing the live loads from the top of the bridge to the arch barrel as well as providing lateral stability to the arch barrel. According to Betti et al. (Betti et al. 2008) not only the presence of the fill, but the characteristics can also heavily affect the collapse load, as a model with non-resistant fill gives a collapse load only 35% of the collapse load of a model with resistant fill. This study examined the relationship between the weight of the backfill and the arch load carrying capacity. As the weight of the backfill is reduced, the pre-compression in the arch reduces making it more vulnerable to the upward and lateral forces introduced during flooding. The bridge is analysed for three backfill density values: 0 (no backfill), 10 kN/m^3 (dry low density fill) and 20 kN/m^3 (saturated fill).
- *Arch ring thickness*: Altunisik et al. (Altunisik et al. 2015) studied the effect of varying the arch thickness on the structural behaviour of masonry arch bridges with finite element models. They found that the maximum displacements increased when the ring thicknesses were decreased and vice versa. Also tensile stresses in the arch increased when the ring thickness was decreased and vice versa. While these results are as one would expect through intuition, the exact numerical relationships are essential to enable engineers to identify the risk to a certain bridge and to thereby take protection measures as necessary. The ring thicknesses chosen for this study are 125mm, 150mm, 215mm, 328mm and 440mm. These are selected based on suitable ring thicknesses for a 5m span. The thicknesses of 215mm, 328mm and 440mm are chosen as they are standard arch ring thicknesses. A ring 215mm thick can be obtained by laying a single layer of bricks as headers. A ring 328mm thick can be achieved by the addition of a brick header and the short end of a brick. A ring 440m can be achieved by laying two bricks head to head. Smaller thicknesses were used to infer the structural behaviour as it approaches zero.

Other critical parameters include the spandrel and wing walls of the masonry arch as well as the foundations and the span/rise ratio all of which have been subject of conducted investigations in literature that define their effect on the capacity of masonry arch bridges. These additional parameters are not investigated in this study but are part of the plans for the future.

2.3.2 Flood-induced forces

When a bridge is partially or fully submerged in a flood, it is subject to upward and lateral forces. These forces that can be broken down into hydrostatic and hydrodynamic components, which are a result of the corresponding static and dynamic pressures on the bridge.

1. *Hydrostatic pressure*: The static pressure exerted by water on the bridge. This increases in proportion to depth from the surface as calculated below:

$$p_{static} = \rho g h \tag{1}$$

where ρ is the fluid density, g is the acceleration due to gravity and h is the height of the water column at the point at which the pressure is computed. Knowledge of the free surface profile is necessary to determine h and hence to compute the pressure.

2. *Hydrodynamic pressure*: The pressure exerted on the bridge from fluids in motion as described below:

$$p_{dynamic} = \frac{1}{2}\rho V^2 \tag{2}$$

where ρ is the fluid density and V is the velocity of the flowing water. Knowledge of the flow velocity V is essential to determine the dynamic pressure.

The sum of these two components gives the total pressure at any given point on the bridge. The CFD results can be used to determine both the components as the simulation provides both the profile of the free surface and the flow velocities across the fluid domain.

Consequently the forces computed from CFD predictions of pressures are then applied to the numerical model of the bridge in the following manner. The pressures vary in both time and space dimensions. To begin with, we approach the problem as a static phenomena and apply only the peak pressures determined from CFD. The surface area of the bridge that is exposed to water, namely the front face and the underside of the arch are sub-divided into smaller area portions. This is shown in Figure 4. The front face is divided into 8 zones and the underside of the arch into 6 zones. The pressures are integrated over these zones and the resultant forces are applied to the model.

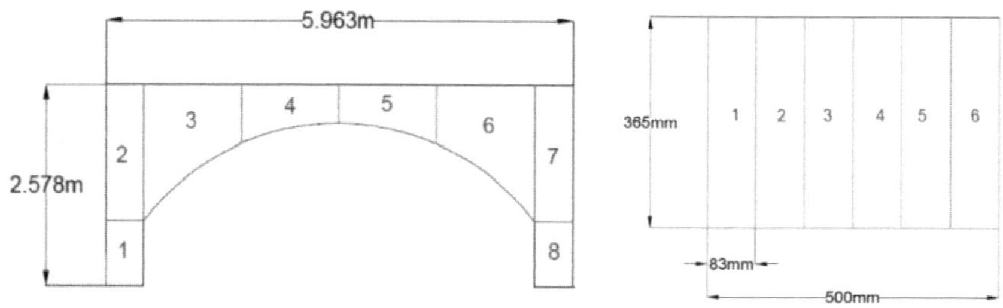

Figure 4. The division of the arch bridge model into zones for applying flood-induced forces; left: the front face, and right: the underside of the arch

2.4 Results

The masonry arch model in Figure 3 under the load combination of dead load (due to self-weight from masonry and from backfill) and the maximum flood-induced loads. Stress results for a backfill density of 20 kN/m^3 and an arch ring thickness of 0.38 m are shown in Figure 4. Regions of high tensile stress are observed near the crown of the arch ring (indicated by node 142 in Figure 5).

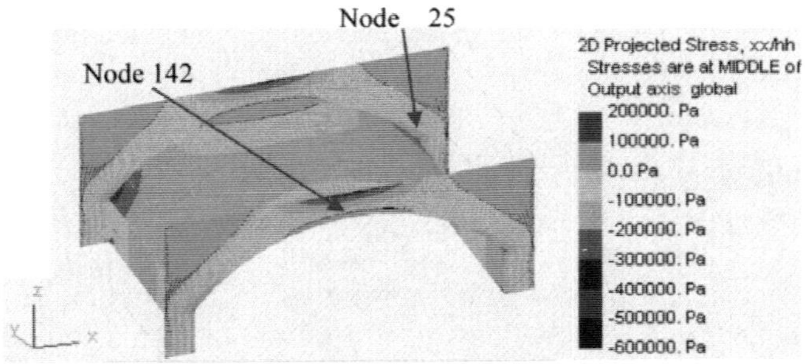

Figure 5. Stress distribution in the arch under dead load and flood-induced loads

Maximum deformations found at the centre of the arch decreases with increasing arch ring thickness and deformations plateau when the arch ring thickness reaches 328 mm. The presence of backfill also shows significant effect on the magnitude of deformation. The deformations are over 50% for the scenario with no backfill when compared to the scenario with 20 kN/m^2 backfill density.

The maximum tensile stress in the bridge model decreases as the arch ring thickness is increased. The maximum values are at the central section of the arch barrel on the front wall of the bridge for all scenarios. As the arch ring thickness is increased, the maximum compressive stress decreases exponentially (not shown in the paper for brevity) until plateauing for arch ring thickness values greater than 328mm. The presence of backfill also gives a significant increase to magnitude of maximum compressive stress, averaging a 50% increase in maximum compressive stress with the introduction of backfill at a density of 20 kN/m^3.

Lastly, for all evaluated scenarios, the tensile stress experienced in the arch ring is smaller than the tensile strength of mortar. This indicates that the arch ring is unlikely to fail for the modelled flood condition. The deflections are also small (of the order of 1-3 mm) supporting a linear, elastic analysis. However, this study did not model the effect of scour and its consequence on the support to foundations. Neither did it consider the impact of angles of attack, which can significantly worsen flow velocities around the bridge. These are to be considered in the next stages of this research.

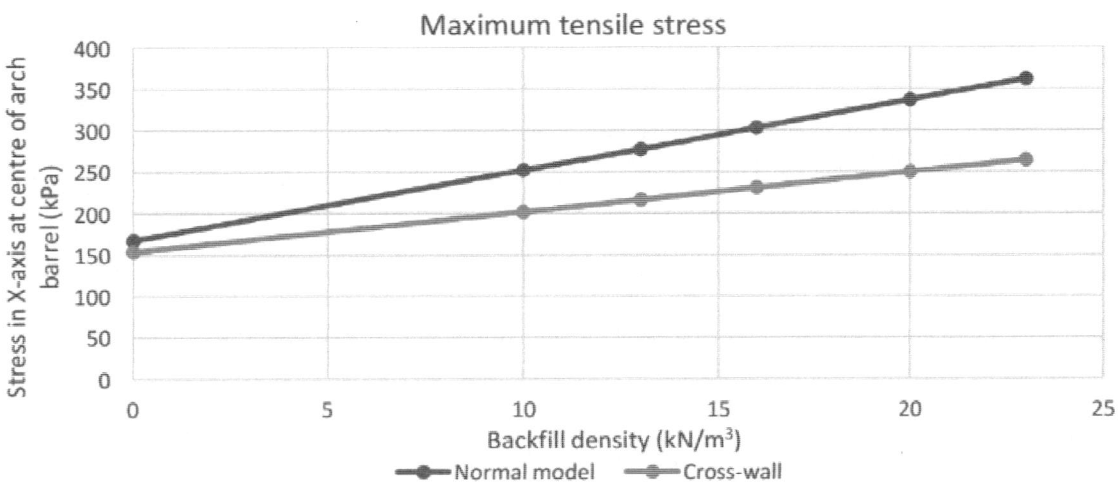

Figure 5. Variation of the maximum tensile stress with backfill density

3. Conclusions

This paper presents preliminary results from the analysis of a masonry arch bridge under flood conditions. The study uses predictions of real hydrodynamic loads on the bridges from CFD models that have been validated using laboratory data. Conclusions of this study are as follows:

- Results show that the considered bridge is not subject to hydraulic pressures that take it beyond its capacity for the modelled flooded condition. This indicates that the bridge is unlikely to collapse for the flood.
- Decreasing the arch ring thickness for a single span arch increases the maximum deformation and stresses experienced by the arch barrel in flooding conditions
- The presence of backfill is influential in spreading the load across the arch barrel in flooding situations. The study of incremental densities of backfill increased the maximum deformation, compressive and tensile stresses experienced by the structure.

Future work will investigate the influence of various structural and hydraulic parameters on structural integrity including the impact of scour at foundations, the forces from debris, angles of attack and channel gradient. All of these are known to affect the structural stability of the bridge in flooding conditions.

Acknowledgments

The research presented in this paper was supported by funding from the UK's Engineering and Physical Sciences Research Council (EPSRC) under grant EP/M017354/1. For further information please refer to the project blog at http://blogs.exeter.ac.uk/ramb/. We would also like to acknowledge the support from all industry partners including Devon County Council who provided details of the Canns Mill Bridge used as a case study in this paper.

References

Altunisik, A.C., Kanbur, B. & Genc, A.F., 2015. The effect of arch geometry on the structural behavior of masonry bridges. *Smart Structures and Systems*, 16(6), pp.1069–1089.

Benn, J., 2013. Railway Bridge Failure during Flood in the UK and Ireland : Learning from the Past. In *Forensic engineeering conference*.

Betti, M., Drosopoulos, G.A. & Stavroulakis, G.E., 2008. Two non-linear finite element models developed for the assessment of failure of masonry arches. *Mecanique*, 336, pp.42–53.

Bridle, R.J. & Sims, F., 2009. The effect of bridge failures on UK technical policy and practice. *Proceedings of the ICE - Engineering History and Heritage*, 162(1), pp.39–49.

Callaway, P., Gilbert, M. & Smith, C.C., 2012. Influence of backfill on the capacity of masonry arch bridges. *Proceedings of the Institution of Civil Engineers - Bridge Engineering*, 165(3), pp.147–157.

Davey, N., 1953. Tests on road bridges. National Building Studies Research Paper No. 16.

Ebrahimi, M. et al., 2016. Hydrodynamic effects of debris blockage and scour on masonry bridges: Towards experimental modelling. In *Scour and Erosion*. Taylor & Francis Group, 6000 Broken Sound Parkway NW, Suite 300, Boca Raton, FL 33487-2742: CRC Press, pp. 743–750.

Ebrahimi, M. et al., 2017. Scour and hydrodynamic effects of debris blockage at masonry bridges: Insights from experimental and numerical modelling. In *IAHR Conference*.

Highways Agency, 2012. The Assessment of Scour and Other Hydraulic Actions at Highway Structures (BD97/12). In *Design Manual for Roads & Bridges*.

Highways Agency, 1994. *The Design of Highway Bridges for Hydraulic Action*,

Hulet, K.M., Smith, C.S. & Gilbert, M., 2006. Load-carrying capacity of flooded masonry arch bridges. *Proceedings of Institution of Civil Engineers: Bridge Engineering*, (BE3), pp.97–103.

Kirby, A.M. et al., 2015. *Manual on scour at bridges and other hydraulic structures* Second edi., London: CIRIA.

McKibbins, L.D. et al., 2006. *Masonry arch bridges: condition appraisal and remedial treatment*, London: CIRIA.
Mountain, A. & Benn, J., 2005. *Safe management of railway structures - flooding and scour risk*, London: RSSB.
Pagliara, S. & Carnacina, I., 2013. Bridge pier flow field in the presence of debris accumulation. *Proceedings of the ICE - Water Management*, 166(4), pp.187–198.
Pagliara, S. & Carnacina, I., 2011. Influence of Wood Debris Accumulation on Bridge Pier Scour. *Journal of Hydraulic Engineering*, 137(2), pp.254–261.
Rowson, J. & Wynne, A., 2009. Cumbrian bridges ripped out by torrential floods. *New Civil Engineer*, pp.6–7.

The 8th International Conference on
Structural Health Monitoring of Intelligent Infrastructure
Brisbane, Australia | 5-8 December 2017

Smart cities to improve resilience of communities

A. Zamani Noori [1], S. Marasco [2], O. Kammouh [3], M. Domaneschi [4], G.P. Cimellaro [5].
[1] *Visiting Researcher, UC Berkeley –USA, Email:* ali.zamani@polito.it
[2] *Visiting Researcher, UC Berkeley –USA, sebastiano.marasco@polito.it*
[3] *PhD Student, Politecnico di Torino –Italy, Email:* omar.kammouh@polito.it
[4] *Assistant Professor, Politecnico di Torino –Italy, Email:* marco.domaneschi@polito.it
[5] *Visiting Professor, UC Berkeley –USA, Email:* gianpaolo.cimellaro@polito.it

Abstract

This paper presents a new approach to predict the potential damage and physical impacts of an earthquake on the built environment. A new methodology for large-scale simulations within a seismic scenario are explored by evaluating multipurpose codes. In particular, a 3-D model of a virtual city is developed for evaluating the seismic effects at increasing intensities. Four different building sectors that provide essential functions to a community, including housing, education, business, and public services are considered. Once the buildings are integrated into the city, parallel simulations are applied to compute the system functionality following a disruptive scenario. Tri-linear elastoplastic backbone curve representative of the global shear behavior of each building is estimated considering the dominant modal shapes and building irregularities. Monte Carlo Simulations (MCS) are applied to take into account the epistemic uncertainties associated with geometry and mechanical properties within the range of observations. The nonlinear dynamic analyses are performed through SAP2000 Application Programming Interface (API) in order to assess the dynamic response of the buildings in an organized and automatic fashion. Accordingly, the city is mapped into different zones representative to the possibility of having different levels of damage (complete, extensive, moderate, and slight). This methodology supports decision-makers to explore the community response to a disruptive events, to develop monitoring and control strategies in urbanized areas, to plan better resilience-building and evacuation strategies.

Keyword: *community disaster resilience, virtual city, smart city, evacuation, monitoring*

1. Introduction

According to the World Bank, disasters killed on the average 58,000 people/year and affected another 225 million people worldwide since 1990. The rising of global populations and the massive economic development in areas prone to disasters have increased the chance of catastrophic incidents. Over the years, community resilience has attracted attention due to the increasing number of natural and man-made disasters. The concept of resilience is multi-dimensional, and therefore involves various subjects of different disciplines. In engineering, resilience is the ability of a system to withstand stress and rebound after severe disasters. Besides the system capacity to absorb the shock, evacuation management right after the event plays a key role in improving the resilience. The evacuation phase subsequently a critical event such as earthquakes and fires following earthquakes has to be monitored by developing suitable methodologies. The use of drones with mounted cameras and wireless networks has been investigated, along with the use of mobile devices and embedded applications. The mobile network, already used by international providers for monitoring roads and traffic, could be used for collecting data in real time for evacuation purpose. Besides, the same mobile devices could be used for driving the crowd along safe paths. The need of approaching such complex problems with rational tools is the object of this research work.

This paper deals with the development of different strategies for monitoring and control the emergency in urbanized areas. The objective is to predict, with the set of large-scale simulations, the potential damage and physical impacts of an earthquake on the built environment. A 3-D virtual city is developed for evaluating the seismic effects at increasing intensities. Four building sectors that provide essential functions to a community including housing (residential building, hotel, shelter), education (school, university, library), business (shopping centers, retail stores, heavy industries), and public services (hospital, police station, churches, airport etc.) are considered.

Given a large number of buildings in a city, the seismic simulation of the buildings cannot be easily implemented without huge computational effort. Thus, several numerical models have been proposed to simulate the seismic damage to buildings in the recent years. Basically, the simulation models are classified into a data-driven method and a physics-driven method. The first is based on statistical data obtained from previous earthquakes. The main limitation is due to the inadequacy and leakage of statistical data for any world areas. To overcome these intrinsic limitations, the physics-driven methods have been progressing in recent years. The buildings seismic damage is based on the structural analysis of an individual building subjected to a given seismic input. Given limited attribute data of each building and the large number of buildings in an urban area, the seismic response prediction model of buildings must be relatively simple to reduce the computational time requested by the analyses.

In this paper, a simplified and efficient physical approach is presented following previous efforts in this direction (Cimellaro et al. 2017). Nonlinear response of a Multi Degree Of Freedom (MDOF) model of each building is obtained considering the dominant modal shapes and irregularities. A trilinear elastoplastic backbone curve representative of global shear behavior of each building is estimated through performing collapse analysis. Monte Carlo Simulations (MCS) are applied in order to take into account the epistemic uncertainties associated with geometry and mechanical properties within the range of observations. For each set of buildings' data, the nonlinear dynamic analysis is performed through SAP2000 application programming interface (API) in order to assess the dynamic response of the buildings in an organized and automatic fashion. The SAP2000 advanced numerical modules permit efficient both pre-analysis and post-analysis computations. In fact, the automated post-processing analysis allows extracting all the response parameters needed to estimate the level of damage to each building.

2. Buildings database

The virtual city model is freely inspired to the buildings stock of the city of Turin, Italy. It has the area of 120.1 km^2 with the total population of 850,000. Four building sectors that provide essential functions to a community, including housing (residential building, hotel, shelter, etc.), education (school, university, library, etc.), business (shopping center, retail store, heavy industry, etc.), and public services (hospital, police station, church, airport, etc.) are considered. Table 1 lists in detail the building sectors supporting the physical, economical, and social dimensions of the virtual city. In total, the virtual city has 30,122 buildings.

The plan dimensions of each building have been gathered through CADMAPPER file for the entire city of Turin. In addition, the numbers of stories have been obtained by the shape-file of "*Carta Tecnica Comunale (CTC)*" of the city of Turin, available at the website http://www.comune.torino.it/geoportale/.

The building inventory of the city is based on the building typology concept already used in many European countries at national and regional levels. However, the lack of information about some buildings makes it difficult to have perfect knowledge of any individual building. For this reason, some building's attributes (e.g. year of construction, type of deck) have been assigned based on known data for the entire city. Six different years of construction categories have been considered according to the principal changing in Italian seismic design codes (Table 2). The numbers of buildings for each year of construction category have been assigned according to Cities on Power (CoP) European program research for the city of Turin (Fracastoro et al., 2013). Corrado et al. (2012) provided typical Italian building construction elements depending on the year of construction. Classification of building construction elements (e.g. deck, wall, etc.) plays a key role in the assessment of mass. Seven different typical deck and three typical external walls have been selected and distributed based on their year of construction (Figure 1).

All the buildings have been divided into groups based on material: concrete, masonry, and steel. Considering limited information in the building attribute data, an accurate determination of the nonlinear structural parameters is rather challenging. The major objectives of this work are to provide a simplified and accurate method for assessing the dynamic response of residential buildings in a generic built environment. For this purpose, the geometric characteristics of the structural elements (e.g. columns and beam sizes) have been defined in order to respect all the technical standards for a given year of construction and for a given seismic hazard scenario (medium or high level).

Table 1. Number of buildings stock and map of the city

Physical Infrastructure	Building	Residential	27830
		Mobile Home	62
	Hospital		17
	Fire Station		3
	Police Station	Carabinieri	18
		Polizia Municipale	11
		Questura	31
	Educational	Elementary School	157
		Middle School	105
		High School	97
		University	70
Social Infrastructure	Hotel		31
	Historical Building		951
	Castel and Palace		18
	Church		176
	Sport		265
	Cinema		48
	Museum		156
	Theater		38
	Library		15
Economic Infrast.	Industrial Build.	Light	321
		Heavy	108
	Commercial	Retail store	25
		Malls	12

Table 2. Year of construction categories

I	II	III	IV	V	VI
< 1916	1916-1937	1938-1974	1975-1996	1996-2008	> 2008

DESCRIPTION		YEAR
	Vault ceiling with solid bricks	< 1900
	Ceiling with wood beams and hollow bricks	< 1900
	Ceiling with reinforced concrete	1900 - 1930
	Vault ceiling with bricks and steel beams	< 1930
	Vault ceiling with hollow bricks and steel beams	1910 - 1940
	Ceiling with reinforced brick-concrete slab	> 1930
	Ceiling with reinforced brick-concrete slab, low insulation	> 1976

(a)

DESCRIPTION		YEAR
	Solid brick masonry (25 cm)	1900 - 1950
	Hollow wall brick masonry (30 cm)	> 1930
	Hollow wall brick masonry with solid and hollow bricks (40 cm)	1930 - 1975

(b)

Figure 1. Typical Italian building's decks (a) and walls (b) used for residential occupancy in different years (Corrado et al., 2012)

3. Nonlinear model

The main component of an urban environment consists in the residential buildings. When a catastrophic event such as an earthquake occurs in a built environment, the consequential structural damage may cause high losses (casualties, repair costs, and repair time). Thus, a concise and methodical approach is needed to estimate the fragility of the building framework. The Nonlinear MDOF shear model is able to satisfactorily capture the nonlinear properties of multi-story buildings, predict the Engineering Demand Parameters (EDPs), and assess a reasonable level of damage (Lu et al. 2014; Xu et al. 2014). In the proposed approach, the inter-story behavior of a regular building is simulated through a trilinear backbone curve (Figure 2). Many studies claim that the trilinear backbone curve model can accurately represent the building's response in terms of inter-story (Vamvatsikos and Cornell 2005; Shi 2014).

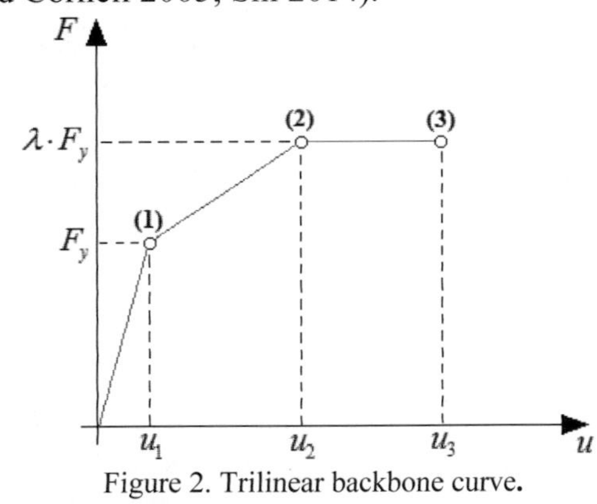

Figure 2. Trilinear backbone curve.

The first point of the trilinear backbone curve (1) indicates the yield point ($\lambda \cdot F_y - u_1$) corresponding to the formation of the first plastic hinge in the weakest base column. After the yield point, the stiffness is significantly reduced until the next point (2), for which the maximum shear base capacity ($\lambda \cdot F_y$) is reached. The ultimate point (3) corresponds to the collapse of the building (complete damage). The evaluation of the shear base and top displacement parameters for each point is discussed in the following subsections.

The three main points of the curve are evaluated using a nonlinear static approach for a MDOF system. To achieve this goal, a MATLAB algorithm has been developed considering the uncertainties on the geometric and mechanical parameters used in the analyses. Variation of the parameters within an acceptable range is considered through an MCS obtaining a set of trilinear backbone curves for each building. The ranges of the main building's parameters are selected according to the knowledge level of the building. In addition, the deterioration of the mechanical properties (such as strength and elastic modulus of concrete) is taken into account through the aging equation proposed by Eurocode 2 (EC2, 2004) according to the year of construction. Once the MATLAB algorithm evaluates the trilinear backbone curve at each step of MCS, the SAP2000 API is used to apply to all the buildings the data set obtained by the algorithm. This automated procedure is capable of reducing the computational time and analyze the dynamic responses dispersion caused by the data uncertainty. Therefore, the mean response and associated dispersion for each building within the virtual city can be estimated. Finally, due to the lack of information about its dynamic behavior, the structure hysteresis behavior is considered according to the Takeda model (Takeda et al., 1970) implemented in SAP2000.

3.1 Elastic parameters

Generally, the geometry of a residential building is mostly regular in plan and elevation, therefore the mass and stiffness can be assumed to be mostly uniformly distributed. In these cases, the evaluation of the response of MDOF system with a nonlinear static procedure is close to the real response of the structure.

Thus, a nonlinear static analysis is performed to assess the base shear and top displacement values corresponding to the formation of the first plastic hinge at the base level (yield point). In order to consider all Degrees Of Freedoms (DOFs), the stiffness matrix of the structure is evaluated considering the building as a bending type system. Since the load patterns are applied in two main directions of the buildings, the modal characteristics are derived by considering the stiffness matrix in the two directions for a 2D system. Thereafter, the static condensation procedures are performed to reduce the number of DOFs to the translational DOFs. Moreover, the model assumes that the mass of each story is concentrated in the center of the mass on its elevation and represented by a mass point (Figure 3).

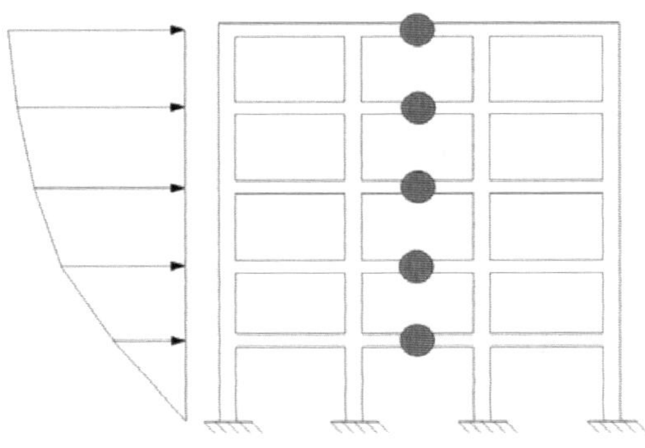

Figure 3. Concept of nonlinear MDOF system

The yielding base shear force is assessed by applying a monotonic load pattern on the building proportional to a given modal shape. A multi-modal approach is carried out to consider all the modal shape contributions, especially for buildings that have geometric irregularities.

$$\Phi_{tot} = \sum_{i=1}^{dof} \{\Phi_i\} \cdot g_i \quad (1)$$

where Φ_{tot} is the modal shape considering all modal contributions (Φ_i). The modal participation factors are represented by the term g_i.

3.2 Post-elastic parameters

Once the structure reaches the yield point, the stiffness is significantly reduced until the point (2) for which the maximum shear base capacity is reached. Thus the top shear base remains constant and the top displacement increases (perfectly plastic behavior) until the ultimate value. The maximum shear base capacity is estimated through the kinematic approach of the limit analysis (Greenberg-Prager theorem). The general global collapse mechanism of a frame subjected to a distribution of horizontal forces (proportional to a given modal shape) is considered (Figure 4).

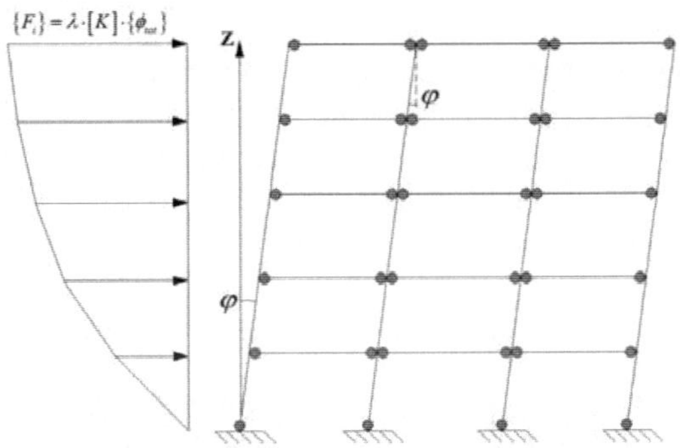

Figure 4. Global collapse mechanism

This approach leads to taking into account the strength contribution of all the structural elements (beams and columns). According to the kinematic theorem of the limit analysis, the global over-strength factor (λ) is assessed by the ratio between the internal and external work of the structural system with equal columns and beams dimensions (Equation (2)).

$$\lambda = \frac{n_c \cdot M_{y,c} + 2 \cdot n_{span} \cdot dof \cdot M_{y,b}}{\sum_{i=1}^{dof} \{F_i\} \cdot \{z_i\}} \tag{2}$$

where $M_{y,c}$ and $M_{y.b}$ are the yielding bending moment for the columns and beams, respectively. The parameters n_c represents the number of columns, n_{span} indicates the number of spans in the considered direction, and *dof* is the number of master DOFs which corresponds to the story number of the building. The external work is given by the denominator expression and it is due to the horizontal load patterns $\{F_i\}$ multiplied by the distance between the considered story and the base at each elevation level $\{z_i\}$. One of the limitations of this procedure consists of the load pattern's shape. In fact, the monotonic horizontal force distribution does not change its shape due to the progressive formation of the plastic hinges in the columns (non-adaptive approach). In addition, a global mechanism has been considered as representative of the collapse mechanism. This hypothesis is reasonable for regular buildings designed according to the seismic standard codes. When a building is not well designed or seismically retrofitted (such as an old building), the collapse can be caused by a local mechanism for which the over-strength factor assumes lower values. In these cases, the possible local mechanisms have to be identified and the over-strength factor will be assumed as the minimum values among all the defined factors. Once the shear base capacity is determined, the top displacements corresponding to points (2) and (3) of the trilinear backbone curve (u_2 and u_3 in Figure 2) have to be assessed. Since the shear base capacity is previously evaluated, the definition of reduction factor (R_μ) can be used to calculate the displacement u_2. The reduction factors depend on ductility (μ), over-strength (λ), damping (ξ) and elastic building characteristics (such as period, T). Several mathematical formulations have been proposed for evaluating the reduction factor. One of the most used expressions is based on the equal energy rule (short period systems, $T<0.5$ s) or equal displacement rule (long period systems, $T>0.5$ s) (Equation (3)).

$$\begin{cases} R_\mu = \sqrt{2\cdot\mu - 1} \quad (T < 0.5s) \\ R_\mu = \mu \quad (T > 0.5s) \end{cases} \tag{3}$$

The ductility parameters are expressed as a ratio between the ultimate displacement and the displacement for which the maximum shear capacity occurs. According to the proposed trilinear backbone curve, the ductility is given by the ratio between displacements u_2 and u_3. Furthermore, the ultimate top displacement is evaluated according to the ultimate capacity of the column section.

4. Analysis implementation and simulation

The proposed approach is capable of applying nonlinear time history analyses to a large number of buildings. The dynamic response of the structural system in a built environment takes into account a considerable amount of parameters. The building inventory, consisting of all the information (such as material, geometry, and mechanical properties), has been developed and allocated to an external server. All this data are accessible by a MATLAB code organized in several functions that manage the seismic input definition, MCS for evaluation of the nonlinear parameters, and SAP2000 API actions. The global behavior of each building has been modeled by using multi-linear plastic link element available in SAP2000. The mechanical characteristics have been defined automatically according to the obtained backbone curves from MCS. Figure 5 depicts the schematic model used for simulating the global shear capacity of each building. The equivalent damping coefficient has been assessed according to the Rayleigh formulation considering the first and second building's period as control periods.

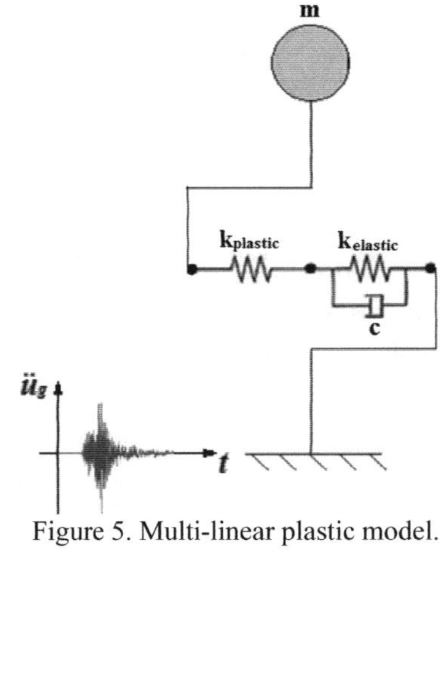

Figure 5. Multi-linear plastic model.

Due to the large number of variables and the time requested for processing, parallel algorithms running on multiple processors are developed with MATLAB.
This approach, with respect to existing ones (Xiong et al. 2017), allows to reduce the number of DOFs in the virtual city model. Consequently, the computational and storage efforts could be consistently decreased with essential benefits for the simulation of complex disaster scenarios (Cimellaro et al. 2017).

4.1 Software architecture

The analysis flow is controlled through an interactive graphical user interface (GUI) that allows for selection of an earthquake scenario in the virtual city (magnitude and epicenter location). Furthermore, the acceleration time history can be selected and processed in both North-South (NS) and East-West (EW) directions. In order to take into account the deamplification of the seismic excitation with the epicenter distance, the shear wave velocity in the upper most 30 m (V_{S30}) for the city of Turin is included in the data. The V_{S30} map has been obtained via USGS website (USGS, 2013) at the link http://earthquake.usgs.gov/hazards/apps/vs30/. The Boore-Atkinson (Boore and Atkinson, 2008) attenuation law is used to estimate the attenuation of the time history's peaks. A MATLAB function is provided for calculating distances between the selected epicenter and the center of the mass of each building. Moreover, the equivalent shear wave velocity is assessed according to the V_{S30} map, and considered in the attenuation model. The main MATLAB function controls the building's data flow, and then the MCSs are carried out to evaluate the backbone curves for each building, considering the epistemic uncertainties in the input model parameters.

4.2 SAP2000 API

The SAP2000 Application Programming Interface (API) is a programming tool that offers efficient access to the analysis and design technology of the SAP2000 structural analysis software. A direct interaction with third-party applications is allowed during run-time analysis. The API software library provides access to a collection of objects and functions capable of remotely controlling the data exchange and setting data in SAP2000. Both pre- and post-processing procedures are managed by a MATLAB language code which mainly provides the two-way data exchange. This procedure is capable of significantly reducing the time needed for data exchange, especially for large data models. Once the MATLAB functions assess the nonlinear parameters (trilinear backbone curve) and the processed seismic input, they are transferred to SAP2000 through API tool. Due to the limited amount of detailed building information, the hysteresis is considered according to the Takeda model. Thus, the nonlinear time history analyses are performed in the SAP2000 environment and the derived output is remotely controlled by MATLAB. Figure 6 shows in detail the software data flow used in the simulations.
According to the maximum drift, the structural damage is assessed for each building and the associated level of damage is evaluated (slight, moderate, extensive, complete). A 3D visualization tool is also provided which shows the dynamic response of the building within the virtual city. This visualization tool can be helpful for monitoring and evacuation management.
The proposed methodology has been applied to a part of the virtual city and the results, in terms of displacements, are shown in Figure 7. A time history recorded during the Central Italy earthquake (Norcia station, PGA=0.42 g), in both horizontal directions, has been used as earthquake scenario for the simulation. The predicted real time response of buildings can help decision makers to monitor and manage the resources during emergencies. This is strictly related the urban development vision to integrate information and communication in a secure approach to manage a city's assets (smart city concept).

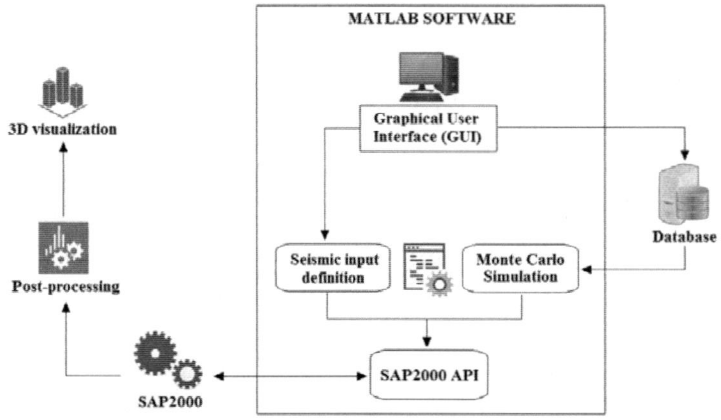

Figure 6. Software data flow

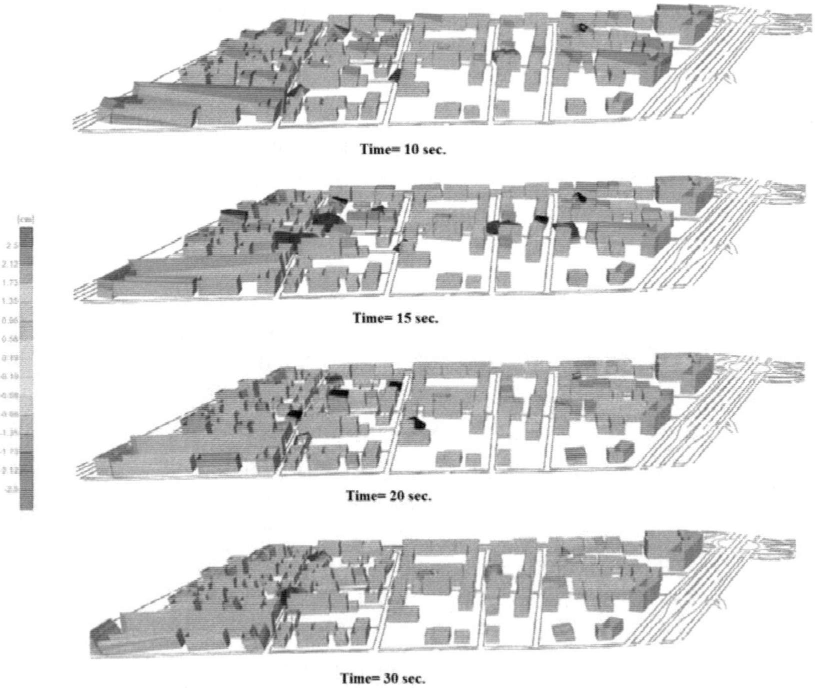

Figure 7. Displacement contours at four different time steps.

5. Conclusions

This paper deals with the development of different strategies for monitoring and control the emergency in urbanized areas. A new methodology to the urbanized systems and large-scale simulations within a seismic scenario are explored, by evaluating multipurpose codes for numerical simulation. A virtual city has been developed for evaluating the seismic effects at increasing intensities. All the characteristics of each building have been collected and organized in a complete database. The dynamic structural response has been simulated through a trilinear backbone curve representative of global behavior of structure through a nonlinear static approach for a MDOF system. The uncertainties on the building geometric and mechanical parameters have been taken into account through a MCS. A MATLAB based software has been developed to control the input data and the SAP2000 workflow through its Application Programming Interface. The post-processing is remotely controlled by MATLAB, and the buildings' damage level is estimated. Finally, a 3D visualization tool is also provided which shows the dynamic response of the building within the virtual city and can be helpful for the evacuation phase. This procedure is capable to significantly reduce the time needed for data exchange, especially for large data models. This methodology supports decision-makers to explore how their community will respond to a disruptive event, to develop different strategies for monitoring and control the emergency in urbanized areas, and to plan better resilience-building and evacuation strategies.

Acknowledgements
The research leading to these results has received funding from the European Research Council under the Grant Agreement n° ERC_IDEal reSCUE_637842 of the project IDEAL RESCUE—Integrated DEsign and control of Sustainable CommUnities during Emergencies.

References

Boore, D.M. & Atkinson, G.M., 2008. Ground-motion prediction equations for the average horizontal component of PGA, PGV, and 5%-damped PSA at spectral periods between 0.01 s and 10.0 s. *Earthquake Spectra*, 24(1), pp.99-138.

Corrado, V., Ballarini, I. & Corgnati, S.P., 2012. National scientific report on the TABULA activities in Italy.

Cimellaro G.P., Domaneschi M., Mahin S., Scutiero, G., 2017, Exploring simulation tools for urban seismic analysis and resilience assessment, COMPDYN 2017, Rhodos, Greece.

EC2 2004. Design of concrete structures-Part 1-1: General rules and rules for buildings. European Committee for Standardization, Bruxelles, BG.

Fracastoro, G. V., Mutani, G., Verda, V. & Vicentini, G., 2013, Analisi del fabbisogno di energia per gli edifici residenziali e della potenzialità del risparmio energetico e delle fonti rinnovabili in ambito urbano, *Progetto Cities on Power.*

Lu, X., Han, B., Hori, M., Xiong, C. and Xu, Z., 2014. A coarse-grained parallel approach for seismic damage simulations of urban areas based on refined models and GPU/CPU cooperative computing. *Advances in Engineering Software*, *70*, pp.90-103.

MATLAB and Statistics Toolbox, *Release 2012b*, The MathWorks Inc., Natick, Massachusetts, United States.

SAP2000, *Version 17.3,* Computer and Infrastructure Inc., Berkeley, CA., United States.

Shi, W., Lu, X., Guan, H. & Ye, L., 2014. Development of seismic collapse capacity spectra and parametric study. *Advances in Structural Engineering*, *17*(9), pp.1241-1255.

Takeda, T., Sozen, M.A. & Nielsen, N.N., 1970. Reinforced concrete response to simulated earthquakes. *Journal of the Structural Division*, *96*(12), pp.2557-2573.

USGS. 2013. Seismic Hazard Analysis tools [Online]. U.S. Geological Survey. Available: http://earthquake.usgs.gov/hazards/designmaps/grdmotion.php.

Vamvatsikos, D. & Cornell, C.A., 2002. Incremental dynamic analysis. *Earthquake Engineering & Structural Dynamics*, *31*(3), pp.491-514.

Xiong C., Lu X.Z., Lin X.C., Xu Z., Ye L.P., 2017, Parameter determination and damage assessment for THA-based regional seismic damage prediction of multi-story buildings, *Journal of Earthquake Engineering*, 21(3), pp. 461-485.

Xu, Z., Lu, X., Guan, H., Han, B. & Ren, A., 2014. Seismic damage simulation in urban areas based on a high-fidelity structural model and a physics engine. *Natural hazards*, *71*(3), pp.1679-1693.

The 8th International Conference on
Structural Health Monitoring of Intelligent Infrastructure
Brisbane, Australia | 5-8 December 2017

Monitoring human emotions during earthquakes

R. Aguilar[1],C. Apostoliti[2], R. Boroshek[3], A. Cardoni[4], D. Cares[5], G.P. Cimellaro[6], M. Domaneschi[7], D. Galdo[8].

[1] Catholic University of Peru, Civil Engineering Department, raguilar@pucp.pe, Politecnico di Torino,[2] Department of Structural, Geotechnical and Building Engineering IT, email: carmelo.apostoliti@studenti.polito.it, *[3] University of Chile, Civil Engineering Department, Santiago, Chile, email:* rborosch@ing.uchile.cl, *[4] Politecnico di Torino, Department of Structural, Geotechnical and Building Engineering IT, email:* alessandro.cardoni@studenti.polito.it, *[5] University of Chile, Civil Engineering Department, Santiago, Chile, email:* david.cares@ug.uchile.cl, *[6] Politecnico di Torino, Department of Structural, Geotechnical and Building Engineering IT, Email:* gianpaolo.cimellaro@polito.it, *[7] Politecnico di Torino, Department of Structural, Geotechnical and Building Engineering IT, Email:* marco.domaneschi@polito.it, *[8] Politecnico di Torino, Department of Structural, Geotechnical and Building Engineering IT, email:* davide.galdo@studenti.polito.it

Abstract

In this research work, the human behaviour during strong seismic events is studied. This aspect is of paramount importance in emergency for being able to develop effective evacuation strategy and increasing the safety level of emergency response personnel and people operating critical facilities. In particular, the quantification of the evacuation and control ability to maintain their stable position of normal people has been initially assessed through shaking table tests. The experimental tests have been performed with the support of virtual reality by replicating indoor and outdoor environments. Vital parameters have been collected through sensors. This work involves at this stage two research groups in Italy and Chile with the aim of evaluating strategies and methods on this topic that is rather new in earthquake engineering. This is supposed to be the first step of further investigations on larger campaigns, and number of individuals as well, in order to achieve a comprehensive statistic.

1. Introduction

The number of casualties caused by an earthquake is usually linked to the collapse of structures in which people cannot properly exit during the event. While the human losses are positively correlated to the structural damage, injuries have also been found to occur even when no damage was present in the structure. Such injuries are the result of the individuals being struck by objects or falling off the staircase while trying to escape from the building. Therefore, the shaking itself is deemed to be a significant cause for injuries and death during earthquakes. In addition some people are required to perform activities during the earthquakes as example critical facility operators that should in some case monitor and actuate certain shutdown procedures, controllers of transport systems that should safely park the vehicles, doctors that should control the patient at the operating room, etc. This preliminary work aims at understanding the ability of normal people to maintain their position during ground shaking. This subject has not been extensively study but some pioneer work has been done by (Takahashi et al. 2004, 2011).

Test on humans due to vibration has been performed in other areas of knowledge, (Griffin 1990, 1998, among many others). The main idea is to evaluate the ability to perform control evacuation or controlled pre-determined tasks. The literature review indicate that this conditions are influence by stability of the individual in standing positions, the motion of individual in siting position, the level and frequency of the vibration at the points of entry to the individual, state of health, gender, age

and most importantly previous experience or practice. Due to the large number of variable and their variability between humans, testing is difficult to capture overall tendencies.
Additionally the monitoring ability to detect the capacity of individuals is difficult due to the spatial motion of the individuals. To understand and initially quantification and evaluate different variables tests were performed on different individuals and human characteristics consider gender, age, height, complexity and state of health. Each person is equipped with sensors to analyze the person's stability and capability to maintain the initial position, heart rate and breathing in order to measure the main vital parameters and anxiety of the person during the shaking.
A six degrees of freedom (3 translations and 3 rotations) shaking table and a one direction shaking table are used to generate artificial earthquakes. In addition, a virtual reality setting is used in order to recreate a more realistic environment. During the experiment, special attention is given to the factor of "surprise" which is necessary to ensure a natural reaction of the individuals.

2. Laboratory facilities

The research groups at Politecnico di Torino and University of Chile perform complementary test. The laboratory facilities and sensors are described below.

2.1 Research at Politecnico di Torino

Virtual Reality Earthquake Simulator
The growing investment by Google, Facebook, Microsoft, Htc and Sony in the Virtual Reality shows that this technology is a reliable solution for many problems such as training, testing and researches. Virtual Reality has been used for years in medicine as surgery training, for motor rehabilitation and psychological treatments (Figure 1). Professional and free of charge developing environment such as Unity 3D, open source modelling software as Blender, cheap virtual reality head mounted display like Google Cardboard or Samsung Gear VR and user tracking/natural interaction sensor like Microsoft Kinect allows the uses of synthetic environment a research opportunity within the reach of any university (Figure 2).
The virtual reality system that has been developed in this research is a client-server application that uses the following hardware equipment:

- Samsung Gear VR 2016
- Samsung Galaxy S8
- Microsoft Kinect V2 Sensor
- A Notebook with Windows 10 and one usb 3.0 port

The software has been developed using Unity3D for the client side, and C# with Kinect V2 SDK for the server side.
The simulation system has to achieve three goals:

- quick development time;
- cost-efficient software/hardware architecture;
- connection between one's movements in real world and those perceived into the virtual environment.

A smartphone-based head mounted display is used to reduce realization costs: the Samsung Gear VR in conjunction with Samsung S8 android smartphone represents a good compromise among costs, computational resources and scalability. It allows the user to move his field of view along the three rotational degrees of freedom. Two virtual environments - indoor and outdoor - have been developed in Unity 3D using mesh modelled with Blender and several free available Unity assets. The virtual environment runs on Samsung S8. The spatial variability of the earthquake motion is reproduced by three sinusoidal temporal histories with random amplitude applied in the three

directions. When the virtual environment is shaken every object in the scene is affected and moves according the input signal: paintings fall off the wall, furniture tilts, in addiction sound effects are used to increase the realism and the emotional response of the user.
Microsoft Kinect sensor allows the system to track the user position along the 3 translational degrees of freedom in this way the third goal is accomplished. The user position is transmitted to the virtual environment by control software via Wi-Fi using a text based message protocol. The transport layer protocol chosen is TCP. The control software is developed using C# and Kinect SDK, it allows choosing which virtual environment to load, to define the maximum displacement and the duration of the experiment.

(a) (b)

Figure 1. VR indoor environment. Before (a) and after (b) the seismic event.

Figure 2. Body monitoring.

Shaking table

The shaking table used in these experimental tests is two degrees of freedom driven by four linear electric actuators. For these preliminary tests, only uniaxial motions have been simulated. The structure itself consists of steel profiles, whereas the upper platform, where specimens can be fixed, is made of aluminium. Basically, there are two parallel tracks located side by side and connected through transversal rectangular sections. Tracks' profiles are 3 meters long and the section's size is 40x100x4 mm. Upon the steel profiles there are aluminium guides in order to allow the motion, along the longitudinal direction, of sliders that support two 600x500x10 mm aluminium platforms. Each track has its own platform, which is moved by a linear electric actuator anchored under it. On the small platforms, other two tracks are fixed. Type and section of the steel profiles are the same of the bottom ones, while the length is shorter (600 mm). The two mobile carriages are positioned at a certain distance along the longitudinal direction in order to receive the upper platform that is

1500x1500x10 mm big (Figure 3a). The other two actuators that accomplish the motion in the transversal direction are fixed under the large platform and between the tracks of the underlying mobile carriages.
The linear electric actuators adopted are manufactured by the company LinMot and each is made of a stator, a slider and a motor. The longitudinal ones have a slider's length of 800 mm and a maximal stroke of 510 mm, whereas the transversal ones have a slider's length of 500 mm and a maximal stroke of 330 mm. The power supply, the two transformers and the four drivers to control the motors are provided by LinMot as well. The drivers are fundamental for the tuning of the motors that is the initial configuration of all the control parameters in order to have a response coherent with the input data. This operation is done through the software LinMot-Talk that is also used to switch on the actuators and to bring them in the home position.
The software used for the activation and control of the shaking table is LabView. A specific code has been written to handle the input and output data. The seismic input is actually sent to the shaking table by a myRIO device provided by National Instruments. This device is physically connected to the motors' drivers and also to an accelerometer, which is located on the platform and allows catching the actual response of the system. Simply, it is possible to connect a USB pen drive containing the seismic signal in terms of displacements to the myRIO device. The LabView code is used to set the input and output sampling rates, to generate a sinusoidal seismic signal or to load a real one, to scale it, to start and stop the motion and finally to compare the data obtained from the accelerometer with the theoretical ones. Figure 3b depicts the target input accelerogram (KOBE 01/16/95 2046, NISHI-AKASHI, 090 – PEER STRONG MOTION DATABASE) in the frequency domain and the recorded one (output) on the table without individuals.

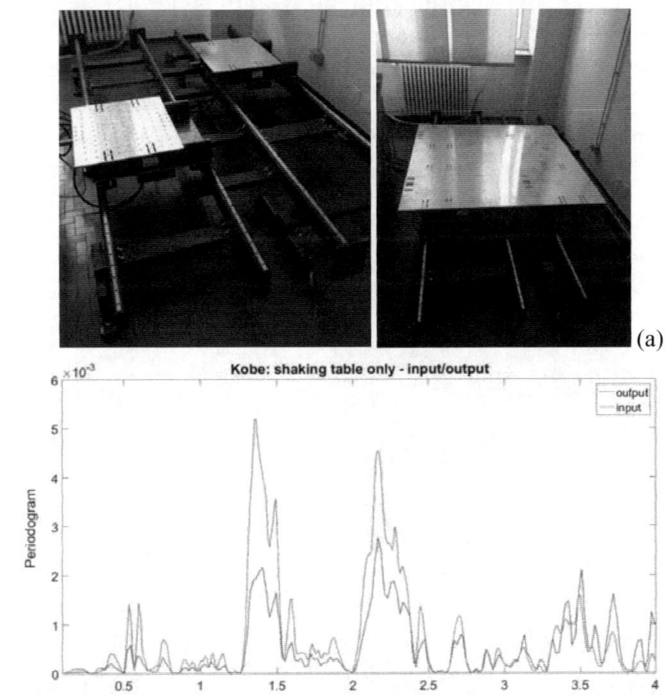

Figure 3. Shaking table (a), input/output in frequency domain without samples on the table (b).

Figure 4 depicts the recorded outputs in terms of acceleration for different performed tests, with and without individuals on the table. In particular, it shows the satisfactory repeatability of the shaking and an essential independency of the output signals to the presence/absence of mounted specimens. It is worth noting that the tuning control parameters of the table remained constant.

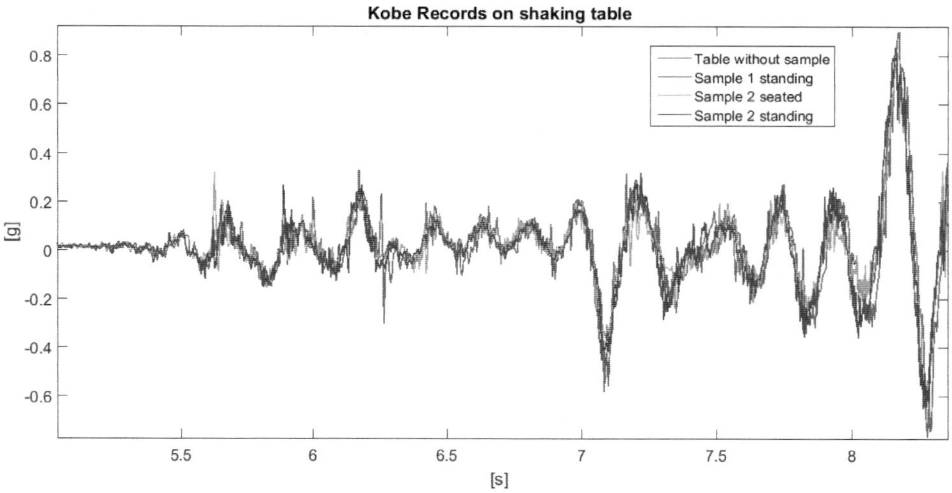

Figure 4. Comparison between shaking table outputs, with and without samples.

Monitoring and sensors

The monitoring systems for ECG (electrocardiogram) and breathing have been applied to the individuals (Figure 5a) in simple configurations. Figure 5b depicts the typical breathing diagram with interpretation details. Individual test was performed on two individuals: 40 and 30 years old, males. First specimen weight was 76 kgf and second one 65 kgf. A lab view code allows to plot the output signals (Figure 6). Finally, Figure 7 depicts standing and seating individual tests on the shaking table, equipped with the VR system and the ECG and breathing sensors. An external laptop allows controlling the whole system and collecting the monitoring data.

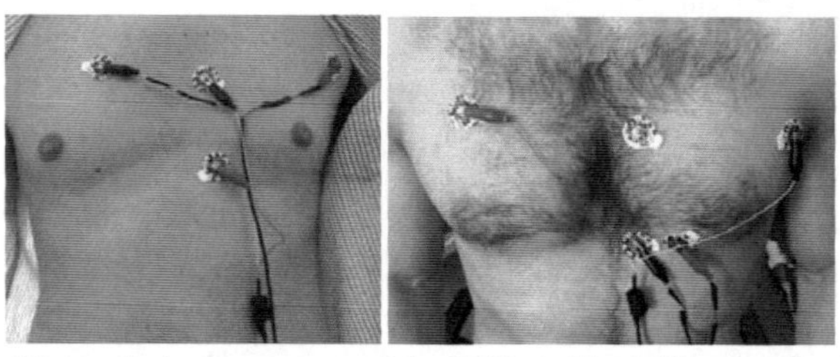
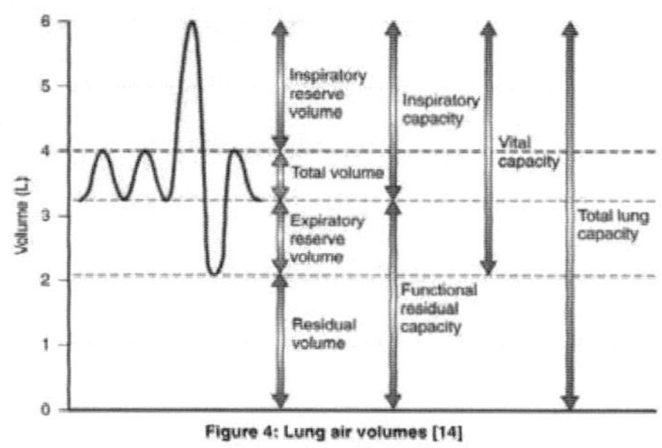

(b)

Figure 5. Arrangements of the ECG and breathing monitoring systems on the evaluated samples (a). Breathing interpretation (b).

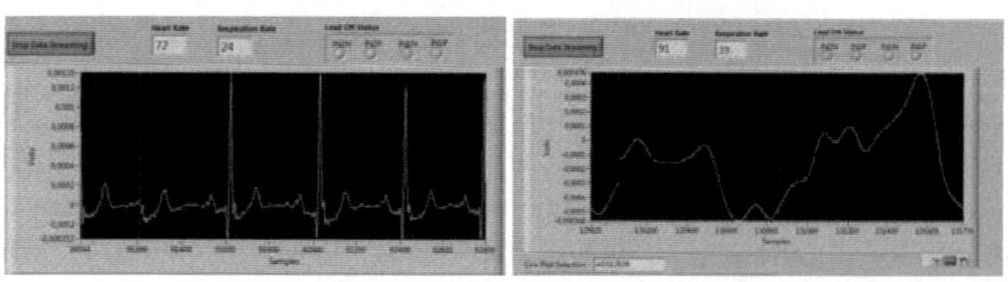
(a)

(b)

Figure 6. ECG and breathing output signals (a). ECG and breathing board (b).

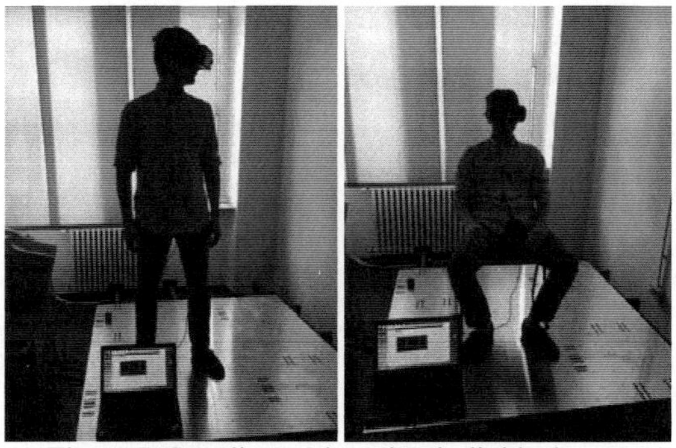

Figure 7. Standing and seating individual tests.

2.2 Civil Engineering Laboratory U. of Chile

Shaking table

The shaking table used at Universidad de Chile has six degrees of freedom. A 2x2 m platform structure was set up for the tested individual (Figure 8). For safety reason the testing area was surrounded by a rail covered with flexible polyethylene foam. Additionally, individuals used a harness to prevent abrupt collapse due to the motions.
The table is capable of reproducing acceleration in the order of 0.6 g and rotations up to 23 degrees. Maximum payload of the system is 1000 kg. Other table nominal capacities are described in Table I. These conditions limits the testing of individuals to environments were the predominant motions are on the low frequency side of the spectrum, in practical terms this corresponds to people inside structures were the predominant motions are below 5 Hz and the acceleration are below 0.6 g.

Figure 8. Shaking table testing system at University of Chile.

Table I. Motion system main features.

Degree of Freedom	Displacement Comb. Motion	Displacement Single DOF	Velocity	Acceleration
Pitch	±22 deg	±21 deg	±30 deg/s	±500 deg/s^2
Roll	±21 deg	±20 deg	±30 deg/s	±500 deg/s^2
Yaw	±23 deg	±22 deg	±40 deg/s	±400 deg/s^2
Heave	±0.18 m (±7.1 in)	±0.18 m (±7.1 in)	±0.30 m/s (±11.8 in/s)	+0.5 G
Surge	±0.27 m (±10.6 in)	±0.25 m (±9.8 in)	±0.50 m/s (±19.7 in/s)	±0.6 G
Sway	±0.26 m (±10.2 in)	±0.25 m (±9.8 in)	±0.50 m/s (±19.7 in/s)	±0.6 G

Sensors

Three accelerometers were mounted on the platform, making a three-dimensional array. Measuring directions were set up according to the initial position of the testing individual: Front-Back (North-South), Side-Side (East-West) and Up-Down. The accelerometers used in test had sensitivity of 0.005 g and a measuring range of ±2g; sampling rate was set to 200 Hz.
A smart bracelet was used to monitor heart rate on the individual's wrist. It was connected to a smart phone via Bluetooth, to have control over the data collection. Every test was recorded and stored for posterior analysis. Sampling rate for this device was 1 Hz.
Video recording was use to monitored testing subject and table motions. Table displacements were monitored using visual targets attach to the table. Individual stability and anxiety was observed visually from videos and from inquiries to the testing subject.
Initial test was performed on three individual ages: 55, 29 and 25, two males, one female. Male weight was 80 kgf and female weight 55 kgf. Testing position was standing with the possibility of walking during testing. Some of the tests were performing using and eye cover to reproduce full blackout during shaking.

3. Results

Some of the initial results are shown in Figures 9 to 11. In this figures the motions of the table, the pulse rate and the individual characteristics are shown. Additionally the testing individuals report the indication of the level of anxiety and stability.
In Figure 9 it is presented (on the top) the pulse rate for a female testing individual. The motion, as recorded on the table, is presented on the bottom three figures, indicating the initial orientation of the individual with respect to the axes of motion (FB Front Back, SS Side to Side and UD vertical). As expected different results are obtained from each individual.

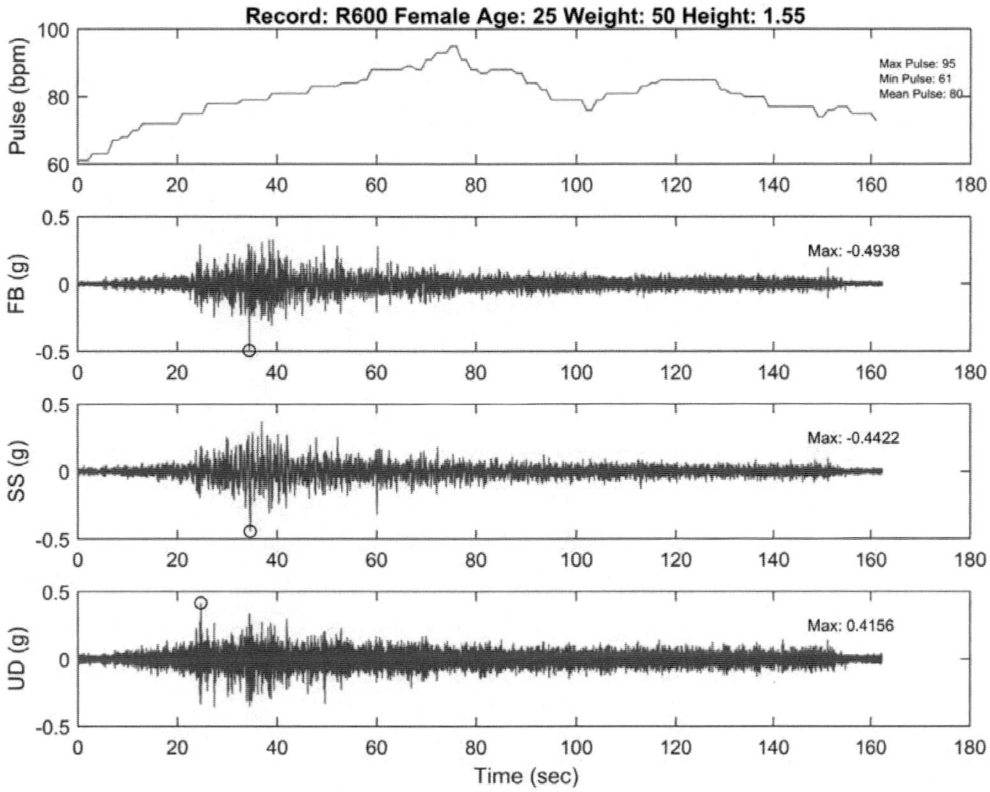

Figure 9. Sample 1 – University of Chile.

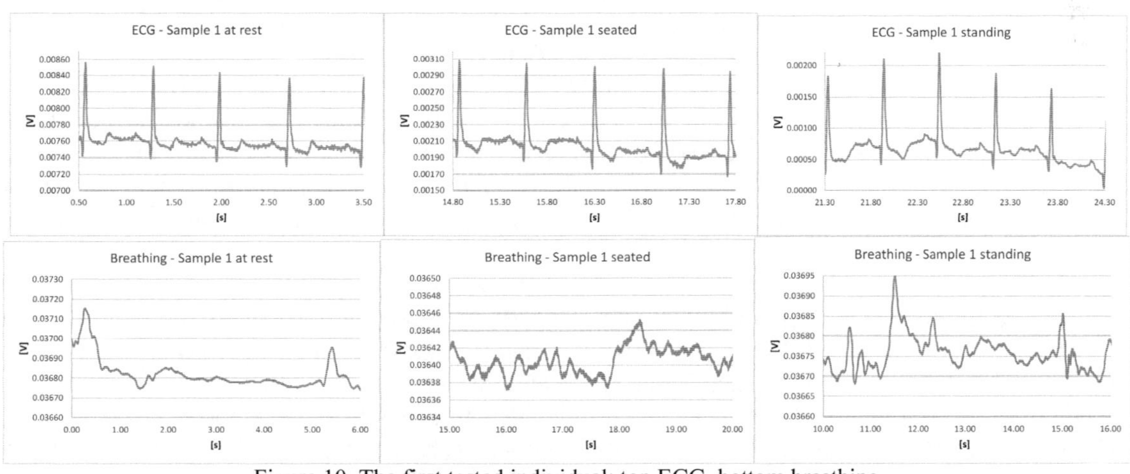

Figure 10. The first tested individual: top ECG, bottom breathing.

Moving to the Politecnico di Torino outcomes, Figure 10, on the first tested individual, highlights increments in the hearth beats from condition "at rest" to "standing" one, passing through "seated" condition. Breathing at rest results steady with regular and clear peaks (inspiration-expiration). On the contrary, when the first tested individual is under shaking (emergency) breathing is disturbed, irregular and uncomplete. Indeed, the breathing sensor is sensible to body movements, arms in particular. When the individual has been shaken, the first body reaction is to lift and opens arms for improving equilibrium.

The second tested individual confirms the first tested one behaviour, even if the breathing "at rest" gives unexpected information owing to arms movements. In fact, as previously discussed, breathing signal can be affected by the chest extension due to breathing but also moving.

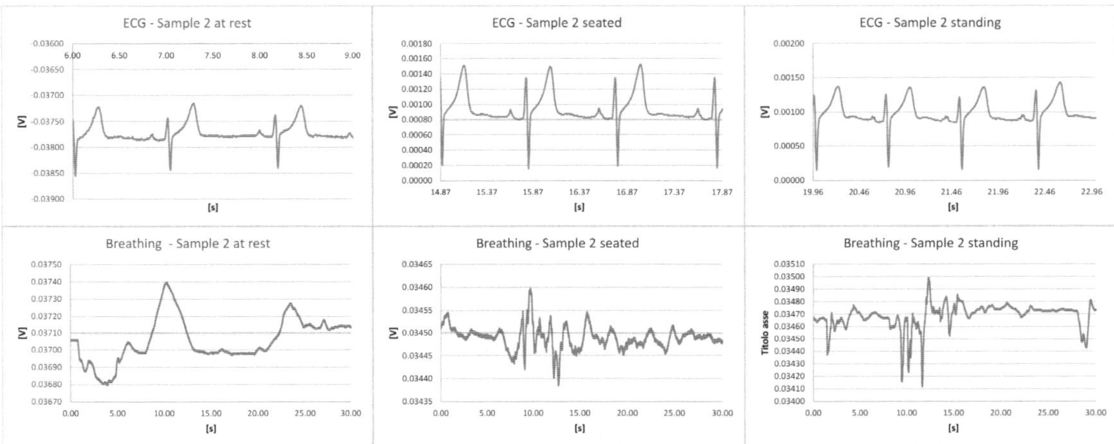

Figure 11. The second tested individual: top ECG, bottom breathing.

4. Conclusions

Several initial conclusions can derived from the performed testing.

Regarding table motions. The results are highly dependent on the number of degrees of freedom of the motions. The standing individual, as he feels instability, try to reposition himself on a more comfortable position. Multiple axis of motion restricts this possibility. Additionally the rotation components augment considerably the instability. Perception of vibrations starts at very low acceleration levels vibration. Literature indicates that sensitivity and reaction are larger in the frequency range from 0.1 Hz up to 80 Hz. Nevertheless inside structures this frequencies are considerable filtered to bands that are lower than 25 Hz. Additionally, for stability, the most critical frequency are lower than 15 Hz. Table motion should be capable of reproducing this motions.

Individual response is highly dependent on people expectations, experience and mental preparation. In Chile due its high rate of seismicity with a magnitude 5 earthquakes, or higher, occurring in average 70 times a year and Mercalli Intensity V occurring every two week, on average in the country, individuals are more trained reacting to earthquake events. Individuals were tested for several different motions and they learned quickly how to reposition themselves on more stable position. Additionally if the motions contained predominant simple periodic motions (like Chile Concepcion 2010 record) the individual quickly detached the motion of his lower part from upper part. High frequency motion produces higher anxiety but not necessarily uncontrollable stability.

Detection of human responses based on the basic sensors used in this stage of the research indicated that they are not as reliable to described stability and anxiety when compare to tested individual verbal description and visual monitoring of overall body position and foot contact with surface.

Increments in the hearth beats from condition “at rest” to “standing” one, passing through “seated” condition are usually detected. Furthermore, breathing at rest results steady with regular and clear peaks (inspiration-expiration). On the contrary, when tested individuals are under shaking (emergency) breathing is disturbed, irregular and uncomplete. Indeed, the breathing sensor is sensible to body movements, arms in particular. When the individual has been shaken, the first body reaction is to lift and opens arms for improving equilibrium.
In general the band wrist sensor shows that in some conditions, especially the long durations of Chilean records, increase of heart rate occurred which was not always consistent with the level of motions. But, as soon the motion was starting to decay the heart rate was higher than at rest conditions.

As expected blindfolded individual, reproducing dark environments, presented much less stability and higher anxiety that individual that had a reference frame. When Virtual devices are use the reference frame is gone and the stability is again diminish indicating the importance of recreating as a real environment as possible for this type o testing.

Acknowledgements

The research leading to these results has received funding from the European Research Council under the Grant Agreement n° ERC_IDEal reSCUE_637842 of the project IDEAL RESCUE— Integrated DEsign and control of Sustainable CommUnities during Emergencies and Civil Engineering Department of University of Chile base research Funds.

References

Takahashi, T, Saito, T, Azuhata, T & Ohtomo, K 2004, ‘Shaking Table Test On Indoor Human Response And Evacuation Action Limit In Strong Ground Motion’, *13th World Conference On Earthquake Engineering*, Vancouver, B.C., Canada, August 1-6, Paper No. 1320.

Takahashi, T, Suzuki, T, Saito, T, Azuhata, T & Morita, K 2011, ‘Shaking Table Test For Indoor Human Response And Evacuation Limit’, *7th International Conference on Urban Earthquake Engineering & 5th International Conference on Earthquake Engineering*, Tokyo, Japan, March 3-5, 187-193.

Griffin, M.J. 1990, *Handbook of Human Vibration*, London, Academic Press.

Griffin, M.J. 1998, ‘A comparison of standardization methods for predicting the hazards of whole-body vibration and repeated shocks’, *Journal of Sound and Vibration*, 215(4), 883-914.

The 8th International Conference on
Structural Health Monitoring of Intelligent Infrastructure
Brisbane, Australia | 5-8 December 2017

Long-term Vibration Monitoring of a Cable-stayed Bridge: Effects of Environmental and Operational Conditions

X.Q. Zhu[1], B. Samali[2], M. Rashidi[2], M.M. Alamdari[3].
[1] *School of Computing, Engineering & Mathematics, Western Sydney University – Australia, Email:* xinqun.zhu@westernsydney.edu.au, [2] *Centre for Infrastructure Engineering, Western Sydney University - Australia,* [3] *DATA61 - Australia*

Abstract

This study is to investigate the effects of environmental and operational conditions on the dynamic behaviour of bridge structures. A long-term monitoring system has been installed on a cable-stayed bridge and there are 24 accelerometers on bridge deck. Monitoring data from two typical days have been analysed and the modal analysis is also carried out to extract modal parameters. Effects of changes in environmental and operational conditions, such as temperature and traffic, on structural dynamic properties are discussed.

1. Introduction

Vibration measurements are generally used with system identification techniques to determine the location and extent of structural damage. In practice, structures are subjected to the changing environmental and operational conditions, and that will affect vibration responses. These ambient variations will mask changes of the system in vibration responses caused by structural damage (Cross et al., 2011). How to separate the changes due to the variations of operational and environmental conditions from the measured responses is a big challenge for structural health monitoring.

The effect of environmental and operational conditions on the structural vibration properties is very complicated. Xia et al. (2006) investigated the variation of frequencies, mode shapes and damping with respect to temperature and humidity changes through the long-term vibration monitoring of a reinforced concrete slab outside of the laboratory. Yuen & Kuok (2010) used the Bayesian spectral density to get trace the variation of modal frequencies with the ambient vibration data from one year measurements of a 22-story reinforced building. A literature review was presented on the statistical correlation between the temperature and frequencies and the variations in material modulus under different temperatures were the major cause of variations in vibration properties (Xia et al., 2012). Westgate et al. (2015) also presented the effect of vehicular loads on the dynamic properties of the bridge. Figueiredo et al. (2014) proposed a Markov-chain Monte Carlo based Bayesian approach to cluster structural responses of bridges into a reduced number of global state conditions through the monitoring data sets from the Z-24 Bridge in Switzerland.

This study is to investigate the effects of environmental and operational conditions on the dynamic behaviour of bridge structures. A long-term monitoring system has been installed on a cable-stayed bridge and there are 24 accelerometers on bridge deck. Monitoring data from two typical days have been analysed and the modal analysis is also carried out to extract modal parameters. Effects of changes in environmental and operational conditions, such as temperature and traffic, on structural dynamic properties are discussed.

2. Long-term monitoring system of a cable-stayed bridge

Figure 1(a) shows the cable-stayed bridge located in Werrington, New South Wales, Australia. The bridge is linked two campuses over the Great Wester Highway with a span 46m. It is a single lane highway bridge with the width of 5m. The bridge is fully instrumented for long-term monitoring. There are total 64 channels in the monitoring system that include 31 channels for accelerations, 31 channels for strain, one temperature and one for optical sensors to indicator vehicle or pedestrian presence. There are total 24 accelerometers on the bridge deck and Figure 1(b) shows the sensor locations.

(a) The cable-stayed bridge

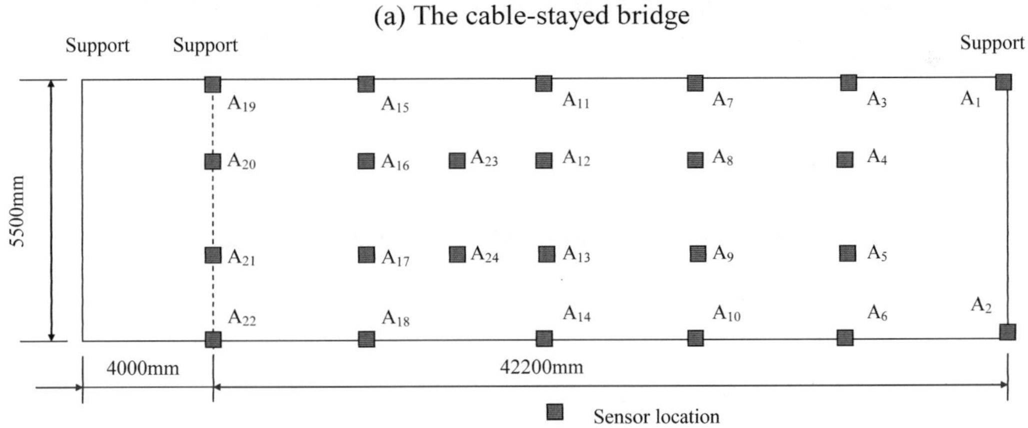

(b) Sensor location

Figure 1 Long-term monitoring of a cable-stayed bridge

Figure 2 shows the data capture diagram from the monitoring system. The data acquisition system continuously records the data from sensors with the sampling rate 600Hz. It creates a file every 2 minutes with the size of 100MB which is saved to the local computer. A data load uploading program is embedded in the local computer to upload the file to the Western Sydney University storage through the fibre optic cable. The data is available to researchers through the internet.

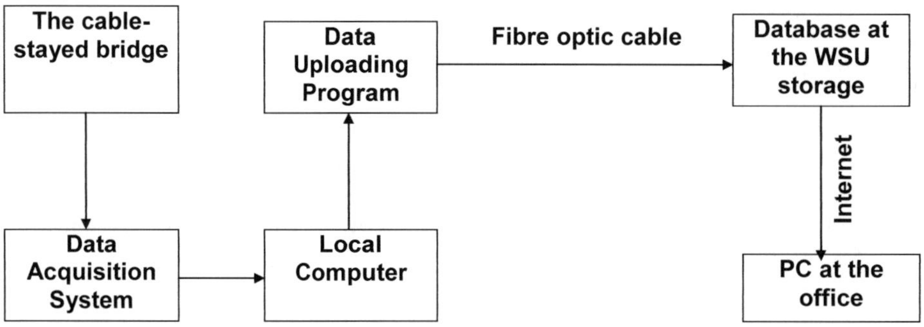

Figure 2 Data capture diagram

3. Results and discussion

3.1 Dynamic responses and their spectrum

Figure 3(a) show a comparison of dynamic responses at Sensor A_5 with or without a passing vehicle and their spectrum are shown in Figure 3(b). The data length is about 300 second. The frequency resolution is 0.003Hz. There is clear impulse response when the vehicle entered the bridge in Figure 3(a) and the high modes are also excited by the passing vehicle as shown in Figure 3(b). This vehicle is a security car to patrol around the campus. From Figure 3(b), the 1st modal frequency is 2.038Hz without the vehicle and it is 2.028Hz with the passing vehicle. So the vehicular load has a big effect on the 1st modal frequency of the bridge.

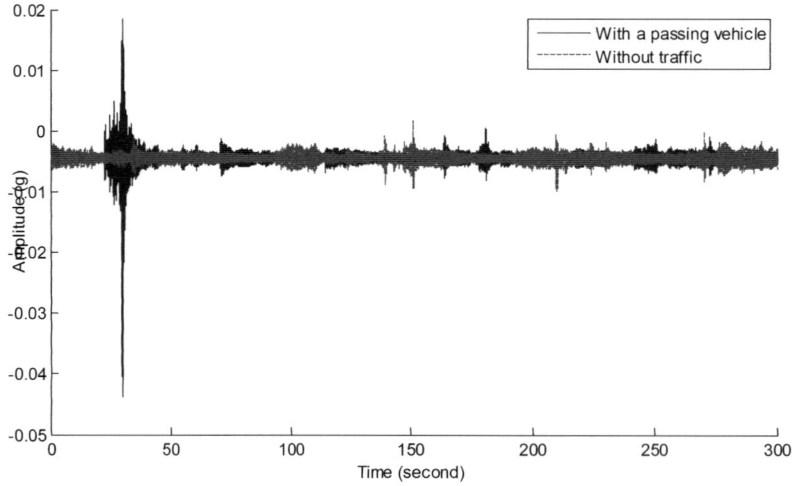

(a) Dynamic response

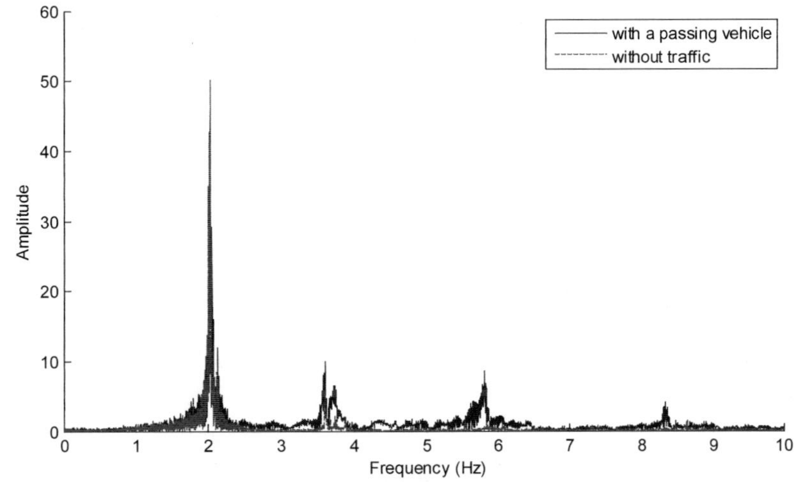

(b) Fourier spectrum

Figure 3 Dynamic responses and their spectrum of the bridge with or without a passing vehicle

3.2 The variation of the 1st modal frequency with the ambient vibration data

Figure 4 shows the variation of the 1st modal frequency with the temperature history on 13 May 2017. From the results, the variation of the 1st modal frequency is very small with the temperature change and it is slightly increased with the temperature increases. There is a big change around 16:00:00. This is due to there is a proof testing and a 6-axle truck was driven over the bridge.

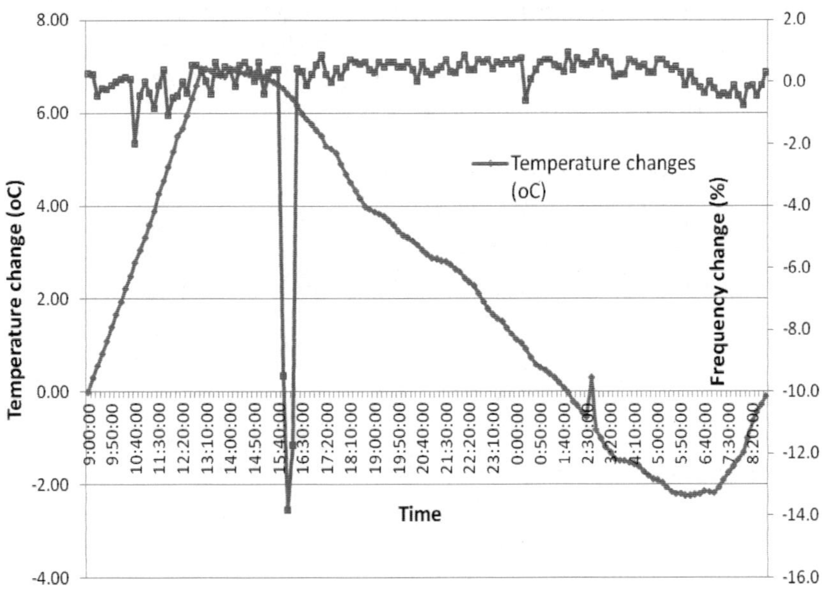

Figure 4 The frequency and temperature on 13 May 2017

Figure 5 show the modal frequencies at two different days. One is on 13 May 2017 and another one is on 11 February 2017. The temperature on 13 May 2017 is from 12°C to 20°C and it is 29°C to 45°C on 11 February 2017. From the results, the mean value of the 1st frequency on 13 May 2017 is 2.033Hz and its standard deviation is 0.036. They are 20.40Hz and 0.095 on 11 February 2017, respective. The result shows the frequency is increased with the temperature increases. The distribution of the frequency pattern is also changed with the temperature. Further study is needed to get the statistical correction between the frequency and the temperature change.

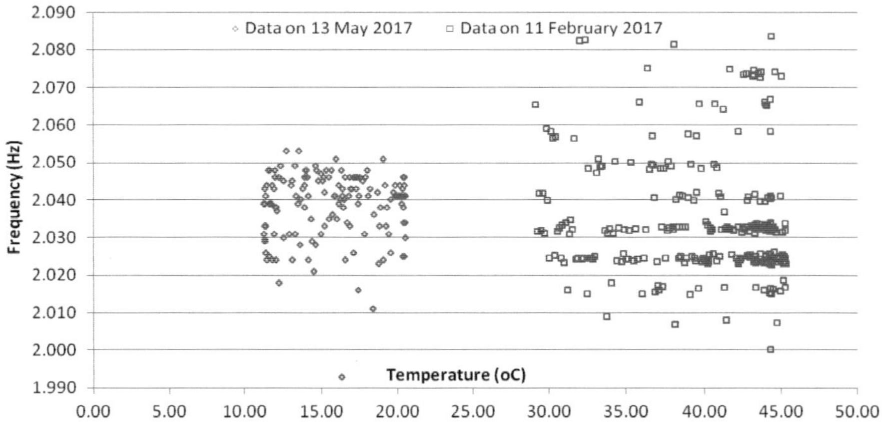

Figure 5 The frequencies of the brdge in two different days

3.3 Comparison of experimental and numerical results

A two-dimensional finite element model has also developed to compare with the results from dynamic measurements using Strand7. The brdige deck and tower are simulated by the beam elements. The bar element is used for cables. Table 1 shows the first five modal frequencies from the numerical modelling and the dynamic measurements. Figure 6 shows the the 1st mode shape from the finite element modelling and the field measurements. The results show that the 1st mode shape from the numerical modelling is agreed well with that from dynamic measurements. The values from the numerical modelling are smaler than that from the dynamic measurements. This is mainly that the compsosite action is not considered in the modelling.

Table 1 Frequencies from filed measurements and the finite element model

Mode No.	Model Frequency (Hz)	Test Frequency (Hz)
1	2.0365	2.038
2	3.2200	3.879
3	4.1935	5.881
4	5.2340	7.007
5	6.2591	9.009

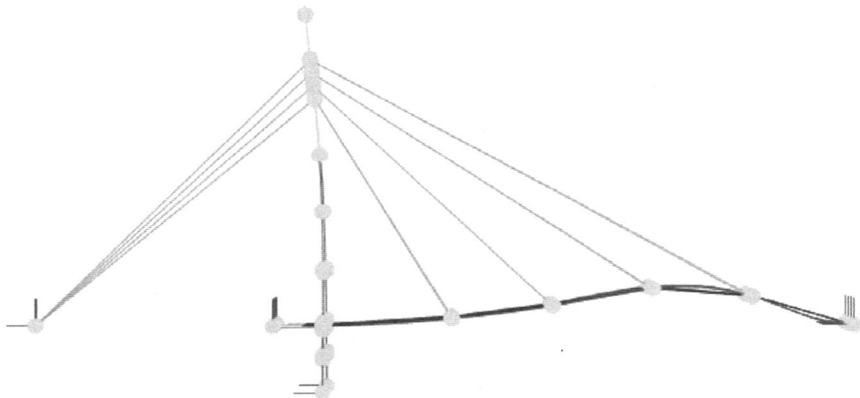

(a) The 1st Mode shape from the finite element model

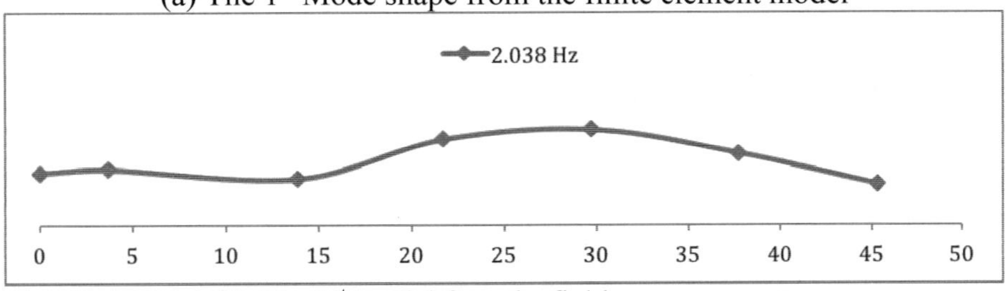

(b) The 1st model from in-field measurements

Figure 6 Comparison of the first mode shapes from measurements and modelling

4. Conclusions

The long-term monitoring system for a cable-stayed bridge has been presented. The monitoring data from two typical days have been analysed. The results show that the 1st modal frequency of the bridge is increased with the temperature increases. The vehicular loads have a big effect on the frequency. It is needed to consider the vehicle-bridge interaction in bridge structural health monitoring. Further study is needed to get the statistical relationship between the frequency and the temperature change.

References

Cross, EJ, Worden, K & Chen, Q 2011, 'Cointegration: a novel approach for the removal of environmental trends in structural health monitoring data', *Proceedings of the Royal Society A: Mathematical, Physical and Engineering Sciences*, Vol. 467, no. 2133, pp. 2712-2732.

Figueiredo, E, Radu, L, Worden, K & Farrar, CR 2014, 'A Bayesian approach based on a Markov-chain Monte Carlo method for damage detection under unknown sources of variability', *Engineering Structures*, Vol. 80, pp.1-10.

Westgate, R, Koo, KY & Brownjohn, J 2015, 'Effect of vehicular loading on suspension bridge dynamic properties', *Structure and Infrastructure Engineering*, Vol. 11, no. 2, pp.129-144.

Xia, Y, Hao, H, Zanardo G & Deeks, A 2006, 'Long term vibration monitoring of an RC slab: temperature and humidity', *Engineering Structures*, Vol.28, no. 3, pp.441-452.

Xia, Y, Chen, B, Weng, S, Ni, YQ & Xu, YL 2012, 'Temperature effect on vibration properties of civil structures: a literature review and case studies', *Journal of Civil Structural Health Monitoring*, Vol. 2, no. 1, pp.29-46.

Yuen, KV & Kuok, SC 2010, 'Ambient interference in long-term monitoring of buildings', *Engineering Structures*, Vol. 32, no. 8, pp.2379-2386.

The 8th International Conference on
Structural Health Monitoring of Intelligent Infrastructure
Brisbane, Australia | 5-8 December 2017

Finite Element Model Updating of Civil Engineering Infrastructures: a Literature Review

H. Moravej[1], S. Jamali[2], T.H.T. Chan[3], A. Nguyen[4]

[1,2,3,4] *School of Civil Engineering & Built Environment, Queensland University of Technology – Australia, Email: h.moravej@qut.edu.au*

Abstract

There are many civil engineering structures in Australia that have experienced different functional deficiencies during their service lives due to the harsh environment, fatigue effect, extreme loads and aging. The current practice of for assessment of structures relies mostly on design conditions. According to a comprehensive review of the literature, current practice in Australian lacks effective techniques to investigate the performance of structures systematically. Assessment of structures is usually performed using simplified methods and design conditions. Besides, code of practice requires revisions such as those made to AS5100.7 to include an additional section for evaluation of existing structures using structural health monitoring (SHM) techniques. Therefore, there is a need to assess the integrity of structures through a holistic approach to satisfy the structural safety. One of the reliable methods to investigate the performance of structures is Finite Element Model Updating (FE model updating) which aims to generate the accurate finite element models (FEM)s of real structures. Availability of reliable FEMs of structures is beneficial in many aspects such as their applications in SHM, evaluation of load carrying capacity, damage detection and performance monitoring of structures in extreme conditions like an earthquake. This paper comprehensively reviewed the previous application of FE model updating in civil structural engineering in Australia and worldwide, clarified the main existing gaps and presented how to make the method more applicable in practice according to the recent advancement in the technology. It is recommended to address further two significant factors, as convergence accuracy and computational intensity mitigation, without sacrificing each other in the future studies. Also, it is proposed to include more of SHM techniques including FE model updating in the assessment of existing structures, and provide a systematic procedure for application of FE model updating in codified approach.

1. Introduction

Australia, as a wide land surrounded by oceans, encounters harsh environment every year especially along the coastal line with the highly urbanised district and numerous infrastructures. Natural hazards such as Cyclone Debbie in Queensland recently or overwhelming Sydney tornado in 2015 cause inevitable damages on structures every year. So after such disasters, are buildings still safe to be reoccupied or bridges are at the same load carrying capacity as before to be still operational? These questions must be addressed.

Thus there is a significant need to evaluate the performance of existing structures during their lifetime. There are a variety of approaches for structural assessment such as visual inspection (Estes 2003), Non-destructive testing (Huston *et al.* 2011), bridge testing using acoustic emissions (Nair 2010), but one increasingly popular approach is the vibration analysis with the aim of assessment and modifying the structural properties with the use of several factors like natural frequencies, mode shape, mode shape curvature, damping ratios, Frequency Response Functions and etc. The vibration analysis would be applied by means of either experimental test on the real structures or by

using analytical approaches to simulate structures to obtain the real structural properties. Among all analytical methods, FE modelling method is possibly the most favorable approach, especially for large and complex structures. Nevertheless, the use of this method is not always easy especially when it comes to the application to assess capacity of existing structures. The inherent deficiencies in FEM in this context could be attributed to false assumptions, simplifications, improper modelling techniques (Jamali *et al.* 2017a) and errors in an approximation of input parameters such as material properties. So in advance of applying FEM in structural assessment, it is essential to adjust its defined parameters with real cases. The most reliable approach to reconciling FEM with real structures is FE model updating with an efficient impression in a wide range of structural fields such as Structural Damage Detection and Health Monitoring (Ng 2014; Chan *et al.* 2009) structural dynamic design functions, Fatigue Analysis and Damage Prognosis, Structural Performance Monitoring (Zanardo *et al.* 2006), Structural Control, Retrofitting & Rehabilitation, nuclear power plants and aeronautic. In light of SHM, bridge assessment is directed recently through FE model updating with the aim of minimizing the discrepancy between analytical and experimental measurements (Jamali *et al.* 2017b; Ding *et al.* 2012)
Recent SHM development at QUT is an example of the innovative application of FE model updating for long-term performance monitoring and assessment of structures in Australia. This SHM system was developed using 8 accelerometers that were synchronised via a cost-effective DAQ system which enables the extraction of modal parameters from ambient excitation (Nguyen *et al.* 2015). Furthermore, modal data of the building are continuously recorded and checked for any anomaly using unsupervised statistical learning methods and application of FE model updating (Nguyen *et al.* 2014; Kodikara *et al.* 2016a; Kodikara *et al.* 2016b). In this study, the model updating process is revisited with application of this framework for the purpose of assessment and monitoring of civil engineering infrastructure. First, the techniques commonly used as auxiliary steps prior to the main model updating task are reviewed, then two main FE model updating approaches as Direct and Iterative are briefly explored before the latter as the most applicable and reliable one for civil engineering applications is discussed in more detail. In the end, the current gaps and the future directions are presented.

2. Initial Comparison between experimental and FEM data

Prior to using any FE model updating method, the consistency of the experimental work should be confirmed which is satisfied by available procedures such as choosing the types, locations and the number of sensors to be used, the sampling rate at which the data to be collected and which size of the data storage should be selected. Also other aspects as normalization of measured data, information condensation, feature extraction and data fusion should be considered as well (Sohn *et al.* 2003). Next, the test data and FE counterpart should be compared to be ensured about the existence of some correlation between their data before conducting the main model updating task.

2.1 Correlation techniques

Correlation indices show how close the experimental data are with FEM's counterpart. These methods comprise assessment of Frequency Response Functions (FRFs), mode shapes and natural frequencies of experimental data and FE data to check the level of agreement.

2.1.1 FRF matching

Generally, a frequency response function is a mathematical representation of the relationship between the input and the output of a system and it is basically obtained from the Eq. (1) (Friswell & Mottershead 2013).

$$H(f) = \frac{Y(f)}{X(f)} \tag{1}$$

In this equation, H(f) is the frequency response function. *Y*(f) and *X*(f) are the output and input ofthe system respectively in the frequency domain.

To compare the results from FE and experimental test, one simple technique would be plotting FRFs for both in one diagram as illustrated in Figure. 1 In this diagram, the vertical axis and horizontal axis stand for magnitude of FRF and Frequency respectively (Chang & Park 1998; D'Ambrogio & Fregolent 1998). As it can be seen, a proper correlation between the FE and experimental FRF results is figured out in Figure. 1(a), on the other hand, Figure 1(b) shows a kind of mismatching between these two counterparts. In such condition, a better matched FEM is required before the main model updating phase can take place.

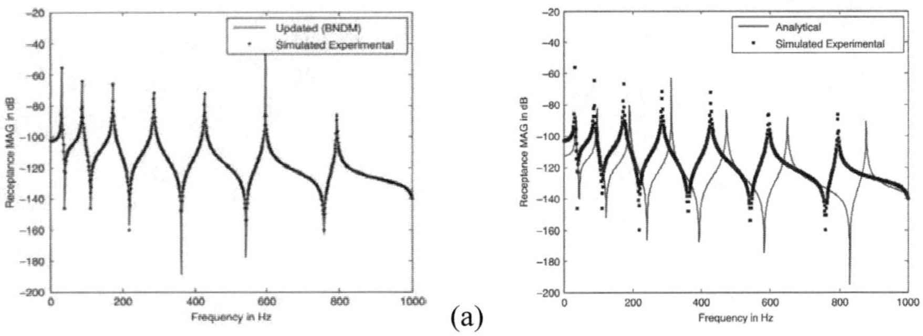

(a) (b)

Figure 1 FRF plots matching. (a) good matching (b) mismatching (Sehgal & Kumar 2016)

2.1.2 Natural frequencies matching

Natural Frequency is another feature which should be considered in comparison between experimental and FE data before updating. In cases the error has a significant percentage, beyond 50%, it is recommended to adjust the current FEM (e.g. by manual tuning) or generate a new FEM before carrying out model updating (Živanović *et al.* 2007; Ewins 1984).

2.1.3 Mode shapes matching

Mode shape as one of the intrinsic dynamic properties of structure that signify the displacement of the structure in each natural frequency. Mode shape depends on material properties and boundary conditions of structures. Overlaying technique is a simple method to compare mode shapes (Sampaio *et al.* 1999). In this method, the divergence between FE and experimental mode shape is visually depicted in a diagram. The significant drawback of this methods is that they are not suitable in automated correlating. Another approach for correlation of mode shape is Modal Scale Factor (MSF) technique which is proper for automated correlating since this method presents a quantitative fitting between two corresponding data. MSF result is transferred on one x–y diagram with mode shape from the experimental test on one axis and FE mode shape on another (Adhikari & Friswell 2010). Perfect correlation between both mode shapes is presented by unity in MSF. In MSF method, the scatter of the x–y graph is not presented, so to overcome the dispersion of data in MSF plot, Modal Assurance Criterion (MAC) is often used which was also defined as mode shape correlation coefficient (Allemang & Brown 1982). MAC can be calculated using Eq. (2).

$$\mathrm{MAC}=\frac{\left|\{\phi_x\}_i^T\{\phi_A\}_j\right|^2}{\left(\{\phi_x\}_i^T\{\phi_A\}_i\right)\cdot\left(\{\phi_A\}_j^T\{\phi_A\}_j\right)} \tag{2}$$

In this equation, $\{\emptyset_X\}_i$ and $\{\emptyset_A\}_j$ represent the ith experimental and j^{th} analytical mode shape respectively; while superscript '*T*' denotes the transpose of the corresponding vector. MAC value would find the perfect matching state between experimental and analytical data when it reaches to unity, and once a MAC value is close to zero explains misfit between two modes (Pandey *et al.* 1991). Figure 2 shows a MAC plot between experimental and FE modes.

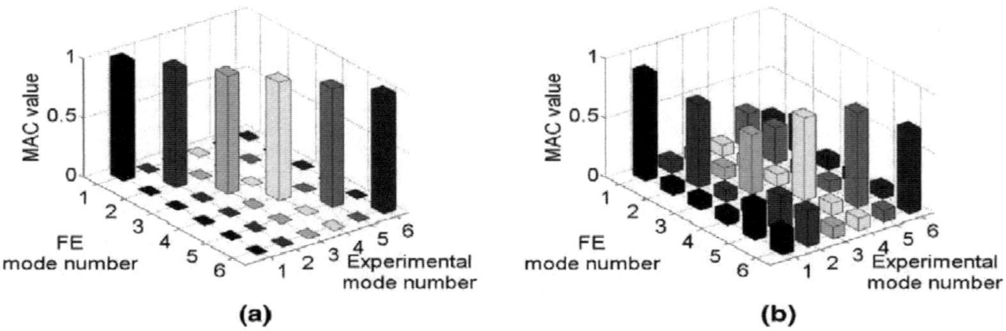

Figure 2 Modes correlation using MAC. (a) Good correlation, (b) poor correlation (Sehgal & Kumar 2016)

It is shown that the diagonal components are equal to one and the others are zero, which demonstrates ideal fitting between two different sets of data. On the other hand, in Figure 2b, a mismatching is illustrated since neither diagonal components are equal to 1, nor off-diagonal ones are equal to zero. Having said that, perfect correlation is not always feasible due to testing condition, corrupted data, and faulty equipment. MAC value has weaknesses in identification of systematic deviances as well as being poor in checking orthogonality due to lack of mass or stiffness matrix in the function. The later drawback can be overcome by applying normalised MAC (NMAC) (Arora *et al.* 2009a). NMAC comprises a weighting matrix [W], which can be substituted by either mass or stiffness matrix. NMAC also has a shortcoming in defining the spatial scattering of correlation, which can be satisfied by using Coordinate MAC (COMAC). COMAC is applicable when number of the correlated mode pairs is found via MAC or NMAC. COMAC value equal to '1' presents excellent correlation at a particular coordinate (Lieven & Ewins 1988).

2.2 Dimension Compatibility Methods

It is common that the degrees of freedom (DOF) measured in an experimental test are generally much smaller than FE's (DOF) as a reason for the lack of a sufficient number of sensors. Also, some DOFs are so challenging to be recognised such as the fixity of support, lateral movement of foundation, local vibration of connecting parts. Accordingly, the dimension of FEM is typically not consistent with their experimental counterparts, whereas in Model Updating, between the two data sets, one point-to-point correlation is necessary. This aim could be satisfied via increasing the experimental data or by decreasing the FEM.

2.2.1 Coordinate Expansion

One approach to achieving the dimension compatibility could be an expansion of measured data sets to reach the same degree of FE's size which is named "coordinate expansion". A simple technique is to replace the unknown components by their FE counterparts. This method is cost-effective in computation, despite sometimes it causes incorrect results in model updating. To conquer such drawback, couple transformation matrix based coordinate expansion should be generated (Arora *et al.* 2009b; Arora *et al.* 2009c). There are several different methods in this category such as System equivalent reduction and expansion processes (SEREP) (Adhikari & Friswell 2010), FE eigenvectors combining with MAC matrix (Arora 2011), Curve fitting (Atalla & Inman 1998) and a method which is made on the assumption that unknown eigenvectors could be considered as a linear combination of measured eigenvectors (Bais *et al.* 2004).

2.2.2 Model Minimization

This approach is mainly on the contrary with the previous method with the aim to reduce the analytical model's size and make its Degree Of Freedom (DOF) close the experimental model counterpart. A basic technique is to cut those unavailable experimental degrees of freedom. So a reduced model is provided. A main lack of this method is that some of mass and stiffness parameters related to the removed DOFs are totally missed and their effects would not be available anywhere (Baruch 1978). Another technique in this area is an approximate method which tries to generate a compact model of a whole structure entirely but approximately through converting the original model. One method of gaining condensed model is static reduction proposed by Baruch & Bar Itzhack (1978) which is mostly known as Guyan reduction method (Guyan 1965). The significant disadvantage of this method is that it would not reflect any of the Eigen parameters of the original full FEM. Improved reduced system (IRS) method is another scheme of model minimization with the aim to reproduce the eigenvalues and eigenvectors closely to the original FEM (Baruch 1978; Bayraktar *et al.* 2011). If reduced model is required to have eigenvalues and eigenvectors exactly the same as that of original FEM then SEREP should be used (Beattie & Smith 1992; Berman 1979).

3. Finite Element Model Updating main frameworks (Direct and Iterative)

The FE model updating methods can be generally categorised into two main groups as Direct and Iterative techniques.

Direct approach is computationally effective since the result would be prepared only in one step with easy converging results. One aspect of this technique is that they reproduce the measured data exactly. The experimental data and their analytical counterpart are rarely to be the same as a result of measurement noise and model deficiency. Updating is fulfilled in one step to optimise the parameters in the model but not with the aim of exactly reproducing the noise. If the updated model exactly reproduces defective measured data, any consequent analysis may be faulty. So, this method requires reliable modal testing and analysis procedures. Direct approach applies matrix methods to calculate the result with performing equation of motion. A major shortcoming of this approach is that the updated stiffness and mass matrices may not be physically sensible. In the other words, while it satisfies the mathematical problem, it is not capable of reflecting the real changes in structural properties (Friswell & Mottershead 2013). This technique is also referred as the technique of reference since one of the three parameters (damping, stiffness and mass matrices) is supposed to be the reference and the other two quantities are adjusted (Baruch & Bar Itzhack 1978; Baruch 1978; Berman 1979). Recently a vibro-acoustic FE model updating technique was offered by Modak (2014). Such method could apply to update the vibro-acoustic FEM of elastic systems. In this technique, stiffness and mass matrix from the structural together with the acoustic components of the model can be updated. Another direct FE model updating method was developed by Bucher and Barun (1993) which was able to handle incomplete known eigensolutions that were tested numerically on a fixed beam and a sequential spring mass system. In 2007, a new method named cross-model-cross mode (CMCM) was established by Hu et al (2007). This method aimed to simultaneously update the physical properties of the system matrices as well as the mass and stiffness matrices. Further, substructure energy direct FE model updating method based on the physical property modification was generated by Fang et al. (2011). The advantage of this technique was that it could deal with insufficient experimental data. Also in this method, the whole structure is separated into several substructures. Thus, just the crucial sub systems are detected and adjusted rather than updating the whole system. Dynamic behaviour of the FEM of structures relies on different parameters, as mentioned before.

In iterative methods, one objective function is generated between input structural parameters and structural responses. Iterative approach aims to update the parameters iteratively in such a way that between FEM's response and the corresponding experimental response minimal difference exists. The cycle of iterations is finished once the responses from FEM and experimental data no longer change in subsequent iterations. Despite being computationally expensive, iterative techniques represent the realistic behaviour of the structure (Berman & Nagy 1983). Since the Iterative method can achieve more reliable results, this technique has further applications in structural engineering by far and many investigations have been conducted to develop the technique which will be addressed in the next part.

4. Finite Element Model Updating based on iterative frameworks

In iterative approach, one objective function is made and it is reduced iteratively to reach the adjusted parameters. These techniques initially started with the development of an iterative eigendata sensitivity technique called inverse eigensensitivity method (IESM) (Collins *et al.* 1974). Lin et al. enhanced the convergence of this technique through applying the FE data together with the experimental responses for assessing sensitivity coefficients (Urn & Du 1995). In IESM, eigen data and damping ratios are performed to construct an objective function. Modal data is provided by modal analysis, so the accuracy in applying the modal analysis has a significant effect on the validity of extracted modal data which is conducted to the model updating results. To circumvent this issue, obtained signal data can directly be transferred for model updating using Response Function Method (RFM) (Lin & Ewins 1990). Since in RFM there is no need for modal extraction to apply to measured data, the chance of errors during modal analysis process decreases. According to a comparative report by Modak et al. (2002), RFM has a better performance than IESM in cases of insufficient experimental data without noise, while IESM is more efficient in the updating cases with the existence of noise; especially when the updating region includes higher number of modes. Furthermore, in RFM, the sensors' positions and sample rate of experimental have significant effects on the convergence of the method. Modak et al. (2005) established a constrained optimization FE model updating method. The method was applied to a fixed-fixed beam and an F-shape frame structure. Despite the technique is computationally more intense than IESM, it can alleviate the effect of variance between the sensitivities of mode shapes and natural frequencies. Atalla and Inman (1998) employed Neural Network (NN) technique in the FE model updating of a flexible frame with training an NN and verified it via a constructed sample of the structure, testing it and comparing the experimental and analytical responses. Such an NN could be exploited to find updated physical parameters by setting experimental responses as inputs. Provided NN model has trained appropriately, its calculation process is rather faster than conventional optimization approaches. Furthermore, NN method is also reliable in the presence of noise (Levin & Lieven 1998a). The main drawback of this technique is that a great amount of training data is necessary.

4.1 Objective function definition in FE model updating

Application of FE model updating in real structures faces a variety of complexities such as errors in measurement, the existence of nonlinearity, damping effect and a large number of updating parameters. Due to such issues, the conventional model updating methods may be unable to converge and minimise the objective functions properly in a global scale. So the application of optimization approaches with the capability to obtain the global optimum is indeed required, such as Genetic Algorithms (GA), Particle Swarm Optimization (PSO) and Simulated Annealing (SA) (Sehgal & Kumar 2016). According to a comparative research by Levin and Lieven (1998)_FE model updating by applying SA leads to a better performance than GA approach. Having said that, the efficiency of FE model updating based on both methods relies on accurately selecting the updating parameters. Marwala (2010) concluded that the application of PSO in FE model updating is so effective in cases with many updating parameters. Further, Mthembu *et al.* (2011) employed PSO in multi-model FE model updating to select the most accurate structural model in the presence of many updated models. For the purpose of damage detection, the residuals should be sensitive to even slight local structural changes. In a research, a sensitivity-based FE model updating is carried out for the purpose of damage detection. The objective function consisting of the modal flexibility residual is formulated and its gradient is derived. The proposed procedure is illustrated with a simulated beam and a laboratory-tested beam with damage. Despite all the elements in the FEM are used as the updating parameters, which is considered as the extreme adverse condition in the FE model updating, the identified damage pattern is comparable. It is demonstrated that the proposed FE updating using the modal flexibility residual is promising for the detection of damaged elements (Jaishi & Ren 2006). In many optimization concepts, single-objective optimization techniques propose reliable algorithms to optimise a single criterion. But this framework is disabled to cope with the optimization cases when different criteria at the same time are considered. In such situations, a multi-objective optimization problem would be evolved (Kim & Park 2004). Perera and Ruiz (2008) proposed a multistage algorithm for damage detection in large-size structures with the application of modal data obtained from experimental tests together with the FE model updating methods applied on simple FEMs. In this method, a multi objective technique was employed to optimise the objective function. Optimization procedure is formulated in a multi-objective context solved by using evolutionary algorithms. The multistage algorithm for damage detection was applied to the I-40 Bridge in New Mexico and it was found that the offered approach is robust, fast and easy to perform. Generally, the results of the several tests on an identical structure would not be same in all cases due to various operational and environmental factors. Consequently, every parameter has an approximate value of a variance and a mean value. Mares *et al.* (2006) included an approximate concept of parameters into FE model updating by applying it to a simulated three degree-of-freedom mass–spring structure and reach the reliable result. The center of the distribution would be defined as the mean value and the size is identified by the variance of test data. Same work was then extended by Mottershead *et al.* (2011) who developed a stochastic FE model updating technique. The framework was constructed based on statistical models with a large database. A Bayesian framework was presented by Wan and Ren (2014) to a Stochastic Model Updating method for measurement of parameters uncertainty which could not be reduced. Bayesian method overcomes the issues of inverse uncertainty quantification in a direct mode, which evades the problems of gradient computation and ill-conditionedness that is related to the optimization process. Some researchers used Bayesian implication to achieve structural reliability updates and to analyse the likelihoods of diverse damage circumstances (Papadimitriou *et al.* 2001; Sohn & Law 2000).

Sun & Betti (2015) developed a hybrid optimization approach for the stochastic model updating of structures. The method is originated as an inverse problem, conducted through a Bayesian framework, and is calculated via a hybrid optimization algorithm. The offered hybrid approach comprises of a global and local assessment operator named Modified Artificial Bee Colony (MABC) and Broyden-Fletcher-Goldfarb-Shanno (BFGS) respectively. Parametric studies and numerical optimization of mathematical functions show the efficiency of the offered method is demonstrated through the numerical and the experimental data sets.

4.2 Reduced Order Techniques in FE model updating

Because of the iterative nature of the FE model updating, this approach would be computationally intensive, especially when it deals with large-scale structures. Guo and Zhang (2004) developed a method to perform the Response Surface methodology (RSM) in FE model updating. The significance of this method is that there is no need to calculate the FE response at each iteration because the FEM is swapped with an approximated mathematical equation, thus the amount of computation is significantly reduced. In this technique, by considering updating parameters as inputs together with the FE responses as outputs, n-dimensional response surfaces are generated. Through applying the response surfaces along with the experimental responses, the model parameters are updated by optimizing the objective function. This type of FE model updating is reliable as the sensitivity-based method, more efficient in computational aspect and faster in convergence aspect (Sehgal & Kumar 2016). Shahidi and Pakzad (2013) applied a time-domain technique in RSM based FE model updating. Since in traditional techniques, measured signals should be converted to response parameters such as mode shape, resources for the training data to generate response surface models are declined. Hence, time-domain results were performed to compensate this issue. One of the significant factors in the FE model updating framework is parameters selection, particularly in the case of large structures. It is obvious that the real structures include so many parameters and considering all of them are updated, the model updating procedure becomes complicated and time-consuming undoubtedly. Moreover, it is likely that the updating problems lead to ill-conditioning as well (Friswell & Mottershead 2013). Recently one parameter selection technique was developed by Kim and Park (2008) which comprises two steps. In the first step, each updating parameter is considered as incorrect. Next two adjacent parameters are combined if they have the same sensitivity signs i.e. both are positive or negative. Thus the number updating parameters is considerably reduced in this way. Such process is continued until all nearby parameters find different sensitivity signs. If the number of updating parameters is less than a limit point, the process will stop; otherwise, the technique is transferred to the second step. In this step, a comparative analysis is applied to the opposite sign sensitivity neighboring groups of the updating parameters, and those two groups are selected to be merged which cause the least decrease in entire sensitivity. Weng et al. (2012) offered a novel substructure model updating approach. In this method, the experimental flexibility matrix for the whole structure is divided into the several substructural matrices. All these substructures matrices are applied as references to update the independent substructures' models. Application of this method is computationally efficient since just the concerned substructure is modified. Furthermore, in this framework, only the measurement of the local area of the focused substructures is needed. The proficiency of the substructure based model updating is numerically illustrated on a large-scale real structure. This method was applied on Guangzhou New TV Tower. and, it was proved that substructure-based procedure is much faster than the traditional global-based procedure.

5. Conlcuding Remarks and Future Research Directions

From the reviewed literature, it is found that structural damping is not considered in the majority of the works, while damping effect significantly affects the structural responses. So it is recommended to investigate the damping effect in the future research to enhance the reliability of the FE model updating as suggested by Friswell & Mottershead (2013). In most of FE model updating cases, the structures are assumed to behave linearly. This assumption could affect the accuracy of FE model updating methods, especially when dealing with assessment of old infrastructures. Response of structures such as buildings and bridges changes over time and in order to reflect these changes, as-is condition of the structure needs to be reflected in the FEM. Material nonlinearity is common in many structures and geometrical nonlinearity to some extent. Such nonlinearities are time-dependant and are affected by various environmental and operational factors. Hence, for assessment purpose, nonlinearity aspect of the structure shall be considered even though the past performance of the structure was satisfactory (Asgarieh *et al.* 2017). Reserved capacity, redundancy, ductility, life cycle and load-distribution are some of the essential information that daily can evaluate when nonlinearity is considered into the modelling. Generally, the nature of FE model updating is the calculation of an inverse problem. It is found that too many updating parameters can lead to an ill-conditioned problem or become computationally costly. So more development in numerical concept is still necessary to converge the inverse problem accurately and with less computation. Major sources of inaccuracy in the FEM are due to physical errors, geometrical errors and modelling error. A systematic comparison with experimental and analytical data using MAC and other modal comparison methods can be effective in alleviating the occurrence of the error. Accurate parameter selection is the most pivotal step in FE model updating which still needs to be improved. Despite a number of techniques have been proposed in this aspect, the parameter selection mostly relies on the engineering judgment and structural response. It is suggested to have a systematic guideline for model updating parameter selection of different structures based on hierarchy level of structural importance. In addition, a generalised process can be formulated to include the application of FE model updating in the routine practices. This includes for buildings, bridges, and any other civil structures that need assessment or retrofitting. The newly released version of AS 5100 Bridge design set (Australian Standards, 2017) is a good example, in which its Part 7 includes the application of SHM for bridge assessment. Such combination of codified approaches with advanced SHM and FE model updating technique would significantly improve the accuracy of the assessment. In turn, the full functionality of the structure is utilised with more confidence in the assessment. Further experimental investigations on benchmark studies such as QUT-SHM benchmark (Nguyen *et al.* 2015) building are needed, which can definitely be helpful in two aspects; to further develop the applicability and accuracy of FE model updating for real structures, and to bridge the gap between the research and the industry.

Acknowledgment

The first author gratefully appreciates the financial support for his research from Queensland University of Technology (QUT).

References

Estes, AC & Frangopol, DM 2003, 'Updating bridge reliability based on bridge management systems visual inspection results', *Journal of Bridge Engineering, 8*(6), 374-382.

Huston, D, Cui, J, Burns, D, & Hurley, D 2011, 'Concrete bridge deck condition assessment with automated multisensor techniques', *Structure and Infrastructure Engineering, 7*(7-8), 613-623.

Nair, A., & Cai, C. 2010. Acoustic emission monitoring of bridges: Review and case studies. *Engineering structures, 32*(6), 1704-1714.

Jamali, S, Chan, THT, Thambiratnam, DP, & Nguyen, A 2017b, 'Comparative study of grillage analogy and finite element method for bridge heavy load assessment', *Proceedings of Austroads Bridge Conference*, Melbourne, Australia, 3-6 April, pp. 1-10.

Ng, C.-T. 2014. 'Bayesian model updating approach for experimental identification of damage in beams using guided waves'. *Structural Health Monitoring, 13*(4), 359-373.

Chan, T., Li, Z., Yu, Y., & Sun, Z. 2009. 'Concurrent multi-scale modeling of civil infrastructures for analyses on structural deteriorating—Part II: Model updating and verification'. *Finite Elements in Analysis and Design, 45*(11), 795-805.

Zanardo, G., Hao, H., Xia, Y., & Deeks, A. J. 2006. 'Stiffness assessment through modal analysis of an RC slab bridge before and after strengthening'. *Journal of Bridge Engineering, 11*(5), 590-601.

Jamali, S, Chan, THT, Thambiratnam, DP, Pritchard, R, & Nguyen, A 2017a, 'Pre-test finite element modelling of box girder overpass-application for bridge condition assessment', *Proceedings of Australian Structural Engineering Conference (ASEC)*, Brisbane, Australia, 23-25 November, pp. 1-8.

Ding, L., Hao, H., Xia, Y., & Deeks, A. J. 2012. 'Evaluation of bridge load carrying capacity using updated finite element model and nonlinear analysis'. *Advances in structural engineering, 15*(10), 1739-1750.

Nguyen, T., Chan, T. H., Thambiratnam, D. P., & King, L. 2015. 'Development of a cost-effective and flexible vibration DAQ system for long-term continuous structural health monitoring'. *Mechanical Systems and Signal Processing, 64*, 313-324.

Nguyen, T., Chan, T. H., & Thambiratnam, D. P. 2014. 'Field validation of controlled Monte Carlo data generation for statistical damage identification employing Mahalanobis squared distance'. *Structural Health Monitoring*, 13(4), 473-488.

Kodikara, K., Chan, T., Nguyen, T., & Thambiratnam, D. 2016a. 'Model updating of real structures with ambient vibration data'. *Journal of Civil Structural Health Monitoring, 6*(3), 329-341.

Kodikara, K. T. L., Chan, T. H., Nguyen, A., & Thambiratnam, D. P. 2016b. 'Model updating incorporating measured response uncertainties and confidence levels of tuning parameters'. *International Journal of Lifecycle Performance Engineering, 2*(1-2), 61-78.

Sohn, H., Farrar, C. R., Hemez, F. M., Shunk, D. D., Stinemates, D. W., Nadler, B. R., & Czarnecki, J. J. 2003. 'A review of structural health monitoring literature: 1996–2001'. *Los Alamos National Laboratory*.

Friswell, M., & Mottershead, J. E. 2013. 'Finite element model updating in structural dynamics' (Vol. 38): *Springer Science & Business Media.*

Chang, K.-J., & Park, Y.-P. 1998. 'Substructural dynamic modification using component receptance sensitivity'. *Mechanical systems and signal processing, 12*(4), 525-541.

D'Ambrogio, W., & Fregolent, A. 1998. 'On the use of consistent and significant information to reduce ill-conditioning in dynamic model updating'. *Mechanical Systems and Signal Processing, 12*(1), 203-222.

Sehgal, S., & Kumar, H. 2016. 'Structural dynamic model updating techniques: A state of the art review'. *Archives of Computational Methods in Engineering, 23*(3), 515-533.

Živanović, S., Pavic, A., & Reynolds, P. 2007. 'Finite element modelling and updating of a lively footbridge: The complete process'. *Journal of Sound and Vibration, 301*(1), 126-145.

Ewins, D. J. 1984. 'Modal testing: theory and practice' Vol. 15: Research studies press Letchworth.

Sampaio, R., Maia, N., & Silva, J. 1999. 'Damage detection using the frequency-response-function curvature method' *Journal of Sound and Vibration, 226*(5), 1029-1042.

Adhikari, S., & Friswell, M. 2010. 'Distributed parameter model updating using the Karhunen–Loève expansion'. *Mechanical Systems and Signal Processing, 24*(2), 326-339.

Allemang, R. J., & Brown, D. L. 1982. 'A correlation coefficient for modal vector analysis'. In *Proceedings of the 1st international modal analysis conference* (Vol. 1, pp. 110-116): SEM, Orlando.

Pandey, A., Biswas, M., & Samman, M. 1991. 'Damage detection from changes in curvature mode shapes'. *Journal of sound and vibration, 145*(2), 321-332.

Arora, V., Singh, S., & Kundra, T. 2009a. 'Finite element model updating with damping identification'. *Journal of Sound and Vibration, 324*(3), 1111-1123.

Lieven, N., & Ewins, D. 1988. Spatial correlation of mode shapes, the coordinate modal assurance criterion (COMAC). *In Proceedings of the sixth international modal analysis conference* (Vol. 1, pp. 690-695).

Arora, V., Singh, S., & Kundra, T. 2009b. 'Damped model updating using complex updating parameters'. *Journal of Sound and Vibration, 320*(1), 438-451.

Arora, V., Singh, S., & Kundra, T. 2009c. 'Comparative study of damped FE model updating methods'. *Mechanical Systems and Signal Processing, 23*(7), 2113-2129.

Arora, V. 2011. 'Comparative study of finite element model updating methods'. *Journal of Vibration and Control, 17*(13), 2023-2039.

Atalla, M. J., & Inman, D. J. 1998. 'On model updating using neural networks'. *Mechanical Systems and Signal Processing, 12*(1), 135-161.

Bais, R. S., Gupta, A., Nakra, B., & Kundra, T. 2004. 'Studies in dynamic design of drilling machine using updated finite element models'. *Mechanism and machine theory, 39*(12), 1307-1320.

Guyan, R. J. 1965. 'Reduction of stiffness and mass matrices'. *AIAA journal, 3*(2), 380-380.

Baruch, M. 1978. 'Optimization procedure to correct stiffness and flexibility matrices using vibration tests'. *AIAA journal, 16*(11), 1208-1210.

Baruch, M., & Bar Itzhack, I. Y. 1978. 'Optimal Weighted Orttiogonalization of Measured Modes'. *AIAA journal, 16*(4), 346-351.
Bayraktar, A., Sevim, B., & Altunişik, A. C. 2011. 'Finite element model updating effects on nonlinear seismic response of arch dam–reservoir–foundation systems'. *Finite elements in analysis and design, 47*(2), 85-97.
Beattie, C., & Smith, S. 1992. 'Optimal matrix approximants in structural identification'. *Journal of Optimization Theory and Applications, 74*(1), 23-56.
Berman, A. 1979. 'Mass matrix correction using an incomplete set of measured modes'. *AIAA journal, 17*(10), 1147-1148.
Modak, S. 2014. 'Direct matrix updating of vibroacoustic finite element models using modal test data'. *AIAA journal, 52*(7), 1386-1392.
Bucher, I., & Braun, S. 1993. 'The structural modification inverse problem: an exact solution'. *Mechanical Systems and Signal Processing, 7*(3), 217-238.
Hu, S.-L. J., Li, H., & Wang, S. 2007. 'Cross-model cross-mode method for model updating'. *Mechanical Systems and Signal Processing, 21*(4), 1690-1703.
Fang, H., Wang, T. J., & Chen, X. 2011. 'Model updating of lattice structures: a substructure energy approach'. *Mechanical Systems and Signal Processing, 25*(5), 1469-1484.
Berman, A., & Nagy, E. 1983. 'Improvement of a large analytical model using test data'. *AIAA journal, 21*(8), 1168-1173.
Collins, J. D., Hart, G. C., Haselman, T., & Kennedy, B. 1974. 'Statistical identification of structures'. *AIAA journal, 12*(2), 185-190.
Urn, M., & Du, H. 1995. 'Improved inverse eigensensitivity method for structural analytical model updating'.
Lin, R., & Ewins, D. 1990. 'Model updating using FRF data'. In *Proceedings of the 15th international seminar on modal analysis* (pp. 141-162).
Modak, S., Kundra, T., & Nakra, B. 2002. 'Comparative study of model updating methods using simulated experimental data'. *Computers & structures, 80*(5), 437-447.
Modak, S., Kundra, T., & Nakra, B. 2000. 'Model updating using constrained optimization'. *Mechanics Research Communications, 27*(5), 543-551.
Modak, S., Kundra, T., & Nakra, B. 2005. 'Studies in dynamic design using updated models'. *Journal of sound and vibration, 281*(3), 943-964.
Levin, R. I., & Lieven, N. 1998a. 'Dynamic finite element model updating using neural networks'. *Journal of sound and vibration, 210*(5), 593-607.
Levin, R., & Lieven, N. 1998b. 'Dynamic finite element model updating using simulated annealing and genetic algorithms'. *Mechanical Systems and Signal Processing, 12*(1), 91-120.
Marwala, T. 2010. 'Finite-element-model Updating Using Particle-swarm Optimization'. *Finite-element-model Updating Using Computional Intelligence Techniques: Applications to Structural Dynamics*, 67-84.
Mthembu, L., Marwala, T., Friswell, M. I., & Adhikari, S. 2011. 'Finite element model selection using Particle Swarm Optimization'. In *Dynamics of Civil Structures, Volume 4* (pp. 41-52): Springer.
Jaishi, B., & Ren, W.-X. 2006. 'Damage detection by finite element model updating using modal flexibility residual'. *Journal of sound and vibration, 290*(1), 369-387.
Kim, G.-H., & Park, Y.-S. 2004. 'An improved updating parameter selection method and finite element model update using multiobjective optimisation technique'. *Mechanical Systems and Signal Processing, 18*(1), 59-78.
Perera, R., & Ruiz, A. 2008. 'A multistage FE updating procedure for damage identification in large-scale structures based on multiobjective evolutionary optimization'. *Mechanical Systems and Signal Processing, 22*(4), 970-991.
Mares, C., Mottershead, J., & Friswell, M. 2006. 'Stochastic model updating: part 1—theory and simulated example'. *Mechanical Systems and Signal Processing, 20*(7), 1674-1695.
Mottershead, J. E., Link, M., & Friswell, M. I. 2011. 'The sensitivity method in finite element model updating: a tutorial'. *Mechanical systems and signal processing, 25*(7), 2275-2296.
Wan, H.-P., & Ren, W.-X. 2014. 'Parameter selection in finite-element-model updating by global sensitivity analysis using Gaussian process metamodel'. *Journal of Structural Engineering, 141*(6), 04014164.
Papadimitriou, C., Beck, J. L., & Katafygiotis, L. S. 2001. 'Updating robust reliability using structural test data'. *Probabilistic Engineering Mechanics, 16*(2), 103-113.
Sohn, H., & Law, K. H. 2000. Bayesian probabilistic damage detection of a reinforced-concrete bridge column'. *Earthquake Engineering Structural Dynamics, 29*(8), 1131-1152.
Sun, H., & Betti, R. 2015. 'A hybrid optimization algorithm with Bayesian inference for probabilistic model updating'. *Computer-Aided Civil and Infrastructure Engineering, 30*(8), 602-619.
Guo, Q., & Zhang, L. 2004. 'Finite element model updating based on response surface methodology'. In *Proceedings of the 22nd international modal analysis conference*.

Shahidi, S. G., & Pakzad, S. N. 2013. 'Generalised response surface model updating using time domain data'. *Journal of Structural Engineering, 140*(8), A4014001.
Kim, G.-H., & Park, Y.-s. 2008. 'An automated parameter selection procedure for finite-element model updating and its applications'. *Journal of Sound and Vibration, 309*(3), 778-793.
Weng, S., Xia, Y., Zhou, X.-Q., Xu, Y.-L., & Zhu, H.-P. 2012. 'Inverse substructure method for model updating of structures'. *Journal of Sound and Vibration, 331*(25), 5449-5468.
Jensen, H., Millas, E., Kusanovic, D., & Papadimitriou, C. 2014. 'Model-reduction techniques for Bayesian finite element model updating using dynamic response data'. *Computer Methods in Applied Mechanics and Engineering, 279*, 301-324.As
Asgarieh, E., Moaveni, B., Barbosa, A. R., & Chatzi, E. 2017. 'Nonlinear model calibration of a shear wall building using time and frequency data features'. *Mechanical Systems and Signal Processing, 85*, 236-251.
Australian Standards. 2017. 'Bridge Design- Part 7: Bridge Assessment (AS 5100.7)'. Sydney, Australia.

The 8th International Conference on
Structural Health Monitoring of Intelligent Infrastructure
Brisbane, Australia | 5-8 December 2017

Smart-based Monitoring of Epoxy Using Piezoelectric Transducers

Y. Y. Lim [1], S. T. Smith [2], I.Izadgoshasb [3]
[1, 2, 3] *School of Environment, Science and Engineering, Southern Cross University – Australia,*
Email: yee.lim@scu.edu.au

Abstract

Structural adhesives are an important component of strengthening and repair techniques for structures. Adequate and proper curing of such materials is essential to achieve its desired strength and durability. In this study, the curing process of epoxy type Sikadur 330 is monitored using the electromechanical impedance (EMI) technique employing smart piezoelectric material, namely Lead Zirconate Titanate (PZT). Experimental results show that the PZT transducer embedded into the epoxy in liquid state is effective in monitoring the initial setting and the subsequent hardening process of a lab scale epoxy coupon. Rightward movement of both the PZT resonance peaks and the structural resonance peaks indirectly reflects the curing process. This study shows that the EMI technique is capable of monitoring the curing process of the structural adhesive. Future study shall focus on developing theoretical model capable of parametrically relate the electrical signatures to the mechanical stiffness and strength.

1. Introduction

Civil engineering discipline has recently witnessed a rise in the use of structural adhesives, in particular Fibre-Reinforced Polymer (FRP) reinforcements in strengthening structural systems. Two-part structural epoxy adhesives are the most commonly used adhesive to bond FRP to other structural elements due to their advanced mechanical and chemical properties (Lettieri and Frigione 2012). The FRP materials can be bonded to various structural components including concrete, steel, and timber for repair and strength enhancement reasons.

In FRP strengthened joints materials, the mechanical performance of the joint is directly related to the strength and the stiffness of the epoxy adhesive, particularly during the early stages of installation while the epoxy is still curing and its mechanical properties are evolving. The performance of the FRP joint is strongly dependent on adequate preparation, application and curing of the epoxy resin.

Adhesives undergo various chemical transformations during the curing process. The freshly mixed liquid adhesive converts into a rubber (gelation) and then into a solid glass (vitrification). As the adhesive cures there is increased cross-linking between monomers, developing a progressively more complex polymeric network. For a structural adhesive to reach its targeted bond strength, the necessary curing time is highly dependent on the environmental conditions, particularly temperature and moisture (Czaderski et al 2012). In general, lower curing temperature results in slower curing process, resulting in decelerated rate of development of the epoxy resins mechanical properties. For practical and economic considerations, cold-curing adhesives are more commonly used in the civil engineering discipline.

Monitoring the curing process of adhesive is therefore significant to ensure structural integrity. In this paper, a novel study is conducted to investigate the feasibility of using the smart piezoelectric based electromechanical impedance technique (EMI) in monitoring the curing process of structural adhesive. The advent of smart based health monitoring technique, such as the EMI technique, which employs piezoelectric transducer (PZT) as active sensor shows distinct superiority over its conventional counterpart. It has the potential to provide autonomous, real-time and remote monitoring at relatively low cost with large scale application. The PZT patches can be conveniently attached to various structures as they are generally non-intrusive. With appropriate protection such as using silicone rubber, the PZT transducer is fairly durable and remains operational for years (Yang, Lim & Soh 2008). The PZT patches used in the EMI technique can be installed at various critical structural members for monitoring the structural health. This technique utilizes the transduction capability of PZT to convert electrical energy to mechanical energy and vice versa. With the PZT transducer attached on a host structure, an alternating voltage of varying frequency is applied by an impedance analyser across the poling direction of the transducer. The transducer can thus be excited (converse effect of piezoelectricity) and correspondingly actuates the host structure. On the other hand, the structural vibratory response interacts with the PZT transducer, thus modulating the current passing through the transducer (direct effect of piezoelectricity).
The modulated current, in terms of complex electrical admittance, can simultaneously be measured by the impedance analyser. In other words, the admittance signatures contain information pertinent to the modes of vibration of the host structure. Any changes in the vibratory behaviour of the host structure caused by curing of concrete (Tawie and Lee 2011; Lim et al. 2016), presence of damage (Lim and Soh 2011; Soh and Bhalla 2000), change in stress (Lim and Soh 2013) or other changes in mechanical properties can therefore be reflected in the admittance signatures.

2. Experimental Setup

The curing process of a commercially available two parts epoxy, namely Sikadur 330 (Sika, 2014) was monitored in this experiment. Epoxy was first mixed according to the specification and placed into standard mould. Three identical coupon specimens were prepared. The specimens were denoted as S330_1, S330_2 and S330_3, respectively. One PZT patch, dimensioned 10mm x 10mm x 0.3mm, was immediately embedded into the epoxy while it is still in its liquid state. A typical epoxy coupon is shown in Figure 1(a). All specimens were cured inside the laboratory with stable temperature (24^0C).

Acquisition of signatures was immediately initiated after the PZT patches were embedded to start monitoring the curing process. An impedance analyser (Hioki IM3570) as shown in Figure 1(b) was used to apply a 5V alternating voltage across the PZT transducer and simultaneously record its electrical admittance. A wide range of frequency (5 kHz – 900 kHz) was considered to capture the changes in vibratory behaviour of both the host structure and the PZT patch. Signatures acquisition was performed periodically, one PZT patch at a time. In the first 24 hours, signatures were acquired every hour as the epoxy was undergoing rigorous curing. The interval was reduced to once per day for the next 7 days as the progress of curing retarded considerably. After 7 days, signatures were recorded every other day up to 21 days. The changes in signatures has practically halted after 21 days as the epoxy was fully cured and no further changes in electrical signatures were observed.

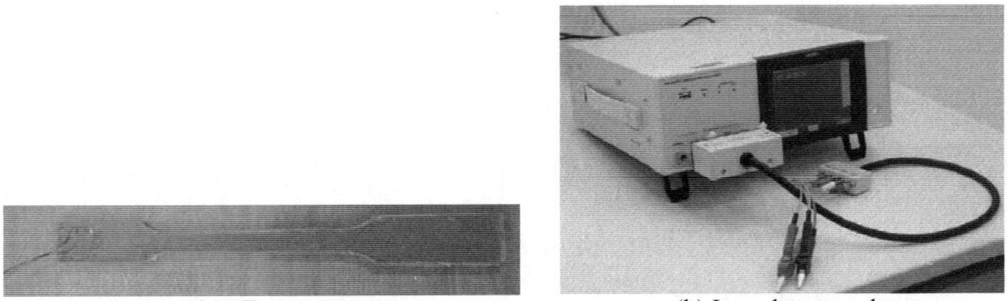

(a) Epoxy coupon (b) Impedance analyser

Figure 1. Pictorial representation of experimental setup for curing monitoring of epoxy using EMI technique.

3. Repeatability of signatures – freely suspended PZT patches

Typical plot of conductance signatures (real component of electrical admittance) versus frequency spectrum of four freely suspended PZT patches under sinusoidal actuation is schematically presented in Figure 2. Note that the four dominant peaks represent the resonances of the PZT patch. Overlapping of the resonance peaks throughout the entire frequency range suggests that the repeatability of the signatures obtained from different PZT patches is very high.

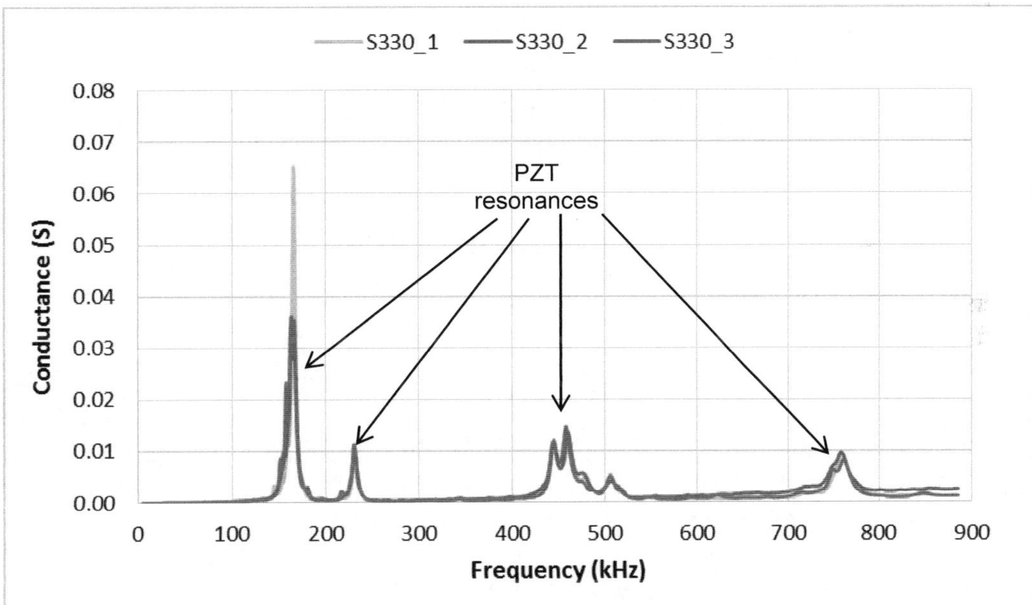

Figure 2. Conductance signatures versus frequency acquired from 4 freely suspended PZT patches (0 – 900 kHz).

4. Monitoring of Curing Process of Epoxy

4.1 PZT Peaks as Indicator

4.1.1 Monitoring of admittance signatures: 1 - 6 hours

The first PZT resonance peak (between 100 and 200 kHz) of Specimen S330_2 is herein examined. In the first 6 hours, the magnitude of the peak reduces rapidly, as shown in Figure 3. The resonance frequency is also gradually reducing, as shown from the leftward movements of the peak with time.

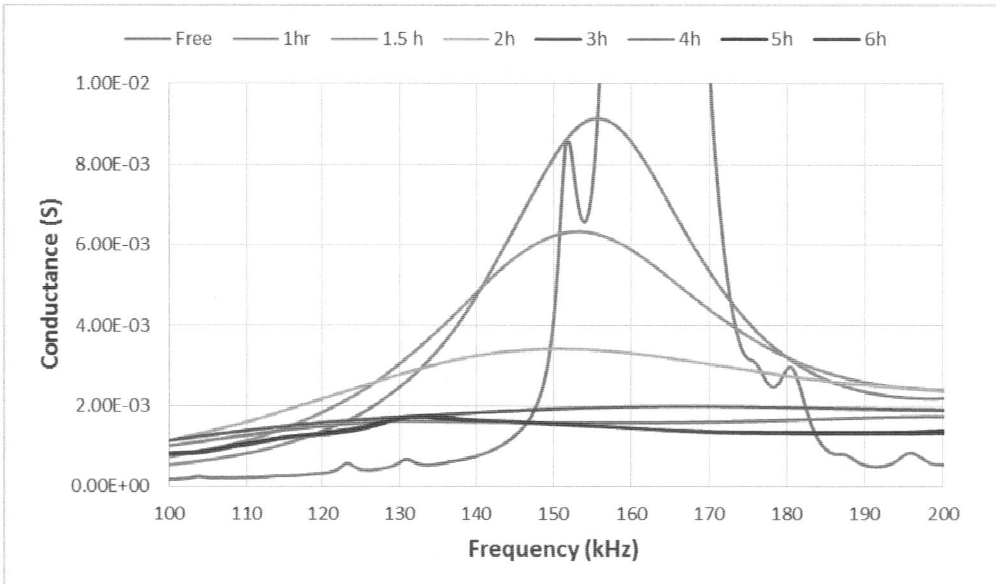

Figure 3: Conductance signatures versus frequency acquired from specimen S330_2 in the first 6 hours (300 – 700 kHz).

This phenomenon can be easily related to the physical process. When the PZT patch is first immersed into the epoxy, the epoxy is in its liquid state. The magnitude of the resonance peak of the PZT patch is significantly dampened by the viscosity of the liquid. The resonance frequency of the peak is also reduced indicating that the vibration of the PZT patch is being gradually restrained, indicating that it can no longer vibrates independently (Yang, Lim & Soh 2008).

This process continues to occur within the first 4 hours, to an extent that the peak almost disappears after 3 hours. The PZT's peak, however, reappears after 5 hours.

It is also worth mentioning that, unlike in the first 5 hours, the magnitudes of all the PZT peaks thereafter remain largely constant, indicating that the damping effect caused by the epoxy has reached a stable state.

4.1.2 Monitoring of admittance signatures: 6 – 24 hours

Figure 4(a) depicts the changes of the first PZT resonance peak from 6 hours to 24 hours and thereafter on Specimen S330_2. From 6 to 8 hours, there is a rapid rightward movement of the resonance peak, indicating that the host structure is gaining significant stiffness. Gain in stiffness can be indirectly related to increase in strength. The rightward movement of the resonance peak continues to occur after 8 hours, but at a much lower rate, as summarised in Figure 5a. From the shape of the curve, one could find that all three specimens exhibit very similar trend of increase.

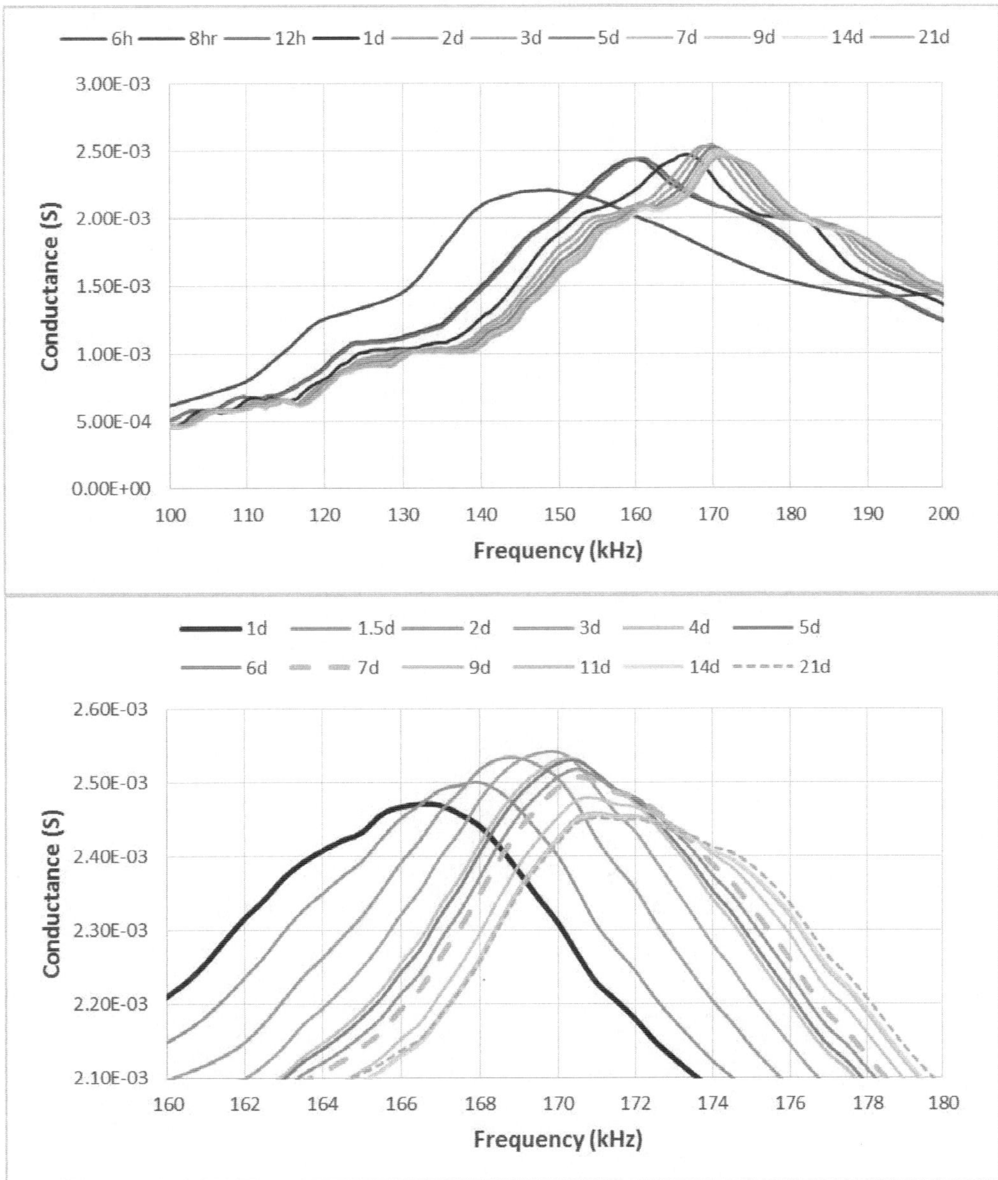

(a)

(b)

Figure 4: Conductance signatures versus frequency acquired from specimen S330_2 after the first 6 hours
(a) 100 – 300 kHz (b) 160 – 180 kHz

4.1.3 Monitoring of admittance signatures: 1 – 21 days

After one day of curing, the rate of rightward movement of the peak further decreases, as shown in Figure 4(a), where all the peaks after 1 day are nearly overlapping each other. Movement of the peaks can still be clearly identified by zooming into a narrower frequency range (160 – 180 kHz), as shown in Figure 4(b). The change in resonance frequency of the first PZT peak for all three specimens are summarised in Figure 5(b). Moderate increase in frequency occurs from day 1 to 4, and tapers thereafter. Note that peak movement from day 14 to 21 can be considered negligible, indicating that the curing process practically ends at approximately 14 days.

Once again, the trend of increase is consistent among all three specimens but the absolute values are different. This is likely to be caused by the slight variations in geometrical shape of the specimens and boundary conditions of the surrounding mould, as the admittance signatures are extremely sensitive to these local variations (Lim and Soh, 2012).

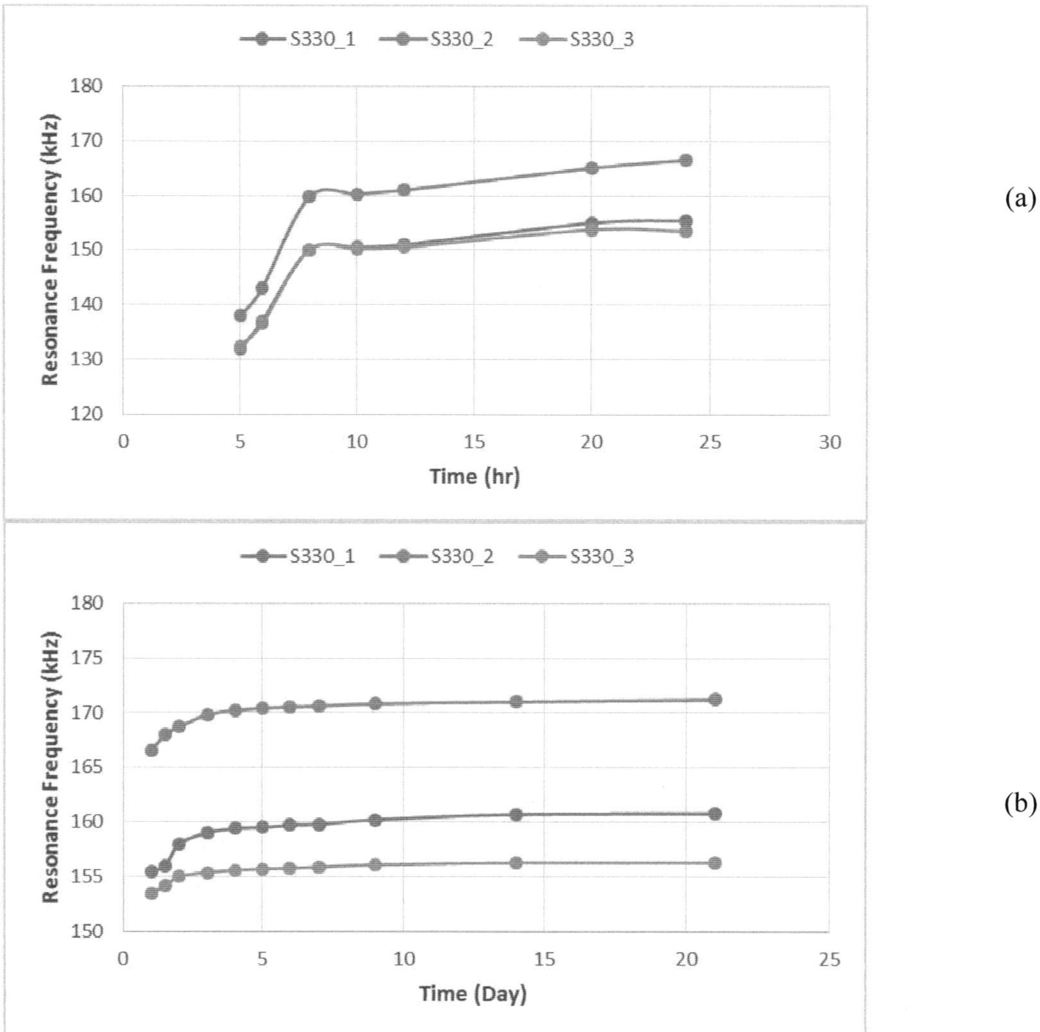

Figure 5: Resonance frequency of first PZT peak versus time of specimen S330_1, S330_2 and S330_3 after the first 6 hours

(a) first 5 – 24 hours (b) 1 – 21 days

4.2 Host Structural Peaks as Indicator

Previous study (Lim et al. 2011) shows that the densely spaced resonance peaks occurring in the lower frequency range (<100 kHz) are better representation of the vibratory behaviour of the host structure. These peaks are often used to observe the variation of host structural resonances (vibratory behaviour) for the purpose of health monitoring.

In this study, no obvious structural peak can be observed in the first 4 hours, as shown in Figure 6. This indirectly reflects that the epoxy is still in a "liquid" form as significant vibration of the host structure is non-existence. Note that the change in slope of the curves are mainly caused by the movement of the first PZT peaks, which does not affect the host structural behaviour.

However, after 5 hours, trace of resonance peaks can be found, as shown from the curves of "5hr" and "6hr", indicating that the epoxy is hardening and transforming into solid. This observation agrees well with its counterpart – PZT's peak (Section 4.1). In addition, referring to the manufacturer's product data sheet (Sikadur 330, 2014), the "tack free time" of Sikadur 330 is approximately 4 – 5 hours, after which, the epoxy will gain significant stiffness. This outcome proves that the EMI technique is very sensitive to the hardening process and can be a useful indicator of the initial hardening.

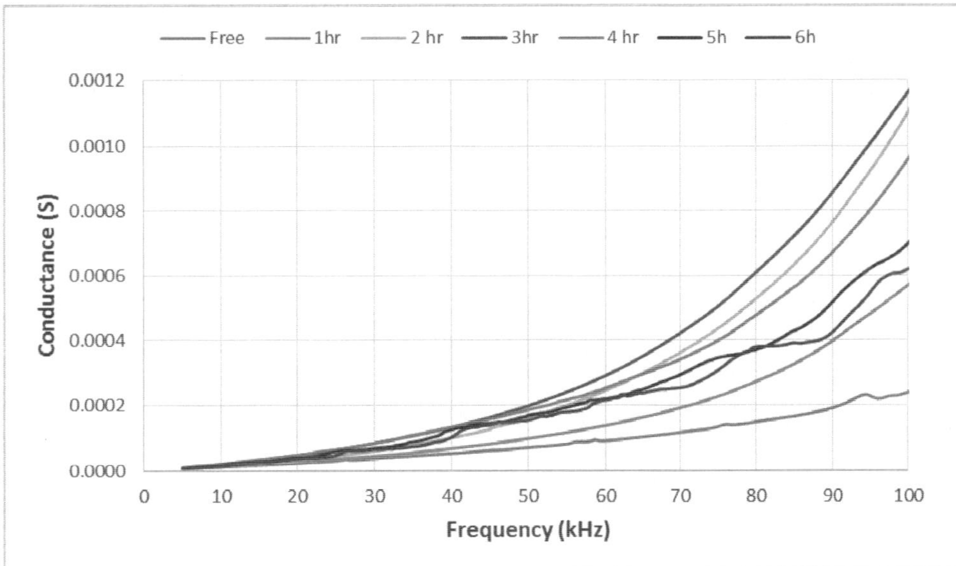

Figure 6: Conductance signatures versus frequency acquired from specimen S330_2 in the first 6 hours (0 – 100 kHz).

From 8 hours onwards, stronger and clearer structural peaks are formed which can be visually identified, as shown in Figure 7(a). Similar to the behaviour of the PZT peak as presented in previous section, the structural resonance peaks move progressively to the right (increase in resonance frequency) as curing progresses. The rate of movement, once again, indirectly reflects the rate of curing, as shown in Figure 7(b).

Thus, both structural resonance peaks and PZT resonance peaks are useful indicator of the curing process of epoxy. The PZT peaks are especially useful in characterising the early curing (first 1-6 hours), including the initial setting of the epoxy.

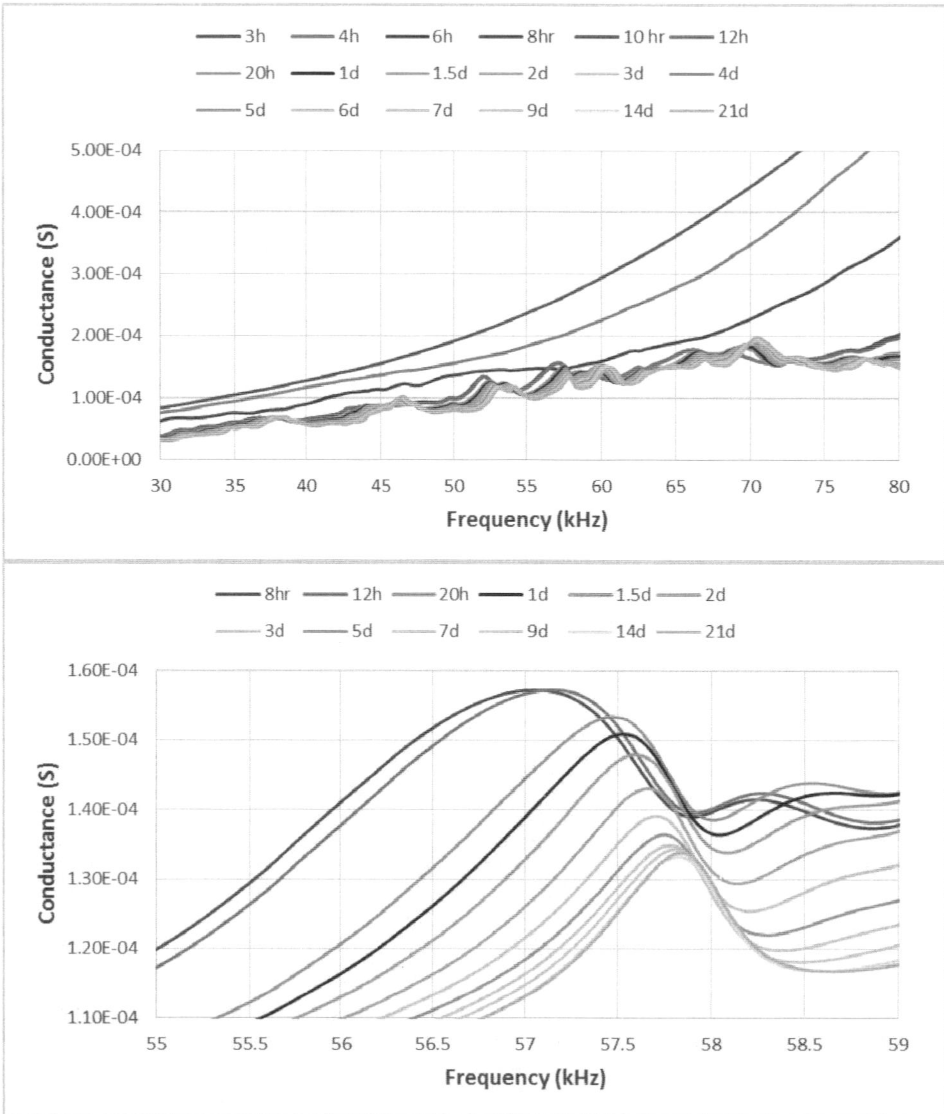

(a)

(b)

Figure 7: Conductance signatures versus frequency acquired from specimen S330_2 showing the structural peaks
(a) 30 – 80 kHz (b) 55 – 59 kHz

5. Conclusion

In this study, the curing process of epoxy Sikadur 330 is monitored using the EMI technique. Experimental results show the PZT transducer is effective in monitoring the initial setting and the subsequent curing process of lab scale epoxy coupons. Rightward movements of both PZT resonance peaks and structural resonance peaks are good representation of the hardening process. Future study can focus on correlating the changes in electrical signatures to the strength or mechanical properties of the epoxy, such as Poisson's ratio and elastic modulus for real-life application. Development of theoretical model is highly desirable. The use of statistical quantification approaches such as Root Mean Square Deviation (RMSD) can also be attempted.

6. References

Czaderski, C, Martinelli, E, Michels, J, & Motavalli, M 2012, 'Effect of curing conditions on strength development in an epoxy resin for structural strengthening', *Composites: Part B,* vol. 43, pp. 398-410.

Lettieri, M, & Frigione, M 2012, 'Effects of humid environment on thermal and mechanical properties of a cold-curing structural epoxy adhesive', *Journal of Construction and Building Materials*, vol. 30, pp. 753-760.

Lim, YY, Kwong, KZ, Liew, WYH & Soh, CK 2016, 'Non-destructive concrete strength evaluation using smart piezoelectric transducer—a comparative study', *Smart Materials and Structures*, vol. 25, no. 8, 085021.

Lim, YY & Soh, CK 2011, 'Fatigue life estimation of a 1D aluminum beam under mode-I loading using the electromechanical impedance technique', *Smart Materials and Structures*, vol. 20, no. 12, 125001.

Lim, YY & Soh, CK 2012, 'Effect of varying axial load under fixed boundary condition on admittance signatures of electromechanical impedance technique', *Journal of Intelligent Material Systems and Structures*, vol. 23, no. 7, pp. 815-826.

Lim, YY & Soh, CK 2013, 'Damage detection and characterization using EMI technique under varying axial load', *Smart structures and systems*, vol. 11, no. 4, pp. 349-364.

Moussa, O, Anastasios, P, de Castro, J & Keller, T 2012, 'Time-temperature dependence of thermomechanical recovery of cold-curing structural adhesives', *International Journal of Adhesion and Adhesives*, vol. 35, pp. 94-101.

Sika 2014, *Product Data Sheet - Sikadur 330 US*, Edition 9.23.

Soh, CK, Tseng, KK, Bhalla, S & Gupta, A 2000, 'Performance of smart piezoceramic patches in health monitoring of a RC bridge', *Smart materials and Structures*, vol. 9, no. 4, pp. 533.

Stewart, A 2012, *Study of Cement-Epoxy Interfaces, Accelerated Testing, and Surface Modification*, Ph.D. Dissertation, University of Florida, Gainesville, FL.

Tawie, R & Lee, HK 2011, 'Characterization of cement-based materials using a reusable piezoelectric impedance-basedsensor', *Smart Materials and Structures*, vol. 20, no. 8, 085023.

Yang, Y, Lim, YY & Soh, CK 2008, 'Practical issues related to the application of the electromechanical impedance technique in the structural health monitoring of civil structures: I. Experiment', *Smart Materials and Structures*, vol. 17, no. 3, 035008.

The 8th International Conference on
Structural Health Monitoring of Intelligent Infrastructure
Brisbane, Australia | 5-8 December 2017

Monitoring Blasting Events in an Underground Mine with Artificial Intelligence Techniques

Linqi Huang[1], Jun Li [2], Hong Hao [3], Xibing Li [4]
[1] Central South University – China [2] Curtin University - Australia, Email: junli@curtin.edu.au
[3] Curtin University – Australia [4] Central South University – China

Abstract

This paper proposes to use Convolutional Neural Network (CNN) to identify the Time Delay of Arrival (TDOA) and subsequently the source location of micro-seismic events. For any two sensor waveforms recorded from the same event, the cross wavelet transform power and phase spectra, and the corresponding output function values can be obtained. They will be treated as the input and output of the CNN model for training. The measured data from eight blasting tests in an underground mine are used to test the trained CNN model and identify the location of the conducted blasting test. The exact locations of these in-field blasting tests are available in prior and will be taken as the references for demonstrating the accuracy of the proposed approach. Results demonstrate the accuracy of using the proposed approach in identifying the in-field blasting event locations.

1. Introduction

Identifying the Time Delay of Arrival (TDOA) in the measured sensor waveforms accurately under micro-seismic (or blasting) events and the source locations in underground mines is challenging since the P-waves may not be clearly observed in weak signals with a low signal-to-Noise Ratio, and the first arrival wave may not necessarily be the P-wave. Tradition methods to identify TDOA include absolute and relative methods. The difference between the determined time instants of P-waves of two sensor waveforms is taken as TDOA in absolute methods (Akram and Eaton 2016). The relative methods measure the similarity of two waveforms to obtain TDOA (Earle and Shearer 1994).

Artificial Neural Networks (ANN) can also be used for identifying the arrival time instants based on the training with a number of data samples (Wang and Luo 2006, Kaur et al. 2013). The global and local characteristics of seismic signals can be used as the input features to the neural networks. Recently, deep learning based neural networks have attracted significant attention and been applied in various research areas in computer science, i.e. speech recognition, image processing and language processing, etc. Krizhevsky et al. (2012) trained a deep Convolutional Neural Networks (CNN) for image processing.

This paper proposes using CNN to identify TDOA and the source location of micro-seismic events. The power and phase spectra of cross wavelet transform between measured waveforms due to micro-seismic events are used as input data to the CNN model. The output would be TDOA between two waveforms. Monitoring data on-site will be used to train the designed CNN model. The identified TDOA will be used to calculate the source locations. In-field blasting tests will be conducted to identify TDOA and detect the source locations by using the proposed approach.

2. Cross Wavelet Transform

The discrete wavelet transform of a time domain signal *x(t)* can be expressed as the convolution of the signal with a mother wavelet function

$$W^x(u,s)=\sqrt{\frac{\delta t}{s}}\sum_{t=1}^{N}x(t)\psi_0\left[(t-u)\frac{\delta t}{s}\right] \tag{1}$$

in which u and s denote the translation and scale parameters, respectively; δt is the time step and N is the number of data points in the signal *x(t)*. Based on the above formulation, the cross wavelet transform of two signals *x(t)* and *y(t)* can be calculated as

$$W^{xy}(u,s)=W^x(u,s)W^{y*}(u,s) \tag{2}$$

where the symbol '*' indicates the complex conjugate. The cross wavelet spectrum power is defined as $\left|W^{xy}(u,s)\right|^2$. The complex argument $\arg(W^{xy})$ is considered as the local relative phase between *x(t)* and *y(t)* in the time frequency space.
Morlet wavelet with ω_0=6 is chosen as the mother wavelet function, which is expressed as

$$\psi_0(\eta)=\pi^{-1/4}e^{i\omega_0\eta}e^{-\frac{1}{2}\eta^2} \tag{3}$$

where ω_0 and η represent the frequency and time scales, respectively. When using cross wavelet transform to identify TDOA between two signals *x(t)* and *y(t)*, which are measured by two sensors S_1 and S_2 due to a blasting event in underground mines, their cross wavelet spectrum power shall reach the maximum between signals x_{i+k} (a delay with $k*\delta t$ time instants on the original signal) and y_i if TDOA between signals x_i and y_i is of k sampling steps (Huang *et al.* 2016).

3. Convolutional Neural Networks

Similar as a traditional ANN, CNN consists of multiple layers. Traditional neural networks define the full connectivity between neurons such that a high dimensionality with a large amount of weights is involved in the training. An outstanding advantage of CNN compared with ANN is that the concepts of local connectivity and shared weights are defined to significantly reduce the number of weights to be trained. These properties enable CNN to achieve a much higher efficiency in training the weights and lowering the requirement of computational memory for training and running the network (Haykin 2007).

3.1 Introduction

A typical architecture of CNN includes a number of layers that transform the input into the output. Three commonly defined types of layers are convolutional layer, down-pooling layer and fully connected layer. The convolutional layer consists of a set of learnable filters, which have a small receptive field but extend through the full range of the input. The number of neural network parameters for training can be significantly reduced by sharing the weights by all the neurons within an output feature map. Therefore the training process is more efficient. The role of the convolutional layer is mapping the input to the output feature map by computing the dot product between the entries of the filter and the input. The function of down-pooling layer is to further reduce the dimension of feature maps as well as the amount of parameters and computational load in the network. This can be used to control the over-fitting. The average or maximum function can be used in the down-pooling process. The final layer of a CNN model is usually a fully connector layer to have a final output.

3.2 Main Idea of the Proposed Approach

The cross wavelet transform spectrum power and phase of any two measured sensor waveforms under the blasting events in the underground mines are taken as the input of the CNN model. Feature extraction and compression are performed by the network. The architecture of the proposed CNN model is designed to extract the potential features of measured waves and recognise the pattern and similarity by examining the cross wavelet transform spectrum to identify TDOA. The flowchart of the proposed approach is shown in Fig. 1. Supervised training is employed in this study by using a feedforward/back propagation training scheme. The learning process is iteratively running until the maximum epoch number is reached to 100 or the validation error is smaller than 1.0×10^{-6}. The final obtained network parameters are used to calculate the output.

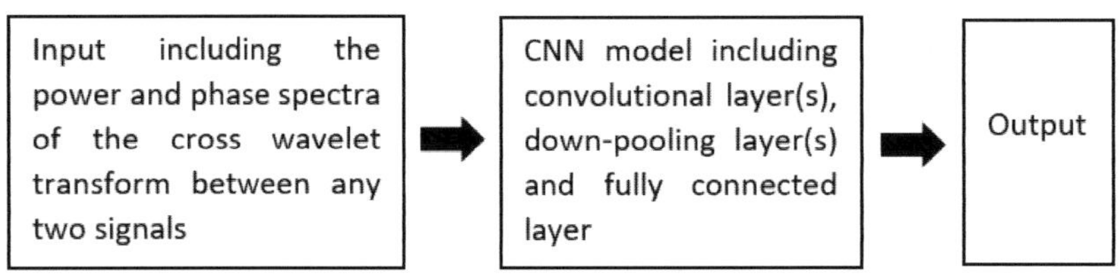

Fig. 1. The procedure of the proposed approach

4. Verfications with In-field Tests

To investigate the performance of using CNN for identifying TDOA and the source location of the blasting events, in-situ experimental blasting tests are conducted. TDOA and source locations are identified with the proposed approach.

4.1 In-field Site and Blasting Tests

In-situ blasting tests are conducted at the Yongshaba mine in Guizhou, China. The surface elevation of the phosphate mine is +1350 meters above the sea level, and the mining depth has reached to 700 m. The existence of more than twenty intensive faults and three dikes lead to the poor stability of Yongshaba mine, especially in the mining area under the Jin Yang highway, which is one of the key areas for micro-seismic monitoring. A digital micro-seismic monitoring system with 32 channels was installed in this underground mine. Institute of Mine Seismology (IMS) 14Hz borehole geophones with a frequency bandwidth of 8-2000Hz were used.
The site conditions, the placed sensors and blasting test locations are shown in Fig. 2. Two of them are tri-axial sensors, denoted as green T1 and T2, and the other twenty-six are single-axis sensors, which are numbered from 1 to 26. The exact coordinates of these sensors are available.

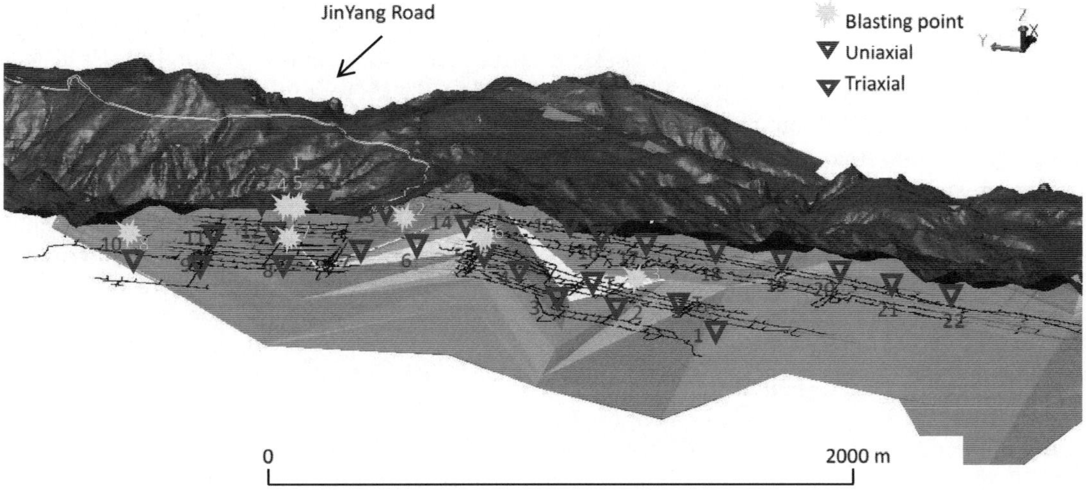

Fig. 2 The locations of placed sensors and blasting events

The recorded data from 10 micro-seismic events are used to train the network. The data from the conducted blast tests are used as the testing case for the network to obtain TDOA between sensor signals. TDOA based source location method is then used to identify the location of the blast tests.

4.2 CNN Model Design

The designed CNN model for this problem is shown in Fig. 3. The architecture of the CNN model includes 5 layers, namely, 2 convolutional layers, 2 down pooling layers and 1 fully connected layer. Convolutional layers are defined to extract the feature maps from the input data, and the down pooling layers are used to reduce the dimensions of extracted features and control the over-fitting. The fully connected layer is used to transform the final high abstract feature vector to the final output. The feature maps from the second down pooling layer will be re-organised as a feature vector to the fully connected layer.

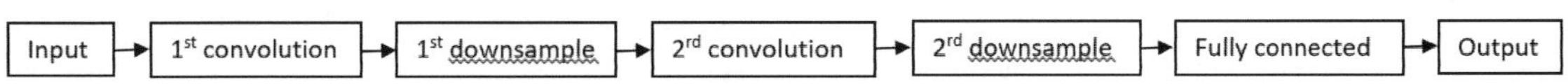

Fig. 3 Designed CNN model in this study

4.3 Network Training

Monitoring data of 10 micro-seismic events recorded by 10 sensors, i.e. 1, 2, 3, 4, 8, 9, 12, 17, 18 and 22, are used for training the network. For two signal segments from any two sensors (defined as segment 1 and segment 2), it should be noted that segment 1 is set as constant and segment 2 is moved with five cases of pre-determined step lengths in a specific range from -50/6000s to +50/6000s. This range is defined based on the pre-pick up statistics results covering the true TDOA between two waveforms. By defining a pair of signal segments from any two sensors, a number of input data samples can be obtained with the defined different sliding windows of different sliding time step lengths.
The labelled output is defined as

$$f(t)=\exp(-\rho|t-T_D|) \tag{4}$$

where t is the sliding time step lengths, T_D is the baseline TDOA which is extracted by the specialist at Central South University. ρ is an empirical parameter, and is defined as 1000 in this study. A larger value ρ will make the shape of the function $f(t)$ sharper with a higher resolution, but it will be more difficult to train the network model.
For any two waveforms recorded from the same event, if delay t is given, the cross wavelet transform power and phase spectra, and corresponding $f(t)$ values can be obtained. They will be treated as the input and output of the CNN model. Regarding the testing stage with in-field data, for any two waveforms recorded from the same event, the curve of $f(t)$ can be obtained as the output of the CNN model with a series of time delays. T_D is the time delay which corresponds to the maximum of the output.

4.4 Investigation with In-field Data

The measured data from these eight blasting tests in an underground mine are used to test the trained CNN model and identify the location of the conducted blasting test. The exact locations of these in-field blasting tests are available in prior and will be taken as the references for demonstrating the accuracy of the proposed approach. Identified TDOA between any two measured sensor waveforms under the in-field blasting tests will be used to predict the location of the blasting events. The errors between the identified and the exact locations will be calculated to evaluate the performance of the proposed approach. Table 1 shows the absolute errors in the source locating accuracy of the in-field blasting tests. Results demonstrate the accuracy and performance of using the proposed approach in identifying the in-field blasting event locations.

Table 1. Absolute errors (m) in the identified blasting event locations

No.	The proposed approach
1	5.62
2	9.85
3	9.16
4	12.23
5	12.68
6	7.23
7	10.55
8	6.94

5. Conclusions

In this paper, identification of TDOA and blasting event locations in an underground mine is conducted by using CNN. The architecture of the designed CNN model is able to extract the potential features of measured waves and recognise the pattern and similarity by examining the cross wavelet transform spectrum to identify TDOA. In-field blasting tests at a phosphorite mine in Guizhou, China are conducted to investigate the accuracy of using CNN for identifying TDOA and the source location. A good accuarcy in locating these blasting events is achieved.

Acknowledgements

The work described in this paper was supported by National Natural Science Foundation of China projects No. 11472311, 41272304 and 51504288.

References

Akram J & Eaton DW 2016, 'A review and appraisal of arrival-time picking methods for downhole microseismic data', *Geophysics*, vol. 81, no. 2, pp. KS71-KS91.

Earle PS & Shearer PM 1994, 'Characterization of global seismograms using an automatic-picking algorithm', *Bulletin of the Seismological Society of America*, vol. 84, no. 2, pp. 366-376.

Haykin S. 2007, Neural Networks: A Comprehensive Foundation (3rd Edition). Prentice-Hall, Inc.

Kaur K, Wadhwa M & Park EK 2013, 'Detection and identification of seismic P-Waves using Artificial Neural Networks', *The 2013 International Joint Conference on Neural Networks*, Dallas, TX, USA, pp.1-6.

Huang L, Hao H, Li X & Li J (2016), 'Micro-seismic monitoring in mines based on cross wavelet transform', *Earthquakes and Structures*, vol. 11, no. 6, pp. 1143-1164.

Wang JF & Luo SX (2006), 'The improved BP neural network and its application in seismic first break picking', *Computing Techniques for Geophysical and Geochemical Exploration*, vol. 28, no. 1, pp. 14-17.

Krizhevsky A, Sutskever I & Hinton GE (2012), 'ImageNet classification with deep convolutional neural networks', *Advances in Neural Information Processing Systems*, pp. 1097-1105

The 8th International Conference on
Structural Health Monitoring of Intelligent Infrastructure
Brisbane, Australia | 5-8 December 2017

Crack Detection via Salient Structure Extraction from Textured Background

F. Nayyeri [1], L. Hou [2], J. Zhou [1], H. Guan [2], A.W.C. Liew [1].
[1] *School of Information and Communication Technology, Griffith University, Nathan – Australia* [2] *School of Engineering, Griffith University, Gold Coast – Australia, Email:* lei.hou@griffith.edu.au

Abstract

A reliable crack detection system is essential for automatic safety inspection of infrastructures such as roads and bridges. Queensland's population is increasing, putting greater pressure on our already aging civil infrastructure. To get the most out of government investment in our State's thousands of bridges, we need to find quicker, cheaper and more reliable ways to assess and maintain them. In this paper, we propose a novel method for crack detection via salient structure extraction from textured background. This method contains two key steps. In the first step, we extract strong edges and distinguish them from strong textures in a local neighborhood via a relative total variation approach. In the second step, the spatial distribution of texture features are calculated so as to detect cracks as salient structures that are not widely spread across the whole image. The outputs from these two steps are fused to calculate the final structure saliency map which is then binarised to generate the crack masks. This method was evaluated on a crack dataset with images collected from the Internet. Comparison with several alternative approaches shows the superior performance of our method.

1. Introduction

Currently, bridge inspections and investigations are conducted manually; requiring inspectors to get within touching distance of all components and manually record the condition using paper-based methods. This is expensive, time consuming and can be dangerous. However, in a country still adjusting to changes in economic drivers and limited capital expenditure for new assets, the Government needs to maximise its investment in public works while reducing risks (Alderman & Yemtsov 2014). Apart from risks, over 75% of bridge costs are incurred during the operations and maintenance phase. Improving current asset management processes can have a significant impact on infrastructure spending. Based on lessons learnt from the Pyrmont Bridge project in Sydney (Chan et al. 2016), the industry is in need of changes towards more autonomous and intelligent inspections system. Migrating to digital processes has driven a more proactive approach to asset management (Kassem et al. 2014). This highlights the cost savings that may be afforded through adopting 'image acquisition' technologies to analyse and detect cracks for a large asset inventory. The ability to efficiently, effectively and consistently track the rate of deterioration of bridges and similar structures will radically change the state of bridge inspectors - engineers will be able to more effectively identify the rate and extent of deterioration. To address the need, this paper is to introduce greater efficiency and address the increasing backlog of inspections and maintenance activities.

2. Literature Review

Cracks are significant linear structures over textured surfaces of infrastructures made of brick, concrete, or asphalt. Reliable crack detection is essential for the safety inspection of roads and bridges. Although understanding and differentiating cracks from the background is an effortless process for human, it is much more challenging for the computer.

During the past years, some efforts have been made on crack detection by exploring various image processing and pattern recognition techniques. Earlier work attempted to use curvilinear features, for example Sobel algorithm (Ayenu-Prah & Attoh-Okine 2008) and Canny edge detector (Zhao et al. 2010), to extract edge like features. Some methods extracted more advanced features, such as wavelet (Zhou et al. 2006) and saliency (Xu et al. 2013). Oliveira and Correia (2013) proposed an unsupervised crack detection method based on learning from some samples to classify non-overlapping image blocks as either containing crack pixels or not. Then they characterized the type of crack according to the Portuguese Distress Catalog. However, this method cannot deal well with thin cracks. Avila et al. (2014) and Amhaz et al. (2016) used the minimum path to detect cracks if this minimal path follows crack pixels. A combined 2D/3D road crack detection approach was presented in (Medina et al. 2014) which used both visual appearance and geometrical information for crack detection. The first deep learning based method was introduced by Zhang et al (2016) who classified each image patch as a part of crack or background. This deep learning framework provides better crack detection performance compared with traditional feature extraction methods.

In this paper, we propose a novel method for crack detection via salient structure extraction from textured background. This method contains two key steps. In the first step, we extract strong edges and distinguish them from strong textures in a local neighbourhood via a relative total variation approach. In the second step, the spatial distribution of texture features are calculated so as to detect cracks as salient structures that are not widely spread across the whole image. The outputs from these two steps are fused to calculate the final structure saliency map which is then binarised to generate the crack masks.

The novel contribution of this paper comes from two aspect. First, we present crack detection as a structure extraction problem. The goal is to treat cracks as prominent foreground structures and distinguish them from background surface textures. Second, we propose to model the global distribution of textures. The background textures are those widely distributed over the whole image. This will leads to a saliency map that can be combined with the local structure distribution for more accurate crack detection.

The rest of this paper is organized as follows. Section 3 introduces the proposed method with details on two key modelling steps. Section 4 presents the experiments on a road dataset and comparison with several alternative approaches. Finally, conclusions are drawn in Section 5.

3. Crack Detection Method

In this section, we describe the proposed method in details. The structure and texture features are extracted at each pixel via a sliding window. Their local property and global distribution are then calculated.

3.1 Local Structure Extraction

The goal of this step is to extract strong structures or edges and distinguish them from background textures. A total variation method can be used for this purpose, which preserves strong edges and suppresses background noises (Rudin et al. 1992). This method was later extended by Xu et al. to cope with the ambiguity between structure and strong textures (Xu et al. 2012) via inherent variation

and relative total variation (RTV) measures. Motivated by this work, we propose to use RTV to calculate the structure feature map for cracks.
Given a crack image I of N pixels, for each pixel $\mathrm{I}(i)$, a local neighbourhood patch N_i can be extracted for local texture analysis. The local total variation measures can be calculated in both horizontal (h) and vertical (v) directions as follows

$$TV_h(i) = \sum_{\mathrm{I}(j)\in\mathrm{N}_i} s(i,j)\,|\frac{\partial\mathbf{I}(j)}{\partial x}| \tag{1}$$

$$TV_v(i) = \sum_{\mathrm{I}(j)\in\mathrm{N}_i} s(i,j)\,|\frac{\partial\mathbf{I}(j)}{\partial y}| \tag{2}$$

$s(i,j)$ is an affinity function defined on the spatial Euclidean distance $ds(i,j)$ between pixels i and j in N_i , such that

$$s(i,j) = \exp\left(-\frac{ds(i,j)^2}{2\sigma^2}\right) \tag{3}$$

where σ controls the contribution of the distance function. Total variation takes the direction of gradients into account. In order to capture the coincident of gradients, inherent variation can be defined as

$$IV_h(i) = |TV_h(i)| \tag{4}$$

$$IV_v(i) = |TV_v(i)| \tag{5}$$

Combining the total variation and the inherent variation, the RTV can be calculated as

$$RTV(i) = \frac{TV_h(i)}{IV_h(i)+\epsilon} + \frac{TV_v(i)}{IV_v(i)+\epsilon} \tag{6}$$

where ϵ is introduced to avoid zero inherent values. Then the foreground structure map $\mathbf{S}$ can be estimated by optimizing the following objective function

$$\arg\min \sum_i (\mathbf{S}_i - \mathbf{I}_i)^2 + \lambda RTV(i) \tag{7}$$

The solution to this optimization problem is provided in (Xu, Yan et al. 2012). The output of this step is a structure map that shows the extracted cracks and suppressed background texture.

3.2 Global Texture Distribution

In this step, we aim to calculate the spatial distribution of textures across the image. The motivation is that background textures, i.e., those formed by road materials, will be widely distributed over the whole image, but the cracks only appear in some parts of the image.
In order to calculate the distribution of textures, we first quantize the texture patterns using Bag-of-Words model. Given vectorised local patches $\mathbf{n}_i(i = 1,\cdots,P)$ at pixel i, where P is the total number of pixels, we use the K-means clustering algorithm to cluster the vectorised image patches into K clusters. Then the probability that a patch $\mathbf{n}_i$ is assigned to cluster center $\mathbf{c}_k(k = 1,\cdots,K)$ is calculated as

$$p(\mathbf{c}_k \mid \mathbf{n}_i) = \frac{1}{Z}\exp(-\frac{dv(\mathbf{c}_k,\mathbf{n}_i)^2}{2\sigma^2}) \tag{8}$$

where $dv(\cdot,\cdot)$ is the distance between two vectors, and Z is a normalization factor. In this way, we get K texture distribution maps.
After that, we calculate the horizontal and vertical spatial variances of pixels for each texture distribution map. The horizontal spatial variance $V_h(k)$ and the vertical spatial variance $V_v(k)$ of texture distribution map k is calculated as follows

$$V_h(k) = \frac{1}{N}\sum_{i} p(\mathbf{c}_k \mid \mathbf{n}_i)x_i \tag{9}$$

$$V_v(k) = \frac{1}{N}\sum_{i} p(\mathbf{c}_k \mid \mathbf{n}_i)y_i \tag{10}$$

where x_i and y_i are the horizontal and vertical coordinates of the pixel i, respectively. The spatial variance of map k is

$$V(k) = V_h(k) + V_v(k) \tag{11}$$

which is then normalized to [0, 1] via

$$V(k) \leftarrow \frac{V(k) - \min_k V(k)}{\max_k V(k) - \min_k V(k)} \tag{12}$$

Finally, the global texture spatial distribution feature $\mathbf{F}(i)$ is defined as a weighted sum

$$\mathbf{F}(i) \propto \sum_{i} p(\mathbf{c}_k \mid \mathbf{n}_i)(1 - V(k)) \tag{13}$$

The spatial distribution map **F** is also normalized to the range [0, 1].

3.3 Fusion Step

Given the generated local foreground structure map **S** and the global texture spatial distribution map **F**, the final saliency map for the cracks can be defined as

$$\mathbf{C} = (1 - \alpha)\mathbf{S} + \alpha\mathbf{F} \tag{14}$$

where α is a parameter that controls the contribution of two components. When α=0, the final saliency map is identical to the local foreground structure map **S**. On the contrary, when α=1, the final saliency map is purely contributed by the global texture spatial distribution map **F**. The best saliency detection can be achieved when both **S** and **F** make appropriate contribution by setting α to be a value between 0 and 1. We will discuss the influence of this parameter in the experiment section.

As the final step, we convert the saliency map into a binary crack map. Considering that the saliency map contains both foreground salient cracks and some noisy background pixels, we use Otsu's method (Otsu 1975) to calculate an optimal threshold for the binarisation. The goal is to minimize the within-class variance and maximize the between-class variance.

3.4 Implementation Details

In processing the crack images, we first convert colour images into grayscale images. In calculating the local structure information and clustering texture features, we extracted a spatial 3×3 neighbourhood of each pixel. According to our experiments, this neighbourhood size generates the best crack detection outcome.

In the global texture distribution calculation step, the background textures are mostly grouped into the same clusters. Based on the assumption that cracks shall be salient features appearing in limited spatial locations, we can save computational cost by removing those clusters with many assigned pixels if hard assignment is adopted. For this reason, we add a hard assignment step, and remove those clusters from further consideration if the total number of pixels assigned to the cluster is larger than a threshold.

In Fig. 1, we show 10 texture spatial distribution maps generated from an example crack image. It can be seen that in some maps, the cracks information are prominent, while several maps mainly show the distribution of background textures.

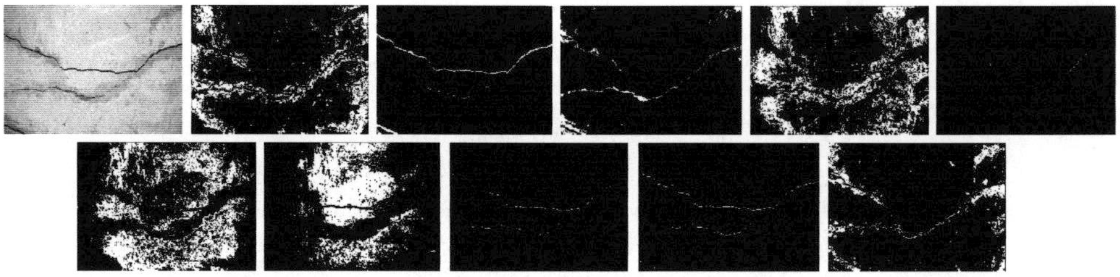

Fig. 1. An example crack image and its 10 texture spatial distribution maps.

Fig. 2 illustrates the results of several key steps in our method on a sample image.

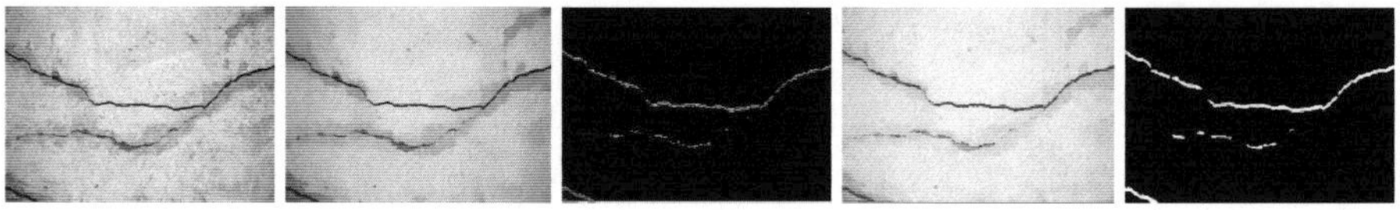

(a) (b) (c) (d) (e)

Fig. 2. (a) Sample crack image; (b) output of local structure extraction step; (c) output of global texture distribution step; (d) result after fusion; (e) binarisation result.

4. Experimental Results

First of all, we analysed the effect of α in equation (14). This parameter controls the contribution from both local and global features. Figure 3 shows that the best performance is achieved when $\alpha = 0.5$.

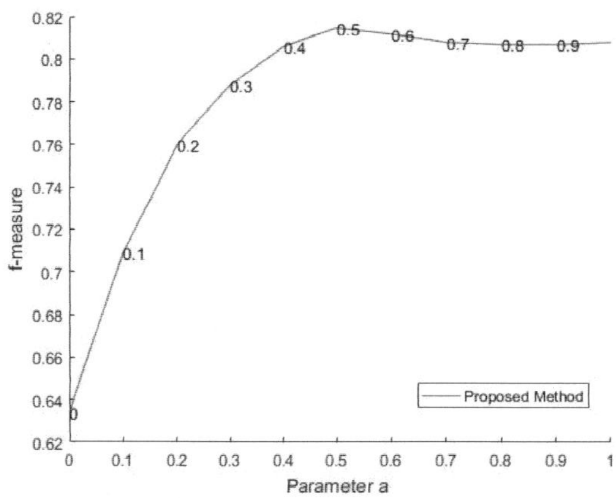

Fig. 3. Influence of the fusion parameter.

This chart shows that the highest f-measure is acquired by when both global and local features make equal contribution to the final saliency map calculation. This is natural as the global and local features characterise two distinctive properties of the distribution of cracks in the image. The global feature tells that crack cannot widely present everywhere in the image as the background concrete texture does. The local feature extracts strong structures or edges, which are very different from the background road textures. They work together to produce reliable detection of cracks.

In this section, we present the experimental results of our method on a road crack dataset with images collected from the Internet. The dataset contains 115 images in different resolutions and aspect ratios, and contains roads made of concrete or asphalt. We manually labelled the location of cracks for the quantitative evaluation purpose. Some sample images and their ground truth can be found in the first and second rows of Figure 4.

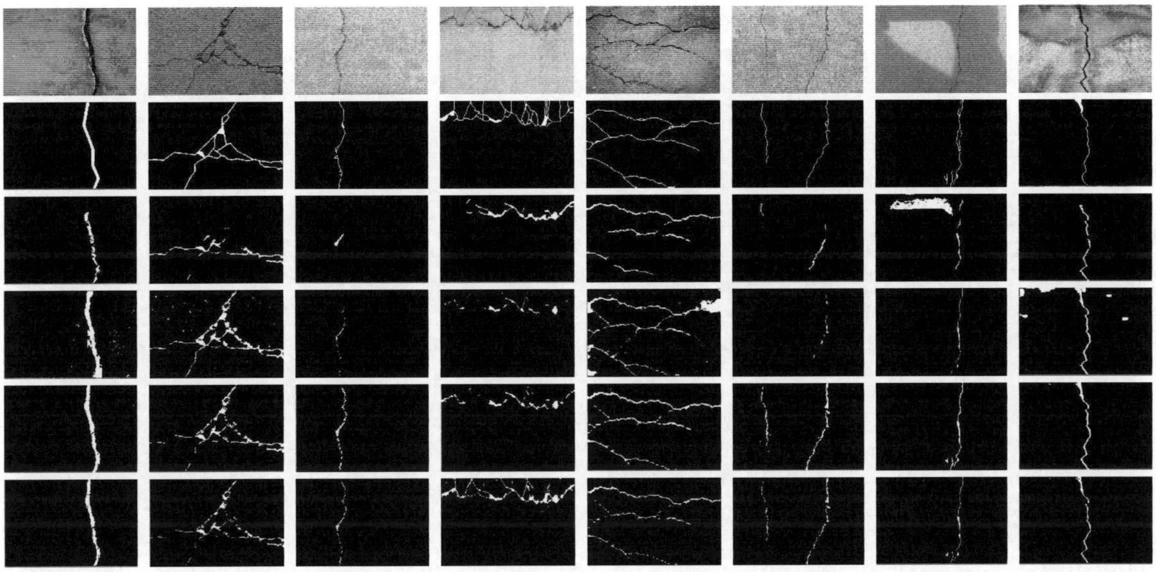

Fig. 4. Crack detection results. From up to down: original image, ground truth, results from saliency detection method (Xu et al. 2013), results from local structure extraction (Xu, Yan et al. 2012), results from global texture distribution, proposed method.

We compared our method with several alternative methods in the literature. These include the saliency detection method (Xu et al. 2013), the local structure extraction method (Xu, Yan et al. 2012) used as the first key step of our method with direct binarisation of structure output, and the global texture distribution method used as the second step of our method with direct binarisation of saliency map. For the methods in (Xu et al. 2013) and (Xu, Yan et al. 2012), we used the original codes provided by the authors for experiments. To evaluate the methods, we calculated precision, recall, and f-measure for each method.

We present the quantitative evaluation results in Table 1. This table shows that the proposed method has demonstrated superior performance than the alternative methods in all three evaluation criteria. The global texture distribution method generated the second best performance. This implies that the spatial distribution of textures is a very important feature to distinguish cracks from road textures. Textures generated by the road materials tend to appear across the whole image, therefore, cracks can be effectively detected by removing those widely distributed textures. Combining both local and global features, the cracks can be more accurately detected.

Table 1. Qualitative results of methods under comparison.

	Precision	Recall	F-measure
Saliency detection [4]	0.814	0.663	0.697
Local structure [11]	0.823	0.536	0.596
Global texture	0.855	0.751	0.790
Proposed method	0.859	0.817	0.815

We also show the qualitative results of eight images in Figure 3. It can be seen in the figure that saliency based method sometimes mistakenly extracts regions that are different from their neighbourhood, for example in the 5th image where road texture is slightly different from its neighbours, or in the 7th image where shadow presents. This method also misses many weak cracks. The local structure extraction method tends to include small noisy features or strong textures in the detection results. The results from global texture distribution method and the final fusion methods are quite close. The latter performs better in most cases by including more complete or thicker cracks.

5. Conclusions

We have presented a novel approach for crack detection. This method models cracks as structures in the textured background. The structure information is calculated via combined local and global information. The local structure extraction step distinguishes cracks from textured background. The global texture distribution step removes road textures that are widely distributed across the image. This methods is compared with several baseline approaches and proved to be very effective.

References

Alderman, H & Yemtsov, R 2014, 'How can safety nets contribute to economic growth?', *The World Bank Economic Review,* 28, 1-20.

Chan, B, Guan, H, Hou, L, Jo, J, Blumenstein, M & Wang, J 2016, 'Defining a conceptual framework for the integration of modelling and advanced imaging for improving the reliability and efficiency of bridge assessments', *Journal of Civil Structural Health Monitoring,* 6, 703-714.

Kassem, M, Kelly, G, Dawood, N, Serginson, M & Lockley, S 2015, 'BIM in facilities management applications: a case study of a large university complex', *Built Environment Project and Asset Management,* 5, 261-277.

Ayenu-Prah, A & Attoh-Okine, N 2008, 'Evaluating pavement cracks with bidimensional empirical mode decomposition', *EURASIP Journal on Advances in Signal Processing*, vol. 2008, no. 1, pp. 861701.

Zhao, H, Qin, G & Wang, X 2010, 'Improvement of canny algorithm based on pavement edge detection', *Image and Signal Processing (CISP), 2010 3rd International Congress on*, IEEE.

Zhou, J, Huang, P & Chiang, F 2006, 'Wavelet-based pavement distress detection and evaluation', *Optical Engineering*, vol. 45, no. 2, pp. 027007-027007-027010.

Xu, W, Tang, Z, Zhou, J & Ding, J 2013, 'Pavement crack detection based on saliency and statistical features', *2013 IEEE International Conference on Image Processing*, IEEE.

Oliveira, H & Correia, PL 2013, 'Automatic road crack detection and characterization', *IEEE Transactions on Intelligent Transportation Systems*, vol. 14, no. 1, pp. 155-168.

Avila, M, Begot, S, Duculty, F & Nguyen, T 2014, '2D image based road pavement crack detection by calculating minimal paths and dynamic programming', *Image Processing (ICIP), 2014 IEEE International Conference on*, IEEE.

Amhaz, R, Chambon, S, Idier, J & Baltazart, V 2016, 'Automatic Crack Detection on Two-Dimensional Pavement Images: An Algorithm Based on Minimal Path Selection', *IEEE Transactions on Intelligent Transportation Systems*, vol. 17, no. 10, pp. 2718-2729.

Medina, R, Llamas, J, Zalama, E & Gómez-García-Bermejo, J 2014, 'Enhanced automatic detection of road surface cracks by combining 2d/3d image processing techniques', *Image Processing (ICIP), 2014 IEEE International Conference on*, IEEE.

Zhang, L, Yang, F, Zhang, Y & Zhu, Y 2016, 'Road crack detection using deep convolutional neural network', *Image Processing (ICIP), 2016 IEEE International Conference on*, IEEE.

Rudin, LI, Osher, S & Fatemi, E 1992, 'Nonlinear total variation based noise removal algorithms', *Physica D: Nonlinear Phenomena*, vol. 60, no. 1-4, pp. 259-268.

Xu, L, Yang, Q, Xia, Yang & Jia, J 2012, 'Structure extraction from texture via relative total variation', *ACM Transactions on Graphics (TOG)*, vol. 31, no. 6, pp. 139.

Otsu, N 1975, 'A threshold selection method from gray-level histograms', *Automatica*, vol. 11, no. 285-296, pp. 23-27.

The 8th International Conference on
Structural Health Monitoring of Intelligent Infrastructure
Brisbane, Australia | 5-8 December 2017

Output-only modal testing and monitoring of civil engineering structures: Instrumentation and test management

A. Nguyen[1], T.H.T. Chan[1], D.P. Thambiratnam[1], K.T.L Kodikara[1], N.T. Le[1], S. Jamali[1]
[1] *School of Civil Engineering and Built Environment, Queensland University of Technology – Australia,*
Email: a68.nguyen@qut.edu.au

Abstract

Development of output-only modal testing and monitoring systems of real civil engineering structures albeit very appealing have been very challenging to engineers of any progression. This paper analyses some of the most critical issues that are needed for the success of implementing this approach to civil engineering structures. Illustrations are made based mainly on the applications developed by the present authors when conducting their research at the Queensland University of Technology (QUT). For illustration purposes, two representative test structures are selected to reflect different characteristics of structures. This paper then focuses in detail on instrumentation aspects since these are amongst the most confusing issues. All necessary components of complete instrumentation systems are analysed with different scenarios and structures. Last shared are some key notes on the test management process which is especially relevant to long-term modal-based monitoring applications.

1. Introduction

In recent years, output-only modal testing and monitoring have become more and more popular for use with civil engineering structures. The main advantages for this experimental approach lie in the fact that it can be used when the structure is under normal service conditions utilising natural ambient and operational loadings as the excitation source. Hence, normal operation of the structure can be fully maintained during the test, while the measured data recorded under in-service situations tend to better reflect the actual behaviour of structures. All of these make the output-only test approach a very convenient and cost-effective tool for emerging engineering areas such as structural health monitoring (SHM) and life-cycle management (LCM) of key civil engineering structures (Ansari and Karbhari, 2009; Frangopol, 2011).
However, developing a quality output-only modal testing or monitoring system for civil engineering structures is not straightforward for the engineer of any single profession. First, there are many general challenges that require the knowledge and skills from other engineering areas such as sensor and measurement, data acquisition (DAQ) and signal processing, rather than just structural engineering. The output-only modal test approach poses some distinct difficulties such as low-amplitude and lengthy acquisition of signal requiring higher care in transducer selection and channel synchronisation. Test structures in practice also vary considerably in their sizes and vibration characteristics and therefore may require different types of transducers and DAQ systems. This paper is therefore dedicated to providing in-depth analysis of some of the key techniques especially in the modal instrumentation and test management aspects because these tend to be the most challenging issues. Illustrations are made based on the applications made by the present authors when conducting their research at the Queensland University of Technology (QUT) with some dating back to 2010. A large number of examples are based on hardware and software modules from National Instruments (NI, www.ni.com) since NI systems are amongst the most

popular platforms for test and measurement applications. Nonetheless, there is an effort made to generalise the information as much as possible, so that an application using another platform is feasible.

The outline of the rest of the paper is as follows. Section 2 provides an overview of general test structure cases and the two illustrative structure cases. Detailed in Section 3 are the necessary components of complete instrumentation systems to suit different scenarios. Section 4 shares three key notes in the test management before concluding remarks are made.

2. Overview of two test structures

As mentioned before, understanding the main features of test structures and applications is crucial for the success in development and implementation of modal testing and monitoring systems. The most critical factors in this regard include (1) the scale or size of the structure, (2) the frequency range of interest, and (3) the type of system required for the structure (i.e. long-term monitoring or short-term testing). The last factor can be decided based on the importance of the structure e.g. located at an important public venue, or fast approaching the end of its service life. To further illustrate these points in the next sections, two different test structures are selected – one is a large-scale building with a long-term monitoring system while the other is a small-scale structure undergone a short-term modal testing series in the laboratory. More details of these two structures are presented below.

Representing real large-scale civil structures is the P block building located in the Gardens Point Campus of QUT, which is a very popular public venue in the city of Brisbane. Arguably the first benchmark real structure for long-term SHM in Australia, this building has been well used in recent SHM studies (Kodikara et al., 2017; Kodikara et al., 2016a; Kodikara et al., 2016b; Nguyen, 2014; Nguyen et al., 2014c; Nguyen et al., 2015). Structurally, the P block is a reinforced concrete (RC) frame structure with post-tensioned slabs and the total height of just over 40m from its first basement. The building has four semi-underground levels the footprint of which are approximately 75m x 65m. With some setback, the six upper floor levels have a smaller area of approximate dimensions of 65m x 45m. As an important public venue, P block has been instrumented with three permanent monitoring subsystems (Nguyen et al., 2015), but for the scope of this paper only components of the vibration monitoring system will be described further in the succeeding sections. The three main challenges of developing this monitoring system include (1) low frequency characteristics, (2) the large scale of the structure, and (3) demanding low-level ambient and operational excitation. Fig. 1 shows a view of P block building with the transducer positions and four selected measured modes of vibration (out of a total of seven that can be estimated). More details of these modes can be found elsewhere (Nguyen et al., 2015).

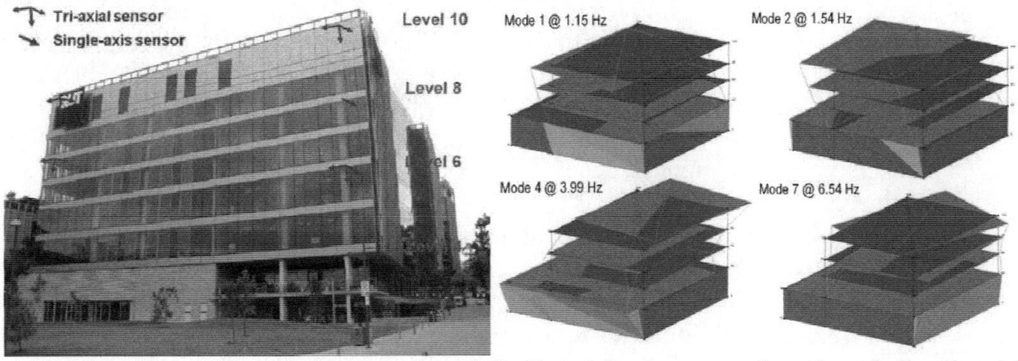

Fig. 1 P block view with transducer positions (left) and four measured modes of vibration (right)

Quite different from the large first test structure, the second test structure is a simply supported steel beam which can be considered to be a representative of small- or medium-scale structures. Owing to the ease of setup as well as introducing damage, this type of beams has been frequently used as a laboratory test structure to validate modal-based damage identification methods developed at QUT. With the span of 2.475 m and the channel section of 75x40x4 mm, this simply supported beam would have the first mode at around 15-20 Hz and therefore belong to the medium frequency characteristics associated with small- medium-scale structures such as short-span bridges or low-rise buildings. Using the instrumentation system chosen as presented in Section 3 (see later), eight modes can be estimated from this test model with modal frequencies ranging from 16.5 Hz (mode 1) to 705 Hz (mode 8). Fig. 2 shows diagrams of the steel beam with other setup details; as well as some selected experimental modal results. The main challenges with this test structure (as well as any other simple beam models) is a suitable modal testing system to estimate a reasonable number of modes (usually 5-10) across a broad frequency range.

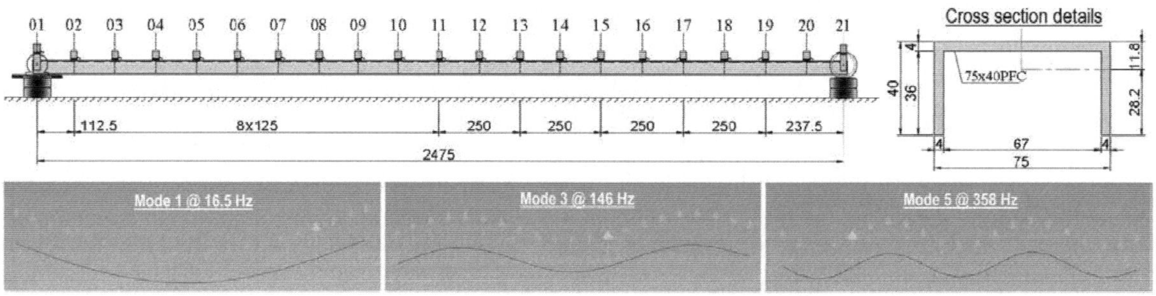

Fig. 2 Diagrams of beam test set up (top) and three measured modes of vibration (bottom)

3. Instrumentation

An instrumentation system for modal testing and monitoring applications often consists of the following main subsystems: (1) modal transducer system, (2) DAQ system, and (3) host control and data storage system. The following subsections will discuss each subsystem in detail and make comparisons where there are multiple options available. Not only are issues in wired instrumentation systems covered, but the wireless sensor system platform is also reviewed in the last subsection.

3.1 Modal transducer

3.1.1 Modal transducer type

The main transducer for modal testing and monitoring is the accelerometer which is to record analogue acceleration response signal from the structure. Accelerometers can be in either AC-response type (e.g. piezoelectric accelerometers) or DC-response type (e.g. piezo-resistive, capacitive or force-balance accelerometers). One of the most critical differences between these two transducer classes is only DC-response accelerometers can capture very low frequency responses (i.e. from 1 Hz downward to zero), making this type of sensors suited for the structures that have such low fundamental frequencies. Other than that, piezoelectric accelerometers may be preferred due to their excellent characteristics such as low noise, broad frequency range and high sensitivity. In the transducer selection stage, it is therefore important to predict the fundamental frequency value and one way to do this is to perform numerical modal analysis via the finite element (FE) model of the structure. The next critical characteristics of accelerometers to consider are sensitivity, input measurement range and frequency response. For the first two parameters, there is an inverse relationship: high sensitivity – low input range and vice versa; and very often, high sensitivity is prioritised for the case when the transducer is to be operated in ambient and operational loading conditions, as opposed to artificial excitation scenarios. The frequency response of the selected accelerometer should accommodate the measured frequency of interest for the test structure. Other characteristics of accelerometers such as noise density and non-linearity are also important especially when assessing inexpensive transducers such as those made by micro-electro-mechanical system (MEMS) technologies.

Implementing the above points to the P block building, the initial FE model developed by the design firm indicated that the first fundamental frequency of the building was around 0.7 Hz while the result from a more detailed FE model constructed based on as-built drawings showed the value of approximately 1 Hz. The DC-response accelerometer was therefore deemed necessary. To record vibration responses under ambient and operational loading conditions, the high sensitivity and low noise MEMS capacitive accelerometer modules 2460-002 (tri-axial) and 2210-002 (single-axis) from Silicon Designs, Inc. were finally chosen – both having 2V/g sensitivity, 2g input range and 13 μg/√Hz noise density. This selection has also considered the fact that the frequency response of the two sensor modules (0 to 300/400 Hz) is sufficiently broad to cover at least the first ten global modes or more of the structure. As can be seen in Fig. 1, the measured fundamental frequency was 1.15 Hz, which is very close to the as-built model's prediction and indicates that DC accelerometer is an appropriate choice.

For the case of the steel beam which has medium frequency characteristics, common AC-response accelerometers such as piezoelectric transducers are the most suitable transducers. Herein, the Kistler piezoelectric accelerometer model 8630B5 with reasonable sensitivity (1V/g), large input range (5g) and broad frequency response (0.5-2000 Hz) was chosen for this test. It is apparent that the use of this broad frequency response has enabled as many as eight modes to be estimated, as reported in Section 2. Although not the case of this test, the lightweight of the transducer (only 4.5 g) makes it very suitable for small test structures especially those that require transducer roving e.g. for detailed modal displacement measurement. Under this circumstance, the impact of transducer mass change due to the roving can be significant for any small structures.

3.1.2 Transducer placement and mounting

Modal transducers are usually placed on main bearing structural elements and should be distributed spatially to capture global modes of vibration. The minimum number of transducers needed in each cross section depends on whether the cross section of the structure can be considered as a rigid body. If this assumption holds, number of measured DOFs can be made at three, equivalent to three single-axis sensors, for the in-plane measurement of each cross section (Nguyen *et al.*, 2014b; SVS, 2011). Transducers should be firmly mounted on the test structure. Any significant looseness or self-vibration of the transducer unit against its mounting surface on the structure will affect the accuracy of measured frequency response or dampen high frequency transmissibility. Popular mounting methods include using studs, magnets, adhesives or mounting blocks; and different methods can affect the high frequency response of the transducer in different ways. In all methods, the preparation of the mounting surface is a key factor as it should be flat and clean.

With the P block building case, the numerical modal animation confirmed the validity of the assumption for most of the stiff concrete floors. Eight transducers were installed using the mounting block to instrument the four main floor levels (i.e. 4th, 6th, 8th and 10th, see Fig. 1) of the upper part of the building. This selection was made because all of the lower levels are semi underground and larger in size effectively acting as a base for the upper parts where the transducers were installed. In the case of the steel beam, the transducer placement strategy is simple and straightforward. As the modes of interest were vertical bending, all transducers were placed vertically using magnet mounting method. The 8630B5 transducer was sticked to the top of the accompanied magnet by using a very thin hot glue layer before the magnet was attached to the top 'deck' of the beam (Fig. 2).

3.2 DAQ system

There are various DAQ system platforms and components in both hardware and software aspects. In terms of DAQ hardware, a DAQ system unit often has (i) input/output (I/O) modules providing connector, signal conditioning and analogue-digital converter (ADC) for transducers or actuators; (ii) a chassis that houses modules; (iii) communication, synchronisation and data transfer medium; among others. On the other hand, the DAQ software has only two components i.e. driver software and application software. The following subsections focus on the main factors governing the development of appropriate DAQ systems for different types of structures.

3.2.1 DAQ system platform

In general, DAQ systems can be broadly categorised into two main groups (i) centralised systems and (ii) distributed systems. A centralised DAQ system normally has one central DAQ chassis which has all I/O modules plugged into it in one place (Fig. 3 left). As the coverage is limited by the length of transducer cable on each side, this type of system is often suitable with small-scale structures such as low-rise buildings or short-span bridges. This was evident for the steel beam herein and the centralised DAQ system was easily established by using one single cDAQ-9172 chassis of NI with a USB connection to computer. With eight I/O module slots, this system can cater up to 32 single-axis accelerometer measurement channels (Fig. 3 right).

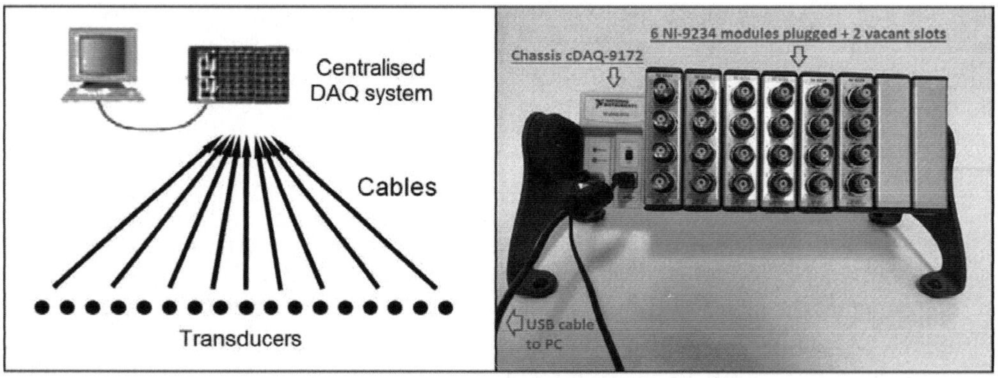

Fig. 3 Centralised DAQ system: schematic diagram (left) and cDAQ-9172 system (right)

On the other hand, a distributed system consists of multiple DAQ chassis nodes connected with the host computer often by Ethernet cables. The use of distributed Ethernet DAQ nodes means the latter can form a much larger network and cater for large-scale structures such as multi-span bridges and multi-storey buildings including the P block building at QUT. For this test structure, the compact reconfigurable I/O (cRIO) platform was adopted and the cost-optimised model cRIO-9074 was chosen to realise the distributed DAQ system. Fig. 4 shows a schematic diagram, actual transducer and DAQ node at one location in the P block as well as a close-up of the cRIO-9074 and its module.

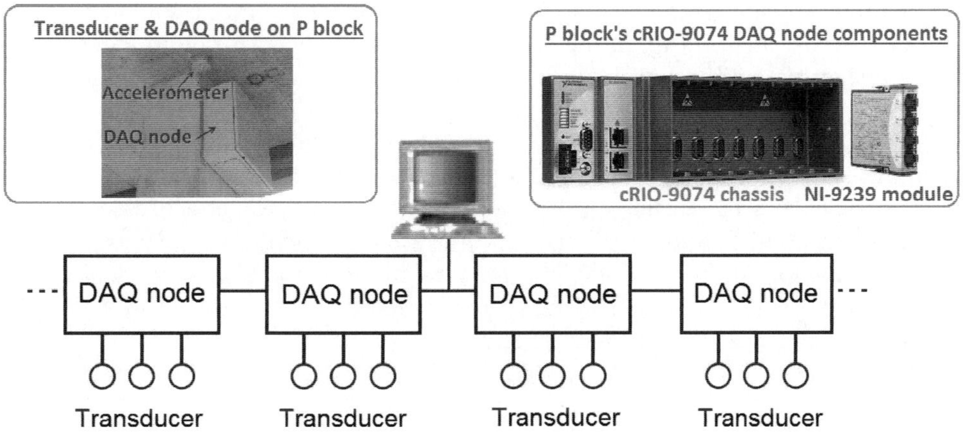

Fig. 4 Distributed DAQ system: schematic diagram (bottom), actual transducer and DAQ node arrangement in P block (top left) and cRIO-9074 DAQ node components (top tight)

3.2.2 DAQ application software

DAQ application software can be classified into three main categories: turnkey, configurable and programmable (National Instruments, 2016b). Turnkey software requires very little user interaction and allows the user to complete quickly all DAQ setup steps. Configurable software expands the functionality to allow the user to configure different aspects of their DAQ application without programming. Programmable software is the most powerful software solution providing the user with the widest scope of functionality and hardware control. Because of the need of a versatile system platform that could be used for both short-term testing and long-term monitoring application, LabVIEW a popular programmable DAQ language of NI has been adopted by the present authors to develop their most recent modal testing and monitoring applications including those presented herein. The use of LabVIEW programming system not only maximizes the hardware and software functions but also allows the builder of standalone executable applications which can be installed on non-LabVIEW computers for use on student research projects. LabVIEW is a graphical programming language which uses icons instead of lines of text to create application programs (National Instruments, 2005). LabVIEW programs are called virtual instruments (VIs) because their appearance and operation imitate physical instruments, such as oscilloscopes and multi-meters. Interested readers can easily find more details of LabVIEW on the NI website (www.ni.com/labview) or other sources.

3.2.3 I/O module

The main I/O module concerned in modal test applications is analogue input (AI) module which is to provide necessary signal conditioning (transducer excitation, signal amplification, isolation and alias filtering) and ADC functions for modal transducers. Desired features for a quality output-only modal testing and monitoring system include simultaneous sampling, high-resolution measurements, and robust noise mitigation. Hence, modules with high-resolution delta-sigma ADC, high sample rate, low input voltage and low input noise are often preferred. Another important note in choosing AI module is to ensure that the type of AI module is compatible and supports the type of transducer as presented in 3.1.1. For example, piezoelectric accelerometers often need AI module supporting integrated electronics piezoelectric (IEPE) signal conditioning including a correct constant current to power the amplifier and AC coupling (to filter out DC component and

improve AC amplitude resolution) amongst others. This is the case of the second test structure – the steel beam and, as illustrated in Fig. 2, the AI module NI-9234 was chosen to meet the excitation current requirements of the Kistler piezoelectric accelerometer model 8630B5 as well as other desired features.
Regarding the DAQ system of P block, since the transducers SDI 2460-002 and 2210-002 are non-IEPE accelerometers, the AI module for these transducers should be selected from general-purpose voltage input modules and the focus of selection lies with the key criteria of ADC as previously mentioned. Considering all criteria together, the voltage input module NI-9239 as shown in Fig. 3 was finally selected amongst those available at the time of purchase, due to many advantages such as high ADC resolution (24 bit), considerably low voltage range (±10 V) in comparison with transducer output voltage (±4 V), and low input noise (70 μVrms).

3.2.4 Synchronisation

Tight synchronisation across different transducer channels is a very important issue in modal testing and modal-based damage identification (Nguyen *et al.*, 2014b; Nguyen *et al.*, 2014a). Even though the synchronisation across the channels within one AI module is almost always automatically enabled, synchronisation across different modules is not a trivial task especially when using non-programmable application software systems. However, with programmable application software system like LabVIEW, precise hardware-based synchronisation across different modules in a centralised DAQ system such as the cDAQ-9172 (Fig. 3) is relatively simple. The only task that needs to be done for this is to assign all channels into the same so-called virtual channel VI (see Fig. 5) so that one sample clock (often the fastest one) from available modules can be elected as a common timing source for all the modules. LabVIEW programming can also allow the selection of a sample clock source of preference via implementing some timing source filtering technique.

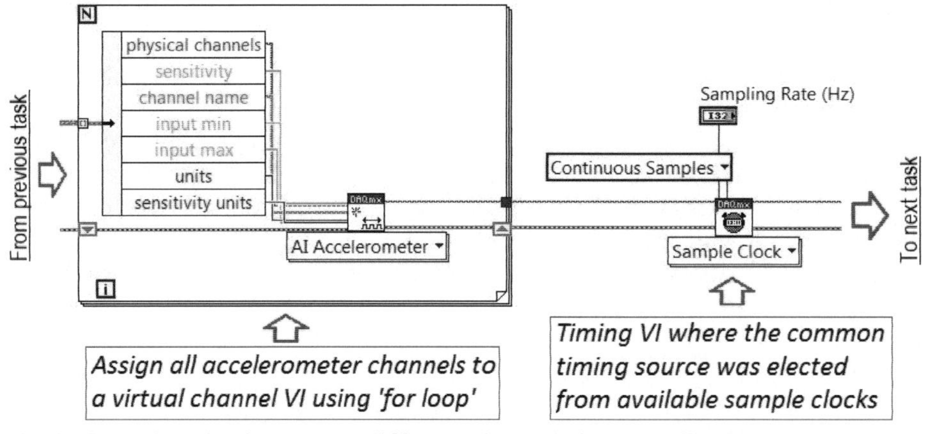

Fig. 5 Method of synchronisation across different channels in centralised DAQ system

For distributed DAQ systems like cRIO, there are generally two main synchronisation approaches, i.e. hardware-based methods and software-based methods. The first possible method in the hardware-based approach is using GPS synchronisation module, but this is not always stable (depending on weather conditions) and not applicable for building or underground structures. Another hardware-based method is the synchronisation via exporting timing signal from a master cRIO unit to slave cRIO nodes via synchronisation modules such as NI-9469 and network cables (CAT5) and. This dedicated synchronisation bus can be expanded to connect multiple cRIO units by using the daisy chain topology thereby spanning large coverage, but it can be quite costly and

requires extra room for cabling. Readers interested in hardware-based synchronisation methods could easily find more details elsewhere (National Instruments, 2016a).
For monitoring applications using cRIO platform, software-based methods can be a very cost-effective choice especially when considering the system scalability, since there is no added cost of synchronisation for additional DAQ nodes. There are two main software-based approaches – one is to implement field-programmable gate arrays (FPGA) timing correlation and the other is to use a common trigger to start the data acquisition and logging before relying on on-module clock in each cRIO node to control the acquisition timing. The second approach has been successfully implemented into the P block monitoring system by making use of the advanced TCP/IP communication medium of cRIO platform to improve the accuracy of the start acquisition timestamps of different cRIO nodes as well as to limit the accumulated synchronisation misalignment due to the use of individual sample clocks. Interested readers could refer to the related published paper for more details (Nguyen et al., 2015).

3.3 Host control and data storage system

3.3.1 PC-based versus controller-based control system

Traditionally, a PC is often used as the host device to control the whole DAQ process via the application software. Simple and straight forward, this control system is often suited for short-term and non-autonomous testing applications such as the laboratory modal tests at QUT. For long-term monitoring applications in real civil structures, embedded control systems such as cRIO controllers or, the newer platform, cDAQ controllers are preferred since they have small form factor and superior ruggedness making them more suitable to be operated autonomously in a confined space and/or harsh conditions. These are also the main reasons for using the cRIO-9074 controller for the P block building. Although not required to run the controller-based system, a PC can still be connected to cRIO or cDAQ controller to save and view data locally when needed.

3.3.2 Local versus server data storage system

As previously mentioned, a PC may be used for both PC-based and controller-based systems; hence local hard disks of the PC including removable ones can be utilised to locally store the acquired data. However, as hard disk space is limited, this local data storage system is often used for short-term modal testing applications and this is also the case of all QUT laboratory tests. Local data storage may be used for long-term monitoring, provided that there is some effective mechanism to transfer data to the users and free up the storage space. Where possible, using file storage server is recommended for long-term continuous monitoring applications due to numerous benefits such as the ability to securely store and simultaneously share a large amount of recorded data with multiple users.

3.3.3 Data logging setup parameters

The first important parameter that needs to be set up for data logging is sample rate. In theory, sample rate is required to be twice the maximum frequency of interest. However, in practice the multiplication factor should be at least 2.5, due to the impact of antialiasing filter typically reducing effective frequency band to 80%. If the user is not completely sure of the maximum frequency of interest, it is recommended to use a relatively high sample rate especially in the case of short-term testing to avoid test repetition just for high frequency estimates. It is worth noting that the sample rate of the recorded data can easily be reduced, for instance, by applying a decimation function in the post-processing process. Using unnecessarily large sample rates is often of a greater concern in continuous monitoring applications. In some cases, the lowest sample rate is bound by the rate range of AI module especially those that use delta-sigma technology. In programmable platforms, an efficient way to overcome this problem is to use point-by-point functions to enable online data decimation (National Instruments, 2001).

The second important parameter in the data logging process is the undisrupted logging duration. A minimum undisrupted logging duration can be estimated from the maximum correlation time in the response and, therefore, inversely proportional to the product of fundamental frequency and damping ratio, $T=10/(f_{min}\varsigma)$ (Brincker and Ventura, 2015). Consequently if one assumes most civil structures have damping ratios of no more than 0.5-1%, they can draw an estimation that is the undisrupted logging length should be around 1000–2000 times the fundamental vibration period. The test engineer of course can always refine this length for later tests, after a reasonable damping estimation can be made from the initial data.

3.4 Wireless sensor system

Besides wired DAQ systems as previously presented, the recent years have witnessed a boom in the development of wireless sensor systems (WSSs) including field implementation for modal testing and monitoring. The main advantage of most WSSs over wired sensor systems is that the removal of the cable connection - transducer is embedded on DAQ node while the DAQ node exchanges data and other information with the host system by a radio signal protocol (e.g. ZigBee, Wi-Fi or Bluetooth); and this makes WSSs very portable. However due to limited space in each sensor node, transducer and DAQ components often do not have the same performance as their counterparts in wired systems. For example, WSSs have poorer ADC resolution and data synchronisation; and are often not capable of long-term continuous simultaneous measurements due to limited storage and/or power supply. Hence, most of the contemporary WSSs are mostly suitable for short-term modal testing, and more enhancement is needed to make them a complete substitute for wired sensor systems in long-term monitoring applications.

4. Some notes on test management

4.1 Excitation issues

While the output-only modal testing and monitoring approach is applicable with many civil structures, the quality of recorded data might vary significantly from one structure to another, even if they have similar vibration characteristics and use the same instrumentation system. One main reason for this symptom could be due to a non-trivial difference between excitations of the structures at the testing time. Understanding the profile and characteristics of excitation is therefore crucial to enable an effective data logging plan. A civil structure is generally subjected to two main excitation sources namely (1) human-related activities such as traffic on a bridge, or movement of occupants in a building, and (2) non-human factors such as wind load and tremor. Amongst these, wind excitation can be significant for flexible structures like long-span bridges or tall buildings, but that may not be the case with stiffer structures like medium rise building or short span bridges. A similar pattern can be seen with traffic-induced excitation on different bridges. A structure can also experience different levels of excitation at different times of the day, e.g. the traffic volume tends to become marginal during night times. In the case of P block building, the time for best excitation was found to be during normal working hours between 9am to 5pm, meaning that human-related activities on the building can be amongst the most significant sources of excitation.

4.2 Management of long-term monitoring system status

Long-term monitoring is mostly done in an unattended manner. To ensure a smooth system operation and limit loss of monitoring data, it is imperative to have a timely reporting mechanism on the system status. For P block monitoring system, an autonomous report channel was established through re-utilising the TCP/IP communication medium. If the system has normal operation, only one report email per day is sent to the management team at a specified time. However, if there is a fault in the system that prevents the normal logging process (e.g. loss of network drive connection), an error email message will be broadcasted immediately to notify the team. Fig. 6 shows two reporting emails – the first was sent to notify a lost connection error whilst the email on the right was set up immediately when the error was fixed to report the normal status of the monitoring system. Due to privacy reasons, the email addresses of the receivers are hidden in Fig. 6.

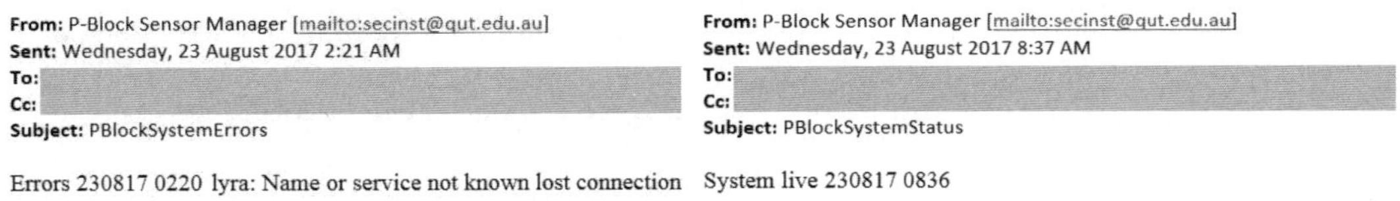
From: P-Block Sensor Manager [mailto:secinst@qut.edu.au]
Sent: Wednesday, 23 August 2017 2:21 AM
To:
Cc:
Subject: PBlockSystemErrors

Errors 230817 0220 lyra: Name or service not known lost connection

From: P-Block Sensor Manager [mailto:secinst@qut.edu.au]
Sent: Wednesday, 23 August 2017 8:37 AM
To:
Cc:
Subject: PBlockSystemStatus

System live 230817 0836

Fig. 6 Reporting monitoring system status through emails

4.3 Detection of strong transient or extreme events

Output-only modal testing and monitoring is usually employed with an expectation that the acquired response data is adequately stationary and approximates a Gaussian sequence so that it can be analysed by operational modal analysis techniques. Hence, it is imperative that strong transient or extreme events such as earthquakes be correctly detected so that appropriate non-transient datasets can be selected for accurate post-event assessment of the structure.

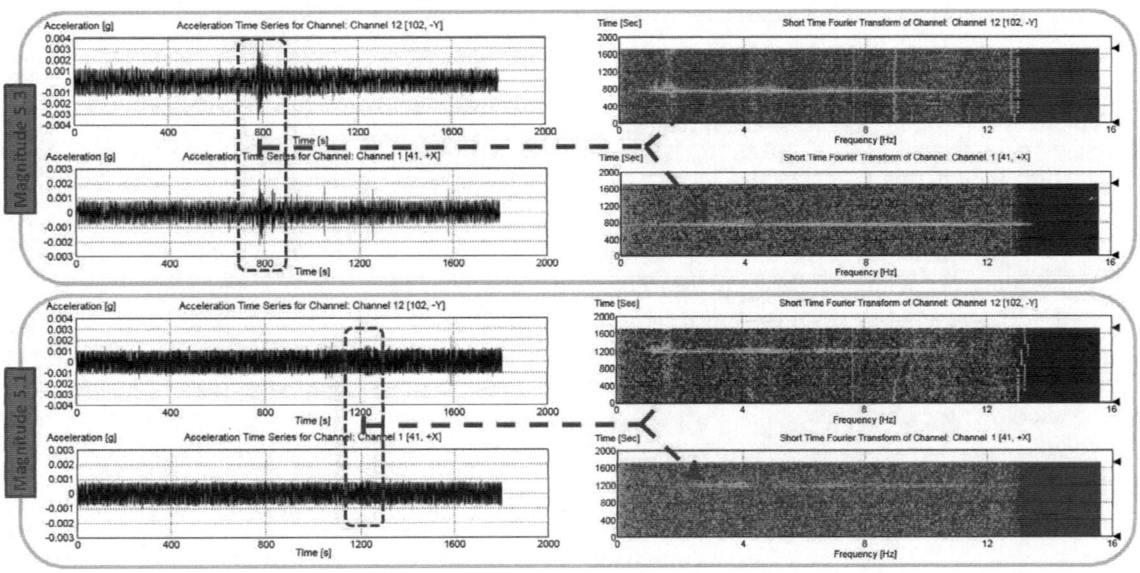

Fig. 7 Detection of earthquake occurrence through time history screening and STFT

As an example, Fig. 7 shows a combination of both time history screening and short-term Fourier transform (STFT) to identify the occurrence of two earthquakes through locating signal singularity in response data. The earthquakes described herein are the two moderate earthquakes with magnitudes of 5.3 and 5.1 (on the Richter scale) that occurred in Queensland in 2015. As shown in the figure, two representative channels are selected – one at level 10 (channel 12) and the other at level 4 (channel 1). As can be seen from the plots, there are perfect agreements in timings of the event occurrences between the two methods. Note that the 5.3 magnitude earthquake is twice as powerful as the 5.1 magnitude earthquake in terms of initial energy release.

5. Conclusions

This paper has analysed some of the most critical issues pertaining to the successful implementation of the output-only modal testing and monitoring approach into civil engineering structures. On the instrumentation aspect, it analysed in detail are all necessary components of complete instrumentation systems covering from modal transducer to each element of DAQ and host system such as appropriate AI module, synchronization and data storage solutions. Through real examples, the paper has shown that the desired type of transducers and DAQ systems depends on the key characteristics of the structure and test application such as the structure size, fundamental frequency as well as test mode. For applications to monitor real large-scale structures with low frequency characteristics, it can be concluded that DC-response accelerometer and distributed DAQ system platform tend to be more suitable candidates. In contrast, AC-response accelerometer and centralised DAQ system can be good options for use with small- to medium-scale structures. On the test management aspect, three key notes have been made on how to cope with adversary situations such as non-uniform excitation, system failure and presence of a strong transient event. Strategies have been given with an emphasis on effective monitoring management and maintenance practices. It is the hope of the present authors that the paper will benefit existing and future developers of output-only modal testing and monitoring applications.

References

Ansari, F. and Karbhari, V. M. 2009. *Structural health monitoring of civil infrastructure systems*. Boca Raton, FL: Woodhead Publishing.

Brincker, R. and Ventura, C. 2015. *Introduction to operational modal analysis*. Chichester, West Sussex: John Wiley and Sons, Inc.

Frangopol, D. M. 2011. Life-Cycle performance, management, and optimisation of structural systems under uncertainty: Accomplishments and challenges. *Structure and Infrastructure Engineering* 7 (6):389-413.

Kodikara, K., Chan, T., Nguyen, A., et al. 2017. Useful tips for automated model updating of medium rise buildings.

Kodikara, K. A. T. L., Chan, T. H. T., Nguyen, T., et al. 2016a. Model updating of real structures with ambient vibration data. *Journal of Civil Structural Health Monitoring* 6 (3):329-341.

Kodikara, K. T. L., Chan, T. H., Nguyen, A., et al. 2016b. Model updating incorporating measured response uncertainties and confidence levels of tuning parameters. *International Journal of Lifecycle Performance Engineering* 2 (1-2):61-78.

National Instruments. 2001. Getting Started with LabVIEW Point-By-Point VIs: National Instruments Corporation, Austin, TX.

National Instruments. 2005. LabVIEW fundamentals: National Instruments Corporation, Austin, TX.

National Instruments. 2016a. Choosing a CompactRIO Synchronization Technology: National Instruments.

National Instruments. 2016b. Choosing the Right Data Logging Software from National Instruments: National Instruments.

Nguyen, T. 2014. SHM through flexible vibration sensing technologies and robust safety evaluation paradigm, PhD Thesis by Publication, Queensland University of Technology.

Nguyen, T., Chan, T. H. T. and Thambiratnam, D. P. 2014a. Effects of wireless sensor network uncertainties on output-only modal-based damage identification. *Australian Journal of Structural Engineering* 15 (1):15-25.

Nguyen, T., Chan, T. H. T. and Thambiratnam, D. P. 2014b. Effects of wireless sensor network uncertainties on output-only modal analysis employing merged data of multiple tests. *Advances in Structural Engineering* 17 (3):319-329.

Nguyen, T., Chan, T. H. T. and Thambiratnam, D. P. 2014c. Field validation of controlled Monte Carlo data generation for statistical damage identification employing Mahalanobis squared distance. *Structural Health Monitoring* 13 (4):473-488.

Nguyen, T., Chan, T. H. T., Thambiratnam, D. P., et al. 2015. Development of a cost-effective and flexible vibration DAQ system for long-term continuous structural health monitoring. *Mechanical Systems and Signal Processing* 64-65:313-324.

SVS. 2011. *ARTeMIS Extractor, Release 5.3, User's Manual*: Structural Vibration Solutions A/S, Aalborg East, Denmark. http://www.svibs.com/products/ARTeMIS_Extractor.aspx.

AUTHOR INDEX

AUTHOR INDEX

AUTHOR INDEX

AUTHOR INDEX

AUTHOR INDEX

International Society for Structural Health Monitoring of Intelligent Infrastructure
University of Manitoba
Agricultural & Civil Engineering Bldg.
A250 - 96 Dafor Road
Winnipeg, Manitoba R3T 2N2
Canada

ISBN 978-1-5108-6457-3

9 781510 864573

CURRAN ASSOCIATES INC.
proceedings.com

8th International Conference on Structural Health Monitoring of Intelligent Infrastructure (SHMII 2017)

Structural Health Monitoring in Real-World Application

Brisbane, Australia
5-8 December 2017

Volume 3 of 3

Editors:

Tommy Chan **Saeed Mahini**

ISBN: 978-1-5108-6457-3

Printed from e-media with permission by:

Curran Associates, Inc.
57 Morehouse Lane
Red Hook, NY 12571

Some format issues inherent in the e-media version may also appear in this print version.

Copyright© (2017) by International Society for Structural Health Monitoring of Intelligent Infrastructure
All rights reserved.

Printed by Curran Associates, Inc. (2018)

For permission requests, please contact International Society for Structural Health Monitoring of Intelligent Infrastructure at the address below.

International Society for Structural Health Monitoring of Intelligent Infrastructure
University of Manitoba
Agricultural & Civil Engineering Bldg.
A250 - 96 Dafor Road
Winnipeg, Manitoba R3T 2N2
Canada

Fax: 204-474-7519

central@ishmii.org

Additional copies of this publication are available from:

Curran Associates, Inc.
57 Morehouse Lane
Red Hook, NY 12571 USA
Phone: 845-758-0400
Fax: 845-758-2633
Email: curran@proceedings.com
Web: www.proceedings.com

TABLE OF CONTENTS

VOLUME 1

KEYNOTE LECTURES

REGULAR SESSIONS

RS1: SMART AND OTHER ADVANCED SENSORS

RS2: WIRELESS AND OTHER ADVANCED SENSOR NETWORKS

RS3: DATA ACQUISITION, PROCESSING AND MANAGEMENT, INCLUDING BIG DATA

VOLUME 2

SUB-TOPIC 3: STRUCTURAL CONDITION ASSESSMENT OF BRIDGES

SUB-TOPIC 4: CONDITION ASSESSMENT OF BUILDING STRUCTURES

RS6: MODEL UPDATING, SAFETY EVALUATION AND RELIABILITY FORECAST

RS7: DAMAGE CONTROL, REPAIR AND STRENGTHENING

RS8: LIFE-CYCLE PERFORMANCE-BASED DESIGN AND PERFORMANCE MONITORING

RS10: SMART MATERIALS, STRUCTURES AND REHABILITATION

MINI-SYMPOSIA

MS1: RECENT SHM ADVANCES IN AUSTRALIA

VOLUME 3

MS2: DATA-DRIVEN STRUCTURAL DAMAGE IDENTIFICATION AND PERFORMANCE ASSESSMENT

MS3: TEMPERATURE BEHAVIOUR OF STRUCTURES

SPECIAL SESSIONS

SS1: RECENT ADVANCES IN RAILWAY SYSTEM MONITORING IN CHINA

SS2: CASE STUDIES AND NEW METHODS FOR MODAL-BASED SHM OF CIVIL STRUCTURES

SS3: OPTICAL BASED MONITORING TECHNIQUES FOR INFRASTRUCTURAL MAINTENANCE AND SAFETY

The 8th International Conference on
Structural Health Monitoring of Intelligent Infrastructure
Brisbane, Australia | 5-8 December 2017

Monitoring of Corrosion Process in Structural Adhesive using Smart Materials

Y. Y. Lim [1], I. Izadgoshasb [2], S. T. Smith [3]
[1, 2, 3] *School of Environment, Science and Engineering, Southern Cross University – Australia,*
Email: yee.lim@scu.edu.au

Abstract

The use of structural adhesives, such as epoxy, for the strengthening and repair of civil infrastructure is gaining popularity in the engineering community. Of particular concern to the viability of intervention measure is the degradation of strength and durability of the epoxy over time, especially when exposed to aggressive environments. Monitoring the corrosion process of epoxy is thus essential to ensure safety of the host structure. In this paper, the electromechanical impedance (EMI) technique that employs a smart piezoelectric transducer is used to monitor the degradation process of commercially available epoxy when exposed to corrosive environments. Standard epoxy coupons with surface bonded piezoelectric transducer are prepared and then immersed in either acidic or saline conditions in the laboratory. Experimental results show that the change in frequency of resonance peaks in the admittance signatures versus the frequency spectrum is a good indicator of the corrosion process. This study proves that the EMI technique is effective in characterising the severity of corrosion inflicted on the epoxy. The capability of the EMI technique in providing autonomous, real-time, remote monitoring with quick response using non-intrusive active sensors could potentially replace its conventional counterparts.

1. Introduction

The use of structural adhesives for repair and strengthening of structure are gaining popularity in the aerospace, mechanical and civil engineering disciplines due to their excellent mechanical and physical properties. Adhesive connections possess various advantages such as the ease of joining different materials and components of complicated forms within short period of time.

In the early years, application of structural adhesives in civil engineering has been focusing on bridge construction, especially for connections between bridge decks and steels girders (Messler 2004).

The use of structural adhesive in the civil engineering discipline has thus far been impeded by the harsh environmental conditions. In depth studies into the influences of environmental factors on the performance of the adhesive joints is necessary, before real life application in the industry and potential replacement of conventional methods of connections such as bolting is possible.

The performance of structural adhesives can often be degraded by hostile environment such as temperature, moisture, salt, acidity. The severity of degradation can be aggravated with a combination of these conditions.

If epoxy-based adhesives is in contact with water, their mechanical performance is altered by the absorption of water into the polymer chains. In addition to weakening the physical and chemical properties of the structural adhesive, the presence of moisture will also degrade the strength of the interfacial bonding between the adhesive and the adherend (Banea and da Silva 2009). For metallic adherend, water will diffuse into the joint through the adhesive. The interface between the adhesive and adherend will also be intruded by the water with time. In general, the weakening of a joint is predominantly caused by the degradation of the interface between the adhesive and adherend rather than the weakening of the adhesive.

Lapique and Redford (2002) conducted an ageing test on commercially available structural adhesive, Araldite 2014 exposed to water vapour over 36 days in a controlled environment at 40^0C. The experiment showed that the adhesive was soften in the presence of moisture. The elastic modulus reduced and the plastic deformation increased correspondingly.

Prolongo and Urena (2007) found that an epoxy-aluminium joint immersed in a saline environment experienced degradation with time. It was also discovered that the durability of the adhesive joints under such environment can be considerably enhanced with the provision of appropriate surface treatments.

Recently, Wong (2017) investigated the feasibility of using the electromechanical impedance (EMI) technique for detecting the wall thickness reduction in pipeline due to corrosion. It was found that the resonance frequency range shifted when the thickness of the wall was reduced. Talakokula et al. (2014) studied the corrosion process of steel reinforcing bars using the EMI technique. The experimental results indicated that the equivalent structural parameters derived from the mechanical impedance of the steel structure were effective in detecting and quantifying the corrosion process.

Thus far, limited study has been conducted on the application of EMI technique on structural adhesive. In this paper, a novel study is presented, in which the feasibility of adopting the smart piezoelectric (PZT) based EMI technique in monitoring the corrosion process of structural adhesives is investigated. The EMI technique possesses various advantages over other conventional health monitoring techniques as the PZT transducers are low cost, non-intrusive, respond rapidly, adaptable to large scale application, with autonomous, real-time and remote monitoring capability. The PZT patches can be installed at various locations on critical structural members in order to monitor local damage or overall structural integrity.

The principal of application of the EMI technique (Lim and Soh 2011) is the piezoelectric effect, in which the PZT is capable of converting electrical energy to mechanical energy and vice versa. When an alternating voltage of varying frequency is applied across the poling direction of a PZT transducer that is mechanically attached to a host structure, the transducer can be mechanically excited (known as the converse effect of piezoelectricity) and subsequently actuates the attached host structure. At the same time, the vibratory response of the host structure will interact with the PZT transducer, thus altering the corresponding electric current.

The current, in the form of complex electrical admittance, can be measured by an impedance analyser. Subsequent alterations in the vibratory behaviour of the host structure as a result of damage (Lim and Soh 2014), change in stress, boundary condition or in mechanical properties (Lim et al. 2016; Kong and Song 2017) can therefore be reflected in the admittance signatures.

2. Experimental study

2.1 Experimental Setup

In this study, commercially available two-part thermosetting epoxy, Sikadur 30 (Sika 2014), was used for the corrosion tests. The liquid epoxy was first mixed and placed into standard "dog bone" shape mould according to the British Standard (BS EN, 1996). Four identical epoxy coupon were produced. They were allowed to fully cure at a constant room temperature of 24^0C for two weeks. One PZT patch, dimensioned 10mm x 10mm x 0.3mm, was surface bonded at the centre of the epoxy coupon. A typical "dog bone" coupon specimen is shown in Figure 1(a).

An impedance analyser (Hioki IM3570) was used to apply an alternating voltage (5V) across the PZT transducer and to also simultaneously record its electrical admittance as shown in Figure 1(b).

The frequency range considered was 5 kHz – 900 kHz, while the electrical signatures were first acquired at a healthy or undamaged stage. The specimens were then subjected to various corrosive environments as summarised in Table 1. Note that the specimens were not fully immersed into the respective solution in order to avoid damaging the PZT patch. Instead, sponges were used to wrap around the specimens which were then dipped into the solution, so that the solution was in contact with the specimens but not the PZT patches. Specimen Ctrl serves as control, which is placed alongside other specimens inside the same laboratory with controlled temperature (24^0C) throughout the period of monitoring.

The admittance signatures were acquired once per day for 11 days. Each time before the signature acquisition started, the specimens were removed from the solution and placed on a dry cloth for 10-15 minutes. This ensured the signatures are acquired under the same condition.

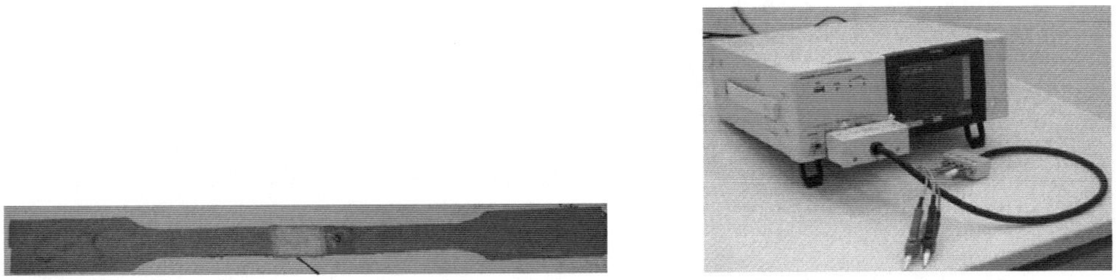

(a) Epoxy coupon with surface-bonded PZT patch

(b) Impedance analyser

Figure 1. Experimental setup for EMI based corrosion monitoring of epoxy.

Table 1: Summary of test specimens under various corrosive environment

Specimen Label	Environment
Ctrl	Control
Wtr	Water
NaCl	Sodium chloride of 10% concentration
H2SO4	Sulphuric acid of 10% concentration

2.2 Admittance Signatures at Healthy Stage

The admittance signatures acquired from the PZT surface bonded on Specimen Ctrl is depicted in Figure 2. The result shows a wide frequency range from 0 – 900 kHz which reflects several large peaks representing the PZT patch resonances. Figure 2(b), on the other hand, offers a magnified view of the first 70 kHz which is characterised by densely spaced structural peaks, representing the resonances of the epoxy coupon. These peaks are a useful indicator of the vibratory behaviour of the host structure (Lim et al. 2006), which will be altered when subjected to structural changes such as corrosion.

2.3 Repeatability of signatures – Control Specimen

Referring again to Figure 2, it can be shown that the repeatability of signatures on Specimen Ctrl can is satisfactorily high throughout the period of monitoring, as shown from the overlapping of signatures throughout the period of monitoring. This indicates that no sign of degradation is shown on this specimen, which is placed inside the laboratory throughout the period of monitoring. In other words, any changes in signatures shown on other specimens would be caused by the corrosion process.

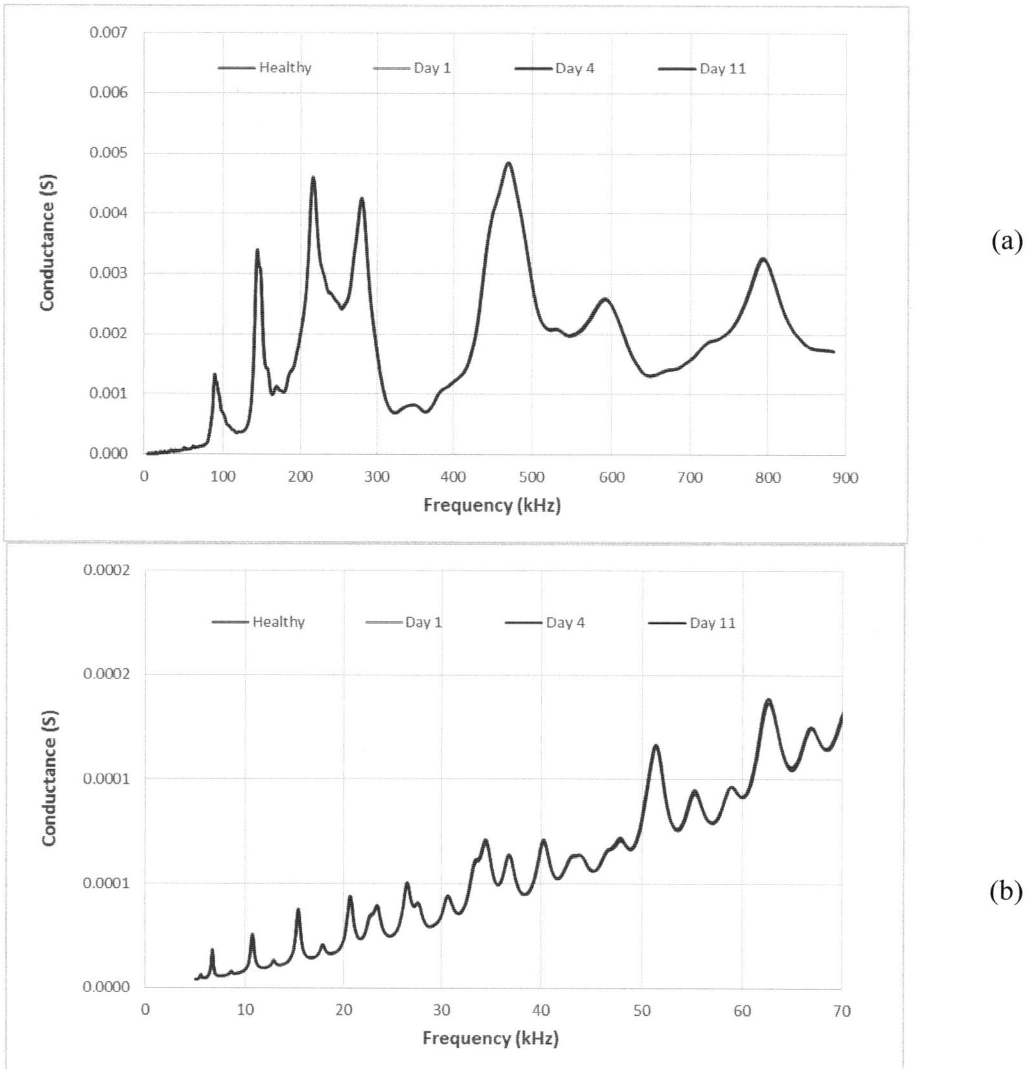

Figure 2. Conductance signatures versus frequency acquired from PZT patch surface bonded on Specimen Ctrl
(a) 0 – 900 kHz (b) – 70 kHz

2.4 Unprotected PZT Patch in 10% Sulphuric Acid

Despite the fact that the PZT patches are not fully immersed into the solution, they are unprotected and hence the acid caused considerable damage to the PZT patch on Specimen H2SO4. This can be reflected from Figure 3, where the signatures acquired after the specimen had been exposed to the corrosive acid for one day deviate significantly from the healthy stage, indicating that the PZT patch itself is damaged. Signs of corrosion can be visually observed on the PZT patch, confirming the changes on the electrical signatures. This observation is similar to the study reported by Yang, Lim and Soh (2008).

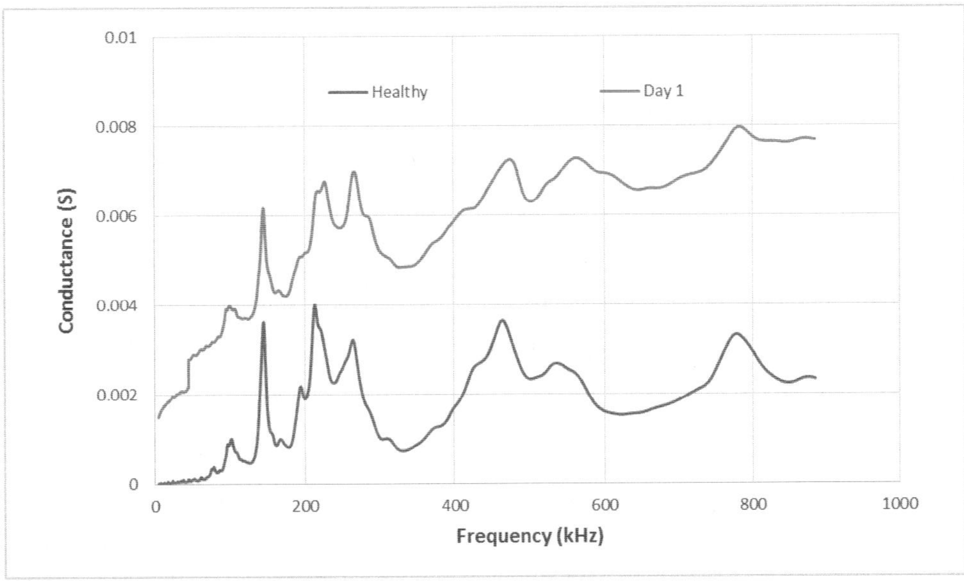

Figure 3. Conductance signatures versus frequency acquired from PZT patch surface bonded on Specimen H2SO4 (without protection)

This is considered as a premature failure and it is undesirable as the PZT patch failed to perform after the first day of monitoring. The test was thus repeated with the PZT patch being protected by applying 2 layers of quick set epoxy over the surface before it was placed in the corrosive environment.

3. Monitoring of Corrosion Process

To monitor the process of corrosion, peaks representing the host structural resonances in the lower frequency range (<70 kHz) were adopted. Figure 4 shows a plot of admittance signatures within a selected frequency range (20 – 40 kHz) for Specimen H2SO4 throughout the period of monitoring. The signatures change continuously throughout that period but the shape of the curve remains essentially the same.

Magnifying the frequency range of 23 – 26 kHz shows one typical structural resonance peak within the lower range spectrum. It is observed that there is an abrupt decrease in resonance frequency (leftward movement) and magnitude after one day. The peak continuously moves to the left at a reduced rate thereafter. Ignoring the change in magnitude, which is less representative of damage (Lim, Bhalla and Soh 2006), the decrease in resonance frequency reflects a reduction in stiffness of the host structure. This observation is in line with the physical corrosion process, where the stiffness of the coupon is expected to reduce when it is attacked by the acid.

The corresponding changes in resonance frequency across the entire period of this particular peak is schematically summarised in Figure 5. Observing the gradual reduction in resonance frequency, it can be confidently concluded that the change in resonance frequency is a good indicator of the corrosion process. On the other hand, observing the gradient of the curve, it can be shown that the corrosion process is significant in the first day and is reduced after 3 days. Note that a similar conclusion can be drawn from other resonance peaks within the first 70 kHz.

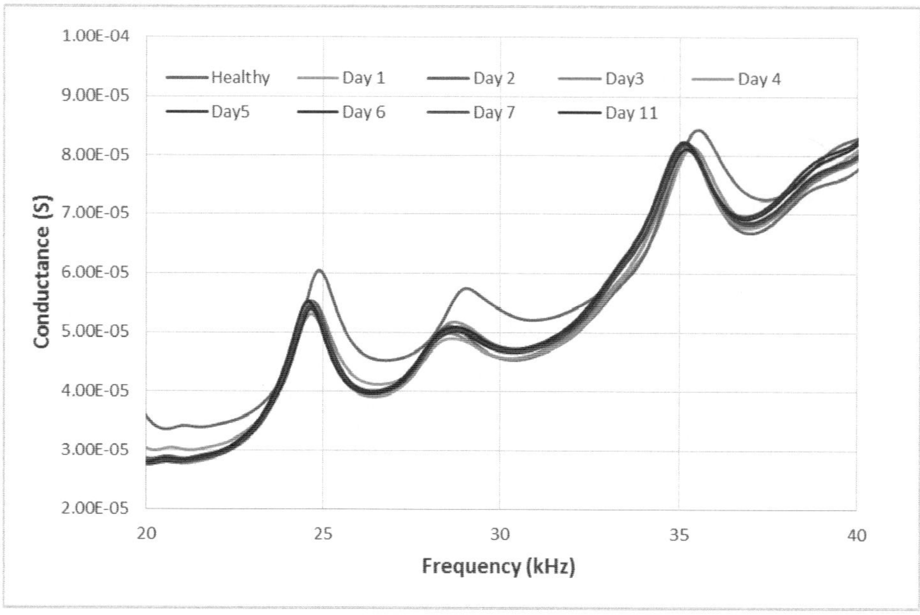

(a)

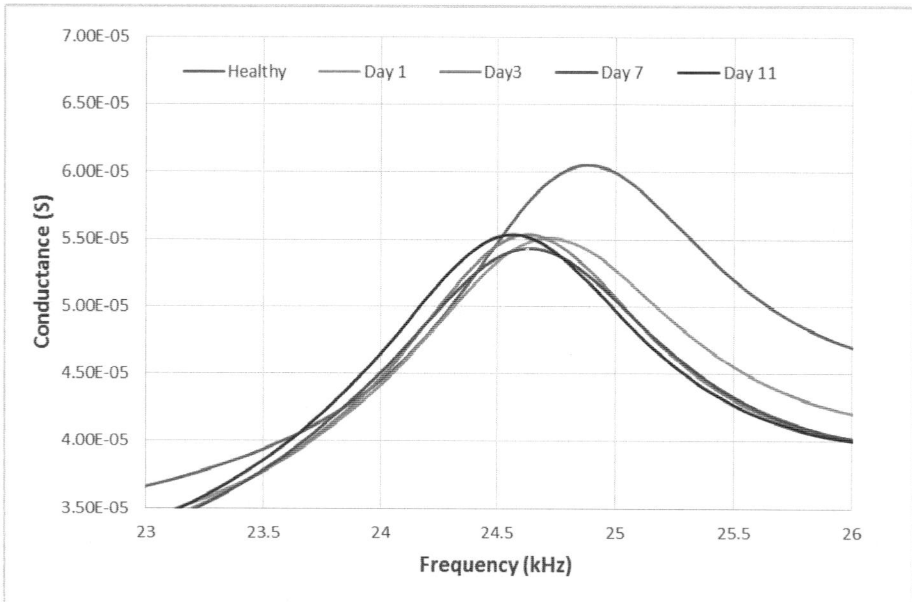

(b)

Figure 4. Conductance signatures versus frequency acquired from PZT patch surface bonded on Specimen H2SO4
(a) 20 – 40 kHz (b) 23 – 26 kHz

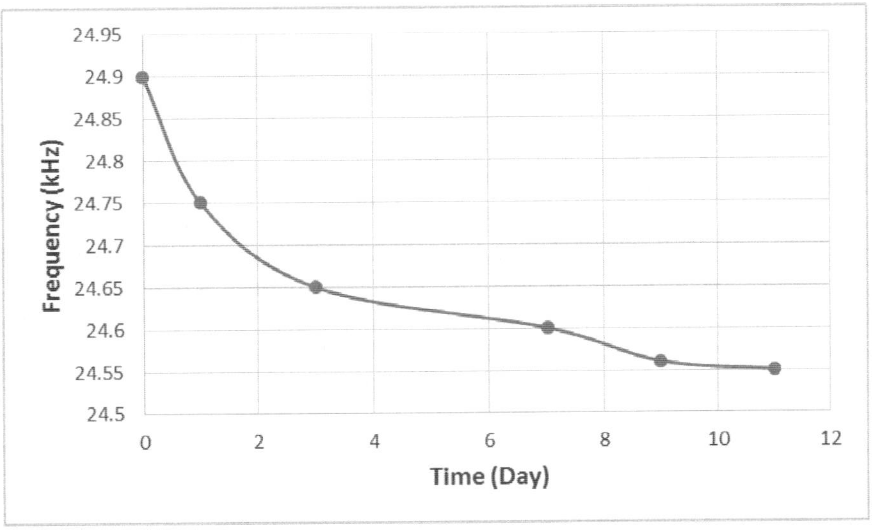

Figure 5. Resonance frequency acquired from PZT patch surface bonded on Specimen H2SO4

Similar observation can be found from Specimen NaCl, which is exposed to 10% sodium chloride, as presented in Figure 6. In this case, a resonance peak occurring at 50.65 kHz is selected as damage indicator. In this case, however, the total reduction in resonance frequency across 11 days is 0.15 kHz, in comparison to a total of 0.35 kHz in Specimen H2SO4. This clearly reflects the fact that the corrosion inflicted on Specimen H2SO4, which is exposed to acidic solution, is much more severe than Specimen NaCl which is exposed to saline solution.

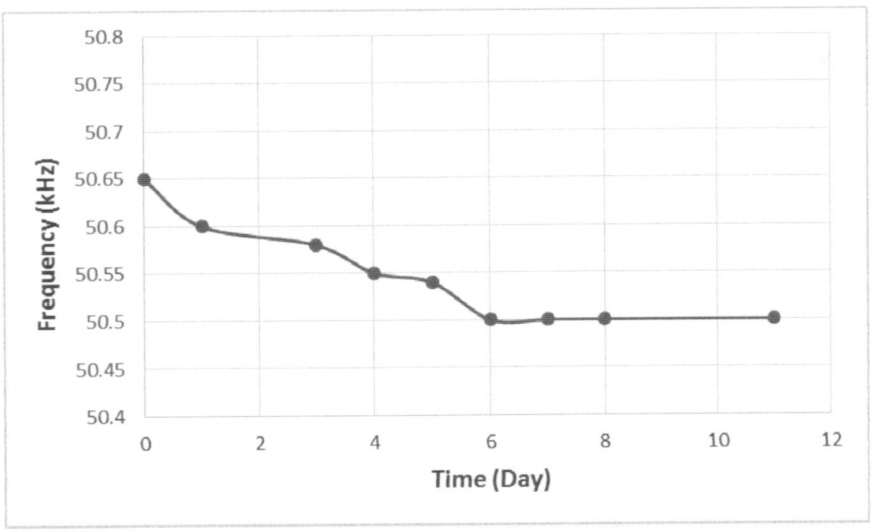

Figure 6. Resonance frequency acquired from PZT patch surface bonded on Specimen NaCl

Finally, observation is also made on Specimen Wtr, which is exposed to tap water. In this case no significant change in resonance frequency is observed, apart from some minor variations in magnitude. This indicates that water is not corrosive and has not caused significant reduction in stiffness of the host structure in such a short term of exposure. However, future study may focus on longer term monitoring to observe any possible changes with time.

4. Conclusion

This paper investigates the feasibility of employing the PZT based EMI technique in monitoring the corrosion process of epoxy. Experimental results show that the changes in electrical admittance acquired can be correlated to the corrosion process. The reduction of resonance frequency of selected structural peaks is found to reflect the stiffness reduction process of the host structure as corrosion takes place. The severity of corrosion on specimen exposed to acid in comparison to sodium chloride is also reflected from the extent of alteration of the resonance frequency, which shows the robustness of the proposed technique in detecting and characterising the corrosion process. Future studies can focus on developing parametric or statistical quantification approaches to monitor the actual severity of the corrosion process. The capability of the EMI technique in providing autonomous, real-time, remote monitoring with quick response using non-intrusive active sensors could potentially replace its conventional counterparts.

5. References

Banea, MD & da Silva LFM 2009, 'Adhesively bonded joints in composite materials: an overview', *Proceedings of the Institution of Mechanical Engineers Part 1, Journal of Materials: Design and Applications*, vol. 223, pp. 1-18.

BS EN ISO 527, 1996, '*Plastics-determination of tensile properties*', UK, British Standards.

Kong, Q & Song, G 2017, 'A Comparative Study of the Very Early Age Cement Hydration Monitoring Using Compressive and Shear Mode Smart Aggregates', *IEEE Sensors Journal*, vol. 17 no. 2, pp. 256-260.

Lapique, F & Redford, K 2002, 'Curing effects on viscosity and mechanical properties of a commercial epoxy resin adhesive', *International Journal of Adhesion & Adhesives*, vol. 22, pp. 337-346.

Lim, YY, Bhalla, S & Soh, CK 2006, 'Structural identification and damage diagnosis using self-sensing piezo-impedance transducers', *Smart Materials and Structures*, vol. 15, no. 4, pp. 987.

Lim, YY, Kwong, KZ, Liew, WYH & Soh, CK 2016, 'Non-destructive concrete strength evaluation using smart piezoelectric transducer—a comparative study', *Smart Materials and Structures*, vol. 25, no. 8, 085021.

Lim, YY & Soh, CK 2011, 'Fatigue life estimation of a 1D aluminum beam under mode-I loading using the electromechanical impedance technique', *Smart Materials and Structures*, vol. 20, no. 12, 125001.

Lim, YY & Soh, CK 2014, 'Electro-mechanical impedance (EMI)-based incipient crack monitoring and critical crack identification of beam structures', *Research in Nondestructive Evaluation*, vol. 25 no. 2, pp. 82-98.

Messler, RW 2004, 'Joining composite materials and structures: some thought provoking possibility', *J Thermoplast Compos Mater*, vol. 17, *pp.* 51-75.

Na, WS 2017, 'Possibility of detecting wall thickness loss using a PZT based structural health monitoring method for metal based pipeline facilities', *NDT & E International*, vol. 88, pp. 42-50.

Prolongo, S & Urena, A 2007, 'Durability of aluminium adhesive joints bonded with a homopolymerised epoxy resin', *Journal of Adhesion*, vol. 83, pp. 1-14.

Sika 2014, *Product Data Sheet - Sikadur 30 US*, Edition 9.23.

Talakokula, V, Bhalla, S, & Gupta, A 2014, 'Corrosion assessment of reinforced concrete structures based on equivalent structural parameters using electro-mechanical impedance technique', *Journal of Intelligent Material Systems and Structures*, vol. 25, no. 4, pp. 484-500.

Yang, Y, Lim, YY & Soh, CK 2008, 'Practical issues related to the application of the electromechanical impedance technique in the structural health monitoring of civil structures: I. Experiment', *Smart Materials and Structures*, vol. 17, no. 3, 035008.

The 8th International Conference on
Structural Health Monitoring of Intelligent Infrastructure
Brisbane, Australia | 5-8 December 2017

Extraction of Bridge Modal Parameters using a Moving Instrumented Vehicle by SSI Technique

J.T. Li[1], X.Q. Zhu, B. Samali[2]

[1, 2] *Centre for Infrastructure Engineering, School of Computing, Engineering and Mathematics, Western Sydney University, Penrith, NSW 2751 - Australia, Email:* jacky.li@westernsydney.edu.au

Abstract

Research on indirect bridge structural identification from the dynamic response of a moving instrumented vehicle has been very attractive in recent years. Unlike conventional bridge structural health monitoring methods that many sensors are installed directly on a bridge to obtain structural dynamic responses, the indirect bridge monitoring utilises an instrumented vehicle to capture the bridge dynamic information during its moving over the bridge. The bridge response often exhibits an inherent randomness due to the road surface roughness as well as the traffic excitation. It is suggested stochastic analysis should be performed. In this paper, the stochastic subspace identification (SSI) method is used for the structural dynamic parameter identification from the vehicle dynamic response only. To reduce the blurring effect of road surface roughness on the indirect structural model identification, two connected identical vehicles are adopted. The white noise support excitation is added to briefly simulate the traffic excitation to the bridge. Cases studies show that the improved SSI using the subtraction residual of the two connected vehicles' responses is very effective. The bridge frequencies can be successfully identified from dynamic responses of the instrumented vehicles.

1. Introduction

The bridge infrastructures are playing a critical role in the modern transportation system. They are inevitably subjected to continuous degradation due to the natural and human factors. The bridge health monitoring is essential since it can provide the health information and give early warning for the potential risk of bridge structures. It has been decades since the conduct of vibration-based bridge health monitoring (Doebling et al. 1998; Carden & Fanning 2004; Abdeljaber et al. 2017). Generally, the structure damage is related to the change of structural physical properties. The changes of the physical properties (i.e. mass, damping and stiffness) will cause variations in the modal properties (i.e. frequencies, mode shapes and modal damping) (Malekjafarian et al. 2015). Accurate modal parameter identification is one of the critical tasks for bridge health monitoring. Conventional approaches use the dynamic responses of the bridge by installing sensors directly on the bridge to extract the bridge dynamic properties. This poses a number of practical challenges, such as the need of the bridge closure, involuntary damage of installed equipment, the high initial and recurring cost of the sensory system etc. Yang et al. (2004) first proposed an indirect approach to extract natural frequencies of bridge structures by analysing the acceleration response of the vehicle during its passing over the bridge. The sensors were installed on the axle of a vehicle instead of the bridge deck. Through further investigation, Yang and Lin (2005) presented that the vehicle response in a vehicle-bridge interaction system contained three groups of frequencies, i.e., the driving frequency, vehicle frequency, and bridge frequency. Among these frequency components, the bridge frequency is that of interest in bridge health monitoring. Obviously, the indirect approach is more portable, convenient and cost-effective than the direct approach. Stochastic subspace identification has widely been used for operational modal analysis to extract the structural dynamic parameters from the system responses only (Bodeux & Golinval 2003; Hong, Ubertini & Betti 2012; Yang & Chen 2015). SSI methods take advantage of robust numerical techniques such as QR factorisation and singular value decomposition by using geometric tools such as the oblique projections of the row space of matrices (Marchesiello et al. 2009).

The dynamic responses recorded from the sensors on the moving vehicle are sensitive to the bridge road surface roughness which would induce a high level of amplitude of excitation at the vehicle-related frequencies (Yang et al. 2012a). This has the blurring effect to the identification of bridge dynamic parameters and damage detection. For practical application of indirect bridge system identification, one of the main challenges for the successfully extraction of bridge modal parameters is to eliminate or reduce the effect of road surface roughness (Zhu & Law 2015). To reduce the effect of the road surface roughness, Yang et al. (2012a) proposed to use two identical vehicles and subtract the synchronized vehicle responses from each other. In this paper, an improved stochastic subspace identification method (SSI) is proposed to identify the bridge frequencies from the residual response. When two identical vehicles pass through the bridge with the same roughness profile, their response amplitudes associated with roughness are assumed as the same. The numerical results show that the bridge frequencies can be identified accurately from the residual response. The roughness effect can be reduced by subtracting one response from the other.

2. Vehicle-bridge Interaction model

2.1 The equation of motion for the vehicle

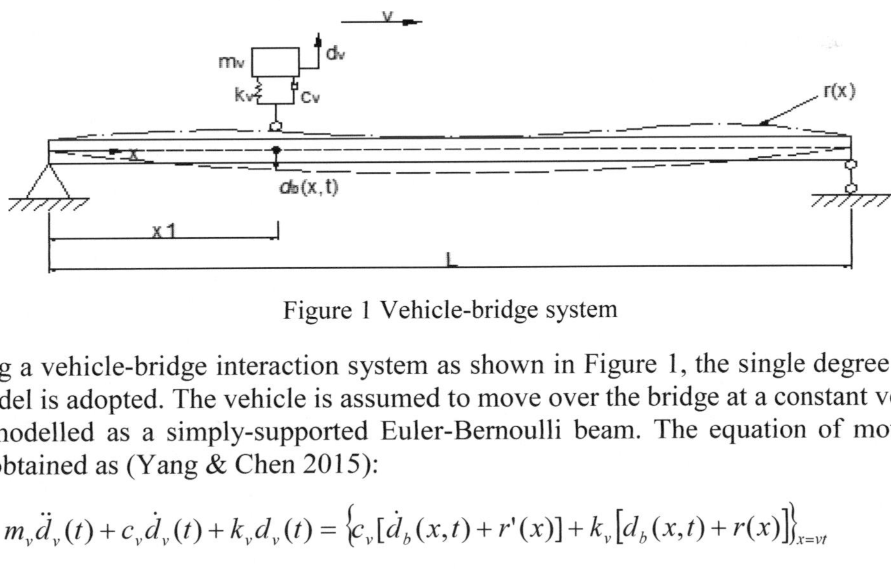

Figure 1 Vehicle-bridge system

Considering a vehicle-bridge interaction system as shown in Figure 1, the single degree of freedom vehicle model is adopted. The vehicle is assumed to move over the bridge at a constant velocity. The bridge is modelled as a simply-supported Euler-Bernoulli beam. The equation of motion for the vehicle is obtained as (Yang & Chen 2015):

$$m_v\ddot{d}_v(t)+c_v\dot{d}_v(t)+k_vd_v(t)=\left\{c_v[\dot{d}_b(x,t)+r'(x)]+k_v[d_b(x,t)+r(x)]\right\}_{x=vt} \tag{1}$$

where m_v is the body mass of the vehicle; k_v is the stiffness of the tyre; c_v represent the damping of the tyre; d_v is the vertical displacement of the axles; v is the speed of moving vehicle, $d_b(x,t)$ is the vertical displacement of the beam bridge at the contact point; $r(x)$ is the surface roughness function.

2.2 The equation of motion for the bridge

According the Euler-Bernoulli beam theory, the governing equation for flexural vibrations of the bridge under a moving vehicle can be written as:

$$EI\frac{\partial^4 d_b(x,t)}{\partial x^4}+\rho\frac{\partial^2 d_b(x,t)}{\partial t^2}+\mu\frac{\partial d_b(x,t)}{\partial t}=-F(x,t)\delta(x-vt) \tag{2}$$

where E is Young's modulus of the beam; I is the moment of inertia of the cross-section; EI is the flexural rigidity of the beam; ρ is the mass of the beam per unit length; μ is the damping coefficient per unit length; $F(x,t)$ is the interaction force between the bridge and vehicle; $d_b(x,t)$ is the bridge's vertical displacement at time t; δ is the function of Dirac.

Using the mode superposition method for a one-dimensional Euler-Bernoulli beam under moving load, the displacement of the beam can be expressed as (Clough & Penzien 2003):

$$d_b(x,t)=\sum_{i=1}^{\infty}\phi_i(x)Y_i(t) \tag{3}$$

Substituting Eq. (3) into Eq. (2) and applying the orthogonality of the mode of vibration, Eq. (2) becomes:

$$\ddot{Y}_n+2\zeta_n\omega_n\dot{Y}_n+\omega_n^{2}Y_n=-F(t)\phi_n \tag{4}$$

where $\phi_n(x)=\sqrt{\frac{2}{\rho L}}\sin\frac{n\pi x}{L}$ is the nth mode of vibration; $Y_n(t)$ is the modal coordinate. $2\zeta_n\omega_n=\mu/\rho$, $\omega_n^{2}=\frac{EI}{\rho}\left(\frac{n\pi}{L}\right)^4$, $F(t)=m_v g+m_v\ddot{d}_v$.

Combining Eqs. (1) and (4), the equation for the vehicle-bridge coupled interaction system can be obtained,

$$\begin{bmatrix}1&0&\cdots&0&m_v\phi_1\\0&1&\cdots&0&m_v\phi_2\\\vdots&\vdots&\cdots&0&\vdots\\0&0&\cdots&1&m_v\phi_N\\0&0&0&0&m_v\end{bmatrix}\begin{bmatrix}\ddot{Y}_1\\\ddot{Y}_2\\\vdots\\\ddot{Y}_N\\\ddot{d}_v\end{bmatrix}+\begin{bmatrix}2\zeta_1\omega_1&0&\cdots&0&0\\0&2\zeta_2\omega_2&\cdots&0&0\\\vdots&\vdots&\cdots&0&\vdots\\0&0&\cdots&2\zeta_N\omega_N&0\\-c_v\phi_1&-c_v\phi_2&\cdots&-c_v\phi_N&c_v\end{bmatrix}\begin{bmatrix}\dot{Y}_1\\\dot{Y}_2\\\vdots\\\dot{Y}_N\\\dot{d}_v\end{bmatrix}$$
$$+\begin{bmatrix}\omega_1^2&0&\cdots&0&0\\0&\omega_2^2&\cdots&0&0\\\vdots&\vdots&\cdots&0&\vdots\\0&0&\cdots&\omega_N^2&0\\-c_v\dot{\phi}_1-k_v\phi_1&-c_v\dot{\phi}_2-k_v\phi_2&\cdots&-c_v\dot{\phi}_N-k_v\phi_N&k_v\end{bmatrix}\begin{bmatrix}Y_1\\Y_2\\\vdots\\Y_N\\d_v\end{bmatrix}=\begin{Bmatrix}-\phi_1 m_v g\\-\phi_2 m_v g\\\vdots\\-\phi_N m_v g\\k_v r(x)+c_v\dot{r}(x)\end{Bmatrix} \tag{5}$$

Eq. (5) is transformed into matrix form as:

$$\mathbf{M}\ddot{\mathbf{d}}+\mathbf{C}\dot{\mathbf{d}}+\mathbf{K}\mathbf{d}=\mathbf{F} \tag{6}$$

where M, C and K are mass, damping and stiffness matrices of the vehicle-bridge system, respectively. The above equation can be solved using the Newmark-beta method. The dynamic responses of the vehicle calculated from the above will be used as the basis for extracting the bridge frequencies.

3. Extraction of Modal Parameters using SSI technique

3.1 Stochastic state-space model

An nth order linear time-invariant dynamic system under a stochastic excitation can be represented in the discrete form as

$$\mathbf{x}_{k+1} = \mathbf{A}\mathbf{x}_k + \mathbf{w}_k \tag{7a}$$

$$\mathbf{y}_k = \mathbf{C}\mathbf{x}_k + \mathbf{v}_k \tag{7b}$$

where $\mathbf{x}_k \in \Re^n$ and $\mathbf{y}_k \in \Re^l$ are the state and output vectors at the *k*th time step, respectively. The matrices $\mathbf{A} \in \Re^{n\times n}$ and $\mathbf{C} \in \Re^{l\times n}$ are the system and observation matrices, respectively. The vectors $\mathbf{w}_k \in \Re^n$ and $\mathbf{v}_k \in \Re^l$ represent the process and measurement noises, respectively. The procedures of using SSI algorithm to obtain bridge frequencies are given in the following section.

3.2 Procedure of the SSI algorithm

SSI method is widely used for the output only system identification. It identifies the state space matrices by constructing a block Hankel matrix using the output data and by the application of numerical techniques such as singular value and QR decomposition. The problem of order determination is solved by constructing a stabilisation diagram for increasing model order. The dynamic parameters of the structure are obtained from the state space matrices. The procedure of the SSI algorithm is given as (Van Overschee & De Moor 2012):

1) The Block Hankel matrix

   The output measurements $\{\mathbf{y}_k, k = 0,1,\cdots s\}$, are gathered in a block Hankel matrix H with $2i$ block rows and j columns

$$Y_{0|2i-1} \overset{\text{def}}{=} \begin{pmatrix} y_0 & y_1 & \cdots & y_{j-1} \\ y_1 & y_2 & \cdots & y_j \\ \cdots & \cdots & \cdots & \cdots \\ y_{i-1} & y_i & \cdots & y_{i+j-2} \\ \hline y_i & y_{i+1} & \cdots & y_{i+j-1} \\ y_{i+1} & y_{i+2} & \cdots & y_{i+j} \\ \cdots & \cdots & \cdots & \cdots \\ y_{2i-1} & y_{2i} & \cdots & y_{2i+j-2} \end{pmatrix} \overset{\text{def}}{=} \left(\frac{Y_{0|i-1}}{Y_{i|2i-1}}\right) \overset{\text{def}}{=} \left(\frac{Y_p}{Y_f}\right) \tag{8}$$

where $j = s - 2i + 1$; the matrices Y_p and Y_f are the past and future output matrix, respectively.

2) Calculate the projection
Projections play a major role in stochastic subspace identification. The projection of the row space of the future outputs into the row space of the past outputs is defined as

$$O_i \overset{def}{=} Y_f / Y_p = Y_f Y_p^T (Y_p Y_p^T)^{\dagger} Y_p \tag{9}$$

where † represents the pseudo inverse. The QR decomposition is used to calculate this orthogonal projection.

3) Calculate the SVD of the weighted projection
Two weighting matrices $W_1 \in \Re^{li \times li}$ and $W_2 \in \Re^{j \times j}$ are introduced. Then, calculate the singular value decomposition to the weighted projection $W_1 O_i W_2$:

$$W_1 O_i W_2 = \begin{pmatrix} U_1 & U_2 \end{pmatrix} \begin{pmatrix} S_1 & 0 \\ 0 & 0 \end{pmatrix} \begin{pmatrix} V_1^T \\ V_2^T \end{pmatrix} = USV^T \tag{10}$$

where $W_1 = \Phi_{[Y_f, Y_f]}^{-1/2}$, $W_2 = I_j$; $\Phi_{[Y_f, Y_f]} = \frac{1}{j} Y_f Y_f^T$. Select a range of different system order n to obtain the corresponding U_1, U_2, S_1.

4) Determine the extended observability matrix Γ_i as:

$$\Gamma_i = W_1^{-1} U_1 S_1^{1/2} \tag{11}$$

where the extended observability matrix Γ_i is defined as $\Gamma_i \overset{def}{=} \begin{pmatrix} C \\ CA \\ CA^2 \\ \vdots \\ CA^{i-1} \end{pmatrix} \in \Re^{li \times n}$

5) Determine the bridge frequencies
The state matrix $\mathbf{A}$ of the bridge contains all the properties of the bridge. It is determined by

$$\mathbf{A} = \underline{\Gamma_i}^{\dagger} \overline{\Gamma_i} \tag{12}$$

where $\underline{\Gamma_i}$ denotes Γ_i without the last l (number of outputs) rows; $\overline{\Gamma_i}$ denotes Γ_i without the first l rows. The state matrix $\mathbf{A}$ is given in discrete form. It can be transformed to continuous form by the logarithmic matrix operation

$$\mathbf{A_c} = \log(\mathbf{A}) / \Delta t \tag{13}$$

By the eigen decomposition, the following is obtained:

$$\mathbf{A_c} = \Psi \Lambda \Psi^{-1} \tag{14}$$

where Ψ =eigenvector, Λ=diagonal matrix containing the eigenvalues λ_k.

The real and imaginary parts of the eigenvalues are related to natural frequencies ω_k as

$$\omega_k = \sqrt{Re(\lambda_k)^2 + Im(\lambda_k)^2} \tag{15}$$

6) Establish the stabilization diagram
   Using the calculated frequency values corresponding to different selected system orders, the stabilization diagram of the frequency can be established. A frequency is considered successfully identified when it continues to exist with an increasing number of system orders without much variation in value.

4. Numerical Simulations

4.1 Case 1: single vehicle moving on smooth or rough road surface

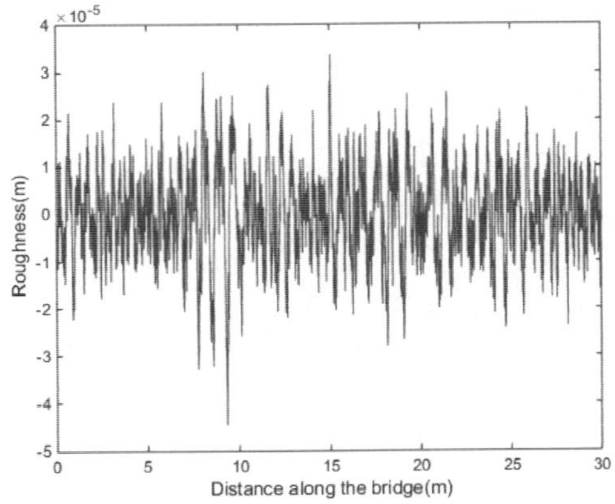

Figure 2 Class A roughness profile along the bridge surface

The vehicle is simulated as a SDOF quarter-car model and the bridge as a simply-supported Bernoulli-Euler beam. Applying the same system parameters given in (Yang & Chen 2015), the parameters of the bridge are: $L = 30m, \rho = 10700kg/m, I = 0.494m^4, E = 29.43GPa$. The first three natural frequencies of the bridge are 2.03, 8.12 and 18.3Hz, respectively. The properties for the vehicle are: $m_v = 100kg,\ k_v = 170kN/m$. The weight ratio between the vehicle and bridge is 3.1×10^{-4} . The natural frequency of the vehicle is 6.5Hz. The vehicle speed is 2 m/s. The damping of the vehicle and bridge are not considered in this study. To study the effect of the road surface roughness, the Class A road surface defined in (Yang et al. 2012a) is considered , as shown in Figure 2. For comparison, the acceleration responses of the vehicle are normalised by dividing their corresponding peak acceleration values. Figure 3 shows the normalized acceleration responses of the vehicle and its corresponding frequency spectrums when the vehicle moves over a smooth or rough bridge deck. In Figure 3(b), when the bridge deck is smooth, there are three dominant frequencies in the spectrum and the identified frequency values are 2.07, 8.20 and 18.27Hz. These frequencies are very close to those first three frequencies of the bridge. So when there is no roughness, the bridge frequencies can be obtained from the response spectrum directly with a light vehicle. As shown in Figure 3(b), when the roughness of the bridge surface is considered, only the vehicle-related frequency dominates the spectrum of vehicle response. No bridge related frequencies are visible in the figure. Therefore, the effect of road surface roughness makes it difficult to identify the bridge frequencies from vehicle responses.

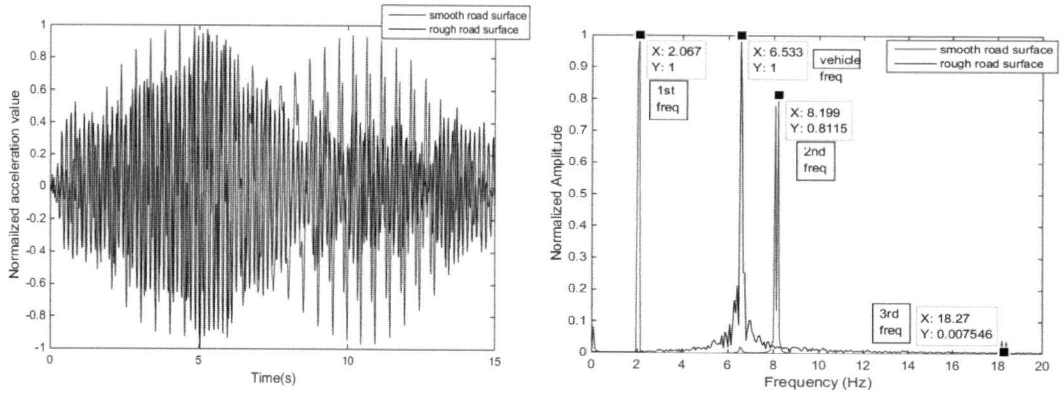

(a) The normalized acceleration of the vehicle (b) The normalized frequency spectrum

Figure 3 The normalised vehicle responses moving on smooth and rough bridge surfaces and their spectrums

4.2 Case 2: two identical cars moving on rough road surface

To reduce the effect of the road surface roughness, Yang et al. (2012b) proposed a method using two connected identical vehicles. Those two identical vehicles with the same parameters as given in Case 1 are used in this study, as shown in Figure 4. The two vehicles move at the same speed with a constant distance of 3m. Ongoing traffic excitation is simulated as the white-noise excitation to the bridge supports. The white noise support excitation is shown in Figure 5. When the Class A roughness is considered here, the dynamic responses of the two vehicles and their frequency spectrums are shown in Figure 6. The time lag between the two vehicle responses is eliminated by shifting the responses to the same start point. Only the vehicle frequency can be seen from the frequency spectrum in the figure. Applying the SSI to the response of the front vehicle, the stabilization diagram is obtained as shown in Figure 7. The results show that only the vehicle-related frequency is clearly identified.

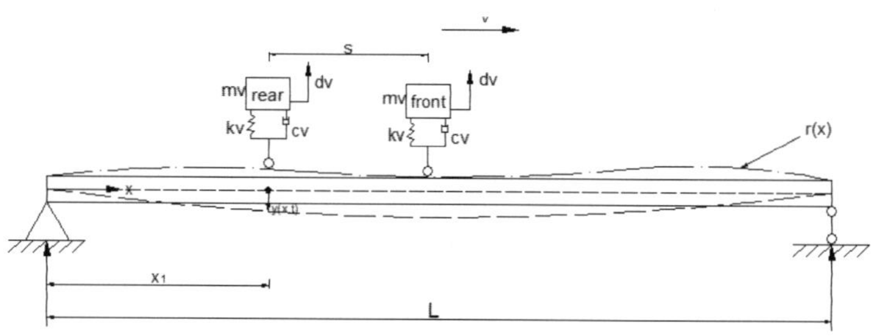

Figure 4 Two SDOF vehicles and bridge model

By subtracting the acceleration time series of one vehicle to that of the other vehicle, the dynamic response residual is obtained. The frequency spectrum of the response residual is given in Figure 8. The bridge related frequencies are invisible in the figure. Applying the SSI to the residual, Figure 9 shows the stabilization graph. From the figure, in addition to the vehicle frequency and the driving frequency, other three frequencies are identified and they are corresponding to those first three frequencies of the bridge. The results show that the bridge frequencies can be identified from the residual response of two identical vehicles.

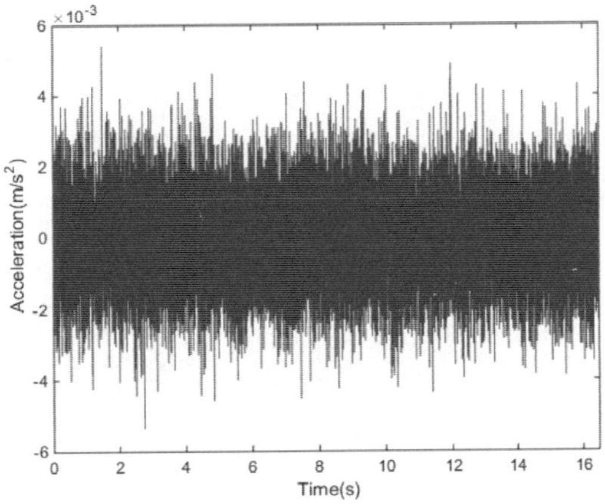

Figure 5 White noise support acceleration

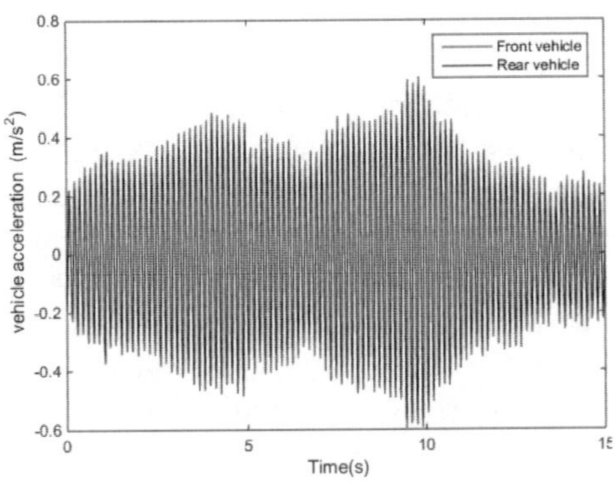

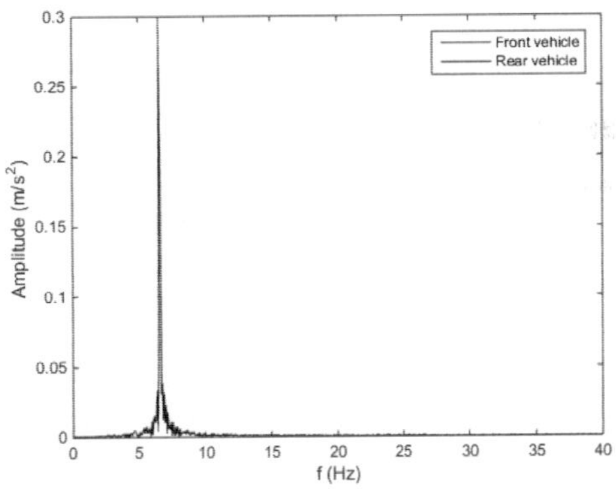

(a) The acceleration of the vehicle (b) The frequency spectrum

Figure 6 Responses of two identical vehicles

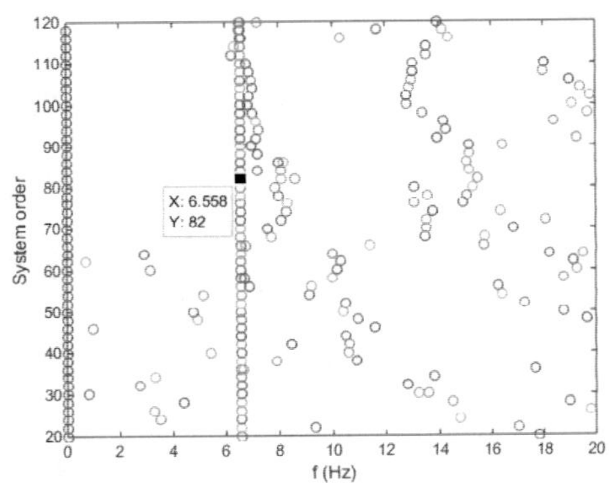

Figure 7 Stabilization graph of the front vehicle response

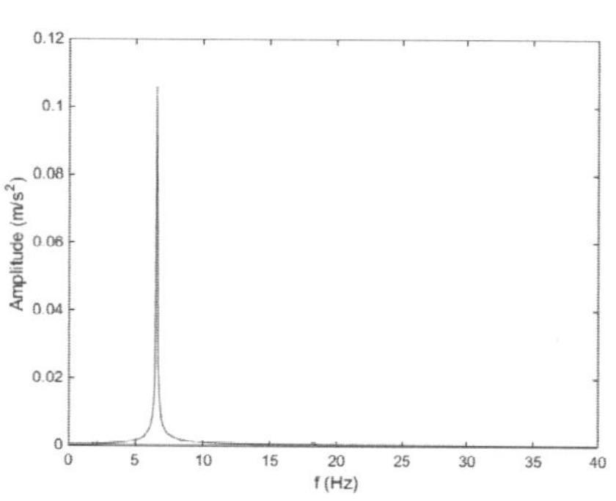

Figure 8 the frequency spectrum of the response residual

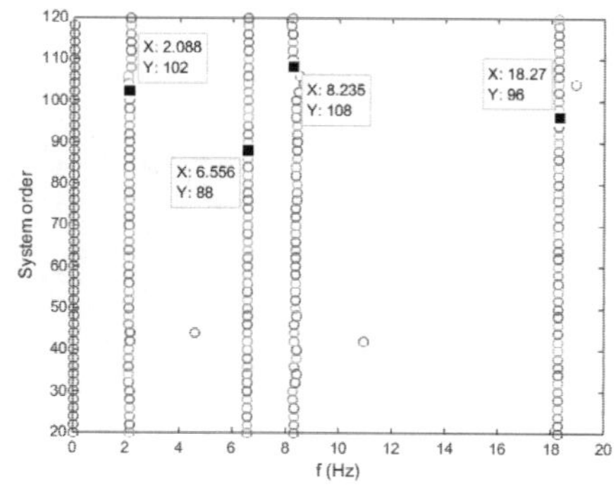

Figure 9 Stabilization graph of the residual

5. Conclusions

An improved stochastic subspace approach has been presented to identify the frequencies of the bridge from the response of the passing vehicles. The bridge frequencies can be obtained directly by inspecting of the power spectrum of the vehicle response when the surface of the bridge is smooth. The road surface roughness in the vehicle-bridge system impairs the ability of indirect bridge modal identification using vehicle responses. The vehicle frequency dominates the spectrum, which makes the bridge frequencies supressed or invisible. By using two identical vehicles to reduce the effect of roughness and using SSI method, the first three bridge frequencies are clearly identified from the residual response. It should be noted that in using of SSI, the assumption of stationary white-noise excitation to the system should be fulfilled. The light vehicle is used to capture the response of the bridge under traffic excitations. Further work will be carried out on the study for the time-variant and non-stationary features of the vehicle responses when the vehicle mass is non-negligible in the vehicle-bridge system.

References

Abdeljaber, O, Avci, O, Kiranyaz, S, Gabbouj, M & Inman, DJ 2017, 'Real-time vibration-based structural damage detection using one-dimensional convolutional neural networks', *Journal of Sound and Vibration*, vol. 388, pp. 154-70.

Bodeux, JB & Golinval, JC 2003, 'Modal identification and damage detection using the data-driven stochastic subspace and armav methods', *Mechanical Systems and Signal Processing*, vol. 17, no. 1, pp. 83-9.

Carden, EP & Fanning, P 2004, 'Vibration based condition monitoring: a review', *Structural Health Monitoring*, vol. 3, no. 4, pp. 355-77.

Clough, RW & Penzien, J 2003, *DYNAMICS OF STRUCTURES*, Third Edition edn, Computers & Structures, Inc., Berkeley, CA 94704 USA.

Doebling, SW, Farrar, CR & Prime, MB 1998, 'A summary review of vibration-based damage identification methods', *Shock and vibration digest*, vol. 30, no. 2, pp. 91-105.

Hong, AL, Ubertini, F & Betti, R 2012, 'New stochastic subspace approach for system identification and its application to long-span bridges', *Journal of Engineering Mechanics*, vol. 139, no. 6, pp. 724-36.

Malekjafarian, A, McGetrick, PJ & OBrien, EJ 2015, 'A review of indirect bridge monitoring using passing vehicles', *Shock and Vibration*, vol. 2015.

Marchesiello, S, Bedaoui, S, Garibaldi, L & Argoul, P 2009, 'Time-dependent identification of a bridge-like structure with crossing loads', *Mechanical Systems and Signal Processing*, vol. 23, no. 6, pp. 2019-28.

Van Overschee, P & De Moor, B 2012, *Subspace identification for linear systems: Theory—Implementation—Applications*, Springer Science & Business Media.

Yang, YB & Chen, WF 2015, 'Extraction of bridge frequencies from a moving test vehicle by stochastic subspace identification', *Journal of Bridge Engineering*, vol. 21, no. 3, p. 04015053.

Yang, YB, Li, YC & Chang, KC 2012a, 'Effect of road surface roughness on the response of a moving vehicle for identification of bridge frequencies', *Interaction and multiscale mechanics*, vol. 5, no. 4, pp. 347-68.

Yang, YB, Li, YC & Chang, KC 2012b, 'Using two connected vehicles to measure the frequencies of bridges with rough surface: a theoretical study', *Acta Mechanica*, pp. 1-11.

Yang, YB & Lin, CW 2005, 'Vehicle–bridge interaction dynamics and potential applications', *Journal of Sound and Vibration*, vol. 284, no. 1, pp. 205-26.

Yang, YB, Lin, CW & Yau, JD 2004, 'Extracting bridge frequencies from the dynamic response of a passing vehicle', *Journal of Sound and Vibration*, vol. 272, no. 3, pp. 471-93.

Zhu, XQ & Law, SS 2015, 'Structural health monitoring based on vehicle-bridge interaction: accomplishments and challenges', *Advances in Structural Engineering*, vol. 18, no. 12, pp. 1999-2015.

The 8th International Conference on
Structural Health Monitoring of Intelligent Infrastructure
Brisbane, Australia | 5-8 December 2017

3D Finite Element Prediction of Scattering and Mode Conversion of Lamb Waves at Delaminations in Composite Laminates

C.T. Ng [1], G.T Pudipeddi [1], A. Kotousov [2]
[1] *School of Civil, Environment & Mining Engineering, The University of Adelaide – Adelaide, Australia,*
Email: alex.ng@adelaide.edu
[2] *School of Mechanical Engineering, The University of Adelaide – Adelaide, Australia,*

Abstract

Detection of damage using Lamb wave has attracted significant attention in the last decade. It has been shown that Lamb wave is sensitive to most types of damages, efficient in detecting small and subsurface damage, and able to inspect large area. However, the capability and sensitivity of Lamb wave on detecting delaminations in composite laminates, especially the mode conversion effect, have not yet been fully investigated, which limits quantitative characterization of the delaminations. This study presents a three-dimensional (3D) finite element (FE) simulation of the fundamental anti-symmetric mode (A_0) Lamb wave scattering and mode conversion from A_0 to fundamental symmetric mode (S_0) Lamb wave at the delamination in composite laminates. The 3D FE simulation is first validated using experimentally measured data, such as group velocity and amplitude of the incident Lamb wave in different wave propagation directions. There is good agreement between the FE calculated and experimentally measured results. The 3D FE model is then employed to predict the A_0 scattered Lamb wave and mode conversion from A_0 to S_0 Lamb wave, which provides understanding about the scattering and mode conversion phenomenon of Lamb wave at the delaminations.

1. Introduction

Fibre-reinforced composite laminates have been widely employed in civil, mechanical, aerospace, and automobile industries due to their light weight, increased toughness and anti-corrosion characteristics. The composite laminates are fabricated by assembling layers of fibrous composite materials. The flexural strength of a laminate is highly dependent on the shear transfer between each of its layers. However, delamination is one of the challenging issues in the applications of this material. It is a separation of adjacent subsurface laminae with no obvious visual evidence on the surface, and hence, it cannot be detected through visual inspection. A small delamination could significantly reduce the stiffness of the composite laminate. Therefore, it is important to provide an early detection of the delaminations.

In the last decade, different non-destructive evaluation (NDE) techniques have been developed to provide safety inspection of composite laminates. Among these NDE techniques, Lamb wave has been widely recognized as one of the promising approaches for detecting the delaminations in the composite laminates. The interactions of Lamb wave with various types of damages, such as hole, crack and corrosion, have been studied for both isotropic plates (Fromme & Sayir 2002, Ratassepp *et al.* 2008, Moreau *et al.* 2012) and composite laminates (Veidt & Ng 2011, Ng *et al.* 2012). Early research in this field focused on the interaction of individual Lamb wave mode with the delaminations, and the nature of the reflected waves. Guo and Cawley (1993) studied the interaction of the fundamental symmetric mode (S_0) Lamb wave with delaminations in a cross-ply composite laminate. They investigated the reflection of Lamb waves from the delaminations located at various interfaces in the composite laminate. It was found that there was no reflection of waves from the delaminations located at interfaces where the shear stress was zero. Demčenko *et al.* (2006) and Ng *et al.* (2012) studied the interaction of the fundamental anti-symmetric mode (A_0) Lamb wave with delaminations. The latter study numerically and experimentally calculated the scattering directivity patterns (SDPs) of A_0 Lamb wave at the delaminations with various sizes and at various locations. Murat *et al.* (2016) also used numerical models to study the effects of delamination shape, size and location on the SDPs for different sizes of delaminations. It was observed that the delamination width has the most significant influence on the scattering directivity while the depth affects the magnitude of the scattered waves.

Ramadas *et al.* (2009) and Ramadas *et al.* (2010) showed that when an incident A_0 Lamb wave interacts with the entrance of the delamination, it generates scattered S_0 Lamb wave that travels within the sub-laminates at the delamination region. Each of these scattered S_0 Lamb waves as well as the incident A_0 Lamb wave further interact at the exit of the delamination and induce mode conversion effect. The mode conversion effect of Lamb wave at the delaminations was studied for both symmetric (Ramadas *et al.* 2009) and asymmetrically (Ramadas *et al.* 2010) located delaminations. It was observed that in the case of symmetrically located delaminations, the mode converted A_0-S_0 Lamb wave is confined to the sub-laminates. In the case of asymmetrically located delaminations, one is able to observe the mode converted S_0 Lamb wave in both the reflected side and the transmitted side of the composite laminates. This phenomenon was studied numerically using a two-dimensional (2D) finite element (FE) model. The main topic of interest for this paper is the phenomenon of mode-conversion that occurs when A_0 Lamb wave interacts with the delaminations. This study employs a three-dimensional (3D) FE model to predict the A_0 Lamb wave and mode converted S_0 Lamb wave at the delaminations.

The structure of the paper is as below. Section 2 describes the 3D FE simulation, which includes the excitation of the Lamb wave and modelling of the delamination. Section 3 presents the experimental results. Section 4 investigates the scattered A_0-A_0 and mode converted A_0-S_0 Lamb wave at the delamination. Finally, conclusions are drawn in Section 5.

2. Three-dimensional finite element simulation of Lamb wave in composite laminates

In this study, a 3D explicit FE method is used to simulate Lamb wave propagation, scattering and mode conversion at the delaminations in an eight-ply $[45/\text{-}45/0/90]_S$ quasi-isotropic composite laminate. A commercial software, ABAQUS, is used in the FE simulations. The composite laminate is made by lamina of Epoxide EHM-32 unidirectional carbon/epoxy prepreg tape with 0.53 fibre volume fraction. The density and thickness of the lamina are 1300 kg/m^3 and 0.2 mm, respectively. The elastic properties of the lamina are calculated using micromechanics theory (Chamois 1989) and they are shown in Table 1. The total thickness of the quasi-isotropic composite laminate is 1.6 mm. In the FE model, each lamina is modeled by eight-noded 3D solid brick element (C3D8I) with full integration, in which each node has three translation degrees-of-freedom (DoFs). Each element has an in-plane square dimension of 0.5×0.5 mm^2.

Table 1: Elastic properties of the Epoxide EHM-32 prepare lamina

E_{11} (GPa)	E_{22} (GPa)	E_{33} (GPa)	G_{12} (GPa)	G_{13} (GPa)	G_{23} (GPa)	ν_{12}	ν_{13}	ν_{23}
111	7.16	7.16	3.52	3.52	2.5	0.33	0.33	0.44

The excitation signal is a 120 kHz narrow-band 5-cycle sinusoidal tone burst pulse modulated by a Hanning window. The A_0 Lamb wave is generated by applying the pressure in the out-of-plane direction to the surface FE nodes covered by a 10 mm diameter half circular transducer area. Based on the cylindrical coordinate system shown in Figure 1, the half circular transducer is located at r = 180 mm and θ =180°. The wavelength of the A_0 and S_0 Lamb wave at this excitation frequency are 9.55 mm and 49.02 mm, respectively. Therefore, at least around 20 FE nodes exist in each wavelength for the A_0 Lamb wave, which ensures the accuracy of the simulation. A dalamination is modeled at the centre of the composite laminate (the origin of the cylindrical coordinate system) as shown in Figure 1. The delamination is modelled by untying of nodes within the area of the delamination, which behaves as a zero volume split delamination (Ramadas *et al.* 2009). The Lamb wave signals are calculated at measurement points at shown in Figure 1. The radius of the circular measurement point is at r = 100 mm, with θ =10° step increment. As this study focuses on the mode conversion from A_0 to S_0 Lamb wave and the wavelength of the S_0 is relative large, therefore, absorbing layers by increasing damping (ALID) (Rajagopal *et al.* 2012) are applied to three edges of the composite laminate as shown in Figure 1. The use of the ALID could avoid the boundary reflection, and hence, the size of the FE model can be minimised to reduce the computational cost.

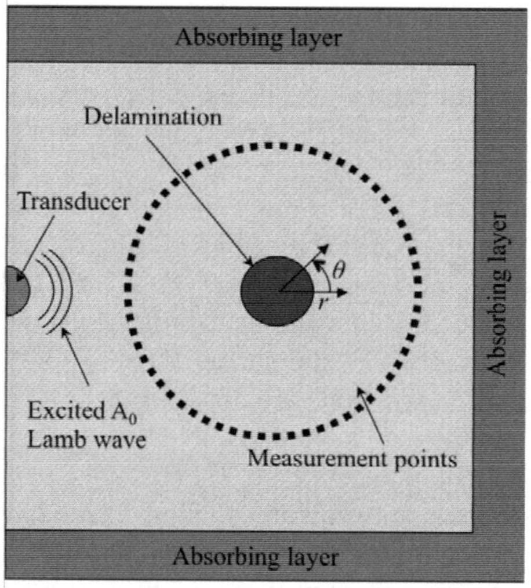

Figure 1. Schematic diagram of the FE model

3. Experimental verification

An 800×800×2 mm^3 eight-ply [45/-45/0/90] s quasi-isotropic composite laminate was used in the experiment. The composite laminate was made by Eporite EHM-32 lamina material, which is the same as the FE model in Section 2. A 10 mm diameter and 2 mm thick piezoceramic transducer was used to excite the A_0 Lamb wave. A 3 mm thick brass disc was used as a backing mass to increase the excitability of the A_0 Lamb wave. The transducer was surface mounted to the composite laminate to excite the Lamb wave. The out-of-plane displacement fields of the A_0 and S_0 Lamb wave were measured using laser vibrometer. Group velocities are calculated for excitation frequencies from 100 to 140 kHz in steps of 10 kHz. Table 2 shows the results of the FE calculated and experimentally measured group velocities at different frequencies. There is good agreement between the FE calculated and experimentally measured group velocities. Figure 2 shows the incident A_0 Lamb wave amplitude at different angles. Circular measurements were taken round the delamination at $0° \leq \theta < 360°$ with $\theta = 10°$ step increment. Reasonable agreement is found between the FE calculated and experimentally measured incident A_0 Lamb wave. Therefore, the 3D FE model is able to provide a reasonable prediction of the Lamb wave in the composite laminates. Due to the fiber direction of the laminate, there are relatively large magnitude of the A_0 Lamb wave at the direction of θ = 45°, 135°, 225° and 315°. This has been demonstrated by Veidt and Ng (2011) in the literature.

Table 2. FE calculated and experimentally measured group velocities

Frequency (kHz)	FE simulation	Experiment
100	1340.3 m/s	1338.7 m/s
110	1345.1 m/s	1344.1 m/s
120	1369.7 m/s	1351.4 m/s
130	1371.5 m/s	1355.0 m/s
140	1374.8 m/s	1356.5 m/s

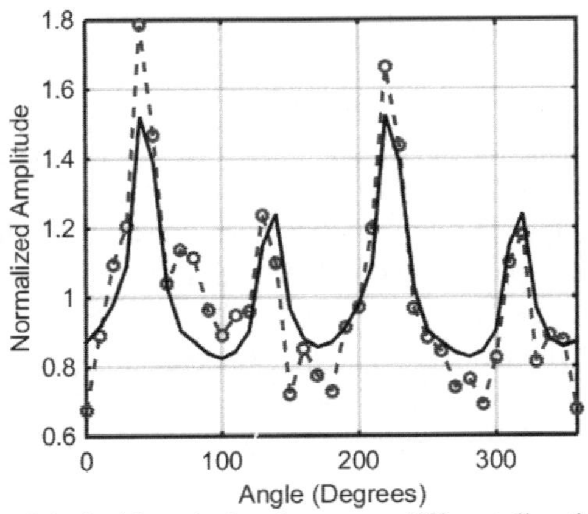

Figure 2. Normalized amplitude of the incident A_0 Lamb wave at different directions (solid line: FE calculated results, dashed line with circles: experimentally measured results)

4. Scattering and mode conversion of A_0 Lamb wave at delaminations

This section presents the FE prediction of the scattered A_0 Lamb wave and mode conversion from A_0 to S_0 Lamb wave. The same 3D FE model described in Section 2 is used in this study. The excitation signal is a 120 kHz narrow-band 5-cycle sinusoidal tone burst pulse modulated by a Hanning window. A 20 mm diameter is modeled at the origin of the cylindrical coordinate system as shown in Figure 1. The delamination is located at third and fourth layers of the laminate. Figure 3 shows an out-of-plane displacement field calculated at the top surface of the composite laminate. The location is at r = 100 mm and θ =180°. In the figure, the first wave package is the incident A_0 Lamb wave. The second and the third arrived wave packages are the mode converted S_0 Lamb wave (A_0-S_0) and scattered A_0 Lamb wave (A_0-A_0) from the delamination. The arrival time of the mode converted S_0 Lamb wave is the time of the A_0 incident wave propagate from the transducer to the delamination and the A_0 Lamb wave is mode converted to S_0 Lamb wave. Since the S_0 Lamb wave propagates faster than the A_0 Lamb wave, A_0-S_0 wave package arrives earlier than the A_0-A_0 wave package as shown in Figure 3.

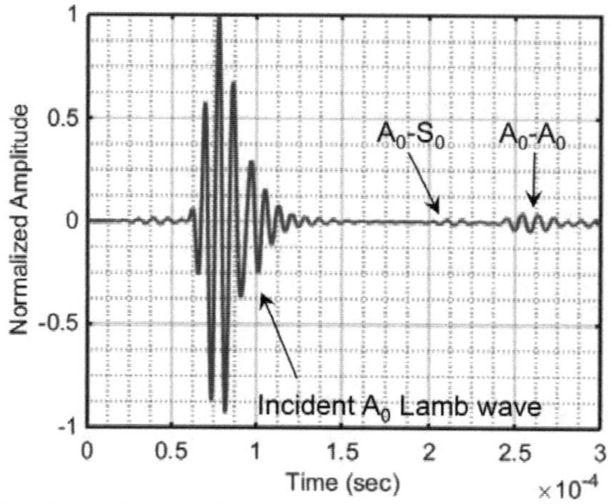

Figure 3. FE simulated time-domain signal at $r = 100$ mm and $\theta = 180°$ of the composite laminate

To extract the scattered wave signals, baseline subtraction is used (Ng & Veidt 2011). Two FE simulations are carried out, one without and the other one with the delamination in the FE model. Both FE models have identical configuration, such as meshing, element size, and excitation and measurement locations. The only difference is the FE model with the delamination, in which the delamination is modeled by untying of nodes within the area of the delamination. In addition the out-of-plane displacement fields at the top and bottom surface of the composite laminate are calculated. The scattered A_0-A_0 wave is obtained by superimposing the out-of-plane displacement fields at the top and bottom surface and then divided by two. However, for the scattered A_0-S_0 wave is obtained by subtracting the out-of-plane displacement field at the top surface by that at the bottom surface, and then divided by two. This process can fully extract the scattered A_0-A_0 and A_0-S_0 wave even they are overlapped in time-domain signals.

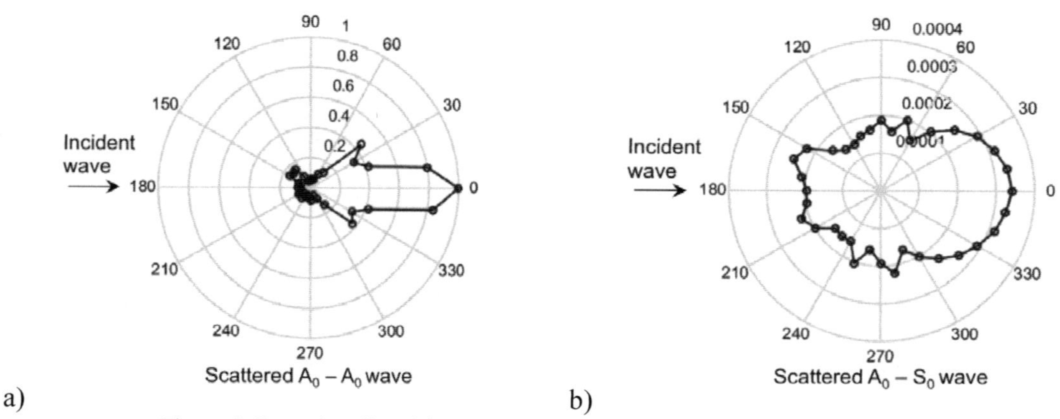

Figure 4. Scattering directivity pattern of the A_0-A_0 and A_0-S_0 Lamb wave at 120 kHz

Figures 4a and 4b show the SDP of the scattered A_0-A_0 and A_0-S_0 Lamb wave at the delamination. The results show that the SDP of both scattered A_0-A_0 and A_0-S_0 wave are not with respect to $\theta =$ 0° direction, which indicate they are both affected by the stacking sequence of the fiber. It should be noted that the magnitude of the scattered A_0-S_0 wave is obtained from the out-of-plane displacement field, which is due to the Poisson effect of the S_0 Lamb wave. If the in-plane displacement field is used, the magnitude of the scattered A_0-S_0 wave will be larger. The results show that the scattered A_0-A_0 wave amplitude is dominated in the forward scattering direction while the scattered A_0-S_0 wave amplitude is evenly distributed in the forward and backward scattering direction.

5. Conclusions

This paper has presented a study of Lamb wave scattering at the delaminations in the quasi-isotropic composite laminates. The 3D FE model has been proposed and verified by experimentally measured data. The results show that there is good agreement between the results of the FE simulation and experimental data. The 3D FE model has then been employed to predict the scattered A_0-A_0 and mode converted A_0-S_0 Lamb wave from the delamination. The results show that the incident A_0 Lamb wave, scattered A_0-A_0 and mode converted A_0-S_0 Lamb wave are affected by the stacking sequence of the composite laminates. The findings of this study have provided further understanding about the scattering and mode conversion phenomenon at the delaminations in the composite laminates

Acknowledgement

This work was supported by Australian Research Council under Grant Number DE130100261. The support is greatly appreciated.

References

Chamis, CC 1989, 'Mechanics of composite materials: Past, present and future', *Journal of Composite Technology and Research*, vol. 11, pp. 3-14.

Demčenko, A, Žukauskas, E, Kažys, R & Voleišis, A 2006, 'Interaction of the A0 Lamb wave mode with a delamination type defect in GLARE3-3/2 composite material', *Acta Acustica united with Acustica,* 92, pp. 540-548.

Fromme, P & Sayir, MB 2002, 'Measurement of the scattering of a Lamb wave by a through hole in plate', *The Journal of Acoustical Society of America*, 111, pp. 1165-1170.

Guo, N & Cawley, P 1993, 'The interaction of Lamb waves with delaminations in composite laminates', *The Journal of Acoustical Society of America*, 94, pp. 2240–2246.

Moreau, L, Velichko, A & Wilcox PD 2012, 'Accurate finite element modeling of guided wave scattering from irregular defects', *NDT & E International*, 45, pp. 46-54.

Murat BI, Khalili P & Fromme P 2016, 'Scattering of guided waves at delaminations in composite plates', *The Journal of Acoustical Society of America*, 139, pp. 3044-3052.

Ng, CT & Veidt, M 2011, 'Scattering of the fundamental anti-symmetric Lamb wave at delaminations in composite laminates', *The Journal of Acoustical Society of America*, 129, pp. 1288-1296.

Ng, CT & Veidt, M 2012, 'Analytical and finite element prediction of Lamb wave scattering at delamination in quasi-isotropic composite laminates', *Journal of Sound and Vibration*, 331, pp. 4870-4883.

Rajagopal, P, Drozdz, M, Skelton, EA, Lowe, MJS & Craster, RV 2012, 'On the use of absorbing layers to simulate the propagation of elastic waves in unbounded isotropic media using commercial available finite element packages, *NDT & E International*, vol. 51, pp.30-40.

Ramadas, C, Balasubramaniam, K, Joshi, M & Krishnamurthy CV 2009, 'Interaction of primary anti-symmetric Lamb mode with symmetric delaminations: numerical and experimental studies', *Smart Materials Structures*, 18, 085011.

Ramadas, C, Balasubramaniam, K, Joshi, M & Krishnamurthy, CV 2010, 'Interaction of guided Lamb waves with an asymmetrically located delamination in a laminated composite plate', *Smart Materials Structures*, 19, 065009.

Ratassepp, M, Lowe, MJS, Cawley, P & Klauson A 2008, 'Scattering of the fundamental shear horizontal mode in a plate when incident at a through crack aligned in the propagation direction of the mode', *The Journal of Acoustical Society of America*, 124, pp. 2873-2882.

Veidt, M & Ng CT 2011, 'Influence of stacking sequence on scattering characteristics of the fundamental anti-symmetric Lamb wave at through holes in composite laminates', *The Journal of Acoustical Society of America*, 129, pp. 1280-1287.

The 8th International Conference on
Structural Health Monitoring of Intelligent Infrastructure
Brisbane, Australia | 5-8 December 2017

Modal-Based Damage Detection of Shear Connectors in Steel-Timber Composite (STC) Beams

A.A. Chiniforush [1], U. Dackermann [1], H. Valipour [1], A. Akbarnezhad [1].
[1] *Centre for Infrastructure Engineering and Safety (CIES), University of New South Wales (UNSW) – Australia,*
Email: a.a.chiniforush@gmail.com

Abstract

Steel-Timber Composite (STC) floors are newly developed sustainable flooring systems that aim to provide light-weight and sustainable flooring systems. The structural behavior of STC beams is greatly dependent on the stiffness and strength of the shear connectors, joining the timber panels to the steel sections. Any failure of shear connectors reduces the composite action between timber panels and steel sections; and consequently, causes a loss in the expected stiffness. Therefore, to assure the structural performance of the new STC flooring systems, the development of a reliable detection strategy for shear connection failures is critical. This study investigates the feasibility and efficiency of using vibration-based damage detection methods for the identification of loose shear connectors in STC flooring systems. Investigated techniques analyse natural frequencies and modes shapes using the Modal Assurance Criterion (MAC), the Coordinated Modal Assurance Criterion (COMAC) and the Modal Flexibility Method. In the experimental study, an STC beam is dynamically tested with three states of shear connector failure. The response of a composite beam excited with a hammer at three different excitation points is recorded by a network of accelerometers and the results are transformed to the frequency domain. Operational modal analysis is executed to extract the modal parameters of the structure with unknown excitation force. To identify matching modes of the different beam shear connector states, the MAC is utilized and four reliable modes are selected. Thereafter, three vibration-based damage detection approaches are implemented analysing the natural frequencies and mode shapes of the selected modes to detect damaged shear connectors and identify their locations. The results show promising identification of the severity and location of loose shear connectors, especially using the COMAC approach.

1. Introduction

Construction and building industry is responsible for about 30% of carbon emission incurred annually (Akbarnezhad & Xiao 2017). A major contributor to construction industry's carbon footprint is the cradle to gate embodied carbon of construction materials which accounts for carbon emissions during various stages of production and processing of materials. With this in mind, reducing the embodied carbon of buildings is a major focus of worldwide research with particular attention to identifying alternative low carbon building materials to replace less sustainable conventional materials including concrete – the most widely used construction material and one of the biggest contributor to embodied carbon of buildings. Apart from its widespread use, the high carbon footprint of concrete is mainly associated with considerable carbon emissions associated with production of Portland cement which accounts for 5 to 7% of global emissions (Akbarnezhad & Xiao 2017). Timber has been advocated widely as a sustainable replacement for concrete mainly due to its considerably lower embodied carbon and recyclability, among other environmental and social benefits. Timber has been also praised for its ease and less labour intensive construction process, which itself contributes to lowering its associated life cycle costs and environmental impacts. However, despite the growing interest in timber in the construction industry, driven by above benefits alongside timber's high strength to weight ratio, widespread adoption of timber as a durable and more sustainable replacement for concrete requires overcoming a number of drawbacks including uncertainties of material properties, moisture dependent expansion and low stiffness. Various efforts to overcome such issues have led to development of reliable timber products including Cross Laminated Timber (CLT) panels (Brandner et al. 2016) with factory assurance quality as well as innovative composite structural systems including Steel-Timber Composite (STC) beams. In particular, STC beams show a significant increase in the load-carrying capacity over long spans with a composite efficiency up to 90% in comparison to bare steel (Hassanieh et al. 2017; Hassanieh et al. 2016; Loss & Davison 2017). In an STC beam, CLT panels are connected to steel girders using shear connectors such as screws and bolts. The shear connectors play a crucial role in ensuring composite action. Accordingly, any damage in shear connectors will be accompanied with a loss of composite action between CLT panels and the steel girder, leading to a loss of stiffness and load-carry capacity of the element. Therefore the availability of a reliable and easy to use monitoring method to detect damage in shear connectors during service life of a STC beam is highly demanding.

STC beams can be considered as a structural dynamic system consisting of mass, damping, and stiffness. Therefore, any damage in shear connectors is expected to alter the dynamic behavior of STC beam. Therefore, monitoring the dynamic behaviour of STC beam can provide a reliable basis for detecting the damage as well as its location, and severity. A variety of vibration-based damage identification methods have been developed for analysing various dynamic parameters for damage detection including changes in natural frequencies (Salawu 1997), mode shapes (Görl & Link 2003), mode shape curvature (Pandey et al. 1991), flexibility matrix (Pandey & Biswas 1994), modal strain energy (Yan et al. 2012; Dackermann et al. 2010; Stubbs et al. 1992), frequency response function (FRF) (Maia et al. 2003; Lee & Shin 2002; Dackermann et al. 2013), and power spectral density function (Liberatore & Carman 2004; Xu & Wu 2007). The viability of a number of these methods for detection of damage in shear connectors used in concrete structures has been previously investigated and promising results have been reported (Li et al. 2014; Li et al. 2015; Xia et al. 2007; Xia et al. 2008; Ren et al. 2008). However, to the best of our knowledge, there is a lack of literature on application of vibration based non-destructive monitoring techniques to detect damage in shear connectors used in STC beams. Due to its unique composite structure, STC beam is expected to possess a unique dynamic response to vibration in presence and absence of damage in shear connectors.

In this study, three types of shear connector damage scenarios were introduced in an experimental STC beam with pre-tensioned bolt shear connectors. The dynamic response of an intact and three damaged composite beams was monitored using a network of 21 accelerometers. The operational modal analysis was performed to extract the natural frequencies and mode shapes from a beam with unknown excitation force, thus providing a continuous structural health monitoring strategy. Four reliable mode shapes were identified and selected to study the viability of Modal Assurance Criterion (MAC), Coordinated Modal Assurance Criterion (COMAC) and the Modal Flexibility methods for detecting damaged or loosened shear connectors.

2. Experimental Setup

For the experimental study, a simply supported STC beam on both ends with 5.8 m span (6.0 m length) connected by pre-tensioned M16 bolt shear connectors was tested (Fig 1 and Fig 2). A standard 310UB32 steel section and two 3000 × 1000 × 120 mm CLT panels were used to fabricate the composite beam structure (Fig 1 and Fig 2). Each panel was connected to the steel section by means of 12 pairs of pretension bolts. The longitudinal and transverse spacing of the bolts were 50 mm and 75 mm, respectively. In practice, the loss of composite action in STC beams occurs due to failure of shear connectors and consequent slip between CLT panel and steel section. This relative slip causes a reduction in the load carry capacity, which can have serious consequences. For this study, to simulate the loss of composite action in the STC beam, a number of the connection bolts were removed. Three damage scenarios with different level of severity (number of removed bolts) were simulated. Table 1 lists the identification numbers of the removed bolts.

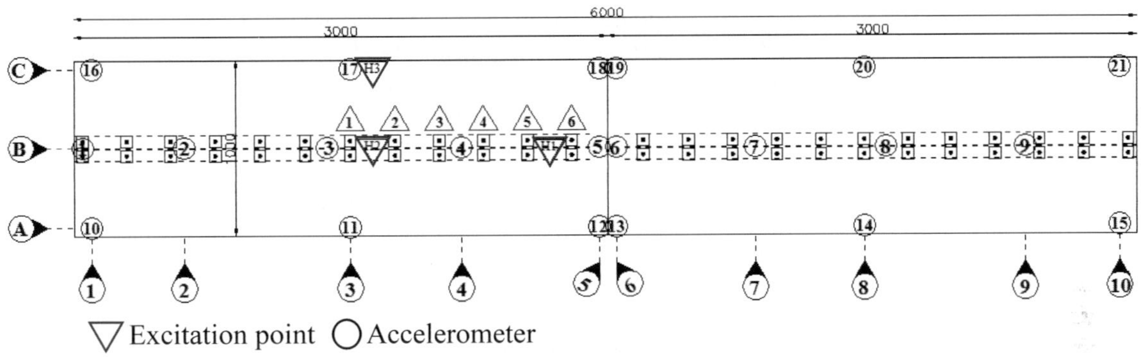

Fig 1. Schematic plan view of STC beam and location of accelerometers (modal nodes)

Fig 1 shows the location of the 21 accelerometers as well as the three hammer excitation points (H1, H2 and H3). The accelerometers were firmly attached to the top surface of the CLT panels using bolted steel plates and magnets. The outer accelerometers (numbers '10' to '21') were mounted with an offset of 5 cm from the edges of panels. For each investigated state of structure (intact and three damage scenarios), three repeated hammer impact tests were performed at each hammer excitation point, giving a total of 36 tests. The National Instrument's NI PXIe-1073 data acquisition system with a 4kHz resolution was used to record the acceleration responses of the structure from the different hammer hits. A hammer with a plastic tip was used for excitation and the duration of the vibration measurements for each test was 10 s.

Fig 2. Plan view of STC beam

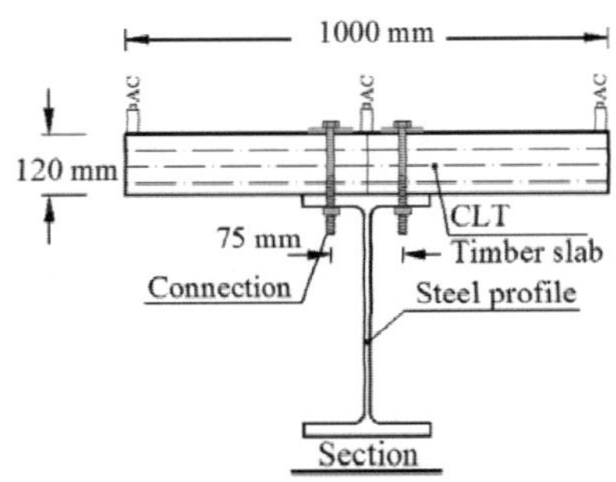

Fig 3. Cross-section of STC beams

Table 1. ID of removed shear connectors (Fig 1)

Damage Scenario	Removed shear connectors
DS.1 (minor)	5, 6
DS.2 (intermediate)	3, 4, 5, 6
DS.3 (severe)	1, 2, 3, 4, 5, 6

3. Damage Detection Methods

Vibration-based damage detection techniques are based on the principle that damage alters both the physical properties of a structure (e.g. mass, stiffness, and damping) as well as its dynamic characteristics (e.g. natural frequencies, damping ratios and mode shapes). Hence, by analysing a structure's dynamic properties from structural vibration, any damage, including its location and severity, can be identified. The following gives an overview of three damage detection methods investigated in this study for their applicability to identify the failure of shear connectors in STC beams.

3.1 Modal Assurance Criterion

The Modal Assurance Criterion (MAC) analyses changes in mode shapes caused by a damage. It is a measure of the similarity of two mode shapes. The MAC gives a single number that varies from 0 to 1. A MAC value of 1 is a perfect match and a value of 0 means the mode shapes are completely dissimilar. For the mode shape vectors of ϕ^u and ϕ^d (e.g. mode shapes of the undamaged and damaged state) the MAC is defined as

$$MAC(\phi^u, \phi^d) = \frac{\left|\sum_{i=1}^{n} \phi_i^u \phi_i^d\right|^2}{\sum_{i=1}^{n} (\phi_i^u)^2 \sum_{i=1}^{n} (\phi_i^d)^2} \tag{1}$$

with n being the modal degrees of freedom (Allemang 2003). In the current study, the MAC was utilized to identify matching mode shapes of the different beam states (intact and three damage states) as well as damage in the element by comparing mode shapes of the damaged states with the intact state.

3.2 Coordinate Modal Assurance Criterion

The Coordinate Modal Assurance Criterion (COMAC) differs from the MAC definition in the way that it gives a pointwise measure of the difference between two sets of mode shapes. It investigates the correlation of modal displacement in modal nodes at modal coordinate p. The COMAC value is defined as:

$$COMAC(\phi^u, \phi^d, p) = \frac{\left[\sum_{i=1}^{nm} (\phi_i^u)_p (\phi_i^d)_p\right]^2}{\sum_{i=1}^{nm} (\phi_i^u)_q^2 \sum_{i=1}^{nm} (\phi_i^d)_q^2} \tag{2}$$

with n_m being the number of considered modes (Lieven & Ewins 1988). Similar to the MAC, the value of COMAC varies from 0 to 1 with a low COMAC value indicating discordance at point p and thus presents a possible damage location indicator.

3.3 Modal Flexibility

For normalized mode shapes with $\phi^T M \phi = 1$, the flexibility (F) and stiffness (K) matrix are related to the modal data as:

$$K = M\phi\Omega\phi^T M = M\left(\sum_{i=1}^{n_m} \omega_i^2 \phi_i \phi_i^T\right) M, \quad \text{and} \quad F = \phi\Omega^{-1}\phi^T = \sum_{i=1}^{n_m} \frac{1}{\omega_i^2}\phi_i\phi_i^T \tag{3}$$

where M is mass matrix, $\phi = [\phi_1, \phi_2, \ldots, \phi_n]$ is the mode shape matrix, $\Omega = \text{diag}(\omega_i^2)$ is the modal stiffness matrix and ω_i is the ith modal frequency (Pandey & Biswas 1994). As can be seen in Eq.3, in the flexibility matrix, the effect of lower natural frequencies is larger compared to higher frequencies. In practice, the measurement of lower natural frequencies is simpler and more reliable; therefore, using the flexibility matrix can lead to more desired results. The concept can be applied to measure and compare the flexibility matrix of an undamaged (F_u) and a damaged (F_d) structure with each column of F representing the nodal displacements due to applying the unit load in one of the freedom degrees. By establishing change in flexibility matrix, $\Delta = F_u - F_d$, a modal-flexibility-based damage detection criterion can be defined as follow:

$$\bar{\delta_j} = max|\delta_{ij}| \tag{3}$$

where δ_{ij} are elements of Δ. A larger value of $\bar{\delta_j}$ indicates damage in the vicinity of modal point j.

4. Results & Discussions

To analyse the vibration response of different damage states of the composite beam structure, acceleration measurements were transformed to the frequency domain and the sum of spectral densities of each test was analysed in a frequency range from 0 to 600 Hz to extract the modal parameters of different modes. The operational modal analysis software ARTeMIS (Structural vibration solutions A/S n.d.) was used in this process. For each vibration test, 24 different modes were initially chosen based on peak-picking using the Frequency Domain Decomposition technique as shown in Fig 4.

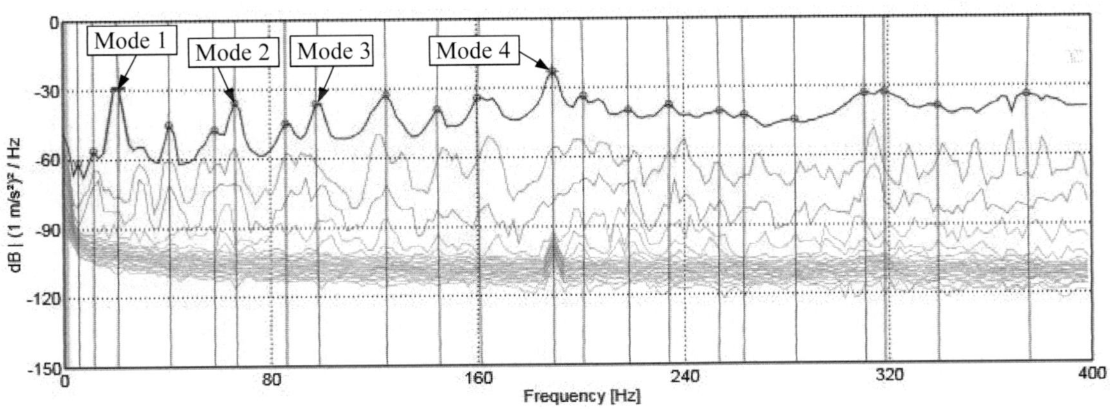

Fig 4. Spectral densities of intact STC beam with hammer excitation at location H1.

To select reliable and matching modes from the four different beam states (intact and three shear connector damage states), the MAC was calculated comparing the 24 mode shapes of the intact structure with hammer impact at location H1(indicated in Fig 1) with all other mode shapes from the various damage states (remaining 35 tests). A MAC value of 0.75 was adopted as an acceptable level for the similarity of two mode shapes. As a result, four modes out of the initial 24 modes were selected for damage detection purposes (see Table 2 and Fig 5).

4.1 Natural Frequencies and MAC Indicator

As it can be seen in Table 2 and Fig 6, the counterpart modes of the damaged beams have lower frequencies than the undamaged beam, as it is expected. Further, the MAC values are decreasing with an increase in faulty shear connectors (see Fig 7). As illustrated in Fig 6, the first and the second natural frequencies were more affected by the removal of the shear connectors. The frequency changes for the third and fourth selected modes were in a similar range. As depicted in Fig 7, the MAC indicator, defined as 1-MAC, shows that the change in mode shapes was maximum for the second mode. The third and fourth mode experienced again a similar level of mode shape change. Overall, for the first mode, only small mode shape changes occur, however, a significant shift in the natural frequency occurs as a function of damage intensity. Also, the effect of damage scenario 1 (DS.1) is very minor such that this small loss of shear connectors does not have any effect on the first and fourth selected natural frequencies and the MAC values. To summarize, the natural frequencies and the MAC values are good measures to indicate the presence and overall severity of shear connector failures in STC beam structures.

Table 2. Average (9 tests) natural frequencies and MAC values of intact and damaged STC beam states.

	Intact beam		Damaged beam						
Mode No.	Frequency (Hz)	MAC (UN)	Frequency (Hz)	MAC (DS1)	Frequency (Hz)	MAC (DS2)	Frequency (Hz)	MAC (DS3)	Ave MAC
1	21.48	0.999	21.48	0.995	20.51	0.900	19.97	0.892	0.947
2	68.14	0.987	66.41	0.918	65.59	0.721	62.61	0.397	0.756
3	99.61	0.998	98.52	0.981	97.98	0.882	96.46	0.623	0.871
4	189.45	1.000	189.45	0.994	187.50	0.950	194.82	0.689	0.908

Fig 5 illustrates the modal displacements of the four selected mode shapes. While all four mode shapes present a combination of flexural and torsional modes, the first, second and third mode shapes are primarily flexural modes while the fourth mode shape is primarily a torsional mode.

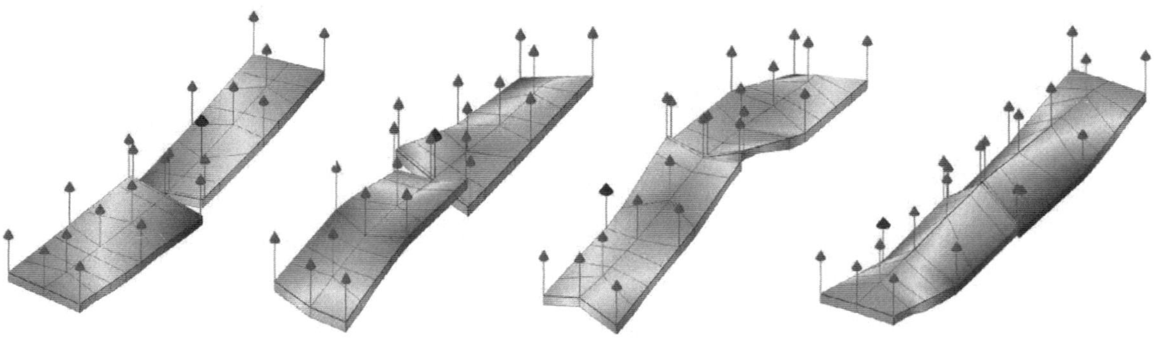

a) 1st mode (21.48Hz) b) 2nd mode (68.14Hz) c) 3rd mode (99.61Hz) d) 4th mode (189.45Hz)

Fig 5. Selected mode shapes of intact STC beam.

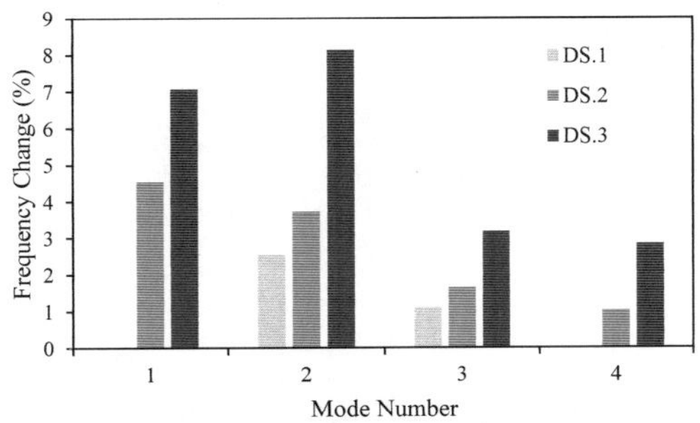

Fig 6. Frequency change (%) of different damaged beam states compared to intact STC beam.

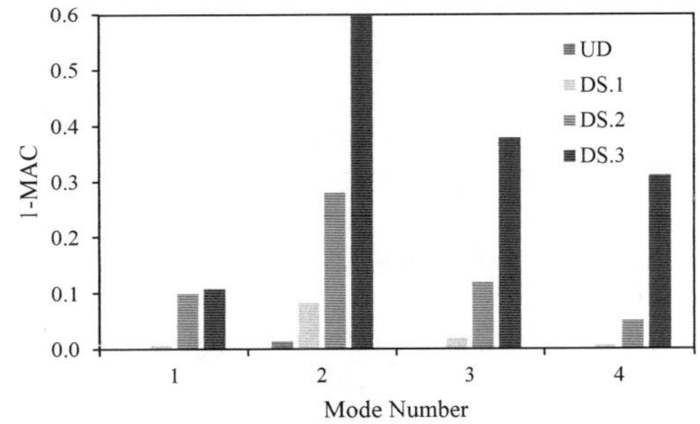

Fig 7. MAC indicator (1-MAC) of four selected modes for different STC beam states.

4.2 COMAC Indicator

Fig 8 displays the averaged COMAC indicator, defined as 1-COMAC, for the measurements at the STC beam centreline, which presents the location of the shear connectors. For the intact structure (Fig 8.UD), the averaged COMAC indicators of all 9 tests (3 repeats at 3 different hammer excitation points) are close to zero for all modal nodes indicating the independence and consistency of the COMAC indicators to the excitation location. For the minor damage scenario DS.1 (Fig 8.DS.1), the averaged COMAC indicator is maximum at location B5 which is in close vicinity to the removed shear connectors (shear connectors ‘5’ and ‘6’). The averaged COMAC indicator values of DS.1 for locations B2 to B8 are 0.0020, 0.0001, 0.0100, 0.0009, and 0.0030, respectively. Compared to the intact structure, these values are 2.92, 0.04, 53.25, 5.51, 16.07 and 17.85 times higher. This shows a clear indication of a structural change (loose or damaged shear connectors) close to location B5. For the intermediate damage scenario DS.2 (Fig 8.DS.2), the average COMAC indicator value for the same points (B2 to B8) are 42.64, 45.49, 29.03, 96.76, 29.05, 17.2 and 6.4 times higher compared to the intact structure, indicating again the damage location and an increase in damage intensity. The same damage indices’ pattern is observed for severe damage scenario DS.3 (Fig 8.DS.3) with averaged COMAC values being 192.03, 145.90, 232.25, 908.48, 264.14, 175.46 and 97.62 times higher in comparison to the intact STC beam. It is noted that COMAC values obtained from accelerometers positioned around the edges of the timber panels were disregarded in this study since the focus was on detecting damage of the shear connectors, which are located on the centreline of the composite beam structure. It is also mentioned here that measurements from accelerometer A1, situated close to the support of the beam, revealed comparatively very high COMAC indicator values suggesting false damage at this location. This phenomenon was previously reported in the literature (Li et al. 2014; Li et al. 2015) and is believed to be caused by singularity issues since mode shape values close to the support are close to zero. Hence it was decided to disregard these measurements.

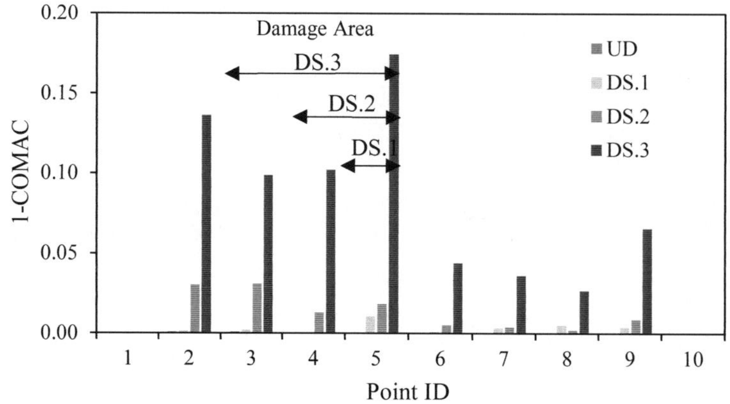

Fig 8. Averaged COMAC values of intact and damaged STC beams at centreline.

4.3 Modal Flexibility

For the modal flexibility index, the calculated differences of the averaged flexibility matrices of all four structural states (δ_{ij}) were normalized to the maximum value of δ_{ij} of all the 36 tests. The corresponding Normalised Difference Modal Flexibility (NDMF) index is illustrated for measurement points at the beam centreline in Fig 9. From the figures, damage is indicated to be primarily in the vicinity of point B4, which is contrary to the results of the COMAC method. Also, an inconsistency is observed with the NDMF values and the correlating damage severity.

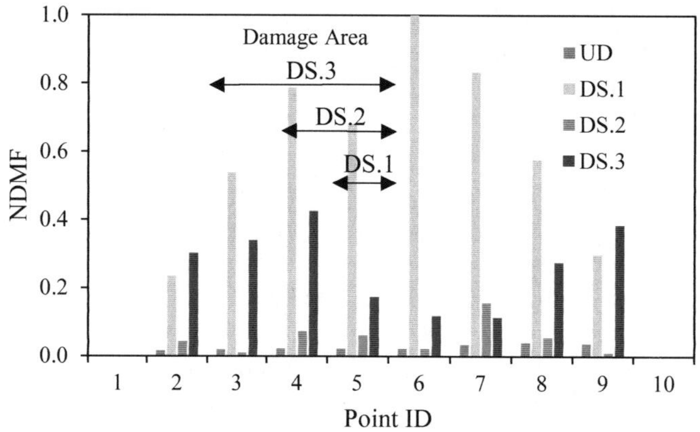

Fig 9. NDMF values of intact and damaged STC beams at centreline.

5. Conclusions

The feasibility of using changes in modal parameters for the detection and localisation of loose or damaged shear connectors in an STC beam structure was experimentally investigated. In particular, changes in natural frequencies and mode shapes were studied and MAC, COMAC, and modal flexibility techniques were evaluated for their applicability to identify shear connection failure in STC beams. The following summarises the study's findings.

- The natural frequencies and MAC values are good measures to indicate the overall presence and severity of shear connector failures in STC beams. In the study, both indicators decreased as a function of damage severity for the selected modes. While for the first mode, only a small decrease in MAC values occurred, a significant shift in the natural frequency was present for an increased number of damaged shear connectors. Further, the effect of the minor damage scenario 1 (DS.1) was very small such that this minor loss of shear connectors had only a very small impact on the first and fourth selected natural frequencies and the MAC values. Hence, it is important to analyse a number of modes since different mode shapes have varying sensitivities to different kind and severities of damage.
- The COMAC indicator revealed very good results in detecting and locating the removed shear connectors. It correctly indicated an increase in damage severity as well as it located the main damage area at position B5, which is in close vicinity of the removed shear connectors.
- For the modal flexibility, the NDMF indicated the general area of damage with some error - the exact location did not coincide with the maximum NDMF values. Also, the NDMF indicator gave an incorrect correlation with the damage severity.
- For mode shape values of accelerometer A1, situated close to the support of the beam, COMAC and NDMF values indicated false damage at this location. This phenomenon was previously reported in the literature and is believed to be caused by singularity issues since mode shape values close to the support are close to zero. Hence these measurements were disregarded.

References

Akbarnezhad, A. & Xiao, J., 2017. Estimation and Minimization of Embodied Carbon of Buildings: A Review. *Buildings*, 7(1), p.5.

Allemang, R.J., 2003. The modal assurance criterion–twenty years of use and abuse. *Sound and vibration*, 37(8), pp.14–23.

Brandner, R. et al., 2016. Cross laminated timber (CLT): overview and development. *European Journal of Wood and Wood Products*, 74(3), pp.331–351.

Dackermann, U., Li, J. & Samali, B., 2010. Dynamic-based damage identification using neural network ensembles and damage index method. *Advances in Structural Engineering*, 13(6), pp.1001–1016.

Dackermann, U., Li, J. & Samali, B., 2013. Identification of member connectivity and mass changes on a two-storey framed structure using frequency response functions and artificial neural networks. *Journal of Sound and Vibration*, 332(16), pp.3636–3653.

Görl, E. & Link, M., 2003. Damage identification using changes of eigenfrequencies and mode shapes. *Mechanical Systems and Signal Processing*, 17(1), pp.103–110.

Hassanieh, A., Valipour, H.R. & Bradford, M.A., 2017. Experimental and numerical investigation of short-term behaviour of CLT-steel composite beams. *Engineering Structures*, 144, pp.43–57.

Hassanieh, A., Valipour, H.R. & Bradford, M.A., 2016. Experimental and numerical study of steel-timber composite (STC) beams. *Journal of Constructional Steel Research*, 122, pp.367–378.

Lee, U. & Shin, J., 2002. A frequency response function-based structural damage identification method. *Computers & Structures*, 80(2), pp.117–132.

Li, J. et al., 2015. Damage assessment of shear connectors with vibration measurements and power spectral density transmissibility. *Structural Engineering and Mechanics*, 54(2), pp.257–289.

Li, J., Hao, H. & Zhu, H.-P., 2014. Dynamic assessment of shear connectors in composite bridges with ambient vibration measurements. *Advances in Structural Engineering*, 17(5), pp.617–638.

Liberatore, S. & Carman, G.P., 2004. Power spectral density analysis for damage identification and location. *Journal of Sound and Vibration*, 274(3–5), pp.761–776.

Lieven, N.A.J. & Ewins, D.J., 1988. Spatial correlation of mode shapes, the coordinate modal assurance criterion (COMAC). In *Proceedings of the sixth international modal analysis conference*. pp. 690–695.
Loss, C. & Davison, B., 2017. Innovative composite steel-timber floors with prefabricated modular components. *Engineering Structures*, 132, pp.695–713.
Maia, N.M.M. et al., 2003. Damage detection in structures: from mode shape to frequency response function methods. *Mechanical Systems and Signal Processing*, 17(3), pp.489–498.
Pandey, A.K. & Biswas, M., 1994. Damage detection in structures using changes in flexibility. *Journal of Sound and Vibration*, 169(1), pp.3–17.
Pandey, A.K., Biswas, M. & Samman, M.M., 1991. Damage detection from changes in curvature mode shapes. *Journal of Sound and Vibration*, 145(2), pp.321–332.
Ren, W.-X. et al., 2008. Damage identification of shear connectors with wavelet packet energy: laboratory test study. *Journal of Structural Engineering*, 134(5), pp.832–841.
Salawu, O.S., 1997. Detection of structural damage through changes in frequency: a review. *Engineering Structures*, 19(9), pp.718–723.
Structural vibration solutions A/S, *ARTeMIS® (2015) Software for operational modal analysis*, NOVI Science Park.
Stubbs, N., Kim, J.T. & Topole, K., 1992. An efficient and robust algorithm for damage localization in offshore platforms. In *Proceedings of the ASCE 10th Structures Congress*. pp. 543–546.
Xia, Y. et al., 2008. Condition assessment of shear connectors in slab-girder bridges via vibration measurements. *Journal of Bridge Engineering*, 13(1), pp.43–54.
Xia, Y., Hao, H. & Deeks, A.J., 2007. Dynamic assessment of shear connectors in slab–girder bridges. *Engineering Structures*, 29(7), pp.1475–1486.
Xu, Z.D. & Wu, Z., 2007. Energy damage detection strategy based on acceleration responses for long-span bridge structures. *Engineering Structures*, 29(4), pp.609–617.
Yan, W.-J., Ren, W.-X. & Huang, T.-L., 2012. Statistic structural damage detection based on the closed-form of element modal strain energy sensitivity. *Mechanical Systems and Signal Processing*, 28, pp.183–194.

The 8th International Conference on
Structural Health Monitoring of Intelligent Infrastructure
Brisbane, Australia | 5-8 December 2017

On the Influence of Structural Complexity on Damage Detection

C.J.L. Cowled[1], D.P. Thambiratnam [1], T.H.T. Chan [1].
[1] *Queensland University of Technology – Australia, Email:* craig.cowled@qut.edu.au

Abstract

Operational bridges range in complexity from relatively simple structures, such as beam bridges, up to incredibly complex structures, such as long-span suspension bridges or through-truss cantilever bridges. While damage detection (DD) research has progressed in tightly controlled numerical and experimental studies, there have been very few examples in the literature of successful DD on operational bridges (*e.g.*, Brownjohn *et al.*, 2011). Environmental and operational variability have previously been identified as creating uncertainty (*e.g.*, Sohn, 2007), but the influence of structural complexity on uncertainty has been treated as almost axiomatic (*e.g.*, Moon & Aktan, 2006) rather than being demonstrated. This paper argues that structural complexity does, in fact, influence the outcome of damage detection studies as a result of uncertainty. The argument begins with a definition of what is meant by complexity and a review of methods of quantifying complexity. The most promising measures are modified to reduce the effect of degeneracy and then used to evaluate the complexity of structures in damage detection studies. A statistical analysis reveals that the structural complexity of experimental structures is negatively correlated with achieved level of DD (p=0.041). A logical argument establishes the causal link between complexity and uncertainty.

1. Introduction

Intuition supports the notion that as a system increases in complexity the uncertainties of that system should accumulate making the damage detection task more difficult.
Moon and Aktan (2006) explicitly state that structural complexity is a primary influence on uncertainty. They provide no evidence to support this assertion, perhaps relying on Zadeh's incompatibility principle which states that "*the complexity of a system and the precision with which it can be analyzed bear a roughly inverse relation to one another*," (Zadeh, 1975, p.201). If the assertion of Moon and Aktan (2006) is correct, then it is reasonable to expect that this relationship will affect the results of damage detection (DD) studies in the literature.
To study the influence of structural complexity on DD, it is necessary to define what is meant by complexity and how it can be quantified.

2. Complexity

2.1 Defining Complexity

An ever-present issue in the complexity sciences has been that of defining complexity and formalising methods of quantifying complexity. A recent publication celebrating the establishment of a European hub for the study of complexity makes the point that defining complexity is notoriously difficult (Schuster, 2017). In her 2009 introductory book on complexity titled,

"Complexity: A Guided Tour," Professor Melanie Mitchell recounts a humorous argument that broke out during a 2004 panel discussion at the Santa Fe Institute, a global centre for the study of complexity, when the expert panel was asked to define complexity. The experts could not even agree on a definition of the very thing they had been studying for two decades. Apparently, the study of complexity is complex!

There is no formal definition of the word *'complexity'* in the sciences and engineering (ElMaraghy *et al.*, 2012). As discussed in Frei and Di Marzo Serugendo (2011), the term is used in many ways with individual researchers focusing on different aspects of complexity. In the field of structural health monitoring (SHM), as argued by Farrar, Worden and Park (2010), *'complexity'* has a deliberately vague meaning. Both Cowled *et al.* (2015a) and Haynes *et al.* (2012) adopted a narrow definition of complexity, limited to the physical geometry of structures.

The English word *'complexity'* has its origins in the Latin words *'com'*, meaning *'together'*, and *'plecto'*, meaning *'plait'* or *'twine'* (Mahoney, 2016). The compound Latin word can be interpreted, therefore, as *'intertwined'* or *'braided together'*. The Oxford dictionary defines *'complexity'* as, *"The state or quality of being intricate or complicated"* (Complexity, 2016).

The above definition recalls the primary characteristic of complexity; that is, the coming together, or interlinking, of many individual parts. It is possible to *objectively* quantify the magnitude of complexity of the resulting amalgam by counting its parts and links, but this doesn't seem to capture the basic meaning of complexity to a human observer.

When a system is described as intricate, complicated or complex, there is an implication for human observers that goes beyond the interlinking of many parts. From the *subjective* viewpoint of the human observer, the primary characteristic of complexity is the degree of difficulty associated with forecasting the properties of interest. Indeed, Sussman (2005) notes that, as the complexity of a system increases, its behaviour becomes increasingly difficult to predict, even though the individual parts or subsystems may be easily predictable.

For this study, complexity will have its dictionary definition, namely, *"The state or quality of being intricate or complicated"* (Complexity, 2016). As argued here, complexity should be objectively quantifiable in terms of the number of parts and the linkages between those parts as well as in terms of the subjective difficulty of predicting the behaviour of the system. As such, it should be possible to plot the complexity of a given system on a continuum from simple to complex and compare it to the complexity of other systems.

2.2 Review of Methods for Quantifying Complexity

Perhaps the most common methods for quantifying the complexity of a system are based on Shannon's entropy (Shannon, 1948). Shannon's entropy was introduced by Rashevsky (1955) to molecular biology as a means of measuring the complexity of graphs. Rashevsky's work was formalised by Trucco (1956) and optimized by Mowshowitz (1968).

Bonchev and Trinajstić (1977) sought to identify an index that would capture differences in the branching characteristics of molecules. They developed two indices, termed information quantities, whose performance was shown to be better than that of five existing indices.

Balaban (1982) reviewed topological indices (TIs) which use the adjacency matrix $\boldsymbol{A}$ and distance matrix $\boldsymbol{D}$. He noticed that, at the time, all the TIs based on $\boldsymbol{D}$ were plagued by degeneracy. Being a chemist, Balaban defined degeneracy as two or more chemical constitutions having the same index value. Balaban proposed a TI using distance sums $\boldsymbol{s_i}$ where $\boldsymbol{D}$ is summed by row or column.

Dehmer (2008) introduced a method to compute the structural information content (*i.e.*, entropy) of a graph. Dehmer's method seeks to assign a probability to each vertex of a set of vertices and compute the structural information content using a version of Shannon's entropy theorem. Dehmer's novel approach was to develop three different *'information functionals'* that are based on certain metrical properties of graphs. As such, the entropy of a given graph can be expressed in three different ways that capture important structural information in a meaningful way.

De Biagi (2013) developed the only known complexity index used in the field of structural engineering. He focused on measuring the redundancy in a structural system and correlated his structural complexity index with the robustness of a structure. As such, his method was a measure of the load paths in a structure rather than being a measure of the geometric complexity of the structure. De Biagi's method was certainly robust in its formulation from a structural engineering perspective; however, it required the analyst to identify all the combinations of determinate substructures within the whole indeterminate structure and calculate the relative work performed by the applied loads. This makes the method very tedious if the analyst wishes to compare many different structures.

Sinha (2014) introduced an index of complexity based on the energy of a system (*i.e.*, the Hamiltonian of a system). Sinha's index captures the complexity of 1) individual nodes in a system, 2) the interaction between nodes, and 3) the topology of the system. These characteristics make it a very useful method for measuring the complexity of structural systems.

Sinha's quantitative measure of structural complexity, C, is the sum of component complexity, C_1, and interface complexity, $C_2 C_3$:

$$C = C_1 + C_2 C_3$$

$$\therefore C = \sum_{i=1}^{n} \alpha_i + \left(\sum_{i=1}^{n} \sum_{j=1}^{n} \beta_{ij} A_{ij} \right) \left(\frac{E(A)}{n} \right) \tag{1}$$

where n is the number of components, α_i is the complexity of the ith component, β_{ij} is the pair-wise interaction complexity of the ith and jth components, A is the binary adjacency matrix or graph energy matrix (*i.e.*, **1** if the ith and jth components are connected, otherwise equal to zero), and $E(A)$ is the sum of the singular values of the adjacency matrix (known as the trace norm of A).

The C_3 term is a scaling factor that is sensitive to the topology of the system. This term allows for different configurations of the components to have quantifiably different values of complexity.

Sinha and de Weck (2016) recently validated this measure of complexity in a convincing experimental study.

Efatmaneshnik and Ryan (2016) present a generalised framework for measuring complexity, which they claim is applicable to any domain. The truly innovative contribution that they make in this paper is to incorporate a subjective element into the measure of complexity. Their general measure of system complexity, also termed total complexity or combination complexity, is:

$$\mathcal{K}(S) = \mathcal{F}(\mu(S), \mathcal{D}(SR)) \tag{2}$$

where, S is the observed system, μ is the objective complexity, $\mathcal{D}$ is the subjective complexity, and $\mathcal{K}$ is system complexity which is a function of μ and $\mathcal{D}$. $\mathcal{D}(SR)$ is a measure of the distance between the system S and a reference system R, which is selected by the observer as an example of a simple system.

The intuition behind the subjective complexity function relies on the assumption that the observer knows what a simple system, R, looks like. Hence, the subjective complexity can be quantified by measuring the comparative difference in the complexity of S and R.

2.3 Quantifying the Complexity of Structures

The previous sub-section of this paper identified Sinha's (2014) method of quantifying complexity as a useful method. Sinha's measure of complexity will be examined in more detail here and modified to improve its applicability in the context of engineered structures. The framework of Efatmaneshnik and Ryan (2016) will be used because it makes it possible for the subjective aspect of complexity to be incorporated.

There are six graphs shown in Figure 1. Intuitively, the complexity of these graphs increases from G_1 to G_6. G_2 is more complex than G_1 because it has a vertical member, whereas G_1 consists of three horizontal members. G_3 is more complex than G_2 because it has an additional member. G_4 is more complex than G_3 because it has a diagonal member. G_5 is more complex than G_4 because it has an additional member and G_6 is more complex than G_5 for the same reason.

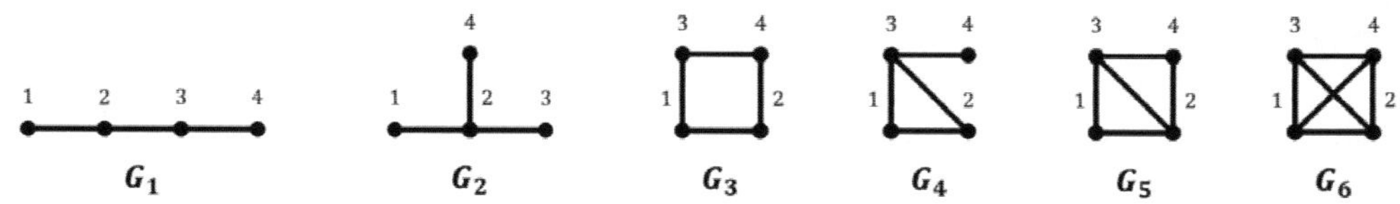

Figure 1. Six graphs having **4** vertices.

The ***A*** matrices for these six graphs are:

$$A_{G_1} = \begin{bmatrix} 0 & 1 & 0 & 0 \\ 1 & 0 & 1 & 0 \\ 0 & 1 & 0 & 1 \\ 0 & 0 & 1 & 0 \end{bmatrix} \quad A_{G_2} = \begin{bmatrix} 0 & 1 & 0 & 0 \\ 1 & 0 & 1 & 1 \\ 0 & 1 & 0 & 0 \\ 0 & 1 & 0 & 0 \end{bmatrix} \quad A_{G_3} = \begin{bmatrix} 0 & 1 & 1 & 0 \\ 1 & 0 & 0 & 1 \\ 1 & 0 & 0 & 1 \\ 0 & 1 & 1 & 0 \end{bmatrix}$$

$$A_{G_4} = \begin{bmatrix} 0 & 1 & 1 & 0 \\ 1 & 0 & 1 & 0 \\ 1 & 1 & 0 & 1 \\ 0 & 0 & 1 & 0 \end{bmatrix} \quad A_{G_5} = \begin{bmatrix} 0 & 1 & 1 & 0 \\ 1 & 0 & 1 & 1 \\ 1 & 1 & 0 & 1 \\ 0 & 1 & 1 & 0 \end{bmatrix} \quad A_{G_6} = \begin{bmatrix} 0 & 1 & 1 & 1 \\ 1 & 0 & 1 & 1 \\ 1 & 1 & 0 & 1 \\ 1 & 1 & 1 & 0 \end{bmatrix}$$

Letting all values of α**=1** and β**=1** [1]in equation **(1)**, the components of complexity and the total complexity for each of these graphs is tabulated in Table 1. These results are counter-intuitive (*i.e.*, degenerate) because they result in the C_3 term of both G_2 and G_3 being evaluated as *less* complex than that of G_1. Note also that the C_1 term is equal to the number of nodes of the graph. The interaction complexity C_2 term is essentially a measure of the number of interactions between nodes. Since a member interacts with two nodes, the C_2 term is equal to twice the number of members.

Table 1. Complexity values, after Sinha (2014), of six graphs having **4** vertices.

GRAPH	C_1	C_2	C_3	C
G_1	4	6	1.12	10.71
G_2	4	6	0.87	9.20
G_3	4	8	1.00	12.00
G_4	4	8	1.24	13.92
G_5	4	10	1.28	16.81
G_6	4	12	1.50	22.00

[1] The α and β terms can have different values. Individual members may be more complex than others in terms of materials, cross-section, structural behaviour *etc*. The α term can be varied to reflect the relative complexity of individual members. Interactions between specific members (*i.e.*, nodes) and boundary conditions may vary in complexity. The β term can be varied to reflect the relative complexity at node points.

In addition to the problem highlighted above, Bonchev and Trinajstić (1977) point out that using the adjacency matrix can lead to a situation where two different graphs produce the same result. This makes the adjacency matrix less than ideal as a measure for differentiating topologies. They proposed using the distance matrix of a graph, denoted ***D***, instead of the adjacency matrix.
The ***D*** matrices for these six graphs are:

$$D_{G_1} = \begin{bmatrix} 0 & 1 & 2 & 3 \\ 1 & 0 & 1 & 2 \\ 2 & 1 & 0 & 1 \\ 3 & 2 & 1 & 0 \end{bmatrix} \quad D_{G_2} = \begin{bmatrix} 0 & 1 & 2 & 2 \\ 1 & 0 & 1 & 1 \\ 2 & 1 & 0 & 2 \\ 2 & 1 & 2 & 0 \end{bmatrix} \quad D_{G_3} = \begin{bmatrix} 0 & 1 & 1 & 2 \\ 1 & 0 & 2 & 1 \\ 1 & 2 & 0 & 1 \\ 2 & 1 & 1 & 0 \end{bmatrix}$$

$$D_{G_4} = \begin{bmatrix} 0 & 1 & 1 & 2 \\ 1 & 0 & 1 & 2 \\ 1 & 1 & 0 & 1 \\ 2 & 2 & 1 & 0 \end{bmatrix} \quad D_{G_5} = \begin{bmatrix} 0 & 1 & 1 & 2 \\ 1 & 0 & 1 & 1 \\ 1 & 1 & 0 & 1 \\ 2 & 1 & 1 & 0 \end{bmatrix} \quad D_{G_6} = \begin{bmatrix} 0 & 1 & 1 & 1 \\ 1 & 0 & 1 & 1 \\ 1 & 1 & 0 & 1 \\ 1 & 1 & 1 & 0 \end{bmatrix}$$

Changing the C_3 term of **(1)** to use the ***D*** matrix:

$$\therefore C = \sum_{i=1}^{n} \alpha_i + \left(\sum_{i=1}^{n} \sum_{j=1}^{n} \beta_{ij} A_{ij} \right) \left(\frac{E(D)}{n} \right) \tag{3}$$

The results from Table 1 are updated accordingly and tabulated in Table 2. These results are also counter-intuitive because the topological complexity C_3 term decreases by **42%** from graph G_1 to G_6. There is one anomaly that bucks the trend where the C_3 term of G_4 is greater than that of G_3. Aside from the degenerate result of G_2, the remaining total values are monotonically increasing.

Table 2. Updated complexity values using the ***D*** matrix in the C_3 term.

GRAPH	C_1	C_2	C_3	C
G_1	4	6	2.58	19.49
G_2	4	6	2.32	17.94
G_3	4	8	2.00	20.00
G_4	4	8	2.05	20.40
G_5	4	10	1.78	21.81
G_6	4	12	1.50	22.00

Neither the ***A*** matrix nor the ***D*** matrix seem to capture topological complexity in an appropriate manner for structures. The maximum possible distance between two nodes on a graph with ***n*** vertices is ***n*–1**, being for the specific graph where all nodes are in a row. Such a graph represents the simplest arrangement of members. It makes sense in formulating the topological complexity C_3 term to attribute a minimum value for large distances and to attribute a maximum value for small distances. A new matrix, being the difference between ***n*–1** and ***D*** is proposed:

$$B = (n-1)(J-I) - D \tag{4}$$

where ***I*** is the identity matrix and ***J*** is the unity matrix.
Using the ***B*** matrix in the C_3 term and halving the magnitude of the C_2 term to reflect the number of members, **(1)** can be reformulated as:

$$\therefore C = \sum_{i=1}^{n} \alpha_i + \frac{1}{2}\left(\sum_{i=1}^{n}\sum_{j=1}^{n} \beta_{ij} A_{ij}\right)\left(\frac{E(B)}{n}\right) \quad (5)$$

The $\boldsymbol{B}$ matrices for these six graphs are:

$$B_{G_1} = \begin{bmatrix} 0 & 2 & 1 & 0 \\ 2 & 0 & 2 & 1 \\ 1 & 2 & 0 & 2 \\ 0 & 1 & 2 & 0 \end{bmatrix} \quad B_{G_2} = \begin{bmatrix} 0 & 2 & 1 & 1 \\ 2 & 0 & 2 & 2 \\ 1 & 2 & 0 & 1 \\ 1 & 2 & 1 & 0 \end{bmatrix} \quad B_{G_3} = \begin{bmatrix} 0 & 2 & 2 & 1 \\ 2 & 0 & 1 & 2 \\ 2 & 1 & 0 & 2 \\ 1 & 2 & 2 & 0 \end{bmatrix}$$

$$B_{G_4} = \begin{bmatrix} 0 & 2 & 2 & 1 \\ 2 & 0 & 2 & 1 \\ 2 & 2 & 0 & 2 \\ 1 & 1 & 2 & 0 \end{bmatrix} \quad B_{G_5} = \begin{bmatrix} 0 & 2 & 2 & 1 \\ 2 & 0 & 2 & 2 \\ 2 & 2 & 0 & 2 \\ 1 & 2 & 2 & 0 \end{bmatrix} \quad B_{G_6} = \begin{bmatrix} 0 & 2 & 2 & 2 \\ 2 & 0 & 2 & 2 \\ 2 & 2 & 0 & 2 \\ 2 & 2 & 2 & 0 \end{bmatrix}$$

The results from Table 1are updated accordingly and tabulated in Table 3. These results accord with intuition as the magnitude of the C_3 term is monotonically increasing from G_1 to G_6.

Table 3. Updated complexity values using the $\boldsymbol{B}$ matrix in the C_3 term and halving the C_2 term.

GRAPH	C_1	C_2	C_3	C
G_1	4	3	2.29	10.86
G_2	4	3	2.30	10.91
G_3	4	4	2.50	14.00
G_4	4	4	2.54	14.14
G_5	4	5	2.77	17.83
G_6	4	6	3.00	22.00

This index passes the degeneracy test but fails to capture a key characteristic of complexity for structural engineering systems. Many structural systems make use of repeated patterns to simplify the design, manufacture and construction of physical structures. These patterns, hereafter referred to as subgraphs, simplify the system and ought to, therefore, capture the effect of simplification. The interaction and topology terms (C_2,C_3) of the objective complexity $\mu(S)=C$, should be calculated for all $\boldsymbol{m}$ subgraphs of a structure and their product summed:

$$\mu(S) = C_1 + \sum_{k=1}^{m} (C_2 C_3)_k \quad (6)$$

The subjective complexity $\mathcal{D}(SR)$, is a measure of the difference between the objective complexity of a system and a subgraph of that system that is determined by the engineer as being an example of a simple system. The maximum common subgraph $MCS(\cdot)$ is the subgraph, common to two graphs, that has the maximum possible number of nodes (Bunke & Shearer, 1998). The reference graph $\boldsymbol{R}$ is chosen as an example of a simple system. Using the objective complexity metric $\mu(\cdot)$, the distance between the observed system $\boldsymbol{S}$ and the reference system $\boldsymbol{R}$ is:

$$\mathcal{D}(SR) = 1 - \frac{\mu(MCS(S,R))}{\mu(S) + \mu(R) - \mu(MCS(S,R))} \quad (7)$$

This can be simplified if ***R=MCS***:

$$\mathcal{D}(\boldsymbol{SR}) = \mathbf{1} - \frac{\boldsymbol{\mu}(\boldsymbol{R})}{\boldsymbol{\mu}(\boldsymbol{S})} \tag{8}$$

The subjective measure of complexity $\mathcal{D}(\mathbf{SR})$ varies between $\mathbf{0}$ and $\mathbf{1}$. A value approaching $\mathbf{1}$ indicates that the observed system is maximally complex with respect to the reference system. A value approaching $\mathbf{0}$ indicates that the observed system is relatively simple. The reference system may, in some circumstances, be the same as the repeating subgraph used in the calculation of objective complexity.

2.4 Worked Example – QUT Benchmark Structure (Cowled *et al.*, 2015)

The QUT benchmark structure in Cowled *et al.* (2015) is shown in Figure 2.

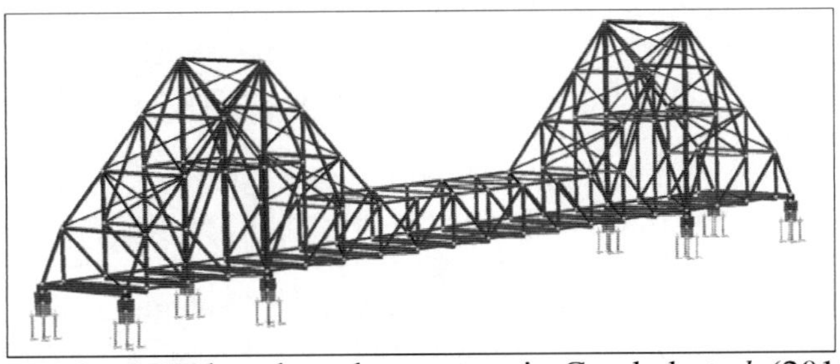

Figure 2. QUT benchmark structure in Cowled *et al.* (2015).

This structure has eight substructures consisting of three different types as shown in Figure 3.

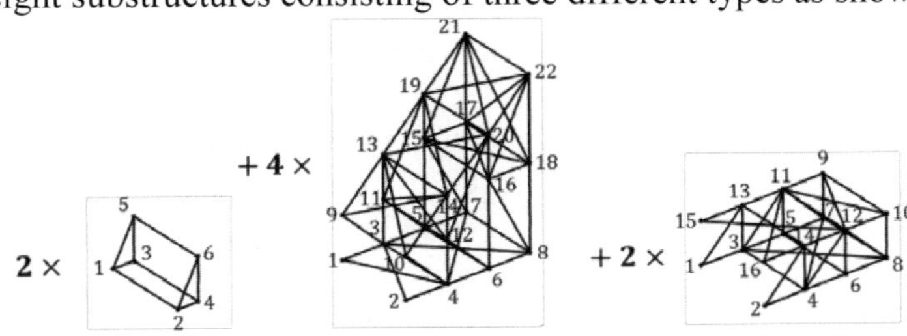

Figure 3. Three types of substructures of the QUT benchmark structure.

The adjacency matrix ***A*** and the derived matrix ***B*** are not shown. From **(6)**:

$$\begin{aligned}\boldsymbol{\mu}(\boldsymbol{S}) &= \boldsymbol{C}_1 + \sum_{\boldsymbol{k}=1}^{\mathbf{8}} (\boldsymbol{C}_2\boldsymbol{C}_3)_{\boldsymbol{k}} \\ &= \mathbf{100} + \big(\mathbf{2} \times (\mathbf{9} \times \mathbf{6}) + \mathbf{4} \times (\mathbf{61} \times \mathbf{36.202}) + \mathbf{2} \times (\mathbf{36} \times \mathbf{24.633})\big) \\ &= \mathbf{10\,815}\end{aligned}$$

The reference subgraph, which represents a simple structure, is taken as one bay of the **16** node substructure in Figure 3. As such, the reference subgraph has **8** nodes and **14** edges:

$$\mu(R) = C_1 + C_2 C_3 = 8 + 14 \times 9.506 = 141$$

$$\mathcal{D}(SR) = 1 - \frac{\mu(R)}{\mu(S)} = 1 - \frac{141}{10\,815} = 0.987$$

$$\mathcal{K}(S) = \mu(S) \times \mathcal{D}(SR) = 10\,815 \times 0.987 = 10\,674$$

3. Complexity of Structures in Damage Detection Studies

The term "truss" was used as a filter to search the journal *Structural Health Monitoring* and **45** studies from **31** papers were selected. An additional **145** studies from **109** papers were selected from *Structural Control and Health Monitoring* using the same search term. The total complexity of all **190** structures was calculated and tabulated in Cowled (2017) along with references. Approximately half of the studies, **93** in total, succeeded in detecting damage; **51** of which achieved a maximum DD level of 3, **26** achieved a maximum DD level of 2, and **16** achieved a maximum DD level of 1. Notably, there are only eight studies of real structures that succeeded in detecting damage, none of which achieved level 3.

To conduct a statistical test of the relationship between complexity and DD level, the data is rearranged and shown as two boxplots in Figure 5. Importantly, the samples shown in DD level 3 are not repeated in DD level 1 & 2 even though they have also achieved DD level 1 & 2. The hypothesis being tested is that the median value of studies that achieved DD level 3 is less than the median value of studies that did not (*i.e.*, studies in DD level 1 & 2). The null hypothesis, then, is that the median value of studies that achieved DD level 3 is greater than or equal to the median value of studies that did not achieve DD level 3. This statistical test is known as a one-sided Wilcoxon rank sum test. The MatLab function '*ranksum*' is used to test the null hypothesis at the **5%** significance threshold. Results from the ranksum test indicate that the null hypothesis is not rejected for numerical structures; however, the null hypothesis is rejected for experimental structures with a p-value of **.041** (the statistical power of this result is better than **99%**).

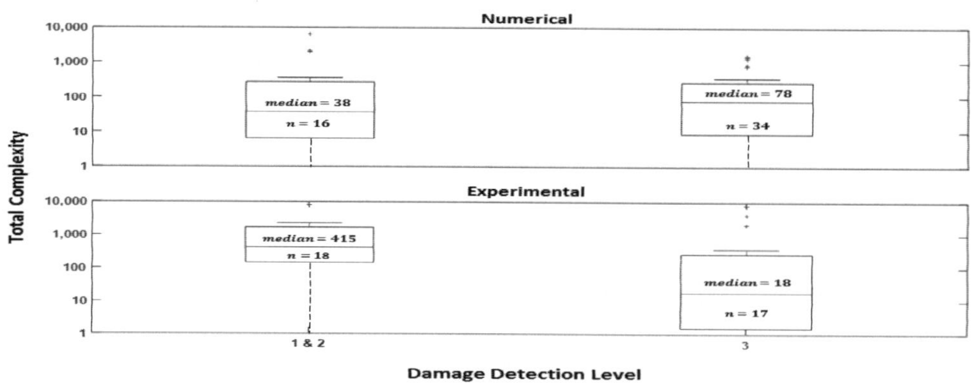

Figure 4. Boxplot of the complexity of **85** structures, all of which achieved some level of damage detection, in *Structural Control and Health Monitoring* and *Structural Health Monitoring* by type of structure and damage detection level.

There is no statistically significant relationship between complexity and achieved DD level for numerical studies. This is most likely because, unlike experimental studies, uncertainties can be managed much better in a numerical study. The spatial resolution and noise levels in numerical studies can be controlled, which has the effect of reducing the uncertainty. Also, the fact that the complexity of most structures in numerical studies is quite low and more tightly clustered than that of laboratory structures may have made it more difficult to differentiate between the two groups (*i.e.*, those that achieved DD level 3 and those that did not).

The result for studies on experimental structures appears to support the intuition that as the complexity of test bed structures increases the uncertainty increases also, making it more difficult to quantify the severity of damage on a structure. But, as the saying goes, "*Correlation does not imply causation.*"

4. Philosophical Rationale for Causation from Structural Complexity to Uncertainty

The famous French scholar Pierre-Simon, marquis de Laplace, introduced a fascinating thought experiment to the world in his 1814 essay titled, "*A philosophical essay on probabilities*." The thought experiment is based on the concept of determinism, a philosophy which requires that we "*regard the present state of the universe as the effect of its anterior state and as the cause of the one which is to follow*," (Laplace, 1902, p.4). Laplace imagined a being, commonly called Laplace's demon, with the intelligence to know the state of every particle in the universe and all the interacting forces and with the capability to compute all this data moment by moment.[2] If Laplace's demon existed, it could determine the past and future states of the universe, from the motion of galaxies right down to the state of the smallest particle. Uncertainty would not exist for such an entity. This reductionist approach to understanding the material world seems, on the face of it, to be entirely reasonable.

Does this all mean that reality is fundamentally deterministic and that all uncertainties can, in theory at least, be reduced to nothing? Not necessarily.

The theory of quantum mechanics (QM), formulated a century after Laplace, brought uncertainty back into the picture. The Copenhagen interpretation of QM holds that physical systems do not have definite properties; rather, they are more appropriately described as a set of probabilities known as the wave function. When a measurement is made, the physical system is observed to have one of the possible values of the wave function. Many physicists find this explanation of QM disturbing on a philosophical level and have formulated their own interpretations of the various QM phenomena. A recent paper by Hollowood (2016) makes the point that an authentic 1920's understanding of QM (not the 1950's interpretation by Werner Heisenberg) coheres comfortably with modern insights from experimental physics of the decoherence phenomenon. Hollowood's reconstruction of the original Copenhagen interpretation sees classical physics as an emergent result of the fundamentally stochastic nature of the real world. Whether this view is correct or not is a question for physicists to argue over. It is known, however, that stochastically random events do occur at the quantum level; hence, uncertainty at the fundamental level of reality is resurrected. Even Laplace's demon may have to cope with uncertainty.

Even if reality was entirely deterministic, Zadeh's incompatibility principle still ensures that finite computational resources will limit our ability to analyse complex systems with precision (Zadeh, 1975). "*Complexity typically induces uncertainty when a process that may be deterministic is so*

[2] Although Laplace is commonly credited with this thought experiment, van Strien (2014) points out that variations of the idea were widespread amongst French philosophers of the time going back half a century before Laplace's essay was first published.

complex that its outcome cannot be predicted with confidence," (Lovell, 1995, p.345). The reductionist modelling approaches, such as the finite element method so favoured by structural engineers, place an exponentially increasing burden on computational resources as the complexity of a system increases. Precision is sacrificed in favour of relevance, allowing uncertainties to creep into the model.

Non-model based methods of analysis, such as the modal analysis at the root of so many DD studies, are also affected by increasing complexity which amplifies the influence of natural stochastic variation, imprecise / incomplete measurement, and computational errors.

Section 3 of this paper demonstrated a statistically significant positive correlation between the complexity of structures in the laboratory and the uncertainty of damage detection outcomes. This section of the paper provides a philosophical grounding for causation in the relationship between complexity and uncertainty. In other words, it has been argued here that increasing the complexity of systems necessarily leads to an increase in overall uncertainty. The argument is complete in that section 3 showed a correlation and section 4 provides a philosophical rationale for causation.

4. Conclusion

This paper has quantifiably shown, with statistical significance (p=0.041), that the structural complexity of test-bed structures in the damage detection literature does influence the outcomes of damage detection. The more complex a structure is, the greater are the uncertainties and, hence, the lesser the likelihood of achieving higher levels of damage detection. A modified index of structural complexity was presented. This index can be used by researchers who may be interested in investigating the link between structural complexity and any other phenomenon of interest.

References

Brownjohn, JMW, De Stefano, A, Xu, Y-L, Wenzel, H, & Aktan, AE 2011, 'Vibration-based monitoring of civil infrastructure: challenges and successes', *Journal of Civil Structural Health Monitoring*, vol. 1, no. 3-4, pp. 79-95.

Sohn, H 2007, 'Effects of environmental and operational variability on structural health monitoring', *Philosophical Transactions of the Royal Society A: Mathematical, Physical and Engineering Sciences*, vol. 365, no. 1851, pp. 539-560.

Moon, FL & Aktan, AE 2006, 'Impacts of epistemic (bias) uncertainty on structural identification of constructed (civil) systems', *The Shock and Vibration Digest*, vol. 38, no. 5, pp. 399-420.

Zadeh, LA 1975, 'The concept of a linguistic variable and its application to approximate reasoning – I', *Information Sciences*, vol. 8, no. 3, pp. 199-249.

Schuster, P 2017, 'Definitions of complexity are notoriously difficult', In S. Thurner (Ed.) *43 Visions for Complexity* (pp. 61-62), World Scientific Publishing Co. Pte. Ltd., Singapore.

Mitchell, M 2009, '*Complexity: A Guided Tour*', Oxford University Press, Oxford, U.K.

ElMaraghy, W, ElMaraghy, H, Tomiyama, T & Monostori, L 2012, 'Complexity in engineering design and manufacturing', *CIRP Annals – Manufacturing Technology*, vol. 61, no. 2, pp. 793-814.

Frei, R & Di Marzo Serugendo, G 2011, 'Concepts in complexity engineering', *International Journal of Bio-Inspired Computation*, vol. 3, no. 2, pp. 123-139.

Farrar, C, Worden, K & Park, G 2010, 'Complexity: A new axiom for structural health monitoring?' In F. Casciati, & M. Giordano (Eds.), *Proceedings of the Fifth European Workshop on Structural Health Monitoring* (pp. 882-888). DEStech Publications, Inc., Lancaster, PA.

Cowled, CJL, Thambiratnam, DP, Chan, THT, & Tan, ACC 2015, 'Structural complexity in structural health monitoring: Design of laboratory model and test plan', In W. Lee, B. Choi, L. Ma, & J. Mathew (Eds.), *Proceedings of the 7th world congress on engineering asset management (WCEAM 2012)* (pp. 171-181). Lecture Notes in Mechanical Engineering. Springer, Cham, Switzerland.

Haynes, C, Todd, MD, Flynn, E & Croxford, A 2012, 'Statistically-based damage detection in geometrically-complex structures using ultrasonic interrogation', *Structural Health Monitoring*, vol. 12, no. 2, pp. 141-152.

Mahoney, KD 2016, 'Latdict: Latin dictionary & grammar resources', July 5, 2016, www.latin-dictionary.net.

Complexity 2016, In '*Oxford Dictionaries*', July 5, 2016, www.oxforddictionaries.com/definition/english/complexity.

Sussman, JM 2005, 'The new transportation faculty: The evolution to engineering systems', In *Perspectives on Intelligent Transportation Systems (ITS)* (pp. 101-116). Springer, Boston, MA.

Shannon, CE 1948, 'A mathematical theory of communication', *Bell System Technical Journal*, vol. 27, no. 3, pp. 379-423.

Rashevsky, N 1955, 'Life, information theory, and topology', *The Bulletin of Mathematical Biology*, vol. 17, no. 3, pp. 229-235.

Trucco, E 1956 'A note on the information content of graphs', *The Bulletin of Mathematical Biophysics*, vol. 18, no. 2, pp. 129-135.

Mowshowitz, A 1968, 'Entropy and the complexity of graphs: I. An index of the relative complexity of a graph', *The Bulletin of Mathematical Biophysics*, vol. 30, no. 1, pp. 175-204.

Bonchev, D & Trinajstić, N 1977, 'Information theory, distance matrix, and molecular branching', *The Journal of Chemical Physics*, vol. 67, no. 10, pp. 4517-4533.

Balaban, AT 1982, 'Highly discriminating distance-based topological index', *Chemical Physical Letters*, vol. 89, no. 5, pp. 399-404.

Dehmer, M 2008, 'Information processing in complex networks: Graph entropy and information functionals', *Applied Mathematics and Computation*, vol. 201, no. 1-2, pp. 82-94.

De Biagi, V 2013, '*Complexity and robustness of structures against extreme events*', (PhD Thesis). Politecnico Di Torino, Turin, Italy.

Sinha, K 2014, '*Structural complexity and its implications for design of cyber-physical systems*', (PhD Thesis). Massachusetts Institute of Technology, Cambridge, MA.

Sinha, K & de Weck, OL 2016, 'Empirical validation of structural complexity metric and complexity management for engineering systems', *Systems Engineering*, vol. 19, no. 3, pp. 193-206.

Efatmaneshnik, M & Ryan, MJ 2016, 'A general framework for measuring system complexity', *Complexity*, vol. 21, no. S1, pp. 533-546.

Cowled, CJL 2017, '*On the influence of structural complexity on the global vibration characteristics of steel truss bridges*', (PhD Thesis under examination). Queensland University of Technology.

Laplace, P-S 1902, '*A philosophical essay on probabilities*', (F.W. Truscott & F.L. Emory, Trans.). John Wiley & Sons, London, U.K. (Original work published 1814).

Hollowood, TJ 2016, 'Copenhagen quantum mechanics', *Contemporary Physics*, vol. 57, no. 3, arXiv:1511.01069 [quant-ph].

Lovell, BE 1995, '*A taxonomy of types of uncertainty*', (PhD Thesis). Portland State University.

van Strien, M 2014, 'On the origins and foundations of Laplacian determinism', *Studies in History and Philosophy of Science*, vol. 45, pp. 24-31.

The 8th International Conference on
Structural Health Monitoring of Intelligent Infrastructure
Brisbane, Australia | 5-8 December 2017

Dynamic analysis of Vehicle-bridge systems based on Explicit Form of Newmark-β Method

Solmaz Pourzeynali[1], Xinqun Zhu[2], Bijan Samali[3], Maria Rashidi[4]
[1,2,3,4] *Centre for Infrastructure Engineering, School of Computing, Engineering and Mathematics, Western Sydney University, Penrith, NSW 2751, Australia. Email: S.pourzeynali@westernsydney.edu.au*

Abstract

In the present study, a new method is proposed to analyse dynamic responses of the bridge subjected to a moving vehicle. Analysing the vehicle-bridge interaction system is useful for moving load identification and bridge structural damage detection. In this study, the algorithm based on explicit form of Newmark-β is proposed for dynamic analysis of the vehicle-bridge system. A three-span continuous bridge is used as a numerical example to investigate the efficiency of the proposed method and the results are compared with the existing results by the conventional state-space method. From the results, the proposed method is reliable and efficient for the vehicle-bridge interaction analysis.
Keywords: vehicle-bridge system, explicit form of Newmark-β method, Interaction force.

1. Introduction

The bridge infrastructures are always subjected to degradation, due to aging, operational environment, and excess loading, placing occupants and users at risk. Nowadays, it has become a big worldwide concern that a large part of bridge infrastructures are approaching their design life, and the number of bridges which are in a poor condition is increasing (Engineers Australia, 2010; American Society of Civil Engineers, 2017). All these make the civil engineering community to come up with developing methods for continuous and more accurate monitoring of bridge infrastructures. New methods need to be developed, as the existing ones are time consuming, and labour intensive in practice as well as they are not accurate enough.

Among all the existing approaches, vibration based methods of structural health monitoring of bridges have attracted the attention of many researchers over three decades (Abdeljaber et al., 2017, Carden and Fanning, 2004, Doebling et al. , 1998). As it is known, repeated traffic loads excitation over time can damage bridge structures and the accuracy of identified moving load can hugely affect the accuracy of damage detection (Law and Li, 2010). Some of the time domain force identification methods have been discussed by Law and Zhu (2011). Law et al. (1997) proposed an explicit integration method for force identification. There are also methods based on sensitivity method (Lu and Law, 2006), zeros order system Markov parameters (Law et al., 2005), and Kalman filter for force identification (Liu et al., 2000). While these methods are developed for force identification, Feng et al. (2015) proposed a method based on state-space method to identify moving load and structural parameters together. Analysis of the vehicle-bridge system based on the state-space method, suffers from a large discretization error and long sampling duration (Liu et al., 2014).

Newmark (1959) presented a general step-by-step method for the solution of problems in structural dynamics. This method is validated for any structures with a group of concentrated masses supported on a deformable structure and for any type of dynamic loading, such as impact, earthquake motion, vibration or blast, and for those of any degree of complication with elastic or inelastic behaviour or plastic behaviour. Liu et al. (2014) presented a method to transform the implicit form of Newmark-β method to its explicit form and the accuracy of this method is investigated for force identification. The results were compared with state-space method and it was concluded that the state-space method is comparatively less accurate when the sampling frequency is low. In the present study, analysis of the vehicle-bridge system based on explicit form of Newmark-β method is suggested and the reliability of the new developed algorithm is investigated by a numerical example.

2. Modelling of the vehicle-bridge interaction system

In this part, the explicit form of Newmark-β algorithm is developed for solving the vehicle-bridge interaction (VBI) system. As shown in Fig.1, a vehicle is modelled as a four-degree-freedom system (half vehicle model), and it is passing over the bridge at speed v. In the figure, m_v and I_v are the body mass and the pitch moment of inertia, respectively; m_{t1} and m_{t2} are the front and the rear axle mass, respectively. Furthermore, k_{s1}, k_{s2}, C_{s1} and C_{s2} are the linear suspension stiffness and the viscous damping parameters of the front and rear axles, respectively. Also, k_{t1}, k_{t2}, C_{t1}, and C_{t2} are the linear tire stiffness and the viscous damping parameters, respectively. a_1 and a_2 are the axle distances with respect to the gravity centre; ρ is the mass per unit length; EI is the flexural stiffness of the bridge, a product of Young's modulus E and the moment of inertia I.

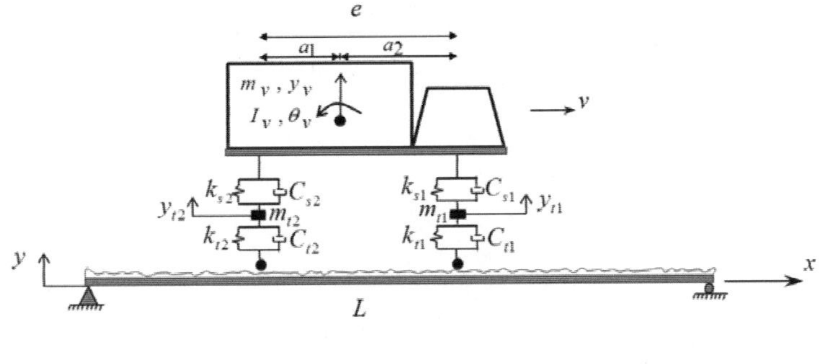

Fig. 1 Vehicle-bridge interaction system

At first, the equation of motion of both vehicle and bridge will be explained. Then, the two equations will be combined at their contact point and solved using the explicit form of the Newmark-β algorithm to get all the responses together in each time step.

2.1 Equation of motion of the vehicle

The equation of motion of the vehicle can be written as below (Feng *et al.*, 2015):

$$\begin{bmatrix} M_{v1} & 0 \\ 0 & M_{v2} \end{bmatrix}\{\ddot{y}_v\}+\begin{bmatrix} C_{v11} & C_{v12} \\ C_{v21} & C_{v22} \end{bmatrix}\{\dot{y}_v\}+\begin{bmatrix} K_{v11} & K_{v12} \\ K_{v21} & K_{v22} \end{bmatrix}\{y_v\}=p_v \tag{1}$$

where M_{v1}, M_{v2}, C_{v11}, C_{v12}, C_{v21}, C_{v22}, K_{v11}, K_{v12}, K_{v21}, and K_{v22} are respectively the vehicle mass, damping and stiffness submatrices. y_v, $\dot{y}_v$, and $\ddot{y}_v$ are the vehicle displacement, velocity and acceleration vectors which have to be calculated. More details are as follows:

$$y_v = \begin{Bmatrix} y_v \\ \theta_v \\ y_{t1} \\ y_{t2} \end{Bmatrix} \qquad P_v = \begin{Bmatrix} 0 \\ 0 \\ F_1(t) \\ F_2(t) \end{Bmatrix} - \begin{Bmatrix} 0 \\ 0 \\ W_1 \\ W_2 \end{Bmatrix} \tag{2}$$

$$F_{\text{int}} = \begin{Bmatrix} F_1(t) \\ F_2(t) \end{Bmatrix} \qquad W_v = \begin{Bmatrix} W_1 \\ W_2 \end{Bmatrix} \tag{3}$$

$$W_1 = (m_v \times a_2 / a + m_{t1})g \qquad W_2 = (m_v \times a_2 / a + m_{t2})g \tag{4}$$

$F_1(t)$ and $F_2(t)$ are the interaction forces happening between the vehicle axle loads and the bridge surface; W_1 and W_2 are the static loads at the front and the rear wheel locations. The vehicle mass, damping and stiffness submatrices in Equation (1) can be written as:

$$M_{v1} = \begin{bmatrix} m_v & 0 \\ 0 & I_v \end{bmatrix}, \quad M_{v2} = \begin{bmatrix} m_{t1} & 0 \\ 0 & m_{t2} \end{bmatrix}, \quad C_{v11} = \begin{bmatrix} C_{s1} + C_{s2} & C_{s1}a_1 - C_{s2}a_2 \\ C_{s1}a_1 - C_{s2}a_2 & C_{s1}a_1^2 + C_{s2}a_2^2 \end{bmatrix}$$

$$C_{v12} = \begin{bmatrix} -C_{s1} & -C_{s2} \\ -C_{s1}a_1 & C_{s2}a_2 \end{bmatrix}, \quad C_{v21} = \begin{bmatrix} -C_{s1} & -C_{s1}a_1 \\ -C_{s2} & C_{s2}a_2 \end{bmatrix}, \quad C_{v22} = \begin{bmatrix} C_{s1} & 0 \\ 0 & C_{s2} \end{bmatrix}$$

$$K_{v11} = \begin{bmatrix} K_{s1} + K_{s2} & K_{s1}a_1 - K_{s2}a_2 \\ K_{s1}a_1 - K_{s2}a_2 & K_{s1}a_1^2 + K_{s2}a_2^2 \end{bmatrix}, \quad K_{v12} = \begin{bmatrix} -K_{s1} & -K_{s2} \\ -K_{s1}a_1 & K_{s2}a_2 \end{bmatrix}$$

$$K_{v21} = \begin{bmatrix} -K_{s1} & -K_{s1}a_1 \\ -K_{s2} & K_{s2}a_2 \end{bmatrix}, \quad K_{v22} = \begin{bmatrix} K_{s1} & 0 \\ 0 & K_{s2} \end{bmatrix} \tag{5}$$

2.2 Equation of motion of the bridge

The bridge is modelled as a beam divided into $NJ-1$ equally spaced elements with NJ nodes. Each node has two degrees of freedom including rotational and vertical translations. The equation of motion of the bridge can be written as follows:

$$[M_b]\{\ddot{y}_b\} + [C_b]\{\dot{y}_b\} + [K_b]\{y_b\} = N_b \times F_{\text{int}} \tag{6}$$

where M_b, C_b, and K_b are the bridge mass, damping, and stiffness matrices, and y_b, $\dot{y}_b$, and $\ddot{y}_b$ are the nodal displacement, velocity and acceleration vectors of the bridge. N_b is a 2NJ×2 matrix that distributes the moving interaction forces into nodal forces. It includes Hermitian shape function vectors at the degrees of the freedom relating to the beam elements on which $F_1(t)$ and $F_2(t)$ are acting and it's all other entries are zero.

$$N_b = \begin{Bmatrix} 0 & \ldots & 0 & \ldots & Nb_1((x_1(t),t) & \ldots & 0 & \ldots & 0 \\ 0 & Nb_2((x_2(t),t) & \ldots 0 & \ldots 0 & & \ldots 0 & \ldots & & 0 \end{Bmatrix}^T \tag{7}$$

where $Nb_n (n = 1,2)$ relates to the n th interactive force on the m th finite element and the shape function of Nb_m can be written in the global coordinate as:

$$Nb_m = \begin{Bmatrix} 1 - 3\times(\frac{\hat{x}_n(t)-(m-1)\times l}{l})^2 + 2\times(\frac{\hat{x}_n(t)-(m-1)\times l}{l})^3 \\ (\hat{x}_n(t)-(m-1)\times l)\times(\frac{\hat{x}_n(t)-(m-1)\times l}{l}-1)^2 \\ 3\times(\frac{\hat{x}_n(t)-(m-1)\times l}{l})^2 - 2\times(\frac{\hat{x}_n(t)-(m-1)\times l}{l})^3 \\ (\hat{x}_n(t)-(m-1)\times l)\times(\frac{\hat{x}_n(t)-(m-1)\times l}{l}-1)^2 - (\frac{\hat{x}_n(t)-(m-1)\times l}{l}) \end{Bmatrix} \tag{8}$$

where $(m-1)\times l \le \hat{x}_n(t) \le ml$.

2.3 Equation of motion for the vehicle-bridge system

By combining Equations (1) and (6) with the compatibility at the contact point between the vehicle wheels and the bridge, the following equation can be obtained

$$\begin{bmatrix} M_b & 0 & N_b\times M_{v2} \\ 0 & M_{v1} & 0 \end{bmatrix}\{\ddot{Y}\} + \begin{bmatrix} C_b & N_b\times C_{v21} & N_b\times C_{v22} \\ 0 & C_{v11} & C_{v12} \end{bmatrix}\{\dot{Y}\} + \begin{bmatrix} K_b & N_b\times K_{v21} & N_b\times K_{v22} \\ 0 & K_{v11} & K_{v12} \end{bmatrix}\{Y\} = -\begin{Bmatrix} N_b W_v \\ 0 \end{Bmatrix} \tag{9}$$

where $\ddot{Y} = \begin{Bmatrix} \ddot{y}_b \\ \ddot{y}_v \end{Bmatrix}$, $\dot{Y} = \begin{Bmatrix} \dot{y}_b \\ \dot{y}_v \end{Bmatrix}$, $Y = \begin{Bmatrix} y_b \\ y_v \end{Bmatrix}$.

This equation can be solved through the explicit form of Newmark-β algorithm.

3. Dynamic analysis of Vehicle-bridge systems based on Explicit Form of Newmark-β Method

The main difference between the Newmark-β method and the ones had been developed before is that they do not include the resisting force variation during the time interval and their results are not as accurate as the Newmark-β method. In this method, the quantities are functions of both of position on the structure and of time, and the aim is to find out displacements, velocities, accelerations, and forces at a time t_{n+1}, which differs from t_n by the time interval h. Two parameters γ and β, were defined to show the share of the acceleration at the end of the interval h in the relations of displacement and velocity at the end of the interval. It is also shown that the best amount for γ is 1/2 and the best amount for β is in the range from 1/6 to 1/4 than in the range below 1/8. Other amounts for γ would introduce a spurious damping. Furthermore, through taking a time interval of the order of 1/5 to 1/6 of the shortest natural period of vibration, the convergence would be rapid and the errors would be small enough for all practical work.

The Newmark-β method is based on the assumed variation of acceleration. The two equations below define the variation of acceleration between two time instants and they are often used for better accuracy and stability of the computed acceleration (Liu et al., 2014).

$$\ddot{x} = (1-\gamma)\ddot{x}_i + \gamma\ddot{x}_{i+1} \qquad 0 \le \gamma \le 1 \tag{10}$$

$$\ddot{x} = (1-2\beta)\ddot{x}_i + 2\beta\ddot{x}_{i+1} \qquad 0 \le \beta \le 0.5 \tag{11}$$

Integrating the acceleration between t_i and t_{i+1} in Fig. 2, the displacement and velocity at t_{i+1} can be obtained as:

$$\dot{x}_{i+1} = \dot{x}_i + \Delta t\ddot{x} \tag{12}$$

$$x_{i+1} = x_i + \Delta t\dot{x}_i + \frac{1}{2}\Delta t^2\ddot{x} \tag{13}$$

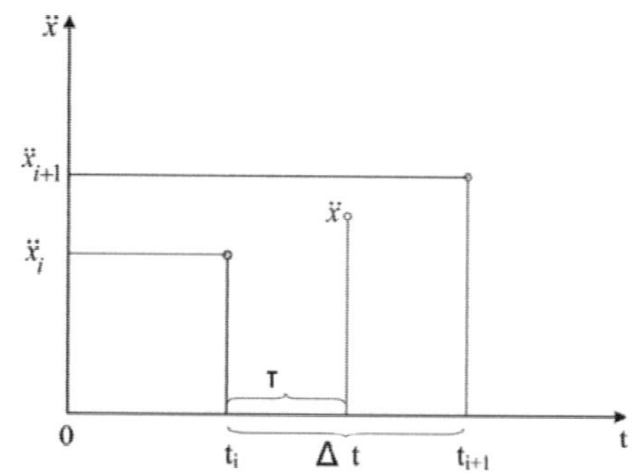

Fig. 2 Time discretisation of the acceleration response

With substituting Equation (10) into Equations (12) and (11) into Equation (13), the followings will be obtained

$$\dot{x}_{i+1} = \dot{x}_i + (1-\gamma)\Delta t\ddot{x}_i + \gamma\Delta t\ddot{x}_{i+1} \tag{14}$$

$$x_{i+1} = x_i + \Delta t\dot{x}_i + (\frac{1}{2}-\beta)\Delta t^2\ddot{x}_i + \beta\Delta t^2\ddot{x}_{i+1} \tag{15}$$

Rewriting Equation (15), the acceleration at t_{i+1} is derived as Equation (16). Substituting Equation (16) into Equation (14), the velocity at t_{i+1} is derived as Equation (17).

$$\ddot{x}_{i+1} = \frac{1}{\beta\Delta t^2}(x_{i+1} - x_i) - \frac{1}{\beta\Delta t}\dot{x}_i - (\frac{1}{2\beta} - 1)\ddot{x}_i \tag{16}$$

$$\dot{x}_{i+1} = \frac{\gamma}{\beta\Delta t}(x_{i+1} - x_i) + (1 - \frac{\gamma}{\beta})\dot{x}_i - (1 - \frac{\gamma}{2\beta})\Delta t\ddot{x}_i \tag{17}$$

The dynamic equation at time t_{i+1} is as follows:

$$M\ddot{x}(t_{i+1}) + C\dot{x}(t_{i+1}) + Kx(t_{i+1}) = N_b(t_{i+1}) \times F_{\text{int}}(t_{i+1}) \tag{18}$$

Substituting Equations (16) and (17) into Equation (18) and rewriting it to get:

$$x_{i+1} = A_0 N_{b\,i+1} F_{\text{int}(i+1)} + A_d x_i + A_v \dot{x}_i + A_a \ddot{x}_i \tag{19}$$

where

$$\hat{K} = K + \frac{1}{\beta\Delta t^2}M + \frac{\gamma}{\beta\Delta t}C$$

$$A_0 = \hat{K}^{-1}, \quad A_d = \hat{K}^{-1}\left[\frac{1}{\beta\Delta t^2}M + \frac{\gamma}{\beta\Delta t}C\right], \quad A_v = \hat{K}^{-1}\left[\frac{1}{\beta\Delta t}M + (\frac{\gamma}{\beta} - 1)C\right],$$

$$A_a = \hat{K}^{-1}\left[(\frac{1}{2\beta} - 1)M + \frac{\Delta t}{2}(\frac{\gamma}{\beta} - 2)C\right]$$

Substituting Equations (19) into Equation (17) and rewriting it to get:

$$\dot{x}_{i+1} = B_0 N_{b\,i+1} F_{\text{int}(i+1)} + B_d x_i + B_v \dot{x}_i + B_a \ddot{x}_i \tag{20}$$

where

$$B_0 = \frac{\gamma}{\beta\Delta t}\hat{K}^{-1}, \quad B_d = \frac{-\gamma}{\beta\Delta t}\hat{K}^{-1}K, \quad B_v = \frac{\gamma}{\beta\Delta t}\hat{K}^{-1}\left[\left(\frac{\beta\Delta t}{\gamma} - \Delta t\right)K + \frac{1}{\gamma\Delta t}M\right],$$

$$B_a = \frac{\gamma}{\beta\Delta t}\hat{K}^{-1}\left[\left(\frac{\beta\Delta t^2}{\gamma} - \frac{\Delta t^2}{2}\right)K + (\frac{1}{\gamma} - 1)M\right]$$

It can be rewritten further as

$$\ddot{x}_{i+1} = C_0 N_{b\,i+1} F_{\text{int}(i+1)} + C_d x_i + C_v \dot{x}_i + C_a \ddot{x}_i \tag{21}$$

where

$$C_0 = \frac{\gamma}{\beta\Delta t^2}\hat{K}^{-1}, \quad C_d = \frac{-1}{\beta\Delta t^2}\hat{K}^{-1}K,$$

Combining Equations (19), (20) and (21), we have

$$\begin{bmatrix} x_{i+1} \\ \dot{x}_{i+1} \\ \ddot{x}_{i+1} \end{bmatrix} = \begin{bmatrix} A_0 \\ B_0 \\ C_0 \end{bmatrix}_{i+1} N_{b(i+1)} F_{\text{int}} + \begin{bmatrix} A_d & A_v & A_a \\ B_d & B_v & B_a \\ C_d & C_v & C_a \end{bmatrix}_{i+1} \begin{bmatrix} x_i \\ \dot{x}_i \\ \ddot{x}_i \end{bmatrix} \tag{22}$$

The general recursive relation can be written as:

$$\begin{bmatrix} x_i \\ \dot{x}_i \\ \ddot{x}_i \end{bmatrix} = \begin{bmatrix} A_0 \\ B_0 \\ C_0 \end{bmatrix}_i N_{b(i)} F_{\text{int}} + \begin{bmatrix} A_d & A_v & A_a \\ B_d & B_v & B_a \\ C_d & C_v & C_a \end{bmatrix}_i \begin{bmatrix} x_{i-1} \\ \dot{x}_{i-1} \\ \ddot{x}_{i-1} \end{bmatrix}$$

$$\begin{bmatrix} x_{i-1} \\ \dot{x}_{i-1} \\ \ddot{x}_{i-1} \end{bmatrix} = \begin{bmatrix} A_0 \\ B_0 \\ C_0 \end{bmatrix}_{i-1} N_{b(i-1)} F_{\text{int}} + \begin{bmatrix} A_d & A_v & A_a \\ B_d & B_v & B_a \\ C_d & C_v & C_a \end{bmatrix}_{i-1} \begin{bmatrix} x_{i-2} \\ \dot{x}_{i-2} \\ \ddot{x}_{i-2} \end{bmatrix}$$

$$\begin{bmatrix} x_1 \\ \dot{x}_1 \\ \ddot{x}_1 \end{bmatrix} = \begin{bmatrix} A_0 \\ B_0 \\ C_0 \end{bmatrix}_1 N_{b(1)} F_{\text{int}} + \begin{bmatrix} A_d & A_v & A_a \\ B_d & B_v & B_a \\ C_d & C_v & C_a \end{bmatrix}_1 \begin{bmatrix} x_0 \\ \dot{x}_0 \\ \ddot{x}_0 \end{bmatrix} \tag{23}$$

and the response at time t_i can be written as:

$$
\begin{bmatrix} x_i \\ \dot{x}_i \\ \ddot{x}_i \end{bmatrix} = \begin{bmatrix} A_0 \\ B_0 \\ C_0 \end{bmatrix}_i N_{bi} F_{\mathrm{int}(i)} +
$$

$$
\begin{bmatrix} A_d & A_v & A_a \\ B_d & B_v & B_a \\ C_d & C_v & C_a \end{bmatrix}_i \begin{bmatrix} A_0 \\ B_0 \\ C_0 \end{bmatrix}_{i-1} N_{bi-1} F_{\mathrm{int}(i)} + \begin{bmatrix} A_d & A_v & A_a \\ B_d & B_v & B_a \\ C_d & C_v & C_a \end{bmatrix}_i \begin{bmatrix} A_d & A_v & A_a \\ B_d & B_v & B_a \\ C_d & C_v & C_a \end{bmatrix}_{i-1} \begin{bmatrix} A_0 \\ B_0 \\ C_0 \end{bmatrix}_{i-2} N_{bi-2} F_{\mathrm{int}(i)} + \dots
$$

$$
+ \begin{bmatrix} A_d & A_v & A_a \\ B_d & B_v & B_a \\ C_d & C_v & C_a \end{bmatrix}_i \begin{bmatrix} A_d & A_v & A_a \\ B_d & B_v & B_a \\ C_d & C_v & C_a \end{bmatrix}_{i-1} \dots \begin{bmatrix} A_d & A_v & A_a \\ B_d & B_v & B_a \\ C_d & C_v & C_a \end{bmatrix}_2 \begin{bmatrix} A_0 \\ B_0 \\ C_0 \end{bmatrix}_1 N_{b1} F_{\mathrm{int}(i)} +
$$

$$
\begin{bmatrix} A_d & A_v & A_a \\ B_d & B_v & B_a \\ C_d & C_v & C_a \end{bmatrix}_i \begin{bmatrix} A_d & A_v & A_a \\ B_d & B_v & B_a \\ C_d & C_v & C_a \end{bmatrix}_{i-1} \dots \begin{bmatrix} A_d & A_v & A_a \\ B_d & B_v & B_a \\ C_d & C_v & C_a \end{bmatrix}_2 \begin{bmatrix} A_d & A_v & A_a \\ B_d & B_v & B_a \\ C_d & C_v & C_a \end{bmatrix}_1 \begin{bmatrix} x_0 \\ \dot{x}_0 \\ \ddot{x}_0 \end{bmatrix} \tag{24}
$$

Equations (22) and (24) are new and explicit form of the Newmark-β method in two different forms, where the responses $x, \dot{x}, \ddot{x}$ are solved together at each time step.

Equation (9) for vehicle-bridge interaction system can be solved by Equation (22) or (24), to obtain the responses of the bridge and vehicle simultaneously.

4. Numerical Examples

In this section, an example is studied to verify the proposed method. The example is a three span continuous bridge with the mid-span of 50 meters and side spans of 30 meters (Fig. 3). The half vehicle model is passing over this bridge at the speed of 20 m/s. The flexural stiffness of the side spans is $EI_{1,3}=1.5*10^{10}$ Nm^2 and of the mid-span is $EI_2=2.5*10^{10}$ Nm^2. Mass per unit length is $\rho = 5\times10^3\ kg/m$, and the damping ratio of all modes is assumed to be $\zeta = 2\%$. The bridge is divided into 55 equal Euler-Bernoulli beam elements with 112 DOFs. The vehicle model information is in Table.4.1.

This example is already solved by Feng *et al.*, (2015) by state-space method and the dynamic responses of both vehicle and bridge were obtained. The results were compared with measurements from seven points of the bridge S1-S7, which are considered as a reference in this paper. In this paper, the dynamic responses of the bridge and the vehicle are obtained by explicit form of both state-space method and Newmark-β method to compare the results. The results of displacement and acceleration responses of the bridge and interaction forces can be seen in Fig. 4 and Fig. 5.

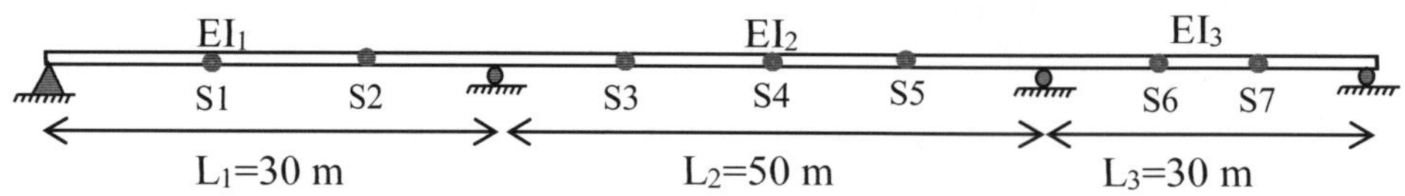

Fig. 3 Three span bridge model

Table.1 The vehicle model information

$m_v = 16200kg$	$m_{t1} = 700\ kg$	$m_{t2} = 1100\ kg$
$I_v = 9.5 \times 10^5\ Nm^2$	$k_{s1} = 0.4 \times 10^6\ N/m$	$k_{s2} = 1.00 \times 10^6\ N/m$
$a_1 = 2.42\,m$	$k_{t1} = 1.75 \times 10^6\ N/m$	$k_{t2} = 3.5 \times 10^6\ N/m$
$a_2 = 1.85\,m$	$C_{s1} = 1.00 \times 10^4\ N/m/s$	$C_{s2} = 2.00 \times 10^4\ N/m/s$
$v = 20\,m/s$	$C_{t1} = 3.90 \times 10^3\ N/m/s$	$C_{t2} = 4.30 \times 10^3\ N/m/s$

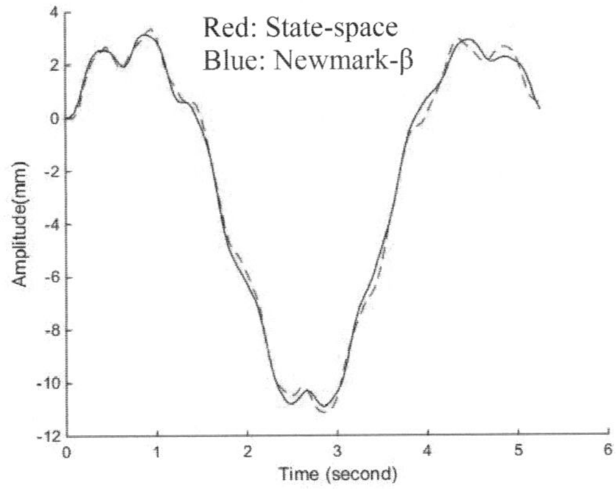

(a) Mid span

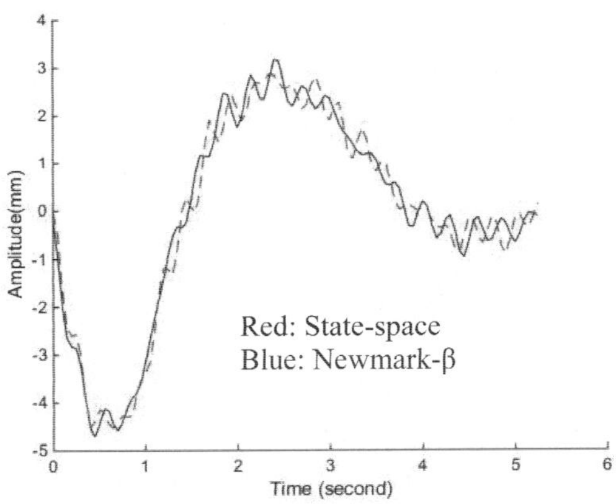

(b) Left span

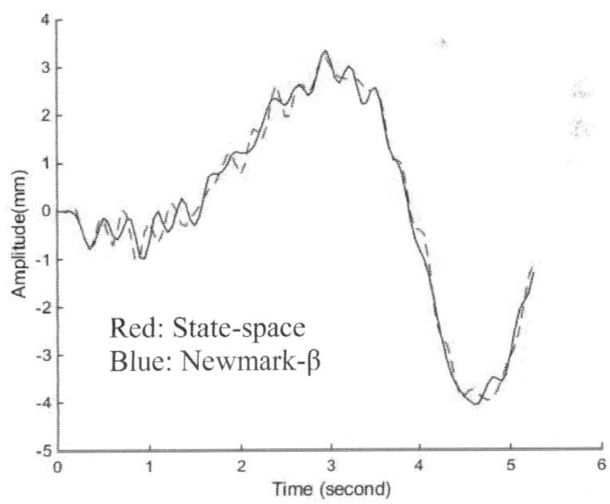

(b) Right span

Fig. 4 Prediction of mid-span displacement of all three spans

Figure 4 shows the displacements by the proposed method that are compared the results by the state-space method from Feng *et al.*, (2015). The results agree well and that shows the proposed method is verified.

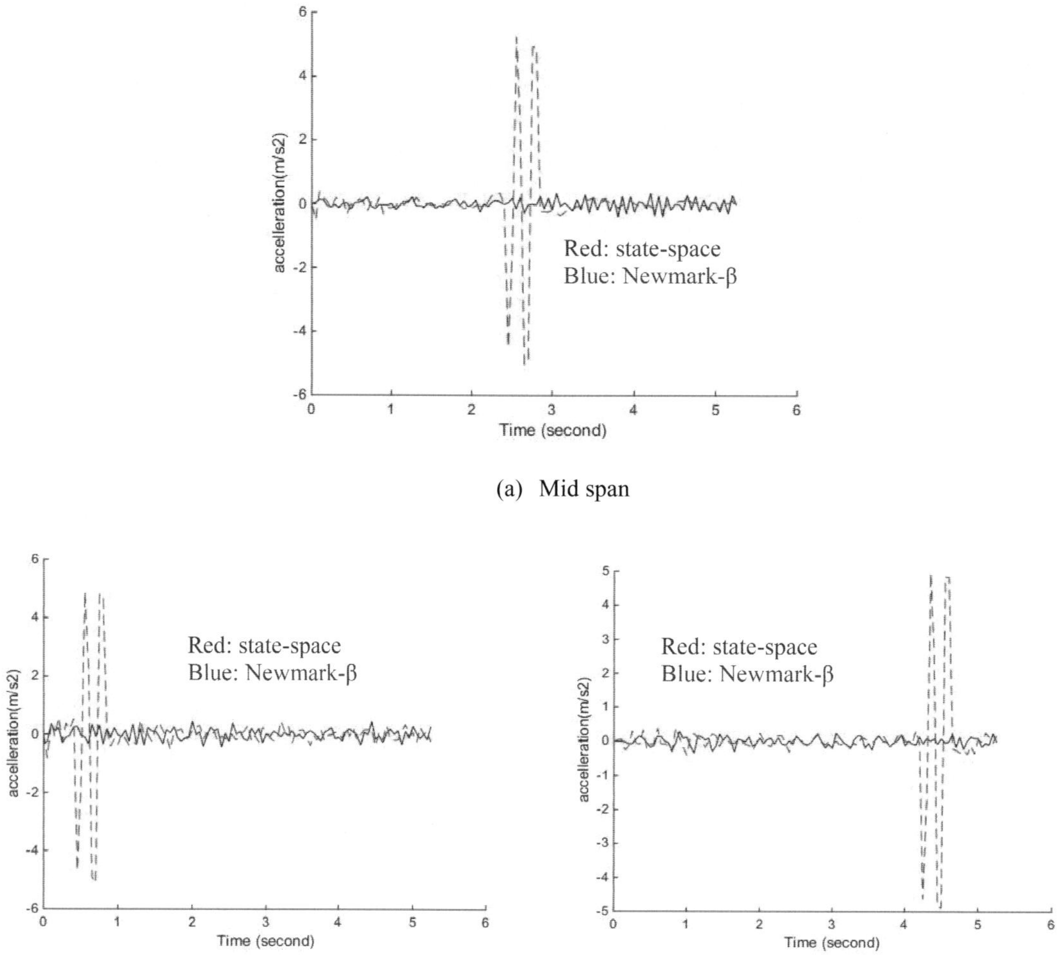

(a) Mid span

(b) Left span

(c) Left span

Fig. 5 Prediction of mid-span acceleration of all three spans

Figure 5 shows the accelerations by the proposed method and the state-space method. The results by the state-space method have a large oscillation around the support and that shows the method cannot be used to calculate the acceleration for continuous bridges. However, the proposed method can get the good results.

5. Conclusions

In this study, the explicit form of Newmark-β algorithm is developed for solving the vehicle-bridge interaction (VBI) system. To verify the results, a numerical example has been studied and the results by the proposed method have been compared with the existing results using the state-space method. It is concluded that a) the proposed method is able to predict dynamic responses of the bridge subjected to moving vehicles; b) the proposed method is more stable around supports in the acceleration response and can provide good results for continuous bridges.

References

Engineers Australia 2010, Engineers Australia Report Card.

American Society of Civil Engineers 2017, Report Card for American Infrastructure.

Abdeljaber, O, Avci, O, Kiranyz, S, Gabbouj, M & Inman, DJ 2017, 'Real-time vibration-based structural damage detection using one-dimensional convolutional neural networks', *Journal of Sound and Vibration,* Vol. 388, pp. 154-170.

Carden, EP & Fanning, P 2004, 'Vibration based condition monitoring: a review', *Structural health monitoring,* Vol. 3, pp. 355-377.

Doebling, SW, Farrar, CR & Prime, MB 1998, 'A summary review of vibration-based damage identification methods', *Shock and vibration digest,* Vol. 30, pp. 91-105.

Law, SS & Li, J, 2010, 'Updating the reliability of a concrete bridge structure based on condition assessment with uncertainties', *Engineering Structures,* Vol. 32, pp. 286-296.

Law, SS & Zhu, XQ 2011, 'Moving load identification-problems and identification', Taylor and Francis, London .

Law, SS, Chan, THT & Zeng, QH 1997, 'Moving force identification: A time domain method. *Journal of Sound and Vibration',* Vol. 201, pp. 1-22.

Lu, ZR & Law, SS 2006, 'Forced identification based on sensitivity in time domain', *Journal of Engineering Mechanics,* Vol. 132, pp. 1050-1056.

Law, SS, Bu, JQ & Zhu, XQ, 2005, 'Time varying wind load identification from structural responses', *Engineering Structures,* Vol. 27, pp. 1586-1598.

Liu, JJ, Ma, CK, Kung, IC & Lin, DC 2000, 'Input force estimation of a cantilever plate by using a system identification technique', *Computer Methods in Applied Mechanics and Engineering,* Vol. 190, pp. 1309-1322.

Liu, K., Law, SS, Zhu, XQ & Xia, Y, 2014, 'Explicit form of an implicit method for inverse force identification', *Journal of Sound and Vibration,* Vol. 333, pp. 730-744.

Newmark, NM 1959, *'A method of computation for structural dynamics',* American Society of Civil Engineers.

Feng, D, Sun, H, & Feng, MQ 2015, 'Simultaneous identification of bridge structural parameters and vehicle loads', *Computers & Structures,* Vol. 157, pp. 76-88.

The 8th International Conference on
Structural Health Monitoring of Intelligent Infrastructure
Brisbane, Australia | 5-8 December 2017

Numerical Investigation of a Linkage Modelling Technique for Damage Identification Using FRF-Based Model Updating

V.V. Nguyen[1], J. Li[2], E. Erkmen [3].
[1] *University of Technology Sydney – Australia, Email:* Van.V.Nguyen-1@student.uts.edu.au [2] *University of Technology Sydney – Australia,* [3] *University of Technology Sydney – Australia*

Abstract

This paper presents a novel method of identifying the location and severity of damage in a bridge component via model updating where the number of degrees of freedom (DOF) in the finite element (FE) model far exceeds the DOFs measured on a structure. First, the FE model is divided into partitions, each with a predefined Young's modulus. These Young's moduli are used as the updating parameters where the reduction in the Young's moduli is used to indicate damage. The mass and stiffness matrices of the FE model is reduced to a linkage model using the System Equivalent Reduction Expansion Process (SEREP). Then the measured DOFs of the bridge component is expanded to the DOF of the linkage model using the mass and stiffness matrices of the linkage model. Based on the expanded modal data, interpolated FRFs are synthesised and used to form the FRF sensitivity matrix, which is iteratively used to calculate the values of the updating parameters until convergence is achieved. The resulting Young's moduli are used to calculate the damage index for each partition of the FE model. The proposed method is explained and verified using a numerical study. This method has the potential to locate and quantify damage in a structure where sensor instrumentation is otherwise inadequate due to the limitation in sensor availability and access for sensor installation.

1. Introduction

In the past couple of decades, Structural Health Monitoring (SHM) has received significant attention in the field of asset management, especially for ageing infrastructures. SHM systems generally aim to provide information on the health state of a structure, including detecting, localising and quantifying damage and deterioration. Numerous damage detection methods have been developed over the years (Doebling et al. 1998; Sohn et al. 2003). One such method includes model updating, which reduces the discrepancy between the results obtained from a finite element (FE) model and the results measured from the physical structure by adjusting the modelling parameters until the analytical and measured results are in agreement.

An important issue in SHM is the limited sensor arrays. Ideally, there would be one sensor corresponding to each degree of freedom (DOF) in the FE model. However, this is rarely possible in practice, due to the limited sensor availability and the inaccessibility of certain areas for sensor placement. Researchers (Mousavi & Gandomi 2016; Hansen et al. 2017) have addressed this issue by reducing the FE model DOF to the measured DOF. Modal expansion has also been considered to solve this issue (Au et al. 2003; Entezami 2017). However, this can be problematic when the number of DOF in an FE model is significantly larger than the measured DOF. A possible solution is to reduce the FE model to the linkage model that can be used to expand to measured DOF.

This paper proposes the use of FE model updating using the FRF sensitivity method to localise and quantify damage in a replicated jack arch structure of the Sydney Harbour Bridge via a linkage model. Details of prior studies of the jack arch structure can be found in Mustapha et al. (2015) and Nguyen et al. (2015). The parameters defined for the model updating procedure include the Young's modulus of 10 divided sections of the specimen and three additional parameters relating to the boundary conditions of the jack arch. The full FE model contains 5,706 DOFs, including DOFs in the vertical and horizontal directions, and is reduced to a linkage model containing 60 vertical DOFs using SEREP. The mode shapes determined experimentally using 11 accelerometers are then expanded to 60 DOFs using the mass and stiffness matrices of the linkage model. The receptance FRFs are synthesised based on the expanded mode shapes and used to form the FRF sensitivity matrix and FRF residual vectors. The linear least squares method incorporating the trust region reflective algorithm is then used to update the parameters. This procedure is iteratively repeated until convergence in the parameters is achieved. The damage location and severity is then determined based on the updated parameters. The procedure is verified numerically.

2. Addressing the Limited Sensor Issue Using a Linkage Model

In real structures, it is impractical to place one sensor at each location corresponding to all DOFs in the FE model. These restrictions are due to the limited number of available sensors for instrumentation and inaccessibility of certain areas of the structure. This study proposes the use of a linkage model that reduces the DOFs of a 3D FE model to a linkage model while expanding the measured DOFs to the same model. In this investigation, SEREP (O'Callahan et al. 1989) is used for model reduction. The procedure requires partitioning of the mass ($\mathbf{M}$) and stiffness ($\mathbf{K}$) matrices according to the expression in Eq. (1).

$$\mathbf{M} = \begin{bmatrix} \mathbf{M}_{\mathrm{mm}} & \mathbf{M}_{\mathrm{ms}} \\ \mathbf{M}_{\mathrm{sm}} & \mathbf{M}_{\mathrm{ss}} \end{bmatrix} \text{ and } \mathbf{K} = \begin{bmatrix} \mathbf{K}_{\mathrm{mm}} & \mathbf{K}_{\mathrm{ms}} \\ \mathbf{K}_{\mathrm{sm}} & \mathbf{K}_{\mathrm{ss}} \end{bmatrix} \tag{1}$$

where the subscript m refers to the master DOF that will be retained and s is the slave DOF that will be removed after the reduction procedure. The transformation matrix, $\mathbf{T}$, used to reduce the FE model can be deduced using Eq. (2)

$$\mathbf{T} = \begin{bmatrix} \boldsymbol{\Phi}_{\mathrm{m}} \\ \boldsymbol{\Phi}_{\mathrm{s}} \end{bmatrix} \boldsymbol{\Phi}_{\mathrm{m}}^{\ +} \tag{2}$$

where $\boldsymbol{\Phi}$ is the matrix containing the mass normalised eigenvectors of the FE model obtained using eigenvalue analysis. The superscript + refers to the pseudo inverse of a matrix. The transformation matrix is then used to reduce the full system matrices to the linkage model system matrices using Eq. (3).

$$\mathbf{M}_{\mathrm{l}} = \mathbf{T}^{\mathrm{T}}\, \mathbf{M}\, \mathbf{T} \text{ and } \mathbf{K}_{\mathrm{l}} = \mathbf{T}^{\mathrm{T}}\, \mathbf{K}\, \mathbf{T} \tag{3}$$

where $\mathbf{M}_{\mathrm{l}}$ and $\mathbf{K}_{\mathrm{l}}$ are the mass and stiffness matrices of the linkage model. The system matrices of the linkage model are then used to expand the measured DOFs to the linkage model DOFs. Friswell et al. (2013) provide the procedure for this method. Consider the expression in Eq. (4).

$$\left(-\omega_{\mathrm{j}}^{2} \begin{bmatrix} \mathbf{M}_{\mathrm{l_{mm}}} & \mathbf{M}_{\mathrm{l_{ms}}} \\ \mathbf{M}_{\mathrm{l_{sm}}} & \mathbf{M}_{\mathrm{l_{ss}}} \end{bmatrix} + \begin{bmatrix} \mathbf{K}_{\mathrm{l_{mm}}} & \mathbf{K}_{\mathrm{l_{ms}}} \\ \mathbf{K}_{\mathrm{l_{sm}}} & \mathbf{K}_{\mathrm{l_{ss}}} \end{bmatrix}\right) \begin{Bmatrix} \boldsymbol{\Phi}_{\mathrm{mj}} \\ \boldsymbol{\Phi}_{\mathrm{sj}} \end{Bmatrix} = \begin{Bmatrix} \mathbf{0} \\ \mathbf{0} \end{Bmatrix} \tag{4}$$

Where $\boldsymbol{\phi}_{\mathrm{mj}}$ and $\boldsymbol{\phi}_{\mathrm{sj}}$ respectively represent the mode shape at the measured and unmeasured DOFs of the jth mode and $\boldsymbol{\omega}_{\mathbf{j}}^{\mathbf{2}}$ is the corresponding eigenvalue. The bottom half of this equation can be rearranged to Eq. (5).

$$\boldsymbol{\phi}_{\mathrm{sj}} = -\left(-\omega_{\mathrm{j}}^{2}\mathbf{M}_{\mathbf{l}_{\mathrm{ss}}} + \mathbf{K}_{\mathbf{l}_{\mathrm{ss}}}\right)^{-1}\left(-\omega_{\mathrm{j}}^{2}\mathbf{M}_{\mathbf{l}_{\mathrm{sm}}} + \mathbf{K}_{\mathbf{l}_{\mathrm{sm}}}\right)\boldsymbol{\phi}_{\mathrm{mj}} \tag{5}$$

which can be used to calculate the interpolated mode shapes for the unmeasured DOFs using the mass and stiffness matrices of the linkage model and the measured eigenvectors. The interpolated mode shapes can be used to synthesise the receptance FRFs at the unmeasured DOFs using a formulation provided by Maia and Silva (1998) expressed in Eq. (6).

$$\alpha_{\mathrm{jk}}(\omega) = \sum_{\mathrm{r}=1}^{\mathrm{N}}\left(\frac{\phi_{\mathrm{jr}}\phi_{\mathrm{kr}}}{\mathrm{i}\omega - \mathrm{s}_{\mathrm{r}}} + \frac{\phi_{\mathrm{jr}}^{*}\phi_{\mathrm{kr}}^{*}}{\mathrm{i}\omega - \mathrm{s}_{\mathrm{r}}^{*}}\right) \tag{6}$$

where there is N number of modes used in the FRF synthesis, $\boldsymbol{\phi}_{\mathbf{jr}}$ represents the modal displacement at response j for mode r, $\boldsymbol{\phi}_{\mathbf{kr}}$ represents the modal displacement at the force input k for mode r. The eigenvalues $\mathbf{s}_{\mathbf{r}}$ are the poles for mode r. The residuals in Eq. (6) are normalised by multiplying the numerators with Eq. (7).

$$\mathrm{a}_{\mathrm{r}} = \frac{1}{2\mathrm{i}\mathrm{m}_{\mathrm{r}}\omega_{\mathrm{dr}}} \tag{7}$$

where $i = \sqrt{-1}$, $\mathbf{m}_{\mathbf{r}}$ is the modal mass and $\boldsymbol{\omega}_{\mathbf{dr}}$ is the damped natural frequency of mode r. The synthesised FRFs can be used in the model updating procedure.

3. Damage Identification Using FRF Sensitivity Model Updating

The location and severity of damage is identified in this study by updating FE models using FRF sensitivities. The method is based on the iterative formulation of the FRF sensitivities derived by Lin and Ewins (1994). In each iteration, this method uses the Linear Least Squares Solver "lsqlin" in MATLAB to minimise the squared 2-norm of the residual expressed in Eq. (8).

$$\min_{\Delta\mathbf{p}} \frac{1}{2}\|\mathbf{S}\,\Delta\mathbf{p} - \Delta\boldsymbol{\alpha}\|_2^2 \tag{8}$$

where $\Delta\mathbf{p}$ is the vector of updating parameters and $\Delta\boldsymbol{\alpha}$ is the vector of FRF residuals calculated using Eq. (9).

$$\Delta\boldsymbol{\alpha} = \begin{Bmatrix} \{\boldsymbol{\alpha}_X(\omega_1)\}_j - \{\boldsymbol{\alpha}_A(\omega_1)\}_j \\ \vdots \\ \{\boldsymbol{\alpha}_X(\omega_n)\}_j - \{\boldsymbol{\alpha}_A(\omega_n)\}_j \end{Bmatrix} \tag{9}$$

where $\{\boldsymbol{\alpha}_X(\omega_i)\}_j$ refers to the vector of the experimentally obtained receptance FRFs at the ith measured frequency with each row of the vector corresponding to the DOF of the measured response with the input applied at coordinate j and $\{\boldsymbol{\alpha}_A(\omega_i)\}_j$ being the analytical counterpart. $\mathbf{S}$ is the FRF sensitivity matrix and can be calculated using Eq. (10).

$$\mathbf{S}=\begin{bmatrix} -[\boldsymbol{\alpha}_A(\omega_1)]\left[\frac{\partial \mathbf{Z}_A(\omega_1)}{\partial p_1}\right]\{\boldsymbol{\alpha}_X(\omega_1)\}_j & \cdots & -[\boldsymbol{\alpha}_A(\omega_1)]\left[\frac{\partial \mathbf{Z}_A(\omega_1)}{\partial p_{Np}}\right]\{\boldsymbol{\alpha}_X(\omega_1)\}_j \\ \vdots & \ddots & \vdots \\ -[\boldsymbol{\alpha}_A(\omega_n)]\left[\frac{\partial \mathbf{Z}_A(\omega_n)}{\partial p_1}\right]\{\boldsymbol{\alpha}_X(\omega_n)\}_j & \cdots & -[\boldsymbol{\alpha}_A(\omega_n)]\left[\frac{\partial \mathbf{Z}_A(\omega_n)}{\partial \mathbf{p}_{Np}}\right]\{\boldsymbol{\alpha}_X(\omega_n)\}_j \end{bmatrix} \tag{10}$$

where $\mathbf{Z}_A(\omega)$ is the dynamic stiffness matrix expressed in the frequency domain in Eq. (11).

$$\mathbf{Z}_A(\omega)=\mathbf{K}+i\omega\mathbf{C}-\omega^2\mathbf{M} \tag{11}$$

where **K** is the stiffness matrix, **C** is the damping matrix and **M** is the mass matrix of the FE model. The formulation of the sensitivity matrix **S** requires the partial derivatives of the dynamic stiffness matrix with respect to the updating parameters to be calculated. The dynamic stiffness matrix derivatives can be reduced to the linkage model using the transformation matrix calculated in Eq. (2). Also, the numerical conditioning of Eq. (8) can be improved by normalising the updating parameters so that the current parameter estimate is unity. The choice of updating parameters depends on the purpose of the model updating procedure. For damage detection purposes, this is typically the Young's modulus of an element or a partition of the model. A decrease in the stiffness of the material can indicate damage in a structure. Dividing the FE model into partitions and using the Young's modulus of each partition as an updating parameter can be used as an approach for damage detection via model updating. The decision of DOFs to retain after model reduction can be tailored around the updating parameters chosen. An example of this is provided in the next section of this paper. The updated parameters can be used to form a damage index (DI) for each partition of the structure based on the overall stiffness reduction of each partition as expressed in Eq. (12).

$$\mathrm{DI}=\frac{E_h-E_d}{E_h} \tag{12}$$

where E_h and E_d refer to the Young's modulus of a partition of the FE model in its healthy state and damaged state respectively. Figure 1 summarises the damage detection procedure.

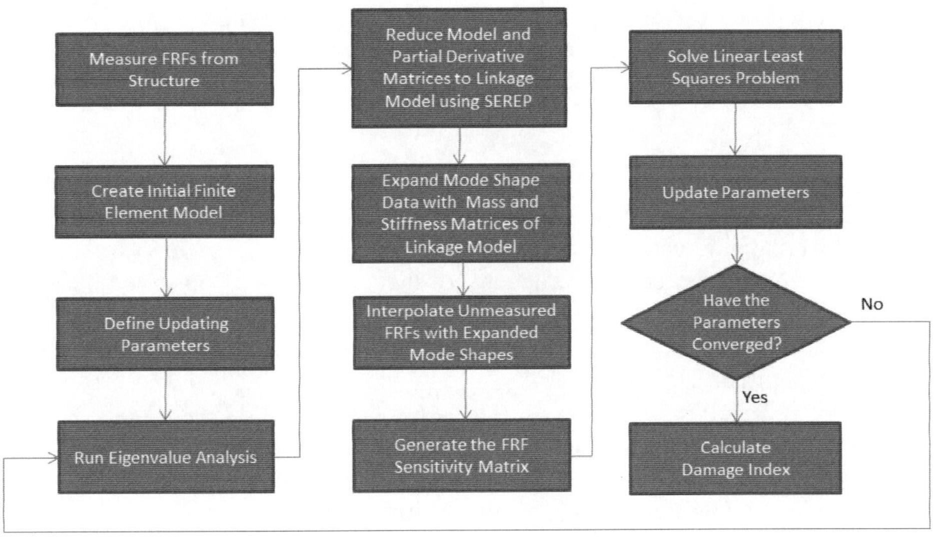

Fig. 1. Flow chart summarising the damage detection procedure.

4. Case Study – Experimental Testing

An experimental specimen replicating a jack arch component from the bus lane of the Sydney Harbour Bridge, shown in Figure 2, was used as the case study for this investigation. The jack arch component was 375 mm in height, 1,000 mm in width and 2,000 mm in length and had an internal steel I-Beam embedded along it. The cantilever structure was adhered to a clamp at the fixed end using plaster and was additionally supported at 1,000 mm from the front of the specimen.

Fig. 2. Experimental jack arch specimen and its boundary condition setup.

Experimental modal analysis (EMA) was conducted on the specimen to extract its dynamic characteristics. An impact hammer was used to excite the structure 25 times to generate multiple data samples and eleven accelerometers (model PCB 352C34), A1 to A11, were used to measure the acceleration response of the structure in the vertical direction. Figure 3 shows the EMA testing set up. The data was collected with a sampling rate of 20 kHz captured over 2 seconds for each impact. A1 was treated as the driving point FRF in this investigation. The accelerance FRFs for each impact sample were calculated and the outliers were removed. The remaining samples were averaged to emphasise the repeatable features of the FRFs while softening the non-repeatable features. The receptance FRFs were then calculated by dividing the accelerance FRFs by $-\omega^2$ at each measured frequency.

The next step of this investigation was to introduce damage to the specimen. Therefore, a static load was applied at the front of the specimen using a hydraulic jack as indicated in Figure 2. The specimen was treated as an overhanging cantilever beam in this configuration. The load was applied until an immediate reduction in the load carrying capacity of the structure was observed. After the loading was stopped, the hydraulic jack was removed and the dynamic test was repeated. A crack was observed at 1,300 mm measured on the left and 1,400 mm measured on the right of the specimen from its front with a depth of 275 mm.

The scope of this conference paper covers the numerical validation of the proposed method only. Simulation of damage is explained in the next section of this paper. An experimental investigation is currently in preparation and will be covered in an upcoming journal paper.

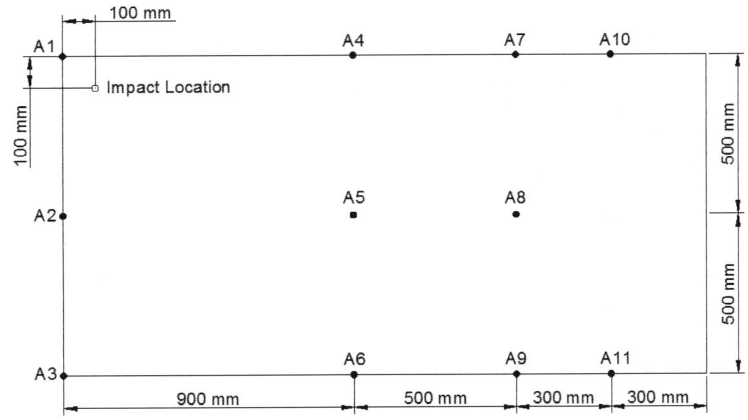

Fig. 3. Schematic for dynamic testing showing the locations of the hammer impact and the accelerometers A1 to A11 measured in the vertical direction.

5. Numerical Validation

An FE model was constructed in ANSYS APDL using 20 noded SOLID 186 elements to exhibit quadratic displacement behaviour in the model. This model is illustrated in Figure 4. The updating parameters in the FE model were defined by dividing the model into ten partitions along the structure with each partition being 200 mm long. This was done to detect the location and severity of the transverse crack that was expected to occur as a result of loading the specimen at its tip in the overhanging cantilever beam configuration. A Young's modulus value was defined for each of these partitions. These parameters are denoted as P1 to P10 in Figure 4. The initial Young's modulus for the concrete specimen was set to 30 GPa. The Young's modulus of the adhesives connecting the specimen to the supports were also considered as updating parameters, as they had a significant influence on the dynamic properties of the structure. They are denoted as BC1 to BC3 in Figure 4. The partial derivatives of the mass and stiffness matrices with respect to the updating parameters were numerically determined by perturbing the updating parameters by a value of 1 MPa and calculating the differences in the perturbed and unperturbed system matrices.

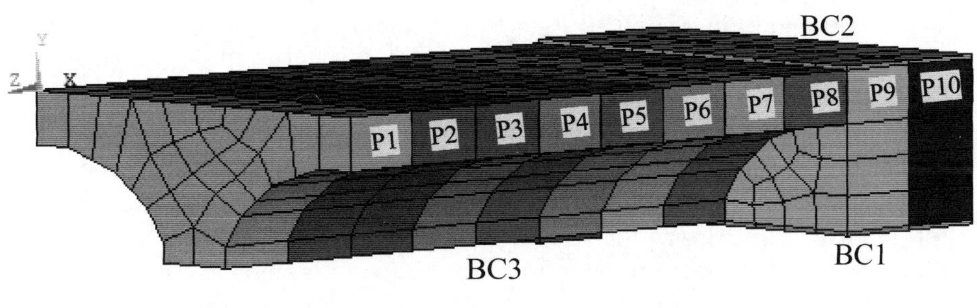

Fig. 4. Finite element model of jack arch specimen.

Smeared cracking was used to simulate damage by reducing the Young's modulus of P7 to 20 GPa, as the crack identified on the specimen was located within the region covered by this parameter. Eleven receptance FRFs were synthesised based on the experimental setup of the modal test shown in Figure 3. Rayleigh damping was assumed in the FRF synthesis and the FE model, using the arbitrary values of $\alpha = 48$ and $\beta = 2 \times 10^{-8}$. The eleven receptance FRFs were then used as the measured data in the numerical study. In the initial iteration of the model updating procedure, the mass and stiffness matrices of the FE model were reduced from the 5,706 DOF shown in Figure 4 to the 60 DOF linkage model shown in Figure 5 using SEREP. The DOFs of the linkage model were chosen based on the updating parameters of the FE model. The partitions of the FE model covered a length of 200 mm along the specimen. Hence, 200 mm was chosen as the distance between the DOFs along the linkage model. The spacing of 250 mm across the specimen was chosen to retain a good aspect ratio for the grid and to ensure that there was a node in the centreline of the structure. This resulted in 55 DOFs for the linkage model. An additional 5 DOFs were added to include measured DOFs, resulting in a total of 60 DOFs for the linkage model. Only the vertical DOFs were considered for the linkage model. The mass and stiffness matrices of the linkage model were then used to expand the measured mode shapes to match the DOFs of the linkage model using Eq. (5). Examples of expanded mode shapes are shown in Figure 6.

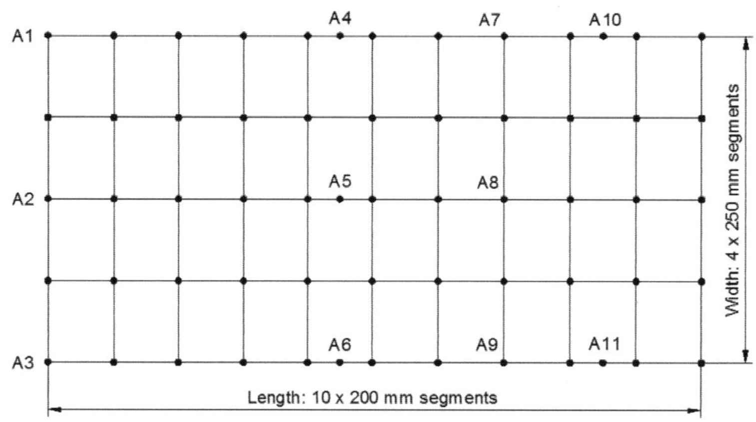

Fig. 5. Linkage FE model.

The interpolated receptance FRFs used in the updating algorithm were then synthesised using the expanded mode shapes with the expression in Eq. (6). The choice of measured frequencies was considered in the updating process. Sufficient measured frequencies were needed to ensure that there was adequate information on the dynamic properties of the structure. However, there was a cut-off where more frequencies points were no longer beneficial as the number of iterations needed to reach convergence was significantly increased (Imregun et al. 1995). The frequency range chosen for the updating procedure was 0 Hz to 1000 Hz with an increment of 1 Hz. The next step of the damage detection method was to produce the FRF sensitivity matrix using Eq. (10). The sensitivity matrix consisted of 13 columns corresponding to the updating parameters and 120,120 rows to accommodate the 1,001 measured frequency points, 60 measurement locations including the measured and interpolated DOFs and the splitting of the real and imaginary components of the FRF. The linear least-squares problem in Eq. (8) was then solved using the linear least-squares solver function in MATLAB with the trust region reflective algorithm. The lower bounds and upper bounds for P1 to P10 were set to 1 GPa and 30 GPa, respectively. A value of 1 GPa was used as a lower limit to allow the Young's Modulus of each partition to drop significantly to indicate its damage severity. 30 GPa was used as the upper limit, as it was understood that the Young's modulus of each partition would not rise above its healthy value. This procedure was repeated iteratively with the linkage model system matrices and interpolated FRFs changing with each iteration according to the adjustments in the updating parameters. The iterative procedure was stopped after 20 iterations. The comparison of the FRFs before and after updating is shown in Figure 7 and the parameter convergence is shown in Figure 8. The finalised updating parameters were used to calculate the damage index using Eq. (12). The results are summarised in Figure 9.

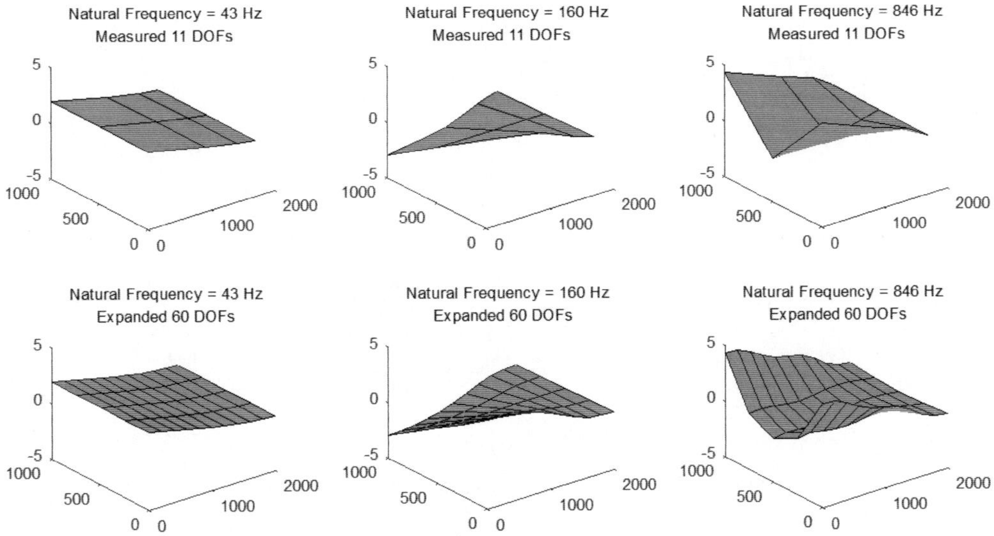

Fig. 6. Examples of mode shapes expanded from 11 DOFs to 60 DOFs.

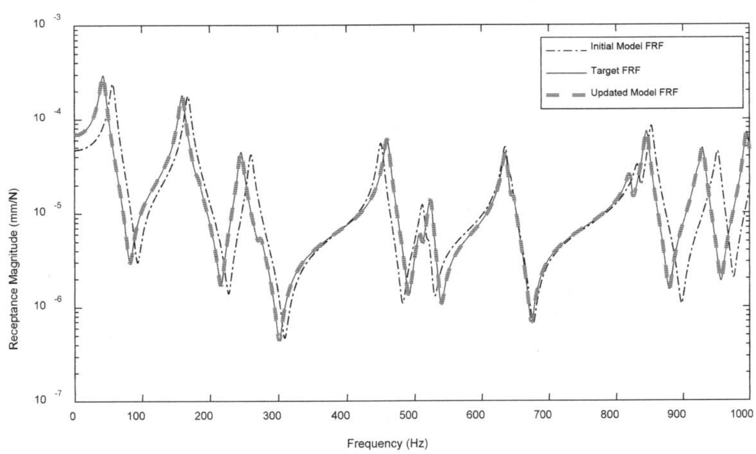

Fig. 7. Driving point FRF before and after updating.

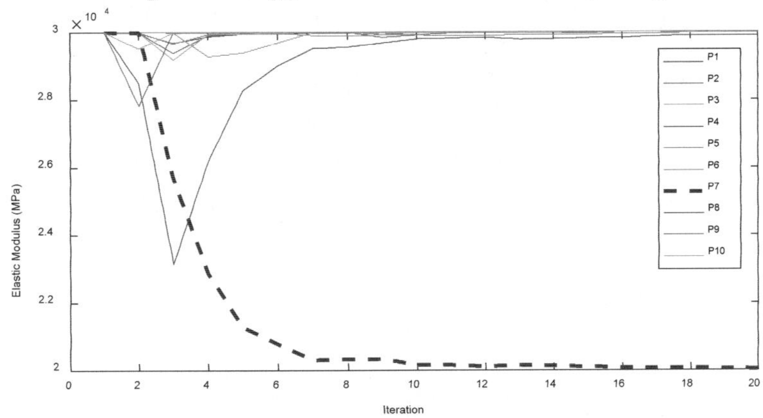

Fig. 8. Convergence of updating parameters for the elastic modulus of specimen partitions.

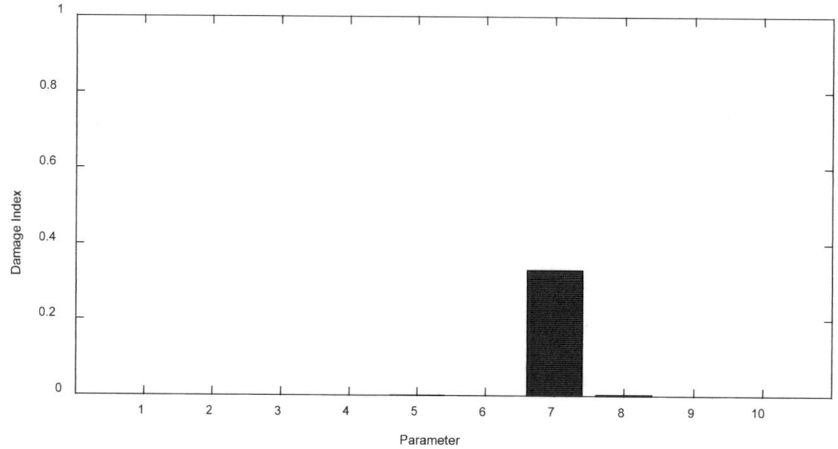

Fig. 9. Damage index for numerical investigation.

The comparison illustrated in Figure 7 showing the driving point FRFs of the FE model and simulated measured data indicates a significant reduction in the discrepancies between the model FRF and the measured FRF after updating. Figure 8 shows that the convergence of the parameters was achieved after 20 iterations. The Young's modulus of each partition remained within 0.4% of the original value except for P7, which dropped down to 20 GPa. The damage index calculated using Eq. (12) revealed that the region covered by P7 had an overall stiffness reduction of 33.2% as shown in Figure 11. The results are summarised in Table 1.

Table 1 Summary of numerical simulated study.

Parameter	Initial Parameter (MPa)	Updated Parameter (MPa)	Damage Index	Measured Parameter (MPa)	Error
P1	30,000	30,000	0.0%	30000	0.0%
P2	30,000	30,000	0.0%	30000	0.0%
P3	30,000	29,999	0.0%	30000	0.0%
P4	30,000	29,998	0.0%	30000	-0.0%
P5	30,000	29,984	0.1%	30000	-0.1%
P6	30,000	30,000	0.0%	30000	0.0%
P7	30,000	20,032	33.2%	20000	0.2%
P8	30,000	29,895	0.4%	30000	-0.4%
P9	30,000	30,000	0.0%	30000	0.0%
P10	30,000	30,000	0.0%	30000	0.0%
BC1	50	69.36	-	70	-0.9%
BC2	50	40.85	-	40	2.1%
BC3	50	7.90	-	8	-1.3%

8. Conclusions

The presented study investigated a method to detect the location and severity of damage in a jack arch specimen by updating a linkage FE model using the FRF sensitivity method. In the developed approach, the specimen was first divided into partitions and then Young's modulus of each partition was updated to identify the stiffness reduction in each section. In the provided numerical study, a

5,706 DOF FE model was reduced to a 60 DOF linkage model using SEREP. The mode shapes determined using the 11 measured DOFs of each identified mode were then expanded to the 60 DOFs of the linkage model using its mass and stiffness matrices. The expanded mode shapes were then used to synthesise the interpolated receptance FRFs from which the FRF sensitivity matrix was formed. The solution of the linear least-squares method was used to update the Young's modulus of the partitions of the specimen and this procedure was repeated for 20 iterations, during which values of the updating parameters reached convergence. The damage index was calculated based on the overall stiffness reduction of each updating parameter. The method was able to accurately update the Young's modulus of each partition and the boundary conditions simultaneously. The method is a practical approach to interpolating data to identify the location and severity of damage for real SHM systems with limited numbers of measurement sensors.

Acknowledgements

This research is supported by an Australian Government Research Training Program Scholarship. The authors also wish to thank the Road and Maritime Services (RMS) in New South Wales, Australia and NICTA for provision of the support and testing facilities for this research work. NICTA is funded by the Australian Government through the Department of Communications and the Australian Research Council through the ICT Centre of Excellence Program. CSIRO's Digital Productivity business unit and NICTA have joined forces to create digital powerhouse Data61.

References

Doebling, S.W., Farrar, C.R. and Prime, M.B., 1998. 'A summary review of vibration-based damage identification methods', *Shock and vibration digest*, vol. 30, no. 2, pp.91-105.

Sohn, H., et al., 2002. *A review of structural health review of structural health monitoring literature 1996-2001*(No. LA-UR-02-2095). Los Alamos National Laboratory.

Mousavi, M. and Gandomi, A.H., 2016. 'A hybrid damage detection method using dynamic-reduction transformation matrix and modal force error', *Engineering Structures*, vol. 111, pp.425-434.

Hansen, J.B., et al., 2017. 'A new scenario-based approach to damage detection using operational modal parameter estimates', *Mechanical Systems and Signal Processing*, vol. 94, pp.359-373.

Au, F.T.K., et al., 2003. 'Structural damage detection based on a micro-genetic algorithm using incomplete and noisy modal test data', *Journal of Sound and Vibration*, vol. 259, no. 5, pp.1081-1094.

Entezami, A., Shariatmadar, H. and Sarmadi, H., 2017. 'Structural damage detection by a new iterative regularization method and an improved sensitivity function', *Journal of Sound and Vibration*, vol. 399, pp.285-307.

Mustapha, S., et al., 2015. 'Pattern recognition based on time series analysis using vibration data for structural health monitoring in civil structures', *Electronic Journal of Structural Engineering*, vol. 14, no. 1, pp.106-115.

Nguyen, V.V., et al., 2015. 'Damage identification of a concrete arch beam based on frequency response functions and artificial neural networks', *Electronic Journal of Structural Engineering*, vol. 14, no. 1, pp.75-84.

Imregun, M., Visser, W.J. and Ewins, D.J., 1995. 'Finite element model updating using frequency response function data: I. Theory and initial investigation', *Mechanical Systems and Signal Processing*, vol. 9, no. 2, pp.187-202.

The 8th International Conference on
Structural Health Monitoring of Intelligent Infrastructure
Brisbane, Australia | 5-8 December 2017

Fatigue cracking identification using nonlinear Lamb waves for FRP-reinforced steel plates

Y. Wang[1], R. Guan [2], Y. Lu [3], Wenhui Duan[4]
[1, 2, 3, 4] *Department of Civil Engineering, Monash University – Australia, Email:* ye.lu@monash.edu

Abstract

Bonding carbon fibre reinforcement polymer (CFRP) laminates to steel structures has attracted a lot of attention for extending the service life of the steel structure. To ensure the functionality of the strengthened structures, an appropriate monitoring method is required. Nonlinear Lamb-wave based method for fatigue crack detection of CFRP-reinforced steel plates was presented in this study. Numerical and experimental studies were carried out for a plate with the CFRP laminates, where a fatigue crack was generated by the servo-hydraulic testing machine to achieve contact acoustic nonlinearity (CAN). The symmetric wave mode S1 was used as the excitation and when it came across the fatigue crack in specimens, S2 mode would be induced. After processing the received signals, the damage-induced Lamb wave nonlinearities were identified with the generation of second harmonic and it was found that the nonlinearities in the CFRP-reinforced steel plate were weaker when the sensing path is further from the fatigue crack. With the correlation coefficient of benchmark and signal with damage at double frequency in the time domain, an imaging method was introduced to locate the fatigue crack in CFRP-reinforced steel plates.

1. Introduction

Bonding carbon fibre reinforcement polymer (CFRP) laminates to ageing steel structures for reinforcement has attracted considerable attention for the extension of structural service life because of outstanding properties of CFRP such as high modulus, high strength-to-weight ratio and excellent resistance to environmental degradation (Zhao and Zhang 2007). To ensure the functionality and integrity of retrofitted structures, long-term monitoring of structural integrity, in particular damage initiation and propagation, is required.

Structural health monitoring (SHM) techniques based on ultrasonic guided waves have proved successful in identifying various types of damage in many engineering structures. Guided waves at relatively low frequencies (kilohertz) can propagate over a reasonably long distance while covering the entire cross-section of thin-walled structures such as metallic plates, where guided waves are also known as Lamb waves. In this respect, Lamb-wave-based SHM techniques have been successfully applied to locate and quantify defects in plate-like metallic structures (Diligent et al. 2002, Lowe et al. 2002). However, guided-wave-based methods using linear wave features are often limited by the wavelength of a selected wave mode, such that they can only detect macro damage of dimensions comparable to the wavelength. Compared with its linear counterpart, nonlinear Lamb waves have shown more sensitivity to microscopic defects and are capable of identifying incipient damage.

Among different nonlinear methods, second-harmonic generation has been employed by many researchers to characterise fatigue crack in metallic structures (Dutta et al. 2009, Hong et al. 2014, Pruell et al. 2009, Su et al. 2014). An experimental procedure (Pruell et al. 2009) was developed using nonlinear Lamb waves to characterise fatigue damage in aluminium plates. An experimental study (Dutta et al. 2009) was also conducted to detect fatigue crack in aluminium and steel structures using nonlinear Lamb waves. Compared with aluminium plates, less effect of nonlinearity was observed in the case of steel structures, due to the size and stiffness of the steel specimens. A modelling technique (Hong et al. 2014) was established for comprehending nonlinear characteristics of Lamb waves travelling in a fatigued medium with the consideration of material, geometric, plasticity-driven and contact acoustic nonlinearity (CAN). Nevertheless, limited studies have been done in detecting fatigue crack initiation with nonlinear Lamb waves in FRP-retrofitted metallic structures used in civil engineering, where concealed crack growth due to fatigue loading is more difficult to monitor because of the bonding of CFRP laminates.

In this study, nonlinear Lamb waves were deployed for detecting and locating fatigue crack in CFRP-reinforced steel plates using surface-bonded piezoelectric sensors. This paper is organised as follows: the theoretical basis and methodology in this research are introduced in Section 2. Section 3 describes the numerical modelling of nonlinear Lamb wave propagation in CFRP-reinforced steel plates. The experimental setup is elaborated in Section 4 to verify the results from the simulation. An imaging method to locate the position of fatigue crack in the CFRP-reinforced steel plate by fusing the signals obtained from experiments is addressed in Section 5, followed by the conclusion in Section 6

2. Methodology

When ultrasonic waves reach an imperfect contact surface, the compressional part of the wave closes the crack and the wave transmits through the crack, whereas the tensile part of the wave opens the crack, blocking such waves (Hong et al. 2014), as illustrated in Figure 1. The repetition of collisions between the contact surfaces generates higher harmonic responses. This effect is known as CAN (Jiao et al. 2014). CAN has attracted increasing attention and has been applied to characterise closed defects and imperfect interfaces (Donskoy et al. 2001, Hong et al. 2014, Jiao et al. 2014, Pecorari 2003, Solodov 1998).

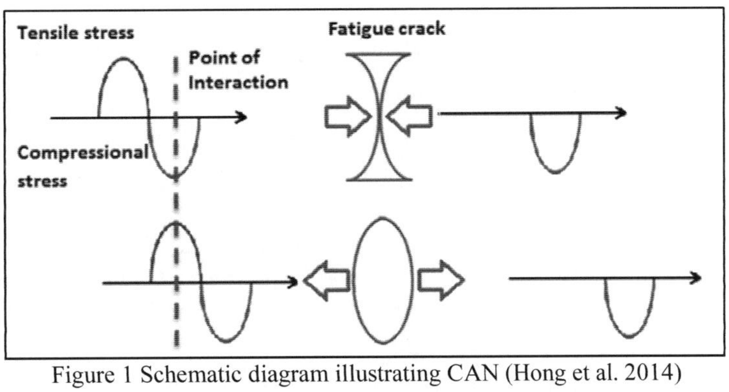

Figure 1 Schematic diagram illustrating CAN (Hong et al. 2014)

To generate enhanced and cumulated second harmonics, two conditions need to be satisfied, i.e. synchronism and non-zero power flux (Rose 2014). To satisfy these conditions in this study, a proper wave mode pair (S1, S2), both propagating at the same phase and group velocities, was selected from a dispersion curve for a 10 mm thick steel plate, which was the specimen to be investigated. As shown in Figure 2, the excitation frequency of S1 mode was selected at 3.77 MHz·mm (377 kHz for the 10 mm plate), which could induce the S2 mode at 7.54 MHz·mm (754 kHz for the 10 mm plate) after interacting with a fatigue crack.

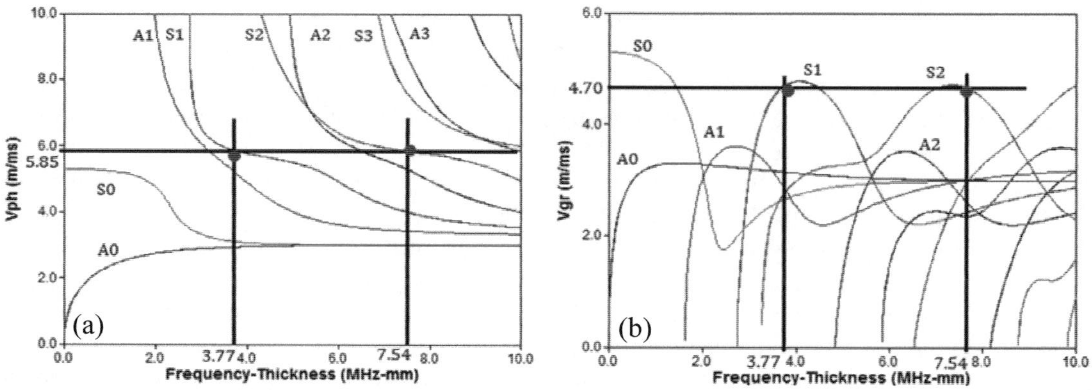

Figure 2 Dispersion curves of (a) phase velocity and (b) group velocity for a 10 mm thick steel plate

3. Numerical simulation

3.1 Modelling of CFRP-reinforced steel plate

Abaqus Dynamic/Explicit® was used to simulate nonlinear Lamb waves propagating in the CFRP-reinforced steel plate. The dimensions of the CFRP-reinforced steel plate with a through-thickness hole and initial notches in the middle of the steel plate are shown in Figure 3 and the material properties are given in Table 1. CFRP laminates were modelled and tied to both sides of the steel plate. The model was meshed using eight-node brick elements (C3D8R) and at least eight elements per wavelength and four nodes in the thickness direction were simulated to achieve accurate solutions (Su and Ye 2009).

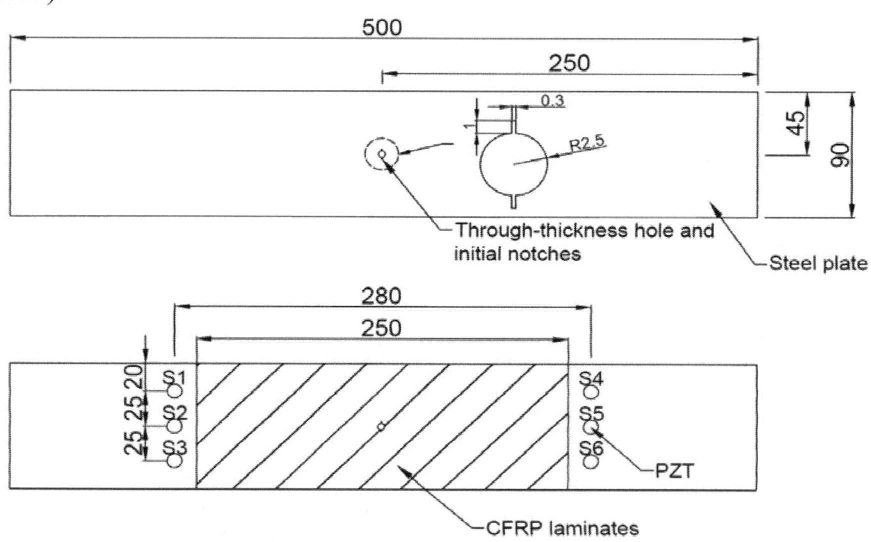

Figure 3 Configuration of CFRP-reinforced steel plate

Table 1 Material properties of steel and CFRP laminates

	Steel	MBRACE^{TM} Laminate 210/3300
Density (kg/m^3)	7850	1900
Tensile strength (MPa)	497	2882
Tensile modulus (GPa)	200	177
Thickness (mm)	10	1.57

A piezoelectric (PZT) transducer was modelled as a disc partitioned out of the steel plate model with the same diameter as the one used in the following experiment. A 15.5-cycle Hanning-windowed sinusoidal tone burst at 377 kHz was excited by applying uniform in-plane radial concentrated forces on the nodes along the periphery of the actuator to simulate Lamb wave excitation. An Abaqus built-in function "seam" was imposed to each surface of the corresponding elements at the end of initial notch to simulate a fatigue crack (Hong et al. 2014). A surface-to-surface contact with small sliding and associated properties was imposed on the crack interfaces. This enabled the "breathing" behaviour of the crack when waves travelled through it and thereby induced the CAN effect.

3.2 Simulation results

A typical benchmark signal (undamaged) and a current signal (with a 0.5 mm seam crack in length) in the time domain acquired in the CFRP-reinforced steel plate model via path S2-S5 are presented in Figure 4(a). As can be seen, these two signals are almost identical in the time domain, indicating that there no apparent waves were scattered by such a small crack. After processing these signals through a fast Fourier transform into the frequency domain, a weak but significant energy was observed in the current signal at twice the excitation frequency of that in the benchmark signal, indicating the existence of a seam crack, as shown in Figure 4(b).

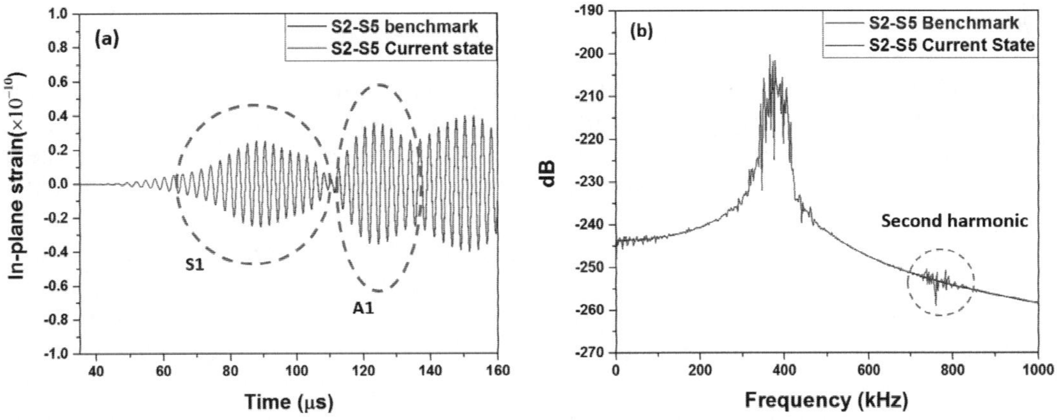

Figure 4 Received signals a) in the time domain and b) in the frequency domain for the CFRP-reinforced steel plate model

A Hanning-windowed finite impulse response bandpass filter was subsequently designed to extract transmitted waves within the frequency range of interest while excluding wave components in other frequency bands. The upper and lower thresholds of the frequency range was selected as ± 50 kHz centred by both fundamental and double frequencies. The amplitudes of wave components at the fundamental and double frequency bands were therefore extracted in terms of time, as shown in Figure 5.

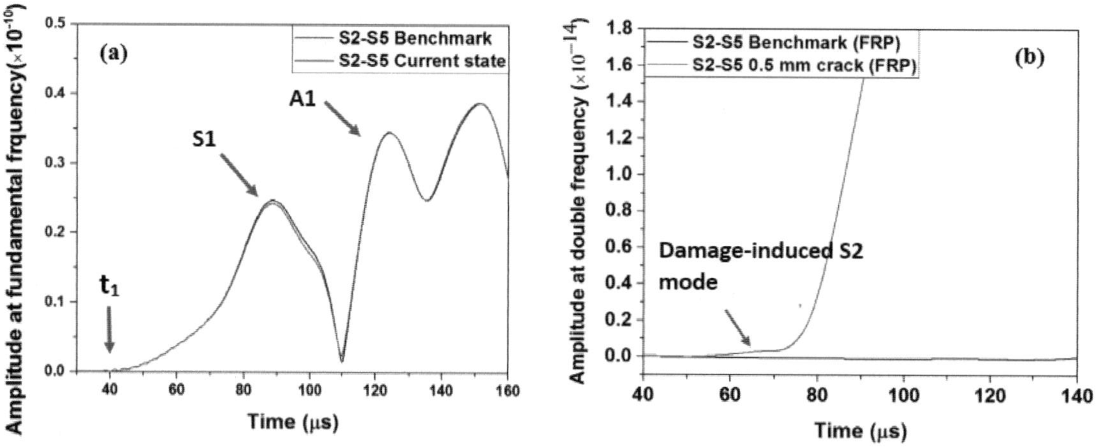

Figure 5 (a) Amplitude profiles at the fundamental frequency, where S1 mode is the first arrival mode at time t1 and (b) amplitude profiles at the double frequency, where damage-induced S2 mode arrives at time t2

As seen in Figure 5(a), the first peak denotes the first arrived transmitted S1 mode at time t1, followed by the A1 mode. The amplitude profiles of the benchmark and current state signals at the fundamental frequency are almost identical. For the wave components at double frequency, the amplitude profiles of the benchmark and the signal with damage are quite different, where an additional peak starting at time t2 can be observed, which is denoted as S2 mode generated by S1 mode when it encountered the seam crack.

4. Experimental evaluation

4.1 Experiment setup

Laboratory-based experiments were conducted on the CFRP-reinforced steel plate with the same dimensions and material properties as those in numerical simulation. 6 piezoelectric wafers (PZT) provided by PI® were surface-bonded on the CRFP-reinforced steel plate, as shown in Figure 6. The CFRP laminates were bonded to both sides of the steel plate by mixing Araldite 420 A/B epoxy adhesive.

The excitation signal was a 15.5-cycle Hanning-windowed tune burst at the central frequency of 377 kHz with a peak-to-peak output voltage of 6 $V_{p\text{-}p}$, which was then amplified to 600 $V_{p\text{-}p}$ by a Ciprian power amplifier. The excited signal was applied on one PZT on the left side, and the wave signals were acquired by all the PZTs on the right side through a Tektronix 4034B digital signal oscilloscope with a sampling frequency of 25 MHz after 128 samples had been averaged.

To generate fatigue cracking, an Instron 8802 servo-hydraulic testing machine was used on the specimen, with sinusoidal cyclic load ranging from 13.5 kN to 135 kN with a frequency of 20 Hz. Since it was impossible to measure the cracked zone due to the presence of the CFRP laminates, the cycle loading was stopped when the signal in the middle sensing path (S2-S5) began to deviate from the benchmark signal, which was after about 280,000 cycles. To confirm there is no disbondings between the CFRP laminates and the steel plate after fatigue loading, an Olympus ultrasonic flaw detector (EPO600) was used to pinpoint the bonding condition via scanning the surface of the sample. The results showed there was no disbondings between the CFRP laminates and the steel plate. In addition, the sample was heated to remove the CFRP laminates from the steel plate after the experiment. The fatigue crack was clearly observed by a digital microscope in the steel plate with lengths around 1mm and 2mm at two notches respectively.

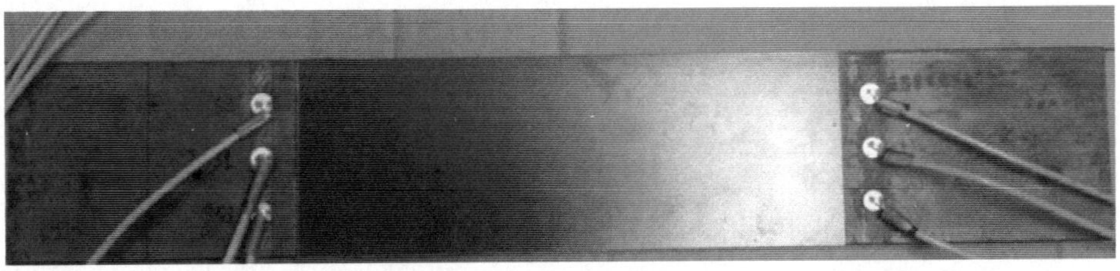

Figure 6 CFRP-reinforced steel plate specimen for experiment

4.2 Experiment results

One of the typical received signals from the middle path S2-S5 and signals from other sensing paths at double frequency in the time domain after the designed filter are shown in Figure 7 and Figure 8. Compared with the benchmark signal, it is noted that the amplitude of the signal at double frequency with cracking is greater than that of benchmark signal, indicating the additional nonlinearity caused by the fatigue crack, as in the simulation results. In Figure 8 it is evident that the nonlinearity of signal becomes weaker when the sensing path is further from the crack. However, it is noted that second harmonics exist for all benchmark signals, which may contribute to the material and geometric nonlinearities of the intact specimen, as well as the inherent nonlinearity from the experimental system. It was therefore difficult to determine the arrival time of the damage-induced S2 mode to locate the position of fatigue crack.

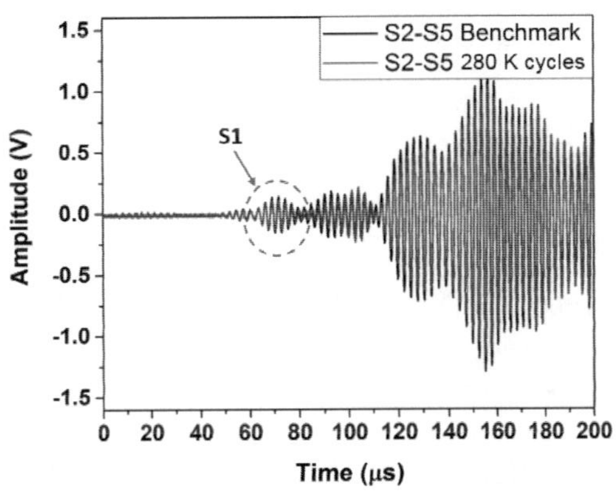

Figure 7 Benchmark and current received signal of path S2-S5 in the time domain

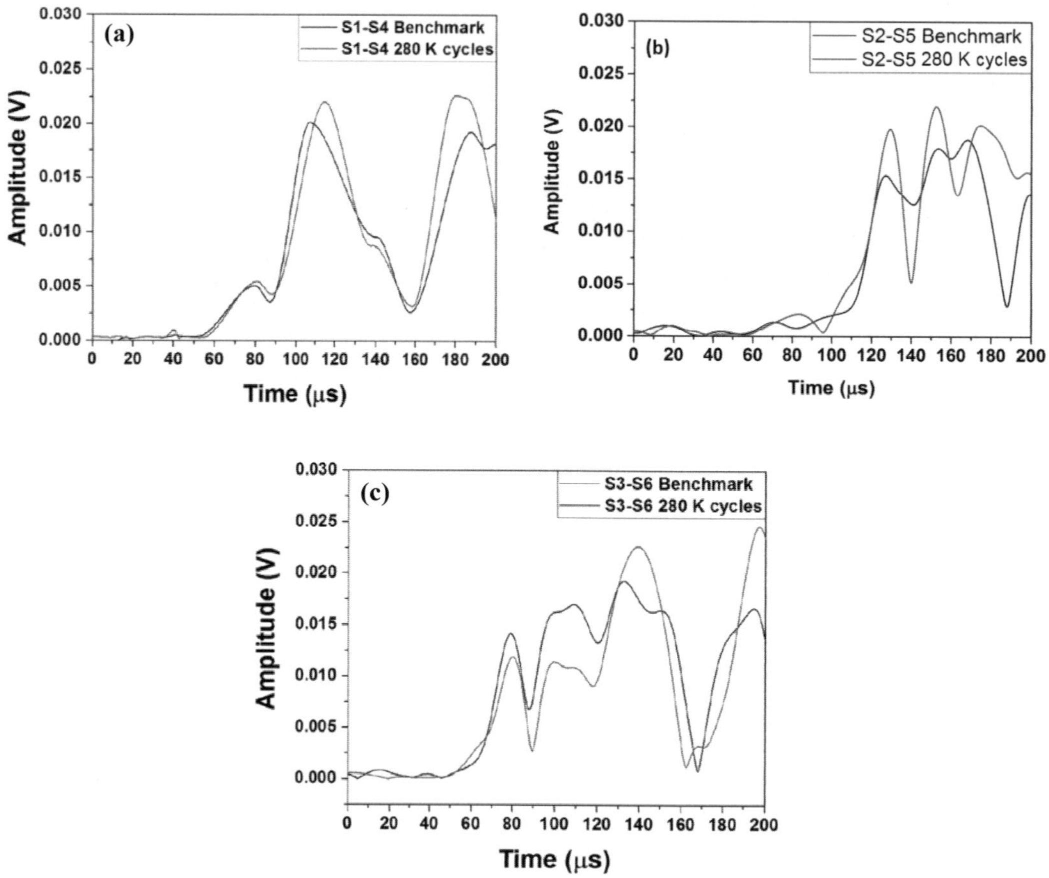

Figure 8 Energy profiles of benchmark and current received signals at double frequency captured via sensing path (a) S1-S4, (b) S2-S5, (c) S3-S6.

5. Probability based imaging method

Because of the unavoidable signal nonlinearity from other sources, which complicated the extraction of genuine wave nonlinearity caused by damage, an imaging method based on correlation coefficients ρ_n, indicating the relationship between the benchmark and the signal with damage at the double frequency, was adopted to locate the crack position. It is understood that the greater the difference between the benchmark and the signal with damage, the smaller the value of the coefficient, $1-\rho_n$, which was defined as a damage index to represent the probability of the presence of damage along one sensing path. The correlation coefficients for the CFRP-reinforced steel plate are summarised in Table 2.

Table 2 Typical damage index for sensing paths in steel and CFRP-reinforced steel plate

CFRP-reinforced steel plate			
Damage index	S1	S2	S3
S4	0.9611	0.9656	0.6303
S5	0.7211	0.2581	0.6934
S6	0.6695	0.9366	0.5221

On each sensing path, the coefficient of each pixel on the inspected area enclosed by the sensors was multiplied by a normal distribution function which was related to the distance from that pixel to the sensing path (Mustapha et al. 2012):

$$f(x,y) = \frac{1}{\sqrt{2\pi}\sigma} \exp\left(-\frac{(l_{ij}-\mu)^2}{2\sigma^2}\right), \tag{1}$$

where l_{ij} is the relative distance of the pixel (x, y) to the nth path, defined as (Wang et al. 2010)

$$l_{ij} = \frac{l_{pa} + l_{ps}}{l_n}, \tag{2}$$

where l_{pa} and l_{ps} are the distances from the pixel (x, y) to the actuator and the sensor, respectively. l_n is the distance from the actuator to the sensor of the nth path. The probability of the presence of damage at position (x, y) for the nth path can be described as

$$P_n(x,y) = (1-\rho_n)\frac{1}{\sqrt{2\pi}\sigma} \exp\left(-\frac{(l_{ij}-\mu)^2}{2\sigma^2}\right), \tag{3}$$

where μ and σ were selected as 0 and 0.08, respectively. This function provided an ellipse zone around each sensing path, and the superposition of the ellipse zones from all valid sensing paths in the specimen as data fusion results in a final image which shows the location of the fatigue crack. After the area of the plate had been meshed into pixels with a density of 1 mm, the final image plotted according to the proposed method above-described for the CFRP-reinforced steel plate is shown in Figure 9. It was observed that the zone with the highest probability of crack occurrence highlighted by the proposed method was quite close to the actual location of the fatigue crack It should also be noted that there were actually two fatigue cracks, one at each side of the hole in the specimen, while the imaging method had relatively low sensitivity to these two cracks, since the higher probability values were always in the direct path between the actuator and sensor for each ellipse zone. These two cracks were therefore detected as one, at an approximate location.

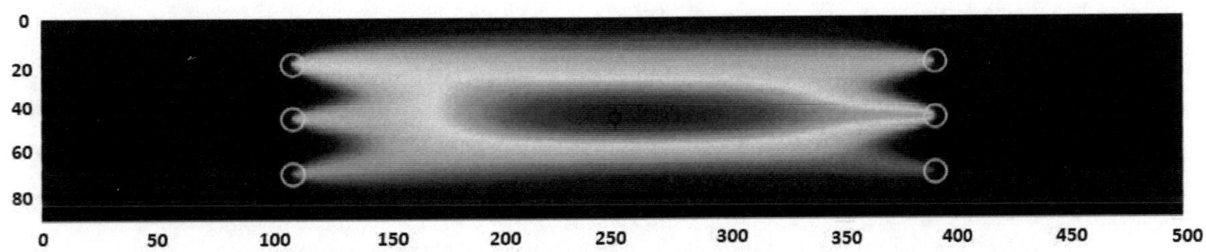

Figure 9 Image identifying the position of fatigue crack in CFRP-reinforced plate

6. Conclusion

Nonlinear Lamb waves were used to detect fatigue crack in the CFRP-reinforced steel plate in this study. Both numerical simulation and experimental method were used to investigate the nonlinear Lamb wave propagation in the specimen. A "seam" crack was introduced in the numerical model to simulate a fatigue crack which could cause CAN when the selected Lamb wave mode interacted with it. The second-harmonic generation indicating the nonlinearities induced by the fatigue crack was successfully observed in the frequency domain and was further verified by the experimental testing with fatigue cracking. On the other hand, signal nonlinearities from other sources were also identified in the benchmark signal in experiments, which complicated extraction of the arrival time of the S2 mode induced by the damage. A cost-effective imaging method based on correlation coefficients was therefore deployed in this study to construct probabilistic distributions of damage occurrence in the specimen, correctly highlighting the location of fatigue crack.

References

Zhao, X.-L. and Zhang, L. 2007, 'State-of-the-art review on FRP strengthened steel structures', *Engineering Structures,* vol. 29, no. 8, pp. 1808-1823.

Diligent, O., Grahn, T., Boström, A., Cawley, P. and Lowe, M. J. 2002, 'The low-frequency reflection and scattering of the S0 Lamb mode from a circular through-thickness hole in a plate: Finite Element, analytical and experimental studies', *The Journal of the Acoustical Society of America*, vol. 112, no. 6, pp. 2589-2601.

Lowe, M. J., Cawley, P., Kao, J. and Diligent, O. 2002, 'The low frequency reflection characteristics of the fundamental antisymmetric Lamb wave a 0 from a rectangular notch in a plate', *The Journal of the Acoustical Society of America*, vol. 112, no. 6, pp. 2612-2622.

Dutta, D., Sohn, H., Harries, K. A. and Rizzo, P. 2009, 'A nonlinear acoustic technique for crack detection in metallic structures', *Structural Health Monitoring*, vol. 8, no. 3, pp. 251-262.

Hong, M., Su, Z., Wang, Q., Cheng, L. and Qing, X. 2014, 'Modeling nonlinearities of ultrasonic waves for fatigue damage characterization: theory, simulation, and experimental validation', *Ultrasonics*, vol. 54, no. 3, pp. 770-778.

Pruell, C., Kim, J.-Y., Qu, J. and Jacobs, L. J. 2009, 'Evaluation of fatigue damage using nonlinear guided waves', *Smart Materials and Structures*, vol. 18, no. 3, 035003.

Su, Z., Zhou, C., Hong, M., Cheng, L., Wang, Q. and Qing, X. 2014, 'Acousto-ultrasonics-based fatigue damage characterization: Linear versus nonlinear signal features', *Mechanical Systems and Signal Processing*, vol. 45, no. 1, pp. 225-239.

Jiao, J.-P., Liu, W., He, C., Wu, B. and Zhang, J. 2014, 'Nonlinear acoustic interaction of contact interfaces', *Experimental Mechanics*, vol. 54, no. 1, pp. 63-68.

Donskoy, D., Sutin, A. and Ekimov, A. 2001, 'Nonlinear acoustic interaction on contact interfaces and its use for nondestructive testing', *NDT & E International*, vol. 34, no. 4, pp. 231-238.

Pecorari, C. 2003, 'Nonlinear interaction of plane ultrasonic waves with an interface between rough surfaces in contact', *The Journal of the Acoustical Society of America*, vol. 113, no. 6, pp. 3065-3072.

Solodov, I. Y. 1998, 'Ultrasonics of non-linear contacts: propagation, reflection and NDE-applications', *Ultrasonics*, vol. 36, no. 1, pp. 383-390.

Rose, J. L. 2014, *Ultrasonic guided waves in solid media*, Cambridge University Press, New York.

Su, Z. and Ye, L. 2009, 'Identification of damage using Lamb waves: from fundamentals to applications', *Springer.*

Mustapha, S., Ye, L., Wang, D. and Lu, Y. 2012, 'Debonding detection in composite sandwich structures based on guided waves', *AIAA journal*, vol. 50, no.8, pp. 1697-1706.

Wang, D., Ye, L., Su, Z., Lu, Y., Li, F. and Meng, G. 2010, 'Probabilistic damage identification based on correlation analysis using guided wave signals in aluminum plates', *Structural Health Monitoring*, vol. 9, no. 2, pp. 133-144.

The 8th International Conference on
Structural Health Monitoring of Intelligent Infrastructure
Brisbane, Australia | 5-8 December 2017

Case Study: Structural Health Monitoring of Timber Bridges using Dynamic Frequency Analysis (DFA)

M. Rashidi[1]*, B. Samali[1], X. Zhu [1], A. Azad [1], M. Ghodrat[1]
[1] Western Sydney University – Australia, *Email: m.rashidi@westernsydney.edu.au

Abstract

Australia has numerous ageing timber bridges in service. The structural efficiency of the majority of them have been impaired over time and their capacity is not known. There is, therefore, a critical need to develop new methods that will enable the load carrying capacity of these bridges to be estimated. In this paper, an established testing method known as Dynamic Frequency Analysis (DFA) and its application for monitoring of bridges is described throughout a case study (Mongarlowe Bridge). The testing procedure involves the attachment of sensors (accelerometers) underneath the bridge. The natural frequency and vibration response of the bridge superstructure is measured when a "calibrated sledgehammer" is used to excite the unloaded deck, and again with a relatively small mass applied at mid-span. The difference in response allows load carrying capacity of the bridge to be estimated. Furthermore, by regular monitoring of the bridge and its natural frequency, the potential degradation of the structure and the existing load bearing capacity can be determined.

1. Introduction

Local Government in Australia is responsible for the asset management of over 20,000 bridges. More than 70% of these bridges include aging timber bridges, the load capacity and structural adequacy of many of which have been impaired over time. A major challenge facing Local Government nationally is to develop effective strategies for the maintenance and management of the extensive timber bridge stocks which form a key element of the road network under its control (Rashidi, et al., 2016a). Raising the reliability and efficiency of bridge maintenance practices of local government has the potential not only to minimise costly emergency repairs, but also to reduce the overall maintenance and repair costs, whilst improving the operational efficiency of its road network (Samali, et al., 2003).
The purpose of the bridge inspection is to identify and quantify the current and projected deterioration of the structure to evaluate the capacity of the structure, and to recommend appropriate management decisions (Rashidi, et al., 2017). Inspection of the bridge components and assessment of its condition are conducted according to the procedure and standard condition rating system as defined by inspection manual (Rashidi, et al., 2016b). The condition rating system reflects the performance, integrity and durability of the structure and its major components (Rashidi, et al., 2016).
This paper presents a case study on structural health monitoring of Mongarlowe bridge using dynamic frequency analysis. The primary test has been undertaken in 2008 in order to estimate the load bearing capacity of the bridge and create a base-line for further analysis. Eight years later, in 2016, another test was conducted to assure the capacity of the bridge has not been changed. Mongarlowe bridge is situated on Little River Rd, in Palerang Shire Council. The bridge is composed of six spans and four girders per span.

2. Basic Concepts Underpinning the Proposed Testing Procedure

The proposed bridge assessment procedure includes the attachment of accelerometers underneath the bridge girders and the measurement of the vibration response of the bridge superstructure unloaded and with one or more loads (such as, water tanker, truck, grader, concrete blocks, etc, of known weight) applied at midspan. The excitation is generated by a modal impact hammer. The resulting dynamic responses are measured with uniaxial accelerometers which are simple and robust to attach. The data is logged and the bridge deck properties evaluated, using dynamic signal analyser or a standard computer with special software.

Two sets of bending frequencies are measured for the bridge, 'as is', and when loaded by the extra weight. By loading the bridge, the bending frequency of the bridge declines. From the resulting frequency shift due to added load, flexural stiffness of the bridge can be estimated.

User friendly software has also been developed which allows the estimation of bridge load carrying capacity from predicted stiffness, adopting a statistically based approach. The proposed test does not necessitate the accurate measurement of deformations, as is the case for static load tests. It is also much faster to conduct compared with proof loading, and hence more affordable than load testing. It is also safer than load testing, especially with respect to old bridges where applying a large load may further jeopardise the integrity of the structure.

3. Instrumentation, Setup and Procedures for Primary Field Testing

Twelve accelerometers with high sensitivity and low frequency-range were selected to record the accelerations of the timber bridge structure. A large 12 lb Modally Tuned ICP Sledge Hammer was used to excite the bridge. Two dynamic signal analysing recorders were employed to record hammer force and acceleration response signals during the tests.

A MATLAB program was developed for data processing and Frequency Response Function (FRF) calculations.

Four accelerometers were located at the bottom face of four girders at mid span to record the dynamic response of the girders at that point. Special base plates were also manufactured for mounting the sensors onto the bridge structure.

The field tests included two sets of tests.

Set 1 - No mass test

The bridge 'as is', without additional mass, is impacted by the modal hammer at 1) the centre of the midspan of each of the six spans.

2) top of the selected girders at the midspan of each of the six spans.

Set 2 – Added mass test

The bridge, carrying additional mass, is impacted by the modal hammer similar to no mass test.

Two types of mass distribution were used:

1) Added mass is distributed uniformly at midspan of the bridge on each of the six spans, one at the time.

2) Added mass is concentrated on selected girders of the bridge on each of the six spans, one at the time.

4. Data Analysis

From recorded dynamic response time histories and using Fast Fourier Transform (FFT), the FRF can be estimated. A computer programme was developed using MATLAB for this purpose. This software offers much greater flexibility when processing the test data. It produces the required Frequency Response Functions at a given bandwidth with good resolutions. The software is being further developed to incorporate more functions such as data acquisition and probabilistic models so that it can become an 'all-round' software for structural assessment (Ariyaratne & Crews, 2003). Advanced Modal Analysis software (Diamond) was also employed in the analysis stage where highly nonlinear and coupled dynamic modes occur, for which normal methods are no longer valid.

5. Results of the Primary Test

As direct results of the modal analysis, dynamic properties of the tested bridge, such as natural frequencies, mode shapes and damping can be obtained. However, the proposed dynamic method requires only the first flexural natural frequency for both with and without added mass cases. Figures 1, 2, 3, 4, 5 and 6 show the comparison of F R Fs with and without added mass for span 1,span 2, span 3, span 4, span 5 and span 6 of the tested bridge, respectively.

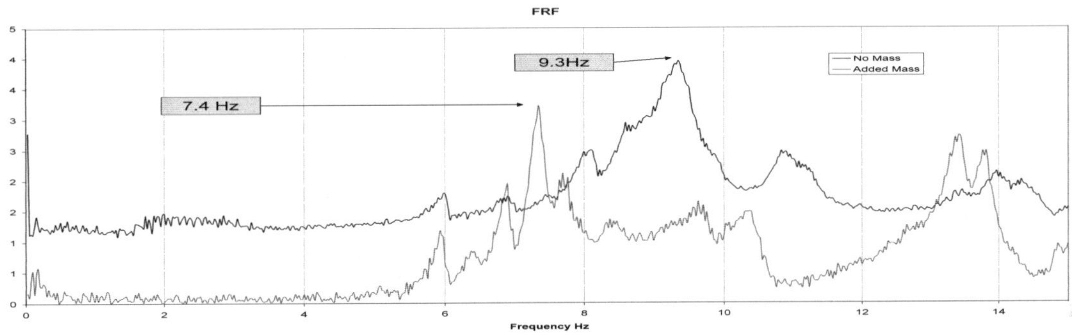

Figure 1 Comparison of sum FRFs for span 1 of the subject bridge with and without added mass

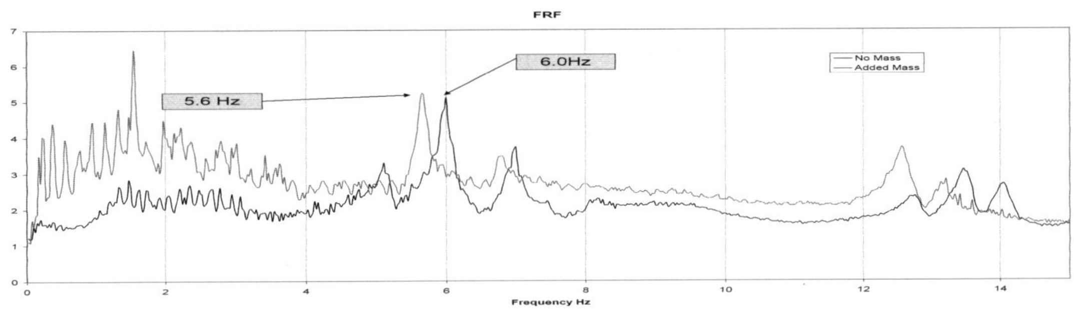

Figure.2 Comparison of sum FRFs for span 2 of the subject bridge with and without added mass

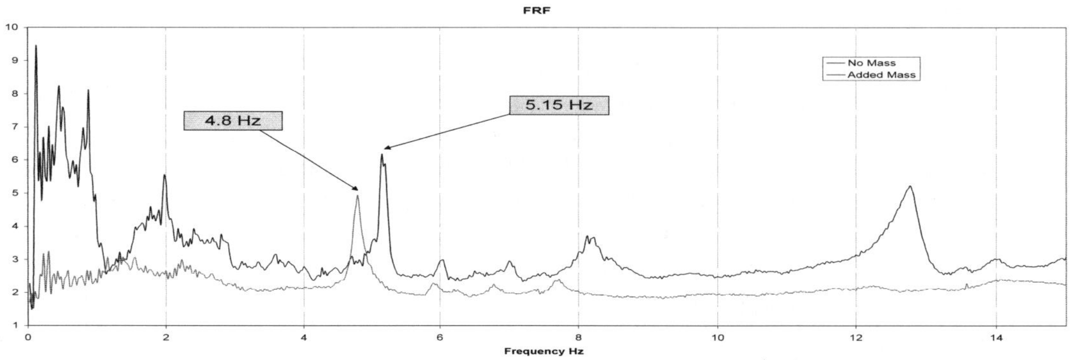

Figure.3 Comparison of sum FRFs for span 3 of the subject bridge with and without added mass

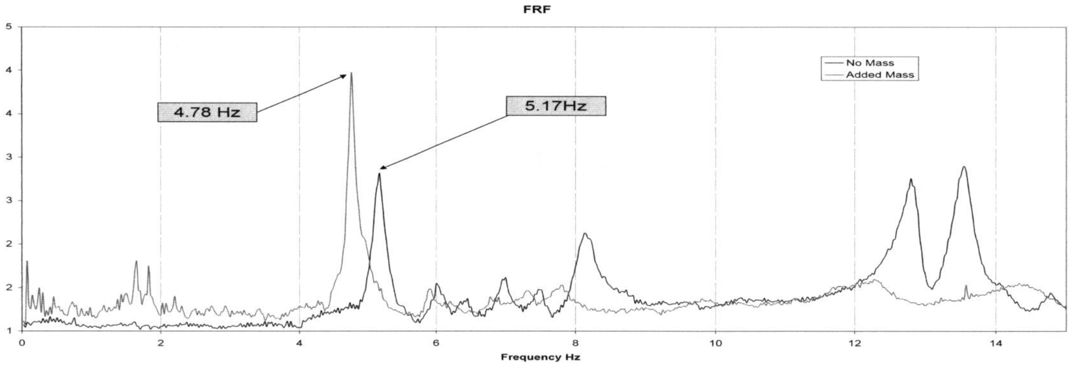

Figure.4 Comparison of sum FRFs for span 4 of the subject bridge with and without added mass

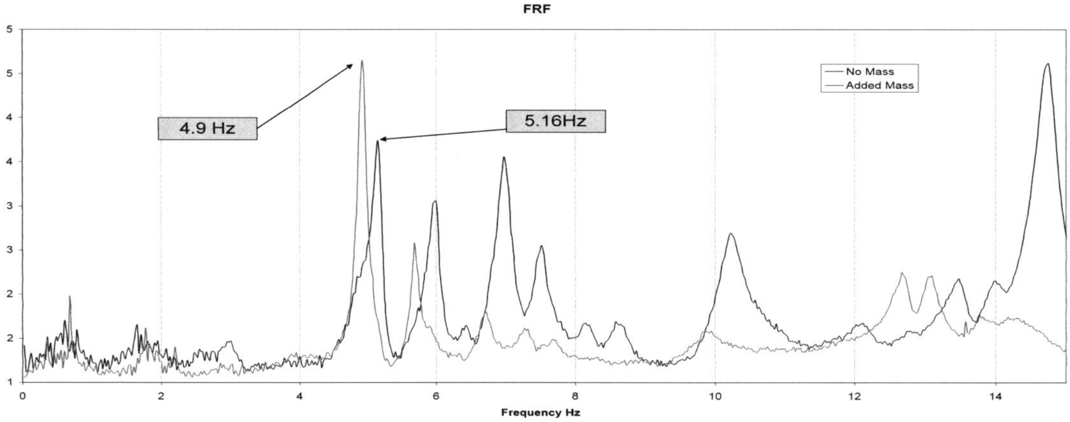

Figure.5 Comparison of sum FRFs for span 5 of the subject bridge with and without added mass

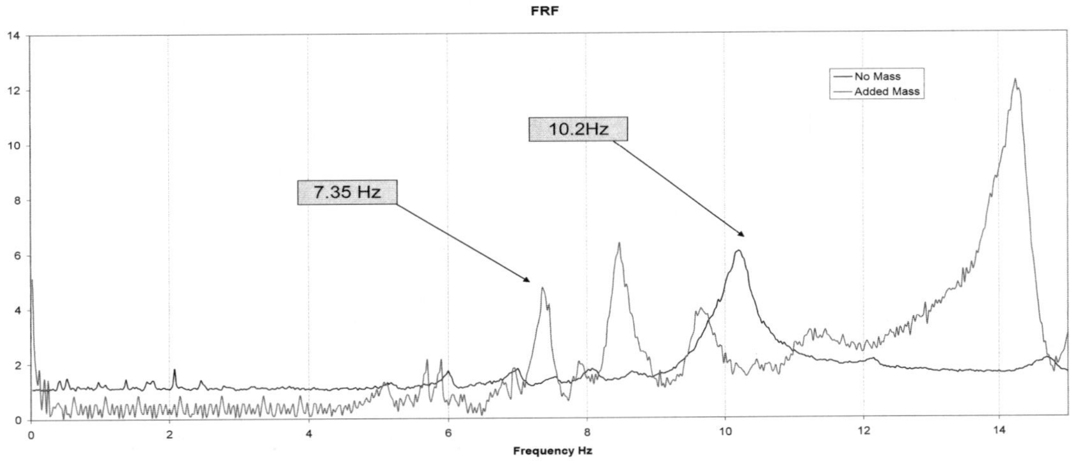

Figure.6 Comparison of sum FRFs for span 6 of the subject bridge with and without added mass

6. Flexural Stiffness of the Tested Bridge

With added mass given, and frequencies with and without mass known, the flexural stiffness of the tested bridge can be estimated. Table 1 demonstrates the amount of mass added to the first natural frequency with and without mass and predicted stiffness of tested bridge.

Table 1 - Results of predicted stiffness using the proposed dynamic method

Span No.	Mass Added (tonnes)	First Natural Frequency (Hz)	Predicted stiffness (kN/mm)
1	0	9.3	
	3	7.4	12.5
2	0	6.05	
	3	5.6	14.2
3	0	5.15	
	3	4.8	11.6
4	0	5.17	
	3	4.78	10.4
5	0	5.16	
	3	4.9	16.1
6	0	10.2	
	3	7.35	9.4

7. Strength Prediction and Live Load Factor Estimation

The determination and prediction of capacity and strength of in-service bridge girders is extremely difficult, unless the girder is broken and loading pattern and the failure load is known. Current "best practice" in Australia generally assumes that the fibre strength of any girder is 80 to 100 MPa. Bending capacity is predicted by multiplying the assumed fibre strength by assumed section modulus "Z". Whilst previous work undertaken by the RMS has involved full scale destructive testing of girders, the basis of current systems is essentially reliant upon testing of "small clear" specimens cut from these structural members, resulting in the assumption of above mentioned fibre stresses (Samali, et al., 2000).

Proof loading of timber bridges is costly and inherently risky, since it is a well-established fact that high load levels cause irrecoverable damage to the wood fibres, which may result, in subsequent failure of a timber girder at load levels considerably less than that indicated by the proof test. Therefore most rating of timber bridges has been based on applying serviceability load levels (such as from a water tanker), measuring the deflections in order to calculate the stiffness and then use an assumed relationship between stiffness and strength to predict the load carrying capacity of each girder (Pesterev, et al., 2002).

The relationship between stiffness and strength used in current load assessment methods is based on the assumed relationship between Modulus of Elasticity (MOE) and Modulus of Rupture (MOR) defined in the Australian Timber Structures Code, AS1720.1. However, it is not commonly understood that this relationship is both theoretical and idealised.

Figure 7, illustrates the problems associated with this approach. This chart presents a plot of MOR vs MOE data achieved from full scale testing of round timbers extracted from service with an average life of 30 years and also compares this with the AS1720 relationship. It can be concluded from the linear regression co-efficient for the test data that the relationship between stiffness and strength for aged girders is not statistically substantial. Furthermore, the theoretical relationship introduced in AS1720 is not reliable for these timbers, since many round timbers have rupture strength considerably lower than that predicted by the Code relationship (Samali, et al., 2003).

In order to evaluate the strength of timber bridge girders with any degree of reliability, it is essential to develop strength models, which reflect the actual bending capacity of timber, taking into account the uncertainties related to determination of the geometric section properties and the actual strength properties. Such a model has been developed to create the basis of the new load testing system. Using data obtained from extensive testing of full scale round timbers, a relationship between actual bending capacity and actual measured stiffness (EI) has been derived, with correlation coefficients in the range 0.4 to 0.5 for lower bound 5th percentiles (Ariyaratne & Crews, 2003).

Using a probabilistic approach, this relationship can be employed in reliability-based models to predict the load bearing capacity of a deck from the stiffness data obtained from the dynamic method, with reliable and transparent degrees of uncertainty.

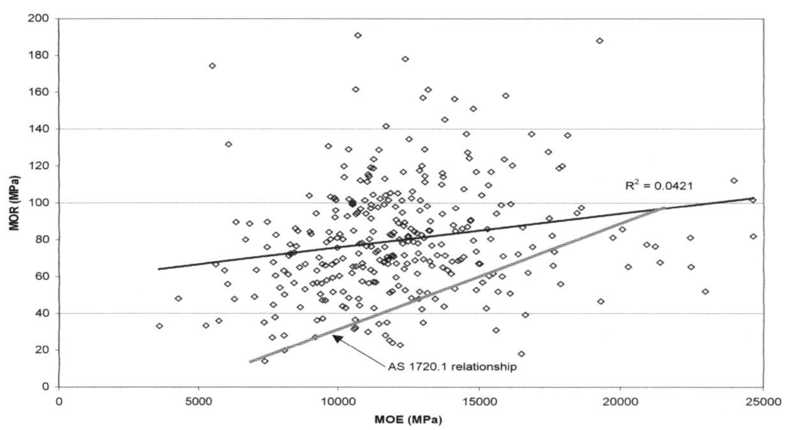

Figure 7 – MOR vs MOE for round timbers (Samali, et al., 2003)

7.1 Live Load Factor Estimation

The purpose of the analysis of the bridge is to evaluate the structure load carrying capacity expressed in terms of a standard truck, eg T44 by calculating the Live Load Factor (LLF) from the Equation (1) (Samali, et al., 2003):

$$LLF = [\text{Ø} \times M^* - 1.2 \times MDL - 2 \times MSDL] / [MLL(1 + DLA) \times LMF] \quad (1)$$

Where:

Ø	Capacity reduction factor
M^*	Ultimate moment capacity of critical girder
M_{DL}	Dead load moment in critical girder
M_{SDL}	Superimposed dead load moment in critical girder
M_{LL}	Live load moment of critical girder due to standard truck from analysis
DLA	Dynamic load allowance (first flexural frequency > 6 Hz) - see AS5100 / AUSTROADS Bridge Design Code
LMF	Load modification factor for two lane loading

Based on the values of the above mentioned parameters the Live Load Factor LLF can be estimated. It can be seen that LLF is the ratio of the factored ultimate capacity of the critical girder minus the factored dead load moments and the adjusted live load moments considering the appropriate Load Modification Factor and Dynamic Load Allowance.
Applying the above mentioned probabilistic approach, the estimated live load factors for the subject bridge are presented in Table 2.

Table 2 – Live load factor(s) for different vehicle types for the subject bridge

VEHICLE TYPE	LLF for Span 1	LLF for Span 2	LLF for Span 3	LLF for Span 4	LLF for Span 5	LLF for Span 6
T44	1.5	1.9	1.5	1.3	1.9	1.5
R20	2.1	3.0	2.3	2.0	3.0	2.3
ST42.5	2.0	2.6	2.0	1.7	2.6	1.9
B-Double (59t)	2.0	2.7	2.0	1.8	2.7	2.0

8. Assessment of the Bridge as Tested

The summary of the assessment of the structural condition of the subject bridge is given in Table 3, based on the field testing and structural analysis described in this paper. The capacity of a bridge to carry live load is generally expressed in terms of the Live Load Factor (LLF), which is estimated using equation (1). A Live Load Factor equal to 1.0 represents the scenario where the bridge superstructure can safely carry a fully loaded standard vehicle positioned on a span to develop the maximum load effect – but without any factor of safety for overloaded vehicles that exceed legal limits.
The maximum bending moment due to traffic load is calculated employing software that "moves" each of the vehicles across each span of the bridge to identify the worst case load effect for the particular axle configuration and axle loads for each vehicle type.

9. Recommendations

The recommended capacities for this bridge, for each of the designated Gross Mass Vehicles (GMV), are presented in Table 3.

Table 3 – Summary of Load Assessment

Vehicle type	GMV	LLF	Recommended load rating
5 axle – T44 truck (9.1m long)	44 tonnes	**1.3**	44 tonnes
Rigid body - 3 axle truck (4.9m long)	20 tonnes	**2.0**	20 tonnes
6 axle semi trailer – general access vehicle (11m long)	42.5 tonnes	**1.7**	42.5 tonnes
8 axle B-Double – restricted access (18.9m long)	59 tonnes	**1.8**	59 tonnes

The load ratings are indicative of the bridge load bearing capacity at the time of assessment. It must be noted that the damage due to overloaded vehicles and biological agents may occur at some point after the date that this test has been undertaken. It is the responsibility of the bridge owner to carry out routine inspections, to identify any potential damage or degradation that would warrant a reduction in the load rating. In accordance with best practice for asset management and health monitoring of timber bridges, it is strongly recommended that a Level 1 visual inspection be undertaken every 6 months and a detailed Level 2 inspection be conducted at least every 2 years and ideally, annually.

A point to consider is that although this methodology is capable of assessing the load carrying capacity of bridges, the load is dynamic in nature and its impact on the bridge is not just a function of the magnitude of the load. A major contributing factor is the surface condition of the bridge deck. A smooth surface will allow a much larger load to be carried safely due to low dynamic amplification attributable to smooth decks, but a similar level load may cause much distress and damage to the bridge if the surface is not sufficiently smooth.

10. The results of secondary test- 8 years later

In order to monitor the structural capacity of the bridge, another DFA test was conducted 8 years later. The first natural frequency of the two of the spans was measured and compared with the initial test. The results of the first natural frequency on span 1 and span 2 (measured through the dynamic testing) can be seen in Table 4.

Table 4- Results of the first natural frequency for the secondary test

Test	Span 1 (Hz)	Span 2 (Hz)
Test 1	9.32	5.69
Test 2	9.46	5.85
Test 3	9.32	5.69
Test 4	9.32	5.69
Test 5	9.32	5.85
Test 6	9.46	5.69
Average	9.36	5.74

Table 5 compares the first natural frequency measured through the initial inspection in 2008 and the recent test (January 2016). As can be seen in the table, the percentage variations are less than 5% and there are no considerable changes between the results. Therefore, the predicted flexural stiffness of the tested bridge remains unchanged. It should be noted that, for timber bridges, some variation in stiffness (and therefore frequency) will naturally occur due to changes in moisture content. The changes reported in Table 5 are not considered to be significant. Hence the load bearing capacity of the bridge is going to be the same.

Table 5- Results of predicted stiffness using the proposed dynamic method

Span No.	First Natural Frequency (Hz) Initial Test (2008)	First Natural Frequency (Hz) Second Test (2016)	Percentage Variation (%)	Predicted Stiffness (kN/mm) (2016)
1	9.3	9.36	0.06%	12.5
2	6.05	5.74	0.31%	14.2

10. Conclusions

In this case study, the application of Dynamic Frequency Analysis (DFA) for estimation of flexural stiffness and structural capacity of timber bridges is demonstrated. The results pertaining to a timber bridge with six spans, along with underlying methodology, is reported. The results clearly show that the dynamic procedure is a cost effective and reliable tool to assess the structural adequacy and efficiency of the timber bridge assets. This method is also a very reliable method for long term health monitoring of bridges. In this case study on Mongarlowe Bridge, the result of two DFA tests conducted in 2008 and 2016 have been analysed. There are neglectable changes between the results and the percentage variations are less than 5%. Hence the bridge can be considered in good condition without any noticeable changes in stiffness and load bearing capacity. The proposed methodology, following the extensive testing program, has now reached a level of improvement, which can be applied to most of timber bridges.

References

Ariyaratne, W. & Crews, K., 2003. *Design and Asset Management of Bridges,* Sydney: Post Graduate Lecture Notes-Road and Traffic Authority of NSW & Faculty of Engineering, University of Technology, Sydney.

Pesterev, A. et al., 2002. *Some recent results in moving load problems with application to highway bridges.* Saitama, Japan, s.n., pp. 20-27.

Rashidi, M. et al., 2017. Remedial modelling of steel bridges through application of analytical hierarchy process (AHP). *Applied Sciences,* 7(2).

Rashidi, M., Samali, B., Azad, A. & Hatamian, H., 2016. *Asset management of steel bridges.* Perth, W.A, CRC Press, pp. 187-192.

Rashidi, M., Samali, B. & Sharafi, P., 2016a. A new model for bridge management: Part A: Condition assessment and priority ranking of bridges. *Australian Journal of Civil Engineering,* 14(1), pp. 35-45.

Rashidi, M., Samali, B. & Sharafi, P., 2016b. A new model for bridge management: Part B: Decision support system for remediation planning. *Australian journal of civil engineering,* 14(1), pp. 46-53.

Samali, B. et al., 2000. *To develop cost effective assessment techniques to facilitate the management of local government bridge assets,* Sydney: Centre for built infrastructure research, UTS.

Samali, B., Bakoss, S., Crews, K. & Saleh, A., 2003. *Field testing and assesment of the Redbank Creek Bridge,* Sydney: Research Report, Centre for Built Infrastructure Research, UTS.

Samali, B. et al., 2003. *Assesing the load carrying capacity of timber bridgesusing dynamicmethods.* Perth, W.A, s.n.

The 8th International Conference on
Structural Health Monitoring of Intelligent Infrastructure
Brisbane, Australia | 5-8 December 2017

Time-Series Coefficient-Based Deterioration Detection Algorithm

B. Monavari[1], T.H.T. Chan[2], A. Nguyen[3], D.P. Thambiratnam[4].
[1, 2, 3, 4] *School of Civil Engineering & Built Environment, Queensland University of Technology – Australia,*
Email: b.monavari@qut.edu.au

Abstract

Regardless how well structures were designed, all existing buildings deteriorate over time due to environmental effects, varying service loads and aging. Hence, it is importent to continuously track the deterioration condition of structures. Structural health monitoring (SHM) based assessment procedures can be employed to evaluate this. In this regard, vibration-based methods are amongst the most effective ones as they can be used in ambient vibration and operational loading conditions. Each building has a unique set of vibration characteristics which change with accumulated deterioration and damage. However, the changes due to deterioration are generally subtler than changes due to damage; consequently, more difficult to detect. Therefore, deterioration detection procedures need to be more accurate and sensitive to these changes. This paper presents a nonparametric statistical test to capture changes in dynamic characteristics of structures using SHM data. First, coefficients of an autoregressive (AR) time-series model are made to represent the vibration response data of a structure. Then, statistical hypotheses of Two-sample t-tests are performed and a function of the resulting *P*-values is used to detect changes in the structure due to deterioration. A novel AR model order estimation procedure was proposed in order to enhance the sensitivity of the method. The proposed vibration-based deterioration identification method was successfully verified utilizing sensor data from an experiment carried out on a bookshelf structure. Results show that the proposed methodology can clearly detect deterioration.

1. Introduction

All infrastructures deteriorate at a slow-pace progressive process due to varieties of challenges in their lifespan such as aging, environmental effects, and varying service loads. Due to the changing characteristics of buildings, SHM data is crucial to evaluate deterioration condition, to maintain their safety, to increase their life expectancy and to reduce their costs of maintenance and repairs (Brownjohn, 2007, Nguyen et al., 2015). Many existing structures are in great need of repair and maintenance, but it is hard to detect deterioration condition prior to damage. On the other hand, most of the current vibration-based methods are not suitable for use to identify deterioration of structures.

Each structure has a unique set of vibration characteristics which changes due to accumulated deterioration and damage. Some researchers defined deterioration as a continuous loss of cross-sectional area in time (Okasha and Frangopol, 2010). It is worth noting that the changes owing to damage are generally greater than changes due to deterioration. Du et al. (2013) concluded that undamaged surface of a concrete structure does not confirm its healthy condition. In other words, deterioration is much more difficult to detect and needs a more accurate and sensitive method than damage detection methods. Farrar et al. (1999) asserted that dynamic properties of a structure alter due to changes in the structure's mass, stiffness or energy dissipative characteristics.

The key role of using SHM data has been to identify damage in structures. Most studies so far in SHM area pertain to damage detection, and not deterioration identification. Doebling et al. (1998), Carden and Fanning (2004) and Chan and Thambiratnam (2011) reviewed a large number of SHM data based methods. The literature shows that not much research has been done for deterioration assessment, especially identifying deterioration using vibration response data. Deterioration has been evaluated in other approaches such as reliability based algorithms (Huang and Chen, 2015). To identify deterioration, the existing vibration based assessment methods need to be enhanced in regard to their sensitivity. Therefore, these assessment methods which have the potential to be developed for deterioration detection will be reviewed here.

Vibration-based damage detection (VBDD) methods have been investigated in the past three decades (e.g. Shih et al., 2009, Alvandi and Cremona, 2006). Some researchers such as Mosavi et al. (2012) asserted that time-series analysis modelling for ambient vibration over the usual frequency domain methods has some advantages. Based on these reviews, time-series analysis, which estimates mathematical models using statistical tools and analyze data such as signals, seems to be more capable of developing deterioration features. Time-series models such as AR model have been used extensively among VBDD methods to detect and locate damage in structures (Gul and Catbas, 2009, Omenzetter and Brownjohn, 2006, Zhang, 2007, Carden and Brownjohn, 2008, Gul and Catbas, 2011, Nguyen et al., 2014a). Statistical time-series methods compare two main different conditions of a considered structure called baseline and assessment phases. The former is defined as the reference or healthy state of the structure and the latter is defined as the assessment state. In each phase, statistical time-series methods use random excitation and/or response signals from the structure to describe and characterize its health state.

The main concept of these methods is that a newly estimated time-series of a damaged and/or deteriorated structure differs from the time-series of the baseline state of that structure (Zhang, 2007, Carden and Brownjohn, 2008, Ling et al., 2009, Lei et al., 2003). Carden and Brownjohn (2008) developed a coefficient-based damage detection method. Their method was capable of identifying damage in structures. The differences between coefficients in baseline and assessment phases indicated changes in the vibration characteristics of the structure due to accumulated deterioration and/or damage.

In real-world structures, environmental and operational (E&O) variations such as external loading and temperature may produce dire effects on structural health assessment and mask subtler structural changes. Assessment features, especially deterioration-sensitive features, are often sensitive to these E&O changes. Therefore, in order to achieve successful deterioration assessment, it is vital to develop data normalization techniques to distinguish and to remove the effects E&O variations. Recently, the damage detection problem has been addressed from a statistical perspective. Nair et al. (2006) developed a novel procedure to detect and locate damage in structures by implementing t-test on AR coefficients of time-series models. Wang and Ong (2015) presented a group of nonparametric statistical formulations and used a function of *P*-values to detect damage. They used three types of statistical hypothesis of two-sample Kolmogorov-Smirnov test, Mann-Whitney test, and Mood test. Nguyen et al. (2014a, b) developed an enhanced Mahalanobis distance method which was validated with SHM data collected from a highway bridge and a large-scale building.

In this study, in order to remove the E&O variations effect on the extracted features, a data normalization procedure including the following two steps are employed. First, data standardization is applied to remove E&O effects on measured acceleration response data. Second, a low-pass Chebyshev filter is employed to mitigate high-frequency content requiring complex time-series models with high model orders to well estimate the data. Then, the acceleration data recorded by SHM systems are modeled as AR time-series. The statistical hypothesis of two-sample t-test is utilized on coefficients of AR models to capture changes in dynamic characteristics of structures due to deterioration. The feature in this approach is defined as a function of the resulting *P*-values. In order to ensure sensitivity of the method, a novel framework called the best model order (BMO) algorithm is developed to estimate the ideal model order satisfying both minimal residual and simplicity of the model. The time-series model orders are unknown values which are required to be estimated with care to be generalized to a wide range of data sets and to be high enough to capture the dynamic characteristics of the structures. These are the two key factors of optimal model orders to be addressed in the present study.

In the following sections, first, the health evaluation framework based on time-series models is discussed. Then, the novel algorithms for structural deterioration detection are presented. Details of the experimental investigation of the proposed method are next summarized, and deterioration detection method is verified using the experimental data. Finally, deterioration detection results are presented before the conclusion is made.

2. Deterioration Detection Algorithm

The AR model was chosen to detect deterioration assuming that the structural response is stationary. In order to remove the E&O effects of variations, data normalization procedure is designed as follows. First, data is collected from the structure using each sensor and standardized as follows:

$$\hat{x}_i = \frac{x_i - \overline{x}}{\sigma} \tag{1}$$

where, x_i denotes amplitude of measured acceleration response data; $\overline{x}$, σ and $\hat{x}_i$ are the mean, standard deviation (STD) and standardized signal of x_i, respectively. Second, the data is filtered with a twelve-order Chebyshev type II low-pass filter with a cutoff frequency of 50. This filter removes high-frequency content. The primary attribute of this filter is its speed. More information can be obtained from Smith (1997). Third, assuming the structural response as stationary, an AR model is fitted to the data:

$$x_k = \sum_{i=1}^{p} \Phi_i^x x_{k-i} + e_k^x \tag{2}$$

where, p is the model order, x_{k-i} represents the $(k-i)^{th}$ previous response, Φ_i^x is the ith AR coefficient of the corresponding previous response, and e_k^x is the residual error of the model.

To increase the sensitivity of time-series analyses, a new framework named the BMO algorithm was developed to estimate an ideal model order satisfying both minimum residual and simplicity of the model. This method estimates the best-fit model to the data considering its complexity. As the time-series models are well fitted to the data, the residual against baseline become very small and so close to zero. The best-fit model order is the one with the least residual and suitable complexity. This technique enables deterioration sensitive features to be detected even with slight changes of vibration characteristics of the deteriorated buildings. It is worth noting that the current procedures are mostly suitable for detecting damage. As the changes in the response of structures due to deterioration are much smaller than those caused by damage, the current techniques for estimating model orders which are widely used in damage detection, cannot be directly used. This is one of the basic concepts for obtaining sensitive features to identify anomalies and deterioration in buildings. If a number of time-series models from the structure in the baseline state are available, the BMO algorithm runs according to the following steps. As this algorithm tries to estimate the most sensitive model order to the changes in the data, a number of data sets in the reference state are required to obtain a good estimation.

Step 1: Obtain AR models using different model orders (limit p equal to 20 for this study) for the first data.

Step 2: Feed another data and predict the data using the obtained AR models in step 1.

Step 3: Calculate residuals of time-series models in step 2.

Step 4: Calculate STD of residuals in step 3.

$$FR_{(i,j)} = \sigma(e_{(i,j)}) \tag{3}$$

where, $j = 1,2,...,m$; and m is the number of data sets in reference state.

Step 5: Calculate C parameter using the following equation to obtain the changes ratio in residuals of different models and data sets.

$$C_{(i,j)} = \frac{FR_{(i,j)} - FR_{(i,1)}}{FR_{(i,1)}} \tag{4}$$

Step 6: Repeat steps 2-5 for the number of data sets in reference state (m). C is the $n \times m$ matrix.

Step 7: Use the following equations and calculate $\boldsymbol{\beta}$ criterion. The minimum value of this parameter corresponds to the model order with a higher sensitivity to changes in data, including changes in data due to deterioration in structures.

$$\boldsymbol{\beta} = \boldsymbol{M} + 2 \times \boldsymbol{S} \tag{5}$$

where, $\boldsymbol{S}$ and $\boldsymbol{M}$ are STD and mean of $\boldsymbol{C}_i$, respectively. $\boldsymbol{C}_i$ is a vector of C parameter for different data sets obtained by using the same model order.

Step 8: The best model order is equal to the model order corresponding to the minimum value of $\boldsymbol{\beta}$ criterion having fit greater than 95% in order to ensure AR models capture the dynamic characteristics of structures.

In this study, a novel coefficient-based deterioration detection method was developed. In this method, statistical hypotheses of two-sample t-test were conducted on coefficients of time-series analyses, which were estimated in the baseline and assessment states of structures, as a deterioration detector. For this hypothesis formulation, a function of the resulting *P*-values is used to define a deterioration feature. The relevant details can be found in statistics literature such as that by Gibbons and Chakraborti (2003).

3. Description of Test Structure and Data

In this study, an experimental three-story bookshelf structure was used to verify the proposed deterioration detection method. The experimental data was downloaded from the website of the Los Alamos National Laboratory (LANL), USA (Figueiredo et al., 2009). The three-story building structure (Figures 1 and 2) was used as accumulated-deterioration detection test bed. Lateral excitation was imposed to the base floor along the center line of the structure by using an electrodynamic shaker. In order to measure the system's response, four accelerometers were attached at the center line of each floor on the opposite side from the excitation source.

Acceleration time histories (time-series or sample records) for various structural states were collected as shown in Table 1 along with information that describes the different states. The structural state conditions were categorized into nine states. The first state (State#1) is the baseline condition. The other states (States #2-#9) are related to the structure when the mass or stiffness of the structure is slightly changed. Real-world structures have E&O variations, which create difficulties in detecting and identifying structural deterioration. Such variations often manifest themselves in changes of the mass or stiffness of a structure. In order to simulate these variations, tests were conducted with different stiffness and mass conditions (States #2-#9). For example, the state condition labelled "State #4" means there was an 87.5% stiffness reduction in the column located between the base and 1st floor at the intersection of plane B and D (abbreviated as 1BD, other abbreviations can be identified in a similar way) (Figure 1). The stiffness reduction consists of replacing the corresponding column by another one with lower stiffness in the direction of shaking. It is assumed that the structure is deteriorated in each of the considered states. In other words, each state shows the condition of the deteriorated structure after a specific time. For instance, it is supposed that in State #4, column 1BD in the first story is deteriorated (87.5% stiffness reduction). It should be noted that each state must be studied individually.

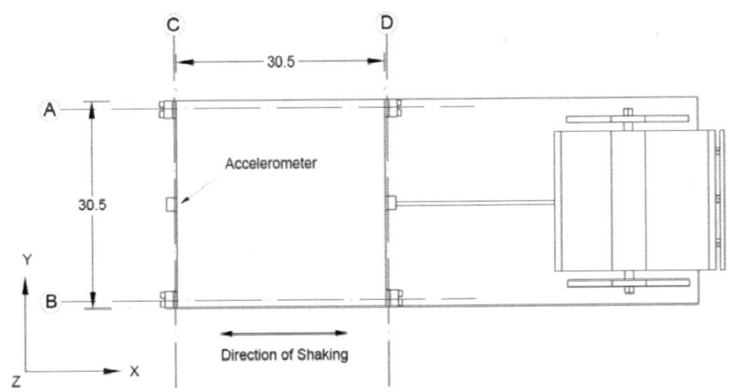

Figure 1. Basic dimensions of the building model (dimensions are in cm) (Figueiredo et al., 2009)

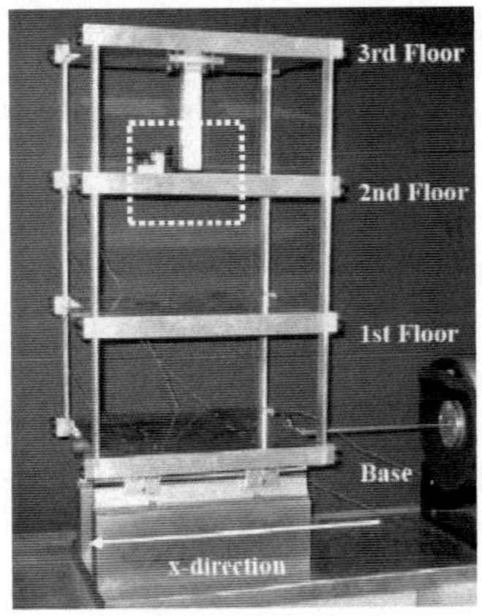

Figure 2. Three-story building model (Figueiredo et al., 2009)

Table 1. Data labels of structural state conditions

State No.	Record No.	Description
State #1	1~50	Baseline condition
State #2	51~100	Mass = 1.2 kg at the base
State #3	101~150	Mass = 1.2 kg on the 1st floor
State #4	151~200	87.5% stiffness reduction in column 1BD
State #5	201~250	87.5% stiffness reduction in column 1AD and 1BD
State #6	251~300	87.5% stiffness reduction in column 2BD
State #7	301~350	87.5% stiffness reduction in column 2AD and 2BD
State #8	351~400	87.5% stiffness reduction in column 3BD
State #9	401~450	87.5% stiffness reduction in column 3AD and 3BD

While no research has been conducted on deterioration assessment of structures, it is known that simulating deterioration in structures is not easy and in some cases impractical. As a result, these data sets, which can be considered as accumulated deterioration, were used to validate the proposed deterioration detection method. A total of 450 records (nine states each of which have 50 records) of 8192 data points with a sampling frequency of 320 Hz is used in each level.

4. Analysis and Results

The mentioned experimental data sets were utilized as input data. In this method, a data normalization procedure including the following two steps were employed. First, data standardization was applied. Second, a low-pass Chebyshev filter was employed. Then, the BMO algorithm was utilized to estimate the best model order. Finally, statistical hypothesis of two-sample F-test was conducted on coefficients of time-series models and the *P*-values of the hypothesis test were used to define a deterioration indicator. This deterioration detection procedure was carried out using MATLAB.

4.1 Model Order Selection

The first step to assess the deterioration using the proposed methods was estimating time-series model orders. To obtain the best model order, 12 datasets in the reference state were used utilizing the novel BMO algorithm. In this case study, AR time-series model was used and the maximum value for model orders (p) is considered to be 20 as the higher model orders result larger β criterion. Figure 3 shows β criterion in the BMO algorithm in which the horizontal axis indicates the model orders. The best model order is equal to the model order corresponding to the minimum value of β criterion having fit greater than 95%. For instance, in this case study, the best model order was 7; as for the first 6 model orders, the fit ratio was less than 95%.

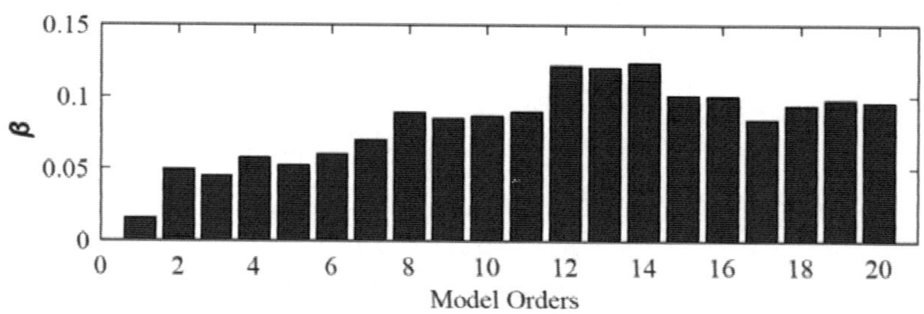

Figure 3. β criterion in BMO algorithm

4.2 Deterioration Detection

In this paper, coefficient-based deterioration detection method is carried out with the experimental data sets. Each state is related to a specific health state of the structure. For instance, record number 401 to 450 is related to deterioration of two columns in the third story. In each state, the response of the structure is recorded at the same time in all the three stories. Figure 4 shows the results of the proposed method. In this method, *P*-values are used as deterioration indices. It can be seen that deterioration was detected clearly and there were no false alarms. Record number 201 to 250 is related to the State #5 which corresponds to deterioration of two columns in the first story.

Deterioration was detected from the recorded response in the all three stories. However, the higher P-values at the first story (Figure 4(a)) indicate that the first story has deteriorated. In other words, it should be noted that deterioration in each story can also be detected from other stories. Furthermore, the results showed that the more deterioration, the higher P-values. For instance, record 351 to 400 is related to the State #6 and record 401 to 450 is related to State #7. State #6 indicates that one column of the third story has deteriorated, and State #7 points out that two columns of the third story have deteriorated. Figure 4(c) also shows larger P-values in State #7 than State #6.

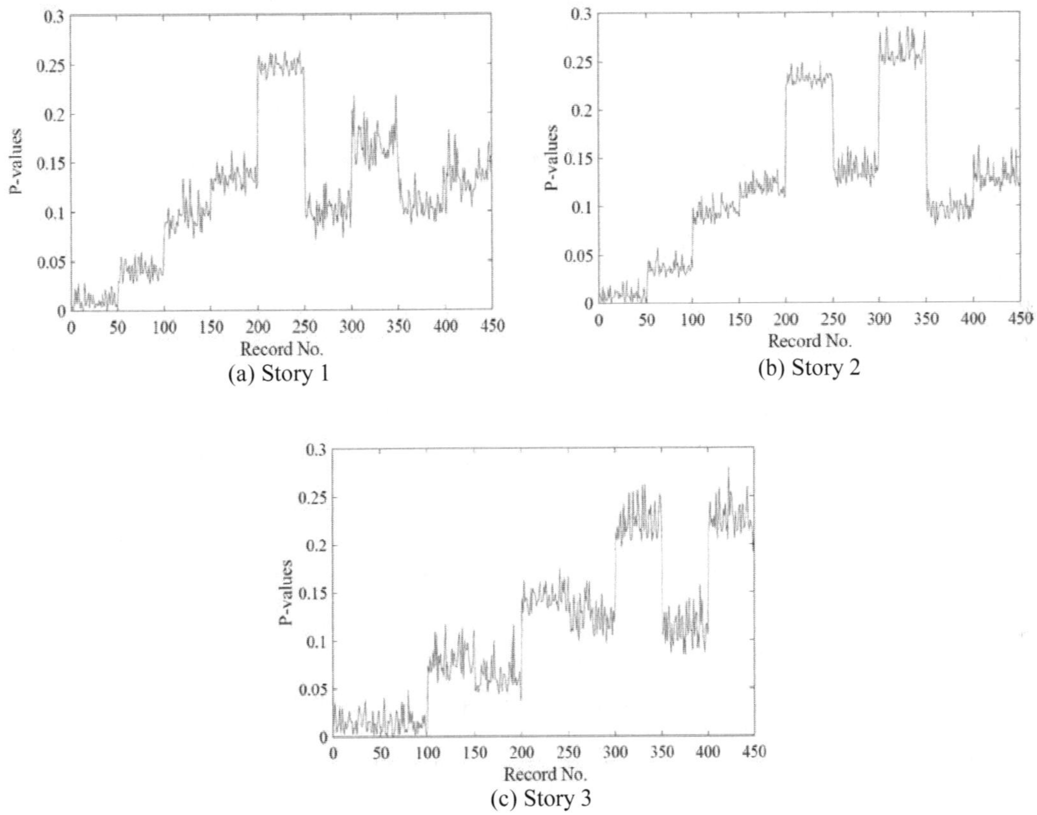

Figure 4. Deterioration detection results

5. Conclusions

All structures, including newly built ones start to deteriorate at a progressive slow-pace process due to environmental effects, varying service loads and ageing. Dynamic characteristics of buildings change due to deterioration. There is no doubt that it is highly crucial to assess the deterioration status of structures to maintain cost-effective maintenance and to extend their life expectancy. This paper presented a deterioration sensitive feature using enhanced AR model coefficients. In this method, statistical hypothesis of two-sample F-test for equal variances was conducted on coefficients of time-series analyses and the P-values of the hypothesis test were used to define a deterioration indicator. In order to increase the sensitivity of the developed method, the BMO algorithm was developed. This novel algorithm estimates the best model order. The effectiveness of the proposed method was shown using an experimental case study. The results showed that the proposed method can clearly detect deterioration. In this procedure, the highest *P*-values in each story corresponds to the deterioration of that story. Moreover, deterioration in a story can be detected in adjacent stories. For instance, the deterioration of the first story can be detected with the second story data. It can be concluded that this procedure may be capable of being developed for locating deterioration in structures, but more work needs to be done to address the cross-level detection problem. This issue will be addressed in the future publications of the authors.

References

Alvandi, A., & Cremona, C. 2006, 'Assessment of vibration-based damage identification techniques', *Journal of Sound and Vibration,* vol. 292, no. 1, 179-202.

Brownjohn, J. M. W. 2007, 'Structural health monitoring of civil infrastructure', *Philosophical Transactions of the Royal Society of London A: Mathematical, Physical and Engineering Sciences,* vol. 365, no. 1851, pp. 589-622.

Carden, E. P., & Brownjohn, J. M. W. 2008, 'ARMA modelled time-series classification for structural health monitoring of civil infrastructure', *Mechanical Systems and Signal Processing,* vol. 22, no. 2, pp. 295-314.

Carden, E. P., & Fanning, P. 2004, 'Vibration based condition monitoring: a review', *Structural Health Monitoring,* vol. 3, no. 4, pp. 355-377.

Chan, T. H. T., & Thambiratnam, D. P. 2011, '*Structural health monitoring in Australia*: Nova Science Publishers.

Doebling, S. W., Farrar, C. R., & Prime, M. B. 1998, 'A summary review of vibration-based damage identification methods', *Shock and vibration digest,* vol. 30, no. 2, pp. 91-105.

Du, Y. G., Chan, A. H. C., Clark, L. A., Wang, X. T., Gurkalo, F., & Bartos, S. 2013, 'Finite element analysis of cracking and delamination of concrete beam due to steel corrosion', *Engineering Structures,* vol. 56, pp. 8-21.

Farrar, C. R., Duffey, T. A., Cornwell, P. J., & Doebling, S. W. 1999, 'Excitation methods for bridge structures', Paper presented at the Society for Experimental Mechanics, Inc, *17th International Modal Analysis Conference.*

Figueiredo, E., Park, G., Figueiras, J., Farrar, C., & Worden, K. 2009, 'Structural health monitoring algorithm comparisons using standard data sets', *Los Alamos National Laboratory (LANL)*, Los Alamos, NM (United States).

Gibbons, J. D. & Chakraborti, S. 2003, 'Nonparametric statistical inference', *4th Edition*, New York: Marcel Dekker.

Gul, M., & Catbas, F. N. 2009, 'Statistical pattern recognition for Structural Health Monitoring using time series modeling: Theory and experimental verifications', *Mechanical Systems and Signal Processing,* vol. 23, no. 7, pp. 2192-2204.

Gul, M., & Catbas, F. N. 2011, 'Structural health monitoring and damage assessment using a novel time series analysis methodology with sensor clustering', *Journal of Sound and Vibration,* vol. 330, no. 6, pp. 1196-1210.

Huang, X., & Chen, J. 2015, 'Time-Dependent Reliability Model of Deteriorating Structures Based on Stochastic Processes and Bayesian Inference Methods', *Journal of Engineering Mechanics,* vol. 141, no. 3, pp. 04014123.

Lei, Y., Kiremidjian, A. S., Nair, K. K., Lynch, J. P., Law, K. H., Kenny, T. W., Carryer, E. D. & Kottapalli, A. 2003, 'Statistical damage detection using time series analysis on a structural health monitoring benchmark problem', *Paper presented at the Proceedings of the 9th international conference on applications of statistics and probability in civil engineering.*

Ling, Q., Tian, Z., Yin, Y., & Li, Y. 2009, 'Localized structural health monitoring using energy-efficient wireless sensor networks', *Sensors Journal, IEEE,* vol. 9, no. 11, pp. 1596-1604.

Mosavi, A. A., Dickey, D., Seracino, R., & Rizkalla, S. 2012, 'Identifying damage locations under ambient vibrations utilizing vector autoregressive models and Mahalanobis distances', *Mechanical Systems and Signal Processing,* vol. 26, pp. 254-267.

Nair, K. K., Kiremidjian, A. S., & Law, K. H. 2006, 'Time series-based damage detection and localization algorithm with application to the ASCE benchmark structure', *Journal of Sound and Vibration,* vol. 291, no. 1-2, pp. 349-368.

Nguyen, T., Chan, T. H. T., & Thambiratnam, D. P. 2014a, 'Controlled Monte Carlo data generation for statistical damage identification employing Mahalanobis squared distance', *Structural Health Monitoring,* vol. 13, no. 4, pp. 461-472.

Nguyen, T., Chan, T. H. T., & Thambiratnam, D. P. 2014b, 'Field validation of controlled Monte Carlo data generation for statistical damage identification employing Mahalanobis squared distance', *Structural Health Monitoring*, vol. 13, no. 4, pp. 473-488.

Nguyen, T., Chan, T. H. T., Thambiratnam, D. P., & King, L. 2015, 'Development of a cost-effective and flexible vibration DAQ system for long-term continuous structural health monitoring', *Mechanical Systems and Signal Processing*, vol. 64, pp. 313-324.

Okasha, N. M., & Frangopol, D. M. 2010, 'Time-variant redundancy of structural systems', *Structure and Infrastructure Engineering,* vol. 6, no. 1-2, pp. 279-301.

Omenzetter, P., & Brownjohn, J. M. W. 2006, 'Application of time series analysis for bridge monitoring', *Smart Materials & Structures,* vol. 15, no. 1, pp. 129-138.

Shih, H. W., Thambiratnam, D. P., & Chan, T. H. T. 2009, 'Vibration based structural damage detection in flexural members using multi-criteria approach', *Journal of Sound and Vibration,* vol. 323, no. 3-5, pp. 645-661.

Smith, S. W. 1997, 'The scientist and engineer's guide to digital signal processing',

Zhang, Q. W. 2007, 'Statistical damage identification for bridges using ambient vibration data', *Computers & Structures,* vol. 85, no. 7-8, pp. 476-485.

Wang, V.Z. & Ong, K.C.G. 2015, 'Nonparametric statistical formulations for structural health monitoring', *Computers & Structures*, vol. 148, pp. 63-74

The 8th International Conference on
Structural Health Monitoring of Intelligent Infrastructure
Brisbane, Australia | 5-8 December 2017

Effect of Changing Stress on Damage Diagnostic with Guided Waves

M. Mohabuth [1], A. Kotousov [2], C-T Ng [3], L. R. F. Rose [4].

[1] *School of Mechanical Engineering, The University of Adelaide, SA 5005 – Australia, Email:* munawwar.mohabuth@adelaide.edu.au, [2] *School of Mechanical Engineering, The University of Adelaide, SA 5005 – Australia,* [3] *School of Civil, Environmental and Mining Engineering, The University of Adelaide, SA 5005 – Australia,* [4] *Aerospace Division, Defence Science and Technology, VIC 3207 - Australia*

Abstract

SHM systems based on guided waves commonly utilise a signal subtraction approach to detect damage using a network of embedded or permanently attached sensors/transducers. In this approach, the presence of damage is normally determined by analysing the residual signals from the structure after subtracting the damage-free reference data. It is well-known that variations in temperature can compromise this approach and lead to false alarms or masking of the damage. Consequently, several temperature compensation strategies have been proposed. In this paper, we demonstrate that changing stress conditions can produce a similar effect, comparable with modest temperature variations. This analysis provides a basis to compensate for the effect of applied or thermally-induced stresses on the detectability of damage. This is particularly important for self-constrained or partially constrained structures, such as rails or reinforced concrete, where the stresses are very difficult to evaluate or control.

1. Introduction

A typical guided wave (GW) based SHM system incorporates a number of permanently bonded or embedded transducers, which can normally operate in the excitation or receiving modes. One of the transducers is usually excited with a tone burst of a few cycles, generating a stress wave that propagates along the structure. The time-domain responses from the transmitter and the receiving transducers are then recorded. This process is then repeated using different transducers as transmitters. The signal remaining after subtraction from damage-free reference data which exceeds the background noise is assumed to be a defect. One of the main contributions to the background noise in real-world situations arises from variations in the temperature of the structure. Ambient temperature variations are arguably the main reason why SHM systems, which have been developed in the last two decades in laboratory conditions, often fail to prove their efficiency in the real-world environment. For example, Croxford et al. (2007) demonstrated that even in the presence of modest temperature fluctuations, the number of sensors required for damage detection can be prohibitively high. Therefore, some form of compensation for changing environmental and operational conditions is essential for GW based SHM systems. Consequently, a number of temperature compensation strategies have been proposed (Aryan et al. 2016).

In this paper, we demonstrate that another factor, changing stress, has a similar effect on noise generation as changing temperatures. The adopted approach is based on the theory of acoustoelasticity (Destrade & Ogden 2012) and a simplified model for the evaluation of the signal noise caused by temperature fluctuations (Croxford et al. 2007). In particular, we provide selected results demonstrating the effect of the applied stress on the phase velocity of the fundamental Lamb wave modes for plates made of different materials. The amplitude of the noise signal is then evaluated in terms of a coefficient relating changes in the phase velocity to the applied stress. A similar coefficient is commonly utilised to compute the background noise amplitude due to fluctuating temperatures (Croxford et al. 2007). Finally, for the fundamental anti-symmetric mode, we investigate the frequencies at which the effect of the applied stress on the phase velocity is negligible. These frequencies might be of interest in the future development of stress compensation strategies for SHM using GW.

2. Phase Velocity Changes due to Applied Stress

In an earlier paper, we derived dispersion equations for Lamb wave propagation in pre-stressed plates by utilising a specialised form of the strain energy function, which accounts for weakly nonlinear elasticity (Mohabuth et al. 2016). The deriviation and the final equations are relatively complex and will not be presented here. Instead, we present numerical results obtained by solving the derived dispersion equations for plates made of different materials and subjected to uniaxial stress. The materials considered in this study and their respective elastic properties are listed in Table 1, where λ and μ are the Lame constants, ρ is the density and l, m and n are the third–order, or Murnaghan constants (Murnaghan 1937).

Table 1: Elastic Properties for the materials considered in the investigation

Material	λ [GPa]	μ [GPa]	l [GPa]	m [GPa]	n [GPa]	ρ [kg/m^3]
Aluminum	54.3	27.2	-281.5	-339.0	-416.0	2704
Hecla Steel	111	82.1	-461	-636	-708	7823
Polystyrene	2.89	1.38	-18.9	-13.3	-10.0	1040

Figures 1 (a-f) show the change in the phase velocity of the fundamental symmetric (S0) and anti-symmetric (A0) modes, compared to the unstressed state, as a function of the magnitude of the uniaxial tensile stress up to the yield stress. The direction of propagation of the waves was considered to be parallel to the applied load, as it was shown to exhibit the highest sensitivity of the wave velocity to the applied stress in the case of longitudinal waves (Egle & Bray 1976). It can be seen that the relative change in the phase velocity is negative for positive (tensile) stresses except for the A0 mode at relatively small values of the frequency-thickness product ($f \times h$). This finding is consistent with the results of bulk wave acoustoelasticity which predicts that a tensile stress causes a decrease in the wave velocity (Hughes & Kelly 1953). A further validation of the analytical model and a comparison with published experimental results was provided in Mohabuth et al. (2016).

It can be also observed from Figure 1 that higher magnitudes of the applied stress result in larger changes of the phase velocity, particularly in the lower $f \times h$ region. However, at higher values, the change in the phase velocity tends to a constant value. This is not surprising since the wave velocity of these modes converge to the Rayleigh wave velocity at higher frequencies (Rose 1999). Compressive stresses have an opposite effect on the change of the phase velocity but the results are not shown here for reasons of clarity.

3. Simplified Noise Generation Model

To quantify the effect of stress on damage diagnostics, we will extend a simplified analytical model recently developed by Croxford et al. (2007). This model was derived for a Hanning-windowed toneburst, which is widely employed in damage detection utilising Lamb waves. It was demonstrated that in the presence of temperature variations, δT, the algebraic signal subtraction of the time trace affected by temperature from the reference data results into the following expression for the residual signal (noise):

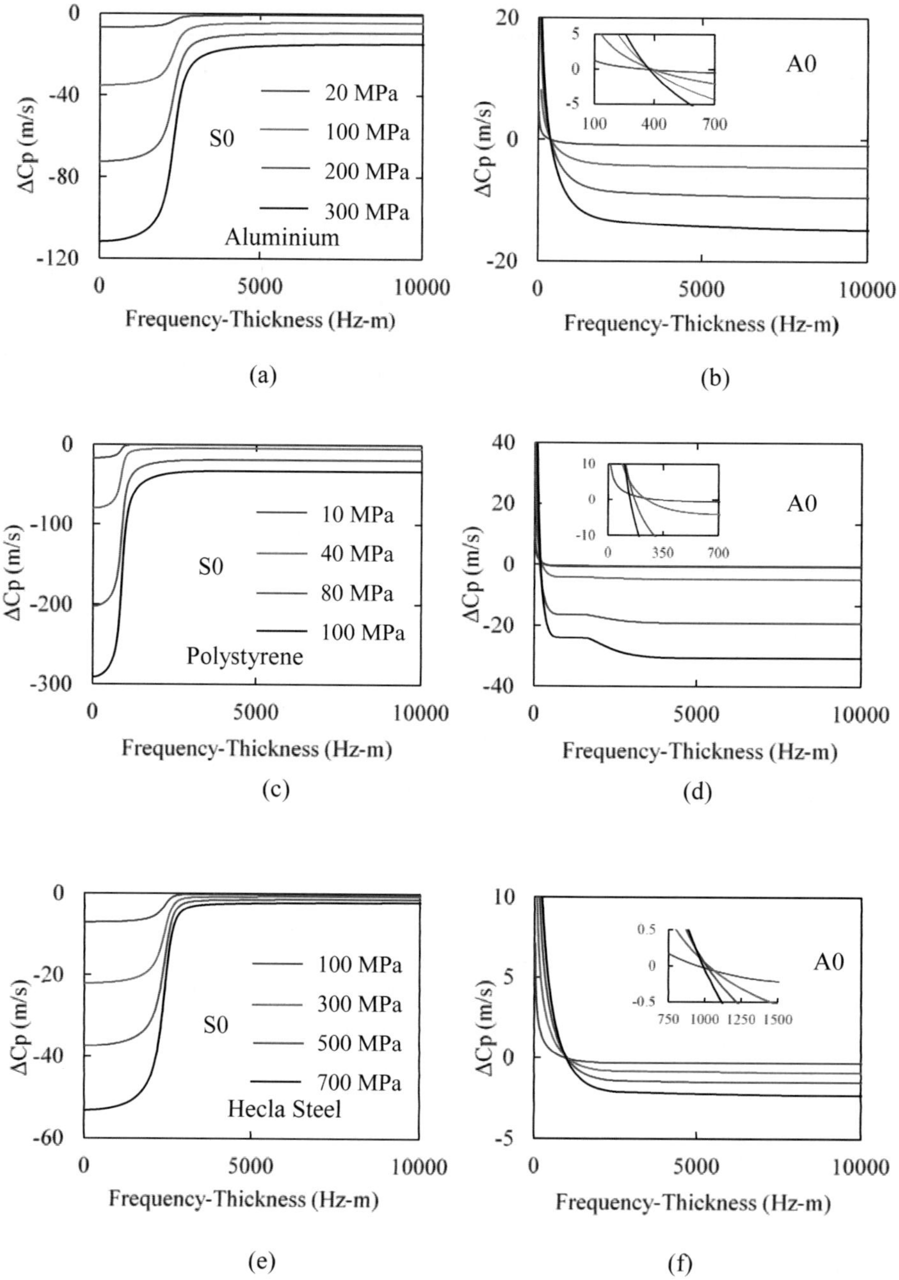

Figure 1. Effect of uniaxial stress on the phase velocity of symmetric (a, c, e) and anti-symmetric (b, d, f) fundamental modes in Aluminum (a, b), Polysterene (c, d) and Steel (e, f) plates.

$$u_{noise} = 2u_0\pi f\frac{d}{v_{ph}}\left(\alpha - \frac{k_{ph}^{T}}{v_{ph}}\right)\delta T \qquad (1)$$

where d is the distance between the transmitter and the sensor; v_{ph} is the phase velocity; α is the coefficient of thermal expansion and k_{ph}^{T} is the coefficient relating changes in v_{ph} to temperature, $k_{ph}^{T} = \delta v_{ph}/\delta T$. The presence of d in the numerator indicates that the time shift associated with changing temperature is proportional to the distance between the transmitter and the sensor. If the signal-to-noise ratio is specified, then the maximum distance or the minimum density of the sensor/transmitter array can be evaluated from equation (1). The latter equation also suggests that faster modes with higher phase velocities are less affected by temperature than slower modes.

The second term in brackets, k_{ph}^{T}/v_{ph}, is typically one or two orders of magnitude larger than α, so the change in the phase velocity with temperature dominates the noise generation and, the thermal expansion is the secondary phenomenon (Croxford et al. 2007). Therefore, the noise generated by the temperature change can effectively be written as

$$u_{noise} \approx -2u_0\pi fd\frac{k_{ph}^{T}}{v_{ph}^{2}}\delta T \ . \qquad (2)$$

Similar considerations can be applied to the evaluation of the signal noise due to changing stress, which also affects the phase velocity in a similar way to temperature. Thus, the following equation can be written:

$$u_{noise} \approx -2u_0\pi fd\frac{k_{ph}^{\sigma}}{v_{ph}^{2}}\delta\sigma \ , \qquad (3)$$

where $k_{ph}^{\sigma} = \delta v_{ph}/\delta\sigma$, which is in this case a function of the applied uniaxial stress, σ, as well as frequency, f.

Table 2 summarises the values of k_{ph}^{T} (provided in Croxford et al. 2007) and range of values of k_{ph}^{σ} as evaluated from the present calculations for Aluminium, see Figure 1 (a) and (b).

Table 2. Modal temperature and stress parameters for Aluminium

	A0	S0
$\mathbf{k_{ph}^{T}}$ (m s^{-1} ^{0}C^{-1})	-0.38	-1.24
$\mathbf{k_{ph}^{\sigma}}$ (m s^{-1} MPa^{-1})	-0.05 – 0[1]	-0.35 – -0.06

[1]This range excludes small values of the frequency×thicknes range where k_{ph}^{σ} is positive.

4. Discussion

From Table 2, it can be seen that the S0 mode is more sensitive to stress as well as to temperature fluctuations. The effect of the applied or thermally-induced stress is significant and comparable to the temperature effect. To demonstrate this point let us consider a constrained plate in one direction and suppose this plate is free from stress state at a reference temperature, T_0. A temperature rise by 1°C will generate a uniaxial thermally-induced stress of approximately $\alpha E \times 1^{o}C \approx 1.5$ MPa. In the low frequency-thickness range and in the case of the S0 mode ($f \times h < 2500$ Hz-m leading to $k_{ph}^{\sigma} \approx$ -0.35 m s^{-1} MPa^{-1}), the amplitude of the signal noise will be 60% of the signal noise amplitude caused by the same temperature increment. This illustrates that temperature and applied stress have a very similar influence on the change in the phase velocity and consequently, on the generation of the signal noise. Obviously, the variation of the applied stress can be caused factors other than temperature fluctuations, and the effect of such variations can be analysed within the current framework using Equation (3).

Figure 1 suggests that the S0 mode, has a threshold value of $f \times h$ at which the effect of the stress becomes relatively small. For example, for Aluminium this threshold value is approximately 2500 Hz-m. As mentioned earlier, this is because the phase velocity converges to the Rayleigh wave propagation speed at higher frequencies. In the case of the A0 mode, a critical value of $f \times h$ can be identified, in the close vicinity of which the effect of the applied stress on the phase velocity is almost negligible, see Figure 1 (b, d, f). For Aluminium this critical value corresponds to $f \times h \approx 380$ Hz m. It is interesting to note that all the materials considered exhibit a very similar dependence of the phase velocity on the applied stress and frequency. Therefore, the simplest strategy for stress compensation would be to utilise the frequency range (if this is possible) above the threshold value when the S0 mode is employed for damage diagnostic. In the case when Lamb wave toneburst is dominated by the A0 mode, the applied or thermally induced stress can be effectively compensated by selecting the excitation frequencies of the toneburst close to the critical value, which will essentially negate any change in stress.

5. Conclusions

In this paper we analysed the effect of uniaxial stresses on the propagation of the fundamental modes of Lamb waves as these modes are normally used for damage detection. We presented results based on dispersion equations derived previously for the change of the phase velocity as a function of the applied stress, frequency and material properties (Mohabuth et al. 2016). Using a signal subtraction approach, we demonstrated that the effect of stress on the phase velocity can be as significant as the effect of temperature fluctuations, and it has to be considered when a SHM system using guided waves is developed for the real-world applications.

It is interesting to note that the temperature change and thermally induced stresses have an opposite effect on the phase velocity of the fundamental symmetric and anti-symmetric modes. For example, a positive temperature change, $\Delta T > 0$, normally induces a compressive (negative) stress, $\Delta\sigma < 0$. The coefficients relating changes to temperature and stress have the same sign, $k_{ph}^{T} < 0$ and $k_{ph}^{\sigma} < 0$, it means that the stress effect can partially compensate the temperature effect on the phase velocity. All materials considered in this study demonstrated very similar dependences of the change in the phase velocity on the applied stress and wave frequency. For all the materials considered, there exists a threshold value above which the effect of the stress diminishes significantly for the S0 mode. We found that there exists a critical value of the frequency-thickness product for the A0 mode at which the influence of stress is almost negligible. These acoustoelastic features can provide a foundation for the development of effective stress compensation strategies for future SHM systems.

Although this paper considered a simple case of uniaxial loading and phase velocity changes in the direction parallel to the applied stress, it should be possible to generalise the current framework and apply it to investigate the effect of bi-axial stress state on wave propagation along an arbitrary direction and evaluate the associated amplitude of the signal noise or conduct the similar analysis for other waveguides, such as circular rods or shells. However, these and other developments will be the focus of future works.

6. Acknowledgement

The work was supported by Department of Further Education, Employment, Science & Technology, Government of South Australia, under Catalyst Research Grant Program and Australian Research Council, project DP160102233. The support is greatly appreciated.

References

Aryan P, Kotousov A, Ng CT & Wildy S 2016, 'Reconstruction of baseline time-trace under changing environmental and operational conditions', *Smart Materials and Structures*, vol. 25, no 3, pp. 035018-1-035018-10.

Croxford AJ, Wilcox PD, Drinkwater BW & Konstantinidis G 2007, 'Strategies for guided-wave structural health monitoring', *Proceedings of the Royal Society A,* vol. 463, pp. 2961–2981.

Destrade M & Ogden RW 2013, 'On stress-dependent elastic moduli and wave speeds', *IMA Journal of Applied Mathematics,* vol. 78, pp. 965-997.

Egle DM & Bray DE 1974, 'Measurement of acoustoelastic and third-order elastic constants for rail steel', *Journal of the Acoustial Society of America*, vol. 60, pp. 741–744.

Hughes, D & Kelly, J 1953, 'Second-Order Elastic Deformation of Solids', *Physical Review*, vol. 92, pp. 1145-1149.

Mohabuth, M., Kotousov, A., & Ng, C 2016, 'Effect of uniaxial stress on the propagation of higher-order Lamb wave modes' *International Journal of Non-Linear Mechanics*, vol. 86, pp. 104-111.

Murnaghan, FD 1937, 'Finite Deformations of an Elastic Solid', *American Journal of Mathematics*, vol. 59, pp. 235-260.

Rose, JL 1999, *Ultrasonic Waves in Solid Media*, Cambridge University Press, Cambridge, United Kingdom.

The 8th International Conference on
Structural Health Monitoring of Intelligent Infrastructure
Brisbane, Australia | 5-8 December 2017

Output-only Structural Damage Detection based on Hard-threshold Method: Theoretical Study

C.D. Pan[1], L. Yu[1, 2].
[1] School of Mechanics and Construction Engineering, Jinan University, Guangzhou 510632, China
[2] MOE Key Lab of Disaster Forecast and Control in Engineering, Jinan University, Guangzhou 510632, China.
Email: lyu1997@163.com

Abstract

Structural damage detection (SDD) with a vehicle-bridge system has been widely studied, since traffic excitations contribute to live load components in bridge design code. As an attempt work, this paper aims to propose an output-only SDD method based on hard-threshold strategy with unknown moving forces. Firstly, a transfer matrix is formulated between any two sensors for a healthy bridge structure. Singular value decomposition (SVD) technology is performed for the transfer matrix, and the corresponding singular vectors are used for decomposing two measured responses based on hard-threshold method. Then the decomposed results are used for estimating some singular values. Finally, a damage indicator (DI), which is defined for classifying status of structures, is calculated by comparing the differences between the estimated and true singular values. In order to assess the accuracy and feasibility of the proposed method, a simply-supported bridge subjected to one moving force is taken as an example for numerical simulations. The illustrated results show that the proposed method has a potential ability for solving the SDD problem. Some drawbacks of the proposed method are also discussed as well.

1. Introduction

Structural damage detection (SDD) is a significant problem in the field of structure health monitoring (SHM), and it has been widely studied in the past few decades. Many effective SDD methods have been proposed (Farrar et al., 2001; Fan & Qiao, 2011). Existing SDD methods can be roughly divided into two categories, i.e. model-based method and signal-based method. Herein, the first category usually performs with considering a structural finite element model (FEM), while the second category always solves the SDD problem via advanced signal processing technology without considering a structural FEM. In fact, one advantage of the model-based SDD methods is that the relationship between structural inputs and outputs is known; therefore, one can easily design a SDD program if the external excitations are known. However, it is hard to obtain the inputs of civil structures in practical application. Hence, it is beneficial to develop some SDD methods with considering unknown structural inputs (Zhu & Law, 2015; Feng & Feng, 2016).
Existing output-only based SDD methods can be roughly divided into two categories such as direct method and iterative method. Herein, the direct method means that the structural damage can be identified without an iterative process, while the iterative method means that iteration is needed for damage detection. Some examples are given in the following contexts. Based on the virtual distortion method (VDM), Zhang et al. (2010) proposed a method for simultaneous identification of external forces and structural damages. Wherein, the structural damage was converted into some virtual forces and then added into a traditional force reconstruction equation. The biggest advantage of VDM-based method is that the identified equation belongs to a classical force reconstruction

problem, which can be solved directly. Lu & Law (2007) focused on identifying both the system parameters and input forces based on dynamic response sensitivity. This method employed a two-step iteration procedure, in which the forces and structural parameters were updated separately. As an extended work, Zhang et al. (2011) presented a probabilistic damage identification approach with uncertainties under unknown inputs. Zhu et al. (2014) proposed a SDD method under unknown inputs by combining with the transmissibility matrix concept. The basic idea based on iteration also has been studied by Sun et al. (2015) and Pan & Yu (2014).

The methods mentioned above aim to estimate the damage extent. They actually can be used for solving the SDD problem for a bridge structure with unknown vehicle-bridge interaction forces. As an exploring attempt to the SDD problem for a bridge structure, this paper aims to propose a novel SDD method with unknown vehicle-bridge interaction forces. Herein, singular value decomposition (SVD) technology is performed for the structural transfer matrix, and the corresponding singular vectors are used for decomposing the measured responses based on hard-threshold method. Both the corresponding singular values and response decomposed results are then used for formulating a damage indicator (DI), which is defined for classifying status of structure. Some numerical simulations are also carried out for assessing the accuracy and feasibility of the proposed method in this paper.

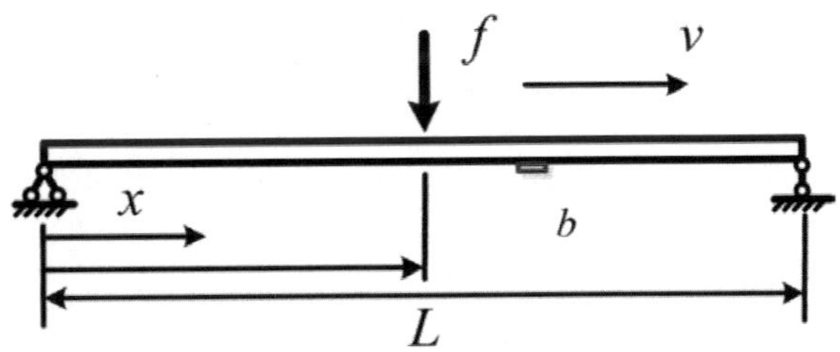

Fig. 1 Simply-supported bridge subjected to moving force.

2. Theoretical background

2.1 Response analysis in time domain

A simply-supported bridge subjected to a moving force, as shown in Fig. 1, is taken as an example for the following analysis. Herein, the bridge is modeled as a discrete one according to the FEM theory. According to the linear vibration theory, the basic motion equation of the bridge can be expressed as:

$$\boldsymbol{M}\ddot{\boldsymbol{y}} + \boldsymbol{C}\dot{\boldsymbol{y}} + \boldsymbol{K}\boldsymbol{y} = \boldsymbol{F} \tag{1}$$

where $\boldsymbol{M}$, $\boldsymbol{C}$ and $\boldsymbol{K}$ are the structural mass, damping and stiffness matrices, respectively. $\ddot{\boldsymbol{y}}$, $\dot{\boldsymbol{y}}$ and $\boldsymbol{y}$ are the acceleration, velocity and displacement vectors, respectively. $\boldsymbol{F}$ means the external forces. Eq. (1) belongs to a second-order differential equation and it can be solved by many numerical methods, such as mode superposition method, Newmark-β method and so on. It should be noted that the above response calculation method is suitable for different types of the external force; however, the relationship between structural inputs and outputs is not intuitive as a linear mapping form. To make it more intuitive, the moving force $f(t)$, $t\in[0,T)$, can be approximately expanded in time domain as:

$$f(t) \approx \sum_{i=0}^{N-1} f_i \times \varphi_i(t)$$
$$\varphi_i(t) = \begin{cases} 1 & t \in [iT/N,\ (i+1)T/N) \\ 0 & others \end{cases} \tag{2}$$

where $\varphi_i(t)$ is the i-th basis function, and f_i is the coefficient of the i-th basis function. N means that the total time range T is divided into N segments. According to the superposition principle, the structural responses of each sensor can be expressed in time domain as:

$$\boldsymbol{b} = [\boldsymbol{h}_0, \boldsymbol{h}_1, \cdots, \boldsymbol{h}_{N-1}][f_0, f_1, \cdots, f_{N-1}]^T = \boldsymbol{Hf} \tag{3}$$

where $\boldsymbol{b}$ is the response vector. $\boldsymbol{H}$ is the system matrix, in which the i-th column $\boldsymbol{h}_i$ means the responses caused by the unit i-th basis function $\varphi_i(t)$. $\boldsymbol{f}$ is the discrete vector of the moving force. It should be noted that $\boldsymbol{h}_i$ can be calculated by solving Eq. (1).

2.2 Transmissibility concept

For any two sensors, the responses obtained from the first sensor are noted as $\boldsymbol{b}_1$ and the other ones are noted as $\boldsymbol{b}_2$. According to the Eq. (3), the responses of two sensors can be expressed as:

$$\boldsymbol{b}_1 = \boldsymbol{H}_1 \boldsymbol{f} \tag{4}$$

$$\boldsymbol{b}_2 = \boldsymbol{H}_2 \boldsymbol{f} \tag{5}$$

where $\boldsymbol{H}_1$ and $\boldsymbol{H}_2$ are system matrices of the first sensor and the second sensor, respectively. According to Eq. (4), the moving force $\boldsymbol{f}$ can be approximately estimated as:

$$\boldsymbol{f} = \boldsymbol{H}_1^+ \boldsymbol{b}_1 \tag{6}$$

where $\boldsymbol{H}_1^+$ is the Moore-Penrose inverse matrix of $\boldsymbol{H}_1$. Substituting Eq. (6) into Eq. (5) yields:

$$\boldsymbol{b}_2 = \boldsymbol{H}_2 \boldsymbol{H}_1^+ \boldsymbol{b}_1 = \boldsymbol{T}_{21} \boldsymbol{b}_1 \tag{7}$$

where $\boldsymbol{T}_{21}$ is a transmissibility matrix from the first sensor to the second sensor. Eq. (7) means that the responses of the second sensor can be estimated from the responses of the first sensor.

3. Basic idea for SDD based on hard-threshold method

3.1 Singular value decomposition of transmissibility matrix

Singular value decomposition (SVD) is a typical matrix decomposition technique in linear algebra. According to the SVD method, the transmissibility matrix $\boldsymbol{T}_{21}$ can be rewritten as:

$$\boldsymbol{T}_{21} = \boldsymbol{U\Sigma V}^{\mathrm{T}} \tag{8}$$

where $\boldsymbol{U}$ and $\boldsymbol{V}$ are two unitary matrices, and $\boldsymbol{\Sigma}$ is a diagonal matrix containing the singular values. It should be noted that the matrix $\boldsymbol{\Sigma}$ is dependent on the status of structure, sampling frequency, sampling time and speed of the moving forces. However, it is independent to the time history of the moving forces. Hence, the singular values have a potential ability for developing a scheme for SDD without considering the input time history.

3.2 Estimation of singular values

Substituting Eq. (8) into Eq. (7) yields:

$$\boldsymbol{U}^{\mathrm{T}} \boldsymbol{b}_2 = \boldsymbol{\Sigma V}^{\mathrm{T}} \boldsymbol{b}_1 \Rightarrow \boldsymbol{\beta}_2 = \boldsymbol{\Sigma \beta}_1 \tag{9}$$

where $\boldsymbol{\beta}_1 = \boldsymbol{V}^{\mathrm{T}}\boldsymbol{b}_1$ and $\boldsymbol{\beta}_2 = \boldsymbol{U}^{\mathrm{T}}\boldsymbol{b}_2$ are the response decomposed results of $\boldsymbol{b}_1$ and $\boldsymbol{b}_2$, respectively. Since the matrix $\boldsymbol{\Sigma}$ is a diagonal matrix, the i-th singular value can easily be estimated as:

$$\sigma_i = \frac{\beta_2(i)}{\beta_1(i)} \tag{10}$$

where σ_i means the i-th singular value, and $\boldsymbol{\beta}(i)$ means the i-th element in vector $\boldsymbol{\beta}$. The estimated singular value obtained from Eq. (10) may be quiet different to the true one if measured noises are considered. In this case, Eq. (10) can be rewritten as:

$$\sigma_i = \frac{\beta_2(i)}{\beta_1(i)+\varepsilon_1} + \frac{\varepsilon_2}{\beta_1(i)+\varepsilon_1} \tag{11}$$

where ε_1 and ε_2 are the measured noise components for $\boldsymbol{\beta}_1$ and $\boldsymbol{\beta}_2$, respectively. Eq. (11) means that the singular value can be accurately estimated only when the measured noise satisfies $\varepsilon \ll \boldsymbol{\beta}(i)$.

3.3 Hard-threshold method and damage indicator

As mentioned in the above section, the singular value can be accurately estimated only when the measured noise levels are small. Hence, it is better to neglect some response components, which have low signal energy. To achieve this goal, the hard–threshold selection method can be applied as:

$$\bar{\beta}_1(i) = \begin{cases} \beta_1(i) & |\beta_1(i)| \geq \delta_1 \\ 0 & |\beta_1(i)| < \delta_1 \end{cases} \qquad \bar{\beta}_2(i) = \begin{cases} \beta_2(i) & |\beta_2(i)| \geq \delta_2 \\ 0 & |\beta_2(i)| < \delta_2 \end{cases} \tag{12}$$

where δ is a threshold value. Eq. (12) means that only the response components, which signal energy is bigger than δ^2, are extracted for the following calculation. According to the hard-threshold selected results, some singular values can be estimated by using a revised formulation as:

$$\sigma_i = \begin{cases} \frac{|\bar{\beta}_2(i)|}{|\bar{\beta}_1(i)|} & |\bar{\beta}_1(i)| > 0 \; and \; |\bar{\beta}_2(i)| > 0 \\ skip & others \end{cases} \tag{13}$$

where *skip* means that the signal components are not used for estimating the singular values. Herein, the absolute sign is used to avoid a possible negative value for σ_i. As the singular values for a healthy structure can be obtained by using the SVD method, a damage indicator (DI) can be easily defined by comparing the estimated singular values to the true ones as:

$$DI = \frac{1}{N}\sum_{i=1}^{N} \frac{|\sigma_i - \bar{\sigma}_i|}{\bar{\sigma}_i} \tag{14}$$

where σ_i and $\bar{\sigma}_i$ are the estimated and true singular values, respectively. N represents the number of the estimated singular values. It can be noted that the DI is defined by calculating an average value of relative errors.

3.4 Basic steps for SDD

As mentioned above, the basic process for the proposed method can be summarized as follows:

- **Step 1.** For any two sensors, three matrices $\boldsymbol{U}$, $\boldsymbol{V}$ and $\boldsymbol{\Sigma}$ are firstly created from a healthy structure.
- **Step 2.** For a testing structure, the measured responses from any two sensors are decomposed by using the matrices $\boldsymbol{U}$ and $\boldsymbol{V}$, respectively.

- **Step 3.** Some effective components with high signal to noise ratio (SNR) are extracted from the decomposed results according to hard-threshold method. Then these effective components are used for estimating the singular values for the testing structure.
- **Step 4.** A damage indicator (DI) is defined by comparing the singular values between the testing and healthy structures.
- **Step 5.** All the DIs are together used for estimating the status of the testing structure.

The basic idea of the proposed strategy is to compare the singular values between the testing and healthy structures without considering the external moving forces. However, it should be noted that the estimated singular values for a testing structure are not necessary to be the true ones. This is mainly because that only the unitary matrices $\boldsymbol{U}$ and $\boldsymbol{V}$ calculated from the health structure, but not ones obtained from the testing structure, are used for estimating singular values. More details are discussed in the following section.

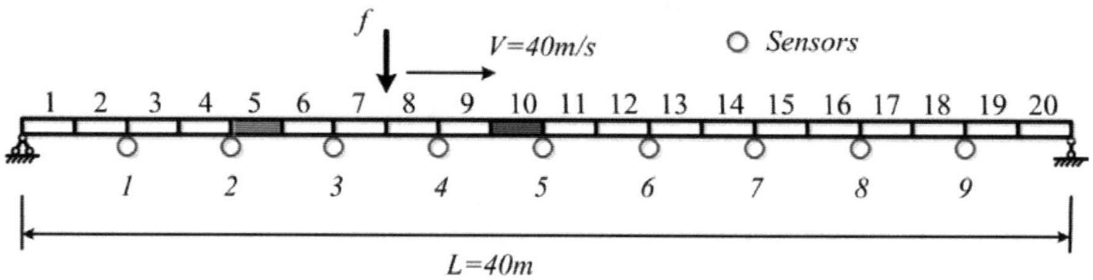

Fig. 2 Finite element model for numerical simulations

4. Numerical simulations for damage detection

To illustrate the proposed method more clearly, several numerical simulations on a simply-supported bridge subjected to a moving force, as shown in Fig. 2, are carried out. Herein, the bridge with a span 40m is equally modeled by 20 elements. The bridge materials are selected as follows: flexural rigidity EI=1.274916×10^{11}N•m^2, density of unit length ρA=12000kg•m^{-1}, the first three damping ratios are $\xi_1=\xi_2=0.02$, $\xi_3=0.04$. The moving force with a speed 40m•s^{-1} is simulated as (Law et al., 1997):

$$f(t)=40\left[1+0.1\sin(10\pi t)+0.05\sin(40\pi t)\right]\text{ kN} \tag{15}$$

The total time is T= 1.1s, and the sampling frequency is f_s = 200Hz. The first three models are used for simulating the structural responses. Nine sensors for bending moment response at L/10, 2L/10, 3L/10, 4L/10, 5L/10, 6L/10, 7L/10, 8L/10 and 9L/10 are simulated. White noises are added to simulate the noise polluted responses as:

$$\boldsymbol{b}_n=\boldsymbol{b}+lev\times\frac{1}{N}\sum_{i=1}^{N}|\boldsymbol{b}_i|\times \boldsymbol{rand} \tag{16}$$

where $\boldsymbol{b}_n$ and $\boldsymbol{b}$ are the responses for noise and noiseless, respectively. N is the element number in vector $\boldsymbol{b}$. lev is the noise level. $\boldsymbol{rand}$ is a standard normal distribution vector.

Only the reduction of element stiffness is considered for simulating the structural damage. Hence, the element stiffness matrix of the damaged structure can be expressed as:

$$\boldsymbol{K}_d=\boldsymbol{K}_h-\sum_{i=1}^{N}\alpha_i\boldsymbol{K}_i \tag{17}$$

where $\boldsymbol{K}_d$ and $\boldsymbol{K}_h$ are the stiffness matrix of the damaged and healthy structures, respectively. α_i is a damaged coefficient for the i-th element with an element stiffness matrix $\boldsymbol{K}_i$. In the following sub-sections, the threshold value δ_j for the j-th sensor is selected as $\delta_j=0.1\times\max\{|\boldsymbol{\beta}_j(i)|\}$. It is beneficial to note again that the threshold value used in the proposed method is to extract the signal components with high SNR.

4.1 Cases study

The noise level is simulated as 1% in this section. Different damage extents, i.e. 0%, 5%, 10% and 15%, at element 5 are studied. All DIs calculated from each sensor to the 2nd sensor are plotted in Fig. 3. From Fig. 3, it can be seen that the DI value increases with an increasing damage extent among the considering damage cases, which means that these DI values are sensitive to the damage occurred at element 5. The above phenomenon shows that the proposed DI has a potential ability for detecting the structure damage. Furthermore, it can be seen that the estimated DI values from the 2nd sensor to the 2nd sensor are much smaller than other DIs. This is mainly because that the transmissibility matrix from i-th sensor to i-th sensor is approximately equal to a unit matrix, and the matrix $\boldsymbol{U}$ is approximately equal to matrix $\boldsymbol{V}$.

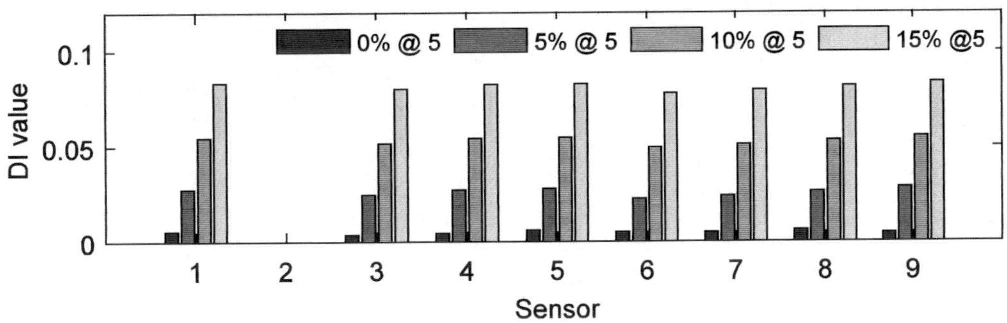

Fig. 3 DIs from each sensor to 2nd sensor (damage element: @5, noise level: 1%).

Two single damage cases, i.e. 5% damage at element 5 and 5% damage at element 10, are calculated, and the DI values are plotted in Fig. 4. From Fig. 4(a), it can be seen that all DIs associated to the 2nd sensor, except DI_{22}, are much bigger than other DI values. Herein, DI_{ij} means the DI value is calculated from the j-th sensor to the i-th sensor. This is mainly because the 2nd sensor is close to the damage element, therefore, its measured responses are more sensitive to the structural damage. This result can also be seen from Fig. 4(b). Hence, the distribution of the DIs shows that the proposed method can be used for detecting the damage location in these two cases. Furthermore, a case that both elements 5 and 10 have 5% damage extent is carried out. The corresponding results are plotted in Fig. 5. From Fig. 5, it can be seen that the proposed method can identify the damage location in this two-damage pattern.

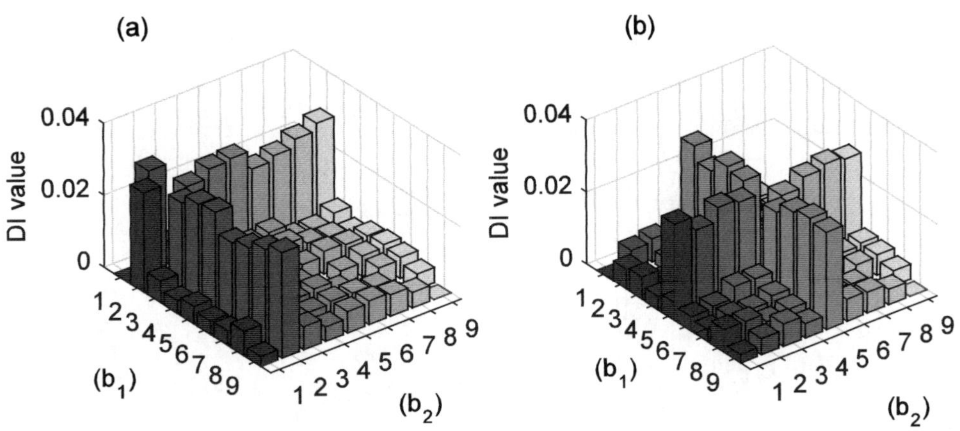

Fig. 4 DIs for two single damage cases with 1% noise level. (a) Damage 5%@5. (b) Damage 5%@10.

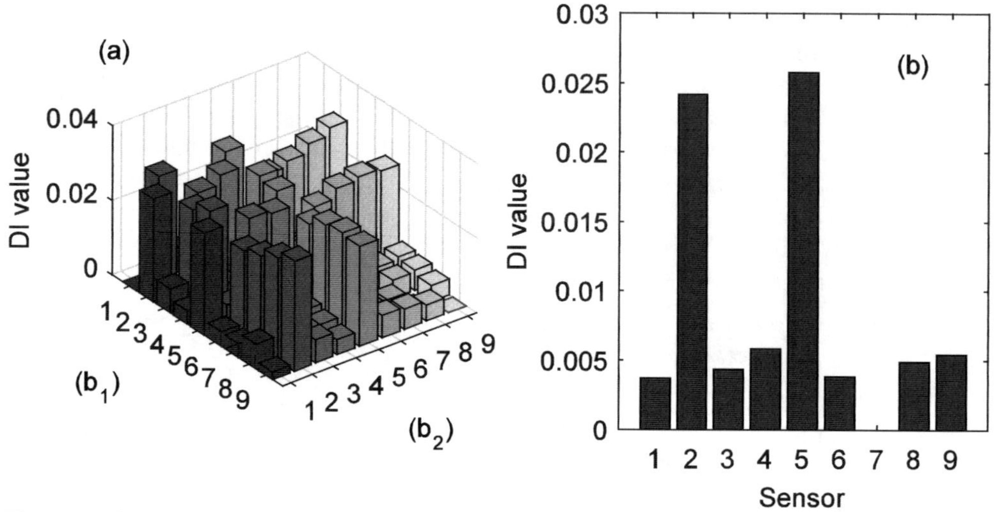

Fig. 5 DIs for two-damage case as 5%@5 and 5%@10. (a) All DIs. (b) DIs from each sensor to 7th sensor.

4.2 Effect of different noise levels

Two damage extents, i.e. 0% and 5%, at element 5 are discussed in this section. Each damage pattern is calculated with four noise levels as 0%, 1%, 3% and 5%. Fig. 6 shows the DIs from each sensor to the 7th sensor for all the cases. From Fig. 6(a), it can be seen that the DI value increases with an increasing noise level for the healthy structure. This is mainly because that the measured noise can influence the accuracies of the estimated singular values, as shown in Eq. (11). This actually brings trouble to the application of the proposed method because a baseline DI value is hard to be selected. From Fig. 6(b), it can be seen that the DI_{72} is much bigger than other DIs when the noise level is less than or equal to 1%. In other word, the proposed method can detect the damage location only when the noise level is less than or equal to 1%. The above phenomenon shows that the proposed method has a weak robustness to the noise, therefore, some effective de-noise methods should be firstly applied for processing of the original signal.

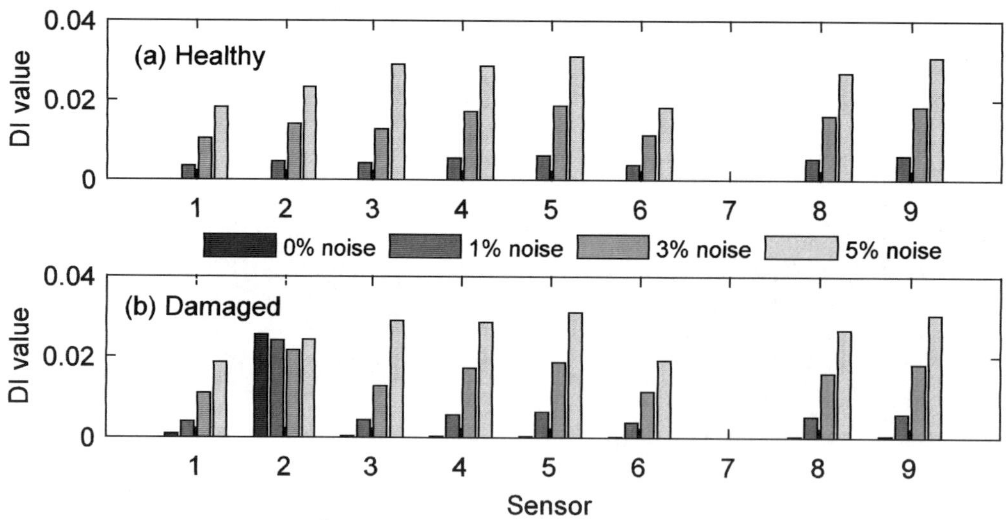

Fig. 6 DIs from each sensor to 7th sensor with different noises. (a) Healthy structure. (b) Damaged structure.

The case with damage extent 5% at element 5 and 5% noise level is carried out for comparison. Herein, all the measured responses are firstly filtered by a traditional moving average filter with a window length 15 sample points. Fig. 7 shows the DIs from each sensor to the 7th sensor for cases with and without considering de-noising method. From Fig. 7(a), it can be seen that the value DI_{27} is not obviously greater than other DI values. However, the value DI_{72} is bigger than other ones if the moving average filter is applied, as shown in Fig. 7(b). This phenomenon shows that the proposed method can used for estimating the damage location in this case with considering the de-noising method.

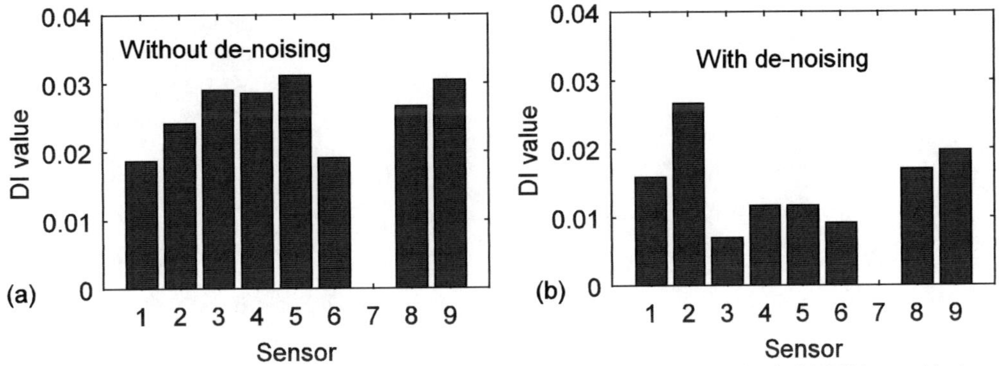

Fig. 7 DIs from each sensor to 7th sensor. (a) Without considering de-noising method. (b) With considering de-noising method.

4.3 Effect of different sampling frequencies

One damage extent as 5% at element 5 is discussed. Four sampling frequencies such as 200Hz, 500Hz, 800Hz and 1000Hz are simulated for comparison. All the cases are performed by using the same hardware and software environments. Fig. 8 shows the DIs from each sensor to the 7th sensor. From Fig. 8, it can be seen that the DI values are not sensitive to the sampling frequency. This is mainly because that the bending moment responses sampled at 200Hz are enough to contain the main information. Moreover, comparison on the CPU time ratios for different cases is shown in Fig. 9. Herein, the CPU time of each case divided by CPU time of 200Hz is used for calculating the CPU time ratio. From Fig. 9, it can be seen that the CPU time ratio increases with an increasing sampling frequency. This is mainly because that the dimension of the transmissibility matrix increases with an increasing sampling frequency; therefore, more CPU time is needed for solving the SVD program for a higher dimension matrix. This phenomenon actually brings trouble to the practical application because the data samples in practice are usually much higher than the ones used in this case.

5. Conclusions

As an attempt work, this paper aims to propose an output-only based method for detecting the structural damage. Herein, singular value decomposition (SVD) method is firstly performed for the transfer matrix between any two sensors. The measured responses are decomposed by using the singular vectors and hard-threshold method, and the decomposed results are used for estimating the singular values. A damage indicator (DI) is defined by comparing the true and estimated singular values. To assess the feasibility of the proposed method more clearly, a simply-supported bridge subjected to one moving force is taken for numerical simulations. The illustrated results show that:

1) The proposed method has a potential ability for detecting the structural damage, because the DI value increases with an increasing damage extent when the noise level is low.
2) The proposed method can be used for detecting the damage location in the simulated cases when the noise level is low.
3) The changes of the singular values can be used for estimating the structural damage location even thought the proposed method has some drawbacks such as weak robustness to noise and time-consuming. Further studies are needed to improve the performance of the proposed method.

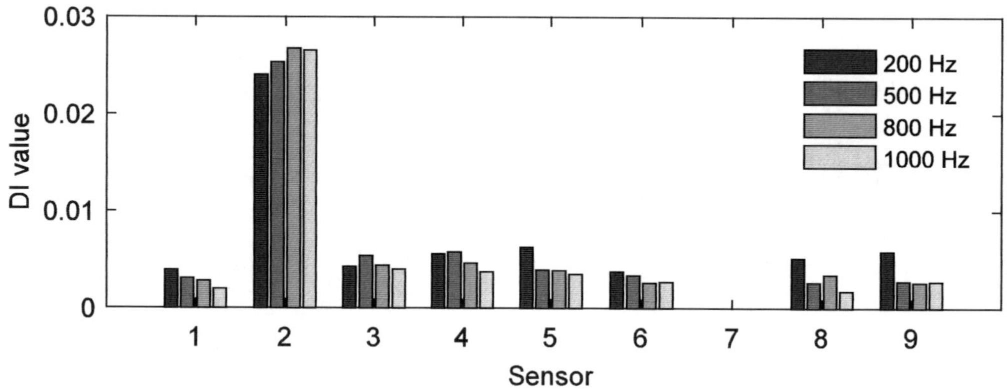

Fig. 8 DIs from each sensor to the 7th sensor for different sampling frequencies.

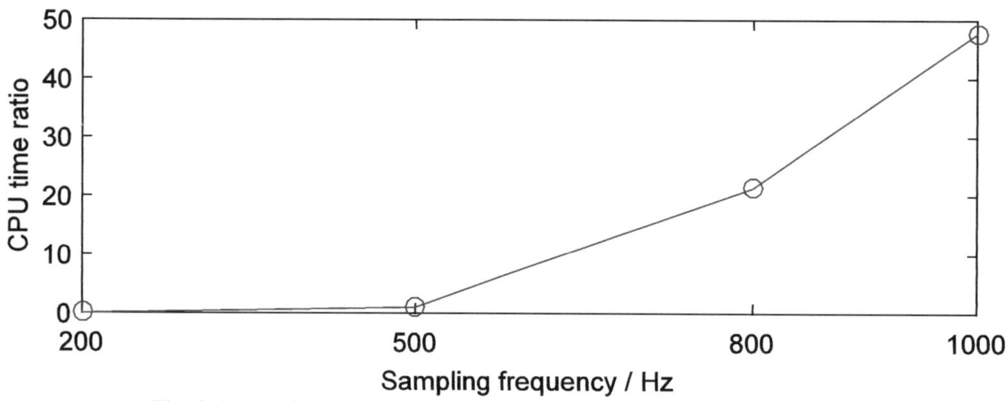

Fig. 9 Comparison on CPU time ratios for different sampling frequencies.

Acknowledgments

This work was jointly supported by the National Natural Science Foundation of China with Grant Numbers 51678278 and 51278226.

References

Farrar, CR, Doebling, SW & Nix, DA 2001, 'Vibration–based structural damage identification', *Philosophical Transactions of the Royal Society of London A: Mathematical, Physical and Engineering Sciences*, vol. 359, no. 1778, pp. 131-149.

Fan, W & Qiao, P 2011, 'Vibration-based damage identification methods: a review and comparative study', *Structural Health Monitoring*, vol. 10, no. 1, pp. 83-111.

Zhu, XQ & Law, SS 2015, 'Structural health monitoring based on vehicle-bridge interaction: accomplishments and challenges', *Advances in Structural Engineering*, vol. 18, no. 12, pp. 1999-2015.

Feng, D & Feng, MQ 2016, 'Output-only damage detection using vehicle-induced displacement response and mode shape curvature index', *Structural Control and Health Monitoring*, vol. 23, no. 8, pp. 1088-1107.

Zhang, QX, Jankowski, L & Duan, ZD 2010, 'Identification of coexistent load and damage', *Structural and Multidisciplinary Optimization*, vol. 41, no. 2, pp. 243-253.

Lu, ZR & Law, SS 2007, 'Identification of system parameters and input force from output only', *Mechanical Systems and Signal Processing*, vol. 21, no. 5, pp. 2099-2111.

Zhang, K, Li, H, Duan, ZD & Law, SS 2011, 'A probabilistic damage identification approach for structures with uncertainties under unknown input', *Mechanical Systems and Signal Processing*, vol. 25, no. 4, pp. 1126-1145.

Zhu, HP, Mao, L & Weng, S 2014, 'A sensitivity-based structural damage identification method with unknown input excitation using transmissibility concept', *Journal of Sound and Vibration*, vol. 333, no. 26, pp. 7135-7150.

Sun, H, Feng, D, Liu, Y & Feng, MQ 2015, 'Statistical regularization for identification of structural parameters and external loadings using state space models', *Computer-Aided Civil and Infrastructure Engineering*, vol. 30, no. 11, pp. 843-858.

Pan, CD & Yu, L 2014, 'Simultaneous identification of structural damages and moving forces from output-only measurements', *Proceedings of the 13th International Symposium on Structural Engineering (ISSE-13)*, Hefei, Anhui, China, pp. 1732-1741.

Law, SS, Chan, THT & Zeng, QH 1997, 'Moving force identification: a time domain method', *Journal of Sound and Vibration*, vol. 201, no. 1, pp. 1-22.

The 8th International Conference on
Structural Health Monitoring of Intelligent Infrastructure
Brisbane, Australia | 5-8 December 2017

Application of Sparsity in Structural Damage Detection

Z.P. Chen [1] and L. Yu [1, 2].

[1] *School of Mechanics and Construction Engineering, Jinan University, Guangzhou 510632, China,* [2] *MOE Key Lab of Disaster Forecast and Control in Engineering, Jinan University, Guangzhou 510632, China, Email:* lyu1997@163.com

Abstract

Structural damage detection (SDD) is one of the major subjects in the field of structural health monitoring (SHM) and various methodologies have been proposed and achieved great accuracy in such field. Inspired by the concept of sparsity, applications of sparsity in SDD are explored in this paper. First of all, an adaptive dictionary is proposed based on intrinsic mode function (IMF), which is utilized to reconstruct compressed acceleration excited by random load. The compressed accelerations are acquired under the application of compressive sensing (CS) technique. Then, as structural damages generally appear locally, SDD problem is characterized by sparsity. Sparse regularization technique is employed to solve SDD by defining two objective functions, to which different sparsity penalty items are introduced. Two iteration procedures inspired by Newton-Gauss algorithm are also presented to solve the problem. Simply-supported beam with different damage degrees are taken as numerical simulations example. The illustrated results show that the proposed method can give an accurate SDD and is robustness to noise. Some related issues are discussed as well.

1. Introduction

Structural damage detection (SDD) is a major research in the field of structural health monitoring (SHM). SHM is a monitoring system which is established to collect different types of responses of structures and expected to provide early warming before serious damage happens (Yan et al., 2007). The target of SDD is to define indices for damage description and detect their variance. Response measurement and signal processing are two major procedures in SDD.

Responses such as acceleration, strain, frequency responses function (FRF) are commonly employed in SDD. For a real world SHM system, the amount of measured responses from sensors are extremely large, requiring a great quantity of energy for transmission and storage for saving. Compressive Sensing (CS), a very recent developed technique, has been popularly studied as its ability to acquire signal at a quite lower sampling frequency than that of Nyquist's theorem (Donoho, 2006). CS is an important application of sparsity and it requires that the signal has a sparse representation in some dictionaries, which are also named bases. It has been popularly applied to SHM. Cortial (Cortial et al., 2007) firstly exploited CS for SHM. He developed a sensor network with a high spatial resolution for the SHM of F-16 via CS techniques, and found the method effective to locate and identify damage to the F-16. Bao *et al* (Bao et al., 2010) developed CS for acceleration signal sampling in SHM, both wavelet and Fourier orthogonal bases are examined. Experiments on Shandong Binzhou Yellow River Highway Bridge showed conclusion that the

values of compression ratios achieved via CS are not high because the bases used are insufficient to sparsely represent signals. But in the works of Mascarenas *et al* (Mascarenas et al., 2013), they found higher compression ratios can be achieved under sinusoidal excitation and Fourier basis is sufficient for satisfying accuracy. However, continuing works about CS application on SHM can be found in References (Yang and Nagarajaiah, 2013, Jayawardhana et al., 2017, Ganesan et al., 2017, Huang et al., 2016). As described above, it is found only impulse and sinusoidal excitation are investigated with application of CS theory. Therefore, in order to extend CS application in SHM, the reconstruction of compressive sampling acceleration excited by random load is studied in this paper.
In the other hand, as structural damages generally appear locally in structures, SDD problem is characterized by sparsity. Most of methodologies dealing with SDD treat it as an optimization problem by minimizing the discrepancy between measured and calculated responses (Pan et al., 2016). Typical algorithms to solve SDD optimization problem include sensitivity-based finite element model updating approach and intelligent algorithm (Lu et al., 2017, Yu and Li, 2014). Both methods have to define an objective function firstly; then the sensitivity-based approach updates the solution and the fitness values of objective function based on sensitivity matrix while the intelligent algorithm updates them based on heuristic iteration formulas. In order to improve the accuracy of identified results when responses are subject to noise, regularization methods such as Tikhonov and consistent regularization (Weber et al., 2009) have been developed to gain an accurate solution. The damage factor of SDD is sparse; therefore, it is essential to determine a sparse solution for SDD. The work of Bao *et al* (Zhou et al., 2013) confirmed that substructure-based sensitivity analysis combined with sparse constraints regularization helps to improve accuracy of SDD. In this paper, sparsity is introduced to objective functions by defining l_1 norm penalty items. Two iteration procedures inspired by Newton-Gauss iteration procedures are also proposed.
In summary, two major works are presented in this paper. The first one proposes IMF dictionary to ensure acceleration responses sparse under random load. The second work presents two iteration procedures based on Newton-Gauss algorithm to solve SDD with sparsity penalty items. Numerical simulations of a simply-supported beam are studied to access the feasibility of the proposed methods.

2. Theoretical background

2.1 Compressive sensing

CS is a technique that deals with signal reconstruction from under-sampled signals. The sparsity of signal is an essential prerequisite which ensures the under-sampled signal to be reconstructed accurately. If a signal $x(t)$ is sparse in time domain with K-sparsity, meaning the number of non-zero element of $x(t)$ is K, a measurement matrix Φ, $M{\times}N$ matrix, is introduced to generate compressive sensing signal y which is formulated as,

$$y = \mathbf{\Phi} x \tag{1}$$

where M is far less than N. The measurement matrix $\mathbf{\Phi}$ should satisfied the restricted isometry property (RIP) (Jayawardhana et al., 2017). $\mathbf{\Phi}$ is commonly defined by sampling independent and identically distributed entries ϕ_{ij} from normal distribution with zero mean and $1/M$ variance.
If $x(t)$ is not sparse in time domain, a dictionary $\mathbf{\Psi}$ that translates x into a sparse representation should be prior determined. And then the compressive sensing signal y can be written as,

$$y = \mathbf{\Phi\Psi}\alpha \quad (2)$$

In the real civil engineering, compressive signals are acquired based on Eq.(2), and the following convex optimization is solved to reconstruct the signals,

$$\alpha_{\text{opt}} = \arg\min \|\alpha\|_1 \;\; s.t.\; \mathbf{\Phi\Psi}\alpha = y \quad (3)$$

where $\|\alpha\|_1$ is the l_1 norm of α. Translating the above optimization problem into unconstrained optimization problem as below,

$$\alpha_{\text{opt}} = \min_{\alpha} \left\{ \|\mathbf{\Phi\Psi}\alpha - y\|_2 + \lambda \|\alpha\|_1 \right\} \quad (4)$$

λ is a regularization parameter. Eq.(4) is a typical *Least absolute shrinkage and selection operator* (LASSO) problem and MATLAB function *lasso* is employed to solve it. The regularization parameter $\lambda$is determined via Bayesian information criterion (BIC), which is formulated as,

$$\text{BIC} = \ln(n)k + n\ln\left(\frac{RSS}{n}\right) \quad (5)$$

where n is the length of y, k is the sparsity of α. RSS is the abbreviation of residual sum of squares, and it is calculated as $RRS = \|\mathbf{\Phi\Psi}\alpha\text{-}y\|_2^2$.

2.2 Sparse representation of responses

In order to make CS technique available in SDD, a dictionary that promises a sparse representation of original responses is required. Based the knowledge of vibration theory, vibration responses of structures under external forces without initial condition includes two major parts. One reflects the tendency of input excitation, named forced vibration, and another reveals modal features of structures, named transient vibration. Under sinusoidal excitation, the transient vibration declines exponentially because of structural damp and the remaining vibration is sinusoidal vibration with same frequency of the excitation, named steady-state vibration; therefore, Fourier basis is sufficient to translate responses to sparse form (Mascarenas et al., 2013). As for impulse responses, only transient vibration is active and it declines exponentially, leading to only coefficients of small scale wavelet components closed to the location of impulse will not be zero; therefore, impulse responses can be translated into a sparsity representation under wavelet basis (Jayawardhana et al., 2017, Ganesan et al., 2017).

When random excitation is applied, it can be seen as a series impulses applied to the structural nodes. The acceleration responses is non-periodic and do not decline over time. In this case, both Fourier and wavelet basis are not suitable for sparsely describing the original responses. Despite the traditional Fourier basis signal decomposition and wavelet basis signal decomposition, empirical mode decomposition (EMD) is also widely used in signal decomposition. EMD method was originally presented to overcome the problem of predetermined bases which not always guarantee a proper sparse decomposition. EMD is employed to find a basis adaptively from the measured signal in an empirical way (Hou and Shi, 2013). The EMD results in decomposition of signal $x(t)$ into n IMFs and a residue $r_n(t)$, which can be formulated as follow,

$$x(t)=\sum_{i=1}^{n}c_i(t)+r_n(t) \tag{6}$$

where $c_i(t)$ is the ith component of IMFs. The IMFs are defined as follow (Huang et al., 1998):

- The number of extreme and the number of zero crossings are equal or differ at most by one on the whole data set;
- The mean value of the envelopes defined by local maxima and local minima is zero at any point.

Inspired by EMD, an adaptive dictionary containing most of possible IMFs is proposed. Assuming that the original measured acceleration responses are acquired from sensors instrumented in structure, a segment of acceleration responses with N continuous sample points are cut from them, written as $\ddot{v}(t)\in R^{N\times m}$ where m represents the number of sensors. $\ddot{v}_i(t)$ means acceleration collected from the ith sensor. Applying EMD method, $\ddot{v}_i(t)$ is decomposed into several IMFs. Because the excitation is random load, vibration performance shows difference in different time intervals. Statistic characteristics of responses are assumed stable during a quite long time monitoring. By this way, a rectangular window with length l is applied on the responses and shifts with a fixed step for p times. Therefore, a testing set with p elements is formulated, which can be described as $\ddot{v}_{iwin}=\left[\ddot{v}_i^1 \quad \ddot{v}_i^2 \quad \ldots \quad \ddot{v}_i^p\right], \ddot{v}_i^j\in R^{l\times 1}, j=1,\ldots,p$. Decomposing each component of $\ddot{v}_{iwin}$ utilizing EMD, a set of IMFs is gained. Arranging all these IMFs in a matrix, then a dictionary is formed and can be expressed as follows,

$$\mathbf{\Psi}=\left[IMF_i^1(1) \quad \cdots \quad IMF_i^1(n_1) \quad \cdots \quad IMF_i^p(1) \quad \cdots \quad IMF_i^p(n_p)\right] \tag{7}$$

where $IMF_i{}^j(n_k)$ means the n_k IMF of the jth section of $\ddot{v}_i(t)$. $\mathbf{\Psi}\in R^{l\times\sum_{j=1}^{p}n_j}$ means that the number of components of dictionary $\mathbf{\Psi}$ is $\sum_{j=1}^{p}n_j$. After $\mathbf{\Psi}$ is defined adaptively based on the measured responses, components of $\ddot{v}_{iwin}$ all have sparsely representation according to the dictionary. It should also be noticed that if the $\sum_{j=1}^{p}n_j$ is big enough, every section of $\ddot{v}_i(t)$ with length l can be sparsely represented under the dictionary. Inspired by mode superposition theory, the responses of structures is a linear combination of the first several mode responses, it is assumed that the acceleration responses from different sensors can be described in a same dictionary $\mathbf{\Psi}$ and they can be express as follows,

$$\ddot{v}_i(t)=a_{i1}\Psi_1+\cdots+a_{i1}\Psi_n,\quad \ddot{v}_j(t)=a_{j1}\Psi_1+\cdots+a_{j1}\Psi_n \tag{8}$$

where Ψ_i is the ith component of $\mathbf{\Psi}$. The reliability of the assumption will be confirmed in the following numerical simulations.

2.3 Sparsity in SDD

The structural damage can be defined as reduction of structural stiffness, $\Delta\mathbf{K}=\sum_{i=1}^{Nele}\alpha_i\mathbf{K}$, where α_i, ranging from 0 to 1,is damage factor of the ith element. As damages always occur in a few elements, SDD is characterized by sparsity. In order to apply sensitivity-based method, the objective function is required firstly. In this paper, two objective functions are defined as follows,

$$J_1(\boldsymbol{\alpha})=\arg\min_{\boldsymbol{\alpha}}\left\|\Delta\mathbf{r}(\boldsymbol{\alpha}^0)+\mathbf{S}(\boldsymbol{\alpha}-\boldsymbol{\alpha}^0)\right\|_2^2+\lambda\|\boldsymbol{\alpha}\|_1 \tag{9}$$

$$J_2(\boldsymbol{\alpha}) = \arg\min_{\boldsymbol{\alpha}} \left\| \Delta\mathbf{r}(\boldsymbol{\alpha}^0) + \mathbf{S}(\boldsymbol{\alpha} - \boldsymbol{\alpha}^0) \right\|_2^2 + \lambda \left\| \boldsymbol{\alpha} - \boldsymbol{\alpha}^0 \right\|_1 \tag{10}$$

where $\boldsymbol{\alpha}^0$ is a selected initial solution, $\boldsymbol{r}(\boldsymbol{\alpha})$ is a vector containing modal frequencies and mode shapes when damage factor equals to $\boldsymbol{\alpha}$, $\Delta\boldsymbol{r}(\boldsymbol{\alpha})$ means that the calculated responses with damage factor $\boldsymbol{\alpha}$ minus the measured ones , $\|\boldsymbol{\alpha}\|_1$ is the l_1 norm of $\boldsymbol{\alpha}$. $\mathbf{S}$ is the sensitivity matrix when damage factor is $\boldsymbol{\alpha}^0$, and λ is a regularization parameter. The difference between the two objective functions is the definition of sparse item. The objective functions $J_1(\boldsymbol{\alpha})$ considers $\boldsymbol{\alpha}$ to be sparse while $J_2(\boldsymbol{\alpha})$ considers the change of $\boldsymbol{\alpha}$, written as $\Delta\boldsymbol{\alpha}$, to be sparse. Iteration process inspired by Newton-Gauss algorithm is proposed. And the pseudo-codes of the iteration of the two objective functions are shown as follows,

Pseudo-codes of $J_1(\alpha)$:
Step 1: giving an initial solution $\boldsymbol{\alpha}^0$

Step 2: solving $\boldsymbol{\alpha}^{k+1} = \arg\min_{\boldsymbol{\alpha}} \left\| \mathbf{S}\boldsymbol{\alpha} - \left(\mathbf{S}\boldsymbol{\alpha}^k - \Delta\mathbf{r}(\boldsymbol{\alpha}^k) \right) \right\|_2^2 + \lambda^k \left\| \boldsymbol{\alpha} \right\|_1$ at the kth iteration.

Step 3: presetting an allowable error ε, if $\left\| \boldsymbol{\alpha}^{k+1} - \boldsymbol{\alpha}^k \right\| \le \varepsilon$ is satisfied, stop iteration; otherwise, setting k=k+1 and returning to Step 2.

Pseudo-codes of $J_2(\alpha)$:
Step 1: giving an initial solution $\boldsymbol{\alpha}^0$

Step 2: solving $\Delta\boldsymbol{\alpha}^k = \arg\min_{\boldsymbol{\alpha}} \left\| \mathbf{S}(\boldsymbol{\alpha} - \boldsymbol{\alpha}^k) - \left(-\Delta\mathbf{r}(\boldsymbol{\alpha}^k) \right) \right\|_2^2 + \lambda^k \left\| \Delta\boldsymbol{\alpha} \right\|_1$ at the kth iteration.

Step 3: presetting an allowable error ε, $\boldsymbol{\alpha}^{k+1} = \boldsymbol{\alpha}^k + \Delta\boldsymbol{\alpha}^k$, if $\left\| \boldsymbol{\alpha}^{k+1} - \boldsymbol{\alpha}^k \right\| \le \varepsilon$ is satisfied, stop iteration; otherwise, setting k=k+1 and returning to Step 2.

It can be seen that the updating of damage factor vector is different. The MATLAB function *lasso* is used to solve the above optimization problem at each iteration. The regularization parameter λ is determined via Bayesian information criterion (BIC) in Eq.(5).

3. Numerical simulations

To check on the correctness and effectiveness of the proposed method, a simply-supported beam is modeled and the diagram is shown in Fig. 1. The parameter of the beam are as follow: elastic modulus E=210GPa, structural density ρ=7800kg/m^3, length of the beam l=20m, damping ratio ξ=0.02, area of the beam cross section A=0.3×0.6m^2, moment of inertia of the cross beam section I= 0.3×0.6^3/12 m^4. The beam is divided into 20 elements with equal length, and each element has four degrees of freedoms (DOFs). The consistent random excitations are applied at y-direction of nodes 2 to 21. Newmark method is adapted to calculate the acceleration responses of the structure. The sampling frequency is 1024Hz and sample time is 6s.

White noise is added to the calculated responses of the beam to simulate the polluted measurements, and noise level is set to 5% which is formulated as $\ddot{v} = \ddot{v}_{cal}\left(1 + E_p\mathbf{N}_{oise}\right)$, where $\ddot{v}$ and $\ddot{v}_{cal}$ are the acceleration with and without noise respectively, E_p is the noise level, and $\mathbf{N}_{oise}$ means standard normal distribution vector with zero mean value and unit standard deviation.

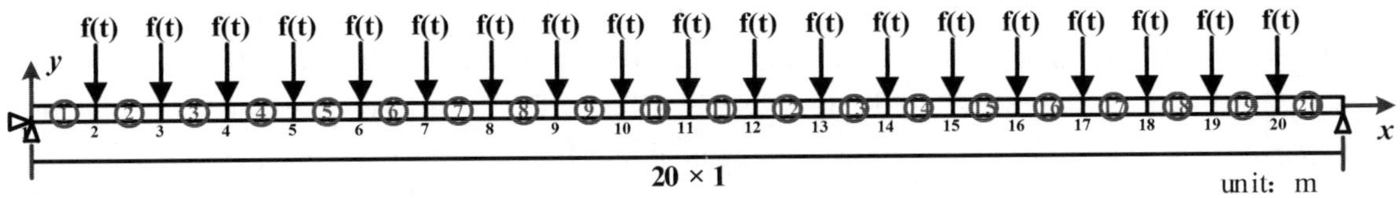

Fig. 1 Diagram of simply-supported beam

3.1 Acceleration reconstruction based on proposed dictionary

Acceleration responses at node 7 is chosen to generate the dictionary as it is not a vibration node of any first five mode shapes. The first four-minute data is divided into 200 segments with equal length, 2048 sample points. Therefore, each component of the dictionary is a $R^{2048\times1}$ vector. The calculated results reveal that $\Psi\in R^{2048\times1629}$. The late two-minute data is used to verify whether sparse representation is existed under the generated dictionary. Coherences in both time domain and frequency domain are used to evaluate the reconstruction accuracy. And they are defined as follows,

$$Coh_{time}=\frac{x_o(t)^T x_{re}(t)}{\sqrt{x_o(t)^T x_o(t) x_{re}(t) x_{re}(t)^T}},\ Coh_{freq}=\frac{X_o(\omega)^T X_{re}(\omega)}{\sqrt{X_o(\omega)^T X_o(\omega) X_{re}(\omega) X_{re}(\omega)^T}}$$

where, $x_o(t)$ and $x_{re}(t)$ are the original and reconstructed responses in time domain respectively, $X_o(\omega)$ and $X_{re}(\omega)$ are the original and reconstructed amplitude of responses in frequency domain respectively.

Table 1 Sparsity of acceleration translated into the proposed dictionary

Node	2	3	4	5	6	7	8	9	10	11
Sparsity	430	428	374	313	404	303	306	325	359	374
Node	**12**	**13**	**14**	**15**	**16**	**17**	**18**	**19**	**20**	
Sparsity	351	328	276	311	400	371	406	396	439	

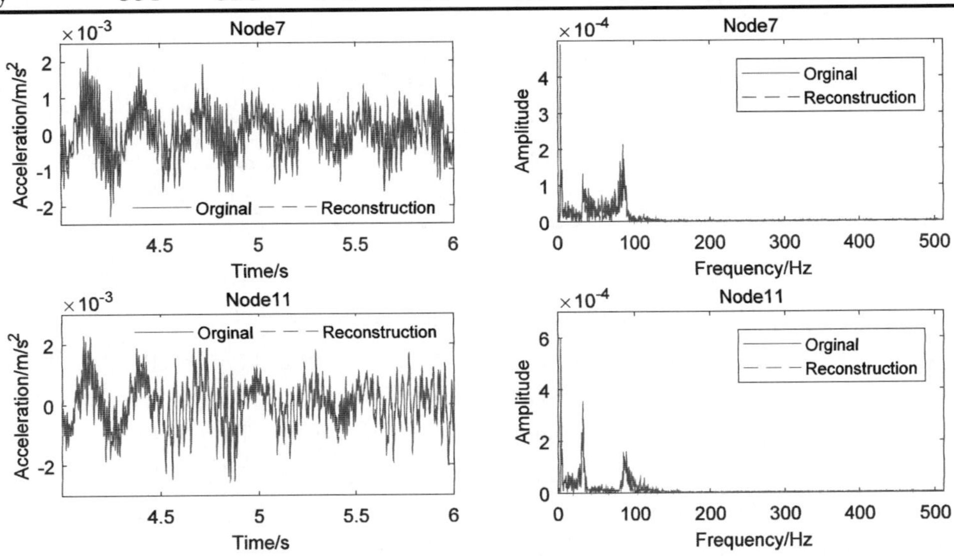

Fig. 2 Comparison on original and reconstructed accelerations at nodes 7 and 11 both in time and frequency domains

Applying l_1 regularization method, the sparsest representation of each node acceleration in the proposed dictionary can be determined and their values of sparsity are listed in Table 1. All values of sparsity are smaller than a quarter of the value of 2048, the original length of each node acceleration; therefore, the proposed dictionary is verified efficient to translated node acceleration into a sparser representation. Fig. 2 shows the original and the reconstructed accelerations from the proposed dictionary at nodes 7 and 11. Both in time domain and in frequency domain, the reconstructed acceleration is coincident with the original one. The values of coherence at different nodes in Fig. 3 show that the coherence values of all nodes are greater than 0.94, which illustrates that the reconstructed acceleration is accurate in the proposed dictionary. Based on the above discussion, conclusions can be made that the proposed dictionary is helpful to translate the node acceleration into a sparser one and to reconstruct the original acceleration accurately.

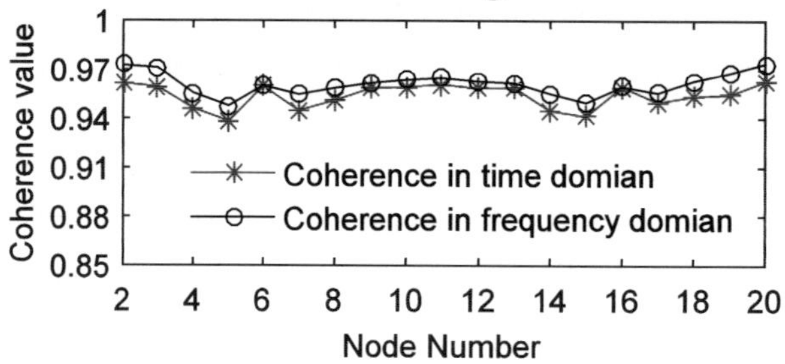

Fig. 3 Comparison on coherence values between original and reconstructed accelerations at different nodes

Most of the time, the element of measured matrix $\mathbf{\Phi}$ is selected to be sampled independently and identically distributed from normal distribution with zero mean and $1/M$ variance. Whether the compressive sensing signal can be reconstructed accurately is greatly depended on the value of M; therefore, a suitable value of M for accurate reconstruction is investigated. Five possible values of M are presented ranging from 512 to 1024.

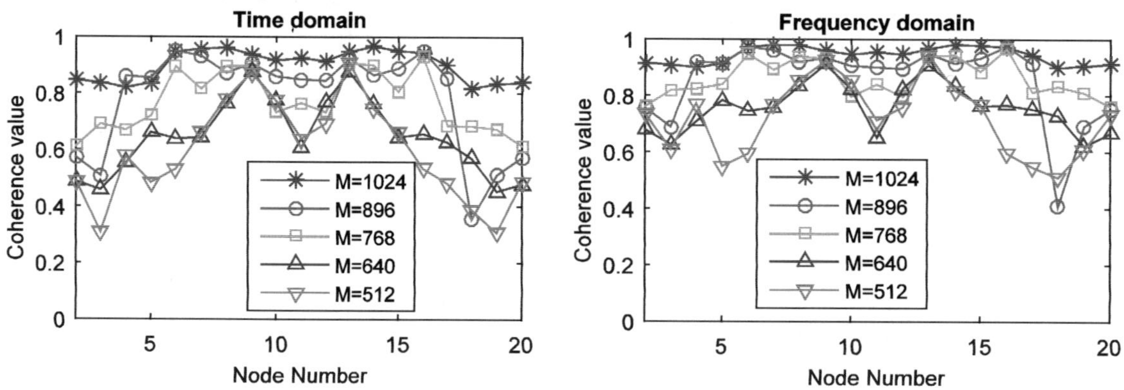

Fig. 4 Coherence value for different node under different M value

In order to evaluate the effect of compression, an index named compression ratio (CR) is defined as dividing l, the length of original signal, by M. By this definition, CRs for preset values of M are 4, 3.20, 2.67, 2.29 and 2 respectively. Fig. 4 shows coherence values of reconstructed accelerations from compressive sampling and original accelerations respectively, both in time domain and frequency domain. It can be seen that the accuracy of reconstruction declines as the value of M becomes smaller. When the M is equal to 512, only few node accelerations can be reconstructed with a satisfied accuracy, such as nodes 9 and 13. While M increases to be 1024, almost all node accelerations can be reconstructed accurately.

Therefore, the original acceleration can be well constructed from CS signals. And the proposed dictionary is available in the process of recovering CS signals. However, the *CR* is limited as only *CR*=2 can give an accurate reconstruction. The proposed dictionary still need to be improved.

3.2 Structural damage detection based on sensitivity method

As it is hard to accurately acquire mode shapes from structural responses, numerical mode parameters contaminated by different level of noises are simulated here. A simply-supported beam with the same structural parameters described above is employed for simulation. Four damage cases are simulated and the details are listed in Table 2. White noise is added to the calculated mode parameters, which can be described as $\mathbf{r}=\mathbf{r}_{cal}\left(1+E_p\mathbf{N}_{oise}\right)$, where $\mathbf{r}$ and $\mathbf{r}_{cal}$ are the mode parameters, containing the first five modal frequencies and mode shapes, with and without noise respectively. E_p is the noise level, which is set to 0%, 1% and 3% respectively. $\mathbf{N}_{oise}$ is a standard normal distribution vector with zero mean value and unit standard deviation.

Table 2 Damage cases

	Case 1		Case 2		Case 3		Case 4	
Element No.	10	-	10	-	10	15	10	15
Damage factor	0.2	-	0.4	-	0.2	0.2	0.2	0.4

To make a comparison with traditional l_2 regularization method, the identification results by the Tikhonov regularization method, in which the *l*-curve method is used to determine its regularization parameter, are also shown. The initial solution is set to zero vector, which represents an intact condition of the beam.

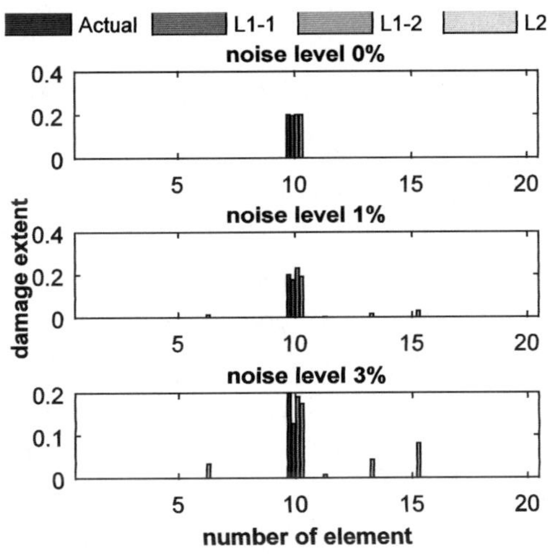

Fig. 5 SDD results in case 1

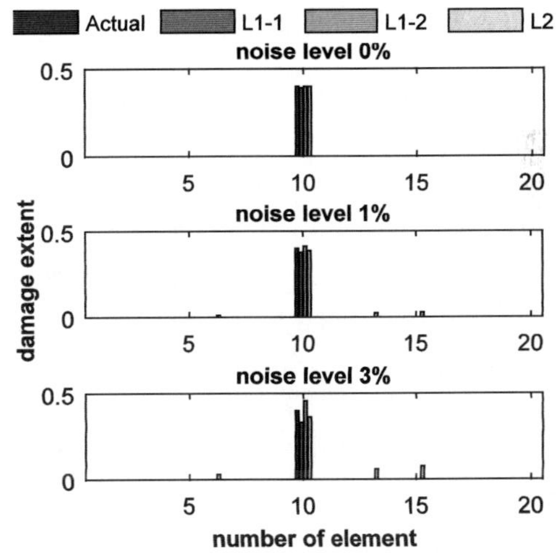

Fig. 6 SDD results in case 2

Fig. 5 to Fig. 8 show the identified results for the whole simulated cases, where 'L1-1' and 'L1-2' represent methods based on $J_1(\boldsymbol{\alpha})$ and $J_2(\boldsymbol{\alpha})$ respectively. If there is no noise, the l_1 regularization method and the l_2 regularization method can both locate the damage with high accuracy. While noise level increases, SDD accuracy declines. All identified results in Figs.5 to 8 also reveal that the l_1 regularization methods are superior to l_2 regularization as fewer misjudgment elements occur by applying the l_1 regularization methods, and the typical case is noise level 3% in case 1. Despite the accuracy of damage extent, the l_1 regularization methods both locate the damage elements, which illustrates that the penalty item of l_1 regularization methods is helpful to gain the sparsest solution for SDD problem. Comparing the two iteration procedures for l_1 regularization, it can be also found that the iteration procedure 2 for l_1 regularization performs better than the iteration procedure 1. Although both iteration procedures can locate damages exactly under the assumed damage cases, the iteration procedure 2 give a higher accurate damage extent than the iteration procedure 1.

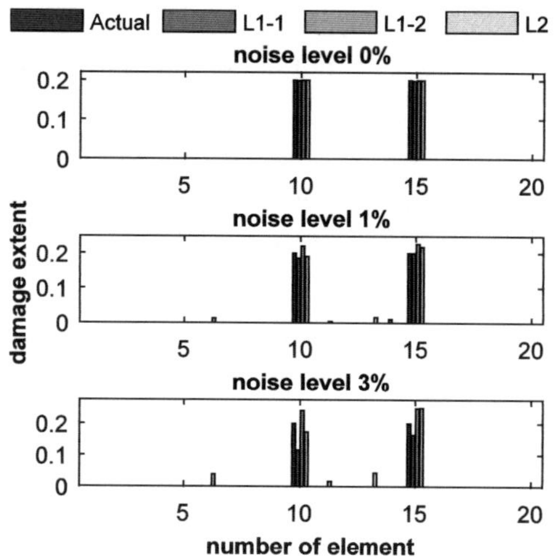

Fig. 7 SDD results in case 3

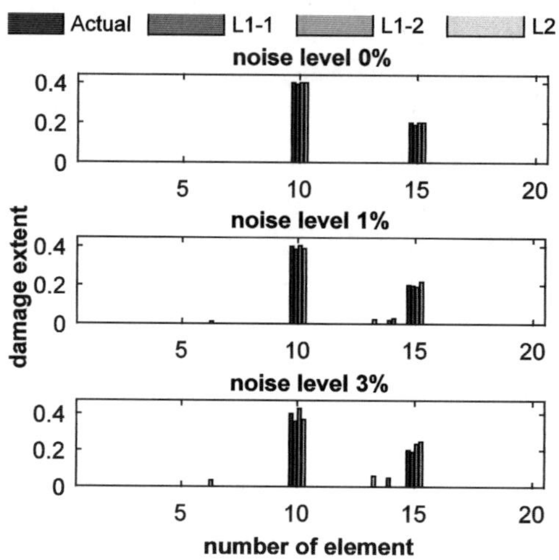

Fig. 8 SDD results in case 4

4. Conclusions

In this paper, structural damage detection (SDD) associated to sparsity is investigated. First of all, the sparse representation of accelerations excited by random load is achieved by proposing a new adaptive dictionary based on empirical mode decomposition (EMD). The compressed accelerations via compressive sensing (CS) technique are well reconstructed using sparse recovery algorithm. Then the SDD is also accomplished accurately applying sparse regulation method by defining two different objective functions. Some conclusions can be summarized as follows,

1) A new adaptive dictionary is proposed to make the acceleration excited by random load sparse, and it is validated accurate to recover the accelerations acquired via compressive sensing. However, the compression ratio is still need to be improved for a better application of the proposed method.
2) Two sparse iteration procedures are presented and compared with traditional Tikhonov method. Numerical simulations show that the sparse iteration outperforms the traditional Tikhonov method. And a more accurate solution can be gained by employing sparse iteration procedure in which less misjudgments occur. Also, the proposed sparse iteration procedure two performs better than the procedure 1.

Acknowledgements

The project is jointly supported by the National Natural Science Foundation of China with grant numbers 51678278 and 51278226 respectively.

References

Yan, YJ, Cheng, L, Wu, ZY & Yam, LH 2007, 'Development in vibration-based structural damage detection technique', *Mechanical Systems and Signal Processing*, vol. 21, no. 5, pp. 2198-2211.

Donoho, DL 2006, 'Compressed sensing', *IEEE Transactions on Information Theory*, vol. 52, no. 4, pp. 1289-1306.

Cortial, J, Farhat, C, Guibas, LJ & Rajashekhar, M 2007, 'Compressed sensing and time-parallel reduced-order modeling for structural health monitoring using a DDDAS', *7th International Conference on Computational Science(ICCS 2007)*, Beijing, China, 27-30 May, pp. 1171-1179.

Bao, YQ, Beck, JL & Li, H 2010, 'Compressive sampling for accelerometer signals in structural health monitoring', *Structural Health Monitoring*, vol. 10, no. 3, pp. 235-246.

Mascarenas, D, Cattaneo, A, Theiler, J & Farrar, C 2013, 'Compressed sensing techniques for detecting damage in structures', *Structural Health Monitoring*, vol. 12, no. 4, pp. 325-338.

Yang, YC & Nagarajaiah, S 2013, 'Blind modal identification of output-only structures in time-domain based on complexity pursuit', *Earthquake Engineering & Structural Dynamics*, vol. 42, no. 13, pp. 1885-1905.

Jayawardhana, M, Zhu, XQ, Liyanapathirana, R & Gunawardana, U 2017, 'Compressive sensing for efficient health monitoring and effective damage detection of structures', *Mechanical Systems and Signal Processing*, vol. 84, pp. 414-430.

Ganesan, V, Das, T, Rahnavard, N & Kauffman, JL 2017, 'Vibration-based monitoring and diagnostics using compressive sensing', *Journal of Sound and Vibration*, vol. 394, no., pp. 612-630.

Huang, Y, Beck, JL, Wu, S & Li, H 2016, 'Bayesian compressive sensing for approximately sparse signals and application to structural health monitoring signals for data loss recovery', *Probabilistic Engineering Mechanics*, vol. 46, pp. 62-79.

Pan, CD, Yu, L, Chen, ZP, Luo, WF & Liu, HL 2016, 'A hybrid self-adaptive Firefly-Nelder-Mead algorithm for structural damage detection', *Smart Structures and Systems*, vol. 17, no. 6, pp. 957-980.

Lu, ZR, Lin, XX, Chen, YM & Huang, M 2017, 'Hybrid sensitivity matrix for damage identification in axially functionally graded beams', *Applied Mathematical Modelling*, vol. 41, pp. 604-617.

Yu, L & Li, C 2014, 'A Global Artificial Fish Swarm Algorithm for Structural Damage Detection', *Advances in Structural Engineering*, vol. 17, no. 3, pp. 331-346.

Weber, B, Paultre, P & Proulx, J 2009, 'Consistent regularization of nonlinear model updating for damage identification', *Mechanical Systems and Signal Processing*, vol. 23, no. 6, pp. 1965-1985.

Zhou, SM, Bao, YQ & Li, H 2013, 'Structural Damage Identification Based on Substructure Sensitivity and l1 Sparse Regularization', *2013 SPIE Conference on Sensors and Smart Structures Technologies for Civil, Mechanical, and Aerospace Systems 2013*, San Diego, California, USA, 10-14 March, pp. 86923N

Hou, TY & Shi, Z 2013, 'Data-driven time–frequency analysis', *Applied and Computational Harmonic Analysis*, vol. 35, no. 2, pp. 284-308.

Huang, NE, Shen, Z, Long, SR, Wu, MC, Shih, HH, Zheng, Q, Yen, NC, Tung, CC & Liu, HH 1998, 'The empirical mode decomposition and the Hilbert spectrum for nonlinear and non-stationary time series analysis', *Proceedings of the Royal Society of London A: mathematical, physical and engineering sciences* vol. 454, no. 1971, pp. 903-995.

The 8th International Conference on
Structural Health Monitoring of Intelligent Infrastructure
Brisbane, Australia | 5-8 December 2017

RIP-based Subspace Projection Method for Structural Damage Detection

Z.W. Luo [1], L. Yu [1, 2].
[1] *School of Mechanics and Construction Engineering, Jinan University, Guangzhou 510632, China*
[2] *MOE Key Lab of Disaster Forecast and Control in Engineering, Jinan University, Guangzhou 510632, China.*
Email: lyu1997@163.com

Abstract

Structural damage detection (SDD) of large span structures needs to deal with a large number of data produced by thousands of sensors. SDD costs vast amounts of calculating time in a structural health monitoring (SHM) system. To overcome this problem, a new method of SDD is proposed in this paper. A measurement matrix satisfying restricted isometry property (RIP) is used to reduce the numbers of sensitivity equation. It projects the sensitivity matrix into the subspace. In the meantime, it also projects the differences of modal parameters between damaged and undamaged structure into this subspace and reduces the dimensions of differences of modal parameters. Then, the information in lower dimensions can be used for SDD. Finally, due to the sparsity of damage coefficients, sparse regularization method can be used to solve the sensitivity equation in subspace and improve the ill-posedness of SDD. In order to assess capabilities of the proposed method, a plane truss is taken as an example for numerical simulations. Different damage scenarios and the effects of measurement noises on the proposed method are considered. The illustrated results show that the proposed method can effectively detect damage location(s) and quantify damage degrees.

1. Introduction

As one of the main research directions in the field of structural health monitoring (SHM), structural damage detection (SDD) has been paid more attention by scholars in recent years and many SDD methods have been proposed (Zhou et al., 2015; Zhou et al., 2016; Mascarenas et al., 2013; Jayawardhana et al., 2017; Pan & Yu. 2016).

On the one hand, as a new sampling technology, compressed sensing (CS) has been proposed (Donoho, 2006), which can collect and compress signals at the same time. Through optimization, the sparsity of a signal can be exploited to recover it from far fewer samples than that required by the Shannon-Nyquist sampling theorem. Therefore, it can save signals in a lower dimensional space (Huang et al., 2014). In the filed of SHM, some scholars have introduced CS into SDD (Mascarenas et al., 2013; Jayawardhana et al., 2017) but they only considered the influence of reconstructed signals for SDD. On the other hand, in order to improve the ill-posedness of SDD, some scholars (Zhou et al., 2015, Zhou et al., 2016) have introduced sparse regularization into it. And the identified results are better than the results without sparse regularization.

In this study, a new SDD method is proposed. Considering first-order sensitivity analysis, a measurement matrix of CS is introduced into it to reduce the dimensions of sensitivity equation. This matrix is required to satisfy a criterion called restricted isometry property (RIP) (Candès, 2008). It projects the sensitivity matrix into the subspace. At the same time, it also projects the differences of modal parameters between damaged and undamaged structure into this subspace and reduces the dimensions of differences of modal parameters. So the information in lower dimensions can be used for SDD. And sparse regularization is used to solve the sensitivity equation in subspace.

To assess the feasibility of the proposed approach, a plane truss is used in numerical simulations. Different damage scenarios and the effects of measurement noises on the proposed method are considered as well.

2. Theoretical background

2.1 Sensitivity analysis

In the service life of structures, cracks inevitably generate and generally occur in partial positions. Based on these features, only changes in stiffness are considered and changes in mass are ignored in this study.

In order to detect damage locations and quantify damage extents of structures, the structure is divided into multiple elements by using finite element method (FEM) so that it can detect the structural damage according to changes in element stiffness. In this study, reduction of element elasticity modulus is used to describe damage extents of elements. Considering the relation between elasticity modulus and stiffness, global stiffness matrix of structures can be expressed as follows:

$$\mathbf{K} = \sum_{i=1}^{N} (1 - \alpha_i) \mathbf{K}_i \left(0 \le \alpha_i \le 1\right) \tag{1}$$

where $\mathbf{K}_i$, α_i represent element stiffness matrix and damage reduction factor of the ith element, respectively. N is the total number of finite elements.

The first-order derivative of eigenvalues and mode shapes with respect to parameter α_i are as follows (Zhou et al., 2016):

$$\frac{\partial \lambda_r}{\partial \alpha_i} = -\boldsymbol{\varphi}_r^T \mathbf{K}_i \boldsymbol{\varphi}_r \tag{2}$$

$$\frac{\partial \boldsymbol{\varphi}_r}{\partial \alpha_i} = \sum_{k=1, k \ne r}^{n} \frac{\boldsymbol{\varphi}_k^T \mathbf{K}_i \boldsymbol{\varphi}_r \boldsymbol{\varphi}_k}{\lambda_k - \lambda_r} \tag{3}$$

where λ_r, φ_r ($r = 1, 2, \ldots, n$) are the rth eigenvalue and corresponding commode shapes of structures, respectively. n is the total degrees of freedom (DOFs) of structures.

The first-order sensitivity analysis can be expressed as follows:

$$\begin{bmatrix} \frac{\partial \lambda_1}{\partial \alpha_1} & \cdots & \frac{\partial \lambda_1}{\partial \alpha_N} \\ \vdots & \ddots & \vdots \\ \frac{\partial \lambda_m}{\partial \alpha_1} & \cdots & \frac{\partial \lambda_m}{\partial \alpha_N} \\ \frac{\partial \boldsymbol{\varphi}_1}{\partial \alpha_1} & \cdots & \frac{\partial \boldsymbol{\varphi}_1}{\partial \alpha_N} \\ \vdots & \ddots & \vdots \\ \frac{\partial \boldsymbol{\varphi}_l}{\partial \alpha_1} & \cdots & \frac{\partial \boldsymbol{\varphi}_l}{\partial \alpha_N} \end{bmatrix} \Delta\boldsymbol{\alpha} = \mathbf{S}\Delta\boldsymbol{\alpha} \approx \Delta\mathbf{f} = \begin{Bmatrix} \Delta\mathbf{f}_\lambda \\ \Delta\mathbf{f}_\varphi \end{Bmatrix} \tag{4}$$

where, $\Delta\mathbf{f}_\lambda = \{\Delta\lambda_1, \Delta\lambda_2, \ldots, \Delta\lambda_m\}^T$ is difference in eigenvalues between undamaged and damaged structures. $\Delta\mathbf{f}_\varphi = \{\Delta\boldsymbol{\varphi}_1, \Delta\boldsymbol{\varphi}_2, \ldots, \Delta\boldsymbol{\varphi}_l\}^T$ is difference in mode shapes between undamaged and damaged structures. m and l are the orders of eigenvalues and mode shapes for computation, respectively. The vector $\Delta\boldsymbol{\alpha} = \{\Delta\alpha_1, \Delta\alpha_2, \ldots, \Delta\alpha_N\}^T$ is a change in damage reduction factors. $\mathbf{S}$ is the closed form of first-order sensitivity matrix.

Equation (4) is the approximation relation between $\mathbf{S}\Delta\boldsymbol{\alpha}$ and $\Delta\mathbf{f}$, so that the sensitivity matrix is updated iteratively in this paper to obtain an accurate solution of $\Delta\boldsymbol{\alpha}$.

2.2 Restricted Isometry Property

According to CS theory, two conditions are required for reconstructing a signal. The first one is sparsity which requires the signal to be sparse in some domain. The second one is incoherence which is applied through the isometric property which is sufficient for sparse signals.
Supposing that we observe

$$\mathbf{y} = \mathbf{\Phi x} \tag{5}$$

where $\mathbf{x} \in \mathbf{R}^P$ is an object we wish to reconstruct, $\mathbf{y} \in \mathbf{R}^Q$ are available measurements, and $\mathbf{\Phi} \in \mathbf{R}^{Q\times P}$ is a known measure matrix.
The measurement matrix $\mathbf{\Phi}$ of CS is used to compress signals so that it can reduce the dimensions of signals. In order to reconstruct signals with good accuracy, the measurement matrix $\mathbf{\Phi}$ should satisfy RIP, which is defined as follows:
It is supposed that signal $\mathbf{x}$ is known to be sparse or nearly sparse in the sense that it depends on a smaller number of unknown parameters. $\mathbf{x}$ has at most k nonzero entries. For each integer k = 1, 2, …, define the isometry constant δ_k of the matrix $\mathbf{\Phi}$ as the smallest number such that (Candès, 2008)

$$\left(1-\delta_k\right)\|\mathbf{x}\|_2^2 \le \|\mathbf{\Phi x}\|_2^2 \le (1+\delta_k)\|\mathbf{x}\|_2^2 \tag{6}$$

holds for all k-sparse vectors, and $Q < P$. $\|\mathbf{\Phi x}\|_2^2$ is the energy of available measurements $\mathbf{y}$ and $\|\mathbf{x}\|_2^2$ is the energy of original signal $\mathbf{x}$.

2.3 Subspace Projection of Sensitivity Function

In CS, the measurement matrices of random generation have two types. They are dense random matrix and sparse random matrix, respectively. The sparse random matrix is better than dense random matrix because of lower computational complexity, convenient update and less storage capacity during compression and reconstruction.
The changes of energy between original signal and available measurements should be in the range mentioned in Eq. (6). So it can reconstruct original signal with good accuracy from the available measurements $\mathbf{y}$. That is to say, the space to which available measurements $\mathbf{y}$ belongs retains enough information of original signal.
Therefore, the measurement matrix $\mathbf{\Phi}$ which satisfies the RIP can project the original signal into the subspace and retain enough information.
In this study, sparse random matrix with fixed column sparsity is chosen as measurement matrix $\mathbf{\Phi}$ which is $Q \times P$ random matrix. It can be generated as follows:
The nonzero elements of the matrix are positive real numbers. The number of nonzero elements in each column is consistent and each column in $\mathbf{\Phi}$ has f ($f < Q$) nonzero elements. The smallest and largest nonzero elements in the matrix are a and b, respectively. In this study, $a = b = 1, f = 4$ (Wu, 2012).
It is possible to recover $\mathbf{x}$ using a Q measurements following the relation:

$$Q \ge Ck\log\left(\frac{P}{k}\right) \tag{7}$$

where C is a constant that has empirically been found to approximately equal 4.0 (Mascarenas et al., 2013).
Because the measurement matrix $\mathbf{\Phi}$ can project the original signal into the subspace and can retain enough information, in this study, this subspace can be introduced into sensitivity equation by

considering $\mathbf{\Delta f}$ is sparse in some domain. Then, by remultiplying $\mathbf{\Phi}$ at both sides of Eq. (4), the following equation can be obtained:

$$\hat{\mathbf{S}}\mathbf{\Delta\alpha} \approx \mathbf{\Delta}\hat{\mathbf{f}} \tag{8}$$

where, $\hat{\mathbf{S}}=\mathbf{\Phi S}$ and $\mathbf{\Delta}\hat{\mathbf{f}}=\mathbf{\Phi\Delta f}$.

2.4 Sparse regularization

Least absolute shrinkage and selection operator (Lasso) regularization algorithm was proposed by Tibshirani (Tibshirani, 1996). It is called l_1-norm regularization for short. Through adding l_1-norm punishment for regression coefficient vector, l_1-norm regularization can compress regression coefficients. Sparseness of regression coefficients can be realized because of setting the minor absolute values of regression coefficients to be zero. $\mathbf{\Delta\alpha}$ is sparse due to damage generally occur in structural partial positions. For the reasons given above, combining sensitivity analysis with l_1-norm regularization can improve the ill-posedness of SDD problems and obtain damage identification results with good accuracy. Adding l_1-norm punishment to Eq. (8) as follows:

$$\min \left\|\mathbf{\Delta\alpha}\right\|_1 \text{ subject to } \left\|\hat{\mathbf{S}}\mathbf{\Delta\alpha} - \mathbf{\Delta}\hat{\mathbf{f}}\right\|_2 \le \varepsilon \tag{9}$$

where, ε is a constant and related to noise in the process of measurement.
Equation (9) can be equivalent to an l_1-norm regularization problem:

$$J = \underset{\mathbf{\Delta\alpha}\in\mathbf{R}^P}{\operatorname{argmin}} \left\{ \frac{1}{2} \left\|\hat{\mathbf{S}}\mathbf{\Delta\alpha} - \mathbf{\Delta}\hat{\mathbf{f}}\right\|_2^2 + \beta \left\|\mathbf{\Delta\alpha}\right\|_1 \right\} \tag{10}$$

where, P is the dimension of sparse vector $\mathbf{\Delta\alpha}$ and β ($\beta > 0$) is the regularization parameter.
In this study, in order to solve the sparse optimization problem as shown in Eq. (10), Sparse Learning with Efficient Projections (SLEP) (Liu et al., 2009) is used. SLEP is based on proximal gradient method and its variants.
For Eq. (10), the value of β is important for stability of solutions. If β is too small, posedness of SDD cannot be improved by regularization; on the contrary, if β is too large, the obtained approximate problem is different with original problem, this is a poor approximation.
To select the optimal regularization parameter, Akaike information criterion (AIC) (Burnham& Anderson, 2002) is used so that reasonable values of $\mathbf{\Delta\alpha}$ can be obtained. AIC is a measure of relative quality of statistical models for a given set of data.
The AIC is formally defined as follows:

$$AIC = 2A - 2\ln\left(\hat{L}\right) \tag{11}$$

where, $\hat{L}$ is the maximized value of the likelihood function. A is the number of nonzero element in $\mathbf{\Delta\alpha}$.
Under the assumption that the model errors are independent and identically distributed with the normal distribution, Eq. (11) can be modified to:

$$AIC = 2A + P\ln\left[RSS(\beta)/P\right] \tag{12}$$

where, $RSS(\beta)$ is the residual sum of squares of $\mathbf{S\Delta\alpha}$ (β)-$\mathbf{\Delta f}$.
The flowchart of SDD using RIP-based subspace projection method is shown in Fig.1.

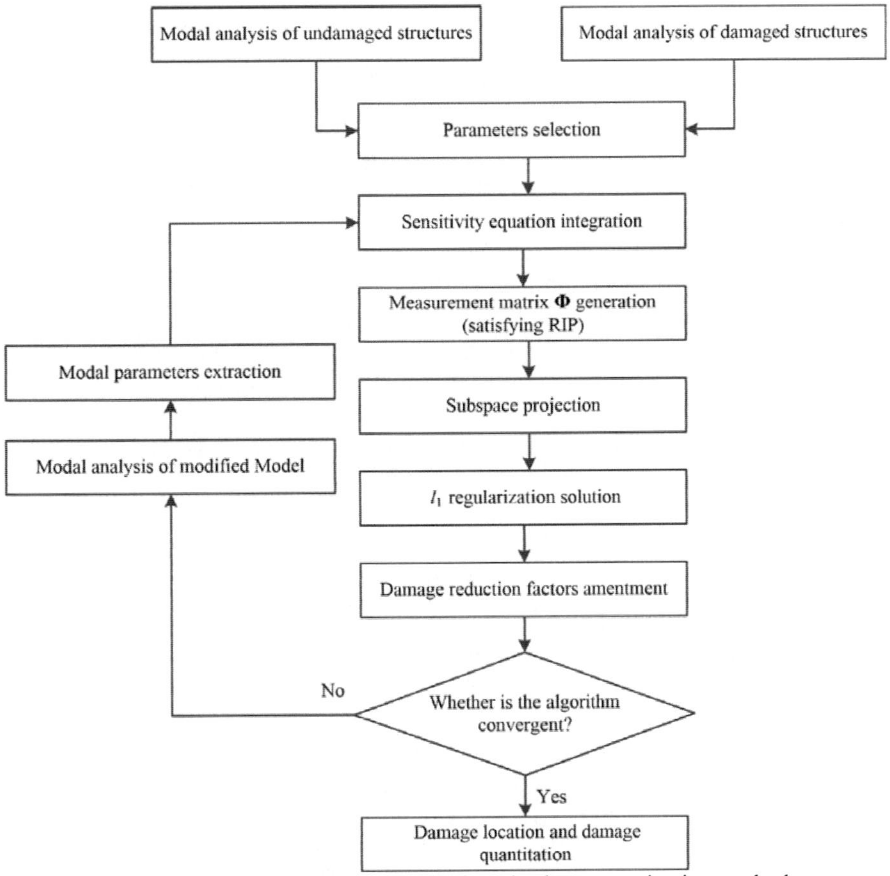

Fig. 1 Flowchart of SDD using RIP-based subspace projection method

3. Numerical simulations

Some numerical simulations of SDD on a plane truss are used to verify the applicability of the proposed method in this study. The plane truss is shown in Fig. 2.

The total length and height of the truss are 5m and 0.5m, respectively. The truss is divided into 40 elements. Numbers in the circles represent the element number and others represent node number in Fig. 2. The elastic modulus, density and cross sectional area of each element are 210GPa, 7800kg/m^3 and 0.003m^2, respectively.

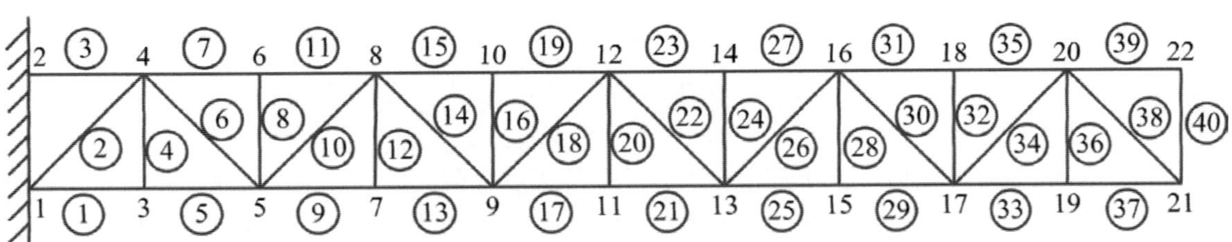

Fig. 2 40-element plane truss with element numbers and node numbers

In practice, frequencies and mode shapes generally contain noise. For describing random measurement errors arising in practice, energy of signals is suitable to represent it. (Bauer & Lukas, 2011) It is considered that the energy of frequency noise and mode shape noise are related to their changes in this study. Frequencies and mode shapes containing noise are as follows:

$$\upsilon_i = \upsilon_i^a + \varepsilon_\upsilon R_i(\frac{\|\mathbf{\upsilon}_s - \mathbf{\upsilon}_e\|_2}{m}) \quad (13)$$

$$\varphi_{ij} = \varphi_{ij}^a + \varepsilon_\varphi R_{ij}(\frac{\|\mathbf{\varphi}_s - \mathbf{\varphi}_e\|_2}{lN}) \quad (14)$$

where, m, l are the orders of frequencies and mode shapes for computation, respectively; N is length of each mode shape vector; υ_s and υ_e are vectors of frequencies in undamaged and damaged structures, respectively; φ_s and φ_e are rearranged vectors of mode shapes in undamaged and damaged structures, respectively; υ_i^a and υ_i (i = 1, 2, ..., m) represent the ith calculated and measured frequencies, respectively; φ_{ij}^a and φ_{ij} (i = 1, 2, ..., l; j = 1, 2, ..., N) are the jth element of the ith calculated and measured mode shapes, respectively; R_i and R_{ij} are subject to the standard normal distribution with zero mean value and unit standard deviation; ε_υ and ε_φ are noise level of frequencies and mode shapes, respectively.

The relative percentage error (RPE) between detected and actual changes of damage reduction factor is calculated as follows:

$$\text{RPE}=\frac{\|\Delta\alpha_c^i - \Delta\alpha_a^i\|_2}{\|\Delta\alpha_a^i\|_2}\times 100\% \quad (15)$$

where, $\Delta\alpha^i{}_c$ and $\Delta\alpha^i{}_a$ are the detected and the actual values of the ith element of $\mathbf{\Delta\alpha}$, respectively.

The first ten frequencies and mode shapes of truss are used for SSD. It is assumed that k is equal to 5 in this study. As shown in Table 1, twelve damage scenarios are considered to verify applicability of the proposed method. To study the influence of different positions with different damage extents, the damages are evenly distributed in the truss. Three noise levels are added to each scenario. By comparing identification results in the same scenario, the robustness of the proposed method is studied. The results of SDD are shown in Tables 2 to 4 and Figs. 3 to 5.

Table. 1 Damage scenarios for plane truss

Number of damage(s)	Scenario no.	Damaged element no. (Damage Extent=x %)
1	1	E7=5%
	2	E10=1%
	3	E16=20%
	4	E21=10%
2	5	E15=20%, E30=10%
	6	E14=15%, E26=10%
	7	E5=5%, E19=15%
	8	E8=5%, E20=10%
3	9	E15=20%, E24=5%, E30=10%
	10	E11=20%, E20=10%, E26=15%
	11	E5=10%, E15=5%, E20=15%
	12	E14=10%, E23=20%, E26=15%

By analyzing the SDD results, the following conclusions are obtained:

The identified errors are stable in given scenarios. For single-damage scenarios, the proposed method is able to detect the single damage location and quantify its damage extent effectively under different noise levels. The SDD results of scenarios 1, 2 and 4 show that the identification accuracy decreases as noise level increases. However, the identified errors in scenario 3 fluctuate as noise level increases. In each scenario and each noise level, measurement matrix is different and it causes different energy loss of **Δf**, therefore the measurement matrix is one of the reason to affect SDD results.

Table. 2 Damage cases and SDD results for single damage location

Scenario no.	Damaged element no.	Damage Extent (%)	Predicted Damage (%)			RPE (%)			Iteration times		
			5% Noise	*10% Noise*	*15% Noise*	*5% Noise*	*10% Noise*	*15% Noise*	*5% Noise*	*10% Noise*	*15% Noise*
1	7	5	5.0037	5.0160	4.9526	0.0744	0.3209	0.9476	6	6	5
2	10	1	1.0002	0.9980	0.9973	0.0185	0.1979	0.2648	7	5	6
3	16	20	19.9784	20.0319	19.9817	0.1080	0.1597	0.0913	7	8	5
4	21	10	10.0061	10.0128	10.0307	0.0613	0.1277	0.3068	5	6	7

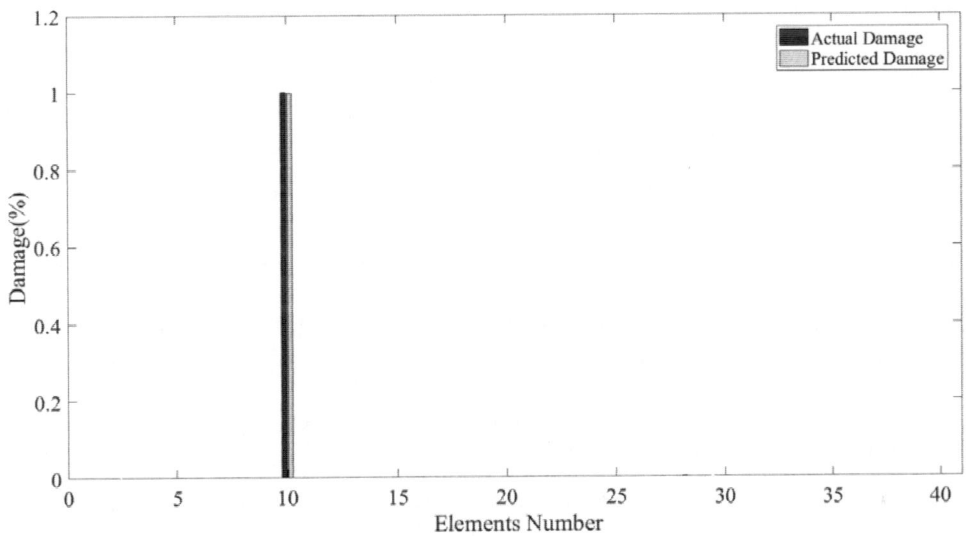

Fig. 3 SDD results under scenario 2 with noise 10%

In scenarios 5 to 8, the damage locations are misjudged sometimes, but the proposed method can still detect the damage locations and quantify their damage extents under different noise levels. Comparing different damages in scenarios 5 to 8, the distributions of damage locations do not normally affect the SDD results. The method can effectively detect damage locations and quantify damage extents even though the difference of each damage extent is large. However, identification accuracy of small damage decreases obviously as damage extent increased. This new SDD approach has good applicability.

Table. 3 Damage cases and SDD results for double damage locations

Scenario no.	Damaged element no.	Damage Extent (%)	Predicted Damage (%)			RPE (%)			Iteration times		
			5% Noise	*10% Noise*	*15% Noise*	*5% Noise*	*10% Noise*	*15% Noise*	*5% Noise*	*10% Noise*	*15% Noise*
5	15	20	20.0048	20.0431	20.1182	0.0238	0.2154	0.5911	6	5	4
	30	10	9.9968	10.0184	10.0468	0.0323	0.1844	0.4684			
6	14	15	15.0090	15.0564	15.1031	0.0598	0.3758	0.6875	7	6	6
	26	10	10.0290	10.0100	9.9150	0.2903	0.1001	0.8499			
7	5	5	4.7834	4.9103	5.0094	4.3319	1.7943	0.1872	6	5	6
	19	15	15.0202	15.0039	15.1376	0.1349	0.0257	0.9174			
8	8	5	4.9931	4.9991	4.9381	0.1385	0.0187	1.2372	7	11	5
	20	10	10.0080	10.0196	10.0335	0.0804	0.1961	0.3346			

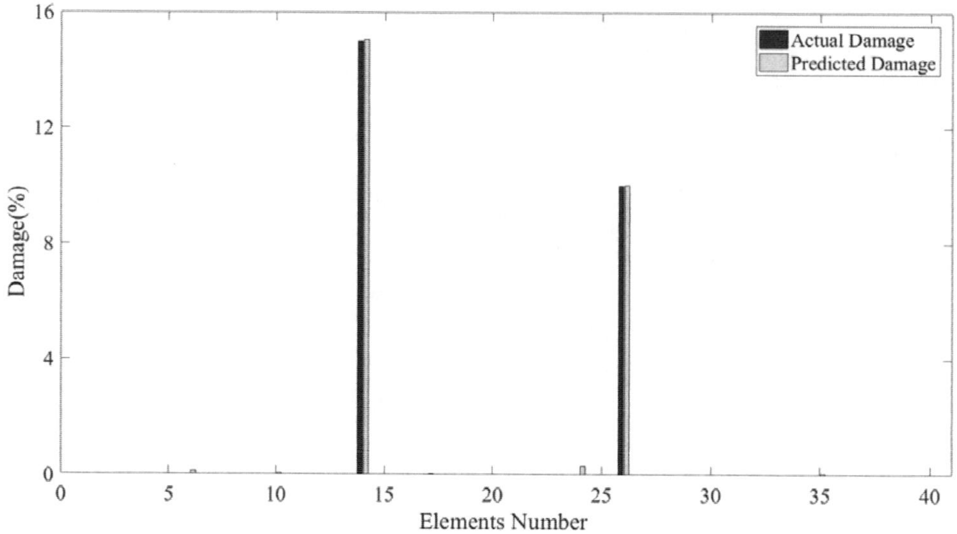

Fig. 4 SDD results under scenario 6 with noise 10%

For three damages, it can be found that identification accuracy of small damage still decreases as damage extent increased. RPE does not increase when noise level increases, that represents RPE is not only effected by noise, but also by randomness of measure matrix. At the same time, for two or three damages, RPEs are obviously larger than ones in single damage scenarios.

Table. 4 Damage cases and SDD results for three damage locations

Scenario no.	Damaged element no.	Damage Extent (%)	Predicted Damage (%)			RPE (%)			Iteration times		
			5% Noise	*10% Noise*	*15% Noise*	*5% Noise*	*10% Noise*	*15% Noise*	*5% Noise*	*10% Noise*	*15% Noise*
9	15	20	20.0390	19.9889	20.0812	0.1951	0.0554	0.4061	7	9	5
	24	5	4.9713	4.9920	4.9077	0.5737	0.1606	1.8468			
	30	10	10.0363	10.0244	9.9196	0.3627	0.2445	0.8041			
10	11	20	20.0351	20.1160	19.8779	0.1753	0.5802	0.6104	7	6	5
	20	10	9.8718	9.5851	9.7185	1.2821	4.1491	2.8151			
	26	15	15.0025	15.0478	15.0733	0.0170	0.3185	0.4890			
11	5	10	10.0070	10.0435	9.9714	0.0700	0.4351	0.2862	5	5	5
	15	5	5.0074	4.9922	4.9636	0.1489	0.1564	0.7273			
	20	15	15.0179	15.2003	15.0376	0.1194	1.3351	0.2505			
12	14	10	10.0328	10.2459	10.0653	0.3279	2.4586	0.6531	5	5	6
	23	20	19.9747	20.2221	19.8855	0.1263	1.1103	0.5726			
	26	15	15.0358	15.1279	15.0480	0.2389	0.8528	0.3201			

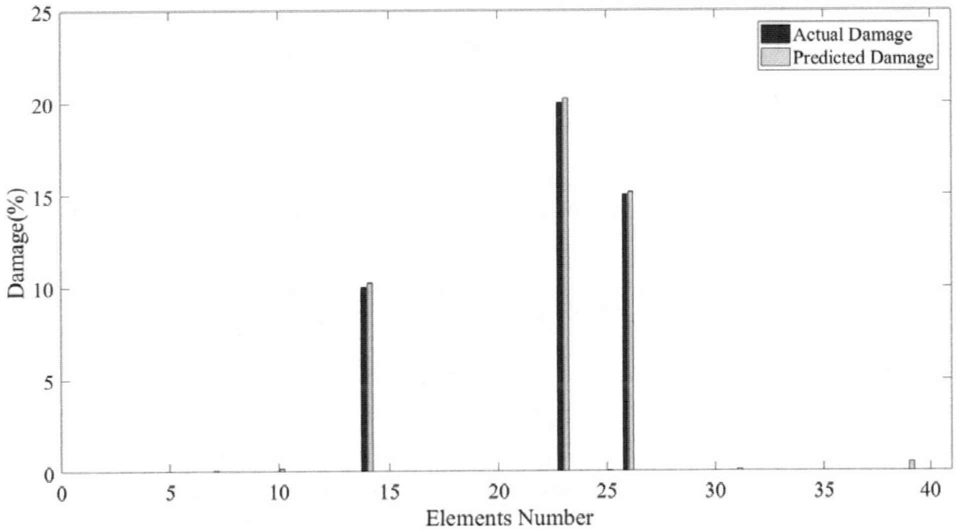

Fig. 4 SDD results under scenario 12 with noise 10%

By comparing all given scenarios, some conclusions are obtained. The proposed method has strong robustness to noise. By adding l_1-norm punishment, it can improve the ill-posedness of sensitivity analysis. Only few iteration times are needed to obtain SDD results with good accuracy in the proposed method, so this method has stable iteration times in different scenarios. Damage locations only have little effect on SDD. Almost all scenarios have no notable difference on identification accuracy except individual scenarios affected by measurement matrix.

4. Conclusions

In this study, a new RIP-based subspace projection method is proposed for the SDD problem. Subspace projection based on RIP is used to reduce the number of equations in sensitivity analysis and it reduces the computation space of SDD. For subspace projection, a measurement matrix $\mathbf{\Phi}$ which satisfies RIP is introduced into sensitivity equation. The obtained subspace according to the measurement matrix $\mathbf{\Phi}$ can retain enough information of vector in original space so that it can reconstruct the original vector with good accuracy. Because of the sparsity of structural damages, sparse regularization approach is used to solve sensitivity equation in the subspace. Enough damage information is retained in the subspace so that it can detect damage locations and quantify damage extents accurately. Some SDD numerical simulations on plane truss are adopted to assess the rationality, applicability and robustness of the proposed approach. Some conclusions can be made as follows:

1) The proposed approach can detect damage locations and quantify damage extents accurately in given scenarios.

2) Both accuracy and robustness to noises of this SDD method are well considered simultaneously in this study. Almost all scenarios have no notable difference on identification accuracy. This indicates the new approach has good applicability with good accuracy and strong robustness to noises in SDD.

3) This new method reduces the number of equations in sensitivity analysis and the computation space of SDD; at the same time, it can locate structural damages and quantifies damage extents well because enough information is retained in the subspace. This has obvious advantage by using this method to not only large scale structure but also online SDD in the field of SHM.

Acknowledgments

This work was jointly supported by the National Natural Science Foundation of China with Grant Numbers 51678278 and 51278226.

References

Zhou, XQ, Xia, Y & Weng, S 2015, 'L1 regularization approach to structural damage detection using frequency data', *Structural Health Monitoring*, vol. 14, no. 6, pp. 571-582.

Zhou, SM, Bao, YQ & Li, H 2016, 'Structural damage identification based on structural sensitivity analysis and sparse restrains optimization', *Journal of Vibration and Shock*, vol. 35, no 9, pp. 135-140. (In Chinese)

Mascarenas, D, Cattaneo, A, Theiler, J & Farrar CR 2013, 'Compressed sensing techniques for detecting damage in structures', *Structural Health Monitoring*, vol. 12, no. 4, pp. 325-338.

Jayawardhana, M, Zhu, XQ, Liyanapathirana, R & Gunawardana, U 2017, 'Compressed sensing for efficient health monitoring and effective damage detection of structures', *Mechanical Systems and Signal Processing*, vol. 84, pp. 414-430.

Pan, CD & Yu, L 2016, 'Simultaneous identification of moving forces and structural damages by L1/2 regularization method', *Proceedings of Mechanics of Structures and Materials: Advancements and Challenges (ACMSM24),* Australia, 6-9 December, pp. 1441-1446.

Donoho, DL 2006, 'Compressed sensing', *IEEE Transactions on information theory*, vol. 52, no. 4, pp. 1289-1306.

Huang, Y Beck, JL & Wu, S, et al 2014, 'Robust Bayesian compressive sensing for signals in structural health monitoring', Computer-Aided Civil and Infrastructure Engineering, vol. 29, no. 3, pp. 160-179.

Candès EJ 2008, 'The restricted isometry property and its implications for compressed sensing', *Comptes Rendus Mathematique*, vol. 346, no. 9, pp. 589-592.

Wu, Y 2012, 'Research on measurement matrix for compressed sensing', Xidian Univesity, Xian. (In Chinese)

Tibshirani, R 1996, 'Regression shrinkage and selection via the lasso', *Journal of the Royal Statistical Society. Series B (Methodological)*, pp. 267-288.

Liu, J, Ji, S & Ye, J 2009, 'SLEP: Sparse learning with efficient projections', Arizona State University.

Burnham, KP & Anderson, DR 2002, 'Model selection and multi-model inference: a practical information-theoretic approach', *Journal of Wildlife Management,* vol. 67, no. 3, pp. 175-196.

Bauer, F & Lukas, M, A 2011, 'Comparing parameter choice methods for regularization of ill-posed problems', *Mathematics and Computers in Simulation*, vol.81, no.9, pp. 1795-1841.

Th[...]rnational Conference on
Structural Health Monitoring of Intelligent Infrastructure
Brisbane, Australia | 5-8 December 2017

Structural Modal Frequency-Environmental Condition Relation Development by Bayesian Learning Approach

H. Q. Mu[1,2], K. V. Yuen[3].

[1]*Associate Professor, School of Civil Engineering and Transportation, South China University of Technology, Guangzhou 510640, P.R. China (corresponding author). E-mail: cthqmu@scut.edu.cn*

[2]*Associate Professor, State Key Laboratory of Subtropical Building Science, South China University of Technology, Guangzhou 510640, P.R. China*

[3]*Professor,Faculty of Science and Technology, University of Macau, Macao, P.R. China.*

Abstract

Structural modal frequency is a well-accepted indicator for the purpose of structural health assessment. However, it is known that this indicator is sensitive to changing environmental conditions (such as temperature and humidity). Thus, development of modal frequency-environmental condition relation is important for making reliable inference. This paper introduces a Bayesian learning approach for modal frequency-environmental conditionrelation development based on the one-year monitoring dataset of a reinforced concrete building. By introducing a sophisticated hyperparameterizaion strategy, Bayesian learning appraoch automatically search the optimal model during hyperparameter optimazation.The appraoch is then utilized for health assessment of a reinforced concrete building based on one-year daily measurements.

1. Introduction

Modal parameteris an important indicator of the overall health status of buildings (Farrar and Worden, 2007). As structures are operated in varying environmental conditions, previous studies have been devoted to develop the relation between modal frequency and environmental conditions. For instance, Alampalli (1998) monitored the fluctuating sensitivity of the modal parameters of a reinforced concrete bridge caused by environment factors and artificial damages. The experimental results showed that the environmental factors under bridge operating conditions could lead to significant fluctuations in bridge modal parameters. Peeters and De Roeck(2001) compared the influence of environmental factors and structural damage on the structural modal parameters. Liu and De Wolf (2007) studied the temperature effects on a curved post-tensioned concrete bridge. Some researchers (Farrar et al. 1997;Cornwell et al. 1999; Sohn et al. 1999; Cross et al. 2013) studied the impact to modal parameters by the rainfall. Zhang et al. (2016)studied the correlation between the structural modal frequency and the ambient temperature, wind speed and vehicle load.Varying environmental conditions can cause significant variation in structural modal parameters, and simply ignoring the environmental conditions in structural health monitoring will lead to misleading health assessment results.

This paper introduces a Bayesian learning approach for modal frequency-environmental condition relation development based on the one-year monitoring dataset of a reinforced concrete building. The learning algorithm, by introducing a sophisticated hyperparameterizaion strategy,is capable to automatically search the optimal model during hyperparameter optimazation(Mu and Yuen, 2016; Mu and Yuen, 2017; Mu et al., 2017).

2.Monitoring dataset

The monitoring structure is a 22-storyreinforced concrete residential building. The floor plan of the building is in asymmetric L-shape, with the building height and the length of the two spans being 64.70m, 51.90m, and 61.75m respectively. The monitoring period is one year between 2 May 2008 to 1 May 2009.Two accelerometers, operated based on standard exploration geophone mass-spring systems with 50V/g sensitivity, were installed on the 18th floor with two orthogonal directions. Based on the ten-minute acceleration time history of each day, the Bayesian spectral density approach (Yuen et al., 2002; Yuen and Beck, 2003; Yuen, 2010) is applied to identify the modal frequencies of the building. The monitoring dataset includes the first two modes (ω_1^2, ω_2^2) and measurements of the temperature (T) and the relative humidity (H).
Previous researches have shown that there is statistically significant dependence of structural modal frequency and temperature along with relative humidity (Xia et al., 2006; Yuen and Kuok, 2010a, 2010b; Mu et al., 2016). In this study, the extracted feature vector $\boldsymbol{x} = [1, T_n, H_n, T_n^2, T_n H_n, H_n^2, T_n^3, H_n^3, T_n^4, H_n^4]^{\mathrm{T}}$, where the normalized temperature $T_n = T/T_{max}$and the normalized relative humidity$H_n = H/H_{max}$ are obtained by considering the rescale of T and H by the maximum value $T_{max} = \max(T)$ and $H_{max} = \max(H)$, receptively. The output vector is the identified squared frequency of the first two modes$\boldsymbol{y} = [\omega_1^2, \omega_2^2]$.

3. Relation development by Bayesian learning approach

The relation between the m-th ($m = 1, \ldots, M$) output vector and input matrix is:

$$\mathbf{y}_m = \mathbf{X}\mathbf{b}_m + \boldsymbol{\varepsilon}_m \tag{1}$$

where $\mathbf{y}_m \in \mathbb{R}^{N\times1}$ is the output vector for the m-th mode of the identified squared modal frequencies; $\mathbf{X} = [\mathbf{X}_1, \mathbf{X}_2, \ldots, \mathbf{X}_{N_h}]$ ($N_h = 10$ for the full model) is the $N \times N_h$input/designmatrix; $\mathbf{b}_m \in \mathbb{R}^{N_b\times1}$ is the m-th weight parameter vector to be trained; and$\boldsymbol{\varepsilon}_m \in \mathbb{R}^{N\times1}$ follows a Gaussian vector with zero mean and covariance matrix$\sigma_m^2\mathbf{I}_N$ with $\mathbf{I}_N$ being $N \times N$ identity matrix. Assume that different components of the output matrix $\mathbf{Y} = [\mathbf{y}_1, \mathbf{y}_2, \ldots, \mathbf{y}_m]$are independent, the likelihood function for the data $\mathrm{D} = \{\mathbf{X}, \mathbf{Y}\}$can be expressed as a product of the Gaussian distribution(Mackay, 1992):

$$p(\mathbf{Y}|\mathbf{X}, \mathbf{b}, \boldsymbol{\sigma}^2) = \prod_{m=1}^{M} (2\pi\sigma_m^2)^{-N/2} \exp\left[-\frac{1}{2\sigma_m^2}(\mathbf{y}_m - \mathbf{X}\mathbf{b}_m)^{\mathrm{T}}(\mathbf{y}_m - \mathbf{X}\mathbf{b}_m)\right] \tag{2}$$

where $\mathbf{b} = [\mathbf{b}_1^{\mathrm{T}}, \mathbf{b}_2^{\mathrm{T}}, \ldots, \mathbf{b}_m^{\mathrm{T}}]^{\mathrm{T}}$ and$\boldsymbol{\sigma}^2 = [\sigma_1^2, \ldots, \sigma_M^2]$. In contrast to the original automatic relevance determination (ARD) prior for the scale output case (Mackay, 1992; Tipping 2004), the ARD prior for the vector outputs is adopted

$$p(\mathbf{b}|\boldsymbol{\alpha}) = \prod_{m=1}^{M} \mathrm{G}(\mathbf{b}_m|\mathbf{0}, [\mathbf{A}(\boldsymbol{\alpha})]^{-1}) \tag{3}$$

where $\mathbf{A}(\boldsymbol{\alpha}) = \mathrm{diag}\{\alpha_1, \alpha_2, \ldots, \alpha_{N_h}\}$ is the precision matrix and $\boldsymbol{\alpha} = [\alpha_1, \alpha_2, \ldots, \alpha_{N_h}]^{\mathrm{T}}$ is the hyperparameter vector parameterizing the prior probability density function (PDF) of the weight parameter vector $\mathbf{b}$.By using Bayes' theorem, the posterior PDF for the uncertain regression parameter $\mathbf{b}$ is given by:

$$p(\mathbf{b}|D,\boldsymbol{\alpha},\boldsymbol{\sigma}^2) = \prod \mathrm{G}(\mathbf{b}_m|\hat{\mathbf{b}}_m,\hat{\boldsymbol{\Sigma}}_m) \tag{4}$$

where the posterior mean vector $\hat{\mathbf{b}}_m$ and the posterior covariance matrix $\hat{\boldsymbol{\Sigma}}_m$ are

$$\hat{\mathbf{b}}_m = \sigma_m^{-2}\hat{\boldsymbol{\Sigma}}_m\mathbf{X}^{\mathrm{T}}\mathbf{y}_m \tag{5}$$

$$\hat{\boldsymbol{\Sigma}}_m = (\sigma_m^{-2}\mathbf{X}^{\mathrm{T}}\mathbf{X}+\mathbf{A})^{-1} \tag{6}$$

It can be observed that the relevancy of each regression term, as well as $\hat{\mathbf{b}}$ and $\hat{\boldsymbol{\Sigma}}$are controlled by the optimal value of $\boldsymbol{\alpha}$ of the ARD prior and $\boldsymbol{\sigma}^2$. In order to achieve the optimal value$\hat{\boldsymbol{\alpha}}$ and $\hat{\boldsymbol{\sigma}}^2$, consider the joint posterior PDF of $\boldsymbol{\alpha}$ and $\boldsymbol{\sigma}^2$by Bayes' theorem: $\mathrm{p}(\boldsymbol{\alpha},\boldsymbol{\sigma}^2|\mathrm{D}) \propto \mathrm{p}(\mathrm{D}|\boldsymbol{\alpha},\boldsymbol{\sigma}^2)\mathrm{p}(\boldsymbol{\alpha},\boldsymbol{\sigma}^2)$, where $\mathrm{p}(\mathrm{D}|\boldsymbol{\alpha},\boldsymbol{\sigma}^2)$ is the evidence (sometimes called the marginal likelihood) conditional on $\boldsymbol{\alpha}$ and $\boldsymbol{\sigma}^2$, and $\mathrm{p}(\boldsymbol{\alpha},\boldsymbol{\sigma}^2)$ is the prior PDF of $\boldsymbol{\alpha}$ and $\boldsymbol{\sigma}^2$. As the maximum a posteriori (MAP) estimation can be well approximately by maximum evidence estimation (Mackay 1992; Bishop 2006), the optimal value for $\boldsymbol{\alpha}$ and $\boldsymbol{\sigma}^2$ can be determined by considering maximization of the logarithm of the evidence $\mathfrak{L}(\boldsymbol{\alpha},\boldsymbol{\sigma}^2) \equiv \ln p(\mathrm{D}|\boldsymbol{\alpha},\boldsymbol{\sigma}^2)$:

$$\mathfrak{L}(\boldsymbol{\alpha},\boldsymbol{\sigma}^2) = -\frac{1}{2}\prod_{m=1}^{M}[N\ln 2\pi + \ln|\mathbf{C}_m| + \mathbf{Y}_m^{\mathrm{T}}\mathbf{C}_m^{-1}\mathbf{Y}_m] \tag{7}$$

where

$$\mathbf{C}_m(\boldsymbol{\alpha},\boldsymbol{\sigma}^2) = \sigma_m^2\mathbf{I}_N + \mathbf{X}^{\mathrm{T}}\mathbf{A}^{-1}\mathbf{X} \tag{8}$$

By using a separation strategy (Bishop, 2006; Mu and Yuen, 2016; Mu and Yuen, 2017), $\mathfrak{L}(\boldsymbol{\alpha},\boldsymbol{\sigma}^2)$ can be separated into two parts:

$$\mathfrak{L}(\boldsymbol{\alpha},\boldsymbol{\sigma}^2) = \mathfrak{L}(\boldsymbol{\alpha}_{-i},\boldsymbol{\sigma}^2) + \mathfrak{l}(\alpha_i,\boldsymbol{\sigma}^2) \tag{9}$$

The first term $\mathfrak{L}(\boldsymbol{\alpha}_{-i},\boldsymbol{\sigma}^2)$ is the log-evidence with removing α_i (and thus $\mathrm{b}_{m,i}$ and $\mathbf{X}_i$), and the second term $\mathfrak{l}(\alpha_i,\boldsymbol{\sigma}^2)$is given by (Thayananthan et al., 2006):

$$\mathfrak{l}(\alpha_i,\boldsymbol{\sigma}^2) = \frac{1}{2}\sum_{m=1}^{M}\left[\ln\alpha_i - \ln(\alpha_i + S_{mi}) - \frac{Q_{mi}^2}{\alpha_i + S_{mi}}\right] \tag{10}$$

with

$$S_{mi} = \frac{\alpha_i s_{mi}}{\alpha_i - s_{mi}} \tag{11}$$

$$Q_{mi} = \frac{\alpha_i q_{mi}}{\alpha_i - s_{mi}} \tag{12}$$

$$s_{mi} = \sigma_m^{-2}\mathbf{X}_i^{\mathrm{T}}\mathbf{X}_i - \sigma_m^{-4}\mathbf{X}_i^{\mathrm{T}}\mathbf{X}\hat{\boldsymbol{\Sigma}}_m\mathbf{X}^{\mathrm{T}}\mathbf{X}_i \tag{13}$$

$$q_{mi} = \sigma_m^{-2}\mathbf{X}_i^{\mathrm{T}}\mathbf{y}_m - \sigma_m^{-4}\mathbf{X}_i^{\mathrm{T}}\mathbf{X}\hat{\mathbf{\Sigma}}_m\mathbf{X}^{\mathrm{T}}\mathbf{y}_m \tag{14}$$

where $\mathbf{X}_i \in \mathbb{R}^{N\times 1}$ is the i-th column of the design matrix $\mathbf{X}$.As the i-th hyperparaemeter α_i has been isolated from the log-evidence $\mathfrak{L}(\boldsymbol{\alpha}, \boldsymbol{\sigma}^2)$, the optimal value of α_i can be obtained by considering$\partial l(\alpha_i, \boldsymbol{\sigma}^2)/\partial\alpha_i = 0$. Similarly, the optimal value $\hat{\sigma}^2$can be obtained by considering $\mathfrak{L}(\boldsymbol{\alpha}, \boldsymbol{\sigma}^2)/\partial\sigma_m^2 = 0$

$$\hat{\sigma}_m^2 = \frac{\left[(\mathbf{Y} - \mathbf{X}\hat{\mathbf{b}})^{\mathrm{T}}(\mathbf{Y} - \mathbf{X}\hat{\mathbf{b}})\right]}{N - N_b + \sum_{i=1}^{N_b} \hat{\alpha}_i \hat{\mathbf{\Sigma}}_{m,ii}} \tag{15}$$

where $\hat{\mathbf{\Sigma}}_{m,ii}$ is the i-th diagonal element of the m-th mode posterior covariance matrix $\hat{\mathbf{\Sigma}}_m$ based on $\mathbf{A}(\hat{\boldsymbol{\alpha}}) = \mathrm{diag}\{\hat{\alpha}_1, \hat{\alpha}_2, \ldots, \hat{\alpha}_{N_b}\}$.

Note that iterations for α_i are required in practical computation. During this process, it can be found that many α_is tend to infinity, leading to that the corresponding $\hat{\mathbf{b}}_{mi}$ are equal to zero. As a result, these regression terms are pruned out from the input-output relation. Iteration completes when the difference of the evidence value of two adjacent iterations is smaller than the prescribed threshold.

4. Results

Based on the Bayesian learning approach, the optimal relation for frequency and environmental condition is:

$$[\omega_1^2, \omega_2^2] = [1, T_n, T_n^3, T_n^4, H_n^4]\begin{bmatrix} 2.00 & 2.69 \\ -0.33 & -0.19 \\ 0.43 & 1.01 \\ 0.03 & -0.70 \\ 0.06 & 0.09 \end{bmatrix} \tag{16}$$

Compared to the full extracted feature vector $\boldsymbol{x} = [1, T_n, H_n, T_n^2, T_nH_n, H_n^2, T_n^3, H_n^3, T_n^4, H_n^4]^{\mathrm{T}}$, only five features are retained in the optimal model. Recall that the term '1' in the extracted feature represents the constant term. Both the temperature and relative humidity exist, showing that these two ambient factors influence should be considered simultaneously in modal frequencies prediction. Figure 1and Figure 2 show the prediction curves of the squared modal frequency-temperature and the squared modal frequency-relative humidity, respectively. The squared modal frequency-temperature curve is plotted with the relative humidity evaluated at its sample mean (67.9%); and the squared modal frequency-relative humidity curve is plotted with the temperature with its sample mean (24.7°C). The increasing trends for both curves can be observed. Figure 3 shows the normalized residual plot for the two modes. The normalized residuals satisfy $|\bar{\varepsilon}_m| \leq 1.96$, where '1.96' corresponds to the absolute value of the 95% confidential bound for a standard Gaussian error. The results of the normalized residuals confirm that the temperature and relative humidity are the primary environmental effects in frequency prediction for the monitored structure.

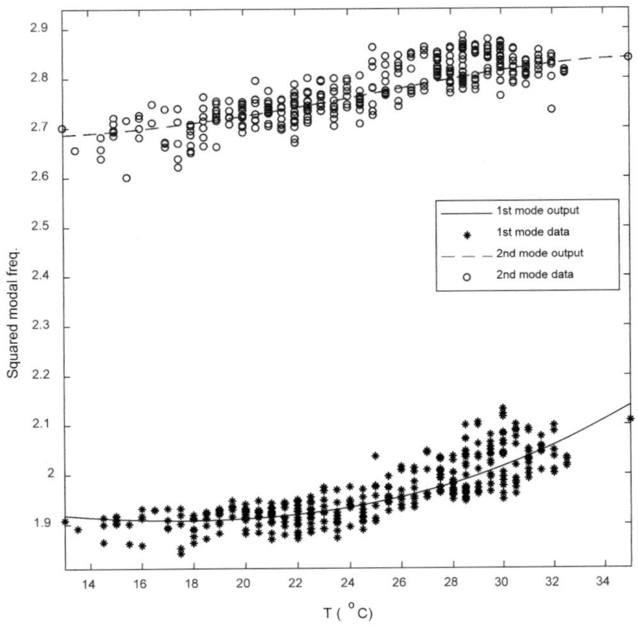

Figure 1. Squared modal frequency-temperature:
1. Lower curve (1st mode); 2. Upper curve (2nd mode)

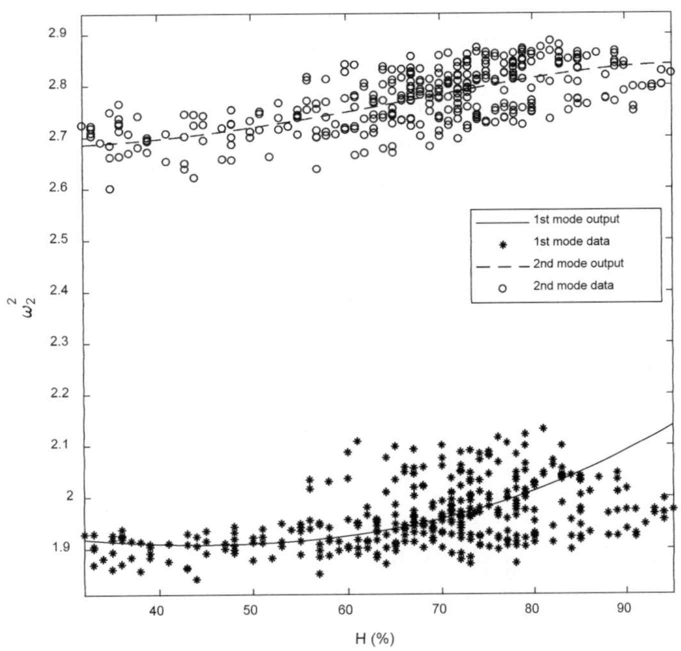

Figure 2. Squared modal frequency-relative humidity:
1. Lower curve (1st mode); 2. Upper curve (2nd mode)

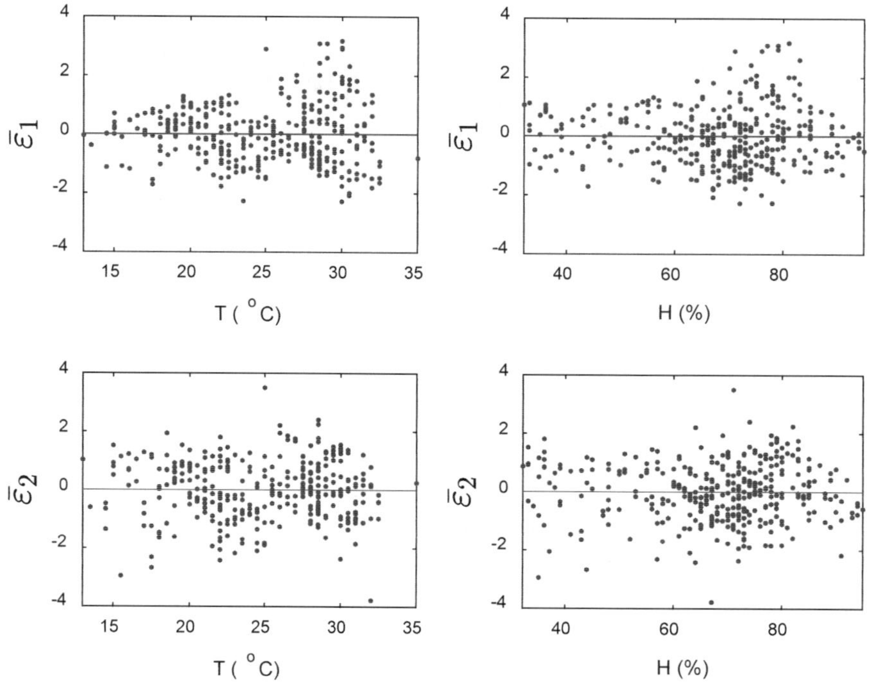

Figure 3. Normalized residual plots of the two modes

5. Conclusions

In this paper, a novel Bayesian learning algorithm is introduced to automatically search for the optimal model in modal frequency-ambient condition pattern recognition based on monitoring data. The learning algorithm is capable for conducting continuous optimal model searching by introducing a sophisticated hyperparameterization on the weighting parameter vector in the prediction model. In contrast to the original ARD prior for the scale output case (Mackay, 1992; Tipping 2004), the ARD prior for the vector outputs is adopted in this study. During the optimization process, the irrelevance components of the weight parameter vector as well as the corresponding features are automatically pruned out while the relevance terms are retained. The analysis results show that both the temperature and relative humidity are related to frequency prediction. The acceptable results of the normalized residual plots show that these two factors are the primary environmental effects in frequency prediction for the monitored structure

Acknowledgement

This work was supported by the National Natural Science Foundation of China (51508201), the State Key Laboratory of Subtropical Building Science, South China University of Technology (2016ZB26). This generous support is gratefully acknowledged.

References

Farrar, CR & Worden, K 2007,'An introduction to structural health monitoring', *Philosophical Transactions of the Royal Society of London A: Mathematical, Physical and Engineering Sciences*, vol. 365, no, 1851, pp. 303-315.

Alampalli, S 1998, 'Influence of in-service environment on modal parameters', *Proceedings of the SPIE the International Society for Optical Engineering*.

Peeters, B & De Roeck, G 2001,'One-year monitoring of the Z24-Bridge: environmental effects versus damage events', *Earthquake Engineering & Structural Dynamics*, vol. 30, no, 2, pp. 149-171.

Liu, CY & De Wolf, JT 2007,'Effect of temperature on modal variability of a curved concrete bridge under ambient loads', *Journal of Structural Engineering-ASCE*, vol. 133, no, 12, pp. 1742-1751.

Farrar, CR, Doebling, SW, Cornwell, PJ & Straser, EG 1997,'Variability of modal parameters measured on the Alamosa Canyon Bridge', *Proceedings of the SPIE The International Society for Optical Engineering*.

Cornwell, P, Farrar, CR, Doebling, SW, & Sohn, H 1999, 'Environmental variability of modal properties', *Experimental Techniques*, vol. 23, no, 6, pp. 45-48.

Zhang, Y, Kurata, M & Lynch, JP 2016, 'Long-term modal analysis of wireless structural monitoring data from a suspension bridge under varying environmental and operational conditions: system design and automated modal analysis', *Journal of Engineering Mechanics*, vol. 04016124.

Sohn, H, Dzwonczyk, M, Straser, EG, Kiremidjian, AS, Law, KH & Meng, T 1999,'An experimental study of temperature effect on modal parameters of the Alamosa Canyon Bridge',*Earthquake Engineering & Structural Dynamics*, vol. 28, no, 8, pp. 879-897.

Cross, EJ, Koo, KY, Brownjohn, JMW & Worden, K 2013,'Long-term monitoring and data analysis of the Tamar bridge',*Mechanical Systems and Signal Processing*, vol. 35, no, 1-2, pp. 16-34.

Mu, HQ &Yuen, KV 2016,'Ground motion prediction equation development by heterogeneous Bayesian learning', *Computer-Aided Civil and Infrastructure Engineering*, vol. 31, no, 10, pp. 761-776.

Mu, HQ& Yuen, KV 2017,'Novel sparse Bayesian learning and its application to ground motion pattern recognition', *Journal of Computing in Civil Engineering (ASCE)*, vol. 04017031.

Mu, HQ, Yuen, KV& Ng, IT 2017,'Prediction on uniaxial compressive strength of rock: Bayesian inference and outlier analysis', *Proceedings of the 12th International Conference on Structural Safety & Reliability (ICOSSAR)*, Wien, Austria.

Yuen, KV, Katafygiotis, LS & Beck, JL 2002,'Spectral density estimation of stochastic vector processes', *Probabilistic Engineering Mechanics*, vol. 17, no, 3, pp. 265-272.

Yuen, KV & Beck, JL 2003,'Updating properties of nonlinear dynamical systems with uncertain input',*Journal of Engineering Mechanics*, vol. 129, no, 1, pp. 9-20.

Yuen, KV 2010,*Bayesian methods for structural dynamics and civil engineering*. John Wiley & Sons.

Xia, Y, Hao, H, Zanardo, G & Deeks, A 2006,'Long term vibration monitoring of an RC slab: temperature and humidity effect', *Engineering Structures*, vol. 28, no, 3, pp. 441-452.

Yuen, KV & Kuok, SC 2010a,'Ambient interference in long-term monitoring of buildings',*Engineering Structures*, vol. 32, no, 8, pp. 2379-2386.

Yuen, KV & Kuok, SC 2010b,'Modeling of environmental influence in structural health assessment for reinforced concrete buildings', *Earthquake Engineering and Engineering Vibration*, vol. 9, no, 2, pp. 295-306.

Mu, HQ, Yuen, KV& Kuok, SC 2016,'Relevance feature selection of modal frequency-ambient condition pattern recognition in structural health assessment for reinforced concrete buildings', *Advances in Mechanical Engineering*, vol. 8, no, 8, pp. 1687814016662228.

MacKay, DJ 1992,'Bayesian interpolation', *Neural computation*, vol. 4, no, 3, pp. 415-447.

Tipping, ME 2004,*Bayesian inference: An introduction to principles and practice in machine learning*. Springer.

Bishop, CM 2006,*Pattern recognition and machine learning*. Springer.

Thayananthan, A, Navaratnam, R, Stenger, B, Torr, P & Cipolla, R 2006,'Multivariate relevance vector machines for tracking',*Proceedings of Computer Vision–ECCV 2006*, pp. 124-138.

The 8th International Conference on
Structural Health Monitoring of Intelligent Infrastructure
Brisbane, Australia | 5-8 December 2017

Exploratory investigation of monitoring data obtained from an in-service pre-stressed concrete bridge incorporating a distinct half-joint layout

Y. Gunawardena[1], F. Aslani[2], Jun Li[3]
[1]*The University of Western Australia, Australia* [2]*The University of Western Australia, Australia*
[3]*Centre for Infrastructural Monitoring and Protection, Curtin University, WA. Email:* junli@curtin.edu.au

Abstract

Measurements obtained from a structural health monitoring (SHM) system currently installed on a three-span pre-stressed concrete bridge are analysed for condition assessment in this paper. The middle span of the bridge contains eight suspended I girders which are supported on half-joints, located at the ends of cantilevered lengths of the side girders. The half-joints also contain external strengthening. The objective of the work is to conduct a preliminary monitoring data based investigation to identify whether and what indices can be derived from the data which could be used to identify the condition or changes in the condition of the bridge.

1. Introduction

Structural health monitoring (SHM) of bridges is an area of continuing interest to many engineers and researchers. Even though such monitoring is mostly done for large iconic bridges, SHM systems have also been installed on short to medium span bridges over the years (Seo et al., 2016). Various indices have been chosen by different researchers, and changes of these indices, which are identified based on the measured data, are used to detect any damage occurring to the bridge (Li and Hao 2016, Li et al., 2014).

In vibration-based SHM, the measured values are usually the accelerations and the indices used are mostly the natural frequencies and mode shapes (Gomez, 2011, Brownjohn et al., 2003). In strain-based SHM, the measured values are strains of the structure at various locations. Strain-based SHM is usually carried out with known loads as the measured strain is dependent on the applied loading. These measured strains can be used to obtain the 'best-fit' parameters of a FE model, so that the measured strains agree with the analytical values (Sanayei et al., 2012). In addition to directly using the measured strains, many researchers have also considered strain based indices such as the neutral axis position, transverse distribution factors and unit influence lines (Wang, 2005, Gangone et al., 2011). In addition to model-based approaches for damage detection referred to above, purely data-based non-model approaches have also been published in the literature. For example Christopher et al. (2014) proposed a method in which the survival distribution functions of maximum measured strains, obtained from a duration of 6 months, were used together with the bootstrapping method to define a 'bridge signature' which was then used as a basis for damage detection.

The study described in this paper has been carried out as part of an ongoing body of work, where the objective is to use SHM data, obtained from a pre-stressed concrete bridge with a unique half-joint arrangement, to understand the behaviour and condition of the bridge and in particular the half-joint. The focus of this work is to use the data in a meaningful manner using simple and established data analysis techniques.

2. Instrumentation system and data measurement methodology

The bridge that has been considered in this paper is a three-span (14m-24m-14m) pre-stressed post-tensioned concrete I girder beam and slab bridge owned by Main Roads Western Australia

(MRWA). The central span of the bridge is made up of eight 17m long suspended I-girders supported on half-joints located at the ends of 3.5m long cantilevered lengths of the side girders. The girder section is solid at the half-joint locations. The half-joints also contain external vertical steel strengthening rods at either side of the half joint as well as horizontal strengthening rods joining the two sides of the joint. The schematic arrangements of the suspended girders of the central span and the half-joint are shown in Figure 1. The half-joint arrangement is different to the typical joint arrangement as there is no bearing between the suspended and supporting nibs while the joint is also post-tensioned by an internal tendon crossing the joint.

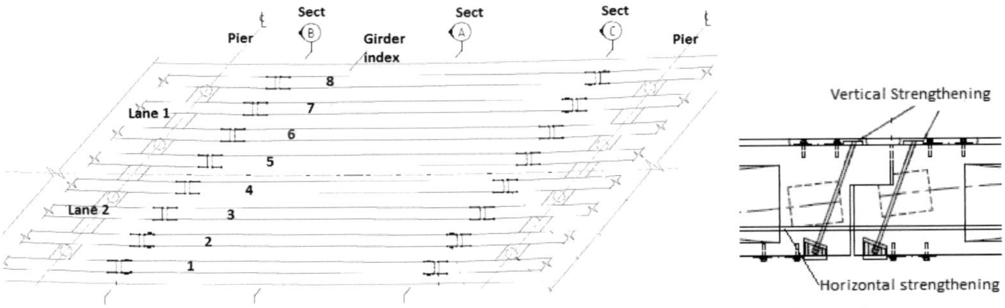

Figure 1. Schematic arrangement of half joint and suspended girders of central span (Continuous internal pre-stressing tendon crossing the joint is not shown for clarity)

An instrumentation system was installed by MRWA on the afore-mentioned bridge in August 2014. The mid-span section of the girders (Section A in Figure 1) was instrumented with 16 strain rings (SR1-16) measuring the strain in the concrete at the lower faces of the top and bottom flanges, as shown in Figure 2. Two tri-axial accelerometers (A1, A2) were also installed on the soffits of girders 1 and 8 as shown at mid-span. In addition, at Sections B and C, as shown, eight strain gauges each (SG1-8, 9-16) were installed on the vertical strengthening rods of the half-joints (on the suspended girder side). This paper discusses the data measured through these strain rings/gauges and accelerometers. The system was setup so that when a vehicle passes over the bridge, if it causes a response above a pre-defined threshold in any of the strain rings (SR1-16), then a two-minute data window (approx. 1 min before to 1 min after the maximum) of all the instrumentation would be recorded. In addition, images taken from a fixed elevated camera showing the vehicle(s) causing the response would also be recorded. MRWA made available 123 such two-minute long 'event data' to the authors recorded during five different months (3, 29 and 1 event(s) in April, September and October 2015 & 40 and 50 events in February and March 2016 respectively).

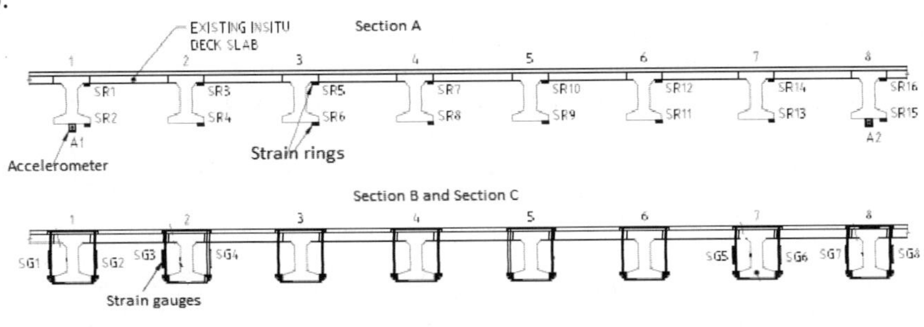

Figure 2. Locations of strain rings/gauges and accelerometers on bridge girders

These event data were recorded normalized with respect to the background measurements and hence contained responses due to the heavy vehicles alone. 18 event data were not used by the authors as those data were found to contain errors and hence this paper discusses the remaining 105 data sets. The data were recorded at a sampling frequency of 130Hz. When evaluating the measured strains, in order to extract the equivalent static response from the measured response, the measured strain signal was filtered by calculating the Fourier transform of the response, then zeroing out all components of the Fourier spectrum above a frequency of 1.25Hz and then taking the inverse Fourier transform to obtain the 'filtered' signal. The cut-off was determined by inspection of the spectrum in a similar manner to that presented by Zaurin and Catbas (2010). Figure 3 shows the unfiltered and filtered responses of the recorded strain signal of one of the sensors due to one event.

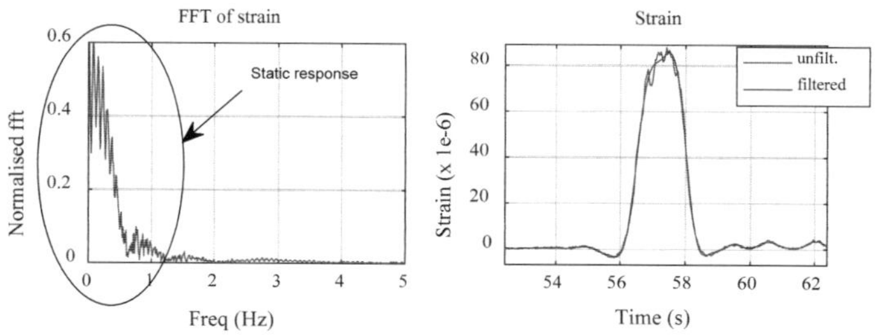

Figure 3. Filtering of measured strain signals to obtain static response

The maximum measured responses from each strain ring/gauge in sections A, B and C were considered in the evaluation. For section A, the tensile and compressive maximums were considered separately. For each event, at the instant when the maximum compressive strain was measured, the corresponding moments (and distribution factors) carried by each girder were calculated by considering the co-existing strains at that instant, assuming no-tension linear elastic behaviour for concrete and linear elastic behaviour for steel. The reinforcement and tendon arrangement were taken from the as-built drawings supplied by the bridge owner. The co-existing strains at the instant of compressive maximum were used so as to minimize any effect of measurement error. The frequency spectra of the vertical acceleration data (negligible in other directions) were obtained using the operational modal analysis (Brincker and Ventura, 2015) technique of Complex Mode Indicator Function (CMIF). In addition to these 'event data', the background response is also recorded by the system, together with environmental conditions such as temperature and wind speed, at 30min intervals. However that data were not used for the work described in this paper.

3. Results and discussion

Based on the measured strains at the mid-span section (Section A), the distributions of the maximum tensile and compressive strains were obtained and are shown in Figure 4. The equivalent static responses due to the passing vehicles were considered for this evaluation. As can be seen in Figure 4, which also differentiates the data by month of measurement, there is considerable scatter in the measured strains. The means and standard deviations of the data, for all the data as well as by month are listed in Table 1. When considering the statistics by month, the events of April and October 2015 were not considered due to inadequate data numbers.

Table 1. Primary statistical properties of measured maximum strains (in micro-strain)

Type	Property	All months	Month 2	Month 4	Month 5
Tensile	Mean	102	104	103	101
	Standard deviation	12.4	11.7	13.3	12.4
Compressive	Mean	-20	-20	-20	-19
	Standard deviation	3.7	3.6	4.0	3.4

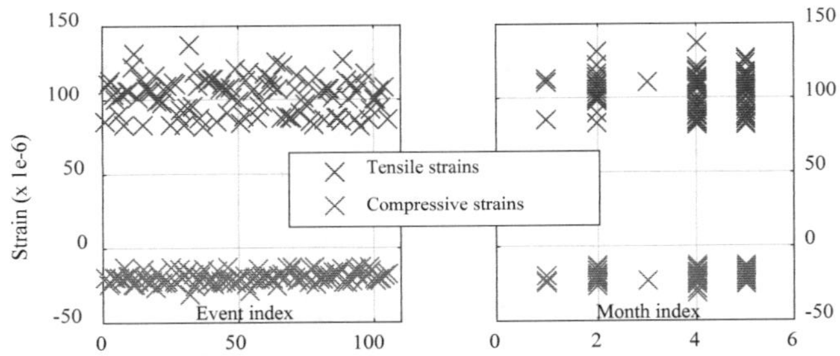

Figure 4. Distribution of maximum strains by event index and by month

There appears to be no significant difference in the primary statistical parameters of the maximum tensile and compressive strains, for the months considered above. Potentially these statistical properties and their variations with time could possibly be used as an indicator of any changes that may occur to the condition of the bridge, in a manner similar to that done by Christopher et al. (2014). However in order to be able to identify any variations, base values of these parameters would have to be established first using a much larger number of event data than the 105 events considered in this paper. Christopher et al. (2014) used strains caused by 1670 heavy vehicle events over a six-month period in order to establish a base-line survival distribution function, which formed the basis of their 'bridge signature' which was used for damage detection.

It should be noted that the maximum tensile and compressive strains did not necessarily always occur in the same girder. The maximum tensile and maximum compressive strains occurred either in the same girder or at girders adjacent to each other or spaced with a girder in between. If the load is primarily carried by bending one would expect the maximum compressive and tensile strains to occur in the same girder, but this apparent discrepancy could be due to differences in the location of the operative elastic neutral axis of each girder. This could be due to the effective operative cross-section, contributing to the stiffness, being different between the eight girders. This difference in effective cross section could be due to a multitude of reasons, such as the skewed nature of the bridge, differences in post-tensioning losses, localized damage occurring in some girders etc., which cannot be established purely based on the measured data. Establishing the reasons behind these differences would require the use of the data together with structural modelling and analysis.

The measured responses inherently depend on the type of the vehicle causing the response and especially on the weights and spacing of the vehicle axles. Since the vehicle type and axle distribution could be qualitatively obtained from the recorded images, it was judged that it would be more representative to consider the distribution of the maximum responses for each vehicle type separately. The distributions of the measured maximum strains by identified vehicle type are shown in Figure 5. Vehicle types for which there were only a very limited number of event data, specifically four and seven-axle mobile cranes and nine and eleven-axle double road trains, together with events for which no image data were available are not shown in Figure 5 (totalling 25 out of the 105 events).

In the absence of sufficient numbers of event data to satisfactorily define base values of the statistical properties of the maximum responses considered previously, it would potentially be possible to use vehicle-type specific maximum data as an indicator to detect any changes occurring to the bridge condition. When using the distribution of all the maximum responses as a condition indicator, the data will be sensitive to any changes in seasonal travel patterns of heavy vehicles, but vehicle type-specific distributions will not have that inherent sensitivity. For vehicle types such as mobile-cranes, which have limited variability in their axle weight and spacing configuration, considering the distribution of the maximum measured responses by vehicle type will also allow greater sensitivity in detecting potential changes. From the data given in Figure 5, it can also be observed that the maximum tensile or compressive strains did not occur in the same girder for all events. For example, it is clear that the majority of the maximum compressive strains were recorded in either girder 3 or girder 5, while most of the maximum tensile strains were recorded in girders 3 or 6.

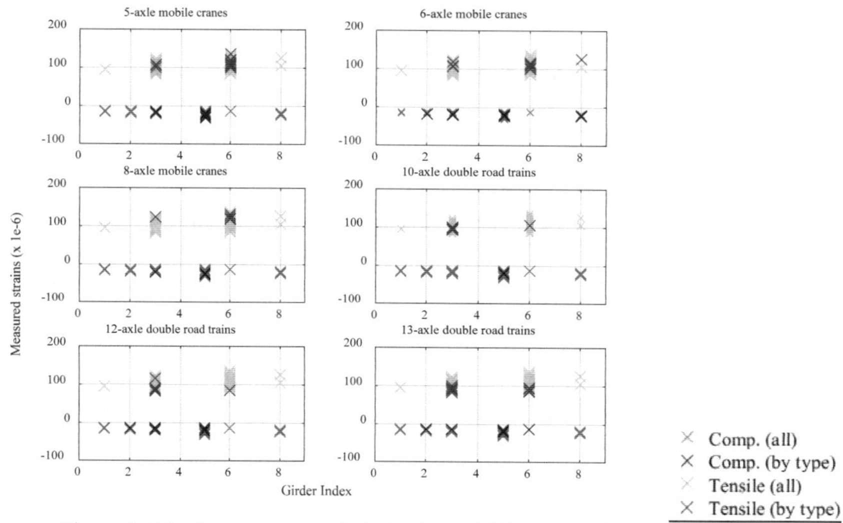

Figure 5. Maximum measured strains by vehicle type and measured girder

The strains measured at the mid-span were also used to calculate the transverse distribution of sectional moment in the girders as described in the methodology. The calculated distribution factors are shown in Figure 6. The distributions have been grouped based on the girder in which the largest measured tensile strain occurred, which is representative of the lane (or lanes) in which the vehicle (or vehicles in the case of two vehicles side by side) causing the response was (were) travelling in.

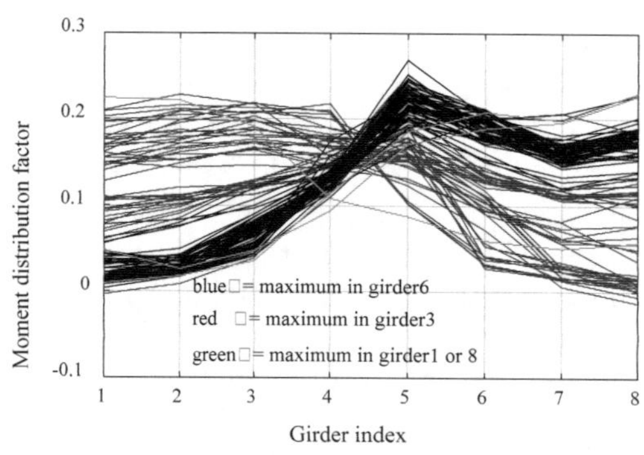

Figure 6. Transverse distribution of moment at mid-span section of suspended span

The transverse distribution of the load could also be used as an indicator of changes occurring to the bridge as has been previously considered by other researchers (Cardini and DeWolf, 2009, Gangone et al., 2011). As can be seen from Figure 6, there is a clearly identifiable general shape of the moment distribution which appears to be dependent on the lane in which the vehicles causing the response were travelling in. However as can be seen from the distributions given in Figure 6, the transverse distributions of moment show far greater variability between the different events. Even for vehicles travelling in the same lane, there is significant variations in the values of the transverse distribution factors. This suggests that the distribution of sectional moment at the mid-span section is quite sensitive to the transverse position of the vehicle within the carriageway lane and hence appears to be less viable as a change indicator.

The distributions of the largest responses measured by the strain gauges attached to the vertical strengthening rods at the half joints (SG1-16) were also considered and are shown in Figure 7. As can be seen from the magnitudes of the strains of the distributions, apart from the maximum (tensile) strains measured by gauge SG1, the measured strains are quite small. Compared with the maximum range of measurement of the strain gauges, which is a strain of 2% (20,000 micro-strain), these small strain values possibly fall within the measurement error of the gauges. Therefore it is demonstrated that it would not be possible to use the strains measured in the strengthening rods as indicators of any changes occurring to the bridge. The maximum strain measured in SG1 corresponds to a stress of 8.4MPa which is 6% of the tension capacity of the rod, which suggests that the utilization of the strengthening rods is low. This in turn suggests that the load-transfer through the half-joint is occurring without significant participation of the rods at present. However the fact that SG1 has non-negligible strains compared to the other gauges suggest that the half-joint corresponding to SG1 may be behaving differently to the other joints.

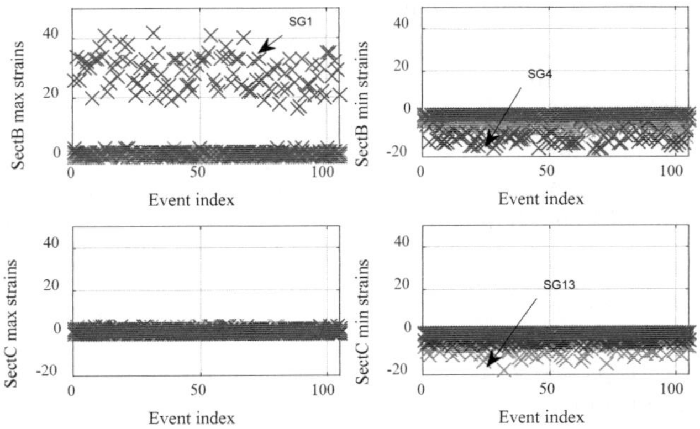

Figure 7. Distributions of maximum responses measured from gauges attached to strengthening rods (in micro-strain)

It is possible in this manner to obtain a qualitative understanding of the half-joint behaviour through the measured rod strains, though for further investigation analysis together with a structural model would be necessary. However, it can be expected that if there is some deterioration of the joint, larger strains would be measured in the gauges attached to the strengthening rods.

The frequency spectra obtained by analysing the measured vertical acceleration responses at the soffits of the edge girders at mid-span for the 105 events are shown in Figure 8. From the spectra as shown in Figure 8, it is observed that there exists a lowest natural frequency of the bridge at approximately 5.8Hz (a number of closely spaced peaks exists in this region as seen in the obtained spectrums) which is in the range of natural frequencies which are generally expected for concrete girder bridges with similar span lengths. For example Brownjohn et al. (2003) obtained through testing that the natural frequency of a 18m span pre-cast pre-tensioned inverted T-beam and slab simply supported bridge was 5.4Hz. If it is considered that the half-joint of the bridge under consideration is an effective simple support, then the suspended central girders can be considered as 17m long simply supported spans and hence the measured value of 5.8Hz compares well with the result obtained by Brownjohn et al. (2003).

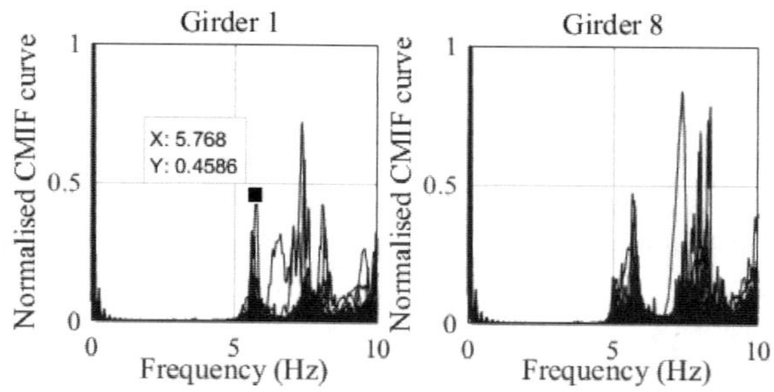

Figure 8. Frequency spectrums obtained from measured accelerations due to 105 events

The peak at approximately 5.8Hz was observed for all the events that have been considered in this paper. This suggests that there may have been no significant changes in the bridge condition, which would affect the vibration characteristics, i.e. frequencies of the bridge, during the time period corresponding to the data. Therefore, it is potentially possible to use the natural frequency of the bridge as a condition indicator, though it is not possible to deduct, using the measured data alone, what changes (and the corresponding sensitivities) in condition could be detected using the measured natural frequency From the work of previous researchers, such as that of Farrar and Doebling (1997), significant condition changes may need to occur for any noticeable changes to occur to the measured frequency spectrum. Since the condition of the half-joint is of particular interest in the bridge that has been considered in this paper, the sensitivity of the measured spectrum to the joint condition needs to be established through structural modelling of the bridge. The measured frequencies could also be used to aid in updating such a model of the bridge.

4. Limitations and Further work

It is also potentially possible to consider a control-chart based approach, as done by Brent et al. (2013) where the relationship between pairs of sensor measurements are considered to detect any anomalies corresponding to any 'changes'. However this would need a significantly more number of event data to establish the control charts than the 105 events used in this paper. A further index which could be used as a condition indicator is the unit influence line of the measurement. Influence lines and specifically changes in the shape of the influence line have been considered by a number of researchers (Wang, 2005, Catbas and Aktan, 2002, Zhu et al., 2014) for condition and damage detection of bridges, though these have mainly been for steel bridges. Influence line based methods (Chen et al., 2016) are attractive as it is dependent on the entire length of the bridge and hence can potentially detect changes which are not localized to the measurement location. However in order to derive the unit influence lines, knowledge of the axle weights and spacings of the specific vehicles causing the response as well as the respective speeds are required. However for the bridge discussed in this paper, complete information on such specifics is not available though it may be possible to use a vehicle-based influence line (which is essentially the measured pulse shaped response shown in Figure 3) as a condition indicator instead, based on the responses due to vehicle types with low variability, such as mobile cranes.

5. Conclusions

This paper has discussed the preliminary investigations carried out on how the measured strains and accelerations, of the suspended central span of a three-span pre-stressed post-tensioned concrete bridge with half-joints, could potentially be used to identify any changes occurring to the condition of the bridge. This work is the initial work of an ongoing overall effort, to use measurement data to better understand the behaviour of the half-joints of the bridge, which are of a unique arrangement. The distributions of the maximum measured girder strains, the moment distribution across the section of measurement, the measured strains in the strengthening rods and the frequency spectra obtained from the acceleration measurements were considered as condition indices. It was clear that except for the measurements obtained from the strengthening rods, the others could potentially be used as indicators to detect changes occurring to the bridge. In some cases, it was also possible to infer a qualitative understanding of the bridge behaviour using the measurements. However it was clearly ascertained that, using the measured data and corresponding data indices alone, it would not be possible to obtain the locations of any changes that would be detected nor would it be possible to obtain the sensitivity of the measured data to any changes that may occur. However, using the data together with a finite element model of the bridge is likely to provide a better physical understanding of the bridge condition and of the half-joints in particular.

Acknowledgements

The support extended to the authors by Main Roads Western Australia (MRWA) through an ARC Linkage project is gratefully acknowledged. The authors also thankfully acknowledge the permission granted by MRWA to publish the information contained in this paper.

References

BRENT, P., PING, L., TERRY, W., LOWELL, G. & JUNWON, S. 2013. Field Validation of a Statistical-Based Bridge Damage-Detection Algorithm. Journal of Bridge Engineering, 18, 1227-1238.

BRINCKER, R. & VENTURA, C. 2015. Introduction to Operational Modal Analysis, Somerset, Somerset : Wiley.

BROWNJOHN, J., MOYO, P. & OMENZETTER, P. 2003. Assessment of highway bridge upgrading by dynamic testing and finite-element model updating. Journal of Bridge Engineering, 8, 162-172.

CARDINI, A. J. & DEWOLF, J. T. 2009. Long-term Structural Health Monitoring of a Multi-girder Steel Composite Bridge Using Strain Data. Structural Health Monitoring, 8, 47-58.

CATBAS, F. N. & AKTAN, A. E. 2002. Condition and damage assessment: Issues and some promising indices. Journal of Structural Engineering, 128, 1026-1036.

CHEN, Z.W., Cai, Q. L. & LI, J. 2016, Stress influence line identification of long suspension bridges installed with structural health monitoring systems, International Journal of Structural Stability and Dynamics, 16, 1640023.

CHRISTOPHER, W. F., MASOUD, S., BRIAN, R. B. & RICHARD, M. V. 2014. Statistical Bridge Signatures. Journal of Bridge Engineering, 19, 04014022.

FARRAR, C. R. & DOEBLING, S. W. 1997. Lessons learned from applications of vibration-based damage identification methods to a large bridge structure. IWSHM, Stanford.

GANGONE, M. V., WHELAN, M. J. & JANOYAN, K. D. 2011. Wireless Monitoring of a Multispan Bridge Superstructure for Diagnostic Load Testing and System Identification. CACIE, 26, 560-579.

GOMEZ, H. C. 2011. System Identification of Highway Bridges using Long-Term Vibration Monitoring Data. 3491285, University of California, Irvine.

LI, J. & HAO, H. 2016. A review of recent research advances on structural health monitoring in Western Australia, Structural Monitoring and Maintenance, 3, 33-49.

LI, J., HAO, H. & ZHU, H.P. 2014. Dynamic assessment of shear connectors in composite bridges with ambient vibration measurements, Advances in Structural Engineering, 17, 617-638.

SANAYEI, M., PHELPS, J. E., SIPPLE, J. D., BELL, E. S. & BRENNER, B. R. 2012. Instrumentation, nondestructive testing, and finite-element model updating for bridge evaluation using strain measurements. Journal of Bridge Engineering, 17, 130-138.

SEO, J., HU, J. W. & LEE, J. 2016. Summary Review of Structural Health Monitoring Applications for Highway Bridges. Journal of Performance of Constructed Facilities, 30, 04015072.

WANG, X. 2005. Structural condition assessment of steel stringer highway bridges. In: SWANSON, J. A. (ed.). ProQuest Dissertations Publishing.

ZAURIN, R. & CATBAS, F. N. 2010. Integration of computer imaging and sensor data for structural health monitoring of bridges. Smart Materials and Structures, 19, 015019.

ZHU, S., CHEN, Z., CAI, Q., LEI, Y. & CHEN, B. 2014. Locate Damage in Long-Span Bridges Based on Stress Influence Lines and Information Fusion Technique. Advances in Structural Engineering, 17, 1089-1102.

The 8th International Conference on
Structural Health Monitoring of Intelligent Infrastructure
Brisbane, Australia | 5-8 December 2017

Damage Identification based on Monte Carlo Method using Likelihood Estimates

B. Wen [1], C.F. Wan[1], R. Huang [1], L.Y. Huang [1], T. Sato [1], Y. Wang [1], L.Y. Xie [2], H.S. Tang [2].

[1] *Southeast University, Key Laboratory of concrete and prestressed concrete structure of Ministry of Education – China, Email:* wan_seu@163.com [2] *Research Institute of Structural Engineering and Disaster Reduction, Tongji University - China*

Abstract

Structural health monitoring of civil engineering structures is a fundamental issue for structural safety and integrity, due to the fact that these structures will deteriorate after they are built and put into services. Within which, damage identification will be the most important problem. In this paper, a structural damage identification method based on Monte Carlo Method, will therefore be introduced. The identification will be made by studying the Likelihood Estimates. The application of the proposed method to a lumped mass shear model structure will be studied. The parameters of the stiffness and damping will be identified and analyzed. Simulations are conduct to verify the effectiveness of the proposed approach. Final identification results show that the proposed method can do the work very well and could be a promising approach for real applications.

1. Introduction

As the most important part of structural health monitoring, damage identification includes four aspects, namely detecting the existence, the location, the severity of structural damage and useful life of this structure. Most damage identification judge the existence and degree of damage by comparing structural parameters' (such as stiffness and damping) change rules or difference before and after the damage occurs (Casas 1994, Doebling 1996). This method offers a useful solution for the problems in structural dynamic process.
At present, there are many structural damage identification methods. The most famous one would be the vibration based method (Farrar, 2001). Also, damages are often identified by the parameter identification algorithms (Wan, 2013). In this paper, a Monte Carlo based Method using response likelihood will be introduced. Monte Carlo Method means when the problem is about the probability of some random events or the expected value of random variables, we can estimate the probability of these random events by the frequency of the event in some "experiments" or get some numerical characteristics of the random variable as the solution (James 1980, Yuen 2004). Through this process, statistics comes to be essential (Feng 2005). In this paper, the damage identification of a single degree of freedom structure using Maximum Likelihood Estimates, which means identifying structural parameters through the whole time domain and then choosing samples with the biggest sum of likelihood values as the result, will be introduced.

2.Single and multiple degree of freedom structure model

Single-degree-of-freedom (SDOF) linear system is the most basic and simplest system. On one hand, many practical problems can be simplified to SDOF linear system. On the other hand, single-degree linear system is also the basis of multi-degree system, continuous system and nonlinear system to conduct vibration analysis. A structure with SDOF linear system is shown in Fig.1.

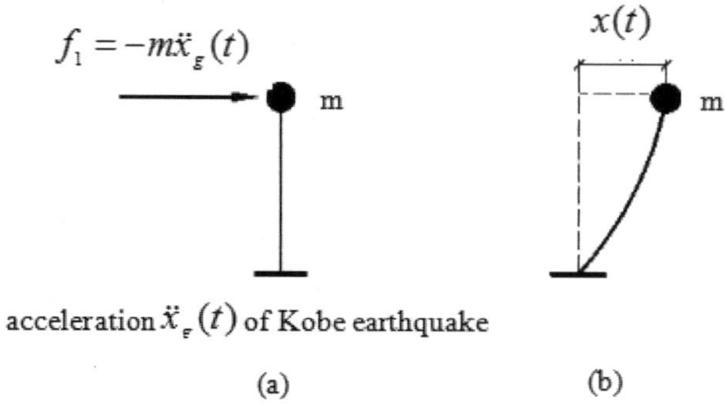

Fig.1 Lumped-mass single degree of freedom structure under translational motion

The structure shown in Fig.1 is founded upon the basic assumptions that the foundation is rigid and the structure is elastic, so the horizontal motion of the whole structure is consistent. During earthquake, horizontal ground motion will drive the building's foundation and cause the vibration of the upper structure, leading to particles' displacement $x(t)$, velocity $\dot{x}(t)$ and acceleration $\ddot{x}(t)$. $\ddot{x}_g(t)$ is the horizontal ground acceleration of Kobe earthquake.

Assuming the initial parameters of the structure shown in Fig.1 as: mass m_0=400kg, stiffness k_0=2500N/m, damping ratio ξ =0.05, then damping coefficient is supposed to be $c_0 = 2mw_n\xi = 2\sqrt{mk}\xi = 2\times 0.05\times\sqrt{400\times 2500} = 100N\cdot s/m$. At the beginning, the displacement and velocity of the structure are both zero. Using $\ddot{x}_g(t)$ as the input power and its time-history curve is illustrated in Fig.2. The effective earthquake force $f(t)$ equals to $-m\ddot{x}_g(t)$.

The motion equation of the structure is shown in Eq.(1)

$$m\ddot{x}(t) + c\dot{x}(t) + kx(t) = f(t) \tag{1}$$

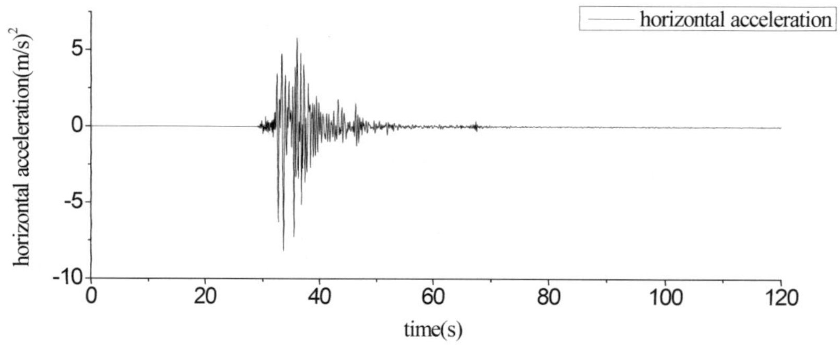

Fig.2 Time-history curve of Kobe earthquake's acceleration

Considering the earthquake motion at the base, dynamic responses can be calculated by Eq.(1). Fig.3 shows the structural acceleration responses subjected to Kobe earthquake excitation at the base, with no noises considered at present.

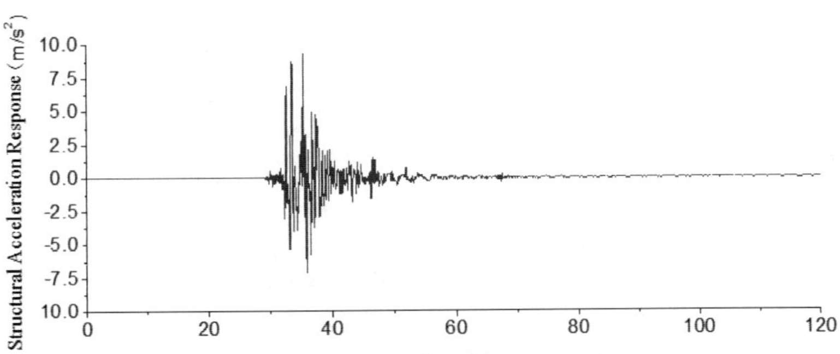

Fig.3 Structural acceleration responses under Kobe earthquake excitation

However, the actual structures are commonly with multi-degree systems. In this paper, multi-degree structures is considered as the following model shown in Fig.4.

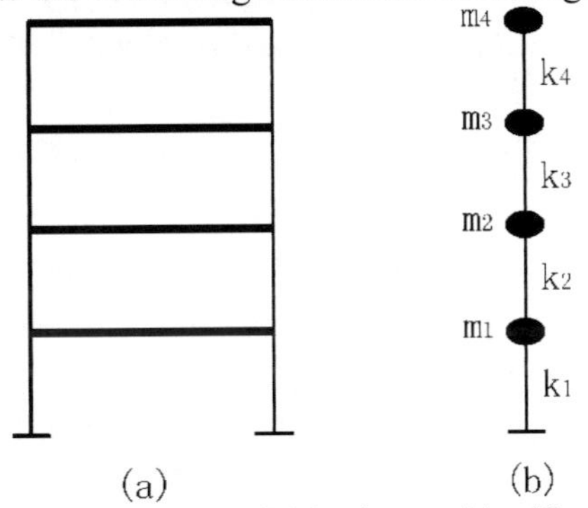

Fig.4 Multi-degree model for damage identification

3.Structural parameters identification without noise

When identifying the parameters (stiffness k and damping c), we should firstly assume the distribution range and predict the interval of these two parameters. In this section, we take n groups of stiffness k_i and damping c_i randomly according to Eq.(2) and Eq.(3)

$$k_i \sim rand_i\left(n,[1500,3000]\right) \quad i=1, 2\ldots n \tag{2}$$

$$c_i \sim rand_i\left(n,[20,120]\right) \quad i=1, 2\ldots n \tag{3}$$

Among the equation, randi is a function in matlab which can produce a uniform distribution of pseudo random integers. When n=1000, structural parameters' (stiffness k_i and damping c_i) distribution of initial predicted sample points is shown in Fig.5.

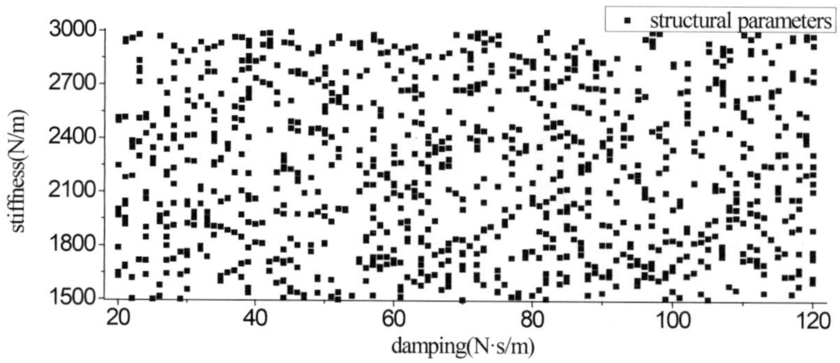

Fig.5 structural parameters' distribution of initial forecast sample points

According to Eq.(4), the SDOF motion equation, we can get n groups of time-history curve of displacement response $x_i(t)$, velocity response $\dot{x}_i(t)$ and acceleration response $\ddot{x}_i(t)$ from n groups of different k_i and c_i. Regarding 0.02s as the time step.

$$m\ddot{x}(t) + c_i\dot{x}(t) + k_i x(t) = -m\ddot{x}_g(t) \tag{4}$$

From initial parameters k_0 and c_0, time-history curve of displacement response $x_0(t)$, velocity response $\dot{x}_0(t)$ and acceleration response $\ddot{x}_0(t)$ can also be achieved according to Eq.4 and considered as observed value.

Through the maximum likelihood function in statistics, likelihood value of each sample point can be calculated in the time interval t and the sample with the maximum likelihood value tends to be the predicted sample. That also means we screen out k_i and c_i nearest to k_0 and c_0.

As shown, maximum likelihood values can be get from Eq.(5), Eq.(6) and Eq.(7)

$$L(x_i(t)) = \max \sum_{i=1}^{n} \frac{1}{\sqrt{2\pi}\sigma} \exp[-\frac{(x_i(t) - x_0(t))^2}{2\sigma^2}] \tag{5}$$

$$L(\dot{x}_i(t)) = \max \sum_{i=1}^{n} \frac{1}{\sqrt{2\pi}\sigma} \exp[-\frac{(\dot{x}_i(t) - \dot{x}_0(t))^2}{2\sigma^2}] \tag{6}$$

$$L(\ddot{x}_i(t)) = \max \sum_{i=1}^{n} \frac{1}{\sqrt{2\pi}\sigma} \exp[-\frac{(\ddot{x}_i(t) - \ddot{x}_0(t))^2}{2\sigma^2}] \tag{7}$$

By using three equations above, the likelihood value of 1000 samples can be calculated separately on the basis of displacement, velocity and acceleration response. Fig.6 shows the likelihood value calculated from acceleration response.

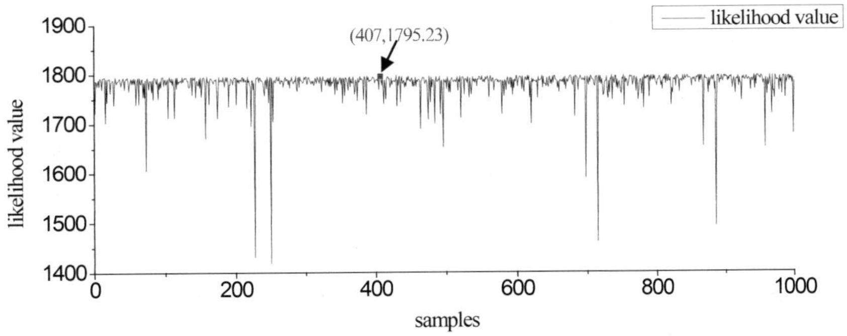

Fig.6 Likelihood value of samples (calculated from acceleration response)

The marked point in Fig.6 is supposed to be the point with maximum likelihood value. Abscissa corresponds to the sample number and ordinate corresponds to the maximum likelihood value.

Final results of noiseless single-degree-of-freedom structural parameters (stiffness k and damping c) based on maximum likelihood estimates are shown in Table.1

Table.1 Damage identification results of the noiseless SDOF structure (n=1000)

Parameters to be identified	Parameters of true value	Identification results	Identification error
$k(N/m)$	2500	2494	0.24%
$c(N\cdot s/m)$	100	102	2%

According to these parameters identified, dynamic time-history curve of identified parameters (stiffness k_s and damping c_s) and parameters of true value (stiffness k_0 and damping c_0) can be drawn.

If we choose more samples, for example, n=2000, final results of noiseless single-degree parameters (stiffness k and damping c) based on maximum likelihood estimates are shown in Table.2

Table.2 Damage identification results of the noiseless SDOF structure (n=2000)

Parameters to be identified	Parameters of true value	Identification results	Identification error
$k(N/m)$	2500	2497	0.12%
$c(N\cdot s/m)$	100	102	2%

Comparing Table.1 and Table.2, we can find the more samples we choose, the higher accuracy this identification method would have and the longer time it would take.

For multi-degree structure, assuming real parameters as following:

Mass matrix: $$[M]=\begin{bmatrix} m_1 & & & \\ & m_2 & & \\ & & m_3 & \\ & & & m_4 \end{bmatrix}=\begin{bmatrix} 400 & & & \\ & 400 & & \\ & & 400 & \\ & & & 400 \end{bmatrix}kg$$

Stiffness matrix: $$[K]=\begin{bmatrix} k_{11} & k_{12} & k_{13} & k_{14} \\ k_{21} & k_{22} & k_{23} & k_{24} \\ k_{31} & k_{32} & k_{33} & k_{34} \\ k_{41} & k_{42} & k_{43} & k_{44} \end{bmatrix}=\begin{bmatrix} 5000 & -2500 & 0 & 0 \\ -2500 & 5000 & -2500 & 0 \\ 0 & -2500 & 5000 & -2500 \\ 0 & 0 & -2500 & 2500 \end{bmatrix}N/m$$

Damping matrix: $[C]=\begin{bmatrix} 200 & -100 & 0 & 0 \\ -100 & 200 & -100 & 0 \\ 0 & -100 & 200 & -100 \\ 0 & 0 & -100 & 100 \end{bmatrix} N\cdot s/m$

n=4^8 samples are chosen and final results are shown in Table.3.

Table.3 Damage identification results of the noiseless MDOF structure (n=4^8)

Parameters to be identified		Parameters of true value	Identification results	Identification error
$k(N/m)$	K_1	2500	2608	4%
	K_2	2500	2419	3%
	K_3	2500	2380	5%
	K_4	2500	2417	3%
$c(N\cdot s/m)$	C_1	100	104.2	4.2%
	C_2	100	96.5	3.5%
	C_3	100	95.9	5.9%
	C_4	100	93.8	6.2%

Considering multi-degree structures, the identification error tends to become larger than single-degree structure. The multi-degree structures are more complicated than single-degree ones, so the simulation analysis is more different from the actual structures, making the results error larger.

4.Structural parameters identification with gaussian noise

Gauss noise is one kind of common noise whose probability density function corresponds to Gaussian distribution. Adding noise to the data in order to simulate the error of the analytical model and the effect of noise on the actual measurement vectors enhances the reliability of the proposed method. Gaussian noise with a mean value of zero and standard deviation equal to one standard deviation of the measured data is added to the input vectors. We add noise with different levels to the true values as the observed signal to study their effect. Fig.7 shows the Observed acceleration signal with 30% Gaussian noise.

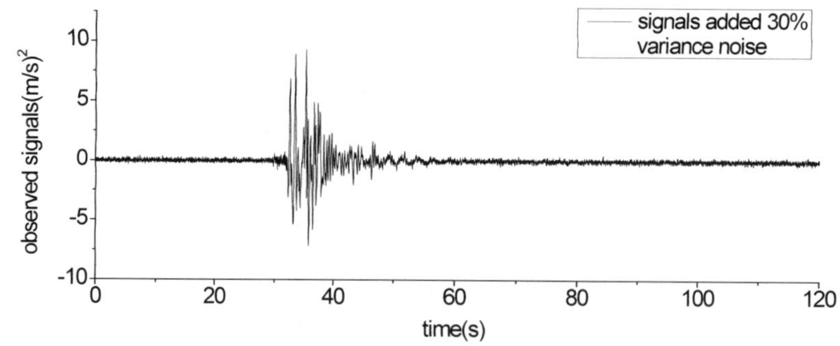

Fig.7 Observed acceleration signal with 30% Gaussian noise

Identifying structural parameters (stiffness k and damping c) under standard deviation Gauss noise of different levels and then comparing the identification results. The comparison is illustrated in Fig.8 and Fig.9.

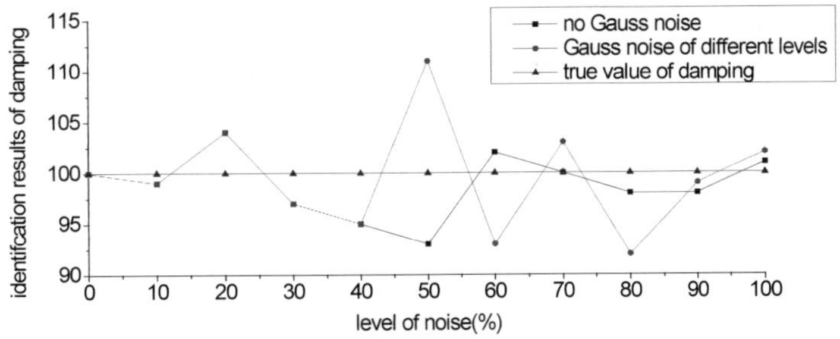

Fig.8 Identification results of damping under Gaussian noise of different levels

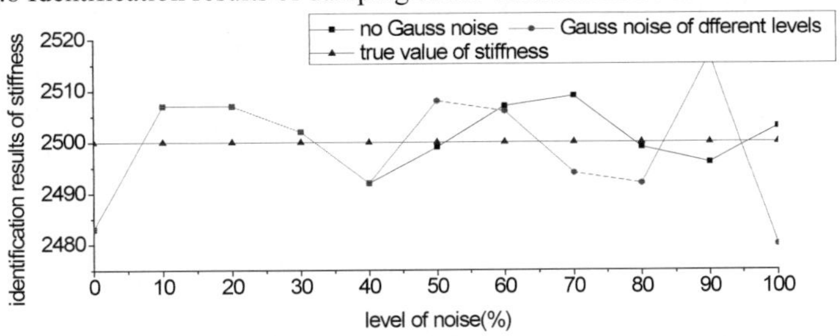

Fig.9 Identification results of stiffness under Gaussian noise of different levels

As shown in Fig.8 and Fig.9, the addition of Gaussian noise does not make obvious effect on the identification of stiffness and damping. This is because all the samples are generated randomly while the identification result will be influenced by samples. However, on the whole, it shows that this method is not so sensitive to the Gaussian noise, even the noise level becomes fairly large, the parameters still can identified without very large errors.

For multi-degree structure, we choose n=4^8 samples and final results are shown in Table.4.

Table.4 Damage identification results of the MDOF structure under Gaussian noise (n=4^8)

Parameters to be identified		Parameters of true value	Identification results	Identification error
$k(N/m)$	K_1	2500	2382	4.7%
	K_2	2500	2520	0.8%
	K_3	2500	2544	1.8%
	K_4	2500	2412	3.5%
$c(N\cdot s/m)$	C_1	100	104.3	4.3%
	C_2	100	97.2	2.8%
	C_3	100	101.3	1.3%
	C_4	100	101.5	1.5%

5.Conclusions

In this paper, parameters identification of SDOF and MDOF structures under no noise and Gauss noise of different levels is studied and discussed. Through comparisons, it can be found that the proposed method can help to identify the structural parameters. With the increase of sampling number, structural parameter identification error will decrease accordingly. It is also found that structural parameter identification based on likelihood estimates has favorable anti-noise performance. Unless the level of noise reaches a certain degree, the identification results seem to be robust to the noise. Sometimes when the structure is faced with damage, the identification results can also be certainly fine using likelihood method.

Acknowledgements

This work was supported by the National Natural Science Foundation of China (No. 51578140), a Project Funded by the Priority Academic Program Development of Jiangsu Higher Education Institutions (PAPD, No. CE02-2-8), Postgraduate Research & Innovation Program of Jiangsu Province and Fundamental Research Funds for Southeast University (No. KYLX15_0084).

References

Casas, JR & Aparicio, AC 1994, 'Structural damage identification from dynamic-test data', *Journal of Structural Engineering*, vol. 120, no. 8, pp. 2437-2450.

Doebling, SW, Farrar, CR, Prime, MB & Shevitz, DW 1996, 'Damage identification and health monitoring of structural and mechanical systems from changes in their vibration characteristics: a literature review', *Shock & Vibration Digest*, vol. 30, no. 11, pp. 2043-2049.

Farrar, CR, Doebling, SW & Nix, DA 2001, 'Vibration–based structural damage identification', *Philosophical Transactions of the Royal Society of London A: Mathematical, Physical and Engineering Sciences*, vol. 359, no. 1778, pp. 131-149.

Wan, CF, Sato, T, Wu, ZS & Zhang, J 2013, 'Damage identification using chaotic excitation', *Smart Structures and Systems*, vol. 11, no. 1, pp. 87-102.

James, F 1980, 'Monte Carlo theory and practice', *Reports on Progress in Physics*, vol. 43, no. 9, pp. 1145-1189.

Yuen, KV, Beck, JL & Au, SK 2004, 'Structural damage detection and assessment by adaptive Markov chain Monte Carlo simulation', *Structural Control and Health Monitoring*, vol. 11, no. 4, pp. 327-347.

Feng, X, Li, G & Zhou, J 2005, 'State-of-the-art of statistical identification for structural health diagnosis in civil engineering', *Earthquake Engineering and Engineering V ibration*, vol. 25, no. 2, pp. 105-113.

The 8th International Conference on
Structural Health Monitoring of Intelligent Infrastructure
Brisbane, Australia | 5-8 December 2017

Structural damage detection using two sensors measured responses with cross-correlation method

Z.H. Nie[1,2]
[1] *School of Mechanics and Construction Engineering, Jinan University, Guangzhou, Guangdong, China.*
Email: Nzhenhua83@gmail.com.
2 *State Key Laboratory for Manufacturing Systems Engineering, Xi'an Jiaotong University, Xi'an, Shanxi, China.*

Abstract

In this paper, a novel damage detection method that use only two sensors to detect the structural damage of beam bridge subjected to a moving mass is proposed. In this method, a moving mass is simulated as a moving vehicle pass through the bridge, and the structure vibration responses at two random positions are measured. A moving window is defined with a certain length determined by the sampling frequency and fundamental frequency of the response. The windowed pair time series cut out from the two measured responses are used to calculate their cross-correlation, and the damage index is defined using the cross-correlation coefficient. In order to demonstrate the feasibility and reliability of the method, a simple supported beam bridge subjected to a moving mass is simulated. The numerical results indicate that the proposed method can identify the damage location accurately which prove that the new index K is a good candidate in structural damage detection.

1. Introduction

Structural damage detection using the vibrational responses has been a hot research topic in the field of structural health monitoring. A direct and fast method is to evaluate and localize damage using the changes in responses between the intact and damaged states of the structure.

Traditional method such as modal-based approaches need enough sensors to measure the response to make sure the accuracy of structural modes (Pandy,1944; Timothy,2008). Therefore, the high cost using these methods in real engineering can not satisfy the requirements of the bridge management. In recent years, many researchers have sought to use a small amount of sensors to detect structural damage, which becomes a hot issue in the research. A novel approach using a single sensor to measure the response at a certain position of the structure subjected to a moving load to identify the damage location was put forward early in 2006 by Zhu and Law (Zhu, 2006). The authors proposed a method using the response obtained at a single measuring point of a beam structure and analyzed by Continuous Wavelet Transform (CWT) and the location of the cracks was estimated. Roveri (Roveri, 2012) proposed a HHT based method using a single point measurement to identify the presence and the location of the damage. Nie (Nie, 2017) proposed a novel damage detection method based on the reconstructed phase space of vibration signals using a single sensor for a moving loaded beam bridge.

The idea of above references (Zhu, 2006; Roveri, 2012; Nie, 2017) is using some local properties change of the single measured response when moving vehicle pass across the damage location of the bridge to locate the damage. This paper proposes another method using only two sensors to locate the damage. The main idea is that using two sensors installed at random two positions (not the same) of bridge measure the responses, the cross-correlation of this two response is analyzed. Using the local cross-correlation singularity of the two responses when the moving vehicle pass cross the damage position, the damage of the bridge is localized. Wang (Wang, 2013) proposed the inner product vector (IPV) as a damage detection algorithm which uses cross correlation functions between vibration responses under band pass white noise excitation. Yang (Yang,2007) proposed an approach to detecting structure damage using the cross correlation function amplitude vector (CorV) of the measured vibration responses. Actually, to the best of author knowledge, the articles with the idea that using the response measured with only two sensors to locate structural damage are relatively rare in structural health monitoring.

2. Methodology

2.1 Dynamic of the beam subjected to a moving mass

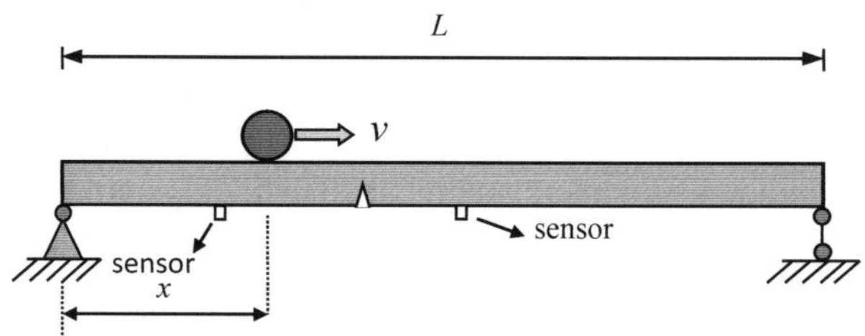

Fig. 1 Beam bridge model subjected to a moving mass.

The uniform Euler-Bernoulli beam with moving mass is considered as shown in Fig.1. The governing differential equation for vibration of the beam within the domain $x \in [0, L]$, neglecting the damping, can be written as:

$$m(x)\frac{d^2 w(x,t)}{dt^2} + Dw(x,t) = p(x,t), \tag{1}$$

where m is the mass of per unit length of the beam, $w(x, t)$ denote the deflection with x and t representing the position of a point in the beam and the time, respectively, and

$$D = EI\frac{\partial^4}{\partial x^4}, \tag{2}$$

$$p(x,t) = m_m\left(g - \frac{d^2 w(x,t)}{dt^2}\right)\delta(x - vt), \tag{3}$$

where m_m is the mass of the moving mass. The displacement $w(x, t)$ can be obtained with Eq.(1-3). In this paper, two points displacement responses are used to locate the damage of the beam.

2.2 Damage detection method using two sensors

This paper proposed a novel approach that using the correlation of two sensors measured responses to locate the damage of beam bridge. The main idea of the method is that when the moving vehicle pass through the bridge, the local correlation of the two measured responses is analyzed. So that the correlation coefficient of the two responses at different time along with the vehicle pass through the bridge are applied to locate the damage. However, if we analyze the whole time series of the two response, we can obtain the correlation coefficient which only indicates the correlation of the two whole responses, and the coefficient can not reflect the local correlation information of the time series. A moving window is defined to cut out a segment of time series with a certain length from the two response respectively. The two coinstantaneous time series sections are used for cross-correlation analysis.

Cross-correlation can be explained as a single number which describes the extent of relationship between two variables. The relationship between these two variables is described through a single value, which is the coefficient. Correlation coefficient is the number that represents the level of relationship between two individual variables. The cross-correlation of the continuous functions $x(t)$ and $y(t)$ is expressed by the formula:

$$R_{xy}(\tau)=\int_{-\infty}^{+\infty} x(t)y(t+\tau)dt \tag{4}$$

where τ is the lag. A positive value of τ actually means that $y(t+\tau)$ leads $y(t)$. Similarly, for discrete signal $x(m)$ and $y(m)$, the cross-correlation is defined as:

$$R_{xy}[n]=\sum_{m=-\infty}^{\infty} x[m]y[m+n] \tag{5}$$

where n is the number of sampling points in the lag. Here a moving window is defined with a length for l to cut out the two measured responses coinstantaneous. The length l can be determined as:

$$L=2\frac{f_s}{f_1} \tag{6}$$

where f_s is the sampling frequency, f_1 is the fundamental frequency of the response. Using the moving window scan the two responses from t=0 to the end, N-l synchronous pairs time series are obtained, where N is the length of the response. Cross-correlation coefficient of the two section signals in the window at ith step is calculated as:

$$R_{x_iy_i}[n]=\sum_{t=1}^{L} x_i[m]y_i[m+n],\quad i=(1\cdots N-l) \tag{7}$$

where n has the length for l. The windowed damage index K_i at ith step can be defined as:

$$K_i=\frac{\sqrt{\sum\left\|R^d_{x_iy_i}\right|-\left|R^u_{x_iy_i}\right\|}}{\sum\left|R^u_{x_iy_i}\right|}, \tag{8}$$

where $R^d_{x_iy_i}$ is the windowed cross-correlation coefficient vector of the two measured responses on the damage beam, while $R^u_{x_iy_i}$ is the undamaged cross-correlation coefficient vector. Due to the window has N-l steps moving through the response, the index K has the length for N-l.

3 Numerical simulation and results

To demonstrate the proposed method, a simple supported beam as shown in Fig. 1 is employed as an example. The physical-geometric parameters of example beam model are as follows: length of beam is L=20m, height is h=0.2m, width b=0.1 m, Young's modulus is E=200GPa, material density is $\rho=7850\ kg\cdot m^{-3}$. The moving mass with the mass for 400kg is used to simulate a passing vehicle. A damage with the ratio equal to 10%, 20%, 30% , 40% and 50% at the position of L_d (L_d=x_d/L, where x_d is the damage position) equal to 0.6 from the left support, as shown in Figure 1, is assumed in the beam in numerical simulations. Displacement responses at random two measurement points are calculated and used in the analysis for damage identification (here we use the 1/4 and mid span points responses).The velocity of the moving mass is 1m/s and the sampling frequency is 200 Hz.

Ignored the influence of gravity, the displacement responses of different damage degrees of the mid-span point are shown as Fig.2-a. As shown, the damaged displacement curves deviate from undamaged one and the deviation increasing with the damage level developing. However, the damage location can not be identified. Then the proposed mothed is tried to implement the damage localization task. Firstly, the moving window should be defined using the FFT spectrum to determine the window length. The Fig.2-b shows the FFT spectrum of the response at 1/4 length of the beam for 10% damaged scenario. As shown, the fundamental frequency of the response is 1.12 Hz, which can be used to calculate the window length with the result for 2×200/1.12=357.

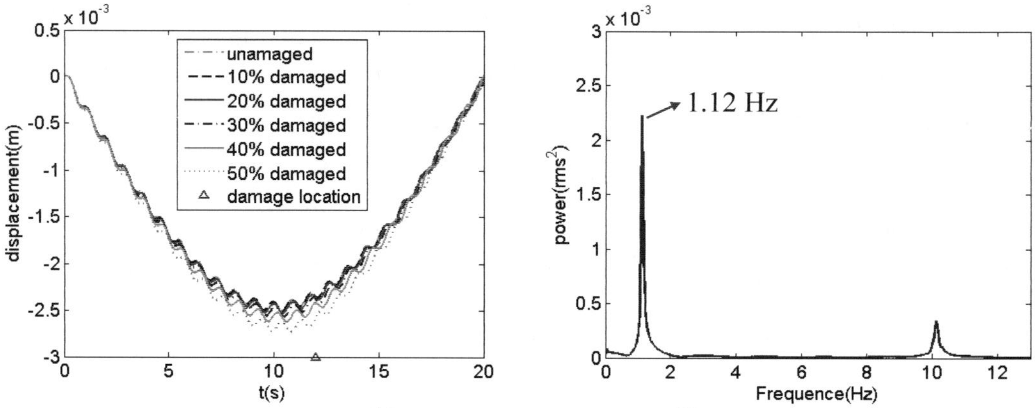

(a) (b)

Fig.2 Displacement responses and the FFT spectrum: (a) Displacement responses of different damage degrees; (b) FFT spectrum of the response at 1/4 length of the beam for 10% damaged scenario.

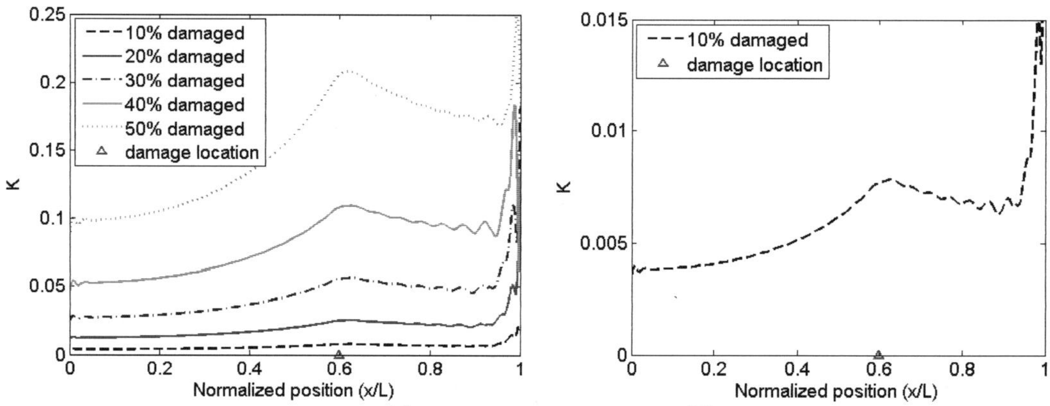

(a) (b)

Fig.3 The results of the damage index K: (a) the curves of K for different damage degree; (b) the curve of K for 10% damaged scenario.

Using the window scan the responses, the results are shown as Fig.3-a. The results indicate that each curve has an obvious peak when the mass pass across the position for x/L=0.6 except for the 10% damaged scenario. Nevertheless, plotting the individual result of 10% damaged scenario, an obvious peak also occur at the normalized position equal to 0.6. This position is exactly where the damage was set in the numerical simulation. Besides, with the damage severity increasing, the value of K also increase. The phenomenon indicates that the proposed method can accurately locate the damage.

4 Conclusions and discussion

In this paper, a novel damage detection method that using only two sensors to detect the structural damage of beam bridge subjected to a moving mass presents is proposed. In this method, a moving mass is simulated as a moving vehicle pass through the bridge, and the structure vibration responses at two random positions are measured. A moving window is defined with a certain length determined by the sampling frequency and fundamental frequency of the response. The windowed pair time series cut out from the two measured responses are used to calculate they cross-correlation, and the damage index is defined using the cross-correlation coefficient. In order to demonstrate the feasibility and reliability of the method, a simple supported beam bridge subjected to a moving mass is simulated. The numerical results indicate that the proposed method can identify the damage location accurately which prove that the new index K is a good candidate in structural damage detection.

In this paper, the main innovation is that only using two sensors to location the damage, which is a leap in sensors consumption compared with some traditional methods such as model-based method. However, there are some deficiencies in this method. The displacement responses are used to extract the damage index. But the acceleration responses are mostly measured in current real structural health monitoring system, while rare displacement responses are measured. Besides, in this method, the undamaged responses are used as the baseline data, nevertheless the baseline data is difficult to be collected. In the future, the main work is to develop a new indicator using only one or two sensors to locate the damage using acceleration without baseline.

Acknowledgements

The authors acknowledge the financial supports from the National Natural Science Foundation of China under grant number (11402098), State Key Laboratory Open Fund of China(SV2014-KF-16), Science and Technology Plan Project of Guangdong Province, China (2013B021500008).

References

Pandy, AK & Biswas, M 1994, 'Damage detection in structures using changes in flexibility', *Journal of Sound and Vibration*, vol.169, no.1, pp. 3-17.

Whalen, TM 2008, 'The behavior of higher order mode shape derivatives in damaged beam-like structures', *Journal of Sound and Vibration*, vol. 309, no.3-5, pp. 426–464.

Zhu, XQ & Law, SS 2006, 'Wavelet-based crack identification of bridge beam from operational deflection time history', *International Journal of Solids and Structures*, vol. 43, no.7-8, pp. 2299-2317.

Roveri, N & Carcaterra, A 2012, 'Damage detection in structures under traveling loads by Hilbert-Huang transform', *Mechanical Systems and Signal Processing*, vol. 28, pp. 128-144.

Nie, ZH, Ngo, T & Ma, HW 2017, 'Reconstructed phase space-based damage detection using a single sensor for beam-like structure subjected to a moving mass', *Shock and Vibration*, vol. 2017, pp.1-20.

Wang, L & Yang, ZC 2013, 'Effect of response type and excitation frequency range on the structural damage detection method using correlation functions of vibration responses', *Journal of Sound and Vibration*, vol. 332, pp.645-653.

Yang, ZC, Yu, ZF & Sun, H 2007, 'On the cross correlation function amplitude vector and its application to structural damage detection', *Mechanical Systems and Signal Processing*, vol. 21, pp. 2918-2932.

T rnational Conference on
Structural Health Monitoring of Intelligent Infrastructure
Brisbane, Australia | 5-8 December 2017

Transducer Array Optimisation for Guided Wave Testing of Pipes Using Finite Element Numerical Simulations and Experimental Studies

X.Niu[1], H.R. Marques[2], H.P. Chen[3].
[1,3]*Department of Engineering Science, University of Greenwich, Chatham Maritime, Kent ME4 4TB, UK,*
*Email:*h.chen@gre.ac.uk
[1,2]*TWI Ltd.Granta Park, Cambridge, Cambridgeshire CB21 6AL, UK.*

Abstract

Oil and gas pipelines are one of the major applications where the structural health monitoring (SHM) technology is particularly attractive.Ultrasonic guided wave (UGW) testing as an advanced non-destructive testing (NDT) technique enables the identification of defects and its location on the pipe circumference in use. In general, a circumferential array of piezoelectric transducers clamped around the surface of the pipe is employed. The sensitivity of the NDT operation is 3 - 9 %of cross–sectional area reduction, depending on the signal-to-noise ratio. The operation frequency range is typically from 20 *kHz* to 100 *kHz*. Optimisation of the transducer array design increases performance in pipe inspection without significant increase in manufacturing and testing costs, which is an area of interest for the industry. This paper presents research on the characterisation of circumferential array design using finite element analysis (FEA) models.The excitation of the fundamental torsional wave mode T(0,1) with flexural wave modes F(n,2) familyis used for pipe inspections. The related experimental testinghas been undertaken to verifythe FEA models. A range of notch dimensions have been introduced to the pipe model, and the T(0,1) with F(n,2) family interaction with these simulated defects is analysed. A series of parameters will be evaluated including transducer excitation variability, number of transducers per array and a gap between the start and end transducers of the array in a practical tool for pipe inspections. The characterisation of multiple transducer array design parameters and its performance could be obtained from the research results.

1. Introduction

Pipeline transport as a traditional transportation method has been widely used in industry. Nowadays, there are over 3 million kilometres of pipeline distributed in the world. Many of them are used to carry oil or natural gas. Structural failuresof pipeline have occurred owing to loss of pipe wall thickness caused by fatigue cracks or corrosion. To control the human health and environmental risk, some advanced NDT or SHM systems have been used to reduce these serious failures, such as ultrasonic testing (UT), X-ray inspection, and guided wave (GW) testing. In these commonly usedtechniques, GW testing has many advantages over other methods for pipeline inspection. It is typically operated lower frequency elastic waves when compared with UT, from 20 *kHz* to 100 *kHz*(Mudge, 2001). Guided waves are able to propagate over a long distance with a high susceptibility to interference on a propagation path(Tse & Wang, 2009). Under an ideal struture condition and surrounding environment, GW testing can travel many tens of metres with minimal attenuation for pipeline screening.Sharan et al (2015)described that GW testing has a high sensitivity (3 -9% of cross – sectional area) depending on the signal-to-noise ratio. Piezoelectric transducers are commonly used in GW pipe inspection system as thesedevices convert electrical signals into vibrations and vice-versa. They are attached to pipes in groups (transducer arrays) working together to manipulate the transmission and reception of ultrasonic guided waves (Niu et al.,2017).

In this paper, the first steps towards optimisation of transducer arrays for pipe inspection are investigated from the study of GW propagation on a steel pipe modelled using finite element numerical simulations. A simulated pipe inspection setup in a real tool with 3 rings and each ring including 24 transducers array is adopted for exciting the torsional wave mode T(0,1). However, T(0,1) propagated on the pipewith a flexural wave modes F(*n*,2) family because of a gap of 33 degrees between two transducers (No.1 and No.24) of the array. An array of 24 transducers equally spacing around the pipe and operating as receivers was modelled and then verified through empirical validation using a real tool. Finally, the defect detection sensitivity of a real tool to various sizes of notch in the pipe is also investigated.

2. Frequency range and wave mode characteristics

Løvstada&Cawley(2011)described that a selection of excitation GW mode and frequency range is essential for an effective function of ultrasonic guided waves in sizing defect. The relation between GW modes and frequency is commonly presented in so-called dispersion curves. Ultrasonic GWs have been widely-employed to measure at frequencies ranging approximately between 20 *kHz* and 100 *kHz*. In general, there are a large number of GW modes available with various displacement characteristics.A nomenclature of **X**(x, y)is often introduced to describe these modes, where**X** represents a letter (L, T, orF, for longitudinal, torsional or flexural, respectively) that relates to characteristics of the individual GW mode (x is the circumferential order, x = 1, 2, 3…; and y is themode order, y = 1, 2, 3…)(Silk and Bainton, 1979).

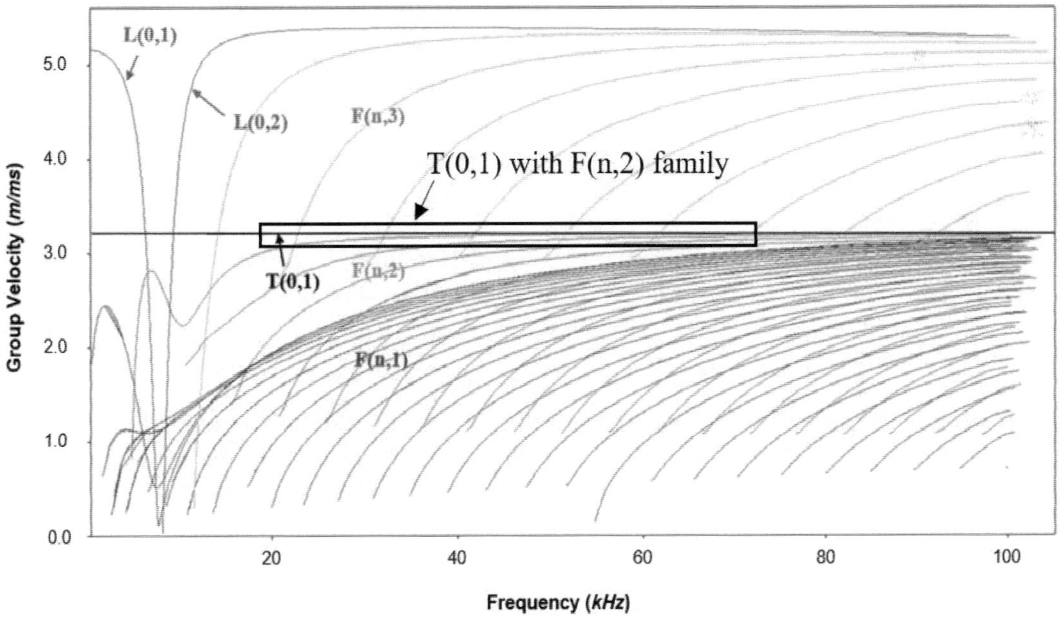

Figure 1 Group velocity dispersion curves for an8-inch(219.1*mm*outer diameter), schedule 40 (8.18*mm* wall thickness) steel pipe

Figure 1 shows the group velocity dispersion curves for an8-inch (219.1*mm* outer diameter) and schedule 40 (8.18*mm* wall thickness) steel pipe. These dispersion curves were generated by

DISPERSE(Pavlakovic, et al., 1997)version 2.0.20a.The rectangular part in the middlerepresentsthe torsional wave mode T(0,1) associated with a family of flexural modesF(n,2). The number of n could be 1, 2, 3..., but lower frequency will be with a less number of F(n,2) family. From 20 *kHz* to 40 *kHz*, the T(0,1) wave modeis highly associated with wave mode F(1,2), comparing with other flexural wave modes in the F(n,2) family.

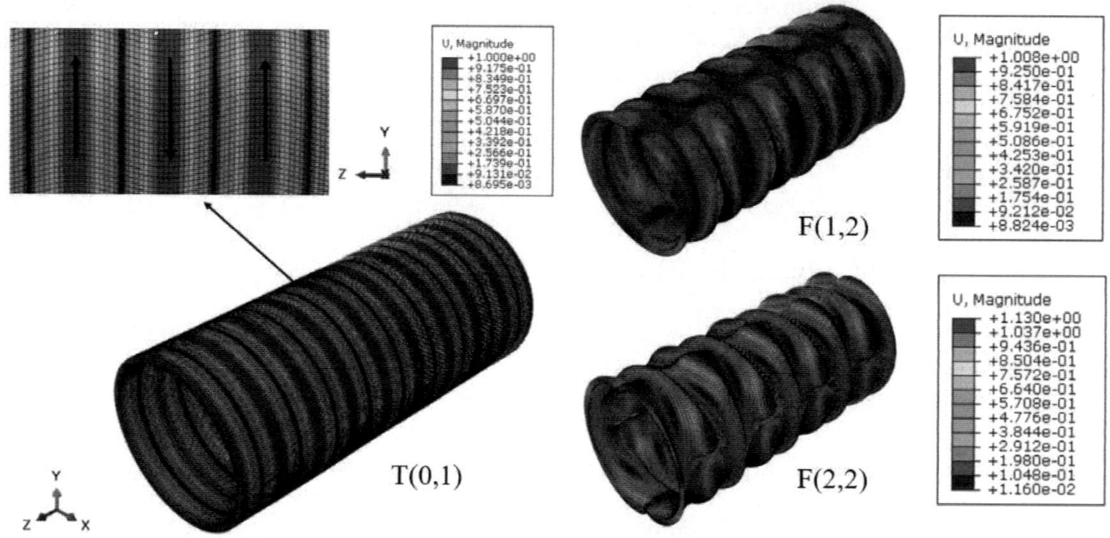

Figure 2 Selection of displacement shapes of guided wave modes in an 8-inch (219.1 mm outer diameter), schedule 40 (8.18 mm wall thickness) steel pipe

Figure 2 shows displacement shapes of torsional mode T(0,1), flexural mode F(1,2) and flexural mode F(2,2). The T(0,1) is an axisymmetric wave mode,butthe F(1,2) and F(2,2) are non-axisymmetric wave modes. If a transducer array has equal spacing for transducers, a pure axisymmetric wave mode, such as one single T(0,1) can be excited using a circumferential array of transducers equally spaced around a pipe. However, if the transducer array has unequal spacing (e.g. a gap of 33 degrees between the start and end transducers of the array), an axisymmetric wave mode will be excited with non-axisymmetric wave modes. Therefore, the F(n,2)will follow T(0,1) when a pure T(0,1) is attempted to be excited by an array of transducers. The effective characteristics of displacement shapes are generatedusing ABAQUS software.

3. Theoretical background

The excitation of T(0,1) torsional wave mode was selected at 35 *kHz*. The mode is axisymmetric, causing in-plane angular displacements. A narrow bandwidth signal of 10 cycles is used as a selective excitation to prevent excitation of dispersive modes. The array excitation was undertaken by a Hanning windowed tone burst signal given by

$$U(t) = \frac{1}{2}\left[1 - \cos\left(\frac{2\pi f t}{n}\right)\right]\sin(2\pi f t) \qquad (1)$$

where$t$ is time,f is the centre frequency of the wave mode, and n is the number of the signal cycles.

Figure 3 shows a channel of transducer collar in a real tool for pipe inspection. Ring 1(R1), Ring 2 (R2) and Ring 3 (R3) are located at front, middle and back of the tool, respectively. There are 24 channels with 30 *mm* spacing for each ring of the tool.

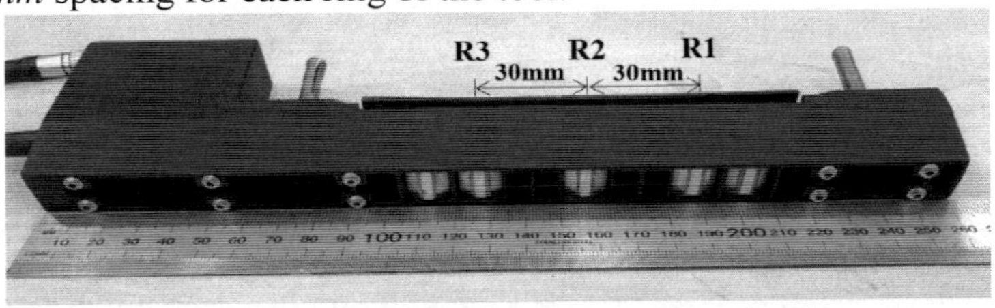

Figure 3 A channel of transducers collar of the testing tool

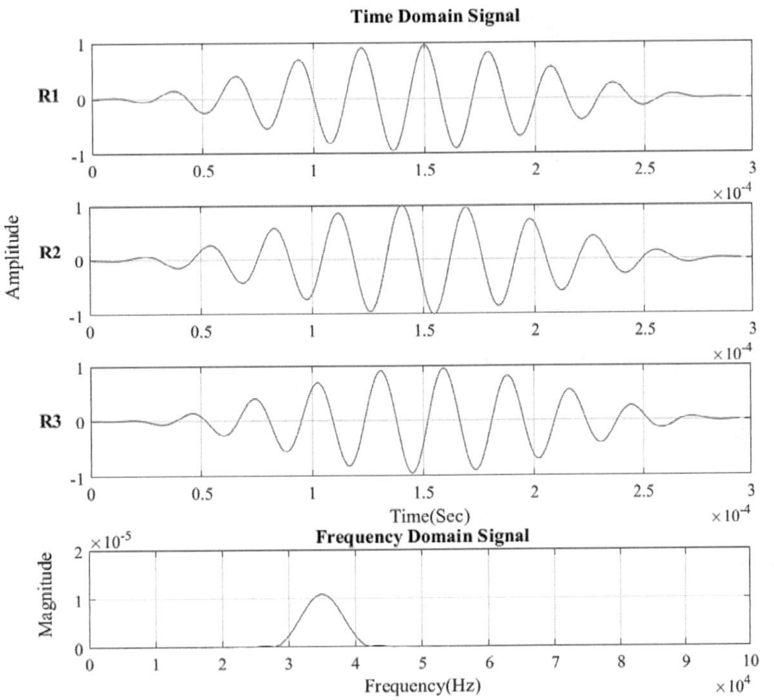

Figure 4 Ultrasonic guided wave excitationsignal: (a) time record; (b) frequency spectrum

Figure 4 shows a 35 *kHz*, 10-cycle Hann windowed pulse in the time and frequency domain. The wave mode T (0,1) was excited and propagated along a pipe with the F(*n*,2). In the tool system, the transducers array of R1 was excited by the signal given in Eq. 1. The signal of R2 was inverted for undesired signal cancellation, and the R3 is time-delayed by the ring spacing of 30 *mm* divided by phase velocity of torsional wave mode T(0,1). The assumed particle displacement components (u_r,u_θ andu_z) are expressed (Rose, 2014)

$$u_r = A_r(r)\cos(m\theta)\, e^{i(kz-\omega t)},$$
$$u_\theta = A_\theta(r)\sin(m\theta)\, e^{i(kz-\omega t)}, \quad (2)$$
$$u_z = A_z(r)\cos(m\theta)\, e^{i(kz-\omega t)},$$

where the terms u_r,u_θ, andu_z are the components of displacement in the radial, circumferential and axial directions, respectively;the terms of $A_r(r)$, $A_\theta(r)$ and $A_z(r)$ are the corresponding displacement amplitudes composed of Bessel functions in the out-of-plane;*m* (0, 1, 2, 3, …) is the circumferential order; and*kz* is the angular wavenumber in the axial (*z*) direction.

4. Finite element numerical modelling

4.1 Transducer array modelling

To suppress the undesired GW modes, a finite element model wasadopted to simulate wave propagationfrom a transducers collar (3 rings, 24 transducers in each ring, 33-degree gap on the top of the pipe in the circumferential direction) as transmitters. This is a common effective technique used to reduce the number of undesired GW modes and simultaneously increase the amplitude of desired GWs in the simulated tool system.

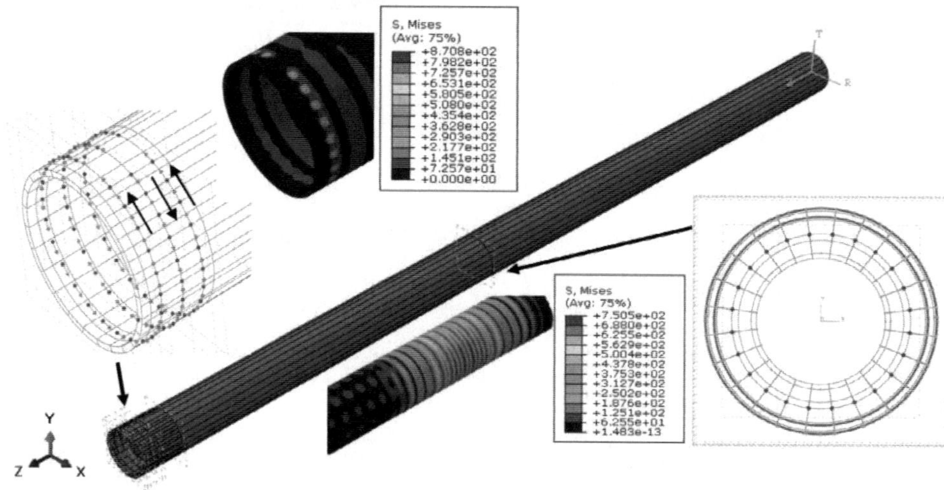

Figure 5 Pitch-catch setup and GW propagation of FEA model for an8-inch (219.1*mm* outer diameter), schedule 40 (8.18*mm* wall thickness) steel pipe

Figure 5shows a pitch-catch setup and GW modes propagation of FEA model for an8-inch (219.1*mm* outer diameter), schedule 40 (8.18*mm* wall thickness) steel pipe. The simulated transducers collaris located on the left pipe end with a 33-degree gap between start and end transducers of the array, and 24 receivers are placed 2 *m* away from the left pipe end. For the steel pipe model, a mass density (ρ) of 7850 kg/m^3, a Young's modulus (E) of 210 *GPa* and a Poisson's ratio (υ) of 0.3 was selected. A total number of 829830 elements were generated for the FE mesh. A mesh size of 3.2*mm* and a step time of 155*ns* were used. From the results, the GW mode T(0,1)propagates along the steel pipe, with a significantly high amplitude comparing with other GW modes. However, the family of flexural modes F(n,2) can also be found at the back of the T(0,1) GW mode.From the FEA results at the predetermined time window, the data was collected with the default Cartesian coordinate system; i.e.X (horizontal), Y (vertical), and$Z$ (out-of-plane). They were then transformed to cylindrical coordinate system, i.e.R (radial), T (circumferential) and$Z$ (axial), and analysed using MATLAB software. These FEA results are verified with a related experiment later.

The results in Figure 6 indicate that a slightly distorted torsional wave mode of T(0,1) was generated by the transducers collar with a gap of 33 degrees between start and end of the array, when excited by a 10-cycle Hann windowed sinusoidal tone burse with the centre frequency of 35 *kHz*. In Figure 6(a), there is a slight problem related to qualityto suppress undesired GW modes. There is one or some other flexural modes followed at the back of T(0,1) wave mode. Figure 6(b) shows a polar plot of the maximum values for displacement in the circumferential direction at each of the transducer in the array of receivers. It should be noted that the maximum values of displacement are reduced at approximately 90 degrees and have a slight increase at 270 degreeswhen no defect exists in the structure. The reason of that is the T(0,1) propagated along the pipe with the F(n,2) owing to a gap of 33 degrees in the array at 90 degrees. Thus, the distribution of displacement amplitudes presents a superposition of wave modes (torsional mode T(0,1) and flexural modes F(n,2))for 24 receivers at 2 m away from the left pipe end.

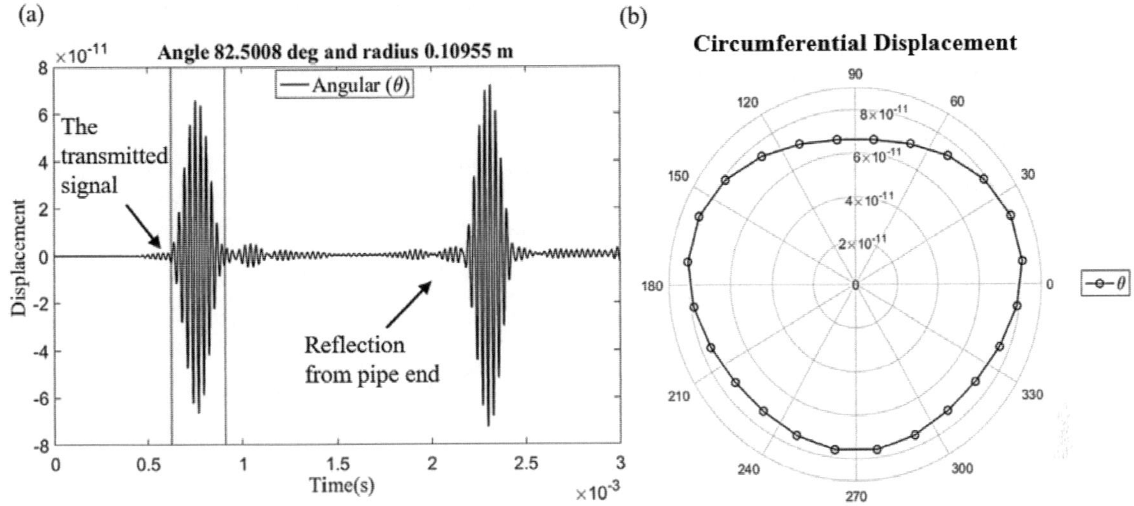

Figure 6 FEA output results for transducers collar transmitted along the pipe (a) displacement/time plot (b) polar plot of all maximum displacements in the circumferential direction measured at each of the 24 receivers

4.2 Numerical modelling with circumferential notch

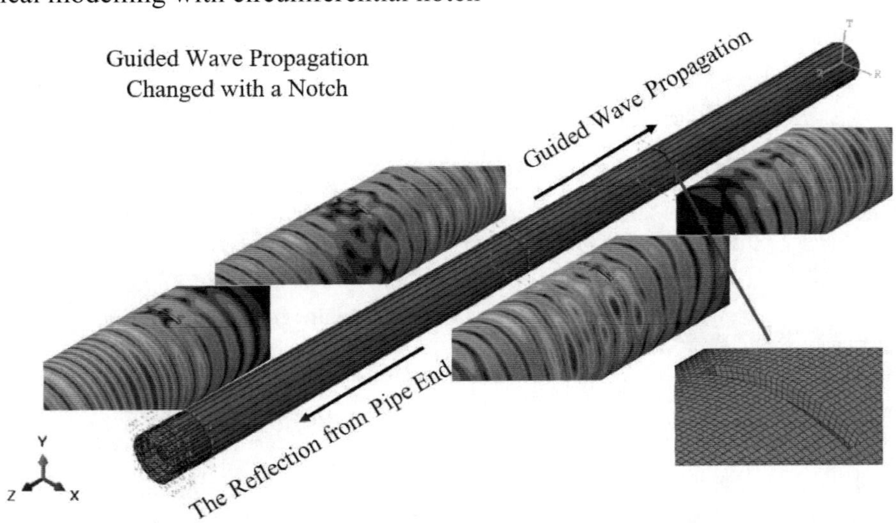

Figure 7 FEA model of steel pipe with a notch in the circumferential direction

A numerical model of a pipe with a circumferential notch is established for finite element analysis. The FEA model is for an8-inch (219.1mm outer diameter), schedule 40 (8.18mm wall thickness) steel pipe with a notch 86.1 mm(arc length)×10mm(width along the axis direction) × 8.18 mm (notch thickness) at the top of the pipe in the circumferential direction. The notch is located 1.0m away from the array of 24 receivers.A total number of 832305 elements were generated with amesh size 3.2mm. A step time of 155$ns$was used.

The FEA model with the notch location and the simulated T(0,1) wave mode with flexural wave modes F(n,2) family interaction with a notch are shown in Figure 7. The left part of pipe modelshows the details of the T(0,1) with F(n,2)propagation and its interaction with notch. The right part of pipe model shows the wave reflection from right pipe end and its interaction with notch. A total of three notch sizes were simulated and their details are described in Table 1.

Table 1: Simulated 3D defects for different notch sizes

Section profile of the notch	Surface profile of the notch
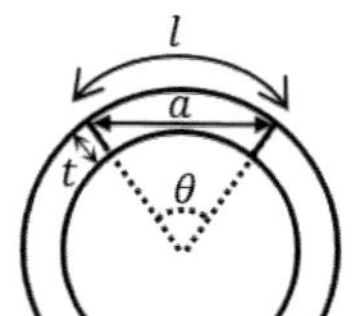	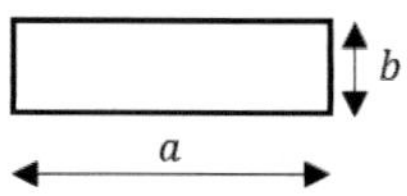 $l_1 = 28.7\text{mm}, \theta_1 = 15^\circ, b_1 = 10\text{mm}, t_1 = 8.18\text{mm};$ $l_2 = 86.1\text{mm}, \theta_2 = 45^\circ, b_2 = 10\text{mm}, t_2 = 8.18\text{mm};$ $l_3 = 143.5\text{mm}, \theta_3 = 75^\circ, b_3 = 10\text{mm}, t_3 = 8.18\text{mm};$

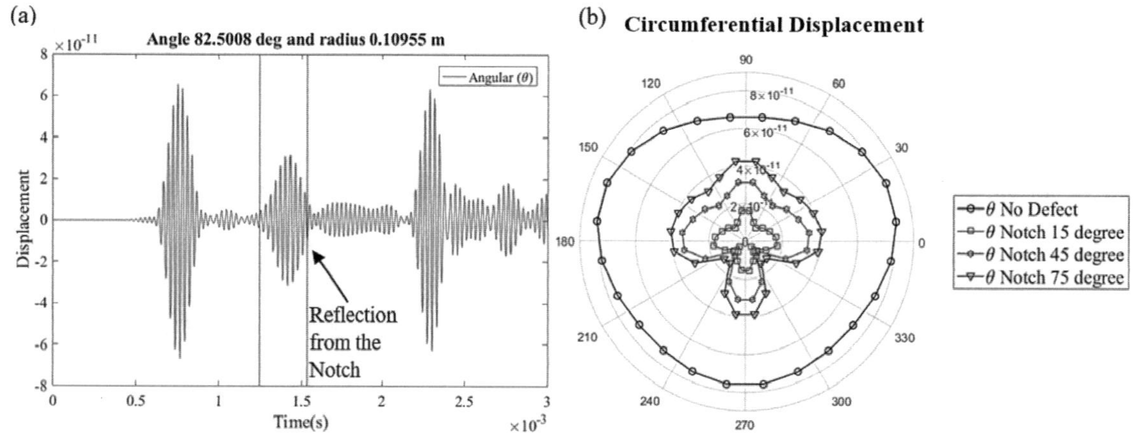

Figure 8 Reflection from the notch (a) displacement/time plot (b) polar plot of all maximum displacements in the circumferential direction measured at each of the 24 receivers

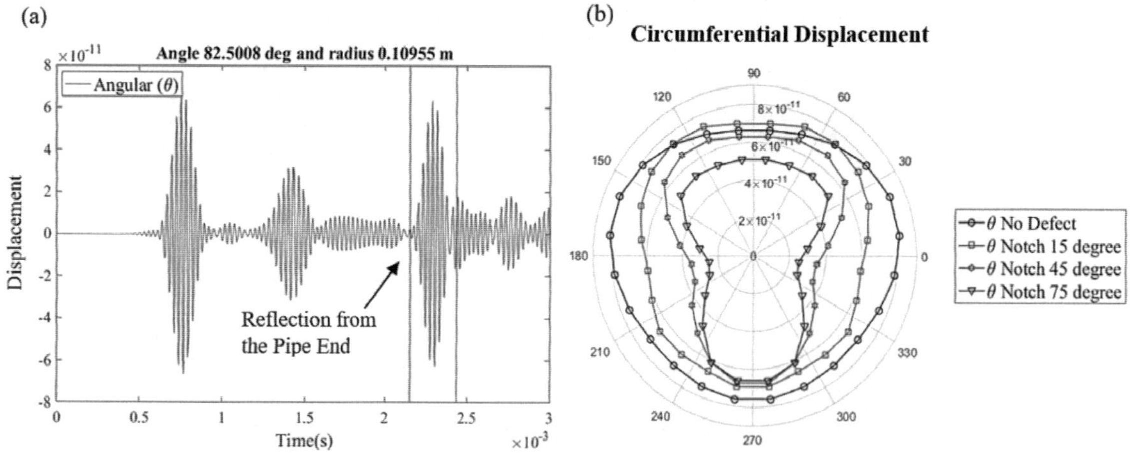

Figure 9 Reflection from the pipe end(a) displacement/time plot (b) polar plot of all maximum displacements in the circumferential direction measured at each of the 24 receivers

Figures 8 and 9 show reflections from the notch and right pipe end, respectively. The amplitude of circumferential displacements in the pitch-catch operation is given in Figure 8(b)and Figure 9(b), respectively. From the results, the patterns of maximum amplitudes of T(0,1) wave mode with F(n,2) have different amplitude distributions with respect to different notch sizes. In the pitch-catch operation, the results of reflection from the notch indicate that the angular displacement amplitude around the pipe increases as the notch grows in the circumferential direction, as shown in Figure 8(b).At approximately 30 degrees, 90 degrees, 150 degrees and 270 degrees, the maximum displacement amplitudes increase as a function of notch growth in the circumferential direction at a low-frequency of 35 *kHz*. In Figure 9(b), the amplitudes reduce at approximately 0 degree, 90 degrees, 180 degrees and 270 degrees as a function of circumferential notch growth.

5. Experimental study

To verify the FEA modelling, the related experiment is undertaken set up at this section. Torsional wave mode T(0,1) of ultrasonic GW with F(n,2) family can be excited by a real tool. A 4.45 *m* long, 8-inch (219.1 *mm* outer diameter), schedule 40 (8.18 *mm* wall thickness) steel pipe is used in this empirical work. The FEA model for this pipe with the transducers collars is validated empirically using experimental measurements. The CLV 3D laser vibrometer system was used to measure the proportions of displacements at 24 receivers around the pipe against frequency in the axial (Z), circumferential (θ) and out-of-plane (R) directions, respectively. In this experiment, the related FEA model is verified where the propagation of guided wave modes is generated against various frequencies in a series of steps (e.g. 20 *kHz*,25 *kHz*, 30 *kHz* 35 *kHz*, 37 *kHz*, 40 *kHz*, 50 *kHz*, 60 *kHz*) using an array of transducers with a gap of 33 degrees between start and end transducers.

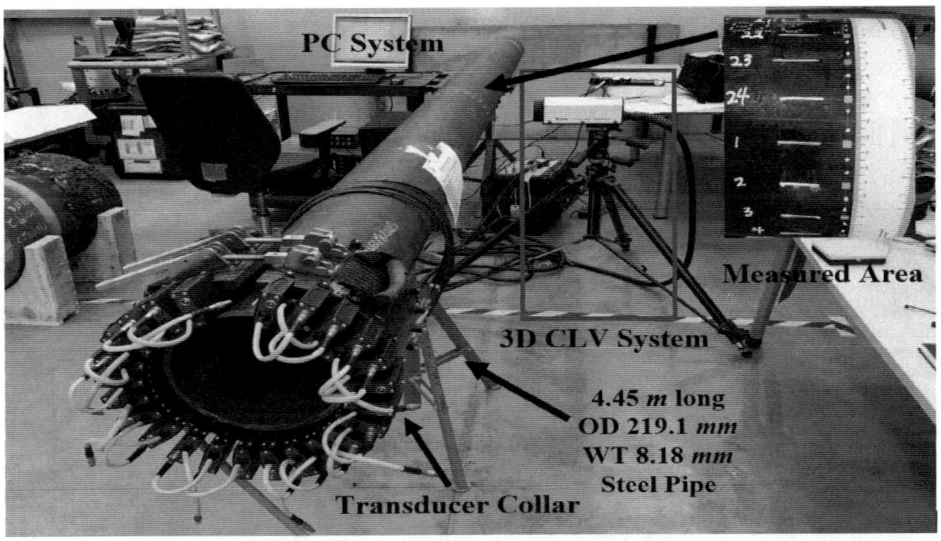

Figure 10: Set up of vibrometer pipeexperiment

A detailed sketch of this experiment has been described, as shown in Figure 11. A 4450 *mm*long, 8-inch (219.1 *mm* outer diameter, schedule 40 (8.18 *mm* wall thickness) steel pipe is used in this experiment. The steel pipe is placed on roller supports. These two supports are located at 1000 *mm* away from the pipe ends. Piezoelectric transducers collar is used to generate torsional wave mode T(0,1), propagating on the pipe with a flexural wave mode 'family' F(n,2) in the circumferential direction. There are 24 points to be measured on the external surface of the pipe, 2000 *mm* away from the transducers collar. The piezoelectric transducers collar was excited by a TelctestFOCUS system. The position of 24 measured points using 3D laser vibrometer equally spaced on the pipe in the circumferential direction.

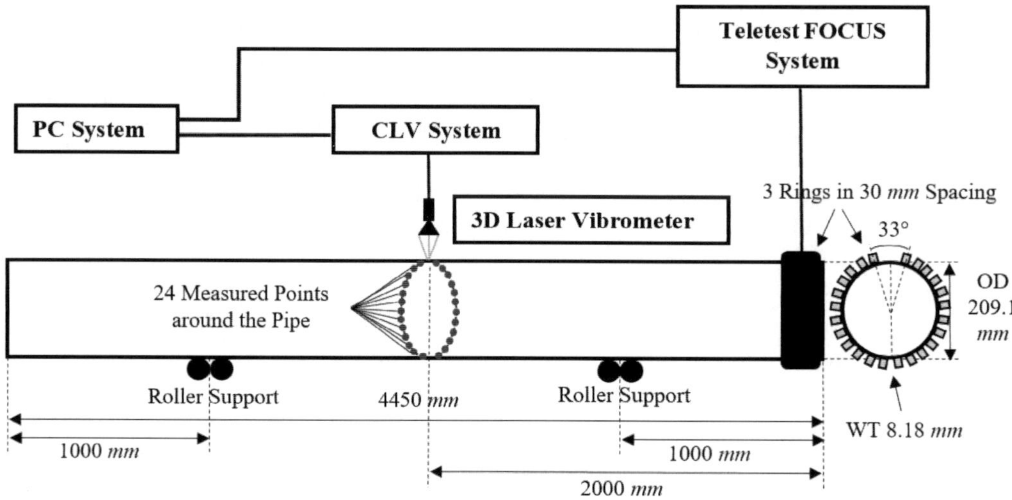

Figure 11 A detailed sketch of experiment set-up

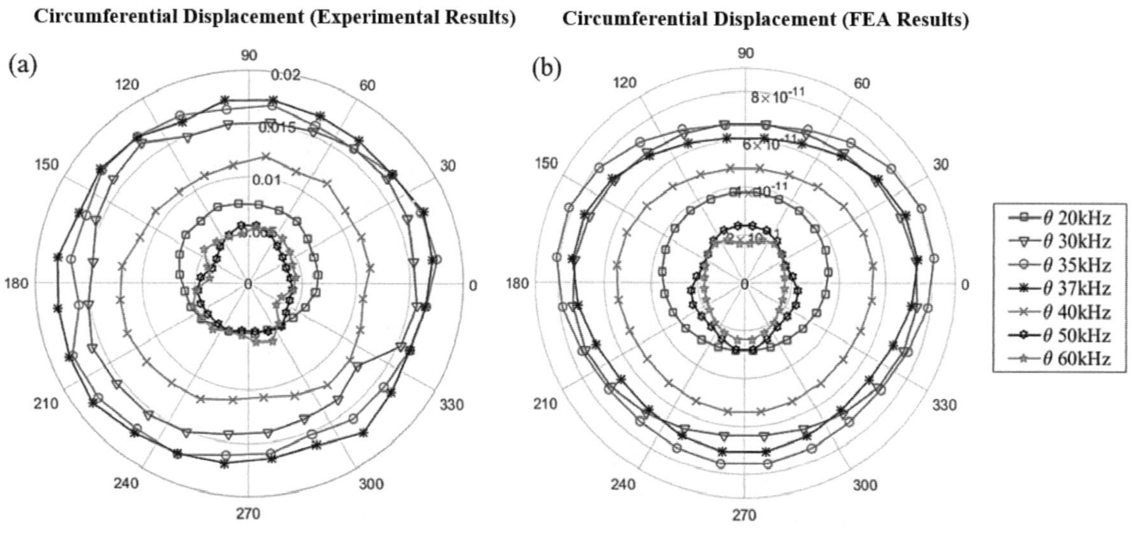

Figure 12 Angular displacement in experimental and FEA results

Figure 12 shows a comparison between experimental results and FEA results in polar plots for maximum amplitude value of displacementat the 24 receivers around the pipe. When the values of each excited load were not considered here, there is a good agreement between the experimental results and the simulated FE results for the distribution proportion of polar plots. Based on these results, The FE models are verified and further research can use this finite element model with torsional wave excitation. For the selection of frequency range, 37 *kHz* is an appropriate testing frequency but 35 *kHz* is more suitable for FEA modelling. An ideal circle shape from polar plot of maximum displacement cannot be achieved, since there is a gap between start and end transducers of the array. The amplitude of T(0,1) with F(1,2)has more effective behaviour than other flexural wave modes in the circumferential direction.

6. Conclusions

This paper presents the characterisation of multiple circumferential transducer arrays using finite element modelling and experimental work. With an excitation of a torsional wave mode T(0,1) with F(n,2), useful results have been simulated and verified at wider frequency range from 20 *kHz* to60 *kHz*. A notched defect simulation has also been undertaken at the frequency of 35 *kHz*. The models are based on a4.45*m* long 8-inch schedule 40 (outer diameter 219.1*mm*, wall thickness 8.18*mm*)steel pipe. From the numerical and experimental studies, following conclusion are noted: (1) A gap of transducer array can influence the quality ofpurity mode for torsional wave.(2) FEA is an effective method for simulating multiple transducer arrays, such as use of 3 rings of transducers collar. (3) The guided wave torsional wave mode has shown high sensitivity to notch growth in the circumferential direction. (4) An appropriate frequency of 37 *kHz* is verified for pipe testing, and 35 *kHz* is an appropriate frequency for FEA modelling without considering the transducer property. Further work is needed to simulate transducer array using numerical linear superposition analysis, and to study effect of the reduced gap between start and end transducers of the array, and to undertake further experiments for empirical validation.

7. Acknowledge

This publication was made possible by the sponsorship and support of TWI Ltd. and University of Greenwich. The work was enabled through, and undertaken at, the National Structural Integrity Research Centre (NSIRC), a postgraduate engineering facility for industry-led research into structural integrity established and managed by TWI through a network of both national and international Universities.

References

Løvstada, A & Cawley, P 2011,'The reflection of the fundamental torsional guided wave from multiple circular holes in pipes'. *Journal of NDT & E International*, vol. 44, no. 7, pp. 553–562.

Mudge, P 2001,'Field application of the Teletest long-range ultrasonic testing technique'. *Insight - Non-Destructive Testing and Condition Monitoring*, vol. 43, no. 2, pp. 74-77.

Niu, X, Marques, HR & Chen, HP 2017,'Coupled piezoelectric transducer array optimization through simulation techniques for guided wave testing of cylindrical structures'. *Proceedings of the 8th ECCOMAS Thematic Conference on Smart Structures and Materials (SMART 2017)*, Madrid, Spain, 5-8 June, pp. 424-435.

Rose, JL 2014,*Ultrasonic Guided Waves in Solid Media*.Cambridge University Press, Cambridge.

Sharan, PK, S, S, Chaitanya, SK & Maddi, HK 2015,'Long Range Ultrasonic Testing - Case Studies'. http://www.nde2015.com/papers-1/paper-83.pdf.Accessed 17 11 2016.

Silk, MG & Bainton, KF 1979,'The propagation in metal tubing of ultrasonic wave modes equivalent to lamb waves'. *Ultrasonics*, vol.17,no. 1, pp. 11-19.

Tse, PW & Wang, X 2009,'Semi-quantitative Analysis of Defect in Pipelines through the use of Technique of Ultrasonic Guided Waves'. *Key Engineering Materials*, vol. 413-414, pp. 109-116.

The 8th International Conference on
Structural Health Monitoring of Intelligent Infrastructure
Brisbane, Australia | 5-8 December 2017

Aerodynamic Performance and Fatigue Damage Assessment of Wind Turbine Composite Blades using Corrected BEM Method

C. Zhang[1], H.P. Chen[2].
[1] *Department of Engineering Science, University of Greenwich, Chatham Maritime, Kent, ME4 4TB, UK;*
[2] *Department of Engineering Science, University of Greenwich, Chatham Maritime, Kent, ME4 4TB, UK,*
Email: h.chen@gre.ac.uk

Abstract

This paper presents a method to estimate the fatigue life of composite blades of horizontal axis wind turbines, and an example of NREL 5 MW wind turbine blade is used to verify this approach. The distributions of aerodynamic loads are analysed by blade element momentum theory, which affects the performance of the blade under various wind speed. The wind speed spectrum over an hour time is generated by Weibull distribution based on the offshore wind parameters in this study. Then, the maximum stress caused by the wind forces generated by aerodynamic load is calculated by finite element model in ANSYS. The stress cycle is established by Rainflow counting in the load spectrum which converts the time-stress domain to fatigue cycles. Fatigue damage performance and the fatigue life of wind turbine blade are predicted by the Goodman diagram and S-N curve for each cycle during the service life of wind turbine blade. Finally, safe working life is predicted by applying the Miner's law for linear fatigue damage accumulation. A numerical example of NREL 5 MW wind turbine blade is investigated by proposed method to estimate fatigue life under one-hour wind speed generated by Weibull distribution in MATLAB. From the results, the proposed method can provide an effective tool for evaluating the fatigue damage and assessing structural performance of the wind turbine blades within a time period containing uncertain wind speed.

1. Introduction

With the increasing energy consumption in the world, renewable and clean energy such as wind resource has been considered as an alternative way to resolve the energy crisis. Energy crisis and greenhouse effect have led to an increasing demand on clean energy. Among them, wind is the most cost-effective and feasible energy, and thus it has been a very attractive option for utilities, independent power producers and companies (Chen *et al.* 2017, Zhang & Chen 2016). In the wind turbine system, the most critical component is the wind turbine blades since the manufacturing cost of the blades is approximately 15-20% of the total manufacturing cost and their maintenance cost is about 25-30% of the wind turbine production cost (Florian & Sørensen 2015). In order to improve the performance of blades in offshore wind turbines, layered fibre-reinforced polymer composite materials are usually adopted for the large blades since these materials have better mechanical properties such as higher fatigue resistance and lighter weight, compared with traditional homogeneous materials i.e. metals (Zhang *et al.*, 2016, Zhang & Chen 2016). One challenge in structural damage estimation is that the wind turbine blade is under complex load conditions, such as aerodynamic load, gravitational load, inertial loads and operational loads (Wang *et al.* 2014). Among these loads, the aerodynamic load is significantly affected by wind speed in environmental condition since the spectrum of wind speed is in high frequency and amplitude. Wind speed could be generated by deterministic distribution, and the Weibull distribution has been used for fitting the wind speed (Epaarachchi & Clausen 2006, Hu *et al.* 2016a, 2016c). It is important to develop a method for analysing the aerodynamic loads in order to obtain the performance of wind turbine blade. From the results for performance assessment, the fatigue analysis can be used to optimise the maintenance strategies, which may prevent the failure of the entire wind turbine.

For blade aerodynamic load calculation, the blade element momentum (BEM) method is simple, and is still widely used for practical engineering application of wind turbine blade at present (Pratumnopharat & Leung 2011, Dai *et al.* 2011, Tahani *et al.* 2017, Wang *et al.* 2014, Wen *et al.* 2017). Most ongoing research on fatigue phenomena uses load spectra obtained by digital sampling or finite element (FE) model of wind turbine blades, which estimated the numerical solution at a location near the root of the blade (Bazilevs *et al.* 2011a, 2011b). Then, to obtain the cycles, traditional counting methods such as Rainflow counting is widely used in recent research (Shokrieh & Rafiee 2006, Epaarachchi & Clausen 2006, Grujicic *et al.* 2010). Based on Goodman diagram and S-N curve, the equivalent load spectra for each cycle can be estimated (Mandell *et al.* 2003, Marin *et al.* 2009). This calculated stress is then utilised to estimate fatigue damage in the blade using Miner's rule (Shokrieh & Rafiee 2006, Grujicic *et al.* 2010), and then the predicted fatigue life is estimated for optimal maintenance strategies.

This study presents a method for investigating the fatigue damage of a composite wind turbine blade. The BEM method is used for calculating aerodynamic loads for different wind speeds, and then the forces used in finite element model can be derived. The Ranflow Counting is used here to obtain the load cycles, based on the results of FE model solution and random wind speeds. The Miner's law is used to predict the fatigue damage during the designed service life of wind turbine by both the S-N curve and Goodman diagram. A numerical example of the NREL 5 MW wind turbine blades (Kim *et al.* 2011, Jonkman *et al.* 2009, Resor 2013) is investigated to estimate fatigue damage under randomised wind speeds, on the basis of Weibull distribution describing the wind speeds in the offshore environment. Results show that FE model can take the variable wind speeds with other blade loads into account and the fatigue cycles can be obtained. The proposed method can be further used for determining the resistance of the composite blades based on S-N curves. The fatigue damage of composite blades can be predicted by Miner's law based on fatigue famage and life cycles in each time period. The fatigue life can be calculated before the cumulative damage exceeds unity. From the results, the proposed method can provide a useful tool for evaluating the fatigue damage and assess structural performance of the wind turbine blades under the uncertain offshore environment.

2. Aerodynamic loads on blades

Blade element momentum (BEM) method is often used to estimate the aerodynamic loads. Unfortunately, these loads could not be used for stress analysis in FE model. Therefore, the edgewise force, flapwise force and pitching moment along with global coordinate of the blade need to be calculated for the FE model based on the results of BEM method.

2.1 Blade element momentum method

BEM method is a typical approach to analyse the performance of wind turbine aerodynamics on the blades. The BEM theory treats the revolving rotor as an actuator disc that combines blade element theory and one-dimensional momentum theory in both the rotor axial and tangential direction. Both rotor axial and tangential induction factors that describe the airflow speed change are introduced in the one-dimensional momentum method to calculate the wind thrust and rotor motivation torque (Dai *et al.* 2011). The blade element theory separates the blade into several elements and ignores the mutual influence between two adjacent elements. The aerodynamic loads on each element are dependent on its local airfoil characteristics, i.e. the lift and drag coefficients.

The BEM method is widely used in current research for wind turbine blade (Pratumnopharat & Leung 2011, Dai *et al.* 2011, Tahani *et al.* 2017, Wang *et al.* 2014, Wen *et al.* 2017). From the one-dimensional momentum method, the thrust dT and torque dQ for each element are expressed, respectively, as (Manwell *et al.* 2010)

$$dT = 4\pi r \rho a(1-a){V_0}^2 f dr \tag{1}$$
$$dQ = 4\pi r^3 \rho a'(1-a)V_0 \Omega f dr \tag{2}$$

where V_0 is free stream velocity, r is radial location along the blade length, Ω is angular velocity, a and a' are axial and tangential induction factor, respectively.

From the blade element theory, the thrust dT and torque dQ for each element are determined by

$$dT = \frac{1}{2} B\rho {V_{vel}}^2 (C_l cos\varphi + C_d sin\varphi) c dr \tag{3}$$
$$dQ = \frac{1}{2} B\rho {V_{vel}}^2 (C_l sin\varphi - C_d cos\varphi) c dr \tag{4}$$

where B is the number of the blades, ρ is the air density, C_l and C_d are lift and drag coefficients at the local angle of attack α, respectively, V_{vel} is local resultant air velocity, and c is chord.
The local angle of attack α is given by

$$\alpha = \varphi - \theta \tag{5}$$

where the θ is the local pitch of the blade, and φ is the angle between the plane of rotation and the relative velocity, which can be expressed as

$$tan\varphi = \frac{(1-a)Vo}{(1+a')\Omega r} \tag{6}$$

The torque and thrust forces depend on the tangential and axial induction factors. To evaluate them it is necessary to implement the momentum and angular momentum conservation equations. From the conservation of the momentum in the axial direction, it is possible to obtain two further expressions for the thrust force dT and the torque dQ. By using Equations (1) and (2) and the corresponding Equations (3) and (4), gives

$$a = \frac{1}{\frac{4sin^2\varphi}{\sigma(C_l cos\varphi + C_d sin\varphi)} + 1} \tag{7}$$

and

$$a' = \frac{1}{\frac{4sin\varphi cos\varphi}{\sigma(C_l sin\varphi - C_d cos\varphi)} + 1} \tag{8}$$

where σ is the rotor solidity, defined as

$$\sigma = \frac{Bc}{2\pi r} \tag{9}$$

Now, all necessary equations for the BEM model have been derived, and the algorithm can be summarised as seven steps below. Since the different control volumes are assumed to be independent, each element can be treated separately and the total force can be accumulated for the whole blade (Pratumnopharat & Leung 2011), namely

Step (1) Initialise a and a', typically $a = a' = 0$.
Step (2) Compute the flow angle φ using Equation (6).
Step (3) Compute the local angle of attack using Equation (5).
Step (4) Read the C_l (α), C_d (α) and $C_{M1/4}$ (α) from the airfoil data.
Step (5) Calculate a and a' from Equations (7) and (8).
Step (6) If a and a' has changed more than a certain tolerance, go to step (2) or else finish.
Step (7) Compute the local loads on the segment of the blades.

Finally, the power P of the wind turbine blade is expressed as $P=\int \Omega dQ=\Omega Q$.

2.2 Loads in finite element model

In this study, the finite element model is introduced for ANSYS software for fatigue analysis under various wind speeds to get the maximum stress. In previous section, the aerodynamic loads have been investigated. However, these loads could not be used for stress analysis in FE model. Therefore, the edgewise force, flapwise force and pitching moment along with global coordinate of the blade need to be calculated, typically shown in Figure 1 (Pratumnopharat 2015).

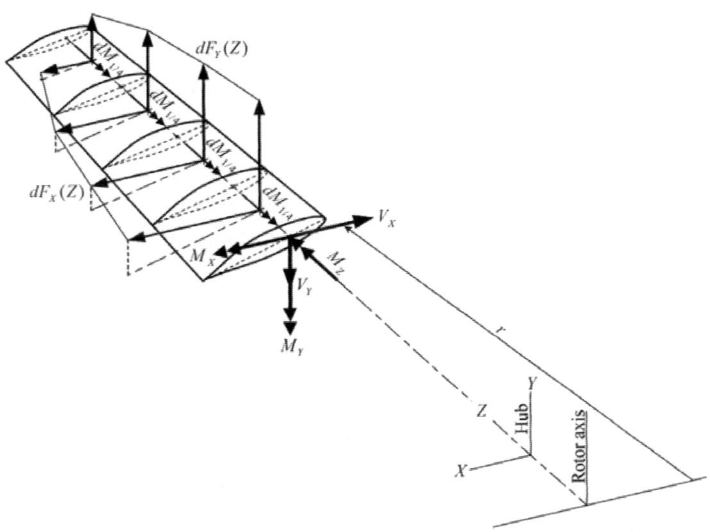

Figure 1: External forces in a global coordinate of wind turbine blade

These loads can be imported into FE model by ANSYS structural software for stress analysis. For simplicity, centrifugal and gravitational loads are ignored in stress analysis in this study. By calculating lift force, drag force, and pitching moment at each station along the blade span, the desired loads can be determined in a global coordinate of the blade in FE model. Then, the solution of the stress for fatigue analysis can be investigated in ANSYS.
The lift force at each station by BEM method is

$$dL = \frac{1}{2}\rho c {V_{vel}}^2 C_l dr \quad (10)$$

The drag force at each station is

$$dD = \frac{1}{2}\rho c {V_{vel}}^2 C_d dr \quad (11)$$

Therefore, the total aerodynamic force in each element is

$$d{F_{aero}}^2 = dL^2 + dD^2 \quad (12)$$

and pitching moment at each station is

$$dM_{1/4} = \frac{1}{2}\rho c^2 {V_{vel}}^2 C_{M1/4} dr \quad (13)$$

By the principle of axes rotation, edgewise and flapwise forces at each blade element are given as, for edgewise force

$$dF_x = -dLsin\varphi + dDcos\varphi \quad (14)$$

for flapwise force

$$dF_y = dLcos\varphi + dDsin\varphi \quad (15)$$

and the pitching moment is

$$dM_z = dM_{1/4} \quad (16)$$

By importing these loads into the FE model via ANSYS commands, the solution of the FE model of wind turbine blades can be inputted and the fatigue damage is therefore to be analysed.

3. Fatigue analysis

Fatigue analysis has been considered as a critical part of the design process of wind turbine blades. According to the BEM method, the forces for FE model of wind turbine blades are established for different of wind speeds. Then the maximum stress in blade is investigated by the results of FE model of wind turbine blades. Fatigue life was estimated by Miner's law for accumulating the damage (Zhang & Chen 2017). By combining these methods, a dynamic structural problem for the fatigue can be converted into a static structural problem, which is easier to be solved.

3.1 Distribution of wind speed

According to existing studies (Epaarachchi & Clausen 2006, Hu *et al.* 2016a, 2016b), the Weibull distribution is adopted to describe the wind speeds. Therefore, the probabilistic density function *f(v)* and cumulative distribution function *F(v)*, respectively, are

$$f(v) = \frac{K}{C}(\frac{V}{C})^{K-1} e^{-(\frac{V}{C})^K} \quad (16)$$

$$F(v) = 1 - e^{-(\frac{V}{C})^K} \quad (17)$$

where K is the shape parameter and C is the scale parameter of the distribution.

3.2 Rainflow counting

When a time-load spectrum is recorded over a measured period, procedures such as the Rainflow counting method are used to find the cycles for fatigue analysis. Within this procedure, the first step involves converting the original load signal into a sequence of the load with points of peaks and valleys. Within the Rainflow counting method, separate counting of load half-cycles is carried out for the ones starting from the peaks and the ones starting from the valleys.

A half-cycle then starts from each peak and end when one of the following three criteria is met (Grujicic *et al.* 2010): (a) when the end of the signal is reached; (b) when the half-cycle in question runs into a half-cycle which originated earlier and which is associated with a higher peak value; and (c) when the half-cycle in question runs into another half cycle which originated at a later time and which is associated with a higher value of the peak.

3.3 Goodman diagram

Constant life diagrams are graphical representations of the safe regime of constant amplitude loading for a given specified life, e.g. the endurance limit or infinite life (Grujicic *et al.* 2010). These diagrams can be drawn in some ways, depending on which parameters are selected to describe the constant amplitude cyclic loading, such as Goodman diagram (Marin *et al.* 2009). Constant amplitude cyclic loading can be defined by specifying the following parameters, i.e. maximum stress and minimum stress. The Goodman diagram gives the equivalent stress under different mean stress. A simple Goodman diagram without considering the stress ratio can be expressed as

$$\sigma_{eqa} = \frac{\sigma_a}{\sigma_m - \sigma_u}(\sigma_{eqm} - \sigma_u) \quad (17)$$

where σ_{eqa} is the equivalent stress in this cycle when the equivalent mean stress σ_{eqm} is given, σ_a is the amplitude stress, and σ_m is the mean stress in this cycle, σ_u is the ultimate stress for the material.

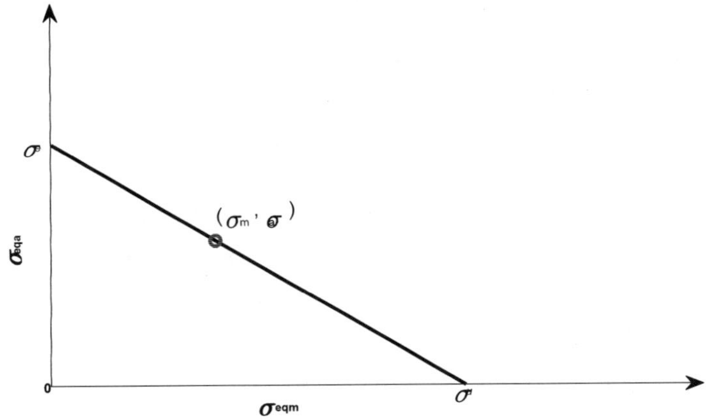

Figure 2: An example of Goodman diagram showing relationship between amplitude stress σ_{eqa} and mean stress σ_{eqm}

Figure 2 gives an illustrative Goodman diagram used in this study. The fatigue stress should use the equal amplitude stress when the average stress is null (σ_m=0) by Goodman diagram for the wind turbine blade, and this amplitude stress is expressed as σ_e.

3.3 S-N curve

To calculate the damage of composite blades by S-N curve, it is necessary to know the fatigue life cycles under constant equal amplitude loading obtained by Goodman diagram. Fatigue properties of materials are usually expressed in the form of S-N curves according to the experimental data, i.e. stress amplitude versus the number of cycles to failure. According to existing studies (Mandell *et al.* 2003, Marin *et al.* 2009), a typical S–N curve model can be expressed as

$$\sigma_e = A - BlgN \ \ or \ \ ratio = \frac{\sigma_e}{\sigma_u} = a - blgN \tag{18}$$

where $ratio=\sigma_e/\sigma_u$ is the stress ratio, A and B are parameters depending on the material, which also can be expressed as $a\sigma_u$ and $b\sigma_u$. Figure 3 represents the relationship between the stress ratio and the stress cycles to calculate the fatigue damage.

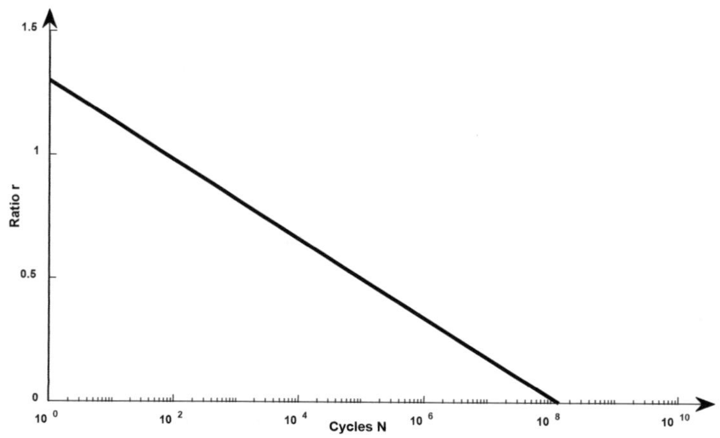

Figure 3: An example of the S-N curve showing linear relationship between the fatigue cycles N and constant ratio r

The fatigue life cycles N can be obtained in this graph given an amplitude stress σ_e. The damage in this cycle is *1/N* by assuming the total damage is a unit.

3.3 Miner's law

The fatigue damage due to each cycle can be accumulated by Miner's rule. According to Miner's rule, the structure is defined as failure when the cumulative damage D exceeds unity (Wang *et al.* 2014), where D is defined as

$$D = \sum_i \frac{n_i}{N_i} \tag{19}$$

where n_i is the load cycle number, and N_i is the allowable load cycle number.

From the method mentioned above, the damage per hour can be generated and the total hour of fatigue life can be estimated. To achieve the design code of wind turbine concerning the total service life, some maintenance strategies may be necessary to increase the performance of the wind turbine blade and prevent the structural failure.

4. Numerical example

In order to assess the accuracy of the performance analysis of the above method, the 5 MW blade developed by the National Renewable Energy Laboratory (NREL) is used as the reference model, and the performance of the blade is analysed by FE model. This composite blade considered in this study is the same geometry as the NREL 5 MW reference wind turbine composite blade (Kim *et al.* 2011, Jonkman *et al.* 2009, Resor 2013). The NREL 5 MW reference wind turbine is a conventional three-bladed upwind turbine based on the NREL report (Jonkman *et al.* 2009). The gross properties and control strategies of this turbine are shown in Table 1, which are the designed data for structural analysis (Wen *et al.* 2017).

Table 1: The gross properties of NREL 5 MW wind turbine blade

Parameters	Value
Rating	5 MW
Blades No.	3
Rotor diameter, hub diameters and height	126m, 3m, 90
Cut in, designed and cut out wind speed	5m/s, 11.4m/s, 30m/s
Rotor speed	12.0 rpm
Pitch angle	3 degree
Company	NREL

Table 2: Aerodynamic properties of 5 MW NREL wind turbine blade

Section	Position [m]	Chord [m]	Twist [°]	Airfoil
1	2.87	3.542	13.308	Cylinder1
2	5.60	3.854	13.308	Cylinder1
3	8.33	4.167	13.308	Cylinder2
4	11.75	4.557	13.308	DU40_A17
5	15.85	4.652	11.480	DU35_A17
6	19.95	4.458	10.162	DU35_A17
7	24.05	4.249	9.011	DU30_A17
8	28.15	4.007	7.795	DU25_A17
9	32.25	3.748	6.544	DU21_A17
10	36.35	3.502	5.361	DU21_A17
11	40.45	3.256	4.188	NACA_64_618
12	44.55	3.01	3.125	NACA_64_618
13	48.65	2.764	2.319	NACA_64_618
14	52.75	2.518	1.526	NACA_64_618
15	56.17	2.313	0.863	NACA_64_618
16	58.9	2.086	0.37	NACA_64_618
17	61.5	1.419	0.106	NACA_64_618

The aerodynamic properties of this model, e.g., airfoil type, chord length, and twist angle, are is listed in Table 2 (Kim *et al.* 2013, Resor 2013). Also, in order to verify the method developed in this research, the specifications for the reference model are introduced as same input parameters for the 5 MW blade design data given above.

In order to obtain static power curve of the blade for comparing the results under same conditions, aerodynamic data of airfoils in Jonkman *et al.* (2009) are used. With these data, aerodynamic performance analysis under normal conditions by BEM method is carried out. Figure 4 shows the aerodynamic coefficients of the different airfoils under the all attack angles for the airfoils in NREL 5 MW wind turbine blade.

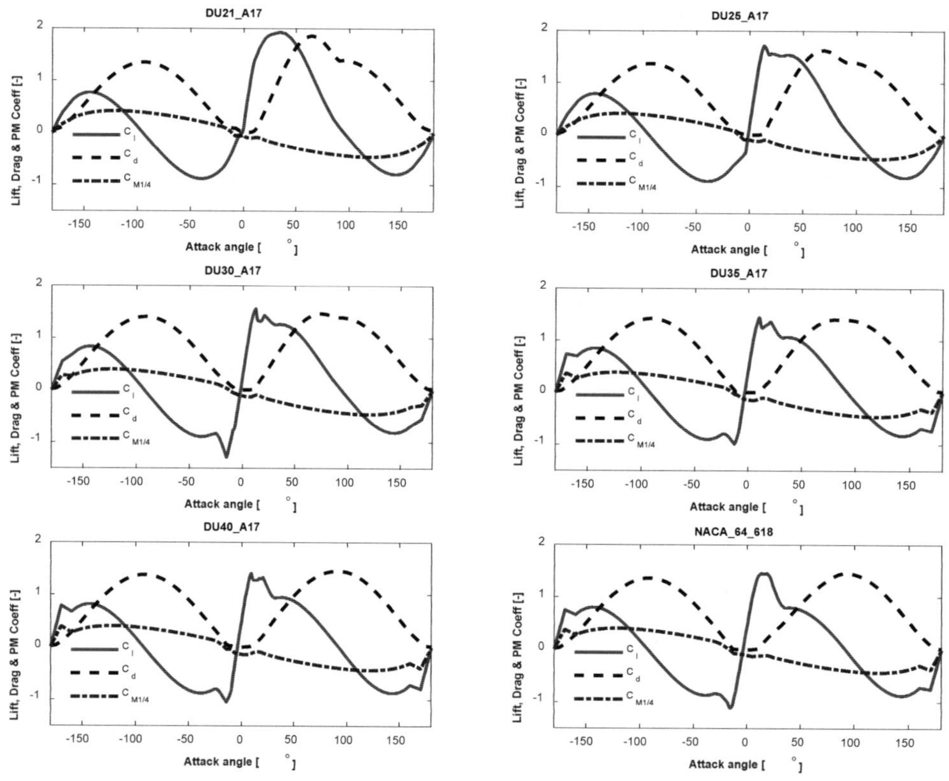

Figure 4: Aerodynamic coefficients of the airfoils in NREL 5 MW wind turbine blade

A blade geometry model was generated by connecting 17 airfoils, which smooth the transition from section to section and reduce the stress concentration, as shown in Figure 5. The 3D blade model can be imported into ANSYS after generating blade geometry model. The FE mesh for the composite blade model is generated by ANSYS 18.0, which adopts layered shell 281 elements for the model for structural analysis.

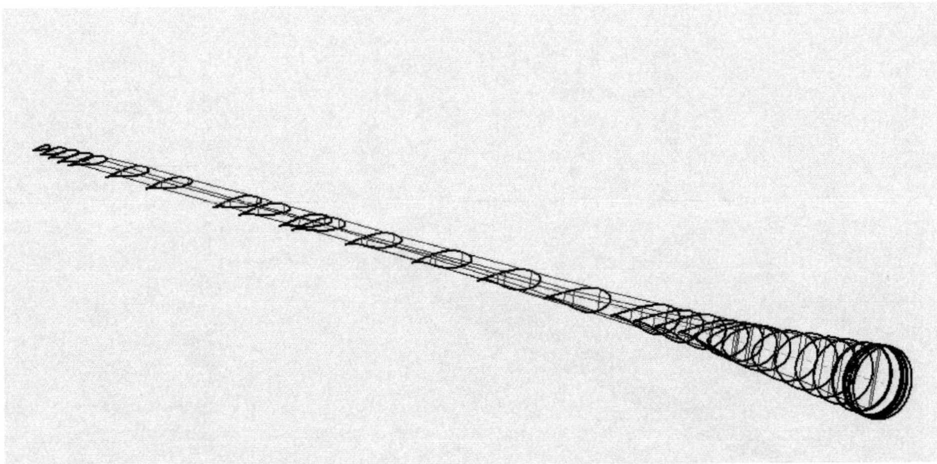

Figure 5: Structural design of NREL 5 MW wind turbine blade based on the NREL report

Since the cut-in wind speed is 5 m/s and the cut-out wind speed is 30 m/s, the classic BEM theory is used in the NREL 5 MW wind turbine blade in order to obtain the aerodynamic loads. And then the edgewise force, flapwise force and pitching moment along with global coordinate of the blade are calculated in each element of wind turbine blade for different wind speeds. Figure 6 shows an example of loads for NREL 5 MW wind turbine blades when the wind speed is 13 m/s.

Figure 6: FE model of NREL 5 MW wind turbine blade in ANSYS with applied loads at wind speed 13 m/s

By applying the blades forces into the model, the maximum stress can be analysed in FE model by ANSYS. In this study, the maximum stress is assumed as the fatigue stress in these loads to calculate the fatigue damage in this certain wind speed. Figure 7 shows the typical maximum stress diagram in ANSYS for NREL 5 MW wind turbine blades when the wind speed is 13 m/s.

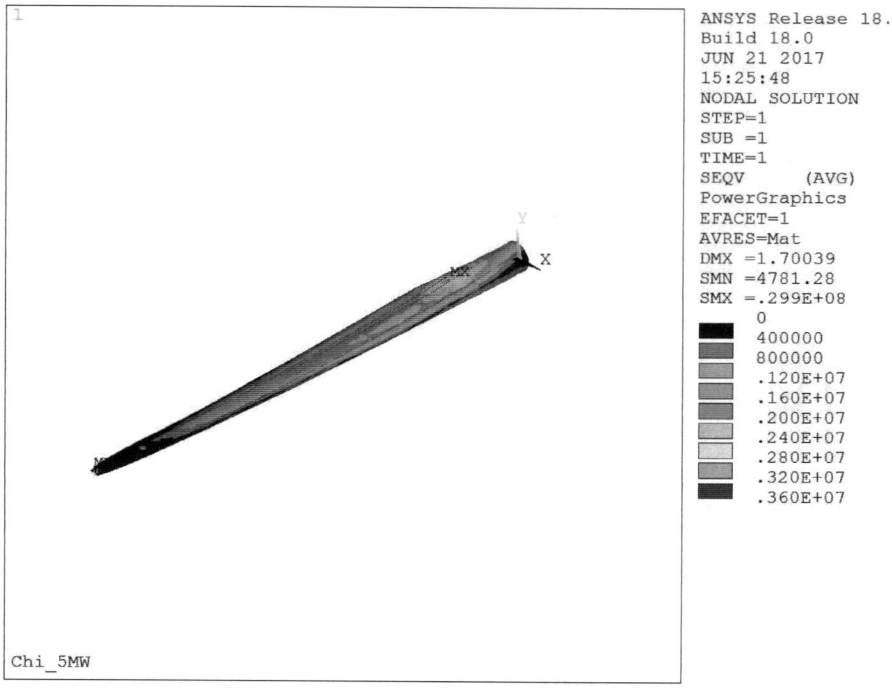

Figure 7: Maximum stress results of NREL 5 MW wind turbine blade in ANSYS at wind speed 13 m/s

The maximum stress in the wind turbine blade depends on the wind speeds shown in Figure 8, according to the results of FE model of wind turbine blade from cut-in wind speed to cut-out wind speed. Based on these results of FE model, the stress-wind speed curve can also be expressed in Figure 8.

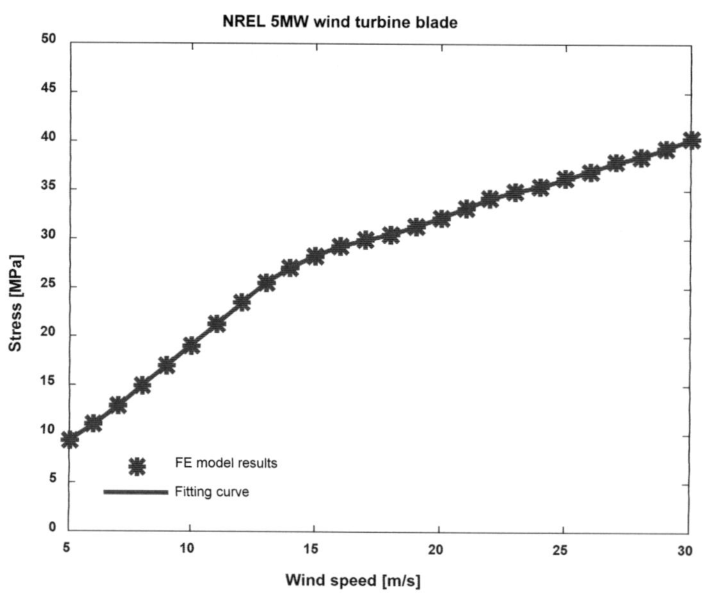

Figure 8: The maximum stress and the curve in various wind speed of FE model

As the wind speed increases, the stress also increase. However, after reaching the designed wind speed, the maximum stress becomes stable. When the wind speed exceeds the designed wind speed for the wind turbine, the attack angle decreases, leading to the reducing wind surface. The wind turbine keeps the stable angular velocity to produce the stable power when the turbulence of wind speed is high (Li *et al.* 2003, Zhang & Chen 2017).

The Weibull distribution of the wind speed of European offshore wind farm over the whole year is estimated, where the shape parameter K=2.0 and the scale parameter C=15 m/s are used in this study. Therefore, the wind field simulation per minute for 1 hour can be randomised in MATLAB, as shown in Figure 9. The spectrum of maximum stress to time diagram for this given hour by wind speed simulation is also shown in Figure 9.

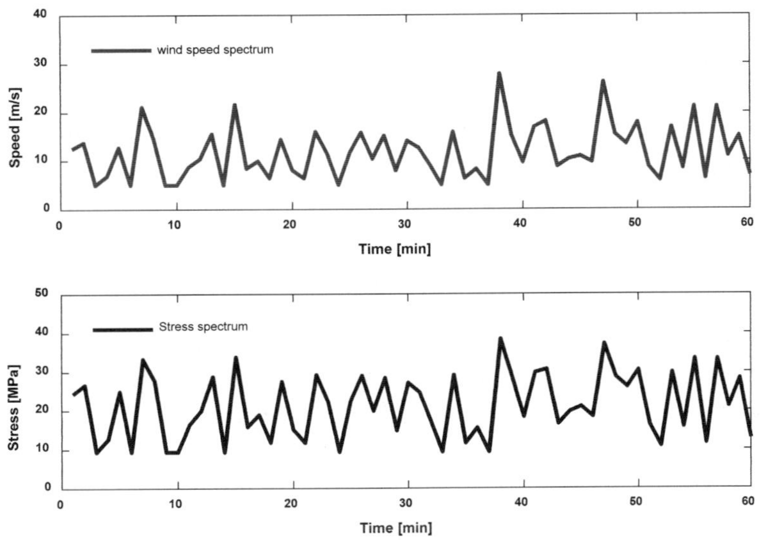

Figure 9: Wind speed generated by Weibull distribution and results of maximum stress to time spectrum in one hour

The number of the cycles can be obtained by Rainflow counting method. Each cycle has the maximum stress σ_{max} and minimum stress σ_{min}, and therefore the mean stress σ_m and amplitude σ_a for this cycle can be calculated. For the glass-fibre reinforced composite material, assuming the ultimate stress σ_u in Goodman diagram σ_u=100MPa, the equivalent stress σ_e at σ_{eqm}=0 can be obtained. The equivalent stress to time spectrum is shown in Figure 10.

According to S-N curve, the life cycles N_i at σ_e can be calculated. The coefficients a and b for the composite material is assumed to be 1.3 and 0.16, respectively. The Miner's law accumulates damage via the cycles by Rainflow counting method, and the failure will occur when the damage accumulates to a certain critical value of unity.

To get the cycle of the fatigue stress in this measuring time, the Rainflow counting method is used to obtain the force cycles during one hour, which is 28 cycles per hour from Figure 10. The fatigue stress cycles can be used to analyse the fatigue damage for the wind turbine blades.

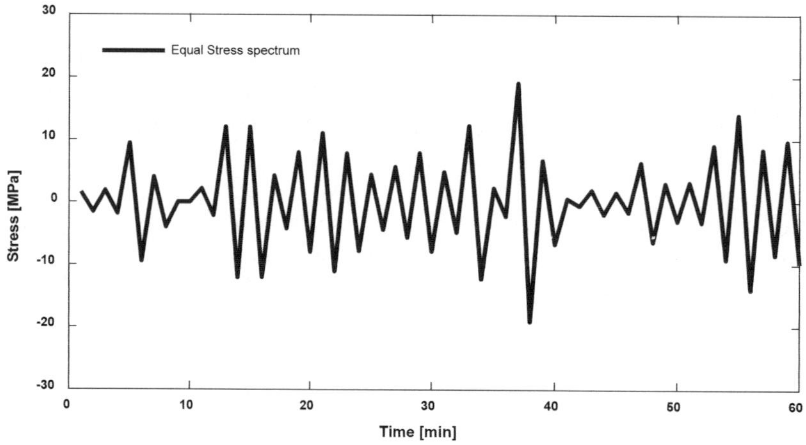

Figure 10: The equivalent maximum stress to time spectrum in one hour

According to S-N curve, the total cycles for wind turbine blades is around 5 million times when the *ratio* is approximately 0.23, and the hourly damage D equals 56×10^{-7}. Therefore, the estimated fatigue life of this NREL 5 MW wind turbine blade is

$$Y = \frac{1}{\sum D\times24\times365} = 20.38 \text{ year} \tag{19}$$

Since the design code expects the service time of offshore wind turbine blades is 25 years, which is slightly longer than the estimated service time by proposed method. Therefore, some proper maintenance strategies should be taken during the design process.

5. Conclusions

This study presents a method for investigating the fatigue damage of the NREL 5 MW wind turbine blades in FE model. The BEM method is used for calculating aerodynamic loads for different wind speeds, and then the forces in FE model can be derived. The Ranflow counting is used here to obtain the load cycles, based on the results of FE model solution and random wind speeds. The Miner's law is used to predict the fatigue damage during the designed service life of wind turbine by both the S-N curve and Goodman diagram. The calculation of the load spectrum and the fatigue assessment of wind turbine blade is performed by analysing the blade loads in FE model using the deterministic method.

From the sample spectrum data during short period operation, the data were rearranged fatigue cycles by Rainflow counting method and equivalent stress by Goodman diagram, and then reordered as life cycle and damage for each cycle by S-N curve. Based on the Miner's law, the accumulation of fatigue damage can be obtained during the sample measuring time. The total service time of the wind turbine blade is estimated as 20.38 years without maintenance.

Further work is needed to determine the optimum repair strategy for the composite blades of an offshore wind turbine according to the obtained results.

References

Bazilevs, Y, Hsu, MC, Akkerman, I, Wright, S, Takizawa, K, Henicke, B, Spielman, T & Tezduyar, TE 2011a, '3D simulation of wind turbine rotors at full scale. Part I: Geometry modeling and aerodynamics', *International Journal for Numerical Methods in Fluids*, vol. 65, no. 1, pp. 207-235.

Bazilevs, Y, Hsu, MC, Kiendl, J, Wüchner, R & Bletzinger, KU 2011b, '3D simulation of wind turbine rotors at full scale. Part II: Fluid–structure interaction modeling with composite blades', *International Journal for Numerical Methods in Fluids*, vol. 65, no. 1, pp.236-253.

Chen, HP, Zhang, C & Huang, TL 2017, 'Stochastic modelling fatigue crack evolution and optimum maintenance strategy for composite blades of wind turbines', *Structural Engineering and Mechanics,* accepted for publication.

Dai, JC, Hu, YP, Liu, DS & Long, X 2011, 'Aerodynamic loads calculation and analysis for large scale wind turbine based on combining BEM modified theory with dynamic stall model', *Renewable Energy*, vol. 36, no. 3, pp. 1095-1104.

Epaarachchi, JA & Clausen, PD 2006, 'The development of a fatigue loading spectrum for small wind turbine blades', *Journal of wind engineering and industrial aerodynamics*, vol. 94, no. 4, pp. 207-223.

Florian, M & Sørensen, JD 2015, 'Wind turbine blade life-time assessment model for preventive planning of operation and maintenance', *Journal of Marine Science and Engineering*, vol. 3, no. 3, pp.1027-1040.

Grujicic, M, Arakere, G, Subramanian, E, Sellappan, V, Vallejo, A & Ozen, M 2010, 'Structural-response analysis, fatigue-life prediction, and material selection for 1 MW horizontal-axis wind-turbine blades', *Journal of materials engineering and performance*, vol.19, no. 6, pp. 790-801.

Hu, W, Choi, KK & Cho, H 2016a, 'Reliability-based design optimization of wind turbine blades for fatigue life under dynamic wind load uncertainty', *Structural and Multidisciplinary Optimization*, vol. 54, no. 4, pp. 953-970.

Hu, W, Choi, KK, Zhupanska, O & Buchholz, JH 2016b, 'Integrating variable wind load, aerodynamic, and structural analyses towards accurate fatigue life prediction in composite wind turbine blades', *Structural and Multidisciplinary Optimization*, vol. 53, no. 3, pp. 375-394.

Jonkman, J, Butterfield, S, Musial, W & Scott, G 2009, 'Definition of a 5-MW reference wind turbine for offshore system development', *National Renewable Energy Laboratory, Golden, CO, Technical Report No. NREL/TP-500-38060.*

Kim, B, Kim, W, Lee, S, Bae, S & Lee, Y 2013, 'Development and verification of a performance based optimal design software for wind turbine blades', *Renewable energy*, vol. 54, pp. 166-172.

Li, DY, Ye, ZQ & Bao, NS 2003, 'Load spectrum and fatigue life analysis of the blade of horizontal axis wind turbine', *Wind Engineering*, vol. 27, pp. 495-506

Manwell, JF, McGowan, JG & Rogers, AL 2010, *Wind energy explained: theory, design and application*, John Wiley & Sons, UK

Mandell, JF, Samborsky, DD, Wang, L & Wahl, NK 2003, 'New fatigue data for wind turbine blade materials', *Journal of solar energy engineering*, vol. 125, pp. 506-514.

Marin, JC, Barroso, A, Paris, F & Canas, J 2009, 'Study of fatigue damage in wind turbine blades', *Engineering failure analysis*, Vol. 16, no. 2, pp. 656-668.

Pratumnopharat, P 2015, *Fatigue Data Editing for Blades of Horizontal Axis Wind Turbines*, LAP LAMBERT Academic Publishing, German

Pratumnopharat, P & Leung, PS 2011, 'Validation of various windmill brake state models used by blade element momentum calculation', *Renewable energy*, vol.36, no. 11, pp. 3222-3227.

Resor, BR 2013, 'Definition of a 5MW/61.5 m wind turbine blade reference model', *Albuquerque, New Mexico, USA, Sandia National Laboratories, SAND2013-2569 2013.*

Shokrieh, MM & Rafiee, R 2006, 'Simulation of fatigue failure in a full composite wind turbine blade', *Composite Structures*, vol. 74, no. 3, pp. 332-342.

Tahani, M, Kavari, G, Masdari, M & Mirhosseini, M 2017, 'Aerodynamic design of horizontal axis wind turbine with innovative local linearization of chord and twist distributions', *Energy*, vol. 131, pp. 78-91.

Wang, L, Liu, X, Renevier, N, Stables, M & Hall, GM 2014, 'Nonlinear aeroelastic modelling for wind turbine blades based on blade element momentum theory and geometrically exact beam theory', *Energy*, vol. 76, pp. 487-501.

Wen, B, Wei, S, Wei, K, Yang, W, Peng, Z & Chu, F 2017, 'Power fluctuation and power loss of wind turbines due to wind shear and tower shadow', *Frontiers of Mechanical Engineering*, vol. 1, pp. 1-12.

Zhang, C, Chen, HP & Huang, TL 2016, 'Stochastic modelling of lifecycle delamination damage evolution of composite blades of wind turbines', *International Conference on Smart Infrastructure and Construction (ICSIC 2016)*, Cambridge, UK.

Zhang, C & Chen, HP 2016, 'A stochastic model for damage evolution of mode II delamination fatigue of composite wind turbine blades', *The Fifth International Symposium on Life-Cycle Civil Engineering (IALCCE)*, Delft, Netherlands.

Zhang, C & Chen, HP 2017, 'Fatigue damage model of wind turbine composite blades under uncertain wind speed', *2nd International Conference on Uncertainty Quantification in Computational Sciences and Engineering (UNCECOMP 2017)*, Rhodes Island, Greece.

The 8th International Conference on
Structural Health Monitoring of Intelligent Infrastructure
Brisbane, Australia | 5-8 December 2017

SA-based Damage Process Stress Monitoring and Verification of RC Columns Subjected to Reversed Cyclic Loading

H.B. Zhang[1], S. Hou[2], J.P. Ou[1]
[1] *Shenzhen Graduate School, Harbin Institute of Technology – China, Email:* zhanghb01@gmail.com
[2] *School of Civil & Transportation Engineering, South China University of Technology - China*

Abstract

The stress history is a critical index to assess the health status of reinforced concrete (RC) structures under seismic loading. Piezoelectric smart aggregate (SA) provides an innovative method for stress monitoring in concrete structure. However, the reliability of SA for stress monitoring in RC structure especially in nonlinear stage is still unknown, due to the factors such as crack contact effect, reinforcement constraint, and compressive and shear stress coupled effect. In this study, the performance of SA for seismic stress monitoring in RC structure during the full damage process was evaluated. Three identical full-scale rectangular RC columns of normal strength concrete with a total of twenty SAs and four strain gauges embedded for concrete stress and reinforcement strain monitoring in the plastic hinge zone in each specimen were employed. The specimens were tested under combined reversed cyclic lateral displacement excursions and axial load by using MTS hydraulic machine. The section moment was calculated by using the monitored data and MTS data, respectively. The accuracy of SA was evaluated by comparing these section moments. It can be found that the monitoring error was only about 5% in nonlinear stage, which indicates that the SA was reliable for seismic stress monitoring in RC structure especially in nonlinear stage.

1. Introduction

Seismic monitoring of structural systems constitutes an integral part of the National Earthquake Hazard Reduction Program in the United States and similar hazard reduction strategies in seismically active regions of the world (Garevski 2013). For reinforced concrete (RC) structures, since the damage mostly occurs at some key local locations during earthquake such as the bottom of the column or shear wall, the ends of the beam, and joint of beam and column, sensors that provide direct local stress/ strain information help supplement existing seismic monitoring systems.

As a type of embedded sensor, the piezoelectric smart aggregate (SA) meets the requirement of seismic stress monitoring including high strength, water-proof, and high sensitivity. The SA composing of a pair of granite blocks as shelter instead of a cement matrix, has been used for compressive and shear stress monitoring (Hou *et al.* 2012, Hou *et al.* 2013). With a specially designed charge amplifier, the modified SA was able to measure stresses of up to 100 MPa with a low frequency bound of 0.05 Hz (Zhang *et al.* 2016). The feasibility and effectiveness of modified SA for stress monitoring in normal strength and high-strength plain concrete were elaborated (Hou *et al.* 2016, Zhang *et al.* 2017). As a matter of fact, concrete is usually combined with steel reinforcement as a form of RC structure to improve the tensile and ductile characteristic. There are three differences between plain concrete and reinforced concrete from the view of seismic stress monitoring.

Firstly, RC structures are subject to multiple loading/unloading cycles during seismic loading, which result in cracks opening and closing many times. The opposing surfaces of these cracks contain multiple asperities that can come into contact under combinations of shear stress (or say 'aggregate interlock') (ACI-ASCE Joint Task Committee 426 1973, Paulay and Loeber 1974) and compressive stress normal to the crack surface (or say 'crack closure behavior') (Nouailletas *et al.* 2015, Jefferson and Mihai 2015), which will result in the stress concentration in the region of sharp corner and variation of the stress field (He and Ou 2006). Therefore, the effects of these factors on the local stress monitoring by SA should be investigated. Secondly, the stirrup in RC structure supplies the hoop restraint, but it will not influence the compressive stress monitoring by SA due to its insensitivity to hoop stress (Hou et al. 2012). Finally, the longitudinal reinforcement in RC structure shares the compressive stress with concrete, the local compressive stress field of concrete around the longitudinal reinforcement could be affected due to the difference of Young's modulus between longitudinal reinforcement and concrete. This effect on the reliability of stress monitoring by SA should also be figured out.

The objective of this paper is to verify the feasibility and effectiveness of SA-based stress monitoring in RC structure. Three identical full-scale RC columns of normal strength with a total of twenty SAs and four strain gauges embedded for concrete stress and longitudinal reinforcement strain monitoring in the plastic hinge zone for each specimen were employed. These specimens were tested under lateral cyclic reversed loading combined with axial loading. The local information including concrete stress and longitudinal reinforcement strain, and global information including lateral force and displacement and axial force, were recorded by data acquisition system at the same time. To verify the reliability of the stress monitored by SA, the section moment monitoring method by using local stress/strain information was proposed. The section moment obtained by local stress/strain monitoring were compared with theoretical value calculated by global load/displacement. It can be concluded that the SA is reliable for seismic stress monitoring in RC structure especially in nonlinear stage.

2. The validation for the reliability of stress monitoring

To validate the reliability of stress monitoring, the Moments of the bottom cross-section of the RC column, which were calculated by using the local stress/strain monitoring method and global load/displacement monitoring method respectively, were compared.

2.1 The moment calculated by local stress/strain monitoring method

The SA consists of a piece of PZT (lead zirconate titanate) patch which is sandwiched between a pair of granite cubes through epoxy. The dimension of the SA after fabricated is 25 mm × 25 mm × 25 mm. The detailed composition of the SA and its feasibility for seismic stress monitoring is introduced in the literature (Hou et al. 2012, Hou et al. 2016, Zhang et al. 2017).

The principle of local stress/strain monitoring method is presented in Figure 1. The concrete stress distribution of the column's section can be monitored by distributed SAs. The strain of the longitudinal reinforcement can be monitored by strain gauges, and the stress of longitudinal reinforcement can be calculated by using its stress-strain relationship. Therefore, the section moment can be calculated by using the following equation.

$$M = \sum (\sigma_{sti} A_s + \sigma'_{sti} A'_s) d_s + \sum_{i=1}^{n} \sigma_i \frac{A_c}{n} d_i \tag{1}$$

where M is the section moment of the RC columns. σ_{sti} and σ_{sti}' is the tensile and compressive stress of the longitudinal reinforcements, respectively. A_s and A_s' is the cross-section area of the tensile and compressive longitudinal reinforcement, respectively. d_s is the distance between the center of longitudinal reinforcement and the vertical center of the section. σ_{i-1} is stress monitored by the $(i-1)^{th}$ SA. d_i is the distance between the center of the ith SA and the vertical center of the section. A_c is the area of the cross-section of the column. n is the quantities of the SA along the direction of the applied moment.

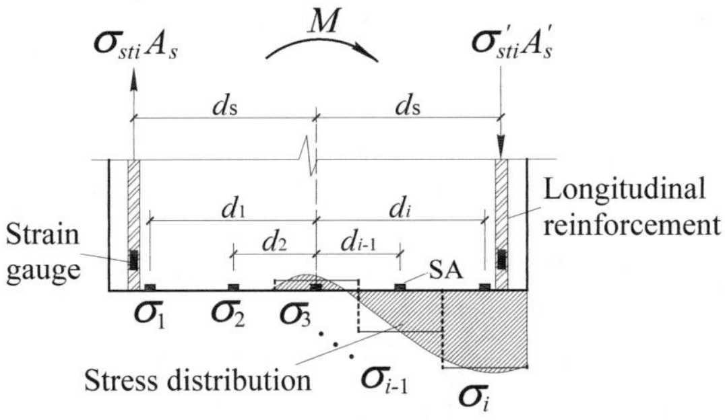

Figure 1 Schematic diagram of local stress/strain monitoring method.

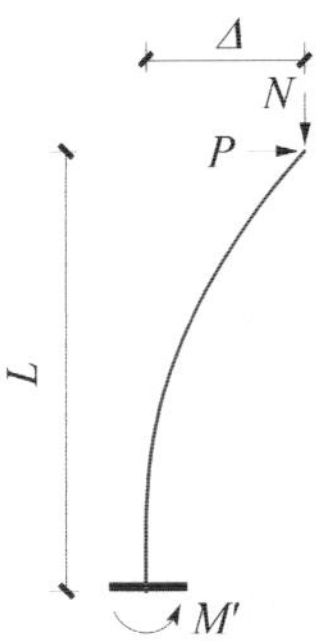

Figure 2 Schematic diagram of global load/displacement monitoring method.

2.2 The moment calculated by global load/displacement monitoring method

The principle of the section moment monitoring by using global load/displacement monitoring method was presented in Figure 2. The axial and lateral load can be monitored by using the load cell located on the loading equipment, respectively. The corresponding displacement of the loading end can be measured by linear variable differential transformer (LVDT). Therefore, the bottom section moment of the RC column can be calculated by equation (2)

$$M' = N\Delta + PL \tag{2}$$

Where M' is the bottom section moment that SA located at. N and P is the axial and lateral load, respectively. Δ and L is the axial and lateral displacement of the loading end, respectively.

3. Experimental setup

3.1 Specimen details

In order to verify the feasibility and effectiveness of seismic stress monitoring of RC columns by using SAs, combination of cyclic reversed displacement excursions and axial loads was applied on RC columns with SAs and strain gauges embedded in the plastic hinge zone. Three RC columns denoted Col-1, Col-2 and Col-3 with identical dimensions and reinforcement layout were involved. The details of these specimens are shown in Figure 3. Each specimen consisted of a 1600 mm × 400 mm × 400 mm column cast integrally with a 1500 mm × 650 mm × 600 mm stub at the bottom, and a 700 mm × 650 mm × 400 mm cap at the top for convenient loading. The clear concrete cover to the stirrups was 30 mm. The column represents the part of a RC frame between the section of maximum moment and the point of contra flexure. The stub represents a discontinuity, such as a beam column joint or a footing. These columns had an effective length of 1400 mm and a span-to-depth ration (L/D) of 3.5.

Eight HRB 400 steel bars (Hot-rolled Ribbed Bar, f_{yk} = 400 N/mm^2) with a diameter of 22 mm were used as longitudinal reinforcement, and HRB 335 steel bars 10mm in diameter (f_{yk} = 335 N/mm^2) were used as stirrups with a spacing of 100 mm in the densely-distributed region. The concrete strength for all specimens was designed to be grade C30. Concrete cubes cured in the same environment as the specimens were also taken to monitor the strength, and the average measured cube strength was 33.5 MPa with a coefficient of variation of 0.13 after 28 days of curing. The mean yield strength and Young's modulus of the longitudinal reinforcement was 460 MPa and 210 GPa, respectively, and the yield strain is about 2200 $\mu\varepsilon$.

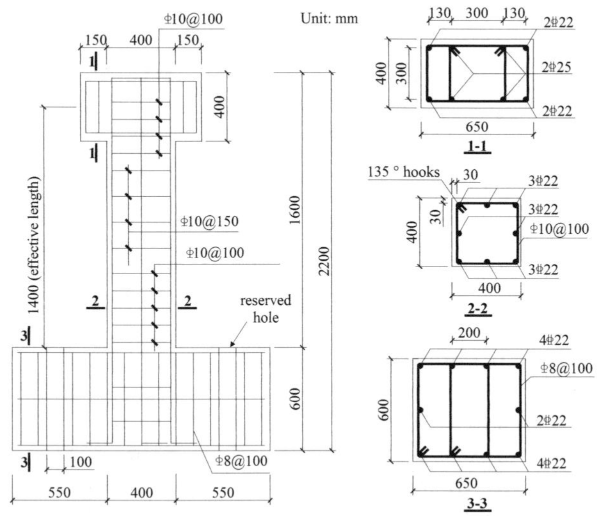

Figure 3 Schematic diagram of the test specimens.

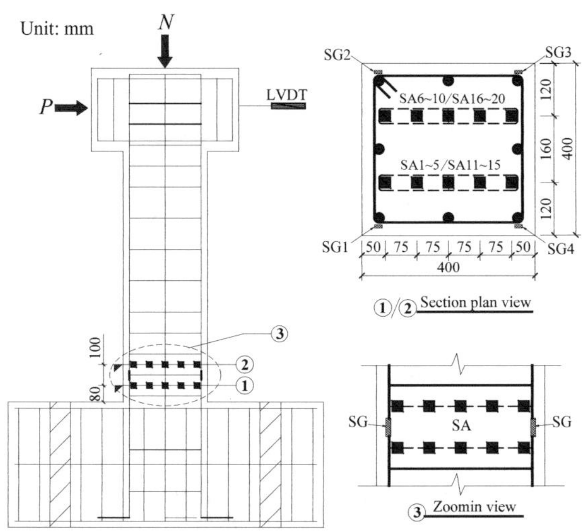

Figure 4 Schematic diagram of the layout of sensors.

3.2 Layout of the sensors

The details of the sensors layout were shown in Figure 4. The section stress at five locations along the loading direction in the plastic hinge zone will be monitored by SAs, which were the center of the section, 75 mm and 150 mm away from the center on each side. Since the monitoring accuracy of SA can be improved by increasing the quantities of SA (Hou et al. 2016), four SAs were employed at each location in the section that 80 mm and 180 mm away from the bottom of the column. The distance between the lines of SA at each section was 160 mm. The average stress of four SAs for each location was calculated to be the nominal stress of the location. A total of twenty SAs were involved, which were denoted from SA1 to SA20. Four stain gauges (SG for abbreviation) which denoted as SG1 to SG4, were deployed on the longitudinal reinforcements of four corners in the cross-section between two layers of SAs.

3.3 Experimental set-up

All the columns were tested under combined axial and lateral load in a testing frame, as shown in Figure 5(a). An initial axial force of 800kN (axial load ratio 0.22) was applied on the top surface of the column cap using a hydraulic jack combined with a hinge support and traveling block. The cyclic reversed lateral displacement was applied on the specimens through a servo-hydraulic test machine with a load capacity of 500 kN (MTS hydraulic actuator, Model 244.41). A pair of spacing rollers were employed to limit the out-plane displacement of the column.

As presented in Figure 5 (b), the generated signals from the SAs and the SGs were converted to a voltage signal by using a charge amplifier (HK9209, Hengke Tech. Co., China) and a dynamic strain indicator (HK9304, Hengke Tech. Co., China), respectively. Data was collected by using a data acquisition system (NI 4472 module, National Instruments) which also recorded the time-history of applied load and displacement at the same time.

(a) Test setup (b) Data acquisition system

Figure 5 The test system.

3.4 Loading scheme

A series of grading cosine waveforms were applied on the specimens at each level of displacement. The initial displacement corresponded to Δ_1=1mm, which was preloading displacement. Subsequent displacements were increased to 4mm; 8mm; 12mm, … , until 56mm. The loading period for each grading was 20s (0.05Hz). It should be noted that the maximum loading displacement of Col-1 was 36 mm due to the unexpected termination. The loading displacement of the other columns reached to 56 mm as expected.

As shown in Figure 6, the measured lateral displacements of hydraulic actuator of the three columns almost coincide, which not only indicates the good consistence of the applied load but also implies the rational design and fabrication of the specimen.

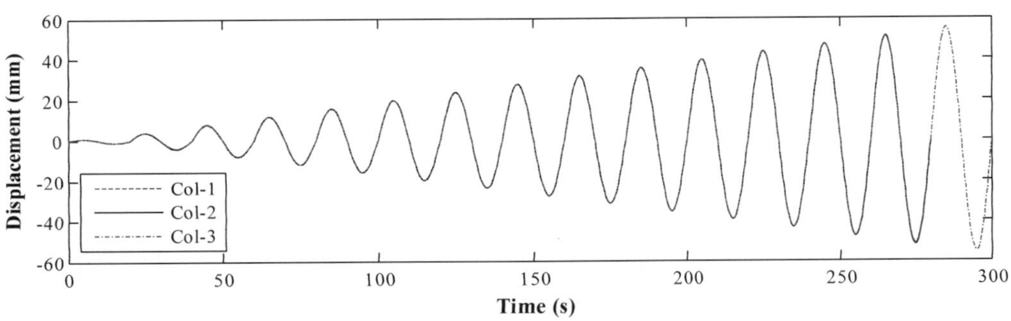

Figure 6 Lateral displacement of the hydraulic actuator.

4. Experimental results and discussion

4.1 Response of the specimen

As presented in Figure 7, it can be found that the axial load was varying with a period of 10s along with the loading process proceeding. The maximum axial load reached to 1500kN (axial load ratio 0.4), which is about twice times than the initial axial load.

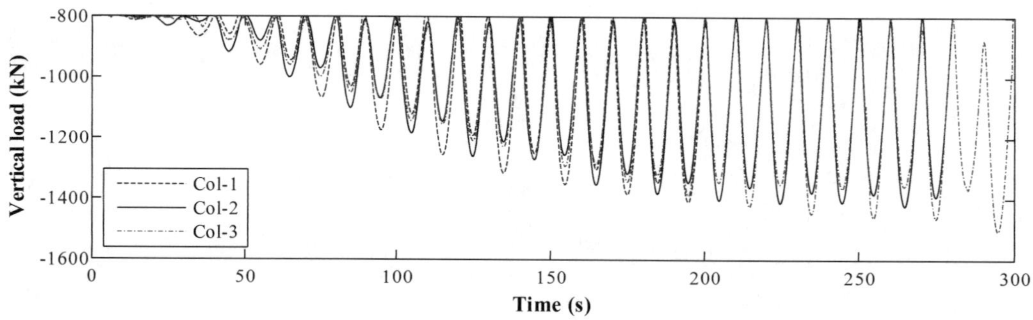

Figure 7 The time-history of axial load.

As shown in Figure 8, the lateral loads of three columns were almost the same. The variation trend of the peak load for each loading period was firstly increasing, then keeping constant, finally decreasing, which was corresponding to the stage of longitudinal reinforcement: not yield, yield, and concrete crushed. The maximum absolute value of the peak and valley lateral load was basically the same, which was about 250kN.

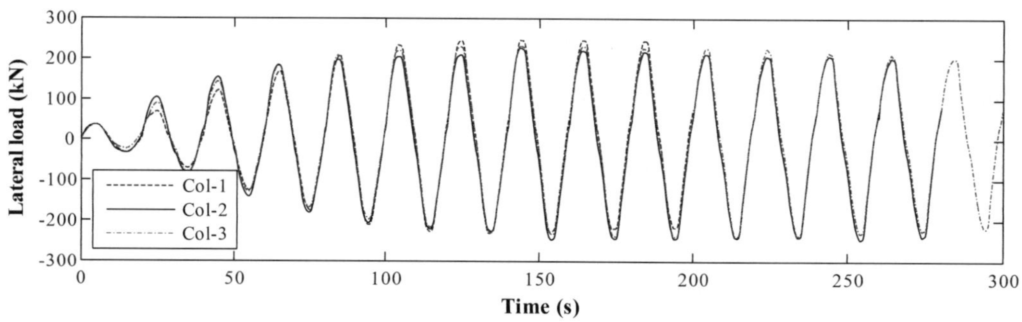

Figure 8 The time-history of the lateral load.

4.2 Time-history of concrete stress

Utilizing the sensitivity curve for the normal strength concrete obtained in the literature (Hou et al. 2016), the output voltage of SA can be converted into concrete stress. The mean stress time-history at each of the five locations along the loading direction were shown in Figure 9. For the concrete stresses of 150mm away from the center line in each column (See Figure 9(a) and (b)), the variation trend was nearly the same. With the increasing of the lateral displacement, the amplitude of compressive stress gradually increases until reaches to concrete strength of 35 MPa. The corresponding tensile stresses ranged from 3 to 4.5 MPa, which was close to the tensile strength of normal strength concrete. In regard to the concrete stresses of 75mm away from the center line as shown in Figure 9(c) and (d), the compressive range was from 10 to 15 MPa, which was lower than the stress at the location of 150mm as expected. As presented in Figure 9(e), the concrete stress at the center line of the cross-section ranged from -2 to 2 MPa, indicating the concrete was still at the linear stage.

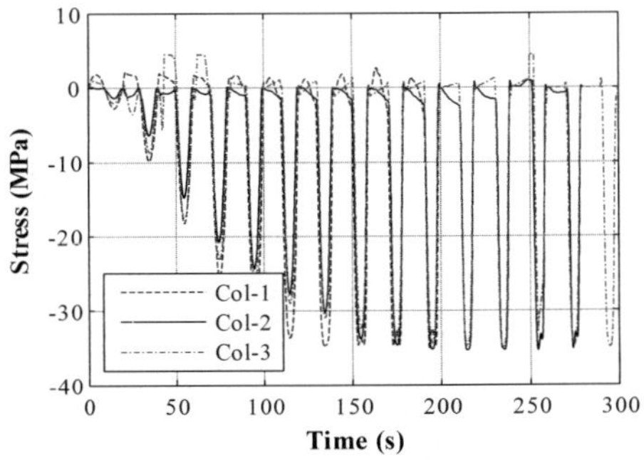

(a) 150mm away from the center line (left)

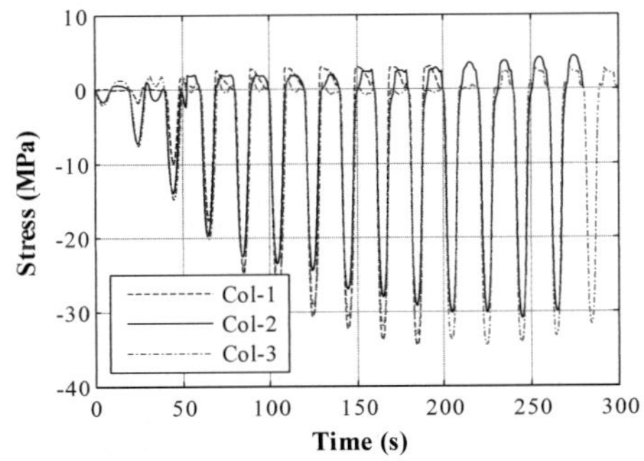

(b) 150mm away from the center line (right)

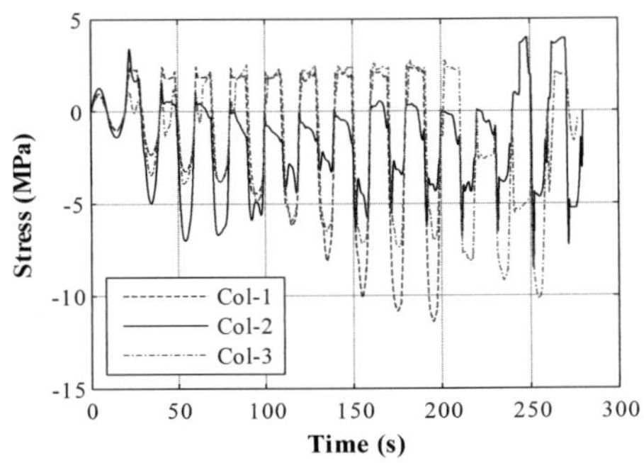

(c) 75mm away from the center line (left)

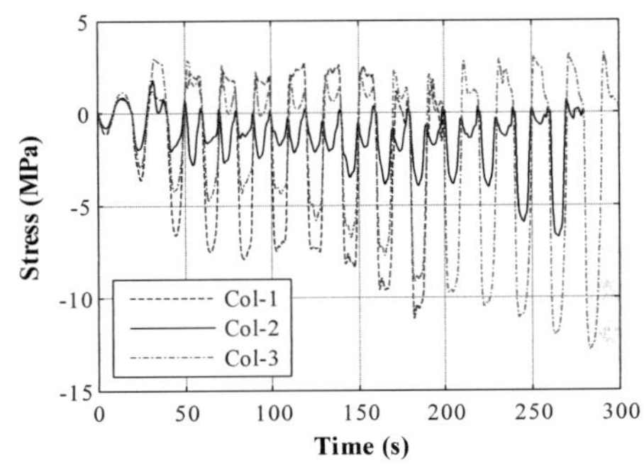

(d) 75mm away from the center line (right)

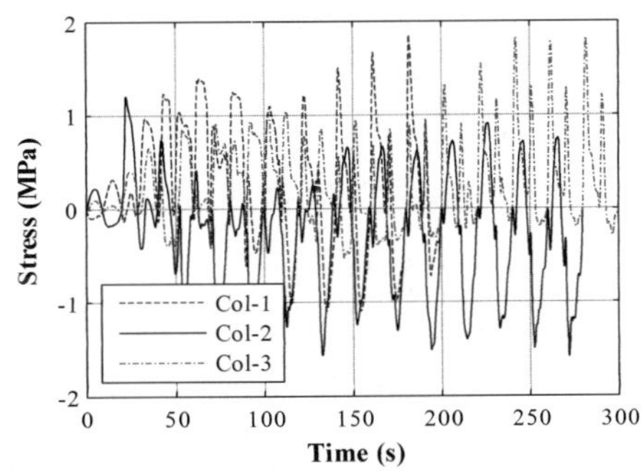

(e) Center line of the cross-section

Figure 9 The stress-histories of five locations along the loading direction.

4.3 Strain of the longitudinal reinforcement

The mean strain history of the longitudinal reinforcement in the same side was shown in Figure 10. It can be found that the amplitude of the strain increases with the passing of the time for three specimens. The moment of the longitudinal reinforcement yield was about 145s or 155s, corresponding to the drift ratio of 1/50. It should be noted that the stress of the longitudinal reinforcement will not exceed the yield strength of 460MPa when the strain is larger than yield strain.

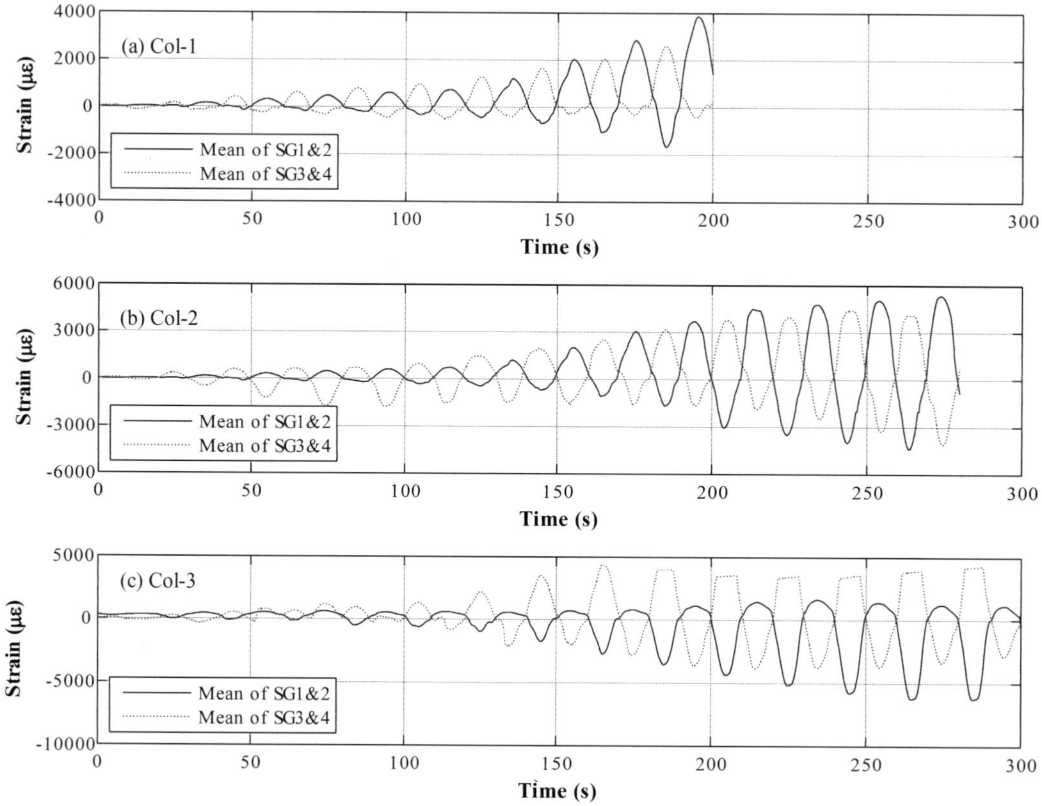

Figure 10 The time-history of the longitudinal reinforcement strain.

4.4 Comparison of the moment monitored by global load/displacement and local stress/strain

The moments respectively calculated by using equation(1) and (2) were compared as shown in Figure 11. The amplitude of the moment firstly increases with the increasing of the lateral displacement, then almost keep constant, indicating the process of the longitudinal reinforcement yielded and concrete crashed. The moment calculated by global load/displacement monitoring method was basically the same with that obtained by local stress/strain monitored method.
In order to depict the error more exactly, the monitoring error can be calculated by equation (3).

$$Error = \frac{|\Delta_p| + |\Delta_N|}{2|A_{\max}|} \times 100\% \tag{3}$$

where $|\Delta_p|$ and $|\Delta_N|$ are the difference between the amplitude of the global load/displacement monitored moment and local stress/strain monitored moment in positive and negative loading direction, respectively. $|A_{\max}|$ is the amplitude of global load/displacement monitored moment.
As shown in Figure 12, the monitoring error reduces with the increasing of drift ratio in the range of 0 to 1/200. Three reasons may lead to the larger monitoring error in this range, such as the gap between the specimen and the loading frame, the initial defect in concrete, and the minor value of displacement amplitude. It should be noted that the error will not be attributed to the effect of aggregate interlock and crack closure behavior, because these effects will increase the monitoring

error until reaches to stability with the increasing of drift ratio. The monitoring error tends to be about 5% when the drift ratio is larger than 1/200, indicating the accuracy of stress monitored by SA in RC structure, and the effect of aggregate interlock, crack closure behavior and restraint of reinforcement on the stress monitoring in RC structure is slight in nonlinear stage.

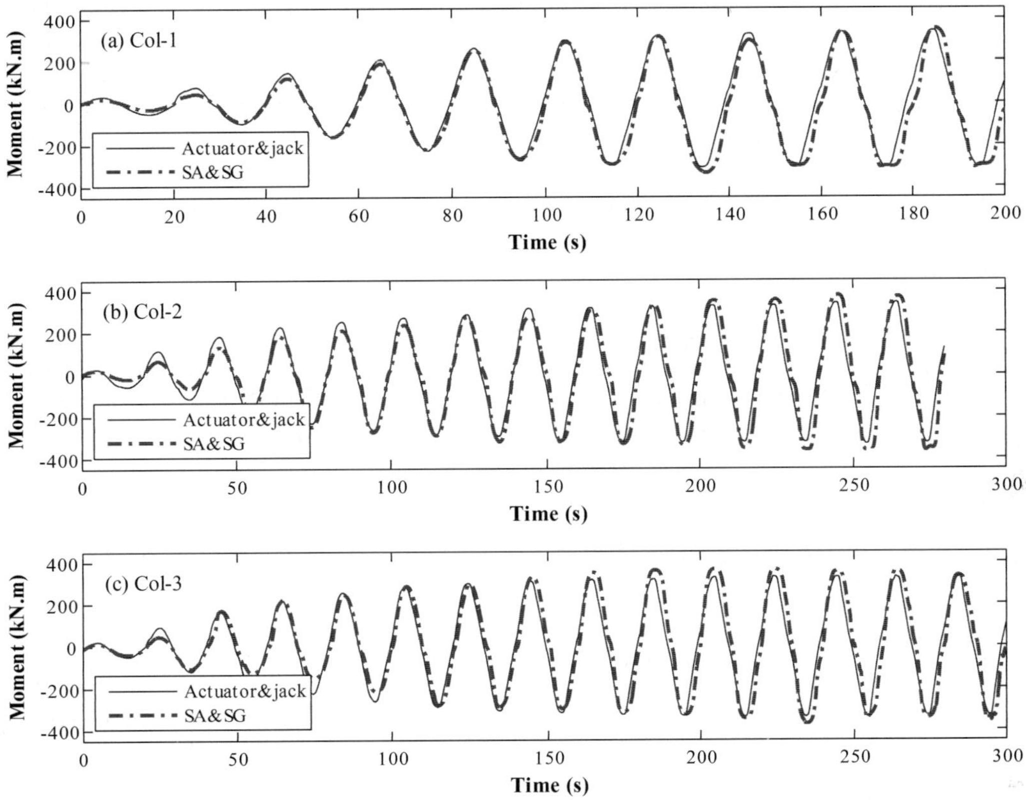

Figure 11 Comparison of the moment between global load/displacement method and local stress/strain method.

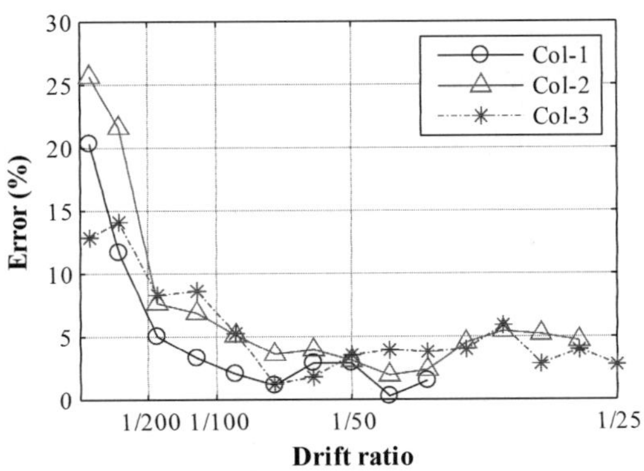

Figure 12 Variation of the monitoring error.

5. Conclusions

This paper presents the feasibility and effectiveness of SA-based concrete seismic stress monitoring in RC structure. Cyclic reversed loading combined with axial loading were conducted on three identical RC columns with twenty SAs and four strain gauges embedded in the plastic hinge zone for each specimen. The performance of local stress/strain monitoring was tested experimentally during the full damage process of the specimens. To verify the accuracy of SA, the section moment monitoring by using local stress/strain method was proposed. The conclusions are summarized as follows.

Throughout the full damage process of RC structure, SAs and strain gauges worked properly and reliably measured the concrete stress and longitudinal strain, respectively. The monitored compressive and tensile strength of concrete is reasonable. The section moment obtained by local stress/strain monitoring method were compared with global load/displacement monitoring method. The Monitoring error was only about 5% when the drift ratio was larger than 1/200 (nonlinear stage), indicating the effects of aggregate lock, crack closure behavior and the restraint of reinforcement on the SA seismic stress monitoring is slight especially in nonlinear stage.

Acknowledgments

Joint financial support from the Key Lab of Structures Dynamic Behavior and Control (Harbin Institute of Technology), Ministry of Education (granted no. FA29100011), National Nature Science Foundation of China (grant no. 51678250) and National Basic Research Program of China (973 Program, granted no. 2015CB057704), are gratefully acknowledged.

References:

Garevski, M 2013, *Earthquakes and Health Monitoring of Civil Structures*, Springer Environmental Science and Engineering, Netherlands.

Hou, S, Zhang, HB & Ou, JP 2012, 'A PZT-based smart aggregate for compressive seismic stress monitoring', *Smart Materials and Structures,* vol. 21, no. 10, pp. 105035.

Hou, S, Zhang, HB & Ou, JP 2013, 'A PZT-based smart aggregate for seismic shear stress monitoring', *Smart Materials and Structures,* vol. 22, no. 6, pp. 065012.

Zhang, HB, Hou, S & Ou, JP 2016, 'Smart aggregate-based seismic stress monitoring system using a specially designed charge amplifier', *Journal of Intelligent Material Systems and Structures,* vol. 27, no. 3, pp. 418-426.

Hou, S, Zhang, HB & Ou, JP 2016, 'SA-based concrete seismic stress monitoring: a case study for normal strength concrete', *Smart Materials and Structures,* vol. 25, no. 9, pp. 095041.

Zhang, HB, Hou, S & Ou, JP 2017, 'Feasibility of SA - based concrete seismic stress monitoring for high-strength concrete', *Journal of Intelligent Material Systems and Structures,* doi: 10.1177/1045389X17689935.

ACI-ASCE Joint Task Committee 426 1973, 'The Shear Strength of Reinforced Concrete Members', *Journal of the Structural Division,* vol. 99 no. 6, pp. 1091-1187.

Paulay, T & Loeber, PJ 1974, 'Shear transfer by aggregate interlock', *ACI Special publication,* no. 42, pp. 1-16.

Nouailletas, O, Borderie, CL, Perlot, C, Rivard, P & Ballivy, G 2015, 'Experimental study of crack closure on heterogeneous quasi-Brittle material", *Journal of Engineering Mechanics,* vol. 141, no. 11, pp. 04015041.

Jefferson, AD & Mihai, IC 2015, 'The simulation of crack opening - closing and aggregate interlock behaviour in finite element concrete models', *International Journal for Numerical Methods in Engineering,* vol. 104, no.1, pp.48-78.

He, Z & Ou, JP 2006, *Nonlinear analysis on reinforced concrete structure*, Harbin Institute of Technology, Harbin, China.

The 8th International Conference on
Structural Health Monitoring of Intelligent Infrastructure
Brisbane, Australia | 5-8 December 2017

Monitoring collapse process of a demolished building using piezo-electric smart aggregate

S. Hou, LL. Cui[1].
[1] *South China University of Technology – China, Email:* cthous@scut.edu.cn

Abstract

Piezo-electric smart aggregate (SA) is capable of full range dynamic stress monitoring for concrete structures. In this study, the collapse processes of a six-story demolished masonry building with inner RC frame at the first floor were monitored by using SAs. Total 12 SAs were installed at the positions of top, bottom and joint area of the two columns of the inner RC frame. Both the normal and the shear stress were monitored in this test. The monitored stresses came from different excitation mechanisms with multiple stress levels, including stresses of a few kPa from ambient vibration, stresses near 1 MPa from wrecking ball collision on the exterior walls of the building, and stresses larger than 10MPa caused by direct collision on the column. Under the ambient excitation, the SAs showed an extraordinary high sensitivity capturing the structural fundamental frequency of the building from the stress signals. Under the vibration by the exterior wall collision, the several former vibration frequencies of the building and their attenuation can be observed and the changes of the fundamental frequency correlate the damage processes of the building. Under direct impacts on a column near the SA, an increase of stress up to magnitude of material strength can be monitored and the redistribution process within the inner frame can be observed. Through this test, the reliability of the SA-based structure damage process monitoring had been demonstrated with positive results.

1 Introduction

Structural Health Monitoring (SHM) is an innovative method for monitoring the structural status and performance without affecting the structure itself. It involves understanding the behavior of structure under different hazardous conditions, reducing the subjection to damage by taking precautions and assisting in the maintenance and repair work of structure.(Modares & Waksmanski, 2013) So an efficient and reliable sensor for monitoring is very necessary. In recent years, a kind of smart sensor made by piezoceramic materials have been successfully employed in SHM of concrete structures due to its advantages of active and passive sensing, low cost, high sensitivity, assorted shapes and simple implementation. Those sensors were usually embedded in concrete before casting and can be treated as aggregates without influence on structures, so they are also called SAs (Smart Aggregates). (Song *et al.*, 2005) By use of the direct and indirect piezoelectricity of this material, researchers have applied SAs in testing the early strength of concrete (Gu, Song, Dhonde, Mo, & Yan, 2006), monitoring the collisions of over-height vehicles with bridges (SongOlmi & Gu, 2007), and surveying the structural health of other concrete structures (Gu, Moslehy, Sanders, Song, & Mo, 2010; Laskar, Gu, Mo, & Song, 2009; Moslehy, Gu, Belarbi, Mo, & Song, 2010; Yan *et al.*, 2009). For all the studies just mentioned, the core theory to identify damage is the assessment of the wave propagation in concrete. The application of SAs was also been extended to monitor seismic stress including compressive and shear stresses. (Hou, Zhang & Ou, 2012, 2013, 2016) With proper charge amplifier, SAs can fulfil the demands of high stress up to 100 MPa and low frequency down to 0.05 Hz, (Zhang, Hou & Ou, 2016) which is satisfied for civil infrastructure seismic monitoring. In this paper, the monitoring system based on SAs proposed by Hou and his team were utilized to detect the dynamic responses of a six-story building structure under impact load and ambient excitations. During the experiment, the building was hit by a heavy swinging steel ball hung from a crane continuously until it collapsed, and the ambient excitation was created by the moving of crane beside the building. At the same time, the embedded SAs which were installed before testing were monitoring the dynamic inner stresses. Through the test, the structure was demolished by the steel ball piece by piece and the whole process was recorded by the monitored stresses. Based the level of severity, 6 damage state was proposed. Within each state, the data directly presents the vibration of structure from which people can obtain the vibration modes and frequencies. With further signal processing, the energy distribution of different modes along time was revealed. Plotting the fundamental frequency of each damage state together, the result shows a decreasing tendency as damage level increases. Besides, the global and local stress redistributions were also detected when major damage happened. Generally speaking, the suggested monitoring system based on SAs provides an easy and feasible method to detect the dynamic characteristics of the structure and can help to recognize the damage to some extent.

2 Building detail and monitoring scheme

A six-story building which was built in 1980s and located in Dalian northeastern China is under investigation in this paper. The first floor has the cast-in-place R.C. slab supported mainly by R.C. frames inside and brick walls outside. And there are also two brick walls around the stairwell, shown in Figure 2 (a). The height of the first and the upper floors are 3.5m and 3m respectively. Upper stories are 3.0 meters high, with precast R.C. slabs, supported by only brick walls, mainly used for residence, shown in Figure 2 (b). The front, back and inner view of the building are shown in Figure 1. The thickness of the exterior wall and the inner partition wall are 370mm and 240mm, respectively. The inner frame is composed of four columns with the same cross section, denoted by C1~C4. The compressive strength of concrete of the inner RC frame was tested to be 20.5 MPa through core sampling method on locations where the least load was carried. The tensile strength of the longitudinal rebar is 410 MPa. The diameter of the stirrup within columns and beams is 6 mm with spacing of 200 mm. The tensile strength of stirrup is 210 MPa.

(a) (b) (c)

Figure 1 Building pictures: (a) front view, (b) back view and (c) inner structures

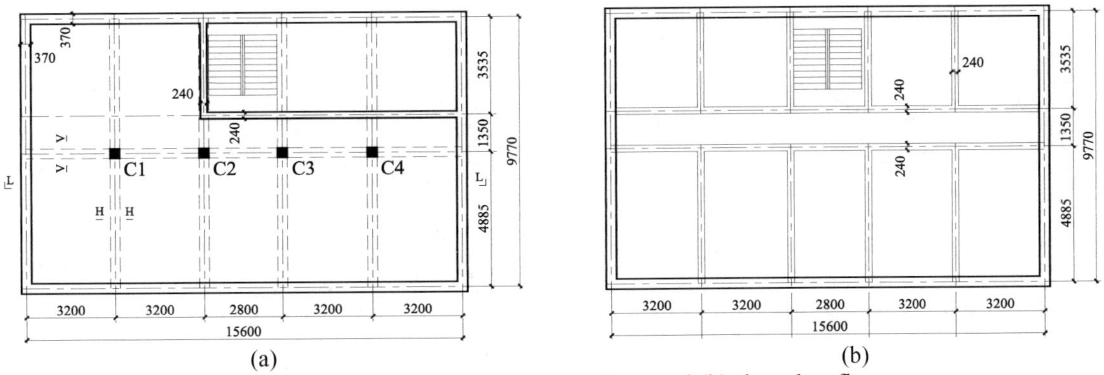

Figure 2 Plan views of (a) the first floor and (b) the other floors

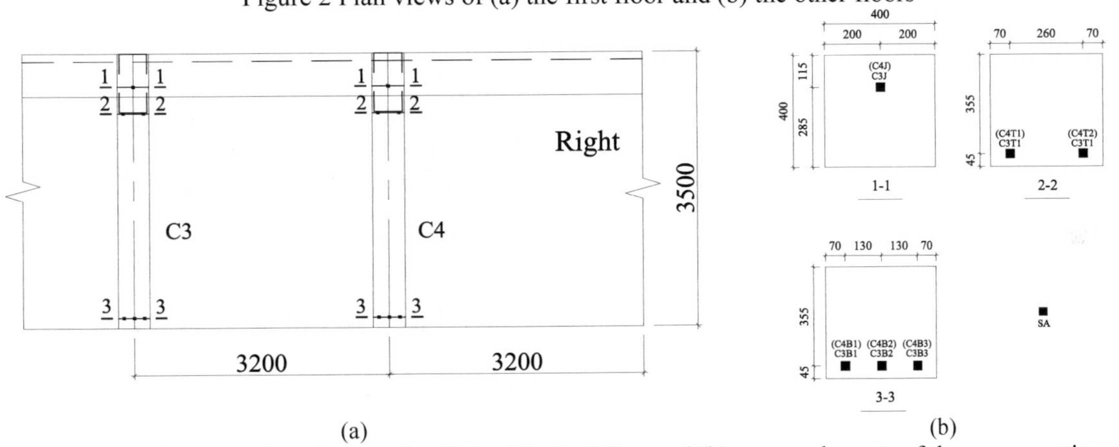

Figure 3 (a) A part of vertical section L-L of the building and (b) sensors layouts of the cross-sections

Before the experiment, the measuring sensors which are smart aggregates (SAs) shown in Figure 4 (a) were installed in the targeted locations of the experimental building by the special instrument like it shows in Figure 4 (b). There are two types of SAs but the same appearance and they were used to monitor the normal and shear stresses respectively. The sensor details had been presented in literature of (Hou, Zhang & Ou, 2012, 2013, 2016; Zhang, Hou & Ou, 2016). In this building, column C3 and C4 that will be damaged in the very late stage during the demolition process were instrumented with the stress sensors. The instrumentation schemes for these two columns are the same. SAs were embedded at three heights in each column where were the beam-column joint, the top and the bottom of the column as shown in Figure 3 (a). At the beam-column joint, the SAs are designed to monitor shear stress, and except them, all the other SAs are designed to monitor normal stress. In this test, total 12 SAs were used and their layouts in the cross-sections are shown in Figure 3 (b).

During the experiment process, the building was subjected to a series of collisions by a wrecking ball hanging by a steel rope which is attached on a crane, shown in Figure 1 (b). The impacting was carried out from left side, back side and right side successively, resulting in a global collapse of the building. Meanwhile, the inner stress data were transferred to the outer data acquisition system in Figure 5 (b). The monitoring scheme is shown in Figure 5 (a).

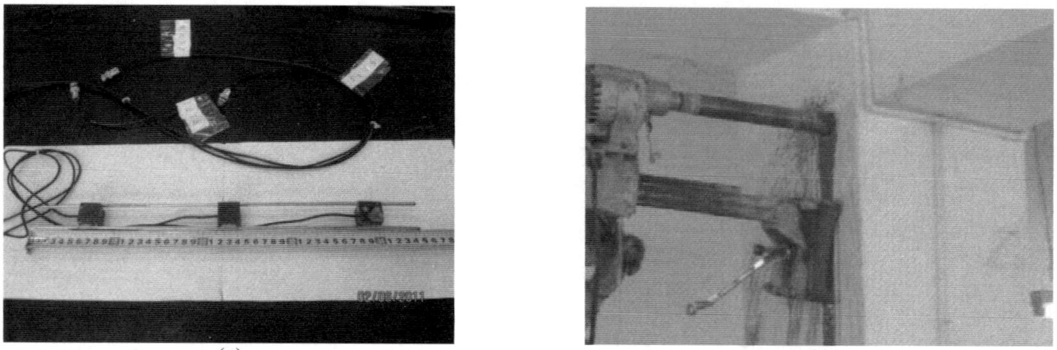

(a) (b)

Figure 4 (a) SAs and (b) installation of SAs

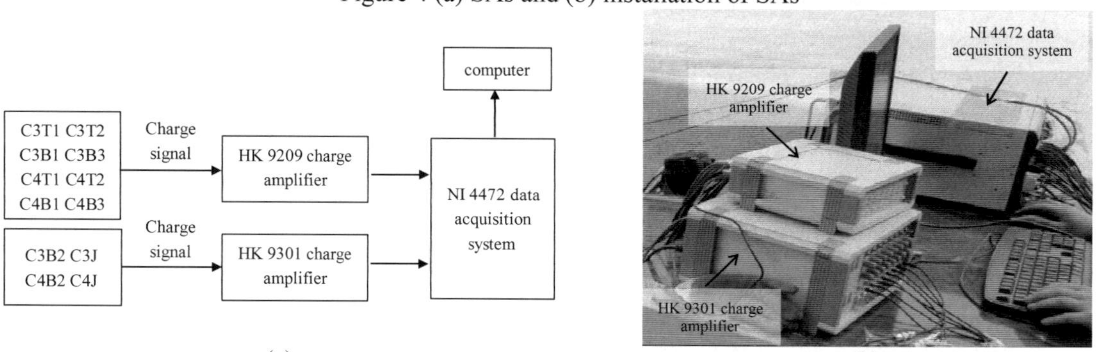

(a) (b)

Figure 5 (a) Monitoring scheme and (d) data acquisition system

3 Structure damage process monitoring

3.1 Environmental vibration monitoring

The experiment wasn't intentionally to perform the ambient excitation. However, in the interim, the moving of crane hanging the steel ball stimulated a series of small vibrations which were just monitored by the sensitive SAs. Figure 6 shows an example of the signal of SA_C3B4 under the ambient excitation during the pristine stage of the building. Note that the amplitude of the monitored stress is less than 2 kPa which is far less than the induced stress caused by impact, like Figure 8 (a). The FFT(Fast Fourier Transformation) spectrum of the SA signal is presented in Figure 7, of which the first three dominant vibration frequencies are identified to be 3.63Hz, 10.84 Hz and 13.72 Hz.

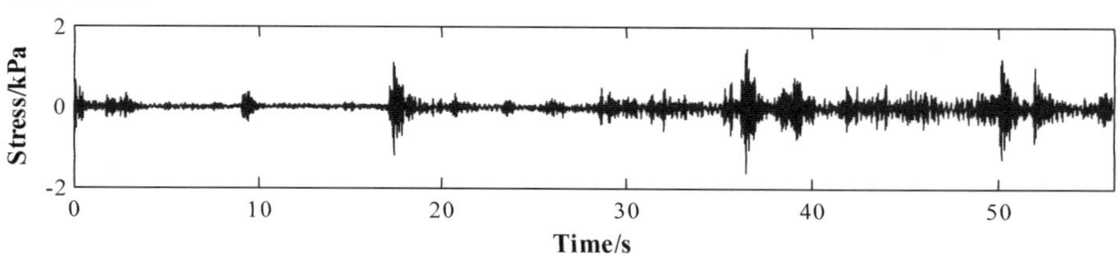

Figure 6 Environmental vibration obtained by SAC4B2

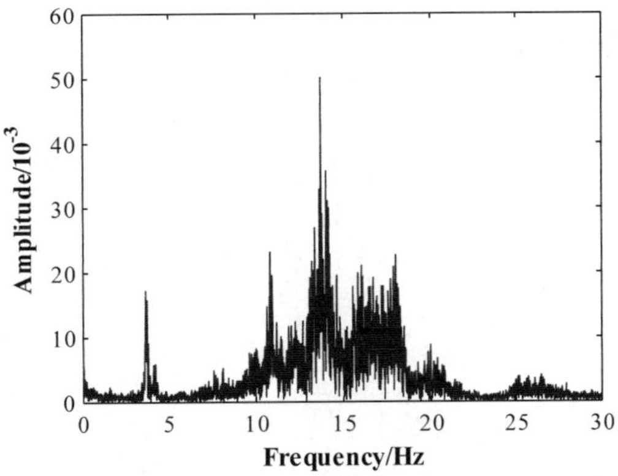

Figure 7 The FFT spectrum of signal of SA C3B2

3.2 Impact process monitoring

The building was demolished place by place in a sequence of left, back and right side, ending with the global collapse. At each place, there existed a gradually wrecking process of those building components, such as walls, beams and columns, from their intact state to serious damaged state, during which the inner stresses were being monitored. As an example, Figure 8 (a) exhibits the time history of SA_C4B2, the stress signal of SA of C4B2, during the pristine state. The figure depicts the whole attenuation process of structural vibration stimulated by impact. Figure 8 (b) shows the time-frequency joint analysis result of the time history. The analysis result represents the energy distribution in both the time and frequency domain so that one can see the energy dissipation process of different frequency components during the vibration. It is obvious that in the beginning of this vibration, a wide frequency band of the building was excited, among which the three apparent frequencies are around 3.5Hz, 10Hz and 15.5Hz which are closed to the frequencies obtained from the ambient excitation. The high frequency contents contained more energy, one can tell by the bright colors, however, the result shows they intend to vanish quickly leaving the building in a free vibration state controlled by the lowest frequency.

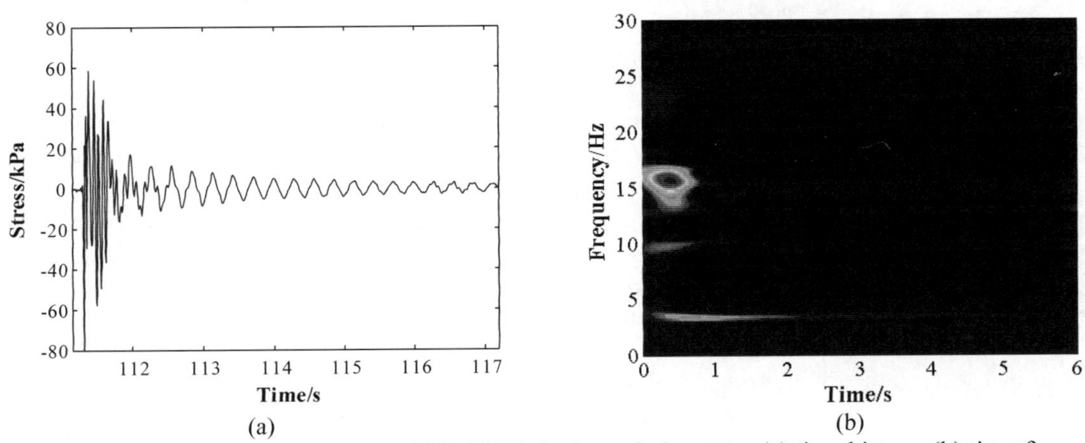

Figure 8 An example impact response of SA_C3B2 during pristine state: (a) time history, (b) time-frequency joint analysis result

In order to make it easier to evaluate the variation of the vibration frequency during the demolishing process, five typical damage states are determined. The five damage stages as well as the pristine state are defined as DS0~DS5 respectively. The descriptions of each damage state are listed in Table 1 and the correspondent photos of each damage state are presented in Figure 9.

Table 1 Description of each damage state

Damage state	Description
DS0	Pristine state
DS1	Left side exterior wall was destroyed;
DS2	More of left side exterior wall was damaged; Column C1 was damaged severely and lose loading capacity;
DS3	Staircase of the building at the first and the second story was destroyed; Column C2 was damaged completely as C1;
DS4	Back side exterior wall at the first and the second story was damaged seriously Concrete floor at the second story was also damaged seriously.
DS5	Column C3 was damaged completely;

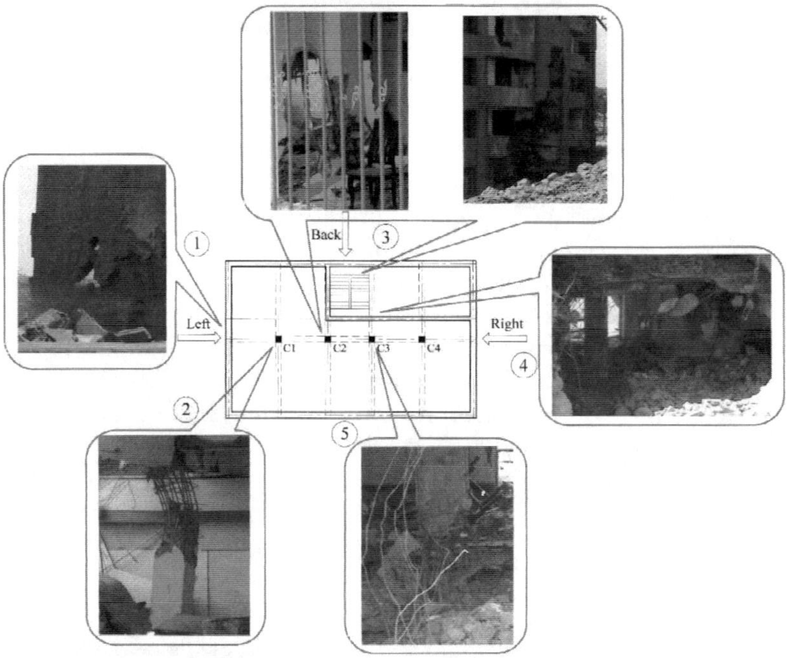

Figure 9 Typical damage states during the demolishing process

Because of the low amplitude of the stress responses in the columns under impacts, noise influences on the stress signal detected by sensors with charge amplifier HK9209 are significant and it's hard to filter. In the case of studying structural free vibration, only the stress signals detected by SA-C3B2, SA-C4B2, SA-C3J, and SA-C4J with charge amplifier HK9301 are studied since their signal are rather clean. The monitored stress signals of SAs of C3J, C4J, C3B2 and C4B2 under one specific impact from different damage states were processed by computer, Figure 10 exhibits an example of DS0, the distinctive feature is that they are all the latter loose part intercepted from a whole vibration process like Figure 8 (a). The reason for doing this is to preserve only the lowest frequency vibration part which is simple to study.

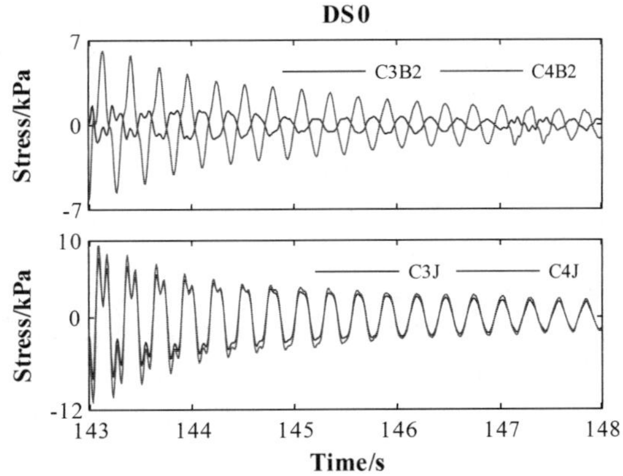

Figure 10 Monitored stresses at positions of C3B2, C4B2, C3J and C4J during free vibration

Table 2 reports the frequencies obtained from FFT, note that in DS5, C3 was broken, the building was on the verge of collapse and the sensors of C3B2, C3J, C4J were all out of order, only C4B2 was still working. Figure 11 shows the fundamental frequency variations of the structure. As was expected that as the damage degree increases, the fundamental frequency of the structure decreases. This is mainly due to the deterioration of structural stiffness which caused by the damage of main structural members and non-structural members.

Table 2 The frequencies in each damage state

	Sensors	Damage State					
		0	1	2	3	4	5
Frequencies	C3J	3.60	3.50	3.45	3.30	3.20	none
	C4J	3.60	3.50	3.45	3.30	3.20	none
	C3B2	3.60	3.50	3.45	3.30	3.10	none
	C4B2	3.60	3.50	3.45	3.30	3.20	2.90

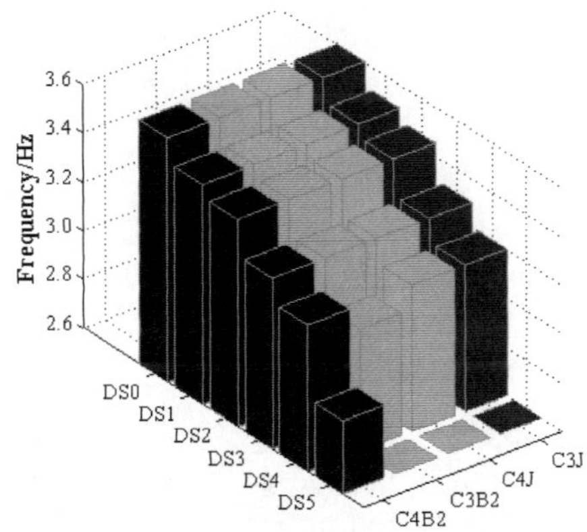

Figure 11 The fundamental frequency variations of the structure

A good designed structure should possess enough safety redundancy and multiple load transformation path to prevent sudden collapse under the circumstances of huge accidental loads and broken of structure members. The building in this experiment is simple in structure. Its vertical bearing capacity and lateral stiffness are mainly provided by four columns, so destroying the column is the key to fell the building. Just before damage state 2 (DS2) that C1 was destroyed, there existed some stress jumps within C3 and C4, shown in Figure 12. The two signals belong to

C3B2 and C4B2. The positive sign means the increase of compression, so the inner compression stress of C3 increased about 15 KPa or more for each impact, while C4 only increased about 2KPa and 5KPa respectively in the last two impacts. This phenomenon is the so called global stress redistribution of structures. When C1 was broken, the shared load originally borne by C1 was released to other bearing structures or columns to achieve new balance, then caused the abrupt jump in stress of C3 and C4. And since C3 was closer to the damaged C1, it increased more. Luckily, the loss of C1 didn't cause the final collapse, which means that the bearing capacity provided by the remained three columns seems to be big enough to hold the building. In the damaging process of C2, the stress redistribution also happened, but the figures won't be included in this paper.

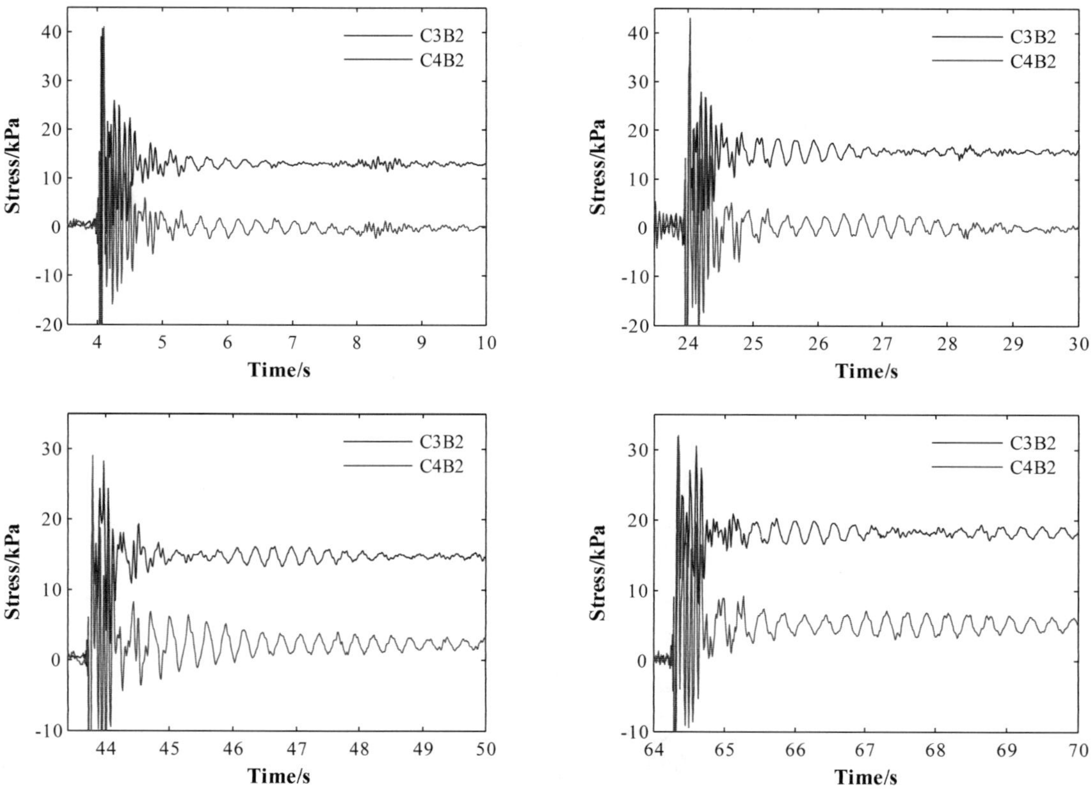

Figure 12 Examples of load redistribution in damage state 2

3.3 Material damage monitoring

After C1 and C2 were severely wrecked, the building was still standing hold by C3 and C4 as well as the brick walls. Figure 13 displays an impact response of all the sensors between DS4 and DS5 when the steel ball angled fell onto the ruined upper floor near C3. The huge kinetic energy was transmitted to the structure members nearby instantaneously. Observing the signals of SAs embedded within C3 in Figure 13, all the vertical compression stresses changes are increased to a much larger level than that shown in Figure 8 (a). Considering the huge extra loads obtained from the broken two columns, the inner stress of concrete was very likely to have reached its bearing capacity. Among these signals of C3, the stress of C3T1 and C3J steps up and steps down for about 13MPa and 0.15MPa respectively, which visibly indicated the local inner stress redistribution of loading sections when concrete was damaged. Besides, for concrete under such violent cyclic loadings, the accumulated plastic deformation was increased and the longitudinal crakes was extended, which dissipated a large amount of energy from impact. This explains why the stress vibration of all the sensors vanished so quickly, which can differentiate the global and local stress redistribution. Looking at the stress changes of C4 shown in Figure 13, the amplitudes are far less than C3 which means most of the kinetic energy was transferred to C3. However, for the less damaged C4, the stress signals still maintain a basic pattern of vibration, which is different from the severely damaged C4. This section offers an indirect method to identify the damage of concrete material.

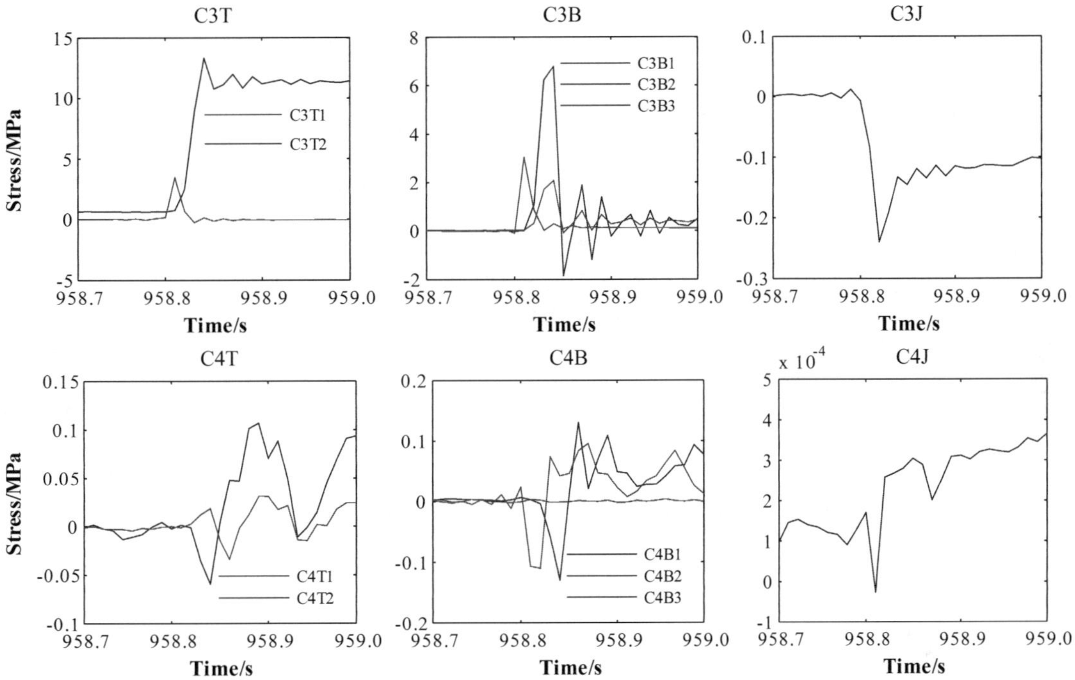

Figure 13 Stress signals of all sensors in C3 and C4 under an impact between DS4 and DS5

4 Conclusions

Keeping aware of the inner stress responses of a concrete structure under dynamic loadings is a crucial part for identifying and evaluating its damage situation. In the experiment of this paper, the applied SAs showed a great feasibility in not only monitoring the strong stress vibration under impact but also catching the small ambient stimulations, which at the meantime reflected the high sensitivity and stability of SAs. Based on the original stress signals, researchers get an access to directly study the internal stress distribution features when the structure was damaged and by analyzing the processed signals, some dynamic characteristics of structure can be revealed, such as the natural frequency decreases as the damage level increases and the energy dissipation process of different vibration modes. However, there remains some improvements to be made, for examples, the SAs should be deployed before concrete casting, installing from outside restricts the number and locations of SAs; it is preferable to build a finite element model to simulate and verify the experiment results. Overall, the experiment exhibits the great potential of SAs to be applied to infrastructures as internal stress transducers for health monitoring and accident notification.

Acknowledgement
Financial supports from the National Science Foundation of China (Grant no. 51678250) and National Basic Research Program of China (973 Program, granted No. 2015CB057704) are gratefully acknowledged.

References
Gu, H., Moslehy, Y., Sanders, D., Song, G., & Mo, Y. L. 2010, 'Multi-functional smart aggregate-based structural health monitoring of circular reinforced concrete columns subjected to seismic excitations', *Journal of Smart Materials and Structures,* vol.19, no.6, 065026.

Gu, H., Song, G., Dhonde, H., Mo, Y. L., & Yan, S. 2006, 'Concrete early-age strength monitoring using embedded piezoelectric transducers', *Journal of Smart Materials and Structures,* vol.15, no.6, pp. 1837-1845.

Hou, S., Zhang, H. B., & Ou, J. P. 2012, 'A PZT-based smart aggregate for compressive seismic stress monitoring', *Journal of Smart Materials and Structures,* vol. 21, no. 10, 105035.

Hou, S., Zhang, H. B., & Ou, J. P. 2013, 'A PZT-based smart aggregate for seismic shear stress monitoring', *Smart Journal of Materials and Structures,* vol. 22, no. 6, 65012.

Hou, S., Zhang, H. B., & Ou, J. P. 2016, 'SA-based concrete seismic stress monitoring: A case study for normal strength concrete', *Journal of Smart Materials and Structures,* vol. 25, no. 9, 50419.

Laskar, A., Gu, H., Mo, Y. L., & Song, G. 2009, 'Progressive collapse of a two-story reinforced concrete frame with embedded smart aggregates', *Journal of Smart Materials and Structures,* vol. 18, no. 7, 075001.

Modares, M., & Waksmanski, N. 2013, 'Overview of structural health monitoring for steel bridges', *Journal of Practice Periodical on Structural Design and Construction,* vol. 18, no. 3, pp. 187-191.

Moslehy, Y., Gu, H., Belarbi, A., Mo, Y. L., & Song, G. 2010, 'Smart aggregate based damage detection of circular RC columns under cyclic combined loading', *Journal of Smart Materials and Structures,* vol. 19, no. 6, pp. 140-158.

Song, G., Gu, H., Mo, Y. L., Hsu, T., Dhonde, H., & Zhu, R. R. H. 2005, 'Health monitoring of a concrete structure using piezoceramic materials', *Proceedings of SPIE - The International Society for Optical Engineering*, San Diego, CA, United states, 7-10, March, pp. 108-119.

Song, G., Olmi, C., & Gu, H. 2007, 'An overheight vehicle-bridge collision monitoring system using piezoelectric transducers', *Journal of Smart Materials and Structures,* vol. 16, no. 2, pp. 462-468.

Yan, S., Sun, W., Song, G., Gu, H., Huo, L., Liu, B., & Zhang, Y. 2009, 'Health monitoring of reinforced concrete shear walls using smart aggregates', *Journal of Smart Materials and Structures,* vol. 18, no. 4, 047001.

Zhang, H., Hou, S., & Ou, J. 2016, 'Smart aggregate-based seismic stress monitoring system using a specially designed charge amplifier', *Journal of Intelligent Material Systems and Structures,* vol. 27, no. 3, pp. 418-426.

The 8th International Conference on
Structural Health Monitoring of Intelligent Infrastructure
Brisbane, Australia | 5-8 December 2017

Mixed-dimensional Coupling Method of Box Section Member Based on Shear Stress Distribution Pattern

W. Lu [1], Y. Cui [2], J. Teng [3,4].

[1] *Harbin Institute of Technology Shenzhen Graduate School - China, Email:* lu.wei@hit.edu.cn [2] *Harbin Institute of Technology Shenzhen Graduate School – China,* [3] *Harbin Institute of Technology Shenzhen Graduate School – China,* [4] *Fujian University of Technology - China*

Abstract

The existing mixed-dimensional coupling method based on the energy conservation is commonly used. However, the assumption of shear stress distribution at the connected interface is relatively rough for the member with box section, which causes the stress near the connected interface to be discontinuous and causes a local stress concentration. The reliability of multi-scale model is in a low level. A simulation method of box section's shear stress distribution pattern is proposed to realize the mixed-dimensional coupling of box section member. A stress distribution calculation model is established to capture the real multipoint stresses at the connected interface firstly. Then the real stress distribution pattern is determined using the multipoint stresses and the surface fitting technology. Finally, the concurrent multi-scale model with high accuracy is established using the multipoint constraint equation. Three multi-scale models, cantilever beam with box section and beam-solid elements, were established respectively to validate the effectiveness of the proposed method, where these multi-scale models were based on the proposed method, one existing method and the CERIG method which is provided by ANSYS®, respectively. Comparing with one existing method and CERIG method, the displacement and stress responses of the multi-scale model based on the proposed method were closer to those of the full solid element model. The proposed method can effectively improve the stress distortion near the connected interface, which can also ensure that the performance analysis of structural local detail be real and effective.

1. Introduction

The application of large-scale space steel structure is gradually extensive. In order to comprehensively understand the structural behaviour, establishing the structural finite element model is a common means. Concurrent multi-scale modelling can simultaneously analyse model based on the different dimensional elements by coupling technique. This kind of model not only can understand the behaviour of the whole structure and the key parts, but also can take into account the computational efficiency and calculation accuracy. A reasonable mixed-dimensional coupling is essential preparation for establishing the concurrent multi-scale model, where the multipoint constraint (MPC) equation based on the energy conservation is the common method for mixed-dimensional coupling (McCune et al 2000; Li et al 2007; Yu et al 2012; Wang et al 2014). This method were developed by McCune (1998) and Monaghan (1998). Afterwards, Hailu et al (2005) emphasized that the constraint equations based on the energy conservation were only as good as the accuracy of the assumed stress distribution at the connected interface.

Because the two principal axes of the box section have a large radius of gyration, this section has good flexural and torsional performance. Comparing with the solid-web member, the box section also has many advantages, such as light weight and less material, so the box section is widely used in bridges and large-scale space steel structures. However the box section under shear force and torque has the complex stress distribution, for this, the existing studies often use the simplified shear stress formulae, which are obtained from the material mechanics, instead of the real shear stress distribution of the box section. For example, Yu et al (2012) used the shear stress of closed thin-walled section under torque and the shear stress of I-shaped section under shear force to derive the constraint equations of beam-shell elements of box section. For the simplified formulae, the variation of shear stress along the thickness of section isn't considered, and the shear stress formula from different section type is adopted. So the shear stress distribution of the box section is rather rough, which causes the stress near the connected interface discontinuously and local stress concentration. Therefore, it is necessary to improve the existing MPC equation method based on the energy conservation, in order to establish a more real and credible structural multi-scale model. In this paper, a simulation method of shear stress distribution pattern of box section is proposed to realize the mixed-dimensional coupling of box section member; the multi-scale analysis of cantilever beam with box section and beam-solid elements is performed to validate the effectiveness of the proposed method.

2. Mixed-dimensional coupling of the box section member

The multi-scale model of cantilever beam with box section, BEAM188 and SOLID185 elements is established using ANSYS® software. Material adopts the isotropic elastic material. Elastic modulus is 2×10^5 N/mm^2. Poisson ratio is 0.3. The specific section size of the multi-scale model of cantilever beam is shown in Figure 1.

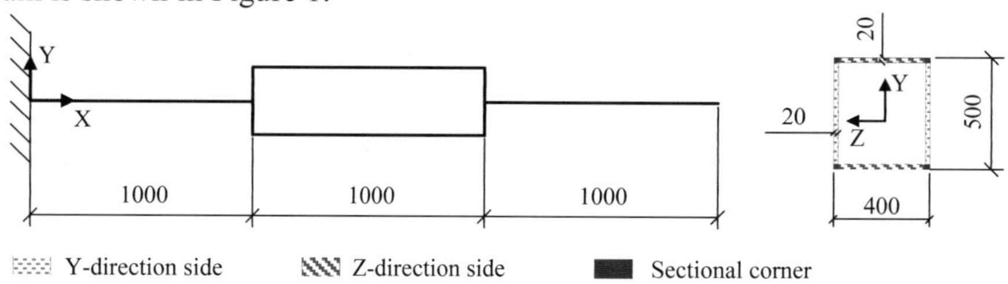

Figure 1. Specific section size of the multi-scale model of cantilever beam (unit: mm)

2.1 Establishment of stress distribution calculation model

A stress distribution calculation model is established firstly, as shown in Figure 2.

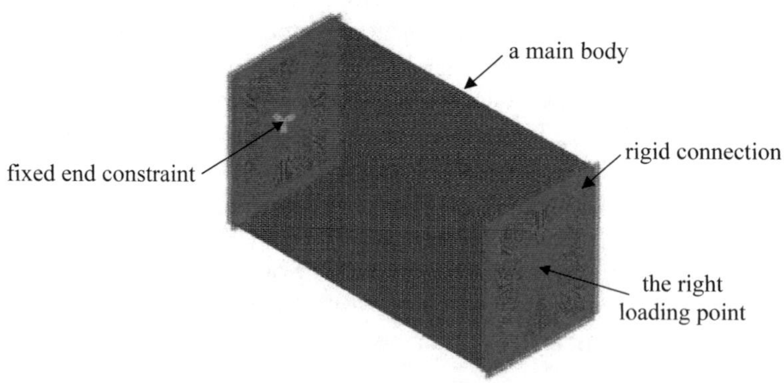

Figure 2. Stress distribution calculation model of box section

This calculation model consists of a main body based on the solid elements, two loading points locate at the centroid of the end of the main body, and the rigid connection between loading point and main body. Which is equivalent to a microcosm of the higher-dimensional part of the multi-scale model, so the cross section size of this calculation model should be consistent with the higher-dimensional part. In addition, According to the Saint-Venant's Principle, the stress distribution will be affected by the boundary effect when the distance between the stress calculation point and the boundary is too small (approximately equal to the member's transverse dimension), so the length of calculation model should not be less than the member's transverse dimension. In this paper, the length of calculation model is taken about twice of the member's transverse dimension. Because the stress distribution is affected by the mesh size, the mesh size should be as small as possible under the premise of ensuring the accuracy of model calculation. Then the fixed end constraint is applied to the left loading point, the unit concentrated force F_x, F_y, F_z and unit moment M_x, M_y, M_z are applied to the right loading point respectively. Because the stress distribution which far away from the boundary is more real and reasonable, the required nodes' stresses are extracted from the internal section of this calculation model, simultaneously, the corresponding coordinate values of these nodes along the y and z axis are extracted.

2.2 Determination of the shear stress distribution pattern

According to the extracted nodes' stress and the coordinate information, the shear stress distribution along the cross section can be plotted, as shown in Figures 3 and 4. The features of shear stress distribution can be preliminarily grasped.

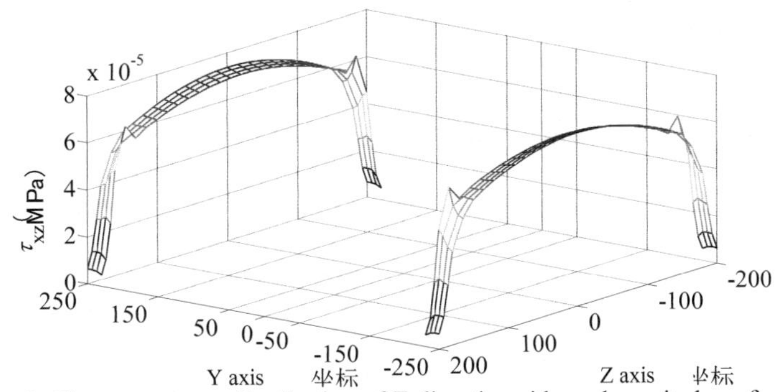

Figure 3. Shear stress τxz distribution of Z-direction side under unit shear force Fz

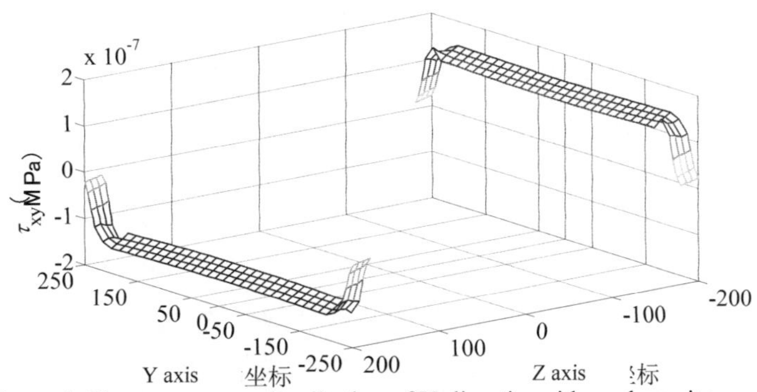

Figure 4. Shear stress τxy distribution of Y-direction side under unit torque

It can be seen from Figure 3 that shear stress τ_{xz} along the Z-direction side of box section are nearly parabola distribution, and which along the thickness direction are nearly linear distribution, while the distribution of shear stress in sectional corner is not the case, the distribution patterns of shear stress along the sectional side and corner should be determined respectively. Similarly, in Figure 4, shear stress τ_{xy} along the Y-direction side of box section are nearly uniform distribution, and which along the thickness direction are nearly linear distribution. According to these features, the functional relationship between the extracted nodes' stresses and the coordinates is established using the surface fitting technology; the real shear stress distribution pattern at the connected interface is finally determined based on the coefficient of determination R^2. Some fitted shear stress distribution patterns are shown in Eqs. (1) - (3). The index R^2 for characterizing the goodness of fit is greater than 0.99.

Shear stress distribution pattern of sectional corner under the unit shear force F_z

$$\begin{aligned}\tau_{xz} &= 0.0035565 - 2.816812\times10^{-5}z + 5.825117\times10^{-6}y \\ &\quad + 5.0682\times10^{-8}z^2 - 3.0362\times10^{-8}zy\end{aligned} \qquad \begin{pmatrix} 180 < z < 200 \\ -250 < y < -230 \end{pmatrix} \tag{1}$$

Shear stress distribution pattern of Z-direction side under the unit shear force F_z

$$\tau_{xz} = 5.706909\times10^{-5} + 6.435298\times10^{-8}y - 5.700108\times10^{-10}z^2 \qquad \begin{pmatrix} -180 < z < 180 \\ 230 < y < 250 \end{pmatrix} \tag{2}$$

Shear stress τ_{xy} distribution pattern under the unit torque

$$\tau_{xy} = 1.1666895\times10^{-7} - 1.3301265\times10^{-9}z \qquad \begin{pmatrix} 180 < z < 200 \\ -230 < y < 230 \end{pmatrix} \tag{3}$$

2.3 Calculation of the coefficient matrix of MPC equations

Substituting the fitted shear stress distribution patterns and the corresponding element shape functions into Eq. (4), the coefficient matrix of MPC equations can be obtained. And then, the multi-scale model of the cantilever beam with box section and beam-solid elements is established, as shown in Figure 5.

$$\boldsymbol{C} = \sum_{J=1}^{K} \left(\int_{A_J} \begin{pmatrix} \sigma_{x,F_x=1}\cdot\boldsymbol{N}_J \\ \tau_{y,F_y=1}\cdot\boldsymbol{N}_J \\ \tau_{z,F_z=1}\cdot\boldsymbol{N}_J \\ \left(\tau_{y,M_x=1}\cdot\boldsymbol{N}_J^{\mathrm{T}}, \tau_{z,M_x=1}\cdot\boldsymbol{N}_J^{\mathrm{T}}\right)^{\mathrm{T}} \\ \sigma_{x,M_y=1}\cdot\boldsymbol{N}_J \\ \sigma_{x,M_z=1}\cdot\boldsymbol{N}_J \end{pmatrix} dA \right)^{\mathrm{T}} \tag{4}$$

Where σ_{x,F_x}, τ_{y,F_y} and τ_{z,F_z} are the stresses of solid elements along x, y and z axis, when the right loading point of the calculation model is subjected to the unit concentrated force; τ_{y,M_x}, τ_{z,M_x}, σ_{x,M_y} and σ_{x,M_z} are the stresses of solid elements along x, y and z axis, when the right loading point of the calculation model is subjected to the unit moment; K is the total number of solid elements at the connected interface; J is the solid element J; A_J is the area of solid element J; $\boldsymbol{N}_J$ is the column vector of shape functions of the solid element J; dA is the symbol of area integral.

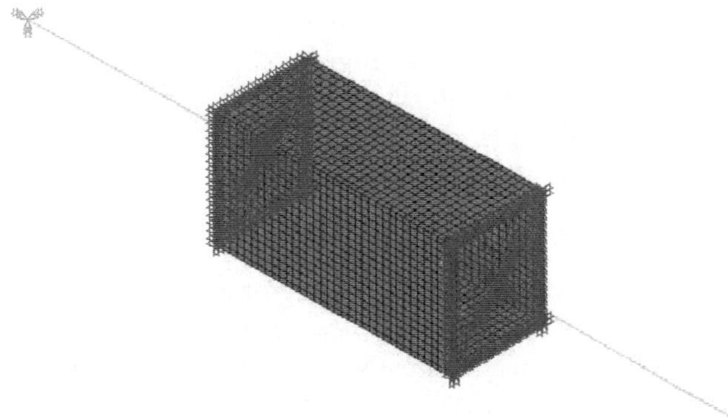

Figure 5. Multi-scale model of the cantilever beam with box section

3. Multi-scale analysis of the box section member

When 10 kN·m torque is applied to the right end of the established multi-scale model, the results of Von Mises stress are shown in Figure 6.

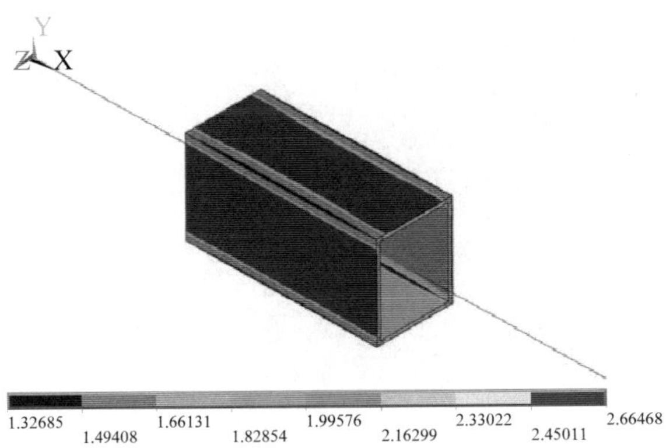

(a) Multi-scale model based on the proposed method

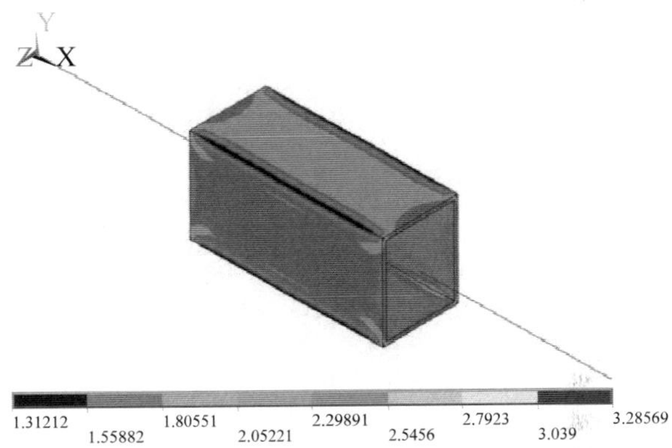

(b) Multi-scale model based on the CERIG method

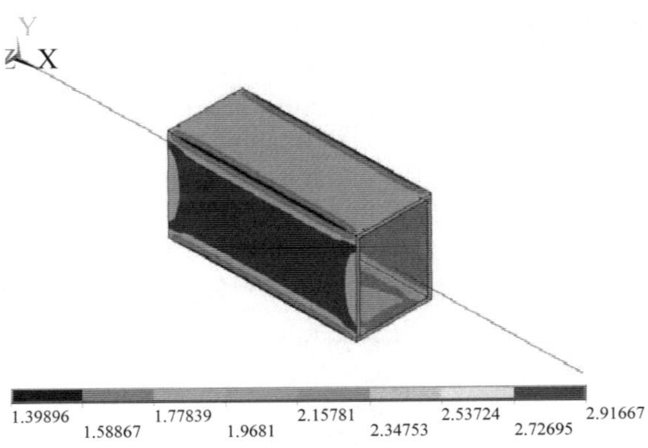

(c) Multi-scale model based on the Yu's method

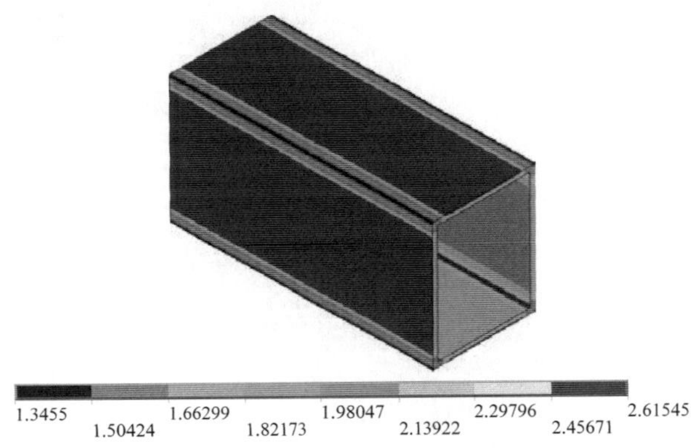

(d) Full solid element model

Figure 6. Von Mises stress contour (unit: MPa)

Von Mises stress of the full solid element model is regarded as the reference value, and then compared with the multi-scale models based on the proposed method, CERIG method and Yu's method (2012) respectively. It can be seen from Figure 6 that the difference between the reference values and the results of multi-scale models, which are based on the CERIG method and Yu's method, is obvious. The relative errors of maximum Von Mises stress are 25.626% and 11.517%, respectively. Moreover, there exists the obvious stress distortion near the interface. The results obtained from the multi-scale model based on the proposed method are in good agreement with the reference value, and the relative error of maximum Von Mises stress is 1.882%. So the proposed method can reflect the stress distribution of structural member more real.

Here, the effective range of local stress of multi-scale model is discussed by the comparative analysis of the node stress of multi-scale model and full solid element model at the same locations. Ten groups of nodes are selected as the stress calculation points, the specific locations are shown in Figure 7, where the distance between each group of nodes and left connected interface increases from 0 mm to 540 mm with 60 mm step. The relative errors of stress are shown in Figure 8.

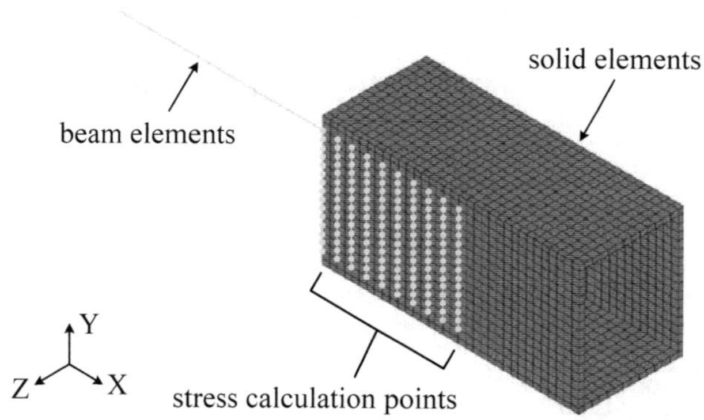

Figure 7. Specific locations of stress calculation points

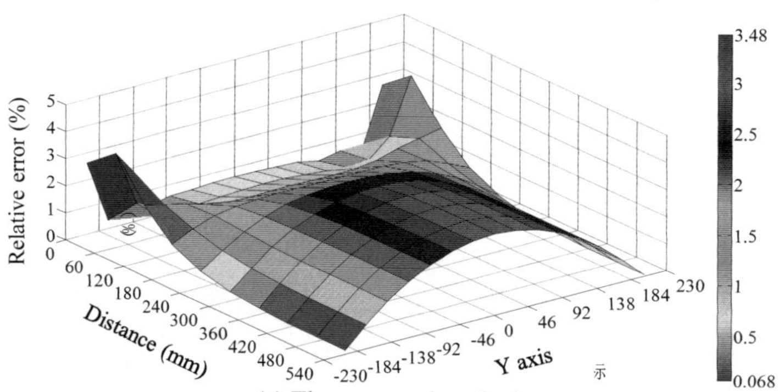

(a) The proposed method

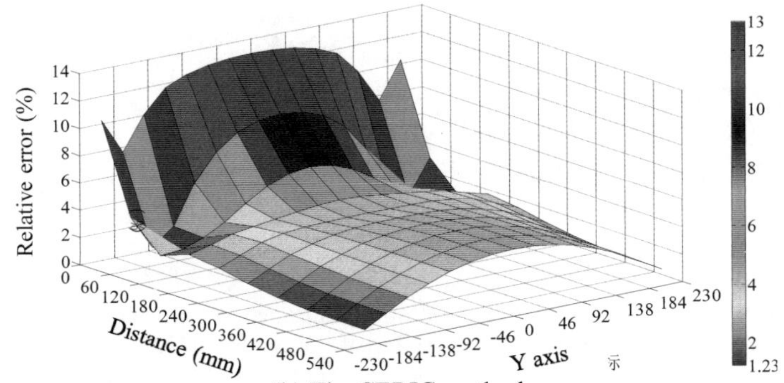

(b) The CERIG method

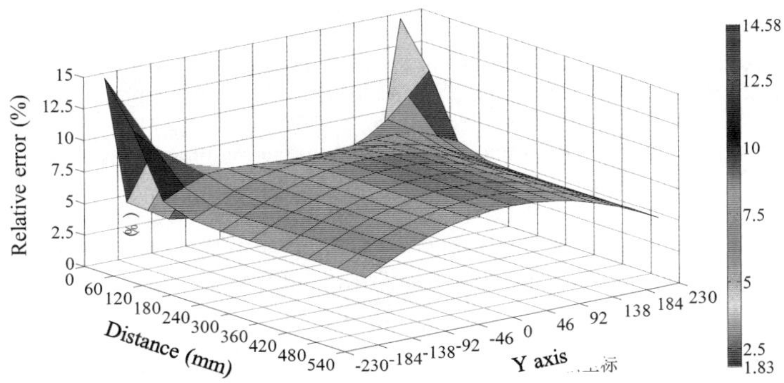

(c) The Yu's method

Figure 8. Relative errors of stress in Y-direction side

It can be known from the relationship between the relative error of node stress and the distance from connected interface:

(1) For the multi-scale model based on the proposed method, the relative errors of stress are all less than 5% in the entire study region. So it can be considered that the multi-scale model based on the proposed method has almost no local stress distortion.

(2) For the multi-scale model based on CERIG method, the relative errors of stress are all less than 5% when the distance from the connected interface is larger than 120 mm. So it can be considered that the distortion region of local stress is about 120 mm.

(3) For the multi-scale model based on the Yu's method, most of the relative errors of stress are larger than 5% in the entire study region, and the relative error is mainly distributed in the range of 5.545%~8.943%. So there is a certain degree of stress distortion in the range of 540 mm.

4. Conclusions

In this paper, a simulation method of shear stress distribution pattern of box section was proposed to solve the problem that the assumption of shear stress distribution for the existing coupling method based on the energy conservation was relatively rough, and realize the mixed-dimensional coupling of box section member, thereby improving the calculation accuracy of the multi-scale model with box section. Three multi-scale models of cantilever beam with box section and beam-solid elements were established, which were respectively based on the proposed method, the Yu's method and the CERIG method. Through the multi-scale analysis, the effectiveness of the proposed method is validated. The proposed method can not only ensure the deformation compatibility and stress continuity at the connected interface simultaneously, but also can reduce the distortion region of local stress effectively, and improve the reliability of multi-scale model.

Acknowledgments

This research is supported by the National Science Foundation of China (Grant No. 51678201, 51308162), and the Fujian National Science Foundation (2014J01171).

References

McCune, RW, Armstrong, CG & Robinson, DJ 2000, 'Mixed-dimensional coupling in finite element models', *International Journal for Numerical Methods in Engineering*, vol. 49, no. 6, pp. 725-750.

Li, ZX, Zhou, TQ & Chan, THT 2007, 'Multi-scale numerical analysis on dynamic response and local damage in long-span bridges', *Engineering Structures*, vol. 29, no. 7, pp. 1507-1524.

Yu, Y, Chan, THT & Sun, ZH 2012, 'Mixed-dimensional consistent coupling by multi-point constraint equations for efficient multi-scale modeling', *Advances in Structural Engineering*, vol. 15, no. 5, pp. 837-853.

Wang, FY, Xu, YL & Qu, WL 2014, 'Mixed-dimensional finite element coupling for structural multi-scale simulation', *Finite Elements in Analysis and Design*, vol. 92, pp. 12-25.

McCune, RW 1998, 'Mixed dimensional coupling and error estimation in finite element stress analysis', PhD thesis, The Queen's University of Belfast, March.

Monaghan, DJ, Doherty, IW & Court, DM 1998, 'Coupling 1D beams to 3D bodies', *Proceedings of the 7th International Meshing Roundtable*, *Sandia National Laboratories*, Michigan, America, pp. 285-293.

Hailu, D, Zekaria, A & Kinde, S 2005, 'A multiple constraint approach for finite element analysis of moment frames with radius-cut RBS connections', pp. 1-10.

The 8th International Conference on
Structural Health Monitoring of Intelligent Infrastructure
Brisbane, Australia | 5-8 December 2017

Data Fusion Based General Kalman Filter with Unknown Inputs

Y. Lei[1], S.J Luo[2] and Y. Su[2].
[1]*School of Architecture and Civil Engineering, Xiamen University- China; E-mail:* ylei@xmu.edu.cn
[2]*School of Architecture and Civil Engineering, Xiamen University- China*

Abstract

The classical Kalman filter (KF) provides a particularly practical and efficient state estimation algorithm with partial measurements of structural responses, but it is requested that all external inputs are known. So far, some methodologies of Kalman filter with unknown inputs (KF-UI) have been proposed. However, previous KF-UI approaches using limited acceleration measurements are inherently unstable which leads poor tracking and fictitious drifts in the identified structural displacements and unknown inputs in the presence of measurement noises. Moreover, it is necessary to have the measurements of acceleration responses at the locations where unknown inputs applied, i.e., with collocated acceleration measurements. In this paper, it is aimed to extend the classical KF approach to circumvent the above limitations for general real time estimation of structural state and unknown inputs without using collocated acceleration measurements. Based on the scheme of the classical KF, an improved KF-UI and without collocated acceleration measurements is derived. Then, data fusion of displacement and acceleration measurements is used to prevent the drifts in the identified structural state and unknown inputs in real time. Finally, the addition of dummy-measurements of accelerations at the locations where unknown inputs applied is proposed to bind the high-frequency oscillation in the identified external inputs. Some numerical examples are used to demonstrate the effectiveness and versatilities of the proposed approach.

1. Introduction

The identification of structural dynamic systems using the measurements of structural vibration data is essential for structural health monitoring and vibration control (Li and Chen 2013; Yuen and Mu 2015). As it is impractical to measure all structural responses, structural identification using only partial measurements of structural responses have received great attentions (Xu and Jia 2012; Xu *et al.* 2015; Lei *et al.* 2014, 2015). In this regard, the Kalman filter (KF), which was proposed by R.E. Kalman in the early sixties (Kalman 1960), provides a particularly practical and efficient state estimation algorithm with partial measurements of structural responses. Moreover, KF has the ability to inherently take the uncertainty in the model into account, which is not possible in the deterministic approaches (Yuen *et al.* 2013; Azam *et al.* 2015; Naets *et al.* 2015). However, in the classical KF approach, it is requested that all external inputs are known (measured).
To circumvent the limitation of the classical KF approach, many improved methodologies have been proposed for Kalman filter based identification of joint structural state and external inputs, e.g., an iterative identification procedure consisting of the least-squares identification technique and a modification process between each iterative step (Chen and Li 2004); the unbiased minimum-variance input and state estimation with direct feed through (Gillijns and Moor 2007); a Kalman filter with unknown inputs approach derived by the weighted least-squares estimation method (Pan *et al.* 2010); a two-stage Kalman estimator in which the classical Kalman filter is first adopted to establish a regression model between the residual innovation and then a recursive least-squares

estimator is proposed to identify the input excitation forces (Ma *et al.* 2003; Wu *et al.* 2009); an augmented Kalman filter (AFK) for force identification in structural dynamics, in which the unknown forces are included in the state vector and estimated in conjunction with the states (Lourens *et al.* 2012); an average acceleration discrete algorithm with regularization (Ding et al. 2013) and implicit Newmark-β algorithm with regularization (Liu *et al.* 2014); a weighted adaptive iterative least-squares estimation with incomplete measured excitations (Xu *et al.* 2015); Kalman estimator with unknown inputs (Lei *et al.* 2014) and a two-stage and two-step algorithm (Lei *et al.* 2015). However, it has been demonstrated that previous Kalman filter with unknown input (KF-UI) approaches using limited number of acceleration measurements are inherently unstable which leads poor tracking and so-called drifts in the estimated unknown external inputs and structural displacements (Azam et al. 2015; Naets *et al.* 2015). Although regularization approaches (Ding *et al.* 2013; Liu *et al.* 2015) or post-signal processing schemes (Lei *et al.* 2014, 2015) can be used to treat the drift in the identified results, these treatments prohibit on-line and real-time identification of coupled structural state and unknown inputs.

Recently, the authors have proposed an improved Kalman filter with unknown inputs based on data fusion of partial acceleration and displacement measurements for real time estimation of joint structural states and the unknown inputs (Liu *et al.* 2016). However, like other previous Kalman filter with unknown input (KF-UI), it is necessary to have the measurements of acceleration responses at the locations where unknown inputs applied, i.e., collocated acceleration measurements at the locations of unknown inputs are requested.

In this paper, it is aimed to extend the classical KF approach and the drawbacks of existing KF-UI approaches for real time estimation of structural states and unknown inputs without using collocated acceleration measurements. Compared with the recent KF-UI (Liu *et al.* 2016), an improved Kalman filter with unknown excitations (KF-UI) and without collocated acceleration measurements is derived. Since accelerations and displacements contains high and low frequencies vibration characteristics, respectively (Ay and Wang 2014; Kim and Sohn 2014), data fusion of displacement and acceleration measurements is used to prevent the low-frequency drifts in the identified structural state vector and unknown external inputs in real time. Finally, the addition of dummy-measurements of accelerations at the locations where unknown inputs applied is proposed to bind the high-frequency oscillation in the identified unknown inputs. Numerical examples of the identification of joint structural state and unknown inputs of a multi-story sharer type building and a plane truss are used to demonstrate the effectiveness and versatilities of the proposed algorithm.

2. Improved KF-UI without Collocated Acceleration Measurements

To further circumvent the limitation of the existing KF-UI approaches, it is proposed to extend the recent KF-UI for the identification of joint structural states and unknown inputs without using collocated acceleration measurements.

Instead of using the conventional zero-order Holder (ZOH) discretization (constant interpolated) method, in which the unknown input force vector in a sampling period is assumed to be constant, the first-order holder (FOH) discretization is used herein (Ding *et al.*, 2013). This discretization within a sampling period is treated by the linear discrete (linear interpolated) method, in which the discrete data is interpolated as:

$$\boldsymbol{f}^{u}(\mathrm{t}) = \boldsymbol{f}_{k}^{u} + \frac{\boldsymbol{f}_{k+1}^{u} - \boldsymbol{f}_{k}^{u}}{\Delta t}(t - k\Delta t) \quad ; \quad \mathrm{k}\Delta t \le \mathrm{t} \le (\mathrm{k}+1)\Delta t \tag{1}$$

Then, the equation of motion of a linear structure under unknown inputs can be converted into the following state equation as follows,

$$X_{k+1} = A_k X_k + B_k f_k^u + B_k f_{k+1}^u + w_k \tag{2}$$

In this paper, it is considered that the acceleration responses at the location of external input are not measured. Therefore, the discrete observation equation is described by:

$$Y_{k+1} = C_{k+1} X_{k+1} + v_{k+1} \tag{3}$$

Analogously, $\tilde{X}_{k+1|k}$ is first predicted as,

$$\tilde{X}_{k+1|k} = A_k \hat{X}_{k|k} + B_k \hat{f}_{k|k}^u + B_k \hat{f}_{k+1|k+1}^u \tag{4}$$

Then, the measurement updated (corrected) $\hat{X}_{k+1|k+1}$ can be obtained by

$$\hat{X}_{k+1|k+1} = \tilde{X}_{k+1|k} + K_{k+1}(Y_{k+1} - C_{k+1}\tilde{X}_{k+1|k}) \tag{5}$$

Under the condition that the number of response measurements (sensors) is no less than the number of unknown external excitations, $\hat{f}_{k+1|k+1}^u$ can be estimated by minimizing the error vector $\boldsymbol{\Delta}_{k+1}$ defined by

$$\Delta_{k+1} = y_{k+1} - C_{k+1}\hat{X}_{k+1|k+1} \tag{6}$$

By inserting the expression of $\hat{X}_{k+1|k+1}$ and $\tilde{X}_{k+1|k}$ in Eq.(5) and Eq.(4) into the above error vector, Δ_{k+1} can be expressed by

$$\Delta_{k+1} = (\mathbf{I} - C_{k+1}K_{k+1})\left[Y_{k+1} - C_{k+1}\left(A_k\hat{X}_{k|k} + \mathbf{B}_k\hat{\mathbf{f}}_{k|k}^u\right)\right] - (\mathbf{I} - C_{k+1}K_{k+1})C_{k+1}\mathbf{B}_k\hat{f}_{k+1|k+1}^u \tag{7}$$

Then, $\hat{f}_{k+1|k+1}^u$ from can be estimated Eq.(7) based on least-squares estimation as

$$\hat{f}_{k+1|k+1}^u = S_{k+1}(C_{k+1}B_k)^{\mathrm{T}}(I - C_{k+1}K_{k+1})\left[Y_{k+1} - C_{k+1}\left(A_k\hat{X}_{k|k} + B_k\hat{f}_{k|k}^u\right)\right] \tag{8}$$

in which

$$S_{k+1} = \left[(C_{k+1}B_k)^{\mathrm{T}}(I - C_{k+1}K_{k+1})(C_{k+1}B_k)\right]^{-1} \tag{9}$$

The error of state estimation defined as $\hat{e}_{k+1|k+1}^X = X_{k+1} - \hat{X}_{k+1|k+1}$ can be derived from Eq.(3) and Eq.(5) as

$$\hat{e}_{k+1|k+1}^X = (I - K_{k+1}C_{k+1})\tilde{e}_{k+1|k}^X - K_{k+1}v_{k+1} \tag{10}$$

where $\tilde{e}_{k+1|k}^X$ is error of state prediction defined as $\tilde{e}_{k+1|k}^X = X_{k+1} - \tilde{X}_{k+1|k}$. From Eqs.(3-4), it is known that

$$\tilde{e}^X_{k+1|k} = A_k \hat{e}^X_{k|k} + B_k \hat{e}^f_{k+1|k+1} + B_k \hat{e}^f_{k|k} + w_k \tag{11}$$

where $\hat{e}^f_{k+1|k+1}$ is error of estimated $\hat{f}^u_{k+1|k+1}$ defined as $\hat{e}^f_{k+1|k+1} = f^u_{k+1} - \hat{f}^u_{k+1|k+1}$. By inserting Y_{k+1} in Eq.(3) into Eq.(8), $\hat{e}^f_{k+1|k+1}$ can be derived by

$$\hat{e}^f_{k+1|k+1} = -S_{k+1}(C_{k+1}B_k)^{\mathrm{T}}(I - C_{k+1}K_{k+1})\left[C_{k+1}\left(A_k \hat{e}^X_{k|k} + B_k \hat{e}^f_{k|k}\right) + C_{k+1}w_k + v_{k+1}\right] \tag{12}$$

From Eq.(10), the error covariance matrix $\hat{P}^X_{k+1|k+1}$ is estimated as

$$\hat{P}^X_{k+1|k+1} = (I - C_{k+1}K_{k+1})\tilde{P}^X_{k+1|k}(I - C_{k+1}K_{k+1})^{\mathrm{T}} + K_{k+1}R_{k+1}K^{\mathrm{T}}_{k+1} \tag{13}$$

To minimize the error covariance matrix $\hat{P}^X_{k+1|k+1}$, K_{k+1} should be selected as

$$K_{k+1} = \tilde{P}^X_{k+1|k} C^T_{k+1}(C_{k+1}\tilde{P}^X_{k+1|k} C^T_{k+1} + R_{k+1})^{-1} \tag{14}$$

Then, $\hat{P}^X_{k+1|k+1}$ in Eq.(13) can be simplified as:

$$\hat{P}^X_{k+1|k+1} = (I - C_{k+1}K_{k+1})\tilde{P}^X_{k+1|k} \tag{15}$$

The error covariance matrix $\hat{P}^f_{k+1|k+1}$ can be estimated from Eq.(12) as

$$\hat{P}^f_{k+1|k+1} = M_{k+1}C_{k+1}\left[A_k \; B_k\right]\begin{bmatrix} \hat{P}^X_{k|k} & \hat{P}^{Xf}_{k|k} \\ \hat{P}^{fX}_{k|k} & \hat{P}^f_{k|k} \end{bmatrix}\begin{bmatrix} A^{\mathrm{T}}_k \\ B^T_k \end{bmatrix}(M_{k+1}C_{k+1})^{\mathrm{T}} + M_{k+1}R_{k+1}M^{\mathrm{T}}_{k+1} + M_{k+1}C_{k+1}Q_k(M_{k+1}C_{k+1})^{\mathrm{T}} \tag{16}$$

So the error covariance matrix $\tilde{P}^X_{k+1|k}$ can be expressed as:

$$\begin{aligned}\tilde{P}^X_{k+1|k} &= (I - B_k M_{k+1}C_{k+1})[A_k \; B_k]\begin{bmatrix} \hat{P}^X_{k|k} & \hat{P}^{Xf}_{k|k} \\ \hat{P}^{fX}_{k|k} & \hat{P}^f_{k|k} \end{bmatrix}\begin{bmatrix} A^{\mathrm{T}}_k \\ B^T_k \end{bmatrix}(I - B_k M_{k+1}C_{k+1})^{\mathrm{T}} \\ &+ B_k M_{k+1}R_{k+1}M^{\mathrm{T}}_{k+1}B^T_k + (I - B_k M_{k+1}C_{k+1})Q_{k+1}(I - B_k M_{k+1}C_{k+1})^{\mathrm{T}}\end{aligned} \tag{17}$$

and the error covariance matrix $\hat{P}^{Xf}_{k+1|k+1}$ can be written as

$$\begin{aligned}\hat{P}^{Xf}_{k+1|k+1} &= -(I - C_{k+1}K_{k+1})(I - B_k M_{k+1}C_{k+1})[A_k \; B_k]\begin{bmatrix} \hat{P}^X_{k|k} & \hat{P}^{Xf}_{k|k} \\ \hat{P}^{fX}_{k|k} & \hat{P}^f_{k|k} \end{bmatrix}\begin{bmatrix} A^{\mathrm{T}}_k \\ B^T_k \end{bmatrix}(M_{k+1}C_{k+1}) \\ &+ \left[(I - C_{k+1}K_{k+1})B_k M_{k+1} + K_{k+1}\right]R_{k+1}M^{\mathrm{T}}_{k+1} - (I - C_{k+1}K_{k+1})(I - B_k M_{k+1}C_{k+1})Q_{k+1}(M_{k+1}C_{k+1})^{\mathrm{T}}\end{aligned} \tag{18}$$

and

$$\hat{P}^{fX}_{k+1|k+1} = \left(\hat{P}^{Xf}_{k+1|k+1}\right)^{\mathrm{T}} \tag{19}$$

However, $\boldsymbol{M}_{k+1}$ in Eq.(6) still contains $\boldsymbol{K}_{k+1}$which is also a function of $\boldsymbol{M}_{k+1}$ as indicated by Eq.(14) and Eq.(17). The direct estimation of $\boldsymbol{M}_{k+1}$ needs to be derived.
By inserting the expression of $\boldsymbol{K}_{k+1}$in Eq.(14) into the term$(\boldsymbol{I}-\boldsymbol{C}_{k+1}\boldsymbol{K}_{k+1})$, it is noted that

$$(\boldsymbol{I}-\boldsymbol{C}_{k+1}\boldsymbol{K}_{k+1})=\boldsymbol{I}-\boldsymbol{C}_{k+1}\tilde{\boldsymbol{P}}^X_{k+1|k}\boldsymbol{C}^T_{k+1}(\boldsymbol{C}_{k+1}\tilde{\boldsymbol{P}}^X_{k+1|k}\boldsymbol{C}^T_{k+1}+\boldsymbol{R}_{k+1})^{-1}=\boldsymbol{R}_{k+1}\left(\boldsymbol{C}_{k+1}\tilde{\boldsymbol{P}}^X_{k+1|k}\boldsymbol{C}^T_{k+1}+\boldsymbol{R}_{k+1}\right)^{-1} \quad (20)$$

From Eq.(11) and Eq.(14),

$$(\boldsymbol{I}-\boldsymbol{C}_{k+1}\boldsymbol{K}_{k+1})=\boldsymbol{R}_{k+1}\left(\boldsymbol{C}_{k+1}\tilde{\boldsymbol{P}}^X_{k+1|k}\boldsymbol{C}^T_{k+1}+\boldsymbol{R}_{k+1}\right)^{-1}=\boldsymbol{R}_{k+1}\left(2\boldsymbol{R}_{k+1}+\boldsymbol{C}_{k+1}\boldsymbol{Q}_{k+1}\boldsymbol{C}^{\mathrm{T}}_{k+1}\right)^{-1} \quad (21)$$

and

$$\boldsymbol{M}_{k+1}=\left[(\boldsymbol{C}_{k+1}\boldsymbol{G}_{k+1})^{\mathrm{T}}\boldsymbol{R}_{k+1}\left(2\boldsymbol{R}_{k+1}+\boldsymbol{C}_{k+1}\boldsymbol{Q}_{k+1}\boldsymbol{C}^{\mathrm{T}}_{k+1}\right)^{-1}(\boldsymbol{C}_{k+1}\boldsymbol{G}_{k+1})\right]^{-1}(\boldsymbol{C}_{k+1}\boldsymbol{G}_{k+1})^{\mathrm{T}}\boldsymbol{R}_{k+1}\left(2\boldsymbol{R}_{k+1}+\boldsymbol{C}_{k+1}\boldsymbol{Q}_{k+1}\boldsymbol{C}^{\mathrm{T}}_{k+1}\right)^{-1} \quad (22)$$

In practice, accelerometers are often used in structural dynamics applications. However, previous KF-UI approaches based only on sparse noisy acceleration measurements are inherently unstable which leads to the so-called spurious drifts in the estimated unknown inputs and structural displacement. The reason can be considered from a physical point of view. At any given time it is not clear whether acceleration is the effect of external forces or from the elastic restoring force due to a certain position. Although either regularization scheme or post-signal processing can be used to treat the drift problem. These approaches prohibit the on-line identification of coupled structural state and unknown inputs. In this paper, it is proposed to add partial measured displacements to the measured accelerations since accelerations and displacements contains high and low frequencies vibration characteristics, respectively. Data fusion of partially measured displacement and acceleration responses is used in the observation equation (Ay and Wang, 2014; Kim and Sohn 2014). Moreover, the addition of dummy-measurements of accelerations at the locations where unknown inputs applied is used to bind the oscillation in the identified unknown inputs (Nates et al. 2015). The dummy measurements have measurement equations as:

$$\boldsymbol{dd}(\boldsymbol{C}'_{k+1}\boldsymbol{X}_{k+1}+\boldsymbol{D}_{k+1}\boldsymbol{f}^u_{k+1})+\boldsymbol{v}_{dm,k+1}=0 \quad (23)$$

where $\boldsymbol{dd}$ is a matrix associated with the locations of applied unknown input vector,$\boldsymbol{C}'_{k+1}$ and $\boldsymbol{D}_{k+1}$are two matrices associated with the state vector and unknown input vectors, respectively. This equation states that the position of the DOFs is zero with an uncertainty v_{dm} with covariance R_{dm}. The entries of this covariance matrix R_{dm} can be used for the tuning of the KF-UI to bind the oscillation in the identified results. The covariance R_{dm} can be estimated from the measured acceleration nearby.

3. Numerical validations of the proposed KF-UI

Two numerical examples are used to validate the proposed KF-UI for the identification of joint structural states and unknown inputs. The theoretically computed displacement and acceleration responses are superimposed with corresponding white noises to consider the influence of measurement noises. These polluted responses are treated as "measured responses" for the identification problem.

3.1 Identification of multi-story shear building and unknown input

A twenty-story shear building is used as an example. Parameters of the building are assumed as: floor mass $m_i = 60kg$, floor stiffness $k_i = 1.2 \times 10^6 N/m$, floor damping $c_i = 1000Ns/m$ (i=1,2,..,20), respectively. An input of wide-banded white noise is applied to the top floor of the building. First, only three acceleration measurements at the 6th, 14th and 19th story floors are used. Measurement noises with 5% noise- to- signal ratio in root mean square (rms) are considered. As shown in Fig.1, significant drifts occur in the identified displacement and oscillation in the identified unknown input.

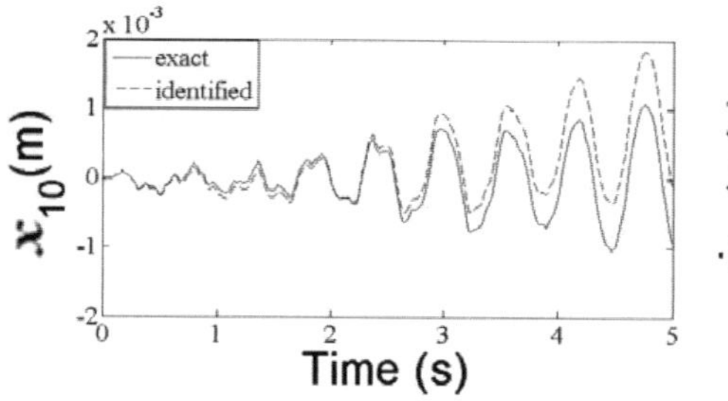

(a)Drift in the identified displacement

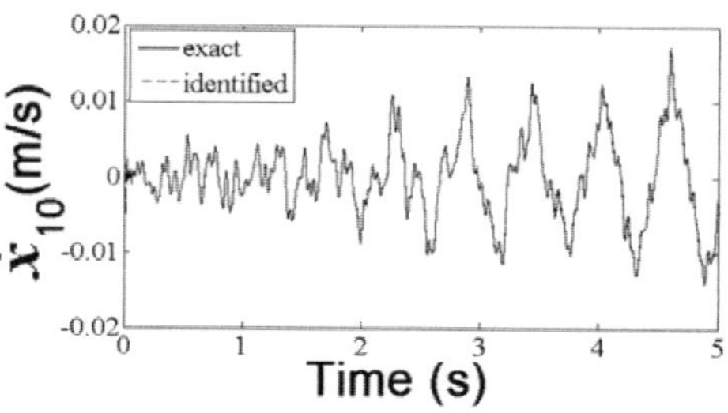

(b) Identified velocity

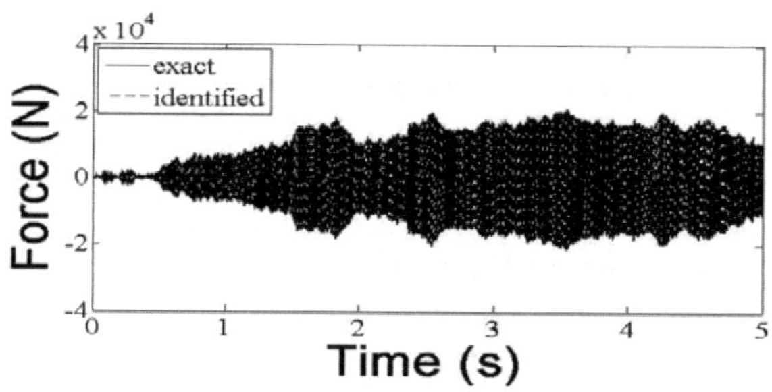

(c) Oscillation in the identified unknown input

Fig.1 : Comparisons of identified structural state and input with noisy acceleration measurements

To circumvent the spurious low-frequency drift problem, partial measured displacements are added in the improved KF-UI. For this relatively simple structural model, the displacement at the 2nd floor of the shear building is added in the measured signals. Data fusion of the measured displacement and the above three accelerations are used. Also, a dummy-measurement of accelerations at the 20th floor of the building is added in the observation equation.
Fig. 2 shows the comparisons of identified structural state and input with their exact values. It is clearly demonstrated that the so-called drift can be avoided by the proposed KF-UI with data fusion of 5% noisy acceleration and displacement measurements.

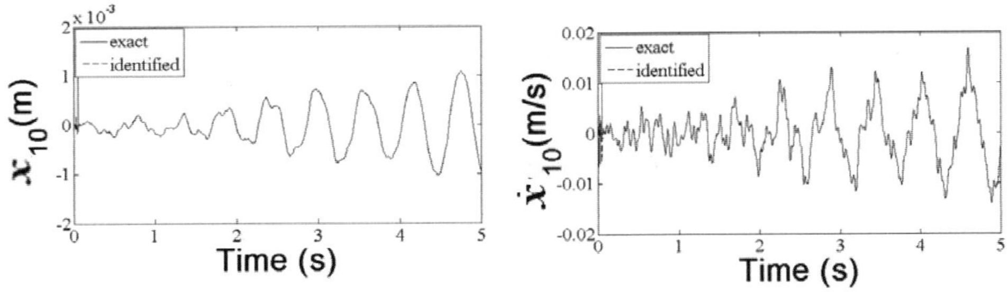

(a) Comparison of identified displacement (b) Comparison of identified velocity

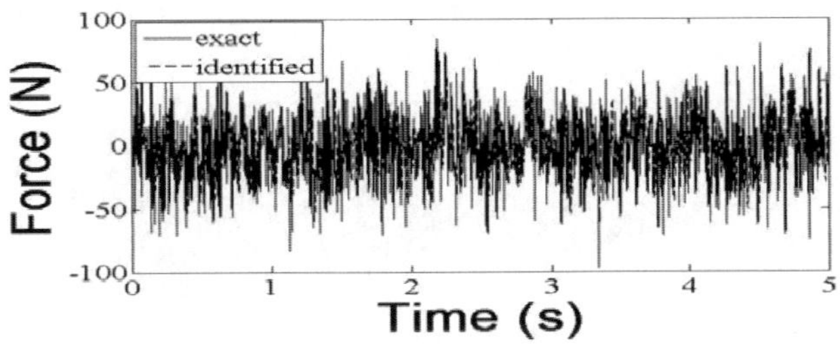

(c)Comparison of identified unknown input

Fig.2 : Comparisions of identified results with data fusion of 5% noisy measuremnts

3.2 Identification of a truss-structure and unknown input

To validate the proposed KF-UI for the identification of other type structures with unknown input, the identification of a plane truss and unknown input is studied. As shown in Fig.3, the truss consists of 11 uniform members. The length of each horizontal and inclined bar are 2m,$\sqrt{2}$m, respectively. Other parameters of the truss are: cross section area A=$7.854 \times 10^{-5} m^2$, Young's module E= 2×10^{11} pa, mass density of truss member = $7.8 \times 10^3 kg/m^3$ and the mass is concentrated on each node. The truss is subjected to an unknown input in the vertical direction at node 4. In this example, Rayleigh damping $C = \alpha M + \beta K$ is employed with α=0.6993 and β=0.0011.

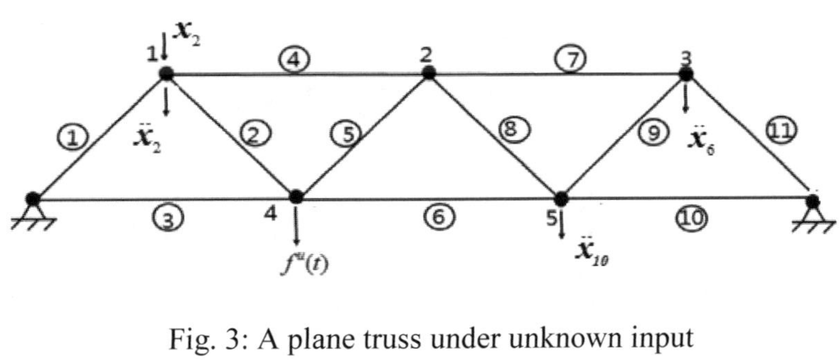

Fig. 3: A plane truss under unknown input

As indicted in Fig.3, acceleration responses in the vertical directions of nodes 1, 4 and 5 are measured. If only these four measured noisy acceleration responses are used in the identification by the KF-UI, significant drifts in the identified displacement and unknown input are shown in Fig. 4 where measured accelerations contains noises with a 5% noise- to- signal ratio in root mean square.

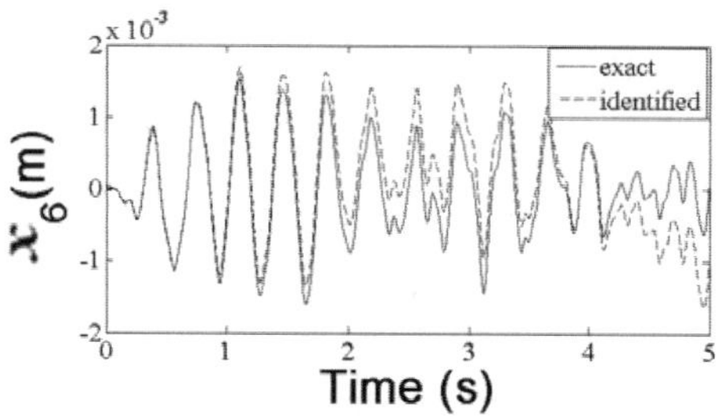

(a) Drift in the identified displacement

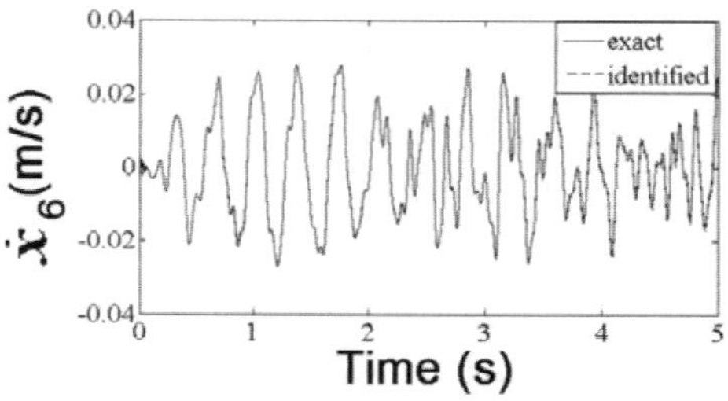

(b) Identified velocity

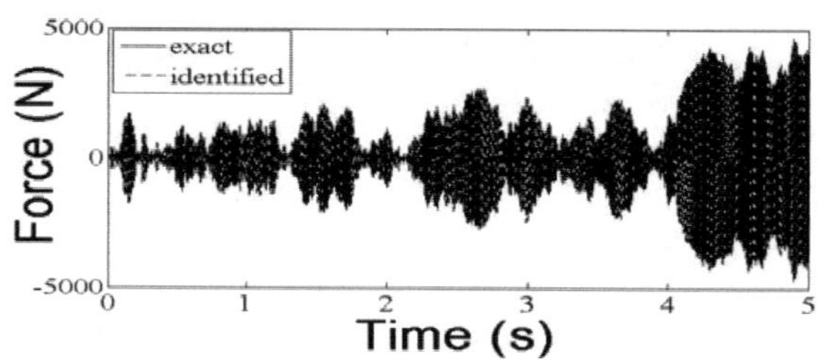

(c) Oscillation in the identified unknown input

Fig.4 : Comparisions of identified structural state and input with noisy acceleration measuremnts

To prevent the above drift problem, partial measured displacements are added in combination with acceleration measurements for the joint identification problem by the proposed KF-UI. For this relatively small size structural model, the displacement in the vertical direction at node 2 in Fig. 3 is added. Data fusion of this measured displacement and the above four accelerations are used and a dummy-measurement of acceleration in the vertical direction of node 4 is added in the observation equation. As shown by the comparisons of identified structural state and input with their exact values in Fig. 5, it is clearly demonstrated that the so-called drifts in estimated structural state and input are avoided by the improved KF-UI.

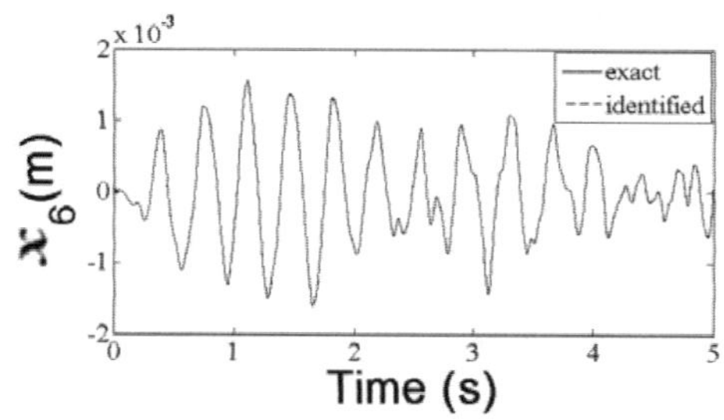

(a) Comparison of identified displacement

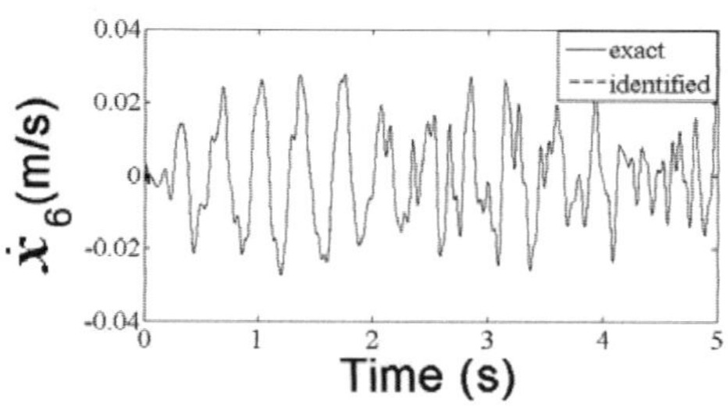

(b) Comparison of identified velocity

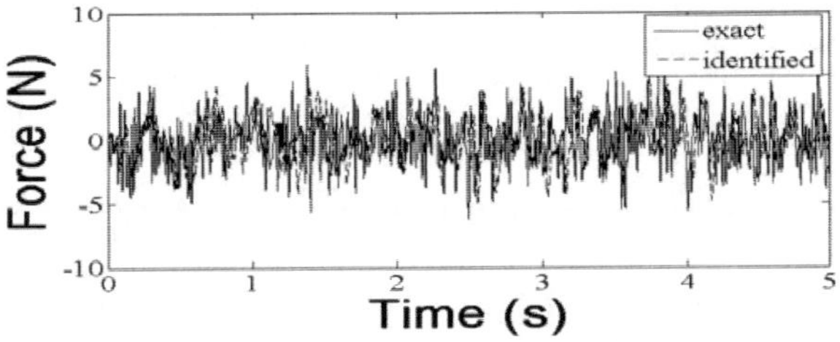

(c) Comparison of identified unknown input

Fig.5 : Comparisions of identified results with data fusion of 5% noisy measuremnts

4. Conclusions

Previous KF-UI approaches based solely on acceleration measurements are inherently unstable which leads the drifts in the estimated unknown inputs and structural displacements. Moreover, it is necessary to have the measurements of acceleration responses at the locations where unknown inputs applied for the recursive estimation of unknown inputs. In this paper, an algorithm is proposed to circumvent these limitations for the estimation of joint structural states and unknown inputs without using collocated acceleration measurements. Based on the scheme of the classical KF, an improved Kalman filter with unknown excitations (KF-UI) and without collocated acceleration measurements is derived. Then, data fusion of displacement and acceleration measurements is used to prevent the drifts in the identified structural state vector and unknown external inputs in real time. Finally, the addition of dummy-measurements of accelerations at the locations where unknown inputs applied is proposed to bind the high-frequency oscillation in the identified external inputs

The proposed algorithm is not available in the literature and the advantages of the proposed algorithm over previous approaches are obvious. Therefore, it provides an efficient algorithm of real time estimation of joint structural states and the unknown inputs, which is important for decentralized structural vibration control as it is essential to estimate the unknown interaction forces between substructures without the requirement of measured responses at the substructure interfaces.

Some numerical examples have demonstrated the effectiveness and versatilities of the proposed approach. More numerical example demonstrations by complex structural configurations and experimental validations are needed. Such investigations are undertaken by the authors.

Acknowledgement

This research is supported by the National Natural Science Foundation of China (NSFC) through Grant No. 51678509 and the Scientific Research Project No.16DZ1200102 of Shanghai Municipal science and Technology Committee.

References

Ay, A.M.,Wang, Y. 2014 'Structural damage identification based on self-fitting ARMAX model and multi-sensor data fusion'. *Structural Health Monitoring*.vol.13, no.4, pp.445-460.

Azam, S.E., Chatzi, E., Costas Papadimitriou, C. 2015, 'A dual Kalman filter approach for state estimation via output-only acceleration measurements'. *Mechanical Systems and Signal Processing,* vol.60-61, pp.866–886.

Chen, J. and Li, J. 2004. 'Simultaneous identification of structural parameters and input time history from output-only measurements', *Computational Mechanics*, vol. 33, no. 5, pp.365-374.

Ding Y., Law S.S., Wu B., Xu G. S., Lin Q., Jiang H. B., Miao Q.S. 2013, 'Average acceleration discrete algorithm for force identification in the state space', *Engineering Structures*, vol.56, pp.1880-1892.

Gillijns S., Moor B. D. 2007, 'Unbiased minimum-variance input and state estimation for linear discrete-time systems with direct feed through', *Automatica,*vol. 43,pp. 934-937.

Kalman, R.E., 1960, 'A new approach to linear filtering and prediction problems', *Journal of Basic Engineering*, vol.82,no.1, pp. 35–45.

Kim, J., Kim, K., Sohn, H. 2014, 'Autonomous dynamic displacement estimation from data fusion of acceleration and intermittent displacement measurements', *Mechanical Systems and Signal Processing*, vol.42,no.1-2, pp.194-205.

Lei, Y. Liu C. and Liu, L.J. 2014, 'Identification of multistory shear buildings under unknown earthquake excitation using partial output measurements: numerical and experimental studies'. *Structural Control & Health Monitoring*, vol.21,no.5, pp.774-783.

Lei, Y., Chen, F. and Zhou H. 2015, 'A Two-Stage and Two-Step Algorithm for the Identification of Structural Damage and Unknown Excitations: Numerical and Experimental Studies'. *Smart Structures and Systems*, vol.15,no.1, pp. 57-80.

Li, Y. Y.; Chen, Y., 2013, 'A review on recent development of vibration-based structural robust damage detection', *Structural Engineering and Mechanics*, vol.45, no.2, pp.159-168.

Liu K., Law S.S., Zhu X.Q., Xia Y. 2014, 'Explicit form of an implicit method for inverse force identification', *Journal of Sound and Vibration*. Vol.333, pp. 730-744.

Liu, L.J., Su, Y., Zhu, J.J., Lei, Y. 2016, 'Improved Kalman filter with unknown inputs based on data fusion of partial acceleration and displacement measurements', *Smart Structures and Systems,*vol.17, no.6, pp.903-915.

Lourens E., Reynders E., Roeck G. De, Degrande G., Lombaert G., 2012, 'An augmented Kalman filter for force identification in structural dynamics', *Mechanical Systems and Signal Processing*, vol.27, pp.446-460.

Ma, C.K. Chang J.M. and Lin D. C., 2003, 'Input forces estimation of beam structures by an inverse method', *Journal of Sound and Vibration*, vol.259, no.2, pp. 387–407.

Naets F., Cuadrado J., Desmet W., 2015, 'Stable force identification in structural dynamics using Kalman filtering and dummy-measurements', *Mechanical Systems and Signal Processing* , vol.50-51, pp. 235-248.

Pan S.W., Su H.Y., Wang H., Chu J., 2010, 'The study of input and state estimation with Kalman filtering', *The Institute of Measurement and Control*, vol.33, no.8, pp.901-918.

Wu, A.L., Loh, C.H., Yang J.N., Weng, J.H., Chen, C.H., Ueng T. S., 2009, 'Input force identification: application to soil-pile interaction', *Structural Control and Health Monitoring*, vol.16, pp. 223-240.

Xu, B., He, J., Rovekamp, R., Dyke, S.J., 2012, 'Structural parameters and dynamic loading identification from incomplete measurements: Approach and validation'. *Mechanical System and Signal Processing*, vol.28, pp.244-257.

Xu, B., He, J., 2015, 'Substructural parameters and dynamic loading identification with limited observations', *Smart Structures and Systems*, vol.15, no.1, pp.169-189.

Yuen, K. V., Mu, H. Q., 2015, 'Real-Time System Identification: An Algorithm for Simultaneous Model Class Selection and Parametric Identification', *Computer-Aided Civil and Infrastructure Engineering*, vol.30, no.10, pp. 785-801.

Yuen, K. V., Liang, P. F., Kuok, S. C., 2013, 'Online estimation of noise parameters for Kalman filter', *Structural Engineering and Mechanics,* vol.47, no.3, pp. 361-381.

The 8th International Conference on
Structural Health Monitoring of Intelligent Infrastructure
Brisbane, Australia | 5-8 December 2017

Nonparametric identification for model-free structural nonlinearities and systems using partial measured structural responses

L.J. Liu[1], Y. Lei [2], M.Y.He [2]
[1] Department of Civil Engineering, Xiamen University-China, Email: liulj214@xmu.edu.cn
[2] Department of Civil Engineering, Xiamen University-China

Abstract

Structural nonlinearity cloud be used as the indicator of the development of structural damage under dynamic excitation. It is necessary to identify structural nonlinearities based on structural dynamic responsments for damage detection and performance evaluation of structures. However, identification of structural nonlinearity is still a difficult task, especially when proper mathematical models for structural nonlinear behaviors are not available. In prior studies on identification of model-free structural nonlinearities, the locations of structural nonlinearities are usually assumed known and all structural responses are measured. In this paper, an identification method is proposed for locating and identifying model-free structural nonlinearities based on limited structural responses measurements.Firstly, equivalent linear structural systems are established and identified by the extended Kalman filter for the identification of the locations of structural nonlinearities. Then, the model-free structural nonlinear restoring forces are approximated by power series polynomial models and are identified by the unscented Kalman filter. Both numerical simulation examples of a multi-story shear building are used to validate the proposed algorithm.

1. Introduction

Structural nonlinearity could be used as the indicator of the development of structural damage under dynamic excitation (Yan *et al.* 2008, Lee *et al.* 2010). So, it is necessary to localize and identify structural nonlinearities using structural dynamic response measurements for damage detection, performance evaluation and remaining service life forecasting of engineering structures (Lu *et al.* 2013, Xia *et al.* 2014). However, identification of nonlinear dynamics structures is much more difficult established than that of linear structures. For some complex nonlinear structural systems, the mathematical forms of models are not clear due to the lack of the knowledge of the patterns of nonlinear performances (Masri *et al.*1982, 2006)

Nonparametric identification approaches for model-free structural nonlinear systems have been proposed. The most popular procedure is the restoring force surface (RFS) approach (Masri and Caughey 1979, Masri *et al.* 1982, Crawley and Aubert 1986). Further extensions and implementation strategies of RFS based methodologies were also investigated (Al-Hadid and Wright 1992, Masri *et al.* 2005, 2006, Hernandez-Garcia *et al.* 2010). In the RFS approaches, the polynomial-basis model is a non-parametric model in which structural nonlinear restoring force is approximated by a linear combination of the power-series expansion of the system displacement and velocity. In recent years, Xu *et al.* (2012) and He *et al.* (2012) presented a power series polynomial modeling approach involving the instantaneous values of system displacement and velocity to approximate structural nonlinear restoring force. However, in prior studies on nonparametric identification, it is prerequisite that the locations of structural nonlinearities are known and complete measurements of structural responses are available. If the locations of structural nonlinearities are not clear, the expansions of all structural nonlinear restoring forces lead to tremendous undermined coefficients in the polynomial series expansions.

In practice, it is impossible to measure all response outputs of a structural system (Lei *et al.* 2013). The extended Kalman filter (EKF) provides a particularly practical and efficient structural identification with incomplete structural responses (Hoshiya and Saito 1984, Yang *et al.* 2006, Lei *et al.* 2013, Su *et al.* 2014). Although EKF can be extended for the identification of nonlinear structures (Yang *et al.* 2006, Lei *et al.* 2012, Lei and He, 2013), the degree of accuracy of the EKF relies on the validity of the linear approximation and is not effective to identify strong nonlinear structural systems (Wu and Smyth 2007). In this regard, the unscented Kalman filter (UKF) (Julier *et al.* 2004, Mariani and Ghisi 2007, Omrani *et al.* 2013) is superior to the EKF in the identification of strong nonlinear structural systems. The UKF is based on the unscented transform of a deterministic sampling scheme consisting of the so-called "sigma points". In each step, the UKF generates a set of sigma points and updates the prediction formulas based on these sigma points. Therefore, if the UKF is directly applied to the identification of a model-free nonlinear structure in which the locations of structural nonlinearities are not prerequisite, it is difficult or even impossible to identify so many undetermined coefficients in the polynomial series expansions.

In this paper, an identification algorithm is proposed for locating and identifying model-free structural nonlinearities using incomplete measurements of structural responses. First, an equivalent linear structural system is established for an original nonlinear system. The equivalent linear system is identified by EKF and the locations of structural nonlinearities can be identified. Then, the model-free structural restoring forces at the locations of identified structural nonlinearities are assumed to be approximated by power series polynomials. The polynomials and the nonlinear structural system are identified by the UKF. Some numerical simulation examples with different nonlinear restoring force models and locations are used to validate the effectiveness of the proposed algorithm.

2. The Identification problem

The equation of motion for a nonlinear structure can be written as

$$\boldsymbol{M}\ddot{\boldsymbol{x}}(\mathrm{t}) + \boldsymbol{C}\dot{\boldsymbol{x}}(\mathrm{t}) + \boldsymbol{K}\boldsymbol{x}(\mathrm{t}) + \boldsymbol{F}_n[\boldsymbol{x}(\mathrm{t}), \dot{\boldsymbol{x}}(\mathrm{t})] = \boldsymbol{B}\boldsymbol{f}(\mathrm{t}) \tag{1}$$

where ***M, C, K*** are the linear mass, damping and stiffness matrices of the structure, respectively; $\boldsymbol{x}(\mathrm{t})$, $\dot{\boldsymbol{x}}(\mathrm{t})$ and $\ddot{\boldsymbol{x}}(\mathrm{t})$ are the vectors of the displacement, velocity and acceleration responses, respectively; $\boldsymbol{F}_n[\boldsymbol{x}(\mathrm{t}), \dot{\boldsymbol{x}}(\mathrm{t})]$ is a vector of the nonlinear restoring force, $\boldsymbol{B}$ and $\boldsymbol{f}(\mathrm{t})$ are the external force influence matrix and external force vector, respectively. Usually, the mass of a structure can be estimated accurately, so ***M*** is assumed known in this paper.

For general complex nonlinear systems, it is hard to establish proper mathematical form of the nonlinear restoring force models due to the lack of the knowledge of the pattern of nonlinear performances. The power series polynomials involving the instantaneous values of the relative displacement and velocity responses of the structural system are utilized to approximate the total nonlinear restoring forces in the identification problem (Masri *et al.* 2005, 2006, 2010, Xu *et al.* 2012), i.e.,

$$F_{n,i,i-1}[\dot{\boldsymbol{x}}(\mathrm{t}), \boldsymbol{x}(\mathrm{t})] \approx F_{n,i,i-1}[s_{i,i-1}, v_{i,i-1}, c^{i,i-1}] \approx \sum_{k=0}^{p}\sum_{j=0}^{q} c_{j,k}^{i,i-1} s_{i,i-1}^{j} v_{i,i-1}^{k} \quad (k+j \neq 0) \tag{2}$$

in which $F_{n,i,i-1}[\dot{\boldsymbol{x}}(\mathrm{t}), \boldsymbol{x}(\mathrm{t})]$ denotes the nonlinear restoring force between *i*-th and (*i*-1)-th degrees of freedom; $c_{j,k}^{i,i-1}$ are the coefficients of power series polynomials; *p* and *q* are the highest power of

polynomials, which are depended on the level of nonlinearity; $s_{i,i-1}$ and $v_{i,i-1}$ represents the relative displacement and relative velocity between i-th and (i-1)-th degrees of freedom, respectively.
When the external excitation is weak, the nonlinear restoring forces can be neglected and the equation of motion of the structural system reduces to

$$\boldsymbol{M}\ddot{\boldsymbol{x}}_l(\mathrm{t}) + \boldsymbol{C}\dot{\boldsymbol{x}}_l(\mathrm{t}) + \boldsymbol{K}\boldsymbol{x}_l(\mathrm{t}) = \boldsymbol{B}\boldsymbol{f}(\mathrm{t}) \tag{3}$$

where $\boldsymbol{x}_l(\mathrm{t})$, $\dot{\boldsymbol{x}}_l(\mathrm{t})$ and $\ddot{\boldsymbol{x}}_l(\mathrm{t})$ are the vectors of the displacement, velocity and acceleration responses of the linear structure, respectively.
The stiffness and damping parameters of the above linear structural system can be identified by the EKF using only partial measurements of the structural responses under weak excitation.
The extended state vector of the above linear structure is defined as $\boldsymbol{X}_l = \left\{\boldsymbol{x}_l^T, \dot{\boldsymbol{x}}_l^T, \boldsymbol{\theta}_l^T\right\}^T$ with $\boldsymbol{\theta}_l$ being a vector of linear stiffness and damping parameters. Then, the equation for the extended state vector of the equivalent linear structure can be derived as

$$\dot{\boldsymbol{X}}_l = \left\{\begin{array}{c} \dot{\boldsymbol{x}}_l \\ \boldsymbol{M}^{-1}\left[\boldsymbol{B}\boldsymbol{f}(t) - \boldsymbol{K}\boldsymbol{x}_l - \boldsymbol{C}\dot{\boldsymbol{x}}_l\right] \\ 0 \end{array}\right\} = \boldsymbol{g}(\boldsymbol{X}_l, \boldsymbol{f}) \tag{4}$$

Usually, it is easy to deploy some accelerometers on structures to measure structural partial acceleration responses. Therefore, the discrete form of equation for measured acceleration vector can be expressed as:

$$\boldsymbol{y}_{l,k+1} = -\boldsymbol{D}\boldsymbol{M}^{-1}\left(\boldsymbol{C}\dot{\boldsymbol{x}}_{l,k+1} + \boldsymbol{K}\boldsymbol{x}_{l,k+1} + \boldsymbol{B}\boldsymbol{f}_{k+1}\right) + \boldsymbol{v}_{k+1} = \boldsymbol{h}\left(\boldsymbol{X}_{l,k+1}, \boldsymbol{f}_{k+1}\right) + \boldsymbol{v}_{k+1} \tag{5}$$

where $\boldsymbol{y}_{l,k+1}$ is the measured linear acceleration response vector at time $t = (k+1) \times \Delta t$ (Δt is the sampling time step), $\boldsymbol{D}$ denotes the location of accelerometers, $\boldsymbol{v}_{k+1}$ is the measurement noise assumed as Gaussian white noise.
Based on the EKF, the extended state vector can be identified as

$$\hat{\boldsymbol{X}}_{l,k+1|k+1} = \tilde{\boldsymbol{X}}_{l,k+1|k} + \boldsymbol{K}_{k+1}[\boldsymbol{y}_{l,k+1} - \boldsymbol{h}(\tilde{\boldsymbol{X}}_{l,k+1|k}, \boldsymbol{f}_{k+1})] \tag{6}$$

$$\tilde{\boldsymbol{X}}_{l,k+1|k} = \hat{\boldsymbol{X}}_{l,k|k} + \int_{t_k}^{t_{k+1}} \boldsymbol{g}(\hat{\boldsymbol{X}}_l, \boldsymbol{f})\, dt \tag{7}$$

where $\hat{\boldsymbol{X}}_{l,k+1|k+1}$ is the estimated state vector $\boldsymbol{X}_l$ at time $t = (k+1) \times \Delta t$, $\boldsymbol{K}_{k+1}$ is the Kalman Gain matrix at the time instant (Hoshiya and Saito 1984, Lei *et al.* 2013).
Therefore, the linear structural parameters can be identified using the EKF with the partial measurements of structural responses under weak excitation.
Under strong external excitation, structural nonlinear effect becomes significant. For the identification of the model-free nonlinear system described by Eq.(1), some algorithms such as the UKF can be utilized. However, when the locations of structural nonlinearities are unknown, it is difficult or even impossible to identify so many undermined coefficients $c_{j,k}^{i,i-1}$ if the UKF is directly applied to the whole nonlinear system. Since structural nonlinearities only exist in local areas of a whole structure, it is better to firstly identify the locations of structural nonlinearities. A novel identification approach is proposed for this purpose in this paper.

2.1 Locating the nonlinearity in nonlinear structural systems

For the above nonlinear structural system under strong external excitation, an equivalent linear structural system is introduced with the following equation of motion

$$\boldsymbol{M}_e\ddot{\boldsymbol{x}}_e(\mathrm{t}) + \boldsymbol{C}_e\dot{\boldsymbol{x}}_e(\mathrm{t}) + \boldsymbol{K}_e\boldsymbol{x}_e(\mathrm{t}) = \boldsymbol{Bf}(\mathrm{t}) \tag{8}$$

in which $\boldsymbol{x}_e(\mathrm{t})$, $\dot{\boldsymbol{x}}_e(\mathrm{t})$ and $\ddot{\boldsymbol{x}}_e(\mathrm{t})$ are the vectors of the displacement, velocity and acceleration responses of the equivalent linear system, respectively; $\boldsymbol{M}_e$,$\boldsymbol{C}_e$ and $\boldsymbol{K}_e$ are the equivalent linear mass, damping and stiffness matrices, respectively; Usually, the mass matrix is hardly changed when nonlinear behavior appears, so $\boldsymbol{M}_e$=$\boldsymbol{M}$.

Analogously, the state vector and the equivalent linear stiffness and damping parameters can be identified by the EKF using the limited measurements of the nonlinear structural responses.

$$\hat{\boldsymbol{X}}_{e,k+1|k+1} = \tilde{\boldsymbol{X}}_{e,k+1|k} + \boldsymbol{K}_{e,k+1}[\boldsymbol{y}_{k+1} - \boldsymbol{h}(\tilde{\boldsymbol{X}}_{e,k+1|k}, \boldsymbol{f}_{k+1})] \tag{9}$$

$$\tilde{\boldsymbol{X}}_{e,k+1|k} = \hat{\boldsymbol{X}}_{e,k|k} + \int_{t_k}^{t_{k+1}} \boldsymbol{g}(\hat{\boldsymbol{X}}_e, \boldsymbol{f})\, dt \tag{10}$$

where $\hat{\boldsymbol{X}}_{e,k+1|k+1}$ is the estimated state vector $\boldsymbol{X}_e$ at time $t=(k+1)\times\Delta t$ and $\boldsymbol{X}_e = \left\{\boldsymbol{x}_e^T, \dot{\boldsymbol{x}}_e^T, \boldsymbol{\theta}_e^T\right\}^T$ with $\boldsymbol{\theta}_e$ being the parametric vector of the equivalent linear structural system.

Then, the locations of structural nonlinearity can be identified by comparing the differences between the identified structural parameters of equivalent linear system and those of the linear structure. Also, the extents of differences indicate the levels of structural nonlinearities.

2.2 Identification of the nonlinear restoring forces and system

After the identification of the locations of structural nonlinearities, the UKF is adopted herein for the identification of the model-free nonlinear system described by Eq.(1)-Eq.(2); however, only the nonlinear forces at the locations of identified structural nonlinearities are approximated by the power series polynomials in which the orders of polynomials p and q are selected based on the levels of identified structural nonlinearities indicated by the extents of differences between the identified structural parameters of the equivalent linear system and those of the linear one in the above section.

In the recursive identification process by UKF, the selection of proper initial values of power series polynomials is important for the identification convergence. This can be accomplished based on the identified equivalent linear structural system by the EKF in the above section.

Based on the identified acceleration, velocity and displacement responses of the equivalent linear structure, the algebraic coefficients used to approximate the total nonlinear restoring forces can be identified by means of least-square techniques. These estimated coefficients are used as the proper initial values of power series polynomials.

Then, the UKF is utilized for the identification of nonlinear system described by Eq.(1)-Eq.(2). The extended state equation and observation equations are nonlinear equations expressed as:

$$\dot{\boldsymbol{X}} = \boldsymbol{g}(\boldsymbol{X}, \boldsymbol{f}) \tag{11}$$

$$\boldsymbol{y}_{k+1} = \boldsymbol{h}(\boldsymbol{X}_{k+1}, \boldsymbol{f}_{k+1}) + \boldsymbol{v}_{k+1} \tag{12}$$

in which $\boldsymbol{X} = \left\{\boldsymbol{x}^T, \dot{\boldsymbol{x}}^T, \boldsymbol{\theta}^T\right\}^T$ with $\theta_i = c_{j,k}^{i,i-1}$

Based on the unscented transform, $2N$+1 deterministic sample points $\chi_{i,k|k}$ can be computed as (Julier *et al*. 2004)

$$\boldsymbol{\chi}_{i,k|k} = \begin{cases} \hat{\mathrm{X}}_{k|k} & , \quad i=0 \\ \hat{\mathrm{X}}_{k|k} + \left(\sqrt{(N+\lambda)\hat{\boldsymbol{P}}_{\boldsymbol{X},k|k}}\right)_i & , \quad i=1,\ldots N \\ \hat{\mathrm{X}}_{k|k} - \left(\sqrt{(N+\lambda)\hat{\boldsymbol{P}}_{\boldsymbol{X},k|k}}\right)_i & , \quad i=N+1,\ldots 2N \end{cases} \tag{13}$$

where N is the dimension of $\boldsymbol{X}$, $\hat{\mathrm{X}}_{k|k} = E\{\boldsymbol{X}_k\}$, $\hat{\boldsymbol{P}}_{\boldsymbol{X},k|k} = E\left\{\left(\boldsymbol{X}_k - \hat{\boldsymbol{X}}_{k|k}\right)\left(\boldsymbol{X}_k - \hat{\boldsymbol{X}}_{k|k}\right)^T\right\}$, λ is a scaling parameter, and $(\cdot)_i$ is the i-th column of the matrix.
First, the sample points and extended state vector are predicated as:

$$\boldsymbol{\chi}_{\mathrm{i},k+1|k} = \boldsymbol{\chi}_{\mathrm{i},k|k} + \int_{k\Delta t}^{(k+1)\Delta t} g\left(\boldsymbol{X}_{t,k}, \boldsymbol{f}\right) dt \; ; \; \tilde{\boldsymbol{X}}_{k+1|k} = \sum_{i=0}^{2N} W_i^m \boldsymbol{\chi}_{i,k+1|k} \tag{14}$$

with the predicted error covariance expressed as:

$$\tilde{\boldsymbol{P}}_{X,k+1|k} = \sum_{i=0}^{2N} W_i^c (\boldsymbol{\chi}_{i,k+1|k} - \tilde{\boldsymbol{X}}_{k+1|k})(\boldsymbol{\chi}_{i,k+1|k} - \tilde{\boldsymbol{X}}_{k+1|k})^T + \boldsymbol{Q}_{k+1} \tag{15}$$

where W_i^m and W_i^c are the weights for the mean and covariance, respectively (Julier *et al*. 2004)
Then, the measurement update steps are as follows:

$$\hat{\boldsymbol{X}}_{k+1|k+1} = \tilde{\boldsymbol{X}}_{k+1|k} + \boldsymbol{K}_{k+1}(\boldsymbol{y}_{k+1} - \hat{\boldsymbol{y}}_{k+1|k+1}) \tag{16}$$

$$\tilde{\boldsymbol{y}}_{i,k+1|k+1} = \boldsymbol{h}\left(\boldsymbol{\chi}_{i,k+1|k}, \boldsymbol{f}_{k+1}\right); \quad \hat{\boldsymbol{y}}_{k+1|k+1} = \sum_{i=0}^{2N} \mathrm{W}_i^m \tilde{\boldsymbol{y}}_{i,k+1|k} \tag{17}$$

$$\hat{\boldsymbol{P}}_{y,k+1} = \sum_{i=0}^{2N} W_i^c (\hat{\boldsymbol{y}}_{k+1|k+1} - \tilde{\boldsymbol{y}}_{i,k+1|k+1})(\hat{\boldsymbol{y}}_{k+1|k+1} - \tilde{\boldsymbol{y}}_{i,k+1|k+1})^T + \boldsymbol{R}_{k+1} \tag{18}$$

$$\hat{\boldsymbol{P}}_{\boldsymbol{X}y,k+1} = \sum_{i=0}^{2n} W_i^c \left\{\boldsymbol{\chi}_{i,k+1|k} - \tilde{\boldsymbol{X}}_{k+1|k}\right\}\left\{\hat{\boldsymbol{y}}_{k+1|k} - \tilde{\boldsymbol{y}}_{i,k+1|k}\right\}^T \tag{19}$$

$$\boldsymbol{K}_{k+1} = \hat{\boldsymbol{P}}_{\boldsymbol{X}y,k+1}(\hat{\boldsymbol{P}}_{y,k+1})^{-1}; \; \hat{\boldsymbol{P}}_{\boldsymbol{X},k+1|k+1} = \tilde{\boldsymbol{P}}_{X,k+1|k} - \boldsymbol{K}_{k+1}\boldsymbol{P}_{y,k+1}\boldsymbol{K}_{k+1}^T \tag{20}$$

After the identification of the state vector and the coefficients of the power series polynomials used to represent the total nonlinear restoring forces, the behaviors of the model-free of nonlinear restoring forces are identified.

3. Numerical Validations

To validate the proposed identification algorithm, numerical identification examples of an 8-story nonlinear shear building with different nonlinear restoring force models and locations are used. The building is subject a white noise excitation at the top floor of the building. The numerical calculated acceleration responses are treated as "measured responses" for the identification problem. To consider partial response measurements, only the acceleration responses at the 1st, 2nd, 4th, 5th and 7th floor levels are used. Also, the calculated acceleration responses are added with the corresponding white noise with 5% noise-to-signal ratio in root mean square (RMS) to consider the influences of measurement noises.

The linear structural parametric values are selected as: mass of floor m_i=500kg, story stiffness k_i=200kN/m, story damping c_i=0.20kN·s/m (i=1,2,…,8). These parameters can be identified using the EKF with the partial acceleration responses of the building under weak external excitation. The identified linear stiffness parameters $k_{l,i}$ and linear damping values $c_{l,i}$ are listed in Table 1.

3.1 Identification of nonlinear restoring forces with Dahl model

In this study a modified Dahl model (Dahl 1976) is employed, which can capture many commonly observed types of nonlinear behavior of MR dampers. The modified Dahl model is described by the following equation for nonlinear force as

$$F_{ni}(t) = k_{0i}u_i(t) + c_{0i}\dot{u}_i(t) + f_{di}z_i(t) + f_{0i} \tag{21}$$

in which, k_{0i}=the stiffness coefficient, c_{0i} = the viscous damping coefficient, f_{di}=the adjustable coulomb friction, f_{0i}=the initial force, u_i=the displacement of the i-th damper, and z_i=a dimensionless parameter which describes the coulomb friction by the equation as:

$$\dot{z}_i = \sigma_i \dot{u}_i \cdot (1 - z_i \cdot \mathrm{sgn}(\dot{u}_i)) \tag{22}$$

When the 8-story building is subject to a strong white noise excitation, structural nonlinear effect needs to be considered. In this example, it is assumed that story nonlinear restoring force with Dahl model occurs in both the 1st and the 5th floors with the parameters as: k_{01}= 0.05kN/m; c_{01}=0.30kN/m; f_{d1}=0.20kN; f_{01}=0.0kN; σ_1=50000; k_{05} = 0.05kN/m; c_{05}=0.30kN/m; f_{d5}=0.15kN; f_{05}=0.0kN; σ_5 =40000 .However, the location of the nonlinear restoring force and the nonlinear force parameters are only used in generating the nonlinear structural responses.

The identification results are shown and compared with those of the linear building in Table 1. From the comparisons, it is noted that there are large differences between the identified equivalent linear parameters and the linear structure parameters in both the 1st floor and the 5th floors as marked by values in bold, which also indicates the locations of the two structural nonlinearities.

Table 1 Parameters of the equivalent linear and linear building

Story No.	$k_{e,i}$ (kN/m)	$k_{l,i}$ (kN/m)	$c_{e,i}$ (kN·s/m)	$c_{l,i}$ (kN·s/m)
1st	198.87	200.73	**0.70**	**0.21**
2nd	201.38	199.96	0.17	0.20
3rd	199.18	200.09	0.21	0.20
4th	200.41	199.79	0.20	0.21
5th	201.23	200.16	**0.63**	**0.18**
6th	200.32	199.76	0.20	0.19
7th	199.58	200.25	0.22	0.21
8th	200.93	200.69	0.20	0.19

Based on the levels of structural nonlinearities indicated by the extents of differences of parameters in Table 1, the orders of power series polynomial for the approximation of nonlinear restoring force at the 1st floor and the 5th floors are selected as p=q=3, so there are 18 undermined coefficients in the power series polynomial. Then, the unscented Kalman filter (UKF) is utilized to identify the nonlinear building. The initial values of the coefficients in the power series polynomials for approximation of the two nonlinear restoring forces at the 1st and the 5th floors are again estimated by the least-square technique based on the above identification results of the equivalent linear structure. In Table 2, all the identified structural parameters and coefficients of polynomial are also shown.

Table 2 Identification parameters of the nonlinear building

Story No.	1st		2nd	3rd	4th	5th	6th	7th	8th
k_i (kN/m)	199.3		201.0	200.0	200.5	200.5	199.8	200.0	199.4
c_i(kN·s/m)	0.23		0.19	0.21	0.20	0.22	0.22	0.20	0.19
Coefficients of polynomial in the 1st story									
	$c_{1,0}$	$c_{0,1}$	$c_{2,0}$	$c_{1,1}$	$c_{0,2}$	$c_{3,0}$	$c_{2,1}$	$c_{1,2}$	$c_{0,3}$
Initial	379.1	623.4	147.3	-102.7	3.2	-5917.9	43.6	-147.7	-81.6
Identified	303.7	500.9	117.9	-49.9	-10.1	-4734.3	35.0	-119.2	-21.0
Coefficients of polynomial in the 5th story									
	$c_{1,0}$	$c_{0,1}$	$c_{2,0}$	$c_{1,1}$	$c_{0,2}$	$c_{3,0}$	$c_{2,1}$	$c_{1,2}$	$c_{0,3}$
Initial	273.6	531.2	702.0	-0.4	3.7	-7789.5	-481.7	-161.7	-48.7
Identified	219.3	427.0	561.6	11.6	20.8	-6231.7	-384.8	-118.3	-26.3

Figs.1 show the comparisons of the identified structural displacement responses with the exact structural responses. In Fig.2, the identified nonlinear restoring force at the 1st floor and the 5th floors of the building is compared with the exact force. It is demonstrated that the identification results are in good agreement with the exact ones.

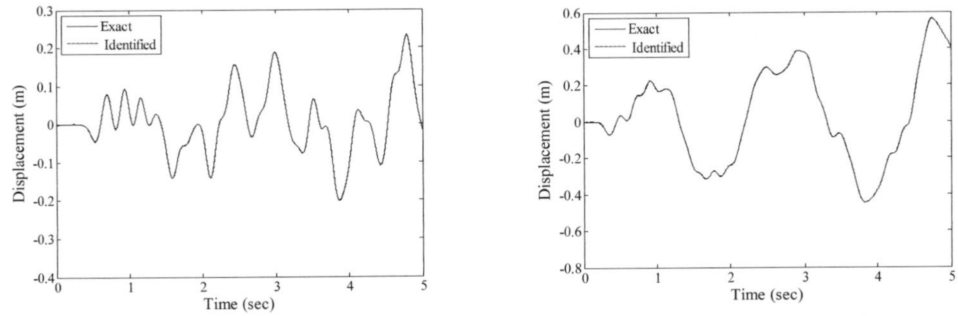

(a) Displacements of the 1st story (b) Displacements of the 5th story

Fig. 1 Comparisons of the displacement responses at the 1st and 5th stories (Dahl models)

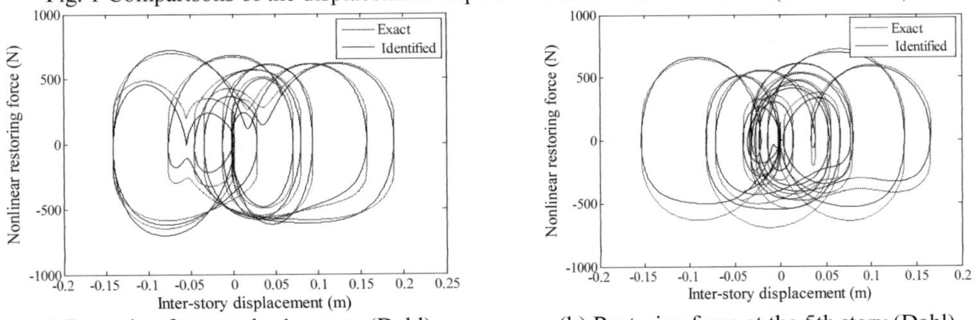

(a) Restoring force at the 1st story (Dahl) (b) Restoring force at the 5th story (Dahl)

Fig. 2 Restoring forces at the 1st and the 5th stories (Dahl models)

3.2 Identification of the nonlinear restoring force with Bingham model

Bingham model is another simple and useful model to describe the nonlinear behavior of MR damper (Woo *et al.* 213). The nonlinear restoring force with Bingham model is described by:

$$F_{ni} = f_{ci} \cdot \mathrm{sgn}(\dot{x}_i) + c_{0i}\dot{x}_i + f_{0i} \tag{23}$$

in which f_{ci}, c_{0i} and f_{0i} are model parameters and $\dot{x}_i$ is the velocity of the i-th damper, respectively. In this numerical example, it is assumed that story nonlinear restoring force with Bingham model in the 1st story of the building and the Bingham model parameters are selected as: f_{ci}=0.02kN , c_{01}=0.15kNs/m , $f_{01}=0$.

Analogously, an equivalent linear structure is established and identified by EKF using the partial measurements of structural nonlinear responses. The identification results of equivalent linear parameters are shown and compared with linear ones in Table 3.

Table 3: Parameters of the equivalent linear and linear building

Story No.	k_{ei} (kN/m)	k_{li} (kN/m)	c_{ei} (kN·s/m)	c_{li} /(kN·s/m)
1st	199.47	200.73	**0.60**	**0.21**
2nd	200.16	199.96	0.20	0.20
3rd	199.92	200.09	0.21	0.20
4th	200.25	199.79	0.20	0.21
5th	200.34	200.16	0.19	0.18
6th	199.78	199.76	0.19	0.19
7th	200.11	200.25	0.21	0.21
8th	200.71	200.69	0.21	0.19

From the comparisons of the identified equivalent linear parameters with the linear ones in Table 3, the location of the structural nonlinearity can be identified. Then, the UKF is utilized to identify the nonlinear building. Table 4 summaries the identified structural parameters and coefficients of the power series polynomial.

Table 4: Identification parameters of the nonlinear building

Story No.	1st	2nd	3rd	4th	5th	6th	7th	8th
k_i (kN/m)	199.8	200.2	200.1	200.1	200.0	200.0	199.9	199.6
c_i(kN·s/m)	0.24	0.20	0.20	0.20	0.20	0.20	0.20	0.18

Coefficients of polynomial in the 1st story

	$c_{1,0}$	$c_{0,1}$	$c_{2,0}$	$c_{1,1}$	$c_{0,2}$	$c_{3,0}$	$c_{2,1}$	$c_{1,2}$	$c_{0,3}$
Initial	-295.7	559.5	464.5	5.2	-4.5	26675	3330.0	287.5	-177.1
Identified	-248.4	482.3	394.8	11.3	-11.9	22674	2830.5	240.8	-151.0

Figs.3-4 show the identified displacement responses and the nonlinear restoring force of the 1st story. The identification results are in good agreements with the exact responses and force.

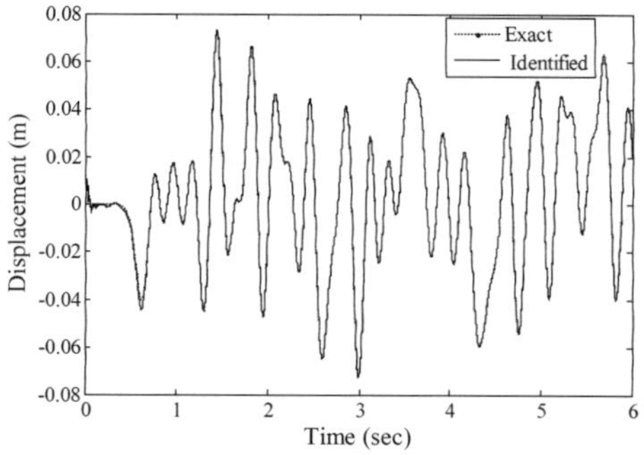

Fig. 3 Displacement of the 1st story

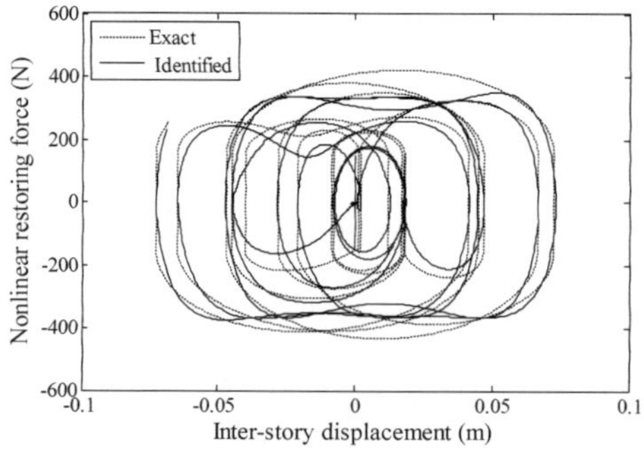

Fig.4 Restoring force at the 1st story (Bingham)

4. Conclusions

In this paper, an identification algorithm is proposed for locating and identifying model-free nonlinear restoring forces and systems using only partial measurements of structural responses. First, an equivalent linear structural system is established for a nonlinear structure. The equivalent linear structure is identified by the EKF. The locations of structural nonlinearities can be identified by comparing the differences between the identified the equivalent linear structural parameters and those of the structure under weak excitation. Also, these differences indicate the levels of structural nonlinearities. Then, model-free nonlinear restoring forces at the locations of identified structural nonlinearities are approximated by power series polynomials with the orders depended on the level of structural nonlinearities. The UKF is utilized to identify the nonlinear structural restoring forces and system with the initial values of power series polynomials estimated by least-square techniques with the identified values of equivalent linear structure.

Compared with prior approaches of direct identification of the nonlinear structures, the proposed algorithm greatly simplifies the identification difficulties. Some numerical simulation examples with different nonlinear models and locations have validated the effectiveness of the proposed algorithm. It is shown that the proposed algorithm is feasible for the identification of model-free nonlinear restoring forces and nonlinear structural systems using only partial measurements of structural responses. Although only shear-type structures are used in the numerical examples of the manuscript, the proposed algorithm can also be used for the identification of other types of nonlinear structures.

Acknowledgement

This research is supported by the National Natural Science Foundation of China (NSFC) through Grant No. 51378445 and the Scientific Research Project No.16DZ1200102 of Shanghai Municipal science and Technology Committee.

References

Al-Hadid, M. A., and Wright, J. R.,1992, 'Estimation of mass and modal mass in the identification of non-linear single and multi degree of freedom systems using the force-state mapping approach', *Mech. Sys. Signal Processing*, vol.6, pp.383-401.

Crawley, E.F., and Aubert, A.C.,1986, 'Identification of nonlinear structural elements by force-state mapping', *AIAA Journal*, vol.24,no.1, pp.155-162.

Dahl, P.R.,1976, 'Solid friction damping of mechanical vibrations', *AIAA J.* vol.14, no.12, pp.1675-1682.

Feng, M. Q., 2009, 'Application of structural health monitoring in civil infrastructure', *Smart Structures and Systems*, vol.5, no.4, pp.469-482.

He, J., Xu, B., Masri, S. F., 2012, 'Restoring force and dynamic loadings identification for a nonlinear chain-like structure with partially unknown excitations', *Nonlinear Dynamics*, vol.69, no.1-2, pp.231-245.

Hernandez-Garcia, M.R., Masri, S. F., Ghanem, R., 2010, 'A structural decomposition approach for detecting, locating, and quantifying nonlinearities in chain-like systems', *Structural Control & Health Monitoring*, vol.17, no.7, pp.761-777.

Hoshiya, M. and Saito, E.,1984, 'Structural identification by extended Kalman filter', *Journal of Engineering Mechanics (ASCE)*, vol.110,no.12, pp.1757-1771.

Julier, S.J., Uhlmann, J.K.,2004, 'Unscented filtering and nonlinear estimation', *Proceedings of the IEEE,* vol. 92, no.3, pp.401-422.

Kerschena, G., Wordenb, K., Vakakis, A.F, Golinvala, J.C., 2006, 'Past, Present and Future of Nonlinear System Identification in Structural Dynamics', *Mechanical Systems and Signal Processing*, vol.20, pp.505-592.

Lee, Y.S., Vakakis, A.F, McFarland, D.M., Bergman, L.A.,2010, 'A Global-local Approach to Nonlinear System Identification: A Review', *Structural Control & Health Monitoring*, vol.17, no.7, pp.742-760.

Lei, Y., Wu, Y., Li, T., 2012, 'Identification of Nonlinear Structural Parameters Under Limited Input and Output Measurements', *International Journal of Non-Linear Mechanics*, vol.47, no.10, pp.1141-1146.

Lei, Y., Liu, C., Jiang, Y.Q. and Mao, Y.K., 2013, Substructure based structural damage detection with limited input and output measurements, *Smart Structures and Systems*, vol.12, no.6, pp.619-640.

Lei, Y. and He, M.Y. 2013, 'Identification of the nonlinear properties of rubber-bearings in base-isolated buildings with limited seismic response data', *Science China-Technological Sciences*, vol.56,no.5, pp.1224-1231.

Lu, W., Teng, J., Xu, Y.L., 2013, 'Identification of damage in dome-like structures using hybrid sensor measurements and artificial neural networks', *Smart Materials and Structures*, vol.22, no.10

Mariani, S., Ghisi, A.2007, 'Unscented Kalman filtering for nonlinear structural dynamics', *Nonlinear Dyn.* Vol.49, no.1-2, pp.131-150.

Masri, S. F., and Caughey, T. K. 1979, 'A nonparametric identification technique for nonlinear dynamic problems', *J. Appl. Mech., ASME*, vol.46, pp.433-447.

Masri, S.F., Sassi, H., Caughey, T.K.,1982, 'A nonparametric identification of nearly arbitrary nonlinear systems', *J. Appl. Mech.* vol.49, pp.619-628.

Masri, S.F., Miller, R.K., Saud, A.F., *et al.*,1987, 'Identification of nonlinear vibrating structures: part II –Application', *Journal of Applied Mechanics*, vol.109, pp.923-929.

Masri, S.F., Caffery, J.P., Caughey, T.K., Smyth, A.W., Chassiakos, A.G., 2005, 'A general data-based approach for developing reduced-order models of non-linear MDOF systems', *Nonlinear Dyn.* vol.39, pp.95-112.

Masri, S.F., Tasbihgoo, F., Caffery, J.P., Smyth, A.W., Chassiakos, A.G., 2006, 'Data-based model-free representation of complex hysteretic MDOF systems', *Struct. Control Health Monit.* vol.13, pp.365-387.

Omrani, R., Hudson, R. E., Taciroglu, E.,2013, 'Parametric Identification of Nondegrading Hysteresis in a Laterally and Torsionally Coupled Building Using an Unscented Kalman Filter', *Journal of Engineering Mechenics-ASCE*, vol.139, no.4, pp.452-468.

Su, J.Z., Xia Y., Xu Y.L., 2014, 'Settlement Monitoring of a Supertall Building Using the Kalman Filtering Technique and Forward Construction Stage Analysis', *Advances In Structural Engineering*, vol.17,no.6, pp.881-893.

Wu, M. and Smyth, A.W., 2007, 'Application of the unscented Kalman filter for real-time nonlinear structural system identification', *Structural Control and Health Monitoring*, vol.14, no.7, pp.971-990.

Xia Y., Chen B, Zhou X.Q., 2013, 'Field monitoring and numerical analysis of Tsing Ma Suspension Bridge temperature behavior' , *Structural Health Control & Health Monitoring*, vol.20, no.4, pp.560-575.

Xu B., He, J., Masri, S. F., 2012, 'Data-based Identification of nonlinear restoring force under spatially incomplete excitations with power series polynomial model', *Nonlinear Dynamics*, vol.67, no.3, pp.2063-2080.

Yan, A.Z., Teng, Jun., Lu, Z.X.,2008, 'Effect of geometrical nonlinearity on the SAVIS device', *Advances in Structural Engineering*, vol.11, no.1, pp.91-105.

Yang, J.N., Lin, S., Huang, H.W. and Zhou, L., 2006, 'An Adaptive Extended Kalman Filter for Structural Damage Identification', *Structural Control and Health Monitoring*, vol.13, no.4, pp. 849-867.

The 8th International Conference on
Structural Health Monitoring of Intelligent Infrastructure
Brisbane, Australia | 5-8 December 2017

A Quantum-Inspired Genetic Algorithm-based Optimization Method for Mobile Impact Test Data Integration

W.J. Zhao [1], S.L. Guo [1], Y. Zhou [2], J. Zhang [1].
[1] *Jiangsu Key Laboratory of Engineering Mechanics (Southeast University), Nanjing, 210096, China, Email:* 230159405@seu.edu.cn [2] *College of Civil Engineering (Hunan University), Changsha, 410082, China*

Abstract

The traditional impact testing can not only identify the basic structural modal parameters, but also identify structural deeply flexibility characteristic. However, it requires a large number of sensors deployed on the entire structure, which cannot meet the requirements of rapid bridge testing. Furthermore, to improve the recognition accuracy and the computational efficiency of the existing mobile impact testing, a new mobile impact test method is proposed by sequentially testing the substructures then integrating the test data of all substructures for flexibility identification of the entire structure. The first novelty of the proposed method is that it doesn't require the reference measurement for integrating the mode shapes of substructures. The second novelty is that the quantum-inspired genetic algorithm (QIGA) is proposed to improve computational efficiency by transforming the scaling factor sign determination problem to an optimization problem. Finally, experimental example of a steel-concrete composite slab is studied which successfully verify the effectiveness of the proposed method.

1. Introduction

Structural health monitoring (SHM) technology has been widely utilized to identify structural characteristics, locate and quantify structural damages (ASCE, 2012). As a widely used solution for SHM, ambient vibration testing, which adopts traffic flows and wind loads etc. to be the natural excitation sources, has been applied in many long span bridges, such as the Humber Bridge (Brownjohn et al. 2010) and the Throgs Neck Bridge (Zhang et al. 2013). Besides, with the finite element (FE) model-driven approach, ambient vibration test data can be used to identify more detailed structural parameters. However, how to decide the single dominant model from many models with high plausibility and uncertainty is still another vital problem.

Multiple reference impact testing (MRIT) used a hammer or other exciters to impact the bridge's deck, during which the impact force and the structural response under the force are recorded for structural identification (Brown & Witter, 2011). Catbas et al. (2004) developed the complex mode indicator function (CMIF) algorithm to process the impact test data and applied it to an aged highway bridge. Zhang et al. (2014) proved that the impact test method with collecting force data has the advantage of identifying the magnitudes of structural frequency response functions (FRFs), and which are necessary for structural flexibility identification.

However, the limitation of the traditional impact test method lies in that it requires a large number of sensors deployed on the entire structure, which usually leads to expensive experimental costs. To overcome the limitation, the idea of mobile impact testing emerges naturally by dividing a structure into substructures then sequentially performing the impact test on substructures. Subsequently, the emerging problem is how to integrate the test data of substructures for modal identification of the entire structure. Although there are many decentralized modal identification algorithms processing the ambient vibration data (Lynch, 2002; Sadhu et al., 2014), they cannot be transplanted directly to solve the data integration problem of the mobile impact test method. In order to address the above challenging problem, Zhang et al. (2015) developed the interface measurement-based approach and the single reference-based approach to integrate all substructures' vibration data for identifying the entire structure's flexibility matrix. However, the requirement of reference nodes greatly reduces test efficiency.

In order to achieve the purpose of rapid impact testing for bridges and optimally improve the testing efficiency, a much more efficient mobile impact test method with reference-free measurement is proposed in this article. In the proposed method, the question of assembling substructures' mode shapes into the global mode shapes of the entire structure is transformed into an optimization problem. There are abundant applications of swarm and evolutionary computations into civil engineering, such as Genetic Algorithm (Han and Kim, 2000; Mencia et al., 2016), Ant Colony Algorithm (Putha et al., 2012). Due to the optimization problem of the mobile impact test data integration studied in this paper falls into the NP-hard binary optimization paradigm, an improved quantum-inspired genetic algorithm (QIGA) is developed because of its excellent performance.

The structure of the paper is as follows: In section 2, the framework of the proposed mobile impact test method is presented, in which the problem of integrating the substructures' mode shapes is transformed into an optimization problem. In section 3, the improved QIGA with dynamic quantum rotation gate and adaptive crossover (mutation) probability rate is developed to solve the optimization problem. In sections 4, the experimental example of a steel-concrete composite slab is studied respectively to verify the effectiveness of the proposed method. Finally, conclusions are drawn.

2. Framework of the proposed mobile impact test method

As shown in Fig. 1, the structure is divided into three substructures and the impact test is independently conducted on each substructure one by one. The FRFs of three substructures are estimated respectively using the recorded input forces and accelerations and their modal flexibility matrices are identified respectively. However, the challenging problem arising is that the identified modal flexibility matrix of the structure are sparse, thus they cannot be used to predict the entire structure's displacements because of unknown elements on the non-diagonal regions of the flexibility matrix.

2.1 Magnitude scaling for mode shapes integrating

The key step to get the full flexibility matrix is integrating the mode shapes of substructures to those of the entire structure. The following operations are implemented to insure all substructures' mode shapes in each mode be at the same scaling level as that of the master substructure: 1) the modal scaling factor Q_r^k of the kth substructure should be the same as Q_r^{master} of the master substructure; 2) adjust the identified mode shape ψ_r^k to $\tilde{\psi}_r^k$ of the kth substructure by using an adjusting factor κ_r^k, $\tilde{\boldsymbol{\psi}}_r^k = \kappa_r^k \boldsymbol{\psi}_r^k$; 3) adjust the identified modal participation factor $\boldsymbol{\Gamma}_r^k$ to $\tilde{\boldsymbol{\Gamma}}_r^k$ by using another adjusting factor δ_r^k, $\tilde{\boldsymbol{\Gamma}}_r^k = \delta_r^k \boldsymbol{\Gamma}_r^k$. Where $\boldsymbol{\psi}_r^k \in \square^{N_o^k \times 1}$ and $\boldsymbol{\Gamma}_r^k \in \square^{N_i^k \times 1}$; N_o^k and N_i^k are the number of output and input of the kth substructure respectively. The subscript r and the superscript k represent the mode order and the substructure number respectively. The symbol "~" represents the variable adjusted.

According to the relations between subset mode shapes consisting of its elements in the indices of driving points and modal participation factors before and after adjusting, ($\mathbf{\Gamma}_r^k = Q_r^k \mathbf{\psi}_{r,drv}^k$, $\tilde{\mathbf{\Gamma}}_r^k = Q_r^{master} \tilde{\mathbf{\psi}}_{r,drv}^k$), and the unchanged residue matrix, which equals to the multiply of mode shape and transpose of modal participation factor, before and after adjusting, ($\mathbf{\psi}_r^k \mathbf{\Gamma}_r^{kT} = \tilde{\mathbf{\psi}}_r^k \tilde{\mathbf{\Gamma}}_r^{kT}$), the two adjusting factors can be calculated by

(1)

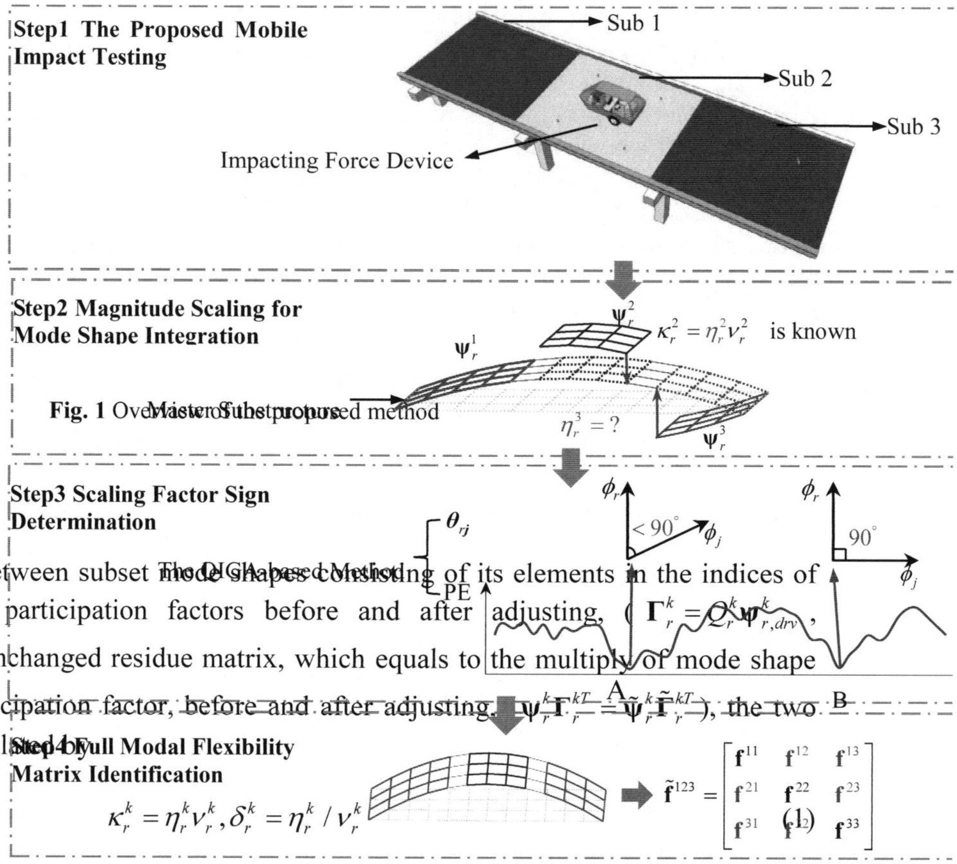

Fig. 1 Overview of the proposed method

where $v_r^k = \sqrt{Q_r^k / Q_r^{master}}$ is the scaling magnitudes which can be calculated, and η_r^k is their undetermined sign factor, i.e. $\eta_r^k = 1$ or -1. If η_r^k is known, the rth mode shape $\tilde{\mathbf{\psi}}_r$ of the entire structure can be assembled from the mode shapes of substructures,

$$\tilde{\mathbf{\psi}}_r = \mathbf{T_r} \mathbf{\psi}_r \tag{2}$$

where $\mathbf{\psi}_r = \{\psi_r^{1T} \quad \cdots \quad \psi_r^{mT}\}^T \in \mathbb{R}^{N_o \times 1}$ is assembled from the substructures' mode shapes which are not at the same scaling level; m is the number of substructures; $\mathbf{T_r} = \mathbf{v}_r \mathbf{\eta}_r$ is the transfer matrix. From the derivation process presented above, it should be noted that the sign factors, η_r^k, are still unknown so far though they are either 1 or -1, which means that the orientation of each substructure's scaled mode shape has not yet been determined. This phenomenon is shown in Sub 3 in the Step 2 of Fig. 1.

2.2 Scaling factor sign determination

A data processing method is proposed to determine the unknown sign factor, η_r^k. It is well known that the accurate displacement of a structure induced by static forces will minimize the structure's potential energy (PE), which can be calculated by

$$\pi_p(\boldsymbol{\eta}_r)=\frac{1}{2}\sum_{r=1}^{N}\left[(\mathbf{F}^{\mathrm{T}}\boldsymbol{\eta}_r\mathbf{v}_r\boldsymbol{\psi}_r)^2\left(\frac{Q_r^{\text{master}}}{\lambda_r^{\text{master}}}+\frac{Q_r^{\text{master*}}}{\lambda_r^{\text{master*}}}\right)\right] \tag{3}$$

where $\mathbf{F}=\left\{\mathbf{F}^{1T} \quad \cdots \quad \mathbf{F}^{mT}\right\}^T \in \mathbb{R}^{N_o\times 1}$ is the static force vector applied on each output node of the entire structure and correspondingly; N is the number of mode orders; The symbol "*" denotes complex conjugate. It is seen that the sign factors can be determined by minimizing the structure's potential energy from Eq. (3). However, when the searching process of the sign factors is carried out, multiple local minimum values even the ill-conditioned minimum that do not obey common sense such as the continuity or basic physical laws due to the noise involved in the measurement. This phenomenon is shown in the Step 3 of Fig. 1. To overcome this problem, the orthogonality of mode shapes is used as constrained conditions of the optimization problem. Mode shapes have the following M-orthogonal: $\tilde{\boldsymbol{\psi}}_r^T\mathbf{M}\tilde{\boldsymbol{\psi}}_j\begin{cases}\neq 0, r=j\\ =0, r\neq j\end{cases}$, where $\mathbf{M}$ is the mass matrix. Assuming the mass distribution is uniform, it can be simplified as: $\tilde{\boldsymbol{\psi}}_r^T\tilde{\boldsymbol{\psi}}_j\begin{cases}\neq 0, r=j\\ =0, r\neq j\end{cases}$. The orthogonal angle between the vectors $\tilde{\boldsymbol{\psi}}_r$ and $\tilde{\boldsymbol{\psi}}_j$ is defined as

$$\theta_{rj}=\arccos\left(\frac{\left|\tilde{\boldsymbol{\psi}}_r^{\mathrm{T}}\tilde{\boldsymbol{\psi}}_j\right|}{\left\|\tilde{\boldsymbol{\psi}}_r\right\|\left\|\tilde{\boldsymbol{\psi}}_j\right\|}\right)(r\neq j) \tag{4}$$

where $\left\|\tilde{\boldsymbol{\psi}}_r\right\|$ is the ℓ_2 norm of the vector $\tilde{\boldsymbol{\psi}}_r$; the notation $|\cdot|$ represents the absolute value of variable. The orthogonal angle θ_{rj} can be used to measure the degree of orthogonality between two different assembled global mode shapes. Therefore, the following constraint conditions are designed for the optimization problem defined in Eq. (3) in order to distinguish the feasible region from infeasible one,

$$\vartheta_r=\min_{j\neq r}\left(\theta_{rj}\right)\geq\vartheta_{\min};\mu_r=\frac{1}{N-1}\sum_{j=1,j\neq r}^{N}\theta_{rj}\geq\mu_{\min};\sigma_r=\sqrt{\frac{1}{N-1}\sum_{j=1,j\neq r}^{N}\left(\theta_{rj}-\mu_r\right)^2}\leq\sigma_{\max} \tag{5}$$

where ϑ_r, μ_r and σ_r are the minimum orthogonal angle, the mean value and the standard deviation of orthogonal angle for the rth mode respectively. $\vartheta_{\min}$, $\mu_{\min}$ and $\sigma_{\max}$ are their corresponding threshold values provided in advance by expertise. In this paper, they are set to be 70°, 85° and 10° respectively.

2.3 Full modal flexibility matrix identification

Once the sign factors are determined by solving the optimization problem described above, the assembled mode shape of the structure can be integrated from the mode shapes of substructures through Eq. (2). In light of the importance of the master substructure, its system pole $\lambda_r^{\text{master}}$ and modal scaling factor Q_r^{master} together with $\tilde{\boldsymbol{\psi}}_r$ are utilized to calculate the full modal flexibility matrix $\tilde{\mathbf{f}}^S$ as followings:

$$\tilde{\mathbf{f}}^S = \sum_{r=1}^{N} \left(\frac{Q_r^{\text{master}} \tilde{\boldsymbol{\psi}}_r \tilde{\boldsymbol{\psi}}_r^{\mathrm{T}}}{-\lambda_r^{\text{master}}} + \frac{Q_r^{\text{master}*} \tilde{\boldsymbol{\psi}}_r^* \tilde{\boldsymbol{\psi}}_r^{*\mathrm{T}}}{-\lambda_r^{\text{master}*}} \right) \tag{6}$$

It is seen that the left question to identify the full modal flexibility matrix of the structure is how to get the sign factors by solving the constrained optimization problem. Because there are $N(m-1)$ sign factors when excluding the ones of the master substructure, there are $2^{N(m-1)}$ possible values for $\pi_p(\boldsymbol{\eta}_r)$ considering all possible combinations of these sign factors. It is obvious that the computation efficiency will be very low if enumerating all values of PE when the scale of the question becomes large. A modified quantum-inspired genetic algorithm (QIGA) is developed in next section to solve the optimization problem.

3 The QIGA-based algorithm for solving the optimization problem

The QIGA is a probabilistic evolutionary algorithm that embeds the quantum computation into the classical genetic algorithm (GA). In QIGA, the basic unit is the quantum bit (Q-bit). It is usually expressed with $[\alpha \ \beta]^T$ or the BraKet notation $|\psi\rangle = \alpha|0\rangle + \beta|1\rangle$, in which α^2 and β^2 are the probability of appearing in the state 0 and the state 1 respectively and they follow the principle of unit normalization $\alpha^2 + \beta^2 = 1$. The position represented by a Q-bit may locate in the state 0, the state 1, or in a linear superposition of both, which is different from classical state.

It is worth to mention that important parameters including the rotation angle, the probabilities of crossover and mutation greatly influence the performance of the QIGA. In order to improve the computational efficiency of the optimization problem, the following works dedicate to design the dynamic rotation angle, the adaptive crossover and mutation probabilities.

3.1 The dynamic quantum rotation gate

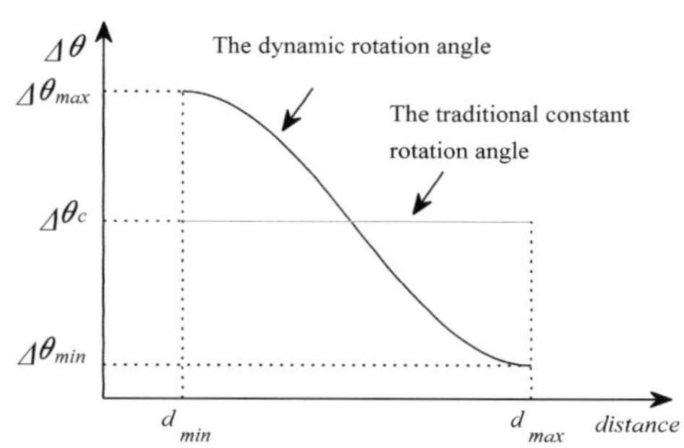

Fig. 2 The dynamic rotation angle line

For conventional genetic quantum algorithm, the rotation angle $\Delta\theta$ is a constant, which reduces the possibility to escape from local optima. In order to overcome this shortcoming, a dynamic rotation angle is designed according to the pattern shown in Fig. 2. It can be seen that this strategy of dynamic rotation angle is able to change dynamically with the fitness of the individuals. Mathematically, if an individual's encoding is far away from the best individual's one, a relative large rotation angle will work to search the optima for the objective function at a large stride on the whole search region. By contrast, when an individual's encoding approaches to the best individual's one, a relative small rotation angle is needed in order to carefully search the local optimal region. The following specific formula just matches with the strategy of the proposed dynamic rotation angle.

$$\Delta\theta_i = a + b \times \sin\left(c + d \times d_i\right) \tag{7}$$

where $a = (\Delta\theta_{max} + \Delta\theta_{min})/2$, $b = -(\Delta\theta_{max} - \Delta\theta_{min})/2$, $c = -\pi(d_{max} + d_{\min})/2(d_{max} - d_{\min})$ and $d = \pi/d_{max} - d_{mean}$. $\Delta\theta_{\max}$ and $\Delta\theta_{\min}$ are the upper and the lower limits of the dynamic quantum rotation angle respectively. In this paper, $\Delta\theta_{max} = 0.05\pi$ and $\Delta\theta_{min} = 0.001\pi$ are used; $d_i = 1 - h_i/L$, where h_i is the Hamming distance between the genes of the individual and the genes of the best one, L is the length of the individual chromosome.

3.2 The adaptive crossover (mutation) strategy

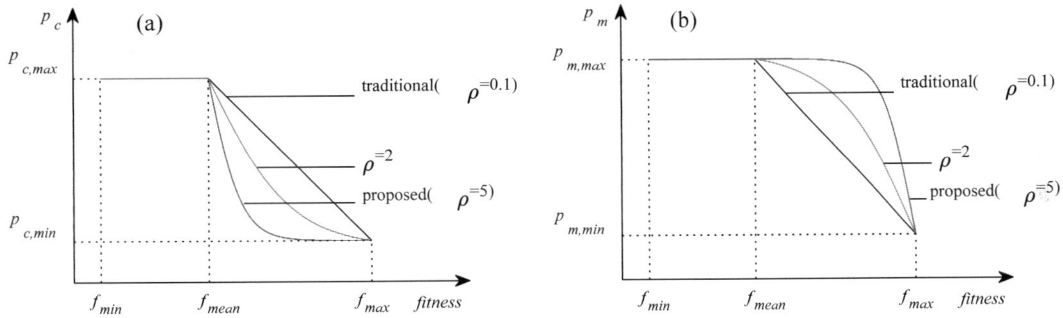

Fig. 3 The variation operators with different bending parameter ρ ; (a) The crossover strategy; (b) The mutation strategy;

Although there are some methods to design the adaptive crossover and mutation probabilities, difference between the traditional and the proposed strategy are illustrated in Fig. 3. It can be clearly seen that the proposed adaptive operators in this paper are not only in response to the fitness values of the solutions adaptively, but also have the exponential smooth transition properties. The adaptive crossover strategy (Fig. 3(a)) can provide the possibility of the diversity of the population during the individual approaches to the best individual in the population. Also the adaptive mutation strategy (Fig. 3(b)) can provide the possibility to jump out the local optimum such that the population can evolve to a better state. The proposed adaptive crossover strategy is defined as,

$$p_{c,i} = \begin{cases} p_{c,max}, \; f_i < f_{mean} \\ p_{c,max} - \dfrac{p_{c,max} - p_{c,min}}{\tanh(\rho)} \times \tanh\left(\dfrac{\rho(f_i - f_{mean})}{f_{max} - f_{mean}}\right), f_i \geq f_{mean} \end{cases} \tag{8}$$

where $p_{c,max}$ and $p_{c,min}$ are the upper and the lower limits of the crossover probability respectively, and $p_{c,max} = 0.9$ and $p_{c,min} = 0.5$ are used in this paper; f_i, f_{mean} and f_{max} are the fitness value of the individual, the best individual and the average fitness value of the population respectively; $\tanh(\cdot)$ is the Hyperbolic Tangent Function, $\rho(\rho > 0)$ is an accelerating factor to regulate the evolution speed of the population, and $\rho = 5$ is used in this paper. It should be noted that a too large value of ρ is prohibited because it may cause the population to evolve in a local optima prematurely. The proposed adaptive mutation strategy is given as,

$$p_{m,i} = \begin{cases} p_{m,max}, f_i < f_{mean} \\ p_{m,min} + \dfrac{p_{m,max} - p_{m,min}}{\tanh(\rho)} \times \tanh\left(\dfrac{\rho(f_{max} - f_i)}{f_{max} - f_{mean}}\right), f_i \geq f_{mean} \end{cases} \quad (9)$$

where $p_{m,max}$ and $p_{m,min}$ are the upper and the lower limits of the mutation probability respectively, and $p_{m,max} = 0.2$ and $p_{m,min} = 0.01$ are used in this paper. The factor ρ in Eq. (9) has the same meaning and value as it in Eq. (8).

3.3 The fitness function

The objective function of the minimum PE subjected to the constraint conditions of the orthogonal angles of the assembled global mode shapes has been proposed in section 2.3. But how to use penalty techniques to transform the constraint conditions into an unconstrained optimization problem is still an essential issue in order to design a reasonable fitness function. This paper proposes a fitness function to determine the sign factors by adopting penalty techniques. According to the constraint conditions in Eq. (5), define the following three penalty factors:

$$p_1 = \sum_{r=1}^{N} \frac{\left[\sin(\vartheta_r)\right]^{\zeta_r}}{2^r}; p_2 = \sum_{r=1}^{N} \frac{\left[\sin(\mu_r)\right]^{\zeta_r}}{2^r}; p_3 = \sum_{r=1}^{N} \frac{\left[\cos(\sigma_r)\right]^{\zeta_r}}{2^r} \quad (10)$$

where ζ_r is the factor to control the degree of punishment. It is obvious that the three penalty factors, p_1, p_2 and p_3 are less than 1. But for the points of the feasible region, their orthogonal angles will be approximately perpendicular; so the three penalty factors will be close to 1. In contrast, for the points of the infeasible region, their orthogonal angles will be far away from perpendicular; thus the three penalty factors will be correspondingly far away from 1. Through this strategy, the feasible and infeasible regions can be discriminated to some extent. Due to the value of PE is usually a negative quantity and yet the fitness value should be better positive, the fitness function is defined as follows,

$$F(\boldsymbol{\eta}_r) = -p_1 p_2 p_3 \pi_p(\boldsymbol{\eta}_r) \quad (11)$$

4 Experiment example of a steel-concrete composite slab

To verify the effectiveness of the proposed method, the mobile impact test of a steel-concrete composite bridge model as shown in Fig. 4 was performed. The model has an overall length of 4.0 m and a width of 2.05 m. Three I-shaped Q235 steel girders were constructed as longitudinal beams, connecting with concrete deck by cheese head studs. The structure was simply supported by three rolling supports (nodes 1, 10, 19) and three fixed supports (nodes 9, 18, 27). A PCB medium size impact hammer was used to impact the structure. Fifteen ICP accelerometers (0.5Hz~7000Hz, acceleration<100g) were mounted on the nodes of the slab, and a DP730 data acquisition system was used to acquire the impact test data.

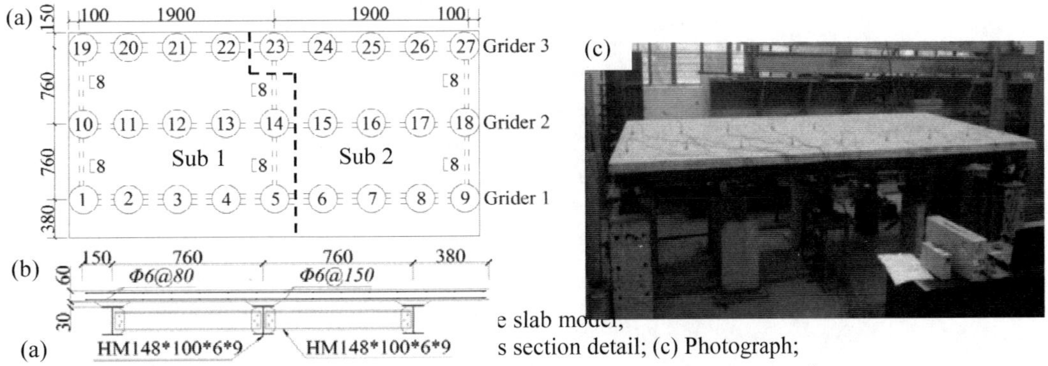

(a) ...e slab model; ...s section detail; (c) Photograph;

As shown in Fig. 4(a), the steel-concrete composite bridge model was subdivided into two substructures. The hammer impact test was sequentially performed on substructures. Fourteen accelerometers were deployed on all nodes of Sub 1, and the impacting forces were applied to all nodes in turn except the rolling supporting nodes 1, 10 and 19. Sub 2 was tested by using thirteen accelerometers deployed on all nodes within this substructure and all nodes except the fixed supporting nodes 9, 18 and 27 were impacted. The technologies of band-width filtering and windowing in time domain were used to reduce noises and leakages of the collected impacting forces and accelerations. The complex mode indicator function (CMIF) method was employed to process the impact test data for structural modal parameter identification.

The frequencies, damping ratios, modal scaling factors, mode shapes and modal participation factors of the two substructures were identified. Choosing Sub 1 as the master substructure, for the first eleven modes identified, there are eleven sign factors needed to be determined. The magnitude adjustment factors and the sign factors of the eleven modes of Sub 2 are listed in Table 1. Modal Assurance Criterion (MAC) values calculated by the assembled global mode shapes and those from the traditional impact test method are also provided in Table 1. Fig. 5 shows the process of assembling substructures' mode shapes of the first and the second mode.

The full modal flexibility matrix of the steel-concrete composite slab model was identified as shown in Fig. 6(a). The computational time of the enumeration method and the proposed QIGA method to solve the optimization problem are compared in Fig. 6(b), which illustrates the high efficiency of the proposed QIGA method.

Table 1 Magnitude adjustment factor and sign factor of the experimental example

Mode No.	v_r^2	η_r^2	MAC	Mode No.	v_r^2	η_r^2	MAC
1	0.84	1	0.9993	7	0.87	1	0.9934
2	1.13	1	0.9986	8	1.18	1	0.9992
3	-1.05	-1	0.9856	9	1.13	1	0.9993
4	0.96	1	0.9988	10	0.88	1	0.9957
5	-1.10	-1	0.9714	11	-0.91	-1	0.9878
6	-1.07	-1	0.9857	--	--	--	--

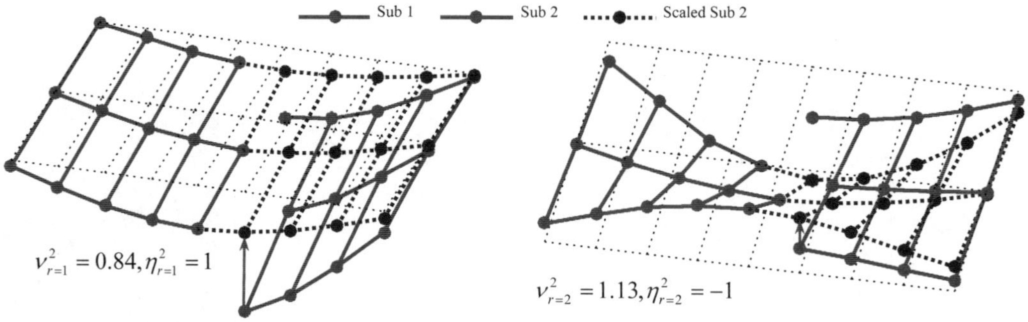

Fig. 5 The process of assembling mode shapes: (a) the first mode; (b) the second mode

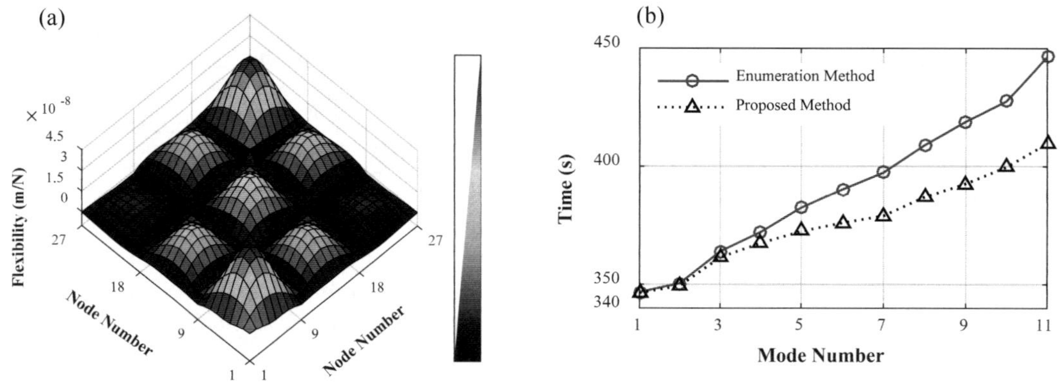

Fig. 6 The identified full modal flexibility matrix and the total time of the enumeration method and the proposed QIGA method

The performance of the QIGA is further compared with that of Quantum Genetic Algorithm (QGA) and GA to verify its effectiveness. In the comparison, 100 independent runs were executed for each algorithm. The probability of crossover and mutation were set to be 0.8 and 0.05 respectively in the GA method. The traditional constant quantum rotation angle strategy was used in the QGA method, in which the rotation angle is $\Delta\theta=0.01\pi$. The results of computations from the proposed QIGA, QGA and GA are shown in Table 2, in which the column of "Evolution No." denotes the average generation converging to the global optimum in the evolving process. It is clearly seen from Table 2 that the QIGA converges to the global optimum with a more fast speed than QGA and GA.

Table 2 Results comparison from the QIGA, QGA and GA methods

Run times	Test case	Algorithm	Fitness value				Evolution No.
			Best	Worst	Mean	Accuracy (%)	
100	case 1	GA	68.52	47.88	65.61	59	40
	10 individuals	QGA	68.52	68.52	68.52	100	28
	100 generations	QIGA	68.52	68.52	68.52	100	11
	case2	GA	68.52	68.52	68	97	18
	50 individuals	QGA	68.52	68.52	68.52	100	10

100 generations	QIGA	68.52	68.52	68.52	100	5

5 Conclusions

This paper presents a new mobile impact testing method for structural flexibility identification, in which the key problem of determining the sign factor is transformed to a constrained optimization problem and it is solved by the developed QIGA method.

The QIGA has been improved by developing the adaptive crossover (mutation) strategy, dynamic quantum rotation gate. Compared to the traditional GA and QGA, the developed QIGA method can explore the search space with a smaller number of individuals and exploit the search space for a global solution within a short span of time, thus it is proved to be much more efficient for determining the sign factors.

An experimental example of a steel-concrete composite slab has been studied, and the results successfully verified the effectiveness of the proposed method by comparing with that of the traditional MRIT method for structural flexibility identification and deflection prediction.

References

ASCE SEI Committee on Structural Identification of Constructed Systems. (2012). "Structural Identification of Constructed Facilities: Approaches, Methods and Technologies for Effective Practice of St-Id.". F. N. Catbas, T. Kijewski-Correa, and A. E. Aktan, eds., ASCE, Reston, VA.

Brownjohn, JM, Magalhaes, F, Caetano, E & Cunha, A 2010, 'Ambient vibration re-testing and operational modal analysis of the Humber Bridge', *Engineering Structures*, 32(8), 2003-2018.

Zhang, J, Prader, J, Grimmelsman, KA, Moon, F, Aktan, AE & Shama, A. 2013, 'Experimental Vibration Analysis for Structural Identification of a Long-Span Suspension Bridge', *Journal of Engineering Mechanics-ASCE*, 139(6), 748-759.

Brown, DL, & Witter, MC 2011, 'Review of Recent Developments in Multiple Reference Impact Testing (MRIT)', *Structural Dynamics*, 45(1), 8-17.

Catbas, FN, Brown, DL & Aktan, AE 2004, 'Parameter estimation for multiple-input multiple-output modal analysis of large structures', *Journal of Engineering Mechanics-ASCE*, 130(8), 921-930.

Zhang, J, Guo, S & Chen, XS 2014, 'Theory of un-scaled flexibility identification from output-only data', *Mechanical Systems and Signal Processing*, 48(1), 232-246.

Lynch, JP 2002, 'Decentralization of wireless monitoring and control technologies for smart civil structures', Ph.D. *Dissertation of Civil and Environmental Engineering*, Stanford University, CA, USA.

Sadhu, A, Narasimhan, S & Goldack, A 2014, 'Decentralized Modal Identification of a Pony Truss Pedestrian Bridge Using Wireless Sensors', *Journal of Bridge Engineering*, 19(6).

Zhang, JZ, Guo, S & Zhang, Q 2015, 'Mobile Impact Testing for Structural Flexibility Identification with only a Single Reference', *Computer-aided Civil and Infrastructure Engineering*, 30(9), 703-714.

Han, K & Kim, J 2000, 'Genetic quantum algorithm and its application to combinatorial optimization problem', *congress on evolutionary computation, IEEE*,2000,pp.1354–1360.

Mencia, R, Sierra, MR, Mencia, C & Varela, R 2016, 'Genetic algorithms for the scheduling problem with arbitrary precedence relations and skilled operators', *Integrated Computer-aided Engineering*, 23(3), 269-285.

Putha, R, Quadrifoglio, L & Zechman, EM 2012, 'Comparing Ant Colony Optimization and Genetic Algorithm Approaches for Solving Traffic Signal Coordination under Oversaturation Conditions', *Computer-aided Civil and Infrastructure Engineering*, 27(1), 14-28.

The 8th International Conference on
Structural Health Monitoring of Intelligent Infrastructure
Brisbane, Australia | 5-8 December 2017

Entropy-based Damage Detection for Three-dimensional Structures

T.K. Lin[1], T.C. Tseng[2], A.G. Laínez[2].
[1]*National Chiao Tung University – Taiwan, Email:* tklin@nctu.edu.tw, [2]*National Chiao Tung University – Taiwan*

Abstract

A three-dimensional structural health monitoring (SHM) system based on multiscale cross-sample entropy (MSCE) is proposed. Ambient vibration signals on the roof of a structure are measured, and vertical MSCE analysis is applied to evaluate the vertical damage location by analyzing individual signals on different floors. Results are quantified using the vertical damage index (DI). The damage orientation of damaged floors is then detected by analyzing the biaxial signals in four directions of each damaged floor through the planar MSCE analysis. Results are physically quantified using the planar DI. According to the results obtained from the evaluation of the proposed system on a three-dimensional structure, the damage condition and elevation were reliably diagnosed. Moreover, average accuracy rates of 93% (vertical) and 91% (planar) were achieved through the proposed DI method. Therefore, structural damage can be reliably detected after major earthquakes.

1. Introduction

Structural health monitoring (SHM) is emerging as a popular research area in civil engineering. SHM techniques can be employed for the periodical inspection of aging structures and for post-disaster damage detection and reinforcement. In past decades, signal processing techniques have been applied in SHM methods to process the measured displacement, velocity, or acceleration signals of structures to obtain dynamic characteristics such as basic vibration frequencyand damping. Thus, damage existence, severity, and possible locations can be detected (Doebling et al. 1998).

The concept of entropy was first introduced by German physicist Clausius in 1865 to evaluate the uncertainty of events in a thermodynamic system. In 1991, an analytical method called "approximate entropy" (ApEn) was developed by Pincus (1991). ApEn can be used to statistically determine regularity in real-world time series. In 2000, Richman and Moorman (2000) proposed a modification of ApEn called "sample entropy" (SampEn). The advantage of SampEn is that the entropy value obtained is not affected by the length of the time series. In 2002, multiscale entropy (MSE) analysis was proposed and subsequently validated through clinical experiments (Costa et al. 2002, 2005). Cross-sample entropy (Cross-SampEn) was developed to evaluate the degree of synchronicity or similarity between a pair of cardiovascular time series (Richman & Moorman 2000). In 2015, Lin and Liang (2015) proposed an SHM system based on multiscale cross-sample entropy (MSCE). Following MSCE analysis, specific locations were determined. In the same year, Rojas *et al.*(2015) developed a damage index based on the entropy distribution using short time wavelet entrropy of the ultrasonic waves to detect artificial damage on a plate-like structure.

According to the results of previous studies, an SHM system was developed in the present study; this system employs the MSE and MSCE methods to analyze the dynamic response signals of a numerically simulated high-rise structure. Moreover, vertical and planar analyses were conducted to diagnose the damage of the whole structure.

2.The Proposed SHM Algorithm

2.1 Cross-Sample Entropy

Cross-SampEn is utilized to evaluate the degree of asynchrony or dissimilarity between two time series derived from the same system.
Let $\{X_i\} = \{x_1, \ldots, x_i, \ldots, x_N\}$ and $\{Y_j\} = \{y_1, \ldots, y_j, \ldots, y_N\}$ represent two individual time series of length N. The signals are segmented into the following templates of length m: $u_m(i) = \{x_i, x_{i+1}, \ldots, x_{i+m-1}\}$, $1 \leq i \leq N - m + 1$ and $v_m(j) = \{y_j, y_{j+1}, \ldots, y_{j+m-1}\}$, $1 \leq j \leq N - m + 1$. The template spaces T_x and T_y are presented as:

$$T_x = \begin{bmatrix} x_1 & x_2 & \cdots & x_m \\ x_2 & x_3 & \cdots & x_{m-1} \\ \vdots & \vdots & \ddots & \vdots \\ x_{N-m+1} & x_{N-m+2} & \cdots & x_N \end{bmatrix} T_y = \begin{bmatrix} y_1 & y_2 & \cdots & y_m \\ y_2 & y_3 & \cdots & y_{m-1} \\ \vdots & \vdots & \ddots & \vdots \\ y_{N-m+1} & y_{N-m+2} & \cdots & y_N \end{bmatrix} \quad (1)$$

The degree of similarity between templates $u_m(i)$ and $v_m(j)$ is defined as $n_i^m(r)$ and is calculated under the criterion of:

$$d[u_m(i), v_m(j)] \leq r, \ 1 \leq j \leq N - m \quad (2)$$

The similarity probability of the templates can be evaluated as follows:

$$U_i^m(r)(v\|u) = \frac{n^m(r)}{(N - m)} \quad (3)$$

The average similarity probability of length m can be calculated using the following equation:

$$U^m(r)(v\|u) = \frac{1}{(N - m)} \sum_{i=1}^{N-m} U_i^m(r)(v\|u) \quad (4)$$

where $U^m(r)(v\|u)$ is the degree of dissimilarity between the two time series.
New template spaces T_x and T_y are created by assembling templates with length $m + 1$, and the average similarity probability $U^{m+1}(r)(v\|u)$ is used to derive the Cross-SampEn values as:

$$\mathrm{CS}_E(m, r, N) = -\ln\left\{\frac{U^{m+1}(r)(v\|u)}{U^m(r)(v\|u)}\right\} \quad (5)$$

2.2 Multi-Scale Entropy

MSE analysis is defined as the process of converting an original signal into signals at different time scales through coarse-graining. After completion of the process, the entropy values for each time scale are calculated using SampEn. The procedure is described as follows: A time series $x_1, x_2, \ldots, x_N$ of length N is segmented into multiple time series with a length of τ points, where τ is the scale factor. Subsequently, each set of data values is averaged, and a new time series $\{y_j^{(\tau)}\}$ is constructed. Each element is calculated according to the following equation:

$$y_j^{(\tau)} = \frac{1}{\tau} \sum_{i=(j-1)\tau+1}^{j\tau} x_i \, , \; 1 \leq j \leq N/\tau \tag{6}$$

SampEn is calculated for each coarse-grained time series $\{y_j^{(\tau)}\}$. The S_E values for each time scale is the MSE of the time series. Finally, the S_E values are plotted as a function of the scale factor ($f(\tau) = S_E$).

2.3 Vertical and Planar Damage Index

For a structure with N stories, the MSCE plot of each floor is expressed as the cross-sample curves of each adjacent floor to the Nth story, where each curve depicts the Cross-SampEn at different scales.

H and D represent the MSCE curves for the healthy and damaged conditions of the structure, respectively. The subscripts depict the analyzed floor; for example, H_1 is the MSCE between the ground and first floor of the healthy structure. After MSCE analysis, the resulting curve illustrates the single-axis vertical characteristics of the first floor, expressed as $H_1 = \{\mathrm{CS}_{E\,H_1}^{1}, \mathrm{CS}_{E\,H_1}^{2}, \mathrm{CS}_{E\,H_1}^{3}, \cdots, \mathrm{CS}_{E\,H_1}^{\tau}\}$, where CS_E is the Cross-SampEn value, the superscript τ is the scale factor and the subscript number is the analyzed floor. Thus, the MSCE of each floor can be expressed as follows:

$$D_F = \{\mathrm{CS}_{E\,D_F}^{1}, \mathrm{CS}_{E\,D_F}^{2}, \mathrm{CS}_{E\,D_F}^{3}, \cdots, \mathrm{CS}_{E\,D_F}^{\tau}\} \tag{7}$$

Subsequently, the following formula can be used to calculate the DI:

$$\mathrm{DI}_F = \sum_{q=1}^{\tau} (\mathrm{CS}_{ED_F}^{q} - \mathrm{CS}_{EH_F}^{q}) \tag{8}$$

where F is the number of the floor to be evaluated for damage.

The DI for a single floor is evaluated by calculating the difference between the MSCE values of the damaged and healthy structures. For a specific floor, a positive DI value indicates the existence of damage on the floor, whereas a negative value indicates a lack of damage.

After the damaged floor has been determined, a planar analysis of the floor can be conducted to further determine the damage direction. The structural model in this study is simulated with bracings on all four sides. Floor damage is represented by the removal of bracings; therefore, planar analysis of the structure was performed in both x- and y-directions. The velocity responses at the center of the north (N), south (S), east (E), and west (W) directions on each floor were thus simulated to reflect the characteristics of the structure.

The two sets of MSCE curves representing the signals of the N, W, S, and E directions on the x- and y-axes of a specific floor to be analyzed in the healthy structure can be expressed in matrices as:

$$\mathrm{MSCE}_{\mathrm{undamaged}} = \{H_N^x, H_W^x, H_S^x, H_E^x\} \qquad \mathrm{MSCE}_{\mathrm{undamaged}} = \{H_N^y, H_W^y, H_S^y, H_E^y\} \tag{9}$$

Similarly, the matrices for the damaged condition of the structure can be expressed as:

$$\mathrm{MSCE}_{\mathrm{damaged}} = \{D_N^x, D_W^x, D_S^x, D_E^x\} \qquad \mathrm{MSCE}_{\mathrm{damaged}} = \{D_N^y, D_W^y, D_S^y, D_E^y\} \tag{10}$$

where H and D represent the healthy and damaged conditions of the structure, respectively; the subscript N, W, S, and E represent the directions of the analyzed floor; and x and y represent the analyzed axes. For example, H_N^x represents the MSCE results of the signal in the N direction of the x-axis of the analyzed floor and the signal of the floor beneath it under healthy conditions. This can be further expressed as $H_N^x = \{CS_{E\,HNx}^{1}, CS_{E\,HNx}^{2}, CS_{E\,HNx}^{3}, \ldots, CS_{E\,HNx}^{\tau}\}$, where in each element, CS_E denotes the MSCE value, the superscript denotes the scale factor τ, the first subscript denotes the healthy condition, the second subscript denotes the analyzed direction, and the third subscript denotes the analyzed axis. Similarly, $D_{P\,axis}$ can be expressed as

$$D_{P\,axis} = \{CS_{E\,DPaxis}^{1}, CS_{E\,DPaxis}^{2}, CS_{E\,DPaxis}^{3}, \ldots, CS_{E\,DPaxis}^{\tau}\} \quad (11)$$

where the subscripts P and $axis$ denote the analyzed direction and axis, respectively.
The dual-axis planar DI can subsequently be calculated using the following formulas:

$$DI_{Px} = \sum_{q=1}^{\tau}(CS_{EDPx}^{q} - CS_{EHPx}^{q})\ DI_{Py} = \sum_{q=1}^{\tau}(CS_{EDPy}^{q} - CS_{EHPy}^{q}) \quad (12)$$

where P is the analyzed direction.
The DI is evaluated as in the vertical analysis. A positive DI value indicates the existence ofdamage on the floor, whereas a negative DI value indicates a lack of damage.

3. SHM Database

In the simulation of the seven-story steel structure, the two sides in the long axis direction of the structure were defined as E and W, and those in the short axis direction were defined as N and S. The four sides of each floor were fitted with a single diagonal brace to support the long and short axis directions. The damaged condition of the floor was denoted by the removal of a brace.The cross section of the columns was set as plate-type of 20 x 75 mm and the beam size was set as 100 x 70 mm.Detachable braces with a cross section defined as L-shaped steel angles measuring 65 x 65 x 6 mm were installed on each face of every floor. An additional mass of 500 kg was mounted on each story.
The details of the damage cases considered for biaxial vertical analysis and those considered for biaxial planar analysis are listed in Table 1. Numbers indicate damaged floors, and N, S, E, and W represent the damage directions.

Table 1. Damage cases for vertical and planar analyses

Case Number	Damage Group	Damaged Floor and Direction	Case Number	Damage Group	Damaged Floor and Direction
V1	Single-story	2W	P1	Single-story	5N
V2		5N	P2	Single-direction	5S
V3		6NES	P3		5W
V4	Two-story	2S7E	P4		5E
V5		3NS4WE	P5	Single-story	3NS
V6		5NE6SW	P6	Multidirectional	4WE
V7	Three-story	1NS4E7NS	P7		6NE
V8		1NES5W7W	P8		1SW
V9		2SW4WE6S	P9	Two-story	3NS4WE
V10		3N4W5S6E	P10	Multidirectional	5NE6SW
V11	Multistory	1NE3NS5SW7N	P11		4N5E6S
V12		2N4W5N6SE7N	P12	Multistory	1NES5W7W
			P13	Multidirectional	1NE3SW5SW7NE
			P14		2N4W5N6SE7N

(a) Diagonal bracing installation

(b) Removal of diagonal bracing

Fig. 1 Bracing configuration diagrams

4. Performance Evaluation

The MSE and MSCE methods were used to diagnose the damage condition of a 3D biaxial structure. After a series of optimization searches, the template length m, threshold r, and signal length N were determined as 3, 0.15 * standard deviation (SD), and 60,000, respectively.

4.1 Vertical Damage Locations

The velocity signals for the X (long) and Y (short) directions were extracted from the center of each floor for every damage condition. Figs. 3(a)-(b) present the MSCE diagrams derived for the healthy condition. Vertical MSCE analysis was performed after acquiring the time history in two directions for every floor. In these figures, G-1F denotes the curve for the first floor, 1F-2F denotes the curve for the second floor, 2F-3F denotes the curve for the third floor, and so forth.

4.1.1 Three-story damage: Case V8-1NES5W7W

The MSCE diagrams for Case V8 are shown in Fig. 3. The curves of the first floor in the Y-direction are all shown to rise as the scale increases. In the X-direction, the curves for the fifth and seventh floors are shown to increase at scales ranging between 1 and 10, whereas the trends of the other floors are shown to be either equal to that of the healthy condition or to decrease only slightly. According to the DI graphs shown in Fig. 4, the first floor in both X- and Y-directions and the fifth and seventh floors in the X-direction were damaged, which is consistent with the MSCE results.

4.1.2 Multistory damage: Case V10-3N4W5S6E

Fig. 5 shows the MSCE diagrams for Case V10. In the Y-direction, the curves for the third and fifth floors increase at scales ranging from 1 to 10. In the X-direction, the curves for the fourth and sixth floors slightly increase compared with those for the healthy condition. All the other curves remain in approximately the same positions as those of the graphs for the healthy condition in both the X- and Y-directions. This indicates that each floor was damaged only in one direction. The DI values shown in Fig. 6 reveal the existence of damage on the third and fifth floors in the Y-direction and on the fourth and sixth floors in the X-direction.The complete results of vertical DI analysis are summarized in Table 2. In the short (Y) direction, all the damaged floors were successfully identified for every analyzed case, whereas in the long (X) direction, one case was misidentified, lowering the accuracy of the analysis. The overall recognition accuracy in both directions was 91%.

4.2 Planar Damage Locations

To conduct planar MSCE analysis, the four signals in the X- and Y-directions of the damaged floor were analyzed by Cross-SampEn. Through the analysis of the discrepancies in all directions, the damage location could be obtained.

4.2.1 Single-story, single-direction damage: Case P1-5N

The MSCE diagrams for Case P1 are presented in Fig. 7. At scales ranging from 2 to 4, the curves gradually separate and differ from those for the healthy condition. From a scale of 5, the curve of the N side is shown to rise to the highest point, and the curves derived for the E, W and S sides are shown to remain lower than those derived for the healthy condition. Thus, damage occurred on the N side in the Y-direction of the structure. In the X-direction, the curves belonging to the four sides can be observed to exhibit no such evident changes, indicating no damage in this direction. Fig. 8 shows the DI results for Case P1. In the Y-direction, only the DI for the N side is observed to be positive, demonstrating that the damage was located only on the N side.

4.2.2 Two-story, multidirectional damage: Case P10-5NE6SW

The MSCE diagrams for case P10 are shown in Fig. 9. In Fig. 9(b), the curve for the N side is shown to rise, whereas that for the S side decreases. In the X-direction of the fifth floor (Fig. 9(a)), the trend

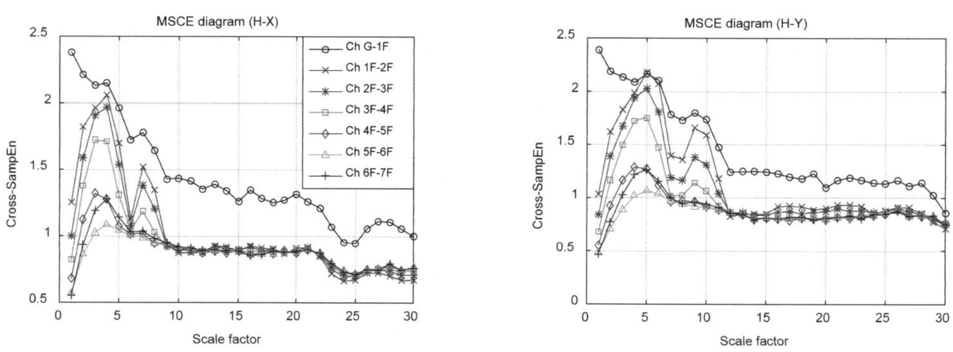

(a) (b)

Figure 2. MSCE diagramsfor the healthy condition: (**a**) X-direction; (**b**) Y-direction.

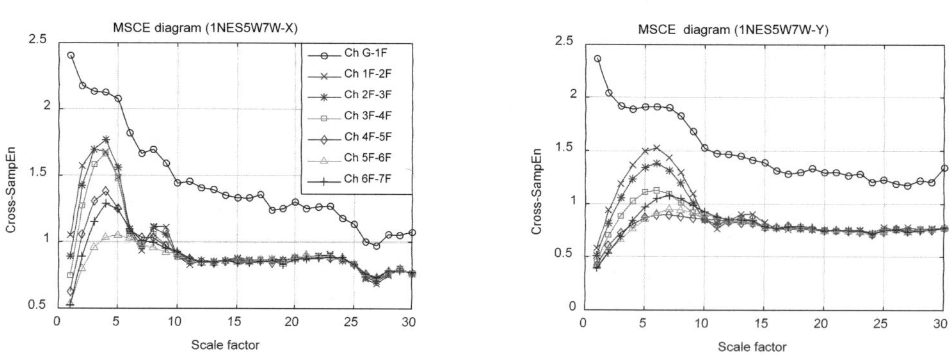

(a) (b)

Figure 3. MSCE diagrams of damage on the N, S, and E sides of the first floor, W side of the fifth floor, and W side of the seventh floor: (**a**) X-direction; (**b**) Y-direction.

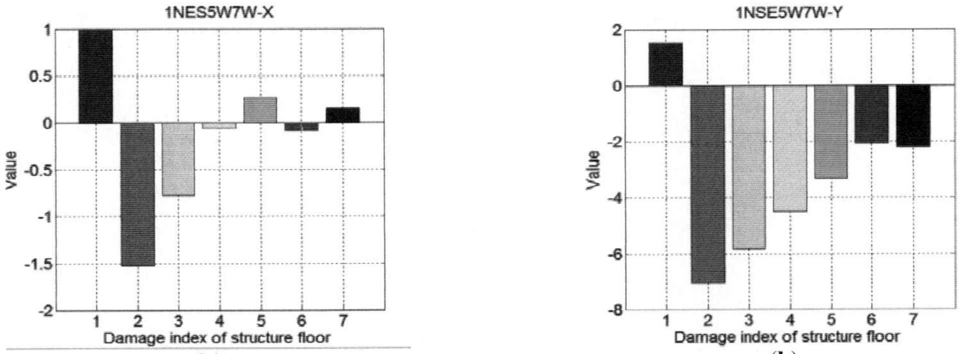

(a) (b)

Figure 4. DI diagram of damage on the N, S, and E sides of the first floor, W side of the fifth floor, and W side of the seventh floor: (**a**) X-direction; (**b**) Y-direction.

of the curve for the E side is significantly higher than those of the curves for the other sides. The analysis results obtained for the sixth floor are presented in Figs. 9(c)-(d). In the Y-direction, the trend of the curve S side is shown to rise sharply, and in the X-direction, the curve derived for the Wside is demonstrated to clearly rise.

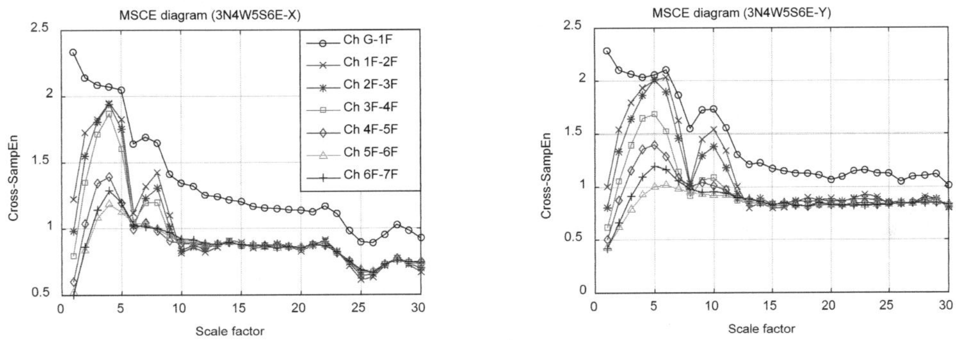

(**a**) X-direction (**b**) Y-direction

Figure 5. MSCE diagrams of damage on the N side of the third floor, W side of the fourth floor, S side of the fifth floor, and E side of the sixth floor.

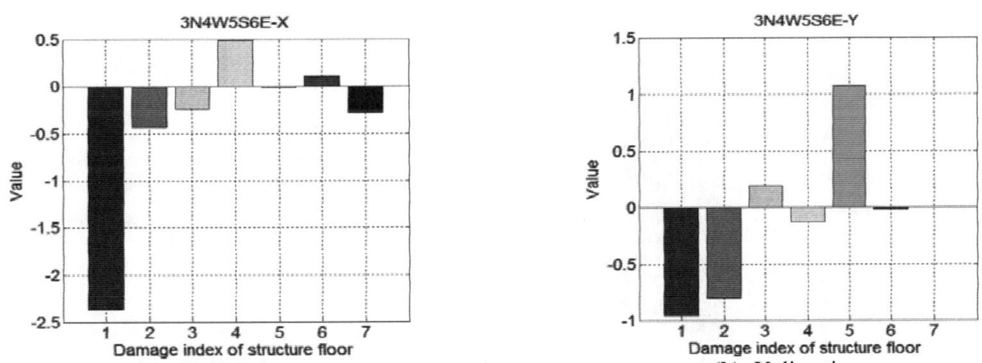

(**a**) X-direction (**b**) Y-direction

Figure 6. DI diagrams of damage on the N side of the third floor, W side of the fourth floor, S side of the fifth floor, and E side of the sixth floor.

Table 2. Identification accuracy of vertical DI analysis.

Case Number	Damaged Floor and Direction	DI (Y-axis)	DI (X-axis)	DI (Both axes)
0	H	✓	✓	✓
V1	5N	✓	✓	✓
V2	2W	✓	✓	✓
V3	6NES	✓	✓	✓
V4	2S7E	✓	✓	✓
V5	3NS4WE	✓	✓	✓
V6	5NE6SW	✓	✓	✓
V7	1NS4E7NS	✓	✓	✓
V8	1NES5W7W	✓	✓	✓
V9	2SW4WE6S	✓	✓	✓
V10	3N4W5S6E	✓	✓	✓
V11	1NE3NS5SW7NE	✓	×(6F&7F)	×(6F&7F)
V12	2N4W5N6SE7NS	✓	✓	✓
	Accuracy (%)	100%	91%	91%

✓ Indicates that damage on all floors was detected; × indicates that damage on some floors was not detected.

According to these results, damage can be inferred to occur on the N and E sides of the fifth floor and on the S and E sides of the sixth floor. The DI analysis results for Case P10 are illustrated in Fig. 10. Through the analysis of the fifth floor, positive DI values were observed for the N side in the Y-direction and E side in the X-direction. On the sixth floor, positive DI values were obtained for the S side in the Y-direction and W side in the X-direction, thereby confirming the MSCE results.

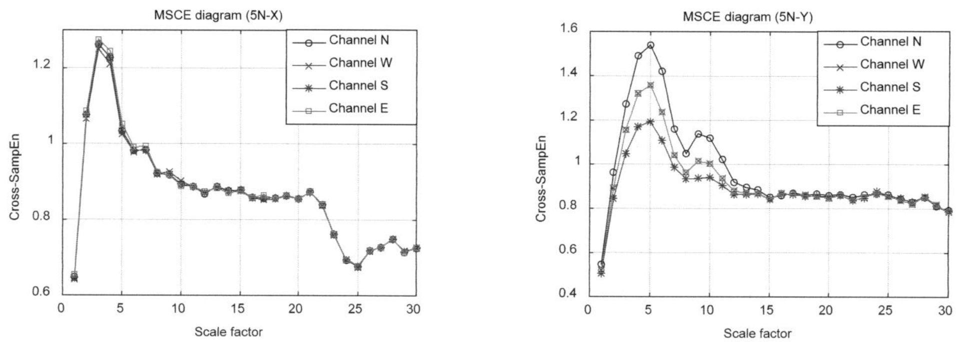

(a) X-direction (b) Y-direction

Figure 7. MSCE diagrams of damage on the N side of the fifth floor.

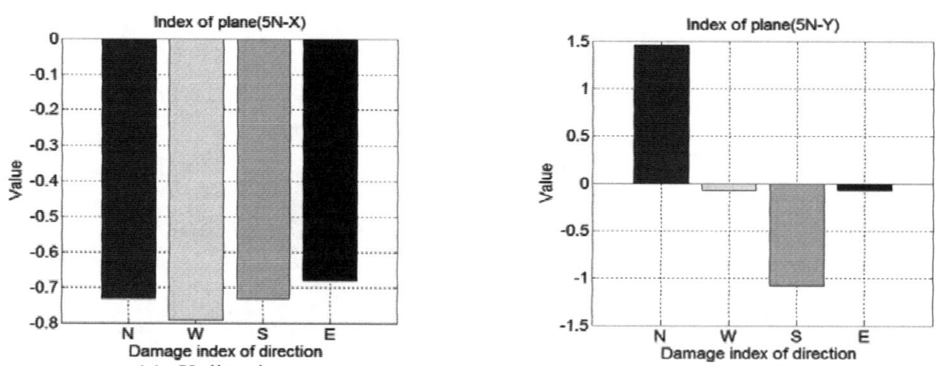

(a) X-direction (b) Y-direction

Figure 8. DI diagrams of damage on the N side of the fifth floor.

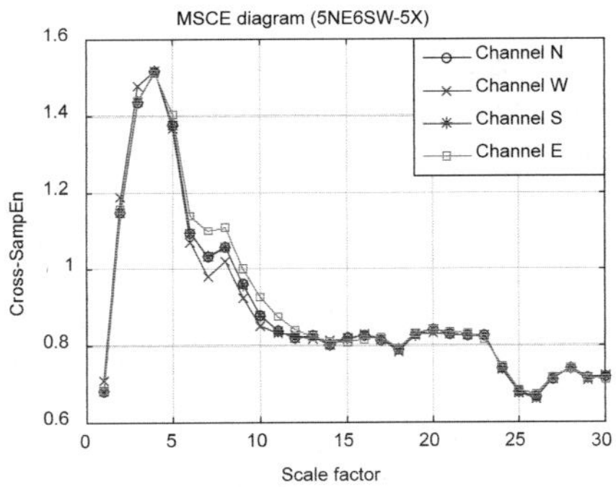

(a) fifth floor, X-direction

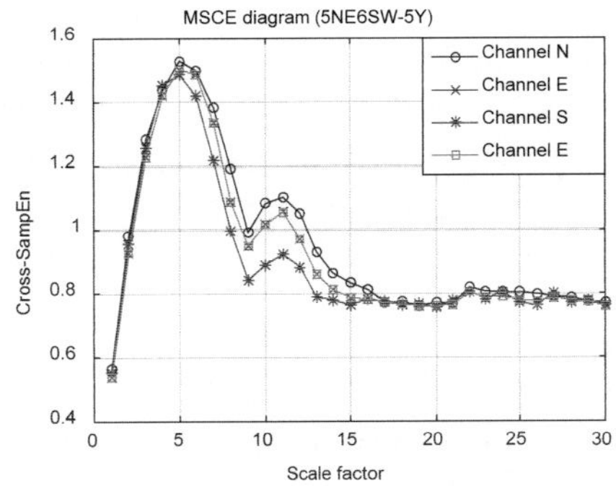

(b) fifth floor, Y-direction

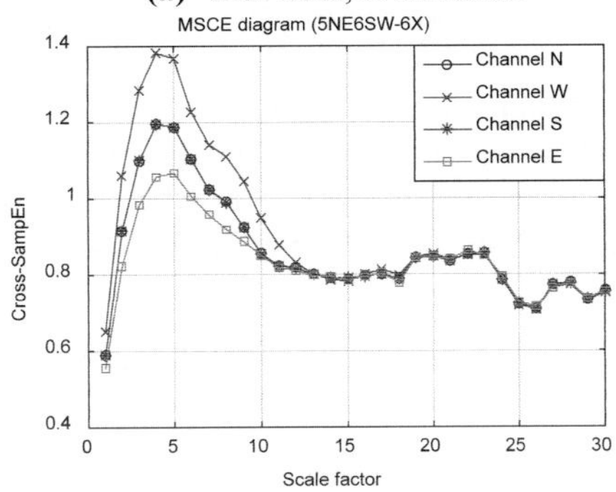

(c) sixth floor, X-direction

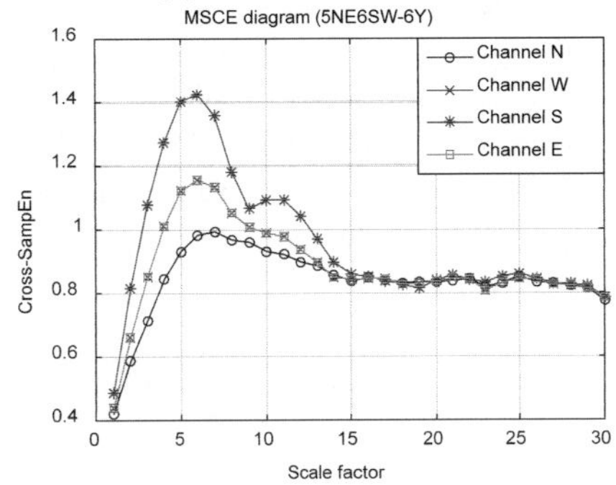

(d) sixth floor, Y-direction

Figure 9. MSCE diagrams for damage on the N and E sides of the fifth floor and S and W sides of the sixth floor.

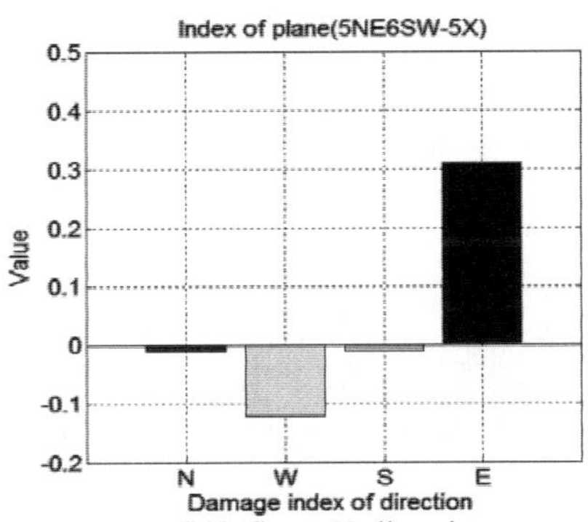

(a) fifth floor, X-direction

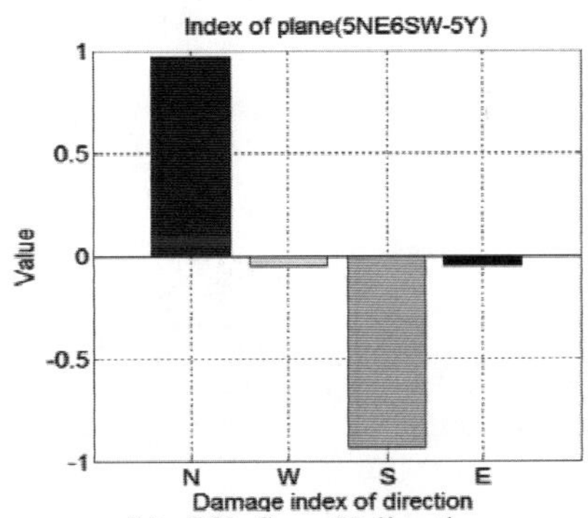

(b) fifth floor, Y-direction

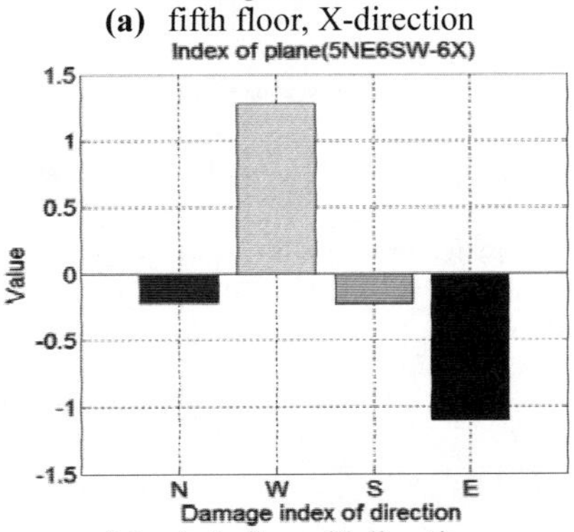

(c) sixth floor, X-direction

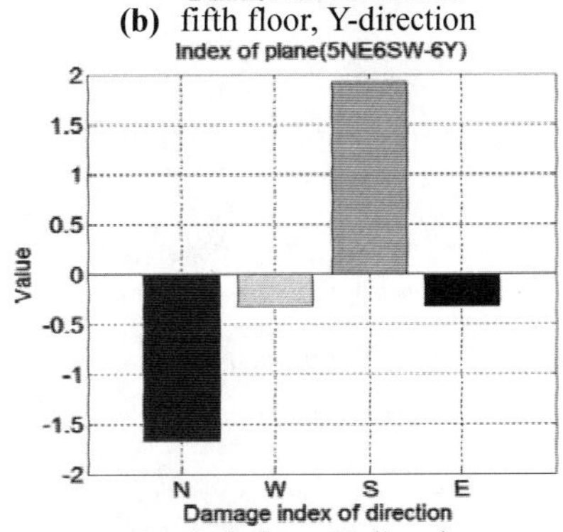

(d) sixth floor, Y-direction

Figure 10. DI diagrams for damage on the N and E sides of the fifth floor and S and W sides of the sixth floor.

Table 3 shows the complete results of planar DI analysis. The identification accuracy levels were 100% in the X-direction and 93% in the Y-direction. Despite there being a misidentified case in the Y-direction, the error only emerged in the second floor results. Therefore, the planar DI can be used to effectively and clearly diagnose the damage direction. The overall identification accuracy of the planar DI was 93%, demonstrating its reliability for practical implementation.

5. Conclusions

In this study, a series of analysis was conducted to highlight and extend the application of MSCE analysis for 3D SHM. A health diagnosis algorithm is proposed, and the reliability and feasibility of the MSCE-based system were verified by analyzing the ambient vibration response signals of a structure. Through the assessment of the complexity of the signals, the damage severity of the structure can be distinguished, indicating that the MSCE-based method is an effective replacement for the more complicated forced vibration response method. In addition, only a set of reference signals from the center of each floor for vertical analysis, as well as biaxial signals in the four directions of each floor for planar analysis is required to initially launch the SHM system; the cost of the structural health diagnosis can be largely reduced.

Case Number	Damaged Floor and Direction	DI (Y-axis)	DI (X-axis)	DI (Both axes)
1	5N	✓	✓	✓
P2	5S	✓	✓	✓
P3	5W	✓	✓	✓
P4	5E	✓	✓	✓
P5	3NS	✓	✓	✓
P6	4WE	✓	✓	✓
P7	6NE	✓	✓	✓
P8	1SW	✓	✓	✓
P9	3NS4WE	✓	✓	✓
P10	5NE6SW	✓	✓	✓
P11	4N5E6S	✓	✓	✓
P12	1NES5W7W	✓	✓	✓
P13	1NE3SW5SW7NE	✓	✓	✓
P14	2N4W5N6SE7N	×(2N)	✓	×(2N)
	Accuracy (%)	93%	100%	93%

Table 3. Identification accuracy of planar DI analysis.

✓ Indicates that damage on all floors was detected; × indicates that damage on some floors was not detected.

References

Doebling, SW, Farrar, CR & Prime, MB 1998, 'A summary review of vibration-based damage identification', *Shock and Vibration Digest*, vol. 30, no. 2, pp. 91-105.

Pincus SM 1991, 'Approximate entropy as a measure of system complexity', Proceedings of the National Academy of Sciences of the United States of America, vol. 88, no. 6, pp. 2297-301.

Richman, JS & Moorman, JR 2000, 'Physiological time-series analysis using approximate entropy and sample entropy', *The American Journal of Physiology - Heart and Circulatory Physiology*, vol. 278, no. 6, pp. H2039-49.

Costa, M, Goldberger, AL & Peng, CK 2002, 'Multiscale entropy analysis of complex physiologic time series', *Physical Review Letters*, vol. 89, no. 6, p. 068102.

Costa, M, Goldberger, AL & Peng, CK 2005, 'Multiscale entropy analysis of biological signals', *Physical Review E*, vol. 71, no. 2, p. 021906.

Lin, TK& Liang, JC 2015, 'Application of multi-scale (cross-) sample entropy for structural health monitoring', *Smart Materials and Structures*, vol. 24, no. 8, p. 085003.

Rojas, E, Baltazar, A & Loh, KJ 2015, 'Damage detection using the signal entropy of an ultrasonic sensor network', *Smart Materials and Structures*, vol. 24, no. 7, p. 075008.

The 8th International Conference on
Structural Health Monitoring of Intelligent Infrastructure
Brisbane, Australia | 5-8 December 2017

Development and Application of Online Monitoring and Safety Evaluation System for Long-span Bridges

H. Jiang[1,2,3], L. X. Wang[1,2,3], K. Yan[1,2,3], X. R. Zhao[1,2,3], H. W. Ma[4], H. Ding[5], Z. H. Nie[4]
[1]*Guangdong Earthquake Agency-China, Email:*stefyk@163.com, [2]*Key Laboratory of Guangdong Province, Earthquake Early Warning and Safety Diagnosis of Major Projects-China,*[3]*Key Laboratory of Earthquake Monitoring and Disaster Mitigation Technology-China,*[4]*Jinan University-China,*[5]*Institute of Industry Technology, Guangzhou & Chinese Academy of Sciences-China*

Abstract

Long-span bridges are the hub of transportation system and occupy an important position in the national economy and society. The damage accumulation and resistance attenuation are unavoidable in the long-term operation of bridge, which reduces its using function and resistance to natural disasters, and leads to damage or destruct in the extreme event, such as earthquake, vehicle and ship collision. Structural health monitoring is an important management tool to realize safe operation and prolong life of structure, and the technical contents of a Online Seismic Safety Monitoring and Evaluation System for Long-span Bridge (OSSMES-LB) and its application are introduced in this paper. The online monitoring and evaluation system includes four aspects of technology: (1) Real-time monitoring technology of strong motion observation array based on Browser/Server (B/S) architecture; (2) Bridge health diagnostic method based on multi-index information fusion technology; (3) Fast numerical simulation technology and seismic performance evaluation method for bridge structures; (4) Integration and visualization system of bridge monitoring and evaluation. The system realizes the integration of real-time acquisition, transmission, analysis, diagnostics, simulation and evaluation of bridge monitoring data. Differences of dynamic characteristics between initial or good conditions and current state are analyzed and diagnosed continuously, and they can provide references for bridge maintenance and reinforcement. When the bridge suffers earthquakes or crash, security status of the bridge is assessed rapidly and the structural abnormalities are alerted. In addition, the intensity information is provided after earthquake, crash or other accidental events.

1. Introduction

Since the reform and opening up, a large number of bridge projects has been built in China, especially in the coastal areas to build a lot of cross-sea, cross-river bridge, which have been used for two or three decades. In the long-term operation process, the bridge has been some health problems in daily loading, corrosion, material aging and other factors. These bridges have been faced with serious threats once they encounter an earthquake or a ship impact. The health monitoring system of long-span bridge has been the hotspot of academic and engineering research at home and abroad and has been applied in some bridges (Huang et al., 2005; Li et al., 2006), such as the Wuhan Yangtze River Bridge, Jiangyin Yangtze River Bridge, Runyang Yangtze River Bridge, Sutong Bridge, Hangzhou Bay Bridge, Hong Kong's Tsing Ma Bridge and other dozens of long-span bridges. Safety monitoring and evaluation system (Du et al., 2015), ship collision monitoring and evaluation system (Ma et al., 2006) for bridges are established by some researchers, and the bridge seismic management and disaster assessment system based on GIS and seismic engineering technology is established (Wang et al., 2003; Chen et al., 2004; Zhao et al., 2006; Xu,

2006). However, due to the complexity of bridge structure and load, the health status and potential danger has not been monitored by the bridge health monitoring and safety assessment system in real time.

Based on the deep research on existing monitoring system at home and abroad, a bridge safety assessment system is developed which integrates the real-time monitoring data acquisition, real-time multi-index fusion analysis of bridge damage, rapid simulation and seismic performance evaluation. The system can not only evaluate the safety status of the bridge in real time, but also evaluate the bridge safety status immediately after the earthquake or ship collision. In addition, according to the seismic environment in site, the system can simulate the possible damage of the bridge under the action of future earthquake, and provide the reference for seismic design and seismic strengthening of the bridge.

2. Overall design of the system

In order to facilitate the management of various functions, the Online Seismic Safety Monitoring and Evaluation System for Long-span Bridge (OSSMES-LB) consists of four subsystems: (1) online monitoring system of strong motion observation array for bridge(referred to as "subsystem 1"); (2) safety evaluation system based on multi index information fusion technology (referred to as "subsystem 2"); (3) rapid simulation and seismic performance prediction system (referred to as "subsystem 3"); (4) integration and visualization system of monitoring, evaluation, simulation, prediction for the bridge (referred to as "subsystem 4"). Through online real-time monitoring of structural vibration and identification of bridge parameters and its evolution characteristics, the prediction and alarm of bridge safety under strong earthquake action are carried out to serve the safe operation of bridges. The main content and relationship of each subsystem is shown in Fig. 1.

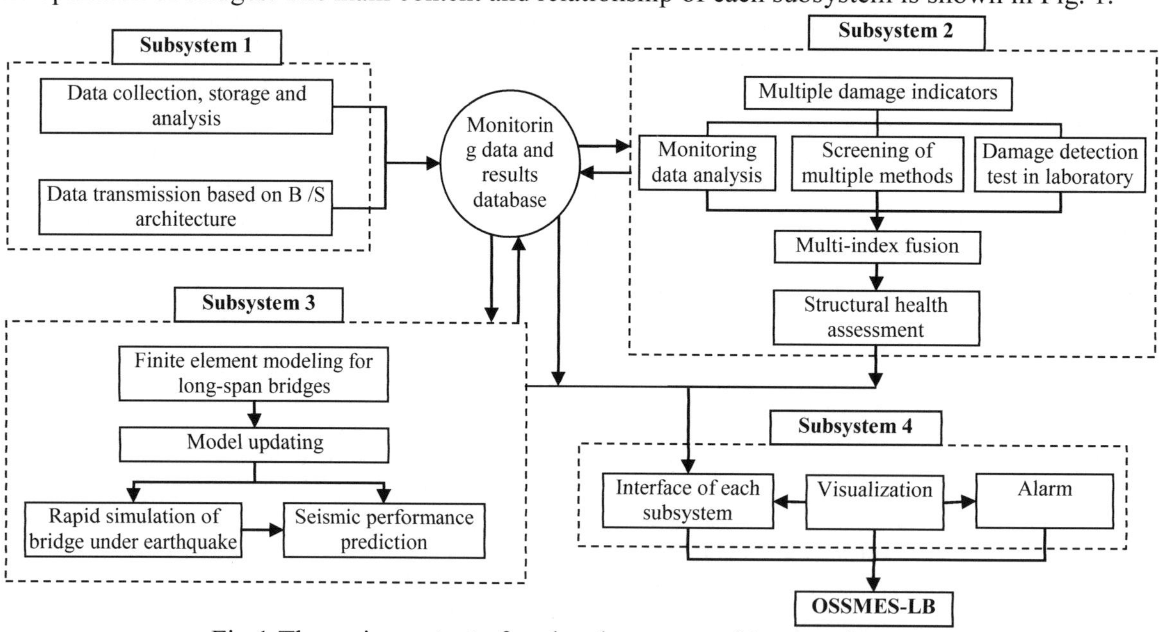

Fig.1 The main content of each subsystem and its associates

3 Monitoring system and safety assessment methods

3.1 Composition of online monitoring system

Online monitoring system mainly includes sensor network, data acquisition and transmission system, which has been applied to different types of bridges such as Jiujiang Bridge of Fokai

Expressway, Zhujiang Huangpu Suspension Bridge and Humen Suspension Bridge. Taking Humen Suspension Bridge as an example, the online monitoring system is introduced.

3.1.1 Sensor network

To monitor real-time vibration and seismic response of the bridge, a total of 36 channels of accelerometer arrays are set up in the critical part of the bridge, such as bridge foundation, pylon and stiffening girder. The sensor arrangement on a long-span suspension bridge is shown in Fig.2.

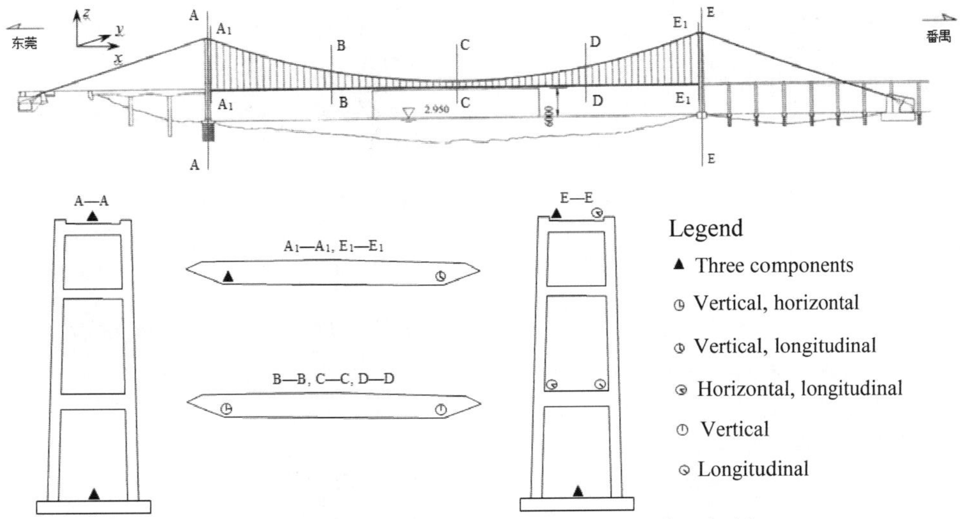

Fig. 2 Sensor arrangement on the suspension bridge

3.1.2 Data acquisition and transmission system

The system is composed of recording data collector and force balance accelerometer. The method of remote monitoring and data storage is adopted in the system, and the operation state of the instrument can be sent back to the seismograph network center regular. In the seismograph network center, remote debugging and data acquisition can be carried out. After the incident is detected, the system will automatically send the relevant parameters and data to the center.

3.2 Safety assessment system

The evaluation method based on four independent damage index is selected in the safety assessment system, which include principal component analysis, wavelet packet energy method, reconstruction phase space method and covariance of covariance (CoC) matrix method (Takens, F, 1981; Li X.Y., etc., 2010) respectively. Each method can be used to reflect the state of the bridge from different aspects, so it is suitable for the quasi real time analysis and evaluation of monitoring data. The first three methods are calculated once every minute, and the CoC matrix method is calculated every two hours. An example of the results of the four methods are shown in Fig. 3.

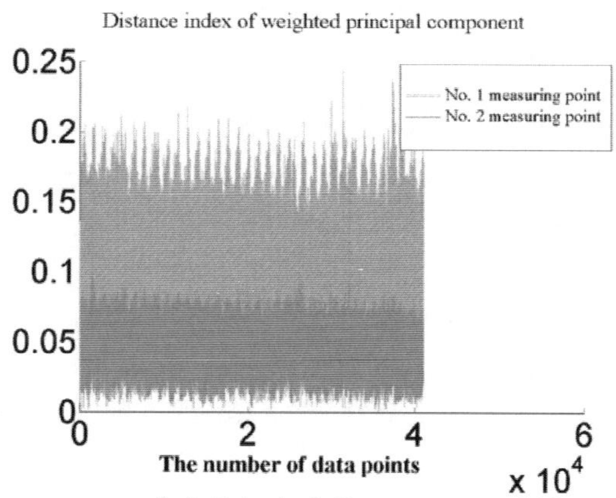

(a) Principal Component methid

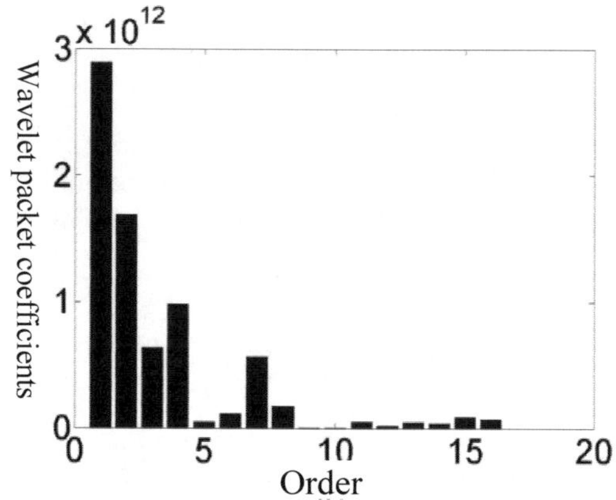

(b) Wavelet packet energy method

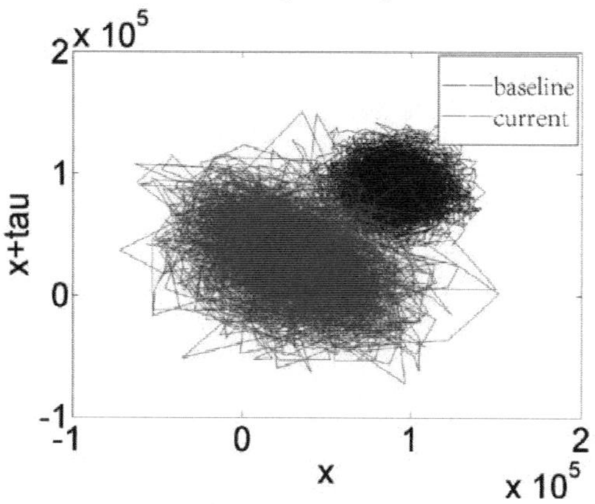

(c) Reconstruction phase space method

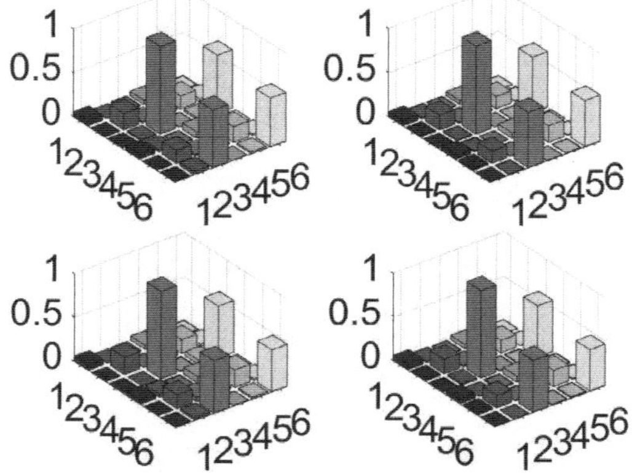

(d) CoC matrix method

Fig.3 Examples of calculation results for the four methods

Because each of these methods is based on a physical characteristic of the bridge, it is difficult to fully reflect the whole information of a single physical feature, and the damage identification method of single index has the possibility of misjudgement and omission, which will reduce the reliability of safety assessment. Therefore, the adaptive weighted least squares method is used to fuse multiple indexes (Nie et al., 2011):

$$\hat{X} = \sum_{i=1}^{n} W_i X_i \tag{1}$$

where W_i is the weight of the corresponding normalized index X_i. The minimum variance of each index is obtained as following:

$$W_k = 1 / \left(\sigma_k^2 \sum_{i=1}^{n} \frac{1}{\sigma_i^2} \right) \tag{2}$$

where σ is the variance of each index.

Then type Eq. (2) into Eq. (1), the fusion index can be obtained. The bridge structure is in normal status when the index value is 1. The measured data are analysed and the results are shown in figure 4. It can be seen that the value of the fusion index fluctuates around 1 during the selected period, which can be inferred that the state of the bridge structure is stable.

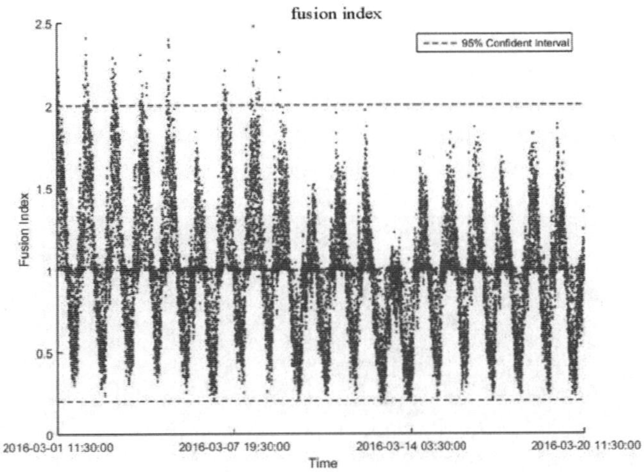

Fig.4 Fusion indicators of bridge health diagnosis

3.3 Simulation and prediction

The essence of fast simulation and prediction of seismic performance is to simulate the dynamic response of the bridge under the action of the load or earthquake. Firstly, the signal abstraction processing of the collected acceleration signal of the bridge is carried out, and the parameters such as the frequency damping ratio of the bridge are calculated by parameter inversion. Then, the finite element model of the bridge is built by the macro element method.

Taking the suspension bridge as an example, the bridge is divided by the macrocell method into 21 synchronization regions, and the finite element model is constructed. The dynamic characteristics of the bridge are calculated by fast simulation system and compared with the calculation results of general finite element software Ansys. It can be seen from Table 1 that the relative errors of the first 10 natural frequencies between the fast simulation system and Ansys are within 1%. The calculation results show that the simulation system can quickly calculate the dynamic characteristics of the long-span bridge under the premise of ensuring the accuracy.

Tab.1 Compare the results of natural frequency of the suspension bridge

Order	Frequency/Hz		Relative error /%	Order	Frequency/Hz		Relative error /%
	Fast simulation system	Ansys			Fast simulation system	Ansys	
1	0.051746573	0.052	0.49	6	0.278132199	0.278	-0.05
2	0.091308454	0.091	-0.34	7	0.322103758	0.322	-0.03
3	0.161949884	0.162	0.03	8	0.388641899	0.388	-0.17
4	0.188875869	0.189	0.07	9	0.39057996	0.389	-0.41
5	0.2158927	0.216	0.05	10	0.421516588	0.421	-0.12

The seismic performance of the bridge is quickly evaluated by finite element simulation. According to the seismic environment in which the bridge is located, the synthetic artificial wave or the natural seismic wave is selected as the seismic input to predict the seismic performance. Seismic load of the bridge is predicted by the dynamic response of the bridge under the action of earthquake, and it is compared with the design load of the bridge to estimate weak part of the bridge. Based on this method, the seismic performance of the bridge can be predicted quickly, and it can provide the basis for bridge maintenance or seismic strengthening.

3.4 Integrated visualization system

The B/S architecture (Browser/Server) which is developed on the basis of J2EE architecture is adopted by the system. The biggest advantage of B/S architecture is that it can be operated anywhere, without the need to install any special software, as long as there is a computer can access the Internet, and the maintenance and update of the system is simple.

This system includes functions as data query, data display, data analysis, decision-making and other functions. The overall interface of the OSSMES-LB is shown in Figure 5, which include 5 menus, "monitoring system", "data browsing", "data analysis", "structural simulation", "output", and so on.

Fig.5 The home page of the system

4 Conclusions

Based on the on-line monitoring of vibration, real-time data analysis and structural simulation of long-span bridges, the Online Seismic Safety Monitoring and Evaluation System for Long-span Bridge (OSSMES-LB) is developed. The system realizes the real-time data acquisition, data storage, processing and interactive analysis of strong vibration monitoring array of bridge through the independent research and development of multi-channel strong vibration data acquisition and its interface software. Based on the multi-index fusion evaluation technique, the comprehensive judgment result of structural damage is obtained, and the judgment of bridge safety condition is more accurate and reliable. The rapid numerical simulation and performance prediction system can predict the performance and the possible damage under earthquake action, and can find the weak parts of the bridge in time and provide a scientific basis for the bridge maintenance and reinforcement.

The research and development of the OSSMES-LB has made a great leap in the strong earthquake observation array technology system, but there are still some problems to be improved, such as: the effectiveness of the assessment method by fusion indexes is not high enough, the method of structural finite element model updating using strong earthquake monitoring data also needs to be studied deeply.

Acknowledgements

This work is supported by the Key Science and Technology Specific Project of Guangdong Province (2012A080102008), Science and Technology Planning Project of Guangdong Province, China (2015A020217007), Science for Earthquake Resilience (XH16031), and the Construction project of Key Laboratory of Earthquake Early Warning and Safety Diagnosis of Major Project, Guangdong Province (2011A060901006), and Independent Project Fund of China Earthquake Administration (CEA) Key Laboratory of Earthquake Monitoring and Disaster Mitigation Technology. Such financial aids are gratefully acknowledged.

References

Chen, M & Hu, SD 2004, Design of seismic analysis of bridge systems based on DDB, *Computer Engineering*, VoL30, No12, pp 165-169.

Du, YL, Su, MB, Liu, YH et al 2015, Study on the long-term health monitoring and safety evaluation system for the Wuhan Yangtze River Bridge, *Journal of the China Railway Society*, Vol.37, No. 4 ,pp.102-110.

Huang, FL, Wang, XM, Chen, ZQ et al 2005, Research progress made on the health monitoring for large-type bridges, *China Railway Science*, Vol. 26 No. 2,pp.1-7.

Li, H, Zhou, WS, Ou, JP et al 2006, A study on monitoring system integration technique of intelligent systems for soundness of long-span bridges, *China Civil Engineering Journal*,vol.39, No.2,pp.46-52.

Ma, J, Liu, S & Wu, TJ 2006, Development of online safety monitoring information management system for bridges in Hangzhou city, *Highway,* Dec. 2006 No. 12, pp.108-114.

Nie, ZH, Cheng, LY & Ma, HW 2011, Visualization method for structural damage detection based on its dynamic characteristics, *Journal of vibration and Shock*, Vol.30, No.12, pp.7-13.

Wang, DS, Zhu, M & Zhong, JR 2003, Aided seismic damage prediction system for bridges using statistical analysis methods, *World Earthquake Engineering*, vol.19, No.3, pp.117-122.

Li, XY & Law, SS 2010. Matrix of the covariance of covariance of acceleration responses for damage detection from ambient vibration measurements. *Mechanical Systems and Signal Processing*, 24(4): 945-956.

Takens, F 1981. Detecting strange attractors in turbulence. In: Rand, D. and Young, L.S.(eds), Dynamical Systems and Turbulence. *Springer Lecture Notes in Mathematics*, New York, NY: Springer, 898:366-381.

The 8th International Conference on
Structural Health Monitoring of Intelligent Infrastructure
Brisbane, Australia | 5-8 December 2017

Heterogeneous Wireless Sensor Deployment for Structural Health Monitoring

C.Y. Liu [1], Z. Peng [1].

[1] *Department of Civil and Environmental Engineering, Harbin Institute of Technology Shenzhen Graduate School, Shenzhen, China, Email*: pengzhen@stmail.hitsz.edu.cn

Abstract

Heterogeneous wireless sensor networks (HWSNs) are widely adopted in structural health monitoring systems due to their potential for performing more complex operations by integrating a diverse set of devices. However, deploying such a HWSN is still in a challenge position due to the heterogeneous nature of data and energy constraints of the network. To respond to these challenges, an optimal deployment framework in terms of both modal information quality and energy consumption is proposed in this study. This framework generates a multi-objective function aimed at maximizing the quality of the modal information identified from heterogeneous data while minimizing the consumption of energy within the network at the same time. Particle swarm optimization algorithm is then implemented to seek solutions to the function effectively. A numerical verification is performed on a four-span prestressed reinforced concrete box-girder bridge. Results show that a set of strategically positioned heterogeneous sensors can maintain a balanced trade-off between the modal information accuracy and energy consumption.

1. Introduction

Traditional Structural health monitoring (SHM) systems rely on wired sensors. The installation and maintenance cost of these sensors takes up a large portion of the total cost of the system. With the advancement of technology, the development of wireless sensor technology in recent years offers new opportunities to the SHM applications (Feltrin & Jalsan 2013). Wireless communication eliminates the cost of wiring and increases the scalability of the SHM systems. Wireless sensor nodes are generally powered by batteries and mounted with multiple sensing devices placed inside and around the structure to periodically collect various types of data. Consequently, wireless sensor networks (WSNs) particularly designed for SHM purpose are evolving into heterogeneous systems (Chin & Rautenberg 2009). The heterogeneous wireless sensor networks (HWSNs) has many advantages. For example, it can potentially provide higher levels of computational power, network densities and lifetimes (Fok & Roman 2012).

As an important issue in SHM, the optimal sensor deployment (OSD) techniques have been demonstrated to be effective in helping to achieve accurate estimation of modal parameters by placing the sensors properly. A handful of works attempted to address the OSD issues for wireless sensor networks. For instance, Bhuiyan et al. proposed a three-phase sensor placement method for SHM that addressed the quality of sensor placements, communication efficiency, and fault tolerance (Bhuiyan & Wang 2012). Onoufriou et al presented a two-step strategy to optimize the number of sensors and their locations to satisfy both specific structural engineering requirements and energy constraint imposed by a WSN (Onoufriou & Soman 2012). Zhou et al. formulated an energy-aware wireless sensor placement framework and developed a hybrid discrete firefly algorithm to solve complex optimization problem (Zhou & Yi 2015,). Fu et al. performed a study to optimize wireless sensors placements for SHM in terms of both the quality of the modal information and network energy consumption (Fu & Ghosh 2013). These OSD algorithms were feasible because most WSNs in these studies were homogeneous (sensors have the same type, storage, processing, battery power, sensing, and communication capabilities). However, in a multi-type sensor network, the orthogonality of the modal vectors cannot be exploited as in single-type sensor networks. Therefore, approaches that do not rely on the orthogonality of the modal vectors are needed.

To the best of the authors' knowledge, none of the existing works attempt to fulfill all the above objectives simultaneously. By employing the ratio of Modal Clarity Index (MCI) and Mode Shape Expansion (MSE), Jalsan et al. took the information quality of the measured data collected from a HWSN into account to represent placement quality of strain and acceleration sensors (Jalsan & Rohan 2014). However, their work did not consider the allocation of data packets along the end-to-end path from a sensor to the base station, and therefore the optimization within HWSNs could be further improved.

In this paper, a framework is proposed to overcome the insufficiency of existing OSD techniques for HWSNs with the consideration of placement quality. Typical heterogeneous devices (accelerometers and strain gauges) are considered herein since they are more commonly available and have proven to be effective in detecting changes of structural properties, particularly in civil structures with low vibration amplitudes. A framework is proposed to optimize the layout of multi-type wireless sensors at critical locations of the structure in terms of the modal information quality and network energy consumption. After the framework is laid out, a numerical verification is performed on a four-span prestressed reinforced concrete box-girder bridge.

2. Optimization Problem Formulation in HWSNs

The optimization for HWSN configuration is aimed to select particular physical locations for the sensors on the monitored structure such that the resulting HWSN exhibits optimal performance. To be more specific, the optimization goal is set to find node locations in order to reliably identify modal information (basis for diagnosing the health of a structure) while consuming minimum energy during data collection. The formulation of optimization goal is described in the following sections.

2.1. Heterogeneous Sensor Placement Quality

In this study, we focus on modal identification performance based on vibration information, specifically acquired through two common types of sensors in SHM – wireless accelerometers and strain gauges. The assessment of sensor placement quality based on modal identification accuracy has been previously studied in homogeneous WSNs. For example, the Modal Assurance Criterion (MAC) which considers the orthogonality of the modal vectors was used to evaluate the sensor placement (Levine-West & Milman 1996). However, in the case of heterogeneous data consisting of acceleration and strain, the orthogonality of the modal vectors is no longer available. Therefore, performance metrics such as the Modal Clarity Index (MCI) and Modal Relative Error (MRE) that do not rely on the orthogonality of the modal vectors need to be employed. A few of heterogeneous sensor placement indices based on these metrics have been proposed in the literature including the weighted sum of MCI and MRE (Ewins 2000), and the ratio of MCI and MRE. In this work, the ratio of MCI and MRE is employed as the placement quality index to quantify the quality of the sensor placement.

The larger the corresponding MRE, the clearer the calculated modal information would be (and vice versa).To better consider the effects of both MRE and MCI, the ratio of the two metrics is defined as the placement quality index to measure the modal identification accuracy.

$$F_M = \frac{I_{MRE}}{I_{MCI}} \quad (1)$$

The ratio of I_{MRE} and I_{MCI} represents a I_{MRE} gain per unit I_{MCI} value. Minimizing Eq. (1) yields the best combination of both measurement types. This formulation overcomes the orthogonality issue arising in heterogeneous sensors deployment and thus is used in this study.

2.2. Network Model

Generally, energy consumption of a typical wireless sensor node in operation mainly occurs in the phases of data acquisition, data processing, data reception, and data transmission(Liu & He 2013). In this study, all of these phases are considered in the energy consumption formulation for acceleration node and strain node separately. Table 1 shows the formulations of energy consumption of each part for different sensor nodes. Table 2 lists the nomenclature of parameters and their typical values used in general WSN platforms for SHM applications(Liu & Fang 2015) and the following equations in this study.

Table 1. The energy consumption functions of each unit about different sensor nodes.

energy consumption	wireless accelerometer node	wireless strain gauge node
data acquisition	$C_{ca} = a_a K$	$C_{cs} = a_s K$
data processing	$C_{pa} = b_a d_a^m K$	$C_{ps} = b_s d_s^m K$
data communication	$C_{ta} = (g_a + d_a d_a^m)K$	$C_{ts} = (g_s + d_s d_s^m)K$
data receiving	$C_{ra} = e_a K$	$C_{rs} = e_s K$

In order to evaluate the energy consumption of HWSNs composed of acceleration and strain nodes, a multi-layer spherical network model with a base station located in the center as shown in Figure 1 is used. Ideal communication without errors and delays is assumed. This network model, presented in the authors' previous work(Liu & Fang 2015), is now extended with heterogeneous sensor nodes deployed. If the nodes are evenly distributed in the network (in other word, obey a uniform distribution), the node density in the spherical space for acceleration p_a and strain p_s can be defined as follows:

$$p_a = \frac{N_a}{V_a}, p_s = \frac{N_s}{V_s} \quad (2)$$

where V_a and V_s are the volume of space covered by the acceleration nodes and strain nodes respectively.

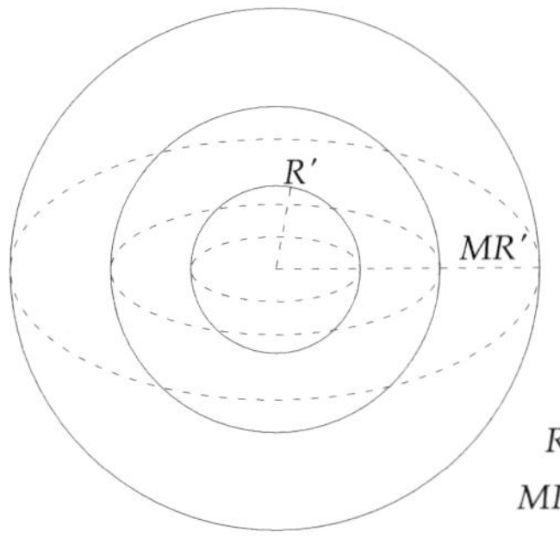

Figure 1. Spherical WSN model.

Table 2. Nomenclature and typical values.

Parameter	Definition	Value
α_a	energy consumption related to data collecting about accelerometer	60×10-9 J/bit
β_a	energy consumption related to data processing about accelerometer	45×10-9 J/bit
γ_a	energy consumption related to the transmission distance about accelerometer	45×10-9 J/bit
δ_a	energy consumption related to the transmission distance about accelerometer	10×10-12 J/bit.m2
ε_a	energy consumption related to data receiving about accelerometer	135×10-9 J/bit
d_a	average distance between all nodes in a network layer to the base station or cluster head node in upper network layer	/
α_s	energy consumption related to data collecting about strain sensor	45×10-9 J/bit
β_s	energy consumption related to data processing about strain sensor	1.35×10-9 J/bit
γ_s	energy consumption related to data transmitting about strain sensor	45×10-9 J/bit
δ_s	energy consumption related to the transmission distance about strain sensor	10×10-12 J/bit.m2
ε_s	energy consumption related to data receiving about strain sensor	135×10-9 J/bit
d_s	average distance between all nodes in a network layer to the base station or cluster head node in upper network layer	/
K	amount of data per second transmitting	200

When considering the pre-defined sensing and routing function of sensor nodes, a multi-layer HWSN contains two main portions of energy consumption. The first portion is the energy consumed by a layer of sensor nodes due to data acquisition, processing, and information transition to the upper level. The other portion is the energy consumed by the same layer of sensor nodes for receiving data from the lower layer and passing them to the upper layer.

In the first portion, the energy consumed is estimated as

$$
\begin{aligned}
F_E(n,n) &= \sum_{i=1}^{N_{na}} F_{node}(V_i)_a + \sum_{i=1}^{N_{ns}} F_{node}(V_i)_s \\
&= \sum_{i=1}^{N_{na}} \{\alpha_a K + \beta_a (d_a)^m K + [\gamma_s + \delta_s (d_s)^m] K\} \\
&+ \sum_{i=1}^{N_{ns}} \{\alpha_s K + \beta_s (d_i)^m K + [\gamma_s + \delta_s (d_i)^m] K\}
\end{aligned}
\tag{3}
$$

where N_{na} is the number of acceleration nodes in the nth layer; and N_{ns} is the number of strain nodes in the nth layer.

Assuming the sensor nodes are uniformly deployed in the network, the number of sensor nodes in each layer can be calculated as

$$
N_{na} = \frac{(3n^2 - 3n + 1)R_a'^3}{R_a^3} N_a \,,\, N_{ns} = \frac{(3n^2 - 3n + 1)R_s'^3}{R_s^3} N_s
\tag{4}
$$

where R_a and R_s are the distance between layers deployed with acceleration nodes and strain nodes, respectively; R'_a and R'_s are the coverage range of all acceleration nodes and strain nodes in the network, respectively.

The expectation of data transmission distance in this process is

$$
E(d_i) = \sum_{i=1}^{N_n} d_i p_i = \int_{(n-1)R}^{nR} \left[x - (n-1)R\right] \frac{4\pi x^2}{4\pi\left\{(nR)^3 - \left[(n-1)R\right]^3\right\}/3} dx = \frac{\frac{3}{2}n^2 - n + \frac{1}{4}}{3n^2 - 3n + 1} R
\tag{5}
$$

Similarly, the energy consumed in the nth layer in the second portion is calculated by

$$
\begin{aligned}
F_E(n,n-1) &= \sum_{i=1}^{N_{na}} (\gamma_a + \delta_a d_a^m + \varepsilon_a) K + \sum_{i=1}^{N_{ns}} (\gamma_s + \delta_s d_s^m + \varepsilon_s) K \\
&= \frac{(3n^2 - 3n + 1)R_a'^3}{R_a^3} N_a \left[\gamma_a + \delta_a \left(\frac{\frac{3}{2}n^2 - n + \frac{1}{4}}{3n^2 - 3n + 1} R_a' \right)^m + \varepsilon_a \right] K \\
&+ \frac{(3n^2 - 3n + 1)R_s'^3}{R_s^3} N_s \left[\gamma_s + \delta_s \left(\frac{\frac{3}{2}n^2 - n + \frac{1}{4}}{3n^2 - 3n + 1} R_s' \right)^m + \varepsilon_s \right] K
\end{aligned}
\tag{6}
$$

Note that the data collected in the nth layer data needs *(n-1)* hops to reach the base station. In this study, we assume the acceleration and strain nodes conform to the same deployment rules, that is

$$R_a' = R_s' = R, R_a = R_s = MR, \tag{7}$$

By summarizing these two portion of energy consumed and its typical values of the corresponding parameters, the HWSN energy consumption per unit time can be expressed by

$$F_E = \sum_{n=1}^{M} \left\{ \begin{array}{l} \dfrac{3n^2-3n+1}{M^3} \times (\dfrac{\frac{3}{2}n^2 - n + \frac{1}{4}}{3n^2-3n+1} R)^2 \times 1.36\times10^{-9} \times (N_a + N_s) \\ \dfrac{3n^2-3n+1}{M^3} \times (1.05\times10^{-7} N_a + 9\times10^{-8} N_s) \end{array} \right\} K$$
$$+ \sum_{n=2}^{M} \left\{ (n-1) \times \frac{3n^2-3n+1}{M^3} \times \left[1.8\times10^{-7} + 10^{-11} (\frac{\frac{3}{2}n^2 - n + \frac{1}{4}}{3n^2-3n+1} R)^2 \right] \times (N_a + N_s) \right\} K \tag{8}$$

3. Objective Function and Solution

There is always a trade-off between modal identification accuracy and energy consumption in WSNs for SHM (Zhou & Yi 2015). Therefore, the objective of the optimization problem herein is to determine the optimal number and location of heterogeneous sensor nodes in order to balance the amount of energy consumed and the modal identification quality in the HWSNs. The following sections describe the formulation of the multi-objective optimization function and its resolving algorithm.

3.1. Multi-objective optimization function

A multi-objective function is formulated to describe the optimization problem as follows:

$$F(F_M, F_E) = f_M w_M F_M(N_a, N_s, R_a, R_s) + f_E w_E F_E(N_a, N_s, R_a, R_s) \tag{9}$$

where f_M is the weight coefficient of modal identification, f_E is the weight coefficient of energy consumption, w_M is the adjustment coefficient of modal identification index, and w_E is the adjustment coefficient of energy consumption. Note that each single objective function (F_M and F_E) has its own numerical dimension, thus adjustment coefficients are utilized to normalize the final calculation results of the multi-object function.

3.2. Particle Swarm Optimization Algorithm

In this study, the particle swarm optimization (PSO) algorithm is employed also due to its unique information diffusion capability and interaction mechanisms which enable PSO to solve complex optimization problem with good performance at low computational cost. Refer to Section 4.2 for the details of the utilization of PSO algorithm in this study.

4. Performance Evaluation

A numerical verification is performed on a four-span continuous reinforced concrete box-girder bridge. The optimization goal is to obtain a HWSN configuration that simultaneously minimize the network energy consumption and maximize the information quality objectives.

4.1. Simulation Setup

The bridge is located in Shenzhen, China, across the Pinghu railway. It is a historical bridge currently being retrofitted. A SHM system composed of wireless acceleration and strain nodes will be implemented on the bridge to monitor its long-term vibration behaviors. The bridge has span length of 42.5m + 2 × 65m + 42.5m as seen from the bridge elevation view in Figure.2.

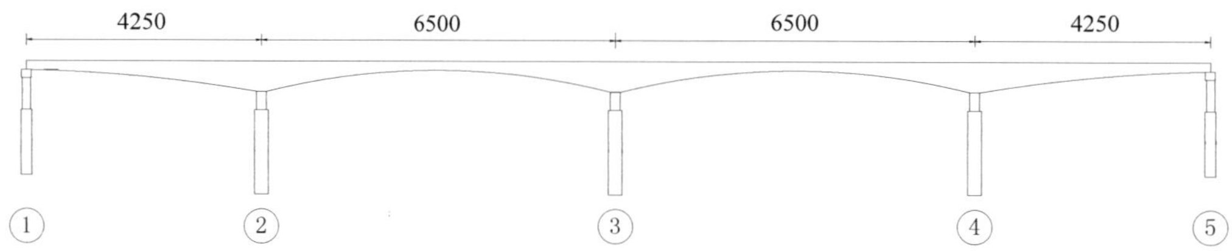

Figure 2. Bridge Elevation View (unit: cm).

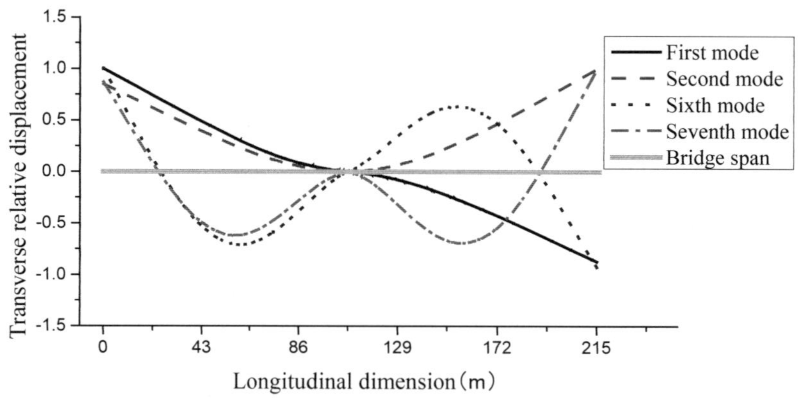

(**a**) Lateral mode of the bridge;

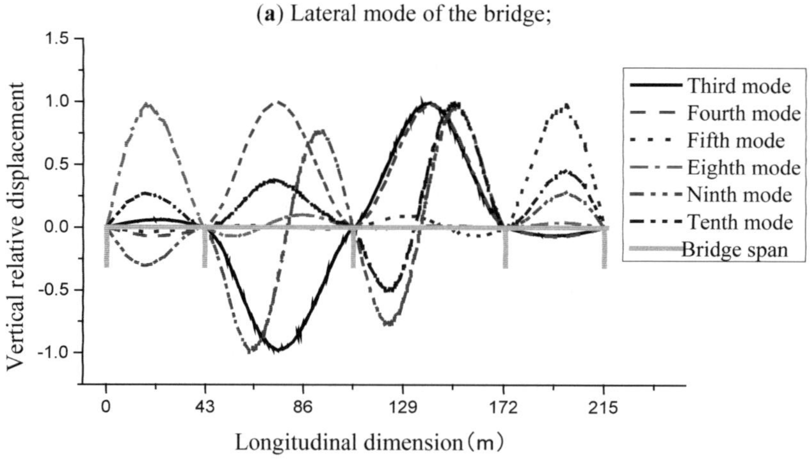

(**b**) Vertical mode of the bridge.

Figure 3. The first ten bending mode shapes of the bridge.

To obtain modal parameters of the bridge, a detailed finite element model was built in ANSYS. The main box girder is modeled with 9854 solid elements and 3354 nodes. In this study, the first 10 mode shapes of the main girder (exhibited in lateral and vertical modes) are considered as the targeted modes and obtained through modal analysis. The first ten modes are shown in Figure. 3.

4.2. Sensor Layout Optimization Results

The proposed OSD techniques are applied to determine both the optimal number of heterogeneous sensors and their best sensing locations. In order to solve the multi-objective function established in section 4.1, it is assumed that the accuracy of the modal parameter identification and network energy consumption for this SHM system has equal significance, meaning that the weight coefficients f_M and f_E in Eq.(9) are both set to be 0.5. Since the two objective values are not in the same scale of magnitude, scale factors $w_R = 1$ and $w_E = 10^6$ are used for normalization. The MATLAB is used to apply the PSO algorithm and carry out the calculations. The PSO parameters of initial population size, maximum number of iterations, and particle velocity, are set to be 300, 150, and [-3, 3], respectively.

The PSO is run for 10 times to generate the network layouts. Herein, the mean values are used as the optimal sensor layout deployed on the bridge plan. The optimal number of accelerometers, strain gauges, and coverage range are 14, 10, and 96.7m, respectively.

The sensor placement layout found from the analysis is shown in Figure. 4, where black circle indicates wireless accelerometers and red rectangle represents the wireless strain gauges, respectively. Numbers 1~22 listed under the bottom of the bridge represents selected cross sections that are equipped with sensors. Numbers listed above the bridge represent longitudinal distance (cm) between sensors. The results showed non-symmetric sensor configurations in the longitudinal direction. Detailed sensor locations on each section are shown subsequently in Figure 5.

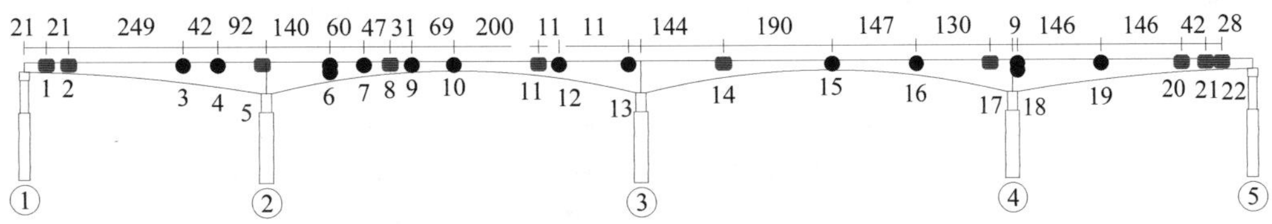

Figure 4. Sensor deployment on the bridge girder.

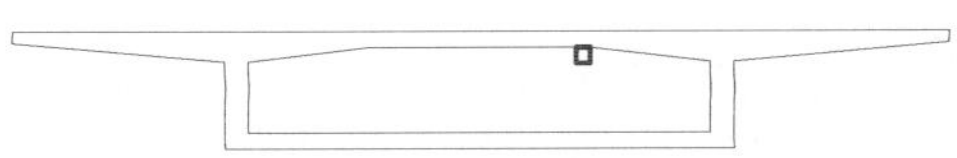

(**a**) Strain sensor deployment in section 1;

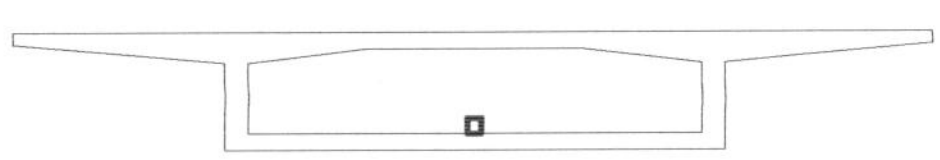

(**b**) Strain sensor deployment in section 2;

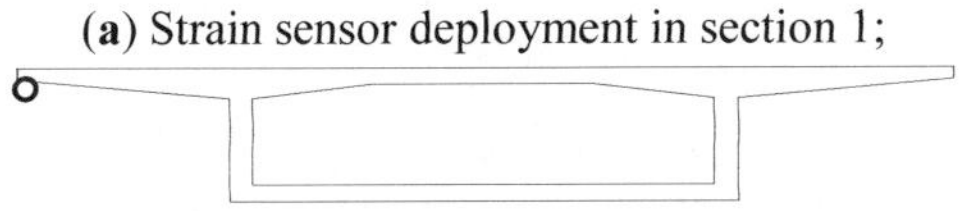

(**c**) Acceleration sensor deployment in 3th section;

(**d**) Acceleration sensor deployment in 4th section;

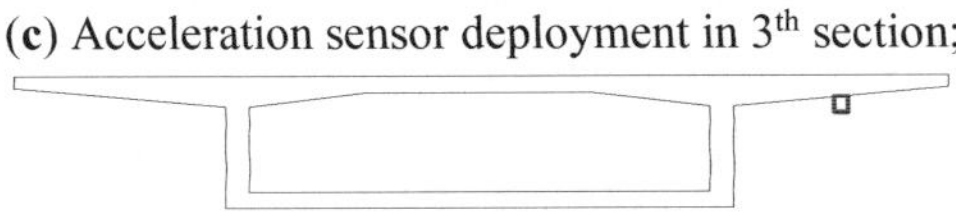

(**e**) Strain sensor deployment in 5th section;

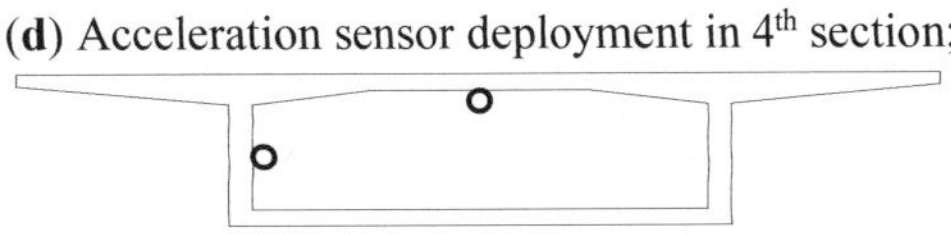

(**f**) Acceleration sensor deployment in 6th section

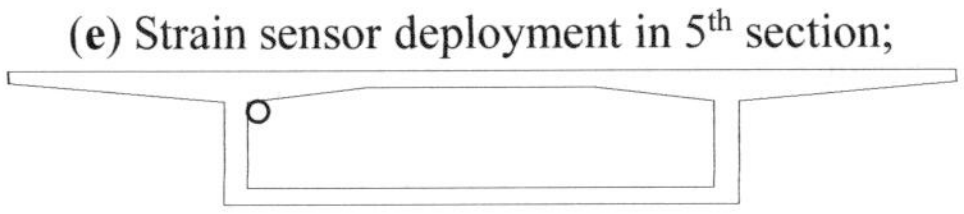

(**g**) Acceleration sensor deployment in 7th section;

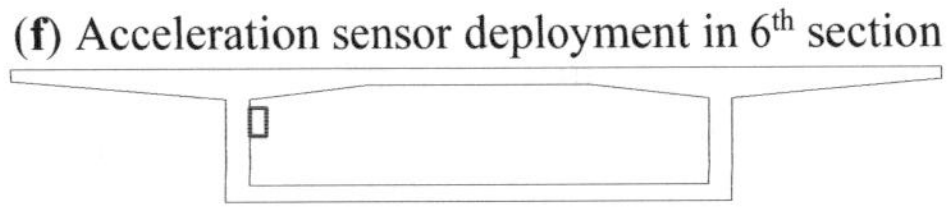

(**h**) Strain sensor deployment in 8th section;

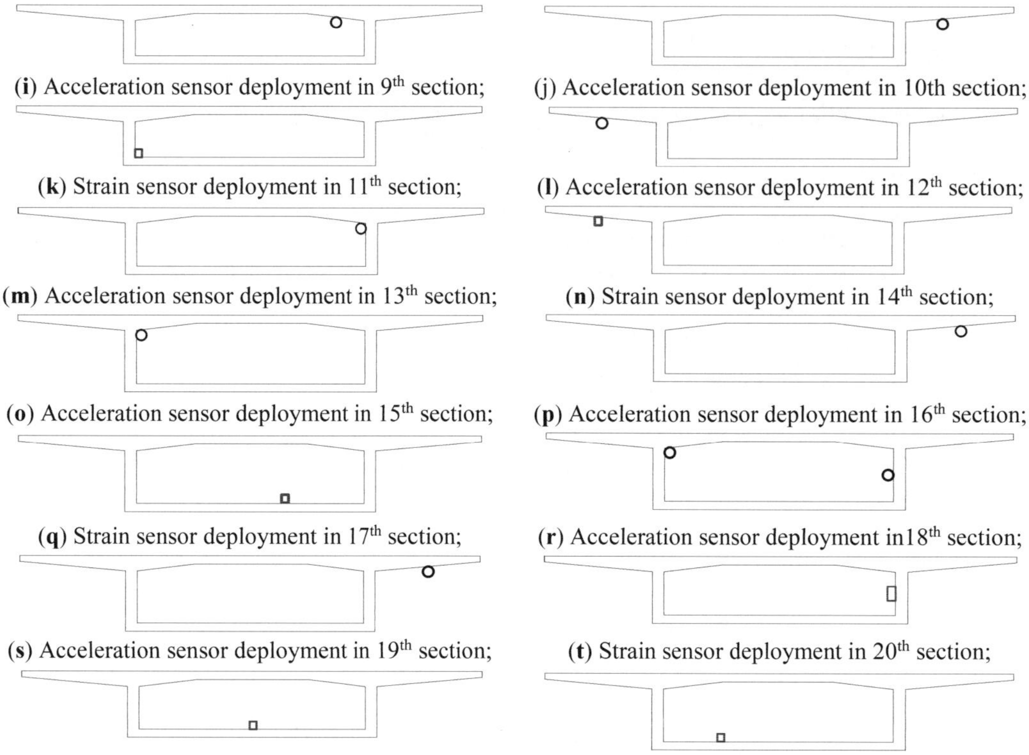

(**i**) Acceleration sensor deployment in 9th section; (j) Acceleration sensor deployment in 10th section;
(**k**) Strain sensor deployment in 11th section; (**l**) Acceleration sensor deployment in 12th section;
(**m**) Acceleration sensor deployment in 13th section; (**n**) Strain sensor deployment in 14th section;
(**o**) Acceleration sensor deployment in 15th section; (**p**) Acceleration sensor deployment in 16th section;
(**q**) Strain sensor deployment in 17th section; (**r**) Acceleration sensor deployment in18th section;
(**s**) Acceleration sensor deployment in 19th section; (**t**) Strain sensor deployment in 20th section;
(**u**) Strain sensor deployment in 21th section; (**v**) Strain sensor deployment in 22th section.
○ Acceleration nodes □ Strain nodes

Figure 5. Sensor deployment on bridge cross sections.

5. Conclusions

The increasing interest of employing HWSNs for SHM applications requires an efficient sensor placement methodology. In this study, a multi-objective layout optimization problem is presented and resolved using the PSO algorithm to determine the trade-off between modal information quality and energy consumption with the consideration of the application requirements. A continuous-span bridge is used as an example to evaluate the performance of the proposed approach for designing a wireless SHM system comprised of acceleration nodes and strain nodes on the bridge.

The proposed optimization approaches can effectively determine the best number and location of heterogeneous wireless sensors for the purpose of SHM. Simulation results demonstrate that a set of strategically positioned heterogeneous sensors can maintain a good balance between the modal information accuracy and energy consumption.

Acknowledgments: This work is financially supported by the National Science & Technology Support Program of China (No. 2014BAL05B06), Shenzhen Science and Technology Projects (No. JCYJ20160525163140206, No.GJHZ20150312114346635), and the authors are grateful to the authorities for their support.

References

Feltrin, G, Jalsan, KE & Flouri, K 2013, 'Vibration monitoring of a footbridge with a wireless sensor network', *Journal of Vibration and Control*, vol. 19, no.15, pp. 2285–2300.

Chin, JC, Rautenberg, JM & Ma, YT 2009, 'An Experimental Low-Cost, Low-Data-Rate Rapid Structural Assessment Network', *IEEE Sensors*, vol. 9, no.11, pp.1361-1369.

Fok, CL, Roman, GC & Lu, CY. Servilla 2012, 'A flexible service provisioning middleware for heterogeneous sensor networks', *Science of Computer Programming*, vol. 77, pp. 663-684.

Bhuiyan, MZA , Wang, G & Cao, JN 2012, 'Sensor deployment with multiple objectives for structural health monitoring', *In WSNs IEEE 14th Int. Conf. on High Performance Computing and Communication & IEEE 9th Int. Conf. on Embedded Software and Systems (HPCC-ICESS).*

Bhuiyan, MZA & Cao, JN 2015, 'Deploying Wireless Sensor Networks with Fault-Tolerance for Structural Health Monitoring', *IEEE Transactions on Computers*, vol. 64, no.2, pp. 382-395.

Onoufriou, T, Soman, RN & Votsis, R 2012, 'Optimization of wireless sensor locations for SHM based on application demands and networking limitations for SHM based on application demands and networking limitations', *Management, Resilience and Sustainability 6th Int. Conf. on Bridge Maintenance.*

Zhou, GD, Yi, TH & Zhang, H 2015, 'Energy-aware wireless sensor placement in structural health monitoring using hybrid discrete firefly algorithm', *Struct. Control Health Monit.*, vol. 22, no.68, pp. 66.

Fu, TS, Ghosh, A, Johnson& Krishnamachari, B 2013, 'Energy-efficient deployment strategies in structural health monitoring using wireless sensor networks', *Struct. Control Health Monit*, vol. 20, no.971, pp. 86.

Jalsan, KE, Rohan, NS & Flouri, K 2014, 'Layout optimization of wireless sensor networks for structural health monitoring', *Smart Struct. Systl*, vol. 14, no.15, pp.39-54.

Levine-West, Milman, M & Kissil, A 1996, 'Mode shape expansion techniques for prediction -experimental evaluation', *AIAA J.*, vol. 34, no.15, pp. 821-829.

Ewins, DJ 2000, *Modal testing, theory, practice and application*; Research Studies Press.

Liu CY & He, XY 2013, 'Wireless sensor integration for bridge model health monitoring', *Vibroeng*, vol. 15, pp. 1028–1040.

Liu, CY, Fang, K & Teng, J, 2015, 'Optimum wireless sensor deployment scheme for structural health monitoring', *Smart Mater. Struct.*, vol. 24, pp. 115034.

The 8th International Conference on
Structural Health Monitoring of Intelligent Infrastructure
Brisbane, Australia | 5-8 December 2017

Thermal Performance Analysis of a Long-span Suspension Bridge with Long-term Monitoring Data

Qi Xia[1], Jian Zhang[1]
[1] *School of Civil Engineering, Southeast University –China, Email:* jian@seu.edu.cn

Abstract

Thermal performance analysis, for instance thermal stress distribution along a section of the bridge girder from measured strains is important for structural safety evaluation. In this study, the structural thermal performance analysis was proposed by processing and analyzing the long-term monitoring data and it was applied to study the thermal and mechanical behavior of a long-span suspension bridge under daily operating conditions. Firstly, statistical analysis of the long-term monitoring data was performed and it found that the measured strain of the main girder may reach 150με under normal operation which is close to the static strain (156με) under the static load testing. Although the two states are similar which produce large strain and deformation, but their mechanisms are totally different, especially on the mechanism between structural stress and measured strain. Secondly, numercial analysis of suspnesion bridge was studeid. Thirdly, the temperature-induced stress on the studied bridge was calcuated. The results show that the effects of temperature on the upper deck could exceed the vehicle loads. The effect of lower deck is mainly subjected to vehicle loads. The influence of temperature-induced stress should be considered in the safety condition of the steel box girder based on structural stress method.
Keywords: Thermal analysis, Thermal stress, self-equilibrating temperature stress, Long-term monitoring

1. Introduction

The thermal effect is widely regarded as one of the most significant and has negative effects on bridges and high-rise structures. The changes of structural temperature and temperature distribution on structure lead to the whole deformations and movements, stresses, cracks, etc. It has been reported that the environmental thermal stress may has a more significant impact on structural behaviours than the external operational loads. (Brownjohn 2007; Xu et al 2012; Yarnold and Moon 2015; Xia et al 2016).
Bridge structures are subject to complex thermal stresses which vary continuously with time. The mechanism between thermal stress and measured strain under temperature loads is in urgent need of research to understand the thermal behaviour of long-span bridges. In terms of finite element simulation, the distribution of temperature and induced strain/stress have been studied by performing finite element analysis of long-span bridges (Xia et al. 2013; Zhou et al 2016).
Static load test has also been performed on the studied bridge, in which the measured static strain of the main girder reached 156με under a truck train load of 17,680kN. In the condition of static test, structural stress can be calculated directly from measured strains by using Hooke's law. It is found from the long-term ambient test data of the studied bridge that the temperature-induced strain reached 150με in operational conditions. It is seen that the strains measured in the static test and in the ambient vibration test are comparable, but it is impossible that their induced stress are in a same

level. Thermal strain will not convert fully to structural stress because it will partially release due to the partial constraints of real bridges. Then, the questions arising are, how to calculate the thermal stress distribution for the real bridges which depends on the boundary condition and the type of temperature load, and what's the role of the thermal stress contributing to structural mechanical performance during evaluating the safety condition of long span bridges.

The objective of this article is to study the thermal stress of a long-span suspension through the long-term monitoring data. The structure of the article is as follows: First, the studied long span suspension bridge and its structural health monitoring system are briefly described. Statistical analysis of the one-year monitoring data is performed. Second, numercial analysis of suspnesion bridge was studeid. Third, the temperature/vehicle -induced stress were calculated and analysed through the long-term monitoring data. Finally, the conclusions are given.

2. The studied bridge and statistical analysis of the long-term monitoring data

2.1 Structural health monitoring (SHM) system of the studied bridge

The studied suspension bridge has a main span length of 1,385 m over the Yangtze River, China (Figure 1). Its main span has a welded streamlined constant depth steel box girder of 3 m height and 36.9 m width which utilized the asphalt concrete pavement, and a navigation clearance of 50 m. The bridge has two reinforced concrete towers of 190 m height, and main cables are anchored in gravity anchorages. The air temperature changes from −6.1°C to 37.2°C in year, and an annual average temperature is about 16.9°C. A SHM system was designed and installed on the bridge in 2005(Ko et al 2005). The sensor layout of the current SHM system includes 170 sensors, such as accelerometers (AS), Fibber Bragg grating sensors (FBG), displacement sensors (DIS), global positioning system (GPS) receivers, shear pins, ultrasonic anemometer etc. Main sensors as shown in Figure 1. A total of 116 FBG were installed on the upper deck and lower deck (Figure 1(b)) at nine equidistant cross sections of main span to measure strain (FBGS) and temperature (FBGT). Other types of sensors were also installed on the bridge as shown in Figure 1(a), but their outputted data were not used in this article. The measured strain and temperature data in the year of 2005 (the first year from installing the SHM system) were used to investigate thermal performance analysis of the studied bridge.

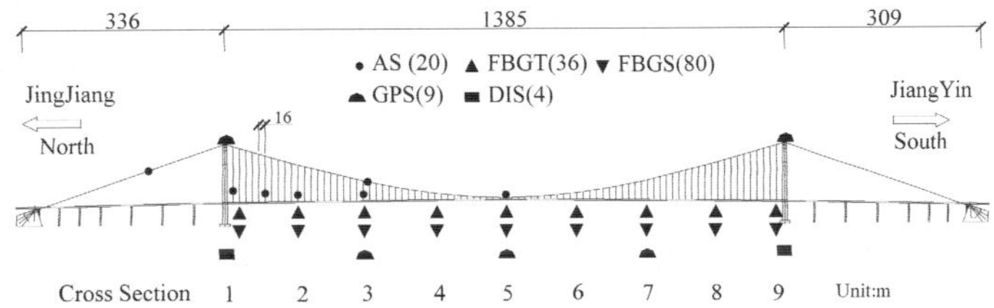

(a)

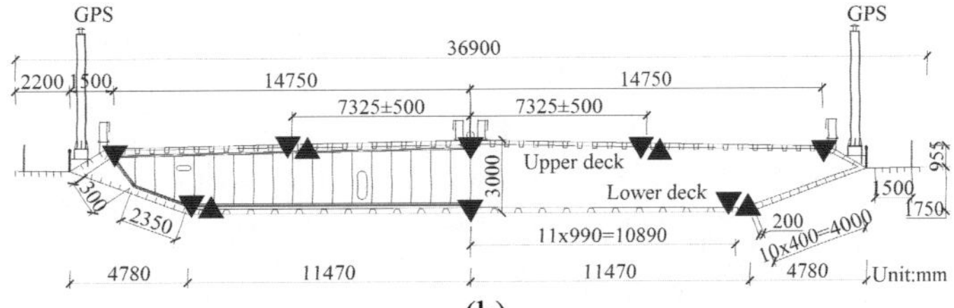

(b)

Figure 1. Sensors layout of the studied bridge: (a) sensors layout; (b) typical cross-section

2.2 Statistical analysis of measured temperatures and strains

Time histories of typical longitudinal strain and temperature responses measured on cross section 5 of the main girder are plotted in Figure 2. The triangular black spots on the cross section indicate the location of FBGS (labelled S-N-#), and the black dots represent positions where strain (labelled by S-N-#) and temperature (labelled by T-N-#) sensors were co-located. For instance, S-5-2 represents the second FBGS on the fifth section of the main girder. Figure 2 shows that strain curves exhibit a similar trend to the temperature curves, and the peaks of the temperature and strain time histories has an around one-hour delay in a day. In the upper surface, the daily measured strain varies from $-100\mu\varepsilon$ to $50\mu\varepsilon$, with a measured temperature from 10°C and 30°C. In the bottom surface, the daily strain varies from $-50\mu\varepsilon$ to $100\mu\varepsilon$, whereas the temperature varies from 10°C to 18°C.

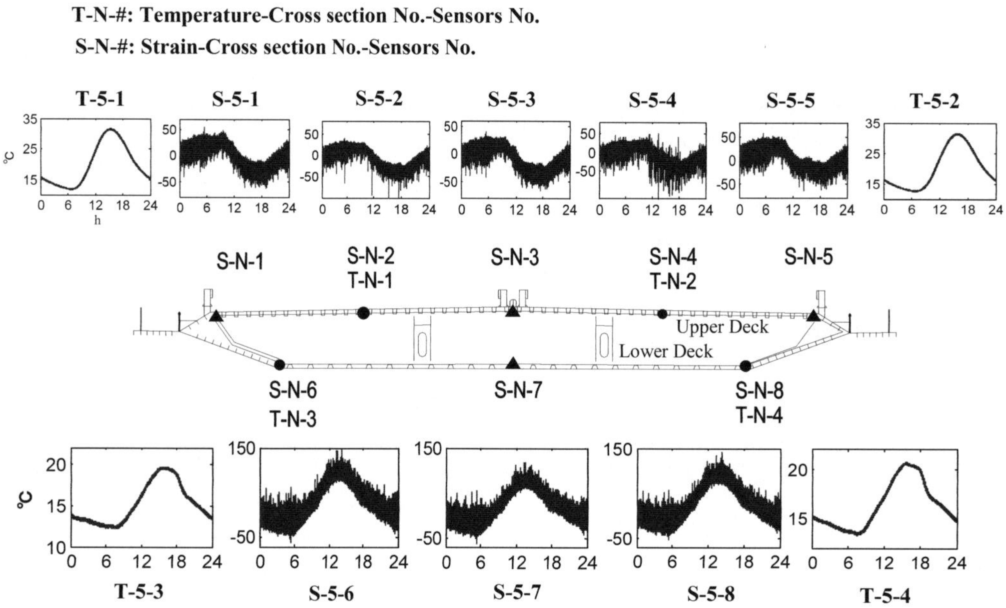

Figure 2. Typical daily strain and temperature time histories on mid-span

The structural strain is an important indicator for condition assessment of bridges. For the studied bridge, the monitored in-service strain in the girder is mainly due to three effects: highway traffic, wind, and temperature. The static strain resulting from initial dead loads is not measurable because the strain gauges were installed after the completion of bridge construction. The statistical analysis of the measured temperature and strain is shown in Figure 3. Figure 3(a) shows the typical strain and temperature time histories on the lower deck, illustrating that the trend of the measured strain is consistent with the temperature change. From the measured data, the temperature-induced strain could exceed $150\mu\varepsilon$. Figure 3(b) shows the typical strain and temperature time histories on the lower deck, displaying that the trend of the measured strain is opposite with the temperature change. Figure 3(c) shows the statistical analysis results of the long-term monitoring data on the upper deck of the mid-span. The upper controlled line (UCL) and the lower controlled line (LCL) in Figure 3(c) represent upper and lower limits of the 0.95 confidence interval of the measured values respectively. The statistical analysis results show that temperature changing is a major cause leading to deformation of long-span suspension bridge. In addition, a controlled static load testing was carried on the studied bridge in July 2014, and the results are also plotted in Figure 3(c). The measured strain is around -20με as represented by circles when no trucks on the bridge in the static test. After that, 52 weighed trucks (52*340kN=17680kN) placed to the mid-span of the bridge, and the measured strain of the girder reached 156με as represented by rectangular in Figure 3(c).

Despite having been investigated for a long time, challenging problems still exist in the field of long-span bridge thermal performance analysis. For instance, during a long-term ambient test, monitoring data indicated that the temperature-induced strain could reached 150με, which is close to the static strain under the static load testing in the Figure 3(c). Although the two states are similar which produce large strain and deformation, but their mechanisms are totally different, especially on the mechanism between structural stress and measured strain. So, how much dose the structural stress caused by temperature load and vehicle load respectively under normal operation?

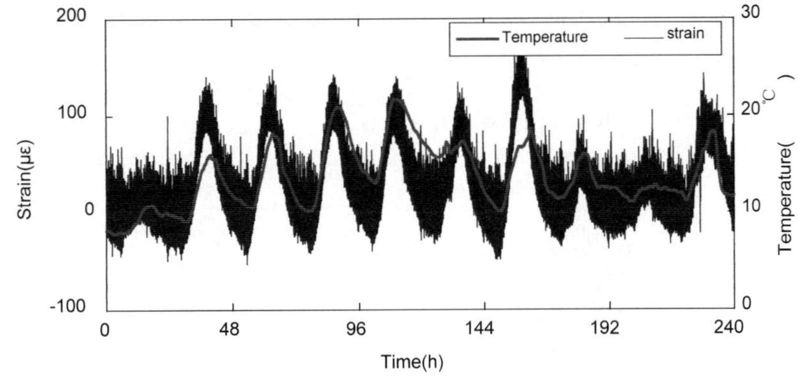

(a)

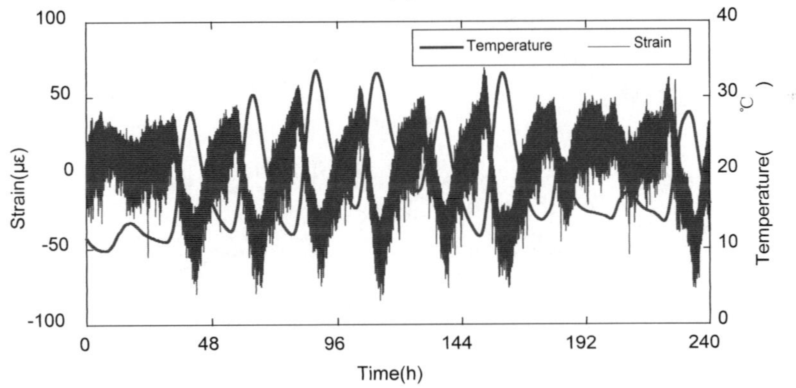

(b)

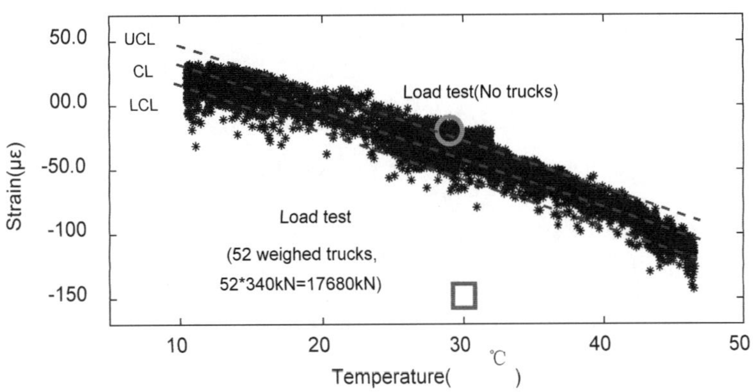

(c)

Figure 3. Statistical analysis results: (a) strain and temperature in the upper deck; (b) strain and temperature in the lower deck; (c) comparison with static test results

2.3 Numerical calculation

Numerical thermal analysis results of the studied bridge were used to validate the proposed method. The bridge was modelled and analysed in the ABAQUS 6.14 software as shown in Figure 4. The steel box girder is simulated by the elastic shell elements 63. The nonlinear elements of link 10 are used in the main cables and hangers. The bridge towers are simulated by the solid elements. The mass density of the girder is $16351.15 kg/m^4$. The vertical and lateral bending moment of inertias are $1.844m^4$ and $93.318m^4$) respectively. The elasticity modulus of the wire rope and the parallel wire strand (PWS) are 1.4×10^5MPa and 2.0×10^5MPa respectively. The temperature load at any time of the day can be expressed by the function and is loaded on the bridge deck surface.

$$T(t)=0.5(T_{max}+T_{min})+0.5(T_{max}-T_{min})\times\sin[(t-9)\times\pi/12] \quad (1)$$

where, T_{max} and T_{min} are the highest and lowest temperatures in a day respectively. t is the time of day. Temperature and strain were extracted on mid-span cross-section as shown in Figure 5.

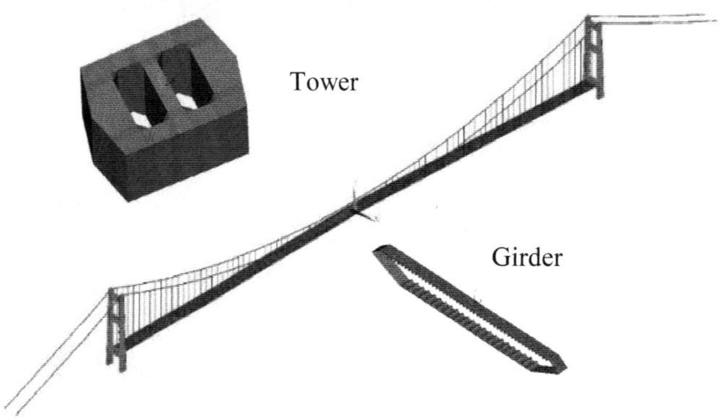

Figure 4. The finite element model of the studied bridge

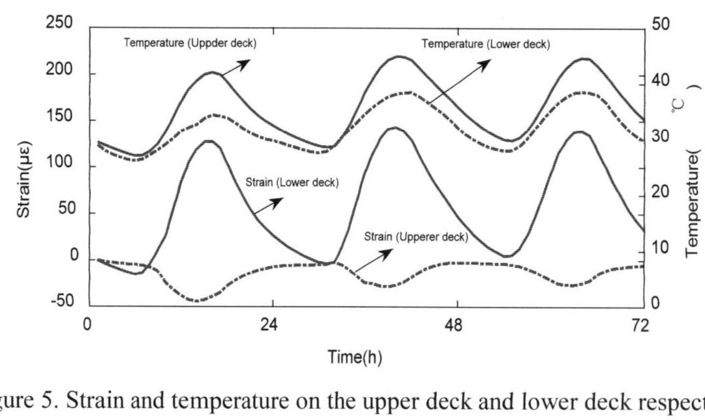

Figure 5. Strain and temperature on the upper deck and lower deck respectively

This section investigates variation of stresses in the bridge members. Considering that temperature tends to cause static strain in a bridge, whereas traffic loadings cause dynamic strain, the hourly mean strain is calculated and variation is regarded as because of temperature changes. The temperature-induced stress σ is calculated from the following equation:

$$\sigma=E(\varepsilon-\alpha\Delta T) \quad (2)$$

where, ε is the measured temperature-induced strain, which separated from measured strain, E is Young's modulus of steel ($2.01*10^{11}$ Pa), and α is the thermal coefficient of linear expansion of steel ($1.1*10^{-5}$/°C). The numerical calculation results and theoretical results are compared in the Figure 6. The classical formula can be used for the calculation of temperature stress.

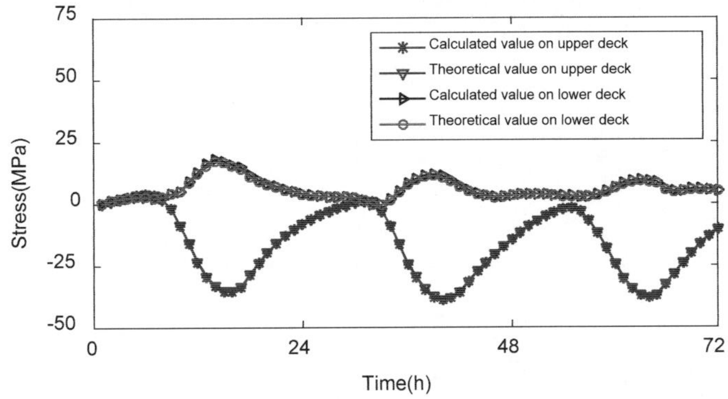

Figure 6. The temperature-induced stress calculation

The stresses time history at the upper/lower deck of mid-span induced by vehicle and temperature as shown in Figure 7. The Figure 7(a) shows the vehicle-induced stress and temperature-induced stress on the upper deck. Vehicle-induced stress is about -15~10MPa under normal operation. Temperature-induced stress has obvious cyclical fluctuations. When the temperature of deck rises to the maximum in a day, maximum temperature-induced stress also appears. During the 0~24h, the change of temperature-induced stress is only -2~4MPa, which is less than the vehicle-induced stress. During the 24~48h, the change of temperature-induced stress is -28~2MPa, which is more than the vehicle-induced stress. When the temperature of upper deck is less than 17°C, the temperature-induced stress will less than vehicle-induced stress. When the temperature of upper deck is above 41°C, the temperature-induced stress may be more than 30MPa. The Figure 7(b) shows the vehicle-induced stress and temperature-induced stress on the lower deck. Vehicle-induced stress is about -7~15MPa and temperature-induced stress is about -12~9MPa. The effect lower deck is mainly subjected to vehicle loads.

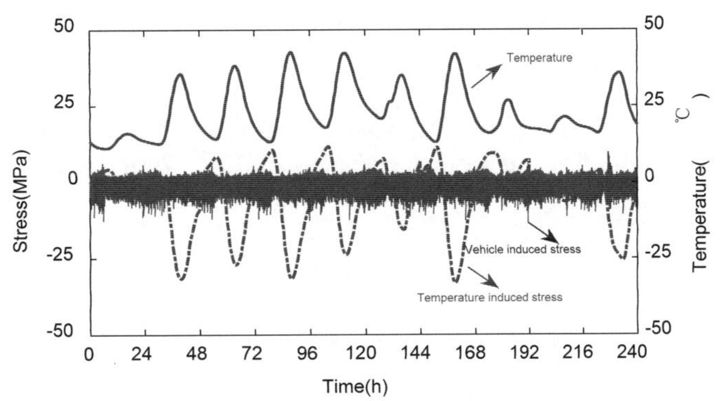

(a)

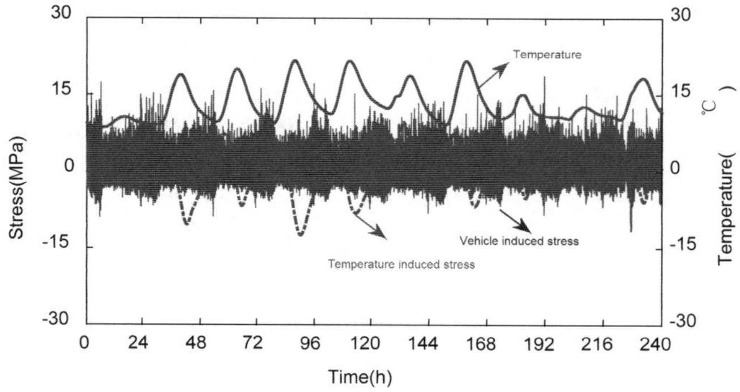

(b)

Figure 7. Temperature-induced stress and vehicle-induced stress time history: (a) stress on the upper deck; (b) stress on the lower deck

It concluded that temperature effects should be caused enough attention in the design of long-span bridges and early warning. The thermal stress is calculated on classical formula, and is actually a temperature-induced stress in the long span suspension bridge. The temperature-induced stress not only conclude the thermal stress, but also the induced stress generated by structural buckling and so on under temperature changing. For example, the trend of the measured strain at upper deck is opposite with the temperature change. The last, most of early warning method based on structural stress only considers the vehicle-induced stress. The influence of temperature will be considered in the early warning method in the following section.

3. Conclusions

The statistical analysis based on one-year monitoring data found that the measured strain of the studied bridge may reach 150με in operation condition, which is equivalent to the strain measurement when 52 trucks (total weight 52*340kN) placed on the steel box girder in static load testing. These two kinds of state of the research are very significant, especially on the study structural stress and reveal mechanics relationship through the monitoring data. The temperature-induced stress and vehicle-induced stress were calculated on the long-term monitoring temperature and strain data at mid-span. Vehicle-induced stress is about -15~10MPa under normal operation. When the temperature of upper deck above 41℃, the temperature-induced stress may be more than 30MPa. When the temperature of upper deck is less than 17℃, the temperature-induced stress will less than vehicle-induced stress. The lower deck is mainly subjected to vehicle loads. Results indicated that the effect of temperature load is greater than the vehicle load effect in the long-span bridge. The performance evaluation method based on structural stress must consider the influence of the temperature induced stress on the long-span bridge.

References

Brownjohn JMW 2007, 'Structural Health Monitoring of Civil Infrastructure', *Phil. Trans. R. Soc,* vol. pp.589-622.

Ko JM & Ni YQ 2005, 'Technology Developments in Structural Health Monitoring of Large-scale Bridges', *Eng Struct*, vol. 27, pp. 1715-1725.

Xia Q, Cheng YY, Zhang J & Zhu FQ 2016, 'In-Service Condition Assessment of a Long-Span Suspension Bridge Using Temperature-induced Strain Data', *J. Bridge Eng.* DOI: 10.1061/(ASCE)BE.1943-5592.0001003.

Xia Y, Chen B, Zhou XQ & Xu YL 2013, 'Field Monitoring and Numerical Analysis of Tsing Ma Suspension Bridge Temperature Behavior', Struct. *Control Health Monit*, vol. 20, pp. 560-575.

Xu YL, Chen B, Ng CL, WongK Y & Chan WY 2012, 'Monitoring Temperature Effect on a Long Suspension Bridge,' *Struct. Control Health Monit*, vol. 17, pp. 632-653.

Yarnold, MT & Moon FL 2015, 'Temperature-based Structural Health Monitoring Baseline for Long-span Bridges', *Eng Struct,* vol. 86, pp. 157-167.

Zhou LR, Xia Y, Brownjohn JMW & Koo K 2016, 'Temperature Analysis of a Long-Span Suspension Bridge Based on Field Monitoring and Numerical Simulation', *J. Bridge Eng.* DOI: 10.1061/(ASCE)BE.1943-5592.0000786

The 8th International Conference on
Structural Health Monitoring of Intelligent Infrastructure
Brisbane, Australia | 5-8 December 2017

3D temperature analysis of a structure subject to nonuniform solar radiation

Fei Gao [1], Pan Chen [1], Shun Weng [1], Hongping Zhu [1], Kailun, Wen [1]
[1] *Huazhong University of Science and Technology – China, Email:* wengshun@hust.edu.cn

Abstract

The influence of the nonuniform solar radiation on a super-tall structure is sometimes as critical as the load the structure bears. The temperature on the irradiation face is much higher than the shade face and the temperature difference has a dynamic changes especially in the sunny day. The nonuniform temperature caused by the solar radiation is a key factor for the change of the structure's strain and lateral displacement. This paper aims at finding the relationship between temperature and strain. Various factors that affect the structure's temperature have been analyzed, such as solar radiation, air temperature, wind velocity. A simplified three-dimensional solid finite element model of Wuhan Yangtze River Shipping Center was established, and the virtual sun is adopted to simulate the actual solar radiation status. Afterwards, the model's whole temperature distribution in 24 hours was calculated to extract the temperature of any node at any time. This method is applied to the Wuhan Yangtze River Shipping Center(about 330m) ,on which a long-term structural health monitoring system including about 256 strain gauges and 64 wireless temperature sensor at different heights will be installed. The monitoring data obtained from the sensors will be used to verify the accuracy of the finite element results and it can also be used to improve the finite element model.

1. Introduction

With the rapid development of social economy, engineering material, more and more super-tall buildings are being constructed. During the construction period, the incomplete structure that exposed to the changing environment has a nonuniform temperature distribution, and it will cause the incline and crack of super-tall structure. The periodic time-variable temperature difference and the low heat conductivity of material will produce nonuniform stress distribution. However, the previous scholars have mainly study the temperature effect in the field of bridge but in building structure (Emerson et al 1977).

There are a variety of factors that affect the strcture's temperature and it's difficult to determine the specific relationship between them. From a large number of measuring data, Zuk et al. found out the solar radiation, air temperature and wind velocity were the main factors that affect the bridge' temperature distribution. Stephenson et al. suggested that the concrete temperature gradient from the outer to the inner should obey index distribution. Numerical analysis and finite element programs have been widely used to solve the problem. However, the finite element models that the scholars used before were mostly two-dimensional plane models (He, Liu et al. 2013). Wang et al. (2016) proposed a three-dimensional solid finite element model that verify the correctness of two-dimensional plane and three-dimensional beam FEM. In terms of finite element thermal analysis in civil engineering, Xu et al. (2009) simulated the actual thermal conditions that vary not only to time

but also to spatial position by establishing a new timing system and using the Ansys auxiliary processor (AUX12).
This study presents a more practical finite element analytical method for calculating a super-tall structure's temperature distribution. The relative position of the sun and the building was determined by establishing the Radiation Calendar timing system. The initial condition used for transient thermal analysis is not a uniform temperature field but a actual nonuniform temperature field. The solid finite element model can simulate the actual solar rotation and apply dynamic boundary condition on the irradiation face automatically. The time-varying temperature of the model at any position can be obtained. This finite-element method was applied to the Wuhan Yangtze River Shipping Center, on which a long-term health monitoring system has been installed as well.

2. The determination of the relative position of the sun

The relative position of the sun and the earth varies in annual periodic. The sun position relative to an any object on the ground can be determined by three parameters: solar altitude angle β, solar azimuth angle α and the distance of the sun from the earth d.

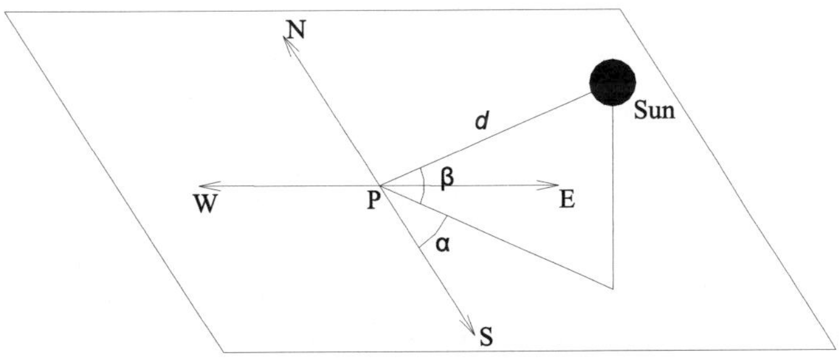

Fig.1 The relative position of the sun and the earth

The conventional calculation formulas to determine the solar altitude and solar azimuth angles is comparatively complex and inadequate. The difference between the global standard time and the local standard time is been taken into consideration but actually it has nearly no influence on the shadow distribution. For simplifying the conventional formulas, a new timing system-the Radiation Calendar is established (Xu et al. 2009).
The Radiation Calendar system sets the Mar. 21th 0:00 as the timing origin. The radiation calendar angle t turns 15 degrees each hour and its maximum value in a year is 15×24×365=131400(24 represents 24 hours, 365 represents 365 days). There is a unique t corresponding to any day's any time. For example, On Mar. 22rd 6:30, the radiation calendar angle t can be calculated as t=1×360+6.5×15=457.5.
By introducing the radiation calendar angle t, the solar altitude and solar azimuth angles can be recalculated, as is shown below:

$$\beta = \frac{1}{A}\arcsin\left\{\sin(A\varphi)\sin[23.5A\sin(\frac{At}{365})] + \cos(A\varphi)\cos[23.5A\sin(\frac{At}{365})]\cos(At)\right\} \quad (1)$$

$$\alpha = \begin{cases} 360 - \dfrac{1}{A}\arccos\left\{\dfrac{\sin(A\beta)\sin(A\varphi) - \sin[23.5A\sin(\frac{At}{365})]}{\cos(A\beta)\cos(A\varphi)}\right\}, 360k < t < 360k+180 \\ \dfrac{1}{A}\arccos\left\{\dfrac{\sin(A\beta)\sin(A\varphi) - \sin[23.5A\sin(\frac{At}{365})]}{\cos(A\beta)\cos(A\varphi)}\right\}, 360k+180 < t < 360k+360 \end{cases} \quad k = 0,1,2...364 \tag{2}$$

where A=*pi*/180, is the multiplier factor for the conversion from radian to angle; φ represents the latitude. The distance of the sun from the earth d varies in annual periadic. The relative change value is only 3.3%, and it can be seen as constant.

3. Theory of thermal analysis

The surface temperature of a concrete structure is related to location, climate, solar radiation, surface orientation, air temperature, wind velocity, etc. For a specific structure, the surface thermal exchange mainly has three ways: solar radiation, convection heat and long wave radiation. Besides, the following assumptions is used to simplify the present study.
(1) The thermal performance parameters of concrete do not change with temperature;
(2) Concrete homogeneity, homogeneity and linear elasticity assumptions are used;
(3) Convective coefficient and wind velocity are linearly related.

3.1 Heat conduction equation

The heat conduction can be defined as the internal energy exchange between the two objects that are completely in contact or between the different parts of an object due to the temperature gradient. It can be expressed by the well-known Fourier heat conduction equation (Elbadry, Ghali et al. 1983):

$$\rho c \frac{\partial T}{\partial t} = k_x \frac{\partial^2 T}{\partial x^2} + k_y \frac{\partial^2 T}{\partial y^2} + k_z \frac{\partial^2 T}{\partial z^2} \tag{3}$$

where ρ represents the density(kg/m^3); c is the specific heat(J/(kg·K)); T is the structure temperature (℃); t is the time; k_x, k_y, k_z (W/(m·K) represent the thermal conductivity of different directions, respectively.

3.2 Initial conditions and boundary conditions

The heat conduction equation establishes the relationship between temperature, time and space. The initial conditions and boundary conditions leads to a specific transient thermal analysis problem. If t=0, the initial temperature distribution of the structure can be seen as a constant.

$$T(x, y, z, 0) = T_0 = \text{constant} \tag{4}$$

Actually, the inner temperature of the structure is usually higher than the outer temperature at the initial time. Here, The temperature at the last time of the previous day is used as the initial temperature distribution.

$$T(x,y,z,0)=T_0(x,y,z) \tag{5}$$

On the other hand, there are usually three types of boundary conditions.
In the first type of boundary condition, the temperature function on the boundary of the object is known to be

$$T|_s = T_b(P,t),(P\in s,t>t_0) \tag{6}$$

where s represents the outer boundary; P=P(x, y, z), is the coordinates of the spatial point; t_0 is the initial time.
In the second type of boundary condition, the heat flux function on the boundary of the object is known to be

$$q_n|_s = -\lambda \frac{\partial T}{\partial n}\bigg|_s = q_b(P,t),(P\in s,t>t_0) \tag{7}$$

where n is the normal direction of the surface.
In the third type of boundary condition, the convection heat function on the boundary of the object is known to be

$$q_n|_s = -\lambda \frac{\partial T}{\partial n}\bigg|_s = h_c(T-T_a),(P\in s,t>t_0) \tag{8}$$

where h_c (W/(m^2·K)) represents the convection coefficient between concrete surfaces and ambient air. T_a is the ambient air temperature around the structure which can be expressed as sinusoidal function (℃).

$$T_a(t)=\frac{T_{max}+T_{min}}{2}+\frac{T_{max}-T_{min}}{2}\sin\frac{(t-9)\pi}{12} \tag{9}$$

When considering the solar radiation, it's difficult to apply the second and third type of boundary conditions on structure surface at the same time. The second type of boundary condition is transferred into the third type of boundary condition. The comprehensive boundary condition can be expressed as

$$(h_c+h_r)(T_a+\frac{q_s-q_{ra}}{h_c+h_r}-T)+k\frac{\partial T}{\partial n}=0 \tag{10}$$

where h_r (W/(m^2·K)) is radiant heat transfer coefficient. q_s is the total heat flux gained from solar radiation at the external boundaries given by equation (11). q_{ra} is the comprehensive long wave radiation heat flux given as equation (12); k is thermal conductivity (W/(m·K)).

$$q_s=\alpha_s(I_{D\phi}+I_{D\beta}+I_f) \tag{11}$$

$$q_{ra}=\frac{1+\cos\gamma}{2}\varepsilon C(1-\varepsilon_a)(273+T_a)^4 \tag{12}$$

where α_s is short wave solar radiation absorption coefficient of the structural outer surface, usually 0.50-0.77; $I_{D\phi}$, $I_{D\beta}$ and I_f (W/m^2), respectively, represent radiation flux of the direct radiation, diffuse radiation and reflected radiation from the surroundings; γ is the surface inclination angle, between the incline and horizontal plane; ε_a is the long wave solar radiation absorption coefficient, usually taken as 0.82; ε is material emissivity, 0.85-0.95; C is Stefan–Boltzmann constant, equals 5.67×10^{-8} W/(m^2·K^4).

4. Finite element model

The size of the Wuhan Yangtze River Shipping Center which has 63 floors in all is about 52.9m × 52.9m × 300m. The model is established by ANSYS. The element type of solid226, which has both the structural and thermal DOFs, is used here to simulate structural and thermal analysis. It's necessary to use different mesh size for different parts because the structure is complex. The solid finite model is shown in Fig.2 and Fig.3.

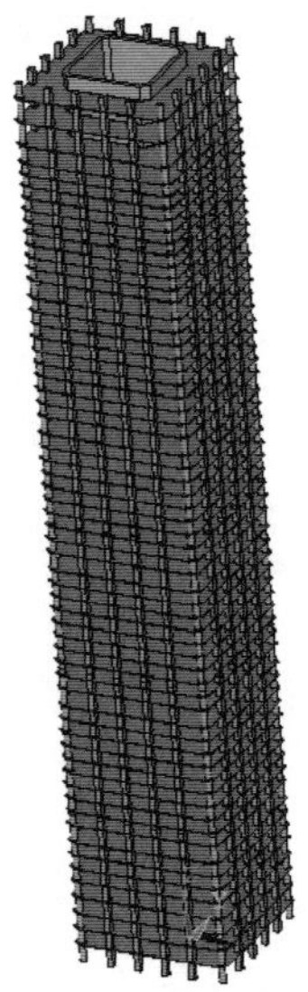

Fig.2 The whole solid finite element model

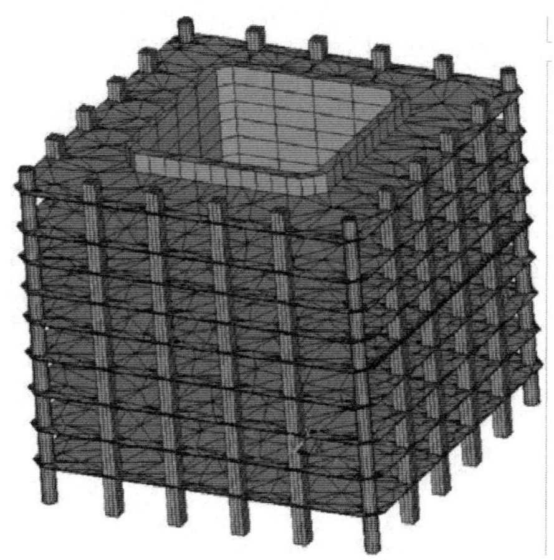

Fig.3 The meshed finite element model(partial)

In order to apply the dynamic solar radiation thermal boundary conditions, the irradiation face is necessary to be separated. A virtual sun is established, which has a same movement trajectory with the real sun.

Fig.4 The virtual sun

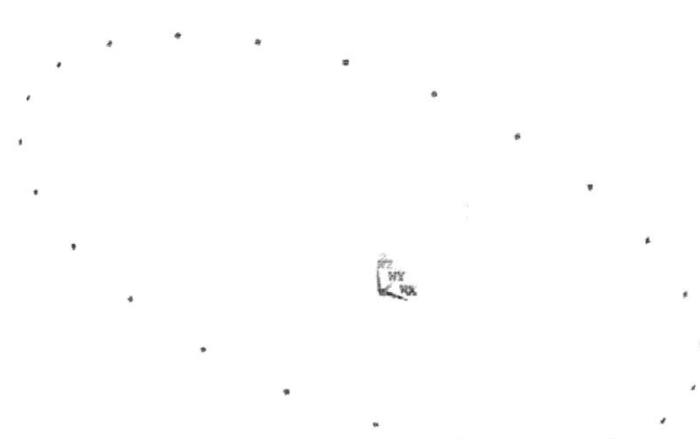

Fig.5 The virtual sun's movement trajectory

For example, the date of March 6 2017 is analyzed. This day's meteorological data is needed, as is shown in table 1.

Table 1 The meteorological data

Time	5:00	6:00	7:00	8:00	9:00	10:00	11:00	12:00	13:00	14:00	15:00	16:00
Air temperature/ ℃	3.4	4.7	5.8	7.1	8.3	9.5	10.5	11.3	11.7	11.6	11.1	10.6
Wind velocity/(m/s)	2.0	2.5	2.0	3.0	2.0	0	0	4.0	0	2.0	1.0	1.0
Solar radiation flux/(W/m^2)	0	41.5	84.0	124.0	442.3	165.0	434.5	757.8	351.3	270.7	318.5	110.1
Time	17:00	18:00	19:00	20:00	21:00	22:00	23:00	0:00	1:00	2:00	3:00	4:00
Air temperature/ ℃	9.9	9.1	8.4	7.7	7	6.4	4.2	4	3.8	3.6	3.5	3.4
Wind velocity/(m/s)	0	0	3.0	1.0	2.0	2.0	2.0	1.0	1.0	2.0	2.0	2.0
Solar radiation flux/(W/m^2)	67.5	33.2	0	0	0	0	0	0	0	0	0	0

The size of the sun element is close to the column element, 0.5m×0.5m×1m. The center of the virtual sun is towards to the center of the structure and the distance between them is 100m. The structure is averagely divided into 6 parts along the height. It is all used for simulating the parallel light. The position of the 24 suns in this day is determined according to the radiation calendar system.

Afterwards, by using the ANSYS auxiliary processor, the sun-structure radiation pair's Angle Coefficient (a radiation energy percentage that emitted from a surface to another surface) can be calculated.

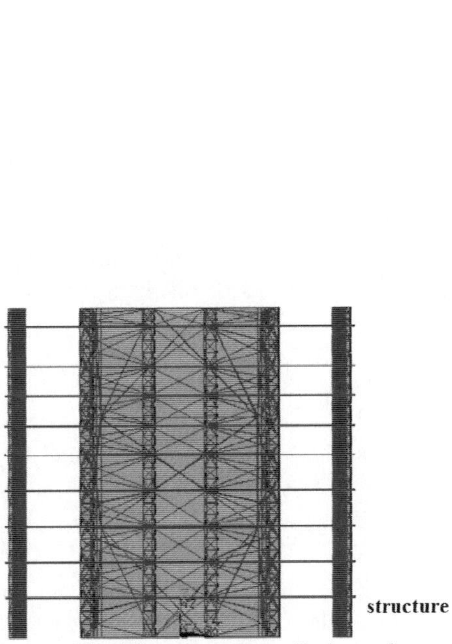

Fig.6 Sun-structure radiation pair

Finally, if the Angle Coefficient is equal to zero, it indicates that the sun cannot "see" the structure element. This element is determined to be shade element. If the Angle Coefficient is not equal to zero, it indicates that the sun can "see" the element. This element is chosen to be irradiation element. The value of the Angle Coefficient can be used to distinguish shade element and irradiation element.

Fig7. The selected irradiation face element(9:00)

The meteorological parameters and material properties were input into ANSYS for transient thermal analysis. The initial temperature field of the structure was assumed to be uniformly distributed as 3.4 ℃ at 5:00 on March 6, 2017. The irradiation element and shade element were applied to different thermal boundary conditions respectively. The boundary conditions were applied every one hour and totally 24 times. The second type of boundary is transferred to the third type of boundary. Therefore, only the radiant heat transfer coefficient and the total heat flux are needed for applying the boundary condition. By solving the program, the temperature of any node at any time can be obtained in a day.

5. Conclusion

The solar radiation has a significant influence on a super-tall structure's temperature distribution. By introducing the radiation calendar system, a new 3D solar radiation thermal analysis method is used to simulate the temperature distribution of Wuhan Yangtze River Shipping Center. The thermal boundary conditions that vary not only to time but also to spatial position is successfully applied to the structure. The temperature on the irradiation face is much higher than the shade face and is closely related to the motion of the sun. It has great significance to more accurate displacement measurements.

6. References

Emerson M. 1976, 'Bridge temperatures estimated from the shade temperature',Transport & Road Research Laboratory (Great Britain), TRRL Report, 1976(696):35~36.

Emerson M. 1973, 'Calculation of the distribution of temperature in bridges', EE/Systems Engineering Today, 1973(LR561): 33.

Emerson M. 1977, 'Temperature differences in bridges: basis of design requirements', ASCE J Struct Div, 1977(765): 39.

Chen C.S., Yan D.H., Cheng H.Q., Tu G.Y., Liu L.T. 2002, 'Analysis of temperature influence on long-span prestress concrete cable-stayed bridge construction control', Gongku Jiaotong Keji/Journal of Highway and Transportation Research and Development, 2002. 19(6): p. 84.

Guo Q.W., Fang Z., Pei B.Z., Deng H., Liu G.D. 2002, 'Temperature effect analysis of concrete cable-stayed bridge', Zhongguo Gonglu Xuebao/China Journal of Highway and Transport, 2002.15(2): p. 48.

Zuk W. 1961, 'Thermal and shrinkage stresses in composite bridges', Journal of America Concrete Institute, 1961, 58(3): 327-340.

Zuk W. 1965, 'Summary review of studies relating to thermal stresses in highway bridges', Highway Research Records, No. 103, 1965.

Zuk W. 1969, 'End movement studies of various types of highway bridges', Highway Research Records, No. 295, 1969.

Priestley M J N, 1978, 'Design of concrete bridges for temperature gradients', ACI Journal Proceedings, ACI, 1978,78(5).

He Y, Ding M, Zhang Q. 2012, 'Numerical Simulation on Temperature Field of CFST Member under Solar Radiation', Advanced Materials Research. Trans Tech Publications, 2012, 354: 1241-1244.

Liu H, Chen Z, Zhou T. 2013, 'Investigation on temperature distribution and thermal behavior of large span steel structures considering solar radiation', International Journal of Advanced Steel Construction, 2013.

Wang Y, Zhan Y, Zhao R. 2016, 'Analysis of thermal behavior on concrete box-girder arch bridges under convection and solar radiation', Advances in Structural Engineering, 2016, 19(7): 1043-1059.

Xu F. 2009, 'Study on the key factors of temperature effect of concrete box-girder bridges', Ph. D. Dissertation, Huazhong University of Science & Technology, Wuhan, China, 2009.

Xu F, Wang B, Zhang H L. 2009, 'A Comparative Research on Thermal Stresses Calculated from Different Element Models of a Concrete Continuous Box-Girder Bridge', Key Engineering Materials. Trans Tech Publications, 2009, 400: 923-928.

Elbadry MM and Ghali, 1983, 'Temperature variations in concrete bridges', Journal of Structural Engineering. 1983,109(10): 2355–2374.

Krkoška L, Moravčík M. 2015, 'The Analysis of Thermal Effect on Concrete Box Girder Bridge', Procedia Engineering, 2015, 111: 470-477.

Kim S H, Cho K I, Won J H, et al. 2009, 'A study on thermal behavior of curved steel box girder bridges considering solar radiation', Archives of Civil and Mechanical Engineering, 2009, 9(3): 59-76.

Saetta A, Scotta R, Vitaliani R. 1995, 'Stress analysis of concrete structures subjected to variable thermal loads', Journal of Structural Engineering, 1995, 121(3): 446-457.

Yarnold M T, Moon F L. 2015, 'Temperature-based structural health monitoring baseline for long-span bridges', Engineering Structures, 2015, 86: 157-167.

The 8th International Conference on
Structural Health Monitoring of Intelligent Infrastructure
Brisbane, Australia | 5-8 December 2017

Some Issues on Temperature Model of Box Girder in HSR

G.L. Dai[1], Y. Tang[1], Y.F. Chen[1,2], J.B. Liang[1], L.H. Yang[1]
[1]. *Central South University – China, Email:* tangyu9433@163.com [2]. *Pennsylvania State University – USA*

Abstract

Solar radiation and ambient temperature will cause deformation in a high-speed rail (HSR) bridge. Such deformation has significant effects on the structural safety and ride comfort. Based on two years' monitoring data of girder-track system on a HSR line, temperature fields for the girder and track slab could be established. Then, by using the probability statistic model asociated with the extreme value theory and an time-series analytical method, the temperature time-dependent law for the struture was studied. The study results show that the distribution of transverse temperatures on the box girder is different between the construction stage and the operation stage, due to the differential track slab's shielding effects. The temperature distribution characteristics of the box girder in different stages were obtained and different temperature calculation models were proposed. In addition, based on the annual temperature data, the temperature could be separated into uniform temperature and fluctuating temperature and the relationship between structure and shade air uniform temperatures was thus established. The concluding results may serve as a basis for the deformation control of a girder-track system.

1. Introduction

High-speed trains require strict track smoothness. However, due to the time lag of thermal transmission in concrete material, the internal and external concrete surfaces of a high-speed rail (HSR) will result in a nonlinear temperature distribution. This is due to the coupling effect of different thermal boundaries, resulting in different temperatures between the top and the bottom of a box girder. Meanwhile, because of the shielding effect of ballastless track on box girder surface, temperatures will be induced horizontally so as to cause the temperature stress and deformation of box girder[1]. The structural annual uniform temperature fluctuation caused by the seasonal change of atmospheric uniform temperatures will lead to the expansion movement of components. The combined effect from the above temperature deformations will no doubt affect the travelling safety and comfort of a high-speed train.

With respect to the various codes/specifications on box girder temperature field, EuroCode 1[2] includes the linear and non-linear methods for establishing box girder's uniform temperatures and the temperature difference calculation. Chinese highway specification is similar to the AASHTO LRFD bridge design specifications [3, 4], providing the distribution of temperatures along the box girder depth with the form of broken lines. The railway bridge and culvert design specification[5] adopts separate exponential functions to describe the horizontal and vertical temperature gradient distributions of a box girder. Nevertheless, all design specifications fail to consider the shielding effect of ballastless track on box girder.

Therefore, aiming to develop a temperature calculation model for HSR's box girders, researchers have attempted to collect field data and conduct relevant research. The experimental researches have mainly focused on the parametric analysis of limited temperature data rather than the long-term field measured data. While the theoretical work has mainly determined the temperature distribution rules under normal days, which cannot be extended to cover the extreme temperature condition that may encounter during the bridge design life. Moreover, previous researches usually considered track and box girder separately and focused on the influence of temperature gradients with insufficient discussion on annual uniform temperatures.

A statistical model was applied to study the temperature pattern for concrete and steel box girders, where the extreme value distribution model was adopted to compute temperatures. However, available researches on the determination of wind and vehicular loads [6] reveal that the results from the short-term data or the extreme value distribution model are poor. Particularly, in the "high tail" portion, a larger error of estimated values is seen[7]. The generalized pareto distribution (GPD) model reduces the samples exceeding a certain threshold value by sorting the samples and then implements a fitting technique for the tail data. This model can better predict the extreme value of sea waves, precipitation, vehicular load, and wind load for certain recurrence periods[8]. However, it has not been used to calculate the extreme temperature load of a box girder.

Based on multiple years' field measured temperature data of box girders, this paper proposes a uniform temperature distribution curve using the time series decomposition method. Meanwhile, the GPD model was used to determine the difference of horizontal temperatures of a box girder for 100 years' recurrence period.

2. Time-history Analysis of Box Girder Temperature Field

2.1 Temperature Field Test

Taking the standard 32m simply supported girder with CRTS I double sleepers ballastless track in a HSR bridge as the example, the azimuth angle of bridge is north by east 9.5°. The cross-section at 15m from the bridge end was selected for the test. The specific arrangement of measure points is shown in Fig. 1. In order to study the influence of track's shielding effect on the distribution of temperatures along the box girder depth, the measure points on sections I and II were selected. They were also chosen for a comparative analysis. The numbering of measure points ranges from A to M.

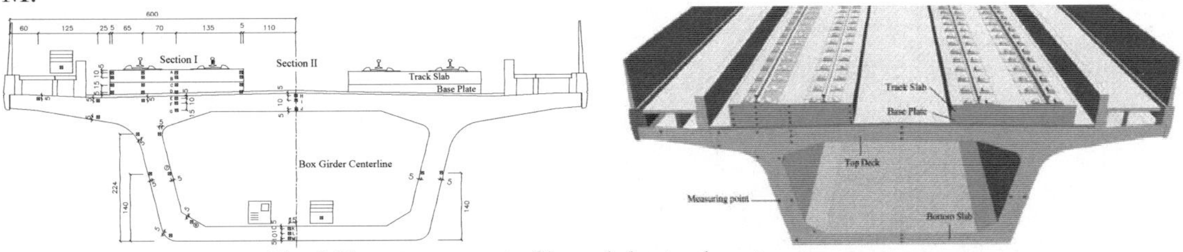

Fig. 1 The arrangement of box girder-track system temperature sensors

The above test connects temperature elements with a data acquisition instrument for automatic data collection and storage for a sampling period of 30 minutes. Later, a 4G wireless remote transmitter was used to download the collected data to a remote computer for long-term data monitoring and collection. Representative test data collected between 0:00 on June 8, 2014 and 24:00 on June 9, 2016 were selected for the temperature field analysis of girder-track system

2.2 Decomposition of Temperature Time Series

Concrete structure is generally affected by three thermal phenomena: sunlight, sudden cooling, and annual temperature change. According to the principle of time series, the time series of temperature monitoring data can be generalized into long-term trend T, annual temperature effect S, sunlight action C, and irregular change I. Thus, the following combination model is valid

$$Y = T + S + C + I \tag{1}$$

However, the concrete structure is mainly influenced by annual temperature variation and solar radiation. So, the above formula can be simplified as

$$T_o = T_u + T_w \tag{2}$$

where T_u is the uniform temperature (°C) and T_w is the fluctuation temperature(°C).

Taking the measuring point A (near the top edge of ballastless track) as an example, the time-history curve of its monitoring temperatures is shown in Fig. 2.

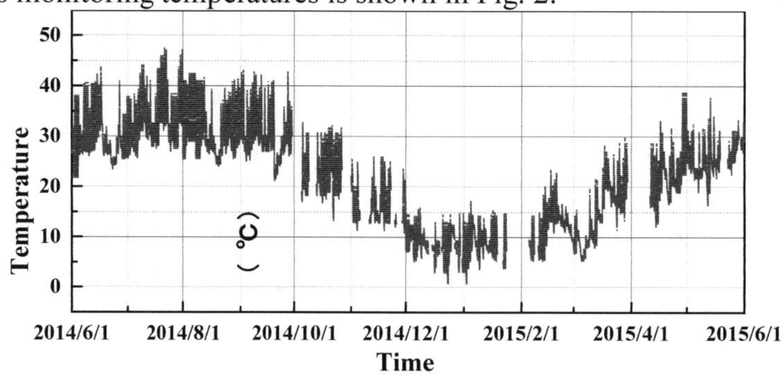

Fig. 2 Time-dependent curve at measuring point A

Decomposition of trend variations and short-term periodic change of monitoring data, with the adoption of time series difference method: Taking day as the cycle, decompose the original data series difference was decomposed and the time-history curve of its trend variations was obtained. Since the uniform processing method was applied to determine the data difference, the time-history curve of temperature variations could be defined as the uniform temperature curve at measuring point A, as shown in Fig. 3.

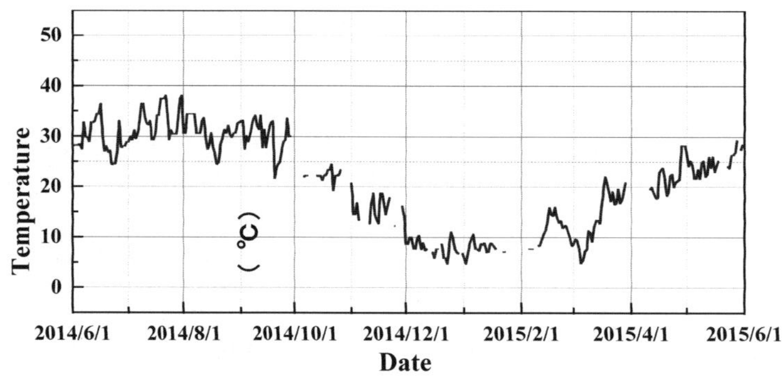

Fig. 3 Time-dependent curve of uniform temperatures at measuring point A

The uniform temperature curve shown in Fig. 3 represents the overall variation trend of the temperatures at measuring point A, displaying generally a sinusoidal variation. It also reflects the following basic seasonable uniform temperature changes: The temperature is higher in summer and lower in winter; the overall temperature variation presents a trigonometric function with obvious annual periodicity.

The separate fluctuation temperature can be obtained by subtrating the original temperature (Fig. 2) from the uniform temperature (Fig. 3). The daily temperature fluctuations due to sunlight are shown in Fig. 4.

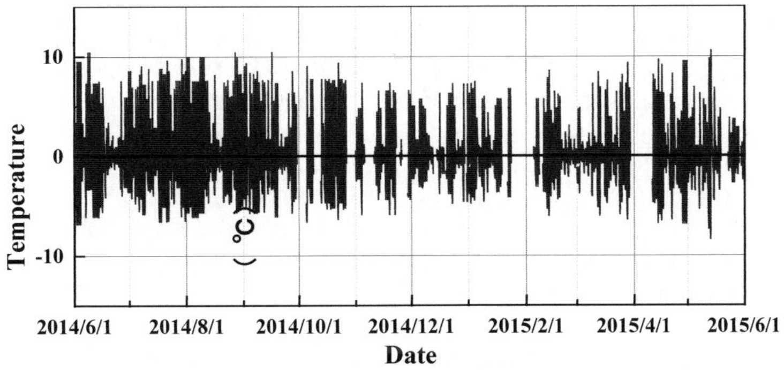

Fig. 4 Time-dependent curve of fluctuating temperature in section A

According to Fig.4, fluctuating temperatures always fluctuate around the zero-temperature line daily, which follows the natural rule of solar radiation changes.

2.3 Uniform Temperature Component

In order to confirm the uniformity of uniform temperatures in a bridge structure, the above method was used to calculate the uniform temperatures at each measurement point of the box girder-track system. The results indicate that the variation rule of uniform temperatures at each measure point is virtually the same, namely that the uniform temperature is mainly affected by the uniform temperature of atmosphere and at lesser degree by the location of measure point. Considering every uniform temperature curve is similar to a Fourier curve, after physically modifying its cycle, using the modified Fourier series fitting technique to fit the curve, the following fitting equation was obtained

$$T_u = a_0 + \sum_{n=1}^{\infty}\left[a_n \cos\left(n\frac{2\pi}{N}t\right) + b_n \sin\left(n\frac{2\pi}{N}t\right)\right] \quad (1)$$

where T_u is the structure uniform temperature (°C), t is the sequence number of time (date, with Jan. 1 as the first day), N is the total days of in a year, and a_0, a_n, and b_n are the fitting parameters (a_0= the mean value of measurement point uniform temperatures).

In doing the fitting, the accuracy of the first order fitting was found to be above 0.9, meeting requirement for fitting accuracy, Eq. (1) can hence be expressed as

$$T_u = a_0 + a_1 \cos\left(\frac{2\pi}{N}t\right) + b_1 \sin\left(\frac{2\pi}{N}t\right) \quad (2)$$

Using the triangle conversion equation, $a\cos(\omega t) + b\sin(\omega t) = \sqrt{a^2+b^2}\cos\left[\omega(t-\varphi)\right]$, the above equation can be further simplified as

$$T_u = a_0 \text{-b} \cos\left(\frac{2\pi}{N}(t-c)\right) \quad (3)$$

where $b = \sqrt{a_1^2 + b_1^2}$ representing the amplitude of the annual variation of uniform temperatures (℃)and c is the sequence number of the date reaching the lowest temperature.

So, according to the fitting of above uniform temperatures at each measurement point, we can obtain the mean value of uniform temperature a and the variation amplitude of uniform temperature as 18.68~20.45 ℃ and 10.68~13.60 ℃, respectively, where the temperature difference is about 2 ℃. The sequence number for the lowest temperature day c varies between 20 and 24, with the maximum difference between the measure points less than 2 days. So, the uniform temperature curve of every measurement point is deemed unchanged. Hence, in practical applications, a single curve may be used to describe the uniform temperature curve rule of a structure. This study takes the mean value from the measure point uniform temperature curve to come up with the time function for describing the annual temperature variation on the structure as

$$T_{u,system} = 19.74 - 11.83\cos\left[\frac{2\pi}{365}(t-20)\right] \quad (4)$$

2.4 Fluctuation Temperature Component

As shown in Fig. 4, when fluctuating temperatures are not affected by the sudden weather change, temperatures will fluctuate near the zero-temperature line in Fourier curve fashion. The daily fluctuation temperature is mainly affected by solar radiation and is also related to rainfall, wind speed, and humidity. Therefore, the temperature fluctuation curve of sunny days was analysed. The measure values at point A form August 15th to 19th, 2015 were selected, as shown in Fig. 5.

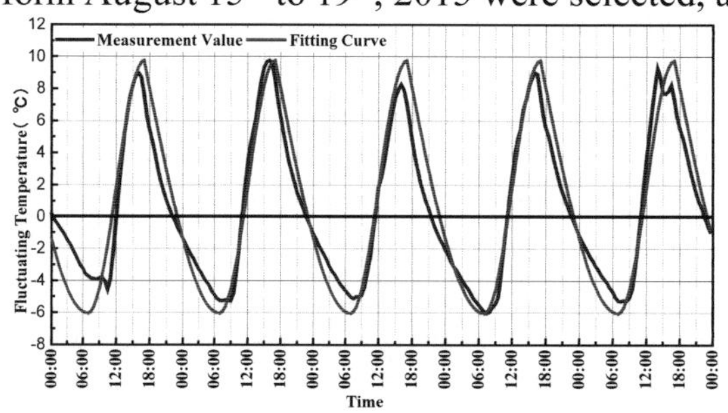

Fig. 5 Time-dependent curve of fluctuating temperatures in 5 days

Fig. 5 shows the positive and negative fluctuating temperature amplitudes are not the same during a day. The positive temperature is actually higher than the negative temperature. In addition, the heating stage is obviously shorter than the cooling stage for concrete structures, indicating that the heating rate of concrete is faster than cooling. To sum up, the fitting of temperature fluctuation curve should be divided into two parts including heating and cooling stages.

3. Extreme Temperature of a Box Girder

3.1 Extreme Value Model

In the extreme value theory, sample data are sorted and when a larger threshold u is exceeded, the rest of sample data satisfies the following equation:

$$P_r(x>u)=D(x)=\begin{cases} 1-\left(1+\xi\dfrac{x-u}{\sigma}\right)^{-1/\xi} & \xi \neq 0 \\ 1-exp\left(-\dfrac{x-u}{\sigma}\right) & \xi = 0 \end{cases} \quad (5)$$

where ξ is the form parameter, σ is the size parameter, and u is a threshold value.

As indicated in the above, the determination of threshold u is critical. Based on the average residual life graph, the minimum squared error (MSE) criterion was used to determine the tail data quantity by weighing the relation between the deviation and the variance as:

$$MSE\left(\hat{\xi}\right) = bias^2\left(\hat{\xi}\right) + var\left(\hat{\xi}\right) \tag{6}$$

where $bias^2\left(\hat{\xi}\right)$ represents the deviation of estimate, calculated by the bootstrap re-sampling method (the number of samplings being determined according to the estimated deviation value without the influence of frequency) and $var\left(\hat{\xi}\right)$ represents the variance of estimate.

In terms of the method for estimating parameters, this study used the generalized method of moments (GMM) to estimate ξ, with the following equation

$$\xi = 0.5\left(1 - \bar{s}^2 / var(s)\right) \tag{7}$$

where, $\{s\} = \{X_{n-k+1} - u, X_{n-k+2} - u, \dots X_n - u\}$, representing the set of D-values between the super threshold samples and the threshold and k is the number of tail data ($=2, 3, \dots, n-2$).

Based on the confirmed threshold u and assuming that the empirical cumulative probability function $f(X_i) = i/(n+1)$ is known ($1 \le i \le n$), the cumulative probability function of random variable can be expressed as:

$$F(x) = \begin{cases} f(x), & x \le u \\ f(u) + \left(1 - f(u)\right) D(x), & x > u \end{cases}, \tag{8}$$

Therefore, according to the empirical cumulative probability function and the above equation, the scatter points can be constructed as:

When $\xi > 0$:

$$\left\{ \left[1 - \frac{j/(n+1) - (n-k)/(n+1)}{1 - (n-k)/(n+1)} \right]^{-\xi} - 1,\ T_j^* \right\} \tag{9}$$

When $\xi \le 0$:

$$\left\{ -ln\left[1 - \frac{j/(n+1) - (n-k)/(n+1)}{1 - (n-k)/(n+1)} \right],\ T_j^* \right\} \tag{10}$$

In Eqs. (9) and (10), $j = n-k+1, \dots, n$ and T_j^* is the temperature difference of corresponding sequence after ranking. By means of the tail data divergence of the upper structure, the linear regression method can be applied to obtain the estimated values of the scale parameter σ.

3.2 Temperature Difference in Girder Section

In order to study the distribution law of temperature gradients in the girder-track system of a HSR, the extreme positive temperature difference occurred on July 21, 2014, with the air temperature of 27~37℃, based on the measured samples of section I and section II in two years. Therefore, the temperature measurement data of this particular day were selected for analysis, as shown in Fig. 6.

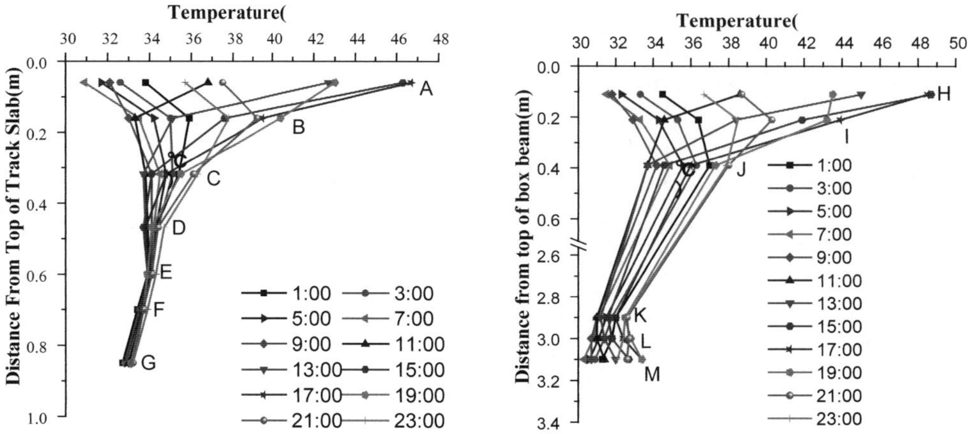

Fig. 6 The vertical temperature gradient distributions of section I and section II

As seen from Fig. 6, the heating from the sun on the girder-track structure is rapid and vertical distribution of temperatures between section I and section II is virtually the same before and after 11:00~21:00. The extreme temperatures of 13.9 C and 17.4 C were reached at 16:00. It can also be seen that the temperature of the covered part of the girder is lower than that of the uncovered part and the temperature of the box girder's roof is higher than the top edge of the base plate due to the thermal insulation of the waterproof layer. At night, the positive temperature gradient began to decrease, the negative temperature gradient appeared before and after 01:00~9:00, and the maximum value was reached at 7:00.

3.3 Extreme Temperature Prediction3.3.1 Extreme Temperature Gradient

First, using the temperatures at measure points A, B, C, D, E, and F in section I to minus the temperatures of measurement points G in same time, accordingly, adopting the temperature of H, I, J, L, and M to minus the temperatures of K. Then, the positive and negative temperature differences of the girder-track system were obtained statistically using the above method. The extreme value theory as described in section 3.1 was later used to estimate the daily extreme values from the statistical samples of each section. The positive and negative temperature difference estimates of cross sections I and II corresponding to the 100-year recurrence period were then obtained. The results are shown in Table 1.

Table.1 Statistic values of temperature differences

Statistical value of temperature difference in Section I					
Distance from the parietal margin of the structure (m)	Number	Measured Value		Estimated Value	
		Max positive	Min negative	Max positive	Min negative
0.06	A-G	15.2	-7	23.36	-12.6
0.16	B-G	7.8	-3.9	12.5	-7.8
0.318	C-G	3.3	-1.9	8.85	-3.9
0.468	D-G	2	-0.8	3.26	-1.83
0.6	E-G	1.3	-0.4	2.56	-1.3
0.7	F-G	0.8	-0.1	1.5	-0.7
Statistical value of temperature difference in Section II					
0.05	H-K	17.4	-4.1	24.4	-8.7
0.15	I-K	11.9	-2.8	17.1	-5.8
0.335	J-K	5.4	-1.4	7.3	-4.9
2.945	L-K	1	-0.5	2.5	-1.7
3.045	M-K	2.9	-1	5.6	-2.1

As indicated in the above table, the maximum positive and negative temperature differences from all measuring points (A-G) of the girder-track system are 15.2 ℃ and ℃, respectively. The GPD model was used to calculate the positive and negative temperature differences corresponding to 100-year recurrence period, giving 23.36℃ and -12.6℃, respectively.

Along the centreline of box girder, the maximum positive and negative temperature differences at H-K are 17.4℃ and -4.1℃, respectively. The estimated values calculated by the GPD model are 24.4℃ and -8.7℃, respectively. From these, we see that the extreme temperature differences calculated by the GPD model can better envelope the maximum temperature differences of the measured points and can predict the vertical temperature gradients under different recurrence periods with different probability demands.

The temperature gradient was fitted by an exponential curve based on the railway code of China.[9] Since the railway specification does not consider the influence of the ground reflection of temperatures on box girder's bottom plate, in doing the fitting a linear variation for the vertical temperature gradient was considered. The fitting results of positive and negative actual measurements and estimated values in 100-year recurrence period for sections I and II are shown in Fig. 7, where the square of fitting correlation factor is above 0.98.

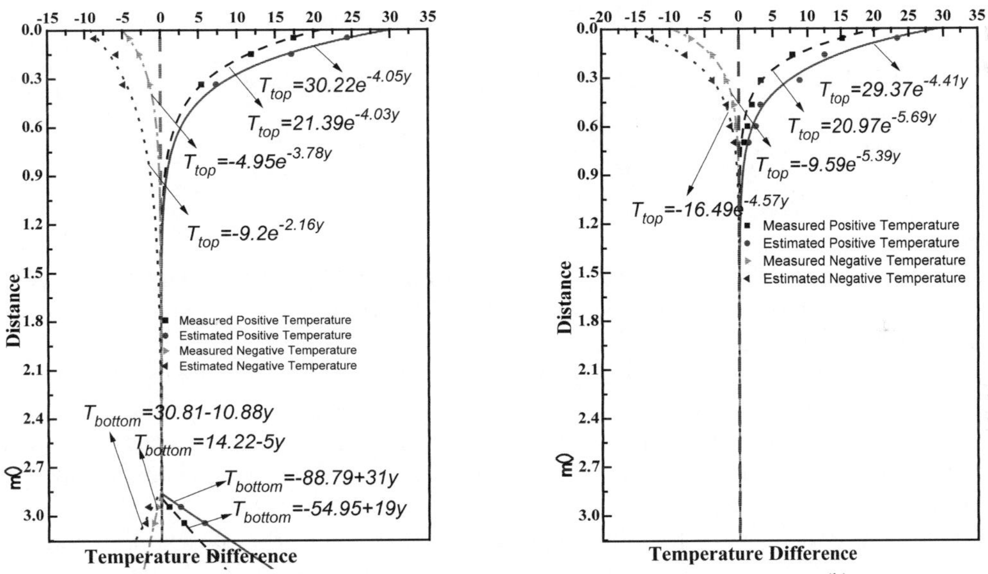

(a) (b)

Fig. 7 Fitting curves of the vertical temperature gradient of section I and section II

From Fig. 7(a), we can conclude that the maximum actual temperature heating pattern for the girder-track as $T_y = 20.97e^{-5.69y}$ and the estimated temperature heating pattern in 100-year recurrence period as $T_y = 29.37e^{-4.41y}$. According to Fig. 7(b), the actual maximum temperature heating pattern for the box girder is $T_y = 21.39e^{-4.03y}$ and the estimated temperature heating pattern in 100-year recurrence period is $T_y = 30.22e^{-4.05y}$. While China's railway specification stipulates that the vertical positive gradient pattern of ballastless two-way bridge is $T_y = 16e^{-10y}$, which indicates a smaller maximum temperature gradient than the measured value.

3.3.2 Extreme Uniform Temperature

Based on annual uniform temperature data of girder-track system, the GPD model was adopted to estimate the extreme uniform temperatures for 100-year recurrence period. The maximum and minimum measured uniform temperatures were calculated, as shown in Fig. 8.

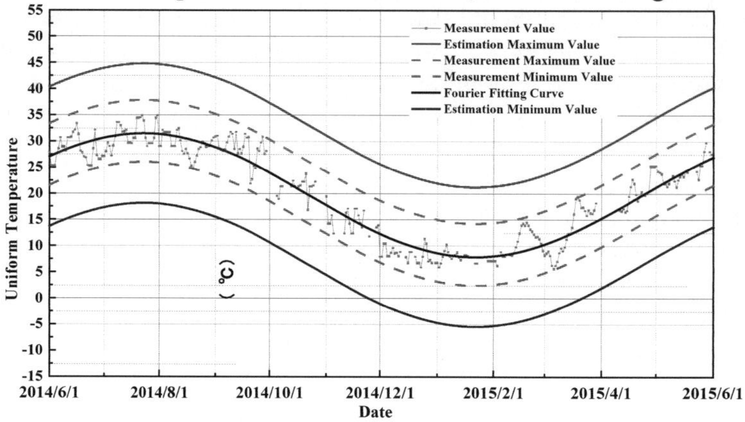

Fig. 8 Fitting curve of uniform temperatures

According to the above figure, the highest extreme uniform temperature estimated by the 100-year recurrence period is 7 ℃ higher than the measured value and the lowest extreme uniform temperature is 7.8 ℃ lower than the measured value.

4. Conclusions

Based on the two-year field data of girder-track system of a HSR and the analysis results of bridge uniform temperatures and extreme temperature differences from the time sequence method and the GPD model , the following conclusions may be made.

(1) The time sequence deposition method may be used to decompose the box girder's temperature as uniform temperature and fluctuating temperature. In addition, a physically meaningful uniform temperature equation for a HSR bridge can be derived using Fourier fitting technique and triangle function transformation. Such equation may be used to find the pattern of yearly temperature variations.

(2) The GPD model provides a reliable theoretical method for determining the maximum temperature load during the design service life of a HSR bridge, especially in the case that the statistical sample size is small and the tail data is sensitive. Through the tail temperature data of daily temperatures, the daily maximum temperature difference at each measuring point can be obtained. At the same time, the maximum temperature difference during the design life can be better enveloped using the extrapolated value of the 100-year recurrence period to the extreme temperature load.

(3) Through the statistical analysis of girder-track system on extremum temperature gradient samples, the GPD model can be utilized to estimate the extreme temperature corresponding to the 100-year recurrence period. The measured and estimated values of maximum positive temperature difference for the covered slab are 15.2 ℃ and 23.36 ℃, respectively and the values for the non-covered area are respectively 17.4 ℃ and 24.4 ℃.

References

[1] Dai, GL & Tang, Y & Liu, Y 2016, 'Numerical Analysis of Solar Radiation Temperature Field of Round End Hollow Tall Pier for High-Speed Railway Bridge', Bridge Construction, Vol.46, No.6, pp.67-72.

[2] European Committee for Standardization 2003, EN 1991-1-5. Eurocode 1: actions on structures – Part 1-5: general actions: thermal actions, Brussels.

[3] American Association of State Highway and Transportation Officials 2013, AASHTO LRFD Bridge Design Specifications, Washington, DC.

[4] CCCC Highway Consultants Co., Ltd. 2015, General Code for Design of Highway Bridge and Culverts, China Communications Press, Beijing.

[5] China Railway Engineering Consulting Group Co., Ltd 2005, Code for design on reinforced and prestressed concrete structure of railway bridge and culvert, China Railway Publishing House, Beijing.

[6] Obrien, EJ & Enright, B & Getachew, A 2010, 'Importance of the Tail in Truck Weight Modeling for Bridge Assessment', Journal of Bridge Engineering, Vol.15, No.2, pp210-213.

[7] Enrique, C & Ali S. Hadi 1997, 'Fitting the Generalized Pareto Distribution to Data'. Journal of the American Statistical Association, Vol 92, No.440, 1609-1620.

[8] Li Y & Cai W 2005, 'Campbell E P. Statistical Modeling of Extreme Rainfall in Southwest Western Australia', Journal of Climate, Vol.18, No.6, pp852-863.

[9] China Railway Design Corporation 2014, Code for design of high speed railway, China Railway Publishing House, Beijing.

The 8th International Conference on
Structural Health Monitoring of Intelligent Infrastructure
Brisbane, Australia | 5-8 December 2017

The experimental study on the concrete hollow slab temperature field in fire

Y. Chen[1,2]; Z. Xu [1,2]; Y. Wang [1,2].

[1] *Guangdong Hualu Transport Technology Co. Ltd., Guangzhou 510550, China; Email : 251735933@qq.com* [2]*Research center for highway traffic safety and emergency security technology and equipment of ministry of transport, Guangzhou 510550,China;*[3]*GuangdongProvincial Communication Group Co. Ltd., Guangzhou 510101, China;*

Abstract

The concrete hollow slab removed from the used bridge was placed in the fire test furnace, and an experimental study on the temperature field for the hollow slab exposed to the standard fire time-temperature curve given in ISO834(1999) was carried out. It showed that the variation of temperature of the hollow slab was nonlinear. The hollow slab was exposed to fire and the time was up to 180min, The temperature of the hollow slab surface exposed to fire was up to 1000℃, and the temperature of the strand position (vertical distance from the fire surface is 56mm) was up to 500℃. Besides, the finite element numerical simulation of the temperature field of the hollow slab was basically consistent with the experimental results. It showed that the thermal parameters which were fitted by a number of correlation studies were suitable for concrete structure member temperature field simulation. The high temperature fire test system, the simulation analysis method for the temperature field of hollow slab, and change law of the temperature field, are not only valuable for fire resistance design of concrete bridge in fire, but also are significant for the detection, maintenance and reinforcement of concrete bridge after fire.

1. Introduction

With the rapid development of society, economy and traffic in China, the concrete hollow slab bridge fire caused by traffic accident occurred frequently(Wu et al. 2015). In fire, heat was transmitted directly to the concrete structure members by means of heat radiation, convection, and conduction. The temperature of the concrete structure members increased gradually. The mechanical properties of concrete and reinforcing bars deteriorated gradually and the safety of the bridge have been seriously affected(Chen,et al. 2016). The studies have shown that the mechanical properties of concrete hollow slab was closely related to its temperature history. The accurate acquisition of temperature field of the hollow slab exposed to fire is the basis of bridge damage assessment(Tong et al. 2012).

There are two main ways to study the temperature field of concrete hollow slab in fire at home and abroad. One way is numerical simulation, the other way is fire test at high temperature. In numerical simulation, Yu et al.(2009)not only analyzed the temperature field of concrete solid slab and hollow slab with the same shape dimension, but also discussed the setting of boundary conditions in the simulation analysis of temperature field of hollow plate. Jing et al.(2016)and Liu et al.(2014).analyzed the temperature field and the mechanism of its bursting of the hollow slab exposed to fire test at high temperature. In the aspect of fire test of structure members, the fire tests of 15 prestressed concrete hollow slabs were carried out to study the fire resistance of prestressed concrete hollow slabs under different loading levels (Xu etal.2013). Hollow slab fire test in the laboratory was carried out to investigate the temperature field, deformation and strain variation in fire(Liu et al.2017). The existing studies focuses on the variation of temperature field and fire resistance of hollow slab under certain fire conditions. However, the parameters of the temperature field simulation model parameters of the hollow slab, the variation of the temperature with time and depth are not thoroughly tested and summarized and verified.

Therefore, according to the calculation principle of fire resistant design and the nonlinear finite element method of concrete structure temperature field, The experimental study and numeric simulation analysis of the temperature field of concrete hollow slab exposed to the standard fire time-temperature curve (ISO834,1999) were carried out. in this paper, the fire test system, the simulation analysis method of the temperature field for hollow slab exposed to fire and the temperature field change laws of hollow slab were valuable for the fire test and calculation of the temperature field of similar bridge structure members, as well as for the inspection, assessment and reinforcement design of similar bridge structure members after fire.

2. Hollow slab fire test

2.1 Hollow slab overview

Hollow slab was removed from the Guangzhou Shenzhen highway interchange ramp bridge in the process of reconstruction and expansion. The bridge was built in 1995 and the superstructure of the bridge was pre-stressed hollow slab. The substructure of the bridge was double column pier, bored pile foundation.The design load was automobile super-20, Trailer-120(CA, JTJ 021-1989). The bridge cross section is shown in Fig.1. The hollow slab and the bridge are shown in Fig. 2.

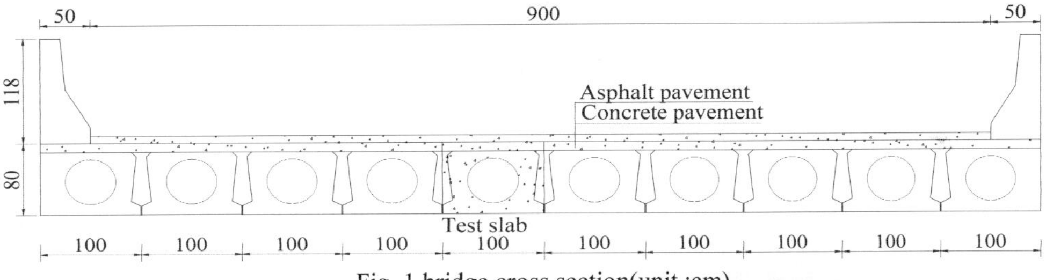

Fig. 1 bridge cross section(unit :cm)

Fig. 2 The hollow slab and the bridge

2.2 Fire test system

The hollow plate fire test system is shown in Fig.3, The whole system were composed of open fire environment system, refractory insulation system, data acquisition system and temperature control system. Among them, the open fire environment system system were mainly composed of furnace, heat insulation baffle, burner console, burner, liquefied gas and air supply equipment.The temperature control was automatically controlled by an industrial computer program. So the standard fire time-temperature curve(ISO834,1999)can be simulated in the furnace. The maximum temperature of the furnace can be up to 1200℃, as shown in Fig. 4. The temperature fluctuation was small and uniform, and the actual heating curve was close to the the standard fire time-temperature curve (ISO834,1999), as shown in Fig. 5.

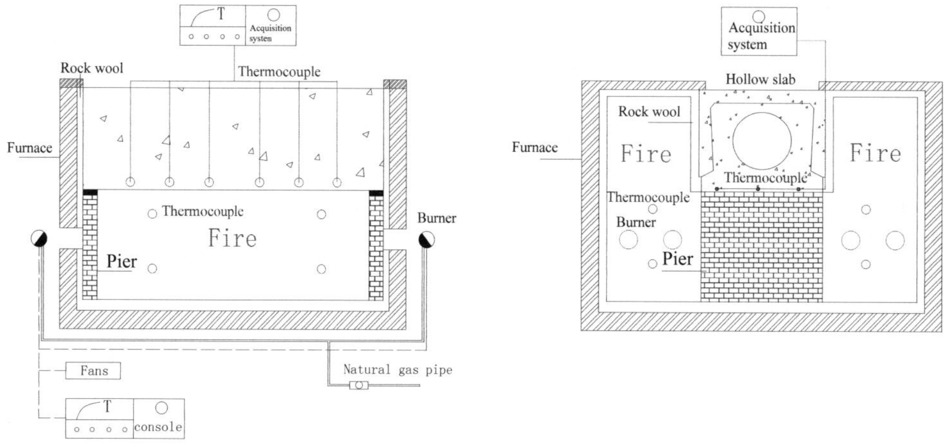

(a) The Elevation view of fire test system of the hollow slab

(b) The side view of fire test system of the hollow slab

Fig. 3 Temperature field test system

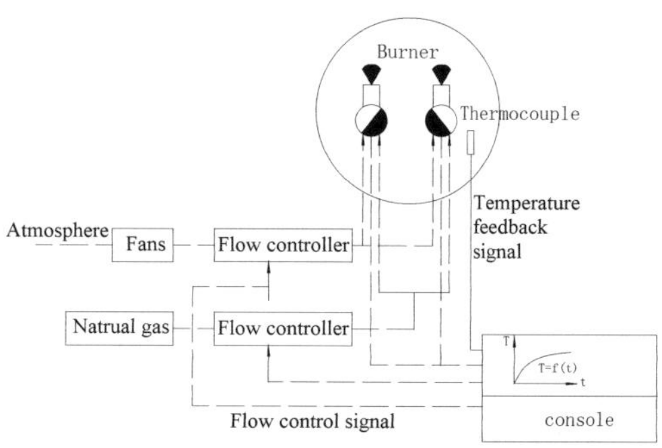

Fig. 4 Temperature control system

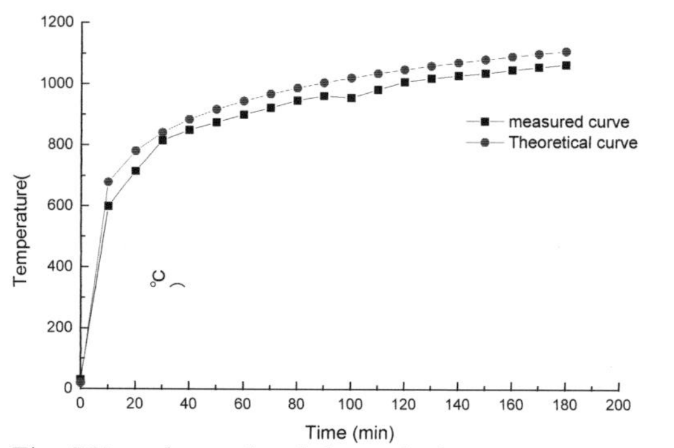

Fig. 5 Experimental and theoretical temperature curve

2.3 The temperature sensor setting

According to the purpose of hollow slab temperature field test, the working conditions of fire test system, the size of hollow slab, the thermocouple performance and so on, The section for temperature measurement was set,The installation of temperature sensors were carried out in accordance with the following principles.

(1)The temperature measuring section setting, fully consider the temperature distribution of the hollow slab in the furnace, when the temperature measuring sections were set up, the complex temperature field near the pier was fully avoided. Along the length of the hollow slab, The test sections were mainly located in the middle part, And 6 sections (section A to section F) were set, the distance between each section was 20cm, the distance from section A to the left pier was 35cm, and the distance from section F to the right pier was 40cm, as shown in Fig. 6.

(2) According to the temperature field of the hollow slab obtained by preliminary numerical simulation, The temperature measuring points were set at different depths at each section, and the

number of temperature measuring points of each section was 3.The setting of the measuring points of each section are shown in Fig.7 and Table 1.

(3) The temperature sensors adopted armored K thermocouple. When they were installed, first of all, the hole with diameter 5mm was drilled in each setting position by the impact drill. And then, In order to make the heat conduct well between the thermocouple and concrete, a small amount of aluminum powder with good thermal conductivity was injected into the hole.Finally, the cement mortar was injected into the hole.

(4)The data acquisition was controlled by computer, and the temperature was collected 1 times every 1min,and the total time was 180min.

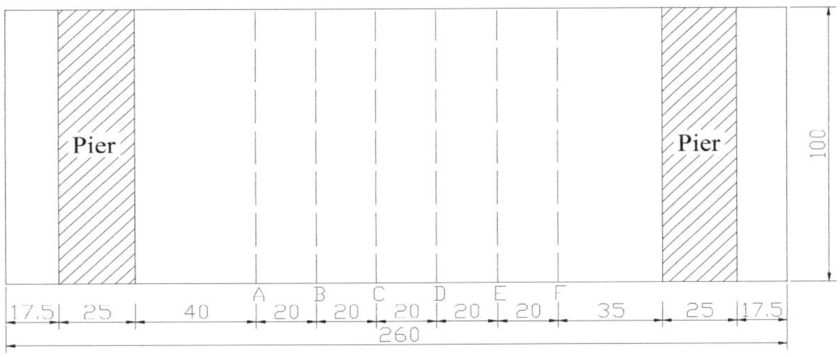

Fig. 6 Temperature field test section layout (unit: cm)

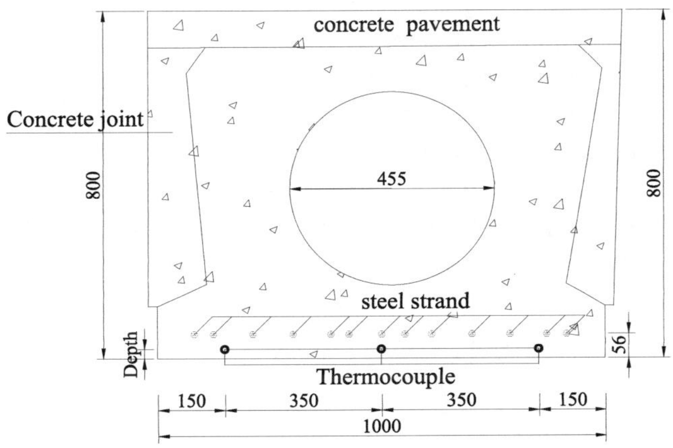

Fig. 7 Temperature measurement point cross-section layout (unit: mm)

Table 1 Temperature sensor embedded depth

Section	A	B	C	D	E	F
Depth (mm)	20	30	42	56	73	100

2.4 The temperature test result

The time temperature curves of each section are shown in Fig. 8. The depth temperature curve of the hollow slab at 30min, 60min, 120min, 150min and 180min are shown in Fig. 9.

(1)The temperature gradually increased with the time increasing, at the same time interval, in the early stage, the temperature increasedrelatively fast, but at a later stage, the rate of temperature rise is relatively slow. In the process of heating, the temperature on the hollow plate fire suface was obviously higher than that of the internal, The experimental results were consistent with the temperature variation mechanism of the concrete members exposed to the standard fire time-temperature curve(ISO834,1999).

(2) At a certain moment, the internal temperature of the hollow slab decreased gradually along the depth direction, and the gradient of decreasing gradually decreased.
(3) When the time of the hollow plate beam in the fire was 180min, the temperature on the hollow slab fire surface was 1061℃, the temperatrue in the strand position (vertical distance from the fire surface is 56mm) was 529℃, and The temperature at depth 90mm (section F) was 312℃.

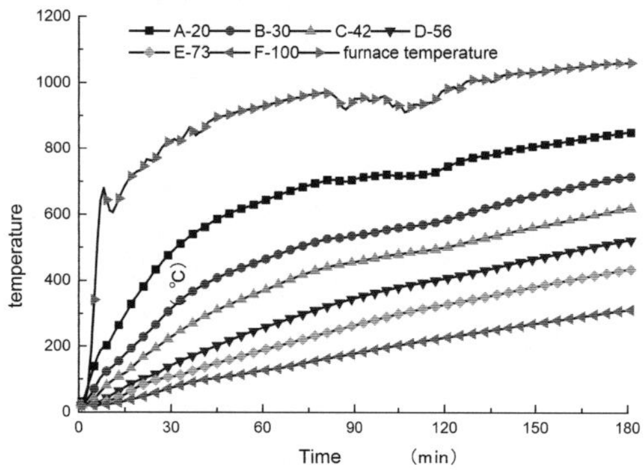

Fig. 8 Temperature-time curve of each section

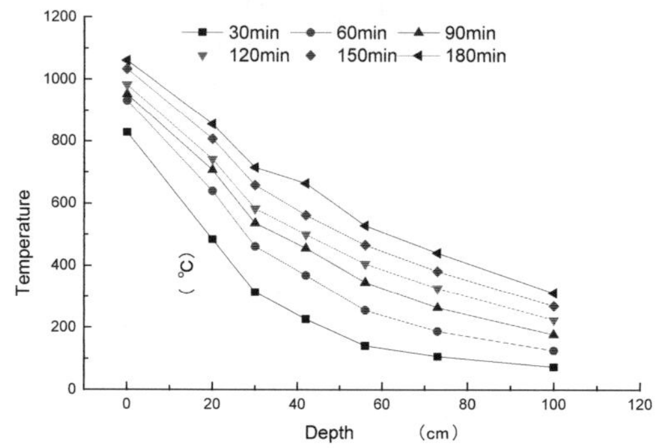

Fig. 9 Temperature-depth curve at different times

According to the time temperature curve of each section of the hollow slab, Further analysis was made on the variation law of the 300℃ and 500℃ isotherms (which determine the deterioration of component performance) with increasing depth. As shown in Fig. 10.
(1)T In the initial stage, with the increase of time, the isothermal temperature at 300℃ and 500℃ moved towards the depth were faster than the latter, In the latter stage, they move deeper into the depth than at the early stage. If the linear relationship is used to fit the function of temperature with time, we can get the following results: the moving average velocity of 300℃ isothermal is 0.5mm/min, and the average speed of 500℃ isothermal moving is 0.3mm/min.
(2)T When the time of the hollow plate exposed to the standard time temperature curve was 180min, the distance between the 300℃ isotherm and the fire surface was 101mm, and the distance between the 500 ℃ isotherm and the fire surface was 62mm.

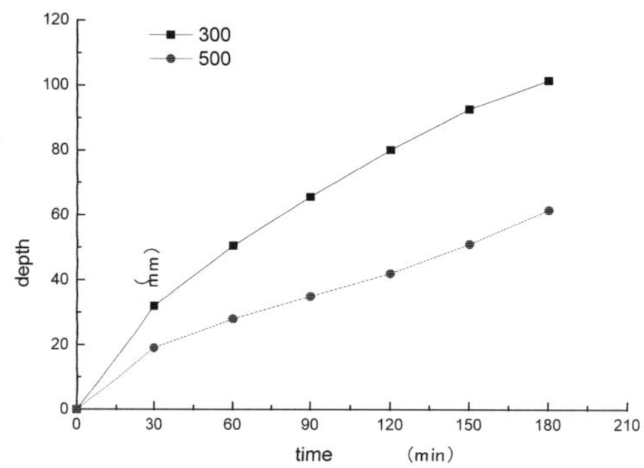

Fig. 10 Depth – time curve of isotherm

3 Simulation of Temperature Field

3.1 Basic Principles of Temperature Field Analysis for Concrete Structures

The heat transfer process of concrete members when it was exposed to fire is real-time. The temperature field in the structure members is transient temperature field. By analysis of the heat of a micro cuboid（Fig.11）, the basic differentia equation of heat conduction can be developed and solution found.

$$\frac{\partial T}{\partial t}=\frac{1}{c\rho}\left[\frac{\partial}{\partial x}\left(\lambda\frac{\partial T}{\partial x}\right)+\frac{\partial}{\partial y}\left(\lambda\frac{\partial T}{\partial y}\right)+\frac{\partial}{\partial z}\left(\lambda\frac{\partial T}{\partial z}\right)\right] \tag{1}$$

ρ—Concrete mass density（kg/m^3）;c—Specific heat capacity（$J/kg\cdot K$）;λ—Thermal Conductivity（$W/m\cdot K$）;t—time（s);T—temperature（K）。

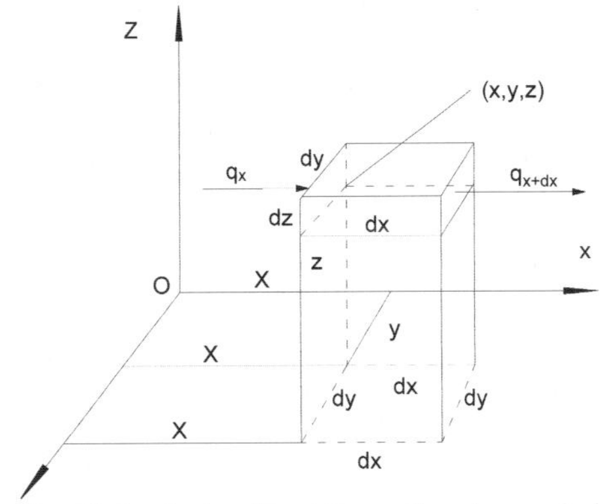

Fig. 11 Analysis of heat flow of a micro-cuboid

When the structure member is subjected to fire, the thermal boundary condition is generally complicated.Because the concrete is subjected to high temperature, its mechanical properties change nonlinearly with the increase of temperature, It is difficult to solve the partial differential equations by analytic method. Usually, the differential equation of heat conduction is converted into equivalent integral form. According to variational principle, A standard finite element formulation for solving the temperature field of a component can be derived.

$$[C]^e\frac{\partial}{\partial t}\{T(t)\}^e+[K]\{T(t)\}^e=\{P\}^e \tag{2}$$

$[C]^e$— the matrix for the special heat capacity of the element,$[K_T]^e$—the matrix for heat conduction of the element,$[K_\beta]^e$—the matrix for heat transfer of the element,$\{P_T\}^e$—the vector for the heat load of the element,$\{P_\beta\}^e$— the vector for the heat transfer load of the element.

By formula (3-2), considering the mode of fire load and thermal analysis boundary condition, a finite element model for the component temperature field analysis is established, and the distribution law of temperature field can be obtained.

3.2 Thermal parameter

According to the calculation principle of temperature field of concrete component, the factors affecting the temperature field of component are not only related to external fire load, heat conduction boundary condition, but also to the thermal parameters of components. Because the concrete mineral composition, mixture ratio, water content and test in the process of heating, heating speed and other random factors, the results of the existing concrete thermal parameters are not exactly the same. In this paper, the average thermal parameters of concrete was adopted, and the function expression of the thermal parameters - temperature were set by polynomial fitting.

According to the results (the concrete societies of UK,1966; Lu,1989; Lie T T,1994; Lu,1995; Li,1991; ENV,1995; chen,2004), the thermal conductivity fitting by polynomial is

$$\lambda_c=1.633-0.111\times(\frac{T}{100})+0.279\times(\frac{T}{1000})^2 \tag{3}$$

According to the results (the concrete societies of UK,1966; Lu,1989; 1994; Lu,1995; Li,1991; ENV,1995)the specific heat capacity fitting by polynomial is

$$C_c = 892.167 + 34.5(\frac{T}{100}) - 1.228(\frac{T}{100})^2 \tag{4}$$

According to the results (the concrete societies of UK,1966; Lu,1989; Hu,2001)the The Mass density fitting by polynomial is

$$\rho_c = 2372 - 30.11 \times (\frac{T}{100}) + 0.256 \times (\frac{T}{100})^2 \tag{5}$$

Table 2 Thermal parameters of concrete structural members

Temperature (℃)	**Thermal conductivity (W/（m.K）)**	**Specific heat capacity (J/（kg.K）)**	**The Mass density (kg/m³)**
20	1.61	900	2363
100	1.52	918	2348
200	1.43	960	2313
400	1.24	1019	2252
600	1.07	1054	2201
800	0.92	1085	2149
1000	0.8	1112	2098
1200	0.71	1134	2047

3.3 The finite element model

According to the theory of nonlinear finite element analysis of concrete structure member temperature field, the finite element model of temperature field of hollow slab was used to analyze the temperature field in the fire by Abaqus program, the model element adopted DC3D8 which is specially used for three-dimensional thermal analysis, The number of model element is 265375, and the overall size of the unit is set to 2cm, as shown in Fig. 12. The thermal boundary conditions of the model are shown in Figure.13

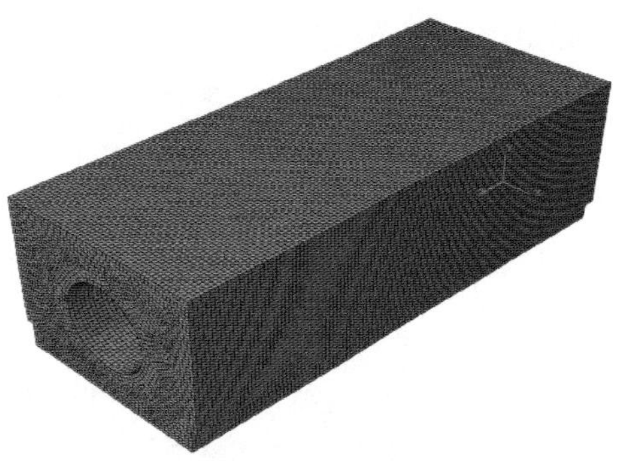

Fig. 12 Finite element model

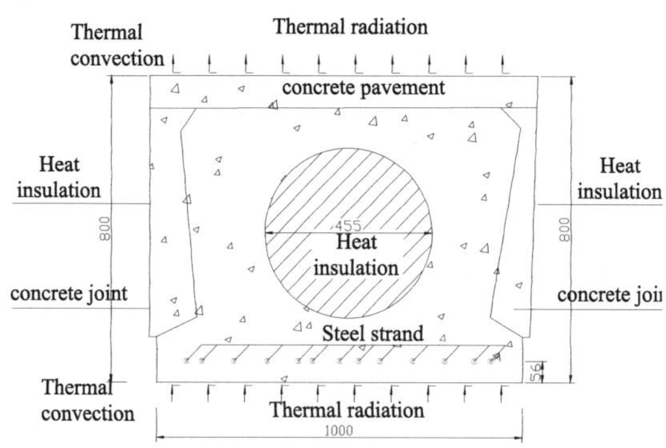

Fig. 13 Boundary conditions of model (unit:mm)

3.4 Numerical simulation analysis of temperature field

The temperature field of the hollow plate at 60min, 120min, 180min is shown in Fig. 14 and the following conclusions can be drawn:

(1) With the increase of time, the high temperature layers at the bottom of the hollow plate gradually moved from outside to inside, and the depth increases gradually.

(2) At a certain time, the temperature gradually decreased with the increase of the depth, and its temperature gradient decreased gradually.
(3) At a certain depth, the rate of temperature rise of the hollow slab at the initial stage is faster than that of the later stage. When the time was 180min, the temperature at the fire surface of the hollow plate was 1030℃, while the temperature of the steel strand position was 548℃ (As shown in Table 3), which was basically consistent with the test results.
the hollow slab was exposed to fire and the time was up to 180min, The results of numerical simulation and test of hollow plate temperature field are listed in Table 4 and plotted in Fig. 15.
(4) The simulation results of the temperature field of the hollow plate are basically consistent with the test. The deviation value is within 50℃, and the deviation ratio is less than 10% , and they are gradually decreasing from inside to outside.
(5) The finite element analysis model can accurately simulate the temperature field changes of structural members which was exposed to the standard temperature time curve (ISO834,1999).

Table 3 The temperature of the hollow slab at the fire surface and the strand (Unit: ℃)

Time	60min	120min	180min
Fire surface	1003	1023	1030
Strand position	306	458	548

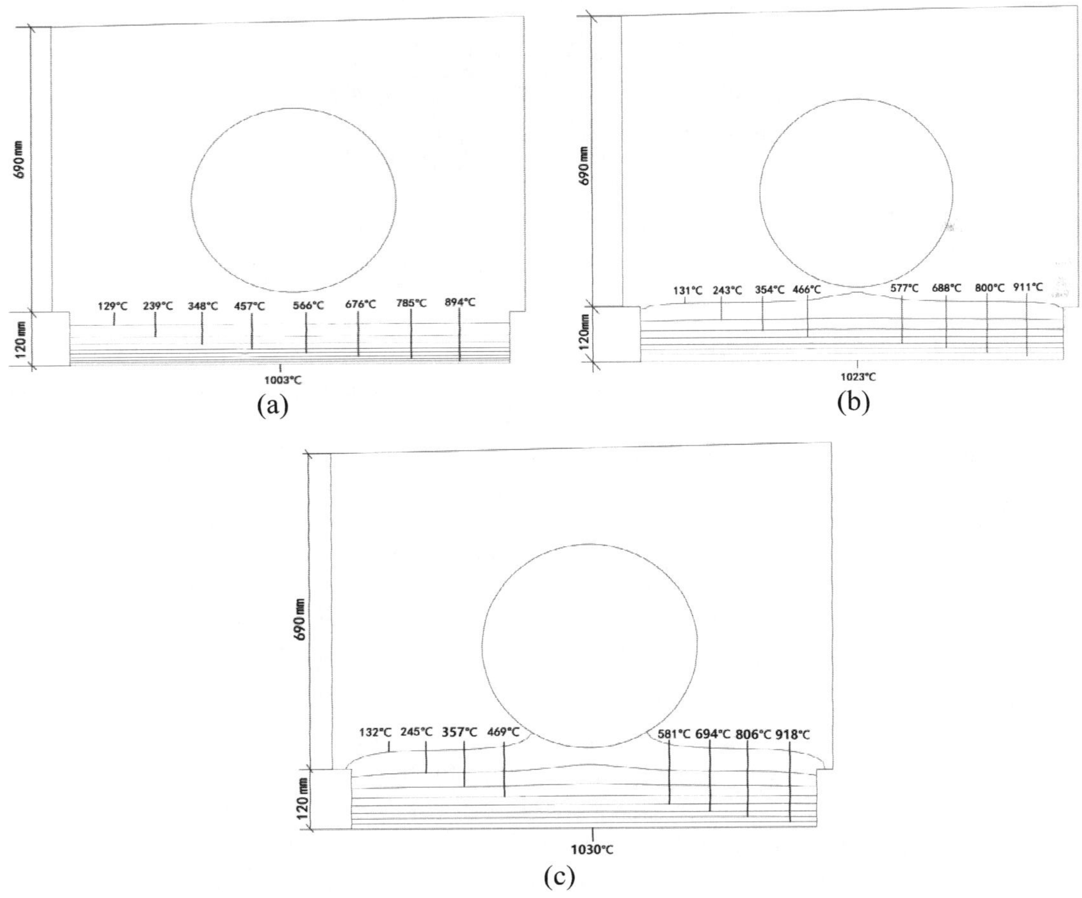

Fig.14 Temperature field of hollow plate (a) 60min (b) 120 min (c) 180min

Table 4 measured the temperature of each section and numerical simulation results

Section	Measured (℃)	Computed (℃)	Deviation(℃)	Deviation Ratio
A-20	855	827	-28	-3.3%
B-30	715	740	25	3.4%
C-42	664	645	-19	-2.9%
D-56	529	548	19	3.7%
E-73	440	452	11	2.6%
F-100	312	339	27	8.7%
Remarks	A-20 represents the A section temperature measurement point depth of 20mm			

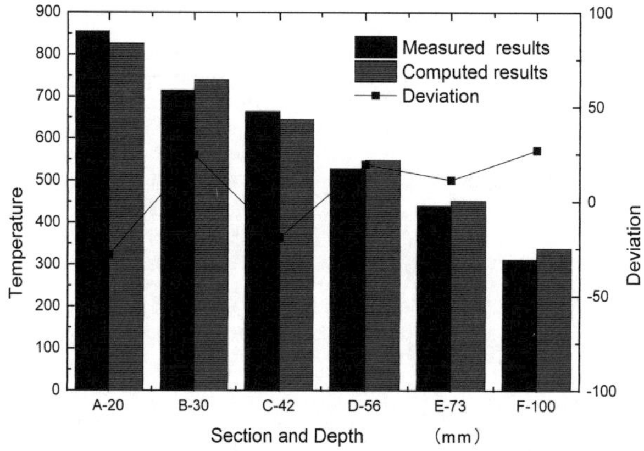

Fig. 15 Temperature measurement and computed results

4. Conclusions

The experimental research and numerical simulation analysis of the temperature field of the concrete hollow slab were carried out. The conclusions are as follows:
(1) With the increase of time, the temperature of the hollow plate gradually increases. At the same time interval, at the initial stage, the rate of temperature rising of the hollow slab was obviously larger than that in the later stage. (2) At a certain moment, the temperature of the hollow slab decreases gradually as the depth increases, and the gradient decreases gradually. The hollow slab was exposed to fire and the time was up to 180min, the temperature on the hollow slab fire surface was up to 1000℃, and the temperature in the strand position (vertical distance from the fire surface is 56mm) was up to 500℃. (3) In the initial stage, with the increase of time, the isothermal temperature at 300℃ and 500℃ moved towards the depth were faster than the latter. The moving average velocity of 300℃ isothermal was 0.5mm/min, and the average speed of 500℃ isothermal moving was 0.3mm/min. When the hollow slab in fire and the time was up to180min, the distance between the 300℃ isotherm and the fire surface was 101mm. The distance between the 500℃ isotherm and the fire surface was 62mm. (4) When the time of the hollow plate exposed to the standard time temperature curve was 180min, the distance between the 300 ℃ isotherm and the fire surface was 101mm, and the distance between the 500 ℃ isotherm and the fire surface was 62mm.

(5) The numerical simulation of the temperature field of the hollow plate was basically consistent with the test results. The absolute deviation was within 50 degrees, and the temperature deviation ratio was less than 10%. It showed that the thermal parameters which were fitted by a number of related references are suitable for concrete member temperature field simulation.

References

ISO 834-1, (1999), 'Fire resistance tests – elements of buildings construction – Part-1 general requirement', International Organization for Standardization, Switzerland.

Wu Jian, Li Bo, & Zhang Lizhi(2015), 'Inspection and evaluation of fire damage of prestressed concrete hollow slab bridge', Journal of Chinese & foreign highway,.vol.35 no.2, pp.209-213.

Chen Yingzhen, Xu Zhaofeng, Wang Yongping,&Lu chan he 'Residual strength of concrete after high temperature (fire) statistical analysis'. Journal of China Concrete and Cement Products. vol.2016 no.9, pp.1-5.

Tong Hao,Ji Xiaoyan,&ZhaoJiajun(2012), 'Fire-damage Evaluation Method of Pretensioned Hollow Platesand Counte rmeasures'.Journal of Highway and Transportation Research and Development, vol.29 no.3, pp.98-103.

Yu Bo,Ye Jianshu&Wen Tianyu(2009), 'Temperature field in concept hollow-slab exposed to fire'.Journal of southeast univesity(natural science edition).vol.39 no.3, pp.536-540.

Jing Duo(2016), 'Numerical analysis on mechanical properties of PC hollow-core slab beambridge exposed to fire', Journal of fire science and technology,Vol.35,no2, pp 205-208

Liu Qi-wei,Wang Chengning&Luo Wenlin(2014), 'Simulation analysis and evaluation of prestressed hollow slab girders subjected to fire disaster',Journal of highway,Vol.1,no1.pp.35-43.

Xu Qingfeng,Han Chongqing&Li Xiangming(2013),'Experimental research on exposed to fire fire endurance of PC hollow-core slab under different load levels'.Journal of Building Structures. Vol.34 No.3 pp.20-27.

Liu Qiwei,Xia Ling&Wang chengming(2017), 'Experimental study of behavior characteristics of PC hollow slab beam exposed to fire'.Journal of Highway and Transportation Research and Development.Vol.34, no.1 67-75.

Standards China (CA)(1989),JTJ 021-89, 'General specifications for design of highway bridges and culverts',Beijing,china.

Wu Bo(2003),The mechanical properties of reinforced concrete structures after fire. Science Press, Beijin.

Shi Xudong&Guo Zhenhai(1996). 'Analysis of the temperature field of reinfored concrete structure.Journal of engineering mechanics'. Journal of engineering mechanics. Vol.13,no.1,pp.35-43.

Eurocode 2(1996), 'Design of concrete structure,part.10:structure fire design'.European Committee for Standardization, Brussels, Belgium.

Lu Zhoudao(1989), 'Experimental study on fire response of simple supported reinforced concrete beams'.ph.D thesis. Department of Civil Engineering, Tongji Uni. Shanghai.

Lie T T&Kodur V K R(1994). Fire Resistance of Circular Steel Columns Filled with Bar-Reinforced Concrete. Journal of Structural Engineering, Vol.120,no.5 pp.30-36.

Lu Chunsen,Qu Lijun&XueWuping(1995), Building Structural fire design, China Architecture & Building PressBeijing.

Li Yinqing,Ma Daozhen&Xu Jian(1991), Resarch on Fire-Endurance Characteristics of Building Materials. China Architecture & Building Press.Beijing.

Eurocode2(2004), 'Design of Concrete Structures, part 1-2: genera rules structural fire design', European Committee for Standardization, Brussels, Belgium.

Chen Ligang, Hu Wenfu&Xin Qingde(2004). 'The temperature distribution of The reinforced concrete slabs under Fire'. Journal of fire safe science,Vol.13,no.4 Oct.2004.

Hu Kexu&Xu Zhaohui(2001).Analysis of Temperature Distribution in Composite Deck Slabs Against Fire.Journal of Tongji university,Vol.29, no.6,pp. 644-646.

The 8th International Conference on
Structural Health Monitoring of Intelligent Infrastructure
Brisbane, Australia | 5-8 December 2017

Structural state monitoring of buried pipelines using distributed fiber optic temperature and strain sensors

X. Feng[1], Y. Han[1], Y. Liu[1], H.F. Liu[1], Z.H. Wang[1], S.L. Gong[1], Jing Zhou[1]
[1] *Dalian University of Technology – China, Email: fengxin@dlut.edu.cn*

Abstract

A method, by using of the distributed fiber optic temperature and strain sensors, is presented to quantitatively assess the structural state for burial pipelines. The monitoring scheme of distributed strain and temperature sensors is proposed according to the pipeline structural analysis, and the assessment approach is built by using of the distributed monitoring data. The time-spatial variation in the bending strain and temperature of the burial pipelines can be monitored by the distributed fiber optic sensors. The structural state of the pipeline can then be assessed with the monitoring data, in conjunction with the pipe parameters and internal pressure. The results from the site application demonstrate that the proposed method can effectively assess the structural state of the burial pipelines with the distributed temperature and strain data.

1. Introduction

The structural deteriorations of pipelines inevitably happen during the life cycle service because of complicated loads, material aging, environmental corrosion and vibration fatigue etc, which may lead to the occurrence of leaks. Regardless of their size, pipeline leaks are a major concern due to the considerable effects that they might have. These effects extend beyond the costs involved by downtime and repair expenses, and can include human injuries as well as environmental disasters. Early detection of pipeline structural deterioration as well as leakage is of great importance for life-cycle maintenance and management of pipeline networks (Zheng & Huang, 2017, Hu et al., 2010).
The application of fiber optic sensing technique has been successfully to provide non-instructive pipeline monitoring. Fiber optic monitoring methods is based on distributed measurements, providing continuous measuring along the entire length of the pipeline. Fiber optic based pipeline monitoring methods are generally divided into two categories: distributed temperature based approach and distributed strain based approach. The former is able to detect and locate leakages along the pipeline given the fact that the fluid leakage may cause the temperature variation in the vicinity of the pipeline (Inaudi & Glisic, 2010; Frings & Walk, 2011); and the latter has the potential to monitor the structural state and structural damage of the pipeline by using of the distributed strain information (Glisic & Yao, 2012). The distributed temperature measurement is commonly based on Roman optical time domain reflectometry (ORTDR). The advantages of ORTDR technique include one-end interrogation, long range measurement, easy installation (no need of bonding on the pipeline), and immune to mechanical disturbance. However the distributed temperature based approach only provides the detection of the leakage events rather that the early warning information of the structural deterioration. The Brillouin optical time domain re reflectometry/analysis (BOTDR/BOTDA) technique achieves the distributed strain measurement along the fiber path. The sensors are both bonded to pipeline and embedded in the soil, in the proximity of pipeline. The distributed strain measurement directly relates to the structural state of

the pipeline. The local novelty of the distributed strain can pinpoint the structural damages in the pipeline, such as corrosion, crack, or winkle. Thus the distributed strain based approach is able to provide the early warning of structural safety before the leakage occurring. Unfortunately the existing research mainly focuses on the detection and localization of the structural damages induced by the natural hazard such as earthquake and landslide and so on. In these cases the structural failures of pipelines are mainly controlled by the bending and shear effects caused by earthquake or landslide induced permanent ground displacements. However the axial thermal load is the control load for oil and gas pipelines under the operation condition. Thus the axial thermal stress should be considered for assessing the structural state of buried oil and pipelines. In fact the axial deformation of pipeline is commonly constrained by the fixed supports. As a result only the thermal stress but no axial thermal strain presents in the pipeline. The challenging issue for the application of distributed fiber optic strain sensor to the oil and gas pipelines is that the axial thermal stress cannot be measured because there is no axial thermal strain.

The objective of this study is to use the distributed fiber optic sensors to monitor the structural state of buried pipelines by considering both the longitudinal bending load and axial thermal load. Based on the theoretical background, a monitoring scheme is presented by combining the distributed temperature measurement and distributed strain measurement. The method is then proposed to assess the structural state of buried pipeline by using of the distributed monitoring data. Then the effectiveness of the proposed approach is validated by the field application of a buried gas pipeline.

2. Monitoring scheme

2.1 Theoretical background

In general, the buried oil or gas pipelines are subjected to the internal pressure, soil pressure and thermal load, and the differential settlement of the foundation leads to the longitudinal bending effect on the pipeline. The axial, circumferential and radial responses form in the pipeline under the combination of the above loads. In the underground environment the temperature distribution of the adjacent soil is uniform along the pipeline. The pipeline is commonly fixed on the support, resulting in that the thermal-induced deformation is constrained in the axial direction. Hence, in the axial direction, only the longitudinal bending deformation presents in the buried oil and gas pipeline.

For pipeline monitoring with distributed fiber optic sensor, the strain sensor is commonly placed in the axial direction. Based on above analysis, the axial placement of distributed strain sensor measures the longitudinal bending deformation of the pipeline. Thus the bending strain/stress can be obtained by the distributed fiber optic sensor. For strain measurement the Brillouin fiber optic sensor is adopted in the monitoring.

According to Kurashima et al. (1990), the Brillouin frequency shift (BFS) Δv_B due to strain and temperature disturbance along the fiber is given by

$$\Delta v_B(x) = \frac{dv_B}{d\varepsilon}\Delta\varepsilon(x) + \frac{dv_B}{dT}\Delta T(x) \tag{1}$$

where x is the location along the fiber; $\frac{dv_B}{d\varepsilon}$ and $\frac{dv_B}{dT}$ are the strain coefficient and temperature coefficient of Brillouin fiber optic sensor, respectively; and $\Delta\varepsilon$ and ΔT are the strain variation and temperature variation, respectively. The strain coefficient and temperature coefficient are obtained by the calibration tests.

Base on the measuring principle, the second term in the right hand of Equation (1) includes two effects: the first one is temperature dependence of the BFS for bare fiber induced by the internal friction with temperature; and the second one is BFS change caused by the thermal strain in the sensor. The previous analysis indicates that the thermal strain of the pipeline is zero. Moreover the Brillouin strain sensor is longitudinally attached on the pipeline. As a result the thermal strain in the sensor is also zero due to the restraint of the monitored pipeline. Then Equation (1) can be rewritten as

$$\Delta v_B(x) = \frac{dv_B}{d\varepsilon}\Delta\varepsilon(x) + \frac{dv_{B,IF}}{dT}\Delta T(x) \tag{2}$$

where $\frac{dv_{B,IF}}{dT}$ is the temperature coefficient induced by the internal friction with temperature.
In order to monitor the bending strain of the buried pipeline, the temperature should be measured independently. Once the temperature variation, ΔT, is obtained, the bending strain of the monitored pipeline, $\Delta\varepsilon_L^b$, can be determined by

$$\Delta\varepsilon_L^b(x) = \frac{\Delta v_B(x) - \frac{dv_{B,IF}}{dT}\Delta T(x)}{\frac{dv_B}{d\varepsilon}} \tag{3}$$

2.2 Monitoring scheme

From equation (3), the distributed strain measurement needs the temperature compensation by using of the distributed temperature measurement. Moreover the distributed temperature monitoring provides the leakage detection for the pipeline. Thus a pipeline monitoring system should include both the distributed strain sensor and the distributed temperature sensor. The monitoring scheme of the buried oil and gas pipeline is schematically shown in Fig. 1. The Raman optical time domain refelectry (RODTR) based sensor is selected to perform the distributed temperature measurement because it is only sensitive to the thermal effect along the fiber path. The Brillouin optical time domain analysis (BOTDA) based sensor is chosen as the distributed strain sensor because it is able to achieve the high measurement accuracy and spatial resolution. Two distributed fiber optic sensors are connected with the ROTDR monitor and BOTDA monitor by the lead optical cables, respectively. The BOTDA strain sensor is bonded on the pipe crown while the ROTDR temperature sensor is placed along the pipeline crown or the pipe invert, which depends on the internal medium is oil or gas. Note that the strain sensor should be adhered to the pipeline but the temperature sensor can be placed along the pipeline without bonding.

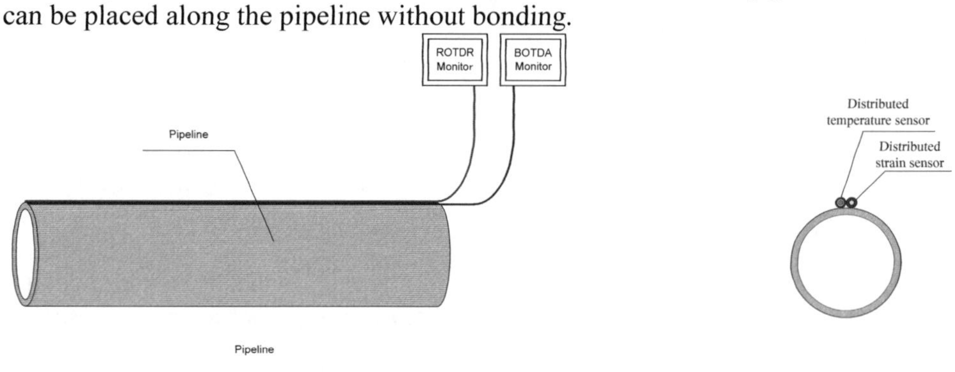

(a) Schematic diagram of distributed monitoring system (b) Cross section view of the placement of distributed FOSs

Figure 1 Monitoring scheme for buried pipeline using distributed fiber optic sensor

The proposed monitoring scheme only needs one distributed strain sensor on the pipe crown in case that the pipeline is not subjected to the lateral loads such as the fault and landslide and so on. The existing methods require at least three parallel sensors along the pipeline to describle the spatial deformation of the monitored pipeline (Glisic & Yao, 2012; Feng et al., 2017). The presented sensor configuration is easily to be applied to the field application. Moreover the existing distributed strain based approaches did not mention the temperature compensation and the thermal stress. But the distributed temperature measurement in the proposed monitoring scheme can provide the temperature compensation for the strain monitoring and the thermal stress monitoring simultaneously. The proposed monitoring scheme is able to monitor both the bending strain and the thermal stress along the pipeline, as well as the temperature variation due to the leakage. Thus the presented methodology is a monitoring tool for the life cycle service of the buried oil and gas pipeline.

2.3 Structural state assessment with distributed monitoring data

By using of the distributed fiber optic sensors, the distributions of the bending strain and the temperature along the pipeline can be obtained in a real time manner. Then the objective of this study is to quantitatively assess the structural state of the pipeline with the distributed monitoring data. Because the proposed method cannot monitor the internal pressure of the pipeline, the internal pressure data from the SCADA system that is extensively adopted in the oil and gas pipeline networks are utilized with the distributed monitoring data.

Under the combinations of operation and environment loads, the stress states in the pipeline are given by

$$\sigma_L(x) = \sigma_a(x) + \sigma_b(x) \tag{4}$$

$$\sigma_a(x) = -E_p \alpha_p \Delta T(x) + v_p p \sigma_h(x) \tag{5}$$

$$\sigma_b(x) = E_p \left[\varepsilon_L^b(x) \right] \tag{6}$$

$$\sigma_h(x) = pD/(2t) \tag{7}$$

where $\sigma_L(x)$ is the longitudinal normal stress along the pipeline; $\sigma_b(x)$ is the longitudinal bending stress; $\sigma_a(x)$ is the axial stress due to the thermal effect and Poisson's effect caused by the internal pressure; and $\sigma_h(x)$ is the hoop stress in the cross section of the pipeline. E_p , α_p and v_p are the Young's modulus, the thermal expansion coefficient and the Poisson ratio of the pipe material, respectively. D and t are the nominal diameter and the wall thickness of the pipeline, respectively. p is the internal pressure, which is commonly measured by the SCADA system of the pipeline.

The distributed strain and temperature variation can be obtained by the proposed monitoring scheme. By performing the temperature compensation with the distributed temperature measurements, the longitudinal bending strain is determined by Equation (3), and then the longitudinal bending stress is calculated by Equation (6). The hoop strain of the pipeline can be obtained by Equation (7) with the internal pressure data. By using of the distributed temperature measurements, the axial stress of the pipeline can be given by Equation (5). As a result the longitudinal normal stress and the hoop stress are obtained with the monitoring data in conjunction with the material parameters of the pipeline. Thus the structural state of the pipeline is totally determined.

During the service the Young's modulus and Poisson's ratio of the pipeline are commonly unchanged and they can be precisely obtained in advance. But the wall thickness of the pipeline may be changed due to the local damages such as corrosion and crack and so on. The local damage leads to the change in the cross-sectional moment of inertial at the damage location, resulting in the local variation in the longitudinal bending strain. The strain irregularity induced by the local damage can be detected by the distributed fiber optic strain sensors (Feng et al., 2013; Feng et al., 2014). Therefore the proposed monitoring scheme can assess the structural state of the pipeline under operation and environment loads as well as the occurrence of the local damage.
Once the longitudinal and hoop stresses are obtained, the structural state of the pipeline is assessed by the following equation.

$$\sigma_c(x) = \sigma_h(x) - \sigma_L(x) < 0.9\sigma_S \tag{8}$$

where $\sigma_c(x)$ is the equivalent stress of the pipeline; and σ_s is the yield stress of the pipe material.
Equations (4) and (8) indicate that the longitudinal bending stress is included for structural state assessment. The bending response is an important behavior due to the overburden loads and the differential settlement of the foundation for buried pipelines. But the existing methods in Chinese and U.S. codes only use the hoop stress and the axial stress because the actual bending load cannot be quantitatively predicted. The proposed method can calculate the actual structural state of the pipeline with the distributed strain and temperature measurements, which overcome the limitations of the existing methods.

3. Field applications

3.1 General description

To investigate the effectiveness of the proposed method, a field application was performed on a buried gas pipeline in Dalian, China. The model of the monitored pipeline is DN300, and the standard grade of the pipe material is 20#. The parameters of the pipeline are listed in Table 1.
The two ends of the monitored pipeline segment pipeline were fixed on two supports. The length of the monitored pipeline is 61.3m. The pipeline was below the traffic road, and the covering depth of the overburden soil was 1.0 m. The width of the pipe trench was 0.9m. The pipeline was installed by welding the multiple pipe segments successively. The standard length of the pipe segment is 12 m.

Table 1 The material and design parameters of the pipeline

Design parameters			Strength parameters		Mechanical parameters		
Internal pressure/ MPa	Outer diameter/ mm	Wall thickness/ mm	Tensile strength/ MPa	Yield strength/ MPa	Young's modulus/ GPa	Poisson's ratio	Thermal expansion coefficient/ (℃$^{-1}$)
1.6	325	8	410	245	206	0.3	1.1×10^{-5}

The distributed fiber optic strain and temperature sensors were placed on the pipe crown in the longitudinal direction according to Figure 1. The distributed temperature sensor used in this study was an armored cable with the loose multi-mode optical fiber inside. The distributed strain sensor was a reinforced cable manufactured by NZSENSING Ltd. The sensing cable consists of one single-mode optical fiber that is reinforced with the spiral steel wires. The cable was selected because of its capacity of the withstanding the harsh installation condition during the pipeline construction, but is still sensitive to strain. The distributed temperature sensor and strain sensor were connected with the ROTDR monitor and BOTDA monitor with the leading cables, respectively. To achieve the high measurement accuracy, the measurement parameters used in the long-term monitoring are listed in Table 2. In the actual conditions the variations in the temperature of the soil round the pipeline is commonly in the order of meter, and thus the presented spatial resolution of the temperature measurement is enough for leakage detection and temperature compensation of strain sensor.

The pipeline started operation in June 24, 2017. Since then the pipeline was periodically monitored using the distributed fiber optic sensors. In this study the monitoring data from June 24 to October 27, 2016 were analyzed for structural state assessment of the pipeline.

Table 2 Monitoring parameters of BOTDA and ROTDR

Monitor type	Spatial resolution/ m	Sampling interval/ cm	Measurement accuracy
BOTDA	0.2	5	±5 με
ROTDR	1	5	±0.1 °C

3.2 Data analysis

3.2.1 Distributed temperature data

The monitoring data from distributed fiber optic temperature sensor is plotted in Figure 2(a). It is shown that the temperature along the pipeline is almost uniform. The uniform temperature indicates that no leakage occurred during the monitoring period. The averaged temperatures (shown in Figure 2(b)) of each monitoring are related to the air temperature with a harmonic trend. The air temperature increased from June, and the temperature of pipeline also increased. The air temperature reached its peak in August 21 and then it reduced, while the pipeline temperature changed accordingly. The lowest air temperature appeared in October 27. The highest and lowest values of pipeline temperature are approximate 27 ℃ and 18 ℃, respectively. In comparison with the air temperature, the pipeline temperature is lower in summer while it is higher in fall and winter. The amplitude of the temperature variation is about 9 ℃ during the monitoring. It is reasonably deduced that the variation amplitude of the pipeline temperature will increase with the full season cycle. The distributed temperature data demonstrate that the thermal load is significant on the buried gas pipeline. The thermal axial stress should be considered in the structural state assessment, and the temperature compensation should be performed for the distributed fiber optic strain sensor in the pipeline monitoring.

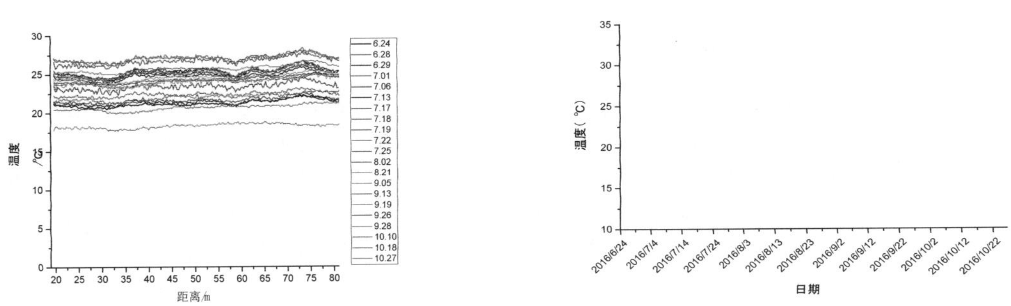

（a）Distributed temperature curves vs. monitoring date （b）Averaged temperature vs. monitoring date

Figure 2 Distributed temperature data

3.2.2 Distributed strain data

The baseline measurements of the distributed strain sensor were carried out after the trench backfill and road pavement were completed. By using the temperature measurements, according to Equation

(2), the temperature compensations were performed for each strain monitoring. The temperature compensated subsequent readings are shown in Figure 3. From Figure 3 the bending responses happend during the monitoring period. The tensile and compression strains appeare alternatively along the pipeline. The ranges and amplitudes of the positive and negative bending strains change with the time. The locations and values of the tensile and compression strains are time-variant. The reason for this phenomenon is that the pipeline construction was not strictly abided by the codes, for example, the solid stuffs below the pipeline, such as rocks and debris, were not removed during the installation. Moreover the differential settlement and soil expansion and compression induced by the temperature changes also led to the bending behavior on the buried pipeline. The maximum and minimum strains on the pipeline crown are 150 με and -320 με, respectively. Accordingly, the counterparts on the pipeline invert are 320 με and -150 με. During the monitoring period, the internal pressure of the pipeline was about 0.4 MPa and hence the hoop strain at the arbitrary cross section of the pipeline was 39.4 με. By comparison with the bending strain, the hoop strain is only 12% of the maximum value of the bending strain. It is indicated that the bending response of the pipeline should be included in the pipeline structural state assessment.

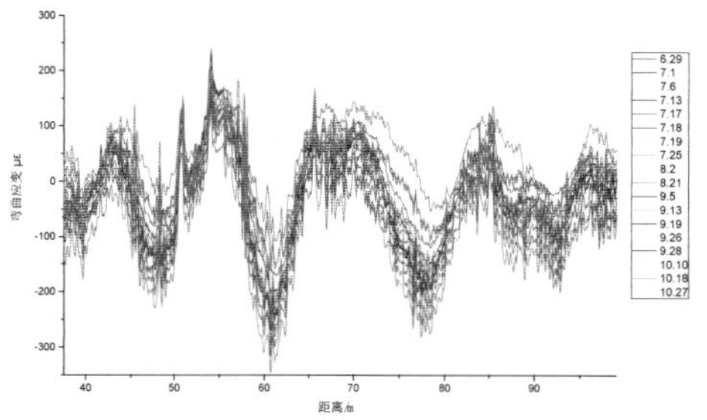

Figure 3 Distributed strain data vs. monitoring date（after performing temperature compensation）

In Figure 3, the time-spatial variations in the bending strains obviously show that the discrete strain sensor at the certain locations cannot capture the peak strain along the pipeline. It is no doubt that only the distributed strain sensor can obtain the complete distribution of the bending response. Therefore the structural state assessment of the actual operation pipelines needs the strain monitoring in the distributed manner rather than the point sensors such as FBG.

3.3 Structural state assessment

By combining the distributed strain, temperature, internal pressure, pipeline material parameters and pipeline geometric parameters, the stress states of the monitored pipeline were calculated according to Equation (5). Figure 4 plots the bending stress along the pipeline. It is shown that the maximum value of the bending compression stress on the pipeline crown is -70 MPa while the peak value of the bending tensile stress is 30 MPa. The counterparts on the pipeline invert are -30 MPa and 70 MPa, respectively.

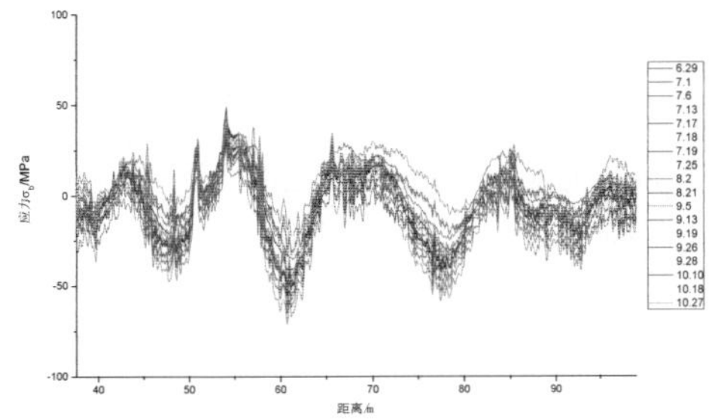

Figure 4 Calculated bending stress on the pipeline crown

Based on SCADA data, the hoop stress caused by the internal pressure is 8.14 MPa. Even in the case that the internal pressure reaches the design value (i.e. 1.6 MPa), the hoop stress in the pipe cross section will be only 32.56 MPa that is less than the half value of maximum bending stress. In this study the monitoring data indicate that the bending stress is the control factor of the stress state of the buried pipeline.

By adding the bending stress to the axial stress induced by temperature and internal pressure, the longitudinal normal stress were obtained by Equation (4). Figure 5 plots the longitudinal normal stresses on the pipeline crown and invert, respectively. The maximum and minimum values of the longitudinal normal stress on the pipeline crown are 45 MPa and -80 MPa, while the counterparts on the pipeline invert are 60 MPa and -40 MPa. From Figure 5 the peak normal stresses both in pipeline crown and invert are much less than the yield strength of the pipeline material.

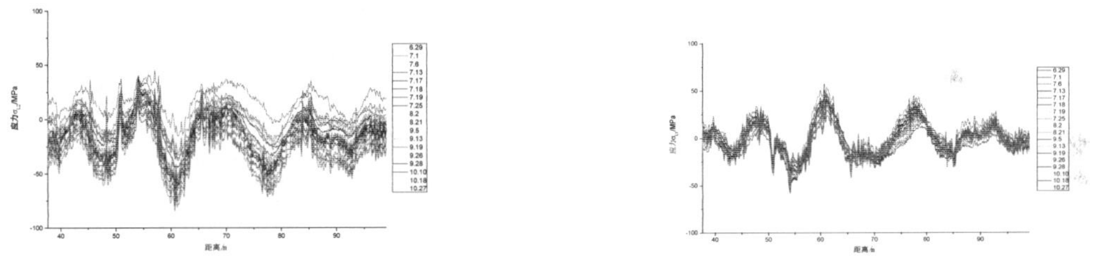

(a) Longitudinal normal stress on pipeline crown (b) Longitudinal normal stress on pipeline invert

Figure 5 Longitudinal normal stress along the pipeline

The maximum compression stress in the monitored pipeline is -80 MPa. According to Equation (7), the equivalent stress of the pipeline is $\sigma_{c,max}$≈88.14 MPa≤0.9σ_s=220.5 MPa. Thus the pipeline was under the safe operation state during the monitoring period.

4. Conclusions

Based on the structural analysis of the buried pipeline, a monitoring scheme is presented by using of distributed fiber optic temperature and strain sensors. The proposed method was applied to monitor a buried gas pipeline at Dalian, China. This paper presents the details of the methodology for pipeline structural state assessment with the distributed temperature and strain measurements. The results from the field applications demonstrate that the proposed monitoring scheme is easy installation during the pipeline construction. The monitoring data from the distributed temperature can provide the calculation parameter of the thermal stress and the temperature compensation of the strain sensor, as well as the leakage detection according to the irregularity in the measured temperature distribution. The data from the distributed strain sensor show that the bending strain along the pipeline has the time-spatial variation characteristics. Further calculation of the bending stress, the thermal stress and the hoop stress from the monitoring data provides the equivalent stress based on the shear strength theory, and thus the structural states of the monitored pipeline can be assessed quantitatively. The limitation of the presented study is that only the restrained straight pipeline is investigated. The further study should be performed on the more complicated conditions, such as the pipelines that cross the fault or have the bends.

References

Zheng, HL & Huang, WH 2017, 'Development status and prospect of safety insurance technologies for oil & gas pipelines and storage and transportation facilities', Oil & Gas Storage and Transportation, vol. 36, no. 1, pp. 1-7.

Hu, ZX, Ma, YB & Tan, DJ, et al 2010, 'Oil Pipeline Sliding Monitoring System Based on Fiber Bragg Grating Sensors', Acta Photonica Sinica, vol.39, no. 1, pp. 33-36.

Inaudi, D & Glisic, B 2010, 'Long-range pipeline monitoring by distributed fiber optic sensing', Journal of Pressure Vessel Technology, vol. 132, no. 1, pp. 1-9.

Frings, J & Walk, T 2011, 'Distributed fiber optic sensing enhances pipeline safety and security', Oil Gas European Magazine, vol. 37, no. 3, pp. 132-136.

Glisic, B & Yao, Y 2012, 'Fiber optic method for health assessment of pipelines subjected to earthquake-induced ground movement', Structural Health Monitoring, vol. 11, no. 6, pp. 696-711.

Kurashima, T, Horiguchi, T & Tateda, M 1990, 'Thermal effects on the Brillouin frequency shift in jacketed optical silica fibers', Applied Optics, vol. 29, no. 15, pp. 2219-2222.

Feng, X, Wu, WJ & Meng, D, et al 2017, 'Distributed monitoring method for upheaval buckling in subsea pipelines with BOTDA sensors', Advances in Structural Engineering, vol. 20, no. 2, pp. 180-190.

Feng, X, Zhou, J & Sun, C, et al 2013, 'Theoretical and experimental investigations into crack detection with BOTDR-distributed fiber optic sensors', Journal of Engineering Mechanics, vol. 139, no. 12, pp. 1797-1807.

Feng, X, Zhang, X & Sun, C, et al 2014, 'Stationary wavelet transform method for distributed detection of damage by fiber-optic sensors', Journal of Engineering Mechanics, vol. 140, no. 4, pp. 1-11.

*** Formatting Issue - Best Available Paper/Graphic ***

The 8th International Conference on
Structural Health Monitoring of Intelligent Infrastructure
Brisbane, Australia | 5-8 December 2017

Stress distribution of Zhuhai Opera House based on temperature variations

Jun Teng[1], Wei Lu[2], Lihang Qiu[3], Kai Huang[4].
[1] *Harbin Institute of Technology Shenzhen Graduate School – China, Fujian University of Technology – China*
[2] *Harbin Institute of Technology Shenzhen Graduate School - China, Email:* lu.wei@hit.edu.cn [3,4] *Harbin Institute of Technology Shenzhen Graduate School – China*

Abstract

The outside structure of Zhuhai Opera House is steel space structure, while the stresses of steel elements changes by environmental temperature variations. In the other hand, the structural health monitoring system of Zhuhai Opera House was implemented, where the temperatures and stresses of the structures are being monitored. The measurements of temperature and stresses of the steel elements are used to analyze the relationship between the stress distribution and the temperature variations. Except the measurements of stresses and temperatures of the structure, the further simulation analysis is carried out to give the detailed relationship between the stress distributions and the temperature variations. The temperature distribution of the structure is simulated by the limited temperature measurements, and the finite element model of the Zhuhai Opera House steel space structure is built to analyze the stress distribution of all the steel elements of the structure. The aims are to reveal the stress distribution of real world project with steel elements based on the temperature variations and to supply the basic database for the optimal construction time of steel space structure closure.

1. Introduction

The structural health monitoring system (SHMS) is widely used in the practical engineering and its technology development is comparatively ripe. The parameter information of each structure members is monitored by the means of the monitoring system, which will play a certain role in the safety management of the structural operation phase (Ko & Ni 2005; Teng, Lu & Wen, et al 2015). The main reason for the cognitive limitations of the structural loads conditions and the overall response status is that the sensors in the SHMS and the data category are limited in number, especially for the space structures in the form of complex. That is why it is particularly important to obtain the structural load information and the response information based on the limited monitoring data. The surfaces of the steel space structure exposed to the environment for extended periods of time, the result for the solar radiation on the surfaces is that the overall temperature field is not uniform. The daily cyclical change of the solar radiation causes that the temperature responses of local elements changes periodically. In order to understand the structural stress response under the actual temperature effect and the extent for the temperature influence on the structural stress distribution, it is necessary to carry out the research on the structural stress distribution under the actual temperature effect.
About the research on the structure temperature field, the current main method of the first part is to obtain the influence of sunshine on structural temperature distribution while considering the solar radiation intensity and the structural material absorption coefficient of solar radiation (GB50019 2003; AITM2-0018 1998). It can also make a simplified calculation on the instantaneous radiation temperature of the elements as stated in the specification (GBJ19-87 2004). The second part is to analyse the heat conduction based on the heat conduction theory by using finite element software

(Kimura 1977; Chen & Krokosky 1976). There are many research have been carried out. The calculation of the steel space structure temperature variations with a traditional method was usually used for those structures which are under the solar radiation (Guo, Tian & Liu 2008). The mode of the structure temperature field under the solar radiation was built and analysed to obtain the reasonable value of the steel space structure closure temperature (Jin, Fan & Li 2012). By using the finite element software, a simplified method was used to simulate the external non-uniform temperature field of the steel space structure under the solar radiation and the distribution regularities of the temperature field were obtained (Xiao, Xu & Liu 2010). In order to simulate the temperature field's time history of the steel arches of a stadium, the transient thermo analysis function of ANSYS was used, and finally the spatial distribution regularities of the most unfavorable temperature field was obtained (Liu, Chen & Wang 2010). About the research on the internal force response of the structures which were under the temperature effect. Many researches were about the uniform temperature field (Tian, Hao & Wang 2012). As for the non-uniform temperature field, the research on the field of concrete structure have already made some progress while the research on the stress monitoring of the steel space structures were still relatively rare (Xiong, Gu & Li 2009). A typical example to combine the temperature field research with SHMS was the national stadium of China. By counting various parameters that were required by the structural calculations, especially for the solar radiation absorption coefficient ρ, and using finite element software to analyse the temperature distribution of the steel structure elements, the optimal construction time of steel space structure closure and the maximum positive/negative temperature difference were determined (Fan, Wang & Tang 2007). The steel structure of the national stadium was subjected to continuous temperature monitoring and the monitoring data was used to divide the temperature effect into uniform temperature field and non-uniform temperature field. In order to study the influence of uniform temperature field and the non-uniform temperature field on the structural stress respectively, the correlation coefficients between the temperature data and the stress data were analysed in detail (Luo, Mei & Shen 2013).

This article was based on the monitoring data of small shell in Zhuhai Opera House. In order to analyze the structural temperature effect and to obtain the elements' stress distribution under different temperature, the structural temperature field was established by using the actual monitoring temperature value. The vibrating wire sensors were installed on the surface of the steel members to monitor the stress data of the members. The digital temperature sensors were installed on the surface of the steel members to monitor the temperature data of the members. The basic thoeries and methods for analyzing the temperature effect and the stress distribution of the steel space structure would be provided by this SHMS.

2. The structural temperature partition methods

2.1 The solar radiation cumulantion

The meshed model was imported into the Ecotect Analysis 2011, as it can be seen in the Figure 1. The whole year's time points of the sunrise/sunset in Zhuhai were counted and set up to the analysis of the incident solar radiation as it was showed in the Figure 2. And the solar radiation cumulantion W of 2014 and 2015 were calculated. With the reason that the Ecotect Analysis can only calculate the period that the time points of the sunrise/sunset are nearly the same, while the time points of the sunrise/sunset in the three months of one season were different, therefore, the average value of time points of the sunrise/sunset were selected to calculate the solar radiation cumulantion. It can be seen in the Figure 3.

*** Formatting Issue - Best Available Paper/Graphic ***

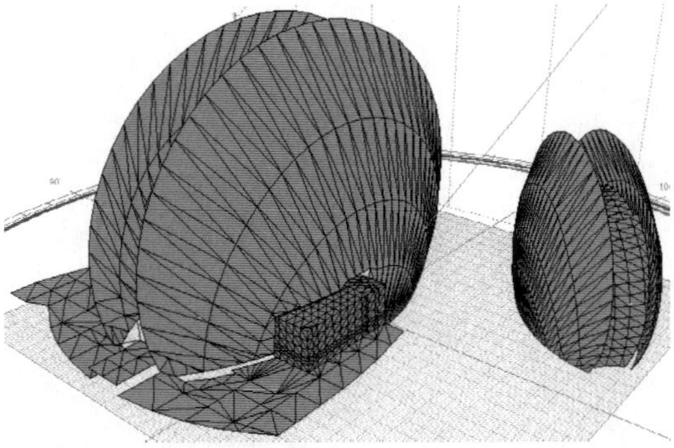

Figure 1 The Ecotect Analysis model.

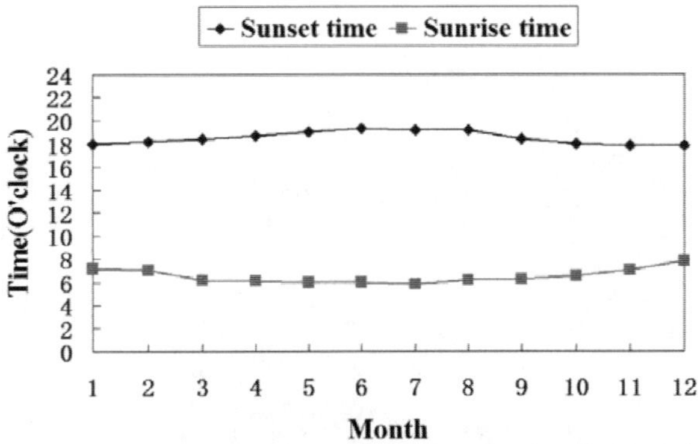

Figure 2 The whole year's sunrise and sunset time in Zhuhai.

a) The first season（1.1—3.31）

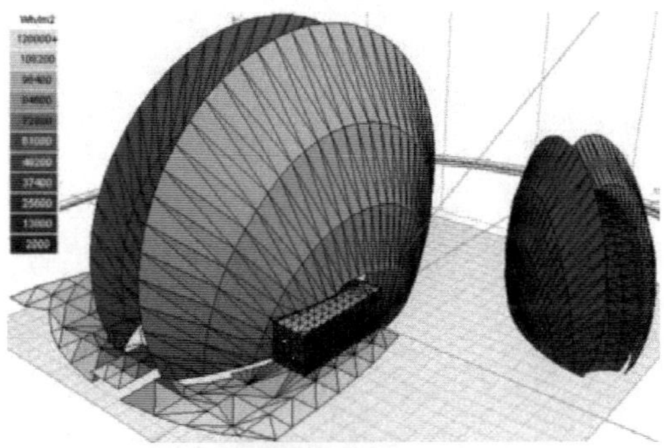

b) The second season（4.1—6.30）

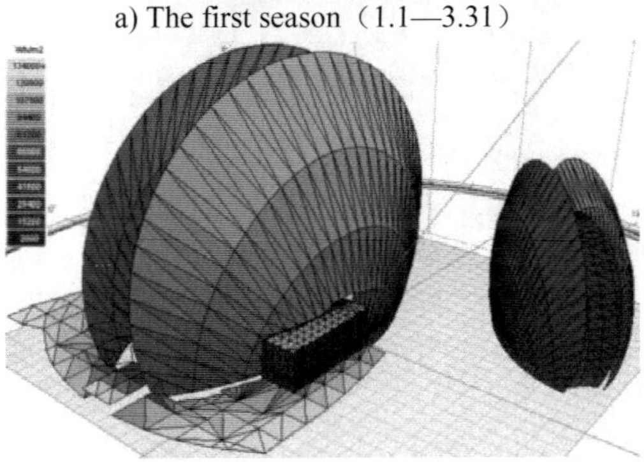

c) The third season（7.1—9.30）

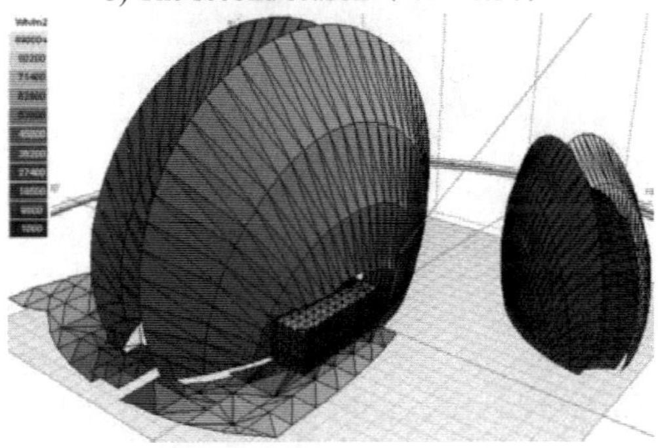

d) The fourth season（10.1—12.31）

Figure 3 The distribution conditions of the solar radiation cumulantion W in each season.

2.2 The structural temperature partitions based on the solar radiation cumulantion value

Accroding to the numerical variation of the solar radiation cumulantion on the surface of the small shell, the areas of which numerical similarity in the cumulantion value were classified as the same partition. There were 4 partitions at all. Sorted the solar radiation cumulantion value from high to low was $W_{\mathrm{I}}>W_{\mathrm{II}}>W_{\mathrm{III}}>W_{\mathrm{IV}}$ as it was showed in the Figure 4.

*** Formatting Issue - Best Available Paper/Graphic ***

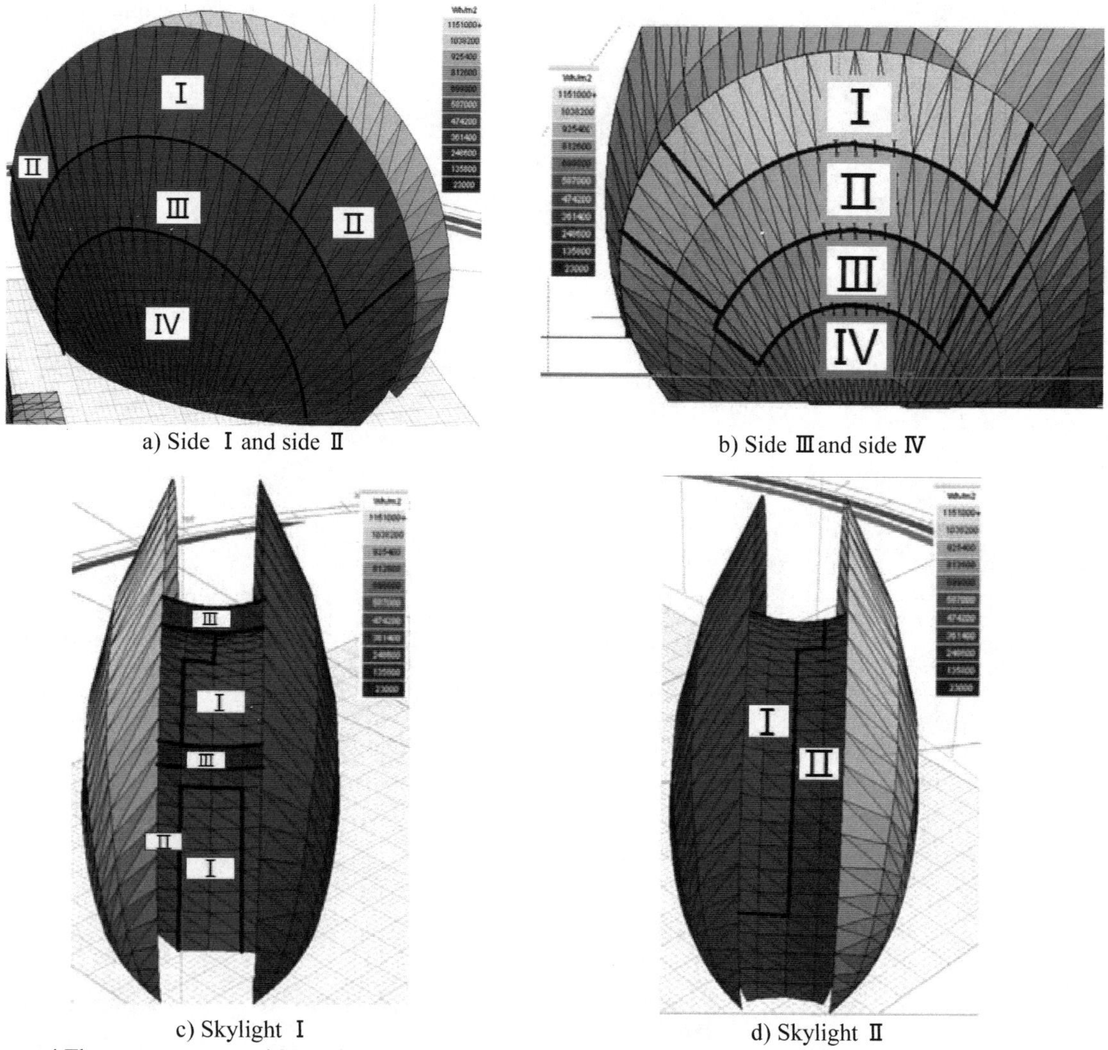

a) Side Ⅰ and side Ⅱ

b) Side Ⅲ and side Ⅳ

c) Skylight Ⅰ

d) Skylight Ⅱ

Figure 4 The temperature partitions of the small shell.

2.3 The standard region and the reference region

Partition Ⅲ, which was installed the most number of sensors within the 4 partitions, was selected as the standard region p. While the reference region q was partition Ⅳ, with the reason that it was the runner-up of the sensors number. The average value of this two partitions' temperature were T_{III}^{M} and T_{IV}^{M} respectively. They were selected as the representative values of the standard region.

3. The calculation methods of the temperature values in the temperature partitions

3.1 Temperature variation radio

The calculation formula of the temperature variation based on the solar radiation cumulantion value was showed as the Formula 1:

$$\Delta T = \frac{\Delta W}{cm} \tag{1}$$

where c means the specific heat capacity, m means the mess, ΔW means the variation value of the solar radiation cumulantion, and ΔT means the variation value of the temperature. According to the formula above, if there was no outside interference, ΔT and ΔW would have a linear relationship when c and m were unchanged. For the structures, if the environment influence on the temperature fields was no considered, the temperature variation value on the surface of the structures and the variation of the solar radiation cumulantion would have a linear relationship in theory. Therefore, the temperature variation radio α_{iq} can be identified as the Formula 2:

$$\alpha_{iq} = \frac{W_i - W_q}{W_p - W_q} = \frac{T_i - T_q}{T_p - T_q} \tag{2}$$

where W_i, W_p and W_q means the solar radiation cumulantion values of partition i, partition p and partition q respectively, while the T_i, T_p and T_q means the temperature values of partition i, partition p and partition q respectively. The solar radiation cumulantion values and the temperature variation radio of esch partition can be seen in the Table 1.

Table 1 The sensors' distribution conditions of each partitions.

Positions	Partitions	The serial number of sensors	W_i (Wh/m^2)	Proportion α_{iq}
Side 1 and 2	I	S1-1	49000	2.33
Side 1 and 2	II	—	39400	1.67
Side 1 and 2	III	S1-2 S3-1 S3-3	29800	1.00
Side 1 and 2	IV	S2-1	15400	0.00
Side 3 and 4	I	S1-4	97000	2.18
Side 3 and 4	II	—	87400	1.59
Side 3 and 4	III	S1-3 S3-2 S3-4	77800	1.00
Side 3 and 4	IV	S2-2	61480	0.00

3.2 The partition temperature representative values

In order to calculate the partition temperature representative values T_i^S, besides the standard region p and the reference region q, the simulative representative values of other partitions i were also needed. It can be seen in the Formula 3:

$$T_i^S = \alpha_{iq}(T_p^M - T_q^M) + T_q^M \tag{3}$$

There were only two monitoring points were arranged on the skylight parts. Therefore, the actual monitoring temperature values were selected to determine the partition temperature representative values, with the reason that the sensors were limited in number. In other words, besides the standard region [illegible] and the reference region IV, the simu [illegible] partitions were calculated by using Formula 3. The actual monitoring temperature values of May 19th afternoon, July 30th morning and October 7th morning in 2014 were selected to calculate the partition temperature representative values, as it can be seen in the Table 2.

*** Formatting Issue - Best Available Paper/Graphic ***

Table 2 The actual monitoring temperature and simulative representative temperature of 3 days.

Positions	Partitions	Dates					
		May 19th afternoon		July 30th morning		October 7th morning	
		AMT (℃)	SRT (℃)	AMT (℃)	SRT (℃)	AMT (℃)	SRT (℃)
Side 1 and 2	I	29.35	29.58	29.65	29.51	29	27.76
Side 1 and 2	II	29.42	29.53	29.3	29.26	27.85	27.23
Side 1 and 2	III	29.48	29.48	29.01	29.02	26.7	26.7
Side 1 and 2	IV	29.4	29.4	28.6	28.6	25.8	25.8
Side 3 and 4	I	29.75	29.56	29.1	31.59	25	28.23
Side 3 and 4	II	29.63	29.53	29.44	30.69	25.98	27.6
Side 3 and 4	III	29.5	29.5	29.78	29.78	26.97	26.97
Side 3 and 4	IV	29.45	29.45	28.25	28.25	25.9	25.9
Skylight	I	29.3	29.3	28.7	28.7	25.8	25.8
Skylight	II	29.8	29.8	29.15	29.15	27.85	27.85
Skylight	III	29.55	29.55	28.93	28.93	26.83	26.83

Note: AMT= Actual Monitoring Temperature; SRT= Simulative Representative Temperature

4. The stress distribution caused by temperature effect

In order to research on the relationship between the overall temperature difference and the response, the members that the tensile stress and the pressure stress in the top 100 were selected, and ΔT =+6℃. The response distribution of the structure was showed in the Figure 5. The different range of T_σ was selected to calculate with the temperature values in different dates, the members' stress distribution can be seen in the Table 3. With the reason that the variation of members' stress positions under different ranges were very small, the members that the tensile stress and the pressure stress in the top 100 were selected as the research objects, in order to observe these variations more conveniently. The distribution of the members can be seen in the Figure 6. Therefore, according to the result of the stress distribution, the members distribution in the morning of May 4th (T_σ=0.11℃) was closed to the distribution in the afternoon of February 23rd (T_σ=0.99℃), while the members distribution in the morning of April 5th (T_σ=2.75 ℃) was closed to the distribution in the afternoon of January 18th (T_σ=2.09 ℃).

Table 3 The temperature massages of each time period.

	May 4th morning	February 23rd afternoon	April 5th afternoon	January 18th afternoon
Mean square error of temperature T_σ (℃)	0.11	0.99	2.75	2.09
Overall temperature T (℃)	22.56	19.14	24.77	21.37
Initial temperature T_0 (℃)	18.97	18.97	18.97	18.97
Temperature difference ΔT (℃)	3.59	0.17	5.80	2.40

*** Formatting Issue - Best Available Paper/Graphic ***

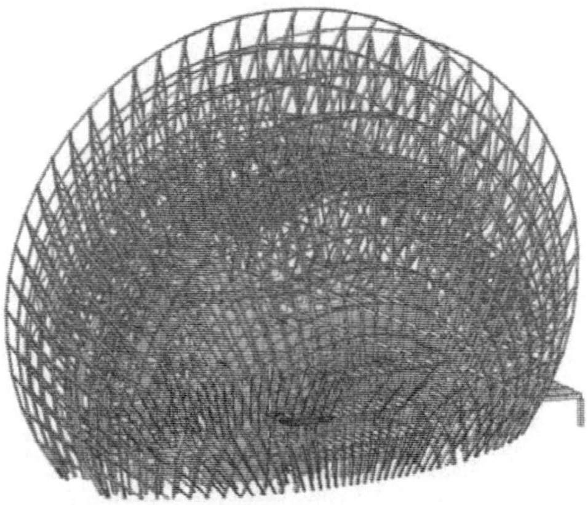

a) Members that the tensile stress in the top 100

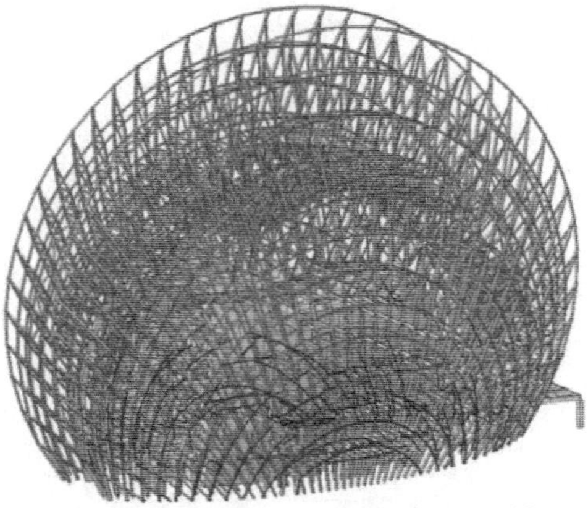

b) Members that the pressure stress in the top 100

Figure 5 The members' stress distribution at ΔT =+6 ℃

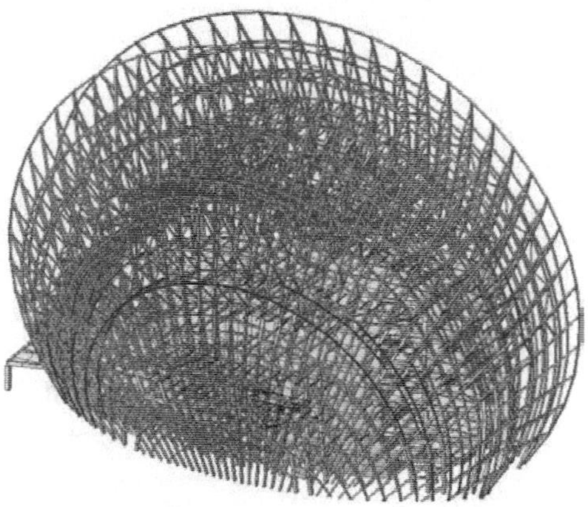

a) Members that the pressure stress in the top 100 in the morning of May 4th

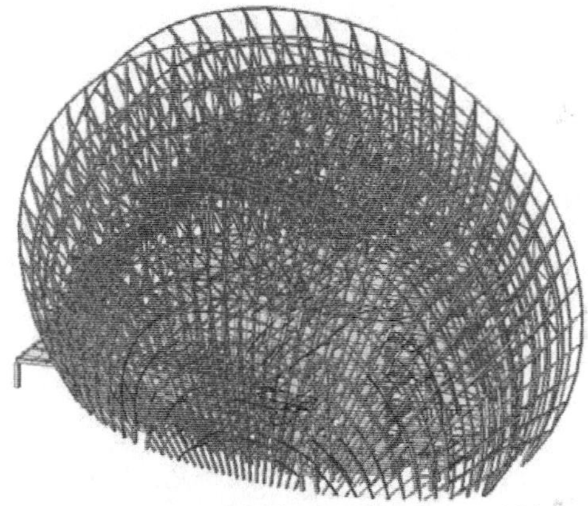

b) Members that the tensile stress in the top 100 in the morning of May 4th

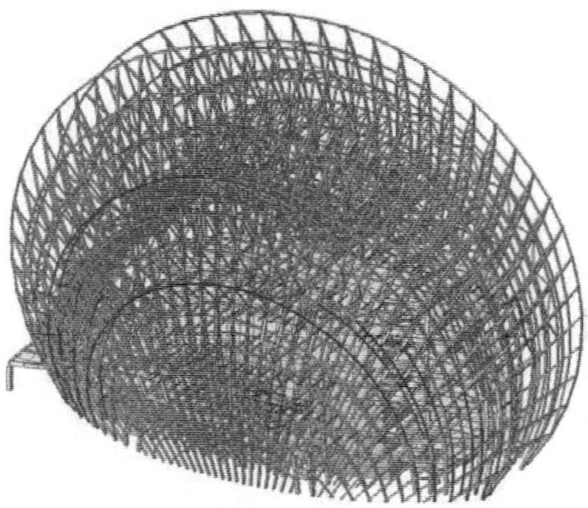

c) Members that the pressure stress in the top 100 in the afternoon of February 23rd

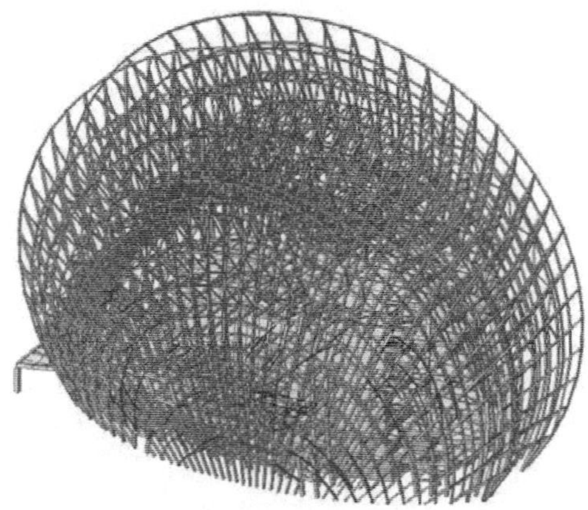

d) Members that the tensile stress in the top 100 in the afternoon of February 23rd

*** Formatting Issue - Best Available Paper/Graphic ***

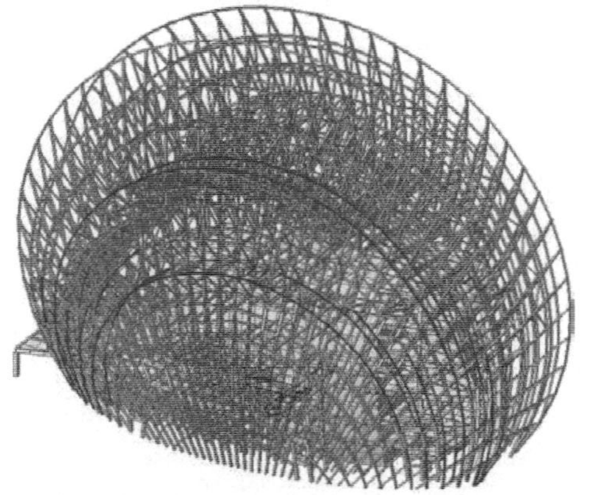

e) Members that the pressure stress in the top 100 in the morning of April 5th

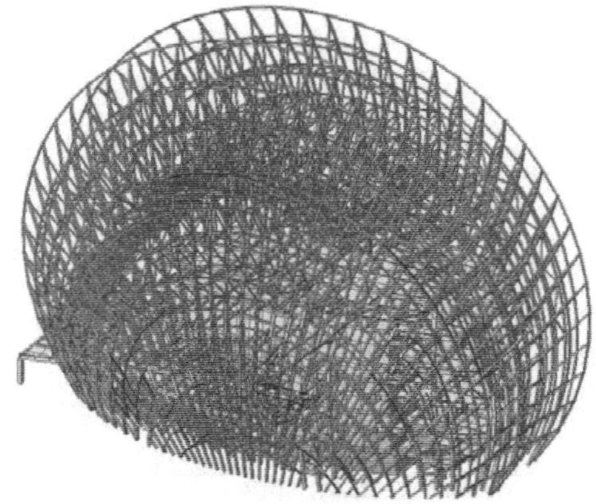

f) Members that the tensile stress in the top 100 in the morning of April 5th

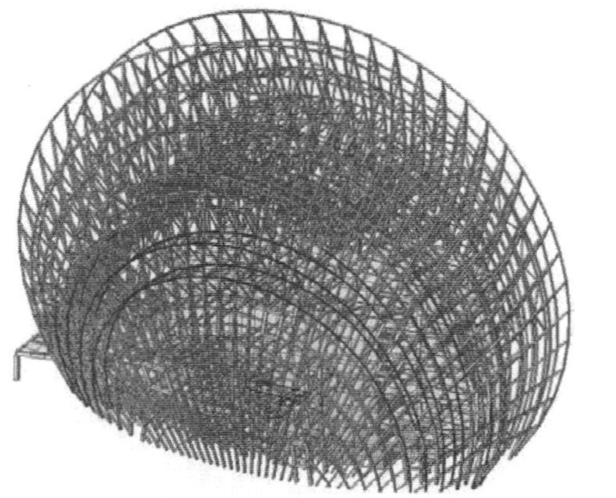

g) Members that the pressure stress in the top 100 in the afternoon of January 18th

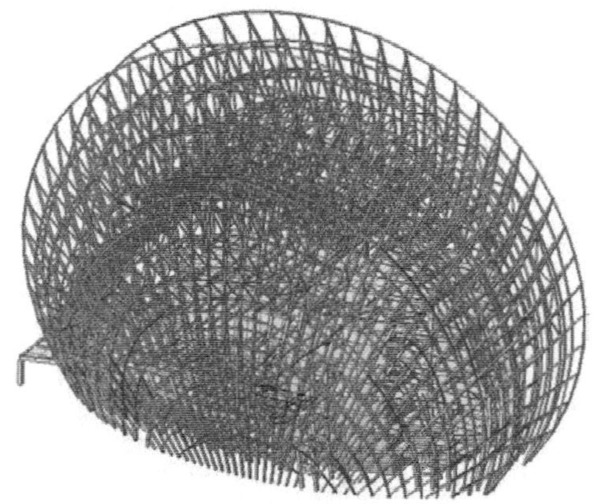

h) Members that the tensile stress in the top 100 in the afternoon of January 18th

Figure 6 The members' stress distribution in the different range of T_σ .

5. Conclusions

The temperature data and the stress data monitored by the sensors were used to analyze the actual temperature effect of the steel space structures. The distribution and the scale of the temperature field and the stress field were studied. The partitions of the structure and the partition load value were studied based on the non-uniform distribution characteristics of the temperature effects. The methods of structural temperature partitions based on the solar radiation cumulanttion were obtained, while the recognition of the structural overall stress field was studied. The Stress distribution of Zhuhai Opera House based on temperature variations was researched. It will not only provide the more convenient, rapid and safe of the early-warning and decision-making for the steel space structural construction process and operation process, but also improve the structural safety and the construction accuracy of the steel space structures.

Acknowledgments

This research is supported by the National Science Foundation of China (Grant No. 51678201, 51308162), and the Fujian National Science Foundation (2014J01171).

*** Formatting Issue - Best Available Paper/Graphic ***

References

Ko JM & Ni YQ 2005, 'Technology Developments in Structural Health Monitoring of Large-Sacle Bridges', *Engineering Structures*, vol. 27, no. 12, pp. 1715-1725.

Teng J, Lu W & Wen RF, et al 2015, 'Instrumentation on Structural Health Monitoring Systems to Real World Structures', *Smart Structures And Systems*, vol. 15, no. 1, pp.151-167.

Department of construction engineering quality & safety supervision ministry of housing and urban-rural development of the P.R.China. '*GB50019-2003.National technical measures for design of civil construction heating, ventilation and air conditioning*', China Planning Press, Beijing, 2003.

'*AITM2-0018.Airbus industries test method solar absorption of paints*', Airbus Industries, 1998.

'*GBJ19-87. Code for design of heating ventilation and air conditioning*', China Planning Press, Beijing, 2004.

Kimura KI 1977, 'Scientific basis of air conditioning', *Applied Science Publishers Ltd.*

Chen CH & Krokosky EM 1976, 'Steady and non steady solar heat transmission through roofs', *Materials and Structures*, vol. 9, no. 1, pp. 19-32.

Guo YL, Tian GY & Liu XW 2008, 'Influence of temperature change on closure process of the cantilever of New CCTV Headquarters', *Construction Technology*, vol. 37, no. 5, pp. 1-4.

Jin XF, Fan F & Li JF 2012, 'Study on the non-uniform temperature load by sunshine of Shanxi Three Gymnasiums', *Spatial Structures*, vol. 18, no. 3, pp. 80-85.

Xiao JC, Xu H & Liu JK 2010, 'The influence of the strong solar radiation on the static performance of the large span spherical reticulated shells', *Chinese Journal of Solid Mechanics*, vol. 31, pp. 75-280.

Liu HB, Chen ZH & Wang XD 2010, 'Temperature effect of suspendome with stacked arch structure considering solar radiation', *Journal of Tianjin University*, vol. 43, no. 8, pp. 705-711.

Tian LM, Hao JP & Wang Y 2012, 'Analysis of temperature effect and determination of temperature upon heating of large-span spatial structures', *China Civil Engineering Journal*, vol. 45, no. 5, pp. 1-7.

Xiong XY, Gu W & Li YM 2009, 'Long-term monitoring and analysis of extra-long PC frame structures', *China Civil Engineering Journal*, vol. 42, no. 2, pp.1-10.

Fan Z, Wang Z & Tang J 2007, 'Analysis on temperature field and determination of temperature upon healing of large-span steel structure of the National Stadium', *Journal of Building Structures*, vol. 28, no. 2, pp. 32-40.

Luo YZ, Mei YJ & Shen YB 2013, 'Field measurement of temperature and stress on steel structure of the National Stadium and analysis of temperature action', *Journal of Building Structures*, vol. 34, no. 11, pp. 24-32.

The 8th International Conference on
Structural Health Monitoring of Intelligent Infrastructure
Brisbane, Australia | 5-8 December 2017

Experimental study on the thermal convection behaviour of concrete material in service environments

Linren Zhou[1], Lan Chen [2], Fengwu Li [3], Yong Xia [4]
[1,2,3] *South China University of Technology – China, Email:* zhoulinren@scut.edu.cn [4] *The Hong Kong Polytechnic University - China*

Abstract

Concrete is a kind of common building material with many advantages and widely used in civil infrastructures. Concrete has the qualities of uneven, anisotropic, large dispersion and thermal inertia, therefore, it is very difficult to accurately determinate its thermophysical properties. The lab testing conditions are much different to the service environments of structures. The inverse method using field measurements affected by so many factors and uncertainties decreases the reliability of results. In this study, the heat convection behaviour of concrete in natural service environments is investigated using specimen experiments. A special testing system is designed and established. It includes a small size thermal insulated cuboid tank, concrete specimens, temperature and wind speed measuring and collection. The heat convection coefficient of concrete in natural environments is investigated taking account of different wind speed and roughness of concrete surface. The results indicate that the roughness of specimen surface has considerable effects on the heat convection coefficient. Based on the measurements, the relationship between heat convective coefficient and wind speed of concrete is regressed using linear equation. Comparing with other literature, the results of this study is reasonable and reliable.

1. Introduction

Concrete is a kind of common building material with many good properties. It has been widely used for civil infrastructures such as buildings and bridges. For high-rise buildings and large long-span bridge, the thermal behaviour is one of the most important issues should be seriously concerned. The analysis of thermal effects on structures is very important for structural design, construction and management. The determination of thermophysical properties of building materials is the basics for the thermal analysis and assessment of structures.

The determination method of concrete surface heat convection coefficient (h_c) can be divided into two types: 1) lab testing using precise instruments and following rigorous procedures; 2) field measurements of structural temperature and inversely calculate the thermal parameters of materials. Zhu (1999) proposed the inversion analysis method for thermal conductivity, heat convection coefficient and solar radiant heat of concrete based on the field measurements of mass concrete temperatures. Song et al (2010) obtained the thermal parameters of concrete through analysing the thermal behaviour of concrete slabs using the measurements of temperature, strain and meteorological data. Liu (2000) proposed the formula for calculating the heat convection coefficient of the outer surface of medium roughness solid and building structure. Zhang and Liu (2006) obtained the convective heat transfer coefficient through fitting the test results from wind tunnel experiments. Kehlbeck (1981) provided a formula for calculating the heat convection coefficient (wind speed less than 5 m/s) using field measurements. Mirambell and Aguado (1990) concluded a formula for calculating the heat convection coefficient of the outer surface of bridge beam. Saetta, et al (1995) concluded a regression formula for heat convection coefficient of concrete structure surface through studying the stress of concrete structure under the action of temperature difference.

Concrete has the qualities of uneven, anisotropic, large dispersion and hot inert materials, therefore, it is very difficult to accurately determinate its thermophysical properties, especially in natural environments. The lab testing conditions are much different to the service environments of structure. The inverse method using field measurements affected by so many factors and uncertainties decreases the reliability of results. In this study, the heat convection coefficient of concrete is investigated using specimen experiments. A testing system is particularly designed and fabricated. The effects of wind speed and concrete surface roughness on the heat convection behaviour are experimented and analysed.

2. The basic principle of the experiment

Usually, the building materials exchange heat with the surrounding environment by way of heat radiation and convection simultaneously. The total exchanged heat Q can be expressed as:

$$Q = Q_c + Q_R \tag{1}$$

where Q_c is the convection heat and Q_R is the radiation heat, respectively.

Thermal convection is the process of energy exchange between the solid surface and the surrounding air. It can be divided into two types: natural convection (atmospheric molecules moved according to thermodynamics) and forced convection (atmospheric molecules moved by environmental wind) (Arpaci, and Larsen, 1984; Bejan, 2013). The convective heat transfer can be expressed using Newton cooling formula:

$$Q_c = h_c (T_w - T_f) A \tag{2}$$

where h_c is the heat convection coefficient; A is the area; T_w and T_f is the temperature of solid and the air temperature, respectively.

The radiation heat transfer can be calculated [6]:

$$Q_R = \varepsilon C_o \left[(T_w / 100)^4 - (T_f / 100)^4 \right] A \tag{3}$$

where ε is the emissivity coefficients of the solid surface; C_0 is the Stefan–Boltzmann constant.

In this experimental study, a specific testing device is designed and fabricated to measuring the total exchanged heat, the temperature of the specimen and surrounding air. The environmental wind is simulated according to real situation and controllable. The convective heat transfer coefficient of concrete is investigated.

3. Experimental study

The heat convection coefficient of concrete is investigated taking account of the effects of wind speed and concrete surface roughness. The experimental system includes testing device, concrete specimen, generating the environmental wind, temperature and wind speed collection, as is shown in Fig. 1.

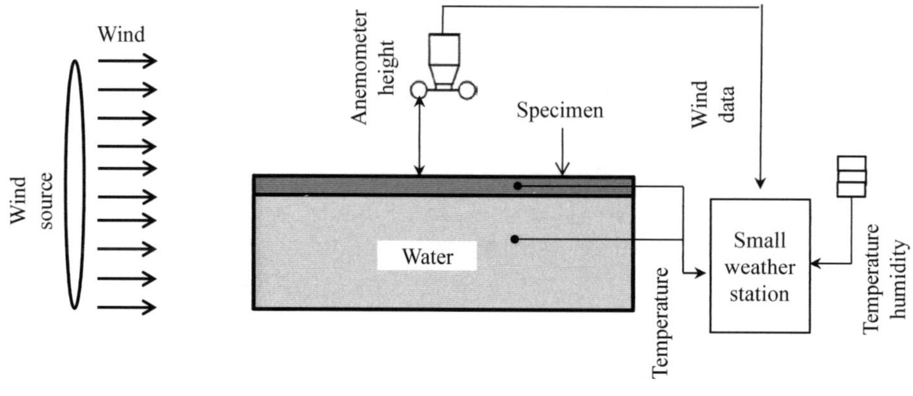

Fig.1. Schematic diagram of the experimental system

3.1 The testing device

A specific testing device is developed, as shown in Fig. 2. A small size thermal insulated cuboid tank is fabricated using insulation material. The upper side of tank can be replaced with the testing specimen. The inside of tank is full of water and using electric heater to warm the water. Besides, a small water pump is placed on the bottom of tank to circulate the water, then, the water temperature is almost even. The specimen is placed on the top of tank with one side exposing to natural environment and the other side immersing in the water of tank. The water can provide a stable heat source and apply a First-type thermal boundary condition on the surface of the testing specimen.

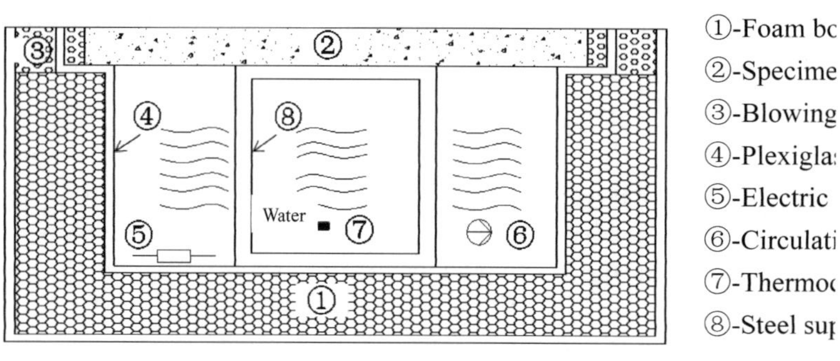

Fig. 2. Profile of thermal insulated cuboid tank

3.2 Measuring the heat loss of the insulation tank system

The insulation tank cannot be absolute heat insulation. In order to accurately calculate the convective heat transferring via the specimen, the amount of heat loss not passing the specimen should be estimated. The cooling tests of the tank are carried out from different initial temperature of water. Three thermocouples are placed at upper, middle and bottom of the tank inside along the depth. The average temperature of the three thermocouples is utilized to calculate the heat loss. The temperature drop curves are shown in Fig. 3. The performance of the heating and insulation of the testing system is good. The heat loss is almost linear and can be easily compensated in the experiments.

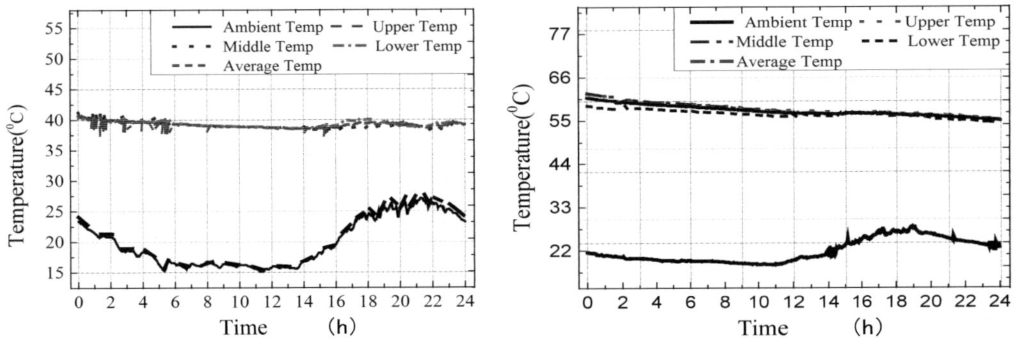

a）Initial temperature of 40 ℃ b）Initial temperature of 60℃

Fig. 3. Temperature drop curves of thermal insulation tank

Two concrete specimens with different surface roughness are fabricated, as shown in Fig. 4. The specimen is 70cm length, 50cm width and 5cm thickness. The concrete axial compressive strength was designed to be 26.8 N/mm^2. The immersing surfaces of the concrete specimen were treated with a flexible waterproof coating to prevent the water penetrating into the concrete slab specimen. The temperature of specimen is measured using T-type thermocouples, and the data is collected by a small automatic weather station. The layout of thermocouples is shown in Fig. 4. Three thermocouples were installed at the center of the specimen (Point-1) along the depth. Another three thermocouples were installed on the specimen surface at the Point-2, 3 and 4, respectively. The surface thermocouples were buried inside specimen with a depth about 1 mm below the concrete surface.

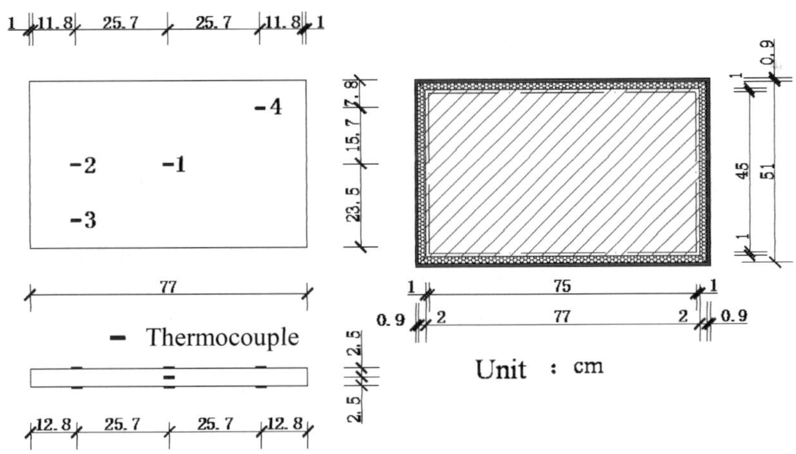

Fig.4. Specimen and the layout of the thermocouples

3.2 Wind field generating

Several industrial fans are used to generate the continuous wind field with controllable wind speed. Figure 5 shows the measured wind speed nearing the specimen surface. It can be seen that the average wind speed of ten minutes is very close to the total average wind speed. The industrial fans can provide stable wind field. Therefore, wind speed in 10 minutes average is used for thermal analysis.

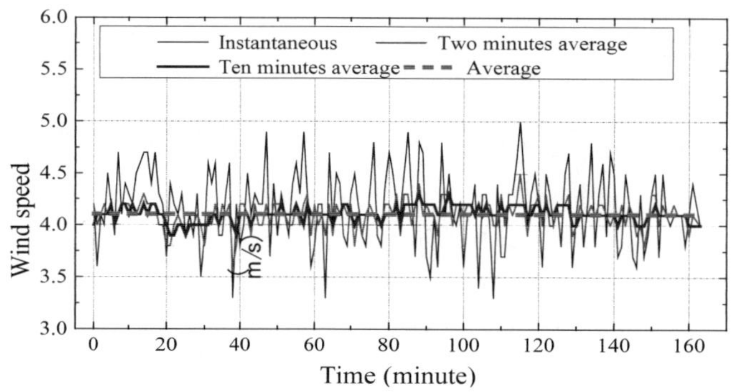

Fig.5 Wind field for experiment

4. Experimental results and analysis

In the natural environment, concrete specimens with different roughness are experimented. The concrete specimen is plastered cement mortar to generate smooth surface. The rough surface is the casted concrete slab without any further processing. The heat convection coefficient h_c of concrete surface is obtained according to different wind speed. The results are listed in Table 1. The difference of h_c between the rough surface and the smooth surface is very small in the natural convective heat transfer condition (wind speed is 0 m/s), however, the difference is significantly increased with the increasing of wind speed.

Table.1. Heat convection coefficients at different wind speed

Wind speed (m/s) / h_c (W/(m²·°C))	0	1.17	3.92	4.11	1.07	4.01
Smooth	5.66	×	×	×	10.09	19.95
Rough	5.31	12.56	20.43	23.20	×	×

Using the least squares method, the relationship between heat convection h_c coefficient and wind speed v of concrete surfaces with different roughness can be fitted as:

$$h_{cS}=3.929v+6.378\ (W/(m^2\cdot°C))\quad (smooth) \tag{4}$$

$$h_{cR}=3.519v+5.941\ (W/(m^2\cdot°C))\quad (rough) \tag{5}$$

The relative difference of heat convection coefficient between rough and smooth surface ($C=(h_{cR}-h_{cS})/h_{cS}$) is shown in Fig. 6. The relative difference rapidly increases to 10% when the wind speed changes from 0 to 4.5 m/s.

The energy exchange of heat convection between the solid and surrounding air is very complicated and deeply affected by many factors. It is not easy to identify the heat convection coefficient (h_c) of concrete using experiments, especially in natural environments. In this study, the discrepancy of the concrete material, the temperature loss of the testing system and other uncertainties will influence the results. The identified heat convection coefficients are compared with the existing conclusions from other literatures, as shown in Fig.7. The rough surface of concrete definitely has the largest h_c since there are more contact area between solid and air. The surface roughness engineering structures is obvious different. It is meaningful taking account of the roughness for thermal analysis of structures.

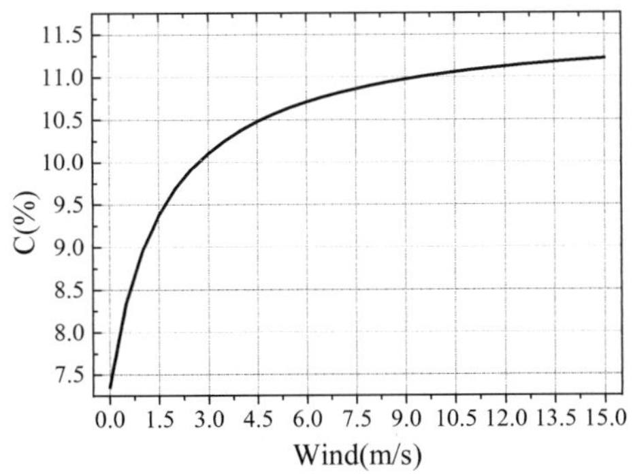

Fig. 6. Discrepancies of h_c between rough and smooth surfaces

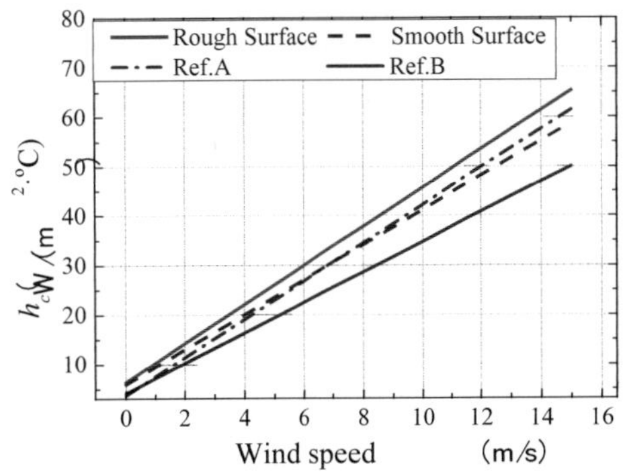

Fig. 7. Comparison of h_c (Ref. A –Bejan, 2013; Ref. B –Mirambell and Aguado, A 1990)

4. Conclusion

In this paper, the heat convection coefficient of concrete in natural environment is investigated. The wind speed and roughness of concrete surface are discussed.

The roughness of concrete surface shows little influence on the natural convection heat transfer process, however, it is significantly impact on the heat convection coefficient of forced convection heat transfer. The relative difference of h_c between rough and smooth rapidly increases to 10% when the wind speed changed from 0 to 4.5 m/s.

Under the condition of forced convection heat transfer, the wind speed has a significant effect on the surface convective heat transfer coefficient. Based on the experimental data, the relationship between the heat convection coefficient h_c and the wind speed v can be regressed as h_c=3.929v+6.378 for rough surfaces and h_c=3.519v+5.941 for smooth surfaces.

Acknowledgments

This research is financially supported by the Technology Planning Project of Guangdong Province (Project No. 2014A020218003), and the Fundamental Research Funds for the Central Universities.

References

Arpaci, VS & Larsen, PS 1984, 'Convection heat transfer', Prentice Hall.

Bejan, A 2013, 'Convection heat transfer', John wiley & sons.

Kehlbeck, F 1981, 'Effect of solar radiation on bridge structure', Trans. X. F. Liu. Chinese Railway Publishing Company, Beijing.

Liu, X 2000, 'Building physics', China Architecture & Building Press, Beijing.

Mirambell, E & Aguado, A 1990, 'Temperature and stress distributions in concrete box girder bridges', Journal of structural engineering, vol.116, no. 9, pp. 2388-2409.

Saetta, A, Scotta, R & Vitaliani, R 1995, 'Stress analysis of concrete structures subjected to variable thermal loads', Journal of Structural Engineering, vol. 121, no. 3, pp. 446-457.

Song, ZW, Xiao, JZ & Zhao, Y 2010, 'Back-Analysis of concrete thermal parameters based on experimental measurements', Journal of Tongji University (Natural Science), vol. 38, no. 1, pp. 35-38.

Zhang, JR & Liu ZQ 2006, 'A study on the convective heat transfer coefficient of concrete in wind tunnel experiment', China Civil Engineering Journal, vol.39, no. 9, pp. 39-42.

Zhu, BF, 1999, 'Mass concrete temperature stress and temperature control', China Electric Power Press, Beijing.

The 8th International Conference on
Structural Health Monitoring of Intelligent Infrastructure
Brisbane, Australia | 5-8 December 2017

Investigation on parametric sensitivity of a transmission tower

B. Chen[1], H.F. Cheng [1], X.H. Tao[2]

[1] *Key Laboratory of Roadway Bridge and Structural Engineering, Wuhan University of Technology – China, Email:* cbsteven@163.com
[2] *Zhanjiang Power Supply Bureau, Guangdong Power Grid Corporation Co. Ltd, – China, Email:* cexhtao@163.com

Abstract

Transmission towers are widely used as electrical power infrastructures throughout the world. It is necessary to evaluate the structural performance of the transmission tower subjected to external dynamic excitations. For a transmission tower with determined parameters, it is troublesome to attain the relationship between dynamic characteristics and changed parameters by numerous recalculation which will be almost impossible for large scale structures. While sensitivity analysis as an alternative approach can take structural parameters as variable and achieve the relationship only with one time analysis, which will dramatically reduce the analytical work especially for large scale structures. The comparative study on frequency sensitivity of a transmission tower is actively carried out in this study. The three dimensional analytical model of a transmission tower is established by using the finite element method. The sensitivity coefficients to natural frequencies are deduced based on the equation of motion of the tower. In addition, the expression of the frequency sensitivity to the Young's modulus, density of material, the cross area of members, torsional stiffness, bending moment inertia is proposed. A real transmission tower constructed in China is taken as an example to examine the feasibility and reliability of the proposed sensitivity computation approach. The work on an example structure indicate that the magnitudes of the sensitivity coefficients of the Young's modulus, the density and the cross area are much larger than those of the torsional stiffness and the bending moment inertia.

1. Introduction

To be a typical spatial steel structure, the transmission tower is widely used as electrical power infrastructures throughout the world. The transmission tower is a high-rise structure with small damping and is prone to strong dynamic excitations, such as earthquakes and wind loadings. It is frequently reported across the world that the excessive vibration of a transmission tower under dynamic excitations may induce the structural damage and even failure. Therefore, it is necessary to evaluate the structural performance of the transmission tower subjected to external dynamic excitations. Many theoretical and experimental investigations have been carried out during the past two decades for examining the performance of the transmission tower. With regard to the approaches and techniques used for performance evaluation, the transmission tower system is conventionally designed and constructed using appropriate design standards without considering the dynamic optimization effects. Therefore, the common approach does not provide deep insights into the structural transient behaviour under strong dynamic excitations, even though the consideration of dynamic effects may be important.
For a transmission tower with determined parameters, it is troublesome to attain the relationship between dynamic characteristics and changed parameters by numerous recalculation which will be almost impossible for large scale structures. While sensitivity analysis as an alternative approach

can take structural parameters as variable and achieve the relationship only with one time analysis, which will dramatically reduce the analytical work especially for large scale structures [16]. The sensitivity analysis is a reasonable and powerful tool for investigating the effects of physical parameters on static and dynamic responses. The sensitivity analysis concerns the relationship between parameters available to the structural responses under dynamic excitations. The dependence of response measures such as displacement, velocity, acceleration, stress, strain, natural frequency, mode shape, and frequency response, is implicitly defined through the governing equations of structural mechanics. Dynamic sensitivity analysis is used to compute the rate of dynamic property change with respect to changes of structural physical parameters.

Current studies on the dynamic assessment of the transmission tower focus on the response computation and mitigation. Up to now, the sensitivity analysis of the transmission tower for dynamic assessment has been investigated rarely. To this end, the comparative study on frequency sensitivity of a transmission tower is actively carried out in this study. A real transmission tower constructed in China is taken as an example to examine the feasibility and reliability of the proposed sensitivity computation approach. The made observations indicate that the magnitudes of the sensitivity coefficients of the Young's modulus, the density and the cross area are much larger than those of the torsional stiffness and the bending moment inertia. The Young's modulus, the density and the cross area are important structural parameters affecting the dynamic properties of the transmission tower.

2. Frequency sensitivity analysis

2.1 Frequency sensitivity

A transmission tower is a typical truss structure constructed by using steel members, which can be commonly modelled by using beam elements based on the finite element (FE) method. The global mass matrix **M** and stiffness matrix **K** of the transmission tower can be expressed respectively as follows

$$\mathbf{M} = \sum_{i=1}^{ne} \mathbf{T}_i^T \mathbf{M}_i^e \mathbf{T}_i \tag{1}$$

$$\mathbf{K} = \sum_{i=1}^{ne} \mathbf{T}_i^T \mathbf{K}_i^e \mathbf{T}_i \tag{2}$$

Where ne is the number of elements of a transmission tower; $\mathbf{T}_i$ is the freedom transform matrix from the local coordinate system (LCS) to the global coordinate system (GCS). $\mathbf{K}_i^e$ and $\mathbf{M}_i^e$ are the element stiffness matrix and mass matrix of the ith member in the LCS.

The eigenvalue equation of a MDOF transmission tower can be expressed as:

$$\left(\mathbf{K} - \omega^2 \mathbf{M}\right)\boldsymbol{\varphi} = 0 \tag{3}$$

Where **M**, **K** and $\boldsymbol{\varphi}$ are the mass matrix, stiffness matrix and modal vector of the transmission tower respectively. The eigenvalue equation of the rth mode vibration is:

$$\left(\mathbf{K} - \omega_r^2 \mathbf{M}\right)\boldsymbol{\varphi}_r = 0 \tag{4}$$

The mass matrix, stiffness matrix, and modal vector are the function of physical parameters. Thus, the first derivative of Equation (14) to the parametric x_i of the ith structural member for the rth mode vibration results in:

$$\left(\frac{\partial \mathbf{K}}{\partial x_i} - \frac{\partial \omega_r^2}{\partial x_i}\mathbf{M} - \omega_r^2 \frac{\partial \mathbf{M}}{\partial x_i}\right)\boldsymbol{\varphi}_r + (\mathbf{K} - \omega_r^2 \mathbf{M})\frac{\partial \boldsymbol{\varphi}_r}{\partial x_i} = 0 \tag{5}$$

In which: ω_r and $\boldsymbol{\varphi}_r$ are the rth circular frequency and modal vector of the system. Since the mass and stiffness matrices are a symmetric matrix, there exists:

$$\boldsymbol{\varphi}_r^T\left(\mathbf{K}-\omega_r^2\mathbf{M}\right)=\left[\left(\mathbf{K}-\omega_r^2\mathbf{M}\right)\boldsymbol{\varphi}_r\right]^T=0 \tag{6}$$

To pre-multiply the vector $\boldsymbol{\varphi}_r^T$ to Equation (15) and considering Equation (16) yield

$$\frac{\partial\omega_r^2}{\partial x_i}=\frac{\boldsymbol{\varphi}_r^T\left(\dfrac{\partial\mathbf{K}}{\partial x_i}-\omega_r^2\dfrac{\partial\mathbf{M}}{\partial x_i}\right)\boldsymbol{\varphi}_r}{\boldsymbol{\varphi}_r^T\mathbf{M}\boldsymbol{\varphi}_r} \tag{7}$$

The relationship between the circular frequency ω_r and natural frequency f_r of the rth vibration mode can be written as

$$\frac{\partial\omega_r^2}{\partial x_i}=2\omega_r\frac{\partial\omega_r}{\partial x_i} \tag{8}$$

The sensitivity of the frequency to the ith structural parameter x_i is

$$\frac{\partial f_r}{\partial x_i}=\frac{1}{8\pi^2 f_r}\cdot\frac{\boldsymbol{\varphi}_r^T\left(\dfrac{\partial\mathbf{K}}{\partial x_i}-4\pi^2 f_r^2\dfrac{\partial\mathbf{M}}{\partial x_i}\right)\boldsymbol{\varphi}_r}{\boldsymbol{\varphi}_r^T\mathbf{M}\boldsymbol{\varphi}_r} \tag{9}$$

Equation (20) provides a way to calculate the sensitivity of the rth natural frequency to the change in the ith structural parameter x_i.

2.3 Sensitivity coefficients of structural parameter

The sensitivity of the rth natural frequency to the structural parameter x_i can be rewritten by using the element stiffness matrix $\mathbf{K}_i^e$ and mass matrix $\mathbf{M}_i^e$ of the ith element in the GCS

$$\frac{\partial f_r}{\partial x_i}=\frac{1}{8\pi^2 f_r}\cdot\frac{\boldsymbol{\varphi}_r^T\left(\sum_{i=1}^{ne}\dfrac{\partial\mathbf{K}_i^e}{\partial x_i}-4\pi^2 f_r^2\sum_{i=1}^{ne}\dfrac{\partial\mathbf{M}_i^e}{\partial x_i}\right)\boldsymbol{\varphi}_r}{\boldsymbol{\varphi}_r^T\mathbf{M}\boldsymbol{\varphi}_r} \tag{10}$$

In which ne denotes the total number of the members of the transmission tower. The sensitivity of the rth natural frequency to the Young's modulus of the ith structural element E_i is given by

$$\frac{\partial f_r}{\partial E_i}=\frac{1}{8\pi^2 f_r}\cdot\frac{\boldsymbol{\varphi}_r^T\left(\sum_{i=1}^{ne}\dfrac{\partial\mathbf{K}_i^e}{\partial E_i}\right)\boldsymbol{\varphi}_r}{\boldsymbol{\varphi}_r^T\mathbf{M}\boldsymbol{\varphi}_r} \tag{11}$$

The sensitivity of the rth natural frequency to the density of the ith structural element ρ_i is

$$\frac{\partial f_r}{\partial \rho_i}=-\frac{f_r}{2}\cdot\frac{\boldsymbol{\varphi}_r^T\left(\sum_{i=1}^{ne}\dfrac{\partial\mathbf{M}_i^e}{\partial \rho_i}\right)\boldsymbol{\varphi}_r}{\boldsymbol{\varphi}_r^T\mathbf{M}\boldsymbol{\varphi}_r} \tag{12}$$

Where A_i and L_i are the cross area and the length of the ith element, respectively. The sensitivity of the rth natural frequency to the cross area of the ith structural element A_i is

$$\frac{\partial f_r}{\partial A_i}=\frac{1}{8\pi^2 f_r}\cdot\frac{\boldsymbol{\varphi}_r^T\left(\sum_{i=1}^{ne}\dfrac{\partial\mathbf{K}_i^e}{\partial A_i}-4\pi^2 f_r^2\sum_{i=1}^{ne}\dfrac{\partial\mathbf{M}_i^e}{\partial A_i}\right)\boldsymbol{\varphi}_r}{\boldsymbol{\varphi}_r^T\mathbf{M}\boldsymbol{\varphi}_r} \tag{13}$$

3. Case study
3.1 Description of an example transmission tower

To examine the frequency sensitivity based on the proposed approach, a real transmission tower constructed in China is taken as the example structure. Figure 1 shows the elevation of a large transmission tower used for electric power transfer. The transmission tower with a height of 31m is located in the southern coastal area in China. The structural members used in the transmission tower are made of Q235 steel with a yielding stress of 235 MPa. The Young's modulus of the steel is 2.01×10^{11} N/m^2 and the density is 7800 kg/m^3. Five platforms are connected to the vertical major members to form the tower body and the skew members are incorporated to increase the vertical and lateral stiffness of the entire tower. Two cross arms are constructed on top of the tower body for the connection of the transmission lines. A three dimensional finite element model is established for the steel transmission tower using a commercial computer package. The movement of all the supports in the three orthogonal directions is restricted. For the sake of convenience in the subsequent discussion, the beam elements in each component of the transmission tower are numbered differently.

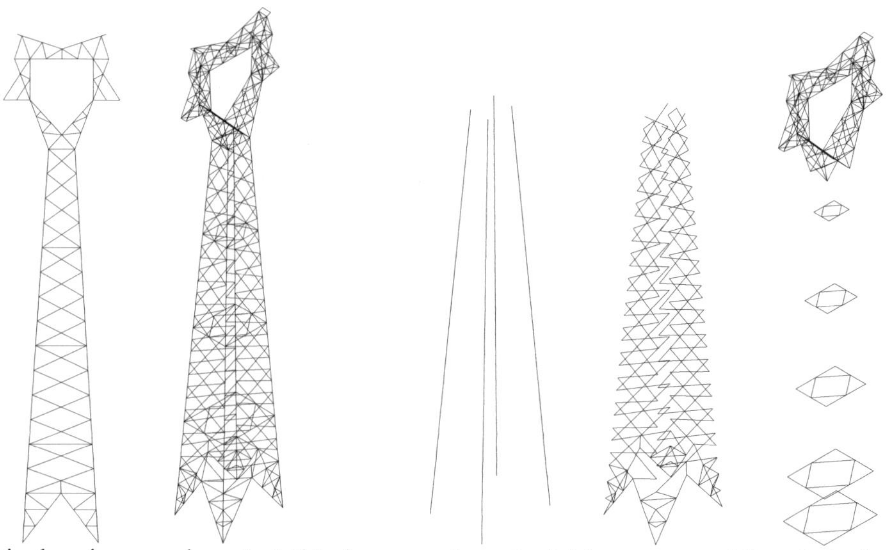

(a) Vertical major members (zc) (b) skew members (xg) (c) cross arm (tt) and platform (pt)

Figure 1 Configuration of the example transmission tower

4.2 Dynamic characteristics of the transmission tower

The dynamic characteristics analysis is conducted based on the established finite element model of the steel transmission tower. The first eight natural frequencies and vibration modes of the tower are depicted in Table 1 and Figure 2. The first natural frequencies of the transmission tower for the in-plane and out-of-plane vibration are 2.174 Hz and 2.744 Hz, respectively. It is seen that the first and second vibration modes are the global vibration mode in the y direction and x direction, respectively. The third vibration mode is a global torsional vibration mode in the x-y plane due to the tower rotation. The fourth to eight vibration modes are the high order translational and torsional vibration modes of the tower. The first eight natural frequencies computed indicate that the natural frequencies of the structure are not closely spaced. The dynamic responses of the first three mode shapes are the major parts of the entire dynamic responses of the transmission tower.

Table 1 Dynamic properties of the transmission tower

No.	Natural frequency (Hz)	Properties of the mode shape
f_1	2.174	1st global vibration mode in the y direction
f_2	2.744	1st global vibration mode in the x direction
f_3	7.437	1st global torsional vibration mode
f_4	9.224	2nd global vibration mode in the y direction
f_5	9.924	2nd global vibration mode in the x direction
f_6	14.614	Local vibration mode of the platform
f_7	17.963	3rd global vibration mode in the y direction
f_8	19.339	3rd global vibration mode in the x direction

(a) 1st (b)2nd (c) 3rd (d) 4th (e) 5th (f)6th (g) 7th (h) 8th

Figure 2 First eight mode shapes of the example transmission tower

4.3 Frequency sensitivity of the transmission tower

Figure 3 shows the sensitivities of the first four natural frequencies to the Young's modulus of each member. It is seen from the first two figures that the first two natural frequencies are more sensitive to the Young's modulus change of members in the vertical major members (zc) than other members. The curves in the third figure indicate that the sensitivity coefficients of the first torsional frequency of the skew members are much larger than those of the other members. The higher natural frequencies of the global vibration are more sensitive to the Young's modulus change of both vertical major members and the skew members. This is consistent with the structural configuration and the modes of vibration: the first two translational modes of vibration are mainly due to the deformation of the vertical major members and the first torsional mode is due to the deformation of the skew members. However, the higher modes of vibration are mainly due to the movement combination of both vertical major members and the skew members.

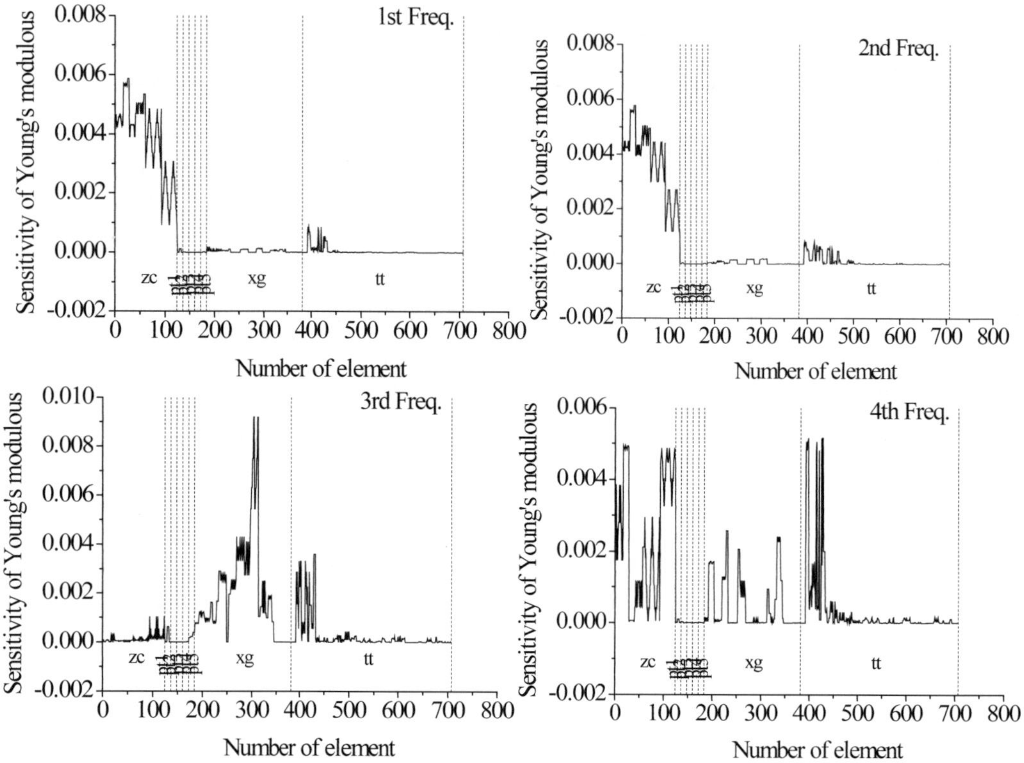

Figure 3 Sensitivity of first eight natural frequencies to the Young's modulus

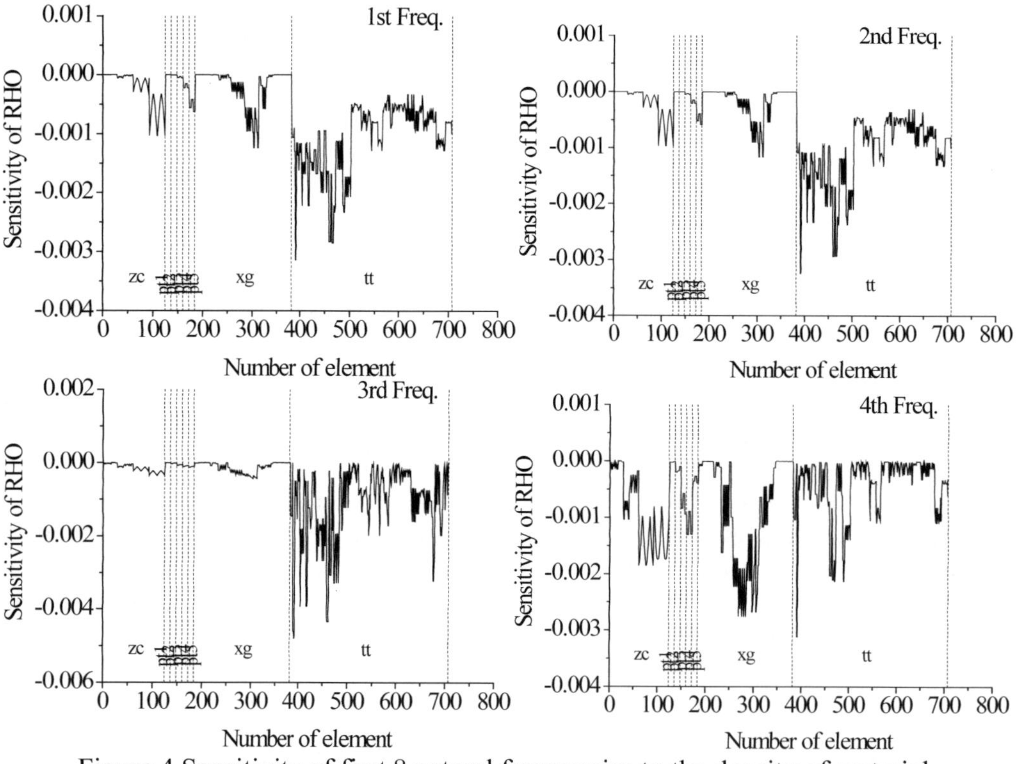

Figure 4 Sensitivity of first 8 natural frequencies to the density of material

Displayed in Figure 4 are the sensitivity coefficients of natural frequencies to the density of the material. It is well known that the natural frequencies decrease with the increasing material density. Therefore, it is indicated from the Figure 4 that the sensitivity coefficients are negative. It is seen that the sensitivity coefficients of the natural frequencies to the density are quite different to those of the Young's modulus. The first two figures demonstrate that the first two natural frequencies are more sensitive to the density change of members in the vertical major members (zc) and cross arms (tt). However, the density sensitivity coefficients of the skew members and platforms are much smaller. The curves in the third figure indicate that the sensitivity coefficients of the first torsional frequency of the two cross arms are much larger than those of the other members. This is because the third mode is the torsional vibration of the two cross arm.

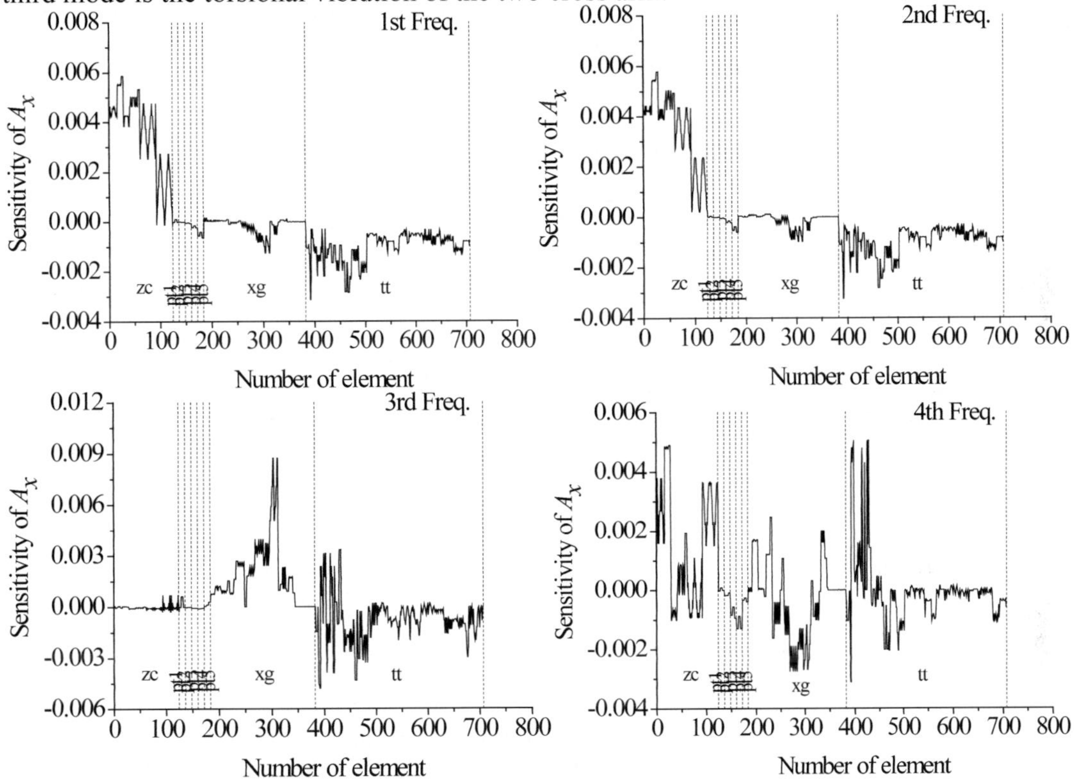

Figure 5 Sensitivity of first eight natural frequencies to the cross area

Figure 5 indicates the sensitivities of the first four natural frequencies to the cross area of each member. It is clear that the sensitivity coefficients of the cross area are quite different to those of the Young's modulus and density. The increase of member cross area does not always cause the increment of the structural natural frequencies. It is seen from the first two figures that the first two natural frequencies are more sensitive to the cross area change of the vertical major members than other members. As far as the first two natural frequencies are concerned, the increase in cross area of most of the vertical major members may induce the increment of the natural frequencies. On the contrary, the increase in the member cross areas of cross arms may lead to the increase of the global mass instead of the stiffness. Therefore, the structural natural frequencies may decrease to some extents. As displayed in the third figure, the skew members can be strengthened by increasing the cross areas and the third natural frequency, corresponding to the first order torsional mode, can be increased. Similar observations can be found from the sensitivity coefficients the other higher natural frequencies.

5. Conclusions

The feasibility of evaluating the parametric effects of a transmission tower based on the frequency sensitivity analysis is conducted in this study. The analytical model of a transmission tower is first constructed by using the FE method. The differential sensitivity analysis approach is presented and the sensitivity coefficients to natural frequencies are deduced based on the equation of motion of the tower. In addition, the expression of the frequency sensitivity to the Young's modulus, density of material and the cross area of members is proposed. A real transmission tower-line system is taken as the example to investigate the effects of the structural parameters on the natural frequency through the detailed parametric study. The made observations demonstrate that the Young's modulus of the vertical major members plays an important role in the first two vibration modes in comparison with all the other vibration modes. The sensitivity coefficients of the skew members are slightly smaller than those of the vertical major members. The sensitivity coefficients of the natural frequencies to the density are different to those of the Young's modulus because the natural frequencies decrease with the increasing material density. The sensitivity coefficients of the cross area are quite different to those of the Young's modulus and density. The increase of member cross area does not always cause the increment of the structural natural frequencies.

Acknowledgement

The writers are grateful for the financial support from the technological project of the Chinese Southern Power Grid Co. Ltd (GDKJXM20161994(030800KK52160004)), the National Natural Science Foundation of China (51678463), and the Chenguang Science and Technology Plan of Wuhan (2016070204010107). Corresponding author is B. Chen.

References

Chen, B, Zheng, J, & Qu, WL. 2009, 'Control of wind-induced response of transmission tower-line system by using magnetorheological dampers', *International Journal of Structural Stability and Dynamics*, vol. 9, no.4, pp. 661-685.

Battista, RC, Rodrigues, RS, & Pfeil, MS. 2003, 'Dynamic behaviour and stability of transmission line towers under wind forces', *Journal of Wind Engineering and Industrial Aerodynamics*, vol. 91, no.8, pp. 1051-1067.

Li, HN, Tang, SY, & Yi, TH. 2013, 'Wind-rain-induced vibration test and analytical method of high-voltage transmission tower', *Structural Engineering and Mechanics*, vol 48, no. 4, pp. 435-453.

Okamura, T, Ohkuma, T, Hongo, E, & Okada, H. 2003, 'Wind response analysis of a transmission tower in a mountainous area'. *Journal of Wind Engineering and Industrial Aerodynamics*, vol. 91, no. 1-2, pp. 53-63.

Chen, B, Zhao, SL, Li, PY. 2014, 'Application of Hilbert-Huang transform in structural health monitoring: a state-of-the-art review', *Mathematical Problems in Engineering*, vol. 2014, Article ID 317954, pp. 1-23.

Chen, B, Xu, YL, & Zhao, X. 2010, 'Integrated vibration control and health monitoring of building structures: a time-domain approach'. *Smart Structures and Systems*, vol. 6, no. 7, pp. 811-833.

Doebling, SW, Farrar, CR, Prime, MB, & Shevitz, DW. 1998, 'A review of damage identification methods that examine changes in dynamic properties', *Shock and Vibration Digest*, vol. 30, no. 2, pp. 91-105.

Lee, PS, & McClure, G. 2007, 'Elastoplastic large deformation analysis of a lattice steel tower structure and comparison with full-scale tests', *Journal of Construction Steel Research*, vol. 63, no. 5, pp. 709-717.

The 8th International Conference on
Structural Health Monitoring of Intelligent Infrastructure
Brisbane, Australia | 5-8 December 2017

Displacement Estimation by Kalman Filter Based Data Fusion

Ke Gao[1], Hongping Zhu [2], Fei Gao [2], Shun Weng [4]
[1,2,3,4] *Huazhong University of Science and Technology – China, Email:* wengshun@hust.edu.cn

Abstract

Single measurement-based displacement estimation methods usually have their own restricts to restore accurate displacement of structure. For instance, acceleration-based method has low-frequency drift due to the double integration error, strain-based technique often brings undesired estimate error caused by its modal assumption. In this paper, a dynamic displacement estimation by data fusion of acceleration sensor and strain gages measurement using Kalman filter is proposed. Virtual work theory is first applied to convert the strain to displacement, without need of strain mode shape. Then Kalman filter enables the data fusion algorithm to fuse acceleration with the strain-estimated displacement. The verification of the proposed method is implemented by a lab scale test of a vibrating cantilever beam. The results and error analysis shows a significant improvement in accuracy compared with strain-based method. Moreover, the data fusion of two measurements can eliminate the precision loss of the peak value of displacement time-history curve, and gives a smooth correction to the estimation result. This approach is advantageous and practical for the displacement estimation of large scale engineering structures, where direct and vision based methods are limited on account of field condition and weather effect.

1. Introduction

Displacement is an intuitive response that directly results from external loads to a structure, therefore it is seen as a critical parameter for the safety monitoring and evaluation of structure (Smyth & Wu 2007). The most common methods of measuring displacement responses of structure are reference-based direct approaches such as linear variable differential transducer (LVDT), laser Doppler vibrometer (LDV), total station, and vision-based systems. These methods, however, are usually limited due to field experiment condition, for instance, occlusion of scaffold, difficulty of sensor installation and weather effect (Park & Sim 2014). As an alternative, indirect methods which are generally reference-free, can be used to reconstruct the displacement by the calculation of other structure responses. Acceleration and strain are the most commonly measured responses in structure monitoring (Kim & Sohn 2016), hence efforts have been made in the application and improvement of these responses in displacement estimation. Acceleration is generally sensitive in measuring dynamic characteristics of a structure, however leads to accumulated integration error when calculating displacement (Park & Sim 2013). Strain measurements can be utilized to obtain displacement with low frequency components (Wang 2014), but the strain-based displacement is vulnerable to measurement noise (Treiber 2008). Strain-displacement methods generally based on modal mapping. Xia (2014) developed a convenient and practical method for displacement estimation of super-tall structures in horizontal plane, this method is based on a cantilever beam assumption and the application of virtual work theory, error analysis shows the derived displacements have higher accuracy than the GPS-measured results, but still at the same magnitude. Thus, data fusion methods that can extract necessary information from different measurements and with high accuracy at the same time are required for displacement estimation (Kim 2013, Cho 2015,

Kim 2016). Cho and Park (2016) proposed a Kalman filter method to estimate displacement on bridges with high accuracy by utilizing acceleration and modal-mapping of strain, however, the modal-mapping approach has difficulty to confirm the neutral axis of large-scale structure which has changing cross sections.

This paper proposes a Kalman filter-based data fusion displacement estimation method using two indirect responses (i.e. acceleration and strain). State-space model and virtual work theory are applied in the proposed method. The performance of this data fusion technique is investigated in a lab scale experiment.

2. Displacement Estimation Method using Kalman Filter

A Kalman based estimation model for the displacement estimation is described in this section. This model provides a combination method of two different measurement data. Acceleration data contains high-frequency dynamic characters of structure responses, displacement with accurate dynamic component can be obtained through the double integration of acceleration, but the double integration leads to serious drift due to the integration error. Strain data can capture the low-frequency and static characters of structure responses, whereas it fails to restore the dynamic components via the strain-displacement formula. Therefore, fuse the information of acceleration and strain can conserve dynamic and static characters, which result in accurate displacement estimation.

In this study, the proposed method which fuse acceleration and strain, is a cooperation of a strain-based displacement estimation technique provided in (Xia 2014), and the conventional Kalman filter (CKF) presented by (Smyth 2007).

2.1 Derivation of Displacement using Strain Measurement

The main body of structure with a large flexibility can be regarded as a beam, such as super-tall buildings, long-span bridges and subway tunnel. A cantilever beam is seen as practical simplified model of a skyscraper according to its large slenderness ratio and its boundary condition. The displacement of the structure can be represented by the lengthwise strain at different heights (Xia 2014). The cantilever beam is divided into n subsection, depend on the amount of lengthwise strain sensors. The length of the i-th subsection is h_i. Assume that the strain between two measurement points is varying linearly, the strain variation at the i-th point is $\Delta\varepsilon_i$, which is the strain difference of opposite sides of the beam.

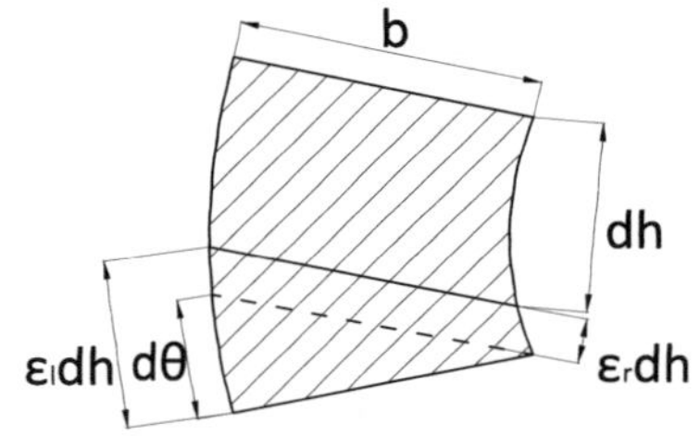

Fig.1 Angular rotation of a micro-element

As shown in Fig.1, the angular rotation of a micro-element is expressed as

$$d\theta = \frac{(\varepsilon_l - \varepsilon_r)dh}{b} = \frac{\Delta\varepsilon dh}{b} \tag{1}$$

where ε_l, ε_r is the longitudinal strain at the element's left and right sides, b is the width of the section.

In order to calculate the horizontal displacement of the beam top (point n), a horizontal virtual unit force is loaded at point n. The bending moment at i-th point is

$$l_i = \sum_{j=i+1}^{n} h_j \,,\; l_n = 0 \tag{2}$$

Assume that the lengthwise deformation of the beam is negligible, then the horizontal displacement is almost due to the bending moment. The displacement at the top point is written as

$$u_n = \int \overline{M}\, d\theta = \frac{1}{b}\int_0^H \overline{M}\, \Delta\varepsilon dh \tag{3}$$

where u_n is the displacement at point n. This formula needs a discretization because of the discrete strain measurements, by utilizing the moment diagram multiplication method which based on the virtual work theory, the discrete form of formula (3) can be rewritten as

$$u_n = \frac{1}{b}\sum_{i=1}^{n} \overline{M}_i\, \Delta\varepsilon_i \cdot h_i = \frac{1}{6b}\sum_{i=1}^{n}\left[h_i(2l_{i-1}\Delta\varepsilon_{i-1} + l_{i-1}\Delta\varepsilon_i + l_i\Delta\varepsilon_{i-1} + 2l_i\Delta\varepsilon_i)\right] \tag{4}$$

where $\overline{M}_i$ is the virtual bending moment at i-th point. Fig.2 shows the bending moment and strain variation diagram under the unit force.

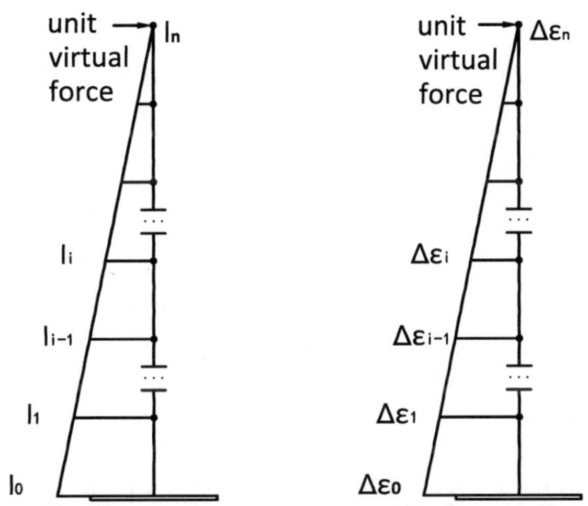

Fig.2 Bending moment and strain variation diagram

2.2 State-Space Model

Note that $x(t)$ is the true displacement at t time, $\dot{x}(t)$ and $\ddot{x}(t)$ are the corresponding velocity and acceleration respectively. Then the state-space model which includes system equation (5) and observation equation (6), can be written as

$$\begin{bmatrix} \dot{x}(t) \\ \ddot{x}(t) \end{bmatrix} = \begin{bmatrix} 0 & 1 \\ 0 & 0 \end{bmatrix} \begin{bmatrix} x(t) \\ \dot{x}(t) \end{bmatrix} + \begin{bmatrix} 0 \\ 1 \end{bmatrix} \ddot{x}(t) + \begin{bmatrix} 0 \\ 1 \end{bmatrix} w(t) \tag{5}$$

$$x_m(t) = \begin{bmatrix} 1 & 0 \end{bmatrix} \begin{bmatrix} x(t) \\ \dot{x}(t) \end{bmatrix} + v(t) \tag{6}$$

where $x_m(t)$, $w(t)$ and $v(t)$ are the measured displacement, system process noise and displacement measurement noise. These noises are Gaussian white noise, with covariance matrix of Q, R. Since the measurements are discrete, so a transform form continuous-time domain to discrete-time domain is required to adjust the filter algorithm. The discrete-time state vector are introduced as

$$X(k) = \begin{bmatrix} x(k) & \dot{x}(k) \end{bmatrix}^T \tag{7}$$

the superscript T means matrix transposition. The state-space model is discretized as

$$X(k) = \begin{bmatrix} 1 & t_a \\ 0 & 1 \end{bmatrix} X(k-1) + \begin{bmatrix} t_a^2/2 \\ t_a \end{bmatrix} \ddot{x}(k) + \begin{bmatrix} 0 \\ 1 \end{bmatrix} w(k-1) = AX(k-1) + B\ddot{x}(k) + w(k-1) \tag{8}$$

$$x_m(k) = \begin{bmatrix} 1 & 0 \end{bmatrix} x(k) + v(k) = Hx(k) + v(k) \tag{9}$$

where t_a is the sampling interval of acceleration, process noise $w(k) \sim N(0, Q)$, measurement noise $v(k) \sim N(0, R)$. The system matrix A, input control matrix B, measurement matrix H, system process noise Q and displacement measurement noise R are defined as

$$A = \begin{bmatrix} 1 & t_a \\ 0 & 1 \end{bmatrix},\ B = \begin{bmatrix} t_a^2/2 \\ t_a \end{bmatrix},\ H = \begin{bmatrix} 1 & 0 \end{bmatrix},\ Q = q \begin{bmatrix} t_a^3/3 & t_a^2/2 \\ t_a^2/2 & t_a \end{bmatrix},\ R = \frac{r}{t_a}.$$

It is worth noting that the process noise Q is an influential factor in the estimation process, therefore it must be properly chosen to achieve an accurate result.

2.3 Two Stages of Kalman Filter

The conventional Kalman filter has two stages, prior estimation process as well as update process. Prior estimation process is based on state-space model in (8) and (9). The main works of estimation process are displacement prior estimation, calculation of estimation error covariance matrix and calculation of Kalman gain. These principle equations are written as

$$\bar{X}(k) = A\hat{X}(k-1) + B\ddot{x}_m(k) \tag{10}$$

$$\bar{P}(k) = A\hat{P}(k-1)A^T + Q \tag{11}$$

$$K(k+1)=\bar{P}(k)H^T\left(H\bar{P}(k)H^T+R\right)^{-1} \quad (12)$$

where $\bar{X}(k)$, $\hat{X}(k)$, $\ddot{x}_m(k)$ are prior state estimates, updated state estimates and acceleration measurement. $\bar{P}(k)$, $\hat{P}(k)$ are prior and updated estimation error covariance matrices (Sohn 2016). The acceleration measurements can be used to estimate prior displacement via (10), the error covariance matrix is a parameter to evaluate the error of estimation, and Kalman gain is regarded as a correction factor or a weight of bias between measurement and estimate.

In the update process, the displacements converted from strain measurements are utilized to correct the displacement estimates. Then updated state and error covariance estimates can be acquired from (13) and (14).

$$\hat{X}(k)=\bar{X}(k)+K(k)\cdot\left(u(k)-H\bar{X}(k)\right) \quad (13)$$

$$\hat{P}(k)=(1-K(k)H)\bar{P}(k) \quad (14)$$

where $u(k)$ is displacement converted by strain via strain-displacement principle (Xia et al 2014).

The updated displacement estimation is the first element of the updated state variable, and the extraction of it is written as

$$\hat{x}(k)=\begin{bmatrix}1 & 0\end{bmatrix}\hat{X}(k) \quad (15)$$

where $\hat{x}(k)$ is the optimal displacement estimation.

In this update process, the fusion of acceleration and strain data is accomplished through an indirect form, and the displacement is the intermediate medium along with desired result.

The accuracy of displacement estimation is quantified by utilizing an error parameter RMSD (root mean square deviation) as follow.

$$RMSD_{est}=\frac{\sqrt{\sum_{n=1}^{N}(x_{est}-x_{ref})^2}}{\sqrt{\sum_{n=1}^{N}(x_{ref})^2}}\times 100\%,\quad RMSD_{strain}=\frac{\sqrt{\sum_{n=1}^{N}(x_{strain}-x_{ref})^2}}{\sqrt{\sum_{n=1}^{N}(x_{ref})^2}}\times 100\% \quad (16)$$

where x_{est}, x_{strain}, x_{ref} are the estimated, strain-based and reference displacement.

3. Experimental Verification

A lab experiment is implemented to validate the feasibility of the proposed method. As shown in Fig.4. The size of cantilever beam is $1000\times 60\times 5$mm, and its first natural frequency is 11.48 according to the free vibration test. The cracks on the beam are considered to simulate the non-uniform rigidity of different sub-elements in a real supertall structure.

Ten resistance strain gages are installed along the length of the beam, with an accordant distance of 10cm between two gages. The electrical property of the gages allow them to measure the dynamic strain of the beam at high sampling rate. Acceleration sensor is placed at the same location as the last strain gage (i.e. 10cm away from the free end of beam).

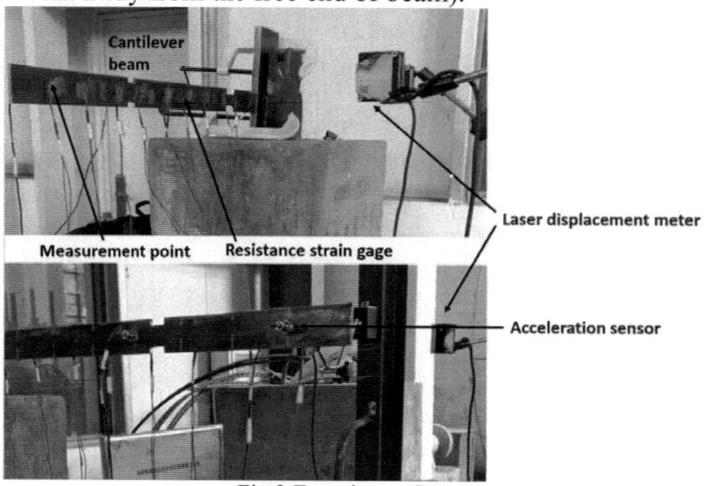

Fig.3 Experiment Setup

The laser displacement sensor is utilized to capture the same point's displacement with high accuracy so that it is used as the reference value for comparison. All the measurements are sampled with the same sampling rates. The measurements at 1000Hz, 500Hz, 200Hz are considered to investigate the effect of sampling frequency.

The beam is excited by an impact load by a hammer. Each impact case is repeated five times to eliminate the sampling uncertainty.

The displacement estimation results of the proposed method at different sampling rates are compared with strain-based method. Fig.5 shows the comparison of two methods' estimation results at a sample rate of 1000Hz. The reference displacement is given by the laser displacement meter. The acc-strain displacement reveals a better agreement with the reference displacement from a contrast of (a) and (b) in Fig.5. The RMS (Root Mean Square) indicator of the estimate error of strain-only method and acc-strain method are 0.2777 and 0.0980 respectively, it exhibits the improvement in accuracy of the proposed method. The close-up for 1.0s to 2.0s shows the smooth effect of this data fusion (see Fig.5 (c)).

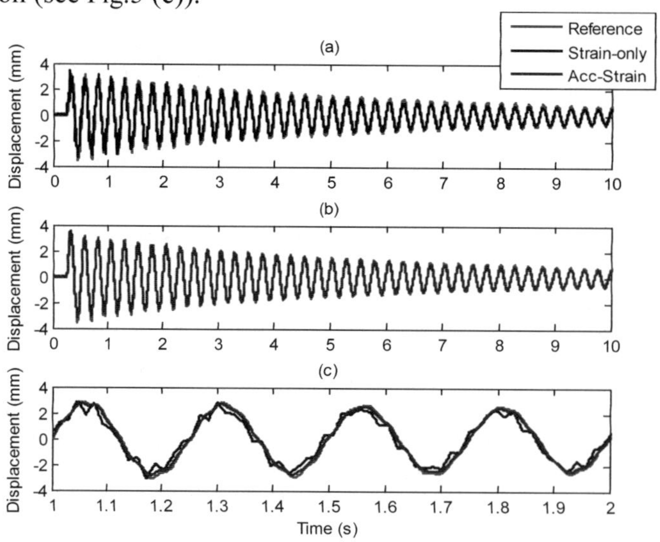

Fig.5 Comparison of the displacement estimated by strain-based method (a) (Xia 2014) and proposed method (b), (c) is the close-up for 1s to 2s, the sampling rate of measurement is 1000Hz.

Fig.6 and Fig.7 show the displacement estimated by two methods at sample rate of 500Hz and 200Hz, respectively. The strain-based method usually loses its precision at every peak of the displacement time-history curve (see (a)), while this defect is eliminated when the strain and acceleration measurement are fused by the proposed algorithm (see (b)). Moreover, the fusion also gives a smoothing correction to the result, which compensate the unsmooth character of strain-based displacement (see (c)). The change of sampling rate doesn't have a significant influence on the estimation results in accordance to these three figures, but further discussions are still needed before a definite conclusion.

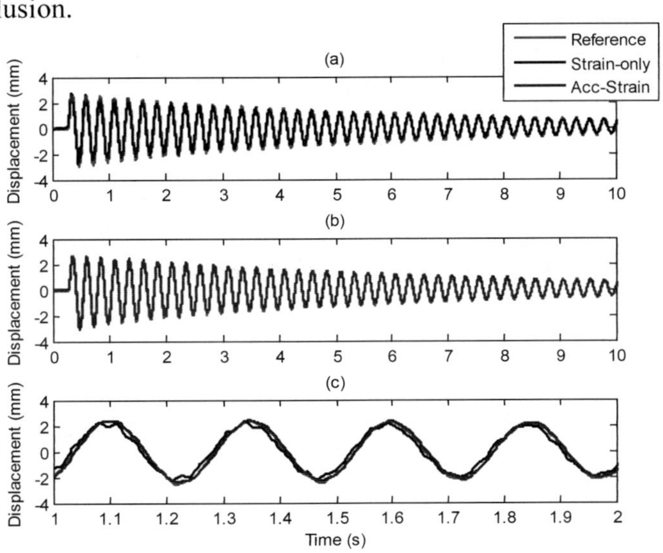

Fig.6 Comparison of the two methods at a sampling rate of 500Hz.

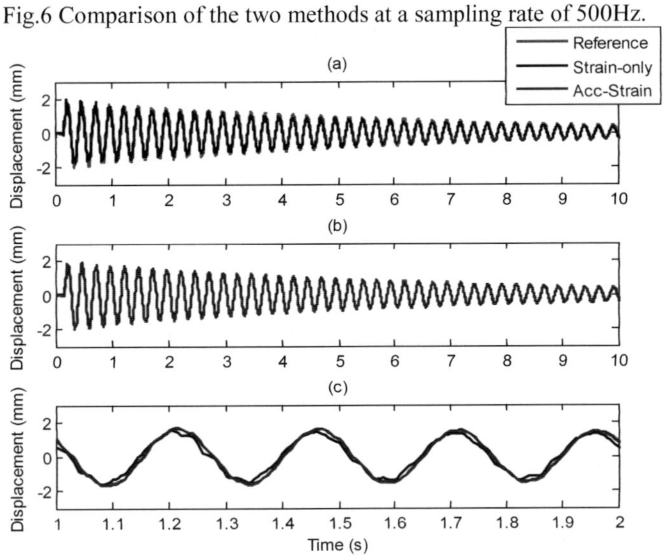

Fig.7 Comparison of the two methods at a sampling rate of 200Hz.

The estimated error of the strain-based method and the proposed method at three different rates are quantified by RMSD percent described in Eq. (16). As shown in Table 1, the average RMSD of the two described methods are 20.82% and 8.02%, respectively. Table 2 shows the RMS indicator comparison of the estimation error of the two methods, the respective average RMS of the two methods are 0.1993 and 0.0779 (unit: mm). It is obviously that the proposed estimate algorithm performs a large improvement of accuracy compared with the strain-based method.

Table 1 Root-Mean-Square-Deviation

RMSD of the strain-based method and the proposed method for 3 sampling rates

Sampling Rate	1000	500	200	Average
Strain-only	22.85%	19.32%	20.31%	20.82%
Acc-Strain	8.06%	7.87%	8.13%	8.02%

Table 2 Root-Mean-Square

RMS of the error of the two methods for 3 sampling rates (unit: mm)

Sampling Rate	1000	500	200	Average
Strain-only	0.2777	0.1775	0.1427	0.1993
Acc-Strain	0.0980	0.0787	0.0571	0.0779

The error shown in Fig.8 is defined as the difference between estimated displacement and reference displacement. Fig.8 illustrates that the error of the proposed estimation algorithm is much smaller than the strain-based algorithm.

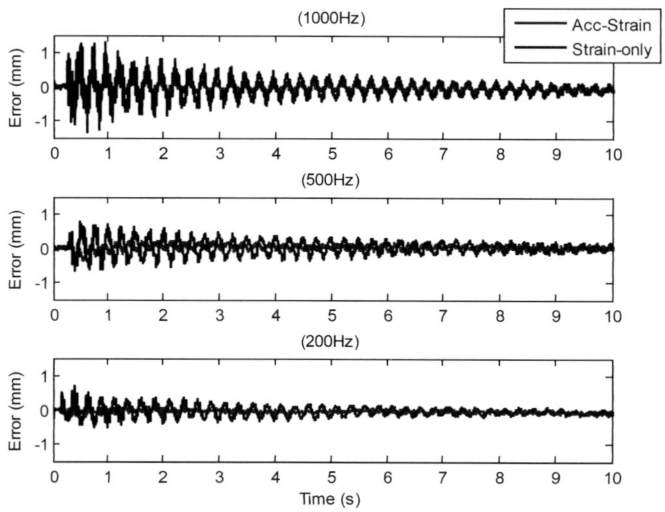

Fig.8 Error of the strain-based method and the proposed method at different sampling rates

4. Conclusions

This paper presents a displacement estimation approach based on acceleration and strain using Kalman filter data fusion algorithm. Firstly, the strain response is converted to displacement using virtual work theory based on a beam model assumption. Meanwhile, it doesn't have to consider the size change of the cross section, because it is already involved in the measurements. Then, the Kalman filter is utilized to combine the displacement and acceleration via a state-space model. The method is verified by a lab experiment of a cantilever beam subjected to impact load at the free end. Based on the results, conclusions can be summarized as follows:

(1) The proposed method can estimate the dynamic displacement of beam structures with 8% RMSD error rate and 0.0779 RMS error level in lab test, which is much improved than the strain-based method.

(2) Fusion with acceleration can not only effectively eliminate the precision loss of the peak value of displacement estimation in strain-based method, but also has a smoothing effect to correct the unsmooth character of strain-based estimation.

(3) The proposed method is expected to apply for large scale engineering structure, such as super-tall structures and subway tunnel where reference point based displacement estimation techniques is unavailable.

References

Smyth, A & Wu, ML 2007, 'Multi-rate Kalman Filtering for the Data Fusion of Displacement and Acceleration Response Measurements in Dynamic System Monitoring', *Mechanical Systems and Signal Processing*, vol. 21, no. 2, pp. 706-723.

Park, JW & Sim, SH 2014 'Wireless displacement sensing system for bridges using multi-sensor fusion', *Smart Materials and Structures*, vol. 23, no. 4, pp. 1-12.

Kim, KY & Sohn, H 2016, 'Dynamic Displacement Estimation by Fusing LDV and LiDAR Measurements via Smoothing Based Kalman Filtering', *Mechanical Systems and Signal Processing*, vol. 82, no. 2, pp. 339-355.

Park, JW & Sim, SH 2013, 'Displacement Estimation Using Multimetric Data Fusion', *IEEE-ASME Transactions on Mechatronics*, vol. 18, no. 6, pp. 1675-1682.

Wang, ZC & Geng, D 2014, 'Strain Modes Based Dynamic Displacement Estimation of Beam Structures with Strain Sensors', *Smart Materials and Structures*, vol. 15, no. 3, pp. 645-663.

Treiber, J & Mueller, UC 2008, 'Filtering techniques in the dynamic deformation estimation using multiple strains measured by FBGs', *Conference on Sensors and Smart Structures Technologies for Civil, Mechanical, and Aerospace Systems2008*, San Diego, CA, 10-13, March, pp. 2A-1–2A-9.

Xia, Y & Zhang, P 2014, 'Deformation Monitoring of A Super-tall Structure Using Real-time Strain Data', *Engineering Structures*, vol. 67, pp. 29-38.

Cho, SJ & Park, JW 2016, 'Reference-Free Displacement Estimation of Bridges Using Kalman Filter-Based Multimetric Data Fusion', *Journal of Sensors*, vol. 2016, pp. 1-9.

Kim, KY & Kim, JH 2013, 'Kalman Filter Based Data Fusion for Dynamic Displacement Estimation Using LDV and LiDAR', *The 2013 World Congress on Advances in Structural Engineering and Mechanics (ASEM13)*, Jeju, Korea, 8-12, September, pp. 1667-1676.

Cho, SJ & Yun, CB 2015, 'Displacement estimation of bridge structures using data fusion of acceleration and strain measurement incorporating finite element model', *Smart Structures and Systems*, vol. 15, no. 3, pp. 645-663.

Kim, KY & Kim, JH 2016, 'Development and full-scale dynamic test of a combined system of heterogeneous laser sensors for structural displacement measurement', *Smart Materials and Structures*, vol. 25, no. 6, pp. 1-15.

The 8th International Conference on
Structural Health Monitoring of Intelligent Infrastructure
Brisbane, Australia | 5-8 December 2017

Using sparse regularization and impedance sensitivity for structural damage detection

Xingyu Fan, Jun Li, Hong Hao
Centre for Infrastructural Monitoring and Protection, School of Civil and Mechanical Engineering, Curtin University, Perth, WA 6102, Australia

Abstract

Electromechanical impedance (EMI) based structural damage detection methods have been widely developed in civil and mechanical engineering community. In EMI-based structural health monitoring techniques, the piezoceramic (PZT) transducer is commonly bonded on the surface of the host structure and is excited by a sweeping voltage. The impedance of the PZT transducer can be measured and used to identify the changes occurred in the host structure by using statistical damage indices, such as RMSD and CC. However, those damage indices are not able to indicate more information of structural damage, such as the location and severity. This paper proposes a new structural damage identification approach based on model updating with impedance sensitivity and using sparse regularization technique to identify the damage location and severity. The sensitivities of impedance responses with respect to the stiffness parameters of the host structure are calculated and used to identify the damage with few number of resonance frequency shifts. The effectiveness and performance of the proposed approach are demonstrated with numerical simulations on an aluminum beam.

1. Introduction

Studies on using smart materials, i.e. piezoelectric ceramic (PZT) transducers, in Structural Health Monitoring (SHM) of civil engineering structures have attracted significant attention in recent decades, because of the light weight, small size, cost-effectiveness and in-sensitive to low-frequency and environmental noise. PZT was first used as both a sensor and an actuator in the health monitoring of aerospace structures (Chaudhry et al. 1995). Liang et al. (1994) performed the coupled electro-mechanical analysis of adaptive material systems using PZT materials as sensor and actuator simultaneously. Later, electromechanical impedance (EMI) based structural health monitoring has been rapidly developed (Sun et al. 1995, Giurgiutiu et al. 1999, Raju et al. 1999, Park et al. 2000, Ong et al. 2002, Park et al. 2006).
In conventional EMI based SHM techniques, the difference between the impedance curves from the current structural state and the baseline state is used to detect whether there is any damage or degradation occurred in the structure. To quantify such difference in impedance curves, various damage indices have been developed. Root Mean Square Deviation (RMSD) was popularly utilized to analyse the impedance signals for structural damage detection (Giurgiutiu & Rogers, 1998). Other statistical damage indicators, such as Covariance Change (CC), Correlation Coefficient Deviation (CCD) and the Mean Absolute Percentage Deviation (MAPD), were compared with RMSD through experimental studies (Zagrai & Giurgiutiu 2001, Tseng & Naidu 2002). The results showed that RMSD is the most sensitive indicator to the new damage occurrence. CC and CCD are more suitable for the detection of the increasing damage extent. Even though these existing impedance based statistical damage indicators can be used to warn the damage occurrence by defining a threshold, they are not able to locate and quantify the structural damage, because the statistical damage index of a

minor damage which is close to the measured location may have the same value as that with a severe damage which is far away from the measured location. Artificial Neural Networks (ANN) as the artificial intelligence based optimization method can be applied to provide a mapping between the inputs and outputs. Pioneering researches have been conducted to enhance the capability of EMI based damage detection methods based on ANN (Min et al. 2012, Na & Lee 2013, He et al. 2014, Fekrmandi et al. 2016, Sevillano 2016). Comparing with the statistical damage indicators that can only detect the existence of structural damage, these studies defined the input as the impedance curves from the single or multiple damage scenarios and the output as the corresponding damage type and severity. However, damage detection by using ANN and EMI based methods can only recognize specific damage type, location and severity from several pre-determined levels and areas.

Both statistical damage indicators and ANN based methods are non-model damage detection methods, in which only the measured data from the structure before and after the damage are required no matter for calculating the damage indicators or training the network. To further explore the capacity of EMI based techniques for accurate structural damage identification, i.e. localization and quantification, model based methods are necessary. Spectral finite element method can be used for structural modelling and impedance calculation, however, it may be only used for modelling structure with simple geometry, i.e. beam and plate structures (Sepehry et al. 2017). On the other hand, finite element modelling can be used to simulate a three-dimensional structure and is more capable than spectral finite element method to simulate structures with complex geometries. The mesh of the finite element modelling shall be fine enough to calculate the impedance accurately. Therefore a significant number of elements are usually involved in the structural modelling. When using the impedance for model-based structural damage identification, there might be a large number of unknown parameters to be identified. This makes the identification equation an underdetermined problem with limited measurement information. To overcome this difficulty, sparse regularization is used in this study for the impedance based finite element model updating and damage identification. To improve the sensitivity of using impedance for damage identification, the frequency shifts are used for the identification.

2. Theoretical Background

To reveal the interaction between the measured electrical impedance of the piezoelectric transducer and the mechanical impedance of the host structure, a semi-analytical model associated with the finite element model (finite element analysis was used to determine the host-structure mechanical impedance which is one part of the impedance formulation) was first proposed by Fairweather & Craig (1998). It demonstrated that finite element modelling techniques can be used for impedance analysis. Efforts have been spent to improve the performance of using finite element modelling for impedance calculation and analysis (Bhalla 2001, Bhalla & Soh 2004, Annamdas & Soh 2007, Liu & Giurgiutiu 2007). Yang et al. (2008) presented a numerical model to calculate the impedance with the commercial finite element analysis software, i.e., ANSYS, and compared the simulation results with experimental results. Simulation results of a single suspended PZT patch and PZT-structure interaction are basically matched with the measured impedance responses. Lim & Soh (2014) updated the parameters used in the previous finite element model and replaced the Rayleigh damping with a hysteretic damping model. This study significantly increased the accuracy of using the finite element modelling to calculate the impedance.

In this study, the updated parameters of the PZT transducer, bonding layer and the host structure in the previous study (Lim & Soh 2014) are used to verify the finite element modelling technique to calculate the impedance and calibrate the finite element models of the PZT patch and host structure. Hysteretic damping is used instead of the Rayleigh damping model which is widely applied in the conventional EMI models. The simulation result will be compared with the experimental result, in order to calibrate the model and achieve an accurate finite element model, which can be used for the subsequent damage detection.

After an accurate finite element model for the electromechanical impedance calculation is obtained, when the model based method, i.e. sensitivity based model updating method is used for structural damage identification, the difference between the calculated impedance response from the numerical model and the measured impedance response from the damaged structure can be obtained. Structural system parameters are identified by solving the ill-posed inverse problem with an iterative procedure. The first-order sensitivity based identification problem can be defined as (Lu & Law 2007)

$$\{\delta_f\} = [S_f]\{\delta_K\} \tag{1}$$

where $\{\delta_f\}$ denotes the shifts in the resonance frequencies, $\{\delta_K\}$ and $[S_f]$ denote the stiffness reduction parameters of structural segments and the sensitivity matrix which is the change of resonance frequency with respect to the stiffness parameter of the host structure. Eq. (1) is usually an ill-posed inverse problem, the conventional damped least-squares method (also called as Tikhonov regularization technique) can be used to solve the identification equation. The ℓ_2-norm regularization term on the solution is added on the objective function of the simple least square method, which can be described as

$$\mathrm{Min}(\left\| S_f \delta_K - \delta_f \right\|^2 + \lambda \left\| \delta_K \right\|_2^2) \tag{2}$$

where λ is the regularization parameter. The identified stiffness changes from Tikhonov regularization technique are usually distributed to most elements of the structure. However, structural damages occur usually at a few locations therefore the damage vector shall be a sparse vector with most of its items are zero or close to zero except the damaged elements with nonzero items. Zhou et al. (2015) presented using ℓ_1-norm regularization approach to overcome this limitation of Tikhonov regularization. To bring this physical understanding into the solution of the inverse problem, ℓ_1-norm regularization technique is applied in this study. Another reason to use the sparse regularization technique is based on the fact that usually a few number of shifts in the impedance resonance frequencies are available for the identification therefore the identification equation may be an underdetermined problem (Hernandez 2014). When using the ℓ_1-norm regularization method, the objective function can be defined as

$$\mathrm{Min}(\left\| S_f \delta_K - \delta_f \right\|_2^2 + \lambda \left\| \delta_K \right\|_1^1) \tag{3}$$

Eq. (3) can be solved by using the spectral projected gradient method (Van den Berg & Friedlander 2011). Sparse regularization, i.e. the l_1-norm regularization technique based sensitivity model updating method will be used for impedance based structural damage detection in this study.

3. Finite Element Model Calibration

A freely suspended PZT patch with the dimension of 10mm×10mm×0.3mm is simulated by using COMSOL Multiphysics version 5.1. Piezoelectric material PIC 151 is used as the type of transducer. The detailed mechanical and piezoelectric property parameters are taken based on a previous study (Lim & Soh 2014) and listed in Table. 1. The mesh size of the finite element model is defined as 0.2 mm, which has been demonstrated as being adequately fine to model the electromechanical impedance behaviours under 800kHz (Makkonen et al. 2001). The simulated impedance signal obtained from the finite element model is compared with the experimental result (Lim & Soh 2014), as shown in Figure 1. A good achievement can be observed from the impedance curves between 1 kHz and 800 kHz.

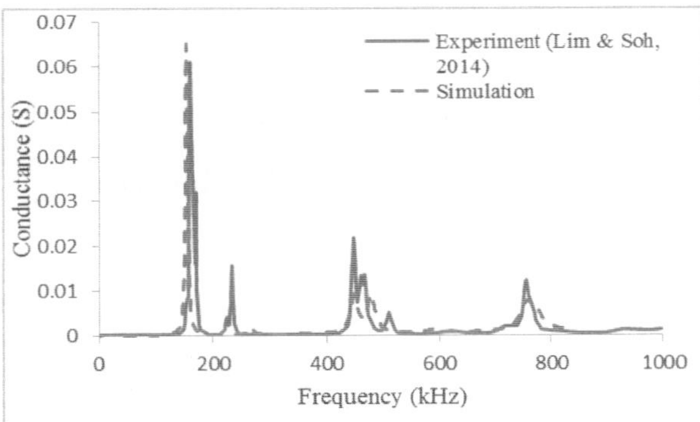

Figure 1. Conductance of a freely suspended PZT patch

Table 1. Material properties of PIC 151 PZT transducer in the finite element analysis

Property	Symbol	Value
Density	ρ	$7760\ kg/m^{-3}$
Compliance	$s_{11} = s_{22}$	$16.8 \cdot 10^{-12} m^2 N^{-1}$
	s_{33}	$19.0 \cdot 10^{-12} m^2 N^{-1}$
	$s_{12} = s_{21}$	$-5.66 \cdot 10^{-12} m^2 N^{-1}$
	$s_{13} = s_{31}$	$-7.11 \cdot 10^{-12} m^2 N^{-1}$
	$s_{23} = s_{32}$	$-7.11 \cdot 10^{-12} m^2 N^{-1}$
	$s_{44} = s_{55}$	$45.0 \cdot 10^{-12} m^2 N^{-1}$
	s_{66}	$45.0 \cdot 10^{-12} m^2 N^{-1}$
Coupling matrix (Strain-charge form)	d_{31}	$-2.08 \cdot 10^{-10} mV^{-1}$
	d_{32}	$-2.08 \cdot 10^{-10} mV^{-1}$
	d_{33}	$4.23 \cdot 10^{-10} mV^{-1}$
	d_{15}	$6.1 \cdot 10^{-10} mV^{-1}$
Relative permittivity	ε_{11}^{S}	1110
	ε_{22}^{S}	1110
	ε_{33}^{S}	852

When building the finite element model to consider the PZT-structure interaction, a free-free aluminum beam with the dimension of 231mm×26mm×4mm is built as the host structure. A

PZT patch with the size of 5mm×5mm×0.3mm is mounted on the structure surface in the middle of the beam. The thickness of the thin bonding layer between the PZT patch and host structure is 0.03 mm. Detailed mechanical properties of the aluminum beam and adhesive layer are shown in Table 2. The mesh size is set as 0.5 mm for both the structure and PZT patch, which will be sufficient for the selected frequency range of 20-60 kHz.
To validate the accuracy and correctness of the finite element modelling technique for calculating the impedance, Figure 2 shows the obtained impedance from this numerical model compared with the experimental result (Lim & Soh 2014). Only the impedance responses in the frequency range from 20-60 kHz are shown, and it is observed that most resonance peaks located in this range are well matched with experimental results. The accuracy in those resonance frequencies is good, and only the magnitudes at some frequencies from the simulated impedance cannot exactly match the values in the experimental study. The possible reason for this phenomenon could be the system uncertainties such as the manufacturing errors and material inhomogeneous. In this paper, only the shifts in the resonance frequencies as observed in the impedance curve will be used for structural damage detection since the accuracy of the obtained resonance frequencies is much higher than the magnitudes and the frequency shifts shall be more sensitive to the structural condition changes.

Table 2 Material properties of the aluminum beam and bonding layer in the finite element modelling

Property	Material	Symbol	Value
Density	Aluminum	ρ	2710 kg/m^{-3}
	Bonding layer		1000 kg/m^{-3}
Young's modulus	Aluminum	Y	70.6 GPa
	Bonding layer		1.4 GPa
Poisson ratio	Aluminum	υ	0.35
	Bonding layer		0.4
Damping ratio	Aluminum	ξ	0.0005
	Bonding layer		0.02

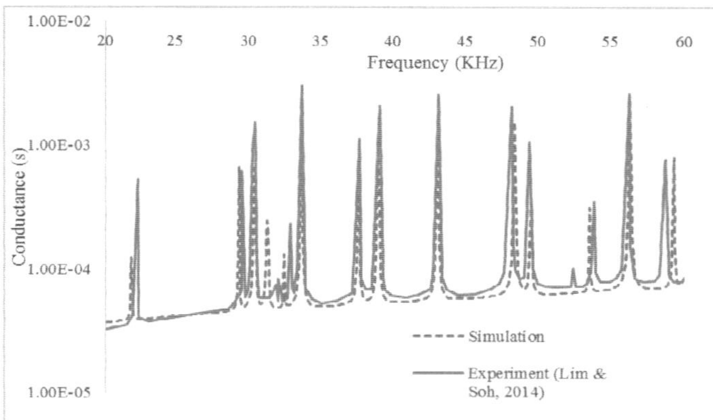

Figure 2. Conductance responses from the finite element model and experimental test

4. Structural Damage Identification

Numerical studies on a finite element model simulating the PZT-structure interaction are conducted for structural damage identification. A PZT transducer is bonded in the middle of an aluminium beam model with a dimension of 220mm×30mm×4mm, as shown in in Fig. 3.

The right side of the host structure is divided into 10 segments. Structural local damage is assumed as the stiffness change on each of these individual segments, i.e. reduction in the elastic modulus (Xia & Hao 2003). D1, D2 to D10 marked on the host structure in Fig. 3 denote the stiffness reduction of each segment. According to the recommendation by (Makkonen et al. 2001), to ensure the accuracy of the finite element simulation to calculate the impedance, there should be three to five nodes within the half wavelength. In this study, since the 8 nodes hexahedral solid elements are used to mesh the model, there should be at least 4 elements per wavelength. The wave velocity travelling in the aluminium plate can be calculated as $c = \sqrt{E/\rho} = 5104.1(m/s)$. The element size is selected as 2.5 mm which is well sufficient to cover the frequency range 20 to 40 kHz used in the following structural damage identification.

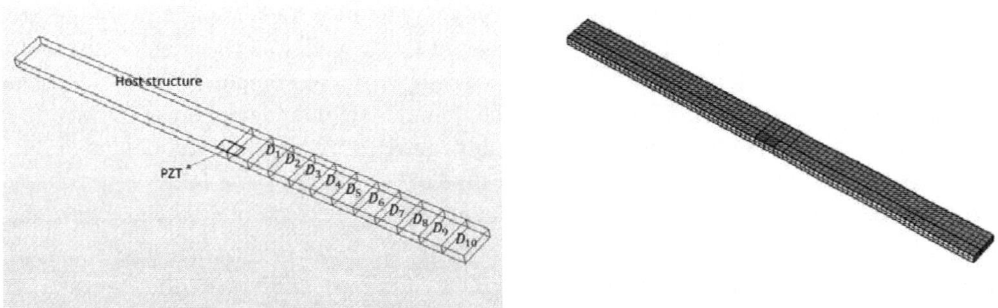

(a) Segments of the host structure

(b) Finite element model of the host structure and PZT

Figure 3. Finite element modelling of the host structure and PZT

A single damage is simulated in the finite element model with 14% stiffness reduction in the third segment. Only 7 resonance frequency shifts located in the frequency range between 20 to 40 kHz are used to identify structural stiffness parameters of those 10 segments. l_1 regularization technique is used to solve this underdetermined inverse problem by using 7 available resonance frequencies to identify 10 unknown parameters. The comparison of identification results by using Tikhonov and l_1 regularization techniques for the single damage identification are shown in Fig. 4. Both the Tikhonov regularization and l_1 regularization can predict the true location of the introduced damage. The predicted results with l_1 regularization technique indicates the largest stiffness reduction is 13.2% on the 3rd segment, which is much closer to the true damage extent than that from the Tikhonov regularization. The solution of using the Tikhonov regularization tends to distribute the damage to more elements around the damage location so that a lower identified damage severity is obtained, which is around 8%. It is only half of the true damage extent. More significant false identifications are observed in the results by using Tikhonov regularization, such as 4% stiffness reduction occurred on the 4th segment and 3.3% on the 9th segment. The false predictions obtained by using the l_1 regularization technique are much less than those from Tikhonov regularization. The largest value in the 10th segment is around 1%. The sparse regularization technique is efficient and accurate in this case to identify the structural damage with only a limited number of impedance resonance frequency shifts before and after damage.

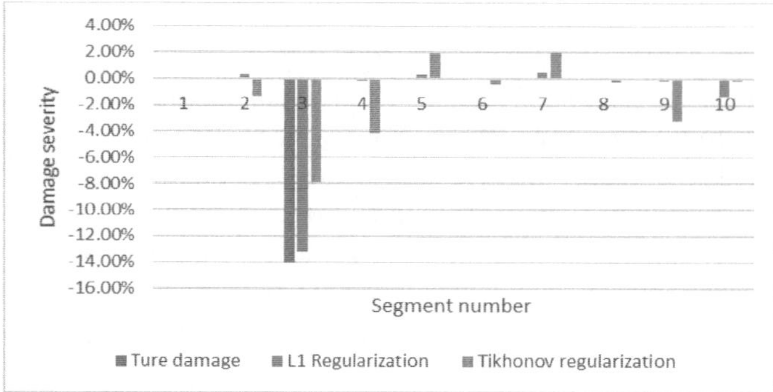

Figure 4. Damage identification results for a single damage scenario

The influence of environmental uncertainty and measurement noise effect is considered in the above case study. Even though that the impedance responses in a high frequency range are used in EMI based SHM techniques, which could avoid the low frequency noise effect, the temperature effect on the PZT material might cause a significant impact on the electrical impedance behaviour (Koo et al. 2009). Gyuhae et al. (1999) have proved that the environmental effect on the impedance response appeared as resonance frequency shifts. According to the experimental study of Sepehry et al. (2011), there was a 0.21% frequency shift of the resonance peak's frequency in the range from 80 kHz to 90 kHz when the temperate was changed from $40\,^{\circ}C$ to $50\,^{\circ}C$. In this study, in order to simulate the effect of environmental uncertainty, i.e. temperate change, a random shift on the resonance frequencies with a mean value of 0.3% of the resonance frequencies is added on the impedance curve calculated from the damaged finite element model. The frequency shifts obtained from the noisy impedance response are used for identification, and the results are shown in Fig. 5. It is observed that the results obtained by using the l_1 regularization can well identify the location and severity of the simulated damage, considering the noise effect on the impedance measurement. The predicted damage is 12.5% stiffness reduction which is lower than the true damage of 14%, but still close to the true damage severity. The false predictions of using the proposed approach are increased in some segments with small values, such as 1% in the 7th segment and 1.5% in the 9th segment.

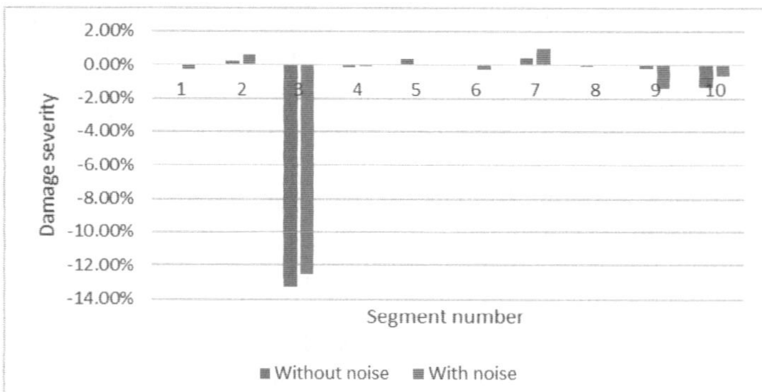

Figure 5. Damage identification results for a single damage scenario with noise effect

5. Conclusion

This paper proposes a structural damage identification approach based on model updating with impedance sensitivity and sparse regularization technique to identify the damage location and severity. The sensitivities of impedance responses with respect to the stiffness parameters of the host structure are calculated and used to identify the damage with a few number of resonance frequency shifts. The effectiveness and performance of the proposed approach are demonstrated with numerical simulations on an aluminum beam. The finite element modelling technique used to calculate the impedance response is verified by calibrating the models of a single PZT patch and PZT-structural interaction. Numerical studies on an aluminum beam attached with a PZT transducer are conducted. The results demonstrate that the proposed approach can well identify both the location and severity of the simulated structural damage, even considering the noise effect.

Acknowledgement

The work described in this paper was supported by an Australian Research Council DECRA Project DE140101741.

Reference

Annamdas, V. G. M. and C. K. Soh 2007, 'Three-dimensional electromechanical impedance model. I: Formulation of directional sum impedance', *Journal of Aerospace Engineering*, vol. 20, no. 1, pp. 53-62.

Bhalla, S. (2001). 'Smart system based automated health monitoring of structures' M. Eng. Thesis, Nanyang Technological University, Singapore.

Bhalla, S. and C. K. Soh 2004, 'Structural health monitoring by piezo-impedance transducers. I: modeling', *Journal of Aerospace Engineering*, vol. 17, no. 4, pp. 154-165.

Chaudhry, Z. A., T. Joseph, F. P. Sun and C. A. Rogers (1995). 'Local-area health monitoring of aircraft via piezoelectric actuator/sensor patches'. *Smart Structures & Materials' 95*, International Society for Optics and Photonics.

Fairweather, J. A. and K. C. Craig (1998). 'Incorporating finite element techniques to simplify the impedance modeling of active structures'. *5th Annual International Symposium on Smart Structures and Materials*, International Society for Optics and Photonics.

Fekrmandi, H., M. Unal, S. R. Neva, I. N. Tansel and D. McDaniel 2016, 'A novel approach for classification of loads on plate structures using artificial neural networks', *Measurement*, vol. 82, no., pp. 37-45.

Giurgiutiu, V., A. Reynolds and C. A. Rogers 1999, 'Experimental investigation of E/M impedance health monitoring for spot-welded structural joints', *Journal of Intelligent Material Systems and Structures*, vol. 10, no. 10, pp. 802-812.

Giurgiutiu, V. and C. A. Rogers (1998). 'Recent advancements in the electromechanical (E/M) impedance method for structural health monitoring and NDE'. *5th Annual International Symposium on Smart Structures and Materials*, International Society for Optics and Photonics.

Gyuhae, P., K. Kabeya, H. H. Cudney and D. J. Inman 1999, 'Impedance-based structural health monitoring for temperature varying applications', *JSME International Journal Series A Solid Mechanics and Material Engineering*, vol. 42, no. 2, pp. 249-258.

He, C., S. Yang, Z. Liu and B. Wu 2014, 'Damage localization and quantification of truss structure based on electromechanical impedance technique and neural network', *Shock and Vibration*, vol. 2014, no.

Hernandez, E. M. 2014, 'Identification of isolated structural damage from incomplete spectrum information using l 1-norm minimization', *Mechanical Systems and Signal Processing*, vol. 46, no. 1, pp. 59-69.

Koo, K.-Y., S. Park, J.-J. Lee and C.-B. Yun 2009, 'Automated impedance-based structural health monitoring incorporating effective frequency shift for compensating temperature effects', *Journal of Intelligent Material Systems and Structures*, vol. 20, no. 4, pp. 367-377.

Liang, C., F. Sun and C. Rogers 1994, 'Coupled Electro-Mechanical Analysis of Adaptive Material Systems—Determination of the Actuator Power Consumption and System Energy Transfer', *Journal of Intelligent Material Systems and Structures*, vol. 5, no. 1, pp. 12-20.

Lim, Y. Y. and C. K. Soh 2014, 'Towards more accurate numerical modeling of impedance based high frequency harmonic vibration', *Smart Materials and Structures*, vol. 23, no. 3.

Liu, W. and V. Giurgiutiu (2007). 'Finite element simulation of piezoelectric wafer active sensors for structural health monitoring with coupled-filed elements'. *The 14th International Symposium on: Smart Structures and Materials & Nondestructive Evaluation and Health Monitoring*, International Society for Optics and Photonics.

Lu, Z. and S. Law 2007, 'Identification of system parameters and input force from output only', *Mechanical Systems and Signal Processing*, vol. 21, no. 5, pp. 2099-2111.

Makkonen, T., A. Holappa, J. Ella and M. Salomea 2001, 'Finite element simulations of thin-film composite BAW resonators', *IEEE transactions on ultrasonics, ferroelectrics, and frequency control*, vol. 48, no. 5, pp. 1241-1258.

Min, J., S. Park, C.-B. Yun, C.-G. Lee and C. Lee 2012, 'Impedance-based structural health monitoring incorporating neural network technique for identification of damage type and severity', *Engineering Structures*, vol. 39, no., pp. 210-220.

Na, S. and H. Lee 2013, 'Neural network approach for damaged area location prediction of a composite plate using electromechanical impedance technique', *Composites Science and Technology*, vol. 88, no., pp. 62-68.

Ong, C.-W., Y. Yang, A. S. Naidu, Y. Lu and C. K. Soh (2002). 'Application of the electro-mechanical impedance method for the identification of in-situ stress in structures'. *SPIE's International Symposium on Smart Materials, Nano-, and Micro-Smart Systems*, International Society for Optics and Photonics.

Park, G., H. H. Cudney and D. J. Inman 2000, 'Impedance-based health monitoring of civil structural components', *Journal of infrastructure systems*, vol. 6, no. 4, pp. 153-160.

Park, G., C. R. Farrar, F. L. di Scalea and S. Coccia 2006, 'Performance assessment and validation of piezoelectric active-sensors in structural health monitoring', *Smart Materials and Structures*, vol. 15, no. 6, pp. 1673.

Raju, V., G. Park and H. H. Cudney (1999). 'Impedance-based health monitoring of composite reinforced structures'. *Ninth International Conference on Adaptive Structures and Technologies*.

Sepehry, N., F. Bakhtiari-Nejad and M. Shamshirsaz 2017, 'Discrete singular convolution and spectral finite element method for predicting electromechanical impedance applied on rectangular plates', *Journal of Intelligent Material Systems and Structures*, no., pp. 1045389X17689931.

Sepehry, N., M. Shamshirsaz and F. Abdollahi 2011, 'Temperature variation effect compensation in impedance-based structural health monitoring using neural networks', *Journal of Intelligent Material Systems and Structures*, vol. 22, no. 17, pp. 1975-1982.

Sevillano, E. 2016, 'Interfacial crack-induced debonding identification in FRP-strengthened RC beams from PZT signatures using hierarchical clustering analysis', *Composites Part B*, vol. 87, no., pp. 322-335.

Sun, F. P., Z. Chaudhry, C. Liang and C. Rogers 1995, 'Truss structure integrity identification using PZT sensor-actuator', *Journal of Intelligent Material Systems and Structures*, vol. 6, no. 1, pp. 134-139.

Tseng, K. K. and A. Naidu 2002, 'Non-parametric damage detection and characterization using smart piezoceramic material', *Smart Materials and Structures*, vol. 11, no. 3, pp. 317.

Van den Berg, E. and M. P. Friedlander 2011, 'Sparse optimization with least-squares constraints', *SIAM Journal on Optimization*, vol. 21, no. 4, pp. 1201-1229.

Xia, Y. and H. Hao 2003, 'Statistical damage identification of structures with frequency changes', *Journal of Sound and Vibration*, vol. 263, no. 4, pp. 853-870.

Yang, Y., Y. Y. Lim and C. K. Soh 2008, 'Practical issues related to the application of the electromechanical impedance technique in the structural health monitoring of civil structures: I. Experiment', *Smart Materials and Structures*, vol. 17, no. 3, pp. 035008.

Zagrai, A. N. and V. Giurgiutiu 2001, 'Electro-mechanical impedance method for crack detection in thin plates', *Journal of Intelligent Material Systems and Structures*, vol. 12, no. 10, pp. 709-718.

Zhou, X.-Q., Y. Xia and S. Weng 2015, 'L1 regularization approach to structural damage detection using frequency data', *Structural Health Monitoring*, vol. 14, no. 6, pp. 571-582.

The 8th International Conference on
Structural Health Monitoring of Intelligent Infrastructure
Brisbane, Australia | 5-8 December 2017

Temperature Behaviour Monitoring of the Canton Tower

J.Z. Su[1], Y. Xia [2].
[1] *College of Civil Engineering, Fuzhou University - China, Email:* jiazhansu@fzu.edu.cn [2] *Department of Civil and Environmental Engineering, The Hong Kong Polytechnic University - China*

Abstract

The thermal action of the 600 m-tall Canton Tower is investigated in this study on the basis of the comprehensive long-term structural health monitoring system installed on the structure and the numerical simulation. The temperature difference between different facades of the inner tube, the temperature difference profile of the outer tube, and the distribution of the temperature difference between the inner and outer tubes along the structural height are presented in detail by using the field monitoring and numberical heat transfer analysis data. The temperature-induced structural resposnes including displacement, stress, and internal forces are then obtained by using the global finite element model of the tower and compared with the measurements. At last, the structural responses (stresses and displacement) obtained from the field monitoring system during a strong typhoon are presented. The purely typhoon-induced quasi-static responses are separated from the total measured responses by subtracting the temperature-induced ones. It shows that the typhoon-induced quasi-static responses of the supertall structure are slightly smaller than the temperature-induced responses in a typical sunny day.

1. Introduction

For a supertall structure, the radiation and daily temperature fluctuations have a significant effect on the overall deflection and stresses, as well as the structural vibration characteristics, of these large-scale structures because of the indeterminacy and nonuniform distribution of temperature. The cross-sectional temperature distribution within an individual structural element consists of four essential components, namely, a uniform temperature component, linearly varying temperature difference components in two transversal directions, and a nonlinear component (Eurocode 2003). The uniform temperature component may cause longitudinal deformation, whereas the linear temperature difference component may cause the curvature of the member. When the deformation is partially or fully constrained, internal forces occur in the member.

In the past decades, investigations of tall buildings are limited due to the uniqueness of the construction site and structural configuration. Since the 1970s, temperature effects on tall TV towers (Pirner and Fischer 1999; Breuer et al. 2008), a steel towers (Tamura et al. 2002), and a stadium (Teng et al. 2014) have been analyzed. A small number of research works have focused on temperature effects on tall structures at the construction stage (Azkune et al. 2007; Li et al. 2012). Although field measurement on some highrise structures has been conducted, the actual temperature distribution of and the associated temperature effects on supertall structures over 300 m have not been investigated. The shortage of field measurement data regarding the thermal action of supertall structures in China leads to only simple specifications in the Chinese national standard (GB 50009-2012 2012) regarding this issue. In particular, the standard only specifies the uniform temperature component but has no clauses on the temperature difference among different members.

2. Canton Tower and the deployment of temperature sensors

2.1 Canton Tower

The Canton Tower is a typical tube-in-tube supertall structure (Figure 1). A global FE model of the Canton Tower has been constructed using the general FE software package SAP2000, to calculate the temperature-induced displacement and stress responses. In this model, four-node and three-node area elements with six degrees of freedom (DOFs) at each node are employed for the shear walls of the inner tube and the floor decks. Two-node 3D beam elements with six DOFs at each node are used to model the outer tube members, the connection girders between the inner and outer tubes, and the antenna mast. All nodes in the basement are fixed in all directions. The full model contains 43,067 elements and 28,305 nodes.

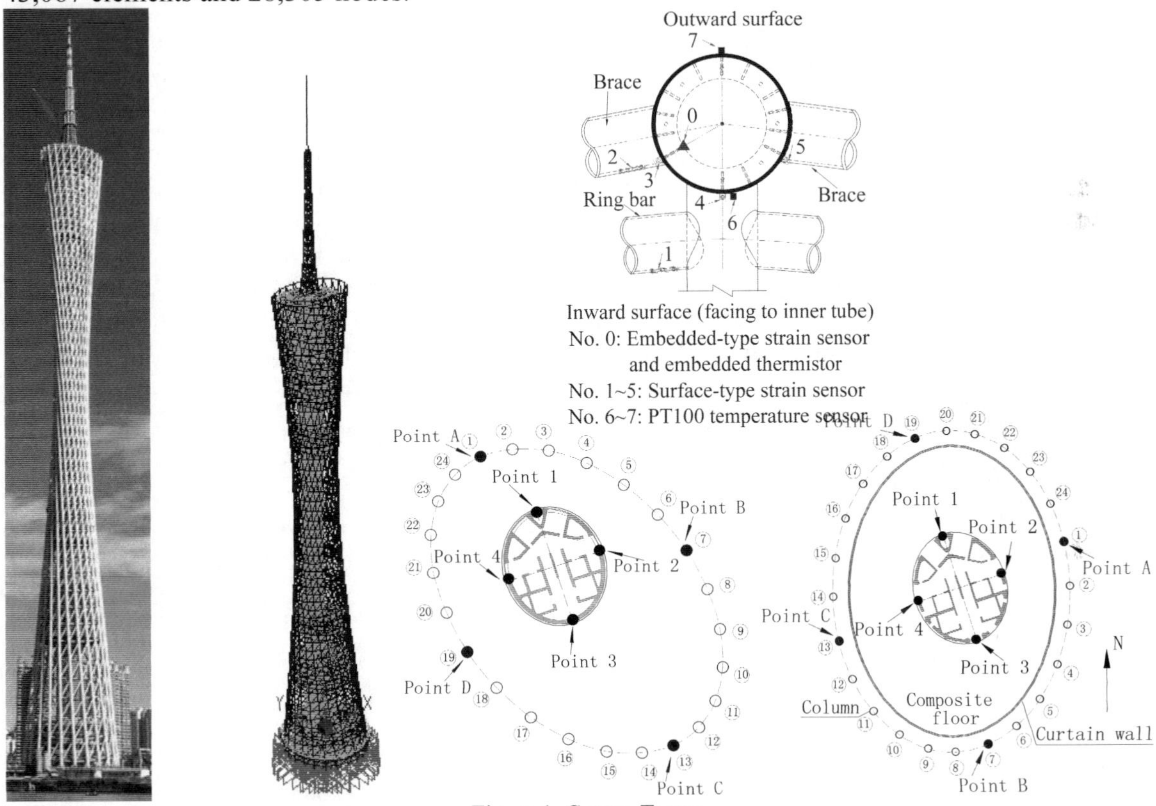

Figure 1. Canton Tower

2.2. Temperature and strain/stress monitoring system

Figures 1 show the layout of the sensors in Sections 3 and 12, respectively. Section 12 has a functional floor enclosed by the curtain wall, whereas Section 3 has no functional floors. The plan locations of the CFT columns in the two sections vary because the columns are inclined in the vertical direction. In the inner tube, Points 1 to 4 are each embedded with three vibrating wire strain gauges and one thermistor in the middle of the concrete core wall. Each vibrating wire incorporates a thermistor such that both strain and temperature at the point are measured. In the outer tube, Points A to D of each section were installed with five vibrating wire strain gauges (denoted as "1" to "5") and two PT100 temperature sensors (denoted as "6" and "7"). From Sections 6 to 12, one vibrating wire was embedded in the concrete of the CFT column (denoted as "0"), which was located at one third of the radius from the column surface. The vibrating wire measures both strain and temperature at the point of the CFT column.

3. Temperature distribution of the Canton Tower

3.1. Effective temperature of the main components

For the Canton Tower, the temperatures at Points 1, 2, 3, and 4 are considered as the effective temperature of the inner core walls facing the north, east, south, and west, respectively, because only one thermistor sensor was installed at the middle thickness of the wall at each point. For the CFT columns, the steel surface and inside the concrete have different temperature distributions. In view of the difficulty in measuring the detailed temperature distribution of the CFT columns by only a few sensors, the temperature distribution can be obtained through a numerical heat conduction analysis. In this regard, the FE model of a typical column section at Point C of Section 8 is constructed in ANSYS. The FE model consists of 400 nodes and 360 two-dimensional Plane55 elements. Plane55 has four in-plane nodes, each with a single DOF of temperature.

Figure 2 shows the temperature contour of the section at 15:00 on 4 August 2011. The temperature is nonuniformly distributed with an approximate 5.3 °C difference between the steel surface and inside the concrete. The numerical results are consistent with the field measurement, with a difference of less than 0.5 °C, which verifies the effectiveness of the heat transfer analysis. The effective temperature of the section is close to the measured concrete temperature. Consequently, the measured temperature inside the column will be considered the effective temperature of the composite section of the column and will be employed in the subsequent analysis.

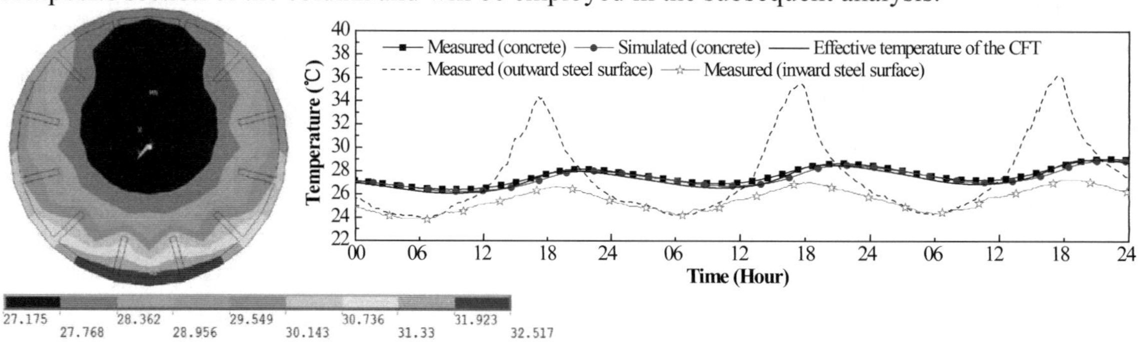

Figure 2. FE model of the CFT column for heat transfer analysis

3.2. Temperature difference between the inner and outer tubes

In general, for Sections 3 to 8 without functional floors, the temperature difference between the inner and outer tubes is relatively small in both seasons. For the other sections with functional floors, the temperature difference between the inner and outer tubes is significant and similar, except for Section 12, as previously explained. Based on the previously presented observations, the envelope of the temperature difference with respect to the height of the structure can be categorized into three groups, as the solid lines shown in Figure 3: (i) the largest temperature difference, $T_a^{+,-}$ for Section 12 and the segment between Sections 11 and 12; (ii) $T_b^{+,-}$ for Sections 1 to 3 and Sections 9 to 11 with functional floors; and (iii) $T_c^{+,-}$ for Sections 3 to 9 without functional floors. The superscripts "+" and "–" denote the positive and negative temperature differences, respectively. In this study, $T_a^+ = 8.0$ °C, $T_b^+ = 5.1$ °C, and $T_c^+ = 3.0$ °C in the summer, whereas $T_a^- = -15.0$ °C, $T_b^- = -9.7$ °C, and $T_c^- = -6.0$ °C in the winter.

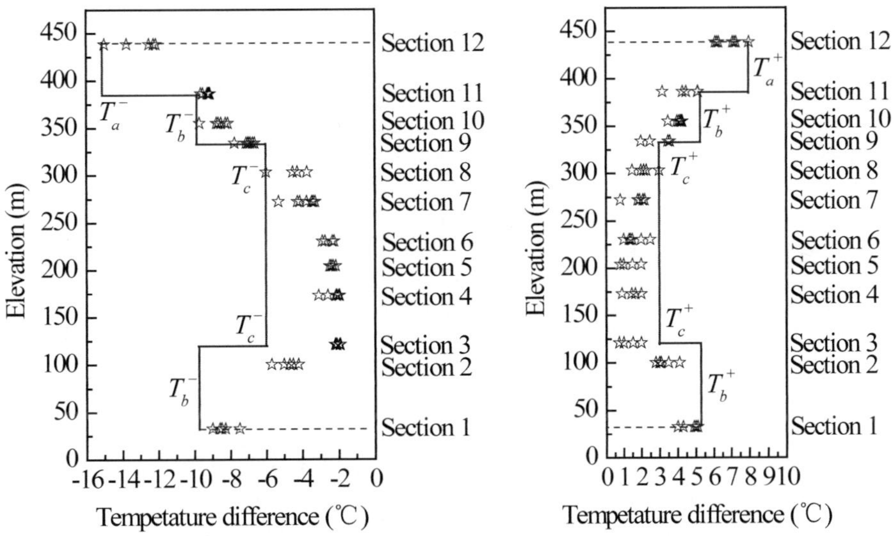

(a) in the winter (b) in the summer

Figure 3. Distribution of temperature differences along the height of the structure

3.3. Temperature difference between the different faces of the structure

3.3.1. Inner concrete wall

Figure 4 illustrates the temperature difference between different facades of Sections 8 and 12 in 2014. The temperature difference fluctuates slightly in the range of −2.0 to 1.5 °C in Section 12, indicating that the inner walls with functional floors enclosed by the curtain walls have a uniform temperature distribution over different facades. Therefore, their temperature difference is negligible.

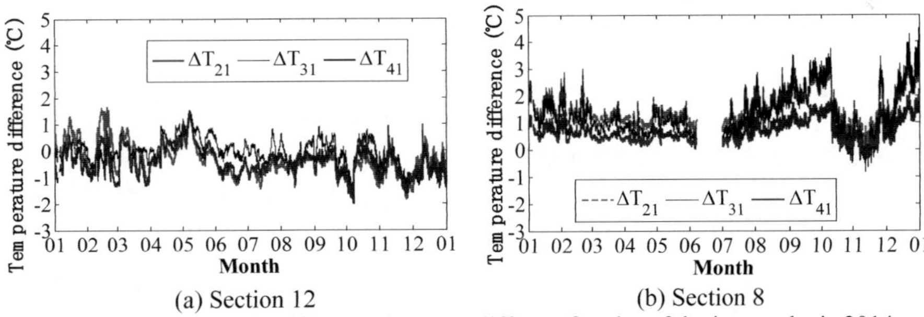

(a) Section 12 (b) Section 8

Figure 4. Temperature difference between different facades of the inner tube in 2014

The temperature difference at Section 8 is almost positive during the entire year, indicating that the temperature of the north facade is the lowest over the year, as expected. Winter has the largest difference of approximately 4.5 °C. Several areas of the exterior surface of the wall are found in the shadow of the CFT columns, thus receiving only a portion of solar radiation. Therefore, the temperature difference between the facades of Section 8 remains at a low level and is slightly higher than that of Section 12.

3.3.2. Outer CFT columns

As the outer columns are inclined in the vertical direction, the monitoring points at different sections rotate in the plan view. The temperature difference of all monitoring points in typical summer and winter days is illustrated in Figure 5, where the alphanumeric code represents the CFT column, for example, "12A" denotes Point A at Section 12. The lowest temperature of the columns is selected as the reference, and the temperature difference is shown in the figure.

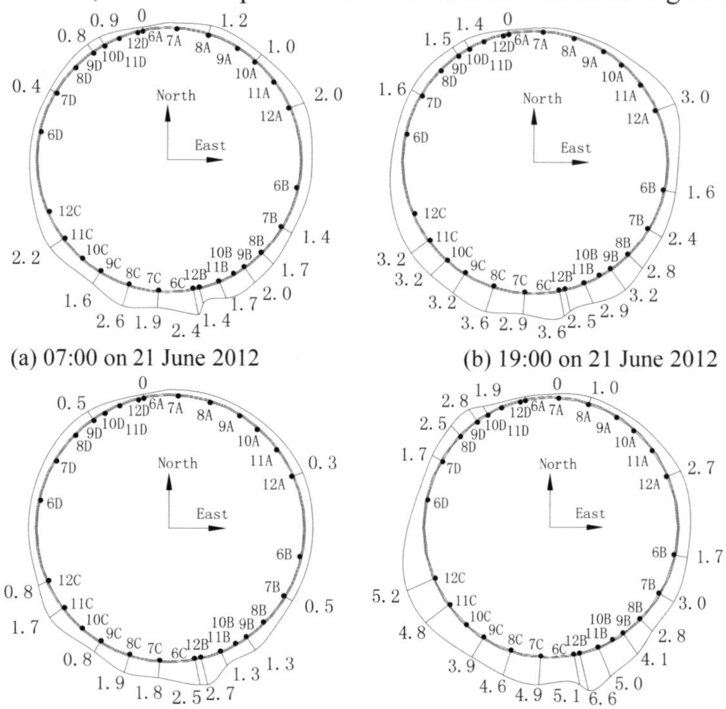

(a) 07:00 on 21 June 2012 (b) 19:00 on 21 June 2012

(c) 08:00 on 22 December 2011 (d) 18:00 on 22 December 2011

Figure 5. Temperature difference profile of the outer tube (unit: °C)

These two typical days are the summer solstice (on 21 June 2012) and the winter solstice (on 22 December 2011), which represent the longest and shortest days of the year in the northern hemisphere, respectively. As expected, the south facade has the highest temperature and the north facade has the lowest. The temperature difference between different facades of the outer tube, which is less than 2.5 °C in the summer and winter, is relatively uniform before sunrise. The difference increases after sunrise and reaches 3.6 °C in the summer and 6.6 °C in the winter. This finding implies that the difference in the winter is more significant than that in the summer.

4. Temperature-induced responses

4.1. Temperature-induced displacement

The measured horizontal displacement track at the top of the tower on four sunny days in different seasons is illustrated in Figure 6. The dates are 14 April 2013, 11 August 2013, 12 October 2013, and 1 January 2014, representing spring, summer, autumn, and winter, respectively. These days all had a significant daily air temperature variation of more than 10 °C. The wind speed of these days was no more than 3.5 m/s. Therefore, the displacement was mainly induced by the thermal load. On 1 January 2014, the tower moved relatively slow before sunrise, moved to the northwest after sunrise, and reached its westernmost position at approximately 12:00. Consequently, the structure bent away from the sun. For the same reason, the temperature of the members in the southwest increased when the sun moved to the southwest in the afternoon, causing the tower to move toward the northeast and reaching its northernmost position at approximately 16:00. After sunset, the temperature difference between the tower members decreased. Therefore, the tower moved back gradually from the north to the south and finally returned to its initial point (at 24:00). The peak-to-peak motion throughout the day was 15.7 cm in the east–west direction and 15.5 cm in the south–north direction, and the maximum displacement was approximately 20 cm.

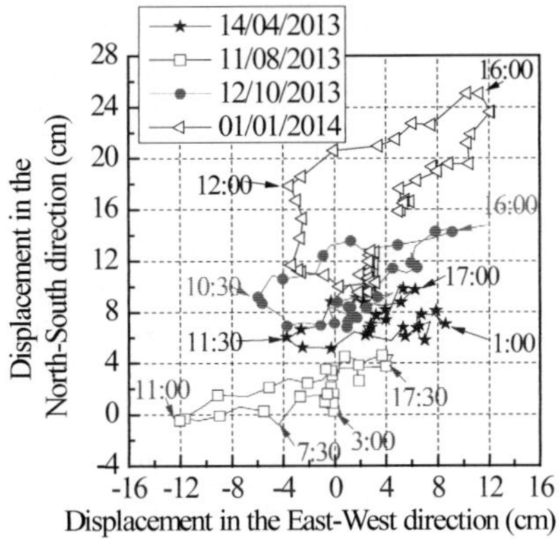

Figure 6. Measured temperature-induced horizontal displacement track at the tower top on four sunny days in different seasons

4.2. Temperature-induced stress

Figure 7 shows the calculated variations in the stresses of Point 2 at different heights between summer and winter, when the structure has the highest temperature in the summer and the lowest temperature in the winter, as compared with the measured counterparts. The variation of stress changed significantly in the transition sections between the zones with and without functional floors (Sections 2, 9, and 11), which is consistent with the measured results. The largest measured variation in stress was observed at Point 2 of Section 2 with approximately −2.8 MPa, accounting for approximately 25% of the total stress. The simulated and measured tendencies of the stress change along the height of the tower are generally consistent, although several discrepancies can be observed in the segments where the inner tube connects with the functional floors.

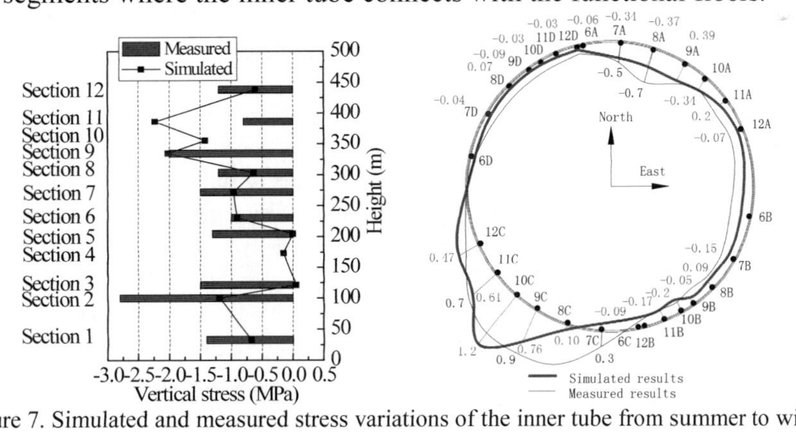

Figure 7. Simulated and measured stress variations of the inner tube from summer to winter

The stresses of the CFT columns on the northeast facade decreased, whereas those on the southwest facade increased. The tower leaned toward the northeast from summer to winter, resulting in increased compression of the columns on the northeast. The stresses of the columns in the southwest significantly increased by 1.2 MPa, which accounted for approximately 11% of the total stress. The largest stress decrease occurred at the column in the northeast.

4.3. Temperature-induced internal forces of beams

Figure 8 illustrates the profiles of the bending moment and shear force induced by the temperature decreasing from summer to winter and the ratio of these variations to the corresponding forces under the dead load. The maximum value among all horizontal girders on each floor of the tower is presented in the figure. The temperature-induced bending moment and shear force of the beams at the highest story and those of the connection girders at the middle of the tower are larger than the forces of other segments. This scenario is different from other loading effects, for example, the wind loading, which generally causes the maximum forces at the bottom of the structure. For the segment without functional floors at the middle of the tower, only four levels of the connection girders link the inner tube to the outer tube. This segment is the weakest along the structural height. These girders are also subject to large thermal actions. The temperature-induced bending moment and shear force can reach 72% and 56% of those induced by the dead load, respectively. Therefore, special attention should be paid to the thermal loading, particularly for this type of complicated supertall structures.

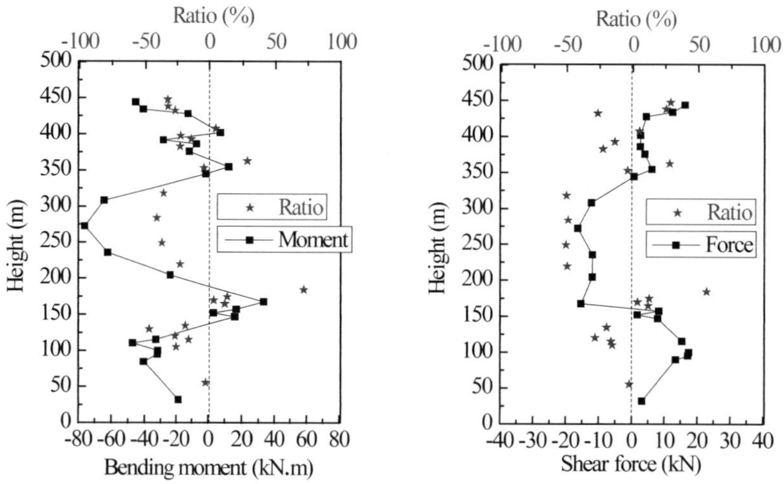

Figure 8. Calculated beam-end bending moment and shear force from summer to winter

5. Quasi-static responses of Canton Tower during typhoon period

Typhoon Usagi, a strong tropical cyclone that struck Guangdong Province from 22 to 23 September, 2013, is selected to investigate the wind-induced quasi-static responses of Canton Tower. The mean wind speed increased to 12.0 m/s at 14:00 on 22 September as the typhoon approached. The speed continued to grow to approximately 18 m/s at 19:00 on 22 September, reaching a maximum of 25.0 m/s at 4:00 on 23 September. The wind speed then decreased to 8.0 m/s as the typhoon continued to move landward. The wind direction was mainly from the north on 21–22 September and turned to the northwest from 0:00 to 4:00 on 23 September, and southwest from 4:00 to 8:00. 21 September was a typical sunny summer day with an air temperature variation of 25.5–31.0 °C and no strong wind. Air temperature then decreased as the typhoon gradually approached. The average air temperature during the typhoon period (from 12:00 on 22 to 12:00 on 23 September) was 4.0 °C lower than that during the non-typhoon period (from 0:00 to 24:00 on 21 September. The structural quasi-static responses during the typhoon period were mainly induced by the combination of temperature changes and mean wind, which must be separated such that those caused by the typhoon solely can be evaluated.

Figure 9 presents the GPS-measured horizontal displacement at the tower top from 21-23 September at ten minute intervals, with the initial position as the origin point. The data were averaged every 10 min period. The tower moved relatively slow before sunrise on 21 September, moving to the west after sunrise and arriving at its westernmost position at approximately 10:00. The tower moved toward the northeast in the afternoon and reached its northernmost position at approximately 17:30. After sunset, the temperature difference between the tower members decreased. The tower gradually moved back from the north to the south and finally returned to the initial point at 24:00. The peak-to-peak motion throughout the day was 17.8 and 8.5 cm in the east-west and south-north direction, respectively.

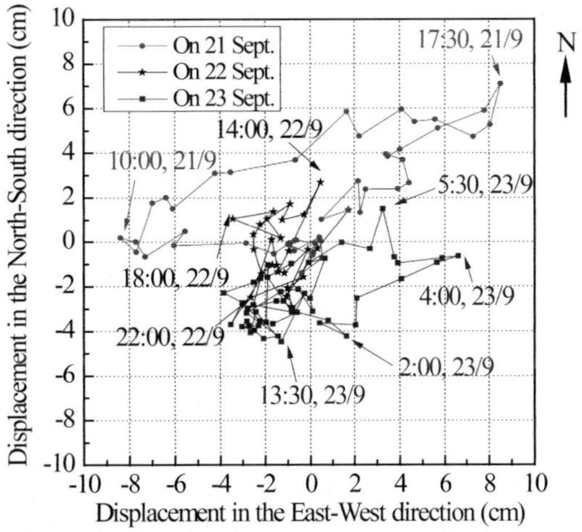

Figure 9. Displacement track at the tower top on 21-23 September 2013

Subsequently, the tower top moved to the south by approximately 7.0 cm from 14:00 on 22 September to 2:00 on 23 September, with the increasing north wind. It then leaned toward the northeast by approximately 6.2 cm from 2:00 to 4:00 on 23 September with the maximum west-northwest wind. Afterward, the tower top returned toward the west as wind speed slowed down from 4:00 to 8:00 on 23 September. The peak-to-peak motion during the typhoon period was roughly 10.1 and 7.0 cm in the east-west and north-south direction, respectively.

Figure 10 shows the vertical stresses of the four façades of the inner tube at an elevation of 303.2 m, from 21 to 23 September 2013. The data were averaged every 10 min period. The stress in this study is the total stress of the structure, including that caused by the dead load.

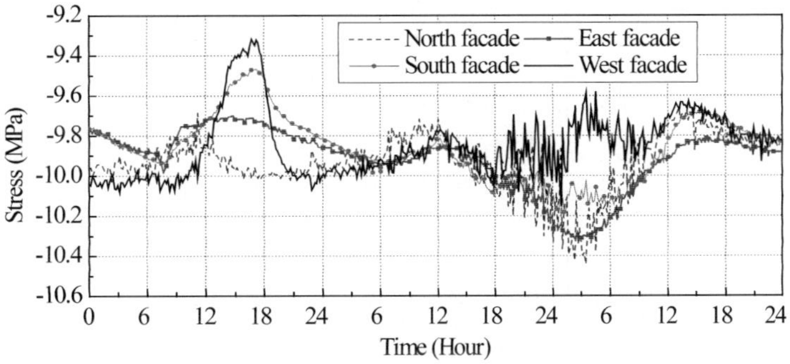

Figure 10. Vertical stress of the inner tube 21-23 September 2013

On 22 September, the stress began to fluctuate significantly from 18:00 to 10:00 on 23 September, when the typhoon weakened. During the period, all façades were subjected to compressive stresses, except the west. The north façade had the largest compressive stress of approximately –0.5 MPa, while the west was under a tensile stress of 0.4 MPa. The wind speed from 0:00 to 24:00 on 21 September was low. The structural responses could be treated as temperature loading effects only. From 2:00 on 22 to 8:00 on 23 September, the wind was strong and temperature variation was significant. Consequently, the structural responses were caused by the combined loading effects. The typhoon-induced quasi-static responses could be separated from the total responses by calculating the temperature-induced responses using FE analysis.

The temperature-induced quasi-static responses of the structure can be calculated effectively from the measured temperature data and FE model. The same approach is applied to calculate the temperature-induced quasi-static responses during the typhoon period. Seven other time instants during the typhoon period on 22-23 September (Table 1) are studied. The temperatures at these instants are inputted into the FE model. The calculated structural responses are regarded as induced by temperature loading solely. Structural displacement and stresses induced by the typhoon only are then determined by subtracting the temperature-induced responses from the measured total responses.

Table 1. Calculated horizontal displacement at the tower top (465 m) induced by temperature variation

	Time instant	Horizontal displacement (cm)	
		East-west	North-south
1	14:00, 22 Sept.	5.0	1.7
2	18:00, 22 Sept.	2.0	1.5
3	22:00, 22 Sept.	0.3	1.3
4	02:00, 23 Sept.	−1.0	1.6
5	04:00, 23 Sept.	−1.2	1.1
6	05:30, 23 Sept.	−1.1	0.5
7	08:00, 23 Sept.	−1.5	−0.1

The maximum east-west displacement is 11.2 cm, occurring at 4:00 on 23 September, whereas the maximum south-north displacement is 4.9 cm, occurring at 2:00 on the same day, as shown in Figure 11. The peak-to-peak horizontal displacement induced by the typhoon is 13.1 and 7.1 cm in the east-west and south-north direction, respectively. The values are larger than those under the combined loading effects (10.1 and 7.0 cm in the east-west and south-north direction, respectively). If the thermal effects are not removed, the typhoon-induced displacement will be underestimated. Nevertheless, the typhoon-induced displacement is smaller than the temperature-induced displacement in a typical sunny day (peak-to-peak displacement of 17.8 and 8.5 cm in the east-west and south-north direction, respectively).

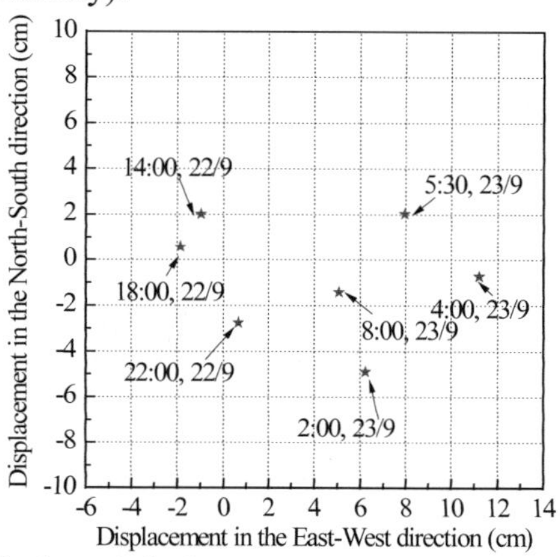

Figure 11. Calculated horizontal displacement at the tower top caused by Typhoon Usagi

The temperature-induced stresses at the elevation of 303.2 m in these different cases are calculated and listed in Table 2. The calculated stress variations of north, east, south, and west façades at the specified elevation at 4:00 on 23 September are approximately −0.10, −0.14, −0.14, and −0.26 MPa, respectively. After eliminating the thermal effect, the typhoon-induced quasi-static stresses of the four façades are approximately −0.30, −0.16, −0.06, and 0.76 MPa, which are comparable to the stresses caused by daily temperature variation. The largest variation in stress induced by the typhoon was roughly 0.76 MPa, accounting for 8% of the total stress approximately.

Table 2. Calculated stresses at an elevation of 303.2 m induced by temperature variation and typhoon

	Time instant	Temperature-induced stress (MPa)				Typhoon-induced stress (MPa)			
		North	East	South	West	North	East	South	West
1	14:00, 22 Sept.	0.02	0.02	−0.02	0.02	0.03	0.02	0.02	0

2	18:00, 22 Sept.	0.01	0.01	0	−0.01	0.01	0.04	0.03	0.01
3	22:00, 22 Sept.	−0.02	−0.02	−0.02	−0.03	−0.06	0.06	0.02	0.06
4	2:00, 23 Sept.	−0.06	−0.08	−0.07	−0.13	−0.28	−0.16	−0.01	0.59
5	4:00, 23 Sept.	−0.10	−0.14	−0.14	−0.26	−0.30	−0.16	−0.06	0.76
6	5:30, 23 Sept.	−0.09	−0.16	−0.15	−0.29	−0.25	−0.13	−0.01	0.51
7	8:00, 23 Sept.	−0.05	−0.11	−0.09	−0.16	−0.12	−0.06	−0.01	0.32

6. Conclusions

The peak-to-peak tower top motion throughout a sunny day in winter was 15.7 cm in the east–west direction and 15.5 cm in the south–north direction, and the maximum displacement was approximately 20 cm. The temperature variations cause the stresses of the different members of the entire structure to redistribute. The temperature difference between the inner and outer tubes produces significant bending moment and shear force in the beams linking the inner tube to the outer tube.

After eliminating the temperature effect, the peak-to-peak horizontal mean displacement at the tower top induced by the strong Typhoon are larger than the combined displacement due to mean wind and temperature. Moreover, the mean wind induced displacements are smaller than those caused by the temperature variation in a typical sunny day. The maximum stress of the inner tube caused by the typhoon is roughly 8% of the total stress of the structure.

Acknowledgments

This research was supported by the Research Grants Council of Hong Kong (Project No. PolyU 152621/16E), NSFC Joint Research Fund for Overseas and Hong Kong and Macao Scholars (Project No. 51629801), and Research Grant of the Fuzhou University (Project No. 510251) and the Education Department of Fujian Province (Project No. 601705).

References

Azkune, M, Puente, I, & Insausti, A 2007, "Effect of ambient temperature on the redistribution of loads during construction of multi-storey concrete structures." *Engineering Structures*, vol. 29, no. 6, pp. 933-941.

Breuer, P, Chmielewski, T, Gorski, P, Konopka, E, & Tarczynski, L 2008, 'The Stuttgart TV Tower-displacement of the top caused by the effects of sun and wind', *Engineering Structures*, vol. 30, no. 10, pp. 2771-2781

Eurocode-1. EN 1991-1-5, 2003 '*Actions on Structures. Part 1–5: General Actions – Thermal Actions*', European Committee for Standardization, Brussels, Belgium.

GB 50009-2012 2012, '*Load Code for the Design of Building Structures*', China Architecture and Building Press, Beijing, China.

Li, DS, Ren, L, Li, HN, & Song, GB 2012, 'Structural health monitoring of a tall building during construction with fiber bragg grating sensors', *International Journal of Distributed Sensor Networks*, 2012, pp. 1-10.

Pirner, M, & Fischer, O 1999, 'Long-time observation of wind and temperature effects on TV towers', *Journal of Wind Engineering and Industrial Aerodynamics*, vol. 79, no. 1-2, pp. 1-9.

Tamura, Y, Matsui, M, Pagnini, LC, Ishibashi, R, & Yoshida, A 2002, 'Measurement of wind-induced response of buildings using RTK-GPS', *Journal of Wind Engineering and Industrial Aerodynamics*, vol. 90, no. 12-15, pp. 1783-1793.

Teng, J, Lu, W, Cui, Y, & Zhang, RG 2014, 'Temperature and displacement monitoring to steel roof construction of Shenzhen Bay Stadium', *International Journal of Structural Stability and Dynamics*, 16, pp. 1-22.

The 8th International Conference on
Structural Health Monitoring of Intelligent Infrastructure
Brisbane, Australia | 5-8 December 2017

Monitoring Technique of High-speed Welded ballastless Turnout on Long Bridge

L.Gao, Y.L. Zhong, G.L. Sun, X.P. Cai, B.W. Hou, C.Y.Zhou
School of Civil Engineering, Beijing Jiaotong University, Beijing, P.R. China,100044, Email: lgao@bjtu.edu.cn

Abstract

Turnout is the most sensitive structure of railways. In order to ensure the operation safety of high speed railway (HSR), it is of great significance to develop a long-term monitoring system dedicated for continuously welded turnout, ballastless track and bridges. This system can realize the online monitoring on the service state of ballastless turnouts laid on bridges in real time and predict/forewarn the diseases that may occur during operation, which can guide the maintenance work of the railway. By taking the monitoring work of the No.42 turnouts on HSR bridges in the Beijing-Shanghai Railway, this paper mainly studies on the monitoring technologies of ballastless turnouts on HSR bridges from the following perspectives: the monitoring contents, monitoring method choices and system construction of monitoring work. The synthetic monitoring technique combining FBG (Fiber Bragg Grating) and VII (Video Intelligent Identification) technology was designed. Given the insufficient of present researches on the data analysis and prediction/decision making assistance, by mining the monitoring data over 3 years, a set of synthetic prediction measures based on the BP Neural Network and Multiple Regression Theory was proposed, which can be used in the forecast of monitoring indexes. Using threshold value and diagnosis of outliers on the fundament of cluster analysis, a forewarning method to abnormal conditions of different ranks was put forward. Furthermore, an information management system for monitoring data was developed, which scientifically directed the maintenance of track system.

1. Introduction

HSR track system is the first receptor of high-speed vehicle load. Its service status directly affects operation safety (Zhai & Zhao 2014). On the one hand, under the coupling effects of train impact, material aging, temperature changes and other factors, the state of track deteriorates continuously. In order to strictly avoid the destruction of the track structure, works department always increase the intensity of maintenance, which may induce the frequent occurance of repairing and overhaul. On the other hand, due to the high utilization rate of HSR, the high density of trafic (Xu 2010) and the limitation of skylight time, it is of great significance to minimize unnecessary maintenance. Therefore, to achieve the scientific and reasonable maintenance of HSR track structure, the comprehensive command of HSR track structure on the basis of real-time monitoring technologies is necessary. Combined with study on the prediction and prewarning of the state of track, more suitable maintenance timing can be made, which can not only correct the bad state of the track timely, but also make it possible to keep the maintenance cycle and cost in a reasonable range.

At present, in terms of HSR track system, it is quite a convention to use manual inspection during skylight time and periodic vehicle inspection (Zhao 2015). However the obtained imformation is often so limited that it can hardly reflect the real and overall service state of the track structure in time. Scholars at home and abroad have used resistance strain gauges, quasi-orbital circuits, vibrating wire, eddy current, magnetic noise, ultrasonic, FBG, VII and other methods of track system monitoring technologies for a certain study (Du & Zhang 2014). H.Y.Tam(H.Y & T & S.L 2007), M. L. Filograno (Filograno & Rodríguez-Barrios 2010) and other researchers have used fiber grating sensors to monitor indicators like vihicle speed and rail acceleration; Pen (Pen & Watson 2014) and other researchers have applied VII techniques to monitor the mechanical behavior of level-crossing track. Domestic scholars like Gao and Cai (Cai & Gao 2015), Jiang (Jiang & Lv 2015) and Wang (Wang 2015) have applied fiber grating technologies into track monitoring. Noway online monitoring of track system has disadvantages like redundancy of measuring points and high cost. As a result, it is not conducive for large-scale promotion and application. Since the functions of mass data storage and information management system are limited, especially with respects of the forecasting, prewarning and alarming, it is difficult to assist the track maintenance work.

In order to solve problems above, a study on the prediction and prewarning methods is conducted. By analysing the real-time monitoring data of the ballastless turnout on a bridge, a data prediction method of track monitoring based on multiple linear regression and BP neural network is developed. A new prewarning scheme on the basis of the combination of threshold and clustering diagnosis is also proposed. For the mass data of monitoring, a data management information system is built to provide reasonable suggestions for HSR maintenance.

2. Monitoring system construction

2.1 Monitoring site

The monitoring site is a seamless turnout on the bridge located in Beijing-Shanghai HSR Tianjin South Station. The track type is CRTSII slab ballastless track. The turnout type is longitudinal slab turnout. There is a NO. 42 turnout set on Tianjin and Shanghai connecting line. The relative position of the turnout and bridge is shown in Figure.1. The turnout is located on a continuous beam of 7-32.7m, and the milepost of the switch frog is DK125+600. The switch frog is 130.8m away from the right end of the continuous beam, and 32.7m away from the fixed bearing. The base slab is longitudinally connected within the turnout area.

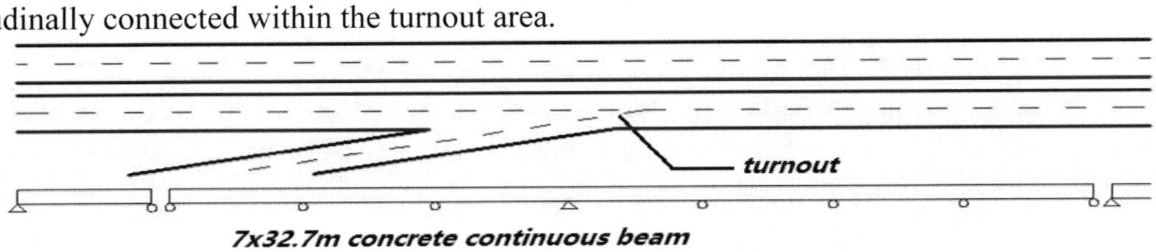

Figure.1 The relative position of the turnout and bridge

2.2 Monitoring sensitive regions

Analysis of the force/deformation sensitive regions and monitor indexes is the scientific basis for the rational layout of monitoring points. It is of great significance to decrease the cost and to reflect the track state precisely.

An HSR vehicle-track-infrastructure spatial coupled model for both static and dynamic analysis is developed, as shown in Figure.2. The vehicle is modeled by multibody dynamics, and turnout and bridge structure models are formed by FEM. Using wheel/rail relationship and turnout-bridge interaction mechanism, the system model is developed (Gao & Tao 2009).

Figure.2 vehicle- track-infrastructure spatial coupled model

On the basis of the model, the temperature loads and the static/dynamic indicators under loads exerted by vehicles are analyzed. To better reflect the state of the track, it is proposed to place monitoring points at regions where the structure deforms intensely or shapely. On the premise that the monitoring requirements are satisfied, the amount of sensors ought to be decreased in order to control the total cost. Additionally, the installation location of sensors should satisfy the actual situation at site and the existence of sensors mustn't influence the normal operation of trains at the same time.

Figure.3(a) indicates the longitudinal stress of rail under the temperature change of 50℃. The coordinate origin is the left end of the 7-span continuous bridge. The stress of turnout rail changes shapely near the restrictor of the point rail. The peak of longitudinal stress appears at the fore-end of the restrictor of the point rail. There are inflexion points at the expansion bearing of the continuous bridge and the fore-end of the nose rail. The ends of beams and fore-ends of point rails should be highlighted by arranging monitoring points.

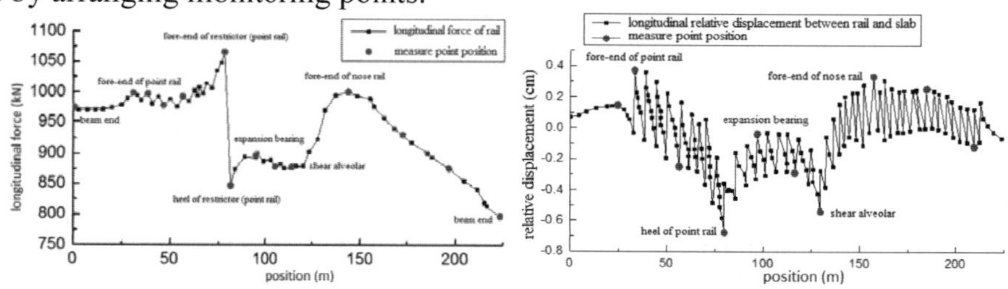

(a) longitudinal force of rail (b) relative displacement between rail and slab

Figure.3 Longitudinal force of rail and relative displacement between rail and slab

Under the temperature change of 50℃, the longitudinal relative placement between rail and slab is also shown in Figure.3(b). Extremums appear at the tip and heel of point rail, expansion bearing, heel of nose rail and the shear alveolar. All of them should be set with monitoring points.

2.3 Monitoring content

The static and dynamic sensitive areas and indicators are carefully analysed, and the specific monitoring contents of rail, ballastless track and bridge are determined, as are shown in Table.1.

Table.1 Contents of long-term monitoring of ballastless turnout on bridge

Content	Location	Amount of monitoring points	Symbol
Longitudinal additional force of turnout	Straight stock rail	4	▲
	Curved stock rail	4	
	Ends of beam	2	
Relative displacement between turnout and slab	Straight stock rail	4	□
	Curved stock rail	1	
	Straight guide rail	1	
Relative displacement between slab and beam	Distributed evenly within switch area	4	▣
Temperature changes	Rail	2	⊛
	Beam of bridge	2	
	Atmosphere temperature	1	☆
	Temperature gradients of ballastless track	9	⊛
Bridge displacement	Longitudinal displacement at the ends of bridge	2	▭
Longitudinal and vertical force of rail	Both ends of beam	4	⬠
Slab stress	In the vicinity of beam joints and fixed bearing	2	⊥
Stress of base slab	In the vicinity of beam joints and fixed bearing	2	⊥
Overall state of turnout area/relative displacement of point rail	Turnout area	1	⊏
Total	/	45	

The actual placement of the monitoring points of the ballastless turnout on the bridge is illustrated in Figure.4.

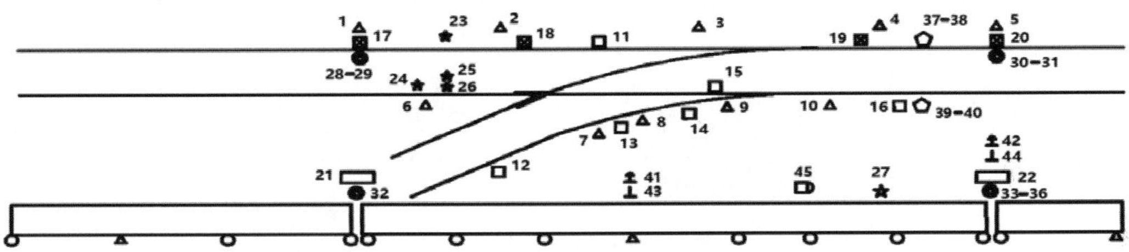

Figure.4 Monitoring points placement of the ballastless turnout on the bridge

2.4 Monitoring methods

The long-term monitoring for HSR track system requires the sensors to be equipped with high security, high stability, corrosion resistance and anti-interference, on the basis of sufficient accuracy (Lin 2015). Among many monitoring methods, fiber grating method has advantages of interference immunity and high precision, and it can meet the need for force, temperature and displacement monitoring (Ren & Jiang 2013). But it is difficult to be applied into the monitoring of some special positions such as the tip of the switch rail and the requirement for installation accuracy is relatively high. VII technology, with its features of non-contact and environmental robustness, is an ideal choice for the monitoring of large displacement or special components.

The above methods have their respective advantages and disadvantages. Thus only using a single method is impossible to meet the needs of comprehensive monitoring on the track structure. In this paper, various means of intelligent monitoring are used, such as FBG and VII technologies.

(1) Fiber Bragg Grating (FBG) technology: By using Finite Element Method (FEM) software, different types of fiber grating strain gauge models are established. The analysis shows that the softer and thinner butterfly strain sensor can better adapt the deformation of the rail.

Conventional displacement tests usually use the equal-strength beam which is unsatisfactory in accuracy and deformation synchronization. As a result, it is difficult to adapt to the needs of fiber grating sensor system. The design uses FBG sensor (shown in Figure.5) to monitor the track displacement, stress and temperature and so on. (Li & Lu 2013).

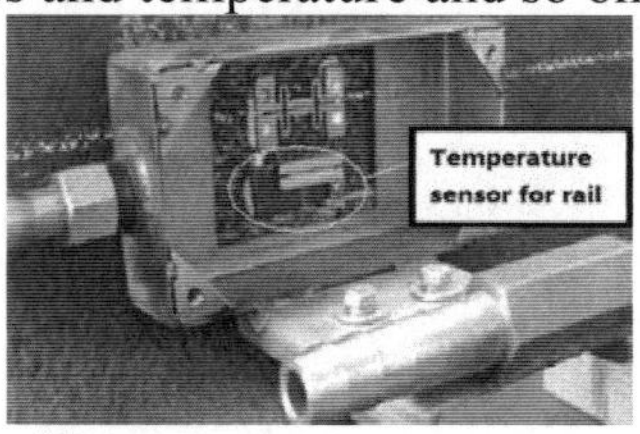

(a) FBG temperature sensor

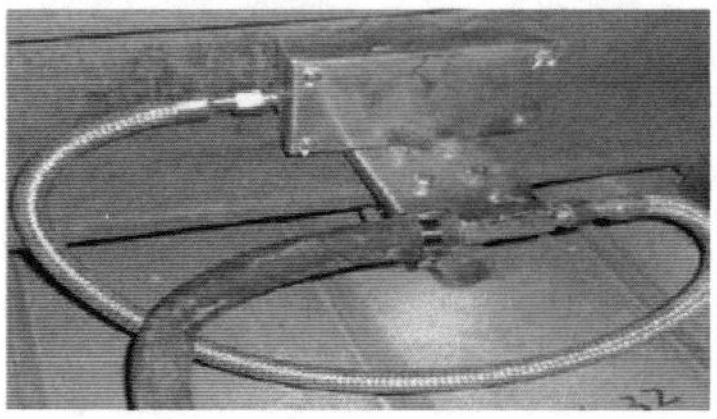

(b) FBG sensor in operation

Figure.5 FBG temperature sensor during installation and FBG sensor in operation

(2) Video Intelligent Identification (VII) technology: By cameras, the image of the relative position of the structure and scale is obtained. Using self-compiled programs, the displacement information carried by images is identified. Consequently the displacement monitoring for large-displacement areas (such as long bridge beam end large displacement) and for special parts where the sensors cannot be installed (such as the tip of the switch rail) is achieved.

Figure.6 is a schematic diagram of recognizing the switch rail telescopic displacement by VII technology. Particularly, for different states of direct and lateral passing, the location of the rail contour line can be used for identification, and then the expansion length of the point rail can be read. And the relative position of the camera and the point rail is used to correct the lateral passing state. In addition, the online monitoring of the track's overall state can be achieved through camera rotation.

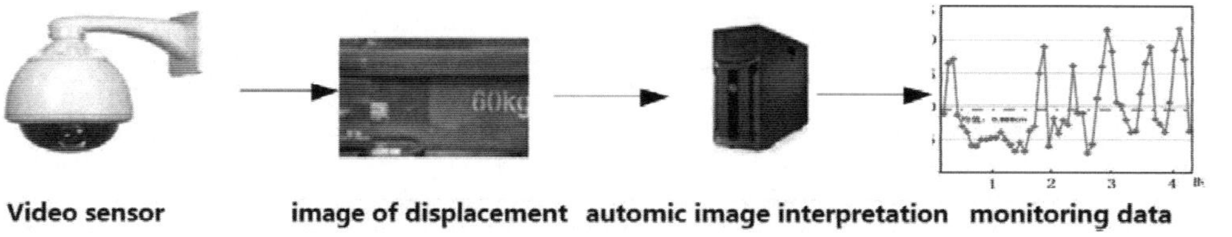

Figure.6 Monitoring of the switch rail extension using VII

3. Monitoring data validation and analysis

3.1 Monitoring data validation

The change of temperature is one of the main causes for the deformation of track system. In order to verify the accuracy of temperature monitoring data, significance tests were carried out by using statistical methods. Comparing the daily maximum temperature monitoring data from August to October in 2014 with the data provided from China Meteorological Administration, it shows that the two kinds of data are consistent, as shown in Figure.7.

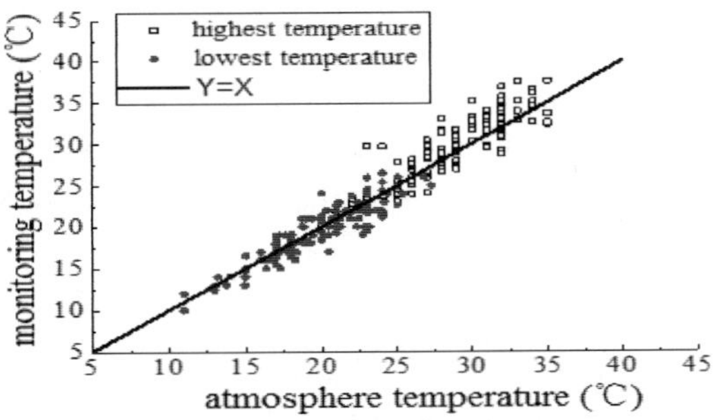

Figure.7 Comparison of monitoring temperature with atmosphere temperature

For a specific seasons the monitoring temperature sample value is subject to the normal distribution, and accords with the overall distribution of temperature in China. Therefore, Smirnov (Смирнов) test method can also be used for significant test. The Smirnov test result shows that the temperature monitoring results have the same distribution as the meteorological data.

In order to verify the accuracy and durability of the monitoring data, authors used the ruler test to compare with the long-term monitoring data at the skylight time in September 17, 2015, as shown in Table.2.

Table.2 Comparison of manual measured values and monitoring values

displacement sensor	manual measured values	monitoring values	displacement sensor	manual measured values	monitoring values
rail displacement11	1mm	0.8mm	base slab displacement 17	-3.5mm	-3.5mm
rail displacement 12	0mm	0.1mm	base slab displacement 18	-3mm	-3.1mm
rail displacement 13	-1mm	-0.7mm	base slab displacement 19	1mm	1.1mm
rail displacement 14	0.5mm	0.5mm	base slab displacement 20	7mm	7.2mm
rail displacement 15	-0.5mm	-0.7mm	beam ends displacement 21	17mm	17.3mm
rail displacement 16	-1mm	-1.2mm	beam ends displacement 22	20mm	19.8mm

The above data show that the monitoring data has a high degree of credibility, the monitoring methods can satisfy the needs, and the system is considerably durable.

3.2 Monitoring data analysis

The track monitoring data reflects the service state of track structure, which is of great importance to the maintenance and design optimization. The analysis of monitoring data is quite necessary.

(1) Distribution feature analysis: Two years' (meaning two periods) samples of different monitoring indicators are taken into account for the distribution analysis. Most samples are obviously subject to normal distribution (shown in Figure.8).

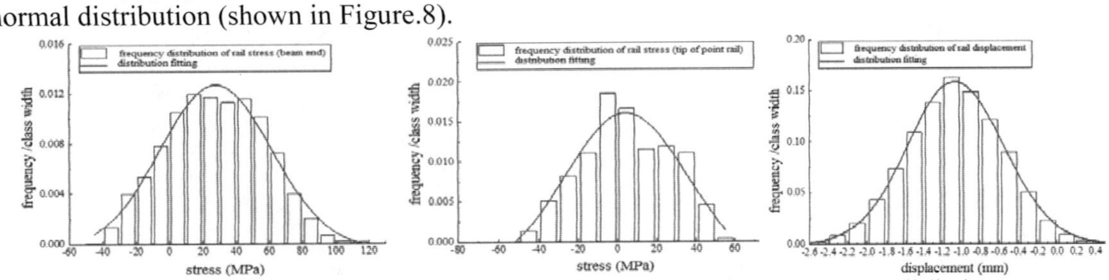

(a) rail stress at beam end (b) rail stress at tip of point rail (c) rail displacement

Figure.8 Distribution histogram of rail stress and displacement

(2) Periodic analysis

The rail temperature, the track slab temperature and the air temperature are monitored. All show obviously periodic characteristics.

To take rail temperature as an example, the data has a short period of one day and a long period of one year, as shown in Figure.9(a). The change trend of the rail temperature is as same as that of the air temperature. The temperature rising range of the rail is larger than that of the air temperature in the day (the maximum difference is about 18℃ in summer), and the rail reaches the peak of temperature 1~2.5 hours earlier than the air temperature dose. When it comes to the time needed for cooling to the lowest temperature, there is no obvious difference between rail and air. The long-term monitoring data shows an annual periodic feature. In a year, the highest rail temperature appears in July and the minimum rail temperature appears in January. During the monitoring, the highest rail temperature reached 53.6 ℃ in July 2015; the minimum rail temperature reached -12.4℃ in January 2016. The temperature difference is 66℃.

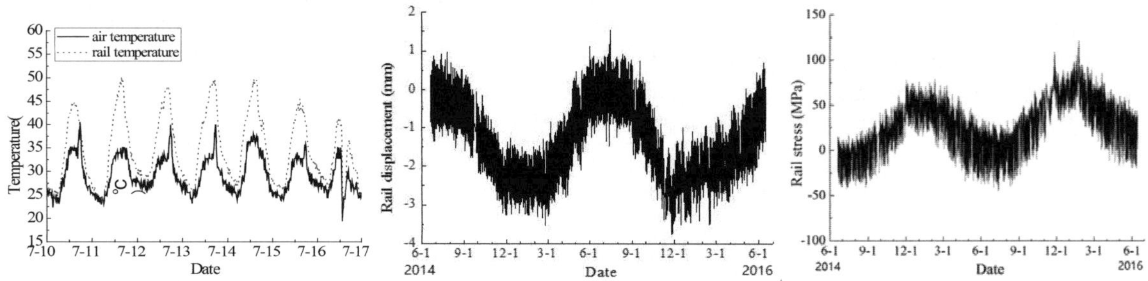

(a) rail temperature (b) rail displacement (c) rail stress

Figure.9 Rail temperature, displacement and stress curve

Similarly, rail displacement and stress also have obvious cyclical changes, also shown in Figure.9(b,c). The tensile and compressive stresses of the switch area are at the highest in July and August. The maximum tensile stress is 13.33MPa, and the maximum compressive stress is 57.73MPa. The maximum compressive stress of the rail in the frog area is greater than that in the bridge beam ends area. As for the maximum tensile stress, however, the situation is reversed. The compressive stress reduces in September and October, but that of the frog area decreases slower. At this point, compressive stress in the frog area is at the peak of 39.60MPa. When November approaches, the force states of all parts, including the switch area and the frog area, change from compress to tension, except the end of beam. The tensile stress of the basic rail in the switch area reaches a maximum of 80.93 MPa in December.

(3) Correlation analysis: Analysis of long-term monitoring data shows that the stress and displacement of the rail and slab (including switch rail) are highly correlated with rail temperature.

As shown in Figure.10, the rail stress is highly correlated with the rail temperature, but there are differences in stress changes at different positions. Similarly, the expansion and contraction changes of the switch rail is in accordance with the change of the rail temperature. With the increase of the rail temperature, the switch rail expands. The regression analysis result shows that the switch rail expansion is about 0.8mm for every 1℃temperature rising. The maximum expansion of the switch rail is 23mm in July, and the maximum contraction is -22mm in December, less than the safety limit of ± 40mm.

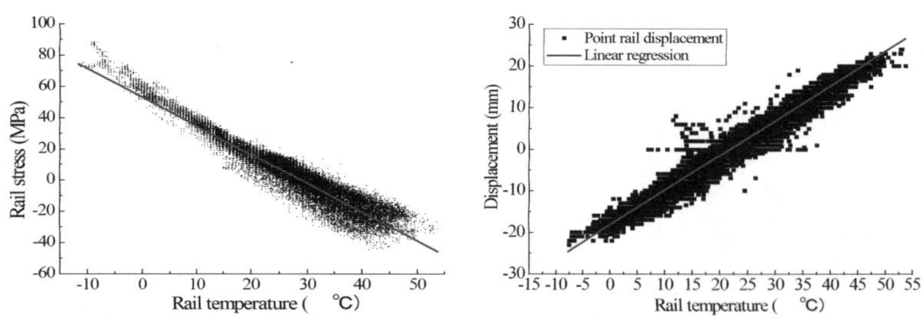

(a) rail stress and temperature (b) rail displacement and temperature

Figure.10 The correlations between the stress/displacement of rail and rail temperature

4. Prediction and forewarning

4.1 Prediction of service state

Considering the high correlation between the track mechanical indexes and the rail temperature, it is quite feasible to adopt the rail temperature to predict track state. Since the air temperature is relatively easy to obtain ahead of time, this paper uses the temperature series as an independent variable to predict the track mechanical state.

Multiple linear regression and BP neural network are used in this study. To exemplify, the BP neural network is introduced as follows.

The BP neural network with single hidden layer was used to predict the temperature, rail stress and the point rail displacement, taking the temperature variation in 12 hours (48 input nodes) into consideration. The historical monitoring data from September 2014 to September 2016 were used as training samples and data in October 2016 were used for tests and verification. Taking the rail stress at the beam end as an example, the prediction results are shown in Figure.11.

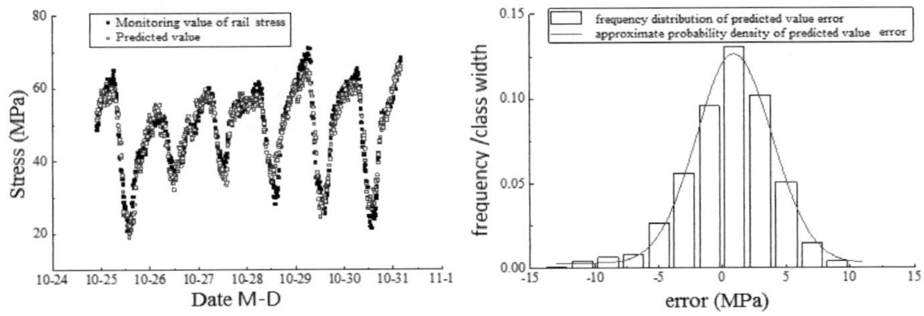

(a) prediction results of rail stress (b) error distribution

Figure.11 Prediction results and error distribution of BP Forecasting Model

The error of the BP prediction model has obvious normal distribution characteristics. After fitting, the error probability density can be expressed as:

$$p(E_i) = 0.123 \times \exp\left(\frac{-(E_i - 0.94)^2}{2 \times 2.97^2}\right) \tag{1}$$

Revise $\hat{Y}_i' = \hat{Y}_i + 1$, using the normal distribution characteristic, there is probability P:

$$P\{Y_i \in (\hat{Y}_i'\text{-}5.95MPa, \hat{Y}_i'+5.95MPa)\} = 95.4\% \tag{2}$$

Similarly, it indicates that the BP prediction method has little error, and the model is reliable.

4.2 Forewarning of breakdown

The limit of rail crawling and switch rails temperature displacement are strictly stipulated in the *HSR Ballastless Track Repair Rules*. Switch rails temperature displacement must be within 40mm, and the basic rail crawling must be within 10mm. Once the monitoring data exceeds the limit, the system needs to make immediate prewarning, reminding the works departments to take maintenance measures.

However, the deterioration of the track state often not only results from a certain indicator's excess. It is actually the joint result of multiple reasons. Therefore, using only one a single static threshold for alarm is often unpractical.

Nine indicators such as air and rail temperature, rail force of different monitoring points are taken into consideration. The different attributive data which is collected at each time point constitute a space vector $\vec{M}$ with 9 degrees of freedom:

$$\vec{M}=[T_1,T_2,T_3,S_1,S_2,S_3,S_4,S_5,S_6] \quad (3)$$

T_1,T_2,T_3—— Atmosphere temperature and temperature gained from point 1,2;

S_1,S_2,S_3,S_4,S_5,S_6—— Displacement of beam ends/frog/nose rail/heel of point rail/bearing/tip of point rail.

The time series constitutes the sample matrix.

$$[M_1, M_2, M_3,\ldots, M_k, \ldots, M_n]^T \ (k=1,2,3, \ldots) \quad (4)$$

The K-means algorithm is used to analyze the sample matrix, and different outlier diagnosis results are obtained by adjusting the cluster amount. The result of the diagnosis is shown in Figure.12. The outlier mainly appears in January 2016, which coincides with the large-scale cold weather Beijing area suffered during the period. The diagnostic result is not only limited to the overrun of a single indicator, but also assesses the overall state of track system.

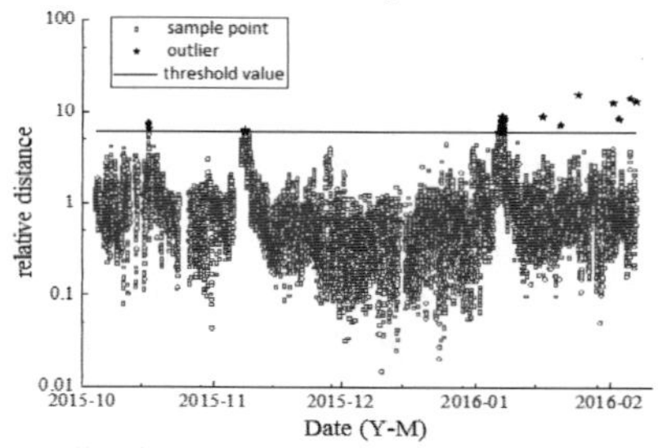

Figure.12 Outlier-identification-Based clustering recognition

Based on the long-term monitoring data, it reveals the changing and deterioration of the track system and strongly supports the study of the mechanical properties of HSR track structure under long-term operating with important data. In addition, the prediction and prewarning system provides a potent guarantee for the safe operation and reasonable maintenance of the HSR track system.

5. Information management system

The information management system is developed to achieve the collection, storage, transmission and analysis of the information, and furthermore to streamline the application and exploitation process for users. Management information system consists of five modules of collection/storage, analysis/processing, prewarning/assisting decision-making and Web browser. As shown in Figure.13, with the Web browser, the information management system provides query and report function according to sensors, measuring points, or monitoring items. It also has functions of data alarming and prewarning. Upon the data is over the threshold, the system can warn the users though popup and alarming sound. According to the analysis of the existing monitoring data change patterns, the neural network prediction model is established to forecast the data development trend, which provides data support for maintenance and optimization design.

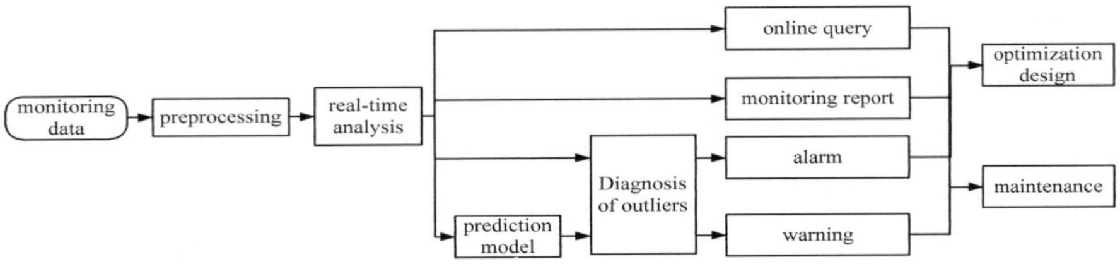

Figure.13 System function

6. Conclusions

Based on the analyses of the static and dynamic characteristics of the track system, the contents of monitoring of the track system and the plan of measuring points layout are put forward. Combined with the high security and other special requirements for HSR, an integration monitoring system based on multi-intelligent methods is developed and the design is optimized. It is very important to establish typical monitoring sites and analyze the massive monitoring data for ensuring the safe operation of the train and the reasonable maintenance of the lines. The specific conclusions are as follows:

(1) A comprehensive monitoring system for turnout on HSR bridges based on the combination of FBG and VII technology is established. The good field application results prove that the system can perfectly meet the high need of safety for HSR and can accomplish the online monitoring function for a long time.

(2) The monitoring data reflect the relationship between air/rail temperature and the force/deformation of the structure. Through the analysis of the monitoring data, a comprehensive method for track state prediction based on linear regression and BP neural network is proposed, which provides assistance for arrangement of the maintenance timing.

(3) An online monitoring information management system for HSR track service status is established. It achieves the online query, prediction, prewarning and alarm function of massive data. The track system is so complicated that its monitoring should meet different needs from outside to inside, from macro-scope to micro-scope, from low frequency to high frequency. Reducing monitoring costs and promoting the urban rail transit is the future direction of development. In addition, image sensors can be further developed to automatically search for typical diseases and structural displacement. It is also desirable to further the study of other non-destructive monitoring technologies.

Acknowledgement

This research was supported by National Natural Science Foundation of China (Grant No. 51578056), Project of Beijing science and technology (Grant No. Z161100001016004) and Project of Science and Technology Research and Development Program of China Railway Corporation (Grant No. 2015G0001-L and 2015G001-B-1).

Reference

Zhai, W.M.& Zhao, C.F. 2014, 'Dynamic evolution of high - speed railway infrastructure and basic scientific problems of service safety', *Science in China: Technology Science*, vol. 07, pp. 645-660

Xu, K.L. 2010, 'Necessity analysis and suggestions on comprehensive maintenance of high - speed railway', *Railway survey*, vol.03, pp.122-125.

Zhao, Y.M. 2015, 'High - speed railway infrastructure service state detection technology', *Railway construction*, vol.10, pp.1-6.

Du, Y.L.& Zhang Y.Z. 2012, 'Construction of Safety Monitoring System for High - speed Railway Line Project', *Journal of Civil Engineering Society*, vol. S2, pp.59-63.

Tam, H.Y. & Lee, T.& Ho S.L. 2007, 'Utilization of Fiber Optic Bragg Grating Sensing Systems for Health Monitoring in Railway Application', Hong Kong: Photonics Research Centre, The Hong Kong Polytechnic University, Hong Kong.
Filograno M L & Rodriguez-Barrios A 2010, 'Real time monitoring of railway traffic using fiber bragg grating sensors', *Proceedings of the 2010 joint rail conference*, New York: Amer Soc Mechanical Engineers, pp.493-50.
Pen L L & Watson G 2014, 'The behaviour of railway level crossings', *Transportation Geotechnics*,vol.1, no.4, pp. 201-213.
Cai X.P. & Gao L. 2015, 'Long - term Monitoring Technology for Track System of Beijing - Shanghai High - speed Railway Elevated Station', *Journal of Railway Engineering Society*, vol.05, pp.35-41.
Jiang J.Z. & Lv G.H 2015, 'Remote Monitoring Data Transmission System for High - speed Railway Track State Based on Fiber Bragg Grating Sensing Technology', *Railway construction*, vol.01, pp.76-79.
Wang Y.Z. & Wang S.R. 2015, 'Monitoring Technology of Ballastless Track in High - speed Railway', *Railway standard design*, vol.08, pp.1-9.
Gao L. & Tao K. 2009, 'Study on Mechanical Properties of Welded Turnout on Passenger Dedicated Line Bridge', *China Railway Science*, vol.30, no.1, pp.29-35.
Lin C. 2015, 'Research on Monitoring Technology of Mechanical Characteristics of Track System of Elevated Station of Beijing - Shanghai High Speed Railway', Beijing Jiaotong University.
Ren L. & Jiang T. 2013, 'Application of Micro FBG Strain Sensor in Dam Model Experiment', *Vibration and testing*, vol.2, pp.277-283+341.
Li W.L. & Lu X.S. 2013, 'Fiber Bragg Grating Monitoring Technology for Viaducts of Viaducts', *Journal of Wuhan University of Technology*, vol.09, pp.133-137+143.

The 8th International Conference on
Structural Health Monitoring of Intelligent Infrastructure
Brisbane, Australia | 5-8 December 2017

Acoustic Emission Monitoring of Switch Rail Detect Based on Wigner-Ville High-order Spectrum and Data Mining Technology

Jingmang Xu [1], Ping Wang [2], Rong Chen [3], Jieling Xiao [4], Weihai Liang [5], Yunhua Hou [6].

[1] *Key Laboratory of High-speed Railway Engineering, Ministry of Education, Southwest Jiaotong University – China,*
[2] *Key Laboratory of High-speed Railway Engineering, Ministry of Education, Southwest Jiaotong University – China, Email:* wping@home.swjtu.edu.cn
[3] *Key Laboratory of High-speed Railway Engineering, Ministry of Education, Southwest Jiaotong University – China,*
[4] *Key Laboratory of High-speed Railway Engineering, Ministry of Education, Southwest Jiaotong University – China,*
[5] *Southwest Jiaotong University Railway Development CO.,LTD - China,*
[6] *Southwest Jiaotong University Railway Development CO.,LTD - China*

Abstract: In order to detect cracks in railroad tracks, various experiments have been examined by Acoustic Emission (AE) method. However, little work has been done on studying switch rail defect detection for railway turnout. Due to the complex constraints and difficult detection of switch rail detect, this paper presents a study on AE detection of switch rail defect for railway turnout based on a large number of field test. Meanwhile, Wigner-Ville high-order spectrum and data mining technology are employed to detect defects. Massive data with and without defects are acquired and characterized based on data mining technology. Wigner-Ville high-order spectrum is applied to achieve the clusters features of switch rail detect by suppressing noise and realizing high time-frequency resolution of signals, and then the subsequent collection of the acoustic emission signal can be classified and identified. The results clearly illustrate that the proposed method can detect switch rail defect for railway turnout effectively.

Keywords: Acoustic emission monitoring; Switch rail detect; Railway turnout; Wigner-Ville high-order spectrum; Data mining technology

1 Introduction

Along with the completion and operation of lots of high-speed railways in China, contradictions between safe transportation and equipment management have become more and more prominent. The establishment of effective monitoring systems and long-term monitoring of fixed equipment can ensure the safe, reliable and effective operation of high-speed railways, especially the monitoring of some key fixed equipment parts such as turnouts, key bridges, tunnels and weak subgrades. Since high-speed turnouts are key parts for providing flexibility to traffic operation (Xu et al. 2017) and have movable members (switch rails and point rails) which are in direct relation to the wheel-rail relationship, the wheel-rail contact state is complicated, the rail sections are variable, the rail's supports and constraint conditions are complicated, many fittings are needed at turnout areas, different parts made of different materials are needed to be installed, engineering-electrification combination is needed for switching --- all the above, definitely, indicate that turnouts are extremely weak parks for fixed equipment of high-speed railways, disasters will be caused once any trouble is not removed in time (Wang et al. 2016, Xu et al. 2016). Fractures are one of the main reasons causing degeneration of rails. The expansion of fractures is an important acoustic emission source, and sound waves caused by them reflect the essence of fractures. Compared with other test methods (ultrasonic, magnetic field tests etc.), the acoustic emission method can be adopted to evaluate the dynamic characteristics of fractures, which is especially suitable for testing the dynamic behaviors of materials and structures and, as an ideal real-time online test method (Al-Dossary et al. 2009), is very important for solving railway safety problems caused by rail fractures, especially those of high-speed turnouts.

The acoustic emission method has been introduced in relevant experimental researches for testing fractures in standard rails, which not only proved the feasibility of the acoustic emission method in testing rail fractures theoretically, but also further verified the effectiveness of the acoustic emission technology in testing rail surface fractures through wheel-rail testing equipment and on-site tests (Zumpano & Meo.2006, Bollas. 2010, Bruzelius & Mba. 2004). The above experimental researches focus on the acoustic emission source of rail surface fractures mainly and no any analysis on acoustic emission source of different fractures in rails has been provided. In order to make a further analysis on the acoustic emission source of fractures in rails, a model for rails and their internal fractures was established based on the finite element method in the researches (Hill et al. 2004, Bartoli et al. 2005), and in the meantime the acoustic emission source of fractures was set at reasonable positions of rails in order to simulate the expansion of fractures, the simulated signals, in relation to those obtained through tests, can provide uniform and ideal data for studying the characteristics of the acoustic emission source of fractures (Zhang et al. 2014, Zhang et al. 2015), however, the acoustic emission method aims to test the fractures in standard rails mainly and relevant researches on testing fractures in turnout switch rails, considering turnout switch rails have weak-constraint variable cross-sections without fasteners, their damage characteristics and the transmission of acoustic emission signals are far more complicated than those of standard rails.

For the indoor test and on-site test presented in this paper, PZT piezoceramic sensors are installed on turnout switch rails for acquiring acoustic emission signals of different types of damaged rails, the Wigner-Ville four-order spectrum is adopted for characterising the acoustic emission signals of damaged rails, and based on which, big data mining and cluster analysis are carried out in order to classify new acoustic emission signals and evaluate the damage to turnout switch rails.

2 Signal acquisition

The phenomenon of transient elastic wave caused by quick release of energy from a local source of a material is called acoustic emission (or stress wave emission). The deformation and fracture expansion of materials under a stress are an important mechanism of structural failure. Such source having direct relation to deformation and fracture mechanism is called acoustic emission source. In recent years, another elastic wave source, caused by liquid leakage, friction, impact and combustion, having no direct relation to deformation and fracture mechanism is called other or secondary acoustic emission source.

Acoustic emission is a common physical phenomenon. The range of acoustic emission signal frequency of different materials is wide - from the Hz class infrasonic frequencies, 20Hz ~ 20KHz audio frequencies to MHz class ultrasonic frequencies; The amplitude range of acoustic emission signals is also wide - from microscopic 10^{-13}m dislocation motion to 1m magnitude seismic waves. A sound can be heard if the stress of acoustic emission meets the requirement. For most materials, the emission can be heard in cases of deformation and fracture, however many have very weak acoustic emission signal intensities that are hard to be heard directly, which can only be detected by sensible electronic instrument. The method for detecting, recording and analysing acoustic emission signals with instrument and testing acoustic emission sources with acoustic emission signals is called acoustic emission technology.

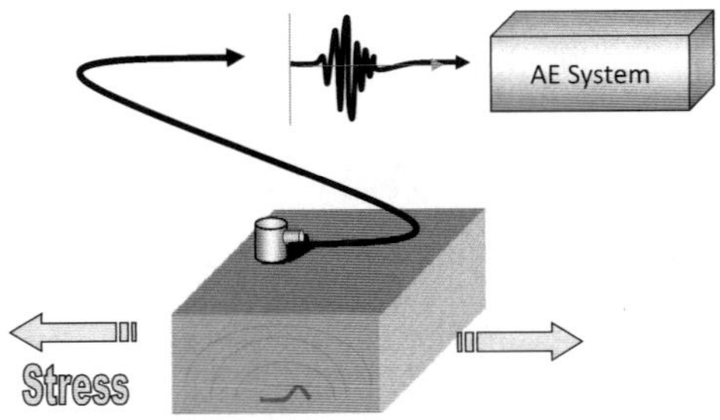

Fig. 1 Acoustic emission signal in case of damage to a material

Acoustic emission sources generate elastic waves that can be transmitted to a tested surface through a certain medium, causing mechanical vibration on the surface. The transient displacement on the surface can be transformed into electrical signals with an acoustic emission sensor. The waveforms or characteristic parameters can be recorded and displayed after the acoustic emission signals are amplified and processed. The characteristics of the acoustic emission sources can be evaluated through data analysis and explanation. The characteristics of acoustic emission signals can be obtained according to the piezoelectricity energy method. The surface of a piezoelectric material generates polarization charge under an external force. The quantity of the electric charge is in proportion to the pressure. Piezoelectric crystals are characterised by low symmetry, the relative displacement of positive and negative ions in crystal cells being subject to deformation under the external force will make the positive and negative charge centres leave from each other, causing macroscopical polarization of the crystals. However the surface density of charge on a surface of a crystal is equivalent to the projection of the polarization intensity on the surface normal, therefore both ends of the piezoelectric material being subject to deformation under a pressure can generate contrary sign charge, whereas when the piezoelectric material polarizes in an electric field, deformation will be caused due to displacement of the charge centre.

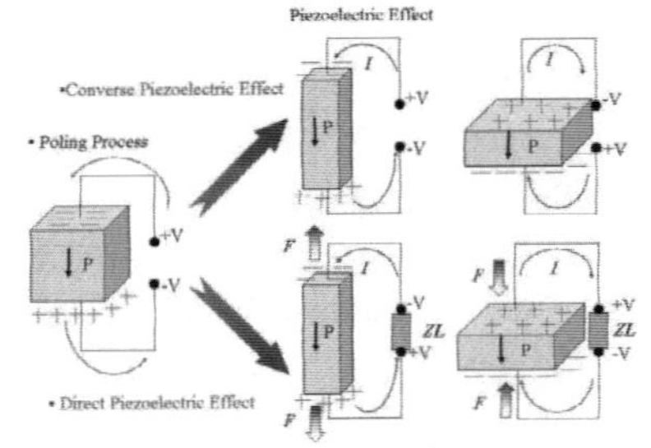

Fig. 2 Schematic of acoustic emission signals during piezoelectric energy test

PZT piezoelectric sensors are characterised by good electromechanical coupling property for realizing conversion between mechanical signals and electric signals. The voltage signals tested with PZT piezoelectric sensors can reflect the conditions of mechanical structures. For the constitutive relation equations, see Eq. (1) ~ (2).

$$\begin{bmatrix} s_1 \\ s_2 \\ s_3 \\ s_4 \\ s_5 \\ s_6 \end{bmatrix} = \begin{bmatrix} S_{11}^E & S_{12}^E & S_{13}^E & 0 & 0 & 0 \\ S_{12}^E & S_{22}^E & S_{23}^E & 0 & 0 & 0 \\ S_{13}^E & S_{23}^E & S_{33}^E & 0 & 0 & 0 \\ 0 & 0 & 0 & S_{44}^E & 0 & 0 \\ 0 & 0 & 0 & 0 & S_{55}^E & 0 \\ 0 & 0 & 0 & 0 & 0 & S_{66}^E \end{bmatrix} \begin{bmatrix} \sigma_1 \\ \sigma_2 \\ \sigma_3 \\ \tau_{23} \\ \tau_{31} \\ \tau_{12} \end{bmatrix} + \begin{bmatrix} 0 & 0 & d_{31} \\ 0 & 0 & d_{32} \\ 0 & 0 & d_{33} \\ 0 & d_{24} & 0 \\ d_{15} & 0 & 0 \\ 0 & 0 & 0 \end{bmatrix} \begin{bmatrix} E_1 \\ E_2 \\ E_3 \end{bmatrix} \tag{1}$$

$$\begin{bmatrix} D_1 \\ D_2 \\ D_3 \end{bmatrix} = \begin{bmatrix} 0 & 0 & 0 & 0 & d_{15} & 0 \\ 0 & 0 & 0 & d_{24} & 0 & 0 \\ d_{31} & d_{32} & d_{33} & 0 & 0 & 0 \end{bmatrix} \begin{bmatrix} \sigma_1 \\ \sigma_2 \\ \sigma_3 \\ \tau_{23} \\ \tau_{31} \\ \tau_{12} \end{bmatrix} + \begin{bmatrix} \varepsilon_{11}^\sigma & 0 & 0 \\ 0 & \varepsilon_{22}^\sigma & 0 \\ 0 & 0 & \varepsilon_{33}^\sigma \end{bmatrix} \begin{bmatrix} E_1 \\ E_2 \\ E_3 \end{bmatrix} \tag{2}$$

In this study, the characteristics of acoustic emission signals are tested through piezoelectric energy. A PZT piezoelectric sensor is installed at the base of a turnout switch rail, the acoustic emission signals of different damages to the switch rail can be acquired through indoor test and an on-site test, providing a reference for the cluster analysis on acoustic emission signals of the damages to the switch rail. For the PZT piezoelectric sensor installed at the base of the turnout switch rail, see Fig. 3.

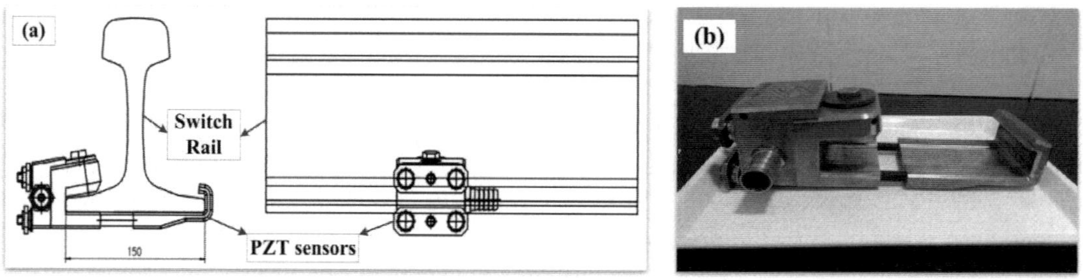

Fig. 3 Installed PZT piezoelectric sensor

3 Methodology

The service state signals of a turnout switch rail are received by a PZT piezoelectric sensor for pretreatment in this paper. The Gaussianity of signal distribution is adopted for confirming whether the received signals are acoustic emission signals; In order to improve the efficiency of the signal analysis algorithm, 1/4 sub-band signals are selected as the characteristic signals through 20KHz highpass filtering and wavelet packet decomposition, and then a mass data can be recharacterised based on the modern signal processing technology and big data mining technology and through the modern signal processing technology and thus the time-frequency high-resolution characterisation of signals can be realised while restraining noise with Wigner-Ville four-order spectrum. Based on the high-resolution characterisation and noise restraint of acoustic emission signals, the big data are mined and clustered by taking the Wigner-Ville four-order spectrum as the characteristics of acoustic emission signals, and the new acoustic emission signals can be classified by taking the clustered mass data as the prior information.

3.1 Data Pre-processing

An acoustic emission signal of a turnout switch rail fracture is similar to a transient impact signal, the emission of energy is transient, and therefore the acoustic emission signal changes with the time. It is a nonstationary random signal since its probability density function changes with the time. Wavelet packet decomposition is helpful for characterising local signals within a time domain and a frequency domain at the same time, which can not only be adopted to describe the spectrum information of signals within a local time interval, but also to indicate the corresponding time domain information of a frequency domain information. In order to improve the efficiency of the signal analysis algorithm, the acoustic emission signal can be decomposed based on the wavelet packet decomposition method, see the following for the details:
Given that a fundamental function is $\psi(t)$, then make

$$\psi_{a,b}(t) = \frac{1}{\sqrt{a}} \psi(\frac{t-b}{a}) \tag{3}$$

Where, both a and b are constants and $a>0$. It is obvious that $\psi_{a,b}(t)$ is obtained when the fundamental function $\psi(t)$ is shifted and then flexed. A family of function $\psi_{a,b}(t)$ can be obtained provided that a and b change continuously. According to the given square integrable signal $x(t)$, i.e. $x(t) \in L^2(R)$, then the wavelet transformation of $x(t)$ can be defined as:

$$WT_x(a,b) = \frac{1}{\sqrt{a}} \int x(t)\psi^*(\frac{t-b}{a})dt = \int x(t)\psi^*_{a,b}(t)dt = \langle x(t), \psi_{a,b}(t)\rangle \tag{4}$$

Where, a is the scale factor, b is the time shifting factor.

Make the spectrum function of the wavelet basis function $\psi(t)$. According to the property of Fourier transform, the spectrum function of the wavelet sequence $\psi_{a,b}(t)$ should be $a^{1/2}\psi(aw)e^{-jab}$. It can be seen that the time factor b only indicates the change of signal phase in the frequency domain, while the scale factor a realises the frequency limit of the signal, i.e. the signal can be divided into different frequency band components. The higher the scale factor is, the lower the frequency will become and the narrower the frequency band will become. Provided that a j scale wavelet analysis will be made for a signal $f(t)$ with a sampling rate $2f_s$, then $f(t) = \sum_{i=1}^{j} D_i + A_j$; Where the band range of A_j is $\left[0, f_s/a^j\right]$, the band range of D_i is $\left[f_s/a^i, f_s/a^{i-1}\right], 1 \le i \le j$.

As shown in Eq. (4), the wavelet sequence function can be considered as a series of window functions, the local analysis on $f(t)$ is made at time b. Suppose that the centre of the wavelet basis function $\psi(t)$ is t^*, the time window width is $2\Delta t$, then Eq. (4) will indicate the local analysis on $f(t)$ within the following time window:

$$\left[at^* + b - a\Delta t, at^* + b + a\Delta t\right] \tag{5}$$

In a similar way, make the centre frequency of the spectrum function $\psi(w)$ of the wavelet basis function $\psi(t)$ be w^* and the band width be $2\Delta w$, then according to the property of Fourier transform, the frequency window in relation to the time window should be:

$$\left[w^*/a - \Delta w/a, w^*/a + \Delta w/a\right] \tag{6}$$

The low scale a is in relation to a high frequency signal, according to Eq. (5) and (6), a small time widow is used in thc timc domain for the wavelet transformation during the local analysis on the function $f(t)$, and big frequency windows are used in the frequency domain of them; While for the analysis on the low frequency signal in relation to the high scale a is quite the reverse. Because the wavelet has variable time / frequency windows, the wavelet transformation is characterised by good localisation in both time and frequency domains, which is quite suitable for analysing those signals with transient transformation.

3.2 Wigner-Ville High-order Spectrum

The Wigner high-order spectrum is the extension of Wigner-Ville distribution, which, based on the good mathematical characteristics of Wigner-Ville distribution, is also provided with the advantages of high-order spectrum analysis. In essence, it reflects the high-order spectrum characteristics of signals on time-frequency planes, i.e. the distribution of signal high-order domain energy on time-frequency planes. The Wigner-Ville high-order spectrum, being also provided with the capacity of high-order spectra in restraining Gaussian noises, keeps the high resolution characterised by the signal transformation domain and realises the synchronous high time-frequency resolution of signals.

The Wigner-Ville distribution is defined as:

$$WVD(t,w)=\int_{-\infty}^{-\infty} x(t+\frac{\tau}{2})x^{*}(t-\frac{\tau}{2})e^{-iw\tau}d\tau \tag{7}$$

The Wigner-Ville high-order spectrum is defined as:

$$WVD_{Order}(t,w_1,\cdots,w_k)=\int_{\tau_1}\cdots\int_{\tau_k} x^{*}(t-\frac{1}{k+1}\sum_{m=1}^{k}\tau_m)\prod_{i=1}^{k}x(t+\frac{1}{k+1}\tau_i-\frac{1}{k+1}\sum_{m=1}^{k}\tau_m)e^{-iw_i\tau_i}d\tau_i \tag{8}$$

When k=4, the Wigner-Ville four-order spectrum can be obtained.

The Wigner-Ville four-order spectrum selected as the basis for extracting the characteristics and cluster of acoustic emission signals of turnout rails is not only helpful for restraining Gauss background noises but also for keeping the high resolution of the signal time-frequency domain. When the Wigner-Ville four-order spectrum is used for processing the acoustic emission signals of rails, the "nice distinction" among different types of acoustic emission signals will not be "neglected" and "blurred", on the contrary, the "nice distinction" among time domains can be put into a time-frequency domain for rearrangement, from this point of view, the Wigner-Ville four-order spectrum realises the effective amplification of the "nice distinction" among different types of acoustic emission signals, laying a solid foundation for the cluster analysis.

3.3 Data Mining Technology

The big data mining technology, in essence, is the combination of signal processing and information processing technology. Information can be mined through signal processing. No any uniform methodology has been provided for big data mining, however, characteristic extraction, classification and clustering are important for the big data mining technology. In order to classify and cluster mass and nonisomorphic data, characterisation is necessary, which can even be deemed as a signal processing technology.

According to the fracture mechanics, it is no doubt that a "distinction" between different acoustic emission signals can be found in a time domain or a transformation domain. Characteristic extraction aims to extract this "distinction" precisely under strong background noises. Characterisation and cluster analysis are the core for acoustic emission signal processing, however characterisation is the foundation for cluster analysis. Theoretically, many methods in processing signals are helpful for characterising acoustic emission signals, but considering the strong background noises at railway turnouts, the Wigner-Ville four-order spectrum can be used to charactrising the acoustic emission signals of damaged turnout switch rails.

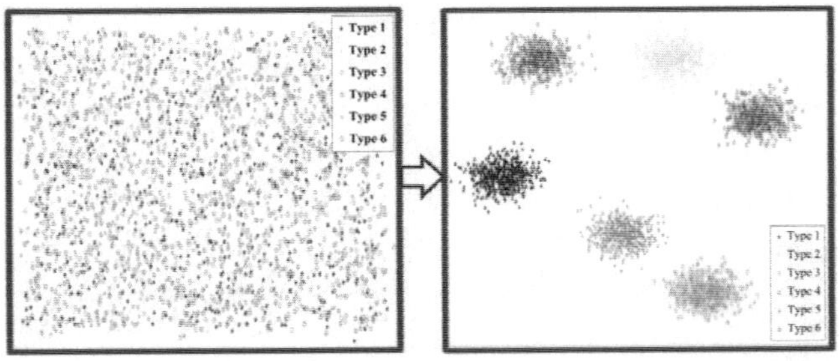

Fig. 4 Acoustic emission signal processing based on big data

Decomposing a set into some subsets according to a uniform standard is called "cluster analysis", each subset is called a cluster, "Birds of a feather flock together" is a typical application for cluster analysis. For cluster analysis, characteristic extraction is the foundation. The uniform standard of clustering comes from a certain transformation of characteristics. From the pure mathematical view, all clusters require the individuals of a subset to be limited within a similar tolerance, while strong subsets are quite different from each other. The acoustic emission characteristics of damaged turnout switch rails are distinguished according to the dynamic clustering method.

4 Acoustic Emission Characteristic

In this paper, the acoustic emission signals of damaged turnout switch rails, including switch rail fracture and spalling of rail foot and rail head, are acquired through an indoor test and an on-site test. The acoustic emission characteristics of different damages to turnout switch rails are obtained according to the characteristic extraction method for acoustic emission signals mentioned in Section 3, providing a theoretical foundation for cluster analysis and online monitoring with regard to damaged turnout switch rails.

①Switch rail fracture

Install a PZT piezoelectric sensor shown in Fig. 3 on the turnout rail of a line, see Fig. 5 (a). Turnouts are laid in marshalling stations with many passing trains every day. The stress environment of turnout rails is poor, and huge economic losses and casualties can be caused in case of rail fractures. When a fracture is caused at an end of a turnout switch rail (Fig. 5 (a)), the time travel curve of acoustic emission signals will be received by the PZT piezoelectric sensor (Fig. 5 (b)). The acoustic emission signal characteristics of the turnout switch rail fracture can be obtained according to the characteristic extraction method for acoustic emission signals mentioned in Section 3, see fig. 5 (c).

Factors such as the diversity of the acoustic emission source mechanism, the complexity of the sonic propagation route, the suddenness and uncertainty of acoustic emission signals as well as the severity of interference noises are of great challenges for acoustic emission signal processing and analysis. Moreover, though the Gauss noises have an impact when acoustic emission signals are acquired at turnout switch rail areas, the characteristic distinction among advantageous signals from different signal sources is minor, therefore a mathematical tool being quite available for restraining Gauss noises and amplifying the "nice distinction" among different types of acoustic emission signals is needed. In this study, the Wigner-Ville high-order spectrum is selected for extracting acoustic emission characteristics. As shown in Fig. 5 (b) and 5 (c), the acoustic signal of the switch rail fracture can be found in both time and frequency domains. The amplitudes of the signal in the time and frequency domains are high within the time axis of 1ms ~ 4ms, indicating that the switch rail is subject to a brittle fracture within the time axis of 1ms ~ 4ms. The signal is "highlighted" at the point of around 90KHz in the frequency domain, which shows the acoustic emission signal characteristic in relation to the switch rail fracture.

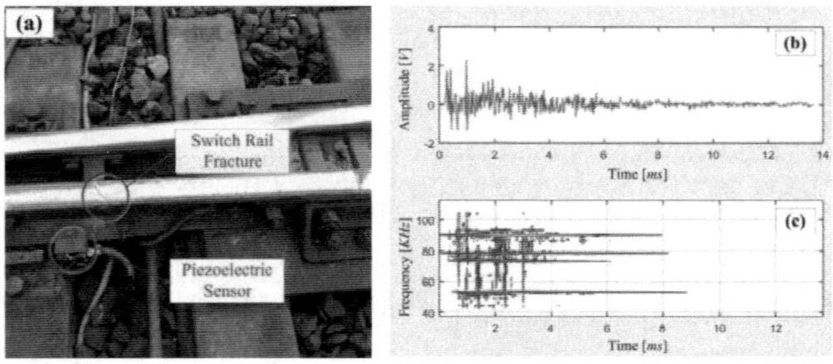

Fig. 5 Characteristic analysis on acoustic emission signals of a fractured site turnout switch rail

② Spalling of rail foot

The experimental study on the acoustic emission signal characteristics of turnout switch rails with spalling of rail foot is carried out by laying a set of switch (Fig. 6 (a)) in a test room to simulate a breaking base piece of a turnout switch rail through breaking a preset base broken piece (Fig. 6 (b)). The characteristics of the acoustic emission signals of spalling of rail foot of the turnout switch rail are obtained based on the time interval curve (Fig. 7 (a)) for the acoustic emission signals received by a PZT piezoelectric sensor and the characteristic extraction method for acoustic emission signals mentioned in Section 3, see Fig. 7 (b).

Fig. 6 Test for turnout switch rail with spalling of rail foot

As shown in Fig. 6 (a) and 6 (b), the acoustic emission signal of spalling of rail foot of the turnout switch rail can be found in both time and frequency domains. The amplitudes of the signal in the time and frequency domains are high within the time axis of 1.5ms ~ 2.5ms, indicating that the switch rail has a spalling of rail foot within the time axis of 1.5ms ~ 2.5ms. The signal is "highlighted" at the point of around 78KHz in the frequency domain, which shows the acoustic emission signal characteristic in relation to the switch rail with spalling of rail foot; In the meantime, it also indicates that the acoustic emission signal can be used to reflect the damage and damaging process in relation to the turnout switch rail quickly.

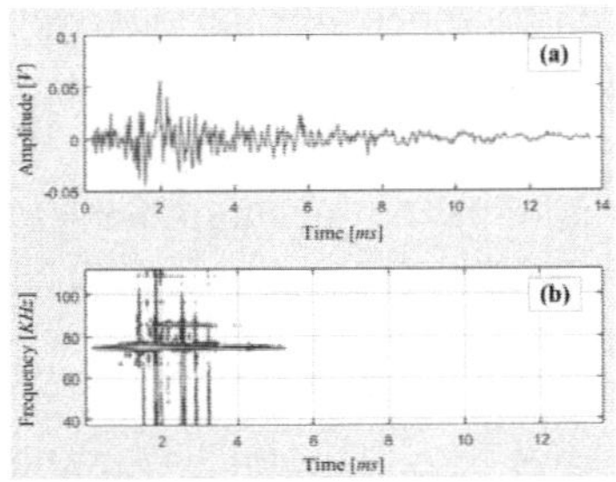

Fig. 7 Characteristic analysis on acoustic emission signals of a turnout switch rail with spalling of rail foot

③ Spalling of rail head

Being similar to those of switch rails with spalling of rail foot, the experimental study on the acoustic emission signal characteristics of turnout switch rails with spalling of rail head is also carried out by laying the same set of switch (Fig. 6 (a)) in a test room to simulate a breaking head piece of a turnout switch rail through breaking a preset spalling of rail head (Fig. 8 (a)). The characteristics of the acoustic emission signals of the spalling of rail head of the turnout switch rail are obtained based on the time interval curve (Fig. 8 (b)) for the acoustic emission signals received by a PZT piezoelectric sensor and the characteristic extraction method for acoustic emission signals mentioned in Section 3, see Fig. 8 (c).

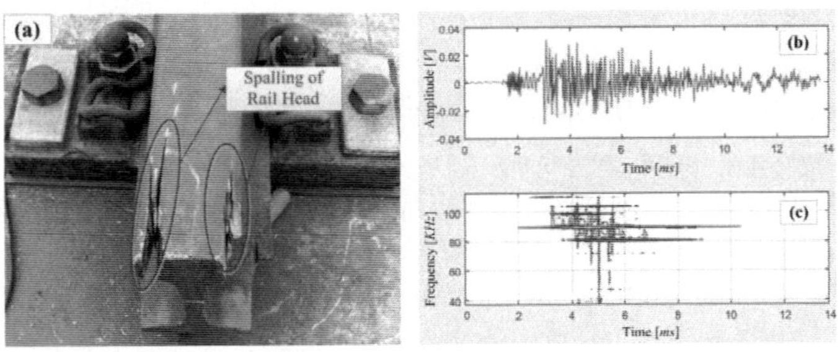

Fig. 8 Characteristic analysis on acoustic emission signals of a turnout switch rail with spalling of rail head

As shown in Fig. 8 (b) and 8 (c), the acoustic emission signal of the spalling of rail head of the turnout switch rail can be found in both time and frequency domains. The amplitudes of the signal in the time and frequency domains are high within the time axis of 3ms ~ 6ms, indicating that the switch rail has a broken head piece within the time axis of 3ms ~ 6ms. The signal is "highlighted" at the point of around 90KHz in the frequency domain, which shows the acoustic emission signal characteristic in relation to the switch rail with spalling of rail head; In the meantime, it also indicates that the acoustic emission signal can be used to reflect the damage and damaging process in relation to the turnout switch rail quickly.

5 Conclusion

Based on the Wigner-Ville high-order spectrum and big data mining technology, this paper presents the method for extracting the characteristics of acoustic emission signals of turnout switch rails with typical damages and analyses the characteristics of acoustic emission signals of damages such as turnout switch rail fracture and spalling of rail foot and rail head. A PZT piezoelectric sensor is installed on a turnout switch rail in an indoor test and an on-site test, when a damage is preset or the site turnout has fractures, the sensor can receive the energy signals. The Gaussianity of signal distribution is adopted for confirming whether the received signals are acoustic emission signals. Acoustic emission signals can be processed through through 20KHz highpass filtering and wavelet packet decomposition, and a sub-band signals are selected as the characteristic signals; The Wigner-Ville four-order spectrum is used for recharacterising mass data in order to realise bid data mining and cluster analysis, laying a foundation for monitoring damages to turnout switch rails with the acoustic emission technology.

References

Xu, J., Wang, P., Ma, X., Xiao, J., & Chen, R. 2017, "Comparison of calculation methods for wheel–switch rail normal and tangential contact" *Proceedings of the Institution of Mechanical Engineers, Part F: Journal of Rail and Rapid Transit*, vol. 231, no.2, pp.148-161.

Wang, P., Xu, J., Xie, K., & Chen, R. 2016, "Numerical simulation of rail profiles evolution in the switch panel of a railway turnout" *Wear*, vol.366, pp.105-115.

Xu, J., Wang, P., Wang, L., & Chen, R. 2016, "ffects of profile wear on wheel–rail contact conditions and dynamic interaction of vehicle and turnout" *Advances in Mechanical Engineering*, vol.8, no.1, pp.1-8.

Clark, R. 2004, "Rail flaw detection: overview and needs for future developments" *NDT & E International*, vol.37, no.2, pp.111-118.

Al-Dossary, S., Hamzah, R. R., & Mba, D. 2009, "Observations of changes in acoustic emission waveform for varying seeded defect sizes in a rolling element bearing" *Applied acoustics*, vol.70, no.1, pp.58-81.

Zumpano, G., & Meo, M. 2006, "A new damage detection technique based on wave propagation for rails" *International journal of solids and structures*, vol.43, no.5, pp.1023-1046.

Bollas, K., Papasalouros, D., Kourousis, D., & Anastasopoulos, A. 2010, "Acoustic emission inspection of rail wheels" *Journal of Acoustic Emission*, vol.28, pp.215-228.

Bruzelius, K., & Mba, D. 2004, "An initial investigation on the potential applicability of Acoustic Emission to rail track fault detection" *NDT & E International*, vol.37, no.7, pp.507-516.

Hill, R., Forsyth, S. A., & Macey, P. 2004, "Finite element modelling of ultrasound, with reference to transducers and AE waves" *Ultrasonics*, vol.42, no.1, pp. 253-258.

Bartoli, I., di Scalea, F. L., Fateh, M., & Viola, E. 2005, "Modeling guided wave propagation with application to the long-range defect detection in railroad tracks" *Ndt & E International*, vol.38, no.5, pp. 325-334.

Zhang, X., Feng, N., Wang, Y., & Shen, Y. 2014, "An analysis of the simulated acoustic emission sources with different propagation distances, types and depths for rail defect detection" *Applied Acoustics*, vol.86, pp. 80-88.

Zhang, X., Feng, N., Wang, Y., & Shen, Y. 2015, "Acoustic emission detection of rail defect based on wavelet transform and Shannon entropy" *Journal of Sound and Vibration*, vol.339, pp.419-432.

The 8th International Conference on
Structural Health Monitoring of Intelligent Infrastructure
Brisbane, Australia | 5-8 December 2017

Construction of Stability Monitoring and Evaluation System for High-speed Railway Roadbed in Cold Regions

Yuzhi Zhang[1], Baochen Sun[1], Weigang Zhao[1], An Wen [1].

[1] *The Key Laboratory for Health Monitoring and Control of Large Structures in Hebei Province, Shijiazhuang Tiedao University, Shijiazhuang, Hebei 050043,China, Email:* donnastin1981@126.com (Y-Z.Zhang)

Abstract

Along with the development of economy, the high-graded transportation line project are extending towards the cold regions gradually. The high-speed railway (HSR) in cold regions requires extreme high stability, thus the rigorous enigineering demands challenge the roadbed stability severely. The construction of the automatic monitoring and evaluation system of the HSR roadbed in cold regions, to master the safety conditions in real time, will guarantee the safe operation of high-speed train strongly. Through the characteristics analysis of the HSR roadbed in cold regions, the monitoring indices, static deformation combined with dynamic responses, are determined. The layout plan of monitoring points, which integrates the transverse discontinuity with horizontal continuity, is put forward. On the basis of conquering the continous high-precision collection sensors and technology suitable for the severe environment in cold regions, the automatic continous monitoring system of the HSR roadbed is developed, and the evaluation method of roadbed stability based on the on-site data information mining and numerical model is built. The monitoring system and stability method have realized the continous monitoring and stability evaluation of the Harbin-Dalian HSR under the water-heat-load combined action. This will be conductive to provide the scientific support for the stability change mechanism, the operation and maintenance of the HSR roadbed in cold regions.

1. Introduction

The area of frozen ground occupies 50% of the land all over the world. Along with the rapid development of railway in China, and the implementation of One Belt and One Road national strategy, many new HSRs are located in severe cold regions. Harbin-Dalian, Harbin-Qiqihar and Shenyang-Dandong HSRs have been constructed and operated successively. Harbin-Jiamusi and Beijing-Shenyang HSRs are being constructed at present. As an important part of Eurasian high-speed transportation corridor, Moscow-Kazan HSR in Russia has become the priority project of China-Russia cooperation. Moscow-Kazan HSR will pass through the Siberian region located in sub frigid zone which has cold weather and very long winter, thus may challenge the HSR design and construction severely.

Although the roadbed construction in cold regions has considered the influence of the frozen ground on the roadbed stability and some relevant engineering measures have been adopted, the operation experiences of railways in frozen soil regions at home and abroad have indicated that the various unexpected troubles and park accidents would occur usually just because of the long-term stability of the roadbed under various natural and human factors. The mutual relations of various influence factors during the HSR freezing process in cold regions are very complicated, the action modes are changeable. Fillings of roadbed endure the cyclic freezing-thawing effect, different

degree damages will occur, thus the conditions of the water infiltration, migration path and the charateristics of water retention will change gradually, and then the frost heaving charateristics will certainly change accordingly (Xu, Wang & Zhang, 2001). But for the HSR roadbed which need high smoothness and stability, the deformation of roadbed overrun will threaten the operation safety severely. Therefore, the research of the condition monitoring and stability evaluation technology of the HSR roadbed will be very important for the roadbed stability maintenances and the guarantee of operation safety. This will be beneficial for the international competitiveness strengthening of Chinese HSR techniques, practically and theoretically.
In this paper, a complete technical and theoretical method of long-term condition monitoring and stability evaluation for the HSR roadbed in cold regions is constructed. The long-term monitoring of Harbin-Dalian high-speed railway has been realized. Thus, the scientific support of roadbed stability rules mining, stability evaluation and development trend predication based on long-term and massive monitoring data is provided.

2. Stability indices and layout plan of the HSR roadbed in cold regions

2.1 Stability characteristics of the HSR roadbed in cold regions

As a new engineering, the HSR roadbed in cold regions has both the frozen ground railway and the HSR characteristics. Based on the research of its performance characteristics, the stability monitoring indices, and then the layout plan, are determined. This is the premise of the long-term monitoring which could provide sufficient and essential information for the roadbed stability evaluation.

2.1.1 Stability characteristics of the ballastless track

The long-term stability and durablity of the roadbed structure should have high properties and applicability because of the ballastless track structure characteristics and the high-speed driving environment. The requirements of the HSR ballstless roadbed is mainly the severe control of the post-construction settlement under the static loads of the roadbed and the foundation and the extral settlement under the long-term dynamic load. According to "Code for Design of High-speed Railway"(TB10321-2014, China) and the relevant research (Hu&Li, 2010), the main stability indices are shown in Table.1.

Table 1 the main stability indices of the ballastless HSR roadbed

Content		Limitation value
Settlement	Post-construction settlement	≤15mm generally
	Differential settlement	≯5mm, the bending angel ≯1/1000
	Additional settlement	≤5mm
Dynamic stability	Critical dynamic stress (CDS)	roadbed surface CDS ≤100kPa; roadbed CDS under the top layer ≤50kPa
	elastic settlement	3.5mm

The settlement monitoring data of the roadbed surface can indicate the roadbed settlement with the time in the late course, thus it could be used to investigate the roadbed static and dynamic stability. The soil stress could be used to analyse the dynamic responses of the roadbed under the train load.

In general, the long-term monitoring indices of the HSR are the settlement of the roadbed surface and the soil stress, which should be put in the relevant position during the construction phase and monitored with the high-precision sensors.

2.1.2 Roadbed stability characteristics in the seasonally frozen regions

The thermal stability and mechanical stability analyses are both very important for the roadbed in frozen soil regions. For the roadbed in permafrost region, the thermal stability is a bit more concerned (Zhang, 2015), but for the HSR roadbed in severe cold regions, whether the emphasis should be put on the thermal stability remains to be further analysed.

The largest challenge of the roadbed brought by the construction of the HSR ballastless track in seasonally frozen regions is how to take effective anti-freezing measures to control its subsequent settlement deformation and ensure its long-term dynamic stability. Due to the intervention of the temperature and the existence of water, and under the influence of phase dramatic change of physical states between solid and liquid caused by the alternation and fluctuation of the positive and negative temperature in soil, a series of special changes have taken place in the engineering properties of roadbed soil and its base layer, which makes the properties and the evaluation method of the roadbed stability in frozen soil regions distinct from those in other regions. The difference of the HSR roadbed between seasonally frozen regions and other regions is that the subsequent settlement of the former always includes frost heaving and thaw settlement caused by seasonal active layer. This deformation is closely related to congenital conditions of the roadbed and foundation, soil temperature and moisture changes, occurs with the alternation of the positive and negative temperature of the ambient environment, and exists throughout all the freezing and thawing cycles from the construction to the operation stage.

Because the deformation of HSR foundation soil consolidated by CFG piles is the compaction deformation mainly, analysis of the HSR roadbed in seasonally frozen regions should be focused on the effects of the seasonal active layer which influences the roadbed stability under the effects of external climate and train load. This is different from the railways in permafrost regions, whose research emphasis is put on the effect of variation of the permafrost upper limit acting on the roadbed stability.

In conclusion, the monitoring indices and technical demands of the HSR roadbed in cold regions are put forward, as shown in table 2.

Table 2 the monitoring indices and monitoring technical demands of the HSR roadbed in cold regions

Monitoring contents	controlling value	measuring range	Testing accuracy
Ground temperature	——	−40℃～+60℃	±0.03℃ in the range of −20℃～+20℃
soil pressure	——	0～300Kpa	±1%FS
moisture content	——	0～100%	±3%
Settlement deformation (including settlement of the roadbed surface and the transition section)	Settlement after construction≤15mm; Additional deformation≤5mm; Bending angle≤1.0×10^{-3}rad	±120mm	±0.2mm

2.1.3 Layout plan of the monitoring points

Considering the geometric environment characteristics and the design of anti-freezing measures of the HSR roadbed in cold regions, a spatial layout plan, including the transverse section and the longitudinal continuity, for the monitoring points of multi fields is designed. The monitoring contents of the roadbed section are the roadbed surface settlement deformation, the ground temperature, the soil stress and water content, four types of monitoring items altogether. In the range of the transition section of different structural types, from the edge of bridge (culvert) frames and along the direction of the railway line, measuring points should be laid every 5m on the roadbed surface of shoulders, forming longitudinal settlement monitoring line and realizing multiparameter tridimensional monitoring. The layout plan of measuring points is shown in figure 1(Zhang, 2015; Sun, Zhang, et al, 2016).

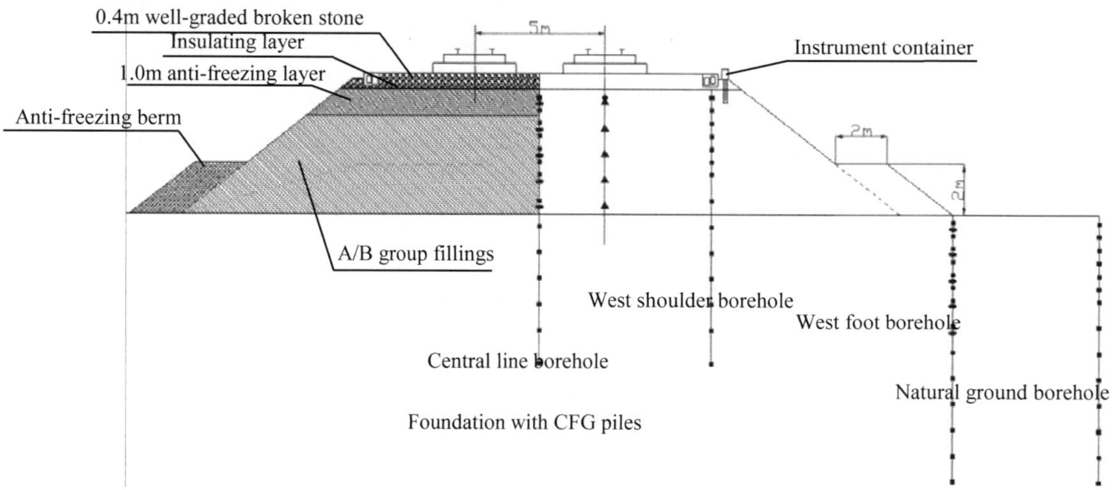

(a) Geometry of the section and positions of the monitoring points

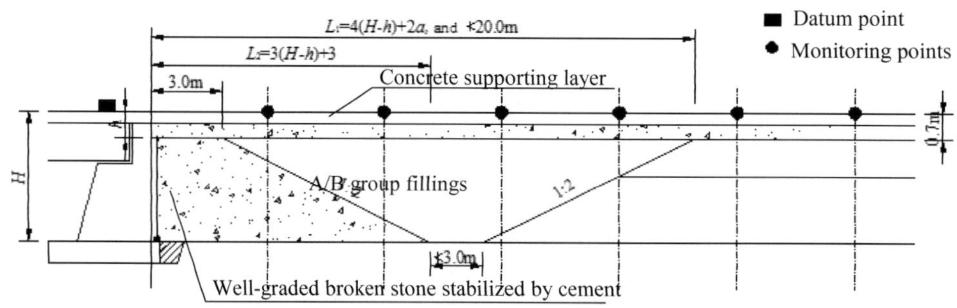

(b) Geometry of the transition section and positions of the monitoring points

Fig.1 Layout plan of monitoring points

3. Establishment of Long-term Automatic Monitoring System for the HSR Stability in Cold Region

At present, the great attention has been paid to the construction monitoring and long-term operation monitoring technology research of railway roadbed all over the world, and a variety of monitoring techniques and instruments have been developed. But there still lacks the technical regulations of the long-term automatic monitoring suitable for the cold climate, low temperature and high-speed conditions. At the same time, the collection instruments are mainly for the laboratory testing, on-site short-term testing. It is difficult for the long-term monitoring under the severe weather, and the testing methods for a single parameter are more than those for multi-parameter centralized testing, thus it cannot realize the long-term, real-time, remote and on-line monitoring; the stability and life-span of the testing instruments are difficult to satisfy the long-term monitoring. Therefore, it is urgent to develop a long-term automatic monitoring technology for multi-parameter and multi-monitoring points of the HSR roadbed condition in severe cold regions so as to improve the real-time and automatic degree of the roadbed monitoring, and then reveal the roadbed health conditions timely, comprehensively and accurately.

The monitoring plan can comprehensively monitor the roadbed stability evaluation indices and its influencing factors. Because the numbers of monitoring variables and measuring points are large, the accuracy and stability requirement is strict; the monitoring system should have the low power consumption, high precision, good reliability, high and low temperatures adaptability and overall coordination performance, and economic requirements as well.

Consequently, a set of long-term automatic monitoring system made up of several on-site monitoring stations and the monitoring center station, as shown in Figure 2. The on-site stations collect various monitoring data automatically and transmit them to the monitoring center through the wireless network. The public network is used to transmit the monitoring data and controlling orders between the on-site monitoring stations and the monitoring center. The system integrates the relatively independent on-site stations, and using the ARM9 core control equipment (CCE) developed in the Windows CE environment. Each station also contains the sensor system, RTT long distance transmission data acquisition terminal (LDTDAT), solar panels, batteries, equipment protection box and other components, as shown in Figure 3(Sun, Zhang, Li & Zhao, 2015).

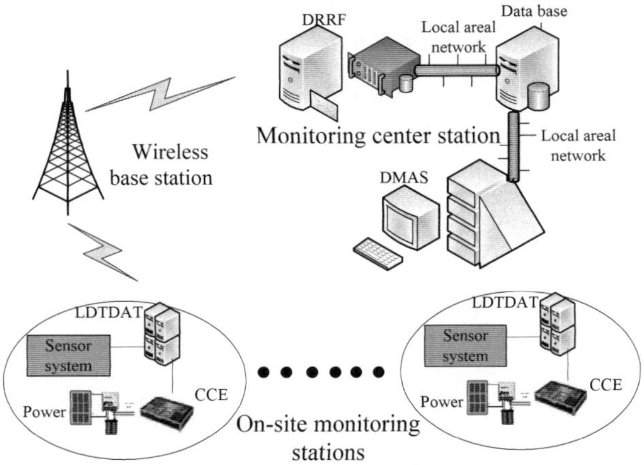

Figure 3 the monitoring system design

The monitoring center functions are implemented through three parts of hardware and software: data reception and resolution facilities (DRRF), monitoring center (Data base), data management and analysis software (DMAS). The on-site station management and monitoring data analysis functions have been designed in the software, by setting different operation rights to distinguish them in order to manage versions and upgrade software easily.

4. Stability Evaluation Method of the HSR roadbed in Cold Regions

The establishment of the long-term monitoring system of the HSR roadbed in the cold regions will inevitably bring about the real-time analysis and processing of the massive monitoring data, the evaluation and prediction of the roadbed stability. The roadbed in cold regions exchanges energy with the outside world. It is a multi-coupled uncertain system of the temperature field, water field, stress field and deformation field. It has the characteristics of the complex huge system, but the on-site monitoring data are usually massive and intermittent. Therefore, more attention should be paid to the on-site monitoring, it is effective to select the appropriate statistical analysis model for the data mining, and then establish a numerical model for the roadbed stability analysis.

4.1 Roadbed temperature estimation model of the cold regions based on the long-term monitoring data varying with time

Based on the long-term temperature monitoring data of the HSR roadbed, considering the average temperature, the temperature amplitude, soil properties and phase difference, in order to overcome the insufficiency of the monitoring data dispersion and partial data missing, the empirical method is adopted to obtain the temperature estimation formula of the roadbed different locations varying with the depth and time continuously.

Taking the phase difference influence into account, and supposing that the soil is homogeneous with immutable state, for a certain roadbed section at a certain region, if the transverse position *x* is determined, $T(z,t)$ at time t and depth z could be estimated as follows (Zhang, Du & Sun, 2015):

$$T(z,t) = T_m + A_s \exp\left(-z\sqrt{\pi/a_u p}\right)\sin\left(2\pi t/p + \varphi - z\sqrt{\pi/a_u p}\right) \tag{1}$$

Where T_m is the annual average temperature at depth *z*, the unit is degree centigrade, °C; A_s is the annual temperature amplitude of roadbed surface; *z* is the depth from the roadbed surface; a_u is the average soil thermal diffusion coefficient ignoring the earth interior heat flux, the unit is m^2/d; p is the vibration period, 365 days; φ is the initial phase angle; *t* is the interval from the starting time, the unit is day; $z\sqrt{\pi/a_u p}$ is the phase lag at the depth *z* from the surface.

We took the average amplitudes of three freezing-thawing cycles as original series to obtain the temperature amplitude functions of the roadbed different parts. According to the time difference between reaching the maximum temperature and starting time, the phase lags were obtained. On the basis of the above work, temperature estimation formulas of different sections could be derived.

4.2 Roadbed temperature field model and analysis model coupling temperature with deformation field

Induced the enthalpy method, the governing differential equation of temperature field is as shown as the equation (2). The method, combing the finite element in spatial domain with finite difference in time domain, is adopted to simulate the temperature field. The results indicate that: the cumulative heat quality of roadbed and the ground under the roadbed during the construction may dissipate ten years later, and then the new thermal equilibrium state will be formed. There will be relatively stable seasonal frost layer and temperature, asymmetric temperature field in the roadbed at last, as shown as Figure 4 (Zhang, Du & Sun, 2014).

$$\frac{\partial H}{\partial t} = \frac{\partial}{\partial x}\left(\lambda\frac{\partial T}{\partial x}\right) + \frac{\partial}{\partial y}\left(\lambda\frac{\partial T}{\partial y}\right) \tag{2}$$

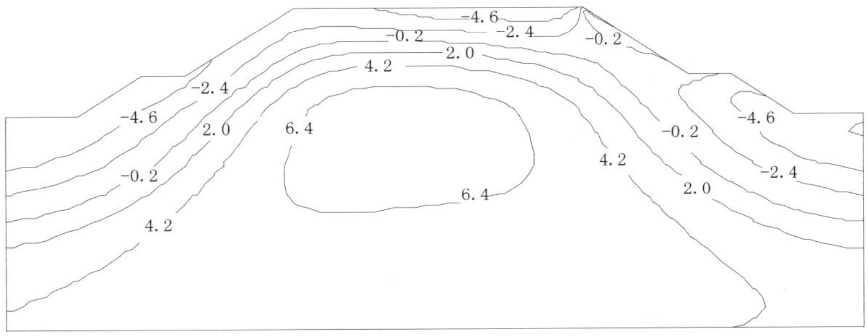

Figure 4 the temperature field at 15th March 2021

Because the indirect coupling model generally adopted cannot satisfy the strict demands of the stability analysis of the HSR roadbed, based on the consideration of the influence of additional stress in phase transformation district, combined with the actual restrict conditions of numerical model, the relation model between the thermal elastic mechanics and the deformation characterization factor of frozen ground is built as shown as the equation (3). Then the governing equation of deformation and stress for the frozen ground is built and the stress and deformation fields are solved at the basis of temperature field, and thus the continuous coupling calculation of temperature and deformation field is realized (Zhang, Du, Sun, Zhang & Han, 2014; Zhang, Zhao, Sun & Bei, 2016).

$$\alpha=\begin{cases} 0 & T>T_u \\ \dfrac{(1-\mu)(T-T_u)}{(1+\mu)(T_l-T_u)}\eta & T_l \le T \le T_u \\ 0 & T<T_l \end{cases} \tag{3}$$

The distribution and evolution laws of roadbed deformation and stress of the HDHSR typical section indicated that: the increase of frost heaving ratio of roadbed filling will make the deformation of roadbed surface larger and the transverse uneven deformation more evident. And as the freezing time continues and the frost depth increases, the uneven deformation will be more evident. The phase transformation action is the direct factor resulted to the roadbed diseases. The longitudinal crack may occur at the roadbed shoulder and the roadbed slope.

4.3 Stability evaluation method of the HSR roadbed in cold regions

In this paper, a roadbed stability evaluation method based on the long-term time varying monitoring data interactive with the theoretical model is proposed. On the basis of the design codes and construction technology standards, the relevant structure parameters in the initial operation phase and the long-term time varying monitoring data of the train load and the environment information are considered. Then the measured variation ranges corresponding to the normal conditions of roadbed structure are determined. These are the evaluation standards of the initial healthy condition parameters of the roadbed structure. Through the temperature monitoring data excavation during a certain operation phase, the actual temperature boundary restriction conditions of the thermal-mechanical model are put forward, then the condition parameters of the HSR roadbed settlements are calculated, the model results are modified based on the monitoring data. Thus, the evaluation standards varied with time on line and the condition evaluation forewarning realized based on the above monitoring data interactive with the theoretical model could be attained. The engineering practices have proved that this method could analyze and evaluate the thermal and mechanical stability of the HSR roadbed in cold regions.

5 Conclusions

(1) The stability monitoring indices, including the settlement, soil stress, soil temperature and water content, and the multifield spatial monitoring point's layout optimization for the HSR roadbed are first put forward. The condition monitoring technological requirements and the indices system is built then.

(2) The long-term automatic monitoring system of the HSR roadbed in cold regions is constructed, and the multiparameter integrated automatic collection, analysis and management have been realized.

(3) The temperature estimation model including the engineering environment and soil properties of the roadbed in frozen regions is established, and the massive monitoring data could be analyzed, excavated and utilized in real time.

(4) The coupled simulation model of the temperature and deformation field of the HSR roadbed is built. The roadbed stability evaluation method based on the long-term time varying monitoring data interactive with theoretical model is put forward, which realizes the precast and evaluation of the roadbed thermal and mechanical stability.

References

Xu, Xiaozu, Wang, Jiacheng & Zhang, Lixin 2001, *Physics of frozen ground, Science Publishing Company*, Beijing, China

Hu, Yifeng, Li, Nufang 2010, *Design principle of high-speed ballastless railway roadbed*, Chinese Railway Publishing Company, Beijing, China

Zhang, Yuzhi 2015, Study on the stability of the high-speed railway roadbed in deeply seasonally frozen region, doctoral thesis, Beijing Jiaotong University, Beijing, China

Sun, Baochen, Zhang Yuzhi, Li Jianzhi, Zhao Weigang 2015, Long term monitoring of high-speed railway roadbed in cold regions, *Journal of Railway Engineering Science*, vol.32, no 1, pp. 22-26.

Zhang, Yuzhi, Du, Yanliang, Sun, Baochen 2015, Temperature distribution analysis of high-speed railway roadbed in seasonally frozen regions based on empirical model, *Cold Regions Science and Technology*, vol. 114, pp. 61-72.

Zhang, Yuzhi, Du, Yanliang, Sun, Baochen 2014, Temperature distribution in roadbed of high-speed railway in seasonally frozen regions, *Chinese Journal of Rock Mechanics and Engineering*, vol. 33, no 6, pp.1286-1295.

Zhang, Yuzhi, Du, Yanliang, Sun, Baochen, Zhang Shaopeng, Han Jing 2014, Roadbed deformation of high-speed railway roadbed due to freezing-thawing process in seasonally frozen regions, *Chinese Journal of Rock Mechanics and Engineering*, vol. 33, no 12, pp. 2546-2553.

Zhang, Yuzhi, Zhao Weigang, Sun, Baochen, Bei Jianghui 2016, Dynamic stability of high-speed railway roadbed in cold regions subjected to train loading, *Chinese journal of underspace and engineering*, vol 12, supp 2, pp.882-888

The 8th International Conference on
Structural Health Monitoring of Intelligent Infrastructure
Brisbane, Australia | 5-8 December 2017

Effective Intrusion Detection for Railway Scene Based on Video Analytics

Y.D. Li [1,2], Y.B. Lin [1], C. Zhang [1], W.G. Zhao [2,3]*

[1] *School of Electronic and Information Engineering, North China University of Technology, Beijing, 100041, China*
[2] *The Key Laboratory for Health Monitoring and Control of Large Structures, Shijiazhuang, 050043, China*
[3] *Institute of Structure Health Monitoring and Control, Shijiazhuang Tiedao University, Shijiazhuang, 050043, China*
E-Mail: zhaoweig2002@163.com

Abstract

Intrusion detection for railway scene is the essential method to avoid railway accidents caused by the invasions of pedestrians, falling rocks, animals, etc. In this paper, an automatic computer-vision based scheme is proposed to detect abnormal events in the track area, and send pre-warning signals to the control center. First, background is estimated by Gaussian mixture models, and the targets are obtained from the difference image between foreground and background combining with threshold, morphological filtering and connected component labeling. Second, mean shift algorithm is used to track the invasive target. To improve the tracking robustness, Histogram of Oriented Gradient (HOG) is chosen as the feature space to establish the target model, which has strong discriminative ability and robustness over illumination variation and small distortion of object. Third, behavior analysis is conducted based on target trajectory to determine whether there exist abnormal events in the track area. If yes, alarm messages with high-definition pictures are sent to control center. The proposed method has been operated on K141 spot of Shuohuang railway. The practical running results show that the intrusion detection system is effective and robust in railway scene.

1. Introduction

Safety is one of the most concerned issues in railway transportation, among which intrusion detection is a fundamental research focus. Presently, the technology of intrusion detection can be categorized into several methods: infrared based, microwave based, laser based and vision based methods(Wang, Shi & Wang, 2009). Among these methods, the microwave technique is sensitive to environmental changes. Guo proposed a laser based scheme in which a 2D laser scanner is used to construct 3D laser curtains to detect intrusion in 3D space(Guo, Zhu & Shi, 2012), but the performance is degraded due to the factors of snowing, raining and fog. Gao proposed a railway fence monitoring system based on Brillouin Optical Time Doman Reflectometer (BOTDR)(Gao, Wang & Li, 2011), which overcomes the drawbacks of traditional perimeter intrusion sensors. Catalano proposed an integrated optical fiber sensing network, composed of FBGs strain sensors, accelerometers, to protect from the rail track accession (Catalano, Bruno & Pisco, 2015).

In the past years, researchers have also expressed interests in intelligent video analysis since vision based method has several advantages of wide view, convenient installation, and large coverage. Shi proposed an automatic intrusion detection system based on machine vision and embedded system, in which the hardware of image acquisition and processing is realized with FPGA and ARM processor(Shi , Chai & Wang, 2015). A fast background updating algorithm based on foreground object distribution is proposed by Guo(Guo ， Yang & Shi , 2016). Besides, Teng proposed a vision railway detection method based on super-pixels rather than pixels(Teng, Liu & Zhang, 2016). Yu came up with a method to detect railway intrusion based on binocular stereo vision(Yu & He ,2011). These vision detection methods have shown some unstable and low precision results in railway scene. In order to overcome the drawbacks, this paper presents a stable and high-precision intrusion detection system based on intelligent video analysis.

2. System scheme design

The workflow of the application software primarily contains the followings. First, original image data are captured from the cameras. Second, image data are passed through the video analysis module for processing. Finally, an alarm signal is sent according to the analysis results(Li, Sun & Ai, 2014).
The key component of the application, i.e., the analysis algorithm, consists of three parts: target extraction, target tracking, and behavior analysis. These parts are described in Sections 3 to 5. The system diagram is shown as Fig.1. In practice, Haikang DS-2CD6233 cameras are used to grab the original image, and industrial PCs are used to conduct the analysis algorithm. An alarm signal is sent by 3G wireless network to the server of control center once an abnormal event is detected.

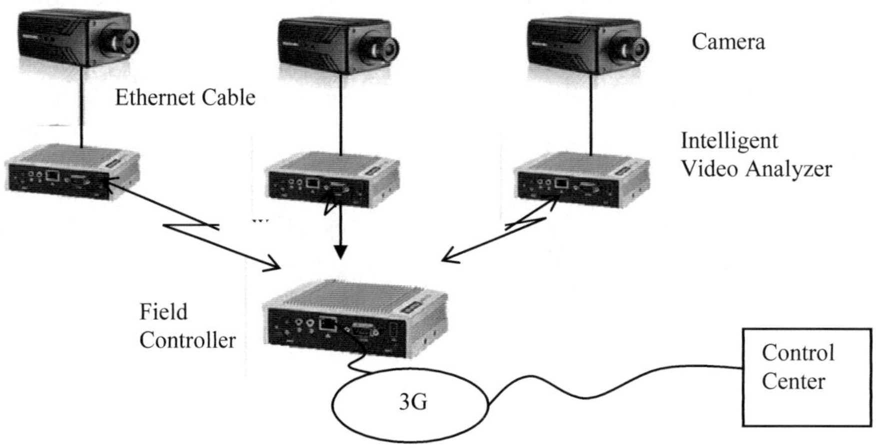

Fig.1 Diagram of railway intrusion detection system.

3. Target extraction

Target extraction is the fundamental unit of the proposed video analysis algorithm. The popular target extraction frameworks include optical flow, frame difference, and background subtraction(Cao, Yun & Zhong, 2016). However, optical flow's computation is too complex, and the concept behind frame difference is simple but its effect is not satisfactory in practice. The background subtraction method divides the scene into background and foreground. The background can be updated to overcome light changes and other disturbances. Among background subtraction methods, the mixture Gaussian models (GMMs) demonstrate the best performance.

3.1 GMMs

The basic principle of GMMs is that a single Gaussian distribution centered at the mean pixel value can describe the values of each pixel with the same location in a video sequence. If the scene remains unchanged, the pixels in the subsequent images will have a high probability of belonging to the background. Otherwise, it may be treated as a foreground point. Thus, we can distinguish foreground points using a suitable threshold value. However, if an object in the scene has multiple surfaces, such as shadows or flickering leaves, a single Gaussian distribution is not enough to describe the distribution. Multiple Gaussian modes are preferred for processing the complex scene.

3.1.1 Principle of mixed Gauss models

The recent history of each pixel, $\{X_1, X_2, ..., X_t\}$, is modeled by a mixture of *K* Gaussian distributions. The probability of the current pixel is defined as:

$$P(X_t) = \sum_{i=1}^{K} \omega_{i,t} * \eta(X_t, \mu_{i,t}, \Sigma_{i,t}) \quad (1)$$

where *K* is the number of GMM, $\omega_{i,t}$ is the weight of ith GMM, and $\mu_{i,t}$ is the mean value of ith GMM at time *t*. $\Sigma_{i,t}$ is the covariance matrix of ith GMM at time *t*. η is the Gaussian probability density function, which is defined as:

$$\eta(X_t, \mu_{i,t}, \Sigma_{i,t}) = \frac{1}{\sqrt{2\pi}|\Sigma|^{\frac{1}{2}}} e^{-\frac{1}{2}(X_t-\mu_t)^T \Sigma^{-1} (X_t-\mu_t)} \quad (2)$$

To decrease computational complexity, let $\Sigma_{k,t} = \sigma_k^{\ 2} I$. The current pixel is tested to determine whether it matches any of the GMMs according to formula (3).

$$\begin{cases} background: & |X_t - \mu_{i,t}| < \lambda\sigma_{i,t} \\ foreground: & |X_t - \mu_{i,t}| \geq \lambda\sigma_{i,t} \end{cases} \quad (3)$$

Then, the weight, mean value, and the covariance of each GMM are updated according to the match result. If the current pixel matches a GMM, then the parameters of the GMM are updated as formulas (4) to (6).

$$\omega_{i,t} = (1-\alpha)\omega_{i,t-1} + \alpha \quad (4)$$

$$\mu_{i,t} = (1-\beta)\mu_{i,t-1} + \beta \quad (5)$$

$$\sigma^2{}_{i,t} = (1-\beta)\sigma^2{}_{i,t-1} + \beta(X_t - \mu_t)^T (X_t - \mu_t) \quad (6)$$

If a GMM is not matched, its weight is updated using formula (7). However, its mean and covariance remain unchanged.

$$\omega_{i,t} = (1-\alpha)\omega_{i,t-1}, \quad (7)$$

where α is the learning factor of the model, $\beta = \alpha\eta(X_t, \mu_{i,t}, \Sigma_{i,t})$.

The GMMs are reordered according to their priority, and only the first *B* distributions are chosen as the background model, where:

$$B = \arg\min_b \left(\sum_{k=1}^{b} \omega_k > T \right) \quad (8)$$

3.1.2 Implementing the Algorithm of GMMs

In this section, detailed steps are presented to describe the target extraction algorithm. Three GMMs are used to estimate the background in our scheme. The procedure can be divided into four

parts: model initialization, model match, model update and target segment.
Step 1: Each pixel is assigned a data structure that includes mean, variance, and the three GMMs. YUV color space is used, and only luminance is selected to establish the model.
Step 2: The parameters of the first GMM are initialized. Let μ_1 = pixel value of first frame, $\omega_1 = 1$, and $\sigma_1 = 20$. The parameters of the other two GMMs are set to zero.
Step 3: The first *B* GMMs according to formula (8) are obtained, where *T*=0.75. The current pixel value is x_i. $Y = \{y_1, y_2, ..., y_n\}$ represents the match results of each pixel, and *n* is the number of pixels. The value of *Y* is determined as follows:

$$y_i = \begin{cases} 255, & |x_t - \mu_i| \geq \lambda\sigma_i \\ 0, & |x_t - \mu_i| < \lambda\sigma_i \end{cases} \quad (9)$$

where $\lambda = 2.5$.
Step 4: If no GMM is matched in Step 3, then the GMM with the least priority is replaced with a new GMM. The mean of the new GMM is set to x_i. Weight is set to an extremely small value (0.05), and its variance is set to a large value (20).
Step 5: The parameters are updated according to formulas (4) to (7), where $\alpha = 0.05$.
Step 6: The weights of the three GMMs are normalized according to formula (10) as follows:

$$\omega_i = \frac{\omega_i}{\sum_{i=1}^{K} \omega_i} \quad (10)$$

Step 7: The GMMs are reordered with respect to their priority ω_i / σ_i.
Step 8: Morphology filtering on binary image *Y* is carried out. All the foreground points are grouped using a connected component operation. Thus, we can set the rectangle around the target.
Step 9: The next frame image is read, and the procedures are repeated starting from Step 3.
The GMMs are trained enough times to estimate the background. We found that the scene must remain unchanged for at least 1 minute before the target appears in the scene.

3.2 The connected component labelling (CCL)

Target extraction is not a simple extraction of each foreground pixel, also need to complete given the same mark on the same target. In order to connect each foreground pixel to form a region, and the region marking information presents each target, we need to connect region labeling. Commonly there are region connecting by four connected region labeling and eight connected region labeling, shown in Fig 2 and 3 respectively.

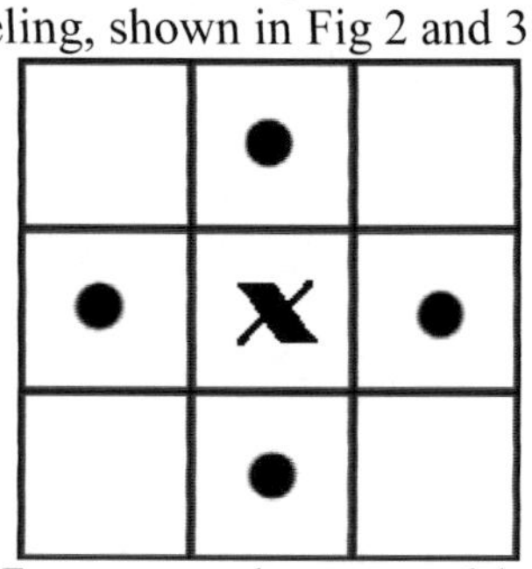

Fig.2 Four connected component labeling.

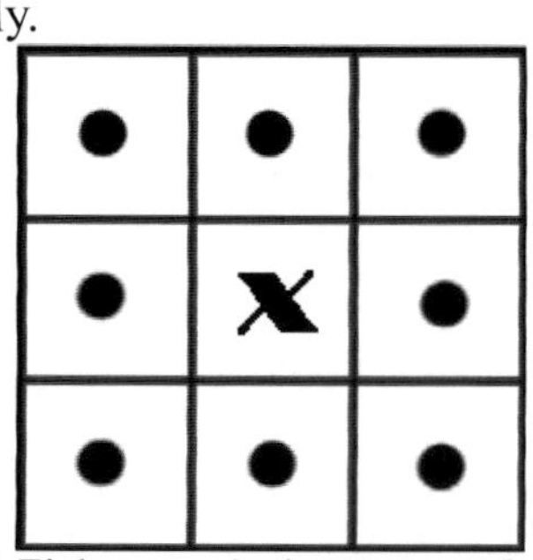

Fig.3 Eight connected component labeling.

The main difference is that the eight connected region mark contains the upper left, the lower left, the upper right and the lower right. For a pixel in an image, the eight connected region can combine

more regions together to form a new region. In this paper, eight connected region labeling is adopted, and the algorithm is implemented as follows:
(1) Scanning the binary image line by line, then we make a sequence of consecutive white pixels in each row that called a run, and record its starting point , end point, and its line number.
(2) If all the run groups except the first row are not overlapped area comparing with previous row, giving it a new label; if it is only one regional area overlapped, the label of the previous row's run will be assigned to it; if it is overlapped area with more than 2 clusters on the line, a connected group minimum label will be assigned to the current run, then it will mark this several runs into equivalence pair, indicating that they belong to a same class.
(3) The equivalence pairs are transformed into equivalent sequences, and each equivalent sequence needs to be given the same label because they are all equivalent. Then giving each equivalent sequence a label starting from (1).
(4) Traversing all the runs , and searching for all equivalent sequences, then giving them new label, and making sure each run's label will be filled into label image.
(5) All targets will be labeled, then drawing the rectangle box on the left, right, top, and bottom of the mark to indicate the boundary of the target.

4. Mean Shift Tracking Using Histogram of Oriented Gradient (HOG)

Target tracking is an essential part of the railway intrusion detection algorithm, which is the basis of behavior analysis. Particle filter and mean shift are the most popular framework algorithms among various tracking algorithms. However, the particle filter method cannot satisfy real-time demands because of its heavy computational payload. Mean shift is widely used because of its non-parametric feature and low computational load. However, tracking stability is affected by factors such as the distinguish capability of the feature space and size change. The distinguish capability of the feature space is the key factor in mean-shift tracking. Color feature space is used in classic mean shift(Jiang & Zhou, 2012). However, disturbances from background objects and light changes have significant effects on it, thus possibly resulting in unstable tracking. To overcome the shortcomings of color space, numerous studies have focused on fusing multiple features by establishing joint histograms of color and other features(Lu, Lei & Hao, 2010; Liu, Tang & Zhao, 2010). A robust mean-shift tracking scheme using a new type of feature space is proposed in our intrusion detector. We use HOG as the feature space instead of RGB histograms. HOG is a feature descriptor based on gradient proposed by Dalal for pedestrian detection(Dalal & Triggs, 2005). HOG is reliable even with illumination changes and slight object deformation.

4.1 Mean Shift tracking with HOG

If HOG is chosen as the feature space, then the HOG of the target is defined as

$$q_u = C_h \sum_{i=1}^{n} M(x_i) k(\| \frac{x_i - y}{h} \|^2) \delta[b(x_i) - u], \tag{11}$$

where $u = 1,...,m$ (m is number of quantizing intervals for the gradient angle), coefficient C_h is the normalization constant, and h is kernel bandwidth. The gradient magnitude of each pixel is defined as

$$M(x_i) = \sqrt{m^2{}_x(x_i) + m^2{}_y(x_i)}, \tag{12}$$

where $m_x(x_i)$ and $m_y(x_i)$ are the horizontal and vertical gradients of the current pixel, respectively. The gradient angle of x_i is defined as $\theta = \arctan(\frac{m_y(x_i)}{m_x(x_i)})$. The gradient angles are mapped into $[0,2\pi]$ and the quantizing interval is 20°, that is, m=18.

Similarly, the HOG of the candidate is defined as

$$p_u(y) = C_h \sum_{i=1}^{n_h} M(x_i) k(\| \frac{x_i - y}{h} \|^2) \delta[b(x_i) - u]. \quad (13)$$

4.2 Work Flow of Mean-Shift Tracking

The rectangle obtained through target extraction can serve as the initial tracking window. The proposed robust mean-shift tracking algorithm using HOG is described as follows.

Step 1: Histogram q of the initial tracking window is calculated.

Step 2: Histogram $p(y_0)$ of the candidate window, which is centered at y_0 in the current frame, is calculated.

Step 3: The Bhattacharyya coefficient is calculated firstly. The mean shift will fail if the Bhattacharyya coefficient is extremely small. Target extraction should be conducted again in this situation.

Step 4: The weight w_i of each pixel is calculated.

Step 5: A new position y_1 is calculated。

Step 6: while $\rho[p(y_1),q] < \rho[p(y_0),q]$, then $y_1 \Leftarrow (y_1 + y_0)/2$.

Step 7: If $\|y_1 - y_0\| < \varepsilon$, then calculation is stopped.

Step 8: $y_0 \Leftarrow y_1$, then proceed to Step 2.

5. Behavior analysis

Target extraction and tracking, which are low-level processes in video analytics, are the bases of behavior analysis. Considering the demand of railway environmental field, the target classification mainly implements to distinguish the people, vehicles and rockfalls from invading the railway tracks.

In this paper, we explore the aspect ratio principle to classify different objects. When the aspect ratio is between 1.5 and 4, it shows the target is human and denoted with '0' on the top of the rectangle of the object. If the aspect ratio is 0.5-1.5, it shows the target is car and donated with '1'. If the width and height of an object are both within a certain size, it is simply determined as a rockfall. We have developed a forbidden area as a warning area by analyzing the target's parameters. The behavior of the moving object is described by two intuitive parameters: persistence time and target center-of-mass coordinate. A rectangle is set as the boundary of forbidden area. If the persistence time of an object exceeds a preset threshold and once the moving target enters the rectangle area with its center-of-mass coordinate sustaining 10 frames in the rectangular area, then an alarm is triggered and the area boundaries will automatically change from yellow to red.

6. Experiments and discusses

In this section, the experimental results are presented to evaluate the performance of the proposed intrusion detection scheme.

(1) Test 1

First, two video sequences are used to verify the performance of the proposed algorithm. Video sequence 1 is taken from PETS2001 library, and the other — video sequence 2, is filmed in a railway scene. Fig. 4 illustrates the immediate results of the detection procedure. The left diagram corresponds to video sequence 1, and the right one is the results of video sequence 2. The mixture Gaussian models (GMMs) are used to estimate the background, and background subtraction is conducted to obtain the foreground targets. The noises are removed by morphological filtering. In order to locate targets, connected labeling is explored to obtain the rectangles surrounding the objects. Fig. 4(a) shows the results of target extraction after background estimation, filtering and connected labeling.

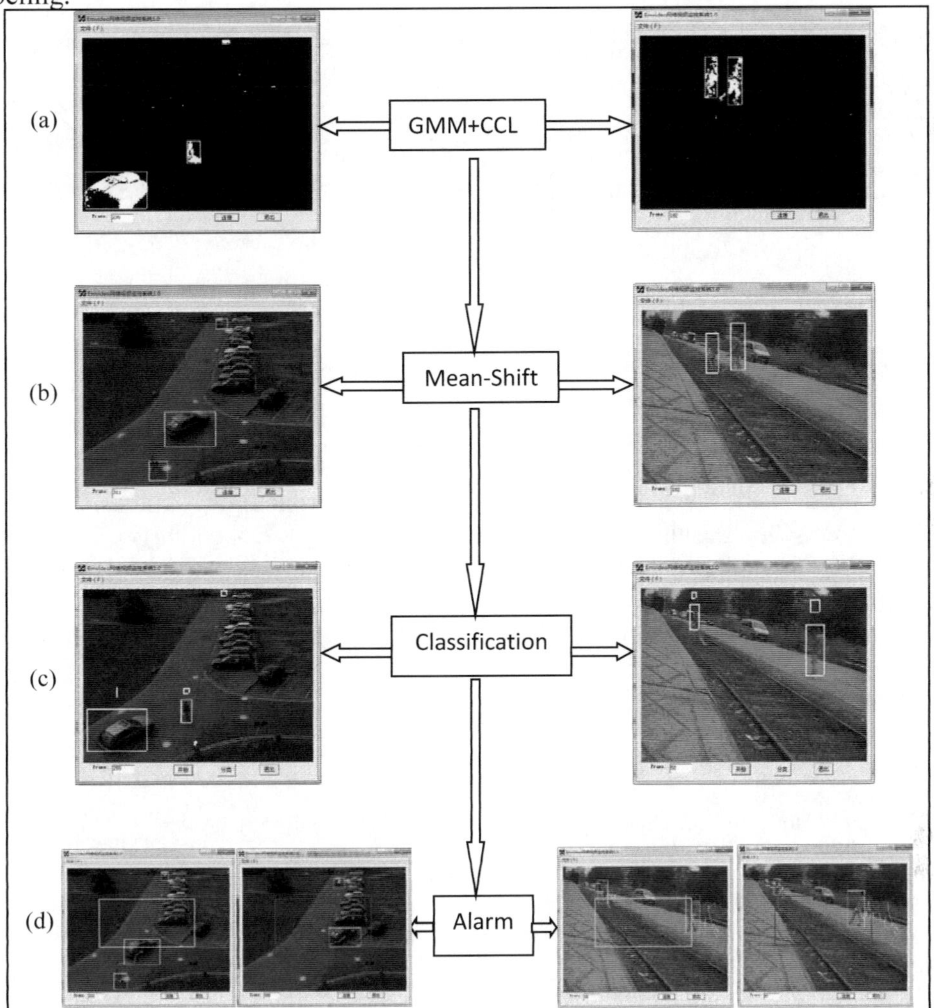

Fig. 4 Test results of video sequence 1 and 2.

Once a valid object is determined, mean shift using HOG is conducted to track the objects. Fig. 4(b) illustrates the tracking results (The forbidden area boundary is yellow and it will turn into red when intrusion is occurred.). The targets are further distinguished by persistence time. Only the targets which appear continuously over a preset period are treated as valid objects, and then classified according to aspect ratio. Fig. 4(c) is classification results. When a valid object enters into the forbidden area, an alarm is triggered and the boundary of forbidden area will change to red which is shown as Fig. 4(d).

(2) Test2

The results of test 2 are taken from the live video surveillance system under a railway environment. In Fig. 5, a person is crossing the tracks. The yellow rectangle is the preset forbidden area. The left pictures of Fig. 5(a), 5(b) and 5(c) are the binary detection results of different positions. In Fig. 5(a), the centroid of the pedestrian hasn't entered into the yellow area, and there is no alarm. But when the person entered into the warning area, the regional boundary will change to red, and trigger an alarm, which is shown in Fig 5(b). When Pedestrian is out of the tracks and far from the warning area, the regional boundary will turn to yellow again, as shown in Fig 5(c).

(a) Before crossing (b) Crossing the track area (c) Out of the track area

Fig. 5 Detection results of crossing tracks.

7. Conclusion

Video surveillance systems are becoming increasingly necessary in protecting the safety of citizens and infrastructure. In this paper, a novel railway intrusion detector using video analytics is presented. The proposed algorithm consists of three parts: target extraction, target tracking, and behavior analysis. The following processes are considered to improve robustness of detection. GMMs are used to estimate the background and the morphological filtering is used to remove noises, then the CCL algorithm is adopted to label each target. Once a target is detected, mean shift based on HOG is employed to track the target. At last, the trajectory of the target is utilized to determine if the target is abnormal and we employ simple aspect ratio for classification. The proposed intrusion detection system has been tested under an outdoor railway environment. Experiments show that the detector is effective and able to adapt to weather changes in outdoor environments.

References

Wang, Tong, Shi, Hong, Wang, Qian 2009, Research Intrusion Monitoring System for PDL, *Railway Computer Application*, Vol.18, No.7, pp. 8-10.

Guo, B, Zhu, L, Shi, H 2012, Intrusion detection algorithm for railway clearance with rapid DBSCAN clustering. *Chinese Journal of Scientific Instrument*, pp. 241-247.

Gao, C, Wang, F, Li Cl 2011,Application of BOTDR in Railway Fence Intrusion Detecting System. *Applied Mechanics Materials*, pp. 4278-4281.

Catalano A, F.A. Bruno, M. Pisco 2015, An intrusion detection system based on the optical fiber technology for the protection of railway assets. *Proceedings of the IEEE*, pp. 1-4.

Shi, Hm, Chai, H, Wang, Y 2015, Study on railway embedded detection algorithm for railway intrusion based on object recognition and tracking. *Tiedao Xuebao/journal of the China Railway Society*, pp. 58-65.

Guo, B , Yang, L , Shi, H 2016,High-speed railway clearance intrusion detection algorithm with fast background subtraction. *Chinese journal of scientific instrument*, vol 37,No 6.

Teng, Z, Liu, F & Zhang, B 2016, *Visual railway detection by super pixel based intracellular decisions*. Kluwer Academic Publishers.

Yu, Z & He, X 2011, Research on intrusion clearance detection system for high-speed railway based on Informatics. *IEEE International Conference on Service Operations* , pp. 532-536.

Li, Y, Sun, C, Ai, J 2014, A robust outdoor intrusion detector using video analytics. *Journal of Computational Information Systems*, pp. 975-982.

Cao, Y, Yun, X, Zhong, T 2016, Research on Vehicle Detection and Tracking Algorithm Based on the Methods of Frame Difference and Adaptive Background Subtraction Difference, *International Conference on Artificial Intelligence and Industrial Engineering*.

Jiang, Zb & Zhou, Sb 2012, An optimized tabu search application on video object tracking, *International Journal of Advancements in Computing Technology*, pp. 258-267.

Lu, X, Lei, H, Hao, Zb 2010, Automatic camshift tracking algorithm based on multi-feature, *Journal of Computer Applications*, pp.650-652.

Liu, Q, Tang, Lb, Zhao, Bj 2010, Infrared target tracking based on adaptive multiple features fusion and mean shift, *Journal of Electronics & Information Technology*, pp. 1137-1141.
N. Dalal & B. Triggs 2005, Histograms of oriented gradients for human detection, *Proceedings of the IEEE Computer Society Conference on Computer Vision and Pattern Recognition*, pp. 886-893.

The 8th International Conference on
Structural Health Monitoring of Intelligent Infrastructure
Brisbane, Australia | 5-8 December 2017

Effective Concrete Surface Crack Detection Based on Extreme Learning Machine

Baoxian Wang[1, 2, 3], Weigang Zhao[1, 2*], Hao Zhang[1, 2], Xinmin Wang[1], Guangyuan Zhang[1, 2, 3]
[1] *Structural Health Monitoring and Control Institute, Shijiazhuang Tiedao University, Shijiazhuang, 050043, China*
E-Mail: zhaoweig2002@163.com
[2] *Key Laboratory for Health Monitoring and Control of Large Structures, Shijiazhuang, 050043, China*
[3] *Collaborative Innovation Center of Large Infrastructure Disaster Prevention and Reduction of Hebei Province*

Abstract

Concrete surface crack detection has been an important field for the safety of civil engineering structure. However, under the background of high-density data acquiring demand, the existing crack detecting methods could not obtain the favorable efficiency and capability. In this work, we present a novel concrete surface crack detector for efficiently dealing with the massive crack images. Specifically, discrete cosine transformation is firstly utilized to extract the frequency domain characteristics of concrete surface images. Then, with the calculated image frequency features, an emergent learning technique (i.e., Extreme Learning Machine) is exploited to find the optimal separate hyper-plane between the crack object and backgrounds. Experiments on the crack images from the railway ballast-less track slab validate the effectiveness of the proposed method.

1. Introduction

With the rapid growth of information technologies, image acquisition system is used to obtain the surface defect information of concrete structure, and the non-contact image non-destructive testing method has become one hot research in the cross research of civil and information engineering (Min, 2015; Rosales, 2009). Due to the huge demand of concrete surface crack detecting, the crack detection research based on image processing has received widespread attention, and various concrete surface crack detecting methods have been proposed.

For the crack region pixel, its value variation is small, and is different from those background contents. Based on the analysis above, the crack detecting model based on the global analysis is presented. Wang *et al.* (Wang, 2010) discussed the distributions of concrete surface noise fields, and presented one image filtering based crack detecting model. Rosales *et al.* (Rosales, 2009) utilized the Sobel and Canny edge detectors for finding the bridge surface crack. Owing to the unsatisfied performance of simply adopted edge detectors, Yamaguchi *et al.* (Yamaguchi, 2006) presented the image crack detection method based on the percolation model. From the point of processing crack images globally, the references above could achieve favourable results only under some certain conditions. However, because of the non-uniform illuminations and various background clutters, the gray values for one same crack object vary widely, and the corresponding detecting results may be bad. Considering that the sparsity of crack contents in the whole image, the crack detectors based on the local analysis are presented. Specifically, the image is divided into many blocks, and the local binary classifier is used to separate the crack blocks from those background ones.

Oliveira *et al.* (Oliveira, 2014) separated the images into non-overlapping blocks, and the crack and non-crack patterns are constructed with the mean and variance features of blocks. Subsequently, Bray *et al.* (Bray, 2006) further computed the histogram features of blocks, and the resultant crack detecting performances are improved. The two methods mentioned above are simple and easy to implement, but the adopted features are both the pixel-level feature. Therefore, they are easily affected by the unknown background disturbance. In addition, owing to the precise crack detection (i.e., 0.3mm width crack can be detected), the acquiring density of concrete surface crack defect is highly boosted. For the resulting massive crack images, it is difficult to obtain the good efficiency and capability via the traditional crack detecting techniques.

In this paper, an emerging learning technique, Extreme learning machine (ELM), is exploited to detect the concrete surface crack defects efficiently with the massive crack images. The ELM proposed by Huang *et al.* (Huang, 2014-2015) is a novel machine learning model for training single layer feed-forward network (SLFN). Compared with traditional methods (Neural network or SVM), ELM's input weight parameters are randomly generated and it only computes the output weights between the hidden layer and the output layer. ELM not only has a faster learning speed but also obtains a better generalization than other traditional learning models.

Fig. 1 illustrates the flow of our proposed crack detecting method. Firstly, we divided the collected crack images into the crack blocks and non-crack ones, which are used as the binary training datasets. Then, the frequency features of concrete surface images are computed via the discrete cosine transformation. Compared with the pixel-level features, the frequency domain characteristics of crack images are more informative, which contribute to the improved performance of crack region detector. Finally, the regularized ELM classifier is used for finding the optimal separate hyper-plane between the crack objects and background ones.

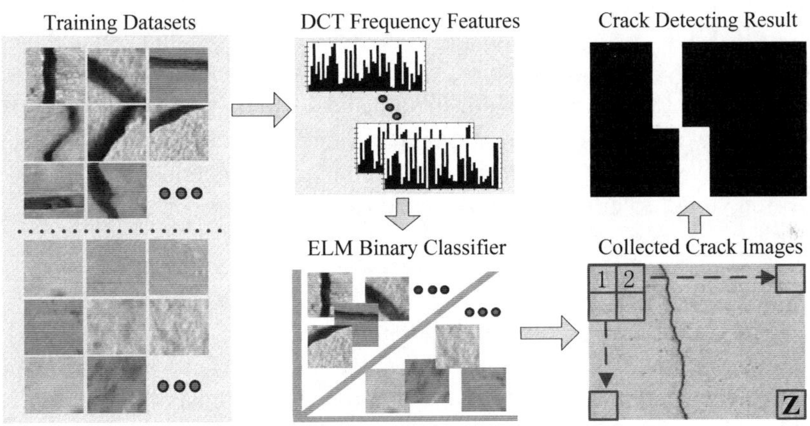

Fig.1. Diagram of the proposed crack detecting method

2. Proposed Method

2.1 Feature representation of crack images

Feature extraction is significant for the performance of crack detector. The existing crack detecting model often utilizes the simple pixel-level features (such as gray mean value, variance, histogram, etc.). The naive and simple feature extraction methods are sensitive to the noise or background disturbances. To obtain the better feature description of crack region, the discrete cosine transformation is used to compute the frequency features of concrete surface images.

Suppose now that the crack region block $\boldsymbol{v} \in \mathbb{R}^{d \times d}$, and we obtain the new vector $\boldsymbol{x}_1 \in \mathbb{R}^{d^2 \times 1}$ by stacking the row data of matrix $\boldsymbol{v}$. Here, the vector $\boldsymbol{x}_1$ can represent the horizontal varying characteristics of matrix $\boldsymbol{v}$. Similarly, we can obtain the other vector $\boldsymbol{x}_2 \in \mathbb{R}^{d^2 \times 1}$ by stacking the column data of matrix $\boldsymbol{v}$, which reflects the vertical changes of matrix $\boldsymbol{v}$. Then, the new vectors $\boldsymbol{x}_1$ and $\boldsymbol{x}_2$ are combined into $f(\boldsymbol{x})=[\boldsymbol{x}_1,\boldsymbol{x}_2]$, and the final image region frequency features are calculated by the following equation.

$$F(\boldsymbol{u})=C(\boldsymbol{u})\sqrt{\frac{2}{N}}\sum_{x=0}^{N-1}f(\boldsymbol{x})\cos\frac{(2x+1)u\pi}{2N} \tag{1}$$

Here, $\boldsymbol{u},\ \boldsymbol{x}=0,1,\cdots,\ N-1$, $(N=2d^2)$. $F(\boldsymbol{u})$ is the DCT coefficient of inputs, and the amplitude of each point denotes the response degree of its corresponding frequency. Fig. 2 shows the DCT coefficients for one crack block and non-crack one.

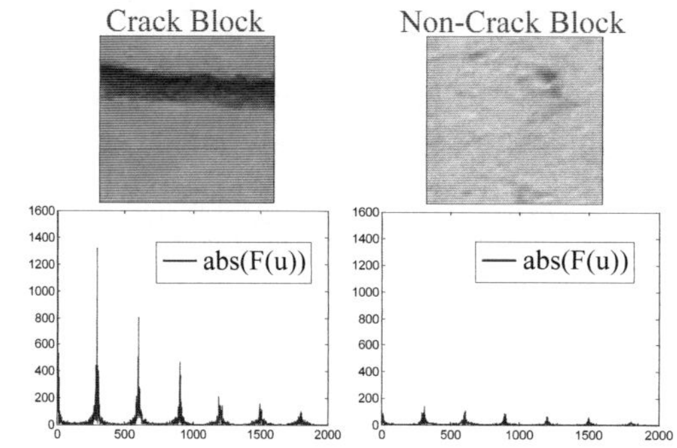

Fig.2 DCT coefficients of a crack block and non-crack one

As illustrated in Fig. 2, owing to the sparse features of crack content in the whole image, the large DCT coefficients of crack object often cluster in some specific frequency domains. In contrast, the noise distribution of non-crack blocks is random, and the resultant DCT coefficients are evenly distributed. In this work, we extract the mean coefficients of eight center frequencies (without the basic frequency) from small to large, which are used as the inputs $\boldsymbol{c}$ for the subsequent crack region detecting model.

2.2 Crack region detecting network

After the crack region feature extraction, the crack detecting model is to maximize the separability between the crack region features and non-crack region ones. At present, the high-density data acquiring often leads to a mass of crack testing images. However, the existing crack detector could not achieve the favorable accuracy and efficiency. To meet the demand of fast and precise maintenance of civil engineering structure, in this section, ELM classifier is utilized for separating the crack object from the backgrounds, which is shown by the following Fig. 3.

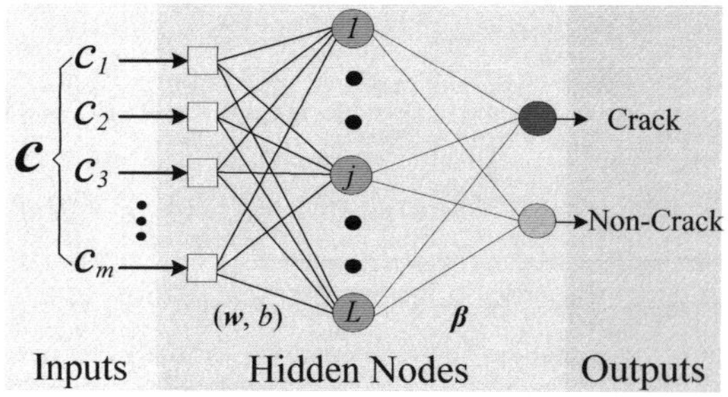

Fig. 3 Efficient crack region detecting network

In Fig. 3, the inputs of crack detecting model are the frequency features of image patches. Let the hidden node number of crack detecting framework be L. The input weights $\boldsymbol{w}_i$ and b_i $(i=1,\ldots,L)$ are randomly assigned. Then, the hidden layer output matrix are calculated by $H=\{h_i(\boldsymbol{c})\}_{i=1}^{L}$. Here, $h_i(\boldsymbol{c})=G(\boldsymbol{w}_i,b_i,\boldsymbol{c})$ is the i-th hidden node output, and $G(\cdot)$ is the activation function. In this work, there are two patterns (i.e., crack and non-crack) which are treated as the outputs $\boldsymbol{T}$ of crack detecting network. Based on above, the goal of ELM-based crack region detector is to minimize the training error as well as the norm of the output weights β

$$\hat{\beta}=\arg\min_{\beta}\left\{\frac{1}{2}\|\beta\|_2^2+\frac{\lambda}{2}\|\boldsymbol{T}-\boldsymbol{H}\beta\|_2^2\right\} \tag{2}$$

Here, λ is a regularization parameter. For simplicity, we rewrite $L_{ELM}=\left(\|\beta\|_2^2+\|\boldsymbol{T}-\boldsymbol{H}\beta\|_2^2\right)/2$. The problem above is commonly known as the regularized least squares or ridge regression. By setting the gradient of L_{ELM} to be zero, we can have the optimal solution as follows

$$\beta=\left(\boldsymbol{H}^{\mathrm{T}}\boldsymbol{H}+\boldsymbol{I}/\lambda\right)^{-1}\boldsymbol{H}^{\mathrm{T}}\boldsymbol{T} \tag{3}$$

And the corresponding crack region discriminate function is

$$J(\boldsymbol{c})=h(\boldsymbol{c})\beta \tag{4}$$

2.3 Crack region testing

According to the expert experiences, and as a trade-off between effectiveness and speed, we divided the collected concrete images into Z non-overlapping image blocks (see Fig. 4(a)). By using the discrete cosine transformation, the frequency features of concrete surface images are calculated. Then, the crack detecting network presented in Section 2.2 are exploited for selecting the crack regions from those candidate blocks. Fig. 4(b) illustrates the crack region detecting results. Specifically, the white blocks represent the crack region, and others denote the non-crack one.

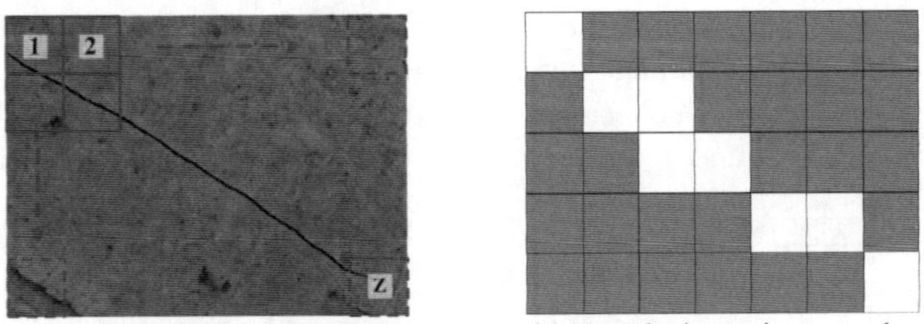

(a) Image block representation (b) Crack detecting results

Fig.4 Illustration of concrete surface crack detecting

3. Experiments and Discussions

To obtain a certain detection ratio of crack defect, the size of image block is set to be 75×75. The hidden node number L is chosen to be 1000; the activation function $G(\cdot)$ is set to be the sigmoid function, i.e., $G(x)=1/(1+e^{-x})$; the regularization parameter λ is 0.1. We collected one hundred representative crack images from the railway ballast-less track slab, and established the ground truth data of crack region via the manual annotation.

In order to evaluate the performance quantitatively, we use two popular evaluation criteria: one is the precision rate (PR), and the other is the recall rate (RE), which are defined as follow.

$$\mathrm{PR} = T_{est} / T_{d}, \quad \mathrm{RE} = T_{est} / T_{g} \tag{5}$$

Here, T_{est} is the number of correctly detected crack blocks, T_{est} is the total number of detected crack blocks, T_g is the number of manually labeled crack blocks.

Fig. 5 shows some representative crack detecting results. In particular, the top row of Fig. 5 (i.e., Fig. 5(a)) exhibits some concrete surface crack images, the second row of Fig.5 (i.e., Fig.5 (b)) includes the ground truth data of crack region via the manual annotation, and the corresponding crack detecting results of our presented method are shown in the last row of Fig. 5 (i.e., Fig. 5(c)).

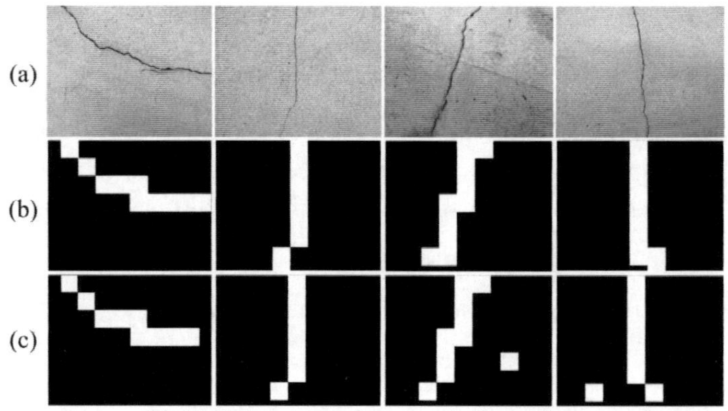

Fig.5 Representative crack detecting results

As shown in Fig. 5 (b-c), compared with the manual labeled ground truth data, the proposed model has achieved the satisfied detecting results as a whole. In this work, we only consider the discriminative attribute of the image block itself. Thus, it is possible that the backgrounds which are similar to the crack contents are incorrectly detected as the crack blocks (See the last two displayed images of Fig. 5).

Furthermore, to validate the effectiveness of proposed model in the case of the massive crack images, we replace ELM classifier with other classical learning technique, i.e., Support Vector Machine (SVM). To give a fair evaluation, the previous feature extraction model is not changed. In this comparison, 4000 divided image blocks are exploited. Here, 2000 image blocks are crack regions, and the rest are non-crack ones. For performance evaluation, the training time, crack detecting precision rate (PR), and crack detecting recall rate (RE) are shown in Table 1.

Table 1 Performance comparisons in terms of training time, PR and RE

Methods	Training time (seconds)	PR (%)	RE (%)
SVM	30s	89.1	90.2
ELM	0.13s	95.3	93.6

For the efficiency aspect, the training of SVM needs to solve a quadratic programming problem, which results in a high computation. In contrast, the parameters of presented crack detecting network can be computed without the iterative calculation. Thus, in case of the massive crack images, our crack detecting method could have a much faster learning speed than SVM model. Moreover, in theory (Huang, 2014), the universal approximation capability of SVM has not been proven. From the optimization point of view, SVM method suffers from the curse of local minima. However, it has been proven that the ELM model has the universal approximation and classification capabilities (Huang, 2014-2015). Therefore, we do not encounter any local minima issues when applying the ELM technique. In this section (See Table 1), we compared the crack detecting PRs and REs between using ELM and SVM, which experimentally demonstrate that the ELM model tends to be preferred in terms of crack detecting accuracy.

4. Conclusions

In the case of high-density data collecting demand, the existing concrete surface crack detectors could not obtain the satisfied detecting results in terms of the efficiency and capability aspects. To address this issue, a novel effective crack region detecting method has been proposed in this work. For feature representation, discrete cosine transformation is used to extract the frequency features of concrete surface images, which are more informative than the pixel-level features. With the calculated features of crack and non-crack training samples, the optimal separate hyper-plane between the crack and backgrounds is fast constructed via the regularized ELM classifier. Experiments on the crack images collected from the railway ballast-less track slab validate the effectiveness of the proposed crack detecting method.

Acknowledgments

This work was supported by the Natural Science Foundation of Hebei Province under Grant No. E2017210113, the Educational Commission Foundation of Hubei Province under Grant No. ZD2017070, and the Collaborative Innovation Center of Large Infrastructure Disaster Prevention and Reduction of Hebei Province.

References

Min Y C, Dyke S J 2015, Vision-Based Automated Crack Detection for Bridge Inspection, *Computer-Aided Civil and Infrastructure Engineering*, vol. 30, no. 10, pp. 759-770.

Rosales M B, Filipich C P, Buezas F S 2009, Crack detection in beam-like structures, *Engineering Structures*, vol. 31, no. 10, pp. 2257-2264.

Wang X J, Qin G F, Zhao H L 2010, Pavement crack detection method based on multilevel denoising model, *Journal of Computer Applications (Chinese)*, vol. 30, no. 6, pp. 1606-1609.

Yamaguchi T, Hashimoto S 2006, Image Processing Based on Percolation Model, *IEICE Transactions on Information & Systems*, vol. 89-D, no. 7, pp. 2044-2052.

Yamaguchi T, Hashimoto S 2006, Automated Crack Detection for Concrete Surface Image Using Percolation Model and Edge Information, *IECON Conference on. IEEE*, pp. 3355-3360.

Zhang D J, Li Q Q, Chen Y 2015, Asphalt Pavement Crack Detection Based on Spatial Clustering Feature, *Acta Automatica Sinica (Chinese),* vol. 42, no. 3, pp. 443-454.

Oliveira H, Correia P L 2014, CrackIT — An image processing toolbox for crack detection and characterization, *IEEE International Conference on Image Processing. IEEE*, pp. 798-802.

Bray J, Verma B, Li X 2006, A Neural Network based Technique for Automatic Classification of Road Cracks, *International Joint Conference on Neural Networks. IEEE*, pp. 907-912.

Huang G B 2014, An Insight into Extreme Learning Machines: Random Neurons, Random Features and Kernels, *Cognitive Computation*, vol. 6, no. 3, pp. 376-390.

Huang G, Huang G B, Song S 2015, Trends in extreme learning machines: A review, *Neural Networks*, vol. 61, pp. 32-48.

Huang G B, Zhou H, Ding X 2012, Extreme learning machine for regression and multiclass classification, *IEEE Transactions on SMC Part B*, vol. 42, no. 42, pp. 513-529.

Huang G B, Bai Z, Kasun L L C 2015, Local Receptive Fields Based Extreme Learning Machine, *IEEE Computational Intelligence Magazine*, vol. 10, no. 2, pp. 18-29.

The 8th International Conference on
Structural Health Monitoring of Intelligent Infrastructure
Brisbane, Australia | 5-8 December 2017

Analysis of Void-disease Detection in Ballastless Track of High-speed Railway based on Directional Filter

Y. Yang[1], W.G. Zhao [2], Y.L. Du [2], X.S. Tian [2].

[1] *School of Electrical and Electronic engineering, Shijiazhuang Tiedao University, Shijiazhuang, China, Email:* sherry_0768@163.com

[2] *The Key Laboratory for Health Monitoring and Control of Large Structures in Hebei Province, Shijiazhuang Tiedao University, Shijiazhuang, China*

Abstract

Owing to the concealment of void disease in cement–asphalt (CA) mortar and the requirements for rapidness, ground penetrating radar (GPR) has a good application prospect in detecting disease on a high-speed railway. Track slab, the covering of the CA mortar layer, is a multi-layer reinforced concrete structure; strong reflection of reinforcement makes disease identification difficult. To solve this problem, directional models were established to study the feature difference of echoes between reinforcement and void disease, and a horizontal filter based on curvelet transform was proposed to identify the void disease in CA mortar. Finally, forward simulation was conducted to validate the performance of the presented algorithm. Results show that the horizontal filter method contributes to the identification of void diseases in the CA mortar layer which were located in the middle of the reinforcement group. This study promotes the application of GPR on disease detection in a high-speed railway.

1. Introduction

Ballastless tracks are widely used in high-speed railways for their high stability, strong structural durability, and easy maintenance[1]. However, due to the repeated action of environmental and train loads, various diseases appear in different degrees.

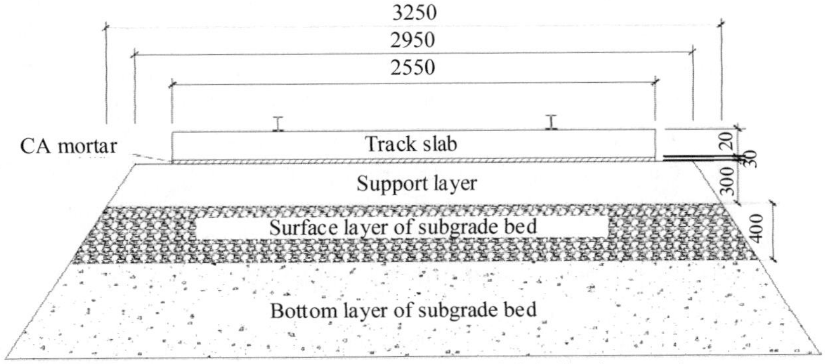

Figure 1. Cross section of CRTS-II slab ballastless track

Fig. 1 shows the structure of CRTS-II slab ballastless track which is composed of with track slab, cement asphalt (CA) mortar, support layer, surface layer of subgrade bed and bottom layer of subgrade bed. The main function of the CA mortar layer is to regulate the lubrication between the track slab and support layer. Warping deformation is affected by the temperature gradient and produced in the track slab. Thus, CA mortar layer is disengaged from track slab. When this situation is further intensified, the void disease of CA mortar is formed. This disease causes the failure of the mortar layer and track slab, weakening of the integrity of ballastless track bed, decrease in the static geometric size and dynamic stability of track, and reduction in the safety of train operation. Therefore, the development of advanced technologies that can help to detect void diseases in CA mortars for high-speed railway health monitoring is of great importance. Due to its fast, continuous, nondestructive, accurate features, ground penetrating radar (GPR) has good application prospects in detecting diseases in high-speed railways. However, in the ballastless track structure, the reinforcement of track plate is densely distributed, and the strong reflection of the reinforced bars on the electromagnetic (EM) wave limits the application of GPR in detecting CA mortar diseases.

On the basis of discussing the main targets in the ballastless track and exploring the difference of different target echoes in characteristics, this study proposes a method of void-disease detection and identification for ballastless track CA mortar.

2. State of the art

Many studies have been developed to help detect diseases in the railway subgrades, mainly concentrating on the ballast tracks. For example, Benedetto et al.[2] set up four different models of railway ballast beds in a laboratory. In addition, relative dielectric permittivity was employed to assess the extent of the “fouling” disease. Khakiev et al.[3–4] proposed a method of assessing the moisture content of a subgrade in different seasons. Anbazhagan et al.[5] built three fouling material models (screened/broken ballast, coal, and iron ore) and advocated the EM wave velocity to identify the degree and type of fouling. Rodés et al.[6] exploited the frequency spectrum of GPR echo signal to analyze the travel time.

Studies on the detection of diseases in ballastless tracks are rare and mainly focus on the settlement after construction[7–8]. Void-disease detection in CA mortar is still at the stage of germination. Dai et al.[9] analyzed the evolution mechanism of interlayer delamination failure for ballastless tracks. Liao et al.[10] built a 2D forward model based on finite difference time domain (FDTD) to simulate void diseases in CA mortars. However, in this article, only two layers of reinforcement were considered and the most densely distributed prestressed bar was ignored.

In a sense, the track slab is regarded as a multi-story reinforced concrete structure; the disease of CA mortar can be attributed to void defects in multi-story reinforced concrete. Scholars in various countries have conducted extensive research on the application of GPR in reinforced concrete. Soldovieri et al.[11] proposed a minimization algorithm to identify the location and diameter of rebar. Chang et al.[12] analyzed the relationship between reflection energy and the diameter of reinforced bar. Zanzi et al.[13] discussed the relationship between the echo spectrum and measurement accuracy of reinforced bars. These studies emphasized the echoes of reinforcements; however, in the current case, a reinforced bar echo is an interference signal. Therefore, further studies should be carried out to explore the characteristics of void-disease echoes. In the present study, directional models of target echoes were established to examine the different characteristics of echoes between reinforced bars and void diseases. Moreover, a horizontal filter based on curvelet transform was proposed to identify the void disease in CA mortar. The remaining parts of this study are organized as follows: Section 3 analyzes the directional features of reinforcement and void-disease echoes, discusses the directional differences between rebar and void echoes, and describes a directional filter based on CT. Section 4 verifies the reliability of the proposed algorithm through forward simulation. Section 5 summarizes our study and presents relevant conclusions.

3. Methodology

3.1. Echo model of main targets in ballastless track

While the measuring line on a track slab is being set up to detect the disease of CA mortars using a GPR, the track slab, which is called covered layer, is in the upper part of the CA mortar layer; the other layers (support layer, surface layer of subgrade bed, and bottom layer of subgrade bed) are in the lower part of the CA mortar. When the sampling window of A-Scan is long enough, GPR echo includes the reflected signals of all layers and their internal targets. The echoes of the support layer and the following layers are removed by adding a time window by the wave velocity estimation. Thus, this study discusses the target echo of track slabs and void diseases in CA mortars.

The track slab is a kind of reinforced concrete structure. The GPR echoes of track slabs are composed of stochastic echoes and reinforcement echoes. Stochastic echoes are mainly caused by uneven mixing of coarse and fine aggregates, whereas reinforcement echoes are provided both in transverse and longitudinal directions. Due to the small distance between rail support platforms in transverse direction, the measuring line is generally parallel to the rail longitudinal axis. In this case, transverse reinforcement is perpendicular to the measuring line, and the cross-sectional area is small, so transverse reinforcement can be regarded as the point target. The longitudinal reinforcement is parallel to the measuring line, and the distances between the steel bar and the antenna are always equal; therefore, the echoes of the longitudinal reinforcement are shown as banded echoes in the B-scan. The banded echoes are suppressed using the mean subtraction method. Therefore, the principal components of the GPR echoes are the following: void-disease echoes in CA mortars, point-target reinforcement echoes, and stochastic echoes.

Given that GPR echo is S, the void-disease echo is V, and the point-target reinforcement and stochastic echoes are R and N, respectively, S can be written as

$$S=V+R+N \tag{1}$$

To simplify the model, this study ignores the stochastic echo N and assumes that the relative permittivity is isotropic. Therefore, Eq. (1) can be expressed as $S=V+R$. In the following, the differences between the void-disease and reinforcement echoes are analyzed.

(1) Void-disease echo model

Void diseases in the CA mortar layers of ballastless tracks have various shapes. This study assumes that the void part has a distinct boundary with the normal part and can be regarded as a rectangle in 2D. According to the EM wave equation, the electromagnetic wave produces reflected echoes at the interface between the concrete and the void disease. As the distances between the antenna and the void disease are different, the two-way travel times of the reflected echo also differ. Therefore, void-disease echoes exhibit different characteristics in various locations as shown in Fig. 2(a).

On both sides of the edge, the phase axis of echo presents a hyperbolic feature; at the center, the phase axis of echo presents a horizontal feature. Given that the width of the void disease is l, the velocity of EM wave is v, and the intersection point between the central symmetry axis and the measuring line is regarded as the coordinate origin, the curve of the phase axis can be expressed as

$$f(x)=\begin{cases}\frac{2}{v}\sqrt{(x+(l/2))^2+(vt_0/2)^2} & x\le -l/2\\ t_0 & -l/2<x<l/2\\ \frac{2}{v}\sqrt{(x-(l/2))^2+(vt_0/2)^2} & x\ge l/2\end{cases} \tag{2}$$

where t denotes the two-way travel time of void disease when the antenna is at $(x, 0)$ and t_0 denotes the two-way travel time of void disease when the antenna is at (0, 0).

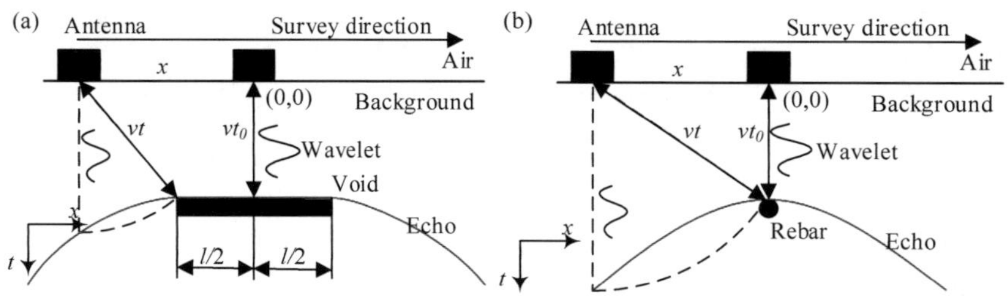

Figure 2. Echo models used in the present study: (a) Void-disease echo model; (b) Reinforcement echo model.

(2) Reinforcement echo model

The transverse reinforcement is perpendicular to the measuring direction. Generally, the diameter φ of reinforcement is much smaller than the width of void disease. Thus, reinforcement can be regarded as a point target, and the phase axis of the reinforcement echo presents a hyperbolic feature, as shown in Fig. 2(b). Similar to Eq. (1), given that the intersection point between the central symmetry axis and the measuring line is regarded as the coordinate origin, the curve of reinforcement echo can be expressed as

$$f(x)=\frac{2}{v}\sqrt{x^2+(vt_0/2)^2} \tag{3}$$

According to the previous analysis, the directions of hyperbolic and horizon are the main features of void-disease echo; the direction of hyperbolic is the main feature of the reinforcement echo. The direction of horizon is the principal difference between the void disease and reinforcement echoes. Thus, building a horizontal filter is a suitable method to suppress the influence of reinforcement echo and identify the void disease.

3.2. Curvelet-based horizontal filter

Curvelet transform (CT)[14–16] is a geometric multiscale analysis (GMA) method developed in the last 10 years. This method can decompose 2D signals in various directions, scales, and translations. Moreover, CT is a tight frame structure, and the original signal can be completely reconstructed by the decomposition coefficients.

Given that the GPR echo is $f(x_1,x_2)$, $\varphi^D_{j,l,k}(x_1,x_2)$ is the "mother" curvelet in time domain; the curvelet coefficient $c^D(j,l,k)$ can be expressed as

$$c^D(j,l,k) = \sum_{0<x_1,x_2 \le N} f(x_1,x_2)\varphi^D_{j,l,k}(x_1,x_2) \tag{4}$$

where j denotes scale, l denotes direction, k denotes translation, and the size of the matrix is $N \times N$. According to the definition of CT, $c^D(j,l,k)$ denotes the decomposition coefficient of $f(x_1,x_2)$ in scale j, direction l, and translation k. When direction l is horizontal, $c^D(j,l,k)$ represents the decomposition coefficient in the horizon.

Given that l_L denotes the lower cutoff direction and l_H denotes the upper cutoff direction, then the curvelet-based horizontal filter can be written as

$$c^D{}'(j,l,k) = \begin{cases} c^D(j,l,k) & l_L < l < l_H \\ 0 & others \end{cases} \tag{5}$$

Passing through the horizontal filter, the GPR echo can be reconstructed as

$$f^D{}'(x_1,x_2) = \sum_{j,l,k} c^D{}'(j,l,k)\varphi^D_{j,l,k}(x_1,x_2) \tag{6}$$

Taking the GPR echo matrix (256×256) as an example, the matrix can be decomposed into five scales: coarse layer, layers 2–4, and finest layer. The coarse layer represents the low frequencies, whereas the finest layer represents the high frequencies. As the directional resolution of the coarse layer and finest layer is 2π, the horizontal filter should not consider these parts. The directional resolution of layer 2 is 2π/16, and those of layers 3–4 are both 2π/32. In the time domain, when l=1, the direction is π/4, and the direction l increases in a counterclockwise direction. Fig. 3 shows the curvelet-based horizontal filter when l_L=2, l_H=3 in layer 2 and l_L=3, l_H=4 in layers 3–4.

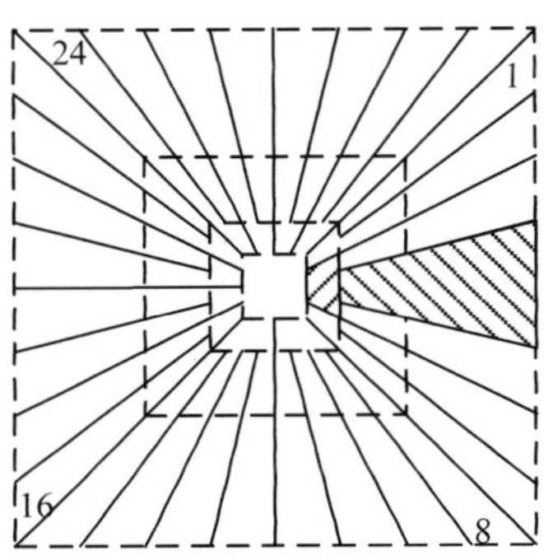

Figure 3. Cartesian-coordinate horizontal filter.

4. Result Analysis and Discussion

To verify the reliability of the curvelet-based horizontal filter on identification of void disease, forward simulation was conducted based on FDTD[17–18].

4.1. Forward model

A CRTS-II ballastless track is shown in Fig.1. The track slab is 6.45 × 2.55 × 0.2 m. In the length direction, the track plate is divided into 10 parts by 10 pairs of rail-bearing plates. Each part is separated by a V-type drain tank. Except both sides of the edge, the interval between drain tanks is 0.65 m. In this study, the section between drain tanks is treated as a model unit as shown in Fig. 4. One unit contains two reinforcement groups; each group is made of three-story reinforcements: reinforced-bar layer I, prestressed-bar layer, and reinforced-bar layer II. Reinforced-bar layers I and II contain two reinforcements with a diameter of φ=8 mm and the same abscissa. Prestressed-bar layer contains three reinforcements with a diameter of φ=10 mm and interval of 40 mm. The reinforcements of the prestressed-bar layer are interlaced with those of the reinforced-bar layers I and II in the abscissa. The height of the CA mortar is 30 mm, whereas that of the support layer is 300 mm. As the void-disease echoes in the CA mortar and reinforcement echoes in the track slab are those considered, the supporting layer is replaced by plain concrete. The software gprMax[19–20] is used to forward simulation, and simulation parameters are shown in Table 1.

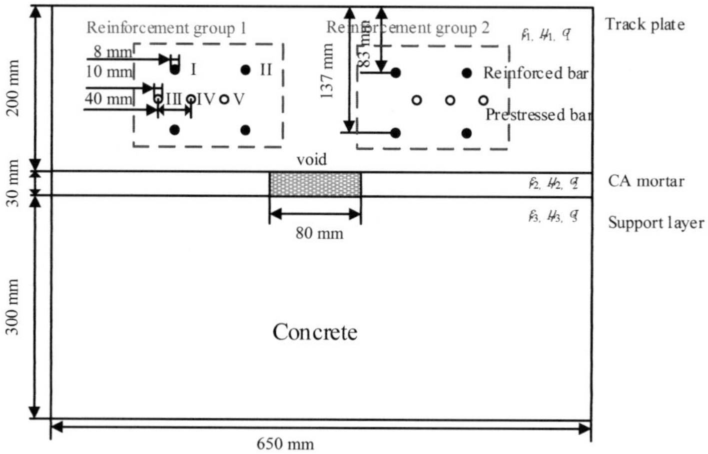

Figure 4. Void disease model of CRTS-II ballastless track.

Table 1. Forward simulation parameters

Parameters	Value	Parameters	Value
time window (ns)	6.6	wavelet	Ricker
samples	1400	trace interval (mm)	2
Yee grid (mm)	(2, 2)	frequency (GHz)	1.5

4.2. Disease-free model echo

To discuss the echoes of reinforcements, forward simulation based on a disease-free model was used. In this case, the relative permittivity of the track slab and support layer and that of CA mortar are close, given that $\varepsilon_{r1}=\varepsilon_{r2}=\varepsilon_{r3}=6$. The simulation result is shown in Fig. 5.

The disease-free echo signals include ground direct wave and reinforcement echo. The upper echo shows the direct wave, which is separated from the reinforcement echo. The lower parts are reinforcement echoes. For dense distribution of reinforcements, the echoes of reinforcements are overlapped.

(1) Echoes of reinforced-bar layer I: The hyperbolic curves of phase axis are clearly visible in the figure. The vertexes are the target positions of the reinforcement, as indicated by the four black dots

in Fig. 5.

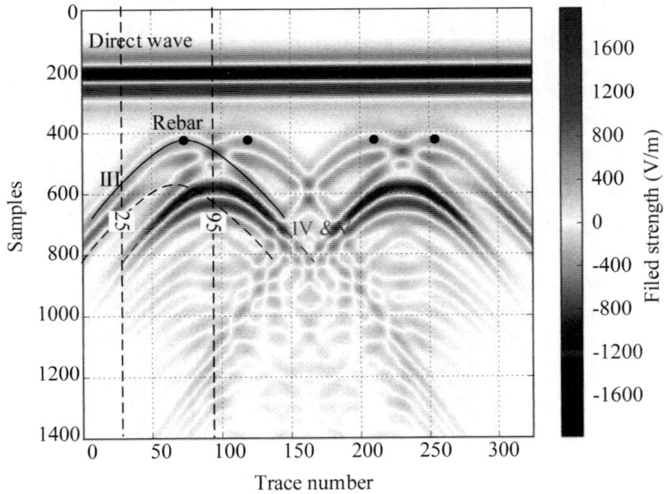

Figure 5. Forward result of disease-free model.

(2) Echoes of prestressed-bar layer: With the echoes of reinforcement group I as an example, in the processing of measurement, the antenna moves from left to right. In the beginning, the antenna is located at the left of the reinforcement group I, the angles between the antenna and reinforcements I and III are both larger, and the echoes of reinforcements I and III are independent. An obvious phase axis appears at the left side, as shown in Fig. 5. As the antenna continues to move, forming a right angle, almost all of the EM waves are reflected by reinforcement I, and some are transmitted to reinforcement III; thus, the vertex is invisible. Fig. 6 shows the time-history curve where the trace number is equal to 25.

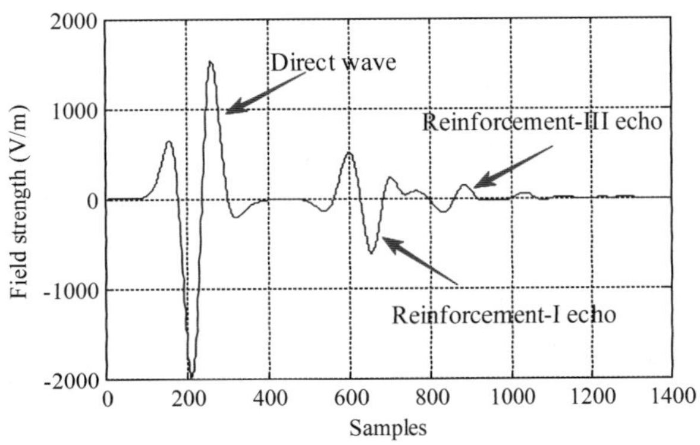

Figure 6. A-Scan echo where trace number=25.

When the antenna continues to move right and reaches the middle of reinforcements I and II, echoes of reinforcements IV and V are received by the antenna. These echoes include two parts: transmitted wave directly from the air-ground surface and reflected wave from reinforcements I and II. As the distance between IV and V shortens, echoes from IV and V are merged into one, as shown in Fig. 5. The snapshots and time-history curve are shown in Fig. 7, where the coordinate of the antenna is at 0.19 and 0 m, trace number=95, and time=1.8 and 2.6 ns.

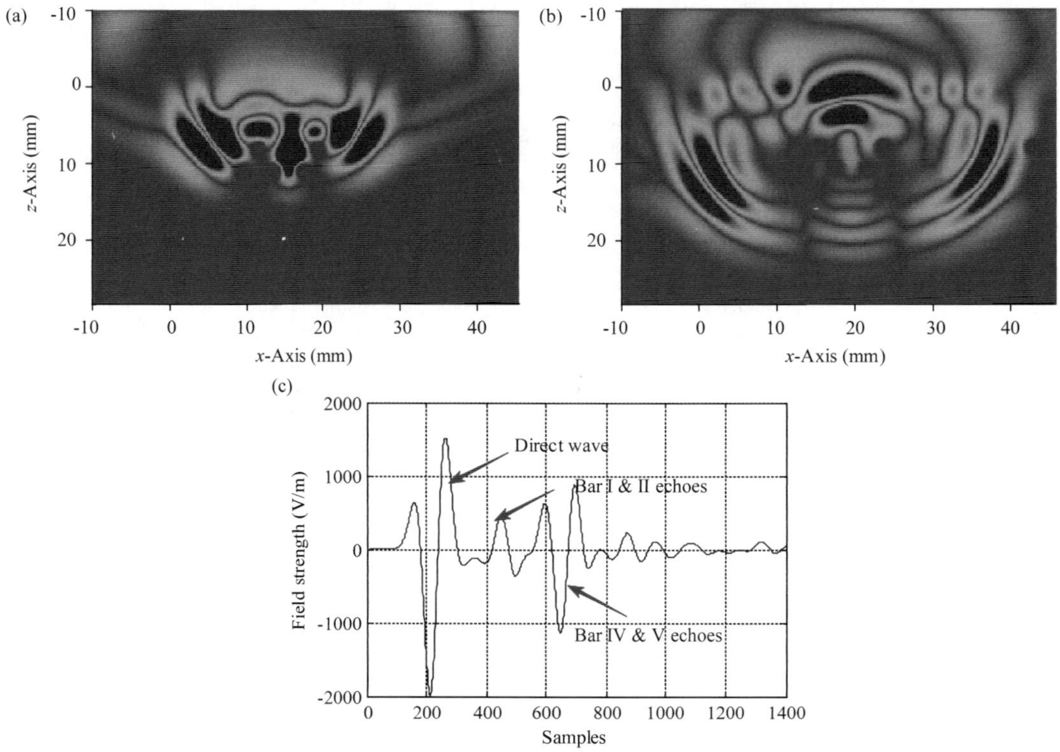

Figure 7. Echoes of reinforcement IV and V: (a) Snapshot when time=1.8 ns; (b) Snapshot when time=2.6 ns; (c) Time travel curve when trace number=95

(3) Echoes of reinforced-bar layer II: The echoes are almost invisible due to the full overlap of the reinforced-bar layers I and II in the abscissa and the shielding influence of the prestressed-bar layer.

4.3. Void disease model echo

A void-disease model of the CA mortar is built to verify the proposed method. The void disease is located at 0.13 and 0.3 m, with length of 80 mm and height of 30 mm. We assume that the void disease is filled with air, where ε_r=1. The simulation result and snapshot (time=2.6 ns) are shown in Fig. 8.

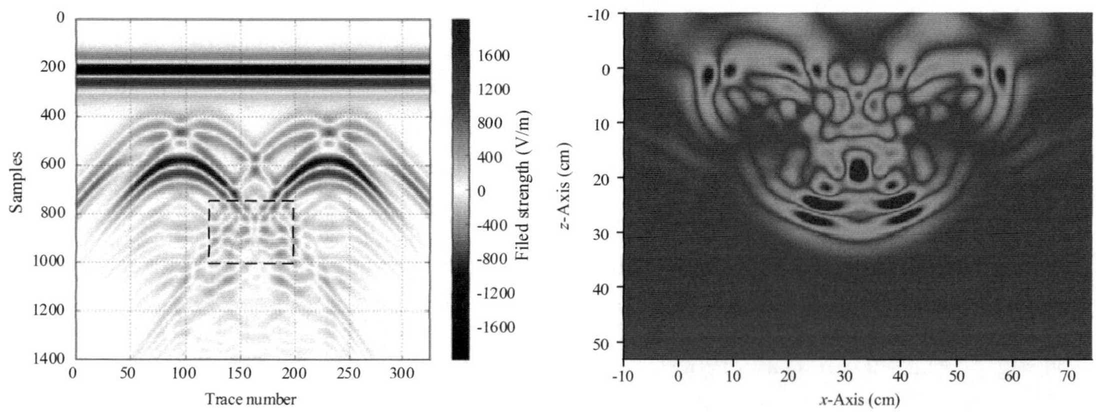

Figure 8. Void-disease echo: (a) Simulation result; (b) Snapshot when time=2.6 ns.

A comparison of Figs. 8(a) and 5 show that a slight difference appears, as indicated in the dash box of Fig. 8(a). The difference is not very significant despite the influence of the reinforcement echoes.

4.4 Processing result

A horizontal filter was conducted to process disease-free and void-disease simulation results. The simulation matrix is 325 × 1400. The CT shows that the matrix is decomposed into six layers. The pass-band directions in different layers are shown in Table 2.

Table 2 Curvelet-based horizontal filter

Coarse layer	Layer 2	Layer 3	Layer 4	Layer 5	Finest layer
{1,1}	{2,3}{10,11}	{3,4}{20,21}	{3,4}{20,21}	{8,9}{40,41}	{1,1}

The processing results are shown in Figs. 9(a) (c). In Fig. 9(c), a horizontal echo exists, where samples ∈ (800, 1000) and trace number ∈ (140, 180). By contrast, the disease-free echo does not exist in the horizontal component.

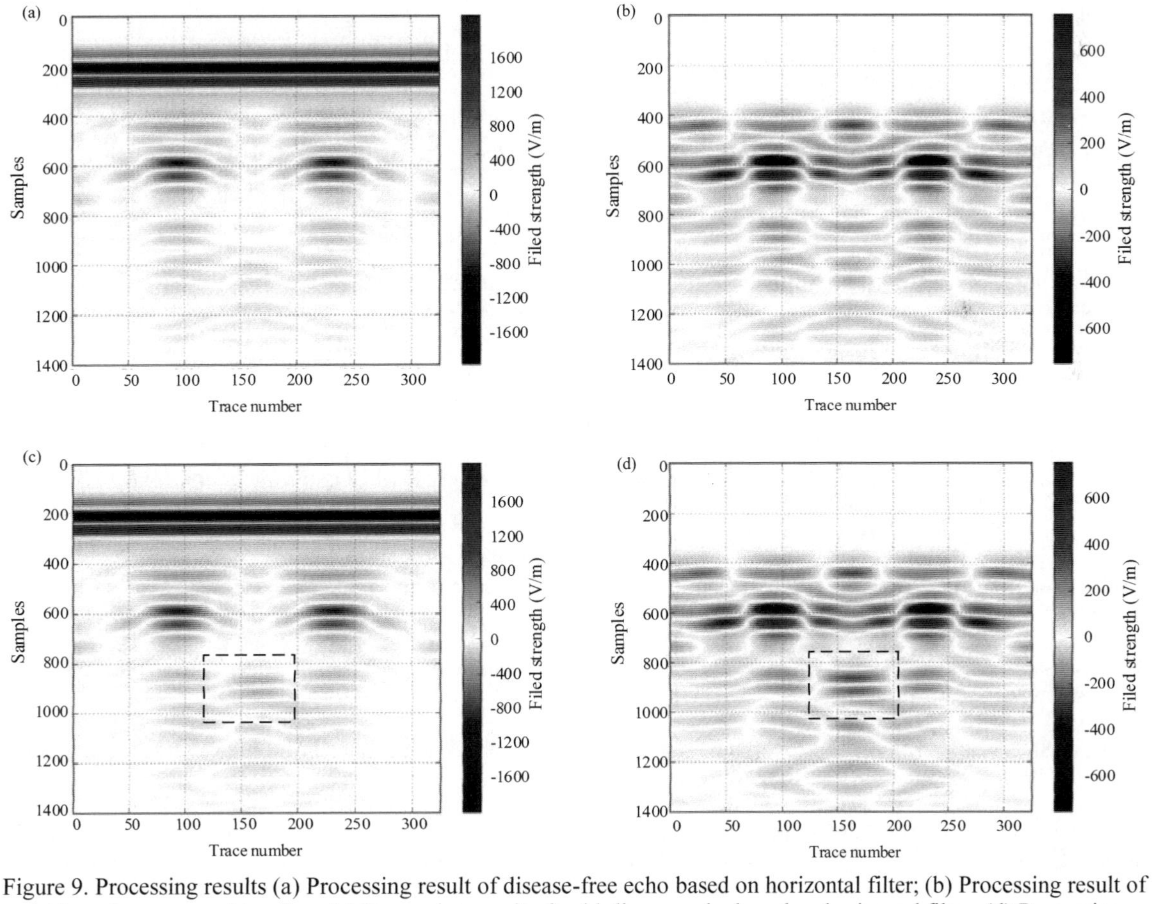

Figure 9. Processing results (a) Processing result of disease-free echo based on horizontal filter; (b) Processing result of (a) based on mean subtraction; (c) Processing result of void-disease echo based on horizontal filter; (d) Processing result of (c) based on mean subtraction.

As the wave moves to the horizontal direction, the horizontal filter cannot suppress the strong direct wave. To enhance the void-disease echo, a method based on mean subtraction is employed. The results are shown in Fig. 9(b) (d). Results show that the void-disease echoes are enhanced.

5 Conclusions

To detect the void disease in the CA mortar and overcome the influence of reinforcement echoes, a horizontal filter was proposed in this study. Forward simulations were conducted to verify the reliability of the algorithm. The following conclusions are drawn:
(1) The strong reflection of reinforcement is the main obstacle to detect the void disease in CA layers. The hyperbolic and horizontal directions are the main directional features of void disease echoes; the hyperbolic direction is the main directional feature of reinforcement echoes. The horizontal direction is the principal difference between the void disease and reinforcement echoes.
(2) The reinforcement echoes of the beam in track slabs are complex. The echoes of reinforcements I and II in reinforced-bar layer I are clearly visible. The echoes of reinforcement III in the prestressed-bar layer are partially visible, whereas the vertex is invisible. Affected by a small interval and reinforcements I and II, the echoes of reinforcements IV and V are combined into one hyperbolic curve. The echoes of reinforced-bar layer II is almost invisible and is influenced by shielding the prestressed-bar layer.
(3) The forward test shows that the curvelet-based horizontal filter can detect and identify the void echo located in the middle of two reinforcement groups.
This result helps us identify the disease of CA mortars in high-speed railways. However, the accuracy of disease identification and diversity of disease distribution need further study.

Acknowledgement

This research was financially supported by National Natural Science Foundation of China (No. 51578349) and the Railway Ministry of Science and Technology Research and Development Program of China (No. 2013G004-A-1).

References

Wheeler, 1976, 'State-of-the-Art of Ballastless Track at-Grade'. *Transportation Engineering Journal*, vol. 102, no. 1, pp.131-145.

Benedetto, Tosti, Ciampoli, 2017, 'Railway ballast condition assessment using ground-penetrating radar – An experimental, numerical simulation and modelling developmen', *Construction & Building Materials*, vol. 140, no. 1, pp. 508-520.

Khakiev, Shapovalov, Kruglikov, 2014, 'Investigation of long term moisture changes in trackbeds using GPR', *Journal of Applied Geophysics*, vol. 110, no. 1, pp.1-4.

Khakiev, Shapovalov, Kruglikov, 2014, 'GPR determination of physical parameters of railway structural layers'. *Journal of Applied Geophysics*, vol. 106, no. 7, pp. 139-145.

Anbazhagan, Dixit, Bharatha, 2016, 'Identification of type and degree of railway ballast fouling using ground coupled GPR antennas', *Journal of Applied Geophysics*, vol. 126, no. 1, pp. 183-190.

Rodés, Pérez, Martínez, 2015, 'Evaluation of the GPR frequency spectra in asphalt pavement assessment', *Construction & Building Materials*, vol. 96, no. 1, pp. 181-188.

Yang, Yang, Fang, 2011, 'Prediction on Red Clay Foundation Settlement of Ballastless Track of High-Speed Railway', *Advanced Materials Research*, vol. 243-249, pp. 2382-2388.

Li, Xu, Chen, 2015, 'Early Warning of Foundation Settlement Deformation for Ballastless High-Speed Railway Tracks', *World Journal of Engineering & Technology*, vol. 03, no. 3, pp. 197-202.

Son, Xiang, 2016, 'Numerical Stimulation of Interface Delamination Failure for Prefabricated Slab Ballastless Track', *Journal of South China University of Technology*, vol. 44, no. 7, pp. 102-108. (in Chinese)

Liao, Zhu, Zan, 2016, 'Detection of Ballastless track disease in high-speed railway based on ground penetrating radar', *Journal of Southwest Jiaotong university*, vol. 51, no. 1, pp. 8-13. (in Chinese)

Soldovieri, Solimene, Monte, 2011, 'Sparse Reconstruction From GPR Data With Applications to Rebar Detection'. *IEEE Transactions on Instrumentation & Measurement*, vol. 60, no. 3, pp. 1070-1079.

Chang, Lin, Lien, 2009, 'Measurement radius of reinforcing steel bar in concrete using digital image GPR', *Construction & Building Materials*, vol. 23, no. 2, pp. 1057-1063.

Zanzi, Arosio, 2013, 'Sensitivity and accuracy in rebar diameter measurements from dual-polarized GPR data', *Construction & Building Materials*, vol. 48, no. 19, pp. 1293-1301.

Donoho, Vetterli, 2003, 'The finite ridgelet transform for image representation', IEEE Transactions on Image *Processing A Publication of the IEEE Signal Processing Society*, vol. 12, no. 1, pp. 16-28.

Dettori, Semler L, 2007, 'A comparison of wavelet, ridgelet, and curvelet-based texture classification algorithms in computed tomography'. *Comput. Biol. Med.*, vol. 37, no. 4, pp. 486-498.

Candès, emanet, Donoho, 2006, 'Fast Discrete Curvelet Transforms', *Multiscale Modeling & Simulation*, vol. 5, no. 3, pp. 861-899.

Yee, Ingham, Shlager, 1991, 'Time-domain extrapolation to the far field based on FDTD calculations', *IEEE Transactions on Antennas & Propagation*, vol. 39, no. 3, pp. 410-413.

Peng, Al-Qadi, 2015, 'Calibration of FDTD Simulation of GPR Signal for Asphalt Pavement Compaction Monitoring', *IEEE Transactions on Geoscience & Remote Sensing*, vol. 53, no. 3, pp. 1538-1548.

Warren, Giannopoulos, Giannakis, 2015, 'An advanced GPR modelling framework: The next generation of gprMax', *International Workshop on Advanced Ground Penetrating Radar*. Florence, Italy : IEEE, pp. 1-4.

Giannakis, Giannopoulos, Warren, 2016, 'A Realistic FDTD Numerical Modeling Framework of Ground Penetrating Radar for Landmine Detection', *IEEE Journal of Selected Topics in Applied Earth Observations and Remote Sensing*, vol. 9, no. 1, pp. 37-51.

The 8th International Conference on
Structural Health Monitoring of Intelligent Infrastructure
Brisbane, Australia | 5-8 December 2017

Field Measurement of Internal Forces Arising in Segmental Lining of Underwater High-speed Railway Shield Tunnel

Kun Feng[1], Zhicheng Dai[1], Wei Wang[1], Chuan He[1,]*, Songyu Cao[1].
[1] *Key Laboratory of Transportation Tunnel Engineering, Ministry of Education, Southwest Jiaotong University, China,*
Email: chuanhe21@163.com

Abstract

The internal forces of segmental lining is subjected to the construction process and complicated geological condition, which also causes the difference of internal forces between actual and theoretical analysis results. This paper presents a field measurement of internal forces arising in segmental lining during the construction of Shiziyang Tunnel in China. Several segments with gauges were installed in typical ground the tunnel passed, the concrete strain data collection began after installing the segments and when the data became stable. Based on the internal forces deduced from the concrete strain data, the variation of the internal forces during the construction phrase, the impact of construction parameters and construction loads, and the influence of ground properties on the internal forces were discussed.

Key words: High-speed railway; underwater shield tunnel; field tests; segmental lining; internal forces

1. Introduction

Segmental lining is commonly applied in shield tunnel as one kind of structure forms, for which processes will be performed such as segment erection, stripping and grouting until the segment lining becomes stable before it turns into effective lining structure. During the process, the load acting on the segments are changing constantly, of which the internal force distribution and the magnitude of it is altered with the change of the load on the segments. In-situ tests are carried out as a useful way to obtain the actual inner force of shield tunnel segments, Lee etc. have proposed one kind of distribution modes for earth pressure based on the long-term monitoring for the earth pressure of shallow tunnels in soft soil foundation (Lee etc., 2001). Mashino, H. etc. have conducted field measurement for shield tunnel in cobble stratum, and have proposed the reasonable analyzing model of segment structure for shield tunnel in cobble stratum by comparing the loads and internal force obtained from in-situ tests and conceptual design method (Mashino, H etc., 2003). He etc. have monitored water and soil pressure and internal forces for shield tunnel in sand stratum in the construction, the results indicated that the load and internal force on the tunnel structure is harshly affected by construction process, and divided water calculation mode of water and soil is preferably applied for the analysis of shield tunnel structure in sand stratum (He etc., 2007). Zhou etc. have conducted in-situ tests for the internal forces of segmental lining structure for shield tunnel in loess stratum, and carried out back-analysis to get the reasonable calculation parameters for the design of the lining structure (Zhou etc., 2011).

According to the results of the existing in-situ tests of shield tunnel, ground condition plays a critical rule for the load on the segmental rings and internal forces of the segmental structure. Based on Shiziyang high-speed railway underwater shield tunnel, the long-term in-situ testing has been conducted to obtain the internal forces of the segmental lining from the construction period to the stabilizing phase, it is summarized for the distribution characteristics and the changing rule of the bending moment, axial forces, and longitudinal stress of the lining structure. Then the theoretical calculated value is compared with the tested value for internal force after the accomplishment of segment erection and in the stable stage, besides, reasonable calculation mode for hydraulic pressure of underwater tunnel is analyzed.

2. Project Background

Shiziyang Tunnel is the first railway underwater tunnel as well as the longest underwater shield tunnel in China. The total length of the tunnel is 10800m, in which the length by shield method accounts for 9340m, shown in Fig.1. The target speed ranks first in the world, and the design running speed is 350km/h. As a matter of fact, the tunnel is recommended as the 'Century tunnel for the railway of China'. Totally 4 slurry balanced shield machines have been applied for the left and right line for the tunnel construction in opposite directions. The internal diameter of the tunnel lining is 9.8m and external diameter is 10.8m. Plate-shaped segments of reinforced concrete and universal wedge-shaped rings with double wedge of 24mm are adopted as the form of lining. Each ring is composed of 7 segments, with thickness of 500 mm and width of 2.0m, the layout of the segmental ring is showed in Fig.2.Authors are recommended to use this template to prepare their papers. This template is an A4 format with 2 cm margins left, right; 3.5 cm margin top and 2 cm margin bottom.

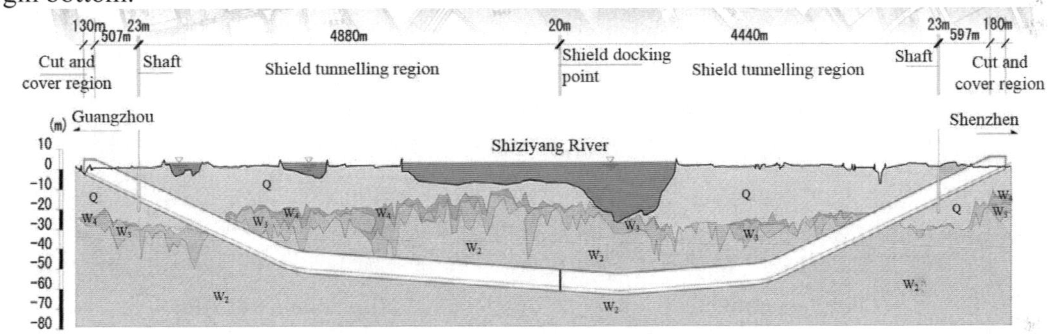

Fig.1 Longitudinal profile of Shiziyang Tunnel

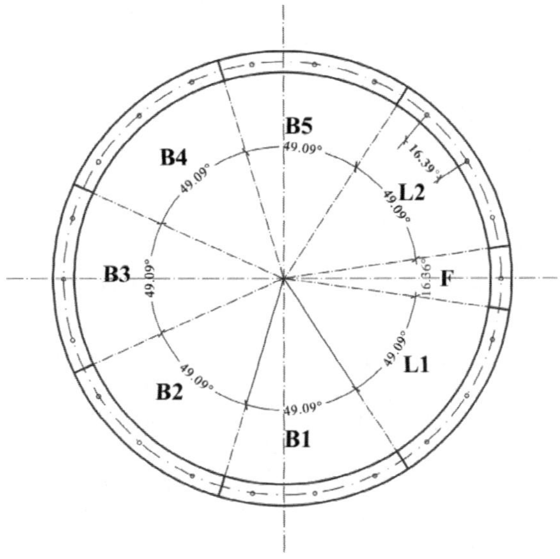

Fig.2 Layout of segmental ring

3. Outline of Field Measurement

Uneven stratum for hardness and softness is chosen for the field measurements, the tunnel is located in compound stratum combined with sand, clay, cobble and intense weathering argillaceous siltstone as shown in Fig.3. The stratum parameters of strata in the testing site is presented in Tab.1.

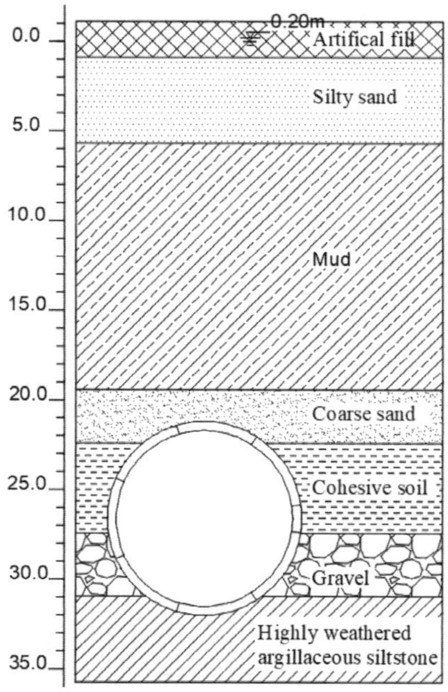

Fig.3 Description of measured tunnel site

Tab.1. Ground parameters of strata in the testing site

Ground type	Thickness (m)	γ ($kN\cdot m^{-3}$)	c (kPa)	φ (°)	SPT-N
Artifical fill soil	2.0	18.0	-	-	-
Silty sand	4.8	19.0	-	30	5.8
Mud	13.7	15.6	8.2	4.3	2.0
Coarse sand	3.0	19.2	-	35	24.8
Cohesive soil	5.0	18.8	23.5	12.6	8.0
Gravel	3.5	20.0	-	40.0	26.9
Highly weathered argillaceous siltstone	-	22.0	-	-	66.1

The internal force of segment structure is obtained through the measurement of concrete strain. 'Vibrating wire type' strain gauge is pre-buried in specific position of the segments. As shown in Fig.4 strain gauges are installed for different placements, first of all, they are placed circumferentially to test the cross-section internal force of the structure. Secondly, placed in pair along the direction of thickness for segments, then placed longitudinally to test the corresponding longitudinal strain, and placed in the center of the point along the segment thickness as well.

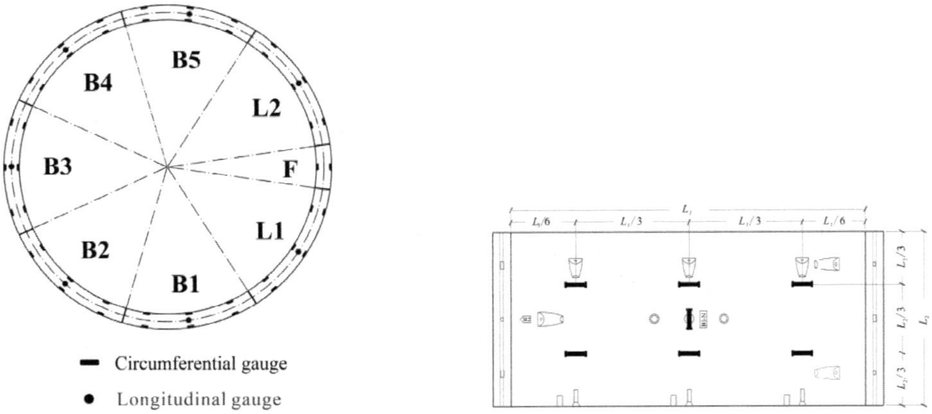

a) Gauge position in the circumferential direction b) Gauge position in the longitudinal direction

Fig.4 Layout of the measurement gauges

The tests begun after the completion of the segments erection, and finished until the data tends to stabilize. At the initial stage of the experiment, as the shield machine tunneling forward, the segments will enter into stratum gradually from the tail of the shield machine, the load on the structure is changing constantly in the stage, therefore, relatively high test frequency is required, however, test frequency can be reduced when certain distance is left between tested ring segments and shield tail as the load on the structure will be stable progressively. Internal force for the cross-section of the tunnel can be obtained through the calculation of the strain gauge for concrete gained by the experiment, the axial force and flexural moment can be calculated by Equation 1 to 2.

$$N_i = \frac{(\varepsilon_{i1}+\varepsilon_{i2})E_c b h}{2} \tag{1}$$

$$M_i = \frac{(\varepsilon_{i1}-\varepsilon_{i2})E_c b h^2}{24(h-2d)} \tag{2}$$

In the equations, ε_i1 is the strain value for intrados, ε_i2 is the strain value for extrados, E_c is the elasticity modulus of concrete, b is the width, h is the thickness of the segments, d is the distance between the center of the tested component and the segment surface. The bending moment is positive when the inner side of the segment is in tension.

4. Measurement Results

As Fig.5 and Fig.6 shown, the values of bending moment were relatively small after the erection of the tested cross-section, and the values increased drastically after the stripping process, then adjustments were applied constantly, and finally the values achieved stability after 11 days (10 rings away from shield tail). The value for bending moment became the largest in stabilization stage, which was 6 times of the value after segment erection. The inner of the top and bottom of the tested ring was in tension after the tested ring enters into stratum, the largest value of positive bending moment came out in B5 at the arch crown, the corresponding value was 279.5kN•m, The axial force increased rapidly after the stripping process, minor decrement for the axial force has been obtained when the stripping process was finished, after 35 days (32 rings away from shield tail), the axial force basically became stable. The values of the axial force after the segment erection were between -315kN and 470kN, and partial region of segments was in tension, uniform distribution for the whole ring of the axial force for the segments was observed, and the value became the largest in the stabilization stage, the largest value of axial force was in the L2, the corresponding value was 8074.6kN.

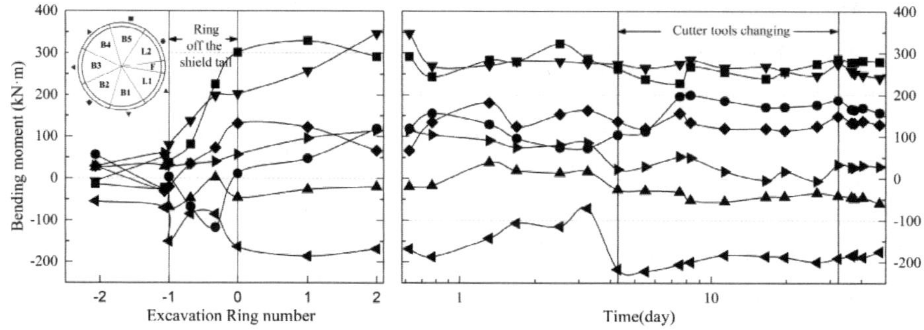

a) Time history of bending moment

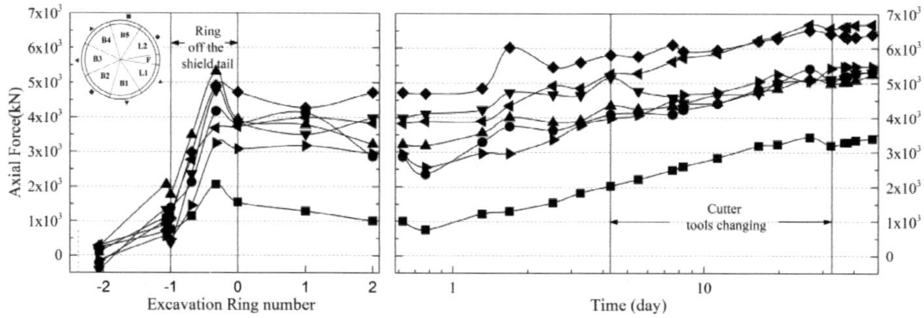

b) Time history of axial force

Fig.5 Time history of internal forces

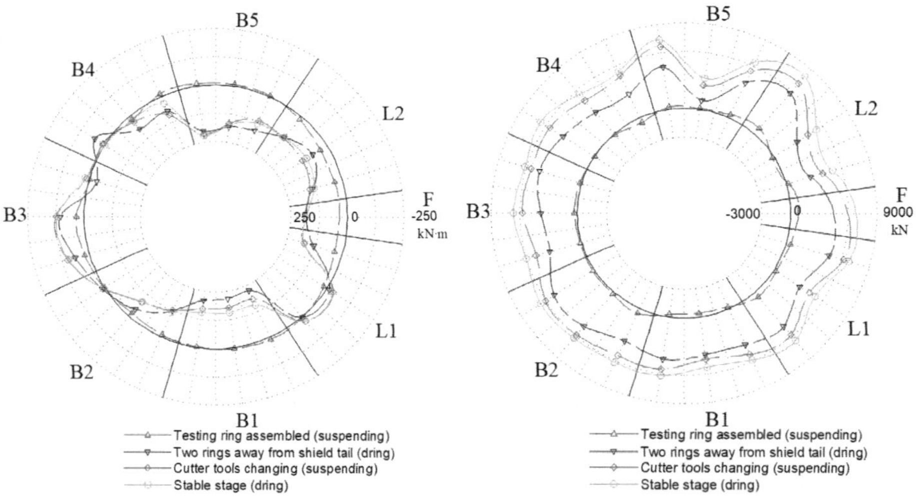

a) Bending moment b) Axial force

Fig.6 Distribution of internal forces in the cross-section

The tested ring was exposed to the constraint force from the jacks after segments erection, the internal forces of the segments were relatively small compared with the stable stage. During the stripping process, due to the extrusion of the shield tail brush and the tail grease, the internal force increased gradually. When the segments fell into stratum, the internal forces were subjected to earth pressure from stratum and synchronous grouting pressure. And due to the situation mainly in soft soil stratum, earth pressure would promptly act on the structure. As a part of strata were of relatively large permeability, grouting pressure dissipated and the grout became solidified progressively, and consequently the axial force would decrease gradually and this trend lasted less than one day. Then the axial force slowly increased, and the axial force of the whole ring tended to increase synchronously, which indicated that the reason for the increment of the axial force was that the hydrostatic pressure on the segments was becoming larger. And that was possibly because of the relatively smaller value of underwater pressure compared with the hydrostatic pressure before construction, under the effect of construction disturbance, the underwater pressure was increasing progressively and finally approached to the value of hydrostatic pressure. The whole process has been completed for about 35 days, and then the axial force of the segments tends to stabilize after the stabilization of the underwater pressure.

As Fig.7 and Fig.8 shown, the longitudinal stress after the segment erected differed a little along the cross-section, the longitudinal stress during excavation process was larger than that before tunnelling. After the segments striped from the shield tail for about 2 to 4 rings, great disturbance for longitudinal stress arose, in which the longitudinal stress for B4 segment increased to 12Mpa. That is possibly because of the uneven thrust of the shield machine on segments in under-constrained ground condition. Stable state has been observed for the longitudinal stress during layoff, and when the testing ring was 30 rings away from the shield tail, it became stable, which indicated that the affected range of the thrust of shield machine was about 30 rings.

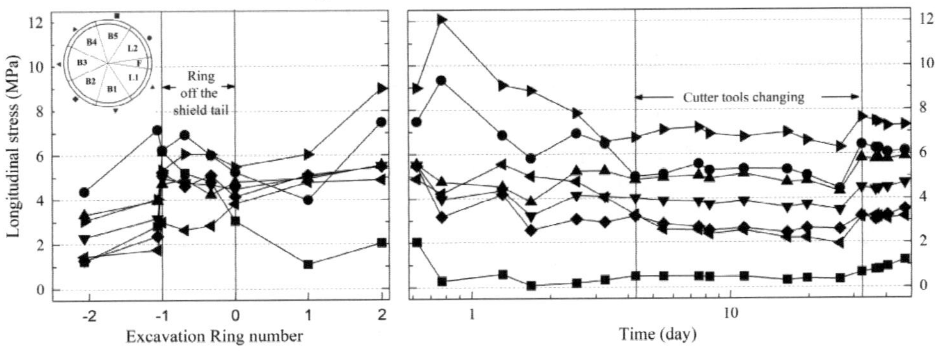

Fig.7 Time history of longitudinal stress

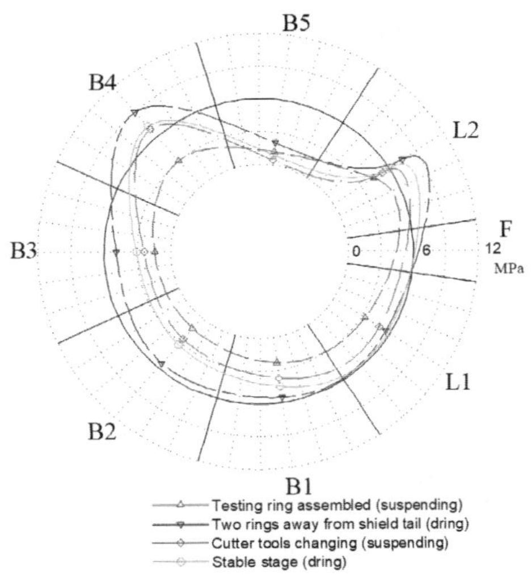

Fig.8 Longitudinal stress distribution in the cross-section

5. Discussion

The beam-spring model applied to calculate the internal forces of the tested cross-section, thus practical cases for the internal force of the structure can be compared with the design phase for that.

5.1 Calculation method

The beam-spring model was applied for calculation as shown in Fig.9, rotational spring elements was adopted for segment joints in the model with a stiffness k_θ, shearing spring elements with a shearing stiffness coefficient k_s which represents the ring joints which connect adjacent segment rings. The springs along radial direction and tangential direction were adopted to simulate the interaction between the segmental structure and the surrounding ground, the corresponding stiffness was k_r 、k_t respectively, and when the radial spring elements were in tension, k_r=0. The relevant parameters are shown for the model as presented in Tab.2.

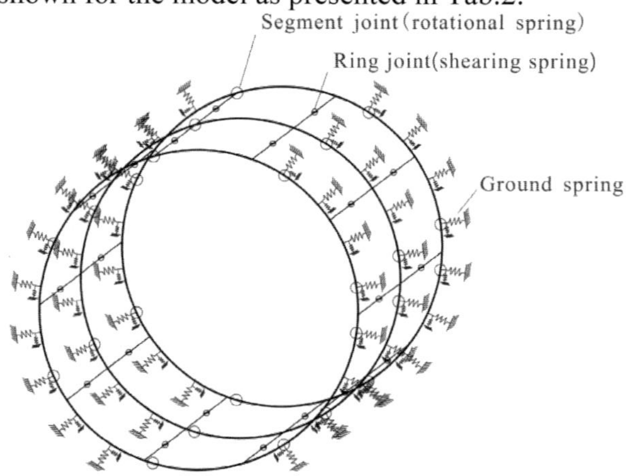

Fig.9 Beam-Spring model

Tab.2 Parameters of the beam-spring model

Tunnel radius *Rc* (m)	5.4	
Thickness of segment *h* (m)	0.5	
Width of segment *w* (m)	2.0	
Elastic modulus of segment *E* (MPa)	34.5×10^3	
Rotational stiffness coefficient of segment joint $k_\theta (MN \cdot m/rad)$	2.5×10^2	
Shearing stiffness coefficient of ring joint $k_s (MN/m)$	1.96	
Lateral earth pressure coefficient λ	Coarse sand	0.5
	Cohesive soil	0.55
	Gravel	0.45
	Highly weathered argillaceous siltstone	0.2
Radial coefficient of ground reaction k_r (MPa)	Coarse sand	13
	Cohesive soil	13
	Gravel	23
	Highly weathered argillaceous siltstone	180
Tangential coefficient of ground reaction k_t	$k_r/3$	

The load mode on the structure in the calculation is presented in Fig.10 and 11. Because it is an underwater tunnel for the project, divided and combined calculation of water and soil pressure are considered respectively. Fig.10 shows the combined water and soil pressure calculation mode and Fig.11 shows the divided water and soil pressure mode. The Terzaghi's theory was applied for the calculation of the top of the earth pressure p_v, as prenseted in Equation 3, the value of γ for the combined calculation of water and soil pressure was taken as the natural unit weight for soil mass, while the value for the divided calculation of water and soil pressure was taken as the floating weight (Zhou etc., 2011).

$$p_{vc} = \frac{B_1(\gamma - c/B_1)}{K_0 \tan\phi}(1 - e^{-K_0 \tan\phi \cdot H/B_1}) + p_0 e^{-K_0 \tan\phi \cdot H/B_1} \quad (3)$$

In the equation, $B_1 = R_0 \cot(\frac{\pi/4 + \phi/2}{2})$,

p_{vc}—the Terzaghi loosen earth pressure,

K_0—the ratio for horizontal earth pressure and vertical earth pressure (usually K_0=1),

ϕ—the angle of internal friction for soil,

P_0—the ground overload,

γ—the unit weight of soil;

c—the adhesion of soil.

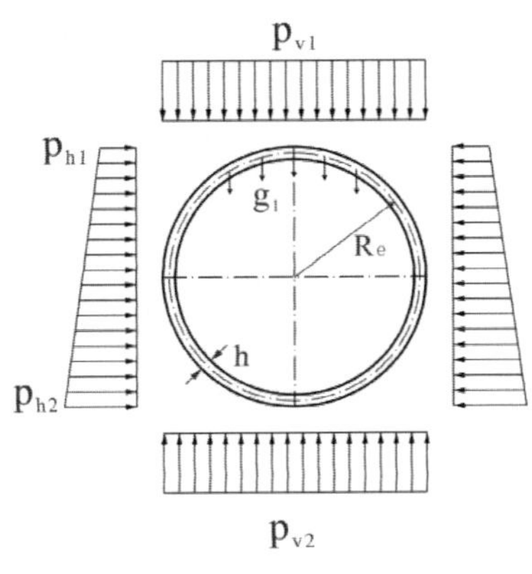

Fig.9 Combined water and soil load mode

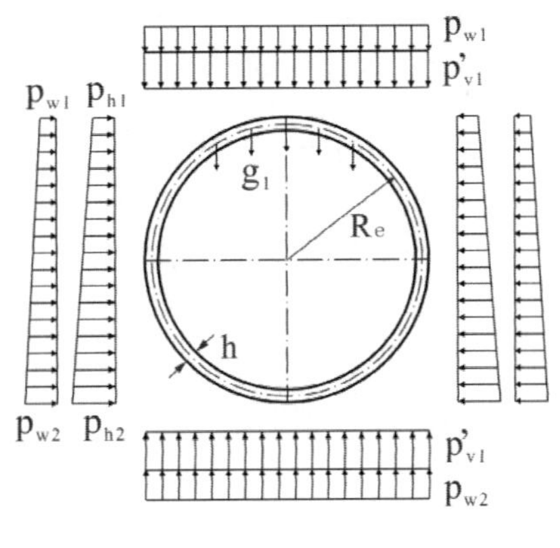

Fig.10 Divided water and soil load mode

5.2 Internal forces in stable phase

For the underwater tunnel, the source of underwater is abundant, the consideration of the underwater effect plays a very important role for structural analysis. Fig.12 shows the comparison between the condition for divided and combined calculation of soil and water pressure for internal force and the practical internal force in stable phase. It can be observed that the results for divided calculation is close to the practical value, the value of structural flexural moment for combined calculation of soil and water is greater than that of practical situation, but the axial force is smaller than the tested one on the contrary. As a result, it is more reasonable to adopt the divided calculation of soil and water pressure as the method for the calculation of structural internal force.

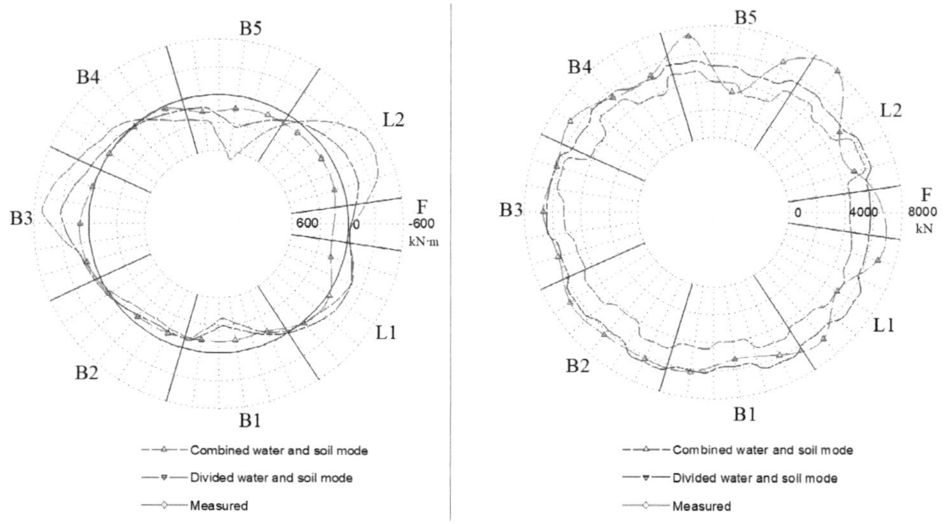

a) Bending moment b) Axial force

Fig.12 Distribution of inner forces in stable phase

5.3 Internal forces after segment assembled

The constraint effect of shield shell acting on the segments should be taken into consideration when it comes to the calculation process for the internal force of segments after the segment erection, $100MN/m^3$ was taken as the value of k_r (Mashino, H etc., 2003), and the results are presented in Fig.13. According to the figure, the practical value for the internal force after the segment erection is larger than the theoretical value of that, and no obvious distributed regularity has been obtained for the tested internal force, which has indicated that the additional load produced by assembling process for the segments plays a more important role than the gravity load. And referring to the preamble for the changes of the segments internal forces, the effect of additional internal forces produced during segment erection decreases after the segments striped from the shield tail.

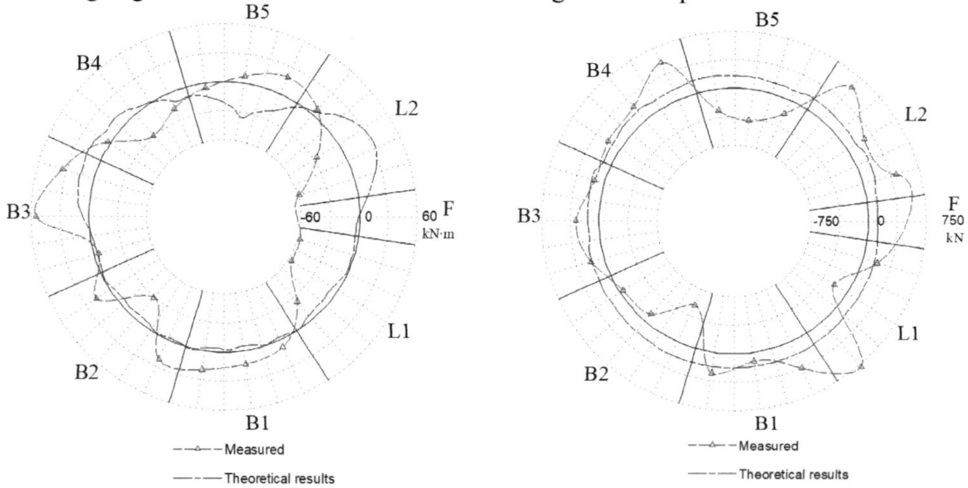

a) Bengding moment b) Axial force

Fig.13 Distribution of inner forces after segment assembled

6. Conclusions

In this study, field measurement and theoretical analysis has been carried out to investigate the internal forces of underwater shield tunnel, conclusion can be summarized as below:

(1) The construction procedure has an obvious effect on the segmental structure for underwater shield tunnel. The bending moment and axial force tend to stabilize after about 35 days. Internal forces become the largest in stabilization stage, greater difference has been recorded in the cross-section for the longitudinal stress, the largest value of that arises after the stripping process. The longitudinal stress is affected by shield machine with an impacting range about 30 rings.

(2) The analysis of internal forces of the segmental structure in the service stage should take the divided water and soil pressure calculation mode for underwater shield tunnel, the design tends to be conservative according to the combined calculation mode of water and soil.

(3) The internal forces of the segmental structure after the segment erection is subjected to the jack thrust force and assembling loads.

Acknowledgements

The authors appreciate the support of the National Natural Science Foundation of China (51578462, U1361210), and the National Key Research and Development Program of China (2016YFC0802202, 2016YFC0802205).

References

He Chuan, Zeng Dong-yang,2007, 'Research on mechanical characteristics of metro shield tunnel segment in sandy strata', *Rock and Soil Mechanics*, vol. 28, no. 5, pp.909-914.

JSCE, 2006, *Standard specifications for tunneling 2006: shield tunnels*, Japan Society of Civil Engineers, Tokyo.

Lee K M, Hou X Y, Ge X W, et al., 2001, 'An analytical solution for a jointed shield-driven tunnel lining', *International Journal for Numerical and Analytical Methods in Geomechanics*, vol. 25, no. 4, p.365-390.

Mashimo H, Ishimura T, 2003, 'Evaluation of the load on shield tunnel lining in gravel', *Tunnelling and underground space technology*, vol. 18, no. 2, pp. 233-241.

Zhou Jimin, He Chuan, Fang Yong, Tan Zhun, 2011, 'Mechanical property testing and back analysis of load models of metro shield tunnel lining in loess strata', *Rock and soil mechanics*, vol. 32, no. 1, pp.165-171.

The 8th International Conference on
Structural Health Monitoring of Intelligent Infrastructure
Brisbane, Australia | 5-8 December 2017

Modal parameter monitoring of the Campobasso's Main Hospital: preliminary results

C. Rainieri [1], D. Gargaro [2], G. Fabbrocino [3].
[1]*University of Molise – Italy, Email:* carlo.rainieri@unimol.it
[2]*University of Molise - Italy,* [3]*University of Molise - Italy*

Abstract

The seismic vulnerability of health facilities may heavily depend on the effects of earthquakes on non-structural components, such as masonry infills, on one hand, and installations and medical equipment, on the other hand. Their failure or malfunctioning might compromise the functionality of the hospital at a time of large demand of medical assistance. As a result, the seismic performance assessment of health facilities requires due consideration of strong motion as well as moderate earthquakes. The primary role of hospitals after hazardous events recommends the development of specific analysis and monitoring strategies for timely damage detection, and management and maintenance of structural as well as non-structural subsystems over time.
In the present paper, some results of an on-going monitoring project for the Main Hospital in Campobasso (Southern Italy) are presented, focusing the attention on the response of the structure in operational conditions and after seismic events. The installed modal based Structural Health Monitoring system is able to provide estimates of the fundamental modal parameters over time on a continuous basis. Preliminary results from modal parameter monitoring are herein presented showing a clear influence of temperature on natural frequency estimates. Moreover, dynamic interaction effects between the two nearby monitored blocks through the joint have been identified. They are mostly influenced by temperature variations, thus remarking the need for compensation of temperature effects in view of effective damage detection. In this perspective, the use of Second Order Blind Identification as an innovative approach to compensation of the environmental influence has been attempted.

1. Introduction

After an earthquake hospitals have to ensure life safety to occupants and assistance to injured people. Thus, they play a key role in the post-earthquake emergency management, as also demonstrated by recent events occurred in Italy and worldwide (Jacques et al. 2014; Masi et al. 2014). The seismic vulnerability of health facilities may depend on the effects of earthquakes on non-structural members, such as masonry infills, on one hand, and installations and medical equipment, on the other hand (Taghavi & Miranda 2003). In fact, their failure and malfunctioning might compromise the functionality of the hospital at a time of large demand of medical assistance. This circumstance has been observed after several earthquakes, such as the 1994 Northridge earthquake (Ayres & Phillips 1998) or the 2010 Chile earthquake (Miranda et al. 2012), or the L'Aquila Earthquake (2009) with several hospitals that were partially or completely inoperable even in the absence of significant structural damage.
According to the definition of the Pan American Health Organization (PAHO), a safe hospital must remain fully operational during and after earthquakes (Pan American Health Organization 2008).

Thus, developing effective seismic protection strategies is critical to assure life protection of occupants and full operation after earthquakes.

The prompt structural health assessment of hospitals after earthquakes by near real time detection and localization of damage definitely represents a promising strategy to enhance their overall safety. Advanced Structural Health Monitoring (SHM) strategies can provide information about the health state of the structure to a remote user in a fully automated way. However, they require the implementation of robust automated data processing procedures in order to provide relevant information about the health conditions of structural as well as non-structural elements. SHM of hospitals can also effectively support the mitigation of administrative and organizational vulnerability by acting on preparedness of staff and supporting management and maintenance, the formulation of disaster mitigation plans and the definition of investment priorities to ensure the overall safety and serviceability (Rainieri & Fabbrocino 2013).

The present paper reports some aspects of the ongoing development of a modal-based SHM system for hospitals located in earthquake prone areas. In particular, attention is focused on the analysis of the dynamic response of a real structure, the Campobasso's Main Hospital (Southern Italy). The installed modal based Structural Health Monitoring is able to provide estimates of the fundamental parameter over time on a continuous basis. Preliminary results from modal parameter monitoring are herein presented, pointing out the influence of the temperature and weak earthquakes on the estimates and the need to compensate it for effective damage detection.

2. Modal parameter monitoring of Campobasso's Main Hospital

An effective SHM requires continuous recording and processing of the vibration response of the monitored structure. In this context, reliable procedures for the analysis of the operational response are of primary concern for structural damage detection (Farrar & Worden 2013, Rainier et al. 2012). For instance, experimental estimates of modal properties can be used for refinement of numerical models in view of seismic analyses under frequent earthquakes (Friswell & Mottershead 1995, Rainieri et al. 2013). The numerical predictions of the seismic response can be compared to vibration limits of equipment sensitive to accelerations. In a similar way, predictions of maximum displacements or interstorey drifts under earthquake loading are useful for vulnerability assessment of non-structural elements sensitive to displacements. Attention is herein focused on continuous monitoring of the dynamic response of the structure for modal parameter estimation and development of pure data driven SHM approaches. This section briefly presents the procedure for automated identification of the modal parameters; the layout of the monitoring system installed at the Campobasso's Main Hospital is described afterwards. Finally, preliminary monitoring results are discussed, highlighting the influence of temperature and weak earthquakes on the fundamental natural frequencies.

2.1 Automated output-only modal identification

An innovative, recently developed automated modal parameter identification procedure, called ARES (acronym for Automated modal paRameter Extraction System), has been used to continuously monitor the modal parameters of the Campobasso's Main Hospital. It is able to provide accurate and precise modal parameter estimates without statically set thresholds and parameters. The key feature of the algorithm is, therefore, the absence of analysis parameters that have to be tuned at each new monitoring application.

The algorithm (Rainieri & Fabbrocino 2015) is based on the combination of different OMA techniques in order to simplify the analysis and interpretation of the stabilization diagram. The Stochastic Subspace Identification (SSI) method (Peeters & De Roeck 1999) is used to estimate

natural frequencies and damping ratios, but a preliminary Blind Source Separation (BSS) (Ans et al. 1985) is operated by the Second Order Blind Identification (SOBI) (Belouchrani et al. 1997, Poncelet et al. 2007) procedure to make easier the identification of the physical poles. In fact, BSS turns the raw data associated to the measured structural response into sources. These can be well-separated (showing the contribution of a single mode to the structural response), not well-separated (noise or minor contributions from other modes could be superimposed to the contribution of the main mode) or noise sources (Poncelet et al. 2007). The auto-correlation of a source can be interpreted as the free decay response of the equivalent Single Degree Of Freedom (SDOF) system corresponding to a mode. Thus, covariance-driven SSI can be applied to extract the associated natural frequencies and modal damping ratios. The approximate separation of the modal contributions by SOBI simplifies the interpretation of the stabilization diagram because the modal information is extracted from the individual sources, which theoretically hold the information about one mode only, and not the multivariate time series of raw data.
The discrimination between physical and spurious poles is based on advanced clustering techniques (Tan et al. 2006). Mode shape estimates are finally obtained from Singular Value Decomposition (SVD) of the output Power Spectral Density (PSD) matrix at the previously estimated frequency of the mode (Brincker et al. 2001).
Effectiveness and accuracy of the algorithm have been demonstrated by a large number of tests based on simulated datasets (Rainieri & Fabbrocino 2015). It is able to ensure a success rate larger than 99% even in the presence of data characterized by low signal-to-noise ratio.

2.2 Layout of the monitoring system

The Main Hospital in Campobasso (Figure 1) consists of a number of reinforced concrete buildings. Most of them were designed and built in the late Seventies, according to outdated seismic design codes.

Figure 1. Aerial view of Campobasso's Main Hospital pointing out the position of the monitored blocks.

A monitoring system has been installed on two joint buildings of the Hospital hosting the inpatient wards. They have overall dimensions of about 78 m x 14 m in plan and 30 m in elevation. One of the buildings covers a rectangular area of about 36 m x 14 m while the other has dimensions of about 42 m x 14 m in plan. The two buildings are denoted as Block A and Block B, respectively (Figure 2).

Figure 2. Sensor layout of the monitoring system installed at the Campobasso's Main Hospital

Sixteen force-balance accelerometers have been installed at the two upper levels of the monitored buildings. The sensors measure the structural response along two orthogonal directions and they are located at opposite corners of the block plans in order to ensure observability of translational as well as torsional modes.
The main features of the accelerometers are: 0.50 g full scale range, 20 V/g sensitivity, 140 dB dynamic range. They ensure high quality measurements in operational conditions. A 16 bit measurement system with on-board anti-aliasing filter completes the system for data acquisition. This is managed by software developed in LabView environment. The collected raw data are stored into a local MySQL database. A sampling frequency of 100 Hz is adopted. ARES continuously processes the acquired data and extract the fundamental modal parameters of the structure.

2.3 Monitoring results

The monitoring system started operating on March 24th, 2016. Sample monitoring results over one month of operation are presented in Figure 3. The fundamental mode shapes of the structure are shown in Figure 4 and can be described as follows:

- the first mode is a global transversal bending mode of the two blocks;
- the second mode is a global longitudinal bending mode of the two blocks;
- the third mode is a global torsional mode involving the two blocks;
- the fourth mode is a torsional mode characterized by counterrotating blocks.

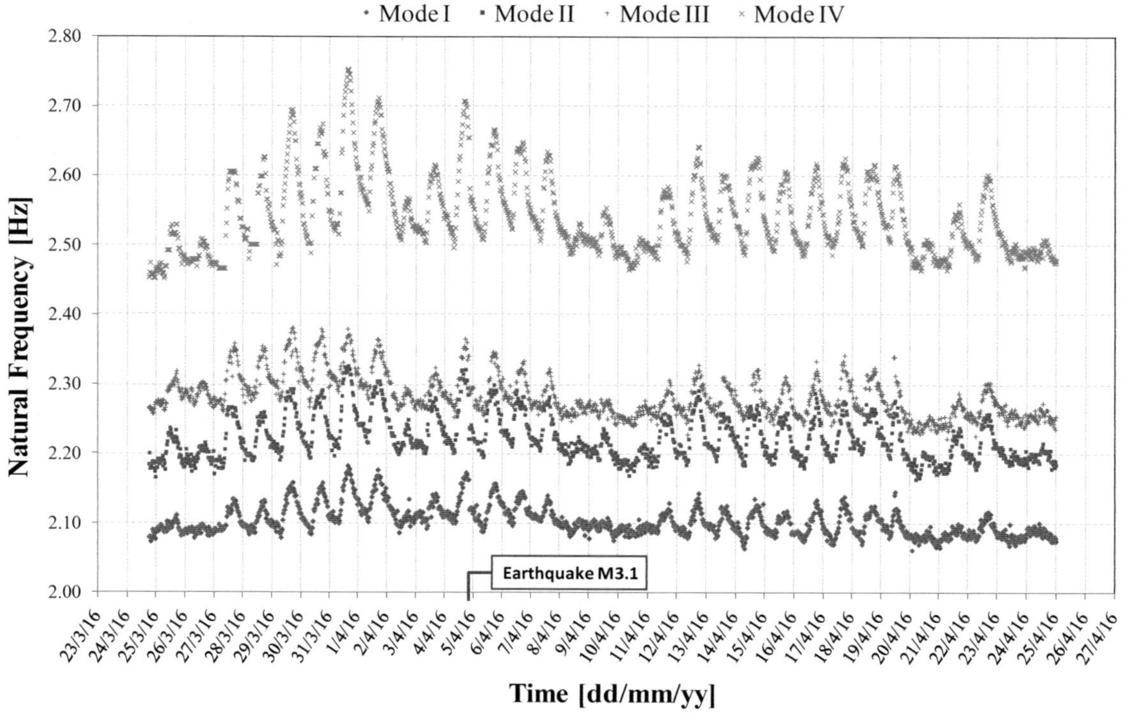

Figure 3. Sample tracking of the fundamental natural frequencies

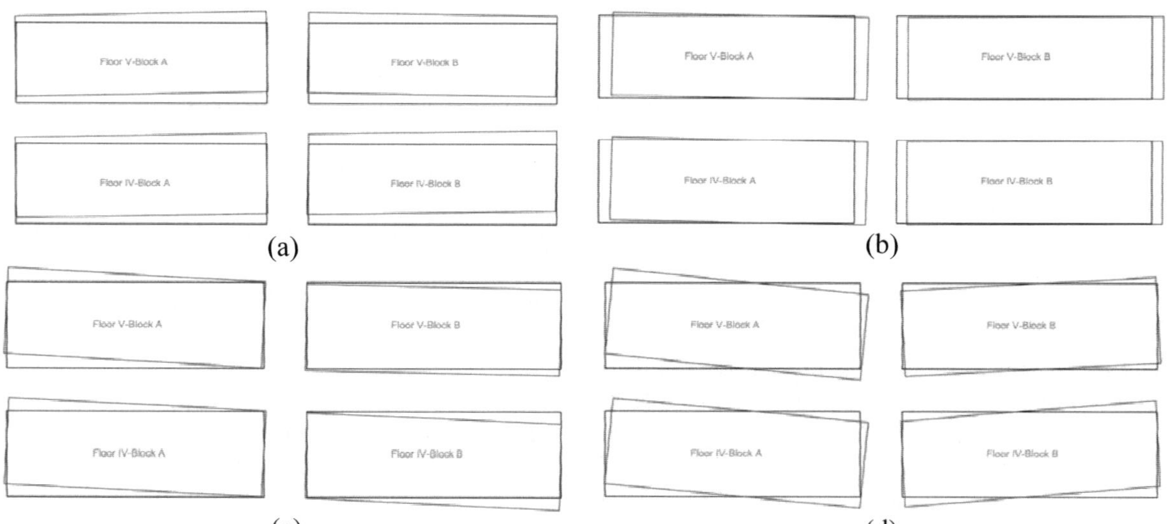

Figure 4. Qualitative representation of mode shapes: mode I (a), II (b), III (c), IV (d)

The mode shapes are influenced by the presence of a small structural joint that divides the two nearby blocks (Figure 5). The small separation and a not perfectly clean space, in fact, make possible some dynamic interactions between the two blocks.

Visual inspection of natural frequency time series in Figure 3 remarks systematic swings occurring every day, with a sudden drop in the night and a gradual increase in the morning until the maximum value reached in the afternoon.

(a)

(b)

Figure 5. Detail of the structural joint between the two monitored blocks: indoor (a) and outdoor view (b).

Comparing the time histories of the fundamental frequency of the block and the local temperature in Campobasso (Figure 6) seems to confirm that the observed variations of the natural frequency estimates can be addressed to environmental factors and, in particular, to thermal effects.

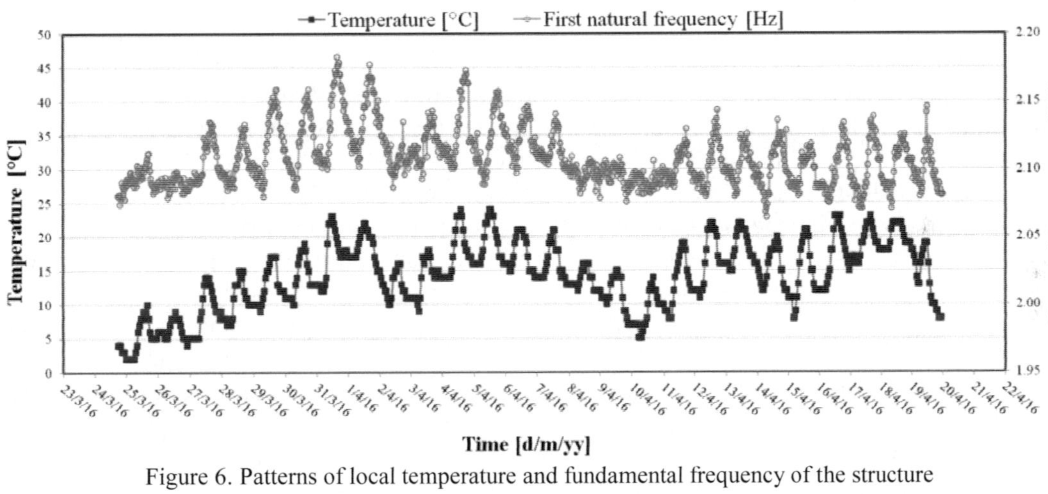

Figure 6. Patterns of local temperature and fundamental frequency of the structure

The increase of the fundamental frequency with the temperature can be addressed to the structural configuration, characterized by predominant longitudinal extension and direct sun radiation on that side, and to the very small joint dividing the two blocks. Because of thermal expansion when the temperature increases, in fact, the distance between the two blocks decreases and the interlocking between the blocks yields some stiffness increase in the longitudinal direction.

A $M_w = 3.1$ earthquake, occurred on April 4th, 2016 (http://cnt.rm.ingv.it/event/6564571), has been also recorded during the considered monitoring period. The epicenter of the earthquake was approximately 1 km far from the hospital. Peak accelerations of 0.015 g and 0.0075 g have been recorded in the longitudinal (Figure 7a) and transversal direction (Figure 7b), respectively, on top of the structure. The occurrence of the earthquake is indicated in Figure 3 and Figure 7c. The latter, in particular, clearly shows a drop of the estimated frequencies because of the occurrence of the earthquake. It is also worth noting that the induced changes of natural frequencies are well within the typical range of variation associated to the daily variation of the temperature. Thus, in view of effective modal based damage detection, the influence of the environmental and operational variables (EOVs) on the natural frequency estimates has to be compensated. The next section describes an innovative application of SOBI for the compensation of the influence of EOVs on modal parameter estimates and the prediction of EOV patterns.

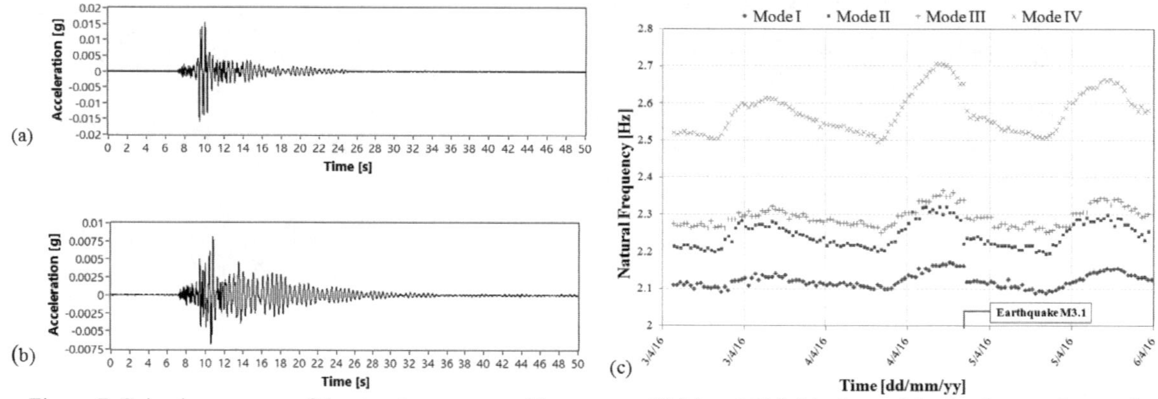

Figure 7. Seismic response of the structure measured by sensors #9 (a) and #10 (b); drop of the fundamental natural frequencies due to the earthquake.

2.4 Analysis of environmental and operational effects

The environmental influence on natural frequency estimates can be modelled following different approaches. They can be broadly classified in two large categories, depending on whether measurements of the EOVs are available or not. When they are available, regression analysis is frequently applied to model the environmental influence. Multiple Linear Regression - MLR - (Hair et al. 2010) allows identifying the relation between some dependent variables (the identified natural frequencies) and several independent variables, the so-called predictors. Once the regression model has been set, the future values of the dependent variables can be predicted from measurements of the predictors only. When MLR is applied in the context of SHM, data have to be collected over a sufficiently long period of time, so that the influence of environmental factors on natural frequencies can be exhaustively characterized. Once the regression model is set, the predicted natural frequencies can be compared to newly collected experimental estimates in order to compensate the environmental influence.

Measurements of the EOVs are not always available. However, statistical methods can still be applied to compensate the influence of EOVs on the estimates. Principal Component Analysis - PCA – (Yan et al. 2005) is probably one the most popular approach in this class of methods. The application of SOBI to remove the effect of EOVs on natural frequency estimates has been also recently proposed (Rainieri et al. 2016a). When SOBI is applied, the natural frequency time series are modelled as a linear combination of sources {s} (the unknown environmental factors) plus the residue {ε}:

$$\{f\}=[A]\{s\}+\{\varepsilon\} \tag{1}$$

The Joint Approximate Diagonalization (Cardoso & Souloumiac 1996) of a set of time-shifted correlation matrices allows computing the mixing matrix [A] and the sources from the collected data.

In the context of SHM, the application of SOBI to remove the influence of EOVs requires a sufficiently large amount of data related to the healthy state of the structure in order to identify the reference mixing matrix [A_{ref}]. Once this matrix is estimated, newly collected data are processed to estimate the corresponding sources, which are recombined by the [A_{ref}] matrix to obtain a model of the variations of the natural frequencies induced by the selected sources. Subtracting the frequency values given by the model from the experimentally obtained values residues are obtained. These are independent of the EOVs so they can be profitably used as damage sensitive features for vibration based SHM (Farrar and Worden 2013).

It is worth noting that the mixing matrix is determined up to a scaling and a permutation of its rows, which prevents the determination of the variances of the identified sources (Rainieri 2014, Poncelet et al. 2007). Thus, an appropriate scaling factor has to be determined to avoid bias when the identified sources are remapped to the original space. It can be computed as the ratio between the first singular value of the reference mixing matrix and that of the mixing matrix estimated from the current data (Rainieri et al. 2016a).

The effectiveness of SOBI in modelling the variability of natural frequency estimates due to individual or multiple EOVs has been discussed in previous papers (Rainieri et al. 2016a, Rainieri et al. 2016b, Rainieri & Magalhaes 2017). However, SOBI is herein applied to the collected data in order to evaluate the possibility of developing a model of natural frequency variations due to EOVs from a short dataset for short-term predictions. Moreover, taking into account that a major advantage of SOBI is the possibility to model the environmental and operational variability of natural frequencies and, at the same time, to trace the pattern of the EOVs up to a scaling factor and an offset, it is used to estimate the temperature pattern in the structure over the monitoring period.

In order to develop the model of natural frequency variations due to EOVs, the dataset shown in Figure 3 is divided in two parts, whose limit is represented by the occurrence of the earthquake. The first dataset is a short dataset consisting of 530 samples, and it is used to develop the model of variation of the natural frequencies due to EOVs for short-term predictions. The model predictions are compared afterwards to the data measured after the occurrence of the earthquake (Figure 8). Predictions are in good agreement with the experimental data even if a very short dataset is used to set the model, thus confirming its effectiveness for short-term predictions. This result yields interesting applicative perspectives for compensation of environmental influence in the short term based on a moving window approach.

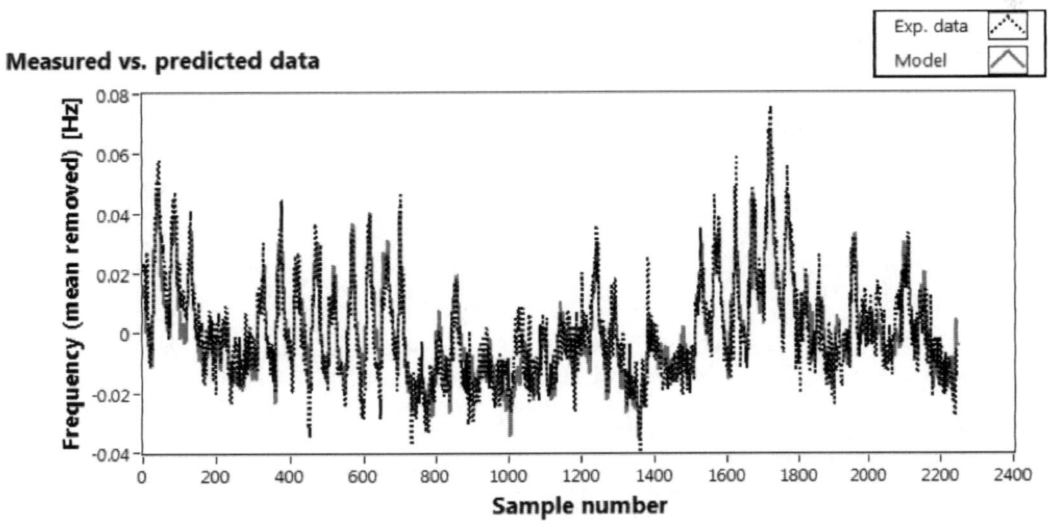

Figure 8. Sample comparison between predicted and measured natural frequencies (mode #1, data are centralized)

The prediction of the temperature pattern provided by SOBI (Figure 9a) is also similar to the local temperature in the area measured by a nearby meteorological station (http://www.paolociraci.it/meteo/campobasso-condizioni-tempo.htm). The differences can probably be addressed to direct sun radiation effects. In addition to temperature, a second source (Figure 9b) has been identified by SOBI as responsible for variations of the natural frequencies. That might represent the effect of the occupants, because recurring peaks can be identified at nearly the same hour of the day. However, further verifications are needed to reach a firm conclusion.

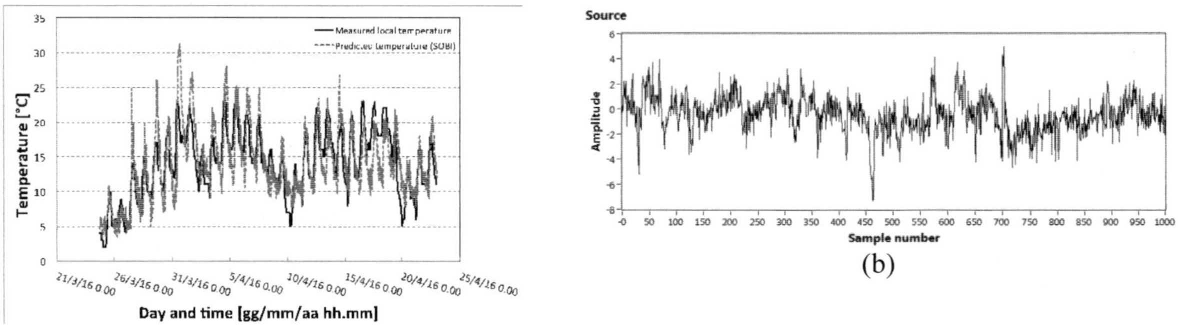

(a) (b)

Figure 9. Estimated sources by SOBI: first source in comparison with air temperature (a), second source

3. Conclusions

Some results of an on-going monitoring project for the main hospital in Campobasso, Italy, have been presented, focusing the attention on the response of the structure in operational conditions and after seismic events. The monitoring system is able to provide estimates of the fundamental modal parameters over time on a continuous basis. The herein presented results showed the influence of the temperature on natural frequency estimates, and the dynamic interaction between the two nearby monitored blocks through the joint. The compensation of the influence of EOVs on natural frequency estimates has been operated by SOBI. Results confirm that a reliable model for short-term predictions can be set from a relatively short dataset.

References

Ans, B, Hérault, J & Jutten, C 1985, 'Adaptive neural architectures: detection of primitives', *Proceedings of COGNITIVA '85*, Paris, France, pp. 593-597.

Ayres, JM & Phillips, RJ 1998, 'Water damage in hospital resulting from the Northridge Earthquake', *ASHRAE Transactions*, vol. 104, pp. 1286–1296.

Belouchrani, A, Abed-Meraim, K, Cardoso, JF & Moulines, E 1997, 'A blind source separation technique using second-order statistics', *IEEE Transactions on Signal Processing*, vol. 45, pp. 434-444.

Brincker, R, Zhang, L & Andersen, P 2001, 'Modal identification of output-only systems using frequency domain decomposition', *Smart Materials and Structures*, vol. 10, pp. 441–445.

Cardoso, JF, Souloumiac, A 1996, 'Jacobi angles for simultaneous diagonalization', *SIAM Journal of Matrix Analysis and Applications*, vol. 17, pp. 161–164.

Farrar, CR & Worden, K 2013, *Structural Health Monitoring: A Machine Learning Perspective*, John Wiley & Sons Ltd., Chichester.

Friswell, MI & Mottershead, JE 1995, *Finite Element Model Updating in Structural Dynamics*, Kluwer Academic Publishers, Dordrecht.

Hair, JF, Black, WC, Babin, BJ & Anderson, RE 2010, *Multivariate Data Analysis*, Pearson Prentice Hall, Upper Saddle River.

Jacques, CC, McIntosh, J, Giovinazzi, S, Kirsch, TD, Wilson, T & Mitrani-Reiser, J 2014, 'Resilience of the Canterbury Hospital System to the 2011 Christchurch Earthquake', *Earthquake Spectra*, vol. 30, no. 1, pp. 533-554.

Masi, A, Santarsiero, G, Gallipoli, MR, Mucciarelli, M, Manfredi, V, Dusi, A & Stabile, TA 2014, 'Performance of the health facilities during the 2012 Emilia (Italy) earthquake and analysis of the Mirandola hospital case study', *Bulletin of Earthquake Engineering*, vol. 12, no. 5, pp. 2419-2443.

Miranda, E, Mosqueda, G, Retamales, R & Pekcan, G 2012, 'Performance of nonstructural components during the February 27, 2010 Chile Earthquake', *Earthquake Spectra*, vol. 28, no. S1, pp. S453–S471.

Pan American Health Organization 2008, *Hospital Safety Index – Guide for Evaluators*, Washington, USA.

Peeters, B & De Roeck, G 1999, 'Reference-based stochastic subspace identification for output-only modal analysis', *Mechanical Systems and Signal Processing*, vol. 13, no. 6, pp. 855-878.

Poncelet, F, Kerschen, G, Golinval, JC & Verhelst, D 2007, 'Output-only modal analysis using blind source separation techniques', *Mechanical Systems and Signal Processing*, vol. 21, pp. 2335-2358.

Rainieri, C 2014, 'Perspectives of Second-Order Blind Identification for Operational Modal Analysis of Civil Structures', *Shock and Vibration*, Article ID 845106.

Rainieri, C & Fabbrocino, G 2013, 'Sustainable design of smart health facilities in seismically prone areas', *Key Engineering Material*s, vol. 569-570, pp. 278-285.

Rainieri, C & Fabbrocino, G 2015, 'Development and validation of an automated operational modal analysis algorithm for vibration-based monitoring and tensile load estimation', *Mechanical Systems and Signal Processing*, vol. 60–61, pp. 512-534.

Rainieri, C, Fabbrocino, G, Manfredi, G & Dolce, M 2012, 'Robust output-only modal identification and monitoring of buildings in the presence of dynamic interactions for rapid post-earthquake emergency management', *Engineering Structures*, vol. 34, pp. 436-446.

Rainieri, C, Fabbrocino, G & Verderame, GM 2013, 'Non-destructive characterization and dynamic identification of a modern heritage building for serviceability seismic analyses', *NDT & E International*, vol. 60, pp. 17-31.

Rainieri, C, Gargaro, D & Fabbrocino, G 2016a, 'On the use of Second Order Blind Identification for the characterization of the environmental influence on modal parameter estimates', *Proceedings of The 8th European Workshop on Structural Health Monitoring*, Bilbao, Spain.

Rainieri, C, Gargaro, D & Fabbrocino, G 2016b, 'Laboratory and field investigations about techniques for compensation of the effects of environmental and operational variables on modal parameter estimates in operational conditions', *Proceedings of The International Conference on Noise and Vibration Engineering ISMA 2016*, Leuven, Belgium.

Rainieri, C & Magalhaes, F 2017, 'Challenging aspects in removing the influence of environmental factors on modal parameter estimates for SHM purposes', *Proceedings of The X International Conference on Structural Dynamics EURODYN 2017*, Rome, Italy (accepted for publication).

Taghavi, S & Miranda, E 2003, *Response assessment of nonstructural building elements*, PEER Report 2003/05, Pacific Earthquake Engineering Research Center, University of California Berkeley, Berkeley.

Tan, P-N, Steinbach, M & Kumar, V 2006, *Introduction to data mining*, Pearson Addison-Wesley, Reading, USA.

Yan, A-M, Kerschen, G, De Boe, P, Golinval, J-C 2005, 'Structural damage diagnosis under varying environmental conditions – Part I: A linear analysis', *Mechanical Systems and Signal Processing*, vol. 19, pp. 847-864.

The 8th International Conference on
Structural Health Monitoring of Intelligent Infrastructure
Brisbane, Australia | 5-8 December 2017

Damage Detection in Arch Bridges using Vibration based Damage Detection Techniques

N. Jayasundara [1], D.P. Thambiratnam[2], T.H.T. Chan[3], A. Nguyen[4].
[1,2,3,4] *School of Civil Engineering & Built Environment, Queensland University of Technology – Australia,*
Email: walpola.jayasundara@hdr.qut.edu.au [1]

Abstract

Most structures are built to have a long life. However, during this long life, they can be incurred damage due to structural deterioration, environmental effects and random actions such as impacts. Early damage assessment and appropriate retrofitting will enable the continued safe and efficient functions of the structures. In this context, vibration based techniques have emerged with the potential for reliable damage assessment. This study develops a dual criteria approach based on the vibration characteristics for detecting and locating damage in arch bridges. Steel arch bridges are one of the aesthetically pleasing bridge types in which damage has been rarely studied. In particular, the arch rib and struts (or columns in deck type bridges) which are important structural components, have received much less attention in damage detection. This study will therefore focus on damage detection in arch bridge structural components using indices based on modified Modal Flexibility (MF) and Modal Strain Energy (MSE) methods. The study is carried out through numerical simulations supported by limited experimental testing. The modelling techniques are first validated through experimental testing of a laboratory scale arch bridge model and this is followed by damage detection studies in this bridge structure. Data obtained from finite element analyses of the healthy and damaged arch bridge models are applied into the modified MF and MSE algorithms for detecting and locating the damage. Results demonstrate that the proposed method is capable of detecting damage in the arch rib and vertical columns of deck type arch bridges.

1. Introduction

Bridges are crucial elements of rail and road infrastructure. But long service periods make bridges more susceptible to everyday traffic loads, fluctuating temperatures and winds, corrosive media, torrential rain, fatigue effects, and material aging. Single or coupling effects of these factors may deteriorate the structural performance and increase the risk of partial or full collapse. Thus the significance of structural integrity, durability and reliability evaluation throughout the life cycle and damage detection prior to unexpected incidents or costly repairs has attracted a great attention over the years. Structural Health Monitoring (SHM) has been identified as a viable technology and an effective approach to ensure safe operations of existing or newly built structures. It is defined by Chan et al (2011) as the use of an on-structure sensing system to monitor structural performance and to evaluate the symptoms of anomalies, deterioration or damages that may affect the operation, serviceability, or safety reliability of a structure.

Precise damage detection is one of the key elements of structural health monitoring. The two basic types of damage detection can be identified as Local methods and Global methods (Doebling et al. 1996). Non-destructive tests such as acoustic emission, ultrasonic, eddy current, magnetic particle inspection and radiography are local damage detection techniques which require prior knowledge of the damaged region (Shih et al. 2011).The limitations of local damage detection techniques were addressed in global methods which examine the changes in vibration properties between the healthy and damaged states of the structure to evaluate the damage. These methods are known as Vibration based Damage Detection Techniques (VBDDT). Comprehensive literature on VBDDT provide evidence of the broad research carried out in this field over the last couple of decades.

A considerable number of studies on the application of SHM and damage detection of different bridge types can be found in the literature (Shih et al. 2011, Wickramasinghe et al. 2016). However, the attention towards structural health monitoring and damage detection of arch bridges has been relatively low compared with other types of bridges. With regards to arch bridges, the bridge deck has been the major concern of all vibration based damage detection (VBDD) studies than the more critical load bearing members such as the arch rib, hangers and struts (columns). This was probably because it is easier to study the deck damage and it might have been envisaged that the deck is the most important part of the bridge where all the human activities occur. But, from a structural engineering point of view, all the other structural components are equally or more important for the safe operations of the bridge. The bridge deck is also the most accessible and visible component in the arch bridge and deck damage would be more easily captured by the bridge inspectors than the damage in the other bridge components. The failure of one cable/hanger or a strut/column is rarely visible and failure of these eventually lead to a collapse of at least a part of the bridge structure and could lead to the progressive collapse of the entire bridge. Therefore this research will focus on developing reliable damage indices for application to arch bridges in order to detect and locate the damage in them.

Vibration based damage index method recognizes structural damage by considering the changes in the vibration properties between healthy and damaged states of the structure. The Model Flexibility (MF) method, first proposed by Pandey & Biswas (1994, 1995), has been used in many damage detection studies due to its accuracy, ease of application and convenient computation (Moragaspitiya et al. 2013; Shih et al. 2009). Another vibration based damage detection technique, Modal Strain Energy (MSE) method was first proposed by Stubbs et al. (1992). This method has been used in further studies by Law et al. (1998) to detect and locate damage in structures with incomplete and noisy measured modal data and by Shi et al. (2000) for damage detection in the structural elements of a single-bay, two-story portal steel frame. A multi-criteria approach incorporating MF and MSE methods was proposed by Shih et al. (2009) for damage detection in beams and slabs. They were able to detect single and multi-damages in the beams and slabs. It was shown that for a single damage, both MF and MSE changes provide accurate results for detecting and locating damage, but for multiple damage cases, the MSE method performed better in predicting the damage locations. However, the application of these methods for damage detection of arch bridges is not evident in the literature.

This paper presents a dual criteria approach which uses damage indices (DIs) based on both the modified MF and MSE changes to detect and locate damage in the main structural components of deck type arch bridge model. Using the 2 DIs simultaneously will enable the results obtained from either DI to complement and supplement the results from the other DI and lead to more accurate prediction of the damage location.The feasibility of the proposed method is demonstrated through its application to a range of damage detection scenarios.

2. Method

2.1 Modal Flexibility Method

Modal flexibility F of a linear structure can be expressed as;

$$[F] = \sum_{i=1}^{m} \frac{1}{\omega_i^2} \emptyset_i \emptyset_i^T \quad (1)$$

i is the mode number considered and ω_i and m are the natural frequency of the structure at mode i and the total number of modes considered, respectively. $\emptyset_i$ is the ith mode shape vector and $\emptyset_i^T$ is its transpose.

According to Pandey and Biswas (1994) the change in flexibility can be expressed as follows where d and h denotes the damage and healthy conditions respectively.

$$MFC = [F]_d - [F]_h = \left[\sum_{i=1}^{n} \frac{1}{\omega_i^2} \phi_i \phi_i^T\right]_d - \left[\sum_{i=1}^{n} \frac{1}{\omega_i^2} \phi_i \phi_i^T\right]_h \quad (2)$$

The diagonal values of MFC matrix are extracted and each term represents the modal flexibility at a particular location along the member.

Damage Index (DI) is obtained by dividing MFC value of a particular location by MF value of that same location at healthy state. Therefore the normalized damage index can be written as follows

$$DI = \frac{\left[\sum_{i=1}^{m} \frac{1}{\omega_i^2} \emptyset_i \emptyset_i^T\right]_D - \left[\sum_{i=1}^{m} \frac{1}{\omega_i^2} \emptyset_i \emptyset_i^T\right]_H}{\left[\sum_{i=1}^{m} \frac{1}{\omega_i^2} \emptyset_i \emptyset_i^T\right]_H} \quad (3)$$

2.2 Modal Strain Energy Method

MSE based method proposed by Stubbs et al. (1995) can be identified as follows.

The strain energy, U of a Bernoulli-Euler beam is given by Equation 4 below. Where EI represents the flexural rigidity of the cross section and d^2y/dx^2 refers to the curvature of the beam in which y is the vertical deflection while x is the distance measured along the beam length.

$$U = \int \frac{EI}{2} \left(\frac{d^2 y}{dx^2}\right)^2 dx \quad (4)$$

With the use of principle of virtual work and modal stiffness derivations, the damage index β_{ji} for the jth element and i^{th} mode of the beam can be provided as follows

$$\beta_{ij} = \frac{k_j}{k_j^*} = \frac{\left(\int_j [\phi_i^{''*}(x)]^2 dx + \int_0^L [\phi_i^{''*}(x)]^2 dx\right) \int_0^L [\phi_i^{''}(x)]^2 dx}{\left(\int_j [\phi_i^{''}(x)]^2 dx + \int_0^L [\phi_i^{''}(x)]^2 dx\right) \int_0^L [\phi_i^{''*}(x)]^2 dx} \quad (5)$$

$$\beta_{ij} = \frac{k_j}{k_j^*} = \frac{\left[(\phi_i^{''*})^2 + \Sigma(\phi_i^{''*})^2\right]\left[\Sigma(\phi_i^{''})^2\right]}{\left[(\phi_i^{''})^2 + \Sigma(\phi_i^{''})^2\right]\left[\Sigma(\phi_i^{''*})^2\right]} \quad (6)$$

For both methods a single indicator is generated by taking several global modes into account as follows.

$$\beta_{ij} \text{ or } DI_{ij} = \frac{\sum_{i=1}^{NM} Num_{ji}}{\sum_{i=1}^{NM} Denom_{ji}} \quad (7)$$

Where NM refers to the number of modes and Num_{ji} is the numerator and $Denom_{ji}$ is the denominator of Equations 3 and 6.

2.3 Modified Modal Flexibility and Modified Modal Strain Energy Methods

The damage indices (DIs) based on the traditional MF and MSE methods were not very effective in detecting and locating damage in the structural members of the deck type arch bridge. This paper therefore develops modified damage detection techniques that considerably improve the results for detecting and locating damage in the arch rib and vertical columns of deck type arch bridges.
In the initial global modes of vibration of deck type arch bridges, the maximum cumulated mass participations occur in the lateral and vertical directions. Thus the DIs using the lateral and vertical components of the mode shapes can provide promising results than those that use the total or resultant mode shapes. To facilitate the use of these component specific features, DIs can be modified as follows.

$$DI_V = \frac{\left[\sum_{i=1}^{m}\frac{1}{\omega_i^2}\phi_i\phi_i^T\right]_{DV} - \left[\sum_{i=1}^{m}\frac{1}{\omega_i^2}\phi_i\phi_i^T\right]_{HV}}{\left[\sum_{i=1}^{m}\frac{1}{\omega_i^2}\phi_i\phi_i^T\right]_{HV}} \tag{8}$$

$$DI_L = \frac{\left[\sum_{i=1}^{m}\frac{1}{\omega_i^2}\phi_i\phi_i^T\right]_{DL} - \left[\sum_{i=1}^{m}\frac{1}{\omega_i^2}\phi_i\phi_i^T\right]_{HL}}{\left[\sum_{i=1}^{m}\frac{1}{\omega_i^2}\phi_i\phi_i^T\right]_{HL}} \tag{9}$$

For MF based damage index, subscripts V and L denote the vertical and lateral component specific DIs using lateral and vertical components of mode shapes respectively.
The MSE based method can be similarly modified and β_{ijV} and β_{ijL} are the vertical and lateral component specific DIs using lateral and vertical components of mode shapes respectively.

$$\beta_{ijV} = \frac{k_j}{k_j^*} = \frac{\left[(\phi_{iV}^{''*})^2+\Sigma(\phi_{iV}^{''*})^2\right]\left[\Sigma(\phi_{iV}^{''})^2\right]}{\left[(\phi_{iV}^{''})^2+\Sigma(\phi_{iV}^{''})^2\right]\left[\Sigma(\phi_{iV}^{''*})^2\right]} \tag{10}$$

$$\beta_{ijL} = \frac{k_j}{k_j^*} = \frac{\left[(\phi_{iL}^{''*})^2+\Sigma(\phi_{iL}^{''*})^2\right]\left[\Sigma(\phi_{iL}^{''})^2\right]}{\left[(\phi_{iL}^{''})^2+\Sigma(\phi_{iL}^{''})^2\right]\left[\Sigma(\phi_{iL}^{''*})^2\right]} \tag{11}$$

3. Application

3.1 Validation of modelling techniques

Before the application of the developed methods for damage detection, a comprehensive laboratory test was conducted on small scale deck type steel arch bridge structure in order to validate the finite element modelling techniques of the bridge models. Subsequently, the validated model was used in damage detection studies.
The test model has a span of 1m and a height of 0.4m. The deck has a width of 0.25m and is made from 2mm thick steel plate and supported by 9, 6mm diameter vertical struts on each side. All members of the bridge are made out of general steel with Young' Modulus, Poisson ratio and density of 2.05×10^{11} Pa, 0.3 and 7870 kg/m^3 respectively. The arch rib is a 12mm x 12mm square section strut bent according to the required shape and the two main arch ribs are connected together via 6mm diameter circular struts, as shown in Figure 1(a). The bridge is fixed to a plate at the base which represents fixed-fixed boundary conditions.

The dynamic testing was performed on the bridge model to obtain vibration parameters of natural frequencies and mode shapes to validate the FE model by comparing the results. The data acquisition system contains single-axial PCB® 393B05 integrated circuit piezoelectric accelerometers to measure vertical and lateral accelerations. These accelerometers can self-calibrate within few seconds and automatically pick up the precise acceleration at the position. Accelerometers were attached to selected nodes in the bridge as shown in Figure 1(b) and each of these indicate the acceleration in the direction of its axis. The modal data was acquired by a National Instruments (NI) data acquisition system including NI cDAQ 9172 chassis, NI 9234 dynamic signal acquisition modules and LabVIEW signal Express software. Since the laboratory model cannot excite naturally, rubber hammer taping was done to excite the structure randomly (Figure 1(c)). Data was gathered for a predetermined time period considering its natural frequencies. Few repeats were carried out to eliminate any random errors and to improve the accuracy of results. All the acceleration data were captured in the time domain and were transferred to the ARTeMIS modal analysis software to obtain modal parameters. The natural frequencies and mode shapes were determined by The Data Driven Stochastic Subspace Identification (SSI-DATA) in ARTeMIS modal analysis software (Nguyen et al. 2014). These natural frequencies and mode shapes of the structure given in Figure will be compared with those from the analysis of finite element model for the validation process.

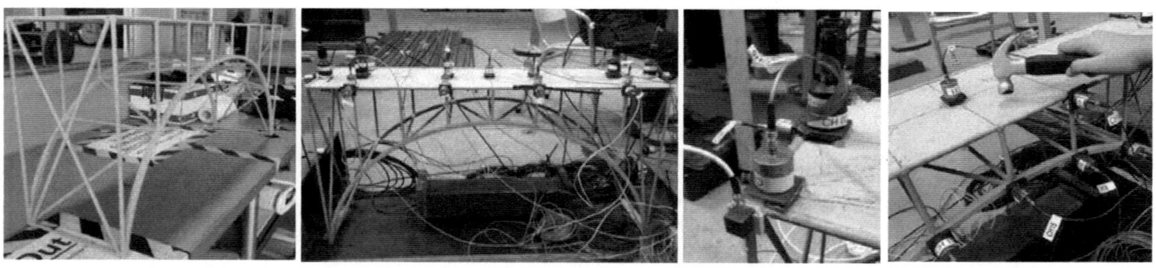

(a) (b) (c)

Figure 1: (a) Bridge model (b) Accelerometer arrangement (c) Random taping

The exact finite element model of laboratory bridge model was created in ANSYS Workbench finite element software which is a commercially available FEM software, facilitated with Graphical User Interface (GUI) for developing complex multibody parts and complex analyses. The geometry of the laboratory bridge was modelled at DesignModeler of ANSYS Workbench as a 3D FE model. Each part (deck, vertical struts, arch rib, and wind bracings) was connected to the other by joint feature in Mechanical module of ANSYS workbench. General mild steel properties for which the Young' Modulus, Poisson ratio and density were 2.05 x10^{11} Pa, 0.3 and 7870 kg/m^3 were assigned to each element of the structure before validation. All elements, except the deck elements, were modelled with beam elements (BEAM 188) and the deck was modelled with shell elements (SHELL 181). Prestressed Modal Analysis was conducted (incorporating the initial stress state if the structure under its self-weight) to obtain natural frequencies and mode shapes of the numerical model of the arch bridge (ANSYS® Academic Research).

Modal updating was performed to tune the structural parameters of the FE model to match with the experimental model. Natural frequencies of the first four modes obtained from the FE model are compared with those from the experiments in Figure 2.

It can be seen that the results from the analysis of the finite element model of the laboratory bridge compare reasonably well with the experimental results and provide confidence in the modelling techniques. These modelling techniques were then used to model the full scale bridge for further analyses.

Mode 1	Mode 2	Mode 3	Mode 4
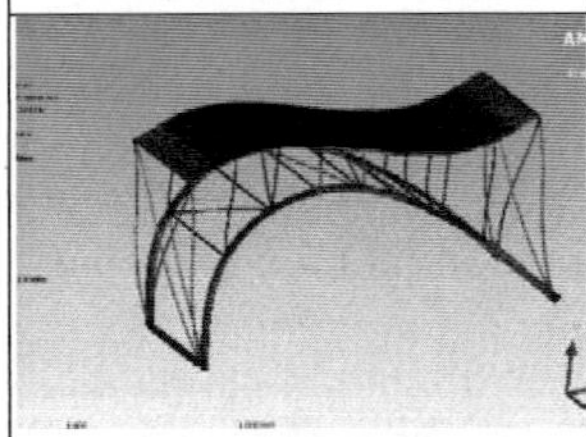	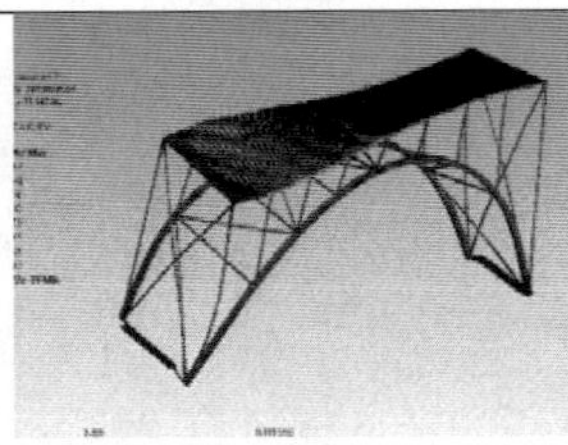	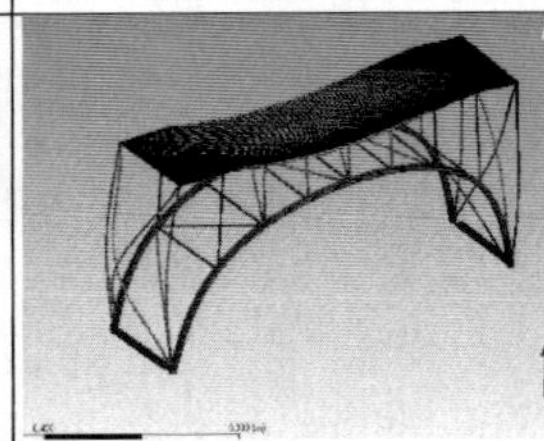	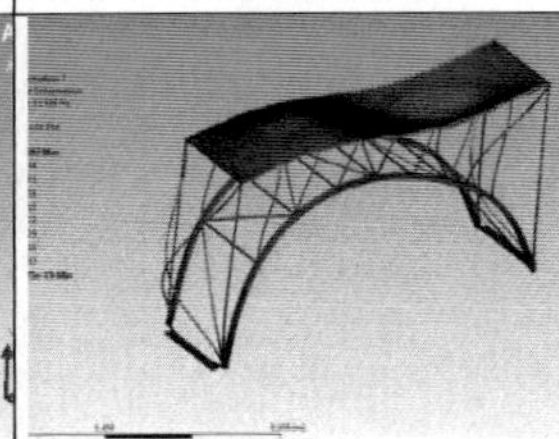
21.51 Hz	34.14 Hz	53.14 Hz	61.95 Hz
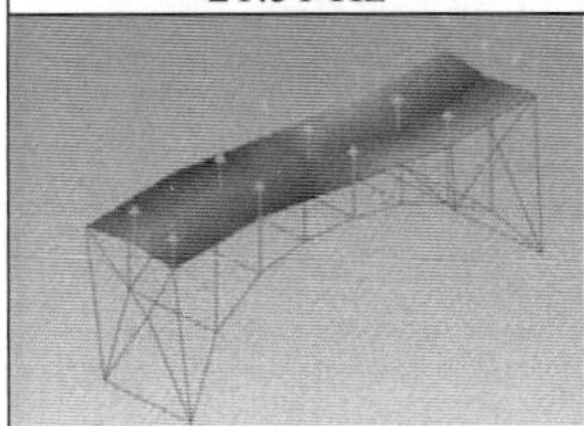	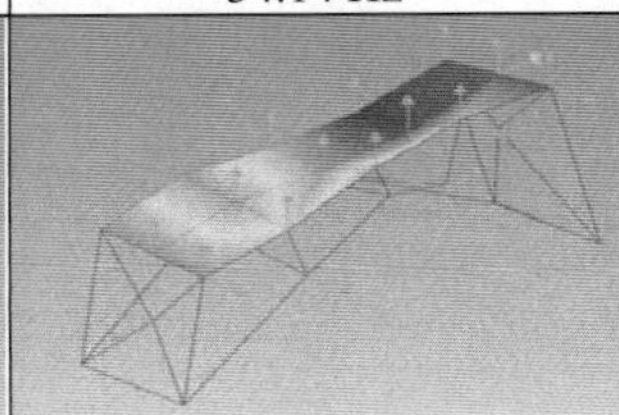	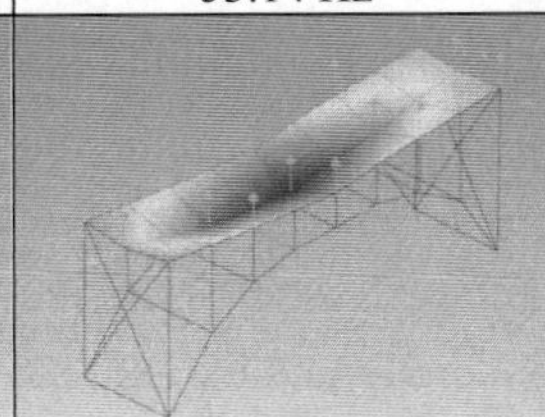	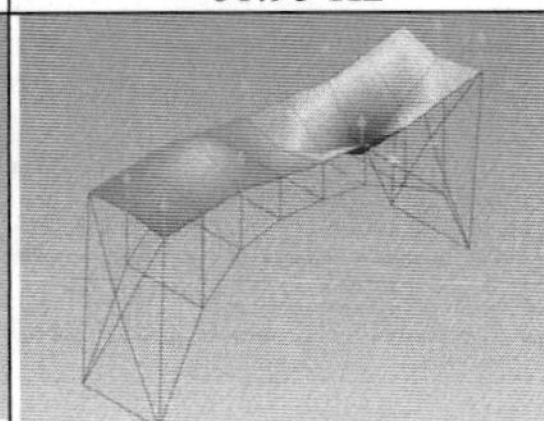
20 Hz	33 Hz	54.25 Hz	64.25 Hz

Figure 2: Numerical (top) and experimental (bottom) results for laboratory scale arch bridge structure

3.2 Damage Detection in arch bridge model

In order to demonstrate the versatility of the damage detection method different damage cases on the arch rib and on the vertical columns were applied to the FE model of the arch bridge. Damage cases are illustrated in Table 1 and Table 2.

Table 1: Damage Scenarios of arch rib

Damage Scenario	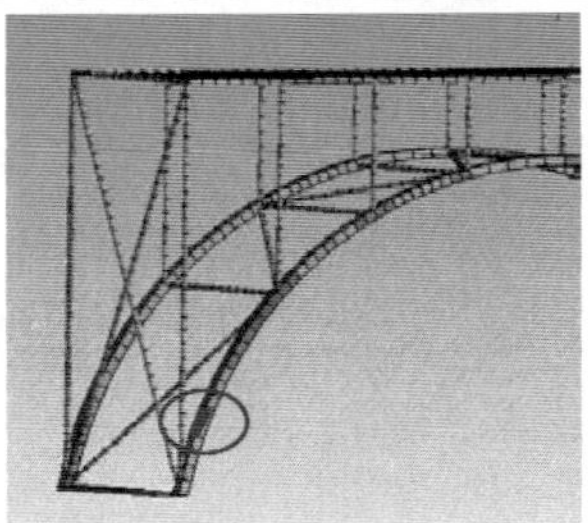	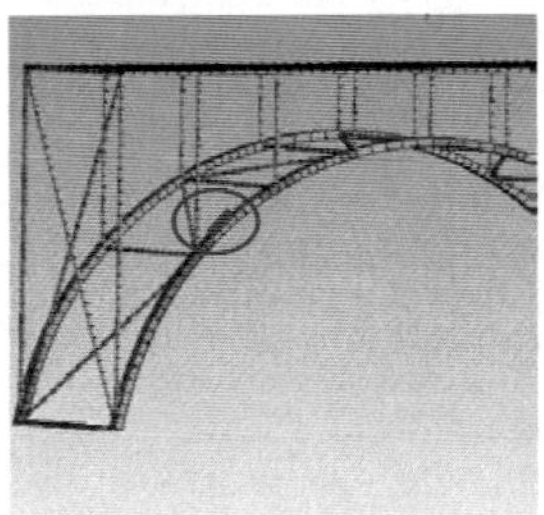	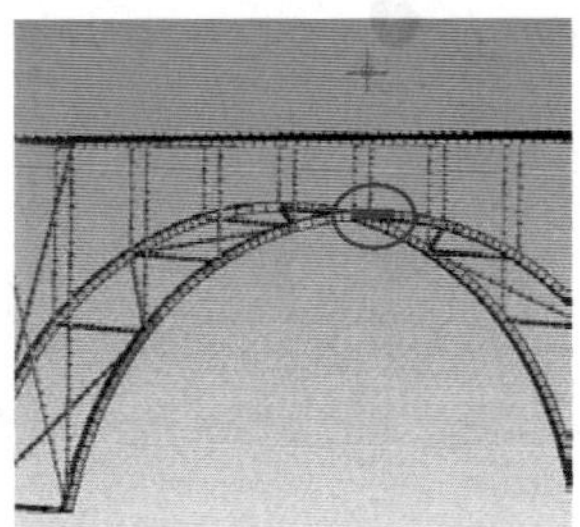
	Rib Damage Case 1	Rib Damage Case 2	Rib Damage Case 3
Damaged Element	Damage at arch spring	Damage at 1/4 span of rib	Damage at crown
Stiffness Reduction	10%	10%	10%

Table 2: Damage Scenarios of vertical columns

Damage Scenario	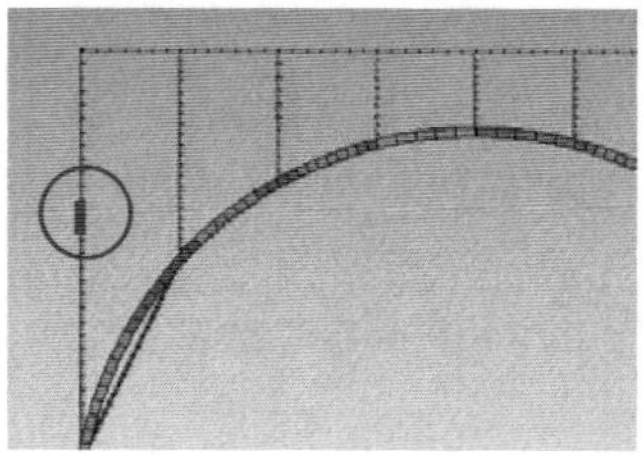	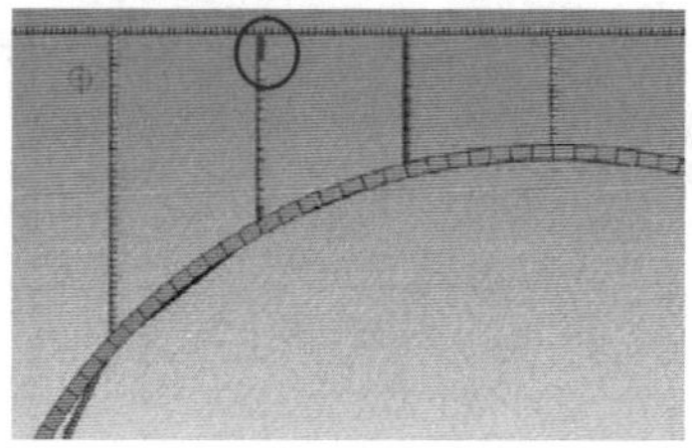
	Vertical column damage Case 1	Vertical column damage Case 2
Damaged Element	Damage at mid of the long column	Damage at the edge of the short column
Stiffness Reduction	10%	10%

Modal parameters such as natural frequencies and mode shapes of healthy and damaged structure were extracted from FE modal analyses and were then used to calculate the proposed damage indices. Mode shape data was extracted from 38 nodal points along the rib and 10-20 nodal points along each vertical column. The proposed modified MF and MSE damage indices are calculated separately for each damage case using the first four global modes of vibration. The peaks of the damage index plots will indicate the location of the damage. The results related to each damage case are presented in Figure 3 – 7.

4. Results and Discussion

- Rib Damage Case 1: Damage at arch spring

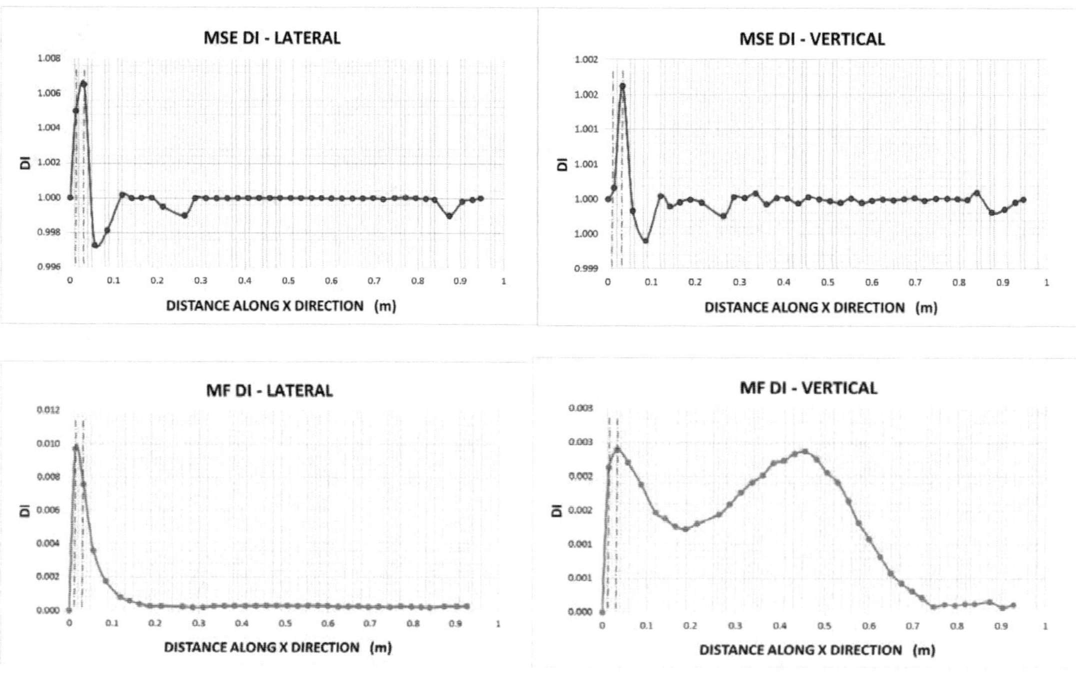

Figure 3: Variation for Lateral and Vertical component specific MSE DIs and MF Dis

- Rib Damage Case 2: Damage at quarter span of the arch rib

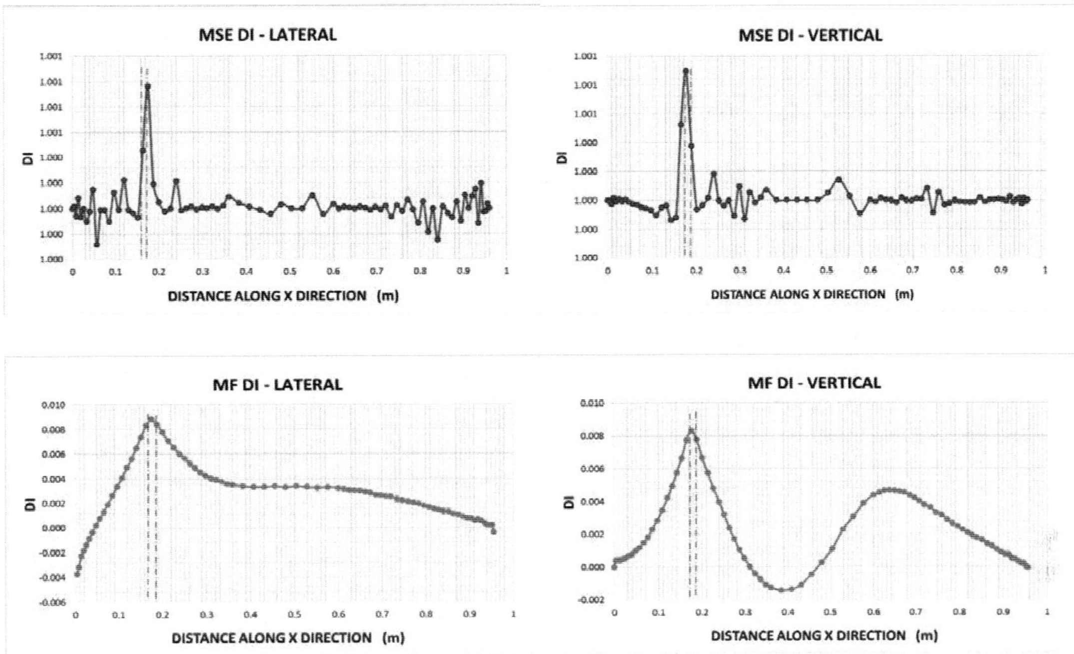

Figure 4: Variation for Lateral and Vertical component specific MSE DIs and MF DIs

- Rib Damage Case 3: Damage at mid span of the arch rib

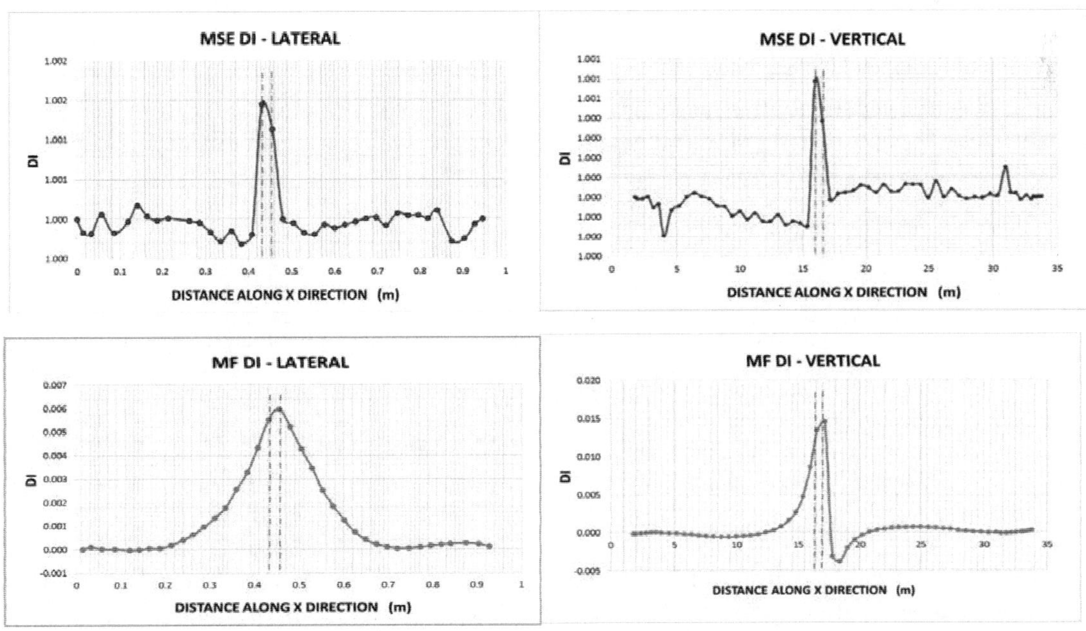

Figure 5: Variation for Lateral and Vertical component specific MSE DIs and MF DIs

- Vertical Column Damage Case1

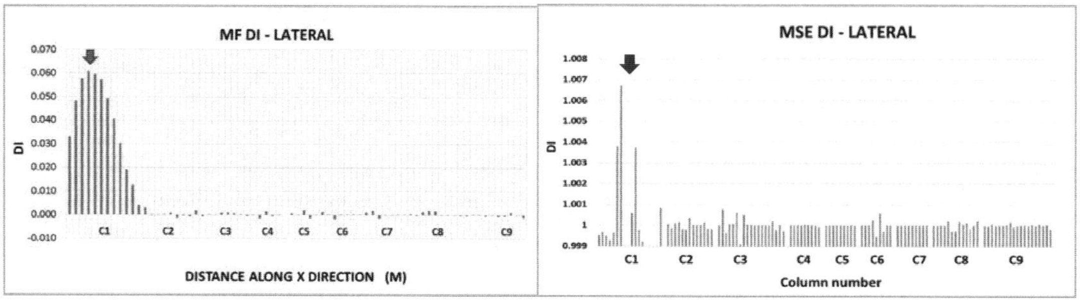

Figure 6: Variation for Lateral and Vertical component specific MSE DIs and MF DIs for the damage at mid of the long column

- Vertical Column Damage Case2

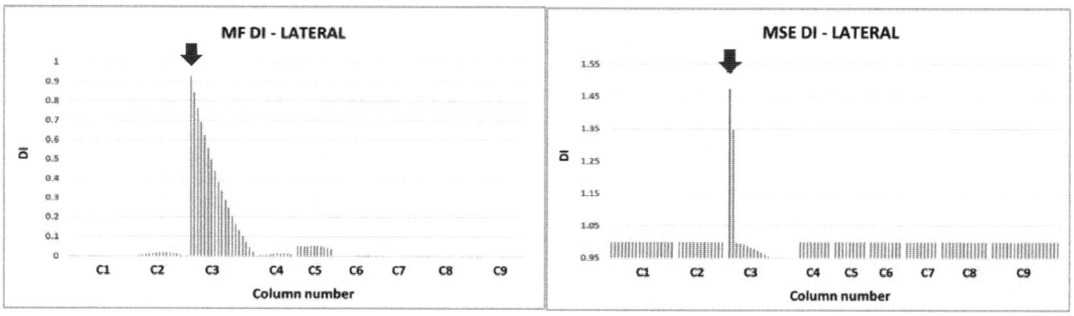

Figure 7: Variation for Lateral and Vertical component specific MSE DIs and MF DIs for the damage at the edge of the short column

The performance of each proposed damage index is summarized in Table 3. Since one damage index cannot reliably identify damage in all scenarios, both MSE and MF methods can be used simultaneously and separately that will complement and supplement the results from the other DI and lead to more accurate prediction of the damage location. This combined evaluation can provide a better chance to cross check and obtain the correct results.

Table 3: The performances of proposed damage indices

Structural component	Damage detection method
Arch rib	Modified MF and Modified MSE methods using both lateral and vertical component specific damage indices. Better indicator is MSE based DI.
Vertical columns	Modified MF and MSE methods with lateral component specific damage index.

5. Conclusion

As a promising method to detect and locate damage in arch bridge structural components, a dual criteria approach based on modified MSE and MF methods is presented in this paper. The traditional MF index and Stubbs' MSE index are modified by decomposing into vertical and lateral indices to capture the damage location more effectively. The feasibility of the proposed procedure is illustrated via its application to identify damage in structural components of an arch bridge model. A total of five damage scenarios were considered related to arch rib and vertical columns. The results clearly show that the both methods are capable of detecting the damage in the arch rib, and vertical columns. But some of the directional DIs are allied with some false alarms. Hence the dual-criteria approach will be favorable in those cases as the results obtained from either DI will complement and supplement the results from the other DI and lead to more accurate prediction of the damage location. This research will further investigate to setup a criteria to refrain from unsafe decisions or unnecessary checking for nonexistence damage, under conflicting results. Finally the procedure developed in this research can be extended to detect and locate damage in other types of arch bridges and will contribute towards the safe and efficient performance of these types of bridge structures.

References

ANSYS® Academic Research, Release 18, Help System, Coupled Field Analysis Guide, ANSYS, Inc.

Chan, T. H. T., Wong, K. Y., A Li, Z. X., & Ni, Y.-Q. 2011. Structural health monitoring for long span bridges: Hong Kong experience & continuing onto Australia Structural Health Monitoring in Australia (pp. 1-32): Nova Publishers.

Doebling, S.W., Farrar, C.R., Prime, M.B. and Shevitz, D.W., 1996. *Damage identification and health monitoring of structural and mechanical systems from changes in their vibration characteristics: a literature review* (No. LA--13070-MS). Los Alamos National Lab., NM (United States).

Law, S., Z. Shi, and L. Zhang, Structural damage detection from incomplete and noisy modal test data. *Journal of Engineering Mechanics*, **1998**. 124(11): p. 1280-1288.

Moragaspitiya, H.P., Thambiratnam, D.P., Perera, N.J. and Chan, T.H., 2013. Development of a vibration based method to update axial shortening of vertical load bearing elements in reinforced concrete buildings. Engineering Structures, 46, pp.49-61.

Nguyen, T., Chan, T. H., & Thambiratnam, D. P. (2014). Effects of wireless sensor network uncertainties on output-only modal analysis employing merged data of multiple tests. *Advances in Structural Engineering, 17*(3), 319-330.

Pandey, A.K. and Biswas, M., 1994. Damage detection in structures using changes in flexibility. Journal of sound and vibration, 169(1), pp.3-17.

Pandey, A., & Biswas, M. 1995. Experimental verification of flexibility difference method for locating damage in structures. Journal of Sound and Vibration, 184(2), 311-328

Shih, H.W., Thambiratnam, D.P. and Chan, T.H., 2009. Vibration based structural damage detection in flexural members using multi-criteria approach. *Journal of sound and vibration*, *323*(3), pp.645-661.

Shi, Z. Y., Law, S. S., & Zhang, L. M. (2000). Damage localization by directly using incomplete mode shapes. Journal of engineering mechanics, 126(6), 656-660.

Shih, H. W., D.P. Thambiratnam and T. H. Chan 2011. Damage detection in truss bridges using vibration based multi-criteria approach. *Structural Engineering and Mechanics* 39(2): 187-206.

Stubbs, N., Kim, J., & Topole, K. 1992. An efficient and robust algorithm for damage localization in offshore platforms. Paper presented at the Proc. ASCE Tenth Structures Congress.

Stubbs, N., Kim, J.T. and Farrar, C.R., 1995, February. Field verification of a nondestructive damage localization and severity estimation algorithm. In *Proceedings-SPIE the international society for optical engineering* (pp. 210-210). SPIE INTERNATIONAL SOCIETY FOR OPTICAL.

Wickramasinghe, W.R., Thambiratnam, D.P., Chan, T.H. and Nguyen, T., 2016. Vibration characteristics and damage detection in a suspension bridge. Journal of Sound and Vibration, 375, pp.254-274.

The 8th International Conference on
Structural Health Monitoring of Intelligent Infrastructure
Brisbane, Australia | 5-8 December 2017

Structural Health Monitoring for Lightweight Footbridges and Pedestrian Vibration Assessment: A Case Study

Liang Wang[1], Ruixiang Song[1] and Yubin Wu[1]
Beijing Municipal Institute of Labour Protection – China, Email: wangliang@bmilp.com

Abstract

Lightweight footbridges subjected to non-vehicular loads have different vibration characteristics from others. In order to take advantage of the development in structural health monitoring (SHM) and investigate its application to lightweight bridges, this paper presents a case study on an aluminium alloy footbridge located in a high pedestrian activity area in Beijing, China. In the test, the structure was excited by the environment with no disturbance to operation. Acceleration time-series at designated structural joints were acquired for a purpose of operational modal analysis (OMA), and the number of pedestrian passage on the structure was also recorded. The identified modal parameters were found to have a proportional variation against the number of pedestrians. These modal parameters were also used to update the structure's numerical model, on which three pedestrian loading models from existing standards were applied to assess human vibration comfort indices. The prototype study has concluded that the involved methods are feasible on data processing, modal identification, and model updating for lightweight footbridges. The updated numerical model may be further used to quantify pedestrian vibration induced by footbridges. The conclusions are informative to SHM initiatives on similar bridges.

1. Introduction

Application of lightweight materials, e.g. stainless steel, aluminium alloy, and fibre-reinforced polymer (FRP) composites, has brought a new challenge to understand proper dynamic responses of footbridges. On one hand, a decreased structural mass tends to render the bridge having low natural frequencies and possibly makes pedestrian uncomfortable to resonance of the structure. On the other hand, many lightweight footbridges have been built in high activity areas and the boundary conditions of the structure may change from time to time with complicated human-structure interaction effects.

As a popular strategy in bridge asset management, today, structural health monitoring (SHM) is able to observe a system's characteristic change over time. The extracted modal data from an SHM system is useful to assess the current state of the subject, detect structural damage, and statistically analyse the asset's reliability. In this manner, many short or long term SHM projects have been motivated to observe dynamic responses and attributes of lightweight footbridges in changing environmental conditions (Occhiuzzi *et al.*, 2008; Huang *et al.*, 2010; and Hu *et al.*, 2013).

In addition to the need of structural integrity, another concern is excessive amount of body vibration received by pedestrians during their passage on footbridges. One example, which later induced a wide discussion on human-structure interaction as well as human body vibration comfort on operational footbridges, was the well-known "wobbly" Millennium Footbridge in London. The bridge experienced serious sideway movements from a synchronous excitation of thousands of people crossing in the opening day. It was the human-structure resonance that resulted in pedestrians passing on the bridge to sway in step, which conversely intensified the vibration of the structure (Dallard, 2001). Similarly, a few other footbridges with various types and spans have also been reported to have the same excessive lateral sway problem (Nakamura, 2003; Brownjohn *et al.*, 2004; Nakamura & Kawasaki, 2006; and Macdonald, 2008). Such problem not only challenges the structural serviceability, it may also lead to possible crowd panic due to insufficient prior knowledge to these severe motions. As one of the deadliest tragedies in recent years, in 2010, a stampede which took hundreds of lives occurred on pedestrians crossing an excessively vibrating footbridge in Cambodia (Hsu & Burkle, 2012). Therefore, it is crucial to assess the vibration comfort of pedestrian occupants on the footbridge.

To extend the practical application of SHM techniques on lightweight footbridges, this work presented a prototype SHM test on an aluminium alloy bridge located in a high pedestrian activity area in Beijing, China. The acquired information was used to identify modal signatures of the structure in different environmental conditions. Monitoring and modal identification results over time were discussed. A finite element (FE) model of the subject was then updated to the experimental results in order to carry pedestrian loading analyses with loading models from three existing standards. Lastly, the pedestrian vibration comfort was assessed with remarks. The prototype study provides practical information for dynamic behaviours of similar lightweight footbridges and draws attention to the suitability of methods to evaluate pedestrian vibration comfort.

2. The case study bridge

An aluminium alloy footbridge connecting the overpassed entrances of two shopping malls at Beijing Xidan Commercial Street (XCS) has been chosen as the subject of the study. The location is known as one of the highest commercial activity areas in Beijing, reported to have a daily flow volume of 145,000 ~ 165,000 passengers (BMBS, 2004). It opened for public in July just before the 2008 Summer Olympic Games, and has soon become a landmark architecture at the south end of XCS. Figure 1(a) shows the location of the case study subject to adjacent shopping centres and a subway interchange station. Figure 1(b) has a street view of the main span of the bridge.

Figure 1 The aluminium alloy footbridge in the case study: bridge location (a); street view of the main span (b).

2.1 Structural information

The footbridge is fabricated with aluminium alloy trusses, whose main span is 38.1m long and 8.0m wide. The clear height underneath the mid-span is 5.1m. The truss members and the deck are made of a relatively new aluminium alloy 6082-T6, whose Young's modulus is 70GPa, yield and ultimate strength is 260MPa, 310MPa respectively. 6082-T6 has been widely used for machining and structural applications among 6000 series due to its high stress capability and excellent corrosion resistance. The designed pedestrian load is 3.6 kPa. A modal test conducted immediately after the completion of the bridge shows that the structure's first natural frequency is 5.72Hz without loading, and 3.18Hz with 1/3 dead load (Yang & Wu, 2011).

2.2 Instrumentation and test process

Uniaxial and triaxial accelerometers were used to form a sensor network for the case study bridge. Specification of sensors is given in Table 1. Prior to the in-situ test, a back-to-back calibration test was carried out against a PCB 301M26 sensor. The result shows that the maximum error at 2Hz of these sensors is 3.23% in amplitude, and 0.97% in phase, which fits the purpose of the case study test (Wang, 2016).

Table 1 Specification of in-situ accelerometers

Accelerometer (Model)	Quantity	DOF	Sensitivity (mV/g)	Amplitude (g)	Freq. Range (Hz)	Mass (g)	Type
PCB 356B18	8	3(X/Y/Z)	1000	±5	0.3~5000	25	ICP
PCB 393B04	10	1	1000	±5	0.05~750	50	ICP

A 36-channel LMS SCADAS Mobile module with the software LMS Test.Lab v14.1 was employed for data acquisition (DAQ). The formed LMS system has an advantageous feature that any signal irregularity during a session can be simultaneously indicated by the DAQ system. In practice, this advantage enhances the in-situ test efficiency and as helps to assure the data quality. In order to identify the eigen-properties for this structure, the data were sampled at 204Hz for an interested range up to 30Hz.

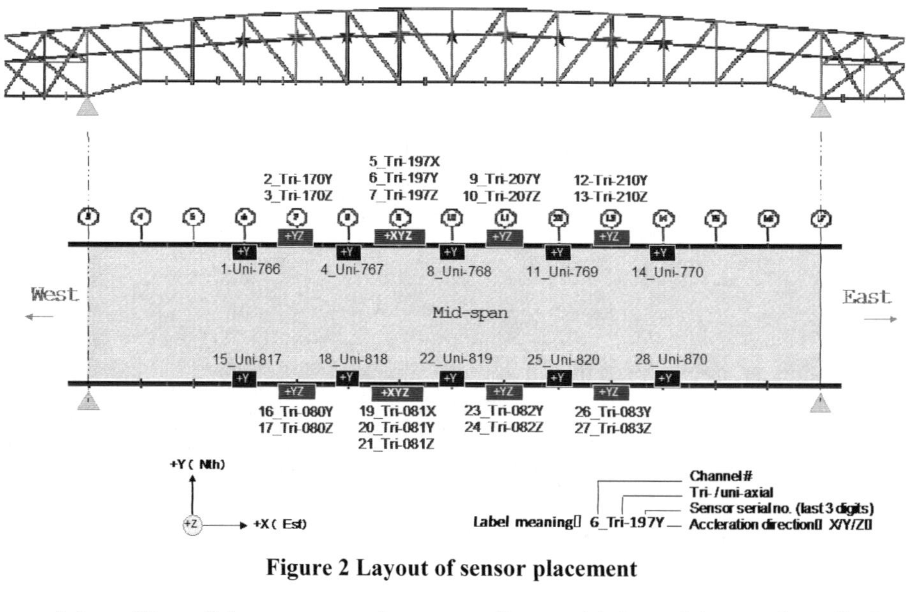

Figure 2 Layout of sensor placement

To capture modal profiles of the structure, 9 sensors from grid-6 to -14 were installed each side of the bridge from the mid-span as described in Figure 2. While the uniaxial sensors placed on even number axes were to measure the lateral sway of the structure, the triaxial sensors on odd number

axes were to record both lateral and vertical movements. Besides, sensors on grid-9 were to reflect the longitudinal vibration of the bridge.
In order to consider different scenarios of the occupant density in weekends and week days, the test began in a Saturday morning at 5.30am and completed in the following Monday at 10.30pm. In total 65 one-hour sessions were obtained from the test.

2.3 Discussion of time-series data

As expected, the vibration magnitudes of the structure vary proportionally against the number of occupants on the bridge. Take the mid-span (grid-10) measurement as an example, Figure 3 shows that the lateral acceleration between 9am and 9pm exceeded 80% of the daily peak value when the number of passengers was high; while between 12.30am to 5.30am, conversely, little vibration was found for the lack of passing activity. It is also found that the vibration in every channel on Monday was weaker than that on weekends.

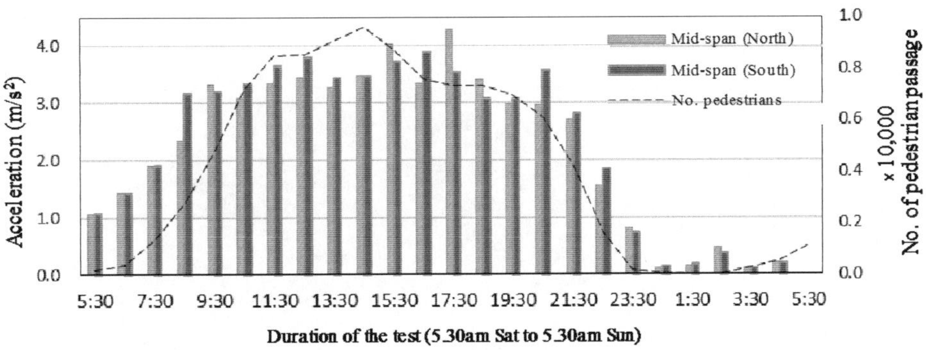

Figure 3 Measured vibration time-history of the lateral acceleration at the mid-span grid-10 (first 24 hours)

It was also observed that the structure vibrated dominantly in the lateral direction in comparison to the vertical direction, given the fact that the measured maximum lateral magnitude is 4.28 m/s^2, which is more than twice of the maximum vertical magnitude. It is perhaps the structure's relatively weak transverse rigidity that causes the natural frequencies of the bridge coincident with the lateral frequency of human walking, especially in a case when the "lock-in" effect happens, which in turn intensifies the vibration to resonance.

3. Modal identification

3.1 Applied modal identification methods

Many frequency or time-domain based output-only methods have been developed as available tools for modal identification in SHM systems. And it is normally assumed that the measured responses of the structure are stochastic. Among them, however, none has yet been recognised as a standalone to yield reliable results in all applications, especially given the fact that the on-line measurements for OMA techniques may contain unknown errors and manual removal of anomalies is often required. In this study, two frequency domain methods, i.e. enhanced frequency domain decomposition (EFDD) and curve-fit frequency domain decomposition (CFDD) were employed, along with three stochastic space identification (SSI) methods in time domain: SSI with Unweighted principal component (UPC), principal component (PC), and canonical variate algorithm (CVA) (Peter & Bart, 1996; Peeters & Roeck, 1999; and Bakir, 2011).

3.2 Signal processing and filtering

A least square fitting method was firstly used to find and remove the drifts in the acquired time-series. In light of the hypothesis mentioned earlier, the data of higher stochasticity were filtered through a Kurtosis-based index which is written in Eq.(1).

$$\gamma^*(x|\mu,\sigma) = \frac{E\left[(x-\mu)^4\right]}{\sigma^4} - 3 \tag{1}$$

where μ and σ is the mean and standard deviation of a series x, respectively. In theory, data following the Gaussian distribution have a Kurtosis value of 3, and here a revised Kurtosis index, γ^*, is used for convenience. Figure 4 depicts a profile of the revised Kurtosis values of both the raw (unfiltered) and filtered data versus time over the entire test. It is observed that the lowest value occurred around 2-6pm in three consecutive days, when the pedestrian activity was high.

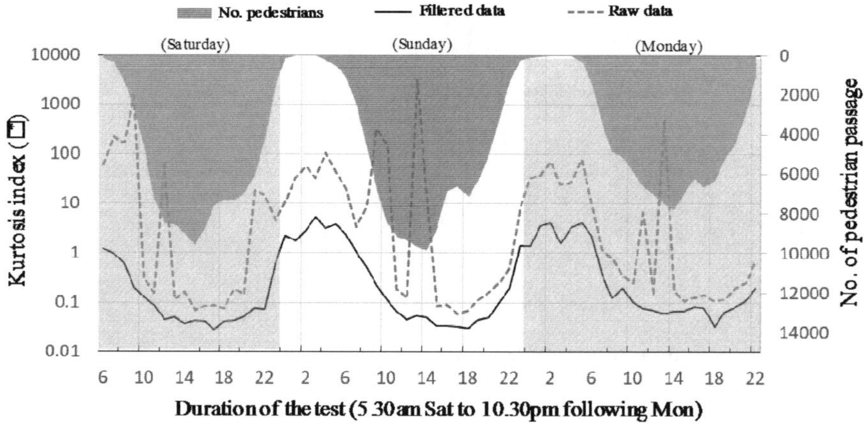

Figure 4 Profile of Kurtosis-based index showing the stochasticity of measurements over time

3.3 Discussion of OMA results

Figure 5 summarises the identified eigen-frequencies of the structure in duration of the test by the aforementioned methods. It is found that, apart from the frequencies at 3.6Hz, 4.8Hz, 5.7Hz, 10.0Hz, and 12.8Hz, the results vary obviously against time. Modes lower than 3Hz are more likely to be captured in high activity hours during the daytime, i.e. from 07:30am to 06:30pm. The mode at 8Hz repetitively occurred from 07:30 to 09:30 in the morning, and a mode at 14Hz was captured but not stable in all hours.

Considering the different scenarios of the footbridge which has numerous (peak hours) or seldom (e.g. after the midnight) occupants, specifically, a comparison of the results can be seen in Figure 6, where the first two modes below 2Hz disappeared after the mid-night.

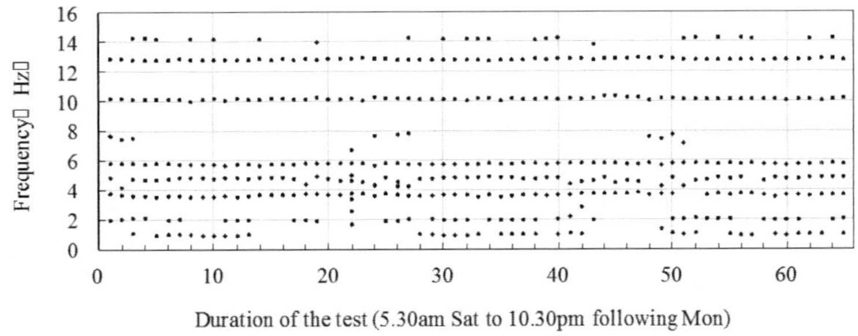

Figure 5 Summary of modal identification results using frequency- and time-domain methods

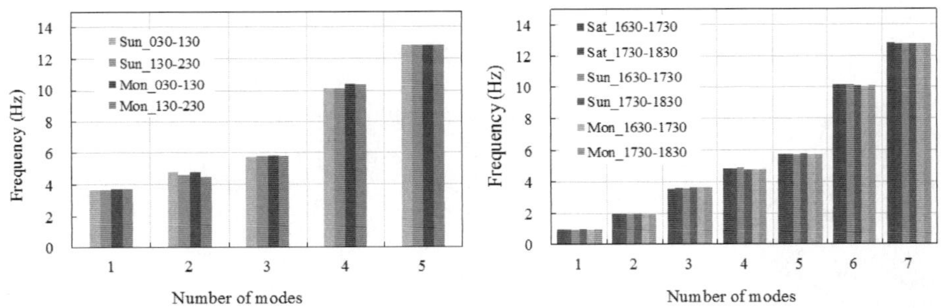

Figure 6 Comparison of modal identification results in quiet and high activity hours

4. Finite element analysis (FEA)

4.1 Finite element modelling

A numeric model of the case study structure was established in SAP2000. The density of the 6082-T6 aluminium alloy is 2.7e3 kg/m^3, the Young's modulus is 77GPa, and the Poisson's ratio is 0.33. In Figure 7, the supports no. 1 to 4, no.7, and 8 are roller in the longitudinal but pinned in the transverse direction, while no. 5 and 6 are pinned in both directions. Figure 8 shows the cross-sections of the truss members of the structure.

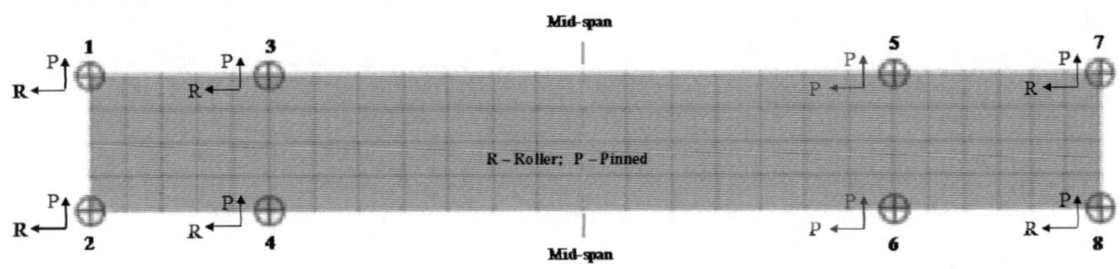

Figure 7 Boundary conditions of the case study footbridge

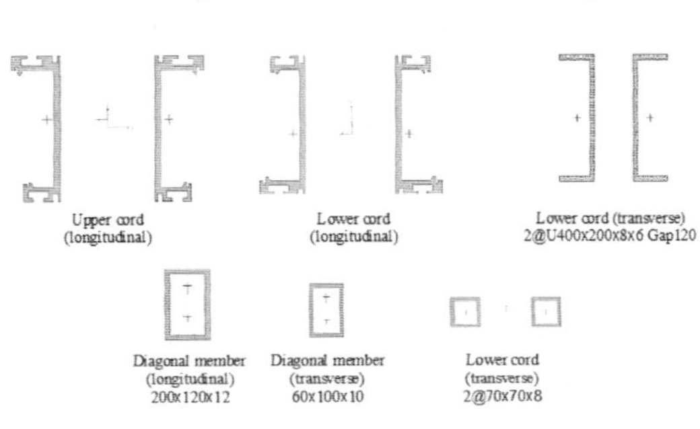

Figure 8 Cross-sections of the case study footbridge

4.2 Model updating

To represent the modal signature obtained from the measurements, the initial model was updated using a sensitivity based method (Wang, 1991). In the calculation of the sensitivity matrix, a set of numerical modes were assumed as a function of the Young's modulus of the structure's elements. The boundary of the change in Young's modulus was up to 20% to the original value. Hence, the sensitivity matrix is calculated as shown in Figure 9(a), and the optimised Young's modulus were sought with a result shown in Figure 9(b). In Table 2, it is evident that the accuracy of the numerical modal analysis has been improved.

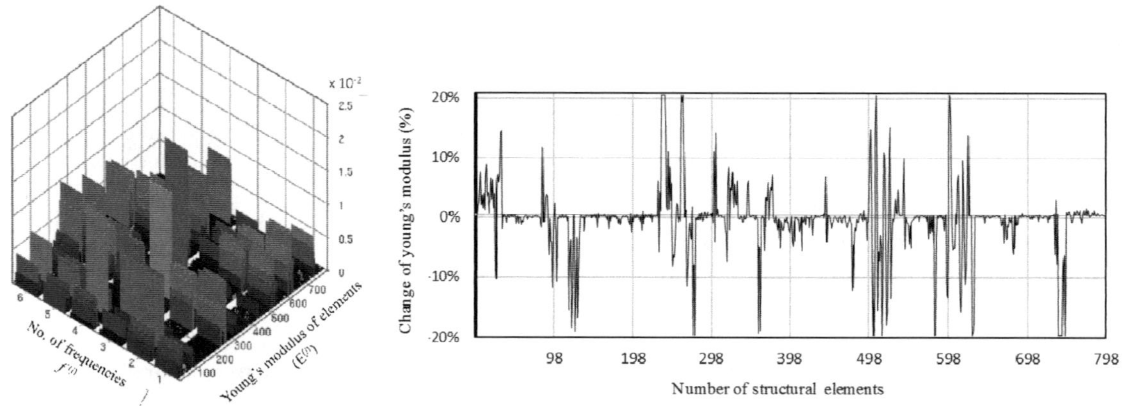

(a) (b)

Figure 9 Model updating of the FE model:
sensitivity matrix to eigen-frequencies (a); change of Young's modulus in elements (b)

Table 2 Comparison of eigen-frequencies before and after model updating

No. mode	Frequency (test)/Hz	Eigen-frequency (before updating) /Hz	Eigen-frequency (after updating) /Hz	Difference (before updating) %	Difference (after updating) %
1	3.62	3.55	3.59	-1.87	-0.75
2	5.76	5.61	5.71	-2.59	-0.92
3	7.57	7.90	7.62	4.35	0.69
4	10.13	10.45	10.18	3.16	0.53
5	12.81	13.14	12.84	2.54	0.21
6	14.19	14.06	14.10	-0.91	-0.64

5. Vibration comfort assessment

5.1 Pedestrian loading models

BS 5400 specifies that the maximum response of the structure shall be calculated by Eq.(2) as a vertical moving load at a speed of $v_f = 0.9f_0$ (m/s), when the first vertical natural frequency is less than 5Hz. Walking frequency f_0 is assumed as the fundamental frequency of the structure without load (BSA, 2006).

$$F_P(t) = \begin{cases} 180\sin(2\pi f_0 t) & f_0 < 4Hz \\ [1-0.3(f_0-4)]\cdot 180\sin(2\pi f_0 t) & 4Hz \le f_0 \le 5Hz \end{cases} \tag{2}$$

The pedestrian loading in Bro 2004 standard is given by a sinusoidal load considering bridge dimensions as shown in Eq.(3). Different from the BS 5400 model, it is a fixed load that results in the maximum vertical structural response. And in this case, it applies at the mid-span of the footbridge (SB, 2004).

$$F_p(t) = k_1 k_2 \sin(2\pi f_p t) \tag{3}$$

The ISO 10137 standard regulates a broader definition of pedestrian loading models. It not only includes a model for the single pedestrian scenario, it also considers another for the crowd. And additionally, both vertical and lateral responses are assessed to evaluate the vibration comfort of pedestrians (ISO, 2007). The employed ISO pedestrian loading models are given in Table 3 and 4 below.

Table 3 Vertical and lateral loading functions of pedestrian passage according to ISO 10137

Scenario No.	Loading freq. f_P (Hz)	Walking freq. specification	Occupant Type	Loading model F_P (N or N/m^2)	Load step Δt (sec)
ISO-1V	3.69	1st bending mode	Single	$750+300\sin(22.5t)+75\sin(45.1t)$	8.70E-3
ISO-2V	1.80	walking freq.	Single	$750+300\sin(11.3t)+75\sin(22.6t)$	1.74E-2
ISO-3V	3.69	1st bending mode	Crowd	$750+300\sin(22.5t)+75\sin(45.1t)$	8.70E-3
ISO-4V	1.80	walking freq.	Crowd	$750+300\sin(11.30t)+75\sin(22.60t)$	1.74E-2
ISO-5V	3.69	1st bending mode	Crowd	$42.0+16.8\sin(22.5t)+4.2\sin(45.1t)$	8.70E-3
ISO-6V	1.80	walking freq.	Crowd	$42.0+16.8\sin(11.30t)+4.2\sin(22.60t)$	1.74E-2
ISO-1L	7.90	1st lateral mode	Single	$75\sin(49.6t)+75\sin(99.2t)$	3.96E-3
ISO-2L	1.00	walking freq.	Single	$75\sin(6.28t)+75\sin(12.56t)$	3.13E-2
ISO-3L	7.90	1st lateral mode	Crowd	$75\sin(49.6t)+75\sin(99.2t)$	3.96E-3
ISO-4L	1.00	walking freq.	Crowd	$75\sin(6.28t)+75\sin(12.56t)$	3.13E-2
ISO-5L	7.90	1st lateral mode	Crowd	$4.2\sin(49.6t)+4.2\sin(99.2t)$	3.96E-3
ISO-6L	1.00	walking freq.	Crowd	$4.2\sin(6.28t)+4.2\sin(12.56t)$	3.13E-2

5.2 Vibration comfort evaluation

Considering single bridge occupant scenarios, relatively large responses are found by using the ISO model than by using the BS and Bro models. For example, the max. vertical acceleration at the mid-span is 25.77 m/s^2. It is probably because of the vertical periodical load in the ISO model that not only has a higher order term, but also has a larger constant factor of 0.4 in the first order term (comparing to 0.256 in BS and 0.2 in Bro). It therefore yields results that are too conservative.
When the bridge is excited by the same ISO loading model but at different frequencies, the acceleration responses at the mid-span are not much different. Under the load applied at the first eigen-frequency (3.59Hz) of the structure, the RMS of the acceleration at the mid-span is 0.781 m/s^2, while under the load at the walking frequency (1.80Hz), it is 0.425 m/s^2.
It is also noticed that the responses of the bridge actuated by a synchronously walking group are way lager than that by a non-synchronous load. The maximum acceleration of the former scenario is as large as 25.77 m/s^2, which is 18.5 times of the counterpart calculated at the non-synchronous frequency. It is in other words a resonance that has occurred to the structure.
It is clearly seen that the comfort evaluation indices among different standards lead to contradictory results. For instance, almost all of the vertical responses from the ISO model are over the limit of the comfort threshold indicated by the standard, the responses using the BS model has met the criterion in the BS standard. Therefore, a comprehensive study of available pedestrian-structure interactive effects is of need in order to provide a reliable evaluation.

Table 4 Evaluation summary of vibration comfort indices using BS, Bro, and ISO models

Scenario	Walking freq. f_P (Hz)	Walking freq. specification	Type of occupants	Case study a_{max} (m/s^2)	Criteria a_{cri} (m/s^2)	Satisfaction
BS 5400-1	3.69	1st bending mode	Individual	0.527	0.96	Yes
BS 5400-2	1.80	walking freq.	Individual	0.003	0.67	Yes
Bro-1	3.69	1st bending mode	Non-sync.	1.741	0.50	No
Bro-2	1.80	walking freq.	Non-sync.	0.327	0.50	Yes
ISO-1V	3.69	1st bending mode□	Individual	0.781	0.34	No
ISO-2V	1.80	walking freq.□	Individual	0.425	0.43	Yes
ISO-3V	3.69	1st bending mode□	Sync. Crowd	25.77	0.34	No
ISO-4V	1.80	walking freq.□	Sync. Crowd	21.84	0.43	No
ISO-5V	3.69	1st bending mode□	Non-sync. Crowd	1.39	0.34	No
ISO-6V	1.80	walking freq.□	Non-sync. Crowd	1.15	0.43	No
ISO-1L	7.90	1st lateral mode□	Individual	0.026	0.61	Yes
ISO-2L	1.00	walking freq.□	Individual	0.012	0.32	Yes
ISO-3L	7.90	1st lateral mode□	Sync. Crowd	3.002	0.61	No
ISO-4L	1.00	walking freq.□	Sync. Crowd	0.049	0.32	Yes
ISO-5L	7.90	1st lateral mode□	Non-sync. Crowd	0.169	0.61	Yes
ISO-6L	1.00	walking freq.□	Non-sync. Crowd	0.012	0.32	Yes

6. Conclusion

This paper discusses a case study of applying the SHM system on an aluminium alloy footbridge towards an evaluation of the pedestrian vibration comfort. Both frequency- and time-domain OMA techniques were implemented and different modal signatures were found over time. Interestingly, more lower modes were observed in high pedestrian activity hours. An FE model was established and updated to represent the dynamic behaviours of the structure. Three pedestrian loading models with respective vibration comfort indices from the BS, Bro, and ISO standards were simulated in numerical analyses. The inconsistent results show that, the BS model of a moving sinusoidal results in smaller values, and it does not include the effect of crowds. The ISO model regards the scenarios of synchronous and non-synchronous crowd as a distributed load on the deck, and yields large responses, particularly for the ISO-3V, -4V, and -3L, which apparently exceeds the given limits.

Acknowledgement

The authors acknowledge the financial aid granted by Beijing Academy of Science and Technology under the Returning Overseas Scholars Grant project (OTP-2014-003) as well as the municipal management authorities in Xicheng District, Beijing, China, for their great support to the test.

References

Bakir PG 2011, 'Automation of the stabilization diagrams for subspace based system identification', *Expert Systems with Applications*, vol.38, no.12, pp. 14390-14397.

BMBS (Beijing Municipal Bureau of Statistics), 2004, *A present survey of passenger flow volume in Xidan Commercial Street*, vol. 19.

Brownjohn, JMW, Fok, P, Roche, M et al. 2004, 'Long span steel pedestrian bridge at Singapore Changi Airport. Part 2: crowd loading tests and vibration mitigation measures', *Structural Engineer*, vol.82, no.16, pp. 28–34.

BSA 2006. *Steel, Concrete and Composite Bridges. Specification for Loads; Appendix C: Vibration Serviceability Requirements for Foot and Cycle Track Bridges*, BS 5400-2:2006. British Standards Association, London, UK.

Dallard, P., Fitzpatrick, AJ, Flint A *et al.* 2001, 'The London Millennium Footbridge', *Structural Engineer*, vol. 79, no.22, pp.17–35.

Hsu, EB & Burkle, FM 2012 'Cambodian Bon Om Touk stampede highlights preventable tragedy', *Prehospital and disaster medicine*, vol.27, no.5, pp. 481-482.

Hu, WH, Caetano, E, Cunha, Á 2013, 'Structural health monitoring of a stress-ribbon footbridge', *Engineering Structures*, vol. 57, pp. 578-593.

Huang, Y, Xu, LT & Li, MS 2010, 'Monitoring of Aluminium Alloy Pedestrian Overpass in Operation', *Building Construction*, vol.32, no.12, pp. 1250-1253.

ISO 2007, *ISO 10137:2007: Bases for design of structures - Serviceability of buildings and walkways against vibrations*, Switzerland.

Macdonald, JHG 2008, 'Pedestrian-induced vibrations of the Clifton Suspension Bridge', *Proceedings of the Institution of Civil Engineers - Bridge Engineering*, vol.161, pp. 69–77.

Nakamura, S 2003, 'Field measurements of lateral vibration on a pedestrian suspension bridge', *Structural Engineer*, vol.81, no.22, pp. 22–26.

Nakamura, S & Kawasaki, T 2006, 'Lateral vibration of footbridges by synchronous walking', *Journal of Constructional Steel Research*, vol. 62, pp. 1148-1160.

Occhiuzzi, A, Spizzuoco, M & Ricciardelli, F 2008, 'Loading models and response control of footbridges excited by running pedestrians', *Structural Control and Health Monitoring*, vol.15, no.3, pp. 349–368.

Peeters B & Roeck GD 1999, 'Reference-based stochastic subspace identification for output-only modal analysis', *Mechanical Systems and Signal Processing*, vol.13, no.6, pp. 855-878.

Peter, VO & Bart, DM 1996, *Subspace Identification for Linear Systems: Theory- Implementation-Applications*, Dordrecht, the Netherlands: Kluwer Academic Publishers.

SB (Svensk Byggtjanst) 2004. *Vagverkets allmanna tekniska beskrivning for nybyggande och forbattring av broar. Svensk Byggtjanst*, Bro 2004, Stockholm, Sverige.

Wang, BP 1991, 'Improved Approximate Methods for Computing Eigenvector Derivatives in Structural Dynamics', *AIAA Journal*, vol.29, no.6, pp. 1018-1029.

Wang, L 2016, *Applying structural health monitoring systems towards pedestrian vibration comfort assessment of aluminium alloy footbridges: the project research report*, Beijing Academy of Science and Technology (in Chinese).

Yang, J & Wu, L 2011, 'Analysis on Load Testing and Carrying Capacity of Aluminium Alloy Pedestrian Overcrossing', *Structural Engineers*, vol.27, pp. 280-284. (in Chinese)

The 8th International Conference on
Structural Health Monitoring of Intelligent Infrastructure
Brisbane, Australia | 5-8 December 2017

Damage quantification in beam-like structures from modal flexibility change

N.T. Le[1,*], D.P. Thambiratnam[2], T.H.T Chan[3], A. Nguyen[4]

[1,2,3,4] *Queensland University of Technology, Australia,* [*] *Email:* ngocthach.le@hdr.qut.edu.au

Abstract

An approach to quantify the severity of damage in Euler-Bernoulli beams from changes in modal flexibility is presented. The approach is built on two pillars. The first is the connection between flexibility relative change (*RFC*) and the damage-induced relative deflection (*DI-RD*). This connection allows one to estimate the *DI-RD* under arbitrary static point loads from the dynamically measured *RFC* matrix. The second pillar is the theoretical relationship between (i) the elemental damage severity and (ii) the *DI-RD* under unit point loads. For statically determinate beams, the scalar function relating these two parameters can be conveniently formulated from the principle of Virtual Work method. It reveals that the scalar function contains only the geometrical information including location of the identified damaged element. Therefore, once the damaged element is located, the damage severity can be calculated. Numerical investigations show that the approach accurately quantifies damage in cantilever and simply supported beams. A merit of this approach is that it has a clear theoretical base and it can be expanded to any statically determinate and indeterminate beam-type structures whose deformation function under point loads can be formulated. As beams can represent various structural components, the proposed approach provides a promising platform targeting the application to real structures.

1. Introduction

Unexpected changes in operational conditions such as exceeding designed loads, improper usage, and extreme natural events; incorporated with age-related deteriorations may cause damage to civil structures such as buildings, bridges and dams. If an initial damage in these structures is not detected, it can lead to increased damage and catastrophic failures. The needs to monitor structural performance and detect the onset of damage in civil structures are thus evident. Structural Health Monitoring (SHM) has emerged as a feasible technique for this purpose by using on-structure, non-destructive sensing systems (Chan et al., 2011). Over the last three or four decades, the research on vibration-based damage identification (VBDI) has received a considerable attention. The fundamental idea of this technique is that damage in a structure can cause changes in its dynamic properties, which then lead to detectable changes in the vibration characteristics. Therefore, by measuring changes in vibration characteristics, one can assess the damage with regard to its presence, location and severity. Extensive literature reviews on this field can be found in Doebling et al. (1998), Sohn et al. (2004), and Fan et al. (2011).

Recently, modal flexibility change (*MFC*) has become one of the feasible damage descriptors due to its high sensitivity to damage. This merit can be explained by the fact that modal flexibility (MF) combines mode shapes and natural frequencies, making it more sensitive to locating damage compared to the methods built on each parameter separately (Zhao et al., 1999). In addition, as MF converges rapidly with increasing natural frequencies, it can be accurately estimated from only a few lower modes of vibration (Pandey et al., 1994). The method was first proposed by Pandey and

Biswas (1994, 1995) and tested for locating damage in beams. It was then explored for locating and quantifying damage in a wide range of structures, such as bridges (Aktan et al., 1994, Doebling et al., 1996, Maeck et al., 2003, Ni et al., 2008, Shih et al., 2013, Wickramasinghe et al., 2015), truss structures (Bernal, 2002, Gao et al., 2002, Duan et al., 2005, Shih et al., 2011), aircraft fuselage (Robinson et al., 1996), shear buildings (Bisht et al., 2006, Koo et al., 2010, Sung et al., 2012), and concrete gravity dams (Le et al., 2016).

This paper presents a new damage quantification approach for Euler-Bernoulli beam structures based on a theoretical relationship between the elemental damage severity and the damage-induced relative deflection (*DI-RD*) under a unit point load. From this theoretical relationship and the estimated *DI-RD* obtained from modal flexibility relative change (*RFC*), the damage severity is interrogated. This article is organised as follows. First, the construction of MF is briefly explained and the formulas relating *MFC* to damage-induced deflection (*DI-D*), and *RFC* to *DI-RD* are given. Second, the *DI-D* and *DI-RD* functions with respect to the damage severity is established, from which a deflection-based damage locating and severity estimation concept is presented. Next, numerical examples are investigated to demonstrate the effectiveness of the proposed method. Finally, the conclusion is followed.

2. Theory

2.1. Relation between modal flexibility change and damage-induced deflection

The modal flexibility matrix F is constructed from mass normalised mode shapes and natural frequencies as follows (Pandey and Biswas, 1994):

$$F_{n\times n} = \Phi \Lambda^{-1} \Phi^T = \sum_{i=1}^{n} \frac{1}{\omega_i^2} \phi_i \phi_i^T \tag{1}$$

where Φ=[ϕ_1 ϕ_2 ... ϕ_n] is the mass-normalised mode shape matrix, Λ=diag(ω_i^2) is a diagonal matrix containing the Eigen-frequencies, ω_i is the ith modal frequency, n is the number of the degrees of freedoms (DOFs) of the system. Equation (1) reveals that the flexibility matrix is inversely proportional to the square of modal frequencies, implying that it can be well approximated with sufficient accuracy by using a few early modes of vibration:

$$F_{n\times n} \cong \sum_{i=1}^{m} \frac{1}{\omega_i^2} \phi_i \phi_i^T \tag{2}$$

where, m is the number of measured or interested modes. After obtaining the flexibility matrices from the undamaged and damage states, F^h and F^d, the difference and the elemental-wise relative difference between these two matrices can be given by:

$$\Delta F_{n\times n} = F^d - F^h \quad ; \tag{3}$$

$$RFC_{n\times n} = \frac{F^d - F^h}{F^h} = \frac{\Delta F}{F^h} \tag{4}$$

where, the superscripts "*d*" and "*h*" denote the damaged and undamaged states, respectively. The deflection change of a structure after damage, or the damage-induced deflection (*DI-D*), and the damage-induced relative deflection (*DI-RD*) under an arbitrary load **f** can be estimated as:

$$\Delta u = \Delta F.\boldsymbol{f} \tag{5}$$

$$\frac{\Delta u}{u^h} = RFC.\boldsymbol{f} \tag{6}$$

2.2. Damage locating and quantification in cantilever beams using *DI-D* and *DI-RD* under unit point load

2.2.1. Formulation of DI-D under unit point load

Consider a cantilever Euler-Bernoulli beam with constant bending stiffness, $(EI)_x=EI$. Under unit point load acting at the tip, the deflection of the undamaged beam can be formulated from the principle of Virtual Work method as:

$$u^h(x) = \int_0^L \frac{M(x_1).m(x_1)}{(EI)_x} dx_1 = \int_0^x \frac{(L-x_1)(x-x_1)}{(EI)_x} dx_1 = \frac{1}{6EI} x^2(3L-x) \tag{7}$$

where, $M(x_1)=L-x_1$ is the bending moment of the beam at $x=x_1$ due to the unit point load at $x=L$, $m(x_1)=x-x_1$ is the virtual bending moment at x_1 due to virtual unit point load at x. In the damaged state, assuming the beam gets damage in the region $a \le x \le a+b$ (Figure 1-a), with the damaged bending stiffness of $(EI)_d=(1-\alpha)EI$, where $0 \le \alpha < 1$ is damage severity coefficient. The *DI-D* of the beam can be formulated from the Virtual Work method as:

$$\Delta u(x) = u^d(x) - u^h(x) = \begin{cases} 0 & , \quad x \le a \\ \beta \dfrac{(x-a)^2}{6EI}(3L-2a-x), & a \le x \le a+b \\ \beta \dfrac{1}{6EI}\big(A + B(x-a-b)\big), & a+b \le x \le L \end{cases} \tag{8}$$

where, the scalars β, A, B are defined as:

$$\beta = \frac{\alpha}{1-\alpha}; \qquad A = b^2(3L-3a-b); \qquad B = 3b(2L-2a-b) \tag{9}$$

Equation (8) characterizes *DI-D* of cantilever beams with respect to the damage severity coefficient at an arbitrary inspection location x. It provides a theoretical base for the damage locating and quantification concepts using flexibility change presented in the following sections.

2.2.2. Locating and quantifying damage in cantilever beams

From equation (8), the *DI-D* is plotted as in Figure 1-b. It can be seen that, compared to the undamaged state, the deflection of the damaged beam remains unchanged in the undamaged region $x \le a$, increases by a third-order curve in the damaged region ($a < x \le a+b$), and linearly increases in the undamaged region $x > a+b$. Therefore, the *DI-D* can be used to identify the damage location (region), which starts from the point where the *DI-D* is positive and ends where the *DI-D* is linearly increased. This provides a theoretical base for indicating damage location using the modal flexibility change reported by Pandey and Biswas (1994).

After the damage location is identified, it is desirable that the damage extent can be interrogated. However, it has been difficult to extract the damage severity from flexibility change (or *DI-D*) due to the inclusion of the unknown stiffness quantity *EI* as can be seen from equation (8). To tackle this end, in this paper, the unknown is eliminated by relating the *DI-D* from equation (8) to the undamaged deflection from equation (7), yielding the damage-induced relative deflection (*DI-RD*) as follows:

$$\frac{\Delta u(x)}{u^h(x)} = \beta \frac{A + B(x-a-b)}{x^2(3L-x)} = \frac{\alpha}{1-\alpha} f(a,b,L,x), \quad (a+b \le x \le L) \tag{10}$$

where, *f(a,b,L,x)* is a scalar function containing only the geometrical entries including damaged element location (*a, b*), beam length (*L*), and the coordinate of the inspection DOFs (*x*):

$$f(a,b,L,x) = \frac{A + B(x-a-b)}{x^2(3L-x)}, \quad (a+b \le x \le L) \tag{11}$$

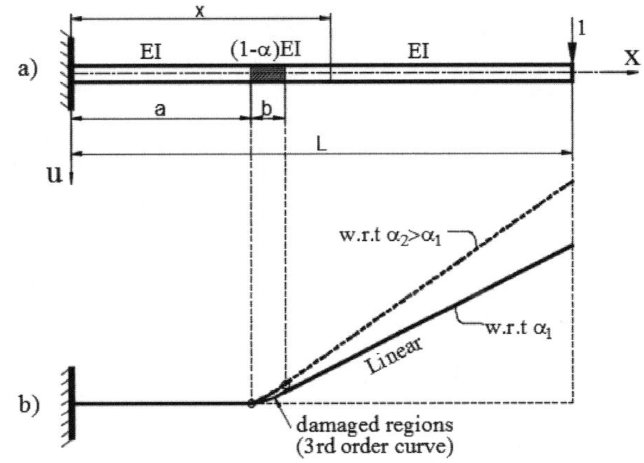

Figure 1. Damage-induced deflection of cantilever beam under unit tip load:
a) Undeformed cantilever beam, b) Damage-induced deflection with different damage severity levels

By simple manipulation of equation (10), the damage severity can be presented as follows:

$$\alpha = \frac{\dfrac{\Delta u(x)}{u^h(x)}}{\dfrac{\Delta u(x)}{u^h(x)} + f(a,b,L,x)}, \quad (a+b \le x \le L) \tag{12}$$

Equations (11) and (12) indicate that when damaged element is identified, its damage severity can be calculated at any inspection point in the undamaged region $a + b < x \le L$. The final damage severity can be obtained by averaging the calculated damage severities at different x locations.

2.3. Damage locating and quantification in simply supported beam using *DI-D* and *DI-RD*

An analogous procedure can be repeated for Euler-Bernoulli simply supported (SS) beams. Consider a SS beam with constant bending stiffness, $(EI)_x=EI$. Under the mid-span unit point load, the undamaged deflection of the beam is formulated by:

$$u^h(x) = \frac{1}{12EI}\left[2\left(x-\frac{L}{2}\right)^3 H\left(x-\frac{L}{2}\right) + x\left(\frac{3}{4}L^2 - x^2\right)\right] \tag{13}$$

where, $H(x-L/2)$ is the Heaviside function which equals to 1 if $x \ge L/2$ and zero if $x<L/2$. The *DI-D*, *DI-RD* and the scalar function are presented as follows:

$$\Delta u(x) = u^d(x) - u^h(x) = \begin{cases} \beta\dfrac{1}{12EI}B_1 x \quad , & x \le a \\ \beta\dfrac{1}{12EI}B_2(L-x)\,, & a+b \le x \le L \end{cases} \tag{14}$$

$$\frac{\Delta u(x)}{u^h(x)} = \frac{\alpha}{1-\alpha} f(a,b,L,x) \tag{15}$$

$$f(a,b,L,x)=\begin{cases}\dfrac{B_1x}{2\left(x-\frac{L}{2}\right)^3H\left(x-\frac{L}{2}\right)+x\left(\frac{3}{4}L^2-x^2\right)}, & x\le a\\ \dfrac{B_2(L-x)}{2\left(x-\frac{L}{2}\right)^3H\left(x-\frac{L}{2}\right)+x\left(\frac{3}{4}L^2-x^2\right)}, & a+b\le x\le L\end{cases} \tag{16}$$

where, the scalars B_1, B_2 are defined as:

$$B_1=\frac{b}{L}[(2a+b)(3L-2a-2b)-2a^2];\quad B_2=\frac{2}{L}[(a+b)^3-a^3] \tag{17}$$

In equation (14), the *DI-D* for $a\le x\le a+b$ is not presented, but it is a third-order function with respect to x. By plotting equation (14) to Figure 2-b, it is evident that the damaged element can be identified as the "peak" region between the two linear lines. This is, again, relevant to what has been reported by Pandey and Biswas (1995) for damage identification in SS beams using MFC. From (15), the damage severity coefficient is calculated using equation (12) at any inspection point. Finally, the damage severity at the identified damaged element is obtained using the averaging technique.

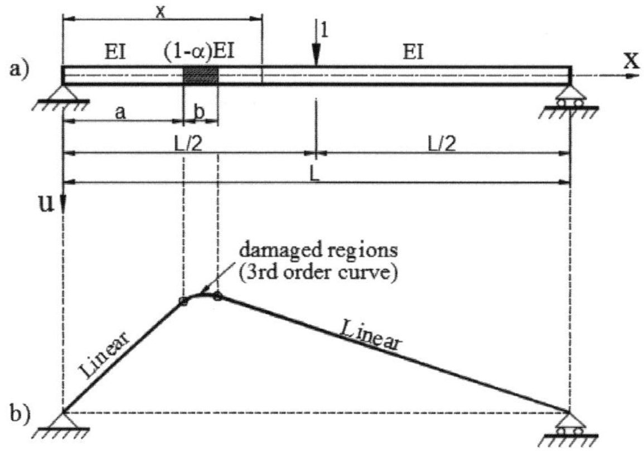

Figure 2. Damage-induced deflection of SS beam under mid-span unit load:
a) Undeformed SS beam, b) Damage-induced deflection

2.4. Damage quantification with *DI-RD* obtained from *RFC* matrix

The process of damage locating and quantification in bending beams using *DI-D* and *DI-RD* obtained from modal flexibility changes can be summarised as follows:

1. Construct modal flexibility matrices in healthy and damaged states: F^h and F^d
2. Determine the flexibility difference (ΔF) and the relative flexibility change (*RFC*) matrices
3. Estimate the *DI-D* vector from ΔF under unit point loads (acting at the free end in cantilever beams, or at mid-span in SS beams) using equation (5).
4. Identify damaged element from the *DI-D* curve
5. Estimate the *DI-RD* vector from *RFC* under the point load in step 3 using equation (6)
6. Calculate damage severity from equations (12)

3. Numerical investigations

3.1. Example 1: cantilever beam with material stiffness reduction

In this example, a horizontal cantilever reinforced concrete beam with the length of 3m and the cross section of 0.2m×0.4m is modelled using frame elements in the finite element (FE) software package Sap2000 (Figure 3). The Young's modulus, Poisson's ratio and mass density for the concrete are 24GPa, 0.2 and 2400kg/m^3 respectively. Two damage scenarios are simulated by reducing the material Young's modulus while retaining the cross section of the damaged elements (Table 1). The first three mode shapes in the undamaged state are presented in Figure 4.

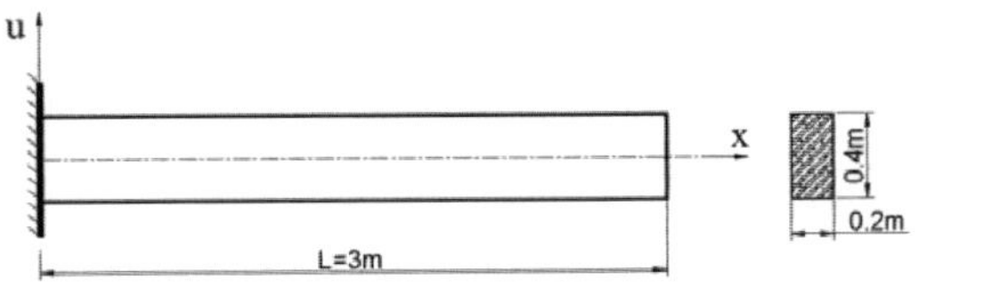

Figure 3. Dimensions of the concrete beam

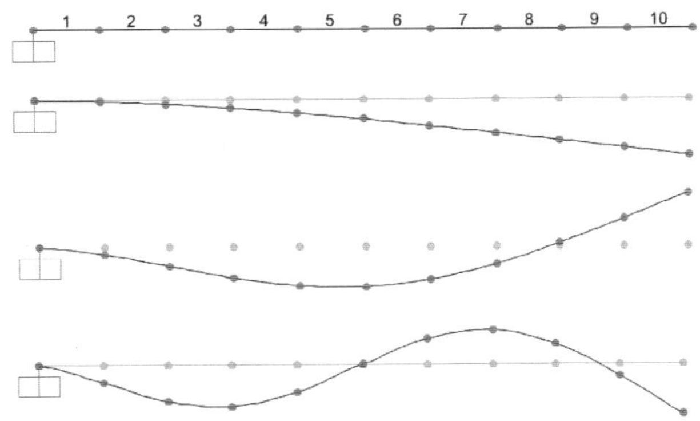

Figure 4. FE model of the beam and its three first mode shapes

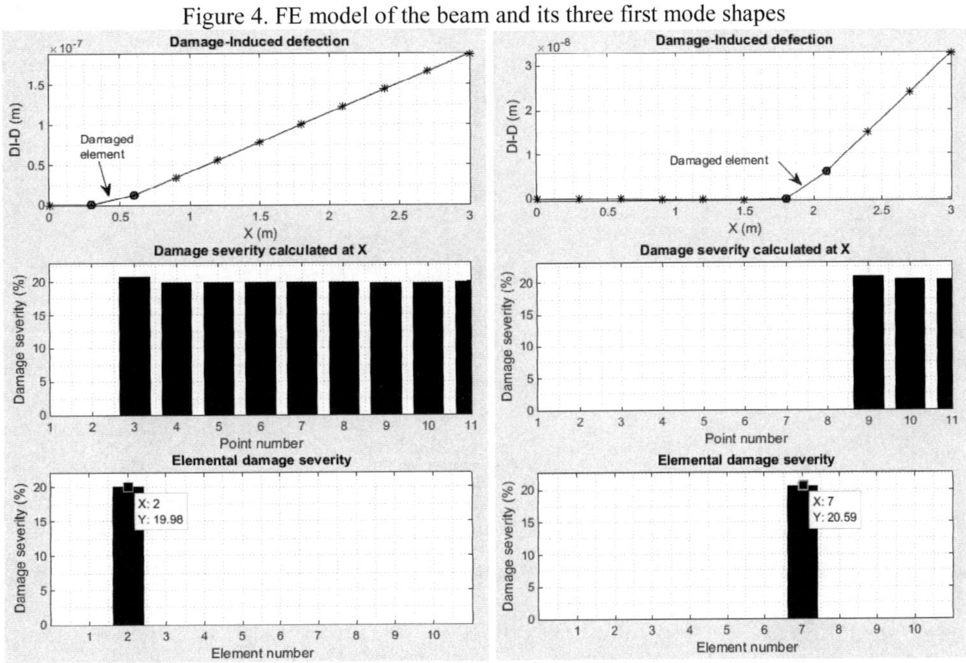

Figure 5. Damage locating and quantification – Case D1.1 (left), Case D1.2 (right)

Damage locating and quantification results using the first 3 modes are illustrated in Figure 5. The *DI-D* is extracted from the last column of the *MFC* matrix, which corresponds to the unit point load acting at the free end of the cantilever. As expected, the flexibility-based *DI-D* follows the theoretical *DI-D* behaviour under unit point load presented in the theory section. From where, the damaged members are correctly identified. From flexibility-based *DI-RD* (not presented in the figures) and equations (11), (12), the damage severity (of the damaged element) is then calculated at different inspection locations. It can be seen that the calculated severities converge rapidly from the second attempt onwards. The final damage severities are finalised by averaging the results, showing very good degree of accuracy (Table 1).

Table 1. Damage scenario definition and damage quantification results for example 1

Damage case	Damaged element	Damage severity	Estimated severity	Error (%)
D1.1	2	20%	19.98%	-0.10%
D1.2	7	20%	20.59%	2.95%

3.2. Example 2: cantilever beam with removing material

A vertical 75PFC steel cantilever with a height of 2.1m is modelled using solid elements in Sap2000 (Figure 6-a). The Young's modulus, Poisson's ratio and mass density of the undamaged material are 200 GPa, 0.3 and 7850 kg/m^3 respectively. The beam is subdivided into 20 segments; each is 105mm in length.

Table 2. Damage scenario definition for example 2

Damage case	a (m)	b (m)	hc (mm)	Damage severity
D2.1	0	0.105	9.5	53.6%
D2.2	0.525	0.105	9.5	53.6%

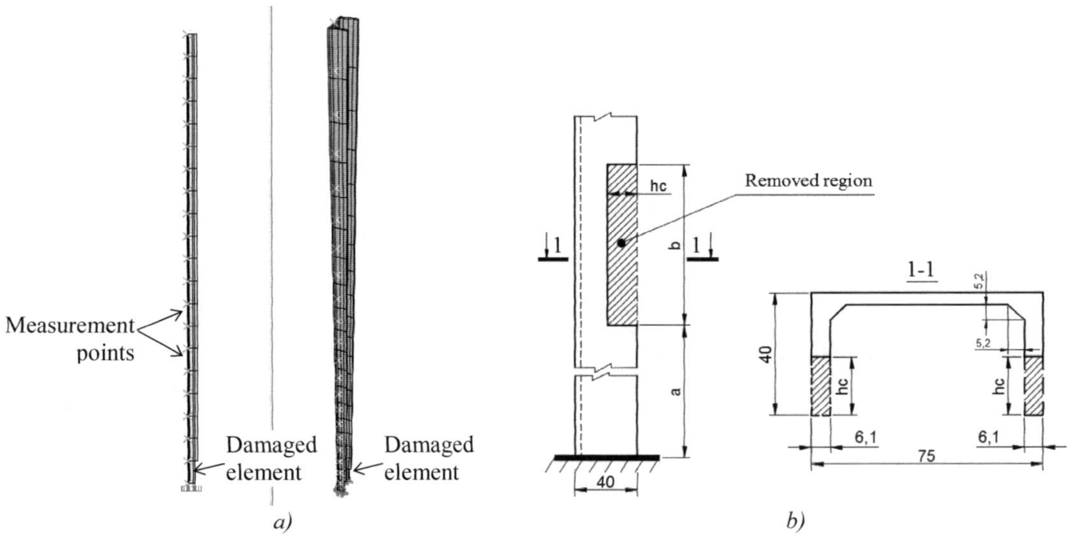

Figure 6. a) FE model of the steel beam using solid elements; b) Dimensions the beam section and the damage

Two damage scenarios are created by cutting the flanges (Figure 6-b), and the resultant sectional stiffness reduction is calculated in Table 2. Figure 7 and Figure 8 illustrate the damage identification results from *DI-D* and *DI-RD* obtained from *MFC* and *FRC* using the first three modes of vibration. Similar to the Example 1, the proposed method can estimate the damage severities of 52.06% (D2.1) and 57.12% (D2.2), with satisfactory accuracy (-2.6% and 6.4% error). These lower accuracy levels compared to Example 1 can be explained by the effect of both stiffness and mass reduction.

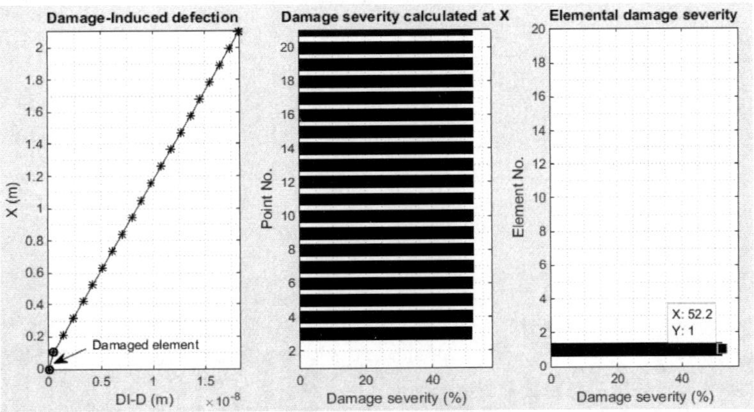

Figure 7. Damage locating and quantification – Case D2.1

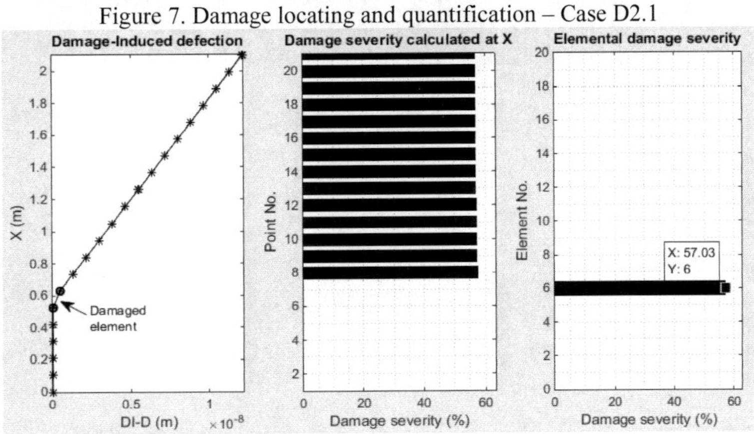

Figure 8. Damage locating and quantification – Case D2.2

3.3. Example 3: simply supported beam with material stiffness reduction

A 6m long SS concrete beam with the cross section of 0.3m×0.5m is modelled using frame elements in using Sap2000 (Figure 9). The Young's modulus, Poisson's ratio and mass density for the concrete are 24GPa, 0.2 and 2400kg/m^3 respectively. Two damage scenarios are simulated by reducing the material Young's modulus (Table 3).

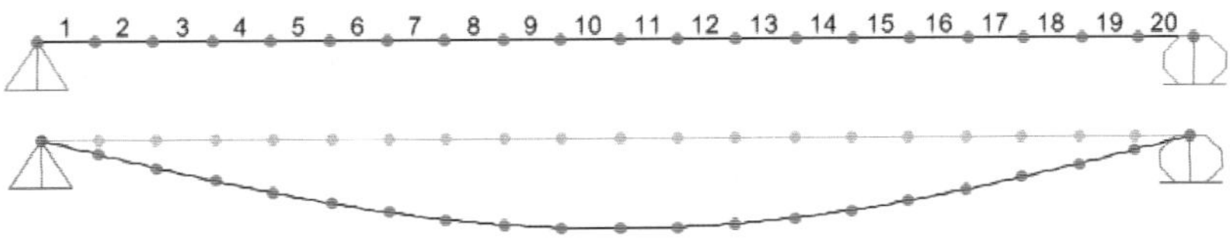

Figure 9. FE model of the SS beam and its first mode shape

Only the first mode is used to construct the MF. As a result, the resultant *DI-D* does not linearly decrease as expected in case D3.1, yet the averaged severity of 19.04% is still very good result compared to the correct value of 20% (Figure 10). For the case D3.2 (damage in the middle region), the proposed method accurately estimates the damage severity of 19.76%, with only 1.2% error.

Table 3. Damage scenario definition for example 3

Damage case	Damaged	Damage

	element	severity
D3.1	5	20%
D3.2	10	20%

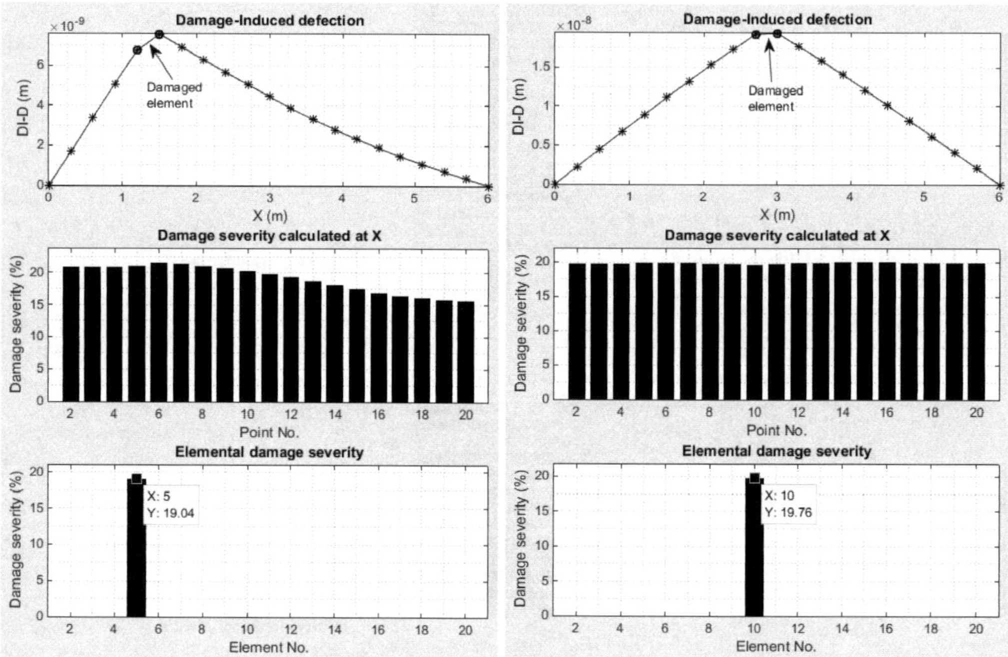

Figure 10. Damage locating and quantification – Case D3.1 (left), Case D3.2 (right)

4. Conclusions

In this paper, a new approach to quantify damage in Euler-Bernoulli beams is presented. It is shown that the established theoretical relationship between the elemental damage severity and the *DI-RD* of the beams under unit point load can be used to estimate the damage severity when the *DI-RD* is extracted from the measured modal flexibility change. Numerical illustrations indicate that the proposed method can accurately quantify various single damage cases in cantilever and simply supported beams by using the first three, or even only the first mode of vibration. This method has a clear theoretical base and can be expanded to other statically determinate and indeterminate beam-type structures whose *DI-D* under unit point load can be formulated. The approach incorporates modal flexibility; hence, mass-normalised mode shapes are required. As beams are common structural components, or can be an idealized model for many civil structures, the proposed approach can be promising for real-world SHM applications.

References

Aktan, A., Lee, K., Chuntavan, C. & Aksel, T. Modal testing for structural identification and condition assessment of constructed facilities. Proceedings-Spie the International Society for Optical Engineering, 1994. 462-462.

Bernal, D. 2002. Load vectors for damage localization. *Journal of Engineering Mechanics,* 128, 7-14.

Bisht, S. & Singh, M. Damage detection and quantification in shear beam buildings. 4th International Conference on Earthquake Engineering, Taipei, Taiwan, 2006.

Chan, T. & Thambiratnam, D. P. 2011. *Structural health monitoring in Australia,* New York, Nova Science Publishers.

Doebling, S. W. & Farrar, C. R. Computation of structural flexibility for bridge health monitoring using ambient modal data. Proceedings of the 11th ASCE Engineering Mechanics Conference, 1996. Citeseer, 1114-1117.

Doebling, S. W., Farrar, C. R. & Prime, M. B. 1998. A summary review of vibration-based damage identification methods. *Shock and vibration digest,* 30, 91-105.

Duan, Z., Yan, G., Ou, J. & Spencer, B. 2005. Damage localization in ambient vibration by constructing proportional flexibility matrix. *Journal of Sound and vibration,* 284, 455-466.

Fan, W. & Qiao, P. 2011. Vibration-based damage identification methods: a review and comparative study. *Structural Health Monitoring,* 10, 83-111.

Gao, Y., Ruiz-Sandoval, M. & Spencer, B. Flexibility-based damage localization employing ambient vibration. 15th ASCE Engineering Mechanics Conference, 2002. Columbia University, New York, NY, 1-8.

Koo, K., Sung, S., Park, J. & Jung, H. 2010. Damage detection of shear buildings using deflections obtained by modal flexibility. *Smart Materials and Structures,* 19, 115026.

Le, N. T., Thambiratnam, D. P., Chan, T. H. T., Nguyen, A. & Huynh, B. K. T. 2016. Damage assessment of concrete gravity dams using vibration characteristics. *In:* ZINGONI, A., ed. The 6th International Conference on Structural Engineering, Mechanics and Computation, SEMC, 2016. 1542 to 1554.

Maeck, J. & De Roeck, G. 2003. Damage assessment using vibration analysis on the Z24-bridge. *Mechanical Systems and Signal Processing,* 17, 133-142.

Ni, Y., Zhou, H., Chan, K. & Ko, J. 2008. Modal Flexibility Analysis of Cable - Stayed Ting Kau Bridge for Damage Identification. *Computer - Aided Civil and Infrastructure Engineering,* 23, 223-236.

Pandey, A. K. & Biswas, M. 1994. Damage detection in structures using changes in flexibility. *Journal of Sound and Vibration,* 169, 3-17.

Pandey, A. K. & Biswas, M. 1995. Experimental verification of flexibility difference method for locating damage in structures. *Journal of sound and vibration,* 184, 311-328.

Robinson, N., Peterson, L., James, G. & Doebling, S. 1996. Damage detection in aircraft structures using dynamically measured static flexibility matrices. Los Alamos National Lab., NM (United States).

Shih, H. W., Thambiratnam, D. & Chan, T. H. 2011. Damage detection in truss bridges using vibration based multi-criteria approach. *Structural Engineering and Mechanics,* 39, 187.

Shih, H. W., Thambiratnam, D. P. & Chan, T. H. T. 2013. Damage detection in slab - on - girder bridges using vibration characteristics. *Structural Control and Health Monitoring,* 20, 1271-1290.

Sohn, H., Farrar, C. R., Hemez, F. M., Shunk, D. D., Stinemates, D. W., Nadler, B. R. & Czarnecki, J. J. 2004. *A review of structural health monitoring literature: 1996-2001,* Los Alamos National Laboratory Los Alamos, NM.

Sung, S., Koo, K., Jung, H. & Jung, H. 2012. Damage-induced deflection approach for damage localization and quantification of shear buildings: validation on a full-scale shear building. *Smart Materials and Structures,* 21, 115013.

Wickramasinghe, W. R., Thambiratnam, D. P. & Chan, T. H. 2015. Use of modal flexibility method to detect damage in suspended cables and the effects of cable parameters. *Electronic Journal of Structural Engineering,* 14, 133-144.

Zhao, J. & DeWolf, J. T. 1999. Sensitivity Study for Vibrational Parameters Used in Damage Detection. *Journal of Structural Engineering,* 125, 410-416.

The 8th International Conference on
Structural Health Monitoring of Intelligent Infrastructure
Brisbane, Australia | 5-8 December 2017

Dynamic testing of a liquid storage tank in view of modal based Structural Health Monitoring

D. Gargaro[1], E. Reynders[2], C. Rainieri[3], G. Fabbrocino[3].
[1] *DiBT Dept., StreGa Lab, University of Molise - Italy, Email:* danilo.gargaro@unimol.it
[2] *Department of Civil Engineering, KU Leuven - Belgium,* [3] *DiBT Dept., StreGa Lab, University of Molise - Italy*

Abstract

Enhancing the knowledge about the dynamic response of tanks or vessels located in critical structures, i.e. hospitals, is of great interest in particular in earthquake prone regions. Their damage can cause explosions and leakage of toxic substances, with the related consequences in terms of pollution and life losses. Previous investigations about the seismic response of tanks have shown that they are very vulnerable if they are not properly anchored. Due to their configuration, liquid and gas storage tanks apply large forces to the supports, which can collapse causing tank toppling. In view of the development of an innovative Structural Health Monitoring (SHM) program for health facilities able to take into account the response of non-structural elements, too, some output-only modal identification tests have been carried out to characterize the dynamic response of a vertical atmospheric water tank under different filling ratios and types of anchoring. This study provides useful hints for the development of modal based SHM systems able to diagnose the health state of similar tanks after seismic events.

1. Introduction

Non-structural components, equipment and installations play a key role to define the operativeness of strategic structures, such as industrial and health facilities (Salzano et al. 2009). These strategic structures should remain undamaged after a seismic event in order to ensure an effective post-earthquake emergency management. However, post-earthquake surveys have shown how real structures are prone to damage (McIntosh et al. 2012, Miranda et al. 2012, Achour et al. 2011, Braga et al. 2011, Ayres & Phillips 1998). In particular, existing hospitals may go out of service because of structural damage, but also because of functional breakdowns (Taghavi & Miranda 2003).

Myrtle et al. (2005) carried out a classification and prioritization of nonstructural systems in hospitals based on the results of extensive surveys of effects of major seismic events (1994 Northridge, 1995 Kobe, 1999 Chi-Chi earthquake and the 1999 Kocaeli earthquake). Surveys included damage surveys, interviews of medical and administrative personnel, and solicitation of expert opinions. This survey pointed out, among the rest, that hospitals cannot operate without electrical services, water, oxygen, and a capacity to dispose waste products. In this contest liquid storage tanks clearly play an important role. Ensuring supply of liquids and medical gases after an earthquake is, in fact, one of the most important tasks to keep the hospital operational.

Unanchored tanks are vulnerable to earthquake loadings. This type of tank is frequently encountered in existing hospitals. Rooftop water tanks, diesel fuel tanks for emergency generators or medical liquid tanks are typical examples of unanchored tanks. Seismic motion can induce large forces at the tanks supports, so that these collapse and the tank overturns, as occurred for instance during the Northridge earthquake (Figure 1) (GeoHazards International, 2009). Strong shaking can also cause rooftop water tanks to break the connections with pipes. Unanchored tanks sliding and breaking the connections with pipes can even fall down from the roof, if they are located close to the edge.

Figure 1 - Tanks failure during Northridge earthquake: tank sliding on its supports (a); overturned medical oxygen tank (b) (GeoHazards International, 2009).

Analyzing the dynamic behavior of those tanks is therefore relevant to improve design procedures and for remote health assessment after an earthquake taking into account the peculiar needs of health facilities (Rainieri & Fabbrocino 2013).
In this paper some results of an ongoing experimental campaign based on extensive dynamic tests on a vertical water tank at different filling ratios and subjected to different boundary conditions are illustrated. They provide useful hints for design of modal based SHM systems for tanks frequently encountered in health facilities.

2. Tank classification

Liquid storage tanks represent relevant non-structural components in hospitals. After FEMA 274 (Federal Emergency Management Agency 1997) non-structural elements have been classified as displacement sensitive and acceleration sensitive. All the components (such as partition walls and pipelines that are rigidly connected to the structure) prone to damage due to an excessive interstory-drift belong to the first group. Acceleration sensitive components are instead vulnerable to shifting or overturning, if their anchorage or bracing is inadequate. This class includes the liquid storage tanks.
The tanks being acceleration sensitive components, their response is influenced by their location along the structure and, as such, the earthquake response of the structure in terms of acceleration represent the input ground motion for the tank. The nuclear code ASCE 4-98 (American Society of Civil Engineers 1998), taking into account that the response of the tank is influenced by the response of the structure, defines a number of criteria to establish whether a coupled or uncoupled analysis structure-tank is needed. These criteria take into account the frequency ratio and the mass ratio between the tank and the structure. Thus, accurate predictions of the dynamic properties of the tanks in operational conditions are definitely fundamental to simulate also their earthquake response.
The amplification of the seismic input throughout the structure can lead to overload at the supports, which could be damaged. Thus, the possibility of remotely recognizing failure at the supports is critical to ensure the hospital efficiency after an earthquake. Taking into account that damage at the supports has an influence on the global stiffness of the tank, this type of damage can be detected by monitoring the variations of the modal parameters of the structure over time (Farrar & Worden 2012).

3. Experimental tests on a water storage tank

A number of dynamic tests have been carried out on a vertical water tank in order to define its dynamic behavior and assess the opportunities of modal based SHM for post-earthquake health assessment. The structure under test (Figure 2a) can be considered as a prototype of atmospheric tanks commonly found in hospitals. It consisted of a cylindrical steel tank welded on three steel, 150 mm high supports. The tank has a capacity storage of about 500 liters and it is characterized by a diameter of 650 mm and a total height of 1540 mm. Two different boundary conditions were considered: in the first test series, each support was anchored to the soil by a 6 mm screw (Figure 2b); in the second test series, free supports were considered to simulate the loss of connection (Figure 2c). In the following, the tank in the first condition is referred to as constrained, while in the second configuration it is referred to as unconstrained.

(a) (b) (c)

Figure 2 - Vertical tank under test (a) and detailed view of supports in the constrained (b) and unconstrained (c) configuration.

For each configuration, different filling ratios were considered: 0%, 25%, 50%, 75%, 90% and 100%, as reported in Figure 3a. In order to apply modal based SHM, the tested structure has to be stationary at least while executing the measurement of its dynamic response. In other words, modal based SHM can be applied to tanks characterized by constant or slowly varying water level.

The ambient vibration response of the structure was recorded by eight accelerometers in three different setups. The sensor layout is schematically reported in Figure 3b. Four sensors (#1, 2, 3 and 4), signed with red *, were fixed while the others were roving sensors. In the first setup the mid section of the tank (section 2) was instrumented, while in the second and third setup sensors were installed also in section 1. The accelerometers were of the piezoelectric type with the following

specifications: ±0.5 g full scale range, 10 V/g sensitivity, 50 g weight. Magnets were used to fix the sensors on the tank. The added mass was about 2% of the total mass of the empty tank.
A measurement device with 24 bit resolution and on-board anti-aliasing filter was used for data acquisition. A sampling frequency of 400 Hz was adopted. The dynamic response of the structure to ambient vibrations, for each setup, was recorded for 90 min. The tank was excited from wind load, traffic induced vibrations and by the passage of the occupants of a nearby structure. Ambient temperature was also monitored during the tests by a RTD sensor. A maximum variation of 5°C was recorded, so it can be expected that the modal identification results are only marginally influenced by the temperature.

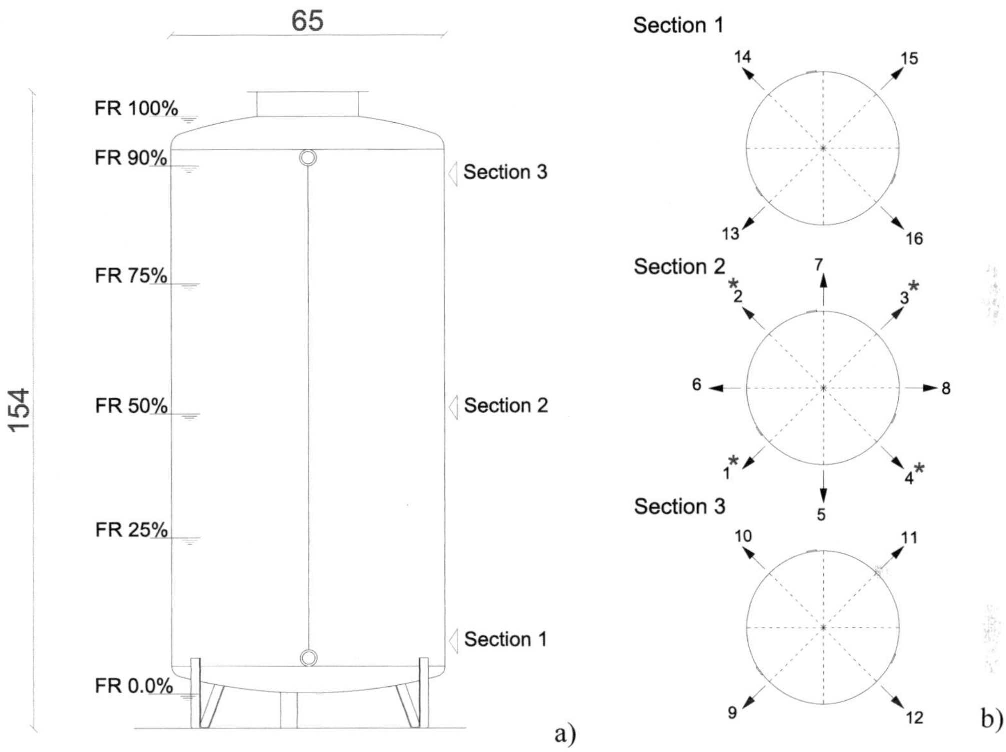

Figure 3 - Front view of the tank (a) and sensor layout (b).

4. Modal identification procedure

Output-only modal parameter estimation has been carried out according to the Covariance Driven Stochastic Subspace Identification (Cov-SSI) (Peeters & De Roeck 1999) and the Frequency Domain Decomposition (FDD) (Brincker et al. 2000, Gade et al. 2005) method. The present section briefly reports the basics of the above-mentioned procedures.
The FDD is based on the Singular Value Decomposition (SVD) of the output Power Spectral Density (PSD) matrix. Structural resonances are identified from the singular value plots by peak picking; the corresponding singular vectors are good estimates of the corresponding mode shapes.
The SSI method works in time domain and it is based on a state space description of the dynamic problem (Peeters 2000, Van Overschee & De Moor 1996). In fact, the second order problem, stated by the differential equation of motion, is converted into two first order problems, defined by the so-called ''state equation'' and ''observation equation''. Such equations, in the output-only case, can be written as follows for the generic discrete time instant $t = k\Delta t$ with $k \in N$, where Δt is the sampling interval:

$$\{x_{k+1}\}=[A]\ \{x_k\}+\{w_k\} \tag{1}$$

$$\{y_k\}=[C]\ \{x_k\}+\{v_k\} \tag{2}$$

$\{x_k\} = \{x(k\Delta t)\}$ is the discrete-time state vector yielding the sampled displacements and velocities, $\{y_k\}$ is the sampled output, [A] is the discrete state matrix, [C] is the discrete output matrix, $\{w_k\}$ is the ''process noise'' due to disturbances and model inaccuracies, $\{v_k\}$ is the ''measurement noise'' due to sensor inaccuracy. These vector signals are both immeasurable. They are assumed to be zero mean Gaussian white noise processes described by their covariance matrices.
In the stochastic framework that characterizes OMA, due to the lack of information about the input, it is implicitly modeled by the noise terms $\{w_k\}$ and $\{v_k\}$. Cov-SSI is based on processing of the output correlations for the estimation of the system matrices.
MACEC software (Reynders et al. 2014) was used for data processing.
Before modal parameter estimation, the recorded time series were pre-processed in order to remove offset and eventual spurious trends. Moreover, they were inspected in order to identify eventual measurement anomalies.
A maximum model order of 100 and a number of block rows i=50 were considered in the construction of the stabilization diagram for the Cov-SSI method. When the FDD method was applied, power spectra were computed according to the Welch procedure by applying the Hanning window and considering a 66% overlap.

5. Test results and discussion

The cylindrical vertical tank tends to vibrate in two directions according to axial wave number *m* and circumferential wave number *n*. Therefore, each mode shape is associated to a couple of values (*m, n*) (Haroun 1983). These mode shapes are the solutions of the theoretical vibration problem and they are employed to study the dynamic behavior of real structures. It is easy to realize how tanks are characterized by several modes. However, attention is herein focused on the first axial mode (m=1) and a maximum of 4 circumferential waves (n=4). The installation of the eight sensors in cross section 2 influences the number of observed circumferential waves.
The first six modes of the structure have been identified from the analysis of the ambient vibration measurements for the constrained as well as unconstrained configuration. Cov-SSI and FDD provided consistent modal identification results.
Results in terms of natural frequencies and damping ratios for different filling ratios and boundary conditions are reported in Table 1, Table 2, Table 3 and Table 4. The 2σ uncertainty bounds are also reported to represent the statistical estimation error for the Cov-SSI method (Reynders et al. 2008). These provide a measure of the accuracy of the estimates. Higher values of the uncertainty bounds are sometimes associated to the first two modes. This variability (observed also in Dooms at al. 2006 and Reyders et al. 2008) depends on the fact that the supports of the structure are not regularly distributed in plan, with slight differences in their relative distances. As a result, the eigenfrequencies of the first two bending modes are similar but not coincident. By the way, a consistent effect of water on the estimated natural frequencies is observed at the different filling ratios. In fact, the difference between the first and second natural frequency is always in the order of 0.4 Hz for the constrained configuration, with the only exception of the configuration characterized by 25% filling ratio. This unexpected result could be due effects of test conditions not explicitly considered in the present study. Thus, further tests and comparisons of results against those provided by numerical models are necessary to confirm the observed natural frequencies in this condition.
A graphic representation of the identified mode shapes is reported in Figure 4, which shows that the first two modes are global bending modes in two orthogonal directions, while the remaining modes

Table 1 - Eigenfrequencies and 2σ confidence intervals for the constrained tank at different filling ratios.

mode	f.r. 0%		f.r. 25%		f.r. 50%		f.r. 75%		f.r. 90%		f.r. 100%	
	f - 2σ [Hz]		f - 2σ [Hz]		f - 2σ [Hz]		f - 2σ [Hz]		f - 2σ [Hz]		f - 2σ [Hz]	
1	25.386	±*0.027*	17.234	±*0.058*	16.201	±*0.021*	12.623	±*0.011*	9.697	±*0.006*	8.386	±*0.016*
2	25.734	±*0.037*	20.050	±*0.018*	16.705	±*0.024*	13.085	±*0.017*	10.045	±*0.014*	8.677	±*0.028*
3	93.104	±*0.012*	77.219	±*0.010*	48.918	±*0.026*	40.386	±*0.045*	39.011	±*0.009*	39.268	±*0.066*
4	94.022	±*0.006*	78.227	±*0.011*	49.982	±*0.019*	38.78	±*0.024*	36.812	±*0.019*	37.040	±*0.110*
5	109.931	±*0.033*	89.390	±*0.009*	58.709	±*0.052*	49.082	±*0.019*	46.522	±*0.071*	44.092	±*0.110*
6	178.479	±*0.006*	131.386	±*0.042*	98.824	±*0.063*	93.052	±*0.026*	91.452	±*0.018*	91.982	±*0.032*

Table 2 - Damping ratios and 2σ confidence intervals for the constrained tank at different filling ratios.

mode	f.r. 0%		f.r. 25%		f.r. 50%		f.r. 75%		f.r. 90%		f.r. 100%	
	ξ - 2σ [%]		ξ - 2σ [%]		ξ - 2σ [%]		ξ - 2σ [%]		ξ - 2σ [%]		ξ - 2σ [%]	
1	1.14	±*0.16*	1.24	±*0.13*	1.12	±*0.11*	1.21	±*0.20*	0.48	±*0.07*	1.06	±*0.21*
2	1.12	±*0.20*	1.43	±*0.15*	0.92	±*0.15*	1.16	±*0.19*	0.84	±*0.18*	0.64	±*0.09*
3	0.06	±*0.01*	0.07	±*0.01*	0.25	±*0.04*	0.58	±*0.09*	0.26	±*0.07*	0.58	±*0.21*
4	0.05	±*0.01*	0.11	±*0.02*	0.19	±*0.03*	0.26	±*0.07*	0.21	±*0.03*	0.47	±*0.19*
5	0.27	±*0.04*	0.09	±*0.01*	0.54	±*0.10*	0.51	±*0.06*	0.82	±*0.09*	0.78	±*0.12*
6	0.07	±*0.01*	0.13	±*0.01*	0.14	±*0.08*	0.05	±*0.04*	0.12	±*0.06*	0.10	±*0.03*

Table 3 - Eigenfrequencies and 2σ confidence intervals for the unconstrained tank at different filling ratios.

mode	f.r. 0%		f.r. 25%		f.r. 50%		f.r. 75%		f.r. 90%		f.r. 100%	
	f - 2σ [Hz]		f - 2σ [Hz]		f - 2σ [Hz]		f - 2σ [Hz]		f - 2σ [Hz]		f - 2σ [Hz]	
1	14.165	±*0.094*	15.660	±*0.033*	14.309	±*0.087*	12.840	±*0.013*	10.058	±*0.015*	8.374	±*0.028*
2	18.482	±*0.086*	19.267	±*0.066*	16.031	±*0.045*	13.413	±*0.017*	10.500	±*0.033*	8.718	±*0.021*
3	92.893	±*0.017*	78.130	±*0.009*	49.162	±*0.015*	39.612	±*0.034*	n.a.	*n.a.*	n.a.	*n.a.*
4	93.968	±*0.031*	79.090	±*0.014*	48.122	±*0.014*	38.939	±*0.020*	36.929	±*0.032*	36.949	±*0.033*
5	109.174	±*0.023*	90.716	±*0.015*	57.398	±*0.124*	49.275	±*0.025*	47.247	±*0.071*	47.609	±*0.046*
6	177.814	±*0.019*	133.034	±*0.062*	98.008	±*0.025*	93.214	±*0.052*	91.515	±*0.047*	91.734	±*0.069*

Table 4 - Damping ratios and 2σ confidence intervals for the unconstrained tank at different filling ratios.

mode	f.r. 0%		f.r. 25%		f.r. 50%		f.r. 75%		f.r. 90%		f.r. 100%	
	ξ - 2σ [%]		ξ - 2σ [%]		ξ - 2σ [%]		ξ - 2σ [%]		ξ - 2σ [%]		ξ - 2σ [%]	
1	2.58	±*0.70*	1.43	±*0.20*	1.60	±*0.23*	0.97	±*0.13*	0.53	±*0.12*	1.55	±*0.46*
2	1.36	±*0.76*	1.83	±*0.57*	1.00	±*0.10*	1.17	±*0.11*	0.63	±*0.17*	1.14	±*0.16*
3	0.10	±*0.01*	0.13	±*0.01*	0.11	±*0.02*	0.07	±*0.13*	n.a.	*n.a.*	n.a.	*n.a.*
4	0.42	±*0.07*	0.09	±*0.02*	0.15	±*0.04*	0.26	±*0.04*	0.21	±*0.07*	0.33	±*0.05*
5	0.10	±*0.02*	0.15	±*0.03*	0.52	±*0.11*	0.51	±*0.04*	0.60	±*0.11*	0.28	±*0.08*
6	0.09	±*0.01*	0.08	±*0.03*	0.17	±*0.05*	0.12	±*0.14*	0.14	±*0.11*	0.20	±*0.09*

affect only the middle section of the tank. Figure 5 and Figure 6 show for the six identified modes how the filling ratio affects their eigenfrequencies and damping ratios. As expected, for a given configuration (constrained or unconstrained), the natural frequency decreases when the overall mass increases, while the damping ratio exhibits a weak increment with the water level that is observed only for the local modes (#3 to #6). The first two modes, that are global in nature, are characterized by a damping ratio higher than 1%; on the contrary, a damping ratio lower than 1% characterizes the local modes.

The mode shapes plotted in Figure 4 do not significantly change when varying the boundary conditions and the filling ratio. However, even if the mode shapes do not vary and, as a consequence, are not informative for the SHM purposes, the eigenfrequencies, especially for the local modes, exhibited significant variations. The differences in terms of eigenfrequencies between the unconstrained and the constrained configuration are reported in Table 5.

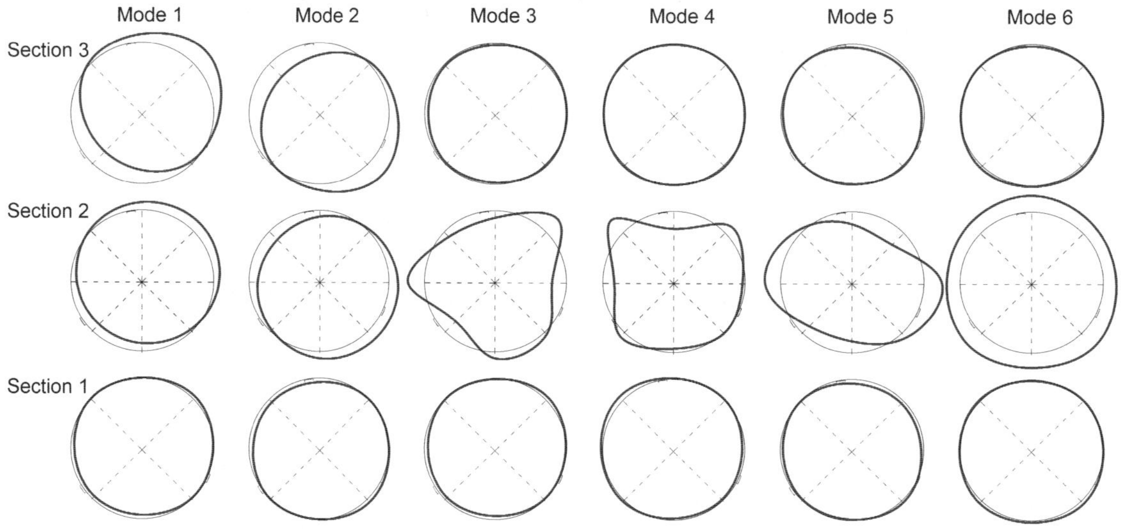

Figure 4 - Graphical representation of the mode shapes.

Table 5 – Eigenfrequency scatters between constrained and unconstrained configuration.

mode	Δf [%]					
	f.r. 0%	f.r. 25%	f.r. 50%	f.r. 75%	f.r. 90%	f.r. 100%
1	-44.20	-9.14	-11.68	6.26	8.28	3.96
2	-28.18	-3.91	-4.04	-1.87	0.13	-3.49
3	-0.23	1.18	0.50	-1.92	n.a.	n.a.
4	-0.06	1.10	-3.72	0.41	0.32	-0.25
5	-0.69	1.48	-2.23	0.39	1.56	7.98
6	-0.37	1.25	-0.83	0.17	0.07	-0.27

Significant differences, in the order of about 44% and 28%, have been observed for the first and second mode, respectively. This percentage variation decreases at a value of about 10% for the 25% filling ratio, and it becomes lower and lower at increasing water levels up to the value of 3% when the tank is full.

Graphic representations of the evolution of the eigenfrequencies and damping ratios are shown in Figure 5 and Figure 6, where red and blue dots denote the values for the unconstrained and constrained configuration, respectively.

While the first two eigenfrequencies are significantly influenced by the boundary conditions, local modes appear to be insensitive to changes at the supports. In the latter case, in fact, the gap between the eigenfrequencies in the two configurations is in the order of 3% or less.

The damping ratios are also insensitive to changes in the boundary conditions: just a small increment has been observed for the first two modes and with a filling ratio up to 50%.

The empty condition of the tank is considered for the sake of completeness. In fact, it is associated to non-operational conditions. Thus, it is of limited interest for practical applications of SHM, but it is relevant from a scientific point of view in order to comprehensively assess the influence of the boundary conditions on the modal properties of the tank.

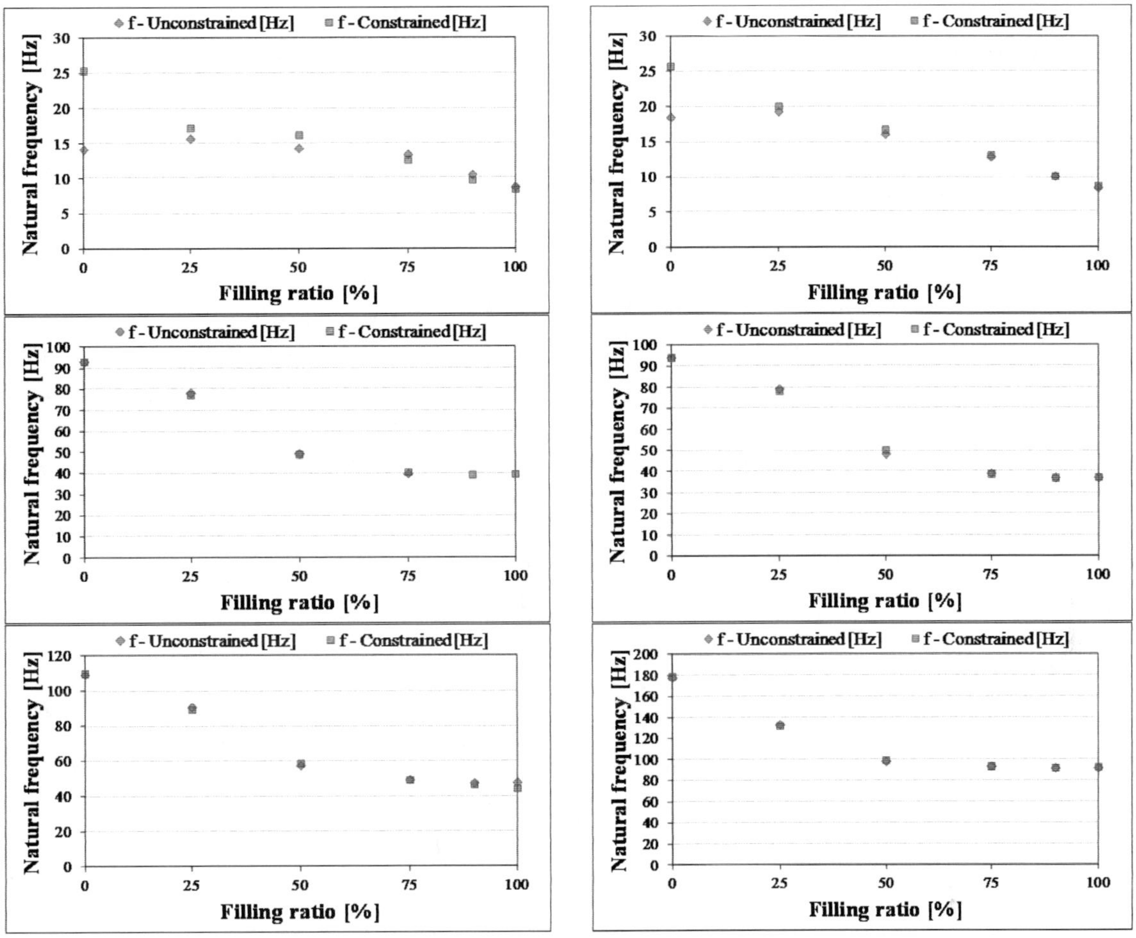

Figure 5 - Evolution of the eigenfrequencies for different filling ratios and boundary conditions.

6. Conclusions

In the framework of a wider research project aimed at implementing effective SHM strategies for health facilities in earthquake prone regions taking also into account the response of non-structural elements, an experimental campaign based on output-only modal identification tests has been carried out to characterize the dynamic response of a vertical atmospheric water tank for different filling ratios and configurations in terms of boundary conditions. The latter have been selected so that the constrained configuration can be referred to as representative of the undamaged condition of the tank, while the unconstrained configuration represents a damaged condition that might occur after a seismic event. The experimental tests have pointed out how the changes in the boundary conditions mainly influence the first two global bending modes of the tank, while little sensitivity has been exhibited by the higher local modes: in fact, the tank acts as a beam which is clamped at the foundations, and support conditions have marginal influence on the local (circumferential) modes. The variation of the dynamic response mainly affects the eigenfrequencies, while the mode shape are basically insensitive to damage at the supports. The changes in the natural frequencies are larger at low filling ratios. Taking into account the reported results, the sensor layout of a SHM system for tanks subjected to damage at the supports can be optimized to observe the fundamental global bending modes only. This is a relevant result for SHM practitioners, since the number of sensors and therefore the costs of the monitoring system can be kept to a minimum; moreover, the sensitivity of the fundamental modes to damage definitely makes the use of (automated) OMA techniques profitable, reducing the risk of missed identification due to poor excitation or limited observability of higher modes.

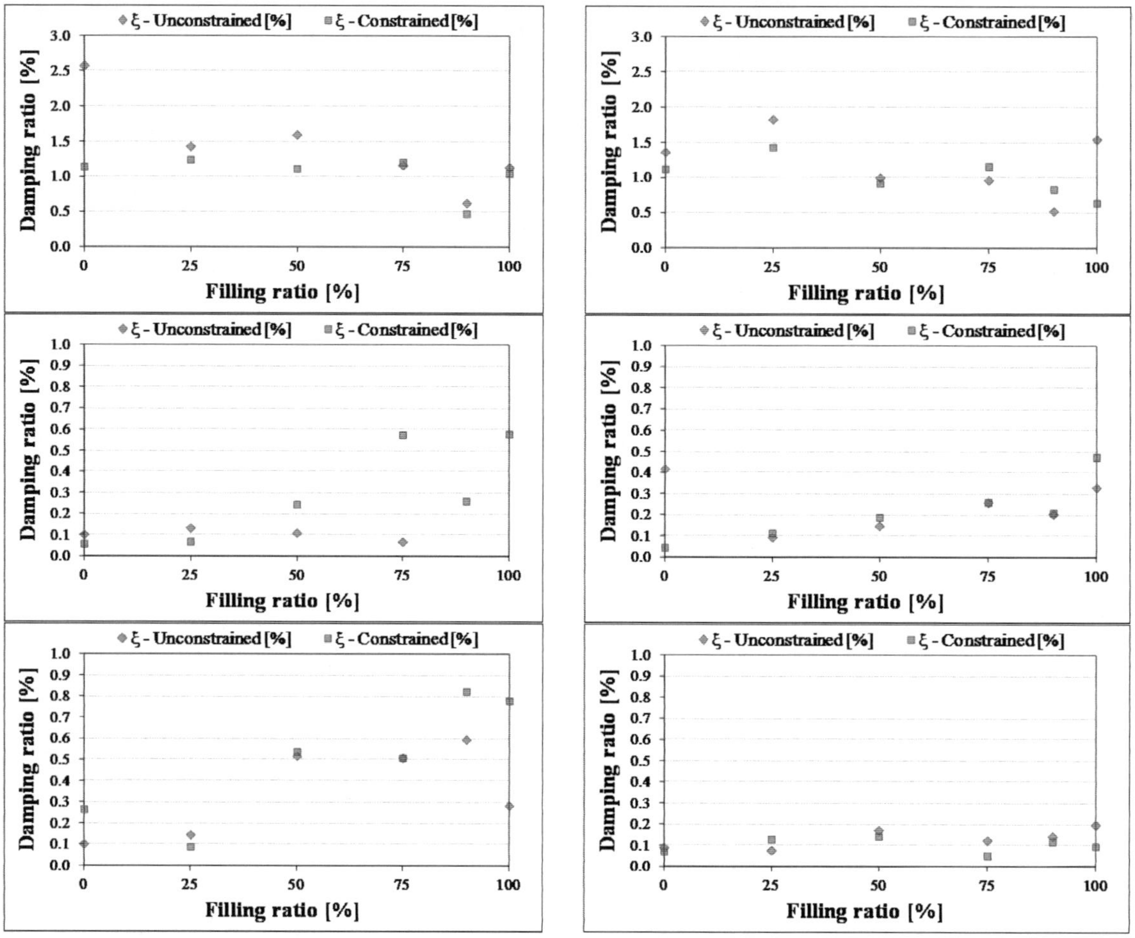

Figure 6 - Evolution of the damping ratios for different filling ratios and boundary conditions.

Acknowledgments

The experimental activities in the present study have been funded the DIBT-IMPACT (Integrated Monitoring of Pressurized and AtmospheriC Tanks) research program at the University of Molise, whose financial support is gratefully acknowledged.

References

Achour, N, Miyajima, M, Kitaura, M & Price A, 2011, 'Earthquake-Induced Structural and Nonstructural Damage in Hospitals', *Earthquake Spectra* 27(3): 617-634.

American Society of Civil Engineers (ASCE), 1998, 'Seismic Analysis of Safety-Related Nuclear Structures and Commentary'.

Ayres, JM & Phillips, RJ, 1998, 'Water damage in hospital resulting from the Northridge Earthquake',*ASHRAE TransPart 1*';SF-98-14-2:1286–96.

Braga, F, Manfredi, V, Masi, A, Salvatori, A & Vona M, 2011, 'Performance of non-structural elements in RC buildings during the L'Aquila, 2009 Earthquake'. *Bulletin of Earthquake Engineering* 2011; 9:307–24.

Brincker, R, Zhang, L & Andersen P, 2000, 'Modal identification from ambient responses using frequency domain decomposition', *The 18th SEM International Modal Analysis Conference*, San Antonio, TX, USA, 625–630.

Dooms, D, Degrande, G, De Roeck, G & Reynders, E, 2006, 'Finite element modelling of a silo based on experimental modal analysis', *Engineering Structures* 28 (4) 532–542.

Farrar, C.R. & Worden, K., 2012, 'Structural health monitoring: a machine learning perspective'. John Wiley and Sons Ltd., Chichester, UK, 631 p.

Federal Emergency Management Agency (FEMA) 274, 1997, 'NEHRP commentary on the guidelines for the seismic rehabilitation of buildings', Washington, D.C.

Gade, S, Møller, N.B, Herlufsen, H & Konstantin-Hansen, H, 2005, 'Frequency domain techniques for operational modal analysis'. *The 1st International Operational Modal Analysis Conference*, Copenhagen, Denmark, 261–271.

GeoHazards International, 2009, 'Reducing Earthquake Risk in Hospitals from Equipment, Contents, Architectural Elements and Building Utility Systems', Menlo Park, CA.

Haroun, MA, 1983, 'Vibration studies and test of liquid storage tanks', *Earthquake Engineering and Structural Dynamics,* vol. 11, 179-206.

McIntosh, JK, Jacques, C, Mitrani-Reiser, J, Kirsch, TD, Giovinazzi, S & Wilson, TM, 2012, 'The Impact of the 22nd February 2011 Earthquake on Christchurch'. Proceedings *of 2012 New Zealand Society for Earthquake Engineering*, New Zealand.

Miranda, E, Mosqueda, G, Retamales, R & Pekcan, G, 2012, 'Performance of nonstructural components during the February 27, 2010 Chile Earthquake'. *Earthquake Spectra* 28(S1): S453–71.

Myrtle, RC, Masri, SF, Nigbor, RL &. Caffrey, JP, 2005, 'Classification and Prioritization of Essential Systems in Hospitals under Extreme Events', *Earthquake Spectra*, Vol. 21, No. 3, pp. 779-802.

Peeters, B, 2000, 'System identification and damage detection in civil engineering', PhD thesis, Katholieke Universiteit Leuven, Leuven, Belgium.

Peeters, B & De Roeck, G, (1999), ' Reference-based stochastic subspace identification for output-only modal analysis' *Mechanical Systems and Signal Processing,* Vol. 13, Issue 6, Pages 855-878.

Rainieri, C & Fabbrocino, G, 2013, 'Sustainable design of smart health facilities in seismically prone areas', *Key Engineering Materials*, Vol. 569-570, 278-285.

Reynders, E, Schevenels, M & De Roeck, G, 2014, 'MACEC 3.3 a Matlab toolbox for experimental and operational modal analysis', Faculty of Engineering Department of Civil Engineering structural Mechanics Section.

Reynders, E, Pintelon, R & De Roeck, G, 2008, 'Uncertainty bounds on modal parameters obtained from stochastic subspace identification'. *Mechanical Systems and Signal Processing* 22 948–969.

Salzano E, Agreda AG, Di Carluccio A & Fabbrocino G, 2009, 'Risk assessment and early warning systems for industrial facilities in seismic zones' *Reliability Engineering & System Safety*, 94 (10): 1577-1584.

Taghavi, S & Miranda, E, 2003, 'Response Assessment of Nonstructural Building Elements', *PEER Report 2003/05*, University of California Berkeley. http://peer.berkeley.edu/publications/peer_reports/reports_2003/reports_2003.html

Van Overschee, P & De Moor, B, 1996, 'Subspace identification for linear systems: theory–implementation–applications'. Kluwer Academic Publishers, Dordrecht, The Netherlands.

The 8th International Conference on
Structural Health Monitoring of Intelligent Infrastructure
Brisbane, Australia | 5-8 December 2017

Continuous Dynamic Monitoring of Offshore Structures

Michele Rizzo[1], Ostilio Spadaccini[2], Andrea Vignoli[3]
[1,2,3] Department of Civil and Environmental Engineering, , Email: mrizzo@dicea.unifi.it
DICeA University of Florence, Florence, Italy

Abstract

Jacket platforms and gravity based structures are located and operating offshore in the Italian sea. The design was implemented to respond to the storms and earthquakes. Different types of analyses have been performed in the design phase: analyses based on a three-step solution where the output describes seabed earthquake motions along with the impedance functions of the rigid foundation, and analyses based on the global (one-step) method including the structures in the computational model of the soil. The output evidences are the time history of earthquake motions (accelerations or displacement) at specified points on the structure or the acceleration response spectra at these specified points. The present paper deals with data collection and the statistical analysis of the structural dynamic response data recorded since 2002; data recorded during storms and seismic events, are used to compare design accelerations of the structures, and to determine the natural frequencies and relevant mode shapes after the events. A new technique based on kernel principal component analysis is developed to eliminate environmental influences on natural frequencies and relevant mode shapes.

1. Introduction

The verification of offshore structures for their use beyond their initial life requires a design of inspection aimed to check-up constantly the structural elements (joints and members). The inspections, their typology and their frequency are a critical issue and inspection planning, were based mainly on probabilistic analysis (Risk Based Inspection, RBI). The inspection of structural damage is in usually hard to perform, taking in account both the water depth and marine growth that hide the structural member.

To overcome these problems the monitoring techniques for damage identification through the analysis of the changes in the modal properties of offshore structures have been developed and the results of previous researches showed indexes capable to detect both offshore damages and mass changes (Betti, Spadaccini & Vignoli 2015).

This research investigates the capacity of monitoring systems to assess possible structural damage in presence of environmental and operational variations, the Principal Components Analysis (PCA) and Novelty Detection are used for this scope. This output-only black box techniques, has been applied for eliminating environmental influences on features such as natural frequencies, which boils down to estimating a relationship between the observed features and unknown environmental factors (Rizzo & Vignoli 2015). An effective method for Structural Health Monitoring (SHM) in changing environmental conditions is the kernel PCA (Reynders, Wursten & De Roeck 2014). In this work the method is improved by using as control parameters, instead of the frequencies, the values of the modal displacements.

As a reference case study the VEGA-A offshore platform, an eight-leg steel fixed jacket platform operating in the Sicily Channel, 25 km offshore, was considered.

2. The Kernel-PCA-modified method

The original method is modified by using a Gaussian RBF (radial basis function) more powerful than the original method, also as control parameters rather than frequencies, they will be used values of the displacement of the first mode shapes. An improved output-only technique to eliminate nonlinear environmental and operational influences on the monitored features is developed and validated on real-life monitoring data; it is based on Gaussian kernel PCA, where the two parameters of the global system model are automatically determined. The first parameter, which represents the standard deviation of the Gaussian (or radial basis function (RBF)) kernel, is chosen in such a way that the matrix of mapped output correlations is maximally informative; in particular a matrix of standard deviations to improve the RBF is used.

If the output of the model is identified as a sequence of vectors $\mathbf{y}_k \in R^{ny}$, k = 1, n_s, the elements of which are damage-sensitive quantities, the RBF kernel is defined as

$$k(\mathbf{y}_i, \mathbf{y}_j) = \exp\left(-\boldsymbol{\delta}_{ij}^t \mathbf{C} \boldsymbol{\delta}_{ij}\right); \quad \boldsymbol{\delta}_{ij} = |\mathbf{y}_i| - |\mathbf{y}_j|; \tag{1}$$

In the vector $\mathbf{y}_k$ are present in the modal displacements ranked in column, each normalized with the respect of the unity norm:

$$\mathbf{y}_k = \begin{Bmatrix} \mathbf{y}_{k,1} \\ \cdots \\ \mathbf{y}_{k,m} \end{Bmatrix}, \tag{2}$$

where $\mathbf{y}_{k,m}$ is the *m-th* modal form obtained in the *k-th* identification (*k-th* sample).

$$\mathbf{C} = \begin{pmatrix} \mathbf{1/2\sigma_1^2} & 0 & 0 \\ 0 & \ldots & 0 \\ 0 & 0 & \mathbf{1/2\sigma_m^2} \end{pmatrix}; \tag{3}$$

In this case we use *m* values of σ, one for each mode shape. The number of principal components to retain, n_u, is then determined by inequality (4);

$$F \le \frac{\sum_{k=1}^{nu} \Sigma_{kk}}{\sum_{k=1}^{nt} \Sigma_{kk}}; \tag{4}$$

where Σ_{kk} denotes the kth diagonal element of Σ, matrix of eigenvalues. It can be seen that K is more informative when σ approach the optimal value which corresponds to the maximum entropy. The parameter σ controls the bandwidth of the inner product matrix K. When σ is very small, this matrix is approximately diagonal. In this case, K provides no information on the correlation among the data points and PCA becomes meaningless. When σ is very large, all elements of the matrix are approximately equal, which makes it again uninformative. An optimal value of σ can be found by requiring that the corresponding inner product matrix is maximally informative as measured, in particular, σ is determined by maximizing the entropy of the inner product matrix K (Shannon 1948) and (Widjaja, Varon, Caicedo & Van Huffel 2012).

3. Case study: VEGA-A platform

In February 1987, the VEGA-A platform was endorsed at a depth of about 122 meters under sea level using a jacket and a steel lattice structure with eight pillars anchored to the seabed by means of 20 piles; on top of these the remaining structural modules hosting production and services plants were subsequently placed. In Figure 1 the VEGA-A platform is shown in the dockyard and in the actual position. The platform is monitored by means of 9 linear accelerometers, 6 linear and 3 rotational, a depth gauge, a current meter and systems for detecting speed and direction of wind. The Department of Civil and Environmental Engineering of the University of Florence is in charge of data analysis for the VEGA platform since 1988 (Spadaccini & Vignoli 1990), when the system was first operated. First in 2001, the monitoring system has been changed, principally with the addition of new accelerometers and a new wave-meter (depth gauge and current meter) and in September 2015, in addition to a second microwave wave-meter.

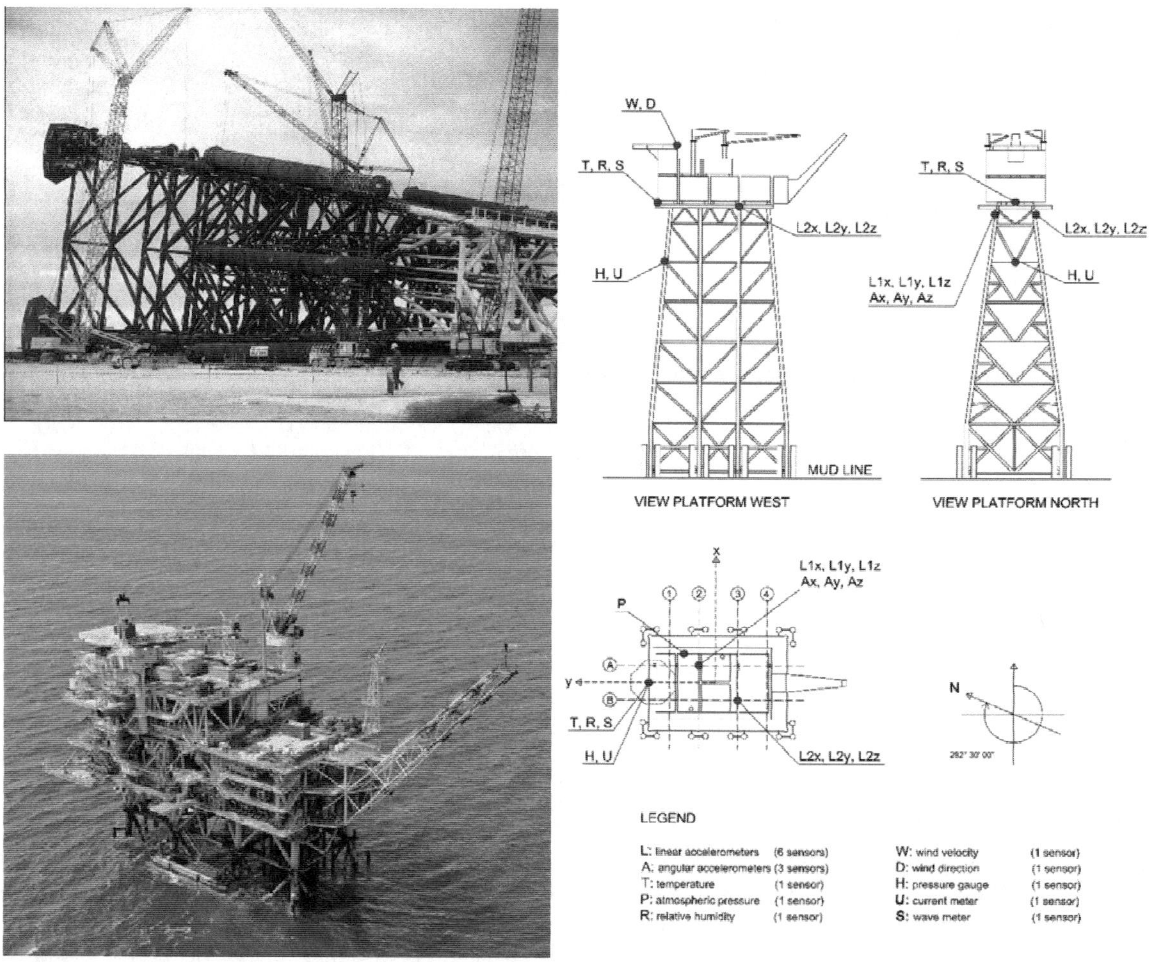

Figure 1. VEGA-A platform, in the dockyard, in service and actual sensor position.

The meteocean and accelerometer signals of the sensors are acquired by an acquisition unit which transmits the data to the control system. The acquisition unit provides to send (email or sms) alarms on the basis of pre-set threshold values.

Actually, the acceleration data are acquired with a sampling frequency of 16 Hz, while all the meteocean data are acquired with a sampling of 2 Hz. For instance data are shown in Figures 2, 3, 4 and 5, where it's possible to point out the recording operated by the monitoring system for the storm of 2016-12-17, with significant wave height of 5.58m. In Table 1 the reduced data (mean on 20') for the year 2016, the height significant and max of the wave, H_s, H_{max}, the up-crossing and significant period, T_z, T_s, the direction of the wave, D_m, and the speed and direction of the wind, W, D_w are shown.

Table 1: Max values for Height significant of the wave Hs, year 2016

month	Day:h	H_s (m)	H_{max} (m)	T_z (s)	T_s (s)	D_m (degN)	W (m/s)	D_w (degN)
January	7:07.00	5.05	8.01	8.14	8.50	218	22.32	287
March	4:03.40	5.65	8.07	8.28	9.45	307	22.93	282
July	15:18.20	3.94	6.87	7.61	8.26	353	17.69	288

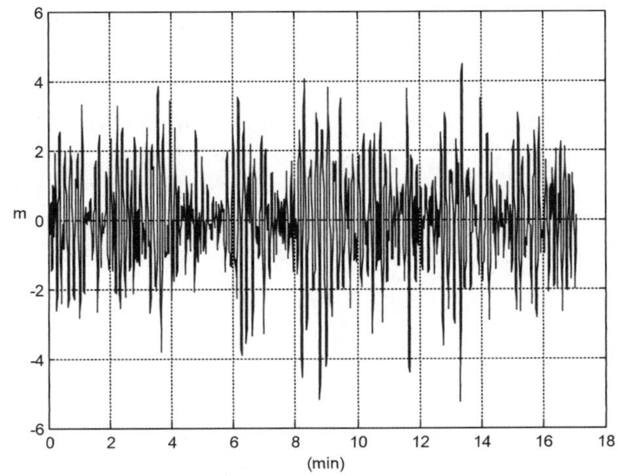
Figure 2. VEGA-A platform, wave elevation, storm of 2016-12-17.

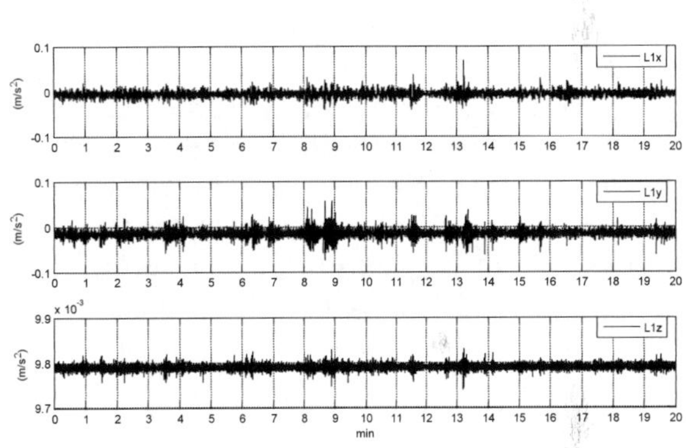
Figure 3. VEGA-A platform, linear acceleration L_1, storm of 2016-12-17.

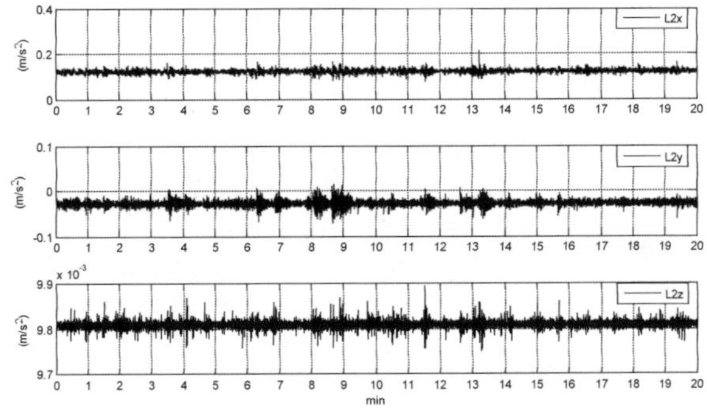
Figure 4. VEGA-A platform, linear acceleration L_2, storm of 2016-12-17.

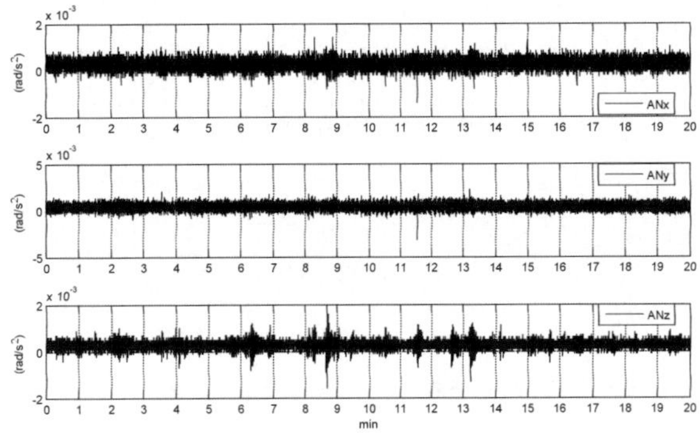
Figure 5. VEGA-A platform, angular acceleration A_N, storm of 2016-12-17.

In Figure 6 we can see, the first three modes identified from the analysis; the first two modes of vibration are characterized by a horizontal translation while the third is a torsional mode shape around the z axis.

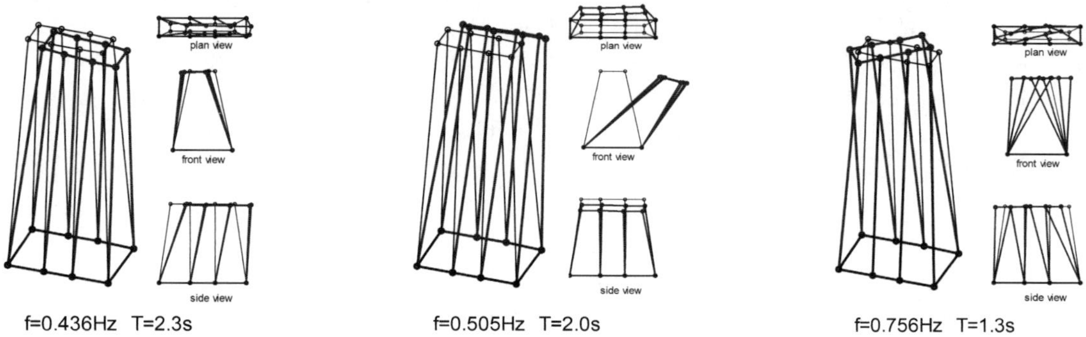

Figure 6. Identified mode shapes of platform, f_1=0.436, f_2=0.505 and f_3=0.756 (H_z).

The modal parameters are obtained through an automated processing implemented on Matlab code; the record used to extract the modal parameters from year 2001 to year 2015 have a duration of 20min at the midnight of every day and the sampling frequency is 10 Hz. One of the objectives of the analysis is to capture the environmental effects, therefore it is considered sufficient that the analyzed time series report an estimation per day. The observation of the evolution of the first three natural frequencies identified in the frequency range 0.4÷0.8 Hz on the basis of the SSI method in the period 01/01/2001 to 31/08/2015, suggest that these frequency estimated are stable; except for the period 2001-2002 in which the derrick on the platform were removed and in the first mounts of 2015 where in the platform were performed heavy maintenance work of the wells.
In Figure 7 is shown the frequency trend.

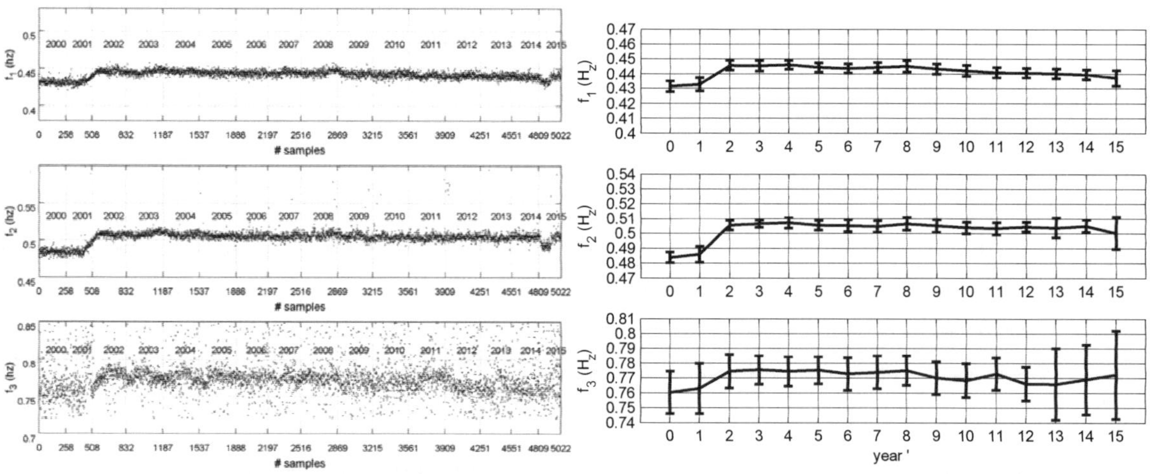

Figure 7. Variation of f_1, f_2 and f_3, mean and standard deviation representation, years 2000-2015.

It can be observed, in addition to the change of mean value due to the removal of derrick in the year 2000-2001 and the variation of 2015 for the work of maintenance, also in that the standard deviation of the third frequency is very large when compared to the other two.

Actually in the monitoring system of the platform, for the health control, is implemented in automatic the linear PCA method. The automated application of linear PCA indicates that the *Novelty Index* from September 2015 to December 2016 remain steadily at the assigned thresholds, so the linear PCA method does not record any substantial changes in the structural behavior.
The Kernel-PCA-modified method is necessary not only because eliminates the environmental effects on modal shapes but because the platform shows nonlinear behavior due to the nature of its foundations. In this case the method can recognize the non-linear behavior of the structure and hence does not identify it as damaging.

3.1 Application of Kernel-PCA-modified method

The proposed Kernel-PCA-modified method, in the section 2, was also used for past data derived from the platform monitoring system. To identify the possible damages, only the first mode shape was taken, then the first two mode shapes and then the first three mode shapes, as shown in Figure 6. In the first case, the entropy is function of one variable σ_1. Therefore, n_t=76 data points are used for training the model (December 2001-February 2002, one identification per day at h.00:00). The number of principal components to retain is then determined with F = 0.99 and his leads to a value n_u = 45. The rest of the data is from March 2002 to august 2015, n=4691 (one identification per day at h.00:00). In Figure 8 it can be seen that K is more informative when σ_1 approach the optimal values 0.13 which corresponds to the maximum entropy.

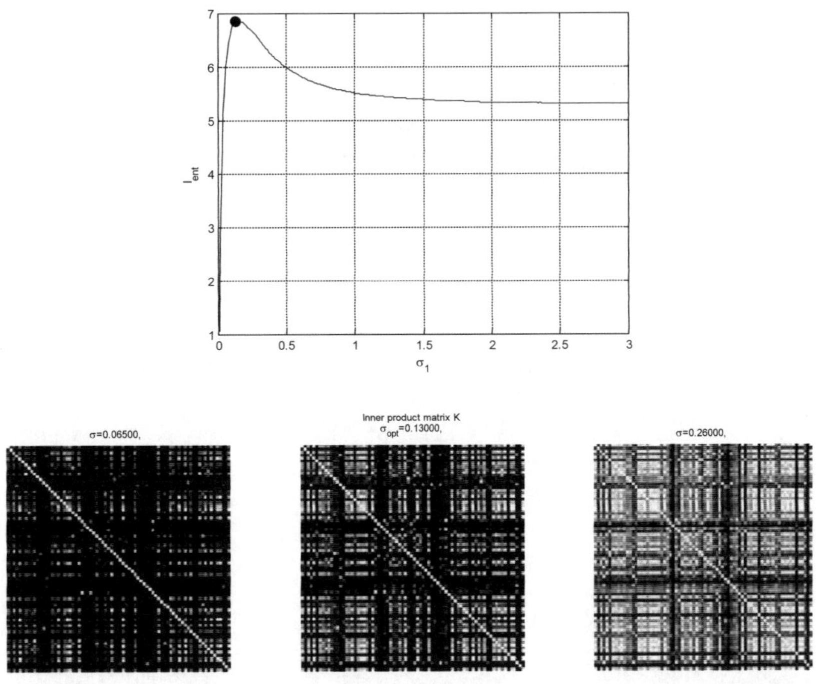

Figure 8. Inner product matrix K plotted as a function of the RBF kernel parameter σ_1, and a grayscale image for three different values of the RBF kernel parameter σ_1.

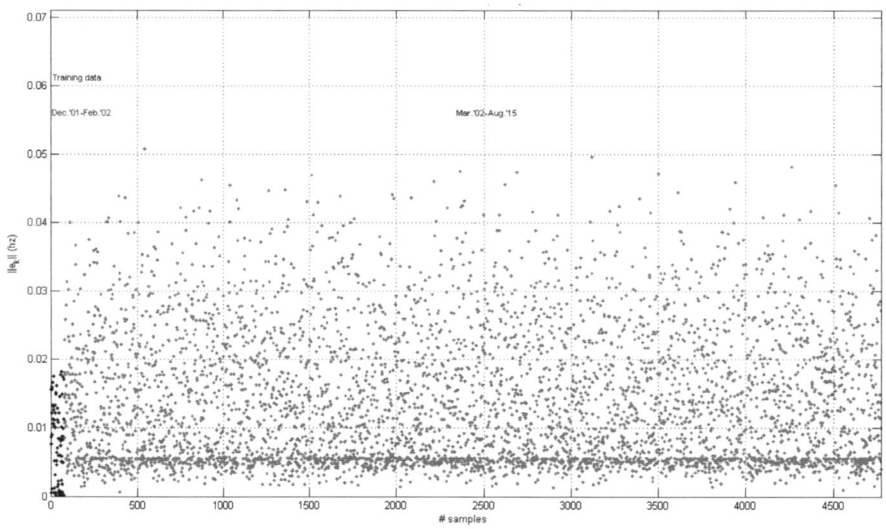

Figure 9. Misfit as a function of the RBF kernel parameter σ_1 with training data December-February 2002.

In the second case (two mode shapes) the entropy is function of one variable σ_1 and σ_2. The number of principal components to retain is then determined with F = 0.99 and his leads to a value $n_u = 51$. The rest of the data is from March 2002 to august 2015, n=4691 (one identification per day at h.00:00). In Figure 10 it can be seen that K is more informative when σ_1 and σ_2 approach the optimal values 0.14 and 0.8 which corresponds to the maximum entropy.

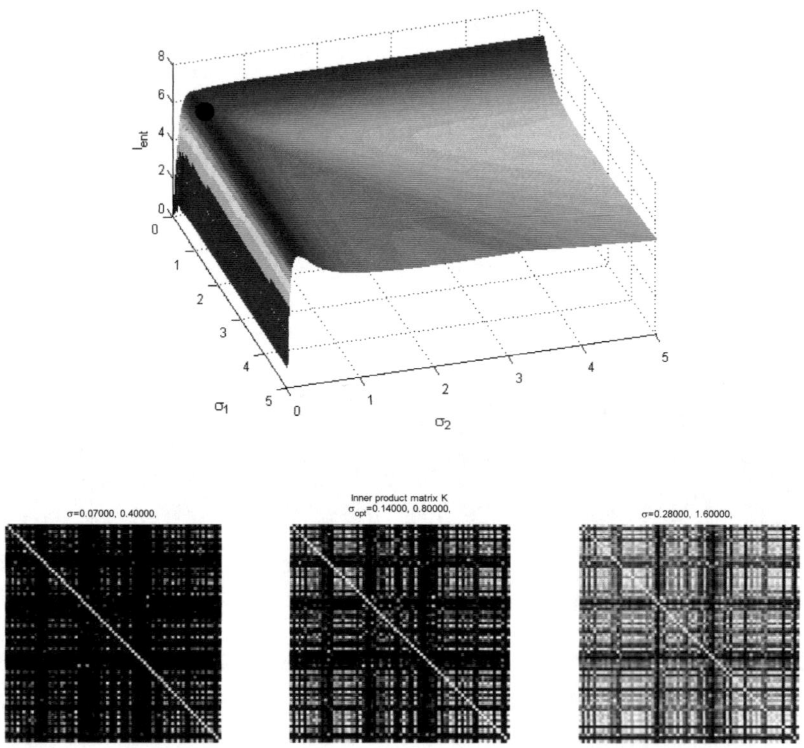

Figure 10. Inner product matrix K plotted as a function of the RBF kernel parameter σ_1 and σ_2, and a grayscale image for three different values of the RBF kernel parameter σ_1 and σ_2.

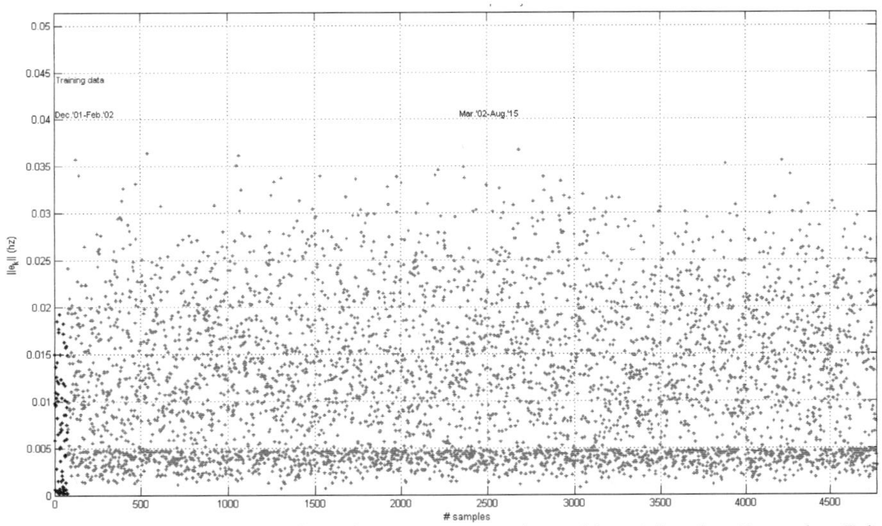

Figure 11. Misfit as a function of the RBF kernel parameter σ_1 and σ_2 with training data December-February 2002.

In the last case the entropy is function of three variables σ_1, σ_2 and σ_3. Therefore, n_t=76 data points are used for training the model (December 2001-February 2002, one identification per day at h.00:00). The number of principal components to retain is then determined with F = 0.99 and his leads to a value n_u = 45. The rest of the data is from March 2002 to august 2015, n=4691 (one identification per day at h.00:00).

In Figure 12 it can be seen that K is more informative when σ_1, σ_2 and σ_3 approach the optimal values 21.8, 4.8 and 0.8 which corresponds to the maximum entropy.

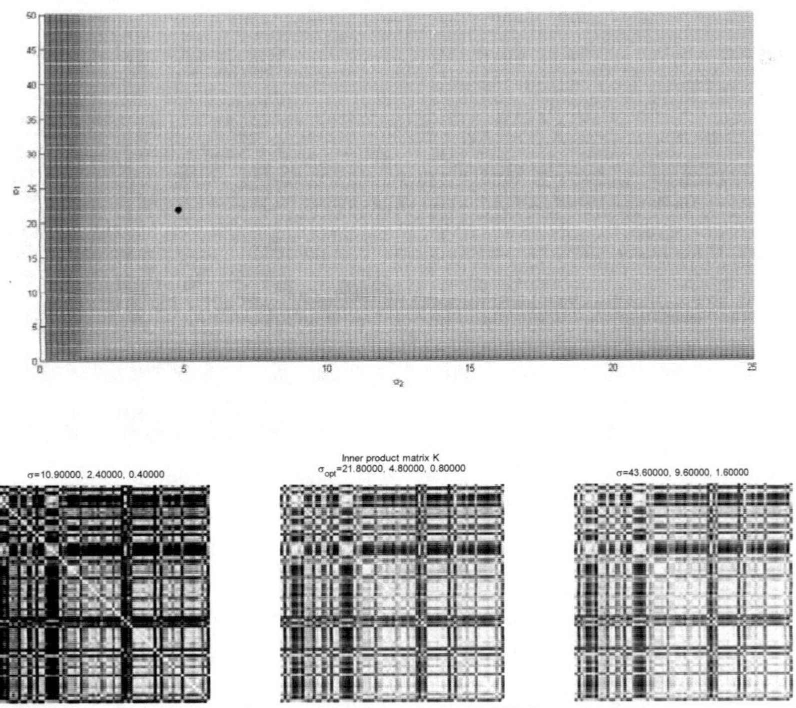

Figure 12. Inner product matrix K plotted as a function of the RBF kernel parameter σ_1, σ_2 and σ_3=0.8, and a grayscale image for three different values of the RBF kernel parameter σ_1, σ_2 and σ_3.

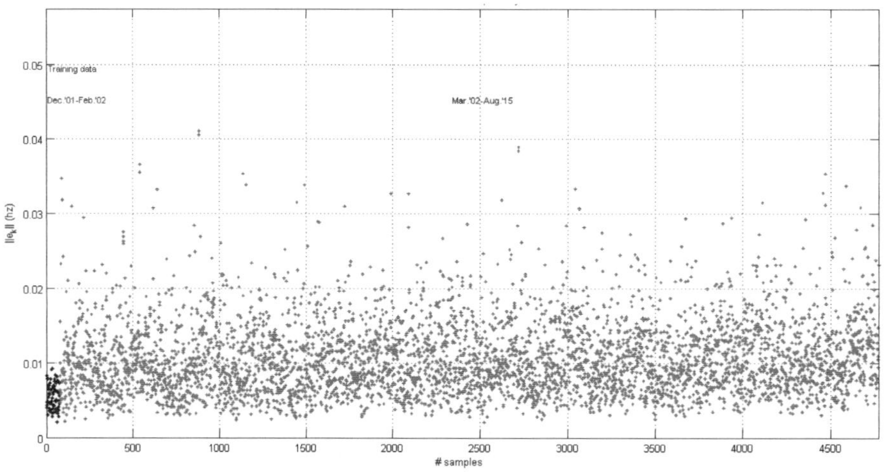

Figure 13. Misfit as a function of the RBF kernel parameter σ_1, σ_2 and σ_3 with training data December-February 2002.

For the identification of the damage it's are not primarily interested in the model parameters but rather in the misfit between the model predictions and the observed behavior of the structure. With the procedure set out above it's possible to evaluate the misfit that is shown in Figures 9, 11 and 13. The results show the effect of the derrick removal (2002) and that no damage occurred on the platform in the following 15 years of activities.

In Figures 14 and 15 can see the results of application of kernel PCA for the data collected in the period 2003-2013 in order to neglect the effects of maintenance works and to focus only on environmental effects.

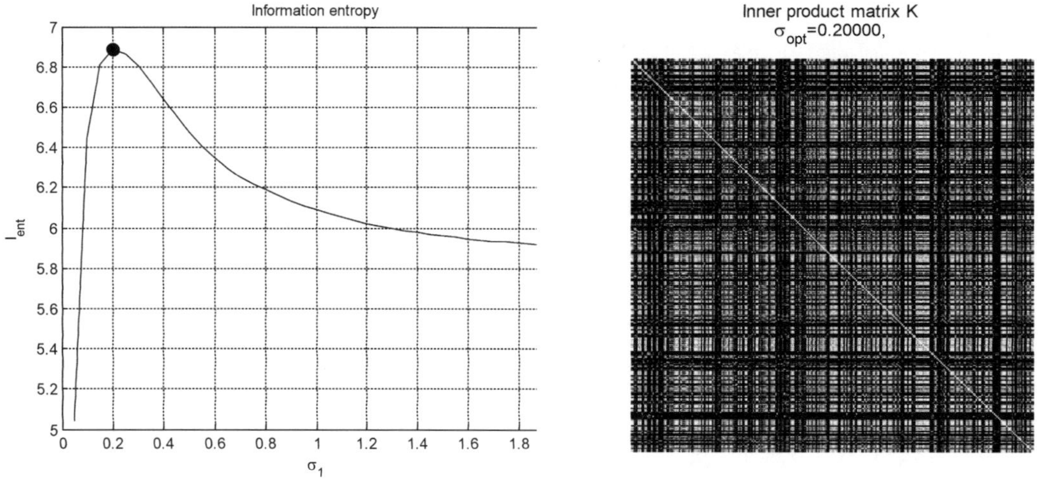

Figure 14. Information entropy and inner product matrix K of the nonlinear output-only model constructed with training data 2003-2013.

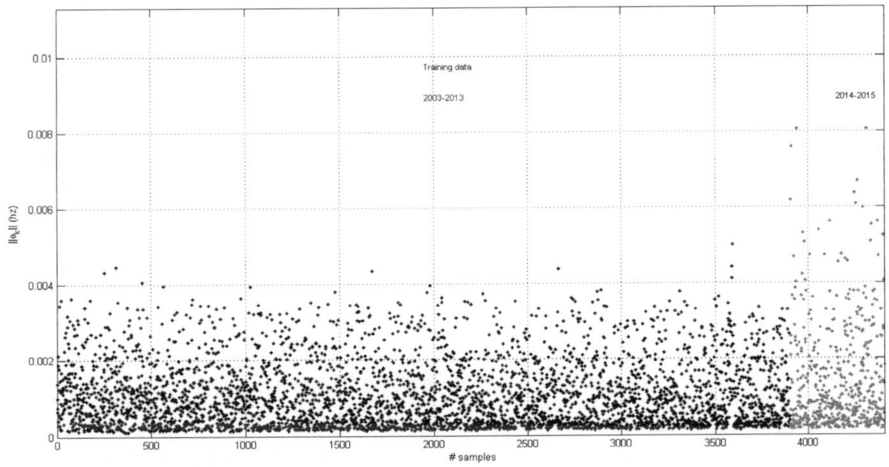

Figure 15. Misfit of the nonlinear output-only model constructed with training data 2003-2013.

4. Conclusions

This paper investigated the dynamic behaviour of the platform VEGA-A, in reference to their mode shapes and their variation over time, from the year 2000 to 2015. The method used, Kernel-PCA-modified, allows the control of the health status of the platform and it's able to detect damages that involve beam of various kinds (vertical legs or diagonal bracings), thanks to the inclusion in the method of displacements related to the mode shapes. A monitoring system of this type provides effective dynamic information, and removing the seasonal effects it can be a valuable tool to detect the structural damage. Application of kernel PCA to data collected in the long period can work as an alert system associated to the verification of vibration serviceability limits and can be a useful support in the risk based inspections.

References

Betti, Spadaccini & Vignoli 2015, 'Offshore Platform Structural Damage Identification Versus Robustness', *Proceedings of the IABSE 2015 Workshop on Safety, Robustness and Condition Assessments of Structures*, Finland, Helsinki, 11-12 February 2015, 86-93.

Rizzo & Vignoli, 2015, 'Monitoring and dynamic identification of Vega-A platform', *Proceedings of the ASME 34th International Conference on Ocean, Offshore and Arctic Engineering*, OMAE2015, May 31-June 5, 2015, St. John's, Newfoundland, Canada.

Reynders, Wursten & De Roeck, 2014, 'Output-only structural health monitoring in changing environmental conditions by means of nonlinear system identification', *Structural Health Monitoring*, Vol 13(1) 82–93.

Shannon 1948, 'A mathematical theory of communication', *The Bell System Technical Journal,Vol. 27: 379–423, 623-656, July, October.*

Widjaja, Varon, Caicedo & Van Huffel, 2012, 'Application of kernel principal component analysis for single-lead-ECG-derived respiration', IEEE T *Bio-Med Eng.*; 59(4): 1169–1176.

Spadaccini & Vignoli, 1990, 'Dynamic identification of a steel jacket platform', *Proceedings of the first European Offshore Mechanics Symposium*, Trondheim, Norway, 20-22 august 1990.

The 8th International Conference on
Structural Health Monitoring of Intelligent Infrastructure
Brisbane, Australia | 5-8 December 2017

Structural vibration identification of bridges by 3D measurement FE analysis and the actual vibration measurement

Kohei YAMAGUCHI[1], Hiroshi MATSUDA[1], Taiki KAWAMURA[1], Takeshi SAIGYO[2], Keisuke KIMOTO[3] and Takafumi NISHIKAWA[1]

[1] *Graduate School of Engineering Nagasaki University, Nagasaki, Japan, Email: kohei@nagasaki-u.ac.jp*
[2] *PAL Corporation, Nagasaki, Japan*
[3] *Keisoku Research Consultant Corporation., Hiroshima, Japan*

Abstract

In this study, short span bridges with a span length less than 15m, which are numerous in local government were no design documents, and the year of construction is often unknown. It was proposed a method to evaluate its safety and risk for those bridges and examined its effectiveness and usefulness by field test. Specifically, a 3D model was created by a relatively inexpensive 3D laser scanner and structural analysis was carried out. After that, it was compared the result with actual vibration measurement data and presented the method of structural vibration identification of bridges by the simplest possible method.

1. Introduction

In Japan, the aging of bridges constructed during the period of high economic growth has occurred frequently in various places. Under such circumstances, it is said that efficient and accurate maintenance management is necessary, but in 2013 a fall accident of the ceiling plate of Sasago tunnel on the Central Expressway occurred. The following year, the Ministry of Land, Infrastructure, Transport and Tourism required the periodic inspection to observe nearly 700,000 road bridges over 2 meters in span at a frequency of once every five years [1)]. In addition, the Cabinet Office has formulated a basic plan to extend the life expectancy of infrastructure. Furthermore, in the SIP: Cross-ministerial Strategic Innovation Promotion Program, "infrastructure maintenance / update / management technology" were taken up as a subject for the next generation infrastructure field. Among them, research and development such as inspection, monitoring, diagnostic technology, robot technology, asset management technology and the like are intensively advanced [2)].

Many bridges managed by the country are well maintained including design documents and others.However, among short span bridges with a span length less than 15m of local government, there are no design documents, and the year of construction is often unknown. In addition to prefectures, many municipalities are absent from technical experts in municipalities. Also, coupled with the shortage of financial resources, the cost of inspection and diagnosis of bridges has become a major issue.

In this study, short span bridges with a span length less than 15m, which are numerous in local government were no design documents, and the year of construction is often unknown. It was proposed a method to evaluate its safety and risk for those bridges and examined its effectiveness and usefulness by field test.

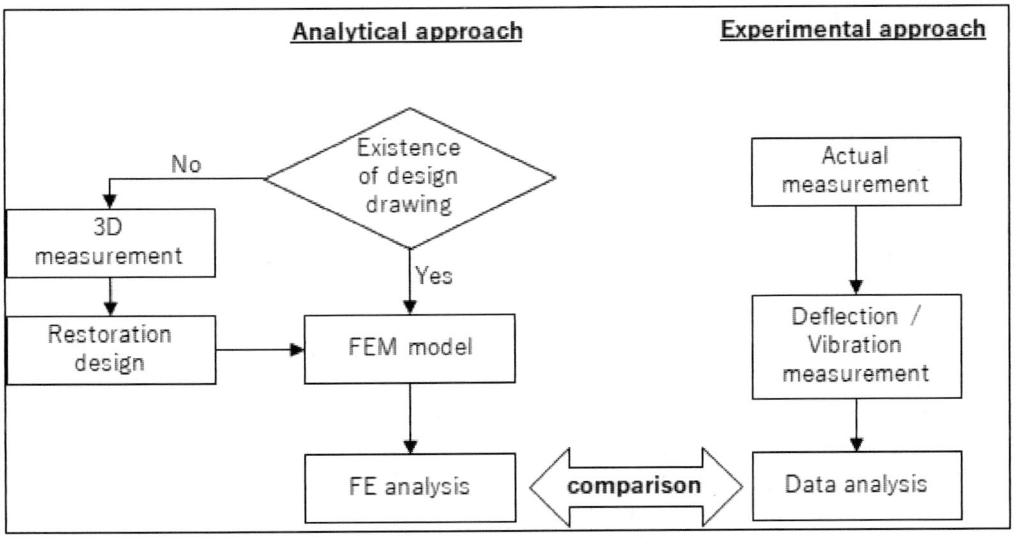

Figure 1 Structural vibration identification

Specifically, a 3D model was created by a relatively inexpensive 3D laser scanner and structural analysis was carried out. After that, it was compared the result with actual vibration measurement data and presented the method of structural vibration identification of bridges by the simplest possible method shown in Figure 1 [3)].

2. Targeted bridges in this study

Among the bridges targeted in this study, medium-sized bridge A is a railway bridge that is not currently in service, the construction year is unknown, simple underpass truss bridge, span length 69.2 m in Nagasaki prefecture. Similarly, the small-scale bridge B is a road bridge managed by Nagasaki Prefecture, 57 years after construction, 2 spans simple posten T girder bridge, span length 16.0 m. Each feature is shown in Table 1.

Table 1 Outline of targeted bridge

	Bridge A	Bridge B
Erection Year	unknown	1960
Bridge type	simple underpass truss bridge	2 spans simple posten T girder bridge
Span length	69.2m	16.0m
Skew	90°	60°
Bearing	steel	rubber
design live load	KS-12	TL-14

3. Overview of 3D Measurement and FE analysis

As shown in Figure 2, the 3D measurement was carried out about 10 minutes once, 9 times for bridge A and 5 times for bridge B so that nearly all point group data of the target bridge could be acquired. The measurement range of the used 3D laser scanner is 0.6m to 330m, the measurement speed is up to 976,000 points / s, the measurement error is ± 2 mm maximum, and the built-in color camera has the maximum resolution 70 Mega pixels.

At first, 3D model shown in Figure 3 was created by converting point group data acquired by 3D measurement into 3D CAD. The size of the created 3D model was measured on the CAD, and an analysis model shown in Figure 4 was created.

For bridge A, an elastic beam element (Bernoulli-Euler) was used. The diagonal material and the secondary member were rigidly jointed, and the material property value of steel was defined as Young's modulus was $2 \times 10^5 N / mm^2$, weight of unit volume was $77.0 kN / m^3$ and Poisson ratio was 0.3. As the analysis parameters, the support condition was set to 4 types of fixed / roller, fixed / fixed, pin / roller and pin / pin. Although it is possible to calculate the size of construction drawing with the acquired point group data, it is not possible to calculate the thickness of the member. Therefore, when

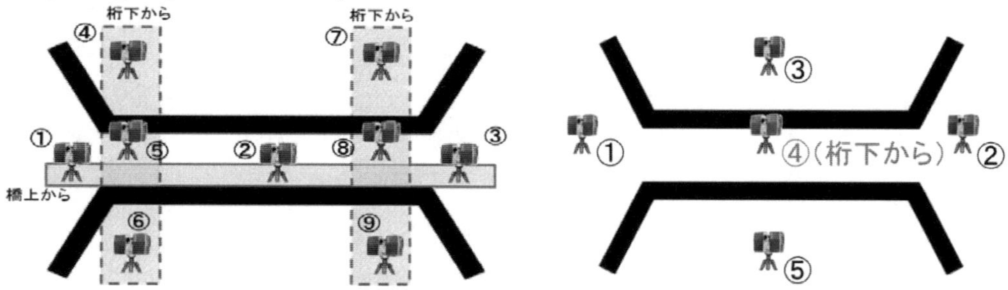

(a) Bridge A (b) Bridge B

Figure 2 Overview of 3D measurement

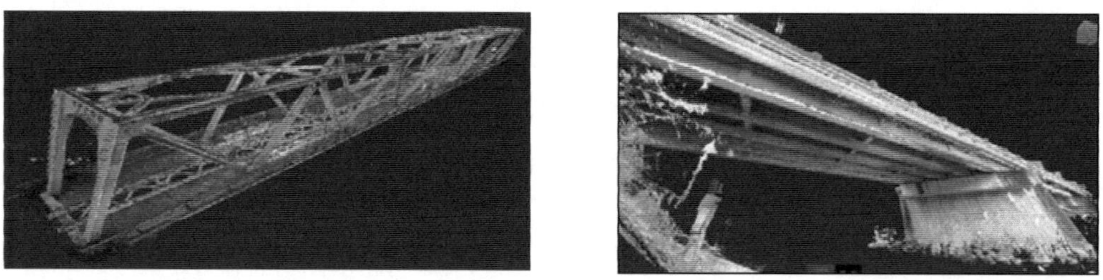

(a) Bridge A (b) Bridge B

Figure 3 3D model

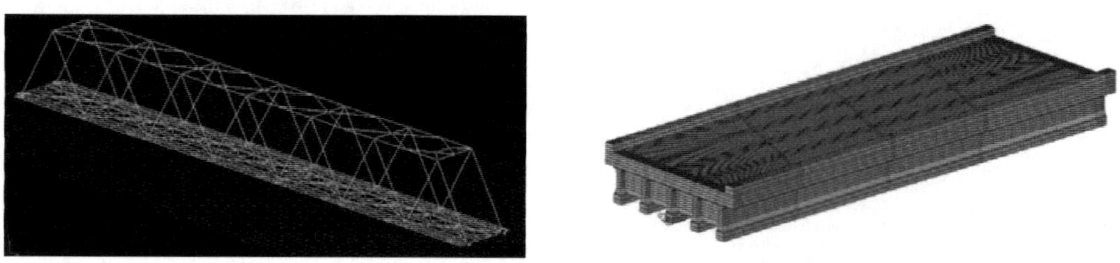

(a) Bridge A (b) Bridge B

Figure 4 3D analysis model

creating the analysis model, only the thickness of the member was determined based on the actual measurement value.

Bridge B is a solid model using isoparametric elements. The material property value of concrete was defined as Young's modulus $2.8 \times 10^4 N / mm^2$, weight of unit volume was $24.5 kN / m^3$ and Poisson ratio was 0.15. As the analysis parameters, two types of support conditions were adopted: simple support pins / rollers and pins / pins supported at both end pins.

4. Outline of vibration measurement

Estimation of the natural frequency was performed using a laser Doppler speed meter. A reflection target was installed and the response speed in the vertical direction was measured. The measurement was carried out under forced oscillation, and the sampling frequency was 1 kHz for bridge A and 500 MHz for bridge B.

Regarding the measurement points, bridge A was located at the lower part of the gusset plate at the joint of the main girder and the transverse girder at the 1/4 span, and bridge B was near the center of the span.

At first, the response speed was measured using a wireless speedometer and LDV for the calculation of natural frequency. Next, FFT analysis was performed on the obtained velocity waveform, and the natural frequency was calculated from the dominant frequency. For bridge A, not only forced oscillation but also calculation of natural frequencies under microtremor, the natural frequency was calculated from the velocity data which was after oscillation had been sufficiently restricted.

5. Results and discussion

5.1 Bridge A

Figure 5 shows the Fourier spectrum and Table 2 shows the natural frequency. The natural frequency was calculated from the outstanding region of this Fourier spectrum. In the Fourier spectrum under microtremor, effective figures were adjusted because the sample data was less than under forced oscillation. Measurement of vibration mode was not done in this measurement. However, the vibration mode was estimated based on the comparison between the frequency obtained by analysis and the frequency obtained by measurement, and the vibration mode obtained by analysis. Originally, since this bridge is erected

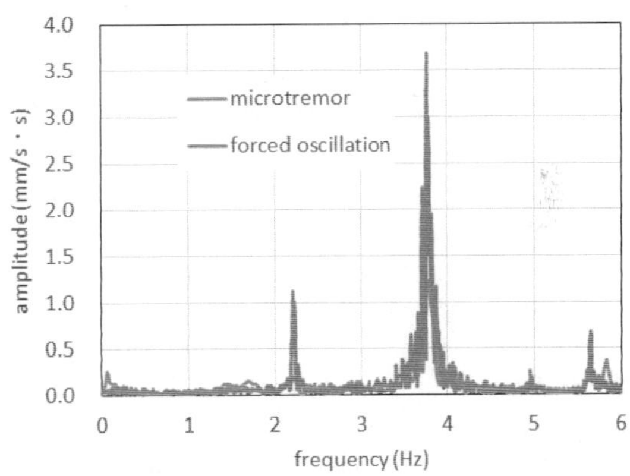

Figure 5 FFT analysis (Bridge A)

Table 2 Comparison of natural frequency (Bridge A)

Support condition	Measurement condition	Horizontal primary mode			Vertical primary mode			Torsional primary mode		
		measured	analysis	mea／ana	measured	analysis	mea／ana	measured	analysis	mea／ana
fixed / roller	forced	2.197	2.247	0.98	3.9	3.900	1.00	5.646	5.654	1.00
	microtremor	2.2	2.247	0.98	3.8	3.900	0.97	5.8	5.654	1.03
fixed / fixed	forced	2.197	2.391	0.92	3.9	4.301	0.91	5.646	5.713	0.99
	microtremor	2.2	2.391	0.92	3.8	4.301	0.88	5.8	5.713	1.02
pin / roller	forced	2.197	2.150	1.02	3.9	3.874	1.01	5.646	5.627	1.00
	microtremor	2.2	2.150	1.02	3.8	3.874	0.98	5.8	5.627	1.03
pin / pin	forced	2.197	2.316	0.95	3.9	4.290	0.91	5.646	5.627	1.00
	microtremor	2.2	2.316	0.95	3.8	4.290	0.89	5.8	5.627	1.03

with pin / roller support, it is presumed that the performance of the bearing has not changed from the time of installation.

From these results, it can be seen that the measured value and the natural frequency of the analysis value are in good agreement with each other in the support condition of the pin / roller.

Next, for each vibration mode, the natural frequency of microtremor and forced oscillation was always nearly the same as the measured value and analytical value. Therefore, it was confirmed that it is possible to calculate the natural frequency even with microtremor. As for the natural frequency of the horizontal primary mode and vertical primary mode, a slight difference of about 2 to 3% was observed between the measurement value and the analysis value. This is thought to be due to omission of accessories and the like which were installed in the bannister and the upper part of the main girder when creating the analysis model.

5.2 Bridge B

Figure 6 shows the comparison of natural frequency. It can be seen that the ratio of the measured value and the analytical value of the vertical primary mode and the torsional primary mode is more matched for both end pin supports than for the simple support condition.

However, from the design books, this bridge was simple support. Here, the actual end of the girder is in the situation as shown in Figure 7, and although it is a rubber bearing, it can be seen that the bearing is close to the state directly placed on the bearing surface. Also, in the past research, it is shown that Coulomb friction acts on the movable bearing part of the actual bridge, so that it often shows no behaviour in the horizontal direction as designed condition [4].

From these facts, it can be judged that the support condition of the main bridge is appropriate both end pin support. Furthermore, by structural identification of the natural frequency, the supporting condition was found to be actually pin support.

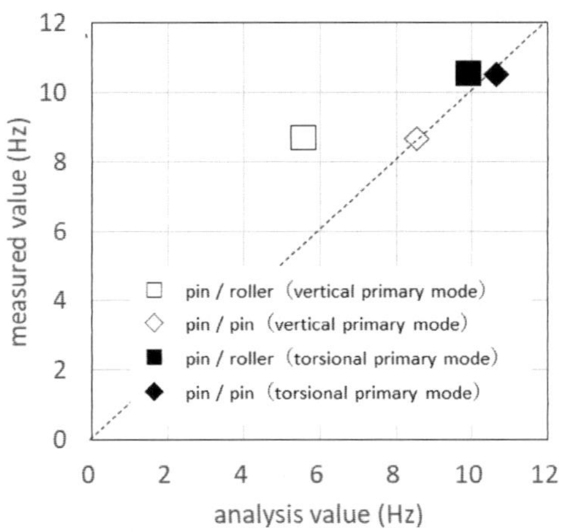

Figure 6 Comparison of natural frequency (Bridge B)

6. Conclusions

The results obtained by this study are shown below.

1) The 3D model obtained from the point group data acquired by the 3D laser scanner can be used as a substitute for the drawing corresponding to the construction drawing.
2) The results of the vibration analysis using the elastic beam element and the solid element well agreed with the actual vibration measurement result.
3) Even with microtremor, results equivalent to forced oscillation can be obtained.

Figure 7 Status of the girder end (Bridge B)

4) It is considered that the support condition can be grasped by structural identification of natural frequency.

From these discussions, it can be confirmed that structural identification is simple and efficient. However, problems such as calculation of member thickness and the like when creating an analysis model from the point group could also be confirmed. In this study, since only support conditions were considered, future studies are required for reduction of the bending stiffness of the girder, decrease in the elastic modulus, decrease in the effective cross section due to cracking and peeling of the deck slab, and the like. By investigating the influence of these variations on natural vibration characteristics, monitoring using simple vibration measurement can be proposed. It is also thought that its effectiveness can be confirmed.

References

Ministry of Land, Infrastructure, Transport and Tourism, Government of Japan: Press Release, http://www.mlit.go.jp/common/001034659.pdf, April. 2014

Cabinet Office, Government of Japan: SIP: Cross-ministerial Strategic Innovation Promotion Program, http://www8.cao.go.jp/cstp/gaiyo/sip/keikaku/7_infura.pdf, April. 2017

Hiroshi Matsuda: Development of efficient and low cost new inspection method using optical measurement method for bridges, http://www.mlit.go.jp/tec/gijutu/kaihatu/josei/162seika.pdf, May. 2016

Masahiro Yoneda: Some Considerations On Damping Characteristics Of Bridge Structures Due To Coulomb Friction Force At Movable Supports, Pp.137～145, 492/Vi-23, *Journal Of Jsce*, June. 1994

The 8th International Conference on
Structural Health Monitoring of Intelligent Infrastructure
Brisbane, Australia | 5-8 December 2017

LiDAR Bridge Inspection Process Analysis and Recommendations for Operation

Haitao Bian[1], P. Sumitro [2], Shenen Chen [3].
[1] *Nanjing Tech University, Nanjing, China, Email:* haitaobian@foxmail.com [2] *Smartsensys, LLC, USA,*
[3]*University of North Carolina at Charlotte, Charlotte, North Carolina, U.S.A.*

Abstract

As a tool for bridge inspection, LiDAR (Light Detection and Ranging) is a critical remote sensing technology with several technical merits including limited disruption to traffic, low labor requirements and providing permanent electronic documentations of the temporal changes of a structure. However, it is also subjected to several environmental effects that may induce data errors during the scanning process. This paper conducts a process analysis of the LiDAR bridge inspection, and points out several issues associated to the application of LiDAR scanning in the inspection process. The challenges in current damage quantification algorithms are listed and possible improvement to the algorithms are suggested to enhance the automation and accuracy of the technique.

1. Introduct.

Both airborne and terrestrial LiDARs have been used extensively in construction management and land development applications. Scanning laser technology for structural monitoring really took off since the 1990s (Fritsch & Kilian, 1994): Because they are non-contact and are able to see from a distance away, scanning lasers have the advantages of limited disruption to traffic, low labor requirements and permanent electronic documentations of the temporal changes of a structure. For infrastructure applications, Al-Turk and Uddin (1999) reported using airborne LiDAR for terrain and roadway mapping with the intent of assessing infrastructure inventory; for structural geometric measurements, Curless and Levoy (1996) reported using laser scanners to construct 3D structural geometry of historical structures; subsequently, several reports described using scanning LiDAR for detection of structural changes (Lichti & Gordon, 2004, Girardeau-Montaut & Roux, 2005, Pieraccini & Parrini, 2007).

For bridge applications, Lefevre (2000) first reported using laser for measuring bridge clearances. By comparing the position change of the scan points at each measurement location, deflection of bridge component can be measured. Fuchs et al. (2004a and 2004b) reported using LiDAR for displacement measurements during several bridge static load tests. However, to monitor multiple lines of a bridge during a load test, their laser needed to be placed at multiple locations manually. There have also been reports of using vehicle-mounted scanning systems for bridge clearance measurements: when traveling at traffic speed (Kim et al., 2008).

Most of these applications remain in validating construction deliverables predominantly in dimension quantification. Describing several applications of scanning laser system for bridge monitoring applications Liu et al. (2011), first reported using scanning LiDAR for surface damage quantification on bridges. As a supplement to visual inspection, scanning LiDAR can calculate area and volume of a damage providing a quantitative measure of a bridge condition. But for damage quantification, there is also a demand for higher standards for scanner resolutions and for processing pointcloud data. Structural damage quantification can include damage area and volume computations, crack length and width measurements and deformation of severely damaged structural elements. Also, because the computed information are linked directly to field inspection and data processing, the data quality control is essential to the final inspection outcomes. Therefore, there is a need to identify possible error sources in LiDAR scans that dictates the quality of laser data as well as the processed damage information.

In this paper, we summarize the outcomes of several studies conducted to assess the accuracy and errors of using scanning lasers for damage quantification. First, a brief description of laser scanner operation is explained and how system errors may be introduced in the process is discussed.

2. Scanning LiDAR Technology and Test Setup

The scanning laser technology described herein can best describe as a mid-range, terrestrial (ground-based) laser scanning system with a single monodyne light beam. The 3D scanning laser or LiDAR (Light detection and ranging) is a static laser that is a close cousin of the airborne LiDAR systems that generates large landscape footprints (Rueger, 1990). Most laser scanners use servo-controlled rotating mirrors to project the laser beams on the target surface. The servo-controller or galvanometers can be either moving-iron or moving-coil types and can be multiple-axial systems (Chen 2012). Laser beams bounce off the mirror and travel to different positions on the target and returns to the same mirror systems. The returned laser beam is then mixed with the reference beam and the distance to the target can be computed based on the Doppler Effect (Drain 1980). There are two approaches to the detection of the position data: 1) time of flight differences between emitted pulse and returned signals and 2) phase differences between the two signals (Jelalian, 1992). Figure 1 shows a schematic of the laser scanner mounted on a tri-pod and the idealized scanning projection, which establishes a hemispheric projection of laser light.

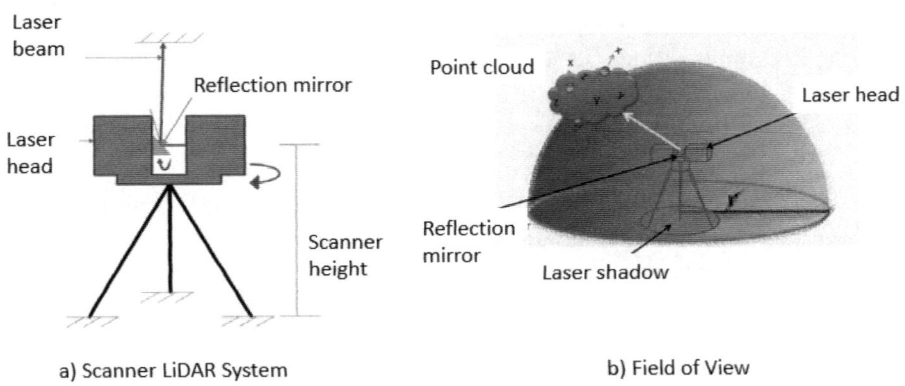

Figure 1 a) Laser scanner schematic and b) the hemispheric projection of laser creating the field of view (FOV)

Limiting our discussion to only terrestrial laser systems that are not mounted on a vehicle, the process of LiDAR scan for bridge inspection involves the inspection team traveling to the bridge site, setting up the laser and collecting the scan results. After getting an overall view of the bridge, the team needs to decide, on the spot, the scan resolution and position to perform the LiDAR scan. Once the decision is make, the team can setup the scanner and conduct a scan. The scanning outcomes are typically 3D pointclouds that can be displayed using graphic processors. Figure 2 shows the 3D reconstruction of a typical LiDAR scan data of a highway bridge and a typical data format consisting of position information in Cartesian coordinates (x, y, z) and reflectivity values.

If properly conducted, LiDAR bridge inspection data can be used for determining the severity of visible damages, which in turn, can provide an accurate assessment of bridge conditions. Such information is lacking in conventional bridge visual inspections.

$$P = \begin{bmatrix} x_1 & y_1 & z_1 \\ x_2 & y_2 & z_2 \\ \vdots & \vdots & \vdots \\ x_n & y_n & z_n \end{bmatrix}, R = \begin{bmatrix} r_1 \\ r_2 \\ \vdots \\ r_n \end{bmatrix}$$

a) Laser Scan Outcomes

b) Data Format: x, y, z, reflectivity (r)

Figure 2 a) Scanner image from pointcloud and b) pointcloud data of position data (x, y, z) and reflectivity value (r)

3. Scanner Error Sources

The error sources from a scan are mostly optical or environmentally related. Environmental effects including fog, smoke, rain, and can be difficult to control. Optical errors can come from insufficient laser energy return to the transceiver – which can result in drop outs in the laser signals. Laser energy returns is functions of the surface conditions of the structure, distance to the structure and power of the laser beam. Typical commercial scanner laser beam has a power of 20 mW, even though military scanners can have much higher powers. The ne-he (helium) lasers work in the infrared range and would work on typical material or painted surfaces. Usually reported in ranging distance, however, for really rough surfaces, poor results may occur due to laser signals bounces off the surface and scattered. Light absorbing surfaces typically would result in poor data quality, however, such studies have not been performed systematically. Figure 3 shows the scanned results of a retaining wall.

Because of the size of the retaining wall, the scans are "stitched" from multiple scan points (Figure 3a) and the effect of energy loss due to distance (Figure 3b) and scattering of laser energy (Figure 3c) are revealed. The light scatter results in background noises and feature recognition techniques will have to be used to remove these noises. These effects ultimately impact damage quantification.

3.1. Structure Surface Reflectivity

Even though LiDAR utilizes a single frequency laser (wavelength = 785 nm), environmental and anthropogenic discoloring (rust or corrosion) or coloring (paint or graffiti) of structure surfaces can have significant effects to reflected laser beams for bridge surface damage evaluation. To understand the effect of surface color on laser reflectivity quality, a first order study using different color paint schemes have been conducted. Figure 4 shows the color scheme (Figures 4a and 4b) and experimental setup (Figure 4c) including placement of the laser scanner. The color scheme included 130 different colors painted on two wood panels and scanned using a terrestrial laser under both dark and normal room lightings.

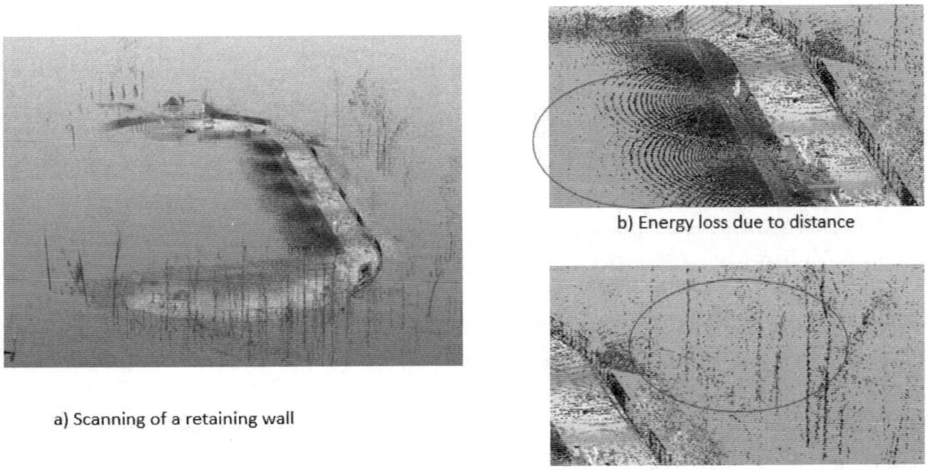

a) Scanning of a retaining wall

b) Energy loss due to distance

c) Energy scatter due to scanning angle

Figure 3 Scanning of a large retaining wall from multiple points: a) Composite laser image of the retaining wall, b) artifacts introduced due to poor energy returns from long distance; c) artifacts due to scattering of laser beams

a) Full Color Panel

b) Partial Color Panel

c) Experimental Setup

Figure 4 Color effects on laser reflectivity (Brigman, 2013)

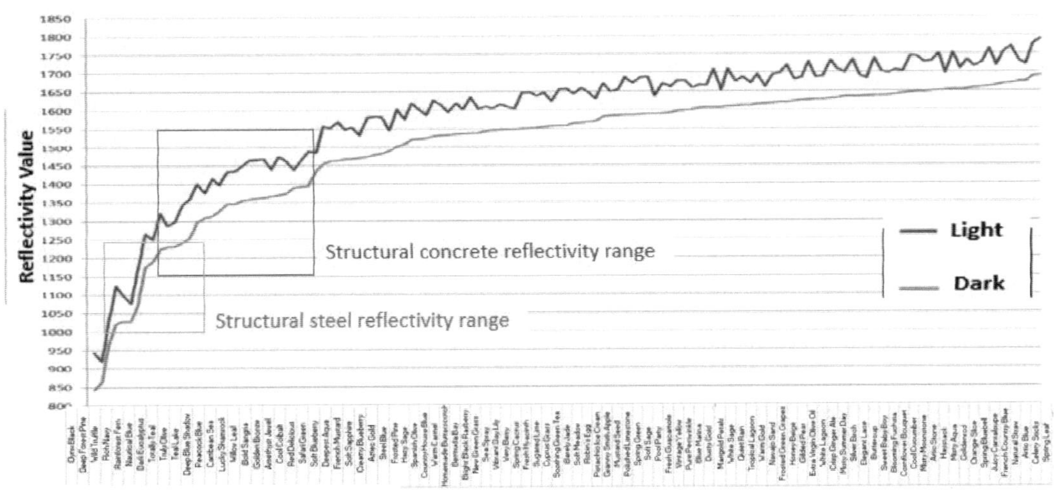

Figure 5 Color effects on laser reflectivity

The reflectivity values from the laser are then compared and evaluated. Figure 5 shows the results of comparison between different color scheme and lighting condition, where dark colors are shown to have much lower reflectivity values than light color schemes. It is also evident that lighting and color brightness (lightness) can influence laser reflectivity and may influence outcomes of dimensional measurements and damage quantification. Also shown in Figure 5 are the range of reflectivity values that are representative of structural steel and structural concrete materials – based on samples of over 100 bridges.

3.2. Scanning Angle and Volumetric Quantification

Because light travels in a straight line, scanning angle and edge of a damage can affect scan quality and significantly influence outcomes of damage quantification. To study the scanning angle effect, a flat panel with six cylindrical indentations is scanned with the LiDAR. The cylindrical indentations are manufactured with different materials representing various surface types. The flat panel was scanned at different locations as shown in Figure 6. The color scheme presented is the coordinate estimate error and is quantified point by point. The non-conforming shape of each location shows deviations to the actual circles. The percentage errors in area and volume quantification are shown in Tables 1 and 2.

Liu (2010) has made a brief analysis of the reason that scanning angle influence happens: The primary factor is that the scans cannot be conducted exactly facing the "damage", as a result, it is difficult to get a clear edge recording. This is because most laser lights touch the damage edge from that direction would scatter rather than return back to the scanner. Therefore, the recorded data will not get sufficient scan points for the edge area. The secondary factor is that the laser light from side scans could be blocked by the damage edge and cannot reach the entire defective surface, thus, the oblique scan cannot capture the entire volume of the damage.

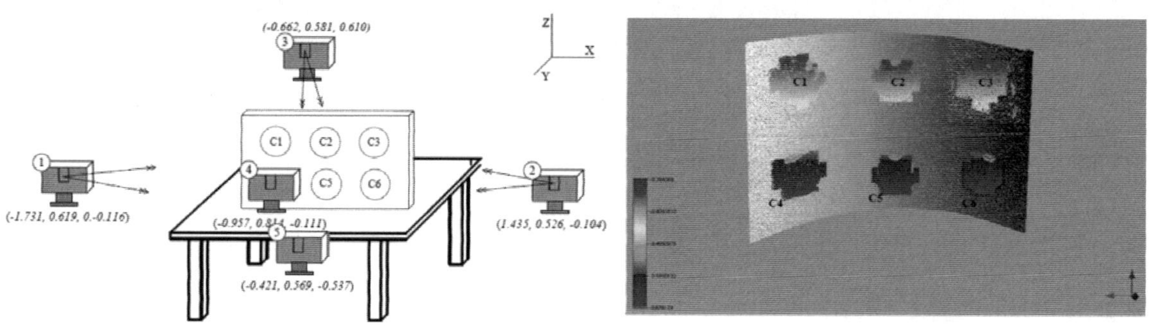

a) Scan angle test setup

b) Scanning results for lower center scanning

Figure 6 Scanning angle test: a) test set up and b) test results for lower center scans

Table 1 Percentage error in area quantification from the scanning angle test

Area	C1	C2	C3	C4	C5	C6	Average	Std.
1. Left	-32.59%	-56.04%	-53.92%	-18.23%	-38.06%	-66.62%	44.75%	0.18
2. Right	-45.58%	-20.58%	1.49%	-29.99%	-37.64%	-24.28%	28.01%	0.15
3. Top	13.21%	3.55%	10.64%	64.37%	-9.26%	-7.41%	18.91%	0.22
4. Center	18.01%	8.77%	12.07%	-1.94%	-2.37%	-8.84%	8.22%	0.06
5. Lower	23.14%	2.75%	19.33%	-3.35%	-5.30%	5.03%	10.74%	0.08

Table 2 Percentage error in volume quantification from the scanning angle test

Volume	C1	C2	C3	C4	C5	C6	Average	Std.
1. Left	-37.62%	-56.04%	-69.23%	-7.67%	-17.93%	-64.66%	42.19%	0.25
2. Right	-55.18%	-20.58%	-6.52%	-17.14%	-17.55%	9.04%	21.00%	0.18
3. Top	11.20%	3.55%	2.31%	61.02%	19.17%	26.73%	20.66%	0.22
4. Center	20.15%	8.77%	9.69%	23.28%	35.06%	23.53%	20.08%	0.10
5. Lower	21.70%	2.75%	7.09%	14.04%	22.83%	32.06%	16.74%	0.11

3.3. Artifacts (line errors) due to passing traffic

At the onset of this paper, we pointed out that one of the benefits of using LiDAR scan for bridge inspection is the minimal disturbance to the ongoing traffic. A LiDAR can be placed at the sidewalk and perform a scan without traffic control. However, another error that can greatly impact LiDAR damage quantification are from the ongoing traffics. Cars traveling at different speeds can pass in front of the LiDAR and may be partially scanned by the LiDAR resulting in random line errors (vertical lines). Passing traffic effect is one of the most common issues when scanning a bridge under heavy traffic, and the line errors created can be difficult to remove during data analysis automation. Figure 7 shows a typical traffic affected scan with several line errors introduced in the image. The color of the line errors is dependent on the vehicle reflection and the scattering of the laser light from the vehicle, which is a function of the vehicle length, size and speed. Thus, the line errors are random in nature.

Figure 7 Passing traffic leaves line errors in the scan

There are also possibilities of the introduction of spurious points due to slow passing vehicles that may complicate damage analysis. However, there has not been any study conducted on this topic.

3.4. Sloped or curved surface analysis

Spurious data points may also be introduced into a LiDAR scan through sloped surfaces – different from scanning angle issue, complexity in object space, such as sloped or curved surfaces, can introduce uncertainty in position point coordinate information. Bian (2012) attempts to address the issue of damage on curved bridge beams and introduced a curvature indexing technique for differentiating point data on a curved surface.

Curvature index is introduced to distinguish if a point in space is part of the bridge surface and to distinguish if the surface contains damage. For each scan point, a curvature index is defined based on the statistical reflectivity value of the point and its surrounding points. The curved surface is determined by comparing the position point to the surrounding points as shown in Figure 8a. The detected surface is then evaluated for damages. Figures 8b and 8c show analysis results indicating damaged areas and non-damaged areas.

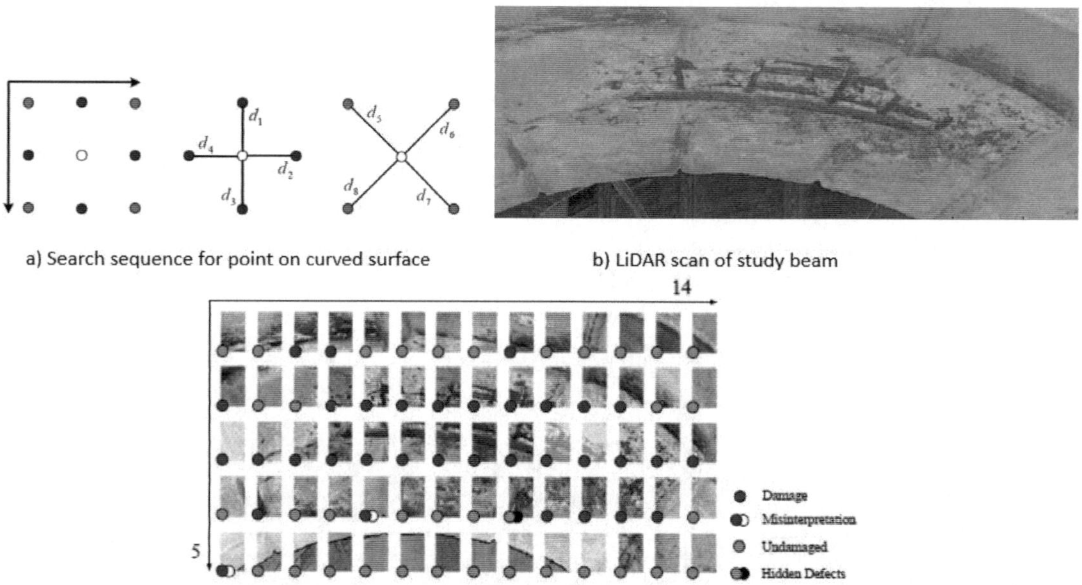

a) Search sequence for point on curved surface

b) LiDAR scan of study beam

c) Detected damaged surface

Figure 8 Case study of damage detection on a curved beam: a) identification of point on a curved surface using curvature index; b) LiDAR scan of a curved beam; and c) analysis results of damage detection on the curved surface.

4. Possible Methods to Eliminate LiDAR Errors

Most of the LiDAR scans are post processed in the lab: Unlike image processing, which has extensive libraries of algorithms, pointcloud data analysis has very limited computational algorithms available for processing the data. Furthermore, the generated pointcloud data can be very large (typical scan results in 200~400 Mbyte data files) and require heavy duty work stations for computation. Hence, there is a demand for effective computation algorithms.

For mass loss-type damage evaluation, pointcloud data analysis differentiates itself from image-based analysis in its ability to utilize directly physical coordinate data. The challenges in automating the damage computation algorithms are in the identification of physical features, in particular, the protruded or concaved surfaces that define damages. Bian et al. (2011) compared two different damage evaluation algorithms for pointcloud data analysis: a) the distance gradient combined method (Liu's Algorithm) – based on two-criterion damage quantifier and requires a precisely selected region for study; and b) the MMSE&T (minimum mean square error and triangulation) Method – based on the adoption of minimum mean square error for reference plane identification and use of Delaunay triangulation for damage quantification. They concluded that both methods rely heavily on the right definition of a reference plane and that MMSE&T method works only on straight reference planes, whereas, the Liu's Algorithm is more flexible for irregularly defined reference planes.

Figure 9 shows a case study on the damage quantification of a bridge pier cap where the deck beams have stressed into the pier cap and caused local concrete mass losses. Using both Liu's Algorithm and the Delaunay's triangulation based MMSE&T Method, less than five percent difference in damaged surface area estimate results. On the other hand, the percentage volume difference estimated by both methods can be as much as 73.4%. This is because the damages are highly irregular and the MMSE&T method underestimated the volumes for both areas 2 and 3. This observation demonstrates the challenges in damage quantification using pointcloud data.

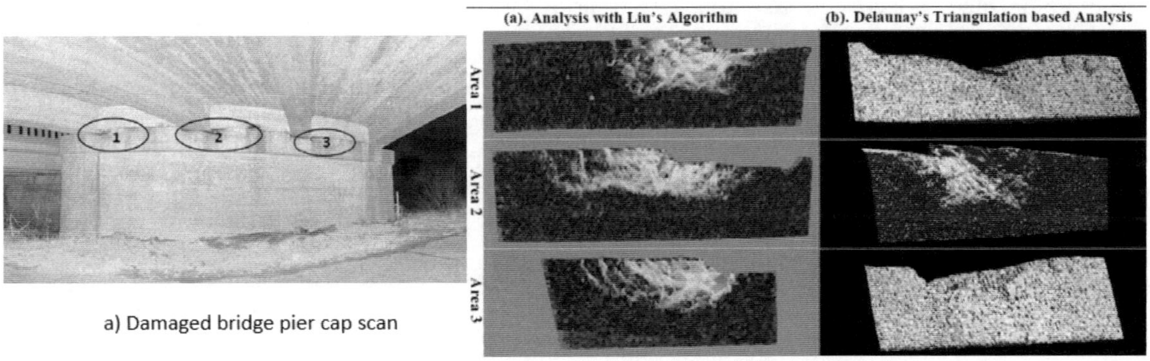

a) Damaged bridge pier cap scan

b) Visualization of Liu's Algorithm and MMSE&T Method

c) Comparison between Liu's Algorithm and MMSE&T Method

Area NO.	Liu's Algorithms Results / Delaunay based Algorithms Results			
	Area (M^2)	**Difference**	**Volume (M^3)**	**Difference**
1	1.38E-01/1.44E-01	4.3%	1.62E-02/1.73E-02	7.5%
2	1.77E-01/1.78E-01	0.6%	1.26E-02/0.62E-02	50.8%
3	1.34E-01/1.38E-01	3.0%	1.24E-02/0.33E-02	73.4%

Figure 9 Comparing Liu's Algorithm and MMSE&T method for damage evaluation on a bridge pier cap: a) LiDAR scan of damaged pier caps with three damage locations; b) damage visualization using both methods; and c) tabulated results from both methods. (Bian 2011)

Figure 9 is based on tests conducted under idealized test conditions, i.e. no rain, fog and ongoing traffic. Both Liu's Algorithm and MMSE&T method are based on the assumption that the test results are devoid of any artefact introduced by environmental or optical elements. Hence, the most logical way to ensure good LiDAR test result is to perform the tests under well controlled test environment. However, this may be difficult to perform under heavy traffic environment. If there is sufficient bridge clearance, then escalating the support height to above the passing vehicles, may be helpful to eliminate the line errors.

Line errors can be eliminated using digital image processing techniques such as Fourier Transformation (to eliminate high frequency components) and image smoothing techniques. Bian (2011) attempted using intensity plots to differentiate the quality of LiDAR scans for damage evaluation. These techniques can be used to help remove errors and artefacts. Post processing of data should be verified with a closer evaluation of the actual scanned surface. Additional measurements onsite may be able to help rectify laboratory processed results.

Finally, many experiences from airborne LiDAR technology can be implemented into terrestrial LiDAR scan practices. However, the LiDAR for bridge evaluation would require higher resolutions and more refined scaling of both laser optics and the damage quantification methodologies beyond typical geometry definition technologies.

5. Conclusions

Terrestrial LiDAR scanners have been gaining tractions to become a remote sensing tool for civil construction monitoring. Recent studies also suggest the technology for bridge damage evaluation, which demands higher precisions and resolutions to geometric measurements. LiDAR (Light Detection and Ranging) is subjected to several environmental effects that may induce data errors during the scanning process. This paper reviewed some of the error issues including effects from structure surface reflectivity, scanning angle, artefacts introduced by passing traffics, sloped or curved surfaces and analysis techniques, etc.

To ensure accuracies in damage detection and quantification, this paper suggests the use of controlled scanning environment and possible applications of advanced data processing algorithms.

ACKNOWLEDGEMENTS

This project is supported by grant DTOS59-07-H-0005 from the United States Department of Transportation (USDOT), Research and Innovative Technology Administration (RITA). The views, opinions, findings and conclusions reflected in this publication are the responsibility of the authors only and do not represent the official policy or position of the USDOT, RITA, or any State or other entity. The authors would like to acknowledge the guidance of Mr. Caesar Singh.

We also acknowledge and appreciate the excellent review and input of our National Advisory Committee: Dr. Sreenivas Alampalli, New York State DOT; Dr. Mrinmay (Moy) Biswas, North Carolina DOT; Dr. Hamad Abu-Harash, Iowa DOT; Mr. Steve Varnedo, American Association of State Highway and Transportation Officials (AASHTO); Dr. K. T. Thirumalai, STI International; Professor Dan Turner, University of Alabama; Dr. Phil Yen, FHWA Turner-Fairbank Research Center and George Connor, Alabama DOT.

We also acknowledge AkzoNobel Company for donating the paint samples for this study.

References

Al-Turk, E & Uddin, W. 1999. Infrastructure Inventory and Condition Assessment Using Airborne Laser Terrain Mapping and Digital Photography, *Transportation Research Records*, TRB, Vol. 1690, pp. 121-125.

Bian, H., Bai, L., Wang, X., Liu, W., Chen, S.E. & Wang, S. 2011. 'Effective LiDAR Damage Detection: Comparing Two Detection Algorithms,' *Structural Engineers*. Vo. 27, No. 137, pp.327-333.

Bian, H., 2012. Advanced Considerations in LiDAR Technology: Application Enhancement, Inspection Workflow Implementation and Data Collection Quality Management, Ph.D. Dissertation, University of North Carolina at Charlotte, Charlotte, NC.

Boehler, W. and Marbs, A. 2002. '3D Scanning Instruments, Proceedings of *CIPA, Heritage Documentation', International Workshop on Scanning for Cultural Heritage Recording*, Corfu, Greece.

Brigman, G. 2013. *Study of LiDAR Reflectivity Effects for Structural Evaluation*, M.S. Thesis, UNC Charlotte, Charlotte, NC.

Chen, S.E. 2012. 'Laser Scanning Technology for Bridge Monitoring,' *Laser Scanner Technology*, InTech Pub., ISBN 979-953-307-265-3.

Curless, B. & Levoy, M. 1996. 'A Volumetric Method for Building Complex Models from Range Images,' *Proceedings of SIGGRAPH,* pp.303-312.

Drain, L.E., 1980. *The Laser Doppler Technique*, Wiley-Interscience, Sussex, England.

Fritsch, D. & Kilian, J.C. 1994. Filtering and Calibration of Laser-Scanner Measurements, *Proceedings of ISPRS Commission III Symposium,* SPIE, Vol. 2357, pp. 227-234.

Fuchs, P.A., Washer, G.A., Chase, S.B. & Moore, M. 2004a. 'Application of laser-based instrumentation for highway bridges.' *ASCE Journal of Bridge Engineering, Vol.* 9, No. 6., pp.541-549.

Fuchs, P.A., Washer, G.A., Chase, S.B. & Moore, M. 2004b. 'Laser-based instrumentation for bridge load testing,' *ASCE Journal of Performance of Constructed Facilities*. Vol. 18, No. 4, pp.213-219.

Girardeau-Montaut, D. & Roux, M. 2005. 'Change detection on points cloud data acquired with a ground laser scanner.' *Proceedings of Workshop Laser scanning 2005,* Enschede, the Netherlands.

Jelalian, A.V. 1992. *Laser Radar Systems*. Artech House, Boston, London.

Kim, Y.R., Hummer, J.E., Gabr, M., Johnston, D., Underwood, B.S., Findley, D.J. & Cunningham, C.M. 2008. *Asset Management Inventory and Data Collection*, Final Report, FHWA/NC/2008-15.

Liu, W.Q., S.E. Chen and Hauser, E. 2011. 'LiDAR-Based Bridge Structure Defect Detection,' *Experimental Techniques, 35(6), 27-34.*

Lefevre, R. J. 2000. 'Radar bridge clearance sensor.' *Proceedings of IEEE 2000 International Radar Conference*, Alexandria, VA, pp. 660-665.

Lichti, D.D. & Gordon, S.J. 2004. 'Error propagation in directly georeferenced terrestrial laser scanner point clouds for cultural heritage recording,' *Proceedings of WSA2 Modeling and Visualization*, Athens, Greece.

Liu, M., Frangopol, D.M. & Kim, S. 2009. Bridge Safety Evaluation Based on Monitored Live Load Effects, *ASCE Journal of Bridge Engineering*, Vol. 14, No. 4, pp. 257-269.

Pieraccini, M. & Parrini, F. et al., 2007. 'Static and dynamic testing of bridges through microwave interferometry.' *NDT&E International*, Vol. 40, pp. 208-214.

Rueger, J.M., 1990. *Electronic Distance Measurements: An Introduction*, Springer-Verlag, New York.

Shinozuka, M. & Rejajaie, S.A. 2000. Analysis of Remotely Sensed Pre- and Post-Disaster Images for Damage Detection, *Smart Structures and Materials*, *Proceedings of SPIE,* Vol. 3988, pp. 307-318.

Smith, D.E., Zuber, M.T., Neumann, G.A. & Lernoine, F.G. 1997. Topography of the Moon from the Clementine LiDAR, *Journal of Geophysical Research,*Vol.102, No. E1, pp. 1591-1611.

The 8th International Conference on
Structural Health Monitoring of Intelligent Infrastructure
Brisbane, Australia | 5-8 December 2017

Development of Self-Temperature Compensated Strain Visualization Sheet

S. Umemoto [1], T. Takaki [2], M. Omachi [1], K. Matsuo [1],
N. Miyamoto [1], I. Ishii [2], T. Aoyama [2].
[1] *Keisoku Research Consultant Co., Japan, Email:* umemoto@krcnet.co.jp
[2] *Hiroshima University, Japan*

Abstract

Health monitoring of rapidly-aging structures has become an urgent issue. A strain visualization device and new strain measurement technology were developed to address this problem. The device, called the "Strain Visualization Sheet," enables quantitative visualization of strain based on the principle of Moiré fringes. It is also possible to obtain extremely accurate strain values from photographed digital images. A power supply is not required, as no electrical elements are used, and durability is also excellent. Thus, this is a suitable sensor for long-term monitoring. The effect of temperature was one problem for field application of the Strain Visualization Sheet. When a strain measurement sensor is attached to the material being measured, temperature changes, which are unrelated to strain due to external force, cause apparent strain as a result of the difference between the linear thermal expansion coefficients of the object material and the strain measurement sensor. Therefore, a self-temperature compensated Strain Visualization Sheet which minimizes this apparent strain was developed and verified for steel and concrete materials. As a result, apparent strain of less than ±1με was obtained in the temperature range of –20°C to +70°C, demonstrating successful self-temperature compensation of the Strain Visualization Sheet. As this temperature compensation range is wider than that of conventional strain gauges, application to structural health monitoring under various environmental temperatures is expected.

1. Introduction

The authors developed a strain visualization device called the "Strain Visualization Sheet" for health monitoring of aging structures. The most important feature of the Strain Visualization Sheet is that quantitative visualization of strain is possible by utilizing the principle of Moiré fringes, without using any electrical elements such as a conventional strain gauge or strain meter. It is also possible to calculate strain values with high accuracy from photographed digital images of the Strain Visualization Sheet. Because no electrical elements are used, a power supply is not required and device life is not limited by electrical failure. Laboratory tests have confirmed measurement accuracy of 50 μm/m by visually reading the strain values and 20 μm/m by digital image analysis.

For application of the Strain Visualization Sheet to steel and concrete structures in the field, the following problems must be solved: (1) Self-temperature compensation suitable for the object of measurement, (2) Improvement of durability and (3) Improvement of ease of installation. This paper discusses the development of a self-temperature compensated type of Strain Visualization Sheet which does not cause apparent strain due to temperature. The principle of self-temperature compensation is discussed, and the developed Self-Temperature Compensated Strain Visualization Sheet is introduced. The results of verification tests are also reported.

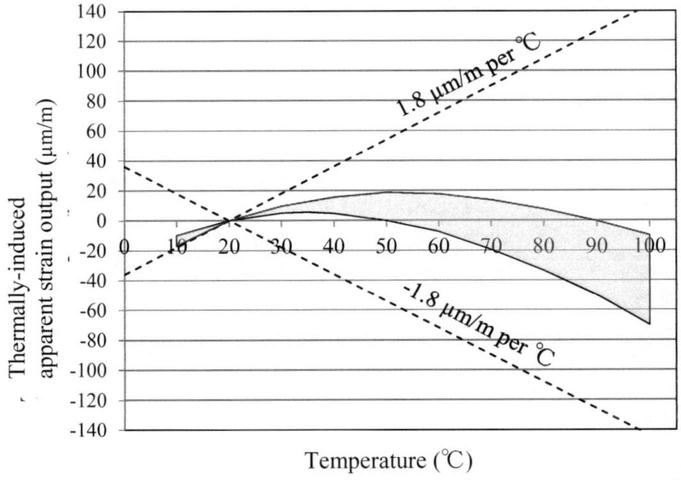

Fig.1 Typical characteristic curve of thermally-induced apparent strain with strain gauges

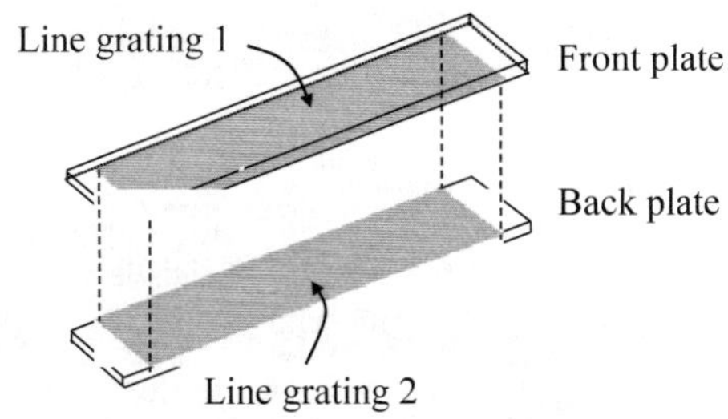

Fig.2 Basic Structure of Strain Visualization Sheet

2. Outline of Self-Temperature Compensated Strain Visualization Sheet

When a strain measurement sensor is installed on a material being measured, temperature changes cause apparent strain in addition to the strain caused by external forces, as a result of the difference in the linear thermal expansion coefficients of the object material and the strain measurement sensor. A self-temperature compensated strain measurement sensor is a type of sensor which minimizes the apparent strain caused by temperature changes. Considering the example of a strain gauge, as shown in Fig. 1, a device which achieves accuracy within ±1.8 μm/m/°C in the prescribed temperature range when installed on a suitable material is called a self-temperature compensated gauge. There are also gauges that achieve ±1.0 μm/m/°C in the 20-40°C range, where they are frequently used. With strain measurement sensors that are self-temperature compensated, the device itself can cancel the apparent strain caused by temperature and output only the strain caused by external force.

3. Principle of Self-Temperature Compensation

3.1 Basic Structure and Measurement Principle of Strain Visualization Sheet

The basic structure of the Strain Visualization Sheet is shown in Fig. 2. Line grating 1 and line grating 2 are drawn on the front plate and back plate, respectively, the two plates are overlaid, and the relative displacement of the two plates in the longitudinal direction is detected by the principle of Moiré fringes. As shown in Fig. 3(1), a striped pattern with a pitch W larger than line gratings 1 and 2 appears due to the superimposition of line grating 1 with pitch p and line grating 2 with pitch $p+\Delta p$, which is Δp ($\ll p$) larger than the pitch of line grating 1. This pattern is called a Moiré fringe. The relationship between the Moiré fringe with the pitch W and line grating 1 with the pitch p is shown in Eq. (1). The display of pitch p can be enlarged visually by $(p+\Delta p) / \Delta p$ times by the Moiré fringe.

$$W = \frac{p + \Delta p}{\Delta p} \cdot p \tag{1}$$

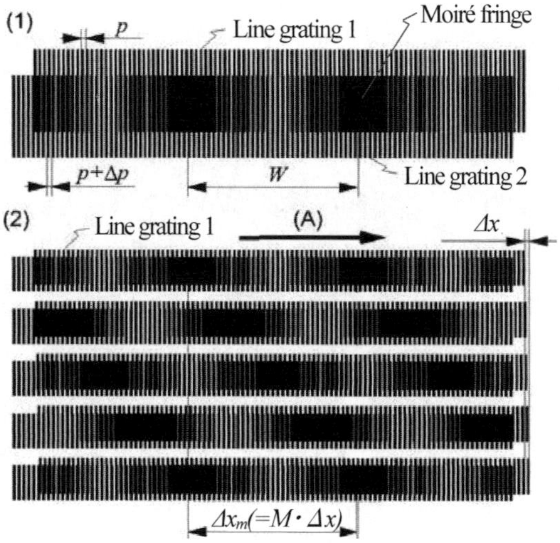

Fig.3 Example of a moiré fringe

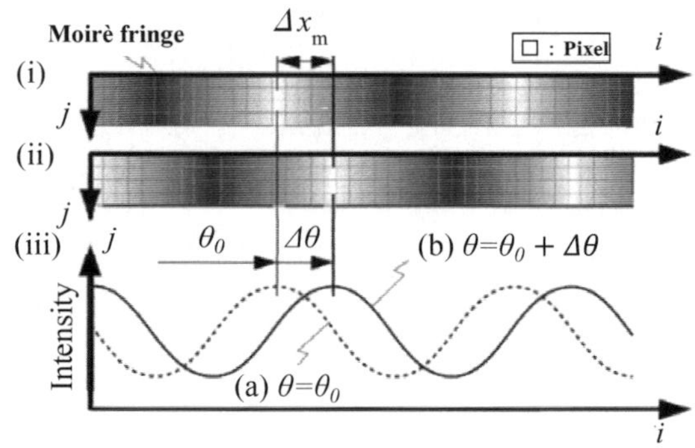

Fig.4 Phase of a moiré fringe

As shown in Fig. 3(2), assuming this magnification factor $(p+\Delta p) / \Delta p$ is M, the amount of shift Δx_m of the Moiré fringe in direction (A) when line grating 1 is moved Δx in direction (A) can be expressed by Eq. (2). In other words, the Strain Visualization Sheet makes it possible to enlarged display of the very small amount of relative displacement of the two plate Δx as the amount of shift Δx_m of the Moiré fringe.

$$\Delta x_m = M \cdot \Delta x \tag{2}$$

Here, we consider the case of that the strain value is calculated by acquiring digital images and performing image processing (Takaki, T. et al. 2012). As shown in Fig. 4, the phase difference $\Delta\theta$ is calculated by a sine curve approximation of the luminance distribution of the Moiré fringe. Further, the amount of shift Δx_m of the Moiré fringe is calculated from the phase difference $\Delta\theta$. and the relationship between Δx_m and $\Delta\theta$ is shown by Eq. (3).

$$\Delta x_m = \frac{W}{2\pi} \Delta\theta \tag{3}$$

Therefore, from Eqs. (1), (2) and (3), the relative displacement of the two plates Δx can be expressed by Eq. (4).

$$\Delta x = \frac{p}{2\pi} \Delta\theta \tag{4}$$

For example, when the amount of relative displacement of the two plate Δx occurs with respect to the reference length L, strain ε can be obtained by Eq. (5).

$$\varepsilon = \frac{p}{2\pi L} \Delta\theta \tag{5}$$

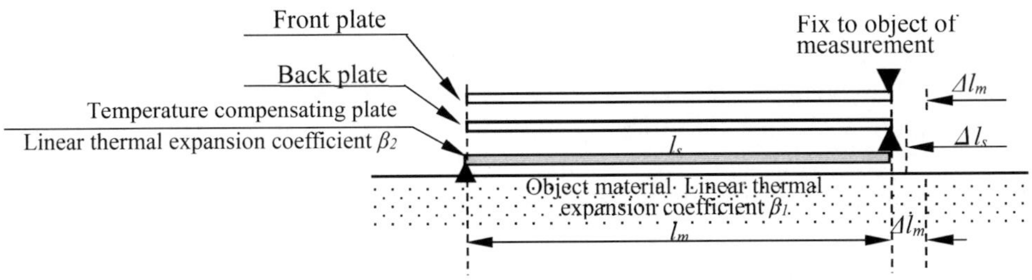

Fig. 5 Proposed self-temperature compensated structure

In addition, in visualization of strain, the Moiré fringe is shown as a numerical value by providing a grating with numerical characters in place of line grating 2, and a numerical value corresponding to the amount of actually occurring strain is displayed by adjusting the magnification factor (Takaki, T. et al. 2012).

3.2 Self-Temperature Compensating Structure

As shown in Fig. 5, the authors conceived a structure in which a temperature compensating plate is incorporated in the basic structure of the Strain Visualization Sheet. When a stress σ acts on an object material which is subjected to a temperature change Δt, the amount of change Δl_m of the length l_m of the object material is equal to the sum of the change of length due to stress σ and the change of length due temperature change Δt. If the elastic modulus of the object material is E and the linear thermal expansion coefficient is β_1, this is expressed by Eq. (6).

$$\Delta l_m = l_m \frac{\sigma}{E} + \beta_1 \cdot \Delta t \cdot l_m \tag{6}$$

If the linear thermal expansion coefficient of the temperature compensation plate is β_2, the amount of change Δl_s in the length l_s of the temperature compensation plate at this time is expressed by Eq. (7).

$$\Delta l_s = \beta_2 \cdot \Delta t \cdot l_s \tag{7}$$

Because the displacement of the front plate is the same as the change Δl_m of the length of the object material, and the displacement of the back plate is the same as the change Δl_s of the length of the temperature compensation plate, the relative displacement Δx of the front plate and back plate can be expressed by the following equation.

$$\Delta x = \Delta l_m - \Delta l_s \tag{8}$$

From Eqs. (6), (7) and (8),

$$\Delta x = l_m \cdot \frac{\sigma}{E} + \beta_1 \cdot \Delta t \cdot l_m - \beta_2 \cdot \Delta t \cdot l_s \tag{9}$$

Assuming $l_m = l_s$ and $\beta_1 = \beta_2$,

$$\Delta x = l_m \cdot \frac{\sigma}{E} \tag{10}$$

Accordingly, equalizing the lengths and linear thermal expansion coefficients of the object material and the temperature compensation plate makes it possible to obtain only the relative displacement Δx of the front plate and back plate caused by acting stress σ, and relative displacement due to temperature change Δt does not occur.

4. Structure of Self-Temperature Compensated Strain Visualization Sheet

The appearance and structure of the developed Self-Temperature Compensated Strain Visualization Sheet are shown in Fig. 6 and Fig. 7, respectively. The device consists of the two glass plates which form the Moiré fringe pattern and a temperature compensating steel plate, and is assembled in a unit form by using a pair of fixing rings. Its external dimensions are 17 mm x 120 mm x 8 mm (thickness), and its gauge length is 105 mm. The Strain Visualization Sheet consists of three types of Moiré fringes. The numbers shown in the upper line are used for strain visualization, the middle line is used for calculation of strain by image processing, and the lower line is a reference moiré for improving the accuracy of image processing. Because the numbers in the upper line are arranged in the form of a scale with an interval of 100 μm/m, strain values can be read directly from this line.

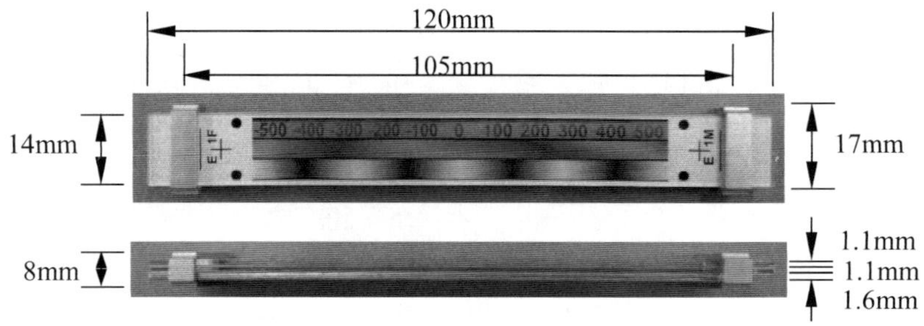

Fig. 6 Appearance of Strain Visualization Sheet

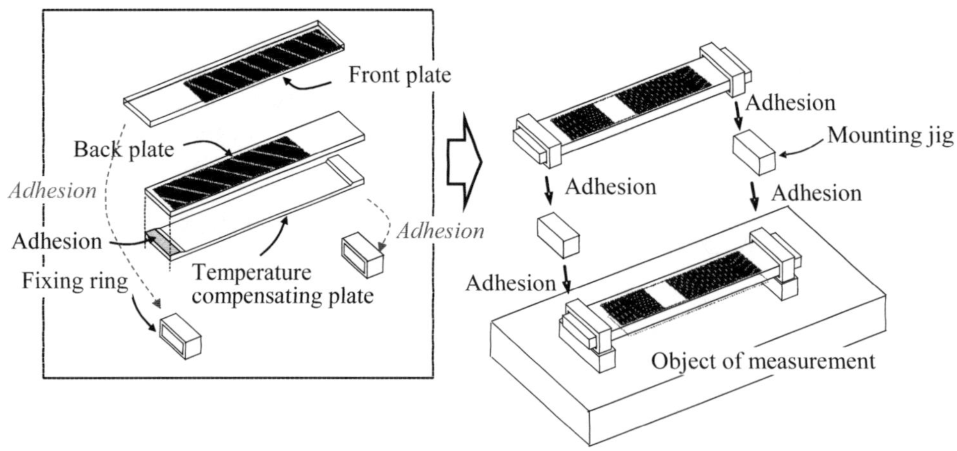

Fig.7 Structure of Strain Visualization Sheet

5. Verification of Self-Temperature Compensated Strain Visualization Sheet

5.1 Test Piece

A steel plate (SPCC, linear thermal expansion coefficient: 11.7 μm/m/°C) with dimensions of 50 mm x 150 mm (t = 1.6 mm) was used as the test piece. The arrangement of sensors on the test piece is shown in Fig. 8, and a photograph of the test piece with the installed sensors is shown in Fig. 9. A conventional strain gauge (gauge length L = 90 mm) was installed on the surface of the steel plate for comparison with the Strain Visualization Sheet. The same type of strain gauge was also installed on the back side of the plate to check for bending of the plate due to temperature. A thermocouple thermometer was installed on the back side of the steel plate to measure the temperature of the plate.

5.2 Verification Method

The test piece was placed in a thermostatic chamber so as to allow free expansion and contraction accompanying temperature changes, and a USB camera was also placed in the chamber for photography (Fig. 10). The USB camera was connected to an external real-time image processing device, and the signal cables of the strain gauges and thermocouple were connected to an external data logger. After measuring the initial strain values, the temperature was changed in the range from –20°C to 70°C in steps of 10°C, and the values of the Strain Visualization Sheet and the strain gauges were recorded when the temperature of the test piece reached the set temperature.

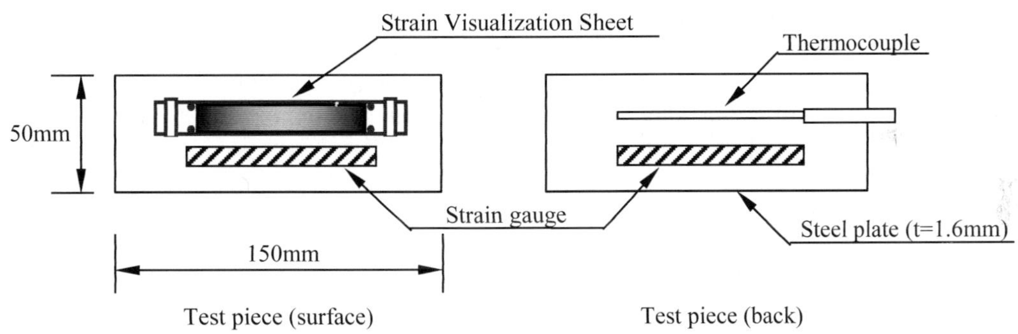

Fig.8 Sensor arrangement

Fig.9 Sensor installation condition

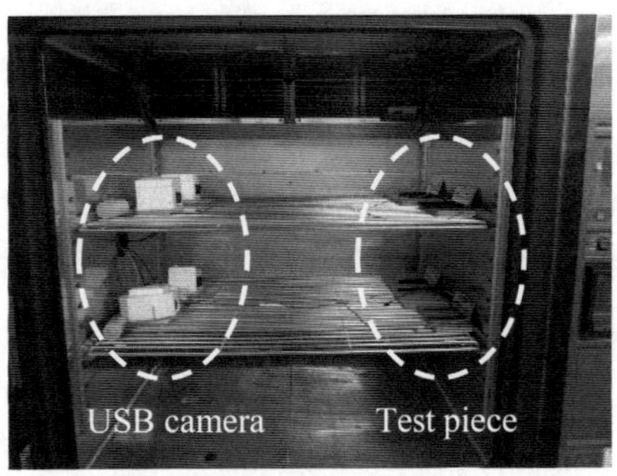

Fig.10 Installation of test piece and USB camera in the thermostatic chamber

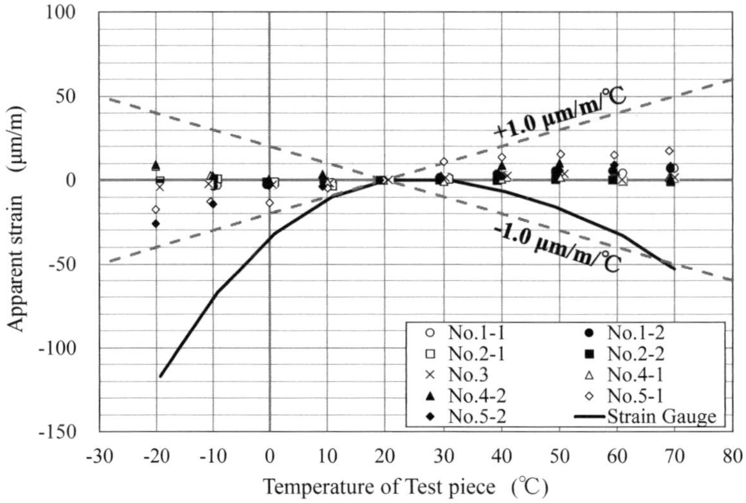

Fig. 11 Relationship between temperature and apparent strain of test piece

5.3 Results of Verification

The relationship between the temperature of a test piece and apparent strain occurred by temperature change is shown in Fig. 11. The apparent strain of the Strain Visualization Sheet was < ±1.0 μm/m/°C, confirming the self-temperature compensation. It is also noteworthy that the Strain Visualization Sheet has a wider self-temperature compensation range than the conventional strain gauge, showing extremely small apparent strain values of approximately 20 μm/m at the high temperature of 70°C and approximately –20 μm/m at –20°C below freezing. The results of this verification test verified the fact that the developed Self-Temperature Compensated Strain Visualization Sheet is a strain measurement sensor with temperature characteristics superior to those of conventional strain gauges.

6. Conclusion

With conventional strain gauges, the self-temperature compensation function is realized by adjusting the temperature coefficient of resistance (TCR) of the strain gauge resistance element material so as to conform to the material being measured. In contrast to this, with the Strain Visualization Sheet discussed in this paper, self-temperature compensation conforming to mild steel and concrete was realized by the constituent materials and structure of the device itself. The development of this new Self-Temperature Compensated Strain Visualization Sheet made it possible to improve applicability in the field. As future work, verification of durability in the field, simplification of the installation method and efforts to realize a low-cost device are planned.

References

Takaki, T. et al., 2012, 'Strain Visualization Sticker using Moiré fringe for remote sensing', *Proceeding of the 6th International Conference on Bridge Maintenance, Safety and Management (IABMAS2012)*, Stresa, Lake Maggiore, Italy, 8-12 July, p. 442.

The 8th International Conference on
Structural Health Monitoring of Intelligent Infrastructure
Brisbane, Australia | 5-8 December 2017

The existing stress measurement of the PC bridges by slit stress relief techniques using the optical full-field measurement method

Masaki YONEMOTO [1], Kenichi HIDA [2], Yukihiro ITO [3], Katuya MITA [3],
Hiroshi MATSUDA [4], Takuji OKAMOTO [1]

[1] *Keisoku Research Consultant CO.,Japan, Email:* yonemoto@krcnet.co.jp [2] *K&T Consultant Co., Japan,*
[3] *Saga University, Japan,* [4] *Nagasaki University, Japan*

Abstract

Tendon of pre-stressed concrete bridge corrodes for damage from chloride attack, and pre-stress decreases. And a bridge falls by corrosion of PC-tendon. It's necessary to grasp the stress which acts on concrete of PC bridges.Therefore authors have developed the existing stress measurement method by slit stress relief techniques using the optical full-field measurement method. This report introduces a measurement method of pre-stress by the existing stress measurement method by slit stress relief techniques using the optical full-field measurement method. The following measured case of pre-stress of PC bridges will be reported. 1)The investigation report of the central hinge bridge which the excessive deformation which passed for about 30 years produced. 2)The strength investigation report of the post tension PCT beam bridge which was damaged by chloride attack. 3)Report of the cause investigation of the abnormal deformation of 3 span continued PC box girder that passed after construction for 38 years

1. Introduction

Concrete structures are believed to be durable permanent structures. However, in recent years, remarkable changes in condition have been observed in some concrete structures. These changes include cracks and corrosion of rebar due to deterioration caused by chloride attack, carbonation, alkali-aggregate reaction, etc.

In post-tensioned prestressed concrete (hereinafter, PC) girders, cases of severe cracking or bridge collapse as a result of corrosion and rupture of PC tendons occurred due to chloride attack or hydrogen embrittlement fracture at points where grout filling was incomplete have been reported. There have also been cases in which prestress was reduced by unexpected creep/shrinkage, and deflection exceeding the design value occurred. Since a decrease in the prestress of a PC structure has a large effect on its load bearing capacity and leads to reduced safety, an accurate grasp of existing stress is necessary for proper maintenance of PC structures. However, the amount of existing prestress is not controlled; only the tensioning force of the PC tendons in the initial stage is controlled. The current methods for measuring the existing stress of PC structures include an estimation method in which strain gauges are placed on 3 axes and core stress is relieved, and a slot stress method developed in France, among others. Because techniques for PC structures have same problems and have not been applied practically very much, a technique that enables accurate estimation is demanded. In order to improve accuracy in comparison with the conventional existing stress measurement methods, in this research, the authors developed an existing stress measurement method for PC structures by the stress relief method which is an existing stress measurement method, and the digital image correlation method. This method has features of highly accurate, noncontact and full-field measurement.

2. Existing stress measurement by slit stress relief technique using optical full-field measurement method

2.1 Overview

If a slit is cut perpendicular to the direction of stress in a concrete member on which uniform compressive stress acts, liberating strain will be generated in the area surrounding the slit. The analytical model of this technique is shown in Fig. 1. Figure 2 shows the relationship between the liberating strain and the distance from the slit center when the slit cutting depth in the model is changed from 10 mm to 20 and 30 mm.

As a feature of the existing stress measurement method by the slit relief technique developed in this research (hereafter, slit stress relief technique), the optical full-field measurement method is used in order to improve accuracy in comparison with the conventional method when measuring the liberating strain caused by slit-cutting.

In the slit stress relief technique, the area around the slit is measured with a full-field strain measurement device before and after cutting the slit, and the distribution of liberating strain is obtained from the measured images by the digital image correlation method, which is one optical full-field measurement method.

By using the optical full-field measurement method, strain measurements can be performed with high accuracy extending to micro areas in the vicinity of the slit. The existing stress of the concrete member is calculated based on the measured liberating stress distribution by inverse analysis by FEM analysis.

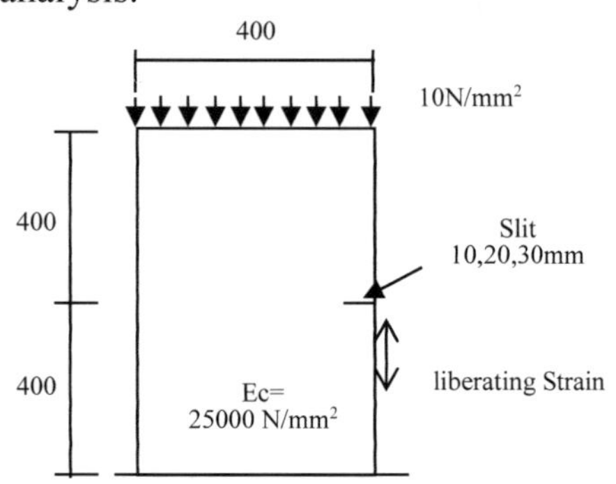

Figure 1. Analytical model

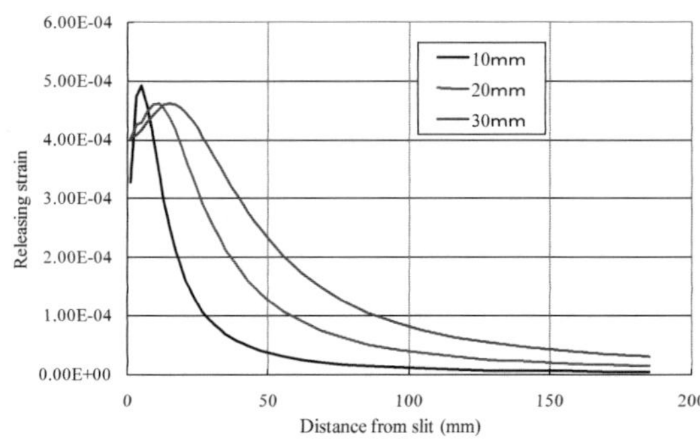

Figure 2. Distribution of liberating strain by slit-cutting

2.2 Optical full-field measurement method

The specifications of the optical full-field strain measurement device are shown in Table 1, and the appearance of the device is shown in Fig. 3. A mechanism that enables highly accurate travel was realized by fixing the drive device for sub-scanning on a high-rigidity frame to improve travel accuracy in the sub-scanning direction, and adopting two moveable linear guide rails for the line sensor and using a stepping motor in a precision ball screw in this part. A linear encoder with resolution of 0.1 μm is used to control the movement distance in the sub-scanning direction. A jig is attached so that the device can be attached easily and firmly to the object of measurement.

Table 1. Specifications

Sensor type	CIS
Greatest manuscript size	210×210mm
Optics resolution(main scanning × vice-scanning)	1200dpi×1200dpi (About0.021mm/pixel)
Source of light	LED
Power supply	DC15V
Electricity consumption	15W
Size (width × depth × height)	387mm×262mm ×90mm
Weight	About 8000g

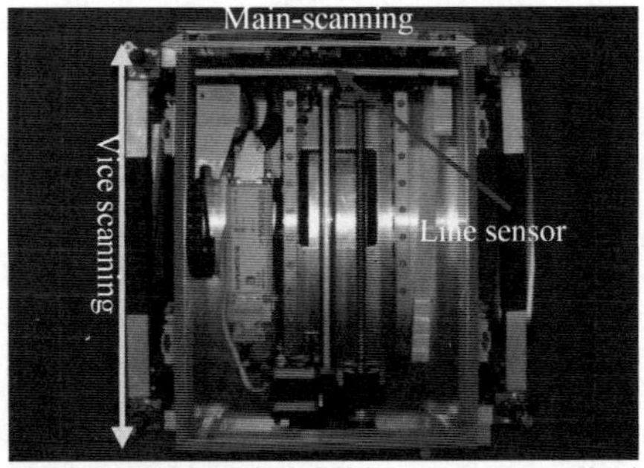

Figure 3. Optical full-field measurement device

2.3 Existing stress measurement by slit stress relief technique

The released strain distribution and change ratio distribution of the distance between the object point is obtained by applying the digital image correlation method to the photographed images. The digital image correlation method, as shown in Fig. 4, is an image analysis technique in which the amount of movement from the position before slit-cutting is obtained based on digital information of the pixel clusters in the initial image, by locating pixel clusters for the same digital information from the image after slit-cutting.

The change ratio distribution of the distance between the object point is obtained from the measured images by the digital image correlation method, and existing stress is then estimated by obtaining a change ratio distribution between object point that approximates the measured change ratio distribution by inverse analysis by FEM analysis.

The change ratio distribution between object point, as shown in Fig. 5, is the ratio of the change in distance between two points, assuming the slit is the center of that distance. Existing stress is estimated from the analysis values obtained by inverse analysis by FEM analysis which coincide with the measured values by using the analytical model shown in Fig. 7 and the mesh diagrams shown in Fig. 8. Figure 6 shows an example of the change ratio distribution of the object point of the analytical values and the measured values at the center of a span. Figure 9 shows the work procedure.

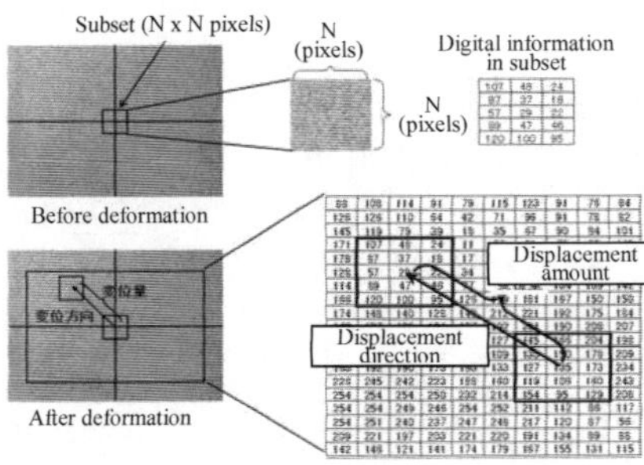

Figure 4. Digital image correlation method

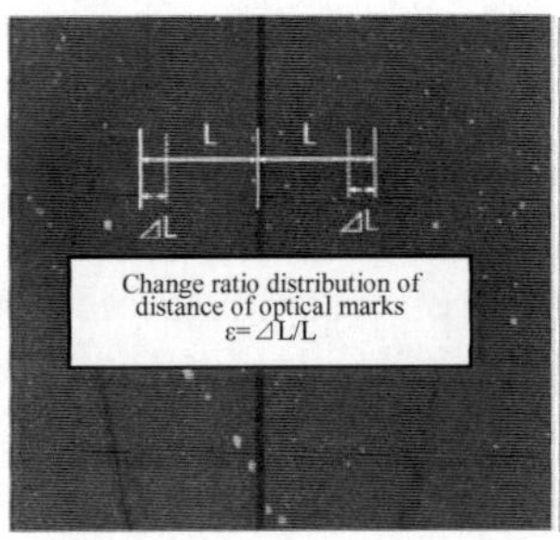

Figure 5. The object point distance rate of change

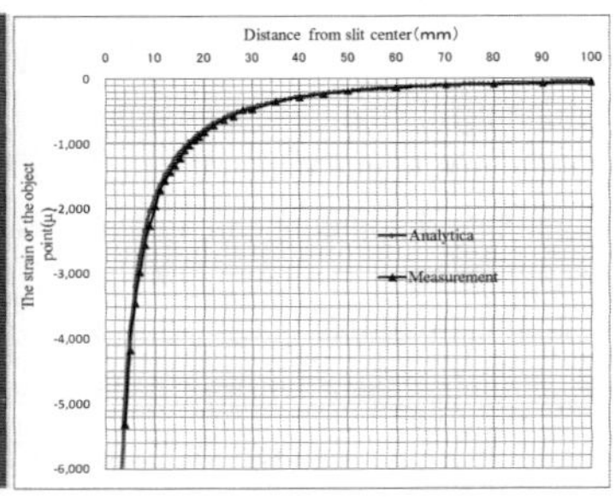

Figure 6. Example of the strain distribution of object point

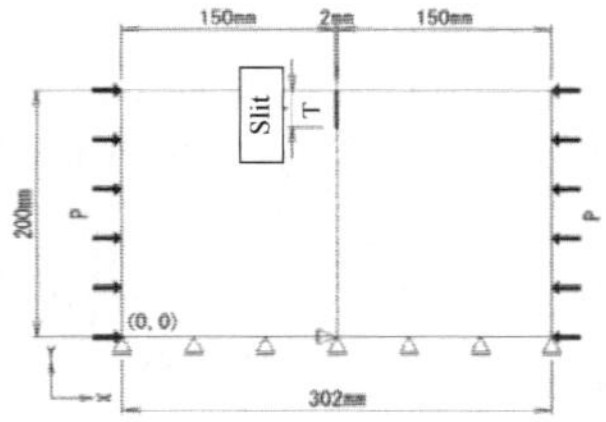

Figure 7. Analytical model

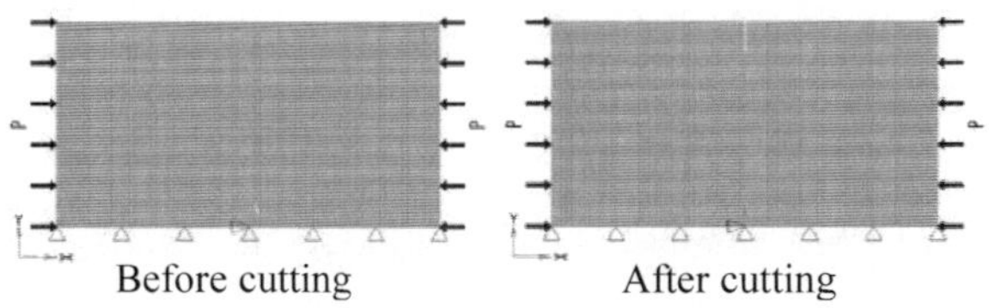

Figure 8. Mesh diagrams

1. Rebar probe Using a reinforcing bar probe, the PC tendon position is confirmed and the measurement position is decided.		4. Initial image The initial image is photographed with a flat scanner.	
2. Surface treatment The surface is smoothed to enable photography of focused images.		5. Slit-cutting slit-cutting is performed with a dedicated cutter.	
3. Surface condition before stress relief A random black and white pattern is attached.	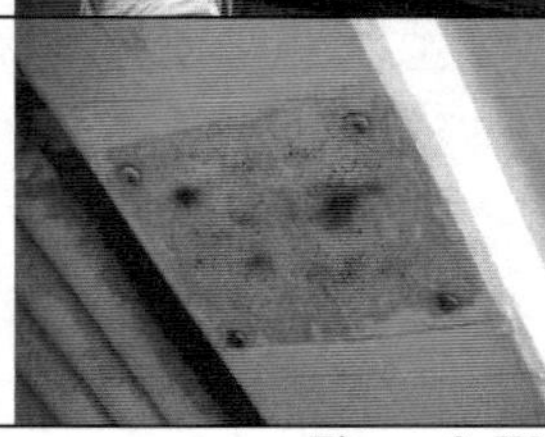	6. Photography after stress relief The image after slit-cutting is photographed with the flat scanner.	

Figure 9. Work procedure

3. Cause investigation of abnormal deformation of central hinge bridge

3.1 Overview

This bridge is a 3-span center hinge PC box girder rigid frame bridge which was constructed in a mountainous area in 1989. Abnormal deflection occurred in the central hinge part. The maximum deflection, as shown in Fig. 10, is approximately 150 mm. Focusing on 1) Lack of Pre-stress in the pier capital part and 2) Abnormality of creep deformation, a study of stress (strain) in the central hinge bridge was carried out to determine the cause of this abnormal deformation. In this stress investigation, the existing strain of the rebar was measured by the slit stress relief technique and the rebar cutting method. The investigation positions are shown in Fig. 11. Cross sections of pier capitals P1 and P2 were taken.

3.2 Measurement results

The measurement results are shown in Table 2. From the results of sample core tests, the elastic modulus of the concrete was assumed to be E_C=22 kN/mm^2. Creep/Shrinkage and elastic strain,

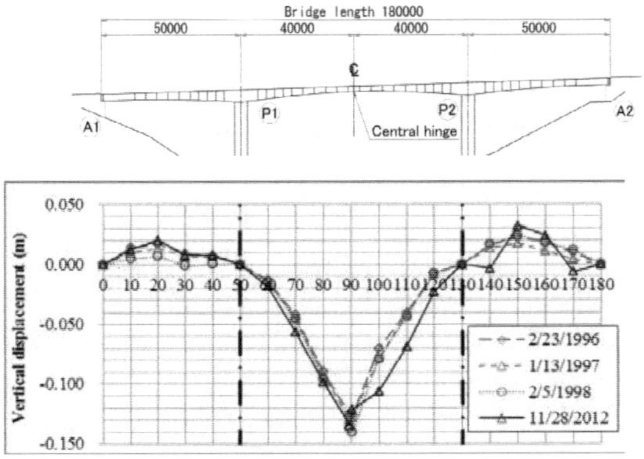

Figure 10. Transition of deflection

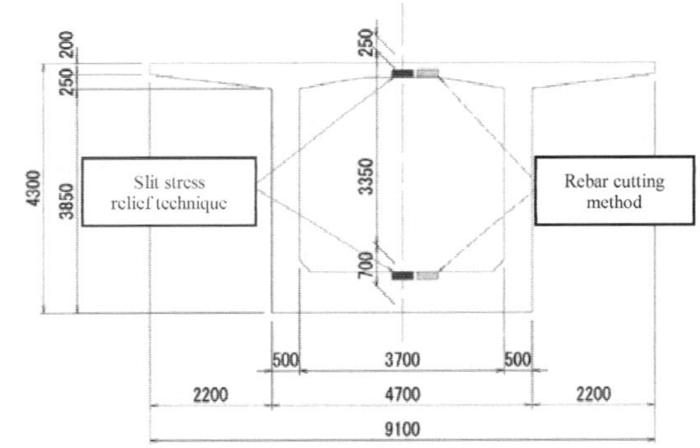

Figure 11. Investigation positions

which were the existing strain at those positions, had occurred in the rebar arranged in the PC bridge.

Table 2. Measurement results

Measurement position		Slit stress relief technique		Rebar cutting method
		Existing stress σ_C	Strain of concrete ε_C	Strain of rebar ε_r
P1	Upper slab	3.89±0.38 N/mm^2	177×10^{-6}	1334×10^{-6}
	Lower slab	5.10±0.59 N/mm^2	232×10^{-6}	1378×10^{-6}
P2	Upper slab	1.73±0.44 N/mm^2	79×10^{-6}	916×10^{-6}
	Lower slab	6.33±0.63 N/mm^2	288×10^{-6}	1328×10^{-6}

That is, the liberating strain which was measured by the rebar cutting method included elastic strain and creep/shrinkage strain. Therefore, if the existing stress of the concrete can be measured, creep/ shrinkage strain can be separated from the liberating strain obtained by the rebar cutting method by using Eq. (1).

$$\varepsilon_r=\varepsilon_l+\varepsilon_{\varphi s}=\sigma_c/E_c+\varepsilon_{\varphi s} \quad (1)$$

where, ε_r: rebar strain, ε_l: elastic strain of rebar ($\varepsilon_l=\sigma_c/E_C$),$\varepsilon_{\varphi s}$: creep/shrinkage strain of concrete, σ_c: concrete stress (N/mm^2),E_c: elastic modulus of concrete (N/mm^2).

3.2 Estimation of cause of abnormal deflection of center hinge

Figure 12 shows the strain distribution diagram of the P1 pier capital for rebar strain (ε_r), which includes concrete strain (ε_c), i.e., a type of elastic strain, and drying shrinkage. ε_r and ε_c show substantially the same slopes. Assuming that shrinkage strain (ε_{sh}) and creep strain (ε_φ) are proportional to the design shrinkage strain and the design creep coefficient, the shrinkage and creep coefficient of the P1 and P2 pier capital parts can be obtained by Eq. (2).

$$\varepsilon_{\varphi s}=\varepsilon_{sh}+\varepsilon_\varphi=n\cdot(\varepsilon_{sh0}+\varepsilon_c\cdot\varphi_0) \quad (2)$$

$$\varepsilon_{sh}=n\cdot\varepsilon_{sh0},\ \varepsilon_\varphi=n\cdot\varepsilon_c\cdot\varphi_0,\ \varphi=n\cdot\varphi_0$$

where, ε_{sh}: shrinkage strain, ε_φ: creep strain, φ: estimated creep coefficient,

ε_{sh0}: design shrinkage (200×10^{-6}), φ_0: design concrete creep coefficient (1.7).

Table 3 shows the estimated creep coefficients and shrinkage. As creep coefficients and shrinkage of P1 and P2 are approximately the same, they are considered to express the creep coefficient and shrinkage of the actual bridge. Figure 13 shows the calculated deflection using the average estimated creep coefficient (φ=3.52) and the results of longitudinal measurement. Because the calculated of deflection using the estimated creep coefficients are in good agreement with the results of longitudinal measurement, it is thought that the main cause of the abnormal deflection in this bridge is creep deformation.

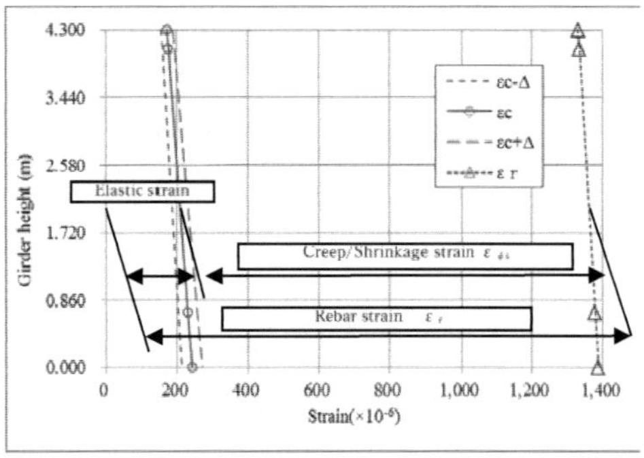

Figure 12. Strain distribution (measured) of P1 pier head11

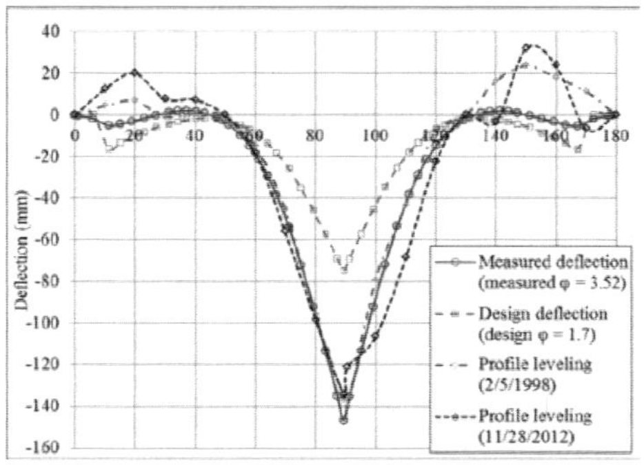

Figure 13. Deflection distribution by estimated creep coefficient

Table 3. Estimated creep coefficients and shrinkage

Measurement position	Girder height (m)	$\varepsilon_r\times10^{-6}$	$\varepsilon_c\times10^{-6}$	$\varepsilon_{\varphi s}\times10^{-6}$	n	$\varepsilon_{sh}\times10^{-6}$	φ
P1	4.300	1330.7	172.7	1158.0	2.3	469.2	3.99
	4.050	1334.0	176.8	1157.2	2.3	462.3	3.93
	0.700	1378.0	231.8	1146.2	1.9	385.9	3.28
	0.000	1387.2	243.3	1143.9	1.9	372.8	3.17
	Average					422.6	3.59
P2	4.300	885.3	63.0	822.2	2.7	535.4	4.55
	4.050	916.0	78.6	837.4	2.5	501.9	4.27
	0.700	1328.0	287.7	1040.3	1.5	301.9	2.57
	0.000	1414.1	331.4	1082.7	1.4	283.6	2.41
	Average					405.7	3.45

4. Confirmation of safety of PC bridge damaged by chloride attack

4.1 Overview

This bridge is a post-tensioned PCT girder bridge which passes over a sea area and had been in service for approximately 40 years since construction. The bridge, as shown in Fig. 14, had suffered progressive deterioration in the forms of loose and peeling scale and corrosion of the rebar, and a reduction of load bearing performance was feared as a result of progress of corrosion of the members due to chloride attack and corrosion of the PC tendons. Therefore, a stress investigation, etc. was carried out to confirm the load bearing capacity of the bridge. In the stress investigation, we measured by the slit stress relief technique and a dynamic load test. The measurement positions are shown in Fig. 15.

Figure 14. Deterioration of main girder

4.2 Measurement results

Table 4 shows the results of measurements by the slit stress relief technique, and Table 5 shows the results of the dynamic load test. As the elastic modulus, the design elastic modules at the time of construction, Ec=35 kN/mm^2 was assumed. In the dynamic load test, a dump truck with a gross weight of 20 t and a 50 t class rafter crane with a gross weight of approximately 40 t were assumed as the load vehicles. One lane was closed to traffic, the load vehicles were stopped in the center of the span on the closed lane, and the vehicles were allowed to travel on the open lane. Measurements were performed under the following three load vehicle patterns.

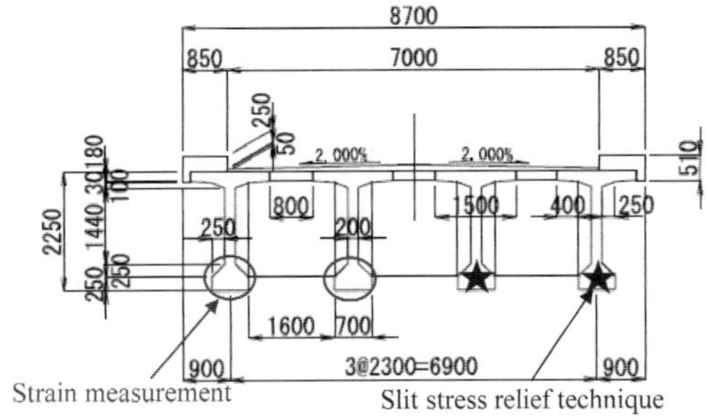

Figure 15. Measurement positions (span center)

Pattern 1: Dump truck stops on closed lane, dump truck travels on open lane.
Pattern 2: Dump truck stops on closed lane, rafter crane travels on open lane.
Pattern 3: Rafter crane stops on closed lane, dump truck travels on open lane.

4.3 Evaluation of load bearing performance

In the PC girder of this bridge, from Table 4, the existing stress, which was stress on the resistance side, has become compressive stress of approximately 3.0 N/mm^2 or more. On the other hand, from Table 5, the maximum tensile stress due to single loading by a dump truck, which seems to have a high possibility of being the actual loading mode, is on the order of 1.0 N/mm^2. The maximum tensile stress due to single loading by a dump truck and parallel loading by dump trucks, which appear to have a small possibility, is about 1.5 N/mm^2, and the maximum

Table 4 Results of measurement by slit stress relief technique

Measurement position	Existing stress (N/mm^2)			Error
G3	3.46	±	0.15	4.30%
G4	3.21	±	0.25	7.70%

Table 5. Results of dynamic load test

		G1	G2	G3	G4
Pattern 1	(A) Dump truck load	-0.2	-0.5	-0.8	-0.8
	(B) Dump truck load (traveling)	-1.1	-0.9	-0.6	-0.3
	(A)+(B)	-1.3	-1.4	-1.4	-1.1
Pattern 2	(A) Dump truck load	-0.3	-0.5	-0.8	-0.9
	(B) Crane load (traveling)	-1.6	-1.4	-0.9	-0.5
	(A) + (B)	-1.9	-1.9	-1.7	-1.4
Pattern 3	(A) Crane load	-0.5	-0.9	-1.3	-1.3
	(B) Dump truck load (traveling)	-1	-0.9	-0.6	-0.4
	(A)+(B)	-1.5	-1.8	-1.9	-1.7
Max. tensile stress	Dump truck single load	-1.1			
	Crane single load	-1.6			
	Dump tuck + dump truck parallel load	-1.4			
	Dump tuck + crane parallel load	-1.9			

Unit: N/mm^2 Compressive (+), tensile (-)

tensile stress due to parallel loading by a dump truck and a rafter crane, which seems to have an extremely small possibility, is about 2.0 N/mm^2.

Therefore this PC bridge is kept enough safe. The reason is even if the heaviest vehicle is loaded, compressive stress acts on a PC bridge.

5. The cause investigation of the abnormal displacement of 3 span continued PC box girder that passed after construction for 38 years.

5.1 Overview

This bridge is 3 span continuous PC box girder bridge located in mountainous region a constructed in 1977. This bridge is a road bridge of bridge lenghth:166.20m and span:40.0m+85.0m+40.0m. (Refer to figure-16 and table-6) About 150 mm of abnormal deformation had caused the central span part as shown in figure-17.So for cause investigation of abnormal deformation a stress was investigated by the central span and th pier capital part.

Figure-17 Displaced situation picture of the span center

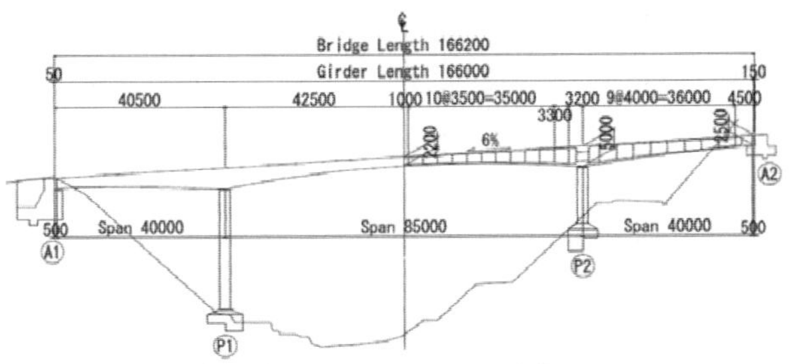

Figure-16 Bridge general figure

Table-6 The bridge specification

A construction year	1977
The bridge form	3 span continuous PC box girder bridge
Bridge length	166.2m
The span length	40.0m+85.0m+40.0m
The width	8.20m roadway 7.00m
Design live load	TL-20
Cross slope	2.0% of both sides
The vertical slope	6.00%
Bevel	90°

We were investigated existing stress by slit stress relief method using the full-field strain measurement equipment and existing rebar strain by rebar cutting method, but we couldn't get a conclusion. Therefore we investigated a factor of abnormal deformation more than the analysis which considered the shrinkage difference in the parts of box section. We report this result.

5.2 Outline of a stress investigation

The existing stress measurement by 1) slit stress relief teachnique and 2) existing rebar strain by rebar cutting method was measured for a stress investigation, like 3 section. The existing stress measurement by slit stress relief teachnique and the existing rebar strain by rebar cutting method location are showed on figure 18.

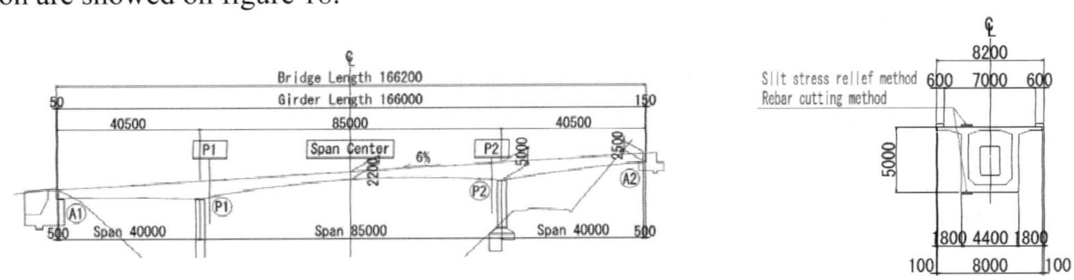

Figure 18 The stress investigation location

5.3 Displacement in the span center by Creep/shrinkage

Creep coefficient and a dry shrinkage were found from the rebar strain, like section 3.Displacement in the span center by Creep/shrinkage was calculated. The displacement in the span center was tens of millimeters of the displaced calculation was big different from deflection 150mm of a measured. Therefore the cause of this bridge abnormal deformation wasn't depended on Creep/shrinkage.

5.4 Cause presumption of an abnormal deformation

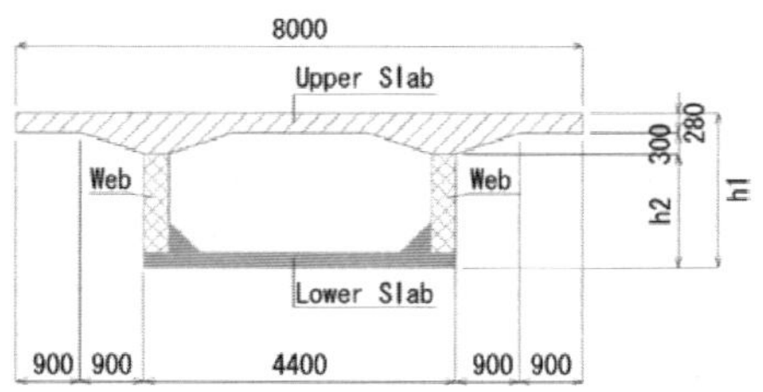

Figure 19 Division section of each part

Design calculation is calculated beam model. (beam analytic theory). Calculated displacement of a beam model is calculated based on a linear Creep law. Or when elucidating long displacement and a mechanism deformation by a beam model, it isn't suitable. Because the statically indeterminate stress which forms isn't considered by the shrinkage difference in the parts of upper deck, web, lower deck, etc. of the box girder section.
So it's divided into upper deck, web, lower deck based on "concrete standard specifications in Japan (2012: The design volume and genus material)" as shown in figure 19, and a concrete shrinkage of each part is established, and the displacement was calculated. Further, based on "concrete standard specifications (the design volume and accessory material)", displacement by the Creep distortion accounted for 60% of the whole amount of displacement and it was supposed that remaining 40% chooses displacement by the dryness shrinkage distortion.
We changed the combination of the shrinkage distortion of the upper and lower deck and calculated displacement using the analysis model on figure 22. As a result we got the displaced calculation result which is 150mm at the shrinkage of the concrete in table 7.
A stress in the measurement location was calculated from design calculation of a beam theoretical analysis using calculated support reaction from the analysis model on figure 20.
Result of this calculation and measured existing stress by slit stress relief technique is in table 10. The calculation value and a measured value are the very close showed table 8.
As a result of the above, abnormal deformation in this bridge has a high possibility it has gained by the concrete shrinkage difference between upper deck and lower deck.

Table 7 The calculation result which considered the concrete shrinkage difference in each parts

Unit mm

	Analysis			Measurement
	Shrinkage (δ s)	Creep (1.5δ s)	Total	
Side Span	12.4	18.6	31	30
Span center	-59.1	-88.7	-148	-150

Table 8 The calculation stress which considered the concrete shrinkage difference in each parts

	The upper (N/mm^2)		The lower (N/mm^2)	
	Exsist stress (measurment)	Analysis	Exsist stress (measurment)	Analysis
P1	4.53±0.54	4.30	6.20±0.76	6.14
Span center	7.06±0.44	7.03	1.80±0.23	2.01
P2	3.10±0.53	4.30	12.37±1.01	6.14

The upper slab shrinkage: ε_U = 230 μ The lower slab shrinkage: ε_L = 280 μ

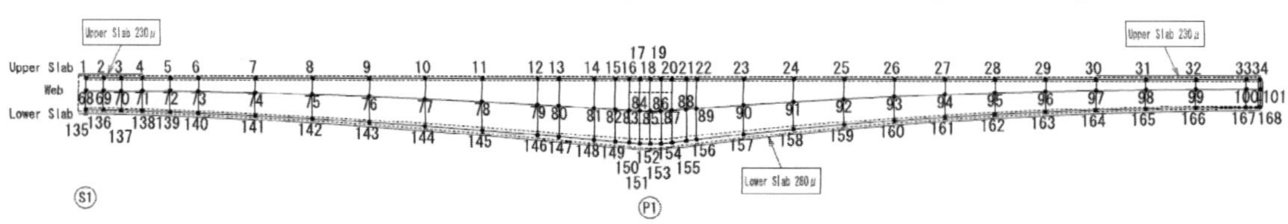

Figure 20 Analysis model

5.5 Consideration

A existing stress by slit stress relief teachnique was measured a lot. The design stress was parallel with a existing stress by slit stress relief teachnique in the statically determinate structure of PC simple girder relatively. But the design stress isn't parallel with a existing stress by slit stress relief teachnique in the statically indeterminate structure of a continuous PC box girder. The calculation stress which considered the shrinkage difference between the upper deck and the lower deck was parallel with the existing stress. And the calculated displacement was also parallel with the measurement displacement. A considered calculation stress possesses the concrete shrinkage difference between upper deck and lower deck at present, the measurement transformable amount was parallel with a stress.

A statically indeterminate structure of a continuous PC box figure has a high possibility that the shrinkage difference in the parts has to consider by this thing. Therefore it'll be expected to put existing stress survey result in the past in order and consider from now on.

6. Conclusions

1) It is possible to separate creep/drying shrinkage strain from concrete stress (strain), which is a type of elastic stress, and estimate the creep coefficient of existing structures by using a combination of the slit stress relief technique and strain measurement of rebar by the rebar cutting method.

2) The load bearing capacity of existing PCT girder bridges, in which a reduction of pre-stress due to deterioration caused by chloride attack is an issue, can be evaluated by a comparative study of the resistance stress obtained by the slit stress relief technique, and the acting stress obtained by strain measurements in the dynamic load test supposing the service load.

3) The amount of reduction of pre-stress can be estimated from the results of the slit stress relief technique, and the proper load bearing performance can be evaluated. The proper amount of strengthening can then be set based on the amount of pre-stress (existing pre-stress) in the current condition.

4) An actual stress in the statically indeterminate structure of a continuous PC box girder was understood from a existing stress result of measurement by slit stress relief teachnique. This stress seems to have a high possibility that we undergo big influence of the concrete shrinkage difference of a part. Therefore we presume a possibility that the actual stress different from the design stress occurs to be very high at the statically indeterminate structure of the continuous PC box girder.

References

Akira Demizu, Kenichi Hida, Yukihiro Ito, Masakazu Uchino, Takuji Okamoto, Hiroshi Matsuda, 'Measurement of existing stress of post-tensioned girders by optical technique and stress release method', *Proceedings of the Annual Conference on Experimental Mechanics, Japanese Society for Experimental Mechanics*, No. 10 GS5-5, pp281-286, 2010.

Kenichi Hida, Yukihiro Ito, Mitsuaki Tsukihara, Haruyuki Koitabashi, 'Report of investigation of central hinge bridge in which abnormal deformation occurred after approximately 30 years in service', *Proceedings of the 22nd Symposium*, pp345-348, 2013.10.

Kenichi Hida, Yukihiro Ito, Akira Demizu, Masakazu Omachi, 'Report on investigation of load bearing capacity of post-tensioned PCT girder bridge with damage by chloride attack', *Proceedings of the 23rd Symposium*, pp649-652, 2014.10.

Kenichi Hida, Sumao Yoshida, Morihiro Horibe, Yukihiro Ito, 'Report of the cause investigation of the abnormal displacement of 3 span continued PC box girder that passed after construction for 38 years', *Proceedings of the 25rd Symposium*, pp531-534, 2016.10.

The 8th International Conference on
Structural Health Monitoring of Intelligent Infrastructure
Brisbane, Australia | 5-8 December 2017

Multipurpose Wireless Sensors for Asset Management and Health Monitoring of Bridges

M. O. Furkan[1], Q. Mao[2], S. Livadiotis[3], M. Mazzotti[4], A. E. Aktan[5], S. P. Sumitro[6], I. Bartoli[7]
[1,2,3,4,5,7] *Drexel University – United States, Email:* mof23@drexel.edu, [6]*SmartSensys – United States*

Abstract

Application of wireless sensors for Structural Health Monitoring (SHM) has been investigated extensively. Key limitations for practical use have been represented by energy requirements, connectivity challenges, and robustness for implementation in harsh field environment. In many field applications, current sensors and sensing systems still rely on wired connectivity for communication and external power source for energy. However, tethered sensors need for extensive cabling throughout a structure, can require longer installation processes, and their maintenance can increase the total cost of the SHM system.
This paper presents a suite of wireless sensors under development that focus on displacement and strain measurements that are rugged, and have long service life and can be used for both rapid structural assessment as well as long term monitoring. The sensors considered are designed by transforming existing, proven, and robust wired sensors into wireless units. In this study, the wireless sensors were tested in laboratory conditions closely resembling field conditions and compared to tethered sensors for calibration and evaluation. This research is part of a broader FHWA sponsored project intended to develop and validate a wireless sensing system on a real, operating structure to account for all the uncertainties, environmental conditions, and operational variability that are encountered in the field.

1. Introduction

Many infrastructures such as bridges, highways, railways, etc. are still in use despite ageing and gradual deterioration. Since operational and environmental loads are inevitable through the service life of structures, serviceability and safety of them are serious concerns for infrastructure stakeholders. In the case of natural disasters or extreme load events, the impact of a crippled infrastructure can be catastrophic and deadly.
The 2015 Federal Highway Administration (FHWA) report (US Federal Highway Administration, 2016) on the condition and performance of bridges in the United States presents a significant picture regarding the necessity of performance monitoring. Based on that in 2012, out of 607,380 bridges 66,749 were rated as structurally deficient and 84,748 were rated as functionally obsolete. That gives us the total bridge deficiency to be approximately 25%. Without proper monitoring and maintenance, lane closures, load postings, and ultimately bridge closures could be experienced. The effect of such limits to especially heavy-traffic transformation networks could be paralyzing.
In the past decades, the need to introduce an alternative strategy to handle the infrastructure system management has been discussed and several studies have presented Structural Health Monitoring (SHM) as a possible approach to that need (Chang et al., 2003, Sohn et al., 2003, Farrar and Worden, 2007). The process of implementing a damage detection and characterization strategy for such systems is referred to as SHM. SHM aims to provide information about the condition of a structure thorough its methods and applications to help with structural maintenance. Visual inspection is still

the most widely used method of inspection to evaluate the conditions of civil infrastructure. However, it can be time-consuming, expensive, labor intensive, and susceptible to error (Li et al., 2015).
Traditional wired sensors have been used successfully for SHM applications for decades. Despite being reliable and proven for a long time, their main disadvantages come from the need for excessive cabling (Celebi, 2002, Brownjohn, 2007), highly time-consuming installation, extensive labor and maintenance efforts, and consequently great costs (Farrar, 2001, Straser et al., 2001). Overall costs of these systems can prevent high-density sensor deployments, and thus limit broader utilization of SHM for civil infrastructures (Hipley, 2001).
Wireless sensors are a potential sensing technology which can advance the abilities of structural engineering field by economically grasping SHM. Without the traditional wired connectivity, wireless sensors can mitigate the disadvantages that are inherent to wired sensors. By removal of wires, wireless sensors provide easier installation and efficient data management which presents a solution for long-term SHM at a lower cost than traditional wired SHM systems (Lynch and Loh, 2006, Kim et al., 2007, Jang et al., 2010)
Various research studies (Doebling et al., 1998, Spencer et al., 2004, Lynch, 2007, Buratti et al., 2009) and applications (Lynch et al., 2004, Lu et al., 2005, Feltrin et al., 2011) have been performed in the field of wireless sensors and SHM systems. However, even more studies need to be done to address the challenges associated with their field application. With the advancements in sensing technology, the demands of the industry can be answered. In this research, the main focus was on the transformation of existing, reliable wired sensors into wireless gages. Through the support of a FHWA project within the Exploratory Advanced Research (EAR) Program, this objective has been investigated by Drexel University and SmartSensys, LLC. In this study, preliminary results from laboratory applications of such transformed sensors are presented and discussed.

2. Drexel University Laboratory Tests

2.1. Validation of Wireless Displacement Gages

The wireless displacement gages tested in this study were transformed from wired Celesco Model PT8510 gages (string potentiometers) that are industrial grade cable-extension transducers. The string potentiometers have a range of 5 in. over an output range of 0 to 10 V, and an accuracy of +/- 0.28% of the full stroke. Those string pots were selected since they are field proven, robust, and durable, showing reliable performance in various field of applications. The wireless connectivity was performed by SmartSensys gateways and wireless transmitters (Figure 1a).
The wireless units were tested alongside the same stringpots (Celesco Model PT8510) with wired connectivity for validation. The wired data acquisition was performed by National Instruments (NI) CompactRIO (cRIO) system. The cRIO system was controlled by the NI LabVIEW software.
Six wireless stringpots were installed on Tinius-Olsen Electromatic Universal Test Machine (Model 398 Display & CMM 496 Controller) along with one wired stringpot for validation as shown in Figure 1a. The wireless displacement sensors were tested simultaneously to observe noise levels, maximum sampling rate, stability of data collection, and accuracy of the data collected. The Tinius-Olsen Test Machine was controlled to impose equal displacement increments of 0.1 in, starting from 0 in and reaching up to 1 in. Then, the same procedure was followed to impose 0.1 in decrements from 1 in to 0 in.

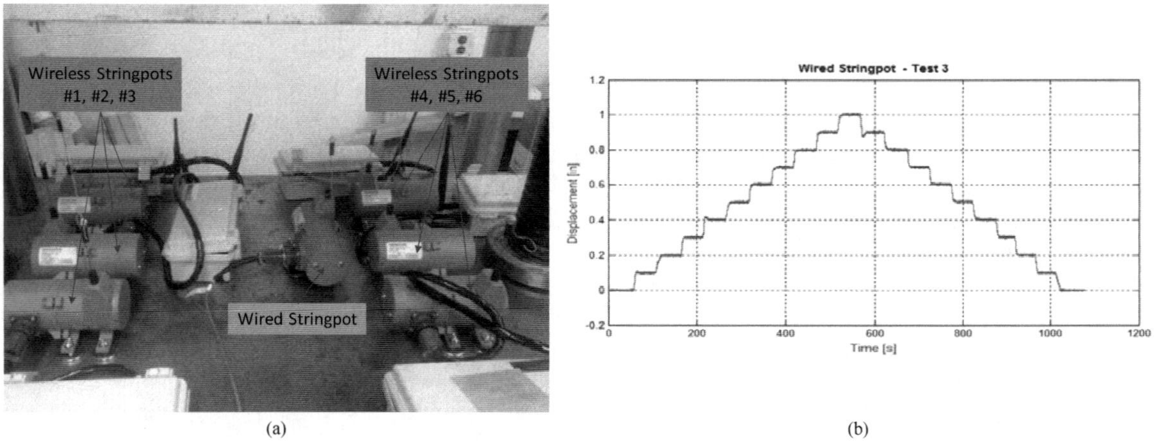

Figure 1 – (a) Wired & wireless stringpots installed on the Tinius-Olsen test machine; (b) Wired displacement time history.

Figure 1b shows the wired displacement sensor time history recorded with a sampling rate of 100Hz. Figure 2 shows the displacement time histories recorded by the six wireless displacement sensors. Some noise can be observed from the plots with outlier data that do not follow the imposed displacement increments and decrements. The noise is induced by the simultaneous recording of data from six separate sensors with one antenna receiver. Increasing the number of receivers can help decreasing such noise. From Figure 2, it can be seen that the overall performance of the wireless stringpots was satisfactory with maximum error of less than 3%.

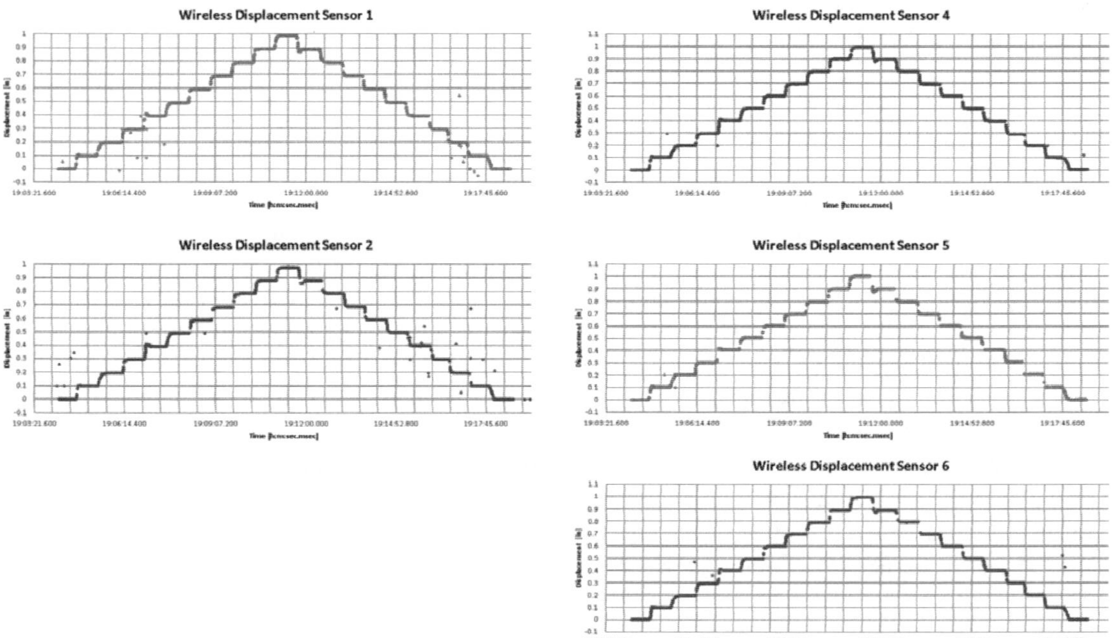

Figure 2 - Wireless displacement time histories

To closely observe the sampling period of each sensing unit, a zoom on the data recorded by the wireless displacement sensor #6 is shown in Figure 3a. The data included in the window shown in Figure 3a highlights how the sampling rate decreases from the highest value of 100Hz to reach minimum sampling of 0.5-1 Hz (1 data/2sec & 1 data/sec). To further investigate the sampling rate variability of each sensing unit, six separate histograms were generated as shown in Figure 3b. Such plots summarize the frequency at which sampling rate range occurs for each sensing unit. For instance, Sensor #1 had approximately 450 events recorded at a rate of 80-90Hz and almost 500 events recorded at sampling frequencies of 90-100Hz. It should be noted that while the loss of data decreases the effective sampling rate of the wireless sensing system, the time synchronization of the six channels is still guaranteed by internal counters within each wireless node.

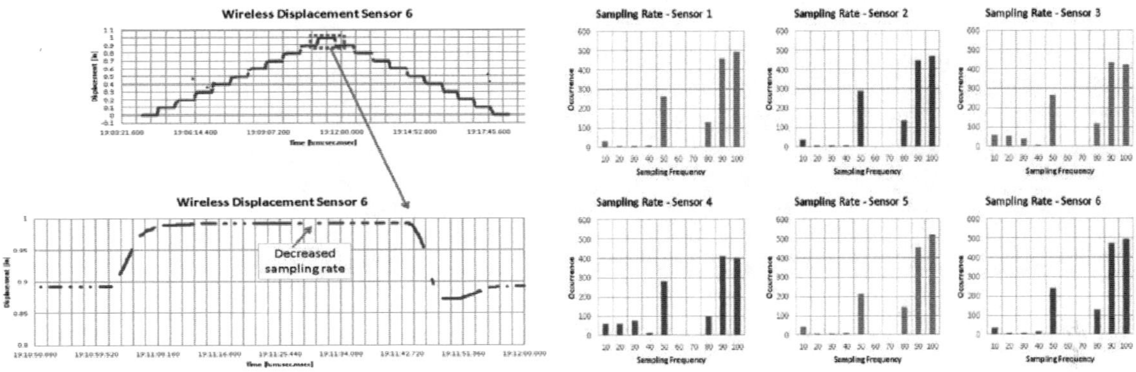

(a) (b)

Figure 3 - (a) Zoomed-in view of wireless stringpot #6 displacement time history; (b) Histograms of wireless data.

2.2. Validation of Wireless Strain Gages

In this test, twelve wired conventional strain gages and twelve wireless strain gages were installed on a 6-foot long specimen (6 ft. x 0.25 in x 2 in). Six of the wired conventional strain gages were from Micro-Measurements (CEA-13-250UW-350/P2), a general-purpose quarter-bridge strain gage with a maximum strain of 5,000 microstrain. Another type of six wired conventional strain gages were chosen to be Omega (SGT-4/1000-FB11), a full-bridge strain gage with a strain limit of 30,000 microstrain. The data from the wired strain gages was obtained by using NI cRIO platform. Besides, another six of the same Omega strain gages (SGT-4/1000-FB11) were transformed into wireless devices by SmartSensys to test the wireless connectivity. Additionally, six magnetic strain gages from Tokyo Sokki (FGMH-2A) were also transformed as wireless units. Those magnetic gages operate through measuring strain caused by friction when the magnet on the gage is activated after pressing down the strain sensing element on the structure. Their installation and removal is fast since they do not require adhesive to bond with the structure which allows repeated, rapid testing. They are full bridge gages with a strain limit of 2,500 microstrain. The wireless connectivity was performed by SmartSensys wireless transmitters and gateways.

The strain gauges were installed at the top, center, and bottom locations on the specimen and on both front and back surfaces. Then, the specimen was placed in the Tinius-Olsen Test Machine (Figure 4a). A closer view in Figure 4b shows setup of each installed strain gage. The same setup was followed at the front and back of the specimen. A uni-axial tensile test was performed on the steel specimen with 1,000 lbs. increments and a maximum load of 6,000 lbs. The loading scenario was repeated for three cycles. The sampling frequency was 100 Hz for the wired gages and one data every twelve seconds for the wireless devices. For the maximum load of 6,000 lbs., the expected theoretical strain value was 414 microstrain.

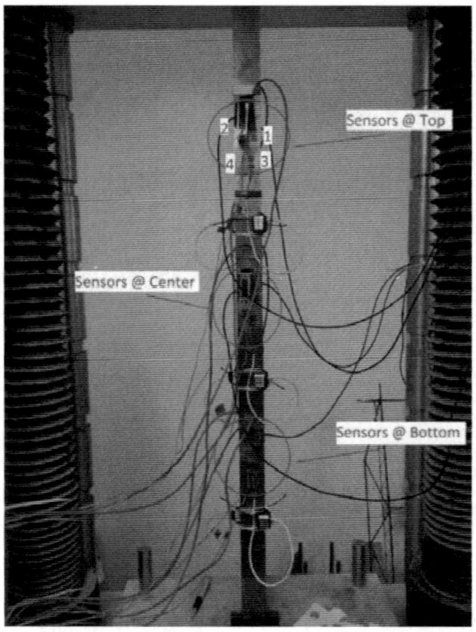

(a)

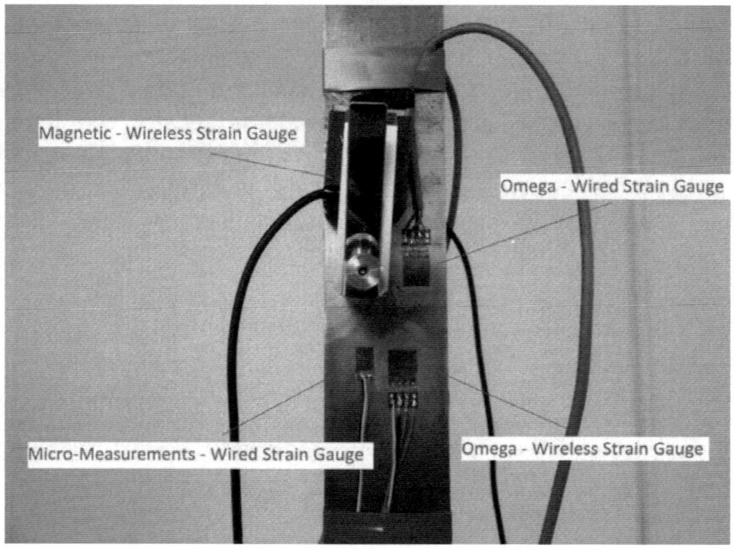

(b)

Figure 4 - (a) The specimen with the wired & wireless strain gages mounted in the Tinius-Olsen test machine; (b) Zoomed-in view of installed strain gages.

Table 1 shows the averaged strain values for each type of sensor and for each location. The strain measurements of the sensors located on opposite sides (front and back) are averaged to eliminate bending effects that can be present in the axial load. It can be seen that the wireless strain gauges performed relatively better than the wired strain gauges. The maximum percentage error for the wireless Omega sensors is around 5%, and 1% for wireless magnetic strain gages. Whereas, the maximum percentage error is around 17% for the wired omega strain gauges, and 1% for the wired Micro-Measurements strain gages. Two of the six wireless magnetic gages (at bottom) malfunctioned during the test and data was not available for those units. The time histories for the center position sensors can be seen in Figure 5. The results are affected by possible misalignment problems which might be due to either misaligned sensors during installation or misalignment of the specimen in the testing machine that can induce bending. Also, different sensor locations could cause minor differences.

Table 1 - Averaged test results for each type of sensor and each location
AVERAGE MAX STRAIN DATA (MICROSTRAIN)

TEST #	Wired MM			Wired Omega			Wireless Omega			Wireless Magnetic	
	Top	Center	Bottom	Top	Center	Bottom	Top	Center	Bottom	Top	Center
1ST CYCLE	-410	-409	-409	-383	-399	-343	-398	-404	-395	-420	-414
2ND CYCLE	-410	-410	-409	-383	-399	-344	-397	-404	-395	-419	-413
3RD CYCLE	-409	-408	-408	-383	-401	-344	-397	-404	-395	-418	-412

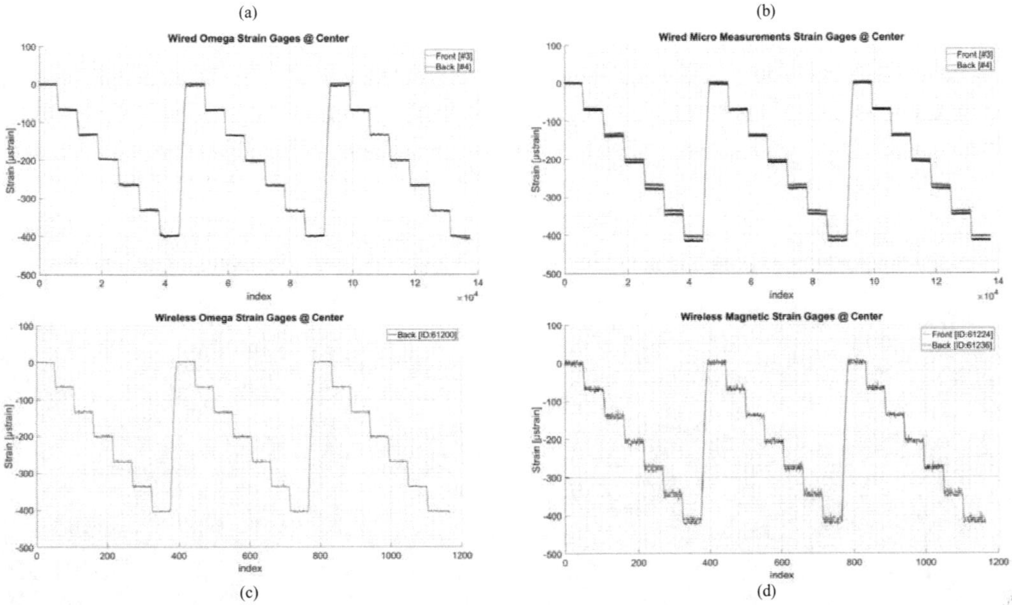

Figure 5 - The time histories from the strain gages in the center: (a) Wired Omega; (b) Wired Micro-Measurements; (c) Wireless Omega; (d) Wireless Magnetic.

3. Tests on a Steel Girder

Following the successful laboratory validation of the wireless sensors, the next step was to test them on a real structure in laboratory conditions. The sensors were tested on a 30-ft steel girder with a precast concrete deck panel grouted on top of the girder with shear studs. A simple representation of the girder can be seen in Figure 6a. The steel girder was loaded by two load frames each supporting an actuator and was supported at both ends (Figure 6b). The tests were performed in three steps; no load, 14.3 kip-load (half-load), and 28.5 kip-load (full-load). As before, the wired sensors were connected to NI cRIO system for data acquisition and the wireless sensors were controlled by SmartSensys software, gateway, and transmitters.

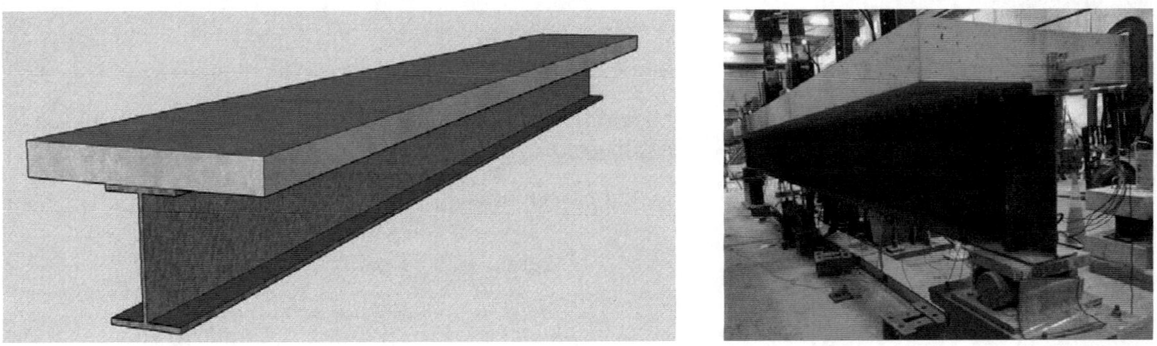

(a) (b)

Figure 6 - The steel girder: (a) model; b) in place.

3.1. Validation of Wireless Displacement Gage

The same wired and wireless displacement gages (Celesco Model PT8510) tested previously were mounted on the girder. In total, three wired and one wireless stringpots were placed under the I-beam for testing as shown in Figure 7. Additionally, the displacements were also measured by an LVDT (Figure 7b) and a FARO Laser Tracker ION (Figure 8a). The laser tracker had a volumetric accuracy of 0.049 mm at 10 m and can measure at a high speed and precision.

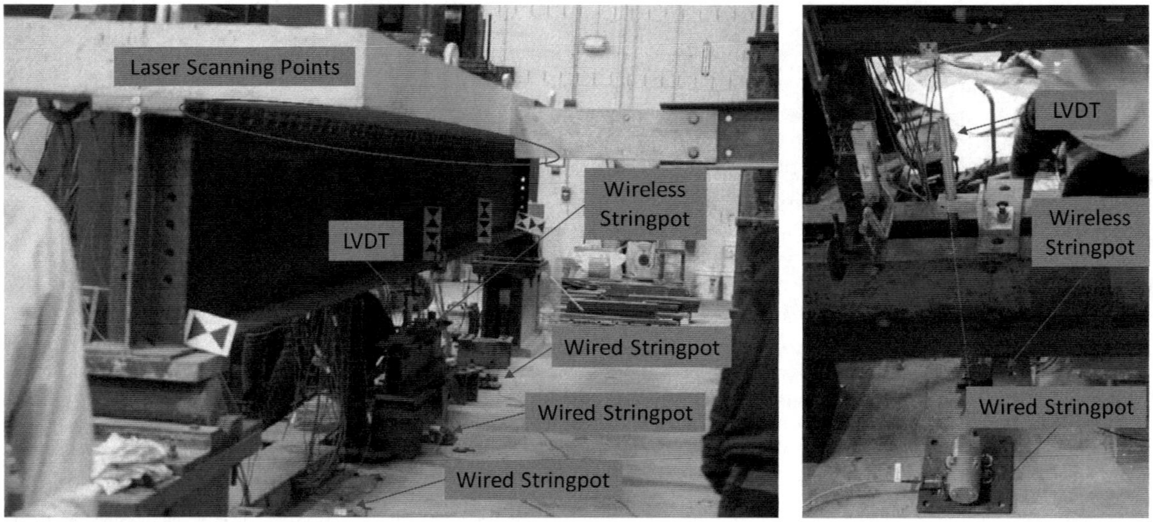

Figure 7 - Setup of wired and wireless displacement gages on the I-beam: (a) front view; (b) side view.

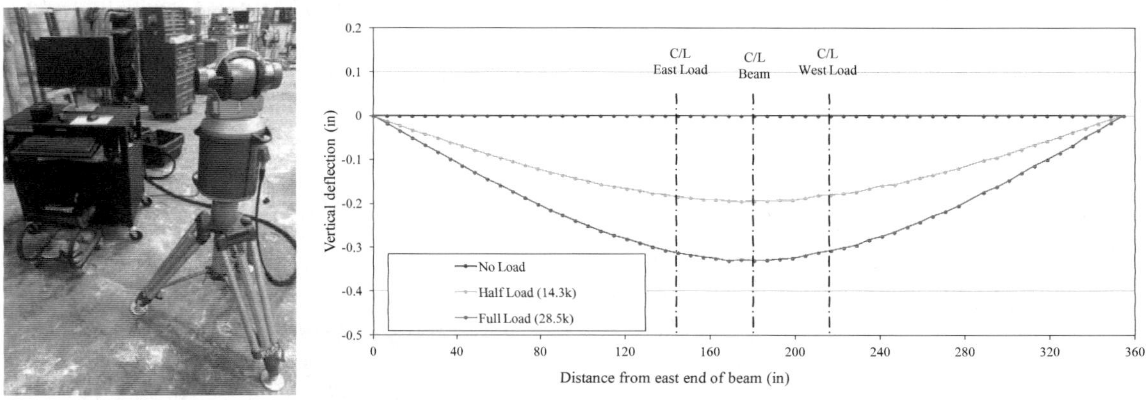

Figure 8 - (a) FARO Laser Tracker ION; (b) Vertical deflection measured by laser tracking.

Data was collected when there was no load, a load of 14.3 kip (half-load), and a load of 28.5 kip (full-load). Figure 8b shows the vertical deflection measured by the FARO laser tracker. The vertical deflection values from each measurement device are given in Table 2 for the mid-section of the girder. Those results are plotted in Figure 9. When we compare laser tracking measurements to the wireless data, the percent error for the wireless stringpot is around 2% at half-load and around 3% for the full-load which is consistent with the previous test results.

Table 2 – Comparison of displacement measurements at mid-section

LOAD [KIP]	LVDT [MM]	LASER SCANNING [MM]	WIRED STRINGPOT [MM]	WIRELESS STRINGPOT [MM]
0	0	0	0	0
14.3	-5.38	-4.93	-4.66	-5.01
28.5	-8.74	-8.38	-8.18	-8.66

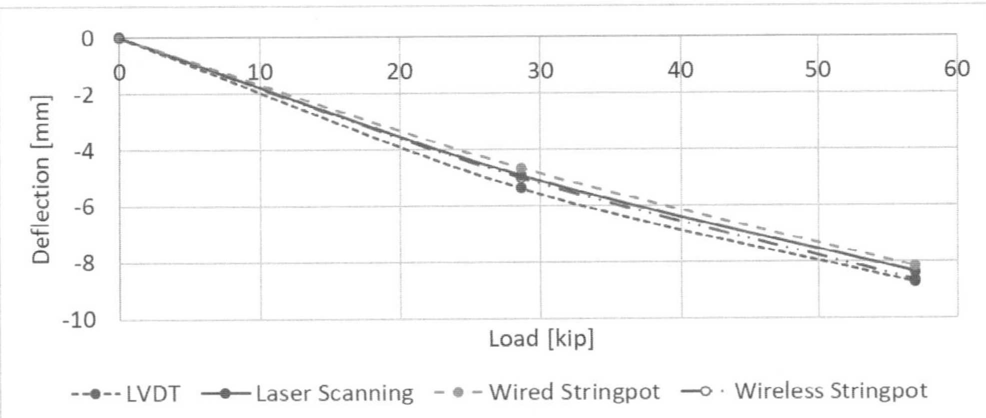

Figure 9 – Comparison of displacement measurements at mid-section

3.2. Validation of Wireless Magnetic Strain Gages

Three wireless magnetic strain gages were mounted on the mid-section of the I-beam and strain levels were measured at the top and mid of the web, and at the bottom of the flange. Additionally, three wired gages were installed at the same locations for comparison. The sensors that were installed on the web can be seen in Figure 10a, and on the flange in Figure 10b.

Figure 10 - Strain gages installed: (a) on the web; (b) on the flange.

The strain gages recorded data during the same loading scenario as before. The results from those sensors for the mid-section of the girder are given in Table 3 and plotted in Figure 11. At the bottom flange, the difference between the wired and wireless strain gages was around 6%. At the top and mid of the web, the difference was between 6-11%. One of the reasons for those differences could be the surface condition of the girder where the magnetic strain gauges were attached. Especially for the top-web, the installation surface was not ideal. Besides, the wired and wireless strain gages were not very close to each other. Another reason could be the ideal attachment type. It was observed that the ideal attachment direction for the wireless magnetic strain was from the bottom. When those gages were installed vertical to the structure, the surface friction may not be enough to hold the sensor in place. Thus, the sensor may slide with time and continuous loading cycles.

Table 3 - Wired & wireless strain levels

	WIRED STRAIN MEASUREMENTS [MICROSTRAIN]			WIRED STRAIN MEASUREMENTS [MICROSTRAIN]		
POSITION	No Load	Half Load	Full Load	No Load	Half Load	Full Load
BOTTOM FLANGE	0	263	494	0	248	466
MID WEB	0	81	176	0	90	190
TOP WEB	0	-77	-86	0	-72	-77

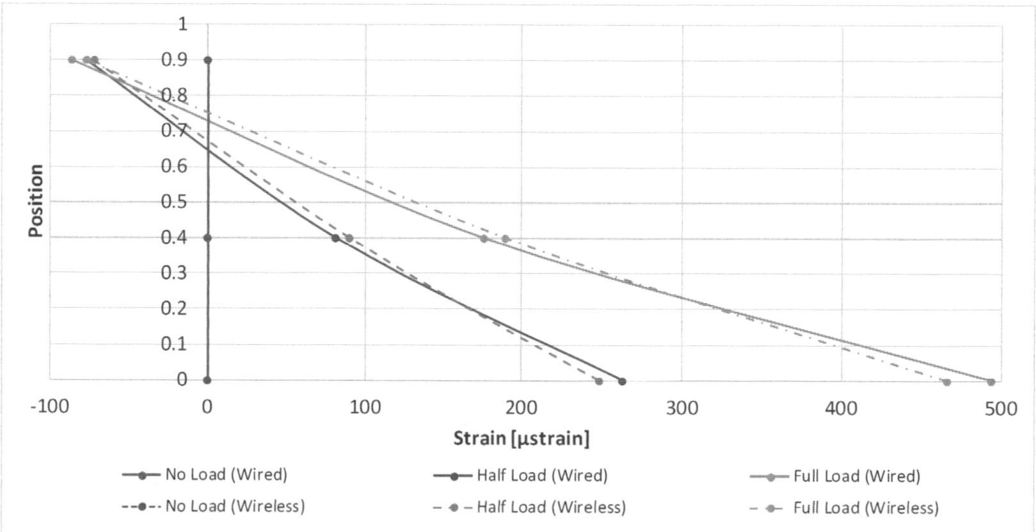

Figure 11 - Comparison of strain measurements

4. Conclusions and Future Work

Several wireless displacement and strain gages that were transformed from existing, reliable wired sensors into wireless devices were tested in two different laboratory environments. The transformation was done by SmartSensys, LLC within a project supported by the FHWA. The developed wireless sensors were tested at Drexel University laboratory for initial calibration and validation. The performance of the wireless devices was mostly satisfactory and comparable with expected results. Upon successful application, the wireless sensors were tested on a real structure, a 30-ft girder. Overall, even though both the wireless displacement and strain gages show promising and satisfactory results, they still need to be improved and upgraded for reliable and robust field application. The wireless sensors will be validated in the field upon performing necessary improvements. Opinions expressed in this paper are those of the authors and do not necessarily reflect the opinions of the sponsors.

References

US Federal Highway Administration 2016. 2015 Status of the Nation's Highways, Bridges, and Transit: Conditions & Performance. Washington, DC.

Chang, P. C., Flatau, A. & Liu, S. 2003. Review paper: health monitoring of civil infrastructure. *Structural health monitoring,* 2, 257-267.

Sohn, H., Farrar, C. R., Hemez, F. M., Shunk, D. D., Stinemates, D. W., Nadler, B. R. & Czarnecki, J. J. 2003. A review of structural health monitoring literature: 1996–2001. *Los Alamos National Laboratory.*

Farrar, C. R. & Worden, K. 2007. An introduction to structural health monitoring. *Philosophical Transactions of the Royal Society of London A: Mathematical, Physical and Engineering Sciences,* 365, 303-315.

Li, J., Mechitov, K. A., Kim, R. E. & Spencer, B. F. 2015. Efficient time synchronization for structural health monitoring using wireless smart sensor networks. *Structural Control and Health Monitoring.*

Celebi, M. 2002. Seismic instrumentation of buildings (with emphasis on federal buildings). *USGS Project, an administrative report, United States Geological Survey, Menlo Park, CA.*

Brownjohn, J. M. 2007. Structural health monitoring of civil infrastructure. *Philosophical Transactions of the Royal Society of London A: Mathematical, Physical and Engineering Sciences,* 365, 589-622.

Farrar, C. R. 2001. Historical Overview of Structural Health Monitoring. *Lecture Notes on Structural Health Monitoring Using Statistical Pattern Recognition.* Los Alamos Dynamics, Los Alamos, NM.

Straser, E. G., Kiremidjian, A. S. & Meng, T. H. 2001. Modular, wireless damage monitoring system for structures. Google Patents.

Hipley, P. Caltran's Current State-of-Practice. Consortium of Organizations for Strong-Motion Observation Systems, 2001.

Lynch, J. P. & Loh, K. J. 2006. A summary review of wireless sensors and sensor networks for structural health monitoring. *Shock and Vibration Digest,* 38, 91-130.

Kim, S., Pakzad, S., Culler, D., Demmel, J., Fenves, G., Glaser, S. & Turon, M. Health monitoring of civil infrastructures using wireless sensor networks. Information processing in sensor networks, 2007. IPSN 2007. 6th international symposium on, 2007. IEEE, 254-263.

Jang, S., Jo, H., Cho, S., Mechitov, K., Rice, J. A., Sim, S.-H., Jung, H.-J., Yun, C.-B., Spencer Jr, B. F. & Agha, G. 2010. Structural health monitoring of a cable-stayed bridge using smart sensor technology: deployment and evaluation. *Smart Structures and Systems,* 6, 439-459.

Doebling, S. W., Farrar, C. R. & Prime, M. B. 1998. A summary review of vibration-based damage identification methods. *Shock and vibration digest,* 30, 91-105.

Spencer, B. F., Ruiz - Sandoval, M. E. & Kurata, N. 2004. Smart sensing technology: opportunities and challenges. *Structural Control and Health Monitoring,* 11, 349-368.

Lynch, J. P. 2007. An overview of wireless structural health monitoring for civil structures. *Philosophical Transactions of the Royal Society of London A: Mathematical, Physical and Engineering Sciences,* 365, 345-372.

Buratti, C., Conti, A., Dardari, D. & Verdone, R. 2009. An overview on wireless sensor networks technology and evolution. *Sensors,* 9, 6869-6896.

Lynch, J. P., Law, K. H., Kiremidjian, A. S., Carryer, E., Farrar, C. R., Sohn, H., Allen, D. W., Nadler, B. & Wait, J. R. 2004. Design and performance validation of a wireless sensing unit for structural monitoring applications. *Structural Engineering and Mechanics,* 17, 393-408.

Lu, K., Wang, Y., Lynch, J., Lin, P., Loh, C. & Law, K. Application of wireless sensors for structural health monitoring and control. Proceedings of KKCNN Symposium on Civil Engineering, Taiwan, 2005.

Feltrin, G., Saukh, O., Bischoff, R., Meyer, J. & Motavalli, M. Structural monitoring with wireless sensor networks: Experiences from field deployments. Proceedings of First Middle East Conference on Smart Monitoring, Assessment and Rehabilitation of Civil Structures SMAR, 2011.

*** Formatting Issue - Best Available Paper/Graphic ***

The 8th International Conference on
Structural Health Monitoring of Intelligent Infrastructure
Brisbane, Australia | 5-8 December 2017

Visualization of Strain and New Strain Measurement Technique

M. Omachi [1], S. Umemoto [1], T. Takaki [2], K. Matsuo [1],
N. Miyamoto [1], I. Ishii [2], T. Aoyama [2]
[1] *Keisoku Research Consultant Co., Japan, Email:* umemoto@krcnet.co.jp
[2] *Hiroshima University, Japan*

Abstract

A strain visualization device and a new strain measurement technique were developed in response to the need for health monitoring of rapidly-aging structures. The device, which is called the "Strain Visualization Sheet," utilizes the principle of Moiré fringes. As a result of repeated improvements to enable application to field measurements, a self-temperature compensated strain visualization sensor was realized. A revolutionary strain measurement technique was also established, which not only enables quantitative visualization of strain, but also simple measurement with an ordinary digital camera by personnel without special skills. Strain visualization resolution of 50μm/m and measurement error of <20μm/m were confirmed in a verification of accuracy in the laboratory with a calibration device. In a comparison with a conventional strain gauge in a tensile test of steel specimens, the same <20μm/m level of error in the strain value was also obtained by this measurement technique, verifying the appropriateness of the measurement technique and the Strain Visualization Sheet as a sensor. Based on these results, Strain Visualization Sheets were applied in two fields. As a result, stable strain data were obtained and applicability to field measurement was verified. With the application of the Strain Visualization Sheet to structural health monitoring, acquisition of strain information on structures are economical and simple. As a result, more advanced regular visual inspections are possible, and strain information on structures can be obtained at low cost.

1. Introduction

Health monitoring of structures is expected to have a variety of benefits, including a higher level of inspection, labor-saving in inspections and early discovery of abnormalities, and also assist in early recovery in disasters and other abnormal cases. In recent years, however, health monitoring been limited to only some structures due to problems such as the condition of public finances, and is also not performed adequately, even in visual inspections. The largest reason for the failure to implement health monitoring of structures is the problem of cost. In addition to the cost of installation, maintenance costs also increase as more sophisticated sensors are used. Moreover, if frequent maintenance of the monitoring equipment itself is necessary, equipment maintenance costs will also increase, and depending on the scale of the structure, this may result in reversing the logical order of things. Therefore, focusing on strain, which is one monitoring item, we developed a device called the Strain Visualization Sheet with the aim of establishing a remote, noncontact strain measurement technique that enables simple measurement by personnel without special skills (Omachi, M. et al. 2015). The Strain Visualization Sheet is a new sensor for use in strain measurements utilizing the principle of Moiré fringes, and has the following distinctive features.

(1) Quantitative visualization of strain is possible.

(2) Because the Strain Visualization Sheet uses absolutely no electrical elements, a power source is not necessary and device life is not determined by electrical failure.
(3) For the same reason, the Strain Visualization Sheet is not affected by electrical noise.
(4) Strain can be measured by a remote, noncontact method with higher accuracy than is possible by visual inspection by using a general digital camera, etc.
(5) Maintenance is easy.

Recently, we developed a self-temperature compensated type of Strain Visualization Sheet for application to field measurements, and after repeated improvement to solve several problems in field measurement, we succeeded in practical application of the new device. This paper presents an outline of the Strain Visualization Sheet, which is a new strain measurement technology. The results of a performance verification are reported, and examples of application of Strain Visualization Sheet in the field are also introduced.

2. Measurement Principle of Strain Visualization Sheet

As shown in Fig. 1(1), when line grating 1 with pitch p and line grating 2 with a pitch $p+\Delta p$ which is Δp ($<<p$) larger than the pitch of line grating 1 are superimposed, a striped pattern with a pitch W larger than that of line gratings 1 and 2 appears. This pattern is called a Moiré fringe. The Strain Visualization Sheet utilizes the principle of the Moiré fringe. The relationship between the pitch of the line gratings and the pitch of the Moiré fringe can be expressed by Eq. (1), and pitch p can be displayed visually enlarged by $(p+\Delta p) / \Delta p$ by the Moiré fringe. This magnification factor is M.

$$W = \frac{p + \Delta p}{\Delta p} \cdot p \tag{1}$$

In order to enhance visibility (legibility), a Moiré fringe with a high magnification ratio is obtained by treating a number n of stripes with pitch p on line grating 2 as one unit, and arranging this at intervals of $np+\Delta p$. The pitch of the Moiré fringe which is formed under this condition can be expressed by Eq. (2), and the magnification factor is $(np+\Delta p) / \Delta p$.

$$W = \frac{np + \Delta p}{\Delta p} \cdot p \tag{2}$$

The amount of shift Δx_m of the Moiré fringe in direction (A) when line grating 1 moves Δx in direction (A), as shown in Fig. 1(2), is expressed by Eq. (3). This means that displacement Δx can be displayed visually enlarged by M times. It is possible to measure the strain level by this principle.

$$\Delta x_m = M \cdot \Delta x \tag{3}$$

In the same manner, if a grating in character form (character grating) is used in place of line grating 1, as shown in Fig. 2, the amount of shift when the character grating is moved in direction (A) can be displayed by changes in the characters, as in (i') to (iii'). In visualization of strain, numbers corresponding to the amount of generated strain are displayed by arranging the character grating in the form of a strain scale and adjusting the magnification factor.

*** Formatting Issue - Best Available Paper/Graphic ***

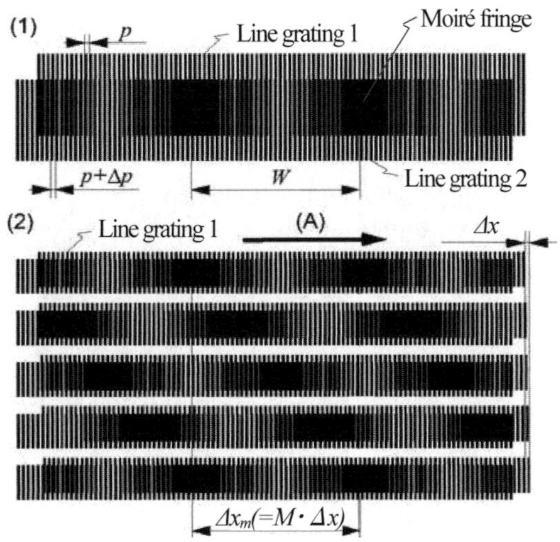

Fig. 1 Example of a moiré fringe

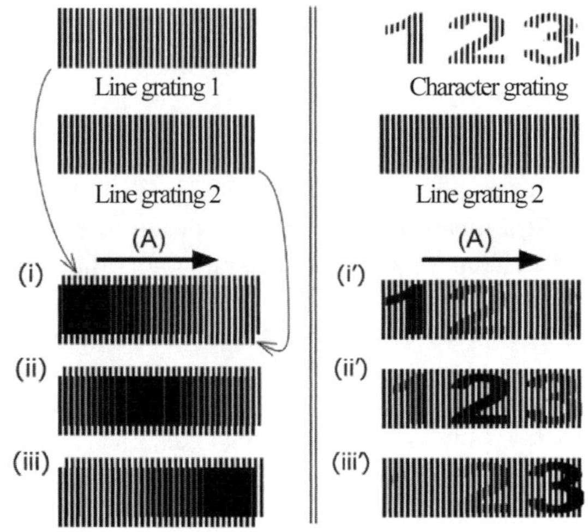

Fig. 2 A moiré fringe with characters

3. Strain Visualization Sheet
3.1 Structure of Strain Visualization Sheet

The structure and appearance of the Strain Visualization Sheet are shown in Fig. 3 and Fig. 4, respectively. The device comprises the two glass plates that form the Moiré fringe pattern and a steel temperature compensation plate for temperature compensation, and is assembled into a unit by using two fixing rings. The length, width and thickness of the Strain Visualization Sheet are 120 mm, 14 mm and 8 mm, respectively, and the gauge length is 105 mm. The Strain Visualization Sheet consists of three fringe types. The Moiré fringes in the top and middle line change according to the strain. The top line shows character patterns in the form of a scale with an interval of 100 μm/m, and allows the user to read strain values directly with the unassisted eye. The middle line is a simple Moiré fringe for strain measurement by image processing, and the bottom line is a reference Moiré fringe for improvement of image cropping accuracy in image processing.

3.2 Specification of Grating Pattern

The pitch p, Δp and number of repetitions n of the grating pattern in the visualization part on the top line are 0.1 mm, 0.01 mm and 73, respectively, and the magnification factor M = 730x. In the image processing part on the middle line, the pitch and number of repetitions of the grating pattern p, Δp and n are 0.07 mm, 0.00025 mm and 1, and M = 280x. Because grating pattern generation technology has improved dramatically, it has become possible to set the pitch p and Δp on the image processing part on the middle line to smaller values than were used until now, and it is also possible to secure the same magnification factor M as in the past, even with the number of repetitions n = 1.

3.3 Structure of Pattern Glass

Fig. 5 shows the structure of the pattern glass. Photomask technology is used in generating the pattern on the glass. Grating pattern 1 is generated on the back side of the front glass. The back glass is first coated with a white undercoat, and grating pattern 2 is then generated. After the patterns are generated a clear overcoat is applied to both line gratings.

*** Formatting Issue - Best Available Paper/Graphic ***

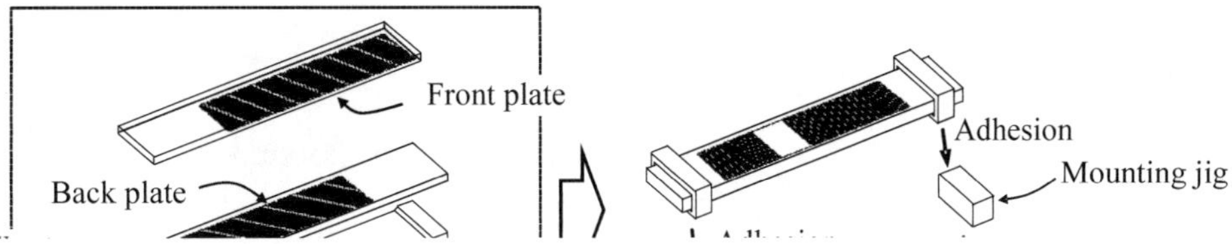

Fig. 3 Structure of Strain Visualization Sheet

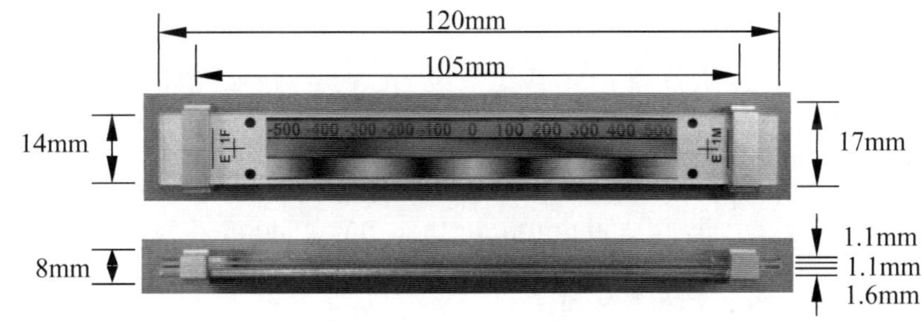

Fig. 4 Appearance of Strain Visualization Sheet

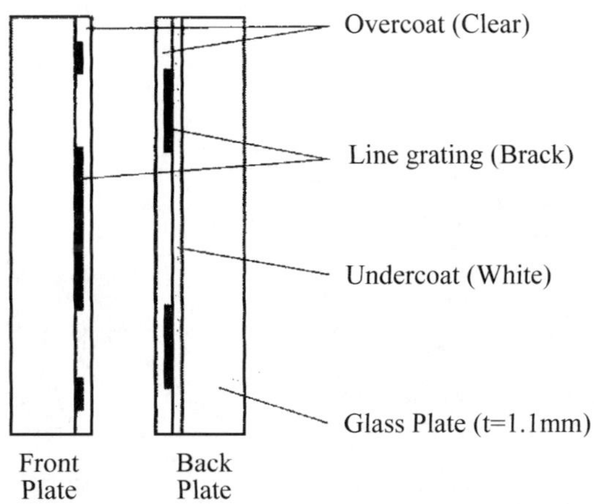

Fig. 5 Structure of Pattern glass

*** Formatting Issue - Best Available Paper/Graphic ***

4. Verification of Accuracy

4.1 Test Device

The appearance of the test device and the displacement control unit are shown in Fig. 6 and Fig. 7, respectively. The test device consists of the displacement stage, laser displacement meter (repeatability: 0.05 μm) and their control PC, a USB camera (resolution: 2 x 10^6 pixels) for use in photographing the Strain Visualization Sheet and a real-time image processing device. The USB camera was set at a height of 90 cm above the Strain Visualization Sheet, and lighting was provided to maintain a constant intensity of illumination.

4.2 Test Method

After the initial value was measured, displacement (strain ε = 45 μm/m equivalent) was applied with a pitch of 5 μm to the Strain Visualization Sheet by the test device, and the Strain Visualization Sheet was photographed continuously with the USB camera. Strain was calculated from the continuous photographed images in real time by the real-time image processing device, and the strain values were recorded at intervals of 5 μm.

4.3 Visual Measurement

Fig. 8 shows the change of the strain values in the visualization part. The strain values were obtained by reading the central value in the numbers that could be recognized visually to a half between indicated numbers. From the verification results, it can be understood that the strain values can be visualized with accuracy of ±50 μm/m for the strain generated at each 90 μm/m.

4.4 Measurement by Image Processing

Fig. 9 shows the results of measurements of the strain generated by the displacement stage obtained by image processing of the Strain Visualization Sheet. The results of measurements by image processing showed extremely good agreement with the actual generated strain. Error was –8 μm/m to +15 μm/m, confirming that strain can be measured with higher accuracy than by measurement by visual inspection.

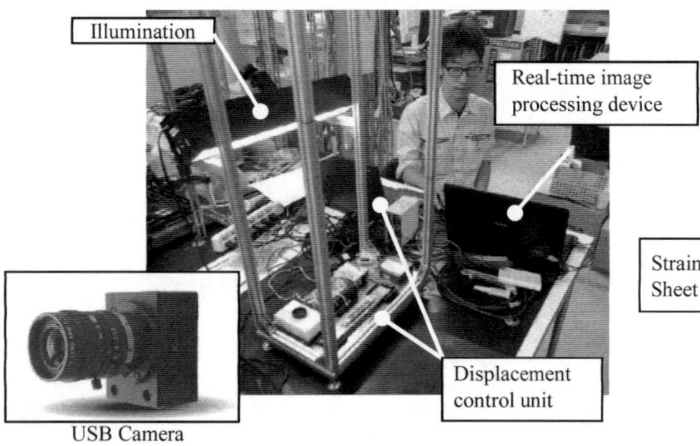

Fig. 6 Appearance of test setup

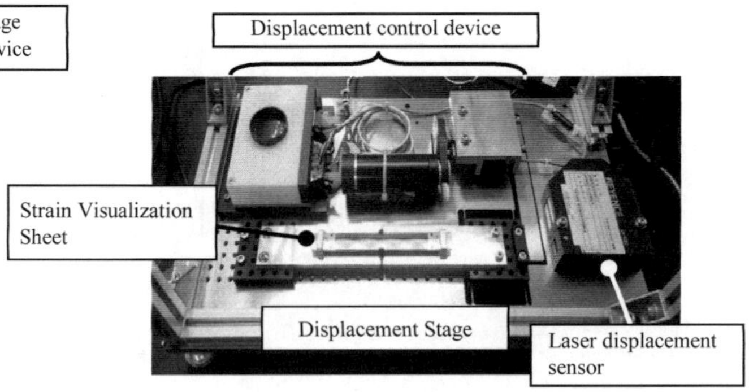

Fig. 7 Displacement control unit

*** Formatting Issue - Best Available Paper/Graphic ***

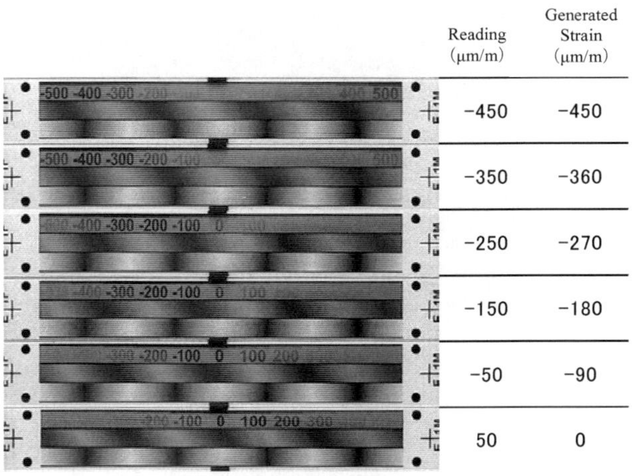

Fig. 8 Change in strain value of visualization part

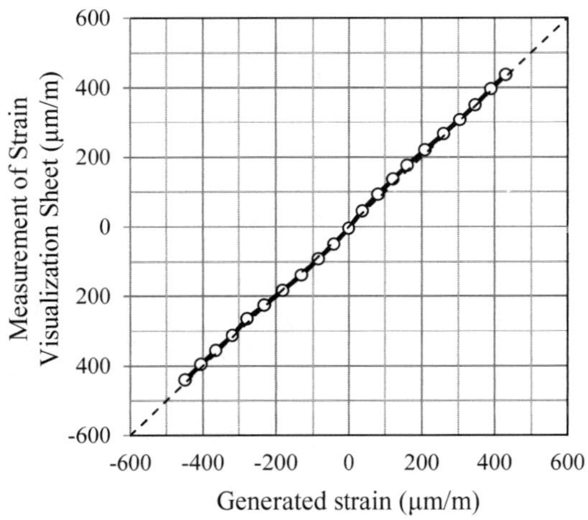

Fig. 9 Image processing result

5. Comparative Verification by Tensile Test

5.1 Test Piece

Fig. 10 shows the tensile test piece (JIS Z2201, No. 14). A Strain Visualization Sheet and a strain gauge as an object of comparison were arranged adjacent to each other on both sides of the test piece at the test piece center. The gauge length of the strain gauge was 90 mm.

5.2 Test Method

The test piece was set in the test machine (rated capacity: 1 000 KN) as shown in Fig. 11, and loading was applied in steps of 20 kN at a constant rate so as not to apply impact to the test piece. Loading was continued to a maximum of 140 kN, after which the load was removed in the same manner. The Strain Visualization Sheets on both side of the test piece were photographed with a general digital camera (Fig. 12) while keeping the load at each 20 kN step. The values of the strain gauges were also recorded at the same time.

The strain calculation application shown in Fig. 13 was used in calculations of the strain values. This application, which was developed for use in field measurements, calculates strain from the static images photographed by the digital camera. Strain is calculated by the procedure of (1) reading the static image, (2) extracting the image of the Moiré fringe, (3) fitting the measured value and the theoretical value of the luminance distribution (sine approximately curve) of the Moiré fringe and (4) calculating the strain value from the phase difference with the original image. In actual operation, the operator simply selects the black circles at the four corners of the Strain Visualization Sheet in order and clicks the analysis button.

5.3 Verification Results

Fig. 14 shows the relationship of load and strain during the loading cycle, and Fig. 15 shows the results during the unloading cycle. The average values of the strain on the two sides of the test piece are shown for both the strain gauges and the Strain Visualization Sheets.

*** Formatting Issue - Best Available Paper/Graphic ***

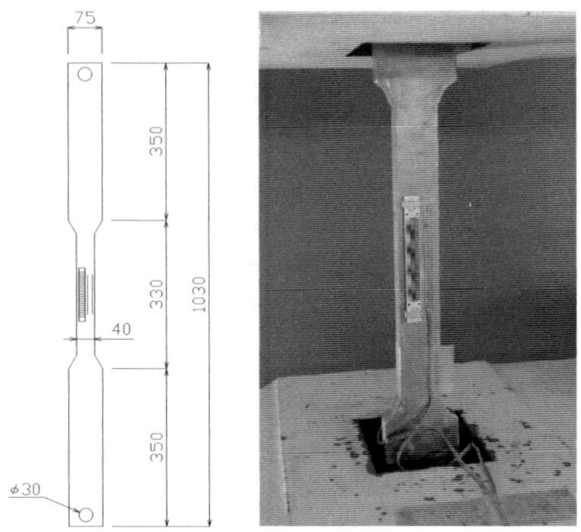

unit : mm   : Strain gauge
: Strain Visualization Sheet

Fig. 10 Test Piece for Tensile Test

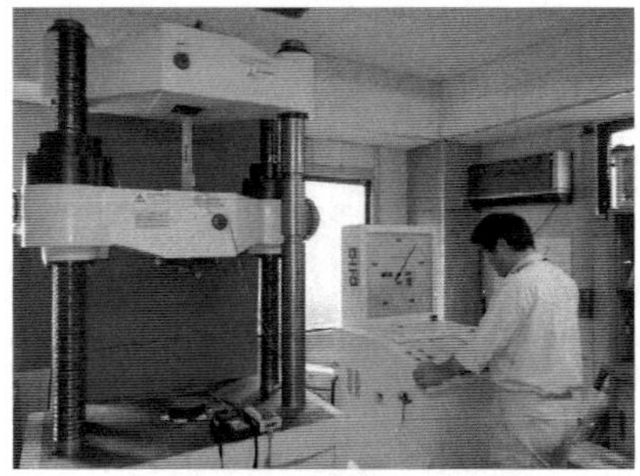

Fig. 11 Testing scene

OLYMPUS STYLUS TG-3 Tough

Fig. 12 Digital camera

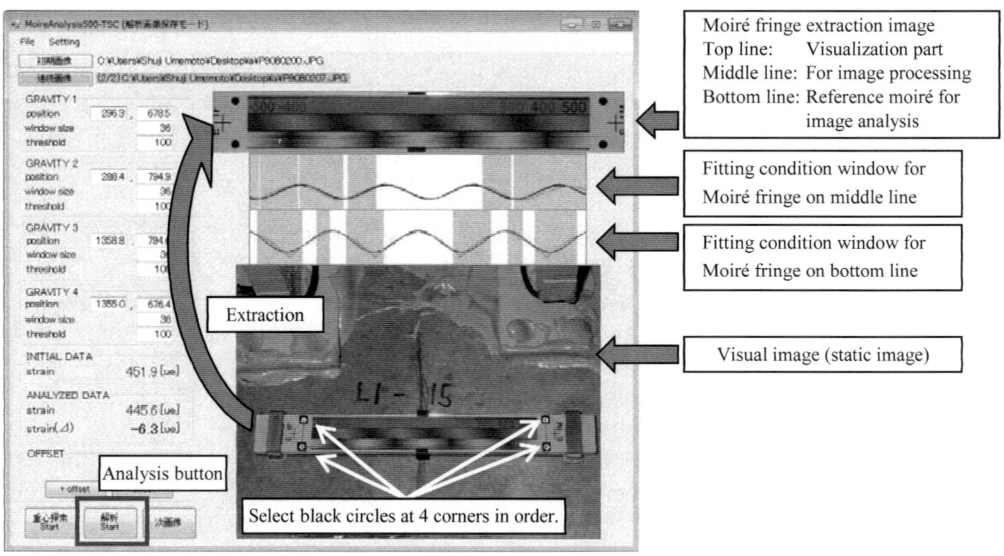

Fig. 13 Strain calculation application

During loading, the values of the Strain Visualization Sheet showed extremely good agreement with the values of the stain gauge, the maximum difference being 14 μm/m. Although the difference with the strain gauge during unloading was somewhat larger than during loading, the average difference was 20 μm/m and the maximum difference was 32 μm/m. Thus, this verification, which assumed field measurement, verified the fact that strain can generally be measured with high accuracy by the Strain Visualization Sheet.

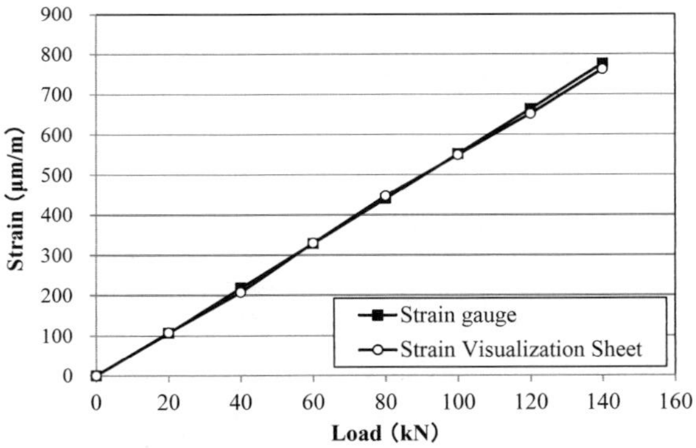

Fig. 14 Relationship of load and strain (loading)

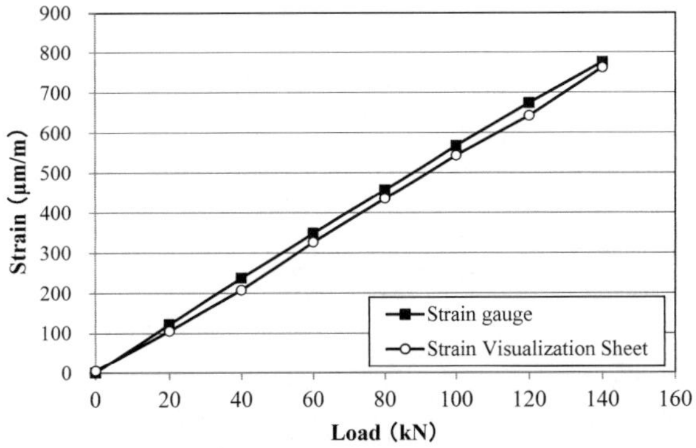

Fig. 15 Relationship of load and strain (unloading)

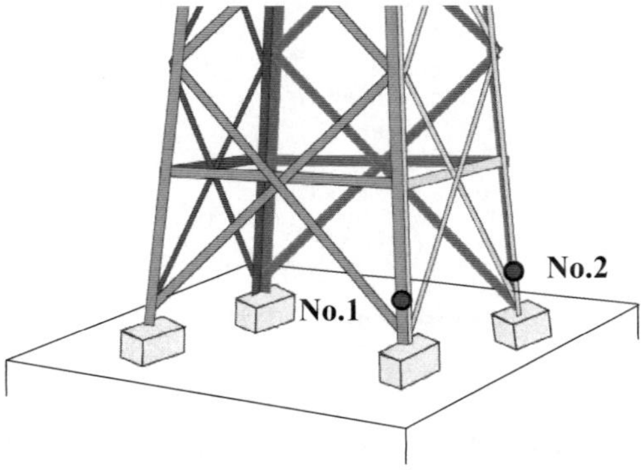

Fig. 16 Measurement positions

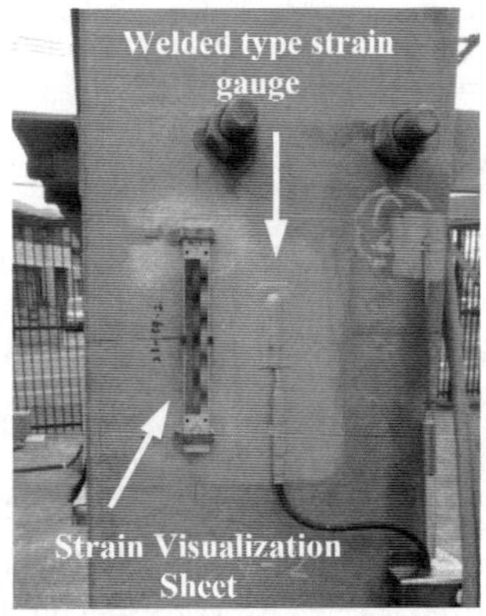

Fig. 17 Condition of measurement device installation

6. Application to Field Measurement

In order to verify the applicability of the Strain Visualization Sheet to field measurement, Strain Visualization Sheets were applied to a steel power transmission tower in Saitama Prefecture, Japan and to a telecommunications tunnel in Tokyo, and strain measurements were performed.

6.1 Strain Measurement of Steel Power Transmission Tower

6.1.1 Outline of Measurement

As shown in Fig. 16, Strain Visualization Sheets were applied to two positions on the legs of the steel transmission tower. In addition to installing these Strain Visualization Sheets on the tower legs, welded type strain gauges (gauge length: 5 mm) were also installed adjacent to the Strain Visualization Sheets for comparison. The condition of installation is shown in Fig. 17.

6.1.2 Measurement Results

The strain measurement results are shown in Fig. 18 (position No. 1) and Fig. 19 (position No. 2). As the measured strain data for the Strain Visualization Sheet, static images were photographed by using digital cameras, and strain was calculated by the strain calculation application software.

Although the amount of change in the strain is slight and the number of measurements is small, the measured values obtained by the Strain Visualization Sheet at the present point in time are trending at a difference of 10-50 μm/m in comparison with the measured values by the welded type strain gauge, and stable measured data are being obtained.

*** Formatting Issue - Best Available Paper/Graphic ***

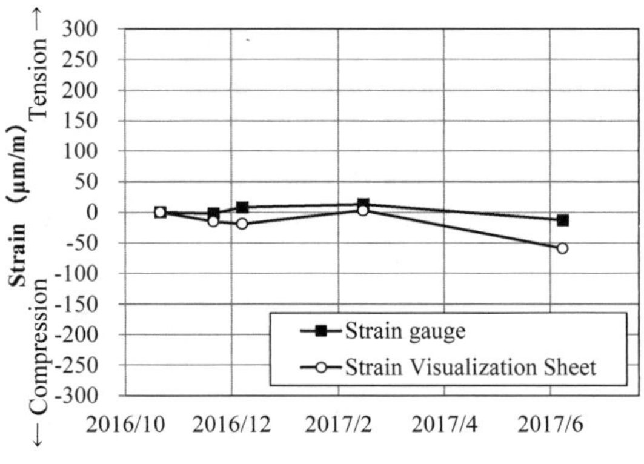

Fig. 18 Strain measurement results (position No. 1)

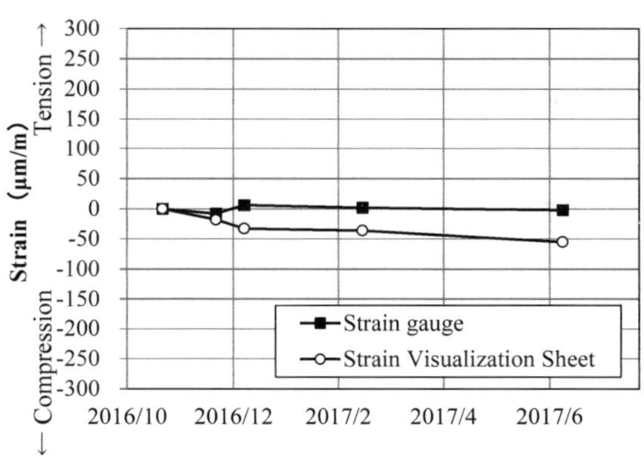

Fig. 19 Strain measurement results (position No. 2)

6.2 Strain Measurement of Telecommunications Tunnel

6.2.1 Outline of Measurement

In this test, the object of measurement was the lining concrete of a telecommunications tunnel. As the measurement locations, two locations with different environmental conditions ((1) Dry condition, (2) Wet condition) were selected.

A Strain Visualization Sheet was installed on the lining concrete, and a strain transducer (gauge length: 100 mm) and a metal base strain gauge were installed near the Strain Visualization Sheet as objects of comparison. The stainless steel type metal base strain gauge was used in order to prevent insulation degradation due to moisture from the concrete. A thermocouple type thermometer was also installed to measure the ambient temperature. Fig. 20 shows the condition of installation of these measuring devices.

6.2.2 Measurement Results

Fig. 21 shows the strain measurement results. The strain data measured by the Strain Visualization Sheet were obtained by photographing static images with a digital camera and calculating the strain values with the strain calculation application software.

At the measurement point under the dry environment, the strain values were approximately in agreement from the start of measurement until about 4 months. However, after 4 months, a maximum difference of approximately 200 μm/m was observed in comparison with the conventional measurement method using the strain gauge and the strain transducer. Since we cannot rule out the possibility that a malfunction occurred in the strain transducer, strain gauge or Strain Visualization Sheet after the measurements in February 2017, we plan to continue checking the trend of the measured data and investigate whether a sensor malfunction of other problem occurred.

At the measurement point under the wet environment, no changes in the strain data were observed after the measurement in February 2017. Although a difference of approximately 100 μm/m occurred between the Strain Visualization Sheet and the strain gauge/strain transducer, the cause is not known at the present time. The strain gauge at the measurement point under the wet environment failed about 1 month after the start of measurements.

*** Formatting Issue - Best Available Paper/Graphic ***

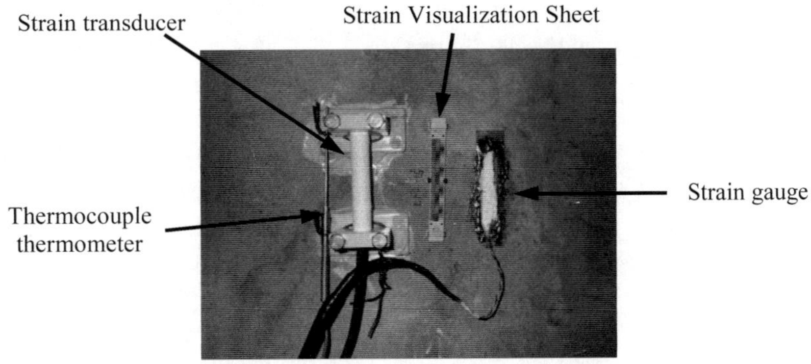

Fig. 20 Condition of installation of measuring devices

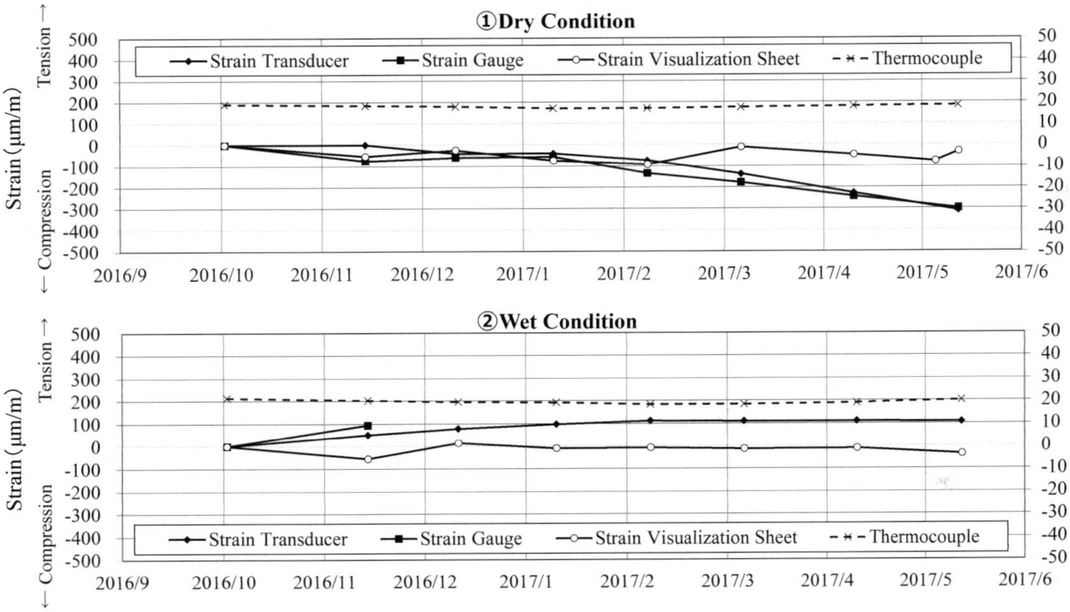

Fig. 21 Strain measurement result

7. Conclusion

To respond to the need for health monitoring of structures, the authors developed and successfully realized practical application of a device called the Strain Visualization Sheet as a new strain measurement technology which enables quantitative visualization of strain. In a verification of measurement accuracy in laboratory tests, highly accurate measurements were possible in both measurement by visual inspection and measurement by image processing. The possibility of measurements at a level that enables practical application was also demonstrated in a tensile test of steel material assuming field measurement. Although there are a few examples of application to actual field measurement, this research showed the possibility of practical application in long-term measurement. In the future, the authors plan to accumulate actual results of practical application in the field and improve the completeness of the Strain Visualization Sheet.

References

Omachi, M. et al., 2015, 'Verification of Strain Visualization Sheet with Aim of Field Application', *Proceeding of the 7th International Conference on Structural Health Monitoring of Intelligent Infrastructure (SHMII-7)*, Torino,Italy, 1-3 July.

The 8th International Conference on
Structural Health Monitoring of Intelligent Infrastructure
Brisbane, Australia | 5-8 December 2017

Development of Bridge Inspection Method without Temporary Scaffolding by Using Optical Measurement Techniques

Keisuke KIMOTO [1], Kohei YAMAGUCHI [2], Toshihiro OKUMATSU [2],
Taiki KAWAMURA [2].Hiroshi MATSUDA [2]
[1] Keisoku Research Consultant CO., Japan, Email: kimoto@krcnet.co.jp [2] Nagasaki University, Japan

Abstract

This paper shows the development of inspecting method about bridge without temporary scaffold, using and applying the optical measurement techniques about Unmanned Aerial Vehicle and Structure from Motion. In late years UAV and SfM progress innovatively, they are becoming common techniques. SfM is a technology for building a 3D model of the target structure only images taken from multiview. By using the UAV, it is possible to acquire an image from a view point which can not be confirmed normally, it can be more effective inspection.
Thereupon authors verified their techniques at the several bridges, and summarized the measurement range, photography method, measurement precision and investigation time for future use.

1. Introduction

Japan has approximately 700,000 bridges with lengths of 2 m or more, including about 500,000 with a length of less than 15 m. In 2014, the Ministry of Land, Infrastructure, Transport and Tourism (MLIT) imposed an obligation to conduct periodic inspections of all of those bridges at a frequency of once in 5 years by proximity visual inspection. However, the reality is that adequate inspection cannot be performed because many of those bridges are managed by municipal governments which are faced with financial problems and shortages of technical personnel accompanying low birth rates, an aging population and population declines. As an additional problem, the percentage of bridges (2 m or more) with ages exceeding 50 years since construction was 20% in 2016, but this is expected to increase rapidly to 44% in 10 years. Thus, the development of more efficient bridge inspection techniques has become an urgent issue. Moreover, in proximity visual inspections by the conventional techniques, it is necessary to approach the structure closely, but this requires a bridge inspection vehicle or lift-type vehicle or temporary scaffolding and special work by rope access, and traffic restrictions must be imposed during such work (Fig. 1). In this case, reduction of inspection costs/labor is an issue.

Therefore, the authors considered the possibility of labor-saving and higher efficiency in inspections by applying SfM (Structure from Motion) technique to bridge investigations. SfM has become a general technique as a result of improvements in personal computer (PC) performance and image analysis technique in recent years. As SfM is a technique for

Fig. 1 Inspection of bridge pier by suspended scaffold and rope access

constructing a 3D model of an object structures by using only images taken from multiple views, it would appear that inspection work is possible on a PC, by indoor work, simply by photographing images with an ordinary digital camera. Although remote measurement is also possible by using a 3D laser scanner, etc., the data acquired by this method is a point group and ultimately only reproduces shape. SfM can reproduce shape, and can also acquire real color information. Since SfM can grasp damage with high accuracy, SfM would seem to be an extremely effective technique for bridge inspections.

The use of unmanned aerial vehicles (UAV) has also spread rapidly in recent years, to the point where UAV have been called an "industrial revolution in the sky." By using UAV to photograph images for SfM analysis, it is possible to acquire images from views that normally cannot be observed, and it is thought that this will enable more effective inspections.

Therefore, we verified the effectiveness of damage investigation using UAV. The object of this verification was a concrete main tower of a cable stayed bridge.

2. Applied Technologies

2.1 SfM (Structure from Motion)

SfM is a technique which makes it possible to construct a 3D model of a photographed object based on images photographed from multiple views.

Unlike conventional photogrammetry, a characteristic point on the surface of the photographic subject is taken from different directions, aiming at an overlap of 80%, and the positions of that point and the camera, in other words, 3-dimensional coordinates (X, Y, Z), are automatically assigned in succession by the stereo method from combinations of the photographed images (Fig. 2). The SfM analysis generates a triangulated irregular network (TIN) of triangular plane elements with those coordinates as the tangent points based on the point group obtained in this manner, and constructs a 3D model. Here, a "characteristic point" means a point on the image and is a point with different luminescence, i.e., color and brightness, compared with its surroundings. For characteristic points p and q in regions (i, j) on two images, SfM calculates the sum total J of their differences of luminescence T compared with the surrounding areas in the respectively regions, that is, the degree of difference in color and brightness compared with the surroundings, and the points are judged to be the same point when the value of J is within a certain judgment value (Fig. 3). Because the photographed images are linked as texture in the generated TIN, a 3D model that faithfully reproduces the photographic subject is constructed. Figure 4 shows an example of the construction of a 3D model that was generated from photographic images by SfM analysis (partial excerpt).

It is also possible to give a scale to the generated TIN by setting targets with clear coordinate values at 3 or more points in the object range, and photographing those targets in the image.

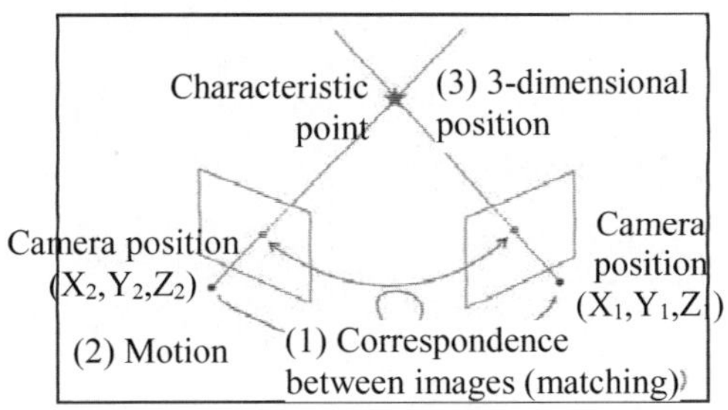

Fig. 2 Method of calculating characteristic point and camera position

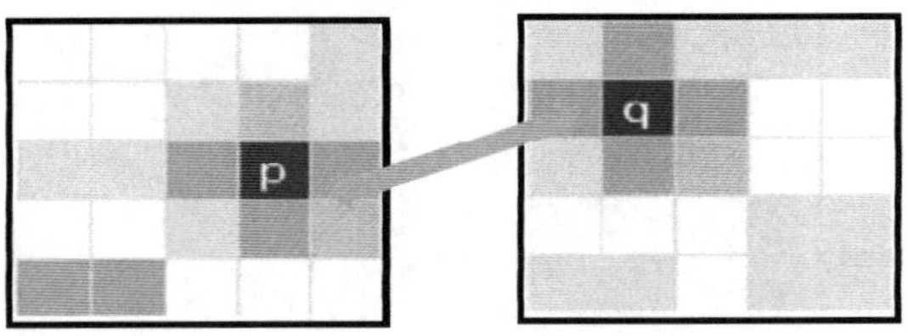

$$J(p,q) = \sum_{i,j} \left| T_p(i,j) - T_q(i,j) \right|^2$$

Fig. 3 Judgment of characteristic point

Images photographed from multiple views

Result of SfM analysis

Fig. 4 Example of construction of 3D model by SfM

2.2 Comparison of Analysis Software

At present, several software programs which perform SfM analysis are widely used. Therefore, we compared the functions of the two types that seem to the most general, S software of A company and P software of B company. The results are summarized in Table 1. We also compared the 3D models generated by the respective programs by using the same photographed images of a concrete structure. As the result of the comparison, P software had many editing functions and high analytical flexibility, but the texture quality was somewhat low. On the other hand, S software had the minimum functions necessary for only generating models, with almost no editing functions, but a texture close with near-photographic resolution could be pasted. Furthermore, the reproducibility of the edges of the structure was also high (Fig. 5) because the TIN is generated by adjusting the edge line when performing automatic analysis. Since the priorities in bridge inspections are the reproducibility of the object structure and the degree of recognition of damage, the verification in this research was performed using the S software.

Table 1 Outline of SfM analysis software programs

Name of program	S	P
Maker	A Co.	B Co.
Principle	SfM	SfM
Camera calibration	No (parameter input is possible)	Yes
Point group editing	No	Yes
Measurement functions	Distance (viewer software)	Distance / Area / Accumulation
Ortho-image output	No	Yes
Edge line confirmation	Yes	No

TIN generated by S software

TIN generated by P software

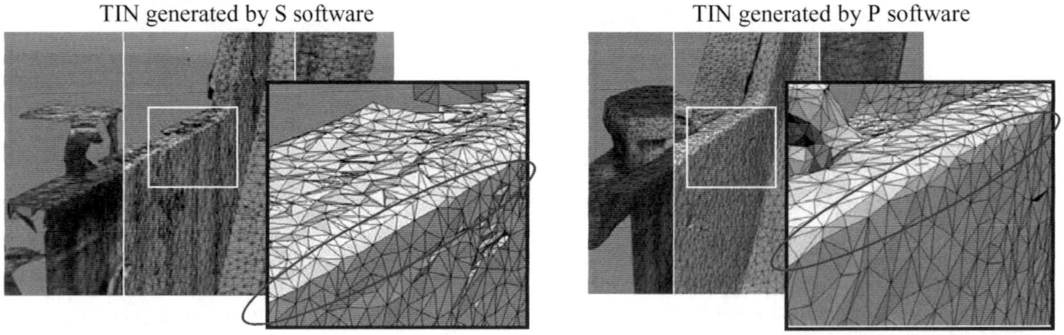

Can adequately distinguish texture.
TIN of structure edge is generated considering edge line.

Quality of texture is somewhat low.
Edge of structure is not accurately expressed.

Fig. 5 Comparison of reproducibility of edge of concrete structure

2.3 UAV (Unmanned Aerial Vehicle)

UAV is a general name for small, unmanned aircraft. UAVs are already firmly established as a measurement technique in the fields of surveying and civil engineering, in addition to aerial photogrammetry, airborne LiDAR and ground measurements. By using UAVs, it is expected to be possible to investigate dangerous locations and the actual site immediately after disasters, which had been difficult in the past, and application to bridge inspections is also amply possible.

As shown in Fig. 6, a GPS antenna, gyro sensor and battery are mounted in a UAV, and automatic flight is possible by setting the flight path in advance. When a camera is mounted on the UAV, the operator manipulates the triaxial gimbal, which enables free vertical and lateral movement of the camera, while constantly monitoring the photographed range from the ground, and the equipment configuration makes it possible to trigger the shutter. Because the gimbal is controlled so as to automatically maintain a horizontal attitude, images that reflect the intention of the photographer can be acquired easily.

Table 2 summarizes the features of the UAV. In comparison with conventional radio-controlled aircraft, the UAV is superior in terms of compact size, light weight and stability, and safer flight is possible. Because it is motor-driven, it is comparatively quiet, but flight time depends on battery capacity and is short, at about 20 min. Although a license is not required in order to operate a UAV, the safety of UAVs has been questioned recently. As this became a social issue, Japan's Civil Aeronautics Act was revised in December 2015 so that it is now necessary to apply in advance to the Ministry of Land, Infrastructure, Transport and Tourism for flights in densely inhabited districts or over an altitude of 150 m, and for flights out of the operator's range of vision.

The verification in this research was carried out giving full consideration to the safety of the UAV, but there are also many restrictions for ensuring safe flight, as UAVs are easily affected by weather, radio waves and the like. Special care is also necessary in flights near roads, railroads, bridges, radio towers and high voltage electric power lines, and an experienced operator is indispensable.

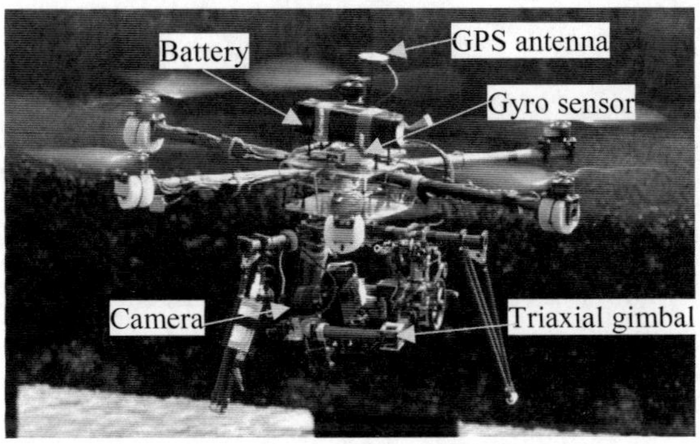

Fig. 6 Typical configuration of UAV (with camera attached)

Table 2 Features of UAV

Advantages	Compact and lightweight; easy to transport
	Operator does not require license.
	Operation is quiet (motor driven).
	Flight is stable due to large number of propeller blades.
	Easy to take aerial photographs
Disadvantages	Short flight time (about 20 min)
	Difficult to use in strong winds (10 m/s or faster)
	Approval is necessary for flights at altitudes higher than 150 m and flights in densely inhabited districts.
	Flight out of the operator's visual range is prohibited.

3. Verification at Long-Span Bridge

3.1 Outline

In bridge inspections in Nagasaki Prefecture, periodic inspections by proximity visual inspection are normally carried out once in 5 years. However, inspections of bridges which are greatly different in bridge scale or structural characteristics and inspections of bridges with a large effect on the region (priority maintenance control bridges) are also performed once a year (hereinafter, annual inspection). In annual inspections, the progress of deterioration of fixed observation points that can be observed from the ground is confirmed.

Accordingly, by using UAV-SfM in annual inspections, it is possible to confirm the condition of damage safely, quickly and at low cost, and at a wider range of points which normally can only be checked by proximity visual inspection using rope access or temporary scaffolding. In addition, since the damage assessments are performed from images, it is also possible to make quantitative judgments that are not dependent on the skill of the inspector, and comparisons over time are also easy by accumulating digital data. Moreover, if UAV-SfM can be established as a new inspection technique in place of proximity visual inspection, application to periodic inspections is also fully possible.

In this research, an investigation using SfM was performed for a concrete main tower (4P) of the Takashima-Hizen Bridge (cable stayed bridge) connecting Hoshika, Hizenmachi, Karatsu City, Saga Prefecture and Kozakimen, Takashima-cho, Matsuura City, Nagasaki Prefecture, as shown in Figs. 7 and 8. The results were compared with the existing inspection results, and the required inspection time was investigated. The following shows the data for the Takashima-Hizen Bridge.

Bridge name : Takashima-Hizen Bridge
Road name: Takashima-Hizen Line (General Prefectural Road)
Bridge length: L = 1251.00 m
Total width: W = 11.50 m
Superstructure type: 5 span continuous steel 2 main girder, 5 span continuous cable stayed bridge, 4 span continuous steel 2 main girder
Substructure: Abutments (2), piers (2)
Year constructed: 2009

Fig. 7 Takashima-Hizen Bridge

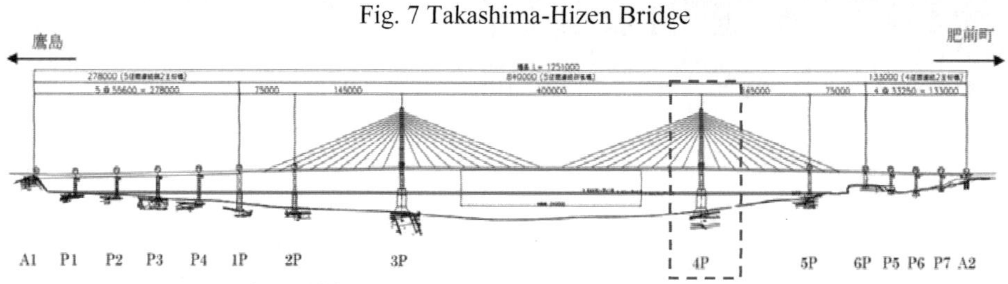

Fig. 8 Side elevation of Takashima-Hizen Bridge

3.2 Equipment Used

The specifications of the UAV and digital camera used in this research are shown in Table 3 and Table 4, respectively. The UAV used here was a Spreading Wings S900 manufactured by DJI with a maximum payload of 2.5 kg. This type was selected because strong winds were assumed above the site of the bridge, as it is possible to mount a somewhat heavy camera on the UAV, and the UAV has wind resistant stability. As the camera, a SONY mirrorless α7R was selected; this camera is equipped with a full-size sensor and enables high pixel photography (36 x 106 pixels). The photograph in Fig. 9 shows the Spreading Wings S900 with the attached α7R camera.

Table 3 Specification of UAV used

DJI Spreading Wings S900	
Payload	2.5 kg
Wind resistant stability	8m/s
Flight time	Upto 15 min
Flight distance	2 km
GPS	1 cycle

Table 4 Specification of camera used

Mirrorless SONY α7R camera	
Pixel number	7360 ×4912(36M),
Sensor size	35 mm full size
Focal length	35mm
Weight	580 g (including lens)

3.3 Content of Measurements and Work Procedure

Fig. 9 UAV used

Using the UAV with the mirrorless SONY α7R camera, the sea side surface and the bridge axis side surface (45° direction) of the main tower were photographed (6 measurement lines shown in Fig. 10). Although photography with a resolution of 2 mm/pix is sufficient for capturing a crack with a 0.2 mm width on a concrete surface, the target resolution was set at 1.5 mm/pix to acquire higher resolution images. (See the statement that "In case of an image with digital resolution of 2 mm/pix, a crack with a width of 0.2 mm can be recognized on a PC monitor" in "Concrete Diagnosis Technique '16.) For photography with constant resolution, it is necessary to maintain a distance of approximately 10 m from the object surface, as shown in Fig. 11. Therefore, the UAV was flown by a 2-person team consisting of a UAV operator and an image checker. While personally confirmed the photographed images transmitted to a tablet in real time, the image checker determined the photographing distance and resolution from the size of the photographed object relative to the photographic angle of view, and gave instructions on the flight position to the operator (Fig. 12). In order to generate a 3D model from the photographed images by SfM analysis, interval photography was performed at 1 frame/sec (fps), assuming 80% overlap of all of the adjacent images. To give a scale to the generated TIN, 3D laser measurements were performed, and the coordinates of arbitrary points were used as the target points. The procedure of the work is shown in Fig. 13.

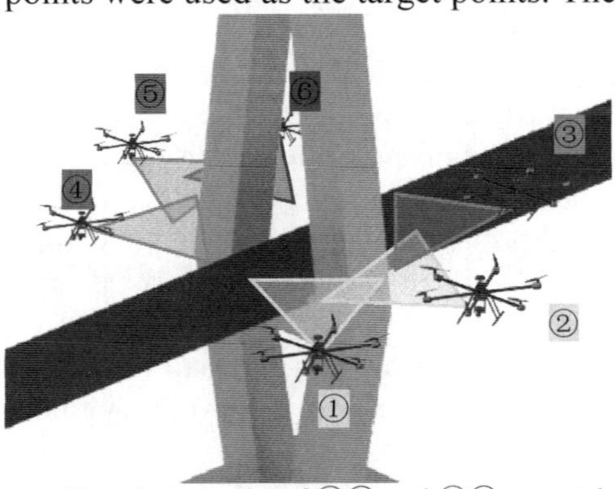

The photographs of ①② and ④⑤ were taken with single flight
Fig. 10 Image of photography (6 measurement lines of outer side of main tower)

To photograph with a fixed focal point, photography is performed so as to keep a constant separation. → Instructions are given while monitoring transmitted images.

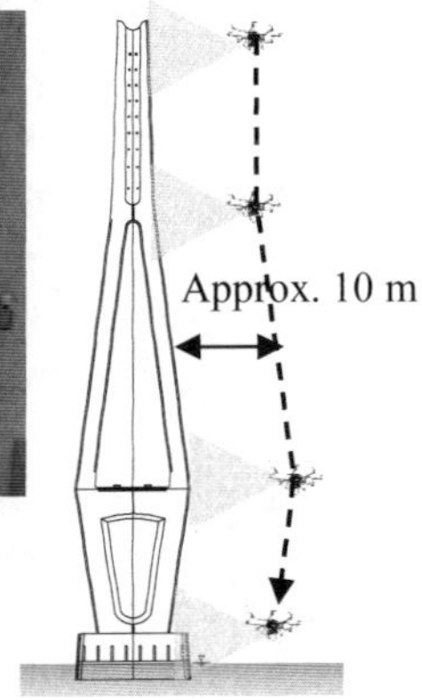

Fig. 11 UAV flight plan

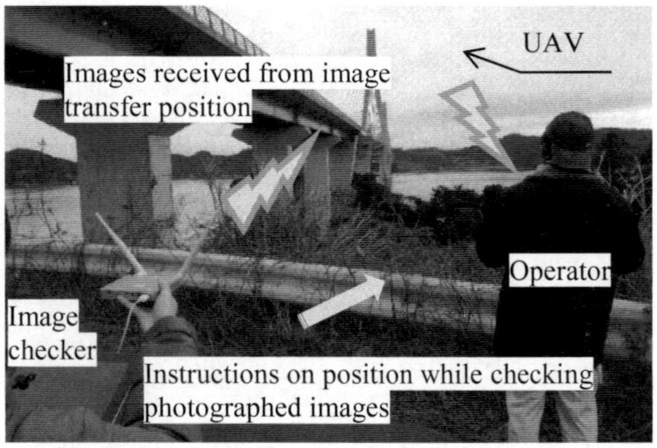

Fig. 12 System during photography

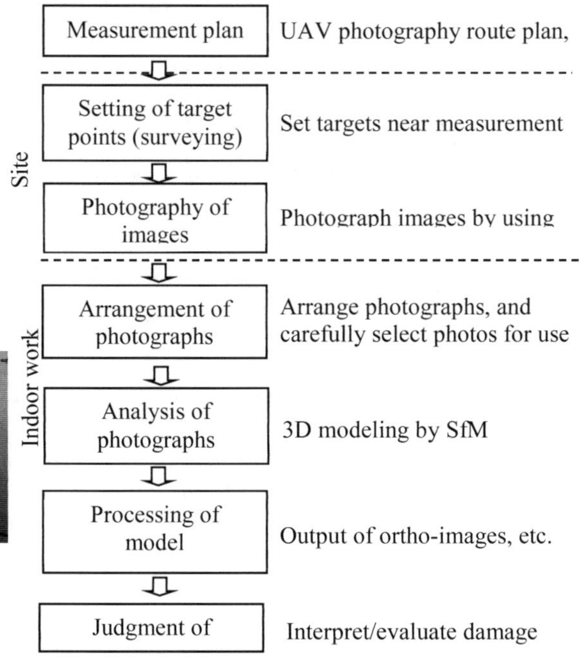

Fig. 13 Work procedure

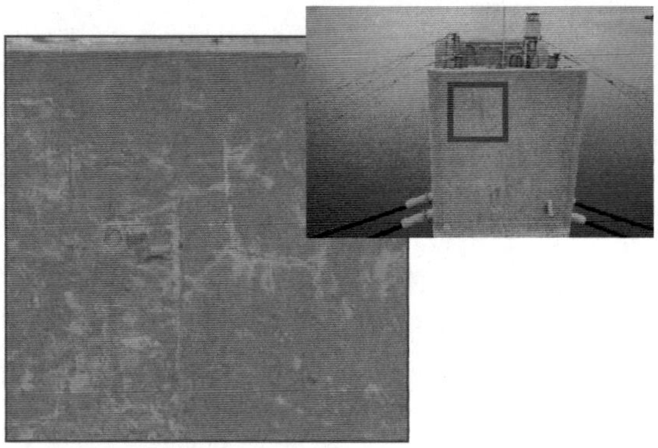
Fig. 14 Example of photographed image

3.4 Results of Photography

A total of 4 flights were carried out to photograph the 6 measurement lines of the sea side surface of the main tower. One flight was completed in about 12 min. Because interval photography was performed at 1 fps, a total of 600 images were acquired. By using the UAV, it was also easy to obtain images of the top part of the tower, which normally can only be checked by rope access, etc., and the condition of the concrete surface could be recognized by enlarging the images. Figure 14 shows an example of a photographed image of the top part of the tower, together with an enlarged view.

3.5 Procedure from SfM Analysis to Generation of Ortho-Image

Conventional photogrammetry was extremely laborious, as it is necessary to perform manual orientation work for the enormous number of images taken by the UAV, and much time was required to match all of the images, but in SfM analysis, substantial labor-saving is possible because automatic orientation is performed using characteristic points. Furthermore, in photogrammetry, faithful reproduction of shapes with depth was difficult, but with SfM, even images photographed from multiple directions can be analyzed easily, and it is also possible to generate models of surfaces with depth.

Figure 15 shows the analysis procedure and results. First, images that are blurred due to camera shaking or are out-of-focus are eliminated from the photographed images. In UAV photography, there are a large number overlapping photographs because interval photography is performed continuously from take-off to landing of the UAV. Therefore, while eliminating poor-quality images, images that secure the proper overlap are carefully selected. Next, these carefully-selected images are analyzed by automatic orientation, and the characteristic points and camera position are calculated. At this time, the rough shape is checked on the PC, and if a shape is not formed, the analysis is repeated several times. A TIN is then constructed based on the calculated characteristic points, and a 3D model is generated by mapping, using the photographed image as the texture. Because this model can be enlarged/reduced, moved and rotated on the PC, it is possible to examine the model from any arbitrary view. The generated model can also be exported to other CG programs (in this research, Autodesk 3dsmax was used), ortho-images can be output, and base drawings for damage investigations can be generated. The time required for generation of a model by SfM is approximately 5 days.

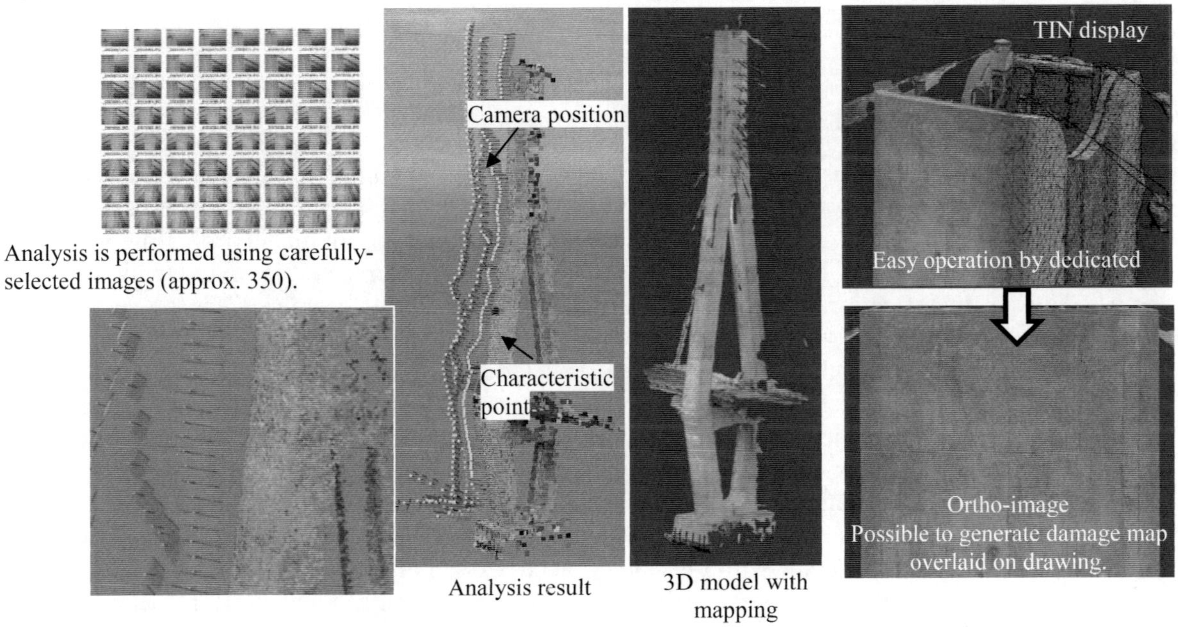

Fig. 15 Procedure from SfM analysis to generation of ortho-image

3.6 Generation of Damage Map

Ortho-images with resolution of 2 mm/pix were generated from a 3D model with mapping, which was created by SfM, and damage mapping was performed for items that could be read from the images. The object items were cracks, exfoliation and exposure of rebars, water leakage and free lime, fallen out parts, and damage of repair/reinforcing materials as shown in Periodic Bridge Inspection Procedure issued by MLIT. Among these, a tree-structure filter is sometimes used in automatic judgments of the position and width of cracks, but in this research, a Crack Position/Width Interpretation and Mapping System utilizing a Crack Index (CI) was used. This system extracts the crack position and width semi-automatically from the correlation between crack width and the total of characteristic values, CI, of the cracked part.

Fig. 16 Condition of damage extraction work

Damage interpretation is performed on multiple screens, as shown in Fig. 16. On the screen at the left, damage is extracted from the ortho-image on the lower screen, using the Crack Position/Width Interpretation and Mapping System, and this damage is described on the screen. In this process, when the image resolution is 2 mm/pix, a range of approximately 3.5 m x 1.5 m is displayed on one screen, and damage is searched in this range. At this time, a detailed interpretation is performed by enlarging the image on the PC so it is approximately on the same level as in visual interpretation. The screen at the right displays the 3D model and is used in checking the object range in 3 dimensions from various views. By searching for damage while observing the 3D model, it is also possible to check for deformation, such as differences in level and swelling, which cannot be seen in 2 dimensions. This is an extremely effective technique, in that the work be performed under calm conditions indoors. It is also possible to convert the damage shown on the screen to CAD data and transfer it to control charts, etc.

3.7 Discussion

No conspicuous damage of the main tower which was the object of this study was found as a result of checking for damage by using the photographed images and 3D model. In this, however, the SfM results were also compared with the results of visual inspection within the range possible.
In the SfM results, as shown in Fig. 17, it was possible to extract approximately the same cracks as in the visual investigation in FY 2014. However, even though fine cracks were extracted in detail in both the visual investigation and the investigation by SfM, slight differences can be seen in the positional relationship. The visual investigation was considered to be inferior in terms of positional accuracy because the results were recorded in sketches, but when using SfM, positional reproducibility was extremely high, as the images were recorded and stored at the same time. Furthermore, while all of the visual inspection results are described as 0.1 mm in width, the SfM results were generally on the order of 0.2 mm, but also included some cracks that were judged to be 0.5 mm. When cracks are extracted by using CI, there are deviations in the results due to the color or soiling of the concrete surface. While it is conceivable that the results of this study were influenced by soiling, evaluation is difficult with only this object range. For the crack width, it is considered necessary to carry out continuing verification work focusing on bridges where clear true values can be obtained, and clarify the relationship between the acquired images and crack width.
In conventional visual inspections, many parts are left to the skill and interpretation of multiple personnel, and quantitative judgment is difficult, for example, in records of the position of damage, etc. However, in investigation by SfM, quantitative judgment is possible because damage interpretation is performed based on images, and it is also possible extract 3D information such the cross section, etc. for arbitrary positions from the 3D model. Moreover, comparisons of changes over time, the progress of deterioration, and other time-related phenomena can also be checked easily by acquiring the same data periodically. Because value-added date such as 3D models, images, etc. is accumulated in this manner, it can be said that this investigation technique has great merit for future operation and maintenance.

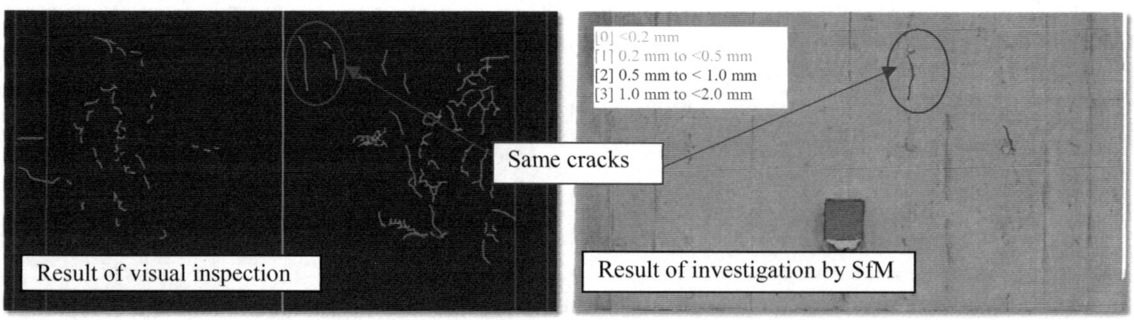

Fig. 17 Comparison of investigation results

3.8 Issues

Although investigation by UAV-SfM is an effective technique, as described above, the following issues also exist.

Adequate consideration of the safety of the UAV is necessary. Particularly when investigating a bridge, as was done in this study, it is important to be aware of weather conditions, etc., as winds blow constantly. Although a large labor-saving and low cost are possible in field work if UAV-SfM is used, the analysis work requires considerable time. Moreover, since the 3D model is constructed based on characteristic points in SfM, certain parts cannot be analyzed, for example, uniformly-painted steel structures in which extraction of the characteristic points is difficult, members with repeated patterns, thin members, etc. Furthermore, items such as lifted defects in concrete, which must be investigated by a contact technique, are difficult to interpret from images. It is thought that this is the reason why investigation by images, in its current state, has been evaluated as unsuitable as a substitute for proximity visual inspection.

4. Conclusion

Use of SfM technique to generate 3D models from images photographed employing a UAV and create damage maps is sufficiently applicable as an inspection technique, and can also realize a large labor-saving and cost reduction in comparison with investigations in which temporary scaffolding is erected. Investigation using images still has not be recognized as a substitute for proximity visual inspection because its crack width detection accuracy is insufficient and it is not possible to detect lifting, which requires investigation by a contact technique. However, a technique using images is considered effective as a primary screening method. Primary screening to select the locations where detailed inspection is necessary makes it possible to allocate time, budgets and personnel appropriately, and thus enables effective, efficient inspections.

Application of UAVs in investigations can be expected in a variety of fields (river surveys, dam investigations, etc.), and use in the rough design of countermeasures during disasters is also amply possible. However, since safety must be secured when using a UAV, from the viewpoints of safety and cost, UAVs are not necessarily effective in all cases. Depending on the required results, an appropriate plan should be prepared, and SfM analysis by manual photography, etc. should also be used effectively. Moreover, because an as-built 3D model that faithfully reproduces the present situation can be constructed by SfM analysis, expansion to BIM and CIM, while continuing the current dissemination of this technique, is considered effective.

In the future, it will be necessary to carry out ongoing verification work and clarify solutions for issues such as clarification of the relationship between crack width and photographic resolution, improvement of detection accuracy by comparison with true values, and shortening of analysis time.

References

Ministry of Land, Infrastructure and Transport 2016.9, 'Annual report on road maintenance', *http://www.mlit.go.jp/road/sisaku/yobohozen/28_3maint.pdf*, p.27.

Keitaro Konuma, Syozo Nishimura, 2014.9, 'Applicability of SfM technology to bridge inspection' *Proceedings of the 69th Annual Conference of the Japan Society of Civil Engineers*, VI-511

Nagasaki Pref. Government 2015.3., Priority Maintenance Bridge Detailed Inspection Report in Nagasaki Pref.

Japan Concrete Institute, 2016, 'Concrete Diagnosis Technique'16', pp.110-111

Syozo Nishimura, et al., 2013.1., 'The Development and Evaluation of Telemetry in Bridge Maintenance' *Journal of Applied Survey Technology* Vol.24, pp.52-61

AUTHOR INDEX

AUTHOR INDEX

AUTHOR INDEX

AUTHOR INDEX

AUTHOR INDEX

International Society for Structural Health Monitoring of Intelligent Infrastructure
University of Manitoba
Agricultural & Civil Engineering Bldg.
A250 - 96 Dafor Road
Winnipeg, Manitoba R3T 2N2
Canada

ISBN 978-1-5108-6457-3

CURRAN ASSOCIATES INC.
proceedings.com